WHO IS WHO
AT THE
EARTH SUMMIT
RIO DE JANEIRO
1992

CO-PUBLISHED BY
VISIONLINK
EDUCATION FOUNDATION
AND
TERRA CHRISTA
COMMUNICATIONS

WHO IS WHO AT THE EARTH SUMMIT

130 Biodome Drive
Waynesville, North Carolina 28786, USA
704 926-2200 FAX: 704 926-9041

Terra Christa Communications
1880 East River Road, Suite 210
Tucson, Arizona 85718, USA
602 577-6476 FAX: 602 577-1195

ISBN 0-9628405-3-X

CONTENTS

INTRODUCTION

VISION STATEMENTS

DEDICATION

DIRECTORY

DIRECTORY / By Country

From the Editor:

A challenge it was. Collecting and sorting of over 30,000 names of participants of the Earth Summit demanded endurance. While most people in "charge" of one or the other databases were open to participate, a few still operated for various reasons under yesterday's syndrome of "state secret", "classified", and other good reasons. Only if we get the Governments, the UN and Institutions to understand that people have a right to information, do we make a step forward towards a unified Earth. This book is meant as a tool for this purpose. Special thanks goes to the dozens of people that have made this directory possible: (please forgive me, if I missed one) Janos Pasztor, Lianne Sorkin Fisher, Sonia Lecca, Joe Sils, Andrey Vavilov, Kerry ten Kate, Monique McClellan, Craig Forte, Hanne Strong, Maurice Strong, Terry Pezzi, Yolanda Kakabadsc, Marjorie Fox, Gary Herbertson, Warwick Manifranto, Sandy McClintock, David Beckett, Brian Smith, Irv Sarnoff, Dean Greenberg, Thomas J. Prendergast, Sally Randall, Don Conroy, Elvie Ruottinen, Shabari Cymerman, Richard Jordan, Charlie Martin, Dick Wilson, Denise Reid, Mark Thomas, Susan Cafferty, Michelle Lappas . . . and last but not least to the home support team of Juri Arusoo and Vanessa Keller.

Hans J. Keller

From Vision . . . . to Reality
Progress in Motion . . . . the vision and dedication of some, in service to the many, in support of us all . . . .
. . . . creating a preferred future of sustainable peace, unifying love and spiritual harmony with well being of all life and our Earth. . . .
. . . . assuring that the world's children and generations to come shall thrive and flourish in the positive environment of hope and all the promise that comes from a world of people existing in the uniting of the spirit of the human family and reflecting the knowledge of our oneness with each other, our earth and God.

Jacqueline Snyder

Design the cover of this important book, we were told. Put it on recycled paper, naturally. Include the logos of the UN and UNCED, the universally recognized symbols of hope and concern for the planet. Feature the pacific blue color of the United Nations – the logical leader in the quest to save our world. And capture the flavor of the host country, where there still exists . . . if only for a while longer . . . the last great jungle wilderness on the face of the earth.

We chose the mighty anaconda as our graphic icon – not only for the potential of its raw power but because in ancient lore the snake could prolong life, restore health to the sick, revive the dead, ensure fertility and counteract poisons . . . in short, a prescription for what ails our planet. To us, it represents the ultimate truth for humanity: that to survive, we must learn to live in harmony with all that surrounds us.

We are proud to be associated with this important undertaking. And priviledged to be a part of a gathering crescendo.

ULTRA DESIGN
Asheville, North Carolina
704.254.7985

Only the United Nations can convene Heads of State or Government to examine complex issues and propose integrated approaches. The Conference on Environment and Development was called to face an immense challenge – how to integrate environment and development at the national, regional and global levels as a prerequisite to movement towards sustainable development; this is the foundation of Agenda 21.

In Rio, as I listened to over 100 Heads of State or Government, I was struck by the universal acceptance of the need for this integrated approach. It was heartening to hear so many leaders of the North putting the emphasis on development and, what is perhaps more significant, the large number of leaders from the South who emphasized the protection of the environment. I consider this a conceptual breakthrough with enormous implications for future world cooperation and for our work in the United Nations system.

After Rio, it is no longer credible to speak of the environment without putting it squarely in the context of economic and social development, nor is it possible to discuss development in isolation from the environment. The Rio Declaration, which places people at the center of our concerns, marks a significant advance: States are now obliged to take global environmental consequences into consideration in their domestic planning decisions. This is a fundamental philosophical undertaking by the world's nations, with potentially far-reaching implications. Agenda 21, as a center-piece of international cooperation, must be followed through and fortified. The establishment of a high-level commission on sustainable development under Article 68 of the Charter – and under the authority of the Economic and Social Council – will be crucial to monitoring and integrating the tasks set by Agenda 21.

There is increasing recognition that deeper international cooperation is both necessary and possible. If we grasp these possibilities with good will and good sense, we will find ourselves on the road to a more humane and successful world.

Dr. Boutros Boutros-Ghali
Secretary-General of the United Nations

Finding solutions to the serious environment and development problems we face as a global community is an enormous challenge, but one we must embrace if we are to secure our common future on this planet. Our success in meeting that challenge, however, will depend upon our ability as nations, as sectors of society and as people, to communicate, to cooperate and to dialogue. Only if we move from confrontation, through dialogue to cooperation can we hope to succeed.
On behalf of the government and people of Brazil, I would like to thank all participants of the "Earth Summit."

Fernando Collor De Mello
President of the
Federative Republic of Brazil

UNCED was an historic event. One hundred eighteen Heads of State and Government attended, making it the largest summit meeting ever, and the first true Earth Summit. Together with other government delegates and representatives from United Nations agencies and intergovernmental organizations, 1,420 non-governmental organizations were accredited to the Conference. I know that this book will be extremely useful to many. It lists participants and will provide a valuable record for all those interested in the Conference and its follow-up. UNCED has been pleased to cooperate in such a worthwhile project.

Maurice F. Strong
Secretary-General of UNCED

By Senator Al Gore

Chairman, U.S. Senate Delegation to the UNCED

As the world turns its attention to the Earth Summit, so too does it turn its attention to the global ecological crisis that has brought the world's leaders together with the world's citizens to work for solutions. We are at a turning point; the relationship between human civilization and the ecological system of the earth has changed dramatically. A new way of thinking is needed. New policies are needed. And, while many of us may be disappointed that the agreements to be signed at the Earth Summit are not stronger, we must recognize that this remarkable meeting represents a beginning. The road to Rio will not end there. The road from Rio demands our attention. The world community has an historic opportunity to move forward in a spirit of global partnership to meet the challenges that will define our common future.

We must take bold and unequivocal action; we must make the rescue of the global environment the central organizing principle for civilization. Whether we realize it or not, we are now engaged in an epic battle to right the balance of our earth, and the tide of this battle will turn only when the majority of people in the world become sufficiently aroused by a shared sense of urgent danger to join in an all out effort.

There is no doubt that with sufficient agreement on our goals, we can achieve the victory we are seeking. Although very difficult changes in established patterns of thought and action will be required, the task of restoring the natural balance of the earth's ecological system is both within our grasp and desirable for other reasons – including our interest in social justice, a democratic government, and free market economics. Ultimately, a commitment to healing the environment represents a renewed dedication to what Jefferson believed were not merely American but universal, inalienable rights: life, liberty, and the pursuit of happiness.

Nowhere is that more clearly defined that in the debate around the Earth Summit where the concerns of the developing nations and the developed nations provide such a contrast. In fact, we are working for the same goals: for a better quality of life; and for a brighter future for our children; for a safe, clean, and sustainable environment.

Along with many others, I had hoped that the Earth Summit would produce meaningful new agreements. Sadly, it will not, in large measure because of opposition to, among other things, specific targets and timetables to control emissions of carbon dioxide, the principal cause of global warming. That disappointment must not discourage progress but rather, must remind us of what still needs to be done.

We must look beyond the Earth Summit to build a new global consensus that will create stronger agreements and real progress. We need to establish a mechanism to carry forward and strengthen the agreements reached at the Earth Summit. For example, I have suggested that the United Nations might consider establishing a Stewardship Council to deal with matters relating to the global environment, just as the Security Council now deals with matters of war and peace. Or, we could establish a tradition of annual environmental summit meetings, similar to the annual economic summits of today, which only rarely find time to consider the environment. Or, an existing U.N. body, such as the U.N. Economic and Social Council could be charged with monitoring the success in implementing Agenda 21 and carrying forward the principles of the Earth Charter. The important point is that we must ensure that there is a mechanism through which nations will remain committed to pursuing sustainable development.

Toward this goal, heads of nations also must commit to improving overseas development practices to ensure that the projects are sustainable. One of the best ways to do this is to increase the transparency of lending practices so that local communities and non-governmental groups have the opportunity to comment on proposals.

Adopting a central organizing principle – agreed to voluntarily – means embarking on an all-out effort to use every policy and program, every law and institution, every treaty and alliance, every tactic and strategy, every plan and course of action – to use, in short, every means to halt the destruction of the environment and nurture our ecological system.

I have proposed a Global Marshall Plan for the environment that combines large-scale, long-term, carefully targeted financial aid to developing nations, massive efforts to design and then transfer to poor nations the new technologies needed for sustained economic progress; a world-wide program to stabilize world population; and binding commitments by industrial nations to accelerate their transition to an environmentally responsible pattern of life.

To work, however, any such effort will also require the wealthy nations to make a transition that in some ways will be more wrenching than that of the Third World, simply because powerful, established patterns will be disrupted. It must emphasize cooperation – in the different regions of the world and globally – while carefully respecting the integrity of individual nation states.

Our strategic goals are obvious; stabilizing world population; developing environmentally appropriate technologies, especially in energy, transportation, agriculture, construction and manufacturing; changing the economic "rules of the road" to a system that accounts for environmental costs; negotiating and approving a new generation of international agreements to make the overall plan a success; and establishing a cooperative plan to educate the world's citizens about the global environment.

Finally, though, what is needed is an ecological perspective that does not treat the earth as something separate from human civilization. We, too, are part of the whole, and looking at the whole ultimately means also looking at ourselves. If we do not see that we are a powerful natural force, like the winds and tides, we cannot see how we threaten to push the earth out of balance.

By Brahma Kumari (Sister) Janki
Spiritual Leader

Since the dawn of history humanity has been fascinated by three puzzles – nature, the soul and God. Countless hours have been spent trying to research, understand, and sometimes master, that which surrounds us (nature and matter), is seemingly most distant from us (God) and that which we are (the soul). The investigation of these three eternal entities has often fallen into the individual domain of different experts, each working within their own boundaries and perspectives and each reaching their own conclusions. When these conclusions have, as so often has been the case, conflicted with each other the result has been to add another twist to the puzzle of life as a whole. With different theories leading to the building of different fractions, and an ever-increasing lack of understanding of the whole, advancement or progress in one area has almost always been at the expense of another. Laws that seemed valid in the realms of matter had no jurisdiction when applied to soul and vice versa. Thus, societies that have achieved materialistic or technological advancement have usually sold their soul to do so, whereas fatalism, or a denial of the reality of matter, has left some communities bereft of the physical wherewithal for a dignified life. But if the puzzle of Life has any solution, and if our existence any meaning or purpose, then its foundation must be based on Truth. If we are to recognize truth when we see it then two of the qualities that may transcend the disguise that it sometimes seems to hide behind are its simplicity and universal applicability. This being so, the path that can lead us through the maze of life should be governed by some simple and universal laws. An understanding of these laws, or the Truth as a whole from which they derive their authority, should therefore enable us to understand these three eternal entities. Further they should allow us to understand the ways in which they interact and also understand each of them when considered in isolation and all of them when taken together.

The Truth of the soul is that human beings are fundamentally spiritual in nature. Inside the material body is the conscient entity of the soul. This eternal being is the energy which gives life to the instrument of the human body. Around the world, nature is reverently referred to as Mother Nature and her very first act of service is to provide the soul with a body. The body and soul together make up a human being. Perceiving ourselves as the immortal spirit, or the soul, our relationship with nature starts with our bodies and every action that we perform affects the physical matter that our bodies are made of. It also has some effect on all other matter around us. This effect or connection can be described by the fundamental law of action (or karma). This law states that it is the conscient entity of the soul that performs actions through the physical medium of the body and that every positive or negative action that it performs will give rise to a correspondingly positive or negative reaction. Thus the state of the matter around us (and the condition of the world as a whole) is a reflection not just of the quality of our actions but also of the state of well-being of the soul. In other words, every human action inevitably has an effect upon matter, and secondly, no one is immune to the effects or consequences of his or her actions.

The law of Nature is a pattern of ceaseless motion and change in which every natural process is irreversible and leads towards a state of decay or disorder, although as a Mother, Nature's tendency is to try to recreate balance and minimize pollution and disorder. However, nearly every action we perform today results in some imbalance and creates further pollution within nature. Nevertheless, scriptures, legends and history abound with references to a former age of happiness and prosperity in which humanity lived in harmony with nature. The fact that the environment is as polluted and abused as it is today is because we have forgotten that we are as intimately connected to nature as is any baby to her mother. Arrogantly treating her as an object for reckless exploitation in the name of our perceived, but often only imaginary needs, as well as desecrating the original beauty of our planet, we have fouled and blocked our own life-giving umbilical cord. As we are confronted with ever greater instances of environmental wounding, the depletion of natural resources, and radio-active and other pollution we, the human family, must accept responsibility for our own wrongful behavior. However, the crisis that this presents need not be interpreted simply as a portent of disaster. Rather it may be seen as a plea, even more desperate, for a fundamental change of course, a change in our way of thinking, a change in our way of life.

The mystery of God need not be shrouded in the veils of uncertainty as we sometimes like to think. For just as we have a natural and intimate relationship with Nature as our Mother so we do with God as our mother and father. To know ourselves as spiritual beings is the first secret that we must master in understanding the truth of our spiritual Mother and Father, in understanding the truth of God. As the parent of all humanity the form of God must be universally recognizable, a form with which all people of all beliefs and traditions can identify. When the soul has been enlightened with the understanding of the self as a being of spiritual light, we can have an easy and direct relationship with God, The Supreme Soul, the Supreme Star of light. Then to see God as the seed of the human family tree is to know and accept ourselves as the creation of the One Creator, the children of the One whose qualities we inherit and share. It has been said that God is truth and truth is God. The truth cannot be denied and so although people may want to deny the existence of God this cannot be proved because God and Truth are inseparable and the Truth is revealed by God. The law of God is love and mercy. When the boat of humanity is about to crash onto the rocks of illusion and falsehood, of greed, arrogance and anger, it is only God, as the merciful and benevolent Teacher, who can remind us of our forgotten identity, of our connection with nature and of our eternal relationship with God.

When we understand the Truth, God and Nature become our friends; they become divine forces which belong to all of us. And when we accept and learn to live by their laws we can start to fulfil an important role in the play of life. Life becomes something sacred and holy; it offers the wonderful opportunity to serve others and help them make life worth living in. We are accustomed to respecting the rule of law which holds us accountable for our actions as individuals within society. But we must also learn to respect the laws of nature. The attention focused from around the world on the need to save our environment is an indication of the fact that these laws are already holding us accountable. However, facing the facts need not just be a reminder of past errors – we can also make it a prescription for the future, the foundation for charting a new course into the next century with confidence and enthusiasm. Let us have faith in our own future and let that faith be founded upon the truth of our understanding of these three eternal entities. For it is from truth that victory emerges.

By Ambassador Tommy Koh, Singapore
Chairman, UNCED Preparatory Committee

The Earth Summit was held just 20 years after the historic United Nations Conference on the Human Environment, which was held in Stockholm in June 1972.

But there is a major difference between the Stockholm Conference and the Earth Summit. At Stockholm, the Conference had a single focus on how to preserve and enhance the human environment. In Rio, the Summit had a double foci: environment and development, and specifically, on how to reconcile mankind's desire for economic progress, on the one hand, with the need to protect our environment, on the other.

Because of its double foci, the Earth Summit was a more complex and difficult conference than the Stockholm Conference. Indeed, some sceptics believe that there is an inherent and irreconcilable conflict between the two objectives. I do not agree with this view. I believe that it is possible to strike a balance between environment and development. Experts call such development, which is environmentally friendly, "sustainable development". The experience of my own country during the last 25 years suggest that sustainable development works, that concern for the environment does not have to impede a country's economic growth.

If the goals of the Earth Summit are ultimately achieved, history will record it as one of the most important conferences of this decade.

There were five principal foci of the Earth Summit:

- An "Earth Charter" containing the principles governing environmental and developmental relations between States and between States and peoples to ensure our common security.
- Two new conventions on climate change and biological diversity.
- A statement of principles on the management and conservation of all types of forests.
- "Agenda 21", containing over 300 programs, including agreements on how the programs will be financed and how the transfer of environmental sound technology from developed to developing countries will be facilitated.
- Agreement on institutional reform.

In recent years, environmental movements in several developed countries in the West have mounted campaigns against the rate at which the rain forests are being felled. The campaigns of the environmental lobbies have had varying degrees of impact on their governments. For example, the Prime Minister of Malaysia has reacted strongly to such campaigns and charges.

In my view, it is better for the developed countries and the developing countries to work together, on the basis of mutual respect, to help the latter pursue the path of sustainable development. Indulging in finger-pointing and mutual recrimination are easy to do, but they are counter-productive. They raise the temperature but produce little light.

Let me suggest a few specific steps:

- The developed countries should try to create a more supportive international economic environment which would reduce the burden of the developing countries as they embrace sustainable development.

- The developed countries should encourage the development of environmentally sound technologies and facilitate the flow of such technologies to Third World countries.
- The developed countries should ensure that there would be adequate financial resources and proper funding mechanisms to implement the various conventions and programs being negotiated.
- The developed countries should refrain from translating their environmental concerns into new trade barriers.
- The developed countries should consider expanding their concept of foreign aid to include subsidizing the costs of environmental improvements by recipient countries.

It is possible for developing countries to reconcile environment with development. To achieve this goal, countries should cooperate with their neighbors and with countries outside the region. I am confident that such cooperation will be broadened and deepened in the coming years.

I would urge dialogue partners to eschew confrontation in favour of cooperation over the environment. Economic progress and care of the environment should unite, not divide us.

By Bella Abzug and Mim Kelber

Founders Of The Women's Environment & Development Organization

We have only one world. Most of us do not plan to colonize the moon or Mars. My dream is that at this historic Earth Summit in Rio, we will agree, without reservations, to work together to make our still beautiful and bountiful planet more accessible to the needs and dreams of all people.

This is a vision shared by women from every region of Mother Earth. In November 1991, fifteen hundred women from 83 countries came together at the World Women's Congress for a Healthy Planet in Miami, Florida, to put together a Women's Action Agenda 21 that would be incorporated into the Earth Summit's Action Agenda.

The women brought with them horror stories, true stories. They told of poisoned rivers, radiation leaks, babies born deformed, of Eastern European children wearing gas masks to school, of toxic dump sites being located in black communities in the United States and toxic wastes being exported to Africa and Asia. They told of crops dying and people going hungry, of huge forests destroyed for cash, of ancient tribes slaughtered because they stood in the way of corporate exploiters, and, in almost every instance, women and children were the chief victims.

They also noted that at the United Nations and in almost every nation of the world, policies and decisions are made by men, and they agreed that this cannot continue. We are the world's population, we are certainly half the world's brains, we are more than half of the working force – even if women don't get paid for much of what they do – and so we are saying, no, we are demanding, let us into the decision-making process.

In our Action Agenda, we said the following:

"As caring women, we speak for the millions of women who experience daily the violence of environmental degradation, poverty and exploitation of their work and bodies. As long as nature and women are abused by a so-called 'free market' ideology and wrong concepts of economic growth, there can be no environmental security."

We said that, "We come together to pledge our commitment to the empowerment of women, the central and powerful force in the search for equity between and among the peoples of the Earth and for a balance between them and the life-support systems that sustain us all."

We pointed out that, "Women are a powerful force for change. In the past two decades, thousands of new women's groups have been organized in every region of the world, ranging from community-based groups to international networks. Everywhere, women are catalysts and initiators of environmental activism. Yet policy-makers continue to ignore the centrality of women's roles and needs as they make decisions about the future of Mother Earth and her children."

We said that we will no longer tolerate the enormous role played by military establishments and industries in making the 20th Century the bloodiest and most violent in all of human history. We declared that it is urgent that resources currently consumed by the military must be redirected to meet the needs of people and our planet.

We declared that a healthy and sustainable environment is contingent upon world peace, respect for human rights, participatory democracy, the self-determination of peoples, respect for indigenous people and their lands, cultures, and traditions, and the protection of all species.

We declared that basic human rights include access to clean air and water, food, shelter, health, education, personal liberty, freedom of choice and freedom of information.

We reaffirmed our right, as women, to bring our values, skills and experiences into policy-making, on an equal basis with men. Not only at the Earth Summit, but on into the 21st Century.

We pledged to undertake our Action Agenda on behalf of ourselves, our families and future generations. And we appealed to all women and men to join in our call for a profound and immediate transformation in human values and activities.

We are pleased that at the recent UNCED PrepCom IV sessions in New York, a Women's Caucus organized by the Women's Environmental & Development Organization was successful in getting many key points of our Women's Action Agenda into the official Action Agenda sent to the Earth Summit. We were able to win recognition of the essential roles women play in environment and development, as well as specific proposals for supporting women's programs and raising their economic, social and political status. There was also recognition of the need for gender balance in international organizations and governments.

Women are making ourselves heard. We are making progress. But the pace is agonizingly slow and the reality frightening. Once again, at the Earth Summit, we were expected to be on the sidelines as this assembly of nations led overwhelmingly by men made "fate of the Earth" decisions that determine the future of men, women and children alike as well as all living species.

Our hope is that world leaders will finally accept the fact that the enormous task of protecting and saving our Earth requires the participation, wisdom, experience and efforts of all humanity. We made a significant contribution in Rio and, one thing we know: women will continue to organize, work, reach out, build our community, national and international networks, and do whatever is necessary to achieve a just and healthy planet.

Note: The Women's Environment and Development Organization (WEDO) is located at 845 Third Avenue, 15th Floor, New York, NY 10022.
Phone: 212-759-7982 Fax: 212-759-8647

By Robert Muller

Former UN Assistant Secretary-General,

Chancellor of the University for Peace in demilitarized Costa Rica

I remember how in the late 1950's, we became very concerned at the UN with the deterioration of the natural resources of this planet. We tried to warn governments, but nobody listened.

I remember how the Government of Sweden came to us at the Economic and Social Council in 1969, complaining that the 1968 UNESCO conference on the biosphere had no effect on governments and on the public, and asked if we could do something. Sweden, one of the most advanced countries on Earth, was already feeling the effects of acid rain. We recommended to use a big drum to alert the public, namely, to hold a world conference. Sweden agreed. Question: a world conference on what?

I remember a meeting was held to find a name. I was for pollution, the only understandable word to the public at the time. Other colleagues were in favor of the more scientific term of biosphere. We compromised on the environment, "what surrounds us humans". The first world conference on the environment thus took place in Stockholm in 1972. Today, environment is a household word all around the globe.

I remember the opposition of the developing countries to the conference. "We want smoke stacks," they said. In a meeting with them and Maurice Strong I suggested that the conference be preceded by regional conferences because conditions were indeed different from continent to continent. The idea was accepted.

I remember recommending in my numerous speeches during the last few years that a second world conference should be held twenty years later to see what progress and regress had taken place. This time the developing countries wanted to be the seat of the conference and Rio de Janeiro was selected.

As a world servant who spent forty years at the UN here are my recommendations for action by the world community at this stage of evolution:

On the environment

1. A new UNCED should be held at the latest ten years from now in 2002;
2. All statistics and data on the environment should be concentrated in a Global Data and Optimum Design Agency to be created within the UN system, equipped with the most powerful and advanced computers, to monitor the situation and issue optimum designs for human life and Earth preservation;
3. A world agency of national parks and nature conservation sites should be established;
4. A long overdue World Consumers Agency should be created;
5. A drastic shift should be effected from world military expenditures to environmental expenditures. The military should be transformed into police, peace and environmental forces;
6. Current economic standards and measurements should be amended to take into account the deterioration and consumption of the Earth capital;

7. During the next eight years, a world conference of scientists should be held in the main fields of environmental concern (atmosphere, seas and oceans, radiation, climate, hydrosphere, immune system, etc.) to tell the world how many years they give us until the planet will become unlivable in their field, if we continue on the present course. There should also be a world conference on garbage; a world conference on advertising, marketing, and obsolescence; and a world conference on simpler and more frugal living.

On development

1. The transformation of the UN Fund for Population Activities into a specialized agency on Population with vastly increased resources;
2. A Marshall Plan for the rapid improvement of the well-being of the countries;
3. A world plan according to which the European Community would care of the development of Africa; North America (US, Canada) would take care of Latin America; and Japan, Australia and New Zealand would take care of the poor areas of Asia;
4. A world immigration policy which would allow people from the world's South to come, work, and repatriate their earnings to their countries, repeating the successful policy followed in Europe for the development of Southern Italy and Spain;
5. Take advantage of the cheaper labour costs in the poor countries only in the field of manufacturing but also in the field of services (e.g., health and geriatric services);
6. A study in depth by the UN of the errors made in transferring development models from the West to poor countries with fundamentally different conditions, and the development of more intelligent development policies for these countries;
7. The holding by ILO of a world conference on small and medium enterprises;
8. The holding of an International Year of the Villages;

On planetics or planetary management

1. Use the fiftieth anniversary of the UN for an unprecedented review where we stand in the world, where we want to go and how to get there;
2. Develop more regional communities along the pattern of the European Community and aim at the creation of, or transformation of the UN into a true world community;
3. Support the Stockholm Initiative to establish a World Group of Eminent Persons to deal with the way this planet should be governed;
4. The UN should establish a Preparatory Committee of all member governments to prepare a celebration of the Year 2000 and our entry into a better, third millennium.

The above recommendations are in addition to the proposals before UNCED II.

With very best wishes to our beautiful planet and to the genial human race.

By Rosalind W. Harris
President, Conference of NGOs

The Conference of NGOS in Consultative Status with the UN Economic and Social Council (CONGO) is committed to facilitating the work of its members in relation to the major issues of concern to the international community. It has, therefore, taken an active role in many aspects of the development and environment debates that have taken place since the time of the Stockholm Conference on environment. Its Development Committees in Geneva and New York have held many discussions, seminars and workshops during the past twenty years which focus on these issues. One of the features of the CONGO process is that it consistently brings together the different aspects of development such as: human rights, women, population, family, youth, and disarmament.

CONGO Members have diverse interests and constituencies. Some focus on a few issues and others have broad programmes which may encompass all of the economic and social aspects of development. Thus they provide a link among the United Nations programs and agencies. The consultative NGOs were in the forefront of urging the connection between the planet's environmental problems and development. In light of their long history of dealing with the problems of poverty and deprivation they supported efforts to relate the issues of the human condition to those of our physical environment.

When the UN General Assembly voted to hold UNCED, CONGO responded with its traditional call to members to form a Planning Committee for UNCED activities. This Committee has monitored all sessions of the UNCED Preparatory Committee and has provided information to CONGO members about the preparations and arrangements for the Rio Conference. More significantly it published *Crosscurrents* during the three substantive UNCED Preparatory Committee sessions and will do so at Rio. This independent newspaper provides commentary on UNCED proceedings, draft documents and activities of various non-governmental groups. In addition the Planning Committee organized forums at the PrepComs which brought together NGO representatives from many countries to further elaborate their common views.

The major UN Conferences of the past decades have been the springboard for NGO activity beyond the actual Conference and UNCED will be no different. Already NGOs are looking at effective programs to maintain the linkages underlined by UCNED material and to review existing programs to see whether they meet the goals of participation in decision-making by all affected sectors of the population; elimination of poverty; environmentally sound development projects.

CONGO members are planning a number of activities, individually and collectively at the Global Forum in Rio. Many also will be fulfilling their consultative status obligations and focusing on the proceedings of UNCED primarily with a view to the future and implementation of decisions which will be made.

The NGO community, especially those represented in CONGO, will continue to encourage members in all countries (together they cover the globe) to adapt UNCED proposals to local conditions and to encourage governments to participate seriously in the work that is required to bring about the desired changes.

During the next few years CONGO will be engaged in activities relating to the Conference on Human Rights (1993), Population (1994) and Women (1995) as well as the Year of Indigenous Peoples and the Year of the Family. All of these are closely linked to the issues before UNCED and NGO involvement will ensure that the ties are made and that the goals of UNCED are carried forward into the other arenas of international debate. A healthy planet is a human right for all: indigenous peoples, women, youth, and most especially children. CONGO members active in the NGO Committee on UNICEF are considering the environment/development issues with UNICEF and in turn are supporting the concept of children's rights and involvement in UNCED.

In undertaking planning for NGO activities related to major conferences CONGO has and will continue to open up participation in its parallel forums to all interested persons and organizations. The networks within and among NGOs are significant factors in bringing issues from the grass-roots to the forefront and then in returning the implementation to the local community. CONGO will continue to find ways to assists its members in this reciprocal process to which they are committed. Ensuring that the concerns identified by the governments of the United Nations become the issues of the people and that more experience and knowledge is contributed to the solution of the problems is the contribution of the consultative NGOs to the UN system.

By Hugh Locke
Director of Only One Earth, Ltd.,
Executive Producer of the 1992 UNEP Global Youth Forum

When the environmental movement first took shape in the late 1960's, it held people's attention for a time before being relegated to the status of a special interest group. When it re-emerged in the late 1980's, concern for the environment struck a new chord as conventional wisdom began to reflect the notion that ecological issues were no longer just local or regional, but global in nature and inseparable from concerns about development. This second wave of the environmental movement – now linked with development – was the engine that powered the Earth Summit. The next wave is currently in the making and is being shaped by youth.

The generation of young people which includes the current youth environment and development movement is distinguished from those preceding it in three important ways. It is the first generation to come to maturity without the debilitating politicization of environmental issues that the Cold War brought; it is the first generation to accept the interconnectedness of one human family occupying a common home as a living reality and not just a conceptual model; and, it is the first generation to have formed environmental opinions based on direct access to the global information system.

It is not so much that young people involved in environment and development are well-organized and able to mount sophisticated lobbying initiatives to influence their governments. As a rule, youth do not work out carefully crafted positions to further their interests, nor do they have the patience to explore the nuance of meaning encoded within the careful wording of official documents. In fact, youthful zeal often puts young people at odds with authorities. But whatever the communication challenges, as a constituency youth constitute almost half the world's population: close to 2 billion of the earth's tenants are 25 years of age or younger. It is not simply that this group will someday inherit the earth – they live here now and are already playing a vital role.

While the Earth Summit represents the highest aspirations of the adult, half of the world, the ideas put forward are not yet seasoned. The natural constituency for the continuance of the Earth Summit spirit is youth, and herein lies their most important contribution. The adult world has in many instances taken the issues as far as they can. In order to translate these truly noble aspirations into action it is necessary to rethink deeply held notions about who we are, how we live together and our relationship to the environment. The courage to make the changes the Earth Summit portends are only possible when you accept in your heart that we live in one world inhabited by one human race whose fate is inextricably linked with the condition of the earth. This is the new reality, the new spirit of our times – and it is youth who are manifesting that spirit as they begin to spearhead a new relationship to our Earth Home.

The Declaration of The Sacred Earth Gathering
Rio de Janiero, 1992

The planet earth is in peril as never before. With arrogance and presumption, humankind has disobeyed the laws of the Creator which are manifest in the divine natural order.

The crisis is global. It transcends all national, religious, cultural, social, political and economic boundaries. The ecological crisis is a symptom of the spiritual crisis of the human being, arising from ignorance. The responsibility of each human being today is to choose between the forces of darkness and the force of light. We must therefore transform our attitudes and values, and adopt a renewed respect for the superior law of Divine Nature.

Nature does not depend on human beings and their technology. It is human beings who depend on Nature for survival. Individuals and government need to evolve "Earth Ethics" with a deeply spiritual orientation or the earth will cleanse itself of all destructive forces.

We believe that the universe is sacred because all is one. We believe in the sanctity and the integrity of all life and life forms. We affirm the principles of peace and non-violence in governing human behavior towards one another and all life.

We view ecological disruption as violent intervention into the web of life. Genetic engineering threatens the very fabric of life. We urge governments, scientists and industry to refrain from rushing blindly into genetic manipulation.

We call upon all political leaders to keep a spiritual perspective when making decisions. All leaders must recognize the consequences of their actions for the coming generations.

We call upon our educators to motivate the people towards harmony with nature and peaceful coexistence with all living beings. Our youth and children must be prepared to assume their responsibilities as citizens of tomorrow's world.

We call upon our brothers and sisters around the world to recognize and curtail the impulses of greed, consumerism and disregard of natural laws. Our survival depends on developing the virtues of simple living and sufficiency, love and compassion with wisdom.

We stress the importance of respecting all spiritual and cultural traditions. We stand for preservation of the habitats and life style of indigenous people and urge restraint from disrupting their communion with nature.

The World Community must act speedily with vision and resolution to preserve the Earth, Nature and humanity from disaster. The time to act is now. Now or never.

DEDICATION

Children are the hope of the future
More correctly, they are the future.
What kind of world will they inherit?
Will the rains bring life sustaining water or an acid bath?
Will the forests be alive with growth, or will they be paved over?
Will birds and fish and wildlife abound,
or will more and more species be extinct?
Will the economy sustain them,
or will poverty and hunger be their plight?
These are the questions that our children face as they grow up.
They are hard questions, urgent questions, life and death questions
that should not be faced by youngsters.
That's why the Earth Summit is so important.
It is a historical beginning to an ongoing process
which will ultimately affect the quality of life on our planet.
It is a co-creative effort to define the problems that face us,
and an agreement to address them together.
This Directory, "Who Is Who At The Earth Summit",
is published by: VisionLink Education Foundation
and Terra Christa Communications, Inc.
To assist all those who have agreed to carry out this mission.
Those in whose hands the future of our planet rests.
It is meant to be a communications tool of the first order.
We dedicate this directory to the children of the world
who will inherit our planet earth –
in the hope that it will be a beacon of light,
containing the love, energy and power of all those listed in it.

Terry Pezzi
Terra Christa Communications
1880 East River Road, Suite 210
Tucson, Arizona 85718 USA
602 577-6476
Fax: 602 577-1195

Hans J. Keller
VisionLink Education Foundation
130 Biodome Drive
Waynesville, NC 28786 USA
704 926-2200
Fax: 704 926-9041

DIRECTORY

GRAND TOTAL ALPHABET

B

C

E

F

Ferreira Barbosa, Alda, Ms., Brazil 150
Ferreira Carneiro, Jaemio, Mr., Brazil 150
Ferreira Da Nova, Abilio, M., Holy See 247
Ferreira Fleitas, Felix Vidal, Paraguay 309
Ferreira Gomes, Ana Lucia, Brazil 112
Ferrer, Maximiliano, Panama 308
Ferrera, Juan, Sr., Honduras 248
Ferreras, Julian Alberto, Mr., Argentina 83
Ferrerira, Valdevino, Brazil 112
Ferrero, Alejandra, Sra., Argentina 83
Ferretti, Janine, Ms., Canada 202
Ferrey, Aura, Switzerland 337
Ferrey, Jo Espinoza, Mr., United Kingdom 351
Ferrez-Weinberg, Marc, Sweden 327
Ferrier, Maria Rosa, Ms., Argentina 83
Ferriera, Marcela, Ms., United States of America 377
Ferro, Nubia De S., Brazil 112
Ferstman, Miriam, Ms., Canada 199
Feruzi, J., Mr., United Republic of Tanzania 354
Festa, Alicia Beatriz, Sra., Argentina 83
Festa, Regina, Brazil 150
Fetnat, Tolba, Egypt 223
Feuer, Andras, Hungary 249
Fevre, Marcel, Mr., French Guyana 236
Fewell, Dorothy, Ms., South Africa 323
Fewell, P.T., Mr., South Africa 323
Feyder, Jean, S.E. M., Luxembourg 286
Fialho, Alaor, Brazil 150
Fialho, Ana Maria, Brazil 150
Fialho, Aurea Vieira, Sra., Brazil 150
Fialho, Maria Elisabeth Arruda, Sra., Brazil 150
Fiander, V. Winston, Mr., Barbados 94
Fibueiredo, Mariza, Portugal 315
Fidalgo, Ana Claudia, Brazil 112
Fidler, Amanda, United Kingdom 347
Fidyka, Leopoldo Javier, Sr., Argentina 83
Fiedler, Eugene G., Mr., United States of America 377
Fiege, Klaus Dieter, Dr., Germany 238
Fielden, Katherine, Australia 89
Fields, Bonnie, Ms., United States of America 377
Fifita, Netatua, Ms., Tonga 342
Figari, Andrea, Ms., Netherlands 300
Figols, Francisca B., Ms., Brazil 150
Figueira, Carlos A. P., Brazil 150
Figueira, Mariza T., Brazil 112
Figueira, Sallel, Switzerland 337
Figueiras, Silvia, Switzerland 337
Figueiredo, Andrea Coelho De, Sra., Brazil 150
Figueiredo, Armenio De, Mr., Portugal 315
Figueiredo, Astrid, Switzerland 337
Figueiredo, Francisco De Assis Gomes De, Sr., Brazil 150
Figueiredo, Licinio Coelho De O., Brazil 196
Figueiredo, Maria Jose S., Brazil 112
Figueiredo, Maria Manuel Reis De, Ms., Portugal 315
Figueiredo, Maria Patricia De L., Brazil 150
Figueiredo, Mariza, Portugal 314
Figueiredo, Newton Carlos, Brazil 112
Figueiredo, Patricia Da Silva, Brazil 112
Figueiredo, Pedro Edson, Brazil 150
Figueiredo, Safira Maria, Brazil 112
Figueiredo, Sergio, Portugal 314
Figueiroa, Americo Magno De Souza, Brazil 150
Figueredo, Anne Elizabeth Grangues, Brazil 150
Figueredo, Luis F.M., Brazil 112
Figueredo, Mauricio De Pavia, Brazil 150
Figueredo, Paulo, Brazil 150
Figueredo, Paulo Fernando De, Brazil 150
Figueroa, Arturo, Mr., Argentina 83
Figueroa, Francisco R., Spain 324
Figueroa, Godofredo, Switzerland 337
Figueroa, Nelson M., Mr., Argentina 83
Figueroa, Ricardo, Mr., Argentina 83
Figuerroa, Juan Augustin, S.E., Chile 207
Fikri, Mohamed, M., Morocco 296
Fila, Raquel, Ms., Brazil 150
Filali, AbDellatif, S.E. M., Morocco 296
Filemyr, Ann, Ms., United States of America 377
Filho, Acyr Ferreira De Souza, Brazil 150
Filho, Amauri Severino Dos Santos, Brazil 150
Filho, Antonio Araujo Fanzeres, Brazil 150
Filho, Antonio Castelo Branco Da C., Brazil 150
Filho, Antonio Gaspar, Brazil 150
Filho, Arlindo, Brazil 150
Filho, Audilio A. C., Brazil 150
Filho, Caetano Scanavino, Brazil 150
Filho, Carlos Vieira Peixoto, Brazil 150
Filho, Durval Santana, Brazil 150
Filho, Eduardo Bruno, Brazil 150
Filho, Eliakim Araujo Parsira, Brazil 112
Filho, Ewaldo Schleder, Brazil 150
Filho, Ezio Sertorio, Brazil 150
Filho, Farid Germano, Brazil 150
Filho, Floriano Dias Viera, Sr., Brazil 150
Filho, Francisco Alves Das Chagas, Brazil 150
Filho, Gerson T. Caffe, Brazil 150
Filho, Guilherme Silva, Brazil 150
Filho, Idalicio Manoel, Brazil 150
Filho, Ivo Carvalho, Brazil 150
Filho, Joao Costa, Brazil 150
Filho, Joao Henrique Da Silva, Mr., Brazil 150
Filho, Joao Zacharias, Brazil 150
Filho, Jose Ataliba Alvares, Brazil 112
Filho, Jose Cabral, Brazil 150
Filho, Jose Domingues De Godoy, Sr., Brazil 151
Filho, Jose Maria Paschoalick, Brazil 151
Filho, Lauro Monteiro, Sr., Brazil 151
Filho, Lourival Santana, Brazil 151
Filho, Lourivaldo Oliveira, Brazil 151
Filho, Mario Elena, Brazil 151
Filho, Mario Honorio Teixeira, Brazil 151
Filho, Masao Goto, Brazil 151
Filho, Murilo Melo, Brazil 151
Filho, Nauile De Barros, Mr., Brazil 129
Filho, Nelson Pompeo, Brazil 151
Filho, Oscar Maron, Brazil 151
Filho, Ovidio De Abreu, Brazil 151
Filho, Paulo Santos, Italy 261
Filho, Paulo Studart, Mr., Brazil 151
Filho, Rubem Didini, Brazil 151
Filho, Rui Machado Lima, Brazil 151
Filho, Sebastiao Leite, Brazil 151
Filho, Sergio Adeodato, Brazil 151
Filho, Sergio Cabral, Deputy, Brazil 196
Filho, Tales Henrique Da C., Brazil 151
Filho, W.D.S. Leal, Dr., Germany 242
Filho Barros, Luiz, Sr., Brazil 151
Filion, J. Paul, Uruguay 396
Filipe, Eduardo, Brazil 151
Filipon, Jane M., Brazil 112
Filippini, Eduardo, Brazil 112
Fillo, Maria De Janea, France 232
Fillus, Acyr Lacerda, Brazil 151
Filmon, Gary, The Hon., Canada 199
Filoni, Fernando, Mgr., Holy See 247
Finamore, Luiz, Brazil 151
Finau, Bishop Patelisio, Tonga 342
Findley, Rober W., Dr., United States of America 377
Fingerle, Jorg, Mr., Brazil 151
Finguerut, Silvia, Brazil 112
Fink, Christopher, Mr., United States of America 377
Finley, Bruce, United States of America 361
Finnbogadottir, Vigdis, H.E. Mrs., Iceland 249
Finney, C., United Kingdom 351
Finnseth, Asle, Norway 305
Fioravanti, Mario, Mr., Brazil 151
Fiorencio, Livia Maria Ferrari, Brazil 112
Fiori, Heloisa Helena Pinheiro, Sra., Brazil 151
Fiorot, Antonio Wilson, Sr., Brazil 151
Firebrace, James, Netherlands 299
Firmino, Hiram, Brazil 112
Firszt, Natalio David, Mr., Argentina 83
Fischer, Alfred-Herwig, Dr. Ing., Germany 238
Fischer, Dietrich, United States of America 356
Fischer, Eric, Mr., Germany 243
Fischer, Gert Roland, Mr., Brazil 151
Fischer, Gudrun, Ms., Germany 243
Fischer, Per, Mr., Denmark 219
Fisher, Frances, Ms., United States of America 377
Fisher, Henrik, Denmark 221
Fisher, Lawrence, United States of America 361
Fisher, Lianne, United States of America 361
Fisher, Mary Pat, Ms., United States of America 377
Fisher, Mike, Mr., United States of America 377
Fisher, Padraic, Mr., United States of America 377
Fisk, Ronald B., Mr., Canada 203
Fisker, Henrik, Denmark 220
Fitch, Brian, Mr., United Kingdom 351
Fitipaldi, Mario, Sr., Argentina 83
Fitter, Perin, Ms., Kenya 284
Fitzgerald, Anne, Ms., United States of America 368
Fitzgerald, John, Mr., United States of America 356
Fitzgibbon, Theresa, Ms., United States of America 377
Fiume, Giorgio Dal, Mr., Italy 261
Fiuza, Guilherme, Brazil 112
Fjallgren, Maille Kristina, Ms., Brazil 151
Fjortoft, Arne, Mr., Sri Lanka 325
Fjortoft, Ragnhild S., Ms., Norway 306
Fjukstad, Sissel Hartmark, Ms., Norway 306
Flache, Stanislas, Dr., Switzerland 333
Flad, Ethan, Mr., United States of America 356
Flaes, R., H.E. Mr., Netherlands 298
Flanders, Stephanie, United States of America 377
Flatow, Ira, United States of America 361
Flavin, Christopher, Mr., United States of America 356
Fleckenstein, Kurt, Dr., Germany 238
Fleming, Tracy Ann, United States of America 377
Flemming, Charles Stephen, H.E. Dr., Saint Lucia 320
Flenley, Clare, Germany 240
Fletcher, Susan, Ms., United States of America 377
Fleury, Maria Lucia Maranhao, Brazil 151
Flor, Gilberto Da Silva, Brazil 151
Florentino, Edilene, Brazil 151
Flores, Celestino, Venezuela 399
Flores, Clara, United States of America 377
Flores, Daniel, Mexico 291
Flores, Gerardo, Mexico 291
Flores, Horacio Calzon, Mr., Argentina 83
Flores, Isabel Cristina B., Brazil 112
Flores, Luis, Sr., Mexico 294
Flores, Maria Ines Huber, Sr., Brazil 151
Flores, Nicolas Nolasco, Sr., Argentina 83
Flores, Pdero Silveira, Brazil 112
Flores, Rene, Mr., Chile 209
Flores, Silvia Maria Pereira, Sra., Brazil 151
Flores, Teresa, Bolivia 98
Florez, Luis Bernardo, Colombia 213
Florian, Dominique Van Egmond, Ms., France 232
Floriano, Mira Ferreira, Brazil 151
Floridi, Jose Aparecido, Mr., Brazil 151
Florin, Raimundo, Sr., Argentina 83
Floter, Udo, Mr., Germany 238
Fluch, Franz, Austria 91
Fluhler, Johannes, Switzerland 333
Foca, Anna, Italy 257
Foer, Paul, United States of America 377
Foguelman, Dina, Ms., Argentina 83
Foley, Kevin, United States of America 361
Folgarait, Alejandra, Argentina 83
Folkes, Anthony, United States of America 394
Folkesson, Helena, Mr., Sweden 328
Fomenti, Ligia H. G., Brazil 151
Fonda, Jane, Switzerland 337
Fondi, Patrizio, Mr., Italy 256
Fonesca, Leticia Silveira, Brazil 151
Fong, Gracie, Ms., Fiji 225
Fonseca, Adolfo Pinto, Brazil 151
Fonseca, Alegria, Sra., Colombia 213
Fonseca, Alexandre Alberto Henklain, Sr., Brazil 151
Fonseca, Ana Maria Paiva Da, Sra., Brazil 151
Fonseca, Carlos, Mr., Colombia 214
Fonseca, Celso, Brazil 112
Fonseca, Claudio Ruy Vasconcelos Da, Sr., Brazil 151
Fonseca, Danielle, Brazil 151
Fonseca, Edmilson, Mr., Brazil 151
Fonseca, Edson Leite Da, Brazil 151
Fonseca, Francisco Carlos Guedes Da, Mr., Brazil 151
Fonseca, Francisco Franco De Assis, Sr., Brazil 151
Fonseca, Geraldo Alberto, Sr., Brazil 151
Fonseca, Gustavo, Mr., Brazil 151
Fonseca, Iva Marques Da, Ms., Brazil 151
Fonseca, Ivan Claret M., Mr., Brazil 151
Fonseca, Joao Carlos Pinheiro Da, Sr., Brazil 151
Fonseca, Joao Mello, Brazil 112
Fonseca, Joberto, Brazil 112
Fonseca, Jorge Carlos Almeida, H.E. Dr., Cape Verde 206
Fonseca, Marcelo, Sr., Brazil 151
Fonseca, Marcelo De Oliveira, Brazil 151
Fonseca, Marcia Prado Prioli, Brazil 112
Fonseca, Marco Antonio, Brazil 112
Fonseca, Maria Do Carmo Monteiro Da, Sra., Brazil 151
Fonseca, Maria Imaculada Pinto, Brazil 151
Fonseca, Maria Leticia Silveira, Brazil 151
Fonseca, Mario Imaculada P., Brazil 112
Fonseca, Randal, Sr., Brazil 151
Fonseca, Reginalda C., Brazil 112
Fonseca, Ricardo Mesquita Da, Mr., Brazil 151
Fonseca, Roberto Lopes Da, Sr., Brazil 151
Fonseca, Silvia, Brazil 112
Fonseca Barrera, Alegria, Colombia 213
Fonseca Brasil, Carlos Avelino, Sr., Brazil 151
Fonseca Gramling, Cristina, Ms., Brazil 151
Fonsecasta, Marcia Prado Prioli, Brazil 151
Font, Eduardo Hugo, Sr., Guatemala 246
Fontan, Alain, Brazil 151
Fontan, Jacques, Mr., France 233
Fontanella, Matia, Mr., Italy 261
Fontanezi, Paulo Jose, Sr., Brazil 151
Fontanilla, Elba Elena Vedoya De, Ms., Argentina 83
Fontenele, Maria Luiza, Sra., Brazil 151
Fonterel, Nathalie, France 233
Fontes, Andrea Cristina Pessoa, Brazil 151
Fontes, Angela, Sra., Brazil 151
Fontes, Angela Maria Belmiro, Ms., United States of America 377
Fontes, Claudia De Fatima, Brazil 112
Fontes, Dora, Uruguay 396
Fontes, Luis Fernando De Meira, Sr., Brazil 151
Fontes, Maria Aparecida, Brazil 112
Fontes, Roldao De Siqueira, Sr., Brazil 151
Fontoura, Lauro Sobrinho, Brazil 112
Fontrel, Nathalie, France 229
Fonzes, Ana, Ms., Brazil 151
Foo, Kim Boon, Singapore 322
Footman, Bill, United Kingdom 351
Foraste, Xavier G., Spain 324
Forbes, Gerald Innes, Mr., United Kingdom 346
Forbes, G.I., Dr., United Kingdom 350
Ford, Carolyn, Ms., United States of America 377
Ford, Mary, Germany 240
Ford, Mary Anne, Ms., United States of America 377
Ford, Simon, United Kingdom 347
Forde, Bjorn, Mr., Denmark 220
Forden, Clark, Mr., United States of America 377
Fordyce, Urma, Switzerland 337
Forero, Enrique, United States of America 356
Forero Monsalve, Luz Marina, Colombia 213
Foresta-Martin, Francesco, Italy 258
Forestier, Hubert Le, France 233
Forkan, Patricia, United States of America 377
Formagini, Elizabeth, Brazil 112
Formenti, Ligia H.G., Brazil 112
Formigli, Ana Lucia Menezes, Sra., Brazil 152
Formoso, Denise Maria Lopes, Sra., Brazil 152
Fornasari Filho, Nilton, Mr., Brazil 152
Forneck, Anselmo, Brazil 152
Forno, Eduardo, Lic., Bolivia 98
Fornos, Werner, Mr., United States of America 356
Forrester, Rosemary, Ms., Brazil 152
Forsberg, Meri Helena, Ms., Sweden 328
Forseth, Elisabeth, Ms., Norway 306
Forster, Maria, Ms., Brazil 152
Forster, Thomas, Mr., United States of America 356
Forsythe, Kathleen, Ms., Canada 203
Fortes, Daniela, Ms., Brazil 152
Fortes, Joao Augusto, Brazil 101
Fortes, Luis, Mr., Brazil 152
Fortes, Marcio, Mr., Switzerland 334
Fortez, Cecilia, Mr., Chile 209
Forteza Martin, Francisco, Cuba 217
Forti, Andrea, Brazil 152
Fortner, Rosanne, Ms., United States of America 377
Forwood, William, Venezuela 399
Foste, Wolfgang, Germany 241
Foster, Peter, Canada 200
Foster, Rebecca, India 250
Foster, Richard, United States of America 361
Foster, Tonia Shoumatoff, Ms., United States of America 377
Foster, Vanira Da Silva, Brazil 152
Foti, Paola Filomena, Italy 258
Fouad, Bitar, Mr., Switzerland 335
Fouin, Joulien, France 233
Fouin, Julien, France 229
Fouque, Antoinette, Ms., France 233
Fouracre, Anthony, Mr., United States of America 377
Fournier, Francine, France 236
Fowlor, Richard, United States of America 377
Fox, Margie, United States of America 356
Fox, Noel, Denmark 221
Fox, Stefan Karl, Germany 241
Fox, Veronica, Germany 241
Foy, Stephen Mark, Australia 90
Frado, Dennis, Mr., Switzerland 333
Fraga, Luiz Felipe Prado, Brazil 112
Fraga, Patricia, Sra., Brazil 152
Fraga Jr., Plinio, Brazil 112
Fragoso, Cristiane Conceicao Maia, Sra., Brazil 152
Fragoso De Abreu, Marcelo Jorge, Brazil 112
Fragoso Pereira, Julio Cesar, Portugal 314
Fraisse, Alberto William, Mr., Brazil 152
Fram, Steven, Switzerland 337
Frampton, Robin, United Kingdom 347
Franca, Antonio Carlos, Mr., Brazil 152
Franca, Jamari, Brazil 112
Franca, Manoel Da Costa, Brazil 152
Franca, Marcelo Piettsch, Brazil 152
Franca, Marines Fransisca De, Ms., Brazil 152
Franca, Martha S.J., Brazil 112
Franca, Mauricio Cintrao, Brazil 112
Franca, Valdo, Sr., Brazil 152
Franca, Wilson, Brazil 152
Franca De Faria Melco, Marcelo, Brazil 100
France, Nicky, United Kingdom 351
Francescato, Grazia, Italy 258
Franceschi, Roberto, Italy 256
Francia, Aldo Gabriel, Mr., Argentina 83
Francis, Diane, Canada 200
Francis, Douglas T., United States of America 356
Francis, Pascale Morand, Mme, Switzerland 331
Francis, Sara K., Ms., Canada 203
Francis, Solveig, Ms., United Kingdom 350
Francis, Tyler, Mr., United States of America 377
Francischetti, Cesar Nascimento, Mr., Brazil 152
Francisco Jose, Eszter, Mr., Argentina 83
Franck, Claude, M., Luxembourg 286
Franck, Mansour Adeoti, Cote d'Ivoire 217
Francke, Ulrika, Ms., Sweden 329
Franco, Carlos R., Brazil 112
Franco, Eleazar, Mexico 291
Franco, Fernando, Mexico 291
Franco, Lucrecia, United States of America 361
Franco, Mandel G., Brazil 152
Franco, Maria Da Penha Silva, Sra., Brazil 152
Franco, Maria Luiza Leite, United States of America 361
Franco, Maria Terza Machado, Sra., Brazil 152
Franco, Marilia Carmem, Ms., Brazil 152
Franco, Mario Ferreira, Brazil 152
Franco, Norberto Mario, Mr., Argentina 83
Franco, Paulo Cesar Campos De Seixas, Sr., Brazil 152
Franco, Paulo Sergio, Brazil 152
Franco, Pedro, Dominican Republic 221
Franco Sanchez, Jose Luciano, Mexico 291
Francoeur, Louis-Gilles, Canada 200
Franeck, Maik, Mr., Germany 243
Frangos, Vassilios, Mr., Greece 245
Frank, Phillip J., United States of America 361
Franke, Tamara Maria, Ms., Brazil 152
Franklin, Bernard B., Dr., United States of America 377
Franqueira, Carlos, Sr., Brazil 152
Franza, Jorge Atilio, Mr., Argentina 83
Franzi, Etienne, M., Monaco 296
Franziska, Erika, Brazil 112
Frare, Jose Luiz, Brazil 112
Fraschini, Angela, Ms., Italy 261
Fraser, Alan, United Kingdom 351
Fraser, C. Gerald, United States of America 361
Fraser, Scott, United States of America 377
Fraser, Whit, Mr., Canada 203
Frassinete, Paula, Brazil 101
Frazao, Mariza Ribeiro, Brazil 152
Frazao, Silvio, Brazil 112
Frazao, Wilson, Brazil 112
Frazer, Garth, Switzerland 332
Frazier, Jeff L., Mr., Argentina 83

G

H

I

J

K

L

Limon, Armida, Sra., Mexico 294
Limon, Dante, Mexico 292
Limon Dominguez, Geraldo Arturo, Sr., Mexico 294
Limonet, Renato, Mr., Italy 262
Lim-Perez, Adelaida, Ms., Philippines 312
Limstrand, Siv, Mr., Switzerland 335
Lin, Ada Wen-Shung Ma, Ms., Taiwan 341
Lin, Edgar J. Y., Mr., Taiwan 341
Lin, Ho, Mr., Taiwan 341
Linares, Felicitas, Ms., Peru 311
Linares, Yamelis, Ms., Venezuela 399
Linares Meneses, Felicitas Estelas, Peru 310
Linck, Ricardo, Brazil 116
Linda, Andy, United States of America 382
Lindahl, Lena, Ms., United States of America 382
Lindahl-Kiessling, Kerstin, Ms., Sweden 329
Lindebotten, John, Norway 306
Linden, Eugene, United States of America 364
Lindenberg, Angela M., Brazil 116
Lindenkamp, Frans, Netherlands 300
Lindevall, Bo, Mr., Sweden 328
Lindgren, Kerstin, Ms., Sweden 329
Lindhe, Jenny, Ms., Sweden 329
Lindne, Jenny, Sweden 329
Lindner, Warren, Mr., Switzerland 338
Lindo Poisson, Rosana Estela, Ms., Argentina 84
Lindor, Deweare Ivann, Haiti 247
Lindovino, Pedro Gerson, Brazil 161
Lindquist, Jan, France 230
Lindquist, Russell, Dr., United States of America 382
Lindsoe, Anne Brit, Ms., Norway 306
Lindstrom, Per, Sweden 328
Lindstrom, U.B., Prof., Finland 226
Lindzvist, Hans, Mr., Sweden 329
Ling, Chee Yoke, Ms., Malaysia 288
Lingen, Carl David Van Der, Mr., South Africa 323
Linggard, Trine, Norway 306
Lingl, Paul, Mr., Canada 204
Lingner, Klaus, Mr., Switzerland 334
Lingscheid, Raines, Mr., Germany 240
Lingsma, Tjitske, Italy 262
Linhares, Ana Felicia De C., Brazil 116
Linhares, Andre Jose Moreira, Brazil 161
Linhart, Ana Maria, Brazil 161
Lini, Fr. Walter, Mr., Vanuatu 398
Lini, Mary, Ms., Vanuatu 398
Lini, Walter H., Mr., United States of America 382
Linn Locher, Monika, Mme, Switzerland 331
Lino, Clayton Ferreira, Brazil 161
Lino, Geraldo Luis S., United States of America 364
Lins, Cristiane, Ms., Brazil 161
Lins, Dalton Estellita, Sr., Brazil 161
Lins, Ivan, Mr., Brazil 161
Lins, Marcelo, Brazil 116
Lins, Selma Rita Severo, Brazil 116
Linz, Helen, Ms., United States of America 382
Linz, Werner, Mr., United States of America 382
Lion, Marco, Sr., Italy 262
Lionel, David, Mr., United States of America 364
Liones, Rocha, Brazil 116
Liphuko, S. D., Mr., Botswana 99
Lipietz, Alain, Mr., France 234
Lipman, Geoffrey H., Mr., United Kingdom 350
Lippert, Merrilee, Ms., United States of America 382
Lippi, Flavia L., Brazil 116
Lippold, Klaus W., Dr., Germany 239
Lippolis, Ignazio, Italy 258
Lira, Ana Maria De Farias, Sra., Brazil 161
Lira, Luiz, Sr., Brazil 161
Lirio, Jose Geraldo, Sr., Brazil 161
Lisboa, Armando, Brazil 161
Lisboa, Edgar, Brazil 161
Lisboa, Marijane, Ms., Brazil 162
Lisboa, Salete, Brazil 116
Lisboa Pacheco, Maria Emilia, Brazil 101
Lisbos, Edgar, Brazil 116
Liskauskas, Suzana F., Brazil 116
Lispi, Ugo, Mr., Italy 262
Li-Teh, Lu, Taiwan 341
Litenski, Chaim, United States of America 382
Lithgow, Adrian, United Kingdom 348
Little, Janice, Ms., Canada 202
Little, Kathryn, Ms., United States of America 382
Little, Paul Elliott, United States of America 364
Littrell, Michael, Mr., Canada 204
Liu, Dizhong, China 212
Liu, Huaqiu, H.E. Mr., China 211
Liu, Jiang, H.E. Mr., China 211
Liu, Jimmy, Mr., Taiwan 341
Liu, Shan, Mr., China 211
Liu, Shuqin, Mr., China 212
Liu, Wen-Chao, Mr., Taiwan 341
Liu, Zhenwu, Mr., China 212
Livengood, David L., Mr., United States of America 382
Lizarraga, Ricardo J., Spain 324
Lizarraga-Reyes, Jose, Kenya 284
Ljungberg, Lars Ove, Mr., Sweden 329
Llanos, Gabriela, Ms., Peru 311
Llanos, Roberto Espinoza, Sr., Peru 311
Llaurado, B. Esther, Spain 324
Lloret, Maria Teresa, Ms., Paraguay 309
Llosa, Gonzalo, Sr., Peru 311
Lloveras, Ramon, Mr., United States of America 382
Lloyd, Judy Amber, Ms., United States of America 382
Lloyd, Robin, United States of America 364
Lo Curto, Aldo, Mr., Italy 262
Loaney, Jenifer, Ms., United States of America 382
Loayza, Juana Bertha Roja, Sra., Brazil 162
Loback, Harald, Sr., Brazil 162
Lobato, Randolpho Marques, Sr., Brazil 116
Lobe, James R., United States of America 364
Lobo, Ana, Brazil 116
Lobo, Beatria Helena B.S., Brazil 116
Lobo, Carlos Freire Ferreira, Brazil 162
Lobo, Claudia Luiza, Brazil 116
Lobo, Cristiana Mendes, Brazil 116
Lobo, Elza Ferreira, Sra., Brazil 162
Lobo, Humberto Haddock, Brazil 162
Lobo, Ines Valladao Campos R., Brazil 116
Lobo, Teresa Cristina, Brazil 162
Lobsack, Lili, Ms., Germany 243
Locke, Brian, United Kingdom 346
Locke, Kevin, Mr., United States of America 382
Locke, Patricia, Ms., United States of America 382
Locke, Waniya, Ms., United States of America 382
Lockhart, C, Switzerland 338
Locklin, Ken, Mr., United States of America 382
Lockne, Gunilla, Sweden 328
Lodi, Leopoldo, Italy 262
Lodico, Yvonne, Ms., United States of America 382
Lody, Raul, Brazil 162
Lodygensky, Andre, Mr., Brazil 162
Loebens, Francisco Gunther, Brazil 162
Loebens, Guenter Francisco, Sr., Brazil 162
Loebsack, Willi, Mr., Germany 239
Loecker, Petra, Mrs., Germany 239
Loewenstein, Louis K., Mr., United States of America 382
Lofgren, Lars, Mr., Sweden 329
Lofonga, Ilanga, Mr., Zaire 401
Loftis, Randy Lee, United States of America 364
Logan, Chris, Brazil 162
Logan, Ronald, Mr., United States of America 382
Logreira, Margaret, Ms., United States of America 382
Lojacomo, Giuseppe, Mr., Italy 262
Lokorio, John, Kenya 283
Lokos, Zoltan, Hungary 249
Lomardi, Gabriel, Mr., Italy 262
Lomas Maldonado, Emilio E., Mexico 292
Lomax, David, Brazil 162
Lombardo, Alina, Italy 258
Lombera, Rocio, Ms., Mexico 294
Londoale, Louya, Mr., Zaire 401
Lonergan, Dennis, United States of America 382
Long, Beverly A., United States of America 357
Long, Donna, Ms., United States of America 382
Long, Suzanne Marie, Sra., Brazil 162
Long, William, United States of America 364
Longa, Regina Do Carmo Avance, Sra., Brazil 162
Longange, Bomina N'Somi, Mr., Zaire 401
Longchamp, Fritz, S.E. M., Haiti 247
Longet, Rene, M., Switzerland 331
Longley, Robert, Mr., United States of America 382
Longo, Justice Giovanni, Mr., Italy 260
Longo, Orlando Celso, Sr., Brazil 162
Longo, Paolo, Italy 258
Longobardi, Maurizio, Brazil 162
Longoria, Martin, Mr., Mexico 294
Longstreth, Molly, Ms., United States of America 382
Longuepee, F., France 234
Lonn, Jan, Mr., Switzerland 334
Lonsa, Kanda, Mr., Zaire 401
Lontra, Paulo Klein, Sr., Brazil 162
Lool, Annie, Ms., Brazil 162
Lool Cheok Hun, , Mr. H.E., Malaysia 287
Looremeta, S., Mr., Kenya 283
Looye, Johanna W., Ms., United States of America 382
Loperena, Demetrio, Mr., Spain 325
Lopes, Alberto, M., Guinea-Bissau 247
Lopes, Alvaro, M., Belgium 95
Lopes, Ana Gabriela Duarte, Brazil 162
Lopes, Ana Maria, Sra., Portugal 315
Lopes, Andre Luis Holanda, Sr., Brazil 162
Lopes, Antonio Batista, Brazil 162
Lopes, Carlos A., Brazil 116
Lopes, Celso Dos Prazeres, Brazil 162
Lopes, Denise Costa, Brazil 162
Lopes, Emilio Mira Y, Mr., Brazil 162
Lopes, Eulinda De Campos, Ms., Brazil 162
Lopes, Geisa Morgado Rego, Ms., Brazil 162
Lopes, Gilberto, United States of America 394
Lopes, Henrique, Mr., Mozambique 298
Lopes, Herminio Fernando Ribeiro, Brazil 162
Lopes, Hugo Jose Guimaraes, Sr., Brazil 162
Lopes, Jaqueline Ricardo, Sra., Brazil 162
Lopes, Joao Humberto Goncalves, Brazil 162
Lopes, Jose Milano, Brazil 116
Lopes, Julio Aurelio Vianna, Sr., Brazil 162
Lopes, Lucia, Ms., Canada 204
Lopes, Marcia R. Antunes, Brazil 116
Lopes, Margarete Vizeu Thomas, Ms., Brazil 162
Lopes, Margarida Maria Do Amaral, Sra., Brazil 162
Lopes, Monica Cristina Almeida, Sra., Brazil 162
Lopes, Ney, Deputy, Brazil 196
Lopes, Patricia, United States of America 394
Lopes, Regina C., Brazil 116
Lopes, Renato Luiz Leme, Sr., Brazil 162
Lopes, Ricardo M., Brazil 162
Lopes, Rita Silva, Ms., Portugal 315
Lopes, Sandra Maria, Ms., Germany 243
Lopes, Sergio Dos Prazeres, Brazil 162
Lopes, Sergio Gabriel, Brazil 116
Lopes, Stella Prata Da Silva, Ms., Brazil 162
Lopes, Valeria R., United States of America 364
Lopes, Vania Goncalves, Brazil 162
Lopes, Victor, Sr., Argentina 84
Lopes, Wania, Brazil 116
Lopes Filho, Oswaldo J., Brazil 116
Lopes Filho, Pedro, Sr., Brazil 162
Lopes Mejia, Ana Isabel, Sra., Colombia 214
Lopes Netto, Andre, Mr., Brazil 162
Lopez, Alfonso, E.S., Colombia 213
Lopez, Argelia Laya, Sra., Venezuela 399
Lopez, Carlos Enrique Diaz, Mr., Argentina 84
Lopez, Domingo, Mexico 292
Lopez, Emilio Mira Y, Mr., Brazil 162
Lopez, Fermina, Ms., Switzerland 335
Lopez, Francisco Sanchez, Ms., Mexico 294
Lopez, Guadalupe, Ms., United States of America 382
Lopez, Hector, Sr., Argentina 84
Lopez, Jean, France 230
Lopez, Jorge Federico S., Chile 208
Lopez, Jose Horacio, Sr., Colombia 213
Lopez, Laura, Mexico 292
Lopez, Leonel, Sr., Guatemala 246
Lopez, Maria Teresa, Sra., Chile 210
Lopez, Noemi Meyer, Ms., Norway 306
Lopez, Olhedyis Zamora, Ms., Costa Rica 216
Lopez, Oscar de Jesus, Sr., Colombia 213
Lopez, Rogelio, Paraguay 309
Lopez, Victoria M., Ms., Philippines 312
Lopez Doriga Velandia, Joaquin, Mexico 292
Lopez Ehavaee, Mario, Switzerland 331
Lopez Jordan, Jesus, Spain 324
Lopez Lorenzo, Lozada, Mexico 292
Lopez Montanez, Wilfredo, United States of America 357
Lopez Nicolas, Luis, Spain 324
Lopez Pastran, Domingo, Mexico 292
Lopez Portillo Vargas, Guillermo, Mexico 292
Lopez Vazquez, Luis, Spain 323
Lopez-Doriga, Joaquin, Mexico 292
Lopezllera, Luis, Sr., Mexico 294
Lopez-Portillo, Guillermo, Mexico 294
Lopinot, Didier, France 234
Lora, Rolando Espinoza, Gen., Bolivia 98
Lorca, Maria Cecilia Rejes, Ms., Brazil 162
Lorch, Andre, Brazil 116
Lord, Arthur, United Kingdom 348
Lording, David Rowland, Mr., Australia 89
Loredo De Izcue, Julioaugusto, Brazil 116
Lorentzen, Erling, Mr., Brazil 196
Lorenzo, Andres B., Sr., Philippines 312
Lorenzo, Aristide De, Italy 262
Lorenzo, Daniel, Argentina 84
Lorenzon, Alino, Mr., Brazil 162
Loreze, Adriana, Brazil 162
Lorina, Luiz Eduardo, United States of America 364
Lorine, Frantz, Haiti 247
Lorio, Margarita, Ms., Nicaragua 303
Lorriman, Donna, Ms., Canada 204
Lorriman, Doug, Mr., Australia 90
Lorud, Christophe, Brazil 162
Losada, Horacio Ruben Perez, Mr., Argentina 84
Lottmann, Jurgen, Mr., Germany 239
Louarn, Jean J., France 230
Louback, Jefferson B., Sr., Brazil 162
Loubser, Jan, Ms., Barbados 94
Louchard, Aimee Gomes, Brazil 116
Loucks, William, Canada 201
Louis, Andre, Mr., United States of America 369
Louis Juffe, Andre, Chile 208
Loukina, Natalya Vladimirovna, Ms., Russian Federation 320
Lounes, Ferhat, Cote d'Ivoire 217
Loureiro, Cesar, Brazil 162
Loureiro, Marcia, Brazil 116
Loureiro, Vania, Sra., Brazil 162
Louren, Jose Seixas, Mr., Brazil 129
Lourencao, Otavio Cunha, Sr., Brazil 162
Lourenco, Antonio Carlos, Brazil 162
Lourenco, Carlos Vitor, Brazil 116
Lourenco, Cesar Augusto Ottoni, Brazil 162
Lourenco, Jose Seixas, Brazil 101
Lourenco, Tarcisio P., Brazil 116
Loures, Maria Aparecida Limongi, Sra., Japan 277
Lourival, Humberto, Brazil 116
Lourival, Reinaldo, Mr., Brazil 162
Lousma, Jack, Mr., United States of America 382
Louven, Mariza, Brazil 116
Louzada, Silvana, Brazil 162
Louzeiro, Andre M, Brazil 162
Loveday, Geoffrey Mark, Mr., Australia 89
Lovejoy, Derek, Prof., United Kingdom 350
Lovejoy, Thomas E., The Hon., United States of America 355
Lovell, Jean, Ms., Switzerland 332
Lovold, Lars, Mr., Norway 306
Low, Patricia, Ms., United States of America 382
Lowcock, Mark, Mr., United Kingdom 345
Lowe, David, Mr., Marshall Islands 289
Lowe, Delia May, Ms., Australia 91
Lowe, Julie, Ms., United Kingdom 352
Lowe, Justin, United States of America 364
Lowenthal, Abrao, United States of America 357
Lowrie, Diane G., Ms., United States of America 357
Lowthian, Douglas, Mr., United States of America 382
Lowthian, Elizabeth, Ms., United States of America 382
Loy, Detlef, Mr., Germany 243
Loya, Yosef, Prof., Israel 255
Loyaga De Ludena, Martha Elena, Ms., Peru 311
Loyola, Cleuler De Barros, Ms., Brazil 129
Lozada, Alejandra Sanchez De, Sra., Bolivia 99
Lozano, Andre, Brazil 116
Lozano, Juan, Sr., Colombia 213
Lozano, Rommel V., Sr., Philippines 312
Lu, Congmin, Mr., China 211
Lu, Jhy-Ming, Dr., Taiwan 341
Lu, Li-Teh, Taiwan 340
Lu, Yucun, Mr., China 212
Luamba, Lua Nsembo, Zaire 401
Lubbers, R.F.M., H.E. Mr., Netherlands 298
Lubi, Elisa Tita, Ms., Philippines 311
Lubin, Flaminia, Italy 258
Lucarati, Guilhermo Hugo, Mr., Argentina 84
Lucas, Francis, Philippines 312
Lucas, Jody Kathleen, Mrs., Australia 89
Lucas, Marco A., Brazil 116
Lucca, Liliana De, United States of America 382
Lucchese, Elizabeth, Brazil 116
Lucchetta, Maria Elisa, Ms., Italy 262
Lucena, Rodolfo De, Brazil 162
Lucena Macedo, Monica Maria, Ms., Brazil 162
Lucena Mendes, Sergio, Mr., Brazil 162
Lucenet, Georges, Mr., France 231
Lucero, Fernando, Sr., Ecuador 222
Lucero Marquez, Alvaro F., Sr., Mexico 294
Lucia, Ricardo Luiz C. Funari Di, Brazil 162
Luciano, Olga, Mr., Dominican Republic 222
Luciolli, Esmeralde, Ms., France 234
Lucis, Frank De, Mr., France 234
Luckett, Dolores Garcia, Ms., United States of America 382
Ludena Luque, Guillermo, Mr., Peru 311
Ludin, Walter, Switzerland 333
Luduena, Manuel Antonio, Mr., Argentina 84
Ludwig, Sayonara Teixeira, Brazil 162
Luedig, Paul, Mr., Estonia 224
Luehring, Karen, Ms., Belgium 97
Luevano, Alejandro, Mr., Mexico 294
Luhnow, David, Mexico 292
Luhulima, Achie Sudiarti, Indonesia 253
Lui, Li-Hui, Taiwan 340
Luigi, Martinelli, Mr., Italy 262
Luik, Heino, Mr., Estonia 224
Luithui, Luingam, India 252
Luiz, Cosme, Brazil 162
Luiz, Nivaldo, Brazil 162
Luiz Carlos, De Almeida, France 228
Lujan, Berta, Sra., Mexico 294
Lujan, Daniel, Brazil 162
Lukabu, Khabouji, Zaire 401
Lukagyu, Bitomwa, Mr., Zaire 402
Lukasu, Dianda, Mr., Zaire 402
Lulic, Shirley J., Ms., United States of America 382
Lumb, John Michael, Dr., Australia 89
Lumbama, Noel, Mr., Kenya 284
Lumsden, Sheila, Ms., Canada 204
Lumuka, Ntonfo, Mr., Zaire 401
Lumuka Manioie, Leon Ntondo, Mr, Zaire 401
Luna, Boris Gomez, Sr., Peru 311
Luna, Henry Gomez, Sr., Peru 311
Luna, Ivanoir Ferreira De Castro, Mr., Brazil 162
Luna, Maria Anita, Ms., Brazil 162
Luna, Roberto Scriviner De, Brazil 162
Luna, Tonatiuh, Mexico 292
Luna D., Jose M., Mexico 292
Luna Espinoza, Ali Iram, Peru 310
Luna Parra, Angelica, Lic. Ma., Mexico 290
Lunardi, Adriana Brasilia, Sra., Brazil 162
Lunceford, Miguel, Mr., United States of America 382
Lund, Alan, Switzerland 338
Lund, H., Austria 92
Lund, Hauk, Mr., Norway 306
Lund, Heidi, Ms., United States of America 382
Lund, Henrik, Mr., Denmark 220
Lundberg, Bjorn, Denmark 220
Lundberg, Hakan, Mr., Austria 92
Lundberg, Hans, Mr., Sweden 329
Lundberg, Jan Christian, Mr., United States of America 382
Lundby, Claus Merrild, Denmark 220
Lunde, Leiv, Mr., Norway 306
Lunden, Jaana, Ms., United Kingdom 352
Lunden, Jan-Ake, Mr., Sweden 329
Lunden, Sven, Mr., Sweden 329
Lundgren, Marten, Mr., Sweden 330
Lundin, Anders, Mr., United States of America 382
Lundin, Michael, Mr., United States of America 382
Lundstol, Ottar, Mr., Norway 306
Lungo, Mario, Mr., Mexico 294
Lunn, Bruce, Mr., Australia 91
Lunz, Heliomar, Mr., Brazil 162
Luo, Guanxing, China 212
Luo, Jibin, Mr., China 211
Luonga, Mutchokv, Mr., Zaire 402
Luraschi, Mario, Mr., Argentina 84
Lusan, Daniel Rodrigues, Brazil 162

M

Mathusse, Hilario, Mr., Mozambique 297
Matias, Ivaci, Brazil 165
Matlack, Carolyn B., Ms., United States of America 364
Matlakala, D., Ms., Botswana 99
Matlock-Meria, Staci, United States of America 383
Matomina, Lusanda, Zaire 401
Matos, Ana Paula Goulart, Brazil 165
Matos, Emilia, Brazil 118
Matos, Francisco Jose De Abreu, Sr., Brazil 166
Matos, Gustavo Gomes De, Sr., Brazil 166
Matos, Jose Santos, Mr., Brazil 166
Matos, Maria Teresa Soares, Sra., Brazil 166
Matos, Mario Jorge Gama, Brazil 166
Matos, Milly, Sra., United States of America 383
Matrianha, Carlos Borges, Brazil 166
Matringe, L., France 234
Matsetse, P., Mr., Botswana 100
Matsuara, Shinici, Japan 277
Matsuba, Kenzo, Mr., Japan 277
Matsubara, Koji, Mr., Japan 265
Matsuda, Keiko, Japan 270
Matsuda, Kimiko, Ms., Japan 277
Matsui, Mitsuma, Japan 267
Matsui, Ryuich, Japan 270
Matsui, Saburo, Dr., Japan 267
Matsumoto, Hideki, Mr., Japan 277
Matsumoto, Masaya, Japan 270
Matsumoto, Tsukio, Mr., Japan 277
Matsumoto, Tsunao, Japan 270
Matsumura, Akio, Mr., United States of America 369
Matsumura, Kunimori, Mr., Japan 277
Matsumura, Maki, Ms., United States of America 383
Matsumura, Makoto, Mr., Japan 277
Matsumura, Takashi, Mr., Japan 265
Matsumura, Tetsuro, Japan 265
Matsuno, Nobuo, Mr., Japan 277
Matsuno, Taro, Dr., Japan 265
Matsuo, Mitsuhiro, Japan 270
Matsushima, Arata, Belgium 96
Matsushima, Tsuneharu, Mr., Japan 277
Matsushita, Kazuo, Switzerland 338
Matsushita, T, Switzerland 338
Matsuura, Shinichi, Japan 270
Matsuyama, Kaori, Japan 270
Matta, Claudinei Da, Brazil 166
Matta, Paulo S., Brazil 118
Matta, Ubiratan Brazil, Brazil 118
Matte-Baker, Andrea, Switzerland 338
Mattedi, Maria Raquel Mattoso, Sra., Brazil 166
Mattesky, Thomas, United States of America 364
Matteuzzi, Mauricio, Italy 259
Matthes, Klaus, Dr., Germany 239
Matthew, Mathai Koyapillil S.J., Italy 259
Matthews, Christopher, Mr., European Economic Community 225
Matthews, Eamonn, United Kingdom 348
Mattioli, Gianni, Hon., Italy 256
Mattos, Carlos Arnaldo Silva, Mr., Brazil 166
Mattos, Cristina, Sra., Brazil 166
Mattos, Grace Dantas, Brazil 118
Mattos, Haroldo, Kenya 284
Mattos, Jean Kleber De Abreu, Sr., Brazil 166
Mattos, Lucia, Brazil 118
Mattos, Luis, Brazil 166
Mattos, Marli M., Sra., United States of America 383
Mattos, Pedro Agilson, Brazil 166
Mattos, Regina Helena Cabral, Sra., Brazil 166
Mattos S., Jorge Luiz, Brazil 118
Maturana, Gonzalo, Mr., Chile 210
Matus Villalobos,, Julio, Chile 208
Matuschka, Mario Graf von, Dr., Germany 239
Matusse, Hilario Manuel Eugenio, Sr., Mozambique 298
Matutes, Abel, European Economic Community 225
Matz, Marcello, Mr., Brazil 166
Matzembacher, Luiz Oscar D., Brazil 118
Matzenbacher, Jaime, Sr., Brazil 166
Maubury-Lewis, Bioru, Mr., Brazil 166
Maud, Isabel, Brazil 118
Maud, Mirval, Ms., French Guyana 236
Maugain, Patricia, France 234
Maughan, Janet, Ms., United States of America 369
Mauldin, Corrie, United States of America 383
Mauleon, Cecelia Oleo, Ms., Peru 311
Maung, Marlar Sein, Miss, Italy 256
Maurer, Kim, Japan 270
Maurice, Antoine, Switzerland 333
Mauricio, Geraldo, Mr., Brazil 166
Mauritzon, Dagmar, Ms., Sweden 329
Maurizi, Mauro, Mr., Italy 262
Mauro, Francesco, Mr., Italy 256
Mauro, Guillermo Javier De, Mr., Argentina 85
Mauroy, Pierre, M., France 228
Maury, Cilulia Maria, Sra., Brazil 166
Mauser, Bruno, Switzerland 332
Mauso, Angela Villarrubia, Brazil 166
Maussan Flota, Jose Jaime, Mexico 292
Mausse, Jaime Zaquere, Mozambique 297
Mautone, Marcelo, Sr., Uruguay 398
Mavoungou, Vincent, M., Gabon 237
Mavoungou Bongo, Donatien, M., Gabon 237
Mavroudi, Maria, Ms., United States of America 383
Mawla, Naif, H.E. Mr., Jordan 282
Mawlawi, Farouk, United States of America 383
Max, Jose Carlos Molina, Sr., Brazil 166
Max, Lisa, Ms., United States of America 383
Maxe, Gote, Mr., Sweden 329
Maximo, Claudio, Dr., Argentina 88
Maximo, Gabriela, Brazil 118
Maxman, Susan, Ms., United States of America 357
Maxx, Suzanne, United States of America 383
May, Charles A., Mr., United Kingdom 352
May, Peter H., Mr., Brazil 101
May, Richard, Mr., United States of America 369
Maybrom, Nis Brummer, Denmark 220
Mayer, Kurt Martin, Austria 91
Mayer, Paul, United States of America 364
Mayer, Stephan Andreas, Germany 241
Mayer, Sylvie, Ms., France 230
Maynard, Edward, Mr., United States of America 364
Mayo, Ed, United Kingdom 346
Mayo, Kristina, United States of America 357
Mayoh, Anne Kathrine, Ms., Denmark 221
Mayr, Heidemarie, Mrs., Germany 239
Mayrink, Jose Maria, Brazil 118
Mayshara, Rurenza, Burundi 198
Mayur, Rashmi, India 250
Maywald, Leila Maria, Sra., Brazil 166
Maywald, Marcelo, Sr., Brazil 166
Maza, Liza, Ms., Philippines 313
Mazairac, L.P.J., H.E. Mr., Netherlands 298
Mazalam, Ricardo, Brazil 166
Mazambani, David, Mr., Senegal 321
Mazambani, David, Mr., Zimbabwe 403
Mazhar, Farhad, Ms., Bangladesh 94
Mazi, Bernardshaw, Mr., Nigeria 304
Mazuncaide, Susana, Ms., Uruguay 398
Mazzarollo, Joanilda, Brazil 118
Mazzei, Alex, Mr., Uruguay 398
Mazzei, Paulo Roberto Britto, Brazil 118
Mazzonna, Claudio, Italy 262
Mazzonna, Marco, Italy 262
Mazzucchelli, Sergio Andre, Mr., Argentina 79
Mazzucco, Roberto M., Brazil 118
Mba, Adrien Nguema, Hon., Gabon 237
Mba Bekale, Serge, S.E. M., Gabon 237
Mba Obame, Andre, M., Gabon 237
Mbaelele, Mankoto-Ma, Mr., Zaire 401
Mbaiherem, Ladukein, Chad 207
Mbala, Kasombo, Mr., Zaire 402
Mbanefo, Sandra, Brazil 166
Mbanga, Wilf, Zimbabwe 403
Mbaya, V. Beaman, Dr., Kenya 284
Mbaye, Babacar Carlos, Mr, Senegal 321
M'bitikon, Raymond, Mr., Central African Republic 207
Mbitikon, Raymond, Mr, Central African Republic 207
Mboho, Kisusu, Mr., Zaire 402
Mbonerane, Albert, Burundi 198
Mboudy, Serge, Gabon 237
Mboya, P., H.E. Mrs., Kenya 283
Mbuba, Labu, Zaire 401
Mbumba, Labu, Mr., Zaire 401
Mbumwae, Louis, Mr., Zambia 402
Mbwana, S., Mr., United Republic of Tanzania 354
Mc Daniel, Robert, Mr., United States of America 383
McAlary, David, United States of America 364
McAlevey, Jane, Ms., United States of America 383
McAlister, Roy E., Mr., United States of America 383
McAllister, Andrew, Brazil 118
McAllister, Ian, Switzerland 332
McAndrew, Wayne, Australia 90
McArthur, Ian, Japan 270
McCabe, Ron, Mr., United States of America 383
McCammon, Antony, Mr., Switzerland 332
McCandliss, John, United States of America 383
McCann, Nancy Wilson, Ms., United States of America 383
McCarthy, Brian, Mr., Ireland 255
McCarthy, John, Italy 259
McCarthy, Michael, United Kingdom 348
McCarthy, Patricia L., Ms., United States of America 383
McCarthy, Thomas, Mr., United States of America 383
McCaughey, Winsome, Ms., Australia 91
McCauley, Mark, United Kingdom 352
McClary, Eric, Mr., United States of America 383
McClean, Oronde, Mr., United States of America 383
McCleary, Joel, Mr., United States of America 357
McClellan, Jo Eva, Sra., United States of America 383
McClellan, Monique, Switzerland 338
McClintock, Sandra, United States of America 364
McCloskey, Maxine E., Ms., United States of America 369
McCloskey, Michael, Mr., United States of America 369
McClugston, Richard, United States of America 383
McCluskey, Ian, United States of America 364
McCollough, Charles, United States of America 357
McComie, Ella Garcia, Ms., United States of America 357
McConnell, Fiona, Ms., United Kingdom 345
McCorry, Denise, Ms., United Kingdom 346
McCoy, Charles, United States of America 364
McCoy, Michael, United States of America 364
McCracken, Sylvia, Ms., Zimbabwe 403
McCrory, Colleen, Ms., Canada 204
McCullough, Edmund, United States of America 364
McCully, Patrick, Mr., Uruguay 396
McCully, Shannon, Ms., United States of America 383
McCurdy, Rev. Archibald, Canada 205
McCusker, Alison, Ms, Italy 256
McDaniel, B, Switzerland 338
McDermott, Lawrence, Mr., Canada 205
McDonagh, Sean, Mr., United Kingdom 352
McDonald, Brian, Mr., European Economic Community 225
McDonald, David, The Hon., Canada 199
McDonald, David C., Mr., United States of America 383
McDonald, Frank, Ireland 255
McDonald, Franklin, Mr., Jamaica 264
McDonald, Greg, United States of America 364
McDonald, Mary S., United States of America 364
McDonald, Norm, Mr., United States of America 384
McDonald, Trevor, United Kingdom 348
McDonough, Lee, Ms., United States of America 357
McDonough, William, United States of America 357
McDougall, Barbara, The Hon., Canada 199
McDowell, Andrew, Mr., Canada 205
McElhone, Jane, Canada 201
McElroy, Susan, Ms., United States of America 384
McFarlane-Taylor, Joann, United States of America 364
McGinnis, Mr.Lloyd, Mr., Canada 205
McGinnis, Mrs.Lloyd, Ms., Canada 205
McGinty, Lawrence, United Kingdom 348
McGlashan, Teresa, Ms., Belgium 96
McGlone, Alistair, Mr., United Kingdom 345
McGonigle, Bridget, Ms., Ireland 255
McGrath, James, United States of America 364
McGrath, Kathleen, Mr., Canada 199
McGraw, Desiree, Mr., Canada 199
McGreal, Shirley, Ms., United States of America 384
McGuinness, Ann, Ms., Ireland 255
McGuinness, John P., Mr., United States of America 355
McGuinness, Patrick, United States of America 357
McGuinty, David J., Mr., United Kingdom 352
McHarg, Allison, United Kingdom 352
McHuch, James T., S.E. Mgr., Holy See 247
McHutchon, Graham, United Kingdom 352
McIlroy, Anne, Canada 201
McIntwre,, United States of America 358
McIntyre, Jeff, Mr., Venezuela 399
McIntyre, Jim, Mr., Venezuela 399
McIvor, Daniel, United States of America 384
McJunkin, James, Mr., United States of America 384
McKay, David, Netherlands 300
McKean, Bruce, Canada 200
McKenney, Joan, Ms., United States of America 384
McKenzie, J.C., Mr., United Kingdom 350
McKeon, Isadora Zoe, Ms., United States of America 384
McLaren, Nadia, Ms., Belgium 96
McLauchlin, David, Canada 201
McLean, Jennifer, Ms., Canada 205
McLoed, Teri, Ms., United States of America 384
McMahon, James P., Mr., United States of America 384
McManus, Doyle, United States of America 364
McNally, Tony, Mr., United Kingdom 352
McNamara, Brendan, Mr., Ireland 255
McNamara, Kristin, United States of America 384
McNamara, Patrick, Mr., United States of America 384
McNealey, Connie, Ms., United States of America 384
McNeil, Mary, Ms., United States of America 384
McNeill, Charles, Dr., United States of America 395
McNeill, Sabine, Ms., United Kingdom 352
McPhail, Ian Roderick, Mr., Australia 89
McPherson, Andrew, United Kingdom 352
McPherson, Janice, Spain 323
McPherson, Mary M., United States of America 364
McPike, Barbara, Ms., United States of America 384
McQuillan, Laurence, United States of America 364
McSweeney, Brenda, Ms., Switzerland 335
Mctiron, Anne, Canada 205
Md Fazlul, Huq, Bangladesh 93
M'Daghri Alaoui, Mustapha, M., Morocco 296
Mdluli, Sethabile E., Ms., Swaziland 327
Mead, A., Ms., New Zealand 302
Mead, Aroha, Ms., New Zealand 302
Mead, Walter, United States of America 364
Meagher, Michael, Canada 201
Mealey, Elisabeth, Brazil 166
Meana, Carlo Ripa di, European Economic Community 225
Meana, Francesco Ripa Di, Sr., Brazil 166
Mecado, Alvaro, Brazil 118
Meda, Gurudutt P., India 250
Medawan, Jean, Ms., United Kingdom 350
Medeiros, Alessandra, Brazil 166
Medeiros, Andrea Da Silva Mury, Brazil 118
Medeiros, Dea Maria Moreira, Ms., Brazil 166
Medeiros, Fatima Beatriz F., Brazil 118
Medeiros, Frederico Freire, Brazil 166
Medeiros, Guilherme Da Motta, Brazil 118
Medeiros, Gustavo Henrique, Brazil 166
Medeiros, Joao, Sr., Brazil 166
Medeiros, Jose Guilnerme, Brazil 118
Medeiros, Lourenco, Portugal 314
Medeiros, Marcelo, Brazil 118
Medeiros, Maria Angelica F., Brazil 166
Medeiros, Monica, Brazil 118
Medeiros, Sonia Maria Gomes De Matos, Brazil 166
Medeiros-Ferandes, Robert, Mr., European Economic Community 225
Medellin, Jorge Alejandro, Mexico 292
Medeng, , M., Gabon 237
Mederos, Jorge Washington, M., United States of America 364
Medina, Ivan, United States of America 395
Medina, Jose Ronaldo De Castro, Mr., Brazil 166
Medina, Nelson, Ecuador 223
Medina, Ofelia, Ms., Mexico 294
Medina, Rodrigo, Mr., Chile 210
Medina, Silvio Jose G, Brazil 166
Medina Acosta, Rolando, Mexico 292
Medina Barranco, Patricia Veronica, Mexico 292
Medvedev, Svyatoslav, H.E. Mr., Kazakhstan 282
Meegan, Anne, Ms., United States of America 384
Meenan, Kyle, United States of America 364
Meersmith, Thomas, United States of America 364
Megee, Pennie, Ms., Brazil 166
Meghar, Hocine, M., Algeria 78
Megret, Alain, France 234
Megyesi, Anna Louise, Sra., Brazil 101
Mehmood, Rao Khalid, Pakistan 308
Mehta, Pradeep S., Mr., India 251
Mehta, Sanat, India 250
Meidlinger, Leo James, United States of America 364
Meier, Gerhard, Mr., Italy 260
Meier Miro-Quesada, Martha E., Ms, Peru 310
Meighan, Matt, Mr., United States of America 384
Mei-Hua, Cheng, Taiwan 340
Meijer, C. M., Ms., Netherlands 301
Meijer-Werner, Eugenia Frota De, Ms., Venezuela 399
Meillon, Jacqueline, France 230
Meimon, Marc, France 230
Meira, Maria Elisa, Brazil 166
Meira, Rita, Sr., Portugal 315
Meireles, Marcos, Brazil 166
Meireles, Sergio, Brazil 118
Meirelles, Domingos Joao, Brazil 166
Meirelles, Julia, Sra., Brazil 166
Meirelles, Marcia, Brazil 166
Meirelles, Nei Flavio, Sr., Brazil 166
Meisen, Peter, Mr., United States of America 384
Meisfjord, Arna, Norway 306
Meisler, Stanley, United States of America 364
Meitin, Acejandro, Argentina 85
Mejia, Alberto Lopez, Brazil 166
Mejia, Ignacio, Colombia 214
Mejia, Louis, United States of America 358
Mejia, Margarita, Ms., Mozambique 298
Mejia, Mario, Mr., Colombia 214
Mekbach Jr., Paulo, Mr., Brazil 166
Mekdad, Fayssal, Mr., Czechoslovakia 218
Melake, Kiflemariam, Ms., Ethiopia 224
Melandez, Gerardo, Colombia 213
Melandri, Elisabetta, Ms., Italy 260
Melandri, Giovanna, Ms., Italy 260
Melantonio, Adriana Ribas, Brazil 166
Melega, Marco, Italy 259
Melendez, Giselle, Switzerland 338
Melendez, Hernan Javier, Colombia 213
Melendez, Pedro, Colombia 213
Melendez, Ricardo, Sr., Colombia 213
Melgarejo-Remus, Marion, Ms., Bolivia 99
Melitsco, Graciela, Sra., Argentina 85
Mellbin, Franz-Michael, Mr., Denmark 219
Mellet, Luiz Ernesto, Sr., Brazil 166
Melli, Frank, United States of America 384
Mello, Carlos, Mr., Brazil 166
Mello, Carlos Eduardo Penzias De, Brazil 166
Mello, Claudia Sorres De, Brazil 166
Mello, Djalma, Brazil 166
Mello, Eduardo, Brazil 166
Mello, Enir Moreira Da Silva, Sra., Brazil 166
Mello, Flavia, Sra., Brazil 166
Mello, Geisa, Brazil 118
Mello, Ivo Da S., Brazil 166
Mello, Joana, Brazil 118
Mello, Luiz Carlos, Brazil 166
Mello, Maria Claudia Cavalcanti S., Brazil 118
Mello, Maria Da Gloria Silveira, Ms., Brazil 166
Mello, Maria De Fatima E. B. De, Brazil 166
Mello, Marta Iara Silveira, Sra., Brazil 166
Mello, Mateo Magarinos De, Mr., Uruguay 398
Mello, Paulo Quintanilha De Mello, Sr., Brazil 166
Mello, Pedro Paulo Thiago De, Brazil 166
Mello, Priscilla L., Brazil 118
Mello, Regina Lara Silveira, Brazil 166
Mello, Solange De Toledo, Ms., Brazil 166
Mellqvist, Lars Erik, Norway 305
Melnick, Ana, Italy 259
Melnicove, Margo, United States of America 364
Melo, Aldair Gomes De, Brazil 166
Melo, Guilherme De Azevedo, Brazil 166
Melo, Jose Caldas Gouveia De, Brazil 166
Melo, Lea Maria C.B., Brazil 118
Melo, Liana, Brazil 118
Melo, Luis Philipe Figueira De, Brazil 166
Melo, Marconi Teodoro, Brazil 166
Melo, Placido Jose Alves De, Brazil 118
Melo, Rosa, Sra., Brazil 166
Melo Filho, Murilo, Brazil 118
Meloche, Francois, Costa Rica 216
Meloi, Ronize Alina De, Brazil 166
Meloye, Mitiku, Mr., Ethiopia 224
Melton, Richard H., The Hon., Brazil 100
Meltzer, Erica, United States of America 384
Melvarg, David, United States of America 364
Melzatamuk, Sinabadam Ka Baqpbo, Ms., Philippines 313
Membreno, Lic. Conny, Nicaragua 303
Memiaghe, Gisele, Mme, Gabon 237
Mena, Ramona Concepcion Alcaraz, Sra., Paraguay 309
Mena, Staci Matlock, United States of America 364
Menagh, Melanie, United States of America 364

Moesch, Alberto Pretto, Sr., Brazil 168
Moglia, Ana M., Ms, Switzerland 331
Moguel, Julio, Mr., Mexico 295
Mohamad Ali, Mr., Malaysia 287
Mohamed, Ali, Mr., Djibouti 221
Mohamed, Brahimi, Mr., Algeria 78
Mohamed, Dileita, Mr., Djibouti 221
Mohamed, N., Mr., New Zealand 302
Mohamed, Son Altesse Royale le Prince Heritier Sidi, Morocco 296
Mohamed Abdul Kadir, Syed Syed, Dr., Malaysia 287
Mohamed Babiker, Elham Elgaily, Sudan 326
Mohamed Fari, Emara, Egypt 223
Mohamed-Cheikh, Biadillah, Morocco 296
Mohammad, Dawood S., Mr., Kuwait 285
Mohammad, Fares Ali, Mr., Iraq 254
Mohammad, Reazuddin, Bangladesh 93
Mohammed, Ahmedullah, India 251
Mohammed, Duri, H.E. Dr., Ethiopia 224
Mohammed Ataur, Rahman, Bangladesh 93
Mohanan, N., Dr., India 252
Mohapatro, Pitabas, India 250
Mohr, Anne Tamara, Ms., Netherlands 299
Mohring, Caroline, Germany 241
Moi, Daniel Toroitich arap, The Hon., Kenya 283
Moiane, Arlindo, Mr., Mozambique 297
Moir, Matthew, Mr., United States of America 384
Moisan, Francois, France 234
Moises, Helvio Nicolau, Sr., Brazil 168
Moita, Carla, Portugal 314
Moitoza, Kaaren, United States of America 384
Mojica, Armando, Mr., Mexico 295
Moka, G.U., Nigeria 304
Mokgothu, B., The Hon., Botswana 100
Mokhtar, Azouz, Dr., Tunisia 343
Mokobi, Ruth F., Ms., Botswana 100
Mokoginta, Dv. H.Ijan, Mr., Indonesia 254
Mokoginta, Jan, Mr, Indonesia 254
Molander, Karin, Ms., Sweden 329
Molander, Sverker, Dr., Sweden 329
Molano, Diego, Mr., Belgium 96
Moldan, Bedrich, Mr., Czechoslovakia 218
Molesti, Romano, Italy 259
Molica, Fernando, Brazil 168
Molica, Jose F., Brazil 119
Molina, Enrique, Sr., Costa Rica 215
Molina, Yolanda, Sra., Venezuela 399
Molinari, Cesar R. S., Brazil 168
Molineaux, Diana, United States of America 364
Molinier, Michel, Mr., Brazil 168
Molinski, Michael, United States of America 364
Moll, Peter, Germany 240
Moller, Lone, Denmark 219
Moller, Per Stig, H.E. Mr., Denmark 219
Molloy, Christina Alves, United States of America 364
Molloy, Tezza, Mr., Australia 91
Molnarne, Eva Csobod, Hungary 249
Mololas, , M., Gabon 237
Molotsi, Prisca, Ms., Zambia 402
Moltke-Leth, Aase, Ms., Denmark 219
Molvaer, Reidulf, Mr., Norway 306
Moma, Fatma E., Ms., Nigeria 304
Mombaur, Peter Michael, Dr., Germany 239
Momendeng, Symphorien Kevin, Cameroon 198
Mometti, Pe. Giovanni, Sr., Brazil 168
Momi, Lingosa, Mr., Zaire 401
Monadjemi, Cyrus, Sr., Brazil 168
Monadjemi, Puran, Sra., Brazil 168
Monaghan, Aileen, United Kingdom 352
Monaghan, Michael, Switzerland 336
Monajad, Abdulhamad, Syrian Arab Republic 340
Monballiu, Jan, Belgium 96
Mondim, Leda, Sra., Brazil 168
Mondino, German Luis, Mr., Argentina 85
Mondonedo, Nora Linda, Ms., Philippines 313
Mondragon, Luis Enrique Sanchez, Mexico 295
Monducci, Delba, Brazil 168
Monediaire, Gerard, Mr., France 234
Monero, Mariucha, Brazil 119
Monetti, Myriam, Ms., United Kingdom 352
Monforte, Carlos, Brazil 168
Mongbe, Rene Valery, S.E.M., Benin 98
Monge, Victor Manuel, Sr., Costa Rica 215
Mongelli, Nicola, Mr., Italy 256
Mongelos, Ricardo Fabian Yorg, Sr., Paraguay 309
Mongenie, Jean Claude, Mr., French Guyana 236
Mongockodji, Pepin, S.E. M., Gabon 237
Monguzzi, Carlo, Mr., Italy 262
Moniaga, Sandra, Ms., Indonesia 254
Monin, Noel, Mr., United States of America 384
Monk, Arlete, Sra., Brazil 168
Monnerat, Monica, Sra., Brazil 168
Monninghoff, Hans, Mr., Germany 243
Monplaisir, Roberte, Mme, Haiti 247
Monsalvo, Bernabe, United States of America 364
Monserrat Filho, Jose, Brazil 119
Monsieur, Christiane, Italy 259
Montalvo, Jaime, Costa Rica 216
Montanari, Pinuccia, Italy 262
Montanes, Cid Jose Francisco, Spain 324
Montanez, Wilfredo, Mr., Switzerland 335
Montano, Jorge, S.E. Dr., Mexico 290
Montant, Francoise, Ms., Switzerland 332
Montautti Ibarra, Americo, Italy 259
Montbrun, Alberto, Sr., Argentina 85
Monte, Ednalva Maia, Ms., Brazil 168
Monte, Gilberto, Mr., United States of America 384
Monte Rosa, Ricardo, Mr, Brazil 100
Monteiro, Anna Claudia Vianna, Brazil 119
Monteiro, Antonio Jose L. C., Mr., Brazil 168
Monteiro, Antonio Mascarenhas, Mr., Cape Verde 206
Monteiro, Edalva Alves, Sra., Brazil 168
Monteiro, Fernando, United States of America 384
Monteiro, Frederico Mai, Brazil 168
Monteiro, Gerson Souza, Jr., Brazil 168
Monteiro, Jose Antonio Pinto, H.E. Mr., Cape Verde 206
Monteiro, Jose Carlos, Brazil 119
Monteiro, Karen, Brazil 168
Monteiro, Liege, Brazil 119
Monteiro, Luiz Antonio C., Brazil 168
Monteiro, Marcia Lima De Azevedo, Sra., Brazil 168
Monteiro, Marco Antonio, Brazil 119
Monteiro, Maria Elisa G. De Castro, Brazil 168
Monteiro, Maria Elizabeth Brea, Brazil 168
Monteiro, Maria Rozane S., Brazil 119
Monteiro, Mario, Brazil 168
Monteiro, Marion, Brazil 119
Monteiro, Roberta De Oliveira, Brazil 119
Monteiro, Rui Miranda, Brazil 168
Monteiro, Simone, Sra., Brazil 168
Monteiro, Valeria, Brazil 119
Montenegro, Armando, E.S., Colombia 213
Montenegro, Marcelo, Brazil 119
Montenegro, Patricia, Ms., Brazil 168
Montenegro, Raul A., Sr., Argentina 79
Montera, Hyacinthe De, France 234
Monteros, Ismael Ramon, Mr., Argentina 85
Monterroso, Luis, Mr., Portugal 315
Monterroso, Tulio, Mr., Guatemala 246
Montes, Egledio Viana, Brazil 168
Montes, Marisa, Sra., Argentina 85
Montes De Oca D., Fernando Jose, Mexico 292
Monteverde, Bernardo, Mr., Brazil 168
Monteverde, Esterzinha, Ms., Brazil 168
Montez, Ceasar, Mr., United States of America 384
Montgomery, Katy, Ms., United States of America 384
Monticelli, Gislene, Sra., Brazil 168
Montiel, Juan Jose, Lic., Nicaragua 302
Montis, Antonella De, Mr., Italy 262
Monto, Laurie, Ms., Canada 202
Montorio Jr., Italo, Sr., Brazil 168
Montoro, Andre Franco, Sr., Brazil 129
Montoussamy, Carmelien, Mr., Martinique 289
Montoya, Ruben, Sr., Venezuela 399
Moodye, Lynda, Sra., Brazil 168
Moon, Young Mee, Ms., Republic of Korea 317
Mooney, Pauline, Switzerland 338
Mooneyhan, Dee, Ms., United States of America 384
Mooneyhan, Wayne, Mr., United States of America 384
Moonstar, Joy, Ms., United States of America 384
Moore, Calvin, Mr., United States of America 384
Moore, Catherine Anne, Mrs., Australia 89
Moore, Christopher T., Dr., United States of America 384
Moore, Esteban, Sr., Argentina 85
Moore, Euna, Ms., Barbados 94
Moore, Howard, Mr., France 236
Moore, Jonathan, The Hon., United States of America 355
Moore, Julia A., Ms., United States of America 369
Moore, Katie, Brazil 168
Moore, Marjorie, Ms., United States of America 384
Moore, Phyllis, United States of America 395
Moore, Rosa Moreno, Ms., Belgium 97
Moore-Kochlacs, Peter, Mr., United States of America 384
Moorhead, Erica S., Ms., United States of America 384
Mooyo-Mhlanga, Khethiwe, Ms., Zimbabwe 403
Mora, Alejandra, Ms., Chile 210
Mora Mora, Rodrigo, Mr., Chile 210
Mora Tavares, Carlos Eduardo, Mexico 292
Moraes, Antonio, Brazil 119
Moraes, Ari, Sr., Brazil 168
Moraes, Catia V., Brazil 119
Moraes, Elza Helena Gimenez, Brazil 119
Moraes, Gustavo Luiz Gripp, Brazil 119
Moraes, Hamilton H. S., Brazil 168
Moraes, Ignez S. Pinto De, Ms., Brazil 168
Moraes, Igor B., Brazil 168
Moraes, Jose Mauro, United States of America 358
Moraes, Julio Cesar De, Brazil 168
Moraes, Lewy Rodrigues De, Brazil 168
Moraes, Luciano De, Brazil 168
Moraes, Luis, Switzerland 338
Moraes, Luzia, Sra., United States of America 384
Moraes, Marco Aurelio Lopes, Brazil 168
Moraes, Maria Aguinaga De, Sra., Brazil 168
Moraes, Maria Do Perpetuo Socorro, Sra., Brazil 168
Moraes, Paulo, Mr., Brazil 168
Moraes, Paulo De, Brazil 168
Moraes, Paulo Roberto S., Brazil 119
Moraes, Renata, Brazil 119
Moraes, Ricardo, Deputy, Brazil 196
Moraes, Rita De Cassia Barboza De, Brazil 168
Moraes, Stella Sodre Vieira De, Brazil 119
Moraes, Susan Eghrari, Sra., Brazil 168
Moraes, Zeni De Souza, Ms., Brazil 168
Moraes Duarte, Maria Cecilia De, Brazil 119
Morais, Eliane De, Sra., Brazil 169
Morais, Marcos Jose De, Sr., Brazil 169
Moral, Javier G., Spain 324
Morales, Esther E., Chile 208
Morales, Freddy, Bolivia 98
Morales, Gunter Morales Arnao, Mr., Peru 311
Morales, Horacio, Mr., Philippines 311
Morales, Juan Manuel, Mr., Argentina 85
Morales Carlos, Gloria Adriana, Sra., Brazil 169
Morales Jr., Horacio R., Mr., Philippines 313
Moran, Stephane, France 234
Morand Francis, Pascale D, Switzerland 331
Morandini, Norma, Spain 324
Morehouse, Ward, Mr., United States of America 369
Moreillon, Jacques, Mr., Switzerland 334
Moreira, Adriana Guimaraes, Brazil 119
Moreira, Alcir Alves, Mr., Brazil 169
Moreira, Antonio, Sr., Portugal 315
Moreira, Antonio Carvalho, Brazil 169
Moreira, Branca, United States of America 395
Moreira, Claudio, Mr., Brazil 169
Moreira, Darcy, Brazil 169
Moreira, Dilciea R. De S., Sra., Brazil 169
Moreira, Dr Neiva, Sr., Brazil 169
Moreira, Eduardo, United States of America 395
Moreira, Elio Raymundo, Sr., Brazil 169
Moreira, Elizabeth, Ms., Brazil 169
Moreira, Gilberto, Brazil 101
Moreira, Gladstone Novaes, Sr., Brazil 169
Moreira, Goretti, Brazil 169
Moreira, Iara Verocal Dias, Ms., Canada 205
Moreira, Jose Luiz Pinheiro, Brazil 169
Moreira, Jose Marcelo Antonez, Brazil 169
Moreira, Jose Roberto, Mr., Brazil 169
Moreira, Lucia Amelia Martins, Brazil 169
Moreira, Luzia Amelia Alres, Brazil 169
Moreira, Maria Da Conceicao Ruffeil, Sra., Brazil 169
Moreira, Maria Goretti, Brazil 169
Moreira, Maria Manuela, Sra., Brazil 169
Moreira, Mario, Brazil 169
Moreira, Nazareth Pinheiro, Sra., Brazil 169
Moreira, Noemi De Souza, Brazil 169
Moreira, Osvaldo, Sr., Argentina 85
Moreira, Paulo Cesar, Ms., Brazil 169
Moreira, Rita De Cassia, Sra., Brazil 169
Moreira, Roberto Munoz, Sr., Nicaragua 303
Moreira, Ronaldo Antunes, Brazil 169
Moreira, Sandra Maria, Brazil 169
Moreira, Teresa Maria, Sra., Brazil 169
Moreira, Tereza, India 252
Moreira, Terezinha Cancella, Brazil 119
Moreira, Vera Cristina Roche, Sra., Brazil 169
Moreira, Veronica, Brazil 169
Moreira Andreatta, Maria Beatris, Brazil 119
Morel, Francoise-Marie, France 230
Morel, Mario, Brazil 119
Morelle, Glaucia, Sra., Brazil 169
Morelli, Ana, Brazil 119
Morelli, Dom Mauro, Mr., Switzerland 335
Morello, Jorge, Brazil 169
Moreno, Carlos, Sr., Argentina 85
Moreno, Hector, Canada 201
Moreno, Icaro, Jr., Mr., Brazil 169
Moreno, Johnny Fernandez, Costa Rica 216
Moreno, Jorge Bastos, Brazil 119
Moreno, Jose Armando Torres, Sr., Brazil 169
Moreno, Julio, Mexico 292
Moreno, R., Italy 264
Moreno, Saulo Pinto, Brazil 119
Moreno, Zoraida, Venezuela 399
Moreno Lacave, Emilio, Spain 324
Moreria, Wilma, Ms., United States of America 384
Moretto, Ana Lucia, Brazil 119
Moretz-Sohn, Claudia, Brazil 119
Moretzsohn, Gerson Ribeiro, Mr., Brazil 169
Moreyra, Sandra Maria, Brazil 119
Morezzi, Ettore, Mr., Italy 262
Morgan, David, United States of America 384
Morgan, Dawn, Ms., Barbados 94
Morgan, Donagh, Mr., Ireland 255
Morgan, Eka, Ms., Netherlands 299
Morgan, Gloria Calero, Ms., Nicaragua 303
Morgan, James, United Kingdom 348
Morgan, Jim, Mr., United States of America 369
Morgan, Marlys, Ms., United States of America 384
Morgan, Maurene M., Ms., United States of America 384
Morgan, Richie, Mr., Switzerland 335
Morgan, Tom, Mr., United States of America 384
Morgan, Victor H., Sr., Costa Rica 215
Morganci, Marcos, Brazil 119
Morgon Colon, Humberto, Sr., Mexico 295
Mori, Kazuaki, Mr., Japan 265
Mori, Kenji, Japan 278
Mori, Miyuki, Brazil 169
Mori, Nobuhito, Japan 270
Mori, Norikazu, Mr., Japan 278
Moria, Jose Carlos Ramalho, Brazil 169
Moriarty, Burke, Mr., United States of America 385
Moriarty, John, Mr., Chile 210
Morich, Horst, Mr., Switzerland 335
Moriconi, Maurizio, Italy 262
Morier, Luiz, Brazil 169
Morikawa, Tohru, Japan 267
Morillo, Marco, Ing., Ecuador 222
Morimoto, Machiko, Japan 278
Morimoto, Sara, Ms., United States of America 385
Morishima, Akio, Japan 267
Morita, , Mr., Japan 278
Morita, Akira, Mr., Japan 278
Morita, Chiharu, Ms., Japan 278
Moriwaki, Kimio, Mr., Japan 267
Moriya, Yoshihiko, Sr., Japan 278
Moriyama, Hiroyuki, Mr., Japan 278
Moriyasu, Tsuneo, Mr., Japan 278
Morna, Colleen, India 251
Moromisato Miasato, Doris, Ms, Peru 309
Morongiu, Antonio, Mr., European Economic Community 225
Morosine, Maria De Fatima, Sra., Brazil 169
Morosini, Liseane, Brazil 169
Morottaja, Matti, Mr., Finland 226
Moroz, Irina, Ms., Australia 91
Morris, Langdon, Mr., United States of America 385
Morris, Mary Lou, Ms., Australia 90
Morrison, Dorothy, Ms., United States of America 385
Morrison, Sabrina, Ms., Guatemala 246
Morrison, Sherman, Mr., United States of America 385
Morrone, Giuliana Faria, Brazil 119
Morselli, L., France 234
Morta Lucia, Nesta D Etacletti, Argentina 79
Mortensen, Pia, Ms., Denmark 219
Mortezayani, Oliver, Sr., Uruguay 398
Mosca, Cataldo, Italy 262
Moscoso, Arturo, Sr., Bolivia 98
Moseley, Jennifer, Ms., United States of America 385
Moses, Resio S., The Hon., Micronesia 296
Moses, Ted, Mr., Canada 205
Moses, Ted, Mr., United States of America 369
Moshe, Nancy, United States of America 358
Mosienyane, Matc Gadibolae, Dr, Botswana 100
Mosley, Marilyn, Ms., United States of America 385
Mosley, Ramaa, Ms., United States of America 385
Mosquea, Rosa Gonzalez, Sra., Dominican Republic 222
Moss, Deborah Kamlani, Ms., United States of America 385
Moss, Douglas E., United States of America 364
Moss, Marcia, Brazil 169
Moss, Vera, Brazil 169
Mostafa, Aisha, Ms., Sudan 326
Mostafa, Intisar, Switzerland 338
Mostapha, El Haiba, Mr., Morocco 297
Mota, Estefania, Brazil 119
Mota, Helena, Mrs., Mozambique 297
Mota, Humberto, Sr., Brazil 169
Mota, Lilian De Oliveira Dantas, Brazil 119
Mota, Maria De Fatima Cantidio, Sra., Brazil 169
Mota, Norma, Brazil 119
Mota, Renata, Sra., Brazil 169
Mota Pereira, Luiz Otavio, Brazil 101
Mota Rodrigues, Neumar, Brazil 119
Motley, Tony, Mr., United Kingdom 352
Motoki, Akiyoshi, Mr., Japan 265
Motonami, Yoshinori, Mr., Japan 278
Motoo, Asakuno, Japan 270
Mott, Richard, United States of America 364
Motta, Aydano A., Brazil 119
Motta, Claudia Dos Reis, Ms., Brazil 169
Motta, Claudio, Brazil 119
Motta, Cristina M. Soares, Brazil 119
Motta, Estefania De Castro, Brazil 169
Motta, Helio, Brazil 169
Motta, Jose Luiz Fernandes Da, Brazil 169
Motta, Jose Machado Da, Brazil 169
Motta, Liane Oliveira Mufarrej, Brazil 169
Motta, Luiz Gonzaga F., Brazil 119
Motta, Marcos Valerio F., Brazil 169
Motta, Paulo Cesar, Sr., Brazil 169
Motta, Paulo De A., Brazil 119
Motta, Paulo Roberto Marins, Brazil 169
Motta, Ricardo Mendes, Brazil 169
Motta Dos S., Hamilton, Brazil 119
Mouasher, Anis, Mr., Jordan 282
Moucani Muesta, Victor, S.E. M., Gabon 237
Mouche, Carol, United States of America 385
Moudud, Hasna, Ms., Bangladesh 94
Moufarrige, Emilio, Mr., Brazil 101
Moufarrige, Lucia Brandao, Sra., Brazil 169
Moulik, T.K., Mr., India 251
Moullade, Pierre, France 234
Moulton, Tim, Sr., Brazil 169
Moumba, Pascal, M., Congo 214
Mouna, Ma Dhyan, Ms., Brazil 169
Mounier, Jean-Francois Le, Brazil 169
Mounir, Amr Ahmed, Mr., Egypt 223
Moura, Abdalaziz De, Brazil 169
Moura, Adriana Maria Magalhaes De, Brazil 169
Moura, Arovel, Sr., Brazil 169
Moura, Carlos, Sr., Portugal 315
Moura, Claudio Carvalho, Brazil 169
Moura, Debora Da Veiga, Brazil 119
Moura, Evelina Brigite Pires De, Sra., Portugal 315
Moura, Fernando Goncalves De, Sr., Brazil 169
Moura, Geraldo De Melo, Sr., Brazil 169
Moura, Jose A., Brazil 119
Moura, Josilda Rodrigues Da Silva De, Sra., Brazil 169
Moura, Luiza, Sra., Brazil 169
Moura, Milton, Sr., Brazil 169
Moura, Miriam, Brazil 119
Moura, Sergio Mascarenhas De, Sr., Brazil 169
Moura, Tereza Ione Souza Filho, Sra., Brazil 169
Moura, Walter Thomas W., Brazil 119

N

O

P

Q

R

S

Sang, Hoon Chun, Mr., Republic of Korea 318
Sang, Stephen, Mr., Kenya 283
Sangare, Yaya, Cote d'Ivoire 217
Sangchul, Park, Mr., Republic of Korea 318
Sang-Hwan, Kim, Mr., Republic of Korea 318
Sangoi, Elizabeth, Sra., Brazil 182
Sangoon, Lee, Mr., Republic of Korea 318
Sangui, Vladimir, Mr., Russian Federation 320
Sanguinetti, Ana, Sra., Mexico 295
Sangweni, Stanislaus S., South Africa 323
Sanhez, Jorge, Argentina 87
Sanielovici, Susanne, Ms., Brazil 182
Sanin De Rubio, Noemi, E.S. Dra., Colombia 213
Sankaran, K., Dr., United States of America 389
Sanna, Lucy, Ms., United States of America 370
Sanousi, Magda M. El, Ms., Sudan 326
Sansone, Marie, Ms., United States of America 389
Santa, David, Denmark 220
Santa Rosa, Paulo, Sr., Brazil 182
Santala, Anna, Ms., Brazil 182
Santamarta Florez, Jose, Spain 323
Santana, Celso Guido De, Brazil 182
Santana, Horacio Gilvan B., Brazil 124
Santana, Jackson, Sr., Brazil 182
Santana, Pedro Franco, Sr., Dominican Republic 222
Santana, Reinaldo Ribeiro De, Sr., Brazil 182
Santana, Romildes Oliveira, Sra., Brazil 182
Santana De Mello, Lucia Maria, Brazil 124
Santana Filho, Lourival, Brazil 124
Sant'Anna, Alexandre, Brazil 182
Sant'Anna, Lucia Maria De Mello, Brazil 182
Sant'Anna, May De Britto, Brazil 124
Santareita, Mariurete, Brazil 182
Santarelli, Eduardo Guilherme, Mr., Brazil 182
Santello, Giovana Vantini, Sra., Brazil 182
Santer, Jacques, S.E. M., Luxembourg 286
Santhakumar, C. Job, India 250
Santiago, Alicia Goncalves, Brazil 124
Santiago, Andre Luis, Brazil 182
Santiago, Carlos, Brazil 124
Santiago, Emiliano, Sr., Brazil 182
Santiago, Patricia Costa, Ms., Brazil 182
Santiago, Ramil B., Sr., Philippines 313
Santiago, Sandra, Brazil 182
Santiago, Susana Irene, Ms., Argentina 87
Santiago, Tiago Xavier, Sr., Brazil 182
Santibanez, Victor, Mr., Argentina 87
Santillan Velazquez, Enrique, Mexico 292
Santini, Maria, Sra., Brazil 182
Santini, Zelia, Ms., Brazil 182
Santo, Alvaro Negrao Do Espirito, Mr., Brazil 182
Santo, Cesar Victor Do Espirito, Sr., Brazil 182
Santo, Joper Padrao Do Espirito, Mr., Brazil 182
Santori, Isa In Fenocchio, Italy 259
Santori, Isabella, Italy 263
Santoro, l' abbe Filippo, M., Holy See 247
Santoro, Luis Fernando, Brazil 182
Santoro, Rosangela, United States of America 389
Santos, Adilson Dos, Brazil 182
Santos, Adriana Dos Santos E, Sra., United States of America 389
Santos, Alberto Do, Brazil 182
Santos, Almir, Sra., Brazil 182
Santos, Aluizio Figueiredo Dos, Brazil 182
Santos, Amilton Nascimento Dos, Brazil 182
Santos, Ana Cristina Lessa, Sra., Brazil 182
Santos, Ana Gaio, Brazil 124
Santos, Ana Lucia Vieira Dos, Brazil 182
Santos, Ana Maria Gaio, Brazil 182
Santos, Anatalia, Ms., Brazil 182
Santos, Angelo, United States of America 389
Santos, Antonio, Sr., Brazil 182
Santos, Antonio Carlos Cordeiro Dos, Brazil 182
Santos, Antonio Carlos Silva, Brazil 124
Santos, Antonio Gabriel De Carvalho, Brazil 182
Santos, Antonio Manoel Pinheiro, Mr., Portugal 315
Santos, Arlene Maria Soares Correia, Sra., Brazil 182
Santos, Arlinda Messias Dos, Ms., Brazil 182
Santos, Aureo, Mr., Switzerland 336
Santos, Carla Regina, Brazil 124
Santos, Carlos A. Ivanir Dos, Sr., Brazil 182
Santos, Carlos Alberto Felicio Dos, Brazil 182
Santos, Carlos Alberto O. Dos, Brazil 182
Santos, Carlos Benedito Dos, Sr., Brazil 182
Santos, Carlos Eduardo Medina Dos, Sr., Brazil 182
Santos, Celio Dos, Brazil 182
Santos, Clarindo Dos, Brazil 124
Santos, Claudia M., Brazil 124
Santos, Claudio De Holanda, Brazil 182
Santos, Claudio Vieira Dos, Brazil 124
Santos, Dalva Maria Vasconcelos Dos, Ms., Brazil 182
Santos, Danillo Augusto, Mr., Brazil 182
Santos, Debora N., Brazil 124
Santos, Denise Marinho Dos, Brazil 182
Santos, Deoclides, Brazil 182
Santos, Edilson, Brazil 182
Santos, Edilson Da Silva, Brazil 182
Santos, Edilson Gomes Dos, Brazil 182
Santos, Edmundo Gonzaga Dos, Brazil 182
Santos, Edson Escouto Dos, Brazil 182
Santos, Eduardo De Oliveira, Brazil 182
Santos, Eliane, Brazil 101
Santos, Eliane Oliveira Dos, Brazil 124
Santos, Elisete Moura, Brazil 182
Santos, Elizabeth, Ms., Brazil 182
Santos, Eloy Dos, Brazil 124
Santos, Ernani Gomes, Brazil 182
Santos, Evandro C., Brazil 182
Santos, Fatima C., Brazil 124
Santos, Fernanda Reznik, Brazil 182
Santos, Fernando Cezar Dos, Brazil 182
Santos, Fernando Dos, Brazil 182
Santos, Fernando Pacheco Dos, Mr., Angola 78
Santos, Filipina Del Rosario, Ms., Philippines 312
Santos, Flavio Augusto Altero Dos, Mr., Brazil 182
Santos, Francelino Jose Dos, Brazil 183
Santos, Francisco Alves, Sr., Brazil 183
Santos, Francisco De Oliveira, Brazil 183
Santos, Gianne Carvalho S. Dos, Brazil 183
Santos, Gilberto Gomes Dos, Mr., United States of America 389
Santos, Helio Dos, Brazil 183
Santos, Ignacio, Sr., Canada 200
Santos, Ilhens Pereira Dos, Brazil 183
Santos, Ilmar Bastos, Sr., Brazil 183
Santos, Ione M.De O., Brazil 124
Santos, Ismael, Brazil 124
Santos, Joao Batista Dos, Sr., Brazil 183
Santos, Joao Francisco, Brazil 124
Santos, Jorge Andre, Brazil 124
Santos, Jose Augusto Correia, Brazil 183
Santos, Jose Eduardo Dos, Brazil 124
Santos, Jose Macena Dos, Brazil 183
Santos, Jose Marcos Tadeu Dos, Brazil 124
Santos, Jose Ricardo Mello Dos, Brazil 183
Santos, Jose Soares Dos, Brazil 183
Santos, Josefina Dos, Ms., Brazil 183
Santos, Jucara Cristina Braga, Brazil 124
Santos, Jurandir S., Brazil 124
Santos, Juscelino Batista Dos, Sr., Brazil 183
Santos, Leila, Brazil 183
Santos, Leila Cordeiro, Brazil 124
Santos, Liana Carvalho, Brazil 183
Santos, Lilian Costa Nabuco Dos, Brazil 183
Santos, Lucia De Araujo A., Brazil 124
Santos, Luis Claudio Dos, Brazil 183
Santos, Luis Feitosa Dos, Brazil 183
Santos, Luis Ricardo, Brazil 183
Santos, Luiz Antonio Dos, Brazil 183
Santos, Luiz Carlos Lessa Dos, Brazil 183
Santos, Luiz Fernando Lima, Brazil 124
Santos, Luiz Henrique Dos, Brazil 183
Santos, Luiz Marques Dos, Brazil 183
Santos, Lydia, Ms., United States of America 389
Santos, Ma.Angela, Brazil 124
Santos, Madalena Rodrigues Dos, Sra., Brazil 183
Santos, Manuel, E.S., Colombia 213
Santos, Manuel Antonio Araujo, Sr., Portugal 315
Santos, Marcelo Kobylinski Dos, Brazil 183
Santos, Marcio A., Brazil 124
Santos, Marcio Alberto Mendonca, Brazil 124
Santos, Marco Aurelio Martins, Brazil 183
Santos, Margaret, Brazil 183
Santos, Maria, Portugal 315
Santos, Maria Cecilia MacDowell Dos, Ms., United States of America 389
Santos, Maria Das Neves, Sra., Brazil 183
Santos, Maria Do Carmo Neves Dos, Ms., Brazil 183
Santos, Maria Francisca M. Dos, Brazil 183
Santos, Maria Isabel Rodrigues Dos, Brazil 183
Santos, Maria Lino Dos, Ms., Brazil 183
Santos, Maria Sirley Dos, Sra., Brazil 183
Santos, Marian Sola, Ms., Spain 325
Santos, Marina, Ms., Brazil 183
Santos, Marli Terezinha Dos, Brazil 183
Santos, Mauricio Rodrigues Dos, Brazil 183
Santos, Moises Soares Dos, Sr., Brazil 183
Santos, Monica Cristina M., Brazil 124
Santos, M.P. Vinicius, Mr., Brazil 183
Santos, Napoleao Bonaparte, Officer, Cape Verde 206
Santos, Neemias Espinola Dos, Brazil 183
Santos, Nelly C. Dos, Sra., Brazil 183
Santos, Nelson J. B., Brazil 183
Santos, Nelson Perez Dos, Brazil 183
Santos, Ney Costa, Sr., Brazil 183
Santos, Nilton, Brazil 124
Santos, Olga De Los, Brazil 183
Santos, Omar Dos, Brazil 183
Santos, Osvaldo J., Brazil 124
Santos, Patricia C., Brazil 124
Santos, Paula Cristina Ventura Dos Sa, Sra., Portugal 315
Santos, Paulo, Sr., Brazil 183
Santos, Paulo Cesar Vieira Dos, Brazil 183
Santos, Paulo Roberto Dos, Brazil 183
Santos, Pedro Manuel Ribeiro Simoes D, Mr., Portugal 315
Santos, Pedro Paulo Ferreira Dos, Sr., Brazil 183
Santos, Raimunda Manuela Dos, Ms., Brazil 183
Santos, Raimundo Nunes, Mr., Brazil 183
Santos, Reinaldo Menezes, Brazil 183
Santos, Renan Cordeiro Dos, Brazil 183
Santos, Renan Dos, Mr., Brazil 183
Santos, Roberto Machado, Brazil 124
Santos, Roman G., Sr., Philippines 313
Santos, Ronaldo Braga Dos, Brazil 183
Santos, Rosangela H. Dos, Brazil 183
Santos, Rubens R., Brazil 124
Santos, Sergio A. G., Brazil 183
Santos, Sidney Jose Dos, Sr., Brazil 183
Santos, Suely Rangel, Brazil 124
Santos, Susana Cristina Nunes C., Sra., Portugal 315
Santos, Suzana Gomes, Brazil 183
Santos, Suzana Maria Da Rocha, Brazil 183
Santos, Tadeu Martins Dos, Brazil 124
Santos, Valerio Cardoso Dos, Sr., Brazil 183
Santos, Valter, Brazil 124
Santos, Vandeval Lourenco Dos, Brazil 183
Santos, Waltencir Goulart Dos, Brazil 183
Santos, Wandreuves Acampora Dos, Brazil 183
Santos, Wilma Resende Araujo, Ms., Brazil 183
Santos, Wilson Franca Dos, Brazil 124
Santos Costa, Carlos Domingos, Brazil 124
Santos Gomes, Suzana, Brazil 125
Santos Moita, Carla Isabel, Portugal 314
Santos Neto, Acylino Jose Dos, Sr., Brazil 183
Santovincenzo, Ivano, Italy 259
Sanvoisin, Marco, Italy 259
Sanz, Luiz Alberto, Brazil 183
Saorio, Petteri, Finland 227
Saouma, Eduardo, Italy 264
Sapia, Cristina Noemi, Ms., Argentina 87
Sappia, Horacio, Mr., Argentina 87
Sapucahy, Mario Lucio, Brazil 184
Sarabhai, Kartikeya Vikram, India 249
Saracino Tagliante, Jeanine, Mme, Cote d'Ivoire 217
Saraiva, Deolinda Cavalcante, Brazil 184
Saraiva, Mauricio, Brazil 125
Saraiva, Sergio R.B., Brazil 125
Sarasana, Cesar, Sr., Peru 310
Sarathy, Partha, Mr., Switzerland 332
Sardar Khan, Malik, Pakistan 308
Sardenberg, Ormi, Mr., Brazil 184
Sardinha, Ocimar, Brazil 125
Sarenas, Patricia M., Ms., United States of America 389
Sargavakian, Ani, Bulgaria 197
Sarikahputi, Yookti, Mr., Thailand 341
Sarkar, Arun Chandra, Mr., Bangladesh 94
Sarli, Luigi, Mr., Italy 263
Sarmento, Elizabeth, Brazil 184
Sarmento, Jair Da Silva, Mr., Brazil 184
Sarmiento, German, Mr., Colombia 213
Sarmiento, Juan Roberto Ramirez-Otarola, Mr., Peru 311
Sarno, Ruy, Brazil 184
Sarquis, Otto, Brazil 125
Sarre, Eustache, S.E.M., Benin 98
Sarrut, Jacques, Mr., France 231
Sarti, Riccardo, Mr., United States of America 389
Sarti, Silvio, Sr., Brazil 184
Sarto, Vilson, Switzerland 339
Sartori, Marcia, Ms., Ireland 255
Sartori, Michele, Ms., Brazil 184
Sartorio, Ronald, Sr., Brazil 184
Sarturi, Linda Cristina Degrazia, Brazil 184
Sarubbo, Mauro Alejandro Baptista, Brazil 184
Saruva, Takaaki, Mr., Japan 279
Sarwer-Foner, Brian, Mr., Brazil 184
Sasada, Naomi, Japan 271
Sasai, Kojiro, Japan 279
Sasaki, Hiroshi, Japan 271
Sasaki, Kiyoshi, Mr., Japan 279
Sasaki, Rica, Ms., Japan 279
Sasaki, Robinson K., Brazil 125
Sasaki, Shuichi, Mr., Japan 266
Sasoon, Shana, Ms., Brazil 184
Sasparovic, Franjo, Dr., Croatia 217
Sass, Luis Alberto, Mr., Brazil 184
Sass, Vera Beatriz, Ms., Brazil 184
Sassano, Marco, Italy 259
Sassi, Carlos Eduardo Lessi, Brazil 184
Sasson, Albert, France 236
Sastre, Vera L., Brazil 125
Satapathy, Annapurna, Ms., India 253
Satayavivad, Jutamaad, Associate Professor Dr., Thailand 341
Satheesh, P.V., Mr., India 251
Sathler, Evandro, Sr., Brazil 184
Sato, Hidenori, Japan 271
Sato, Hiroyuki, Germany 240
Sato, Kazuo, Japan 279
Sato, Kenji, Japan 271
Sato, Marina Yeko, Brazil 184
Sato, Mikio, Japan 272
Sato, Mitsuo, Mr., Japan 279
Sato, Naoko Sasada, Brazil 184
Sato, Noriaki, Japan 272
Sato, Sandra, Brazil 125
Sato, Shuji, Mr., Japan 279
Sato, Takayuki, Japan 272
Sato, Toshio, Japan 272
Sato, Tsutomu, Mr., Japan 279
Sato, Yoshimi, Japan 272
Sato, Yosuke, Mr., Japan 279
Sato, Yousuke, Mr., Japan 279
Satoh, Keiichi, Japan 272
Satouris., Elizabeth, Ms., United States of America 389
Sattar, Danyal, Mr., United Kingdom 350
Satu, Huttunen, Ms., Finland 227
Sauaget, Elza, Brazil 100
Saucero, Jose Pedro, Uruguay 397
Sauer, Adeum Hilario, Mr., Brazil 184
Saugier, Bernard, Mr., France 235
Sauni, Miriama, Mr., United States of America 389
Sautier, Denis, Mr., France 235
Savadogo, , Ms., Cote d'Ivoire 217
Savage, Eleanor W., Ms., United States of America 355
Savage, J.A., United States of America 366
Savaget, Claudio C., Brazil 125
Savaget, Eliza, Brazil 184
Savaget, Elza K.K., Brazil 125
Savaget, Rubens Carvalho, Brazil 125
Savaget, Ruth De Castro, Brazil 125
Savathvong, Silavanh, M., Laos People's Democratic Republic 285
Savio, Roberto, Italy 259
Savioli, Adriano, Brazil 184
Savitsky, Basil, Mr., United States of America 389
Savitt, Chuck, Mr., United States of America 389
Savlia, Paolo, Italy 263
Savoie, Donat, Mr., Canada 205
Savory, Judith Eve T., Canada 201
Saw, Aliou, Mauritania 290
Sawa, Bozena, United States of America 366
Sawa, Yuji, Japan 272
Sawada, Akiyoshi, Mr., Japan 267
Sawada, Isao, Japan 272
Sawama, Yasuo, Mr., Japan 274
Sawaya, J., France 236
Sawhill, John C., Mr., United States of America 395
Sawhney, Kamini, India 251
Sawitri, Isma, Indonesia 254
Sawyer, Donald, Brazil 101
Saxe, Dianne, Canada 201
Saxena, Mahesh Prasad, Dr., India 253
Sayao, Lea, Brazil 125
Sayao, Silvia, Brazil 125
Sayd, Cristiane, Sra., Brazil 184
Sayedazzaman, Md., Mr., Bangladesh 93
Saygilioglu, Nevzat, Mr., Turkey 343
Sayne, Pamela, Ms., Canada 205
Sayres, Sandra, Mr., United States of America 389
Sazanami, Hidehiko, United States of America 395
Sbai, Nafissa, Ms., Morocco 297
S.Benevides, Ana Madalena P., Brazil 181
Sbrenna, Giuseppe, Mr., Italy 263
Scaglione, Cynthia M., Ms., United States of America 389
Scalabrini, Laura, Ms., Italy 263
Scamparini, Ilze Lia, Brazil 125
Scanlon, John, Australia 90
Scannavino Netto, Eugenio, Brazil 101
Scaramuzza, Carlos, Switzerland 339
Scarborough, Katherine Tracy Alvord, United States of America 366
Scarf, Franklin, Mr., Australia 90
Scarvalone, Maria, Germany 241
Schaal, Florence, France 230
Schaback, Susanna, Mrs., Germany 239
Schaffer, Marilyn, Ms., United States of America 389
Schaffer, Wigold, Sr., Brazil 184
Schaffrath, Sara, Ms., United States of America 389
Schafhausen, Franz Josef, Mr., Germany 239
Schak, Sabine, Switzerland 339
Schaller, Andrea, Germany 244
Schaluck, Herman, Rev., Italy 263
Schapira, Catherine, France 235
Scharf, Regina Claudia Gross, Brazil 125
Schartz, Michelle, Ms., United States of America 389
Schatt, Rahera, Ms., Switzerland 336
Schaub, Nancy, Ms., United States of America 389
Schawrnopf, Paulo Inacio, Brazil 184
Schearer, S. Bruce, Mr., United States of America 358
Schechter, Daniel, United States of America 366
Schedin, Sven O., Mr., Sweden 330
Scheele, Huub, Netherlands 300
Scheele, Reinier J., Mr., Italy 263
Scheffler, Marion, Mr., Germany 244
Schei, Peter Johan, Mr., Norway 307
Scheid Lopes, Alfredo, France 229
Scheiner, Paulo Cezar De Lemos, Sr., Brazil 184
Schenk, Maria, Ms., Austria 92
Schenkel, Celso, Mr., Brazil 184
Schepelmann, Philip, Germany 240
Scherbak, Yuri M., H.E. Mr., Ukraine 344
Scherdien, Celia Regina, Brazil 125
Scherr, S Jacob, Mr., United States of America 366
Schettino, Luiz Fernando, Mr., Brazil 184
Scheuer, James H., Congressman, United States of America 396
Scheuermann, Horst, Mr., Germany 239
Scheumann, Waltina, Ms., Germany 244
Schiavo, Marcio, United States of America 389
Schiavo, Marcio Ruiz, Mr., Brazil 184
Schiavo, Patricia, Colombia 213
Schiavone, Giuseppe, Italy 263
Schiavoni, Anna, Ms., Italy 263
Schicker, Elgin Veronica, Sra., Brazil 184
Schiefer, N., United States of America 395
Schielmann, Horst, Germany 244
Schiesser, Walter, Switzerland 333
Schiligting, Sandra, Ms., Brazil 184
Schimke, Fritz, Austria 91
Schiper, Ruth, Sra., Brazil 184
Schipper, Lee, Mr., United States of America 389
Schipulle, Hans-Peter, Dr., Germany 239
Schirmer, Gerson Lopes, Brazil 125
Schlanger, John, Mr., Netherlands 300

U

V

X

Y

Z

NOTES

DIRECTORY

LISTINGS OF GOVERNMENTAL DELEGATIONS TO UNCED

ADDRESSES AND TELEPHONE NUMBERS NOT PROVIDED. PLEASE SEND CORRESPONDENCE TO THE ATTENTION OF THE RESPECTIVE UN MISSIONS IN NEW YORK, WHICH ADDRESS AND TELEPHONE NUMBER YOU CAN FIND IN THE ALPHABETICAL LISTING FOR EACH COUNTRY.

AFGHANISTAN - GOVERNMENT

Husain, Sultan, Mr.
Deputy
Ministry of Planning
Permanent Mission to the United Nations, 866 United Nations Plaza, Suite 520
New York, NY 10017, 754 1191, 1192

Salimi, Mohamad Naim, Mr.
Permanent Mission of Afghanistan
866 United Nations Plaza, Suite 520
New York, NY 10017, Afghanistan
T: 754 1191, 1192

Shams, Abdul Jalil, H.E. Dr.
Acting Deputy Minister
Foreign Affairs
Permanent Mission to the United Nations, 866 United Nations Plaza, Suite 520
New York, NY 10017, United States of America
T: 754 1191, 1192

ALBANIA - GOVERNMENT

Deliu, Gani, Mr.
Expert On Environment
GOVERNMENT DELEGATION
Committee of Environment Protection Ministry of Health
Tirana, Albania
T: 00355 42 28390 F: 00355 42 28696

Koci, Kosta, Mr.
Expert On Environment
GOVERNMENTAL DELEGATION
Committee for Environment Ministry of Health
Tirand, Albania
T: 003554228390 F: 003554228696

Selfo, Birim, Mr.
Chairman Committee For Environment
GOVERNMENTAL DELEGATION
Committe for Environment Ministry of Health
Tirana, Albania
T: 003554228390 F: 003554228696

ALGERIA - GOVERNMENT

Abdelbari, Sid Ali, M.
Chef Bureau des Affaires Scientifiques et Techniques, Ministere des Affaires Etrangeres
Permanent Mission of Algeria
15 East 47th Street
New York, NY 10017, United States of America
T: 750 1960, 1961, 1962 F: 759 9538

Ahmia, Mourad, M.
Troisieme Secretaire
Permanent Mission of Algeria
15 East 47th Street
New York, NY 10017, United States of America
T: 750 1960, 1961, 1962 F: 759 9538

Bachir, Kadik
Directeur General
MINISTERE AGRICULTURE
Ann Jardin D'Essa du Hamma B445 Elanasser
Alger, Algeria
T: 02 67 4072 F: 02 67 4065

Boukhlifa, Khaled, M.
Directeur du Patrimoine Ministere de l'energie
MINISTRY OF ENERGY
80 Avenue Ahmed
Ghermoul, Algeria
T: 2 65 06 49

Daghmoum, Abdelfetah, M.
Secretaire de l'Ambassade d'Algerie au Bresil
Permanent Mission of Algeria
15 East 47th Street
New York, NY 10017, United States of America
T: 750 1960, 1961, 1962 F: 759 9538

Djoghlaf, Ahmed, M.
Sous-Director des Etudes Economiques Ministere des Affaires Estrangeres
Permanent Mission of Algeria
15 East 47th Street
New York, NY 10017, United States of America
T: 750 1960, 1961, 1962 F: 759 9538

Gaid, Abdelkader, M.
Directeur charge de l'Environnement Secretariat d'etat a la Recherche Scientifique
Permanent Mission of Algeria
15 East 47th Street
New York, NY 10017, United States of America
T: 750 1960, 1961, 1962 F: 759 9538

Ghozali, Sid Ahmed, S.E. M.
Premier Ministre
Republique Algerienne Democratique et Populaire
15 East 47th Street
New York, NY 10017, United States of America
T: 750 1960, 1961, 1962 F: 759 9538

Hadid, Rabah, M.
Ministre Plenipotantiaire
Permanent Mission of Algeria
15 East 47th Street
New York, NY 10017, United States of America
T: 750 1960, 1961, 1962 F: 759 9538

Kadik, Bachir, M.
Directeur General de l'Agence Nationale pour la Conservation de la Nature, Ministere de l'Agriculture
Permanent Mission of Algeria
15 East 47th Street
New York, NY 10017, United States of America
T: 750 1960, 1961, 1962 F: 759 9538

Khelladi, Mourad, S.E. M.
Secretaire d'Etat a la Recherche Scientifique
Permanent Mission of Algeria
15 East 47th Street
New York, NY 10017, United States of America
T: 750 1960, 1961, 1962 F: 759 9538

Laala, Mohamad, S.E. M.
Ambassadeur Extraordinaire et Plenipotentiaire au Bresil
Permanent Mission of Algeria
15 East 47th Street
New York, NY 10017, United States of America
T: 750 1960, 1961, 1962 F: 759 9538

Laoussine, Nordine Ait, S.E. M.
Ministre de l'Energie
Permanent Mission of Algeria
15 East 47th Street
New York, NY 10017, United States of America
T: 750 1960, 1961, 1962 F: 759 9538

Meghar, Hocine, M.
Directeur des Relation Economiques et Culturelles Ministere des Affaires Etrangeres
Permanent Mission of Algeria
15 East 47th Street
New York, NY 10017, United States of America
T: 750 1960, 1961, 1962 F: 759 9538

Ouiguin, Kleira, Mme.
Sous-Director des Affaires Culturelles, Scientifiques et Techniques Ministere des Affaires Etrangeres
Permanent Mission of Algeria
15 East 47th Street
New York, NY 10017, United States of America
T: 750 1960, 1961, 1962 F: 759 9538

NGO

Benaboud, Fifi-Halima, Ms.
FEMMES, ENVIRONNEMENT ET DEVELOPPEMENT - FED
17 Rue Shakespeare,
El Mouradia, Algeria

Brahimi, Mohammed
AAPNE
Cite 56 logements, Bt. F, Rue Abane Ramdane
BP 154, 10000 Bouira, Algeria

Derbal, Hammouche
AAPNE
Cite 56 logements, Bt. F, Rue Abane Ramdane
BP 154, 10000 Bouira, Algeria

Sekkal, Zohir, Ms.
President
MOUVEMENT ECOLOGIQUE ALGERIEN
17, rue Shakespeare
Alger, Algeria
T: 213 2 604650 F: 213 2 606382

ORGANIZATIONS

Mohamed, Brahimi, Mr.
As. Algerienne pour la Protection de la Nature et de l'Environnement
Cite 56 logements, Bt. F, Rue Abane Ramdane
BP 154, 10000 Bouira, Algeria

GLOBAL FORUM

Ouared, Akila, Ms.
ASSOCIATION DE DEFENSE ET PROMOTION DES FEMMES
Alger, Algeria
T: 213 2 61 70 45

Younsi, Said, Mr.
TOWIZA ASSOCIATION
18, Rue Hozawon Abdel Aziz
Alger, Algeria
T: 213 2 264-9992

ANGOLA - GOVERNMENT

Abrev Van-Dunen, Cesaltina, Ms
Adviser
PRIME MINISTER'S OFFICE
P.O. Box 2746
Luanda, Angola
T: 391400 - 469441 F: 00244-2-354711

De Rohas, Liliana, Mr
Delegation
MINISTRY OF AGRICULTURE AND DEVELOP Ment Rural
Luanda, Angola
T: 323934

Dos Santos, Jose Eduardo, Mr.
President
Permanent Mission of the Poeple's Republic of Angola
125 East 73rd Street
New York NY 10021, United States of America
T: 212 861 5656 F: 212 861 9295

Guilherme, Victor Hugo, Mr
Delegate
MINISTERIO DO PLANO
B.P. 1205-Largo Do Palacio
Luanda, Angola
T: 333835/392796

Serodio De Almeida, Joao Manuel
Delegado
GOVERNO DE ANGOLA
Faculdade De Ciencias P.O.Box 2110- Luanda
Luanda, Angola
T: 345000/298/RESI34489 F: 00 244 2 345000 EXT

Tati, Jose Amaro
Vice-Ministro / Delegado
MINISTERIO LA AGRICULTURA
Luanda - C.P. No 523
R P, Angola
T: 320552 323065 363091

Teixeira De Carvalho, Victor Da Silva, Ms
Delegate
MINISTERIO DA AGRICULTURA E DES. RU
C.P. 74 Luanda - Angola Instituto De Desgnvolvimento Flores
Luanda, Angola
T: 323934 OR 323581

Theisohn, Thomas A, Mr
Development Programme
RUA MAJOR KANHANGULO 197
PO Box 910
Luanda, Angola
T: 331181 331188 331198 F: 871 150 5360

GLOBAL FORUM

Agaia Russo, Vladimir, Mr.
JUVENTUDE ECOLOGICA
Avenue 4 de Fevereiro no.32 - 10o.Esq.
Luanda, Angola
T: 244 2 33 47 18 F: 244 2 33 40 37

Belo, Jeronimo Otavio Xavier, Sr.
Bibliotecario
ASSOCIACAO ANGOLANA DO AMBIENTE
Caixa Postal 3675
Luanda, Angola
T: 244 2 332001

Cain, Allan, Mr.
Representative
AMIGOS DO AMBIENTE NGOLA KILUANJE
a/c CP 3360
Luanda, Angola
T: 244 2 33024 F: 244 2 393445
TL: 4072

Domingos, Carlos, Mr.
President
AMIGOS DO AMBIENTE NGOLA KILUANJE
a/c CP 3360
Luanda, Angola
T: 244 2 33024 F: 244 2 393445
TL: 4072

Menezes, Eunice, Ms.
Socio Fundador
JUVENTUDE ECOLOGICA ANGOLANA
R. Conego Manuel Das Neves, 87 - A 3rd
Wanda, Angola
T: 345 651

Santos, Fernando Pacheco Dos, Mr.
ADRA
BP 3788
Luanda, Angola
T: 244 2 33 55 89 F: 244 2 33 40 37

Varela, Benjamim Makunzy Simao, Sr.
Secretario Geral
P.H.C.O.- PROGRAMA HUMANITARIO PARA CRIANCAS ORFAS
Caixa Postal 6274
Luanda, Angola
T: 244 2 351765

Viegas, Walter, Sr.
Socio
ADRA
Av. 4 De Fevereiro 30/1- Andar
Luanda, Angola
T: 244 2 332822
TL: 3147 AN

Weba, Fernando, Mr.
JUVENTUDE ECOLOGICA ANGOLANA
Rua Karl Marx ,24
Luanda, Angola

Zage, Paulo Figueira, Mr.
Secretary
AMIGOS DO AMBIENTE NGOLA KILUANJE
a/c CP 3360
Luanda, Angola
T: 244 2 33024 F: 244 2 393445
TL: 4072

UN AND UNCED

Theisohn, Thomas, Mr.
Programme Office
UNDP
CP 010, Rua Majo Uanbauojulo 197
Luanda, Angola
T: 244 2 33 12 49 F: 244 2 871 150 5360

ANGUILLA - GLOBAL FORUM

Banks-Devonish, Linda, Dr.
Clinical Psychologist
WORLD COUNCIL OF CHURCHES
PO Box 85
The Valley, Anguilla

ANTIGUA AND BARBUDA - GOVERNMENT

Ashe, John W., Dr.
Counsellor
Permanent Mission of Antigua and Barbuda
610 Fifth Avenue, Suite 311
New York, NY 10020, United States of America
T: 541 4117 F: 757 1607

Bird, Oscar, Mr.
Chairman Historical Conservation and Environmental Commission
Ministry of Economic Development
Permanent Mission to the United Nations, 610 Fifth Avenue, Suite 311
New York, NY 10020, United States of America
T: 541 4117 F: 757 1607

Cort, Errol, Dr.
Adviser
Ministry of Economic Development and Tourism
Permanent Mission to the United Nations, 610 Fifth Avenue, Suite 311
New York, NY 10020, United States of America
T: 541 4117 F: 757 1607

Hurst, Lionel A., H.E. Mr.
Ambassador Extraordinary and Plenipotentiary
Permanent Mission of Antigua and Barbuda
610 Fifth Avenue, Suite 311
New York, NY 10020, United States of America
T: 541 4117 F: 757 1607

Williams, Candice, Ms.
Adviser to the Minister of Economic Development
Permanent Mission of Antigua and Barbuda
610 Fifth Avenue, Suite 311
New York, NY 10020, United States of America
T: 541 4117 F: 757 1607

Williams, Rodney, H.E. Dr.
Minister of Economic Development
Permanent Mission of Antigua and Barbuda
610 Fifth Avenue, Suite 311
New York, NY 10020, United States of America
T: 541 4117 F: 757 1607

ARGENTINA - GOVERNMENT

Abraham, Elena Maria, Ms
Researcher
IADIZA (Arid Zones Institute) - CONICET
C.C. 507 (5500)
Mendoza, Argentina
T: 54 61 241995 F: 54 61 380370

Bortagaray, Lucia Leontina
Asesor Tecnico
PRESIDENCIA DE LA NACION
Balcarce 24
Buenos Aires, Argentina
T: 343 0738 F: 331 6376 Y 331 1849

Carballo, Yolanda, Ms
Sociologa
FUNDAL.PROMEVA/DELEGACION ARGENTINA
Uruguay Pso.'4101.8'-Basas Argentin
BSAS, Argentina
T: 428364 F: 812 459Q

Claver, Silvia, Ms.
Investigador
IADIZA
C.C.507 - Correo Central
Mendoza 5500, Argentina
T: 54 61 241995 F: 54 61 380370
TL: 55438-CYTME-AR, E: 5500

Curia, Eugenio Maria, Mr
Assistant Legal Adviser
MINISTRY FOREIGN AFFAIRS
Reconquista 1088 #12
Buenos Aires, Argentina
T: 541 8144821 F: 541 3123539

Fabre, Eduardo, Mr
Secretary Of Science & Technology
GOVERMENT OF MENDOZA
Casa De Gobierno
Mendoza 5500, Argentina
T: 54 61 241230 F: 54 61 240385

Gentile, Rodolfo A., Ms
Delegado
MIN. RELAC.EXTER.(ECON DEPT)
Reconquista 1088,4TH Fl. (Dgase)
Capital Federal, Argentina
T: 311 0071 INT 178 9

Guastoni, Monica Cristina
Asesora Tecnica
PRESIDENCIA DE LA NACION
Balcarce 24
Buenos Aires, Argentina
T: 343 0738 F: 331 6376/331 1849

Gutierrez, Tomas, Mr
Vice Presidente
FUNDACION NATURAL(ARGENTINA)
U De Enero 2481
3000 Santa Fe, Argentina
T: 59 42 90795 F: 59 42 50944

Isola, Juan Jose, Ms
Concejal
CONCESO DELIBERANTE
Municipio De Quilmes Libertad 872-Codigo 1878
Quilmes,Buenos Aires, Argentina
T: 541 253 7241 F: 541 253 9003

Jilek, Ricardo Alejandro, Mr
Presidente
CONSEJO FEDERAL DE MEDRO AMBIENTE
Calle 51 Y 18 - 50 Piso (1900)-la Plata-Buenos Aires
la Plata, Argentina
T: 054 021 510530 F: 54 21 515737

Maiztegui, Cristina Elena, Ms
Director Medio Ambiente
MINISTERIO SALUD 1 ACCIONSOCIAL
65 No 423 (1900 la Plata)
Argentina

Maqueda, Eduardo Roberto, Mr
Adviser
MUNICIPALIDAD DE BUENOS AIRES
Esmeralda 740 10 Floor of 1007
(1425) Buenos Aires, Argentina
T: 54 1 393 5940/5947 F: 54 1 393 6046

Massei, Cristina, Ms
Coordinadora Evaluacion Ambiental
SECRETARID DE ENERGIA ELECTRICA
Alsina 1418 - 60 Piso Cap. Federal
Argentina, Argentina
T: 381 3988 / 381 3089 F: 334 7112

Menem, Carlos Saul, Mr.
President
Permanent Mission of Argentina
One United Nations Plaza,25th floor
New York NY 10017, United States of America
T: 212 688 6300 F: 212 980 8395

Morta Lucia, Nesta D Etacietti
Consejera Temp Mujer
Reconquista 1088 70PISO
Buenos Aires, Argentina
T: 541 311 4253

Rolon, Jose Manuel, Mr
Prefecto Mayor
PREFECTURA NAVAL ARGENTINA
Ave Madero N 235
Buenos Aires 1106, Argentina
T: 311 2664 F: 0054133 2876

Rubtno, Mariel
Arquttecta
RUNDATCON PBOMUBYA
Uruguay 950 4 To 8
Buenos Aries, Argentina
T: 42 8364 410547 F: 812 4599

Spinelli, Sara Matilde, Ms
Rectora
UNIVERSIDAD PERITOMORENO FUNDACION
Moreno 1623
Buenos Aires, Argentina
T: 01 361 4609 40 2897 F: 01 361 4609 40 2880

Zebellos De Sisto, Maria Cristina, Ms
Abogada
FUNDACION UNIVERSIDAD PERITO MORENO
Moreno 1621 Salguero 1509.80 (Codigo Postal 1093) (Codigo Postal
Buenos Aires, Argentina
T: 40 2897 40 2715 F: 01 361 4608

NGO

Araujo, Ariel
ACPG
Casilla de Correo 2
3332 Capiovy, Misiones, Argentina

Badaraco De Bevera, Susana
FEDU
Monroe 2142
(1428) Buenos Aires, Argentina

Bianco, Mabel, Sra.
Presidenta
FUNDACION PARA ESTUDIO Y INVESTIGACION DE LA MUJER
Parana 135 - 3 Piso 13
Buenos Aires 1017, Argentina
T: 54 1 476-2763 F: 54 1 274-274
TL: 18969

Di Pace, Maria J.
IIEDAL
Corrientes 2835-60 Piso-Cuerpo A
1193 Buenos Aires, Argentina

Geller, Horacio Sergio
FEDA
Paraguay 2342, Primer Piso
1121 Buenos Aires, Argentina

Gentile, Georgina, Ms.
Board Member
FUNDACION PARA LA DEFENSA DEL AMBIENTE - FUNAM
Avda Elflein 3775, 9 C, Olivos
Buenos Aires 1636, Argentina
T: 54 1 7900051 F: 54 1 3936380

Hernandez, Mario Alberto
FJER
Alsina 1816
1090 Buenos Aires, Argentina

Inet, Inet Jose Maria
ACPG
Casilla de Correo 2
3332 Capiovy, Misiones, Argentina

Mazzucchelli, Sergio Andre, Mr.
Investigator Associado
IIED-AMERICA LATINA
Corrientes 2835 - Cuerpo A - 6to piso 'B'
Buenos Aires 1193, Argentina
T: 54 1 961 3050 F: 54 1 961 1854

Onestini, Maria
Co-Director
CENTRO DE ESTUDIOS AMBIENTALES - CEDEA
C. Correo 116-Suc. 28
1428 Buenos Aires, Argentina

Palos, Claudio
Co-Director
CENTRO DE ESTUDIOS AMBIENTALES - CEDEA
C. Correo 116-Suc. 28
1428 Buenos Aires, Argentina

Perrone, Alejandra, Ms.
Secretary
FUNDACION PARA ESTUDIO E INVESTIGACION DE LA MUJER
Parana 135 - 3o. 13
Buenos Aires 1017, Argentina
T: 54 1 476-2763 F: 54 1 274 274

Risiga, Amilcar Hugo
Presidente
FUMDACON NATURA (ARGENTINA)
4 De Eneo 2481
3000 Santa Fe, Argentina
T: 54 42 40795 61895 F: 54 42 50944

Schroeder, Juan
CETAA
Rivadavia 4097, Marcos Paz
Provincia de Buenos Aires, Argentina

Tabares, Pablo Guillermo
CENTRO DE PROTECCION A LA NATUREZA
San Lorenzo 1582, Casilla Correo 538
(3000) Santa Fe, Argentina
T: 54 42 32425 F: 54 42 52219

Tarak, Pedro
FARN
Monroe 2142-1428
Buenos Aires, Argentina

Trombetta, Claudio Daniel
FEDA
Paraguay 2342, Primer Piso
1121 Buenos Aires, Argentina

Vigil, Carlos
Presidente
CONVOCATORIA PARA LA DEFENSA AMBIENTAL
Casilla de Correo No. 67, Suc. 2 (CP 1402)
Buenos Aires, Argentina

Voloschin, Clarissa
FEDU
Monroe 2142
(1428) Buenos Aires, Argentina

PRESS

Calaf, Rosa Maria
Corresponsal - Directora Para
TELEVISION ESPANHOLA
Viamonte 166 - 2o. piso
Buenos Aires, Argentina
T: 54 1 13 2692/440 6927 F: 54 1 13-2678

Marx, Gary
South American Correspondent
CHICAGO TRIBUNE
1662 Rodriguez Pena #2B
Buenos Aires, Argentina
T: 541 411662 F: 541 3947655

ORGANIZATIONS

Alvarez, Javier Francisco, Sr.
Centro de Proteccion a la Naturaleza (CPNAT)
San Lorenzo 1582, Casilla Correo 538
(3000) Santa Fe, Argentina
T: (54 42) 32 425 F: (54 42) 52 219

Cano, Guillermo J., Mr.
Fundacion Ambiente y Recursos Naturales (FARN)
Monroe 2142-1428
Buenos Aires, Argentina
T: (54 1) 781 6115/781 9171 F: (54 1) 781 6115

Canziani, Osvaldo, Mr.
Fund. Jorge Esteban Roulet/Inst. de Estudios e invest. sobre el Medio Ambiente
Alsina 1816
1090 Buenos Aires, Argentina
T: (54 1) 462 951/451 850

D'Alessandro, Maria, Lic.
Convocatoria para la Defensa Ambiental
Casilla de Correo No. 67, Suc. 2 (CP 1402)
Buenos Aires, Argentina
T: (54 1) 334 4273

Galan, Patricia, Lic.
Federacion Sindicatos Unidos Petroleros del Estado
Ave. Rivadavia 869 Piso 2 (1002)
Buenos Aires, Argentina
T: (54 1) 493 201/495 451 F: (54 1) 405 904

Joseph, Andres Hamilton, Sr.
Presidente
LOS ALGARROBOS
C.C. 96
La Cumbre - Cordoba 5178, Argentina
T: 54 548 52322 F: 54 548 51607
TL: 17661 Estalar

Montenegro, Raul A., Sr.
President
Fundacion Para La Defensa Del Ambiente (FUNAM)
Casilla de Correo 83, Correo Central
5000 Cordoba, Argentina
T: (54 51) 226 252 F: (54 51) 520 260

Pace, Maria Di, Ms.
Instituto Internacional De Medio Ambiente Y Desarrollo (IIED America Latina)
Corrientes 2835-60 Piso-Cuerpo A
1193 Buenos Aires, Argentina
T: (54 1) 961 3050 F: (54 1) 961 1854

Ramos, Lorenzo, Mr.
As. De Comunidades Del Pueblo Guarani
Casilla de Correo 2
3332 Capiovy, Misiones, Argentina
T: (54 752) 38 280 F: (54 752) 36 578

Seferche, Juan, Dr.
Fundacion Hernandiana
Balcarce 73, Villa Ballester
Buenos Aires, Argentina
T: (54 1) 768 0700

Zulkarnaren, M., Mr.
Fundacion Educambiente
Monroe 2142
(1428) Buenos Aires, Argentina
T: (541) 781 9171 F: (541) 781 6115

GLOBAL FORUM

Abaleron, Carlos Alberto, Mr.
Arquiteto
FUNDACION BARILOCHE
Lonquimay, 4102
Bariloche Zip.: 8400, Argentina
T: 54 944 22050 F: 54 944 22050

Abet, Jose Manuel, Sr.
Planificador
CENTRO ENTRERRIANO MEDIO AMBIENTE
Vucetich 186 - Parana
Entre Rios, Argentina
T: 54 043 224415
TL: 328

Abramides, Lic. Elias, Sr.
WORLD COUNCIL OF CHURCHES
Hipolito Yrigoyen 1512
Florida - Bs.As. 1602, Argentina

Acosta, Elba Amelia, Ms.
FUNDACION FRATERNITAS
Mitre 930
Rosario de Santa Fe, Argentina
T: 54 41 71-0663

Acuna, Ruben Marcelo, Mr.
FUNDACION DESARROLLO ECOLOGICO
Corrientes 1762 Piso 1, of. C
Buenos Aires, Argentina
T: 54 1 490099 F: 54 1 490099

Adamich, Carlos Nestor, Mr.
CEAV - CENTRO ESTUDIOS AMBIENTALES VETERINARIA
Fray Luis Beltran, 262
Rosario 2000, Argentina
T: 54 41 55 9714

Agostinelli, Graciela, Sra.
E.L.C.I. BOARD/ TALLER ECOLOGISTA
Casilla de Correo 56, Sucursal 8
Rosario, Argentina
T: 54 41 39-8677 F: 54 41 25-9089

Aiscorbe, Carlota, Sr.
FUNDACION ECOLOGICA UNIVERSAL
Avda. Corrientes 2763/7o. piso
Buenos Aires 1046, Argentina
T: 54 1 962-0237 F: 54 1 952-1535

Aizcorbe, Maria R. Sanchez, Ms.
President
PRO- TIGRE (FUND. ECOL. UNIVERSAL)
Estrada 1112 Tigre
Buenos Aires 1648, Argentina
T: 54 1 749-0800

Alasino, Gustavo, Mr.
Member
FUNDACION PARA LA DEFENSA DEL AMBIENTE - FUNAM
Casilla de Correo 83, Correo Central
Cordoba 5000, Argentina
T: 54 51 690282 F: 54 51 520260

Alberto, Fernadez Luis, Sr.
Estudiante
U.N.A.M
Tucuman 1830
Posadas -Misiones, Argentina
T: 047320711

Albo, Augusto
DIARIO EL CRONISTA
Honduras 5673
Buenos Aires, Argentina
T: 54 1 7753090 F: 54 1 7753694

Alderete, Jorge D., Sr.
Socio Activo
CONVOCJATORIA PARA LA DEFENSA DEL MEDIO AMBIENTE
Casilla de Correo 67, Suv. 2
Capital Federal, Argentina
T: 54 1 334-4273

Alderoky, Silvia, Sra.
Coordinadora Cs. Sociales
SECRE. EDUCACION
Jufre 796
Buenos Aires, Argentina
T: 54 1 773-2131

Alenedo, Regholo Miguel
Journalist
TELEFE
Pavon 24444
Argentina
T: 9413411

Alesandri, Carlos Tomas, Mr.
Intendente
MUNICIPIO EMBALSE
Hipolito Yrigoyen 252
Cordoba, Argentina
T: 54 51 0571-99258

Almiron, Daniel Eduardo, Sr.
Secretario General
MOVIMENTO ARGENTINO DE JUVENTUDE DE NACIONES UNIDADAS
9 de Julio 65 PB
Rosario de Santa Fe 2000, Argentina
T: 54 41 240006 F: 54 41 240006

Alonso, Ana Esther, Ms.
FUNDACION VIDA SILVESTRE ARGENTINA
Yujuy 267
Obera - Misiones 3360, Argentina
T: 59 755 21553

Alonso, Rafael, Mr.
Director
FUNDACION PROYDESA
Sarmiento 930/32
Buenos Aires, Argentina
T: 54 1 35 1888 F: 54 1 35 0150

Alonso, Ricardo Emilio, Sr.
Director
DIREC. CONSERV. SUELOS
Cordoba 1039
San Miguel Tucuman 4000, Argentina
T: 54 81 218150

Alvarez, Marcelo, Mr.
UNIVERSIDAD DE CORDOBA-FACULDAD DE CIENCIAS EXACTAS FISICAS Y NATURALES
Avenida Bariloche 537, Parque Horizonte
Cordoba, Argentina
E: CP 5000

Alvarez, Ruben Ruiz, Mr.
Miembro
MAJNU
9 de Julio 3785
Rosario de Santa Fe 2000, Argentina
T: 54 41 394844

Amat, Anibal, Mr.
Professor
UNIVERSIDAD NACIONAL DE MISIONES
Felix De Azara 1552
Posadas - Misiones 3300, Argentina
T: 54 752 22186

Ambrogetti, Abel, Sr.
Architect
Jujui, 1660
Corrientes, Argentina
T: 54 717 6-0017

Americo, Vianna Pedro, Sr.
COPASEL FEMICAP
Quiroga 772
Salto Encantado, Argentina
T: 54 755 98123

Amitrano, Marcelo, Mr.
Member
FUNDACION PARA LA DEFENSA DEL AMBIENTE - FUNAM
Casilla de Correo 83, Correo Central
Cordoba 5000, Argentina
T: 54 51 690282 F: 54 51 520260

Andia, Eduardo Gonzalez, Mr.
Director
BOSQUE ESMERALDA
42 No. 1520
La Plata 1900, Argentina
T: 54 21 246657

Andrinolo, Dario, Mr.
BIOGAS - MT - UNLP
522 Bis 1276
La Plata - Bs. As., Argentina
T: 54 1 294-4716

Antonini, Barbara, Sra.
Arquiteta
Entre Rios, 1
Corrientes, Argentina
T: 54 717 2-2287

Antoniow, Marta Victoria, Ms.
FUNDACION AMBIENTIS
Rivadacvia 3561- 5. P 27
Buenos Aires 1204, Argentina
T: 54 1 88-0130

Antuna, Juliana, Sra.
Directora Del Instituto Area
AYLLU-TINKUY
Bulnes 1638 #3B
Buenos Aires 1425, Argentina
T: 54 1 825-3692 F: 54 1 463349

Aoki, Monica, Sra.
Member Of Board Of Director
VIDA EN ALERTA
Felipe Boero 2054 - Rosedad
Cordoba 5010, Argentina
T: 54 51 603571 F: 54 51 241405

Aostri, Fernando, Mr.
INSTITUTO DE COOPERACION INTERNACIONAL
Chazbueo 387
San Luis, Argentina
T: 54 652 23088 F: 54 652 23088

Aparicio, Eduardo, Sr.
Vice President
PLANETA NEUSTRO
Dean Funes 1015 / CP 1231
Buenos Aires, Argentina
F: 54 1 342-6541

Apertile, Elsa, Ms.
WORLD YOUNG WOMEN'S CHRISTIAN ASSOCIATION
Humberto 1 2360
Buenos Aires 1229, Argentina
T: 54 1 941-3775
TL: WYMCA 41 23 32

Aprosio, Pablo Benetti, Mr.
DEFENSORIA DEL PUEBLO
Montevideo 2284
Rosario, Argentina
T: 54 41 48742 F: 54 41 66438

Apud, Emilio J., Mr.
Ingeniero
FUNDACON ARGENTINA Y EL MUNDO - FUNAM
Cerrito 836 - 5o. Piso
Buenos Aires 1010, Argentina
T: 54 1 40 4007 F: 54 1 46 9932

Aquino, Maria Cristina, Sra.
FUNDACION PROYDESA
SARMIENTO 930 - 3o. PISO
Buenos Aires, Argentina
T: 54 1 351888 F: 54 1 350150

Arcocha, Carlos Enrique, Mr.
Director Gen.
DEFENSORIA DEL PUEBLO
Montevideo 2284
Rosario, Argentina
T: 54 41 48742 F: 54 41 66438

Arecco, Aurelia Elena Roldan De, Ms.
Professora
ESCUELA NORMAL SUPERIOR 8
Elmensu 38
Iguazu - Misiones, Argentina
T: 54 757 20872

Arecco, Norberto Natalio Angel, Mr.
Director
E.N.E.T NUMERO 1 O.E.A.
Av. Victoria Aguirre 369
Pto Iguazu-Misiones, Argentina
T: 757 20872

Arenas, Joaquin H., Sr.
Member Consejo Administ.
FUNDACION ECOLOGICA UNIVERSAL
Corrientes 2763
Buenos Aires, Argentina
T: 54 1 962-0237 F: 54 1 952-1535
E: 1046

Arias, Elsa Perez, Sra.
Membro
A E FORO VERDE
Diagonal 79, no. 525
La Plata 2900, Argentina
T: 54 21 3-3057

Arias, Marcelo, Sr.
UNIVERSIDAD DE CORDOBA-FACULDAD DE CIENCIAS EXACTAS FISICAS Y NATURALES
Avenida Bariloche 537, Parque Horizonte
Cordoba 5000, Argentina

Ariza, Jose Omar, Mr.
JARDIN BOTANICO
colon 1270
Posadas 3300, Argentina

Arlettaz, Monica Del Carmen Rodriguez De
Dir De Meio Ambiente
CIMARU
Scelzi 378
C.del Uruguay,Ent.Rs 3260, Argentina
T: 54 442 27 623 F: 54 442 23 676

Arman, Raul Marcelo, Mr.
IPA - INVESTIGACIONES Y PROYECTOS AMBIENTALES
Espana 919 - 1o.B
Rosario, Argentina
T: 54 41 48129

Artaza, Antonio M. A., Mr.
FACULTAD CS. EX, QUIMICAS Y NATURALES
Sarmiento 468
Posadas 3300, Argentina

Artigas, Eleonora, Ms.
T E A
la Merced # 527 Ensenada
Buenos Aires, Argentina
T: 54 1 69-2433

Artime, Veronica Viviana, Ms.
Director
TEA
Diag. 107 no. 50
La Plata, Argentina

Aun, Elisa
FUNAM- ENVIROMENT DEFENCE FOUNDANTION
Alvear 1040
Rio Cuarto-Cordoba 5800, Argentina
T: 54 586 26347

Badgen, Irene Rut Wais De, Sra.
Chief
MUSEO ARGENTINO DE CIENCIAS NATURALES
Anasco 1792
Buenos Aires 1416, Argentina
T: 54 1 582-4522 F: 54 1 582-9973

Baldassarri, Daniela Paccosi De, Ms.
Dir Of Chemical Engeneering
EQUIPOS Y CONTROLES SA
Florida 537 - 6o. piso
Buenos Aires, Argentina
T: 54 1 3946281/4995 F: 54 1 3258644
TL: 24745

Baquedano, Maria Helena, Sra.
Teacher And Secretary
UNIVERSIDAD NACIONAL DE LA PLATA
14 n$ 555 - 1$ A
La Plata, Argentina
T: 54 21 24-5845

Barbosa, Helena Beatriz, Ms.
Secretaria Ecologia/Medio Amb.
FED. VEC. FUNES
Av. Tomas de La Torre, 489 bis
Funes 2132, Argentina
T: 54 41 93-1680

Barrandegui, Jorgelina, Ms.
Member
FUNDACION PARA LA DEFENSA DEL AMBIENTE - FUNAM
Casilla de Correo 83, Correo Central
Cordoba 5000, Argentina
T: 54 51 690282 F: 54 51 520260

Barreto, Maria Angela, Sra.
Active Member
ASOC. UNESCO
Chile M28 Dto.D
Corrientes, Argentina

Barreto, Ricardo A., Mr.
CORPORACION DEL DESAROLLO
Solis y Urquiza
Gualeguaycho- E.Rios, Argentina
T: 044626116 F: 044626116

Barrionuevo, Veronica P. Aballay, Ms.
Member
FUNDACION PARA LA DEFENSA DEL AMBIENTE - FUNAM
Casilla de Correo 83, Correo Central
Cordoba 5000, Argentina
T: 54 51 690282 F: 54 51 520260

Barsotti, Omar, Mr.
HONORABLE CONCEJO MUNICIPAL DE ROSARIO
Buenos Aires, 2521 - 7o. 6
Rosario 2000, Argentina
T: 54 41 818999

Basti, Abel Ricardo, Mr.
A.G.N.R.A
Paso 129, 1 andar / Casilla Correo 700
Bariloche, Argentina
T: 54 944 4422607 F: 54 944 4423654

Bazano, Adriana Beatriz, Ms.
ORGANISMO DE DERECHOS CRISTIANOS
La Valleya 1321 Dpto. 150
Cordoba, Argentina
T: 54 51 72 7858

Behrenz, Alfredo Jorge, Mr.
Head Of Ecology Department
M.A.G.I.C
Bv. Pellegrini 3100
Santa Fe 3000, Argentina
T: 54 42 62352

Belis, Claudio A., Sr.
Miembro
AMIDA
Calle 3 no. 1054
La Plata 1900, Argentina
T: 54 21 25-0261

Bello, Maria Julia, Ms.
Asesor.
CAMARA SENADORES
Calle 5 # 569
La Plata - Bs. As. 19001, Argentina
T: 54 21 33007

Bellorio, Dino Luis, Mr.
Member
CONVOCATORIA PARA LA DEFENSA AMBIENTAL
Alvarez Jonte 5084 C.P 1407
Buenos Aires, Argentina
T: 54 1 5669253 F: 54 1 3124891

Benavente, Andres, Sr.
Investigator
FUNDACION 'INTEGRACION'
Sarmiento 440 4o.P
Buenos Aires 1347, Argentina
T: 54 1 394-0275 F: 54 1 394-9227

Bercovich, Patricia, Sra.
President
FUND. NEHUEN
Palacios 474
Neuquen 8300, Argentina
T: 54 943 32492

Bernardello, Luis, Sr.
UNIVERSIDAD DE CORDOBA-FACULDAD DE CIENCIAS EXACTAS FISICAS Y NATURALES
Avenida Bariloche 537, Parque Horizonte
Cordoba 5000, Argentina

Bernardez, Rosa, Sra.
FUNDACION ECOLOGICA UNIVERSAL
Avda. Corrientes 2763/7o. piso
Buenos Aires 1046, Argentina
T: 54 1 962-0237 F: 54 1 952-1535

Berta, Sperber, Ms.
Directora
CENTRO DE ESTUDIO DE LA PERSONA
Arenales 1630 1A
Buenos Aires, Argentina
T: 54 1 416055

Berta, Waldman, Ms.
Directora
CENTRO DE ESTUDIO DE LA PERSONA
Arenales 1630 1A
Buenos Aires, Argentina
T: 54 1 416055

Bertolutti, Amanda, Sra.
RIE -RED INFORMATICA ECOLOGISTA - FEU
Avda. Corrientes 2763/7o. piso
Buenos Aires 1046, Argentina
T: 54 1 962-0237 F: 54 1 952-1535

Bezerra, Jorgelina Rodrigues, Sra.
President
PRO- TIGRE (FUND. ECOL. UNIV.)
Av. Sernambetiba 3600 Bl.6 apt.104 Cond. Atlantico Sul
Rio de Janeiro - RJ, Argentina
T: 55 21 439-1715

Bianchedi, Eugenia G., Sra.
Medical Director
CETAAR
Casilla de Correo 80
Marcos Paz - Bs.As. 1727, Argentina
T: 54 1 2207-2171 F: 54 1 334-7802

Bianco, Delia De, Ms.
Educadora
MUJERES DE NEGOCIOS Y PROFESIONALES - FAMNYA
Pedernera 958
San Luis, Argentina
T: 0652 20854

Biani, Ricardo D., Mr.
Director General Ecology
M.A.G.I.C
Bv. Pellegrini 3100
Santa Fe 3000, Argentina
T: 54 42 62352

Biasco, Hugo Horaciolar, Mr.
Forestal Tech.
MUNICIPALIDAD DE SAN MARTIN DE LA ANDES
The Cual Pery 89
S.M. Andes, Argentina
T: 54 972 27-315 F: 54 972 27334

Bilbao, Jorge, Mr.
FUNDACION AMBIENTE Y RECURSOS NATURALES
Monroe 2142 -1428
Buenos Aires, Argentina
T: 54 1 7816115 F: 54 1 7816115

Bilisich, Miguel, Mr.
Titular
COMUNA VILLA MUGUETA
Santa Fe 252 - Villa Muguena
Santa Fe, Argentina
TL: 2175

Bilmezis, Horacio Rafael, Mr.
Fundador
FUNDACION MADRE SILVA
la Rioja
la Rioja, Argentina
T: 54 822 26807

Birgin, Haydee, Sra.
CEADEL
Juan M. Gutierrez 3998 5th
Buenos Aires 1425, Argentina

Blanc, Graciela, Ms.
Arquitecta
BLOQUE RADICAL CAMARA DEPUTADOS DE E. RIOS
Toriente Gutierez, 124
San Salvador, Jujuy 3218, Argentina
T: 045 910 278 F: 045 910 108

Blebel, Rene M., Ms.
FEDEVI
Florida 868 3th G
Buenos Aires 1005, Argentina
T: 54 1 313 7578

Bluma, Pedro Omar, Sr.
Concejal
HONORABLE CONCEJO MUNICIPAL DE ROSARIO - HCM
Pje Ferranz no. 359 - (Ayacucho 6000)
Rosario 2000, Argentina
T: 54 41 83-3184 F: 54 41 25-8492

Bo, Maria Susana, Sra.
Professor
COLEGIO MAR DEL PLATA DE LAS COLINAS
Strobel 9032
Mar del Plata Bs.As. 7600, Argentina
T: 54 23 73-6828 F: 54 23 51-8422

Bocelli, Jose Maria, Sr.
Advogado
INSTITUTO FEDERAL DE ESTUDIOS PARLAMENTARIOS SENADO DE LA NACION ARGENTINA
Mendoza 281 - 1st piso
Buenos Aires 1428, Argentina
T: 54 1 786 8552

Boivin, Mauricio, Mr.
Director
DIRECCION DE FAUNA
Salta 233 Parana
Entre Rios 3100, Argentina
T: 07 043 211566

Bonazzi, Vicente Angel, Mr.
Professor
INSTITUTO LA SALETTE
Argentina
Buenos Aires 1430, Argentina
T: 54 1 541-5263 F: 54 1 542-2389

Borches, Walberto Manuel
M.E.L.
Albert Schweitzer 371 - Quilmes
Buenos Aires 1878, Argentina
T: 54 1 254 9491

Boscolo, Ana Maria, Ms.
Lawyer
MUNICIPALIDAD CIUDAD BUENOS AIRES
14 de Julio no. 1531
Buenos Aires, Argentina
T: 54 1 552-6049 F: 54 1 311-6840

Boshi, Ana Maria, Sra.
Teacher
FAC. DE TURISMO
Buenos Aires 1400
Neuquen, Argentina
T: 54 943 72537 F: 54 943 84266

Botta, Ricardo A., Sr.
Tecnico Agronomo
DIRECCION DE ECOLOGIA
Larrechea 2441 MB. 41 P.01 - D.04
Rosario de Santa Fe, Argentina
T: 54 41 55-4180

Bottazzi, Luis A., Sr.
CAMARA DE DIPUTADOS - ARGENTINA
Camara de Diputados
La Plata, Argentina
T: 54 21 25115

Brand, Ulrich, Mr.
FRIEDRICH EBERT FOUNDATION
Avenida Santa Fe 3242
Buenos Aires, Argentina

Bregaglio, Arturo E., Mr.
Promocion Y Desarroyo
CECOPAL
Av. Colon 1141
Cordoba 5000, Argentina
T: 54 51 223-528 F: 54 51 223-528

Bregagnolo, Raul Orlando, Mr.
Colaborador
GREENPEACE F.VIDA SILVESTRE
Leandro M Alem 596- A. del Valle
Misiones 3364, Argentina
T: 54 075590397

Breitman, Sara B De, Sra.
INTERNATIONAL COUNCIL OF JEWISH WOMEN
Av. Libertador 3672- P10
Buenos Aires 1425, Argentina
T: 54 1 773-5582 F: 54 1 983-0930

Briziela, Jose Antonio, Mr.
Organizador De Encontro
E C O V A L
R Sarmiento 890
Rio Ceballos 5111, Argentina
T: 54 51 75 2104 F: 54 51 952109

Brodtmann, Lilian, Ms.
Miembro
ASOCIACON UNESCO PAS, AMB. Y DES
Acuna de Figueroa 1511 70B
Buenos Aires, Argentina
T: 54 1 892355 F: 54 1 892355

Bronzoni, Viviana, Ms.
FACULTAD CS. EX, QUIMICAS Y NATURALES
San Lorenzo 280
Posadas 3300, Argentina

Budkin, Yanina, Ms.
Colaboradora
FUNDACION PARA ESTUDIO E INVESTIGACION DE LA MUJER
Parana 135 - 3o. 13
Buenos Aires 1017, Argentina
T: 54 1 476-2763 F: 54 1 274 274

Bulacio, Fabian, Mr.
Profesor Educacao Fisica
C.69 Number 976 - la Plata
Buenos Aires 1900, Argentina
T: 54 021 524531

Burghi, Graciela Del Carmen, Ms.
Fundadora
ALUMNOS DEFENSORES DE LA NATUREZA
Cordoba y Cuffia no. 1707
Armstrong, Argentina
T: 54 61607

Burlando, Jorge Alberto, Mr.
UNIVERSIDAD NACIONAL DEL SUR
Universidad Nacional del Sur
Bahia Blanca, Bs.As. 8000, Argentina
T: 54 91 44987

Bustos, Eduardo, Sr.
CAMARA DE DIPUTADOS - ARGENTINA
Camara de Diputados
La Plata, Argentina
T: 54 21 25115

Bustos, Fabiana, Ms.
GRUPO ARQHITECTURA DE ACCION
Jeronimo Cortez 554
Cordoba, Argentina
T: 54 51 71-9434
E: 5147

Cabouli, Claudia Patricia, Ms.
Coordinacion Museologica
DIRECCION NACIONAL DE MUSEOS
Cervino # 3741/2 B
Buenos Aires 1425, Argentina
T: 54 1 801-0785

Caceres, Claudia Graciela, Sra.
Estudiante
ESCOLA INDUSTRIAL DE PUERTO IGUAZU
Villa Nueva - Urquiza 124
Pto Iguazu 3370, Argentina
T: 075720839

Cafiero, Juan Pablo, Sr.
Diputado De La Nacion
H. CAMARA DE DIPUTADOS DE LA NACION
Riobamba 25-Oficina 1330
Buenos Aires, Argentina
T: 54 1 9541111 F: 54 1 3135793
TL: 17549CEDIPAR

Cajal, Jorge Luis, Sr.
Comite Cientifico
FUCEMA
Diaz Velez 740, 1 P DEP 3
la Lucila, Argentina
T: 799 1847

Caldiero, Victorino, Mr.
Direccion Gobierno
MUNICIPALIDAD
Espana y Urquiza
Cipolletti 8324, Argentina
T: 943-73801 F: 74602

Calero, Daniel, Mr.
Priest
UNITED LUTHERAN CHURCH
Gob. Barreyro 921
Obera - Misiones 3360, Argentina
T: 59 755 21119

Calvino, Luis
Acesor
FUNDACION PROYDESA
SARMIENTO 930 32
Buenos Aires, Argentina
T: 54 1 35-1888

Camarero, Graciela, Ms.
Licenciada En Genetica
FAC. CIENCIAS EXATAS- UNAM
Felix de Azorra 174 -or Ap. Postal 22-3370 -Pto. Iguazu
Posadas-Misiones, Argentina
T: 054 0757 20382

Cambrini, Maria Margarita, Ms.
Vice Chairman
ARC PEACE (ECO-GEO ARGENTINA)
Av. Roca (S) no. 811
Santiago del Estero 4200, Argentina
T: 54 85 214170 F: 54 85 224350

Cambrini, Silvia Beatriz, Ms.
Secretary
ARC PEACE (ECO-GEO ARGENTINA)
Av. Roca (S) no. 811
Santiago del Estero 4200, Argentina
T: 54 85 214170 F: 54 85 224350

Campanella, Marcelo Gustavo, Mr.
CIRCULO ROSARIONA DE ESTUDOS PARAPSIQUICOS
Oatamaroa 3352 P.Alta
Rosario 2000, Argentina
T: 54 41 389322

Campanella, Roberto, Sr.
Asesor Diputado Bermudez
CAMARA DIPUTADO
San Lorenzo 1754
Rosario de Santa Fe, Argentina
T: 54 41 252 454 F: 54 41 252 454

Campilongo, Marcela, Ms.
Medico
FLACAM
Tucuman 1563 / 5 Piso
Rosario de Santa Fe, Argentina

Camps, Sibila
Journalist
DIARIO 'CLARIN'
Piedras 1743
Buenos Aires, Argentina
T: 270061 F: 239003

Camusso, Rosa, Ms.
Polonia 365
Lomas del Mirador, Argentina

Candia, Lilian Natalia, Sra.
Estudiante
ESCOLA INDUSTRIAL DE PUERTO IGUAZU
Barrio Ignacio Abiaru E.-12
Puerto Iguazu 3370, Argentina

Candioti Leiva, Alicia Graciela
Asesora De Educacion
CONCEJO DELIBERANTE - BUENOS AIRES
Peru 130/4 Piso-Anexo
Capital Federal, Argentina
T: 54 1 804 1234

Candugiia Pivorano, Jose Augusto, Sr.
FUNDACION ARGENTINA DE ECOLOGIA - FADE
Avda of Los Italianos 365-501
Buenos Aires 1107, Argentina
T: 54 1 313-2948 F: 54 1 313-2948

Cano, Guillermo S., Sr.
H. President
ASOCIACION INTERNACIONAL DE DERECHO DE AGUAS
Arenales 2040 piso 7-B
Buenos Aires, Argentina
T: 54 1 847292 F: 54 1 781-6115
TL: 22088 CARTE AR

Cantarell, Eduardo, Mr.
Expositor
F.E.U.
Av. Felipe Varela 150 - C.P. 4700
Catamarca, Argentina
T: 054 833 27696

Cape, Graciela, Sra.
Professora
CONSDEJO PROVINCIAL EDUCACION
Villegas 701
S.M. Andes, Argentina
T: 54 972 27070 F: 54 972 28091

Capezio, Daniel, Sr.
Estudiante
ESCOLA INDUSTRIAL DE PUERTO IGUAZU
Esquiu 180
Pto. Iguazu 3370, Argentina
T: 075721273

Cappato, Jorge Alberto, Mr.
Executive Director
FUNDACION PROTEGER
Casilla de Correo 550
Santa Fe 3000, Argentina
T: 54 42 35339 F: 54 42 52293

Carbaio, Vilma, Sra.
FUNDACION ECOLOGICA UNIVERSAL
Avda. Corrientes 2763/7o. piso
Buenos Aires 1046, Argentina
T: 54 1 962-0237 F: 54 1 952-1535

Carballo, Guillermo, Mr.
Vice President
FUNDACION DESARROLLO ECOLOGICO
Corrientas 1762 - 1o. C
Buenos Aires, Argentina
T: 54 1 49-0099 F: 54 1 49-0099

Cardozo, Norberto Gerardo, Mr.
El Ratologo
EMPRESARIO RATOLOGO
Calle 9 No. 837 Piso 6, Dto. 'A'
La Plata 1900, Argentina
T: 54 21 24-4606

Carena, Mauricio, Mr.
Member
FUNDACION PARA LA DEFENSA DEL AMBIENTE - FUNAM
Casilla de Correo 83, Correo Central
Cordoba 5000, Argentina
T: 54 51 690282 F: 54 51 520260

Careri, Elsa, Sra.
Asesora
LEGISLATURA DE LA PROVINCIA DE JUJUY
San Salvador, Jujuy, Argentina

Carlitti, Julio Roberto
Empleado, Locutor Radial Y Eco
RADIO FM 106.9
Buenos Aires, Argentina
T: 54 1 6201833

Carlos, Martinez Salinas,
Journalist
LA RIBERA
770 Correo Central
Rosario -Santa Fe 20, Argentina
T: 54 41 250675 F: 54 41 829450

Carranza, Jeronimo, Mr.
Director
TEA
Calle 510 /1915
La Plata, Argentina

Carranza, Marcelo, Sr.
CONCEJO DELIBERANTE DE LA PLATA
12 y 51
La Plata, Argentina
T: 54 21 36978 F: 54 21 256384

Carrasco, Dino David Parra, Mr.
Director
A.CU.FO.M.
W. Nunez 744
Rivadavia 5577, Argentina
T: 062343077

Carrasco, Marcelo Angel Javier, Sr.
Estudiante
ESCOLA INDUSTRIAL DE PUERTO IGUAZU
Urquiza, 41
Pto Iguazu 3370, Argentina

Carreira, Ariel Emilio, Mr.
GRUPO NATURA
Avda. Eden 385
La Falda - Cordoba 5172, Argentina
T: 54 8 22781

Carreras, Ana Maria, Ms.
Representante De Delegacion
U.N.L.Z.
Lavalle 3597 Remedios de Escalada
Buenos Aires 1826, Argentina
T: 54 1 248-1395

Carreras, Luis Dario, Mr.
Administrative Director
FUNDACION PROTEGER
Casilla de Correo 550
Santa Fe 3000, Argentina
T: 54 42 35339 F: 54 42 52293

Carrero, Diana, Ms.
FUNDACION PROTEGER
Casilla de Correo 550
Santa Fe 3000, Argentina
T: 54 42 35339 F: 54 42 52293

Carrillo, Eduardo Marcelo, Sr.
LEGISLATURA DE LA PROVINCIA DE JUJUY
San Salvador, Jujuy, Argentina

Castelli, Jorge, Mr.
Jornalista
RADIO NACIONAL
Lafinur y Ayacucho
San Luis - SL, Argentina
T: 24771 F: 25539

Castelvitri, Silvina, Ms.
Biology Student
UNIVERSIDEAD NACIONAL DE CORDOBA
Progresso 1850 - Bo.Villa Cabrera CP.5009
Cordoba, Argentina
T: 54 51 807839

Castro, Adrian Adolfo, Mr.
Plancamiente
MUNICIPALIDAD
Espana y Urgniza - Cipolletti
Rio Negro 8324, Argentina
T: 54 944 73 801

Castro, Ma. Mercedes Del Valle, Sra.
LEGISLATURA DE LA PROVINCIA DE JUJUY
San Salvador, Jujuy, Argentina

Catala, Pablo Esteban, Sr.
Asesor En Medio Ambiente
MUNICIPALIDAD DE LA PLATA
Calle 8, 728, piso 5, depto. C
La Plata, Argentina
T: 54 21 3-0862 F: 54 21 255004

Cecchi, Silvina, Sra.
Coordenadora
INSTITUTO DE CALIDAD TOTAL
Paraguay 1345, 8o. piso
Buenos Aires, Argentina
T: 54 1 41-9719-20 F: 54 1 41-9722

Cecchini, Carlos Alberto, Mr.
Head Of Ecology Department
M.A.G.I.C
Bv. Pellegrini 3100
Santa Fe 3000, Argentina
T: 54 42 62352

Cervantes, Raquel Silmara, Sra.
Student
FCEFYU / U.N.C.
Gral Paz 2682
Cordoba 5000, Argentina
T: 54 51 713 420

Cettour, Hugo, Mr.
SENADO/ COMISSION DE SAUDE PUBLICA E MEIO AMBIENTE
Eva Peron 181
Entre Rios, Argentina

Chemin, Pablo Daniel, Mr.
AGRUPACION ECOLOGISTA IBIRA PUCU / E.C.E.N.A.A.
Colon 78 - Parana
Entre Rios 3100, Argentina
T: 54 43 213 248

Chiesa, Alberto, Mr.
Member
CONVOCATORIA PARA LA DEFENSA AMBIENTAL
Alvarez Jonte 5084 C.P 1407
Buenos Aires, Argentina
T: 54 1 5669253 F: 54 1 3124891

Cinquini, Jorge Omar, Mr.
Head Of Ecology Department
M.A.G.I.C
Bv. Pellegrini 3100
Santa Fe 3000, Argentina
T: 54 42 62352

Cinti, Elena, Sra.
Architector
ESCUELA 134 - NEUQUEN
Koessler 2119
Neuquen, Argentina
T: 54 943 27117 F: 54 943 27334

Cirino, Julio, Sr.
Director Area International
FUNDACION 'INTEGRACION'
Sarmiento 440 4o.P
Buenos Aires 1347, Argentina
T: 54 1 394-0275 F: 54 1 394-9227

Clerico, Agustin Vargas, Mr.
Biology Student
UNIVERSIDAD NACIONAL DE CORDOBA
CC. 1023 Correo Central
Cordoba 5000, Argentina
T: 54 51 606061 F: 54 51 241405

Cobas, Gladys
PRODUFE
Viamonte 1145
Buenos Aires PB 10, Argentina
T: 46 0851

Cocconi, Henry Miguel Amadeo, Dr.
Presidente
HONORABLE CONSEJO DELIBERANTE
Crisostomo Alvarez 723
San Miguel de Tucuma, Argentina
T: 54 81 238050 F: 54 81 311524

Cocconi, Monica Cristina Merli De, Sra.
HONORABLE CONSEJO DELIBERANTE
Crisostomo Alvarez 723
San Miguel de Tucuma, Argentina
T: 54 81 238050 F: 54 81 311524

Colagion, Lidia, Sra.
FUNDACION ECOLOGICA UNIVERSAL
Avda. Corrientes 2763/7o. piso
Buenos Aires 1046, Argentina
T: 54 1 962-0237 F: 54 1 952-1535

Colcombet, Patricia Keller, Ms.
Engeneer
FUNDACION VIDA SILVESTRE
C.C. 149 Cae
Misiones, Argentina
T: 54 075121351 F: 54 22000

Contardi, Norma, Mr.
ASOCIACION GAIA ARGENTINA
Almafuerte 1732, Villa Maipu, San Martin
Buenos Aires, Argentina
T: 54 1 752-1549 F: 54 1 752-1549

Cooper, Maria Rosa Cecilia
FREELANCER
Mendoza 310 3A,
Cordoba 5000, Argentina
T: 54 51 242945 F: 54 51 214123

Coppa, Juan Jose, Mr.
Lic En Biologia
DIRECCION DE CULTURA Y TURISMO DE TRES ARROYOS
Pte Roca 786 90. Piso
Rosario de Santa Fe, Argentina
T: 54 41 2588963

Coppola, Gustavo E., Sr.
Concejal
HONORABLE CONCEJO DELIBERANTE DE LA CIUDAD DE BUENOS AIRES
Peru 130, 1er. piso, oficina 108
Capital Federal, Argentina
T: 54 1 342-7051 2224 2226 F: 54 1 334-1823

Corbellini, Jorge Luis, Mr.
VIDA EN ALERTA
25 de Mayo 73, P.A. Oficina 6
Cordoba, Argentina
T: 54 51 37924

Corona, Norma Liliana
Docente
ESCUELA PROVINCIAL N8 'ALMAFUERTE'
La Picada Dpto Parana
Parana - Entre Rios, Argentina

Corra, Lilian Alcira, Ms.
President
FUNDACION PROTEGER
Casilla de Correo 550
Santa Fe 3000, Argentina
T: 54 42 35339 F: 54 42 52293

Corral, Julio Cesar, Sr.
Investigator
FUNDECMA
Cordoba 1090
San Miguel Tucuman 4000, Argentina
T: 54 81 217619 F: 54 81 311524
TL: 3361143

Correa, Nelida Elma, Ms.
INSTITUTO DE YOGA Y DISCIPLINAS ALTERNATIVAS 'OSIRIS'
Oayamarca 9338
Rosario 2000, Argentina
T: 54 41 386686

Cortes, Martin, Sr.
CAMARA DE DIPUTADOS - ARGENTINA
Camara de Diputados
La Plata, Argentina
T: 54 21 25115

Corti, Daniel Juan, Mr.
Iparraguire 203- Ptc. Deryui
Buenos Aires 1635, Argentina
T: 322 85800

Costero, Dario Daniel Garcia, Mr.
Proydesa
FUNDACION PROYECTOS Y DESAROLLOS ARGENTINOS
Sarmiento 930 - 3o. Piso
Buenos Aires 1386, Argentina
T: 54 1 35-1888 F: 54 1 35-0150

Cremaschi, Gustavo, Sr.
CONCEJO DELIBERANTE DE LA PLATA
12 y 51
La Plata, Argentina
T: 54 21 36978 F: 54 21 256384

Crevico, Agustin Vargas
FUNDACION TELEMACO
C.C. 1023 Corres Central
Cordoba 5000, Argentina
T: 54 51 606061 F: 54 51 241405

Cristaldo, Dario Marcelo F., Sr.
Estudiante
ESCOLA INDUSTRIAL DE PUERTO IGUAZU
Felipe Varela 188 (Villa Alta)
Iguazu - Misiones 3370, Argentina

Curto, Erio, Mr.
UNIVERSIDADE NACIONAL CORDOBA
independencia 1225 b. nueva
Cordoba, Argentina
T: 54 51 602209

Cutropia, Alberto Mario, Mr.
UNIVERSIDAD DE MENDOZA
Boulogne Sur Mer 665
Mendonza 5500, Argentina
T: 54 61 24-7017 F: 54 61 24-7017
TL: 18660

D'Errico, Gladis Ines, Sra.
LEGISLATURA DE LA PROVINCIA DE JUJUY
San Salvador, Jujuy, Argentina

Dall' Armellina, Eduardo, Mr.
Assesor
COMUNA V. MUGUETA
Santa Fe 252 - Villa Mugueta
Santa Fe, Argentina
T: 54 4269 97014
TL: 2175

Damian, Abril, Mr.
GALILEO GALILEI (PUNAM)
Alvear - Rio Cuarto
Cordoba, Argentina

Daverio, Sergio, Sr.
Docent
UNIVERSIDAD TECNOLOGICA NACIONAL
Pronunciamiento s/n
Parana - PR, Argentina
T: 54 43 243694 F: 54 43 243589
E: Alma/verte 1033

Del Bianco, Milena, Sra.
Student
FCEFYU / U.N.C.
J.M. Estrada 144 3o. D sol III
Cordoba 5000, Argentina

Del Valle, Guillermo Gustavo, Mr.
Presidente
ASO. DIF. PROG.NACIONES UNIDAS MEDIO AMBIENTE
27 De Abril N. 864
Cordoba 5000, Argentina
T: 54 51 226304 F: 54 51 226304

Delgado, Hilda Felisa, Sra.
Secretaria
LEGISLATURA DE LA PROVINCIA DE JUJUY
San Salvador, Jujuy, Argentina

Della Cha, Cecilia Ines, Sra.
COPADE
Belgrano 370
Neuquen, Argentina
T: 54 943 23727 F: 54 943 22750
E: 9300

Destefanis, Ariana, Ms.
Member
FUNDACION PARA LA DEFENSA DEL AMBIENTE - FUNAM
Casilla de Correo 83, Correo Central
Cordoba 5000, Argentina
T: 54 51 690282 F: 54 51 520260

Devia, Leila, Sra.
INTI
Av. L.N. Alem 1067-70 Piso
Buenos Aires, Argentina
T: 54 1 313-4008 F: 54 1 313-2130

Di Pacce, Alfredo Antonio, Mr.
Oficina Ejecutiva
CENTRAL L. AMERICANA DE TRABAJADORES - CLAT
Calle Alberti 36
Buenos Aires 1082, Argentina
T: 54 1 953 2776 F: 54 1 953 2763

Di Placido, Daniel, Sr.
CONCEJO DELIBERANTE DE LA PLATA
12 y 51
La Plata, Argentina
T: 54 21 36978 F: 54 21 256384

Diaz, Carlos A., Sr.
CAMARA DE DIPUTADOS - ARGENTINA
Camara de Diputados
La Plata, Argentina
T: 54 21 25115

Diaz, Cristina Rojas De, Sra.
Investigator
FUNDECMA
Alsina 2719
San Miguel Tucuman 4000, Argentina
T: 54 81 217619 F: 54 81 311524

Diaz, Daniel, Sr.
UNIVERSIDAD DE CORDOBA-FACULDAD DE CIENCIAS EXACTAS FISICAS Y NATURALES
Avenida Bariloche 537, Parque Horizonte
Cordoba 5000, Argentina

Diaz, Hugo, Mr.
Estanislao Del Campo, 1002 - Florida
Buenos Aires 1602, Argentina
T: 54 1 7606092 F: 54 1 3117500

Diaz Montero, Maria Ines, Ms.
Rel. Publicas - Traductor
CONVOCATORIA PARA LA DEFENSA AMBIENTAL
Casilla Correo 67, Suc. 2
Buenos Aires 1402, Argentina
T: 54 1 334-4273

Dieguez, Damian Sabino, Sr.
Gran Canarias 926
Quilmes - Bs. As. 1878, Argentina

Dignani, Jose Carlos, Sr.
RED LATINOAMERICANA DE ECOSOFIA
Tucuman 1059 Dto. 5
Rosario, Argentina
T: 54 41 21 9676

Dolber, Marisa Rut, Sra.
Tecnico Docente
DIRECCION DE ECOLOGIA
Larrechea 2441 MB. 41 P.01 - D.04
Rosario de Santa Fe, Argentina
T: 54 41 55-4180

Domingo, Alberto Virdo, Mr.
Lawyer
FUNDACION AMBIENTIS
Lavalle 1282/Piso 2/ Oficina 14
Buenos Aires, Argentina
T: 54 1 35-5518 F: 54 1 785-4020

Dotta, Ruben, Mr.
GRUPO NATURA
Volivar 539
Rio III - Cordoba, Argentina

Dovio, Nora Maria, Sra.
Professora
CENTRO PROVINCIAL ENSENANZA MEDIA NO.12
BP Co.Vi. Sal casa No.94
S.M. Andes, Argentina
T: 54 972 28060

Drovetta, Diogenes, Sr.
Miembro
MAJNU
Balcarce 1471
Rosario de Santa Fe 2000, Argentina
T: 54 41 217936

Easdale, Allan, Mr.
Civil Engineer
Casilla de Correo 1
Jujuy, Argentina
T: 54 25077

Easdale, Susana De Jong De, Sra.
DIR. AGUA POTABLE
El Naranjo No. 15 - Barrio Los Perales
S.S. de Jujuy-Jujuy, Argentina
T: 25077

Elgue, Raul Alfredo Vanotti, Mr.
INSTITUTO NACIONAL DE TECHONOLOGIA AGROPECUARIA
Rocamora 55, 3153 Victoria - Entre Rios
Enugu, Argentina
T: 234 4201 F: 234 4201

Elizondo, Ramon Rosas, Mr.
CAMARA DE DIPUTADOS
San Juan, Argentina
T: 1 809 064222750

Els, Erica, Sra.
UNIVERSIDAD DE CORDOBA-FACULDAD DE CIENCIAS EXACTAS FISICAS Y NATURALES
Avenida Bariloche 537, Parque Horizonte
Cordoba 5000, Argentina

Encabo, Matilde, Sra.
Assistent
FAC. DE TURISMO
Buenos Aires 1400
Neuquen, Argentina
T: 54 943 72537 F: 54 943 84266

Epelbaum, Renee, Ms.
Relaciones Exteriores
MADRES DE PLAZA DE MAYO/LINEA FUNDADORA
Larrea 1058
Buenos Aires 1117, Argentina
T: 54 1 835893 F: 54 1 812 3406

Escobar, Susana A. D., Sra.
FEDEVI
La Pampa 1424
Buenos Aires, Argentina
T: 54 1 782-6869

Esparza, Ernesto, Mr.
DEPARTAMENTO DE INGENIERIA AMBIENTAL
C.C. 961 - Bariloche
Rio Negro 8400, Argentina
T: 54 944 22121 F: 54 944 23051

Esper, Mario, Mr.
Medico
MINISTERIO SALUD PUBLICA - MISIONES
San Martin 1654 (Ex-421)
Posadas 3300, Argentina

Esteban, Andrea San
RADIO FM 97
Paraguay 191
Rosario, Santa Fe, Argentina

Estevez, Carlos, Mr.
ASOCIACION GAIA ARGENTINA
Almafuerte 1732 Villa Maipu, San Martin
Buenos Aires, Argentina
T: 54 1 752-1549 F: 54 1 752-1549

Estevez, Diana, Ms.
A E M A B
Calle 15 # 5675- Berazategui
Buenos Aires 1884, Argentina
T: 54 1 256-2032

Estrada, Paula, Sra.
COLEGIO DE ABOGADOS SAN ISIDRO
Billinghurst 1565 8A
Buenos Aires, Argentina
T: 54 1 824-2698

Evangelista, Robert, Sr.
Member
COMITE UCR
Paseo Colon 669
Buenos Aires, Argentina
T: 54 1 334-5114

Fantin, Maria Alejandra
Educacion Sindical
CONFEDARECION DE TRABAJADORES DE LA EDUCACION DE LA REPUBLICA ARGENTINA - CTERA
San Buenaventura Del Monte Alto 481
Resistencia, Argentina

Farias, Oscar Jorge, Sr.
President
ASOC. UNESCO
Roca 13
Corrientes, Argentina
T: 54 717 63800

Farre, Stella Maris, Sra.
Asesora
HONORABLE CONCEJO DELIBERANTE CIUDAD DE BUENOS AIRES
Peru 130, 1er. piso, oficina 108
Capital Federal, Argentina
T: 54 1 342-7051-9 2224 F: 54 1 334-1823

Fasce, Alfredo Luis, Mr.
COMISION ASESORA DE ECOLOGIA Y MEDIO AMBIENTE CONCEJO MUNICIPAL DE ROSARIO
Entre Rios, 1372 - Pso 1 - Bto 9
Rosario 2000, Argentina
T: 54 41 210421

Feijoo, Enrique Albeto, Mr.
AMIGOS DE LA ECOLOGIA
Bolivar 395
Tucuman, Tucuman 4000, Argentina
T: 54 81 214566 F: 54 81 227866

Ferloli, Daniel, Dr.
Member
COMUNIDAD BAHA'I - ARGENTINA
Otamendi 215
Buenos Aires, Argentina
T: 54 1 9011078 F: 54 1 9011076

Fernandez, Carlos R., Mr.
Journalist
RADIO Y PRE
Bolivar
Posadas, Argentina
T: 54 752 30570 F: 54 752 30570

Fernandez, Esther, Sra.
Asesora Tecnica
COPADE
Belgrano 370
Neuquen 8300, Argentina
T: 54 943 23-727 F: 54 943 23-152
TL: 84175 Gonevar

Fernandez, Gustavo, Ms.
PRO- TIGRE (FUND. ECOL. UNIVERSAL)
Estrada 1112 Tigre
Buenos Aires 1648, Argentina
T: 54 1 749-0800

Fernandez, Miguel
DIARIO EL CRONISTA
Honduras 5673
Buenos Aires, Argentina
T: 54 1 7753090 F: 54 1 7753694

Fernandez, Norma C.
REVISTA 'ALTERNATIVA LATINAMERICANA'
Rua Velez Sarsfield 1278
Cordoba 5000, Argentina
T: 51 605707 F: 51 605707

Fernandez, Sergio Antonio, Mr.
Cordinador Area Paisajismo
ASOC. UNESCO
Salta 852
Corrientes, Argentina
T: 54 717 0783-22044

Ferre, Norberto, Mr.
INSTITUTO JUAN MANTOVANI
Garcia del Rio,4559
Buenos Aires, Argentina
T: 54 1 5410888 F: 54 1 393 6767

Ferreras, Julian Alberto, Mr.
FACULTAD CS. EX, QUIMICAS Y NATURALES
Sarmiento 468
Posadas 3300, Argentina

Ferrero, Alejandra, Sra.
Membro
A E FORO VERDE
Diagonal 79, no. 525
La Plata 1900, Argentina
T: 54 21 3-3057

Ferrier, Maria Rosa, Ms.
Fundacion Jardin Botanico
Blas Parera 2526
Posdadas 3300, Argentina
T: 0752-36687

Festa, Alicia Beatriz, Sra.
Architect
FLACAM
Superi 2135 - 2/ 19
Buenos Aires, Argentina
T: 54 1 543-0505 F: 54 1 312-6612

Fidyka, Leopoldo Javier, Sr.
Director Del Instituto Area Ec
AYLLU-TINKUY
Sarmiento 4696 3N
Buenos Aires 1197, Argentina
T: 54 1 862-6718 F: 54 1 463349

Figueroa, Arturo, Mr.
Concejal
HONORABLE CONCEJO DELIBERANTE
24 de Setiembre 346
San Miguel Tucuman, Argentina
T: 54 81 311118

Figueroa, Nelson M., Mr.
APROBA AGRO. PRODUCTORES ORGANICOS DE BUENOS AIRES
Vicente Lopez 2698
Quilmes 1878, Argentina
T: 54 1 254-2222 F: 54 1 253-8727

Figueroa, Ricardo, Mr.
Paisagista-Coordenador
MUNICIPALIDAD DE ROSARIO
Moreno 2300
Rosario de Santa Fe 2000, Argentina
T: 54 41 81-1932

Firszt, Natalio David, Mr.
Architect
UNIVERSIDAD DE BUENOS AIRES
14 de julio no. 1531
Buenos Aires, Argentina
T: 54 1 552-6049 F: 54 1 311-6840

Fitipaldi, Mario, Sr.
CAMARA DE DIPUTADOS - ARGENTINA
Camara de Diputados
La Plata, Argentina
T: 54 21 25115

Flores, Horacio Calzon, Mr.
Member
AMIGOS DE LA TIERRA
Rio de Janeiro 500, 1405
Buenos Aires, Argentina

Flores, Nicolas Nolasco, Sr.
LEGISLATURA DE LA PROVINCIA DE JUJUY
San Salvador, Jujuy, Argentina

Florin, Raimundo, Sr.
Associate
BUSINESS COUNCIL FOR SUSTAINABLE DEVELOPMENT
Maiph 1300, 3o. Piso
Buenos Aires, Argentina
T: 54 1 312-5599 F: 54 1 312-5599

Foguelman, Dina, Ms.
M.A.E.
MOVIMENTO ARGENTINO ECOLOGICO
Camargo 771
Buenos Aires, Argentina
T: 54 1 856-3891 F: 54 1 802-1266

Folgarait, Alejandra
Director
CENTRO DE DIVULGACAO CIENTIFICA/UNIVERSIDAD DE BUENOS AIRES
Catamarca 106 12A
Buenos Aires, Argentina
T: 54 1 76692

Fontanilla, Elba Elena Vedoya De, Ms.
Arquiteta
PROVINCIA DE SALTA - ARGENTINA
Caseros 1523
Salta 4400, Argentina
T: 54 87 218-587

Francia, Aldo Gabriel, Mr.
Lic. De Biologia
DIRECCION DE CULTURA Y TURISMO DE TRES ARROYOS
S. Costa 4482
Tres Arroyos, Argentina
T: 098326016

Francisco Jose, Eszter, Mr.
MUNICIPIO DE LA MATANZA
Arjeta 4935
Peja, Buenos Aires, Argentina
T: 651 0565 F: 651 03 3045

Franco, Norberto Mario, Mr.
Pte. Peron, 1457 - 2 Cuerpo - 1er.piso - OF.17
Buenos Aires 1037, Argentina
T: 54 1 405536

Franza, Jorge Atilio, Mr.
Member
CONVOCATORIA PARA LA DEFENSA AMBIENTAL
Alvarez Jonte 5084 C.P 1407
Buenos Aires, Argentina
T: 54 1 5669253 F: 54 1 3124891

Frazier, Jeff L., Mr.
FUNDACION AMBIENTE Y RECURSOS NATURALES
Esmeralda 933
Buenos Aires, Argentina
T: 54 1 89-0644

Frederick, Pablo, Sr.
Coord. Adjunto
SERPAJ-AL
Mexico 479
Buenos Aires 1997, Argentina
T: 54 1 348266 F: 54 1 3307036

Fuente, Alejandro De La
FUNDACION ECOLOGICA UNIVERSAL
Avda. Corrientes 2763/7o. piso
Buenos Aires 1046, Argentina
T: 54 1 962-0237 F: 54 1 952-1535

Gabbi, Alejandro, Mr.
Director De Parques Y Paseos
MUNICIPALIDAD DE ROSARIO
Moreno 2300
Rosario de Santa Fe 2000, Argentina
T: 54 41 81-1932

Gaitan, Miguel Esteban, Sr.
Estudiante
ESCOLA INDUSTRIAL DE PUERTO IGUAZU
Qto. 69, 1 Piso Av. Misiones (51)
Puerto Iguazu 3370, Argentina
T: 075720591

Gaitan, Norberto N.E., Mr.
Defensa De La Fauna
UNIVERSIDAD NACIONAL RIO CUARTO (CORDOBA)
Sarmiento 1886
Rio Cuarto-Cordoba 5800, Argentina
T: 54 586 34045

Galafassi, Guido, Mr.
A E M A B
Rivadavia 1396- Berazategui
Buenos Aires 1884, Argentina
T: 54 1 256-1901

Galano, Carlos, Sr.
Docente Universitario
RENACER
Colon 2359 Vila Constitucion
Santa Fe, Argentina
T: 54 42 75451 F: 54 42 74668

Galletti, Sylvia, Sra.
SECRETARIA TURISMO PROVINCÍA RIO NEGRO
Av, Da FCO De Viedma
Viedma, Argentina
T: 0920 22150

Gallo, Melibea, Sra.
Student
FCEFYU / U.N.C.
Achayal Rodriguez 10 5B Hueba
Cordoba 5000, Argentina

Galloni, Maria Del Carmen, Ms.
Ambientalista
MINISTERIO DE CULTURA Y EDUCACION
2 Educacion
Buenos Aires, Argentina
T: 54 1 802-9390 F: 54 1 806-1846

Gandolla, Enrique, Mr.
Ingeniero
UNIVERSIDAD NACIONAL DE MISIONES
Morcillo 2692
Posadas 3300, Argentina
T: 54 752 24750.31011 F: 54 752 31011

Garcia, Hugo, Sr.
UNIVERSIDAD TECNOLOGICA NACIONAL
Pronunciamiento s/n
Parana - PR, Argentina
T: 54 43 243694 F: 54 43 243589
E: Alma/verte 1033

Garcia Napo, Osvaldo
Periodista
RADIO SPLENDID
Grai Rodrigues 174 Lamis
Buenos Aires, Argentina
T: 54 1 46-4530

Gari, Norma Edith, Sra.
RETARD
Rivadavia 4097
Marcos Paz - Bs.As., Argentina
T: 54 1 353305

Gaviglio, Luis Angel, Sr.
ASOCIACION MEDIO AMBIENTE SAN FRANCISCO
LeJuy 108
San Fmacisco, Argentina
T: 54 564 23431 F: 54 564 26546

Gay, Patricia, Ms.
President
AMIGOS DE LA TIERRA
Correo Central 1000, Casilla de Correo 1107
Buenos Aires, Argentina
T: 54 1 725-950 F: 54 1 725-950

Genesio, Javier, Mr.
GRUPO NATURA
Bolivar 535
Rio III - Cordoba, Argentina

Gentili, Haydee Cristina, Sra.
Docente
UNIVERSIDEAD NACIONAL DEL COMAHUE
Buenos Aires 1400
Neuquen 8300, Argentina
T: 54 943 23596 F: 54 943 23609
TL: 84266UNCNON

Giliberti, Adriana
FM DEL LAGO - SAN CARLOS DE BARILOCHE
Bariloche, Argentina

Gimbernat, Maria Angelica, Sra.
FEDEVI
Florida 868
Buenos Aires 1005, Argentina
T: 54 1 313-7578

Giordano, Ruben, Mr.
Arquitecto
FACULTAD DE ARQUITETURA
Facundo Zuviria 4676
Santa Fe 3000, Argentina
T: 54 42 32566

Gitto, Santiago G., Sr.
FUNDACION ARGENTINA DE ECOLOGIA - FADE
Avda of Los Italianos 365-501
Buenos Aires 1107, Argentina
T: 54 1 313-2948 F: 54 1 313-2948

Gomes, Nora Gladys Casals, Sra.
Teacher
FAC. DE TURISMO
Buenos Aires 1400
Neuquen, Argentina
T: 54 943 72537 F: 54 943 84266

Gomez, Adrian Alejandro
Cargo Directivo
RENACER
Guiemes 670
Viela Constitucion, Argentina

Gomez, Rodolfo Sixto, Mr.
CONCIENCIA ECOLOGICA
Vieytes 70 - 5'D' - Bahia Blanca
Buenos Aires 8000, Argentina
T: 54 091 33020

Gomez, Rut Noemi, Sra.
Estudante
FAU
Av. Gdor. Ruiz, 2281
Corrientes, Argentina
T: 54 717 2-2575

Gomez, Tito, Sr.
FUNDACION ECOLOGICA UNIVERSAL
Avda. Corrientes 2763/7o. piso
Buenos Aires 1046, Argentina
T: 54 1 962-0237 F: 54 1 952-1535

Gonzales, Hugo, Mr.
Director
BOSQUE ESMERALDA
42 No. 1520
La Plata 1900, Argentina
T: 54 21 246657

Gonzalez, Adriana Gabriela, Ms.
Miembro
FUNDECMA
Sermiento, 2520 - Menuel B. Gonnet
La Plata Cep.: 1897, Argentina
T: 54 21 84-1128 F: 54 21 312-3427

Gonzalez, Francisca Celia, Ms.
Lic Geografia
UNIVERSIDAD NACIONAL DEL SUR - BAHIA BLANCA
Haiti 57
Buenos Aires 8000, Argentina
T: 54 91 30024 F: 54 91 27876
TL: 6H GONZALEZ

Gonzalez, Gabriel Esteban
Reporter
REVISTA GENTE
Azopardo 579
Buenos Aires, Argentina
T: 3314591 F: 3314591

Gonzalez, Nilda, Ms.
Licenciada
INSTITUTO DE ESTUDIOS E INVESTIGACIONES SOBRE EL MEDIO AMBIENTE
Alsina 1816
Buenos Aires 1090, Argentina
T: 54 1 462 951 F: 54 1 476 1850

Gonzalez, Norma Diana, Ms.
TIERRA SIN MAL
Salta Yjunin Vanguaroj 7'C'
Posadas - Misiones 3300, Argentina

Gonzalez Prieto, Pedro, Mr.
Pres. Comision Medio Ambiente
ROTARY CLUB BAHIA BLANCA NORTE
Reousuista 187
Bahia Blanca, Argentina

Goytia, Cristina, Ms.
FEDERACION ARGENTINA MUJERES PROFESIONALES Y DE NEGOCIOS
Mitre 509
Rosario de Santa Fe, Argentina
T: 54 41 42557 F: 54 41 244386

Gozzani, Oscar Marcelo, Sr.
Guarda Parques
ASOCIACION PRO-DESARROLLO INTEGRAL DE LA REGION CENTRAL (BROTE)
Gaonay 25 de Agosto
Unquillo Argentina, Argentina
T: 05 43 88398 F: 051 00 520038

Greenberg, Miguel, Mr.
MUTANTIA
1880 East River Road
Tucson - AZ, Argentina
T: 1 602 577-6476 F: 1 602 577-1195

Gregoret, Maria Silvina, Sra.
Member
FUNAM
Casilla de Correo 83 - Correo Central
Cordoba 5000, Argentina
T: 54 51 690282 F: 54 51 520260

Grinberg, Miguel
Environmental Editor
REVISTA NUEVA /MUTANTIA TV
Esmeralda 343
Buenos Aires 1069, Argentina
T: 54 1 3254500/3258157 F: 54 1 3256114

Groef, Hector Van Der
Periodista
PRIMER PLANO CABLEHOGAR REVISTA
San Lorenzo, 1754
Rosario de Santa Fe, Argentina
T: 54 41 240971 F: 54 41 252459

Gropelli, Eduardo, Mr.
FUNDACION PROTEGER
C.C. 550
Santa Fe 3000, Argentina
T: 54 42 970-298 F: 54 42 52293

Guevara, Irma
FUNCACION PARA LA DEFENSA DELA VIDA - FEDEVI
Florida no. 868 - Piso 3o. DF G
Buenos Aires 1005, Argentina
T: 54 1 313-7578 F: 54 1 393-5792

Guinazu, Angel Hugo, Mr.
Educacion & Ecologia
GRUPO ECOLOGISTA CHACO
Ramon Vasquez 95
Resistencia 3500, Argentina

Gutierrez, Ana Maria, Ms.
Professora
FUNDACION NATURA
Francia 3336
Santa Fe 3000, Argentina
T: 54 42 40795

Gutierrez, Martha, Ms.
DEFENSA DERECHOS DEL ANIMAL
Julian Alvarez 143
Buenos Aires 1414, Argentina
T: 54 1 855-0078

Gyenge, Javier, Mr.
UNIVERSIDADE NACIONAL CORDOBA
leon pinelo 1745 b los naranjos
Argentina

Halac, Raul, Mr.
Director
UNIV. NAC. DE CORDOBA INSTITUTO DEL AMBIENTE HUMANO - FAC. DE ARQUITEC.
BUENOS AIRES 1085 - 1o.
Cordoba, Argentina
T: 54 51 600604

Hamilton, Joseph Andrew, Mr.
President
A E FORO VERDE
Diagonal 79, no. 525
La Plata 1900, Argentina
T: 54 21 3-3057

Hamity, Enrique Juan, Sr.
President
LEGISLATURA DE LA PROVINCIA DE JUJUY
San Salvador, Jujuy, Argentina

Hansen, Pablo, Sr.
Docente
UNIVERSIDAD NACIONAL DE MAR DEL PLATA
CALLE 67 No.3322 NECOCHEA
Buenos Aires 7630, Argentina

Hartkopf, Anamaria, Ms.
Docente
M.E.TAMANDUA
Asuncion S/N Km 10
Eldorado 3380, Argentina
T: 54 075121472

Heredia, Dante, Mr.
Ingeniegro
PARTIDO JUSTICIALISTA DISTRITO CORDOBA
A.M. Bas 481- Capital
Cordoba, Argentina

Heremite, Amalia
FRIENDS OF THE EARTH
Salguero 1863, 20A
Capital Federal, Argentina
T: 541 847383 F: 541 3256389

Hermitte, Amalia, Ms.
FRIENDS OF THE EARTH - FOE
Correo Central 1000, Casilla de Correo 1107
Buenos Aires, Argentina
T: 54 1 725-950 F: 54 1 725-950

Hernandez, Marzia Pilar, Ms.
Professor
CONSEJO PCIAL EDUCACION
Rohde y Fosbery - Depto - 7
Neuquen, Argentina

Hernando, Jorge E., Sr.
Investigator
FUNDECMA
San Martin 951 - 3A
San Miguel Tucuman 4000, Argentina
T: 54 81 222421 F: 54 81 311524

Herrera, Julio, Sr.
UNIVERSIDAD DE CORDOBA-FACULDAD DE CIENCIAS EXACTAS FISICAS Y NATURALES
Avenida Bariloche 537, Parque Horizonte
Cordoba 5000, Argentina

Herrera, Pablo Luis, Mr.
Park Ranger
A.P.N. ARGENTINA
Avenida Victoria Aguirre,66
Pto.Iguazu-Misiones, Argentina
T: 54 07 20-180

Hisas, Liliana, Sr.
Press
F.E.U. (IG.)
Av. Corrientes 2763 - 7o.
Buenos Aires, Argentina
T: 54 1 9120237 F: 54 1 4521535

Hodari, Claudia, Ms.
ECO - RANCH LA ESTANCIA SAN JOSE
Lavalle 3643/7o. p.
Buenos Aires 1009, Argentina
T: 54 1 89-0644

Holguin, Monica, Ms.
FUNDACION NATURA
Francia 3336
Santa Fe 3000, Argentina
T: 54 42 40795

Huisman, Rosa Mirtha Fuentes, Sra.
President
EQUISTICA
H. de la Quintana 33
Rosario 2000, Argentina

Ibarra, Herman Ignacio, Mr.
TIERRA SIN MAL
Posadas (MNES)
Posadas - Misiones 3300, Argentina
T: 54 752 22186

Iglesias, Zulma, Sra.
FACULTAD DE DERECHO Y CIENCIAS SOCIALES
Almafuerte 1732, Villa Maipu, San Martin
Buenos Aires, Argentina
T: 54 1 752-1549 F: 54 1 752-1549

Inon, Javier, Sr.
PROYECTO PEREYRA
Nazarre 2880
Buenos Aires 1417, Argentina
T: 54 1 53-6768 F: 54 1 786-8578

Invernizzi, Arzenio, Mr.
Ica
ALIANZA COOPERATIVA INTERNATIONAL
Moreno 1729 - 2nd P.
Buenos Aires, Argentina
T: 54 1 45 6733 F: 54 1 393 9806

Isasi, Luis Maria, Sr.
Asesor
HONORABLE CAMARA DE DIPUTADOS PROV. DE BUENOS AIRES
Avda Rivadavia y Riobamba
Buenos Aires, Argentina
T: 54 1 954-1111

Jatib, Maria Ines, Ms.
Assessora Camara Diputados De
COMISSION RECURSOS NATURALES
Riobamba 25
Buenos Aires, Argentina
T: 54 1 954-1110 F: 54 1 954-1100

Jonch, Juan Jose, Sr.
CONVOCATORIA PARA LA DEFENSA AMBIENTAL
Alvarez Jonte 5084 C.P 1407
Buenos Aires, Argentina
T: 54 1 5669253 F: 54 1 3124891

Jong De Easdale, Susana, Ms.
GOVIERNO DE JUJUY
Santibanez, 1602
San Salvador, Jujuy, Argentina
T: 54 088225077 F: 54 088229073

Jorge, Raul, Mr.
Director Espacios Verdes
MUNICIPALIDAD S.S. DE JUJUY
San Salvador, Jujuy, Argentina
T: 54 088224728

Jure, Cristian, Mr.
Director
TEA
Calle 510 /1915
La Plata, Argentina

Juricich, Dante
FUNDACION ECOLOGICA UNIVERSAL
Buenos Aires, Argentina

Kadlec, Anibal Francisco, Mr.
Educacion Sindical
CONFEDARECION DE TRABAJADORES DE LA EDUCACION DE LA REPUBLICA ARGENTINA - CTERA
H. Irigoyen 1290
Resistencia, Argentina
T: 26-969

Kanenguiser, Martin Diego
KANENGUISER CHURBA ASOC.
Buenos Aires, Argentina

Kannemann, Oscar Frederico, Sr.
Profesor
ESCOLA INDUSTRIAL DE PUERTO IGUAZU
Av. Victor Aguirre 350
Puerto Iguazu, Argentina
T: 20653

Kaufman, Miriam, Sra.
Coordinator
SECRETARIA DE EDUCACION
Av. el Cano 3409, 8o. B.
Buenos Aires 1426, Argentina
T: 54 1 552-7153 F: 54 1 802-6962

Khairallah, Eva Luz, Sra.
ARC PEACE (ECO-GEO ARGENTINA)
Av. Roca (s) No.811
Santiago del Estero 4200, Argentina
T: 54 85 214170 F: 54 85 224350

Kiczka, Edmundo, Mr.
Sub-Secretary
MUNICIPAL
Rivadavia y San Martin
Posadas, Argentina
T: 54 752 33313

Kohn, Celia, Ms.
Medecin
CONSEJO DELIBERANTE
Alem 68 Rivada Via
Rivadavia-Mondoza 5577, Argentina

Kraft, Siegfrido, Mr.
Professor / Director
UNIVERSIDAD DE RIO CUARTO - UNRC
Panama 180, Banda Norte
Rio Cuarto-Cordoba 5800, Argentina
T: 54 586 28458

Labbate, Gabriel David, Mr.
Humboldt 1667
Buenos Aires, Argentina
T: 54 1 771-9854 F: 54 1 801-5183

Lafuente, Juan Manuel, Mr.
MOVIMIENTO DE SALVACION NATURAL
Rivadavia 1689 - Coronda
Santa Fe, Argentina

Lago, Patricio D., Mr.
BIOGAS - MT - UNLP
C. Bernardi 2246
Alte. Brown - Bs.As., Argentina
T: 54 21 294-4716

Laharrague, Peter, Mr.
Agricultural Engineer
C.C. 64 (7530) Coronel Pringles

Lainocoyi, Silvia, Ms.
Women Representative
ASOCIACION INDIGENA MEGUESOXOCHI
Municipalidad 3705
Castelli - Chaco, Argentina

Lancelle, Anna Irene, Sra.
Coordinator
ASOC. UNESCO
Balboa 2664
Corrientes, Argentina
T: 54 717 0783-67435

Landazuri, Leticia, Ms.
Member
FUNDACION PARA LA DEFENSA DEL AMBIENTE - FUNAM
Casilla de Correo 83, Correo Central
Cordoba 5000, Argentina
T: 54 51 690282 F: 54.51 520260

Laporte, Juan Pablo Gaggino, Mr.
Asociado
FORO VERDE
40 no.33
La Plata, Argentina
T: 54 21 24-5135

Laquidara, Jose Luis, Sr.
Asesor
CONGR. NACIONAL
Avda. Rivadavia y Riobamba
Buenos Aires, Argentina
T: 54 1 954-1111

Lardelli, Miguel Angel
Presidente
COMUNA DE VILLA MUGUETA
Santa Fe 252
Villa Mugueta 2175, Argentina
T: 046997125-97014

Lario, Paula, Ms.
Member
FUNDACION PARA LA DEFENSA DEL AMBIENTE - FUNAM
Casilla de Correo 83, Correo Central
Cordoba 5000, Argentina
T: 54 51 690282 F: 54 51 520260

Larsen, Peter, Mr.
2130, Uriarte 1A
Buenos Aires, Argentina
T: 54 1 771-1234

Lascano, Lilian, Sra.
MODIM
Rivadavia 1367
Buenos Aires, Argentina
T: 54 1 750-0561

Ledesma, Jose Chuza, Sr.
GRUPO ARQUITECTURA DE ACCION
Jeronimo Cortez 554
Cordoba 5001, Argentina
T: 54 51 719-434

Leguizamon, Ruben, Sr.
CONCEJO DELIBERANTE DE LA PLATA
12 y 51
La Plata, Argentina
T: 54 21 36978 F: 54 21 256384

Lencina, Juan Carlos, Mr.
Santiago del Estero, Argentina

Leon, Graciela, Ms.
F L A C A M
sanlorenzo 6776
Rosario, Argentina
T: 54 41 569996

Liberatore, Augusto H., Mr.
Estudiante
PROCOVI - PROMOTORA DE COOP. DE VIV.
Entre Rios 1109 1o. 1
Rosario 2000, Argentina
T: 54 41 815-711

Liliana, Hisas
Correspondent
INFORME GENESIS
Av.Corrientes, 2763
Capital Federal, Argentina
T: 1 541 9620237 F: 1 541 4521535

Lindo Poisson, Rosana Estela, Ms.
MOVIMIENTO EDUCADORES MISIONEROS
SALTA 1857 - 2o.C
Posadas 3300, Argentina

Lopes, Victor, Sr.
President
PLANETA NEUSTRO
Dean Funes 1015 / CP 1231
Buenos Aires, Argentina
F: 54 1 342-6541

Lopez, Carlos Enrique Diaz, Mr.
Concejal
HONORABLE CONCEJO DELIBERANTE
24 de Setiembre 346
San Miguel Tucuman, Argentina
T: 54 81 311118

Lopez, Hector, Sr.
FUNDACION ECOLOGICA UNIVERSAL
Avda. Corrientes 2763/7o. piso
Buenos Aires 1046, Argentina
T: 54 1 962-0237 F: 54 1 952-1535

Lorenzo, Daniel
Press Representant
FRIENDS OF THE EARTH
Salguero 1863, 20A
Buenos Aires, Argentina
T: 54 1 847383 F: 54 1 3256389

Losada, Horacio Ruben Perez, Mr.
Colaborador
Avenida Cramter 2454 5o.A - 1428
Buenos Aires, Argentina

Lucarati, Guilhermo Hugo, Mr.
Vice-President
FUNDACION UNIVERSIDAD 'PERITO MORENO'
Piedras 736
Buenos Aires, Argentina
T: 54 1 361-3700 F: 54 1 361-4604

Luduena, Manuel Antonio, Mr.
INSTITUTO JUAN MANTOVANI
Garcia del Rio,4559
Buenos Aires, Argentina
T: 54108080 F: 54 1 3136767

Luraschi, Mario, Mr.
Jefe Depto Contaminacoes
DEPARTAMENTO G. IRRIGACION
Av. Espanha y Barcala
Mendoza, Argentina
T: 54 61 234 000

Luti, Ricardo, Sr.
Director
CERNAR-UNC
Ave. Velez Sarsfield 299
Cordoba 5000, Argentina
T: 54 51 3-9473 F: 54 51 24-4092
TL: 051822BUCOR, E: 5000 Cordoba,Arg.

Luy, Amalia M., Sra.
FUNDACION CONASED
117 No. 1594
La Plata, Argentina
T: 54 21 40846 F: 54 21 217167

Mabel, Tosi Liliana, Ms.
Fundadora
ADN - ALUMNOS DEFENSORES DE NATUR
Pagani 1441
Armstrong, Argentina
T: 54 61376

Macchiavello, Gabriel, Mr.
Lawyer
ACCION VERDE
Rodriguez Pena
Buenos Aires 1020, Argentina
T: 54 1 40-4308 F: 54 1 40-8830

Machado, Pablo, Sr.
FUNDACION CONASED/FUNDACION DE ESTUDIOS MUNICIPALES Y SOCIALES
13 No. 865
La Plata, Argentina
T: 54 21 40846 F: 54 21 217167

Madar, Irma, Sra.
FEDEVI
Taodoro Garcia 1815- 1 A
Buenos Aires, Argentina
T: 54 1 783-9479

Magnani, Cesar Raul, Mr.
Jefe Area Juridica
DEPARTAMENTO GENERAL IRRIGACION
Av. Espanha y Barcala
Mendoza 5500, Argentina
T: 54 61 234 000

Maini, Mario, Mr.
FUNDACION SUR
Piedras 456/ 6
Buenos Aires 1070, Argentina
T: 54 1 343-8210 F: 54 1 343-8210

Maletti, Ernesto Ruben, Mr.
Presidente Cap. Misiones
FUNDACION VIDA SILVESTRE
Nenquem, 264
Obera, Argentina

Mammini, Susana Elena
NOTICIAS ARGENTINAS
Hipolito Irigoyen 910 1er Piso 'C'
Buenos Aires CF, Argentina
T: 01 3345648 F: 01 3348628

Manjon, Julio, Mr.
FUNDACION VIDA SILVESTRE
Mariano Acosta 116
Buenos Aires, Argentina
T: 54 1 612-9842

Marchetti, Beatriz Dora, Sra.
CEN. EST. AVAN.
Florida 439, 2 piso
Buenos Aires, Argentina
T: 54 1 901-4632 F: 54 1 311-0516
TL: 18694 IBUBA AR

Mare, Clotilde G.B. De, Sra.
FEV
San Martin de Tours 2916 - 5o. piso
Buenos Aires 1425, Argentina
T: 54 1 803-4171

Marelli, Gustavo, Sr.
CAMARA DE DIPUTADOS - ARGENTINA
Camara de Diputados
La Plata, Argentina
T: 54 21 25115

Marena, Antonio, Mr.
Tesorero
COMUNA VILLA MUGUETA
Santa Fe 252
Villa Magueta, Argentina
T: 469 97125

Maria Angelica, Politis, Sra.
Docente
ESCUELA MAR DEL PLUTA
Falucho 2768
Mar del Plata 7600, Argentina
T: 54 23 43305 F: 54 23 518422

Marignac, Veronica Martinez, Ms.
FACULTAD CS. EX, QUIMICAS Y NATURALES
SALTA1743
Posadas 3300, Argentina
T: 54 752 23850

Marques, Hector Pina, Sr.
CAMARA DE DIPUTADOS - ARGENTINA
Camara de Diputados
La Plata, Argentina
T: 54 21 25115

Martinese, Patricia Isabel Virginia, Ms.
Professor
UNIVERSIDAD NAC. DEL COMAHUE
Los Helechos 130
Neuquen, Argentina

Martinez, Adolfo Garcia, Sr.
CONSEJO PROFESIONAL DE INGENIERIA QUIMICA
Bartolome Mitre 1470
Buenos Aires, Argentina
T: 54 1 381-8796 F: 54 1 383-7256

Martinez, Carlos, Sr.
Vocal
FUND. NEHUEN
Palacios 474
Neuquen 8300, Argentina
T: 54 943 32492

Martinez, Miriam, Ms.
Director
BOSQUE ESMERALDA
La Plata, Argentina
T: 54 21 246657

Martini, Daniel Jose, Sr.
Sec. Gen. Juventud
PARTIDO SOCIALISTA POPULAR
Monter gudo 170 6'c'
Buenos Aires 1428, Argentina
T: 54 1 277293

Mascolo, Hector Jose, Mr.
Rel. Publicas/Internacionales
CONVOCATORIA PARA LA DEFENSA AMBIENTAL
Casilla Correo 67, Suc. 2
Buenos Aires 1402, Argentina
T: 54 1 334-4273

Masso, Maria Del Carmen, Ms.
Member
CONVOCATORIA PARA LA DEFENSA AMBIENTAL
Virrey Loreto 2685 Piso 4to. 'A'
Buenos Aires, Argentina
T: 54 1 856-3891 F: 54 1 802-1266

Massola, Maria Victoria, Sra.
ASOCIACION CONSERVACIONISTA DEL SUR - TELLUS
Caronti 126 - Bahia Blanca
Buenos Aires 8000, Argentina
T: 54 1 091-46397

Mastrangelo, Alice, Sra.
INTERNATIONAL WOMEN'S TRIBUNE CENTRE
Avda. Corrientes 2763/7o. piso
Buenos Aires 1046, Argentina
T: 54 1 962-0237 F: 54 1 952-1535

Mastrogiacomo, Miguel A., Sr.
CAMARA DE DIPUTADOS - ARGENTINA
Camara de Diputados
La Plata, Argentina
T: 54 21 25115

Matarasso, Horacio F., Mr.
Presidente
FUNDACION BIOMA
Av. Honorio Pueyrredon 786
Buenos Aires 1426, Argentina
T: 54 1 901-6551 F: 54 1 901-6551

Mauro, Guillermo Javier De, Mr.
Guardafaune
ASOCIACION PARA LA DIFUSION DE LOS PROGRAMAS DE LAS NACIONES UNIDAS PARA EL M.A.
Recta Martinoli 3939 - 8o. Villa Belgrano
Cordoba 5147, Argentina
T: 54 51 20850

Meitin, Acejandro
RADIO UNIVERSIDAD
55 No 681 7B
La Plata, Argentina
T: 253740 F: 253740

Melitsco, Graciela, Sra.
YBF
Pareja 3633
Buenos Aires, Argentina
T: 54 1 9811309 F: 54 1 501-1458

Mendez, Maria Susana, Ms.
Professora
FUNDACION BARILOCHE
Lonquimay, 4102
Bariloche, Argentina
T: 54 944 22050 F: 54 944 22050

Menendez, Andrei
Editor
TELEVISION ESPANHOLA
Viamonte 166 - 2o. piso
Buenos Aires, Argentina
T: 54 1 13 2692/440 6927 F: 54 1 13-2678

Mercado, Silvia
Asesora
COMISSION RECURSOS NATURALES
Riobamba 25
Buenos Aires, Argentina
T: 54 1 954-1111 F: 54 1 954-1113

Mestre, Victor Hugo, Sr.
Bouquet Roldan, 127
Neuquen 23333, Argentina

Micou, Luis A.
EDITORIAL ATLANTICA
Buenos Aires, Argentina

Miguel, Marcelo, Sr.
UNIVERSIDAD DE CORDOBA-FACULDAD DE CIENCIAS EXACTAS FISICAS Y NATURALES
Transito Caceres de Allende 469
Cordoba 5000, Argentina

Miguel Garo, Raul, Mr.
CONSEJO MUNICIPAL ROSARIO COMISSION DE ECOLOGIA
Ituzaingo,915
Rosario, Argentina
T: 54 41 82-0780

Mihura, Enrique Raul, Sr.
Asesor
UNIVERSIDAD
25 de Mayo no.64
Parana, Argentina
T: 54 43 226308 F: 54 43 215462

Minervini, Osvaldo Eugenio, Sr.
Vice-Presidente
FORO VERDE
143 y 487 Gorina- La Plata
Buenos Aires, Argentina

Moccero, Gustavo, Sr.
CAMARA DE DIPUTADOS - ARGENTINA
Camara de Diputados
La Plata, Argentina
T: 54 21 25115

Mondino, German Luis, Mr.
Geologo
SENADO PROVINCIA SANTA FE
Cabal 2123
San Justo - Santa Fe, Argentina
T: 04 98 28010

Montbrun, Alberto, Sr.
Engenheiro
AMPROVE
A.Villanueva, 425
Mendoza 5500, Argentina
T: 54 61 291893 F: 54 61 249365

Monteros, Ismael Ramon, Mr.
Sec. Politico
CAMARA DE DIPUTADOS
Mitre 550
Salta 4400, Argentina
T: 54 87 218239

Montes, Marisa, Sra.
Director
TEA
Calle 15 no. 1775
La Plata, Argentina
T: 54 21 24-6657

Moore, Esteban, Sr.
1st Vice President
PLANETA NEUSTRO
Dean Funes 1015 / CP 1231
Buenos Aires, Argentina
F: 54 1 342-6541

Morales, Juan Manuel, Mr.
La Plata, Argentina
E: 1900

Moreira, Osvaldo, Sr.
CONCEJO DELIBERANTE DE LA PLATA
12 y 51
La Plata, Argentina
T: 54 21 36978 F: 54 21 256384

Moreno, Carlos, Sr.
CONCEJO DELIBERANTE DE LA PLATA
12 y 51
La Plata, Argentina
T: 54 21 36978 F: 54 21 256384

Moyamo De Burt, Maria A., Sra.
Miembro Titular
COCHUNA VIDA SI
Rawson 994
Concepcion-Tucuman 4146, Argentina
T: 54 81 865-22449

Moyano, Amilcar Atilio, Mr.
UNIVERSIDAD DE MENDOZA
Av. Boulogne Sur Mer, 665
Mendonza, Argentina
T: 54 61 24-7017 F: 54 61 31-1100

Munoz, Graciela Elsa, Ms.
Domingo 1431
Neuquen, Argentina
T: 54 43 249-56 F: 54 43 20579

Munuce, Cecilia, Sra.
FUNDACION ECOLOGICA UNIVERSAL
Avda. Corrientes 2763/7o. piso
Buenos Aires 1046, Argentina
T: 54 1 962-0237 F: 54 1 952-1535

Murekian, Noemi G., Sra.
Pca. Arabe Siria 2930, piso 14, dep. C
Buenos Aires, Argentina
T: 54 1 804-4263

Muslera, Irma Munoz, Ms.
Lincenciada
MINISTERIO DE EDUCACION Y CULTURA
Hipolito Irigoyen, 440 - 7P
Buenos Aires 1406, Argentina
TL: 345-2200

Nadal, Margaret
Correspondent
NUEVO DIARIO
Maipi 639, 4220 Las Termas de Rio Hondo
Santiago del Estero, Argentina

Nani, Susana Garcia, Ms.
Member
CONVOCATORIA PARA LA DEFENSA AMBIENTAL
Virrey Loreto 2685 piso 4to. 'A'
Buenos Aires, Argentina
T: 54 1 856-3891 F: 54 1 802-1266

Natalia Paola, Petit Marty, Ms.
FACULTAD CS. EX, QUIMICAS Y NATURALES
Salta 1743
Posafdas 3300, Argentina

Navarro, Fausto German, Sr.
LEGISLATURA DE LA PROVINCIA DE JUJUY
San Salvador, Jujuy, Argentina

Nelly, Plot Beatriz, Ms.
Professora
MINIST. OBRAS PUBLICAS
12-681- DI 4
La Plata - Bs. As. 19001, Argentina
T: 54 21 46447

Neme, Javier, Sr.
Student
FCEFYN / U.N.C.
Colon 1155 6o. B
Cordoba, Argentina

Niederle, Berta Rosa, Sra.
Chefe Delegacion Tucuman
INST. NAC. DE ALIMENTOS
Paso De Los Andes,750 -Dpto. 9
S.M.Tucuman -Tucuman, Argentina
T: 081214153

Nielsen, Gustavo Leslie, Mr.
Soler # 3653, Sto.- A
Buenos Aires, Argentina
T: 54 1 821-6655

Nieto, Ariel, Mr.
Director
TEA
Calle 510 /1915
La Plata, Argentina

Nieto, Diana, Sra.
FUNDACION ECOLOGICA UNIVERSAL
Avda. Corrientes 2763/7o. piso
Buenos Aires 1046, Argentina
T: 54 1 962-0237 F: 54 1 952-1535

Noher De Halac, Rita Ines, Mr.
UNIVERSIDAD CATOLICA DE CORDOBA
Obispo Trejo 323
Cordoba 5000, Argentina
T: 54 51 38389 F: 54 51 695101

Nunez, Emilio Victoriano, Mr.
MODEMA/FLACAM
San Martin, 982
C.del Uruguay,Ent.Rs, Argentina
T: 54 442 27671

Ocampo, Raul Granillo, Sr.
President
FUNDACION INTEGRACION
Sarmiento 440
Buenos Aires, Argentina
T: 54 1 394-0275 F: 54 1 394-9227

Ojeda, Roberto Tomas, Mr.
DEPARTAMENTO DE INGENIERIA AMBIENTAL
C.C. 961 - Bariloche
Rio Negro 8400, Argentina
T: 54 944 22121 F: 54 944 23051

Oliva, Viviana, Ms.
GRUPO NATURA
Avda. Eden 385
La Falda - Cordoba, Argentina

Oliveri, Nestor Jorge, Sr.
INSTITUTO NACIONAL DE TECNOLOGIA AGRIPECUARIA (INTA)
Estacion exp. Agropecuaria Misiones, C.C. 101, 3315 Leandro N. Alem
Misiones, Argentina

Olivier, Altamirano Diego, Mr.
A E FORO VERDE
Diagonal 79, no. 525
La Plata 1900, Argentina
T: 54 21 3-3057

Olivier, Maria Luisa Scorpati De, Sra.
Diretora
A E FORO VERDE
Diagonal 79, no. 525
La Plata 1900, Argentina
T: 54 21 3-3057

Olivier, Santiago R., Mr.
Diretor
A E FORO VERDE
Diagonal 79, no. 525
La Plata 1900, Argentina
T: 54 21 3-3057

Omar, Vassallo, Mr.
Professor
UNIVERSIDAD NAC ROSARIO
Dorrego 1580
Rosario 2000, Argentina
T: 54 41 240896

Orellano, Luis, Sr.
CAMARA DE DIPUTADOS - ARGENTINA
Camara de Diputados
La Plata, Argentina
T: 54 21 25115

Orlando, Angela, Ms.
Education
UNIVERSIDAD NACIONAL DE ROSARIO
Entre Rios 206
Rosario de Santa Fe 2000, Argentina
T: 54 41 249130

Orosco, Otero Mario, Mr.
Jornalista
RADIO NACIONAL
Lafinur y Ayacucho
San Luis - SL, Argentina
T: 652 24771

Ortega, Luis, Mr.
INTENDENCIA SAN MIGUEL
Grar. Sarmiento
Buenos Aires, Argentina
T: 54 1 351888

Oscar, Samuel
Correspondent
NUEVO DIARIO
Maipi 639, 4220 Las Termas de Rio Hondo
Santiago del Estero, Argentina

Paez, Leticia, Ms.
Head Of English Teaching Staff
CENTRO PROVINCIAL ENSENANZA
Moreno 829, San Martin Andes
Neuquen 8370, Argentina
T: 54 972 27387

Paez Allende, Clara, Ms.
Secretaria
AMIGOS DEL ARBOL
Santiago Del Estero 449 - Parana
Entre Rios, Argentina
T: 54 42 222595

Pagnotta, Leonardo
FUNDACION ARGENTINA DE ECOLOGIA - FADE
Rivadavia 4623-3$-Dto 15-501
Buenos Aires 1424, Argentina
T: 54 1 313-2948 F: 54 1 313-2948

Palladino, Nelson Luis, Sr.
TRAMA
Belen 721 D43
Buenos Aires 1407, Argentina

Pangol, Juan
Camarografo
TELEVISION ESPANHOLA
Viamonte 166 - 2o. piso
Buenos Aires, Argentina
T: 54 1 13 2692/440 6927 F: 54 1 13-2678

Panick, Betina, Ms.
FORO ARGENTINO DE BIOTECNOLOGIA
SARMIENTO 930 - 3o.P
Buenos Aires 1022, Argentina
T: 54 1 351888

Paredes, Victor, Dr.
Doctor
INSTITUTO DE ESTUDIOS E INVESTIGACIONES SOBRE EL MEDIO AMBIENTE
Alsina 1816
Buenos Aires 1090, Argentina
T: 54 1 462 951 F: 54 1 476 1850

Parizek, Bernardo, Mr.
Member
FUNDACION PARA LA DEFENSA DEL AMBIENTE - FUNAM
Casilla de Correo 83, Correo Central
Cordoba 5000, Argentina
T: 54 51 690282 F: 54 51 520260

Pascale, Alejandra Alba, Sra.
CEN. EST. AVAN.
Florida 439, 2 piso
Buenos Aires, Argentina
T: 54 1 879-809 F: 54 1 311-0156
TL: 18694 IBUBA AR

Pascuarelli, Cristina Teresa, Ms.
Technitian
DIR. TEN. DE PROGRAMAS PROJECTOS E IMP. EN EL
Rioja 229 - Ministerio de Obras Publicas
Neuquen, Argentina
T: 54 43 34742 F: 54 43 25746-31165

Pastorelli, Martha Gallardo De, Ms.
Legisladora
CAMARA DE DIPUTADO
Falucho 1290 (OESTE)
Argentina

Paura, Aliga Clara, Sra.
INTI
L.N. Aleui 1067
Buenos Aires, Argentina
T: 54 1 313-3966 F: 54 1 313-2130

Pecora, Carolina, Ms.
Member
FUNDACION PARA LA DEFENSA DEL AMBIENTE - FUNAM
Casilla de Correo 83, Correo Central
Cordoba 5000, Argentina
T: 54 51 690282 F: 54 51 520260

Pedroni, Maria Esther, Sra.
C. DEL LITORAL
3 de Febrero 74 - 1o. Piso
Rosario, Argentina
T: 54 41 49939

Pelatia, Adriana, Ms.
Licenciada
PARTIDO JUDICIALISTA DISTRITO CORDOBA
Av. Sarriento 374 - Capital
Cordoba, Argentina

Pelegrini, Valeria, Sra.
FUNDACION ECOLOGICA UNIVERSAL
Avda. Corrientes 2763/7o. piso
Buenos Aires 1046, Argentina
T: 54 1 962-0237 F: 54 1 952-1535

Pellegrini, Florencia, Sra.
FUNDACION ECOLOGICA UNIVERSAL
Avda. Corrientes 2763/7o. piso
Buenos Aires 1046, Argentina
T: 54 1 962-0237 F: 54 1 952-1535

Pellegrini, Valeria, Ms.
Miembro
F.E.U.
Aguilar 2325
Buenos Aires, Argentina
T: 54 1 786 6838

Perez, Jorge, Sr.
CAMARA DE DIPUTADOS - ARGENTINA
Camara de Diputados
La Plata, Argentina
T: 54 21 25115

Perez, Raul Joaquin, Sr.
CONCEJO DELIBERANTE DE LA PLATA
12 y 51
La Plata, Argentina
T: 54 21 36978 F: 54 21 256384

Perlini, Alcira Hoemi, Ms.
FUNDACION ESTUDIOS Y ACCION SOLIDARIA DE LA MUJER
Avda. Montes De Oca 1408 - 5B
Buenos Aires 1271, Argentina
T: 54 1 214935

Pesci, Hugo Eduardo, Sr.
Member
PANAVI
Peru 1058
Alta Gracia, Cordoba 5186, Argentina
T: 54 547 21274 F: 54 51 520860

Pesci, Ruben, Mr.
Director Editorial
EDITORIAL A/MBIENTE
Calle 57, 393
La Plata 1900, Argentina
T: 54 21 214-915 F: 54 21 256-556

Pessoa, Efrain, Mr.
FUNDACION DESARROLLO ECOLOGICO
Corrientes 1762 Piso 1, of.C
Buenos Aires, Argentina
T: 54 1 490099 F: 54 1 490099

Picca, Graciela, Sra.
FEDEVI
Florida 868
Buenos Aires 1005, Argentina
T: 54 1 313-7578

Piedra Gomez, Maria Eugenia, Ms.
Teacher
A.R.E.S.
Rivero Miguel 48 Arguello (C.P. 5147)
Cordoba, Argentina
T: 54 51 3.23663

Pigoni, Anibal, Mr.
INTENDENCIA SAN MIGUEL
San Miguel
Buenos Aires, Argentina
T: 54 1 315-1888

Pina, Carlos, Mr.
UNIVERSIDADE NACIONAL CORDOBA
Adolfo Orma 1815 Cerro de las Rosas
Cordoba, Argentina
T: 54 51 810-230 F: 54 51 244-689

Pineiro, Edgar Antonio, Sr.
Coordinator
ASOC. UNESCO
Jujuy 1360
Corrientes, Argentina
T: 54 717 63800

Pirovano, Jose Augusto Gandugiia, Mr.
F.A.D.E. (FUNDACION ARGENTINA DE ECOLOGIA)
Avda. de los Italianos 365-501 (cod.1107)
Buenos Aires, Argentina
T: 54 1 313-2948 F: 54 1 313-2948

Pirro, Hernan C., Sr.
UNIVERSIDAD TECNOLOGICA NACIONAL
Pronunciamiento s/n
Parana - PR, Argentina
T: 54 43 243694 F: 54 43 243589
E: Alma/verte 1033

Pischedda, Escobar, Mr.
CEISAL
Jujuy 551
Corrientes, Argentina
T: 54 717 078327074

Pithod, Pablo Victor, Mr.
UNIVERSIDAD DE MENDOZA
BOULOGNE SUR MER 665 5o.PISO
Mendonza 5500, Argentina
T: 54 61 24-7017 F: 54 61 31-1100

Poggiese, Hector Aricio, Mr.
Coordinador Area Planif.
FACULDAD LATINOAMERICANA CIENCIAS SOCIALES(FLACSO)
Federico Lacrose 2097
Buenos Aires 1426, Argentina
T: 54 1 772-2407 F: 54 1 775-6937

Politis, Maria Angelica, Sra.
Saavedro 01 Depto. C
Mar del Plata 7600, Argentina
T: 54 23 51-8422

Polotto, Luis Alberto, Mr.
FUNDACION SUR
PIEDRAS, 456 6o. P-A
Buenos Aires 1072, Argentina
T: 54 1 343-8210 F: 54 1 343-8210

Pombo, Claudia, Ms.
INTENDENCIA SAN MIGUEL
San Miguel
Buenos Aires, Argentina
T: 54 1 315-1888

Ponce, Mariana, Sra.
Member
GRUPO NACIMIENTO
Casilla de Correos no. 53, Maipu 1455
Rosario 2000, Argentina
T: 54 41 561436

Presta, Carlos, Sr.
FUNDACION ECOLOGICA UNIVERSAL
Avda. Corrientes 2763/7o. piso
Buenos Aires 1046, Argentina
T: 54 1 962-0237 F: 54 1 952-1535

Quevedo Carrillo, Marcelo, Sr.
Vice President
LEGISLATURA DE LA PROVINCIA DE JUJUY
San Salvador, Jujuy, Argentina

Quillay, Liliana Mabel, Ms.
FM LATINA, FM 97
Avda. Netri 1560
Rosario, Santa Fe, Argentina
T: 555469

Quintana, Maria Matha, Ms.
Biology Student
UNIVERSIDEAD NACIONAL DE CORDOBA
Duarte Quiros 461 7o.B.
Cordoba, Argentina
T: 54 51 87839

Ramirez, Gustavo, Mr.
ASOCIACION GAIA ARGENTINA
Almafuerte 1732 Villa Maipu, San Martin
Buenos Aires, Argentina
T: 54 1 752-1549 F: 54 1 752-1549

Ramos, Rodolfo, Mr.
Profesor
UNIVERSIDAD NACIONAL DE MISIONES
Beron Astrada 8674
Posadas 3300, Argentina
T: 54 752 22186

Ramos, Valentin, Sr.
LEGISLATURA DE LA PROVINCIA DE JUJUY
San Salvador, Jujuy, Argentina

Raventos, Alicia Perez, Sra.
Director
BOSQUE ESMERALDA
53 No. 724 1/2
La Plata, Argentina
T: 54 21 246657

Ravera, Anibal Ruben, Mr.
Director
PARQUE ECOLOGICO CULTURAL G.E. HUDSON
Jujuy 2727
Quilmes 1879, Argentina
T: 54 1 251-1974 F: 54 1 251-2066

Ravera, Ruben, Mr.
PROYECTO AUTOSUFICIENCIA REGIONAL - PAR
Chacabuco 699
Bernal, Argentina
T: 54 259-2125

Real, Patricia, Ms.
Colaboradora
RED LATINOAMERICANA DE ECOSOFIA
Tucuman 1059 Dto. 5
Rosario, Argentina
T: 54 41 21 9676

Recia, Nelida E., Sra.
President
FUNDACION ALBERT SCHWEITZER
Balcare e37 Bis
Rosario 2000, Argentina
T: 54 41 259-547

Reimondi, Ivan, Sr.
FUNDACION CEPA
57 no. 393
La Plata 1900, Argentina
T: 54 21 214915 F: 54 21 256556

Remonda, Luisa B. De, Ms.
Vicepresident
FUNDACION PARA LA DEFENSA DEL AMBIENTE - FUNAM
Casilla de Correo 83, Correo Central
Cordoba 5000, Argentina
T: 54 51 690282 F: 54 51 520260

Reutemann, Graciela Cecilia
RADIO ROSARIO - LT8
Cordoba 1843
2000 Rosario, Argentina
T: 0054 41 245075 F: 0054 41 247882

Reyeros, Alberto Luna, Dr.
Scientist
FUNDACION MIGUEL LILLO
Mendoza 3820
Tucuman, Tucuman, Argentina
T: 54 81 23-2421

Reyes, Nelida, Ms.
Professora
MINISTERIO EDUCACION
Lacroze, 2173
La Plata Zip.: 1897, Argentina
T: 54 21 845-360

Reynoso, Miguel
PRODUFE
Viamonte 1145
Buenos Aires PB 10, Argentina
T: 46 0851

Riccardi, Diana, Ms.
Administ. Sec.
AMIGOS DE LA TIERRA
Correo Central 1000, Casilla de Correo 1107
Buenos Aires, Argentina
T: 54 1 725-950 F: 54 1 725-950

Richard, Daniel, Sr.
UNIVERSIDAD TECNOLOGICA NACIONAL
Pronunciamiento s/n
Parana - PR, Argentina
T: 54 43 243694 F: 54 43 243589
E: Alma/verte 1033

Rinaldi, Sergio, Mr.
E.L.C.I. BOARD/ TALLER ECOLOGISTA
Casilla de Correo 56, Sucursal 8
Rosario, Argentina
T: 54 41 39-8677 F: 54 41 25-9089

Rinaldi, Susana, Ms.
Vocal Sup
MODEMA
San Martin 281
Parana, Argentina

Rissi, Luis Antonio, Mr.
Expositor
FUND. ECOLOGIA UNIVERSAL
La Madrid # 81
La Rioja, Argentina
T: 082224170 F: 082226766

Rivas, Gonzalo Mandagaran, Sr.
CAMARA DE DIPUTADOS - ARGENTINA
Camara de Diputados
La Plata, Argentina
T: 54 21 25115

Roa, Blas Hector, Mr.
FUNDACION YGUAZU
Ayacocho 1406
Posadas, Argentina
T: 54 752 30842

Robotti, Mario, Mr.
Arquitecto
FLACAM
M. T. de Alvear 452 - 3o. C
Buenos Aires 1058, Argentina

Roca, Magdalena, Sra.
CONCIENCIA (ASOCIACION)
Puey Medon 2471 - 7th floor
Eap Fed 1119, Argentina
T: 803 2830 F: 763 5123

Roca, Margarita Gonzalez Del Solar, Ms.
Environment Comittee Director
ASOCIACION CONCIENCIA
Florida 633, 3rd floor
Buenos Aires, Argentina
T: 54 1 3935191 F: 54 1 3937196

Roca, Roberto Daniel, Mr.
Docente
UNIVERSIDAD DE BUENOS AIRES - FACULTAD DE AGRONOMIA
Centeneja 174
Buenos Aires, Argentina

Rocha, Eduardo
DIR. AGUA POTABLE
El Naranjo No. 15 - Barrio Los Perales
S.S. de Jujuy-Jujuy, Argentina
T: 25077

Rodriguez, Claudia, Sra.
Docente
UNIVERSIDAD NACIONAL DE RIO CUARTO
Enlace Rutas 8 y 36
Rio Cuarto-Cordoba 5800, Argentina
T: 54 586 33858 F: 54 586 23980
TL: 54572cunrc-ar

Rodriguez, Daniel
Graphic Reporter
DIARIO CLARIN
Piedras 1743
Buenos Aires, Argentina
T: 270061 F: 239003

Rodriguez, Felix Isla, Mr.
Esmeralda 570 /6 piso of 21
Buenos Aires, Argentina
T: 54 1 394-7607 F: 54 1 394-7607
TL: 17989

Rodriguez, Flavio V., Mr.
Coordinador General
ASO.DIF. PROG. NACIONES UNIDAS MEDIO AMBIENTE
27 De Abril, 864
Cordoba, Argentina
T: 54 51 226304 F: 54 51 226304

Rodriguez, Hector, Sr.
Cerimonial Director
CAMARA DE DIPUTADOS - ARGENTINA
Camara de Diputados
La Plata, Argentina
T: 54 21 25115

Rodriguez, Pablo, Sr.
Director
TEA
Calle 59 no. 511 Piso 2 Dto.2
La Plata, Argentina

Rodriguez, Veronica Maria, Ms.
TIERRA SIN MAL
Av. Mitre 1941
Posadas 3300, Argentina
T: 54 752 22876

Rojo, Sebastian Rubens Y, Mr.
CISV
Luis M. Campos 152 4B
Buenos Aires 1245, Argentina
T: 54 1 7714940

Romani, Andrea, Sra.
UNIVERSIDAD DE CORDOBA - FACULDAD DE CIENCIAS EXACTAS FISICAS Y NATURALES
Avenida Bariloche 537, Parque Horizonte
Cordoba 5000, Argentina

Romero, Carlos, Sr.
FUNDACAO UNIVERSIDADE PATAGONICA
Ruta 258 KM 10.500 C.C 657-CP 8400
S. Carlos De Bariloc, Argentina
T: 54 944 24734

Romero, Jorge Emilio, Mr.
Director
TEA
Calle 510 /1915
La Plata, Argentina

Roque, Juan Carlos, Mr.
President
CONAM (CONSERVACION AMBIENTAL)
Esperanza 1558
Cordoba 5850, Argentina
T: 54 51 23-996 F: 54 51 22-214

Roque, Maria Teresa Ferrero De, Ms.
Secretary
CONAM (CONSERVACION AMBIENTAL)
Esperanza 1558
Cordoba 5850, Argentina
T: 54 571 23-996 F: 54 571 22-214

Rosenthal, Christian Gaston Borges, Sr.
Estudiante
ESCOLA INDUSTRIAL DE PUERTO IGUAZU
Iray Luis Betron 69
Puerto Iguazu 3370, Argentina

Rosenthal, Mario, Mr.
CONVOCATORIA AMBIENTAL
Guruchaga 2126
Buenos Aires 1425, Argentina
T: 54 1 714080

Rossi, Marta, Ms.
ROTARY INTERNACIONAL
ENTRERROS 134 - 6th ANDAR 2000 ROSA RIO
Santa Fe 2000, Argentina
T: 210 800

Rouillet, Hugo, Mr.
Ingeniero
MINISTERIO GOBIERNO
Lacroze, 2173
La Plata Zip.: 1897, Argentina
T: 54 21 845-360

Roze, Janis, Mr.
FUNDACION ELPIS
Monteagudo 20, San Marcos Sierra
Cordoba, Argentina
T: 54 51 96101 F: 54 51 543-7562

Rubel, Diane, Ms.
Biologist
FAC. DE CIENCIAS EXACTAS Y NATURALES
Godoy Cruz 2439-Depto. 15
Buenos Aires 1425, Argentina
T: 54 1 773-4099

Rubilar, Alejandra Andrea, Ms.
Coupe 1548
Neuquen, Argentina
T: 54 43 28830

Ruiz, Alejandro, Mr.
UNIVERSIDAD DE CORDOBA-FACULDAD DE CIENCIAS EXACTAS FISICAS Y NATURALES
Blvd.Chacabuco, 654
Cordoba 5000, Argentina

Russo, Cintia, Ms.
UNIVERSIDAD DE BUENOS AIRES
V. Loreto 3145 Ato 4
Buenos Aires, Argentina
T: 54 1 552 6395

Saba, Nicolas Alberto, Sr.
Biologo
12 de Octobre 1646 - Bo. Providencia - CP 5000
Cordoba, Argentina
T: 54 51 720320

Sabattoli, Carlos, Mr.
C.H. Rodriguez 282
Neuquen 8300, Argentina
T: 54 943 22629

Saeed, Dario, Sr.
Representative
COMUNIDADE BAHA©I
Alvarez Jonte 3861 p. 15 Dpto. F
Buenos Aires, Argentina
T: 54 1 566-3187

Saenz, Fernando, Mr.
DEPARTAMENTO DE INGENIERIA AMBIENTAL
C.C. 961 - Bariloche
Rio Negro 8400, Argentina
T: 54 944 22121 F: 54 944 23051

Saez, Angela Aurelia Frontini De, Sra.
Miembro Titular
COCHUNA VIDA SI
Moreno 1445
Concepcion-Tucuman 4146, Argentina
T: 54 81 865-22449

Sago, Miriam Delia, Ms.
FUNDECMA
Junin 461/9 andar, Cap Fed
Buenos Aires 1026, Argentina
T: 54 1 953-6182 F: 54 1 312-7781

Sala, Sergio Gabriel, Mr.
Expositor
FLACAM
Rio Pilcomayo, 881 - Bella Vista
Buenos Aires 1661, Argentina
T: 54 1 666-1789 F: 54 1 381-1136

Salemme, Monica, Ms.
Researcher
CADIC - CONICET
C.C. 92
Ushuaia 9410, Argentina
T: 1 901 22320 F: 1 901 22318

Salerno, Gustavo, Sr.
Estudiante
UNIVERSIDAD NACIONAL DE MAR DEL PLATA
CALLE 67 No. 3322 NECOCHEA
Buenos Aires, Argentina

Salibian, Alfredo, Dr.
WORLD COUNCIL OF CHURCHES
Helguera 2050
Buenos Aires 1416, Argentina

Sambucetti, Sebastian, Sr.
Periodista
MCDONALDS DE ARGENTINA
H.. Irigoyen 910, 1er Piso 'C'
Buenos Aires, Argentina
T: 54 1 334-5648 F: 54 1 334-8628

Sammartino, Sergio, Sr.
Coordinator
RED ECOLOGICA DEL SUR
C.C.895 - Fitz Roy 53 / Oficina 09
Bahia Blanca 8000, Argentina
T: 54 91 49 898

Sand, Hugo Anibal, Mr.
Ingeniero Agronomo
UNITED LUTHERAN CHURCH
Gobernador Barreiro 947
Obera 3360, Argentina
T: 755 22340

Sanhez, Jorge
Tv Cameraman
REVISTA NUEVA /MUTANTIA TV
Esmeralda 343
Buenos Aires, Argentina
T: 54 1 3254500/3258157 F: 54 1 3256114

Santiago, Susana Irene, Ms.
Arquitecta
PROCOVI - PROMOTORA DE COOPERATIVA VIVENDA
Entre Rios 1109 - 1o. 1
Rosario 2000, Argentina
T: 54 41 40942

Santibanez, Victor, Mr.
Representative In Chile
FUNDACION PARA LA DEFENSA DEL AMBIENTE - FUNAM
Casilla de Correo 83, Correo Central
Cordoba 5000, Argentina
T: 54 51 690282 F: 54 51 520260

Sapia, Cristina Noemi, Ms.
Representante De Delegacion
U.N.L.Z.
Oncativo 1552 - Lanus
Buenos Aires, Argentina
T: 54 1 241-3559

Sappia, Horacio, Mr.
GRUPO NATURA
Bolivar 535
Cordoba, Argentina

Schmee, Carolina, Sra.
FUNDACION ECOLOGICA UNIVERSAL
Avda. Corrientes 2763/7o. piso
Buenos Aires 1046, Argentina
T: 54 1 962-0237 F: 54 1 952-1535

Schnook, Karin Ines, Ms.
F.E.D.A.
Vop Leblon Etamel 2172 4th andar - Bairro Paraguay
Buenos Aires, Argentina
T: 785 7298 F: 962 0062

Schnukler, Edith, Sra.
Investigador
FLACSO
Av. Lacroze 2097
Buenos Aires, Argentina
T: 54 1 772-6407 F: 54 1 703-5727

Schroder, Juan, Mr.
RENACER
Sarmiento 2554-2o
Buenos Aires 1045, Argentina
T: 54 1 951-5739

Sciangula, Cristina Sciangula De, Sra.
Partner
AMIGOS ECOLOGIA
Ituzaingo 929, Yerba Buena
San Miguel de Tucuma, Argentina
T: 54 81 252868

Scilinguo, Enzo
Reporter
FM ROCK AND DOP
Belgrano 270
Buenos Aires, Argentina

Seara, Luis
Somidista
TELEVISION ESPANHOLA
Viamonte 166 - 2o. piso
Buenos Aires, Argentina
T: 54 1 13 2692/440 6927 F: 54 1 13-2678

Sefercheoglou, Juan Horacio, Mr.
Presidente
FUNDACION HERNANDIANA
Balcarce 73 - Villa Ballester
Buenos Aires 1653, Argentina
T: 54 1 7680700 F: 54 1 3136471

Segui, Jorge, Sr.
Asesor
SENADO DE ARGENTINA
Tucuman 3021 / 11o. '7'
Buenos Aires 1189, Argentina

Segurado, Maria Del Carmen, Ms.
BLOQUE DIPUTADOS SOCIALISTA (SANTA FE)
Paraguay 3598
Rosario, Argentina
T: 54 41 84-5178

Serafini, Maria Crsitina, Ms.
Ing.Agronoma
UNIVERRSIDAD NACIONAL DE LUJAN - ARGENTINA
Rutas 5 Y 7- Lujan
Buenos Aires 6700, Argentina
T: 54 1 323-21030 F: 54 1 323-21030

Servi, Aldo Luis, Sr.
Assistant Professor
UNIVERSIDAD NACIONAL DE LA PLATA
48 - 687 Piso 12 - D
La Platà 1900, Argentina
T: 54 21 24-5845

Silva, Lady Rosa, Sra.
Docente
Juan de Rivadeneyro 3885, La France
Cordoba, Argentina
T: 54 51 733357

Silva, Telma Marta, Sra.
Architect
FLACAM
Superi 2135 - 2o. 19
Buenos Aires, Argentina
T: 54 1 543-0505 F: 54 1 312-6612

Simek, Maya Iriondo, Ms.
CONSEJO UNIV. DE GESTION AMBIENTAL
SECRETARIA DE EXTENSION UNIV. - Pak II, CIUDAD UNIVERSITARIA
Buenos Aires 1428, Argentina

Simone, Leonardo, Mr.
Arquitecto
MINSTERIO DE CULTURA Y EDUCACION
Libertador, 4966 PB A
Buenos Aires, Argentina
T: 54 1 772-2350 F: 54 1 806-1846

Simonelli, Fabricio
Tv Assistant
REVISTA NUEVA /MUTANTIA TV
Esmeralda 343
Buenos Aires, Argentina
T: 54 1 3254500/3258157 F: 54 1 3256114

Sisto, Maria Cristina Zeballos De, Ms.
Member
FUNDACION UNIVERSIDAD 'PERITO MORENO'
Piedras 736
Buenos Aires, Argentina
T: 54 1 361-3700 F: 54 1 361-4604

Solf, Italo, Sr.
Active Member
ASOC. UNESCO
Junin 2062
Corrientes, Argentina
T: 54 717 63800

Sonjic, Ana Maria
FUNDACION ECOLOGICA UNIVERSAL/ASSOC. AMIGOS DE LA NATURALEZA
Avda. Corrientes 2763/7o. piso
Buenos Aires 1046, Argentina
T: 54 1 962-0237 F: 54 1 952-1535

Sosa, Enrique Eduardo, Mr.
CECOPAL
Av. Colon 1141
Cordoba 5000, Argentina
T: 54 51 225528 F: 54 51 223528

Sotelo, Jorge, Mr.
F L A C A M
Paraguay 226 Depto. 03- Rosario
Rosario, Argentina

Soto, Sergio Daniel, Mr.
Educacion Sindical
CONFEDARECION DE TRABAJADORES DE LA EDUCACION DE LA REPUBLICA ARGENTINA - CTERA
Barranqueiras M363 PC26 500 Viviendas
Barranqueras, Argentina
E: 3503

Speranza, Vicente, Mr.
INSTITUTO JUAN MANTOVANI
Garcia Del Rio 4559
Buenos Aires, Argentina
T: 54 1 393-6767

Spiaggi, Eduardo, Mr.
CENTRO DE ESTUDIOS AMBIENTALES DE VETERINARIA (CEAV)
Fac. Ciencias Veterinarias - Ov. Lagos - Ruta 33
Casilda Cep.: 2170, Argentina
T: 54 464 23377 F: 54 464 22050

Stefani Irizar, Jose Lucas, Sr.
Estudiante
ESCOLA INDUSTRIAL DE PUERTO IGUAZU
1 de Maio 114
Iguazu - Misiones 3370, Argentina
T: 075721156

Suarez, Cristina Silvia, Ms.
SOC. ECOLOGICA REGIONAL
Casa de La Cultura - El Bolson - Rio Negro
Rio Negro 8430, Argentina
T: 54 944 92204 F: 54 944 92204

Suarez, Esteban, Sr.
Student
UNIVERSIDAD TECNOLOGICA NACIONAL
Pronunciamiento s/n
Parana - PR, Argentina
T: 54 43 243694 F: 54 43 243589
E: Alma/verte 1033

Sunsta, Lee Min, Ms.
FUNDACION ECOLOGICA UNIVERSAL
Avda. Corrientes 2763/7o. piso
Buenos Aires 1046, Argentina
T: 54 1 962-0237 F: 54 1 952-1535

Tarat, Pedro, Mr.
Attorney
FUNDACION AMBIENTE Y RECURSOS NATURALES
Monroe 2142 -1428
Buenos Aires, Argentina
T: 54 1 7816115 F: 54 1 7816115

Temporelli, Berta Leonor, Ms.
COMISION ECOLOGIA DE ROSARIO
Pje. Ferran, 359
Rosario, Argentina

Thibaud, Miguel H, Mr.
SOCIEDAD RURAL ARGENTINA
Florida 460
Buenos Aires, Argentina
T: 54 1 322-3431 F: 54 1 325-8231
TL: 23414SRAAR

Tito, Gustavo Mauricio, Mr.
Miembro
FUNDECMA
47 1763 D2
La Plata Cep.: 1900, Argentina
T: 54 21 208-2186 F: 54 21 209-3075

Tognetti, Mariana, Mr.
FUNDACION UNIVERSIDAD PERITO MORENO
Paraguay 5480/7o. C
Buenos Aires, Argentina
T: 54 1 773-9091

Tomatis, Nelson Luis, Mr.
L. DEL LITORAL
Paraguay 726 - Piso 11 Dt.2
Rosario, Argentina
T: 54 41 263946

Torre, Ricardo Manuel De La, Mr.
UNIVERSIDAD DE MENDOZA
Boulogne Sur Mer, 665 Piso 5
Mendonza 5500, Argentina
T: 54 61 24-7017 F: 54 61 31-1100

Torres, Carolina Cecilia, Sra.
Student
FCEFYN / U.N.C.
Esposos Curie 1850 Barrio Maipu 2a. Seccion
Cordoba 5014, Argentina
T: 54 51 550-350

Torres, Liliana B., Sra.
Estudiante
Aldane y Ortega No. 3684 Patricios M.
Cordoba, Argentina
T: 54 51 725584

Tortora, Maria Isabel, Ms.
Directora Ord. Ambiental
MUNICIPALIDAD SAN FERNANDO
Miguens, 2893 - Victoria San Fernando
San Fernando - Bs.As, Argentina
T: 54 1 745-6009

Tossenberger, Alex, Mr.
Sub-Director
DESBORDAR
Jose M. Paz 2311 - Olivos
Buenos Aires, Argentina
T: 54 1 791-2273

Trevisan, Miryam, Ms.
Invest.
UNIVERSIDAD NACIONAL DE ROSARIO
Italia, 123
Rosario Cep.:2000, Argentina
T: 54 41 251930 F: 54 41 251930

Ureta, Demetrio, Mr.
GRUPO NATURA
San Martin 2120
Huerta Grande,Cordob, Argentina

Urquiza, Alberto Luiz De, Mr.
Delegate
EMAUS INTERNATIONAL
Amenedo 1058
Burzaco, Argentina
T: 54 1 299-1593

Vadell, Edith Susana, Ms.
9 De Julio 3301
Rosario, Argentina
T: 54 41 371355

Vaich, Susana, Sra.
AMIDA
12 no. 1924 'PB'
La Plata 1900, Argentina
T: 54 21 51-9926

Valenzuela, Antonio, Mr.
Hipolito Irigoyen, 365
Puerto Vilelas, Argentina
T: 72281012

Valestreti, Florencia
FUNDACION ECOLOGICA UNIVERSAL
Avda. Corrientes 2763/7o. piso
Buenos Aires 1046, Argentina
T: 54 1 962-0237 F: 54 1 952-1535

Valladares, Alfredo Jose, Mr.
Secretario
INSTITUTO ARGENTINO PARA EL DESARROLLO INTEGRAL DE CIENCAS HIDROGRAFICAS
Reconquista, 1048- 1o. piso
Buenos Aires, Argentina

Valle, Genovesi Gabriela Del, Sra.
Professor
FUNDACION DE ACCION ECOLOGICA
Ricardo Pedroni 1712 Bajo Palerma
Cordoba, Argentina
T: 54 51 80-0265 F: 54 51 241405

Valle, Guillermo Del, Mr.
President
ADNUMA
27 de Abril 864
Cordoba 5000, Argentina
T: 54 51 226304 F: 54 51 226304

Valls, Claudia, Sra.
Sub-Diretora Ambiente
COLEGIO DE ABOGADOS SAN ISIDRO
Billinghurst 1565 8A
Buenos Aires, Argentina
T: 54 1 824-2698

Varela, Damian, Sr.
Coordinator
CIPOCAR
Monroe 5617
Buenos Aires 5643, Argentina
T: 54 1 51-8689

Venchiarutti, Raul Alberto, Mr.
Socio
FUNDACION VIDA SILVESTRE
Av. Libertad, 536
Obera, Argentina

Venturino, Jorge Joaquin, Mr.
Coordinador
DIRECCION FAUNA
Salta 233 Parana
Entre Rios 3100, Argentina
T: 07 043 211566

Vicente, Carlos, Sr.
CENTRO DE TECNOLOGIA APROPRIADA DE ARGENTINA
Palais de Nations
Geneve CH 1211, Argentina
F: 41 22 733-6542

Viguier, Marcelo, Sr.
CONCEJO DELIBERANTE DE LA PLATA
12 y 51
La Plata, Argentina
T: 54 21 36978 F: 54 21 256384

Villalba, Juan, Mr.
FUNDACION DESARROLLO ECOLOGICO
Corrientas 1762 - 1o. 'C'
Buenos Aires, Argentina
T: 54 1 49-0099 F: 54 1 49-0099

Villalonga, Juan Carlos, Mr.
Coordinator
E.L.C.I. BOARD/ TALLER ECOLOGISTA
Casilla de Correo 56, Sucursal 8
Rosario, Argentina
T: 54 41 39-8677 F: 54 41 25-9089

Villar, Hector Ricardo, Sr.
LU 5 RADIO NEUQUEN
Casilla de Correo 204
Neuquen 8300, Argentina
T: 54 943 32772 F: 54 943 32821

Villarroel, Marlein, Ms.
GRUPO NATURA
Bolivar 535
Rio III - Cordoba, Argentina

Viola, Carlos, Mr.
Vice-President
AEROTERRA S.A.
Gorostiaga 2465
Buenos Aires 1426, Argentina
T: 54 1 775 6205 F: 54 1 775 3714

Virasoro, Maria Del Carmen Longa, Sra.
FACULTAD DE DERECHO / UNIVERSIDAD DE BUENOS AIRES
Juramento 3030, 2o. B
Buenos Aires, Argentina
T: 54 1 7850605 F: 54 1 3116086
TL: 22130PASA AR

Virdo, Alberto Nicolas, Mr.
Geographer
FUNDACION AMBIENTIS
lavalle 1282/Piso 2/Oficina 14
Buenos Aires, Argentina
T: 54 1 35-5518 F: 54 1 785-4020

Waldelich, Otto, Sr.
WORLD COUNCIL OF CHURCHES
3371 Colonizacion Andresito - Estafeta Almir. Brown
Misiones, Argentina

Weigandt, Sandra E. Rosas De, Ms.
Architect
Velez Sarsfield 1440-Bo Couemes
Cordoba, Argentina
T: 54 51 69-0916

Weigandt, Wilfredo Alberto, Mr.
Architect
MUNICIPIO EMBALSE
Velez Sarsfield 1440, Barrio Guemes
Cordoba, Argentina
T: 54 51 69-0916

Weisburd, Eloisa, Ms.
MUNICIPALIDAD DE LA PLATA
Calle 12 C/51 Y 53
La Plata, Argentina
T: 54 21 257130 F: 54 21 257130

Welcz, Felisa, Ms.
FUNDACION VIDA SILVESTRE
Cc 149 (Cae)
Eldorado 3380, Argentina
T: 54 751 23507 F: 54 751 22000

Wessels, Astrid
Contributer
ALTERNET
Mendonza 1858 14 Piso 'B'
Buenos Aires 1428, Argentina
T: 1 541 7872761 F: 1 541 8144066

Westerkamp, Angela, Ms.
MOVIMENTO POR LA VIDA Y LA PAZ
Julian Alvarez 2797-3rd
Buenos Aires 1425, Argentina
T: 54 1 8244558 F: 54 1 8044935

Westerkamp, Jose Federico, Mr.
Movip (President)
MOVIMENTO POR LA VIDA Y LA PAZ
Julian Alvarez 2797-3rd
Buenos Aires 1425, Argentina
T: 54 1 8244558 F: 54 1 8044935

Woslawski, Mauricio, Mr.
FUNDACAO UNIVERSIDADE PATAGONICA
Ruta 258 Km10.500 -c.c 657
Sao Carlos Bariloche 8400, Argentina
T: 54 944 24734

Yankelevich, Jeanette S. De, Sra.
General Director
La Rioja 229 - Ministerio de Obras Publicas
Neuquen, Argentina
T: 54 943 34742 F: 54 943 25746

Zamar, Martin Esteban, Sr.
DIR. AGUA POTABLE
El Naranjo No. 15 - Barrio Los Perales
S.S. de Jujuy-Jujuy, Argentina
T: 25077

Zapata, Nelida Carlota, Sra.
Docente
UNIVERSIDEAD NACIONAL DEL COMAHUE
Buenos Aires 1400
Neuquen 8300, Argentina
T: 54 943 23596 F: 54 943 23609
TL: 84266UNCNON

Zaugg, Martha, Ms.
GRUPO NATURA
San Martin 2120
Huerta Grande,Cordob, Argentina

Zerda, Luis De La, Sr.
LEGISLATURA DE LA PROVINCIA DE JUJUY
San Salvador, Jujuy, Argentina

Zimmer, Dieter, Mr.
Chairman
ARC PEACE (ECO-GEO ARGENTINA)
Av. Roca (S) no. 811
Santiago del Estero 4200, Argentina
T: 54 85 214170 F: 54 85 224350

Zin, Claudio
Editor
DIARIO EL CRONISTA
Honduras 5673
Buenos Aires, Argentina
T: 54 1 7753090 F: 54 1 7753694

Zuvanic, Laura, Sra.
Investigador
FLACSO
Av. Federico Lacroze 2097
Buenos Aires, Argentina
T: 54 1 772-2407 F: 54 1 703-2751

PARLIAMENTARIANS / SPIRITUAL LEADERS

Benitez, Alfredo, Sen.
Calle Hipolito Yrigoyen 1835
Capital Federal, Argentina
T: 953 6359
TL: 17100 Conar AR

Gas, Adolfo, Sen.
Calle Hipolito Yrigoyen 1835
Capital Federal, Argentina
T: 953 6359
TL: 17100 Conar AR

Malharro De Torres, Margarita, Sen.
Calle Hipolito Yrigoyen 1835
Capital Federal, Argentina
T: 953 6359
TL: 17100 Conar AR

Maximo, Claudio, Dr.
Senate
Comm. on Pop. & Dev.
Calle Hipolito Yrigoyen 1835
Capital Federal, Argentina
T: 953 6359
TL: 17100 Conar AR

Rabassa, Jorge, Deputy
Calle Hipolito Yrigoyen 1835
Capital Federal, Argentina
T: 953 6359
TL: 17100 Conar AR

Solana, Jorge, Senator
Pres.
Comm. on Pop. & Dev.
Calle Hipolito Yrigoyen 1835
Capital Federal, Argentina
T: 953 6359
TL: 17100 Conar AR

Varela, Eduardo
Dep. V.P.
Latin American Parliament
Calle Hipolito Yrigoyen 1835
Capital Federal, Argentina
T: 953 6359
TL: 17100 Conar AR

Velasquez, Hector Jorge, Sen.
Calle Hipolito Yrigoyen 1835
Capital Federal, Argentina
T: 953 6359
TL: 17100 Conar AR

ARMENIA - GOVERNMENT

Ter-Petrossian, Levon, Mr.
President
Permanent Mission of the Republic of Armenia
630 2nd. Ave.
New York NY 10016, United States of America
T: 212 686 9079 F: 212 686 3934

ARUBA - GOVERNMENT

Curet, Eduardo Armando, Mr
Geologist
DIRECTORATE OF ENERGY AFFAIRS
Caya Betico Croes 90 Oranjestad
Oranjestad, Aruba
T: 297 8 24460 26912 24 F: 297 8 25502

Willemsen, Frans
Legal Adviser
MINISTRY OF JUSTICE, LEGISLATIVE DE
Wilheiminastraat 19
Oranjestad, Aruba
F: 297 8 23134 21702
TL: 297 8 38723

AUSTRALIA - GOVERNMENT

Bayly, Brett Michael, Mr.
Media Adviser
Department of Foreign Affairs and Trade
Permanent Mission to the United Nations, United Nations Plaza
New York, NY 10017, United States of America
T: 212 963 7174 F: 212 963 1921

Bellear, Solomon David, Mr.
Aboriginal and Torress Strait Islander Commission
Permanent Mission of Australia
United Nations Plaza
New York, NY 10017, United States of America
T: 212 963 7174 F: 212 963 1921

Blunn, Anthony Stuart, Mr.
Secretary
Department of the Arts, Sport, the Environment and Territories
Permanent Mission to the United Nations, United Nations Plaza
New York, NY 10017, United States of America
T: 212 963 7174 F: 212 963 1921

Butler, Richard William, Mr.
Ambassador Extraordinary and Plenipotentiary
Permanent Representative to the United Nations, New York
Permanent Mission to the United Nations, United Nations Plaza
New York, NY 10017, United States of America
T: 212 963 7174 F: 212 963 1921

Disano, Jo Anne Giovanna, Mrs.
First Assistant Secretary
Environment Conservation Policy Division, Department of the Arts, Sport, the Environment and Territories
Permanent Mission to the United Nations, United Nations Plaza
New York, NY 10017, United States of America
T: 212 963 7174 F: 212 963 1921

Dunn, Robert Joseph, Mr.
Director, Special Projects
Department of the Arts, Sport, the Environment and Territories
Permanent Mission to the United Nations, United Nations Plaza
New York, NY 10017, United States of America
T: 212 963 7174 F: 212 963 1921

Garrett, Paul David, Mr.
Executive Officer Policy Liaison Section
Department of the Arts, Sport, the Environment and Territories
Permanent Mission to the United Nations, United Nations Plaza
New York, NY 10017, United States of America
T: 212 963 7174 F: 212 963 1921

Gilbert, Shane, Mrs.
Ministerial Staff for the Minister for Tourism and Resources
Permanent Mission to the United Nations, United Nations Plaza
New York, NY 10017, United States of America
T: 212 963 7174 F: 212 963 1921

Grieve, Narelle Ann, Mrs.
Head of Mission Secretary
Permanent Mission of Australia
United Nations Plaza
New York, NY 10017, United States of America
T: 212 963 7174 F: 212 963 1921

Griffiths, Alan Gordon, H.E. Mr.
Minister for Tourism and Resources
Permanent Mission of Australia
United Nations Plaza
New York, NY 10017, United States of America
T: 212 963 7174 F: 212 963 1921

Hamilton, Genevieve Mary, Mrs.
Multilateral Environment Section
Department of Foreign Affairs and Trade
Permanent Mission to the United Nations, United Nations Plaza
New York, NY 10017, United States of America
T: 212 963 7174 F: 212 963 1921

Hegarty, David, Mr.
Ministerial Staff for the Minister for Trade and Overseas Development
Permanent Mission of Australia
United Nations Plaza
New York, NY 10017, United States of America
T: 212 963 7174 F: 212 963 1921

Higgins, Sally, Mrs.
Administrative Assistant
Embassy of Australia, Santiago
Permanent Mission to the United Nations, United Nations Plaza
New York, NY 10017, United States of America
T: 212 963 7174 F: 212 963 1921

Hillier, John Colin Leslie, Mr.
Australian Federal Police
Permanent Mission to the United Nations, United Nations Plaza
New York, NY 10017, United States of America
T: 212 963 7174 F: 212 963 1921

Holesgrove, Roderick Leslie, Mr.
Director, Policy Liaison Section
Department of the Arts, Sport, and Environment and Territories
Permanent Mission to the United Nations, United Nations Plaza
New York, NY 10017, United States of America
T: 212 963 7174 F: 212 963 1921

Keller, Edward Lee, Mr
Director, Air Quality
ENVIRONMENTAL INSTITUTE FOR PETROLE
34 Queens Road, Suite 702
Melbourne, VIC 3004, Australia
T: (03) 820 2377 F: (03) 820 3680

Kelly, Roslyn Joan, H.E. Mrs.
Head of Delegation
Minister for the Arts, Sport, the Environment and Territories
Permanent Mission to the United Nations, United Nations Plaza
New York, NY 10017, United States of America
T: 212 963 7174 F: 212 963 1921

Kerin, John Charles, H.E. Mr.
Minister for Trade and Overseas Development
Permanent Mission of Australia
United Nations Plaza
New York, NY 10017, United States of America
T: 212 963 7174 F: 212 963 1921

Kruk, Robyn Caroline, Mr.
Policy Manager Natural, Resources Branch
The Cabinet Office, NSW
Permanent Mission to the United Nations, United Nations Plaza
New York, NY 10017, United States of America
T: 212 963 7174 F: 212 963 1921

Lambert, Judy, Mrs.
Ministerial Staff
Minister for the Arts; Sport, the Environment and Territories
Permanent Mission to the United Nations, United Nations Plaza
New York, NY 10017, United States of America
T: 212 963 7174 F: 212 963 1921

Leigh, Katherine Grace, Mrs.
Attorney-General's Department
Permanent Mission of Australia
United Nations Plaza
New York, NY 10017, United States of America
T: 212 963 7174 F: 212 963 1921

Lenehan, Susan Mary, Ms.
Minister for the Environment and Planning, South Australia
Permanent Mission of Australia
United Nations Plaza
New York, NY 10017, United States of America
T: 212 963 7174 F: 212 963 1921

Lording, David Rowland, Mr.
Minsterial Staff
Minister for the Arts, Sport, the Environment and Territories
Permanent Mission to the United Nations, United Nations Plaza
New York, NY 10017, United States of America
T: 212 963 7174 F: 212 963 1921

Loveday, Geoffrey Mark, Mr.
Senior Ministerial Adviser
Minister for the Environment and Planning South Australia
Permanent Mission to the United Nations, United Nations Plaza
New York, NY 10017, United States of America
T: 212 963 7174 F: 212 963 1921

Lucas, Jody Kathleen, Mrs.
Administrative Assistant
Embassy of Australia, Brasilia
Permanent Mission to the United Nations, United Nations Plaza
New York, NY 10017, United States of America
T: 212 963 7174 F: 212 963 1921

Lumb, John Michael, Dr.
Victorian Office for the Environment
Permanent Mission of Australia
United Nations Plaza
New York, NY 10017, United States of America
T: 212 963 7174 F: 212 963 1921

McPhail, Ian Roderick, Mr.
Executive Director
Environment Protection Authority, Department of the Arts, Sport, the Environment and Territories
Permanent Mission to the United Nations, United Nations Plaza
New York, NY 10017, United States of America
T: 212 963 7174 F: 212 963 1921

Moore, Catherine Anne, Mrs.
National Parks and Wildlife Service, South Australia
Permanent Mission of Australia
United Nations Plaza
New York, NY 10017, United States of America
T: 212 963 7174 F: 212 963 1921

Pearson,, Warwick Robinson, Mr.
Ambassador
Federative Republic of Brasil
Permanent Mission to the United Nations, United Nations Plaza
New York, NY 10017, United States of America
T: 212 963 7174 F: 212 963 1921

Purnell, Alain Henri Kenneth, Mr.
Director
Environmental Affairs, Business Council of Australia
Permanent Mission to the United Nations, United Nations Plaza
New York, NY 10017, United States of America
T: 212 963 7174 F: 212 963 1921

Rattigan, Carolyn Jane, Mrs.
Director, Land Resources Division
Department of Primary Industry and Energy
Permanent Mission to the United Nations, United Nations Plaza
New York, NY 10017, United States of America
T: 212 963 7174 F: 212 963 1921

Rayner, Peter Brian, Mr.
First Secretary
Embassy of Australia, Brasilia
Permanent Mission to the United Nations, United Nations Plaza
New York, NY 10017, United States of America
T: 212 963 7174 F: 212 963 1921

Redhead, Noreen Elizabeth, Mrs.
Environment Section
Australian International Development Assistance Bureau
Permanent Mission to the United Nations, United Nations Plaza
New York, NY 10017, United States of America
T: 212 963 7174 F: 212 963 1921

Ross, Paul Neville, Mr.
Director
Director, UNCED Coordination
Department of Primary Industry and Energy
Permanent Mission to the United Nations, United Nations Plaza New York, NY, 10017
F: 212 963 7174
TL: 212 963 1921

Samarcq, Bob Michel, Mr.
First Assistant Secretary
Department of Industry, Technology and Commerce
Permanent Mission to the United Nations, United Nations Plaza
New York, NY 10017, United States of America
T: 212 963 7174 F: 212 963 1921

Shogren, Rodney Francis, Mr.
First Assistant Secretary
The Treasury
Permanent Mission to the United Nations, United Nations Plaza
New York, NY 10017, United States of America
T: 212 963 7174 F: 212 963 1921

Simpson, Anthony, Mr.
Australian Conservation Foundation
Permanent Mission of Australia
United Nations Plaza
New York, NY 10017, United States of America
T: 212 963 7174 F: 212 963 1921

Stephen, Ninian Martin, Sir
Ambassador for the Environment
Permanent Mission of Australia
United Nations Plaza
New York, NY 10017, United States of America
T: 212 963 7174 F: 212 963 1921

Tayles, Kathryn Dorothy, Mrs.
Business Council of Australia
Permanent Mission of Australia
United Nations Plaza
New York, NY 10017, United States of America
T: 212 963 7174 F: 212 963 1921

Tilemann, John Alexander, Mr.
Director, Environment Multilateral Section
Department of Foreign Affairs and Trade
Permanent Mission to the United Nations, United Nations Plaza
New York, NY 10017, United States of America
T: 212 963 7174 F: 212 963 1921

Tilley, John William, Mr.
Assistant Secretary
Department of Primary Industry and Energy
Permanent Mission to the United Nations, United Nations Plaza
New York, NY 10017, United States of America
T: 212 963 7174 F: 212 963 1921

Tomkinson, Corinne Denise, Mrs.
First Secretary
Permanent Mission of Australia
United Nations Plaza
New York, NY 10017, United States of America
T: 212 963 7174 F: 212 963 1921

Tsang, Alderman Henry, Mr.
Deputy Lord Mayor
SYDNEY CITY COUNCIL
G.P.O. Box 1591
Sydney NSW 2001, Australia
T: 61 2 265-9706 F: 61 2 265-9416

Walker, Ronald Alfred, Mr.
Ambassador Extraordinary and Plenipotentiary
Permanent Representative to the United Nations, Geneva
Permanent Mission to the United Nations, United Nations Plaza
New York, NY 10017, United States of America
T: 212 963 7174 F: 212 963 1921

Wensley, Penelope Anne, Mrs.
First Assistant Secretary
International Organisations and Legal Division, Department of Foreign Affairs and Trade
Permanent Mission to the United Nations, United Nations Plaza
New York, NY 10017, United States of America
T: 212 963 7174 F: 212 963 1921

Wilks, Eric Anthony, Mr.
Representative
Australian Council of Trade Unions
Permanent Mission to the United Nations, United Nations Plaza
New York, NY 10017, United States of America
T: 212 963 7174 F: 212 963 1921

Willis, Janet Thea, Media Adviser
Department of the Arts, Sport, Environment and Territories
Permanent Mission of Australia
United Nations Plaza
New York, NY 10017, United States of America
T: 212 963 7174 F: 212 963 1921

Worrell, William Alan Vincent, Mr.
Technical Officer
Embassy of Australia, Washington, D.C.
Permanent Mission to the United Nations, United Nations Plaza
New York, NY 10017, United States of America
T: 212 963 7174 F: 212 963 1921

Zillman, John William, Mr.
Director
Bureau of Meterology
Permanent Mission to the United Nations, United Nations Plaza
New York, NY 10017, United States of America
T: 212 963 7174 F: 212 963 1921

NGO

Banks, Joan M., Ms.
International President
SOROPTIMIST INTERNATIONAL
400 Ryrie Street
Geelong, Australia
T: 61 52 213987 F: 61 52 294795

Boer, Ben, Mr.
Professor
LAWASIA/INTERNATIONAL ENVIRONMENTAL LAW CENTER
Faculty of Law, University of Sidney-New South Wales
Sydney - NSW, Australia
T: 61 2 232-5944 F: 61 2 221-5635

Byrne, Peter, Mr.
Vice President
INTERNATIONAL FEDERATION OF SURVEYORS
PO Box 2
Belconnen ACT 2616, Australia

Cass, Daniel, Mr.
Global Change Officer
AUSTRALIAN CONSERVATION FOUNDATION
P.O.Box 2699 Act 2601
Fitzroy, Australia
T: 61 3 416-1455 F: 61 3 416-0767

Cullen, Anthony Maxwell
LAWASIA
G.P.O. Box A 35
Perth 6001, Australia

Davis, Julie, Ms.
Solicitor
NATIONAL ENVIRONMENTAL LAW ASSOCIATION
P.O. Box 913
Dickson ACT 2602, Australia

Downton, Paul
UREA
P O Box 3040, Grenfell Street
Adelaide (Tandanya Bioregion) SA 5000, Australia

Fielden, Katherine
EYAA
PO Box 1504, Collingwood 3066
Victoria, Australia

Goetzberger, Adolf
ISES
PO Box 124
Caulfield East, Victoria 3145, Australia

Gordon, Pauline, Ms.
RICE
PO Box A100
South Sydney N.S.W. 2000, Australia

Greenwood, Garry
ERF
36 Mount View Avenue
Hazelbrook NSW 2779, Australia

Holmes, Raymond
IFS
PO Box 2
Belconnen ACT 2616, Australia

Hoyle, Cherie
UREA
P O Box 3040, Grenfell Street
Adelaide (Tandanya Bioregion) SA 5000, Australia

Hubbard, Frank
EVIA
PO Box 930
Dickson, ACT 2602, Australia

Ilett, Annie
AAEE
P.O. Box 12003
Brisbane Elizabeth Street, Queensland 4002, Australia

Illet, Annie
GA
PO Box 9868
Canberra City 2601, Australia

Inhof, Aviva
EYAA
PO Box 1504, Collingwood 3066
Victoria, Australia

Khol, Brian
WS
1A James Lane
Sydney NSW 2000, Australia

Lorriman, Doug, Mr.
ISES
PO Box 124
Caulfield East, Victoria 3145, Australia

McAndrew, Wayne
UMFA
72-74 Buckingham Street
Surry Hills NSW 2010, Australia

Nancy, Rolfe
RICE
PO Box A100
South Sydney N.S.W. 2000, Australia

Paterson, John
ACM
G.P.O. Box 1469 N
Melbourne, Victoria 3001, Australia

Read, Wallace, Mr.
INTERNATIONAL SOLAR ENERGY SOCIETY
55 Aycmer Street North Balwyn
Aveiro, Australia
T: 351 34 859-3961

Roberts, Brian
GA
PO Box 9868
Canberra City 2601, Australia

Robottom, Ian
AAEE
P.O. Box 12003
Brisbane Elizabeth Street, Queensland 4002, Australia

Roper, Anita, Ms.
Director
AUSTRALIAN CHAMBER OF MANUFACTURES
380 St. Kilda Road
Melbourne 3088, Australia
T: 61 3 698-4111 F: 61 3 699-1729

Scanlon, John
NATIONAL ENVIRONMENTAL LAW ASSOCIATION
P.O. Box 913
Dickson ACT 2602, Australia

Turbayne, David
AUSTRALIAN COUNCIL FOR OVERSEAS AID
GPO Box 1562
Canberra ACT 2601, Australia

Vang, Kevin, Mr.
WILDERNESS SOCIETY
PO Box K29 Haymarket
Sydney 2000, Australia
T: 61 2 267 7929 F: 61 2 264-2673

Whitecross, Peter
ERF
36 Mount View Avenue
Hazelbrook NSW 2779, Australia

Wilks, Tony
UMFA
72-74 Buckingham Street
Surry Hills NSW 2010, Australia

PRESS

Orr, John
Journalist
FREELANCER
51 Amelia Street Coorparoo
Brisbane 4151, Australia
T: 61 7 3944298 F: 61 7 8471268

Pankhurst, Gary
Reporter Radio/Television
RADIO STATION 2NCR-FM
60 Hindmarsh St.
NSW 2480 Australia, Australia

Peacock, Matthew
Correspondent
AUSTRALIAN B'CASTING CORP.
1724 Corroran St., NW, A
Washington, DC 20009, USA, Australia

Philpott, Leanne
Reporter Radio/Television
2 SER FM RADIO
11 Harris St.
Cronula 2230, Australia, Australia

Reeve, Simon
Reporter Radio/Television
BEYOND 2000
5/49 Elizabeth Bay
2011 NSW, Australia, Australia

Ryan, Peter
Correspondent
AUSTRALIAN B'CASTING CORP.
#B, 1717 P St.
Washington, DC 200036 USA, Australia

Sanders, Stafford, Mr.
Reporter Radio/Television
AUSTRALIAN PUBLIC RADIO
213 Wilson St.
Newtown 2042, N.S.W., Australia, Australia

Tate, Alan
Reporter Radio/Television
AUSTRALIAN B'CASTING CORP.
57/4 Ward Avenue
Potts Point, 2011, Australia, Australia

Taylor, Lenore Ruth
Correspondent
AUSTRALIAN
17 Hope St.
Dickson Act 2600 Australia, Australia

Thorman, Pelham
Reporter Radio/Television
2 SER FM RADIO
1 Ormond Street, Bondi
Sydney NSW, Australia, Australia

Worner, Timothy
Editor
BEYOND 2000
1 Cardigan Road
Roseville Chase 2069 NSW, Australia, Australia

ORGANIZATIONS

Begg, Susannah, Ms.
Environmental Youth Alliance Australia, The
PO Box 1504, Collingwood 3066
Victoria, Australia
F: (61 3) 416 0767

Burnup, Chris, Ms.
Business Council of Australia
10 Queens Road, PO Box 7225
Melbourne 3004, Australia
T: (61 3) 867 6288 F: (61 3) 867 7861

Hare, Bill, Mr.
Australian Conservation Foundation
340 Gore Street
Fitzroy, Victoria 3065, Australia
T: (61 03) 416 1455 F: (61 03) 416 0767

Jurd, Karenne, Ms.
Wilderness Society, The
1A James Lane
Sydney NSW 2000, Australia
T: (61 2) 267 7929 F: (61 2) 264 2673

Morris, Mary Lou, Ms.
Environment Institute of Australia
PO Box 930
Dickson, ACT 2602, Australia
T: (61 6) 247 7300 F: (61 6) 247 4994

Nicklas, Michael, Mr.
International Solar Energy Society
PO Box 124
Caulfield East, Victoria 3145, Australia
T: (62 3) 571 7557 F: (62 3) 563 5173

Scarf, Franklin, Mr.
Earth Repair Foundation
36 Mount View Avenue
Hazelbrook NSW 2779, Australia
T: (61 47) 586 393 F: (61 47) 586 904

Shipton, Roger, Mr.
United Nations Association of Australia Inc.
Room 5, Trinity Building 671 Hay Street
Perth WA 6000, Australia
T: (61 9) 321 7030 F: (61 9) 321 8269

Talvitie, Juha, Mr.
International Federation of Surveyors
PO Box 2
Belconnen ACT 2616, Australia
T: (61 6)253 2922 F: (61 6) 253 1741

Worth, David, Mr.
Ideas Centre Inc.
PO Box A100
Sydney South N.S.W. 2000, Australia
T: (61 2) 281 8099 F: (61 2) 281 9639

GLOBAL FORUM

Adam, Sylvia, Ms.
Marine Researcher
MACQUARIE UNIVERSITY
Flat 12, 34 Sturdee Parade, Dee Why
Syongy, Australia
T: 61 2 971 6067

Bin-Sallik, Maryann, Ms.
Delegate
INTERNATIONAL PAN AFRICAN MOVEMENT
15 Baliol St.
College Park - SA 5069, Australia
T: 8 362-1725 F: 8 302-6711

Blainey, Janette, Ms.
Director
EARTH ENVIRONMENTAL TITHING TRUST
Kyle House, 27-31 Macquaire Place
Sydney 2.000, Australia
T: 61 2 241-4203 F: 61 2 241-5662

Bressau, Almir, Jr., Mr.
Presidente
ABEMA
Av. Jeronimo Monteiro, 240 - salas 811/812
Vitoria 29010900, Australia
T: 61 53 322 4193 F: 61 53 222 3725

Caputo, Joe, Mr.
GREENHOUSE ACTION AUSTRALIA
PO Box 16
North Melbourne, Australia
T: 61 3 3290633 F: 61 3 3284445

Cato, Leigh, Ms.
Managing Diretor
BUSINESS OF ECOLOGY
254 SWAN St.
Melbourne - VIC 3000, Australia
T: 61 3 4296577 F: 61 3 4299902

Clark, Geoff, Mr.
Coordinator
NATIONAL FEDERATION OF ABORIGINAL LAND COUNCILS - AUSTRALIA
Franungham Aboriginal Trust
Purnim 3279, Australia
T: 61 55 671-003 F: 61 55 671-298

Colin, O'Neill, Graeme
Science Reporter
THE AGE
250 Spenceer Street
Melbourne 3000, Australia
T: 61 3 6012647 F: 61 3 6707514

Cookson, Michel, Mr.
GAA
GREENHOUSE ACTION AUSTRALIA
6 Banfield Square - MT Waverley
Melbourne 3149, Australia
T: 61 3 8074857 F: 61 3 565-2948

Corbett, Helen, Ms.
National Chairperson
NATIONAL COMMITTEE TO DEFEND BLACK RIGHTS / ABORIGINAL COOPORATION
PO Box 498 Broadway
New South Wales, Australia
T: 61 9 244-3197 F: 61 9 244-3197

Coyle, Gary, Mr.
PROUTIST UNIVERSAL
PO Box 713, Burwood 2134
Sydney, Australia
T: 61 2 742-5390

Crossley, Louise, Ms.
GREENHOUSE ACTION AUSTRALIA
PO BOX 754, Rosny Park
Tasmania 7018, Australia
T: 61 02 674570

Cullen, Tony, Mr.
ENVIRO REPORT GROUP
38 Robertson Rd. Scotland Island
Sydney NSW 2105, Australia
T: 61 2 997-8536 F: 61 2 2699116

Davis, Carole Rae, Ms.
Consultant
GREENGAUGE
93 Addison Rd- Manly
Sydney - NSW 2095, Australia
T: 61 2 977-2696 F: 61 2 977-0975

Duarte, Fernanda, Ms.
Reporter/Interpreter
AUSTRALIAN PUBLIC RADIO
3/18-A Carr St.
Coogee 2034, Australia
T: 61 2 6652230 F: 61 2 3303099

Engelhart, Dan
Filming
BEYOND 2000
34 Hotham Parade, Altarmon 2064
New South Wales, Australia
T: 61 2 4385155 F: 61 2 4396549

Falla, Natalie, Ms.
RAINFOREST INFORMATION CENTRE
4, Orchard Rd - Farnborough
NR. Orpington -Kent Kent BR67BP, Australia
T: 44 689 854004 F: 44 689 5226

Foy, Stephen Mark
Filming
BEYOND 2000
34 Hotham Parade, Altarmon 2064
New South Wales, Australia
T: 61 2 4385155 F: 61 2 4396549

Gadd, Diana, Ms.
WORLD YOUNG WOMEN'S CHRISTIAN ASSOCIATION
RMB 257i Wilson
Jingellic 2642, Australia
T: 61 60 371 377 F: 61 03 482 3661
TL: WYMCA 41 23 32

Garran, Robert, Mr.
Reporter
AUSTRALIAN FINANCIAL REVIEW
GPO Box 506
Sidney 2001, Australia

Gilleatt, Brad, Mr.
Reporter
2SER-FM / GREEN LEFT MAGAZINE
3 Lauderdale Ave., Fairlight 2094
Sydney, Australia
T: 61 2 9487290 F: 61 2 3303099

Greer, Pam, Ms.
P.O. Box 1151
Grafton-NSW 2460, Australia
T: 0 66 494-531 F: 0 66 561-8709

Greyem, Gabrielle, Ms.
Student
UNIVERSITY OF NEW SOUTH WALES
21 Prospect St. - Waverley
Sydney 2024, Australia
T: 61 2 389-7724

Hannocks, Christopher John, Mr.
P.O.Box 1282 North Sydney
Sydney NSW 2059, Australia
T: 61 2 4382217 F: 61 2 9543786

Hannocks, Sharyn Sinclair, Ms.
Environmental Management Offic
SYDNEY CITY COUNCIL
Envnmt'l Managememt Unit, Corp Division, Sydney City Council, Box 1591
Sydney NSW 2001, Australia
T: 61 2 265-9717 F: 61 2 265-9780

Harley, Jonathan
Journalist
RADIO 2UE
8/55-57 Wrights Road
Drummonyne -NSW 204, Australia
T: 61 02 812405 F: 61 02 3303

Heap, Eliza, Ms.
WILDERNESS SOCIETY
1 A James Lane
Sydney 2000, Australia
T: 61 2 267-7929 F: 61 2 264-2678

Hegarty, Therese
Production Manager
BEYOND 2000
34 Hotham Parade, Altarmon 2064
New South Wales, Australia
T: 61 2 4385155 F: 61 2 4396549

Henderson, Judy, Ms.
National Chair
AUSTRALIAN COUNCIL FOR OVERSIAS AID
164 Waterworks Rd.
Dynnyrne 7005, Australia
T: 61 2 343-870 F: 61 2 241489

Hitchman, Kylien, Ms.
Reporter
2 ABC RADIO / 2SER-FM
3 Lauderdale Ave., Fairlight 2094
Sydney, Australia
T: 61 2 9487290

Holland, Browner Elizabeth
Correspondent
FRIENDS OF THE EARTH CHAIN REACTION
33 Pirie St
Adelaide South Aust, Australia
T: 612 307394 F: 612 3137984

Holland, Geoff
Media Correspondent
PUBLIC RADIO 2 SER - FM 107.3
PO Box 123 Broadway
New South Wales 200, Australia
T: 02 3303000 F: 02 3303099

Hubacek, K, Mr.
UNIVERSITY
J. Hayohg 25
Breiterlerum, Australia

Imhof, Aviva, Ms.
Environment Officer
UNIVERSITY OF NSW STUDENT'S UNION
PO Box 1
Kensington NSW2033, Australia
T: 61 2 663 0461 F: 61 2 662 1702

Isaacs, Clarence, Mr.
ABORIGINAL GOVERNMENT OF AUSTRALIA
c/o Rainforst Information Centre, P. O. Box 368
Lismore - N.S.W. 2480, Australia
T: 61 66 218505 F: 61 66 222339

Jefferys, Dean, Mr.
Freelancer
PUBLIC TELEVISION
15 Prospect Waverley
Sidney 2024, Australia
T: 61 2 387-8290

Keller, Amanda Rose
Filming
BEYOND 2000
34 Hotham Parade, Altarmon 2064
New South Wales, Australia
T: 61 2 4385155 F: 61 2 4396549

Kelly, Gabrielle
Producer
ARCHANGEL AUSTRALIA
60 the Esplanade
Semaphore, Australia
T: 453518 F: 456467

Kohl, Brian, Mr.
WILDERNESS SOCIETY
1A James Lane
Sydney 2000, Australia
T: 61 2 267 7929 F: 61 2 264-2673

Langston, Marcia Lynne, Ms.
MALQUARIE UNIVERSITY
School of Behavioural Sciences, Malquarie University Sydney
Sydney NSW 2109, Australia
T: 61 2 805-8077 F: 61 2 805-5062

Lowe, Delia May, Ms.
Secretaria
NATIONAL COALITION ABORIGINAL ORGANIZATION
13 Mansfield St.
Sydney - NSW 2000, Australia
T: 61 2 6603444

Lunn, Bruce, Mr.
AUSTRALIAN AUTOMOBILE ASSOCIATION
P. O. Box 1555
Canberra ACT 2601, Australia
T: 61 6 247-7311 F: 61 6 257-5320

Maddigan, Tania, Ms.
BHP - ENVIRONMENTAL AFFAIRS
Floor 47 BHP Tower, Bourke Place 600 Bourke St.
Melbourne 3000, Australia
T: 61 3 609-3534 F: 61 3 609-3048

Marsden, Val, Ms.
Convenor
NATIONAL WOMEN'S CONSULTATIVE COUNCIL
3-5 National Circuit- Barton
Canberra 2600, Australia
T: 61 62 2715723 F: 61 62 2715751

Martins, Nilo, Mr.
Gerente
ABEMA
Av. Jeronimo Monteiro, 240 - salas 811/812
Vitoria, Australia
T: 61 53 322 4193 F: 61 53 222 3725

Mathews, Fiona Amanda
BEYOND 2000/GREENHOUSE ACTION
34 Hotham Parade, Altarmon 2064
New South Wales, Australia
T: 61 2 4385155 F: 61 2 4396549

McCaughey, Winsome, Ms.
Chief Executive
GREENING AUSTRALIA
P.O. Box 9868
Canberra 2601, Australia
T: 61 6 81-8585 F: 61 6 81-8590

Mitsos, James, Mr.
Vice President
BRUNSWICK VALLEY PEOPLE FOR PEACE
60 Armstrong St.
Byron Bay NSW 2481, Australia
T: 61 66 853026 F: 61 66 856962

Molloy, Tezza, Mr.
Facilitator
EARTH TRAIN/WORLDLINK/GLOBAL YOUTH SUMMIT
20 Nevis St.
Camberwell - Vic. 3124, Australia
T: 61 3 509-2628

Moroz, Irina, Ms.
BRUNSWICK VALLEY PEOPLE FOR PEACE
60 Armstrong St.
Byron Bay NSW 2481, Australia
T: 61 66 853026 F: 61 66 856962

Muller, Jeff, Mr.
Board Member
TREE TECHNOLOGY INTERNATIONAL, INC.
440 Park Avenue South 13th floor
New York - NY, Australia
T: 1 212 685-8400 F: 1 212 689-3701

Noyce, Philip J., Mr.
Director
GLOBAL 500 ACTION FORUM
P.O. Box 16
Melbourne 3051, Australia
T: 61 3 329-0633 F: 61 3 328-4445

O'Connor, Mark, Mr.
Vice President
AESP - AUSTRALIANS FOR AN ECOLOGICALLY SUSTAINABLE POPULATION
8 Banjine St. O'Connor
O'Connor 2601, Australia
T: 61 6 247-3341

Orszanski, Roman, Mr.
FOE AUSTRALIA
PO Box 131
Marden 5070, Australia
T: 8 61 261-6050 F: 8 61 223-4075

Pain, Nicola, Ms.
Convenor
INTERNATIONAL ENVIRONMENTAL LAW CENTRE
Suite 82, 280 Pitt Street
Sydney 2000, Australia
T: 61 2 261-3599 F: 61 2 267-7548

Quinn, Mark, Mr.
Member
BEYOND THE EARTH SUMMIT
C/ Erika Harms P.O. Box 3702-1000
San Jose, Australia
T: 50 6 32-6783 F: 50 6 21-9127

Richardson, Benjamin James, Mr.
INTERNATIONAL ENVIRONMENTAL LAW CENTRE
c/o Prof Ben Boer, Faculty Law, Univ Sydney, 173-175 Phillip St,
Sydney 2000, Australia
T: 61 2 957-4060 F: 61 2 221-5635

Rodger, Allan, Mr.
Professor
INTERNATIONAL UNION OF ARCHITECTS
Dept Architecture & Building, Univ of Melbourne
Parkville - Victoria 3052, Australia
T: 61 03 344-6454/347-7292

Rolfe, Nitya, Ms.
RAINFOREST INFORMATION CENTRE
PO Box 368 - NSW
Lismore, Australia
T: 61 55 218505 F: 61 55 222339

Rounsefell, Vanda, Ms.
ECOCO
Adelaide, Australia

Ruhlmann, Danny Peter
Filming
BEYOND 2000
34 Hotham Parade, Altarmon 2064
New South Wales, Australia
T: 61 2 4385155 F: 61 2 4396549

Shaw, Terri, Ms.
199 Canterbury Rd West, St. Kilda
Melbourne - Victoria 3182, Australia
T: 61 3 534-8719 F: 61 3 670-6030

Simons, Max J., Mr.
Managing Director
ECLECTECH DESIGN AND DEVELOPMENT
2/375 pacific highway
Artarmon 2064, Australia
T: 61 2 439-8236 F: 61 2 439-7538

Simpson, Tony, Mr.
INTERNATIONAL ENVIRONMENTAL LAW CENTRE
Suite 82, 280 Pitt Street
Sydney 2000, Australia
T: 61 2 261-3599 F: 61 2 267-7548

Sinclair-Hannocks, Sharin, Ms.
SIDNEY CITY COUNCIL
Box 1591
Sydney, Australia
T: 61 2 265.9717

Smith, Adam, Mr.
Environmental Scientist
AGC WOODWARD-CLYDE
P. O. Box 12635, Elizabeth St.
Brisbane 4002, Australia
T: 61 7 393-1533 F: 61 7 391-8019

Smyth, Joseph, Mr.
Founder
ECOCO
71 Fullarton Road Itent Town 5067
Adelaide 5067, Australia
T: 61 8 332-1099

Sykes, Roberta B., Ms.
Delegate
INTERNATIONAL PAN AFRICAN MOVEMENT
PO Box 1022, Strawberry Hills
New South Wales 2012, Australia
T: 2 699-1201 F: 2 699-1201

Thorman, Rob, Mr.
MANIDIS ROBERTS CONSULTANTS
Level 5, 88-90 Foveaux Street, Surry
Sydney 2010, Australia
T: 61 2 281-5199 F: 61 2 281-9406

Villis, Peter
ECOCO
Adelaide, Australia

Wallman, Peter
ECOCO/BECAP
Sydney, Australia

Williams, Nick Hart
ARCHANGEL AUSTRALIA
60 the Esplanade
Semaphore, Australia
T: 453518 F: 456467

Worner, Tomthy Geoffrey Redmond
Filming
BEYOND 2000
34 Hotham Parade, Altarmon 2064
New South Wales, Australia
T: 61 2 4385155 F: 61 2 4396549

Yeomans, Neville Thomas, Dr.
Consultant
PETFORD ABORIGINAL TRAINING FARM
19 Cedar Street
Yungaburra 4872, Australia
T: 61 70 953120

Yeomans, Quan, Mr.
Liaison
PETFORD ABORIGINAL TRAINING FARM
19 Cedar Street
Yungaburra 4872, Australia
T: 61 70 953120

UN AND UNCED

Koeyers, Johannes, Mr.
UNAAS
Room 5, Trinity Building 671 Hay Street
Perth WA 6000, Australia

Murray, Kaye
UNAAS
Room 5, Trinity Building 671 Hay Street
Perth WA 6000, Australia

ADVISORS

Ellyard, Peter, Dr.
Executive Director, Preferred Futures Pty. Ltd.
Melbourne 3000, Australia
T: 61 3 521 3803 F: 61 3 521 3804

PARLIAMENTARIANS / SPIRITUAL LEADERS

Chaney, Frederick Michael, Mr.
MP
The Senate, Parliament House
Canberra, A. C. T. 2600, Australia
T: 72 12 11
TL: AA62326

AUSTRIA - GOVERNMENT

Abdulai, Yesufu Seyyid Momoh
Director-General
OPEC FUND FOR INTERNATIONAL DEVELOP
P.O. Box 995 Parkring 8
Vienna A-1011, Austria
T: 51564 0 F: 513 92 38

Daniel, I. Muki
ENVIRONMENT COORDINATION UNIT UNIDO
Vienna Int'L Center P.O. Box 309 Room D2123
A-1400 Vienna, Austria
T: 431 21131 3778 3779 F: 431 2304779

Manahl, Christian
FEDERAL CHANCELLORY
Ballhausplatz
Vienna 1014, Austria
T: 53 115 4474 F: 53 115 3619

Miremadi, Assadollah
Head Energy Section
OPEC
Obere Donaustrasse 93
Vienna A1020, Austria
T: 43 1 21112205 F: 43 1 264320

Subroto,
Secretary General
ORGANIZATION OF PETROLEUM
Exporting Countries (OPEC) Obere Donaustrasse 93
Vienna A1020, Austria
T: (43-1) 21112 F: (43-1) 264320

Vranitsky, Franz, Mr.
Chancellor
Permanent Mission of Austria
809 United Nations Plaza 7th floor
New York NY 10017, United States of America
T: 212 949 1840 F: 212 953 1302

NGO

Cayford, James, Mr.
INTERNATIONAL UNION OF FORESTRY
IUFRO Secretariat
A-1131 Wien-Schonbrunn, Austria

De Freitas, Amantino R.
IUFRO
IUFRO Secretariat
A-1131 Wien-Schonbrunn, Austria

De Maeyer, Marc
ICSW
Koestlergasse 1/29
A-1060 Vienna, Austria

Heilig, Gerhard, Mr.
INTERNATIONAL INSTITUTE FOR APPLIED SYSTEMS ANALYSIS
Schlossplatz 1
Laxenburg A-2361, Austria
T: 43 223671521 F: 43 223671313
TL: 079137iiasa a

Lazaretti, Alfredo
IUSY
Neustiftgasse 3, A-1070 Vienna
Austria

Lutz, Wolfgang, Mr.
INTERNATIONAL INSTITUTE FOR APPLIED SYSTEMS ANALYSIS
Schlossplatz 1
Laxenburg A-2361, Austria
T: 43 223671521 F: 43 223671313
TL: 079137iiasa a

Oberg, Sture, Mr.
Professor
INTERNATIONAL INSTITUTE FOR APPLIED SYSTEMS ANALYSIS
Schlossplatz 1
Laxenburg A-2361, Austria
T: 43 223671521 F: 43 223671313
TL: 079137iiasa a

Ricard, Torrell
IUSY
Neustiftgasse 3, A-1070 Vienna
Austria

Russel, Robert I.
ICSW
Koestlergasse 1/29
A-1060 Vienna, Austria

Selle, Ronaldo
IUSY
Neustiftgasse 3, A-1070 Vienna
Austria

Turid, Birkeland
IUSY
Neustiftgasse 3, A-1070 Vienna
Austria

PRESS

Cumberbatch, Nigel
Correspondent
OPEC NEWS AGENCY (OPECNA)
Ave. Jose Maria Vargas,Apt.111
Santa Fe Norte, Caracas, Venezuela, Austria

Fluch, Franz
Reporter Radio/Television
ORF
Gallitzinstrasse 77/7/4
1160 Vienna, Austria, Austria

Koessler, Franz
Correspondent
AUSTRIAN TV
1206 Eton Ct. NW
Washington, DC USA, Austria

Mayer, Kurt Martin
Editor
KURIER
StrauBengasse 7/5
A-1050 Wien, Austria, Austria

Narr, Monika
Editor
WIENER ZEITUNGA
Hockegasse 17-A-1180
Vienna, Austria, Austria

Schimke, Fritz
Reporter Radio/Television
MCTV
Hermammgasse 14/9
1070 Vienna, Austria, Austria

Schreiberhuber, Hermine
Editor
AUSTRIA PRESS AGENCY
Josefstadterstr 5l/4/21
Vienna, Austria, Austria

Seinitz, Kurt
Editor
KRONEN-ZEITUNG
Weihburggasse 5/8
A-1010 Wien, Austria, Austria

Simburger, Franz
Reporter Radio/Television
ORF-AUSTRIAN RADIO
A-1070 Bandgasse 30/15
Vienna, Austria, Austria

Werner, Christian
DER STANDARD
R. Riachuelo, Apt. 2402
90010 Porto Alegre-RS, Brazil, Austria

Wilk-Sanatani, Sangita
Correspondent
SOL
A-1190 Vienna, Huleschgasse
13/B-3 Austria, Austria

ORGANIZATIONS

Baum-Breuer, Elisabeth, Ms.
International Council On Social Welfare
Koestlergasse 1/29
A-1060 Vienna, Austria
T: (43 1) 587 81 64 F: (43 1) 587 99 51

Burley, J., Dr.
International Union of Forestry Research Organizations (IUFRO)
IUFRO Secretariat
A-1131 Wien-Schonbrunn, Austria
T: (43 1) 820 151

Schenk, Maria, Ms.
European Union Of Women*
Korntnerstrass 51
A-1010 Wien, Austria
T: (43 222) 515 21 326 F: (43 222) 512 24 68

Shaw, Roderick, Mr.
International Institute For Applied Systems Analysis
A-2361 Laxenburg, Austria
T: (43 2236) 715 210 F: (43 2236) 713 13

GLOBAL FORUM

Carneiro, Sergio De Alvim, Sr.
Castellezg. 26/22
Vienna, Austria
T: 43 1 216-4715

Fedra, Kurt, Mr.
INT'L INSTITUTE FOR APPLIED SYSTEMS ANALYSIS (IIASA)
Schlossplatz 1
Laxenburg A-2361, Austria
T: 43 2236 715210 F: 43 2236 73148

Hausknost, Daniel, Mr.
GLOBAL 2000
Lindauergasse 5/14
Vienna 1238, Austria
T: 43 1 8891801 F: 43 1 3105023

Hoegh, Hans, Mr.
Assis. Secr. General
UNITED NATIONS
G_UN
Vienna, Austria

Ilg, Andreas
SALTO
Kaiserstrasze 67
Vienna, Austria
F: 43 1 5236885

Kaspar, Martin, Mr.
GLOBAL 2000
Hahngasse 15/14
Vienna 1090, Austria
T: 43 1 3104077 F: 43 1 3105023

Kindl, Monica, Ms.
Schwarzspanier Str. 10/8
Vienna 1090, Austria

Kornexl, Werner, Mr.
STUDENT REPRESENTATIVE
Augasse 2_6
Vienna, Austria

Lundberg, Hakan, Mr.
INTERNATIONAL UNION OF SOCIALIST YOUTH
Neustiftgasse 3
Vienna, Austria
T: 46 8 714-4800 F: 46 8 7149508

Maeyer, Marc De
Correspondent
INTERNATIONAL COUNCIL ON SOCIAL WELFARE
Koestiergasse 1/29
Vienna 1060, Austria
T: 43 15 878164 F: 43 15 879951

Mikula, Boguslaw, Mr.
INTERNATIONAL INSTITUTE FOR APPLIED SYSTEM ANALYSIS
Schlossplatz 1, A-2361 Laxemburg
Laxenburg, Austria
T: 43 223686591538 F: 43 2236

Schreyerhuber, Hermine
Editor
AUSTRIA PRESS AGENCY
A - Gunoldstr. 14
Vienna 1190, Austria
T: 3605143 F: 3605224

Shaw, Myria, Ms.
INTERNATIONAL INSTITUTE FOR APPLIED SYSTEMS ANALYSIS
Scholossplatz 1
Laxenburg A-2361, Austria
T: 43 2236715210 F: 43 223671313
TL: 079 137 iiasa a

Strobl, Alice, Ms.
GLOBAL 2000
Hahngasse 15/14
Vienna 1090, Austria
T: 43 1 222-3104078 F: 43 1 222-3105023

Weaver, Paul, Dr.
Director
INTERNATIONAL INSTITUTE FOR APPLIED SYSTEMS ANALYSIS
Schlossplatz 1
Laxenburg A-2361, Austria
T: 43 2236715210 F: 43 223673147
TL: 079137iiasa a

Weilharter, Ulrike, Ms.
FRAUENSOLIDARITAT
Weyrgasse 5/1
Vienna 1030, Austria
T: 43 1 222 713 35 94 F: 43 1 222 713 35 9

Wiltrand, Englaender, Ms.
WORLD CONGRESS ALTERNATIVES AND ENVIRONMENT
Richard-Straussstr. 12
Salzburg A-5020, Austria
T: 646-354
TL: 632-400

Zanini, Ernst, Dr.
Amld. Shesk. Landesregierung RA 6 Learmeliterpl. 2
Graz 8044, Austria
T: 0043316 877-3181

UN AND UNCED

Billand, K.
United Nations Industrial Development Organization
Vienna International Centre, P.O. Box 300
A-1400 Vienna, Austria
T: 43-22 2111310 F: 43-22 232156

Chambaiu, B.
United Nations Industrial Development Organization
Vienna International Centre, P.O. Box 300
A-1400 Vienna, Austria
T: 43-22 2111310 F: 43-22 232156

Daniel, Muki
United Nations Industrial Development Organization
Vienna International Centre, P.O. Box 300
A-1400 Vienna, Austria
T: 43-22 2111310 F: 43-22 232156

Lund, H.
United Nations Industrial Development Organization
Vienna International Centre, P.O. Box 300
A-1400 Vienna, Austria
T: 43-22 2111310 F: 43-22 232156

Nakazawa, Kenji
Associate Expert
ENVIRONMENT COORDINATION UNIT/UNIDO
Vienna International Centre PO Box 300 A-1400
Vienna, Austria
T: 43 1 21131 3367 F: 43 1 2307449

Pedrosa, W.
United Nations Industrial Development Organization
Vienna International Centre, P.O. Box 300
A-1400 Vienna, Austria
T: 43-22 2111310 F: 43-22 232156

Pembleton, Peter
United Nations Industrial Development Organization
Vienna International Centre, P.O. Box 300
A-1400 Vienna, Austria
T: 43-22 2111310 F: 43-22 232156

Siazon, Domingo
United Nations Industrial Development Organization
Vienna International Centre, P.O. Box 300
A-1400 Vienna, Austria
T: 43-22 2111310 F: 43-22 232156

Tcheknavorian, A.
United Nations Industrial Development Organization
Vienna International Centre, P.O. Box 300
A-1400 Vienna, Austria
T: 43-22 2111310 F: 43-22 232156

Williams, Robert
United Nations Industrial Development Organization
Vienna International Centre, P.O. Box 300
A-1400 Vienna, Austria
T: 43-22 2111310 F: 43-22 232156

PARLIAMENTARIANS / SPIRITUAL LEADERS

Keppelmueller, Peter, Dr.
MP
Bundesrat, Parlamentsgebaude, Dr. Karl-Renner-Rin 3
A-1017 Wien, Austria
T: (222) 4804
TL: 115231 parld a

Langthaler, Monika, Mrs.
MP
Bundesrat, Parlamentsgebaude, Dr. Karl-Renner-Rin 3
A-1017 Wien, Austria
T: (222) 4804
TL: 115231 parld a

Schweitzer, Karl
MP
Bundesrat, Parlamentsgebaude, Dr. Karl-Renner-Rin 3
A-1017 Wien, Austria
T: (222) 4804
TL: 115231 parld a

Steinbauer, Heribert
MP
Bundesrat, Parlamentsgebaude, Dr. Karl-Renner-Rin 3
A-1017 Wien, Austria
T: (222) 4804
TL: 115231 parld a

AZERBAIJAN - GOVERNMENT

Hajiev, Ibrahim, Mr.
Director
Department of Culture and Echology, Ministry of Foreign Affairs
Permanent Mission to the United Nations, Permanent Mission to the United Nations, 136 E. 67th St.
New York, NY 10021, United States of America
T: 212 949 0150

Huseinov, Ragim, H.E. Mr.
Prime Minister
Azerbaidjani Republic
136 E. 67th St.
New York, NY 10021, United States of America
T: 212 949 0150

Mansurov, Arif Enver-Ogly, Mr.
Chairman
Committee for Environment Protection
Permanent Mission to the United Nations, 136 E. 67th St.
New York, NY 10021, United States of America
T: 212 949 0150

GLOBAL FORUM

Atababyev, Elchin, Mr.
AZERBAIJAN GREENS MOVEMENT
Sumgait, Azerbaijan
T: 873 140-2631

Bahar, Hajyzadeh, Ms.
AZERBAIJAN GREEN MOVEMENT
370000 Azerbaijan, Baku, Kirov Prospekt 14, - 9
Baku, Azerbaijan
T: 89 22 93 09 51 F: 89 22 98 73 03
TL: 142172BAKINSU

Elchin, Atababayev, Mr.
AZERBAIJAN GREEN MOVEMENT
Institute of Genetics and Selection, Lenin Prospekt, 155
Baku 370606, Azerbaijan
T: 89 22 62 97 95 F: 89 22 62 95 02
TL: 142172 BAKIN 14

Hajyzadeh, Bahar, Mr.
AZERBAIJAN GREENS MOVEMENT
Baku, Azerbaijan
T: 7 892 140-2631

Mamed, Efendiyeva Effet, Ms.
MOVEMENT OF AZERBAYZHANIAN GREENS
Alekperova St. 97, Floor 22
Baku 370141, Azerbaijan
T: 7 892 395281 F: 7 892 13370

BAHAMAS - GOVERNMENT

Carey, Sandra, Ms.
Second Secretary
Permanent Mission of Bahamas
East 46th Street
New York, NY 10017, United States of America
T: 421 6925, 6926, 6929 F: 759 2135

Carter, Charles, The Hon.
Minister
Health (Head of Delegation)
Permanent Mission to the United Nations, East 46th Street New York, NY, 10017
F: 421 6925, 6926, 6929
TL: 759 2135

Cooper, Donald Leon, Dr.
Chief Public Analyst
Ministry of Health
P.O. Box N8903
Nassau Np, Bahamas
T: 809 322 2845 F: 809 323 3863

Stubbs, Michael, Mr.
Senior Meteorologist
Meteorological Department
N-8330 Nassau Bahamas
Nassau, Bahamas
T: 809 325 4049 809 325 F: 809 327 5275

NGO

Larson, Gary
BNT
P.O. Box N4105
Nassau, Bahamas

Larson, Susan
BNT
P.O. Box N4105
Nassau, Bahamas

BAHRAIN - GOVERNMENT

Al-Arrayed, Jawad Salem, H.E. Mr.
Minister Of Health
Ministry of Health
P.O. Box 145
Manama, Bahrain
T: (973) 252605 F: (973) 252569

Al-Kuwari, Zahwa Mohamed, Miss
Engineer
Environmental Protection Committee
Permanent Mission to the United Nations, United Nations Plaza
New York, NY 10017, United States of America
T: 212 963 7174 F: 212 963 1921

Al-Rumaihi, Nabil Mohamed, Mr.
Secretary
Minister of Health
Permanent Mission to the United Nations, United Nations Plaza
New York, NY 10017, United States of America
T: 212 963 7174 F: 212 963 1921

Al-Zayani, Faisal, Mr.
Councellor
Ministry of Foreign Affairs
Permanent Mission to the United Nations, United Nations Plaza
New York, NY 10017, United States of America
T: 212 963 7174 F: 212 963 1921

Fakhro, Khalid Mohamed, Mr.
Vice Chairman
Environmental Protection Committee
Permanent Mission to the United Nations, United Nations Plaza
New York, NY 10017, United States of America
T: 212 963 7174 F: 212 963 1921

BANGLADESH - GOVERNMENT

Ahmed, Risalat, Mr.
Director General
Department of Environment
Government of Bangladesh Road No.2,House No.16,Dhanmondi Ra
Dhaka-1209, Bangladesh
T: 812461

Al-Noman, Abdullah, H.E. Mr.
Minister
Enviroment and Forests
Permanent Mission to the United Nations, United Nations Plaza, 8th Floor
New York, NY 10017, United States of America
T: 867 3434, 3435, 3436

Babu, Sayedazzaman
Second Secretary
EMBASSY OF BANGLADESH
Brazilia - Df, Brazil
Bangladesh,
T: BANGLADESH 312544

Chowdhury, Iftekhar Ahmed, Dr.
Director General
Ministry of Foreign Affairs
Permanent Mission to the United Nations, United Nations Plaza, 8th Floor
New York, NY 10017, United States of America
T: 867 3434, 3435, 3436

Chowdhury, S. H. M. B. Nur, Mr.
Minister
Embassy of Bangladesh, Brasilia
Permanent Mission to the United Nations, United Nations Plaza, 8th Floor
New York, NY 10017, United States of America
T: 867 3434, 3435, 3436

Huq, Md. Fazlul, Mr.
Joint Secretary
Ministry of Environment
Permanent Mission to the United Nations, United Nations Plaza, 8th Floor
New York, NY 10017, United States of America
T: 867 3434, 3435, 3436

Islam, Mohammad Anwarul, Mr.
Deputy Director
Department of Environment
House-2 Road-16 Dhanmondi R/A
Dhaka-1209, Bangladesh
T: 318682 328676

Kabir, Humayun, Dr
Senior Assistant Secretary
Ministry of Environment & Forests
Room 1303, Bldg 6, Bangladesh Secretariat, Dhaka-1000
Dhaka, Bangladesh
T: 88 02 246607 834156

Katebi, Muhammad Nurul Alam, Mr.
Chief Conservator Of Forests
Forest Department
33 Bailey Road
Dhaka, Bangladesh
T: 605923 401260

Khan, M. Sultan Mahmud
Deputy Secretary
MINISTRY OF ENVIRONMENT AND FOREST
Government of Bangladesh Bangladesh Secretariat
Dhaka, Bangladesh
T: 880 2 246607

Khan, Z. N. Tahmida, Dr.
Professor
Department of Botany, University of Dhaka
Dhaka-1000, Bangladesh
Dhaka, Bangladesh
T: 500545 (R) 506378

Majumdar, Shamsuzzaman, Dr.
Division Chief
Planning Commission
Permanent Mission to the United Nations, United Nations Plaza, 8th Floor
New York, NY 10017, United States of America
T: 867 3434, 3435, 3436

Md Fazlul, Huq
Joint Secretary
MINISTRY OF ENVIRONMENT & FOREST
Govt of Bangladesh
Dhaka, Bangladesh
T: 606704

Mohammad, Reazuddin
Deputy Director
DEPARTMENT OF ENVIRONMENT
3/2 Secretary Hostel
Shangshad Bhavan, Bangladesh
T: 323575

Mozumder, Shams Uz Zaman, Dr.
Division Chief
PLANNING COMMISSION
Govt of Bangladesh Block-17, Room-21
Sher-E-Banaka Nagar Dhaka, Bangladesh
T: 0088 02: 814724

Rahman, A.S.M. Mustafizur, H.E. Mr.
Minister
Foreign Affairs
Permanent Mission to the United Nations, United Nations Plaza, 8th Floor
New York, NY 10017, United States of America
T: 867 3434, 3435, 3436

Rashid, Mohammad Abdur, Dr.
Secretary
Ministry of Environment & Forests
Flat 13-C Eastern Housing Apt Sidheswari
Dhaka 1217, Bangladesh
T: 008802 833 782

Reazuddin, Md., Mr.
Deputy Director
Department of Environment
Permanent Mission to the United Nations, United Nations Plaza, 8th Floor
New York, NY 10017, United States of America
T: 867 3434, 3435, 3436

Salam, Abdus, H.E. Major Gen.
Ambassador-designate Bangladesh to Brazil
Permanent Mission of Bangladesh
United Nations Plaza, 8th Floor
New York, NY 10017, United States of America
T: 867 3434, 3435, 3436

Sayedazzaman, Md., Mr.
Second Secretary
Embassy of Bangladesh, Brasilia
Permanent Mission to the United Nations, United Nations Plaza, 8th Floor
New York, NY 10017, United States of America
T: 867 3434, 3435, 3436

Siddiqui, Kamaluddin, Dr.
Secretary to the Prime Minister
Permanent Mission of Bangladesh
United Nations Plaza, 8th Floor
New York, NY 10017, United States of America
T: 867 3434, 3435, 3436

Sobhan, Farooq, Mr.
Additional Foreign Secretary
Permanent Mission of Bangladesh
United Nations Plaza, 8th Floor
New York, NY 10017, United States of America
T: 867 3434, 3435, 3436

Syed, M.a., Mr.
Alternate Executive Director
World Bank
Permanent Mission to the United Nations, United Nations Plaza, 8th Floor
New York, NY 10017, United States of America
T: 867 3434, 3435, 3436

NGO

Alam, A.K.M. Jahangir
BNP
GPO Box 4046
Ramna, Dhaka, Bangladesh

Ali, Hashem
BNP
GPO Box 4046
Ramna, Dhaka, Bangladesh

Bano, Selina
UC
GPO Box 2296
Dhaka 1000, Bangladesh

Bashida,
NGOF
4/6, Block E
Lalmatia, Dhaka-1207, Bangladesh

Hoque, Mozammel
UC
GPO Box 2296
Dhaka 1000, Bangladesh

Hossain, Sanovar
POUSH
5/7 Sir Syed Road, Block A
Mohammadpur, Dhaka 1207, Bangladesh

Huq, Saleemul, Mr.
Executive Director
BANGLADESH CENTRE FOR ADVANCED STUDIES
620 Road 10A (New) Dhanmondi, GPO Box 3971
Dhaka 1205, Bangladesh

Khan, Abul Haseeb, Mr.
RESOURCE INTEGRATION CENTRE (RIC)
3/9 Block-C, Lalmatia
Dhaka 1207, Bangladesh
T: 88 2 31.1839 F: 88 2 81.3014
TL: 671166 ICIL BJ

Khan, Adilur Rahman, Mr.
RESOURCE INTEGRATION CENTRE (RIC)
3/9 Block-C, Lalmatia
Dhaka 1207, Bangladesh
T: 88 2 31.1839 F: 88 2 81.3014
TL: 671166 ICIL BJ

Mohammed Ataur, Rahman
CESR
68/1 Purana Pattan
Dhaka-1000, Bangladesh

Rahman, A. Atiq, Dr.
BANGLADESH CENTRE FOR ADVANCED STUDIES
Palais de Nations
Geneve CH 1211, Bangladesh
F: 41 22 733-6542

Rahman, Mustafizur
BMS
Natun Bazar
Parbatipur Dinajpur 5250, Bangladesh

Rakib, Abdur
BMS
Natun Bazar
Parbatipur Dinajpur 5250, Bangladesh

Sabur, Mohammad Zia-Us, Mr.
NGO FORUM FOR DRINKING WATER SUPPLY AND SANITATION
4/6 Block E LalMatia
Dhaka, Bangladesh
T: 88 2 894346

PRESS

Chowdhury, Shahadat
Editor
BICHITRA
30 Hatkhola Road, Tikatooly
Dhaka 1203, Bangladesh, Bangladesh

Yagub, Nerun
Editor
THE BANGLADESH TIMES
1/10 Block C, 1st Flr, Lamatia
Dhaka 1207, Bangladesh, Bangladesh

ORGANIZATIONS

Haque, Mozammel, Mr.
United Club
GPO Box 2296
Dhaka 1000, Bangladesh
T: (880 2) 233 873/282 011 F: (880 2) 863 357/863 861

Hug, Saleemel, Dr.
Bangladesh Centre for Advanced Studies
620 Road 10A (New) Dhanmondi, GPO Box 3971
Dhaka 1205, Bangladesh
T: (8802) 315 793/310 538 F: (8802) 863 379/863 170

Milon, Imdadul Haque, Mr.
Bohumukhi Milon Shangha
Natun Bazar
Parbatipur Dinajpur 5250, Bangladesh
T: 363 (PARBATIPUR) F: (880 2) 833 297/833 306

Mirza, M. Monirul Qader, Mr.
Centre for Environmental Studies and Research
68/1 Purana Pattan
Dhaka-1000, Bangladesh
T: (88 02) 254 936
TL: 671060 SLIMA BJ

Rashid, Haroun Er, Mr.
POUSH
5/7 Sir Syed Road, Block A
Mohammadpur, Dhaka 1207, Bangladesh
T: (8802) 812 430 F: (8802) 863 060

Revington, John, Mr.
Resource Integration Centre
3/9 Block-C
Lalmatia 1207, Bangladesh

GLOBAL FORUM

Abed, Fazle Hasan, Mr.
ADAB-ASSN. OF DEVELOPMENT AGENCIES IN BANGLADESH
House No. 1/3, Block-F, Lalmatia
Dhaka 1207, Bangladesh
T: 88 2 313318 F: 88 2 813095

Absar, MD, Mr.
Delegate
DAYEMI COMPLEX BANGLADESH
42/2, Azimpur Road,
Dhaka 1205, Bangladesh

Adhikari, Susanta, Mr.
Executive Director
CHRISTIAN COMISSION FOR DEVELOPMENT IN BANGLADESH (CCDB)
88,Senpara Parbatta , Mirpur-10
Dhaka 1216, Bangladesh
T: 88 2 801971-73 F: 88 2 2803556
TL: 632429 CCDB BJ

Ahmad, Nilufar
DEVELOPMENT ALTERNATIVES WITH WOMEN FOR A NEW ERA (DAWN)
c/o WAND, UWI School of Continuing Studies, Pinelands
St Michael, Bangladesh
T: 1 809 4269288 F: 1 809 4294854

Ahmed, Qazi Faruque, Mr.
Chairman
ADAB-ASSN. OF DEVELOPMENT AGENCIES IN BANGLADESH
House No. 1/3, Block-F, Lalmatia
Dhaka 1207, Bangladesh
T: 88 2 313318 F: 88 2 813095
TL: 642940 ADAB BJ

Ahmed, Yasmin, Ms.
ADAB-ASSN. OF DEVELOPMENT AGENCIES IN BANGLADESH
House No. 1/3, Block-F, Lalmatia
Dhaka 1207, Bangladesh
T: 88 2 313318 F: 88 2 813095

Akhter, Farida, Ms.
UBINIG/RESISTANCE NETWORK
5/3 Barabo Mahanpm Ring Road, Shaymoli
Dhaka 1207, Bangladesh
T: 88 2 811465 F: 88 2 813065

Chowdhury, Mohamed Bahauddin, Mr.
SEA RESOURCE LTD
Shadarghat, CPA Building
Chittagong, Bangladesh
T: 880 31 22 69 63

D'Rosario, Benedict
Father
CARITAS BANGLADESH
Box 994
Bangladesh, Asia, Bangladesh
T: 4037060754
TL: 642940

Debnath, Krishnadulal, Mr.
Group Member
IEDS - FOE
29, Purana Paltan - PO Box 4222
Dhaka, Bangladesh
T: 88 2 23-1464 F: 88 2 86-3797
TL: 632170

Farruque, Musharraf A., Mr.
WORLD CONSTITUTION AND PARLIAMENT ASSOCIATION
26 Lalit Mohon Das Lane, P.O. Peel Khana 1205
Dhaka 1205, Bangladesh
T: 88 2 40 8081 F: 88 2 83474
TL: 671440

Ferdous, Abdur Rahman, Mr.
WORLD CONSTITUTION AND PARLIAMENT ASSOCIATION
26 Lalit Mohon Das Lane, P.O. Peel Khana 1205
Dhaka 1205, Bangladesh
T: 88 2 40 8081 F: 88 2 83 4741
TL: 671440

Haque, Nasimul, Mr.
Research Officer
BANGLADESH CENTRE FOR ADVANCED STUDIES
#620, Road 10 A Dhanmondi R.A.
Dhaka 1205, Bangladesh
T: 880 2 315793 F: 880 2 863170

Hasan, F.R. Mahmood, Mr.
ADAB-ASSN. OF DEVELOPMENT AGENCIES IN BANGLADESH
House No. 1/3, Block-F, Lalmatia
Dhaka 1207, Bangladesh
T: 88 2 313318 F: 88 2 813095

Hossain, Anwar, Mr.
WORLD CONSTITUTION AND PARLIAMENT ASSOCIATION
26 Lalit Mohon Das Lane, P.O. Peel Khana 1205
Dhaka 1205, Bangladesh
T: 88 2 40 8081 F: 88 2 83474
TL: 671440

Hussein, Mian Akbar, Mr.
Coordinator
SERVICE CIVIL INTERNATIONAL
GPO Box 3254
Dhaka 1000, Bangladesh
T: 88 2 313-623 F: 88 2 863-145
TL: 642940

Ibrahm, Muhammad, Mr.
CMES
37C Asad Avenue
Dhaka, Bangladesh
T: 88 2 811-898

Inu, Hasanul Haq, Mr.
Chairman
BANGLADESH FRONT INC.
257/4 Elephant Road
Dhaka, Bangladesh
T: 88 2 506834

Iqbal, Mohammad, Mr.
Member
SCI-PROJECT COMMITTEE
House 78/H, Road 7, Banani
Dhaka 1212, Bangladesh
T: 88 2 607-604 F: 88 2 863-412

Islam, Kazi Monirul, Mr.
Vice President
HOPE '87 BANGLADESH
56, Satish Sarker Road, Gandaria
Dhaka 1204, Bangladesh
T: 88 2 236 886 F: 88 2 863797

Kabir, Khushi, Ms.
Vice Chairperson
ADAB-ASSN. OF DEVELOPMENT AGENCIES IN BANGLADESH
House No. 1/3, Block-F, Lalmatia
Dhaka 1207, Bangladesh
T: 88 2 812353 F: 88 2 813095
TL: 642940 ADAB BJ

Karim, Mohammed Rezaul
Secretary General
DEVELOPMENT ASSOCIATION FOR YOUTH
16 Haricharan Roy Road, Faridabad
Dhaka 1204, Bangladesh
T: 88 2 252678 F: 88 2 863797

Khan, Abdul M., Dr.
Director
BAGLADESH CENTRE FOR ADVANCED STUDIES
620 Dhanmond Road 10A
Dhaka, Bangladesh
T: 88 2 815829 F: 88 2 863379
TL: 642940

Khandkar, Mir Manjurul Huq, Mr.
Delegate Member
DAYEMI COMPLEX BANGLADESH
42/2 Azsimpur Road,
Dhaka 1205, Bangladesh
T: 88 2 503-445 F: 88 2 863-057
TL: 632458 MEGNA BJ

Khandker, Ahjmed Ul Huq, Mr.
Delegate Member
DAYEMI COMPLEX BANGLADESH
42/2 Azsimpur Road,
Dhaka 1205, Bangladesh
T: 88 2 503-445 F: 88 2 863-057
TL: 632458 MEGNA BJ

Majid, Anwaruc, Mr.
President
WORLD CONSTITUTION AND PARLIAMENT ASSOCIATION
26 Lalit Mohon Das Lane, P.O. Peel Khana 1205
Dhaka, Bangladesh
T: 88 2 40 8081 F: 88 2 83474
TL: 671440

Mazhar, Farhad, Ms.
UBINIG
Palais de Nations
Geneve CH 1211, Bangladesh
F: 41 22 733-6542

Miah, Md. Abdul Latif, Mr.
Delegate Member
DAYEMI COMPLEX BANGLADESH
42/2 Azsimpur Road,
Dhaka 1205, Bangladesh
T: 88 2 503-445 F: 88 2 863-057
TL: 632458 MEGNA BJ

Moudud, Hasna, Ms.
GLOBAL 500 AWARD LAUREATE/CARDMA
159,Gulshar Ave.
Dhaha, Bangladesh
T: 600076 F: 880 2-865547

Nabi, Mohd Nur, Mr.
Delegate Member
DAYEMI COMPLEX BANGLADESH
42/2 Azsimpur Road,
Dhaka 1205, Bangladesh
T: 88 2 503-445 F: 88 2 863-057
TL: 632458 MEGNA BJ

Naser, Moinuddin
Special Correspondent
HOLIDAY
30 Tejgron Industrial Area North of Moghbalar Railway Crossing
Dhaka, Bangladesh
T: 88 2 329163/329786 F: 88 2 8802 832903

Omar, Ikhtyar, Dr.
Professor
BANGLADESH UNIVERSITY OF ENG. & TECH.
CHEM. ENG. DEPAT. bUET DHAKA 1000
Dhaka, Bangladesh
T: 88 2 501664 F: 88 2 863026

Rahman, Maswoodur, Mr.
Chairman
SERVICE CIVIL INTERNATIONAL
GPO Box 3254
Dhaka 1000, Bangladesh
T: 88 2 313-623 F: 88 2 863-145
TL: 675847 BCIC BJ

Rob, Abdur, Mr.
General Secretary
IEDS - FOE
29, Purana Paltan - PO Box 4222
Dhaka, Bangladesh
T: 88 2 23-1464 F: 88 2 86-3797
TL: 632170

Sarkar, Arun Chandra, Mr.
Program Officer
SERVICE CIVIL INTERNATIONAL
GPO Box 3254
Dhaka 1000, Bangladesh
T: 88 2 313-623 F: 88 2 863-145
TL: 642940

Shahjahan, M., Dr.
Professor
BANGLADESH UNIVERSITY OF ENGINEERING AND TECHNOLOGY
Dhaka 1000, Bangladesh
T: 88 2 861666 F: 88 2 863046

Shelley, Mizanur Rahman, Dr.
President
SERVICE CIVIL INTERNATIONAL
House 89-A, Road 8-A (new) Dhanmandi R.A.
Dhaka 1209, Bangladesh
T: 88 2 313 623 F: 88 2 863145
TL: 642940

Sultana, Shireen, Ms.
Campaign Coordinator
IEDS - FOE
29, Purana Paltan - PO Box 4222
Dhaka, Bangladesh
T: 88 2 23-1464 F: 88 2 86-3797
TL: 632170

Uddin, Mohammad Dabir, Mr.
Secretary
DEVELOPMENT ASSOCIATION FOR YOUTH
16, Haricharan Roy Road, Faridabad
Dhaka 1204, Bangladesh
T: 88 2 252678 F: 88 2 863797

Zohora, Fatema, Ms.
Secretary General
SOCIETY FOR INTERNATIONAL DEVELOPMENT
17 Elephant Rd. PO Box 5007, New Market
Dhaka, Bangladesh
T: 88 2 508097 F: 88 2 813466
TL: 671054FRCBJ

WISDOMKEEPERS

Ahmed, Benjir
OISCA Int'l
House 121, Rd. #13
Banani, Dhaka Dhaka, Bangladesh
T: 8802-865332 or 863521 F: 8802-863521

BARBADOS - GOVERNMENT

Betts, Johnny, Mr.
Honourary Consul, Barbados
Permanent Mission of Barbados
800 Second Ave. 18th Floor
New York, NY 10017, United States of America
T: 212 867 8431 F: 212 869 1030

Bulbulia, David, Mr.
Foreign Service Officer
Ministry of Foreign Affairs
Permanent Mission to the United Nations, 800 Second Ave. 18th Floor
New York, NY 10017, United States of America
T: 212 867 8431 F: 212 869 1030

Clarke, Carl, The Hon. Dr.
Minister of Labour
Consumer Affairs and the Environment
Permanent Mission to the United Nations, 800 Second Ave. 18th Floor
New York, NY 10017, United States of America
T: 212 867 8431 F: 212 869 1030

Edwards, Pearl, Mrs.
Senior Executive Secretary
Prime Minister's Office
Permanent Mission to the United Nations, 800 Second Ave. 18th Floor
New York, NY 10017, United States of America
T: 212 867 8431 F: 212 869 1030

Griffith, Mark, Dr.
Environmental Officer
Ministry of Labour, Consumer Affairs and Environment
Permanent Mission to the United Nations, 800 Second Ave. 18th Floor
New York, NY 10017, United States of America
T: 212 867 8431 F: 212 869 1030

Headley, Oliver St Clair, Prof.
Adviser UNCED Professor At Work
University of West Indies
Cove Hill Campus PO Box 64
Bridgetown, Barbados
T: 809 425 1310 F: 809 425 1327

Sandiford, , The Rt.Hon. L. Erskine
Prime Minister
Barbados
800 Second Ave. 18th Floor
New York, NY 10017, United States of America
T: 212 867 8431 F: 212 869 1030

Sandiford, Erskine, Mr.
Prime Minister
Permanent Mission of Barbados
800 Second Ave. 18th floor
New York NY 10017, United States of America
T: 212 867 8431 F: 212 986 1030

NGO

Burke, Reginald, Mr.
CARIBBEAN CONSERVATION ASSOCIATION
The Garrison
St Michael, Barbados
T: 1 809 426-9635 F: 1 809 429-8483

Flander, V. Winston, Mr.
CCA Advisor
CARIBBEAN CONSERVATION ASSOCIATION
Savannah Lodge, The Garrison
St Michael, Barbados
T: 1 809 426-2456 F: 1 809 429-8483

Morgan, Dawn, Ms.
DEVELOPMENT ALTERNATIVES WITH WOMEN FOR A NEW ERA (DAWN)
c/o WAND, UWI School of Continuing Studies, Pinelands
St Michael, Barbados
T: 1 809 4269288 F: 1 809 4294854

PRESS

Thompson, Wendy, Ms.
Reporter Radio/Television
CARIBBEAN NEWS AGENCY
25B Graeme Hall Park
Barbados, W.I., Barbados

ORGANIZATIONS

Antrobus, Peggy, Mrs
General Coordinator
Development Alternatives With Women For A New Era (DAWN)
c/o Wand, University of the West Indies, School of Continuing Studies,
Pinelands, St. Michael, Barbados
T: (1 809) 426 9288 F: (1 809) 429 4854

Howell, Calvin, Mr.
Caribbean Conservation Association
Savannah Lodge, The Garrison
St. Michael, Barbados
T: (1 809) 426 5373/426 9635 F: (1 809) 429 8483

GLOBAL FORUM

Baulu, Jean, Mr.
Director
BARBADOS PRIMATE RESEARCH AND WILDLIFE RESERVE
Farley Hill
St Peter, Barbados
T: 1 809 422 8826 F: 1 809 422 8946
TL: 2203BetBru

Dow, Jocelyn, Ms.
DEVELOPMENT ALTERNATIVES WITH WOMEN FOR A NEW ERA (DAWN)
c/o WAND, UWI School of Continuing Studies, Pinelands
St Michael, Barbados
T: 1 809 4269288 F: 1 809 4294854

Ellis, Pat
ICAE
12 Rajaji Marg
New Delhi 110011, Barbados
T: 91 11 301-4309

Emberson-Bain, Atu, Ms.
Researcher
DEVELOPMENT ALTERNATIVES WITH WOMEN FOR A NEW ERA (DAWN)
c/o WAND, UWI School of Continuing Studies, Pinelands
St Michael, Barbados
T: 1 809 4269288 F: 1 809 4294854

French, Joan, Ms.
CARIBBEAN POLICY DEV. CENTRE
c/o CCC PO Box 616
Bridgetown, Barbados

Loubser, Jan, Ms.
Research Coordinator
DEVELOPMENT ALTERNATIVES WITH WOMEN FOR A NEW ERA -DAWN
c/o Wand, UWI School of Continuing Studies Pinelands
St Michael, Barbados
T: 1 809 426-9288 F: 1 809 429-4854

Mahon, Susan, Ms.
BARBADOS ENVIRONMENTAL ASSOCIATION
P.O.Box 132,
Bridgetown West Indies, Barbados
T: 1 809 427-0619 F: 1 809 432-6446

Mason, S. Conrad, Mr.
CCC
CNR George Street and Bannister Land
St Michael, Barbados

Moore, Euna, Ms.
DEVELOPMENT ALTERNATIVES WITH WOMEN FOR A NEW ERA (DAWN)
c/o WAND, UWI School of Continuing Studies, Pinelands
St Michael, Barbados
T: 1 809 4269288 F: 1 809 4294854

ADVISORS

Barrow, Nita, Dame
Governer General, Barbados
St. Michael, Barbados
T: 1 809 429 2646/2962 F: 1 809 429 6652

BELARUS - GOVERNMENT

Dorofeev, Anatoli M., H.E. Dr.
Minister of Environment (Deputy Head of Delegation)
Permanent Mission of Belarus
United Nations Plaza
New York, NY 10017, United States of America
T: 212 963 7171 F: 212 963 1921

Kravchanka, Petr K., H.E. Dr.
Minister for Foreign Affairs (Deputy Head of Delagation)
Permanent Mission of Belarus
United Nations Plaza
New York, NY 10017, United States of America
T: 212 963 7171 F: 212 963 1921

Mozhoukhov, Alexie A., Dr.
Head of the Department
Ministry of Foreign Affairs
Permanent Mission to the United Nations, United Nations Plaza
New York, NY 10017, United States of America
T: 212 963 7171 F: 212 963 1921

Ryabkov, Oleg P., Mr.
Ministry of Foreign Affairs
Permanent Mission to the United Nations, United Nations Plaza
New York, NY 10017, United States of America
T: 212 963 7171 F: 212 963 1921

Shushkevich, Stanislav S., H.E. Dr.
Chairman of the Supreme Soviet
Permanet Mission of the Republic of Belarus
United Nations Plaza
New York, NY 10017, United States of America
T: 212 963 7171 F: 212 963 1921

BELGIUM - GOVERNMENT

Aguero, Maria Elena, Ms
Consultant
INTERAMERICAN DEVELOPMENT BANK
2 Val des Becasses
Brussels 1150, Belgium
T: 322 770 5500 F: 322 772 7508

Antoine-Gregoire, Jean-Louis, Mr
Administrateur Principal
PARLEMENT EUROPEEN
89-91 Rue Belliard
1040 Bruxelles, Belgium
F: 32 2 230 12 14

Auraujo, Sebastiao
Attache a la Delegation
Permanent Mission of Belgium
809 United Nations Plaza, 2nd Floor
New York, NY 10017, United States of America
T: 212 599 5250 F: 212 599 6843

Baeyens, Willy, M.
Chef de Cabinet Adjoint de la Ministre de la Sante Publique de l'Environnement et de l'Integration Sociale
Permanent Mission of Belgium
809 United Nations Plaza, 2nd Floor
New York, NY 10017, United States of America
T: 212 599 5250 F: 212 599 6843

Bastin, Patricia, Ms
Peintre
ASSOCIATION DES ARTISTES POUR L'ECL
47 Rue des Franchimontois
Liege, Belgium
T: 32 2 216 76 28

Baudru, Andre, M.
Consulat General de Belgique a Rio de Janeiro
Permanent Mission of Belgium
809 United Nations Plaza, 2nd Floor
New York, NY 10017, United States of America
T: 212 599 5250 F: 212 599 6843

Cartuyvels, Pierre, M.
Charge de Mission au Cabinet du Ministre de l'Environment
Square De Meeus 35
Bruxelles 1040, Belgium
T: 021515 88 11 F: 021511 36 15

Claeys, Mirielle, Mme.
Directeur Service des Nations Unies
Ministere des Affaires Etrangeres
Permanent Mission to the United Nations, 809 United Nations Plaza, 2nd Floor
New York, NY 10017, United States of America
T: 212 599 5250 F: 212 599 6843

Cordeiro, Robson, M.
Attache a la Delegation
Permanent Mission of Belgium
809 United Nations Plaza, 2nd Floor
New York, NY 10017, United States of America
T: 212 599 5250 F: 212 599 6843

Cuyvers, Jo, M.
Premier Vice-President de la Commission pour Sante Publique et Environnement
Permanent Mission of Belgium
809 United Nations Plaza, 2nd Floor
New York, NY 10017, United States of America
T: 212 599 5250 F: 212 599 6843

Czetwertynski, Michel, S.E. M.
Ambassadeur Extraordinaire et Plenipotentiaire de Belgique au Bresil
Permanent Mission of Belgium
809 United Nations Plaza, 2nd Floor
New York, NY 10017, United States of America
T: 212 599 5250 F: 212 599 6843

Dartois, Philippe, M.
Consul general de Belgique
Rio de Janeiro
Permanent Mission to the United Nations, 809 United Nations Plaza, 2nd Floor
New York, NY 10017, United States of America
T: 212 599 5250 F: 212 599 6843

Debrabandere, Jan, Maurits
Charge de Mission Services de Programmation de la Politique Scientifique
Permanent Mission of Belgium
809 United Nations Plaza, 2nd Floor
New York, NY 10017, United States of America
T: 212 599 5250 F: 212 599 6843

Decaddt, Brigitte, Mme.
Charge de Mission Services de Programmation de la Politique Scientifique
Permanent Mission of Belgium
809 United Nations Plaza, 2nd Floor
New York, NY 10017, United States of America
T: 212 599 5250 F: 212 599 6843

Dehaene, Jean-Luc, S.E. M.
Prime Ministre
Royaume de Belgique
809 United Nations Plaza, 2nd Floor
New York, NY 10017, United States of America
T: 212 599 5250 F: 212 599 6843

Delfosse, Michel, M
Ministre Plenipotentiaire
Chef du Service Nations Unies, Ministere des Affairs Etrangeres
Permanent Mission to the United Nations, 809 United Nations Plaza, 2nd Floor
New York, NY 10017, United States of America
T: 212 599 5250 F: 212 599 6843

Demuelenaere, Donald, M.
Conseil Central de l'economie
Permanent Mission to the United Nations, 809 United Nations Plaza, 2nd Floor
New York, NY 10017, United States of America
T: 212 599 5250 F: 212 599 6843

Derycke, Erik, M.
Secretaire d'Etat a la Cooperation au Development
Permanent Mission of Belgium
809 United Nations Plaza, 2nd Floor
New York, NY 10017, United States of America
T: 212 599 5250 F: 212 599 6843

Dielenant, Achille, M.
President de la Commission pour la Sante Publique et l'Environnement
Permanent Mission of Belgium
809 United Nations Plaza, 2nd Floor
New York, NY 10017, United States of America
T: 212 599 5250 F: 212 599 6843

Driessens, Jacquelina, Mme.
Attache a la Delegation
Permanent Mission of Belgium
809 United Nations Plaza, 2nd Floor
New York, NY 10017, United States of America
T: 212 599 5250 F: 212 599 6843

Gedopt, Mark, M.
Secretaire d'Ambassade
Representation Permanente de la Belgique aupres de l'Office des Nations Unies
Permanent Mission to the United Nations, 809 United Nations Plaza, 2nd Floor
New York, NY 10017, United States of America
T: 212 599 5250 F: 212 599 6843

Geel, Lydia Van
Attache a la Delegation
Permanent Mission of Belgium
809 United Nations Plaza, 2nd Floor
New York, NY 10017, United States of America
T: 212 599 5250 F: 212 599 6843

Georges, Koutzoukos, Dr
Conseiller Energie Et Questions
REPRESENTATION PERMANENTE
Av. De Cortenberg 71,B1040
Bruxelles, Belgium
T: 322 7395688 F: 322 73 55 979

Ghekiere, Dany
Head Of Multilateral Department
MINISTRY FOR AFFAIRS AND DEVELOPMEN
Marsveldpleiu 5-Box 57 1050 Brussels
Brussels, Belgium
T: 02/519 05 13 F: 02/519 05 85

Giet, Thierry, M.
Secretaire de Cabinet de la Ministre de la Sante Publique et de l'Environnement
Permanent Mission of Belgium
809 United Nations Plaza, 2nd Floor
New York, NY 10017, United States of America
T: 212 599 5250 F: 212 599 6843

Gillet, Ronald
President du Conseil l'Environnement de la Region de Bruxelles-Capitale
Permanent Mission of Belgium
809 United Nations Plaza, 2nd Floor
New York, NY 10017, United States of America
T: 212 599 5250 F: 212 599 6843

Gots, Franz, M.
Conseiller Adjoint
Administration de la Trasorerie, Ministere des Finances
Permanent Mission to the United Nations, 809 United Nations Plaza, 2nd Floor
New York, NY 10017, United States of America
T: 212 599 5250 F: 212 599 6843

Gouze, Nadine, Mme.
Charge de Mission au Cabinet de la Ministre de la Sante Publique et de l'Environnement
Permanent Mission of Belgium
809 United Nations Plaza, 2nd Floor
New York, NY 10017, United States of America
T: 212 599 5250 F: 212 599 6843

Hens, Luc, M.
Mimbre du Conseil de l'Environnement de la Region Flamande-Chef du Department Ecologie Humaine a la V.U.B.
Permanent Mission of Belgium
809 United Nations Plaza, 2nd Floor
New York, NY 10017, United States of America
T: 212 599 5250 F: 212 599 6843

Hoornaert, Chris, M.
Attache d'Ambassade Direction d'Administration des Relations Economiques Multilaterales
Permanent Mission of Belgium
809 United Nations Plaza, 2nd Floor
New York, NY 10017, United States of America
T: 212 599 5250 F: 212 599 6843

Janssens, Jean-Claude
Attache a la Delegation
Permanent Mission of Belgium
809 United Nations Plaza, 2nd Floor
New York, NY 10017, United States of America
T: 212 599 5250 F: 212 599 6843

Kuijken, Eckhard, M.
Directeur de l'Institut pour la Conservation de la Nature de la Region Flamande
Permanent Mission of Belgium
809 United Nations Plaza, 2nd Floor
New York, NY 10017, United States of America
T: 212 599 5250 F: 212 599 6843

Lambert, Marcel
Direction Celcule Environment
Boulevard Pacheco,19
Bruxelles, Belgium
T: 32 2 2104543 F: 32 2 2104704

Leunse, Maurice, M.
Conseil Central de l'economie
Permanent Mission to the United Nations, 809 United Nations Plaza, 2nd Floor
New York, NY 10017, United States of America
T: 212 599 5250 F: 212 599 6843

Leynse, Nick
Conseil Central de l'economie
Permanent Mission of Belgium
809 United Nations Plaza, 2nd Floor
New York, NY 10017, United States of America
T: 212 599 5250 F: 212 599 6843

Libert, Jose, M.
Conseil Central de l'economie
Permanent Mission to the United Nations, 809 United Nations Plaza, 2nd Floor
New York, NY 10017, United States of America
T: 212 599 5250 F: 212 599 6843

Lopes, Alvaro, M.
Attache a la Delegation
Permanent Mission of Belgium
809 United Nations Plaza, 2nd Floor
New York, NY 10017, United States of America
T: 212 599 5250 F: 212 599 6843

Mahoux, Philippe, M.
Deuxieme Vice-President de la Commission pour Sante Publique et Environnement
Permanent Mission of Belgium
809 United Nations Plaza, 2nd Floor
New York, NY 10017, United States of America
T: 212 599 5250 F: 212 599 6843

Michiels, Liliane, Mme.
Attache a la Delegation
Permanent Mission of Belgium
809 United Nations Plaza, 2nd Floor
New York, NY 10017, United States of America
T: 212 599 5250 F: 212 599 6843

Nieuwejaers, Robert, M.
Ingenieur
Chef de Service a l'Administration pour l'Environnement la Conservation de la Nature et l'Amenagement
Permanent Mission to the United Nations, 809 United Nations Plaza, 2nd Floor
New York, NY 10017, United States of America
T: 212 599 5250 F: 212 599 6843

Oeyen, Brigitte, Mme.
Attache de Presse au Cabinet du Ministre de l'Environnement de la Region Wallone
Permanent Mission to the United Nations, 809 United Nations Plaza, 2nd Floor
New York, NY 10017, United States of America
T: 212 599 5250 F: 212 599 6843

Onkelinx, Laurette, S.E. Mme.
Ministre de la Sante Publique de l'Environnment et de l'Integration Sociale
Permanent Mission of Belgium
809 United Nations Plaza, 2nd Floor
New York, NY 10017, United States of America
T: 212 599 5250 F: 212 599 6843

Parent, Ginette, Mme.
Conseil Central de l'economie
Permanent Mission to the United Nations, 809 United Nations Plaza, 2nd Floor
New York, NY 10017, United States of America
T: 212 599 5250 F: 212 599 6843

Peeters, Kris, M.
Conseil Central de l'economie
Permanent Mission to the United Nations, 809 United Nations Plaza, 2nd Floor
New York, NY 10017, United States of America
T: 212 599 5250 F: 212 599 6843

Pendeville, Bernard, M.
Conseiller au Cabinet de la Ministre de la Sante Publique et de l'Environnement
Permanent Mission of Belgium
809 United Nations Plaza, 2nd Floor
New York, NY 10017, United States of America
T: 212 599 5250 F: 212 599 6843

Pierrerd,, Hughes, M.
Attache a la Delegation
Permanent Mission of Belgium
809 United Nations Plaza, 2nd Floor
New York, NY 10017, United States of America
T: 212 599 5250 F: 212 599 6843

Smedt, Jan de, M.
Conseil National d'avis sur le climat l'environement et le developpement
Permanent Mission to the United Nations, 809 United Nations Plaza, 2nd Floor
New York, NY 10017, United States of America
T: 212 599 5250 F: 212 599 6843

Struye, Dominique, M.
Conseiller d'Embassade
Conseiller Diplomatique au Cabinet du Premier Ministre
Permanent Mission to the United Nations, 809 United Nations Plaza, 2nd Floor
New York, NY 10017, United States of America
T: 212 599 5250 F: 212 599 6843

Thiers, Godfried, Mr
Director
Institut d'Hygiene et d'Epidemiologie
Rue J. Wytsman 14
Brussels B-1050, Belgium
T: 00 32 2 642 51 11 F: 00 32 2 642 50 01

Trio, Wendel, M.
Conseil National d'avis sur le climat l'environment et let developpement
Permanent Mission to the United Nations, 809 United Nations Plaza, 2nd Floor
New York, NY 10017, United States of America
T: 212 599 5250 F: 212 599 6843

Van Driessche, Christian, M.
Conseiller d'Ambassade
Ambassade de Belgique au Bresil
Permanent Mission to the United Nations, 809 United Nations Plaza, 2nd Floor
New York, NY 10017, United States of America
T: 212 599 5250 F: 212 599 6843

Van Eevercoren, Dirk, M.
Conseil Central de l'economie
Permanent Mission to the United Nations, 809 United Nations Plaza, 2nd Floor
New York, NY 10017, United States of America
T: 212 599 5250 F: 212 599 6843

Van Ypersele De Strihou, Jean Pascal
Conseil National d'avis sur le climat l'environnement et le development
Permanent Mission of Belgium
809 United Nations Plaza, 2nd Floor
New York, NY 10017, United States of America
T: 212 599 5250 F: 212 599 6843

Vanbarel, Nancy, Mme.
Attache de Presse au Cabinet de la Sante Publique et de l'Environnement
Permanent Mission to the United Nations, 809 United Nations Plaza, 2nd Floor
New York, NY 10017, United States of America
T: 212 599 5250 F: 212 599 6843

Vanderstraten, Martine, M.
Attache Services de Programmation de la Politique Scientifique
Permanent Mission of Belgium
809 United Nations Plaza, 2nd Floor
New York, NY 10017, United States of America
T: 212 599 5250 F: 212 599 6843

Vankeerbergen, Annic, Mme.
Attache a la Direction Generale des Ressources Naturelle et de l'Environnement de la Region Wallonne
Permanent Mission of Belgium
809 United Nations Plaza, 2nd Floor
New York, NY 10017, United States of America
T: 212 599 5250 F: 212 599 6843

Vassilopoulos, Miltiades, Mr.
Cancelor
Permanent Delegation of Greece to the European Communities
71,Av Cortenberg 1040 Bryxelles Belgium
Cortenberg, Belgium
T: 00322/739 5611 F: 00 322 7355979

Verschaeve, Marc, M.
Counseil central de l'economie
Permanent Mission to the United Nations, 809 United Nations Plaza, 2nd Floor
New York, NY 10017, United States of America
T: 212 599 5250 F: 212 599 6843

Vershueren, Karl, M.
President de la Commission pour Cooperation au Developpement
Permanent Mission of Belgium
809 United Nations Plaza, 2nd Floor
New York, NY 10017, United States of America
T: 212 599 5250 F: 212 599 6843

Vyghen, Roger, M.
Chancellier, Vice-consul
Consulat General de Belgique a Rio de Janeiro
Permanent Mission to the United Nations, 809 United Nations Plaza, 2nd Floor
New York, NY 10017, United States of America
T: 212 599 5250 F: 212 599 6843

Waregne, Jean Marie, M.
Attache de la Communaute Francaise de Belgique a Rio de Janeiro
Permanent Mission of Belgium
809 United Nations Plaza, 2nd Floor
New York, NY 10017, United States of America
T: 212 599 5250 F: 212 599 6843

Willems, Stephane, M.
Charge de Mission au Cabinet de la Ministre de la Sante Publique et de l'Environnement
Permanent Mission of Belgium
809 United Nations Plaza, 2nd Floor
New York, NY 10017, United States of America
T: 212 599 5250 F: 212 599 6843

Wispelaere, Frank de, M.
Attache de Cabinet du Secretaire d'Etat a la Cooperation au Developement
Permanent Mission of Belgium
809 United Nations Plaza, 2nd Floor
New York, NY 10017, United States of America
T: 212 599 5250 F: 212 599 6843

NGO

Albers-Schonberg, G.
GLOBE
50 rue du Taciturne
1040 Bruxelles, Belgium

Bento Xavier, Raimundo
FIMARC
Rue Africaine, 92
B-1050 Bruxelles, Belgium

Carro, Fernando, Mr.
President
AIESEC INTERNATIONAL
40 rue Washington
Brussels 1050, Belgium
T: 32 2 646-2420 F: 32 2 646-3764

Casswell, Tim
UIA
Rue Washington 40
1050 Bruxelles, Belgium

Condray, James Ronald, Mr.
CEFIC
Av. E. van Nieuwenhuyse 4, Bte. 1
B-1160 Brussels, Belgium

Dunlop, Nicholas, Mr.
Internatiol Coordinator
EAN/EDRC
26 Boulevard Louis Schmidt
Brussels 1040, Belgium
T: 32 2 736-8052 F: 32 2 735-8895

Gatta, Mario
CEFIC
Av. E. van Nieuwenhuyse 4, Bte. 1
B-1160 Brussels, Belgium

Green, Reg G.
ICFTU
Rue Montagne aux Herbes Potageres 37-41
1000 Bruxelles, Belgium

Hammond, Thomas, Mr.
CAREI
Boulevard du Regent 58
1000 Brussells, Belgium

Harrison, James J.
CEPCEO
Avenue de Tervueren 168, Bte. 11
B-1150 Brussels, Belgium

Hazelwood, Peter, Mr.
CARE
c/o Boulevard du Regent 58/10
Brussels, Belgium
T: 32 2 5024333 F: 32 2 5028202
TL: 20253 CAREB

Heller, Jean Marc
IFHR
27 rue Jean Dolent
75014 Paris, Belgium

Jordan, Lisa, Ms.
GLOBE
50 rue du Taciturne
1040 Bruxelles, Belgium

Krauss, Kit
ICAI
Rue Amidie Lynen, 8
B-1030 Brussels, Belgium

Makhurane, Lesedi
AIESEC
40 rue Washington, B-1050,
Brussels, Belgium

McGlashan, Teresa, Ms.
EDRC
26 Boulevard Louis Schmidtlaan
1040 Brussels, Belgium

McLaren, Nadia, Ms.
UIA
Rue Washington 40
1050 Bruxelles, Belgium

Molano, Diego, Mr.
VP Programmes
AIESEC INTERNATIONAL
40 rue Washington
Brussels 1050, Belgium
T: 32 2 646-2420 F: 32 2 646-3764

O'Gorman, Fergus
EEB
Rue de la Victoire 26, Bte 12
B 1060 Bruxelles, Belgium

Ramos Mariano, Maria De Lourdes
IYCW
Rue Plantin 11
1070 Brussels, Belgium

Rankin, Elaine
EBCD
Rue de la Science, 9
B-1040 Brussels, Belgium

Reinaud, Guy
CEFIC
Av. E. van Nieuwenhuyse 4, Bte. 1
B-1160 Brussels, Belgium

Rodrigues De Souza, Antonio
IYCW
Rue Plantin 11
1070 Brussels, Belgium

Sertic, Viktor, Dr.
CARE INTERNATIONAL
c/o Boulevard du Regent 58/10
Brussels, Belgium
T: 32 2 5024333 F: 32 2 5028202
TL: 20253 CAREB

Skreta, Marek
AIESEC
40 rue Washington, B-1050,
Brussels, Belgium

Smet, Jan De
SGIP
Breughelstraat 31-33
B-2018 Antwerpen, Belgium

Van Ermen, Raymond
EEB
Rue de la Victoire 26, Bte 12
B 1060 Bruxelles, Belgium

Williams, Lynn R.
ICFTU
Rue Montagne aux Herbes Potageres 37-41
1000 Bruxelles, Belgium

Winteler, Ursula
President
INSTITUTE OF CULTURAL AFFAIRS INT'L
Rue Amidie Lynen, 8
B-1030 Brussels, Belgium

Wright, Michael
ICFTU
Rue Montagne aux Herbes Potageres 37-41
1000 Bruxelles, Belgium

Zaunberger, Karin
GLOBE
50 rue du Taciturne
1040 Bruxelles, Belgium

PRESS

Cogels, Olivier
Correspondent
DE MORGEN
Chaussef DF Huy 230
1325 Chaumont-Gistoux, Belgium, Belgium

De Vroey, Hubert
Reporter Radio/Television
BRT-N RADIO 1
Pieter de Smedtstraat 24
2140 Antwerpen, Belgium, Belgium

Dresser, Christa Leila
Editor
POTSDAMER NEUESTEN NACHRICHTEN
Rosa-Luxemburg-Str. 39
D-0-1820, Belgium, Belgium

Gree, Reg
Correspondent
FREE LABOUR WORLD (ICFTU)
Kouterstraat 90
3090 Overijse, Brussels, Belgium, Belgium

Hopper, Ronald
Correspondent
GAS MATTERS
7 Stag Lane, Buckhurst Hill
Essex IG95TD, Belgium

Matsushima, Arata
Correspondent
TELEVISION TOKYO CHANNEL 12, LTD
98 Ave du Bois de la Cambre
1050 Brusseles, Belgium
T: 32 2 5120138 F: 32 2 5111749

Menten, Luc
Editor
BELGA
Dorpsstraat 25
B-3440 Zoutleeuw, Belgium, Belgium

Meskens, Joelle
Editor
LE SOIR
58, av. P. Deschanel
1030 Bruxelles, Belgium, Belgium

Monballiu, Jan
Editor
VTM-NEWS
Acacia Laan 3
2620 Hemiksem, Belgium, Belgium

Nieuwenhuys, Benedicte
Correspondent
SENEPART
Avenue de la Clairiere 7
B-1050 Brussels, Belgium, Belgium

Quiring, Holger
Correspondent
VWD FINANCIAL NEWS
R. Plaine des Sports B-1970
Wembeek, OPPEm, Belgium, Belgium

Thiran, Yves
Reporter Radio/Television
RTBF TELEVISION
Avenue Paul Beschanel, 19/7
B-1030 Brussels, Belgium, Belgium

Van Goethem, Anne
Editor
EUROPE INFORMATION SERVICE
15 Brameschhof
Luxemburg, Belgium, Belgium

Van Haver, Kristien
Editor
DE FINANCIEE
TH Coppensstraat 261
1731 Zellik, Belgium, Belgium

Wouters, Antoen
Editor
DE STANDAARD
Koning Albertlaan 53
B-3010 Leuven (Kessel-Lo), Belgium, Belgium

ORGANIZATIONS

Beulink, Anne, Ms.
Liaison Committee Of Development Ngos To The European Communities
62 Avenue de Cortenbergh
B-1040 Bruxelles, Belgium
T: (32 2) 736 4087 F: (32 2) 732 1934

Bridel, Renee, Mrs
International Association Of Democratic Lawyers*
263 Avenue Albert
1180 Brussels, Belgium

Dixson-Decleve, Sandrine, Ms.
GLOBE - Global Legislators Organisation For A Balanced Environment
50 rue du Taciturne
1040 Bruxelles, Belgium
T: (32 2) 230 6589 F: (32 2) 230 9530

Ermen, Raymond Van, Mr.
European Environmental Bureau
Rue de la Victoire 26, Bte 12
B 1060 Bruxelles, Belgium
T: (32 2) 539 0037 F: (32 2) 539 0921

Gigacz, Stefan, Mr.
International Young Christian Workers*
Rue Plantin 11
1070 Brussels, Belgium
T: (32 2) 521 6983 F: (32 2) 521 6944

Gillet, P., Mr.
International Collective In Support Of Fishworkers
65 rue Gretry
B-1000 Brussels, Belgium
T: (32 2) 218 1538

Holschuh, Lenhard, Mr.
International Iron And Steel Institute
Rue Colonel Bourg, 120
B-1140 Brussels, Belgium
T: (32 2) 735 89 75 F: (32 2) 735 8012
TL: 22639

Judge, Anthony J.N., Mr.
Union of International Associations
Rue Washington 40
1050 Bruxelles, Belgium
T: (32 2) 640 4109/640 1808 F: (32 2) 649 3269

Lebrun, Marc, Mr.
International Union For The Scientific Study Of Population*
Rue des Augustins, 34
4000 Liege, Belgium
T: (32 41) 22 40 80

Lever, H.H., Mr.
European Chemical Industry Council (CEFIC)
Av. E. van Nieuwenhuyse 4, Bte. 1
B-1160 Brussels, Belgium
T: (32 2) 676 7211 F: (32 2) 676 7300

Mallaart, Jeannette Kungham, Ms.
Environment And Development Resource Centre
26 Boulevard Louis Schmidtlaan
1040 Brussels, Belgium
T: (32 2) 736 8050 F: (32 2) 735 8895

Persson, Bengt, Mr.
Council Of European National Youth Committees
8 Avenue des Courses
B-1050 Brussels, Belgium

Pirson, Joseph, Mr.
Federation Internationale Des Mouvement D'adultes Ruraux Catholiques
Rue Africaine, 92
B-1050 Bruxelles, Belgium
T: (32 2) 538 78 42

Radcliffe, George, Mr.
CARE International
Boulevard du Regent 58
1000 Brussells, Belgium
T: (32 2) 502 4333 F: (32 2) 502 8202

Rogler, J. Van Der Stichelen, Mr.
As. Of The Coal Producers Of The European Community (CEPCEO)
Avenue de Tervueren 168, Bte. 11
B-1150 Brussels, Belgium
T: (32 2) 771 9974 F: (32 2) 771 4104

Sterner, Sven Erik, Mr.
International Confederation Of Free Trade Unions
Rue Montagne aux Herbes Potageres 37-41
1000 Bruxelles, Belgium
T: (32 2) 217 8085 F: (32 2) 218 8415

Stocker, Simon, Mr.
Eurostep (european Solidarity Towards Equal Participation Of People)
115 rue Stevin
1040 Brussels, Belgium
T: (32 2) 231 1709

Thorpe, Victor, Mr.
International Federation Of Chemical, Energy & General Worker's Unions*
Avenue Emile de Beco 109
B-1050 Brussels, Belgium
T: (32 2) 647 02 35 F: (32 2) 648 43 16
TL: 20847 ICEFBX

Veys, Frances, Ms.
Bureau International De La Recuperation
Rue du Lombard 24, Bte 14
B-1000 Bruxelles, Belgium

GLOBAL FORUM

Barsamian, Stephanie, Ms.
2, Place St. Nicolas
Antwerp 2000, Belgium
T: 03 233 7776

Baumann, Heiner, Mr.
Delegate
AIESEC INTERNATIONAL
40 rue Washington
Brussels 1050, Belgium
T: 32 2 646-2420 F: 32 2 646-3764

Bettomo, Virginio, Mr.
Member Of E.P. (greens)
EUROPEAN PARLIAMENT
97 Rue Belliard
Bruxelles - B-8 1040, Belgium
T: 32 2 284-5197 F: 32 2 284-9197

Bjoerkman, Malin, Ms.
Delegate
AIESEC INTERNATIONAL
40 rue Washington
Brussels 1050, Belgium
T: 32 2 646-2420 F: 32 2 646-3764

Boes, Marc, Mr.
INSTITUT VOOR MILIEURECHT K.U. LEUVEN, COLLEGIUM FALCONIS
Tiensestraat 41
Leuven 3000, Belgium

Cailnut, Bridget, Ms.
Delegate
AIESEC INTERNATIONAL
40 rue Washington
Brussels 1050, Belgium
T: 32 2 646-2420 F: 32 2 646-3764

Cantarella, Eric, Mr.
FONDS DE COOPERATION AU DEVELOPPEMENT
Boulevard de l'Empereur 15/4
Brussels B1000, Belgium
T: 32 2 513-7545 F: 32 2 5428816

Canto, Fabricio Martins, Mr.
Delegate
AIESEC INTERNATIONAL
40 rue Washington
Brussels 1050, Belgium
T: 32 2 646-2420 F: 32 2 646-3764

Castro, Viviane, Ms.
Office Manager
EARTHACTION
26 Blvd. Louis Schmidt
Brussels 1040, Belgium
T: 3 22 736-8052 F: 32 2 735-8895

Corral, Luis, Mr.
Delegate
AIESEC INTERNATIONAL
40 rue Washington
Brussels 1050, Belgium
T: 32 2 646-2420 F: 32 2 646-3764

Courtmanche, Chris, Mr.
Delegate
AIESEC INTERNATIONAL
40 rue Washington
Brussels 1050, Belgium
T: 32 2 646-2420 F: 32 2 646-3764

Csicsery-Ronay-Smith, Elizabeth, Ms.
Place De la Liberte, 7
Brussels 1000, Belgium
T: 32 2 218-7975

Decadt, Brigitte, Ms.
Program Manager Remote Sensine
SCIENCE POLICY OFFICE - BELGIUM
Wetenschapsstraat 8
Brussels, Belgium
T: 32 2 30-6074

Dechamp, Eric, Mr.
Project Coordinator
OXFAM
39 rue du Conseil
Brussels 1050, Belgium
T: 32 2 512-9990 F: 32 2 511-8919
TL: 63939 OXFAM

Dupriez, Patrick, Mr.
Representant
ECOLO
Rue de Thy, 12
Genappe 1475, Belgium
T: 32 67 780079 F: 32 67 217149

Dureau, Glenna, Ms.
Delegate
AIESEC INTERNATIONAL
40 rue Washington
Brussels 1050, Belgium
T: 32 2 646-2420 F: 32 2 646-3764

Freudenberg, Andres, Mr.
Delegate
AIESEC INTERNATIONAL
40 rue Washington
Brussels 1050, Belgium
T: 32 2 646-2420 F: 32 2 646-3764

Gast, Leontine, Ms.
Delegate
AIESEC INTERNATIONAL
40 rue Washington
Brussels 1050, Belgium
T: 32 2 646-2420 F: 32 2 646-3764

Goplen, Line, Mr.
Delegate
AIESEC INTERNATIONAL
40 rue Washington
Brussels 1050, Belgium
T: 32 2 646-2420 F: 32 2 646-3764

Grundmann, Michael, Mr.
Delegate
AIESEC INTERNATIONAL
40 rue Washington
Brussels 1050, Belgium
T: 32 2 646-2420 F: 32 2 646-3764

Gucht, Robert Van Der, Mr.
Director
CENTRE DECUMENIQUE DE LIAISSONS INTERNATIONALES
Rue Kessesls, 66
Brussels B-1030, Belgium
T: 32 2 241-4116

Guliner, Pascale, Ms.
Delegate
AIESEC INTERNATIONAL
40 rue Washington
Brussels 1050, Belgium
T: 32 2 646-2420 F: 32 2 646-3764

Heller, Thomas, Mr.
Delegate
AIESEC INTERNATIONAL
40 rue Washington
Brussels 1050, Belgium
T: 32 2 646-2420 F: 32 2 646-3764

Hernandez, Sarah, Ms.
Delegate
AIESEC INTERNATIONAL
40 rue Washington
Brussels 1050, Belgium
T: 32 2 646-2420 F: 32 2 646-3764

Hirsch, Robert, Mr.
Delegate
AIESEC INTERNATIONAL
40 rue Washington
Brussels 1050, Belgium
T: 32 2 646-2420 F: 32 2 646-3764

Hooghe, Marc R.
Correspondent
BELGIAN RADIO & TELEVISION
Blvd. Reyers 52 3 L27
Brussels B1043, Belgium
T: 1 322 7415656 F: 1 322 7358273

Jim, Daniel, Mr.
Vp Programmes
AIESEC
40, Rue Washington
Brussels 1050, Belgium
T: 32 2 646-2420 F: 32 2 646-3764

Johnstone, Diana, Ms.
Press Attachee
GROUP DES VERTS AU PARLEMENT EUROPEAN
97, Rue Belliard
Brussels 1047, Belgium
T: 32 2 284.5157 F: 32 2 284.9157

Jonasson, David, Mr.
Delegate
AIESEC INTERNATIONAL
40 rue Washington
Brussels 1050, Belgium
T: 32 2 646-2420 F: 32 2 646-3764

Kok, Ping Soon, Mr.
Delegate
AIESEC INTERNATIONAL
40 rue Washington
Brussels 1050, Belgium
T: 32 2 646-2420 F: 32 2 646-3764

Kroll, Lucien, Mr.
President
AUAI-SPRL
Av. L-Berlaimont,20
Brussels 1160, Belgium
T: 32 2 673 3539 F: 32 2 673 8927

Kuhnle, Sabine, Ms.
Delegate
AIESEC INTERNATIONAL
40 rue Washington
Brussels 1050, Belgium
T: 32 2 646-2420 F: 32 2 646-3764

Kuyken, Eckhart, Dr.
Professor
INSTITUTE OF NATURE CONSERVATION - MINISTRY OF THE FLEMISH COMMUNITY
Kiewitdregf
Hasselt B-3500, Belgium
T: 32 11 210 110 F: 32 11 242 262

Labrecque, Laura Lee, Ms.
Delegate
AIESEC INTERNATIONAL
40 rue Washington
Brussels 1050, Belgium
T: 32 2 646-2420 F: 32 2 646-3764

Larsen, Svein, Mr.
Delegate
AIESEC INTERNATIONAL
40 rue Washington
Brussels 1050, Belgium
T: 32 2 646-2420 F: 32 2 646-3764

Lassenius, Malin, Ms.
Delegate
AIESEC INTERNATIONAL
40 rue Washington
Brussels 1050, Belgium
T: 32 2 646-2420 F: 32 2 646-3764

Lienard, Pierre, Mr.
18 R.E. Montulet,5053 St. Germain
St. Germain 5053, Belgium
T: 081 512493

Luehring, Karen, Ms.
Delegate
AIESEC INTERNATIONAL
40 rue Washington
Brussels 1050, Belgium
T: 32 2 646-2420 F: 32 2 646-3764

Moore, Rosa Moreno, Ms.
Official Delegate
ECOLO-BELGIUM
Rue Elise 79
Brussels 1050, Belgium
T: 32 2 648 9574 F: 32 2 217 5290

Nieuwejaers, Jozef, Mr.
Engineer
MINISTERIE VLAAMSE GEMEENSCHAP
Bellionohtrast 14-18
Brussels 1040, Belgium
T: 32 2 5076797 F: 32 2 5073005

Okimura, Tadashi, Mr.
Delegate
AIESEC INTERNATIONAL
40 rue Washington
Brussels 1050, Belgium
T: 32 2 646-2420 F: 32 2 646-3764

Oldenburg, Claes, Mr.
Delegate
AIESEC INTERNATIONAL
40 rue Washington
Brussels 1050, Belgium
T: 32 2 646-2420 F: 32 2 646-3764

Page, Karyn, Ms.
Delegate
AIESEC INTERNATIONAL
40 rue Washington
Brussels 1050, Belgium
T: 32 2 646-2420 F: 32 2 646-3764

Paladella, Miguel De, Mr.
Delegate
AIESEC INTERNATIONAL
40 rue Washington
Brussels 1050, Belgium
T: 32 2 646-2420 F: 32 2 646-3764

Poznanski, Marek, Mr.
COLLECTIF STRATEGIES ALIMENTAIRES
9 Quai du Commerce
Brussels, Belgium
T: 32 2 223 0381 F: 32 2 217 6078

Premal, Desai, Mr.
Delegate
AIESEC INTERNATIONAL
40 rue Washington
Brussels 1050, Belgium
T: 32 2 646-2420 F: 32 2 646-3764

Roncerel, Annie Bonnin, Ms.
STOCKHOLM ENVIRONMENT INSTITUTE
44 Rue du Taciturne
Brussels 1040, Belgium
T: 32 2 231-0180 F: 32 2 230-5713

Rydberg, Erik
Reporter - Environmental Desk
LA CITE
26 Rue Saint Laurent
Brussels, Belgium
T: 32 2 2172390 F: 32 2 2176995

Salinas, Viviane Castro, Ms.
Office Manager
EARTHACTION
26 Boulevard Louis Schmidt
Brussels 1040, Belgium
T: 32 2 736-8052 F: 32 2 735-8895

Salle, Silvia, Ms.
Delegate
AIESEC INTERNATIONAL
40 rue Washington
Brussels 1050, Belgium
T: 32 2 646-2420 F: 32 2 646-3764

Sancy, Mary, Ms.
Professeur
27A av des Ombrages
Brussels 1200, Belgium

Scholes, Rossana, Ms.
Enviroment Consultant
RIS LTD.
Ave E. Mounier 83 Bte 1
Bruxelles 1200, Belgium
T: 32 2 7728070 F: 32 2 7727237

Schulte, Ilka, Ms.
Delegate
AIESEC INTERNATIONAL
40 rue Washington
Brussels 1050, Belgium
T: 32 2 646-2420 F: 32 2 646-3764

Silva, Victor A. Da, Sr.
Director
AIESEC INTERNATIONAL
40 rue Washington
Brussels 1050, Belgium
T: 32 2 646-2420 F: 32 2 646-3764

Sperrer, Gerhard, Mr.
Delegate
AIESEC INTERNATIONAL
40 rue Washington
Brussels 1050, Belgium
T: 32 2 646-2420 F: 32 2 646-3764

Stell, Marc, Mr.
Delegate
AIESEC INTERNATIONAL
40 rue Washington
Brussels 1050, Belgium
T: 32 2 646-2420 F: 32 2 646-3764

Strandqvist, Joakim, Mr.
Delegate
AIESEC INTERNATIONAL
40 rue Washington
Brussels 1050, Belgium
T: 32 2 646-2420 F: 32 2 646-3764

Subramaniam, Narendran, Mr.
Delegate
AIESEC INTERNATIONAL
40 rue Washington
Brussels 1050, Belgium
T: 32 2 646-2420 F: 32 2 646-3764

Tegenmark, David, Mr.
Delegate
AIESEC INTERNATIONAL
40 rue Washington
Brussels 1050, Belgium
T: 32 2 646-2420 F: 32 2 646-3764

Vanderstraetem, Martine, Ms.
Attache
SCIENCE POLICY OFFICE - BELGIUM
Wetenshapsstraat 8
Brussels 1040, Belgium
T: 32 2 238-3411 F: 32 2 230-5912

Vanypersele, Jean-Pascal, Mr.
Dr.
UNIVERSITE CATHOLIQUE DE LOUVAIN
ASTR/UCL,2 Chemin du Cyclotron
Louvain-la-Neuve B-1348, Belgium
T: 32 10 473296 F: 32 10 474722
TL: 59065 IPL - B

Ven, Kari, Mr.
Director
B.I.I.
Desguinlei 206
Antwerp 2018, Belgium
T: 39 165 237-2519 F: 39 165 248-3717

Ven, Livien, Mr.
Ing.
B.I.I.
Desguinlei 206
Antwerp 2018, Belgium
T: 32 3 2372519 F: 32 3 2483717

Vila, Jose, Mr.
Delegate
AIESEC INTERNATIONAL
40 rue Washington
Brussels 1050, Belgium
T: 32 2 646-2420 F: 32 2 646-3764

Vogel, Michael, Mr.
Delegation Manager
AIESEC - INTERNATIONAL ASSOCIATION OF STUDENTS IN MANAGEMENT AND ECONOMICS
40 Rue Washington
Brussels 1050, Belgium
T: 32 2 646-2420 F: 32 2 646-3764

Wirnsberger, Michael, Mr.
Delegate
AIESEC INTERNATIONAL
40 rue Washington
Brussels 1050, Belgium
T: 32 2 646-2420 F: 32 2 646-3764

Woehler, Jens, Mr.
Delegate
AIESEC INTERNATIONAL
40 rue Washington
Brussels 1050, Belgium
T: 32 2 646-2420 F: 32 2 646-3764

BELIZE - GOVERNMENT

Belisle, Richard, Mr.
Principal Forestry Officer in charge of Conservation
Permanent Mission of Belize
820 Second Avenue, Suite 922
New York, NY, United States of America
T: 599 0233, 0286 F: 599 3391

Godfrey, Glenn, Hon.
Attorney General
Minister of Tourism and the Environment
Permanent Mission to the United Nations, 820 Second Avenue, Suite 922
New York, NY, United States of America
T: 599 0233, 0286 F: 599 3391

Gonzalez, Victor, Dr.
Permanent Secretary
Ministry of Tourism and the Environment
Permanent Mission to the United Nations, 820 Second Avenue, Suite 922
New York, NY, United States of America
T: 599 0233, 0286 F: 599 3391

Marin, Florencio, Hon.
Deputy Prime Minister, Minister and Head of Delegatiom Natural Resources
Belize
820 Second Avenue, Suite 922
New York, NY, United States of America
T: 599 0233, 0286 F: 599 3391

Rogers, Carl Lindbergh, Mr.
Ambassador Extraordinary and Plenipotentiary
Permanent Mission of Belize
820 Second Avenue, Suite 922
New York, NY, United States of America
T: 599 0233, 0286 F: 599 3391

ORGANIZATIONS

O'Hara, Jennifer, Ms.
Programme For Belize
PO Box 749
Belize City, Belize
T: (501 2) 75616/75617/75634 F: (501 2) 75635

GLOBAL FORUM

Engleton, Lucia, Ms.
Chairperson
ASSOCIATION OF NAT'L DEVELOPMENT AGENCIES
#20 Stann Creek Street, P.O. Box 175
Belmopan, Belize
T: 501 8 23-383 F: 501 8 23-383

Gamero, Maria, Sra.
RED DE MUJERES Y MEDIO AMBIENTE - CENTRO AMERICA Y CARIBE
P. O. Box 196
Belmapan, Belice
T: 501 3 11663

Panton, Bernard, Mr.
Chairperson
BELIZE YOUTH CONSERVATION CORPS.
PO Box 666, 55 Eve Street
Belize City, Belize
T: 501 2 32348 F: 501 2 32347

BENIN - GOVERNMENT

Ahlonsou, Epiphane D., M.
Chef
Division Climatologie, l' Asecna-Benin
Permanent Mission to the United Nations, 4 E. 73rd Street New York, NY, 10021
F: 212 249 6014
TL: 212 734 4735

Ahoyo, Roger, S.E.M.
Directeur de Cabinet du President de la Republique
Permanent Mission of Benin
4 E. 73rd Street
New York, NY 10021, United States of America
T: 212 249 6014 F: 212 734 4735

Akplogan, Zacharie, M.
Directeur des Organisations Internacionales
Permanent Mission of Benin
4 E. 73rd Street
New York, NY 10021, United States of America
T: 212 249 6014 F: 212 734 4735

Biaou, Rogatien, M.
Premier Conseiller
Mission Permanent aupres des Nations Unies
Permanent Mission to the United Nations, 4 E. 73rd Street
New York, NY 10021, United States of America
T: 212 249 6014 F: 212 734 4735

Boko, Marcel Andre, Mr
Directeur
SERVICE METEOROLOGIQUE
BP 379
Cotonou, Benin
T: 301413 301202

Holo, Theodore, S.E.M.
Ministredes Affaires Estrangeres et de la Cooperation
Permanent Mission of Benin
4 E. 73rd Street
New York, NY 10021, United States of America
T: 212 249 6014 F: 212 734 4735

Houeto, Damien, M.
Directeur de l' Environnement
Permanent Mission of Benin
4 E. 73rd Street
New York, NY 10021, United States of America
T: 212 249 6014 F: 212 734 4735

Mongbe, Rene Valery, S.E.M.
Ambassadeur Extraordinaire et Plenipotentiaire, Representant Permanent aupres des Nations Unies,New York
Permanent Mission of Benin
4 E. 73rd Street
New York, NY 10021, United States of America
T: 212 249 6014 F: 212 734 4735

Otchou, Barthelemy, M.
Directeur des Forets et Ressources Naturelles
Permanent Mission of Benin
4 E. 73rd Street
New York, NY 10021, United States of America
T: 212 249 6014 F: 212 734 4735

Sarre, Eustache, S.E.M.
Ministre de l' Environnement de l' Habitat et de l' Urbanisme
Permanent Mission of Benin
4 E. 73rd Street
New York, NY 10021, United States of America
T: 212 249 6014 F: 212 734 4735

Soglo, Nicephore, S.E.M.
President
Republique du Benin
Permanent Mission of the Republic of Benin, 4 E. 73rd Street
New York, NY 10021, United States of America
T: 212 249 6014 F: 212 734 4735

Tonoukouin, Lucien, S.E.M.
Ambassadeur Directeur du Protocole d' Etat
Permanent Mission of Benin
4 E. 73rd Street
New York, NY 10021, United States of America
T: 212 249 6014 F: 212 734 4735

NGO

Badjagou, Pascal O.
AO
BP 80
Dassa-Zoume, Benin

Ogawin, Solange J.
AO
BP 80
Dassa-Zoume, Benin

PRESS

Ahounou, Alfred
Correspondent
BENIN TV
BP 366
Cotonou, Benin

Brathier, Leon
Editor
BENIN PRESS
Ortb - BP 366
Cotonou, Benin

Tidjani, Sikirou
Correspondent
BENIN TV
BP 366
Cotonou, Benin

ORGANIZATIONS

Badjagou, P., Mr.
As. Orukutuko
BP 80
Dassa-Zoume, Benin

Djogbenou, Paul C., Mr.
As. Pour La Sauvegarde De L'environnement Et Le Developpement
Boite Postale 06-1269
Cotonou, Benin

GLOBAL FORUM

Sotinkon, Leonie, Ms.
FOVAD-LONLON-BEHA
B.P. 04-511
Cotonou 1221, Benin
T: 223 321321 F: 300778
TL: 1221

BERMUDA - PRESS

Hayward, Stuart
Correspondent
DE FONTES B'CASTING
10 A Perkeley Road
Pembroke, Bermuda, Bermuda

GLOBAL FORUM

Junos, Leyoni, Ms.
20-A Paddock Drive.Warwick WK 04
Bermuda

Richardson-Campbell, Janine, Ms.
1 Brac Lane
Penbroke HM09, Bermuda
T: 2727

BHUTAN - GOVERNMENT

Dorji, Kinlay, Mr.
Deputy Secretary
Ministry of Agriculture and Environment
Permanent Mission to the United Nations, Two United Nations Plaza, 27th Floor New York, NY, 10017
F: 212 826 1919
TL: 212 826 2998

Tashi, Tshering, Mr.
Head, National Environment Secretariat
Planning Commission
Permanent Mission to the United Nations, Two United Nations Plaza, 27th Floor New York, NY, 10017
F: 212 826 1919
TL: 212 826 2998

Tshering, Ugyen, Mr. H.E.
Ambassador Extraordinary and Plenipotentiary
Permanent Mission of Bhutan
Two United Nations Plaza, 27th Floor
New York, NY 10017, United States of America
T: 212 826 1919 F: 212 826 2998

Wangchuck, Princess Sonam Choden, H.R.R.
Special Envoy, Representative of His Majesty and Cabinet Minister Ministry of Agriculture and Environment
Bhutan
Two United Nations Plaza, 27th Floor
New York, NY 10017, United States of America
T: 212 826 1919 F: 212 826 2998

Wangchuck, Sonam Choden, Princess
Permanent Mission of the Kingdom of Bhutan
Two United Nations Plaza, 27th floor
New York NY 10017, United States of America
T: 212 826 1919 F: 212 826 2998

Wangchuk, Yanki, Mrs.
Deputy Auditor General
Permanent Mission of Bhutan
Two United Nations Plaza, 27th Floor
New York, NY 10017, United States of America
T: 212 826 1919 F: 212 826 2998

BOLIVIA - GOVERNMENT

Araujo, Heriberto, H.
Honorable Senador
Permanent Mission of Bolivia
211 E. 43rd Str., 8th Floor
New York, NY 10017, United States of America
T: 212 682 8132 F: 212 687 4642

Baudoin, Mario, Dr.
Director de Areas Protegidas
Secretaria General del Medio Ambiente
Permanent Mission to the United Nations, 211 E. 43rd Str., 8th Floor New York, NY, 10017
F: 212 682 8132
TL: 212 687 4642

Brockmann, Ericka, Lic.
Secretaria General de la Presidencia
Permanent Mission of Bolivia
211 E. 43rd Str., 8th Floor
New York, NY 10017, United States of America
T: 212 682 8132 F: 212 687 4642

Burela, Guillermo Aponte, H.
Honorable Senador
Permanent Mission of Bolivia
211 E. 43rd Str., 8th Floor
New York, NY 10017, United States of America
T: 212 682 8132 F: 212 687 4642

Calpineiro, Emigdio Flores, H.
Vice Presidente de la Comision del Medio Ambiente y Recursos Naturales de la Honorable Camara de Diputados
Permanent Mission of Bolivia
211 E. 43rd Str., 8th Floor
New York, NY 10017, United States of America
T: 212 682 8132 F: 212 687 4642

Camberos Bolanos, Carlos, Coronel DESP
Director Nacional de la Policia Forestal y Medio Ambiente de Bolilia
Permanent Mission of Bolivia
211 E. 43rd Str., 8th Floor
New York, NY 10017, United States of America
T: 212 682 8132 F: 212 687 4642

Claros, Angel Zannier, Ing.
Embajador ante la Republica
Federativa del Brasil
Permanent Mission to the United Nations, 211 E. 43rd Str., 8th Floor New York, NY, 10017
F: 212 682 8132
TL: 212 687 4642

Cortez, Jorge, Lic.
Director de Politicas
Secretaria General del Medio Ambiente
Permanent Mission to the United Nations, 211 E. 43rd Str., 8th Floor New York, NY, 10017
F: 212 682 8132
TL: 212 687 4642

Cronembold, Ricardo, Dr.
Director de Asuntos Juridicos
Secretaria General del Medio Ambiente
Permanent Mission to the United Nations, 211 E. 43rd Str., 8th Floor New York, NY, 10017
F: 212 682 8132
TL: 212 687 4642

Cuellar, Oscar Serrate, Ing.
Enbajodor
Representante Permanente de las Naciones Unidas
Permanent Mission to the United Nations, 211 E. 43rd Str., 8th Floor
New York, NY 10017, United States of America
T: 212 682 8132 F: 212 687 4642

Cusicanqui, Jorge Escobari, H.
Honorable Senador
Permanent Mission of Bolivia
211 E. 43rd Str., 8th Floor
New York, NY 10017, United States of America
T: 212 682 8132 F: 212 687 4642

Elio, Rene Beotheguy, Lic.
Director General de Organismos Internacionales
Ministerio de Relaciones Exteriores y Culto
Permanent Mission to the United Nations, 211 E. 43rd Str., 8th Floor New York, NY, 10017
F: 212 682 8132
TL: 212 687 4642

Enriquez, Juan Carlos, Ing.
Director de Calidad Ambiental
Secretaria General del Medio Ambiente
Permanent Mission to the United Nations, 211 E. 43rd Str., 8th Floor New York, NY, 10017
F: 212 682 8132
TL: 212 687 4642

Forno, Eduardo, Lic.
Gerente Tecnico del Fondo Nacional del Medio Ambiente
Permanent Mission of Bolivia
211 E. 43rd Str., 8th Floor
New York, NY 10017, United States of America
T: 212 682 8132 F: 212 687 4642

Hurtado, Neysa Roca, H.
Presidente de la Comision del Medio Ambiente y Recursos Naturales de la Honorable Camara de Diputados
Permanent Mission of Bolivia
211 E. 43rd Str., 8th Floor
New York, NY 10017, United States of America
T: 212 682 8132 F: 212 687 4642

Lora, Rolando Espinoza, Gen.
Consul General de Bolivia en Rio de Janeiro
Permanent Mission of Bolivia
211 E. 43rd Str., 8th Floor
New York, NY 10017, United States of America
T: 212 682 8132 F: 212 687 4642

Maclean Abaroa, Ronald, Lic.
Ministro de Relaciones Exteriores y Culto
Permanent Mission of Bolivia
211 E. 43rd Str., 8th Floor
New York, NY 10017, United States of America
T: 212 682 8132 F: 212 687 4642

Martin, Fernando San, Sr.
Director de Ceremonial
Ministerio de Relaciones Exteriores y Culto
Permanent Mission to the United Nations, 211 E. 43rd Str., 8th Floor New York, NY, 10017
F: 212 682 8132
TL: 212 687 4642

Obleas, Jorge Torrez, H.
Diputado Nacional
Permanent Mission of Bolivia
211 E. 43rd Str., 8th Floor
New York, NY 10017, United States of America
T: 212 682 8132 F: 212 687 4642

Paz Zamora, Jaime, Lic.
Presidente Constitucional
Republica de Bolivia Jefe de la Delegacion
211 E. 43rd Str., 8th Floor
New York, NY 10017, United States of America
T: 212 682 8132 F: 212 687 4642

Quintela, Carlos, Lic.
Director Ejecutivo del Fondo Nacional para el Medio Ambiente
Permanent Mission of Bolivia
211 E. 43rd Str., 8th Floor
New York, NY 10017, United States of America
T: 212 682 8132 F: 212 687 4642

Quintela Vaca Diez, Senador Enrique, Hon.
Presidente de la Comicion de Ecologia y Medio Ambiente del Senado Nacional de Bolivia
Permanent Mission of Bolivia
211 E. 43rd Str., 8th Floor
New York, NY 10017, United States of America
T: 212 682 8132 F: 212 687 4642

Rivero, Wigberto, Lic.
Director Ejecutivo del Institutó Indigenista de Bolivia
Permanent Mission of Bolivia
211 E. 43rd Str., 8th Floor
New York, NY 10017, United States of America
T: 212 682 8132 F: 212 687 4642

Rocha, Ramon, Sr.
Asesor de la Presidencia
Permanent Mission of Bolivia
211 E. 43rd Str., 8th Floor
New York, NY 10017, United States of America
T: 212 682 8132 F: 212 687 4642

Turcat, Benoit, Dr.
Asesor Internacional de la Presidencia de la Republica (Coordinador del fondo Indigena)
Permanent Mission of Bolivia
211 E. 43rd Str., 8th Floor
New York, NY 10017, United States of America
T: 212 682 8132 F: 212 687 4642

Ulloa, Gerardo Aguirre, Ing.
Secretario General del Medio Ambiente
Permanent Mission of Bolivia
211 E. 43rd Str., 8th Floor
New York, NY 10017, United States of America
T: 212 682 8132 F: 212 687 4642

Zamora, Jaime Paz, Mr.
President
Permanent Mission of Bolivia
211 E. 43rd. Str., 8th floor
New York NY 10017, United States of America
T: 212 682 8132 F: 212 687 4642

NGO

Flores, Teresa
LIDEMA
Ave. 20 de Octubre No. 1763
Casilla No. 11237 La Paz, Bolivia

Moscoso, Arturo, Sr.
LIDEMA
Av. 20 de Outubro 1763
Casilla No. 11237 La Paz, Bolivia
T: 591 2 356249 F: 591 2 392321

Rodrigo Gainza, Luis Alberto
LIDEMA
Ave. 20 de Octubre No. 1763
Casilla No. 11237 La Paz, Bolivia

Sanchez De Lozada, Alexandra
LIDEMA
Ave. 20 de Octubre No. 1763
Casilla No. 11237 La Paz, Bolivia

PRESS

Alvarez, Nataniel
Editor
RADIO SANTA CRUZ
Antenor Vasques 258
Santa Cruz, Bolivia, Bolivia

Gruenberger, Maria Jenny
Editor
PRESENCIA
Casilla 1064
La Paz, Bolivia, Bolivia

Morales, Freddy
Correspondent
AGENCIA NOTICIAS ERBOL
Graci 6788
La Paz, Bolivia, Bolivia

Paniagua F., Saul
Correspondent
HOY
Beni 653
Santa Cruz de la Sierra, Bolivia, Bolivia

Valcarce, Alfonso P.
Editor
BOLIVIA PRESS
Francisco Bedregal 540
La Paz, Bolivia, Bolivia

ORGANIZATIONS

Apaza, Julio Tumiri, Mr.
Centro de Accion Social Union Achiri
Calle Collana s/n/, Urbanizacion Union Achiri,
El Alto, Casilla 12043, La Paz, Bolivia

GLOBAL FORUM

Agreda Lema, Carlos, Mr.
Coordinador Nacional
COORDINADORA DE CENTROS
Cochabamba, Bolivia
T: 591 42 23603 F: 591 42 41176

Antelo, Ramiro, Sr.
FORO BOLIVIANO-CER-DET
Coson 977
Tarja 83, Bolivia
T: 66 24120 F: 665865

Aramayo, Silvio, Mr.
FORUM BOLIVIANO
1RO. De Maio No.544
Camiri, Bolivia

Astorga, Alberto, Mr.
Director Executivo
C E B O L I D
La Paz, Bolivia
T: 591 2 812824

Augstburger, Franz, Mr.
Casilla 1836
Cochabamba, Bolivia
T: 591 42 46283 F: 591 42 88381

Ayala, Wilma, Sra.
CASA DE LA MUJER
21 de Mayo 541
Santa Cruz, Bolivia
T: 59 133 324931 F: 59 133 332257

Balcazar, Freddy Luis Heinrich, Dr.
ESTUDIO JURIDICO
La Paz, Bolivia

Baracatt, Gabriel, Mr.
FORO BOLIVIANO
Daniel Campos 707
Tarija-Tarija, Bolivia
T: 66 25 865 F: 66 25 865
TL: P.O.B.O. 59

Bedregal, Teresa Flores, Ms.
Mgr. Communication
LIGA DE DEFENSA DEL MEDIO AMBIENTE
PO BOX 499 Cas Av. 20 de Octubre 1763
La Paz, Bolivia
T: 591 2 782130 F: 591 2 392321

Cappelletti, Peirlorenzo, Sr.
ITALCOWSULT
9 De Febrero
Cobija, Bolivia

Cardenas, Daniel Angel Vildozo, Sr.
PROYECTO HORTICULTURA Y FRUTICULTURA LURIBAY
Casilla 5816
La Paz, Bolivia

Castro, Miguel, Mr.
FORO BOLIVIANO
Avenida Abaroa E-549
Tarija 83, Bolivia
T: 591 066 24120

Claure, Eduardo, Mr.
Director
(COPLA) - FORO BOLIVIANO DESARROLLO Y MEDIO AMBIENTE
6 De Octubre # 6269
Oruro 1072, Bolivia
T: 591052 55838 F: 591 052 60624

Curi, Marianela, Ms.
Cordinadora Proyecto
CIEC
Av. Ecuador 2459
La Paz, Bolivia
T: 591 2 360583 F: 591 2 328933

Czerwenka, Jurgen, Mr.
INSTITUTO DE ECOLOGIA LA PAZ
Cota Cota Calle 17
La Paz, Bolivia
T: 591 2 797511 F: 591 2 797511

De Noe, Carmen Pereira, Ms.
CENTRAL DE MUJER INDIGENA
C. Jose Bopi
Trinidad - Beni, Bolivia
T: 21575

Delgado Burgoa, Freddy
UNIV.MAYOR DE SAN SIMON (AGRUCO) FORO BOLIVIANO
Av. Petrolera KM. 4 1/2 Carretera Antigua A Sta. Cruz
Cochabamba, Bolivia
T: 591 42 27485 F: 591 42 45613

Enriquez, Juan Carlos, Mr.
Tecnico
ASOCIACION PRO-DEFENSA DE LA NATURALEZA
Cassila 7884
La Paz, Bolivia
T: 591 2 361647 F: 591 2 369304

Espinosa, Anita Banegas, Ms.
President
COMITE CIVICO FEMENINO
Canada Strongest #70
S.Cruz de la Sierra 2630, Bolivia
T: 591 3 341594 F: 591 3 341812

Fabricano, Marcial Noe, Mr.
Presidente
CENTRAL DE PUEBLOS INDIGENA BENI-CPIB
Casilla 58, Trinidad-Beni,
Beni - Trinidad, Bolivia
T: 591 46 21575

Gainza, Luis Alberto Rodrigo, Sr.
LIGA DE DEFENSA DEL MEDIO AMBIENTE
Palais de Nations
Geneve CH 1211, Bolivia
F: 41 22 733-6542

Galindo, Leon, Mr.
Assistant
PRODENA
Casilla 11065
La Paz, Bolivia
T: 591 2 79-6340

Garbellini, Paolo, Mr.
Responsable De Proyectos
ACRA
Casilla 10424
La Paz, Bolivia
T: 591 2 357932 F: 591 2 357932

Gonzalez, Nicanor Diaz, Mr.
Coordinador
CULTURAL SURVIVAL
Casilla 14259, Calle Murillo, Edificio 1060, Ofic #5 Mezz
La Paz, Bolivia
T: 591 2 392301 F: 591 2 392301

Gutierrez, Sergio R. Antelo, Mr.
Calle Arenales 3682
Santa Cruz, Bolivia
T: 59 133 331197
E: 2249

Gutierrez, Vladimir, Mr.
FORO BOLIVIANO-CEDEC
J. Mostaso 125
Sucre, Bolivia
T: 591 64 25008 F: 591 64 32628

Haquim, David, Mr.
Asesor
AIPE - UNITAS
La Paz, Bolivia

Helguero, Freddy Arce, Dr.
CENTRO DE INVESTICACIONES DE TECNOLOGIA INTEGRAL - INTI
Casilla 13139
La Paz, Bolivia

Houston, Patricia, Ms.
Member
WEDO - WOMEN & ENVIRONMENT
Casilla 657
Cochabamba, Bolivia
T: 591 42 42-43685

Ino, Vicente, Mr.
Cacique
CENTRAL INDIGENA SIRIONO
Trinidad - Correo Central
Trinidad - Beni, Bolivia

Lozada, Alejandra Sanchez De, Sra.
LIGA DE DEFENSA DEL MEDIO AMBIENTE
Palais de Nations
Geneve CH 1211, Bolivia
F: 41 22 733-6542

Luther, Sheila, Ms.
Agrologist
FORO BOLIVIANO
Avenida 16 Octubre
Oruro, Bolivia
T: 591 52 55038

Melgarejo-Remus, Marion, Ms.
CENTRO DE LUZ SOL
Casilla 5651
Santa Cruz, Bolivia
T: 59 133 360649

Merida, Gonzalo, Mr.
Planifilador Ambiental
PAAB
La Paz 1508, Bolivia
T: 591 2 353583 F: 591 2 392936

Miranda, Victor, Mr.
Sec. Relaciones
C S U T C B FORO BOLIVIANO
Piso G4a
La Paz, Bolivia
T: 591 2 364975

Navia, Carlos, Mr.
Director
CIDDEBENI - FORO BOLIVIANO
Fabian Monesterio Claure #25
Trinidad - Beni, Bolivia
T: 591 46 22824

Ortiz, Carlos Calvimontes, Mr.
Foro Boliviano
CEDEMABT
Av, Central VV P De Abril #8- San Borja
Beni, Bolivia
T: 671200

Padilla, Beatriz Guerra, Sra.
Directora
JUNTA NACIONAL DE SOLIDARIDAD Y DESARROLLO SOCIAL
Presidencia de la Republica en Prov. Gran Chcaco
Tarija, Bolivia
T: 2063

Perez, Jenny Gruenberger, Ms.
LIGA DE DEFENSA DEL MEDIO AMBIENTE
Av. Bollivian 939 - Calacoto
La Paz, Bolivia
T: 591 2 790644

Pierola I., Wilhelm, Mr.
C.D.R.
Cobida Ii 64
Poootosi, Bolivia
T: 06225559 F: 062 22372

Ponce, Rosa Maria Quiroga De, Sra.
JUNTA NACIONAL DE SOLIDARIDAD Y DESARROLLO SOCIAL
Presidencia de la Republica en Prov. Gran Chcaco
Tarija, Bolivia
T: 2063

Quiroga, Juan Carlos, Mr.
Director
C U M A T
Av. Bush 1870
La Paz, Bolivia

Ramalho, Paulo Sergio De Moraes, Sr.
Economy Analist
FONPLATA
Espana #74 PO Box 47
Sucre, Bolivia
T: 591 64 30322 F: 591 64 32997
TL: 25882260

Ramirez, Juan Marcel, Mr.
AIPE - FORO BOLIVIANO
1206 C De Agusto
La Paz, Bolivia
T: 591 2 327272 F: 591 2 392242

Reynaga, Maria Luisa La Fuente De, Ms.
LIMPAL
Potosi,856
Santa Cruz, Bolivia
T; 59 133 329888 F: 59 133 332988

Richard, Annas, Mr.
Ecologist
CIDDE BENI
Casilla 159
Trindad - Beni, Bolivia
T: 046 28246

Rodriguez, Loida, Ms.
FORO BOLIVIANO SOBRE MEDIO AMBIENTE Y DESARROLLO
Abdon Saavedra 2223
La Paz, Bolivia
T: 591 2 53048 F: 591 2 391365

Russo, Hugo Salas, Mr.
Director
IPDRE - FORO BOLIVIANO
C/ BAURES No.10
Trinidad - Beni, Bolivia
T: 591 40 20943
TL: 305

Schollaert, Alain, Mr.
BOLIVIAN FORUM
Calle Argentina
Potosi 287, Bolivia
T: 591 62 32066 F: 591 62 32070

Soruco M., Saul, Mr.
Planificacion
SEMTA/FORO BOLIVIANO
Alfredo Ascarrunz 2675
La Paz 15041, Bolivia
T: 591 2 34-2600 F: 591 2 39-1458

Szwagrzak, Andrzej, Mr.
PRODENA
Casilla 4422
La Paz, Bolivia
T: 591 2 785115 F: 591 2 392321

Tamo, Ernesto Noe, Mr.
CENTRAL PUEBLO INDIGENA
C/ Jose Bopi
Trinidad - Beni, Bolivia
T: 591 40 21575

Torrico, Licy T. De, Ms.
WORLD YOUNG WOMEN'S CHRISTIAN ASSOCIATION
Manual Ignacio Salvatierra no. 482
Santa Cruz, Bolivia
T: 59 133 324011 F: 59 133 1977
TL: WYMCA 41 23 32

Trigo, Maria Soledad Quiroga, Ms.
FORO BOLIVIANO M.A. Y DESARROLLO
Av. 20 De Octubre 1763
La Paz 11237, Bolivia
T: 591 2 356249 F: 591 2 392321

Velarde, Richard Suarez, Sr.
Representante
CEDERNAT
C/Tajija #81- Trinidad
Boni, Bolivia
T: 21585

Villarroel Garcia, Martin, Mr.
Director
P A A C (FORO BOLIVIANO)
San Lorenzo no. 231
Cochabamba, Bolivia
T: 591 42 80517

Visinoni, Andrea, Mr.
Prof. Ecologia
UNIVERSIDAD NUR - FORO BOLIVIANO
Av. Banzer 100- Casilla 3273
S. Cruz, Bolivia
T: 0059103 333939 F: 331850

BONAIRE - GOVERNMENT

Newton, Eric, Mr
Island Government Environmental Off
PLANNING AND ENVIRONMENT OFFICE
Bonaire Island Government
Bonaire
T: 599 7 5330 8708 H F: 599 7 8416

BOTSWANA - GOVERNMENT

Bhandari, V. T., Dr.
Deputy Director
Botswana Technology Center
Permanent Mission to the United Nations, 103 E. 37th Street
New York, NY 10016, United States of America
T: 212 889 2277 F: 212 725 5061

Bhawdari, Prof Dr Vinod Kumar
Technology Dev Manager
BOTSWANA TECHNOLOGY CENTRE
Private BAG 0082 Gaborone
Gaborone, Botswana
T: 267 314161 F: 267 374677

Chiepe, Gaositwe K. T., The Hon.
Minister for External Affairs
Permanent Mission of Botswana
103 E. 37th Street
New York, NY 10016, United States of America
T: 212 889 2277 F: 212 725 5061

George, S. C., Mr.
Chief of Protocol
Permanent Mission of Botswana
103 E. 37th Street
New York, NY 10016, United States of America
T: 212 889 2277 F: 212 725 5061

Hunter, N., Mr.
Director of Wildlife and National Parks
Ministry of Commerce and Industry
Permanent Mission to the United Nations, 103 E. 37th Street
New York, NY 10016, United States of America
T: 212 889 2277 F: 212 725 5061

Kwerepe, Raymond Moemedi, Mr
Senior Rangeland Ecologist
Ministry of Agriculture Government
P.O. Box 710
Gaborone, Botswana
T: 267 350511 267 37339 F: 267 356027 267 35993

Legwaila, Legwaila J. M. J., H.E. Mr.
Ambassador Extraordinary and Plenipotentiary
Permanent Representative to the United Nation
Permanent Mission to the United Nations, 103 E. 37th Street
New York, NY 10016, United States of America
T: 212 889 2277 F: 212 725 5061

Liphuko, S. D., Mr.
Executive Secretary
National Resources Agency
Permanent Mission to the United Nations, 103 E. 37th Street
New York, NY 10016, United States of America
T: 212 889 2277 F: 212 725 5061

Masire, Ketumile, Mr.
President
Permanent Mission of Botswana
103 E. 37th Str.
New York NY 10016, United States of America
T: 212 889 2277 F: 212 725 5061

Matlakala, D., Ms.
Assistant Attorney General
Attorney General's Chambers
Permanent Mission to the United Nations, 103 E. 37th Street
New York, NY 10016, United States of America
T: 212 889 2277 F: 212 725 5061

Matsetse, P., Mr.
Permanent Secretary (Development)
Office of the President
Permanent Mission to the United Nations, 103 E. 37th Street New York, NY, 10016
F: 212 889 2277
TL: 212 725 5061

Mokgothu, B., The Hon.
Assistant Minister of Local Government
Lands and Housing
Permanent Mission to the United Nations, 103 E. 37th Street
New York, NY 10016, United States of America
T: 212 889 2277 F: 212 725 5061

Mosienyane, Matc Gadibolae, Dr
Director of Veterinary Services
Ministry of Agriculture
P/BAG 0032
Gaborone, Botswana
T: 350616 267 F: 356027 267

Ntakhwana, Z. J., Mr.
External Affairs Officer
Protocol
Permanent Mission to the United Nations, 103 E. 37th Street
New York, NY 10016, United States of America
T: 212 889 2277 F: 212 725 5061

Outlune, Samuel O., Mr.
First Secretary
Permanent Mission of Botswana
103 E. 37th Street
New York, NY 10016, United States of America
T: 212 889 2277 F: 212 725 5061

Ramothwa, G., Ms.
Director of Meteorological Services, Ministry of Works Transports and Communications
Permanent Mission to the United Nations, 103 E. 37th Street
New York, NY 10016, United States of America
T: 212 889 2277 F: 212 725 5061

Sebele, B. K., H.E. Mr.
Ambassador to the United States of America
Permanent Mission of Botswana
103 E. 37th Street
New York, NY 10016, United States of America
T: 212 889 2277 F: 212 725 5061

Sekwahe, Moremi, Mr.
Director Of Water Affairs
Ministry of Mineral Resources and Water Affairs
Caborone, Botswana
T: 267 351601 F: 267 374372

Selepeng, M. L., Mr.
Permanent Secretary (Political)
Office of the President
Permanent Mission to the United Nations, 103 E. 37th Street
New York, NY 10016, United States of America
T: 212 889 2277 F: 212 725 5061

Sentle, B. K., Mr.
Senior Private Secretary to the President
Permanent Mission of Botswana
103 E. 37th Street
New York, NY 10016, United States of America
T: 212 889 2277 F: 212 725 5061

Taolo, D., Mr.
Factory Inspector
Ministry of Labour and Home Affairs
Permanent Mission to the United Nations, 103 E. 37th Street
New York, NY 10016, United States of America
T: 212 889 2277 F: 212 725 5061

Tuogelang, Gabolekwe Lesole, Mr.
Chief Economist
Ministry of Finance and Development Planning
P/BAG 008
Gabarone, Botswana
T: 350100 350272 F: 356086

NGO

Garraway-Stayers, Deborah
BTC
P Bag 0082, Gaborone
Botswana

Vinod, Bhandari
BTC
P Bag 0082, Gaborone
Botswana

PRESS

Sibiya, Victor
Correspondent
MMEGI THE REPPORTER
P.O. Box 993 Gaborone
Botswaga, Botswana

GLOBAL FORUM

Mokobi, Ruth F., Ms.
BOTSWANA YWCA
PO Box 359
Caborone, Botswana
T: 353681

Mqhum, Agnes, Ms.
Teacher
BOTSWANA YWCA
PO Box 556
Mochudi, Botswana
T: 353-681

Segodi, R., Mr.
THUSANO LEFATSHENG
Palais de Nations
Geneve CH 1211, Botswana
F: 41 22 733-6542

Stayers, Deborah, Ms.
BOTSWANA TECHNOLOGY CENTER
Palais de Nations
Geneve CH 1211, Botswana
F: 41 22 733-6542

PARLIAMENTARIANS / SPIRITUAL LEADERS

Balopi, Patrick, Min.
National Assembly, P.O. Box 240
Gaborone, Botswana
T: 55681
TL: 2414 BD

BRAZIL - GOVERNMENT

Barbouth, Carlos Alberto, Mr
Correspondent
JORNAL DE ESPANA
Rua Do Quvidor 90, CEP 20040
Rio De Janeiro, Brazil
T: (5521) 224-5422 F: (5521) 221-9264

Belfort Simoes, Cecilia, Ms
Producer Television
GLOBO ECOLOGIA PROGRAMME
Roa General Glicerio 58/104 Larawjeiras CEP.22.251
Rio De Janeiro, Brazil
T: 021 2856099 F: 021 2246112

Billand, Klaus G K
Unido Country Director Brazil
UNIDO
Quadra 2, Lote B Scn
FDF10 Brazilia Df, Brazil
T: 061 223 8101 F: 061 225 9339

Blajberg, Israel, Mr
Coordinating Engineer
BNDES/ ENVIRONMENT DEPT
P. O. Box 1910
Rio De Janeiro, Brazil
T: 55 21 277 7888 F: 220 7397

Braga, Ana Cecilia
Special Advisor For UNCED
GOVERNO DO ESTADO DO RIO DE JANEIRO
Rua Pacheco Leao,915
Rio De Janeiro,, Brazil
T: 021 294 7722 F: 021 294 9897

Chediak, Nadsa
Delegation From Guam
GOVERNOR
Rua Humalta 282 707 Bloco 2
Rio De Janeiro, Brazil
T: 226 1480

Clarke, Philip, Mr.
Director Of Operations In Brazil
WORLD FOOD PROGRAMME
Scn Quadra 02 Bloco B Terreo 70710
Brazilia, Brazil
T: 226 7872 F: 322 2080

Colagrossi, Fernanda, Sra.
Presidente
ASSOCIACAO AMIGOS DE PETROPOLIS-APANDE/COMITE BRASILEIRO DO ICOMOS
Rua 13 de Maio 236 - Centro
Petropolis - RJ 25685, Brazil
T: 55 242 430-674 F: 55 21 267-4756

Collor de Mello, Fernando, Mr.
President
Permanent Mission of Brazil
747 third Ave, 9th floor
New York NY 10017, United States of America
T: 212 832 6868 F: 212 371 5716

D'Agostino, Gustavo Alberto, Mr
Correspondent-Film Maker
DIARIO AMBITO FINANCIERO
Jose Cubas 4390 CP.1419 Capital Federal-Argentina
Argentina, Brazil
T: 566 5269 501 3295 F: 568 4555

De Mancilha, Maria Lucia, Ms
Secretaria
GTN
Rua Nascimento Silva 36 Apt 201 B Ipanema
Rio De Janeiro, Brazil
T: 021 227 7298

De Sa'Earp Hertz, Alfredo, Mr
Director
AAPPP - APANDE
87 Av Barau Rio Branco 25 680
Petropolis RJ, Brazil
T: 0242 42 6159

Dogbe, Komi Ruben
First Secretary
EMBASSY OF TOGO
SHIS QI11 Conj G Casa 10
Brazilia Df 71600, Brazil
T: (061) 248 4209 F: (61) 248 4752

Drummond, Maria - Claudia, Ms
Legislative Aide
FEDERAL SENATE-BRAZIL
Hla Das Comissoes-Sala 7-Subsolo
Federal-Brazillia-Df-70165, Brazil
T: 061 311 3179 3180 F: 061 321 7333

Franca De Faria Melco, Marcelo
External Relations Assistant
GTN GRUPO DE TRABALLIO NACIUAL
Ruo Cousellelro Lafayette 23/202
Rio De Faveiro 22081, Brazil
T: 021 521 4067

Garcia, Santiago, Mr
F.A.A. Senior Representative
FAA
Amcongen-Rio Unit 3501 APO AA 34030
Rio De Jonero, Brazil
T: 292 7117 XT 2501 F: 240 9853

Goes, Zaque
Cameraman
GLOBO ECOLOGIA PROGRAMME
RUA SANTA ALEXANDRINA 343/C-3 RIOCOMPRIDO - 20261
RIO DE JANEIRO, Brazil
T: 021 5025451 F: 021 2739081

Marcia, Pereira
Production Assistant For Radio
DPI PRESS
Rua Esteves Junior Fo 104 Laruljenras CP22231
Rio De Janerio, Brazil
T: 245 35 35

Melton, Richard H., The Hon.
Ambassador Extraordinary and Plenipotentiary
Federative Republic of Brazil
The Capitol
Washington, D.C. 20510, United States of America
T: 202 224 3121

Menezes, Waldson
Journalist
GLOBO ECOLOGIA PROGRAMME
Rua Marques De Abrantes 88/501 Flamengo UP-22.230
Rio De Janeiro, Brazil
T: 021 5524085 F: 021 2739087

Monte Rosa, Ricardo, Mr
Director
EMA VIDEO
Scn 708/9 Bl-E Lj 10 Fofyo
Brazila, Brazil
T: 061 274 6683 F: 061 347 3352

Munhoz, Tania Maria
Assesseur
IPEA/MINISTCOW DA ECOMONIC/BRAZIL
Sbs Ed.Bndes 11 Brasima/Df 70000 Brazil
Brastmid, Brazil
T: (061) 226 4289 F: (061) 223 5282

Murriello, Sandra Elena, Ms
Correspondent
UNIVERSIDAD NACIONAL DE LA PLATO
Calle 62 No 218 1/2-D To E la Plato-(1900)-Pasa BSAS
Argentina, Brazil
T: 01 6420145 F: 01 5684555

Nehrer, Fla'Vio
Journalist
BRAZIL UPDATE
Rua General Urquza 242/208 Leblon 22431 040
Rio De Janeiro, Brazil
T: 021 2596379 F: 021 2739087

Pereira,
Proputor
NNN PROTETO VISUAL
Rua Marques De Sao Ncente 423 501 22451
Rio De Janero, Brazil
T: 2743003 2742318

Pinheiro, Theresa Teixeira Maetins
Assistant
PRESIDNCY OF THE CONFERENCE - UNCED
Av. Niemeyer 179 C/102
Rio De Janeiro 22450RY, Brazil
T: 021 274 1105 5121776

Sauaget, Elza
Director
GLOBO ECOLOGIA PROGRAMME
Rua Santa Alexandrina 343/C-3 Riocomprido - 20260
Rio De Janeiro, Brazil
T: 021 5025451 F: 021 2739087

Veuthey, Michel, Mr.
Head Of Divis.
INTERNATIONAL COMMITEE OF THE RED CROSS (ICRC)
C/O ICRC SHIS QI 06 Conj. 09, Casa 16 - Lago Sul
Brasilia - DF, Brazil
T: 55 61 248-0250 F: 55 61 248-5173

NGO

Abella, Emilio Miguel
FBNU
Rua Manoel da Nobrega, 456
04001 Sao Paolo/SP, Brazil

Abreu Mendes, Carlos Henrique
PV
Rua da Assembleia 93/801
CEP 20011, Rio de Janeiro RJ, Brazil

Acselrad, Henri
IBASE
Rua Vicente de Souza, 29
22251 Botafogo, Rio de Janeiro JR, Brazil

Aduan, Wanda
APRV
R. Sao Francisco Xavier, 524 SL. 12037
Bloco F, CEP 20550, Rio de Janeiro, Brazil

Afonso Da Silva, Luis Virgilio
CAXI
Rua Riachuelo 194, Centro CEP 01007
Sao Paulo SP, Brazil

Alegrette, Mary
FBNU
Rua Manoel da Nobrega, 456
04001 Sao Paolo/SP, Brazil

Allegretti, Mary, Ms.
INSTITUTO DE ESTUDOS AMAZONICOS
Rua Monte Castalo, 380
Curitiba, Parana 82500, Brazil

Amaral, Weber Antonio Do, Sr.
FBNU
Rua Manoel da Nobrega, 456
04001 Sao Paolo/SP, Brazil

Andrade, Marcello
IBPEA
Rua Monte Alegre, 374
Rio de Janeiro RJ 20240, Brazil

Andujar, Claudia, Ms.
Coordenadora
CCYP
Rua Manoel da Nobrega, 111, cj. 32
04001 Sao Paulo S.P., Brazil

Aragon, Luis Eduardo
UNAMAZ
C. Postal 558
Belem-Para 66040, Brazil

Araujo Evangelista, Helio
ISMECN
Fazenda Lagoa, Caixa Postal 6
Monte Belo 37132 MG, Brazil

Aroucha, Mauricio Lins
FBNU
Rua Manoel da Nobrega, 456
04001 Sao Paolo/SP, Brazil

Asquini, Lidia Izabel Da Luz
CPISP
Rua Ministro Godoy, 1484, S/ 56-57
CEP 05015 Sao Paulo sp,, Brazil

Barbosa, Renata Egito
IBPEA
Rua Monte Alegre, 374
Rio de Janeiro RJ 20240, Brazil

Bezerra, Agostino F.
FGE
SAS Quadra 06, Bloco 'L', Edificio ASCB
Sala 801-A, 70070 Brasilia DF, Brazil

Bombona, Maria Celia
AIPA
Fazenda Sao Miguel, Caixa Postal 83
CEP 13300, Itu, SP, Brazil

Bonato, Sergio
FBNU
Rua Manoel da Nobrega, 456
04001 Sao Paolo/SP, Brazil

Born, Rubens
FBNU
Rua Manoel da Nobrega, 456
04001 Sao Paolo/SP, Brazil

Bosco Senra, Joao
FBNU
Rua Manoel da Nobrega, 456
04001 Sao Paolo/SP, Brazil

Braga, Marilita Gnecco De Camargo, Sra.
INTERNATIONAL FEDERATION OF PEDESTRIANS
Rua Cosme Velho, 315 / Bloco 2 / Ap. 508
Rio de Janeiro - RJ 22241, Brazil
T: 55 21 245-3066 F: 55 21 290-6626

Bueno Arruda, Moacir
FBNU
Rua Manoel da Nobrega, 456
04001 Sao Paolo/SP, Brazil

Buhler Alves, Norma Elena
FEDAPAM
Av. Brigadeiro Luiz Antonio 2576
CEP 01402 Sao Paulo SP, Brazil

Cabixi, Daniel
CIT
S.C.S.Ed. Palacio do Comercio, Sala 1103
CEP 70 372, Brasilia/DF, Brazil

Castro, Giselda Escosteguy, Sra.
Vice-President
ADFG / AMIGOS DA TERRA
Rua Miguel Tostes, 694
Porto Alegre - RS 90420, Brazil
T: 55 512 332-8884 F: 55 512 332-8884

Chacon De Assis, Jose
AFEA
Av. Roberto Silveira, 245
Niteroi, Rio de Janeiro, Brazil

Chagas Martorelli, Marco Aurelio
CAXI
Rua Riachuelo 194, Centro CEP 01007
Sao Paulo SP, Brazil

Coelho, Franklin
FBNU
Rua Manoel da Nobrega, 456
04001 Sao Paolo/SP, Brazil

Confalonieri, Ulisses
ETB
Rua Almirante Alexandrino, 2161
20241 Rio de Janeiro, Brazil

Cortez Costa, Jairo
FBCN
R. Miranda Valverde 103, Botafogo 22281
Rio de Janeiro RJ, Brazil

Costa Ribeiro, Wagner
FBNU
Rua Manoel da Nobrega, 456
04001 Sao Paolo/SP, Brazil

Couto Soares, Maria Clara
IBASE
Rua Vicente de Souza, 29
22251 Botafogo, Rio de Janeiro JR, Brazil

Czapski, Silvia Virginia
AIPA
Fazenda Sao Miguel, Caixa Postal 83
CEP 13300, Itu, SP, Brazil

Darcy De Oliveira, Rosiska
IAC
Av. Ataulfo de Paiva 566/305
22440 Rio de Janeiro, Brazil

De Barros, Nilson
SODEPAN
Ave. Americo Carlos da Costa 320, Parque de Exposicoes Laucidio Coelho, Campo Grande-MS, Brazil

De Barros Loyola, Cleuler
IBAM
Largo IBAM, 1
22282 Rio de Janeiro, Brazil

De Brito Pereira, Sonia Regina
MBDV
Rua Carvalho de Mendonca No. 36/802
Copacabana, Rio de Janeiro RJ CEP 22020, Brazil

De Mendonca Filho, W.F.
BBSEP
Caixa Postal 3591, Ave. Pasteur,404/3 andar (CPRM)
Urca-CEP 22290 Rio de Janeiro RJ, Brazil

De Oliveira, Marco Antonio
FBNU
Rua Manoel da Nobrega, 456
04001 Sao Paolo/SP, Brazil

Della Vedova Ribeiro Dantas, Jose Carlos
CASSC
Rua Paulo Barreto, No. 23, Botafogo
CEP 22280, Rio de Janeiro RJ, Brazil

Diniz, Nilo
FBNU
Rua Manoel da Nobrega, 456
04001 Sao Paolo/SP, Brazil

Do Prado Franco, Albano
CBI
Av. Nilo Pecanha, 50 - 34$ andar
CEP 20044 Rio de Janeiro RJ, Brazil

Dourado, Ana
FBNU
Rua Manoel da Nobrega, 456
04001 Sao Paolo/SP, Brazil

Dubois, Jean C. L., Mr.
Presidente
REDE BRASILEIRA AGROFLORESTAL - REBRAF
Rua Visconde de Piraja 111/713
Rio de Janeiro - RJ 22410-010, Brazil
T: 55 21 521-7896 F: 55 21 521-1593

Espirito Santo, Cesar Victor
FU
SCLN 107, Ed. Gemini Center 11, Bloco B
Salas 201/10, CEP 70743 Brasilia D.F., Brazil

Fernandes, Marlene, Sra.
INSTITUTO BRASILEIRO DE ADMINISTRACAO MUNICIPAL - IBAM
Edificio Diogo Lordello de Mello. Largo IBAM no.1. Humaita
Rio de Janeiro - RJ 22282, Brazil
T: 55 21 266-6622 F: 55 21 537-1262
TL: 22638 INBM BR

Fernandes Botelho, Paulo Roberto
MBDV
Rua Carvalho de Mendonca No. 36/802
Copacabana, Rio de Janeiro RJ CEP 22020, Brazil

Fernandes Sampaio, Alvaro
FBNU
Rua Manoel da Nobrega, 456
04001 Sao Paolo/SP, Brazil

Fortes, Joao Augusto
PR
Rua Uruguaiana, 10-20 Andar
CEP 20050, Rio de Janeiro RJ, Brazil

Frassinete, Paula
FBNU
Rua Manoel da Nobrega, 456
04001 Sao Paolo/SP, Brazil

Giannasi, Fernanda
CONTREN
Rua Tamandare, 348;
01525 Liberdade, Sao Paulo, Brazil

Glantz, Michael H., Mr.
Scientist
FUNDACAO GRUPO ESQUEL
SAS Quadra 06, Bloco 'L', Edificio ASCB
Sala 801-A, 70070 Brasilia DF, Brazil

Glauco Nascimento, Jorge
FBNU
Rua Manoel da Nobrega, 456
04001 Sao Paolo/SP, Brazil

Goncalvez Barbosa, Joao Manoel
CONTREN
Rua Tamandare, 348;
01525 Liberdade, Sao Paulo, Brazil

Gordon, Flavio, Sr.
Presidente
ACAO ECOLOGICA - ASSOCIACAO CAMPINEIRA DE ACAO ECOLOGICA
Rua Maestro Florence 30
Campinas - SP 13070, Brazil
T: 55 192 421312

Groeneveld, Willem
IPAE
Caixa Postal 585
78 900 Porto Velho, Rondonia, Brazil

Harmon, Mavy, Mr.
BRAZILIAN RED CROSS
Praca Cruz Vermelha 10/1o.
Rio de Janeiro - RJ, Brazil
T: 55 21 232-6325 F: 55 21 242-6760
TL: 2130532

Hossmann, Elizabeth De O.
PROVENT
Rua Uruguaiana, 10-20 Andar
CEP 20050, Rio de Janeiro RJ, Brazil

Ivern, Francisco, Sr.
SBE
Rua Bambina 115 CEP 22251
Rio de Janeiro, Brazil

Joly, Carlos Alfredo
SUC
Caiza Postal 1170
13100 Campinas SP, Brazil

Jose Trinidade, Olympio Serra
FMV
SCS-Ed Palacio do Comercio - Sala 1.103
70.300 Brasilia DF, Brazil

Jost, Niara
FMV
SCS-Ed Palacio do Comercio - Sala 1.103
70.300 Brasilia DF, Brazil

Karaja, Idjarruri, Mr.
CIT
S.C.S.Ed. Palacio do Comercio, Sala 1103
CEP 70 372, Brasilia/DF, Brazil

Klabin, Luiz Roberto
SMAF
Rua Manoel da Nobrega 456, Cep 04001
Sao Paulo, SP, Brazil

Krieger Merico, Luis Fernando
FEEC
Caixa Postal 628, 89160 Rio do Sul SC
Brazil

Leis, Hector
FBNU
Rua Manoel da Nobrega, 456
04001 Sao Paolo/SP, Brazil

Leme Machado, Paulo Affonso
SBDMA
Rua Barao de Piracicamirim No.1783
CEP 13.400 Piracicaba SP, Brazil

Lima Jr., Aurelio Vianna
CEDDEI
Rua Santo Amaro 129
22211 Rio de Janeiro RJ, Brazil

Lisboa Pacheco, Maria Emilia
FOASE
Rua Bento Lisboa, No. 58
Catete, Rio de Janeiro CEP 22221, Brazil

Lourenco, Jose Seixas
UNAMAZ
C. Postal 558
Belem-Para 66040, Brazil

Magalhaes, Antonio R.
FGE
SAS Quadra 06, Bloco 'L', Edificio ASCB
Sala 801-A, 70070 Brasilia DF, Brazil

Magalhaes, Maria Luiza
SBDMA
Rua Barao de Piracicamirim No.1783
CEP 13.400 Piracicaba SP, Brazil

Magalhaes, Solange Maria
APRV
R. Sao Francisco Xavier, 524 SL. 12037
Bloco F, CEP 20550, Rio de Janeiro, Brazil

Maia, Katia Drager
FBNU
Rua Manoel da Nobrega, 456
04001 Sao Paolo/SP, Brazil

Mangabeira, Nancy
SBE
Rua Bambina 115 CEP 22251
Rio de Janeiro, Brazil

Mantovani, Ivolin, Sr.
Presidente
FUNDACAO AMAZONAS FOREVER GREEN
Av. Giovani Gronchi 6186 Morumbi
Sao Paulo - SP 05724, Brazil
T: 55 11 521-3963 F: 55 11 246-8092

Martins, Jefferson
FBNU
Rua Manoel da Nobrega, 456
04001 Sao Paolo/SP, Brazil

May, Peter H., Mr.
Projector Coordinator - Proeco
INTERNATIONAL SOCIETY FOR ECOLOGICAL ECONOMICS
CPDA-UFRRJ - Av.Presidente Vargas, 417 - 8 Andar
Rio de Janeiro - RJ 20071, Brazil
T: 55 21 252-3001 F: 55 21 246-8003

Megyesi, Anna Louise, Sra.
Assistente Administrativa
ECOTEC
Rua Vigario Tenorio, 194 s/301 - Bairro do Recife
Recife - PE 50030, Brazil
T: 55 81 224 2863 F: 55 81 224 4654
TL: 81..1367

Mesquita, Afranio Rubens
IO
Cidade Universitaria - Butanta
CEP 05508 - Sao Paulo SP, Brazil

Miller, Carlos, Mr.
Director
FUNDACAO VITORIA AMAZONICA
Av. Darcy Vargas 520
Manaus - AM 69050, Brazil
T: 55 92 642-1336 F: 55 92 236-3257

Moreira, Gilberto
FO
Travessa da Ajuda, 01, Edif. Martins Catharino, Sa
Sala 707, 40020 Salvador, Bahi, Brazil

Mota Pereira, Luiz Otavio
ABESA
Av. Beira-Mar, 216-13
Andar-20021, Rio de Janeiro, RJ, Brazil

Moufarrige, Emilio, Mr.
Gerente
ASSOC. PALAS ATHENA DO BRASIL CENTRO ESTUDOS FILOSOFICOS
Rua Leonciode Carvalho 99
Sao Paulo - SP 04003, Brazil
T: 55 11 288-7356

Muller, Frederico
UNAMAZ
C. Postal 558
Belem-Para 66040, Brazil

Oliveira, Antonio
FO
Travessa da Ajuda, 01, Edif. Martins Catharino, Sa
Sala 707, 40020 Salvador, Bahi, Brazil

Otero Amazonas, Patricia
FBCN
R. Miranda Valverde 103, Botafogo 22281
Rio de Janeiro RJ, Brazil

Padrao Do Espirito Santo, Joper
ABESA
Av. Beira-Mar, 216-13
Andar-20021, Rio de Janeiro, RJ, Brazil

Paginoccheschi, Bruno
ISPN
Caixa Postal 02-0444
CEP 70001, Brasilia DF, Brazil

Paranhos Do Rio Branco, Roberto
FAFG
Av. Giovanni Gronchi, 6186
CEP 05724, Morumbi, Sao Paulo SP, Brazil

Pawlowicz, Basilio
APAB
Rua Leoncio de Carvalho, 99
Paraiso, Sao Paulo, CEP 04003, Brazil

Pedrosa, Trudy De Andrade
ECOTEC
Rua Vigario Tenorio No. 194, Sala 301
50.030 Recife PE, Brazil

Piedade, Edilis A.
IPAE
Caixa Postal 585
78 900 Porto Velho, Rondonia, Brazil

Pinage De Lima, Luiz Carlos
FMV
SCS-Ed Palacio do Comercio - Sala 1.103
70.300 Brasilia DF, Brazil

Pinheiro Sombra, Wilson
FBNU
Rua Manoel da Nobrega, 456
04001 Sao Paolo/SP, Brazil

Pinon Frias,
GEDEBAM
GEDEBAM, Nucleo Monte Alegre, Rua Raimundo
Uchoa, 171, Monte Alegre-Para, CEP 68 220, Brazil

Pinto De Oliveira, Nilson
UNAMAZ
C. Postal 558
Belem-Para 66040, Brazil

Prochnow, Miriam
APREMAVI
Caixa Postal 628, 89.160 Rio do Sul SC,
Brazil

Quresh, Ata
FGE
SAS Quadra 06, Bloco 'L', Edificio ASCB
Sala 801-A, 70070 Brasilia DF, Brazil

Ramos, Pedro
FBNU
Rua Manoel da Nobrega, 456
04001 Sao Paolo/SP, Brazil

Ribeiro, Gustavo Lins, Dr.
Vice Presidente
INSTITUTO DE ESTUDOS AMAZONICOS E AMBIENTAIS
UNE - Colina - Bloco J, ap. 501
Brasilia - DF 70510, Brazil
T: 55 61 273-9030

Rodrigues, Waldemar
FBNU
Rua Manoel da Nobrega, 456
04001 Sao Paolo/SP, Brazil

Rogerio Da Silva, Alberto
GEDEBAM
GEDEBAM, Nucleo Monte Alegre, Rua Raimundo
Uchoa, 171, Monte Alegre-Para, CEP 68 220, Brazil

Saavedra Durao, Jorge Eduardo
FOASE
Rua Bento Lisboa, No. 58
Catete, Rio de Janeiro CEP 22221, Brazil

Sahihi Pezeshk, Venus
NSABRA
AEN dos Bahais do Brasil, Escritorio de Meio
Ambiente, Caixa Postal 7035, 71619 Brasilia- DF, Brazil

Santos, Eliane
CIT
S.C.S.Ed. Palacio do Comercio, Sala 1103
CEP 70 372, Brasilia/DF, Brazil

Sawyer, Donald
Coordenador Executivo
INSTITUTO SOCIEDADE POPULACAO E NATUREZA - ISPN
Caixa Postal 02-0444
CEP 70001, Brasilia DF, Brazil

Scannavino Netto, Eugenio
CASSC
Rua Paulo Barreto, No. 23, Botafogo
CEP 22280, Rio de Janeiro RJ, Brazil

Schweizer, Jorge
SODEPAN
Ave. Americo Carlos da Costa 320, Parque de Exposicoes Laucidio Coelho, Campo Grande-MS, Brazil

Scovillos, Michael, Mr.
President
EUROPEAN ENVIRONMENTAL BUREAU
Ambassade de France au Bresil SES Av. das Nacoes Lote 4
Brasilia - DF 70404, Brazil
T: 55 61 3215015 F: 55 61 2269021
TL: 614688

Serricchio, Claudio
FEDAPAM
Av. Brigadeiro Luiz Antonio 2576
CEP 01402 Sao Paulo SP, Brazil

Silveira, Ricardo
PV
Rua da Assembleia 93/801
CEP 20011, Rio de Janeiro RJ, Brazil

Souza Dias, Ines
SMAF
Rua Manoel da Nobrega 456, Cep 04001
Sao Paulo, SP, Brazil

Suely, Anderson
REBRAF
Rua Visconde de Piraja, 111-Sala 713
CEP 22410, Ipanema, Rio de Janeiro RJ, Brazil

Tarcsay, Ronaldo M.
AFEA
Av. Roberto Silveira, 245
Niteroi, Rio de Janeiro, Brazil

Thais, Rodrigues Corral
IAC
Av. Ataulfo de Paiva 566/305
22440 Rio de Janeiro, Brazil

Theiss, Gilberto
ASSOCIACAO DE PRESERVACAO DO MEIO AMBIENTE DO ALTO VALE DO ITAJAI
Caixa Postal 628, 89.160 Rio do Sul SC, Brazil

Tommasi, Luiz Roberto
IO
Cidade Universitaria - Butanta
CEP 05508 - Sao Paulo SP, Brazil

Valente De B. Barre, Fernando
ACAE
Rua Maestro Florence, No. 30
Jd. Chapadao, Campinas, Sao Paulo - CEP 13.070, Brazil

Venturini, Maria Helena
AIPA
Fazenda Sao Miguel, Caixa Postal 83
CEP 13300, Itu, SP, Brazil

Vieira, Andre Tavares
FEEC
Caixa Postal 628, 89160 Rio do Sul SC
Brazil

Vieira, Liszt
FBNU
Rua Manoel da Nobrega, 456
04001 Sao Paolo/SP, Brazil

Vieira Do Nascimento, Jose
CBI
Av. Nilo Pecanha, 50 - 34$ andar
CEP 20044 Rio de Janeiro RJ, Brazil

Vogt, Carlos Alberto
SUC
Caiza Postal 1170
13100 Campinas SP, Brazil

Weber, Bertoldo
UPAN
Caixa Postal 189
CEP 93001, Sao Leopoldo, RS, Brazil

Yanomani, Davi Kopenawa
CCYP
Rua Manoel da Nobrega, 111, cj. 32
04001 Sao Paulo S.P., Brazil

Zuquim Antas, Paulo De Tarso
FU
SCLN 107, Ed. Gemini Center 11, Bloco B
Salas 201/10, CEP 70743 Brasilia D.F., Brazil

PRESS

Abbott, Maria Luiza
Correspondent
JORNAL DO BRASIL
SHIS QI 2I Cj.2 c/17
Brasilia, D.F., Brazil, Brazil

Abott, Maria Luiza
Editor
GLOBO
SCS Ed.Oscar Niemeyer-13 andar
Brasilia-DF, Brazil, Brazil

Abrahao, Rosangela De Oliveira
Editor
RADIO MINISTERIO DE EDUCACAO
Tr.Vasconcelos,11/302
Andarai, Rio de Janeiro, RJ, Brazil

Abramo, Beatriz
Editor
FOLHA DE SAO PAULO
Al. Barao de Limeira, 425
Sao Paulo-SP, Brazil, Brazil

Abranches, Sergio H. H.
Editor
TV EDUCATIVA
R. Raimundo de Magalhaes, 22
Rio de Janeiro, Brazil, Brazil

Abrao, Bernadette S.
Editor
GLOBO CIENCIA
Rua Aimbere, 1807 Apt.21
01258 Sao Paulo-SP, Brazil, Brazil

Abreu, Elaine G.
Editor
FLUMINENSE
Rua Nisia Floresta, 89
Andarai, Rio de Janeiro, Brazil, Brazil

Abreu, Teresa
Editor
ROLF PRESS
R.Tomas Coelho 26/201
Rio de Janeiro, Brazil, Brazil

Abritta, Alexandra Benevenuti
Reporter Radio/Television
TV EDUCATIVA
Pr. do Flamengo, 98/602
Rio de Janeiro, Brazil, Brazil

Absalao, Carlos De Lima
Editor
GLOBO TV
Rua Von Martius, 22 sala 328
Rio de Janeiro, Brazil, Brazil

Adeodato Filho, Sergio
Editor
GAZETA MERCANTIL
R. Mj. Quedinho, 90 - 4 andar
Sao Paulo, Brazil, Brazil

Agatha, Iris
Editor
GLOBO RADIO
R. Vinte de Abiel, 28
Rio de Janeiro-RJ, Brazil, Brazil

Aguas, Jose Eduardo Lafon
Editor
REDE OM BRASIL DE TELEVISAO
R. Prdeo Ludovico, 24
Rio de Janeiro, Brazil, Brazil

Aguiar, Antonio Jose
Reporter Radio/Television
MANCHETE TV
Rua Cacapava, 321 Fundos
Andarai, Brazil, Brazil

Aguiar, Marcelo
Editor
JORNAL DO COMMERCIO
R.Vol. da Patria, 266/307
Botafogo, RJ, Brazil, Brazil

Aguiar, Ricardo O.Gomes
Editor
AGENCIA ESTADO
R. Constante Ramos, 105/801
Copacabana, Rio de Janeiro-RJ, Brazil, Brazil

Aguiar, Sergio Ricardo Soares
Reporter Radio/Television
SUPER RADIO TUPI
Rua Gomes Serpa, 431
20740, Rio de Janeiro-RJ, Brazil, Brazil

Aguilera, Maria Veronica
Editor
REVISTA ENGENHARIA/RODO CARGA
R. Camaragibe, 9/803
Rio de Janeiro, Brazil, Brazil

Ahmed, Marcelo Bruno
JORNAL TRIBUNA DA IMPRENSA
Rua do Lavradio, 98
20230 Rio de Janeiro-RJ, Brazil, Brazil

Aiello, Cristiane
Editor
VISNEWS
Rua Cardoso Quintao, 413
Rio de Janeiro, Brazil, Brazil

Ajara, Norma Maselli
MANCHETE TV
R.General Esp.Sto.Cardoso 200
Apt.407, Tijuca, RJ, Brazil, Brazil

Ajuz, Christine B.
Editor
DIA
R. Barao Reis, 93
Sta. Teresa, Rio de Janeiro-RJ, Brazil, Brazil

Alarcon, Cesar A.S.
NEON RIO/21ST CENTURY NEWS
Rua Bento Lisboa 89/204
22221 Rio de Janeiro-RJ, Brazil, Brazil

Albano, Valeria
Editor
JORNAL DO COMMERCIO
R. Mal. Coelho, 288/502,K-11
Nova Iguacu, RJ, Brazil, Brazil

Albuquerque, Benedito G. De M.
Editor
GLOBO TV
Rua Von Martius, 22 sala 328
Rio de Janeiro, Brazil, Brazil

Albuquerque, Carlos Otavio Mendes
Reporter Radio/Television
O GLOBO
Av. Atlantica 4066/303
Copacabana, Rio de Janeiro, RJ,Brazil, Brazil

Albuquerque, Cleomar Da Conceicao
Editor
JORNAL DO PARLAMENTAR
Rua Senador Dantas 45/506
Rio de Janeiro, Brazil, Brazil

Albuquerque, Ruth
Reporter Radio/Television
TV BUZIOS
Rua Alice 139
Laranjeiras Rio de Janeiro-RJ, Brazil, Brazil

Albuquerque, Severino
Correspondent
FOLHA DA MANHA
R.Ribeiro de Almeida 38/601
Laranjeiras, Rio de Janeiro-RJ, Brazil, Brazil

Alcantara, Euripedes
Editor
VEJA
Av.Caxingui, 175 Apt.84
05579 Sao Paulo-SP, Brazil, Brazil

Alcantara, Jose Luiz
Editor
DIA
Rua Aparana, 38/205
Leblon, Rio de Janeiro-RJ, Brazil, Brazil

Alcantara, Marcos J. Coelho De
Reporter Radio/Television
TV EDUCATIVA
R. Padre Achotegui, 51/504
Rio de Janeiro, Brazil, Brazil

Alcouloumbre Jr., Albert
Editor
GLOBO
R. Barao da Torre, 489/503
Rio de Janeiro-RJ, Brazil, Brazil

Alegria, Victor
Editor
MEIO AMBIENTE
SQN 105 Bl.K apt.402
Brasilia-DF, Brazil, Brazil

Alencar, Aristoteles Torres Filho
Reporter Radio/Television
RADIO NOVA DIMENSAO
Dr. Cesar 1227 Apt. 163
Sao PAulo, Brazil, Brazil

Alexandria, Ana Maria Capri
Correspondent
JORNAL DOS CONSTRUTORES
Pca. Nilo Pecanha 97
NILOPOLIS, RJ, Brazil, Brazil

Alfradique, Marcelo Fernandez
Editor
GLOBO TV
Rua Von Martius, 22 sala 328
Rio de Janeiro, Brazil, Brazil

Allam, Lise
Reporter Radio/Television
RADIO ALVORADA
Rua Pareto, 20/201
Rio de Janeiro, Brazil, Brazil

Almeida, Fernando
Editor
GLOBO RADIO
Av. Afonso de Taunay, 143
Barra da Tijuca, Rio de Janeiro, Brazil, Brazil

Almeida, Liara De Oliveira Avelar
Editor
RADIO MINISTERIO DA EDUCACAO
R.Humaita,282 Bl.2/1008
Rio de Janeiro, RJ Brazil, Brazil

Almeida, Lucio P.
Reporter Radio/Television
RADIOBRAS
R.Dr. Souza Lopes 23/202
Rio de Janeiro, Brazil, Brazil

Almeida, Marcus Vinicius De R.
Editor
GLOBO TV
Rua Von Martius, 22 sala 32l
Rio de Janeiro, Brazil, Brazil

Almeida, Ney Bianchi De
Jornalista
REVISTA MANCHETE
Rua do Russel, 804
Rio de Janeiro - RJ, Brazil
T: 55 21 2652012

Almeida, Raquel De Q.
Reporter Radio/Television
TV EDUCATIVA
R.Haddock Lobo 35/503
Rio de Janeiro, Brazil, Brazil

Almeida, Ronaldo
Correspondent
FOLHA DO MEIO AMBIENTE
Rua Catete, 534/304
30.480 Belo Horizonte, MG, Brazil, Brazil

Almeida, Sergio Braga
Reporter Radio/Television
RADIO MANCHETE
R.Jorge Rudge 110 c/17
Rio de Janeiro, Brazil, Brazil

Almeida, Tania
Editor
VEJA
Est.Velha da Tijuca, 400/403
Alto da Tijuca, Rio de Janeiro, Brazil, Brazil

Almeida, Wilson Ferraz De
Editor
NOSSO AMIGUINHO/VIDA E SAUDE
Pca. Joao Casemiro, 324
Tatui, SP, Brazil, Brazil

Almeida Neves, Marina Isabel
Editor
PROVIDEO
R.Prof.Arthur Ramos, 48/202
Leblon, Rio de Janeiro, RJ, Brazil, Brazil

Alonso, George
Editor
FOLHA DE SAO PAULO
Al. Barao de Limeira, 425
Sao Paulo-SP, Brazil, Brazil

Alpha, Monica
Reporter Radio/Television
REDE OM BRASIL DE TELEVISAO
Rua do Bispo, 295/405
Tijuca, RJ, Brazil, Brazil

Alvarenga, Marcos A.
Reporter Radio/Television
MANCHETE TV
R.Marques de Valenca, 146 C/5
Brazil, Brazil

Alvarenga, Tales
Editor
EDITORA ABRIL
R.da Passagem, 123/8 andar
Rio de Janeiro, RJ, Brazil
T: 021 5468282 F: 021 2759347

Alvarenga Arnold, Eliane
Correspondent
AGENCIA DE NOTICIAS ACOPI
R. Firmino Pires da Silva, 245
06800 Sao Paulo-SP, Brazil, Brazil

Alvares, Gisela Marietta
Reporter Radio/Television
JORNAL DO BRASIL-RADIO AM
Rua Dez, Quadra 30, Lote 66
Niteroi, Rio de Janeiro, Brazil, Brazil

Alvares Filho, Jose A.
Editor
RADIOBRAS
Almte.Pereira Guimaraes 40/401
Rio de Janeiro, Brazil, Brazil

Alvarez, Gloria
Editor
JORNAL DO BRASIL
R. Alberto Campos, 10/1703
Ipanema, Rio de Janeiro-RJ, Brazil, Brazil

Alvarez, Maria Regina Costa
Reporter Radio/Television
FOLHA DE SAO PAULO
SHIN QI 4 Lago Norte
Brasilia, DF, Brazil, Brazil

Alves, Carlos Alberto
Reporter Radio/Television
RADIO SOCIEDADE PASSOS
R.Tiradentes 208
Passos, Minas Gerais, Brazil, Brazil

Alves, Cristina
Editor
GLOBO
R. Marques de Olinda, 61/803
Botafogo, Rio de Janeiro-RJ, Brazil, Brazil

Alves, Eli Siqueira
Reporter Radio/Television
RADIO MARANATA
Av. Braz de Pina,2051
Vista Alegre, RJ, Brazil

Alves, Francisca Stella Faga
Editor
GAZETA MERCANTIL
R. Mj. Quedinho, 90/ 4 andar
Sao Paulo, Brazil, Brazil

Alves, Jose Henrique
Editor
GLOBO
Rua Conde de Bonfim, 648/703
20520-Rio de Janeiro-RJ, Brazil, Brazil

Alves, Luis Antonio
Editor
RADIOBRAS
SQN 404 Bl. I Ap.302
Brasilia, D.F., Brazil, Brazil

Alves, Maria Fernanda Esteves
Reporter Radio/Television
GLOBO TV
Rua Von Martius, 22 sala 328
Rio de Janeiro, Brazil, Brazil

Alves, Milton
GLOBO
Av. Cidade Jardim, 400
Brazil, Brazil

Alves, Paulo Cesar Nogueira
Editor
JORNAL METRO PRESS
Rua Carmela Dutra, 97
Rio de Janeiro-RJ, Brazil, Brazil

Alves, Ramiro
Editor
GLOBO
Rua Benjamin Constant, 134/205
20241 Rio de Janeiro,Brazil, Brazil

Alves Da Silva, Rosani
EDITORA AZUL/REVISTA CONTIGO
Rua Ademar Gonzaga, 210
Jacarepagua, Rio de Janeiro, Brazil, Brazil

Alves Tinoco, Helio
Reporter Radio/Television
MANCHETE RADIO
Rua do Oro 185 Campo Grande
Rio de Janeiro, Brazil, Brazil

Amado, Roberto Kovalick
Reporter Radio/Television
GLOBO TV
Rua Von Martius, 22 sala 321
Rio de Janeiro, Brazil, Brazil

Amaral, Elisa Amelia F. Nunes Do
Reporter Radio/Television
SBT TV
Rua Domingo Ferreira, 63/1107
22050 Rio de Janeiro-RJ, Brazil, Brazil

Amaral, Francoise Vernot
Editor
GLOBO TV
Rua Von Martius, 22 sala 32l
Rio de Janeiro, Brazil, Brazil

Amaral, Irene Cristina Gurgel
Reporter Radio/Television
TV EDUCATIVA
Av. Pres. Joao Goulart 401
Rio de Janeiro, Brazil, Brazil

Amaral, Paulo Roberto
Reporter Radio/Television
GLOBO TV
Rua Von Martius, 22 sala 32l
Rio de Janeiro, Brazil, Brazil

Amaral, Sergio De Paula
Reporter Radio/Television
SBT TV
HCGN 714 Bloco 'A'
Brasilia-DF, Brazil, Brazil

Amaral, Sergio Ferreira
Correspondent
DIARIO DO POVO
Rua Antonio Zaine 30
Sao Paulo, Brazil, Brazil

Amaro, Aurezia Da Silva
Reporter Radio/Television
RADIO UNIVERSITARIA
Av.Oeste 1700 Bl. 3A/201
Setor Norte Ferr. Goiania, GO, Brazil, Brazil

Amauri, Ribeiro Jr.
Editor
FOLHA DE SAO PAULO
Al. Barao de Limeira, 425
Sao Paulo-SP, Brazil, Brazil

Ambrosio, Marcelo
JORNAL DE BRASILIA
Rua General Glicerio 114/402
Laranjeiras, Rio de Janeiro, Brazil, Brazil

Amorim, Aramis
Editor
AMAZONAS ILUSTRADA
R. Conde Bonfim 119/602
Rio de Janeiro, Brazil, Brazil

Amorim, Paulo Henrique
Correspondent
GLOBO TV
254 E. 68th St.apt.15A
New York, N.Y.10021 U.S.A., Brazil

Amorim Jr., Galeno
Correspondent
AGENCIA ESTADO
R. Joao Bim, 1.125 Bl.4 Apt.21
Ribeirao Preto, Sao Paulo, Brazil, Brazil

Anastacio, Sueli De Souza Barbosa
Editor
Rio de Janeiro, RJ Brazil
R. Enaldo Cravo Peixoto 35/606
Brazil

Anderson, Carter
Reporter
JORNAL O GLOBO
Rua Irineu Marinho, 35
Rio de Janeiro - RJ, Brazil
T: 55 21 2722000

Andrade, Alexandre M.
Editor
DIA
R.Conselheiro Lafaiete, 29/402
Copacabana, Rio de Janeiro-RJ, Brazil, Brazil

Andrade, Arlindo
Editor
JORNAL PATROPI - ZONA OESTE
Rua Amanajo 66 - Bangu
Rio de Janeiro, Brazil, Brazil

Andrade, Olivar
Reporter Radio/Television
TV EDUCATIVA
R. Sta. Rosa 60/306 Bl.ZA
Rio de Janeiro, Brazil, Brazil

Andrade, Rui Manoel Pizarro
Reporter Radio/Television
JORNAL DO BRASIL-RADIO AM
Rua Sto. Amaro, 33 Cob. 01
Gloria, Rio de Janeiro-RJ, Brazil, Brazil

Andries, Elisa
Reporter
JORNAL O GLOBO
Rua Irineu Marinho, 35
Rio de Janeiro - RJ, Brazil
T: 55 21 2722000

Angelim, Ana Cristina S.
FOLHA DA MANHA
Av. Presidente Vargas 509/6
Rio de Janeiro, Brazil, Brazil

Angelo, Fatima Lucia
Editor
FLUMINENSE
Rua Teodoro da Silva,316/127
Vila Isabel, Rio de Janeiro, Brazil, Brazil

Angelo, Maria Madalena Lourenco
Correspondent
JORNAL DOS CONSTRUTORES
Santo 333
Rio de Janeiro, Brazil, Brazil

Angert, Maria
NEON RIO LTDA
Rua do Oriente, 172
Sta.Teresa 20.240 Rio de Janeiro,Brazil, Brazil

Annoni, Jose William G.
Editor
RADIOBRAS
Ave.Sernambetiba 5800
Rio de Janeiro, Brazil, Brazil

Antonelli, Diana F.
Correspondent
AGENCIA ESTADO
SQN 216 Bl.K Apt.207
Brasilia-DF, Brazil, Brazil

Antunes, Aureclydes Ponce De Leon
Director
RADIO BANDEIRANTES
Rua Pilot Saldanha 147, Cob 401 Lagoa
Rio de Janeiro, Brazil
T: 55 21 2205824

Antunes, Claudia
Editor
JORNAL DO BRASIL
R. Marques de Sao Vicente
29/30, Gavea, Rio de Janeiro-RJ, Brazil, Brazil

Antunes, Danielle P.L.
Reporter Radio/Television
BANDEIRANTES RADIO
R.Prof.Saldanha 147/401
Rio de Janeiro, Brazil, Brazil

Antunes, Elizabete
Editor
VEJA
R. Alfredo do Vale Cabral,210
Realengo, Rio de Janeiro, Brazil, Brazil

Antunes, Jesus Soares
Editor
JORNAL DOS IMOVEIS
Av. H. Dumont, 126/302
Rio de Janeiro, Brazil, Brazil

Antunes, Laura De J.
Editor
GLOBO
Rua Conde de Bonfim, 1132
Tijuca, Rio de Janeiro-RJ, Brazil, Brazil

Antunes, Marcelino Ribeiro
Editor
O NORTE DE PARANA
R.Rio Grande do Sul 190
Siqueira Campos,Parana, Brazil

Antunes, Sergio Ferreira
Editor
RADIO GUANABARA/BANDEIRANTES
R.Prof.Saldanha 147/cob 401
Jardim Botanico, Rio de Janeiro,Brazil, Brazil

Antunes, Zanoni A.
Reporter
AGENCIA ESTADO
Av. Caetano Alvares, 55
Sao Paulo -SP 02598, Brazil
T: 55 11 8562210 F: 55 11 2652297

Aquino, Guilherme A.G. Thoaz De
Reporter Radio/Television
TV RIO
Av. Dr. Acurcio Torres, #406
Piratininga, Niteroi, RJ, Brazil, Brazil

Aquino, Wilson Mendonca
Editor
O DIA
R.Riachuelo 359
Rio de Janeiro, Brazil, Brazil

Aranha, Hubert
Reporter Radio/Television
GLOBO TV
Rua Lopes Quintas, 303
Rio de Janeiro-RJ, Brazil, Brazil

Araripe, Ana P. C., Reporter
DIA
Rua Maria Quiteria, 53/301
Ipanema, Rio de Janeiro-RJ, Brazil, Brazil

Araripe, Sonia
Editor
JORNAL DO BRASIL
R.Raquel Prado 130/102
Ilha Governador, Rio de Janeiro, Brazil, Brazil

Araujo, Ana Paula Peceico
Reporter
SISTEMA GLOBO DE RADIO
Rua do Russel, 434, Gloria
Rio de Janeiro, Brazil
T: 55 21 2058282 F: 55 21 2054096

Araujo, Beatriz F.
Reporter Radio/Television
MANCHETE TV
Av.Sernambetiba, 3500/1703
Brazil, Brazil

Araujo, Catarina L.
Correspondent
DIARIO DE PERNAMBUCO
R.Jose Nunes da Cunha,5024
Candeias, Jaboatao-PE, Brazil, Brazil

Araujo, Celso Pires
Editor
JORNAL DE BRASILIA
Quadra 1, Conjunto B1, Casa 5
Sobradinho, DF, Brazil, Brazil

Araujo, Iza Antunes
Editor
MEIO AMBIENTE
SQS 112 Bl.G Apt.204
70375 Brasilia-DF, Brazil, Brazil

Araujo, Ledice
Editor
GLOBO
Rua Carolina Santos, 113-C
Meier, Brazil, Brazil

Araujo, Mauro Fontainha De
Producer
CNN SPANISH
R. Lopes Quintas 732 Jardim Botanico
Rio de Janeiro -RJ, Brazil
T: 55 21 5115139 F: 55 21 2397977

Araujo, Paulo M.
GLOBO
Ave. Paula e Souza, 46/302
Rio de Janeiro-RJ, Brazil, Brazil

Araujo, Rosa Maria Faria
Reporter
RADIOBRAS
Q1 10 Conj. Z casa 3 Guara I
Brasilia, D.F., Brazil, Brazil

Araujo, Vera
Editor
JORNAL DO BRASIL
R. Noemia Nunes, 636
Olaria, Rio de Janeiro-RJ, Brazil, Brazil

Araujo De R., Marcelo
Reporter Radio/Television
TV EDUCATIVA
Rua Viuva Dantas, 229 c114
Rio de Janeiro, Brazil, Brazil

Arbache, Marcia
Editor
GLOBO
Rua Mariz e Barros, 184/603
Icarai-Niteroi-RJ, Brazil, Brazil

Arbex, Jose Mauro
Editor
GAZETA MERCANTIL
R. Mj. Quedinho, 90 - 4 andar
Sao Paulo, Brazil, Brazil

Arco E Flexa, Rodrigo
Editor
JORNAL DA USP
Rua Bela Cintra, 1760/34
Sao Paulo-SP, Brazil, Brazil

Arkader, Rosane
Reporter Radio/Television
MANCHETE TV
Rua Aperana, 84/401
Leblon, Rio de Janeiro-RJ, Brazil, Brazil

Arruda, Edgar A.
Editor
GLOBO
Rua Maxwell, 400/303
Vila Isabel, Rio de Janeiro-RJ, Brazil, Brazil

Arruda, Lilian A.
Correspondent
FOLHA DA MANHA
Av.Marechal Rondon,477 C/44
Rio de Janeiro-RJ, Brazil, Brazil

Arruda, Roldao Oliveira
Editor
ESTADO DE SAO PAULO
R. Azevedo Junior 185/33
Bras, Sao Paulo-SP, Brazil, Brazil

Arthur, Gisele
Reporter Radio/Television
RADIOBRAS
HCGN 713 Bl.H Ap. 109
Brasilia, D.F., Brazil, Brazil

Ascenso, Maria Isabel
Editor
JORNAL DA TARDE
R. Cel.Arthur de Godoy 228
V.Mariana, Sao Paulo-SP, Brazil, Brazil

Assad, Jotair Elias
Editor
GLOBO TV
Rua Von Martius 22 sala 328
Rio de Janeiro, Brazil, Brazil

Assis, Denise
Editor
ISTO E SENHOR
Rua Ipiranga, 44 casa 14
Sao Paulo-SP, Brazil, Brazil

Assuncao, Sonia
Reporter Radio/Television
RADIO TV BANDEIRANTES
Rua Cayona 1260, ap. 196
Sao Paulo, Brazil, Brazil

Astigarraga, Rene Antonio Correa
Editor
GLOBO TV
Rua von Martius. 22, sala 321
Rio de Janeiro, Brazil, Brazil

Auler, Marcelo
Editor
VEJA
R.Prof.Gastao Baiana 495
Apt.1805, Copacabana-RJ, Brazil, Brazil

Azem, Antonio Carlos
Editor
SAO PAULO STATE GOVERNMENT
R.Alves Guimaraes 461
Sao Paulo, S.P., Brazil, Brazil

Azevedo, Andre Luiz
Reporter
REDE GLOBO
R. Lopes Quintas 303
Rio de Janeiro -RJ 2, Brazil
T: 55 21 2947732

Azevedo, Cristina H.
Reporter Radio/Television
REVISTA GEOGRAFICA UNIVERSAL
Rua Salomao Filho 577 bl.6/507
Rio de Janeiro, Brazil, Brazil

Azevedo, Eliane
Editor
VEJA
Rua Gonzaga Bastos, 419 C/10
Vila Isabel, Rio de Janeiro, Brazil, Brazil

Azevedo, Katia R.
Reporter Radio/Television
RADIO ELDORADO FM
R.Alfredo Inacio Trindade55/21
Sao Paulo-SP, Brazil, Brazil

Azevedo, Rosamaria Urbaneto De
Editor
GLOBO TV
Rua Von Martius, 22 sala 32l
Rio de Janeiro, Brazil, Brazil

Babo, Roberta Campos
Editor.
JORNAL TRIBUNA DA IMPRENSA
Rua do Lavradio, 98
20230 Rio de Janeiro-RJ, Brazil, Brazil

Baggio, Marcos Antonio
Editor
RADIO GAUCHA
R. Goncalves Diss. 96
90060 Porto Alegre, Brazil, Brazil

Bahe, Bernardette Duarte
Reporter Radio/Television
TV EDUCATIVA
Av. T. de Castro 77/302
Bonsucesso, RJ, Brazil, Brazil

Baker Ribeiro, Mario Arnaldo
Editor
JORNAL MONITOR MERCANTIL
Rua Benjamin Cosntant, 149/503
Rio de Janeiro-RJ, Brazil, Brazil

Baptista, Sergio
Editor
JORNAL SUL DAS GERAES
R.Afonsina Guimaraes Cobra 119
Pouso Alegre, Minas Gerais, Brazil, Brazil

Baptista, Zulmira M.Castro
Redatora & Editora
JORNAL PERSONA de Brasilia
R.Prudente de Moraes, 1440/304
Rio de Janeiro - RJ, Brazil
T: 55 21 2391433 F: 55 21 2397636

Barata, Luiz Cezar
Reporter Radio/Television
VIDEO CIENCIA PRODUCOES LTDA.
Rua Prud.de Morais,1441/505
22420 Rio de Janeiro-RJ, Brazil, Brazil

Barbanti, Olympio
Correspondent
FOLHA DE SAO PAULO
Av. Higienopolis 741, 401
Sao Paulo, SP, Brazil, Brazil

Barbieri, Edison, Mr.
REVISTA CIDADE NOVA - MOVIMENTO DOS FOCOLARES
Via Bandeirantes, Km47 Cx.Postal 30
Vargem Grande Paulis, Brazil
T: 0055 11 7900638 F: 0055 11 7900

Barbosa, Ana Carla L.
Reporter Radio/Television
GOVERNO MATO GROSSO SUL
R.Maracaju 773/06
Campo Grande, Mato Grosso Sul, Brazil, Brazil

Barbosa, Ana Claudia
Editor
GLOBO
SCS Ed. Oscar Niemeyer
Brasilia-DF, Brazil, Brazil

Barbosa, Celsiane
Reporter Radio/Television
MANCHETE RADIO
Trav. Mario de Almeida 112
Sao Goncalo, Rio de Janeiro, Brazil, Brazil

Barbosa, Joao Paulo S.
Editor
JORNAL DO COMERCIO
R.Nelson de Sa Earp 153/704
Rio de Janeiro, Brazil, Brazil

Barbosa, Lustosa Walkiria
Editor
BUZIOS TV
Rua Principado De Monaco,68/10
Rio de Janeiro-RJ, Brazil, Brazil

Barbosa, Marcia Maria Carvalho
Reporter Radio/Television
BLUE LIGHT PROD. E PROM. LTDA
R. Conde de Baependi, 78/910
Rio de Janeiro, Brazil, Brazil

Barbosa, Marcio S.V.
Editor
JORNAL DE SERVICO
Est. do Tindiba 1477
Rio de Janeiro-RJ, Brazil, Brazil

Barbosa, Marise De Paula
Reporter Radio/Television
JORNAL DO BRASIL-RADIO AM
Rua Jose Higino, 241/201
Rio de Janeiro-RJ, Brazil, Brazil

Barbosa, Mirian Cristina Carneiro
Reporter Radio/Television
MANCHETE TV
Av.Ns.Copacabana, l24l/l2l9
Copacabana, Rio de Janeiro, RJ, Brazil, Brazil

Barbosa, Regina Maura Villela
Reporter Radio/Television
TV-BANDEIRANTES-RIO
Rua Magalhaes Couto, 758/705
Rio de Janeiro-RJ, Brazil, Brazil

Barbosa, Rodrigo F.
Correspondent
FOLHA DA MANHA
R.Conde de Baependi,78/910
Laranjeiras, Rio de Janeiro-RJ, Brazil, Brazil

Barbosa, Veronica Maria Moreira
JORNAL TRIBUNA DA IMPRENSA
Rua do Lavradio, 98
20230 Rio de Janeiro-RJ, Brazil, Brazil

Barbosa, Vilma
Reporter Radio/Television
BUZIOS TV
R.Principado de Monaco,68/101
Rio de Janeiro, RJ, Brazil, Brazil

Barbosa, Zilda Martins
Editor
RADIO MINISTERIO DA EDUCACAO
R.Barao da Torre,133/201
Ipanema Rio de Janeiro RJ Brazil, Brazil

Barboza, Luiz Felipe Da S.
Reporter Radio/Television
REDE OM BRASIL TV
R.Sa Ferreira 44/Cob.2
Rio de Janeiro, Brazil, Brazil

Barca, Lacy
Editor
VIDEO CIENCIA PRODUCOES LTDA.
Rua Dona Mariana, 205 c/02
22280 Rio de Janeiro-RJ, Brazil, Brazil

Barcellos, Geraldo
Reporter Radio/Television
MANCHETE TV
Rua Luiza Bairota 734 C/1
Padre Miguel, Rio de Janeiro-RJ, Brazil, Brazil

Barcelos, Claudio
Reporter Radio/Television
GLOBO TV
Rua Von Martius 22 sala 328
Rio de Janeiro, Brazil, Brazil

Bardanachvili, Eliane
Editor
JORNAL DO BRASIL
R. General Polidoro, 58/305
Botafogo, Rio de Janeiro-RJ, Brazil, Brazil

Barreto, Edson Rodrigues
Reporter Radio/Television
TV EDUCATIVA
N.S. Da Guia 155
Mairer, RJ, Brazil, Brazil

Barreto, Francisco Americo T.
Editor
GLOBO TV
Rua Von Martius 22 sala 32l
Rio de Janeiro, Brazil, Brazil

Barreto, Mauricio P.S. Menna
Editor
TV RIO
R. Barao da Torre, 527, #101
Ipanema, Rio de Janeiro, Brazil, Brazil

Barreto, Nelson De Sa
Reporter Radio/Television
RADIO MARANATA
Av. Copacabana,666
Rio de Janeiro, Brazil, Brazil

Barreto, Rita De C.
Reporter Radio/Television
MANCHETE TV
Rua Galvao 192 C/8
Barreto, Niterio-RJ, Brazil, Brazil

Barreto, Valdir
GLOBO TV
Rua Von Martius 22 sala 32l
Rio de Janeiro, Brazil, Brazil

Barriocanal, Mercedes
Reporter Radio/Television
RADIO CARDINAL AM-STEREO
Mayor Bullo, 667
Asuncion, Paraguay, Brazil

Barros, Ary F.
Editor
CORREIO FLUMINENSE
R. Lagoinha 226
Rio de Janeiro, Brazil, Brazil

Barros, Claudia De Souza De Camargo
GAZETA MERCANTIL
Rua Major Quedinho 90
Sao Paulo - SP, Brazil
T: 55 11 256 3133

Barros, Cristiano
Correspondent
RADIOBRAS
SQN 2l1A apt.301
Brasilia, D.F., Brazil, Brazil

Barros, Flavio Henrique
Editor
BRAZIL TURISMO
Alvarez de Acevedo 89/1001
Niteroi, Rio de Janeiro, Brazil, Brazil

Barros, Guilherme S.
Editor
REVISTA EXAME
Rua 8 lote 31 Q.5
Niteroi, Rio de Janeiro-RJ, Brazil, Brazil

Barros, Jefferson
Editor
A CLASSE OPERARIA
Rua Lima e Silva 1704/18
Porto Alegre, RS, Brazil, Brazil

Barros, Jorge Luis
Reporter Radio/Television
MANCHETE TV
Rua Teixeira Campos 427 L.1
C/53 Brazil, Brazil

Barros, Marcus
Editor
MANCHETE
SQS 216 Bl.H Apt.505
Brasilia-DF, Brazil, Brazil

Barros, Maria Lucia M.
Editor
PREFEITURA RIO DE JANEIRO
R.Voluntarios Patria 166/901
Rio de Janeiro, Brazil, Brazil

Barros, Marinete De A. De Campos
Reporter Radio/Television
TV EDUCATIVA
R. C. de Vasconcelos, 44/202
Rio de Janeiro, Brazil, Brazil

Barros, Odilon Guimaraes
Editor
REVISTA GLOBO RURAL
Rua do Curtume, 665/5 andar
05065 Sao Paulo-SP, Brazil, Brazil

Barros Pinto, Marcus De
Editor
O GLOBO
R. Coelho Neto, 52/1008
Laranjeiras, RJ, Brazil, Brazil

Barros Santana, Horacio G.
Reporter Radio/Television
TV EDUCATIVA
P. M. da Silva 255/302
Taquara, Rio de Janeiro, Brazil, Brazil

Barsante, Cassio E.
Editor
MANCHETE
Rua Aristides Espinola, 59/301
22440-Rio de Janeiro-RJ, Brazil, Brazil

Basile, Sidnei
Editor
GAZETA MERCANTIL
R. Mj.Quedinho, 90 -4 andar
Sao Paulo, Brazil CEP 01050, Brazil

Bastos, Ana
ECABE PRESS
Rua Rodrigo Otavio, 205/302
Rio de Janeiro-RJ, Brazil, Brazil

Bastos, Claudia
Editor
BELEM DO PARA PRESS
R.Jose Higino 254/101
Rio de Janeiro, Brazil, Brazil

Bastos, Luiza G. N.
Editor
TV CULTURA DO PARA
Av. Amirante Barroso, 735
Para, Brazil, Brazil

Bastos, Marisa T.
Reporter Radio/Television
MANCHETE TV
R.Senador Vergueiro, 203/811
Brazil, Brazil

Bastos, Solange A.
Reporter Radio/Television
MANCHETE TV
R.Paissandu, 283 Bl.B Apt.203
Rio de Janeiro-RJ, Brazil, Brazil

Bastos, Sonia C.
Editor
ECABE PRESS
Rua General Rabelo, 65/404
Rio de Janeiro-RJ, Brazil, Brazil

Batista, Jose Messias
Editor
TV RIO
R.Ipanambeirs 224, #404
Jacarepagua, Rio de Janeiro, Brazil, Brazil

Batista, Sirlei
Reporter Radio/Television
RADIOBRAS
SQN 303 Bl. A Ap. 402
Brasilia, D.F., Brazil, Brazil

Batista, Tarlis
Editor
MANCHETE
R. Barao de Itambi, 42
22232-Rio de Janeiro-RJ, Brazil, Brazil

Batista, Vera Lucia De Souza
JORNAL TRIBUNA DA IMPRENSA
Rua do Lavradio, 98
20230 Rio de Janeiro-RJ, Brazil, Brazil

Batista Filho, Ismaelino P.
Editor
TV CULTURA DO PARA
Av. Atlantica, 3040/102
Rio de Janeiro-RJ, Brazil, Brazil

Batista Sena, Alberto De
Editor
ESTADO DE MINAS
R. Dep. A. Sales, 453/702
Sto. Antonio, Belo Horizonte-MG, Brazil, Brazil

Bauer, Carlos Eduardo Da Nobrega
Editor
TV EDUCATIVA
R. Piratininga, 16/102
Gavea, RJ, Brazil, Brazil

Beck, Leda
Editor
ESTADO DE SAO PAULO
R. Irmao Goncalo 74/21
05439 Sao Paulo-SP, Brazil, Brazil

Becker Filho, Alfredo
Editor
JORNAL EXPRESSAO
Av. Atlantica, 2736/402
Copacabana, Rio de Janeiro-RJ, Brazil, Brazil

Beckert, Roberto
Reporter Radio/Television
TV EDUCATIVA
R. Eduardo Santos, 171
Sta. Teresa, Rio de Janeiro, Brazil, Brazil

Beirao, Niriando L.
Editor
ESTADO DE SAO PAULO
Al.Casa Branca, 822/23
Sao Paulo-SP, Brazil, Brazil

Belchior, Fatima Lucia R.
Editor
GAZETA MERCANTIL
R. Mj. Quedinho, 90 - 4 andar
Sao Paulo, Brazil, Brazil

Belem, Claudia
Editor
GLOBO
Rua Itacuraja 26, Apt. 710
Brazil, Brazil

Belem, Miriam De Freitas
Editor
REDE GLOBO
R. Lopes Quintas 303
Rio de Janeiro -RJ 2, Brazil
T: 55 21 2947732

Belfort, Cecilia S.
Editor
PROGRAMA GLOBO ECOLOGIA
R. Gen. Glicerio, No. 58
Laranjeiras, Rio de Janeiro-RJ, Brazil, Brazil

Belmonte, Gecy N.
Correspondent
AGENCIA ESTADO
SQN 202 Bl.F Apt.103
Brasilia-DF, Brazil, Brazil

Belmonte, Roberto Villar
Reporter Radio/Television
RADIO GAUCHA
R. Augusto Melechi, 61,ap.204
90060 Porto Alegre, Brazil, Brazil

Beltrao, Lislaine Andrea M.
Editor
GLOBO TV
Rua Vono Martius 22 sala 32l
Rio de Janeiro, Brazil, Brazil

Bemfeito, Fabiola Damato
Editor
PANROTAS EDITORA LTDA.
R. Santa Clara, 86/807
Copacabana, Rio de Janeiro, Brazil, Brazil

Benante, Daniela Santos
Correspondent
UNIVERSIDADE FED MATO GROSSO SUL
R. das Gracas, 1716 Bl. B/33
Campo Grande, MS, Brazil, Brazil

Benedicto, Marcos Carvalho De
Editor
MOCIDADE/DECISAO
R. Jose Del Fiol, 125
Tatui, SP, Brazil, Brazil

Benjamin, Cid
Editor
GLOBO
Rua Sao Joao Batista, 47/303
22270 Rio de Janeiro-RJ, Brazil, Brazil

Benzi, Liliam D.
Editor
F&C EMBALAGEM-EDITORA CRISOL
R.Cajaiba, 520
05025 Sao Paulo-SP, Brazil, Brazil

Bergamaschi, Mara A.
Correspondent
AGENCIA ESTADO
SQS 305 Bl.B Apt.103
Brasilia-DF, Brazil, Brazil

Bergamo, Monica
Editor
VEJA
R.Deputado Lacerda, 494
05418 Sao Paulo-SP, Brazil, Brazil

Berio, Cristina Luisa
Reporter Radio/Television
CHICO JR & ASSOCIADOS
R. Dr. Julio Otoni 254
Rio de Janeiro, Brazil, Brazil

Berlinck, Deborah C.
Correspondent
GLOBO
107 Rue de Lausane, Ap. 62
1202-Geneve, Suisse, Brazil

Berman, Deborah
Editor
MANCHETE
Rua Joao Afonso, 10/304
Humaita, Rio de Janeiro-RJ, Brazil, Brazil

Bernardes, Betina Dos S.
Editor
ETHOS COMUNICACAO
Rua Maxwell, 149
Tijuca, Rio de Janeiro-RJ, Brazil, Brazil

Bernardes, Fatima Gomes
Reporter Radio/Television
GLOBO TV
Rua Von Martius 22 sala 32l
Rio de Janeiro, Brazil, Brazil

Bernardo, Gilmara Leite
Correspondent
UNIVERSIDADE FED MATO GROSSO SUL
R. Liberdade, 59 V. Carvalho
Campo Grande, MS, Brazil, Brazil

Berrondo Da Silva, Vania Maria
Reporter Radio/Television
RADIO FEDERAL
R. Condessa Belmonte, 228/201
Rio de Janeiro, Brazil, Brazil

Beting, Joelmiri Jose
Reporter Radio/Television
GLOBO TV
Rua Von Martius 22 sala 328
Rio de Janeiro, Brazil, Brazil

Beting, Lucila Zioni
Editor
GLOBO TV
Rua Von Martius, 22 sala 321
Rio de Janeiro, Brazil, Brazil

Bexiga, Maria Paula
Reporter Radio/Television
RECORD TV
R. Angelica Pereira, 67
Imirim, Sao Paulo, SP, Brazil, Brazil

Bezerra, Maria Helena F.
REDE OM BRASIL DE TELEVISAO
Rua Gustavo Augusto de Rezende
Bl.04 Apt.102 Ilha do Gov., RJ, Brazil, Brazil

Bezerra, Mucio Luiz
GLOBO
R. N.S das Mercis, Fonseca
Niteroi-RJ, Brazil, Brazil

Bezerra De Menezes, Maria Celia
Editor
RADIOBRAS
SCLN 410 Bl. E Ap. 101
Brasilia, D.F., Brazil, Brazil

Bial, Pedro
Reporter Radio/Television
GLOBO TV
Oval Road, Camden Lock
London NWI 7 EP England, Brazil

Biancarelli, Aureliano
Editor
FOLHA DE SAO PAULO
Al. Barao de Limeira, 425
Sao Paulo-SP, Brazil, Brazil

Bieging, Elisabeth
Reporter Radio/Television
MANCHETE TV
Rua da Praca do Flamengo, 2
Rio de Janeiro-RJ, Brazil, Brazil

Bigaton, Lucia Helena Boldrini Lacava
Editor
FOLHA DE SAO PAULO
Al. Barao de Limeira, 425
Sao Paulo, Brazil, Brazil

Bigaton, Ulisses L.
Reporter Radio/Television
RADIOBRAS
Sqn 407 K-107
Brasilia, D.F., Brazil, Brazil

Biggi, Walcyr Nogueira
Reporter Radio/Television
GLOBO RADIO
R. Fagundes Varela, 149
Rio de Janeiro, Brazil, Brazil

Bignetti, Leda Maria
Editor
A.N.EDITORA/PRODUTORA
Rua Oscar Freire 515
Sao Paulo, S.P., Brazil, Brazil

Bissio, Beatriz
Editor
ECOLOGIA E DESENVOLVIMENTO
R.Cosme Velho 639 Bl.2 Apt.140
22241 Rio de Janeiro-RJ, Brazil, Brazil

Bittencourt, Carlos Eduardo De Almeida F.
Reporter Radio/Television
RADIO FEDERAL
Praca Eugenio Jardim, 55
Rio de Janeiro, Brazil, Brazil

Bizzi, Luiz Souza
Correspondent
NEW BORDER FILMES
R.Visconde Inhauma 58/801
Rio de Janeiro, Brazil, Brazil

Blanc, Valeria
Editor
GLOBO
Av. Almirante Ary Parreiras
Icarai, Niteroi, RJ, Brazil, Brazil

Blanco, Maria Dolores R.
Correspondent
BANDEIRANTES TV
Av. S. Ines, 813/41
Sao Paulo, SP, Brazil, Brazil

Blasifera, Bianca Maria
Reporter Radio/Television
RADIO ALVORADA FM
R. Gen.Ribeiro da Costa 56/501
Leme, Rio de Janeiro-RJ, Brazil, Brazil

Bocayuva, Luis
Editor
GLOBO
R. I. Simoes Lopes, 62/202
Rio de Janeiro-RJ, Brazil, Brazil

Bochichi, Regiane
Reporter Radio/Television
RADIO ELDORADO
R.Sebastiao Jorge Fco. 222
Sao Paulo, S.P., Brazil, Brazil

Boechat, Claudia
Editor
JORNAL DO BRASIL
R. Barao da Torre, 309/401
Leblon, Rio de Janeiro-RJ, Brazil, Brazil

Boechat, Raquel Macedo
Reporter Radio/Television
MANCHETE TV
Av. Visconde de Piraja 28/802
Rio de Janeiro, Brazil, Brazil

Bomfim, Octavio
Correspondent
RADIOBRAS
HIGS 714 Bl. O c/08
Brasilia, D.F., Brazil, Brazil

Bonalume, Neto Ricardo
Editor
FOLHA DE SAO PAULO
Al. Barao de Limeira, 425
Sao Paulo-SP, Brazil, Brazil

Bonas, Ariovaldo
Editor
ESTADO DE SAO PAULO
R. Artur 246
02467 Sao Paulo-SP, Brazil, Brazil

Bonchristiano, Ana Cristina
Editor
FOLHA DE SAO PAULO
Al. Barao de Limeira, 425
Sao Paulo-SP, Brazil, Brazil

Bonelli, Ariete
Editor
A CRITICA
SQN 216 Bl.F Apt.305
70875 Brasilia-DF, Brazil, Brazil

Bonemer Jr., Willian
Reporter Radio/Television
GLOBO TV
Rua Von Martius 22 sala 32l
Rio de Janeiro, Brazil, Brazil

Borges, Gustavo
Correspondent
JORNAL METRO PRESS
Rua da Tranquilidade, 36
Rio de Janeiro-RJ, Brazil, Brazil

Borges, Joao
Editor
GLOBO
SCS. Ed. Oscar Niemeyér
Brasilia-DF, Brazil, Brazil

Borges, Leticia
AGENCIA ESTADO
SQS 305 Bl.I Apt.604
Brasilia-DF, Brazil, Brazil

Borges, Marisa De R.
Editor
ROLF PRESS
R.Conde Afonso Celso 131/207
Rio de Janeiro, Brazil, Brazil

Borges, Paulo Marcos
Reporter Radio/Television
TV EDUCATIVA
Itaperu 155 Bl Q - 807
Catumbi, RJ, Brazil, Brazil

Bortolot, Carlos Eduardo R.
Editor
GOVERNO MATO GROSSO SUL
R.Dom Aquino 2274
Campo Grande, Mato Grosso Sul, Brazil, Brazil

Bottari, Elenilce V.
Editor
GLOBO
Rua Rainha Guilhermina, 187
Leblon-RJ, Brazil, Brazil

Bottini, Paulo Rogerio Neubern
Editor
SBT TV
R. Prof. Paul Hugon
98 Agua Fria, Brazil, Brazil

Boyago, Claudia
Reporter Radio/Television
RECORD TV
R. Tuim 484, #152
Sao Paulo, SP, Brazil, Brazil

Boyd, Aristides Dos Santos
Editor
REDE GLOBO
R. Lopes Quintas 303
Rio de Janeiro -RJ 2, Brazil
T: 55 21 2947732

Braga, Edson R.
Editor
GLOBO
Rua Pacheco Leao 320
Rio de Janeiro-RJ, Brazil, Brazil

Braga, Jabes Teixeira
Editor
JORNAL DO IMIGRANTE
R.M.S. Bom Conselho 320
Rio de Janeiro, Brazil, Brazil

Braga, Joao
Editor
VEJA
Rua Prof.Ortis Monteiro,132
Laranjeiras, Rio de Janeiro, Brazil, Brazil

Braga, Tuleu Ruth
Editor
JORNAL DO IMIGRANTE
Rua N.Sra. Bom Conselho 320
Sao Paulo, Brazil, Brazil

Branco, Paulo Cesar
Reporter Radio/Television
TV-BANDEIRANTES-RIO
Rua Anita Garibaldi, 9/202
Copacabana, Rio de Janeiro-RJ, Brazil, Brazil

Brandao, Herminia
Editor
FOTOGRAMA
R. Fernao Dios 264/82
Pinheiros, SP, Brazil, Brazil

Brandao, Marcia
Reporter Radio/Television
GLOBO RADIO
R. Araujo Raitao, 607
Rio de Janeiro-RJ, Brazil, Brazil

Brandao, Nilson De Souza
Editor
GLOBO
Rua Jorge Rudge 29/206
Vila Isable, Rio de Janeiro-RJ, Brazil, Brazil

Brandao, Priscila J.
Editor
GLOBO TV
Rua Von Martius 22 sala 32l
Rio de Janeiro, Brazil, Brazil

Brasil, Cristina Indio
Reporter Radio/Television
RADIOBRAS
Ql 31 Bl 07 Ap. 519 Guara II
Brasilia, D.F., Brazil, Brazil

Brasiliense, Jose Ronaldo
Editor
JORNAL DO BRASIL
Av. Alcindo Cacela, 995/402
Centro, Belem, PA, Brazil, Brazil

Braun, Maria Helena A.
Editor
GLOBO TV
Rua Von Martius 22 sala 32l
Rio de Janeiro, Brazil, Brazil

Braun, Silvio
Editor
TV EDUCATIVA
Av. P. D. Cardoso, 800 Bl 3,
608 Barra, RJ, Brazil, Brazil

Breitinger, Jaqueline
Editor
ISTO E SENHOR
Av.Sete de Setembro,234/705
24230 Niteroi-RJ, Brazil, Brazil

Brito, Francisco Jose
Reporter Radio/Television
GLOBO TV
Rua Von Martius 22 sala 328
Rio de Janeiro, Brazil, Brazil

Brito Filho, Manoel
Correspondent
JORNAL DO BRASIL S/A
2111 Wisconsin Ave. N.W.723
Washington, D.C. 20007, USA, Brazil

Britto, Carlos Alfredo
Reporter Radio/Television
GOVERNO DO DISTRITO FEDERAL
QI 16 Bl. G/308
Brasilia, D.F., Brazil, Brazil

Britto Jr., Hilton
Reporter Radio/Television
GLOBO TV
Rua Von Martius 22 sala 328
Rio de Janeiro, Brazil, Brazil

Bruno, Ricardo Alberto De Oliveira
Editor
TV EDUCATIVA
R. Gumaraes Natatal 16 -306
Rio de Janeiro, Brazil, Brazil

Bruno, Roberto Da S.
REVISTA FLUMINENTE
Rua Sen.Vergueiro 45/1201
Rio de Janeiro, Brazil, Brazil

Bueno, Eduardo
Editor
ZERO HORA
Rua Cel. Bordini, 815
Porto Alegre-RS, Brazil, Brazil

Bulhoes, Leticia R.K.
Reporter Radio/Television
GLOBO RADIO
Rua da Gloria 214/1406
Rio de Janeiro, Brazil, Brazil

Burger Jr., Darcy
Editor
MANCHETE TV
Rua Visconde Piraja 66/705
Rio de Janeiro, Brazil, Brazil

Burnier, Jose Roberto
Reporter Radio/Television
GLOBO TV
Rua Von Martius 22 sala 328
Rio de Janeiro, Brazil, Brazil

Buzzo, Ivany T.
Reporter Radio/Television
RECORD TV
Fradique Coutinho 216/112
Sao Paulo, Brazil, Brazil

Cabral, Graca
Reporter Radio/Television
RADIO ELDORADO FM
R.Sen.C.Lacerda Vergueiro 596
Apt.72 Sumarezinho, Sao Paulo-SP, Brazil, Brazil

Cabral, Luiza Barbosa Da Silva
Editor
CIRANDA DOS BAIRROS
Estrada da Cacuia, 231
Rio de Janeiro, Brazil, Brazil

Cafaro, Pedro
Editor
ESTADO DE SAO PAULO
R. Apiacas 570/102
Sao Paulo-SP, Brazil, Brazil

Caifman, Lucia Ribeiro
Editor
GLOBO TV
Rua Von Martius 22 sala 32l
Rio de Janeiro, Brazil, Brazil

Cajaty, Paula Romero
Correspondent
JORNAL DOS IMOVEIS
Rua Marechal Trompouski,42
Rio de Janeiro-RJ, Brazil, Brazil

Cajaty, Pedro Paulo
Editor
JORNAL DOS IMOVEIS
R. Mal. Trompowski, 42
Tijuca, Rio de Janeiro-RJ, Brazil, Brazil

Caldas, Maria S.
Editor
AGENCIA ESTADO
R. Barao de Lucena,124/401
Botafogo, Rio de Janeiro-RJ, Brazil, Brazil

Calheira, Zoraya
Redator
ECOLOGIA E DESENVOLVIMENTO
Rua da Gloria 122 gr 105/106
Rio de Janeiro - RJ, Brazil
T: 55 21 242 1957/222 1370

Calheiros, Pedro
Reporter Radio/Television
JORNAL DO BRASIL - RADIO
R. Senador Vergueiro, 26/503
Rio de Janeiro, Brazil, Brazil

Calife, Jorge
Editor
JORNAL DO BRASIL
R. Magda, 60/102
Bonsucesso, Rio de Janeiro-RJ, Brazil, Brazil

Callado, Tessy
Reporter Radio/Television
TV EDUCATIVA
R. Rubens Berardo, 175/807
Rio de Janeiro, Brazil, Brazil

Calmon, Cristina
Editor
SISTEMA JORNAL DO BRASIL
Av. Brasil 500
Rio de Janeiro -RJ, Brazil
T: 55 21 5854422 F: 55 21 5854428

Calmon Alves, Rosental
Editor
JORNAL DO BRASIL
R. Desembargador Joao Claudio
22795, Rio de Janeiro-RJ, Brazil, Brazil

Camacho, Marcelo Goncalves
Editor
EDITORA O DIA
Av.Sernambetiva 6600/1705
Rio de Janeiro, Brazil, Brazil

Camargo, Marcelo De Albuquerque
Editor
RADIOBRAS
Rua Maracana, 1334/101
Rio de Janeiro, Brazil, Brazil

Camargo, Pedro Geraldo
Editor
ANO ZERO
R. Voluntarios da Patria
22270-Rio de Janeiro-RJ, Brazil, Brazil

Campanella, Marco Antonio
Editor
HORA DO POVO
SQN 403 BLO Apt.207
Brasilia-DF, Brazil, Brazil

Campello, Lilia De Lima
Reporter Radio/Television
MANCHETE TV
R.Pintora Djanira 156/102
Rio de Janeiro, Brazil, Brazil

Camphora, Ana Lucia
Reporter Radio/Television
TV-BANDEIRANTES-RIO
Av. Padre Leonel Franca
Gavea, Rio de Janeiro-RJ, Brazil, Brazil

Campos, Alexandre D.
Correspondent
POVO
Rua Mariz e Barros, 396/303
24.230-Rio de Janeiro-RJ, Brazil, Brazil

Campos, Ivanildo Ferreira
Correspondent
FOLHA CRISTA
Rua Anadia, 82
Rio de Janeiro-RJ, Brazil, Brazil

Campos, Luiz Sergio
Reporter Radio/Television
RECORD TV
R.Q.I. J. De Oliveira 1127/c20
Brazil

Campos, Rolland Gianotti
JORNAL TRIBUNA DA IMPRENSA
Rua do Lavradio, 98
20230 Rio de Janeiro-RJ, Brazil, Brazil

Canellas, Marcelo P.
Reporter Radio/Television
GLOBO TV
Rua Von Martius 22 sala 32l
Rio de Janeiro, Brazil, Brazil

Canfran Mas, Vera Lucia
Reporter Radio/Television
RADIOBRAS
HIGS 710 Bl. Q Casa 36
Brasilia, D.F., Brazil, Brazil

Cangucu, Angela C.
Editor
RADIO & TV BANDEIRANTES
R.Caiova, 869/702
Sao Paulo, S.P., Brazil, Brazil

Canton, Katia
Correspondent
ISTO E SENHOR
303 E 71th St. (#4C)
New York, NY 10021, USA, Brazil

Capaz, Andrea Ferretti
Reporter Radio/Television
JORNAL DO BRASIL-RADIO AM
Rua Delgado de Carvalho, 65
Tijuca, Rio de Janeiro-RJ, Brazil, Brazil

Capriglione, Laura
Editor
FOLHA DE SAO PAULO
Al. Barao de Limeira, 425
Sao Paulo-SP, Brazil, Brazil

Capriles, Rene
Editor
REVISTA ECORIO
Rua Antonio Vieira, No. 22
22010, Rio de Janeiro-RJ, Brazil, Brazil

Carden, Elizabeth Flanigen
Correspondent
ABIM AGENCY
2429 McCormick Rd.
Rockville, MD 20850 USA, Brazil

Cardim, Roberto
Correspondent
FUNDACAO ROBERTO MARINHO
Est.das Canoas, 1476 c/36
S.Conrado, Rio de Janeiro-RJ, Brazil, Brazil

Cardoso, Beatriz Teixeira
Reporter
CONTEXTO
Av Calogeras 12/63 Centro
Rio de Janeiro - RJ, Brazil
T: 55 21 2204653 F: 55 21 2204653

Cardoso, Carlos
Reporter Radio/Television
RADIO ALVORADA FM
R. Eng.Nazare, 56/102
Abolicao, Rio de Janeiro-RJ, Brazil, Brazil

Cardoso, Cesar
Editor
REDE GLOBO
R. Lopes Quintas 303
Rio de Janeiro -RJ 2, Brazil
T: 55 21 2947732

Cardoso, Ivo Da Silva
GLOBO TV
Rua Von Martius 22 sala 328
Rio de Janeiro, Brazil, Brazil

Cardoso, Jorge Luiz Pereira
Reporter Radio/Television
GLOBO RADIO
R. Otavio Carneiro, 126/402
Niteroi, RJ, Brazil, Brazil

Cardoso, Luciana
Reporter Radio/Television
GLOBO TV
Ave. Alvorada, 2600
Tijuca, Rio de Janeiro-RJ, Brazil, Brazil

Cardoso, Manoel Simao
Reporter Radio/Television
RADIO CLUBE DE PERNAMBUCO
R. Neto Mendoca 25, #07
Aflitos - Recife - Pernambuco, Brazil, Brazil

Cardoso, Marco Valerio
Editor
RADIOBRAS
SQN 310 Bl. A Ap. 105
Brasilia, D.F., Brazil, Brazil

Cardoso, Mauricio
Editor
RADIOBRAS
SQS 205 Bloco A Ap. 403
Brasilia, D.F., Brazil, Brazil

Cardoso, Monica Goncalves
Reporter Radio/Television
RADIOBRAS
Q. 505 Bl. B/401
Brasilia, D.F., Brazil, Brazil

Cardoso, Vera Lucia
Reporter Radio/Television
MANCHETE TV
Rua Coronel Costa, 109/301
Brazil, Brazil

Cardoso Castilho, Marcelo
Reporter Radio/Television
RADIO FEDERAL
R. Pontes Correa, 34/401
Rio de Janeiro, Brazil, Brazil

Cardozo Neto, Sylvio Dos Santos
Editor
EDITORA GLOBO
R.Retiro Artistas 607/905/Bl.5
Rio de Janeiro, Brazil, Brazil

Carlini, Monica
Reporter Radio/Television
TV-BANDEIRANTES-RIO
Rua Conde de Bonfim, 890/205
Tijuca, Rio de Janeiro-RJ, Brazil, Brazil

Carlos, Joao Carlos Gonzalez
Editor
RADIOBRAS
SQS 103 Bl. E Ap. 112
Brasilia, D.F., Brazil, Brazil

Carmo, Augusto Jose Do
Reporter Radio/Television
GLOBO RADIO
R. Antonino, 37
Miguel Couto, RJ, Brazil, Brazil

Carmo, Marcia Eliane M.
Correspondent
JORNAL DO BRASIL
SNS 205 Bl.B, apt. 406
Brasilia, D.F., Brazil, Brazil

Carneiro, Adriana Cristina Sampaio
Reporter Radio/Television
GLOBO RADIO
R. Pres. Carlos Campos, 244
Rio de Janeiro, Brazil, Brazil

Carneiro, Claudio Nelson F.
Reporter
SISTEMA GLOBO DE RADIO
Rua do Russel, 434
Rio de Janeiro, RJ, Brazil
T: 021 2058282 F: 021 2054096

Carneiro, Helio P.
Editor
MANCHETE
Rua Gomes Carneiro, 49/601
22071-Rio de Janeiro-RJ, Brazil, Brazil

Carneiro, Lucia De Oliveira Castro
Reporter Radio/Television
RADIO & TV BANDEIRANTES
SHIS QI 28 -conjunto 15/c 18
Brasilia, DF, Brazil, Brazil

Carneiro, Luiz Orlando Q.
Editor
JORNAL DO BRASIL
SHIS, QI-28, Conj. 15
Brasilia, DF, Brazil, Brazil

Carneiro, Marcelo
DIA
R. Voluntarios da Patria, 452
22270, Rio de Janeiro-RJ, Brazil, Brazil

Carneiro, Sonia Maria De Abranches
Correspondente Brasilia
SISTEMA JORNAL DO BRASIL
Av.Brasil, 500
Rio de Janeiro - RJ, Brazil
T: 55 21 5854422

Carneiro Fonseca, Ana Maria
Editor
JORNAL DO COMMERCIO
R. Sao Clemente, 185/1208 B.1
Botafogo, Rio de Janeiro, Brazil, Brazil

Carnevale, Yolanda Maria Mieli
Reporter Radio/Television
TV EDUCATIVA
R. Oliveira Botelho 33/201
Rio de Janeiro, Brazil, Brazil

Caropreso, Alvaro A.
Editor
AGENCIA ESTADO
R. Mourato Coelho, 1430/142
05417 Sao Paulo-SP, Brazil, Brazil

Carvalho, Alberto A.
Reporter Radio/Television
MANCHETE
Rua Russel 804/8 and.
Rio de Janeiro, Brazil, Brazil

Carvalho, Antonio Jose Lemos De
Editor
GAZETA DE NOTICIAS
R. Dona Zulmira 21-503
Brazil

Carvalho, Bernardo De A.
Editor
RADIO & TV BANDEIRANTES
R. Avanhandava, 40/206
Bairro Bela Vista, Sao Paulo, Brazil, Brazil

Carvalho, Cintia Neves
Reporter Radio/Television
TV EDUCATIVA
Av. Gomes Freire, 474
Rio de Janeiro, Brazil, Brazil

Carvalho, Jorge De Azevedo
Editor
RADIOBRAS
Itapiru, 155 bl. R/1207
Rio de Janeiro, Brazil, Brazil

Carvalho, Liana
Reporter Radio/Television
REDE BRASIL
Rua Marques de Abrantes 199
Rio de Janeiro, Brazil, Brazil

Carvalho, Ligia Melo
Reporter Radio/Television
REDE OM BRASIL DE TELEVISAO
Rua Barao da Torre, 521/403
Rio de Janeiro, Brazil

Carvalho, Maria C.
Editor
GLOBO
TV Oriente, 111 S/301
20.240, Rio de Janeiro-RJ, Brazil, Brazil

Carvalho, Maria De Fatima G.
Reporter Radio/Television
GLOBO RADIO
Rua da Gamboa, 199 C/04
Saude, Rio de Janeiro-RJ, Brazil, Brazil

Carvalho, Maria Luiza
Correspondent
SAO PAULO ZONA SUL
Mario Pederneira 51/3011
Rio de Janeiro, Brazil, Brazil

Carvalho, Nivia
JORNAL O GLOBO
Rua Mariz e Barros, 156, Ap504
Niteroi-RJ, Brazil, Brazil

Carvalho, Patricia Mauricio
Reporter Radio/Television
JORNAL DO BRASIL-RADIO AM
Rua Jorge Rudge, 89
Vila Isabel, Rio de Janeiro-RJ, Brazil, Brazil

Carvalho, Ronald Dias Pinto De
Editor
GLOBO TV
Rua Von Martius 22 sala 32l
Rio de Janeiro, Brazil, Brazil

Casado, Jose Aurino
Editor
GAZETA MERCANTIL
R. Mj. Quedinho, 90 -4 andar
Sao Paulo, Brazil, Brazil

Casali, Romuno Henrique Bartoly
Reporter Radio/Television
RADIOBRAS
R. M. de Abrantes, 178/1505
Rio de Janeiro, Brazil, Brazil

Case, Rafael O.
Editor
GLOBO TV
Rua Von Martius 22 sala 32l
Rio de Janeiro, Brazil, Brazil

Casotti, Bruno
Editor
JORNAL DO BRASIL
R. Jardim Botanico, 227/403
20470, Rio de Janeiro-RJ, Brazil, Brazil

Casoy, Boris
Reporter
TVSBT
Campo de S.Cristovao, 105
Rio de Janeiro - RJ, Brazil
T: 55 21 5803688 / 5800678 F: 55 21 5809685

Cassilly, Michael C.
Editor
ECABE PRESS
Rua Barao da Torre, 639
Ipanema, Rio de Janeiro-RJ, Brazil, Brazil

Castellani, Marisa P.
Editor
AGENCIA ESTADO
R. Alves de Brito, 31/201
Tijuca, Rio de Janeiro-RJ, Brazil, Brazil

Castelo Branco, Adriana
Editor
JORNAL DO BRASIL
Rua Humaita, 104/202
22261, Rio de Janeiro-RJ, Brazil, Brazil

Castelo Branco, Jose
Correspondent
DIARIO DE PERNAMBUCO
Av.Beira Mar 1312
Piedade, Jaboatao-PE, Brazil, Brazil

Castilho, Ines
Editor
REVISTA VISAO
Rua Alves Guimaraes, 139/36
05410 Sao Paulo-SP, Brazil, Brazil

Castro, Daniele De O.
Editor
A NOTICIA
Barao da Torre, 521/302
Rio de Janeiro, Brazil, Brazil

Castro, Marcos Almeida M.
Editor
RIO DE JANEIRO PRESS
Av.Edson Passos 1023/102
Rio de Janeiro, Brazil, Brazil

Castro, Thereza C.
Editor
PROGRAMA GLOBO ECOLOGIA
R. Voluntarios da Patria, 431
Botafogo, Rio de Janeiro-RJ, Brazil, Brazil

Castro Alves, Janet
Editor
PREFEITURA RIO DE JANEIRO
R.Gonzaga Bastos 20/303
Rio de Janeiro, Brazil, Brazil

Castro Neves, Jose Octavio
Reporter Radio/Television
TV EDUCATIVA
R. B. da Torre, 450/501
Rio de Janeiro, Brazil, Brazil

Cavagnari, Victor
Editor
GLOBO
R. Marquis de S. Vicente, 351
Brazil, Brazil

Cavalcante, Itamar
Correspondent
JORNAL DA USP
Rua Itarare, 214
Sao Paulo-SP, Brazil, Brazil

Cavalcanti, Fernanda
Correspondent
JORNAL PLANTAS MEDICINAIS
R. Voluntarios da Patria, 127
Rio de Janeiro-RJ, Brazil, Brazil

Cavalcanti, Karla V.
Reporter Radio/Television
TV EDUCATIVA
R.Joaquim Tavora 74
Rio de Janeiro, Brazil, Brazil

Cavalcanti, Roberto Luiz L.
Editor
GLOBO TV
Rua Von Martius 22
Rio de Janeiro, Brazil, Brazil

Cavalcanti, Vanda Leston
Reporter Radio/Television
TV-BANDEIRANTES-RIO
Rua Francisco Xavier, 342/507
Tijuca, Rio de Janeiro-RJ, Brazil, Brazil

Cavalheiro, Lilian Nascimento
Editor
REDE GLOBO
R. Lopes Quintas 303
Rio de Janeiro -RJ 2, Brazil
T: 55 21 2947732

Cavalleiro, Teresa Cristina
Editor
GLOBO TV
Rua Von Martius 22 sala 328
Rio de Janeiro, Brazil, Brazil

Caversan, Luiz Carlos
Editor
FOLHA DA MANHA
R.Redentor, 63 Cobertura
Rio de Janeiro-RJ, Brazil, Brazil

Celestino, Helena
Editor
GLOBO
Igarapava, 35/403, Leblon
Rio de Janeiro-RJ, Brazil, Brazil

Centeno, Ayrton J.
Correspondent
RADIO ELDORADO FM
R.Dr.Barcelos, 909/601
91900 Tristeza, Porto Alegre-RS, Brazil, Brazil

Cerqueira, Sofia Silva
Reporter
JORNAL O GLOBO
Rua Irineu Marinho, 35
Rio de Janeiro, RJ 2, Brazil
T: 021 2722000

Cerri, Claudio Emilio
Editor
REVISTA GLOBO RURAL
Rua do Curtume, 665/5 andar
05065 Sao Paulo-SP, Brazil, Brazil

Chabassus, Luiz Augusto V.
Editor
GLOBO TV
Rua Von Martius 22 sala 32l
Rio de Janeiro, Brazil, Brazil

Chacel, Cristina
Editor
GLOBO
Rua Sao Clemente, 398
Rio de Janeiro-RJ, Brazil, Brazil

Chagas, Acy Porto
Editor
REDE GLOBO
R. Lopes Quintas 303
Rio de Janeiro -RJ 2, Brazil
T: 55 21 2947732

Chagas, Adelia Maria
Correspondent
AGENCIA DE NOTICIAS ACOPI
R. Jose Pinto de Moura, 51
13070, Sao Paulo-SP, Brazil, Brazil

Chagas, Helena
Editor
GLOBO
SCS. Ed. Oscar Niemeyer, 13
Brasilia-DF, Brazil, Brazil

Chagas, Ivone B.
Correspondent
ZERO HORA
SCLN-Q. 307 Bloco D, Apt. 214
70.000 Brasilia-DF, Brazil, Brazil

Chagas, Marcos A.S.
Editor
RADIOBRAS
SAIN CAESB Casa 04 Asa Norte
Brasilia, D.F., Brazil, Brazil

Chalita, Jose Marcio Paiva
Reporter Radio/Television
REDE OM BRASIL DE TELEVISAO
R. Pereira da Silva,568/905
Laranjeiras, RJ, Brazil, Brazil

Chamilete, Jose
Editor
JORNAL DO COMERCIO
R.Gustavo Sampaio, 598/404
Leme, Rio de Janeiro-RJ, Brazil, Brazil

Chapot, Luis R.
Reporter Radio/Television
MANCHETE TV
R.das Dalias, 115 C/9
Brazil, Brazil

Charam, Isaac
Correspondent
INSTITUTO BRASILEIRO DE MEDICINA
Praca Serzedelo Correa 15/703
Rio de Janeiro, Brazil, Brazil

Chayb, Lucia
Editor
ECORIO
R. Antonio Vieira, 22/601
Leme, Rio de Janeiro, Brazil, Brazil

Chervezon, Maria Sirlene
Reporter Radio/Television
GLOBO RADIO
Rua Garibaldi, 133
Tijuca, Rio de Janeiro-RJ, Brazil, Brazil

Chiaretti, Daniela
Editor
FOLHA DE SAO PAULO
Al. Barao de Limeira, 425
Sao Paulo-SP, Brazil, Brazil

China, Julio T.
Editor
BRASIL SEIKYO PUBL.HOUSE
R. Ten.Gomes Ribeiro 58/131
Vila Clementino, SP, Brasil 04038, Brazil

Chor, Osmar Fabio
Editor
GLOBO TV
Rua Von Martius 22 sala 32l
Rio de Janeiro, Brazil, Brazil

Cichowski, Cecilia
Editor
JORNAL DO COMMERCIO
Av. Osvaldo Cruz, 86/301
Flamengo, RJ, Brazil, Brazil

Ciucci, Jose Augusto
Reporter Radio/Television
TV CULTURA-REPORTER ECO
R. Pedroso Xavier, 420
02732 Sao Paulo-SP, Brazil, Brazil

Ciuffreda, Giuseppina
IL MANIFESTO
Rua Girassol 1317/44
Sao Paulo - SP, Brazil
T: 55 11 815 4528 F: 55 11 825-7861

Clark Bueno, Walter
Editor
TV EDUCATIVA - RJ
R. P. Gastao Bahiana, 615/1601
Rio de Janeiro, Brazil, Brazil

Claudio, Manoel
GLOBO TV
Ave. Alvorada, 2600
Tijuca, Rio de Janeiro-RJ, Brazil, Brazil

Clela, Alssa Susi
Editor
FOLHA DE SAO PAULO
Al. Barao de Limeira, 425
Sao Paulo-SP, Brazil, Brazil

Clement, Paulo Julio Moraes
Editor
GLOBO
Rua do Bispo, 337/401
Rio Comprido, Rio de Janeiro-RJ, Brazil, Brazil

Clemente, Fabio
Editor
JORNAL DO COMMERCIO
R. Reseda, 26/201
Fonte da Saudade, RJ, Brazil, Brazil

Cochrane, Marilia Gabriela
Reporter Radio/Television
RADIO TV BANDEIRANTES
Rua Radiantes, 13, Morumbi
Sao Paulo, Brazil, Brazil

Coelho, Cosme
Reporter Radio/Television
GLOBOSAT
R.Mario Portela 161 Bl./1102
Rio de Janeiro, Brazil, Brazil

Coelho, Darlene Mara Mello
Correspondent
PANROTAS EDITORA LTDA.
R. Bambina 17/201, Botafogo
22251 Rio de Janeiro, Brazil, Brazil

Coelho, Isabelle Da R.
Editor
DIA
Estr. do Cabucu, 108
Campo Grande, Rio de Janeiro-RJ, Brazil, Brazil

Coelho, Virginia Marques
Editor
GLOBO TV
Rua Von Martius 22 sala 32l
Rio de Janeiro, Brazil, Brazil

Colla, Marcia
Correspondent
RADIO ELDORADO FM
R.Lucio Marques, 104 Apt.02
11.900 Registro, Sao Paulo, Brazil, Brazil

Coluccion, Ana Maria Fuster
Editor
JORNAL SAO PAULO ZONA SUL
Ave. Senador Casemira da Rocha
04047-Sao Paulo-SP, Brazil, Brazil

Continentino, Dudu
NEON RIO LTDA
Rua Sta.Crisitina, 89
Sta.Teresa 20.241 Rio de Janeiro,Brazil, Brazil

Copstein, Jayme
Reporter Radio/Television
RADIO GAUCHA
Av. Venancio Aires, 495, Ap.41
90000 Porto Alegre, Brazil, Brazil

Cordeiro, Antonio G.
Editor
RADIO UNIVERSIDADE
R.Mariz e Barros 272/809
Rio de Janeiro, Brazil, Brazil

Cordeiro, Antonio Roberto Martins
Editor
REDE GLOBO
R. Lopes Quintas 303
Rio de Janeiro -RJ 2, Brazil
T: 55 21 2947732

Cordeiro, Ligia E.C.R.
Editor
REVISTA VISAO
Rua Felipe Camarao 75 c/15
Rio de Janeiro, Brazil, Brazil

Cordeiro, Renato J.S.
Correspondent
POVO
Rua Barao de Mesquita, 647/702
20.540-Rio de Janeiro-RJ, Brazil, Brazil

Corradini, Marcia Raposo
Editor
GAZETA MERCANTIL
R. Mj. Quedinho, 90/ 4 andar
Sao Paulo, Brazil, Brazil

Correa, Arlindo Lopes
Reporter Radio/Television
TV EDUCATIVA
Av. Fernando Mattos 122
Rio de Janeiro, Brazil, Brazil

Correa, Douglas
Reporter Radio/Television
RADIOBRAS
R. dos Carijos, 55/301
Rio de Janeiro, Brazil, Brazil

Correa, Elena Rosecler B.
Reporter Radio/Television
REDE OM BRASIL DE TELEVISAO
Rua Alberto de Campos, 258/106
Ipanema, Rio de Janeiro-RJ, Brazil, Brazil

Correa, Geraldo Pereio
Reporter
REVISTA VISAO
R.Dr.Almeida Lima, 1384
Sao Paulo - SP, Brazil
T: 55 11 2914866 R.476 F: 55 11 2927438

Correa, Gilmar
Editor
NOTICIA
Rua Demetrio Ribeiro, 14/12
Centro, Florianopolis, SC, Brazil, Brazil

Correa, Marcos Sa
Editor
DIA
R. Antenor Rangel, 124
Gaveo, Rio de Janeiro-RJ, Brazil, Brazil

Correa, Martha N.
Reporter Radio/Television
MANCHETE TV
R.Mario Portela 161/603 Bl.B
Laranjeiras, Rio de Janeiro-RJ, Brazil, Brazil

Correa, Mauricio
Correspondent
AGENCIA ESTADO
SHIN-QI 11 Conj.11 Casa 12
Brasilia-DF, Brazil, Brazil

Correa, Paulo Cesar
Editor
GLOBO TV
Rua Von Martius 22 sala 321
Rio de Janeiro, Brazil, Brazil

Correa Farias, Ana Maria
Editor
EDITORA AZUL/REVISTA CONTIGO
R. das Laranjeiras, 192 Bl.L
Rio de Janeiro-RJ, Brazil, Brazil

Correao, Claudio Bernardo
Reporter Radio/Television
TV EDUCATIVA
Couto 29
Penha, RJ, Brazil, Brazil

Cortes, Celina
Reporter
SISTEMA JORNAL DO BRASIL
Av. Brasil 500
Rio de Janeiro -RJ, Brazil
T: 55 21 5854422 F: 55 21 5854428

Cosme, Luiz
Reporter
TVSBT
Campo de S.Cristovao, 105
Rio de Janeiro - RJ, Brazil
T: 55 21 5803688 / 5800678 F: 55 21 5809685

Costa, Antonio Carlos
Reporter Radio/Television
GLOBO RADIO
R.Domingos Ferreira 188
Rio de Janeiro, Brazil, Brazil

Costa, Carlos Alberto Silveira Da
Reporter Radio/Television
TV EDUCATIVA - RJ
R. Eurico Cruz 48/202
Rio de Janeiro, Brazil, Brazil

Costa, Claudia Lima
Reporter
O DIA
R. Riachuela 359
Rio De Janeiro, RJ 2, Brazil
T: 55 21 272-8000 F: 55 21 507-1228

Costa, Joao Carlos
Reporter Radio/Television
RADIO UNIVERSIDADE/UFSM
Av. N.S. Medianeira 1780
Centro, Sta. Maria, RS, Brazil, Brazil

Costa, Jose Eduardo
Reporter Radio/Television
RADIO ITATIAIA DE BELO HORIZONTE
R. Itatiaia 117
Belo Horizonte, Minas Gerais, Brazil, Brazil

Costa, Julio C.
Reporter Radio/Television
MANCHETE TV
Av.N.S de Copacabana 1099/901
Rio de Janeiro-RJ, Brazil, Brazil

Costa, Luiz Antonio V.
Editor
MANCHETE TV
R.Altinopolis 369
Ilha Governador, Rio de Janeiro, Brazil, Brazil

Costa, Marcelo
Reporter Radio/Television
REDE OM BRASIL DE TELEVISAO
Rua Sao Clemente, 114/201 Bl.2
Botafogo, Rio de Janeiro-RJ, Brazil, Brazil

Costa, Mari-Angela Heredia Da
Editor
RADIO& TV BANDEIRANTES
SQN 215 - Bloco A- Apto. 615
Brasilia Brazil, Brazil

Costa, Maria Cecilia
Editor
GLOBO
R. Barao da Torre, 175 c/3
Rio de Janeiro-RJ, Brazil, Brazil

Costa, Maria Rosa
Correspondent
AGENCIA ESTADO
SQN 206 Bl.B Apt.302
Brasilia-DF, Brazil, Brazil

Costa, Maria Teresa
Editor
CORREIO POPULAR
R. Bern. de Campos, 646 Apt.2
13.013 Campinas-SP, Brazil, Brazil

Costa, Mary Diana Da
Editor
RADIOBRAS
HCGN 706 Bl. B Casa 30
Brasilia, D.F., Brazil, Brazil

Costa, Mauro Jose Da
Editor
SBT TV
R. Pres. Pedreira, 97/1001
24210, Rio de Janeiro-RJ, Brazil, Brazil

Costa, Nilaz Botteon
IRIS FOTO
R. Jacucaim 67
04563 Sao Paulo, SP, Brazil, Brazil

Costa, Paula Gasparini A.
Reporter Radio/Television
ABRIL TV
Av. Sao Gualter 1578
Alto de Pinheiros, Sao Paulo, Brazil, Brazil

Costa, Renata Affonso Marques Da
Reporter Radio/Television
SBT TV
Rua Visconde do Itamarati
20550, Rio de Janeiro-RJ, Brazil, Brazil

Costa, Sergio
Editor
GLOBO
Rua Bom Pastor, 89 C/20
Rio de Janeiro-RJ, Brazil, Brazil

Costa, Sergio Luiz Alves Da
Reporter Radio/Television
TV BANDEIRANTES
Rua Anibal Benevolo 185 c/3
Rio de Janeiro, Brazil, Brazil

Costa, Sergio Queiroz
Editor
DIA
P. da Bandeira, 461, I.do Gov.
Rio de Janeiro-RJ, Brazil, Brazil

Costa Filho, Joao
Editor
REVISTA VISAO
R.Alm.Alexandrino, 3780
20241 Rio de Janeiro-RJ, Brazil, Brazil

Couceiro, Leila
Editor
VEJA
Rua Dois de Dezembro 25, 202
Rio de Janeiro, Brazil, Brazil

Coutinho, Henrique Jose B.
Editor
GLOBO TV
Rua Von Martius 22 sala 323
Rio de Janeiro, Brazil, Brazil

Coutinho, Paulo Sergio Bento
Editor
GLOBO TV
Rua Von Martius 22 sala 32l
Rio de Janeiro, Brazil, Brazil

Coutinho, Vanda De C.
Editor
GLOBO TV
Rua Von Martius 22 sala 328
Rio de Janeiro, Brazil, Brazil

Couto, Luis A.
MANCHETE
Est. dos Bandeirantes 8325/107
Bloco 7 - Brazil, Brazil

Crim, Ofelia Alvarenga
Reporter
O DIA
R. Riachuela 359
Rio De Janeiro, RJ 2, Brazil
T: 55 21 272-8000 F: 55 21 507-1228

Crisostomo, Simone
Reporter Radio/Television
GLOBO RADIO
Rua Prof. Valadares, 119/202
Grajau, Rio de Janeiro-RJ, Brazil, Brazil

Cruz, Alberico De Souza
Editor
REDE GLOBO
R. Lopes Quintas 303
Rio de Janeiro -RJ 2, Brazil
T: 55 21 2947732 F: 55 21 2947247

Cruz, Cristina
Reporter Radio/Television
GLOBOSAT
Rua Redentor 279/201
Rio de Janeiro, Brazil, Brazil

Cruz, Mario Ribeiro
Reporter Radio/Television
MTV MUSIC TV
F. Francisco Sampaio 213
Sao Paulo, Brazil, Brazil

Cruz, Paulo Matos Da
Reporter Radio/Television
RADIO CONTEMPORANEA AM
Rua do Bispo, 323/404
Tijuca, RJ, Brazil, Brazil

Cruz, Sandra B.
Editor
GLOBO
SCS Ed.Oscar Niemeyer-13 andar
Brasilia-DF, Brazil, Brazil

Cruz, Simone Marques Da
Reporter Radio/Television
SBT TV
Rua pedro de Carvalho, 439/902
20721, Rio de Janeiro-RJ, Brazil, Brazil

Cruz E Sousa, Jose Da
Editor
JORNAL DO BRASILIA
SQN 106-Bloco C, Apt. 107
Brasilia, DF, Brazil, Brazil

Cruz Filho, Antonio C.B.
Reporter Radio/Television
GLOBO TV
Rua Von Martius, 22 sala 32l
Rio de Janeiro, Brazil, Brazil

Cunha, Angela
Editor
JORNAL DO BRASIL
R. Pereira da Silva, 622/201
22221, Rio de Janeiro-RJ, Brazil, Brazil

Cunha, Clarice Passos
Editor
GLOBO TV
Rua Vono Martius 22 sala 32l
Rio de Janeiro, Brazil, Brazil

Cunha, Gelcio Antonio
Editor
GLOBO RADIO
Ave. Marechal Rondon, 300
Rio de Janeiro-RJ, Brazil, Brazil

Cunha, Luiz Antonio
Editor
JORNAL O LIDER
Tr.Deoclides Almeida 230
Altamira, Paraiba, Brazil, Brazil

Cunha, Marcio Jose Parente
Editor
TV EDUCATIVA - RJ
R. das Laranjeiras, 280/202
Rio de janeiro, Brazil, Brazil

Cunha, Margareth B. De A.
Editor
GLOBO TV
Rua Von Martius 22 sala 32l
Rio de Janeiro, Brazil, Brazil

Curado, Maria Olga
Editor
GLOBO TV
Rua Von Martius 22 sala 32l
Rio de Janeiro, Brazil, Brazil

Curry, Andreia
Editor
JORNAL DO BRASIL
Rua Leopoldo Miguez, 19/102
22060, Rio de Janeiro-RJ, Brazil, Brazil

Cury, Adamir Marun, Jr.
Reporter Radio/Television
TELENEWS SERVICE LTDA.
R. Humaita 46/902
Botafogo, Rio de Janeiro, Brazil, Brazil

D'Angelo, Paulo
Editor
SAO PAULO ZONA SUL
Rua Massanduba 19
Tabaquara, Sao Paulo, Brazil, Brazil

D'Angelo, Wagner Luiz
Editor
JORNAL SAO PAULO ZONA SUL
Rua Morato Coelho, 1427
05417-Sao Paulo-SP, Brazil, Brazil

D'Avila, Cristiane
Editor
JORNAL DO COMMERCIO
R. Gomes Carneiro, 161/503
Ipanema, RJ, Brazil, Brazil

D'Avila Pizzolante, Maria Lucia
Editor
JORNAL PERSONA
SHIS QI 23, Cj.05 c/02
Brasilia, D.F., Brazil, Brazil

D'Oliveira, Moises
Reporter Radio/Television
GLOBO TV
Rua Macembu, 186
Rio de Janeiro-RJ, Brazil, Brazil

D'Orleans, Jannie
Editor
RIO EM FOCO
R.Visc. Piraja 306/202
Rio de Janeiro, Brazil, Brazil

Da Costa, Aylton F.
Editor
RADIO GUANABARA/BANDEIRANTES
Rua Belford Roxo, 284/402
Copacabana, Rio de Janeiro-RJ, Brazil, Brazil

Da Costa, Cassio
Reporter Radio/Television
MANCHETE TV
Av.Henrique Valadares,02/15
Centro, Rio de Janeiro-RJ, Brazil, Brazil

Da Cruz, Carmen Cleci
Correspondent
CORREIO BRAZILIENSE
QE 28 Conj.T C/10 GUARA II
Brasilia-DF, Brazil, Brazil

Da Cruz, Sergio Siqueira
Editor
HORA DO POVO
R. Rio Grande, 50 Apt. 83
Sao Paulo-SP, Brazil, Brazil

Da Cunha, Eurico Carvalho
Reporter Radio/Television
GLOBO TV
R. Prof. Ferreira da Rosa, 161
Rio de janeiro, Brazil, Brazil

Da Luz, Candido O.
Reporter Radio/Television
RADIO UNIVERSIDADE/UFSM
Rua Visconde de Pelotas,505
97.050 Sta.Maria, RS, Brazil, Brazil

Da Rocha, Jorge, Jr.
Reporter Radio/Television
GLOBO RADIO
Rua Sao Cristovao, 516
Rio de Janeiro-RJ, Brazil, Brazil

Da Rosa, Elbio
Editor
HORA DO POVO
Rua Voluntarios da Patria,3533
Santana,SP, Brazil, Brazil

Da Rosa, Marilia Mattos
Correspondent
IMAGEM
Rua Uruguai 205/204
Rio de Janeiro-RJ, Brazil, Brazil

Da Silva, Adriana B.
Editor
DIARIO COMERCIO IND. (DCI)
Rua Visconde Silva, 44/403
22281 Rio de Janeiro-RJ, Brazil, Brazil

Da Silva, Adriana Nunes
Reporter Radio/Television
GLOBO RADIO
Rua Cordovil, 1240
Rio de Janeiro-RJ, Brazil, Brazil

Da Silva, Artur Luiz Andrade
Correspondent
PANROTAS EDITORA LTDA.
Siqueira Campos 143/1203, B1
Copacabana, Rio de Janeiro, RJ, Brazil, Brazil

Da Silva, Clara Elisabeth Alves
Reporter Radio/Television
TRIBUNA DA IMPRENSA
Rua das Laranjeiras, 103/604
Rio de Janeiro, Brazil, Brazil

Da Silva, Claudio O.
Editor
ETHOS COMUNICACAO
Av. Jameiro, 175
Jacarepagua, Rio de Janeiro, Brazil, Brazil

Da Silva, Darcy M.
Editor
DIA
R. Dois de Fevereiro, 938
Rio de Janeiro-RJ, Brazil, Brazil

Da Silva, Joao C.
Reporter Radio/Television
MANCHETE TV
R.Carlos Alves de Oliveira1422
Brazil, Brazil

Da Silva, Joao Elias A.
Reporter Radio/Television
A.N.EDITORA/PRODUTORA
R.Almte. Alexandrino 1772-B
Rio de Janeiro, Brazil, Brazil

Da Silva, Joel
Editor
REVISTA DEFATO
Rua Marino Ruch, 66/206
205 Rio de Janeiro-RJ, Brazil, Brazil

Da Silva, Jorge Eduardo Marques
Editor
JORNAL A NOTICIA
R. Barata Ribeiro, 228/503
Rio de Janeiro, Brazil, Brazil

Da Silva, Jose Pereira
Correspondent
JORNAL DO COMERCIO
R.Taubate, 184
Trindade, Rio de Janeiro, Brazil, Brazil

Da Silva, Lilian Claudia M.
Editor
FERROVIARIA
Rua Americo Rocha,1542 Apt.102
Honorio Gurgel, Rio de Janeiro, Brazil, Brazil

Da Silva, Marcia V.
Editor
FLUMINENSE
Rua Barao do Amazonas, 02/1003
Centro, Niteroi, Rio de Janeiro, Brazil, Brazil

Da Silva, Maria Aparecida
Correspondent
JORNAL DE ESPANA
Rio de Janeiro, Brazil, Brazil

Da Silva, Marita B.
Editor
AGENCIA ESTADO
R. H.de Barros, 130/201
Sta. Tereza, Rio de Janeiro-RJ, Brazil, Brazil

Da Silva, Orlando
Editor
GAZETA DE NOTICIAS
R. Viscoude de Sepetibe 162
20280 Niteroi, Rio de Janeiro, Brazil, Brazil

Da Silva, Paulo Cesar
Correspondent
O PROGRESSO
Rua Luiz Domingues 20188
Imperatriz, Ma, Brazil, Brazil

Da Silva, Paulo O.
Editor
DIA
Av. 28 de Setembro, 264/510
Rio de Janeiro-RJ, Brazil, Brazil

Da Silva, Renato Ribeiro
Reporter Radio/Television
JORNAL DO BRASIL
Rua Garibaldi, 120/505
Tijuca, Rio de Janeiro-RJ, Brazil, Brazil

Da Silva, Ricardo Ferreira
Reporter Radio/Television
GLOBO RADIO
Rua Conselmeiro 10 B/M
Rio de Janeiro-RJ, Brazil, Brazil

Da Silva, Rogerio Alves
Editor
RADIOBRAS
QE 32 Bl. B Casa 33 Gr.II
Brasilia, D.F., Brazil, Brazil

Da Silva, Sergio V.
Editor
AGENCIA ESTADO
R. Iperoig, 527/131
05016 Perdizes, Sao Paulo-SP, Brazil, Brazil

Da Silva, Sonia
Correspondent
FOLHA DA MANHA
R.Pires de Almeida, 65/101
Laranjeiras, Rio da Janeiro-RJ, Brazil, Brazil

Da Silva, Tania Regina
GLOBO
Rua Antonio Dias, 48
Campo Grande, Rio de Janeiro-RJ, Brazil, Brazil

Da Silva, Teodomiro B.
Correspondent
JORNAL DO BRASIL
529 14th St., NW, Suite 1286
Washington, DC 20045, USA, Brazil

Da Silva, Vanderlei Novaes
Reporter Radio/Television
RADIO ROQUETTE PINTO
R.Andre Rocha, 372/903
Rio de Janeiro, Brazil, Brazil

Da Silva, Wagner Rodrigues
Reporter Radio/Television
MANCHETE TV
R.Comm.Tarque Horta, 42
Rio de Janeiro-RJ, Brazil, Brazil

Da Silva, Yara Eugenia
Editor
RECORD TV
Av. Santo Amaro, 1932, #06
V. Olimpia, Rio de Janeiro, Brazil, Brazil

Da Silva Amorim, Carlos
Editor
GLOBO TV
Rua Vopn Martius, 22 sala 328
Rio de Janeiro, Brazil, Brazil

Da Silva Gomes, Braz
Editor
GLOBO TV
Rua Von Martius, 22 sala 321
Rio de Janeiro, Brazil, Brazil

Da Silva Osorio, Paulo Mauricio
Editor
GLOBO TV
Rua Von Martius, 22 sala 321
Rio de Janeiro, Brazil, Brazil

Da Silva Passos, Maria H.
Correspondent
FOTOGRAMA
R. Iperoig 229
Soa Paulo, Brazil, Brazil

Da Silveira, Adriana O.
Editor
ESTADO DE MINAS
Av. do Contorno, 5048/101
Funcionario, Belo Horizonte-MG, Brazil, Brazil

Da Silveira, Walter
Reporter Radio/Television
RADIO GUAIBA
Rua dos Andradas, 221/501
Porto Alegre, RS, Brazil, Brazil

Da Veiga, Rita De Cassia Ferreira
Reporter Radio/Television
TV-EDUCATIVA
R.Mundo Novo,376/03b
Botafogo, Rio de Janeiro, Brazil, Brazil

Daher, Leila
Editor
RADIOBRAS
SQS 212 Bl. B Ap.201
Brasilia, D.F., Brazil, Brazil

Damato, Marcelo
Editor
FOLHA DE SAO PAULO
Al. Barao de Limeira, 425
Sao Paulo-SP, Brazil, Brazil

Dantas, Andrea
Editor
FOLHA DE SAO PAULO
Al. Barao de Limeira, 425
Sao Paulo-SP, Brazil, Brazil

Dantas, Edna
Editor
FOLHA DE SAO PAULO
Al. Barao de Limeira, 425
Sao Paulo-SP, Brazil, Brazil

Dantas, Jose Fernando Vianna
Reporter Radio/Television
TV-BANDEIRANTES-RIO
Estrada Velha da Pavuna, 4539
Rio de Janeiro-RJ, Brazil, Brazil

Dantas, Sergio
Editor
A NOTICIA
N. S. de Copacabana, 777/304
Rio de Janeiro, Brazil, Brazil

Dante, Gastaldoni
Editor
ETHOS COMUNICACAO
Rua Mapendi, 359
Jacarapagua, Rio de Janeiro, Brazil, Brazil

Danziger, Rolf B.
Reporter Radio/Television
MANCHETE TV
R.Voluntarios da Patria248/701
Rio de Janeiro-RJ, Brazil, Brazil

Das Chagas, Renata Monteiro
Reporter Radio/Television
BUZIOS TV
R.Ocidental 186/102
Rio de Janeiro, Brazil, Brazil

David, Lillian Ben
Editor
GAZETA MERCANTIL
R. Mj. Quedinho, 90 - 4 andar
Sao Paulo, Brazil, Brazil

De Abreu, Beatriz Nogueira
Editor
AGENCIA ESTADO
San 210 E 206
Brasilia, DF, Brazil, Brazil

De Abreu, Rosani S.
MANCHETE TV
R.Mario Portela, 161 Bl.A/1102
Brazil, Brazil

De Aguiar, Leonel Azevedo
Editor
REVISTA VISAO
Rua Alm.Alexandrino 3780
20241 Rio de Janeiro-RJ, Brazil, Brazil

De Almeida, Antonio Cesar
Correspondent
MANCHETE TV
Rua Dr. Bernardino 280
Rio de Janeiro, Brazil, Brazil

De Almeida, Cynthia
Editor
VEJA
Av. B.de Campos Gerais, 35
05684 Sao Paulo-SP, Brazil, Brazil

De Almeida, Ma.Lucia
Reporter Radio/Television
TV CULTURA-REPORTER ECO
R.Maria Helena, 81
02914 Pequeri, Sao Paulo-SP, Brazil, Brazil

De Almeida, Roberto
Reporter Radio/Television
TV EDUCATIVA - RJ
R. Barao de Cotegipe 139/206
Rio de Janeiro, Brazil, Brazil

De Almeida, Simone Fernandes Figueiredo
Reporter Radio/Television
GLOBO RADIO
Marica 74
Rio de Janeiro, Brazil, Brazil

De Almeida, Soraya
Correspondent
CIENCIA HOJE
Rua Alvaro Ramos, 235/102
22280 Rio de Janeiro-RJ, Brazil, Brazil

De Almeida, Sylvio Dalgado
Reporter Radio/Television
TV EDUCATIVA - RJ
Edgar Werneck 588
Jacarepagua, RJ, Brazil, Brazil

De Andrade, Miriam Lopes
Editor
JORNAL A NOTICIA
R. Senador Vergueiro, 38/409
Rio de Janeiro, Brazil, Brazil

De Araujo, Alexandre T. B. E.
Reporter Radio/Television
GLOBO RADIO
Rua Pedro Telles, 049/301
Rio de Janeiro-RJ, Brazil, Brazil

De Araujo, Amadeu Perreira
Reporter Radio/Television
RADIO CLUB DE PERNAMBUCO
R. Desembargador P. Andre 78
Tejipio-Recife-Pernambuco, Brazil, Brazil

De Araujo, Carlos A.
Reporter Radio/Television
MANCHETE TV
Praia do Botafogo 154/215
Rio de Janeiro-RJ, Brazil, Brazil

De Araujo, Hilda
Editor
JORNAL DA USP
Rua N.Sousa Queiroz, 759/137
04105 Sao Paulo-SP, Brazil, Brazil

De Araujo, Joao D.
Correspondent
AGENCIA ESTADO
Cond. Vivendas Paraiso,150
Conj.A Lote 1, Brasilia-DF, Brazil, Brazil

De Assis, Olga
Editor
HORA DO POVO
Rua Carolina Santos,119/201
Rio de Janeiro-RJ, Brazil, Brazil

De Azevedo, Ana
GLOBO
Rua Siqueira Campos,143
Bloco A/401, Brazil, Brazil

De Azevedo, Dirceu Ezequiel
Editor
JORNAL DO COMERCIO
Rua Feguereido de Magalhaes
Rio de Janeiro-RJ, Brazil, Brazil

De Barros, Armando Filho
Correspondent
TV EDUCATIVA
Rua Silva Vale 952, casa 01
Rio de Janeiro, Brazil, Brazil

De Barros, Celia Ribeiro
Reporter Radio/Television
JORNAL DO COMMERCIO
Av.M. Rondon, 1155,Bl.3/402
Rocha, Rio de Janeiro, Brazil, Brazil

De Barros, Teresa C.
Reporter Radio/Television
TV CULTURA-REPORTER ECO
R. Luisiania, 47 C/11
04560 Sao Paulo-SP, Brazil, Brazil

De Beaurepaire, Lucila
Editor
GLOBO
Rua Barao da Torre, 100/402
Ipanema, Rio de Janeiro-RJ, Brazil, Brazil

De Brito, Maria Eugenia
Reporter Radio/Television
GLOBO TV
Ave. Alvorada, 2600
Tijuca, Rio de Janeiro-RJ, Brazil, Brazil

De Camargo, Roberto Penteado F.
Editor
FOLHA DA EMBRAPA
ML 12 Norte Lago Norte
71500 Brasilia, Df, Brazil, Brazil

De Carli, Irineu
RADIO TV BANDEIRANTES
R. Ricardo Lonadelli 32
Brazil

De Carvalho, Felix
Editor
AGENCIA ESTADO
R. Amalfi, 26
Lins, Rio de Janeiro-RJ, Brazil, Brazil

De Carvalho, Roberto Barros
Correspondent
CIENCIA HOJE
Rua Nicaragua 56 Apt.106 S10N
30330 Belo Horizonte, MG, Brazil, Brazil

De Carvalho Amazonas, Marcio
Correspondent
TV BUZIOS
R.Joao Lira 166/101
Rio de Janeiro, Brazil, Brazil

De Faria, Antonio C.
Correspondent
FOLHA DA MANHA
R.Senador Vergueiro, 170/1407
Flamengo, Rio de Janeiro-RJ, Brazil, Brazil

De Faria, Claude Bornel
Correspondent
MONITOR MERCANTIL
Rua Camerino, 26 Apt.202
Centro, Rio de Janeiro-RJ, Brazil, Brazil

De Farias, Wellington A.
Correspondent
SISTEMA CORREIO DE COMUNICACAO
Ed. M. Pires #901, P. S. de
Lucena N. 205 - J. Pessoa, PA, Brazil, Brazil

De Freitas, Claudia Queiroz
Editor
JORNAL DE HOJE
R.Maria Fernandes 50 N. Iguacu
Rio de Janeiro, Brazil, Brazil

De Freitas, Dylza Goncalves
Reporter Radio/Television
RADIO EL DORADO
Conde dee Bonfin 615/608
Tijuca, RJ, Brazil, Brazil

De Freitas, Nivaldo
Editor
INTERFACE
QSA 01 LTS 14/15 Apt.305
Taguatinga,Brasilia-DF, Brazil, Brazil

De Gabriel, Joao
Editor
VEJA
R.Prof. Martins, 59
04513 Sao Paulo-SP, Brazil, Brazil

De Godoy, Vanessa
Editor
ESTADO DE SAO PAULO
R. Madre Rita Amada de Jesus79
Granja Julieta, Sao Paulo, Brazil, Brazil

De Goes, Denise
Editor
AGENCIA ESTADO
R. Heitor Penteado, 1577
Sao Paulo-SP, Brazil, Brazil

De Jesus, Regina
Editor
RADIO ALVORADA FM
R. Teixeira Jr., 383/302
S. Cristovao, Rio de Janeiro-RJ, Brazil, Brazil

De Lima, Gilberto Teixeira
Correspondent
JORNAL 'S O S TERRA'
R. Monte Libano 67
Teresopolis, Brazil

De Lima, Paulo C.
Reporter Radio/Television
MANCHETE TV
Praca Euvaldo Lodi, 15/204
Brazil, Brazil

De Lima, Ronaldo D.
Correspondent
FOLHA DA MANHA
R.Mario Portela, 161/1003 Bl.B
Laranjeiras, Rio de Janeiro-RJ, Brazil, Brazil

De Luccia, Marina F.
Editor
RADIO ELDORADO FM
R.Princesa Isabel 1252 Apt.131
Brazil

De Lucena, Rodolfo
Editor
FOLHA DE SAO PAULO
Al. Barao de Limeira, 425
Sao Paulo-SP, Brazil, Brazil

De Matos, Ronize Aline
Correspondent
JORNAL DOS SPORTS
Rua Alvaro Ramos, 405 Bl.1
Rio de Janeiro-RJ, Brazil, Brazil

De Mello, Kathia L.
Reporter Radio/Television
MANCHETE TV
R.Pres.Pedreira, 38/301 Bl.A
Inga, Niterio-RJ, Brazil, Brazil

De Melo, Paulo F. Queiros
Reporter Radio/Television
JORNAL FLUMINENSE
Jean Valentin de M. 364-a
Varza da Mocas, Niteroi, RJ, Brazil, Brazil

De Menezes, Mauricio M.
Editor
AGENCIA ESTADO
R. Souza Cruz, 128/503
Tijuca, Rio de Janeiro-RJ, Brazil, Brazil

De Menezes Telles, Katia A.
Editor
GLOBO TV
Rua Von Martius, 22
Rio de Janeiro, Brazil, Brazil

De Menezez Rodrigues, Dario J.
Editor
GLOBO TV
Rua Von Martius, 22 sala 328
Rio de Janeiro, Brazil, Brazil

De Miranda, Evaristo
Editor
AGENCIA ESTADO
R.Ant.Augusto de Almeida,1178
13083 Campinas, Sao Paulo, Brazil, Brazil

De Moraes, Luciano
Editor
GLOBO TV
Rua Von Martius, 22
Rio de Janeiro, Brazil, Brazil

De Moraes, Marcelo Caetano
Editor
O GLOBO
R. Cosme Velho, 315/201 Bl. 2
Rio de Janeiro, Brazil, Brazil

De Oliveira, Adriana Porto
Reporter Radio/Television
RADIO FEDERAL
Rua Joaquim Nabuco 98/c-01
Rio de Janeiro, Brazil, Brazil

De Oliveira, Castro Carneiro Teresa
Reporter Radio/Television
MANCHETE TV
SQN 105, Bloco J, apto. 601
Brasilia, DF, Brazil, Brazil

De Oliveira, Claudio
Editor
GLOBO
R. Real Grandeza, No. 59
Botafogo, Rio de Janeiro-RJ, Brazil, Brazil

De Oliveira, Eucimar
Editor
JORNAL DO BRASIL
Rua Pinheiro Guimaraes, 77
22281, Rio de Janeiro-RJ, Brazil, Brazil

De Oliveira, Fabiola
Editor
INST. NACIONAL DE PESQUISAS ESPACIAIS
B. de Oliveira e Silva 148
12213 S. J. dos Campos, S. Paulo Brazil, Brazil

De Oliveira, Helen R.
Reporter Radio/Television
TV CULTURA-REPORTER ECO
R. Paula Ney, 137
04107 Sao Paulo-SP, Brazil, Brazil

De Oliveira, Neper Dimas
Editor
LAMOUNIER CINE VIDEO
R. Prof. Gabizo 210/601
Tijuca, Rio de Janeiro, Brazil, Brazil

De Oliveira, Nielmar
Reporter Radio/Television
RADIOBRAS
Trav.Guimaraes Natal 9/303
Rio de Janeiro-RJ, Brazil, Brazil

De Oliveira, Ribarmar
Editor
AGENCIA ESTADO
SHIN QI 03 Conj.12 Casa 20
Brasilia-DF, Brazil, Brazil

De Oliveira, Roberto
RADIO TV BANDEIRANTES
Rua Radiante 13, Morumbi
Sao Paulo, Brazil, Brazil

De Oliveira Alencar, Auri
Editor
GLOBO TV
Rua Von Martius, 22
Rio de Janeiro, Brazil, Brazil

De Oliveira Filho, Manuel
Correspondent
MONITOR MERCANTIL
Av.28 de Setembro, 177/703
Vila Isabel, Rio de Janeiro-RJ, Brazil, Brazil

De Oliveira Garcia P., Jose Marcio
Editor
GLOBO TV
Rua Von Martius, 22 sala 328
Rio de Janeiro, Brazil, Brazil

De Oliveira Goncalves, Eermando
Editor
NOTICIAS EM DESTAQUE
Senador Vergueiro 56, Apt.103
22230 Rio de Janeiro, Brazil, Brazil

De Oliveira Jr., Oswaldo
Editor
POVO
Rua Sao Gabriel, 1110/Apt.201
20.781-Rio de Janeiro-RJ, Brazil, Brazil

De Oliveira Ramos, Marcos
Reporter Radio/Television
MANCHETE RADIO
R. Guimaraes Rosa 143/1303
Rio de Janeiro, Brazil, Brazil

De Oliveira Rodrigues, Marcelo
Editor
GLOBO TV
Rua Von Martius, 22
Rio de Janeiro, Brazil, Brazil

De Oliveira Santos, Eduardo
Editor
GLOBO TV
Rua Von Martius, 22 sala 321
Rio de Janeiro, Brazil, Brazil

De Paula, Jana
Editor
JORNAL DO COMMERCIO
R. Alfredo Chaves, 39
Botafogo, RJ, Brazil, Brazil

De Sa, Francisco
Editor
FOLHA DE SAO PAULO
Al. Barao de Limeira, 425
Sao Paulo-SP, Brazil, Brazil

De Sa, Jose Jaime Matos
REVISTA GLOBO RURAL
Rua do Curtume, 665/5 andar
05065 Sao Paulo-SP, Brazil, Brazil

De Sa, Jose Ribamar Coelho
Editor
ECOBRASIL
Rua Conde de Bonfim, 460/402
Rio de Janeiro-RJ, Brazil, Brazil

De Salles, Luzia Elisa
EDITORA AZUL/REVISTA CONTIGO
Rua Angelica Motta, 440/201
Olaria, Rio de Janeiro-RJ, Brazil, Brazil

De Santana, Marcos Valerio Goncalves
Editor
MANCHETE TV
R.Renato Rebechi, 470
Bangu, Rio de Janeiro,RJ, Brazil, Brazil

De Siqueira, Carla V.
ECABE PRESS
Rua Ipanema, 197 Apt.1504
22631 Rio de Janeiro-RJ, Brazil, Brazil

De Souza, Andreia Salles
Editor
JORNAL A NOTICIA
R. Pe. Achotegui, 25/102
Rio de Janeiro, Brazil, Brazil

De Souza, Ciro Dias
Editor
VOZ DE PORTUGAL
R.Combu, 50/302
Rio de Janeiro, Brazil, Brazil

De Souza, Elaine
Editor
JORNAL DE HOJE
R. Ipameri,105 N. America
Rio de Janeiro, Brazil, Brazil

De Souza, Eliana
Editor
JORNAL PLANTAS MEDICINAIS
R.Riachuelo 221 COB 03
20030 Rio de Janeiro-RJ, Brazil, Brazil

De Souza, Francisco
Editor
RADIOBRAS
SQS 108 Bl.H Apt.506
Brasilia-DF, Brazil, Brazil

De Souza, Jorge Carlos Santos
Editor
OS CAMINHOS DA TERRA
Av. Nacoes Unidas, 5777
Sao Paulo, Brazil, Brazil

De Souza, Jorge M.
Reporter Radio/Television
RADIO ELDORADO FM
R.Azevedo Jr, 185 Bl.2 Apt.101
Bras, Sao Paulo-SP, Brazil, Brazil

De Souza, Juracy Rabelo
Reporter Radio/Television
TV CULTURA DO PARA
Av.Augusto Monte Negro, b3/l05
Belem, Para, Brazil, Brazil

De Souza, Lia
Editor
TV CULTURA-REPORTER ECO
R.Cesar Lacerda Vergeiro, 373
05435 Sao Paulo-SP, Brazil, Brazil

De Souza, Luiz Carlos
Correspondent
DIARIO COMERCIO IND. (DCI)
Rua Babilonia, 22
20511 Rio de Janeiro-RJ, Brazil, Brazil

De Souza, Maria Francisca Alves
Editor
ECO NOTICIAS
Av.Henrique Dumont 126/102
Rio de Janeiro, Brazil, Brazil

De Souza, Marise Rodrigues
Reporter Radio/Television
RADIO CLUBE DE PERNAMBUCO
Rua Sta.Maria da Boa, 162
Artun Lundgren II, Paulista, PE,Brazil, Brazil

De Souza, Patricia Costa Coelho
Editor
HORA DO POVO
R. Magdalena, 124
Villa Madalena, Sao Paulo, Brazil, Brazil

De Souza, Reis
Editor
AGENCIAS DE NOTICIAS BRASILIA
Centro Comercial Cruzeiro,4
70.640 Brasilia-DF, Brazil, Brazil

De Vives, Vera
Editor
FLUMINENSE
Rua Prudente de Moraes,462/402
Ipanema, Rio de Janeiro-RJ, Brazil, Brazil

Degrazia S., Linda Cristina
Correspondent
DIARIO DA MANHA
Luciana de Abreu 74/90570
Moinhos de Vento, Porto Alegre, Brazil, Brazil

Dehrends, Herbert Dilson
Editor
REDE GLOBO
R. Lopes Quintas 303
Rio de Janeiro -RJ 2, Brazil
T: 55 21 2947732

Delmanto, Renato
Editor
JORNAL DA TARDE
R. Tanabi 233 Apt.06
05002 Sao Paulo-SP, Brazil, Brazil

Deus, Corban Giovannini
Reporter Radio/Television
RADIOBRAS
SQS 104 Bl. C Ap.306
Brasilia, D.F., Brazil, Brazil

Dias, Alberto Carlos Osorio
Correspondent
POLICIA EM ACAO REVISTA
Rua Senador Dantas, 117/1611
20034 Rio de Janeiro-RJ, Brazil, Brazil

Dias, Ariadne G.
Editor
FLUMINENSE
Rua Conde de Bonfim, 119/601
Rio de Janeiro-RJ, Brazil, Brazil

Dias, Aurelio Gomes
Producer / Editor
EBU - EUROPEAN BROAD. UNION
Rua Clodomird Amazonas, 960/171 B 04537-002 SP
Sao Paulo, Brazil
T: 005511 820 0908 F: 005511 820 4150

Dias, Jose Luiz Fernandes
Editor
RADIO IMPRENSA
R. Atila de S. Peixoto, 192
Realengo, Rio de Janeiro, Brazil, Brazil

Dias, Juliana De P.
Editor
ECOLOGIA E DESENVOLVIMENTO
R. Prof. Gabizo 30/202
Tijuca, Rio de Janeiro-RJ, Brazil, Brazil

Dias, Luciano Sebastiao Rodrigues
Editor
EDITORA IMAGEM
Rua Mateus Silva 259
Rio de Janeiro, Brazil, Brazil

Dias, Mauricio
Editor
ISTO E SENHOR
Rua Efiginio Sales,100/302
22241 Cosme Velho, Brazil, Brazil

Dias, Mauro
Editor
AGENCIA ESTADO
R. Paulo Alves 110 Bl.A
Inga, Niteroi, Rio de Janeiro, Brazil, Brazil

Dias, Rogerio Ferreira
Reporter Radio/Television
GLOBO RADIO
Ladeira do Tabajara, 301/101
Rio de Janeiro, Brazil, Brazil

Diaz, Juan Jose Modesto
Correspondent
EDITORA BRASIL ENERGIA
Praca Barao de Taquara
Rio de Janeiro, Brazil, Brazil

Didini, Rubem
Editor
TV RIO
Av. Canal de Marapendi, 2800
#310, Barra da Tijuca, RJ, Brazil, Brazil

Didonet, Marcos, Mr.
Editor
BUZIOS TV
Rua Edson Passo, 1481/201
Rio de Janeiro-RJ, Brazil, Brazil

Diegoli, Vera Lucia
Editor
TV CULTURA-REPORTER ECO
R.Rafael Correia, 16-V
05043 Romana-SP, Brazil, Brazil

Diniz, Henrique Augusto De O.
Editor
A NOTICIA
Gal. Goes Monteiro, 8 Bl.G/601
Rio de Janeiro, Brazil, Brazil

Diniz, Luis Eduardo Garcia
Editor
JORNAL DO BRASIL-RADIO AM
Rua Nascimento E Silva
Ipanema, Rio de Janeiro-RJ, Brazil, Brazil

Diocleclo, Luz
Editor
FOLHA DO MEIO AMBIENTE
Caixa Postal 08582
70312 Brasilia, DF, Brazil, Brazil

Diogo, Maria Da Gloria
Editor
VIDEO CLIPPING PRODUCOES LTDA.
R. M.Portelo 161-Bl.B,Apt.2003
22241, Rio de Janeiro-RJ, Brazil, Brazil

Do Amaral, Sergio P.
GLOBO
Rua Conde de Bonfim, 517
Rio de Janeiro-RJ, Brazil, Brazil

Do Carmo, Antonio Jose
Correspondent
AGENCIA ESTADO
Av. Bandeirantes, 2895
Sao Paulo-SP, Brazil, Brazil

Do Nascimento, Argemiro Do Carmo
Editor
ECOBRASIL
Rua Visconde de Niterio,298
Rio de Janeiro-RJ, Brazil, Brazil

Do Nascimento, Jorge Batista
RADIO FEDERAL
Rua das Laranjeiras 478/801
Rio de Janeiro, Brazil, Brazil

Do Nascimento, Josinaldo
Reporter Radio/Television
MANCHETE TV
Est.do Camboata,3083 R.B C/26
Brazil, Brazil

Do Nascimento, Maria A.
Editor
TV GAZETA-CANAL 11
R.Alm.Marques de Leao, 638/141
Bl.A 01330 Sao Paulo-SP, Brazil, Brazil

Doimo, Morgana C.
Reporter Radio/Television
MANCHETE TV
R.Zamenhof, 62 Apt.901
Rio de Janeiro-RJ, Brazil, Brazil

Domingues, Daria Francisca Brum
Reporter Radio/Television
SUPER RADIO TUPI
Rua Alice 51/101
Laranjeiras, Rio de Janeiro-RJ, Brazil, Brazil

Domingues, Fernanda
Editor
GLOBO
Conselheipo Brotero, 1378
Sao Paulo-SP, Brazil, Brazil

Domingues, Paulo Dunlop
Reporter Radio/Television
TV EDUCATIVA - RJ
R. Imbuia, 180 B. da Tijuca
Rio de Janeiro, Brazil, Brazil

Donato, Ana Maria
Editora
PANROTAS EDITORA
Av.Jabaquara, 1761
Sao Paulo - SP, Brazil
T: 55 11 2750211 F: 55 11 2761602

Donegana, Costanzo
Redator Revista
REVISTA CIDADE NOVA - MOVIMENTO DOS FOCOLARES
Via Bandeirantes, Km47 Cx.Postal 30
Varre-Sai - RJ, Brazil
T: 55 249 7900638 F: 55 249 7900017

Dorea Filho, Gomercindo Rocha
Correspondent
AGENCIA DE NOTICIAS ACOPI
R. Eca de Queiroz, 720/105
V. Mariana, Sao Paulo-SP, Brazil, Brazil

Dorneles, Ana
Editor
MANCHETE TV
Av.Bartolomeu Mitre 1079/205
Rio de Janeiro, Brazil, Brazil

Dos Reis, Sebastiao P.
Editor
AGENCIA ESTADO
R. Engenheiro Richard,273
Grajau, Rio de Janeiro, Brazil, Brazil

Dos Santos, Alfredo Rodriguez
Editor
FOLHA CRISTA
Rua Petrolina, 72
Honorio Gurgel, Rio de Janeiro, Brazil, Brazil

Dos Santos, Carlos A.
Editor
RADIO ALVORADA FM
R. Aquidabam, 874/604
Brazil, Brazil

Dos Santos, Francisca M.
Reporter Radio/Television
TV BUZIOS
Av. Bento Ribeiro Dantas,265
Cabo Frio, Rio de Janeiro, Brazil, Brazil

Dos Santos, Jose Gabriel
Correspondent
ESTADO DE MINAS
R. Goias, 36
Belo Horizonte, Brazil, Brazil

Dos Santos, Juan Luiz Louzada
Reporter Radio/Television
TV EDUCATIVA
Rua Daniel Carneiro 17/301
Rio de Janeiro, Brazil, Brazil

Dos Santos, Rosangela H.
Editor
AGENCIA ESTADO
R. Prof. Clark 413/201
Vila da Penha, Rio de Janeiro-RJ, Brazil, Brazil

Dos Santos, Sandro Roberto
Editor
A NOTICIA
R.Barao de Itapagipe, 401Bl.B
Rio de Janeiro-RJ, Brazil, Brazil

Dos Santos Freitas, Itamar
Reporter Radio/Television
MANCHETE RADIO
R. M. Joao Alberto 445/101
Rio de Janeiro, Brazil, Brazil

Doti, Andrea E.
Editor
PREFEITURA RIO DE JANEIRO
R.Desemb.Izidro 129/302
Rio de Janeiro, Brazil, Brazil

Dourado, Leonardo B.
Editor
TELENEWS SERVICE
Av. Sernambetiba, 3300/1505
Rio de Janeiro-RJ, Brazil, Brazil

Duarte, Edson G.
Reporter Radio/Television
RADIO E JORNAL DE JUAZEIRO
Ave. Flaviano Guimaraes, 37
Juazeiro-Bahia, Brazil, Brazil

Duarte, Ibere Baena
Editor
FM RAIZES
Rua Olavo Bilac, Capivari
Sao Paulo, Brazil, Brazil

Duarte, Jose Luis L.
Editor
A NOTICIA
Rua Jorge Rudge, 29/105
Vila Isabel, Rio de Janeiro-RJ, Brazil, Brazil

Duarte, Luiz A.
Editor
ZERO HORA
SQS 103 Bloco D, Apt. 108
Brasilia-DF, Brazil, Brazil

Duarte, Roberto
Reporter Radio/Television
TV BUZIOS
R.Barao de Mesquita,1025/404
Rio de Janeiro,R.J.,Brazil, Brazil

Duarte, Simone
Editor
GLOBO TV
Rua Pinheiro Machado, 69/401
Laranjeiras, Rio de Janeiro-RJ, Brazil, Brazil

Duarte, Solange
Editor
GLOBO
Rua Magnolia Brasil Travessa
Niteroi, Rio de Janeiro-RJ, Brazil, Brazil

Duchesne, Andrea
Reporter
JORNAL DO COMMERCIO
R.Pacheco Leao,704, c 6
Jardim Botanico, RJ, Brazil, Brazil

Duffrayer, Luciano
Editor
ECORIO
R. Pedra Branca, 130
Taquara, RJ, Brazil, Brazil

Dunningham, Andrea
Editor
GLOBO
Rua Nascimento Silva, 350
Cobertura, Rio de Janeiro-RJ, Brazil, Brazil

Duque, Andrea Paula O.
Correspondent
JORNAL INTEGRACAO
Trav.Soledada 25/416
Rio de Janeiro, Brazil, Brazil

Duque, Jose Ronaldo Lopes
Correspondent
PROVIDEO CINEMA E TV
SHIS QI 29 Conj.6 c/7
71600 Brasilia-DF, Brazil, Brazil

Durao, Vera Ligia H.N. Saavedra
Editor
GAZETA MERCANTIL
R. Mj. Quedinho, 90 - 4 andar
Sao Paulo, Brazil, Brazil

Dutra, Gisele
Reporter Radio/Television
RADIOBRAS
SQN 303 Bl. A Ap.504
Brasilia, D.F., Brazil, Brazil

Dutra, Isabela R.
Editor
GLOBO TV
Rua Von Martius 22 sala 32l
Rio de Janeiro, Brazil, Brazil

Dutra, M. Ligia
Reporter Radio/Television
MANCHETE TV
R. Piratininga, 16/102
Rio de Janeiro, Brazil, Brazil

Dutta-Roy, Amitava
Correspondent
RNT
Rockefeller Center
New York, NY 10185, USA, Brazil

Eich, Nuri V.
Correspondent
FOLHA DA MANHA
R.Felicio dos Santos 14/301
Sta.Tereza, Rio de Janeiro-RJ, Brazil, Brazil

Eleuterio, Regina M.
GLOBO
Rua Maxwell, 400, Apto. 303
Rio de Janeiro-RJ, Brazil, Brazil

Eller, Osmerio
Reporter Radio/Television
TV EDUCATIVA
Alameda Salano Salles,335
Rio de Janeiro-RJ, Brazil, Brazil

Elliot, Stela Marina
Editor
PREFEITURA RIO DE JANEIRO
R.Barao Guaratiba 71/301
Rio de Janeiro, Brazil, Brazil

Emolo, Francisco
Editor
JORNAL DA USP
Rua Vitoria, 9
Jardim Julieta, Itapevi, SP, Brazil, Brazil

Eosta, Erancisco O.
Editor
AGENCIA ESTADO
E. Morais e Silva 97/104
Tijuca, Rio de Janeiro-RJ, Brazil, Brazil

Erlanger, Luis
Editor
GLOBO
Rua Irineu Marinho, 35
Rio de Janeiro-RJ, Brazil, Brazil

Escobar, Otavio G.
Editor
PROVIDEO
Av. Delfin Moreira, 514/410
Rio de Janeiro-RJ, Brazil, Brazil

Espinola, Marcio Raimundo
Correspondent
MANCHETE TV
A. P. F. Moreira, 46/701
Rio de Janeiro, Brazil, Brazil

Esteves, Jaine R.
Editor
PREFEITURA RIO DE JANEIRO
R.Aristides Lobo 237302
Rio de Janeiro, Brazil, Brazil

Esteves, Nelma M.
Reporter Radio/Television
MANCHETE TV
Rua Morro do Pinheiro, 66
Brazil

Estrada, Ma. Ignez
Editor
CIENCIA HOJE
Rua Getulio das Neves 16 c/6
22461 Rio de Janeiro-RJ, Brazil, Brazil

Euloterio, Marcia Manga
Reporter Radio/Television
RADIO ROQUETE PINTO
R. B. Constant, 82/701
Gloria, Rio de Janeiro, Brazil, Brazil

Fadul, Sergio Luis
Editor
JORNAL DO COMMERCIO
R. Sambaiba, 613/202
Leblon, RJ, Brazil CEP 22450, Brazil

Faerman, Marcos
Correspondent
JORNAL DA TARDE
R. Tacito de Almeida 119
Sao Paulo-SP, Brazil, Brazil

Faga, Marcelo J.W.
Editor
ESTADO DE SAO PAULO
Alameda Jau, 600/102
Sao Paulo-SP, Brazil, Brazil

Fajardo, Elias
ECOLOGIA E DESENVOLVIMENTO
R. das Laranjeiras 531/66A
Rio de Janeiro-RJ, Brazil, Brazil

Falcao, Catia Gama
Reporter
REDE BANDEIRANTES DE TELEVISAO
R.Alvaro Ramos, 492
Rio de Janeiro - RJ, Brazil
T: 55 21 2122072 F: 55 21 2957391

Falcao, Joaquim
Correspondent
FUNDACAO ROBERTO MARINHO
Rua Frederico Eyer, 93
Gavea, Rio de Janeiro-RJ, Brazil, Brazil

Falcao, Raimundo
Editor
AGENCIA ESTADO
R. Eduardo Nadruz, 216
Ilha do Governador, RJ-RJ, Brazil, Brazil

Falceta Jr., Walter
Editor
ESTADO DE SAO PAULO
R. Florentinos 48
Vila Sta. Izabel, Brazil, Brazil

Falco Ortiz, Pedro Henrique
Correspondent
AGENCIA DE NOTICIAS ACOPI
Al. dos Guainumbis, 549
04067 Sao Paulo-SP, Brazil, Brazil

Farah, Sergio Danilo
Editor
GAZETA MERCANTIL
Rua Major Quedinho, 90 - 4 andar
Sao Paulo, Brazil
T: 55 11 256 3133

Faria, Patricia
Editor
GLOBO
Rua Aires Sadanha, 72/306
Copacabana, Rio de Janeiro-RJ, Brazil, Brazil

Farias, Ma.De Lourdes C.
Correspondent
POVO
Rua das Laranjeiras, 95/501
22.240-Rio de Janeiro-RJ, Brazil, Brazil

Farias, Paulo Cesar
Correspondent
JORNAL PATROPI-ZONA OESTE
R.Julio Reis 56 - Bangu
Rio de Janeiro, Brazil, Brazil

Farias, Sebastiao C.
Editor
SUNDAY NEWS
R.Pardal 279
Osasco, Sao Paulo, Brazil, Brazil

Farruggia, Alessandro
Editor
POLIPRESS
Via Donatello 67/d - 00196
Roma, Italy, Brazil

Faustino, Jose
Reporter Radio/Television
GLOBO RADIO
R. B. Constant, 10/211
Rio de Janeiro, Brazil, Brazil

Favaro, Julio Ricardo
Editor
SBT TV
R. Adolfo Gordo 166
Apt. 21, Campos Eliseos, Brazil, Brazil

Favilla De Oliveira Lima, Clara Lucia
Editor
RADIOBRAS
SHIS QI 27 Conj. 07 C.07
Brasilia, D.F., Brazil, Brazil

Feital, Irapuan Portugal
Reporter Radio/Television
TV EDUCATIVA - RJ
R. Sorocaba, 518/203
Rio de Janeiro, Brazil, Brazil

Feldman, Eliane I.
Editor
MANCHETE TV
Rua Sao Clemente, 272/201 B1
Brazil, Brazil

Felice, Douglas Guarino De
Reporter Radio/Television
JORNAL DO BRASIL-RADIO AM
Rua Edgard Werneck, 131
Freguesia, Rio de Janeiro-RJ, Brazil, Brazil

Felippe, Maria Bernardete Pinzon
Reporter Radio/Television
RADIO ROQUETE PINTO
Av. N.S. Copacabanam 664/902
Rio de Janeiro, Brazil, Brazil

Feres, Roberto Pereira
Editor
SUPER RADIO TUPI
Rua Martins Ribeiro, 35
Flamengo, Rio de Janeiro-RJ, Brazil, Brazil

Fernandes, Adriana Campos Gomes
Editor
RADIOBRAS
SHIS QI 27 Conj.07 C. 9
Brasilia, D.F., Brazil, Brazil

Fernandes, Claudia Maria
Correspondent
MONITOR MERCANTIL
Rua Getulio 114/201, Meier
Rio de Janeiro, Brazil, Brazil

Fernandes, Cristina Da S.
Reporter Radio/Television
RADIOBRAS
Rua Vistula 156
Ilha Governador, Rio de Janeiro, Brasi, Brazil

Fernandes, Gleise S.C.
Editor
ESTADO DE SAO PAULO
R.Prof.Romilde Nogueira Sa 854
Imirim, Sao Paulo Brazil, Brazil

Fernandes, Laerte
Editor
AGENCIA ESTADO
S. Vicente de Paulo 501/506
Rio de Janeiro, Brazil, Brazil

Fernandes, Luiz Marcos
Editor
FOLHA DO TURISMO
R.Dezesseis de Marco 17/803
Petropolis, Rio de Janeiro, Brazil, Brazil

Fernandes, Ma. Cristina
Editor
VEJA
R. M. Francisco, 334 Apt.1204
01226 Sao Paulo-SP, Brazil, Brazil

Fernandes, Marcelo Luiz R.
Reporter Radio/Television
GLOBO TV
Rua Von Martius 22 sala 328
Rio de Janeiro, Brazil, Brazil

Fernandes, Nelson
Editor
GLOBO TV
Ave. Alvorada, 2600
Tijuca, Rio de Janeiro-RJ, Brazil, Brazil

Fernandes, Rita De Cassia. A.
Editor
GLOBO TV
Rua Von Martius 22 sala 328
Rio de Janeiro, Brazil, Brazil

Fernandes, Rodolfo
Editor
GLOBO
SCS Ed.Oscar Niemeyer-13 andar
Brasilia-DF, Brazil, Brazil

Fernandes, Simone
Reporter Radio/Television
GLOBO RADIO
Rua Marca, 74
Rio de Janeiro-RJ, Brazil, Brazil

Fernandes Filho, Helio
Editor
JORNAL TRIBUNA DA IMPRENSA
Rua do Lavradio, 98
20230 Rio de Janeiro-RJ, Brazil, Brazil

Fernandes G., Julio Cesar
Reporter Radio/Television
TV EDUCATICA
Av. P. Roosevelt 1021
Vista Alegre, SG, Brazil

Fernandes Jr., David
Correspondent
REVISTA QUESTAO AMAZONICA
Trav.9 Janeiroooo 2332
Belem, Para, Brazil, Brazil

Fernandez Y Garcia, Hector
Correspondent
NOSSA VOZ
R. Copesquintas, 390/104
Rio de Janeiro-RJ, Brazil, Brazil

Ferrari, Cynthia Regina
Correspondent
RADIO & TV BANDEIRANTES
R. Siqueira Campos 945/82
Santo Andre, Sao Paulo, Brazil, Brazil

Ferrari, Jose
Editor
PROGRAMA GLOBO ECOLOGIA
Av. Paulo de Frontim, 568
Rio Comprido-RJ, Brazil, Brazil

Ferrari, Luis Carlos
Editor
GLOBO
Rua das Laranjeiras, 531
Rio de Janeiro-RJ, Brazil, Brazil

Ferraz, Ivone
MANCHETE TV
Av. Rodrigo Otavio, 226/104
Gavea, Rio de Janeiro-RJ, Brazil, Brazil

Ferraz, Maria Emilia Figueira
Reporter Radio/Television
TV EDUCATIVA - RJ
R. D. Mariana, 29/103
Rio de Janeiro, Brazil, Brazil

Ferraz, Patricia
Editor
JORNAL DA TARDE
R. Ame. Alves Pereira Filho 57
Apt.22, Morumbi, Sao Paulo-SP, Brazil, Brazil

Ferraz, Paulo
Editor
FOLHA DE SAO PAULO
Al. Barao de Limeira, 425
Sao Paulo-SP, Brazil, Brazil

Ferraz Tavares, Ana M.
Editor
ESTADO DE SAO PAULO
R.Claudino Inacio Joaquin 81
Tucuruvi, Sao Paulo Brazil, Brazil

Ferreira, Antonio C.
Reporter Radio/Television
MANCHETE TV
Alameda 28, Ticos No.133
Pavuna, Rio de Janeiro-RJ, Brazil, Brazil

Ferreira, Argemiro Humberto
Editor
TV EDUCATIVA - RJ
Av. Gomes Freire 474 Centro
Rio de Janeiro -RJ, Brazil
T: 55 21 2426736 F: 55 21 2426740

Ferreira, Cleusa Maria
Editor
JORNAL DO BRASIL
R. Miguel Pereira, 59/301
Rio de Janeiro-RJ, Brazil, Brazil

Ferreira, Edna
Editor
VIDEO CIENCIA PRODUCOES LTDA
Est.Velha da Pavuna, 4800
20761 Inhauma, Rio de Janeiro-Brazil, Brazil

Ferreira, Edson Luis
Editor
GLOBO
SCS Ed.Oscar Niemeyer-13 andar
Brasilia-DF, Brazil, Brazil

Ferreira, Eduardo Band
Correspondent
CAMERA PRODUCTIONS
R. Conde Afonso Celso 131
Rio de Janeiro, Brazil, Brazil

Ferreira, Maria Teresa Carneiro
Reporter Radio/Television
GLOBO RADIO
Conde de Bonfim, 171
Tijuca, Rio de Janeiro-RJ, Brazil, Brazil

Ferreira, Marines Da Fonseca
Correspondent
FOLHA DO MEIO AMBIENTE
Rua Belmiro Braga, 374
Belo Horizonte, MG, Brazil, Brazil

Ferreira, Monica
Editor
GLOBO
Rua Maia Lacerda, 244/303
Estacio, Rio de Janeiro-RJ, Brazil, Brazil

Ferreira, Nadia
Editor
DIA
R. Giulhermina Guinle, 296/205
Botafogo, Rio de Janeiro-RJ, Brazil, Brazil

Ferreira, Nucia Maira Da S.
Editor
A NOTICIA
Francisco Moratoria, 46
Rio de Janeiro, Brazil, Brazil

Ferreira, Roberto P.
Editor
A NOTICIA
Prof. Fausto Moreira, 46/1705
B. da Tijuca, Rio de Janeiro, Brazil, Brazil

Ferreira, Sergio Vitor Martins
Reporter Radio/Television
TV EDUCATIVA - RJ
Av. Canal Marapendi 2980 - 303
Rio de Janeiro, Brazil, Brazil

Ferreira, Valdevino
Reporter Radio/Television
TV EDUCATIVA - RJ
Av. Darcy B. Costa Q.1 B.4/403
Rio de Janeiro, Brazil, Brazil

Ferreira, Vera Aparecida
Correspondent
FOLHA DO MEIO AMBIENTE
R.Araujo Leitao, 545 Bl.A/304
Rio de Janeiro-RJ, Brazil, Brazil

Ferreira Gomes, Ana Lucia
Reporter Radio/Television
MANCHETE RADIO
Rua Girua, 98 Anchieta
Rio de Janeiro, Brazil, Brazil

Ferrerira, Valdevino
Reporter Radio/Television
TV EDUCATIVA - RJ
Av. Darcy B. Costa Q.1 B.4/403
Rio de Janeiro, Brazil, Brazil

Ferro, Nubia De S.
Editor
GLOBO
SCS Ed.Oscar Niemeyer-13 andar
Brasilia-DF, Brazil, Brazil

Fidalgo, Ana Claudia
RIO LIFE
Est da Gavea 611/Bl 2/304 Sao Conrado
Rio de Janeiro - RJ, Brazil
T: 55 21 322 2290 F: 55 21 322 2990

Figueira, Mariza T.
Editor
GLOBO
Rua Cupertino Durao, 139
Leblon, Rio de Janeiro-RJ, Brazil, Brazil

Figueiredo, Maria Jose S.
Reporter Radio/Television
GLOBO TV
Rua Von Martius 22 sala 321
Rio de Janeiro, Brazil, Brazil

Figueiredo, Newton Carlos
Reporter Radio/Television
TV-BANDEIRANTES-RIO
Rua Doutor Julio Ottoni, 215
Rio de Janeiro-RJ, Brazil, Brazil

Figueiredo, Patricia Da Silva
Reporter Radio/Television
RECORD TV
R. Bom Conselho 2A
Guarulhos, Sao Paulo, SP, Brazil, Brazil

Figueiredo, Safira Maria
Correspondent
RADIOBRAS
SQN 312 BIK Ap.608
Brasilia, D.F. Brazil, Brazil

Figueredo, Luis F.M.
Editor
PROGRAMA GLOBO ECOLOGIA
R. Barata Ribeiro, 92
Copacabana-RJ, Brazil, Brazil

Filho, Eliakim Araujo Paraira
Correspondent
MANCHETE TV
Jose da Brito 35
Rio de Janeiro, brazil, Brazil

Filho, Jose Ataliba Alvares
RADIOBRAS/TV NACIONAL
R. Botafogo, 852/804
Rio de Janeiro, Brazil, Brazil

Filipon, Jane M.
Correspondent
RADIO ELDORADO FM
Av. Ganzo, 445/402
90060 Porto Alegre-RS, Brazil, Brazil

Filippini, Eduardo
Reporter Radio/Television
BANESER
R. Alvaro Martins Ferreira,85
Sao Paulo, Brazil, Brazil

Finguerut, Silvia
Editor
PROGRAMA GLOBO ECOLOGIA
Av. Paulo de Frontim, 568
Rio Comprido-RJ, Brazil, Brazil

Florencio, Livia Maria Ferrari
Editor
GAZETA MERCANTIL
R. Mj. Qudinho, 90 - 4 andar
Sao Paulo, Brazil, Brazil

Firmino, Hiram
Editor
ESTADO DE MINAS
R. Pedra Bonita, 790/602
30.000-Belo Horizonte-MG, Brazil, Brazil

Fiuza, Guilherme
Editor
JORNAL DO BRASIL
R. A. Gomes Pereira, 130/301
Rio de Janeiro-RJ, Brazil, Brazil

Flores, Isabel Cristina B.
Reporter Radio/Television
RADIO FLUMINENSE
Rua Tirol 861/Bl.23/304
Rio de Janeiro, Brazil, Brazil

Flores, Pdero Silveira
Correspondent
GOVERNO ESTADO RIO GRANDE DO SUL
R.Ramiro Barcelos 2255/52
Porto Alegre, Rio Grande do Sul, Brazil, Brazil

Fonseca, Celso
Editor
JORNAL DA TARDE
R. Bom Jesus de Pirapora 112
13.200 Jundiai, Sao Paulo, Brazil, Brazil

Fonseca, Joao Mello
Editor
GLOBO TV
Rua Von Martius 22 sala 328
Rio de Janeiro, Brazil, Brazil

Fonseca, Joberto
Correspondent
FUNDACAO ROBERTO MARINHO
Rua Serafim Valandro, 19
Botafogo, Rio de Janeiro-RJ, Brazil, Brazil

Fonseca, Marcia Prado Prioli
Reporter Radio/Television
SBT TV
Rua Cap. Cesar Andrade,168/704
22431, Rio de Janeiro-RJ, Brazil, Brazil

Fonseca, Marco Antonio
Editor
MEZCLA LATINA JOURNAL
R. Guararema 145 c/2
Rio de Janeiro, Brazil, Brazil

Fonseca, Mario Imaculada P.
Editor
GLOBO TV
Rua Von Martius 22 sala 321
Rio de Janeiro, Brazil, Brazil

Fonseca, Reginalda C.
Reporter Radio/Television
TV GAZETA-CANAL 11
R.Arlindo Veiga Santos, 25/91
04671 Sao Paulo-SP, Brazil, Brazil

Fonseca, Silvia
GLOBO
Rua Joaquin Tavora, 122/705
Niteroi, Rio de Janeiro, Brazil, Brazil

Fontes, Claudia De Fatima
Editor
JORNAL METRO PRESS
Rua Piaui, 409/403
Rio de Janeiro-RJ, Brazil, Brazil

Fontes, Maria Aparecida
Editor
GLOBO
SCS. Ed. Oscar Niemeyer, 13
Rio de Janeiro-RJ, Brazil, Brazil

Fontoura, Lauro Sobrinho
Reporter Radio/Television
TV & RADIO BANDEIRANTES
Alameda Ribeirao Preto, 783
Stna. Paraiba, S.P. Brazil, Brazil

Formagini, Elizabeth
Editor
TVT REDE DE COMUNICACAO
R. Ipiranga, 107
Rio de Janeiro, Brazil, Brazil

Formenti, Ligia H.G.
Correspondent
ESTADO DE SAO PAULO
R. Conselheiro Brotero 1.417
Sao Paulo-SP, Brazil, Brazil

Fraga, Luiz Felipe Prado
Editor
RADIOBRAS
SQN 216 Bl.K/610
Brasilia, D.F., Brazil, Brazil

Fraga Jr., Plinio
Correspondent
FOLHA DA MANHA
R.Belfort Roxo, 283/704
Copacabana, Rio de Janeiro-RJ, Brazil, Brazil

Fragoso De Abreu, Marcelo Jorge
Editor
JORNAL DO COMMERCIO
R. Setubal 1662, Apt. 801
Rio de Janeiro, RJ, Brazil, Brazil

Franca, Jamari
Editor
JORNAL DO BRASIL
R. Heraclito Graca, 347/402
Vasconcelos, Rio de Janeiro-RJ, Brazil, Brazil

Franca, Martha S.J.
Editor
ESTADO DE SAO PAULO
R.J.Antonio Coelho 661 Apt.42
04011 Sao Paulo-SP, Brazil, Brazil

Franca, Mauricio Cintrao
Editor
REVISTA VISAO
R.Dr.Almeida Lima, 1384
Sao Paulo - SP, Brazil
T: 55 11 2914866 R.476 F: 55 11 2927438

Franco, Carlos R.
Editor
AGENCIA ESTADO
Av. Nsa.Copacabana 872 Ap$.803
Copacabana, Rio de Janeiro-RJ, Brazil, Brazil

Franziska, Erika
Reporter Radio/Television
VIDEO CIENCIA PRODUCOES LTDA
Rua do Oriente, 367/201
20240 Rio de Janeiro-RJ, Brazil, Brazil

Frare, Jose Luiz
Editor
REVISTA NOVA ESCOLA
R.Haddock Lobo 403
Sao Paulo, S.P., Brazil, Brazil

Frazao, Silvio
Reporter Radio/Television
RADIO ALVORADA FM
R. Barao, 207/202
Jacarepagua, Rio de Janeiro-RJ, Brazil, Brazil

Frazao, Wilson
Correspondent
JORNAL A RAZAO
R.Joao Paulo I 387/301
Rio de Janeiro, Brazil, Brazil

Freire, Edmilson De Souza
Reporter Radio/Television
TV EDUCATIVA
Rua Omar Fontoura 98/204
Rio de Janeiro, Brazil, Brazil

Freire, Gustavo E.
Correspondent
ZERO HORA
SQN 315 Bloco D, Apt. 401
Brasilia-DF, Brazil, Brazil

Freire, Vera Lucia C.
Editor
JORNAL DA TARDE
R.Oncioneiro de Evora,174/100
Chacara St.Antonio, Sao Paulo, Brazil, Brazil

Freire, Vinicios Torres
Editor Assistant
FOLHA DE SAO PAULO
Alameda Barao de Limeira 425
Sao Paulo SP, Brazil
T: 55 11 874 2608/874 2602 F: 55 11 2231644

Freitas, Armando C.
Editor
GLOBO
R. Almte Sadock de Sa, 120/103
22411, Rio de Janeiro-RJ, Brazil, Brazil

Freitas, Consuelo Lage
Reporter Radio/Television
TV EDUCATIVA - RJ
R. C. de Vasconcelos, 57/103
Rio de Janeiro, Brazil, Brazil

Freitas, Gloria
Editor
GLOBOSAT
R.Barao Oliveira Castro 17/102
Rio de Janeiro, Brazil, Brazil

Freitas, Igor Lima De
Reporter
SISTEMA JORNAL DO BRASIL
Av.Brasil, 500
Rio de Janeiro - RJ, Brazil
T: 55 21 5854422

Freitas, Jose Itamar
Editor
GLOBO TV
Rua Von Martius 22 sala 321
Rio de Janeiro, Brazil, Brazil

Freitas, Luiz Antonio Signates
Reporter Radio/Television
RADIO UNIVERSITARIA
R. Serra dos Javaes,Q9, L.6
Sonho Verde, Goiania, GO, Brazil, Brazil

Freitas, Marise Lugullo De
Reporter Radio/Television
JORNAL DO BRASIL-RADIO AM
Rua Benjamin Constat, 397/404
Niteroi, Rio de Janeiro, Brazil, Brazil

Freitas, Monica
Reporter Radio/Television
GLOBO RADIO
R.Uruguai 545/503
Rio de Janeiro, Brazil, Brazil

Frigeri, Rosane
Correspondent
CORREIO DO POVO (EJCJ)
Sarmento Leite, 860/202
Porto Alegre, RS, Brazil, Brazil

Fruet, Luiz H.
Editor
GLOBO CIENCIA
Av. Dr.Altino Arantes,120 c/4
04042 Sao Paulo-SP, Brazil, Brazil

Fumagalli, Marcia E.
Reporter Radio/Television
TV EDUCATIVA - RJ
r. Almirante Alexandrino 660
Sta. Teresa, RJ, Brazil, Brazil

Furtado, Marba R.
Correspondent
ZERO HORA
SQN 104 Bloco H, Apt. 101
Brasilia-DF, Brazil, Brazil

Furtado, Teresa Urban
Editor
ISTO E SENHOR
Rua Brigadeiro Franco,549
80405 Curitiba, PR, Brazil, Brazil

Fusco, Tania Mara
Editor
ISTO E SENHOR
Av. Atlantica, 1536/602
Copacabana, Rio de Janeiro-RJ, Brazil, Brazil

Fuser, Igor
Editor
VEJA
R.Itapicuru, 643 Apt.84
Sta. Cecilia, Sao Paulo-SP, Brazil, Brazil

Gabeira, Fernando P.
Correspondent
FOLHA DA MANHA
R.Lopes Quintas, 340 C/113
Jardim Botanico, Rio de Janeiro, Brazil, Brazil

Galazi, Jovaci Lea
Editor
AGENCIA ESTADO
R. Conde de Itaquai, 610/101
Tijuca, Rio de Janeiro-RJ, Brazil, Brazil

Galeazzi, Marlene
Correspondent
AGENCIA ESTADO
HIGN 716, Bl.G Apt.119
Brasilia-DF, Brazil, Brazil

Gallo, Emilio
Reporter Radio/Television
BLUE LIGHT PROD. E PROM. LTDA
R. Campinas, 133/504
Rio de Janeiro, Brazil, Brazil

Galvao, Maria Angela Botelho
Reporter Radio/Television
TV EDUCATIVA - RJ
Efigenio de Salles 138/202
Cosme Velho, RJ, Brazil, Brazil

Galves, Monica S.
Editor
GLOBO TV
Rua Von Martius 22 sala 328
Rio de Janeiro, Brazil, Brazil

Gamboa, Aldo H
Redator
ECOLOGIA E DESENVOLVIMENTO
Rua da Gloria 122 gr 105/106
Rio de Janeiro RJ 20, Brazil
T: 55 21 242 1957/222 1370

Gamboa, Claudia G.
Editor
ECOLOGIA E DESENVOLVIMENTO
R. Conde de Lages, 68/201
20241 Gloria, Rio de Janeiro-RJ, Brazil, Brazil

Gancia, Barbara
Editor
FOLHA DE SAO PAULO
Al. Barao de Limeira, 425
Sao Paulo-SP, Brazil, Brazil

Garambone, Sidney
Correspondent
GLOBO
Rua Figueiredo Magalhaes, 581
Rio de Janeiro-RJ, Brazil, Brazil

Garcia, Alexandre
Reporter Radio/Television
GLOBO TV
Rua Von Martius 22 sala 328
Rio de Janeiro, Brazil, Brazil

Garcia, Mauro Alves
Reporter Radio/Television
TV EDUCATIVA - RJ
Carvalho Alvim 583
Tijuca, RJ, Brazil, Brazil

Gaspar Filho, Antonio
Editor
ESTADO DE SAO PAULO
R.Padre Francisco Liberman 162
Butanta, Sao Paulo-SP, Brazil, Brazil

Gay, Lygia Maria
Editor
GLOBO TV
Rua Von Martius 22 sala 321
Rio de Janeiro,Brazil, Brazil

Gazio, Anderson Coelho
Editor
GLOBO TV
R. Von Martius, 22
Jardim Botanico, Rio de Janeiro, Brazil, Brazil

Gentil, Ermelinda Rita De Sao Thiago
Reporter Radio/Television
GLOBO RADIO
Barao de Itapagipe, 245
Rio de Janeiro-RJ, Brazil, Brazil

Genuncio, Elen
Correspondent
POVO
Rua Maria Amalia, 547/204
Tijuca, Rio de Janeiro-RJ, Brazil, Brazil

Germano Filho, Farid
Reporter Radio/Television
RADIO GAUCHA
R. Demetrio Ribeiro, 1168
Ap. 61, 90000 Porto Alegre, Brazil, Brazil

Gertel, Luis Carlos
Reporter Radio/Television
RADIO & TV BANDEIRANTES
R. Levon Apovian, 60
Sao Paulo-SP, Brazil, Brazil

Gertel, Vera
Editor
SECRETARIA ESTADUAL MEIO AMBIENTE
R.Benjamim Batista 14201
Rio de Janeiro, Brazil, Brazil

Gherman, Naum
Correspondent
MANCHETE TV
Rua Barata Ribeiro 622/901
Rio de Janeiro-RJ, Brazil, Brazil

Ghivelder, Joel
Correspondent
GLOBO TV
Avda. Sernambetiba 3300
Rio de Janeiro, Brazil, Brazil

Giannini, Alessandro
Correspondent
JORNAL DA TARDE
R. da Consolacao 2984/81
01416 Jardim Paulista, SP Brazil, Brazil

Giannini, Mirian Clark
Editor
NOTICIAS POPULARES SA
O. Tarquinio de Souza 1289/101
Sao Paulo, SP, Brazil, Brazil

Gil, Luiz Fernando
Reporter Radio/Television
MANCHETE RADIO
R. Teodoro da Silva, 332/202
Rio de Janeiro, Brazil, Brazil

Gilio, Paulo Sergio Bittencourt
Reporter Radio/Television
TV EDUCATIVA - RJ
R. Paulo Alves, 73/15 Inga
Rio de Janeiro, Brazil, Brazil

Gillet, Edson
Correspondent
RADIO LIBERAL
Av.25 Setembro 883
Belem, Para, Brazil, Brazil

Giovanini, Dener
Reporter Radio/Television
RADIO TRES RIOS LTDA
R. Barao do Rio Branco, 255
25800 Rio de Janeiro-RJ, Brazil, Brazil

Giusti, Andre Luis De Almeida
Reporter Radio/Television
TV-BANDEIRANTES-RIO
Rua Leopoldo 215
Rio de Janeiro-RJ, Brazil, Brazil

Godoy, Marcelo L.
Editor
GLOBO TV
Rua Von Martius 22 sala 328
Rio de Janeiro, Brazil, Brazil

Goes, Hercules
Editor
JORNAL ECOTURISMO
R.Argentina, 16
Porto Velho, Rondonia, Brazil, Brazil

Goes, Severino De
Director
AGENCIA ESTADO
Av. Caetano Alvares, 55
Sao Paulo -SP 02598, Brazil
T: 55 11 8562210 F: 55 11 2652297

Gois, Ancelmo
Editor
EDITORA ABRIL
R.da Passagem, 123/8 andar
Rio de Janeiro, RJ, Brazil
T: 021 5468282 F: 021 2759347

Gomes, Ana Helena Nogueira R.
Producer
REDE GLOBO
R. Lopes Quintas 303
Rio de Janeiro -RJ 2, Brazil
T: 55 21 2947732

Gomes, Andrea Dos Reis
Reporter Radio/Television
TV RIO
R. Morro Dois Irmaos, 131
Jacarepagua, Rio de Janeiro, Brazil, Brazil

Gomes, Clarice Maria Abdalla
Coordenadora De Pauta
SISTEMA JORNAL DO BRASIL
Av.Brasil, 500
Rio de Janeiro - RJ, Brazil
T: 55 21 5854422

Gomes, Claudio C.
Correspondent
JORNAL DO COMERCIO
R. Jose Carvalheira 203
Apt.1302, Recife-PE, Brazil, Brazil

Gomes, Eusebio
Reporter Radio/Television
MANCHETE TV
R.Assis Carneiro,404
Rio de Janeiro-RJ, Brazil, Brazil

Gomes, Gloria
Editor
MANCHETE RADIO
Rua da Justica 351 V. da Penha
Rio de Janeiro, Brazil, Brazil

Gomes, Lea
Editor
GLOBO
Rua Desembargador Toledo Pizza
Niteroi, Rio de Janeiro-RJ, Brazil, Brazil

Gomes, Luiz
Correspondent
CORREIO BRAZILIENSE
Rua Mariz e Barros no.38/502
Niteroi, Rio de Janeiro, Brazil, Brazil

Gomes, Marcelo Simoes
Editor
TV RIO
Av. Pio Borges, 1015-bl.A,#104
Pita-Sao Goncolo Niteroi, RJ, Brazil, Brazil

Gomes, Marcos E.
Editor
ESTADO DE SAO PAULO
R. Anast.de Souza Pinto 333/83
02926 Freguesia, Sao Paulo-SP, Brazil, Brazil

Gomes, Paulo
Editor
REVISTA FLUMINENSE
Rua Republica Libano 61/1120
Rio de Janeiro, Brazil, Brazil

Gomes, Rogerio
Editor
GLOBO TV
Ave. Alvorada, 2600
Tijuca, Rio de Janeiro-RJ, Brazil, Brazil

Gomes, Rose Almenteiro
Editor
GLOBO TV
R. Von Martius, 22 S.321
Rio de Janeiro, Brazil, Brazil

Goncalves, Ana Paula De Miranda
Editor
REDE GLOBO
R. Lopes Quintas 303
Rio de Janeiro -RJ 2, Brazil
T: 55 21 2947732

Goncalves, Francisco
Correspondent
JORNAL DO BRASIL
SQS 214 Bl.C apt.405
Brasilia, D.F., Brazil, Brazil

Goncalves, Jose Esmeraldo
Editor
REVISTA MANCHETE
Rua Russel 804/8 and.
Rio de Janeiro, Brazil, Brazil

Goncalves, Luis Eduardo Soares
Editor
NOTICIAS EM DESTAQUE
Senador Verqueiro 56, Apt. 103
22230 Rio de Janeiro, Brazil, Brazil

Goncalves, Maria Lucia
Reporter Radio/Television
TV EDUCATIVA - RJ
Marquez de Parana 51 A, 2 -308
Niteroi, RJ, Brazil, Brazil

Goncalves, Roberto
Editor
GLOBO TV
Rua Von Martius 22 sala 328
Rio de Janeiro, Brazil, Brazil

Goncalves, Rosa Maria Magalhaes
Editor
GLOBO TV
Rua Von Martius, 22 S. 321
Rio de Janeiro, Brazil, Brazil

Goncalves, Sergio B.
Editor
RADIOBRAS
R.Gilberto Cardoso 230/1503
Leblon, Rio de Janeiro-RJ, Brazil

Gondim, Abnor Duarte De Souza
Reporter
FOLHA DE SAO PAULO
Alameda Barao de Limeira 425
Sao Paulo SP, Brazil
T: 55 11 874 2608/874 2602 F: 55 11 2231644

Gontijo, Angelica Maria De Carvalho
Reporter Radio/Television
RADIOBRAS
SHIN Q I 04 Conj. 06 C. 8
Lago Norte, Brasilia, D.F., Brazil, Brazil

Gontijo, Mariyana T.
Editor
ESTADO DE MINAS
R. Niquel, 237/802
Serra-Belo Horizonte-MG, Brazil, Brazil

Gonzaga, Douglas
Editor
TV EDUCATIVA
R. Cenno Sbreg, 37
Sao Paulo, Brazil, Brazil

Gonzaga, Heloisa E.
Editor
GLOBO TV
Rua Von Martius 22 sala 321
Rio de Janeiro, Brazil, Brazil

Gonzaga, Luiz
Reporter Radio/Television
TV EDUCATIVA - RJ
Rua B 95, Ilha da Conceicao
Niteroi, Rio de Janeiro, Brazil, Brazil

Gonzales, Amelia
Editor
GLOBO
Rua Mario Vianna, 380
Niteroi-Rio de Janeiro, Brazil, Brazil

Gonzales, Nora Virginia Celmi
GAZETA MERCANTIL
Rua Major Quedinho 90
Sao Paulo - SP, Brazil
T: 55 11 256 3133

Gonzales, Ursula Rey
Reporter Radio/Television
MANCHETE TV
Rua Comendador Siqueira, 218
Brazil, Brazil

Gonzalez, Nora V.C.
Editor
GAZETA MERCANTIL
Rua Molliere 106/32
Sao Paulo - SP, Brazil

Gonzalez, Silvio J.
Reporter Radio/Television
MANCHETE TV
Rua Joao Pessoa, 936 C/2
Brazil, Brazil

Gorgulho, Jose Silvestre
Editor
FOLHA DO MEIO AMBIENTE
QI 15 Conj.07 C/22
Brasilia-DF, Brazil, Brazil

Gosselin, Loick
Correspondent
VISNEWS
Rua Baraó de Guaratiba, 218
Rio de Janeiro, Brazil, Brazil

Goulart, Antonio Silveira
Editor
ASSESSORIA DE IMPRENSA RGS
R. Jose do Patrocinio 737/601
Porto Alegre, Brazil, Brazil

Goulart, Maria Antonieta P.
RADIO TV BANDEIRANTES
Rua SQM, 310 Bloco L
Brasilia, Brazil, Brazil

Goulart, Rubeny S.
Editor
REVISTA EXAME
Rua Borda do Mato, 106/306
Grajau, Rio de Janeiro-RJ, Brazil, Brazil

Goulart, Silvino Santiago Fernandez
Editor
EXPRESSAO
Rua Jose do Vale Pereira 181
Coqueiros, Brazil

Graca Netto, Marcia
Editor
PROGRAMA GLOBO ECOLOGIA
R. Marques de Olinda, No. 61
Botafogo, Rio de Janeiro-RJ, Brazil, Brazil

Gracie, Marcela
Reporter Radio/Television
RADIOBRAS
SQS 315 Bl. C Ap. 104
Brasilia, D.F., Brazil, Brazil

Gralha, Ana Catarina
Editor
RADIOBRAS
SQN 411 Bl.I apt.108
Brasilia, D.F., Brazil, Brazil

Gramatico, Daurea
Reporter Radio/Television
GLOBO RADIO
R. Gal. Gois Monteiro
Botafogo, Rio de Janeiro-RJ, Brazil, Brazil

Granato, Fernando
Correspondent
JORNAL DA TARDE
R. Monte Caseros 313 Apt.23
Butanta, Sao Paulo-SP, Brazil, Brazil

Granetto, Ederson
Editor
TV EDUCATIVA-CANAL 4
R.Joao Akamine, 690
Campo Grande, Mato Grosso do Sul,Brazil, Brazil

Gravina, Regina C.
Editor
AGENCIA ESTADO
R. Moraes e Silva 139 Apt.602
Maracana, Rio de Janeiro-RJ, Brazil, Brazil

Greenlees, Andrew Stuart
Editor
FOLHA DE SAO PAULO
Al. Baro de Limeira 425
Sao Paulo, Brazil, Brazil

Gregati, Ana Lucia
Reporter Radio/Television
TV EDUCATIVA - RJ
R. Aires Saldanha, 72/504
Copacabana, RJ, Brazil, Brazil

Grillo, Klvia De Almeida Pereira
Reporter Radio/Television
TV EDUCATIVA - RJ
R. Miguel Lemos, 124/604
Rio de Janeiro, Brazil, Brazil

Grillo, Ma. Cristina
Editor
GLOBO
Rua Prof. Gabizo, 38, Apt. 803
Tijuca, Rio de Janeiro-RJ, Brazil, Brazil

Gronquist, Greger
Reporter Radio/Television
OY VLEISRADIO AB
Punavuorenk 16 D 64
00150 Helsinki 15, Finland, Brazil

Grunewald, Marcia
Reporter Radio/Television
UNIVERSIDADE CANDIDO MENDES
Mascarenhas de Morais 258/502
Rio de Janeiro, Brazil, Brazil

Grunfeld, Selma Dias
Reporter Radio/Television
REDE OM BRASIL DE TELEVISAO
R. Gustavo de Sampaio, 395/704
Leme, Rio de Janeiro, Brazil, Brazil

Grynberg, Halina
Reporter Radio/Television
TV-BANDEIRANTES-RIO
Estrada dos Canoa 1476
22610-Rio de Janeiro-RJ, Brazil, Brazil

Gryzinski, Vilma
Editor
VEJA
R.Caiowa, 1951 Casa 10
01258 Sao Paulo-SP, Brazil, Brazil

Guedes, Leonor
Editor
ORBE-PRESS
Praia do Botafogo, 416/1208
22250 Rio de Janeiro-RJ, Brazil, Brazil

Guedes, Niedja Maria
Reporter Radio/Television
O TROMBONE
QD B-1 Lote 02, Ouro Preto
Olinda, PE, Brazil, Brazil

Guedes, Octavio
Editor
JORNAL DO BRASIL
R. Mariz e Barros, 54/801
Niteroi, Rio de Janeiro-RJ, Brazil, Brazil

Guedes, Wilson
Correspondent
TRIBUNA DO RIBEIRA
Rua Shitiro Maeji, 133
11.900 Registro, SP, Brazil, Brazil

Gueiros, Fernando Da S.
Editor
GLOBO TV
Rua Von Martius 22 sala 328
Rio de Janeiro, Brazil, Brazil

Guerra, Antonio Jose Teixeira
Journalist
Rua do Russel 434, Gloria
Rio de Janeiro, Brazil
T: 55 21 2058282

Guerreiro, Marcia I.
Correspondent
ESTADO DE SAO PAULO
Av. Casa Verde 2901/114
Sao Paulo-SP, Brazil, Brazil

Guidalli, Roberto Jose
Editor
GLOBO TV
R. Von Martius, 22 S.321
Jardim Botanico, RJ, Brazil, Brazil

Guilherme, Vilma Goulart S.
Editor
RECORD TV
R. Maociri. 95
Roo de Janeiro, Brazil, Brazil

Guilhermino, Luiz A.
Editor
AGENCIA ESTADO
R. Barao do Flamengo 04
Apt.108 Flamengo, RJ-RJ, Brazil, Brazil

Guilhermino, Maria Lucia
Editor
MANCHETE
Rua Barao do Flamengo, 04
Flamengo, Rio de Janeiro-RJ, Brazil, Brazil

Guimaraens, Gonzalo
Editor
AGENCIA BOA IMPRENSA (ABIM)
Eduardo Acevedo, 1404 P7
Montevideo, Uruguay, Brazil

Guimaraes, Alexandre
Editor
JORNAL DE HOJE
R.Piragibe, 115 B
Inhauma, Rio de Janeiro, Brazil, Brazil

Guimaraes, Ana Luiza Pereira
Editor
REDE GLOBO
R. Lopes Quintas 303
Rio de Janeiro -RJ 2, Brazil
T: 55 21 2947732

Guimaraes, Antonio C.C.
Correspondent
POVO
Rua Ferreira Sampaio, 18
Abolicao-Rio de Janeiro-RJ, Brazil, Brazil

Guimaraes, Emir Rodriguez
Editor
ROLAND HENZE PRODUCOES
Rua das Laranjeiras, 115/203
22240 Rio de Janeiro-RJ, Brazil, Brazil

Guimaraes, Eneida Maria C.
Correspondent
JORNAL SOS TERRA
Lad.Tabajaras 126/907
Rio de Janeiro, Brazil, Brazil

Guimaraes, Maria Mirtes
Correspondent
O DIA
R. Riachuelo, 221/1011
Rio de Janeiro, RJ, Brazil

Guimaraes, Patricia Boueri
Reporter Radio/Television
GLOBO RADIO
R.Carvalho Alvim 333/502
Rio de Janeiro, Brazil, Brazil

Guimaraes, Regina
Editor
TV GAZETA-CANAL 11
R.Dr.Hoemen de Mello, 189/101
05007 Sao Paulo-SP, Brazil, Brazil

Guimaraes, Renato, Jr.
Editor
DIARIO DO POVO
R.Dr. Francisco Toledo 501
Campinas, S.P., Brazil, Brazil

Guimaraes, Vania Cardoso
Reporter Radio/Television
TV EDUCATIVA
Av. Atlantica no. 554/1202
Rio de Janeiro, Brazil, Brazil

Gut, Flavio De Araujo
Editor
AGENCIA ESTADO
Trv da R. Arnoldo Gut Jr.240
Varzea Paulista, Sao Paulo-SP, Brazil, Brazil

Guy-Bernard, Sorman
Correspondent
AGENCIA ESTADO
13 rue d'Uzes
7500 Paris, France, Brazil

Gvedegeoson, Ricardo
Reporter Radio/Television
TV EDUCATIVA
Sergio de Oliveira 37 c 04
Rio de Janeiro, Brazil, Brazil

Hadba, Gustavo
Editor
GLOBO TV
Ave. Alvorada, 2600
Tijuca, Rio de Janeiro-RJ, Brazil, Brazil

Hamu, Mariangela
Editor
GLOBO TV
SQN 314 Bl.B/513
Brasilia, D.F., Brazil, Brazil

Harasymowicz, Tania R.M.
Editor
AGENCIA ESTADO
SQS 103, Bloco I. 601
Brasilia-DF, Brazil, Brazil

Heilborn, Luis E.
Reporter Radio/Television
MANCHETE TV
R.Pres.Carlos de Campos,115
Bl.2 Apt.201 - Rio de Janeiro-RJ, Brazil, Brazil

Hermano, Ricardo Icassatti
Editor
CBIA
SHIS QI,Conjunto 20,casa 6
Brasilia, DF,Brazil, Brazil

Hermes, Gilmar Adolfo
Correspondent
RADIO SAO LEOPOLDO
R.C. Colombo 117 B. Rio Branco
Novo Hamburgo, RS, Brazil, Brazil

Herszenhut, Marcia
Reporter Radio/Television
TV EDUCATIVA - RJ
Paraia do Flamengo 328/501
Rio de Janeiro, Brazil, Brazil

Hippertt, Joyce B.
Editor
PREFEITURA RIO DE JANEIRO
Av.Seis 244 - Itaipu
Niteroi, Rio de Janeiro, Brazil, Brazil

Hlebnikian, Arturo
Correspondent
AGENCIA BOA IMPRENSA (ABIM)
J. Benito Blanco, 657
Montevideo, Uruguay, Brazil

Hollander, Claudia
Editor
FOLHA DE SAO PAULO
Rua Traipu 1298
01235 Sao Paulo, Brazil, Brazil

Horta, Beatriz
Editor
MANCHETE
R. Nina Rodrigues 49-101
22461-Rio de Janeiro-RJ, Brazil, Brazil

Horta, Nelio Jorge T.
Editor
GLOBO TV
Rua Von Martius 22 sala 328
Rio de Janeiro, Brazil, Brazil

Hossell, Johnny
Editor
BRAZIL TODAY
P.O. Box 341356
Los Angeles, CA 90034, USA, Brazil

Hugen, Artur
Editor
RADIO CLUBE DE CANELA
Horacio Cardoso, s/n
Gramado, RS, Brazil, Brazil

Ilha, Sonia Fonseca
Correspondent
JORNAL DO COMMERCIO
Av. N.S.Copacabana,1058/602
Copacabana, RJ, Brazil, Brazil

Inacio, Manoel
Reporter Radio/Television
TV EDUCATIVA - RJ
R. Conrado B. Souza 200/1804
Niteroi, RJ, Brazil, Brazil

Iorio, Vitor
Editor
PREFEITURA RIO DE JANEIRO
Av.Rodrigo Otavio 145/303
Rio de Janeiro, Brazil, Brazil

Ishii, Hisatoshi
Editor
TOKYO BROADCASTING SYSTEM
Praia Botafogo 528 Bl.A.1207
Rio de Janeiro, Brazil, Brazil

Ivanissevich, Alicia
Editor
JORNAL DO BRASIL
R. joaquim Murtinho, 886/S-501
Rio de Janeiro-RJ, Brazil, Brazil

Izidoro, Marcia Beatriz L.
Editor
PREFEITURA RIO DE JANEIRO
R.Alte. Ary Rangel 291
Rio de Janeiro, Brazil, Brazil

Izique, Maria Claudia P.
Editor
GAZETA MERCANTIL
R. Mj. Quedinho, 90 - 4 andar
Sao Paulo, Brazil, Brazil

Jacobskind, Mario Augusto
JORNAL TRIBUNA DA IMPRENSA
Rua do Lavradio, 98
20230 Rio de Janeiro-RJ, Brazil, Brazil

Jannotti, Dulce
Editor
JORNAL DO BRASIL
R. Sambaiba, 91/402
Rio de Janeiro-RJ, Brazil, Brazil

Jardim, Lauro
Editor
GLOBO
Rua Paulo Barreto 46/605
Rio de Janeiro-RJ, Brazil, Brazil

Jatahy, Ana Claudia
Correspondent
POVO DO AMAZONAS
Av.Sernambetiba 3300/2304
Rio de Janeiro, Brazil, Brazil

Jesus, Angela Teresa Cottas De
Editor
REDE GLOBO
R. Lopes Quintas 303
Rio de Janeiro -RJ 2, Brazil
T: 55 21 2947732

Jesus, Eliziario Alexandrino De
Editor
SBT TV
R. M. Lourenco Fernandes 144
Jd.Ma. Rosa Taboao da Serra, Brazil, Brazil

Jobim, Nelson Franco
Editor
JORNAL DO BRASIL
Rua Faro 51/202
Rio de Janeiro, Brazil, Brazil

John, Liana
Editor
AGENCIA ESTADO
R.Ant. Augusto de Almeida 1178
13083 Campinas, Sao Paulo, Brazil, Brazil

Joory, Eva
Correspondent
FOLHA DA MANHA
R.Bartolomeu Mitre, 119/201
Leblon, Rio de Janeiro-RJ, Brazil, Brazil

Judice, Fabio
Reporter Radio/Television
TV EDUCATIVA - RJ
R. S. Fco. Xavier, 278/504
Rio de Janeiro, Brazil, Brazil

Juliao, Ana Cristina De A. Lemgruber
Editor
JORNAL DO COMMERCIO
R. Vitor Meireles, 188/201
Rio de Janeiro, Brazil, Brazil

Kacelnik, Alexandre
Editor
EMBRATUR
Rua Gal. Gois Monteiro, 156
apt.307 Rio de Janeiro, Brazil, Brazil

Kaiser, Delorgel
Editor
SECRETARIA DO PLANEJAMENTO-RGS
Rua Demetrio Ribeiro, 718/602
90010 Porto Alegre, RS, Brazil, Brazil

Kappel, Fernando Hecker
Editor
OFFICIAL PRESS ESTADO RS
R. Villamil, 41
Porto Alegre, RS, Brazil, Brazil

Karam, Elisabeth
Editor
JORNAL DA USP
Rua Bueno de Andrade, 391/46
01526 Rio de Janeiro-RJ, Brazil, Brazil

Kaskus, Marcia
Editor
VOZ DE PORTUGAL
R.Antonio Vieira, 17/802
Rio de Janeiro, Brazil, Brazil

Kaye, Gustavo Jose Campos
Editor
REDE GLOBO
R. Lopes Quintas 303
Rio de Janeiro -RJ 2, Brazil
T: 55 21 2947732

Keller, Wera
Reporter Radio/Television
RADIOBRAS
HIGS 712 Bl. M. Casa 13
Brasilia, D.F., Brazil, Brazil

Keppe, Sandra Ingrid
Correspondent
JORNAL DE PSICANALISE
Av. Dr. Arnaldo, 2217
Sao Paulo, Brazil, Brazil

Kinjo, Celso Iwao
Editor
JORNAL DA TARDE
R. Luiziania 120 C/26
04560 Brooklin, Sao Paulo-SP, Brazil, Brazil

Kobelus, Olga De Mello
Editor
GLOBO
Rua Barao da Torre 15/203
Rio de Janeiro-RJ, Brazil, Brazil

Kolandra, Andre
Reporter Radio/Television
RADIOBRAS
Rua do Bispo, 94/203 Bl. 2
Rio Comprido, Rio de Janeiro, Brazil, Brazil

Koin, Sergio
Reporter Radio/Television
TV JOVEM PAN
Praca Roosevelt 178/52
Sao Paulo-SP, Brazil, Brazil

Korman, Rachel
Editor
RADIOBRAS
B. da Torre, 583/3
Rio de Janeiro, Brazil, Brazil

Kosak, Carmen
Correspondent
AGENCIA ESTADO
SQS 408 Bl.B Apt.312
Brasilia-DF, Brazil, Brazil

Koury, Jorge Eduardo G.
Correspondent
AGENCIA BOA IMPRENSA (ABIM)
R.Antonio Basilio, 555/504
Rio de Janeiro-RJ, Brazil, Brazil

Kraiker, Thomas
Editor
BANESER
R. Cel. N. de Azevedo, 84
Sao Paulo, Brazil, Brazil

Krause, Friedrich Helmut Gottlier
Editor
RADIOBRAS
SHCGN Bl.E Casa 05
Brasilia, D.F., Brasil CEP 70750, Brazil

Krengiel, Marcia Rodrigues De Brito
Reporter Radio/Television
TV EDUCATIVA - RJ
Fonte da Saudade, 293/1002
Rio de janeiro, Brazil, Brazil

Kruger, Rudi Augusto
Correspondent
RADIO NOVA DIMENSAO
Uruguai 545/203
Tijuca, RJ, Brazil, Brazil

Kruger,, Wilma Lucia Leitao
Correspondent
RADIO NOVA DIMEMSAO
Rua Uruguai, 545/203-Tijuca
Rio de Janeiro - RJ, Brazil

Kubrusly, Mauricio A.
Reporter Radio/Television
GLOBO TV
Rua Von Martius 22 sala 328
Rio de Janeiro, Brazil, Brazil

Kuck, Claudio Renato
GAZETA MERCANTIL
Rua Major Quedinho 90
Sao Paulo - SP, Brazil
T: 55 11 256 3133

Kuntz, Rolf N.
Editor
ESTADO DE SAO PAULO
R. Sao Jose 860
Santo Amaro, Sao Paulo, Brazil, Brazil

L'Abelle, Jair Guerra
Editor
JORNAL DOS SPORTS
Rua Fadel, 84 Apt. 704
Rio de Janeiro-RJ, Brazil, Brazil

La Camera, Andrea
Reporter Radio/Television
RADIO UNIVERSIDADE
R.Guilherme Marconi 67/602
Rio de Janeiro, Brazil, Brazil

Labre, Romario Soares
Reporter Radio/Television
MANCHETE TV
R.Francisco Guedes, 49 B2/206
Brazil, Brazil

Lacerda, Fernando
Editor
JORNAL DO BRASIL
R.Bueno Brandao 398
Belo Horizonte, Minas Gerais, Brazil, Brazil

Lacerda, Rita
Reporter Radio/Television
RADIO FLUMINENSE
R.Alvares de Azevedo 77/1207
Niteroi, Rio de Janeiro, Brazil, Brazil

Lacerda Gertel, Eloi
Editor
AGENCIA ESTADO
Av.Ministro Godoy 860 Apt.3131
05015 Perdizes, Sao Paulo-SP, Brazil, Brazil

Ladeira, Francisco Rorigues
Reporter Radio/Television
TV EDUCATIVA - RJ
R. C 30
Miguel Couto, RJ, Brazil, Brazil

Lago, Fernando M.S.R.
Editor
POVO
Estrada dos Bandeirantes, 8325
Jacarepagua, Rio de Janeiro-RJ, Brazil, Brazil

Lago, Rudolfo
Editor
GLOBO
SCS Ed.Oscar Niemeyer-13 andar
Brasilia-DF, Brazil, Brazil

Lagoa, Ana Mascia
Editor
REVISTA NOVA ESCOLA
Rua Haddock Lobo 403-D
Sao Paulo, S.P., Brazil, Brazil

Lamounier, Saulo Morete
Editor
RADIOBRAS
CNB 05 Lote 02 Ap. 504
Taguatinga Norte, D.F., Brazil, Brazil

Landeira, Ines Dominguez
Editor
GLOBO
Est.Velha da Tijuca, 511/201
Alto da Boa Vista, RJ, Brazil, Brazil

Laplace, Monica Dias
Reporter
GLOBO RADIO
Rua Rino Levi 375
Tijuca, Rio de Janeiro-RJ, Brazil, Brazil

Largman, Ricardo
Editor
ISTO E SENHOR
Rua Theodor Herzl, 50/101
22260 Brazil, Brazil

Larmer, Brook
Correspondent
NEWSWEEK
Av. S. Sebastiao 157 Urca 444 Madison Ave. NY-NY 10017 USA
Rio de Janeiro -RJ 2, Brazil
T: 55 21 5419697 F: 55 21 5426583

Lassance, Gelson
Editor
VIDEO CIENCIA PRODUCOES LTDA
Rua Oscar Valdetaro 176/706
22700 Rio de Janeiro-RJ, Brazil, Brazil

Latge, Luis Claudio C.
Editor
GLOBO TV
R.Von Martius 22 sala 321
Rio de Janeiro, Brazil, Brazil

Lau, Fabio Pinheiro
Editor
DIA
R. Arquias Cordeior, 518/404
Meier, Rio de Janeiro-RJ, Brazil, Brazil

Laukenicas, Sandra Almada
Reporter Radio/Television
GLOBO TV
Rua Von Martius, 22 sala 328
Rio de Janeiro, Brazil, Brazil

Lavigne, Jose
Editor
REDE GLOBO
R. Lopes Quintas 303 J.Botanico
Rio de Janeiro RJ 22, Brazil
T: 55 21 2947732

Lazaretti, Mariella
Correspondent
JORNAL DA TARDE
R. D.Germaine Burchard 585 C.7
Perdizes, Sao Paulo-SP, Brazil, Brazil

Leal, Celia
Reporter Radio/Television
VISNEWS
Av.Rio Branco, 25
Rio de Janeiro, Brazil, Brazil

Leal, Daniela Farina
Reporter
O DIA
R. Riachuela 359
Rio De Janeiro, RJ 2, Brazil
T: 55 21 272-8000 F: 55 21 507-1228

Leal, Francisco
Editor
TRIBUNA DA SERRA
Rua Bolivar Felix 279
Valenca, Est. Rio de Janeiro, Brazil, Brazil

Leal, Karla
Editor
PROGRAMA GLOBO ECOLOGIA
R. Haddock Lobo, No. 209
Rio de Janeiro-RJ, Brazil, Brazil

Leal, Lucia Ma.
Editor
TV CULTURA-REPORTER ECO
R.Alcino Braga, 207 Apt.52
Paraiso, Sao Paulo-SP, Brazil, Brazil

Leal, Paulo Cesar
Reporter Radio/Television
MANCHETE TV
Rua Barao, 207 Praca da Seca
Brazil, Brazil

Leal, Ricardo
Correspondent
CACHOEIRAS JORNAL
Rua Padre Andre Moreira,305
Rio de Janeiro-RJ, Brazil, Brazil

Leeuwen, Paulo Alceu Van
Reporter Radio/Television
REDE OM BRASIL DE TELEVISAO
R. D. Cardoso Bl.II/101
Barra da Tijuca, RJ, Brazil, Brazil

Leitao, Miriam
Editor
GLOBO
R. Raimunao de Magalhaes, 22
Gavea, Rio de Janeiro-RJ, Brazil, Brazil

Leite, Eulalia F.J.
Editor
PREFEITURA RIO DE JANEIRO
R.Indiana 49
Rio de Janeiro, Brazil, Brazil

Leite, Geraldo
Editor
JORNAL DO BRASIL-RADIO AM
Rua Sambaiba, 699
Leblon, Rio de Janeiro-RJ, Brazil, Brazil

Leite, Jose Antonio Bicalho
Editor
JORNAL DO COMMERCIO
Rua Benjamin Constant, 104
Gloria, Rio de Janeiro, Brazil, Brazil

Leite, Jose Pedro M.
Reporter Radio/Television
SAO PAULO STATE GOVERNMENT
Palacio Bandeirantes
Sao Paulo, S.P., Brazil, Brazil

Leite, Luis Ricardo
Reporter Radio/Television
MANCHETE TV
Av. Pasteur, 395
Rio de Janeiro-RJ, Brazil, Brazil

Leite, Marcal Alves
Editor
JORNAL DO COMERCIO
R. Olinda 236/202
Rio de Janeiro, Brazil, Brazil

Leite, Marcelo
Editor
FOLHA DE SAO PAULO
Al. Barao de Limeira, 425
Sao Paulo-SP, Brazil, Brazil

Leite, Otavio Soares
Editor
FLUMINENSE
Rua Capitao Resende, 408 C.43
Meier, Rio de Janeiro, Brazil, Brazil

Leite, Virginie C.
Editor
VEJA
Rua Perereira Nune, 347/201
Vila Isabel, Rio de Janeiro-RJ, Brazil, Brazil

Leite Vieira, Cassio
Editor
CIENCIA HOJE
R.Joaqim Martinho, 886 S-501
20241 Rio de Janeiro-RJ, Brazil, Brazil

Leme, Luiz Antonio Padovani
Editor
SBT TV
Rua Rodrigues de Campos Leite
Sao Paulo-SP, Brazil, Brazil

Lemos, Ana Amelia
Correspondent
ZERO HORA
SHIN QI 08 Conj.08, Casa 05
71.500 Brasilia-DF, Brazil, Brazil

Lemos, Clautenis D.
Correspondent
ZERO HORA
QI 23 Lote 12, Apt. 414
Guara II, Brasilia-DF, Brazil, Brazil

Lemos, Regina
Reporter Radio/Television
GLOBO RADIO
R. Constante Ramos, 161/202
Rio de Janeiro, Brazil, Brazil

Lenard, Magda
Correspondent
DIARIO DA TARDE
Uruguai 302/510
Rio de Janeiro, Brazil, Brazil

Lencastre, Carla
Editor
GLOBO
Rua Eurico Aragao, 86
24350 Niteroi-RJ, Brazil, Brazil

Lengsfeld, Richard
Editor
ECO RADIO
Rainha Elizabeth 571, 501
22081 Rio de Janeiro-RJ, Brazil, Brazil

Lessa, Ana Cecilia
Editor
EDITORA ABRIL/CAPRICHO
R. General Asdrubal da Cunha
1118, 055565, Sao Paulo, SP, Brazil, Brazil

Lessa, Jefferson Pinheiro Mariano
Editor
JORNAL DO COMMERCIO
Av. B. de Medeiros, 2545/1004
Lagoa, Rio de Janeiro, Brazil, Brazil

Levi, Elifas
Reporter Radio/Television
MANCHETE TV
Rua Morro do Amorim, 153
Brazil, Brazil

Levinsohn, Claudia
GLOBO RADIO
R. General Urquisa 326
Rio de Janeiro-RJ, Brazil, Brazil

Liberado, Casia
Reporter Radio/Television
RADIO ELDORADO DE SAO PAULO
SCRN 706/7 bloco G
Brasilia, DF, Brazil, Brazil

Liborio, Antonio J.F.
Correspondent
POVO
Rua Pinheiro Guimaraes, 149
22.281-Rio de Janeiro-RJ, Brazil, Brazil

Liedke, Enoi
Editor
SECRETARIA DO PLANEJAMENTO-RGS
Av. Venancio Aires, 70/404
Porto Alegre, RS, Brazil, Brazil

Lilenbaum, Denise
Reporter Radio/Television
REDE OM BRASIL DE TELEVISAO
Rua Ronald Carvalho, 154/10
Rio de Janeiro-RJ, Brazil, Brazil

Lima, Ana Catarina
Editor
GLOBO
SCS Ed.Oscar Niemeyer-13 andar
Brasilia-DF, Brazil, Brazil

Lima, Antonio Jose
Editor
JORNAL FAMTOUR
Av. Rio Branco 12-/501
Rio de Janeiro, Brasi, Brazil

Lima, Carlos Gustavo Trindade
Reporter
EDITORA MULTIMIDIA
Rua Riachuelo, 221 Cob.03
Rio de Janeiro - RJ, Brazil
T: 55 21 2223192 / 2215894

Lima, Clara
Reporter Radio/Television
RADIOBRAS
SHIS QI 27 Conj.07 c 7
Brasilia, D.F., Brazil, Brazil

Lima, Darcilia Pereira De
Reporter
SISTEMA GLOBO DE RADIO
Rua do Russel, 434
Rio de Janeiro - RJ, Brazil
T: 55 21 2058282 F: 55 21 2054096

Lima, Edvaldo Pereira
Editor
UNIVERSIDAD DE SAO PAULO
rua Barao de Morenos, 88
05453 Sao Paulo, SP Brazil, Brazil

Lima, Elza
Correspondent
SECRETARIA DA CULTURA ESTADO DO PARA
Av. Conselheiro Furtado 1698
Belem, Para, Brazil, Brazil

Lima, Eucy Da Silva
Reporter Radio/Television
JORNAL DO BRASIL - RADIO
R. Anibal Porto, 730 Bl. 3/102
Rio de Janeiro, Brazil, Brazil

Lima, Everton Silva
Editor
DIA
Estr. da Agua Grande, 1.525
Iraja, Rio de Janeiro-RJ, Brazil, Brazil

Lima, Gloria Maria De Heraclito
Editor
TV EDUCATIVA - RJ
R. Dr. Julio Otoni, 455
Rio de janeiro, Brazil, Brazil

Lima, Joao Luiz Alves De
Editor
SBT TV
Av. Imirim
1070 Imirim, Brazil, Brazil

Lima, Leila Pinage
Reporter Radio/Television
RADIO ROQUETE PINTO
R. Dom Bosco, 55 casa 2
Jardim Icarai, Niteroi, RJ, Brazil, Brazil

Lima, Ma.Da Consolacao
Editor
GLOBO
SCS Ed.Oscar Niemeyer-13 andar
Brasilia-DF, Brazil, Brazil

Lima, Marcia G. S.
Reporter Radio/Television
SUPER RADIO TUPI
Rua Silva Rabelo, 91
Meier, Rio de Janeiro-RJ, Brazil, Brazil

Lima, Marco Aurelio
Editor
MANCHETE
Rua Alvaro Ramos, 405/Bl. 02
Botafogo, Rio de Janeiro-RJ, Brazil, Brazil

Lima, Maria Da Consolacao
Correspondent
O GLOBO
Sqs 205 J 401
Brasilia, DF, Brazil

Lima, Mauricio Goncalves
Reporter Radio/Television
GOVERNOO ESTADO SAO PAULO
R.Maria Borba 75/52
Sao Paulo, S.P., Brazil, Brazil

Lima, Priscila Anders De Souza
Editor
GLOBO TV
Rua Von Martius 22 sala 321
Rjo de Janeiro, Brazil, Brazil

Lima, Rosa
Reporter
DIA
Praca Radial Sul, 25/1001
22260-Rio de Janeiro-RJ, Brazil, Brazil

Lima, Sonilda Ferreira
Reporter Radio/Television
RADIO UONIVERSIDADE
R. Raul Pompeia 131601
Rio de Janeiro, Brazil, Brazil

Lima, Valter Angelo
Reporter Radio/Television
RADIOBRAS
QNL 12 Conj. G Casa 21
Taguatinga, D.F., Brazil, Brazil

Lima, Viviane G.
Reporter Radio/Television
MANCHETE TV
R.Cel Moreira Cesar 330
Apt.701 Bl.A Brazil, Brazil

Lima Filho, Ruy M.
Editor
GLOBO TV
Rua Von Martius 22 sala 321
Rio de Janeiro, Brazil, Brazil

Linck, Ricardo
Correspondent
DIA
Ave. Rui Barbosa, 762
Flamengo, Rio de Janeiro-RJ, Brazil, Brazil

Lindenberg, Angela M.
Reporter Radio/Television
GLOBO TV
Rua Von Martius 22 sala 321
Rio de Janeiro, Brazil, Brazil

Linhares, Ana Felicia De C.
RADIOBRAS
SQN 406 Bl.H Apt.302
Brasilia-DF, Brazil, Brazil

Lins, Marcelo
Editor
GLOBOSAT
R.Moura Brasil, 74/801
Rio de Janeiro-RJ, Brazil, Brazil

Lins, Selma Rita Severo
Editor
SBT TV
Rua Oscar Freire, 1218/114
Sao Paulo-SP, Brazil, Brazil

Liones, Rocha
Correspondent
JORNAL INDUSTRIA E COMERCIO DO PARANA
QSA 1 Casa 01
Taguatinga, Brasilia-DF, Brazil, Brazil

Lippi, Flavia L.
Editor
TV CULTURA-REPORTER ECO
Av.Cotovia 141/44
04517 Sao Paulo-SP, Brazil, Brazil

Lisboa, Salete
Editor
DIA
R. Ribeiro Guimaraes, 220/1102
Tijuca, Rio de Janeiro-RJ, Brazil, Brazil

Lisbos, Edgar
Editor
JORNAL DE BRASILIA
SHIS Qi 13
Conjunto 2, Casa 5, Brasilia, DF,Brazil, Brazil

Liskauskas, Suzana F.
Editor
DIA
Rua Soldado Paiva, 66, Penha
Rio de Janeiro-RJ, Brazil, Brazil

Lobato, Randolpho Marques, Sr.
President
ASSOCIACAO BRASILEIRA DE PREVENCAO A POLUICAO DO AR - ABPPOLAR
Avenida Tres Poderes 303
Sao Paulo - SP 05514, Brazil
T: 55 11 212-5544 F: 55 11 65 09 30

Lobo, Ana
Editor
RADIOBRAS
R.Visc. Figueiredo 42/202
Rio de Janeiro, Brazil, Brazil

Lobo, Beatria Helena B.S.
Editor
PREFEITURA RIO DE JANEIRO
R.Proc.Luiz O.Cardoso 440
Rio de Janeiro, Brazil, Brazil

Lobo, Claudia Luiza
Editor
MANCHETE
Rua Lopez Trovao, 294
Niteroi, Rio de Janeiro-RJ, Brazil, Brazil

Lobo, Cristiana Mendes
Correspondent
GLOBO
SHIS QI 07 Conj.05 C/07
Brasilia-DF, Brazil, Brazil

Lobo, Ines Valladao Campos R.
Reporter Radio/Television
RADIO ROQUETE PINTO
R. Morro do Amorim, 153
Jacarepagua, Rio de Janeiro, Brazil, Brazil

Lopes, Carlos A.
Editor
ECOLOGIA E DESENVOLVIMENTO
R. das Laranjeiras 531 Apt.34
Rio de Janeiro-RJ, Brazil, Brazil

Lopes, Jose Milano
Editor
FOLHA DO MEIO AMBIENTE
QI 25 Conj.7 C/25
71.600 Brasilia-DF, Brazil, Brazil

Lopes, Marcia R. Antunes
Editor
O JORNAL DO PARANA
R.Minas Gerais 643
Siqueira Campos,PR, Brazil, Brazil

Lopes, Regina C.
Editor
MANCHETE TV
R.Min.Raul Fernandes 180/507
Brazil, Brazil

Lopes, Sergio Gabriel
Reporter Radio/Television
RADIO & TV BANDEIRANTES
Pca. Coronel Lisboa 58
Sto. Amaro, Sao Paulo, SP, Brazil, Brazil

Lopes, Wania
Reporter Radio/Television
REDE OM BRASIL DE TELEVISAO
Rua Almirante Goncalves,23/801
Copacabana, Rio de Janeiro-RJ, Brazil, Brazil

Lopes Filho, Oswaldo J.
Reporter Radio/Television
RADIO & TV BANDEIRANTES
R.Belo Horizonte 240/A
Manaus, Amazonas, Brazil, Brazil

Lorch, Andre
Reporter Radio/Television
TV ABRIL
Pamplona 1551/602
Sao Paulo 01405, Brazil, Brazil

Loredo De Izcue, Julioaugusto
Correspondent
REVISTA CATOLICISMO
Cons. Pedro Luiz 313
Santana, Sao Paulo, Brazil, Brazil

Louchard, Aimee Gomes
Reporter
O DIA
R. Riachuela 359
Rio De Janeiro, RJ 2, Brazil
T: 55 21 272-8000 F: 55 21 507-1228

Loureiro, Marcia
Reporter Radio/Television
REDE OM BRASIL DE TELEVISAO
Rua M.Pires Ferreira, 10/408
Cosme Velho, Rio de Janeiro-RJ, Brazil, Brazil

Lourenco, Carlos Vitor
Editor
RADIOBRAS
R.Pacheco Leao 1235 c/37
Rio de Janeiro, Brazil, Brazil

Lourenco, Tarcisio P.
Correspondent
DIARIO CATARINENSE
R. Pascoal Simone No.69, Apt18
Coqueiros, Florianopolis-SC, Brazil, Brazil

Lourival, Humberto
Editor
REDE OM BRASIL DE TELEVISAO
R. Alvares de Azevedo, 32/202
Icarai, Rio de Janeiro, Brazil, Brazil

Louven, Mariza
Editor
GLOBO
R. Saboia Lima, 1L/101
Tijuca, Rio de Janeiro-RJ, Brazil, Brazil

Lozano, Andre
Editor
FOLHA DE SAO PAULO
Al. Barao de Limeira, 425
Sao Paulo-SP, Brazil, Brazil

Lucas, Marco A.
Reporter Radio/Television
MANCHETE TV
Av. N.S. de Fatima, 74/702
Brazil, Brazil

Lucchese, Elizabeth
Reporter Radio/Television
GLOBO RADIO
R.Sampaio Viana 339 c/14
Rio de Janeiro, Brazil, Brazil

Luz, Benedito Paulo Da
Editor
TVSBT
Campo de S.Cristovao, 105
Rio de Janeiro - RJ, Brazil
T: 55 21 5803688 / 5800678 F: 55 21 5809685

Luz, Cristina R.M.
Reporter Radio/Television
REDE OM BRASIL TV
Av.S.Sebastiao 66/301
Rio de Janeiro, Brazil, Brazil

Luz, Leonardo Pimentel Da
Editor
MANCHETE
Rua Santa Sofia 115/102
20540-Rio de Janeiro-RJ, Brazil, Brazil

Luz, Sergio Pereira
Editor
JORNAL TRIBUNA DA IMPRENSA
Rua do Lavradio, 98
20230 Rio de Janeiro-RJ, Brazil, Brazil

Luz, Tercio
Editor
MANCHETE TV
SQS 208 Bl.E/102
Brasilia, D.F., Brazil, Brazil

Lyra, Paulo Shayer, Mr.
Correspondent
WWF-FUNDO MUNDIAL PARA A NATUREZA
SQS 213 Bl.J Apt.208
70292 Brasilia-DF, Brazil

Macedo, Antonio Carlos
Reporter Radio/Television
RADIO GAUCHA
R. Pedro Ivo. 206, ap. 203
90000 Porto Alegre, Brazil, Brazil

Macedo, Antonio Sergio
Reporter
FREELANCER
Rio de Janeiro - RJ, Brazil
T: 55 21 2247130 F: 55 21 2528455

Macedo, Eldio
Reporter Radio/Television
RADIO GAUCHA
Av.Nossa Sra.de Copacabana 903
Ap. 1102, 21060 Rio de Janeiro, Brazil, Brazil

Macedo, Lourival Antonio
Reporter Radio/Television
RADIOBRAS
SCRN 710/711 Bl. E Ap. 201 E40
Brasilia, D.F., Brazil, Brazil

Macedo, Wilton
Editor
GLOBO TV
Rua Von Martius, 22
Rio de Janeiro, Brazil, Brazil

Macedo Freire, Joao Batista
Editor
MANCHETE
R. Almirante Tamandare, 10/904
Flamengo, Rio de Janeiro-RJ, Brazil, Brazil

Machado, Alexandre
Reporter Radio/Television
TV EDUCATIVA - RJ
R. Sampaio Vidigal, 564
Sao Paulo, Brazil, Brazil

Machado, Altino
Correspondent
RADIO ELDORADO FM
Conj. M.Juliao, Q.10 Casa 11
69.900 Rio Branco, Acre, Brazil, Brazil

Machado, Alvaro Soledade
NOTICIAS EM DESTAQUE
Praia do Flamento 374, Apt.703
22210 Rio de Janeiro, Brazil, Brazil

Machado, Gisele
Editor
GLOBO TV
Rua Von Martius, 22
Rio de Janeiro, Brazil, Brazil

Machado, Marcia
Reporter Radio/Television
ONE WORLD ENVIRONMENTAL TV
218 E. 82nd
New York, 10028, USA, Brazil

Machado, Renata Fraga
Editor
DIA
R. Arthur Menezes, 15
20271-Rio de Janeiro-RJ, Brazil, Brazil

Machado, Renato
Reporter Radio/Television
MANCHETE TV
Rua Eng.Penna Chaves, 63/101
Rio de Janeiro-RJ, Brazil, Brazil

Machado, Rodolpho Villanova
Reporter Radio/Television
TV EDUCATIVA
Rua Marques de Abrantes 115
Rio de Janeiro, Brazil, Brazil

Machado De Souza, Carlos Alberto
Editor
FOLHA DE SAO PAULO
Al. Barao de Limeira, 425
Sao Paulo-SP, Brazil, Brazil

Machado Fernandes, Aroldo
Reporter Radio/Television
GLOBO TV
Rua Von Martius, 22 sala 328
Rio de Janeiro, Brazil, Brazil

Macieira De Souza, Maria Teresa
Editor
RADIOBRAS
SQN 408 Bl. M/201
Brasilia, D.F., Brazil, Brazil

Maciel, Candido Dos Santos De Abreu
Reporter Radio/Television
TV EDUCATIVA - RJ
R. Leopoldo, 1057 casa 3
Rio de Janeiro, Brazil, Brazil

Maciel, Ricardo Zarzur Burlini
Reporter Radio/Television
GLOBO RADIO
R. Frei Inocencio, 77 Bl 11
Rio de Janeiro, Brazil, Brazil

Madeira, Loane Barbosa P.
Editor
DIA
Av. Marechal Rondon, 2777/102
Rio de Janeiro-RJ, Brazil, Brazil

Madureira, Paulo Sergio De Jesus
Reporter Radio/Television
MANCHETE RADIO
Rua Jau, 4/302 E. Novo
Rio de Janeiro, Brazil, Brazil

Magalhaes, Heloisa Teixeira
Editor
GAZETA MERCANTIL
R. Mj. Quedinho, 90 - 4 andar
Sao Paulo, Brazil, Brazil

Magalhaes, Leila
Editor
JORNAL DO BRASIL
R. Leopoldina Rego, 372/Casa24
21021, Rio de Janeiro-RJ, Brazil, Brazil

Magalhaes, Maria Francecirley A.
Correspondent
DIFUSORA RADIO CAJAZEIRAS
R.Emidio Assis 159
Cajazeiras, Paraiba, Brazil, Brazil

Magalhaes, Paulo Francisco
Editor
GAZETA DE NOTICIAS
R. Firmino do Amaral 341
Freguesia, Jacarepagua, RJ, Brazil, Brazil

Magalhaes, Renato Silva De
Reporter Radio/Television
RADIO ROQUETE PINTO
Av. Mal. Fontenele, 3740/407
Sulacap, Rio de Janeiro, Brazil, Brazil

Magalhau, Andrea
Reporter Radio/Television
GLOBO RADIO
Rua Xavier Leal 15
Rio de Janeiro-RJ, Brazil, Brazil

Magnavita, Monica C.
Editor
REVISTA EXAME
Rua Felix da Cunha, 38/406
Tijuca, Rio de Janeiro-RJ, Brazil, Brazil

Maia, Alvaro Avila
Reporter Radio/Television
GLOBO RADIO
R. Barao., 717 Bl. E/101
Rio de Janeiro, Brazil, Brazil

Maia, Ana Claudia
Reporter Radio/Television
RADIO GUANABARA/BANDEIRANTES
Rua Antonio Badajos, 141
Madureira, Rio de Janeiro-RJ, Brazil, Brazil

Maia, Andrea Quintiere
Editor
RADIOBRAS
SQS 216 Bl. D Ap. 603
Brasilia, D.F., Brazil, Brazil

Maia, Katia Fabiana Chaves
Reporter Radio/Television
RADIOBRAS
SQN 215 Bl. G Ap.405
Brasilia, D.F., Brazil, Brazil

Maia, Marcio L.S.
Editor
RADIO SOCIEDADE PASSOS
R.Antonio Carlos 268
Passos, Minas Gerais, Brazil, Brazil

Makita, Takayoshi
Correspondent
KIODO NEWS SERVICE
Praia do Flamengo, 168/701
Rio de Janeiro - RJ, Brazil
T: 55 21 2852412 F: 55 21 2852270

Maklouf, Luiz C.
Correspondent
JORNAL DA TARDE
R. Prof. Djalma Bento 206/101
Sao Paulo-SP, Brazil, Brazil

Malaman, Conceicao
Reporter
LEGIAO DA BOA VONTADE
Av.Rudge, 700
Sao Paulo - SP, Brazil
T: 55 11 2202599 F: 55 11 2205196

Malbergier, Sergio
Editor
FOLHA DE SAO PAULO
Al. Barao de Limeira, 425
Sao Paulo-SP, Brazil, Brazil

Malheiros, Tania M.
Editor
AGENCIA ESTADO
R. Silveira Martins 24 Apt.401
Flamengo, Rio de Janeiro-RJ, Brazil, Brazil

Mallot, Christian
Correspondent
NEW BORDER FILMES
R. Nascimento Silva 312
Rio de Janeiro, Brazil, Brazil

Malta, Cynthia Bueno De Oliveira
GAZETA MERCANTIL
Rua Major Quedinho 90
Sao Paulo - SP, Brazil
T: 55 11 256 3133

Malta, Dacio
Editor
JORNAL DO BRASIL
R. Almeida Godinho, 26
22471, Rio de Janeiro-RJ, Brazil, Brazil

Maluf Filho, Aziz
Correspondent
FOLHA DA MANHA
R.Benjamin Constant, 66/407
Gloria, Rio de Janeiro-RJ, Brazil, Brazil

Mancini, Luciana
Editor
TELEPRESS LATINOAMERICA
Av. Paulista, 1159 Conj.1015
Sao Paulo-SP, Brazil, Brazil

Mancuso Negrao, Maria Antonieta
Editor
RADIOBRAS
SHIN QI05 Conj. 06 Casa 17
Brasilia, D.F., Brazil, Brazil

Manhaes, Nathercia Motta
Reporter Radio/Television
SBT TV
Rua Leite Ribeiro, 57/802
24120, Rio de Janeiro-RJ, Brazil, Brazil

Manso, Jussara
Editor
VISNEWS
Rua Italia D Incu 191
Rio de Janeiro, Brazil, Brazil

Mansur, Alexandre
Editor
ANO ZERO
R.Viuva Lacerda 128/201
22261 Rio de Janeiro-RJ, Brazil, Brazil

Manzur, Jose Alberto
Editor
GLOBO TV
Rua Von Martius, 22
Rio de Janeiro, Brazil, Brazil

Maranini, Nicolau Jose Carvalho
Reporter Radio/Television
JORNAL DO BRASIL-RADIO AM
Rua Visconde de Sta. Isabel
Rio de Janeiro-RJ, Brazil, Brazil

Marcal, Neise Martins
Reporter Radio/Television
GLOBO RADIO
Rua Prof. Edmundo March, 2
Niteroi, Rio de Janeiro-RJ, Brazil, Brazil

Marcelino, Francisco Altino
Reporter Radio/Television
MANCHETE TV
R.Gen.Ribeiro da Costa 230/203
Brazil, Brazil

Marcello, Jorge Guilherme P.
Editor
GLOBO RADIO
Est.Gavea Pequena 952 c/112
Rio de Janeiro, Brazil, Brazil

Marcelo, Madureira
Reporter Radio/Television
GLOBO TV
Ave. Alvorada, 2600
Tijuca, Rio de Janeiro-RJ, Brazil, Brazil

Marchand, Carlos Alberto Luiz
Reporter Radio/Television
BUZIOS TV
R.Alceu Amoroso Lima, 65/702
Rio de Janeiro, RJ, Brazil, Brazil

Marcolin, Neldson J.
Editor
ESTADO DE SAO PAULO
R.Dr.Pinto Ferraz 155/161
Vila Mariana, Sao Paulo-SP, Brazil, Brazil

Marcovici, Micaela R.
Editor
MARCO DE COMUNICACOES SC LTDA
Praca Vilaboim 78 Apt.51
Higienopolis, Sao Paulo-SP, Brazil, Brazil

Maria, Helen Martinho
Reporter Radio/Television
RBS TV
Rua Felipe de Oliveira, 292
Port Alegre, RS, Brazil, Brazil

Maria, Jose Carlos
Editor
GLOBO TV
Rua Von Martius, 22 sala 321
Rio de Janeiro, Brazil, Brazil

Mariano, Nilson
Editor
ZERO HORA
Av. Getulio Vargas,143 Ap. 305
Menino Deus, Porto Alegre-RS, Brazil, Brazil

Marinho, Jose Ricardo
Reporter Radio/Television
TV EDUCATIVA - RJ
R. Hadock Lobo 309/104
Tijuca, RJ, Brazil, Brazil

Marino, Alexandre
Correspondent
AGENCIA ESTADO
SQN 407 Bl.G Apt.311
Brasilia-DF, Brazil, Brazil

Marmori, Margareth E.
Correspondent
CIENCIA HOJE
SCLR 703 Bl.I 57 Asa Norte
Brasilia-DF, Brazil, Brazil

Marotta, Eduardo L.
Editor
MANCHETE TV
Rua Carolina Santos 113 C/3
Brazil, Brazil

Marques, Fabricio
Editor
EDITORA ABRIL
R.da Passagem, 123/8 andar
Rio de Janeiro, RJ, Brazil
T: 021 5468282 F: 021 2759347

Marques, Francisco Henrique Moura
Reporter Radio/Television
RADIOBRAS
SQS 102 Bl. D Ap. 111
Brasilia, D.F., Brazil, Brazil

Marques, Jose Reinaldo B.
Editor
AFRONEWS COMUNICACOES
Rua Petrocochino, 7/203
Rio de Janeiro-RJ, Brazil, Brazil

Marques, Joseti
Editor
REDE OM BRASIL DE TELEVISAO
R. H. Osvwald, 211/903
Copacabana, RJ, Brazil, Brazil

Marques, Jota Carlos
Reporter Radio/Television
JORNAL DO BRASIL-RADIO AM
Rua Espumas, 148/201
Rio de Janeiro-RJ, Brazil, Brazil

Marques, Marcia
Editor
FOLHA DE SAO PAULO
Al. Barao de Limeira, 425
Sao Paulo-SP, Brazil, Brazil

Marques, Randau
Editor
JORNAL DA TARDE
R.Isabel de Castela 105
05445 Sao Paulo-SP, Brazil, Brazil

Marques, Raul Beiriz
Editor
JORNAL DO COMMERCIO
R. M. Gaffre, 16/701
Grajau, Rio de Janeiro, Brazil, Brazil

Martan, Aroldo
Reporter Radio/Television
RADIO GUACU DE TOLEDO LTD.
Rua Santo Angelo, 587
Toledo, Parana, Brazil, Brazil

Martinez, Gisela L.
Correspondent
FOLHA DO TURISMO
R.Sen.Correa 33/401
Rio de Janeiro, Brazil, Brazil

Martinez, Jorge Daniel
Reporter Radio/Television
LBV-RADIO
Anhanquera, 704
Barra Funda, Sao Paulo-SP, Brazil, Brazil

Martinez, Monica
Editor
UNIVERSIDAD DE SAO PAULO
Rua Barao de Morenos, 106
05453 Sao Paulo -SP, Brazil, Brazil

Martins, Ana Maria
Correspondent
EMBRATUR
R.Lopes Quintas 340
Rio de Janeiro, Brazil, Brazil

Martins, Elaine Duim
Editor
FLUMINENSE
Travessa das Flores, 382
Mangueira, Rio de Janeiro, Brazil, Brazil

Martins, Elza P.
Correspondent
AGENCIA ESTADO
SHIN QI 05 Conj. 03 Casa 16
Brasilia-DF, Brazil, Brazil

Martins, Jorge De Azevedo
Reporter Radio/Television
TV EDUCATIVA - RJ
Ipinambes 340 - 126
Jacarepagua, RJ, Brazil, Brazil

Martins, Jose Pedro
Editor
CORREIO POPULAR
R. D.Presciliana Soares, 192
Apt.102, 13025 Campinas-SP, Brazil, Brazil

Martins, Lasier Costa
Reporter Radio/Television
RADIO GAUCHA
Av. Getulio Vargas, 1351
Ap. 708, 90060 Porto Alegre, Brazil, Brazil

Martins, Lincoln
Jornalista
REVISTA MANCHETE
Rua do Russel, 804
Rio de Janeiro - RJ, Brazil
T: 55 21 2652012

Martins, Luiz Carlos
Correspondent
AGENCIA ESTADO
R. dos Cristais, 58
17500 Marilia, Sao Paulo, Brazil, Brazil

Martins, Marcio Andre
Editor
DIA
R. Prof. Gabizo, 332/402
Rio de Janeiro-RJ, Brazil, Brazil

Martins, Sebastiao
Editor
RADIO JORNAL DO BRASIL
R.Joao da Mata, 160/301
Rio de Janeiro, Brazil, Brazil

Martins Pinto, Francisco Jr.
Editor
AG. DE COMMUNICACAO LTDA.
R. Lopes Quintas, No. 340
22460, Rio de Janeiro-RJ, Brazil, Brazil

Massaran, Luisa
Journalist
CIENCIA HOJE
Av Venceslau Bras 71 Fundos Casa 27
Rio de Janeiro, Brazil
T: 55 21 2954846 F: 55 21 5415342

Matheus, Fernando Leonardo
Reporter Radio/Television
TV EDUCATIVA - RJ
R. Dona Amelia 47
Andarai, RJ, Brazil, Brazil

Matos, Emilia
Reporter Radio/Television
TV EDUCATIVA - RJ
Jose Eulalio 691
Rio de Janeiro, Brazil, Brazil

Matta, Paulo S.
Editor
GLOBO
Rua Ataulfo de Paiva, 983
Leblon, Rio de Janeiro-RJ, Brazil, Brazil

Matta, Ubiratan Brazil
Editor
JORNAL DA TARDE
Rua Apinages, 1268/505
Sao Paulo, SP, Brazil, Brazil

Mattos, Grace Dantas
Editor
JORNAL DO BRASIL
R. M. F. de Moura, 149/301
Rio de Janeiro-RJ, Brazil, Brazil

Mattos, Lucia
Editor
PREFEITURA RIO DE JANEIRO
Rua Rita Ludolf 87/603
Rio de Janeiro, Brazil, Brazil

Mattos S., Jorge Luiz
Reporter Radio/Television
TV EDUCATIVA - RJ
R. Pedro Americo, 244/1009
Rio de Janeiro, Brazil, Brazil

Matzembacher, Luiz Oscar D.
Editor
ASSESSORIA DE IMPRENSA RGS
R.24 de Outubro, 1281/508
Porto Alegre, Brazil, Brazil

Maud, Isabel
Editor
GLOBO
Rua Clarice l'Indio do Brazil
22230 Botafogo, Rio de Janeiro-RJ,Brazil, Brazil

Maximo, Gabriela
Editor
JORNAL DO BRASIL
R. Bartolomeu Portela, 18/404
Botafogo, Rio de Janeiro-RJ, Brazil, Brazil

Mayrink, Jose Maria
Editor
JORNAL DO BRASIL
Rua Aimberi, 945
Sao Paulo-SP, Brazil, Brazil

Mazzarollo, Joanilda
Editor
GLOBO TV
Rua Von Martius, 22 sala 321
Rio de Janeiro, Brazil, Brazil

Mazzei, Paulo Roberto Britto
Editor
JORNAL BRAZIL TURISMO
Av. Nova York, 364/201
Bonsucesso, Rio de Janeiro-RJ, Brazil, Brazil

Mazzucco, Roberto M.
Editor
ESTADO DE SAO PAULO
R. Sabara 75 apt.101
Higienopolis, Sao Paulo-SP, Brazil, Brazil

McAllister, Andrew
Reporter Radio/Television
BUZIOS TV
R.Xavier da Silveira,99/702
Rio de Janeiro, RJ, Brazil, Brazil

Mecado, Alvaro
Editor
GAZETA DE NOTICIAS
Prais do Flamengo 374, Apt.703
22210 Rio de Janeiro, Brazil, Brazil

Medeiros, Andrea Da Silva Mury
Editor
JORNAL DO BRASIL-RADIO AM
Rua Toneleiros, 370/501
Copacabana, Rio de Janeiro-RJ, Brazil, Brazil

Medeiros, Fatima Beatriz F.
Reporter Radio/Television
RADIO JORNAL DO BRASIL
R.Bambina 50/403
Rio de Janeiro, Brazil, Brazil

Medeiros, Guilherme Da Motta
Editor
SBT TV
R. Guaracininga
162 Vila Guilherme, Brazil, Brazil

Medeiros, Jose Guilnerme
Reporter Radio/Television
TV EDUCATIVA - RJ
Av. Ataulfo de Paiva 386
Apt. 406, Rio de Janeiro, RJ, Brazil, Brazil

Medeiros, Marcelo
MTV
Rua Albuquerque Lins 565
Sao Paulo, Brazil, Brazil

Medeiros, Monica
Editor
GLOBO
SCS. Ed. Oscar Niemeyer, 13
Brasilia-DF, Brazil, Brazil

Meireles, Sergio
Reporter Radio/Television
GLOBO RADIO
R.Gal.Cristovao Barcelos 55
Rio de Janeiro, Brazil, Brazil

Mello, Geisa
Editor
AGENCIA ESTADO
SQN 314 Bl.E Apt.614
Brasilia-DF, Brazil, Brazil

Mello, Joana
Editor
EDITORA ABRIL/ELLE
R. Visconde de Caravelas98/907
22271 Rio de Janeiro-RJ,Brazil, Brazil

Mello, Maria Claudia Cavalcanti S.
Editor
RADIOBRAS
SHIS QI 16 Conj. 2 casa 6
Brasilia, D.F., Brazil, Brazil

Mello, Priscilla L.
Correspondent
GLOBO
Rua Noronha Torrezao, 395
Niteroi-RJ, Brazil, Brazil

Melo, Lea Maria C.B.
Editor
EMBRATUR
Rua Sa Ferreira, I38 apt.I03
Rio de Janeiro, Brazil, Brazil

Melo, Liana
Editor
GLOBO
Senador Vergueiro, 197/601
Rio de Janeiro-RJ, Brazil, Brazil

Melo, Placido Jose Alves De
Editor
GLOBO RADIO
Estrada dos Bandeirantes 8325
Rio de Janeiro-RJ, Brazil, Brazil

Melo Filho, Murilo
Editor
MANCHETE
Rua do Russel, 804
22210-Rio de Janeiro-RJ, Brazil, Brazil

Mendes, Agnaldo Soares
Reporter Radio/Television
SUPER RADIO TUPI
Rua Silveira Martins, 146
Rio de Janeiro-RJ, Brazil, Brazil

Mendes, Andre Trigueiro
Reporter
SISTEMA JORNAL DO BRASIL
Av.Brasil, 500
Rio de Janeiro - RJ, Brazil
T: 55 21 5854422

Mendes, Armando N.
Reporter
AGENCIA ESTADO
Av. Caetano Alvares, 55
Sao Paulo -SP 02598, Brazil
T: 55 11 8562210 F: 55 11 2652297

Mendes, Jorge R.
Correspondent
JORNAL FLOR DA MANGA
Trav.Retiro Saudo 247
Sao Goncalo, Rio de Janeiro, Brazil, Brazil

Mendes, Lucas
Correspondent
RECORD TV
2 E 12St. 3,
New York NY USA, Brazil

Mendonca, Jair Jayme Cardoso De
Editor
SBT TV
Travesso Jose Higino, 144
Tijuca, Rio de Janeiro, Brazil

Mendonca, Jose Carlos
Editor
AGENCIA ESTADO
Rua Carlos Vasconcelos, 116
Tijuca, Rio de Janeiro-RJ, Brazil, Brazil

Mendonca, Luiz Augusto Guimaraes
Reporter Radio/Television
RADIOBRAS
SHIN QL 11 Conj. 3 Casa 15
Brasilia, D.F., Brazil, Brazil

Mendonca, Magno Da S.
Editor
GLOBO ECOLOGIA
Rua Lucidio Lago 325409
Rio de Janeiro, Brazil, Brazil

Mendonca, Martha M.
Editor
DIA
Av. Maracana, 1320/404
Tijuca, Rio de Janeiro-RJ, Brazil, Brazil

Mendonca, Silvia
Editor
ROLF PRESS
R.Benjamim Constant 52/802
Rio de Janeiro, Brazil, Brazil

Mendonca, Stella Maris Cermeno
Reporter
RADIO NEDERLAND (FREELANCER)
R. Miguel de Resende 520 c/3 Santa Teresa
Rio de Janeiro -RJ 2, Brazil
T: 55 21 5022271

Menegato, Nestor Navarro
Reporter Radio/Television
TVT REDE DE COMUNICACAO
R. E. Luis C. Barbante, 197
Sao Paulo, Brazil, Brazil

Menezes, Carlos Alberto Cabral
Correspondent
TV BUZIOS
Rua Visc. de Albuquerque 324
Rio de Janeiro, Brazil, Brazil

Menezes, Guilherme
Reporter Radio/Television
MANCHETE TV
SQS 502 Bl.B Apt.105
Brasilia-DF, Brazil, Brazil

Menezes, Janice
Editor
JORNAL DO BRASIL
Av.Joao Brasil, 366/Bl.1/301
Niteroi, Rio de Janeiro, Brazil, Brazil

Menezes, Marcia
Editor
GLOBOSAT
R. Paula Freitas, 19/202
Copacabana, RJ, Brazil, Brazil

Menezes, Paula Cabral De
JORNAL TRIBUNA DA IMPRENSA
Rua do Lavradio, 98
20230 Rio de Janeiro-RJ, Brazil, Brazil

Menezes, Paulo
Editor
FEIRAS E CONGRESSOS
R.Marechal Tronpowsky,20
Cobertura 02, Rio de Janeiro-RJ,Brazil, Brazil

Menezes Neto, Pedro Vieira
Editor
CBIA
Av.Barao de Maruim,441/50
Acaraju, SE, Brazil, Brazil

Menezes Rocha, Julius
Editor
GLOBO
R.Eduardo Santos 161/301
Rio de Janeiro, Brazil, Brazil

Mengel, Rodolfo
Editor
JORNAL DA USP
Rua Bom Pastor, 350 c/6
04203 Sao Paulo-SP, Brazil, Brazil

Merces, Abimael
Reporter Radio/Television
TV EDUCATIVA - RJ
Rua Brasil 110/402
Rio de Janeiro, Brazil, Brazil

Mesquita, Rodrigo L.
Editor
AGENCIA ESTADO
R. Belgica, 107
01448 Sao Paulo-SP, Brazil, Brazil

Messias, Alex Barbosa
MANCHETE
R. do Russel 804
Rio de Janeiro -RJ 2, Brazil
T: 55 21 2652012 F: 55 21 2855148

Michahelles, Kristina
Editor
JORNAL DO BRASIL
R. Corcovado, 253/302
22460, Rio de Janeiro-RJ, Brazil, Brazil

Migiaccio, Marcelo B.
Correspondent
FOLHA DA MANHA
R.Ozorio de Almeida, 29/102
Urca, Rio de Janeiro-RJ, Brazil, Brazil

Miglioli, Leopoldo
Reporter Radio/Television
RADIO SENTINELA DO VALE LTDA
Rua do Riachuelo, 54 Gr.702
20.230 Ibirama, Sta. Catarina, Brazil, Brazil

Miguez, Ana Cristina
Editor
DIA
R. Candido Oliveira, 143/101
Rio Comprido, Rio de Janeiro-RJ, Brazil, Brazil

Milesi, Lauci
Reporter Radio/Television
GLOBO RADIO
R.Desembargador Izidro 36/201
Rio de Janeiro, Brazil, Brazil

Militao, Francisco
Reporter Radio/Television
RADIO UNIVERSIDADE
Trav.Antonio Costa 85
Niteroi, Rio de Janeiro, Brazil, Brazil

Mineiro, Ana Silvia
Editor
MANCHETE TV
R. Sao Clemente 365/202
Rio de Janeiro, Brazil, Brazil

Miranda, Alvaro
Editor
GLOBO
Rua Marechal Rondon, 477
Rio de Janeiro-RJ, Brazil, Brazil

Miranda Filho, Ricardo
Correspondent
JORNAL DO BRASIL
SHIN QI 02 Cj.06 c/08
Brasilia, D.F., Brazil, Brazil

Miranda Theodoro, Gerson
Editor
MAIORIA FALANTE
R. Venancio Ribeiro, 109/1001
Rio de Janeiro, Brazil, Brazil

Mmehta, Pradeep S.
Correspondent
RAJASTHAN PATRIKA
110 Lake Terrace
Calcuta, India, Brazil

Mocarzel, Evaldo
Editor
AGENCIA ESTADO
R. Newton Prado, 85
St.Rosa, Niteroi, Rio de Janeiro, Brazil, Brazil

Modernell, Renato
Editor
GLOBO CIENCIA
R.Fradique Coutinho, 1692-23
05416 Sao Paulo-SP, Brazil, Brazil

Molica, Jose F.
Correspondent
FOLHA DA MANHA
R.Cacapava, 20 Apt.504
Grajau, Rio de Janeiro-RJ, Brazil, Brazil

Monero, Mariucha
Editor
JORNAL DO BRASIL
R. Pio Correa, 80/401
22461, Rio de Janeiro-RJ, Brazil, Brazil

Monserrat Filho, Jose
Editor
CIENCIA HOJE
Rua Joaquim Murtinho,772/404
20.241 Rio de Janeiro-RJ, Brazil, Brazil

Monteiro, Anna Claudia Vianna
Sub-Editora De Cidade
SISTEMA JORNAL DO BRASIL
Av.Brasil, 500
Rio de Janeiro - RJ, Brazil
T: 55 21 5854422

Monteiro, Jose Carlos
Editor De Politica
SISTEMA JORNAL DO BRASIL
Av.Brasil, 500
Rio de Janeiro - RJ, Brazil
T: 55 21 5854422

Monteiro, Liege
Correspondent
JORNAL EXPRESSAO
Farme de Amoedo 152, Apt.401
Rio de Janeiro-RJ, Brazil, Brazil

Monteiro, Marco Antonio
Reporter Radio/Television
GAZETA MERCANTIL
Av. 28 de Setembro 288/703
Rio de Janeiro, Brasil 20551, Brazil

Monteiro, Maria Rozane S.
Editor
DIA
Av. pres. Rosewel, 396
Niteroi, Rio de Janeiro-RJ, Brazil, Brazil

Monteiro, Marion
Reporter Radio/Television
REDE OM BRASIL DE TELEVISAO
R. Santo Amaro, 51/903
Gloria, Rio de Janeiro, Brazil, Brazil

Monteiro, Roberta De Oliveira
Reporter Radio/Television
JORNAL DO BRASIL RADIO
Rua Conde de Bonfim,549/704
Tijuca, Rio de Janeiro-RJ, Brazil, Brazil

Monteiro, Valeria
Reporter
REDE GLOBO
R. Lopes Quintas 303
Rio de Janeiro -RJ 2, Brazil
T: 55 21 2947732

Montenegro, Marcelo
Editor
ECOLOGIA E DESENVOLVIMENTO
R. Silveira Martins 48 Apt.104
22221 Flamengo, Rio de Janeiro, Brazil, Brazil

Moraes, Antonio
SISTEMA GLOBO DE RADIO
R. de Russel 434
Rio de Janeiro RJ, Brazil
T: 55 21 2058282 F: 55 21 2054096

Moraes, Catia V.
Reporter Radio/Television
MANCHETE TV
R.Jangadeiro, 28
Rio de Janeiro-RJ, Brazil, Brazil

Moraes, Elza Helena Gimenez
Reporter Radio/Television
GLOBO RADIO
Rua Tavares Basios, 29
Rio de Janeiro-RJ, Brazil, Brazil

Moraes, Gustavo Luiz Gripp
Correspondent
TV EDUCATIVA
R.Andrade Neves 94
Nova Friburgo, R.J., Brazil, Brazil

Moraes, Paulo Roberto S.
FOLHA DA MANHA
Ave.Presidente Vargas 509/6and
Rio de Janeiro, Brazil, Brazil

Moraes, Renata
Editor
JORNAL DO BRASIL
R. Saint Roman, 182
Copacabana, Rio de Janeiro-RJ, Brazil, Brazil

Moraes, Stella Sodre Vieira De
Reporter
O DIA
R. Riachuela 359
Rio De Janeiro, RJ 2, Brazil
T: 55 21 272-8000 F: 55 21 507-1228

Moraes Duarte, Maria Cecilia De
Reporter Radio/Television
GLOBO RADIO
Av. Oswaldo Cruz, 81-1702
Rio de Janeiro-RJ, Brazil, Brazil

Moreira, Adriana Guimaraes
JORNAL TRIBUNA DA IMPRENSA
Rua do Lavradio, 98
20230 Rio de Janeiro-RJ, Brazil, Brazil

Moreira, Terezinha Cancella
Reporter Radio/Television
TVSBT -CHANNEL 11
Av.N.S.Copacabana, 115/II06
Rio de Janeiro, Brazil, Brazil

Moreira Andreatta, Maria Beatris
Editor
REVISTA QUALIDADE BRASIL
R. Carlos Spera, 52
S. Bernardo do Campo, SP, Brazil, Brazil

Morel, Mario
Editor
TV EDUCATIVA - RJ
R. General Glicerio, 176/702
Rio de janeiro, Brazil, Brazil

Morelli, Ana
Editor
RADIOBRAS
SQN, 316 Bl.I Ap. 208
Brasilia, D.F. Brazil, Brazil

Moreno, Jorge Bastos
Editor
O GLOBO
SCS Rd. Oscar Niemeyer 139
Brasilia, Brazil, Brazil

Moreno, Saulo Pinto
Editor
RADIOBRAS
Ql 23 Lote 09/11 Ap.306
Guara II, D.F. Brazil, Brazil

Moretto, Ana Lucia
Reporter Radio/Television
RADIO ELDORADO FM
R.Haddock Lobo, 403/16-B
Sao Paulo-SP, Brazil

Moretz-Sohn, Claudia
Editor
REVISTA EXAME
Rua Benjamim Constant, 149/301
Gloria, Rio de Janeiro-RJ, Brazil, Brazil

Moreyra, Sandra Maria
Editor
GLOBO TV
Rua Von Martius, 22
Rio de Janeiro, Brazil, Brazil

Morganci, Marcos
Editor
SBT TV
R. Frei Vital 246
Apt. 21, Santos, SP, Brazil, Brazil

Morrone, Giuliana Faria
Reporter Radio/Television
SBT TV
SQS 112 Bloco 'E'
Brasilia-DF, Brazil, Brazil

Mota, Estefania
Reporter Radio/Television
RADIO ALVORADA FM
Rua J.de Menezes 240/402
Vista Alegre, Rio de Janeiro-RJ, Brazil, Brazil

Mota, Lilian De Oliveira Dantas
Reporter Radio/Television
BUZIOS TV
Roberto Rowley Mendes 120/1010
Niteroi, RJ, Brazil, Brazil

Mota, Norma
Reporter Radio/Television
RADIOBRAS
R. Visc. Rio Branco, 763/105
Niteroi, RJ, Brazil, Brazil

Mota Rodrigues, Neumar
Editor
GLOBO TV
Rua Von Martius, 22 sala 321
Rio de Janeiro, Brazil, Brazil

Motta, Aydano A.
Editor
GLOBO
Rua Dona Mariana, 182, Apt.904
Botafogo, Rio de Janeiro-RJ, Brazil, Brazil

Motta, Claudio
Editor
GLOBO
Rua Tirol No. 153
Rio de Janeiro-RJ, Brazil, Brazil

Motta, Cristina M. Soares
Reporter Radio/Television
BLUE LIGHT PROD. E PROM. LTDA
Av. Sao Sebastiao, 275
Rio de Janeiro, Brazil, Brazil

Motta, Luiz Gonzaga F.
EMA VIDEO
SHIS QI.6 Cj.1 c/18
Brasilia, D.F., Brazil, Brazil

Motta, Paulo De A.
Editor
PROGRAMA GLOBO ECOLOGIA
R. Marques de Olinda, No. 61
Botafogo, Rio de Janeiro-RJ, Brazil, Brazil

Motta Dos S., Hamilton
RADIDOBRAS
Teodoro da Silva, 328/302
Rio de Janeiro, Brazil, Brazil

Moura, Debora Da Veiga
Produtora
PROVIDEO CINEMA E TV
SHIS Q1 21 Conj 7 casa 15 Lago Sul
Brasilia - DF, Brazil
T: 55 61 366 1851 F: 55 61 366 3182

Moura, Jose A.
Reporter Radio/Television
A.N.EDITORA/PRODUTORA
Rua Senador Pompeu 188/23
Rio de Janeiro, Brazil, Brazil

Moura, Miriam
Editor
AGENCIA ESTADO
SHIS -QI 21 Conj.2 Casa 17
Brasilia-DF, Brazil, Brazil

Moura, Walter Thomas W.
Reporter Radio/Television
MANCHETE RADIO
Praia do Flamengo 72/112
Rio de Janeiro, Brazil, Brazil

Mourad, Mohamad Said
Editor
JORNAL AL URUBAT
R.Fonseca da Costa 329
Sao Paulo, S.P., Brazil, Brazil

Moussali, Sergio Emilio
Correspondent
RADIO & TV BANDEIRANTES
R.Frei Caneca 218/12
Sao Paulo, S.P., Brazil, Brazil

Mouteira, Bonanca
Reporter
JORNAL DO COMMERCIO
Rua do Livramento, 189/4 andar
Rio de Janeiro, Brazil
T: 55 21 2336562 F: 55 21 2337690

Muggiati, Anna
Editor
JORNAL DO BRASIL
R. Epitacio Pessoa, 2142/1003
Lagoa, Rio de Janeiro-RJ, Brazil, Brazil

Muggiati, Roberto Fernando
Editor
MANCHETE
R. Real Grandeza, 82 Casa 3
22281-Rio de Janeiro-RJ, Brazil, Brazil

Muhana, Leticia V.
Editor
GLOBOSAT
Est.Tamba 401 Bl.3 Bl.111
Rio de Janeiro, Brazil, Brazil

Muller, Cristina
Editor
AGENCIA ESTADO
R. Francisco Leitao 163
Sao Paulo-SP, Brazil, Brazil

Munhoz, Claudio Goncalves
Editor
JORNAL PONCHE VERDE, RADIO SULINA
Av. Rui Barbosa, 818
Dom Pedrito - Rio Grande do Sul, Brazil

Muniz, Magda Lucia
Reporter Radio/Television
RADIO GUANABARA/BANDEIRANTES
Av. Atlantica 734/501
Leme, Rio de Janeiro-RJ, Brazil, Brazil

Muniz, Marise De Souza
Correspondent
CIENCIA HOJE
Rua Marechal Hermes 694/102
Belo Horizonte, MG, Brazil, Brazil

Muniz Fernandes, Fabio
Reporter Radio/Television
TV & RADIO BANDEIRANTES
R.R. Tavares km.13.5 Torre
Cap. Dantibes 34, Sao Paulo, Brazil, Brazil

Murphy, Thomas, Mr.
Correspondent
KNIGHT-RIDDER NEWS
Av. Paulista 2001/Sala 412
Sao Paulo - SP, Brazil
T: 55 11 287-3137 F: 55 11 289-0444
TL: 1137550

Nabuco, Helio d'Andrea Filho
Reporter Radio/Television
MANCHETE TV
Rua Canavieiras 207, 101
Rio de Janeiro, Brazil, Brazil

Nadai, Elvira C. De O.
Editor
REVISTA NOVA ESCOLA
Rua Haddock Lobo 403/D
Sao Paulo, S.P., Brazil

Nader, Carlos A.
Reporter Radio/Television
A.N.EDITORA/PRODUTORA
R.Antonio A. Rebello, 912
Sao Paulo, S.P., Brazil, Brazil

Nakamae, Ivan Jun
Editor
REVISTA GLOBO RURAL
Rua do Curtume, 665/5 andar
05065 Sao Paulo-SP, Brazil, Brazil

Nardelli, Rodrigo Pereira
Editor
RADIOBRAS
SQS 206 Bl. J Ap. 401
Brasilia, D.F., Brazil, Brazil

Nascimento, Antenor
Editor
VEJA
R.Joao Moura,434 Apt.3
05412 Sao Paulo-SP, Brazil, Brazil

Nascimento, Arcanjo Antonio Do
Reporter
O DIA
R. Riachuela 359
Rio De Janeiro, RJ 2, Brazil
T: 55 21 272-8000 F: 55 21 507-1228

Nascimento, Carlos
Reporter Radio/Television
GLOBO TV
Rua Von Martius, 22 sala 328
Rio de Janeiro, Brazil, Brazil

Nascimento, Daisy Castro Do
Reporter Radio/Television
RADIOBRAS
OAS 07 Bl. B Ap. 512 Octogonal
Brasilia, D.F., Brazil, Brazil

Nascimento, Denise Ramalho
Reporter
REDE BANDEIRANTES DE TELEVISAO
R.Alvaro Ramos, 492
Rio de Janeiro - RJ, Brazil
T: 55 21 2122072 F: 55 21 2957391

Nascimento, Gilberto Pereira
Editor
PREFEITURA DE NOVA IGUACU
R. Bagda, 345
Rio de Janeiro, Brazil, Brazil

Nascimento De Azeredo, Ricardo
Reporter Radio/Television
RBS TV
Rua Mucioa Teixeira, 585
Porto Alegre, RS, Brazil, Brazil

Nass, Hermann
Editor
GLOBO
R. Dezenove de Fevereiro, 146
Rio de Janeiro-RJ, Brazil, Brazil

Nassar, Jose Roberto
Editor
EDITORA GLOBO
Rua Ministro Godoi, 657/111
05015 Sao Paulo-SP, Brazil, Brazil

Navarro Filho, Carlos
Editor
RADIO ELDORADO FM
Av.Paulo VI, Q.D, Bl.A Apt.201
Pituba, Salvador-BA, Brazil, Brazil

Nazario, Maisa De Lacerda
Editor
REVISTA CONTACTO
R.Joao Moura 1373/11
Sao Paulo, S.P., Brazil, Brazil

Negreiros, Mario Luis Paiva
Reporter Radio/Television
JORNAL DO BRASIL-RADIO AM
Rua Timoteo da Costa, 444
Leblon, Rio de Janeiro-RJ, Brazil, Brazil

Negromonte, Marcia F.
Editor
AGENCIA ESTADO
R.Dr. Alfredo Ellis 262/02
01322-Sao Paulo-SP, Brazil, Brazil

Nehrer, Ricardo
Editor
PROGRAMA GLOBO ECOLOGIA
R. Gen. Urquisa, No. 242
Leblon-RJ, Brazil, Brazil

Nelo, Liana
Editor
GLOBO
Senador Vergueiro, 197/601
Rio de Janeiro-RJ, Brazil, Brazil

Nemes, Nemer
Reporter Radio/Television
RADIO METROPOLITANA
R. Prof. Gabiso, 38/103
Rio de Janeiro, Brazil, Brazil

Nemes, Wilson
Editor
RADIO METROPOLITANA
R. Prof. Gabiuso, 38/103
Rio de janeiro, Brazil, Brazil

Neres, Luciana
Editor
JORNAL DO COMMERCIO
R. Sabino Barroso, 318
Vila Valqueire, RJ, Brazil, Brazil

Neto, Dermeval Coutinho
Reporter Radio/Television
TV EDUCATIVA - RJ
R. Gaviao Peixoto, 288/601
Brazil

Neto Carneiro De Moraes, Geneton
Editor
GLOBO TV
Rua Von Martius, 22 sala 328
Rio de Janeiro, Brazil, Brazil

Netto, Jose Carlos
Editor
GAZETA DE NOTICIAS
R. Marques de Aracati, #116
Iraja 21231 Rio de Janeiro, Brazil, Brazil

Netto, Patricia
Reporter Radio/Television
GLOBOSAT
Av.Alexandre Ferreira 300/302
Rio de Janeiro, Brazil, Brazil

Neusstadt, Monica N.
Reporter Radio/Television
RADIO UNIVERSIDADE
R.Haddock Lobo 379/302
Rio de Janeiro, Brazil, Brazil

Neves, Edmilson
Editor
TV JOVEM PAN
Rua Ed.Juventino Fontes,180
Vila Prudente, Sao Paulo, Brazil, Brazil

Neves, Maria Heloisa Cardoso
Reporter Radio/Television
FOLHA DE SAO PAULO
S.QS 108, Bloco A, apto. 505
Brasilia, DF, Brazil, Brazil

Neves, Marina Isabel
Editor
PROVIDEO
R. Prof. Arthur, 48/202
Leblon, RJ, Brazil, Brazil

Neves, Napoleao De Castro
Reporter Radio/Television
RADIO CLUB DE PERNAMBUCO
R. Dom Rinaldi, 50
Centro-Carpina-Pernambuco, Brazil, Brazil

Neves, Rogerio Ferreira Das
Editor
SBT TV
R. Silvio Dante Bertach
1059 Vila Sonia, Brazil, Brazil

Neves, Rubens Ruy Silva
Reporter Radio/Television
MANCHETE TV
Rua 15 de Novembro 175 C/5
Brazil, Brazil

Neves De Assis, Luiz Henrique
Correspondent
HORA DO POVO
Padre Roma 246/202
Sao Paulo, Brazil, Brazil

Ninomiya, Phillip H.
Reporter Radio/Television
MANCHETE TV
Est.da Urussanga, 225/202
Freguesia, Jacarepagua, RJ, Brazil, Brazil

Nitrini, Dacio
Editor
TVSBT
Campo de S.Cristovao, 105
Rio de Janeiro - RJ, Brazil
T: 55 21 5803688 / 5800678 F: 55 21 5809685

Nogueira, Elisabete De F.
Reporter Radio/Television
RADIO UNIVERSIDADE
R. Pascal 580/301
Rio de Janeiro, Brazil, Brazil

Nogueira, Gabriel
Editor
AGENCIA ESTADO
R. Bartolomeu Portela 28/203
Botafogo, Rio de Janeiro-RJ, Brazil, Brazil

Nogueira, Gilberto Da Conceicao De S.
Reporter Radio/Television
TV CULTURA DO PARA
Cidade 5, WE 6E, 10-91
Belem, Para, Brazil, Brazil

Nogueira, Paulo Ricardo Passos
Reporter Radio/Television
SBT TV
Rua Rainha Guilhermina
22441, Rio de Janeiro-RJ, Brazil, Brazil

Nogueira, Paulo Roberto De P.
Reporter Radio/Television
RADIOBRAS
SQN 403 m-204
Brasilia, DF, Brazil, Brazil

Nogueira, Rui
Coordinator
FOLHA DE SAO PAULO
Alameda Barao de Limeira 425
Sao Paulo - SP, Brazil
T: 55 11 874 2608/874 2602 F: 55 11 2231644

Nolasco, Patricia
Editor
DIA
Rua Gal. Severiano, 205/404
Botafogo, Rio de Janeiro-RJ, Brazil, Brazil

Nonato, Estrella
Editor
GLOBO TV
Ave. Alvorada, 2600
Tijuca, Rio de Janeiro-RJ, Brazil, Brazil

Norberto, Natalicio
Editor
FOLHA DO RIO
Praia Botafogo 416/1104
Rio de Janeiro, Brazil, Brazil

Novaes, Viviane
Reporter Radio/Television
MANCHETE TV
R.Pinneiro Machado,60/104
Rio de Janeiro, Brazil, Brazil

Novaes, Washington
Editor
JORNAL DO BRASIL
SHIS, QI5 Conj. 18 c.5 71635
Brasilia, D.F., Brazil, Brazil

Novis, Carlos Henrique
Reporter
MTV - RIO
Praca Nossa Senhora da Gloria 46
Rio de Janeiro -RJ, Brazil
T: 55 21 2855946 F: 55 21 2055394

Nunes, Angelina S.
GLOBO
Av. 28 de Setembro, 269
Vila Isabel, Rio de Janeiro-RJ, Brazil, Brazil

Nunes, Leticia
GLOBO
Rua Antonio Silva, 54, Apt.301
Niteroi, Rio de Janeiro-RJ, Brazil, Brazil

Nunes, Luciana
DIA
Rua Cinco de Junho, 335/803
Copacabana, Rio de Janeiro-RJ, Brazil, Brazil

Nunes, Ma. Clara M.
MANCHETE TV
Estrada da Gavea, 847/303
Rio de Janeiro-RJ, Brazil, Brazil

Nunes, Maria Alice
Editor
REVISTA TOUR IN RIO
Av.Copacabana 960/502
Rio de Janeiro, Brazil, Brazil

Nussenzveig, Micheline
Editor
CIENCIA HOJE
Rua Hilario de Gouveia 81/602
22040 Rio de Janeiro-RJ, Brazil, Brazil

Obliziner, Alfredo
Editor
CORREIO BRAZILIENSE
SHIN-QI-4 Con.6 C/5
Brasilia-DF, Brazil, Brazil

Ohta, Noriko
Editor
GLOBO
Rua Marques, 17, Apt. 302
Botafogo, Rio de Janeiro-RJ, Brazil, Brazil

Oinegue, Eduardo
Editor
VEJA
SQS 108 Bl.K Apt.405
Brasilia-DF, Brazil, Brazil

Olavo Junqueira Dantas, Fernando
Reporter Radio/Television
GAZETA MERCANTIL
R. Aperana, 57/203
Rio de Janeiro, Brazil, Brazil

Oliveira, Adriana
Editor
GLOBO
Ave. Editacio Dessoa 2780/810
Rio de Janeiro-RJ, Brazil, Brazil

Oliveira, Auderi Alencar De
Editor
REDE GLOBO
R. Lopes Quintas 303
Rio de Janeiro -RJ 2, Brazil
T: 55 21 2947732

Oliveira, Barbara
Correspondent
JORNAL DA TARDE
Largo Sta Cecilia 158 Apt.74
Sao Paulo-SP, Brazil, Brazil

Oliveira, Carla Cristina Pereira De
Redatora
SISTEMA JORNAL DO BRASIL
Av.Brasil, 500
Rio de Janeiro - RJ, Brazil
T: 55 21 5854422

Oliveira, Carlos
Correspondent
ECABE PRESS
Av.Ricardo Medina Filho, 697
S.Paulo-SP, Brazil, Brazil

Oliveira, Danielle
Reporter Radio/Television
GLOBO RADIO
Rua Dias da Rocha, 20/1004
Copacabana, Rio de Janeiro-RJ, Brazil, Brazil

Oliveira, Denise Ribeiro
Correspondent
CACHOEIRAS JORNAL
Rua Uruguay 228/402
Rio de Janeiro, Brazil, Brazil

Oliveira, Diogenes Lopes De
Editor
RADIO METROPOLITANA
R. Prof. Gabiso, 38/103
Rio de Janeiro, Brazil, Brazil

Oliveira, Edmundo
Editor
JORNAL DA TARDE
R. Antonio Gomide 238
Sao Paulo-SP, Brazil, Brazil

Oliveira, Fernando O.
MANCHETE TV
Av. Afr.Melo Franco 422/104-B
Alto Teresopolis, RJ, Brazil, Brazil

Oliveira, Francisco G. C.
Reporter Radio/Television
DIFUSORA RADIO CAJAZEIRAS
R.Emidio Assis 159
Cajazeiras, Paraiba, Brazil, Brazil

Oliveira, Helen Regina M.
Reporter Radio/Television
TVN
Alameda Jaiaperi 1083/51
Sao Paulo, S.P., Brazil, Brazil

Oliveira, Iris A.
Editor
GLOBO RADIO
R.Vinte Abril 28/204
Rio de Janeiro, Brazil, Brazil

Oliveira, Jose Carlos Da Silva
Reporter Radio/Television
JORNAL DO BRASIL-RADIO AM
Rua Maria Calmon, 91/212
Meier, Rio de Janeiro-RJ, Brazil, Brazil

Oliveira, Luciana R.
Editor
RADIO UNIVERSIDADE
R. Barao Mesquita 70/502
Rio de Janeiro, Brazil, Brazil

Oliveira, Luciano Barroso De
Reporter Radio/Television
RADIOBRAS
Quadra 02 Conj. A Casa 17
Candangolandia, D.F., Brazil, Brazil

Oliveira, Luiz Alberto De
Reporter Radio/Television
RADIOBRAS
Q 1.311 Bl. H Ap. 307
Cruzeiros Novo, D.F., Brazil, Brazil

Oliveira, Luiz Carlos De
Reporter Radio/Television
TV-BANDEIRANTES-RIO
Rua Marques de Abrantes
Rio de Janeiro-RJ, Brazil, Brazil

Oliveira, Ma. Aparecida
Editor
ESTADO DE SAO PAULO
R.Costa Aguiar 875/12
04204 Ipiranga, Sao Paulo, Brazil, Brazil

Oliveira, Maria Claudia De
Editor
O GLOBO
R. G. Marconi, 61/804
Rio de Janeiro, Brazil, Brazil

Oliveira, Maria Helena Alves
Reporter Radio/Television
REDE OM BRASIL DE TELEVISAO
Condominio Green Park,328
Itaipu, Niteroi, RJ, Brazil, Brazil

Oliveira, Nilandi C.
Editor
A NOTICIA
B. de Petropolis, 721,casa24
Rio de Janeiro, Brazil, Brazil

Oliveira, Nilson De
Editor
FOLHA DE SAO PAULO
Alameda Barao de Limeira 425
Sao Paulo SP, Brazil
T: 55 11 874 2608/874 2602 F: 55 11 2231644

Oliveira, Paulo Augusto Santos
Editor
JORNAL DO COMMERCIO
Av. Princesa Isabel, 214/902
Copacabana, RJ, Brazil, Brazil

Oliveira, Rozane Ma.
Editor
TRIBUNA DA IMPRENSA
R.Prof.Cantanhede 311/403
Laranjeiras, Rio de Janeiro-RJ, Brazil, Brazil

Oliveira, Walter Clemente De
GAZETA MERCANTIL
Rua Major Quedinho 90
Sao Paulo - SP, Brazil
T: 55 11 256 3133

Oliveira Duarte, Neide
Reporter Radio/Television
GLOBO TV
Rua Von Martius, 22 sala 328
Rio de Janeiro, Brazil, Brazil

Oliveira E Souza, Sheila Maria De
Reporter Radio/Television
JORNAL HORA DO POVO
R. Costa 41/32
Rio de Janeiro, Brazil, Brazil

Oliveira Filho, Idalicio M.
Editor
PREFEITURA RIO DE JANEIRO
R.Teodoro Silva 917 c/12
Rio de Janeiro, Brazil, Brazil

Oliveira Luz, Nina Rosa De
Reporter Radio/Television
TV EDUCATIVA - RJ
R. Borges de Medeiros 2415/302
Rio de Janeiro, Brazil, Brazil

Onofre, Jose
Editor
ZERO HORA
Rua Pareci, 518
Porto Alegre-RS, Brazil, Brazil

Ortman, Carlos Oscar A. De Souza
Reporter Radio/Television
TV EDUCATIVA - RJ
R. Itapema 127/101
Engenho de Dentro, RJ, RJ, Brazil, Brazil

Ota, Hajime
Diretor Correspondente
NHK
Av Nilo Pecanha 50 Grupo 1208
Rio de Janeiro - RJ, Brazil
T: 55 21 262 9777 F: 55 21 262 9514

Outeda, Carmen S.
Reporter Radio/Television
TV GAZETA-CANAL 11
Av.Lins de Vasconcelos 1961/86
01537 Sao Paulo-SP, Brazil, Brazil

Ozorio, Adriana De Almeida
SISTEMA GLOBO DE RADIO
R. de Russel 434
Rio de Janeiro RJ, Brazil
T: 55 21 2058282 F: 55 21 2054096

Ozorio, Rodrigo Otavio De A.
Reporter Radio/Television
GLOBO RADIO
Rua Cosme Velho 1041/501
Rio de Janeiro, Brazil, Brazil

Pacola, Jose F.
Editor
RADIO ELDORADO FM
R.Barreto Leme, 915 Apt.43
Campinas, Sao Paulo, Brazil, Brazil

Padilha, Laura Beatriz Duque De Magalha
Reporter Radio/Television
Radiobras
SQS 114 Bloco G 103
Brasilia,Brazil, Brazil

Padinha, Antonio De Padua Silva
Editor
DIARIO DE FRANCA
R. Maria C. Archetti 2877
Franca, Sao Paulo, Brazil, Brazil

Padovani, Renato
Reporter Radio/Television
BUZIOS TV
R.Pereira da Silva,444/c04
Laranjeiras, Rio de Janeiro,RJ, Brazil, Brazil

Padua, Tarcisio
Editor
INTERFACE
Quadra 10 con. H, casa 05
Sobradinho, Brasilia-DF, Brazil, Brazil

Paes, Luciene Soares
Reporter Radio/Television
JORNAL DO BRASIL-RADIO AM
Rua Marques de Olinda
Botafogo, Rio de Janeiro-RJ, Brazil, Brazil

Paglia, Ernesto George
Reporter Radio/Television
GLOBO TV
Pca. Marechal Deodoro 356
Sao Paulo, Brazil, Brazil

Pahim, Fabio
Editor
JORNAL DA TARDE
Av. Arruda Botelho 490 Apt.102
05466 Sao Paulo-SP, Brazil, Brazil

Paiva, Anabela
Correspondent
JORNAL DO BRASIL S/A
211 Winsconsin Ave, NW Ap.723
Washington, D.C. USA, Brazil

Paiva, Fernando
Editor
EDITORA ABRIL/ELLE
R.Dr.Veiga Filho 176/41
Sao Paulo-SP, Brazil, Brazil

Palatnik, Ilana
MANCHETE TV
Rua Igarapava, 80/301
Brazil, Brazil

Paldes, Renata Maria A.
Reporter Radio/Television
RADIO UNIVERSIDADE
R.Pinheiro Cunha 110/201
Rio de Janeiro, Brazil, Brazil

Paletta, Mauricio
Executive Producer
GLOBO ECOLOGIA PROGRAMME
Rua Fadel Fadel 54/101 Leblon
Rio De Janeiro, Brazil
T: 021 2395257 & 021 28 F: 021 2739087

Palhares, Joanna B.
Editor
JORNAL DE TURISMO
Av. Ores. A. Carlos, 25/602
Rio de Janeiro,Brazil, Brazil

Palomino, Joao
Reporter Radio/Television
SBT TV
Rua Joao Passalacqua, 140/52
Sao Paulo-SP, Brazil, Brazil

Pamplona, Walmor M.
Editor
GLOBO
Rua Pedra da Gavea, 50
Rio de Janeiro-RJ, Brazil, Brazil

Pan Chacon, Monica F.
Correspondent
JORNAL DO BRASIL-RADIO AM
Gontrischstrasse, 4
3006 Bern, Switzerland, Brazil

Panayotis, Courcouroupoulos
Reporter Radio/Television
GLOBO TV
Rua Von Martius, 22 sala 328
Rio de Janeiro, Brazil, Brazil

Pannunzio, Fernanda
Reporter Radio/Television
TV JOVEM PAN
R.Itapetininga, 243
Sorocaba Sao Paulo-SP, Brazil, Brazil

Pantazopoulos, Achiles
Reporter Radio/Television
RADIOBRAS
SQN 216 Bl. K Ap. 504
Brasilia, D.F., Brazil, Brazil

Paranhos Fernandes, Marcelo
Reporter Radio/Television
GLOBO TV
Av.Alvorada 2600-Barra
Rio de Janeiro, Brazil, Brazil

Parreira, Adriana De Mendonca
Editor
TV EDUCATIVA - RJ
R. Gilberto Cardoso, 230/2503
Leblon, RJ, Brazil, Brazil

Passos, Juarez M.
Reporter Radio/Television
MANCHETE TV
Ladeira dos Trabajaras 126/302
Brazil, Brazil

Passos, Marcelo A.
Reporter Radio/Television
MANCHETE TV
Rua Amianto 388
Belo Horizonte, Minas Gerais, Brazil, Brazil

Paul, Gustavo
Editor
VEJA
SQN 105 Bl.H Apt.305
Brasilia-DF, Brazil, Brazil

Paula, Adriana Augusta N.
Reporter Radio/Television
REDE OM BRASIL TV
Av.Sernambetiba 3500 II/1103
Rio de Janeiro, Brazil, Brazil

Paula, Regis Adriano De
Editor
JORNAL DO BRASIL-RADIO AM
Rua Julio Horta Barbosa, 309
Rio de Janeiro-RJ, Brazil, Brazil

Paula, Vera Lucia
Correspondent
JORNAL DOS CONSTRUTORES
Dulce 858, Olinda
Nilopolis, RJ, BRAZIL, Brazil

Paulino, Antonio Carlos
REVISTA OPODER
R. Quirino dos Santos, 355/202
11ha doGovernador, RJ, RJ, Brazil, Brazil

Paulino Neto, Fernando
Reporter Radio/Television
GAZETA MERCANTIL
R. Sao Salvador, 99/703
Rio de Janeiro, Brazil, Brazil

Paulo, Ivani Flora
Reporter Radio/Television
TELENEWS SERVICE LTDA.
Rua Barao da Torre 464/101
Ipanema, Rio de Janeiro, Brazil, Brazil

Paulo, Jose Wilson
Reporter Radio/Television
CIRANDA DOS BAIRROS
R.Visc.Cabo Frio 15/104
Rio de Janeiro, Brazil, Brazil

Pecego, Rodrigo
Correspondent
VIA URBANA
R. Sambaiba, 479 Apt.101
Lebon, Rio de Janeiro-RJ, Brazil, Brazil

Pecorelli, Maria
Editor
MANCHETE
Rua do Russell, 766
22210-Rio de Janeiro-RJ, Brazil, Brazil

Pedreira, Antonio Dos S.
Editor
REVISTA OPODER
R. Henrique Lacombe, 344/101
11he do Governador, RJ, RJ, Brazil, Brazil

Pedreira, Mauricio
Editor
VIDEO CIENCIA PRODUCOES LTDA.
Rua Vis. de Caravelas, 41/404
22271 Rio de Janeiro-RJ, Brazil, Brazil

Pedrosa, Fernanda
Editor
JORNAL DO BRASIL
Rua 61, Casa 55
24350, Rio de Janeiro-RJ, Brazil, Brazil

Peixoto, Ari Chaveut
Reporter
REDE GLOBO
R. Lopes Quintas 303
Rio de Janeiro -RJ 2, Brazil
T: 55 21 2947732

Peixoto F. Vieira, Carlos
Editor
GLOBO TV
Rua Von Martius, 22 sala 321
Rio de Janeiro, Brazil, Brazil

Peleias, Sandra Regina Dos S.
Editor
PREFEITURA RIO DE JANEIRO
R.Candido Mendes 129/202
Rio de Janeiro, Brazil, Brazil

Pelissari, Jose H.
Editor
TV GAZETA-CANAL 11
R.Alm.Marques de Leao, 638/141
Bl.A 01330 Sao Paulo-SP, Brazil, Brazil

Pellegrini, Mayriuce Velloso Villela
Reporter Radio/Television
RADIOBRAS
SQN 416 Bl. J Ap. 107 Norte
Brasilia, D.F., Brazil, Brazil

Penha, Gerson
Correspondent
ESTADO DE SAO PAULO
R.Rodrigues de Campos Leite100
Sao Paulo-SP, Brazil, Brazil

Penna, Andrea Monteiro
Editor
HORA DO POVO
Rua Visconde de Piraja,180/901
Ipanema, Rio de Janeiro-RJ, Brazil, Brazil

Penna, Marcia Mello
Editor
CONTEXTO AG. DE FOTOJORNALISMO
R. Newton Prado 32, Apt. 801
Niteroi, Rio de Janeiro, Brasil 24240, Brazil

Pereira, Amandio B.
Reporter Radio/Television
TV EDUCATIVA - RJ
Pincesa Isabel 350 - 1207
Rio de Janeiro, Brazil, Brazil

Pereira, Ana Albertina Evelin
Editor
RADIOBRAS
SQS 306 Bl. G/102
Brasilia, D.F., Brazil, Brazil

Pereira, Antonio Jr.
Editor
PROGRAMA GLOBO ECOLOGIA
Av. Marica, 255
Jd. Alcantara-SG, RJ, Brazil, Brazil

Pereira, Citene G.
Editor
GLOBO
P.Heloi Sa Camargo, 803
Sao Paulo-SP, Brazil, Brazil

Pereira, Edmar
Editor
JORNAL DA TARDE
R. Teodoro Herzl 49/41
Sao Paulo-SP, Brazil, Brazil

Pereira, Elson Lino
Editor
TRIBUNA DA IMPRENSA
Av. Min.Edgard Romero, 715/308
Madureira, RJ, Brazil, Brazil

Pereira, Joaquim Jose Ribeiro
Editor
GLOBO RADIO
Rua Casemiro de Abreu, 60
Rio de Janeiro-RJ, Brazil, Brazil

Pereira, Maria R.L.
Editor
DIA
Rua Palmira Goncalvez Maia
Tijuca, Rio de Janeiro-RJ, Brazil, Brazil

Pereira, Pablo
Editor
FOLHA DE SAO PAULO
Al. Barao de Limeira, 425
Sao Paulo-SP, Brazil, Brazil

Pereira, Paulo F. De Camargo
Editor
ESTADO DE SAO PAULO
R. Visconde de Ouro Preto 138
Apt.23 Consolocao, Sao Paulo-SP, Brazil, Brazil

Pereira, Paulo G.
Correspondent
DIARIO DE PERNAMBUCO
Av.Mario Melo, 612/05
50.050 Sto.Amaro, Recife-PE, Brazil, Brazil

Pereira, Paulo Renato C.
JORNAL TELENEGOCIOS
Rua Humaita, 207-906
Botafogo, Rio de Janeiro-RJ, Brazil, Brazil

Pereira, Robson
Editor
AGENCIA ESTADO
R. Mapendi 359 Taquara
Jacarepagua, Rio de Janeiro, Brazil, Brazil

Pereira, Sergio Leo De Almeida
Editor
O GLOBO
SCS. Ed. Oscar Nyemeyer 13
Brasilia, DF, Brazil, Brazil

Pereira, Sergio R.
Editor
VEJA RIO
Rua Candido Mendes, 279/502
Gloria, Rio de Janeiro-RJ, Brazil, Brazil

Pereira, Waldir Muniz
ORBE-PRESS
Rua Carlos Pimenta, 446
24400 Rio de Janeiro-RJ, Brazil, Brazil

Pereira A. Rodrigues, Margarida Maria
Editor
GLOBO TV
Rua Von Martius, 22 sala 321
Rio de Janeiro, Brazil, Brazil

Pereira Dos Santos, Eliane
Reporter Radio/Television
RADIO ELDORADO
Tr. Manoel Martinho, 58
Sao Paulo, Brazil, Brazil

Pereira Filho, Eliakim Araujo
Reporter Radio/Television
MANCHETE TV
Rua Jose de Brito 35
Rio de Janeiro, Brazil, Brazil

Pereira Lisboa, Ricardo
Editor
GLOBO TV
Rua Von Martius, 22 sala 321
Rio de Janeiro, Brazil, Brazil

Perelberg, Katia
Reporter Radio/Television
JORNAL DO COMERCIO
R.Marquesa dos Santos,77/805
Laranjeiras, Rio de Janeiro, Brazil, Brazil

Peres, Giovani De Almeida
Reporter
ESTADO DE MINAS
R. Goias 36 Centro
Belo Horizonte -MG 3, Brazil
T: 55 31 2732322 F: 55 31 2268070

Perez, Fabbio
Reporter Radio/Television
GLOBO TV
Rua Von Martius, 22 sala 321
Rio de Janeiro, Brazil, Brazil

Perez, Regina
Correspondent
FERROVIARIA E VIA URBANA
Av.Atlantica, 792 apt.401
Leme, Rio de Janeiro-RJ, Brazil, Brazil

Perin, Katia
Correspondent
JORNAL DA TARDE
Av. Francisco Matarazzo 108/63
Agua Branca, Sao Paulo, Brazil, Brazil

Perin, Orivaldo
Editor
JORNAL DO BRASIL
Rua Pereira Nunes, 90
24230, Rio de Janeiro-RJ, Brazil, Brazil

Pessanha, Ana Lucia F. V.
Reporter Radio/Television
TV-BANDEIRANTES-RIO
Rua Pinheiro Machado, 103/801
Rio de Janeiro-RJ, Brazil, Brazil

Pessano, Jorge
Editor
TV EDUCATIVA
Rua 24 Lt.6 Q.79
Itapu, Niteroi, RJ, Brazil

Pessoa, Isa Therezinha E.S.
Editor
GLOBO TV
R.Carlos Gois 476/201
Rio de Janeiro, Brazil, Brazil

Petraglia, Claudia Guimaraes
Editor
TV-BANDEIRANTES-RIO
Rua Alvaro Ramos, 492 B
Rio de Janeiro-RJ, Brazil, Brazil

Petropolis, Ibsen Spantacus Dias
Reporter
DCI - DIARIO COMERCIO e INDUSTRIA
Av.Rio Branco, 177/6 andar
Rio de Janeiro - RJ, Brazil
T: 55 21 2621016 F: 55 21 2200805

Petry Fragoso, Luiz Alberto
Editor
GLOBO TV
Rua Von Martius, 22 sala 328
Rio de Janeiro, Brazil, Brazil

Petti, Carin
Editor
JORNAL DA TARDE
R. Madalena 317
05434 Vila Madalena-SP, Brazil, Brazil

Pfeifer, Claudia Escobar
Reporter Radio/Television
TV EDUCATIVA - RJ
R. Nascimento Silva, 451/4-1
Rio de Janeiro, Brazil, Brazil

Piasentini, Maria Cristina
Editor
GLOBO TV
Rua Von Martius, 22 sala 328
Rio de Janeiro, Brazil, Brazil

Pinheiro, Francisco De A.
RADIO TV BANDEIRANTES
Rua Radiantes, 13
Sao Paulo, Brazil, Brazil

Pinheiro Machado, Paulo De Tarso
Editor
OFFICIAL PRESS ESTADO RS
Av. Independencia, 986/1103
Rio Grande do Sul, Brazil, Brazil

Pinho, Elizabeth Fernandez
REVISTA EXPRESSAO
Rua Retiro dos Armisama 607
Rio de Janeiro-RJ, Brazil, Brazil

Pino Toro, Manuel Alejandro
Correspondent
UNIVERSIDAD FEDERAL DE MS
R. Julio Anffe, 249
Campo Grande, MS Brazil, Brazil

Pinto, Andre R.
NEON RIO LTDA
Rua Felinto de Almeida, 382
22.241 Rio de Janeiro-RJ, Brazil, Brazil

Pinto, Bonifacio Filho Clausem
Correspondent
THE COUSTEAU SOCIETY
Sqn 308, Blk 602
Brasilia, DF, Brazil, Brazil

Pinto, Fernando Augusto
Editor
MANCHETE
SQN 407 Bl.F Apt.105
Brasilia-DF, Brazil, Brazil

Pinto, Gilson A.
Reporter Radio/Television
MANCHETE TV
Est.Velha da Pavuna,4800 Bl.11
Apt.101, Brazil, Brazil

Pinto, Iris M. Fernandes
Editor
GAZETA DE NOTICIAS
R. Flack, #170 Casa 03
Rio de Janeiro, Brazil, Brazil

Pinto, Luis Antonio
Editor
VEJA
AOS 08 Bl.E Apt.5DS
70.350 Brasilia-DF, Brazil, Brazil

Pinto, Pericles Santana
Reporter Radio/Television
RADIO ROQUETE PINTO
Rua das Camelias, 571
Rio de Janeiro, Brazil, Brazil

Pinto, Solange Amado Da Costa
Editor
TV EDUCATIVA - RJ
R. Engenho da Rainha, 316 C3-1
Rio de Janeiro, Brazil, Brazil

Pinto Neto, Francisco
Correspondent
A UNIAO
Rua Carneiro da Cunha, 1289
Joao Pessoa, Brazil, Brazil

Piquet, Jorge Renato Dos Santos
Reporter Radio/Television
GLOBO RADIO
R. Lavini, 366
Realengo, RJ, Brazil, Brazil

Pires, Janete L.
Correspondent
ZERO HORA
Ql 02 Bloco T, Apt. 101
Guara I, Brasilia-DF, Brazil, Brazil

Pires, Lucia Helena S.M.
RADIOBRAS
Rua Radmaker 37 c/2
Rio de Janeiro, Brazil, Brazil

Pires Moreira, Laura
Editor
GLOBO TV
Rua Von Martius, 22 sala 321
Rio de Janeiro, Brazil, Brazil

Pompeu, Sonia
Reporter Radio/Television
MANCHETE TV
R.Prof.Manoel Ferreira, 88/501
Brazil, Brazil

Ponde, Eduardo C.
Reporter Radio/Television
RADIO GUANABARA/BANDEIRANTES
Av.Borges de Medeiros 3693/302
Lagoa, Rio de Janeiro-RJ, Brazil, Brazil

Pontes, Jorge Guilherme Marcello
Reporter Radio/Television
GLOBO RADIO
Estrada da Gavea Pequena
Rio de Janeiro-RJ, Brazil, Brazil

Pontes, Marcelo
Editor
JORNAL DO BRASIL
R. Ministro Waldemar Falcao,81
Tijuca, Rio de Janeiro-RJ, Brazil, Brazil

Pontes, Ruy
Editor
RADIOBRAS
SQS 305, Bloco I
Brasilia, DF, Brazil, Brazil

Pontual, Angela C.
MANCHETE TV
Rua Sambaiba, 404/301
Brazil, Brazil

Pontual Faure, Jorge Alexandre
Editor
GLOBO TV
Rua Von Martius, 22 sala 328
Rio de Janeiro, Brazil, Brazil

Portela, Maria C.
Editor
DIA
R. Antonio Fernandes, 16/201
Niteroi, Rio de Janeiro-RJ, Brazil, Brazil

Porto, Continentino
Editor
RADIOBRAS
R.Miguel de Frias,230/1005
Icarai,Niteroi, RJ, Brazil, Brazil

Portocarrero, Vera Lucia
O ESTADO
Av.Vitor Konder 103
Rio de Janeiro, Brazil, Brazil

Prado, Virginia Silviano Do
Reporter Radio/Television
RADIO ITATIAIA DE BELO HORIZONTE
Rua Itatiaia no. 117, Bonfim
Rio de Janeiro, Brazil, Brazil

Prandini, Paula
Editor
TV CULTURA-REPORTER ECO
R.Medeiros de Albuquerque, 130
05436 Sao Paulo-SP, Brazil, Brazil

Praxedes, Wanderley
Editor
GLOBO TV
Rua Von Martius, 22
Rio de Janeiro, Brazil, Brazil

Preard, Paulo Virgilio
Editor
RADIOBRAS
Rua da Passagem, 114/603
Rio de Janeiro-RJ, Brazil, Brazil

Propato, Valeria
Reporter
O DIA
R. Riachuela 359
Rio De Janeiro, RJ 2, Brazil
T: 55 21 272-8000 F: 55 21 507-1228

Puga, Monica De C.
Reporter Radio/Television
SBT TV
Rua Caruaru 485
Rio de Janeiro, Brazil, Brazil

Pugliese, Sergio
Editor
DIA
R. Almte Alexandrino, 447/201
Sta. Tereza, Rio de Janeiro-RJ, Brazil, Brazil

Puppin, Violeta Mafalda G.
Correspondent
CORREIO DA REGIAO
Av. Rui Barbosa, 636/1403
22.250-020 Rio de Janeiro-RJ, Brazil, Brazil

Quaino, Lilian
Editor
GLOBO
Rua Fadel Fadel, 186/101
Rio de Janeiro-RJ, Brazil, Brazil

Queiroz, Alvaro Goncalves
Reporter Radio/Television
RADIO ROQUETE PINTO
Est. Bandeirantes,8325, Bl.1
Apo.705 B.Tijuca,Rio de Janeiro,Brazil, Brazil

Queiroz, Luiz R.
Editor
ESTADO DE SAO PAULO
R. Pintassilgo, 59 Apt.91-A
Sao Paulo-SP, Brazil, Brazil

Quintanilha, Luiz Renato R.
Editor
PREFEITURA RIO DE JANEIRO
R.Gonzaga Bastos 209/B/304
Rio de Janeiro, Brazil, Brazil

Quintao, Ayle-Salassie F.
Editor
JORNAL MOMENTO POLITICO
SHIN Ql 04 Cj.10 c/16
Brasilia, D.F.,, Brazil

Rabelo, Clovis I.
Editor
TV GAZETA-CANAL 11
Alm dos Nhambiquaras, 1645/402
04090 Sao Paulo-SP, Brazil, Brazil

Rabinovici, Moises
Correspondent
AGENCIA ESTADO
9215 Bardon Rd.
Bethesda, MD 20814, USA, Brazil

Rabinovitch, Judith
Correspondent
BRAZILIAN GAZETTE
Rua Desembargador Burle, 73
Rio de Janeiro, Brazil, Brazil

Ramalho, Cristiane
Editor
JORNAL DO BRASIL
Av. Epitacio Pessoa, 2142/1003
22471, Rio de Janeiro-RJ, Brazil, Brazil

Ramos, Leo Borges
Editor
MANCHETE
R. Marques de Abrantes 119/505
Flamengo, Rio de Janeiro-RJ, Brazil, Brazil

Ramos, Margarida
Correspondent
FUNDACAO ROBERTO MARINHO
Av.Epitacio Pessoa, 604/204
Lagoa, Rio de Janeiro-RJ, Brazil, Brazil

Ramos, Monica Ribeiro
Reporter Radio/Television
JORNAL DO BRASIL - RADIO
Trav. Cruz, 8/302
Tijuca, RJ, Brazil, Brazil

Ramos, Paulo Cezar
Editor
HORA DO POVO
Rua Barao de Macabuba, 21/301
Botafogo, Rio de Janeiro-RJ, Brazil, Brazil

Ramos Filho, Jose
Correspondent
JORNAL DO BRASIL
SQN 410 Bl.A, Apto.307
Brasilia, D.F., Brazil, Brazil

Rangel, Renata
Editor
FOLHA DE SAO PAULO
Al. Barao de Limeira, 425
Sao Paulo-SP, Brazil, Brazil

Rangel Do Nascimento, Maria Lucia
Editor
GLOBO TV
Rua Von Martius, 22 sala 328
Rio de Janeiro, Brazil, Brazil

Ranney-Gumaer, Sally
Editor
WILD SIDE JOURNAL
8550 E. Davies Pl.
Englewood, Colo 800112, USA, Brazil

Ranzani, Elya Marina
Reporter Radio/Television
TV JOVEM PAN
Av. Invernada, 432/103
Sao Paulo-SP, Brazil, Brazil

Ranzolin, Armindo Antonio
Radio Reporter
RADIO GAUCHA
Av Ipiranga 1075
Porto Alegre - RS 90060, Brazil
T: 55 512 223 6600 F: 55 512 2336600 r 57

Rapello, Nelson
Editor
JORNAL A HORA
R.Guapui, 15
Rio de Janeiro, Brazil, Brazil

Raposo, Patricia
Correspondent
JORNAL DO COMMERCIO
R. Francisco Lisboa s/no.Bl.18
500010 Recife, PE, Brazil, Brazil

Ray, Dixy Lee
Correspondent
ABIM AGENCY
600 3rd Ave., Fox Island
Washington 98333, USA, Brazil

Raymundo, Ricardo H. De Souza
Editor
JORNAL DO BRASIL-RADIO AM
Rua Timoteo da Costa, 600
Leblon, Rio de Janeiro-RJ, Brazil, Brazil

Reali, Elpidio
Correspondent
AGENCIA ESTADO
Rue Ranelagh 52
Paris 16, France, Brazil

Rebello, Joao B.
Editor
AGENCIA ESTADO
SHIN QL 04 Conj.01 Casa 17
70.000 Brasilia-DF, Brazil, Brazil

Rebello, Octacio F.V.
Editor
PREFEITURA RIO DE JANEIRO
R.Buarque Macedo 46/803
Rio de Janeiro, Brazil, Brazil

Rech, Marcelo
Editor
ZERO HORA
Rua Barbedo 640, Apt. 608
Menino Deus, Porto Alegre-RS, Brazil, Brazil

Reganelli, Wilson Roberto
Editor
FM RAIZES
Av. Julia Conceicao Alves 422
Sao Paulo, Brazil, Brazil

Regis Gouvea, Marcia
Reporter Radio/Television
GLOBO TV
Rua Von Martius, 22 sala 321
Rio de Janeiro, Brazil, Brazil

Rego, Lucia
Editor
MANCHETE
R. Gilberto Amado, 915/201
Barra, Rio de Janeiro-RJ, Brazil, Brazil

Reis, Andhrea Dias Lopes Tavares
Reporter Radio/Television
RADIO NACIONAL DE BRASILIA
SQS 109 Bl. D Ap. 210
Brasilia, D.F., Brazil, Brazil

Reis, Claudia
Editor
DIA
R. A. Alexandrino, 345/501
Sta. Tereza, Rio de Janeiro-RJ, Brazil

Reis, Cleber Cabral
Reporter
DCI - DIARIO COMERCIO e INDUSTRIA
Av.Rio Branco, 177/6 andar
Rio de Janeiro - RJ, Brazil
T: 55 21 2621016 F: 55 21 2200805

Reis, Diocelio Lopes
Editor
JORNAL DO VALE
R. Bom Jesus, 580
Centro, Pouso Alegre, MG, Brazil, Brazil

Reis, Jose Antonio Dos
Reporter Radio/Television
GLOBO RADIO
R. Cabo Rocha, 155
Vila da Penha, RJ, Brazil, Brazil

Reis, Lea Maria
Editor
GLOBO TV
Rua Des. Alfredo Russel 773
Rio de Janeiro, Brazil, Brazil

Reis, Leila Cristina Maia Dos
Reporter Radio/Television
TV-BANDEIRANTES-RIO
Rua Sta. Cristina, 144/102
Rio de Janeiro-RJ, Brazil, Brazil

Reis, Marcelo De Jesus
Editor
JORNAL DOS SPORTS
R.Maia Lacerda 222/603
Rio de Janeiro, Brazil, Brazil

Reis, Marco Antonio
Reporter Radio/Television
SUPER RADIO TUPI
Rua Machado de Assis, 74
Flamengo, Rio de Janeiro-RJ, Brazil, Brazil

Reis, Mario Jose Rodrigues
Editor
RADIOBRAS
SQN 416 Bl. N/302
Brasilia, D.F., Brazil, Brazil

Renault, Marcia
Reporter Radio/Television
TV EDUCATIVA - RJ
Av. Gomes Freire 474
Rio de Janeiro, RJ, Brazil, Brazil

Resende, Helena L.
Reporter Radio/Television
MANCHETE TV
R.Bar.de Oliveira Castro, 48
Apt.402 Brazil, Brazil

Rezende, F.L. Vania
Editor
JORNAL DO BRASIL
Rua Guimaraes Rose 143/1301
Rio de Janeiro, Brazil, Brazil

Rezende, Luiz Eduardo
Editor
JORNAL DO BRASIL
R. Tonelero, 380/702
Copacabana, Rio de Janeiro-RJ, Brazil, Brazil

Rezende, Maria Helena Malta
Editor
MANCHETE
Rua Peri, 251
Jardim Botanico-RJ, Brazil, Brazil

Rezende, Sidney Nolasco De
Reporter Radio/Television
TV-BANDEIRANTES-RIO
Rua Barao de Oliveira Castro
Rio de Janeiro-RJ, Brazil, Brazil

Rezende, Sonia Ma.
Editor
GLOBO
R. Pinheiro Guimaraes 115
Rio de Janeiro-RJ, Brazil, Brazil

Ribas, Carlos H. Goncalves Camargo
Editor
RADIOBRAS/TV NACIONAL
SQS 103 Bl. E/312 Sul
Brasilia, D.F., Brazil, Brazil

Ribeire De Souza, Efren
Editor
FOLHA DE SAO PAULO
Al. Barao de Limeira, 425
Sao Paulo-SP, Brazil, Brazil

Ribeiro, Alvaro Pereira
Reporter Radio/Television
GLOBO RADIO
Av. Sete Setembro, 28/1204
Icarai, RJ, Brazil, Brazil

Ribeiro, Antonio Couto
Reporter
SISTEMA JORNAL DO BRASIL
Av.Brasil, 500
Rio de Janeiro - RJ, Brazil
T: 55 21 5854422

Ribeiro, Arnor Da Silva
Correspondent
UNIVERSIDADE FED. DE MATO GROSSO SUL
R.Ouro Branco, 1064
Campo Grande, MS, Brazil, Brazil

Ribeiro, Carlos Alberto Couto
Reporter Radio/Television
JORNAL DO BRASIL-RADIO AM
Rua Getulio, 249
Rio de Janeiro-RJ, Brazil, Brazil

Ribeiro, Damiao Cesar
Correspondent
JORNAL DOS SPORTS
Rua Monte Alegre, 468
Rio de Janeiro-RJ, Brazil, Brazil

Ribeiro, Denise Maria Dos Santos
Reporter
SISTEMA GLOBO DE RADIO
Rua do Russel, 434
Rio de Janeiro - RJ, Brazil
T: 55 21 2058282 F: 55 21 2054096

Ribeiro, Gerardo Lucas Da C.L.
Reporter Radio/Television
TV EDUCATIVA - RJ
R. Sao Clemente 185
Apto. 90, Rio de Janeiro, RJ, Brazil, Brazil

Ribeiro, Helton
Editor
GLOBO
Rua Bela Cintra, 85/142
Sao Paulo-SP, Brazil, Brazil

Ribeiro, Ines V.
Reporter Radio/Television
MANCHETE TV
Rua Lopes Quintas, 243/201
Rio de Janeiro-RJ, Brazil, Brazil

Ribeiro, Jose Augusto
Editor
ECOLOGIA E DESENVOLVIMENTO
Rua da Gloria 122 gr 105/106
Rio de Janeiro RJ 20, Brazil
T: 55 21 242 1957/222 1370

Ribeiro, M. Cristiane M. De Avila
Reporter Radio/Television
RADIOBRAS
R. Amancio da Silva, 620
Rio de Janeiro, Brazil, Brazil

Ribeiro, Marcos
Editor
MANCHETE TV
R.Barao Oliveira Castro 48/402
Rio de Janeiro, Brazil, Brazil

Ribeiro, Mehanne
Editor
GLOBO
Rua Xingu, 73/102
Jacarepagua, Rio de Janeiro-RJ, Brazil, Brazil

Ribeiro, Nelson Pinto Da F., Reporter Radio/Television
MANCHETE
R. do Russel 804
Rio de Janeiro -RJ 2, Brazil
T: 55 21 2652012 F: 55 21 2855148

Ribeiro, Pedro A.
Editor
JORNAL DA TARDE
R.Dr.Teophilo R.de Andrade 157
95466 Sao Paulo-SP, Brazil, Brazil

Ribeiro, Renato
Editor
GLOBO TV
Rua Von Martius, 22 sala 321
Rio de Janeiro, Brazil, Brazil

Ribeiro, Vicente De Paula Vetromille
Reporter Radio/Television
TV EDUCATIVA - RJ
R. Tirol, 861 Bl. 09/203
Rio de Janeiro, Brazil, Brazil

Ribeiro Jr., Descartes N.
RADIOBRAS
R.Pinheiro Guimaraes 115/1501
Rio de Janeiro, Brazil, Brazil

Ribeiro P., Maria Cristina
Correspondent
O DIA
Antonio Fernandes 16/201
Niteroi, Rio de Janeiro, Brazil, Brazil

Ricci, Cesar Arnaldo
Editor
VEJA
R.Paulo VI, 600/20E
22230 Rio de Janeiro-RJ, Brazil, Brazil

Richers, Leila De Moura
Editor
MANCHETE TV
R. Joaquim Mourtinho, 549
Santa Teresa, RJ, Brazil, Brazil

Rigueira Jr., Itamar
Editor
EMBRATUR
Rua Miguel Frias, 23 apt.1502
Niteroi, Rio de Janeiro, Brazil, Brazil

Rimoli, Laerte Lima
Editor
GLOBO RADIO
R. Senador Vergueiro 35/1109
Rio de Janeiro, Brazil, Brazil

Rinski, Wolf Lobo
Editor
ACADEMIA/ORDEM JORNALISTAS BRASIL
R.Araujo Porto Alegre 71
Rio de Janeiro, Brazil, Brazil

Rio, Elianne De Andrade P. Do
Editor
TV EDUCATIVA - RJ
Av. Gomes Freire 474 Centro
Rio de Janeiro -RJ, Brazil
T: 55 21 2426736 F: 55 21 2426740

Rivera, Clara
INTERNATIONAL TELEVISION NEWS
Av. Rio Branco, 25/9 andar
Rio de Janeiro RJ, Brazil
T: 55 21 2335480

Rizzotto, Rodolfo
Editor
JORNAL DO COMMERCIO
Av.Atlantica 3514/701
Rio de Janeiro, Brazil, Brazil

Rocha, Antonio Carlos
Correspondent
JORNAL DOS CONSTRUTORES
Praca Nilo Pecanha 97
Rio de Janeiro, Brazil, Brazil

Rocha, Caius Lucicius Burche
Reporter
CORREIO BRAZILIENSE
Rua Cubato, 436
Sao Paulo, Brazil
T: 55 11 8858235 F: 55 11 8851849

Rocha, Jaqueline Silveira
Correspondent
TV EDUCATIVA
Av.Sao Luiz Gonzaga c/1
Nova Friburgo, Rio de Janeiro, Brazil, Brazil

Rocha, Joao Batista Torres
Editor
JORNAL DO BRASIL-RADIO AM
Rua Cupertino Durao, 219/1402
Leblon, Rio de Janeiro-RJ, Brazil, Brazil

Rocha, Jose M.
Reporter Radio/Television
MANCHETE TV
Av.Mons.Felix, 1059 Rua A
No.145-1/602, Iraja, RJ, Brazil, Brazil

Rocha, Jose Ricardo
Editor
MANCHETE TV
R.Gal. Glicerio 32/1002
Rio de Janeiro, Brazil, Brazil

Rocha, Marcia De Araujo
Reporter Radio/Television
TV EDUCATIVA - RJ
Mundo Novo 376 - 3
Botafogo, RJ, Brazil, Brazil

Rocha, Maria Lucia M.
Correspondent
FOLHA DO MEIO AMBIENTE
Rua Saldanha Marinho,689/601
80.410 Curitiba, PR, Brazil, Brazil

Rocha, Nilton Araujo
Editor
SBT TV
R. Embiras 339
Jardim Popular Penha, Brazil, Brazil

Rocha, Patricia Pereira Da
Reporter Radio/Television
GLOBO RADIO
Rua do Russel 344
Rio de Janeiro, Brazil, Brazil

Rodrigues, Agnes Consula J.
33RD PRODUCTIONS/NEON RIO
Largo das Neves 11 Sta.Teresa
Rio de Janeiro, Brazil, Brazil

Rodrigues, Altenir Santos
Editor
TV EDUCATIVA - RJ
Av. Gomes Freire 474 Centro
Rio de Janeiro -RJ, Brazil
T: 55 21 2426736 F: 55 21 2426740

Rodrigues, Azelma
Editor
RADIOBRAS
SRTVS, Radio Nacional
Brasilia, D.F., Brazil, Brazil

Rodrigues, Claufe
REDE GLOBO
R. Lopes Quintas 303 J.Botanico
Rio de Janeiro RJ 22, Brazil
T: 55 21 2947732

Rodrigues, Cristiane De Freitas
Correspondent
ECABE PRESS
Av.Sernambetiba, 3500/bl1/306
Rio de Janeiro, Brazil, Brazil

Rodrigues, Danubio
Reporter
S/A CORREIO BRAZILIENSE
R. do Livramento 158/4 andar
Rio de Janeiro -RJ, Brazil
T: 61 3212123 F: 061 3213864

Rodrigues, Elaine R.
Editor
GLOBO
Rua Marques de Caxias, 21
Rio de Janeiro-RJ, Brazil, Brazil

Rodrigues, Enilton, Jr.
MANCHETE TV
R.Barao de SF,245 Bl.G Apt.310
Vila Isabel, Rio de Janeiro-RJ, Brazil, Brazil

Rodrigues, Ernesta De Amorim
Editor
TV CULTURA - SP
R.Amoroso Lima, 100/5032
Cidade Nova, RJ, Brazil, Brazil

Rodrigues, Genesio
Correspondent
REVIST OPODER
R. Uruquai, 545/304
Tijuca, Rio de Janeiro, RJ, Brazil, Brazil

Rodrigues, Jacira Lucas
Reporter Radio/Television
TV EDUCATIVA
Rua Rodmaker 33/302
Rio de Janeiro-RJ, Brazil, Brazil

Rodrigues, Jose Fernando
Editor
FOLHA DE SAO PAULO
Al. Barao de Limeira, 425
Sao Paulo-SP, Brazil, Brazil

Rodrigues, Marco Antonio Ortiz
Reporter Radio/Television
TV EDUCATIVA
Rua Acau, no. 30, Engenho Novo
Rio de Janeiro, Brazil, Brazil

Rodrigues, Maria De Fatima
Editor
JORNAL DO BRASIL
R. Sta. Clara, 431/204
Copacabana, Rio de Janeiro-RJ, Brazil, Brazil

Rodrigues, Ricardo De Souza
Reporter Radio/Television
GLOBO RADIO
Rua Noronha Torrezao, 407
Niteroi, Rio de Janeiro-RJ, Brazil, Brazil

Rodrigues, Ricardo Lessa
Editor
ISTO E SENHOR
Praia do Botafogo, 252/702
22250 Rio de Janeiro-RJ, Brazil, Brazil

Rodrigues, Roberto Maia E Carmem
Correspondent
FUNDACAO BRASIL 2000
Brazil

Rodrigues, Sebastiao C.
MANCHETE TV
Praia de Botafogo, 484/310
Rio de Janeiro-RJ, Brazil, Brazil

Rodrigues, Vivian
Editor
TV-BANDEIRANTES RIO
Rua Bogari 126/103
Rio de Janeiro, Brazil, Brazil

Rodrigues Carneiro, Ernesto
Editor
GLOBO TV
Rua Von Martius, 22 sala 321
Rio de Janeiro, Brazil, Brazil

Rodriguez, Ana Lucia M.
Editor
RADIO MEC
Rua Joaquim Caetano, 10
Urca, Rio de Janeiro-RJ, Brazil, Brazil

Rodriguez, Bienvenido Gomez
Reporter Radio/Television
TV EDUCATIVA - RJ
R. Sen. Nabuco, 252/202A
Rio de Janeiro, Brazil, Brazil

Rodriguez, Sergio L.
Editor
GAZETA MERCANTIL
Rua Major Quedinho, 90/4 andar
Sao Paulo-SP 01050 Brazil, Brazil

Rodriguez, Valeria S.
Editor
DIA
R. Araucaria, 65/301
Rio de Janeiro-RJ, Brazil, Brazil

Roger Sauer, Jules
Correspondent
BRAZILIAN GAZETTE
Av. Atlantica, 270/1601
Rio de Janeiro, Brazil, Brazil

Romanholli, Luiz H.
Editor
GLOBO
Rua Gen. Roca, 443/302, Tijuca
20.521-Rio de Janeiro-RJ, Brazil, Brazil

Romero, Cristiano
Editor
RADIOBRAS
Sqn 210 A 301
Brasilia, D.F., Brazil, Brazil

Roque, Antonio J.
Editor
SUNDAY NEWS
R. Tita Ruffo 879
Sao Paulo, S.P., Brazil, Brazil

Rosa, Ana Cristina
Reporter Radio/Television
RADIO GUAIBA
Rua Liberal, 3110/272
Cavalhada, Porto Alegre, RS, Brazil, Brazil

Rosa, Carlos S.M.
Editor
REVISTA NOOVA ESCOLA
Rua Haddock Lobo 403/D
Sao Paulo, S.P., Brazil, Brazil

Rosa, Claudio Roberto
Editor
SAO PAULO GOVERNMENT PRESS
R.Mura 180
Sao Paulo, S.P., Brazil, Brazil

Rosa, Ricardo M.
Editor
EMA VIDEO
EPPN Km4 - Chacara Ipanema
Brasilia, D.F., Brazil, Brazil

Rosa, Ronaldo Da S.
Reporter Radio/Television
MANCHETE TV
Rua Aarao Reis, 37/201
Sta.Tereza, Rio de Janeiro-RJ, Brazil, Brazil

Rosa, Rosalva N.
Correspondent
ZERO HORA
SQN 405 Bloco A, Apt. 201
Brasilia-DF, Brazil, Brazil

Rosa, Vera Lucia
Correspondent
JORNAL DA TARDE
Av. Nove de Julho 4706 Apt.13
Sao Paulo-SP, Brazil, Brazil

Rosario, Marta Crisostomo
Editor
RADIOBRAS
SQN 203 Bl. D Ap. 302
Brasilia, D.F.., Brazil, Brazil

Rosenthal, Adolfo
Editor
MANCHETE TV
Av. P. Dulcidio Cardoso 800
Bl. s/303 Rio de Janeiro, Brazil, Brazil

Rossi, Angelo
Correspondent
JORNAL DO COMMERCIO
R. Alte. Sadock de Sa 7/901
Ipanema, RJ, Brazil, Brazil

Rotenberg, Elsie L.
Editor
EDITORA ABRIL/ELLE
R.Visconde de Porto Seguro1105
04642 Sao Paulo-SP, Brazil, Brazil

Rothenburg, Denise
Editor
GLOBO
SCS Ed.Oscar Niemeyer-13 andar
Brasilia-DF, Brazil, Brazil

Roxo, Heliete Dos S.
Correspondent
CORREIO FLUMINENSE
R. Marques Abrantes 177/302
Rio de Janeiro, Brazil, Brazil

Roxo, Jose D.
Editor
CORREIO FLUMINENSE
R. Marques Abrantes 177/302
Rioi de Janeiro, Brazil, Brazil

Rubin, Rosane
Editor
JORNAL DO BRASIL
R. Maria Angelica, 428/301
Ipanema, Rio de Janeiro-RJ, Brazil, Brazil

Ruiz, Simone
Editor
DIA
Av. Atlantica, 538/901
Rio de Janeiro-RJ, Brazil, Brazil

Sa, Lilian Saback Nogueira De
Reporter
TV EDUCATIVA - RJ
Av. Gomes Freire 474 Centro
Rio de Janeiro -RJ, Brazil
T: 55 21 2426736 F: 55 21 2426740

Sa, Wlamir
RADIOBRAS
Sampaio Vidal 75, S. Goncalo
Rio de Janeiro, Brazil, Brazil

Saboia, Napoleao
Editor
AGENCIA ESTADO
35 R. de Garches, 92210
St. Clout, France, Brazil

Saint-Clair, Clovis
Editor
VEJA
Rua Barata Ribeiro, 622/602
Copacabana, Rio de Janeiro-RJ, Brazil, Brazil

Saldanha De Araujo, Jorge
Editor
JORNAL BRAZIL TURISMO
Av. Marca 255, #703
RJ, Alcantara S. Gancalo, Brazil, Brazil

Salem, Helena
Editor
JORNAL DO BRASIL
R. Pres. C. de Campos, 183/804
Laranjeiras, Rio de Janeiro-RJ, Brazil, Brazil

Sales Da Silva, Carmen
Reporter Radio/Television
TV EDUCATIVA - RJ
Av. Gomes Freire 474
Rio de Janeiro, RJ, Brazil, Brazil

Salomon, Marta M.
Correspondent
AGENCIA ESTADO
SQN 211 Bl.C Apt.106
70.000 Brasilia-DF, Brazil, Brazil

Salvador, Marli O.
Editor
RADIO ELDORADO FM
R.Nilo Pecanha, 110 Apt.114
09540 S.Caetano do Sul-SP, Brazil, Brazil

Sampaio, Joaquim
Reporter Radio/Television
TV EDUCATIVA
Travessa Joaquim Sampaio 67
Sao Pedro da Aldeia, Brazil, Brazil

Sampaio, Porfirio V.
Editor
PREFEITURA RIO DE JANEIRO
R.Buarque Macedo 36
Rio de Janeiro, Brazil, Brazil

Sanches, Sandra
Editor
GLOBO
Trav. Olano Bastos, 31/201
Niteroi-RJ, Brazil, Brazil

Sanches, Valdir
Editor
JORNAL DA TARDE
R.Caetano D'Andrea 10
Guarulhos, Sao Paulo, Brazil, Brazil

Sanchez, Ana Maria T.
Editor
REVISTA NOVA ESCOLA
Rua Haddock Lobo 403/D
Sao Paulo, S.P., Brazil, Brazil

Sanchez, Consuelo Martins F.
Correspondent
IMAGEM
Rua Catalao, 10
Rio de Janeiro-RJ, Brazil, Brazil

Sant'Anna, May De Britto
Editor
RADIO MEC
Rua Conde de Bonfrim 892
Rio de Janeiro-RJ, Brazil, Brazil

Santana, Horacio Gilvan B.
Reporter Radio/Television
TV EDUCATIVA - RJ
P. Moreira da Silva 255 - 302
Taguara, RJ, Brazil, Brazil

Santana De Mello, Lucia Maria
Editor
GLOBO TV
Rua Von Martius, 22 sala 321
Rio de Janeiro, Brazil, Brazil

Santana Filho, Lourival
Editor
ESTADO DE SAO PAULO
Av. Jurema 147 Apt.172
Moema, Sao Paulo-SP Brazil, Brazil

Santiago, Alicia Goncalves
Reporter Radio/Television
RADIOBRAS
CRS 515 Bl. B Ent. 22 Ap.305
Brasilia, D.F., Brazil, Brazil

Santiago, Carlos
Editor
MONITOR MERCANTIL
Rua Paratinga, 948
Vista Alegre, Rio de Janeiro, Brazil, Brazil

Santos, Ana Galo
Editor
MANCHETE
Rua Ypiranga, 13/ap. 102
Laranjeiras, Rio de Janeiro-RJ, Brazil, Brazil

Santos, Antonio Carlos Silva
Reporter Radio/Television
RADIOBRAS
CSB 8 Lote 3/4 Ap.405
Taguatinga Sul D.F., Brazil, Brazil

Santos, Carla Regina
Editor
JORNAL BRAZIL TURISMO
Av. Marica 255 Bl.6 Apt.703
Jardim Alcantara, Rio de Janeiro,Brazil, Brazil

Santos, Clarindo Dos
Editor
RODOCACEA ENG : EM REVISTA
Rua Sehador Dantas, 7A - 12 andar
Rio de Janeiro RJ 2, Brazil
T: 55 21 2622591

Santos, Claudia M.
Reporter Radio/Television
RADIOBRAS
R.Columbia 125
Rio de Janeiro, Brazil, Brazil

Santos, Claudio Vieira Dos
Reporter
O DIA
R. Riachuela 359
Rio De Janeiro, RJ 2, Brazil
T: 55 21 272-8000 F: 55 21 507-1228

Santos, Debora N.
Editor
RADIO & TV BANDEIRANTES
R.Dr. Renato P.Barros 550/113
Rio de Janeiro, Brazil, Brazil

Santos, Eliane Oliveira Dos
Correspondent
UNIVERSIDADE FED MATO GROSSO SUL
R. Jorn. Belizario Lima, 336
Campo Grande, MS, Brazil, Brazil

Santos, Eloy Dos
Reporter Radio/Television
TV EDUCATIVA - RJ
R. Columbia, 125
Rio de Janeiro, RJ, Brazil, Brazil

Santos, Fatima C.
Editor
GLOBO
Rua Sao Salvador, 59
Rio de Janeiro-RJ, Brazil, Brazil

Santos, Ione M.De O.
Editor
PREFEITURA RIO DE JANEIRO
R.Felix da Cunha 42/102
Rio de Janeiro, Brazil, Brazil

Santos, Ismael
Reporter Radio/Television
SBT TV
Av.Copacabana 115/1106
Rio de Janeiro, Brazil, Brazil

Santos, Joao Francisco
Correspondent
FOLHA DA MANHA
R.Lopes Trovao, 294/1304 Bl.1
Icarai, Rio de Janeiro-RJ, Brazil, Brazil

Santos, Jorge Andre
Editor
AGENCIA ESTADO
R.Waldir dos Santos, 4021
Galo Branco, Sao Goncalo-RJ, Brazil, Brazil

Santos, Jose Eduardo Dos
Editor
REVISTA CICLO MAGAZINE
R. Claudio Ghirelli, 140 A
02841 - Sao Paulo, Brazil

Santos, Jose Marcos Tadeu Dos
Reporter Radio/Television
RADIO GLOBO
Rua Heitor Penteado, 220
Sao Paulo, Brazil, Brazil

Santos, Jucara Cristina Braga
Editor
SJPMRJ
R. Laurinda S.Lobo, 52/301
Rio de Janeiro, Brazil, Brazil

Santos, Jurandir S.
Reporter Radio/Television
CORREIO DO POVO/RADIO GUAIBA
R.Caldasl Junir 219
Porto Alegre, Rio Grande Sul, Brazil, Brazil

Santos, Leila Cordeiro
Reporter Radio/Television
MANCHETE TV
Rua Jose de Brito, 35
Rio de Janeiro, Brazil, Brazil

Santos, Lucia De Araujo A.
Correspondent
JORNAL METRO PRESS
R.Ana Leonida, 270/102
Engenho de Dentro 20.730 RJ, Brazil, Brazil

Santos, Luiz Fernando Lima
Reporter Radio/Television
SBT TV
Rua voluntarios da Patria
22270, Rio de Janeiro-RJ, Brazil, Brazil

Santos, Ma.Angela
Editor
TV CULTURA-REPORTER ECO
R. das Rosas, 299/06
04048 Sao Paulo-SP, Brazil, Brazil

Santos, Marcio A.
Reporter Radio/Television
MANCHETE TV
Est.Jose Rucas, 462/102
Penha, Rio de Janeiro-RJ, Brazil, Brazil

Santos, Marcio Alberto Mendonca
Reporter Radio/Television
RADIO E TELEVISAO RIBAMAR
Av. Alcantara Machado, 3
Maranhao, Brazil, Brazil

Santos, Monica Cristina M.
Reporter Radio/Television
RADIO ITATIAIA DE BELO HORIZONTE
R. Itatiaia 117
Belo Horizonte, Minas Gerais, Brazil, Brazil

Santos, Nilton
Reporter Radio/Television
TV EDUCATIVA - RJ
R. do Riachuelo 221/515
Rio de Janeiro, Brazil, Brazil

Santos, Osvaldo J.
Editor
JORNAL DA USP
R.Pres.Bernardes, 42
Freguesia, SP, Brazil, Brazil

Santos, Patricia C.
Reporter Radio/Television
MANCHETE TV
R.Vaparaiso, 82 Apt.603
Tijuca, Rio de Janeiro-RJ, Brazil, Brazil

Santos, Roberto Machado
Reporter Radio/Television
TVT Rede De Comunicacao
R. dos Ingleses, 347
Sao Paulo, Brazil

Santos, Rubens R.
Editor
JORNAL VERDE
Manoel Tourinho 880
Sao Paulo, Brazil, Brazil

Santos, Suely Rangel
Editor
CEFET-RJ
Rua Cel.Moreira Cesar,331/204
Icarai, Niteroi, Rio de Janeiro, Brazil, Brazil

Santos, Tadeu Martins Dos
Editor
RADIOBRAS
ONN 26 Conj. F Casa 54
Taguatinga Norte, D.F., Brazil, Brazil

Santos, Valter
Reporter Radio/Television
TV JOVEM PAN
R.Maria Cecilia, 36
Sao Paulo-SP, Brazil, Brazil

Santos, Wilson Franca Dos
JORNAL TRIBUNA DA IMPRENSA
Rua do Lavradio, 98
20230 Rio de Janeiro-RJ, Brazil, Brazil

Santos Costa, Carlos Domingos
Reporter Radio/Television
TV EDUCATIVA - RJ
R. Cabo Reis, 22
Rio de Janeiro, Brazil, Brazil

Santos Gomes, Suzana
Editor
JORNAL DO COMMERCIO
R.Idume, 131
Braz de Pina, RJ, Brazil, Brazil

Saraiva, Mauricio
Reporter Radio/Television
RADIO GUAIBA
Rua Duque de Caxias, 638/1003
Porto Alegre, RS, Brazil, Brazil

Saraiva, Sergio R.B.
Editor
JORNAL DO COMERCIO
R. Sta. Teresinha 395
Apt. 21, Rio de Janeiro, Brazil, Brazil

Sardinha, Ocimar
Correspondent
JORNAL FLOR DA MANGA
R.Felix Antonio Da Silva,16
Engenho Pequeno,Sao Goncalo/RJ Brazil, Brazil

Sarquis, Otto
Editor
GLOBO
SCS. Ed. Oscar Niemeyer, 13
Brasilia-DF, Brazil, Brazil

Sasaki, Robinson K.
Editor
REVISTA NOVA ESCOLA
Rua Haddock Lobo 403-D
Sao Paulo, S.P., Brazil, Brazil

Sastre, Vera L.
Editor
AGENCIA ESTADO
R. Marques de Olinda, 64
Botafogo, Rio de Janeiro-RJ, Brazil, Brazil

Sato, Sandra
Correspondent
AGENCIA ESTADO
Sqn 210
Brasilia, Brazil, Brazil

Savaget, Claudio C.
Editor
PROGRAMA GLOBO ECOLOGIA
R.Conselheiro Lampreia,No.326
Cosme Velho, Rio de Janeiro-RJ, Brazil, Brazil

Savaget, Elza K.K.
Editor
PROGRAMA GLOBO ECOLOGIA
R.Conselheiro Lampreia, No.326
Cosme Velho, Rio de Janeiro-RJ, Brazil, Brazil

Savaget, Rubens Carvalho
Editor
GLOBO ECOLOGIA
R.Santa Clara 316 c/01
Rio de Janeiro, Brazil, Brazil

Savaget, Ruth De Castro
Editor
ROLAND HENZE PRODUCOES
Rua Humaita, 234 Bl. 2/204
22261 Rio de Janeiro-RJ, Brazil, Brazil

Sayao, Lea
Correspondent
Jornal do Estado do Rio de Janeiro
Barata Ribeiro 74/202
Rio de Janeiro 20000, Brazil

Sayao, Silvia
Editor
GLOBO TV
Rua Von Martius, 22 sala 328
Rio de Janeiro, Brazil, Brazil

Scamparini, Ilze Lia
Reporter Radio/Television
GLOBO TV
Rua Von Martius, 22 sala 328
Rio de Janeiro, Brazil, Brazil

Scharf, Regina Claudia Gross
Editor
GAZETA MERCANTIL
R. Mj. Quedinho, 90 - 4 andar
Sao Paulo, Brazil, Brazil

Scherdien, Celia Regina
Editor
RADIOBRAS
SQS 205 Bl. B Ap. 411
Brasilia, D.F., Brazil, Brazil

Schirmer, Gerson Lopes
Editor
REVISTA EXPRESSAO
Est. Geral do Cuato da Lagoa
1502 Florianopolis, SC, Brazil, Brazil

Schieder, Ewaldo F.
Editor
BOLETIM ECOPRENSA
Simao Bolivar 903
Curitiba, PR, Brazil, Brazil

Schmidt, Selma Regina
Editor
GLOBO
Rua Aristides Espinola 107
Leblon, Rio de Janeiro-RJ, Brazil, Brazil

Schmitt, Renata C.
Editor
DIA
R. Gal Goes Monteiro, 8
Botafogo, Rio de Janeiro-RJ, Brazil, Brazil

Schreiner, Sandro Garcia
Reporter Radio/Television
RADIO BANDEIRANTES
Dr.V. de Paula Dutra 47/303
Porto Alegre, RS, Brazil, Brazil

Schroder, Carlos Henrique
Editor
REDE GLOBO
R. Lopes Quintas 303
Rio de Janeiro -RJ 2, Brazil
T: 55 21 2947732 F: 55 21 2947247

Schuffner, Claudia
Editor
JORNAL DO COMMERCIO
Av. A. de Paiva, 900/302
Leblon, RJ, Brazil CEP 22440, Brazil

Schwob, Wiliane
GLOBO
Rua Rubens Berardo Bl.1/125
Gavea, Rio de Janeiro-RJ, Brazil, Brazil

Scofield Jr., Gilberto
Editor
JORNAL DO BRASIL
R.Padre Roma, 525/201
Grajau, Rio de Janeiro-RJ, Brazil, Brazil

Seabra, Leonardo Dos Santos
Reporter Radio/Television
GLOBO RADIO
R. Plinio Bastos, 420
Olaria, RJ, Brazil, Brazil

Seara, Berenice
Reporter
JORNAL O GLOBO
Rua Irineu Marinho, 35
Rio de Janeiro - RJ, Brazil
T: 55 21 2722000

Sebastiao, Ana A.
Editor
ECOLOGIA E DESENVOLVIMENTO
Estrada do Cafunda 1757 Apt702
22725 Rio de Janeiro-RJ, Brazil, Brazil

Seixas, Luciano Rocha
Reporter Radio/Television
RADIOBRAS
SQS 402 Bl. R Ap. 103
Asa Sul, Brasilia, D.F.,Brazil, Brazil

Seleme, Ascanio A.
Editor
GLOBO
SCS. Ed. Oscar Niemeyer, 13
Brasilia-DF, Brazil, Brazil

Senna, Benedito Rocha
Editor
REDE OM BRASIL DE TELEVISAO
R.Francisco S. Melo, 2/105
Rio de Janeiro, Brazil, Brazil

Sequeira, Monica M.P.
Editor
PREFEITURA RIO DE JANEIRO
R.Paissandu 279/103
Rio de Janeiro, Brazil, Brazil

Serejo, Barbara
Reporter Radio/Television
RADIO CULTURA
Es.9B Rua 5, casa 14
Sr. Mansoes, Sobradinho, DF, Brazil, Brazil

Serr, Silvia
GLOBO RADIO
Rua General Gaizon, 28
Rio de Janeiro-RJ, Brazil, Brazil

Serra, Veronica A.
Editor
EDITORA ABRIL
R.Antonio Gouveira Giudice 737
Sao Paulo, S.P., Brazil, Brazil

Serra, Wilson
Editor
GLOBO TV
Rua Von Martius, 22 sala 328
Rio de Janeiro, Brazil, Brazil

Serra Ferreira, Cristina
Editor
GLOBO TV
Rua Von Martius, 22 sala 321
Rio de Janeiro, Brazil, Brazil

Serradeira, Carlos Lazaro Vianna
Reporter Radio/Television
GLOBO RADIO
Av. 28 Setembro, 74 c 5/202
Rio de Janeiro, Brazil, Brazil

Serrao, Jorge Fernando B.
Editor
PREFEITURA RIO DE JANEIRO
R.Joao Pessoa 338/304
Niteroi, Rio de Janeiro, Brazil, Brazil

Sforza, Ricardo
Editor
VIDEO CIENCIA PRODUCOES LTDA.
Rua Faro, 76/404
22461 Rio de Janeiro-RJ, Brazil, Brazil

Shackleton, Andreas
Editor
PROGRAMA GLOBO ECOLOGIA
R. Min. Viveiro de Castro, 66
Copacabana, Rio de Janeiro-RJ, Brazil, Brazil

Shiguti, Aldo Watao
Editor
JORNAL PAULISTA
R.Angelica A Pro.Machado,50
Vila Clara, SP, Brazil, Brazil

Shiv, Khemka
MANCHETE
25 Eaton Square, Flat H
London SW1, United Kingdom, Brazil

Sholl, Daniella
Editor
JORNAL DO BRASIL
R. A. Alexandrino, 447/201
Sta. Teresa, Rio de Janeiro-RJ, Brazil, Brazil

Sigiliano, Renato
Editor
JORNAL A NOTICIA
Lad.Frei Orlando 51
Rio de Janeiro, Brazil, Brazil

Silva, Aida Lobo
Editor
DIARIO MERCANTIL
R.Prof.Saboia Ribeiro 47/1103
22430 Rio de Janeiro-RJ, Brazil, Brazil

Silva, Ana Maria Carvalho
Reporter Radio/Television
JORNAL DO BRASIL-RADIO AM
Praca Vereador Rocha Leao
Copacabana, Rio de Janeiro-RJ, Brazil, Brazil

Silva, Antonio Cesar G. Da
Editor
RADIO CONTEMPORANEA AM
R. Dr. E. Cordeiro, 28
Rio Bonito, RJ, Brazil, Brazil

Silva, Antonio De Padua
Editor
DIARIO DE FRANCA
R. M.C.Archetti, 2877
Franca, SP, Brazil, Brazil

Silva, Antonio Gilson Braga Da
Reporter Radio/Television
GLOBO RADIO
R. Dr. Roberto Freire, 133
Rio de Janeiro, Brazil, Brazil

Silva, Artur Pereira
Correspondent
JORNAL DO BRASIL
Qda.12 Cj.A-12 c/15
Cruzeiro Velho, Brasilia, D.F., Brazil, Brazil

Silva, Carla Ramos E
Reporter Radio/Television
TV EDUCATIVA - RJ
Est. Do Boubanville, 442
Barra da Tijuca, RJ, Brazil, Brazil

Silva, Carlos
Reporter Radio/Television
TV JOVEM PAN
Rua Piava, 224
C.A.E. Carvalho Sao Paulo, Brazil, Brazil

Silva, Carlos Borges
Editor
TV EDUCATIVA
Rua Sao Gabriel 315/407
Rio de Janeiro-RJ, Brazil, Brazil

Silva, Carlos Roberto Guimaraes
Reporter Radio/Television
MANCHETE TV
Rua Francisca Peixoto 47, casa
Rio de Janeiro, Brazil, Brazil

Silva, Cesar
Editor
JORNAL VEGAN
R.Riachuelo 220/406
Rio de Janeiro, Brazil, Brazil

Silva, Clara Elizabeth Alves Da
JORNAL TRIBUNA DA IMPRENSA
Rua do Lavradio, 98
20230 Rio de Janeiro-RJ, Brazil, Brazil

Silva, Edileide Goncalves M.
Reporter Radio/Television
TV-BANDEIRANTES-RIO
Rua Sao Salvador, 59/608
Rio de Janeiro-RJ, Brazil, Brazil

Silva, Edilson
Editor
MANCHETE TV
Alm. Gomes Pereira, 76/303
Rio de Janeiro, Brazil, Brazil

Silva, Ely
Reporter Radio/Television
GLOBOSAT
Av. Epitacio Pessoa,1834/302
Rio de Janeiro, Brazil, Brazil

Silva, Fernando G.O.
Reporter Radio/Television
TV EDUCATIVA
R.Gal. Glicerio 74/303
Rio de Janeiro, Brazil, Brazil

Silva, Genilson De Araujo
Reporter Radio/Television
JORNAL DO BRASIL-RADIO AM
Rua Quiririm, 1523
Jacarepagua, Rio de Janeiro-RJ, Brazil, Brazil

Silva, Ilimar F.
Correspondent
ZERO HORA
SQN 403, Bloco J, Apt. 201
Brasilia-DF, Brazil, Brazil

Silva, Ingrid Marzocchi
Reporter Radio/Television
RADIO UNIVERSIDADE
R.Alzira Brandao 295/403
Rio de Janeiro, Brazil, Brazil

Silva, Isaias
Correspondent
RADIO NOVA DIMENSAO
R. Don Carlos F.Ferraz,442
Sao Paulo, Brazil, Brazil

Silva, Joelson R.
Reporter Radio/Television
MANCHETE TV
R.Coelho de Souza L.9 Q.5
San Goncalo, Rio de Janeiro-RJ, Brazil, Brazil

Silva, Jorge Luiz
Correspondent
ECABE PRESS
Rua General Rabelo, 65/404
22451 Rio de Janeiro-RJ, Brazil, Brazil

Silva, Jorge Ribeiro
Correspondent
ORBE-PRESS
Estr. Bandeirantes, 8325 Bl.3
Rio de Janeiro, Brazil, Brazil

Silva, Jose P.
Editor
AGENCIA ESTADO
R. Uruguai 247 C.04,Tijuca
Rio de Janeiro-RJ, Brazil, Brazil

Silva, Linei Zago
Reporter Radio/Television
RADIO GUAIBA
Rua Anita Garibaldi, 1877/218
Porto Alegre, RS, Brazil, Brazil

Silva, Luci Braga
Reporter Radio/Television
SUPER RADIO TUPI
Rua Joaquim martins, 395
Piedade, Rio de Janeiro-RJ, Brazil, Brazil

Silva, Lucia Ramos E
Reporter Radio/Television
TV EDUCATIVA - RJ
Av. Gomes Freire, 474
Rio de Janeiro, Brazil, Brazil

Silva, Luciano Garrido
Reporter Radio/Television
GLOBO RADIO
Rua Visconde de Ouro Preto
Rio de Janeiro-RJ, Brazil, Brazil

Silva, Luiz Carlos Barroso C.
Editor
MANCHETE TV
R. Evilasio Silva, 86
Niteroi, RJ, Brazil, Brazil

Silva, Marcos Gomes Da
Editor
OFFICIAL PRESS EST. RIO DE JANEIRO
R. Itaguai,m 140/101
Niteroi, RJ, Brazil, Brazil

Silva, Maria Aparecida
Correspondent
JORNAL DE ESPANA
Rua Teodoro da Silva, 312
Rio de Janeiro, Brazil, Brazil

Silva, Maria Arlete Ferreira
Editor
JORNAL MAIORIA FALANTE
R.Francisco Vieira Souza 20
Rio de Janeiro, Brazil, Brazil

Silva, Maria Das Gracas
Editor
REVISTA TOUR IN RIO
R.Joaquim Murtinho 273
Rio de Janeiro, Brazil, Brazil

Silva, Maria De Fatima Vilas
Editor
LIGHT SERVICOS DE ELETRICIDADE S.A.
R. H. de M. Costa, 200 cob.01
Rio de Janeiro, Brazil, Brazil

Silva, Messias Da
Correspondent
O ESPECTADOR
Rua Werner Krause, 228
Araruama, Brazil, Brazil

Silva, Nelson Carlos De Souza
Reporter Radio/Television
MANCHETE TV
Rua Santa Rosa, 57 C/3
Bento Ribeiro, RJ, Brazil, Brazil

Silva, Pedro Leobino Da
Reporter Radio/Television
TV EDUCATIVA
Av. Gomes Freire 474, Centro
Rio de Janeiro, Brazil, Brazil

Silva, Peoro
Editor
GLOBO TV
Rua Von Martius, 22
Rio de Janeiro, Brazil, Brazil

Silva, Procopio Mineiro Da
Reporter Radio/Television
RADIO ROQUETE PINTO
Av. Sao Paulo, lote 34 Q. 39
Vilar dos Teles, RJ, Brazil, Brazil

Silva, Roberto
Reporter Radio/Television
GLOBO TV
Ave. Alvorada, 2600
Tijuca, Rio de Janeiro-RJ, Brazil, Brazil

Silva, Robson Jose M.
Reporter Radio/Television
RADIO & TV RIO NEGRO
Rua 17 No.31 Nucleo 05
Manaus, Amazonas, Brazil, Brazil

Silva, Rosane M.
Editor
AGENCIA ESTADO
R.Engenheiro Gama Lobo 463/404
Rio de Janeiro-RJ, Brazil, Brazil

Silva, Sebastiao Claudio
Reporter Radio/Television
TV EDUCATIVA - RJ
Rafael Rodrigues 32 C-1
Rio de Janeiro, Brazil, Brazil

Silva, Sergio
Reporter Radio/Television
RADIOBRAS
R. Silvio Polis, 201/1101
Rio de Janeiro, Brazil, Brazil

Silva, Sergio M.
Editor
GLOBO
Rua Assuncao, 140
Botafogo, Rio de Janeiro-RJ, Brazil, Brazil

Silva, Sergio R.
Editor
MANCHETE RURAL
R.Princesa Januaria 45/201
Flamengo, Rio de Janeiro, 22050,Brazil, Brazil

Silva, Silvianne Ma. T.
Correspondent
TV CULTURA DO PARA
Av. Atlantica, 3040/02
Rio de Janeiro-RJ, Brazil, Brazil

Silva, Vanderlei Novaes Da
Reporter Radio/Television
RADIO ROQUETE PINTO
R. Andre Rocha, 372 Bl4/903
Jacarepagua, RJ, Brazil, Brazil

Silva, Zileide
Reporter
TVSBT
Campo de S.Cristovao, 105
Rio de Janeiro - RJ, Brazil
T: 55 21 5803688 / 5800678 F: 55 21 5809685

Silva A., Roberto
Reporter Radio/Television
TV EDUCATIVA - RJ
Rua Tirol 876/105
Rio de Janeiro, Brazil, Brazil

Silva Alves F., Manoel
Editor
GLOBO TV
Rua Von Martius, 22 sala 321
Rio de Janeiro, Brazil, Brazil

Silva Froes, Iara Da Cruz
Editor
GLOBO TV
Rua Von Martius, 22 sala 321
Rio de Janeiro, Brazil, Brazil

Silva Matta, Gloria Maria
Reporter Radio/Television
GLOBO TV
Rua Von Martius, 22 sala 328
Rio de Janeiro, Brazil, Brazil

Silva Ortis, Delis
Correspondent
GLOBO TV
Rua Von Martius, 22 sala 328
Rio de Janeiro, Brazil, Brazil

Silva Rodrigues, Amauri
Reporter Radio/Television
RADIO TRES RIOS LTDA
Av. Heitor Zanatta, 129
25800 Rio de Janeiro-RJ, Brazil, Brazil

Silveira, Ana Paula
Editor
DIA
R. Joaquim Murtinho, 802
Rio de Janeiro-RJ, Brazil, Brazil

Silveira, Andrea Christine Laureiro
Reporter Radio/Television
RADIOBRAS
Daniel Carneiro 17/301
Rio de Janeiro, Brazil, Brazil

Silveira, Evanildo
Editor
JORNAL DO BRASIL
R. Paulistania, 46
Sao Paulo-SP, Brazil, Brazil

Silveira, Joao
Editor
JORNAL COPACENTRO
R. Domingos Ferreira 102
Apt. 804, Copacabana, RJ, Brazil, Brazil

Silveira, Mauro
Editor
MANCHETE
Rua Cicero Valadares, 100
Sao Paulo-SP, Brazil, Brazil

Silveira De Almeida, Emilia Rosa
Editor
GLOBO TV
Rua Von Martius, 22 sala 321
Rio de Janeiro, Brazil, Brazil

Silverio, Marcia Luisa
Editor
A NOTICIA
Rua Rocha Pita, 149 C/9
Cachambi, Rio de Janeiro, RJ, Brazil, Brazil

Simao, Fernando
Reporter Radio/Television
JOVEM PAN TV
R.Padre Galanti, 28
Jardim da Gloria 04114 Sao Paulo,Brazil, Brazil

Simao, Ronaldo
Reporter
TROFEU MARINER
Praia de Botafogo, 324/1007
Rio de Janeiro, Brazil, Brazil

Simoes, Maria M.P.S.
Editor
POVO
Rua Amaral, 114, Casa 05
Tijuca, Rio de Janeiro-RJ, Brazil, Brazil

Simoes, Ricardo J.
Correspondent
DIARIO DE PERNAMBUCO
R.Sete de Setembro, 464/706
Boa Vista, Recife-PE, Brazil, Brazil

Simoes, Rodolfo
Editor
DIA
Rua Cosme Velho, 639/404
Rio de Janeiro-RJ, Brazil, Brazil

Simquevitz Neto, Carlos
Correspondent
JORNAL EXPRESSAO
Av. Atlantica 2736/402
Rio de Janeiro, Brazil, Brazil

Sinelli, Monica
Editor
EDITORA PINI
R.Gal.Glicerio, 107/701
Rio de Janeiro, Brazil, Brazil

Sinicco, Sandra
Editor
ECOPRESS
R. Tete 59
Sao Paulo, Brazil, Brazil

Siniscalchi, Veronica
Reporter Radio/Television
TV CULTURA-REPORTER ECO
R.Bernardo Jose, 214
02345 Sao Paulo-SP, Brazil, Brazil

Siqueira, Carmem Lucia Muller
Reporter Radio/Television
TVT REDE DE COMUNICACAO
Rua dos Ingleses, 347
Sao Paulo, Brazil, Brazil

Siqueira, Jair
Correspondent
DIARIO CATARINENSE
R. Desembargador Pedro Silva
2745 Apt. 103, Florianopolis-SC,Brazil, Brazil

Siqueira, Priscila D.
Correspondent
RADIO ELDORADO FM
Alameda Santana, 426
11.600 Sao Sebastiao, Sao Paulo, Brazil, Brazil

Siqueira Campos De Barros, Marcelo
Editor
MANCHETE
Rua Grajau, 88
Grajau, Rio de Janeiro-RJ, Brazil, Brazil

Slezynger, Yvonne B.
Editor
SUNDAY NEWS
R.Eron Domingues 45
Sao Paulo, S.P., Brazil, Brazil

Smelli, Lameh
Editor
JORNAL AL URUBAT
R.Dr.Genesio Pereira 64
Sao Paulo, S.P., Brazil, Brazil

Soares, Elizabeth Sanz
Reporter Radio/Television
RADIOBRAS
R. Prof. S. Ribeiro, 83/1004
Rio de Janeiro, Brazil, Brazil

Soares, Kathia Pompeu
Editor
MANCHETE
Rua Decio Vilares, 180/402
Copacabana, Rio de Janeiro-RJ, Brazil, Brazil

Soares, Lucia H.
Reporter Radio/Television
TV CULTURA-REPORTER ECO
R. Homem de Mello, 804/22
05007 Sao Paulo-SP, Brazil, Brazil

Soares, Lucila
Editor
GLOBO
Av. Epitacio Pessoa, 4720/102
Rio de Janeiro-RJ, Brazil, Brazil

Soares, Luiz Antonio G.
Editor
RECORD TV
R. Siqueira Bueno, 217
Bairro do Belenzinho, Brazil

Soares, Nelson F.
Editor
AGENCIA ESTADO
R. Zeferino de Assis 16
Ramos, Rio de Janeiro-RJ, Brazil, Brazil

Soares, Rubens Valente
Correspondent
UNIVERSIDADE FED. MATO GROSSO SUL
R. Manoel Rosa, 110
Campo Grande, MS, Brazil, Brazil

Soares, Sara Benvinda
Editor
SBT TV
Av. Heitor Baltrao, 6/502
Tijuca, Rio de Janeiro-RJ, Brazil, Brazil

Soares, Terezinha Maria Muniz
Editor
TV EDUCATIVA - RJ
R. Sao Leobaldo, 369
Rio de janeiro, Brazil, Brazil

Soares, Vera Lucia Ferreira
JORNAL DO COMMERCIO
R. Francisco Muratori, 15/301
Sta.Teresa, Rio de Janerio-RJ, Brazil, Brazil

Sobral, Mauro Lopes
Editor
SBT TV
R. Cesario Motta Jr., 476
Apt. 1107 B1.B V. Buarque, Brazil, Brazil

Sobral Leite Barbosa, Maria Helena
Editor
GLOBO TV
Rua Von Martius, 22 sala 328
Rio de Janeiro, Brazil, Brazil

Sobrinho Cunha Da Silva, Denise
Editor
GLOBO TV
Rua Von Martius, 22 sala 321
Rio de Janeiro, Brazil, Brazil

Socile A., Mercedes
Correspondent
BRAZIL TURISMO
Vitor Meirelles 505
Rio de Janeiro, Brazil, Brazil

Solano, Jaime Biaggio
Editor
MANCHETE RADIO
R. Gal. Urquiza 63/1002,Leblon
Rio de Janeiro, Brazil, Brazil

Solberg, Ruy C.
Reporter Radio/Television
TV EDUCATIVA
Rua Eng. Alfredo Duarte,215
Rio de Janeiro-RJ, Brazil, Brazil

Somogyi, Flavio
NEON RIO LTDA.
Rua Sta.Cristina, 89
Sta.Teresa 20.241 Rio de Janeiro,Brazil, Brazil

Sotero, Paulo
Correspondent
AGENCIA ESTADO
5526 Greystone St.
Chevy Chase, MD 20815, USA, Brazil

Sousa, Ana Paula
Editor
GLOBO
SCS Ed.Oscar Niemeyer-13 andar
Brasilia-DF, Brazil, Brazil

Sousa, Emerson
Editor
EMBRATUR
SQN 108 Bl.6/406
Brasilia, D.F., Brazil, Brazil

Sousa, Jessie
Editor
ECOLOGIA E DESENVOLVIMENTO
R. Dr. Sardinha 155, St. Rosa
Niteroi, Rio de Janeiro, Brazil, Brazil

Souto, Veronica F.
Correspondent
DIARIO DE PERNAMBUCO
R.Guilherme Pinto, 66/704
52.011 Gracas, Recife-PE, Brazil, Brazil

Souza, Armando Carlos P.M.
Correspondent
JORNAL CORREIO FLUMINENSE
R.Marques Abrantes 168/711
Rio de Janeiro, Brazil, Brazil

Souza, Claudio Medeiros De
Reporter Radio/Television
GLOBO RADIO
R. Porto das Pedras, 81/302
Rio de Janeiro, Brazil, Brazil

Souza, Denival Barbosa De
Reporter Radio/Television
TV EDUCATIVA - RJ
R. Agostinho Barbalho 321 c1
Rio de Janeiro, Brazil, Brazil

Souza, Francisco Clovis Pinto De
Editor
RADIOBRAS
SQS 108 Bl.H/506 Sul
Brasilia, D.F., Brazil, Brazil

Souza, Gilberto De
Editor
FUNDACAO ROBERTO MARINHO
R. Gustavo Sampaio 630
Apt. 1101, Rio de Janeiro, RJ, Brazil, Brazil

Souza, Guilherme
Reporter Radio/Television
GLOBO RADIO
Rua do Russel 434
Rio de Janeiro, Brazil

Souza, Joel
Reporter Radio/Television
RADIO BANDEIRANTES
R. Luiz Cardamon, 123
Sao Paulo, Brazil, Brazil

Souza, Luiz Otavio
Reporter Radio/Television
RADIOBRAS
R.Almte. Guilhem 332/706
Rio de Janeiro, Brazil, Brazil

Souza, Ma. Alice M.
Editor
CONTEXTO AG. DE FOTOJORNALISMO
R. Xavier da Silveira, 15/1101
22.061 Rio de Janeiro-RJ, Brazil, Brazil

Souza, Maria Angelica Borges De
Editor
RADIOBRAS
HIGS 713 Bl.Z Casa 12
Brasilia, D.F., Brazil, Brazil

Souza, Maria De Fatima O.
Reporter Radio/Television
RADIO & TV BANDEIRANTES
Rua Luiz Noveletto 20
Sao Paulo, S.P., Brazil, Brazil

Souza, Maria Fernanda Da Silva
Reporter Radio/Television
RADIO CONTEMPORANEA AM
R. J. Miranda da Motta,396
Rio Bonito, RJ, Brazil, Brazil

Souza, Milton
Editor
JORNAL MOVIMENTACAO
Rua Senado 230/320
Rio de Janeiro, Brazil

Souza, Murta Katemary
Reporter Radio/Television
RADIO BANDEIRANTES
Av. Beira Mar 406
Rio de Janeiro, Brazil, Brazil

Souza, Nelson L.F.
Correspondent
POVO
Rua Comandante J. de Moura
Tijuca, Rio de Janeiro-RJ, Brazil, Brazil

Souza, Nilson F.
FOLHA DA MANHA
Av.Presidente Vargas 509/6and.
Rio de Janeiro, Brazil, Brazil

Souza, Randolpho De
Reporter Radio/Television
RADIO ROQUETE PINTO
R. Thomas Guimaraes, 221/202
Cachambi, RJ, Brazil, Brazil

Souza, Roberto P.
Editor
JORNAL DA TARDE
Av.Jurema 143 Apt.103
04079 Moema, Sao Paulo-SP, Brazil, Brazil

Souza, Teresa Felix
Reporter Radio/Television
MANCHETE
R. J. Conde de Bonfim, 526/704
Tijuca, RJ, Brazil, Brazil

Souza, Vera Lucia De
Reporter Radio/Television
VIDEO CLIPPING PRODUCOES LTDA.
Rua Senador Vergueiro 138
Rio de Janeiro-RJ, Brazil, Brazil

Souza Barreto Paes, Marcello
Editor
TV EDUCATIVA - RJ
R. Dois de Dezembro 15/601
Rio de Janeiro, RJ, Brazil, Brazil

Souza Dos Santos, Silvio
Reporter Radio/Television
TV EDUCATIVA - RJ
R. Gal. Savaget 419/301
Rio de Janeiro, Brazil, Brazil

Souza Neto, Helcio
MANCHETE TV
Rua Dois de Dezembro 62/104
Brazil, Brazil

Souza Silva, Sirlene Maria Chervezan
Reporter Radio/Television
GLOBO RADIO
R.Garibaldi 133/1301 Bl.B
Rio de Janeiro, Brazil, Brazil

Spencer-Chagas, B.
Editor
DIA
Rua Carlos Gois, 355/402
Leblon, Rio de Janeiro-RJ, Brazil, Brazil

Spinello, Sonia Matilde Bridi
Reporter Radio/Television
GLOBO TV
Rua Von Martius, 22 sala 321
Rio de Janeiro, Brazil, Brazil

Spinola, Rodolfo
Reporter Radio/Television
RADIO ELDORADO FM
R.Fausto Cabral, 1021
Fortaleza, Ceara, Brazil, Brazil

Stabec, Claudio
Reporter Radio/Television
MANCHETE TV
R.Voluntario da Patria166/1101
Botafogo, Rio de Janeiro-RJ, Brazil, Brazil

Stallbaum, Imara
Correspondent
DIARIO CATARINENSE
Av. Desembargador Pedro Silva
No. 1166, Florianopolis-SC, Brazil, Brazil

Stigger, Ivo
Editor
ZERO HORA
R. Vieira de Castro,481,Apt402
Santana, Porto Alegre-RS, Brazil, Brazil

Strauss, Joao Vitor
Editor
ESTADO DE SAO PAULO
R.Com.Miguel Calfat 303 C.4
04537 Vila Olimpia, Sao Paulo-SP, Brazil, Brazil

Stycer, Daniel
Editor
GLOBO
Rua Gen. Gois Monteiro, 54/702
Botafogo, Rio de Janeiro-RJ, Brazil, Brazil

Stycer, Mauricio Jose
Editor
FOLHA DE SAO PAULO
Al.Barao de Limeira, 425/4
Sao Paulo, SP, Brazil, Brazil

Suassuna, Luciano
Editor
ZERO HORA
SHIS QI 17, Conj. 08, Casa 04
Lago Sul, Brasilia-DF, Brazil, Brazil

Sukman, Hugo
Editor
JORNAL DO COMMERCIO
R. Caning, 19/501
Ipanema, RJ, Brazil, Brazil

Sussekind, Helio Carlos
Editor
MANCHETE
Rua Araucaria, 178
Jardim Botanico-RJ, Brazil, Brazil

Suzuki, Geraldo
Director
AMERICA DO SUL PRODUCOES DE DOCUMENTARIOS LTDA (NAMBEI TSUSHIN)
Av. Paulista 509 /11 andar conj.1104 Cerqueira Cesar
Sao Paulo -SP 01311, Brazil
T: 55 11 2886141 F: 55 11 2512065

Sydenstricker, Angela H.
Editor
MANCHETE TV
Rua Paulo Barreto, 46/1006
Brazil, Brazil

Sztajn, Andre
Editor
MANCHETE TV
R.Frei Leandro, 42/401
Rio de Janeiro, Brazil, Brazil

Tachinardi, Maria Helena
Editor
GAZETA MERCANTIL
R. Mj. Quedinho, 90 - 4 andar
Sao Paulo, Brazil, Brazil

Takeuti, Rene
Editor
BRASIL SEIKYO PUBL.HOUSE
Rua Pires da Mota 595
Sao Paulo-SP, Brazil, Brazil

Tambasco Penedo, Francisco
Editor
GLOBO TV
Rua Von Martius, 22
Rio de Janeiro, Brazil, Brazil

Tarcitano, Terezinha Lucia A.
Editor
EMBRATUR
Av. Paulo Frontin, 167 apt.505
Rio de Janeiro, Brazil, Brazil

Tardin, Marcos
Editor
GLOBO
Praia de Botafogo, 528/203-A
Rio de Janeiro-RJ, Brazil, Brazil

Tartaglia, Cesar A.
Editor
GLOBO
Rua Haddock Lobo, 375
Tijuca, Rio de Janeiro-RJ, Brazil, Brazil

Taublib, Janete A.
Editor
PREFEITURA RIO DE JANEIRO
R.Dr. Satamini 91/307
Rio de Janeiro, Brazil, Brazil

Tavares, Eduardo Diogo
Editor
JORNAL CORREIO DA BAHIA
Av. Garibaldi, 707
Salvador, Bahia, Brazil, Brazil

Tavares, Gerson
Editor
TV BANDEIRANTES
R.Barata Ribeiro 59/804
Rio de Janeiro, Brazil, Brazil

Tavares, Helena
Editor
ECABE PRESS
Rua General Rabelo, 65/404
22451 Rio de Janeiro-RJ, Brazil, Brazil

Tavares, Rita Cristina
Correspondent
JORNAL DO BRASIL
SQS 402 Bl.G apt.414
Brasilia, D.F., Brazil, Brazil

Tavares, Suzane
Reporter Radio/Television
MANCHETE TV
Rua Grao Para, 295/402
Brazil, Brazil

Taves, Rodrigo J.
Editor
GLOBO
Rua Miguel Couto, 371
Niteroi-RJ, Brazil, Brazil

Tedesco, Jose Carlos
Editor
MONITOR MERCANTIL E BEAUTIFUL WEEK
Jose Linhares 244/1002
Rio de Janeiro, Brazil, Brazil

Teich, Daniel H.
Editor
ESTADO DE SAO PAULO
R. Dr.Licinio Maragliano 149
04753 Sao Paulo-SP, Brazil, Brazil

Teixeira, Aldo Mussi
Reporter Radio/Television
TV BUZIOS
R.Sao Salvador 53/203
Rio de Janeiro, Brazil, Brazil

Teixeira, Aloy J.
Editor
GLOBO
Rua Miguel Cervantes, 143
Rio de Janeiro-RJ, Brazil, Brazil

Teixeira, Amauri
Editor
GLOBO
SCS Ed.Oscar Niemeyer-13 andar
Brasilia-DF, Brazil, Brazil

Teixeira, Ana Lucia De Castro
Editor
RADIOBRAS
SQB 115 Bl. G Ap. 103
Brasilia, D.F., Brazil, Brazil

Teixeira, Andre De Melo
Editor
UNISYS ELETRONICA
R.Lauro Muller 128/202
Rio de Janeiro, Brazil, Brazil

Teixeira, Eloisa Q.
Editor
GLOBO ECOLOGIA
R.Humaita 77/902
Rio de Janeiro, Brazil, Brazil

Teixeira, Flavio
Correspondent
SECRETARIA ESTADO MATO GROSSO SUL
Rua Palmital, 199
Campo Grande, MS, Brazil

Teixeira, Mario Luiz
Correspondent
POVO
Rua Manoel Carneiro, 24
20241 Rio de Janeiro-RJ, Brazil, Brazil

Teixeira, Regina Marcia
Reporter Radio/Television
MONITOR MERCANTIL
R.Conselheiro Autran,36
Vila Isabel, Rio de Janeiro,RJ,Brazil, Brazil

Teixeira Filho, Mario Honorio
Editor
ECORIO
Av. Ataulfo de Paiva,1151/802
Rio de Janeiro-RJ, Brazil, Brazil

Teixeira Leite, Luciana Savaget
Editor
GLOBO TV
Rua Von Martius, 22 sala 328
Rio de Janeiro, Brazil, Brazil

Teixeira Neubarth, Liliane
Reporter Radio/Television
GLOBO TV
Rua Von Martius, 22 sala 328
Rio de Janeiro, Brazil, Brazil

Temer, Cristian
Correspondent
MANCHETE TV
R. Coronel Moreira Cezar
279 Rio de Janeiro-RJ, Brazil, Brazil

Tenorio, Roberto L.
BLOCH EDITORES
Rua do Russel, 804
Rio de Janeiro, Brazil, Brazil

Teodoro Do Nascimento, Paulo Fernando
IRIS FOTO
R. Jacucaim 67
04563 Sao Paulo, SP, Brazil, Brazil

Terra, Patricia
Editor
ECOLOGIA E DESENVOLVIMENTO
R. Joaquim Murtinho 1042/301
Sta Tereza, Rio de Janeiro-RJ, Brazil, Brazil

Tetti, Laura Maria Regina
Reporter Radio/Television
GLOBO TV
Rua Von Martius, 22 sala 328
Rio de Janeiro, Brazil, Brazil

Tetu, Odilon
Director
TELENEWS SERVICE LTDA
Av Princesa Isabel 500/1613
Rio de Janeiro - RJ, Brazil
T: 55 21 546 6153 F: 55 21 5424223

Theme, Ana Claudia
Reporter Radio/Television
SBT TV
Rua Vinte Quatro de Maio
20951, Rio de Janeiro-RJ, Brazil, Brazil

Theodoro, Gerson Miranda
Editor
JORNAL MAIORIA FALANTE
R.Venancio Ribeiro 109/1001
Rio de Janeiro, Brazil, Brazil

Thielmann Monteiro Da Silva, Beatriz Helena
Reporter Radio/Television
GLOBO TV
Rua Von Martius, 22 sa al 321
Rio de Janeiro, Brazil, Brazil

Thomas, Erika
Editor
JORNAL O DIA
Av.Rui Barbosa 364/501
Rio de Janeiro, Brazil, Brazil

Tognolli, Claudio Julio
Correspondent
FOLHA DE SAO PAULO
R.Eca de Queiros, 257/1
Paraiso, Sao Paulo, SP, Brazil, Brazil

Toledo, Jose Luciano Cardoso
Correspondent
TV EDUCATIVA
R.Visconde Inhauma 25
Rio de Janeiro, Brazil, Brazil

Toribio, Lucia
Editor
GLOBO
SCS. Ed. Oscar Niemeyer, 13
Brasilia-DF, Brazil, Brazil

Torreao, Nelson
Editor
GLOBO
SCS. Ed. Oscar Niemeyer, 13
Brasilia-DF, Brazil, Brazil

Torres, Flavia De Almeida
Reporter Radio/Television
JORNAL DO BRASIL-RADIO AM
Av. Rui Barbosa, 520/1401
Flamengo, Rio de Janeiro-RJ, Brazil, Brazil

Torres, Sergio M.
Correspondent
FOLHA DA MANHA
R.Joaquim Tavora, 122/705 Bl.B
Niteroi-Rio de Janeiro, Brazil, Brazil

Tosta, Wilson
Editor
GLOBO
Ave. Suburbana, 7397/202
Rio de Janeiro-RJ, Brazil, Brazil

Tostes, Octavio Hermanny
Editor
GLOBO TV
Rua Von Martius, 22 sala 321
Rio de Janeiro, Brazil, Brazil

Totti, Paulo I.
Editor
GLOBO
Rua Visconde Silva, 102
Botafogo, Rio de Janeiro-RJ, Brazil, Brazil

Toutz, Carlos
Correspondent
ECOLOGIA E DESENVOLVIMENTO
R. Eduarod Novdruz 574
Ilha do Governador, RJ-RJ, Brazil, Brazil

Townes, Nelson
Correspondent
RADIO ELDORADO FM
Av.Carlos Gomes, 1952 Casa 1
Porto Velho, Est. de Rondonia, Brazil, Brazil

Trolles, Rosangela
Editor
ECOLOGIA E DESENVOLVIMENTO
R. do Riachuelo 32 Apt.607
20230 Lapa, Rio de Janeiro-RJ, Brazil, Brazil

Tsuguo, Murano
Correspondent
JORNAL PAULISTA
Mansion Nine 301 1-2-12
Igusa, Suginami-Ku, Tokyo, Japan, Brazil

Tupynamba, Jose Paulo
Editor
GLOBO
SCS. Ed. Oscar Niemeyer, 13
Brasilia-DF, Brazil, Brazil

Turazi, Deigma Maria Silva
Editor
RADIOBRAS
SHIN QL-02 Conj.11 c/06 Norte
Brasilia, D.F., Brazil, Brazil

Turcato, Marcia
Correspondent
AGENCIA ESTADO
QSB 6 Casa 27
Taguatinga, Brasilia-DF, Brazil, Brazil

Turela, Claudia
Editor
GLOBO TV
Rua Von Martius, 22 sala 321
Rio de Janeiro, Brazil, Brazil

Turibio, Ivany Buzzo
Reporter Radio/Television
RECORD TV
Rua Fradique Coutinho,316/112
Sao Paulo-SP, Brazil, Brazil

Uchida, Rubens Minoru
Correspondent
AGENCIA BOA IMPRENSA (ABIM)
Rua Victor Meirelles, 164
20.951 Rio de Janeiro-RJ, Brazil, Brazil

Uchoa, Joao
Editor
TV BUZIOS
Av.J.B. Ribeiro, 265
Cabo Frio, RJ, Brazil, Brazil

Uchoa, Julio
Editor
TV BUZIOS
Rua Visconde de Caravelas,116
Rio de Janeiro-RJ, Brazil, Brazil

Uchoa, Lygia Maria A.
FOLHA DA MANHA
Av.Presidente Vargas 509/6and.
Rio de Janeiro, Brazil, Brazil

Uday, Khemka
Correspondent
MANCHETE
25 Eaton Square,FlatH
London SW1 United Kingdom, Brazil

Vala, Sandro A.
Editor
AGENCIA ESTADO
R. Armando Colaferri 354
13200 Jundiai, Sao Paulo, Brazil, Brazil

Vale, Antonia Marcia Bezerra
Reporter Radio/Television
RADIOBRAS
SQS 410 Bl.M Entr. B Ap. 102
Brasilia, D.F., Brazil, Brazil

Valenca, Ribeiro Andreia
Reporter Radio/Television
GLOBO RADIO
Rua Tirol 861
Jacarepagua, Rio de Janeiro-RJ, Brazil, Brazil

Valente, Luize Mendes Pinheiro
Reporter Radio/Television
TV-BANDEIRANTES-RIO
Rua Paissandu, 283/302
Rio de Janeiro-RJ, Brazil, Brazil

Valerio, Marcos
Editor
GLOBO
Ave. Epitacio Pessoa 4720/102
Rio de Janeiro-RJ, Brazil, Brazil

Valla, Maria Augusta
Editor
GAZETA MERCANTIL
R. Mj. Quedinho, 90 - 4 andar
Sao Paulo, Brazil, Brazil

Valle, Luisa
Reporter Radio/Television
MTV
Rua Sambaiba 250/101
Rio de Janeiro, Brazil, Brazil

Valporto, Oscar De A.
Editor
GLOBO
Rua D. Mariana, 182/904
Botafogo, Rio de Janeiro-RJ, Brazil, Brazil

Varejao, Claudia
GLOBO
Rua Zaco Parana 126/104
Rio de Janeiro-RJ, Brazil, Brazil

Vargas, Francisco
Editor
DIA
Estr. de Jacarepagua, 2434
Rio de Janeiro-RJ, Brazil, Brazil

Varjao, Guilherme
Editor
RADIOBRAS
R. B. de Ipanema, 32/404
Rio de Janeiro, Brazil, Brazil

Vasconcellos, Claudio Carvalho A.
Editor
MANCHETE
Estrada Velha da Tijuca, 511
Rio de Janeiro-RJ, Brazil, Brazil

Vasconcellos Cattoni, Bruno
Editor
GLOBO TV
Rua Von Martius, 22 sala 321
Rio de Janeiro, Brazil, Brazil

Vasconcelos, Bianca
Reporter Radio/Television
TVN
R.Purpurina 162/93
Sao Paulo, S.P., Brazil, Brazil

Vasconcelos, Carlos H.
Editor
DIA
Av. Benjamin Constant, 497/102
Niteroi, Rio de Janeiro-RJ, Brazil, Brazil

Vasconcelos, Fabio A.
Reporter Radio/Television
MANCHETE TV
Rua Lopes Quintas 200/205
Rio de Janeiro, Brazil, Brazil

Vasconcelos, Palmerio D.
Editor
DIARIO DO PARA
Rua O de Almeida 1102 Casa 35
Belem-PA, Brazil, Brazil

Vasconcelos, Sonia Maria A.
Editor
EDITORA GLOBO
R.Retiro Artistas 607/5/105
Rio de Janeiro, Brazil, Brazil

Vasconselos, Roberto Barreira
Editor
MANCHETE
Rua do Russell, 766
22210-Rio de Janeiro-RJ, Brazil, Brazil

Vasilcousky, Carlos
Editor
REVISTA ALAGOASTUR
Rua Anita Garibaldi, 37/1001
Rio de Janeiro-RJ, Brazil, Brazil

Vaz, Andrea Resende
Reporter Radio/Television
GLOBO TV
Avenida Alvorada, 2600
Tijuca, Rio de Janeiro-RJ, Brazil, Brazil

Vaz, Elida Mattos
Editor
GLOBO
Rua Luiz Tomas 270/302
Rio de Janeiro-RJ, Brazil, Brazil

Vaz Martins, Antonio Carlos
Editor
GLOBO TV
Rua Von Martius, 22 sala 328
Rio de Janeiro, Brazil, Brazil

Veiga, Maria Lucia Moreira A.
Editor
GOVERNO DO DISTRITO FEDERAL
Palacio do Buriti, Brazil

Veiga, Rita De Cassia Ferreira Da
Reporter Radio/Television
TV EDUCATIVA
Rua Mundo Novo 376, Botafogo
Rio de Janeiro, Brazil, Brazil

Velasco, Ma. Daniella
Editor
FLUMINENSE
Rua Barao do Amazonas, 02/1204
Centro, Niteroi, Rio de Janeiro, Brazil, Brazil

Velloso, Eliane B.
GLOBO
Rua Jose Bonifacio, 42
Niteroi-RJ, Brazil, Brazil

Velloso, Marcelo Fernandes
Reporter Radio/Television
JORNAL DO BRASIL-RADIO AM
Rua Sa Viana, 117 Casa 5
Rio de Janeiro-RJ, Brazil, Brazil

Veloso, Suzane T.
Editor
GLOBO
R.Moura Brito, 189/106
Rio de Janeiro, Brazil, Brazil

Velyanovsky, Naum
Correspondent
GLOBO TV
Rua Von Martius, 22 sala 321
Rio de Janeiro, Brazil, Brazil

Ventura, Luzia Christina
Editor
TV EDUCATIVA - RJ
R. Franco Job 21
Monero Ilha do Governador, RJ, Brazil, Brazil

Ventura, Zuenir Carlos
Editor
JORNAL DO BRASIL
Rua Redentor, 99/101
Rio de Janeiro - RJ, Brazil

Verdini, Liana
Editor
GLOBO
Rua Senador Vergueiro, 238
Flamengo, Rio de Janeiro-RJ, Brazil, Brazil

Vianna, Claudio Besserman
REDE GLOBO
R. Lopes Quintas 303 J.Botanico
Rio de Janeiro RJ 22, Brazil
T: 55 21 2947732

Vianna, Marilena Cavalcanti
Reporter Radio/Television
RADIO ROQUETE PINTO
R. Campos Sales, 25/1104
Tijuca, Rio de Janeiro, Brazil, Brazil

Vianna, Rosangela S.
Editor
BLOCH EDITORES
R. Sa Ferreira 227/707
Rio de Janeiro, Brazil, Brazil

Victorino, Gustavo A.G.
Reporter Radio/Television
RADIO GUAIBA
Rua Visconde de Piraja,525/211
Rio de Janeiro-RJ, Brazil, Brazil

Vidal, Ney Filho
Correspondent
ZERO HORA
R. Antenor de Moraes, No. 128
Florianopolis-SC, Brazil, Brazil

Vidal, Paulo Sergio M.
Reporter Radio/Television
RADIO FLUMINENSE
R.Miguel Couto 344/302
Niteroi, Rio de Janeiro, Brazil, Brazil

Vidal Campos, Sueli
Reporter Radio/Television
GLOBO RADIO
Rua Senador Verqueiro, 73
Rio de Janeiro-RJ, Brazil, Brazil

Vieira, Agostinho
Editor Local
JORNAL O GLOBO
Rua Irineu Marinho, 35
Rio de Janeiro - RJ, Brazil
T: 55 21 2722000

Vieira, Felipe Garcia
Reporter Radio/Television
RADIO GAUCHA
R. Demetrio Ribeiro, 247/205
90000 Porto Alegre-RS, Brazil, Brazil

Vieira, Jose Carlos
Editor
JORNAL DE ESPANA
Rua dos Artistas, 225
Rio de Janeiro, Brazil, Brazil

Vieira,, Ma. Candida
Editor
DIARIO COMERCIO IND. (DCI)
R.Heitor Penteando 2009/604
Sumarezinho, Sao Paulo, Brazil, Brazil

Vieira, Marcelo Martins
Reporter Radio/Television
TV RIO
R. Lineu Machado, 826, #404
Jardim Botanico, Rio de Janeiro, Brazil, Brazil

Vieira, Marceu
Editor
DIA
Rua Gago Coutinho, 89/602
Rio de Janeiro-RJ, Brazil, Brazil

Vieira, Marcia
Editor
VEJA RIO
Rua Sao Salvador, 99/703
Laranjeiras, Rio de Janeiro-RJ, Brazil, Brazil

Vieira, Paulo Sergio M.
Editor
GLOBO
R. Sao Salvador 84-502
Rio de Janeiro-RJ, Brazil, Brazil

Vieira, Wilma Lucia
Editor
A NOTICIA
R.Gurindiba 134/S-102
Rio de Janeiro, Brazil, Brazil

Vieira De Souza, Gerson
EDITORA AZUL/REVISTA CONTIGO
Rua Arnaldo Quintela, 10/402
Botafogo, Rio de Janeiro-RJ, Brazil, Brazil

Vignais, Alain
Reporter Radio/Television
GLOBO TV
Canal Marapendi 2980
Rio de Janeiro, Brazil, Brazil

Vilanova, Candida Roberta Couto
Reporter
DIARIO DO PARA
Av. Gaspar Viana, 773
Belem - Para, Brazil
T: 55 91 2227388 F: 55 91 2243162

Villanova, Rodolpho
Reporter Radio/Television
TV EDUCATIVA
Rua Marques de Abrantes,115
Rio de Janeiro-RJ, Brazil, Brazil

Villas Boas, Marcos
Reporter Radio/Television
GLOBO RADIO
SHIS QI 6/8 Bl. A
Brasilia-DF, Brazil, Brazil

Vinhaes, Paulo
Reporter Radio/Television
REDE OM BRASIL DE TELEVISAO
R. Buarque de Macedo, 37/506
Flamengo, RJ, Brazil, Brazil

Vinhal, Sara M.
Reporter Radio/Television
MANCHETE TV
Rua Clemencenu 126
Bonsucesso, RJ, Brazil, Brazil

Visoni, Claudia
Editor
EDITORA ABRIL/CAPRICHO
R. Brasilio Machado 47/112
01230 Sao Paulo, Brazil, Brazil

Vita, Rachel Beatriz Faria
Editor
DIA
Rua Maxwell, 40
Vila Isabel, Rio de Janeiro-RJ, Brazil, Brazil

Vitta, Joao
Editor
GLOBO RADIO
Rua Georgino Rego Barros,185
23.075 Rio de Janeiro-RJ, Brazil, Brazil

Waack, William
Correspondent
AGENCIA ESTADO
Rusternallee 19
1000 Berlin 19, Germany, Brazil

Wakimn, Roberto Goncalves
Reporter Radio/Television
GLOBO RADIO
R. Barbacena, 104
Rio de Janeiro, Brazil, Brazil

Walcacer, Ma. Theresa
Editor
PROGRAMA GLOBO ECOLOGIA
R. Maria Eugenia, No. 210
Humaita, RJ, Brazil, Brazil

Waldvoguel, Monica De Lourdes
Reporter Radio/Television
GLOBO TV
Rua Von Martius, 22 sala 328
Rio de Janeiro, Brazil, Brazil

Walsh, Lionel
Correspondent
MANCHETE TV
F3 Les Rives de l'Yerres
Yerres, Paris, Brazil

Wambier, Telmo Luiz Meira
Editor
AGENCIA ESTADO
R. Assis Brasil 143, Bl.1
Copacabana, Rio de Janeiro-RJ, Brazil, Brazil

Wamburg, Jorge Cesar Bellez
Reporter Radio/Television
RADIOBRAS
SQS 316 Bl. K Ap. 208
Brasilia, D.F., Brazil, Brazil

Weber, Leila Paixao
Editor
JORNAL DO COMERCIO
R. Gen. Cipriano Ferreira 551
Apt. 07, Rio de Janeiro, Brazil, Brazil

Weichert, Marcio Armin
Editor
MANCHETE
Rua 21 de Abrix, 30
Niteroi, Rio de Janeiro-RJ, Brazil, Brazil

Weingrill, Carmen Lucia Visconti
Reporter
CIENCA HOJE
Av Wenceslan Bids 71 Casa 27
Rio de Janeiro, Brazil
T: 55 21 2954846 F: 55 21 5415342

Weissemberg, Ma.Cecilia B.
MANCHETE TV
Rua Lauro Muller 26/504
Brazil, Brazil

Wenzel, Jose Alberto
Reporter Radio/Television
EMISSORAS REUNIDAS
R.Andrade Neves 104
Santa Cruz do Sul, RS, Brazil, Brazil

Wesley, Carlos
Editor
JORNAL DO BRASIL
R. Timoteo da Costa, 1033/203
Leblon, Rio de Janeiro-RJ, Brazil, Brazil

White, Christine De Oliveira
Reporter Radio/Television
JORNAL DO BRASIL-RADIO AM
Rua Alfredo Backer, 536
Niteroi, Rio de Janeiro, Brazil, Brazil

Wilson, Ana Teresa Robusti
Redatora
SISTEMA JORNAL DO BRASIL
Av.Brasil, 500
Rio de Janeiro - RJ, Brazil
T: 55 21 5854422

Woeltje, Gert Rodolfo
Editor
AMAZONAS MULTIMEDIA STOCK
Rua das Angelicas, 173
Manaus, Brazil, Brazil

Wynn-Jones, Elizabeth
Editor
RIOLIFE
R. Prof. Brandao 60/401
Lebon, 22450, Rio de Janeiro-RJ, Brazil, Brazil

Xavier, Dieny Dayse Cordeiro
Reporter Radio/Television
RADIO CONTEMPORANEA AM
R. Sigisfredo Bravo, 340
Bacaxa, Saquarema, RJ, Brazil, Brazil

Xavier, Jose Messias
Reporter
REVISTA VISAO
R.Dr.Almeida Lima, 1384
Sao Paulo - SP, Brazil
T: 55 11 2914866 R.476 F: 55 11 2927438

Xavier, Rui Cardoso
Editor
DIA
Rua Barao da Torre, 19/101
Ipanema, Rio de Janeiro-RJ, Brazil, Brazil

Xisto, Domingos
Reporter Radio/Television
REDE OM BRASIL DE TELEVISAO
R. Henrique Oswaldo, 211/903
Copacabana, RJ, Brazil, Brazil

Yanakiew, Monica C.
Correspondent
GLOBO
Av. Franklin Roosevelt 182/3
Belgica, Bruxelas, Brazil

Yusim Salles, Liliane
Editor
GLOBO TV
Rua Von Martius, 22 sala 321
Rio de Janeiro, Brazil, Brazil

Zacconi, Carla
Editor
GLOBO
Rua Barao de Itambi, 34/1503
Botafogo, Rio de Janeiro-RJ, Brazil, Brazil

Zanettini, Paulo E.
Editor
JORNAL DA TARDE
R.Belgica 245
01448 Sao Paulo-SP, Brazil, Brazil

Zappa, Regina
Editor
JORNAL DO BRASIL
R. Sambaiba, 479/801
22450, Rio de Janeiro-RJ, Brazil, Brazil

Zocatelli, Ricardo R.
Editor
FLUMINENSE
Rua Barata Ribeiro, 334/1004
Copacabana, Rio de Janeiro-RJ, Brazil, Brazil

Zorzo, Rosangela
Correspondent
ZERO HORA
Ql 23, Bloco F, Apt. 157
70.000 Guara II, Brasilia-DF, Brazil, Brazil

Zucchi, Luiz
Reporter Radio/Television
JORNAL EXPRESSAO
R. Humaita 207/403
Rio de Janeiro, Brazil, Brazil

Zygmunt, Sulistrowski
Editor
ORGANIZACAO CINEMETOGRAFICA BRAS.
Monsenhor Coutinho, 527
Manaus, Amazonas, Brazil, Brazil

ORGANIZATIONS

Aunvalle, Angela Escostiguy, Ms.
Acao Democratica Feminina Gaucha./friends Of The Earth, Brazil
Rua Miguel Tostes, 694
90 000 Porto Alegre, Brazil
T: (55 512) 32 88 84 F: (55 512) 31 31 05

Aveline, Carlos, Mr.
Uniao Protetora do Ambiente Natural (UPAN)
Caixa Postal 189
CEP 93001, Sao Leopoldo, RS, Brazil
T: (55 512) 927 933 F: (55 512) 926 617

Bonofaz, Neptali, Mr.
Institute for Ethnobiology of the Amazon
Alameda III, Casa 1A, Conjunto Maguari
Belem-PA 66000, Brazil

Bruni, Dorival Correia, Mr.
Biosfera-brazilian Society Of Environmental Promotion
Caixa Postal 3591, Ave. Pasteur,404/3 andar (CPRM)
Urca-CEP 22290 Rio de Janeiro RJ, Brazil
T: (55 11) 295 4347/295 0032 F: (55 11) 533 2342/262 5946

Candotti, Ennio, Mr.
Sociedade Brasileira para o Progresso da Ciencia
Caixa Postal 11008
CEP 05499, Sao Paulo, SP, Brazil
T: (55 11) 211 0933/211 5008 F: (55 11) 212 1376

Capobianco, Joao Paulo, Mr.
Forum Of Brazilian Ngos For UNCED 92
Rua Manoel da Nobrega, 456
04001 Sao Paolo/SP, Brazil
T: (55 11) 887 1195 F: (55 11) 885 1680

Costa, Jairo, Sr.
Fundacao Brasilliera Para E Conservacao Da Natureza (FBCN)
R. Miranda Valverde 103, Botafogo 22281
Rio de Janeiro RJ, Brazil
T: (55 21) 226 2654 F: (55 21) 226 1345

Czapski, Julian D., Mr.
As. Ituana de Protecao Ambiental (AIPA)
Fazenda Sao Miguel, Caixa Postal 83
CEP 13300, Itu, SP, Brazil

De Almeida, Talita, Ms.
Brazilian Organization for Special Projects (OBRAPE)
Rua Marques de Sao Vicente, 355
Gavea, Rio de Janeiro, RJ, CEP 22451, Brazil
T: (51 21) 274 6593

De Lemos Monteira Da Silva, Alberto Mr.
As. Brasiliera De Ecologia
Av. Nilo Pecanha 12, Grupos 801/825
Centro
Rio de Janeiro RJ, Brazil
T: (55 21) 242 4374

De Oliveira, Rosiska Darcy, Ms.
Instituto de Acao Cultural
Av. Ataulfo de Paiva 566/305
22440 Rio de Janeiro, Brazil
T: (55 21) 511 0142 F: (55 21) 511 0142

Diskin, Lia, Ms.
As. Palas Athena do Brazil
Rua Leoncio de Carvalho, 99
Paraiso, Sao Paulo, CEP 04003, Brazil
T: (55 11) 288 7356

Eghrari, Iradj Roberto, Mr.
National Spiritual Assembly of the Bahais of Brazil (NSA-BRA)
AEN dos Bahais do Brasil, Escritorio de Meio Ambiente, Caixa Postal 7035, 71619 Brasilia- DF, Brazil
T: (55 61) 248 4718 F: (55 61) 248 4321

Filho, Naulle De Barros, Mr.
Sociedade De Defensa Do Pentanal (SODEPAN)
Ave. Americo Carlos da Costa 320, Parque de Exposicoes Laucidio Coelho, Campo Grande-MS, Brazil
T: (55 67) 721 1891

Friaes, Pinon, Mr.
Group for Study & Defense of Ecosystems of the Lower & Middle Amazon Region
GEDEBAM, Nucleo Monte Alegre, Rua Raimundo Uchoa, 171, Monte Alegre-Para, CEP 68 220, Brazil
T: (55 91) 533 1437 F: (55 91) 533 1437

Goree, Langston James, IV, Mr.
Institute Of Pre-history, Anthropology And Ecology
Caixa Postal 585
78 900 Porto Velho, Rondonia, Brazil
T: (55 69) 223 3945 F: (55 69) 223 3945

Gross, Anthony, Mr.
Centro Ecumenico de Documentacao e Informacao
Rua Santo Amaro 129
22211 Rio de Janeiro RJ, Brazil
T: (55 21) 224 6713 F: (55 21) 242 8847

Hosmann, Elizabeth, Ms.
Pro-rio 92
Rua Uruguaiana, 10-20 Andar
CEP 20050, Rio de Janeiro RJ, Brazil
T: (55 21) 242 4618 F: (55 21) 252 4273

Lacerda, Leonard, Mr.
Fundacao Vitoria Amazonica
Avenida Darcy Vargas, 520 Chapada
Manaus, Amazonas, 69 050 Brazil, Brazil
T: (55 92) 642 1336 F: (55 92) 642 2255

Louren, Jose Seixas, Mr.
As. of Amazonian Universities (UNAMAZ)
C. Postal 558
Belem-Para 66040, Brazil
T: (55 91) 229 5666/229 4339 F: (55 91) 229 9677/229 7288

Loyola, Cleuler De Barros, Ms.
Instituto Brasileiro de Administracao Municipal
Largo IBAM, 1
22282 Rio de Janeiro, Brazil
T: (55 21) 266 6622 F: (55 21) 537 1262

Montoro, Andre Franco, Sr.
Instituto Latino Americano
Rua Colombia, 66- Jardim America
CEP 01438 Sao Paulo, Brazil
T: (55 11) 852 7799 F: (55 11) 640 179

Netto, Eugenio Scannavino, Mr.
Center for Advanced Studies in Social Care/Health and Happiness Project
Rua Paulo Barreto, No. 23, Botafogo
CEP 22280, Rio de Janeiro RJ, Brazil
T: (55 21) 266 7897/226 1519 F: (55 21) 226 3235

Padua, Maria Teresa Jorge, Ms.
Funatura
SCLN 107, Ed. Gemini Center 11, Bloco B
Salas 201/10, CEP 70743 Brasilia D.F., Brazil
T: (55 61) 274 5449 F: (55 61) 274 5324

Paioli, Waldemar, Mr.
As. Mundial De Ecologia (AME)
Rua Senador Feijo, 338-20, Caixa Postal 108
CEP 06700 Sao Paulo, Brazil
T: (55 11) 493 2636 F: (55 11) 493 5679/492 2209

Pereira, Sonia Regina Brito, Sra
Brazilian Movement in Defense of Life (MBDV)
Rua Carvalho de Mendonca No. 36/802
Copacabana, Rio de Janeiro RJ CEP 22020, Brazil
T: (55 21) 542 1385

Perrier, Denis, Mr.
Ecotropic - Brazil
Rua Almirante Alexandrino, 2161
20241 Rio de Janeiro, Brazil
T: (55 21) 221 7997 F: (55 21) 242 8939

Pessis, Anne-Marie, Ms.
Fundacao Museu Do Homem Americano
Rua Abdias neves 551
64770 Sao Raimundo Nonato, Piaui, Brazil
T: (55 86) 582 1612

Saavedra, Jorge Eduardo, Sr.
Federacao de Orgaos para Assistancia Social e Educacional
Rua Bento Lisboa, No. 58
Catete, Rio de Janeiro CEP 22221, Brazil

Silva, Ricardo, Mr.
Pro Vita
Rua da Assembleia 93/801
CEP 20011, Rio de Janeiro RJ, Brazil
T: (55 21) 221 4761 F: (55 21) 222 1569

Terena, Marcos, Mr.
Comite Intertribal - 500 Anos De Resistencia
S.C.S.Ed. Palacio do Comercio, Sala 1103
CEP 70 372, Brasilia/DF, Brazil
T: (55 61) 224 3547 F: (55 61) 321 4377

Turtelli, Armando, Jr., Mr.
State University Of Campinas
Caiza Postal 1170
13100 Campinas SP, Brazil
T: (55 192) 39 1301 F: (55 192) 39 4717

Vamos, Robert K., Mr.
Instituto Brasileiro De Pesquisas E Estudos Ambientais, Pro-natura
Rua Monte Alegre, 374
Rio de Janeiro RJ 20240, Brazil
T: (55 21) 252 1191/224 3994 F: (55 21) 222 6742

Weyland, Maria Christina, Ms.
Instituto Sul-Mineiro de Estudios e de Conservacao da Natureza
Fazenda Lagoa, Caixa Postal 6
Monte Belo 37132 MG, Brazil

GLOBAL FORUM

Abdala, Fabio De Andrade, Mr.
UNIVERSIDADE ESTADUAL DE CAMPINAS
NEPAM UNICAMP Caixa Postal 1170
Campinas - SP 13081, Brazil
T: 55 192 398151 F: 55 192 394717
TL: 1150

Abe, Toyofica, Sra.
IECAM
Dr. Moreira 250
Manaus - AM 69000, Brazil

Abrantes, Jerry, Mr.
INTERNATIONAL UNION OF SOCIALIST YOUTH
Neustiftgasse 3
Vienna A 1070, Brazil
T: 43 1 931-267 F: 43 1 5231-24385
TL: 75312469 SJOE

Abreu, Cecilia Rita De
PROVENT
Rua Uruguiana, 10 2 andar
Rio de Janeiro, Brazil
T: 55 21 2424618

Abreu, Claudette M., Sra.
Presidente
ASSOCIACAO MULHERES NEGOCIOS PROFISSIONAIS-LONDRINA
Rua Alagoas 1439
Londrina-PR 86020, Brazil
T: 222754

Abreu, Jose Agostinho Almada De, Sr.
Coordenador Comissao Vmb/Eco92
VISAO MUNDIAL BRASIL
Rua Antonio de Albuquerque, 788
Belo Horizonte - MG 30110, Brazil
T: 55 31 221-8944 F: 55 31 227-2995
TL: 31.6052

Abreu, Leila, Sra.
Secretary
CMB
Rua Avanhadava no. 756
Sao Paulo - SP 01306, Brazil
T: 55 11 2564354 F: 55 11 2577648

Abreu, Marcelo, Sr.
Reporter
AUCUBA PESQUISA E DOCUMENTACAO
R. Ricardo Hardman 105 Aflitos
Recife - PE 52050, Brazil
T: 55 81 268-2715

Abreu, Marcia Braga De, Sra.
Recepcionista
IBFAN
Rua Arco-Ires 148 Freguesia/Jacarepagua
Rio de Janeiro - RJ, Brazil
T: 55 21 342-3563

Abreu, Patricia, Sra.
PREFEITURA DE ANTONINA - PARANA
R. Voluntarios da Patria 616a
Curitiba - PR 80000, Brazil
T: 55 41 225 5423

Abreu, Sandra Maria De Paula
FORUM DO FUTURO
Rio de Jaeiro, Brazil

Abreu, Silvana Rocha, Sra.
INSTITUTO INTERNACIONAL DE INFORMACOES ECOLOGICAS
Av. Treze de Maio, 23 Grupo 423 e 424
Rio de Janeiro - RJ 20031, Brazil
T: 55 21 220-5548 F: 55 21 240-7472
E: 1442

Abrunhosa, Orlando
Fotografo
REVISTA MANCHETE
Rua do Russel, 804
Rio de Janeiro - RJ, Brazil
T: 55 21 2652012

Abud, Mylene
ESTADO DE SAO PAULO
Av. Rio Branco 151 - Sobre - Loja sala 201 a 204
Rio de Janeiro RJ, Brazil
T: 55 21 2217111

Abugoche, Veronica, Sra.
Biologa
INSTITUTO DE MEIO AMBIENTE DO ACRE
Rua Rui Barbosa, 450
Acre, Brazil
T: 55 68 224 5497 F: 55 68 224 5694
TL: 682692

Acha, Marcos Rogerio, Mr.
Eng. Flor.
I T C F
Rua Alexandre Belot, S/N
Mimoso do Sul - ES, Brazil
T: 55 27 555-1025

Ache, Lina Maria, Sra.
Geografa
SECRETARIA DE MEIO AMBIENTE DO ESTADO DE SAO PAULO
Av. Irai, 1337
Sao Paulo - SP 04082, Brazil
T: 55 11 543 9640

Acherman, Luiza, Ms.
LIGA FEMININA ISRAELITA DO BRASIL
Av. Copacabana 690 sl.1001
Rio de Janeiro - RJ 22050, Brazil

Acioli, Claudia
Jornalista
REVISTA MANCHETE
Rua do Russel, 804
Rio de Janeiro - RJ, Brazil
T: 55 21 2652012

Acosta, Adao, Mr.
Eng. Agronomo
COTRIJUI'
Cx. Postal 111
Ijui-RS 98700, Brazil
T: 55 55 332-2400

Acosta, Fernando Tomas, Sr.
President Director
INSTITUTO DE O.
Rua Miguel Pereira, 44 - A -
Rio de Janeiro - RJ, Brazil
T: 55 21 226-3733

Acosta, Tomas Diaz
PRENSA LATINA
R. Mascarenhas de Moraes 121/203 Copacabana
Rio de Janeiro -RJ 2, Brazil
T: 55 21 2567259

Adario, Paulo, Mr.
ASSOCIACAO CIVIL GREENPEACE
Rua Mexico 21, Grp 1301, Centro
Rio de Janeiro - RJ 20031, Brazil
T: 55 21 262-7318 F: 55 21 240-1690

Advincula, Aluizio, Mr.
SOCIEDADE NORDESTINA DE ECOLOGIA
C.Postal 7807
Recife - PE 50739, Brazil

Affonso, Cleber V.
Technician
REDE GLOBO
R. Lopes Quintas 303
Rio de Janeiro -RJ 2, Brazil
T: 55 21 2947732

Affonso, Luiz Carlos, Mr.
C C C M
Av. Amaral Peixoto, 334/801 - Centro
Niteroi - RJ 24020, Brazil
T: 55 21 717 0112

Afonso, Cleyde
Producer
IBASE
R.Vicente de Souza, 29 Botafogo
Rio de Janeiro -RJ, Brazil
T: 55 21 2265412 F: 55 21 2860541

Afonso, Dilson Bacarat
Journalist
FREELANCER
Rua Benjamin Constant, 47/702
Rio de Janeiro, Brazil
T: 55 21 2229849

Afonso, Luiz Carlos
Sound Operator
RADIO TUPI
Rua do Livramento 189/8th Floor
Rio de Janeiro - RJ, Brazil
T: 55 21 203 2421 r 34 F: 55 21 2637323

Afonso, Neila Maria Batista, Sra.
Assistente Social
FUNDO CRISTAO PARA CRIANCAS
Rua Curitiba 689 - 5o. andar
Belo Horizonte - MG 30170, Brazil
T: 55 31 201-4955 F: 55 31 212-6957

Aguiar, Chico, Sr.
Presidente
ASSOCIACAO ECOLOGICA E PROTETORA DO PARQUE NACIONAL DA TIJUCA - ECO TIJUCA
Rua Carlos Vasconcelos 155 - Gr.402
Rio de Janeiro - RJ 20521, Brazil
T: 55 21 284-8551

Aguiar, Ivalnilda De Souza, Ms.
ISER/SOFIA
Caixa Postal 16011
Rio de Janeiro - RJ 22201, Brazil
T: 55 21 265 3747

Aguiar, Leonel Azevedo De
Jornalista
REVISTA VISAO
Av.Rio Branco, 177/6 andar
Rio de Janeiro - RJ, Brazil
T: 55 21 2621085 F: 55 21 2200805

Aguiar, Ludmilla, Ms.
CONSERVATION INTERNATIONAL
Rua Bueno Brandao 393
Belo Horizonte - MG 31010, Brazil
T: 55 31 226-5145 F: 55 31 222-8429

Aguiar, Luis Alexandre
LEGIAO DA BOA VONTADE
Av. Rudge 700 Bom Retiro
Sao Paulo -SP 01134, Brazil
T: 55 21 2338381/R:223 F: 55 21 2634025

Aguiar, Rene, Sr.
Chefe Nucleo Ctl Ambiental
INSTITUTO DO MEIO AMBIENTE DO AMAZONAS - IMA/AM
Rua Recife 3280 - Parque 10
Manaus - AM 69050, Brazil
T: 55 92 236-2309 F: 55 92 236-2535

Aguzzoli, Enio, Sr.
Medico
ROTARY
R. Prof. Annes Dias, 154/1203
Porto Alegre - RS 90020, Brazil
T: 55 051 226 5743

Ahouagi, Tania
CONFEDERACAO NACIONAL DA INDUSTRIA
Rio de Janeiro, Brazil

Aissa, Susi Cleia
Reporter
FOLHA DE SAO PAULO
Alameda Barao de Limeira 425
Sao Paulo SP, Brazil
T: 55 11 874 2608/874 2602 F: 55 11 2231644

Aizenman, Renato
Reporter
EDITORA ABRIL
R.da Passagem, 123/8 andar
Rio de Janeiro - RJ, Brazil
F: 55 21 2759347

Akhavan, Elena Martin, Sra.
COMUNIDADE BAHA'I
Rua Bela Cintra, 1413/112
Sao Paulo - SP 01415, Brazil

Akhavan, Said
COMUNIDADE BAHA'I
Rua Bela Cintra 1413/112
Sao Paulo - SP 01415, Brazil

Alajmo, Marco, Mr.
MLAL
c/o Bask - Rua Vicente de Souza 12
Rio de Janeiro - RJ, Brazil
T: 55 21 286-6161 F: 55 21 286-0541

Alambert, Zuleika, Ms.
Assessora Especialpresidenta
CONSELHO ESTADUAL DA CONDICAO DA MULHER
Av. Paulista 1776-17o Andar
Sao Paulo - SP 01310, Brazil
T: 55 11 287-4332 F: 55 11 284-1667

Albano, Adriana, Ms.
ASSOCIACAO CIVIL GREENPEACE
Rua Mexico 21, Grp 1301, Centro
Rio de Janeiro - RJ 20031, Brazil
T: 55 21 262-7318 F: 55 21 240-1690

Albano, Angelus R., Dr.
Cientista
SOCIEDADE BRASILEIRO DE GEOGRAFIA
Rua Senador Dantas, 117 Sala 824
Rio de Janeiro - RJ 20031, Brazil

Albano, Maria Tereza, Sra.
PREFEITURA MUNICIPAL DE PORTO ALEGRE
Av. Borges de Medeiros 2244
Poa - RS 90000, Brazil
T: 55 51 221-0366

Albin, Ricardo, Mr.
MUSEU DA IMAGEM E DO SOM
Av.Atlantica,1260/1001
Rio de Janeiro - RJ, Brazil
T: 55 21 542-1162 F: 55 21 295-0742

Albino, Ozias Jose
Technician
REDE GLOBO
R. Lopes Quintas 303 Jardim Botanico
Rio de Janeiro -RJ 2, Brazil
T: 55 21 2947732

Albino Neto, Osvaldo Marcos, Mr.
Conselheiro De Saude Do Trabal
HOT-RJ
Av. Presidente Vargas 502/18 andar Centro
Rio de Janeiro - RJ 20100, Brazil
T: 55 21 252 5817

Albuquerque, Sandra
Reporter
LEGIAO DA BOA VONTADE
Av.Rudge, 700
Sao Paulo - SP, Brazil
T: 55 11 2202599 F: 55 11 2205196

Albuquerque, Sidney Monteiro, Mr.
Pesquisador
MERCK S/A
Estrada dos Bandeirantes, 1099 - Jacarepagua
Rio de Janeiro - RJ, Brazil
T: 55 21 3424646-204

Alcantara, Everaldo Siqueira, Sr.
Diretor Superintendente
ECOSS-CONSUTORIA E PLANEJAMENTO AMBIENTAL
Av. Trancredo Neves, 274, Bl-B,, Salas 320/325
Salvador - BA 40100, Brazil
T: 55 71 359-1589
TL: (071)1896

Alcantara, Julio Sergio B. De
Photographer
AGENCIA ESTADO
Av. Caetano Alvares, 55
Sao Paulo -SP 02598, Brazil
T: 55 11 8562210 F: 55 11 2652297

Alcantara, Sandra Cristina, Sra.
Representativ
FUNDACAO PROJETO SANITARISTA
Rua Batataes, 333 Apto 11 Jardim Paulista
Sao Paulo - SP 01423, Brazil
T: 55 11 8879735

Alechandre, Andrea, Sra.
Professora/Tecnica
UNIVERSIDADE FEDERAL DO ACRE
Univ. Federal Do Acre
Rio Branco - AC, Brazil

Aleksandrowicz, Vanessa Ronit, Ms.
Av. Delfim Moreira 12/302
Rio de Janeiro - RJ, Brazil
T: 55 21 239 2057

Alemeida, Rubem, Mr.
ASSOCIACAO CIVIL GREENPEACE
Rua Mexico 21, Grp 1301, Centro
Rio de Janeiro - RJ 20031, Brazil
T: 55 21 262-7318 F: 55 21 240-1690

Alencar, Jesse Claudio Fontes De, Sr.
Coordenador Do Gt De Ma/UNCED
CONFEDERACAO NACIONAL DA INDUSTRIA
Av. Nilo Pecanha, 50 sala 3018
Rio de Janeiro - RJ 20044, Brazil
T: 55 21 262 0181 F: 55 21 262 1465
TL: 22634

Alencar, Jose Macedo De
Correspondent
REDE JUL BAHAI DE COMMICACAO
Pca da Independencia, 244, 2 andar
Teixeira de Freitas, Brazil
T: 55 73 2913090 F: 55 73 2912333

Alessi, Renato Neves, Sr.
Membro
APOENA
Rua 7 de Setembro, 1633
Presid. Prudente-SP 19015, Brazil
T: 55 182 22-5371

Alexandre, Marta, Sra.
CMB
Rua Avanhadava no. 756
Sao Paulo - SP 01306, Brazil
T: 55 11 2564354 F: 55 11 2577648

Alexandrisky, Davy
Director
FOTOGRAFOS ASSOCIADOS
R. Mariz e Barros 1058 Box 12
Rio de Janeiro -RJ, Brazil
T: 55 21 2645435 F: 55 21 2645435

Alfinito, Fernanda De O., Sra.
Rua Reverendo Armando Ferreira 174
Niteroi - RJ 24230, Brazil
T: 55 21 711-9381

Alfonsin, Ana Isabel, Ms.
ASSOCIACAO DAS MULHERES PAPELEIRAS
Rua General Rondon 349
Porto Alegre - RS 91900, Brazil
T: 55 512 249-2743

Algarra, Luiz
TVT REDE DE COMUNICACAO DOS TRABALHADORES
Sao Paulo, Brazil

Alger, Keith, Mr.
Regional Researcher
FULBRIGHT FOUNDATION REGIONAL RESEARCH
Horta Zoo Botanico, Caixa Postal 7, Km 22 Rod. Ilheus-Itabuna, Itabuna BA
Ilheus - BA 45600, Brazil
F: 55 73 231-8801

Algranti, Nicole
Photographer
FREELANCER
Rua Senador Vergueiro 81/1103 Flamengo
Rio de Janeiro - RJ, Brazil
T: 55 21 2451597/2654516

Aliandro, Andre Luiz
Techinician
AGENCIA ESTADO
Av. Caetano Alvares, 55
Sao Paulo -SP 02598, Brazil
T: 55 11 8562210 F: 55 11 2652297

Allamo, Valeria Dias
Reporter
JORNAL DO COMMERCIO
Rua do Livramento, 189/4 andar
Rio de Janeiro, Brazil
T: 55 21 2336562 F: 55 21 2337690

Allegretti, Jose Raul, Sr.
ASSOCIACAO COMERCIAL DO RIO DE JANEIRO
Rio de Janeiro - RJ, Brazil

Allegretti, Luiz Fernando, Sr.
Chef.Do Escrit.De Brasilia
INSTITUTO DE ESTUDOS AMAZONICOS E AMBIENTAIS
SCLN 210 ED. Center Bloco C Sala 211 e 212
Brasilia - DF 70862, Brazil
T: 55 61 347-9874

Allen, Claudia, Ms.
ROYAL COMMISSION FOR JUBAIL & YANBU
R. Sambaiba, 190/601
Rio de Janeiro - RJ 22450, Brazil

Allison, Charles Gary
TWENTY FIRST CENTURY INC.
Rio de Janeiro, Brazil

Almeida, Aldo Roberto De
Technician
REDE GLOBO
R. Lopes Quintas 303
Rio de Janeiro -RJ 2, Brazil
T: 55 21 2947732

Almeida, Antonio Sergio Alves De
Technician
REDE GLOBO
R. Lopes Quintas 303
Rio de Janeiro -RJ 2, Brazil
T: 55 21 2947732

Almeida, Candido Mendes De, Sr.
President
INTERNATIONAL SOCIAL SCIENCE COUNCIL
Conj. Univ. Candido Mendes, Praca 15 de Novembro 101
Rio de Janeiro - RJ, Brazil
T: 55 21 2310648 F: 55 21 2324782

Almeida, Celso Goncalves De
Technician
REDE GLOBO
R. Lopes Quintas 303
Rio de Janeiro -RJ 2, Brazil
T: 55 21 2947732

Almeida, Cosme Coelho De
Editor
GLOBO
Rua Itapiru 1209 - Rio Comprido
Rio de Janeiro, Brazil
T: 55 21 5023535

Almeida, Derblay De
LEGIAO DA BOA VONTADE
Av. Rudge 700 Bom Retiro
Sao Paulo -SP 01134, Brazil
T: 55 21 2338381/R:223 F: 55 21 2634025

Almeida, Dom Luciano Mendes De, Sr.
President
C.N.B.B. - CONFERENCIA NACIONAL DOS BISPOS DO BRASIL
S.E./ Sul - Quadra 801 - Conjunto B
Brasilia - DF 70401, Brazil
T: 55 61 2252955
TL: 1104, E: 13.2067

Almeida, Fernando Monteiro, Sr.
Repres.
MISSAO NOVAS TRIBOS DO BRASIL
Rua Principado de Monaco, 24/102 Botafogo
Rio de Janeiro - RJ 22281, Brazil
T: 55 21 286 1450 F: 55 21 768-1037
TL: 213-8879

Almeida, Francisca Mesquita De, Sra.
Assessora
HIDROMET
Shin Ql 10, conj.5, casa 11
Brasilia - DF 71535, Brazil
T: 55 61 5774932

Almeida, Francisco Antonio De
ROLAND HENZE
Rio de Janeiro, Brazil

Almeida, Gizelle Tonin De
LEGIAO DA BOA VONTADE
Av. Rudge 700 Bom Retiro
Sao Paulo -SP 01134, Brazil
T: 55 21 2338381/R:223 F: 55 21 2634025

Almeida, Ivani De
Technician
REDE GLOBO
R. Lopes Quintas 303
Rio de Janeiro -RJ 2, Brazil
T: 55 21 2947732

Almeida, Jorge Soares De, Mr.
Professor Pesquisador
UNIVERSIDADE DE CUIABA
Av. Beira Rio, 3100 - Jardim Europa
Cuiaba - MT 78015-480, Brazil
T: 55 065 321 4488 F: 55 065 624 6444

Almeida, Joseney Farias, Sra.
Rua 5 de Julho 125/202
Rio de Janeiro - RJ 22051, Brazil
T: 55 21 236 7697

Almeida, Jozimar Paes De
Professor
FREELANCER
Universidade Estadual de Londrina Campus Universitario C.P. 600L
Londrina -SC, Brazil
T: 1 432 212000 F: 432 276932

Almeida, Liege De
JORNAL DE ESPANA
Rio de Janeiro, Brazil

Almeida, Lincoln Ricardo F.De
Diretor
JORNAL CANAL DIRETO
R.Santa Maria, 21
Colatina - ES, Brazil
T: 55 27 222203 F: 55 27 7222203

Almeida, Marcia De Lourdes Silva, Sra.
INTERNACIONAL (AMD)
Av. Henrique Lott B1 apt 1302
Rio de Janeiro - RJ, Brazil
T: 55 21 431-1990

Almeida, Marco Antonio Accioly De
MANCHETE
R. do Russel 804
Rio de Janeiro -RJ 2, Brazil
T: 55 21 2652012 F: 55 21 2855148

Almeida, Maria Romilda Padvan De
Journalist
FREELANCER
Rua Claudia N. Siqueira 250 Centre
Birtiba Mirim, Brazil
T: 55 11 4621252 F: 55 11 4621354

Almeida, Mariza Gomes De
OPERATOR FOTOJORNALISMO
Rio de Janeiro, Brazil

Almeida, Paulo Cezar Carrazedo De, Sr.
Director
TERRA VIVA PROJ. ALT.
Rua Araujo Lima 108
Rio de Janeiro - RJ 20541, Brazil
T: 55 21 288-9035

Almeida, Pedro
LEGIAO DA BOA VONTADE
Av. Rudge 700 Bom Retiro
Sao Paulo -SP 01134, Brazil
T: 55 21 2338381/R:223 F: 55 21 2634025

Almeida, Roberto De
Executive Producer
TV EDUCATIVA
Av Gomes Freire 474 Centro
Rio de Janeiro, Brazil
T: 55 21 2920012

Almeida, Roberval Tavares, Sr.
FAUNE
Rua Rio Solimoes 231 Areias
Recife - PE, Brazil
T: 55 81 251 0452

Almeida, Ronaldo Luiz Monteiro
Reporter
HOJE EM DIA
Rua Padre Rolim 462
Belo Horizonte RJ, Brazil
T: 55 31 2734296 F: 55 31 2731080

Almeida, Sandra
ICAE
Rede Mulher C.P.4651
Sao Paulo SP 01061, Brazil
T: 55 11 627050 F: 55 11 8713457

Almeida, Silvio Vieira
Tecnico De Audio
SISTEMA GLOBO DE RADIO
Rua do Russel, 434
Rio de Janeiro - RJ, Brazil
T: 55 21 2058282 F: 55 21 2054096

Almeida, Simone Fernandes De
Reporter
SISTEMA GLOBO DE RADIO
Rua do Russel, 434
Rio de Janeiro - RJ, Brazil
T: 55 21 2058282 F: 55 21 2054096

Almeida, Soraya Araujo De
Journalist
REVISTA CIENCIA HOJE
Av Venesian Bras, 71 Casa 27
Rio de Janeiro, Brazil
T: 55 21 2954846

Almeida, Talita De, Sra.
Diretora
ORGANIZACAO BRASILEIRA DE PROJETOS ESPECIAIS - OBRAPE
Mqs de Sao Vicente 355 - Gavea
Rio de Janeiro - RJ 22451, Brazil
T: 55 21 512-1356/274-6593 F: 55 21 511-5270

Almeida, Tatiana Lea Constant De
Photographer
AGENCIA ESTADO
Av. Caetano Alvares, 55
Sao Paulo -SP 02598, Brazil
T: 55 11 8562210 F: 55 11 2652297

Almeida, Terezinha Wiggers, Ms.
Pesquisadora
NUPES
Rua Da Assembleia N. 100 Sala 502
Rio de Janeiro - RJ 20011, Brazil
T: 55 21 2224-8622 F: 55 21 --254--

Almeida, Vanderlei
Photograpaher
REUTERS NEW SERVICE
Rua Sete de Setembro 99/4th floor Centro
Rio de Janeiro - RJ, Brazil
T: 55 21 5074151 F: 55 21 5072120

Almeiteros, Antonio
Producer
GRUPO NOVO DE CINEMA E TV
Rua Morechel Niemeyer 24
Rio de Janeiro - RJ, Brazil
T: 55 21 2663637 F: 55 21 2250435

Alonso, Helio, Sr.
Presidente
FACULDADE HELIO ALONSO
Rua Barata Ribeiro 264 - Copacabana
Rio de Janeiro - RJ 22040, Brazil
T: 55 21 256-7898

Altafef, Paul, Sr.
IECAM
Dr. Moreira 250
Manaus - AM 69000, Brazil

Altberg, Julia De Abrev
Editor
DIADEMA PRODUCOES
Rua Visconde Piraja 596, sala 206
Rio de Janeiro, Brazil
T: 55 21 2594782 F: 55 21 5110754

Altberg, Marco
Director
DIADEMA PRODUCOES
Rua Visconde Piraja 596, sala 206
Rio de Janeiro, Brazil
T: 55 21 2594782 F: 55 21 5110754

Alvarenga, Paulino
Assistente Cinematografico
PROVIDEO CINEMA E TV
SHIS Q1 21 Conj 7 casa 15 Lago Sul
Brasilia - DF, Brazil
T: 55 61 366 1851 F: 55 61 366 3182

Alvarenga, Telma
Editor Assistente
EDITORA ABRIL
R.da Passagem, 123/8 andar
Rio de Janeiro - RJ, Brazil
F: 55 21 2759347

Alvarenga, Wania, Ms.
Artista
TERRA E TRAJE
Rua Pana Tempo 39 Apto. 201
Belo Horizonte - MG 30330, Brazil
T: 55 31 221-4615

Alvarenga, Zulmira Galvao, Sra.
MOVIMENTO DE SAUDE DA ZONA LESTE
Rua Joao Ramalho, 001 - Perdizes
Sao Paulo - SP 05008, Brazil
T: 55 11 262-9407 F: 55 11 871-3457

Alvarez, Marta, Sra.
Executive Secretary
SERVICO UNIVERSITARIO MUNDIAL
Rua Cardoso de Almeida 990 (Caixa Postal 14631)
Sao Paulo - SP 03698, Brazil
T: 55 11 263.0211 F: 55 11 263.5804

Alves, Alfredo
Coordenador Produtora De Video
IBASE
Rua Vicente de Souza 29 Botafogo
Rio de Janeiro - RJ, Brazil
T: 55 21 226 5412 F: 55 21 286 0541

Alves, Arnaucy Ferreira Leite, Sra.
Rua Sao Clemente, 127/208 B
Rio de Janeiro - RJ 22260, Brazil
T: 55 21 286-0109

Alves, Carlos Bernardo M., Sr.
CONSERVATION INTERNATIONAL
c/o Sonia Rigueira, Rua Bueno Brandao, 393 - Santa Tereza
Belo Horizonte - MG 31010, Brazil
T: 55 31 226-5145 F: 55 31 222-8429

Alves, Cyrene Dos Santos, Sra.
IECAM
Dr. Moreira 250
Manaus - AM 69000, Brazil

Alves, Dircilene Pimentel
ESTADO DE SAO PAULO
Av. Rio Branco 151 - Sobre - Loja sala 201 a 204
Rio de Janeiro RJ, Brazil
T: 55 21 2217111

Alves, Eduardo Antonio, Sr.
Expositor
CIEE
Av. Maracana, 1524/1528 - Tijuca
Rio de Janeiro - RJ 20511, Brazil
T: 55 21 268-6447 F: 55 21 288-0748
TL: 34763

Alves, Eliones Oliveira, Sr.
ASSOCIACAO DA UNIAO ESTE BRAS. DOS ASD
Rua do Matoso, 97 - Praca da Bandeira
Rio de Janeiro - RJ 20270, Brazil
T: 55 21 273-5422 F: 55 21 273-5422

Alves, Fernando Henrique, Sr.
TELE-SALES REPRESENTACAO LTDA. / REVISTA TIME
Av. Rio Branco 151 gr. 403
Rio de Janeiro-RJ 20040, Brazil
T: 2420264 F: 2320198

Alves, Heloisa Stopatto Da Cruz, Ms.
UNITED METHODIST CHURCH OF BRAZIL
Rua Urujuai 530 apto 203, Tijuca
Rio de Janeiro - RJ 20510, Brazil

Alves, Joao Cicero De Sousa, Sr.
Coordenador De Projetos
NOVA PESQUISA ASS EM EDUCACAO
Barao do Flamengo, 22 Sala 803
Rio de Janeiro - RJ, Brazil
T: 55 21 205-6219

Alves, Jorge
Technician
REDE GLOBO
R. Lopes Quintas 303
Rio de Janeiro -RJ 2, Brazil
T: 55 21 2947732

Alves, Junior Francisco
MANCHETE
R. do Russel 804
Rio de Janeiro -RJ 2, Brazil
T: 55 21 2652012 F: 55 21 2855148

Alves, Leonidio, Sr.
Rua Acarau 25 F
Rio de Janeiro - RJ 21210, Brazil
T: 55 21 351 1800 F: 55 21 521 1987

Alves, Liane C. A.
Coordinator
FUNDACAO SOS MATA ATLANTICA
R. Manuel da Nobrega 456
Sao Roque - SP, Brazil
T: 55 11 8871195 F: 55 11 8851680

Alves, Marcus Vinicius P., Sr.
Rua Dr. Oscar Pimentel 48/303
Rio de Janeiro - RJ 20260, Brazil
T: 55 21 248-7309

Alves, Maria Cristina, Ms.
GOVERNO DO ESTADO DA BAHIA - SEC. DE EDUCACAO
Ilheus - BA, Brazil
F: 55 73 212-2195

Alves, Maria Da Penha Rodrigues, Ms.
Assistente
FINTEC S.A.
Rua Lauro Muller 116/4204 Botafogo
Rio de Janeiro - RJ, Brazil
T: 55 21 542-1426 F: 55 21 542 1426

Alves, Maria De Jesus, Sra.
Assistente Social
Rua Candido Mendes, 279/405 Gloria
Rio de Janeiro - RJ 20241, Brazil
T: 55 21 252-5190

Alves, Mario Jorge De Oliveira
Technician
REDE GLOBO
R. Lopes Quintas 303
Rio de Janeiro -RJ 2, Brazil
T: 55 21 2947732

Alves, Orane Falcao De Souza, Ms.
Professora
UNIV. FEDERAL DA BAHIA
R. Mons. Gaspar Sadock, 292/101 A - Costa Azul
Salvador - BA 41760, Brazil
T: 55 71 230 1869

Alves, Raimunda De Oliveira, Sra.
Presidenta
CLUBE DE MAES
Rua Congonhas no.355 - Parque Erasmo Assuncao
Santo Andre - SP 09270, Brazil
T: 55 11 415-3474

Alves, Ricardo Marcio Martins, Sr.
Economista
ENGEVIX ENGENHARIA S.A.
Av Afonso Pena 1500 Andar 13
Belo Horizonte - MG 30133, Brazil
T: 55 31 201 5100 F: 55 31 273 5639

Alves, Roberto De Brito
Photographer
AMAZON FILMS
Correa Sutra 59/401
Rio de Janeiro, Brazil
T: 55 21 2456779

Alves, Robson
MANCHETE
Rio de Janeiro, Brazil

Alves, Rosental Calmon
Editor
SISTEMA JORNAL DO BRASIL
Av. Brasil 500
Rio de Janeiro -RJ, Brazil
T: 55 21 5854422 F: 55 21 5854428

Alves, Sebastiao Francisco, Mr.
Planejamento Ambiental
INSTITUTO JONES DOS SANTOS NEVES
Av. Cezar Hilal, 437 - Praia do Sua 29
Vitoria - ES 29050, Brazil
T: 55 27 227 5044 F: 55 27 227 6067

Alves, Vanilda De Fatima Bongiorno, Ms.
U.S.P.
Rua Carlos Ribeiro Das Chagas 291
Sao Carlos - SP 19560, Brazil
T: 55 162 2723910 F: 55 162 2723910

Alves, Wandette Aboim Andrade, Sra.
Representative
SINDICATO DOS MEDICOS DO DISTRITO FEDERAL - SMDF
SBS - Ed. Seguradoras, 9o. andar
Brasilia - DF 70072, Brazil
F: 55 61 225-9694

Alves, Wilmar Antonio
Diretor
RADIO UNIVERSITARIA
Universidade Federal de Goias - Setor Oeste Av Anhanguera com Alameda das Rosas
Goiania - GO, Brazil
T: 55 62 233-4473/6896

Alves De Brito, Ana Maria, Ms.
Relacoes Publicas
HOPE OF THE FUTURE
Rua Hugo Leal, 67 - Ilha do Governador
Rio de Janeiro - RJ, Brazil
T: 55 21 393-0129

Alves Filho, Aristophio Andrade, Sr.
Delegate
UNIVERSALA ESPERANTO-ASOCIO
Quadra B55, no.85 Janoleiro do Martins
Maceio - AL 21931, Brazil

Alves Reis, Natalia, Ms.
WOMEN'S CLUB
Rua Real Grandeza # 99
Rio de Janeiro - RJ, Brazil

Alvim, Paulo T., Mr.
President
FUNDACAO PAU BRASIL
Horto Zoobotanico/Ceplac Caixa Postal 7
Ibatuna - Ba 45600, Brazil
T: 55 073 2143002 F: 55 073 2143020

Amaral, Beatriz, Sra.
Coordenadora Administrativa
FEDERACAO DAS BANDEIRANTES DO BRASIL
Av. Marechal Camara 186 - 2 Andar
Rio de Janeiro - RJ 20020, Brazil
T: 55 21 240-9220

Amaral, Claudio Castro Do
Technician
REDE GLOBO
R. Lopes Quintas 303
Rio de Janeiro -RJ 2, Brazil
T: 55 21 2947732

Amaral, Jefferson Gontijo Do, Mr.
UFG
Av. Para, 303 Setor Campinas
Goiania - GO 75110, Brazil
T: 55 62 233-1403

Amaral, Luiz Fernando, Mr.
Diretor
FUNDACAO S.O.S. PIRACICABA
Rua Ipe , 10 Bairro Laranjeiras
Joao Monlevade - MG 35930.384, Brazil
T: 851 5553

Amaral, Maria Tereza, Sra.
Diretora De Departamento
SECRETARIA DE CULTURA
Tereza Guimaraes 111 361
Rio de Janeiro - RJ, Brazil

Amazonas, Marcio, Sr.
Gerente De Meio Ambiente
COCA COLA
Rua Joao Lira 166, apto 101
Rio de Janeiro - RJ 22430, Brazil
T: 55 21 546 4077

Amazonas, Patricia O., Sra.
Tecnica Em Educ. Ambiental
FUNDACAO BRASILEIRA PARA A CONSERVACAO DA NATUREZA
Rua Miranda Valverde 103
Rio de Janeiro - RJ 22281, Brazil
T: 55 21 266-5008 F: 55 21 266-1345
TL: 213-7984

Amdrade, Rui Mandel Pizzaro
Reporter
RADIO JORNAL DO BRASIL
Av Brasil, 500 Sala 728
Rio de Janeiro, Brazil
T: 55 21 5805898 F: 55 21 5854466

Amore, Luiz, Mr.
INTERNATIONAL HOLISTIC UNIVERSITY
P.O. Box 09521
Brasilia - DF, Brazil
T: 55 61 553-1885 F: 55 61 553-1202

Amorim, Adilson Barbosa De
Fotografo
A STOP FOTOS & VIDEO
Rua Alvaro Alvim 37/1422-1 Cinelandia
Rio de Janeiro - RJ, Brazil
T: 55 21 240 1192 F: 55 21 240 4469

Amorim, Gil Soutellinho
Technician
MANCHETE
Rua do Russel, 804 Botafogo
Rio de Janeiro, Brazil
T: 55 21 2056898

Amorim, Marcos Alexandre Caires
Cinegrafista
REDE BANDEIRANTES DE TELEVISAO
R.Alvaro Ramos, 492
Rio de Janeiro - RJ, Brazil
T: 55 21 2122072 F: 55 21 2957391

Amorim, Mario Luiz De Freitas
Technician
REDE GLOBO
R. Lopes Quintas 303
Rio de Janeiro -RJ 2, Brazil
T: 55 21 2947732

Amorim, Paulo Cesar Forreira
Image Director
UFRJ/TVE
Avienda Nossa Senhora de Fatima, 74/604
Rio de Janeiro, Brazil
T: 55 21 2424284

Amurelli, Vera Frncisca F. Mussi, Sra.
Procuradora
PROCURADORIA GERAL DO DISTRITO FEDERAL
Higs 704 Bloco B Casa 42
Brasilia - DF 70331, Brazil
T: 55 61 223-3133 F: 55 61 223-6661

Anand Premraj, SW
OSHO SPIRITUAL HEALTH ORGAN. - OSHO
Rua Celestino Basilio 26
Rio de Janeiro - RJ, Brazil
T: 55 21 437-7488 F: 55 21 322-0327

Ancillotti, Lusardo Novo De Almeida
Tecnhician
SISTEMA GLOBO DE RADIO
Rua do Russel
Rio de Janeiro RJ, Brazil
T: 55 21 2058282

Andersen, Sigrid
Vice President
INSTITUTO GAIA DO BRASIL
Rua Mateus Leme 240/301
Curitiba - PR, Brazil
T: 55 41 223 9381/254 8153

Anderson, Anthony B., Mr.
Program Officer
THE FORD FOUNDATION
Praia Do Flamengo 100/12ANDAR
Rio de Janeiro - RJ 22210, Brazil
T: 55 21 2658252 F: 55 21 2851250

Anderson, Elza Suely, Ms.
Diretora Executiva
REDE BRASILEIRA AGROFLORESTAL - REBRAF
Rua Visconde de Piraja 111/713 - Ipanema
Rio de Janeiro - RJ 22410-010, Brazil
T: 55 21 521-7896 F: 55 21 521-1593

Anderson, Hugh G., Mr.
First Secretary
SOUTH AFRICAN EMBASSY
Ses Av Das Nacoes Lote 6
Brasilia - DF 70406, Brazil
T: 55 61 223 4873

Andrade, Ana Clara Marinho De
Reporter
ADMINISTRACAO DO DISTITO ESTUDUAL DE FERNANDO DE NARONHA
Rua Barrao de Sao Borfa 243 Boa Vista
Recife PE, Brazil
T: 55 81 2317570 F: 55 81 2311594

Andrade, Aroldo Souza, Sr.
Membro
GAMBA
Rua Renato Laport 99
Ubitaba, Brazil
T: 073 230 1072

Andrade, Artur Luiz
Jornalista
PANROTAS EDITORA
Av.Jabaquara, 1761
Sao Paulo - SP, Brazil
T: 55 11 2750211 F: 55 11 2761602

Andrade, Claudio Fernando De F.
Technician
REDE GLOBO
R. Lopes Quintas 303
Rio de Janeiro -RJ 2, Brazil
T: 55 21 2947732

Andrade, Coneico
Journalist
FREELANCER
Veopasiano Grande
Belo Horizonte, Brazil
T: 55 21 6211177

Andrade, Ernani, Sr.
Presidente
ADEJA
Av. Itapoama - Caixa Postal 110.801
Saquarema - RJ 28993, Brazil

Andrade, Fernando Antonio Thome, Sr.
Director
FUNDACAO PRO-NATUREZA (FUNATURA)
SCLN 107, Bloco B, Salas 201/17
Brasilia - DF 70743, Brazil
T: 55 61 274-5449 F: 55 61 274-5324

Andrade, Fernando Cesar Cerqueira De, Sr.
Expositor
CIEE
Av. Maracana, 1524/1528 - Tijuca
Rio de Janeiro - RJ 20511, Brazil
T: 55 21 268-6447 F: 55 21 288-0748
TL: 34763

Andrade, Filippe Augusto Vieira De, Dr.
PREFEITURA MUNICIPAL
Praca Maua s/n 1 andar
Santos - SP 11010, Brazil
T: 55 132 342-2093 F: 55 132 345-809
TL: 131048

Andrade, Heliton Figueiredo
FREELANCER
Rua Silveira Martins, 40 Ap 1008, Flamengo
Rio de Janeiro, Brazil
T: 55 21 2451447

Andrade, Hilda Fadiga De, Sra.
REDE MULHER
Rua Joao Ramalho, 991
Sao Paulo - SP 05008, Brazil
T: 55 11 262-9407 F: 55 11 871-3457

Andrade, Jose, Sr.
SIEMESP
Av.Paulista
Sao Paulo - SP, Brazil
T: 55 11 542-4222

Andrade, Jose Paulo, Sr.
MONSANTO
Rua Paes Leme 524
Sao Paulo - SP 05424, Brazil
T: 55 11 817-6206

Andrade, Luiz Carlos De
Tecnico De Audio
SISTEMA GLOBO DE RADIO
Rua do Russel, 434
Rio de Janeiro - RJ, Brazil
T: 55 21 2058282 F: 55 21 2054096

Andrade, Moises Henrique, Sr.
Diretor
FLAMA
Rua Marques de Abrantes, 55 - Flamengo
Rio de Janeiro - RJ 22230, Brazil
T: 55 21 265-5690

Andrade, Monica Costa, Sra.
Recepcionista
SIND. IND. ESTAMPARIA DE METAIS DE S.P.
Rua Wilson de Araujo, 650 - Xerem
Duque de Caxias 25400, Brazil

Andrade, Raquel Soares, Ms.
Rua Espumas 85
Rio De Janeiro -RJ, Brazil

Andrade, Rasemary M.R., Ms.
Professora
UNIV. SANTA URSULA
Rua: Jornalista Orlando Dantas, 59 - Botafogo
Rio de Janeiro - RJ 22231, Brazil
T: 55 21 552 5422 F: 55 21 551 6446

Andrade, Sebastiao Carlos De Oliveira, Mr.
Diretor Tecnico
AMIGOS PARQUE CIPO
Rua Runel, 388/801
Rio de Janeiro - RJ 22210, Brazil
T: 55 21 245 9419

Andrade, Virgilio Da Silva, Mr.
Presidente
CENTRO MINEIRO PARA CONSERVACAO DA NATUREZA
Universidade Federal de Vicosa-CMCN
Vicosa - MG 36570, Brazil
T: 031 891-2254

Andrade, Wallace Gomes De
Technician
REDE GLOBO
R. Lopes Quintas 303
Rio de Janeiro -RJ 2, Brazil
T: 55 21 2947732

Andre, Marcia, Ms.
Coordenadora De Projetos
ASHOKA
Rua do Russel 450/401, Gloria
Rio de Janeiro - RJ, Brazil
T: 55 21 285-7553 F: 55 21 205-5087

Andrea, Rita
CEMINA
R.Barao do Flamengo, 22/304
Rio de Janeiro - RJ, Brazil
T: 55 21 2056297 F: 55 21 5563383

Andreata, Jose Vanderli, Sr.
Prof. De Ciencias Biologicas
UNIVERSIDADE SANTA URSULA
Rua Fernando Ferrari, 75
Rio de Janeiro - RJ 22231, Brazil
T: 55 21 551-5542 F: 55 21 551-6446

Andreatta, Maria Beatriz Maurice
Editor
REVISTA QUALIDADE
Rua Ernesta Peosini, 22 sala 220
Sao Paulo, Brazil
T: 55 11 4480041 F: 55 11 4586169

Andreotti, Durvalino, Sr.
Embassador / Delegate
ORDEM DOS CAVALEIROS DA CONCORDIA
Av. Rui Barbosa no.1654 - Gracas
Recife - PE 52011, Brazil
T: 55 81 3391486 F: 55 81 3395338

Andres, Maria Helena, Ms.
Artista Plastica
Rua Santa Rita Dulcio 358
Belo Horizonte - MG 30146, Brazil
T: 55 31 221-2248

Andrews, Susan, Ms.
Coordinator
ASSOCIACAO NEO HUMANISTA
Rua Maria Angelica, 534, casa
Rio de Janeiro - RJ 224661, Brazil
T: 55 21 286-9908

Angelo, Alex Garcia Smith, Sr.
Membro
APOENA
Rua Cel.Albino, 1489
Presid. Prudente-SP 19023, Brazil
T: 55 182 22-3147

Angulo, Antonia, Sra.
ASOCIACION PARA LA SALUD INTEGRAL COMUNITARIA
Rua Eubanque, 78
Rio de Janeiro - RJ, Brazil

Angus, Jones, Mr.
Liason Off.
DEFENCE MAPPING AGENCY
Av. Presidente Wilson 147
Rio de Janeiro - RJ, Brazil
T: 55 21 292-7117

Anisio, Jose
MAIORIA FALANTE
Rua da Lapa 200 s/808
Rio de Janeiro RJ 2, Brazil
T: 55 21 5071462

Anjos, Hermano Odilon Dos, Sr.
Decano Do Centro De Ciencias H
UNIVERSIDADE SANTA URSULA
Rua Fernando Ferrari, 75
Rio de Janeiro - RJ 22231, Brazil
T: 55 21 551-5542 F: 55 21 551-6446

Anjos, Jose, Sr.
Geologo
CENTRO DE RECURSOS AMBIENTAIS
Av. Cardeal Da Silva, Ed. Studio Avant Gard - Ap 201
Salvador - BA CEP: 40.220, Brazil
T: 55 71 312-7199

Anjos, Wagner Figueiredo, Sr.
Analista De Sis.
EMBRATEL
Rua Marques de Leao, 32 / 1102 Engenho Novo
Rio de Janeiro - RJ 20780, Brazil
T: 55 21 581 2394

Ansell, Sam, Mr.
MUSEU DA IMAGEM E DO SOM
Av: Atlantica, 1260/1001
Rio de Janeiro - RJ, Brazil
T: 55 21 542-1162 F: 55 21 295-0742

Antas, Paulo De Tarso Zuquim, Sr.
Presidente
FUNDACAO PRO-NATUREZA (FUNATURA)
SCLN 107, Bloco B, Salas 201/17
Brasilia - DF 70743, Brazil
T: 55 61 274-5449 F: 55 61 274-5324

Antolini, Jorge Luiz, Sr.
INSTITUTO HAHNEMANNIANO DO BRASIL
Rua Frei Caneca 94 - Centro
Rio de Janeiro - RJ, Brazil
T: 55 21 221-1831

Antonio, Silva
Photographer
ASSOCIATED PRESS
Av. Brasil 500
Rio de Janeiro, Brazil
T: 55 21 5804442 F: 55 21 5802309

Antunes, Jose, Mr.
Advogado
SIND.GAR. DO TAPAJOS
Hugo Mendonca 293
Itaituba - PR, Brazil
T: 55 91 518-1162 F: 55 91 518-1162

Antunes, Marlene, Sra.
Rel.Publicas
INSTITUTO MEDICINA NATURAL
Av.Praia Belas, 422/1005
Porto Alegre - RS 90040, Brazil
T: 55 512 226-8298

Anunciacao, Ana Lucia Gomes
Reporter/Assessora
COMUNIDADE BAHA'I DO BRASIL
Rua Olegario Maciel 2403 Bairro Saraiva
Uberlandi, Brazil
T: 215 4320 F: 236 9225

Aoki, Hideyo, Mr.
INSTITUTO FLORESTAL
R. Horto S/N
Sao Paulo - SP, Brazil
T: 55 11 952 8555

Aparicio, Teresa, Sra.
Diretora Esecutiva
IAMA
Rua Turi no 16
Sao Paulo - SP 05443, Brazil
T: 55 11 210-1301 F: 55 11 210-1338

App, Dietmar
COORDENACAO INTERNACIONAL CONTRA OS PERIGOS DA BAYER
Rua Monte Alegre 482/205 - Santa Teresa
Rio de Janeiro - RJ 20242, Brazil
T: 55 21 232-0967

Appio, Idivar Francisco, Sr.
Deputado
ASSEMBLEIA LEGISLATIVA
Riveira 281-apt. 703
Porto Alegre - RS, Brazil
T: 55 512 226-8035

Apruda, Maria Teresa De
FREELANCER
Rio de Janeiro, Brazil

Aquino, Carlos Pessoa
Secretario Adjunto
SECRETARIA DO MEIO AMBIENTE
Centro Administrativo 2 Bloco 4 Andar Jaguaribe
Joao Pessoa - PB 58000, Brazil
T: 55 83 222-2851

Aquino, Daniela Cosendey De
FOX FILMES DO BRASIL
Rio de Janeiro, Brazil

Aquino, Marcia, Ms.
Biologa
SUDEMA/PB
Av. Monsenhor Walfredo Leal 181 Tambia
Joao Pessoa - PB 58.023, Brazil
T: 55 83 222-4663

Aquino, Welson Klein De, Mr.
Coordenador
MOVIMENTO GBM
R. Ernesto Nazareth, bl. 74,med 7, apto. 325, Estancia
Recife - PE 50771, Brazil
T: 55 81 455-2774 1

Aragao, Maria Helena Franca Moniz, Sra.
Assessoria
PASTORAL DO MENOR
Rua Umari, no.50 - Laranjeiras
Rio de Janeiro - RJ 22221, Brazil
T: 55 21 269-0914 F: 55 21 221-8093

Arantes, Jose Valadao, Jr., Mr.
Technical Director
CENTRO DE PESQUISAS DO MAR (CEPEMAR)
Rua Mexico, 31/Gr. 501- Centro
Rio de Janeiro - RJ, Brazil
T: 55 21 240-3822 F: 55 21 240-9516

Arantes, Paulo Henrique Carrara
Photographer
FREELANCER
Rua Prudente do Morris, 1440 - 304 Ipanema
Rio de Janeiro, Brazil
T: 55 21 2321433

Araripe, Alex
Producer
FUND. CULTURAL DE RADIODIFUSAO EDUCATIVA DA REGIAO DOS LAGOS
Av. Jose Bento Ribeiro Dantas 265 Manguinhos 28905
Armacao dos Buzios -, Brazil
T: 55 246 2753848 F: 55 21 2956494

Araujo, Abel Luiz De
Technician
REDE GLOBO
R. Lopes Quintas 303
Rio de Janeiro -RJ 2, Brazil
T: 55 21 2947732

Araujo, Alcione Cavalcante De
Reporter
LEGIAO DA BOA VONTADE
Av.Rudge, 700
Sao Paulo - SP, Brazil
T: 55 11 2202599 F: 55 11 2205196

Araujo, Aloizio
Reporter
REDE GLOBO
R. Lopes Quintas 303
Rio de Janeiro -RJ 2, Brazil
T: 55 21 2947732

Araujo, Amadeu Ferreira De
RADIO CLUBE DE PERNAMBUCO
Rio de Janeiro, Brazil

Araujo, Anderson Netto Da Cruz
Technician
REDE GLOBO
R. Lopes Quintas 303
Rio de Janeiro -RJ 2, Brazil
T: 021 2947732

Araujo, Anisio Andrade De
JORNAL MAIORIA FALANTE
Rio de Janeiro, Brazil

Araujo, Armando Goncalves De
EDITORA AZUL
Rio de Janeiro, Brazil

Araujo, Daniel, Sr.
Presidente
ASSOCIACAO COLATINENSE DE DEFESA ECOLOGICA
Rua Hilario Delacqua 136 Caixa Postal 100
Colatina - ES, Brazil
T: 55 27 722-1687 F: 55 27 722-2203

Araujo, Denis Ferreira De
Technician
REDE GLOBO
R. Lopes Quintas 303
Rio de Janeiro -RJ 2, Brazil
T: 55 21 2947732

Araujo, Edilson Divino De, Sr.
Biologist
UNIVERSIDADE FEDERAL DE GOIAS
Av. 20-A No. 88 - Centro
Goiania - GO 74000, Brazil
T: 55 62 261-7766 F: 55 62 251-1694

Araujo, Francisco Ferreira De
AGENCIA GILLVAN
Av Treze de Maio 47/1110 Cinelandia
Rio de Janeiro RJ 2, Brazil
T: 55 21 2205400

Araujo, Helenice
ESTADO DE SAO PAULO
Av Rio Branco 151
Rio de Janeiro, Brazil
T: 55 21 2217111

Araujo, Jacirema Bernardo, Ms.
Diretora
CENTRO JOSUE DE CASTRO
R. Dom Bosco, 770
Recife - PE 50070, Brazil
T: 55 81 222 1906 F: 55 81 231 2765

Araujo, Janderson Netto Da Cruz
Technician
REDE GLOBO
R. Lopes Quintas 303
Rio de Janeiro -RJ 2, Brazil
T: 55 21 2947732

Araujo, Joao D. De
Reporter
AGENCIA ESTADO
Av. Caetano Alvares, 55
Sao Paulo -SP 02598, Brazil
T: 55 11 8562210 F: 55 11 2652297

Araujo, Jorge
Photographer
FOLHA DE SAO PAULO
Alameda Barao de Limeira 425
Sao Paulo SP, Brazil
T: 55 11 874 2608/874 2602 F: 55 11 2231644

Araujo, Jose Antonio Pessoa De, Sr.
Rua Barao Da Torre 42 - Apt. 401
Rio de Janeiro - RJ, Brazil

Araujo, Jose Cortes Pereira De, Sr.
SOCIEDADE POTIGUAR DE EDUCACAO - SPEL
Rua Floriano Peixoto, 295, Petropolis
Natal - RN 59020, Brazil
T: 55 84 222-2314 F: 55 84 221-6255

Araujo, Jose Reduzino De
Photographer
CONTEXTO
Av Calogeras 12/63 Centro
Rio de Janeiro - RJ, Brazil
T: 55 21 2204653 F: 55 21 2204653

Araujo, Lea Sayao Carvalho, Ms.
ORDEM DO MERITO CULTURAL BERMARDO SAYAO
Rua Barata Ribeiro, 74, apto. 202
Rio de Janeiro - RJ, Brazil
T: 55 21 275-0603

Araujo, Leonardo Sampaio A.
MANCHETE
R. do Russel 804
Rio de Janeiro -RJ 2, Brazil
T: 55 21 2652012 F: 55 21 2855148

Araujo, Lourice Neves
Photographer
JORNAL GAZETA DE NOTICIAS
R. Leandro Martins 72 Centro
Rio de Janeiro -RJ, Brazil
T: 55 21 2630435 F: 55 21 2630584

Araujo, Luis Roberto
Techinician
AGENCIA ESTADO
Av. Caetano Alvares, 55
Sao Paulo -SP 02598, Brazil
T: 55 11 8562210 F: 55 11 2652297

Araujo, Marialdo
SISTEMA JORNAL DO BRASIL
Av. Brasil 500
Rio de Janeiro -RJ, Brazil
T: 55 21 5854422 F: 55 21 5854428

Araujo, Marialva Mendes De, Sra.
2o.Tesoureiro
AMNPR
Av. Cons. Aguiar, 3384 apt.1002 - B. Viagem
Recife - PE, Brazil
T: 55 81 325-0315 F: 55 81 224-7994

Araujo, Marlene
Editor
FREELANCER
R. Barata Ribeiro 70 /1301 Copacabana
Rio de Janeiro -RJ, Brazil
T: 55 21 5421823

Araujo, Monica De Azevedo, Ms.
ASSOCIACAO PROJETO RODA VIVA
R. Mario Portela 261/1301-A Laranjeiras
Rio de Janeiro - RJ 22241, Brazil

Araujo, Neverton Carlos De
Technician
REDE GLOBO
R. Lopes Quintas 303
Rio de Janeiro -RJ 2, Brazil
T: 55 21 2947732

Araujo, Paulo Cesar Santos
JORNAL O FLUMINENSE
Niteroi, Brazil

Araujo, Paulo Roberto
Reporter
JORNAL O GLOBO
Rua Irineu Marinho, 35
Rio de Janeiro - RJ, Brazil
T: 55 21 2722000

Araujo, Sandra Kelly De, Sra.
UNIV. FED. RONDONIA
Av. Pres. Dutra 2965 - CENTRO
Porto Velho - RO 78.900, Brazil
T: 55 69 2218188 F: 55 69 2215444
TL: (69)2152

Araujo, Selma Maria De, Ms.
Professor
UNIVERSIDADE FEDERAL DA PARAIBA
Rua Aprigio Veloso 882
Campina Grande - PB 58100, Brazil
T: 55 83 333 1000 F: 55 83 333 1080

Araujo, Suely Mara Vaz Guimaraes De, Sra.
Vice Presidente
INSTITUTO DE ARQUITETOS DO BRASIL
SHIS QL 14 Conj. 8, casa 9 - Lago Sul
Brasilia - DF 71600, Brazil
T: 55 61 248-3579 F: 55 61 225-3459

Araujo, Tadeu R., Sr.
Rua Joaquim Nabuco, 236/101
Rio de Janeiro - RJ 20080, Brazil
T: 55 21 247-0297

Araujo, Tania Goncalves De, Ms.
FUNDACAO CULTURAL AVATAR
Rua Min. Octavio Kelly 352 c.6
Niteroi - RJ 24220, Brazil
T: 55 21 611-2489

Araujo, Washington Luiz De, Sr.
Assessor De Imprensa
FUNDACAO SOS MATA ATLANTICA
Rua Manoel da Nobrega, 456
Sao Paulo - SP 04001, Brazil
T: 55 11 887-1195 F: 55 11 885-1680

Araujo Caldas, Jose Geraldo, Mr.
Vice-Pres.
MOVIMENTO VERDE DE PARACATU
Rua Rubens Bittencourt, 77
Paracatu cep: 38.600, Brazil
T: 55 61 671-6234

Araujo Silva, Fernando Antonio, Sr.
Engenheiro
CENTRO DE RECURSOS AMBIENTAIS
Rua Tupinambas, 504 - Ap. 202 - Rio Vermelho
Salvador - BA CEP: 41.910, Brazil
T: 55 71 240-3863

Arcangeli, Cristiana, Ms.
President
PHYTOERVAS IND. COSM. NATURAIS LTDA.
Al. Araguaia 2179
Barveri, SP 06400, Brazil
T: 421-1188 F: 709-1397

Archer, Marcio, Mr.
Constante Ramos, 110/1501
Rio de Janeiro - RJ 22250, Brazil

Arcuri, Regina Celia, Sra.
Rua Pedro Bolato, 80/202
Rio de Janeiro - RJ 22621, Brazil
T: 55 21 493-9048

Areas, Silvia Fittipaldi
Producer
MANCHETE
Rua do Russel, 766 6 andar
Rio de Janeiro, Brazil
T: 55 21 2850057 F: 55 21 2651570

Argento, Mauro Sergio Fernandes, Sr.
Coordenador
CENTRO DE ESTUDOS E PESQUISAS AMBIENTAIS - CEPAM
Rua Ibitiuva, 151
Rio de Janeiro - RJ 21710, Brazil
T: 55 21 331-3022 216 F: 55 21 331-3024

Argolo, Jose Amaral
HELIO ALONSO
Rua Muniz Barreto, 51
Rio de Janeiro RJ, Brazil
T: 55 21 5515448

Arikawa, Seiichiro, Mr.
Av. Nilo Pecanha, 50 s/2110
Rio de Janeiro - RJ, Brazil
T: 55 21 532-2455 F: 55 21 220-9916

Arnaud, Edy Paola, Ms.
C G I L - PROGETTO SVILUPPO
Sao Paulo - SP, Brazil

Arnold, Eliane Alvarenga
AGENCIA DE NOTICIAS ACOPI
Sao Paulo, Brazil

Arnt, Clarice, Ms.
Assessora Internacional
ABES
Av. Beira Mar 216/13o. andar
Rio de Janeiro - RJ 20021, Brazil
T: 55 21 210-3221 F: 55 21 262-6838
TL: 31902 ABST

Arnt, Ricardo
Correspondente
IL MANIFESTO
Rua Girassol 1317/44
Sao Paulo - SP, Brazil
T: 55 11 815 4528 F: 55 11 825-7861

Arozo, Maria Amalia Soares, Sra.
CBCISS
Rua Jardim Botanico 438/201
Rio de Janeiro - RJ 22461, Brazil
T: 55 21 226-3579

Arreu, Leandro Pimentel
Photographer
MANCHETE
R. do Russel 804 Gloria
Rio de Janeiro RJ 2, Brazil
T: 55 21 2454816 F: 55 21 2056196

Arruda, Abiatar
Technician
REDE GLOBO
R. Lopes Quintas 303
Rio de Janeiro -RJ 2, Brazil
T: 55 21 2947732

Arruda, Andre
SISTEMA JORNAL DO BRASIL
Av. Brasil 500
Rio de Janeiro -RJ, Brazil
T: 55 21 5854422 F: 55 21 5854428

Arruda, Dalton, Sr.
IECAM
Dr. Moreira 250
Manaus - AM 69000, Brazil

Arruda, Frederico, Sr.
IECAM
Dr. Moreira 250
Manaus - AM 69000, Brazil

Arruda, Moacir Bueno, Sr.
Coordenador Geral
REEA-DF REDE E ECOLOGIA
Sclrn 710, Bl. E, Cs. S1
Brasilia - DF 70750, Brazil
T: 55 61 226-6371

Arruda, Silvia De Castro, Sra.
National Coordenator
IBFAN - INTERNATIONAL BABY FOOD ACTION NETWORK
Rua Apeninos 336, apto 52 /Aclimacao
Sao Paulo - SP 01533, Brazil
T: 55 11 279-1898 F: 55 11 832-6889

Arruda, Sineide Aparecida, Ms.
TRABALHADORAS RURAIS
Rua Tamandare, 667
Sao Paulo - SP, Brazil

Arruda, Zelia, Sra.
Director
CMB
Rua Avanhadava no. 756
Sao Paulo - SP 01306, Brazil
T: 55 11 2564354 F: 55 11 2577648

Asbec, Jose Carlos
PALMARES PRODUCOES E JORNALISMO LTDA
Rio de Janeiro, Brazil

Ascoli, Maria Isabel L.
PROVENT
Rua Uruguiana, 10 2 andar
Rio de Janeiro, Brazil
T: 55 21 2424618

Asmus, Alexandre Rios, Mr.
Rua Silva Da Rocha Pozzis, 431/302
Rio de Janeiro - RJ CEP: 20.973, Brazil
T: 55 21 240-3605 F: 55 21 220-0972

Assaad, Farhah, Mr.
Ph.D
Av.Ministro Rocha Azevedo 482
Sao Paulo - SP 10410, Brazil
T: 55 11 282-3755 F: 55 11 282-3755

Assef, Ricardo, Mr.
A. Assistant
WORLD FOOD PROGRAMME / UN
SCN Quadra 02 Bloco 'B' Terreo
Brasilia - DF 70710, Brazil
T: 55 61 226-7872 F: 55 61 322-2080
TL: 61-1243

Assis, Alair, Mr.
Secretario Municipal
SECRETARIA MEIO AMBIENTE MOJI GUACU
Rua Apolinario, 189 - Centro
Moji Guacu - SP 13840, Brazil
T: 0192614063 F: 0192614934

Assis, Geise, Sra.
ILDES/FES
Rua. Lisboa 191
Sao Paulo - SP 05413, Brazil
T: 55 11 282-3900 F: 55 11 280-1916
TL: 1180616

Assis, Luiz Rogerio Pena
Photographer
AGENCIA ESTADO
Av. Caetano Alvares, 55
Sao Paulo -SP 02598, Brazil
T: 55 11 8562210 F: 55 11 2652297

Assis, Olga
EDITORA HORA DO POVO S.A.
Rio de Janeiro, Brazil

Assiz, Sergio, Mr.
Diretor
ICI BRASIL
Rua Verbo Divino 1356
Sao Paulo - SP 04719, Brazil
T: 55 11 525-2008 F: 55 11 525-7932

Assuncao, Marlene Felippe, Sra.
Assessora Administrativa
UNAMAZ
Trav. Mauriti, 3275/804 - Marco
Belem - PA 66240, Brazil
T: 55 91 229-9310 F: 55 91 229-4339
TL: 1013

Asuiguti, Luiz Sadayuki, Sr.
Rua Paraiba 36
Rio de Janeiro - RJ, Brazil
T: 55 21 273-8346 F: 55 21 293-3894

Ataide, Eva Aparecida R. De, Ms.
OSHO SPIRITUAL HEALTH ORGANIZATION
Rua Barao Do Flamengo 50/606
Rio de Janeiro - RJ 22220, Brazil
T: 55 21 2455479

Athayde, Simone Ferreira
Biologo
SOC. DE PESQ. EM VIDA SELVAGEM E EDUCACAO AMBIENTAL
Rua Gutemberg 345
Curitiba - PR 80420-030, Brazil
T: 55 41 222-3425 F: 55 41 222-3425

Atkinson, David B.
Repres. Do Bid No Brazil
BCO. INTERAMERICANO DE DESENVOLVIMENTO
PO Box 08738
Brasilia - DF 70312, Brazil
T: 55 61 321-7848 F: 55 61 321-3112

Atkinson, Martha Luz Diaz De
Esposa Do Rep. No Brazil
BCO. INTERAMERICANO DE DESENVOLVIMENTO
Caixa Postal 08738
Brasilia - DF 70312, Brazil
T: 55 61 321-7848 F: 55 61 321-3112
TL: 61-1204

Attie, Cynthia
Director
INSTITUTO DE ARQUITETOS DO BRASIL
Rua do Pinheiro, 10 Flamengo
Rio de Janeiro, Brazil
T: 55 21 2853480

Audi, Antonio, Mr.
CESP
Rua da Consolacao, 1875, 9 andar, Cerq. Cesar
Sao Paulo - SP 01301, Brazil
T: 55 11 258-2165 F: 55 11 256-2101

Augusti, Maria Teresa, Ms.
Presidenta
CONSELHO ESTADUAL DA CONDICAO DA MULHER
Av. Paulista 1776-17o Andar
Sao Paulo - SP 01310, Brazil
T: 55 11 287-4332 F: 55 11 284-1667

Augusto, Carlos, Mr.
TERRA CHRISTA
1880 East River Road
Tucson - AZ, Brazil
T: 1 602 577-6476 F: 1 602 577-1195

Augusto, Celso
Image Editor
Manchete
Rua Gal Gois Monteiro, 184/904
Rio de Janeiro, Brazil
T: 55 21 2850033

Augusto, Vera Maria Haj Mussi, Sra.
CONSELHO ESTADUAL DA CONDICAO FEMININA
Rua Ebano Pereira 240,Rua Jose De Alencar 1941
Curitiba - PR 80040, Brazil
T: 55 41 2322583 F: 55 41 234 4839

Avakian, Raffi V. V., Sr.
Av. Cesario De Melo 2363 Apto 201
Rio de Janeiro - RJ 23050, Brazil
T: 55 21 322 4893

Avelar, Lucila Vasconcelos De
Super Indendent
FUNDACAO ROQUETTE-PINTO
Prasa da Republica, 141 A Centro
Rio de Janeiro, Brazil
T: 55 21 2522203

Avelleda, Ana Maria
92 GLOBAL FORUM
Rio de Janeiro - RJ, Brazil

Avila, Carlos Frederico De Brito, Sr.
AMAZONIA FOREVER GREEN
Av. Pe.A.J. dos Santos 921 - Brooklin
Sao Paulo - SP, Brazil
T: 55 11 240-2266

Avila, Dagoberto Nunes, Sr.
COLEGIO N.S. DO BOM CONSELHO
R. Ramiro Barcelos, 996
Porto Alegre - RS 90210, Brazil
T: 55 512 226-0522

Avila, Fabio Augusto De Brito
AMAZONIA FOREVER GREEN
Av. Pe.A.J. dos Santos 921 - Brooklin
Sao Paulo - SP, Brazil
T: 55 11 240-2266

Avila, Jose Ubiratam, Sr.
Coordenador De Centro
INSTITUTO PRINCIPIO UNICO
Rua Presidente Backer 140
Niteroi - RJ 24220, Brazil
T: 55 21 714-6306

Avila, Maria Betania, Ms.
Researcher
SOS CORPO
Rua do Hospicio, 859, Boa Vista
Recife - PE 50050, Brazil
T: 55 81 221-3018 F: 55 81 221-3947

Avila, Maria Elizabeth De Brito, Sr.
AMAZONIA FOREVER GREEN
Av. Pe.A.J. dos Santos 921 - Brooklin
Sao Paulo - SP, Brazil
T: 55 11 240-2266

Aymone, Carlos, Sr.
Secretario
ASOC. LATINO AMERICANA MULTIDICIPLINARIA PLANEACION EST. MEDIO AMBIENTE
Av. Da Igreja, 346 - DEP. Municipal De Meio Ambiente
Tramandai - RS CEP: 95.590, Brazil
T: 55 51 684-1222 F: 55 51 661-1762

Ayres, Alvaro Tourinho Junqueira, Dr.
Director
CRUZ VERMELHA BRASILEIRA
Praca Cruz Vermelha 10/12 - 1o. andar
Rio de Janeiro - RJ, Brazil
T: 55 21 232-6325 F: 55 21 242-6760

Ayres, Esther Cristina, Sra.
REDE MULHER
Rua Jose Ramalho, 991 - Perdizes
Sao Paulo - SP 05008, Brazil
T: 55 11 262-9407 F: 55 11 871-3457

Ayumi, Sandra
Editor
YOMIURI SHIMBUN
Av Oswaldo Cruz 149/1201
Rio de Janeiro, Brazil
T: 55 21 5514500 F: 55 21 5522279

Azambuja, Heloisa, Sra.
Tecnica
CONFEDERACAO NACIONAL DA INDUSTRIA
Av. Nilo Pecanha, 50 sala 3018
Rio de Janeiro - RJ 20044, Brazil
T: 55 21 262 0181 F: 55 21 262 1465
TL: 22634

Azaziel, Marcus, Sr.
Presidente
COOPERATIVA DOS PROFISSIONAIS LIBERAIS - MULTICOOP / SAIBA
Av. Beira Mar 406/1002
Rio de Janeiro - RJ 20021, Brazil
T: 55 21 240-7923 F: 55 21 240-0465

Azem, Milton, Sr.
Diretor
SCHOELLER PLAST HOLDING
Rua Cunha Gago, 800 - Pinheiros
Sao Paulo - SP 05421, Brazil
T: 55 11 212-9719 F: 55 11 212-8169

Azeredo, Edson Bedim De, Sr.
Presidente
MOVIMENTO RESSURGENCIA
Rua Monte Castelo 163, Roca Velha
Arraial Do Cabo 28930-000, Brazil
T: 55 246 222472

Azeredo, Maril, Sra.
Escultora
PRISMART CULTURA
Av. Atlantica 554 - Leme
Rio de Janeiro - RJ 22010, Brazil
T: 55 21 275-3303

Azevedo, Adalberto
ARTE RIO
Rio de Janeiro, Brazil

Azevedo, Alfredo Silveira De
Technician
REDE GLOBO
R. Lopes Quintas 303
Rio de Janeiro -RJ 2, Brazil
T: 55 21 2947732

Azevedo, Ana Lucia
Reporter
JORNAL O GLOBO
Rua Irineu Marinho, 35
Rio de Janeiro - RJ, Brazil
T: 55 21 2722000

Azevedo, Andre Louis Tenuta, Sr.
Associado
FUNDACAO SOS RIO PIRACICABA
Ipatinga - MG 35160, Brazil
T: 55 31 821-2897

Azevedo, Celso R. A. De
ECO TV
Parati, Brazil

Azevedo, Christina De
MANCHETE
Rua do Russel 766/804
Rio de Janeiro RJ, Brazil
T: 55 21 2652012 F: 55 21 2059998

Azevedo, Dirceu Ezequiel
BRASILTURIS JORNAL
Sao Paulo, Brazil

Azevedo, Fernando Cesar De
Camera
TVSBT
Campo de S.Cristovao, 105
Rio de Janeiro - RJ, Brazil
T: 55 21 5803688 / 5800678 F: 55 21 5809685

Azevedo, Gennyson C., Sr.
Secretario Executivo
ASSOCIACAO BRASILERIA DE TELECOMUNICACOES
Av. Pasteur 383 - Urca
Rio de Janeiro - RJ 22290, Brazil
T: 55 21 295-4432 F: 55 21 542-4092
TL: 23057

Azevedo, Isabel Cristina Michelan De, Sra.
Coordenadora Do Dep. Cultural
ASSOCIACAO PALAS ATHENA
Rua Dr. Rafael de Barros, 252 - apto 42
Sao Paulo - SP 04003, Brazil
T: 55 11 288 7356

Azevedo, Isaias Andrade De
Technician
REDE GLOBO
R. Lopes Quintas 303 Jardim Botanico
Rio de Janeiro -RJ 2, Brazil
T: 55 21 2947732

Azevedo, Jandrira Gomes De
FORUM DO FUTURO
Rio de Jaeiro, Brazil

Azevedo, Jose Carlos
Reporter
REDE GLOBO
R. Lopes Quintas 303
Rio de Janeiro -RJ 2, Brazil
T: 55 21 2947732

Azevedo, Josias Andrade De
MANCHETE
R. do Russel 804
Rio de Janeiro -RJ 2, Brazil
T: 55 21 2652012 F: 55 21 2855148

Azevedo, Luiz Claudio Lucas De
Reporter
REDE GLOBO
R. Lopes Quintas 303
Rio de Janeiro -RJ 2, Brazil
T: 55 21 2947732

Azevedo, Luiz Eduardo Barros
Technician
REDE GLOBO
R. Lopes Quintas 303
Rio de Janeiro -RJ 2, Brazil
T: 55 21 2947732

Azevedo, M. Augusto
INSTITUTO DO MEIO AMBIENTE
Rua Mj. Cicero De Goes Monteiro 2197
Maceio - AL, Brazil
T: 55 82 221 6747

Azevedo, Maria Do Rosario Rates, Ms.
Associada
FUNDACAO SOS RIO PIRACICABA
R. Monte Alegre, 40, apto. 12 - Serra
Belo Horizonte - MG 30240, Brazil
T: 55 31 227-6125

Azevedo, Maria Lucia De
Producer
MANCHETE
Rua Afonso Pena, 1500 16andar
Belo Horizonte MG, Brazil
T: 55 31 2733900 F: 55 31 2240900

Azevedo, Maria Margarida Figueiredo, Ms.
CIA VALE DO RIO DOCE
Av. Graca Aranha, 26
Rio de Janeiro - RJ 20.030, Brazil
T: 55 21 272 4401 F: 55 21 220 0593

Azevedo, Mauricio Pereira De
TV RIO
Rio de Janeiro, Brazil

Azevedo, Patricia De
ABVP
Rua Treze de Maio 489, Bela Vista
Sao Paulo SP 01327, Brazil
T: 55 11 2847862

Azevedo, Paulo
Cameraman
MTV - RIO
Praca Nossa Senhora da Gloria 46
Rio de Janeiro -RJ, Brazil
T: 55 21 2855946 F: 55 21 2055394

Azevedo, Pedro Paulo, Dr.
GOVERNO DO ESTADO DO RIO DE JANEIRO
Rio de Janeiro - RJ, Brazil

Azevedo, Saias Andrade De
Technician
REDE GLOBO
R. Lopes Quintas 303 Jardim Botanico
Rio de Janeiro -RJ 2, Brazil
T: 021 2947732

Azevedo, Sarah, Ms.
Psicologa Pesquisadora
RUA 1o. DE JANEIRO 1164
Altamira PA 68320, Brazil
T: 55 91 515-1070

Azoury, Ricardo
Reporter Fotografico
AGENCIA N IMAGENS
Av.Brig.Luis Antonio, 2724 Casa 6
Sao Paulo - SP, Brazil
T: 55 11 8870847 F: 55 11 8853694

B.Teixeira, Aloy Jupiara
Reporter
JORNAL O GLOBO
Rua Irineu Marinho, 35
Rio de Janeiro, RJ 2, Brazil
T: 021 2722000

Babuy, Carlos Alberto
Technician
REDE GLOBO
R. Lopes Quintas 303
Rio de Janeiro -RJ 2, Brazil
T: 55 21 2947732

Bacchin, Jose Antonio, Mr.
Fellow
ASHOKA
a/c Rua do Russel 450 sala 401
Rio de Janeiro - RJ, Brazil
T: 55 21 285-7553 F: 55 21 205-5087

Bacellar, Alzimara, Sra.
Director
CMB
Rua Avanhadava no. 756
Sao Paulo - SP 01306, Brazil
T: 55 11 2564354 F: 55 11 2577648

Backus, Edward, Mr.
CONSERVATION INTERNATIONAL
Rua Bueno Brandao 393
Belo Horizonte - MG 31010, Brazil
T: 55 31 226-5145 F: 55 31 222-8429

Bacon, Glen, Ms.
Linguist
SUMMER INSTITUTE OF LINGUISTICS
SAI/NO, Lote 'D', Bloco 03
Brasilia - DF 70770, Brazil
T: 55 61 272-1200

Badiani, Rita
SOCIEDADE CIVIL DO BEM-ESTAR FAMILIAR NO BRASIL -BEMFAM
Av. Republica do Chile 230/ 17-andar
Rio de Janeiro -RJ, Brazil
T: 55 21 2102448 F: 55 21 2204057

Baggio, Adriana
Reporter
DCI - DIARIO COMERCIO e INDUSTRIA
Av.Rio Branco, 177/6 andar
Rio de Janeiro - RJ, Brazil
T: 55 21 2621016 F: 55 21 2200805

Bahia, Evier Pereira, Sr.
UNIAO ESTE BRASILEIRA DAS IASD
Av. Sete de Setembro, 69
Niteroi - RJ Cep.: 24.300, Brazil
T: 55 21 711 1062 F: 55 21 611 1571

Bahiana, Luiz Carlos, Sr.
V. Presidente Cons. Adm.
TELEBRASIL
Av. Pasteur 383 Urca
Rio de Janeiro - RJ 22290, Brazil
T: 55 21 295-4432 F: 55 21 542-4092
TL: 2123057

Bahiense, Jose Ferreira
Reporter
REDE GLOBO
R. Lopes Quintas 303
Rio de Janeiro -RJ 2, Brazil
T: 55 21 2947732

Baia, Carlos Alberto
FORUM DO FUTURO
Rio de Janeiro, Brazil

Baia, Eduardo De
Correspondent
ASSOCIATED PRESS
Av Brasil 500/847 /Caixa Postal 72-ZC-00-CEP 20001
Rio de Janeiro RJ 20, Brazil
T: 55 21 580 4442/585 460 F: 55 21 580230

Baldez, Sandra Mota, Sra.
Auxiliar
MATA ATLANTICA/UF - VICOSA / POLYGLOTT USA
Rua Gomes Barbosa 84
Vicosa, MG. 36570, Brazil
T: 31 899-2139

Balhestero, Andre, Sr.
Diretor
FRATERNIDADE PAX UNIVERSAL
Rua Dom Domingos de Silos 315
Sao Paulo - SP 02526, Brazil
T: 55 11 290-3002 F: 55 11 950-2726

Balhestero, C. Lucia, Sra.
Organisation Founder
FRATERNIDADE PAX UNIVERSAL
Rua Dom Domingos de Silos 315 Jardim Sao Bento
Sao Paulo - SP 02526, Brazil
T: 55 11 2903002 F: 55 11 9502726

Balhestero, Vera, Sra.
Administration Manager
FRATERNIDADE PAX UNIVERSAL
Rua Dom Domingos de Silos 315 Jardim Sao Bento
Sao Paulo - SP 02526, Brazil
T: 55 11 2903002 F: 55 11 9502726

Baltazar, Wilson S., Mr.
J.D. TAIT ASSOCIATES CANADA
Rua Carlos De Carvalho, 851
Curitiba - PR, Brazil
T: 55 41 263 1672 F: 55 41 224 2641

Balthazar, Neusa Costa, Ms.
UNICEF
SBS Edificio Seguradoras 13o Andar
Brasilia - DF, Brazil
T: 55 61 224-7145 F: 55 61 224-5035

Bananeira, Rui Moura, Mr.
Coord. De Unidades De Cons.
INSTITUTO DO MEIO AMBIENTE DO AMAZONAS - IMA
Rua Recife, 3280 - Parque 10
Manaus - AM 69050, Brazil
T: 55 92 236-2309 F: 55 92 236-2535

Bandeira, Milton Leite, Mr.
Diretor
ASSDAK
Av. Almirante Barroso, 6 - sala 1902
Rio de Janeiro-RJ 20031, Brazil
T: 220-8837

Bandeira, Ricardo
Journalist/Producer
RADIO ALVORADA -PROGRAMA MOMENTO DA TERRA
Av. Getulio Vargas 668
Belo Horizonte -MG, Brazil
T: 55 31 2732525 F: 55 31 2733535

Baracho, Marcelo Jose Peres
Reporter
REDE GLOBO
R. Lopes Quintas 303
Rio de Janeiro -RJ 2, Brazil
T: 55 21 2947732

Barahona, Antonio
ECOTROPIC
1880 East River Road
Rio de Janeiro - RJ, Brazil
T: 55 21 577-6476 F: 55 21 577-1195

Barata, Eduardo De Souza
Redator
SISTEMA JORNAL DO BRASIL
Av.Brasil, 500
Rio de Janeiro - RJ, Brazil
T: 55 21 5854422

Baratta, Teresa Cristina, Sra.
INSTITUTO BRASILEIRO DE ADMINISTRACAO MUNICIPAL - IBAM
Edificio Diogo Lordello de Mello. Largo IBAM no.1. Humaita
Rio de Janeiro - RJ 22282, Brazil
T: 55 21 266-6622 F: 55 21 537-1262
TL: 22638 INBM BR

Barbosa, Abadia Elizete Silva Leal, Ms.
Secretaria
INSTITUTO BRASILEIRO DE CIENCIAS PENAIS
Av. Goias, 310 - Conj. 408 - Centro
Goiania - GO 74110, Brazil
T: 55 62 224-0239 F: 55 62 225-9333

Barbosa, Alberto Pires, Sr.
PREFEITURA MUNICIPAL
Praca Maua s/n 1 andar
Santos - SP 11010, Brazil
T: 55 132 342-2093 F: 55 132 345-809
TL: 131048

Barbosa, Amaury, Sr.
Vice-Presidente
ORG. COOP. BRASILEIRAS
SQS, Ed. Baracat - 4. andar
Brasilia - DF Cep.: 70309-900, Brazil
T: 55 61 225 8766 F: 55 61 226 8766
TL: 1879

Barbosa, Deise Juventina
Recepcionista
SIND. IND. ESTAMPARIA DE METAIS DE S.P.
Rosa Da Fonseca, 41 Fundos Casa 17 Manguinhos
Rio de Janeiro - RJ 21041, Brazil

Barbosa, Everaldo Torres, Mr.
Professor
UNIVERSIDADE FEDERAL DA PARAIBA - CAMPOS 1
Av. Sao Paulo,75/103 B. dos Estados
Joao Pessoa - PB 58030, Brazil
T: 55 83 244 1918

Barbosa, Francisco Das Chagas P.
REDE GLOBO
R. Lopes Quintas 303 J. Botanico
Rio de Janeiro - RJ, Brazil
T: 55 21 5292000 F: 55 21 2942042

Barbosa, Joao Manoel Goncalves, Sr.
CONTREN
Rua Tamandare, 348
Liberdade, SP 01525, Brazil
T: 55 11 279-3811 F: 55 11 268-3811

Barbosa, Lazaro Martins, Sr.
Chefe Do Servico De Intercambi
IDESP
Av. Nazare 871
Belem - PA 06035, Brazil
T: 55 91 224-4411 F: 55 91 225-3414
TL: 091 6905

Barbosa, Licinio Leal, Sr.
President
INSTITUTO BRASILEIRO DE CIENCIAS PENAIS
Av. Goias, 310, Conj. 408 - Centro - Ed. Villa Boa
Goiania - GO 74110, Brazil
T: 55 62 224-0239 F: 55 62 225-9333

Barbosa, Marco Antonio, Mr.
Reasercher
ASFEEMA
Rua Fonseca Teles 58 - Sao Cristovao
Rio de Janeiro - RJ 20.940, Brazil
T: 55 21 284-2772

Barbosa, Mario Dos Santos
TVT REDE DE COMUNICACAO DOS TRABALHADORES
Sao Paulo, Brazil

Barbosa, Mariuce, Sra.
Director
CMB
Rua Avanhadava no. 756
Sao Paulo - SP 01306, Brazil
T: 55 11 2564354 F: 55 11 2577648

Barbosa, Monica Maria Gomes
Editor
MANCHETE
Rua do Russel, 804 - 4 andar
Rio de Janeiro, Brazil
T: 55 21 2850033 x383

Barbosa, Renata, Sra.
Diretora Executiva
REYNOLDS/LATASA
Av. Presidente Wilson,231/2403
Rio de Janeiro - RJ, Brazil
T: 55 21 282 1382

Barbosa, Rizio, Sr.
FUNDACAO ESCOLA DE SERVICO PUBLICO RJ - FESP
Av. Carlos Peixoto, 64
Rio de Janeiro - RJ 22290, Brazil

Barbosa, Roberto Santos, Sr.
SOUZA CRUZ/ELDOCOR
Av. Rio Branco 125/21o.
Rio de Janeiro - RJ, Brazil
T: 55 21 227-0044
TL: 34858

Barbosa, Sonia Regina Da Cal Seixas G., Ms.
Researcher
UNIVERSIDADE ESTADUAL DE CAMPINAS
NEPAM UNICAMP Caixa Postal 1170
Campinas - SP 13081, Brazil
T: 55 192 398151 F: 55 192 394717
TL: 1150

Barbosa, Walkiria, Sra.
Diretora Executiva
MEDIA INFORMATION CENTER
Rua Frei Leandro 35 - Lagoa
Rio de Janeiro - RJ 22470, Brazil
T: 55 21 246-7791 F: 55 21 286-1191

Barbosa Neto, Domingos, Mr.
Gerente Controle
CETREL
Km 19 da BR 345 - Estrada do Coco / Polopetroquimico
Camacari - BA 42810, Brazil
T: 55 71 832-1186 F: 55 71 832-2567

Barcat, George, Sr.
Office Of Tibet - 02
ASSOCIACAO PALAS ATHENA DO BRASIL CENTRO DE ESTUDOS FILOSOFICOS
Rua Leoncio de Carvalho 99 - Paraiso
Sao Paulo - SP 04.003, Brazil
T: 55 11 288-7356/283-0867

Barcellos, Gilson Da Gama
Editor
SISTEMA GLOBO DE RADIO
Rua do Russel, 434 - 4 andar
Rio de Janeiro, Brazil
T: 55 21 2850448

Barcelos, Marco Antonio Narvaez
Cinegrafista
VIDEO CLIPPING
Av.Rio Branco, 181 Sl.910
Rio de Janeiro - RJ, Brazil
T: 55 21 2204002 / 2629240 F: 55 21 2204002

Barens, Nathalie Braum, Ms.
Filer
JOVEM TURISMO
SCLN 202 Bloco B Lojas 7/11
Brasilia - DF, Brazil
T: 55 61 224-5526 F: 55 61 321-0345
TL: 61-4705

Barg, Nelson
Palestrante
UNIAO ESTE BRASILEIRA DAS IASD
Av. Sete de Setembro, 69
Niteroi - RJ Cep.: 24.300, Brazil
T: 55 21 711 1062 F: 55 21 611 1571

Barki, Alain
Photographer
JAIME CAMARA
Jornal de Brasilia SIG, Trecho 1 Lot.585/645
Brasilia -DF, Brazil
T: 61 2252515 F: 61 2266735

Barkov, M.A., Mr.
President
ZARUBEZHTSVETMET
Rua Araujo Porto Alegre, 36 - sala 902
Rio de Janeiro - RJ 20030, Brazil
T: 55 21 262-4442 F: 55 21 262-4709
TL: 34957

Barreira, Roberto
Jornalista
REVISTA MANCHETE
Rua do Russel, 804
Rio de Janeiro - RJ, Brazil
T: 55 21 2652012

Barreiros, Elizabeth
Coorden. De Eventos
BCO. INTERAMERICANO DE DESENVOLVIMENTO
Caixa Postal 08738
Brasilia - DF 70312, Brazil
T: 55 61 321-7848 F: 55 61 321-3112
TL: 61-1204

Barreiros, Maria Da Graca
Recepcionista
BCO. INTERAMERICANO DE DESENVOLVIMENTO
Caixa Postal 08738
Brasilia - DF 70312, Brazil
T: 55 61 321-7848 F: 55 61 321-3112
TL: 61-1204

Barreto, Ana Marina
ASSOCIACAO POTIGUAR DE EDUCACAO E CULTURA
Rua Floriano Peixoto, 295 Petropolis
Natal - RN 59020, Brazil
T: 55 84 221-6255 F: 55 84 221-6255

Barreto, Juracy Mendes
Technician
REDE GLOBO
R. Lopes Quintas 303
Rio de Janeiro -RJ 2, Brazil
T: 55 21 2947732

Barreto, Mizael Araujo, Sr.
ASSOCIACAO POTIGUAR DE EDUCACAO E CULTURA
Rua Floriano Peixoto, 295 Petropolis
Natal - RN 59020, Brazil
T: 55 84 221-6255 F: 55 84 221-6255

Barreto, Reinaldo Leite Paes, Sr.
Secretary General
INSTITUTO HERBERT LEVY
Av Presidente Wilson 164 - 9o. andar
Rio de Janeiro - RJ 20030, Brazil
T: 55 21 262 5482 F: 55 21 220 2764

Barreto, Rosalee Cunha
Gerente De Comunicacao
PRO-NATURA INSTITUTO BRASILEIRO PESQ.EST.AMBIENTAIS
Rua Monte Alegre, 374
Rio de Janeiro - RJ, Brazil
T: 55 21 2243994 F: 55 21 2226742

Barreto, Suzanne, Sra.
INSTITUTO PRO NATURA
Rua Monte Alegre 374 - Santa Teresa
Rio de Janeiro - RJ 20240, Brazil
T: 55 21 224-39904 F: 55 21 224-3994

Barrett, Mary Ellin, Ms.
Senior Research Associate
COMMONWEAL
PO Box 316
Brasilia - DF, Brazil
T: 55 61 8680970 F: 55 61 8682230

Barretto, Lucia, Ms.
R. Siqueira Campos 12/1003
Rio de Janeiro - RJ 22031, Brazil
T: 55 21 2560035

Barroca, Karla, Ms.
WORLD YOUNG WOMEN'S CHRISTIAN ASSOCIATION
Eng. Bittencourt Vasc. 76 Cidade Universidade
Recife - PE 50741, Brazil
T: 55 81 271-4697
TL: WYMCA 41 23 32

Barroio, Cecilia Dos Santos
Reporter
JORNAL DO COMMERCIO
Rua Livramento, 189 4 andar
Rio de Janeiro, Brazil
T: 55 21 2538370 F: 55 21 2337690

Barros, Arquimedes Luiz James
CONFEDERACAO NACIONAL DA INDUSTRIA
Rio de Janeiro, Brazil

Barros, Aulton Rodrigues De
Audio Operator
L.A.S. CINEVIDEO
Rua de Matriz 30 Botafogo
Rio de Janeiro, Brazil
T: 55 21 2866998

Barros, Claudia
Fotografa
UNIVERSIDADE ASPERJ
Av.N.S.Copacabana, 787/402
Rio de Janeiro - RJ, Brazil
T: 55 21 2572726 F: 55 21 2572726

Barros, Djalma De Almeida
REDE RECORD
Av. Miruna 713
Sao Paulo _SP 04084, Brazil
T: 55 11 5429000 F: 55 11 612228

Barros, Edite Geraldo, Ms.
PROJETO FOLHA VERDE
Rua Carlos Finlay
Sao Paulo - SP, Brazil

Barros, Flavia Lessa De, Ms.
UNIVERSIDADE DE BRASILIA
SQS 304- Bloco A/102
Brasilia - DF 70337, Brazil
T: 55 61 226-3711

Barros, Francisco Anselmo Gomes De, Sr.
Presidente
FUNDACAO PARA CONSERVACAO DA NATUREZA
av. tamandare, 1818
Campo Grande - MS 79025, Brazil
T: 55 67 383-2528

Barros, Jorge, Sr.
CENTRO DE ART. POP. MARGINAL - CEAP
Rua da Lapa, 200 - S.805
Rio de Janeiro - RJ 20021, Brazil
T: 55 21 252-2302 F: 55 21 205-3099

Barros, Jose De Arimatea
Reporter
REDE GLOBO
R. Lopes Quintas 303
Rio de Janeiro -RJ 2, Brazil
T: 55 21 2947732

Barros, Jose Ribamar Magalhaes
MANCHETE
R. do Russel 804
Rio de Janeiro -RJ 2, Brazil
T: 55 21 2652012 F: 55 21 2855148

Barros, Katia Maria, Sra.
C. De Cont. Fisc.
INSTITUTO DE MEIO AMBIENTE DO ACRE
Rua: Rui Barbosa,450 - Centro
Rio Branco - AC 69.900, Brazil
T: 55 68 224 2857 F: 55 68 224 5694
TL: 68.2692

Barros, Luciana Rocha De Magalhaes
Editor
Rua Republica do Peru, 36/6 apto. 10
Rio de Janeiro, Brazil
T: 55 21 2850033

Barros, Luiz Mendes De, Mr.
Rua S. Pedro, 90 - Guarcia Torta
Maceio - AL 57000, Brazil
T: 55 82 2311053

Barros, Marcelo
Jornalista
REVISTA MANCHETE
Rua do Russel, 804
Rio de Janeiro - RJ, Brazil
T: 55 21 2652012

Barros, Martha Maria Machado, Ms.
ELETROBRAS
Av, Presidente Vargas, 642/10
Rio de Janeiro - RJ 20079, Brazil
T: 55 21 263-3137 F: 55 21 233-3248

Barros, Moacir R.
Technician
REDE GLOBO
R. Lopes Quintas 303
Rio de Janeiro -RJ 2, Brazil
T: 55 21 2947732

Barros, Ricardo Luis Peixoto De, Sr.
Tecnico
CONFEDERACAO NACIONAL DA INDUSTRIA
Av. Nilo Pecanha, 50 sala 3018
Rio de Janeiro - RJ 20044, Brazil
T: 55 21 262 0181 F: 55 21 262 1465
TL: 22634

Barros, Tania Andrade De Souza
INFORMATIVO 'BRASIL-MEIO AMBIENTE E DESENVOLVIMENTO'
Rio de Janeiro, Brazil

Barros, Valtemir De
MANCHETE
R. do Russel 804
Rio de Janeiro -RJ 2, Brazil
T: 55 21 2652012 F: 55 21 2855148

Barros Cruz, Luiz Eduardo Monteiro De, Mr.
Biologo
FUNAI
Rua francisco Otaviano 86/704 Copacabana
Rio de Janeiro - RJ 22080, Brazil
T: 55 21 247 9432

Barroso, Fabio Goncalves
Photographer
CACHOEIRAS JORNAL
Av. Lord Baden Powell, 62 s.102
Cachoeiras de Macacu, Brazil
T: 649 2692 F: 649 2692

Barroso, Felipe, Sr.
KEY MARKETING BRASIL
Av, Ataulfo de Paiva135 gr. 710
Rio de Janeiro - RJ 22440, Brazil

Barroso, Luiz Carlos Da S.
Technician
REDE GLOBO
R. Lopes Quintas 303
Rio de Janeiro -RJ 2, Brazil
T: 55 21 2947732

Barsanti, Hugo, Mr.
ASSOCIACAO CIVIL GREENPEACE
Rua Mexico 21, Grp 1301, Centro
Rio de Janeiro - RJ 20031, Brazil
T: 55 21 262-7318 F: 55 21 240-1690

Bartar, Tauran Golestani, Sra.
Treasurer
COMUNIDADE BAHA'I DO BRASIL
Av. Higienopolis, 938 - apto. 21
Sao Paulo - SP 01238, Brazil
T: 55 11 66-9828

Basho, Kyoto, Sr.
Diretor Do Instituto
INSTITUTO PERFECT LIBERTY DE PESQUISAS BOTANICAS
Av. PL do Brasil 8000
Aruja - SP 07400, Brazil
T: 55 11 466-0188 F: 55 11 466-0168

Basilio, Maria Carmen De Souza, Sra.
TRIBUNAL JUSTICA DF
SQS 308 Bloco B ap. 308
Brasilia - DF 20354, Brazil
T: 55 61 243-8393

Basilio, Ronaldo, Sr.
Director
SQUALO
Ladeira Felipe Neri NO. 7/301 - Pc.Maua
Rio de Janeiro - RJ, Brazil
T: 55 21 233-4108

Bastos, Cesar, Mr.
TRIBO COSMICA
Caixa Postal 1030
Rio de Janeiro - RJ 20001-970, Brazil
T: 55 21 233-7434
TL: 1030

Bastos, Clea Margarida Maciel, Sra.
Administradora
IBAMA-SUPES-BA
Av. Juraci Magalhaes Jr 608 Rio Vermelho
Salvador - BA 41930, Brazil
T: 55 71 240-7322 F: 55 71 240-7913

Bastos, Guilherme
Fotografo
JORNAL O GLOBO
Rua Irineu Marinho, 35
Rio de Janeiro - RJ, Brazil
T: 55 21 2722000

Bastos, Jonathan, Sr.
C D O / M J
Saia 07, Lt. 23, Sps Asa Sul
Brasilia - DF 70610, Brazil
T: 55 61 245-7425

Bastos, Katja, Ms.
TRIBO COSMICA
Rua Buenos Aires No. 2 - 1402
Rio de Janeiro - RJ 22970, Brazil
T: 55 21 233-7454

Bastos, Paulo Cesar O.
FBCN - JORNAL INFORMATIVO
Rio de Janeiro, Brazil

Bastos, Silas Pereira
Cinegrafista
TV RIO
R. Miguel de Frias, 57 Estacio
Rio de Janeiro-RJ 20, Brazil
T: 55 21 2939888 F: 55 21 2938626

Bastos, Tonia
UNIVERSIDADE GAMA FILHO
Rua Manuel Vitorino, 625
Rio de Janeiro, Brazil
T: 55 21 2697272

Batalha, Antonio Jose
Fotografo
AGENCIA NEXUS
Av.Beira Mar, 406/402
Rio de Janeiro - RJ, Brazil
T: 55 21 2402573 F: 55 21 2200094

Batisse, Michel
CONSERVATION INTERNATIONAL
Rua Bueno Brandao 393
Belo Horizonte - MG 31010, Brazil
T: 55 31 226-5145 F: 55 31 222-8429

Batista, Adauta Luz, Sra.
Asst. Social
ASSOCIACAO COMUNITARIA ARY CESAR - ACACE
R. Antonio Gomes de Souza, no. 221
Luciara - MT 78660, Brazil
T: 529-1113

Batista, Alberto De Sena
Editor
ESTADO DE MINAS
R. Goias 36 Centro
Belo Horizonte -MG 3, Brazil
T: 55 31 2732322 F: 55 31 2268070

Batista, Augusto, Sr.
GULF PROMOCOES
Rua do Carmo 11/ 802 Centro
Rio de Janeiro - RJ, Brazil
T: 55 21 224-5173

Batista, Camilo
TELEVISA
Av. Sao Sebastiao 157 URCA
Rio RJ 22291, Brazil
T: 55 21 295-6075

Batista, Luiz Carlos, Sr.
Decano D Centro De Arquitetura
UNIVERSIDADE SANTA URSULA
Rua Fernando Ferrari, 75
Rio de Janeiro - RJ 22231, Brazil
T: 55 21 551-5542 F: 55 21 551-6446

Batista, Marcia Aguiar, Ms.
Conselheira
IAB - RIO DE JANEIRO
av. oswaldo cruz 115/105
Rio de Janeiro - RJ 22250060, Brazil
T: 55 21 5519095

Batista, Pedro Afonso Rodrigues, Sr.
Presidente
CAMARA MUNICIPAL DE GOIANIA
Rua 4 # 515- Ed. Parthenon Center, 9 Andar- Centro
Goiania - GO 74020, Brazil
T: 55 62 225-1368 F: 55 62 225-4773
TL: 62-1095

Batista, Roberto Mauricio Costa, Mr.
Counselor
CREA-PE
Agamenom Magalhaes 2978 Espinheiro
Recife - PE, Brazil
T: 55 81 2315253 F: 55 81 2315170
TL: 1199

Batistela, Evandro Marcelo, Sr.
Vice President
ASSOCIACAO CIENTIFICA ECOLOGICA 'AUGUSTO RUSCHI' - ACEAR
Cerquilho SP 18520, Brazil
T: 55 015 284-1274 F: 55 015 284-1205

Battey, Michael, Mr.
CHIKYU NO TOMO KANAZAWA
Turistico Hotel Ladeira da Gloria 30, Gloria #123
Rio de Janeiro - RJ, Brazil

Bauler, Denise Haddad
Journalist
CAIEME
Matto Grosso 78030, Brazil
T: 3215820

Baum, Cristina, Ms.
Educadora
SACTS
Rua: Frei Mansueto, 1077
Brazil

Baumfeld, Carlos Minc, Sr.
Vicepresident
DEFENSORES DA TERRA
Rua Paulo Cesar de Andrade 222/802 Pq.Guinle
Rio de Janeiro - RJ 242-9353, Brazil
T: 55 21 2920111 F: 55 21 2429353

Baumgartner, Bill, Mr.
Co-Director
CENTRO ESPIRITA LEON DENIS
Rua Abilio dos Santos 137
Bento Ribeiro - RJ 21331, Brazil
T: 54 3571846

Bauquedel, Angela, Ms.
Relacoes Publicas
SEC. DE MINAS E ENERGIA
R. Luiz Leopoldo Fernandes Pinheiro 517, sala 1403
Niteroi - RJ 24030, Brazil
T: 55 21 719-7171

Bautista, Hortensia Pousada, Ms.
Biologa
Av. Paulo VI, no. 2609, apt.1302 - Itaigara
Salvador - BA 41810, Brazil
T: 55 71 258-5128 F: 55 71 359-8054
TL: 071-1918

Bavaresco, Frederico, Sr.
Vice-Presidente
ORG. COOP. BRASILEIRAS
SQS, Ed. Baracat, 4. andar
Brasilia - DF Cep.: 70309-900, Brazil
T: 55 61 225 0275 F: 55 61 226 8766
TL: 1879

Beaudet, Normand, Mr.
Counselor For Latin America
CSN - CONFEDERATION DES SYNDICATS NATIONAUX
Caixa Postal 1759, Centro
Sao Paulo - SP 01059, Brazil
T: 55 11 270-1257

Beaurepaire, Lucila
Reporter
JORNAL O GLOBO
Rua Irineu Marinho, 35
Rio de Janeiro - RJ, Brazil
T: 55 21 2722000

Bebiano, Alexandre, Sr.
Divulgador
INSTITUTO BRASILEIRO DE ANALISES SOCIAIS E ECONOMICAS - IBASE
Rua Vicente de Souza 29 - Botafogo
Rio de Janeiro - RJ 22251, Brazil
T: 55 21 286-0348 F: 55 21 286-0541
TL: 2136466 BASE BR

Becker, Beatriz
MANCHETE
Rio de Janeiro, Brazil

Begossi, Alpine, Mr.
UNIVERSIDADE ESTADUAL DE CAMPINAS
NEPAM UNICAMP Caixa Postal 1170
Campinas - SP 13081, Brazil
T: 55 192 398151 F: 55 192 394717
TL: 1150

Belchior, Murilo Bastos, Mr.
Vice-President
CRUZ VERMELHA BRASILEIRA
Praca Cruz Vermelha 10/12 - 1o. andar
Rio de Janeiro - RJ, Brazil
T: 55 21 232-6325 F: 55 21 242-6760

Belfort, Fernando, Jr
Cinegrafista
GLOBO ECOLOGIA
Rua Santa Alexandrina 343/c3 Rio Comprido
Rio de Janeiro RJ, Brazil
T: 55 21 502-5451 F: 55 21 273-9087

Belfort, Lilia
Operador De Audio
GLOBO ECOLOGIA
Rua Santa Alexandrina 343/c3 Rio Comprido
Rio de Janeiro - RJ, Brazil
T: 55 21 502-5451 F: 55 21 273-9087

Bellicanta, Maria Lucia, Sra.
Sociologist
CERIS
Rua Dr. Julio Ottoni, 571/35
Rio de Janeiro - RJ 20241-400, Brazil
T: 55 21 265-5177 F: 55 21 285-7276

Bello, Flavio Bello
Technician
REDE GLOBO
R. Lopes Quintas 303
Rio de Janeiro -RJ 2, Brazil
T: 55 21 2947732

Bello, Susan Mintz, Ms.
ECO-ART
SQS 406 Bloco M apt.103
Brasilia - DF 70255, Brazil
T: 55 61 243-6610

Belloc, Beth
92 GLOBAL FORUM
Rio de Janeiro - RJ, Brazil

Belluomini, Roberta Vaz
Tecnica
TVSBT
Campo de S.Cristovao, 105
Rio de Janeiro - RJ, Brazil
T: 55 21 5803688 / 5800678 F: 55 21 5809685

Belotti, Paola, Sra.
HOSPITAL PIO X
Praca Pio X M1
Ceres - Goias 76300-000, Brazil

Beltrao, Jimena Felipe, Sra.
Coordenadora Editorial
UNAMAZ - ASSOCIACAO DE UNIVERSIDADES AMAZONICAS
Caixa Postal 558
Belem - PA 66.059, Brazil
T: 55 91 229-9310/229-2088 F: 55 91 229-4339
TL: 5591-1013UPBCBR

Benatti, Claudia De Aquino, Sra.
SERVICO MULHER MARGINALIZADA
R. Guilherme Maw, 64, casa 20 - Luz
Sao Paulo - SP 01105, Brazil
T: 55 11 2284955

Benazzi, Marcia Geogina
Journalist
MANCHETE
Rua do Russel, 804/ 6 andar
Rio de Janeiro, Brazil
T: 55 21 2850033

Benedicto, Nair
Reporter Fotografica
AGENCIA N IMAGENS
Av.Brig.Luis Antonio, 2724 Casa 6
Sao Paulo - SP, Brazil
T: 55 11 8870847 F: 55 11 8853694

Benevides, Mara Rubia Gomes, Ms.
Eng. Florestal
INSTITUTO DO MEIO AMBIENTE DO AMAZONAS
Rua Recife no. 3280
Manaus 69055, Brazil
T: 092 2362064 F: 092 2362535

Beni, Franco De, Sr.
Diretor Agropecuario
AGIP DO BRASIL SA
Av Paulista 2073 - 2o. terraco
Sao Paulo - SP 01311, Brazil
T: 55 11 288 8573 F: 55 11 288 7483
TL: 21573

Bening, Carsten, Mr.
GRUPO AMBIENTALISTA DA BAHIA - GAMBA
Rua Borges dos Reis, 46 - sala J
Salvador - BA 41910, Brazil
T: 55 71235-5836 R F: 55 71235-5836

Benites, Marcello
REVISTA CIDADE NOVA
Vargem Grande Paulis, Brazil

Bennett, Sergio Jose Martins
REDE GLOBO
R. Lopes Quintas 303 J. Botanico
Rio de Janeiro - RJ, Brazil
T: 55 21 5292000 F: 55 21 2942042

Benoliel, Antonio Cezar C., Mr.
Presidente
UNIAO BRASILEIRA DE ENGENHARIA DE SEGURANCA
rua General Aristides Athayde Jr. 588
Curitiba - PR 80730-370, Brazil
T: 55 41 273 6262 F: 55 41 272 6448
TL: 41-5847, E: 7087

Beraldo Guimaraes, Luiz Fernando, Mr.
Conselheiro
CREA - SAO PAULO
Rua Jacutinga 610 apto 23
Sao Paulo - SP 04515, Brazil
T: 55 11 532 0328

Berbem, Dionisia Pimenta, Ms.
UNIAO DOS MOVIMENTOS POPULARES E MEIO AMBIENTE
Rua dos Menestreis, 10
Sao Paulo - SP, Brazil

Bergamo, Carmem, Ms.
CONFEDERACAO NACIONAL IND.
Nilo Pecanha,50/andar 38
Rio de Janeiro - RJ 22250, Brazil
T: 55 21 552-0099 F: 55 21 552-9392

Berggravf, Berit, Ms.
Jr. Executive Officer
NORWEGIAN DELEGATION TO UNCED
Real Consulado de Noruega, Caixa Postal 436
Rio de Janeiro - RJ 20000, Brazil
T: 55 21 552-1540 F: 55 21 551-9091

Berna, Vilmar, Sr.
President
SOS TERRA (FREELANCER)
Rua Lemos Cunha, 545 apt.102 - Icarai
Niteroi - RJ 24230, Brazil
T: 55 21 2920111 F: 55 21 2429353

Bernades, Renata C. P. Raposo
FREELANCER
R. Correa de Sa, 5 Santa Teresa
Rio de Janeiro -RJ, Brazil
T: 55 21 2243574 F: 55 21 2529370

Bernandes, Sidney Correia
Technician
REDE GLOBO
R. Lopes Quintas 303
Rio de Janeiro -RJ 2, Brazil
T: 55 21 2947732

Bernardes, Aline Tristao, Sra.
Representative
FUNDACAO BIODIVERSITAS
Rua Bueno Brandao 393
Belo Horizonte - MG 31010, Brazil
T: 55 31 226-5145 F: 55 31 222-8429

Bernardes, Cicero, Sr.
GREENPEACE BRASIL
Rua Mexico 21, Grupo 1301 A/B
Rio de Janeiro - RJ 20031, Brazil
T: 55 21 262-7318 F: 55 21 240-1690

Bernardes, Mauro Roberto
MANCHETE
Rio de Janeiro, Brazil

Bernardo, Christianne, Ms.
Diretora
GECA - CRUPO DE ESTUDO E CONSCIENTIZACAO AMBIENTAL
R. General Artigas no. 85/201 - Leblon
Rio de Janeiro - RJ 22450, Brazil
T: 55 21 239-2418 F: 55 21 274-1601

Bernardo, Maristela Bezerra, Sra.
Conselheira
INSTITUTO DE ESTUDOS AMAZONICOS E AMBIENTAIS
SCLN 210 Ed. Center, Bloco C, Salas 211 e 212
Brasilia - DF 70862, Brazil
T: 55 61 347-9874 F: 55 61 347-9874
TL: AX.1EA

Bernstein, Ida Baumfeld, Ms.
Membro
CENTRO DORJE JIGJE
Av Borges De Medeiros 3535 Apto. 1006- Lagoa
Rio de Janeiro - RJ 22470, Brazil
T: 55 21 286-6331

Berti, Maria Jose Ferreira
Enga. Sanitarista
FEMA
Rua K-Z - No. 45
Sinop - MT 78550, Brazil
T: 065 531-1408

Bessa, Marcos Antonio Teixeira
Technician
REDE GLOBO
R. Lopes Quintas 303
Rio de Janeiro -RJ 2, Brazil
T: 55 21 2947732

Bettencourt, Flavio Estevao, Sr.
Professor
FACULDADE JOAO PAULO II
Rua Benjamin Constant, 23
Rio de Janeiro - RJ 20241, Brazil
T: 55 21 242-4552

Bezerra, Alexandre Soarez
REDE GLOBO
Rio de Janeiro, Brazil

Bezerra, Ligia Maria Demelo, Sra.
I B G E
PQ Julio Cesar, Ed. Servio, ap. 801 - Pituba
Salvador - BA 41350, Brazil

Bezerra, Manoel Aristarcho De B.
CONFEDERACAO NACIONAL DA INDUSTRIA
Rio de Janeiro, Brazil

Bezerra, Mauro, Mr.
Prom.Just.
MINISTERIO PUBLICO DO AMAZONAS
Rua: 24 De Maio,321 - Centro
Manaus - AM, Brazil
T: 55 92 233-5547

Bezerra, Osvaldo Da, Mr.
SIND. DOS TRABALHADORES NAS IND. QUIM. E FARMACEUTICAS DE S.PAULO
Rua Tamandare 348 - Liberdade
Sao Paulo - SP 01525, Brazil
T: 55 11 279-3811 F: 55 11 279-0662

Bezerra, Regina
Jornalista
REVISTA ECORIO
Av.Pres.Antonio Carlos, 54 Gr.802
Rio de Janeiro - RJ, Brazil
T: 55 21 5330069 F: 55 21 2402658

Bezzoco, Francisco
Produtor
TV Rio
R. Miguel de Frias, 57 Estacio
Rio de Janeiro - RJ, Brazil
T: 55 21 2939888 F: 55 21 2938626

Biar, Paulo, Sr.
Menbro Do Conselho Curador
FUNDACAO XAPURI
Rua Da Assembleia # 10 Sala 1621- Centro
Rio de Janeiro - RJ 20119, Brazil
T: 55 21 221-9424 F: 55 21 224-7213

Biasi, Marta De, Ms.
Translator
CENTRO DE PESQUISAS DO MAR (CEPEMAR)
Rua Mexico, 31/Gr. 501- Centro
Rio de Janeiro - RJ, Brazil
T: 55 21 240-3822 F: 55 21 240-9516

Bicalho, Deusiana Dias, Sra.
Vice-President
GRUPO ECOLOGICO NATI-VERDE
Rua cel. Alipio dias, 863
Sao Jose do Rio Pard 13720, Brazil
T: 55 196 61-2836

Bicalho, Paulo Barreto
Reporter
REDE GLOBO
R. Lopes Quintas 303
Rio de Janeiro -RJ 2, Brazil
T: 55 21 2947732

Biderman, Maria Tereza, Ms.
President
GLOBAL HARMONY FOUNDATION
Rua Cardeal ArcoVerde 99, casa 2 Pinheiros
Sao Paulo - SP, Brazil
T: 55 11 8812360

Biderman, Rachel, Sra.
Conselheira
HARMONIA GLOBAL
Av. Dr. Cardoso de Mello, 1750-7o. andar
Sao Paulo - SP 04548, Brazil
T: 55 11 829-0043 F: 55 11 829-3743

Biella, Luiz Carlos, Mr.
CESP
Rua da Consolacao, 1875, 9 andar, Cerq. Cesar
Sao Paulo - SP 01301, Brazil
T: 55 11 258-2165 F: 55 11 256-2101

Bifleveld, May, Ms.
UNIVERSALA ESPERANTO-ASOCIO
Rua Santa Alexandrina 481/606 - Rio Comprido
Rio de Janeiro - RJ 20261, Brazil
T: 55 21 273-8108

Biggs, Gonzalo
Oficial De Relacoes Externas
BCO. INTERAMERICANO DE DESENVOLVIMENTO
Caixa Postal 08738
Brasilia - DF 70312, Brazil
T: 55 61 321-7848 F: 55 61 321-3112
TL: 61-1204

Bilesky, Andre
BRASIL ENERGIA
Rio de Janeiro, Brazil

Billo, Daniel
Camera Assistant
MTV - RIO
Praca Nossa Senhora da Gloria 46
Rio de Janeiro -RJ, Brazil
T: 55 21 2855946 F: 55 21 2055394

Bisogno, Francisco Paulo
Tecnical Operator
RADIO GAUCHA
Av Ipiranga 1075
Porto Alegre - RS 90060, Brazil
T: 55 512 223 6600 F: 55 512 2336600 r 57

Bissacot, Raquel, Sr.
Membro
APOENA
Rua Cuiaba,1-19
Presid. Prudente-SP 19470-000, Brazil
T: 55 182 21-4222 F: 55 182 22-6177

Bittancurt, Oswaldo Cassio Vieira, Sr.
Seguranca
BANCO AMERICA DO SUL YASUDA
Av. Brigadeiro Luiz Antonio, 2020 - 5o andar
Sao Paulo - SP 01318-911, Brazil
T: 55 11 285-1411 F: 55 11 285-1426

Bittar, Paulo, Sr.
ASSOCIACAO ECOLOGICA DE TREKKING
Rua Irineu Marinho 120-Bom Pastor
Juiz de Fora - MG 36021-580, Brazil
T: 55 32 211138

Bittencourt, Julia Helena Guimaraes
FREELANCER
Rio de Janeiro, Brazil

Bittencourt, Noemi
BRASIL ENERGIA
Rio de Janeiro, Brazil

Bittencourt, Sybil
Hotel Regente, Av. Atlantica 3716
Rio de Janeiro - RJ, Brazil
T: 55 21 287-4212

Bizarria, Raugeston, Mr.
Assessor Parlamentar
ASSEMBLEIA LEGISL.
Av. Pedro Alvares Cabral s/n
Sao Paulo - SP, Brazil
T: 55 11 884-2748 F: 55 11 881-2748

Bizinover, Ana Lucia
Journalist
HILL AND KNOWLTON
Av. Afranio de Melo Franco 135 /5 andar Leblon
Rio de Janeiro -RJ 2, Brazil
F: 55 21 2394697

Blumenschein, Markus, Sr.
Student
UNIVERSIDADE DE TUBINGEN
Rua Dias da Rocha 20, apto. 1006, Copacabana
Rio de Janeiro - RJ 22051, Brazil

Blumenshein, Raquel Naves, Ms.
Architect
P & R LTD
S.Q.S. 107 Bl. E apto. 304
Brasilia - DF 70346, Brazil
T: 55 61 225-4359

Boas, Jose Henrique, Mr.
Filiado
GERMEN/IBGE
Rua Rio De Janeiro 591/1004- Pituba
Salvador - BA 41830, Brazil
T: 55 71 248-3431

Bocage, Sergio Menezes Du
Reporter
TV EDUCATIVA - RJ
Av. Gomes Freire 474 Centro
Rio de Janeiro -RJ, Brazil
T: 55 21 2426736 F: 55 21 2426740

Bocayuva, Christina
Photographer
FOLHA DE SAO PAULO
Alameda Barao de Limeira 425
Sao Paulo SP, Brazil
T: 55 11 874 2608/874 2602 F: 55 11 2231644

Bock, Udo
Oficial Mobilizacao Social
UNICEF
SBS Ed.Seguradoras, 13 andar
Brasilia - DF, Brazil
T: 55 61 2247145 F: 55 61 2245035

Boclin, Carolina
MACACO PREGO
1880 East River Road
Rio de Janeiro - RJ, Brazil
T: 55 21 577-6476 F: 55 21 577-1195

Boeira, Sergio Luis
Journalist
FREELANCER
R. Delminda Silveira 235/E-402
Florianopolis, SC 88, Brazil
T: 1 482 335-151 F: 1 478 222-497

Boff, Frei Leonardo
FRANCISCANS AT THE UNITED NATIONS
SINFRAFUPE do Brasil, CP 90174
Petropolis - RJ 25655, Brazil
T: 55 242 42 52 47

Bombardelli, Helio Marino, Sr.
Assesor Regionais
FEPAM
Av. A.J. Renner, 10
Porto Alegre - RS 90245, Brazil
T: 55 512 342-0224 F: 55 512 342-0224

Bona, Paulo
Journalist
FREELANCER
R. Santa Amelia 50 Bl/2 apt. 1008 Catumbi
Rio de Janeiro RJ 2, Brazil
T: 55 21 2738726

Bonder, Cintia, Sra.
Consultora
CONSELHO INTERNACIONAL DE BEM ESTAR SOCIAL
Ramiro Barcelos 1522/13
Porto Alegre - RS 90210, Brazil
T: 55 512 221 3217

Bonduki, Maria India, Ms.
Jornalista
DIALOGOS
Rua Juri 195
Sao Paulo - SP 05443, Brazil
T: 55 11 815-5782

Bonilla, Javier Eduardo
Correspondent
EL PAIS
Plaza Cagancha 1162
Montevideo, Brazil
T: 59 82 920115 F: 59 82 920632

Bonis, Rodolfo De
Reporter
O DIA
R. Riachuela 359
Rio De Janeiro, RJ 2, Brazil
T: 55 21 272-8000 F: 55 21 507-1228

Bonnekamp, Ulli
TWENTY FIRST CENTURY INC.
Rio de Janeiro, Brazil

Bonner, Willian
Reporter
REDE GLOBO
R. Lopes Quintas 303
Rio de Janeiro -RJ 2, Brazil
T: 55 21 2947732

Boorhem, Roberto Leal, Sr.
Medico E Professor
INSTITUTO DE ACUPUNTURA DO RIO DE JANEIRO
Travessa Pepe 81 Botafogo
Rio de Janeiro - RJ 22290, Brazil
T: 2951481

Borba, Ivaldo, Mr.
Expositor
EDITORA TERCEIRO MUNDO
Rua da Gloria, 122/101
Rio de Janeiro - RJ, Brazil

Borba, Nestor Da Costa, Mr.
Director
FUNDACAO BRASIL FRANCA PARA O DESENVOLVIMENTO
SCLN 402, Bloco E, Sala 109
Brasilia - DF 70834, Brazil
T: 55 61 226.8173

Borba, Rufino Nogueira
Photographer
JORNAL GAZETA DE NOTICIAS
R. Leandro Martins 72 Centro
Rio de Janeiro -RJ, Brazil
T: 55 21 2630435 F: 55 21 2630584

Borba, Virginia Guimaraes, Sra.
Relacoes Publicas
AMNPF
Felipe Schmidt 96- Centro
Florianopolis - SC, Brazil
T: 55 482 226084

Borges, Ana, Ms.
WORLD YOUNG WOMEN'S CHRISTIAN ASSOCIATION
Rua Manoel Barreto 52-apt.201 Graca
Salvador - BA 40230, Brazil
T: 55 71 247-6882
TL: WYMCA 41 23 32

Borges, Claudio, Mr.
GOVERNO DE SAO PAULO
Sao Paulo - SP, Brazil

Borges, Edson Benedito Roffe, Sr.
Economista
BANCO DA AMAZONIA S/A
Av. Pres. Vargas, 800 - 7o. andar - GERHU-Ditre
Belem - PA, Brazil
T: 55 91 216-3359 F: 55 91 216-3407
TL: 1093

Borges, Maria Jose, Sr.
ARQUIPELAGO DE FERNANDO DE NORONHA
Rua Barao de Sao Borja, 243 - Boa Vista
Recife - PE 50070, Brazil
T: 55 81 221-3217 F: 55 81 231-1594
TL: 81.2533

Borges, Ray Taylon Alves
Technician
REDE GLOBO
R. Lopes Quintas 303
Rio de Janeiro -RJ 2, Brazil
T: 55 21 2947732

Borgeth, Pedro
Photographer
MANCHETE
R. do Russel 804 Gloria
Rio de Janeiro RJ 2, Brazil
T: 55 21 2454816 F: 55 21 2056196

Borghi, Gertrudes Silva Nogueira, Sra.
Dir.De Estudos E Projetos
DEPARTAMENTO DE RECURSOS MINERAIS
Rua Marechal Deodoro, no.351 - Centro
Niteroi - RJ 24030, Brazil
T: 55 21 722-7074 F: 55 21 719-6846

Born, Gemina Cirilo Cabral, Sra.
IECAM
Dr. Moreira 250
Manaus - AM 69000, Brazil

Borriero, Jose Luiz, Mr.
Diretor
SIND. DOS TRABALHADORES NAS IND. QUIM. E FARMACEUTICAS DE S.PAULO
Rua Tamandare 348 - Liberdade
Sao Paulo - SP 01525, Brazil
T: 55 11 279-3811 F: 55 11 279-0662

Borrmann, Luciane
Reporter
FREELANCER
Caixa postal 82
Paulinia - SP 13140, Brazil
T: 55 192 741909 F: 55 192 741909

Bosso, Arnaldo
Tecnico
LEGIAO DA BOA VONTADE
Av.Rudge, 700
Sao Paulo - SP, Brazil
T: 55 11 2202599 F: 55 11 2205196

Botsaris, Alexandros Spiros, Sr.
Medico E Professor
INSTITUTO DE ACUPUNTURA DO RIO DE JANEIRO
Travessa Pepe 81 Botafogo
Rio de Janeiro - RJ 22290, Brazil
T: 55 21 295-1481

Botton, Andre De, Sr.
Administrator
Rua do Passeio, 56
Rio de Janeiro - RJ 20021, Brazil

Bouchardet, Carlos Alberto, Mr.
CONSERVATION INTERNATIONAL
Rua Bueno Brandao 393
Belo Horizonte - MG 31010, Brazil
T: 55 31 226-5145 F: 55 31 222-8429

Bouvier, Antoine, Mr.
M.Legal Affairs Div.
INTERNATIONAL COMMITEE OF THE RED CROSS (ICRC)
C/O ICRC, SHIS QL06 Conj. 09, Casa 16 - Lago Sul
Brasilia - DF, Brazil
T: 55 61 248-0250 F: 55 61 248-3840

Boyum, Richard, Mr.
Eto
U S I S
Al. Femao Cardim 377/141
Sao Paulo - SP 01403, Brazil
T: 55 11 287-6927 F: 55 11 852-1395

Braga, Alceste, Sra.
CLUBE SOROPTIMISTA INTERNACIONAL
SQN 106 BL. 1 ap. 406
Brasilia - DF 70742, Brazil
T: 55 61 273-5806 F: 55 61 321-4320

Braga, Calarice Jucafa
SECRETARIA DE ESTADO DE MINAS
Rua Sao Christavao 1200
Rio de Janeiro, Brazil
T: 55 21 2340942 F: 55 21 2340942

Braga, Clarice Girafa, Sra.
SECRETARIA DE ESTADO DE MINAS E ENERGIA DO RIO DE JANEIRO
Rua Luis Leopoldo Fernandes Pinheiro, 517, sala 1003
Niteroi - RJ 24016, Brazil
T: 55 21 234-0942 F: 55 21 234-0942
TL: 2132079

Braga, Guirlanda Maria Maia Benevide, Sra.
Diretora Administrativa Financ
IPREDE
Rua Prof. Carlos Lobo, 15 Cidade dos Funcionarios
Fortaleza - CE 60815, Brazil
T: 55 85 239-1300

Braga, Helena Liva Ribeiro, Ms.
Eng. Florestal
Travessa D, 12/1101
Cuiaba - MT CEP:78.070, Brazil
T: 55 125 624-4899 F: 55 125 624-4899

Braga, Lais Martins
Researcher
TV EDUCATIVA
av Gomes Freire, 474 Centro
Rio de Janeiro, Brazil
T: 55 21 2920012 x169

Braga, Marcia, Sra.
KEY MARKETING BRASIL
Av. Atraulfo de Paiva 135 Gr.710
Rio de Janeiro - RJ 22440, Brazil

Braga, Maria Christina De Almeida, Sra.
Antropologa
PAU-BRASIL COMUNICACOES
R. Mator Rubens Vaz, 702/1102 - Gavea
Rio de Janeiro - RJ, Brazil
T: 55 21 274 8824 F: 55 21 342 8567

Braga, Ricardo Correa Braga, Sr.
FLAMA
Rua Marques de Abrantes, 55
Rio de Janeiro - RJ 22250, Brazil

Braga, Roberto Seljan
CONFEDERACAO NACIONAL DA INDUSTRIA
Rio de Janeiro, Brazil

Braga, Samuel
Vereador
CAMARA MUNICIPAL DE FORTALEZA
Rua Antonele Bezerra
Fortaleza - CE 60.000, Brazil
T: 55 85 221-1791

Braga, Teodomiro
SISTEMA JORNAL DO BRASIL
Av. Brasil 500
Rio de Janeiro -RJ, Brazil
T: 55 21 5854422 F: 55 21 5854428

Braga, Vilson, Sr.
GOVERNO DO ESTADO DO RIO DE JANEIRO
Rio de Janeiro - RJ, Brazil

Braganca, Alcides Viana
Tecnico De Audio
SISTEMA GLOBO DE RADIO
Rua do Russel, 434
Rio de Janeiro - RJ, Brazil
T: 55 21 2058282 F: 55 21 2054096

Braia, Ana Lucia S., Sra.
Director
CMB
Rua Avanhadava no. 756
Sao Paulo - SP 01306, Brazil
T: 55 11 2564354 F: 55 11 2577648

Braile, Pedro Marcio, Sr.
Tecnico
CONFEDERACAO NACIONAL DA INDUSTRIA
Av. Nilo Pecanha, 50 sala 3018
Rio de Janeiro - RJ 20044, Brazil
T: 55 21 262 0181 F: 55 21 262 1465
TL: 22634

Branco, Adriana Castelo
Reporter
SISTEMA JORNAL DO BRASIL
Av. Brasil 500
Rio de Janeiro -RJ, Brazil
T: 55 21 5854422 F: 55 21 5854428

Branco, Joanna Lucia Mota, Ms.
Tecnico (assist. Social)
IDESP
Av. Nazare 817
Belem - PA 66035, Brazil
T: 55 91 224-4411 F: 55 91 225-3415
TL: 6905

Branco, Jorge, Sr.
CENTRO COMUNITARIO DOS PEQUENOS NEGOCIOS
Rua Abilio dos Santos 137
Bento Ribeiro - RJ 21331, Brazil
T: 55 21 357-1846

Branco, Maria Da Graca B. Castello, Sra.
Diretora
CENTRO BRASILEIRO DE COOPERACAO E INTERCAMBIO DE SERVIDORAS SOCIAIS
R. Sta. Luzia 685/206
Rio de Janeiro - RJ 20030, Brazil
T: 55 21 2208174

Branco, Maria Luiza Castelo
Reporter
AGENCIA 355
Rua Cosme Velho, 355
Rio de Janeiro RJ 2, Brazil
T: 55 21 2251813 F: 55 21 2422878

Brandalise, Ana Maria, Sra.
Assesor Diretoria
CONSELHO ESTADUAL DA CONDICAO FEMININA
R. Ebano Pereira,240
Curitiba - PR 80410240, Brazil
T: 55 41 232 2583 F: 55 41 234 4835
TL: 041 6245

Brandani, Lucineide Pereira L., Ms.
Eng. Florestal
FEMA - FUNDACAO ESTADUAL MEIO AMBIENTE
Centro Politico Administrativo
Cuiaba - MT 78080, Brazil
T: 55 065 313-2901

Brandao, Claudia, Ms.
ASSOCIACAO CIVIL GREENPEACE
Rua Mexico 21, Grp 1301, Centro
Rio de Janeiro - RJ 20031, Brazil
T: 55 21 262-7318 F: 55 21 240-1690

Brandao, Cristina De Jesus Botelho
Jornalista/Redatora
JORNAL MUSEU AO VIVO
Rua das Palmeiras, 55
Rio de Janeiro - RJ, Brazil
T: 55 21 2862097 F: 55 21 2860845

Brandao, Eduardo Barbosa, Sr.
Diretor
FLAMA
Rua Marques de Abrantes, 55 - Flamengo
Rio de Janeiro - RJ 22230, Brazil
T: 55 21 265-5690

Brandao, Leticia Domingues, Ms.
Biologa
INSTITUTO FLORESTAL DE SAO PAULO
Caixa Postal 1322
Sao Paulo - SP 01059, Brazil
T: 55 11 952-8555 F: 55 11 204-8067
TL: 11-22877BR

Brandao, Luiz Roberto De P.
Technician
REDE GLOBO
R. Lopes Quintas 303
Rio de Janeiro -RJ 2, Brazil
T: 55 21 2947732

Brandao, Margarida Luiza Ribeiro, Sra.
Professora
NUCLEO DE ESTUDOS SOBRE A MULHER (NEM) PUC/RJ
Rua Pacheco Leao 506, casa 42
Rio de Janeiro - RJ 22460, Brazil
T: 55 21 294-5294 F: 55 21 274-0238

Brandao, Maria Inez Alves
Journalist
HILL AND. KNOWLTON
Av. Afranio de Melo Franco 135 /5 andar Leblon
Rio de Janeiro -RJ 2, Brazil
F: 55 21 2394697

Brandao, Marilia Lopes, Sra.
Vice-Coordinator
INSTITUTO DE ECOLOGIA HUMANA
Av. Dom Manuel de Medeiros, s/n - Ed. Newton Banks 3o. andar
Recife - PE 52071, Brazil
T: 55 81 441-3931 F: 55 81 441-4697

Brandao, Mario Augusto
Reporter
LEGIAO DA BOA VONTADE
Av.Rudge, 700
Sao Paulo - SP, Brazil
T: 55 11 2202599 F: 55 11 2205196

Brandao, Zilah, Sra.
Diretoria
ASSOCIACAO MULHERES NEGOCIOS PROFISSIONAIS-LONDRINA
Rua Alagoas, 1439
Londrina PR 86020, Brazil
T: 55 0432222754

Brandat, Sandra
FREELANCER
Rio de Janeiro, Brazil

Brandi, Luiz Bernardo Moretosonn
Audio
FREELANCER
Rua Othon Bezerro de Melo, 154
Horto Florestal RJ, Brazil
T: 55 21 2206349

Brasileiro, Tula Viera, Ms.
Ass. Pedagogica
FUNDACAO FE E ALEGRIA
Rua Sao Clemente, 226 - Botafogo
Rio de Janeiro - RJ 22260, Brazil
T: 55 21 537-2621

Brasiliense, Ronaldo
SISTEMA JORNAL DO BRASIL
Av. Brasil 500
Rio de Janeiro -RJ, Brazil
T: 55 21 5854422 F: 55 21 5854428

Braun, Maria Cecilia, Sra.
Sec. Municipal
PREFEITURA MUNICIPAL NOVO HAMBURGO
Praca da Bandeira, 806
Novo Hamburgo - RS 93300, Brazil
T: 055 051 595-2929

Braune, Jorge Augusto Melo, Sr.
PARLAMENTO MUNDIAL DE ESTADO PARA SEGURANCA E PAZ
Rua Gomes Carneiro 86-702 - Ipanema
Rio de Janeiro - RJ 22422, Brazil
T: 55 21 521-7963 F: 55 21 521-7963
E: Cx. Post. 70128

Brauniger, Jorge, Sr.
Vice-Presidente
BANK OF AMERICA NT & SA
Av. Rio Branco, 110 - 39o. Andar
Rio de Janeiro - RJ 20040, Brazil
T: 55 21 224-7882 F: 55 21 231-0929
TL: 21500

Bravin, Adriana
MANCHETE
Rio de Janeiro, Brazil

Braz, Francisco Clemylton De
Photographer
ASSOCIACAO PELOS DIREITOS DE TODOS
R Coronel Camissao 1286 Cordovil
Rio de Janeiro - RJ, Brazil
T: 55 21 354 2205

Braz, Rosaura Maria, Ms.
ASSOCIACAO BRASILEIRA DE VIDEO POPULAR
Rua Joaquim Murtinho 700/102
Rio de Janeiro - RJ 21240, Brazil
T: 55 21 221 0395

Braz, Vera Nobre, Ms.
Professora
UNIVERSIDADE FEDERAL DO PARA
Rua Oliveira Belo 488
Belem - PA 66000, Brazil
F: 55 91 229-1458

Brazil, Cristina Indio
Reporter
RADIOBRAS
Brasilia, Brazil

Brazil, Eri, Sr.
Assessor De Projetos/Ce
VISAO MUNDIAL BRASIL
Tv. Para, 12 - Ed. Sulamerica - 10o. andar - Centro
Fortaleza - CE 60025, Brazil
T: 55 85 251-1331

Brazil, Luiz Eduardo
OSHO SPIRITUAL HEALTH ORGANIZATION
Rua Pinheiro Guimaraes 149 - Casa 129
Rio de Janeiro - RJ 22.281, Brazil
T: 55 21 266-5881

Brazil, Nilza Azevedo Vital, Sra.
Associada
FAMNEP
Ave Horizonte 214 J. Perola
Guaratinguera SP, Brazil
T: 012522 4181

Brazil, Ricardo Assis
Photographic Reporter
PERFIL COMUNICACAO
Rua Senador Dantas, 71/40 andar
Rio de Janeiro, Brazil
T: 55 21 2206446 F: 55 21 2407827

Breckenfeld, Silvio Guilherme Beltrao, Mr.
MISTER ORGANIZATION SUNDAE
SHLS-QL8 conj.5 casa 15
Brasilia - DF 71600, Brazil
T: 55 61 248-1123

Bredariol, Marcio
Ass.Camera
IBASE
R.Vicente de Souza, 29 Botafogo
Rio de Janeiro -RJ, Brazil
T: 55 21 2265412 F: 55 21 2860541

Breder, Celia, Sra.
Bibliotecaria
LIBRARY OF CONGRESS OFFICE
Av. Presidente Wilson, 147 2o. andar
Rio de Janeiro - RJ 20030, Brazil
T: 55 21 292-7117 F: 55 21 262-0218

Brenacher, Maria Helena, Sra.
Coordenadora
FUNDACAO PARA CONSERVACAO DA NATUREZA
av. tamadare', 1818
Campo Grande - MS 79025, Brazil
T: 55 67 3832528

Brennard, Joao Carlos
Technician
REDE GLOBO
R. Lopes Quintas 303
Rio de Janeiro -RJ 2, Brazil
T: 55 21 2947732

Breton, Binka Le
Environmental Writer
FREELANCER
Fazenda Iracambi Rosario da Limeira
Muriare MG 36887, Brazil
F: 55 32 721 3351

Briones, Mario Hugo Tapia
Technician
REDE GLOBO
R. Lopes Quintas 303
Rio de Janeiro -RJ 2, Brazil
T: 55 21 2947732 F: 55 21 2947247

Brisset, Carolyn, Sra.
KEY MARKETING BRASIL
Av. Ataulfo de Paiva 135 gr. 710
Rio de Janeiro - RJ, Brazil

Britmeier, Helmut, Mr.
Ph.D.
UNIVERSIDADE DE TUBIGEN
Rua dias da Rocha 20, apto 1006, Copacabana
Rio de Janeiro - RJ 22051, Brazil

Brito, Alberto Campos
MADE FOR TV
Rio de janeiro, Brazil

Brito, Cristovao
WTN
Rio de Janeiro, Brazil

Brito, Lazaro Jose De, Sr.
TELEBRASIL
Av. Pasteur 383 Urca
Rio de Janeiro - RJ 22290, Brazil
T: 55 21 295-4432 F: 55 21 542-4092
TL: 2123057

Brito, Orlando
Photographer
EDITORA ABRIL
R.da Passagem, 123/8 andar
Rio de Janeiro, RJ, Brazil
T: 021 5468282 F: 021 2759347

Brito, Robson Nonato Macedo De
Technician
REDE GLOBO
R. Lopes Quintas 303
Rio de Janeiro -RJ 2, Brazil
T: 55 21 2947732

Brito, Sandra Maria Linares De, Sra.
FLAMA
Rua Marques de Abrantes, 55 - Flamengo
Rio de Janeiro - RJ, Brazil
T: 55 21 265-5690

Brito, Vanessa Maria De
Jornalista
COMUNIDADE BAHA'I DO BRASIL
Rua Bela Cintra 1413/112
Sao Paulo - SP, Brazil
T: 55 11 215 4320 F: 55 11 236 9225

Britto, Evandro Rodrigues De, Mr.
Environment Coordin.
ABES
Av. Beira Mar, 216 13andar
Rio de Janeiro - RJ 20000, Brazil
T: 55 21 210.3221 F: 55 21 262.6838

Britto, Francisco Eugenio Calmon De
FUNDACAO ROBERTO MARINHO
Rio de Janeiro RJ, Brazil

Britto, Gloria Maria, Sra.
Presidente
ASSOCIACAO BRASILEIRA DE ANALISTAS DE SIMBOLOS
R. Silveira Martins, 30/401
Rio de Janeiro - RJ 22221, Brazil
T: 55 21 225-6744

Britton, Walter, Sr.
ADRA
Caixa Postal 0-2600
Brasilia - DF 70279, Brazil
T: 55 61 244-0066 F: 55 61 244-0705
TL: 1612

Brizola, Dra. Neuza, Sra.
GOVERNO DO ESTADO DO RIO DE JANEIRO
Rio de Janeiro - RJ, Brazil

Brouillard, Denise, Ms.
Latin American Project
CSN - CONFEDERATION DES SYNDICATS NATIONAUX
Caixa Postal 1759, Centro
Sao Paulo - SP 01059, Brazil
T: 55 11 270-1257

Brugioni, Raul, Sr.
Teacher
SUSILA - LIGHT TOUCH
Rua Araucaria 90/201
Rio de Janeiro - RJ 22461, Brazil
T: 55 21 286-8373 F: 55 21 512-1912

Brum, Daria Francisca
Reporter
RADIO TUPI
Rua do Livramento 189/8th Floor
Rio de Janeiro - RJ, Brazil
T: 55 21 203 2421 r 34 F: 55 21 2637323

Bruno, Helcio
Editor
REVISTA FLUMINENSE
Rua Republica do Libano, 61 Sala 1120, Centro
Rio de Janeiro RJ, Brazil
T: 55 21 2429809

Bruns, Jam
Camera
2DF-TV ALEMA
Av.Delfim Moreira, 662/704
Rio de Janeiro - RJ, Brazil
T: 55 21 2947434 F: 55 21 5122120

Brussolo Jr., Joao Batista
Com.Social
OSHO SPIRITUAL HEALTH ORGANIZATION
Rua Paissandu, 261/304
Rio de Janeiro - RJ, Brazil
T: 55 21 552-0022

Buarque, Cristovam, Sr.
President
SERVICO UNIVERSITARIO MUNDIAL
Rua Cardoso de Almeida 990 (Caixa Postal 14631)
Sao Paulo - SP 03698, Brazil
T: 55 11 263.0211 F: 55 11 263.5804

Bucci, Edson, Mr.
GOVERNO DE SAO PAULO
Sao Paulo - SP, Brazil

Buck, Alfred, Sr.
UNIVERSIDADE DE TUBINJEN
Rua Dias da Rocha 20, apto.1006
Rio de Janeiro - RJ 22051, Brazil
T: 55 21 225-0232

Buege, Tania Regina, Ms.
ABEMA
Av. Jeronimo Monteiro, 240/S. 811/812
Vitoria - ES 29010-900, Brazil
T: 55 27 322-4198 F: 55 27 222-3725

Bueno, Elza, Sra.
Presidente
ASSOCIACAO MANTENEDORA DE ORFANATOS E ASILOS
Rua Imbui 192/Fundos - Jacarepagua - Tanque
Rio de Janeiro - RJ 20060, Brazil
T: 55 21 423-1962

Bueno, Ivo
Technician
REDE GLOBO
R. Lopes Quintas 303
Rio de Janeiro -RJ 2, Brazil
T: 55 21 2947732

Bueno, Jeronymo Coimbra, Sr.
President
F.C.B. - FUNDACAO COIMBRA BUENO
Praia do Flamengo, 262 - 12o. Cobertura
Rio de Janeiro - RJ 22210, Brazil
T: 55 21 5522288

Bulcao, Joao Luiz M.
Photographer
EDITORA ABRIL
R.da Passagem, 123/8 andar
Rio de Janeiro, RJ, Brazil
T: 021 5468282 F: 021 2759347

Buneo, Julio Cesar Marcondes
Technician
REDE GLOBO
R. Lopes Quintas 303
Rio de Janeiro -RJ 2, Brazil
T: 55 21 2947732

Burgess, Chris, Ms.
GREENPEACE
Brazil
T: 55 21262-7318 F: 55 21240-1690

Burhoe, Robin Knight, Sr.
Diretor Comercial
SATELLITE SYSTEMS DO BRASIL
Av. Rio Branco, 12/ 4 Andar
Rio de Janeiro - RJ 20090, Brazil
T: 55 21 233-3788 F: 55 21 263-7047

Burlan, Paulo Procopio, Sr.
Auxiliar De Pesquisa
NEPO - UNICAMP
Caixa Postal 6166
Campinas - SP 13081, Brazil
T: 55 192 39 4000

Burnoe, Robin Knight, Sr.
Diretor
SATELLITE SYSTEMS INTERNATIONAL
Av. Rio Branco, 12/4o. andar
Rio de Janeiro - RJ 20090, Brazil
T: 55 21 847-7800 F: 55 21 263-7067

Busto, Roberto Muchuli Marques, Sr.
Expositor
LASERE - UFF
R. Almirante Ingram 105 Vila da Penha
Rio de Janeiro - RJ 21210, Brazil

Butembendur, Maria Cecilia, Sra.
Secretaria
MOVIMENTO EM DEFESA DA VIDA
Av. Guaiba 316
Porto Alegre - RS 91700, Brazil
T: 55 512 248-4008 F: 55 512 343-6671

Buttendorff, Arno Roberto, Sr.
Supervisor De Servicos De Mkt
Leao Junior S/A
Av. Getulio Vargas, 253
Curitiba - Pr 80230, Brazil
T: 55 41 233-7822 F: 55 41 222-0028
TL: 5134

Byington, Maria
ABVP
Rua Treze de Maio 489, Bela Vista
Sao Paulo SP 01327, Brazil
T: 55 11 2847862

C.Monteiro, Luiz Antonio
Assessor De Imprensa
PRO-NATURA INSTITUTO BRASILEIRO PESQ.EST.AMBIENTAIS
Rua Monte Alegre, 374
Rio de Janeiro, RJ 2, Brazil
T: 021 2243994 F: 021 2226742

Cabral, Adilson Vaz
ABVP
Rua Treze de Maio 489, Bela Vista
Sao Paulo SP 01327, Brazil
T: 55 11 2847862

Cabral, Ana Maria
Secretaria
BCO. INTERAMERICANO DE DESENVOLVIMENTO
Caixa Postal 08738
Brasilia - DF 70312, Brazil
T: 55 61 321-7848 F: 55 61 321-3112
TL: 61-1204

Cabral, Ana Maria Rezende, Ms.
Professor
UFMG - UNIVERSIDADE FEDERAL DE MINAS GERAIS
Rua Rovigo 470
Belo Horizonte - MG 31310, Brazil
T: 55 31 441-3976

Cabral, Gina Maria Da Molta Maia
ECABE PRESS
Rio de Janeiro, Brazil

Cabral, Ines Tavares, Sra.
Arquiteta
FLAMA
Rua Marques de Abrantes, 55 - Flamengo
Rio de Janeiro - RJ 22230, Brazil
T: 55 21 265-5690

Cabral, Jose Augusto, Sr.
IECAM
Dr. Moreira 250
Manaus - AM 69000, Brazil

Cabral, Luiz Alberto Caetano
MANCHETE
R. do Russel 804
Rio de Janeiro -RJ 2, Brazil
T: 55 21 2652012 F: 55 21 2855148

Cabral, Luiz Antonio, Sr.
APOENA
R. Dr. Gurgel 860
Presidente Prudente 19015-140, Brazil
T: 018221 4222 F: 22 6177

Cabral, Maria Regina Martins, Sra.
Tecnica Em Eduacacao
RAAAB - CEAAL
Rua Das Horas 134, Centro
Sao Luiz - MA 65000, Brazil
T: 55 98 221-5428 F: 55 98 232-1421

Cabral, Oscar Vieira
Photographer
EDITORA ABRIL
R.da Passagem, 123/8 andar
Rio de Janeiro, RJ, Brazil
T: 021 5468282 F: 021 2759347

Cabral Mattos, Regina Helena, Sra.
Chefe Secao Museo
PARQUE ZOOBOTANICO GETULIO VARGAS
Amilcar Falcao, 03/101 - Morro do Gato
Salvador - BA 40160, Brazil
T: 55 71 237-6008

Cabrera, Evilasio
ESTADO DE SAO PAULO
Av. Rio Branco 151 - Sobre - Loja sala 201 a 204
Rio de Janeiro RJ, Brazil
T: 55 21 2217111

Cadamuro, Maria Aparecida
OSHO SPIRITUAL HEALTH ORGANIZATION
Shcgn 713 Bl. F Ap. 104
Brasilia - DF 70.760, Brazil

Calabro, Enzo, Sr.
Cooperante
CROCEVIA
R. Barao de Lucena 140
Rio de Janeiro - RJ, Brazil
T: 55 21 537-2638

Caldas Filho, Gil Luiz, Sr.
President
FAMA
Rua Gustavo Correa Guimaraes, 8no.
Benvindos - PA 68795, Brazil
T: 55 91 225-0019 F: 55 91 225-0019

Caldos, Dulce Tupy, Sra.
GOVERNO DO ESTADO DO RIO DE JANEIRO
Rio de Janeiro - RJ, Brazil

Call, Ricardo
Editor
MANCHETE
R do Russel 804 Gloria
Rio de Janeiro - RJ, Brazil
T: 55 21 205 4527 F: 55 21 2855148

Calixto, Fernando De Moura
Reporter
REDE GLOBO
R. Lopes Quintas 303
Rio de Janeiro -RJ 2, Brazil
T: 55 21 2947732

Callegari, Giorgio, Mr.
Sacerdote
CEPE
Praca Da SE, 158
Sao Paulo - SP, Brazil
T: 55 11 36.95.71

Camara, Andre Luis
Journalist
IBASE
R.Vicente de Souza, 29 Botafogo
Rio de Janeiro -RJ, Brazil
T: 55 21 2265412 F: 55 21 2860541

Camara, Breno Pimentel
ABVP
Rua Treze de Maio 489, Bela Vista
Sao Paulo SP 01327, Brazil
T: 55 11 2847862

Camara, Henrique Dutra, Sr.
UNICEF
SBS Edificio Seguradoras 13o Andar
Brasilia - DF, Brazil
T: 55 61 224-7145 F: 55 61 224-5035

Camara, Leonor Moreira, Ms.
CEDEC
Rua Airosa Galvao 64
Sao Paulo - SP 05002, Brazil
T: 55 11 871-2966 F: 55 11 871-2123

Camargo, Ademir De Almeida
Technician
REDE GLOBO
R. Lopes Quintas 303
Rio de Janeiro -RJ 2, Brazil
T: 55 21 2947732

Camargo, Claudio
Editor
FOLHA DE SAO PAULO
Alameda Barao de Limeira 425
Sao Paulo - SP, Brazil
T: 55 11 874 2608/874 2602 F: 55 11 2231644

Camargo, Elizabeth
Editora Assistente
REVISTA FERROVIARIA
Rua Mexico, 41 Sl.904
Rio de Janeiro - RJ, Brazil
T: 55 21 2401747/1248 e 5320260 F: 55 21 2400139

Camargo, Fernando Ferreira De, Mr.
CESP
Rua da Consolacao, 1875, 9 andar, Cerq. Cesar
Sao Paulo - SP 01301, Brazil
T: 55 11 258-2165 F: 55 11 256-2101

Camargo, Jose Carlos
Reporter
MTV - RIO
Praca Nossa Senhora da Gloria 46
Rio de Janeiro -RJ, Brazil
T: 55 21 2855946 F: 55 21 2055394

Camargo, Sergio
Assistente Tecnico
UNIVERSIDADE ASPERJ
Av.N.S.Copacabana, 787/402
Rio de Janeiro - RJ, Brazil
T: 55 21 2572726 F: 55 21 2572726

Camargo Filho, Joao Manoel, Sr.
1o. Secretario
GRUPO ECOLOGICO SENTINELA DOS PAMPAS
Rua Morom, 2032 - Centro
Passo Fundo - RS 99050, Brazil

Camargo-Moro, Fernanda De, Sra.
President Director
ICMA - ICOM/MOV
Rua Professor Brandao Filho no.60-101
Rio de Janeiro - RJ 22450, Brazil
T: 55 21 2741366 F: 55 21 2595499

Camaz, Adir Dos Santos
Technician
REDE GLOBO
R. Lopes Quintas 303
Rio de Janeiro -RJ 2, Brazil
T: 55 21 2947732

Cambraia, Rosana Passos, Ms.
Representante
CAMARA MUNICIPAL MURIAE'-MG
Alameda Sao Jose, 43 Apto. 504
Muriae Cep.: 36.880, Brazil
T: 032 721-2862

Camera, Andrea La
Jornalista
RADIO UNIVERSITARIA FM
Rua Paula Frassinete, 42
Rio de Janeiro - RJ, Brazil
T: 55 21 2933847 F: 55 21 2930592

Camisao, Cristina, Sra.
FLAMA
Rua Marques de Abrantes, 55 - Flamengo
Rio de Janeiro - RJ 22230, Brazil
T: 55 21 265-5690

Camisao, Olga
IBAMA
Rio de Janeiro, Brazil

Cammarota, Jocell, Ms.
PREFEITURA MUNICIPAL DE PETROPOLIS
Rua Barao Do Rio Branco, 279
Petropolis - RJ, Brazil
T: 55 242 31-2121

Campello, Myriam, Sra.
Maria Angelica 451-301 / Jardim Botanico
Rio de Janeiro - RJ 22461, Brazil
T: 55 21 226-8499

Campos, Ana Prodan, Sra.
Presidente
INSTITUTO INTERNACIONAL DE COSMOBIOLOGIA
Rua Senador Dantas 75/2705
Rio de Janeiro - RJ 20031, Brazil
T: 55 21 220-7583

Campos, Elisa Guerra Malta, Ms.
ELETROBRAS
Av. Presidente Vargas, 642/10
Rio de Janeiro - RJ 20079, Brazil
T: 55 21 263-3137 F: 55 21 233-3248

Campos, Ezio
Reporter
FOLHA METROPOLITANA
Rua General Osorio, 60, Centro
Guarulhos, SP 0702, Brazil
T: 55 11 9689199 F: 55 11 2094787

Campos, Franklin De S.
Chefe De Reportagem
TV EDUCATIVA - RJ
Av. Gomes Freire 474 Centro
Rio de Janeiro -RJ, Brazil
T: 55 21 2426736 F: 55 21 2426740

Campos, Ivonice Aires, Ms.
CENTRAIS ELECTRICAS DO SUL DO BRASIL S.A.
Rua Deputado Antonio Edu Vieira, 353
Florianopolis - SC 88048, Brazil
T: 55 482 317105 F: 55 482 343434

Campos, Jose Olinto De Arruda, Sr.
Partner
PINHEIRO NETO ADVOGADOS
Av. Rio Branco, 131 - 9o. andar
Rio de Janeiro - RJ 20040, Brazil
T: 55 21 221 5877 F: 55 21 242 3292
TL: 23220

Campos, Leandro
TV RIO
Rio de Janeiro, Brazil

Campos, Luiza Maria De
EDITORA SARAIVA
Campinas SP, Brazil

Campos, Mariusi, Ms.
UFPA/UNAMAZ
Av. Cons. Furtado 958 - apto 705
Belem - PA 66000, Brazil
T: 55 91 229 2088

Campos, Narcisa
Reporter
FOTOGR. & DESIGNERS ASSOC
Rua Mariz e Barros, 1058, Box 12 Tijuca
Rio de Janeiro, Brazil
T: 55 21 2645435 F: 55 21 2546124

Campos, Rosanita Monteiro, Sra.
Presidente
CONFEDERACAO DAS MULHERES DO BRASIL - PRESIDENTE
Rua Avanhadava no. 756
Sao Paulo - SP 01306, Brazil
T: 55 11 2564354 F: 55 11 2577648

Campos, Sergio Jorge De
Technician
REDE GLOBO
R. Lopes Quintas 303
Rio de Janeiro -RJ 2, Brazil
T: 55 21 2947732

Campos, Vanessa
Reporter
RADIO ROQETE PINTO
Rio de Janeiro, Brazil

Campos Almeida, Marilene, Ms.
Tecnologa
UNIVERSIDADE FEDERAL DO ACRE
Campus Universitario - BR 364 Km 4
Rio Branco - AC Cep.: 69.900, Brazil
T: 55 68 226-1641

Canapini, Anita, Ms.
MISSIONARY BENEDICTINE
Rua 19 Fevereiro 172
Rio de Janeiro - RJ 22280, Brazil
T: 55 21 542-0293

Candido, Amilton, Sr.
Membro
APOENA
Rua Cuiaba, 1-19
Presid. Prudente-SP 19470-000, Brazil
T: 55 182 21-4222 F: 55 182 22-6177

Candido, Francisco, Sr.
Vice-Presidente
ORG. COOP. BRASILEIRAS
SCS, Ed. Baracat, 4. andar
Brasilia - DF Cep.: 70309-900, Brazil
T: 55 61 225 0275 F: 55 61 226 8766
TL: 1898

Candido, Sidnei
Technician
REDE GLOBO
R. Lopes Quintas 303
Rio de Janeiro -RJ 2, Brazil
T: 55 21 2947732

Canedo, Eneida V.S. Ostria, Sra.
Representante
CONFEA
Rua Parintiks Qd. 'B' - casa 4 - Parque Amazonas
Sao Luiz - MA 65030, Brazil
T: 55 69 222-6660

Caneiro, Eneraldo De Oliveira
Photographer
IMAGENS DA TERRA
R. das Laranjeiras 203/202
Rio de Janeiro RJ 2, Brazil
T: 55 21 5562502 F: 55 21 2204653

Canettiere, Marcos Barreira
INTERNATIONAL NGOs AND SOCIAL MOVEMENTS FORUM
Rio de Janeiro, Brazil

Cannabrava, Benita Beatriz Accioli, Sra.
REDE MULHER
Rua Joao Ramalho, 991 - Perdizes
Sao Paulo - SP 05008, Brazil
T: 55 11262-9407 R F: 55 11841-3457

Cannone, Karen Harris, Sra.
Training Coord.
INSTITUTO DE ASSUNTOS CULTURAIS
Av. Graca Aranha 416/1116
Rio de Janeiro - RJ 20030, Brazil
T: 55 21 2424445

Cantore, Jacqueline
Director
MTV - RIO
Praca Nossa Senhora da Gloria 46
Rio de Janeiro -RJ, Brazil
T: 55 21 2855946 F: 55 21 2055394

Capalbo, Lucio
Journalist
COMUNIDADE BAHA'I DO BRASIL
Rua Olegario Maciel 2403 Bairro Saraiva
Uberlandia MG 38400, Brazil
T: 55 34 215 4320 F: 55 34 236 92

Capato, Paulo
JOVEM PAN TV
Sao Paulo, Brazil

Capistrano, Lilian, Sra.
Coordenadora Do Projeto
PROJETO CETACEOS
Rua Miranda Valverde 103 - Botafogo
Rio de Janeiro - RJ 22281, Brazil
T: 55 21 226-5008 F: 55 21 226-1345
TL: 2137984

Capitoni, Flavio
Reporter
REDE GLOBO
R. Lopes Quintas 303
Rio de Janeiro -RJ 2, Brazil
T: 55 21 2947732

Capobianco, Elizabeth Maria S.
CEG - COMPANHIA ESTADUAL DE GAS DO RIO DE JANEIRO
Rio de Janeiro, Brazil

Caprario, Francisco X., Sr.
Biologo
UNIVERSIDADE FEDERAL DE SANTA CATARINA
Rua Frederico Veras, 36 - Pantanal
Florianopolis - SC 88020, Brazil
T: 55 482 332871

Caran, Maly
IECAM
Dr. Moreira 250
Manaus - AM 69000, Brazil

Caravello, Flavia Mantuano, Sra.
INSTITUTO PRO NATURA
Rua Monte Alegre 374 - Santa Teresa
Rio de Janeiro - RJ 20240, Brazil
T: 55 21 224-39904 F: 55 21 224-3994

Caravello, Jose Luiz Rocha
WDR GERMAN TV CHANNEL
Rua Eca de Queiroz, 479 casa 12
Sao Paulo - SP, Brazil
T: 55 11 5721262 F: 55 11 5728959

Cardeman, David, Sr.
Diretor De Inf. Gerenciais
IPLANRIO
R.Gago Coutinho, 52 / 5o. andar
Rio de Janeiro - RJ, Brazil
T: 55 21 556-1863 F: 55 21 285-2098

Cardoso, Carlos Prestes, Sr.
President
ASSOCIACAO BRASILEIRA DE ENSINO DE ENGENHARIA - ABENGE
Rua Gastao Ruch - 15
Niteroi - RJ 24220, Brazil
T: 55 21 711 5122 F: 55 21 717 4446
TL: 02121619

Cardoso, Fernando Rodrigues
Technician
FUND. CULTURAL DE RADIODIFUSAO EDUCATIVA DA REGIAO DOS LAGOS
Av. Jose Bento Ribeiro Dantas 265 Manguinhos 28905
Armacao dos Buzios -, Brazil
T: 55 246 2753848 F: 55 21 2956494

Cardoso, Gustavo, Mr.
UNESCO
SCN -Quadra 2 Lote B
Brasilia - DF 70710, Brazil
T: 55 61 223-8684 F: 55 61 322-4261

Cardoso, Marcia, Mr.
GOVERNO DE SAO PAULO
Sao Paulo - SP, Brazil

Cardoso, Maria Irony Bezerra, Sra.
Coordenadora
CEDAC
Rua Benjamin Constant, 108 Gloria
Rio de Janeiro - RJ 20241, Brazil
T: 55 21 242-9693 F: 55 21 222-2527

Cardoso, Maria Isabel Lima
APCEF - RJ
Rio de Janeiro, Brazil

Cardoso, Osvaldo, Ms.
EXPO 2000 PLANNING GROUP
Planckstr. 2
Hannover 3000, Brazil
T: 49 511 1206991 F: 49 511 1202670

Cardoso, Vania Andrade
Reporter
FUNTEVE - RADIO MEC
Praca da Republica, 141
Rio de Janeiro RJ, Brazil
T: 55 21 2204466

Cardoso Filho, Gerson
92 GLOBAL FORUM
Rio de Janeiro - RJ, Brazil

Cardozo, Euclides Lara, Mr.
Representante Da Entidade
PROMAM
Rua Rio de Janeiro 2671 a/c Milton Menancin
Medianeira - PN 85870, Brazil
T: 55 452 642144

Caresia, Gislaine, Sra.
Director
CMB
Rua Avanhadava no. 756
Sao Paulo - SP 01306, Brazil
T: 55 11 2564354 F: 55 11 2577648

Carmello, Adriane
Photographer
FREELANCER
Rua Marques de Olinda, 61/703 - B Botafogo
Rio de Janeiro, Brazil
T: 55 21 2663203

Carmo, Antonio J. Do
Reporter
AGENCIA ESTADO
Av. Caetano Alvares, 55
Sao Paulo -SP 02598, Brazil
T: 55 11 8562210 F: 55 11 2652297

Carmo, Luiz Paulo Mesquita Do
Reporter
REDE GLOBO
R. Lopes Quintas 303
Rio de Janeiro -RJ 2, Brazil
T: 55 21 2947732

Carmo, Quintino Manuel, Sr.
CREA-RJ
AV. RIO BRANCO 133 - 21o. ANDAR
Rio de Janeiro - RJ, Brazil

Carmo, Roberto Luiz De, Sr.
Research Assistant
NEPO-UNICAMP
Av. Sta. Izabel, 1.125, casa B6
Campinas - SP 13083, Brazil
T: 55 192 39-4000

Carmuto, Ana Luiza Leica Texeira De
Photographic Producer
BLOCH EDITORES
Rua do Russel, 804 Gloria
Rio de Janeiro, Brazil
T: 55 21 5554000

Carnaval, Marcelo
Fotografo
JORNAL O GLOBO
Rua Irineu Marinho, 35
Rio de Janeiro - RJ, Brazil
T: 55 21 2722000

Carneiro, Evilasio Paraense Soares
Reporter
REDE GLOBO
R. Lopes Quintas 303
Rio de Janeiro -RJ 2, Brazil
T: 55 21 2947732

Carneiro, Leonardo, Sr.
L T T
Rua Barao De Jaguaribe 199/4 Andar
Rio de Janeiro - RJ, Brazil
T: 55 21 521-5588 F: 55 21 521-6611

Carneiro, Maria Aparecida Campos, Sra.
PREFEITURA MUNICIPAL
Praca Maua s/n 1 andar
Santos - SP 11010, Brazil
T: 55 132 342-2093 F: 55 132 345-809
TL: 131048

Carneiro, Roberto, Sr.
INTERNATIONAL INSTITUTE FOR LAW, SCIENCE AND TECHNOLOGY
Av. Rio Branco, 4-18o. andar
Rio de Janeiro - RJ, Brazil
T: 55 21 253-7676 F: 55 21 253-1422

Carneiro, Tereza De Oliveira
MANCHETE
Rio de Janeiro, Brazil

Carnevale, Rosangela
OPERATOR FOTOJORNALISMO
Rio de Janeiro, Brazil

Carradas, Cassiana, Ms.
Biologa
COMP. DESENV. LIMEIRA
Rua Henrique Forster, 47
Limeira - SP 13480, Brazil
T: 0194417348

Carreira, Edimar, Sr.
MINERACOES BRASILEIRAS REUNIDAS SA MBR
Rua Rio Doce, 26 - Sao Lucas
Belo Horizonte - MG 30240, Brazil
T: 55 31 5471102-203

Carreira Filho, Orion, Mr.
SUPERINTENDENCIA DE DESENVOLVIMENTO INDUSTRIAL E COMERCIAL
Br 324 KM 18,5, Ed. Pres. Costa E Silva
Salvador - BA, Brazil
T: 55 71 594-9400 F: 55 71 594-9194
TL: 711015

Carstens, Renato Irineu, Mr.
Organizador
ASSOCIACAO RIO DE JANEIRO IASD BOTAFOGO
Rua da Matriz, 16
Rio de Janeiro - RJ, Brazil

Carvacho, Edelmira Castillo, Ms.
Nat.Gen.Secret.
YWCA DO BRASIL
AV. PAULISTA, 2073-19o. ANDAR
Sao Paulo - SP, Brazil
T: 55 11 283-3715

Carvalho, Aloisio, Sr.
Training Coord.
INSTITUTO DE ASSUNTOS CULTURAIS
Av. Graca Aranha 416/1116
Rio de Janeiro - RJ 20030, Brazil
T: 55 21 2424445

Carvalho, Cacilda Nascimento De, Sr.
Professora
UNIVERSIDADE FEDERAL FLUMINENSE
Rua Francisco Gois Calmon 420
Rio de Janeiro - RJ, Brazil
T: 55 21 393-5449

Carvalho, Carlos F. A., Sr.
FREELANCER
R.Almte.Alexandrino, 2966
Rio de Janeiro - RJ, Brazil
T: 55 21 2654990 F: 55 21 2654990

Carvalho, Celia Maria Dias De, Ms.
Av. Oswaldo Cruz 96 - Apt. 1105
Rio de Janeiro - RJ 22250, Brazil
T: 55 21 552-2469

Carvalho, Cesar
JIJI PRESS NEWS AGENCY
Av Paulista 854 - 13o. andar cj 133 Bela Vista
Sao Paulo - SP, Brazil
T: 55 11 285 0025/287 9526 F: 55 11 285 3816

Carvalho, Claudia Maria De
INSTITUTO DE ARQUITETOS DO BRASIL
Rua do Pinheiro, 10 Flamengo
Rio de Janeiro, Brazil
T: 55 21 2853480

Carvalho, Cristina Cavalcanti De, Sra.
Public Relations
INSTITUTO GAIA
Rua da Assembleia 10 s/ 3619
Rio de Janeiro - RJ, Brazil
T: 55 21 252-9420

Carvalho, Denise
TVT REDE DE COMUNICACAO DOS TRABALHADORES
Sao Paulo, Brazil

Carvalho, Domingues Ailton Ribeiro
Reporter
CIDADE SOL FM
Rua Cel Joao Borges, Edif. Multicenter, Sala 3
Brazil
T: 73 5251499

Carvalho, Edison Durval Ramos, Mr.
UNICAMP/IG/NPCT
Americana - SP 13470, Brazil
T: 55 194 621-228

Carvalho, Edson De, Sr.
Diretor
FLORESTA AMAZONICA
Av. Perimetral OESTE 2001
Alta Floresta - Mt, Brazil
T: 11 258-7258 F: 11 258-5159

Carvalho, Emygdia Maria De, Sra.
Diretora
FLAMA
Rua Marques de Abrantes, 55 - Flamengo
Rio de Janeiro - RJ 22230, Brazil
T: 55 21 265-5690

Carvalho, Feliz O. De
Reporter
AGENCIA ESTADO
Av. Caetano Alvares, 55
Sao Paulo -SP 02598, Brazil
T: 55 11 8562210 F: 55 11 2652297

Carvalho, Fernando
ESTADO DE SAO PAULO
Av. Rio Branco 151 - Sobre - Loja sala 201 a 204
Rio de Janeiro RJ, Brazil
T: 55 21 2217111

Carvalho, Fernando Mauro Mendes De, Sr.
Director
GEMA-GEOLOGIA, ENGENHARIA E MEIO AMBIENTE
R. Francisco Gois Calmon, 420
Rio de Janeiro - RJ 21940, Brazil
T: 55 21 393-5449

Carvalho, Flavio Henrique Lazaro, Mr.
PARTIDO VERDE
Camara Municipal - Praca Floriano s/n$ - gabinete 1103
Rio de Janeiro - RJ, Brazil
T: 55 21 220-5188 F: 55 21 532-1834

Carvalho, George Orlando De, Mr.
Distribuicao Material
BANCO INTERAMERICANO DE DESENVOLVIMENTO
Rua Viscondessa De Pirasununga, 39 - Casa - Estacio
Rio de Janeiro - RJ CEP: 20.211, Brazil
T: 55 21 293-9356

Carvalho, Gilberto Cavalcanti De, Mr.
Diretor
INSTITUTO GAIA
Rua da Assembleia 10 s/ 3619
Rio de Janeiro - RJ, Brazil
T: 55 21 252-9420

Carvalho, Hosana Aparecida Sartorio De, Ms.
Secretary
CENTRO DE PESQUISAS DO MAR (CEPEMAR)
Rua Mexico, 31/Gr. 501- Centro
Rio de Janeiro - RJ, Brazil
T: 55 21 240-3822 F: 55 21 240-9516

Carvalho, Isabel C. M., Sra.
Psychologist
CERIS
Rua Dr. Julio Ottoni, 571/35
Rio de Janeiro - RJ 20241-400, Brazil
T: 55 21 265-5177 F: 55 21 285-7276

Carvalho, Ivanir Da Costa
FREELANCER
Rua Baao de Sao Borja 17/801
Rio de Janeiro, Brazil
T: 55 21 2493103

Carvalho, Izabel
Pesquisadora
IBASE
Rua Vicente de Souza 29 Botafogo
Rio de Janeiro - RJ, Brazil
T: 55 21 226 5412 F: 55 21 286 0541

Carvalho, Joao Luis Homem De, Sr.
Professor
UNIVERSIDADE DE BRASILIA
Higs - 712 - P - 44
Brasilia - DF 70361, Brazil
T: 55 61 245-1516

Carvalho, Joaquim Francisco, Mr.
GOVERNO DE SAO PAULO
Sao Paulo - SP, Brazil

Carvalho, Jose Augusto Salgado Filho
SISTEMA GLOBO DE RADIO
R. de Russel 434
Rio de Janeiro RJ, Brazil
T: 55 21 2058282 F: 55 21 2054096

Carvalho, Jose Carlos, Sr.
Diretor Geral
I E F / M G
Rua Paracatu 304
Belo Horizonte - MG 30180, Brazil
T: 55 31 295-1405 F: 55 31 295-2123

Carvalho, Jose Carlos Rodrigues De
Photographer
EDITORIA MERGULHAR
Rua Prof. Alfredo Gomes, 3 Botafogo
Rio de Janeiro RJ 22, Brazil
T: 55 21 2867445

Carvalho, Jose Luis De, Sr.
Dir. Do Instituto De Ciencias
UNIVERSIDADE SANTA URSULA
Rua Fernando Ferrari, 75
Rio de Janeiro - RJ 22231, Brazil
T: 55 21 551-5542 F: 55 21 551-6446

Carvalho, Jose Ribamar De
CORREIO FLUMINENSE
Rio de Janeiro, Brazil

Carvalho, Manuel Lampreia Da
MADE FOR TV
Rio de Janeiro, Brazil

Carvalho, Marcelo Velloso De
Cinegrafista
FUND. CULTURAL DE RADIODIFUSAO EDUCATIVA DA REGIAO DOS LAGOS
Av. Jose Bento Ribeiro Dantas 265 Manguinhos 28905
Armacao dos Buzios -, Brazil
T: 55 246 2753848 F: 55 21 2956494

Carvalho, Marcia Monteiro De
Journalist
REVISTA FLUMINENSE
Rua Republica do Libano, 61 Sala 1120, Centro
Rio de Janeiro RJ, Brazil
T: 55 21 2429809

Carvalho, Marcio, Mr.
PREFEITURA MUNICIPAL DE APIACA
Rua W-3, 7
Apiacas - MT 78595, Brazil

Carvalho, Marcos Aurelio De
Coordenador Do Projeto Radio
IBASE
Rua Vicente de Souza 29 Botafogo
Rio de Janeiro - RJ, Brazil
T: 55 21 286 6161/8642 F: 55 21 286 0541

Carvalho, Marcus Renato De, Sr.
Cooerdenador
IBFAN - INTERNATIONAL BABY FOOD ACTION NETWORK
Rua Faro 35/101,
Rio de Janeiro - RJ 22461, Brazil
T: 55 21 286-0139 F: 55 21 270-2193

Carvalho, Maria Nazareth Rocha De, Ms.
ASSOCIACAO DE MEIO AMBIENTE DA REGIAO DE BANGU - AMAR-BANGU
Rua Raimundo Sjisnando, 51 - Padre Miguel
Rio de Janeiro - RJ 21710, Brazil
T: 55 21 332-5435 F: 55 21 719-6846

Carvalho, Maria Teresa, Ms.
Tecnica
FUNDACAO IPLANCE
Centro Administrativo Do Governo Do Estado Do Ceara - Cambeba
Fortaleza - CE 60000, Brazil
T: 55 85 274-1519 F: 55 85 274-1285

Carvalho, Othilia, Sra.
Rua Cosme Velho 67/1003
Rio de Janeiro - RJ 22241, Brazil
T: 55 21 265 1742

Carvalho, Priscila Assis Schueler De, Ms.
Recepcionista
IBFAN
Rua Padre Francisco Lana, 96/403 Bl.II
Rio de Janeiro - RJ, Brazil
T: 55 21 288-7771

Carvalho, Raquel Melo De, Mr.
Associada
ASPAN
Caixa Postal 7862
Recife - PE 50739, Brazil
T: 55 81 222-2038 F: 55 81 222-2038

Carvalho, Ricardo Alves
Reporter
REDE GLOBO
R. Lopes Quintas 303
Rio de Janeiro -RJ 2, Brazil
T: 55 21 2947732

Carvalho, Rosemeri De
Assistant
REVISTA FLUMINENSE
Rua Republica do Libano, 61 Sala 1120, Centro
Rio de Janeiro RJ, Brazil
T: 55 21 2429809

Carvalho, Sonia, Ms.
Assessora Para Cooperacao
MOVIMENTO NACIONAL DE MENINOS E MENINAS DE RUA
SDN - CNB 3o. andar, sala L 403
Brasilia - DF 70070, Brazil
T: 55 61 226-9631 F: 55 61 225-1577

Carvalho, Sonia Maria Solano, Sra.
UNIVERSIDADE FEDERAL DO PARA
UNAMAZ Caixa Postal 558
Belem - PA 66059, Brazil
T: 55 91 229-9310 F: 55 91 229-4339
TL: (91)1013 UPBCBR

Carvalho, Valeria, Ms.
Coord.
NUCLEO DE ESTUDOS DA AMAZONIA
UNB - Campus Universitario - Predio Multiuso Bl. A
Brasilia - DF 70000, Brazil
T: 55 61 348-2366

Carvalho, Vitoria Da Riva, Sra.
Director
FLORESTA TOUR
Rua Teodoro Baima 100-112
Sao Paulo - SP 01220, Brazil
T: 55 11 2587258 F: 55 11 2585159
TL: 1131495

Casa Nova, Cora Maria Da Costa, Sra.
Coordenador
MOVIMENTO EM DEFESA DA VIDA
Rua Santa Cecilia # 2201
Porto Alegre - RS 90000, Brazil
T: 55 512 331-9902 F: 55 512 343-6671

Cascino, Fabio, Sr.
Presidente
CLUBE ALPINO PAULISTA
Av. Ramalho Optigao, 235
Sao Paulo - SP 04130, Brazil
T: 55 11 276-0852

Case, Hamilton Paes, Mr.
Engenheiro
PLANEJAMENTO ARQUITETONICO E AMBIENTAL
Ladeira De Nossa Senhora 158 Gloria
Rio de Janeiro - RJ 22211, Brazil
T: 55 21 265-2699 F: 55 21 205-8398

Casilunha, Robson Rodrigues
MANCHETE
Rua do Russel 766
Rio de Janeiro, Brazil
T: 55 21 2850033

Casper, Gastao Vitor, Sr.
R. Barata Ribeiro 628/1001
Rio de Janeiro - RJ 22051, Brazil
T: 55 21 2171685

Cassiano, Laloa De Oliveira
Auxiliary
Av. Gomes Freire 474, Centro
Rio de Janeiro, Brazil
T: 55 21 2920012 x108

Cassio, Simone De, Sra.
INSTITUTO PRO NATURA
Rua Monte Alegre 374 - Santa Teresa
Rio de Janeiro - RJ 20240, Brazil
T: 55 21 224-39904 F: 55 21 224-3994

Castain, Bruno
Photographer
FREELANCER
Rua Ministro Viveiros Castro 51/501
Rio de Janeiro RJ, Brazil
T: 55 21 5411749 F: 55 21 5411749

Castellani, Daruiz
Director
INSTITUTO DE ARQUITETOS DO BRASIL
Rua do Pinheiro, 10 Flamengo
Rio de Janeiro, Brazil
T: 55 21 2853480

Casthro Neto, Geraldo Freire De, Sr.
INSTITUTO PRO NATURA
Rua Monte Alegre 374 - Santa Teresa
Rio de Janeiro - RJ 20240, Brazil
T: 55 21 224-39904 F: 55 21 224-3994

Castiguiola, Concetta
Chefe De Reportagem
TV EDUCATIVA - RJ
Av. Gomes Freire 474 Centro
Rio de Janeiro -RJ, Brazil
T: 55 21 2426736 F: 55 21 2426740

Castilho, Marcelo Cardosa
Reporter
RADIO MANCHETE
Rua do Russel 804 - Gloria
Rio de Janeiro, Brazil
T: 55 21 2855997 F: 55 21 2857078

Castilho, Paulo Jose, Mr.
APOENA
Rua Aquidaban 12
Presidente Prudente 19020, Brazil
T: 27 12 22-1178

Castorini, Marcio Antonio
Audio
RADIO MANCHETE
Rua do Russel 766 Gloria
Rio de Janeiro RJ, Brazil
T: 55 21 4501484 F: 55 21 2057078

Castra, Loreta Pinheiro De
Reporter
RADIO ROQETE PINTO
Rio de Janeiro, Brazil

Castrioto, Renee, Ms.
WOMEN'S CLUB DE PETROPOLIS
Rua Locarno 275 - Morim
Petropolis - RJ 25630, Brazil
T: 55 242 42-9483

Castro, Agostinho Vicente De
Tecnico
TVSBT
Campo de S.Cristovao, 105
Rio de Janeiro - RJ, Brazil
T: 55 21 5803688 / 5800678 F: 55 21 5809685

Castro, Aloysio De, Mr.
PREFEITURA DE PARATY
Rua Dr. Samuel Costa 29
Paraty - RJ 23970, Brazil
T: 55 243 71-1266 F: 55 243 71-2105

Castro, Carla Di
RADIO UNIVERSITARIA
Universidade Federal de Goias - Setor Oeste Av Anhanguera com Alameda das Rosas
Goiania - GO, Brazil
T: 55 62 233-4473/6896

Castro, Cassia Cristina De
Editor
JORNAL DO MOVIMENTO CONSERVACIONISTA TERESOPOLITANO
R Barao de Sao Francisco 245/1108 Bloco H Vila Isabel
Rio de Janeiro RJ 20, Brazil
T: 55 21 238 9259

Castro, Ewerton Pery De
MANCHETE
Rio de Janeiro, Brazil

Castro, Haroldo, Mr.
CONSERVATION INTERNATIONAL TV PRODUCTIONS
Rua Bueno Brandao 393
Belo Horizonte - MG 31010, Brazil
T: 55 31 226-5145 F: 55 31 222-8429

Castro, Heloisa N. De, Ms.
Oper. Da Reconstrucao
ASSOC.DE JORNALISTAS E ESCRITORES DE BRASIL
R. Guarara, 117 - apto 72
Sao Paulo - SP 01425, Brazil
T: 55 11 887 1704

Castro, Ivo Lubrina
Conselheiro
USAGAL
Hugo Mendonca, 295
Itaituba - PR 68180, Brazil
T: 55 91 518-1702 F: 518-1162

Castro, Jorge, Mr.
Consultor
IMATEL
Rua Oito, 345 - Recanto dos Passaros
Cuiaba - MT, Brazil
T: 55 065 386 1389 F: 55 065 321 7009
TL: 78100

Castro, Josue De, Sr.
Tesoureiro
IGREJA ADVENTISTA DO SETIMO DIA/VOZ DA PROFECIA
Rua da Matriz, 16 - Botafogo
Rio de Janeiro - RJ 22260, Brazil
T: 55 21 273-5422 F: 55 21 273-5422

Castro, Lauro Pereira De
Techinician
AGENCIA ESTADO
Av. Caetano Alvares, 55
Sao Paulo -SP 02598, Brazil
T: 55 11 8562210 F: 55 11 2652297

Castro, Manoel
REDE GLOBO
R. Lopes Quintas 303 J.Botanico
Rio de Janeiro RJ 22, Brazil
T: 55 21 2947732

Castro, Marcia Correa E
ABVP
Rua Treze de Maio 489, Bela Vista
Sao Paulo SP 01327, Brazil
T: 55 11 2847862

Castro, Pedro Mendes, Sr.
Engenheiro
C E M I G
AV. BARBACENA 1200 - 20o. B1U
Belo Horizonte - MG 30161, Brazil
T: 55 31 349-4555

Castro, Raimir De, Sr.
Driver
WORLD FOOD PROGRAMME
SCH Quadra 02 Bloco 'B' Terreo
Brasilia - DF 70710, Brazil
T: 55 61 226-7872 F: 55 61 322-7080

Castro, Ronaldo Souza De, Sr.
Professor
UNIVERSIDADE FEDERAL DO RIO DE JANEIRO
Rio de Janeiro - RJ, Brazil
T: 55 21 396-4265

Castro, Samuel De, Mr.
Programme Assistant
WORLD FOOD PROGRAMME / UN
SCN Quadra 02 Bloco 'B' Terreo
Brasilia - DF 70710, Brazil
T: 55 61 226-7872 F: 55 61 322-2080
TL: 61-1243

Castro Coimbra, Jose, Mr.
Deputado
ASSEMBLEIA LEGISLATIVA DE SAO PAULO
Av. Pedro Alvares Cabral, S/N
Sao Paulo - SP 04097, Brazil
T: 55 11 884-4446 F: 55 11 884-4760

Castro Costa, Aparecida Bisconsin, Sra.
Assessora De Gabinete
PREFEITURA MUNICIPAL
Av. Castelo Branco 1046 Centro
Pimenta Bueno 78984000, Brazil
T: 069 451-2417

Castro Sobrinho, Geraldo, Mr.
Professor-Diretor
SINDICATO DOS PROFESSORES
Av. Ruy Barbosa 22 Madre De Deus
Sao Luiz - MA 65000, Brazil
T: 55 69 222 7540

Cataldo, Sergio Luiz Do Valle
Tecnico De Audio
SISTEMA GLOBO DE RADIO
Rua do Russel, 434
Rio de Janeiro - RJ, Brazil
T: 55 21 2058282 F: 55 21 2054096

Catao, Sergio
ECOTROPIC
1880 East River Road
Rio do Sul - SC, Brazil
T: 55 478 577-6476 F: 55 478 577-1195

Catao Aguiar, Sergio, Mr.
Diretor
CENDAE
Rua Praia do Flamengo Q.01-L.7
Lauro Frestas - BA 42700, Brazil
T: 55 71 379-0525 F: 55 71 379-0525

Catarino, Marcio Dos Santos
REDE RECORD
Av. Miruna 713
Sao Paulo _SP 04084, Brazil
T: 55 11 5429000 F: 55 11 612228

Cavalcante, Amaro Jose
Diretor
JORNAL SOL VIAGEM TURISMO
Caixa Postal 172
Maua - SP, Brazil
T: 55 011 450 4854 F: 55 011 477 4881 (Su

Cavalcante, Antonio Maria Filgueiras, Sr.
Professor
UNIVERSIDADE FEDERAL DO PARA
Rua Diogo Moia 407
Belem - PA 66030, Brazil
T: 55 91 222-3980

Cavalcante, Guido
Editor De Video
2DF-TV ALEMA
Av.Delfim Moreira, 662/704
Rio de Janeiro - RJ, Brazil
T: 55 21 2947434 F: 55 21 5122120

Cavalcante, Orlando Ferreira
Technician
REDE GLOBO
R. Lopes Quintas 303
Rio de Janeiro -RJ 2, Brazil
T: 55 21 2947732

Cavalcante, Sandra Maria, Sra.
VISAO MUNDIAL
Rua do Carmo, 27/601
Rio de Janeiro - RJ 20011, Brazil
T: 55 21 221-1772

Cavalcante, Wagner Jorge Silva
MANCHETE
R. do Russel 804
Rio de Janeiro -RJ 2, Brazil
T: 55 21 2652012 F: 55 21 2855148

Cavalcanti, Alcyr
SISTEMA JORNAL DO BRASIL
Av. Brasil 500
Rio de Janeiro -RJ, Brazil
T: 55 21 5854422 F: 55 21 5854428

Cavalcanti, Guarani De Hollanda, Sr.
Rua Barao de Guaratiba 108/102 Gloria
Rio de Janeiro - RJ 22211, Brazil
T: 55 21 265-2532

Cavalcanti, Jose Tupinamba Vidal
Photographer
AGENCIA ESTADO
Av. Caetano Alvares, 55
Sao Paulo -SP 02598, Brazil
T: 55 11 8562210 F: 55 11 2652297

Cavalcanti, Marco Antonio
SISTEMA JORNAL DO BRASIL
Av. Brasil 500
Rio de Janeiro -RJ, Brazil
T: 55 21 5854422 F: 55 21 5854428

Cavalcanti, Maria Jose Marques, Ms.
Arquiteta
FIDEM
Rua Bebinho Salgado 60 Apipucos
Recife - PE 52071, Brazil
T: 55 81 268-5549

Cavalcanti, Maria Suzanete Da Silva, Ms.
SIND DE PURIFICACAO DE AGUA E ESGOTO - STIPDASE
Av. Carneiro da Cunha, 1036 - Torre
Joao Pessoa - PB 58030, Brazil
T: 55 83 224 8531

Cavalcanti, Matilde, Sra.
Professora
Rua Fernando Osorio, 2 Ap 34 Flamengo
Rio de Janeiro - RJ 22.230, Brazil
T: 55 21 552-4022

Cavalcanti, Ronaldo Camara, Sr.
Assessor De Meio Ambiente
CHESF
Rua Bebinho Salgado , 60 Apipucos
Recife - PE 52071, Brazil
T: 55 81 268 5549

Cavalin, Paulo Fernando, Sr.
FLAMA
Rua Marques de Abrantes, 55 - Flamengo
Rio de Janeiro - RJ 22230, Brazil
T: 55 21 265-5690

Cavallo, Claudia
CHICO JR. & ASSOCIADOS
Rio de Janeiro, Brazil

Cechin, Antonio, Sr.
Assessor
ASSOCIACAO DE MULHERES PAPELEIRAS TRABALHADORAS EM GERAL
Porto Alegre - RS, Brazil

Cechin, Matilde, Sra.
Professora
ASSOCIACAO DE MULHERES PAPELEIRAS TRABALHADORAS EM GERAL
R. Coronel Vicente 444 /130
Porto Alegre - RS 90030, Brazil
T: 55 512 2219021 F: 55 512 2275229

Celano, Sandra, Sra.
Assessora
INSTITUTO DE TECNOLOGIA INTUITIVA E BIO-ARQUITETURA
Rua Ingles de Souza, 296
Rio de Janeiro - RJ, Brazil
T: 55 21 274-1762 F: 55 21 295-0580

Cepeda, Renan
SISTEMA JORNAL DO BRASIL
Av. Brasil 500
Rio de Janeiro -RJ, Brazil
T: 55 21 5854422 F: 55 21 5854428

Cepelowicz, Luciana
BBC TV
Sao Paulo, Brazil

Cepollina, Elio, Mr.
SINDICATO DAS INDUSTRIAS DE ESTAMPARIA DE METAIS DO ESTADO DE SAO PAULO
Av Paulista 1313 - 8o. andar, C 804, Bela Vista
Sao Paulo - SP 01311, Brazil
T: 55 11 2892856 F: 55 11 2843971

Cerqueira, Cristina F. M. De, Sra.
P N U D
Scn Q-2 Bl-B
Brasilia - DF 70710, Brazil
T: 55 61 321-5758 F: 55 61 225-9339

Cerqueira, Joao
SISTEMA JORNAL DO BRASIL
Av. Brasil 500
Rio de Janeiro -PJ, Brazil
T: 55 21 5854422 F: 55 21 5854428

Cerqueira, Roberto De Azevedo G.
Photographer
OPERATOR FOTOJORNALISMO
R. Humaita 18/105
Rio de Janeiro -RJ 2, Brazil
T: 55 21 2463853

Cesar, Aluisio
92 GLOBAL FORUM
IPC Staff
Brazil

Cesario, Manuel, Mr.
Consultant
JOVEM TURISMO
SCLN 202 Bloco B Lojas 7-11 - Asa Norte
Brasilia - DF 70832, Brazil
T: 55 61 224-5526 F: 55 61 321-0345

Chacon, Monica Fitzpatrick Da C.Pan
Correspondente Internacional (
SISTEMA JORNAL DO BRASIL
Av.Brasil, 500
Rio de Janeiro - RJ, Brazil
T: 55 21 5854422

Chagas, Behula Spencer
Reporter
O DIA
R. Riachuela 359
Rio De Janeiro, RJ 2, Brazil
T: 55 21 272-8000 F: 55 21 507-1228

Chagas, Denise Machado, Sra.
Secretaria
UNIAO ESTE BRASILEIRA DAS IASD
Av. Sete de Setembro, 69
Niteroi - RJ Cep.: 24.300, Brazil
T: 55 21 711-1062 F: 55 21 611-1571

Chagas, Elizabeth Sheila Campos
JORNAL O FLUMINENSE
Niteroi, Brazil

Chagas, Francisco Silva
Technician
REDE GLOBO
R. Lopes Quintas 303
Rio de Janeiro -RJ 2, Brazil
T: 55 21 2947732

Chagas, Marcelo
Editor
FUND. CULTURAL DE RADIODIFUSAO EDUCATIVA DA REGIAO DOS LAGOS
Av. Jose Bento Ribeiro Dantas 265 Manguinhos 28905
Armacao dos Buzios -, Brazil
T: 55 246 2753848 F: 55 21 2956494

Chagas, Maria Helena S. C., Sra.
Bibliotecaria
ELETRONORTE
Scn-Q06, Conj.A, Bloco C-Venancio 3000, Sala 313
Brasilia - DF 70000, Brazil
T: 55 61 212-6157 F: 55 61 225-9755
TL: 1279

Chagas, Mauro, Mr.
GEOSPHRE PROJECT
Av. 9 de Julho 5049/11o.
Sao Paulo - SP 01407, Brazil
T: 55 11 881-6135 F: 55 11 893-938

Chagas, Renata Monteiro Das
Director
FUND. CULTURAL DE RADIODIFUSAO EDUCATIVA DA REGIAO DOS LAGOS
Av. Jose Bento Ribeiro Dantas 265 Manguinhos 28905
Armacao dos Buzios -, Brazil
T: 55 246 2753848 F: 55 21 2956494

Chaloub, Regina, Sra.
Gerente
CEPPD - O SEGUNDO BRASIL
Av. Rio Banco 43/5o. andar
Rio de Janeiro - RJ 20090, Brazil
T: 55 21 263-7798

Chapada, Guarim De Lorna
Photographer
FREELANCER
Av. Itaoca 1908 Travessa Monterey 211
Inhauma -RJ, Brazil
T: 55 21 2802972 F: 55 21 2802972

Chatawy, Mark
HILL AND KNOWLTON
Rio de Janeiro, Brazil

Chaves, Miriam, Ms.
REDES AT
R. Timoteo da Costa 197/401
Rio de Janeiro - RJ, Brazil
T: 55 21 274-7850 F: 55 21 511-2724

Chaves, Ricardo De Leoni
Photographer
AGENCIA ESTADO
Av. Caetano Alvares, 55
Sao Paulo -SP 02598, Brazil
T: 55 11 8562210 F: 55 11 2652297

Chaves, Rogerio, Sr.
Coordenador De Distribuicao
INSTITUTO CAJAMAR
Via Anhanguera, Km 46,5
Cajamar - SP 07750, Brazil
T: 55 11 437-3098 F: 55 11 437-4343

Chaves, Sandra Vasconcellos
CHICO JR. & ASSOCIADOS
Rio de Janeiro, Brazil

Chaves, Sergio Augusto De Miranda, Sr.
President
CEDRE DO BRASIL
Rua Benjamin Constant 120
Rio de Janeiro - RJ 20241, Brazil
T: 55 21 553-3645 F: 55 21 239-2705

Chaves, Tiana Sento Se, Mr.
COPACOBANA PALACE
Rio de Janeiro - RJ, Brazil

Chaves Soares, Geralda, Ms.
CENTRO DE DOCUMENTACAO
Rua Tiradentes, 2564 Sala 7 Bairro Industrial
Contagem, Brazil
T: 031 3337683

Cheok Hun, Looi, Mr.
Ambassador
AMBASSADOR OF MALAYSIA
Brasilia
Brazil

Cheregati, Roberto Brigido
TWENTY FIRST CENTURY INC.
Rio de Janeiro, Brazil

Chermont, Claudia, Ms.
UNDP/UNESCO
Estrada da Gavea 655/1601
Rio de Janeiro - RJ 22450, Brazil
T: 55 21 322-4971

Chervezan, Sirlene
Reporter
SISTEMA GLOBO DE RADIO
Rua do Russel, 434
Rio de Janeiro - RJ, Brazil
T: 55 21 2058282 F: 55 21 2054096

Chetan Luiz, Sw, Mr.
OSHO SPIRITUAL HEALTH ORGAN. - OSHO
R. Celestino Basilio 26
Rio de Janeiro - RJ, Brazil
T: 55 21 437-7488

Chiapeti, Maria Isabel, Sra.
FEPAM
Av. A.J. Renner, 10
Porto Alegre - RS 90250, Brazil
T: 55 512 343-5799 F: 55 512 342-0224

Chiara, Jose Felipe De Paiva, Sr.
Presidente
ASSOC. BRAS. PREV. ACID
Av. Rio Branco, 37 18 andar
Rio de Janeiro - RJ 20090, Brazil
T: 55 21 233 9033 F: 55 21 263 0763
TL: 2139071

Chiaranda, Roberto, Mr.
Professor
UFMT
Av. Fernando Correa da Costa s/n
Cuiaba - MT 78060-900, Brazil
T: 55 065 315 8630 F: 55 065 361 1119

Chicarino, Antenor Placido Carvalho, Sr.
Deputado
ALESP
Av. Pedro Alvarez Cabral S/#, Ibirapuera
Sao Paulo - SP 04097, Brazil
T: 55 11 884-4423

Chindler, Ana, Sra.
Diretora
CENTRO DA MULHER BRASILEIRA - GALERIA PE DE BOI
Rua Ipiranga 55 - Laranjeiras
Rio de Janeiro - RJ, Brazil
T: 55 21 285-4395 F: 55 21 224-1035

Chip, Molley
HILL AND KNOWLTON
Rio de Janeiro, Brazil

Cho, Sei Young, Sr.
Director
SUNKYONG
Av. Paulista, 2073 Horsa II - Cj. 44
Sao Paulo - SP 01311, Brazil
T: 55 11 283-34308 F: 55 11 284-6050
TL: 38095

Christiano, Raul, Sr.
Secretario Do Ma
PREFEITURA MUNICIPAL DE CUBATAO
Praca dos Emancipadores s/no. - Centro
Cubatao 11500, Brazil
T: 55 613131 F: 55 611873
TL: 131278

Chveidel, Henrique
Journalist
FREELANCER
R. Santa Amelia 50 Bl/2 apt. 1008 Catumbi
Rio de Janeiro RJ 2, Brazil
T: 55 21 2738726

Cicliato, Ronaldo Natalino, Sr.
Membro
APOENA
Rua Dr.Gurgel, 860
Presid. Prudente-SP 19015-140, Brazil
T: 55 182 21-4222 F: 55 182 22-6177

Cintra, Heitor B., Mr.
SOCIEDADE BRASILEIRA DE ESPELEOLOGIA
Rua Joao Juliao 296/11A
Sao Paulo - SP 01323, Brazil
T: 55 11 289-6441 F: 55 11 289-8580

Civitate, Hector, Sr.
Secretario Geral
ASSOCIACAO CRISTA DE MOCOS
Rua da Lapa 86
Rio de Janeiro - RJ, Brazil
T: 55 21 242-2074

Clajus, Gerhard, Sr.
Capelao
UNIAO ESTE BRASILEIRA DAS IASD
Rodovia para Ibiuna, Km 59,5
Ibiuna Cep.: 18.140, Brazil
T: 55 011 425 7100

Clapp, Jorge
ASSOCIACAO COMERCIAL DO RIO DE JANEIRO
Rio de Janeiro, Brazil

Clares, Jose Augusto Bernardino
Operador De Vt
REDE BANDEIRANTES DE TELEVISAO
R.Alvaro Ramos, 492
Rio de Janeiro - RJ, Brazil
T: 55 21 2122072 F: 55 21 2957391

Clavia, Gilmar Da Fonseca
FOLHA VERDE
Av. Basil 1831 sala 606
Belo Horizonte -MG, Brazil
T: 55 31 2263700 F: 55 31 2251386

Clutilde, Maria, Sra.
FUNDACAO ESCOLA DE SERVICO PUBLICO RJ - FESP
Av. Carlos Peixoto, 64
Rio de Janeiro - RJ 22290, Brazil

Cobas, Veronica Pereira Garcia
AGENCIA TEXTO E ARTE
Rio de Janeiro, Brazil

Codes, Jose De, Dr.
Contry Representative
PATHFINDER INTERNATIONAL / BRAZIL
Rua Alagoinhas, 363, Rio Vermelho
Salvador - BA 41910, Brazil
T: 55 71 245-3255 F: 55 71 247-9446

Coelho, Alfredo Da Silva, Sr.
PREFEITURA MUNICIPAL
Praca Maua s/n 1 andar
Santos - SP 11010, Brazil
T: 55 132 342-2093 F: 55 132 345-809
TL: 131048

Coelho, Carlos Alberto Almeida
FORUM DO FUTURO
Rio de Jaeiro, Brazil

Coelho, Eliomar, Sr.
Vereador
CAMARA MUNICIPAL
Pca. Floriano, S/N Sala 705
Rio de Janeiro - RJ 20031, Brazil
T: 55 21 2620789

Coelho, Elizabeth, Ms.
Tec.Nivel Apoio
SERVICO SOCIAL DO COMERCIO
Rua Voluntarios Da Patria, 169
Rio de Janeiro - RJ 22270, Brazil
T: 55 21 286-5152 F: 55 21 286-2638

Coelho, Helder Marques De, Sr.
PREFEITURA MUNICIPAL
Praca Maua s/n 1 andar
Santos - SP 11010, Brazil
T: 55 132 342-2093 F: 55 132 345-809
TL: 131048

Coelho, Luiz Alberto De Oliveira, Sr.
R Moreira Cesar 220 - Lava Pes
Paraiba do Sul - RJ, Brazil
T: 55 242 631421

Coelho, Maria Das Gracas, Sra.
Tec. Ed. Amb.
SOCIEDADE POTIGUAR DE EDUCACAO - SPEL
Rua Floriano Peixoto, 295, Petropolis
Natal - RN 59020, Brazil
T: 55 84 222-2314 F: 55 84 221-6255

Coelho, Marinete Dos Santos, Sra.
Therapist
Rua Antonio Parreiras, 28/302 Boa Viagem
Niteroi - RJ 24210, Brazil
T: 55 21 722-1064

Coelho, Ricardo Mendes
FREELANCER
Rio de Janeiro, Brazil

Coelho, Silvia Helena, Ms.
Associada
ASSOCIACAO DOS AMIGOS DA TERRA
Rua Jaime Calado L.2Q.4 - Lagoinha - Alcantara
Rio de Janeiro - RJ, Brazil
T: 55 21 463-2645

Coelho, Vitoria, Sra.
Assessora
SECRETARIA DO MEIO AMBIENTE / SEMAM / PR
Esplanada dos Ministerios -Bl. B- 5o. andar
Brasilia-DF 70000, Brazil
T: 061 2230129 F: 2264345
TL: 1548

Coelho, Wander Oliveira Duarte, Mr.
PREFEITURA DE PARATY
Rua Dr. Samuel Costa 29
Paraty - RJ 23970, Brazil
T: 55 243 71-1266 F: 55 243 71.2105

Coeli, Regina Medina
CONFEDERACAO NACIONAL DA INDUSTRIA
Av. Nilo Pecanha 50 - 34$ andar
Rio de Janeiro - RJ 20044, Brazil
T: 55 21 292-7766 F: 55 21 262-1495
TL: (5521) 22634

Coello, Maria Mercedes
Produtora
WDR GERMAN TV CHANNEL
Rua Eca de Queiroz, 479 casa 12
Sao Paulo - SP, Brazil
T: 55 11 5721262 F: 55 11 5728959

Coffin, Corinne, Sra.
Presidente
REFRIGERANTES NITEROI S/A
Rua Dr. Luiz Palmier, 280 - Barreto
Niteroi - RJ 24110, Brazil
T: 55 21 717-5476 F: 55 21 622-1126
TL: 32120

Cognolato, Rosana, Sra.
Geografa
ESPACO - FORMACAO
Rua Osvaldo Caetano Tosi, 183 - Cidade Dutra
Sao Paulo - SP 04810, Brazil
T: 55 11 522-3238 F: 55 11 512-8646

Cogo, Simone, Sra.
CENTRO DE TERAPIAS HOLISTICAS - SUSILA
Estrada Muriqui, 360
Niteroi - RJ, Brazil
T: 55 21 616-2342

Cohen, Deborah, Ms.
C.E.D.I
R. Viuva Lacerda 433/501
Rio de Janeiro - RJ 22261, Brazil
T: 55 21 266-0360

Coimbra, Carlos Antonio, Sr.
Diretor
REFRIGERANTES NITEROI S/A
Rua Dr. Luiz Palmier, 280 - Barreto
Niteroi - RJ 24110, Brazil
T: 55 21 717-5476 F: 55 21 622-1126
TL: 32120

Coimbra, Custodio
Fotografo
JORNAL O GLOBO
Rua Irineu Marinho, 35
Rio de Janeiro - RJ, Brazil
T: 55 21 2722000

Cola, Lucia Oliveira, Ms.
WOMEN'S CLUB DE PETROPOLIS
Rua Locarno 275 - Morim
Petropolis - RJ 25630, Brazil
T: 55 242 42-9483

Colaco, Maria, Ms.
Visconde de Albuquerque 1350/201
Rio de Janeiro - RJ 22040, Brazil
T: 55 21 274-0435

Colares, Jose Goncalves
Technician
REDE GLOBO
R. Lopes Quintas 303
Rio de Janeiro -RJ 2, Brazil
T: 55 21 2947732

Cole, M., Ms.
UNESCO
SCN -Quadra 2 Lote B
Brasilia - DF 70710, Brazil
T: 55 61 223-8684 F: 55 61 322-4261

Coleman, John
Producer
BBC TELEVISION NEWS
Rua Tavares Bastos 414/66 Catete
Rio de Janeiro RJ 22, Brazil
T: 55 21 2255547 F: 55 21 2853109

Colhaco, Claudia, Ms.
Travel Agent
CENTRO DE PESQUISAS DO MAR (CEPEMAR)
Rua Mexico, 31/Gr. 501- Centro
Rio de Janeiro - RJ, Brazil
T: 55 21 240-3822 F: 55 21 240-9516

Colhaco, Tania, Ms.
CENTRO DE PESQUISAS DO MAR (CEPEMAR)
Rua Mexico, 31/Gr. 501- Centro
Rio de Janeiro - RJ, Brazil
T: 55 21 240-3822 F: 55 21 240-9516

Colonesi, Sergio Rogerio
MANCHETE
R. do Russel 804
Rio de Janeiro -RJ 2, Brazil
T: 55 21 2652012 F: 55 21 2855148

Conceicao, Nilson, Mr.
Porteiro
SERVICO SOCIAL DO COMERCIO
Rua Voluntarios Da Patria, 169
Rio de Janeiro - RJ 22270, Brazil
T: 55 21 286-5152 F: 55 21 286-2638

Concleiro, Flavio, Sr.
Diretor De Planejamento
SERVICO SOCIAL DO COMERCIO-DEPARTAMENTO NACIONAL
Rua Voluntarios da Patria 169 Botafogo
Rio de Janeiro - RJ, Brazil
T: 55 21 286-5152 F: 55 21 286-2638

Consolo, Dirce, Sra.
PAX UNIVERSAL
R. Alcantilado, 13
Sao Paulo - SP 02979, Brazil
T: 55 11 875 9994

Contaeffer, Wilson Da Silva
Tecnico
TVSBT
Campo de S.Cristovao, 105
Rio de Janeiro - RJ, Brazil
T: 55 21 5803688 / 5800678 F: 55 21 5809685

Contente, Lilian De Aguiar
Geografa
IBGE
Sen. Teotonio Vilela, 263 Apto. 902
Salvador - BA 41960, Brazil
T: 55 71 358-8026

Cony, Amauri
FOX FILMES DO BRASIL
Rio de Janeiro, Brazil

Cordani, Umberto G., Sr.
UNIVERSIDADE DE SAO PAULO
Cx. Postal 20899
Sao Paulo - SP 01498, Brazil
T: 55 11 813 8777 F: 55 11 210 4958

Cordeiro, Adilson Barbosa
Technician
REDE GLOBO
R. Lopes Quintas 303
Rio de Janeiro -RJ 2, Brazil
T: 55 21 2947732

Cordeiro, Elmano Augusto
Reporter
S/A CORREIO BRAZILIENSE
R. do Livramento 158/4 andar
Rio de Janeiro -RJ, Brazil
T: 61 3212123 F: 061 3213864

Cordeiro, Flor
Photograpaher
REUTERS NEW SERVICE
Rua Sete de Setembro 99/4th floor Centro
Rio de Janeiro - RJ, Brazil
T: 55 21 5074151 F: 55 21 5072120

Cordeiro, Joao Roberta Ripper B.
Editor
IMAGENS DA TERRA
R. das Laranjeiras 203/202
Rio de Janeiro RJ 2, Brazil
T: 55 21 5562502 F: 55 21 2204653

Cordeiro, Ronaldo De Paula
Reporter
REDE GLOBO
R. Lopes Quintas 303
Rio de Janeiro -RJ 2, Brazil
T: 55 21 2947732

Cornelio, Paulo Fernando Oliveira, Sr.
President
GRUPO ECOLOGICO SENTINELA DOS PAMPAS
Rua Morom, 2032 - Centro
Passo Fundo - RS 99050, Brazil

Corral, Ariel
Assistente
DE VOLKSKRANT/REVISTA ELSEVIER/TV NOS/VARA TV
Rua do Russel 300/701 Gloria
Rio de Janeiro - RJ, Brazil
T: 55 21 205 5049 F: 55 21 205 4951

Corral, Cikissie Bazan, Sra.
Membro
APOENA
Rua 7 de Setembro, 639
Presid. Prudente-SP 19100, Brazil
T: 55 182 22-2958

Corral, Thais
Coordenadora
CEMINA
R.Barao do Flamengo, 22/304
Rio de Janeiro - RJ, Brazil
T: 55 21 2056297 F: 55 21 5563383

Correa, Algemiro Gomes
Technician
REDE GLOBO
R. Lopes Quintas 303
Rio de Janeiro -RJ 2, Brazil
T: 55 21 2947732

Correa, Ana Cristina
Journalist
FASE
Rua Bento Lisboa
Rio de Janeiro, Brazil
T: 55 21 2852998

Correa, Armezinda Rosa, Ms.
Av. Nossa Senhora de Copacabana, 420-710
Rio de Janeiro - RJ 22020, Brazil
T: 55 21 257-3285

Correa, Elidio De Barros
Technician
REDE GLOBO
R. Lopes Quintas 303
Rio de Janeiro -RJ 2, Brazil
T: 55 21 2947732

Correa, Franklin B., Sr.
Director
CONSELHO FISCAL AEVRE
Praca Savio Gama, 53
Volta Redonda - RJ, Brazil
T: 55 243 42-0234

Correa, Jacques De Castro, Sr.
Teacher
ESCOLA DAS NACOES
SHIS QI 21 Area Especial Lote C-01 Caixa Postal 7093
Brasilia - DF 71600, Brazil
T: 55 61 366-1800 F: 55 61 366-3377

Correa, Javro De Araujo
Photographer
JORNAL DO PARLIAMENTAR
Evaristo da Veiga 35 8 andar
Rio de Janeiro, Brazil
T: 55 21 2320711

Correa, Jose Westphalen, Sr.
Deputado
ASSEMBLEIA LEGISTATIVA
Pca Marechal Deodoro s/no.
Porto Alegre - RS 90010, Brazil
T: 55 512 221-4260 F: 55 512 228-6740

Correa, Renata, Ms.
CONFEDERACAO NACIONAL IND.
Nilo Pecanha,50/andar 38
Rio de Janeiro - RJ 22250, Brazil
T: 55 21 552-0099 F: 55 21 552-9392

Correa, Rosmary, Ms.
Deputada Estadual
ASSEMBLEIA LEGISL.
Av. Pedro Alvares Cabral S/#
Sao Paulo - SP, Brazil
T: 55 11 884-2748 F: 55 11 8842748

Correia, Berenice Maria
Technician
FUND. CULTURAL DE RADIODIFUSAO EDUCATIVA DA REGIAO DOS LAGOS
Av. Jose Bento Ribeiro Dantas 265 Manguinhos 28905
Armacao dos Buzios -, Brazil
T: 55 246 2753848 F: 55 21 2956494

Correia, Maria Helena B. C. Arrochelas, Sra.
Diretora
CAALL
Rua Mosela, 289
Petropolis - RJ 25675, Brazil
T: 55 242 42-6332 F: 55 242 42-6332

Correia, Rosimeire De Castro, Ms.
Assessora
CONSELHO NACIONAL DOS SERINGUEIROS
Av. 28 de Setembro,728
Cruzeiro do Sul AC 69980, Brazil
T: 68 3223326 F: 68 3222942

Correia, Yedre De Franca
Journalist
REVISTA MANCHETE
Praia do Russel, 804, 8 andar, Gloria
Rio de Janeiro, Brazil
T: 55 21 2850033 F: 55 21 2056196

Corso, Fernando Del
REDE GLOBO
R. Lopes Quintas 303 J. Botanico
Rio de Janeiro - RJ, Brazil
T: 55 21 5292000 F: 55 21 2942042

Corta, Maria Ines Rodrigues De, Ms.
ASSOCIACAO CIVIL GREENPEACE
Rua Mexico 21, Grp 1301, Centro
Rio de Janeiro - RJ 20031, Brazil
T: 55 21 262-7318 F: 55 21 240-1690

Cortes, Carlos Augusto Angelo
Cinegrafista
TV Rio
R. Miguel de Frias, 57 Estacio
Rio de Janeiro - RJ, Brazil
T: 55 21 2939888 F: 55 21 2938626

Cortes, Iaris Ramalho, Sra.
Representative
ASSOCIACAO BRASILIENSE DE COMBATE A AIDS - GRUPO ARCO-IRIS
Av. W-3 Norte, Q. 701 - Ed. Brasilia Radio Center sala 1130
Brasilia - DF 70710, Brazil

Cortes, Jose Villarino, Mr.
Aguiar de Andrade 14/30
Santos - Paqueta, Brazil
T: 327 111 322621

Cortez Carnauba, Manoel, Sr.
Assessor
SALGEMA S/A
Maceio - AL 57010, Brazil
T: 55 82 221 5422 F: 55 82 221 5209

Cosenza, Emilio
Reporter
ANSA
Av. Pres. Antonio Carlos, 40
Rio de Janeiro, Brazil
T: 55 21 2626574 F: 55 21 2626574

Cosenza, Silvia
Reporter
RADIO ROQETE PINTO
Rio de Janeiro, Brazil

Costa, Ana Maria, Sra.
Diretora
SIND. DOS MEDICOS D.F.
SBS - Ed. Seguradoras, andar 9
Brasilia - DF, Brazil
T: 55 61 2260027 F: 55 61 2259694

Costa, Ana Paula Ferreira Da
HELIO ALONSO
Rua Muniz Barreto, 51
Rio de Janeiro RJ, Brazil
T: 55 21 5515448

Costa, Ana Paula Gouvea, Sra.
Rua Das Laranjeiras 441/702
Rio de Janeiro - RJ 22240, Brazil
T: 55 21 2650795

Costa, Antonio Aurelio Oliveira, Sr.
Diretor De Programas
FUNDO CRISTAO PARA CRIANCAS
Rua Curitiba 689 - 5o. andar
Belo Horizonte - MG 30170, Brazil
T: 55 31 201-4955 F: 55 31 212-6957

Costa, Antonio Jose F.Da
Tecnico
TVSBT
Campo de S.Cristovao, 105
Rio de Janeiro - RJ, Brazil
T: 55 21 5803688 / 5800678 F: 55 21 5809685

Costa, Aparecido Antonio Carvalho, Mr.
Gerente De Marketing
FUNDACAO FLORESTAL
Av. Miguel Stefano 3900
Sao Paulo - SP 04301-903, Brazil
T: 55 11 577-2240 F: 55 11 276-6799

Costa, Aristoteles J. Riani, Mr.
Rua 5
Rio Claro - SP 13500, Brazil
T: 55 195 242706 F: 55 195 717957

Costa, Carla Gomez Da, Ms.
Recepcionista
CERJ
Rua Nascimento Silva 39 - Apt.201
Rio de Janeiro - RJ 20000, Brazil
T: 55 21 287-1662

Costa, Cassio Vieira Da
MANCHETE
R. do Russel 804
Rio de Janeiro -RJ 2, Brazil
T: 55 21 2652012 F: 55 21 2855148

Costa, Cremilda Luiza De Almeida
Educadora
MUSA - CENTRO DE EDUCACAO E SAUDE DA MULHER
Rua Formosa 48
Belo Horizonte - MG 31010, Brazil
T: 55 31 467-5875

Costa, Delaine Martins, Sra.
INSTITUTO BRASILEIRO DE ADMINISTRACAO MUNICIPAL - IBAM
Edificio Diogo Lordello de Mello. Largo IBAM no.1. Humaita
Rio de Janeiro - RJ 22282, Brazil
T: 55 21 266-6622 F: 55 21 537-1262
TL: 22638 INBM BR

Costa, Edna, Sra.
Director
CMB
Rua Avanhadava no. 756
Sao Paulo - SP 01306, Brazil
T: 55 11 2564354 F: 55 11 2577648

Costa, Elaine
INFORMATIVO 'BRASIL-MEIO AMBIENTE E DESENVOLVIMENTO'
Rio de Janeiro, Brazil

Costa, Elizabeth Guimaraes
Photographer
FREELANCER
Rua Bias Fortes, 15
Rio de Janeiro, Brazil
T: 55 21 5904906

Costa, Francisco O. A.
Reporter
AGENCIA ESTADO
Av. Caetano Alvares, 55
Sao Paulo -SP 02598, Brazil
T: 55 11 8562210 F: 55 11 2652297

Costa, Geraldo George De Mello
Technician
REDE GLOBO
R. Lopes Quintas 303
Rio de Janeiro -RJ 2, Brazil
T: 55 21 2947732

Costa, Givanildo Ramos, Mr.
UNIVERSALA ESPERANTO-ASOCIO
R. Santa Alexandrina 481/606
Rio de Janeiro - RJ, Brazil
T: 55 21 273-8108

Costa, Guilherme Alves Da
SISTEMA GLOBO DE RADIO
R. de Russel 434
Rio de Janeiro RJ, Brazil
T: 55 21 2058282 F: 55 21 2054096

Costa, Ione
CHICO JR. & ASSOCIADOS
Rio de Janeiro, Brazil

Costa, Jaci De Souza, Ms.
GOVERNO DE SAO PAULO
Sao Paulo - SP, Brazil

Costa, Jurandir Noe Da
Photographer
JORNAL DO COMERCIO
Rua do Lisamento, 189
Rio de Janeiro, Brazil
T: 55 21 2634821

Costa, Lara Moutinho Da, Sra.
Tesoureira
DEFENSORES DA TERRA
Rua Cruz Lima 8/902 - Flamengo
Rio de Janeiro - RJ 22230, Brazil
T: 55 21 551-0035 F: 55 21 2429353

Costa, Leda Leal Da, Sra.
Fiscal
SOZED
R. Getulio das Neves, 26, apto. 101 - Jardim Botanico
Rio de Janeiro - RJ, Brazil
T: 55 21 226-0266

Costa, Leilah Borges, Sra.
Advogada
ESCRITORIO PARTICULAR DE ADVOCACIA
Rua da Quitanda,19 610/11
Rio de Janeiro - RJ 20011, Brazil
T: 55 21 222 6832

Costa, Leonora Pires, Sra.
Biologa
FUNDACAO BIODIVERSITAS
Rua Mariz Vaz de Melo, 71
Belo Horizonte - MG 31250, Brazil
T: 55 31 443-2119 F: 55 31 441-7037

Costa, M., Sra.
Oceanografa
INPE
Av. Ademar de Barros 1136/51
Sao Josedos Campos 22215, Brazil
T: 55 012321-1483

Costa, Manuel Augusto, Sr.
Diretor
CEPPD - O SEGUNDO BRASIL
Av. Rio Banco 43/5o. andar
Rio de Janeiro - RJ 20090, Brazil
T: 55 21 263-7798

Costa, Manuel Da
Correspondent
AGENCIA ANGOLA PRESS (ANGP)
R Assis Brasil 143 b1/508 Copacabana
Rio de Janeiro - RJ, Brazil
T: 55 21 220 9439

Costa, Maria Do Socorro, Ms.
SOCIEDADE MARANHAENSE DE DIREITOS HUMANOS
Rua Antonio Rayol 290 - Centro
Sao Luiz - MA 65015-040, Brazil
T: 55 69 221-3838

Costa, Maria Ester
TVT REDE DE COMUNICACAO DOS TRABALHADORES
Sao Paulo, Brazil

Costa, Maria Imaculada Conceicao, Ms.
UNITED METHODIST CHURCH OF BRAZIL
Rua Marechal Deodoro, 1290-Caixa Postal 182
Sto.Ant.Platina 86430, Brazil

Costa, Marilena, Ms.
Secretaria
SERVICO SOCIAL DO COMERCIO
Rua Voluntarios Da Patria, 169
Rio de Janeiro - RJ 22270, Brazil
T: 55 21 286-5152 F: 55 21 286-2638

Costa, Mauro San Rego, Mr.
UNIVERSIDADE LIVRE
Rua Marques de Olinda, 100/503
Rio de Janeiro - RJ Cep.: 22.251, Brazil
T: 55 21 552 7239

Costa, Musa, Ms.
CONSERVATION INTERNATIONAL
Rua Bueno Brandao 393
Belo Horizonte - MG 31010, Brazil
T: 55 31 226-5145 F: 55 31 222-8429

Costa, Orbenes Da Silva, Sr.
Diretor
ASSOCIACAO DOS MORADORES E AMIGOS DO BAIRRO ANAIA
Rua Marly Motta Lobo,90 ANAIA
Sao Goncalo- RJ 24740, Brazil
T: 55 21 701 7287

Costa, Renata Maria Ferraz Da
Technician
REDE GLOBO
R. Lopes Quintas 303 Jardim Botanico
Rio de Janeiro -RJ 2, Brazil
T: 55 21 2947732

Costa, Ricardo Cadila Da
Technician
REDE GLOBO
R. Lopes Quintas 303
Rio de Janeiro -RJ 2, Brazil
T: 55 21 2947732

Costa, Ricardo Guimaraes
Technician
REDE GLOBO
R. Lopes Quintas 303
Rio de Janeiro -RJ 2, Brazil
T: 55 21 2947732

Costa, Robson Da
REDE GLOBO
R. Lopes Quintas 303 J.Botanico
Rio de Janeiro RJ 22, Brazil
T: 55 21 2947732

Costa, Shirley Floreti, Sra.
Diretora
ASSOCIACAO DOS MORADORES DE IPANEMA
Rua Nascimento Silva 120/103
Rio de Janeiro - RJ 22421, Brazil
T: 55 21 287 0969 F: 55 21 287 0969

Costa, Sidney Alves, Sr.
Rua Dr. Figueiredo, 1010
Valenca - RJ, Brazil
T: 55 244 520667

Costa, Silvio Jose Frazoa Da
RADIO ALVORADA FM
Rio de Janeiro, Brazil

Costa, Sueli De Souza Melo Da, Sra.
INSTITUTO DO MEIO-AMBIENTE DO ACRE
Rua Rui Barbosa, 450
Rio Branco - AC 69900, Brazil
T: 55 68 224-5497 F: 55 68 224-5694
TL: 682692

Costa, Suely Carrilho Gomes Da
Photographer
LIP CONSULTORIA DE COMUNICACAOS LTDA
Av Almirante Barroso 63/209
Rio de Janeiro, Brazil
T: 55 21 2203567 F: 55 21 2625401

Costa, Suely Cristina Da, Sra.
Biologa
ASSEMBLEIA LEGISLATIVA DO ESPIRITO SANTO
Rua Japao 147
Marataizes - ES, Brazil
T: 55 27 223-0489

Costa, Tarso Drumond M. Da
ECO TV
Parati, Brazil

Costa, Teresa Brandao
Produtora-Diretora
MTVC
R.Marqwues de Sao Vicente, 429/303
Rio de Janeiro - RJ, Brazil
T: 55 21 274 1553 F: 55 21 239 6905

Costa, Vicente Paim
FREELANCER
Rio de Janeiro, Brazil

Costa Filho, Alvino, Sr.
CIA. DE CIGARROS SOUZA CRUZ
Rua Da Candelaria 66 10ANDAR
Rio de Janeiro - RJ 20092-900, Brazil
T: 55 21 2769090

Costa Filho, David Capistrano Da, Sr.
PREFEITURA MUNICIPAL
Praca Maua s/n 1 andar
Santos - SP 11010, Brazil
T: 55 132 342-2093 F: 55 132 345-809
TL: 131048

Costa Filho, Jose, Sr.
Menbro
GRUPO DE ETNOECOLOGIA
Rua Cel. Sergio Henrique Cardin 115 Bl.A Apto. 304
Recife - PE 51011, Brazil
T: 55 81 465-2526

Costa Neto, Perminio De Castro, Sr.
Prefeito
PREFEITURA MUNICIPAL
Av.. Castelo Branco, 1046, Centro - CP 99
Pimenta Bueno - RO 78984-000, Brazil
T: 55 69 451-2417
TL: (69)7066

Cotrim, Maria Ester, Sra.
Secretaria Da Associacao
ASSOCIACAO BRASILEIRA DE ONGS
Rua Bento Lisboa 58, Catete (Caixa Postal 16151)
Rio de Janeiro - RJ 22222, Brazil
T: 55 21 285-2998 F: 55 21 205-3099
TL: 2134201 FOAS BR

Cotta, Barbara Elisa Babo
Journalist
FREELANCER
R. Pedro de Carvalho 501 /cob.02 Bl.A
Rio de Janeiro -RJ 2, Brazil
T: 55 21 5946813

Courcouroupoulos, Panayotis
Reporter
REDE GLOBO
R. Lopes Quintas 303
Rio de Janeiro -RJ 2, Brazil
T: 55 21 2947732

Couro, Maria Celia
Coordinator
FUNDACAO PLANETA OCEANO
R. Camargo 26
Rio de Janeiro -RJ, Brazil
T: 55 21 2124473 F: 55 21 8204454

Coutinho, Angelica Lourdes De Matos
Professora
CONSELHO ESTADUAL DA MULHER - MINAS GERAIS
Rua Comendador Viana 300
Belo Horizonte - MG, Brazil
T: 55 31 223-7750

Coutinho, Antonio Carlos, Sr.
LAB. DE SENSORIAMENTO REMOTO DA UFF
Instituto Geociencias Av. Citoranga s/n Boa Viagem
Niteroi - RJ 24020, Brazil
T: 55 21 712-4241

Coutinho, Carlos Luiz De Azevedo
Technician
REDE GLOBO
R. Lopes Quintas 303
Rio de Janeiro -RJ 2, Brazil
T: 55 21 2947732

Coutinho, Celia Maria Perdigao, Sra.
Arquiteto
IBPC
Rua Juvenal Galeno 512
Fortaleza - CE 60015, Brazil
T: 55 85 223 0706

Coutinho, Marco Antonio D. Freitas, Mr.
CIRCES INTERNACIONAL CHANCELARIA DO BRASIL
Rua Anfilupiuo De Carvalho 29/216 Centro RJ
Rio de Janeiro - RJ, Brazil
T: 55 21 266-5572 F: 55 21 537-1287

Couto, Christine Seixas Telles
Editor
JORNAL SHOPPING MODAS
Rua Tonelero 130/704
Rio de Janeiro RJ, Brazil
T: 55 21 2362283

Couto, Fernando Dela Rocque, Sr.
IECAM
Dr. Moreira 250
Manaus - AM 69000, Brazil

Couto, Ilzani Jacira S.Rocha
Camera
TVSBT
Campo de S.Cristovao, 105
Rio de Janeiro - RJ, Brazil
T: 55 21 5803688 / 5800678 F: 55 21 5809685

Couto, Mauricio
JORNAL BRAZIL EXPORT
Av. Franklin Roosevelt 194 Sobre Loja Conj.202
Rio de Janeiro -RJ 2, Brazil
T: 55 21 2629920 F: 55 21 2404705

Coya, Hugo
Correspondent
UPI
R. Uruguaiana 54/18 andar
Rio de Janeiro -RJ 2, Brazil
T: 55 21 2244194 F: 55 21 2328295

Cozzi, Marco Aurelio, Sr.
Diretor De Meio Ambiente
FEDERACAO NACIONAL DE MEDICINAS ALTERNATIVAS NATURAIS
Rua Professora Iacema Pimenta 135/135 - Sao Bento
Belo Horizonte - MG 30350, Brazil
T: 55 31 344-6433

Crabtree, Brad, Mr.
Consultant
Av. Brigadeiro Luiz Antonio 4442
Sao Paulo - SP 01402, Brazil
T: 55 11 887-9369

Craveiro, Fani De J. Leme Marques, Sra.
Pesquisadora
Rua Novo Horizonte, 613
Campinas - SP 13983, Brazil
T: 55 192 543722

Crema, Stefano Camillo, Mr.
President
TRENT UNIVERSITY - STUDENT UNION
Rua General Urquiza 161/101
Rio de Janeiro - RJ, Brazil
T: 55 21 259-3974

Crespo, Telson Emmanuel Ferreira, Mr.
Coordenador Geral
CONESAN
Avenida Amazonas, 5253, Nova Suica
Belo Horizonte - MG 30410, Brazil
T: 55 31 3328400 R137 F: 55 31 3345042
TL: 313863

Crossin, Judith
Correspondent
REUTERS NEW SERVICE
Rua Sete de Setembro 99/4th floor Centro
Rio de Janeiro - RJ, Brazil
T: 55 21 5074151 F: 55 21 5072120

Cruz, Ana Lucia Da, Sra.
Tecnica
INSTITUTO DE ESTUDOS AMAZONICOS E AMBIENTAIS
SCLN 210 Ed. Center Bloco C salas 211 e 212
Brasilia - DF 70062, Brazil
T: 55 61 347-9874 F: 55 61 347-9874

Cruz, Beatriz Bastos
INSTITUTO DE ARQUITETOS DO BRASIL
Rua do Pinheiro, 10 Flamengo
Rio de Janeiro, Brazil
T: 55 21 2853480

Cruz, Beth
Photograpaher
REUTERS NEW SERVICE
Rua Sete de Setembro 99/4th floor Centro
Rio de Janeiro - RJ, Brazil
T: 55 21 5074151 F: 55 21 5072120

Cruz, Carlos Alberto Da
ARTE RIO
Rio de Janeiro, Brazil

Cruz, Carmem Cleci Da
Reporter
S/A CORREIO BRAZILIENSE
R. do Livramento 158/4 andar
Rio de Janeiro -RJ, Brazil
T: 61 3212123 F: 061 3213864

Cruz, Dirce Maria De Araujo, Ms.
Diretora Do Desa
SECRETARIA DA SAUDE
Av. Conselheiro Furtado 3520/101
Belem - PA 66.050, Brazil
T: 55 91 223-9355

Cruz, Eliane Lobato Alves Da, Sra.
Distribuicao
CEDI
Rua Santo Amaro 129
Rio de Janeiro - RJ 22211, Brazil
T: 55 21 205-5993 F: 55 21 205-5993
TL: 213982CIED.BR

Cruz, Jorio Roberto De Castro, Sr.
President
MIDADE/IMAC/MIDADEN
Av. Santa Catarina, 2000
Barra Velha - SC 89240, Brazil
T: 55 474 460747

Cruz, Jose Ricardo Candido Da
Operador De Vt
TV Rio
R. Miguel de Frias, 57 Estacio
Rio de Janeiro - RJ, Brazil
T: 55 21 2939888 F: 55 21 2938626

Cruz, Juan Santa
Consultor De Meio Ambiente
BCO. INTERAMERICANO DE DESENVOLVIMENTO
Caixa Postal 08738
Brasilia - DF 70312, Brazil
T: 55 61 321-7848 F: 55 61 321-3112
TL: 61-1204

Cruz, Lucia Santa
Photographic Reporter
PERFIL COMUNICACAO
Rua Senador Dantas, 71/40 andar
Rio de Janeiro, Brazil
T: 55 21 2206446 F: 55 21 2407827

Cruz, Luis Antonio Da, Mr.
Director
SOCIEDADE DE AMIGOS DE TIRADENTES
Rua Bueno Brandao 393
Belo Horizonte - MG 31010, Brazil
T: 55 31 226-5145 F: 55 31 222-8429

Cruz, Maria Alves Da, Sra.
Pedagoga
CENTRO DE RECURSOS AMBIENTAIS
Rua Rio Sao Francisco, 1 - Monte Serrat
Salvador - BA CEP: 40.425, Brazil
T: 55 71 312-7192 F: 55 71 312-7198

Cruz, Maria Do Carmo, Sra.
FLAMA
Praia do Flamengo, 221 - Flamengo
Rio de Janeiro - RJ, Brazil
T: 55 21 295-2234

Cruz, Maria Herbenia Lima, Sra.
Rua Do Progresso 426/B/403
Recife - PE 50000, Brazil
T: 55 81 4652247

Cruz, Ramon Bienvenido Martinez
Representante No Brasil De Age
PRENSA LATINA
R Mal. Mascarenhas de Morais 121/203 Copacabana
Rio de Janeiro - RJ, Brazil
T: 55 21 256 7259

Cruz Filho, Vicente, Mr.
PREFEITURA DE PARATY
Rua da Patitiba 12
Paraty - RJ 23970, Brazil
T: 55 243 71-1266 F: 55 243 71-2105

Csillag, Paula, Sra.
Estudante
FUNDACO GETULIO VARGAS
Praca Amadeu Amaral
Sao Paulo - SP 01327, Brazil
T: 55 11 287 0148

Cunha, Alexandre Magno Machado Da
REDE GLOBO
Rio de Janeiro, Brazil

Cunha, Alexia Celeste Da, Sra.
INSTITUTO PRO NATURA
Rua Monte Alegre 374 - Santa Teresa
Rio de Janeiro - RJ 20240, Brazil
T: 55 21 224-39904 F: 55 21 224-3994

Cunha, Ana Paula, Ms.
Av. Maracana, 1063 - Ap. 32
Rio de Janeiro - RJ, Brazil
T: 55 21 288-9471

Cunha, Cassia Cristina De Castro
Editor
JORNAL DO MOVIMENTO CONSERVACIONISTA TERESOPOLITANO
R Barao de Sao Francisco 245/1108 Bloco H Vila Isabel
Rio de Janeiro - RJ, Brazil
T: 55 21 238 9259

Cunha, Cristiana Rodrigues Da
RADIO O DIA FM
Praca Mana, 7 21 andar
Rio de Janeiro, Brazil
T: 55 21 2536725 F: 55 21 2332513

Cunha, Daniel Renout Da, Sr.
Lawyer
Rua Antonio Parreiras, 56/apt. 204-A - Ipanema
Rio de Janeiro - RJ, Brazil
T: 55 21 262-8572

Cunha, Eldis C. Neves Da, Sra.
Diretora
GECA - GRUPO DE ESTUDO E CONSCIENTICAO AMBIENTAL
R. Muniz de Souza 608
Sao Paulo - SP 01534, Brazil
T: 55 11 278-4382

Cunha, Eugenio, Mr.
ASSOCIACAO POTIGUAR DE EDUCACAO E CULTURA
Rua Floriano Peixoto, 295 Petropolis
Natal - RN 59020, Brazil
T: 55 84 221-6255 F: 55 84 221-6255

Cunha, Gertrudes Carneiro Da, Sra.
FLAMA
Rua Marques de Abrantes, 55 - Flamengo
Rio de Janeiro - RJ 22230, Brazil
T: 55 21 265-5690

Cunha, Icaro, Sr.
FUNDACAO NACIONAL DE ACAO ECOLOGICA
Rua Brigadeiro Luiz Antonio 4442
Sao Paulo - SP, Brazil
T: 55 11 885-3424

Cunha, Joaquim Marques, Mr.
PREFEITURA DE PARATY
Largo de Santa Rita s/n
Paraty - RJ 23970, Brazil
T: 55 243 71-1266 F: 55 243 71-2105

Cunha, Jonas Correa
Photographer
AGENCIA ESTADO
Av. Caetano Alvares, 55
Sao Paulo -SP 02598, Brazil
T: 55 11 8562210 F: 55 11 2652297

Cunha, Luis Roberto Lisboa Da, Mr.
Rua Timoteo da Costa, 1001/506, Leblon
Rio de Janeiro - RJ 22450, Brazil
T: 55 21 259-5565

Cunha, Miguel Gastao Da, Sr.
FLAMA
Rua Marques de Abrantes, 55 - Flamengo
Rio de Janeiro - RJ 22230, Brazil
T: 55 21 265-5690

Cunha, Milton Augusto Da
Technician
REDE GLOBO
R. Lopes Quintas 303
Rio de Janeiro -RJ 2, Brazil
T: 55 21 2947732

Cunha, Priscilla, Sra.
PRO-NATURA
Av. Rodrigues Alves, 179, 2o. andar
Rio de Janeiro - RJ 20081, Brazil
T: 55 21 253-8655

Cunha, Renato Paes Da, Sr.
FORUM DAS ONG'S BAIANAS
Rua Borges dos Reis 46, Sala J, Boulevard Rio Vermelho
Salvador - BA, Brazil
T: 55 71 235-5836 F: 55 71 235-5836

Cunha, Ricardo Lisboa Da, Sr.
Diretor
FUNDACAO BRASIL - FRANCA
Av. Beira Mar, 406 - conj 305
Rio de Janeiro - RJ 20021, Brazil
T: 55 21 262 7782 F: 55 21 262 7782

Cunha, Rodrigo Carneiro Da, Sr.
FLAMA
Rua Marques de Abrantes, 55 - Flamengo
Rio de Janeiro - RJ 22230, Brazil
T: 55 21 265-5690

Cunha, Sergio Servulo Da, Dr.
Vice Prefeito
PREFEITURA MUNICIPAL
Praca Maua s/n 1 andar
Santos - SP 11010, Brazil
T: 55 132 342-2093 F: 55 132 345-809
TL: 131048

Cunha, Silvio, Mr.
Expositor
EDITORA TERCEIRO MUNDO
Rua da Gloria 122/101
Rio de Janeiro - RJ, Brazil

Cunha, Sonia Silva Santos, Sra.
SOUZA CRUZ/ELDOCOR
Av. Rio Branco 125/21o.
Rio de Janeiro - RJ, Brazil
T: 55 21 224-0077
TL: 34858, E: 2244

Cunha E Silva, Nuno, Mr.
ARACRUZ CELULOSE S.A.
Rua Lauro Muller, 116 - 22o. andar
Rio de Janeiro - RJ, Brazil
T: 55 21 541-1699 F: 55 21 541-5443

Cunningham, Patricia, Ms.
Phd. Professor
INST. OCEANOGRAFICO USP
Cid. Univ. Butanta
Sao Paulo - SP 05508, Brazil

Curi, Giselle, Ms.
PRISMART CULTURA
Av. Atlantica 14 Edificio Regina Feighl
Rio de Janeiro - RJ 22010, Brazil
T: 55 21 541 3715

Curial, Anne-Claire S., Ms.
Caete 263
Porto Alegre - RS 91.900, Brazil
T: 55 512 249-3151

Curtis, Elaine
Produtora
REDE BANDEIRANTES DE TELEVISAO
R.Alvaro Ramos, 492
Rio de Janeiro - RJ, Brazil
T: 55 21 2122072 F: 55 21 2957391

Cury, Walmir Marun
Technician
REDE GLOBO
R. Lopes Quintas 303 Jardim Botanico
Rio de Janeiro -RJ 2, Brazil
T: 55 21 2947732

Cussate, Fernando
Fotografo
REVISTA MANCHETE
Rua do Russel, 804
Rio de Janeiro - RJ, Brazil
T: 55 21 2652012

Cuvi, Guadalupe, Sra.
Rua Olartinho de Mesquite, 241
Rio de Janeiro - RJ 22620, Brazil
T: 55 21 493-8477

D'Albuquerque, Lucia
Producao
FREELANCER
Caixa Postal 18414
Sao Paulo - SP, Brazil
T: 55 11 543 0566 F: 55 11 534 8234

D'Almeida, Ernani
Fotografo
ZNZ AGENCIA FOTOGRAFICA
Rua Cosme Velho 355 Cosme Velho/
Rio de Janeiro - RJ, Brazil
T: 55 21 225 5805/0066 F: 55 21 265 6567

D'Almeida, Sonia Maria Rodrigues
Fotografa
FREELANCER
Rua Dezoito de Outubro 129/101
Rio de Janeiro - RJ, Brazil
T: 55 21 268 6511

D'Arce, Alvaro Ferraz, Sr.
Membro
APOENA
Rua Guadalajara, 232
Presid. Prudente-SP 19020, Brazil
T: 55 182 22-1765

D'Avila, Justo
Reporter
RADIO TUPI
Rua do Livramento 189/8th Floor
Rio de Janeiro - RJ, Brazil
T: 55 21 203 2421 r 34 F: 55 21 2637323

D'Avila, Roberto, Sr.
GOVERNO DO ESTADO DO RIO DE JANEIRO
Rio de Janeiro - RJ, Brazil

D'Oliveira, Luiz Paulo S., Sr.
Vice-President
ASS. BRAS. EN. ALTERNATIVA E MEIO AMBIENTE
Estrada dos Bandeirantes 5817- Jacarepagua
Rio de Janeiro - RJ 22780, Brazil
T: 55 21 342-5815 F: 55 21 342-5271

D'Oliveira, Marcos Durao
ARTE RIO
Rio de Janeiro, Brazil

D'Oliveira, Teresa Cristina Almeida
Assistant Producer
Av. Gomes Freire 474 Centro
Rio de Janeiro, Brazil
T: 55 21 2920012 x181

Da Mota, Ivan Suarez, Sr.
Pesquisador Cientifico
INSTITUTO FLORESTAL SAO PAULO
Parque Estadual Da Serra Do Mar- Nucleo Caraguatatuba
Caraguatatuba-SP 11660, Brazil
T: 012422-3166

Da Silva, Antonio, Mr.
Expositor
EDITORA TERCEIRO MUNDO
Rua da Gloria, 122/101
Rio de Janeiro - RJ, Brazil

Da Silva, Antonio Carlos Vilas Boas, Sr.
Estatistico
CENTRO DE RECURSOS AMBIENTAIS
Rua Rio Francisco, 1 - Monte Serrat
Salvador - BA 40425, Brazil
T: 55 71 312-7192 F: 55 71 312-9198

Da Silva, Carlos Eduardo, Sr.
INSTITUTO FLORESTAL
r. do horto ,931
Sao Paulo - SP 02377, Brazil
T: 55 11 952-8555 F: 55 11 204-8067

Da Silva, Carlos Roberto Soares, Sr.
Sec. De Planejamento
PREFEITURA MUNICIPAL DE MARECHAL DEODORO
av. Tavares Bastos s/n - Centro - Marechal Deodoro
Alagoas 57160, Brazil
T: 082 263 1365

Da Silva, Claudio Ramos
HOSPITAL ADVENTISTA SILVESTRE
Ladeirados Guararapes, 263 - Santa Tereza
Rio de Janeiro - RJ, Brazil
T: 55 21 265-0212

Da Silva, Delfim Milagres, Sr.
Rua Do Bispo 323/501
Rio de Janeiro - RJ 20.261, Brazil
T: 55 21 234-8309

Da Silva, Dimas Antonio, Mr.
Scientist Recearch
INSTITUTO FLORESTAL - SECR. DO MEIO AMBIENTE
Rua do Horto 931
Sao Paulo - SP, Brazil
T: 55 11 952-8555 F: 55 11 204-8067
TL: 11-22877BR

Da Silva, Jose Dibamar, Sr.
FUNDACAO NACIONAL SAUDE-FNS
Rua Moacir Strauch 85 Praia Do Canto
Vitoria, Brazil
T: 027 227 6411

Da Silva, Leny Pereira, Sra.
Advogada
PROCURADORIA GERAL DO DISTRITO FEDERAL
Q 06 , Conj D , Casa 08
Brasilia - DF 70000, Brazil
T: 55 61 234 2126

Da Silva, Luiz Jose, Mr.
Expositor
EDITORA TERCEIRO MUNDO
Rua da Gloria, 122/101
Rio de Janeiro - RJ, Brazil

Da Silva, Marconi, Mr.
Seguranca
UIJDC
Rua Francisco Otaviano, 42 - Copacabana
Rio de Janeiro - RJ, Brazil

Da Silva, Roberto, Sr.
Agronomo
CENTRO DE RECURSOS AMBIENTAIS
Rua Rio Sao Francisco, 1 - Monte Serrat
Salvador - BA CEP: 40.425, Brazil
T: 55 71 312-7192 F: 55 71 312-7198

Da Silva, Ronaldo Martins, Sr.
Gerente
CENTRO DE RECURSOS AMBIENTAIS
Rua Rio Sao Francisco # 1
Salvador - BA 40000, Brazil
T: 55 71 312-7192 F: 55 71 312-7198

Da Silva, Sidmar, Mr.
Seguranca
UIJDC
Rua Francisco Otaviano, 42 - Copacabana
Rio de Janeiro - RJ, Brazil

Dacorso, Daniela Hallack
Journalist
ECABE PRESS
Rua Gal. Rabelo 65/404 Gavea
Rio de Janeiro, Brazil
T: 55 21 5113107 F: 55 21 2742759

Dacorso, Giovana Hallack
Journalist
ECABE PRESS
Rua Gal. Rabelo 65/404 Gavea
Rio de Janeiro RJ, Brazil
T: 55 21 5113107 F: 55 21 2742759

Dadia, Daniel
Assistente
WDR GERMAN TV CHANNEL
Rua Eca de Queiroz, 479 casa 12
Sao Paulo - SP, Brazil
T: 55 11 5721262 F: 55 11 5728959

Dahlen, Ulrika, Ms.
Participante
UBV
Av Rui Barbosa, 560 # 702
Rio de Janeiro - RJ 22250, Brazil
T: 55 21 552 1521

Damaso, Romualdo
INSTITUTO PRINCIPIO UNICO
Praca Carlos Gomes 60, 1 andar - Liberdade
Sao Paulo - SP 01501, Brazil
T: 55 11 329738 F: 55 11 5773235

Damazio, Heitor
ECOTROPIC
1880 East River Road
Rio de Janeiro - RJ, Brazil
T: 55 21 577-6476 F: 55 21 577-1195

Dana, Vicky, Ms.
HERMANN
Rua Visconde De Piraja 608/910-911
Rio de Janeiro - RJ 22410, Brazil
T: 55 21 511-4045

Daniela, Luise, Ms.
Voluntario Internacional
MLAL
Rua Goias,191
Luciara - MT 78 660000, Brazil

Dansa, Claudia Valeria De Assis, Sra.
Diretora Cultural
A D U F U - S. SINDICAL
R. Laerte Vieira Goncalves 2402
Uberlandia - MG 38400, Brazil
T: 55 34 216-3528

Dantas, Eloizio Henrique H., Sr.
Superintendente
SUDEMA
Rau Walfredo Leal 181 - Tambia
Joao Pessoa - PB 58023, Brazil
T: 55 83 229-3149

Dantas, Grace
Editor
SISTEMA JORNAL DO BRASIL
Av. Brasil 500
Rio de Janeiro -RJ, Brazil
T: 55 21 5854422 F: 55 21 5854428

Dantas, Joao Portela R.
ASSOCIACAO COMERCIAL DO RIO DE JANEIRO
Rio de Janeiro, Brazil

Dantas, Jose Pereira
Reporter
REDE GLOBO
R. Lopes Quintas 303
Rio de Janeiro -RJ 2, Brazil
T: 55 21 2947732

Dantas, Lenira Junqueira, Sra.
Rua Santa Luzia 651 33 Andar
Rio de Janeiro - RJ 20030, Brazil
T: 55 21 2976699 F: 55 21 262 2382

Dantas, Luis Olavo, Sr.
Diretor
SERTEP S.A. ENGENHARIA E MONTAGEM
Rua Santa Luzia 651- 33 Andar
Rio de Janeiro - RJ 20030, Brazil
T: 55 21 297-6699 F: 55 21 262-2382
TL: 21-22489

Dantas, Luiza
Photographic Reporter
FREELANCER
Rua Aracaria No. 12 apto 102 - Jardim Botanico
Rio de Janeiro RJ 2, Brazil
T: 55 21 2460982

Dantas, Paulo Cesar Esteves, Sr.
Prof. De Ciencias Biologicas
UNIVERSIDADE SANTA URSULA
Rua Fernando Ferrari, 75
Rio de Janeiro - RJ 22231, Brazil
T: 55 21 551-5542 F: 55 21 551-6446

Dantas, Paulo Morais De Souza
MADE FOR TV
Rio de Janeiro, Brazil

Dantas, Ricardo, Mr.
Biologist
CENTRO DE PESQUISAS DO MAR (CEPEMAR)
Rua Mexico, 31/Gr. 501- Centro
Rio de Janeiro - RJ, Brazil
T: 55 21 240-3822 F: 55 21 240-9516

Dantas, Roberto Antonio S., Sr.
Secretario
SECRETARIO MUNICIPAL DE MEIO AMBIENTE
Fonte Nova, Antigo 7 Andar Batalhao De Policia
Salvador - BA 40000, Brazil
T: 55 71 243 6158

Dapieve, Arthur
Sub-Editor
EDITORA ABRIL
R.da Passagem, 123/8 andar
Rio de Janeiro - RJ, Brazil
F: 55 21 2759347

Dauch, Angela Andriolo, Sra.
President Of Borors District
FED. DAS BANDEIRANTES DO BRASIL
Av. Marechal Camara, 186/2 andar
Rio de Janeiro - RJ 20020, Brazil
T: 55 21 240-9220

David, Luis Carlos
SISTEMA JORNAL DO BRASIL
Av. Brasil 500
Rio de Janeiro -RJ, Brazil
T: 55 21 5854422 F: 55 21 5854428

Davies, Jim
Environment Reporter
LONDON DAILY EXPRESS/MUSICAL AMERICA
Miguel Lemos 51/604 Copacabana
Rio de Janeiro - RJ, Brazil
T: 55 21 227 7594 F: 55 21 227 7594

Davis, Roger Langdon, Sr.
Teacher
ESCOLA DAS NACOES
SHIS QI 21 Area Especial - Lote C-1 - Caixa Postal 7093
Brasilia - DF 71600, Brazil
T: 55 61 366-1800 F: 55 61 366-3777

De Albuquerque, Jose Eduardo Cavalcanti, Sr.
COMISSAO ESPECIAL LEI 1849/92
Rio de Janeiro - RJ, Brazil
T: 55 21 273 9442

De Almeida, Luiz Carlos P., Mr.
INTERNATIONAL REAL ESTATE FEDERATION
888 Av. Faria Lima 8o.
Sao Paulo - SP 05613, Brazil
T: 55 11 211-9811 F: 55 11 210-0958

De Araujo, Eyer, Mr.
Rep. De Com.
CIER
Rua Maria Angelica, 416 Apto. 201
Rio de Janeiro - RJ, Brazil
T: 55 21 232 7381 F: 55 21 233 4334
TL: (21)22329 CELB

De Assis, Jarbas Ribeiro, Jr., Sr.
Secretario De Meio Ambiente
SECRETARIA DO ESTADO DO ESPIRITO SANTO P/ ASSUNTOS DE MEIO AMBIENTE E DESENVOLVIMENTO
Av. Princesa Isabel, 629 - 6$ andar - Centro
Vitoria - ES 29045, Brazil
T: 55 27 223-4022 F: 55 27 222-7908

De Blasis, Cristina De La Luz Guzman, Ms.
ASSOC. MUN. OESTE
Rua Maria Angelica 326/403
Rio de Janeiro - RJ, Brazil
T: 55 21 266-1409

De Campos, Francisco Jose Alcantara, Sr.
Coordenador
INDEA MATO GROSSO
RUA Porto Esperidiao Q 69 No02
Cuiaba - MT 78070, Brazil
T: 55 065 341 2805

De Campos, Jose Marcio, Sr.
COMISSAO ESPECIAL LEI 1849/92
Rio de Janeiro - RJ, Brazil
T: 55 21 273 9442

De Campos, Tatiana Pitta, Ms.
Rio de Janeiro - RJ 22020, Brazil
T: 55 21 275-7682

De Darmas, Nilda, Sra.
Leopoldo Miguez Via 81 apto 106 - Copacabana
Rio de Janeiro - RJ, Brazil
T: 55 21 2357646

De Lira, Rafael Ruis, Mr.
Coordenador
I C I
Ses - Av. Das Nacoes Lote 44
Brasilia - DF, Brazil
T: 55 61 244-2121

De Medeiros, Tibucio, Sr.
Geologo
CENTRO DE RECURSOS AMBIENTAIS
Engenheiro Celso Torres, 16 A Ap. 602 - Ed. Vale Verde
Salvador - BA CEP: 40.150, Brazil
T: 55 71 245-2485

De Mello, Ney, Sr.
Geografo
CENTRO DE RECURSOS AMBIENTAIS
Rua Rio Sao Francisco, 1 - Monte Serrat
Salvador - BA CEP: 40.425, Brazil
T: 55 71 312-7192 F: 55 71 312-7198

De Moraes, Marcia, Sra.
COMISSAO ESPECIAL LEI 1849/92
Rio de Janeiro - RJ, Brazil
T: 55 21 273 9442

De Oliveira, Raphaela, Ms.
Artur Araripe 53 Apto 701 Gavea
Rio de Janeiro - RJ 22451, Brazil
T: 55 21 2945191

De Oliveira, Walfredo, Mr.
GRUPO QUATRO S.C.
Rua 124, No. 66
Goiania - GO 74000, Brazil
T: 55 62 224-2555 F: 55 62 229-0110

De Paula, Francisco C. F., Mr.
Pesquisador
DEPT DE GEOQUIMICA DA UFF
Morro do Valonguinho, s/n
Niteroi - RJ, Brazil

De Paulo, Carlos Alberto Santos, Sr.
Militante
MOVIMENTO NEGRO
Rua Joarez Correia Coelho
Curitiba - PR 80001, Brazil
T: 55 41 285 2307

De Queiroz, Sandra, Sra.
Biologa
SUREHMA
Rua Colombo 414 Apt 401
Curitiba - PR 80530, Brazil
T: 55 41 225 6283 F: 55 41 225 3411

De Sa, Carlos Alberto Morais, Sr.
Professor
UNIVERSIDADE DO RIO DE JANEIRO
R. Barao Sao Francisco 120/902 - Vila Isabel
Rio de Janeiro - RJ 20560, Brazil
T: 55 21 254 7988 F: 55 21 254 5249

De Sousa, Paulo Sergio, Mr.
Expositor
EDITORA TERCEIRO MUNDO
Rua da Gloria, 122/101
Rio de Janeiro - RJ, Brazil

De Souza, Herbert, Sr.
COMISSAO ESPECIAL LEI 1849/92
Rio de Janeiro - RJ, Brazil
T: 55 21 273 9442

De Valera, Cile, Sra.
Parlamentar
DELEGACAO IRLANDA
Av. Princesa Isabel, 323/1205
Rio de Janeiro - RJ 22011, Brazil
T: 55 21 275 0196 F: 55 21 275 6299

De Vasconcellos, Gerhard Sardo, Sr.
Presidente
GRUPO CAMINHANTE INDEPENDENTE
Rua Lopes Trovao 284/2104 Bloco 01
Niteroi - RJ 24220, Brazil
T: 55 21 711 4665

Debieuvre, Marie Odile, Ms.
ECOTROPIC
Rua Almirante Alexandrino 2161
Rio de Janeiro - RJ, Brazil

Debs, Maria Ausiliadora
Reporter
AGENCIA 355
Rua Cosme Velho, 355
Rio de Janeiro RJ 2, Brazil
T: 55 21 2251813 F: 55 21 2422878

Dechtiar, Jairo Bensussan, Sr.
Member
COMUNIDADE BAHA'I
Caixa Postal 999
Porto Velho - RO 78900, Brazil
T: 55 69 2217461 F: 55 69 2233945

Deitenbach, Armin, Mr.
CMSP, E.V.C., Km 34
Caieiras-SP 07700, Brazil
T: 55 11 4313133 F: 55 11 4312471
TL: 01179669

Delfina, Jorge
Cameraman
MANCHETE
Rua do Russel 804 Gloria
Rio de Janeiro, Brazil
T: 55 21 2850033 x505

Della Cava, Mirka, Sra.
NATURAL RESOURCES DEFENSE COUNCIL
Rua Diamantina, 13/201 Jardim Botanico
Rio de Janeiro - RJ 22461, Brazil
T: 55 21 2260769 F: 55 21 2260769

Delling, Ricardo
Supervisor Of Communications
TV CULTURA (SAO PAULO)
Rua Cenno Sbrig, 378
Sao Paulo, Brazil
T: 55 11 2639111 F: 55 11 2636552

Denghy, John De
Correspondent And Editorial Re
FOREIGN PRESS CLUB
R Nilza Medeiros Martins 279/132 Morumbi
Sao Paulo - SP, Brazil
T: 55 11 255 1815 F: 55 11 2583920

Depner, Uta
Produtora
2DF-TV ALEMA
Av.Delfim Moreira, 662/704
Rio de Janeiro - RJ, Brazil
T: 55 21 2947434 F: 55 21 5122120

Descio, Fernando, Mr.
Chefe Parque Est Da Cantareira
INSTITUTO FLORESTAL - SP
Rua Horto 931
Sao Paulo - SP 02377, Brazil
T: 55 11 9528555

Descomps, Gregeire
A GOTA DE ESPERANCA
Rio de Janeiro, Brazil

Dettmar, Glaucio Bastos
Journalist
RADIOBRAS - BRASILIA
Rua Rezende 103/apt 701 - B. Fatima
Rio de Janeiro, Brazil
T: 55 21 2224449

Deus, Carlos Edegard De, Sr.
Professor
UNIVERSIDADE FEDERAL DO ACRE
Br.364 - KM 4, Campus Universitario (Pz)
Rio Branco - AC 69900, Brazil
T: 55 68 226-1641 F: 55 68 226-1162
TL: 2532

Deva Bodhi, Ma, Ms.
OSHO SPIRITUAL HEALTH ORGAN. - OSHO
Rua Celestino Basilio 26
Rio de Janeiro - RJ, Brazil
T: 55 21 437-7488 F: 55 21 322-0327

Deva Prashanto, Sw, Mr.
OSHO SPIRITUAL HEALTH ORGAN. - OSHO
Rua Paula Mattos, 121, sobrado
Rio de Janeiro - RJ 22251, Brazil
T: 55 21 232-5008

Deva Raji, Ma
OSHO SPIRITUAL HEALTH ORGAN. - OSHO
R. Celestino Basilio 26
Rio de Janeiro - RJ, Brazil
T: 55 21 437-7488

Devantie, Christine, Sra.
SERVICO ALEMAO DE COOPERACAO
Rua Presidente Carlos de Campos, 183/105
Rio de Janeiro - RJ Cep.: 22.231, Brazil
T: 55 21 553 9928

Dhyan Kavita, Ma, Ms.
OSHO SPIRITUAL HEALTH ORGAN. - OSHO
Rua Celestino Basilio 26
Rio de Janeiro - RJ, Brazil
T: 55 21 437-7488 F: 55 21 322-0327

Di Villarosa, Francesco Notarbartolo, Sr.
Responsavel Programas
TERRA NUOVA
Lgo. Sto. Antonio da Barra s/n - Ladeira da Barra
Salvador - BA 40130, Brazil
T: 55 71 247.9707 F: 55 71 2479707

Diana, Ana Claudia Costa
Reporter
FREELANCER
Rua Palas, 706 casa 4
Rio de Janeiro - RJ, Brazil
T: 55 21 3758677

Dias, Adriane Guimaraes
JORNAL O FLUMINENSE
Niteroi, Brazil

Dias, Antonio Jose Da S.
Techinician
AGENCIA ESTADO
Av. Caetano Alvares, 55
Sao Paulo -SP 02598, Brazil
T: 55 11 8562210 F: 55 11 2652297

Dias, Carlos Augusto, Mr.
PREFEITURA MUNICIPAL DE BELO HORIZONTE
Av. Afonso Pena, 1212
Belo Horizonte - MG, Brazil
T: 55 31 238 4257

Dias, Claudete Maria Miranda, Sra.
Professora
UNIVERSIDADE FEDERAL DO PIAUI
Rua David Caldas, 1451/SUL
Teresina, PI 64010, Brazil
T: 86 232-3344 F: 86 222-1012

Dias, Fatima Maria Bessa, Sra.
Rua Belfort Roxo, 169/ 503
Rio de Janeiro - RJ 22020, Brazil
T: 55 21 541-7984

Dias, Fernando
INSTITUTO DE ARQUITETOS DO BRASIL
Rua do Pinheiro, 10 Flamengo
Rio de Janeiro, Brazil
T: 55 21 2853480

Dias, Henrique Roberto
Technician
REDE GLOBO
R. Lopes Quintas 303
Rio de Janeiro -RJ 2, Brazil
T: 55 21 2947732

Dias, Ines De Souza
Coordinator
FUNDACAO SOS MATA ATLANTICA
R. Manuel da Nobrega 456
Sao Roque - SP, Brazil
T: 55 11 8871195 F: 55 11 8851680

Dias, Joao Francisco
REDE GLOBO
Rio de Janeiro, Brazil

Dias, Joao Henrique Pinheiros, Mr.
CESP
Rua da Consolacao 1875, 9 andar, Cerq. Cesar
Sao Paulo - SP 01301, Brazil

Dias, Jose Carlos Azevedo
Produtor
REDE BANDEIRANTES DE TELEVISAO
R.Alvaro Ramos, 492
Rio de Janeiro - RJ, Brazil
T: 55 21 2122072 F: 55 21 2957391

Dias, Jose Fernandes, Sr.
Pesquisador
CEDAC
Rua Benjamin Constante, 108
Rio de Janeiro - RJ 20241, Brazil
T: 55 21 231 0263 F: 55 21 222 2527

Dias, Marcia Dos Santos, Sra.
Participant
IBFAN
Rua Jose Roberto no.28
Rio de Janeiro - RJ 21050, Brazil
T: 55 21 290-3802

Dias, Marcos Henrique De Castro
MANCHETE
R. do Russel 804
Rio de Janeiro -RJ 2, Brazil
T: 55 21 2652012 F: 55 21 2855148

Dias, Margarida, Sra.
Bibliotecaria
LIBRARY OF CONGRESS OFFICE
Av. Presidente Wilson, 147 2o. andar
Rio de Janeiro - RJ 20030, Brazil
T: 55 21 292-7117 F: 55 21 262-0218

Dias, Marizia Biondi, Sra.
President
GRUPO ECOLOGICO NATI VERDE
R. Saturnino Barbosa no.145
Sao Jose do Rio Pard 13720, Brazil
T: 55 196 611791

Dias, Osvannildo
JORNAL O FLUMINENSE
Niteroi, Brazil

Dias, Paulo Roberto Castro, Sr.
Member
ALAMA - ASSOCIACION
Av. da Igreja 346 Demcul
Tramandai - RS 95590, Brazil
T: 55 51 6611322 F: 55 51 6611762

Dias, Rodrigo
Editor De Fotografia
AGENCIA SPID LDA
Rua Cardeal D Sebastiao Leme 350/301 Sta Tereza
Rio de Janeiro - RJ, Brazil
T: 55 21 242 2878 F: 55 21 242 2878

Dias, Sophie De Panaskhet M., Sra.
Director
ESPACO BRASIL 2001 / LES APPRENTIS DE L'ESPERANCE/MULTICOOP
a/c Maison de France - Rua Antonio Carlos 58 - Castelo
Rio de Janeiro - RJ 20020, Brazil
T: 55 21 210-1272..165 F: 55 21 220-3629

Dias, Zoraida
Photograpaher
REUTERS NEW SERVICE
Rua Sete de Setembro 99/4th floor Centro
Rio de Janeiro - RJ, Brazil
T: 55 21 5074151 F: 55 21 5072120

Dibala, Eduardo
ASSOCIATED PRESS
Rio de Janeiro, Brazil

Didonet, Hermeto, Sr.
Gerente De Projetos Especiais
COOPERATIVA AGROPECUARIA MISTA CANARANA LTDA
Av. Ministro Joao Alberto 5 - Centro
Barra do Garca - MT 78600, Brazil
T: 55 65 446-1640

Diegues, Antonio Carlos, Sr.
Professor
UNIVERSIDADE CATOLICA
Colmeia Favo 6 - Rua do Anfiteatro 181
Sao Paulo - SP 05508, Brazil
T: 55 11 887-5204

Dilger, Robert, Sr.
Coordenador
SUREHMA
Rua Eng. Reboucas 1206
Curitiba - PR 80215100, Brazil
T: 55 41 225 6283 F: 55 41 254 8257

Dinis, Alan
MANCHETE
Rio de Janeiro, Brazil

Diniz, Elizeu, Mr.
Assistente Tecnico
SEC. DE ESTADO DE SAUDE
Av. Sao Luis, 99 - °o. Andar
Sao Paulo - SP, Brazil
T: 55 11 257-7437

Diniz, Iliria Maria, Sra.
Supervisor De Setor
FUNDACAO SOS MATA ATLANTICA
Rua Manoel da Nobrega, 456
Sao Paulo - SP 04001, Brazil
T: 55 11 887-1195 F: 55 11 885-1680

Diniz, Jalcione N. Nuniz, Sra.
CESP
Rua da Consolacao, 1875, 9 andar, Cerq. Cesar
Sao Paulo - SP, Brazil

Diniz, Joao Helder Alves Da Silva, Sr.
Supervisor Da Area/Pe
VISAO MUNDIAL BRASIL
Rua Dom Vital, 587 - 2o. andar - sala 05 - Centro
Petrolina - PE 56300, Brazil
T: 55 81 961-1115

Diniz, Joubert C., Mr.
Socio
BRAZCO
Praia Do Flamengo 66 Bl. B S/202
Rio de Janeiro - RJ 22.610, Brazil
T: 55 21 285-5839 F: 55 21 285-3943

Diniz, Paulo, Sr.
AQUALUNG CONFECCAO LTDA.
Av. Londres, 174 - Bonsucesso
Rio de Janeiro - RJ 21041, Brazil
T: 55 21 270-1955

Dinsmore, Paul Campbell
President
DINSMORE ASSOCIATES/PROJECT MET INST
R da Lapa 180/cob
Rio de Janeiro - RJ, Brazil
T: 55 21 224 5542 F: 55 21 2324932

Ditcheff, Fabia, Sra.
KEY MARKETIG BRASIL
Av. Ataulfo de Paiva 135 gr. 710
Rio de Janeiro - RJ 22440, Brazil

Dixon, Diana, Ms.
Information Assistant
WORLD FOOD PROGRAMME / UN
SCN Quadra 02 Bloco 'B' Terreo
Brasilia - DF 70710, Brazil
T: 55 61 226-7872 F: 55 61 322-2080
TL: 61-1243

Do Valle, Marta, Sra.
OSHO INSTITTUTO BRASIL
24 de outubro 544/61
Porto Alegre - RS, Brazil
T: 55 512 222-9888

Dobberthin, Lutz, Mr.
AMAZON INDS.
Rua Consolacao
Sao Paulo - SP, Brazil
T: 55 11 264 9712

Dobrovolski, Ricardo Luiz, Sr.
Coordenador
FEPAM-RS
Av. A. J. Renner 10
Porto Alegre - RS 90250, Brazil
T: 55 512 3435799 F: 55 512 3420224

Dominges, Irma, Ms.
MISSIONARY BENEDICTINE
Rua 19 Fevereiro, 172
Rio de Janeiro - RJ 22280, Brazil
T: 55 21 542-0293

Domingos, Irene B., Sra.
Rua Paes Leme no. 1081 apt.103
Londrina PR 86100, Brazil

Domingues, Eloisa, Sra.
I B G E
R, Paula Fernandes 24/1 Andar
Rio de Janeiro - RJ, Brazil
T: 55 21 2736582

Domingues, Rachel Bertol
Reporter
PETROBRAS
Rio de Janeiro, Brazil
T: 55 21 5343226 F: 55 21 220527

Domingues, Zilna Hoffmann, Sra.
Diretora
ITCF
Desembargador Motta 3384
Curitiba - PR 80430, Brazil
T: 55 41 234-1611 F: 55 41 222-2850
TL: 415899

Dominici, Maria Celeste Macedo, Ms.
Tec. Planejamento
CODEPLAN - GDF
SCN ED. CODEPLAN 2o. ANDAR
Brasilia - DF 70.775, Brazil
T: 55 61 322-4119 F: 55 61 226-1941

Domitra, Michael, Sr.
ILDES/FES
Rua. Lisboa 191
Sao Paulo - SP 05413, Brazil
T: 55 11 282-3900 F: 55 11 280-1916
TL: 1180616

Donati, Humberto, Sr.
ASSOCIACAO COMERCIAL DO RIO DE JANEIRO
Rio de Janeiro - RJ, Brazil

Donida, Andrea, Ms.
CANADA
SHIS 0115 Canjos Casa 13
Brasilia - DF, Brazil

Doria, Mauro Zambrotti
Camera
TVSBT
Campo de S.Cristovao, 105
Rio de Janeiro - RJ, Brazil
T: 55 21 5803688 / 5800678 F: 55 21 5809685

Dos Anjos, Miguel Pereira, Sr.
SINDICATO DOS TRABALHADORES DAS INDUSTRIAS PLASTICAS DE SAO PAULO
Rua Nicolau De Souza Queiroz 141
Sao Paulo - SP 04755, Brazil
T: 55 11 575-6914

Dos Passos, Claudia Vieira, Sra.
Tecnica
CONFEDERACAO NACIONAL DA INDUSTRIA
Av. Nilo Pecanha, 50 sala 3018
Rio de Janeiro - RJ 20044, Brazil
T: 55 21 262 0181 F: 55 21 262 1465
TL: 22634

Dos Santos, Hosana, Sra.
Sociologa
CENTRO DE RECURSOS AMBIENTAIS
Rua Rio Sao Frfancisco No.1 - Monte Serrat
Salvador - BA 40425, Brazil
T: 55 71 312-7192 F: 55 71 312-7198
TL: (071)7148Ramb

Dos Santos, Margaret Ferreira, Sra.
Pos-Graduado
EESC-USP
R 9 De Julho 1407
Sao Carlos - SP 13560, Brazil
T: 55 162 719 538

Dos Santos, Raimundo Jose Pereira, Sr.
Deputado Estadual
ASSEMBLEIA LEGISLATIVA
Praca D. Pedro Ii, # 130
Belem - PA 66000, Brazil
T: 55 91 225-3355

Dos Santos, Roberto Machado
TVT REDE DE COMUNICACAO DOS TRABALHADORES
Sao Paulo, Brazil

Dost, Paulo, Sr.
Tecnico
CONFEDERACAO NACIONAL DA INDUSTRIA
Av. Nilo Pecanha, 50 sala 3018
Rio de Janeiro - RJ 20044, Brazil
T: 55 21 262 0181 F: 55 21 262 1465
TL: 22634

Dourado, Ana Cristina Dubeux, Ms.
SOS CORPO
Rua do Hospicio, 859, Boa Vista
Recife - PE 50050, Brazil
T: 55 81 221-3018 F: 55 81 221-3947

Dourado, Joao Henrique
Journalist
HILL AND KNOWLTON
Av. Afranio de Melo Franco 135 /5 andar Leblon
Rio de Janeiro -RJ 2, Brazil
F: 55 21 2394697

Dourado, Maria Cristina Cezar De O. C., Sra.
Professor
UNIVERSIDADE FEDERAL DO PARA
Trav. Oliveira Belo 238/902
Belem - PA 66030, Brazil
T: 55 91 241-3702

Dourado, Miriam B. De F., Sra.
SOS CORPO
Rua do Hospicio 859 - Boa Vista
Recife - PE 50050, Brazil
T: 55 81 221-3018 F: 55 81 221-3947

Dourojeanni, Marc
Chefe Dept. De Meio Amb.
BCO. INTERAMERICANO DE DESENVOLVIMENTO
Caixa Postal 08748
Brasilia - DF 70312, Brazil
T: 55 61 321-7848 F: 55 61 321-3112
TL: 61-1204

Dressel, Gerhard
UNIVERSIDADE ABERTA - UNIVERTA
R: Monte Alegre, 35
Rio de Janeiro - RJ 20240, Brazil
T: 55 21224-1454 R

Drummond, Fabiana, Sra.
BRITISH EMBASSY
Setor de Embaixadas Sul Quadra 801 Conjunto K
Brasilia - DF, Brazil
T: 55 61 225-2710
TL: 0611360

Drumond, Yvens Lucio Do Amaral, Sr.
President
AMDA
Rua Silva Jardim 472 - Vila Isolino Almeida
Macae - RJ 28700, Brazil
T: 55 247 621766 F: 55 0247625110

Drumond, Zenilda
GAIA - ASSOC.CONS.M.AMBIENTE E PROD.INTEGRADA ALIM.AMAZONIA
Arse-51 - 39
Palmas, TO 77000, Brazil
T: 063 8621553/1108

Duarte, Andrea, Sra.
CONSTRUTORA NORBERTO ODEBRECHT
Rua Voluntarios da Patria, 89 7o. 3
Rio de Janeiro - RJ 22271, Brazil
T: 55 21 226 3842 F: 55 21 286 5495

Duarte, Jeanne
FREELANCER
Rua do Russel, 680/11 Gloria
Rio de Janeiro, Brazil
T: 55 21 2657097 F: 55 21 2200497

Duarte, Lilian Cristina Burlamaqui, Sra.
Student
PONTIFICIA UNIVERSIDADE CATOLICA DO RIO DE JANEIRO - PUC
Rua Marques de S.Vicente 225 - Vila dos Diretorios, Casa XIX - Gavea
Rio de Janeiro - RJ 22453, Brazil
T: 55 21 529-9493 F: 55 21 274-1296
TL: 31048

Duarte, Lucia Maria Alvarenga, Sra.
COMISSAO ESPECIAL LEI 1849/92
Rio de Janeiro - RJ, Brazil
T: 55 21 273 9442

Duarte, Maria Cecilia De Moraes
Reporter
SISTEMA GLOBO DE RADIO
R do Russel 434 /401 Gloria
Rio de Janeiro - RJ, Brazil
T: 55 21 205 8282 r 410/429 F: 55 21 2054096

Duarte, Maria Jose Ribeiro
Producer
CALANDRA FM
Av. Sete de Setembro 1015
Bage -RS 96200, Brazil
T: 532 424341 F: 532 422242

Duarte, Neide De Oliveira
Reporter
REDE GLOBO
R. Lopes Quintas 303
Rio de Janeiro -RJ 2, Brazil
T: 55 21 2947732

Duarte, Renato Barros, Sr.
Pesquisador
JARDIM BOTANICO DE RIO DE JANEIRO
Rua Francisco Otaviano 49/503 - Copacabana
Rio de Janeiro - RJ 22080, Brazil
T: 55 21 227-2181

Duarte, Ricardo, Mr.
Delegate
GOVERNMENT OF CANADA
AV. PAULISTA 854 - 5o. ANDAR
Sao Paulo - SP 01310, Brazil
T: 55 11 287-2122

Duarte, Ruy, Sr.
ASSOCIACAO COMERCIAL DO RIO DE JANEIRO
Rio de Janeiro - RJ, Brazil

Dumont, Annette, Sra.
Vice President
SOLIDARIEDADE FRANCA-BRASIL
Rua Uniao, 12 Santo Cristo
Rio de Janeiro - RJ 20220, Brazil
T: 55 21 253-1168 F: 55 21 220-1279

Dumont, Didier, Mr.
Professor
SOCIEDADE FRANCO BRASILEIRA DE ENSINO
728 Rua Pereira da Silva - Laranjeiras
Rio de Janeiro - RJ, Brazil
T: 55 21 285-3946

Dunbar, Richard
Media Coordinator
INTERNATIONAL NGO FORUM
Hotel Florida
Rio de Janeiro RJ, Brazil
T: 55 21 2855242

Dunin, Claudie, Sra.
Presidente
SOCIEDADE ZOOFILA EDUCATIVA
Av. Delfin Moreira 1044/701 - Leblon
Rio de Janeiro - RJ 22441, Brazil
T: 55 21 259-3926 F: 55 21 275-4846

Dunjwa-Blajberg, Jennifer, Ms.
Member
AFRICAN NATIONAL CONGRESS (SOUTH AFRICA)
Rua Buarque De Macedo 5/101-A, Flamengo
Rio de Janeiro - RJ 22221, Brazil
T: 55 21 205-1501 F: 55 21 267-6818

Dunker, Hans Joachim, Mr.
Consul General
CONSULADO GENERAL DA ALEMANHA
Rua.: Presidente C. de Campos, 417 - Laranjeiras
Rio de Janeiro - RJ, Brazil
T: 55 21 553 6777

Duran, Tulio H. F. Arvelo, Sr.
Professor
UNIVERSIDADE SANTA URSULA
Rua Jornalista Orlando Dantas, 59
Rio de Janeiro - RJ 22231, Brazil
T: 55 21 552-5422 F: 55 21 551-6446

Durans, Daniel Da Silva
TWENTY FIRST CENTURY INC.
Rio de Janeiro, Brazil

Durao, Bruno Ribeiro
ARTE RIO
Rio de Janeiro, Brazil

Durao, Jorge Eduardo Saavedra, Sr.
Diretor Coordenador Nacional
FASE
Rua Bento Lisboa 58
Rio de Janeiro - RJ 22221, Brazil
T: 55 21 285 2998 F: 55 21 205 3099
TL: 2134201

Dusek, Andre
Photographer
AGENCIA ESTADO
Av. Caetano Alvares, 55
Sao Paulo -SP 02598, Brazil
T: 55 11 8562210 F: 55 11 2652297

Dusilek, Darci, Sr.
Executive Director
VISAO MUNDIAL BRASIL
Rua Antonio Albuquerque 788 - Funcionarios
Belo Horizonte - MG 30110, Brazil
T: 55 31 221 8944 F: 55 31 227 2995
TL: 316052

Dutra, Alcides, Sr.
Coordenador
UFSC / PROJETO LARUS
Campus Universitario
Florianopolis - SC 88049, Brazil
T: 55 482 31 9302 F: 55 482 34 4069

Dutra, Claudia Glasser
Journalist
FREELANCER
Av. Afranio de Mello Franco 353/402
Rio de Janeiro RJ, Brazil
T: 55 21 2949401

Dutra, Pedro Paulo Loureiro, Mr.
CIA VALE DO RIO DOCE
Av. Graca Aranha, 26
Rio de Janeiro - RJ 20.030, Brazil
T: 55 21 272 4401 F: 55 21 220 0593

Earp, Ney Affonso De Sa, Sr.
Professor
DIOCESE DE PETRoPOLIS
Rua Dom Pedro I, 306
Petropolis - RJ 25610, Brazil
T: 55 242 42-6620

East, May, Sra.
OMAME - GAIA FOUNDATION
Rua Atlantica 771
Sao Paulo - SP 01440, Brazil
T: 55 11 852-2817 F: 55 11 852-2817

Eastland, Todd, Mr.
Telecommunications Officer
DEPARTMENT OF STATE
US Consulate Unit 3501
Rio de Janeiro - RJ 34030, Brazil

Eckert, Betina, Sra.
Rua Ludovico Geronasso 717/14
Curitiba - PR 82.500, Brazil
T: 55 41 256-3700

Eckhardt, Ulrike, Ms.
Eng. Agronoma
S A C T E S
Rua Cesar Leitao SIN
Cajazeiras - PB 58900, Brazil
T: 085 531-2133

Eghrari, Mahruv Vahdat, Ms.
COMUNIDADE INTERNACIONAL BAHA'I
Rua Nathalia Wolfgang 120
Araguari - MG 38440, Brazil
T: 55 34 241-5626

Eiras, Isabel Cristina
INSTITUTO DE ARQUITETOS DO BRASIL
Rua do Pinheiro, 10 Flamengo
Rio de Janeiro, Brazil
T: 55 21 2853480

Eksterman, Eleny Victoria
OSHO SPIRITUAL HEALTH ORGANIZATION
Av. Epitacio Pessoa, 1196/201
Rio de Janeiro - RJ 22.471, Brazil
T: 55 21 227-6236

El-Deir, Soraya, Ms.
Recife, Brazil
T: 45 081341-1465

Eleilson, Antonio, Sr.
Coordenador Do Setor De Distr.
CENTRO ECUMENICO DE DOCUMENTACAO E INFORMACAO - CEDI
Av. Higienopolis 983
Sao Paulo - SP 01238, Brazil
T: 55 11 825-5544 F: 55 11 825-7861
TL: 1126561 ECUM BR

Elliott, Joy
Senior Correspondent
EARTH SUMMIT TIMES
Luxor Regente, Av. Alantica
Rio de Janeiro RJ, Brazil
T: 55 21 2874212 F: 55 21 5218899

Emely, Akasaka
Correspondent
YOMIURI SHIMBUN
Av Oswaldo Cruz 149/1201
Rio de Janeiro, Brazil
T: 55 21 5514560 F: 55 21 5522279

Emidio, Adriana, Sra.
PAX UNIVERSAL
Av. Bras Leme, 2428
Sao Paulo - SP 02526, Brazil
T: 55 11 2997673

Endresen, H.E. Sigurd
Ambassador Brasilia
NORWEGIAN DELEGATION TO UNCED
Embaixada Real da Noruega, Caixa Postal 07 0670
Brasilia - DF 70351, Brazil
T: 55 61 243-8720 F: 55 61 242-7989

Engel, Wanda
Director
ASSOC. PROJETO RODA VIVA
R.Sao Francisco Xavier, 524/12037-F
Rio de Janeiro - RJ, Brazil
T: 55 21 2645196 F: 55 21 2645196

Engelhardt, Marc, Mr.
Estudante
Rua 2 de Dezembro 351/503
Rio de Janeiro - RJ 22220, Brazil

Ennes, Eduardo Luiz, Sr.
Rua Belizario Augusto 79/1001
Niteroi - RJ 24230, Brazil
T: 55 21 714-3906

Erb, Marilia Sa
PROVENT
Rua Uruguiana, 10 2 andar
Rio de Janeiro, Brazil
T: 55 21 2424618

Ernestino, Helio
Operador De Som
RADIO ROQUETE PINTO
Av. Erasmo Braga 118/ 11andar Castelo
Rio de Janeiro -RJ 2, Brazil
T: 55 21 2247727 F: 55 21 2524584

Escobar, Mauricio
Cinegrafista
WDR GERMAN TV CHANNEL
Rua Eca de Queiroz, 479 casa 12
Sao Paulo - SP, Brazil
T: 55 11 5721262 F: 55 11 5728959

Escobar, Omar Bula-, Sr.
Assistant Project Officer
WORLD FOOD PROGRAMME
SCH Quadra 02 Bloco 'B' Terreo
Brasilia - DF 70710, Brazil
T: 55 61 226-7872 F: 55 61 322-7080

Escosteguy, Angela, Ms.
Assessora
CONS.REG.MED.VET.
Rua Eudoro Berlink 920 Ap. 403
Porto Alegre - RS 90420, Brazil
T: 55 512 331-2360 F: 55 512 331-2360

Escudeiro, Marcelo Lavoyer, Mr.
Rua Professor Hernani Melo 80/702 - S. Domingos
Niteroi - RJ 24210, Brazil
T: 55 21 7178847

Eskinazi, Marcia Cristina S. A. V.
Editor
LIP CONSULTORIA DE COMUNICACAOS LTDA
Av Almirante Barroso 63/209
Rio de Janeiro, Brazil
T: 55 21 2203567 F: 55 21 2625401

Espindola, Albimarcia Das Neves
Biologa
FUNDACAO ESTADUAL DO MEIO AMBIENTE - MT
Av. Cel Escolastica, 515 Apto 704
Cuiaba - MT 78020, Brazil
T: 55 065 624-5227 F: 55 065 321-2171
TL: 652512

Espindola, Jaly, Mr.
M A E
R. Desembargador Westelabin 116
Curitiba - PR 8000, Brazil
T: 55 41 63-1367

Espinola, Ana Lucia Queiroz, Sra.
Coord. Educacao Ambiental
SUDEMA
Rua Monsenhor Walfredo Leal 181 - Tambia
Joao Pessoa - PB 58000, Brazil
T: 55 83 222-4663

Estery, Joao Leonel B, Sr.
GOVERNO DO ESTADO DO RIO DE JANEIRO
Rio de Janeiro - RJ, Brazil

Esteves, Joacir Lyra
Director
INSTITUTO DE ARQUITETOS DO BRASIL
Rua do Pinheiro, 10 Flamengo
Rio de Janeiro, Brazil
T: 55 21 2853480

Esteves, Jose Jorge Paz
Technician
REDE GLOBO
R. Lopes Quintas 303
Rio de Janeiro -RJ 2, Brazil
T: 55 21 2947732

Esteves, Maria Aparecida
ESTADO DE SAO PAULO
Av Rio Branco 151
Rio de Janeiro, Brazil
T: 55 21 2217111

Estides, Mirian F., Ms.
Sociologa
SQS 311 BL. I Apto 503
Brasilia - DF, Brazil
T: 55 61 244-8207

Ettori, Ligia De Castro, Ms.
Researcher
INSTITUTO FLORESTAL DE SAO PAULO
Caixa Postal 1322
Sao Paulo - SP 01059, Brazil
T: 55 11 952-8555 F: 55 11 204-8067
TL: 228-77BR

Eudima, Paulo
JOVEM PAN TV
Sao Paulo, Brazil

Ewart-White, Fraser
92 GLOBAL FORUM
Rio de Janeiro - RJ, Brazil

F.A.Carvalho, Carlos
FREELANCER
R.Almte.Alexandrino, 2966
Rio de Janeiro, RJ 2, Brazil
T: 021 2654990 F: 021 2654990

F.Stein, Richard
Camera
2DF-TV ALEMA
Av.Delfim Moreira, 662/704
Rio de Janeiro, RJ, Brazil
T: 021 2947434 F: 021 5122120

Faccioli, Cesar
GAZETA MERCANTIL
Rua Major Quedinho 90
Sao Paulo SP, Brazil
T: 55 11 256 3133

Fainguelernt, Estela Kaufman, Sra.
Directora Pos-Graduacao
UNIVERSIDADE SANTA URSULA
Rua Fernando Ferrari, 75
Rio de Janeiro - RJ 22231, Brazil
T: 55 21 551-5542 F: 55 21 551-6446

Faissal, Rogerio
Fotografo
EDITORA ABRIL
R.da Passagem, 123/8 andar
Rio de Janeiro - RJ, Brazil
F: 55 21 2759347

Falcao, Antonia Hestia Guimaraes
Coord. De Est. Ambientais
SUDEMA GOVERNAMENTAL - PB
Av. Walfredo Leal, 18
Joao Pessoa - PB 58000, Brazil
T: 55 83 222-4663 F: 55 83 224-0068

Falcao, Denis Adriano Holanda, Mr.
Representante
PREFEITURA MUNICIPAL DE SAO JOAO DO ARAGUIAIA
Praca Inacio Moura S/N
Para, Brazil

Falcao, Flavio
CHICO JR. & ASSOCIADOS
Rio de Janeiro, Brazil

Falcao, Heloisa Helene, Sra.
Raul Pompeia 9/401
Rio de Janeiro - RJ 22080, Brazil
T: 55 21 267-9083

Falcao, Raquel, Sra.
Director
CMB
Rua Avanhadava no. 756
Sao Paulo - SP 01306, Brazil
T: 55 11 2564354 F: 55 11 2577648

Falcao Filho, Horacio Arruda, Mr.
L E M N A
Av. Lineu de Paula Machado 104/102- Jardim Botanico
Rio de Janeiro - RJ 22470, Brazil
T: 55 21 259-7646 F: 55 21 756-4643

Fanias, Maria De
Reporter
REVISTA MULHER DE HOJE
Rua do Russel, 604/ 8 andar
Rio de Janeiro, Brazil
T: 55 21 5554225 F: 55 21 2059998

Fanny, Rubio, Mr.
Sociologa
S A C T E S
R Almeida Cunha,364 - Boa Vista
Recife - PE 50050, Brazil
T: 55 81 221 0075

Faraco, Marcia
IBAMA
Rio de Janeiro, Brazil

Farage, Fabiola Beatriz Resende, Ms.
Jornalista
Rua dos Timbiras # 1633
Belo Horizonte - MG 30140, Brazil
T: 55 31 224-0782

Farah, David, Mr.
Area Director
SUMMER INSTITUTE OF LINGUISTICS
SAI/NO, Lote 'D', Bloco 03
Brasilia - DF 70770, Brazil
T: 55 61 272-1200

Farah, Ivete Mello Calil, Sra.
Arquiteta
FLAMA
Rua Marques de Abrantes, 55 - Flamengo
Rio de Janeiro - RJ 22230, Brazil
T: 55 21 265-5690

Faria, Adalberto Borges, Sr.
Supervisor De Area/Mt
VISAO MUNDIAL BRASIL
R. Campo Grande, 427 - Galeria Medalhao - Centro
Cuiaba - MT 78020, Brazil
T: 55 125 321-8642

Faria, Alcides B., Sr.
Presidente
ECOA - ECOLOGIA E ACAO
Rua Petunia 219 Bairro Cidade Jardim
Campo Grande - MS, Brazil
T: 55 67 382 9065

Faria, Alvaro Jose De Souza
INSTITUTO DE ARQUITETOS DO BRASIL
Rua do Pinheiro, 10 Flamengo
Rio de Janeiro, Brazil
T: 55 21 2853480

Faria, Ana Dalva
Assessora De Imprensa
PRO-NATURA INSTITUTO BRASILEIRO PESQ.EST.AMBIENTAIS
Rua Monte Alegre, 374
Rio de Janeiro - RJ, Brazil
T: 55 21 2243994 F: 55 21 2226742

Faria, Andrea Alice Da Cunha, Ms.
Agronoma
REDE BRASILEIRA AGROFLORESTAL - REBRAF
Rua Visconde de Piraja 111/713 - Ipanema
Rio de Janeiro - RJ 22410-010, Brazil
T: 55 21 521-7846 F: 55 21 521-1543

Faria, Arlete Bonelli H. De
Reporter
JORNAL A CRITICA
Rede Calderaro de Com. - SCS Ed. Denasa Sala 304
Brasilia - DF, Brazil
T: 55 61 2256451 F: 55 61 2264705

Faria, Marcos Henrique De
Reporter-Fotografico
JORNAL A CRITICA
Rede Calderaro de Com. - SCS Ed. Denasa Sala 304
Brasilia - DF, Brazil
T: 55 61 2256451 F: 55 61 2264705

Faria, Maria Do Carmo Bettencourt De, Sra.
Reitora
UNIVERSIDADE SANTA URSULA
Rua Fernando Ferrari, 75
Rio de Janeiro - RJ 22231, Brazil
T: 55 21 551-5542 F: 55 21 551-6446

Faria, Tereza Helena Oliveira De
Fotografa
REVISTA VISAO
Av.Rio Branco, 177/6 andar
Rio de Janeiro, RJ, Brazil
T: 021 2621085 F: 021 2200805

Farias, Adelia Oliveira, Ms.
TRABALHADORA RURAL
Rua Santa Cruz, 281
Sao Paulo - SP, Brazil

Farias, Ana Maria Correa
EDITORA AZUL
Rio de Janeiro, Brazil

Farias, Ben Hur, Sr.
ADFG / AMIGOS DA TERRA
Rua Miguel Tostes, 694
Porto Alegre - RS 90420, Brazil
T: 55 512 332-8884 F: 55 512 332-8884

Farias, Celeide De Queiroz, Sra.
President
CONSELHO MUNICIPAL DOS DIREITOS DA MULHER - CAMPINA GRANDE/PB
Rua de Campina Grande / Desembargador Trindade 435
Campina Grande - PB 58000, Brazil
T: 55 83 341-2917

Farias, Jorge Antonio De, Sr.
Engenheiro Florestal
ASS. FUMICULTORES BRASIL
Rua Julio de Castilhos, 1031
Santa Cruz do Sul-RS 96800, Brazil
T: 55 51 713 2722 F: 55 51 713 2109
TL: 510021

Farias, Luiz Heron Bueno De, Mr.
Coordenador
MOVIMENTO EM DEFESA DA VIDA
Av. Guaiba 316
Porto Alegre - RS 91700, Brazil
T: 55 512 248-4008 F: 55 512 343-6671

Farias, Marcelo Cesar Bastos De
92 GLOBAL FORUM
Rio de Janeiro - RJ, Brazil

Farinatti, Silviana Maria Veras
Journalist
BLOCH EDITORES
Rua Anita Gaibaldi, 38/apt 302
Rio de Janeiro, Brazil
T: 55 21 2566527

Farjalha, Luciana
Journalist
SOCIEDADE CIVIL DO BEM-ESTAR FAMILIAR NO BRASIL -BEMFAM
Av. Republica do Chile 230/ 17-andar
Rio de Janeiro -RJ, Brazil
T: 55 21 2102448 F: 55 21 2204057

Fassini, Sonia
CEMINA
R.Barao do Flamengo, 22/304
Rio de Janeiro - RJ, Brazil
T: 55 21 2056297 F: 55 21 5563383

Fattore, Antonio, Mr.
Cooperante
PROGETTO SVILUPPO- CGIL
Rua Dona Brigida, 299 - Vila Mariana
Sao Paulo - SP 04111, Brazil
T: 55 11 575 5155

Faustino, Renato, Sra.
Professor
PREFEITURA MUNICIPAL
Praca Maua s/n 1 andar
Santos - SP 11010, Brazil
T: 55 132 342-2093 F: 55 132 345-809
TL: 131048

Favero, Luiz A., Mr.
Assessor Tecnico
FUNDACAO PARQUE TECNOLOGICO - PB
Rua Floriano Peixoto, 715 3rd andar
Campina Grande - PB 58100, Brazil
T: 55 83 341 2210 F: 55 83 321 7424

Fayet, Augusto Cesar, Sr.
Tecnico
INSTITUTO DE ESTUDOS AMAZONICOS E AMBIENTAIS
SCLN 210 Ed. Center Bloco C salas 211 e 212
Brasilia - DF 70062, Brazil
T: 55 61 347-9874 F: 55 61 347-9874

Fechado, Luiz Eduardo Garrido, Sr.
BRADESCO VISA
Av. Rio Branco 131 3$
Rio de Janeiro - RJ 20021, Brazil
T: 55 21 224-0031

Feiden, Alberto, Sr.
C E A - CAMPO / IAFA / SINDASPP
Rua Alagoas 504
Mal. Cdo. Rondon-PR 85960, Brazil
T: 55 0452 542032 F: 55 0452 542032

Feijo, Ateneia
Reporter/Editor
TEXTO E ARTE AM (FREELANCER)
R. Prof. Brandao Filho 60/404 Leblon
Rio de Janeiro- RJ, Brazil
T: 55 21 2391478 F: 55 21 2391478

Feitosa, Roberto, Sr.
GULF PROMOCOES
Rua Do Carmo 11/802 Centro
Rio de Janeiro - RJ, Brazil
T: 55 21 577-6476 F: 55 21 577-1195

Feitosa, Wallace
Fotografo
PANROTAS EDITORA
Av.Jabaquara, 1761
Sao Paulo - SP, Brazil
T: 55 11 2750211 F: 55 11 2761602

Felden, Adriana
Photographic Reporter
FREELANCER
Ladeira Ari Barroso, No. 4 ap. 101, Leme
Rio de Janeiro RJ, Brazil

Feldhaus, Guilherme, Sr.
Superintendente
CLUBE ECOLOGICO INT. DA AMAZONIA
Av. Jose Bonifacio 1927
Belem - PA 66000, Brazil
T: 55 91 229-9362

Felgueiras, Jose Ayres
Technician
REDE GLOBO
R. Lopes Quintas 303
Rio de Janeiro -RJ 2, Brazil
T: 55 21 2947732

Felice, Gisele Daltrini, Sra.
Professora
UNIVERSIDADE DO RIO GRANDE
Caixa Postal 1015 Cassimo
Cassimo-RG 96200, Brazil
T: 55 053232-3200 F: 55 053262-1456

Felinto, Valdo, Sr.
INSTITUTO TIBA
Rua Ingles de Souza 296
Rio de Janeiro - RJ, Brazil
T: 55 21 274-1762

Felipe, Bernadete
Reporter
RADIO ROQUETE PINTO
Av. Erasmo Braga 118/ 11andar Castelo
Rio de Janeiro -RJ 2, Brazil
T: 55 21 2247727 F: 55 21 2524584

Felix, Joao Carlos Da Rosa
HELIO ALONSO
Rua Muniz Barreto, 51
Rio de Janeiro RJ, Brazil
T: 55 21 5515448

Felix, Selma Abreu, Ms.
GOVERNO DE SAO PAULO
Sao Paulo - SP, Brazil

Felix, Teresa Cristina
92 GLOBAL FORUM
Rio de Janeiro - RJ, Brazil

Fenerich, Carlos Henrique
Photographer
AGENCIA ESTADO
Av. Caetano Alvares, 55
Sao Paulo -SP 02598, Brazil
T: 55 11 8562210 F: 55 11 2652297

Fernandes, Annemarie
FUNAI
Rua Francisco Otaviano, 86 apto 704 Copacabana
Rio de Janeiro - RJ 22080, Brazil
T: 55 21 247 9437

Fernandes, Cristina De Fatima
LEGIAO DA BOA VONTADE
Av. Rudge 700 Bom Retiro
Sao Paulo -SP 01134, Brazil
T: 55 21 2338381/R:223 F: 55 21 2634025

Fernandes, Cristina Maria
Editor
EDITORA ABRIL
R.da Passagem, 123/8 andar
Rio de Janeiro, RJ, Brazil
T: 021 5468282 F: 021 2759347

Fernandes, Fabio De Borba, Mr.
Engenheiro Florestal
UFMT
Av. Fernando Correa Da Costa s/n - Coxipo
Cuiaba - MT, Brazil
T: 55 065 315 8638 F: 55 065 361 1119

Fernandes, Glaucia Mendes, Ms.
FUNDACAO BIODIVERSITAS
Rua Maria Vaz de Melo 71
Belo Horizonte - MG 31250, Brazil
T: 55 31 443-2119

Fernandes, Helio
TRIBUNA DA IMPRENSA
Rio de Janeiro, Brazil

Fernandes, Iara Regina De Azevedo, Sra.
Rua Bonsucesso 446/502
Rio de Janeiro - RJ 21041, Brazil
T: 55 21 2808934

Fernandes, Inacia Assis, Ms.
Assistente/Executive Secretary
REDE BRASILEIRA AGROFLORESTAL - REBRAF
Rua Visconde de Piraja 111/713
Rio de Janeiro - RJ 22410-010, Brazil
T: 55 21 521-7896 F: 55 21 521-1593

Fernandes, Ivone Silva, Ms.
SAUDE E VIDA
Rua Desembargador Alfredo Russel, 19 - A
Sao Paulo - SP, Brazil

Fernandes, Jose Boaventura
Technician
REDE GLOBO
R. Lopes Quintas 303
Rio de Janeiro -RJ 2, Brazil
T: 55 21 2947732

Fernandes, Luciana Cristina Matsuda, Sra.
Membro
APOENA
Av. das Americas, 472
Alvares Machado 19160, Brazil
T: 055 018273-1580

Fernandes, Lucio Ribeiro
MANCHETE
Rio de Janeiro, Brazil

Fernandes, Manlio
ECOTROPIC
1880 East River Road
Rio de Janeiro - RJ, Brazil
T: 55 21 577-6476 F: 55 21 577-1195

Fernandes, Marcos Inacio, Mr.
EMATER/ACRE
Jardim Solar -Q-D-Conj.19
Rio Branco - AC 69.900, Brazil
T: 55 68 226-4897

Fernandes, Ottoni, Jr
Editor
GAZETA MERCANTIL
Rua Major Quedinho 90
Sao Paulo SP, Brazil
T: 55 11 256 3133

Fernandes, Rosangela De Jesus
Radialista/Jornalista
IBASE
Rua Vicente de Souza 29 Botafogo
Rio de Janeiro - RJ, Brazil
T: 55 21 286 6161/8642 F: 55 21 286 0541

Fernandes, Sandra, Ms.
Tec.Nivel Apoi
SERVICO SOCIAL DO COMERCIO
Rua Voluntarios Da Patria, 169
Rio de Janeiro - RJ 22270, Brazil
T: 55 21 286-5152 F: 55 21 286-2638

Fernandes, Valeria Dantas
FREELANCER
Rua Marechal Camoro, 210/ 1 andar
Rio de Janeiro, Brazil
T: 55 21 2207779 F: 55 21 2207779

Fernandes, Valmira Goncalves, Sra.
Geografa
INSTITUTO GAIA
Rua Barao de Contendas, 50 apt.301 - Aflitos
Recife - PE 52050, Brazil
T: 55 81 231-0287

Feron, Jean-Jacques
INTERNATIONAL TELEVISION NEWS
Av. Rio Branco, 25/9 andar
Rio de Janeiro RJ, Brazil
T: 55 21 2335480

Ferrao, Cristina, Sra.
EDITORA INDEX / EDITORA LIBRIS / ISLAND PRESS
Av. Rio Branco 45 / 1707
Rio de Janeiro - RJ 20000, Brazil
T: 55 21 516-2336 F: 55 21 253-3507

Ferrao, Jose Cardoso, Mr.
Interpreter
CHRISTIAN ENVIRONMENT ALLIANCE
R. Alcides Carneiro,40
Passa Quatro - MG 37460-000, Brazil
T: 55 35 3711137

Ferrarese, Claudio, Sr.
Representative
FONDAZIONE SIPEC / ITALIA
Av. Copacabana 945/1001
Rio de Janeiro - RJ 22060, Brazil
T: 55 21 255-3352 F: 55 21 242-9871
TL: 32628

Ferrari, Antenor, Mr.
Socio Fundador
FUNDACAO BRASIL-FRANCA
Av. Beira-Mar 406/305
Rio de Janeiro - RJ 20021, Brazil
T: 55 21 2202644 F: 55 21 2627782

Ferrari, Josemar
SISTEMA JORNAL DO BRASIL
Av. Brasil 500
Rio de Janeiro -RJ, Brazil
T: 55 21 5854422 F: 55 21 5854428

Ferrari, Robisson Gomes
Engineer
AMERICA DO SUL PRODUCOES DE DOCUMENTARIOS LTDA (NAMBEI TSUSHIN)
Av. Paulista 509 /11 andar conj.1104 Cerqueira Cesar
Sao Paulo -SP 01311, Brazil
T: 55 11 2886141 F: 55 11 2512065

Ferrari, Sergio
Technician
REDE GLOBO
R. Lopes Quintas 303
Rio de Janeiro -RJ 2, Brazil
T: 55 21 2947732

Ferraz, Ana Cristina Novaes, Ms.
Tecnica
SEPLAN - PE
Av. 17 De Agosto 892/304 - Casa Forte
Recife - PE 52.060, Brazil

Ferreira, Amauri Costa
LEGIAO DA BOA VONTADE
Av. Rudge 700 Bom Retiro
Sao Paulo -SP 01134, Brazil
T: 55 21 2338381/R:223 F: 55 21 2634025

Ferreira, Antonio Sergio, Sr.
Desenhista
INSTITUTO FLORESTAL
Rua do Horto 931
Sao Paulo - SP 02377, Brazil

Ferreira, Aurea, Mr.
ASSOCIACAO CIVIL GREENPEACE
Rua Mexico 21, Grp 1301, Centro
Rio de Janeiro - RJ 20031, Brazil
T: 55 21 262-7318 F: 55 21 240-1690

Ferreira, Carlos Alberto, Sr.
Rua Cura D'ars 64/01 Prado
Bolo Hte - MG 20410, Brazil
T: 334 9084

Ferreira, Cesar
Technician
REDE GLOBO
R. Lopes Quintas 303
Rio de Janeiro -RJ 2, Brazil
T: 55 21 2947732

Ferreira, Damy, Mr.
Religiao Batista
CONVENCAO BATISTA CARIOCA
Rua Senador Furtado no.12
Rio de Janeiro 20270, Brazil
T: 2840988

Ferreira, Eliana, Sra.
Assessora Internacional
ORG. COOP. BRASILEIRAS
SQS, Ed. Baracat, 4. andar
Brasilia - DF Cep.: 70309-900, Brazil
T: 55 61 225 0275 F: 55 61 226 8766
TL: 1879

Ferreira, Evangelista
Fotografo
LEGIAO DA BOA VONTADE
Av.Rudge, 700
Sao Paulo - SP, Brazil
T: 55 11 2202599 F: 55 11 2205196

Ferreira, Helio
Technician
REDE GLOBO
R. Lopes Quintas 303
Rio de Janeiro -RJ 2, Brazil
T: 55 21 2947732

Ferreira, Helio Divino, Mr.
Advogado
OAB/GO UFG
R Bh Q 28 L 07 B Capuava
Goiania - GO 74450, Brazil
T: 55 62 271 6208

Ferreira, Iedo
JORNAL MAIORIA FALANTE
Rio de Janeiro, Brazil

Ferreira, Isabelle Martins Dos Santos, Ms.
Assistente Da Delegacao
CIRCES INTERNACIONAL DELEGACAO DO BRASIL
Araujo Lima,47 - apto 401
Rio de Janeiro - RJ, Brazil
T: 55 21 288 4247

Ferreira, Jairo Roserto
Audio Journalist
REDE MANCHETE
Rua do Russel, 804 Gloria
Rio de Janeiro, Brazil
T: 55 21 2850033

Ferreira, Joao Flavio
MANCHETE
Rio de Janeiro, Brazil

Ferreira, Joao Marcelo
Reporter
BLOCH EDITORES
Rua do Russel 804, 8 andar
Rio de Janeiro, Brazil
T: 55 21 2050295 F: 55 21 2059998

Ferreira, Jorge
FORUM DO FUTURO
Rio de Janeiro, Brazil

Ferreira, Jorge Antonio De Paula, Mr.
IBAM
Rio de Janeiro - RJ, Brazil

Ferreira, Jorge De Faria Vieira
Editor
JORNAL O DIA
Rua Riachuelo
Rio de Janeiro, Brazil
T: 55 21 2728034

Ferreira, Leila Da Costa, Ms.
Professor
UNIVERSIDADE ESTADUAL DE CAMPINAS
NEPAM UNICAMP Caixa Postal 1170
Campinas - SP 13081, Brazil
T: 55 192 398151 F: 55 192 394717
TL: 1150

Ferreira, Luiz Roberto Alves, Sr.
Presidente
CONDEMA-OSASCO
Rua Narciso Sturlini, 111 sala 103
Osasco, SP 06010, Brazil

Ferreira, Manoel Eduardo Tavares
JORNAL PAU BRASIL
Ribeirao Preto, Brazil

Ferreira, Maria Da Gloria Nin, Sra.
Diretoria
CBCISS
Rua Fonte da Saudade 47/103
Rio de Janeiro - RJ 22471, Brazil
T: 55 21 226-2833

Ferreira, Maria De Fatima, Sra.
Assistente Social
IBFAN
Rua Joaquim Palhares 267/503 Pca da Bandeira
Rio de Janeiro - RJ, Brazil
T: 55 21 293-0064

Ferreira, Maria De Fatima De Araujo, Sra.
Socia
SOCIEDADE NORDESTINA DE ECOLOGIA
Rua Pessoa de Melo, 355 - Madalena
Recife - PE 50710, Brazil
T: 55 81 227-2487 F: 55 81 222-3361

Ferreira, Maria De Lourdes, Ms.
Rua Correa Dutra 117/202 - Catete
Rio de Janeiro - RJ 22210, Brazil
T: 55 21 245-4109

Ferreira, Mario
Camera
2DF-TV ALEMA
Av.Delfim Moreira, 662/704
Rio de Janeiro - RJ, Brazil
T: 55 21 2947434 F: 55 21 5122120

Ferreira, Marly Prata, Sra.
CENTRO DORJE JIG JE
Rua Nascimento Silva 76/304
Rio de Janeiro - RJ 22421, Brazil
T: 55 21 521 4600

Ferreira, Nelio
Assistant Photographer
GRUPO NOVO DE CINEMA E TV
Rua Morechel Niemeyer 24
Rio de Janeiro - RJ, Brazil
T: 55 21 2663637 F: 55 21 2250435

Ferreira, Renato, Mr.
Gerente
PREFEITURA MUNICIPAL PORTO ALEGRE-GUAIBA VIVE
Usina Do Gasometro
Porto Alegre - RS CEP: 90.010, Brazil
T: 55 512 228-8769

Ferreira, Rosa Monica, Sra.
Director
CMB
Rua Avanhadava no. 756
Sao Paulo - SP 01306, Brazil
T: 55 11 2564354 F: 55 11 2577648

Ferreira, Rosangela Sales, Ms.
CIA VALE DO RIO DOCE
Av. Graca Aranha, 26
Rio de Janeiro - RJ 20.030, Brazil
T: 55 21 272 4401 F: 55 21 220 0593

Ferreira, Rosely
BRASIL ENERGIA
Rio de Janeiro, Brazil

Ferreira, Tania, Ms.
UNDP/UNESCO
Av Joao Ribeiro,497 Casa 06
Rio de Janeiro - RJ 20760, Brazil
T: 55 21 591 9432

Ferreira, Valdir
Reporter
REDE GLOBO
R. Lopes Quintas 303
Rio de Janeiro -RJ 2, Brazil
T: 55 21 2947732

Ferreira, Vera Lucia Da Silva
PREFEITURA DO RIO DE JANEIRO
Av Presidente Vargas 817 17 andar, Centro
Rio de Janeiro RJ, Brazil
T: 55 21 2525422 F: 55 21 2244973

Ferreira Barbosa, Alda, Ms.
Prof.
MINISTERIO DA EDUCACAO MUSEU NACIONAL-RJ
Rua das laranjeiras,21/704
Rio de Janeiro - RJ, Brazil
T: 55 21 265 8440

Ferreira Carneiro, Jaemio, Mr.
FUNDACAO UNIVERSO E VIDA-UNIDA
Almirante Barroso, 410 - Centro
Joao Pessoa - PB 58020, Brazil
T: 083 221 8581

Festa, Regina
ICAE
Rede Mulher C.P.4651
Sao Paulo SP 01061, Brazil
T: 55 11 627050 F: 55 11 8713457

Fialho, Alaor
SISTEMA JORNAL DO BRASIL
Av. Brasil 500
Rio de Janeiro -RJ, Brazil
T: 55 21 5854422 F: 55 21 5854428

Fialho, Ana Maria
Director
INSTITUTO DE ARQUITETOS DO BRASIL
Rua do Pinheiro, 10 Flamengo
Rio de Janeiro, Brazil
T: 55 21 2853480

Fialho, Aurea Vieira, Sra.
President
ZONTA CLUBE - RIO
Rua Tonelero 295/702
Rio de Janeiro - RJ 22030, Brazil
T: 55 21 2573477

Fialho, Maria Elisabeth Arruda, Sra.
Tecnica
CONFEDERACAO NACIONAL DA INDUSTRIA
Av. Nilo Pecanha, 50 sala 3018
Rio de Janeiro - RJ 20044, Brazil
T: 55 21 262 0181 F: 55 21 262 1465
TL: 22634

Figols, Francisca B., Ms.
JAMA
Rua Juri 16, Vila Madalena
Sao Paulo - SP 03443, Brazil
T: 55 11 210-1301 F: 55 11 210-1301

Figueira, Carlos A. P.
Technician
REDE GLOBO
R. Lopes Quintas 303
Rio de Janeiro -RJ 2, Brazil
T: 55 21 2947732

Figueiredo, Andrea Coelho De, Sra.
Biologa
ELETRONORTE
Scn-Q06, Conj.A, Bloco C-Venancio 3000, Sala 313
Brasilia - DF 70000, Brazil
T: 55 61 212-6157 F: 55 61 225-9755
TL: 1279

Figueiredo, Francisco De Assis Gomes De, Sr.
Counseler
GRUPO CONSCIENCIA ECOLOGICA
Rua Afonso Cesar Siqueira, 212 ap. 18
S.Jose dos Campos-SP 12243, Brazil
T: 55 123 234291

Figueiredo, Maria Patricia De L.
CONFEDERACAO NACIONAL DA INDUSTRIA
Rio de Janeiro, Brazil

Figueiredo, Pedro Edson
MANCHETE
R. do Russel 804
Rio de Janeiro -RJ 2, Brazil
T: 55 21 2652012 F: 55 21 2855148

Figueiroa, Americo Magno De Souza
Reporter
REDE GLOBO
R. Lopes Quintas 303
Rio de Janeiro -RJ 2, Brazil
T: 55 21 2947732

Figueredo, Anne Elizabeth Grangues
Editor
TVE - JORNAL DA ASDUER
Rua Deznove de Fereira, 56 CASA 11 Botafogo
Rio de Janeiro, Brazil
T: 55 21 2662143

Figueredo, Mauricio De Pavia
Editor
JORNAL DOS SPORTS
Rua Temente Possale, 15 Centro
Guaz Vermelho, Brazil
T: 55 2428592

Figueredo, Paulo
TVE
Rua Generol Venancio Flores 580/204 Leblon
Rio de Janeiro, Brazil
T: 55 21 2943067

Figueredo, Paulo Fernando De
Reporter
JORNAL DOS SPORTS
Rua Temente Possale, 15 Centro
Guaz Vermelho, Brazil
T: 55 2428592

Fila, Raquel, Ms.
Eng. Florestal
I T C F
Rua Fernando de Noronha,631 Ed. Castelo de Viena
Londrina PR 86020, Brazil
T: 55 0432232631 F: 55 0432223026

Filho, Acyr Ferreira De Souza
Technician
REDE GLOBO
R. Lopes Quintas 303
Rio de Janeiro -RJ 2, Brazil
T: 55 21 2947732

Filho, Amauri Severino Dos Santos
Reporter
SISTEMA GLOBO DE RADIO
Rua do Russel, 434
Rio de Janeiro - RJ, Brazil
T: 55 21 2058282 F: 55 21 2054096

Filho, Antonio Araujo Fanzeres
Technician
REDE GLOBO
R. Lopes Quintas 303
Rio de Janeiro -RJ 2, Brazil
T: 55 21 2947732

Filho, Antonio Castelo Branco Da C.
Reporter
REDE GLOBO
R. Lopes Quintas 303
Rio de Janeiro -RJ 2, Brazil
T: 55 21 2947732

Filho, Antonio Gaspar
Editor
ESTADO DE SAO PAULO
Av. Engenheiro Caetano Alvares 55
Sao Paulo -SP 02598, Brazil

Filho, Arlindo
Fotografo
LEGIAO DA BOA VONTADE
Av.Rudge, 700
Sao Paulo - SP, Brazil
T: 55 11 2202599 F: 55 11 2205196

Filho, Audilio A. C.
Techinician
ESTADO DE SAO PAULO
Av. Engenheiro Caetano Alvares 55
Sao Paulo -SP 02598, Brazil

Filho, Caetano Scanavino
Director
CENTRO DE ESTUDOS AVANCADOS EM PROMOCAO SOCÍAL
Av. Borges Leal 2284
Santarem PA, Brazil
T: 55 91 5225090 F: 55 91 5231083

Filho, Carlos Vieira Peixoto
Editor
REDE GLOBO
R. Lopes Quintas 303
Rio de Janeiro -RJ 2, Brazil
T: 55 21 2947732

Filho, Durval Santana
MANCHETE
R. do Russel 804
Rio de Janeiro -RJ 2, Brazil
T: 55 21 2652012 F: 55 21 2855148

Filho, Eduardo Bruno
Assistant Editor
REVISTA FLUMINENSE
Rua Republica do Libano, 61 Sala 1120, Centro
Rio de Janeiro RJ, Brazil
T: 55 21 2429809

Filho, Ewaldo Schleder
Editor Boletim Eco Prensa
INSTITUTO GAIA DO BRASIL
Rua Mateus Leme 240/301
Curitiba - PR, Brazil
T: 55 41 223 9381/254 8153

Filho, Ezio Sertorio
Techinician
AGENCIA ESTADO
Av. Caetano Alvares, 55
Sao Paulo -SP 02598, Brazil
T: 55 11 8562210 F: 55 11 2652297

Filho, Farid Germano
Reporter
RADIO GAUCHA
Av Ipiranga 1075
Porto Alegre - RS 90060, Brazil
T: 55 512 223 6600 F: 55 512 2336600 r 57

Filho, Floriano Dias Viera, Sr.
SINDICATO DOS URBANITARIOS - RJ
Av. Alda Garrido No.240 Barra Da Ti Fuca
Rio de Janeiro - RJ, Brazil

Filho, Francisco Alves Das Chagas
Reporter
FRILAS SERVICOS JORNALISTICOS
Rua Anfilofio de Carvalho, 29/407
Rio de Janeiro RJ 2, Brazil
T: 55 21 2205444

Filho, Gerson T. Caffe
Editor
EDITORIA MERGULHAR
Rua Prof. Alfredo Gomes, 3 Botafogo
Rio de Janeiro RJ 22, Brazil
T: 55 21 2867445

Filho, Guilherme Silva
MANCHETE
Rio de Janeiro, Brazil

Filho, Idalicio Manoel
COORDENADORIA DE COM. SOCIAL-PREFEITURA DO RIO DE JANEIRO
R. Afonso Cavalcante 455/3/1520 Cidade Nova
Rio de Janeiro -RJ 2, Brazil
T: 55 21 2931744 F: 55 21 2737633

Filho, Ivo Carvalho
Journalist
FREELANCER
R. Almirante Alexandrino 3780 Bl/G2 Apt. S102 Santa Tereza
Rio de Janeiro -RJ 2, Brazil
T: 55 21 2656479

Filho, Joao Costa
Reporter
REVISTA VISAO
Av.Rio Branco, 177/6 andar
Rio de Janeiro - RJ, Brazil
T: 55 21 2621085 F: 55 21 2200805

Filho, Joao Henrique Da Silva, Mr.
CIA VALE DO RIO DOCE
Av. Graca Aranha, 26
Rio de Janeiro - RJ 20.030, Brazil
T: 55 21 272 4401 F: 55 21 220 0593

Filho, Joao Zacharias
Editor
MOMENTO SINDICAL EMEUE EDITORA
R. do Riachuelo 330/GR 214 Centro
Rio de Janeiro -RJ 2, Brazil
T: 55 21 2522676 F: 55 21 2221914

Filho, Jose Cabral
Sound Operator
RADIO TUPI
Rua do Livramento 189/8th Floor
Rio de Janeiro - RJ, Brazil
T: 55 21 203 2421 r 34 F: 55 21 2637323

Filho, Jose Domingues De Godoy, Sr.
GRUPO DE ESTUDOS DOS RECURSOS NATURAIS NAO RENOVAVEIS E ENERGETICOS
Caixa Postal 998
Cuiaba - MT 78000, Brazil
T: 55 65 361-5570
TL: 1371

Filho, Jose Maria Paschoalick
JORNAL PAU BRASIL
Ribeirao Preto, Brazil

Filho, Lauro Monteiro, Sr.
Presidente
ABRAPIA
R. Pinheiro Machado 39 - Laranjeiras
Rio de Janeiro - RJ 22231, Brazil
T: 55 21 5538181 F: 55 21 5525829

Filho, Lourival Santana
Reporter
ESTADO DE SAO PAULO
Av. Engenheiro Caetano Alvares 55
Sao Paulo -SP 02598, Brazil

Filho, Lourivaldo Oliveira
Reporter
MANCHETE LTDA
Rua do Russel 804 (andar
Rio de Janeiro, Brazil
T: 55 21 2850033 F: 55 21 2089098

Filho, Mario Elena
Operador De Som
RADIO ROQUETE PINTO
Av. Erasmo Braga 118/ 11andar Castelo
Rio de Janeiro -RJ 2, Brazil
T: 55 21 2247727 F: 55 21 2524584

Filho, Mario Honorio Teixeira
REVISTA ECORIO
Rio de Janeiro - RJ, Brazil

Filho, Masao Goto
Photographer
AGENCIA ESTADO
Av. Caetano Alvares, 55
Sao Paulo -SP 02598, Brazil
T: 55 11 8562210 F: 55 11 2652297

Filho, Murilo Melo
Jornalista
REVISTA MANCHETE
Rua do Russel, 804
Rio de Janeiro - RJ, Brazil
T: 55 21 2652012

Filho, Nelson Pompeo
FBCN - JORNAL INFORMATIVO
Rio de Janeiro, Brazil

Filho, Oscar Maron
Journalist
FREELANCER
Barao de Ipanema 102 -1000
Rio de Janeiro -RJ, Brazil
T: 55 21 2573167

Filho, Ovidio De Abreu
MADE FOR TV
Rio de Janeiro, Brazil

Filho, Paulo Studart, Mr.
SQS, 209 Bloco I Apto 306
Brasilia - DF 70272, Brazil
T: 55 61 223 2673

Filho, Rubem Didini
Diretor
TV RIO
R. Miguel de Frias, 57 Estacio
Rio de Janeiro-RJ 20, Brazil
T: 55 21 2939888 F: 55 21 2938626

Filho, Rui Machado Lima
Editor
REDE GLOBO
R. Lopes Quintas 303
Rio de Janeiro -RJ 2, Brazil
T: 55 21 2947732

Filho, Sebastiao Leite
Technician
REDE GLOBO
R. Lopes Quintas 303
Rio de Janeiro -RJ 2, Brazil
T: 55 21 2947732

Filho, Sergio Adeodato
GAZETA MERCANTIL
Rua Major Quedinho 90
Sao Paulo - SP, Brazil
T: 55 11 256 3133

Filho, Tales Henrique Da C.
FREELANCER
Rio de Janeiro, Brazil

Filho Barros, Luiz, Sr.
Ass. Tecnico
IMAIAM
Rua Recife no.3280
Manaus-AM 69000, Brazil

Filipe, Eduardo
ESTADO DE SAO PAULO
Av. Rio Branco 151 - Sobre - Loja sala 201 a 204
Rio de Janeiro RJ, Brazil
T: 55 21 2217111

Fillus, Acyr Lacerda
Reporter
REDE GLOBO
R. Lopes Quintas 303
Rio de Janeiro -RJ 2, Brazil
T: 55 21 2947732

Finamore, Luiz
GRAFITTO GRAFICA EDITORA
Rio de Janeiro, Brazil

Fingerle, Jorg, Mr.
Rua Mundau 36, Bairro Aerolandia
Fortaleza - CE 60830, Brazil

Fioravanti, Mario, Mr.
Antropologo
Ladeira da Capela 9, Xerem, Duque de Caxias
Rio de Janeiro - RJ 25250, Brazil
T: 55 21 779 1503

Fiori, Heloisa Helena Pinheiro, Sra.
Diretora
ABES
Av. Beira Mar, 216 - 13o. andar
Rio de Janeiro - RJ 20021, Brazil
T: 55 21 210-3221 F: 55 21 262-6838
TL: 21-31902

Fiorot, Antonio Wilson, Sr.
FUNDACAO DA ARVORE E DA CRIANCA / G.L.M.E.E.S.
Rua Padre Antonio Ribeiro Pinto 38 - Apt. 110
Vitoria - ES 29000, Brazil
T: 55 27 2270317

Fischer, Gert Roland, Mr.
Vice Presidente
APREMA-SC
Rua Marinho Lobo 80, sala 605
Joinville-SC, Brazil
T: 55 474 224874 F: 55 474 224874

Fjallgren, Maille Kristina, Ms.
Vice Consul
SWEDISH CONSULATE GENERAL
Praia do Flamengo, 344/9o. andar
Rio de Janeiro - RJ, Brazil
T: 55 21 552-2422 F: 55 21 551-9091
TL: 2121192 EFINEDR

Fleury, Maria Lucia Maranhao
Reporter
FOLHA DO MEIO AMBIENTE
Saldanha Marinho, 689 apt. 601 Centro
Curitiba -PR 80416, Brazil
T: 41 2222370

Flor, Gilberto Da Silva
Technician
REDE GLOBO
R. Lopes Quintas 303
Rio de Janeiro -RJ 2, Brazil
T: 55 21 2947732

Florentino, Edilene
LEGIAO DA BOA VONTADE
Av. Rudge 700 Bom Retiro
Sao Paulo -SP 01134, Brazil
T: 55 21 2338381/R:223 F: 55 21 2634025

Flores, Maria Ines Huber, Sr.
Coordenador Regional
FBCN/RS
Rua da Republica 292/13
Porto Alegre - RS, Brazil
T: 55 512 24-7014

Flores, Silvia Maria Pereira, Sra.
Professora
UNIVERSIDADE FEDERAL DO PARA
jeronimo pimentel 900 apto. 502
Belem - PA 66055, Brazil
T: 55 91 2252305

Floriano, Mira Ferreira
MANCHETE
Rua do Russel 766
Rio de Janeiro, Brazil
T: 55 21 2850033

Floridi, Jose Aparecido, Mr.
Ass. Diretoria
INSTITUTO FLORESTAL
Rua Do Horto 931
Sao Paulo - SP 02377, Brazil
T: 55 11 952-8555 F: 55 11 204-8067
TL: 22877

Fomenti, Ligia H. G.
Reporter
ESTADO DE SAO PAULO
Av. Engenheiro Caetano Alvares 55
Sao Paulo -SP 02598, Brazil

Fonesca, Leticia Silveira
FREELANCER
Rua Gorceix 18/401
Rio de Janeiro RJ 2, Brazil
T: 55 21 2874723

Fonseca, Adolfo Pinto
TV RIO
Rio de Janeiro, Brazil

Fonseca, Alexandre Alberto Henklain, Sr.
Vice-Presidente
FUNDACAO DO MEIO AMBIENTE E TECNOLOGIA DE RORAIMA
Av. Ville Roy,1630 - W
Boa Vista _ RR 69300, Brazil
T: 55 095 2249454 F: 55 095 2242228

Fonseca, Ana Maria Paiva Da, Sra.
Diretora Geral
FUNDACAO BIODIVERSITAS
Rua Maria Vaz de Melo 71, Bairro Da. Clara
Belo Horizonte - MG 31250, Brazil
T: 55 31 4432119 F: 55 31 4417037

Fonseca, Claudio Ruy Vasconcelos Da, Sr.
President
FUNDACAO VITORIA AMAZONICA
Av. Darcy Vargas 520
Manaus - AM 69050, Brazil
T: 55 92 642-1336 F: 55 92 236-3257

Fonseca, Danielle
ARQUIPELAGO DE FERNANDO DE NORONHA
Av. Beira Mar 216 GR. 504
Rio de Janeiro - RJ 20.021, Brazil
T: 55 21 262-8011 F: 55 21 240-8005
TL: 21-34849

Fonseca, Edmilson, Mr.
Tesoureiro
ABES
Av. Beira Mar, 216 - 13o. andar
Rio de Janeiro - RJ 20021, Brazil
T: 55 21 210-3221 F: 55 21 262-6838
TL: 21-31902

Fonseca, Edson Leite Da
TV RIO
Rio de Janeiro, Brazil

Fonseca, Francisco Carlos Guedes Da, Mr.
CIA VALE DO RIO DOCE
Av. Graca Aranha, 26
Rio de Janeiro - RJ, Brazil
T: 55 21 272 4401 F: 55 21 220 0593

Fonseca, Francisco Franco De Assis, Sr.
Counselor
BIODIVERSITAS
R. Arthur Bernardes, 43/806 - Catete
Rio de Janeiro - RJ 22220, Brazil
T: 55 21 265-7240

Fonseca, Geraldo Alberto, Sr.
Eng.Quimico
CENTRO DE RECURSOS AMBIENTAIS
Rua Rio Sao Francisco, 1 - Monte Serrat
Salvador - BA CEP: 40.425, Brazil
T: 55 71 312-7192 F: 55 71 312-7198

Fonseca, Gustavo, Mr.
CONSERVATION INTERNATIONAL
Rua Bueno Brandao 393
Belo Horizonte - MG 31010, Brazil
T: 55 31 226-5145 F: 55 31 222-8429

Fonseca, Iva Marques Da, Ms.
FUNDACAO CULTURAL AVATAR
Rua Min. Octavio Kelly 352 c.6
Niteroi - RJ 24220, Brazil
T: 55 21 611-2489

Fonseca, Ivan Claret M., Mr.
Presidente
A MAO QUE SEMEA
Rua Pocos de Caldas 165
NANUQUE- Mg 39860, Brazil
T: 11506

Fonseca, Joao Carlos Pinheiro Da, Sr.
Editor
TELEBRASIL
Av. Pasteur 383 Urca
Rio de Janeiro - RJ 22290, Brazil
T: 55 21 295-4432 F: 55 21 542-4092
TL: 2123057

Fonseca, Marcelo, Sr.
Estudante
Rua Dom Joao VI, 129 c/9
Nova Friburgo - RJ Cep.: 28620, Brazil

Fonseca, Marcelo De Oliveira
Photographer
IMAGENS DA TERRA
R. das Laranjeiras 203/202
Rio de Janeiro RJ 2, Brazil
T: 55 21 5562502 F: 55 21 2204653

Fonseca, Maria Do Carmo Monteiro Da, Sra.
Expositor
IBASE
Rua Vicente de Souza 12- Botafogo
Rio de Janeiro - RJ 22251, Brazil
T: 55 21 286-6161 F: 55 21 286-0541
TL: 21-36466

Fonseca, Maria Imaculada Pinto
Editor
REDE GLOBO
R. Lopes Quintas 303
Rio de Janeiro -RJ 2, Brazil
T: 55 21 2947732

Fonseca, Maria Leticia Silveira
Ass.Imprensa
FREELANCER
R. Gorceix 18 apt. 401 Ipanema
Rio de Janeiro -RJ 2, Brazil
T: 55 21 2874723

Fonseca, Randal, Sr.
ARQUIPELAGO DE FERNANDO DE NORONHA
Rua Barao de Sao Borja, 243 - Boa Vista
Recife - PE 50070, Brazil
T: 55 81 221-3217 F: 55 81 231-1594
TL: 81.2533

Fonseca, Ricardo Mesquita Da, Mr.
Filer
JOVEM TURISMO
SCLN 202 Bloco B Lojas 7/11
Brasilia - DF, Brazil
T: 55 61 224-5526 F: 55 61 321-0345
TL: 61-4705

Fonseca, Roberto Lopes Da, Sr.
Coordenador
C P V - CENTRO DE DOCUMENTACAO E PESQUISA VERGUEIRO
RUA PROFESSOR SEBASTIAO SOARES DE FARIA, 27 2o. ANDAR
Sao Paulo - SP 01317, Brazil
T: 55 11 258-6288

Fonseca Brasil, Carlos Avelino, Sr.
CONSELHO ESTADUAL SAUDE
Av. Borges Medeiros 1501/5 Andar-S/3 Alanorte Chergs
Porto Alegre - RS 90060, Brazil
T: 55 512 226-3100 F: 55 512 226-3309

Fonseca Gramling, Cristina, Ms.
IDAC
Rua Jose Rangel 10
Rio de Janeiro - RJ 21931, Brazil
T: 55 21 393-3645

Fonsecasta, Marcia Prado Priolli
Reporter
TVSBT
Campo de S.Cristovao, 105
Rio de Janeiro - RJ, Brazil
T: 55 21 5803688 / 5800678 F: 55 21 5809685

Fontan, Alain
Editor
LETTRE DU BRESIL
Av.Presidente Antonio Carlos 58/10andar Castelo
Rio de Janeiro -RJ, Brazil
T: 55 21 2201015

Fontanezi, Paulo Jose, Sr.
Presidente
MOVER - MOVIMENTO ECOLOGICO DE RESENDE
Rua Vila Adelaide no. 458
Resende - RJ 27500, Brazil
T: 55 243 545503

Fontenele, Maria Luiza, Sra.
Deputada Federal
CAMARA DOS DEPUTADOS
Gabinete 582, Anexo Iii
Brasilia - DF 70160, Brazil
T: 55 61 311-5582 F: 55 61 311-9582

Fontes, Andrea Cristina Pessoa
FRILAS SERVICOS JORNALISTICOS
Rio de Janeiro, Brazil

Fontes, Angela, Sra.
INSTITUTO BRASILEIRO DE ADMINISTRACAO MUNICIPAL - IBAM
Edificio Diogo Lordello de Mello. Largo IBAM no.1. Humaita
Rio de Janeiro - RJ 22282, Brazil
T: 55 21 266-6622 F: 55 21 537-1262
TL: 22638 INBM BR

Fontes, Luis Fernando De Meira, Sr.
Tecnico
CONFEDERACAO NACIONAL DA INDUSTRIA
Av. Nilo Pecanha, 50 sala 3018
Rio de Janeiro - RJ 20044, Brazil
T: 55 21 262 0181 F: 55 21 262 1465
TL: 22634

Fontes, Roldao De Siqueira, Sr.
Presidente
FUMBRASIL
Av. Caxanga 2200- Parque de Exposicoes
Recife - PE 50000, Brazil
T: 55 81 227-2427 F: 55 81 465-2043

Fonzes, Ana, Ms.
ASSOCIACAO CIVIL GREENPEACE
Rua Mexico 21, Grp 1301, Centro
Rio de Janeiro - RJ 20031, Brazil
T: 55 21 262-7318 F: 55 21 240-1690

Formigli, Ana Lucia Menezes, Sra.
Assessora
CEAS
Rua Aristides Novis, 101 Federacao
Salvador - BA 40010, Brazil

Formoso, Denise Maria Lopes, Sra.
Enga. Agronoma
INSTITUTO ESTADUAL DE FLORESTAS - MG
Rua Paracatu # 304 Sala 403
Belo Horizonte - MG 30180, Brazil
T: 55 31 295-1655

Fornasari Filho, Nilton, Mr.
Av. Eduardo Prado 838 - Apt.211
Sao Paulo - SP 01218, Brazil
T: 55 11 661018

Forneck, Anselmo
JORNAL PORANTIM -CONSELHO INDIGENISTA MISSIONARIO
SCS Ed. Venancio III S/311 C.P.03679
Brasilia -DF 70084, Brazil
T: 61 2259457 F: 61 2259401

Forrester, Rosemary, Ms.
THE GRAY LINE
Av. Niemeyer, 121 Terreo
Rio de Janeiro - RJ 22610, Brazil
T: 55 21 274-7146 F: 55 21 259-5847

Forster, Maria, Ms.
Humberto De Campos 480/2001
Rio de Janeiro - RJ 22430, Brazil
T: 55 21 2594630

Fortes, Daniela, Ms.
CONFEDERACAO NACIONAL IND.
Nilo Pecanha,50/andar 38
Rio de Janeiro - RJ 22250, Brazil
T: 55 21 552-0099 F: 55 21 552-9392

Fortes, Luis, Mr.
CONFEDERACAO NACIONAL IND.
Nilo Pecanha, 50/andar 38
Rio de Janeiro - RJ 22250, Brazil
T: 55 21 552-0099 F: 55 21 552-9392

Forti, Andrea
RAI
Hotel Gloria
Rio de Janeiro - RJ, Brazil

Foster, Vanira Da Silva
Editor
RADIOBRAS
Brasilia, Brazil

Fraga, Patricia, Sra.
Prof. Com. Social
PETROLEO BRASILEIRO S/A - PETROBRAS
Av. Republica do Chile 65/20 And.-S/2052
Rio de Janeiro - RJ 20035, Brazil
T: 55 21 534-3256 F: 55 21 220-5052

Fragoso, Cristiane Conceicao Maia, Sra.
SALGEMA INDUSTRIAS QUIMICAS S/A
Av. Assis Chateaubriand 5260 -Pontal Da Barra
Maceio - AL 57010, Brazil
T: 55 82 2216458 F: 55 82 2215209

Fraisse, Alberto William, Mr.
Rua Figueiredo De Magalhaes, 285/803
Rio de Janeiro - RJ 22.031, Brazil
T: 55 21 240-3605 F: 55 21 220-0572

Franca, Antonio Carlos, Mr.
CESP
Rua da Consolacao 1875, 9 andar, Cerq. Cesar
Sao Paulo - SP 01301, Brazil

Franca, Manoel Da Costa
Technician
REDE GLOBO
R. Lopes Quintas 303
Rio de Janeiro -RJ 2, Brazil
T: 55 21 2947732

Franca, Marcelo Piettsch
MADE FOR TV
Rio de Janeiro, Brazil

Franca, Marines Fransisca De, Ms.
UFRPE
Av. Alfredo Bandeira De Melo, 290
Igarassu -Pe 53600, Brazil

Franca, Valdo, Sr.
IECAM
Dr. Moreira 250
Manaus - AM 69000, Brazil

Franca, Wilson
TRIBUNA DA IMPRENSA
Rio de Janeiro, Brazil

Franclschetti, Cesar Nascimento, Mr.
Gen. Dionisio 40/103
Rio de Janeiro - RJ 22271, Brazil
T: 55 21 226-4645

Franco, Mandel G.
Reporter
REVISTA VISAO
Av. Rio Branco 177/6 andar
Rio de Janeiro RJ, Brazil
T: 55 21 2621016 F: 55 21 2200805

Franco, Maria Da Penha Silva, Sra.
Presidente
CENTRO BRASILEIRO DE COOPERACAO E INTERCAMBIO DE SERVIDORAS SOCIAIS
R. Sta. Luzia 685/206
Rio de Janeiro - RJ 20030, Brazil
T: 55 21 2208174

Franco, Maria Terza Machado, Sra.
Asses.Imprensa
ASSOCIACAO BRASILEIRA DE AEROSOIS
Rua Ana Catharina Randi, 25
Sao Paulo - SP 04637, Brazil
T: 55 11 240-5466 F: 55 11 240-5528

Franco, Marilia Carmem, Ms.
Asses.Pres
FUNDACAO BRASIL-FRANCA
Av. Beira Mar 406/305
Rio de Janeiro - RJ, Brazil
T: 55 21 220 2644 F: 55 21 262 7782

Franco, Mario Ferreira
Technician
REDE GLOBO
R. Lopes Quintas 303
Rio de Janeiro -RJ 2, Brazil
T: 55 21 2947732

Franco, Paulo Cesar Campos De Seixas, Sr.
Diretor
GEONEX DO BRASIL SERVICOS TECNICOS LTDA
Rua Estacio Coimbra no. 50
Rio de Janeiro - RJ, Brazil
T: 55 21 286-7036 F: 55 21 266-1944

Franco, Paulo Sergio
Technician
REDE GLOBO
R. Lopes Quintas 303
Rio de Janeiro -RJ 2, Brazil
T: 55 21 2947732

Franke, Tamara Maria, Ms.
Professora
PUC - RS
Av. Ipiranga, 6.600 - Predio 7
Porto Alegre - RS, Brazil
T: 55 512 339-1511

Franqueira, Carlos, Sr.
AQUALUNG CONFECCAO LTDA.
Av. Londres, 174 - Bonsucesso
Rio de Janeiro - RJ 21041, Brazil
T: 55 21 270-1955

Frazao, Mariza Ribeiro
Researcher
JORNAL O LIBERAL
Av. Perimetral S/N
Belem PA, Brazil
T: 091 2417996

Freaza, Iza De Salles
EXPRESSO DE PORTUGAL
R. Jose Linhares 14/109 Leblon
Rio de Janeiro -RJ, Brazil
T: 55 21 2593160 F: 55 21 2391495

Freire, Ana Maria
ICAE
12 Rajaji Marg
New Delhi 110011, Brazil
T: 91 11 301-4309

Freire, Beatiz Muniz, Ms.
Sintrasef
SINTRASEF
rua treze maio 13/10andar
Rio de Janeiro - RJ, Brazil
T: 55 21 533-0569

Freire, Joao Batista Macedo
Jornalista
REVISTA MANCHETE
Rua do Russel, 804
Rio de Janeiro - RJ, Brazil
T: 55 21 2652012

Freire, Lilia Zanetti
TV EDUCATIVA
Av Gomes Freire 474
Rio de Janeiro RJ 2, Brazil
T: 55 21 2920012

Freire, Liz Helena Costa Varella, Ms.
Diretora
ESCOLA AMIGOS DO VERDE
Av. Cristovao Colombo 3437
Porto Alegre - RS 90440, Brazil
T: 55 512 3420589 F: 55 512 343 1125

Freire, Luiz Rodrigues, Sr.
CREA-RJ
AV. RIO BRANCO 133/21o. ANDAR
Rio de Janeiro - RJ, Brazil

Freire, Marcus
Assistente Tecnico
UNIVERSIDADE ASPERJ
Av.N.S.Copacabana, 787/402
Rio de Janeiro - RJ, Brazil
T: 55 21 2572726 F: 55 21 2572726

Freire, Octactlio
Coordinator
COORDENADORIA DE COM. SOCIAL-PREFEITURA DO RIO DE JANEIRO
R. Afonso Cavalcante 455/3/1520 Cidade Nova
Rio de Janeiro -RJ 2, Brazil
T: 55 21 2931744 F: 55 21 2737633

Freire, Paulo
ICAE
12 Rajaji Marg
New Delhi 110011, Brazil
T: 91 11 301-4309

Freire, Ricardo, Sr.
Executive Director
ASSOCIACAO BRASILEIRA DE CACA E CONSERVACAO - ABC
R. Mourato Coelho, 1372
Sao Paulo - SP 05417, Brazil
T: 55 11 813 8238 F: 55 11 212 8501

Freitag, Heiko, Mr.
Professor
UNIVERSIDADE FEDERAL DE SERGIPE
Caixa Postal 363
Sao Cristovao - SE 49100, Brazil
T: 55 79 224-1331 F: 55 79 241-3995

Freitas, Aparecido De
TVT REDE DE COMUNICACAO DOS TRABALHADORES
Sao Paulo, Brazil

Freitas, Braz Jose, Mr.
Eng. Ambiental
PETROFERTIL
Praca Mahatma Gandhi 14 Sala 1302
Rio de Janeiro - RJ 20031, Brazil
T: 55 21 217 5329 F: 55 21 212 3880

Freitas, Joao Fialho, Mr.
Professor
MOVIMENTO VOLUNTARIOS INTER BRASILEIROS PARA AMAZONIA -VIBRA JOAO XXIII
Travessa do Alenquer 123
Belem - PA 66010, Brazil
T: 55 91 224-1965 F: 55 91 223-4855

Freitas, Jorge Genovencio De, Sr.
Biologo
UFSC / PROJETO LARUS
Univ.Fed. de S.C.
Florianopolis - SC 88030, Brazil
T: 55 482 31 9302 F: 55 482 34 4069
TL: 240

Freitas, Jorge Gloria De Azevedo De
Technician
REDE GLOBO
R. Lopes Quintas 303
Rio de Janeiro -RJ 2, Brazil
T: 55 21 2947732

Freitas, Jorge Roberto Martins De
GAZETA MERCANTIL
Rua Major Quedinho 90
Sao Paulo - SP, Brazil
T: 55 11 256 3133

Freitas, Marina
Assistente
TROFEU MARINER
Praia de Botafogo 324/1007
Rio de Janeiro - RJ, Brazil
T: 55 21 552 5473 F: 55 21 5525473

Freitas, Regina Helena, Sra.
SUDECAP
Rua Paraiba 890 - Funcionarios
Belo Horizonte - MG 30130, Brazil
T: 55 31 201-6433 F: 55 31 273-4228

Freitas, Roberto De
Technician
REDE GLOBO
R. Lopes Quintas 303
Rio de Janeiro -RJ 2, Brazil
T: 55 21 2947732

Freitas, Rosa Maria De
Assistente De Producao
GLOBO ECOLOGIA
Rua Santa Alexandrina 343/c3 Rio Comprido
Rio de Janeiro - RJ, Brazil
T: 55 21 502-5451 F: 55 21 273-9087

Freitas, Roseman, Sr.
KLABIN CELULOSE
Av. Suburbana 5000
Rio de Janeiro - RJ, Brazil

Freitas, Vera Lucia Chalegre De, Sra.
Pesquisadora
INSTITUTO DE ECOLOGIA HUMANA
Av. Dom Manuel de Medeiros, s/n - Ed. Newton Bancks 3o. andar
Recife - PE 52071, Brazil
T: 55 81 441-3931 F: 55 81 441-4697

Freixo, Leila Ambros
TV RIO
Rio de Janeiro, Brazil

Freneaux, Pascal
A GOTA DE ESPERANCA
Rio de Janeiro, Brazil

Fresta, Aldo Cezar, Sr.
Diretor Cultural
AMDA
R. 'E', 80, Loteamento Bela Vista
Macae - RJ 28700, Brazil
T: 55 247 62-5110

Frias, Lena, Sra.
Rua Jorge Rudge 120, Bl. I, Apt. 1001, Vila Isabel
Rio de Janeiro - RJ, Brazil
T: 55 21 228-7619

Friederich, Dr Nelton, Sr.
GOVERNO DO ESTADO DO RIO DE JANEIRO
Rio de Janeiro - RJ, Brazil

Friedrich, Romeu, Sr.
Ass. Tecnico
PREFEITURA MUNICIPAL DE IMBE
Av. Paraguacu 1607
Imbe - RS 95625, Brazil
T: 55 666-2655

Frigoes, Francisca, Ms.
Researcher
IAMA
Rua Turi no 16
Sao Paulo - SP 05443, Brazil
T: 55 11 210-1301 F: 55 11 210-1338

Froes, Vera, Sra.
Presidente
INSTITUTO DE ESTUDOS DA CULTURA AMAZONICA
Rua Barros de Alarcao 816 Pedra de Guaratiba
Rio de Janeiro - RJ 23025, Brazil
T: 55 21 395-3586 F: 55 21 252-7779
TL: 21-940

Frota, Francisco Ivaldo, Sr.
Engenheiro
ELETRONORTE
Scn-Q06, Conj.A, Bloco C-Venancio 3000, Sala 313
Brasilia - DF 70000, Brazil
T: 55 61 212-6157 F: 55 61 225-9755
TL: 1279

Frurtado, Viviane Dorneles, Ms.
Biologa E Orientadora
AMNP-CA
Barao Do Rio Branco 1580
Cruz Alta 98100, Brazil
T: 322-3215 F: 055 322-4577

Fuimairae, Jorge Luiz Sauti
President
JORNAL ECOLOGIA GLOBAL
Rua Prof Julio do A. Carvalho, No 20
Rosa Maria, Sao Paul, Brazil
T: 55 11 3595326 F: 55 71 3583113

Fuji, Roberto
Ass. Director
AMERICA DO SUL PRODUCOES DE DOCUMENTARIOS LTDA (NAMBEI TSUSHIN)
Av. Paulista 509 /11 andar conj.1104 Cerqueira Cesar
Sao Paulo -SP 01311, Brazil
T: 55 11 2886141 F: 55 11 2512065

Fujiwar, Fumio
Writer/Photographer
YOMIURI SHIMBUN
Av Oswaldo Cruz 149/1201
Rio de Janeiro - RJ, Brazil
T: 55 21 551 6187 F: 55 21 552 2279

Fullgraf, Frederico, Sr.
President
INSTITUTO TERRA - ESTUDOS ECOLOGICOS, EDUCACAO AMBIENTAL
Av. Paulo de Frontim 581, casa B - Rio Comprido
Rio de Janeiro - RJ 20260, Brazil
T: 55 21 273-9190 F: 55 21 551-5908

Furlan, Nello, Sr.
Gerente Administrativo
ASSOCIACAO BRASILEIRA DE ENTIDADES DE MEIO AMBIENTE - ABEMA
Ave. Jeronimo Monteiro, 240 - Centro
Vitoria - ES 29000, Brazil
T: 55 27 322-4193 F: 55 27 222-3725

Furriella, Fernando Nabais De, Sr.
Counselor
HARMONIA GLOBAL
Av. Dr. Cardoso de Mello 1750/7o. andar
Sao Paulo - SP 04548, Brazil
T: 55 11 829-0043 F: 55 11 829-3743

Furtado, Mariangela
Assistent
IBASE
R.Vicente de Souza, 29 Botafogo
Rio de Janeiro -RJ, Brazil
T: 55 21 2265412 F: 55 21 2860541

Furusho, Maria Yoshie, Ms.
A L L B E
Rua Midibu 140, Vila Romana / 0
Sao Paulo - SP 05049, Brazil
T: 55 11 262-1994

Fusco, Mauro Enrico, Sr.
Vendor
INSTITUTO CAJAMAR
Via Anhanguera Km.46,5
Cajamar - SP 07750, Brazil
T: 55 11 437-3098 F: 55 11 437-4343

G., Carol Britton, Sra.
Consul
CONSULADO GENERAL DA COSTA RICA
Rua Brigadeiro Tobias, 577 - 9$ andar - sala 906
Sao Paulo - SP 01032, Brazil
T: 55 11 229-8722

Gabriel, Joao
Editor
EDITORA ABRIL
R.da Passagem, 123/8 andar
Rio de Janeiro, RJ, Brazil
T: 021 5468282 F: 021 2759347

Gabriel, Luiza Teixeira
Photographer
REVISTA FLUMINENSE
Rua Republica do Libano, 61 Sala 1120, Centro
Rio de Janeiro RJ, Brazil
T: 55 21 2429809

Gabriel Jose De, Souza, Sr.
Praia do Flamengo 60/805
Rio de Janeiro - RJ 22210, Brazil
T: 55 21 245-2977

Gabrielle, Marcelo Farias
Technician
REDE GLOBO
R. Lopes Quintas 303
Rio de Janeiro -RJ 2, Brazil
T: 55 21 2947732

Gaio, Ana
Jornalista
REVISTA MANCHETE
Rua do Russel, 804
Rio de Janeiro - RJ, Brazil
T: 55 21 2652012

Gaitan, Paula
Director
GRUPO NOVO DE CINEMA E TV
Rua Morechel Niemeyer 24
Rio de Janeiro - RJ, Brazil
T: 55 21 2663637 F: 55 21 2250435

Galan, Mary
Correspondent
NEW YORK TIMES
R visconde de Piraja 550/2012 Ipanema
Rio de Janeiro RJ, Brazil
T: 55 21 259 4146 F: 55 21 274419

Galante, Jacintho, Sr.
Rua Euclides da Cunha, 94 Ap. 03
Santos - SP 11060, Brazil
T: 55 132 46837

Galanternick, Mary
NEW YORK TIMES
Rio de Janeiro, Brazil

Galeno, Ricardo Aguilar, Mr.
Colaborador
YMCA
Rua Matins Cardoso, 165 - Sto. Agostinho
Belo Horizonte - MG, Brazil
T: 55 31 337-3032 F: 55 31 337-1153

Galindo, Carlos R.
PM STUDIO FOTOGRAFICO
Rio de Janeiro, Brazil

Galle, Werner Rene
Wdr Brazilian Correpondent
WDR GERMAN TV CHANNEL
Rua Eca de Queiroz 479 casa 12
Sao Paulo - SP, Brazil
T: 55 11 572 1262 F: 55 11 572-1702

Gallo, Berenice Maria Gomes, Sra.
Gerente Setorial De Meio Ambie
PROJETO LINHA VERDE - FUNDART/UBATUBA
Pca. Anchieta No. 38 - Centro
Ubatuba - SP 11680, Brazil
T: 55 124 324584 F: 55 124 321292

Galuzzi Jr., Mario, Sr.
CIOSL - ORIT
Rua 24 de Maio, no 77 cj 902
Sao Paulo - SP 01041, Brazil
T: 55 11 223-2174 F: 55 11 223-2153

Galvan, Carlos Matias
AGENCIA DE NOTICIAS ACOPI
Sao Paulo, Brazil

Galvao, Claudia Valeria, Sra.
Gerente Administrativa
FUNDACAO MOVIMENTO ONDAZUL
Travessa da Ajuda 1 - Sala 707 - Centro
Salvador - BA 40020, Brazil
T: 55 71 241-3046 F: 55 71 243-9518

Galvao, Gilberto Nascimento
Journalist
TV BANDEIRANTES
Rua dos Radiautes, 13
Sao Paulo SP, Brazil
T: 55 11 8423011

Gama, Fernando, Mr.
GREEN PEACE
Rua do Russel # 724 Apto. 1002
Rio de Janeiro - RJ, Brazil
T: 55 21 245-2824

Gama, Jose Arnaldo De Souza, Mr.
CIA VALE DO RIO DOCE
Av. Graca Aranha, 26
Rio de Janeiro - RJ 20.030, Brazil
T: 55 21 272 4401 F: 55 21 220 0593

Gama, Maria Eloisa De Oliveira, Ms.
CIA VALE DO RIO DOCE
Av. Graca Aranha, 26
Rio de Janeiro - RJ 20.030, Brazil
T: 55 21 272 4401 F: 55 21 220 0593

Gancalves, Gloria Regina Almeida
LINEA PROPAGANDA S/C LTDA
Rio de Janeiro, Brazil

Gandia, Luiz Antonio, Sr.
JUNDILAB
Jundiai
Sao Paulo - SP 13200, Brazil
T: 55 11 437-9033

Ganter, Bernardus, Mr.
Medical Officer
WORLD HEALTH ORGANIZATION
S.E.M., Lote 19
Brasilia - DF 70800, Brazil
T: 55 61 3211200 F: 55 61 2251551

Garcia, Alan Gomes
Technician
REDE GLOBO
R. Lopes Quintas 303
Rio de Janeiro -RJ 2, Brazil
T: 55 21 2947732

Garcia, Antonio Correia
ESTADO DE SAO PAULO
Av Rio Branco 151
Rio de Janeiro, Brazil
T: 55 21 2217111

Garcia, Barbara
Reporter
FOLHA DE SAO PAULO
Alameda Barao de Limeira 425
Sao Paulo - SP, Brazil
T: 55 11 874 2608/874 2602 F: 55 11 2231644

Garcia, Cesar, Mr.
ASSOCIACAO DOS SERVIDORES DA SUDENE
R. Antonio Curado, s/n - Engenho do Meio
Recife - PE 51030, Brazil
T: 55 81 271.1994 F: 55 81 271.1594

Garcia, Gabriela Mauro, Sra.
Membro
APOENA
Rua Dr. Gurgel, 1099
Presid. Prudente-SP 19015, Brazil
T: 55 182 22-2097

Garcia, Hector Oswaldo Fernandez Y
NOSSA VOZ
Rua do Catete, 243
Rio de Janeiro RJ 2, Brazil

Garcia, Iolanda
OSHO SPIRITUAL HEALTH ORGAN. - OSHO
Av. Atlantica 3056/302
Rio de Janeiro - RJ, Brazil
T: 55 21 255-5753 F: 55 21 322-0327

Garcia, Jerubal Eliel
Cameraman
AMERICA DO SUL PRODUCOES DE DOCUMENTARIOS LTDA (NAMBEI TSUSHIN)
Av. Paulista 509 /11 andar conj.1104 Cerqueira Cesar
Sao Paulo -SP 01311, Brazil
T: 55 11 2886141 F: 55 11 2512065

Garcia, Jorge
PM STUDIO FOTOGRAFICO
Rio de Janeiro, Brazil

Garcia, Marcos Cotrin, Sr.
Administrador
ELETRONORTE
Scn-Q06, Conj.A, Bloco C-Venancio 3000 Sala 313
Brasilia - DF 70000, Brazil
T: 55 61 212-6157 F: 55 61 225-9755
TL: 1279

Garcia, Mariana Fellows, Ms.
Diretora
SIMBIO
Rua Siqueira Campos 43/Sala 926
Rio de Janeiro - RJ 22031, Brazil
T: 55 21 236-4999

Garcia, Martin, Mr.
Advogado
BINDER GARCIA ADVS.
Rua Sorocaba 782/501
Rio de Janeiro - RJ 22271, Brazil
T: 55 21 262 6901 F: 55 21 240 6776

Garcia, Rosa De Lourdes Barros, Sra.
Sqn 416 A - 308
Brasilia - DF 70879-010, Brazil
T: 55 61 273-4477

Garcia Cid, Clysia, Ms.
Diretoria
A.M.N.P.LONDRINA
Rua Alagoas, 1439
Londrina PR 86020, Brazil
T: 55 222754 F: 55 274129

Garciduenas, Alfredo, Sr.
Project Officer
WORLD FOOD PROGRAMME
SCH Quadra 02 Bloco 'B' Terreo
Brasilia - DF 70710, Brazil
T: 55 61 226-7872 F: 55 61 322-7080

Garrido, Juan Carlos Pinto De, Sr.
Engenheiro
INPE
Av. dos Astronautas, 1758
S.Jose dos Campos-SP 12201, Brazil
T: 55 123 418977

Gasbarra, Livia, Sra.
Presidente
CLUBE ECOLOGICO INT. DA AMAZONIA
Av. Jose Bonifacio 1927
Belem - PA 66000, Brazil
T: 55 91 229-9362

Gasparini, Paula
Producer
MTV - RIO
Praca Nossa Senhora da Gloria 46
Rio de Janeiro -RJ, Brazil
T: 55 21 2855946 F: 55 21 2055394

Gato, Fernando
Director
INSTITUTO DE ARQUITETOS DO BRASIL
Rua do Pinheiro, 10 Flamengo
Rio de Janeiro, Brazil
T: 55 21 2853480

Gaudenzi, Luiz Americo
Vice President
INSTITUTO DE ARQUITETOS DO BRASIL
Rua do Pinheiro, 10 Flamengo
Rio de Janeiro, Brazil
T: 55 21 2853480

Gauderio, Antonio
Photographer
FOLHA DE SAO PAULO
Alameda Barao de Limeira 425
Sao Paulo SP, Brazil
T: 55 11 874 2608/874 2602 F: 55 11 2231644

Gauer, Wolf Ruediger
Reporter
WDR GERMAN TV CHANNEL
Rua Eca de Queiroz, 479 casa 12
Sao Paulo - SP, Brazil
T: 55 11 5721262 F: 55 11 5728959

Gaurio, Jalten De Paula
Journalist
FUNDACAO MUSEU DA IMAGEM E SOM
Rua Barbosa No. 1
Rio de Janeiro, Brazil
T: 55 21 2620309

Gauz, Valeria, Ms.
Chefe Divilgacao
FUNDACAO BIBLIOTECA NACIONAL
Av. Rio Branco 219
Rio de Janeiro - RJ 20040-008, Brazil
T: 55 21 220-0890 F: 55 21 220 4173

Gava, Joao Paulo, Sr.
Gerente Area
SOUZA CRUZ
Av. Suburbana #2066
Rio de Janeiro - RJ 21050, Brazil
T: 55 21 281-6122 F: 55 21 201-7395
TL: (021) 31187

Gaviao, Fabio Ferreira
Correspondent
YOMIURI SHIMBUN
Av Oswaldo Cruz 149/1201
Rio de Janeiro, Brazil
T: 55 21 5514560 F: 55 21 5522279

Gay, Marco Antonio
Forum de ONGs Brasileiras
Museu de Republica, Rua do catete
Rio de Janeiro, Brazil
T: 55 21 2851212 F: 55 21 2057537

Gazio, Adalberto Jose
REDE GLOBO
Rio de Janeiro, Brazil

Gazio, Analder Lopes
Technician
REDE GLOBO
R. Lopes Quintas 303
Rio de Janeiro -RJ 2, Brazil
T: 55 21 2947732

Gazzaneo, Luiz Manoel Cavalcanti, Mr.
Rua Xavier da Silveira
Rio de Janeiro - RJ 22061, Brazil
T: 55 21 256-7003

Geber, Helida, Sra.
ASSOCIACAO COMERCIAL DO RIO DE JANEIRO
Rio de Janeiro - RJ, Brazil

Gelert, Heleni
Representative
LIGHT INSTITUTE OF GALISTEO
R. Nascimento Silva 555/501
Rio de Janeiro - RJ, Brazil
T: 2594190 F: 5110535

Gelman, Marco Leao, Sr.
Coordenador Cursos Extra Curri
FACULDADES HELIO ALONSO
R. Barata Ribeiro 264 cob.01
Rio de Janeiro - RJ 22040, Brazil
T: 55 21 256-7898

Genari, Alziro
Fotografo
LEGIAO DA BOA VONTADE
Av.Rudge, 700
Sao Paulo - SP, Brazil
T: 55 11 2202599 F: 55 11 2205196

Genio, Durvalina Da Costa Di, Ms.
UNICEF
SBS Edificio Seguradoras 13o Andar
Brasilia - DF, Brazil
T: 55 61 224-7145 F: 55 61 224-5035

Gentile, Lilliam
Photographer
NEWSWEEK
Av. S. Sebastiao 157 Urca 444 Madison Ave. NY-NY 10017 USA
Rio de Janeiro -RJ 2, Brazil
T: 55 21 5419697 F: 55 21 5426583

Gentille, Paola
GOVERNO DO ESTADO DE SAO PAULO
Sao Paulo, Brazil

Genuncio, Ellen
O POVO
Rio de Janeiro, Brazil

Germano, Ana Lucia
SECRETARIA DE ESTADO DE CULTURA DO RIO DE JANEIRO
Rio de Janeiro, Brazil

Germano, Ines
A GOTA DE ESPERANCA
Rio de Janeiro, Brazil

Germano, Sonia, Sra.
Director
CMB
Rua Avanhadava no. 756
Sao Paulo - SP 01306, Brazil
T: 55 11 2564354 F: 55 11 2577648

Geronimo, Jose Do Carmo
Sao Paulo, Brazil

Gershon, Ron
BLOCH
Rua do Russel 766/804 Gloria
Rio de Janeiro, Brazil
T: 55 21 2652012

Gianotti, Rolland
TRIBUNA DA IMPRENSA
Rio de Janeiro, Brazil

Gil, Aldo Oliveira, Mr.
Assessor
CAMARA DOS DEPUTADOS
Praca Tres Poderes - Anexo Iii, Gab. 64
Brasilia - DF 70160, Brazil
T: 55 61 318-6745

Gil, Olga Maria De Oliveira, Ms.
Rua Cristovao Diniz, 67/201
Sao Paulo - SP 01426, Brazil
T: 55 11 282-8043

Gilderis, Magrin, Ms.
Engenheira
PREFEITURA M. PORTO ALEGRE
Av. Borges de Medeiros, 2244 6andar
Porto Alegre - RS 90060, Brazil
T: 55 512 2288769

Gimenez, Aurelio Vanderlei Batista
Reporter
FREELANCER
R.Barao de Itapagipe 385/307 Tijuca
Rio de Janeiro -RJ, Brazil
T: 55 21 2545531

Gindry, Reny, Mr.
Diretoria
FUNDACAO AMAZONAS FOREVER GREEN
Av. Giovani Gronchi 6186 - Morumbi
Sao Paulo - SP 05724, Brazil
T: 55 11 523-2993 F: 55 11 246-8092

Glembotzki, Jurgen
ECOTROPIC
1880 East River Road
Rio de Janeiro - RJ, Brazil
T: 55 21 577-6476 F: 55 21 577-1195

Gobel, Peter, Sr.
Assessor
GTZ
Av. da Nacoes L 25
Brasilia - DF 70339, Brazil
T: 55 61 244-1118 F: 55 61 243-4991
TL: 061 1687

Godinho, Marco, Mr.
Presidente
A.A.G.G.
Cx. Postal
Vitoria - ES 29000, Brazil
T: 55 27 227-0357

Godoi Filho, Jose Domingues De, Sr.
Coordenacao
GRUPO DE ESTUDOS DOS RECURSOS NATURAIS RENOVAVEIS E ENERGETICOS - GERNE
Caixa Postal 998
Cuiaba - MT 78000, Brazil
T: 55 125 361-5570
TL: 651371

Godoy, Vanessa De
Reporter
ESTADO DE SAO PAULO
Av. Engenheiro Caetano Alvares 55
Sao Paulo -SP 02598, Brazil

Godoy, Vicente De Freitas
Technician
REDE GLOBO
R. Lopes Quintas 303
Rio de Janeiro -RJ 2, Brazil
T: 55 21 2947732

Goes, Denise T. De
Reporter
AGENCIA ESTADO
Av. Caetano Alvares, 55
Sao Paulo -SP 02598, Brazil
T: 55 11 8562210 F: 55 11 2652297

Goes, Liliane Lopes
Editora
JORNAL PACOTIM
Av. Brasil 248/sl.603 Santa Efigenia,
Belo Horizonte-MG 30, Brazil
T: 55 31 2228085 F: 55 31 2272959

Goes, Manoel
Editor
GLOBO ECOLOGIA
Rua Santa Alexandrina 343/c3 Rio Comprido
Rio de Janeiro - RJ, Brazil
T: 55 21 502-5451 F: 55 21 273-9087

Goes, Ruy De, Mr.
ASSOCIACAO CIVIL GREENPEACE
Rua Mexico 21, Grp 1301, Centro
Rio de Janeiro - RJ 20031, Brazil
T: 55 21 262-7318 F: 55 21 240-1690

Goes, Yara Ferraz De, Sra.
Owner
HARAS-PIQUEROBY
Rua Campo Belo, 148 - Laranjeiras
Rio de Janeiro - RJ 22221, Brazil
T: 55 21 245-8476

Goffman, Roseli, Sra.
Assessora
INSTITUTO TIBA
Rua Ingles de Souza 296
Rio de Janeiro - RJ, Brazil
T: 55 21 274-1762

Goldberg, Hetty
Director
INSTITUTO DE ARQUITETOS DO BRASIL
Rua do Pinheiro, 10 Flamengo
Rio de Janeiro, Brazil
T: 55 21 2853480

Goldberg, Ricardo, Mr.
UNESCO
SCN -Quadra 2 Lote B
Brasilia - DF 70710, Brazil
T: 55 61 223-8684 F: 55 61 322-4261

Goldberg, Rolf, Mr.
UNESCO
SCN -Quadra 2 Lote B
Brasilia - DF 70710, Brazil
T: 55 61 223-8684 F: 55 61 322-4261

Goldfarb, Solano
Fotografo
REVISTA MANCHETE
Rua do Russel, 804
Rio de Janeiro - RJ, Brazil
T: 55 21 2652012

Goldstein, Arthur, Mr.
ATLANTIC
Rua Marques De Sao Vicente, 30 - 1602
Rio de Janeiro - RJ 22451, Brazil

Goldstein, Donna, Ms.
Anthropologist
UNIVERSITY OF CALIFORNIA BERKELEY
Rua Almirante Alexandrino 3179 Apto 202 / Santa Teresa
Rio de Janeiro - RJ 20241, Brazil
T: 55 21 245-2956 F: 55 21 245-2956

Gollo, Nestor Jose
Journalist
FUNDACAO CULTURAL RIOGRANDENSE
R.General Sampaio 189
Caxias do Sul -RS 9, Brazil
T: 54 22 62222

Gomberg, Denis, Sr.
GOVERNO DO ESTADO DO RIO DE JANEIRO
Rio de Janeiro - RJ, Brazil

Gomes, Adriana
CEMINA
R.Barao do Flamengo, 22/304
Rio de Janeiro - RJ, Brazil
T: 55 21 2056297 F: 55 21 5563383

Gomes, Albano Rita, Sr.
UNAMAZ
BR 316 Conj. Jardim Ananindema, cs. 100
Ananindema - PA 63000, Brazil
T: 55 91 2354849 F: 55 91 2294339
TL: 1013

Gomes, Alcio Luiz De Jesus, Sr.
Medico E Professor
INSTITUTO DE ACUPUNTURA DO RIO DE JANEIRO
Travessa Pepe 81 Botafogo
Rio de Janeiro - RJ 22290, Brazil
T: 55 21 295-1481

Gomes, Carmen
Secretaria Executiva
SOCIEDADE CIVIL DO BEM-ESTAR FAMILIAR NO BRASIL -BEMFAM
Av. Republica do Chile 230/ 17-andar
Rio de Janeiro -RJ, Brazil
T: 021-2102448 F: 021-2204057

Gomes, Celisa Altivo, Sra.
Paisagista
ARTES E PLANTAS
R. Pires Da Mota 250 / 504 Bloco C Vila Imperial
Belo Horizonte - MG, Brazil
T: 55 31 329-2508

Gomes, Elzaguirre Paranhos
REDE GLOBO
Rio de Janeiro, Brazil

Gomes, Ely Dominques, Sr.
National Director
FUNDO CRISTAO PARA CRIANCAS
Rua Curitiba, 689/5o. andar
Belo Horizonte - MG 30170, Brazil
T: 55 31 201-4955 F: 55 31 212-6957

Gomes, Gerson Toller
Editor
REVISTA FERROVIARIA
Rua Mexico, 41 Sl.904
Rio de Janeiro - RJ, Brazil
T: 55 21 2401747/1248 e 5320260 F: 55 21 2400139

Gomes, Hildon Carrapito
Cinegrafista-Editor
UNIVERSIDADE ASPERJ
Av.N.S.Copacabana, 787/402
Rio de Janeiro - RJ, Brazil
T: 55 21 2572726 F: 55 21 2572726

Gomes, Joao Carlos
CORREIO FLUMINENSE
Rio de Janeiro, Brazil

Gomes, Jose Carlos
MANCHETE
R. do Russel 804
Rio de Janeiro -RJ 2, Brazil
T: 55 21 2652012 F: 55 21 2855148

Gomes, Jose Fernando, Mr.
Ger De Programas E Avaliacao
BADESUL
Rua Sete de Setembro 666/8o Andar
Porto Alegre - RS 90010, Brazil
T: 55 512 289455 F: 55 512 272221

Gomes, Luiz Carlos Amaral, Sr.
CENTRO DE ARTICULACAO DE POPULACOES MARGINALIZADAS - CEAP
Rua da Lapa, 200 - S.805
Rio de Janeiro - RJ 20021, Brazil
T: 55 21 252-2302 F: 55 21 205-3099

Gomes, Luiz Fernando
Photographic Reporter
ZERO HORA
Ipiranga 1075
Porto Alegre, Brazil
T: 55 51 2234300

Gomes, Luiz Fernando De Oliveira
Reporter
O DIA
R. Riachuela 359
Rio De Janeiro, RJ 2, Brazil
T: 55 21 272-8000 F: 55 21 507-1228

Gomes, Luiz Henrique Lopes
Technician
REDE GLOBO
R. Lopes Quintas 303
Rio de Janeiro -RJ 2, Brazil
T: 55 21 2947732

Gomes, Marco Antonio
Secretario Geral De Redacao
GAZETA MERCANTIL
Rua Major Quedinho 90
Sao Paulo SP, Brazil
T: 55 11 256 3133

Gomes, Maria De Fatima, Ms.
Assistant
GLOBAL FORUM OF SPIRITUAL AND PARLIAMENTARY LEADERS ON HUMAN SURVIVAL
Rua Alvaro Ramos 405 Bl. II apto 306 Botafogo
Rio de Janeiro - RJ 22280, Brazil
T: 55 21 295-0325 F: 55 21 325-2993

Gomes, Mercio Pereira, Sr.
GOVERNO DO ESTADO DO RIO DE JANEIRO
Rio de Janeiro - RJ, Brazil

Gomes, Nilo Sergio
Radialista/Jornalista
IBASE
Rua Vicente de Souza 29 Botafogo
Rio de Janeiro - RJ, Brazil
T: 55 21 286 6161/8642 F: 55 21 286 0541

Gomes, Sandra Lima, Sra.
Director
CMB
Rua Avanhadava no. 756
Sao Paulo - SP 01306, Brazil
T: 55 11 2564354 F: 55 11 2577648

Gomes, Sonia Regina
Jornalista
REVISTA PORTOS E NAVIOS
Av.Rio Branco, 185/1914
Rio de Janeiro - RJ, Brazil
T: 55 21 2203578

Gomes, William Ribeiro
REDE RECORD
Av. Miruna 713
Sao Paulo _SP 04084, Brazil
T: 55 11 5429000 F: 55 11 612228

Gomes Oliveira, Silvia Cristina, Ms.
Biologa
CIA DESENVOLVIMENTO DE LIMEIRA
R. Teresa Cosci Fontana, 125 - Jd. Nova Italia
Limeira - SP 13480, Brazil
T: 01 94 512293

Gomma, Carlota, Ms.
Praia do Botafogo 74/901
Rio de Janeiro - RJ 22250, Brazil
T: 55 21 551-7816

Goncalves, Alcindo Fernandes, Sr.
Diretor Presidente - Prodesan
PREFEITURA MUNICIPAL
Praca Maua s/n 1 andar
Santos - SP 11010, Brazil
T: 55 132 342-2093 F: 55 132 345-809
TL: 131048

Goncalves, Alvaro Orlando
REDE RECORD
Av. Miruna 713
Sao Paulo _SP 04084, Brazil
T: 55 11 5429000 F: 55 11 612228

Goncalves, Antonio
WTN
Rio de Janeiro, Brazil

Goncalves, Clesio Oliveira, Mr.
Diretor
FUNDACAO S.O.S. PIRACICABA
Elvio Bastieri, 10 Bairro Jose de Alencar
Joao Monlevade -MG 35930.284, Brazil
T: 031 8515130

Goncalves, Dorival, Jr., Sr.
GRUPO DE ESTUDOS DOS RECURSOS NATURAIS NAO-RENOVAVEIS E ENERGETICOS
Caixa Postal no. 998
Cuiaba - MT 78000, Brazil
T: 55 125 3615570 F: 55 125 3158751

Goncalves, Edmilson Ferreira
Technician
REDE GLOBO
R. Lopes Quintas 303
Rio de Janeiro -RJ 2, Brazil
T: 55 21 2947732

Goncalves, Elaine Blach, Ms.
Foneticista Terapeuta
TRYBO COSMICA
Alameda Franca 1331-62
Sao Paulo - SP 01422, Brazil
T: 55 11 853 7154

Goncalves, Fabio Correia, Mr.
Eng. Florestal
ITCF-ES/AABRI
Cachoeiro-ITCF
Cachoeiro de Itapeme 24300, Brazil
T: 027 522-2386

Goncalves, Fernando De Almeida
Reporter
REDE GLOBO
R. Lopes Quintas 303
Rio de Janeiro -RJ 2, Brazil
T: 55 21 2947732

Goncalves, Henrique Lobo
FUNSEC
Av. Brasil, 2920, 4 andar- Centro
Gov. Valladares - MG 35010, Brazil
T: 55 332 271-1333 F: 55 332 271-6769
TL: 332169

Goncalves, Ivan Rodrigues
Technician
REDE GLOBO
R. Lopes Quintas 303
Rio de Janeiro -RJ 2, Brazil
T: 55 21 2947732

Goncalves, Jacinto Fatima
Cinegrafista
REDE BANDEIRANTES DE TELEVISAO
R.Alvaro Ramos, 492
Rio de Janeiro - RJ, Brazil
T: 55 21 2122072 F: 55 21 2957391

Goncalves, Joao Marcelo Fiorezi, Mr.
Deputado Estadual
ASSEMBLEIA LEGISLATIVA DO ESTADO DE SAO PAULO
Av. Pedro Alvares Cabral, s/#- Sala 3007
Sao Paulo - SP 04097, Brazil
T: 55 11 884-4209 F: 55 11 884-2230
TL: 22273

Goncalves, Lourdes, Sra.
ASSOCIACAO COMERCIAL DO RIO DE JANEIRO
Rio de Janeiro - RJ, Brazil

Goncalves, Mara Nelia Da Silva
HELIO ALONSO
Rua Muniz Barreto, 51
Rio de Janeiro RJ, Brazil
T: 55 21 5515448

Goncalves, Marcio F. De Souza
Technician
REDE GLOBO
R. Lopes Quintas 303
Rio de Janeiro -RJ 2, Brazil
T: 55 21 2947732

Goncalves, Marco Antonio
Reporter
REDE GLOBO
R. Lopes Quintas 303
Rio de Janeiro -RJ 2, Brazil
T: 55 21 2947732

Goncalves, Mauricio
SINDICATO DOS TRABALHADORES IND. QUIM. FARM. DE S.PAULO
Sao Paulo, Brazil

Goncalves, Mauro Henrique B.
Photographer
IMAGENS DA TERRA
R. das Laranjeiras 203/202
Rio de Janeiro RJ 2, Brazil
T: 55 21 5562502 F: 55 21 2204653

Goncalves, Paulo Jorge Barros
REDE GLOBO
R. Lopes Quintas 303 J.Botanico
Rio de Janeiro RJ 22, Brazil
T: 55 21 2947732

Goncalves, Pedro
REDE GLOBO
R. Lopes Quintas 303 J.Botanico
Rio de Janeiro RJ 22, Brazil
T: 55 21 2947732

Goncalves, Rachel Mader, Sra.
President
ASSOC. CRISTA FEM.
Visconde do Rio Branco - 1800 ap. 802
Curitiba - PR 80420, Brazil
T: 55 41 222-0238

Goncalves, Rita De Cassia
MANCHETE
R. do Russel 804
Rio de Janeiro -RJ 2, Brazil
T: 55 21 2652012 F: 55 21 2855148

Goncalves, Sergio Da Silva
TV RIO
Rio de Janeiro, Brazil

Goncalves, Wagner
SISTEMA GLOBO DE RADIO
R. de Russel 434
Rio de Janeiro RJ, Brazil
T: 55 21 2058282 F: 55 21 2054096

Gondek, Silvia, Sra.
Tecnica
CONFEDERACAO NACIONAL DA INDUSTRIA
Av. Nilo Pecanha, 50 sala 3018
Rio de Janeiro - RJ 20044, Brazil
T: 55 21 262 0181 F: 55 21 262 1465
TL: 22634

Gondim, Alexandre Luis Morais, Mr.
Coordenador - Rio 92
SOCIEDADE BRASILEIRA DE EUBIOSE
Rua Barao de Mesquita, 850/808 Bloco I
Rio de Janeiro - RJ 20540, Brazil
T: 55 21 278-3571

Gondim, Marilene, Ms.
TAKE IT BACK
Av. Canal De Marapendi 2500 Bl. 3/1402
Rio de Janeiro - RJ 22.631, Brazil
T: 55 21 326-2494 F: 55 21 326-2494

Gonzalez, Eduardo
Correspondent
EL PAIS
Plaza Cagancha 1162
Montevideo, Brazil
T: 59 82 920115 F: 59 82 920632

Gonzalez, Gerardo
Correspondent
AGENCIA EFE S/A
Avenida Rio Branco 25/13 andar
Rio de Janeiro -RJ, Brazil
T: 55 21 2534465 F: 55 21 2534861

Gopfert, Edmauro, Sr.
CONTREN
Av. Angelica no. 589 - Higienopolis
Sao Paulo - SP, Brazil
T: 55 11 211-6011 F: 55 11 279-0662

Gorini, Ronaldo
TRIBUNA DA IMPRENSA
Rio de Janeiro, Brazil

Gostoli, Renzo
Photograpaher
REUTERS NEW SERVICE
Rua Sete de Setembro 99/4th floor Centro
Rio de Janeiro - RJ, Brazil
T: 55 21 5074151 F: 55 21 5072120

Gotardi, Rogerio
JOVEM PAN TV
Sao Paulo, Brazil

Gottschalk, Joachim, Sr.
ILDES/FES
Rua. Lisboa 191
Sao Paulo - SP 05413, Brazil
T: 55 11 282-3900 F: 55 11 280-1916
TL: 1180616

Gottsfritz, Giselda Tozzi, Ms.
Dir. Cult.
PROJETO LINHA VERDE - FUNDART/UBATUBA
Pca Anchieta, 38
Ubatuba - SP 11.680, Brazil
T: 55 124 32-4584 F: 55 124 32-2982

Goulart, Aida Eghrari, Sra.
Doctor
COMUNIDADE INTERNACIONAL BAHA'I
Av. Nicolau Dorazio 555
Araguari 38440, Brazil
T: 55 34 241-5785

Goulart, Jose Silveira
SINDICATO NACIONAL DOS AEROVIARIOS
Rio de Janeiro, Brazil

Gouvea, Elio, Sr.
Professor
SOBEU
Rua Vereador Pinho de Carvalho, 267 - Centro
Barra Mansa - RJ 27330, Brazil
T: 55 243 22-0222 F: 55 243 22-3690
TL: 22-3640

Gouvea, Giana Raquel Rosa, Ms.
Dep. Ed. Ambiental
ASS. PRO PARQUE NACIONAL DO ITATIAIA
Rua Pro Parque Nac. do Itatiaia
Itatiaia-RJ 27580, Brazil
T: 55 024352-1606

Gouvea, Johnson Barbosa
Reporter
REDE GLOBO
R. Lopes Quintas 303
Rio de Janeiro -RJ 2, Brazil
T: 55 21 2947732

Gouveia, Carla
RADIO NACIONAL - PROGRAMA GINCANA NACIONAL
Rio de Janeiro, Brazil

Gouveia, Luis
JORNAL PORANTIM -CONSELHO INDIGENISTA MISSIONARIO
SCS Ed. Venancio III S/311 C.P.03679
Brasilia -DF 70084, Brazil
T: 61 2259457 F: 61 2259401

Gouveia, Vera Maria, Sra.
Estudante
UFMT FENF
Rua Caceres 251B - ALvorada
Cuiaba - MT 78000, Brazil
T: 55 065 322 8205

Gouver, Alberto Alves
Editor
REDE MANCHETE/TV GLOBO
Rua do Russel, 804 Gloria
Rio de Janeiro, Brazil
T: 55 21 2850033

Goz, Raimundo, Sr.
SENBES
Rua Barbosa 230
Rio de Janeiro - RJ 21350, Brazil
T: 55 21 2696249

Graca, Gilena, Sra.
Professora
U S P
Cidade Universitaria - Iee
Sao Paulo - SP 05508, Brazil
T: 55 11 8152423 F: 55 11 815-4272

Graca, Lucio, Sr.
Especialista Gis
IBM BRASIL
Av. Rio Branco 1 / 14 Andar
Rio de Janeiro - RJ 20090, Brazil
T: 55 21 271-4553 F: 55 21 271-4030

Graca, Paulo
MACACO PREGO
1880 East River Road
Rio de Janeiro - RJ, Brazil
T: 55 21 577-6476 F: 55 21 577-1195

Gralow, Hainsi, Sr.
Diretor-Presidente
ASS. FUMICULTORES BRASIL
Rua Julio de Castilhos, 1031
Santa Cruz do Sul-RS 96800, Brazil
T: 55 51 713 2722 F: 55 51 713 2109
TL: 510021

Granado, Elisabeth Tavares, Sra.
Medica
FAMNEP
Rua Durval Rocha 255 Vila Paraiba
Guaratingueta SP, Brazil
T: 012522-4587

Grazioli, Nadia Regina, Sra.
PROJETO LINHA VERDE - P.M. CARAGUATATUBA
Prefeitura Municipal de Sao Sebastiao, Secretaria de Cultura e Turismo
Sao Sebastiao - SP, Brazil

Green, Reg, Mr.
CIOSL - ORIT
Rua 24 de Maio, no 77 cj 902
Sao Paulo - SP 01041, Brazil
T: 55 11 223-2174 F: 55 11 223-2153

Gregorio, Orlamar Teixeira, Mr.
Promotor De Justica
CURADORIA DO MEIO AMBIENTE/CONSELHO MUNICIPAL DO MEIO AMBIENTE DE DOURADOS
Dourados
Dourados - MS, Brazil

Greilier, Patrice
MTVC-TV Francesa FR3
R.Tenente Marcio Pinto, 96
Rio de Janeiro, RJ 2, Brazil
T: 021 322 52 80 F: 021 322 1181

Grill, Jorge Alberto Duarte, Sr.
Deputado Estadual
ASSEMBLEIA LEGISTATIVA
Rua Joao Manoel 582 ap.22
Porto Alegre - RS 90010, Brazil
T: 55 512 228-6261

Grillo, Renato, Sr.
President
AGIP DO BRASIL SA
Av Paulista 2073 - 2o. terraco
Sao Paulo - SP 01311, Brazil
T: 55 11 288 8573 F: 55 11-288 7483
TL: 21573

Grimaldi Filho, Gabriel, Sr.
Pesquisador
FUNDACAO OSWALDO CRUZ
Av. Brasil 4365 - Manguinhos
Rio de Janeiro - RJ 21045, Brazil
T: 55 21 280-1480 F: 55 21 280-1589

Gropsert, Edmauro, Sr.
COMISSAO NACIONAL DOS TRABALHADORES EM ENERGIA NUCLEAR
Rua Tamandare, 348 - Liberdade
Sao Paulo - SP, Brazil
T: 55 11 279-3811 F: 55 11 279-0662

Grossman, Elio, Sr.
Designer
IBFAN - INTERNATIONAL BABY FOOD ACTION NETWORK
Travessa Poti 41
Rio de Janeiro - RJ 20240, Brazil
T: 55 21 242-1605 F: 55 21 242-1645

Grubbe, Helmut, Sr.
Executivo
MERCEDES - BENZ DO BRASIL S.A.
Rua Lauro Muller 116 - 3201
Rio de Janeiro - RJ 22290, Brazil
T: 55 21 275-1444 F: 55 21 542-4794

Grudice, Diego
ASSOCIATED PRESS
Rio de Janeiro, Brazil

Gruzman, Zina
OSHO SPIRITUAL HEALTH ORGANIZATION
Rua Correa Dutra, 119/508
Rio de Janeiro - RJ 22.210, Brazil
T: 55 21 205-7665

Grybowski, Ewa
WIADOMOSCI
Rio de Janeiro, Brazil

Grzybowski, Lourdes, Sra.
Coordenadora Setor Comunicacao
REDE PTA
Bento Lisboa, 58, 3 andar
Rio de Janeiro - RJ 22221, Brazil
T: 55 21 285-2998 F: 55 21 205-3099
TL: 34201 FOAS BR

Guariglia, Rene
Camera Assistant
MTV - RIO
Praca Nossa Senhora da Gloria 46
Rio de Janeiro -RJ, Brazil
T: 55 21 2855946 F: 55 21 2055394

Guatemie, Vicente De Paulo Pedroso
TWENTY FIRST CENTURY INC.
Rio de Janeiro, Brazil

Guazzelli, Maria Jose, Ms.
CAE IPE
CP 21
IPE-RS 95240, Brazil
T: 55 051 330-3890

Gudme, Niels Nikolaj, Mr.
President
ASSOCIACAO NEO HUMANISTA
Rua Maria Angelica, 534, casa
Rio de Janeiro - RJ 224661, Brazil
T: 55 21 286-9908

Guedes, Edson Ricardo L.
Aux. Tecnico
TV EDUCATIVA - RJ
Av. Gomes Freire 474 Centro
Rio de Janeiro -RJ, Brazil
T: 55 21 2426736 F: 55 21 2426740

Guedes, Marcia Denise, Ms.
MUSEU NACIONAL -UFRJ
Rua Pedra Branca. 60 Bairro da Graca - Jacarepagua
Rio de Janeiro - RJ 22715-330, Brazil
T: 55 21 446 6568

Guedes, Maria Cecilia Soares
Reporter
FRILAS SERVICOS JORNALISTICOS
Rua Anfilofio de Carvalho, 29/407
Rio de Janeiro RJ 2, Brazil
T: 55 21 2205444

Guedes, Sidney Pereira
SISTEMA GLOBO DE RADIO
R. de Russel 434
Rio de Janeiro RJ, Brazil
T: 55 21 2058282 F: 55 21 2054096

Gueron, Sergio Jacques, Sr.
Vice-Reitor De Planejamento
UNIVERSIDADE SANTA URSULA
Rua Fernando Ferrari, 75
Rio de Janeiro - RJ 22231, Brazil
T: 55 21 551-5542 F: 55 21 551-6446

Guerra, Eduardo Teixeira, Sr.
Delegate
ORDEM SOBERANA E MILITAR DO TEMPLO DE JERUSALEM
Av. Rui Barbosa no. 1654 - Gracas
Recife - PE 52011, Brazil
T: 55 81 2685092 F: 55 81 3395338

Guerra, Euclides
Coordenador Comunicacoes Green
GREENPEACE
Rua Mexico, 21/1301 Gr.A e B
Rio de Janeiro - RJ, Brazil
T: 55 21 2401690 F: 55 21 2627318

Guerra, Gema Galgani, Ms.
Professora/Advogada
FAPEMIG
R. Guajajaras 600/602
Belo Horizonte - MG 30180, Brazil
T: 55 31 226-2927

Guerreiro, Isabel Maria Estudante, Sra.
Nutricionista
Rua 56 Lote 05 Quadra 67 - Piratininga - Cafuba
Niteroi-RJ 24350, Brazil
T: 021 7091076

Guerrero, Claudio, Sr.
Diretor
ABES
Av. Beira Mar, 216 - 13o. andar
Rio de Janeiro - RJ 20021, Brazil
T: 55 21 210-3221 F: 55 21 262-6838
TL: 21-31902

Guia, Cecilia Fraga Mares, Sra.
BRAHMA KUMARIS WORLD SPIRITUAL UNIVERSITY
Rua Dr. Estevao de Almeida 53
Sao Paulo - SP 05014, Brazil
T: 55 11 864-3694 F: 55 11 872-7838

Guilherme, Marcia Lucia, Sra.
Coordenadora Do Gt Rio-92
INSTITUTO DE ARQUITETOS DO BRASIL - IAB
Rua Bento Freitas 306 - 4 Andar
Sao Paulo - SP 01220, Brazil
T: 55 11 259-9897 F: 55 11 259-6597

Guilhon, Madalena
CEMINA
R.Barao do Flamengo, 22/304
Rio de Janeiro - RJ, Brazil
T: 55 21 2056297 F: 55 21 5563383

Guimaraes, Christiane Barreto, Ms.
Organizadora
PROJETO QUADRILATERO DE PORTO SEGURO TRANCOSO
Povoado de Trancoso
Porto Seguro 40000, Brazil
T: 071867 1115

Guimaraes, Claudia, Ms.
WORLD WILDLIFE FUND - WWF
Rue du Mont Blanc
Gland, Brazil
T: 41 22 64 95 25 F: 41 22 64 58 79

Guimaraes, Eduardo Santos, Sr.
Tesoureiro
ONCA
R. Soldado Jose Alves De Abreu 193
Cacapava 12280, Brazil
T: 55 122 521830

Guimaraes, Euberlandio, Sr.
Advogado
C.A.R.
Av. A. C. Magalhaes 2671, Sala 505, Itaicara
Salvador - BA 41960, Brazil
T: 55 71 359-5850

Guimaraes, Iolanda, Sra.
NUCLEO DE ESTUDOS E ORGANIZACAO DAS MULHERES
Rua 1 Quadra 1 Casa 6
Cuiaba - MT 78000, Brazil
T: 55 125 361-2884

Guimaraes, Iveraldo, Sr.
Diretor Presidente
ASSOCIACAO POTIGUAR DE EDUCACAO E CULTURA
Rua Floriano Peixoto, 295 Petropolis
Natal - RN 59020, Brazil
T: 55 84 221-6255 F: 55 84 221-6255

Guimaraes, Jose Carlos, Sr.
Diretor Tecnico
TRANSMAR CONSULTORIA
Rua Henrique Moscoso, 1019 s/310
Vila Velha - ES 29100, Brazil
T: 55 27 329-0129 F: 55 27 329-1006

Guimaraes, Jose Iveraldo, Sr.
Tec. Ed. Amb.
SOCIEDADE POTIGUAR DE EDUCACAO - SPEL
Rua Floriano Peixoto, 295, Petropolis
Natal - RN 59020, Brazil
T: 55 84 222-2314 F: 55 84 221-6255

Guimaraes, Jussara Cerdeira, Sra.
MOVIMENTO ECOLOGICO AMIGOS DA NATUREZA
Rua Aracuai - 50 Jacarepagua
Rio de Janeiro - RJ 21341, Brazil
T: 55 21 359 6192

Guimaraes, Luiz Daniel
Technician
REDE GLOBO
R. Lopes Quintas 303
Rio de Janeiro -RJ 2, Brazil
T: 55 21 2947732

Guimaraes, Maracy
Journalist
FORUM DE ONGs BRASILEIRAS
Museu da Republica, Rua do Catete, 153
Rio de Janeiro RJ 2, Brazil
T: 55 21 2851212 F: 55 21 2057537

Guimaraes, Maria Benita Fernandes, Ms.
A D I C O
Av Rio Branco, 181, 80
Rio de Janeiro - RJ, Brazil
T: 55 21 222 3981

Guimaraes, Marilia De Lacerde
Producer
FUNDACAO ROQUETTE-PINTO
Prasa da Republica, 141 A Centro
Rio de Janeiro, Brazil
T: 55 21 2522203

Guimaraes, Paulo Ricardo
Operador De Vt
TV RIO
R. Miguel de Frias, 57 Estacio
Rio de Janeiro-RJ 20, Brazil
T: 55 21 2939888 F: 55 21 2938626

Guimaraes, Renato De Paiva
Jornalista
ASSOC. PROJETO RODA VIVA
R.Sao Francisco Xavier, 524/12037-F
Rio de Janeiro - RJ, Brazil
T: 55 21 2645196 F: 55 21 2645196

Guimaraes, Rose, Sra.
Director
INSTITUTO TIBA
Rua Ingles de Souza 296
Rio de Janeiro - RJ, Brazil
T: 55 21 274-1762

Guimaraes, Sergio, Mr.
UNICEF
SBS Edificio Seguradoras 13o Andar
Brasilia - DF, Brazil
T: 55 61 224-7145 F: 55 61 224-5035

Guimaraes, Viviane Maria De Castro, Sra.
Estudante
UNIVERSIDADE FEDERAL GO
Rua 83 # 487 Setor Sul
Goiania - GO 74000, Brazil
T: 55 62 223-2594 F: 55 62 251-1694

Guimaras, Rosangela Aguiar De
Reporter
FOLHA DIRIGIDA
Rua Maharma Gaiandi, No. 2 Cinelandia
Rio de Janeiro, Brazil
T: 55 21 2624122 F: 55 21 2409474

Guimares, Robert, Sra.
Gerente
ECOMERCADO
R. Barao de Itambi, 119C - Botafogo
Rio de Janeiro - RJ, Brazil
T: 55 21 553-5777

Guinle, Nina
Producer
MTV - RIO
Praca Nossa Senhora da Gloria 46
Rio de Janeiro -RJ, Brazil
T: 55 21 2855946 F: 55 21 2055394

Guitton, Serge Claude
Reporter-Cameraman
MTVC-TV Francesa FR3
R.Tenente Marcio Pinto, 96
Rio de Janeiro - RJ, Brazil
T: 55 21 322 52 80 F: 55 21 322 1181

Guizelini, Alberto Costa, Sr.
Biologist
UNIVERSIDADE FEDERAL DE GOIAS - UFG
Rua C 136 No. 346 - Jardim America, Qd. 562, Lt. 08
Goiania - GO 74275, Brazil
T: 55 62 251-1694 F: 55 62 251-1694

Gurgel, Jose Eduardo Souza
MANCHETE
Rio de Janeiro, Brazil

Gurjao, Jose
ESTADO DE SAO PAULO
Av. Rio Branco 151 - Sobre - Loja sala 201 a 204
Rio de Janeiro RJ, Brazil
T: 55 21 2217111

Gursola, Marilena
Coordenator
ASSOC. PROJETO RODA VIVA
R.Sao Francisco Xavier, 524/12037-F
Rio de Janeiro - RJ, Brazil
T: 55 21 2645196 F: 55 21 2645196

Gusmao, Cyntia
Operadora De Vt
COMULHER - COMUNICACAO MULHER
Rua Rocha, 119/504
Sao Paulo - SP, Brazil
T: 55 11 2515626

Gussonato, Vanderlei
ICAE
Rede Mulher C.P.4651
Sao Paulo SP 01061, Brazil
T: 55 11 627050 F: 55 11 8713457

Gutierrez, Edda, Sra.
MULHERANDO
Roberto Dias Lopez 100/1003 Leme
Rio de Janeiro - RJ, Brazil

Gutman, Eunice
Director
FREELANCER
R. Ataulo de Paiva 50 Bl/C2 apt.801
Rio de Janeiro -RJ, Brazil
T: 55 21 2591114

Guttmann, Marcos
Fotografo
REVISTA ANO ZERO
Rua do Russel 450/802
Rio de Janeiro - RJ, Brazil
T: 55 21 205 4907 F: 55 21 285 6096

H., Margarete
Assistente Operacional
ABIA -ASSOCIACAO BRASILEIRA INTERDISCIPLINAR DE AIDS
R. Lopes Quintas 576 J. Botanico
Rio de Janeiro -RJ 2, Brazil
T: 55 21 2315171 F: 55 21 2945602

Hacidume, Leo Ryoske
NHK
Av Nilo Pecanha 50 Grupo 1208
Rio de Janeiro - RJ, Brazil
T: 55 21 262 9777 F: 55 21 262 9514

Halabi, Adnan Hamadani, Mr.
Encarregado De Negocios
CONSULADO GERAL DA SIRIA
Av. Nossa S. de Copacabana, 162 / 6o. andar
Rio de Janeiro - RJ 22020, Brazil
T: 55 21 541-2992 F: 55 21 541-2948
TL: 32873

Hallan, Thomas Patrick
Foreign Correspondent
CREATION MAGAZINE & S.E.O.P.
Rua Paulino Afonso. 120
Petropolis RJ 25680, Brazil
T: 55 0242437774

Hallmam, Stephan
Correspondent
2DF-TV ALEMA
Av.Delfim Moreira, 662/704
Rio de Janeiro - RJ, Brazil
T: 55 21 2947434 F: 55 21 5122120

Hamburger, Diana Sarita, Sra.
Geografa
FUNDACAO SOS MATA ATLANTICA
Rua Manoel da Nobrega, 456
Sao Paulo - SP 04001, Brazil
T: 55 11 887-1195 F: 55 11 885-1680

Hamond, Beatrice, Ms.
SERVICO SOCIAL DO COMERCIO
Rua Voluntarios Da Patria, 169
Rio de Janeiro - RJ 22270, Brazil
T: 55 21 286-5152 F: 55 21 286-2638

Handa, Rosangela M., Sra.
Instr. Ens. 2-Grau
SENAI - PR
Rua Chile 1678
Curitiba - PR 80220, Brazil
T: 55 41 233-0322 F: 55 41 233-0322
TL: 4135141

Handfas, Anita, Sra.
Director
CMB
Rua Avanhadava no. 756
Sao Paulo - SP 01306, Brazil
T: 55 11 2564354 F: 55 11 2577648

Handler, Bruce
Director In Brazil
ASSOCIATED PRESS
Av Brasil 500/847 /Caixa Postal 72-ZC-00-CEP 20001
Rio de Janeiro - RJ, Brazil
T: 55 21 580 4442/585 4607 F: 55 21 5802309

Hansen, John
Reporter
FREELANCER
R.das Camelias H/5 Cond.Bleu et Vert Bairro Braunes
Nova Friburgo - RJ, Brazil
T: 55 245 220192 F: 55 245 220192

Hansford, Frances, Sra.
Training Coord.
INSTITUTO DE ASSUNTOS CULTURAIS
Av. Graca Aranha 416/1116
Rio de Janeiro - RJ 20030, Brazil
T: 55 21 2424445

Harada, Noboru, Sr.
Guia
Avinida Rio Branco, 15615,.1115
Rio de Janeiro - RJ 20040, Brazil
T: 55 21 210-32634 F: 55 21 220-9854
TL: 30605

Haro, Ursula, Ms.
Rua Marquesa De Santos 53/903
Rio de Janeiro - RJ, Brazil
T: 55 21 265-0697

Harres, Antonio Carlos
URANUS
Rio de Janeiro, Brazil

Harris, Sandra, Ms.
Diretora
SIMBIO
Rua Siqueira Campos 43 Sala 926
Rio de Janeiro - RJ 22031, Brazil
T: 55 21 236-4999

Hassan, Saleh Haidar, Mr.
Interpreter
CONSULADO GERAL DA SIRIA
Av. Nossa S. de Copacabana, 162 / 6o. andar
Rio de Janeiro - RJ 22020, Brazil
T: 55 21 541-2992 F: 55 21 541-2948

Hassett, David M., Sr.
Coordenador
SANTUARIO ECOLOGICO DE PIPA
Rua Acu 495 - Petropolis
Natal - RN 59020, Brazil
T: 55 84 222-6626 F: 55 84 222-7991

Hauer, Margit, Ms.
INSTITUTO DE TERRAS, CARTOGRAFIA, FLORESTAS
Desembargador Motta, 3384
Curitiba - PR 80410, Brazil
T: 55 41 234 1611

Hauer, Xenia, Sra.
Biologa
SOC. DE PESQ. EM VIDA SELVAGEM E EDUCACAO AMBIENTAL
Rua Guttemberg 345
Curitiba - PR 80420-030, Brazil
T: 55 41 222-3425 F: 55 41 222-3425

Haug, Steffen, Mr.
NATURFREUNDEJUGEND DEUTSCHLAND
Praia do Flamengo, 98/806
Rio de Janeiro - RJ 22210, Brazil
T: 55 21 225-0322

Hautin-Guiraut, Dennis
Correspondent
LE MONDE
R.Jose Linhares 44-501
Rio de Janeiro -RJ, Brazil
T: 55 21 5112815 F: 55 21 2593460

Hawrylyshyn, Olgha Dierdre Tibytiba
92 GLOBAL FORUM
Rio de Janeiro - RJ, Brazil

Haynes, Thomas Stanley, Mr.
MARGARET MEE
Praca Cardeal Arco Verde 25/902 Copacabana
Rio de Janeiro - RJ, Brazil
T: 55 21 235-1840 F: 55 21 2363786

Hegdorni, Eronildo
Technician
REDE GLOBO
R. Lopes Quintas 303
Rio de Janeiro -RJ 2, Brazil
T: 55 21 2947732

Hehl, Ana Paola, Sra.
Gerente
ECOMERCADO
R. Barao de Itambi, 119C - Botafogo
Rio de Janeiro - RJ, Brazil
T: 55 21 553-5777

Heller, Eliane, Ms.
ASSOCIACAO CIVIL GREENPEACE
Rua Mexico 21, Grp 1301, Centro
Rio de Janeiro - RJ 20031, Brazil
T: 55 21 262-7318 F: 55 21 240-1690

Hemsley, Peter, Sr.
Rua Jorge Rudge 120, Bl. I, Apt. 1001
Rio de Janeiro - RJ, Brazil
T: 55 21 228-7619

Heng, Wu Yong
Diretor P/ O Brazil
AGENCIA NOTICIOSA XINHUA
Rua Paissandu,7 apto.401
Rio de Janeiro - RJ, Brazil
T: 55 21 285 0590 / 285 06 03

Henrique, Roberto
Technician
REDE GLOBO
R. Lopes Quintas 303
Rio de Janeiro -RJ 2, Brazil
T: 55 21 2947732

Henrique, Vanderlei, Sr.
SECRETAIA ESTADUAL DE SAUDE
Rua N 210 Bairro Cajus
Jucimeira - MT, Brazil
T: 55 65 412-1182

Henze, Roland Werner
ROLAND HENZE
Rio de Janeiro, Brazil

Heras, Txomin Las
Correspondent In Brazil
UENPRES- AGENCIA VENEZUELANA DE NOTICIAS
Alameda Surutaia 73/26 Jardim Paulista
Sao Paulo - SP, Brazil
T: 55 11 885 6858

Hermeto, Maria Cristina, Sra.
Biologa
SHIN QL 15 Conj.04 casa 08
Brasilia - DF 71500, Brazil
T: 55 61 577-4353

Hernkenhoff, Fernando, Sr.
Consultor Florestal
Leao Junior S/A
Av. Getulio Vargas, 253
Curitiba - Pr 80230, Brazil
T: 55 41 233-7822 F: 55 41 222-0028
TL: 5134

Herrman, Wallace
EDITORIA MULTIMIDIA
Rio de Janeiro, Brazil

Higuchi, Yasuhiro, Sr.
Researcher
INSTITUTO PERFECT LIBERTY DE PESQUISAS BOTANICAS
Avenida Paulista do Brasil no.8000
Aruja - SP 07400, Brazil
T: 55 11 466 0188

Hilber, Anton, Mr.
Coordinator
PROJETO LAT
CP 2001
Natal - RN 59085, Brazil
T: 55 84 236-3536 F: 55 84 219-3366

Hirano, Hajimu
Director
AMERICA DO SUL PRODUCOES DE DOCUMENTARIOS LTDA (NAMBEI TSUSHIN)
Av. Paulista 509 /11 andar conj.1104 Cerqueira Cesar
Sao Paulo -SP 01311, Brazil
T: 55 11 2886141 F: 55 11 2512065

Hirose, Haruki
JIJI PRESS NEWS AGENCY
Av Paulista 854 - 13o. andar cj 133 Bela Vista
Sao Paulo - SP, Brazil
T: 55 11 285 0025/287 9526 F: 55 11 285 3816

Hirota, Marcia Makiko, Sra.
Responsavel Pela Documentacao/
FUNDACAO SOS MATA ATLANTICA
Rua Manoel da Nobrega, 456
Sao Paulo - SP 04001, Brazil
T: 55 11 887-1195 F: 55 11 885-1680

Hirtsch, Christina David
Photographer
MANCHETE
R. do Russel 804 Gloria
Rio de Janeiro RJ 2, Brazil
T: 55 21 2454816 F: 55 21 2056196

Hockstein, Beatriz
Produtora
2DF-TV ALEMA
Av.Delfim Moreira, 662/704
Rio de Janeiro - RJ, Brazil
T: 55 21 2947434 F: 55 21 5122120

Hofmann, Stefan
Assistente De Camera
WDR GERMAN TV CHANNEL
Rua Eca de Queiroz, 479 casa 12
Sao Paulo - SP, Brazil
T: 55 11 5721262 F: 55 11 5728959

Hogetop, Juliana, Sra.
Chefe Do Ramo / Guia
GRUPO ESCOTEIRO ALBERT SCNCHEITZER
Rua Lusitana, 1261 - Higienopolis
Porto Alegre - RS 90520, Brazil
T: 55 512 342-8581

Hoisel, Evelina, Ms.
Professora
UFBA
Monsenhor Gaspar Sadoc 48, Jardim Ala
Salvador - BA 41750, Brazil
T: 55 71 231 8989

Holanda, Marcia
RADIO NACIONAL - PROGRAMA GINCANA NACIONAL
Rio de Janeiro, Brazil

Holanda, Marcio Vitor Paixao, Mr.
Estudante
UNESPA
Av Magalhaes Barata 92/507
Belem - PA 66000, Brazil
T: 55 91 225-0038

Holanda, Roosevelt Nogueira De
FREELANCER
R. Xavier da Silveira 104 apt. 601 Copacabana
Rio de Janeiro -RJ, Brazil
T: 55 21 2569985

Hollanda, Kristina De, Ms.
VBV - COOPERACAO TECNICA SUECA
Rua Dona Anunciada 188, Gracas
Recife - PE, Brazil
T: 55 81 222-0810

Hollanda, Jr., Lamartine, Sr.
Delegado Chefe
FEDERATION INTERNATIONAL DES CORPS ET ASSOCIATION CONSULAIRES
Av Rui Barbosa 1654 - Gracas
Recife - PE, Brazil
T: 55 81 2686185/2685092 F: 55 81 221-2178

Holmquist, Fredrik, Mr.
VBV - COOPERACAO TECNICA SUECA
Av. Rui Barbosa 560 apto 702
Rio de Janeiro - RJ, Brazil
T: 55 21 552-1521

Holtwisk, Yneke
Correspondente Para America La
DE VOLKSKRANT/REVISTA ELSEVIER/TV NOS/VARA TV
Rua do Russel 300/701 Gloria
Rio de Janeiro - RJ, Brazil
T: 55 21 205 5049 F: 55 21 205 4951

Holzman, Flora, Sra.
BRITISH EMBASSY
Setor de Embaixadas Sul Quadra 801 Conjunto K
Brasilia - DF, Brazil
T: 55 61 225-2710
TL: 0611360

Holzschuh, Sebastian, Sr.
Rua Nobrega 194
Niteroi - RJ 24220, Brazil

Honey, Martha
Greenpeace Usa
GREENPEACE
Rua Mexico, 21/1301 Gr.A e B
Rio de Janeiro - RJ, Brazil
T: 55 21 2401690 F: 55 21 2627318

Hora, Jorge Expedito S. Da
Technician
REDE GLOBO
R. Lopes Quintas 303
Rio de Janeiro -RJ 2, Brazil
T: 55 21 2947732

Horn, Dan, Sr.
Diretor
SATELLITE SYSTEMS INTERNATIONAL
Av. Rio Branco, 12/4o. andar
Rio de Janeiro - RJ 20090, Brazil
T: 55 21 847-7800 F: 55 21 263-7067

Horn, Marcelo
Photographer
MANCHETE
R. do Russel 804 Gloria
Rio de Janeiro RJ 2, Brazil
T: 55 21 2454816 F: 55 21 2056196

Horsay, Marcos Antonio
REDE GLOBO
R. Lopes Quintas 303 J. Botanico
Rio de Janeiro - RJ, Brazil
T: 55 21 5292000 F: 55 21 2942042

Horta, Bernardo
REVISTA ANO ZERO
Rua do Russel 450/802
Rio de Janeiro - RJ, Brazil
T: 55 21 205 4907 F: 55 21 285 6096

Houelleur, Yann Le
FREELANCER
R.Iquiririm, 464 Vila Indiana Butanta
Sao Paulo - SP, Brazil
T: 55 11 8143382 F: 55 11 2523299

Hovoruski, Alexandre
Coordenador De Programacao
SISTEMA JORNAL DO BRASIL
Av.Brasil, 500
Rio de Janeiro - RJ, Brazil
T: 55 21 5854422

Howard, Jonhy
Photographer
GRUPO NOVO DE CINEMA E TV
Rua Morechel Niemeyer 24
Rio de Janeiro - RJ, Brazil
T: 55 21 2663637 F: 55 21 2250435

Hozmann, Elizabeth
Diretora
BRASIL MEIO AMBIENTE E DESENVOLVIMENTO
Rua Uruguaiana 10 - 2o. andar
Rio de Janeiro - RJ, Brazil
T: 55 21 242 9733/242 4618 F: 55 21 252 4273

Hua, Yang Jing
Fotografo
AGENCIA NOTICIOSA XINHUA
Rua Paissandu,7 apto.401
Rio de Janeiro - RJ, Brazil
T: 55 21 285 0590 / 285 06 03

Hueen, Artur
Reporter
RADIO CLUBE CANELA
Av Julio de Cratilhos 238
Canela, Brazil
T: 55 54 2822000

Hug, Alfons, Sr.
Director
CP 08619
Brasilia - DF 70312, Brazil
T: 55 61 224-6773 F: 55 61 224-6522

Huhn, Joehen
Correspondent
WDR GERMAN TV CHANNEL
Rua Eca de Queiroz 479 casa 12
Sao Paulo - SP, Brazil
T: 55 11 572 1262 F: 55 11 572-1702

Hundt, Reed
TWENTY FIRST CENTURY INC.
Rio de Janeiro, Brazil

Hungria, Carlos Luis
Reporter Fotografico
REVISTA ISTO E
R.William Speers,1000/1and. Cj.39
Sao Paulo - SP, Brazil
T: 55 11 2600533 F: 55 11 2611440

Hutchinson, Chuck, Mr.
CONSERVATION INTERNATIONAL
Rua Bueno Brandao 393
Belo Horizonte - MG 31010, Brazil
T: 55 31 226-5145 F: 55 31 222-8429

Hyodo, Adalgisa
JIJI PRESS NEWS AGENCY
Av Paulista 854 - 13o. andar cj 133 Bela Vista
Sao Paulo - SP, Brazil
T: 55 11 285 0025/287 9526 F: 55 11 285 3816

Iashalom, , Mr.
FAR
Av. Epitacio Pessoa 3964 CO. 03
Rio de Janeiro - RJ, Brazil
T: 55 21 2865686

Ide, Sandra, Sra.
Membro
APOENA
Rua Professor Hugo Mieli, 159
Presid. Prudente-SP 19100, Brazil
T: 55 182 21-4222

Iglesias, Enrique
Presidente Do Bid
BCO. INTERAMERICANO DE DESENVOLVIMENTO
Caixa Postal 08738
Brasilia - DF 70312, Brazil
T: 55 61 321-7848 F: 55 61 321-3112
TL: 61-1204

Ikeda, Masakasu
Photographer
YOMIURI SHIMBUN
Av Oswaldo Cruz 149/1201
Rio de Janeiro - RJ, Brazil
T: 55 21 551 6187 F: 55 21 552 2279

Ilha, Andre Silva, Sr.
COMISSAO ESPECIAL LEI 1849/92
Rio de Janeiro - RJ, Brazil
T: 55 21 273 9442

Imada, Kunio, Mr.
Fundador
LIFE WATCH NETWORK - JAPAN
R.Prof. Cristiano Volkart, 54 - Centro
Valinhos - SP 13270, Brazil
T: 55 192 71-33292

Imaguire, Nobor, Sr.
Professor
UNIVERSIDADE FEDERAL DO PARANA
Jardim Das Americas
Curitiba - PR 80040, Brazil
T: 55 41 252-1590

Impallari, Luis Fernando C., Sr.
Rua Padre Champagnat, 21/20M
Rio de Janeiro - RJ 20511, Brazil
T: 55 21 254-5202

Inacio, Antonio Rogerio Da Silva, Sr.
Director
ABES
Av. Beira Mar, 216 - 13o. andar
Rio de Janeiro - RJ 20021, Brazil
T: 55 21 210-3221 F: 55 21 262-6838
TL: 21-31902

Ines, Hernan, Mr.
Estudiante
ESCOLA INDUSTRIAL DE PUERTO IGUAZU
Av. das Cataratas 507
Foz Iguacu - PR, Brazil

Ingberg, Ismar
SISTEMA JORNAL DO BRASIL
Av. Brasil 500
Rio de Janeiro -RJ, Brazil
T: 55 21 5854422 F: 55 21 5854428

Inocencio, Nilo
Technician
REDE GLOBO
R. Lopes Quintas 303
Rio de Janeiro -RJ 2, Brazil
T: 55 21 2947732

Intrator, Simone, Sra.
Expositor
IBASE
Rua Vicente de Souza 12- Botafogo
Rio de Janeiro - RJ 22251, Brazil
T: 55 21 286-6161 F: 55 21 286-0541
TL: 21-36466

Inukai, Noboru, Mr.
Diretor
Av. Presidente Antonio Carlos, 697 - 4th andar Castelo
Rio de Janeiro - RJ, Brazil
T: 55 21 2245582 F: 55 21 2226035

Ioruba, Togo
JORNAL MAIORIA FALANTE
Rio de Janeiro, Brazil

Ioselli, Ana Cristina
Journalist
FREELANCER
Rua Irineu Marinho 35 Centro
Rio de Janeiro, Brazil
T: 55 21 2922000 x700

Ipiranga, Cicero Moreira
GAIA - ASSOC.CONS.M.AMBIENTE E PROD.INTEGRADA ALIM.AMAZONIA
Arse-51 - 39
Palmas, TO 77000, Brazil
T: 063 8621553/1108

Irmao, Jose Rufino
MANCHETE
R. do Russel 804
Rio de Janeiro -RJ 2, Brazil
T: 55 21 2652012 F: 55 21 2855148

Ishida, Yasurido, Mr.
Diretor
JETRO
Praia de Botafogo, 228, Setor B, sala 809
Rio de Janeiro - RJ 22250, Brazil
T: 55 21 551-1196 F: 55 21 552-1196

Ishihara, Francisco
Engineer
AMERICA DO SUL PRODUCOES DE DOCUMENTARIOS LTDA (NAMBEI TSUSHIN)
Av. Paulista 509 /11 andar conj.1104 Cerqueira Cesar
Sao Paulo -SP 01311, Brazil
T: 55 11 2886141 F: 55 11 2512065

Isidoro, Cristiana
Photographer Reporter
AGENCIA SPID R
R Cardeal D Sebastiao Leme 350/301 Santa Tereza
Rio de Janeiro - RJ, Brazil
T: 55 21 242 2878 F: 55 21 2422878

Isidoro, Elsson Batista Campos
Photography Reporter
JORNAL MUNDO PORTUGUES
R do Rezende 96/201
Rio de Janeiro - RJ, Brazil
T: 55 21 242 2356

Israel, B.
Free Lance
Rua Visc.Cabo Frio, 25
Rio de Janeiro - RJ, Brazil
T: 55 21 2683078 F: 55 21 2201397

Issa, Marcos
Fotografo
JORNAL O GLOBO
Rua Irineu Marinho, 35
Rio de Janeiro - RJ, Brazil
T: 55 21 2722000

Itoh, Osamu, Mr.
Vice-Diretor De Divulgacao
CEFLAM - CENTRO DE ESTUDOS FLORESTAIS DA AMAZONIA
a/c ATS Rua Mandurucus, 1826, Bairro Batista Campo
Belem - PA 66010, Brazil
T: 55 91 241-1099

Iullianeli, Jorge Atilio, Mr.
Publishing Dept.
WORLD COUNCIL OF CHURCHES
S:T Sigfrids Folkhogskola
Vaxjo 35590, Brazil

Iwakami, Shoichi, Sr.
Gerente Geral De Seguranca
BANCO AMERICA DO SUL YASUDA
Av. Brigadeiro Luiz Antonio, 2020 - 5o andar
Sao Paulo - SP 01318-911, Brazil
T: 55 11 285-1411 F: 55 11 285-1426

Iwasa, Jo
Photographer
YOMIURI SHIMBUN
Av Oswaldo Cruz 149/1201
Rio de Janeiro - RJ, Brazil
T: 55 21 551 6187 F: 55 21 552 2279

Izumi, Jaques Fernandes
Technician
REDE GLOBO
R. Lopes Quintas 303
Rio de Janeiro -RJ 2, Brazil
T: 55 21 2947732

Jablonka, Jayme, Mr.
Relacoes Publicas
RELACOES PUBLICAS
Av. N.S.Copacabana 680/Cobertura
Rio de Janeiro - RJ, Brazil
T: 55 21 256-1395

Jaccarino, Claudio
Periodista
TELEVISA
Av. Sao Sebastiao 157 URCA
Rio RJ 22291, Brazil
T: 55 21 295-6075

Jacob, Arnaldo Cesar R.
Ass.Editor
EDITORA ABRIL
R.da Passagem, 123/8 andar
Rio de Janeiro, RJ, Brazil
T: 021 5468282 F: 021 2759347

Jacobi, Pedro Roberto, Mr.
Researcher
CEDEC
Rua Airosa Galvao 64
Sao Paulo - SP 05002, Brazil
T: 55 11 871-2966 F: 55 11 871-2123

Jacobs, Casimiro Medeiros, Sr.
UNISINOS
Rua Riachuelo, 1290/804
Porto Alegre - RS, Brazil
T: 55 512 222-8429

Jacobson, Steve, Mr.
Public Affairs
SUMMER INSTITUTE OF LINGUISTICS
SAI/NO, Lote 'D', Bloco 03
Brasilia - DF 70770, Brazil
T: 55 61 272-1200

Jacques, Geny Coelho Jeanes
CONFEDERACAO NACIONAL DA INDUSTRIA
Rio de Janeiro, Brazil

Jaffet, Edmea E.
ICAE
12 Rajaji Marg
New Delhi 110011, Brazil
T: 91 11 301-4309

Jahara, Virginia De Souza, Sra.
Coordenadora
IIU LEONARD ORR
Rua Raimundo Correa 43/302
Rio de Janeiro - RJ 22040, Brazil
T: 55 21 256-8901

Jakobskind, Mario Augusto
TRIBUNA DA IMPRENSA
Rio de Janeiro, Brazil

Jamal, Leila, Sra.
ORGANIZACAO PARA LIBERTACAO DA PALESTINA
SHIS QI 09, Conjunto 06, casa 02
Brasilia - DF 71625, Brazil
T: 55 61 2484760 F: 55 61 2485879

Janer, Arianne, Ms.
Director
BIONAUT
R. Gabriel Garcia Moreno, 498
Rio de Janeiro - RJ 22610, Brazil

Janer, Bengt, Mr.
President
INSTITUTO BIODOMUS
Av. Antonio Carlos, 51/601
Rio de Janeiro - RJ 20020, Brazil
T: 55 21 240.4761 F: 55 21 262.1103

Janot, Luiz Fernando
Presidente
INSTITUTO DE ARQUITETOS DO BRASIL
Rua do Pinheiro, 10 Flamengo
Rio de Janeiro, Brazil
T: 55 21 2853480

Janotti, Dulce
Reporter
SISTEMA JORNAL DO BRASIL
Av. Brasil 500
Rio de Janeiro -RJ, Brazil
T: 55 21 5854422 F: 55 21 5854428

Jardelino, Jose Rubens, Mr.
WARC
Rua Maranhao, Apto. 32
Sao Paulo - SP 01240, Brazil

Jardim, Virginia, Ms.
Arquiteta
PREFEITURA M. PORTO ALEGRE
Av. Borges De Medeiros, 2244 - 6 Andar
Porto Alegre - RS 90060, Brazil
T: 55 512 228-8769

Jares, Paulo Roberto M.
Photographer
EDITORA ABRIL
R.da Passagem, 123/8 andar
Rio de Janeiro, RJ, Brazil
T: 021 5468282 F: 021 2759347

Jassus, Carlos Henrique
ARTE RIO
Rio de Janeiro, Brazil

Jauregui, Jorge Mario, Mr.
Arquiteto
PLANEJAMENTO ARQUITETONICO E AMBIENTAL
Ladeira De Nossa Senhora 158 Gloria
Rio de Janeiro - RJ 22211, Brazil
T: 55 21 265-2699

Jensen, Elisabeth
Journalista
FUND. CULTURAL DE RADIODIFUSAO EDUCATIVA DA REGIAO DOS LAGOS
Av. Jose Bento Ribeiro Dantas 265 Manguinhos 28905
Armacao dos Buzios -, Brazil
T: 55 246 2753848 F: 55 21 2956494

Jesus, Joaquim Santana De
Audio Operator
Rua Barao s. Felix, No 26 Centro
Rio de Janeiro, Brazil
T: 55 21 2638272

Jesus, Marco Antonio De
Reporter
SISTEMA GLOBO DE RADIO
Rua do Russel, 434
Rio de Janeiro - RJ, Brazil
T: 55 21 2058282 F: 55 21 2054096

Jesus, Matlvaldo Gomes De
REDE RECORD
Av. Miruna 713
Sao Paulo _SP 04084, Brazil
T: 55 11 5429000 F: 55 11 612228

Jesus, Miguel Pires De
MANCHETE
Rua do Russel, 804 Gloria
Rio de Janeiro, Brazil
T: 55 21 2850033

Jesus, Nilton S. De
Aux. Tecnico
TV EDUCATIVA - RJ
Av. Gomes Freire 474 Centro
Rio de Janeiro -RJ, Brazil
T: 55 21 2426736 F: 55 21 2426740

Jesus, Regina Coeli Silva De
RADIO ALVORADA FM
Rio de Janeiro, Brazil

Jiquirica Filho, Reinaldo, Sr.
Director
INSTITUTO INTERNACIONAL DE INFORMACOES ECOLOGICAS
Av. Treze de Maio, 23 Grupo 423 e 424
Rio de Janeiro - RJ 20031, Brazil
T: 55 21 220-5548 F: 55 21 240-7472
E: 1442

Jiyama, Masashi
Correspondent
YOMIURI SHIMBUN
Av Oswaldo Cruz 149/1201
Rio de Janeiro, Brazil
T: 55 21 5514560 F: 55 21 5522279

Joaquim, Rui Paulo
Technician
REDE GLOBO
R. Lopes Quintas 303
Rio de Janeiro -RJ 2, Brazil
T: 55 21 2947732

Joffily, Ruth
Fashion Editor
REVISTA DESFILE
Praia do Russel, 804/90
Rio de Janeiro RJ, Brazil
T: 55 21 5554271

Jofily, Vera, Ms.
Delegada Com. Meio Ambiente
OAB/RJ
Rua Mexico 70/1001
Rio de Janeiro - RJ 20031, Brazil
T: 55 21 220-8625 F: 55 21 262-7301

Johanson, Bo
Cinegrafista
IBASE
R.Vicente de Souza, 29 Botafogo
Rio de Janeiro -RJ, Brazil
T: 55 21 2265412 F: 55 21 2860541

Johnson, Don, Mr.
SUMMER INSTITUTE OF LINGUISTICS
SAI/NO, Lote 'D', Bloco 03
Brasilia - DF 70770, Brazil
T: 55 61 272-1200

Johnson, Eduardo Adolfo
Cinegrafista
WDR GERMAN TV CHANNEL
Rua Eca de Queiroz, 479 casa 12
Sao Paulo - SP, Brazil
T: 55 11 5721262 F: 55 11 5728959

Joner, Genera Antonio
Journalist
ZERO HORA
Av. Ipinanga 1075
Ponto Alegre Rio Gra, Brazil
T: 55 51 2234300 F: 55 51 2234265

Jones, Justyn
Media Coordinator
INTERNATIONAL NGO FORUM
Hotel Florida
Rio de Janeiro RJ, Brazil
T: 55 21 2855242

Jongh, Helena Nussbaum De, Sr.
Prefeito Municipal
PREFEITURA MUNICIPAL DE IMBE
Av. Paraguacu 1607
Imbe - RS 95625, Brazil
T: 55 666-2655

Joras, Sonia Cardoso Ryer, Sra.
Vicepresidente
ASSOCIACAO BRASILEIRA DE EDUCACAO MONTESSORIANA
Rua Marques de Sao Vicente, 355 - Gavea
Rio de Janeiro - RJ 22451, Brazil
T: 55 21 274-4420 F: 55 21 511-5270

Jordan, Juan Carlos
Jornalista
REVISTA ECORIO
Av.Pres.Antonio Carlos, 54 Gr.802
Rio de Janeiro - RJ, Brazil
T: 55 21 5330069 F: 55 21 2402658

Jordao, Eduardo Pacheco, Sr.
Diretor Tecnico
SEAERJ
Rua do Russel NO. 01 - Gloria
Rio de Janeiro - RJ, Brazil
T: 55 21 205-6797 F: 55 21 205-2795

Jorge, Thais De Mendonca
Assessora De Imprensa
BCO. INTERAMERICANO DE DESENVOLVIMENTO
Caixa Postal 08738
Brasilia - DF 70312, Brazil
T: 55 61 321-7848 F: 55 61 321-3112
TL: 61-1204

Jose, Alvaro Patrocionio
MANCHETE
R. do Russel 804
Rio de Janeiro -RJ 2, Brazil
T: 55 21 2652012 F: 55 21 2855148

Jose, Claudia, Sra.
UNIVERSIDADE FEDERAL GO
11. Avenida Esquina C/ Rua 206 N.747
Vila Velha - ES 74000, Brazil
T: 55 27 2615781 F: 55 27 2511694

Josua, Leila Marcia Leal
CORREIO FLUMINENSE
Rio de Janeiro, Brazil

Juchem, Peno, Sr.
Coordenador
SUREHMA
Rua Engenheiro Reboucas 1206
Curitiba - PR 80215-100, Brazil
T: 55 41 225 3411 F: 55 41 225 3411

Julio, Jorge De Azevedo
MANCHETE
R. do Russel 804
Rio de Janeiro -RJ 2, Brazil
T: 55 21 2652012 F: 55 21 2855148

Julio, Vanderlei Teixeira
Technician
REDE GLOBO
R. Lopes Quintas 303
Rio de Janeiro -RJ 2, Brazil
T: 55 21 2947732

Jung, Johaann
Assistente De Camera
2DF-TV ALEMA
Av.Delfim Moreira, 662/704
Rio de Janeiro - RJ, Brazil
T: 55 21 2947434 F: 55 21 5122120

Junio, Galeno Amorim
Reporter
AGENCIA ESTADO
Av. Caetano Alvares, 55
Sao Paulo -SP 02598, Brazil
T: 55 11 8562210 F: 55 11 2652297

Junior, Antonio Siqueira
INSTITUTO DE ARQUITETOS DO BRASIL
Rua do Pinheiro, 10 Flamengo
Rio de Janeiro, Brazil
T: 55 21 2853480

Junior, Ariosto Ruiz De C.
Technician
REDE GLOBO
R. Lopes Quintas 303
Rio de Janeiro -RJ 2, Brazil
T: 55 21 2947732

Junior, Ary Preihs
FBCN - JORNAL INFORMATIVO
Rio de Janeiro, Brazil

Junior, Avanir Magalhaes De Oliveira
Photographer
AGENCE FRANCE PRESSE
R. Mexico 21/7
Rio de Janeiro, Brazil
T: 55 21 240-6634 F: 55 21 262-6717

Junior, Carlos Armando S. Ribeiro
FREELANCER
Rio de Janeiro, Brazil

Junior, Celio Alves Rodrigues
Photographer
AGENCIA ESTADO
Av. Caetano Alvares, 55
Sao Paulo -SP 02598, Brazil
T: 55 11 8562210 F: 55 11 2652297

Junior, Dario Rodrigues De Menezes
Editor
REDE GLOBO
R. Lopes Quintas 303
Rio de Janeiro -RJ 2, Brazil
T: 55 21 2947732

Junior, Elpidio Reali
Correspondent
AGENCIA ESTADO
Av. Caetano Alvares, 55
Sao Paulo -SP 02598, Brazil
T: 55 11 8562210 F: 55 11 2652297

Junior, Francisco Das Chagas Barbosa
Technician
REDE GLOBO
R. Lopes Quintas 303 Jardim Botanico
Rio de Janeiro -RJ 2, Brazil
T: 55 21 2947732

Junior, Francisco Martins Pinto
CHICO JR. & ASSOCIADOS
Rio De Janeiro, Brazil

Junior, Francisco Silveira
TVN PRODUCOES ARTISTICAS
Sao Paulo, Brazil

Junior, Gelson Correa Da Silva
INSTITUTO DE ARQUITETOS DO BRASIL
Rua do Pinheiro, 10 Flamengo
Rio de Janeiro, Brazil
T: 55 21 2853480

Junior, Gilberto De Abren
Reporter
FOTOGR. & DESIGNERS ASSOC
Rua Mariz e Barros, 1058, Box 12 Tijuca
Rio de Janeiro, Brazil
T: 55 21 2645435 F: 55 21 2546124

Junior, Helio Toledo De C. M.
Director
AGENCIA ESTADO
Av. Caetano Alvares, 55
Sao Paulo -SP 02598, Brazil
T: 55 11 8562210 F: 55 11 2652297

Junior, Hilton Brito
Reporter
REDE GLOBO
R. Lopes Quintas 303
Rio de Janeiro -RJ 2, Brazil
T: 55 21 2947732

Junior, Ivan Vernon Gomes Torres, Sr.
Coordenador
GLOBAL FORUM OF SPIRITUAL AND PARLIAMENTARY LEADERS ON HUMAN SURVIVAL
Av. Canal de Marapendi 2500 - Bl.4/1904
Rio de Janeiro - RJ 22631, Brazil
T: 55 21 325-2993 F: 55 21 325-2993

Junior, Janir Hollanda
Reporter
BLOCH EDITORES
Rua do Russel 804/ 8 andar
Rio de Janeiro, Brazil
T: 55 21 2850295 F: 55 21 2059998

Junior, Jayme Teixeira Therezo
Technician
REDE GLOBO
R. Lopes Quintas 303
Rio de Janeiro -RJ 2, Brazil
T: 55 21 2947732

Junior, Jorge Da Rocha
Reporter
SISTEMA GLOBO DE RADIO
Rua do Russel, 434
Rio de Janeiro - RJ, Brazil
T: 55 21 2058282 F: 55 21 2054096

Junior, Jose Amarilio Souza De O.
FREELANCER
Rio de Janeiro, Brazil

Junior, Jose Antonio Da Silva
MANCHETE
R. do Russel 804
Rio de Janeiro -RJ 2, Brazil
T: 55 21 2652012 F: 55 21 2855148

Junior, Jose Luciano Dos Santos
Director Of Cultural Journalis
MAPE
Rua de Cesario Mottajr, No 565/12
Sao Paulo, Brazil
T: 55 11 2828837 F: 55 11 2800492

Junior, Laercio Tenorio
Technician
REDE GLOBO
R. Lopes Quintas 303
Rio de Janeiro -RJ 2, Brazil
T: 55 21 2947732

Junior, Luis Carlos Mora
SINDICATO DOS TRABALHADORES IND. QUIM. FARM. DE S.PAULO
Sao Paulo, Brazil

Junior, Luiz Clarismiao Alves
Cameraman
TV CANCAO NOVA
Alto da Bela Vista
Cachoeira Paulista, Brazil
T: 55 125 612400

Junior, Norberto Marques Quirino
Technician
REDE GLOBO
R. Lopes Quintas 303
Rio de Janeiro -RJ 2, Brazil
T: 55 21 2947732

Junior, Orlando Hardt, Mr.
Medico
Rua Comendador Remo Cesaroni 425 - 3 andar
S.Jose dos Campos-SP 12200, Brazil
T: 55 123 231206

Junior, Raimundo Genito Do C.
MANCHETE
R. do Russel 804
Rio de Janeiro -RJ 2, Brazil
T: 55 21 2652012 F: 55 21 2855148

Junior, Renato Guimaraes
ABPOLAR - ASSOC. BRASIL DE PREVENCAS A POLUICAS DO AR
Campinas SP, Brazil
F: 55 0192393799

Junior, Tarcisio Meirelles Padilha, Sr.
Decano Do Centro De Ciencias E
UNIVERSIDADE SANTA URSULA
Rua Fernando Ferrari, 75
Rio de Janeiro - RJ 22231, Brazil
T: 55 21 551-5542 F: 55 21 551-6446

Junior, Vantoeu Pereira
Photographer
ZNZ AGENCIA FOTOGRAFICA
R. Cosme Velho 335
Rio de Janeiro -RJ 2, Brazil
T: 55 21 2251813 F: 55 21 2422878

Junior, Walter Falceta
Editor
ESTADO DE SAO PAULO
Av. Engenheiro Caetano Alvares 55
Sao Paulo -SP 02598, Brazil

Junior, Wilson Cardoso
Reporter
FOTOGR. & DESIGNERS ASSOC
Rua Mariz e Barros, 1058, Box 12 Tijuca
Rio de Janeiro, Brazil
T: 55 21 2645435 F: 55 21 2546124

Junqueira, Eulalia
HOSPITAL ADVENTISTA SILVESTRE
Ladeira dos Guararapes, 263 - Santa Tereza
Rio de Janeiro - RJ, Brazil
T: 55 21 265-0212

Jurgens, Paul
Reporter
EDITORA ABRIL
R.da Passagem, 123/8 andar
Rio de Janeiro - RJ, Brazil
F: 55 21 2759347

K.Stevens, William
NEW YORK TIMES
R visconde de Piraja 550/2012 Ipanema
Rio de Janeiro RJ, Brazil
T: 55 21 259 4146 F: 55 21 2744196

Kac, Abilio, Sr.
CENTRO DE TERAPIAS HOLISTICAS
Rua Anita Garibaldi, 18 Apto,301
Rio de Janeiro - RJ 22041, Brazil
T: 55 21 257-4695

Kacser, Regina, Ms.
Fisioterapeuta
Av. Nossa Senhora de Fatima 50/607
Rio de Janeiro - RJ 20240, Brazil
T: 55 21 232 3975

Kahane, Sandra Perla, Sra.
Rua Leblon # 4
Rio de Janeiro - RJ 22441, Brazil
T: 55 21 294-5660

Kaingang, Iracema Forte, Sra.
APARAI
Rua Jose Bonifacio 915
Porto Velho - RO 78900, Brazil
T: 55 68 227 1586

Kaiser, Wolfgang, Mr.
Av. Nossa Senhora de Copacabana, 2/603
Rio de Janeiro - RJ 22010, Brazil
T: 55 21 542.1706

Kaiya, Hideo, Mr.
CEFLAM - CENTRO DE ESTUDOS FLORESTAIS DA AMAZONIA
a/c ACTA Av. Dionisio Bentes 210
Tome-Acu - PA 66010, Brazil
T: 55 91 734-1316

Kaickmann, Amalia Suzana, Sra.
REDE MULHER
Rua Joao Ramalho, 991
Sao Paulo - SP 05008, Brazil
T: 55 11 262-9407 F: 55 11 871-3457

Kambe, Tamotsu, Mr.
Diretor
ASSOC. DE INTERCAMBIO JAPAO-BRASIL
R. Machado De Assis 715
Sao Paulo - SP 06104, Brazil
T: 55 11 571-4728 F: 55 11 571-4728

Kaminski, Frederico, Sr.
Rua General Urquiza, 98 - 302 Leblon
Rio de Janeiro - RJ 22431, Brazil
T: 55 21 511 2351

Kamizaki, Yasanori
JIJI PRESS NEWS AGENCY
Av Paulista 854 - 13o. andar cj 133 Bela Vista
Sao Paulo - SP, Brazil
T: 55 11 285 0025/287 9526 F: 55 11 285 3816

Kan, Hu Ji
Reporter
AGENCIA NOTICIOSA XINHUA
Rua Paissandu,7 apto.401
Rio de Janeiro - RJ, Brazil
T: 55 21 285 0590 / 285 06 03

Kanazawa, Paulo Shiguenori, Mr.
GOV. MATO GROSSO DO SUL
Parque dos Poderes- Bloco 3
Campo Grande - MS, Brazil
T: 55 67 726-4033 F: 55 67 726-4038
TL: 67-2567

Kaneko, Erica, Sra.
R.Prof. Cristiano Volkart, 54 - Centro
Valinhos - SP 13270, Brazil
T: 55 192 71-33292

Kanto, Kazuhiko, Sr.
Director
KAWASAKI STEEL - COMERCIO E SID. LTDA
Praia de Botafogo, 228 Setor 'B' salas 508/509
Rio de Janeiro - RJ 22250, Brazil
T: 55 21 551-8196 F: 55 21 551-1099

Karabogosian, Lilian, Ms.
VBV - COOPERACAO TECNICA SUECA
Av. Rui Barbosa 560 apto 702
Rio de Janeiro - RJ, Brazil
T: 55 21 552-1521-

Kassow, Isabela
SISTEMA JORNAL DO BRASIL
Av. Brasil 500
Rio de Janeiro -RJ, Brazil
T: 55 21 5854422 F: 55 21 5854428

Kasuno, Yuji
JORNAL PAULISTA
Sao Paulo - SP, Brazil

Kayser, Arno Leandro, Mr.
Coordenador
MOVIMENTO ROESSLER P/ DEFESA AMBIENTAL
C.P. 2115
N. Hamburgo - RS 93500, Brazil

Kazuhiko, Uematsu, Mr.
M M A J- RIO
Av Rio Branco 156 - 1726
Rio de Janeiro - RJ, Brazil
T: 55 21 202 4440

Kehrig, Maria Carolina G., Sra.
Presidente
AMNPF
Rua Sao Francisco 227
Florianopolis - SC, Brazil

Kello, Nuria
Felipe dos Santos Reis 186 - Barra
Rio de Janeiro - RJ 22631, Brazil

Kelly, Andrea Coelho
MANCHETE
Rio de Janeiro, Brazil

Kenzler, Wilhelm, Mr.
Rua Sao Benedito, 1661
Sao Paulo - SP 04735, Brazil
T: 55 11 247-0368

Keppler, Alexandre
Produtor
2DF-TV ALEMA
Av.Delfim Moreira, 662/704
Rio de Janeiro - RJ, Brazil
T: 55 21 2947434 F: 55 21 5122120

Kesselring, Adolpho Luiz Bezerra, Mr.
Consultor
MINISTERIO DA SAUDE
QE 19 Conj. A casa 42 - Guara II
Brasilia - DF Cep.: 71.050, Brazil
T: 55 61 226-1369

Khan Malek, Houtan, Sr.
Director
ELF-AQUITAINE
Av Rio Branco 80 -10o. andar
Rio de Janeiro - RJ 20040, Brazil
T: 55 21 2219755

Khan Malek, Nicole, Ms.
ELF-AQUITAINE
Av Rio Branco 80 -10o. andar
Rio de Janeiro - RJ 20040, Brazil
T: 55 21 2219755

Khury, Paulo Roberto, Mr.
Relacoes Publicas
HOPE OF THE FUTURE
Rua Hugo Leal, 67 - Ilha do Governador
Rio de Janeiro - RJ, Brazil
T: 55 21 393-0129

Kievel, Jair, Sr.
Adviser
AMENCAR
Sao Leopoldo - RS, Brazil
T: 55 512 592-8111

Kikuchi, Tomio, Mr.
Presidentera De Projetos
INSTITUTO PRINCIPIO UNICO
Praca Carlos Gomes, 60/1o. andar - Liberdade
Sao Paulo - SP 01501, Brazil
T: 55 11 329738 F: 55 11 5773239

Kireff, Maria Lopes, Sra.
2nda Vice-Presidente
A.M.N.P. LONDRINA
Rua Alagoas 1439
Londrina PR 86020, Brazil
T: 55 222754

Kishinami, Roberto, Mr.
ASSOCIACAO CIVIL GREENPEACE
Rua Mexico 21, Grp 1301, Centro
Rio de Janeiro - RJ 20031, Brazil
T: 55 21 262-7318 F: 55 21 240-1690

Kiss, Cassia, Sra.
M.Associado
INSTITUTO PRINCIPIO UNICO
R. Alte. Alexandrino 3780, 6o., S. 601
Rio de Janeiro - RJ, Brazil
T: 55 21 225-6953

Klabin, Roberto L., Sr.
President
FUNDACAO SOS MATA ATLANTICA
Rua Manoel da Nobrega, 456
Sao Paulo - SP 04001, Brazil
T: 55 11 887-1195 F: 55 11 885-1680

Klein, Romeu
COLEGIO N.S. DO BOM CONSELHO
R. Ramiro Barcelos, 996
Porto Alegre - RS 90210, Brazil
T: 55 512 226-0522

Kloster, Marcus Vinicius, Sr.
Representative
GRUPO ECOLOGICO DOS CAMPOS GERAIS
Monteiro Lobato, bloco 27A, apto. 12
Ponta Grossa - PR 84100, Brazil
T: 55 422 238852
E: 710

Knobloch, Claudia, Ms.
Estudante
Rua Nobrega 194
Niteroi - RJ 24220, Brazil
T: 55 21 711-9190

Kobayashi, Samuel Shigeru
Cameraman
AMERICA DO SUL PRODUCOES DE DOCUMENTARIOS LTDA (NAMBEI TSUSHIN)
Av. Paulista 509 /11 andar conj.1104 Cerqueira Cesar
Sao Paulo -SP 01311, Brazil
T: 55 11 2886141 F: 55 11 2512065

Kodaira, Hitoshi, Mr.
EMBAIXADA DO JAPAO
Sqn 202 Bl C Apto 306
Brasilia - DF, Brazil
T: 55 61 242 6866

Kodja, Gisela, Sra.
PREFEITURA MUNICIPAL
Praca Maua s/n 1 andar
Santos - SP 11010, Brazil
T: 55 132 342-2093 F: 55 132 345-809
TL: 131048

Koenig, Vladimir, Mr.
Professor
VERJ
Av. Prado Junior 297/ 803
Rio de Janeiro - RJ 22011, Brazil
T: 55 21 2700403

Koganemary, Ayako, Ms.
Interprete - Fotografa
R. Jean Dornauff, 300 - Bairro Socorro
Mogi das Cruzes - SP 87000, Brazil
T: 55 11 468 3949

Kogl, Ildiko Csech, Mr.
Diretora
PROJETO ITATUBA
Rua da Conceicao 188/2605
Niteroi - RJ 24.020-083, Brazil

Kohno, Hiroko
YOMIURI SHIMBUN
Av Oswaldo Cruz 149/1201
Rio de Janeiro - RJ, Brazil
T: 55 21 551 6187 F: 55 21 552 2279

Konrath, Magale, Ms.
Fiscal Sanitaria
PREFEITURA M. DE SAPIRANGA
Av. Joao Correa, 793
Saripanga - RS 93800000, Brazil
T: 051599 1891 F: 051599 3315
TL: Cx.P.32

Korn, Joel, Sr.
Senior Vice President
BANK OF AMERICA NT & SA
Av. Rio Branco 110 - 39 andar
Rio de Janeiro - RJ 20040, Brazil
T: 55 21 224-7882 F: 55 21 231-0929

Kramarz, Keiko Mayma
Correspondent
YOMIURI SHIMBUN
Av Oswaldo Cruz 149/1201
Rio de Janeiro, Brazil
T: 55 21 5514560 F: 55 21 5522279

Kranz, Marcia
Fotografa
FREELANCER
R.Professor Saldanha, 154/S-204
Rio de Janeiro - RJ, Brazil
T: 55 21 2269741

Krause, Maria Amalia, Ms.
Vice- Dir.
ARQUIPELAGO DE FERNANDO DE NORONHA
Alameda Boldro, s/numero
Fernan.de Noronha-PE 53990, Brazil
T: 55 81 610-1255
TL: 814123

Kreisel, Burkard
Cinegrafista
WDR GERMAN TV CHANNEL
Rua Eca de Queiroz, 479 casa 12
Sao Paulo - SP, Brazil
T: 55 11 5721262 F: 55 11 5728959

Kretschmer, Johannes, Mr.
PUC - RIO
R. Joaquim Murtinho, 900
Rio de Janeiro - RJ 20240, Brazil

Kriesel, Stephan, Mr.
FACULDADE DE TEOLOGIA
Rua Sacramento 230
Sao Bernardo, Brazil
T: 055 011 4573733

Kruger, Hans Jurgen, Sr.
Director
GTZ
Av. da Nacoes L 25
Brasilia - DF 70339, Brazil
T: 55 61 244-1118 F: 55 61 243-4991
TL: 061 1687

Kube, Reimund, Sr.
Coo.Amazonia
SACTES
CP 1655
Recife - PE, Brazil

Kuroda, Junko, Ms.
NKK
Av. Afranio de Melo Franco, 43 / 301
Rio de Janeiro - RJ, Brazil
T: 55 21 2246477

Kuroda, Naotaka, Mr.
Diretor
NKK
Rua da Assembleia, 10 / s.3501
Rio de Janeiro - RJ, Brazil
T: 55 21 2246499 F: 55 21 2423928

Kurtz, Frederico, Sr.
Prof. De Ciencias Biologicas
UNIVERSIDADE SANTA URSULA
Rua Fernando Ferrari, 75
Rio de Janeiro - RJ 22231, Brazil
T: 55 21 551-5542 F: 55 21 551-6446

Kuznetsov, V.G., Mr.
Dty President
ROSNEFTEPRODUCT
Rua Araujo Porto Alegre, 36 - sala 902
Rio de Janeiro - RJ 20030, Brazil
T: 55 21 262-4442 F: 55 21 262-4709
TL: 34957

Labao, Mario
WTN
Rio de Janeiro, Brazil

Lacave, Emilio Moreno
ANTENA 3 DE RADIO
Rio de Janeiro, Brazil

Lacerda, Jose Paulo
Photographer
AGENCIA ESTADO
Av. Caetano Alvares, 55
Sao Paulo -SP 02598, Brazil
T: 55 11 8562210 F: 55 11 2652297

Lacomly, Paulo
Photographer
EDITORIA LEGENDA LTDA
Rua Guarana 174
Sao Paulo, Brazil
T: 55 11 2856544 F: 55 11 2832768

Laeao, Valeria Ferreira
Journalist
CENTRO DE ESTUDOS AVANCADOS EM PROMOCAO SOCIAL
Av. Borges Leal 2284
Santarem PA, Brazil
T: 55 91 5225090 F: 55 91 5231083

Laet, Ana, Sra.
Rua Francisco Manuel 91
Rio de Janeiro - RJ 20911, Brazil
T: 55 21 284-6533

Lafleur, James Rudolph, Sr.
Executive Director
ECOTEC
Rua Vigario Tenorio, 194 sala 301 - Bairro do Recife
Recife - PE 50030, Brazil
T: 55 81 224 2863 F: 55 81 224 4654
TL: 81..1367

Lage, Miriam
Editor
SISTEMA JORNAL DO BRASIL
Av. Brasil 500
Rio de Janeiro -RJ, Brazil
T: 55 21 5854422 F: 55 21 5854428

Lagreca, Leonardo
ASSOCIACAO COMERCIAL DO RIO DE JANEIRO
Rua da Candelaria 9/12 andar
Rio de Janeiro RJ, Brazil
T: 55 21 2031229 F: 55 21 2637613

Lagrotaria, Antonio Carlos
Tecnico
TVSBT
Campo de S.Cristovao, 105
Rio de Janeiro - RJ, Brazil
T: 55 21 5803688 / 5800678 F: 55 21 5809685

Lamber, Rafaela Carla, Sra.
Membro
APOENA
Rua Oscar Toledo Cesar, 716
Pirapozinho 19200, Brazil
T: 055 018241-1461

Lambert, Magda Soares, Sra.
CIAT COLOMBIA
SQN 215 Bloco H apto 503
Brasilia - DF 70874, Brazil
T: 55 61 389-3016

Lamberti, Carlos, Sr.
PREFEITURA MUNICIPAL
Praca Maua s/n 1 andar
Santos - SP 11010, Brazil
T: 55 132 342-2093 F: 55 132 345-809
TL: 131048

Lamego, Valeria
REVISTA ESTUDOS FEMINISTAS - U.F.R.J.
Rio de Janeiro, Brazil

Lamontagne, Celine, Ms.
Vice-President
CSN - CONFEDERATION DES SYNDICATS NATIONAUX
Caixa Postal 1759, Centro
Sao Paulo - SP 01059, Brazil
T: 55 11 270-1257

Landin, Helvidio, Sr.
Diretor
ECOLOG
Centro Empresarial Iguatemi - Bloco B - 5830
Salvador - BA 41820, Brazil
T: 55 71 358-9644 F: 55 71 358-8196

Lange Jr., Francisco Lothar, Sr.
Engenheiro
ITCF
Desembargador Motta 3384
Curitiba - PR 80430, Brazil
T: 55 41 234-1611 F: 55 41 222-2850
TL: 415899

Lara, Leonardo Pinho
Reporter
FREELANCER
Rua Joao XXIII No 139
Betem MG 32500, Brazil
T: 55 31 5314086 F: 55 31 5311900

Lara, Odete
Reporter
REVISTA THOT
Largo do Marchado 11 ap. 403
Rio de Janeiro - RJ, Brazil
T: 55 21 5561774

Larangeira, Jose Maria, Mr.
IBAM
Rio de Janeiro - RJ, Brazil

Larissa, Simon, Sra.
KEY MARKETING BRASIL
Av. Ataulfo de Paiva 135 gr. 710
Rio de Janeiro - RJ, Brazil

Larouco, Maria, Sra.
Praia de Botafogo 524/404
Rio de Janeiro - RJ, Brazil
T: 55 21 295-974

Larsson, Lennart
UBV
Av.Rui Barbosa, 560/702
Rio de Janeiro - RJ, Brazil
T: 55 21 2654576 F: 55 21 2860541 Rio

Laskorin, B.N., Mr.
Academician
RUSSIAN ACADEMY OF SCIENCES
Rua Araujo Porto Alegre, 36 - sala 902
Rio de Janeiro - RJ 20030, Brazil
T: 55 21 262-4442 F: 55 21 262-4709
TL: 34957

Laskorin, S.B., Mr.
General Director
ECOTECHNOLOGIA
Rua Araujo Porto Alegre, 36 - sala 902
Rio de Janeiro - RJ 20030, Brazil
T: 55 21 262-4442 F: 55 21 262-4709
TL: 34957

Laskorina, F.G., Ms.
Scientific Secretary
ECOFORUM
Rua Araujo Porto Alegre, 36 - sala 902
Rio de Janeiro - RJ 20030, Brazil
T: 55 21 262-4442 F: 55 21 262-4709
TL: 34957

Laslowski, Adao, Sr.
Advogado
SUREMMA
Rua Professor Dario Vellozo 1025
Curitiba - PR 80320-050, Brazil
T: 55 41 226 6283 F: 55 41 225 3411

Latarjet, Stephane Charles, Mr.
Technician
CERIS
Rua Dr. Julio Ottoni, 571/35
Rio de Janeiro - RJ 20241-400, Brazil
T: 55 21 265-5177 F: 55 21 285-7276

Latini, Sidney, Sr.
ASSOCIACAO COMERCIAL DO RIO DE JANEIRO
Rio de Janeiro - RJ, Brazil

Lauredo, Maria Bethania
LEGIAO DA BOA VONTADE
Av. Rudge 700 Bom Retiro
Sao Paulo -SP 01134, Brazil
T: 55 21 2338381/R:223 F: 55 21 2634025

Lavinas, Lena
REVISTA ESTUDOS FEMINISTAS - U.F.R.J.
Rio de Janeiro, Brazil

Laxe, Mauricio Cortines, Sr.
Membro
ASPAN
Rua Almirante Nelson Fernandes, 185
Recife - PE 51030, Brazil
T: 55 81 341- 3342 F: 55 81 341-3325

Leal, Antonia, Ms.
Chefe Biblioteca
SERVICO SOCIAL DO COMERCIO
Rua Voluntarios Da Patria, 169
Rio de Janeiro - RJ 22270, Brazil
T: 55 21 286-5152 F: 55 21 286-2638

Leal, Cecilia
INSTITUTO DE ESTUDOS DA RELIGIAO - ISER
Ladeira da Gloria, 98 - Gloria
Rio de Janeiro - RJ 22430, Brazil
T: 55 21 2655635 F: 55 21 511-5708

Leal, Luciana Nunes
Reporter
O DIA
R. Riachuela 359
Rio De Janeiro, RJ 2, Brazil
T: 55 21 272-8000 F: 55 21 507-1228

Leal, Luiz Maria Camacho, Sr.
Consultor
ASSOC.BRAS. ENG. SANITARIO E AMBIENTAL
Av. Beira Mar, 216 - 13 Andar
Rio de Janeiro - RJ 20021, Brazil
T: 55 21 210-3221 F: 55 21 262-6838
TL: 21-31902

Leal, Marisa, Mr.
GOVERNO DE SAO PAULO
Sao Paulo - SP, Brazil

Leal, Rolando Pires, Mr.
Interprete
UIJDC
Rua Francisco Otaviano, 42 - Copacabana
Rio de Janeiro - RJ, Brazil

Leao, Antonio Carlos, Sra.
Diretor Vice President
LEAO JUNIOR S/A
Av. Getulio Vargas, 253
Curitiba - PR 80230, Brazil
T: 55 41 233-7822 F: 55 41 222-0028
TL: 5134

Leao, Armando Zurita, Sr.
Diretor Geral
INSTITUTO DE DIVULGACAO DA AMAZONIA
Tv. Apinages 621 - sala 203
Belem - PA 66010, Brazil
T: 55 91 223-8908 F: 55 91 222-7789

Leao, Dario Nascimento
Technician
REDE GLOBO
R. Lopes Quintas 303
Rio de Janeiro -RJ 2, Brazil
T: 55 21 2947732

Leao, Ivan
Diretor De Desenvolvimento
REVISTA VIA URBANA
Rua Sao Jose, 90/2005
Rio de Janeiro - RJ, Brazil
T: 55 21 2320157 / 2523276 F: 55 21 2400139

Leao, Marcia Brandao Carneiro, Ms.
rua ilneus # 151
Sao Paulo - SP 01251, Brazil
T: 55 11 266-5225

Leao, Maria Rita, Sra.
Coord. Area De Saude
AMAR-BANGU
Estrada da Agua Branca 5865 casa 2 Bangu R-J
Rio de Janeiro - RJ 21840, Brazil
T: 55 21 332-2371

Leao, Marilene Regina Caruso, Sra.
Diretora Cultural
SINDICATO TRABALHADORES EM CONSULTORIA
R. Alvaro Alvim 34/501
Rio de Janeiro - RJ, Brazil
T: 55 21 240-6328

Leao, Mauro Sileno Saraiva, Sr.
IBGE
Rua Silvio Valente 65 Apto. 103
Salvador - BA 40850, Brazil
T: 55 71 3580098

Leao, Noemi Vianna, Ms.
Pesquisador
CONVENIO EMBRAPA/WOODS HOLE
Trav. Eneas Pinheiro, S/N
Belem - PA 66240, Brazil
T: 55 91 226 6622

Lebre, Fernando Ribeiro
FUND. CULTURAL DE RADIODIFUSAO EDUCATIVA DA REGIAO DOS LAGOS
Av. Jose Bento Ribeiro Dantas 265 Manguinhos 28905
Armacao dos Buzios -, Brazil
T: 55 246 2753848 F: 55 21 2956494

Ledon, Gilbert Rene, Sr.
Colaborador
UNIVERSAL ESPERANTO ASOCIO
Rua Santana, 73/806
Rio de Janeiro - RJ 20230, Brazil

Ledon, Symilde Schenk, Sra.
Colaborador
UNIVERSALA ESPERANTO ASOCIO
Rua Maria Aparecida Anacleto 111
Sao Paulo - SP 20230, Brazil

Leidgens, Christine, Ms.
Conjunto Jaguarema A10 - Anil
Sao Luis - MA, Brazil
T: 55 69 225-0022

Leipnitz, Itamar Ivo, Sr.
ARQUIPELAGO DE FERNANDO DE NORONHA
Rua Barao de Sao Borja, 243 - Boa Vista
Recife - PE 50070, Brazil
T: 55 81 221-3217 F: 55 81 231-1594
TL: 81.2533

Leira, Jose
Correspondent
AGENCIA EFE S/A
Avenida Rio Branco 25/13 andar
Rio de Janeiro -RJ, Brazil
T: 55 21 2534465 F: 55 21 2534861

Leitao, Sergio Barreira
CHICO JR. & ASSOCIADOS
Rio de Janeiro, Brazil

Leitao, Wilma Marques, Ms.
FUNAI
Travessa Do Pescador 4 - Ilha De Paqueta
Rio De Janeiro 20397-190, Brazil
T: 55 21 397-1015

Leite, Adi
Photographer
FOLHA DE SAO PAULO
Alameda Barao de Limeira 425
Sao Paulo SP, Brazil
T: 55 11 874 2608/874 2602 F: 55 11 2231644

Leite, Ana Alexandre, Sra.
INSTITUTO INTERNACIONAL DE COSMOBIOLOGIA
Rua Senador Dantas 75/2705
Rio de Janeiro - RJ 20031, Brazil
T: 55 21 220-7583

Leite, Ana Maria Ferreira, Sra.
Socia
FAMN E P
Rua Sargento Boracho 155
Guaratingeta - SP 12500, Brazil
T: 55 125 22 2426 F: 55 125 22 2700

Leite, Antonio Carlos Cohen, Sr.
FLAMA
Rua Marques de Abrantes, 55 - Flamengo
Rio de Janeiro - RJ 22230, Brazil
T: 55 21 265-5690

Leite, Antonio Elielson, Sr.
Distribuicao
CEDI
Av Higienopolis 983
Sao Paulo - SP 01238, Brazil
T: 55 11 825-5544 F: 55 11 825-7861
TL: 1126561

Leite, Carlos, Sr.
Assessor De Imprensa
AMPARO AO MENOR CARENTE
Rua Andre Ebling, 234 - C.Postal 433
Sao Leopoldo - RS 93040, Brazil
T: 55 512 592-8111 F: 55 512 592-8623

Leite, Dirceu Cesar, Mr.
Geologist & Engeneer
PETROBRAS - APOSENTADO
Rua Raul Drumond, 252
Salvador - BA 40135, Brazil
T: 55 71 245 1043

Leite, Flavio
Director
ABIA -ASSOCIACAO BRASILEIRA INTERDISCIPLINAR DE AIDS
R. Lopes Quintas 576 J. Botanico
Rio de Janeiro -RJ 2, Brazil
T: 55 21 2315171 F: 55 21 2945602

Leite, Francisberto N., Mr.
IBAM
Rio de Janeiro - RJ, Brazil

Leite, Jorge Wilson Passos
Tecnico
TVSBT
Campo de S.Cristovao, 105
Rio de Janeiro - RJ, Brazil
T: 55 21 5803688 / 5800678 F: 55 21 5809685

Leite, Jurandir Carvalho, Sr.
Sub. Coordenador
PROJETO TERRAS INDIGENAS
Quinta da Boa Vista s/n
Rio de Janeiro - RJ, Brazil
T: 55 21 284-9642

Leite, Leila Pinto Correia, Ms.
ASSOCIACAO DIPLOMADAS UNIVERSITARIAS
Travessa Do Carmo 50
Porto Alegre - RS, Brazil

Leite, Lesley Gasparini, Sra.
Gerente
ORGANIZACAO PREFEITO FARIA LIMA - CEPAM
Av. Prof. Lineu Prestes, 913
Sao Paulo - SP 05508, Brazil
T: 55 11 212-3144 F: 55 11 813-5969
TL: (11)83141

Leite, Luciana Savaget Teixeira
Producer
REDE GLOBO
R. Lopes Quintas 303
Rio de Janeiro -RJ 2, Brazil
T: 55 21 2947732

Leite, Manoel, Mr.
Advogado
UNIVERSIDADE DO RIO GRANDE DO SUL
Travessa Do Carmo 50
Porto Alegre - RS, Brazil

Leite, Marcela Augusta, Ms.
Secretary
CENTRO DE PESQUISAS DO MAR (CEPEMAR)
Rua Mexico, 31/Gr. 501- Centro
Rio de Janeiro - RJ, Brazil
T: 55 21 240-3822 F: 55 21 240-9516

Leite, Romeu, Sr.
Diretor
YAMAGISHI ASSOCIATION
SP.340 Km 138
Jaguariuna - SP 13820, Brazil
T: 0192971173 F: 0192973034

Leite, Sergio Fernando Vieira
Technician
REDE GLOBO
R. Lopes Quintas 303
Rio de Janeiro -RJ 2, Brazil
T: 55 21 2947732

Leite, Washington Zopur, Mr.
Administrativo
IMATEL
Praca Brasilia 830
Denise - MT 78380, Brazil
T: 065 742 1186

Leite, Yuri, Mr.
FUNDACAO BIODIVERSITAS
Rua Maria Vaz de Melo 71
Belo Horizonte - MG 31250, Brazil
T: 55 31 443-2119

Leite De Castro, Carmen C, Sra.
GOVERNO DO ESTADO DO RIO DE JANEIRO
Rio de Janeiro - RJ, Brazil

Lemes, Armando
MANCHETE
R. do Russel 804
Rio de Janeiro -RJ 2, Brazil
T: 55 21 2652012 F: 55 21 2855148

Lemle, Marina, Ms.
WORLD INFORMATION TRANSFER INC
274 Madison Ave, Suite 601
New York - NY 10016, Brazil
T: 1 212 2730253 F: 1 212 3792004

Lemos, Alvaro Alfredo
Photographer
MANCHETE
R. do Russel 804 Gloria
Rio de Janeiro RJ 2, Brazil
T: 55 21 2454816 F: 55 21 2056196

Lemos, Ana Maria, Sra.
ASSOCIACAO COMERCIAL DO RIO DE JANEIRO
Rio de Janeiro - RJ, Brazil

Lemos, Elizabeth Mascarenhas Sanches
Educadora
MUSA - CENTRO DE EDUCACAO E SAUDE DA MULHER
Rua Formosa 48
Belo Horizonte - MG 31010, Brazil
T: 55 31 467-5875

Lemos, Eneida Spolgim, Sra.
Av. Getulio Vargas - 78
Brasopolis - MG 37530, Brazil

Lemos, Fernando
Fotografo
EDITORA ABRIL
R.da Passagem, 123/8 andar
Rio de Janeiro - RJ, Brazil
F: 55 21 2759347

Lemos, Jose Gomes, Sr.
Diretor Conselheiro
INSTITUICAO RELIGIOSA PERFECT LIBERTY
Rua Pirapitingui no.204
Sao Paulo - SP 01508, Brazil
T: 55 11 270 0111 F: 55 11 279 0910

Lemos, Maria Angelica
Jornalista/Diretora
COMULHER - COMUNICACAO MULHER
Rua Rocha, 119/504
Sao Paulo - SP, Brazil
T: 55 11 2515626

Lemos, Vicente, Mr.
Expositor
EDITORA TERCEIRO MUNDO
Rua da Gloria, 122/101
Rio de Janeiro - RJ, Brazil

Lengen, Johan Van, Sr.
Director
INSTITUTO TIBA
Rua Ingles de Souza 296
Rio de Janeiro - RJ, Brazil
T: 55 21 274-1762

Lenox, Roberto, Mr.
Diretor De Meio Ambiente
AEP-DF ASSEMBLEIA ECOLOGICA PERMANENTE / FENAMAN
SQS 314 Bl 'F' 205
Brasilia - DF 70383, Brazil

Lessa, Candace Albertal, Ms.
Director
ASHOKA
Rua do Russel 450/401, Gloria
Rio de Janeiro - RJ, Brazil
T: 55 21 285-7553 F: 55 21 205-5087

Lessa, Maria Jose
SISTEMA JORNAL DO BRASIL
Av. Brasil 500
Rio de Janeiro -RJ, Brazil
T: 55 21 5854422 F: 55 21 5854428

Lessa, Sergio, Dr.
a/c Rua do Russel 450 sala 401
Rio de Janeiro - RJ, Brazil
T: 55 21 285-7553 F: 55 21 205-5087

Levy, Roberta, Sra.
OSHO SPIRITUAL HEALTH ORGANIZATION
Caixa Postal 37036
Rio de Janeiro - RJ 22622, Brazil
T: 55 21 256-4679 F: 55 21 437-7488

Lewan, Todd
Reporter
ASSOCIATED PRESS
Avenida Brasil 500 sala 847
Rio de Janeiro - RJ, Brazil
T: 55 21 580 4442 F: 55 21 580 2309

Libaneo, Maria Lucia Leonadi, Sra.
Diretor Centro Educacional Amb
PREFEITURA DE SAO PAULO
R. Caio Graco, 758, casa 1
Sao Paulo - SP 05044, Brazil
T: 55 11 872-6579 F: 55 11 273-9221

Libanio, Aristoteles Pinheiro
President Director
JORNAL GERACAO
Av Getulio Vargas 375/308 Galeria Kennedy Centro
Volta Redonda - RJ, Brazil
T: 55 243 423 918/431 634

Libardoni, Marlene, Sra.
CENTRO FEM. ESTUDOS E ASSESSORIA - CFEEA
Caixa Postal 153061 - Campus - UNB
Brasilia - DF 70910, Brazil
T: 55 61 273-7353

Lichtenberger, Dirk, Mr.
Scientist
OFFICE FOR EXCEPTIONAL MEASURES
c/o M. da Veiga Greuel, R. de el Salvador 208
Sao Paulo - SP 04910, Brazil
T: 55 11 514-5074

Licinio, Rafael
Tecnico
TVSBT
Campo de S.Cristovao, 105
Rio de Janeiro - RJ, Brazil
T: 55 21 5803688 / 5800678 F: 55 21 5809685

Lie, Eva
Cooperadora Tecinca
UBV
Av.Rui Barbosa, 560/702
Rio de Janeiro - RJ, Brazil
T: 55 21 2654576 F: 55 21 2860541 Rio

Lie, Liong Khing
Techinician
AGENCIA ESTADO
Av. Caetano Alvares, 55
Sao Paulo -SP 02598, Brazil
T: 55 11 8562210 F: 55 11 2652297

Liedtke, Paulo Fernando, Sr.
Relacoes Publicas
UFSC / PROJETO LARUS
Rod. SC 404 no.337 B6 Apt.403
Florianopolis - SC 88030, Brazil
T: 55 482 31 9302 F: 55 482 34 4069
TL: 240

Lima, Adriano M. A. Da Conceicao
MANCHETE
R. do Russel 804
Rio de Janeiro -RJ 2, Brazil
T: 55 21 2652012 F: 55 21 2855148

Lima, Alexandre Paranhos
Technician
REDE GLOBO
R. Lopes Quintas 303
Rio de Janeiro -RJ 2, Brazil
T: 55 21 2947732

Lima, Ana Cristina Siqueira, Sra.
Vice-Presidente
FUMBRASIL
Av. Caxanga 2200
Recife - PE, Brazil
T: 55 81 339-1585 F: 55 81 465-2043

Lima, Ana Maria, Sra.
Asst. Tecnica-Marketing
VISAO MUNDIAL
Rua Antonio de Albuquerque, 788 - Funcionarios
Belo Horizonte - MG 30110, Brazil
T: 55 31 225-3581 F: 55 31 227-2995
TL: 31-6052-VIMU

Lima, Ana Wisa
FREELANCER
Rua do Russel, 680/11 Gloria
Rio de Janeiro, Brazil
T: 55 21 2657097 F: 55 21 2200497

Lima, Artur Vitor Campos
SISTEMA GLOBO DE RADIO
R. de Russel 434
Rio de Janeiro RJ, Brazil
T: 55 21 2058282 F: 55 21 2054096

Lima, Charles A., Sr.
UNIVERSIDADE FEDERAL DO PARA
Greamo s/n
Belem - PA, Brazil
T: 55 91 228 8611

Lima, Clovis Ferreira
Photographer
AGENCIA ESTADO
Av. Caetano Alvares, 55
Sao Paulo -SP 02598, Brazil
T: 55 11 8562210 F: 55 11 2652297

Lima, Dora Maria De Mendonca
MADE FOR TV
Rio de Janeiro, Brazil

Lima, Elizabeth Teixeira, Sra.
Tecnico Pleno Ii
ORGANIZACAO PREFEITO FARIA LIMA - CEPAM
Av. Prof. Lineu Prestes, 913
Sao Paulo - SP 05508, Brazil
T: 55 11 212-3144 F: 55 11 813-5969
TL: (11)83141

Lima, Ilvis Ponciano Araujo, Sra.
Professora Autora
TV EDUCATIVA - CEARA
Rua Fausto Cabral, 1243 - Papicu
Fortaleza - CE 60155, Brazil
T: 55 85 234-1242 F: 55 85 205-5144

Lima, Ismael
LEGIAO DA BOA VONTADE
Av. Rudge 700 Bom Retiro
Sao Paulo -SP 01134, Brazil
T: 55 21 2338381/R:223 F: 55 21 2634025

Lima, Ivaneuza, Sra.
Arquiteta
CENTRO DE RECURSOS AMBIENTAIS
Rua Rodolfo Pimentel, 12 - Brotas
Salvador - BA, Brazil
T: 55 71 244-6125

Lima, Jorge Damasceno
REDE GLOBO
R. Lopes Quintas 303 J.Botanico
Rio de Janeiro RJ 22, Brazil
T: 55 21 2947732

Lima, Jose Alves
WTN
Rio de Janeiro, Brazil

Lima, Jose Soares
Photographer
JORNAL GAZETA DE NOTICIAS
R. Leandro Martins 72 Centro
Rio de Janeiro -RJ, Brazil
T: 55 21 2630435 F: 55 21 2630584

Lima, Juliana Claudia, Sra.
Coord. Cons. Mata Atlantica
SUDEMA
Caixa Postal 3064
Joao Pessoa - PB 58000, Brazil
T: 55 83 245-1821

Lima, Lucia Maria, Ms.
Assessora
REDE ACREANA DE HOMENS E MULHERES
Rua Das Laranjeiras 1436
Pedreiras 65725, Brazil

Lima, Luiz Eduardo Correa, Sr.
Presidente
ONCA
Caixa Postal 105
Cacapava 12280, Brazil
T: 55 122 524695 F: 525855

Lima, Luiz Otavio Mafra, Sr.
Promotor De Vendas
CIA DE CIGARROS SOUZA CRUJZ
Rua Aristides Logo, 80
Rio de Janeiro - RJ 20250, Brazil
T: 55 21 273-4322 F: 55 21 293-0243

Lima, Marcia De Souza
Photographer
AMFOC
Av 13 do Maio, 23 5/2227
Rio de Janeiro, Brazil
T: 55 21 2409466

Lima, Marcos A. Alves, Mr.
SUSILA DHARMA INTERNATIONAL
Av. Prof. Frederico Herman Jr., 345
Sao Paulo - SP CEP: 05489, Brazil
T: 55 11 210-1100

Lima, Maria Tereza F.C.Rosa, Sra.
Educadora
NINHO-MOVIMENTO DA DEFESA DA MULHER
Av. Itaoca,2358 - Villa Levio Tulio 48 - Bonsucesso
Rio de Janeiro - RJ, Brazil
T: 55 21 280 4675

Lima, Miriam
Reporter
LEGIAO DA BOA VONTADE
Av.Rudge, 700
Sao Paulo - SP, Brazil
T: 55 11 2202599 F: 55 11 2205196

Lima, Paulo Gerson De, Sr.
Tec. Ed. Amb.
SOCIEDADE POTIGUAR DE EDUCACAO - SPEL
Rua Floriano Peixoto, 295, Petropolis
Natal - RN 59020, Brazil
T: 55 84 222-2314 F: 55 84 221-6255

Lima, Pedro Cerqueira, Mr.
Consultor Meio Ambiente
CETREL
Rua Augusto Lopes Pontes, 455 Ed. Costa do Sol - Apto. 202
Salvador - BA 41700, Brazil
T: 55 71 230-8666 F: 55 71 832-2567

Lima, Rachel Cristina Linhares De
FUND. CULTURAL DE RADIODIFUSAO EDUCATIVA DA REGIAO DOS LAGOS
Av. Jose Bento Ribeiro Dantas 265 Manguinhos 28905
Armacao dos Buzios -, Brazil
T: 55 246 2753848 F: 55 21 2956494

Lima, Ricardo, Sr.
AQUALUNG CONFECCAO LTDA.
Av. Londres, 174 - Bonsucesso
Rio de Janeiro - RJ 21041, Brazil
T: 55 21 270-1955

Lima, Ricardo Barbosa De
DIRETORIO CENTAL ESTUD. UN. FED. GOIAS - DCE/UFG
R: Campus Samambaia - Estrada de Neropolis KM 13
Goiania - GO 74000, Brazil
T: 55 62205-1000 R

Lima, Roberto Xavier, Sr.
Biologo
SOCIEDADE DE PESQUISA EM VIDA SELVAGEM E EDUCACAO AMBIENTAL
Rua GOTTEMBERG 345
Curitiba - PR 80420, Brazil
T: 55 41 222-3425 F: 55 41 222-3425

Lima, Roni
Reporter
FOLHA DE SAO PAULO
Av Presidente Vargas 509/601
Rio de Janeiro - RJ, Brazil
T: 55 21 292 3929

Lima, Silvia Helena P. Nogueira, Ms.
Pesquizadora
CEPEL
Rua Marques S. Vicente 390/104
Rio de Janeiro - RJ 22451, Brazil
T: 55 21 274-5343

Lima, Sonia Maria, Ms.
ELETROBRAS
Av. Presidente Vargas, 642/10
Rio de Janeiro - RJ 20079, Brazil
T: 55 21 263-3137 F: 55 21 233-3248

Lima, Susana Aranha De, Sra.
ASSOCIACAO EVANGELICA BRASILEIRA
Rua da Conceicao, 188/S. 2807-B - Centro
Niteroi - RJ, Brazil
T: 55 21 717-6017 F: 55 21 717-2296

Lima, Valdete Dos Santos
Journalist
JORNAL MAIORIA FALANTE
R. da Lapa, 200 sala 808
Rio de Janeiro -RJ, Brazil
T: 55 21 5071462

Lima, Vanilson Benjamin Araujo, Sr.
Coordenador
MANTRA - MEIO AMBIENTE ACAO E TRANSCEDENCIA
SCLN 306 Bloco A sl.10
Brasilia - DF 70745, Brazil
T: 55 61 347-5985

Lima, Vera Alves De, Ms.
Director
FUNDO CRISTAO PARA CRIANCAS
Rua Tavares Coutinho, 310 - Bairro Papicu
Fortaleza - CE 60155, Brazil
T: 55 85 234-3133
TL: 852673 BNSA, E: 1223 = E-Mail

Lima Neves, Maria Lucia De Carvalho, Sra.
COMISSAO ESPECIAL LEI 1849/92
Rio de Janeiro - RJ, Brazil
T: 55 21 273 9442

Limabre, Joao Rodrigues De
SISTEMA GLOBO DE RADIO
R. de Russel 434
Rio de Janeiro RJ, Brazil
T: 55 21 2058282 F: 55 21 2054096

Limeira, Carlos Augusto M.
Photographer
AGENCIA ESTADO
Av. Caetano Alvares, 55
Sao Paulo -SP 02598, Brazil
T: 55 11 8562210 F: 55 11 2652297

Lindovino, Pedro Gerson
REDE RECORD
Av. Miruna 713
Sao Paulo _SP 04084, Brazil
T: 55 11 5429000 F: 55 11 612228

Linhares, Andre Jose Moreira
Technician
REDE GLOBO
R. Lopes Quintas 303
Rio de Janeiro -RJ 2, Brazil
T: 55 21 2947732

Linhart, Ana Maria
Coordinator
NUCLEO AUDIOVISUAL DE DOCUMENTACAO/UFRJ
Largo de Sao Francisco, 1 /109
Rio de Janeiro -RJ 2, Brazil
T: 55 21 2229353 F: 55 21 2211470

Lino, Clayton Ferreira
Fotografo
GLOBO ECOLOGIA
Rua Santa Alexandrina 343/c3 Rio Comprido
Rio de Janeiro - RJ, Brazil
T: 55 21 502-5451 F: 55 21 273-9087

Lins, Cristiane, Ms.
Estudante
CANDIDO MENDES
Rua Visconde de Piraja 310/504 Ipanema
Rio de Janeiro - RJ 22410, Brazil
T: 55 21 247-8627

Lins, Dalton Estellita, Sr.
Diretor Gerente
DALTON ESTELLITA LINS ADVOCACIA
Av. 13 de Maio, 13 sala 1714
Rio de Janeiro - RJ 20031, Brazil
T: 55 21 220 1365 F: 55 21 286 8885

Lins, Ivan, Mr.
TAKE IT BACK
Av. Conde De Marapendi, 2500 Bl. 3/1402
Rio de Janeiro - RJ 22.631, Brazil
T: 55 21 326-2494 F: 55 21 326-2494

Lira, Ana Maria De Farias, Sra.
Presidente
CASA DA MULHER DO NORDESTE
Rua Napoleao Laureano 40A Madalena
Recife - PE 50751, Brazil
T: 55 81 227-0531

Lira, Luiz, Sr.
President
UMA TERRA SO
Estrada do Arraial, 4558
Recife - PE, Brazil
T: 55 81 268-9208

Lirio, Jose Geraldo, Sr.
Eng. Agron.
CENTRO DE RECURSOS AMBIENTAIS
Rua Rio Sao Francisco, 1 - Monte Serrat
Salvador - BA CEP: 40.425, Brazil
T: 55 71 312-7191 F: 55 71 312-7198

Lisboa, Armando
Representative
FREELANCER
R. Delminda Silveira 235/E-402
Florianopolis, SC 88, Brazil
T: 1 482 335-151 F: 1 478 222-497

Lisboa, Edgar
Director
JAIME CAMARA
Jornal de Brasilia SIG, Trecho 1 Lot.585/645
Brasilia -DF, Brazil
T: 61 2252515 F: 61 2266735

Lisboa, Marijane, Ms.
ASSOCIACAO CIVIL GREENPEACE
Rua Mexico 21, Grp 1301, Centro
Rio de Janeiro - RJ 20031, Brazil
T: 55 21 262-7318 F: 55 21 240-1690

Loayza, Juana Bertha Roja, Sra.
Assessora
UNAMAZ
Rua Dr.Assis, 834 - Cidade Velha
Belem - PA, Brazil
T: 55 91 229-9310 F: 55 91 229-4339
TL: 1013

Loback, Harald, Sr.
Assessor
SACTES
Rua Dona Mariana 72/1001
Rio de Janeiro - RJ 22280, Brazil
T: 55 21 266-2314 F: 55 21 2622565

Lobo, Carlos Freire Ferreira
Technician
REDE GLOBO
R. Lopes Quintas 303
Rio de Janeiro -RJ 2, Brazil
T: 55 21 2947732

Lobo, Elza Ferreira, Sra.
Presidente
REDE MULHER
Rua Joao Ramalho, 991 - Perdizes
Sao Paulo - SP 05008, Brazil
T: 55 11 262-9407 F: 55 11 871-3457

Lobo, Humberto Haddock
CONFEDERACAO NACIONAL DA INDUSTRIA
Rio de Janeiro, Brazil

Lobo, Teresa Cristina
Reporter
SISTEMA JORNAL DO BRASIL
Av. Brasil 500
Rio de Janeiro -RJ, Brazil
T: 55 21 5854422 F: 55 21 5854428

Lody, Raul
FUNDACAO GILBERTO FREYRE
Rio de Janeiro - RJ, Brazil

Lodygensky, Andre, Mr.
President
FINTEC
Rua Lauro Muller,116 Gr4204
Rio de Janeiro - RJ 22290, Brazil
T: 55 21 542 1426 F: 55 21 542 1426

Loebens, Francisco Gunther
JORNAL PORANTIM -CONSELHO INDIGENISTA MISSIONARIO
SCS Ed. Venancio III S/311 C.P.03679
Brasilia -DF 70084, Brazil
T: 61 2259457 F: 61 2259401

Loebens, Guenter Francisco, Sr.
Secretary General
CIMI
SDS - Ed. Venancio III - Salas 309/304 C.P.03679
Brasilia - DF, Brazil
T: 55 61 225-9457 F: 55 61 225-9401
TL: 614293-CONM

Logan, Chris
MAIL ON SUNDAY (LONDON)
c/o Claus Meyer Barao do Flamengo 22/503
Rio de Janeiro - RJ, Brazil

Lomax, David
Reporter
BBC TELEVISION NEWS
Rua Tavares Bastos 414/66 Catete
Rio de Janeiro RJ 22, Brazil
T: 55 21 2255547 F: 55 21 2853109

Long, Suzanne Marie, Sra.
Membro
APOENA
Av.Washington Luiz, 840
Presid. Prudente-SP 19015, Brazil
T: 55 182 22-2780

Longa, Regina Do Carmo Avance, Sra.
Supervisora De Area/Rj
VISAO MUNDIAL BRASIL
Rua M, 116 - casa 01 - Conjunto Quafa - Bangu
Rio de Janeiro - RJ 21850, Brazil
T: 55 21 338-1556

Longo, Orlando Celso, Sr.
CREA-RJ
AV. RIO BRANCO 133 - 21o. ANDAR
Rio De Janeiro - RJ, Brazil

Longobardi, Maurizio
TVT REDE DE COMUNICACAO DOS TRABALHADORES
Sao Paulo, Brazil

Lontra, Paulo Klein, Sr.
Vice President
PLASTIVIDA
Rua Alexandre Dumas, 2016
Sao Paulo - SP 04707, Brazil
T: 55 11 247-6911 F: 55 11 247-8981
TL: 011/57690, E: 51583

Looi, Annie, Ms.
WIFE OF AMBASSADOR
Brasilia
Brazil

Lopes, Ana Gabriela Duarte
Producer
CNN SPANISH
R. Lopes Quintas 732 Jardim Botanico
Rio de Janeiro -RJ, Brazil
T: 55 21 5115139 F: 55 21 2397977

Lopes, Andre Luis Holanda, Sr.
Rua Conselheiro Tristao 588
Fortaleza - CE, Brazil
T: 55 85 231 6981

Lopes, Antonio Batista
Auxiliar De Camera
TV RIO
R. Miguel de Frias, 57 Estacio
Rio de Janeiro-RJ 20, Brazil
T: 55 21 2939888 F: 55 21 2938626

Lopes, Celso Dos Prazeres
MANCHETE
R. do Russel 804
Rio de Janeiro -RJ 2, Brazil
T: 55 21 2652012 F: 55 21 2855148

Lopes, Denise Costa
PREFEITURA DA CIDADE DO RIO DE JANEIRO
Rio de Janeiro, Brazil

Lopes, Emilio Mira Y, Mr.
IECAM
Dr. Moreira 250
Manaus - AM 69000, Brazil

Lopes, Eulinda De Campos, Ms.
Eng. Sanitarista
FUNDACAO ESTADUAL DO MEIO AMBIENTE
Centro Politico Administrativo
Cuiaba - MT 78000, Brazil
T: 55 065 313-2541 F: 55 065 321-2171
TL: 652512

Lopes, Geisa Morgado Rego, Ms.
Professora
ESCOLA MEXICO
R. Constante Ramos 99/801
Rio de Janeiro - RJ 22051, Brazil
T: 55 21 257-9018

Lopes, Herminio Fernando Ribeiro
Technician
REDE GLOBO
R. Lopes Quintas 303
Rio de Janeiro -RJ 2, Brazil
T: 55 21 2947732

Lopes, Hugo Jose Guimaraes, Sr.
Prof. De Ciencias Biologicas
UNIVERSIDADE SANTA URSULA
Rua Fernando Ferrari, 75
Rio de Janeiro - RJ 22231, Brazil
T: 55 21 551-5542 F: 55 21 551-6446

Lopes, Jaqueline Ricardo, Sra.
Secretaria Executiva
MORENO
Rua Xavier da Silveira, 23
Rio de Janeiro - RJ, Brazil
T: 55 21 521 7272

Lopes, Joao Humberto Goncalves
Tecnico De Audio
SISTEMA GLOBO DE RADIO
Rua do Russel, 434
Rio de Janeiro - RJ, Brazil
T: 55 21 2058282 F: 55 21 2054096

Lopes, Julio Aurelio Vianna, Sr.
Presidente
MOVIMENTO RESISTENCIA ECOLOGICA
R. Cel. Moreira Cesar 168/901
Niteroi - RJ 24230, Brazil
T: 55 21 710-3040

Lopes, Margarete Vizeu Thomas, Ms.
PREFEITURA MUNICIPAL DE PETROPOLIS
Petropolis - RJ, Brazil

Lopes, Margarida Maria Do Amaral, Sra.
FACULDADE DE ADM. / DEPT. ECONOMIA USP
a/c Rua do Russel 450 sala 401
Rio de Janeiro - RJ, Brazil
T: 55 21 285-7553 F: 55 21 205-5087

Lopes, Monica Cristina Almeida, Sra.
Expositor
CIEE
Av. Maracana, 1524/1528 - Tijuca
Rio de Janeiro - RJ 20511, Brazil
T: 55 21 268-6447 F: 55 21 288-0748
TL: 34763

Lopes, Renato Luiz Leme, Sr.
Assessor
CAMARA DOS DEPUTADOS
Sqs 211 B 303
Brasilia - DF 70274, Brazil
T: 55 61 318-6698

Lopes, Ricardo M., Mr.
Business Director
ASSOCIACAO MANTEDORA DE ORFANATOS E ASILOS
Rua Visconde de Rio Branco, 05 sl. 05 Centro
Rio de Janeiro - RJ 20060, Brazil
T: 55 21 2420035 F: 55 21 2374774

Lopes, Sergio Dos Prazeres
MANCHETE
R. do Russel 804
Rio de Janeiro -RJ 2, Brazil
T: 55 21 2652012 F: 55 21 2855148

Lopes, Stella Prata Da Silva, Ms.
Vice-Presidente
AMNP - DF
SQS 308 - Bloco F apto 102
Brasilia - DF 70355, Brazil
T: 55 61 242-5788

Lopes, Vania Goncalves
Reporter
FRILAS SERVICOS JORNALISTICOS
Rua Anfilofio de Carvalho, 29/407
Rio de Janeiro RJ 2, Brazil
T: 55 21 2205444

Lopes Filho, Pedro, Sr.
AMIGOS / PARQUE CIPO
Rua Marques de Olinda, 90/03
Rio de Janeiro - RJ 22551, Brazil
T: 55 21 246-4145

Lopes Netto, Andre, Mr.
Director
ASSOCIACAO LATINO AMERICANA DE ENGENHARIA DE SEGURANCA DO TRABALHO
Av. Rio Branco 185 Grupos 1518/19
Rio de Janeiro - RJ 20040-007, Brazil
T: 55 21 240 6072 F: 55 21 262 5632
TL: 24737-Argentina

Lopez, Emilio Mira Y, Mr.
Medico)pres. Do Cemat)
SECRETARIA DE ESTADO DE SAUDE DO RIO DE JANEIRO
Rua Mexico,128 - Centro
Rio de Janeiro - RJ 20260(20031), Brazil
T: 55 21 297-7133
TL: Ramal 289

Lorca, Maria Cecilia Rejes, Ms.
ICAE
Rua Marques de Lages, 1532 Bloco 2o, Apto 93, Saude
Sao Paulo - SP 04162, Brazil
T: 55 11 946 4801

Lorenzon, Alino, Mr.
UERJ
Rua Sao Francisco Xavier 524, Sl 1010/F
Rio de Janeiro - RJ 20550, Brazil

Loreze, Adriana
SISTEMA JORNAL DO BRASIL
Av. Brasil 500
Rio de Janeiro -RJ, Brazil
T: 55 21 5854422 F: 55 21 5854428

Lorud, Christophe
A GOTA DE ESPERANCA
Rio de Janeiro, Brazil

Louback, Jefferson B., Sr.
Representante
HOPE OF THE FUTURE / KIDDY SHOW
Rua Hugo Leal 67 - Bairro Portuguesa Ilha do Governador
Rio de Janeiro - RJ, Brazil
T: 55 21 393-0129

Loureiro, Cesar
Fotografo
JORNAL O GLOBO
Rua Irineu Marinho, 35
Rio de Janeiro - RJ, Brazil
T: 55 21 2722000

Loureiro, Vania, Sra.
CEAT
Rua: Andrade Pertence, 50 - 305
Rio de Janeiro - RJ Cep.: 22.220, Brazil
T: 55 21 205 3627

Lourencao, Otavio Cunha, Sr.
Coorde. Adm. Fin. De Projetos
FUNDACAO SOS MATA ATLANTICA
Rua Manoel da Nobrega, 456
Sao Paulo - SP 04001, Brazil
T: 55 11 887-1195 F: 55 11 885-1680

Lourenco, Antonio Carlos
LINEA PROPAGANDA S/C LTDA
Rio de Janeiro, Brazil

Lourenco, Cesar Augusto Ottoni
Technician
WASHINGTON POST
Av. Portugal 248/501 Urca
Rio de Janeiro - RJ, Brazil
T: 55 21 5426765 F: 55 21 5421917

Lourival, Reinaldo, Mr.
CONSERVATION INTERNATIONAL
Rua Bueno Brandao 393
Belo Horizonte - MG 31010, Brazil
T: 55 31 226-5145 F: 55 31 222-8429

Louzada, Silvana
TRIBUNA DA IMPRENSA
Rio de Janeiro, Brazil

Louzeiro, Andre M
Fotografo
ECOLOGIA E DESENVOLVIMENTO
Rua da Gloria 122 gr 105/106
Rio de Janeiro - RJ, Brazil
T: 55 21 242 1957/222 1370

Lucena, Rodolfo De
Editor Assistent
FOLHA DE SAO PAULO
Alameda Barao de Limeira 425
Sao Paulo - SP, Brazil
T: 55 11 874 2608/874 2602 F: 55 11 2231644

Lucena Macedo, Monica Maria, Ms.
C P R H
Rua Santana, 367 Casa Forte
Recife - PE, Brazil
T: 55 81 268-1788

Lucena Mendes, Sergio, Mr.
Vice Presidente
SOC. AMIGOS MUSEU MELLO LEITAO
Av. Jose Ruschi 4
Santa Teresa - ES 29650, Brazil
T: 027 259 1182

Lucia, Ricardo Luiz C. Funari Di
Photographic Editor
IMAGENS DA TERRA
R. das Laranjeiras 203/202
Rio de Janeiro RJ 2, Brazil
T: 55 21 5562502 F: 55 21 2204653

Ludwig, Sayonara Teixeira
Journalist
THE GRAYLINE TOURISM
Av. Niemeyer - Hotel Sharaton
Rio de Janeiro - RJ, Brazil
T: 55 21 2941444 F: 55 21 2747146

Luiz, Cosme
Photographer
ZNZ AGENCIA FOTOGRAFICA
R. Cosme Velho 335
Rio de Janeiro -RJ 2, Brazil
T: 55 21 2251813 F: 55 21 2422878

Luiz, Nivaldo
Techinician
AGENCIA ESTADO
Av. Caetano Alvares, 55
Sao Paulo -SP 02598, Brazil
T: 55 11 8562210 F: 55 11 2652297

Lujan, Daniel
Film Director
ASSEMBLEIA ESPIRITUAL NACIONAL DOS BAHAI'S DO BRASIL
R Eng Gama Lobo 267
Rio de Janeiro - RJ, Brazil
T: 55 21 288 9846

Luna, Ivanoir Ferreira De Castro, Mr.
CIA VALE DO RIO DOCE
Av. Graca Aranha, 26
Rio de Janeiro - RJ 20.030, Brazil
T: 55 21 272 4401 F: 55 21 220 0593

Luna, Maria Anita, Ms.
PESCADEIRA DE PERNAMBUCO
Rua Dom Bosco, 779
Recife - SP, Brazil

Luna, Roberto Scriviner De
Technician
REDE GLOBO
R. Lopes Quintas 303
Rio de Janeiro -RJ 2, Brazil
T: 55 21 2947732

Lunardi, Adriana Brasilia, Sra.
Member
GRUPO NATIVA - GOIAS/BR
AOS 06-Bloco A- apt.611
Brasilia - DF 70651, Brazil
T: 55 61 234-1410 F: 55 61 233-8763

Lunz, Heliomar, Mr.
Diretor Pesquisas
FUNTAC
Caixa Postal 395
Rio Branco - AC 69900, Brazil
T: 55 68 226-2304 F: 55 68 226-1667
TL: 68-2508

Lusan, Daniel Rodrigues
Jornalista
COMUNIDADE BAHA'I DO BRASIL
Rua Bela Cintra 1413/112
Sao Paulo SP 01415, Brazil
T: 55 34 215 4320 F: 55 34 236 92

Lutz, Beryl Eirene, Ms.
Biologist
UNIVERSIDADE FEDERAL DE GOIAS - UFG
Rua: 122 Lt45 QdF 43 A Setor Sul
Goiania - GO 74000, Brazil
T: 55 62 241-2346 F: 55 62 251-1694

Luz, Antonio Jose Rodrigues
MANCHETE
R. do Russel 804
Rio de Janeiro -RJ 2, Brazil
T: 55 21 2652012 F: 55 21 2855148

Luz, Dioclecio
Editor
FOLHA DO MEIO AMBIENTE
Ed. Venancio 2000 Ent. B-60 sala 230
Brasilia -DF, Brazil
T: 61 3213765 F: 62 3215809

Luz, Fernando Antonio De Souza Da
Technician
REDE GLOBO
R. Lopes Quintas 303
Rio de Janeiro -RJ 2, Brazil
T: 55 21 2947732

Luz, Leda, Sra.
Eng. Florestal
CONSELHO NACIONAL DE SERINGUEIROS
Conj. Biarritz Bl 8/103 - Est. do Coqueiro
Ananindema - PA 67000, Brazil
T: 55 91 235 4395

Luz, Leonora Hermes, Sra.
Tecnica
CONFEDERACAO NACIONAL DA INDUSTRIA
Av. Nilo Pecanha, 50 sala 3018
Rio de Janeiro - RJ 20044, Brazil
T: 55 21 262 0181 F: 55 21 262 1465
TL: 22634

Luz, Lourdes
Director
INSTITUTO DE ARQUITETOS DO BRASIL
Rua do Pinheiro, 10 Flamengo
Rio de Janeiro, Brazil
T: 55 21 2853480

Luz, Sidney Bonfim Da
REDE RECORD
Av. Miruna 713
Sao Paulo _SP 04084, Brazil
T: 55 11 5429000 F: 55 11 612228

Luzineide, , Sra.
Director
CMB
Rua Avanhadava no. 756
Sao Paulo - SP 01306, Brazil
T: 55 11 2564354 F: 55 11 2577648

Lyare, Miguel Henriquez-Berciano, Mr.
Representative
UNESCO BRASILIA, UNESCO PARIS, AND INTERGOVERNMENTAL OCEANOGRAPHICS COMMISSION
Scn Q 2 Lote B
Brasilia - DF 70710, Brazil
T: 55 61 223-8684 F: 55 61 322-4261

M. Trentin, Vera Regina, Ms.
Assistente Social
PREFEITURA M. SAPIRANGA
Av. Joao Correa, 793
Sapiranga - RS 93800 000, Brazil
T: 051 599 1891 F: 051599 3315

Maani-Hessari, Touba
Rua 23 de Maio no.70 apt.1201 CP1999
Vitoria - ES 29010, Brazil
T: 55 27 223-3802

MacLeod, Sofiah, Ms.
(from Scotland)
Alberto de Campos 247/502
Rio de Janeiro - RJ, Brazil
T: 55 21 267-5004

Macchi Silva, Margareth V., Ms.
Arquiteta/Engenheira
METROPLAN
Av. Ipiranga, 1365/ 7th andar
Porto Alegre - RS, Brazil
T: 55 051 223 1377

Macedo, Amauri Ferreira, Mr.
COPACOBANA PALACE
Rio de Janeiro - RJ, Brazil

Macedo, Ana Heloise, Ms.
CANADA
SHIS Q115 Cory 05 Casa 13
Brasilia - DF, Brazil
T: 55 61 321 2171 F: 55 61 321 4524

Macedo, Carlos Antonio Campos
TEXTO A ARTE
Rua Prof. Gabiza 332/502 Tijuca
Rio de Janeiro RJ, Brazil
T: 55 21 2482037

Macedo, Claudia
Techinician
ASSOC. PROJETO RODA VIVA
R.Sao Francisco Xavier, 524/12037-F
Rio de Janeiro - RJ, Brazil
T: 55 21 2645196 F: 55 21 2645196

Macedo, Dirceu, Sr.
Professor
ANBEM - UFPE
Av. Rosa e Silva, 1847 - Apto. 06
Recife - PE Cep.: 50.000, Brazil
T: 55 81 268-6801

Macedo, Eneas
GOVERNO DO ESTADO DE SAO PAULO
Sao Paulo, Brazil

Macedo, Helder Bonfim De, Mr.
Av. Barao Stuart 1125
Fortaleza - CE, Brazil
T: 55 85 272-1058

Macedo, Maria Ibiapina, Sra.
Bibliotecaria
IDESP
Av. Nazare 871
Belem - PA 06035, Brazil
T: 55 91 224-4411 F: 55 91 225-3414
TL: 091 6905

Machade, Maria De Lurdes Rosa
FORUM DO FUTURO
Rio de Jaeiro, Brazil

Machado, Alberto Henrique, Mr.
Rua Francisco Manuel 91
Rio de Janeiro - RJ 20911, Brazil
T: 55 21 214-6533

Machado, Allan Retilho, Sr.
Member
FLAMA
Praia do Flamengo, 22
Rio de Janeiro - RJ, Brazil
T: 55 21 236-1274

Machado, Andre Fernando F.
REVISTA INCRIVEL
Rio de Janeiro, Brazil

Machado, Anna Virginia Muniz, Ms.
Techn. Advisor
ABES
Av. Beira Mar, 216 13 andar
Rio de Janeiro - RJ 20000, Brazil
T: 55 21 210.3221 F: 55 21 262.6838

Machado, Antonio A.
PM STUDIO FOTOGRAFICO
Rio de Janeiro, Brazil

Machado, Aroldo
Editor
REDE GLOBO
R. Lopes Quintas 303
Rio de Janeiro -RJ 2, Brazil
T: 55 21 2947732

Machado, Berenice Ribeiro, Sra.
Alameda dos Cedros - 376 - Floresta
Juiz de Fora - MG 36070, Brazil
T: 55 32 235-2466

Machado, Cleocides, Mr.
Porteiro
SERVICO SOCIAL DO COMERCIO
Rua Voluntarios Da Patria, 169
Rio de Janeiro - RJ 22270, Brazil
T: 55 21 286-5152 F: 55 21 286-2638

Machado, Daniel De Freitas, Mr.
IBAM
Rio de Janeiro - RJ, Brazil

Machado, Joao Carlos, Sr.
UNISINOS
Rua Riachuelo, 1290/804
Porto Alegre - RS, Brazil
T: 55 512 222-8429

Machado, Jorge Da Costa
MANCHETE
R. do Russel 804
Rio de Janeiro -RJ 2, Brazil
T: 55 21 2652012 F: 55 21 2855148

Machado, Jose Marcus
REDE GLOBO
Rio de Janeiro, Brazil

Machado, Julio Silveira, Sr.
Av. Ataulfo de 50 B2 - 1102 - Leblon
Rio de Janeiro - RJ 22440, Brazil
T: 55 21 234-3987

Machado, Katia Helena Serafina Cruz, Sra.
Assessora P/ Ass Da Amazonia
IECLB - CPT (RO)
6a. Rua no. 1713 CxP 151
Ariquemes - RO 78914, Brazil
T: 55 69 535 5583 F: 55 69 223 1135

Machado, Lucio Antonio, Sr.
SPVS
Rua Petit Carneiro, 352
Curitiba - PR 80240, Brazil
T: 55 41 243 1424

Machado, Luis Fernando, Mr.
ASSOCIACAO CIVIL GREENPEACE
Rua Mexico 21, Grp 1301, Centro
Rio de Janeiro - RJ 20031, Brazil
T: 55 21 262-7318 F: 55 21 240-1690

Machado, Luiz De Oliveira, Sr.
Diretor
TELEBRASIL
Av. Pasteur 383 Urca
Rio de Janeiro - RJ 22290, Brazil
T: 55 21 295-4432 F: 55 21 542-4092
TL: 2123057

Machado, Marcello De A.
TV RIO
Rio de Janeiro, Brazil

Machado, Marco Aurelio A.C., Sr.
Presidente
ABRACAVE
AV. CONTORNO,6321 - 9o. ANDAR
Belo Horizonte - MG, Brazil
T: 55 31 2278811 F: 55 31 2278505

Machado, Maria, Sra.
Ass. Social
IBAMA / SUPRES / BA
Av. Juracy Magalhaes, 608 - Rio Vermelho
Rio Vermelho - BA 41930, Brazil

Machado, Maria Fatima F. Da Silva
CIA DE TURISMO DO ESTADO RO RIO - TURISRIO
Rio de janeiro, Brazil

Machado, Paulo De Tarso Pinheiro, Mr.
Sec. Executivo S.C.P.
GOVERNO DO ESTADO DO RIO GRANDE DO SUL
Borges De MEDEIROS1501
Porto Alegre - RS, Brazil
T: 55 512 511708 F: 55 512 2269722

Machado, Ricardo Bonfim, Mr.
FUNDACAO BIODIVERSITAS
Rua Maria Vaz de Melo 71
Belo Horizonte - MG 31250, Brazil
T: 55 31 443-2119

Machado, Rodolpho Lopes
Fotografo
AGENCIA NEXUS
Av.Beira Mar, 406/402
Rio de Janeiro - RJ, Brazil
T: 55 21 2402573 F: 55 21 2200094

Machado, Sergio Jose De Araujo, Sr.
FLAMA
Rua Marques de Abrantes, 55 - Flamengo
Rio de Janeiro - RJ 22230, Brazil
T: 55 21 265-5690

Machado, Sergio Luis Vicente
Cameraman
AMERICA DO SUL PRODUCOES DE DOCUMENTARIOS LTDA (NAMBEI TSUSHIN)
Av. Paulista 509 /11 andar conj.1104 Cerqueira Cesar
Sao Paulo -SP 01311, Brazil
T: 55 11 2886141 F: 55 11 2512065

Machado, Ziliani De Oliveira, Sra.
Professora
SOBEU
Rua Vereador Pinho de Carvalho, 267 - Centro
Barra Mansa - RJ 27330, Brazil
T: 55 243 22-0222 F: 55 243 22-3690
TL: 22-3640

Maciel, Andrea Gasparini, Ms.
Comunicadora Visual
PROJETO NOVAMERICA
R. Maques de Sao Vicente 331 / 5 andar, Gavea
Rio de Janeiro - RJ 22451, Brazil
T: 55 21 274.2149 F: 55 21 274.2149

Maciel, Clovis Souza
Camera
TVSBT
Campo de S.Cristovao, 105
Rio de Janeiro - RJ, Brazil
T: 55 21 5803688 / 5800678 F: 55 21 5809685

Maciel, Dilson, Sr.
Secretario Geral
ASOC. LATINOAMERICANA MULTIDICIPLINARIA DE PLANEACION ESTRATEGICA DEL MEDIO AMBIENTE
Av. da Igreja 346 - 2$ andar
Tramandai - RS 95.590, Brazil
T: 55 51 661-1322 Ramal-224 F: 55 51 661-1762

Maciel, Jacques Antonio Alvim, Sr.
Representative
FUNDACAO PROJETO SANITARISTA
BR 020, Km 33 Caixa Postal 08273
Planaltina 73301, Brazil
T: 55 61 5684545

Maciel, Rui, Mr.
Administrador
SERVICO SOCIAL DO COMERCIO
Rua Voluntarios Da Patria, 169
Rio de Janeiro - RJ 22270, Brazil
T: 55 21 286-5152 F: 55 21 286-2638

Maciel, Tereza Maria Cristina Bastos, Sra.
Representativ
FUNDACAO PROJETO SANITARISTA
BR 020, Km 33 Caixa Posta 08273
Planaltina 73301, Brazil
T: 55 61 5684545

Maciel Filho, Luiz Anselmo, Sr.
Expositor
IBASE
Rua Vicente de Souza, 12 Botafogo
Rio de Janeiro - RJ, Brazil
T: 55 21 2866161 F: 55 21 2860541
TL: 2136466

Maddelena, Benito
Photographic Reporter
REVISTA RAZOR
Rua Halfelo 414, 9 andar, Caixa Postal 662
Juz de Fora MG, Brazil
T: 55 32 2113587 F: 55 32 2152472

Madeira, Paulo Laercio Soares, Mr.
Presidente
BADESUL
Rua Sete de Setembro 666/8o Andar
Porto Alegre - RS 90010, Brazil
T: 55 512 289455 F: 55 512 272221

Madureira, Marcelo
REDE GLOBO
R. Lopes Quintas 303 J.Botanico
Rio de Janeiro RJ 22, Brazil
T: 55 21 2947732

Madureira, Mario Luiz, Mr.
Director
FUNDACAO BRASIL FRANcA
Rua Vicente de Fontoura 2683/403
Porto Alegre - RS 90630, Brazil
T: 55 512 332-4664 F: 55 512 233-2916

Magahaes, Mauricio Coelho
REUTERS
Rua Sete de Setembro 99-4, Sala 401
Rio de Janeiro RJ, Brazil
T: 55 21 5074151 F: 55 21 5072120

Magalhaes, Andrea Da S.
Reporter
SISTEMA GLOBO DE RADIO
Rua do Russel, 434
Rio de Janeiro - RJ, Brazil
T: 55 21 2058282 F: 55 21 2054096

Magalhaes, Cecilia Maria Da Silva, Ms.
Professora
UNIRIO
Rua Frei Caneca, 94 - Centro Dpto de Ciencias Naturais - 5o. andar
Rio de Janeiro - RJ 20000, Brazil
T: 55 21 232 4104

Magalhaes, Celso Rezende
CREA-RJ
AV. RIO BRANCO 133 - 21o. ANDAR
Rio De Janeiro, Brazil

Magalhaes, Francisco, Sr.
Vice-Presidente
ORG. COOP. BRASILEIRAS
SQS, Ed. Baracat, 4. andar
Brasilia - DF Cep.: 70309-900, Brazil
T: 55 61 225 0275 F: 55 61 226 8766
TL: 1879

Magalhaes, Ines
Photographer
FREELANCER
R. Ataulo de Paiva 50 Bl/C2 apt.801
Rio de Janeiro -RJ, Brazil
T: 55 21 2591114

Magalhaes, Leticia, Sra.
Rua Francisco Manuel 91
Rio de Janeiro - RJ 20911, Brazil
T: 55 21 284-6533

Magalhaes, Maria Da Conceicao Ferreira, Ms.
Procurador
PROCURADORIA GERAL ESTADO DO CEARA
Rua Silva Paulet, 324
Fortaleza - CE 60000, Brazil
T: 55 85 244 9190

Magalhaes, Ricardo Mattos
Technician
REDE GLOBO
R. Lopes Quintas 303
Rio de Janeiro -RJ 2, Brazil
T: 55 21 2947732

Magalhaes, Rogerio T., Sr.
Diretor
VILA RICA CONSULTORIA
Rua Anfilofio de Carvalho, 29/920
Rio de Janeiro - RJ 20030, Brazil
T: 55 21 240-2208 F: 55 21 262-3559

Magalhaes, Rosalvo De, Sr.
Presidente
CENTRO DE ESTUDOS E CONSERVACAO DA NATUREZA
Rua Mal. Rondon 13 - Bairro do Conego Caixa Postal 97411
Nova Friburgo - RJ 28600, Brazil
T: 55 245 225-699

Magalhaes, Sergio Faria
Diretor
ABAPORU EMPREENDIMENTOS
Caixa Postal 11286
Sao Paulo - SP, Brazil
T: 55 11 210 7657

Magalhaes, Sergio Moreira
Tecnico
TVSBT
Campo de S.Cristovao, 105
Rio de Janeiro - RJ, Brazil
T: 55 21 5803688 / 5800678 F: 55 21 5809685

Magalhaes, Theresa
GLOBO ECOLOGIA
Rua Santa Alexandrina 343/c.3 Rio Comprido
Rio de Janeiro, Brazil
T: 55 21 2739087 F: 55 21 5025451

Mage, Alexandre Da Silva
TV RIO
Rio de Janeiro, Brazil

Maglio, Ivan Carlos, Sr.
Assessor De Cooperacao Intern.
SECRETARIA DE MEIO AMBIENTE DO ESTADO DE SAO PAULO
Rua Frederico Hermann Jr. 345 - 9 andar/S. 902
Sao Paulo - SP 05459, Brazil
T: 55 11 814-8639 F: 55 11 813-0227
TL: 80631 CETSBR

Magro, Teresa, Ms.
ESALQ/USP
Av. Padua Dias, 11 LCF
Piracicaba - SP 13400, Brazil
T: 55 194 334-124 F: 55 194 336-081

Mahoney, Patricia
ICAE
12 Rajaji Marg
New Delhi 110011, Brazil
T: 91 11 301-4309

Maia, Fernando
Fotografo
JORNAL O GLOBO
Rua Irineu Marinho, 35
Rio de Janeiro - RJ, Brazil
T: 55 21 2722000

Maia, George Bittencourt Doyle, Sr.
Decano Do Centro De Ciencias
UNIVERSIDADE SANTA URSULA
Rua Fernando Ferrari, 75
Rio de Janeiro - RJ 22231, Brazil
T: 55 21 551-5542 F: 55 21 551-6446

Maia, Margareth Peixoto, Sra.
CETREL
Via Atlantica KM 09 Polo Petroquimico
Camacari 42810, Brazil
T: 55 71 8321186 F: 55 71 8322583

Maia, Maria Tereza
PROVENT
Rua Uruguiana, 10 2 andar
Rio de Janeiro, Brazil
T: 55 21 2424618

Maia, Miriam, Sra.
PREFEITURA MUNICIPAL DE BELO HORIZONTE
Av. Alonso Pena 1212
Belo Horizonte - MG, Brazil
T: 55 31 238-4186

Maia, Nailton De Agostinho
HELIO ALONSO
Rua Muniz Barreto, 51
Rio de Janeiro RJ, Brazil
T: 55 21 5515448

Maier, John
Photographer
TIME MAGAZINE
Rua Prof Gastao Baiana 429 Apt 402 Copacabana
Rio de Janeiro RJ 22, Brazil
T: 55 21 521 3989 F: 55 21 521 5647

Maier, Karl
Student
PREFEITURA MUNICIPAL DE PARACAMBI / TERRA DE EDUCAR
R. Luiz Emilio Carmo 50
Paracambi 26600, Brazil

Maimon, Dalia, Ms.
Diretora Adjunta
FEA/UFRJ
R. Barao da Torre 269/301, Ipanema
Rio de Janeiro - RJ 22411, Brazil
T: 55 21 287-5806 F: 55 21 287-5806

Mainzer-Heyers, Barbara, Ms.
Etnologia/Enfermeira
SACTES
Rua Joaquim Nabuco # 155
Itambe - PE 55920, Brazil
T: 55 81 635-1444

Maki, Ricardo Massao, Mr.
Coordenador De Acumpultura
SECRETARIA DE ESTADO DE SAUDE DO RIO DE JANEIRO
Rua Mexico,128/719
Rio de Janeiro - RJ 20031, Brazil
T: 55 21 297-7133
TL: Ramal 289

Malavasi, Javier Pierola
Camera
TVSBT
Campo de S.Cristovao, 105
Rio de Janeiro - RJ, Brazil
T: 55 21 5803688 / 5800678 F: 55 21 5809685

Malavaze, Aparecida, Sra.
Director
CMB
Rua Avanhadava no. 756
Sao Paulo - SP 01306, Brazil
T: 55 11 2564354 F: 55 11 2577648

Maldonado, Claudia, Sra.
KEY MARKETING BRASIL
Av. Ataulfo de Paiva 135 gr. 710
Rio de Janeiro - RJ, Brazil

Malheiros, Tadeu Fabricio, Sr.
FUNDACAO BRASIL-FRANCA
R. Cardoso De Almeida 668 Apto.111
Sao Paulo - SP 05013, Brazil
T: 55 11 650971

Maliz, Caroline
Photographer
FREELANCER
R. Alberto de Campos 101 apt.603
Rio de Janeiro -RJ, Brazil
T: 55 21 2676060

Malta, Flavio Jose Nery C., Sr.
Colaborador
IAMA
Rue Turi 16, Sila Madeleria
Sao Paulo - SP 05543, Brazil
T: 55 11 6724 F: 55 11 1338

Malta, Ligia Maria De Carvalho
SISTEMA GLOBO DE RADIO
Rio de Janeiro, Brazil

Malta, Maria Helena
Jornalista
REVISTA MANCHETE
Rua do Russel, 804
Rio de Janeiro - RJ, Brazil
T: 55 21 2652012

Mamede, Silvia
92 GLOBAL FORUM
Rio de Janeiro - RJ, Brazil

Mancini, Angelo A. P. Gian-Paolo
INTERNATIONAL PRESS ASSOCIATION
Rio de Janeiro, Brazil

Mancz, Antonio Wagner
MANCHETE
Rio de Janeiro, Brazil

Mandarino, Joao
UNIVERSIDADE GAMA FILHO
Rua Manuel Vitorino, 625
Rio de Janeiro, Brazil
T: 55 21 2697272

Manfredi, Eleara Maria
Editor
JORNAL SOBREVIVENCIA
Rua Lopo Goncalves, 599
Porto Alegre RJ, Brazil
T: 55 51 2287352

Manga, Marcia
Reporter
RADIO ROQUETE PINTO
Av. Erasmo Braga 118/ 11andar Castelo
Rio de Janeiro -RJ 2, Brazil
T: 55 21 2247727 F: 55 21 2524584

Mangolis, Mac
Correspondent
NEWSWEEK
Av. S. Sebastiao 157 Urca 444 Madison Ave. NY-NY 10017 USA
Rio de Janeiro -RJ 2, Brazil
T: 55 21 5419697 F: 55 21 5426583

Manoel, Claudio
REDE GLOBO
R. Lopes Quintas 303 J.Botanico
Rio de Janeiro RJ 22, Brazil
T: 55 21 2947732

Mansur, Douglas
Reporter Fotografico
O SAO PAULO
Av Higienopolis 890
Sao Paulo - SP, Brazil
T: 55 11 669 660 F: 55 11 826 3049

Mansur, Gustavo Teodoro
Photographer
FREELANCER
R Viuva Lacerda 128/201 Humaita
Rio de Janeiro - RJ, Brazil
T: 55 21 286 7757

Mansur, Jorge Alberto
Editor
REDE GLOBO
R. Lopes Quintas 303
Rio de Janeiro -RJ 2, Brazil
T: 55 21 2947732

Mansur, Renata Nascimento Y
PREFEITURA DO RIO DE JANEIRO
Av Presidente Vargas 817 17 andar, Centro
Rio de Janeiro RJ, Brazil
T: 55 21 2525422 F: 55 21 2244973

Mantovani, Mario Cesar, Sr.
Coordenador Geral
FUNDACAO SOS MATA ATLANTICA
Rua Manoel da Nobrega, 456
Sao Paulo - SP 04001, Brazil
T: 55 11 887-1195 F: 55 11 885-1680

Manutchehri, Mayram
ASSEMBLEIA ESPIRITUAL NACIONAL DOS BAHA'IS DO BRASIL
Rio de Janeiro, Brazil

Manutchehri, Shirin
ASSEMBLEIA ESPIRITUAL NACIONAL DOS BAHA'IS DO BRASIL
Rio de Janeiro, Brazil

Manutcheri, Hushmand
Film Crew Director
ASSEMBLEIA ESPIRITUAL NACIONAL DOS BAHAI'S DO BRASIL
R Eng Gama Lobo 267
Rio de Janeiro RJ 20, Brazil
T: 55 21 288 9846

Manzano, Concepcion, Ms.
I C I
Av. Das Nacoes Lote 44
Brasilia - DF, Brazil
T: 55 61 244-2121

Manzano, Ivan
UNIVISION
Rio de Janeiro - RJ, Brazil

Mara, Tania
RADIO NACIONAL - PROGRAMA GINCANA NACIONAL
Rio de Janeiro, Brazil

Maranesi, Fernanda
RAI
Av. Atlantica 1572-201
Rio de Janeiro - RJ 22021, Brazil
T: 55 21 541 -2353 F: 55 21 541-9857

Maranhao, Aluizio
Director
ESTADO DE SAO PAULO
Av. Engenheiro Caetano Alvares 55
Sao Paulo -SP 02598, Brazil

Maranhao, Bruno, Sr.
Coordenador
INSTITUTO DE ESTUDOS POLITICOS MARIO ALVES
Rua Capitao Rabelo, 208
Sao Paulo - SP 02039, Brazil
T: 55 11 267 75 20 F: 55 11 267 75 20

Maranhao, Clenia, Sra.
Directorc
CMB
Rua Avanhadava no. 756
Sao Paulo - SP 01306, Brazil
T: 55 11 2564354 F: 55 11 2577648

Maranhao, Luis Carlos Gomes
Correspondent
S/A CORREIO BRAZILIENSE
R. do Livramento 158/4 andar
Rio de Janeiro -RJ, Brazil
T: 61 3212123 F: 061 3213864

Maranhao, Sydia, Sra.
Director
URBANA: PLANEJAMENTOI E PROJETOS
R. Leao Coroado, 45
Recife - PE 50060, Brazil
T: 55 81 222-0707 F: 55 81 231-7221

Marcelo, Augusto, Mr.
Expositor
EDITORA TERCEIRO MUNDO
Rua da Gloria, 122/ 101
Rio de Janeiro - RJ, Brazil

Marcia, Yamahata
Correspondent
YOMIURI SHIMBUN
Av Oswaldo Cruz 149/1201
Rio de Janeiro, Brazil
T: 55 21 5514560 F: 55 21 5522279

Marcos, Marcia Magda
ASSOCIACAO DOS SERVIDORES DO HOSPITAL DE IPANEMA
Rio de Janeiro, Brazil

Marcos, Paulo
Fotografo
EDITORA ABRIL
R.da Passagem, 123/8 andar
Rio de Janeiro - RJ, Brazil
F: 55 21 2759347

Marcovaldi, Maria Angela, Sra.
Presidente
FUNDACAO PRO TAMAR
Av. Juracy Magalhaes Junior 608
Salvador - BA, Brazil
T: 55 71 835-1150 F: 55 71 876-1045

Maretzki, Gerry
EDITORA NOVA FRONTEIRA
Rio de Janeiro, Brazil

Margalho, Hernandes Espinosa, Sr.
Coordinator
CENTRO ACADEMICO DE DIREITO EDSON LUIZ
Passagem das Flores, 259
Belem - PA 66000, Brazil

Margulies, Andre Jorge
Journalist
FUND. CULTURAL DE RADIODIFUSAO EDUCATIVA DA REGIAO DOS LAGOS
Av. Jose Bento Ribeiro Dantas 265 Manguinhos 28905
Armacao dos Buzios -, Brazil
T: 55 246 2753848 F: 55 21 2956494

Maria, Adriana Maria Magalhais De
Journalist
RADIOBRAS
SQN 205 Bloco G, apto 601
Brasilia 70843, Brazil
T: 55 61 2721911 F: 55 61 3221063

Maria, Cleusa
Reporter
SISTEMA JORNAL DO BRASIL
Av. Brasil 500
Rio de Janeiro -RJ, Brazil
T: 55 21 5854422 F: 55 21 5854428

Mariani, Jose, Sr.
Socio-Diretor
ANDALUZ PRODUCOES CINEMATOGRAFICAS
Praia do Flamengo 66 #710
Rio de Janeiro - RJ, Brazil
T: 55 21 205 5487 F: 55 21 225 8918

Mariano, Ciro
Fotografo
TYBA AGENCIA FOTOGRAFICA
Rua Barao do Flamengo 22/503 Flamengo
Rio de Janeiro - RJ, Brazil
T: 55 21 285 4640/265 4839 F: 55 21 222 6742

Mariano, Claudio Luiz Diniz
Operador De Vt
REDE BANDEIRANTES DE TELEVISAO
R.Alvaro Ramos, 492
Rio de Janeiro - RJ, Brazil
T: 55 21 2122072 F: 55 21 2957391

Mariano, Maria De Lourdes Ramos, Sra.
INTERNATIONAL YOUNG CHRISTIAN WORKERS
Rua Joao Timoteo de Andrade, 352, Bairro Santana
Carlacica - ES 29140, Brazil
T: 55 27 223-6711.

Mariasch, Telma, Sra.
IECAM
Dr. Moreira 250
Manaus - AM 69000, Brazil

Marin, Ana Magdalena
AGENCIA DE NOTICIAS ACOPI
Sao Paulo, Brazil

Marin, Cintra Moraes
Reporter
FUNTEVE - RADIO MEC
Praca da Republica, 141
Rio de Janeiro RJ, Brazil
T: 55 21 2204466

Marinho, Geraldo, Mr.
Coordenador
ARRUAR
R. Dona Anunciada, 188 - Gracas
Recife - PE 52011, Brazil
T: 55 81 222 0810

Marinho, Jailson Gama
HOSPITAL ADVENTISTA SILVESTRE
Ladeira dos Guararapes, 263 - Santa Tereza
Rio de Janeiro - RJ, Brazil
T: 55 21 265-0212

Marinho, Nilson Cezar
Journalist
ZERO HORA
Rua Ipamanga 1075
POA Ponto Alegre, Brazil
T: 55 21 2234300

Marinho, Paulo
Reporter
RADIO ROQETE PINTO
Rio de Janeiro, Brazil

Marinho, Sebastiao
Fotografo
JORNAL O GLOBO
Rua Irineu Marinho, 35
Rio de Janeiro - RJ, Brazil
T: 55 21 2722000

Marino, Horacio Antonio, Sr.
Inv./research Dep. Director
CEPAVE
Av. Venceslao Braz 14-309 - Botafogo
Rio de Janeiro - RJ, Brazil
T: 55 21 295-2271 F: 55 246223-2300

Mariotti, Silvia, Sra.
Coordenadora
AMIGOS DA TERRA-ITALIA
Av. Brigadeiro Luiz Antonio 4442
Sao Paulo - SP, Brazil
T: 55 11 887-8228 F: 55 11 884-2795

Mariotto, Francisco Ercilio
Vt Operator
TV CANCAO NOVA
Alto da Bela Vista
Cachoeira Paulista, Brazil
T: 55 125 612400

Maron, Edison, Sr.
JUNDILAB
jUNDAI
Sao Paulo - SP 13200, Brazil
T: 55 11 434-5622

Marquei, Aida
Producer
MP2 PRODUCOES
Rua Ito Amano, 112
Rio de Janeiro, Brazil
T: 55 21 2528357

Marques, Ayrton Tadeu Alves
Editor
JORNAL VERDADE
R.Braulio Mario Ribas, 1000
Entre-Ijuis, RS 9885, Brazil
T: 55 31 25996 F: 55 3611339

Marques, Gilberto De Souza, Sr.
Av Bartolomeu de Gusmao 34 apt 125
Santos - SP 11040, Brazil
T: 55 132 380242

Marques, Iara Maria De Almeida, Sra.
P N U D
Shin Ql 12 Conj 06 Capa 07
Brasilia - DF 71500, Brazil
T: 55 61 577-3851

Marques, Jose Geraldo W., Sr.
Coordinator
GRUPO DE ETNOECOLOGIA
Rua Jornalista Augusto Jose Filho, 241 - Farol
Maceio - AL 57055, Brazil
T: 55 82 241-1631

Marques, Jose Roque, Mr.
MINISTERIO PUBLICO DO AMAZONAS
Rua: Londres, C/01 Q23 - Campos Eliseos
Manaus - AM, Brazil
T: 55 92 656-4357 F: 55 92 232-3244

Marques, Judith Early, Ms.
Observer
ESCOLA AMERICANA DE BRASILIA
Sqs 215 Bld. Apto. 215
Brasilia - DF, Brazil
T: 55 61 242 0459

Marques, Lucia Benfatti, Sra.
Diretora Tesoureira
ASSOCIACAO PALAS ATHENA
Alameda Lorena, 289 - apto. 902
Sao Paulo - SP 01424, Brazil
T: 55 11 288 7356

Marques, Maria De Fatima C.
Reporter
REVISTA FERROVIARIA
Rua Mexico, 41 Sl.904
Rio de Janeiro - RJ, Brazil
T: 55 21 2401747/1248 e 5320260 F: 55 21 2400139

Marques, Maria Fernanda Cabral, Ms.
Assistant
GLOBAL FORUM OF SPIRITUAL AND PARLIAMENTARY LEADERS ON HUMAN SURVIVAL
Rua S. Joao Batista 47 apto 503 Botafogo
Rio de Janeiro - RJ 22270, Brazil
T: 55 21 246-8046 F: 55 21 325-2993

Marques, Maria Lucia, Sra.
Eng.Quimica
CENTRO DE RECURSOS AMBIENTAIS
Rua Rio Sao Francisco, 1 - Monte Serrat
Salvador - BA CEP: 40.425, Brazil
T: 55 71 312-7191 F: 55 71 312-7198

Marques, Myriam Silva, Ms.
Conselheira
CONSELHO DE DEFESA DIREITOS DA MULHER
Rua Realeza 314
Minas Gerais 35930, Brazil
T: 031 851-3237

Marques, Nelson
ESTADO DE SAO PAULO
Av. Rio Branco 151 - Sobre - Loja sala 201 a 204
Rio de Janeiro RJ, Brazil
T: 55 21 2217111

Marques, Regina De Souza, Ms.
Rua Bento Lisboa, 24/1003
Rio de Janeiro - RJ, Brazil
T: 55 21 225-0600

Marques, Regina Maria, Sra.
FLAMA
Rua Marques de Abrantes, 55 - Flamengo
Rio de Janeiro - RJ 22230, Brazil
T: 55 21 265-5690

Marques, Rogerio Fernandes
TV RIO
Rio de Janeiro, Brazil

Marques, Telma Ferracioli, Sra.
Coordenadora De Projetos
ABRAPIA
Rua Paissandu 155/1101
Rio de Janeiro - RJ 22210, Brazil
T: 55 21 265-7893

Marques, Walter Duton
Reporter
JJN NOTICIAS
Praca Almirante Jaceguai 72 apt. 815
Rio de Janeiro -RJ, Brazil
T: 55 21 282-1292 F: 55 21 240-3073

Marques Filho, Mario Ribeiro, Mr.
GOV. MATO GROSSO DO SUL
Parque dos Poderes Bloco 03
Campo Grande - MS 79031902, Brazil
T: 55 67 726-4033 F: 55 67 726-4038
TL: 67-2567

Marquette, Joe
ASSOCIATED PRESS
Rio de Janeiro, Brazil

Marriel, Nelson, Mr.
WORLD COUNCIL OF CHURCHES
150 route de Ferney
Brazil
F: 41 22 791 0361

Marrocos, Ricardo Paysano, Sr.
Eng. Sanitarista
FUNDACAO NACIONAL DE SAUDE
Sqs 102 Bloco B Apto. 501
Brasilia - DF 70230, Brazil
T: 55 61 224-5840

Marshall, Elvyn, Ms.
Socia Gerente
SOFTLANDING ASSESSORIA DE EMPRESAS LTDA
av. Marechal Camara 271/ grupo 1103
Rio de Janeiro - RJ 20020, Brazil
T: 55 21 262 2286 F: 55 21 262 5946
TL: 213 8806 SFLA

Marteletto, Walkiria, Sra.
Membro
APOENA
Rua Cuiaba, 1-19
Presiden.Epitacio-SP 19470-000, Brazil
T: 055 018221-4222 F: 055 018622-6177

Martes, Eunice, Sra.
Participante
ISER/SOFIA
Caixa Postal 16011
Rio de Janeiro - RJ 22201, Brazil
T: 55 21 265-5797

Martin, Eduardo C. San, Mr.
CESP
Rua da Consolacao, 1875, 9 andar, Cerq. Cesar
Sao Paulo - SP, Brazil
T: 55 11 258-2165 F: 55 11 256-2101

Martinez, Bernard
Photographer/ Journalist
GAMMA
R Joao Lira 140/304 22430
Rio de Janeiro - RJ, Brazil
T: 55 21 239 7086

Martinez, Eliana M.G.
ICAE
12 Rajaji Marg
New Delhi 110011, Brazil
T: 91 11 301-4309

Martinez, Ricardo Horacio, Sr.
Vendedor
NEWSWEEK INTERNATIONAL
Rua Do Russel 450 Sala 801
Rio de Janeiro - RJ 22210, Brazil
T: 55 21 265-4392 F: 55 21 204-6887

Martinez, Simone Caroline V., Sra.
Membro
APOENA
Rua Tiradentes, 836
Presid. Prudente-SP 19200, Brazil
T: 55 182 41-1112

Martini, Nildo, Mr.
Presidente
ECOLOGICAL CAMPAIGN INTERNATIONAL COMMUNITY-ECIC
Rua Senador Dantas 45 - Grupo 902
Rio de Janeiro - RJ 20031, Brazil
T: 55 21 220-4045

Martini, Ronaldo Lucas
VIDEOARTE RIO
Rua Senradom 45
Rio de Janeiro, Brazil

Martiniano, Roberto
Technician
REDE GLOBO
R. Lopes Quintas 303
Rio de Janeiro -RJ 2, Brazil
T: 55 21 2947732

Martins, Aristeu, Jr., Sr.
Tec.Ed. Amb.
SOCIEDADE POTIGUAR DE EDUCACAO - SPEL
Rua Floriano Peixoto, 295, Petropolis
Natal - RN 59020, Brazil
T: 55 84 222-2314 F: 55 84 221-6255

Martins, Beatriz Cintra
ASSOCIACAO DE MORADORES E AMIGOS DE LARANJEIRAS
Rio de Janeiro, Brazil

Martins, Carlos H., Sr.
INSTITUTO PRO NATURA
Rua Monte Alegre 374 - Santa Teresa
Rio de Janeiro - RJ 20240, Brazil
T: 55 21 224-39904 F: 55 21 224-3994

Martins, Claudia De Lima Alves, Ms.
Repr. Comite De Juventude
COMUNIDADE INTERNACIONAL BAHA'I
Rua Wanderley Pinho, 570, apt.101 - Itaigara
Salvador - BA 41840, Brazil
T: 55 71 358-1903

Martins, Daniella Vanessa A., Sra.
Biologa
ELETRONORTE
Scn-Q06, Conj.A, Bloco C-Venancio 3000, Sala 313
Brasilia - DF 70000, Brazil
T: 55 61 212-6157 F: 55 61 225-9755
TL: 1279

Martins, Edgard Da Silva
Technician
REDE GLOBO
R. Lopes Quintas 303
Rio de Janeiro -RJ 2, Brazil
T: 55 21 2947732

Martins, Helio Bittencourt
Editor
SISTEMA CELOSO DE RADIO
Rua do Russel, 439, 4 andar
Rio de Janeiro, Brazil
T: 55 21 2850448

Martins, Idalia, Ms.
MOVIMENTO EM DEFESA DA VIDA
Praca Monsenhor Emilio Lotterman, 96
Porto Alegre - RS 90.000, Brazil
T: 55 512 248-4098 F: 55 512 343-6671

Martins, Laise, Ms.
WORLD YOUNG WOMEN'S CHRISTIAN ASSOCIATION
Av. Epitacio Pessoa, 2664 Bl.B-Apt804 Lagoa
Rio de Janeiro - RJ 22471, Brazil
T: 55 21 287-0376
TL: WYMCA 41 23 32

Martins, Manuel Ferreira
Cinegrafista
REDE BANDEIRANTES DE TELEVISAO
R.Alvaro Ramos, 492
Rio de Janeiro - RJ, Brazil
T: 55 21 2122072 F: 55 21 2957391

Martins, Margarete, Sra.
Secretaria
ASSOCIACAO MACAENSE DE DEFESA AMBIENTAL
Av. Presidente Sodre 316 - Sala 304
Macae - RJ 28700, Brazil
T: 55 247 628-364 F: 55 247 625-110

Martins, Maria Ines
Director
INSTITUTO DE ARQUITETOS DO BRASIL
Rua do Pinheiro, 10 Flamengo
Rio de Janeiro, Brazil
T: 55 21 2853480

Martins, Maria Lucia Silva, Sra.
Tecnica
CONFEDERACAO NACIONAL DA INDUSTRIA
Av. Nilo Pecanha, 50 sala 3018
Rio de Janeiro - RJ 20044, Brazil
T: 55 21 262 0181 F: 55 21 262 1465
TL: 22634

Martins, Maura, Ms.
Membro Da Coordenacao
ASSEMBLEIA ECOLOGICA PERMANENTE - AEP
SQS 314 - Bloco 'F' sala 205
Brasilia - DF, Brazil

Martins, Nelson Luis
Technician
PROVIDEO CINEMA E TV
SHIS Q1 21 Conj 7 casa 15 Lago Sul
Brasilia DF, Brazil
T: 55 61 366 1851 F: 55 61 366 31

Martins, Nyelse Eliane Trench, Sra.
PROJETO LINHA VERDE - P.M. S.SEBASTIAO
Prefeitura Municipal de Sao Sebastiao, Secretaria de Cultura e Turismo
Sao Sebastiao - SP, Brazil

Martins, Paulo Amilcar
Techinician
AGENCIA ESTADO
Av. Caetano Alvares, 55
Sao Paulo -SP 02598, Brazil
T: 55 11 8562210 F: 55 11 2652297

Martins, Rosana C.
OSHO SPIRITUAL HEALTH ORGANIZATION
Rua Fernando Osorio 19/601
Brazil

Martins, Sergio Augusto Pinto, Mr.
Rua Barao De Ipanema 29 - Apt. 404
Rio de Janeiro - RJ 22050, Brazil
T: 55 21 235-5837

Martins, Virginia, Sra.
ORGANIZACAO PARA LIBERTACAO DA PALESTINA
SHIS Ql 09, Conjunto 06, casa 02
Brasilia - DF 71625, Brazil
T: 55 61 2484760 F: 55 61 2485879

Martorelli, Marco Aurelio Chagas
CENTRO ACADEMICO XI DE AGOSTO
Rua Riachuelo, 194
Sao Paulo - SP 01007, Brazil
T: 55 11239-0652 R F: 55 11 37-5300

Marx, Luis Eduardo P.
ECO TV
Parati, Brazil

Mascarenhas, Lousival
Technician
REDE GLOBO
R. Lopes Quintas 303
Rio de Janeiro -RJ 2, Brazil
T: 55 21 2947732 F: 55 21 2947247

Mascarenhas, Maria Ivone Coelho, Sra.
Tecnica
CONFEDERACAO NACIONAL DA INDUSTRIA
Av. Nilo Pecanha, 50 sala 3018
Rio de Janeiro - RJ 20044, Brazil
T: 55 21 262 0181 F: 55 21 262 1465
TL: 22634

Mascarenhas, Teresa, Sra.
COMISSAO ESPECIAL LEI 1849/92
Rio de Janeiro - RJ, Brazil
T: 55 21 273 9442

Mastromatteo, Matteo, Sr.
Commercial Attache
CONSULADO GERAL DA ITALIA
Av. Presidente Antonio Carlos, 40/7o. andar
Rio de Janeiro - RJ 20020, Brazil
T: 55 21 220-3460

Mata, Alba Soares Fonseca
Reporter
RADIO ROQETE PINTO
Rio de Janeiro, Brazil

Mata, Danilo Leonel Da
CONFEDERACAO NACIONAL DA INDUSTRIA
Rio de Janeiro, Brazil

Mathias, Francisco, Sr.
ECOTROPIC
1880 East River Road
Tucson - AZ, Brazil
T: 1 602 577-6476 F: 1 602 577-1195

Mathieu, Hans, Sr.
ILDES/FES
Rua. Lisboa 191
Sao Paulo - SP 05413, Brazil
T: 55 11 282-3900 F: 55 11 280-1916
TL: 1180616

Mathis, Armin, Mr.
Pesquisador
UNIVERSIDADE LIVRE DE BERLIN
Trav. Mauriti 1006/904
Belem - PA 66085, Brazil
T: 55 91 244 0592

Matias, Ivaci
Special Reporter And Editor
GLOBO
TV Globo de Sao Paulo (Globo Rural) R Paraguacu 404/403
Sao Paulo - SP, Brazil
T: 55 11 629912/8234275 F: 55 11 8262112

Matos, Ana Paula Goulart
Technician
REDE GLOBO
R. Lopes Quintas 303 Jardim Botanico
Rio de Janeiro -RJ 2, Brazil
T: 55 21 2947732

Matos, Francisco Jose De Abreu, Sr.
IECAM
dr.moreira 250
Manaus - AM 69000, Brazil

Matos, Gustavo Gomes De, Sr.
Tecnico
CONFEDERACAO NACIONAL DA INDUSTRIA
Av. Nilo Pecanha, 50 sala 3018
Rio de Janeiro - RJ 20044, Brazil
T: 55 21 262 0181 F: 55 21 262 1465
TL: 22634

Matos, Jose Santos, Mr.
Diretor Video
REDE MULHER
Rua Joao Ramalho, 991 Perdizes
Sao Paulo - SP 05008, Brazil
T: 55 11 262 9407 F: 55 11 871 3457

Matos, Maria Teresa Soares, Sra.
Diretora De Apoio Tecnico
IPREDE
Rua Prof. Carlos Lobo, 15 Cidade dos Funcionarios
Fortaleza - CE 60815, Brazil
T: 55 85 239-1300

Matos, Mario Jorge Gama
Technician
REDE GLOBO
R. Lopes Quintas 303
Rio de Janeiro -RJ 2, Brazil
T: 55 21 2947732

Matrianha, Carlos Borges
Editor
FUNDACAO ROQUETTE-PINTO
Prasa da Republica, 141 A Centro
Rio de Janeiro, Brazil
T: 55 21 2522203

Matta, Claudinei Da
Technician
REDE GLOBO
R. Lopes Quintas 303
Rio de Janeiro -RJ 2, Brazil
T: 55 21 2947732

Mattedi, Maria Raquel Mattoso, Sra.
INSTITUTO DE ESTUDOS AMAZONICOS E AMBIENTAIS
Rua Mae Luzia, 582
Macapa - AP 60900, Brazil
T: 55 96 222-1010

Mattos, Carlos Arnaldo Silva, Mr.
Advogado
O A B
Rua Candido Benicio 1201/806 B
Rio de Janeiro - RJ, Brazil

Mattos, Cristina, Sra.
Pesquisadora
EMBRAPA
Av. Dr. Julio Soares de Arruda 803, Pq. S. Quirino
Campinas - SP 13085, Brazil
T: 55 192 52-5977 F: 55 192 54-1100
TL: 197686 EBPABR

Mattos, Jean Kleber De Abreu, Sr.
FUB
Qd 4 Bl U C/8, Cruzeiro Velho
Brasilia - DF 70640, Brazil

Mattos, Luis
ECOTROPIC
1880 East River Road
Rio de Janeiro - RJ, Brazil
T: 55 21 577-6476 F: 55 21 577-1195

Mattos, Pedro Agilson
Reporter Fotografico
REVISTA ISTO E/SENHOR
Av.Almte.Barroso,63/1512
Rio de Janeiro - RJ, Brazil
T: 55 21 2405224 F: 55 21 2402925

Mattos, Regina Helena Cabral, Sra.
Museum Sector Chief
PA. ZOOBOTANICO G. VARGAS
Rua Prof. Amilcar Falcao no. 03, apt.101 Morro do Gato
Salvador - BA 40160, Brazil
T: 55 71 237-6008

Matz, Marcello, Mr.
ROYAL INSTITUTE OF INTERNATIONAL AFFAIRS
Rua Barao da Torre269 B 01
Rio de Janeiro - RJ, Brazil
T: 55 21 2471844 F: 55 21 2875806

Matzenbacher, Jaime, Sr.
TERRA CHRISTA COMMUNICATIONS
Alameda Barao de Limeira 1283 COB #131 Campos Eliseus
Sao Paulo - SP 01202, Brazil

Maubury-Lewis, Bloru, Mr.
FUNDACAO FORD
Praia Do Flamengo 100 / 12 Andar
Rio de Janeiro - RJ 22210, Brazil
T: 55 21 225-0662 F: 55 21 285-1250

Mauricio, Geraldo, Mr.
DENACOOP - M.A.
Rua Frei Leandro 35
Rio de Janeiro - RJ 22470, Brazil
T: 55 21 286-9511 F: 55 21 286-1191

Maury, Cilulia Maria, Sra.
Biologa
FUNDACAO PRO-NATUREZA (FUNATURA)
SCLN 107, Bloco B, Salas 201/17
Brasilia - DF 70743, Brazil
T: 55 61 274-5449 F: 55 61 274-5324

Mauso, Angela Villarrubia
REVISTA 'CONCIENCIA PLANETARIA'
Sao Paulo, Brazil

Max, Jose Carlos Molina, Sr.
Engenheiro Florestal
GOVERNO DO ESTADO DE SAO PAULO / INSTITUTO FLORESTAL
Rua Sebastiao Silva Leite 990
Assis - SP 19800-000, Brazil
T: 55 21 222682

Maywald, Leila Maria, Sra.
Presidente
ASSOCIACAO DOS MORADORES E AMIGOS DO FLAMENGO - FLAMA
Rua Marques de Abrantes, 55 - Flamengo
Rio de Janeiro - RJ 22.230, Brazil
T: 55 21 245-8000

Maywald, Marcelo, Sr.
FLAMA
Rua Marques de Abrantes, 55 - Flamengo
Rio de Janeiro - RJ, Brazil
T: 55 21 265-5690

Mazalam, Ricardo
ASSOCIATED PRESS
Rio de Janeiro, Brazil

Mbanefo, Sandra
92 GLOBAL FORUM
Rio de Janeiro - RJ, Brazil

Mealey, Elisabeth
Greenpeace Communications
GREENPEACE
Rua Mexico, 21/1301 Gr.A e B
Rio de Janeiro - RJ, Brazil
T: 55 21 2401690 F: 55 21 2627318

Meana, Francesco Ripa Di, Sr.
Coordenador P/ Brazil
COOPERACaO ITALIANA EM SAuDE
Av. Sao Rafael 2289 - Pau da Lima
Salvador - BA 41600, Brazil
T: 55 71 399-6138 F: 55 71 393-3563

Medeiros, Alessandra
92 GLOBAL FORUM
Rio de Janeiro - RJ, Brazil

Medeiros, Dea Maria Moreira, Ms.
Rua Antonio Ataide 218
Vila Velha - ES 29100, Brazil
T: 55 27 229-1691

Medeiros, Frederico Freire
Photographer
SERVICO SOCIAL DA INDUSTRIA
Av. Nilo Pecanha 50/2409 Centro
Rio de Janeiro -RJ, Brazil
T: 55 21 2621267 F: 55 21 2204166

Medeiros, Gustavo Henrique
Journalist
FUND. CULTURAL DE RADIODIFUSAO EDUCATIVA DA REGIAO DOS LAGOS
Av. Jose Bento Ribeiro Dantas 265 Manguinhos 28905
Armacao dos Buzios -, Brazil
T: 55 246 2753848 F: 55 21 2956494

Medeiros, Joao, Sr.
Diretor Presidente
PREFEITURA MUNICIPAL DE NITEROI/ ENITUR - EMPRESA NITEROENSE DE TURISMO
Estrada Leopoldo Froes, 773
Niteroi - RJ, Brazil
T: 55 21 710-2727 F: 55 21 710-8489

Medeiros, Maria Angelica F.
92 GLOBAL FORUM
Rio de Janeiro - RJ, Brazil

Medeiros, Sonia Maria Gomes De Matos
Assessora Legislativa
ASSEMBLEIA LEGISLATIVA - PE
Caixa Postal, 7873 - Cidade Universitaria
Recife - PE, Brazil

Medina, Jose Ronaldo De Castro, Mr.
IBAM
Rio de Janeiro - RJ, Brazil

Medina, Silvio Jose G
MANCHETE
Rio de Janeiro, Brazil

Megee, Pennie, Ms.
DENACOOP - M.A.
Rua Frei Leandro 35
Rio de Janeiro - RJ 22470, Brazil
T: 55 21 286-9511 F: 55 21 286-1191

Meira, Maria Elisa
INSTITUTO DE ARQUITETOS DO BRASIL
Rua do Pinheiro, 10 Flamengo
Rio de Janeiro, Brazil
T: 55 21 2853480

Meireles, Marcos
FORUM DO FUTURO
Rio de Janeiro, Brazil

Meirelles, Domingos Joao
Reporter
REDE GLOBO
R. Lopes Quintas 303
Rio de Janeiro -RJ 2, Brazil
T: 55 21 2947732

Meirelles, Julia, Sra.
Coordenadora Tecnica
CONSORCIO MATA ATLANTICA - UNICAMP
Rua Ribeiro do Vale 236
Sao Paulo - SP 04568, Brazil
T: 55 11 530-2359 F: 55 11 39-4717
TL: (019) 1150

Meirelles, Marcia
Camera/Fotografia
COMULHER - COMUNICACAO MULHER
Rua Rocha, 119/504
Sao Paulo - SP, Brazil
T: 55 11 2515626

Meirelles, Nei Flavio, Sr.
ATOBA - MOVIMENTO DE EMANCIPACAO HOMOSEXUAL
Rua Prof. Carvalho de Melo 471
Rio de Janeiro - RJ 21730, Brazil
T: 55 21 3320787 F: 55 21 3311527

Mejia, Alberto Lopez
General Coordinator
ABVP - ASSOCIACAO BRASILEIRA DE VIDEO POPULAR
R treze de Maio 489 Bela Vista
Sao Paulo - SP, Brazil
T: 55 11 284 7862 F: 55 11 2872259

Mekbach Jr., Paulo, Mr.
GOVERNO DE SAO PAULO
Sao Paulo - SP, Brazil

Melantonio, Adriana Ribas
JORNAL ENERGIZANDO
Rio de Janeiro, Brazil

Mellet, Luiz Ernesto, Sr.
Assessor De Comunicacao
ARQUIPELAGO DE FERNANDO DE NORONHA
Rua Barao de Sao Borja, 243 - Boa Vista
Recife - PE 50070, Brazil
T: 55 81 221-3217 F: 55 81 231-1594
TL: 81.2533

Mello, Carlos, Mr.
Vendedor
EDITORA INDEX/LIBRIS/ISLAND PRESS
Av, Rio Branco 45/1707
Rio de Janeiro - RJ 20090, Brazil
T: 55 21 516 2336 F: 55 21 253 3507

Mello, Carlos Eduardo Penzias De
Produtor
MANCHETE
Rua do Russel 804 Gloria
Rio de Janeiro, Brazil
T: 55 21 2850033 x423

Mello, Claudia Sorres De
Editor
GLOBOSAT
Rua Itaperu, 1209
Rio Comprido, Brazil
T: 5023535

Mello, Djalma
Coordenador De Programacao
SISTEMA JORNAL DO BRASIL
Av.Brasil, 500
Rio de Janeiro - RJ, Brazil
T: 55 21 5854422

Mello, Eduardo
INFORMATIVO 'BRASIL-MEIO AMBIENTE E DESENVOLVIMENTO'
Rio de Janeiro, Brazil

Mello, Enir Moreira Da Silva, Sra.
Professora
UFMT/NUEPON
Rua 44, no.41, apto. 304 - Ed. Liege Cristino - Boa Esperanca
Cuiaba - MT 78090, Brazil
T: 55 125 361-2732

Mello, Flavia, Sra.
IBASE
01 B. P. 305, BoBo-Dioulasso 01
Brazil
T: 98-2094 F: 97-0802

Mello, Ivo Da S.
Reporter
SISTEMA GLOBO DE RADIO
Rua do Russel, 434
Rio de Janeiro - RJ, Brazil
T: 55 21 2058282 F: 55 21 2054096

Mello, Luiz Carlos
CHICO JR. & ASSOCIADOS
Rio de Janeiro, Brazil

Mello, Maria Da Gloria Silveira, Ms.
SEC. MEIO AMBIENTE
Rua Antonio Correa Barbosa 2233 - 6o Andar
Piracicaba 13400, Brazil
T: 22-4975 F: 33-9949

Mello, Maria De Fatima E. B. De
Reporter
FUNTEVE - RADIO MEC
Praca da Republica, 141
Rio de Janeiro RJ, Brazil
T: 55 21 2204466

Mello, Marta Iara Silveira, Sra.
REDE MULHER
Rua Joao Ramalho, 991
Sao Paulo - SP 05008, Brazil
T: 55 11 262-9407 F: 55 11 871-3457

Mello, Paulo Quintanilha De Mello, Sr.
Reitor Academico
UNIVERSIDADE SANTA URSULA
Rua Fernando Ferrari, 75
Rio de Janeiro - RJ 22231, Brazil
T: 55 21 551-5542 F: 55 21 551-6446

Mello, Pedro Paulo Thiago De
FUND. CULTURAL DE RADIODIFUSAO EDUCATIVA DA REGIAO DOS LAGOS
Av. Jose Bento Ribeiro Dantas 265 Manguinhos 28905
Armacao dos Buzios -, Brazil
T: 55 246 2753848 F: 55 21 2956494

Mello, Regina Lara Silveira
Diretora
ABAPORU EMPREENDIMENTOS
Caixa Postal 11286
Sao Paulo - SP, Brazil
T: 55 11 210 7657

Mello, Solange De Toledo, Ms.
GOVERNO DE SAO PAULO
Sao Paulo - SP, Brazil

Melo, Aldair Gomes De
MANCHETE
Rua do Russel, 804 Gloria
Rio de Janeiro, Brazil
T: 55 21 2850033

Melo, Guilherme De Azevedo
REDE GLOBO
Rio de Janeiro, Brazil

Melo, Jose Caldas Gouveia De
Fotografo
FREELANCER
R.Sao Clemente, 114/705 Bl.II
Rio de Janeiro - RJ, Brazil
T: 55 21 2464871

Melo, Luis Philipe Figueira De
Director
CONCREMAT ENGENHARIA & TECNOLOGIA
R Euclides da Cunha 106 Sao Cristovao
Rio de Janeiro - RJ, Brazil
T: 55 21 284 4022 F: 55 21 284 2702

Melo, Marconi Teodoro
Technician
REDE GLOBO
R. Lopes Quintas 303
Rio de Janeiro -RJ 2, Brazil
T: 55 21 2947732

Melo, Rosa, Sra.
Director
CMB
Rua Avanhadava no. 756
Sao Paulo - SP 01306, Brazil
T: 55 11 2564354 F: 55 11 2577648

Meloi, Ronize Alina De
JORNAL DOS SPORTS
Rua Tenente Passolo 15/25 Centro
Rio de Janeiro RJ, Brazil
T: 55 21 2328010

Mendanha, Italo
Assistente
TROFEU MARINER
Praia de Botafogo 324/1007
Rio de Janeiro - RJ, Brazil
T: 55 21 552 5473 F: 55 21 5525473

Mendeiros, Roberto
President
ASSOCIACAO BRASILIERA DE JORNALISMO CIENTIFICO
Rua Novo Horiznze, 210 apt. 111 Chicara de Bana 13093680
Campinas, Brazil
T: 55 192 551095 F: 55 192 522099

Mendes, Alberto
Operador De Som
RADIO ROQUETE PINTO
Av. Erasmo Braga 118/ 11andar Castelo
Rio de Janeiro -RJ 2, Brazil
T: 55 21 2247727 F: 55 21 2524584

Mendes, Alvaro Marcelo
Technician
REDE GLOBO
R. Lopes Quintas 303
Rio de Janeiro -RJ 2, Brazil
T: 55 21 2947732

Mendes, Edinaldo
Militar
SISTEMA GLOBO DE RADIO
Rua do Russel
Rio de Janeiro RJ, Brazil
T: 55 21 2058282

Mendes, Frederico
Fotografo
REVISTA MANCHETE
Rua do Russel, 804
Rio de Janeiro - RJ, Brazil
T: 55 21 2652012

Mendes, Maria, Ms.
CSN - CONFEDERATION DES SYNDICATS NATIONAUX
Caixa Postal 1759, Centro
Sao Paulo - SP 01059, Brazil
T: 55 11 270-1257

Mendes, Marly Lucilia, Ms.
Assessora Especial
GOVERNO DO ESTADO DO AMAZONAS
Rua Paulo VI, 500 Apt. 2004
Rio de Janeiro - RJ 22230, Brazil
T: 55 21 551-7205

Mendes, Mauro, Sr.
Gerente De Circulacao
EDITORA TERCEIRO MUNDO
Rua da Gloria 122, 1 andar
Rio de Janeiro - RJ 20241, Brazil
T: 55 21 252-7440 F: 55 21 252-8455
TL: 33054

Mendes, Sandro Da Silva
VIDEOARTE RIO
Rua Senradom 45
Rio de Janeiro, Brazil

Mendes, Vanda Amancio B., Sra.
Principal
ESCOLA DAS NACOES
SHIS QI 21 Area Especial Lote C-1 - Caixa Postal 7093
Brasilia - DF 71600, Brazil
T: 55 61 366-1800 F: 55 61 366-3377

Mendes De Leon, Nicole, Sra.
Rua Eng. Cortes Sigaud 100/402
Rio de Janeiro - RJ 22450, Brazil
T: 55 21 274 5668

Mendes De Leon, Ramon, Mr.
Rua Eng. Cortes Sigaud 100/402
Rio de Janeiro - RJ 22450, Brazil
T: 55 21 559-7343

Mendonca, Abimel Conceicao
MANCHETE
Rio de Janeiro, Brazil

Mendonca, Joao Batista
Motorista
WDR GERMAN TV CHANNEL
Rua Eca de Queiroz, 479 casa 12
Sao Paulo - SP, Brazil
T: 55 11 5721262 F: 55 11 5728959

Mendonca, Marcia Helena, Sra.
Jornalista/Professora
Rua Machado 155/01
Belo Horizonte - MG 31110, Brazil
T: 55 31 442-3217

Mendonca, Maria Antonieta De, Sra.
FLAMA
Rua Marques de Abrantes, 55 - Flamengo
Rio de Janeiro - RJ 22230, Brazil
T: 55 21 265-5690

Mendonca, Paulo Knauss
Photographer
NOVA ESCOLA
Rua Haddock Lobo 403-D Cerqueira Cesar
Sao Paulo, Brazil
T: 55 11 2843733

Mendonca, Rosamaria, Sra.
SOCIEDADE CIVIL BEM-ESTAR FAMILIAR
Av. Chile 230 / 17 Andar
Rio de Janeiro - RJ 20031, Brazil
T: 55 21 533-3898 F: 55 21 220-4057
TL: 21-30634

Mendonca, Vera Maria De, Ms.
Socia
BRAZCO
Praia Do Flamengo, 66 Bl B S/202
Rio de Janeiro - RJ 22.610, Brazil
T: 55 21 285-5839 F: 55 21 285-3943

Mendonca Soares, Tania Lucia, Sra.
ABEMA
Av. Jeronimo Monteiro 240/S.811/812
Vitoria - ES, Brazil
T: 55 27 322-4193 F: 55 27 222-3725

Menezes, Clarice Moreira
Reporter
FUNTEVE - RADIO MEC
Praca da Republica, 141
Rio de Janeiro RJ, Brazil
T: 55 21 2204466

Menezes, Claudia, Sra.
CEPAVEH
Estrada da Fazenda Inglesa 742
Petropolis - RJ 25600, Brazil
T: 55 242 423-807 F: 55 242 423-200

Menezes, Eline Viana, Sra.
CENTRO DE EST. HONESTINO GUIMARAES
Rua das Acacias, 79 - Pituba
Salvador - BA 41810, Brazil
T: 55 0 R

Menezes, Eric, Sr.
PRO-NATURA
Av. Rodrigues Alves, 179, 2o. andar
Rio de Janeiro - RJ 20081, Brazil
T: 1 021 253-8655

Menezes, Fausto Adelino De
PALMARES PRODUCOES E JORNALISMO LTDA
Rio de Janeiro, Brazil

Menezes, Henrique M. Campos De
EDITORA TERCEIRO MUNDO
Rio de Janeiro, Brazil

Menezes, Jose Eugenio De Oliveira, Sr.
AGENCIA ECUMENICA DE NOTICIAS
Equipe de Redacao de Sem Fronteiras / Caixa Postal 55
Taboao da Serra-SP 06751, Brazil
T: 55 11 229-6734 F: 55 11 229-6806

Menezes, Katia Amorim Telles De
Editor
REDE GLOBO
R. Lopes Quintas 303
Rio de Janeiro -RJ 2, Brazil
T: 55 21 2947732

Menezes, Lucia Beckmann De Castro, Sra.
UNIVERSIDADE FEDERAL DO PARA
Av. Gov. Jose Malcher 2271 - apt. 702
Belem - PA 66000, Brazil
T: 55 91 224-9221

Menezes, Maria Celia B. De
Editor
RADIOBRAS
Brasilia, Brazil

Menezes, Maria De Fatima Lins De, Sra.
Coordenadora
SUDEMA
R. Monsenhor Walfredo Leal 181, Tambia
Joao Pessoa - PB 58000, Brazil
T: 55 83 222-4663

Menezes, Mario Assis, Sr.
Tecnico
INSTITUTO DE ESTUDOS AMAZONICOS E AMBIENTAIS
SCLN 210 Ed. Center Bloco C salas 211 e 212
Brasilia - DF 70062, Brazil
T: 55 61 347-9874 F: 55 61 347-9874

Menezes, Maurico M. De
Reporter
AGENCIA ESTADO
Av. Caetano Alvares, 55
Sao Paulo -SP 02598, Brazil
T: 55 11 8562210 F: 55 11 2652297

Menezes, Placido Batista, Sr.
Membro
APOENA
Rua Fernando Costa, 950
Presid. Prudente-SP 19020, Brazil
T: 55 182 21-3314

Menezes, Sidine Antunes De
Cinegrafista
FUND. CULTURAL DE RADIODIFUSAO EDUCATIVA DA REGIAO DOS LAGOS
Av. Jose Bento Ribeiro Dantas 265 Manguinhos 28905
Armacao dos Buzios -, Brazil
T: 55 246 2753848 F: 55 21 2956494

Menezes, Thereza Macedo Soares Dias, Ms.
Socio-Filantr.
WOMEN'S CLUB OF RIO DE JANEIRO
Rua Real Grandeza, 99
Rio de Janeiro - RJ, Brazil
T: 55 21 205-8934

Menezes, Wellington, Mr.
Economista
CETREL
Km 19 DABR 347-Estrada do Coco - Camacari
Salvador - BA 42810, Brazil
T: 55 71 832-1186

Merguihao, Claudia Bezerra
Reporter
FOLHA DA PINHEIRO
R.Silveira martins, 153
Rio de Janeiro - RJ, Brazil
T: 55 21 2957995

Merle, Daniel
Photograpaher
REUTERS NEW SERVICE
Rua Sete de Setembro 99/4th floor Centro
Rio de Janeiro - RJ, Brazil
T: 55 21 5074151 F: 55 21 5072120

Mesquita, Andre Coimbra De
MANCHETE
R. do Russel 804
Rio de Janeiro -RJ 2, Brazil
T: 55 21 2652012 F: 55 21 2855148

Mesquita, Carlos
SISTEMA JORNAL DO BRASIL
Av. Brasil 500
Rio de Janeiro -RJ, Brazil
T: 55 21 5854422 F: 55 21 5854428

Mesquita, Marlene Dantas
INSTITUTO DE ARQUITETOS DO BRASIL
Rua do Pinheiro, 10 Flamengo
Rio de Janeiro, Brazil
T: 55 21 2853480

Mesquita, Mercedes, Sra.
PLANTA DESIGN
Rua DR. Zefenino # 11
Uberaba 38100, Brazil
T: 55 034 333-4449

Mesquita, Ruth Brazil
Coordenadora
INSTITUTO DA LUZ E VIDA
Rua H Jardim Imperial, 01
Salvador - BA Cep:40.000, Brazil
T: 55 71 226 1450

Mesquita, Sonia Lucia Athayde, Ms.
Assessora Comunicacao Social
CRUZ VERMELHA BRASILEIRA
Praca Cruz Vermelha 10/12 - 1o. andar
Rio de Janeiro - RJ, Brazil
T: 55 21 232-6325 F: 55 21 242-6760

Mesquita Neto, Jarbas, Sr.
Professor
UNIRIO
Rua Frei Caneca, 94 - Centro Dpto de Ciencias Naturais - 5o. andar
Rio de Janeiro - RJ 20000, Brazil
T: 55 21 241 3913

Metri, Paulo, Mr.
Conselheiro
CLUBE DE ENGENHARIA
Rua Pompeu Loureiro, 44/402
Rio de Janeiro - RJ 22.061, Brazil
T: 55 21 255-7753

Meyer, Claus C
Fotografo
TYBA AGENCIA FOTOGRAFICA
Rua Barao do Flamengo 22/503 Flamengo
Rio de Janeiro - RJ, Brazil
T: 55 21 285 4640/265 4839 F: 55 21 222 6742

Meyer, Maria Helena, Sra.
Journalist
ASSOCIACAO DE MULHERES PAPELEIRAS E TRABALHADORAS
R. Coronel Aurelio Bittencourt 110/304
Porto Alegre - RS 90410, Brazil
T: 55 512 3317492 F: 55 512 2275229

Mias, Jose Sebastiao Menezes
FREELANCER
av. Brasil 5000 6 andar
Rio de Janeiro, Brazil
T: 55 21 5854621

Miceli, Fernando
Photographer
IBASE
R.Vicente de Souza, 29 Botafogo
Rio de Janeiro -RJ, Brazil
T: 55 21 2265412 F: 55 21 2860541

Michel, Alazet, Sr.
Professor
Rua 19 de Fevereiro , 110/ Cobertura
Rio de Janeiro - RJ 22280, Brazil
T: 55 21 537 3047

Midkiff, Peter, Mr.
ASS. BRAS. DE COMUNIDADES ALTERNATIVAS-ABRASCA/PI
Caixa Postal 516 - Centro
Teresina PI 64000, Brazil
T: 55 0 R

Miessler, Robert J., Sr.
Platonista
COMUNIDADE INTERNACIONAL BAHA'I
Rua Frei Antonio de Padua 1045
Campinas - SP 13073, Brazil
T: 55 192 41-0730 F: 55 192 43-3243

Migliaccio, Ivani
ESTADO DE SAO PAULO
Av. Rio Branco 151 - Sobre - Loja sala 201 a 204
Rio de Janeiro RJ, Brazil
T: 55 21 2217111

Miguel, Sebastiao R.
Technician
REDE GLOBO
R. Lopes Quintas 303
Rio de Janeiro -RJ 2, Brazil
T: 55 21 2947732

Miguel, Victor E Silva, Mr.
Fotografo
UIJDC
Rua Francisco Otaviano,42 - Copacabana
Rio de Janeiro - RJ, Brazil

Milanez, Francisco, Mr.
Presidente
ASSOCIACAO DE JOVENS EMPRESARIOS/ECOFUND
Olavo Barreto Viana, 66 apt.201
Porto Alegre - RS 90460, Brazil
T: 55 512 222-3542 F: 55 512 222-8591

Milani, Feizi Masrour, Dr.
Medico
COMUNIDADE INTERNACIONAL BAHA'I
CX. Postal 7380
Salvador - BA 41831, Brazil
T: 55 71 359-2529

Milani, Guitty Masrour, Sra.
Membro
COMUNIDADE BAHA'i DO BRASIL
SQS 208 Bl. A, apto 403
Brasilia - DF 70254, Brazil
T: 55 61 244-1458 F: 55 61 248-4321

Milani, Razi Abbassi, Sr.
Member
COMUNIDADE BAHA'I DO BRASIL
SQS 208 Bl. A apto 403
Brasilia - DF 70254, Brazil
T: 55 61 244-1458 F: 55 61 248-4321

Milena, Antonio
Photographer
EDITORA ABRIL
R.da Passagem, 123/8 andar
Rio de Janeiro, RJ, Brazil
T: 021 5468282 F: 021 2759347

Miles, Gerald, Mr.
Coordinator
SOUTH PACIFIC ISLANDS
Rua Toneleiro 154
Rio de Janeiro - RJ, Brazil

Milko, Peter, Sr.
CENTRO DE ESTUDOS E ATIVIDADES P/ CONSERVACAO DA NATUREZA - CEACON
Rua Verissimo Gloria 93
Sao Paulo - SP 01252, Brazil
T: 55 11 864-5303/864-8471 F: 55 11 864-4663

Miller, Paula
INSTITUTO DE ARQUITETOS DO BRASIL
Rio de Janeiro, Brazil

Millikan, Brent, Sr.
Representative
INSTITUTO DE ESTUDOS AMAZONICOS E AMBIENTAIS
Rua Alvaro Maia, 1034 Caixa Postal 791
Porto Velho - RO 78900, Brazil
T: 55 68 223-3945

Milward, Sonia Latge
Coordinator
INSTITUTO DE COMUNICACAO SINDICAL
Rua Visconde de Inhauma, 134 sala 715 Centro
Rio de Janeiro, Brazil
T: 55 21 2333516

Mineiro, Procopio
Sub-Editor
ECOLOGIA E DESENVOLVIMENTO
Rua da Gloria 122 gr 105/106
Rio de Janeiro RJ 20, Brazil
T: 55 21 242 1957/222 1370

Mineiros, Elisa Bezerra, Ms.
Professora Universitaria
ASSOCIACAO DOS DOCENTES DA UFPB
Centro De Vivencia - Campus Universitario - Universidade Federal
Joao Pessoa - PB 28.000, Brazil
T: 55 83 224-7200

Ming, Chien Yu
NHK
Av Nilo Pecanha 50 Grupo 1208
Rio de Janeiro - RJ, Brazil
T: 55 21 262 9777 F: 55 21 262 9514

Minowa, Isack, Mr.
Director
YAMAGISHI ASSOCIATION
SP 340 Km 138
Jaguariuna - SP 13820, Brazil
T: 019297 1173

Minowa, Marcelin, Ms.
Colaboradora
YAMAGISHI ASSOCIATION
SP 340 Km 138
Jaguariuna - SP 13820, Brazil
T: 0192971173 F: 0192973034

Miranda, Carlos Aurelio W. De
CHICO JR. & ASSOCIADOS
Rio de Janeiro, Brazil

Miranda, Claudia, Sra.
PREFEITURA DE PARATY
Largo de Santa Rita s/n
Paraty - RJ 23970, Brazil
T: 55 243 71-1266 F: 55 243 71-1205

Miranda, Clovis, Sr.
Universidade Federal Mato Grosso
Caixa Postal 833
Cuiaba - Mt, Brazil
T: 55 125 627 1396 F: 55 125 3611119

Miranda, Daniel
Journalist
REVISTA FLUMINENSE
Rua Republica do Libano, 61 Sala 1120, Centro
Rio de Janeiro RJ, Brazil
T: 55 21 2429809

Miranda, Hely
Sistema Globo De Radio
R. de Russel 434
Rio de Janeiro RJ, Brazil
T: 55 21 2058282 F: 55 21 2054096

Miranda, Itamar Ferreira De
Tecnico
TVSBT
Campo de S.Cristovao, 105
Rio de Janeiro - RJ, Brazil
T: 55 21 5803688 / 5800678 F: 55 21 5809685

Miranda, Jose Luiz Da Silva
Tecnico
TVSBT
Campo de S.Cristovao, 105
Rio de Janeiro - RJ, Brazil
T: 55 21 5803688 / 5800678 F: 55 21 5809685

Miranda, Laucia Milesi De
SISTEMA GLOBO DE RADIO
R. de Russel 434
Rio de Janeiro RJ, Brazil
T: 55 21 2058282 F: 55 21 2054096

Miranda, Luzia, Ms.
Seringueira
MOVIMENTO DE MULHERES DO CAMPO E DA CIDADE
Travessa Tupimanbas, 703
Belem - PA 66010, Brazil
T: 55 91 224-2385

Miranda, Marcia, Ms.
WORLD COUNCIL OF CHURCHES
150 route de Ferney
Brazil
F: 41 22 791 0361

Miranda, Rita K., Ms.
WORLD YOUNG WOMEN'S CHRISTIAN ASSOCIATION
Rua Mozorlandia 95
Feira de Santana-BA 44040, Brazil
T: 55 75 221-31170
TL: WYMCA 41 23 32

Miranda, Umbelina, Sra.
CLUBE SOROPTIMISTA INTERNACIONAL
SHIS QI 4 Conj.2 Casa 9
Brasilia - DF 71600, Brazil
T: 55 61 2481805 F: 55 61 3214320

Mirando, Luciano De
INTERNATIONAL TELEVISION NEWS
Av. Rio Branco, 25/9 andar
Rio de Janeiro RJ, Brazil
T: 55 21 2335480

Mirands, Washington Duarte
Editor
SINDICATO DOS ENGENHEIROS NO ESTADO DO RIO DE JANEIRO
Av. Rio Branco 277/17 andar Cinelandia
Rio de Janeiro, Brazil
T: 55 21 5321398

Mitchell, Martin Philip, Mr.
Director
BRITISH COUNCIL
SCRN 708/709 Bloco F Lojas 1/3
Brasilia - DF 70740, Brazil
T: 55 61 272-3060 F: 55 61 274-5070
TL: 061-1859

Mitsui, Yasuhiro
EMBAIXADA DO JAPAO
Av Da Nacoes Lote 39
Brasilia - DF, Brazil
T: 55 61 242 6866

Miura, Midori, Sra.
Secretaria Executiva
MOREMO
Rua Xavier da Silveira, 23
Rio de Janeiro - RJ, Brazil
T: 55 21 521-7272

Miziara, Rosana, Sra.
Rua Atuau, 46
Sao Paulo - SP, Brazil
T: 55 11 212-0412

Mizuno, Takashi
Journalist
YOMIURI SHIMBUN
Av Oswaldo Cruz 149/1201
Rio de Janeiro - RJ, Brazil
T: 55 21 551 6187 F: 55 21 552 2279

Mochizuki, Jiro, Sr.
Cinefotografo
YASUDA FIRE & MARINE INS. CO., LTD
Rua General Leite De Castro, 706, Jardim Santa Cruz
Sao Paulo - SP, Brazil
T: 55 11 947-0092 F: 55 11 884-7553

Mochizuki, Masaomi
Fotografo
REVISTA MANCHETE
Rua do Russel, 804
Rio de Janeiro - RJ, Brazil
T: 55 21 2652012

Modern, Renata H., Sra.
Aid
BAHA'I - COMUNIDADE INTERNACIONAL
Rua Dr. Melo Alves, 742 / apt. 161
Sao Paulo - SP 01417, Brazil
T: 55 11 852-5905

Moesch, Alberto Pretto, Sr.
Assessor Superior
ASSEMBLEIA LEGISTATIVA
Pca Marechal Deodoro s/no.
Porto Alegre - RS 90010, Brazil
T: 55 512 221-4260 F: 55 512 228-6740

Moises, Helvio Nicolau, Sr.
Gerente
ORGANIZACAO PREFEITO FARIA LIMA - CEPAM
Av. Prof. Lineu Prestes, 913
Sao Paulo - SP 05508, Brazil
T: 55 11 212-3144 F: 55 11 813-5969
TL: (11)83141

Molica, Fernando
Reporter
FOLHA DE SAO PAULO
Alameda Barao de Limeira 425
Sao Paulo SP, Brazil
T: 55 11 874 2608/874 2602 F: 55 11 2231644

Molinari, Cesar R. S.
Technician
REDE GLOBO
R. Lopes Quintas 303
Rio de Janeiro -RJ 2, Brazil
T: 55 21 2947732

Molinier, Michel, Mr.
Dir. Pesquisas
ORSTOM
C. P. 09747
Brasilia - DF 70001, Brazil
T: 55 61 248-5323 F: 55 61 248-5378

Mometti, Pe. Giovanni, Sr.
Presidente
MOVIMENTO VOLUNTARIOS INTER BRASILEIROS PARA AMAZONIA -VIBRA JOAO XXIII
Travessa do Alenquer, 125
Belem 66010, Brazil
T: 091 224-1965 F: 091223.4855

Monadjemi, Cyrus, Sr.
Representative
BAHA'I INTERNATIONAL COMMUNITY
R. Mosteiro 37/303
Belo Horizonte - MG 30350, Brazil
T: 55 31 3446239

Monadjemi, Puran, Sra.
BAHA'I INTERNATIONAL COMMUNITY
R. Mosteiro 37/303
Belo Horizonte - MG 30350, Brazil
T: 55 31 3446239

Mondim, Leda, Sra.
PREFEITURA MUNICIPAL
Praca Maua s/n 1 andar
Santos - SP 11010, Brazil
T: 55 132 342-2093 F: 55 132 345-809
TL: 131048

Monducci, Delba
OSHO SPIRITUAL HEALTH ORGANIZATION
Rua Pinheiro Guimaraes, 149 Casa 129
Rio de Janeiro - RJ 22.281, Brazil

Monforte, Carlos
CONFEDERACAO NACIONAL DA INDUSTRIA
Rio de Janeiro, Brazil

Monk, Arlete, Sra.
Secretary
FLAMA
Rua Marques de Abrantes, 55 - Flamengo
Rio de Janeiro - RJ 22230, Brazil
T: 55 21 265-5690

Monnerat, Monica, Sra.
Gerente De Projetos
IPLANRIO
R.Gago Coutinho, 52 / 5o. andar
Rio de Janeiro - RJ, Brazil
T: 55 21 556-1863 F: 55 21 285-2098

Monte, Ednalva Maia, Ms.
Av. Democraticos 8406 ASM 6, Ap 101
Rio de Janeiro - RJ, Brazil
T: 55 21 280-4705

Monteiro, Antonio Jose L. C., Mr.
PINHEIRO NETO ADVOGADOS
Rua Boa Vista ,254 - 9 andar
Sao Paulo - SP 01014, Brazil
T: 55 11 258 7455 F: 55 11 36-5011
TL: 21391

Monteiro, Edalva Alves, Sra.
Director
SAGMIMA
Rua Joao Galberto
Sao Luiz - MA 55000, Brazil
T: 55 69 222.7174 F: 55 69 221-2009

Monteiro, Frederico Mai
Diretor Geral
RADIO ATIVIDADE
Av.Isac Povoas, 1331 Sl.13
Cuiaba - MT, Brazil
T: 55 125 3221019 F: 55 125 3221667

Monteiro, Gerson Souza, Jr.
TV RIO
Rio de Janeiro, Brazil

Monteiro, Karen
Reporter
FREELANCER
Rua Silva Jardim 994/2002
Curitiba - PR, Brazil
T: 55 41 224 5691/244 6386

Monteiro, Luiz Antonio C.
Assessor De Imprensa
PRO-NATURA INSTITUTO BRASILEIRO PESQ.EST.AMBIENTAIS
Rua Monte Alegre, 374
Rio do Sul - SC, Brazil
T: 55 478 021 2243994 F: 55 478 021 2226742

Monteiro, Marcia Lima De Azevedo, Sra.
Pesquisadora
INSTITUTO DE ECOLOGIA HUMANA
Av. Dom Manuel de Medeiros s/n - Ed. Newton Bancks - 3o. andar
Recife - PE 52071, Brazil
T: 55 81 441-3931 F: 55 81 441-4697

Monteiro, Maria Elisa G. De Castro
FREELANCER
R. Santa Clara 239/901 Copacabana
Rio de Janeiro -RJ, Brazil
T: 55 21 2576419

Monteiro, Maria Elizabeth Brea
Technical Consulter
MUSEUDO INDIO/FUNAI
Rua das Palmeiras, 55
Rio de Janeiro, Brazil
T: 55 21 2867745 F: 55 21 2860845

Monteiro, Mario
Redator
TEXTO E ARTE AM (FREELANCER)
R. Prof. Brandao Filho 60/404 Leblon
Rio de Janeiro- RJ, Brazil
T: 55 21 2391478 F: 55 21 2391478

Monteiro, Rui Miranda
Tecnico De Audio
SISTEMA GLOBO DE RADIO
Rua do Russel, 434
Rio de Janeiro - RJ, Brazil
T: 55 21 2058282 F: 55 21 2054096

Monteiro, Simone, Sra.
CENTRO COMUNITARIO DOS PEQUENOS NEGOCIOS
Rua Abilio dos Santos 137
Bento Ribeiro - RJ 21331, Brazil
T: 55 21 357-1846

Montenegro, Patricia, Ms.
COPACOBANA PALACE
Rio de Janeiro - RJ, Brazil

Montes, Egledio Viana
Technician
REDE GLOBO
R. Lopes Quintas 303
Rio de Janeiro -RJ 2, Brazil
T: 55 21 2947732

Monteverde, Bernardo, Mr.
Presidente
MONTEVERDE ENG. COM. IND.
Rua Evaristo da Veiga , 55 - 3andar
Rio de Janeiro - RJ 20031, Brazil
T: 55 21 240 4747 F: 55 21 220 9457

Monteverde, Esterzinha, Ms.
Diretora
MONTEVERDE ENG. COM. IND.
Rua Evaristo da Veiga, 55 - 3 andar
Rio de Janeiro - RJ 20031, Brazil
T: 55 21 240 4747 F: 55 21 220 9457

Monticelli, Gislene, Sra.
ADFG / AMIGOS DA TERRA
Rua Miguel Tostes, 694
Porto Alegre - RS 90420, Brazil
T: 55 512 332-8884 F: 55 512 332-8884

Montorio Jr., Italo, Sr.
Membro
APOENA
Rua Cuiaba, 1-19
Presiden.Epitacio-SP 19470-000, Brazil
T: 055 018221-4222 F: 055 018222-6177

Moodye, Lynda, Sra.
INTERNATIONAL INSTITUTE FOR LAW, SCIENCE AND TECHNOLOGY
Av. Rio Branco, 4-18o. andar
Rio de Janeiro - RJ, Brazil
T: 55 21 253-7676 F: 55 21 253-1422

Moore, Katie
HILL AND KNOWLTON
Rio de Janeiro, Brazil

Moraes, Ari, Sr.
IECAM
Dr. Moreira 250
Manaus - AM 69000, Brazil

Moraes, Hamilton H. S.
Technician
REDE GLOBO
R. Lopes Quintas 303
Rio de Janeiro -RJ 2, Brazil
T: 55 21 2947732

Moraes, Ignez S. Pinto De, Ms.
Diretora De Divisao
SECRETARIA DE CULTURA
Tereza Guimaraes 111 - 301
Rio de Janeiro - RJ, Brazil

Moraes, Igor B.
Technician
REDE GLOBO
R. Lopes Quintas 303
Rio de Janeiro -RJ 2, Brazil
T: 55 21 2947732

Moraes, Julio Cesar De
Technician
REDE GLOBO
R. Lopes Quintas 303
Rio de Janeiro -RJ 2, Brazil
T: 55 21 2947732

Moraes, Lewy Rodrigues De
Diretor Editor De Fotografia
AGENCIA NEXUS
Av.Beira Mar, 406/402
Rio de Janeiro - RJ, Brazil
T: 55 21 2402573 F: 55 21 2200094

Moraes, Luciano De
Editor
REDE GLOBO
R. Lopes Quintas 303
Rio de Janeiro -RJ 2, Brazil
T: 55 21 2947732

Moraes, Marco Aurelio Lopes
Operador De Vt
REDE BANDEIRANTES DE TELEVISAO
R.Alvaro Ramos, 492
Rio de Janeiro - RJ, Brazil
T: 55 21 2122072 F: 55 21 2957391

Moraes, Maria Aguinaga De, Sra.
ZONTA CLUBE - RIO
Av. Presidente Vargas, 290 2o. andar
Rio de Janeiro - RJ 20091, Brazil
T: 55 21 2332882 F: 55 21 2202305

Moraes, Maria Do Perpetuo Socorro, Sra.
AMIGOS / PARQUE CIPO
Praia de Botafogo, 428/12o.
Rio de Janeiro - RJ 22551, Brazil
T: 55 21 246-4145

Moraes, Paulo, Mr.
Porteiro
SERVICO SOCIAL DO COMERCIO
Rua Voluntarios Da Patria, 169
Rio de Janeiro - RJ 22270, Brazil
T: 55 21 286-5152 F: 55 21 286-2638

Moraes, Paulo De
LEGIAO DA BOA VONTADE
Av. Rudge 700 Bom Retiro
Sao Paulo -SP 01134, Brazil
T: 55 21 2338381/R:223 F: 55 21 2634025

Moraes, Rita De Cassia Barboza De
Radialista/Jornalista
IBASE
Rua Vicente de Souza 29 Botafogo
Rio de Janeiro - RJ, Brazil
T: 55 21 286 6161/8642 F: 55 21 286 0541

Moraes, Susan Eghrari, Sra.
Member
COMUNIDADE BAHA'I
Praca Dom Emanuel no.180 Centro
Catalao - GO 75700, Brazil
T: 55 62 441-3302

Moraes, Zeni De Souza, Ms.
UNESCO
SCN -Quadra 2 Lote B
Brasilia - DF 70710, Brazil
T: 55 61 223-8684 F: 55 61 322-4261

Morais, Eliane De, Sra.
Assessora De Contato
LEGIAO DA BOA VONTADE
Av. Rudge 700 - Bom Retiro
Sao Paulo - SP 01134, Brazil
T: 55 11 222-3199
TL: 1123051 ELBV BR

Morais, Marcos Jose De, Sr.
ARQUIPELAGO DE FERNANDO DE NORONHA
Rua Barao de Sao Borja, 243 - Boa Vista
Recife - PE 50070, Brazil
T: 55 81 221-3217 F: 55 81 231-1594
TL: 81.2533

Morales Carlos, Gloria Adriana, Sra.
Eng.Sanitaria
ASOCIACION PERUANA DE INGENIERIA SANITARIA
Rua Lauro Muller, 56/1007
Rio de Janeiro - RJ 22290, Brazil

Moreira, Alcir Alves, Mr.
Coordinator
SOUTH PACIFIC ISLANDS
Rua Toneleiro 154
Rio de Janeiro - RJ, Brazil

Moreira, Antonio Carvalho
Technician
REDE GLOBO
R. Lopes Quintas 303
Rio de Janeiro -RJ 2, Brazil
T: 55 21 2947732

Moreira, Claudio, Mr.
Adviser
BRASILIAN STANDARDS ORGANIZATION
Rio de Janeiro - RJ, Brazil

Moreira, Darcy
Cheif Reporter
O DIA
R. Riachuela 359
Rio De Janeiro, RJ 2, Brazil
T: 55 21 272-8000 F: 55 21 507-1228

Moreira, Dilclea R. De S., Sra.
Finantial Director
PMVR
Praca Savio Gama, 53
Volta Redonda - RJ, Brazil
T: 55 243 43-2449

Moreira, Dr Neiva, Sr.
GOVERNO DO ESTADO DO RIO DE JANEIRO
Rio de Janeiro - RJ, Brazil

Moreira, Elio Raymundo, Sr.
Sociologist
CERIS
Rua Dr. Julio Ottoni, 571/35
Rio de Janeiro - RJ 20241-400, Brazil
T: 55 21 265-5177 F: 55 21 285-7276

Moreira, Elizabeth, Ms.
WOMEN'S CLUB DE PETROPOLIS
Rua Locarno 275 - Morim
Petropolis - RJ, Brazil
T: 55 242 42-9483

Moreira, Gladstone Novaes, Sr.
Projetos De Meio Ambiente
ARQUITETURA AMBIENTAL
Av. Rio Branco 277 GR1210
Rio de Janeiro - RJ 20040, Brazil
T: 55 21 263-0769

Moreira, Goretti
Fotografa
JORNAL MUSEU AO VIVO
Rua das Palmeiras, 55
Rio de Janeiro - RJ, Brazil
T: 55 21 2862097 F: 55 21 2860845

Moreira, Jose Luiz Pinheiro
REDE GLOBO
R. Lopes Quintas 303 J. Botanico
Rio de Janeiro - RJ, Brazil
T: 55 21 5292000 F: 55 21 2942042

Moreira, Jose Marcelo Antonez
Reporter
FREELANCER
Rua Visconde de Irlabopai, No. 184
Rio de Janeiro, Brazil
T: 55 21 7176220 F: 55 21 7146344

Moreira, Jose Roberto, Mr.
Diretor Exec
BIOMASS USERS NETWORK
Rua Francisco Dias Velho 814
Sao Paulo - SP 04581, Brazil
T: 55 11 543-5494 F: 55 11 543-5494

Moreira, Lucia Amelia Martins
Reporter
FUNTELDA
Av. Almirante 735/4 andar
Belem PA, Brazil
T: 55 91 2281000

Moreira, Luzia Amelia Aires
Redator
ECOLOGIA E DESENVOLVIMENTO
R da Gloria 122/105 Gloria
Rio de Janeiro RJ 20, Brazil
T: 55 21 242 1957 F: 55 21 2528455

Moreira, Maria Da Conceicao Ruffeil, Sra.
UNIVERSIDADE FEDERAL DO PARA
UNAMAZ Caixa Postal 558
Belem - PA 66059, Brazil
T: 55 91 229-9310 F: 55 91 229-4339
TL: (91)1013 UPBCBR

Moreira, Maria Goretti
Fotografa
JORNAL MUSEU AO VIVO
Rua das Palmeiras, 55
Rio de Janeiro, RJ 2, Brazil
T: 021 2862097 F: 021 2860845

Moreira, Maria Manuela, Sra.
Assistente Tecnica
C E M I G
AV. BARBACENA 1200 - 20o.
Belo Horizonte - MG 30161, Brazil
T: 55 31 349-3581

Moreira, Mario
Editor
SISTEMA JORNAL DO BRASIL
Av. Brasil 500
Rio de Janeiro -RJ, Brazil
T: 55 21 5854422 F: 55 21 5854428

Moreira, Nazareth Pinheiro, Sra.
Organizadora
GRUPO ECOSSISTEMA
Rua Juiz da Costa Val no. 641
Belo Horizonte - MG 30240, Brazil
T: 55 31 283-1598

Moreira, Noemi De Souza
Member
FLAMA
Rua Barao do Flamengo, 50/802
Rio de Janeiro - RJ, Brazil
T: 55 21 225-4496

Moreira, Paulo Cesar, Ms.
COPACOBANA PALACE
Rio de Janeiro - RJ, Brazil

Moreira, Rita De Cassia, Sra.
I B G E
r. silvia valente n.65 apto 201
Salvador - BA 41850, Brazil
T: 55 71 259-2911 F: 55 71 359-5909

Moreira, Ronaldo Antunes
Reporter
REDE GLOBO
R. Lopes Quintas 303
Rio de Janeiro -RJ 2, Brazil
T: 55 21 2947732

Moreira, Sandra Maria
Reporter
REDE GLOBO
R. Lopes Quintas 303
Rio de Janeiro -RJ 2, Brazil
T: 55 21 2947732

Moreira, Teresa Maria, Sra.
ABEMA
Av. Jeronimo Monteiro, 240/S. 811/812
Vitoria - ES 29010-900, Brazil
T: 55 27 322-4193 F: 55 27 222-3825

Moreira, Vera Cristina Roche, Sra.
Sociologa
CRUZADA DO MENOR
Rua Barata Ribeiro 193/201 - Copacabana
Rio de Janeiro - RJ 22011, Brazil
T: 55 21 255-2084

Moreira, Veronica
TRIBUNA DA IMPRENSA
Rio de Janeiro, Brazil

Morelle, Glaucia, Sra.
Director
CMB
Rua Avanhadava no. 756
Sao Paulo - SP 01306, Brazil
T: 55 11 2564354 F: 55 11 2577648

Morello, Jorge
Consultor De Meio Ambiente
BCO. INTERAMERICANO DE DESENVOLVIMENTO
Caixa Postal 08738
Brasilia - DF 70312, Brazil
T: 55 61 321-7848 F: 55 61 321-3112
TL: 61-1204

Moreno, Icaro, Jr., Mr.
Superint. Obras
PREFEITURA DO RIO DE JANEIRO
Rua Helade 37
Rio de Janeiro - RJ 20730, Brazil
T: 55 21 289-8367

Moreno, Jose Armando Torres, Sr.
Diretor Tecnico
EMPRESA DE PROJETOS BIODINAMICOS LTDA
Caixa Postal 1250
Recife - PE 50000, Brazil
T: 55 81 268-3856

Moretzsohn, Gerson Ribeiro, Mr.
Advogado
Praia do Flamengo 144/11o. Andar
Rio de Janeiro - RJ, Brazil
T: 55 21 245-9239

Mori, Miyuki
Coordinator
AMERICA DO SÙL PRODUCOES DE DOCUMENTARIOS LTDA (NAMBEI TSUSHIN)
Av. Paulista 509 /11 andar conj.1104 Cerqueira Cesar
Sao Paulo -SP 01311, Brazil
T: 55 11 2886141 F: 55 11 2512065

Moria, Jose Carlos Ramalho
TV RIO - REDE RECORD
Rio de Janeiro, Brazil

Morier, Luiz
SISTEMA JORNAL DO BRASIL
Av. Brasil 500
Rio de Janeiro -RJ, Brazil
T: 55 21 5854422 F: 55 21 5854428

Morosine, Maria De Fatima, Sra.
Coordenadora
SUDEMA
Rua Monsenhor Walfredo Leal 181 - Tambia
Joao Pessoa - PB 58000, Brazil
T: 55 83 222-3149

Morosini, Liseane
JORNAL DO COMMERCIO
Rio de Janeiro, Brazil

Moss, Marcia
CAD DESIGN CLUB
Rio de Janeiro, Brazil

Moss, Vera
Cineasta
PRODUTORA LUZ
R. Visconde de Piraja 494/102 Ipanema
Rio de Janeiro -RJ 2, Brazil
T: 55 21 2393120

Mota, Humberto, Sr.
ASSOCIACAO COMERCIAL DO RIO DE JANEIRO
Rio de Janeiro - RJ, Brazil

Mota, Maria De Fatima Cantidio, Sra.
Tecnica
CONFEDERACAO NACIONAL DA INDUSTRIA
Av. Nilo Pecanha, 50 sala 3018
Rio de Janeiro - RJ 20044, Brazil
T: 55 21 262 0181 F: 55 21 262 1465
TL: 22634

Mota, Renata, Sra.
OSHO SPIRITUAL HEALTH ORGAN. - OSHO
Av. Atlantica, 3056/202
Rio de Janeiro - RJ, Brazil
T: 55 21 255-5753 F: 55 21 322-0327

Motta, Claudia Dos Reis, Ms.
Prof. Ingles/Estud. Psicologia
INSTITUTO DA LUZ E VIDA
Rua Professor Bernardino de Souza, 15
Salvador - BA CEP:40.415, Brazil
T: 55 71 312-1727

Motta, Estefania De Castro
RADIO ALVORADA FM
Rio de Janeiro, Brazil

Motta, Helio
Photographer
MANCHETE
R. do Russel 804 Gloria
Rio de Janeiro RJ 2, Brazil
T: 55 21 2454816 F: 55 21 2056196

Motta, Jose Luiz Fernandes Da
Techinician
AGENCIA ESTADO
Av. Caetano Alvares, 55
Sao Paulo -SP 02598, Brazil
T: 55 11 8562210 F: 55 11 2652297

Motta, Jose Machado Da
IBAMA
Rio de Janeiro, Brazil

Motta, Liane Oliveira Mufarrej
TWENTY FIRST CENTURY INC.
Rio de Janeiro, Brazil

Motta, Marcos Valerio F.
Technician
REDE GLOBO
R. Lopes Quintas 303
Rio de Janeiro -RJ 2, Brazil
T: 55 21 2947732

Motta, Paulo Cesar, Sr.
R. Laerte Vieira Goncalves 2402
Uberlandia - MG 38400, Brazil
T: 55 34 216-3528

Motta, Paulo Roberto Marins
Technician
REDE GLOBO
R. Lopes Quintas 303 Jardim Botanico
Rio de Janeiro -RJ 2, Brazil
T: 55 21 2947732

Motta, Ricardo Mendes
SISTEMA GLOBO DE RADIO
R. de Russel 434
Rio de Janeiro RJ, Brazil
T: 55 21 2058282 F: 55 21 2054096

Moufarrige, Lucia Brandao, Sra.
Diretora Secretaria
ASSOCIACAO PALAS ATHENA
Rua Alencar Araripe, 231
Sao Paulo - SP 04253, Brazil
T: 55 11 63 6304

Moulton, Tim, Sr.
Diretor Cientifico
CEPARNIC
Caixa Postal 43
Cananeia - SP 11990, Brazil
T: 1 805 511163 F: 1 805 511683

Mouna, Ma Dhyan, Ms.
OSHO - SPIRITUAL HEALTH ORGANIZATION
R Dr Lauro Pimentel 968
Campinas - SP 13083, Brazil
T: 55 192 392-681 F: 55 192 411-679

Mounier, Jean-Francois Le
Director For Brazil
AGENCE FRANCE-PRESSE
Rua Mexico, 91/7 andar
Rio de Janeiro - RJ, Brazil
T: 55 21 2406634 F: 55 21 2626717

Moura, Abdalaziz De
REDE PROGRESSO DE COMUNICACAO
Gravata-Pe, Brazil

Moura, Adriana Maria Magalhaes De
Journalist
FUNDACAO ESQUEL
SQN 205 B/G apt.601
Brasilia -DF, Brazil
T: 61 2721911 F: 61 3221063

Moura, Arovel, Sr.
Diretor De Comunicao
IGREJA ADVENTISTA DO SETIMO DIA/ASSOCIACAO DA UNIAO ESTE BRASILEIRA
Rua 7 de Setembro, 69 - Icarai
Niteroi - RJ, Brazil
T: 55 21 273-5422 F: 55 21 273-5422

Moura, Claudio Carvalho
SISTEMA GLOBO DE RADIO
R. de Russel 434
Rio de Janeiro RJ, Brazil
T: 55 21 2058282 F: 55 21 2054096

Moura, Fernando Goncalves De, Sr.
Idealista
Rua Dr. Francisco Portela, 794
Rio de Janeiro - RJ 24400, Brazil
T: 55 21 712-2005

Moura, Geraldo De Melo, Sr.
ESTADO DO ACRE SECRETARIA DE CIENCIA TECNOLOGIA E MEIO AMBIENTE
Rua Rui Barbosa, 450
Rio Branco - AC, Brazil
T: 55 68 224-2857 F: 55 68 224-5694

Moura, Josilda Rodrigues Da Silva De, Sra.
President
DEPARTAMENTO DE RECURSOS MINERAIS
Rua Marechal Deodoro, no.351 - Centro
Niteroi - RJ 24030, Brazil
T: 55 21 722-7074 F: 55 21 719-6846

Moura, Luiza, Sra.
Director
CMB
Rua Avanhadava no. 756
Sao Paulo - SP 01306, Brazil
T: 55 11 2564354 F: 55 11 2577648

Moura, Milton, Sr.
Assessor
CESE
Rua da Graca# 164-Graca
Salvador - BA 40120, Brazil
T: 55 71 336-5457 F: 55 71 336-0733

Moura, Sergio Mascarenhas De, Sr.
Assesor De Imprensa
AMIGOS DA TERRA-ITALIA
R. St. Clara 188/602
Rio de Janeiro - RJ, Brazil
T: 55 21 236-5619

Moura, Tereza Ione Souza Filho, Sra.
Chefe De Biblioteca
IDESP
Av. Nazare 871
Belem - PA 06035, Brazil
T: 55 91 224-4411 F: 55 91 225-3414
TL: 091 6905

Mourao, Jorge
Editor
FOLHA DE TRANCOSO
Prudente de Morais 1440/802
Rio de Janeiro RJ, Brazil
T: 55 21 2667467 F: 55 21 2667467

Mourao, Koki Ayamara
Assistente
MTVC
R.Marqwues de Sao Vicente, 429/303
Rio de Janeiro - RJ, Brazil
T: 55 21 274 1553 F: 55 21 239 6905

Mousasticoshvily, Andrei, Mr.
VIRGINIA COMMONWEALTH UNIVERSITY
Rua Figueiredo Magalhaes 421/301
Rio de Janeiro - RJ 22031, Brazil
T: 55 21 255-6368 F: 55 21 235-1205

Mousasticoshvily, Igor, Mr.
Figueiredo Magalhaes 421/301
Rio de Janeiro - RJ 22031, Brazil
T: 55 21 255-6368 F: 55 21 235-1205

Moya, Gonzalo E., Sr.
Professor
UFRRJ
Km 47 Rio-Sao Paulo
T: 21 682-1149

Mozart, Fernando, Sr.
SHOPPING CULTURAL FUNDICAO PROGRESSO
Rua dos Arcos 28-42
Rio de Janeiro - RJ 20230, Brazil
T: 55 21 220 5022 F: 55 21 240 9432

Mueller, Johannes Georg
Redator
WDR GERMAN TV CHANNEL
Rua Eca de Queiroz, 479 casa 12
Sao Paulo - SP, Brazil
T: 55 11 5721262 F: 55 11 5728959

Mueller, Lourenco, Sra.
Consultor
L.M PLANEJAMENTO CONSULTORIA
R. Humberto De Campos
Salvador - BA 40150, Brazil
T: 55 71 3362235

Muello, Peter
Reporter
ASSOCIATED PRESS
Avenida Brasil 500 sala 847
Rio de Janeiro - RJ, Brazil
T: 55 21 580 4442 F: 55 21 580 2309

Muller, Jackson, Mr.
Biologo
SEMEC-NH/RS
Rua Jose Inacio Plawgg, 66
N.Hamburgo-RS 93510, Brazil
T: 55 051 595-1533

Muniz, Paulo C.
PM STUDIO FOTOGRAFICO
Rio de Janeiro, Brazil

Murad, Jamil, Mr.
Dep. Estadual
ASSEMBLEIA LEGISLATIVA DE SAO PAULO
Av. Pedro Alvares Cabral s/n 4o. Andar, Sala 4008
Sao Paulo - SP 04097, Brazil
T: 55 11 884-2054 F: 55 11 884-2054

Murad, Sergio
Technician
REDE GLOBO
R. Lopes Quintas 303
Rio de Janeiro -RJ 2, Brazil
T: 55 21 2947732

Murano, Tsuguo
JORNAL PAULISTA
Sao Paulo - SP, Brazil

Murauskas, Luiz Carlos
Photographer
FOLHA DE SAO PAULO
Alameda Barao de Limeira 425
Sao Paulo SP, Brazil
T: 55 11 874 2608/874 2602 F: 55 11 2231644

Murayama, Toshio, Mr.
Av. Nilo Pecanha, 50 grupo 2110
Rio de Janeiro - RJ, Brazil
T: 55 21 532-2455 F: 55 21 220-9916

Murta, Fernando Antonio De Oliveira, Sr.
FLAMA
Rua Marques de Abrantes, 55 - Flamengo
Rio de Janeiro - RJ 22230, Brazil
T: 55 21 265-5690

Mutter, Theo, Sr.
Coooperative
SERVICO ALEMAO DE COOPERACAO TECNICA E SOCIAL SACTES
P.O. Box 1655
Recife - PE 50050, Brazil
T: 55 81 221-0075 F: 55 81 222-1959

Myers, Eric, Mr.
Researcher
IMAZON
Tv. Apinages 131/1502
Belem - PA 66000, Brazil
T: 55 91 235-4214 F: 55 91 235-4214

Myers, J. Peterson, Mr.
Director
W. ALTON JONES FOUNDATION
c/o Sonia Rigueira, Rua Bueno Brandao, 393 - Santa Tereza
Belo Horizonte - MG 31010, Brazil
T: 55 31 226-5145 F: 55 31 222-8429

Nabulsi, Fatima, Ms.
Information Assistant
WORLD FOOD PROGRAMME / UN
SCN Quadra 02 Bloco 'B' Terreo
Brasilia - DF 70710, Brazil
T: 55 61 226-7872 F: 55 61 322-2080
TL: 61-1243

Nadall, Carlos, Sr.
ARQUIPELAGO DE FERNANDO DE NORONHA
Rua Barao de Sao Borja, 243 - Boa Vista
Recife - PE 50070, Brazil
T: 55 81 221-3217 F: 55 81 231-1594
TL: 81.2533

Nader, Roxana M.M., Sra.
President
AMNPR
Rua Ana Angelica, 25 /02 Derby
Recife - PE 52010, Brazil
T: 55 81 231-2295 F: 55 81 221-2302

Naegeli, Cristina, Ms.
COPPE/UFRJ
Rua Maria Amalia 106/204
Rio de Janeiro - RJ 20510, Brazil
T: 55 21 258-3587

Nagao, Pedro Seitiro, Sr.
PREFEITURA MUNICIPAL
Praca Maua s/n 1 andar
Santos - SP 11010, Brazil
T: 55 132 342-2093 F: 55 132 345-809
TL: 131048

Nagy, Vanice Rodrigues Rose, Sra.
Associada
FEMNAP
Rua Hugo Soares Fagundes, 135 - Codesg
Guaratingueta - SP, Brazil
T: 0125224236

Nakai, Sussumu, Sr.
Guia
Avinida Rio Branco, 15615,.1115
Rio de Janeiro - RJ 20040, Brazil
T: 55 21 210-32634 F: 55 21 220-9854
TL: 30605

Nakamura, Antenor Eiji, Mr.
Diretor
SIND. DOS TRABALHADORES NAS IND. QUIM. E FARMACEUTICAS DE S.PAULO
Rua Tamandare 348 - Liberdade
Sao Paulo - SP 01525, Brazil
T: 55 11 279-3811 F: 55 11 279-0662

Nakamura, Nobuco, Ms.
Av. Paulista, 1294 - 23o. andar
Sao Paulo - SP 01310-100, Brazil
T: 55 11 289-4988 F: 55 11 288-8058
TL: 31465

Nakano, Maria
Coordenator
IBASE
R.Vicente de Souza, 29 Botafogo
Rio de Janeiro -RJ, Brazil
T: 55 21 2265412 F: 55 21 2860541

Nakao, Yoshito
Engineer
AMERICA DO SUL PRODUCOES DE DOCUMENTARIOS LTDA (NAMBEI TSUSHIN)
Av. Paulista 509 /11 andar conj.1104 Cerqueira Cesar
Sao Paulo -SP 01311, Brazil
T: 55 11 2886141 F: 55 11 2512065

Nakatune, Edi Sachiko
JIJI PRESS NEWS AGENCY
Av Paulista 854 - 13o. andar cj 133 Bela Vista
Sao Paulo - SP, Brazil
T: 55 11 285 0025/287 9526 F: 55 11 285 3816

Nakauchi, Tadafumi, Sr.
Guia
Avinida Rio Branco, 15615,.1115
Rio de Janeiro - RJ 20040, Brazil
T: 55 21 210-32634 F: 55 21 220-9854
TL: 30605

Napolitano, Maria Christina, Ms.
Tecnica
CNI/SESI
Rua Dezoito de Outubro 187/101, Tijuca
Rio de Janeiro - RJ 20530, Brazil

Nascentes, Jose
INSTITUTO DE ARQUITETOS DO BRASIL
Rua do Pinheiro, 10 Flamengo
Rio de Janeiro, Brazil
T: 55 21 2853480

Nascimento, Baubino
JOVEM PAN TV
Sao Paulo, Brazil

Nascimento, Damiao
ASSOCIACAO DE MORADORES E AMIGOS DE LARANJEIRAS
Rio de Janeiro, Brazil

Nascimento, Gediel Jose Do, Sr.
Engenheiro Florestal
DER-PE
Rua Nova Descoberta, 1475 Casa Amarela
Recife - PE, Brazil
T: 55 81 4213484

Nascimento, Joao Pedro Cunha
Tecnico De Audio
SISTEMA GLOBO DE RADIO
Rua do Russel, 434
Rio de Janeiro - RJ, Brazil
T: 55 21 2058282 F: 55 21 2054096

Nascimento, Jose Geraldo Do
REDE GLOBO
R. Lopes Quintas 303 J. Botanico
Rio de Janeiro - RJ, Brazil
T: 55 21 5292000 F: 55 21 2942042

Nascimento, Josinaldo Diniz Do
MANCHETE
R. do Russel 804
Rio de Janeiro -RJ 2, Brazil
T: 55 21 2652012 F: 55 21 2855148

Nascimento, Jovelina Tonello Do, Sr.
REDE MULHER
Rua Joao Ramalho, 991
Sao Paulo - SP 05008, Brazil
T: 55 11 262-9407 F: 55 11 871-3457

Nascimento, Julio Jose Do
President, Jjn Noticias
JJN NOTICIAS
Praca Almirante Jaceguai 72 apt. 815
Rio de Janeiro -RJ, Brazil
T: 55 21 282-1292 F: 55 21 240-3073

Nascimento, Luana
OSHO SPIRITUAL HEALTH ORGAN. - OSHO
Rua Celestino Basilio 26
Rio de Janeiro - RJ, Brazil
T: 55 21 437-7488 F: 55 21 322-0327

Nascimento, Marcia C. De Almeida
EDITORA TERCEIRO MUNDO
Rio de Janeiro, Brazil

Nascimento, Maria Isabel Do, Ms.
COMAFRICA
Rua Senador Dantas, 75/1901-1903
Rio de Janeiro - RJ 20001, Brazil
T: 55 21 220-0781

Nascimento, Maria Nilza R. Do, Sra.
Diretora - Presidente
ASSOCIACAO ECOLOGICA DE VOLTA REDONDA
Rua 539 A - No. 468 Nossa Senhora das Gracas
Volta Redonda - RJ 27295, Brazil
T: 55 243 427-388

Nascimento, Neilton
ESTADO DE SAO PAULO
Av Rio Branco 151
Rio de Janeiro, Brazil
T: 55 21 2217111

Nascimento, Reynaldo Ribeiro Do, Sr.
Environment Desk
REN - BAHA I BRASIL/ ESCRITORIO MEIO AMBIENTE
Av. Hermes Fontes - Cond. Verdes Mares - Bl. G - Ap. 403
Aracaju - SE 49025, Brazil
T: 55 232-1397

Nascimento, Silvane Magali Vale, Ms.
Assessora
SOCIEDADE MARANHAENSE DE DIREITOS HUMANOS
Rua Antonio Rayol 290 - Centro
Sao Luis - MA 65015-040, Brazil
T: 55 98 221-3838

Nascimento, Sonia Satiko, Sra.
INSTITUICAO RELIGIOSA PL
Rua General Barbosa Lima,47 Copacabana
Rio de Janeiro - RJ 22011, Brazil
T: 55 21 255-9252 F: 55 21 256-6578

Nascimento, Volmer, Sr.
Coordenador Estadual
MOVIMENTO NACIONAL DE MENINOS E MENINAS DE RUA
Av. Beira-Mar, 405, Sala 705 - Castelo
Rio de Janeiro - RJ 20021, Brazil
T: 55 21 240-6872 F: 55 21 240-6795

Nascimento, Wilson Roberto Do
REDE RECORD
Av. Miruna 713
Sao Paulo _SP 04084, Brazil
T: 55 11 5429000 F: 55 11 612228

Nascomento, Edison De Oliveira
Reporter
MANCHETE
Rua do Russel 834
Rio de Janeiro, Brazil
T: 55 21 2059895

Nassarala, Jorge Luis Martins
Tecnico
VIDEO CLIPPING
Av.Rio Branco, 181 Sl.910
Rio de Janeiro - RJ, Brazil
T: 55 21 2204002 / 2629240 F: 55 21 2204002

Nations, Jim, Mr.
Vice-President For Lat.Am.
CONSERVATION INTERNATIONAL
Rua Bueno Brandao 393
Belo Horizonte - MG 31010, Brazil
T: 55 31 226-5145 F: 55 31 222-8429

Naves, Plinio Lazaro Faleiro, Mr.
Estudante
UFG
Rua 01 no. 130 S. Norte Ferroviario
Goiania - GO 74000, Brazil
T: 55 62 225-8796

Naves, Viviane A., Ms.
Senior Secretary
WORLD FOOD PROGRAMME / UN
SCN Quadra 02 Bloco 'B' Terreo
Brasilia - DF 70710, Brazil
T: 55 61 226-7872 F: 55 61 322-2080
TL: 61-1243

Needham, Lucy
Producer/Researcher
BBC TELEVISION
R.Gilberto Olimpio Maria, 100 Recreio dos Bandeirantes
Rio de Janeiro - RJ, Brazil
T: 55 21 437 7174 F: 55 21 437 7174

Negrao, Maria A. Mancuso
Editor
RADIOBRAS
Brasilia, Brazil

Negreiros, Gustavo Hees, Sr.
LASERE - UFF
Rua Guaranis , 61 Sao Francisco
Niteroi - RJ 24250, Brazil
T: 55 21 711-4645

Negret, Rafael
Especialista Em Meio Ambiente
BCO. INTERAMERICANO DE DESENVOLVIMENTO
Caixa Postal 08738
Brasilia - DF 70312, Brazil
T: 55 61 321-7848 F: 55 61 321-3112
TL: 61-1204

Negri, Francisco Luiz Negri, Mr.
Diretor
NANCON
Rua Castro Alves, 654
Sao Paulo - SP, Brazil
T: 55 11 277 5540 F: 55 11 464 4446

Neiman, Zysman, Mr.
Diretor
PHYSIS- CULTURA E AMBIENTE
Rua Diogenes Ribeiro De Lima 516 - Alto De Pinheiros
Sao Paulo - SP, Brazil
T: 55 11 212-0952

Neiser, Osmar Goncalves
REDE RECORD
Av. Miruna 713
Sao Paulo _SP 04084, Brazil
T: 55 11 5429000 F: 55 11 612228

Nepomuceno, Ronise Malena Pio
Journalist
RADIO ALVORADA -PROGRAMA MOMENTO DA TERRA
Av. Getulio Vargas 668
Belo Horizonte -MG, Brazil
T: 55 31 2732525 F: 55 31 2733535

Nepomueno, Carlos
Reporter
JORNAL BANCARIO
Av. Presidente Vargas, 502/21 andar
Rio de Janeiro, Brazil
T: 55 21 2234117 x37/38

Nery, Luciana Alvares
Member
FLAMA
Praia do Flamengo, 22
Rio de Janeiro - RJ, Brazil
T: 55 21 476-1219

Neto, Alberto Thiago Da Silva
Technician
REDE GLOBO
R. Lopes Quintas 303
Rio de Janeiro -RJ 2, Brazil
T: 55 21 2947732

Neto, Alfredo Campos
Technician
REDE GLOBO
R. Lopes Quintas 303
Rio de Janeiro -RJ 2, Brazil
T: 55 21 2947732

Neto, Claire
Correspondent
AGENCE FRANCE PRESSE
Rua Mexico 21/7 andar, Centro
Rio de Janeiro, Brazil
T: 55 21 2406634 F: 55 21 2626717

Neto, Dirceu Antunes Camargo
MANCHETE
R. do Russel 804
Rio de Janeiro -RJ 2, Brazil
T: 55 21 2652012 F: 55 21 2855148

Neto, Felicissimo P. Marques, Sr.
Engenheiro
C E M I G
Av Barbacena 1200 20 B1
Belo Horizonte - MG 30310, Brazil
T: 55 31 349 4173

Neto, Francisco Ferreira
GOBBI E FERREIRA IMAGENS
Sao Paulo, Brazil

Neto, Frederico Amandio Sil
MANCHETE
Rio de Janeiro, Brazil

Neto, Geneton Carneiro De M.
Editor
REDE GLOBO
R. Lopes Quintas 303
Rio de Janeiro -RJ 2, Brazil
T: 55 21 2947732

Neto, Guilherme Fonseca
Techinician
AGENCIA ESTADO
Av. Caetano Alvares, 55
Sao Paulo -SP 02598, Brazil
T: 55 11 8562210 F: 55 11 2652297

Neto, Helcio Alves De Souza
MANCHETE
R. do Russel 804
Rio de Janeiro -RJ 2, Brazil
T: 55 21 2652012 F: 55 21 2855148

Neto, Ibsen Costa M.
Techinician
AGENCIA ESTADO
Av. Caetano Alvares, 55
Sao Paulo -SP 02598, Brazil
T: 55 11 8562210 F: 55 11 2652297

Neto, Isias Nunes Pereira
ACIE
Av. Alantica 4240/137 - Rio Palace Hotel
Rio de Janeiro, Brazil
T: 55 21 5213232 x137

Neto, Ivo Leao, Sr.
LEAO JUNIOR S/A
Av. Getulio Vargas 253
Curitiba - PR 80230, Brazil
T: 55 41 233-7822 F: 55 41 222-0028
TL: 5134

Neto, Joao Estavam De Lima
Tecnico
TVSBT
Campo de S.Cristovao, 105
Rio de Janeiro - RJ, Brazil
T: 55 21 5803688 / 5800678 F: 55 21 5809685

Neto, Joao Flor De Oliveira, Sr.
CIA VALE DO RIO DOCE
Av. Graca Aranha 26
Rio de Janeiro - RJ 20.030, Brazil
T: 55 21 2724857 F: 55 21 2206126

Neto, Jose Augusto De Almeida
HELIO ALONSO
Rua Muniz Barreto, 51
Rio de Janeiro RJ, Brazil
T: 55 21 5515448

Neto, Libero
JORNAL MAIORIA FALANTE
Rio de Janeiro, Brazil

Neto, Mario De Souza
Tecnico
VIDEO CLIPPING
Av.Rio Branco, 181 Sl.910
Rio de Janeiro - RJ, Brazil
T: 55 21 2204002 / 2629240 F: 55 21 2204002

Neto, Mario Leite De Carvalho
CORREIO FLUMINENSE
Rio de Janeiro, Brazil

Neto, Nestor Sezefredo Dos Passos
BBC TV
Sao Paulo, Brazil

Neto, Pedro Jose De Oliveira
Technician
PROVIDEO CINEMA E TV
SHIS Q1 21 Conj 7 casa 15 Lago Sul
Brasilia DF, Brazil
T: 55 61 366 1851 F: 55 61 366 31

Neto, Raimundo Alves
FREELANCER
Niteroi RJ, Brazil

Neto, Ricardo Bonalume
Reporter
FOLHA DE SAO PAULO
Alameda Barao de Limeira 425
Sao Paulo - SP, Brazil
T: 55 11 874 2608/874 2602 F: 55 11 2231644

Neto, Rodrigo Trinidade
OSHO SPIRITUAL HEALTH ORGANIZATION
Rua Decio Villares 286/402
Rio de Janeiro - RJ 22.041, Brazil
T: 55 21 236-0104

Neto, Salvador Bizzarro
REDE RECORD
Av. Miruna 713
Sao Paulo _SP 04084, Brazil
T: 55 11 5429000 F: 55 11 612228

Netto, Arturo Piscotti
FREELANCER
Rio de Janeiro, Brazil

Netto, Marcia Graca
Produtora
GLOBO ECOLOGIA
Rua Santa Alexandrina 343/c3 Rio Comprido
Rio de Janeiro - RJ, Brazil
T: 55 21 502-5451 F: 55 21 273-9087

Netto, Marise Baunilha Correa
Technician
REDE GLOBO
R. Lopes Quintas 303
Rio de Janeiro -RJ 2, Brazil
T: 55 21 2947732

Neus, Luciana Andreia Jorge
Reporter
JORNAL DO COMMERCIO
Rua do Livramento, 189/4 andar
Rio de Janeiro, Brazil
T: 55 21 2336562 F: 55 21 2337690

Neves, Alice Martins Baeta
Producer
FUND. CULTURAL DE RADIODIFUSAO EDUCATIVA DA REGIAO DOS LAGOS
Av. Jose Bento Ribeiro Dantas 265 Manguinhos 28905
Armacao dos Buzios -, Brazil
T: 55 246 2753848 F: 55 21 2956494

Neves, Ana Elidia Caffer
Jornalista
O SAO PAULO
Av Higienopolis 890
Sao Paulo - SP, Brazil
T: 55 11 669 660 F: 55 11 826 3049

Neves, Antonio Carlos De Castro
ARQUIPELAGO DE FERNANDO DE NORONHA
Alameda do Boldro s/n
F.de Noronha PE 53990, Brazil
T: 55 81619-1210 R

Neves, Carlos Alberto
Assistant Cameraman
SAVI FILMES E TAPES
Rua Oliveira Fausto 35A Botafogo
Rio de Janeiro, Brazil
T: 55 21 5424544

Neves, Luiz Carlos Baeta, Sr.
DENACOOP - M.A.
Rua Frei Leandro 35 - Lagoa
Rio de Janeiro - RJ 22470, Brazil
T: 55 21 286-1191 F: 55 21 286-1191

Neves, Marcos Humberto Cavalcanti
CONFEDERACAO NACIONAL DA INDUSTRIA
Rio de Janeiro, Brazil

Neves, Marinita Vasconcelos
General Coordinator
AUCUBA PESQUISA E DOCUMENTACAO
R. Ricardo Hardman 105 Aflitos
Recife, PE 52050, Brazil
T: 81 268-2715

Neves, Paulo Sergio Azevedo Das
HOSPITAL ADVENTISTA SILVESTRE
Ladeira dos Guararapes, 263 - Santa Tereza
Rio de Janeiro - RJ, Brazil
T: 55 21 265-0212

Newell, Sachiko, Ms.
Av. Rui Barbosa 348/501- Flamengo
Rio de Janeiro - RJ, Brazil
T: 55 21 551-6335

Newton, Gregg
Photograpaher
REUTERS NEW SERVICE
Rua Sete de Setembro 99/4th floor Centro
Rio de Janeiro - RJ, Brazil
T: 55 21 5074151 F: 55 21 5072120

Ng, Gustavo Emilio
Latin American Correspondent
YOMIURI SHIMBUN
Av Oswaldo Cruz 149/1201
Rio de Janeiro - RJ, Brazil
T: 55 21 551 6187 F: 55 21 552 2279

Nicacio, Jomar Wilson, Sr.
Expositor
IBASE
Rua Vicente De Souza 12- Botafogo
Rio de Janeiro - RJ 22251, Brazil
T: 55 21 286-6161 F: 55 21 286-0541
TL: 21-36466

Nicholson, Brian
Editor
UP DATE (AMERICAN CHAMBER OF COMMERCE)
R Pamplona 1341 /1
Sao Paulo - SP, Brazil
T: 55 11 884 9220 F: 55 11 8849220

Nicolelia, Paulo
SISTEMA JORNAL DO BRASIL
Av. Brasil 500
Rio de Janeiro -RJ, Brazil
T: 55 21 5854422 F: 55 21 5854428

Niemeyer, Naila Costa, Sra.
Farmaceutica
PREFEITURA DO RIO DE JANEIRO
Largo do Humaita No. 207 apt.508 - Humaita
Rio de Janeiro - RJ 22261, Brazil
T: 55 21 293-5549

Nietman, Anneliese, Ms.
WOMEN'S CLUB DE PETROPOLIS
Rua Locarno 275 - Morim
Petropolis - RJ 25630, Brazil
T: 55 242 42-9483

Ning, Zhang
Reporter
AGENCIA NOTICIOSA XINHUA
Rua Paissandu,7 apto.401
Rio de Janeiro - RJ, Brazil
T: 55 21 285 0590 / 285 06 03

Nishinoiri, Yumiko, Ms.
Casa de Estudantes Harmonia Av. Caminho do Mar, 2709
Sao Bernard.Campo-SP, Brazil
T: 55 11 457-9355

Nobre, Miriam, Sra.
PREFEITURA DE PIRACICABA
Rua Dr. Paulo Pinto, 2415
Piracicaba - SP, Brazil
T: 55 194 342 999

Nobre, Reginaldo Francisco
SISTEMA GLOBO DE RADIO
R. de Russel 434
Rio de Janeiro RJ, Brazil
T: 55 21 2058282 F: 55 21 2054096

Noffs, Paulo Da Silva, Mr.
CESP
Rua da Consolacao 1875, 9 andar, Cerq. Cesar
Sao Paulo - SP 01301, Brazil

Nogueira, Claudio
Reporter
JORNAL O GLOBO
Rua Irineu Marinho, 35
Rio de Janeiro - RJ, Brazil
T: 55 21 2722000

Nogueira, Jose Roberto Dias
MANCHETE
R. do Russel 804
Rio de Janeiro -RJ 2, Brazil
T: 55 21 2652012 F: 55 21 2855148

Nogueira, Marcos
Fotografo
JORNAL GERACAO
Av Getulio Vargas 375 sala 308 Galeria Kennedy Centro
Volta Redonda - RJ, Brazil
T: 55 243 423 918/431 634

Nogueira, Maria Antonia Costa, Sra.
FETAEMG
Rua Joao Ramalho, 001 - Perdizes
Sao Paulo - SP 05008, Brazil
T: 55 11 262-9407 F: 55 11 871-3457

Nogueira, Paula Freitas Borges, Sra.
Professora
UNIVERSIDADE FEDERAL DE MINAS GERAIS
R. Monte Alegre 74/101 - Serra
Belo Horizonte - MG 30240-230, Brazil
T: 55 31 227-0604

Nolasco, Mauricio
Jornalista
FREELANCER
R.Sete de Setembro, 67/8 andar
Rio de Janeiro - RJ, Brazil
T: 55 21 2242200

Nonato, Raimundo Estrela
REDE GLOBO
R. Lopes Quintas 303 J.Botanico
Rio de Janeiro RJ 22, Brazil
T: 55 21 2947732

Nonoyama, Yoshihiro, Mr.
Av.Atlantica 3264 Copacabana
Rio de Janeiro - RJ, Brazil
T: 55 21 2558812

Noqueira, Maria Helena, Ms.
Vice Presidente
FED.BRAS.DIFUSAO CULT.E AMIGOS DE MUSEUS
R. Augusta 1819/191
Sao Paulo - SP 01413.000, Brazil
T: 55 11 283 1706

Noroes, Ana Maria T., Sra.
Diretora-Presidente
INSTITUTO DE PREVENCAO A DESNUTRICAO E A EXCEPCIONALIDADE - IPREDE
Rua Professor Carlos Lobo 15 Cidade dos Funcionarios
Fortaleza - CE 60815, Brazil
T: 55 85 239-1300

Nova, Jose Eduardo Vila
JOVEM PAN TV
Sao Paulo, Brazil

Novaes, Lourdes Rego, Sra.
President Director
COMITE BRASILEIRO ICOM
Rua Murtinho Nobre n.89
Rio de Janeiro - RJ 20241, Brazil
T: 55 21 224-5461

Novello, Maria Do Carmo, Sra.
Secretary
PROJETO NOVAMERICA INTERCULTURAL
R. Maques de Sao Vicente 331 / 5 andar, Gavea
Rio de Janeiro - RJ 22451, Brazil
T: 55 21 274.2149 F: 55 21 274.2149

Nowotny, Monica, Sra.
A.P.T.A.
Av. Gil Veloso 464 Apto. 305
Vila Velha - ES, Brazil
T: 55 27 329-3343

Numes, Eduardo
ASSOCIATED PRESS
Rio de Janeiro - RJ, Brazil

Nunes, Alexandra De Araujo
Produtor
REDE BANDEIRANTES DE TELEVISAO
R.Alvaro Ramos, 492
Rio de Janeiro - RJ, Brazil
T: 55 21 2122072 F: 55 21 2957391

Nunes, Altamiro
Fotografo
ASSOCIATED PRESS
Avenida Brasil 500 sala 847
Rio de Janeiro - RJ, Brazil
T: 55 21 580 4442 F: 55 21 580 2309

Nunes, Cairu Teles
ESTACAO ECOLOGICA
Cataguases MG, Brazil

Nunes, Eduardo Albuquerque, Sr.
Sp Project Supervisor
VISAO MUNDIAL BRASIL
Rua Domingos de Moraes 1414 - Vila Mariana
Sao Paulo - SP 04010, Brazil
T: 55 11 549 2196

Nunes, Gloria, Sr.
Assistant
INUIT CIRCUMPOLAR CONFERENCE
Rua Rabino Henrique Lemle, 403
Rio de Janeiro - RJ 22700, Brazil
T: 55 21 437 9130

Nunes, Henrique, Ms.
Professor/Pesq. Ambiental
NEPA / URCAMP
Rua Constabile Hipolito 645
Bage - RS 96400-100, Brazil
T: 0532423967 F: 0532422898
TL: 141

Nunes, Isabella Rosado
CHICO JR. & ASSOCIADOS
Rio de Janeiro, Brazil

Nunes, Joao Mario De Aquino
Photographer
MANCHETE
R. do Russel 804 Gloria
Rio de Janeiro RJ 2, Brazil
T: 55 21 2454816 F: 55 21 2056196

Nunes, Lidio Anibal Lopes, Sr.
Biologo
FEPAM
av. a.j. renner, 10
Porto Alegre - RS 90250, Brazil
T: 55 512 343-5799 F: 55 512 342-0224

Nunes, Maria Clara Marinho
MANCHETE
R. do Russel 804
Rio de Janeiro -RJ 2, Brazil
T: 55 21 2652012 F: 55 21 2855148

Nunes, Nilton
Technician
REDE GLOBO
R. Lopes Quintas 303
Rio de Janeiro -RJ 2, Brazil
T: 55 21 2947732

Nunes, Paulo Ricardo
Assistente
2DF-TV ALEMA
Av.Delfim Moreira, 662/704
Rio de Janeiro - RJ, Brazil
T: 55 21 2947434 F: 55 21 5122120

Nunes, Petrina Portela
INSTITUTO DE ARQUITETOS DO BRASIL
Rua do Pinheiro, 10 Flamengo
Rio de Janeiro, Brazil
T: 55 21 2853480

Nunes, Roberto Da Silva, Sr.
Presidente
C S R M M R - E S
Av. Pelotas 208
Serra - ES, Brazil
T: 55 27 322-4641

Nunes, Stella Matutina Costa, Sra.
Prof. Ensino Medio
Rua Teodoro da Silva, 546, Cob. 02, Vila Isabel
Rio de Janeiro - RJ 20560, Brazil
T: 55 21 208-6700

Nunis, Alexandre Joaquim
Engineer
AMERICA DO SUL PRODUCOES DE DOCUMENTARIOS LTDA (NAMBEI TSUSHIN)
Av. Paulista 509 /11 andar conj.1104 Cerqueira Cesar
Sao Paulo -SP 01311, Brazil
T: 55 11 2886141 F: 55 11 2512065

Nwe, Yin Yin, Sra.
Pjroject Officer-Comm
UNICEF
SBS Edificio Seguradoras 13o Andar
Brasilia - DF, Brazil
T: 55 61 224-7145 F: 55 61 224-5035

O., Ricardo, Sr.
Coordenador
AUCUBA - PESQUISA E DOCUMENTACAO
Rua Ricardo Hardman 105 - Aflitos
Recife - PE 52050, Brazil
T: 55 81 441-2722

O. Roxo, Carlos Alberto, Mr.
ARACRUZ CELULOSE S.A.
Av. Lauro Muller, 116/22o. andar
Rio de Janeiro - RJ 22290, Brazil
T: 55 21 541-2993 F: 55 21 541-5443

O.Ribeiro, Rogerio Antonio
Camera
TVSBT
Campo de S.Cristovao, 105
Rio de Janeiro - RJ, Brazil
T: 55 21 5803688 / 5800678 F: 55 21 5809685

Oberhofer, Maria, Ms.
Gardner
R. General Dionisio 63, Botafogo
Rio de Janeiro - RJ, Brazil
T: 1 021 286-0303

Ochmitt, Renata Catarina
Reporter
O DIA
R. Riachuela 359
Rio De Janeiro, RJ 2, Brazil
T: 55 21 272-8000 F: 55 21 507-1228

Odanara, Helvia Yuri Cafe, Sra.
Membro
APOENA
Rua Tomizo Kawaguti, 500
Presid. Prudente-SP 19013, Brazil
T: 55 182 22-7175

Oestreich, Helena Maria
INSTITUTO DE ARQUITETOS DO BRASIL
Rua do Pinheiro, 10 Flamengo
Rio de Janeiro, Brazil
T: 55 21 2853480

Oitaven, Juan Carlos Martin
Technician
REDE GLOBO
R. Lopes Quintas 303
Rio de Janeiro -RJ 2, Brazil
T: 55 21 2947732

Oiticica, Vera, Sra.
Lucis Trust
FUNDACAO CULTURAL AVATAR
Rua Ministro Octavio Kelly, 352, c-6
Niteroi - RJ 24220, Brazil
T: 55 21 611-2489

Oldemburgo, Cristina
FUND. CULTURAL DE RADIODIFUSAO EDUCATIVA DA REGIAO DOS LAGOS
Av. Jose Bento Ribeiro Dantas 265 Manguinhos 28905
Armacao dos Buzios -, Brazil
T: 55 246 2753848 F: 55 21 2956494

Oldenburg, Laura Couto Guedes
Journalist
HILL AND KNOWLTON
Av. Afranio de Melo Franco 135 /5 andar Leblon
Rio de Janeiro -RJ 2, Brazil
F: 55 21 2394697

Olive, Lea
LUCIS TRUST - FUND. CULT. AVATAR
Min. Otavio Kelly 352 c/6
Niteroi - RJ 24230, Brazil
T: 55 21 611-2489

Oliveira, Aarao Reis
Tecnico
TVSBT
Campo de S.Cristovao, 105
Rio de Janeiro - RJ, Brazil
T: 55 21 5803688 / 5800678 F: 55 21 5809685

Oliveira, Acir Silva, Sr.
Diretor Tecnico
FUNDACAO S.O.S. RIO PIRACICABA
Av. Brasil 35A Bairro Iguacu
Ipatinga - MG 35160, Brazil
T: 55 31 821-4047

Oliveira, Afonso A., Sr.
ASSOCIACAO CIENTIFICA ECOLOGICA 'AUGUSTO RUSCHI' - ACEAR
Cerquilho SP 18520, Brazil
T: 55 15 284-1274 F: 55 15 284-1205

Oliveira, Aluizio De
MANCHETE
Rio de Janeiro, Brazil

Oliveira, Ana Beatriz Rodrigues, Sra.
EDITORA GLOBO S/A.
Rua Do Cortume , 675
Sao Paulo - SP, Brazil
T: 55 11 874-6000

Oliveira, Ana Maria Maia De
CONFEDERACAO NACIONAL DA INDUSTRIA
Rio de Janeiro, Brazil

Oliveira, Ana Paula De
CHICO JR. & ASSOCIADOS
Rio de Janeiro, Brazil

Oliveira, Antonio Marcus De
VIDEOARTE RIO
Rua Senradom 45
Rio de Janeiro, Brazil

Oliveira, Astrid Prange De
Correspondente
DIE TAGESZEITUNG
Av das Americas 411 apt 111 Bl 2
Rio de Janeiro - RJ, Brazil
T: 55 21 389 5561 F: 55 21 389 5561

Oliveira, Auri Alencar De
Editor
REDE GLOBO
R. Lopes Quintas 303
Rio de Janeiro -RJ 2, Brazil
T: 55 21 2947732

Oliveira, Claudomiro Moura De, Mr.
CIA VALE DO RIO DOCE
Av. Graca Aranha, 26
Rio de Janeiro - RJ, Brazil
T: 55 21 272 4401 F: 55 21 220 0593

Oliveira, Condesmar Fernandes De, Sr.
PREFEITURA MUNICIPAL
Praca Maua s/n 1 andar
Santos - SP 11010, Brazil
T: 55 132 342-2093 F: 55 132 345-809
TL: 131048

Oliveira, Dolores, Sra.
CLUBE SOROPTIMISTA INTERNACIONAL
SHIN Q 13 conj. 8 casa 13
Brasilia - DF 71588, Brazil
T: 55 61 577-2357 F: 55 61 321-4320

Oliveira, Donezetti Teixeira De, Sr.
Psicologo
Avenida Joao XXIII 759
Campo Bom - RS 93700, Brazil
T: 55 51 597-1727

Oliveira, Edileia Maria De, Sr.
Diretor De Planejamento
SERVICO NACIONAL DE APRENDIZAGEM COMERCIAL - SENAC
Rua Dona Mariana, 48 - Botafogo
Rio de Janeiro - RJ 22280, Brazil
T: 55 21 266-1522 F: 55 21 286-0645
TL: 2131129

Oliveira, Eduardo Borges, Mr.
Advogado
SOCIEDADE DE DEFESA - DIREITOS HUMANOS
Rua Antonio Rayol 190 - Centro
Sao Luiz - MA 65000, Brazil
T: 55 85 222-8546

Oliveira, Elaina Machado De
Reporter
CORREIO DA BAHIA
Alameda Genova, 377, Pituba
Salvador, Bahia, Brazil
T: 55 3593714

Oliveira, Elmo Da Cunha
Technician
REDE GLOBO
R. Lopes Quintas 303
Rio de Janeiro -RJ 2, Brazil
T: 55 21 2947732

Oliveira, Erundina Barbosa De, Sra.
Ass. Presidencia
FUND. ECOLOGICA VALE DO GUAPORE RONDONIA
Rua Abuna, 2107- Sao Cristovao Caixa Postal 27
Porto Velho - RO 78900, Brazil
T: 55 68 224-3691 F: 55 68 222-5297

Oliveira, Eucimar
Editorial Secretary
SISTEMA JORNAL DO BRASIL
Av. Brasil 500
Rio de Janeiro -RJ, Brazil
T: 55 21 5854422 F: 55 21 5854428

Oliveira, Everaldo R. Ramos De
Photographer
IMAGENS DA TERRA
R. das Laranjeiras 203/202
Rio de Janeiro RJ 2, Brazil
T: 55 21 5562502 F: 55 21 2204653

Oliveira, Flavio
COORDENADORIA DE COM. SOCIAL-PREFEITURA DO RIO DE JANEIRO
R. Afonso Cavalcante 455/3/1520 Cidade Nova
Rio de Janeiro -RJ 2, Brazil
T: 55 21 2931744 F: 55 21 2737633

Oliveira, Francisco De Assis A. De
Technician
REDE GLOBO
R. Lopes Quintas 303
Rio de Janeiro -RJ 2, Brazil
T: 55 21 2947732

Oliveira, Francisco Jose P. De
Technician
REDE GLOBO
R. Lopes Quintas 303
Rio de Janeiro -RJ 2, Brazil
T: 55 21 2947732

Oliveira, Gilberto Duarte De, Mr.
Linguist
SUMMER INSTITUTE OF LINGUISTICS
SAI/NO, Lote 'D', Bloco 03
Brasilia - DF 70770, Brazil
T: 55 61 272-1200

Oliveira, Iara Pietricovsky, Sra.
Assessora
INSTITURO DE ESTUDOS SOCIO-ECONOMICOS
Ed. Venancio 2000 Bloco b-50 s/433
Brasilia - DF 70333, Brazil
T: 55 61 226-8131 F: 55 61 226-8042
TL: 4837

Oliveira, Jenner, Ms.
WORLD COUNCIL OF CHURCHES
150 route de Ferney
Brazil
F: 41 22 791 0361

Oliveira, Joacj Barbosa De
FORUM DO FUTURO
Rio de Jaeiro, Brazil

Oliveira, Joao Fernandes, Mr.
PREFEITURA DE PARATY
Rua da Floresta s/n
Paraty - RJ 23970, Brazil
T: 55 243 71.1193

Oliveira, Joao Pacheco De, Sr.
Coordenador
PROJETO TERRAS INDIGENAS
Quinta da Boa Vista s/n
Rio de Janeiro - RJ, Brazil
T: 55 21 284-9642

Oliveira, Joao Vicente Ganzarolli De
Journalist
FREELANCER
Rua Jardim Botanico 657/302 Jardim Botanico
Rio de Janeiro - RJ, Brazil
T: 55 21 274 4024

Oliveira, Joel Marques De, Mr.
Diretor
COOPSYSTEMS
Praia Flamengo ,66 B Sala 214
Rio de Janeiro - RJ, Brazil
T: 55 21 205 6268

Oliveira, Jorge Gomes De
Technician
REDE GLOBO
R. Lopes Quintas 303
Rio de Janeiro -RJ 2, Brazil
T: 55 21 2947732

Oliveira, Jorge Luiz De
Operador De Vt
REDE BANDEIRANTES DE TELEVISAO
R.Alvaro Ramos, 492
Rio de Janeiro - RJ, Brazil
T: 55 21 2122072 F: 55 21 2957391

Oliveira, Jose Antonio De
Motorista
BCO. INTERAMERICANO DE DESENVOLVIMENTO
Caixa Postal 08738
Brasilia - DF 70312, Brazil
T: 55 61 321-7848 F: 55 61 321-3112
TL: 61-1204

Oliveira, Jose Maurino, Sr.
Tecnico
UNIVERSIDADE POPULAR DO TRABALHO
Rua Pastor Manoel Virgilio De Souza 1310-B
Curitiba - PR 82810-400, Brazil
T: 55 41 267 2007

Oliveira, Jose Robert De
Editor
TV EDUCATIVA
Rua Gomes Frano, 474 Centro
Rio de Janeiro, Brazil
T: 55 21 2920012 x130

Oliveira, Jose Soares
Technician
REDE GLOBO
R. Lopes Quintas 303 Jardim Botanico
Rio de Janeiro -RJ 2, Brazil
T: 55 21 2947732

Oliveira, Julio Cezar Pereira De
Photographer
AGENCE FRANCE PRESSE
Rua Mexico 21/7 andar
Rio de Janeiro -RJ 2, Brazil
T: 55 21 2406634 F: 55 21 2626717

Oliveira, Lea Lobato De Carvalho, Ms.
Tecnico (eng. Agronomo)
IDESP
Av. Nazare 817
Belem - PA 66035, Brazil
T: 55 91 224-4411 F: 55 91 225-6905
TL: 6905

Oliveira, Lucia Leiga De, Ms.
UNITED METHODIST CHURCH OF BRAZIL
Rua Itajuba, 2086 apto. 402
Belo Horizonte - MG 3l.030, Brazil

Oliveira, Luiz Marcos
MANCHETE
Rio de Janeiro, Brazil

Oliveira, Manoel Carlos R. De
Reporter
REDE GLOBO
R. Lopes Quintas 303
Rio de Janeiro -RJ 2, Brazil
T: 55 21 2947732

Oliveira, Marco Antonio Souza De, Sr.
Secretario Geral
ABES
Av. Beira Mar, 216 - 13o. andar
Rio de Janeiro - RJ 20021, Brazil
T: 55 21 210-3221 F: 55 21 262-6838
TL: 21-31902

Oliveira, Marcos Garcia De
GOVERNO DO ESTADO DE SAO PAULO
Sao Paulo, Brazil

Oliveira, Marcus Vinicius, Sr.
Conselheiro
CONFEA
Rua Antunes Frota, 297/104
Fortaleza - CE 60020, Brazil
T: 55 85 243-4269

Oliveira, Maria Aparecida De
Editor
FUNDACAO ROQUETTE-PINTO
Prasa da Republica, 141 A Centro
Rio de Janeiro, Brazil
T: 55 21 2522203

Oliveira, Maria Da Conceicao Maia, Sra.
Coord. Ed. Amb.
INSTITUTO DE MEIO AMBIENTE DO ACRE
Rua Rui Barbosa, 450 - Centro
Rio Branco - AC 69.900, Brazil
T: 55 68 224 5497 F: 55 68 224 5694
TL: 68 2692

Oliveira, Maria Do Ceu F. De, Ms.
Assistente Social
UNIVERSIDADE CATOLICA DO SALVADOR
Av. Cardeal Da Silva 205-Viracom
Salvador - BA, Brazil
T: 55 71 247 7353 F: 55 71 245 8434

Oliveira, Maria Helena De, Sra.
Representante Em Brasilia
GRUPO NATIVA - PROTECAO PESQUISA E INFORMACAO AMBIENTAL
AOS Quadra 6 Bl.A/611
Brasilia - DF 70651, Brazil
T: 55 61 234-1410 F: 55 61 233-8763

Oliveira, Maria Ines Couto De, Ms.
Local Coordinator
INTERNATIONAL BABY FOOD ACTION NETWORK
Rua Faro 35/101
Rio de Janeiro - RJ 22461, Brazil
T: 55 21 286-0139 F: 55 21 270-2193

Oliveira, Marilma Torres Gouveia De, Ms.
ASSOCIACAO BRASILEIRA MULHERES CARREIRA JURIDICA - ABMCJ
Rua Deputado Siloe Tavares - 170 - Farol
Maceio - AL, Brazil
T: 55 82 241-5937

Oliveira, Mario Ignacio
Technician
REDE GLOBO
R. Lopes Quintas 303
Rio de Janeiro -RJ 2, Brazil
T: 55 21 2947732

Oliveira, Mario Ramao, Jr., Mr.
Biologo Chefe
CENTRO DE PESQUISAS DA REPRODUCAO
Rua Miguel de Frias 245 ER.601
Niteroi - RJ 24220, Brazil
T: 55 21 722 5337

Oliveira, Marly Brito Ribeiro De, Sra.
Rua Ajax Correia Rabelo, 195 - Alto das Mangabeiras
Belo Horizonte - MG, Brazil

Oliveira, Mauro Luiz De
Technician
REDE GLOBO
R. Lopes Quintas 303
Rio de Janeiro -RJ 2, Brazil
T: 55 21 2947732

Oliveira, Mayra Jurua, Sra.
rua embaixador carlos taylor 95 101 bl.3
Rio de Janeiro - RJ 22451, Brazil
T: 55 21 274-2834

Oliveira, Moarym Peoreira Gomes De
VIDEOARTE RIO
Rua Senradom 45
Rio de Janeiro, Brazil

Oliveira, Nazare Do S. Leite, Sra.
EDITORA GLOBO S/A.
Rua Cortume, 675
Sao Paulo - SP, Brazil
T: 55 11 874-6000

Oliveira, Orquinezio, Sr.
Presidente
ORDEM DOS ADVOGADOS PRESIDENT NITEROI
Avenida Amaral Peixoto, 507
Niteroi - RJ 24020, Brazil
T: 55 21 719-8470

Oliveira, Paulo Marcio De
MANCHETE
Rio de Janeiro, Brazil

Oliveira, Regina Martins De
TVN PRODUCOES ARTISTICAS
Sao Paulo, Brazil

Oliveira, Ribamar De
Coordinator
AGENCIA ESTADO
Av. Caetano Alvares, 55
Sao Paulo -SP 02598, Brazil
T: 55 11 8562210 F: 55 11 2652297

Oliveira, Rita De Cassia Motta Pires De, Ms.
Gerente
HOTEL F. CALEDONIA INN
Reserva Ecologica Do Pico Da Caledonia S/N
Nova Friburgo - RJ, Brazil
T: 55 245 223775 F: 55 245 223775

Oliveira, Rossano
Assistente De Producao
GLOBO ECOLOGIA
Rua Santa Alexandrina 343/c3 Rio Comprido
Rio de Janeiro - RJ, Brazil
T: 55 21 502-5451 F: 55 21 273-9087

Oliveira, Sandra De Fatima, Ms.
Professora
Al. Xavier De Almeida # 237- Setor Pedro Lodovico
Goiania - GO 74000, Brazil
T: 55 62 241-1477

Oliveira, Sara Soares, Ms.
Auxiliar De Enfermagem
Rua Barata Ribeiro, 90 Apto. 802
Rio de Janeiro - RJ, Brazil
T: 55 21 227-8592

Oliveira, Sebastiao Luiz De, Sr.
Member
GRUPO NATIVA - GOIAS/BR
AOS 06-Bloco A- apt.611
Brasilia - DF 70651, Brazil
T: 55 61 234-1410 F: 55 61 233-8763

Oliveira, Sergio
Editor
JORNAL NOVA ERA
R.Paulino Fernandes, 98/503
Rio de Janeiro - RJ, Brazil
T: 55 21 2957359

Oliveira, Sylvia De, Ms.
Social Assist.
SESC/ARRJ
R. Cincinato Silva 152
Rio de Janeiro - RJ 21341, Brazil
T: 55 21 359-6551

Oliveira, Vanderlei De
Technician
REDE GLOBO
R. Lopes Quintas 303
Rio de Janeiro -RJ 2, Brazil
T: 55 21 2947732

Oliveira, Vera Lucia A., Sra.
INB
Praia do Flamengo 224/311
Rio de Janeiro - RJ, Brazil

Oliveira, Wagner Carvalho De
MANCHETE
R. do Russel 804
Rio de Janeiro -RJ 2, Brazil
T: 55 21 2652012 F: 55 21 2855148

Oliveira, Wilson Barbosa De, Mr.
Ass. Sec. Saude E M. Ambiente
PETROFERTIL S.A.
Praca Mahatma Ghandi, 14 Sala 1304
Rio de Janeiro - RJ 20031, Brazil
T: 55 21 217-5330 F: 55 21 240-4523
TL: 212-3880

Oliveira Filho, Jurandir Pinheiro, Mr.
Coord. Und. Cons.
INSTITUTO DE MEIO AMBIENTE DO ACRE
Rua: Rui Barbosa, 450 - Centro
Rio Branco - AC 69.900, Brazil
T: 55 68 224 5497 F: 55 68 224 5694
TL: 68.2692

Olivet, Lea, Ms.
Assistente Social
Rua Antonio Parreiras 51/403
Rio de Janeiro - RJ 22411, Brazil
T: 55 21 267-3831

Olivieira, Cesar Augusto
Technician
REDE GLOBO
R. Lopes Quintas 303
Rio de Janeiro -RJ 2, Brazil
T: 55 21 2947732

Olivieri, Silvio, Mr.
Vice-President For Science
CONSERVATION INTERNATIONAL
Rua Bueno Brandao 393
Belo Horizonte - MG 31010, Brazil
T: 55 31 226-5145 F: 55 31 222-8429

Omanquim, Sergio Monteiro
Technician
REDE GLOBO
R. Lopes Quintas 303
Rio de Janeiro -RJ 2, Brazil
T: 55 21 2947732

Oporto, Paulina, Ms.
WORLD YOUNG WOMEN'S CHRISTIAN ASSOCIATION / BRASIL
37, Quai Wilson
Geneva 1201, Brazil
T: 41 22 732 31 00 F: 41 22 731 79 38
TL: WYMCA 41 23 32

Ornellas, Erica
FUND. CULTURAL DE RADIODIFUSAO EDUCATIVA DA REGIAO DOS LAGOS
Av. Jose Bento Ribeiro Dantas 265 Manguinhos 28905
Armacao dos Buzios -, Brazil
T: 55 246 2753848 F: 55 21 2956494

Orsi, Ana Lucia, Sra.
Arquiteto
FAU-USP
R Barao De Sta Eulavia, 170/142
Sao Paulo - SP 05685, Brazil
T: 55 11 843 0479

Ortis, Delis
Reporter
REDE GLOBO
R. Lopes Quintas 303
Rio de Janeiro -RJ 2, Brazil
T: 55 21 2947732

Ortiz, Eduardo, Sr.
Representativ
FUNDACAO PROJETO SANITARISTA
Rua Batataes, 333 Apto 11 Jardim Paulista
Sao Paulo - SP 01423, Brazil
T: 55 11 8879735

Ortiz, Francisco, Sr.
Representativ
FUNDACAO PROJETO SANITARISTA
Rua Batataes, 333 Apto 11 Jardim Paulista
Sao Paulo - SP 01423, Brazil
T: 55 11 8879735

Ortiz, Pedro Henrique Falco
AGENCIA DE NOTICIAS ACOPI
Sao Paulo, Brazil

Ortiz, Sonia Regina Carboni, Sra.
Representative
FUNDACAO PROJETO SANITARISTA
Rua Batataes, 333 apto 11 Jd. Paulista
Sao Paulo - SP 01423, Brazil
T: 55 11 8879735

Osawa, Kenji, Sr.
Avinida Rio Branco, 15615,.1115
Rio de Janeiro - RJ 20040, Brazil
T: 55 21 210-32634 F: 55 21 220-9854
TL: 30605

Ota, Sueli Naomi, Sra.
Biologa
SOC. DE PESQ. EM VIDA SELVAGEM E EDUCACAO AMBIENTAL
Rua Gutemberg 345
Curitiba - PR 80420-030, Brazil
T: 55 41 222-3425 F: 55 41 222-3425

Otaki, Masayuki, Sr.
Guia
Avinida Rio Branco, 15615,.1115
Rio de Janeiro - RJ 20040, Brazil
T: 55 21 210-32634 F: 55 21 220-9854
TL: 30605

Otsuki, Yoshinori
AMERICA DO SUL PRODUCOES DE DOCUMENTARIOS LTDA (NAMBEI TSUSHIN)
Sao Paulo, Brazil

Ottoni, Deisi, Sra.
Arquiteta
R. Sacopa, 61 - apto 102
Rio de Janeiro - RJ 22471, Brazil
T: 55 21 246 9244 F: 55 21 226 8540

Outlule, Samuel, Mr.
THE KALAHARI CONSERVATION SOCIETY
Av. Afranio de Mello Franco, 135/5o. Andar - Leblon
Rio de Janeiro - RJ 22430, Brazil
T: 55 21 511-0110 F: 55 21 239-4697

Owa, Gisaburo
Director
AMERICA DO SUL PRODUCOES DE DOCUMENTARIOS LTDA (NAMBEI TSUSHIN)
Av. Paulista 509 /11 andar conj.1104 Cerqueira Cesar
Sao Paulo -SP 01311, Brazil
T: 55 11 2886141 F: 55 11 2512065

Ozaki, Carlos Yuzo
NHK
Av Nilo Pecanha 50 Grupo 1208
Rio de Janeiro - RJ, Brazil
T: 55 21 262 9777 F: 55 21 262 9514

Ozaki, Roberto Shigeo
NHK
Av Nilo Pecanha 50 Grupo 1208
Rio de Janeiro - RJ, Brazil
T: 55 21 262 9777 F: 55 21 262 9514

Paarlberg, Robert, Mr.
Professor
HARVARD UNIVERSITY
1737 Cambridge St.
Cambuquira - MG 02138, Brazil
T: 55 35 4951294

Pacheco, Lia, Sra.
Magisterio
CEPERGS
Rua Castro Alves,1068
Porto Alegre - RS, Brazil
T: 55 512 332-3923

Pacheco, Luzia Santos, Sra.
Psicologa
CONSELHO REGIONAL DE PSICOLOGIA
Rua Paulo Barreto 86
Rio de Janeiro - RJ 20510, Brazil
T: 55 21 208-8590

Pacheco, Mario Poppe De M., Mr.
Presidente
PIELE IND. E COMERCIO DE COSM. NATURAIS LTDA.
Rua Senador Dantas 75 Sala 1803
Rio de Janeiro - RJ 20031, Brazil
T: 55 21 220-1881 F: 55 21 262-5209

Pacheco, Regina Silvia, Ms.
Presidente
NUCLEO DE ESTUDOS REGIONAIS E URBANOS
Rua Consolacao 3497/12
Sao Paulo - SP 01416, Brazil
T: 55 11 852-5401 F: 55 11 440-7198

Pacios, Valentim Manduca, Sr.
Geologo
FUNDACAO BRASIL-FRANCA
R. Jaci Parana' 2758
Porto Velho - RO 78900, Brazil
T: 55 68 2218438

Padilha, Eliseu, Mr.
Prefeito
PREFEITURA DE TRAMANDAI - RS
Av, Da Igreja, 346
Tramandai - RS CEP: 95.590, Brazil
T: 55 51 684-1222 F: 55 51 661-1762

Padilla, Anivaldo, Mr.
WORLD COUNCIL OF CHURCHES
No. 133, Sec 3 Chung-Zong Rd.
Brazil

Padovano, Luca, Sr.
AQUALUNG CONFECCAO LTDA.
Av. Londres, 174 - Bonsucesso
Rio de Janeiro - RJ 21041, Brazil
T: 55 21 270-1955

Padua, Jose Augusto, Mr.
ASSOCIACAO CIVIL GREENPEACE
Rua Mexico 21, Grp 1301, Centro
Rio de Janeiro - RJ 20031, Brazil
T: 55 21 262-7318 F: 55 21 240-1690

Paes, Humberto
LEGIAO DA BOA VONTADE
Av. Rudge 700 Bom Retiro
Sao Paulo -SP 01134, Brazil
T: 55 21 2338381/R:223 F: 55 21 2634025

Paes, Jeniel Pinheiro
MANCHETE
R. do Russel 804
Rio de Janeiro -RJ 2, Brazil
T: 55 21 2652012 F: 55 21 2855148

Paetow, Regina, Ms.
Sociologa
SACTES - PASTORAL DA CRIANCA
Rua Almeida Cunha 364
Recife - PE 50000, Brazil
T: 55 81 221-0075
TL: 1655

Paim, Rosalda, Sra.
Director
CMB
Rua Avanhadava no. 756
Sao Paulo - SP 01306, Brazil
T: 55 11 2564354 F: 55 11 2577648

Paiva, Dinalva Gomes, Sra.
UNIVERSIDADE FEDERAL DE MATO GROSSO
Av. Fernando Correa S/N
Cuiaba - MT 78000, Brazil
T: 55 125 3158304 F: 55 125 3611119

Paiva, Eunice, Dr.
JAMA
Rua Juri 16, Vila Madalena
Sao Paulo - SP 03443, Brazil
T: 55 11 210-1301 F: 55 11 210-1301

Paiva, Gabriel
Photographer
ESTADO DE MINAS
R. Goias 36 Centro
Belo Horizonte -MG 3, Brazil
T: 55 31 2732322 F: 55 31 2268070

Paiva, Maria Cristina Ribeiro, Ms.
FUNDACAO BIODIVERISITAS
Rua Maria Vaz de Melo 71
Belo Horizonte - MG 31250, Brazil
T: 55 31 443-2119

Paiva, Ricardo
Cinegrafista
AUCUBA PESQUISA E DOCUMENTACAO
R. Ricardo Hardman 105 Aflitos
Recife, PE 52050, Brazil
T: 81 268-2715

Paixao, Julio, Mr.
Engenheiro Florestal (ms)
UNIVERSIDADE FERERAL DE VICOSA
Rua Vaz de Mello, 57/202
Vicosa, MG. 36570, Brazil
T: 031891 3024

Paixao, Maria Filomena Da
GAIA - ASSOC.CONS.M.AMBIENTE E PROD.INTEGRADA ALIM.AMAZONIA
Arse-51 - 39
Palmas, TO 77000, Brazil
T: 063 8621553/1108

Palacio, Vanuza F., Ms.
UNESCO
SCN -Quadra 2 Lote B
Brasilia - DF 70710, Brazil
T: 55 61 223-8684 F: 55 61 322-4261

Palatnik, Elchanan, Mr.
Consulting Engineer
Rua Senador Vergueiro 14/1001
Rio de Janeiro - RJ 22230-001, Brazil
T: 55 21 225-9415 F: 55 21 225-9415

Palazzo Jr., Jose Truda, Sr.
Coordenador
IWC/BRASIL
Caixa Postal 5087
Florianopolis - SC 88041, Brazil
T: 55 482 341580 F: 55 482 341580

Palermo, Renata
92 GLOBAL FORUM
Rio de Janeiro - RJ, Brazil

Palese, Blair
Greenpeace Usa
GREENPEACE
Rua Mexico, 21/1301 Gr.A e B
Rio de Janeiro - RJ, Brazil
T: 55 21 2401690 F: 55 21 2627318

Pallaridini, Manoel Horacio Gimenez
Diretor No Rio
ANSA
Av Presidente Antonio Carlos 40 - cobertura
Rio de Janeiro - RJ, Brazil
T: 55 21 262-6574/220-5528 F: 55 21 262-6574

Palmeira, Rosa Cristina
CHICO JR. & ASSOCIADOS
Rio de Janeiro, Brazil

Pandalhi, Iris Bramclao
Reporter
FREELANCER
Av. Ataulfo do Paiva 1389 Apt 302
Rio de Janeiro, Brazil
T: 55 21 5191067

Pandela, Helene Amilie De S.
Coordinator
Praca do Republica, 141 sala 207 Centro
Rio de Janeiro, Brazil
T: 55 21 2522203

Panini, Irma Carmela
FRANCISCANS AT THE UNITED NATIONS
CP 307
Rio do Sul - SC 89160, Brazil
T: 55 478 22 06 14

Panizza, Silvio, Sr.
IECAM
Dr. Moreira 250
Manaus - AM 69000, Brazil

Panol, Mythel, Sra.
Tecnica
CASA DA MULHER DO NORDESTE
Rua Napoleao Laureano 40A Madalena
Recife - PE 50751, Brazil
T: 55 81 227-0531

Paoli, Doris Lisboa De, Ms.
Socia-Filantro
WOMEN'S CLUB OF RIO DE JANEIRO
Rua Real Grandeza, 99
Rio de Janeiro - RJ, Brazil
T: 55 21 325-2451

Paolucci, Luciana
REVISTA CIDADE NOVA
Vargem Grande Paulis, Brazil

Paova, Varicisio Ferraira
Director
INTERFREE PROD. DE VIDEO LTDA
Brazil
T: 61 345435

Paranagua, Ricardo, Sr.
Jornalista
PAU-BRASIL COMUNICACOES
Rua Mator Rubens Vaz, 702/1102 - Gavea
Rio de Janeiro - RJ, Brazil
T: 55 21 274 8824 F: 55 21 342 8587

Paredes, Patricia Mousinho, Sra.
Diretora
SIMBIO
R. Siqueira Campos 43 Sala 926
Rio de Janeiro - RJ 22031, Brazil
T: 55 21 236-4999

Parmegiani, Mariliza Rezende
LEGIAO DA BOA VONTADE
Av. Rudge 700 Bom Retiro
Sao Paulo -SP 01134, Brazil
T: 55 21 2338381/R:223 F: 55 21 2634025

Pasquale, Joaquim C. Andrade, Mr.
Rua Espirito Santo 2404/501
Belo Horizonte - MG 30160, Brazil
T: 55 31 2235814

Passos, Elza G. Doria, Sra.
Coordenadora Nacional Do Ma
FAMN E P
Rua Carlos Hugueney, 90 Jd.Estados
Campo Grande - MS 79015, Brazil
T: 55 67 382-6537 F: 55 67 725-7282

Passos, Maria Helena
Jornalista Colaboradora
GEMINI NEWS SERVICE
R.Cayowaa, 1373 casa 3
Sao Paulo - SP, Brazil
T: 55 11 8728231 F: 55 11 2874489

Passos, Suzana De Castro, Sra.
Coordinator
FED. DAS BANDEIRANTES DO BRASIL
Av. Marechal Camara, 186/2 andar
Rio de Janeiro - RJ 20020, Brazil
T: 55 21 240-9220

Pastor, Mabel Lidia Arias De, Sra.
Palestrista
GRUPO INFORMAL DA CIENCIA CRISTA
Av. Oswaldo Cruz, 87/907
Rio de Janeiro - RJ 22250, Brazil
T: 55 21 552-4979 F: 55 21 240-7472

Pastuk, Marilia, Sra.
Consultora
FRIENDS OF THE EARTH - ITALY
Rua Fonte da Saudade, 47/203
Rio de Janeiro - RJ, Brazil
T: 55 21 246-8003

Patel, Mustak, Mr.
WORLD MUSLIM CONGRESS
Est. do Donde 1692/Apto.201 - Ilha
Rio de Janeiro - RJ, Brazil
T: 55 21 260-8759 F: 55 21 260-8759

Patrocinio, Maria Angelica Torres Do
HOSPITAL ADVENTISTA SILVESTRE
Ladeira dos Guararapes, 263 - Santa Tereza
Rio de Janeiro - RJ, Brazil
T: 55 21 265-0212

Paul, Leonard
Video Producer
WERY PRODUCOES
Rua Diogenes Sampaio 92/301 Humaita
Rio de Janeiro - RJ, Brazil
T: 55 21 246 3272

Paula, Dalva De
ASSOCIACAO POTIGUAR DE EDUCACAO E CULTURA
Rua Floriano Peixoto, 295 Petropolis
Natal - RN 59020, Brazil
T: 55 84 221-6255 F: 55 84 221-6255

Paula, Daniel Monteiro De
Reporter
FREELANCER
R. Barata Ribeiro 70 /1301 Copacabana
Rio de Janeiro -RJ, Brazil
T: 55 21 5421823

Paula, Jurema De
ASSOCIACAO POTIGUAR DE EDUCACAO E CULTURA
Rua Floriano Peixoto, 295 Petropolis
Natal - RN 59020, Brazil
T: 55 84 221-6255 F: 55 84 221-6255

Paula, Paulo De
ASSOCIACAO POTIGUAR DE EDUCACAO E CULTURA
Rua Floriano Peixoto, 295 Petropolis
Natal - RN 59020, Brazil
T: 55 84 221-6255 F: 55 84 221-6255

Paula Fernandes, Nivia Maria, Ms.
Professora
UNIVERSIDADE FEDERAL DO ACRE
Campus Universitario - BR 364 - KM 4
Rio Branco - AC Cep.: 69.900, Brazil
T: 55 68 226-1641

Paula Filho, Fernando Candido De, Sr.
Membro
APOENA
Av.Washington Luiz, 318
Presid. Prudente-SP 19010, Brazil
T: 55 182 22-0187

Pavao, Marilene L. Goncalves, Sra.
PLENUM, ASS. CONS. LTDA.
Av. do Contorno, 5445/601
Belo Horizonte - MG 30110, Brazil
T: 55 31 227-7978

Paveri, Manuel, Mr.
ONU - FAO
Shin Qi 12-4-29
Brasilia - DF 20152, Brazil
T: 55 61 225-2487 F: 55 61 236-4936

Pavia, Jose Antonio Gollo
ETHOS COMUNICACAO
Rio de Janeiro, Brazil

Pavlick, Inez, Sra.
PREFEITURA MUNICIPAL DE PORTO ALEGRE
Av. Vista Alegre 59/302
Poa - RS 91040, Brazil
T: 55 51 341-4604

Paz, Ana, Ms.
Environment Desk
COMUNIDADE BAHA'I
Rua Luis Primola da Silva 57, Bancarios
Joao Pessoa - PB 58000, Brazil
T: 55 235-1565

Pecanha, Ana Lucia Ferreira Viana
Editora
REDE BANDEIRANTES DE TELEVISAO
R.Alvaro Ramos, 492
Rio de Janeiro - RJ, Brazil
T: 55 21 2122072 F: 55 21 2957391

Pecanha, Durval Pereira Gomes
Director
JORNAL COMMUNITARIO 'O POVAO'
Rua Aristides Caire, 19, Olavo Bilar
Duque de Caxias RJ, Brazil
T: 55 7722860

Pecorone, Maria Izabel, Sra.
Rua Pompeu Loureiro, 44 apto 102
Rio de Janeiro - RJ 22061, Brazil
T: 55 21 255-7753

Pecorone, Sonia Penha
Diretoria
CEPS
Rua Cipriano Da Silveira, 96 - Santa Eugenia
Nova Iguacu - RJ 26.285, Brazil
T: 55 21 768-6114

Pedrao, Fernando
ECOTROPIC
1880 East River Road
Rio de Janeiro - RJ, Brazil
T: 55 21 577-6476 F: 55 21 577-1195

Pedrazzi, Antenor Henrique P., Sr.
IECAM
Dr. Moreira 250
Manaus - AM 69000, Brazil

Pedreiro, Carlos Antonio Da Silva
MANCHETE
R. do Russel 804
Rio de Janeiro -RJ 2, Brazil
T: 55 21 2652012 F: 55 21 2855148

Pedro, Claudia Sao
CHICO JR. & ASSOCIADOS
Rio de Janeiro, Brazil

Pedro, Jonathan De Almeida
Interpreter
UNIRHMA
PO Box 275
Barra do Garcas MT, Brazil
T: 55 4463316

Pedrosa, Wilson Expedito
Photographer
AGENCIA ESTADO
Av. Caetano Alvares, 55
Sao Paulo -SP 02598, Brazil
T: 55 11 8562210 F: 55 11 2652297

Pedroso, Marcelo
PREFEITURA MUNICIPAL
Praca Maua s/n 1 andar
Santos - SP 11010, Brazil
T: 55 132 342-2093 F: 55 132 345-809
TL: 131048

Pedroso, Nelson
ASS. COMUNITARIA DE SAO BERNARDO DO CAMPO
R. dos Vianas, 194 - Centro
S. Bernardo do SP 09700, Brazil
T: 55 11448-8000 R

Pedroza, Fabricio, Mr.
Arquiteto
ATA - TECNOLOGIAS ALTERNATIVAS
Sqs 204 Bloco C Apto 204
Brasilia - DF, Brazil
T: 55 61 226-2447

Pedroza, Marilia Sylvia, Ms.
Alvaro Chaves 44 apto 802 Laranjeiras
Rio de Janeiro - RJ 22231, Brazil
T: 55 21 5531007

Pedrozo, Tania Barcellos Rodgrigues L., Sra.
Assessora De Comunicacao Socia
SOCIEDADE PSICANALITICA DO RIO DE JANEIRO
Rua Fernandes Guimaraes 92
Rio de Janeiro - RJ, Brazil
T: 55 21 295-3248 F: 55 21 295-3148

Peixoto, Hugo De Sa
Reporter
REDE GLOBO
R. Lopes Quintas 303
Rio de Janeiro -RJ 2, Brazil
T: 55 21 2947732

Peixoto, Luiz Evangelista, Sr.
Secretario
SECRETARIA MUNICIPAL MEIO AMBIENTE
Praca Sao Goncalo 18
Contagem 32010, Brazil
T: 031 351 2555 F: 351 7144

Peixoto, Ovivido Mota
Editor
RADIO TUPI
Rua Silvino Martins 188 Centro de Sao Joao Meriti
Rio de Janeiro, Brazil
T: 55 21 2032421 F: 55 21 2637323

Peixoto, Saudia Helena Barboza
Reporter
FREELANCER
Rua General Claudio, 227
Rio de Janeiro, Brazil
T: 55 21 2456534

Peixto, Eliane Fernandes Dos Santos
Reporter
JORNAL FOLHA DIRIDA
Rua Candido Mendes 263/503, Cinelandia
Rio de Janeiro, Brazil
T: 55 21 2624122 F: 55 21 2409414

Pelacani, Guiomar, Sra.
CMB
Rua Avanhadava no. 756
Sao Paulo - SP 01306, Brazil
T: 55 11 2564354 F: 55 11 2577648

Pelicioni, Andrea, Sra.
Supervisora
EDITORA GLOBO
Rua do Curtume, 665 - 2$ andar Bloco C
Sao Paulo - SP, Brazil
T: 55 11 874-6010 F: 55 11 864-0271

Pelliccione, Andre Luis
SINDICATO NACIONAL DOS AEROVIARIOS
Rio de Janeiro, Brazil

Pellini, Rodolfo, Sr.
FLAMA
Rua Senador Vergueiro, 52 - Ap. 14 - Flamengo
Rio de Janeiro - RJ 22230, Brazil
T: 55 21 265-1363

Pen, Jeane, Sra.
FUNDACAO S.O.S. MATA ATLANTICA
Avenida 9 de Julho, no. 4877 - 8$ andar
Sao Paulo - SP 01407, Brazil
T: 55 11 853-5905 F: 55 11 885-1680

Pena, Helio De La
REDE GLOBO
R. Lopes Quintas 303 J.Botanico
Rio de Janeiro RJ 22, Brazil
T: 55 21 2947732

Pena, Lidia Maria Goncalves
Assessora De Imprensa
SOCIEDADE CIVIL DO BEM-ESTAR FAMILIAR NO BRASIL -BEMFAM
Av. Republica do Chile 230/ 17-andar
Rio de Janeiro -RJ, Brazil
T: 55 21 2102448 F: 55 21 2204057

Penha, Eri Pimenta Da
FUNDACAO SOS RIO PIRACICABA
Av. Brasil # 35 Iguacu
Ipatinga - MG 35160, Brazil
T: 55 31 821-2897 F: 55 31 821-1994

Penido, Antonio
CHICO JR. & ASSOCIADOS
Rio de Janeiro, Brazil

Penido, Dirce M. Soares, Sra.
Librarian
Av. Padre Anchieta 160 - Bairro Carioca
Cambuquira - MG 37420, Brazil
T: 55 35 251-1482

Penido, Jose Marcio
Editor
REDE GLOBO
R. Lopes Quintas 303
Rio de Janeiro -RJ 2, Brazil
T: 55 21 2947732

Penna, Leonardo Miranda
Technician
REDE GLOBO
R. Lopes Quintas 303
Rio de Janeiro -RJ 2, Brazil
T: 55 21 2947732

Penna, Tadeu De Araujo, Mr.
FLAMA
Praiá do Flamengo, 22
Rio de Janeiro - RJ, Brazil
T: 55 21 265-5044

Pepeira, Carmen Lucia Ribeiro
Radio Roquete Pinto
Rio de Janeiro, Brazil
T: 55 21 2200897

Peppe, Atilio Machado, Mr.
Pesquisador
IBRADES/ CENTRO JOAO XXIII
Rua Bambina 115 - Botafogo
Rio de Janeiro - RJ CEP: 22.251, Brazil
T: 55 21 286-8522

Perego, Jose Neder
Camera
TVSBT
Campo de S.Cristovao, 105
Rio de Janeiro - RJ, Brazil
T: 55 21 5803688 / 5800678 F: 55 21 5809685

Pereira, Alex D. C., Mr.
Geologo
Av. Antonio Carlos Magalhaes, 3840 - 4 andar - Iguatemi
Salvador - BA Cep.: 41.820, Brazil
T: 55 71 359-8677 F: 55 71 359-8054
TL: 711918

Pereira, Andrea Mendes
ASSOCIACAO DOS CORRESPONDENTES DE IMPRENSA ESTRANGEIRA
Rio de Janeiro, Brazil

Pereira, Angela Peixoto, Sra.
Professora Ikebana
FUNDACAO MOKITI-OKADA - IGREJA MESSIANICA MUNDIAL
Rua Itabaiana, 74
Rio de Janeiro - RJ 20561, Brazil
T: 55 21 577-4433 F: 55 21 577-5524

Pereira, Celso Sanchez, Mr.
Estudante
Rua Dona Amelia, 30/202 Andarai
Rio de Janeiro - RJ 20541, Brazil
T: 55 21 571-7189

Pereira, Claudia Basdos
SEC. MEIO AMBIENTE - PA
Rua Jose Higino 254, ap. 101 - Tijuca
Rio de Janeiro, Brazil
T: 55 21 5718301

Pereira, Decio Cesar Cantro
Assistant Director
BLOCH EDITORES
Rua do Russel 804, 8 andar
Rio de Janeiro, Brazil
T: 55 21 2050295 F: 55 21 2059998

Pereira, Dirce
Reporter Fotografica
TEXACO
Rua Candido Mendes 140/107 Gloria
Rio de Janeiro - RJ, Brazil
T: 55 21 232 6366

Pereira, Elizabete Tshcerne, Sra.
Assessora Imprensa
SIND. DOS TRABALHADORES NAS IND. QUIM. E FARMACEUTICAS DE S.PAULO
Rua Tamandare 348 - Liberdade
Sao Paulo - SP 01525, Brazil
T: 55 11 279-3811 F: 55 11 279-0662

Pereira, Elizabeth Mendes
ASSOCIACAO DOS CORRESPONDENTES DE IMPRENSA ESTRANGEIRA
Rio de Janeiro, Brazil

Pereira, Francisca Das Chagas, Ms.
Socia
SINDICATO TRABALHADORES RURAIS DE ESPERANTINA
Rua Mal. Deodoro 668
Esperantina - PI 64280000, Brazil
T: 383-1213

Pereira, Geraldo Jordao, Sr.
Diretor-Executivo
CISAT
Av Nilo Pecanha 155/301
Rio de Janeiro - RJ 20020, Brazil
T: 55 21 240 6306 F: 55 21 240 4755

Pereira, Gilson Gomes
Cameraman
AMERICA DO SUL PRODUCOES DE DOCUMENTARIOS LTDA (NAMBEI TSUSHIN)
Av. Paulista 509 /11 andar conj.1104 Cerqueira Cesar
Sao Paulo -SP 01311, Brazil
T: 55 11 2886141 F: 55 11 2512065

Pereira, Giselda Almeida, Ms.
WORLD COUNCIL OF CHURCHES
150 route de Ferney
Brazil
F: 41 22 791 0361

Pereira, Heloisa, Sra.
Fazenda Pindorama cxp 03
Sao Simao 75890, Brazil

Pereira, Henrique Dos Santos, Sr.
Representative
INSTITUTO DE ESTUDOS AMAZONICOS E AMBIENTAIS
Rua Men de Sa 705 DJ Dom Pedras
Manaus - AM 69040, Brazil
T: 55 92 238-1728

Pereira, Hilda Pacheco De Campos, Dr.
Advogada
ASSOCIACAO DAS DIPLOMADAS BRASILEIRAS
Travessa Do Carmo 50
Porto Alegre - RS, Brazil

Pereira, Hipolito Vieira
Photographer
AGENCIA ESTADO
Av. Caetano Alvares, 55
Sao Paulo -SP 02598, Brazil
T: 55 11 8562210 F: 55 11 2652297

Pereira, Ivanoe Gomes
Photographer
FREELANCER
Praca Floriana 91 28 andar
Rio de Janeiro, Brazil
T: 55 21 2624695

Pereira, Jackeline Braga De B.
SINDICATO NACIONAL DOS AEROVIARIOS
Rio de Janeiro, Brazil

Pereira, Jorge Luiz Da Silva
Editor
BLOCH
Rua do Russel, 766
Rio de Janeiro, Brazil
T: 55 21 2850033

Pereira, Jose Alberto
ESTADO DE SAO PAULO
Av Rio Branco 151
Rio de Janeiro, Brazil
T: 55 21 2217111

Pereira, Jose Carlos
ESTADO DE SAO PAULO
Av. Rio Branco 151
Rio de Janeiro RJ, Brazil
T: 55 21 2217111

Pereira, Jose H.Paulo
Tecnico
TVSBT
Campo de S.Cristovao, 105
Rio de Janeiro - RJ, Brazil
T: 55 21 5803688 / 5800678 F: 55 21 5809685

Pereira, Laerte Willmann, Sr.
Therapist
LABORATORIO DA FLORESTA
Corrego do Maculo, S/N$ - Lumiar - 5$ Distrito
Nova Friburgo - RJ 28635, Brazil

Pereira, Liesio, Mr.
CANADA
Shis Q 115 casa 13
Brasilia - DF, Brazil
T: 55 61 321-2171 F: 55 61 321-4529

Pereira, Luiz Otavio Mota, Mr.
Presidente
ABES
Trafalgar 476 - Barrio Alta Cordoba, C.P. 5001
Cordoba, Brazil
T: 54 51 739851

Pereira, Mandel Filomeno
Reporter
FREELANCER
Praia do Flamengo, 19/505 Flamengo
Rio de Janeiro RJ 22, Brazil
T: 55 21 2455784

Pereira, Marcia Erthal Rocha
JORNAL O FLUMINENSE
Niteroi, Brazil

Pereira, Marcia Lombardi De Mattos
Jornalista
FREELANCER
Rua Esteves Junior 70/104 Laranjeiras
Rio de Janeiro - RJ, Brazil
T: 55 21 245 3535

Pereira, Marco Aurelio, Sr.
Presidente
ASSOCIACAO DOS AMIGOS DA TERRA - ECOFILIA
Rua Jayme Calado - Lote 2 - Casa 4 - Lagoinha
Alcantara - RJ, Brazil
T: 55 21 463-2645

Pereira, Marcos
TWENTY FIRST CENTURY INC.
Rio de Janeiro, Brazil

Pereira, Maria Do Rosario, Ms.
GRUPO ANASTACIA
Rua Joao Manoel Vaz, 11
Sao Paulo - SP, Brazil

Pereira, Maria Regina Peixoto
Editor
BLOCH EDITORES
Rua do Russel 766
Rio de Janeiro, Brazil
T: 55 21 5554271

Pereira, Mariana, Sra.
Director
CMB
Rua Avanhadava no. 756
Sao Paulo - SP 01306, Brazil
T: 55 11 2564354 F: 55 11 2577648

Pereira, Mauri, Mr.
SECRETARIA DO MEIO AMBIENTE (PR)
Rua Eng. Reboucas 2206
Curitiba - PR, Brazil
T: 55 41 224-1864 F: 55 41 225-3411

Pereira, Mauricio Broinizi, Mr.
CIOSL - ORIT
Rua 24 de Maio, no 77 cj 902
Sao Paulo - SP 01041, Brazil
T: 55 11 223-2174 F: 55 11 223-2153

Pereira, Miguel
Director
ABIA -ASSOCIACAO BRASILEIRA INTERDISCIPLINAR DE AIDS
R. Lopes Quintas 576 J. Botanico
Rio de Janeiro -RJ 2, Brazil
T: 55 21 2315171 F: 55 21 2945602

Pereira, Paulo Henrique De Souza, Sr.
Engenheiro
GEONEX DO BRASIL SERVICOS TECNICOS LTDA
Rua Estacio Coimbra no. 50
Rio de Janeiro - RJ, Brazil
T: 55 21 286-7036 F: 55 21 266-1944

Pereira, Regina Francisca, Sra.
I B G E
Rua Fonte Da Saudade 280/103- Lagoa
Rio de Janeiro - RJ 22471, Brazil
T: 55 21 226-1306

Pereira, Roberto Rodrigues
Technician
REDE GLOBO
R. Lopes Quintas 303
Rio de Janeiro -RJ 2, Brazil
T: 55 21 2947732

Pereira, Valeria De Fatima Gomes, Ms.
Chefe Dedivisao
FUNTAC
Av. das Acacias, lote 1 zona a
Rio Branco - AC 69.900, Brazil
T: 55 68 226-1667 F: 55 68 226-1667

Peres, Cecilia Antinori, Ms.
FRATERNIDADE PAX UNIVERSAL
Ri Santo Afonso, 252
Cotia - SP 06700, Brazil
T: 55 11 4926039 F: 55 11 4922507

Perez, Luiz Eduardo
O POVO
Rio de Janeiro, Brazil

Perez, Marcos Godoy
Editor
EDITORA MULTIMIDIA
Rua Riachuelo, 221 Cob.03
Rio de Janeiro - RJ, Brazil
T: 55 21 2223192 / 2215894

Perez-Golzalez, Maria Dolores, Ms.
GRUPO TORTURA NUNCA MAIS
R. Ribeiro de Almeida 46/401
Rio de Janeiro - RJ, Brazil
T: 55 21 285-0733

Periotto, Adalgiza
Reporter
LEGIAO DA BOA VONTADE
Av.Rudge, 700
Sao Paulo - SP, Brazil
T: 55 11 2202599 F: 55 11 2205196

Perozim, Marly Madalena, Sra.
IECAM
Dr. Moreira 250
Manaus - AM 69000, Brazil

Perruci, Arthur Emilio Da Costa, Sr.
Bolsista
INSTITUTO DE ECOLOGIA HUMANA
Av. Dom Manuel de Medeiros, s/n Ed. Newton Bancks-3th floor
Recife - PE 52071, Brazil
T: 55 81 4413931 F: 55 81 4414697

Perruso, Mari, Sra.
Director
CMB
Rua Avanhadava no. 756
Sao Paulo - SP 01306, Brazil
T: 55 11 2564354 F: 55 11 2577648

Perry, Katrina, Ms.
Relacoes Publicas
HOPE OF THE FUTURE
Rua Hugo Leal, 67 - Ilha do Governador
Rio de Janeiro - RJ, Brazil
T: 55 21 393-0129

Peruzini, Luiz Tadeu
Technician
REDE GLOBO
R. Lopes Quintas 303
Rio de Janeiro -RJ 2, Brazil
T: 55 21 2947732

Pestana, Ronaldo
MANCHETE
Rio de Janeiro, Brazil

Peters, Donald Gordon, Mr.
Country Representative
MENNONITE CENTRAL COMMITTEE
Caixa Postal 111
Recife - PE, Brazil
T: 55 81 228-4865

Petersen, Sonia, Ms.
AMERICAN SOCIETY OF RIO DE JANEIRO
Rua: Bolivar, 84 - apto. 603 - Copacabana
Rio de Janeiro - RJ, Brazil
T: 55 21 256 0464

Petr, Emil A., Sr.
ASSOCIACAO POTIGUAR AMIGOS DA NATUREZA
Sitio Nebraska
S.Gonc.do Amarantes RN 59290, Brazil
T: 278-2275 F: 219-3366

Petrone, Ana Ligia Barboza
INTERNATIONAL NGOs AND SOCIAL MOVEMENTS FORUM
Rio de Janeiro, Brazil

Pezeshk, Venus S., Ms.
ASSEMBLEIA ESPIRITUAL NACIONAL DOS BAHA'IS DO BRASIL
Caixa Postal-7035
Brasilia - DF 71619, Brazil
T: 55 61 248 - 4718 F: 55 61 248 -4321
TL: 7366 AEN BRA, E: PA 1008

Pfeiffer, Florian
Cinegrafista
WDR GERMAN TV CHANNEL
Rua Eca de Queiroz, 479 casa 12
Sao Paulo - SP, Brazil
T: 55 11 5721262 F: 55 11 5728959

Phebo, Luciana R., Sra.
Coord.Editoracao
ASSOCIACAO BRASILEIRA MULTIPROFISSIONAL DE PROTECAO A INFANCIA E ADOLECENCIA
Rua Pinheiro Machado, 39 - Laranjeiras
Rio de Janeiro - RJ 22231, Brazil
T: 55 21 553-8181 F: 55 21 552-5829

Philippi, Arlindo, Sr.
IBAMA
Av. Jeronimo Monteiro, 240 - Sala 811/812 Centro
Vitoria - ES 29045, Brazil
T: 55 27 322-4193 F: 55 27 222-3725

Phillips, M.Ann, Ms.
ONTARIO COALITION OF VISIBLE MINORITY WOMEN
Alameda Trete 191 apt. 42
Sao Paulo - SP 01417, Brazil
T: 55 11 646755

Philot, Almir Taveira
CONFEDERACAO NACIONAL DA INDUSTRIA
Rio de Janeiro, Brazil

Philot, Anibal
Fotografo
JORNAL O GLOBO
Rua Irineu Marinho, 35
Rio de Janeiro - RJ, Brazil
T: 55 21 2722000

Picalho, Martha
Photographer
FREELANCER
Rua Visconde de Piraja 259/801
Rio de Janeiro RJ 22, Brazil
T: 55 21 5534648

Picanco, Clodoaldo De Araujo, Sr.
PREF. MUN. DE APIACAS
Av. Principal 15, Centro
Apiacas - MT 78595, Brazil

Pierce, Susan Marie, Ms.
Representante
LEGACY INTERNATIONAL
Fundacao Vitoria Amazonica, Av. Darcy Vargas, Chapada 520
Manaus - AM 69050, Brazil
T: 55 92 642-1336

Pietro, Marcio Di
GAIA - ASSOC.CONS.M.AMBIENTE E PROD.INTEGRADA ALIM.AMAZONIA
Arse-51 - 39
Palmas, TO 77000, Brazil
T: 063 8621553/1108

Pietro, Ricardo
REVISTA PRO NATURA
Rua do Russel 450/802
Rio de Janeiro - RJ, Brazil
T: 55 21 205 4907 F: 55 21 285 6096

Pignato, Vanda, Sra.
Secretaria Executiva
SERVICO UNIVERSITARIO MUNDIAL
Caixa Postal 14.631
Sao Paulo - SP 03698, Brazil
T: 55 11 263-5804 F: 55 11 63-5804

Pimentel, Germana Polimeni, Sra.
Pesquisadora
GRUPO DE MAMIFEROS AQUATICOS NORDESTE
Rua Santo Elias, 404 - 1103 - Espinheiro
Recife - PE 52.020, Brazil
T: 55 81 241-5699

Pinage, Leila
Reporter
RADIO ROQUETE PINTO
Av. Erasmo Braga 118/ 11andar Castelo
Rio de Janeiro -RJ 2, Brazil
T: 55 21 2247727 F: 55 21 2524584

Pinder, Maria Izabel, Sra.
PROJETO LINHA VERDE - P.M. S.SEBASTIAO
Prefeitura Municipal de Sao Sebastiao, Secretaria de Cultura e Turismo
Sao Sebastiao - SP, Brazil

Ping, Fang Chang
Reporter
AGENCIA NOTICIOSA XINHUA
Rua Paissandu,7 apto.401
Rio de Janeiro - RJ, Brazil
T: 55 21 285 0590 / 285 06 03

Pinheiro, Alexandre, Mr.
Filer
JOVEM TURISMO
SCLN 202 Bloco B Lojas 7/11
Brasilia - DF, Brazil
T: 55 61 224-5526 F: 55 61 321-0345
TL: 61-4705

Pinheiro, Ana Maria Dettow
UNIAO EM DEFESA DA NATUREZA
Rua Granja Julieta 345
Sao Paulo - SP 04721, Brazil
T: 55 11 246-9111 F: 55 11 246-9354

Pinheiro, Aparecido Nunes
Journalist
AGENCIA GILLVAN
Av 13 Maio No 475, 1110 Cinelandia
Rio de Janeiro, Brazil
T: 55 21 2205400

Pinheiro, Fatima Lucia Angelo
JORNAL O FLUMINENSE
Niteroi, Brazil

Pinheiro, Flavio
Editor Executivo
EDITORA ABRIL
R.da Passagem, 123/8 andar
Rio de Janeiro - RJ, Brazil
F: 55 21 2759347

Pinheiro, Joao, Jr., Sr.
Advogado
SUREHMA
Rua Engenheiros Reboucas, 1206 - Capanema
Curitiba - PR 80210, Brazil
T: 55 41 225 3411

Pinheiro, Joelma Da Silva
RADIO UNIVERSITARIA
Universidade Federal de Goias - Setor Oeste Av Anhanguera com Alameda das Rosas
Goiania - GO, Brazil
T: 55 62 233-4473/6896

Pinheiro, Jose Ivan Melo
Techinician
REDE GLOBO
R. Lopes Quintas 303
Rio de Janeiro -RJ 2, Brazil
T: 55 21 2947732

Pinheiro, Julia Maria, Sra.
Tecnica
CONFEDERACAO NACIONAL DA INDUSTRIA
Av. Nilo Pecanha, 50 sala 3018
Rio de Janeiro - RJ 20044, Brazil
T: 55 21 262 0181 F: 55 21 262 1465
TL: 22634

Pinheiro, Maria Hilma Abrahim, Sra.
Dir. Cont. Amb.
INSTITUTO DE MEIO AMBIENTE DO ACRE
Rua 16 de Outubro, 428 - Bairro XV
Rio Branco - AC 69.900, Brazil
T: 55 68 224 4202 F: 55 68 224 5694
TL: 68.2692

Pinheiro, Mauricio, Sr.
C D O / M J
Sais 7, Lt 23, SPS Asa Sul
Brasilia - DF 71000, Brazil
T: 55 61 245-7425

Pinheiro, Miguel Angelo De Brito
CONFEDERACAO NACIONAL DA INDUSTRIA
Rio de Janeiro, Brazil

Pinheiro, Vera Lucia, Ms.
Teacher
UFMT/NUEPOM
Rua 01, 640, Boa Esperanca
Cuiaba - MT 78090, Brazil
T: 55 125 361-5316

Pinho, Flavia
CHICO JR. & ASSOCIADOS
Rio de Janeiro, Brazil

Pinho, Roberto Costa, Mr.
Coordenador
PROJETO QUADRILATERO DE PORTO SEGURO TRANCOSO
Povoado de Trancoso
Porto Seguro 45.820, Brazil
T: 071 359 6440

Pinho Filho, Paulo, Sr.
Reasercher
ASFEEMA
Rua Fonseca Teles 58 - Sao Cristovao
Rio de Janeiro - RJ 20.940, Brazil
T: 55 21 284-2772

Pinter, Miguel, Sr.
ASS. CONS. M. AMB. PROD. INTEGR. ALIMEN. AM-GAIA
Arse-51 - 39
Palmas, TO 77000, Brazil
T: 063 8621553/1108

Pinto, Ana Ma. Ferreyra Leite, Sra.
Associada
FAMNEP
Rua Sargento Baracho 155
Guaratingueta -SP 12500, Brazil
T: 22 2426 F: 22 2700

Pinto, Ana Maria Barros, Ms.
Member
ECOFUND
Cel. Genuino, 342/410
Porto Alegre - RS 90010, Brazil
T: 55 512 226.0330 F: 55 512 222.8591

Pinto, Arley, Sr.
RiBaiac de Sagurra
Campinas - SP 13015, Brazil

Pinto, Carlos Luiz
Technician
REDE GLOBO
R. Lopes Quintas 303
Rio de Janeiro -RJ 2, Brazil
T: 55 21 2947732

Pinto, Carmenlu
ECOTROPIC
1880 East River Road
Rio de Janeiro - RJ, Brazil
T: 55 21 577-6476 F: 55 21 577-1195

Pinto, Jorge Lucio De C.
CONFEDERACAO NACIONAL DA INDUSTRIA
Rio de Janeiro, Brazil

Pinto, Maina Dalva Miguel, Sra.
Auxiliar
LIGA DA L. NACIONAL
Rua Andre Cavalcanti, 161
Rio de Janeiro - RJ 20231, Brazil
T: 55 21 252-0188

Pinto, Marcos Andre
Fotografo
JORNAL O GLOBO
Rua Irineu Marinho, 35
Rio de Janeiro - RJ, Brazil
T: 55 21 2722000

Pinto, Maria Conceicao Correa, Sra.
Membro
NUCLEO DE ESTUDOS SOBRE A MULHER (NEM) PUC/RJ
Rua Domingos Ferreira 100/1101
Rio de Janeiro - RJ 22050, Brazil
T: 55 21 255-4815

Pinto, Maria Lucia, Sra.
INSTITUICAO RELIGIOSA PL
Rua General Barbosa Lima 47 Copacabana
Brazil

Pinto, Odilon Ferreira Leite, Sr.
FAMN E P
Rua Sargento Baracho 155
Guaratinguera SP 12500, Brazil
T: 012522 2426 F: 012522 2700

Pinto, Oswaldo Wilson, Sr.
INSTITUTO PRO NATURA
Rua Monte Alegre 374 - Santa Teresa
Rio de Janeiro - RJ 20240, Brazil
T: 55 21 224-39904 F: 55 21 224-3994

Pinto, Paulo Cezar, Sr.
Presidente Conselho Fiscal
AIDIS
Av. Beira Mar, 216 - 13 Andar
Rio de Janeiro - RJ 20021, Brazil
T: 55 21 210-3221 F: 55 21 262-6838
TL: 21-31902

Pinto, Paulo Wander Gerutti, Mr.
Projects Manager
CENTRO DE PESQUISAS DO MAR (CEPEMAR)
Rua Mexico, 31/Gr. 501- Centro
Rio de Janeiro - RJ, Brazil
T: 55 21 240-3822 F: 55 21 240-9516

Pinto, Ricardo Costa, Sr.
Socio
SOCIEDADE NORDESTINA DE ECOLOGIA
Rua Pessoa de Melo, 355 - Madalena
Recife - PE 50710, Brazil
T: 55 81 227-2487 F: 55 81 222-3361

Pinto, Salvador Santos, Sr.
ASSEMBLEIA LEGISLATIVA
Praca Moreira Cabral S/N
Cuiaba - MT 78000, Brazil
T: 55 65 316-8268

Pinto, Wanderley Lisboa, Mr.
Diretor
JETRO
Praia de Botafogo, 228, Setor B, sala 809
Rio de Janeiro - RJ 22250, Brazil
T: 55 21 551-8144 F: 55 21 552-1196

Pinudo, Marcos Samuel
MANCHETE
R. do Russel 804
Rio de Janeiro -RJ 2, Brazil
T: 55 21 2652012 F: 55 21 2855148

Pires, Arnaldo Dias Cardoso, Sr.
Diretor Presidente
SEAERJ
Rua do Russel no.01 - Gloria
Rio de Janeiro - RJ, Brazil
T: 55 21 205-6797 F: 55 21 205-2795

Pires, Jadson De Araujo, Sr.
Conselheiro
ABES
Av. Beira Mar, 216 - 13 Andar
Rio de Janeiro - RJ 20021, Brazil
T: 55 21 210-3221 F: 55 21 262-6838
TL: 21-31902

Pires, Lorena Fornari De Ary, Ms.
Engenheira
ELETROBRAS
SAv. Pres. Vargas 409- 12 Andar Centro
Rio de Janeiro - RJ, Brazil
T: 55 21 232-8344

Pires, Maria America De Carvalho, Sra.
Coord. Geral
CENPLA
Rua Tiradentes, 215-sobrado - INGA
Niteroi - RJ, Brazil
F: 55 21 285-5503

Pires, Maria Conceicao Santana, Sra.
Biologa
IBAMA-SUPES-BA
Av Juracy Maganhaes Jr. 608 - Rio Vermelho
Salvador - BA 41930, Brazil
T: 55 71 240-7322 F: 55 71 240-7913

Pires, Paula Moreira
Editor
REDE GLOBO
R. Lopes Quintas 303
Rio de Janeiro -RJ 2, Brazil
T: 55 21 2947732

Pires, Reinaldo Bouzan
Cameraman
CNN SPANISH
R. Lopes Quintas 732 Jardim Botanico
Rio de Janeiro -RJ, Brazil
T: 55 21 5115139 F: 55 21 2397977

Pires, Rodrigo Fiume Pinheiro
RADIO UNIVERSITARIA
Universidade Federal de Goias - Setor Oeste Av Anhanguera com Alameda das Rosas
Goiania - GO, Brazil
T: 55 62 233-4473/6896

Pires Filho, Ivon, Sr.
Treasurer
SOCIEDADE NORDESTINA DE ECOLOGIA
Rua Pessoa de Melo, 355 - Madalena
Recife - PE 50710, Brazil
T: 55 81 227-2487 F: 55 81 222-3361

Piubel, Maximino Salzmann
Tecnico De Audio
SISTEMA GLOBO DE RADIO
Rua do Russel, 434
Rio de Janeiro - RJ, Brazil
T: 55 21 2058282 F: 55 21 2054096

Pizarro, Angelica
RIO LIFE
Est da Gavea 611/Bl 2/304 Sao Conrado
Rio de Janeiro - RJ, Brazil
T: 55 21 322 2290 F: 55 21 322 2990

Pizarro, James, Mr.
Vereador
CAMARA DE VEREADORES
Rua Gomes Carneiro,409
Santa Maria 97.050, Brazil
T: 55 21 221 5190

Platz, Eraldo
JORNAL MAIORIA FALANTE
Rio de Janeiro, Brazil

Plotkin, Mark, Mr.
CONSERVATION INTERNATIONAL
c/o Sonia kRigueira, Rua Bueno Brandao, 393 - Santa Tereza
Belo Horizonte - MG 31010, Brazil
T: 55 31 226-5145 F: 55 31 222-8429

Pokorny, Bread
Jornalista
COMUNIDADE BAHA'I DO BRASIL
Rua Olegario Maciel 2403 Bairro Saraiva
Uberlandia MG 38400, Brazil
T: 55 34 215 4320 F: 55 34 236 92

Polette, Marcus, Mr.
Oceanografo
SECRETARIA DO MEIO AMBIENTE
Rua: Tabapua, 81
Sao Paulo - SP 04533, Brazil
T: 55 11 883 0766 F: 55 11 280 5468

Pollak, Harrison, Mr.
Researcher
IMAZON
Rua Presidente Vargas, 257
Belem - PA 66000, Brazil
T: 55 91 223-1009 F: 55 91 235-4214

Pompco, Fernanda
Iluminadora
COMULHER - COMUNICACAO MULHER
Rua Rocha, 119/504
Sao Paulo - SP, Brazil
T: 55 11 2515626

Pompeia, Silvia Maria, Sra.
Tecnico Master I
ORGANIZACAO PREFEITO FARIA LIMA - CEPAM
Av. Prof. Lineu Prestes, 913
Sao Paulo - SP 05508, Brazil
T: 55 11 212-3144 F: 55 11 813-5969
TL: (11)83141

Pompeu, Kathia
Jornalista
REVISTA MANCHETE
Rua do Russel, 804
Rio de Janeiro - RJ, Brazil
T: 55 21 2652012

Pontes, Jose Renato Perpetuo
Journalist/Photographer
FREELANCER
R. Marques de Parana, 62 apt. 602 Flamengo
Rio de Janeiro -RJ, Brazil
T: 55 21 5515882

Pontes, Josecelia Rangel, Sra.
Geografa
SUDEMA - GOVERNAMENTAL
Rua Mons Walfredo Leal 181 - Tambia
Joao Pessoa - PB 58000, Brazil
T: 55 83 222-1647 F: 55 83 224-0068

Pontes, Norma Bahia, Sra.
Representante No Rio
SOCIEDADE DE AMIGOS DA AMAZONIA BRASILEIRA
Caixa Postal 44093 - Copacabana
Rio de Janeiro - RJ 22062, Brazil
T: 55 21 257-4594

Pontes, Otavio, Sr.
President
PLASTIVIDA
Rua Alexandre Dumas, 2016
Sao Paulo - SP 04707, Brazil
T: 55 11 247-6911 F: 55 11 247-8981
TL: 011/57690, E: 51583

Pontes, Suzana Pastuk, Sra.
Administradora De Projeto
INTERNATIONAL ASSOCIATION FOR ECOLOGICAL ECONOMICS
CPDA, Av. Presidente Vargas, 417 - 8o. andar
Rio de Janeiro - RJ 20071, Brazil
T: 55 21 252-3001 F: 55 21 252-3001

Pontes Lins, Janizete Rangel, Sra.
Geografa
SUDEMA - GOVERNAMENTAL
Rua Ester Borges Bastos 722 Jaguaube
Joao Pessoa - PB 58000, Brazil
T: 55 83 221-6446 F: 55 83 224-0068

Pontual, Andrea Coutinho, Sra.
U F P E
Av. Beira Rio 650 - Apt. 501 - Madalena
Recife - PE 50610, Brazil
T: 55 81 228-1190

Pontual, Francisco, Mr.
R. Pacheco Leao 1510 Apto. 502 Horto
Rio de Janeiro - RJ 22460, Brazil
T: 55 21 294 7027 F: 55 21 294 5600

Poole, Alan, Mr.
Diretor
ECOTROPIC
Rua Almirante Alexandrino 2161 - Santa Teresa
Rio de Janeiro - RJ 20241, Brazil
T: 55 21 259-7181 F: 55 21 242 8939

Poppovic, Malak El Chichini
PLANETA FEMEA
Hotel Gloria, Sala 462
Rio de Janeiro, Brazil
T: 55 21 2855949 F: 55 21 2054114

Porfirio, Wladimir
Reporter
RADIO ROQETE PINTO
Rio de Janeiro, Brazil

Porphirio, Ismael Thomaz, Sr.
Assessor
TECNOOP INFORMATICA
Rodolfo Coutinho 7/47
Rio de Janeiro - RJ 20231, Brazil
T: 55 21 280-1922

Porpino, Francisca Lucia Botelho, Mr.
CIA VALE DO RIO DOCE
Av. Graca Aranha, 26
Rio de Janeiro - RJ 20.030, Brazil
T: 55 21 272 4401 F: 55 21 220 0593

Porro, Noemi Sakiago Miyasbe, Mr.
Assessora
ASSEMA
Rua das Laranjeiras, 1436
Pedreiras - MA, Brazil
T: 98 642-2061

Porto Neto, Jose Carlos, Mr.
PREFEITURA DE PARATY
Rua Dr. Samuel Costa 29
Paraty - RJ 23970, Brazil
T: 55 243 71-1266 F: 55 243 71-2105

Portugal, Sonia, Sra.
Biologa
ECOSS CONSULTORIA E PLAN. AMBIENTAL
Av. Tancredo Neves, 274 Bl. B Sala 320
Salvador - BA, Brazil
T: 55 71 3591589

Posey, Darrell
GLOBAL VISION
Museu Goeldi
Belem, Para, Brazil
T: 55 91 2480153 F: 55 91 2480601

Posner, Andrew Roy
FREELANCER
Rio de Janeiro, Brazil

Potiguara, Eliane, Sra.
GRUPO MULHER - ED.INDIGENA - GRUMIN
Rua da Quitanda, 185/503
Rio de Janeiro - RJ 20091, Brazil
T: 55 21293-1745 R

Pozzo, Maria Dal, Ms.
Membro
CEQMTR-PR
Rua Almirante Tamandare, 434
Curitiba - PR 80050, Brazil
T: 55 41 2634936

Prado, Barbara, Sra.
IF - SMA
Estrada Do Guarau
Paraiba 11750, Brazil
T: 013295-2053

Prado, Luiz Da Silva
Photographer
AGENCIA ESTADO
Av. Caetano Alvares, 55
Sao Paulo -SP 02598, Brazil
T: 55 11 8562210 F: 55 11 2652297

Prado, Regina, Sra.
Presidente
ASSOCIACAO DE MORADORES / AMIGOS DE LARANJEIRAS
Rua Pinheiro Machado 31 - sobrado
Rio de Janeiro - RJ 22231, Brazil
T: 55 21 553-5941

Prado, Rosane M., Ms.
CIESIN
R. Maestro F. Braga 350/306
Rio de Janeiro - RJ 22041, Brazil
T: 55 21 325 2514

Prado Filho, Jose Francisco, Mr.
Teacher
UNIV. FEA. OURO PRETO
Praca Tiradentes, 20 - Escola de Minas
Ouro Preto-MG, Brazil
T: 55 31 551-2786 F: 55 31 551-1689
TL: 312954

Prasildananda, Dada, Sr.
AMURT: ANANDA MARGA UNIVERSAL RELIEF TEAM
Rua Alfredo Zumkeller 59, Parque Mandaqui
Sao Paulo - SP 02421, Brazil
T: 55 11 204-7954 F: 55 11 298-6571

Prates, Ana Paula Leite, Ms.
Pos-Graduando
EESC-USP
Rua Sergio Fleury 24
Sao Carlos - SP 13560, Brazil
T: 55 162 719 531

Pratginestos, Juan Antonio Gili
Photographer
WWF BRAZIL - WORLD WIDE FUND FOR NATURE
SHIS QI 07 Bl/B Sala 204
Brasilia -DF 71615, Brazil
T: 61 2482899 F: 61 2487176

Preiwitz, Hienz
Photographer
LETTRE DU BRESIL
Av.Presidente Antonio Carlos 58/10andar Castelo
Rio de Janeiro -RJ, Brazil
T: 55 21 2201015

Prem Aja, Ma
OSHO SPIRITUAL HEALTH ORGAN. - OSHO
Estr. das Canoas 1476, casa 14
Rio de Janeiro - RJ 22610, Brazil
T: 55 21 322-1920

Prem Astiko, Ma, Ms.
OSHO SPIRITUAL HEALTH ORGAN. - OSHO
Rua Raimundo Correia 60/901
Rio de Janeiro - RJ 22040, Brazil
T: 55 21 255-0892

Prest, Valeria Inez
Journalist
HILL AND KNOWLTON
Av. Afranio de Melo Franco 135 /5 andar Leblon
Rio de Janeiro -RJ 2, Brazil
F: 55 21 2394697

Preussler, Lara
ASSOCIACAO BRASILEIRA DE COMUNIDADES ALTERNATIVAS (ABRASCA)
Brasilia, Brazil

Prichard, Nancy Beth
TWENTY FIRST CENTURY INC.
Rio de Janeiro, Brazil

Proepper, Heribert
ASSOCIATED PRESS
Rio de Janeiro, Brazil

Protasio, Paulo Manoel, Sr.
ASSOCIACAO COMERCIAL DO RIO DE JANEIRO
Rio de Janeiro - RJ, Brazil

Prudencio, Joao Carlos Agostinho
JORNAL MOVIMENTACAO
Rio de Janeiro, Brazil

Publio, Sergio
SISTEMA JORNAL DO BRASIL
Av. Brasil 500
Rio de Janeiro -RJ, Brazil
T: 55 21 5854422 F: 55 21 5854428

Puerta, Marcos Jesus Boaretti, Sr.
Membro
APOENA
Rua Dr.Gurgel, 860
Presid. Prudente-SP 19015-140, Brazil
T: 55 182 21-4222 F: 55 182 22-6177

Puga, Antonio Augusto
Reporter
O FLUMINENSE
Rua Visconde de Itaborai, 184
Niteroi, Brazil
T: 7178368

Pulcherio, Maria Isabel Capistrano, Sra.
Gerente
ABES
Av. Beira Mar,, 216 - 13 Andar
Rio de Janeiro - RJ 20021, Brazil
T: 55 21 210-3221 F: 55 21 210-3221
TL: 21-31902

Punjabi, Mohini
Journalist
BRAHMA KUMARIS
R. Pinheiro da Cunha 185 Tijuca
Rio de Janeiro -RJ, Brazil
T: 55 21 5715268 F: 55 11 8727838

Puyol, Nadine, Ms.
Filosofia
Av. Pasteur 493/09
Rio de Janeiro - RJ 22290, Brazil
T: 55 21 542 8725

Quartarone, Heloisa
GLOBO ECOLOGIA
Rio de Janeiro, Brazil

Quartin, Francesco, Sr.
Rua Pacheco Leao 1510 - Apt. 502
Rio de Janeiro - RJ, Brazil

Queiros, Lelia Maria Almeida
REVISTA ENFERMAGEM CIENTIFICA
Rio de Janeiro, Brazil

Queiroz, Almir De
Reporter Cinematografico
PROVIDEO CINEMA E TV
SHIS Q1 21 Conj 7 casa 15 Lago Sul
Brasilia - DF, Brazil
T: 55 61 366 1851 F: 55 61 366 3182

Queiroz, Ana Carolina, Sr.
Monitora
UFRJ / INSTITUTO DE CIENCIAS BIOLOGICAS
R. Cosme Velito 412 casa 6
Rio de Janeiro - RJ 22241, Brazil
T: 55 21 265 1965

Queiroz, Antonieta Amaral, Sr.
FLAMA
Praia do Flamengo, 66 bl.B s/1214
Rio de Janeiro - RJ, Brazil
T: 55 21 205-9696

Queiroz, Antonio Carlos Paulino De
Editor
JORNAL PORANTIM -CONSELHO INDIGENISTA MISSIONARIO
SCS Ed. Venancio III S/311 C.P.03679
Brasilia -DF 70084, Brazil
T: 61 2259457 F: 61 2259401

Queiroz, Georgina, Sra.
Director
CMB
Rua Avanhadava no. 756
Sao Paulo - SP 01306, Brazil
T: 55 11 2564354 F: 55 11 2577648

Queiroz, Joao De Deus Freire, Sr.
Representante
CONFEA
Rua Mons. Joao da Mata 29
Natal - RN 59075, Brazil
T: 55 84 231-6118

Queiroz, Jose Eurides P.
Editor
Sindicto Dos Bancarios
Av. Pres. Vargas 502/16 Andar
Rio De Janeiro, Brazil
T: 021 2332113

Queiroz, Renato Olympio De
Technician
REDE GLOBO
R. Lopes Quintas 303
Rio de Janeiro -RJ 2, Brazil
T: 55 21 2947732

Queiroz, Victoria Graca Graun, Sra.
ASSOCIACAO DE MULHERES DE NEGOCIOS E PROFISSIONAIS
Avenida Afonso Pena, 3850
Campo Grande - MS 79010, Brazil
T: 55 67 624-3160

Quenneville, Serge, Mr.
Counselor On Environment
CSN - CONFEDERATION DES SYNDICATS NATIONAUX
Caixa Postal 1759, Centro
Sao Paulo - SP 01059, Brazil
T: 55 11 270-1257

Queroi, Nilton Joaquim
Photograoher
FOLHA METROPOLITANO
Rua Gal Osario, 60
Sao Paulo SP, Brazil
T: 55 11 9689199 F: 55 11 2094787

Quinaud, Evandro, Mr.
Rua Abade Ramos 29, Apto. 302
Rio de Janeiro - RJ 22461, Brazil
T: 55 21 266-7509

Quintela, Maria Fernanda Alves P.
Journalist
FUND. CULTURAL DE RADIODIFUSAO EDUCATIVA DA REGIAO DOS LAGOS
Av. Jose Bento Ribeiro Dantas 265 Manguinhos 28905
Armacao dos Buzios -, Brazil
T: 55 246 2753848 F: 55 21 2956494

Rabello, Fernando Augusto Martins
FREELANCER
Rio de Janeiro, Brazil

Rabello, Joao Bosco B.
Director
AGENCIA ESTADO
Av. Caetano Alvares, 55
Sao Paulo -SP 02598, Brazil
T: 55 11 8562210 F: 55 11 2652297

Rabello, Luiz Henrique Alvarenga
Technician
REDE GLOBO
R. Lopes Quintas 303
Rio de Janeiro -RJ 2, Brazil
T: 55 21 2947732

Rabelo, Alvaro
FOLHA VERDE
Av. Basil 1831 sala 606
Belo Horizonte -MG, Brazil
T: 55 31 2263700 F: 55 31 2251386

Rabelo, Helio
Director
FOLHA VERDE
Av. Basil 1831 sala 606
Belo Horizonte -MG, Brazil
T: 55 31 2263700 F: 55 31 2251386

Rabi, Nidia Albessa, Sra.
INSTITUTO BRASILEIRO DE ADMINISTRACAO MUNICIPAL - IBAM
Edificio Diogo Lordello de Mello. Largo IBAM no.1. Humaita
Rio de Janeiro - RJ 22282, Brazil
T: 55 21 266-6622 F: 55 21 537-1262
TL: 22638 INBM BR

Racy, Marcelo, Sr.
CONTREN
Av. Angelica no. 589 - Higienopolis
Sao Paulo - SP, Brazil
T: 55 11 211-6011 F: 55 11 279-0662

Radwanski, Adriana E.N., Ms.
ECOTROPIC
Rua. Almirante Alexandrino 2161
Rio de Janeiro - RJ 20241, Brazil
T: 55 21 232-3956 F: 55 21 242-8939

Radwanski, Eduardo Nascimento, Mr.
Oceanographer
CENTRO DE PESQUISAS DO MAR (CEPEMAR)
Rua Mexico, 31/Gr. 501- Centro
Rio de Janeiro - RJ, Brazil
T: 55 21 240-3822 F: 55 21 240-9516

Radwanski, Henrique, Dr.
CLINICA IVO PITANGUY
Av. VISCONDE DE ALBUQUERQUE 570 LEBLON
Rio de Janeiro - RJ, Brazil
T: 55 21 274-5776

Ragonha Wehmuth, Evaldo, Dr.
Professor
INSTITUTO DE GEOCIENCIAS E CIENCIAS EXATAS
Av. 24-A No. 1515
Rio Claro - SP 13500, Brazil
T: 55 195 35-1839

Ramalho, Celuta Cardoso, Sra.
Vice President
ORDEM DOS ADVOGADOS VICE PRESIDENTE NITEROI DH
Av. Amaral Peixoto, 507 - 11andar
Niteroi - RJ 24020, Brazil
T: 55 21 719-8470

Ramalho, Luiz, Sr.
Coordenador
SACTES
CP 1655
Recife - PE, Brazil
T: 55 81 2210075

Ramalho, Marisa, Sra.
Director
CMB
Rua Avanhadava no. 756
Sao Paulo - SP 01306, Brazil
T: 55 11 2564354 F: 55 11 2577648

Ramalho, Pedro Alexandre
Technician
REDE GLOBO
R. Lopes Quintas 303
Rio de Janeiro -RJ 2, Brazil
T: 55 21 2947732

Ramid, Joao
Photographic Reporter
DIARIO DO PARA
Av. Gaspar Viana, 773
Belem - Para, Brazil
T: 55 91 2227388 F: 55 91 2243162

Ramirez, Manuel, Sr.
CONSERVATION INTERNATIONAL
c/o Sonia Rigueira, Rua Bueno Brandao, 393 - Santa Tereza
Belo Horizonte - MG 31010, Brazil
T: 55 31 226-5145 F: 55 31 222-8429

Ramos, Adriana Oliveira, Sra.
FUNDACAO AMAZONAS FOREVER GREEN
Av. Giovanni Gronchi 6186 - Morumbi
Sao Paulo - SP 05724, Brazil
T: 55 11 523-2993 F: 55 11 246-8092

Ramos, Alvaro Alfredo
MANCHETE
Rio de Janeiro, Brazil

Ramos, Alzira, Sra.
INSTITUTO DE DIVULGACAO DA AMAZONIA
Trav. Sao Pedro 566 sala 1005
Belem - PA 66010, Brazil
T: 55 91 223-8908 F: 55 91 222-7786

Ramos, Cristiano Viega, Sr.
Bolsista
INSTITUTO DE ECOLOGIA HUMANA
Av. Dom Manuel de Medeiros s/n - Ed. Newton Bancks - 3o. andar
Recife - PE 52071, Brazil
T: 55 81 441-3931 F: 55 81 441-4697

Ramos, Dirceo Torrecillas, Sr.
Governor
LIONS INTERNACIONAL L-4
Av. Ipiranga, 978 - 1o. andar
Sao Paulo - SP 01040, Brazil
T: 55 11 222 1566 F: 55 11 223 9753

Ramos, Gilberto Correa Ferreira, Sr.
WISA COMPANY
R. Lauro Muller ,116 1801/1801 Torre Do Rio Sul Botafogo
Rio de Janeiro - RJ 22290, Brazil
T: 55 21 5418595 F: 55 21 5426345

Ramos, Haroldo
TWENTY FIRST CENTURY INC.
Rio de Janeiro, Brazil

Ramos, Jeanete Marion Ramos, Sra.
Chanceler
AUSU ASSOCIACAO UNIVERSITARIA SANTA URSULA
Rua Jornalista Orlando Dantas, 59
Rio de Janeiro - RJ, Brazil
T: 55 21 552-5422 F: 55 21 551-6446

Ramos, Lucia Waleska L. F.
INTERNATIONAL NGOs AND SOCIAL MOVEMENTS FORUM
Rio de Janeiro, Brazil

Ramos, Wilson Nogueira
Technician
REDE GLOBO
R. Lopes Quintas 303
Rio de Janeiro -RJ 2, Brazil
T: 55 21 2947732

Ramos Filho, Artur, Sr.
SECRETARIO MUNICIPAL DE SAUDE DE JUSCIMEIRA MT
Rua n.210 Casus
Jucimeira - MT, Brazil
T: 55 65 412 1182

Ranali, Virginia Conceicao Amorim, Sra.
Director Dpt. Of Parks And Ga.
PREFEITURA MUNICIPAL DE GUARULHOS
Rua Darci Vargas, no. 80 - 6o. andar - apt.62 / Jardim Zaira
Guarulhos - SP 07090, Brazil
T: 55 11 968-0011 F: 55 11 940-5166

Rangel, Claudio Truccolo, Sr.
Assessor
F.A.F.G.
Av Giovani Gronghi 6186
Sao Paulo - SP, Brazil
T: 55 11 323 2993 F: 55 11 246 8092

Rangel, Maria Do Carmo Martins, Sra.
Coordenadora De Eventos
BCO. INTERAMERICANO DE DESENVOLVIMENTO
Caixa Postal 08 738
Brasilia - DF 70312, Brazil
T: 55 61 321-7848 F: 55 61 321-3112
TL: 61-1204

Ransel, Antonio Jose
Director
OFICINA CULTURAL RIO
Silveira Martins 136/201
Rio de Janeiro, Brazil
T: 55 21 5562761

Raphael, Aura Lopes, Sra.
Ladeira Nossa Senhora 176/52
Rio de Janeiro - RJ 22211, Brazil
T: 55 21 265-4086

Raposo, Claudio A.
OSHO SPIRITUAL HEALTH ORGANIZATION
Rua Paissandu, 385/304
Rio de Janeiro - RJ 22.210, Brazil
T: 55 21 553-2419

Ratton, Eduardo, Sr.
ARQUIPELAGO DE FERNANDO DE NORONHA
Rua Barao de Sao Borja, 243 - Boa Vista
Recife - PE 50070, Brazil
T: 55 81 221-3217 F: 55 81 231-1594
TL: 81.2533

Raviolo, Daniel Gerardo, Sr.
Secretario Executivo
COMUNICACAO E CULTURA
Rua Castro e Silva 121 s/110
Fortaleza - CE 60030, Brazil
T: 55 85 231-6092

Raw, Anthony, Mr.
REBRAF / ECOTEC
Rio de Janeiro - RJ, Brazil

Rebello, Gilson
ASSOCIACAO COMERCIAL DO RIO DE JANEIRO
Rio de Janeiro, Brazil

Rebelo, Cesar De Freitas
Technician
REDE GLOBO
R. Lopes Quintas 303
Rio de Janeiro -RJ 2, Brazil
T: 55 21 2947732

Recca, Lucio
Repr. Do Bid
BCO. INTERAMERICANO DE DESENVOLVIMENTO
Caixa Postal 08738
Brasilia - DF 70312, Brazil
T: 55 61 321-7848 F: 55 61 321-3112
TL: 61-1204

Rech, Deotila Berta, Ms.
Coord. Da Past. Da Crianca
CNBB/SUL 3
R. Gancalo Carvalho 390
Porto Alegre - RS, Brazil
T: 55 512 2284090

Redondo, Odair
TVT REDE DE COMUNICACAO DOS TRABALHADORES
Sao Paulo, Brazil

Rega, Paulo Cavalcanti
Photographic Reporter
JORNAL GRAJAU
Rua Barrao de Mesquita, 779 Anjaret
Rio de Janeiro, Brazil
T: 55 21 26882898

Rego, Claudio Guilherme Soares
MANCHETE
R. do Russel 804
Rio de Janeiro -RJ 2, Brazil
T: 55 21 2652012 F: 55 21 2855148

Rego, Patricia Amorim, Sra.
Ass. Juridica
INSTITUTO DE MEIO AMBIENTE DO ACRE
Rua: Rui Barbosa, 450 - Centro
Rio Branco - AC 69.900, Brazil
T: 55 68 224 2857 F: 55 68 224 5694

Rego, Sergio Ricardo Fonseca, Sr.
BRADESCO VISA
AV. RIO BRANCO 131 - 3o. ANDAR
Rio de Janeiro - RJ 20021, Brazil
T: 55 21 224-0031

Rego, Terezinha De Jesus Almeida S., Ms.
IECAM
Av. AQ. 9 C/3 - Caixa Postal 1129
Calhau - AM 65070, Brazil

Regua, Marcelo
SISTEMA JORNAL DO BRASIL
Av. Brasil 500
Rio de Janeiro -RJ, Brazil
T: 55 21 5854422 F: 55 21 5854428

Regueira, Nadia
ASS. DESENVOLVIMENTO COMUNITARIO CAXAMBU
Rua do Bomfim 24
Pirenopolis - GO 72980, Brazil
T: 062 331-1316

Reichel, Lilian De Souza
Photographer
FREELANCER
R. Miguel Lemos 126 apt.201 Copacabana
Rio de Janeiro -RJ, Brazil
T: 55 21 2575480

Reichel, Sigmar, Sr.
Exec. Secretary
AMENCAR
Sao Leopoldo - RS, Brazil
T: 55 512 592-8111

Reigota, Marcos, Sr.
Researcher - Professor
UNICAMP - NEPAM
Caixa Postal 1170 - Nepam
Campinas - SP 13081, Brazil
T: 55 192 398151 F: 55 192 394717

Reilly, Patricia, Ms.
CONSERVATION INTERNATIONAL
Rua Bueno Brandao 393
Belo Horizonte - MG 31010, Brazil
T: 55 31 226-5145 F: 55 31 222-8429

Reinaud, Axel, Mr.
ICI AGROCHEMICALS
c/p Pro-Natura, Inst. Bras.Pesq.Est.Amb. - Rua Monte Agre, 374 Santa Tereza
Rio de Janeiro - RJ 20240, Brazil
T: 55 21 252-1191 F: 55 21 222-6742

Reinaud, Guy, Mr.
Business Director
ICI AGROCHEMICALS
c/p Pro-Natura, Inst. Bras.Pesq.Est.Amb. - Rua Monte Agre, 374 Santa Tereza
Rio de Janeiro - RJ 20240, Brazil
T: 55 21 252-1191 F: 55 21 222-6742

Reis, Aloisio Clovis Dos, Sr.
Vice-Presidente
ABES
Av. Beira Mar, 216 - 13o. andar
Rio de Janeiro - RJ 20021, Brazil
T: 55 21 210-3221 F: 55 21 262-6838
TL: 21-31902

Reis, Carmem Lucia, Ms.
Diretora
MAE DIVINA IMP./EXP. LTDA
Rua Carvalho Alvin, 624 Casa 7
Rio de Janeiro - RJ 20510-100, Brazil
T: 55 21 278 4288

Reis, Claudio Henrique, Sr.
Expositor
LASERE - UFF
Instituto de Geociencias Av. Litoranea s/n Boa Viagem
Niteroi - RJ 24020, Brazil
T: 55 21 7194241

Reis, Eliana, Sra.
Director
CMB
Rua Avanhadava no. 756
Sao Paulo - SP 01306, Brazil
T: 55 11 2564354 F: 55 11 2577648

Reis, Heloisa Stella Marini Dos, Sra.
Presidente Da Diretoria
SABALEIA
Rua Baltazar da Veiga no. 367 apt.21
Sao Paulo - SP 04510, Brazil
T: 55 11 533-8399 F: 55 11 229-6191

Reis, Jorge
TRIBUNA DA IMPRENSA
Rio de Janeiro, Brazil

Reis, Junior M. Jose R.
Editor
RADIOBRAS
Brasilia, Brazil

Reis, Matheus Alexandre, Sr.
Membro
APOENA
Rua Jose Colnago, 39
Regente Feijo 19670-000, Brazil
T: 055 018242-1845

Reis, Monica Torres Maia Dos, Sra.
Journalist
APOIO MARKETING E EDITORA LTDA.
SAAN - Quadra 03 - Lote 70/80
Brasilia - DF 71220-000, Brazil
T: 55 61 234-8181 F: 55 61 234-8131
TL: 2147

Reis, Nilo Del Cueto De, Mr.
Membro Delegado
ECOFUND
Olavo Barreto Viana, 66 apt.201
Porto Alegre - RS 90460, Brazil
T: 55 512 222-3542 F: 55 512 222-8591

Reis, Roberto Mello
Technician
REDE GLOBO
R. Lopes Quintas 303
Rio de Janeiro -RJ 2, Brazil
T: 55 21 2947732

Reis, Rogerio
Editor
SISTEMA JORNAL DO BRASIL
Av. Brasil 500
Rio de Janeiro -RJ, Brazil
T: 55 21 5854422 F: 55 21 5854428

Reis, Sebastiao P. Dos
Reporter
AGENCIA ESTADO
Av. Caetano Alvares, 55
Sao Paulo -SP 02598, Brazil
T: 55 11 8562210 F: 55 11 2652297

Reis, Solange, Sra.
Director
CMB
Rua Avanhadava no. 756
Sao Paulo - SP 01306, Brazil
T: 55 11 2564354 F: 55 11 2577648

Reksodihardjo, Wertit Soegeng, Dr.
INTERNATIONAL SEED TRADE FEDERATION
Rua Espirito Santo, 1678
Londrina PR 86020, Brazil
F: 55 432247-980

Remigio, Mantovan, Mr.
Member
C.C.A.P.
Rua Ipiranga 25/801
Rio de Janeiro - RJ, Brazil
T: 55 21 265-3713

Renato, Bacci Junio
Editor
RADIOBRAS
Brasilia, Brazil

Renggli, Urs, Mr.
Eng.
P & M EQUIPAMENTOS E TECNOLOGIA
Rua Rio De Janeiro, 591 Ap. 1002 Pituba
Salvador - BA 41.830, Brazil
T: 55 71 240-1176 F: 55 71 392-3369
TL: 07113718

Renno, Marilia Dos Santos Dias, Ms.
Estagiaria
PINHEIRO NETO ADVOGADOS
AV. RIO BRANCO 131 9o. ANDAR
Rio De Janeiro-RJ 20040, Brazil
T: 2215877 F: 2423292
TL: 23220

Rerreira, Renato
Journalist/Reporter
FREELANCER
R.Pompeia 65 apt. 503 Copacabana
Rio de Janeiro -RJ, Brazil
T: 55 21 2279442

Resende, Luis Eduardo
Reporter
SISTEMA JORNAL DO BRASIL
Av. Brasil 500
Rio de Janeiro -RJ, Brazil
T: 55 21 5854422 F: 55 21 5854428

Resende, Marco Antonio
SISTEMA JORNAL DO BRASIL
Av. Brasil 500
Rio de Janeiro -RJ, Brazil
T: 55 21 5854422 F: 55 21 5854428

Resende, Roberto, Dr.
ASSOCIACAO MUNDIAL DE ESPERANTO
pr. Mira 196-44
Moscow 129128, Brazil
T: 7 95 18-70222 F: 7 95 292-6511 684
TL: 411700 6845

Resende, Wilma M.
Reporter
JJN NOTICIAS
Praca Almirante Jaceguai 72 apt. 815
Rio de Janeiro -RJ, Brazil
T: 55 21 282-1292 F: 55 21 240-3073

Rey, Jesus Alvarez
Editor
AGENCIA EFE S/A
Avenida Rio Branco 25/13 andar
Rio de Janeiro -RJ, Brazil
T: 55 21 2534465 F: 55 21 2534861

Rezende, Divaldo Jose Da Costa, Sr.
ASS. CONS. M. AMB. PROD. INTEGR. ALIMEN. AM - GAIA
Caixa Postal no. 84
Palmas - TO 77000, Brazil

Rezende, Jose Perez, Sr.
Sup. Juridico
UNIVERSIDADE SANTA URSULA
Rua Jornalista Orlando Dntas, 59
Rio de Janeiro - RJ 22231, Brazil
T: 55 21 551-5422 F: 55 21 551-6446

Rezende, Maura Lucia M., Sra.
Rua General Glicerio 407/1002
Rio de Janeiro - RJ 22251, Brazil
T: 55 21 225-2402

Rezende, Ranier Nardi, Mr.
Exhibitor
GEA-EXO
Av.Hum, 78
Belo Horizonte - MG 31950, Brazil
T: 55 31 337-8451 F: 55 31 493-3069

Rezende, Vania
Editor
SISTEMA JORNAL DO BRASIL
Av. Brasil 500
Rio de Janeiro -RJ, Brazil
T: 55 21 5854422 F: 55 21 5854428

Ribas, Carlos Eduardo Gurgel
Manager
INSTITUTO DE RADIODIFUSAO EDUCATIVA DA BAHIA
R.Pedro Gama 413 E Alto do Sobradinho - Federacao
Salvador -BA 4230, Brazil
T: 71 2355251 F: 71 2353871

Ribas, Tania
JORNAL ENERGIZANDO
Rio de Janeiro, Brazil

Ribeira, Andrea Rocha
Editor
GLOBOSAT
Rua Itapiru 1209
Rio de Janeiro, Brazil
T: 55 21 5023535 x222

Ribeiro, Adele R. E. Hahlbohm M., Sra.
Secretary General
SOZED
R. Getulio das Neves, 26, apto. 101 - Jardim Botanico
Rio de Janeiro - RJ, Brazil
T: 55 21 226-0266

Ribeiro, Alexandre Dorea, Sr.
AMAZONIA FOREVER GREEN
Av. Pe.A.J. dos Santos 921 - Brooklin
Sao Paulo - SP, Brazil
T: 55 11 240-2266

Ribeiro, Alfredo
Editor
EDITORA ABRIL
R.da Passagem, 123/8 andar
Rio de Janeiro - RJ, Brazil
F: 55 21 2759347

Ribeiro, Amaury, Jr.
Reporter
FOLHA DE SAO PAULO
Alameda Barao de Limeira 425
Sao Paulo SP, Brazil
T: 55 11 874 2608/874 2602 F: 55 11 2231644

Ribeiro, Ana Claudia Ricci
Produtora
TV Rio
R. Miguel de Frias, 57 Estacio
Rio de Janeiro - RJ, Brazil
T: 55 21 2939888 F: 55 21 2938626

Ribeiro, Andrea Valenca
Reporter
SISTEMA GLOBO DE RADIO
R do Russel 434 /401 Gloria
Rio de Janeiro - RJ, Brazil
T: 55 21 205 8282 r 410/429 F: 55 21 2054096

Ribeiro, Angelina Marques, Ms.
Student
RAINBOW OVER SOUTH AMERICA
c/o M. da Veiga Greuel, R. de el Salvador 208
Sao Paulo - SP 04910, Brazil
T: 55 11 514-5714

Ribeiro, Bianca
MACACO PREGO
1880 East River Road
Rio de Janeiro - RJ, Brazil
T: 55 21 577-6476 F: 55 21 577-1195

Ribeiro, Carlos Dimas Martins, Mr.
Assessor Pedagogico
FUNDACAO FE E ALEGRIA
Rua Sao Clemente, 226 - Botafogo
Rio de Janeiro - RJ 22260, Brazil
T: 55 21 537-2621

Ribeiro, Carlos Roberto Griesbach
Director
RAINBOW VIDEO
Rua Aylton Vasconcelos, 290
Rio de Janeiro, Brazil
T: 55 21 3932927 F: 55 21 4631849

Ribeiro, Darcy, Dr.
GOVERNO DO ESTADO DO RIO DE JANEIRO
Rio de Janeiro - RJ, Brazil

Ribeiro, Eduardo Rivail
RADIO UNIVERSITARIA
Universidade Federal de Goias - Setor Oeste Av Anhanguera com Alameda das Rosas
Goiania - GO, Brazil
T: 55 62 233-4473/6896

Ribeiro, Esther Rodrigues Magalhaes, Ms.
ASSOCIACAO CIVIL GREENPEACE
Rua Mexico 21, Grp 1301, Centro
Rio de Janeiro - RJ 20031, Brazil
T: 55 21 262-7318 F: 55 21 240-1690

Ribeiro, Geraldo Do Carmo
O POVO
Rio de Janeiro, Brazil

Ribeiro, Ilka Dos Santos, Ms.
Tecnico (administracao)
IDESP
Av. Nazare 817
Belem - PA 66035, Brazil
T: 55 91 224-4411 F: 55 91 225-3414
TL: 6905

Ribeiro, Jose Roberto Soares, Sr.
Student
PONTIFICA UNIVERSIDADE CATOLICA
Rua Sao Joaquim, 441
Sao Carlos - SP 13560, Brazil
T: 55 162 718025

Ribeiro, Luis Antonio
Technician
REDE GLOBO
R. Lopes Quintas 303
Rio de Janeiro -RJ 2, Brazil
T: 55 21 2947732

Ribeiro, Malu
Reporter
JORNAL URTIGA
Av. Brig. Luiz Antonio 4667 Itu C.P. 83 13300
Sao Paulo -SP 01401, Brazil
T: 55 11 8872423/r.8 F: 55 11 8844835

Ribeiro, Maria Claudia, Sra.
Expositor
CIEE
Av. Maracana, 1524/1528 - Tijuca
Rio de Janeiro - RJ 20511, Brazil
T: 55 21 268-6447 F: 55 21 288-0748
TL: 34763

Ribeiro, Maria De Fatima, Ms.
Professora
CENTRO DE ESTUDOS AMBIENTAIS
Rua 25, Quadra 27 'A' Casa 4
Foz Iguacu - PR 85890, Brazil
T: 55 455 733093

Ribeiro, Milton Roberto Monteiro
Photographer
IMAGENS DA TERRA (AGENCIA)
Rua Joaquim Murtinho 886/204
Rio de Janeiro RJ 2, Brazil
T: 55 21 2222838

Ribeiro, Nilce, Sra.
IECAM
Dr. Moreira 250
Manaus - AM 69000, Brazil

Ribeiro, Paulo Lazaro
Technician
REDE GLOBO
R. Lopes Quintas 303
Rio de Janeiro -RJ 2, Brazil
T: 55 21 2947732

Ribeiro, Pedro Luiz, Sr.
Coordenador De Centro
INSTITUTO PRINCIPIO UNICO
R. Santa Luzia, 405/207
Rio de Janeiro - RJ 20030, Brazil
T: 55 21 262-6306

Ribeiro, Ricardo Pita
Journalist
FREELANCER
R Almirante Alexandrino 3780 Bloco D1 Apt 402 Santa Teresa
Rio de Janeiro - RJ 20241, Brazil
T: 55 21 225 7538 F: 55 21 2651539

Ribeiro, Sidney Dos Santos
Technician
REDE GLOBO
R. Lopes Quintas 303
Rio de Janeiro -RJ 2, Brazil
T: 55 21 2947732

Ribeiro, Tania Regina Santos, Sra.
Associada
GRUPO GERMEM
rua Barros Falcao 79 - Matatu
Salvador - BA 40255, Brazil
T: 55 71 2445241

Ribeiro, Vasti De Moura, Sra.
Gerente
ABES
Av. Beira Mar, 216 - 13 Andar
Rio de Janeiro - RJ 20021, Brazil
T: 55 21 210-3221 F: 55 21 262-6838
TL: 21-31902

Ribeiro Jr, Amaury
Reporter
FOLHA DE SAO PAULO
Alameda Barao de Limeira 425
Sao Paulo SP, Brazil
T: 55 11 874 2608/874 2602 F: 55 11 2231644

Ribeiropinto, Ana Lucia, Dr.
Presidente
CENTRO NACIONAL BERTHA LUTZ
Rua Sto. Afonso, 110 Grs. 605/06 - Tijuca
Rio de Janeiro - RJ, Brazil
T: 55 21 264-2888 F: 55 21 246-9342

Ribemboim, Clara Goldeman, Sra.
Assessora
ARQUIPELAGO DE FERNANDO DE NORONHA
Rua Barao de Sao Borja, 243 - Boa Vista
Recife - PE 50070, Brazil
T: 55 81 221-3217 F: 55 81 231-1594
TL: 81.2533

Ricardo, Nilton
Fotografo
REVISTA MANCHETE
Rua do Russel, 804
Rio de Janeiro - RJ, Brazil
T: 55 21 2652012

Ricardo, Vagner
ASSOCIACAO COMERCIAL DO RIO DE JANEIRO
Rio de Janeiro, Brazil

Riehs, Leonor, Sr.
Sec. Mun. De Educacao & Cultur
PREFEITURA MUNICIPAL DE IMBE
Av. Paraguacu 1607
Imbe - RS 95625, Brazil
T: 55 666-2655

Rigo, Sergio Rocha
Technician
REDE GLOBO
R. Lopes Quintas 303
Rio de Janeiro -RJ 2, Brazil
T: 55 21 2947732

Rigon, Elide, Sra.
Cons. Federal
O A B / M S
Rua 15 de Novembro no. 230, apt. 103
Campo Grande- MS, Brazil
T: 382 9440 F: 3844955
TL: 79100

Rigueira, Itamar, Jr.
Editor
EMBRATUR
Rua Marie e Barros, 13 Pca da Bandeira
Rio de Janeiro, Brazil
T: 55 21 2933042 F: 55 21 2933929

Rigueira, Sonia, Sra.
Representative
CONSERVATION INTERNATIONAL
Rua Bueno Brandao 393
Belo Horizonte - MG 31010, Brazil
T: 55 31 226-5145 F: 55 31 222-8429

Rio, Christina, Ms.
Member Of The Board
ARTISTS UNITED FOR NATURE E.V.
Rio de Janeiro - RJ, Brazil

Rioja, Guillermo, Mr.
CONSERVATION INTERNATIONAL
c/o Sonia Rigueira, Rua Bueno Brandao, 393 - Santa Tereza
Belo Horizonte - MG 31010, Brazil
T: 55 31 226-5145 F: 55 31 222-8429

Rios, Lenimar Goicalves, Sra.
PREFEITURA MUNICIPAL
Praca Maua s/n 1 andar
Santos - SP 11010, Brazil
T: 55 132 342-2093 F: 55 132 345-809
TL: 131048

Rivera, Leonardo
EDITORA TERCEIRO MUNDO
Rio de Janeiro, Brazil

Rme, Guilher
Redacao
GAZETA MERCANTIL
Rua Major Quedinho 90
Sao Paulo SP, Brazil
T: 55 11 256 3133

Robell, Suzanne
Photographer
IL MANIFESTO
Rua Girassol 1317/44
Sao Paulo SP 05433, Brazil
T: 55 11 815 4528 F: 55 11 825-78

Robiyn, , Sr.
Fundador E Diretor Presidente
RENASKIGI
Caixa Postal 106.051
Niteroi - RJ 24232, Brazil
T: 55 21 711-5251 F: 55 21 616-1355

Rocco, Jorge Luiz Silva, Mr.
Eng. Sanitarista
CETESB
Rua Treze de Maio, 35 Apto. 71
Limeira - SP 13480-270, Brazil
T: 0194423777

Rocha, Alvaro Pereira Da
SISTEMA GLOBO DE RADIO
R. de Russel 434
Rio de Janeiro RJ, Brazil
T: 55 21 2058282 F: 55 21 2054096

Rocha, Cesar
Professor
MOVIMENTO ECOLOGICO
Av.Nilo Pecanha, 12 Gr.801
Rio de Janeiro, RJ, Brazil
T: 021 2424374

Rocha, Clayton, Sr.
Assessor De Comunicacao
UNIVERSIDADE FEDERAL DE PELOTAS / PREFEITURA DE STA VITORIA DO PALMAR
Campus Universitario de Capao do Leao - Reitoria
Pelotas-RS, Brazil
T: 55 532 21-5103 F: 55 532 21-5023

Rocha, Clea Carpi Da, Sra.
Vice President
MOVIMENTO EM DEFESA DA VIDA
Av. Independencia, 1184 ap 81
Porto Alegre - RS 90210, Brazil
T: 55 512 221 0550 F: 55 512 227 5229

Rocha, Geraldo Marcos Ferreira Da
MANCHETE
R. do Russel 804
Rio de Janeiro -RJ 2, Brazil
T: 55 21 2652012 F: 55 21 2855148

Rocha, Ibraim Jose Das Merces, Sra.
Coordinator
EXECUTIVA NORTE I DE DIREITO-CADEL-UFPA
YRod. Augusto Correa s/n
Belem - PA 66030, Brazil
T: 55 91 229-2088

Rocha, Jane Do Nascimento Da, Sra.
ARQUIPELAGO DE FERNANDO DE NORONHA
Rua Barao de Sao Borja, 243 - Boa Vista
Recife - PE 50070, Brazil
T: 55 81 221-3217 F: 55 81 231-1594
TL: 81.2533

Rocha, Joao Eugenio Diaz
Journalist
ICS INSTITUTO DE COMUNICACAO SINDICAL
Rua Visconde de Inhauma, 134-Sala 715, Centro
Rio de Janeiro RJ 2, Brazil
T: 55 21 2331430 F: 55 21 2333516

Rocha, Jose Luiz
Redator
WDR GERMAN TV CHANNEL
Rua Eca de Queiroz, 479 casa 12
Sao Paulo, SP 04011, Brazil
T: 011 5721262 F: 011 5728959

Rocha, Jose Raimundo Albuquerque
Editor
PRAIAS DO NORTE JORNAL
Rua Santa Luzia 94
Salvador, Bahia, Brazil
T: 378-1234 F: 249-4514

Rocha, Ligia Moreira, Ms.
FUNDACAO PARA A CONSERVACAO E A PRODUCAO FLORESTAL DO ESTADO DE SAO PAULO
Rua Miguel Stefano 3900
Sao Paulo - SP, Brazil
T: 55 11 952-8555
TL: RAMAL 228

Rocha, Maria Da Conceicao Militao, Sra.
Bibliotecaria
FUNDACAO NACIONAL DO INDIO
Sqs 103, Bloco K - Apto. 205
Brasilia - DF 70342, Brazil
T: 55 61 225-2524

Rocha, Murilo Ferreira Da
FREELANCER
Rio de Janeiro, Brazil

Rocha, Nicola Von Hollanderjose Luiz
Redator
WDR GERMAN TV CHANNEL
Rua Eca de Queiroz, 479 casa 12
Sao Paulo - SP, Brazil
T: 55 11 5721262 F: 55 11 5728959

Rocha, Orlando Garces Da
MANCHETE
R. do Russel 804
Rio de Janeiro -RJ 2, Brazil
T: 55 21 2652012 F: 55 21 2855148

Rocha, Regina P. S. Medina Da, Ms.
LIBRA - LIGA DAS MULHERES ELEITORAS DO BRASIL
Rua Manuel da Nobrega, 627, apt 11, Paraiso
Sao Paulo - SP 04001, Brazil
T: 55 11 885-4369 F: 55 11 363-764

Rocha, Ricardo Luiz Sodre, Sr.
Geologo
DEPARTAMENTO DE RECURSOS MINERAIS
Rua Marechal Deodoro, no.351 - Centro
Niteroi - RJ 24030, Brazil
T: 55 21 722-7074 F: 55 21 719-6846

Rocha, Vania Da, Sra.
Representative
GRUPO ECOLOGICO DOS CAMPOS GERAIS
Monteiro Lobato, bloco 27A, apto. 12
Ponta Grossa - PR 84100, Brazil
T: 55 422 238852
E: 710

Rockstroh, Peter Muller, Mr.
Political Scientist
ASA - PROGBOMM
Rua Prof. Anibal de Matos 290/201 St. Antonio
Belo Horizonte - MG, Brazil
T: 55 31 3440686 F: 55 31 2737888

Rodrigues, Alexandre
Director
FREELANCER
Rua Conde da Bonfim, 702 - 306
Rio de Janeiro, Brazil
T: 55 21 5515448

Rodrigues, Almir Vendas
MANCHETE
R. do Russel 804
Rio de Janeiro -RJ 2, Brazil
T: 55 21 2652012 F: 55 21 2855148

Rodrigues, Amauri Da Silva, Sr.
Environmental Assessor
RADIO TRES RIOS
Av. Heitor Zanatta 129, Monte Castelo
Tres Rios - RJ 25800, Brazil
T: 55 242 523390

Rodrigues, Arnaldo Dos Santos
ASSOCIACAO DE RADIOS LIVRES DE SAO PAULO
Sao Paulo, Brazil

Rodrigues, Aurelio
Fotografo
REVISTA MANCHETE
Rua do Russel, 804
Rio de Janeiro - RJ, Brazil
T: 55 21 2652012

Rodrigues, Elinaldo Jose, Mr.
Press Officer
APAN
Rua Joao da Mata, s/n - Residencia Universitaria -Paraiba
Joao Pessoa - PB 58015, Brazil
T: 55 83 221-0008

Rodrigues, Ernesto Carneiro
Editor
REDE GLOBO
R. Lopes Quintas 303
Rio de Janeiro -RJ 2, Brazil
T: 55 21 2947732

Rodrigues, Evanildo, Sr.
KLABIN CELULOSE
Av. Suburbana 5000 Del Castilho
Rio de Janeiro - RJ, Brazil

Rodrigues, Fernando
Editor
FOLHA DE SAO PAULO
Alameda Barao de Limeira 425
Sao Paulo - SP, Brazil
T: 55 11 874 2608/874 2602 F: 55 11 2231644

Rodrigues, Flavio
Editor
SISTEMA JORNAL DO BRASIL
Av. Brasil 500
Rio de Janeiro -RJ, Brazil
T: 55 21 5854422 F: 55 21 5854428

Rodrigues, Florida M. Acioli, Sra.
Presidente
CENTRO NACIONAL BERTHA LUTZ
Rua Sto. Afonso, 110 Grs. 605/06 - Tijuca
Rio de Janeiro - RJ, Brazil
T: 55 21 264-2888 F: 55 21 246-9342

Rodrigues, Francisco Jose, Sr.
Gerente Financeiro
CETREL
Km 19 da BR 345 - Estrada do Coco Camacari
Camacari - BA 42810, Brazil
F: 55 71 832-2567

Rodrigues, Francisco Paulo Lucena, Sr.
Coord. De Nucleo Tematico
UNIVERSIDADE DO AMAZONAS
Rua Ramos Ferreira 1036
Manaus - AM 69001, Brazil

Rodrigues, Herth
Technician
REDE GLOBO
R. Lopes Quintas 303
Rio de Janeiro -RJ 2, Brazil
T: 55 21 2947732

Rodrigues, Joao Alfredo, Mr.
Assessor
I M A C
Rua : Rui Barbosa, 450
Rio Branco - AC 69.900, Brazil
T: 55 68 224 5497

Rodrigues, Jorge Augusto, Sr.
Pesquisador Cientifico
SOUZA CRUZ
Av. Suburbana # 2066
Rio de Janeiro - RJ 21050, Brazil
T: 55 21 281-6122 F: 55 21 201-7395
TL: (021) 31187

Rodrigues, Jose Lucio
Reporter
REDE GLOBO
R. Lopes Quintas 303
Rio de Janeiro -RJ 2, Brazil
T: 55 21 2947732

Rodrigues, Juan Jose Da
Reporter
MANCHETE
Rua do Russel, 804
Rio de Janeiro, Brazil
T: 55 21 2850033

Rodrigues, Juares Martins, Sr.
APARAI
Rua Jose Bonifacio 915
Porto Velho - RO 78900, Brazil
T: 55 68 227 1586

Rodrigues, Katia Maria De Lima, Ms.
Av. Barao Stuart 1125
Fortaleza - CE, Brazil
T: 55 85 272-2058

Rodrigues, Lino De Souza
EDITORA AZUL
Rio de Janeiro, Brazil

Rodrigues, Marcelo
Editor
REDE GLOBO
R. Lopes Quintas 303
Rio de Janeiro -RJ 2, Brazil
T: 55 21 2947732

Rodrigues, Marco Antonio Leal
MANCHETE
R. do Russel 804
Rio de Janeiro -RJ 2, Brazil
T: 55 21 2652012 F: 55 21 2855148

Rodrigues, Mario Henrique Louzada
Technician
REDE GLOBO
R. Lopes Quintas 303
Rio de Janeiro -RJ 2, Brazil
T: 55 21 2947732

Rodrigues, Maurino, Mr.
N.S. De Copacabana 793
Rio de Janeiro - RJ, Brazil

Rodrigues, Nara Costa, Sra.
Voluntaria Conferencista
MOVIMENTO EM DEFESA DA VIDA
Av. Guaiba 316
Porto Alegre - RS 91700, Brazil
T: 55 512 248-4008 F: 55 512 343-6671

Rodrigues, Paula Isabel Scherner, Sra.
MIDADE - MOVIMENTO INTERNACIONAL DO APOSTOLADO DAS CRIANCAS
Av. Beira Mar no. 48
Barra Velha - SC 88390, Brazil

Rodrigues, Paulo
Editor
SOS TERRA (FREELANCER)
R.Monte Libano, 67 apt. 206 Bairro Centro
Teresopolis -RJ 259, Brazil
T: 55 21 2426237 F: 55 21 2426237

Rodrigues, Paulo Damiao De Oliveira
Reporter
RADIO IMPRENSA FM
Rua Filipe Camarao, 96 Maracana
Rio de Janeiro, Brazil
T: 55 21 5719191

Rodrigues, Pedro
Technichan
BBC TELEVISION NEWS
Rua Tavares Bastos 414/66 Catete
Rio de Janeiro RJ 22, Brazil
T: 55 21 2255547 F: 55 21 2853109

Rodrigues, Roberto, Sr.
Presidente
ORG. COOP. BRASILEIRAS
Rua Correia Dias, 171
Sao Paulo - SP Cep.: 04104, Brazil
T: 55 11 573 6111 F: 55 11 570 0619
TL: 33686

Rodrigues, Sergio
Editor
EDITORA ABRIL
R.da Passagem, 123/8 andar
Rio de Janeiro - RJ, Brazil
F: 55 21 2759347

Rodrigues, Valeria Cristina Da S.
Reporter
O DIA
R. Riachuela 359
Rio De Janeiro, RJ 2, Brazil
T: 55 21 272-8000 F: 55 21 507-1228

Rodrigues, Vladimir Damiao
MANCHETE
R. do Russel 804
Rio de Janeiro -RJ 2, Brazil
T: 55 21 2652012 F: 55 21 2855148

Rodriguez, Carlos
INTERNATIONAL TELEVISION NEWS
Av. Rio Branco, 25/9 andar
Rio de Janeiro RJ, Brazil
T: 55 21 2335480

Rodriguez, Hector
Editor
WDR GERMAN TV CHANNEL
Rua Eca de Queiroz, 479 casa 12
Sao Paulo - SP, Brazil
T: 55 11 5721262 F: 55 11 5728959

Rodriguez, Javier
Jornalista
PRENSA LATINA
Rua Marechal Mascarenhas de Morais 121/203 Copacabana
Rio de Janeiro - RJ, Brazil
T: 55 21 256 7259

Rodriguez, Jose Maria, Sr.
TERRA CHRISTA COMMUNICATIONS
Rua Sao Clemente 373/804 Botafogo
Rio de Janeiro - RJ 22260, Brazil

Rodriguez, Karla Bruner
92 GLOBAL FORUM
Rio de Janeiro - RJ, Brazil

Rodriguez, Waldemar, Sr.
FORUM DAS ONG'S BRASILEIRAS
Palais de Nations
Geneve CH 1211, Brazil
F: 41 22 733-6542

Rodriques, Luis Antonio, Mr.
Presidente
ASSOCIACAO C.U.P.T.
Rua Jurubatuba 1019
S.Bernardo Campo-SP, Brazil
T: 55 11 443-3912

Roempczyk, Elmar, Sr.
Presidente Do Representacao Do
ILDES/FES
Rua. Lisboa 191
Sao Paulo - SP 05413, Brazil
T: 55 11 282-3900 F: 55 11 280-1916
TL: 1180616

Rogick, Valeria
THE MIAMI HERALD
Rio de Janeiro, Brazil

Rohde, Geraldo Mario, Sr.
Programa Tecn. Amb.
CIENTEC
Rua Washington Luiz, 675
Porto Alegre - RS 90010, Brazil
T: 55 512 21-4688 F: 55 51 26-0207
TL: 512009

Rohrig, Eleonora, Ms.
U I P A
Rua Piaui 77 - Apt.71
Sao Paulo - SP, Brazil
T: 55 11 255-6573

Roig, Henrique Llacer, Sr.
Coordenador Do Gaia
UNICAMP / IG / NPCT
Felipe Pattaro 82, Barao Geraldo
Campinas - SP 13083, Brazil
T: 55 192 397552

Roldan, Rosa, Sra.
Secretaria
CONSELHO NACIONAL DOS SERINGUEIROS
Rua Alexandre Farhat 206
Rio Branco - AC 69920, Brazil
T: 55 68 224-9645 F: 55 68 224-3622

Roldao, Alexandre Craveiro
Editor
REDE GLOBO
R. Lopes Quintas 303
Rio de Janeiro -RJ 2, Brazil
T: 55 21 2947732

Rolim, Claudia Celeste
Jornalista
RADIO UNIVERSITARIA FM
Rua Paula Frassinete, 42
Rio de Janeiro - RJ, Brazil
T: 55 21 2933847 F: 55 21 2930592

Rolim, Denise De Moura
O POVO
Rio de Janeiro, Brazil

Rolim, Nestor, Sr.
ASSOCIACAO COMERCIAL DO RIO DE JANEIRO
Rio de Janeiro - RJ, Brazil

Rolim, Wagner, Sr.
DNOCS
Av. Duque De Caxias, 1300
Fortaleza - CE 60000, Brazil
T: 55 85 2241090

Rolla, Maria Edith, Sra.
Bioligist
C E M I G
Av. Barbacena 1200 / 20 Andar B1
Belo Horizonte - MG 31162, Brazil
T: 55 31 349-4271

Rolla, Silvana, Ms.
Assessora
CENTRAL UNICA DOS TRABALHADORES
Rua Sao Bento 405 - 7 Andar
Sao Paulo - SP 01011, Brazil
T: 55 11 255-7500..38 F: 55 11 375-626
TL: 1121543 cutb

Roller, Mirjam, Ms.
Member
LANDESKIRCHLICHE SCHULERINNEN - UND SCHULERARBEIT
Praia do Flamengo 98/APTO 806 i CEP 22.210 Flamengo
Rio de Janeiro - RJ, Brazil
T: 55 21 225-0232

Roman, Hector San, Mr.
CIOSL - ORIT
Rua 24 de Maio, no 77 cj 902
Sao Paulo - SP 01041, Brazil
T: 55 11 223-2174 F: 55 11 223-2153

Romanini, Pedro Umberto, Mr.
CESP
Rua da Consolacao, 1875, 9 andar, Cerq. Cesar
Sao Paulo - SP 01301, Brazil
T: 55 11 258-2165 F: 55 11 256-2101

Romano, Carlos Antonio, Sr.
Diretor Financeiro Adjunto
SEAERJ
Rua do Russel no.01 - Gloria
Rio de Janeiro - RJ, Brazil
T: 55 21 205-6797 F: 55 21 205-2795

Romano, Walter C., Jr., Mr.
Aftn
MEFP
Rua Silvestre Rocha 26/901
Niteroi - RJ 24230, Brazil
T: 55 21 710-2208

Romero, Marcos Augusto, Sr.
Professor
UNIVERSIDADE FEDERAL DA PARAIBA
Rua Jose Peregrino, 251 = Centro
Joao Pessoa - PB 58020, Brazil
T: 55 83 222-2636

Romero, Reinaldo
JORNAL PAU BRASIL
Ribeirao Preto, Brazil

Romero, Susana Elizabeth De Oliveira, Sra.
Nutricionista
PREFEITURA MUNICIPAL DE JOAO PESSOA
Rua Jose Peregrino, 251 = Centro
Joao Pessoa - PB 58020, Brazil
T: 55 83 222-2636

Rong, Wang Chun
Reporter
AGENCIA NOTICIOSA XINHUA
Rua Paissandu,7 apto.401
Rio de Janeiro - RJ, Brazil
T: 55 21 285 0590 / 285 06 03

Roque, Atila
Pesquisador
IBASE
Rua Vicente de Souza 29 Botafogo
Rio de Janeiro - RJ, Brazil
T: 55 21 226 5412 F: 55 21 286 0541

Roque, Javier Rodriguez
PRENSA LATINA
R. Mascarenhas de Moraes 121/203 Copacabana
Rio de Janeiro -RJ 2, Brazil
T: 55 21 2567259

Roque, Maria Jose, Ms.
ELETROBRAS
Av. Presidente Vargas, 642/10
Rio de Janeiro - RJ 20079, Brazil
T: 55 21 263-3137 F: 55 21 233-3248

Roriz, Julio Cesar
INTERNATIONAL TELEVISION NEWS
Av. Rio Branco, 25/9 andar
Rio de Janeiro RJ, Brazil
T: 55 21 2335480

Roriz, Luiz Frederico Alfonso
Reporter
REDE BANDEIRANTES DE TELEVISAO
R.Alvaro Ramos, 492
Rio de Janeiro - RJ, Brazil
T: 55 21 2122072 F: 55 21 2957391

Rosa, Delbio Fernando
EDITORA HORA DO POVO S.A.
Rio de Janeiro, Brazil

Rosa, Fabio Luis De Oliveira, Sr.
Director
CENTRO DE COOPERACAO P/O DESENVOLVIMENTO
Rua Oswaldo Aranha, 1092 - 25
Porto Alegre - RS 90420, Brazil
T: 55 512 221-2089 F: 55 512 221-6629

Rosa, Julio Cesar Portela, Sr.
Student
UNIVERSIDADE FEDERAL DE GOIAS
Rua 207, no. 223 - Vila Nova
Goiania - GO 74000, Brazil
T: 55 62 225-9534 F: 55 62 251-1694

Rosa, M. Antonio
TERRA CHRISTA COMMUNICATIONS
Professor Manuel Ferreira 88/305 Gavea
Rio de Janeiro - RJ 22451, Brazil

Rosa, Maria Fatima, Ms.
UNIAO DE MOVIMENTOS POPULARES E MEIO AMBIENTE
Rua Jose Bezerra Filho, 183
Sao Paulo - SP, Brazil

Rosa, Nelson Benedito, Sr.
COMISSAO NACIONAL DOS TRABALHADORES EM ENERGIA NUCLEAR
Rua Tramandare,348 - Liberdade
Sao Paulo - SP 01525, Brazil
T: 55 11 279-3811.30 F: 55 11 279-3811.33

Rosa, Sylvia, Sra.
TERRA CHRISTA COMMUNICATIONS
Professor Manuel Ferreira 88/305 Gavea
Rio de Janeiro - RJ 22451, Brazil

Rosa, Vania Santa
Journalist
FREELANCER
Av. Armando Hombardi, 370 Ilha Da Gigoia, Casa 14 Barra Da Tijuca
Rio de Janeiro RJ, Brazil
T: 55 21 3998869

Rosa, Vinicius De Magalhaes, Sr.
Teacher
ESCOLA DAS NACOES
SHIS QI 21 - Area Especial - Lote 01
Brasilia - DF 71600, Brazil
T: 55 61 366-1800 F: 55 61 366-3777

Rosa Filho, Oscar De Aguiar, Sr.
Chefe De Gabinete
INSTITUTO DE CIENCIA E TECNOLOGIA DO DF
SRTVS Quadra 701, Ed. Assis Chateaubriand, 7o. andar
Brasilia - DF 70390, Brazil
T: 55 61 321-6947 F: 55 61 226-9477

Rosa Jr., Antonio Fernando Silva, Sr.
Diretor Socia
DEPARTAMENTO ECOLOGICO DA CASA DA CULTURA DE ITAPETININGA
Rua Dr. Virgilio de Rezende 855
Itapetininga - SP 18200, Brazil
T: 55 152 711400 F: 55 152 711161

Rosario, Frederico
Photographer
FOLHA DE SAO PAULO
Alameda Barao de Limeira 425
Sao Paulo SP, Brazil
T: 55 11 874 2608/874 2602 F: 55 11 2231644

Rosario, Jose Manuel
CEG - COMPANHIA ESTADUAL DE GAS DO RIO DE JANEIRO
Rio de Janeiro, Brazil

Rosas, Anna Cristina Macedo, Ms.
Rua Domingos Ferreira 147 Apt.901- Copacabana
Rio de Janeiro - RJ 22050, Brazil
T: 55 21 235-3981

Rosemberg, Diana Carolina
Secretaria
WDR GERMAN TV CHANNEL
Rua Eca de Queiroz, 479 casa 12
Sao Paulo - SP, Brazil
T: 55 11 5721262 F: 55 11 5728959

Rosembluth, Haim
Cinegrafista
WDR GERMAN TV CHANNEL
Rua Eca de Queiroz, 479 casa 12
Sao Paulo - SP, Brazil
T: 55 11 5721262 F: 55 11 5728959

Rosenmann, Maria Cecilia De Leao, Sra.
Acionista/Conselheira
LEAO JUNIOR S/A
Av. Getulio Vargas, 253
Curitiba - PR 80230, Brazil
T: 55 41 233-7822 F: 55 41 222-0028
TL: 5134

Ross, Ricardo Penido Dutt
OSHO SPIRITUAL HEALTH ORGANIZATION
Rua Voluntarios Da Patria, 221/603
Rio de Janeiro - RJ 22.270, Brazil
T: 55 21 246-3500

Rossi, Geoff G.
TWENTY FIRST CENTURY INC.
Rio de Janeiro, Brazil

Rossi, Robert G.
TWENTY FIRST CENTURY INC.
Rio de Janeiro, Brazil

Rossi, Rosangela, Ms.
NUCLEO INTEGRADO DE PSICOLOGIA
Rua Barbosa Lima 284
Juiz de Fora - MG, Brazil
T: 55 32 231-1164 F: 55 32 211-3024

Rotenberg, Marisa, Ms.
PARLAMENTO DA TERRA
Smln-MI09-Conj.04 Casa 25
Brasilia - DF 71500, Brazil
T: 55 61 577-2138

Rotenberg, Ricardo, Sr.
GOVERNO DO ESTADO DO RIO DE JANEIRO
Rio de Janeiro - RJ, Brazil

Roth, Maureen Ellen, Sra.
SOZED
Av. Pasteur 403 #601, Urca
Rio de Janeiro - RJ 22290, Brazil
T: 55 21 245-4054 F: 55 21 295-5179

Rousseau, Leon, Sr.
Member
ABIO - ASSOCIACAO DE AGRICULTORES BIOLOGICOS DO ESTADO DO RIO DE JANEIRO
Rua Voluntarios da Patria, 448 B 61/62 - Humaita
Rio de Janeiro - RJ, Brazil
T: 55 21 5514377

Roxa, Hellete Dos Santos
CORREIO FLUMINENSE
Rio de Janeiro, Brazil

Roxo, Maria Luiza
Diretora De Educacao - Sp
ARTE MAHIKARI
Gabriel Mont da Silva 2000
Sao Paulo - SP 01442, Brazil
T: 55 11 212 3081

Royes, Herson
INTERNATIONAL TELEVISION NEWS
Av. Rio Branco, 25/9 andar
Rio de Janeiro RJ, Brazil
T: 55 21 2335480

Roza, Alfredo
MANCHETE
Rua do Russel 766/804
Rio de Janeiro RJ, Brazil
T: 55 21 2652012 F: 55 21 2059998

Rozemblatt, Flavio
FUNDACAO FE E ALEGRIA
Rio de Janeiro, Brazil

Rubin, Nani
Editor
SISTEMA JORNAL DO BRASIL
Av. Brasil 500
Rio de Janeiro -RJ, Brazil
T: 55 21 5854422 F: 55 21 5854428

Rufino, Olavo
SISTEMA JORNAL DO BRASIL
Av. Brasil 500
Rio de Janeiro -RJ, Brazil
T: 55 21 5854422 F: 55 21 5854428

Ruggero, Waldir, Mr.
Adviser
BRAZILIAN STANDARDS ORGANIZATION
Rio de Janeiro - RJ, Brazil

Russo, Guiomar Barbosa
PREFEITURA MUNICIPAL
Praca Maua s/n 1 andar
Santos - SP 11010, Brazil
T: 55 132 342-2093 F: 55 132 345-809
TL: 131048

Ryff, Tito, Dr.
GOVERNO DO ESTADO DO RIO DE JANEIRO
Rio de Janeiro - RJ, Brazil

S.Benevides, Ana Madalena P.
Coordinator
INSTITUTO DE COMUNICACAO SINDICAL
Rua Visconde de Inhauma, 134 sala 715 Centro
Rio de Janeiro, Brazil
T: 55 21 2333516

S.Lyra, Paulo
Assessor Comunicacao P/Brazil
WWF - FUNDO MUNDIAL PARA A NATUREZA
SHIS QI 07 Bloco B Sala 204
Brasilia - DF, Brazil
T: 55 61 2482899 F: 55 61 2487176

Sa, Anelise, Sra.
Assistant
KEIDANREN
Av. das Americas,489 - Barra
Rio de Janeiro - RJ, Brazil
T: 55 21 493-6184

Sa, Bernardo Jose Soares De
Producer
FOCUS
Rua Manuel de Carvalho, 16 - 9 andar
Rio de Janeiro, Brazil
T: 55 21 5320456 F: 55 21 2625640

Sa, Elaine Maria Pimentel De
FREELANCER
Rio de Janeiro, Brazil

Sa, Elizabeth Regina Dos Santos
Jornalista
FREELANCER
Rua Ernesto Pujol 188 Maria da Graca
Rio de Janeiro - RJ, Brazil
T: 55 21 201 2803

Sa, Evaristo Juliano De
FOLIO FOTOGRAFIA
Rio de Janeiro, Brazil

Sa, Fernando De Almeida
Journalist
IBASE
R.Vicente de Souza, 29 Botafogo
Rio de Janeiro -RJ, Brazil
T: 55 21 2265412 F: 55 21 2860541

Sa, Francisco
Reporter
FOLHA DE SAO PAULO
Alameda Barao de Limeira 425
Sao Paulo SP, Brazil
T: 55 11 874 2608/874 2602 F: 55 11 2231644

Sa, Gabriel, Mr.
Presidente
MOVIMENTO VIVA BUZIOS
rua armindo berioldo 31 centro
Buzios 28905, Brazil
T: 246 232106

Sa, Hebe Maria, Sra.
Professora
UNIVERSIDADE DO RIO DE JANEIRO
Rua Barao de Sao Francisco, 120/902
Rio de Janeiro - RJ 20560, Brazil
T: 55 21 571 2264 F: 55 21 254 5249

Sa, Marcus Aurelio M. De
Technician
REDE GLOBO
R. Lopes Quintas 303
Rio de Janeiro -RJ 2, Brazil
T: 55 21 2947732

Sa, Ney Ribeiro De
Editor/Redator/Assessor De Imp
SENDA COM E EVENTOS /GRUPO AMBIENTALISTA DA BAHIA
Rua Borges dos Reis 46/J Boulevard Rio Vermelho
Salvador - BA, Brazil
T: 55 71 2355836 F: 55 71 2355836

Sa, Tania Machado De, Sra.
PREFEITURA MUNICIPAL
Praca Maua s/n 1 andar
Santos - SP 11010, Brazil
T: 55 132 342-2093 F: 55 132 345-809
TL: 131048

Sa Rego, Eduardo Sa De, Mr.
Rua Leopardo Miguel 101/ apto 503 Copacabana
Rio de Janeiro - RJ 22060, Brazil
T: 55 21 235 0215

Saavedra, Carlos, Mr.
MOTT FOUNDATION
Rua Bueno Brandao 393
Belo Horizonte - MG 31010, Brazil
T: 55 31 226-5145 F: 55 31 222-8429

Sabellaa, Margarette, Ms.
GOVERNO DE SAO PAULO
Sao Paulo - SP, Brazil

Sabia, Irene
REDE MULHER
Dalacete Das Aguas 766 Apto.21
Sao Paulo - SP 11000, Brazil
T: 55 11 2618225 F: 55 11 871 3457

Sabino, Bernardo
Producer
FUND. CULTURAL DE RADIODIFUSAO EDUCATIVA DA REGIAO DOS LAGOS
Av. Jose Bento Ribeiro Dantas 265 Manguinhos 28905
Armacao dos Buzios -, Brazil
T: 55 246 2753848 F: 55 21 2956494

Sabola, Elaine T.
Reporter
AGENCIA ESTADO
Av. Caetano Alvares, 55
Sao Paulo -SP 02598, Brazil
T: 55 11 8562210 F: 55 11 2652297

Saboia, Haroldo, Mr.
Deputado Federal
PARTIDO DOS TRABALHADORES
Av. Grande Oriente Quadra 62 Casa 32 Renascenca
Sao Luiz - MA, Brazil
T: 55 69 227 0652

Sabra, Georgina
92 GLOBAL FORUM
Rio de Janeiro - RJ, Brazil

Sacavino, Susana, Ms.
Coordenadora
PROJETO NOVAMERICA
Rua Marques de Sao Vicente 331 - 5o. Andar, Gavea
Rio de Janeiro - RJ 22451, Brazil
T: 55 21 2742149 F: 55 21 2742149

Saguez, Pierre Louis, Mr.
Cineasta
MATA VIRGEM
116 Rua Taylor - Centro
Rio de Janeiro - RJ 20241, Brazil
T: 55 21 224 2415

Saibert, Luiza, Mr.
Membro
CEQMTR - PR
Rua Almirante Tamandare, 434
Curitiba - PR 80050, Brazil
T: 55 41 2644455

Said, Dulcelina, Ms.
1o.. Secretaria
ASSOCIACAO DOS AMIGOS DA TERRA
Rua Jaime Callado L.2 Q.4 - Lagoinha - Alcantara
Rio de Janeiro - RJ, Brazil
T: 55 21 463-2645

Saidel, Rochelle G
Brazil Correspondent
CATHOLIC NEWS SERVICE
Rua Tiapira 22
Sao Paulo - SP, Brazil
T: 55 11 813 6953 F: 55 11 813 6953

Saito, Marcia Megumi Watanabe, Ms.
Ass. Administ.
M M A J- RIO
Av Rio Branco 156 - 1726
Rio de Janeiro - RJ 20040, Brazil
T: 55 21 262 4440 F: 55 21 262 4440

Sako, Roberto Shiniti, Mr.
Membro
KISSO KOSEI-KAI
Av.Aclimacao 502
Sao Paulo - SP, Brazil
T: 55 11 248-9058

Salazar, Alvaro Posaoas, Mr.
Animal Protector
SOZED
R. Lopes Quintas, 642 Jardim Botanico
Rio de Janeiro - RJ 20000, Brazil
T: 55 21 295.5179

Salcides, Fernando
Tecnical Reporter
RADIO GAUCHA
Av Ipiranga 1075
Porto Alegre - RS 90060, Brazil
T: 55 512 223 6600 F: 55 512 2336600 r 57

Salcides, Jose Ricardo Vieira
Tecnico De Audio
SISTEMA GLOBO DE RADIO
Rua do Russel, 434
Rio de Janeiro - RJ, Brazil
T: 55 21 2058282 F: 55 21 2054096

Saldanha, Beatriz, Sra.
Socia Gerente
ECOMERCADO COMERCIO E CONFECCAO LTDA
Rua Barao de Itambi, loja 119C - Botafogo
Rio de Janeiro - RJ 22231, Brazil
T: 55 21 553-5777

Saldanha, Nelson Alvarenga, Sr.
Director
CENTRO DE PESQUISAS DO MAR (CEPEMAR)
Rua Mexico, 31/Gr. 501- Centro
Rio de Janeiro - RJ 20031, Brazil
T: 55 21 240-3822 F: 55 21 240-9516

Salem, Ricardo, Mr.
Organizador
PROJETO QUADRILATERO DE PORTO SEGURO TRANCOSO
Povoado De Trancoso
Porto Seguro 45820, Brazil
T: 073 8671115

Sales, Claudia Cristina De, Sra.
Rua dos Geraneos Lote 23 Quadra 107 - Pera Flor
Nova Iguacu-RJ 26355, Brazil
T: 0212866325

Sales, Milton Martins
MANCHETE
R. do Russel 804
Rio de Janeiro -RJ 2, Brazil
T: 55 21 2652012 F: 55 21 2855148

Salgado, Gustavo Souto Maior
ASSOCIACAO AMIGOS DO PARQUE NACIONAL DE BRASILIA
HCGN 711, Bloco A, casa 13
Brasilia - DF 70750, Brazil
T: 55 61.274-50-89 F: 55 61 347-2281

Salgado, Tania, Sra.
IECAM
Dr. Moreira 250
Manaus - AM 69000, Brazil

Salgueiro, Lucia Maria Di Tommaso, Sra.
Receptionist
JOVEM TURISMO
SCLN 202-B-Lojas 7/11
Brasilia - DF 70832, Brazil
T: 55 61 224-5526 F: 55 61 321-0345
TL: 614705

Saliba, Edson
Photographer
AGENCIA ESTADO
Av. Caetano Alvares, 55
Sao Paulo -SP 02598, Brazil
T: 55 11 8562210 F: 55 11 2652297

Salles, Fabio Augusto Moreira
Photographer
AGENCIA ESTADO
Av. Caetano Alvares, 55
Sao Paulo -SP 02598, Brazil
T: 55 11 8562210 F: 55 11 2652297

Salles, Luzia Elisa De
EDITORA AZUL
Rio de Janeiro, Brazil

Salles, Valeria C., Ms.
Caixa Postal 1106
Maceio - AL 57000, Brazil
T: 55 82 2310457

Salomao, Andre Nascimento
JORNAL ENERGIZANDO
Rio de Janeiro, Brazil

Salomao, Luis Alfredo, Sr.
GOVERNO DO ESTADO DO RIO DE JANEIRO
Rio de Janeiro - RJ, Brazil

Salomao, Rafael, Sr.
Pesquisador
MUSEU GOELDI
Museu Goeldi
Belem - PA, Brazil

Salome, Dep. Nelson, Sr.
Deputado Estadual Sp
ASSEMBLEIA LEGISLATIVA/SP
Av. Pedro Alvares Cabral, s/n
Sao Paulo - SP 04097, Brazil
T: 55 11 884-1006

Salvador, Elisabete D., Sra.
Engenheiro Agronomo
ENGENHEIRA AGRONOMO
Rua Antonio de Gouveia Giuaice, 1406 Alto de Pinheiros
Sao Paulo - SP 05460, Brazil
T: 55 11 831 2358

Salvador, Luis Otavio
FREELANGER
R. Levi Carneiro 605 Barra da Tijuca
Rio de Janeiro -RJ 2, Brazil
T: 55 21 4332208

Salvador Elvira, Americo, Sr.
Diretor
SUSILA- HOLISTIC THERAPY CENTRE
Praca Cruz Vermelha, 9/ XX - Centro
Rio de Janeiro - RJ 20230, Brazil
T: 55 21 222-8782 F: 55 21 5333643

Salves, Flavio
Correspondent
EUROPEAN CHEMICAL NEWS
Ataulfo de Pavia 135/808
Rio de Janeiro RJ 2, Brazil
T: 55 21 2748974 F: 55 21 2399290

Sampaio, Arnaldo Mesquita, Sr.
Coordenadoria Juridica
MOVIMENTO BOTUCATUENSE VIDA
Rua Capitao Jose Paes de Almeida, 71 Caixa Postal 322
Botucatu-SP 18600, Brazil
T: 55 149 22 64 53 F: 55 149 21 11 61

Sampaio, Francisco De Assis, Sr.
Membro
FUNDACAO PARA CONSERVACAO DA NATUREZA
av. tamandare',1808
Campo Grande - MS 79025, Brazil
T: 55 67 383-2528

Sampaio, Paulo Marcelo, Sr.
Tecnico
CONFEDERACAO NACIONAL DA INDUSTRIA
Av. Nilo Pecanha, 50 sala 3018
Rio de Janeiro - RJ 20044, Brazil
T: 55 21 262 0181 F: 55 21 262 1465
TL: 22634

San Miguel, Luis Alberto Mansilla, Sr.
Rua Paula Freitas 66 apto 707
Rio de Janeiro - RJ, Brazil
T: 55 21 542-0038

Sanches, Carlos Roberto, Mr.
Membro
GERNE
Caixa Postal 998
Cuiaba - MT 78000, Brazil
T: 55 065 315 8751 F: 55 065 361 1119
TL: 651371

Sanchez, Luis Enrique, Sr.
Representative
SOCIEDADE BRASILEIRA DE ESPELEOLOGIA
Rua Visconde da Luz 151/41
Sao Paulo - SP 04587, Brazil
T: 55 11 549-3623 F: 55 11 575-9094

Sanchez, Reginaldo
ESTADO DE SAO PAULO
Av Rio Branco 151
Rio de Janeiro, Brazil
T: 55 21 2217111

Sandri, Sonia Maria, Ms.
Estudante
UFRGS
Jardim Medianeira, Acesso 11 #88
Porto Alegre - RS 90640, Brazil
T: 55 051 233 9539

Sangoi, Elizabeth, Sra.
Coordenadora
MOVIMENTO EM DEFESA DA VIDA
Av. Guaiba 316
Porto Alegre - RS 91700, Brazil
T: 55 512 248-4008 F: 55 512 343-6671

Sanielovici, Susanne, Ms.
Professor
UNIVERSIDADE DE CIENCIAS DE PARIS
a/c ELF-AQUITAINE - Av Rio Branco 80 -10o. andar
Rio de Janeiro - RJ 20040, Brazil
T: 55 21 2219755

Sant'Anna, Alexandre
Fotografo
EDITORA ABRIL
R.da Passagem, 123/8 andar
Rio de Janeiro - RJ, Brazil
F: 55 21 2759347

Sant'Anna, Lucia Maria De Mello
Editor
REDE GLOBO
R. Lopes Quintas 303
Rio de Janeiro -RJ 2, Brazil
T: 55 21 2947732

Santa Rosa, Paulo, Sr.
Doctor
FUNDACAO OSWALDO CRUZ
Escola Nacional De Saude Publica
Rio de Janeiro - RJ, Brazil
T: 55 21 590 3789

Santala, Anna, Ms.
Junior Professional Officer
WORLD FOOD PROGRAMME
SCH Quadra 02 Bloco 'B' Terreo
Brasilia - DF 70710, Brazil
T: 55 61 226-7872 F: 55 61 322-7080

Santana, Celso Guido De
REDE RECORD
Av. Miruna 713
Sao Paulo _SP 04084, Brazil
T: 55 11 5429000 F: 55 11 612228

Santana, Jackson, Sr.
Economista
CENTRO DE RECURSOS AMBIENTAIS
Rua Rio Sao Francisco, 1 - Monte Serrat
Salvador - BA CEP: 40.425, Brazil
T: 55 71 312-7191

Santana, Reinaldo Ribeiro De, Sr.
FLAMA
Rua Senador Vergueiro, 52 - Ap. 14 - Flamengo
Rio de Janeiro - RJ 22230, Brazil
T: 55 21 265-1363

Santana, Romildes Oliveira, Sra.
MINISTERIO PUBLICO DO ESTADO DO MARANHAO
Sao Luiz - MA 65075, Brazil
T: 55 69 227-1171

Santareita, Mariurete
Journalist
QUE FAZER
Rua Dantas (Senador), No. 127 sala 444
Rio de Janeiro, Brazil
T: 55 21 5331169

Santarelli, Eduardo Guilherme, Mr.
CESP
Rua da Consolacao, 1875, 9 andar, Cerq. Cesar
Sao Paulo - SP, Brazil

Santello, Giovana Vantini, Sra.
Membro
APOENA
Rua Norberto Valim, 181
Presid. Prudente-SP 19050, Brazil
T: 55 182 33-3593

Santiago, Andre Luis
ROLAND HENZE
Rio de Janeiro, Brazil

Santiago, Emiliano, Sr.
Consultor
ABES
Av. Cardeal da Silva, 555 / 904
Salvador - BA, Brazil
T: 55 71 2459446

Santiago, Patricia Costa, Ms.
JUVENTUDE DO PSDB
Av.Nove de Julho, 1349
Sao Paulo - SP 01407, Brazil
T: 55 11 872-8177

Santiago, Sandra
ICAE
12 Rajaji Marg
New Delhi 110011, Brazil
T: 91 11 301-4309

Santiago, Tiago Xavier, Sr.
Pesquisador
MESTRADO EM SOCIOLOGIA/LABORATORIO DE PESQUISAS SOCIAIS/IFCS/UFRJ
Rua das Laranjaras, 55/Apto 1301
Rio de Janeiro - RJ 22240, Brazil
T: 55 21 225 0112

Santini, Maria, Sra.
Conrado Hoffmann 610
Santa Maria 97050-130, Brazil

Santini, Zelia, Ms.
Membro
CEQMTR - PR
Rua Almirante Tamandare, 434
Curitiba - PR 80050, Brazil
T: 55 41 2644455

Santo, Alvaro Negrao Do Espirito, Mr.
Assessor Especial
SUDAM
Av.Magalhaes Barata, 1150 casa 95
Belem - PA 66040, Brazil
T: 55 91 226-0615 F: 55 91 226-8595

Santo, Cesar Victor Do Espirito, Sr.
Sup. Adjunto
FUNDACAO PRO-NATUREZA (FUNTARA)
SCLN 107, Bloco B, Salas 201/17
Brasilia - DF 70743, Brazil
T: 55 61 274-5449 F: 55 61 274-5324

Santo, Joper Padrao Do Espirito, Mr.
Superintendent
ABES
Av. Beira Mar, 216 13 andar
Rio de Janeiro - RJ 20000, Brazil
T: 55 21 210.3221 F: 55 21 262.6838

Santoro, Luis Fernando
ICAE
Rede Mulher C.P.4651
Sao Paulo SP 01061, Brazil
T: 55 11 627050 F: 55 11 8713457

Santos, Adilson Dos
MANCHETE
Rua do Russel 804 Gloria
Rio de Janeiro, Brazil
T: 55 21 5554000

Santos, Alberto Do
INTERNATIONAL TELEVISION NEWS
Av. Rio Branco, 25/9 andar
Rio de Janeiro RJ, Brazil
T: 55 21 2335480

Santos, Almir, Sra.
IDEAS
Artigas 3246 30 P. '8'
Brazil

Santos, Aluizio Figueiredo Dos
REDE GLOBO
R. Lopes Quintas 303 J. Botanico
Rio de Janeiro - RJ, Brazil
T: 55 21 5292000 F: 55 21 2942042

Santos, Amilton Nascimento Dos
Technician
REDE GLOBO
R. Lopes Quintas 303
Rio de Janeiro -RJ 2, Brazil
T: 55 21 2947732

Santos, Ana Cristina Lessa, Sra.
Engenheira
ECOSS-CONSULTORIA E PLANEJAMENTO AMBIENTAL
Av. Tancredo Neves, 274, Bl.-8, S. 320/325
Salvador - BA 40100, Brazil
T: 55 71 359-1589

Santos, Ana Lucia Vieira Dos
Director
INSTITUTO DE ARQUITETOS DO BRASIL
Rua do Pinheiro, 10 Flamengo
Rio de Janeiro, Brazil
T: 55 21 2853480

Santos, Ana Maria Galo
MANCHETE
Rua do Russel 766/804
Rio de Janeiro RJ, Brazil
T: 55 21 2652012 F: 55 21 2059998

Santos, Anatalia, Ms.
Conselheira
CONSELHO NACIONAL DOS SERINGUEIROS
Rua Alexandre Farhat 206
Rio Branco - AC 69900-000, Brazil
T: 55 68 224-9645 F: 55 68 224-3622

Santos, Antonio, Sr.
ARQUIPELAGO DE FERNANDO DE NORONHA
Rua Barao de Sao Borja, 243 - Boa Vista
Recife - PE 50070, Brazil
T: 55 81 221-3217 F: 55 81 231-1594
TL: 81.2533

Santos, Antonio Carlos Cordeiro Dos
FREELANCER
Av. Sao Jose, 50
Rio de Janeiro, Brazil
T: 55 21 2328090 x293

Santos, Antonio Gabriel De Carvalho
REDE GLOBO
R. Lopes Quintas 303 J. Botanico
Rio de Janeiro - RJ, Brazil
T: 55 21 5292000 F: 55 21 2942042

Santos, Arlene Maria Soares Correia, Sra.
Membro
APOENA
Rua Dr.Gurgel, 860
Presid. Prudente-SP 19015-140, Brazil
T: 55 182 21-4222 F: 55 182 22-6177

Santos, Arlinda Messias Dos, Ms.
Assessora
ADITTEP
Rua Desembargador Westphalen, 1373
Curitiba - PR 80230, Brazil
T: 55 41 223 3280

Santos, Carlos A. Ivanir Dos, Sr.
Exec. Secret.
CENTRO DE ARTICULACAO DE POPULACOES MARGINALIZADAS - CEAP
Rua da Lapa, 200 - S.805
Rio de Janeiro - RJ 20021, Brazil
T: 55 21 252-2302 F: 55 21 205-3099

Santos, Carlos Alberto Felicio Dos
FOLIO FOTOGRAFIA
Rio de Janeiro, Brazil

Santos, Carlos Alberto O. Dos
Technician
REDE GLOBO
R. Lopes Quintas 303
Rio de Janeiro -RJ 2, Brazil
T: 55 21 2947732

Santos, Carlos Benedito Dos, Sr.
Professor
COLEGIO CIDADE DE ITU
Estrada Barao Rhodia, Km. 15 s/n - Vila Holandia - Barao Geraldo
Campinas - SP 13083, Brazil
T: 55 192 482-5365

Santos, Carlos Eduardo Medina Dos, Sr.
SOCIEDADE POTIGUAR DE EDUCACAO - SPEL
Rua Floriano Peixoto, 295, Petropolis
Natal - RN 59020, Brazil
T: 55 84 222-2314 F: 55 84 221-6255

Santos, Celio Dos
MANCHETE
R. do Russel 804
Rio de Janeiro -RJ 2, Brazil
T: 55 21 2652012 F: 55 21 2855148

Santos, Claudio De Holanda
RADIO NACIONAL - PROGRAMA GINCANA NACIONAL
Rio de Janeiro, Brazil

Santos, Dalva Maria Vasconcelos Dos, Ms.
Tecnico Economista
IDESP
Av. Nazare 817
Belem - PA 66035, Brazil
T: 55 91 224-4411 F: 55 91 225-3414
TL: 6905

Santos, Danilo Augusto, Mr.
Presidente
MOVIMENTO VERDE DE PARACATU
Rua Olegario Maciel,470 - Ap.201
Paracatu CEP: 38.600, Brazil
T: 55 61 671-4397

Santos, Denise Marinho Dos
Editor
CNN SPANISH
R. Lopes Quintas 732 Jardim Botanico
Rio de Janeiro -RJ, Brazil
T: 55 21 5115139 F: 55 21 2397977

Santos, Deoclides
REDE RECORD
Av. Miruna 713
Sao Paulo _SP 04084, Brazil
T: 55 11 5429000 F: 55 11 612228

Santos, Edilson
Desenhista
SERVICO SOCIAL DO COMERCIO-DEPARTAMENTO NAC
Rua Voluntarios Da Patria, 169
Rio de Janeiro - RJ 22270, Brazil
T: 55 21 286-5152 F: 55 21 286-2638

Santos, Edilson Da Silva
MANCHETE
R. do Russel 804
Rio de Janeiro -RJ 2, Brazil
T: 55 21 2652012 F: 55 21 2855148

Santos, Edilson Gomes Dos
Technician
REDE GLOBO
R. Lopes Quintas 303
Rio de Janeiro -RJ 2, Brazil
T: 55 21 2947732

Santos, Edmundo Gonzaga Dos
Tecnico De Audio
SISTEMA GLOBO DE RADIO
Rua do Russel, 434
Rio de Janeiro - RJ, Brazil
T: 55 21 2058282 F: 55 21 2054096

Santos, Edson Escouto Dos
Reporter
REDE GLOBO
R. Lopes Quintas 303
Rio de Janeiro -RJ 2, Brazil
T: 55 21 2947732

Santos, Eduardo De Oliveira
Editor
REDE GLOBO
R. Lopes Quintas 303
Rio de Janeiro -RJ 2, Brazil
T: 55 21 2947732

Santos, Elisete Moura
SINDICATO DOS TRABALHADORES IND. QUIM. FARM. DE S.PAULO
Sao Paulo, Brazil

Santos, Elizabeth, Ms.
WORLD COUNCIL OF CHURCHES
150 route de Ferney
Brazil
F: 41 22 791 0361

Santos, Ernani Gomes
Tecnico De Audio
SISTEMA GLOBO DE RADIO
Rua do Russel, 434
Rio de Janeiro - RJ, Brazil
T: 55 21 2058282 F: 55 21 2054096

Santos, Evandro C.
PM STUDIO FOTOGRAFICO
Rio de Janeiro, Brazil

Santos, Fernanda Reznik
ABVP
Rua Treze de Maio 489, Bela Vista
Sao Paulo SP 01327, Brazil
T: 55 11 2847862

Santos, Fernando Cezar Dos
ROLAND HENZE
Rio de Janeiro, Brazil

Santos, Fernando Dos
FUND. CULTURAL DE RADIODIFUSAO EDUCATIVA DA REGIAO DOS LAGOS
Av. Jose Bento Ribeiro Dantas 265 Manguinhos 28905
Armacao dos Buzios -, Brazil
T: 55 246 2753848 F: 55 21 2956494

Santos, Flavio Augusto Altero Dos, Mr.
CIA VALE DO RIO DOCE
Av. Graca Aranha, 26
Rio de Janeiro - RJ, Brazil
T: 55 21 272 4401 F: 55 21 220 0593

Santos, Francelino Jose Dos
FBCN - JORNAL INFORMATIVO
Rio de Janeiro, Brazil

Santos, Francisco Alves, Sr.
Vice-Presidente
ORG. COOP. BRASILEIRAS
SQS, Ed. Baracat, 4. andar
Brasilia - DF Cep.: 70309-900, Brazil
T: 55 61 225 0275 F: 55 61 226 8766
TL: 1879

Santos, Francisco De Oliveira
Tecnico
TVSBT
Campo de S.Cristovao, 105
Rio de Janeiro - RJ, Brazil
T: 55 21 5803688 / 5800678 F: 55 21 5809685

Santos, Gianne Carvalho S. Dos
Photographer
IMAGENS DA TERRA
R. das Laranjeiras 203/202
Rio de Janeiro RJ 2, Brazil
T: 55 21 5562502 F: 55 21 2204653

Santos, Helio Dos
Cinegrafista
REDE BANDEIRANTES DE TELEVISAO
R.Alvaro Ramos, 492
Rio de Janeiro - RJ, Brazil
T: 55 21 2122072 F: 55 21 2957391

Santos, Ilhens Pereira Dos
Tecnico
TVSBT
Campo de S.Cristovao, 105
Rio de Janeiro - RJ, Brazil
T: 55 21 5803688 / 5800678 F: 55 21 5809685

Santos, Ilmar Bastos, Sr.
Chefe Dept Tecnico/Cientifico
FUNDACAO BIODIVERSITAS
Rua Maria Vaz de Melo 71, Bairro Da. Clara
Belo Horizonte - MG 31250, Brazil
T: 55 31 443-2119 F: 55 31 441-7037

Santos, Joao Batista Dos, Sr.
Desembargador
COMITE ECOLOGICO DE DEFESA DE SAO LUIZ
Rua Beco do Couto 56 - Centro
Sao Luiz - MA, Brazil
F: 55 69 223-3540

Santos, Jose Augusto Correia
Reporter
REDE GLOBO
R. Lopes Quintas 303
Rio de Janeiro -RJ 2, Brazil
T: 55 21 2947732

Santos, Jose Macena Dos
Militar
SISTEMA GLOBO DE RADIO
Rua do Russel
Rio de Janeiro RJ, Brazil
T: 55 21 2058282

Santos, Jose Ricardo Mello Dos
Technician
REDE GLOBO
R. Lopes Quintas 303
Rio de Janeiro -RJ 2, Brazil
T: 55 21 2947732

Santos, Jose Soares Dos
REDE GLOBO
R. Lopes Quintas 303 J. Botanico
Rio de Janeiro - RJ, Brazil
T: 55 21 5292000 F: 55 21 2942042

Santos, Josefina Dos, Ms.
Teacher
MEC
Rua Professor Anibal de Matos 290/201 - St. Antonio
Belo Horizonte - MG, Brazil
T: 55 31 3440686 F: 55 31 2737888

Santos, Juscelino Batista Dos, Sr.
INSTITUTO DE MEIO AMBIENTE DO AMAZONAS
Rua 25 C-522
Manaus - AM 69068, Brazil
F: 55 92 236-2535

Santos, Leila
Fotografa
LEGIAO DA BOA VONTADE
Av.Rudge, 700
Sao Paulo - SP, Brazil
T: 55 11 2202599 F: 55 11 2205196

Santos, Liana Carvalho
Reporter
FREELANCER
Rua Marques de Abrantes 199/104 Botafogo
Rio de Janeiro RJ, Brazil
T: 55 21 5515506 F: 55 21 2288878

Santos, Lilian Costa Nabuco Dos
Editor
FREELANCER
UERF - Unvversidade do Estado do Rio de Janeiro
Rio de Janeiro, Brazil
T: 55 21 2395576

Santos, Luis Claudio Dos
FUND. CULTURAL DE RADIODIFUSAO EDUCATIVA DA REGIAO DOS LAGOS
Av. Jose Bento Ribeiro Dantas 265 Manguinhos 28905
Armacao dos Buzios -, Brazil
T: 55 246 2753848 F: 55 21 2956494

Santos, Luis Feitosa Dos
REDE RECORD
Av. Miruna 713
Sao Paulo _SP 04084, Brazil
T: 55 11 5429000 F: 55 11 612228

Santos, Luis Ricardo
TV EDUCATIVA
Av Gomes Freire 474
Rio de Janeiro RJ 2, Brazil
T: 55 21 2920012

Santos, Luiz Antonio Dos
Editor
SERVICO SOCIAL DA INDUSTRIA
Av. Nilo Pecanha 50/2409 Centro
Rio de Janeiro -RJ, Brazil
T: 55 21 2621267 F: 55 21 2204166

Santos, Luiz Carlos Lessa Dos
Techinician
AGENCIA ESTADO
Av. Caetano Alvares, 55
Sao Paulo -SP 02598, Brazil
T: 55 11 8562210 F: 55 11 2652297

Santos, Luiz Henrique Dos
Image Editor
FREELANCER
Estr. dos tres Rios, 579
Rio de Janeiro, Brazil
T: 55 21 3929923

Santos, Luiz Marques Dos
TVN PRODUCOES ARTISTICAS
Sao Paulo, Brazil

Santos, M.P. Vinicius, Mr.
Membro
GERNE
Caixa Postal 998
Cuiaba - MT 78000, Brazil
T: 55 065 315 8751 F: 55 065 361 1119
TL: 651371

Santos, Madalena Rodrigues Dos, Sra.
Professora
UFMT/NUEPON
Res. dos Trabalhadores - Bl 6, apto. 102 - Res. Sao Carlos
Cuiaba - MT 78000, Brazil

Santos, Marcelo Kobylinski Dos
Technician
REDE GLOBO
R. Lopes Quintas 303
Rio de Janeiro -RJ 2, Brazil
T: 55 21 2947732

Santos, Marco Aurelio Martins
Journalist
FREELANCER
Rua dos Invalidos 138 BL 1 COB 01 Centro
Rio de Janeiro, Brazil
T: 55 21 2329078

Santos, Margaret
Assess.
SERVICO SOCIAL DO COMERCIO-DEPARTAMENTO NAC
Rua Voluntarios Da Patria, 169
Rio de Janeiro - RJ 22270, Brazil
T: 55 21 286-5152 F: 55 21 286-2638

Santos, Maria Das Neves, Sra.
Reg. Planning & Dev.
ASSOCIACAO SERVIDORES SUDENE
Praca Joao Goncalves de Souza s/n - Engenio do Meio
Recife - PE 50000, Brazil
T: 55 81 271-1044 2741

Santos, Maria Do Carmo Neves Dos, Ms.
Geologa
INSTITUTO DO MEIO AMBIENTE DO AMAZONAS
Rua Dr. MACHADO NO. 501-CENTRO
Manaus 69013, Brazil
T: 092 2330800 F: 092 2362535

Santos, Maria Francisca M. Dos
Producer
FUND. CULTURAL DE RADIODIFUSAO EDUCATIVA DA REGIAO DOS LAGOS
Av. Jose Bento Ribeiro Dantas 265 Manguinhos 28905
Armacao dos Buzios -, Brazil
T: 55 246 2753848 F: 55 21 2956494

Santos, Maria Isabel Rodrigues Dos
VIDEOARTE RIO
Rua Senradom 45
Rio de Janeiro, Brazil

Santos, Maria Lino Dos, Ms.
CONSELHO NACIONAL DOS SERINGUEIROS
Rua Alexandre Farhat 206
Rio Branco - AC 69900000, Brazil
T: 55 68 224-1209

Santos, Maria Sirley Dos, Sra.
PREFEITURA MUNICIPAL
Praca Maua s/n 1 andar
Santos - SP 11010, Brazil
T: 55 132 342-2093 F: 55 132 345-809
TL: 131048

Santos, Marina, Ms.
CONSERVATION INTERNATIONAL
Rua Bueno Brandao 393
Belo Horizonte - MG 31010, Brazil
T: 55 31 226-5145 F: 55 31 222-8429

Santos, Marli Terezinha Dos
Assessora De Imprensa
GAIA - ASSOC.CONS.M.AMBIENTE E PROD.INTEGRADA ALIM.AMAZONIA
Arse-51 - 39
Palmas - TO, Brazil
T: 55 63 553/1108

Santos, Mauricio Rodrigues Dos
Coordenador De Jornalismo
RADIO UNIVERSITARIA FM
Rua Paula Frassinete, 42
Rio de Janeiro - RJ, Brazil
T: 55 21 2933847 F: 55 21 2930592

Santos, Moises Soares Dos, Sr.
DEPARTAMENTO DE RECURSOS MINERAIS - DRM
Rua Mal. Deodoro no. 351 Centro
Niteroi - RJ 24030, Brazil
T: 55 21 719 5454 F: 55 21 719 6846

Santos, Neemias Espinola Dos
Technician
REDE GLOBO
R. Lopes Quintas 303
Rio de Janeiro -RJ 2, Brazil
T: 55 21 2947732

Santos, Nelly C. Dos, Sra.
Av. Rui Barbosa 394 /1201
Rio de Janeiro - RJ, Brazil
T: 55 21 551-1444

Santos, Nelson J. B.
Reporter
Av. Rio Branco, 14 13 andar
Rio de Janeiro, Brazil
T: 55 21 2531828

Santos, Nelson Perez Dos
O POVO
Rio de Janeiro, Brazil

Santos, Ney Costa, Sr.
ANDALUZ PRODUCOES CINEMATOGRAFICAS
Praia Do Flamengo 66 Sala 710 Bloco 1
Rio de Janeiro - RJ 22228, Brazil
T: 55 21 205-5497 F: 55 21 225-8918

Santos, Olga De Los
Correspondent
JORNAL LA NACION
Av.Ipiranga, 200 Bl.E
Sao Paulo - SP, Brazil
T: 55 11 2559528

Santos, Omar Dos
Technician
REDE GLOBO
R. Lopes Quintas 303 Jardim Botanico
Rio de Janeiro -RJ 2, Brazil
T: 55 21 2947732

Santos, Paulo, Sr.
CIA VALE DO RIO DOCE
Av. Graca Aranha 26
Rio de Janeiro - RJ, Brazil
T: 55 21 2724857 F: 55 21 2206126

Santos, Paulo Cesar Vieira Dos
Technician
REDE GLOBO
R. Lopes Quintas 303
Rio de Janeiro -RJ 2, Brazil
T: 55 21 2947732

Santos, Paulo Roberto Dos
Photographer
JORNAL MAIORIA FALANSE
Rua da Lapa 200/308
Rio de Janeiro RJ, Brazil
T: 55 21 5071462

Santos, Pedro Paulo Ferreira Dos, Sr.
Diretor De Projetos
SUDECAP
Rua Paraiba 890 - Funcionarios
Belo Horizonte - MG 30130, Brazil
T: 55 31 201-6433 F: 55 31 273-4228

Santos, Raimunda Manuela Dos, Ms.
SOCIEDADE MARANHAENSE DE DIREITOS HUMANOS
Rua Antonio Rayol 290 - Centro
Sao Luiz - MA 65015-040, Brazil
T: 55 69 221-3838

Santos, Raimundo Nunes, Mr.
GRUPO MAE NATUREZA ECOLOGIA VIVA
Av. Presidente Vargas 362
Livramento - BA, Brazil
T: 55 71 444-2119

Santos, Reinaldo Menezes
Reporter
REDE GLOBO
R. Lopes Quintas 303
Rio de Janeiro -RJ 2, Brazil
T: 55 21 2947732

Santos, Renan Cordeiro Dos
Auxiliar De Camera
TV Rio
R. Miguel de Frias, 57 Estacio
Rio de Janeiro - RJ, Brazil
T: 55 21 2939888 F: 55 21 2938626

Santos, Renan Dos, Mr.
Medico (coorden. Homeopatia)
SECRETARIA DE ESTADO DE SAUDE DO RIO DE JANEIRO
Rua Mexico,128 Sala 719 - Centro
Rio de Janeiro - RJ 20260(20031), Brazil
T: 55 21 297-7133
TL: Ramal 289

Santos, Ronaldo Braga Dos
MANCHETE
R. do Russel 804
Rio de Janeiro -RJ 2, Brazil
T: 55 21 2652012 F: 55 21 2855148

Santos, Rosangela H. Dos
Reporter
AGENCIA ESTADO
Av. Caetano Alvares, 55
Sao Paulo -SP 02598, Brazil
T: 55 11 8562210 F: 55 11 2652297

Santos, Sergio A. G.
Technician
REDE GLOBO
R. Lopes Quintas 303
Rio de Janeiro -RJ 2, Brazil
T: 55 21 2947732

Santos, Sidney Jose Dos, Sr.
Distribuicao
CEDI
Av Higienopolis 983
Sao Paulo - SP 01238, Brazil
T: 55 11 825-5544 F: 55 11 825-7861
TL: 1126561

Santos, Suzana Gomes
Reporter
JONAL DO COMMERCIO
Rua do Livramento 176
Rio de Janeiro, Brazil
T: 55 21 2531817 F: 55 21 2532735

Santos, Suzana Maria Da Rocha
CIA DE TURISMO DO ESTADO RO RIO - TURISRIO
Rio de janeiro, Brazil

Santos, Valerio Cardoso Dos, Sr.
Superintendente
IBAMA - SUPES PARAIBA
Av. Pedro II no.3284 - Mata do Buraquinho
Joao Pessoa - PB 58000, Brazil
T: 55 83 224-0068 F: 55 83 224-0068

Santos, Vandeval Lourenco Dos
INSTITUTO DE ARQUITETOS DO BRASIL
Rua do Pinheiro, 10 Flamengo
Rio de Janeiro, Brazil
T: 55 21 2853480

Santos, Waltencir Goulart Dos
Technician
REDE GLOBO
R. Lopes Quintas 303
Rio de Janeiro -RJ 2, Brazil
T: 55 21 2947732

Santos, Wandreuves Acampora Dos
Tecnico De Audio
SISTEMA GLOBO DE RADIO
Rua do Russel, 434
Rio de Janeiro - RJ, Brazil
T: 55 21 2058282 F: 55 21 2054096

Santos, Wilma Resende Araujo, Ms.
Presidente
AMNPPA-FAMNEP
Coroados 750
Porto Alegre - RS, Brazil
T: 55 512 2271961

Santos Neto, Acylino Jose Dos, Sr.
Director
ABES
Av. Beira Mar, 216 - 13o. andar
Rio de Janeiro - RJ 20021, Brazil
T: 55 21 210-3221 F: 55 21 262-6838
TL: 21-31902

Sanz, Luiz Alberto
Programme Coordinator
RADIO MEC
Praca da Republica 141A
Rio de Janeiro, Brazil
T: 55 21 2322123

Sapucahy, Mario Lucio
Photographic Reporter
FREELANCER
Rua Bonda do Mato, 172 apto 304 Grajao
Rio de Janeiro RJ 2, Brazil
T: 55 21 2683422

Saraiva, Deolinda Cavalcante
Reporter
FRILAS SERVICOS JORNALISTICOS
Rua Anfilofio de Carvalho, 29/407
Rio de Janeiro RJ 2, Brazil
T: 55 21 2205444

Sardenberg, Ormi, Mr.
Assistant
CHRISTIAN ENVIRONMENT ALLIANCE
Av. Rui Barbosa, 490
Niteroi - RJ 24250, Brazil

Sarmento, Elizabeth
IBAMA
Rio de Janeiro, Brazil

Sarmento, Jair Da Silva, Mr.
President
FUND. BRASIL - FRANCA PARA ECO
Sqn-304 - bloco. D -Apt. 412
Brasilia - DF 70.736, Brazil
T: 55 61 225-7438 F: 55 61 262-7782

Sarno, Ruy
CONCREMAT ENGENHARIA & TECNOLOGIA
R Euclides da Cunha 106 Sao Cristovao
Rio de Janeiro - RJ, Brazil
T: 55 21 284 4022 F: 55 21 284 2702

Sarti, Silvio, Sr.
IECAM
Dr. Moreira 250
Manaus - AM 69000, Brazil

Sartori, Michele, Ms.
Voluntario
MLAL
Rua Goias, 191
Luciara- MT 78- 660000, Brazil
T: 065 5281163 F: 065 5221133

Sartorio, Ronald, Sr.
CERJ
Rua Nascimento Silva 39 - Apt. 201
Rio de Janeiro - RJ 20000, Brazil
T: 55 21 287-1662

Sarturi, Linda Cristina Degrazia
EMPRESSA JOURNALISTA DIARIO DA MANHA
Av. Sete de Setembro 509
Passofundo, Rio Gran, Brazil
T: 55 54 3121100 F: 55 54 3122692

Sarubbo, Mauro Alejandro Baptista
Cameraman
V.I.P.
Caio de Melo Franco 45
Rio de Janeiro, Brazil
T: 55 21 2469747

Sarwer-Foner, Brian, Mr.
Member
THE EXPLORERS CLUB
Rua Felipe Schmidt no. 735 apt.1103
Florianopolis - SC 88000, Brazil
T: 55 482 84 1242

Sasoon, Shana, Ms.
Coordinator
YMCA OF MINAS GERAIS
Rua Matias Cardoso, 165 Santo Agostinho
Belo Horizonte - MG, Brazil
T: 55 31 337-3032 F: 55 31 337-1153

Sass, Luis Alberto, Mr.
Ass Juridico
UNIAO ERECHINENSE PROTETORA DOS ANIMAIS - UEPA
R. Torres Goncalves, 153/501
Erechim - RS 99700, Brazil
T: 054 321 1329

Sass, Vera Beatriz, Ms.
Diretora
UNIAO ERECHINENSE PROTETORA DOS ANIMAIS - UEPA
R. Torres Goncalves, 153/501
Erechim - RS 99700, Brazil
T: 054 321 1329

Sassi, Carlos Eduardo Lessi
Technician
REDE GLOBO
R. Lopes Quintas 303
Rio de Janeiro -RJ 2, Brazil
T: 55 21 2947732

Sathler, Evandro, Sr.
FUNIVALE
R. Otavio Carneiro, 134/703
Niteroi - RJ 24230, Brazil
T: 55 21 736-5186 F: 55 21 714-6863

Sato, Marina Yeko
Director
AMERICA DO SUL PRODUCOES DE DOCUMENTARIOS LTDA (NAMBEI TSUSHIN)
Av. Paulista 509 /11 andar conj.1104 Cerqueira Cesar
Sao Paulo -SP 01311, Brazil
T: 55 11 2886141 F: 55 11 2512065

Sato, Naoko Sasada
JIJI PRESS NEWS AGENCY
Av Paulista 854 - 13o. andar cj 133 Bela Vista
Sao Paulo - SP, Brazil
T: 55 11 285 0025/287 9526 F: 55 11 285 3816

Sauer, Adeum Hilario, Mr.
GOVERNO DO ESTADO DA BAHIA - SEC. DE EDUCACAO
Ilheus - BA, Brazil
F: 55 73 212-2195

Savaget, Eliza
GLOBO ECOLOGIA
Rua Santa Alexandrina 343/c.3 Rio Comprido
Rio de Janeiro, Brazil
T: 55 21 2739087 F: 55 21 5025451

Savioli, Adriano
ESTADO DE SAO PAULO
Av Rio Branco 151
Rio de Janeiro, Brazil
T: 55 21 2217111

Sayd, Cristiane, Sra.
SANTUARIO ECOLOGICO DE PIPA - SEP
Rua Acu,495
Natal RN 59020, Brazil
T: 55 84222-6656 R
TL: 55 0222-799

Schaffer, Wigold, Sr.
Coordenador Geral
FEDERACAO DE ENTIDADES ECOLOGISTAS CATARINENSES
Caixa Postal 628
Rio do Sul - SC 89160, Brazil
T: 55 478 222-497 F: 55 478 222-996

Schawrnopf, Paulo Inacio
Diretor Tecnico
FUND. CULTURAL DE RADIODIFUSAO EDUCATIVA DA REGIAO DOS LAGOS
Av. Jose Bento Ribeiro Dantas 265 Manguinhos 28905
Armacao dos Buzios -, Brazil
T: 55 246 2753848 F: 55 21 2956494

Scheiner, Paulo Cezar De Lemos, Sr.
Vendedor
NEWSWEEK INTERNATIONAL
Rua Do Russel, 450 Sala 801
Rio de Janeiro - RJ 22210, Brazil
T: 55 21 265-4392 F: 55 21 205-6887

Schenkel, Celso, Mr.
DENACOOP - M.A.
Rua Frei Leandro 35
Rio de Janeiro - RJ 22470, Brazil
T: 55 21 286-9511 F: 55 21 286-1191

Schettino, Luiz Fernando, Mr.
Eng. Florestal
ITCF/ACAPA/CAMARA
R.Jocarly Garcia 582
Castelo 29360, Brazil
T: 027542-1233

Schiavo, Marcio Ruiz, Mr.
Representative
POPULATION COMMUNICATION INTERNATIONAL
Av. das Americas, 3939-Bl-2/209-Barra da Tijuca
Rio de Janeiro - RJ 22631-003, Brazil
T: 55 21 431-1593 F: 55 21 325-7149

Schicker, Elgin Veronica, Sra.
Rua Comendador Remo Cesaroni 425 3rd andar
S.Jose dos Campos SP 12200, Brazil
T: 0123 231206

Schiligting, Sandra, Ms.
Secretaria Nacional
MOVIMENTO NACIONAL DE MENINOS E MENINAS DE RUA
SDN - CNB 3o. andar, sala L 403
Brasilia - DF 70070, Brazil
T: 55 61 226-9631 F: 55 61 225-1577

Schiper, Ruth, Sra.
Palestrante
INSTITUTO INTERNATIONAL DE COSMOBIOLOGIA
Rua Senador Dantas, 75 Sala 2705
Rio de Janeiro - RJ 22031, Brazil
T: 55 21 220 7583 F: 55 21 265 1042

Schlee, Renata Lobato, Ms.
CENTRO DE ESTUDOS AMBIENTAIS (CEA)
Deodoro 1335 Apto. 203
Pelotas-RS 96015, Brazil

Schlenker, Harvey F., Mr.
INSTITUTO DE TERRAS, CATOGRAFIA E FLORESTAS DO PARANA
R. Oes. Motta 3384
Curitiba - PR 80.410, Brazil
T: 55 41 234-1611

Schmall, Hermann, Mr.
EMBRATEL
rua MACEDO SOBRINHO 74/104
Rio de Janeiro - RJ 22271, Brazil
T: 55 21 2265257

Schmidt, Amalia Ruth B., Sra.
Vice President
FAMN E P
Rua Peixoto Gomide 1772 apto 113
Sao Paulo - SP, Brazil
T: 55 11 282-3552 F: 55 11 881-4091

Schmidt, Andrea, Ms.
WORLD YOUNG WOMEN'S CHRISTIAN ASSOCIATION
Caixa Postal 51568
Sao Paulo - SP 01414, Brazil
T: 55 11 283-3725
TL: WYMCA 41 23 32

Schmidt, Hans Christian, Mr.
Assessor
SACTES/DED
Rua Dimores 1297, 4 Andar
Belo Horizonte - MG 30140, Brazil
T: 55 31 226-5193 F: 55 31 224-4088
TL: 1505

Schmidt, Winfried
SOCIEDADE ALEMA DE COOOPERACAO TECNICA LTDA. (GTZ)
Ave. A.J. Renner, 10
Porto Alegre - RS 90250, Brazil
T: 55 512 432 345 F: 55 512 432 345

Schmith, Carla
ENCYCLOPAEDIA BRITANNICA DO BRASIL PUBLICACOES LTDA
Av. Nilo Pecanha, 50/509 Centro
Rio de Janeiro - RJ 20020, Brazil
T: 55 21 252-7638 F: 55 21 242-1002

Schmitz, Heribert, Mr.
Coordenador Geral
SACTES / DED
Rua Aindres 1297/40
Belo Horizonte - MG 30.140, Brazil
T: 55 31 226-5139 F: 55 31 224-4088

Schneider, Ekkehard
REVISTA CIDADE NOVA
Vargem Grande Paulis, Brazil

Schneller, Tania, Ms.
Estudante
Rua Nobrega 194
Niteroi - RJ 24220, Brazil
T: 55 21 711-9190

Schonor, Paulo Alziro
Superint. Marketing & Comunica
LEGIAO DA BOA VONTADE
Av.Rudge, 700
Sao Paulo - SP, Brazil
T: 55 11 2202599 F: 55 11 2205196

Schrage, Clemens A. F., Mr.
GTZ
AV Major Cicero de Goes Monteiro 2197
Maceio - AL 57060, Brazil
F: 55 82 2219299

Schreiber, Waltraud
R. Eliseu Viscoti 366 Santa Tereza
Rio de Janeiro - RJ 20241, Brazil
T: 55 21 222-2979
TL: 061 1687

Schreiber, Yara, Ms.
Rua Albuquerque Lins 1293 Apto. 91
Sao Paulo - SP, Brazil
T: 55 11 67 51 34

Schroder, Karla Ketylin De Lima
Photographic Reporter
FREELANCER
Rua Medina 240 C/2 apt/201 - Meier
Rio de Janeiro, Brazil
T: 55 21 5956094

Schroeder, Christa, Sra.
Vice-President
SOZED
R. Getulio das Neves, 26, apto. 101 - Jardim Botanico
Rio de Janeiro - RJ, Brazil
T: 55 21 226-0266

Schubert, Klaus Juergen, Sr.
ILDES/FES
Rua. Lisboa 191
Sao Paulo - SP 05413, Brazil
T: 55 11 282-3900 F: 55 11 280-1916
TL: 1180616

Schulz, Luis, Sr.
Director
ADRA
Caixa Postal 0-2600
Brasilia - DF 70279, Brazil
T: 55 61 244-0066 F: 55 61 244-0705
TL: 1612

Schumacher, Jorge Machado Fernandes, Sr.
Representante
ASSOCIACAO BRASILEIRA DE CACA E CONSERVACAO
Rua Silva Rosa, 170 apto 501
Rio de Janeiro - RJ 21050, Brazil
T: 55 21 201 31 94

Schuman, Gustavo, Sr.
Tesoureiro
IGREJA ADVENTISTA DO SETIMO DIA/HOSPITAL ADVENTISTA SILVESTRE
Ladeira dos Guararapes, s/n - Sta. Tereza
Rio de Janeiro - RJ, Brazil
T: 55 21 273-5422 F: 55 21 273-5422

Schurink, Jos
Correspondent
ALGEMEEN DAGBLAD
Caixa Postal 137
Sao Roque - SP, Brazil
T: 55 11 425 4775 F: 55 11 4255824

Schwamborn, Friedhelm, Dr.
Diretor
SERVICO ALEMAO DE INTERCAMBIO ACADEMICO
R. Pres. Carlos de Campos 417
Rio de Janeiro - RJ 22231, Brazil
T: 55 21 5533296 F: 55 21 5539261

Schwartz, Vera Lucia
ARTE RIO
Rio de Janeiro, Brazil

Schwarz, Liviu B., Sr.
Executive Director
PLASTIVIDA
Rua Alexandre Dumas, 2016
Sao Paulo - SP 04707, Brazil
T: 55 11 247-6911 F: 55 11 247-8981
TL: 011/57690, E: 51583

Scigarro, Claudio Monteiro
Assistente
WDR GERMAN TV CHANNEL
Rua Eca de Queiroz, 479 casa 12
Sao Paulo - SP, Brazil
T: 55 11 5721262 F: 55 11 5728959

Scipioni, Lamberto
REVISTA PLANETA
Sao Paulo, Brazil

Scorza, Antonio Jose
Photographer
AGENCE FRANCE PRESSE
Rua Mexico 21,7 andar
Rio de Janeiro, Brazil
T: 55 21 240-6634 F: 55 21 262-6717

Scott, Phili Conrad, Sr.
Prof. De Ciencias Biologicas
UNIVERSIDADE SANTA URSULA
Rua Fernando Ferrari, 75
Rio de Janeiro - RJ 22231, Brazil
T: 55 21 551-5542 F: 55 21 551-6446

Scotti, Antonio, Sr.
Diretor
MORENO
Rua Xavier da Silveira, 23
Rio de Janeiro - RJ, Brazil
T: 55 21 287 7643

Scovino, Ronald Jose Terra
Rua Domingues De Sa, 350/902
Niteroi - RJ 24.200, Brazil
T: 55 21 710-3455

Scucato, Ronaldo, Sr.
Vice-Presidente
ORG. COOP. BRASILEIRAS
SQS, Ed. Baracat, 4. andar
Brasilia - DF Cep.: 70309-900, Brazil
T: 55 61 225 0275 F: 55 61 226 8766
TL: 1879

Scussel, Lucia, Sra.
Geografa
Av. 41 No. 535 - Cidade Jardim
Rio Claro - SP 13500, Brazil
T: 55 195 34-8939

Seabra, Marsel, Mr.
INSTITUTO DA CULTURA NOVA
Rua das Laranjeiras, 102/701
Rio de Janeiro - RJ 22241, Brazil
T: 55 21 2450809

Searlatelle, Raquel, Sra.
Director
CMB
Rua Avanhadava no. 756
Sao Paulo - SP 01306, Brazil
T: 55 11 2564354 F: 55 11 2577648

Sebastiao, Loni
AMNP - CA
Rua Joao Manoel 290
Cruz Alta, Brazil
T: 055 3221631

Secundino, Robson
GLOBO ECOLOGIA
Rio de Janeiro, Brazil

Sedane, Paulo Ferreira, Sr.
Seguranca
BANCO AMERICA DO SUL YASUDA
Av. Brigadeiro Luiz Antonio, 2020 - 5o andar
Sao Paulo - SP 01318-911, Brazil
T: 55 11 285-1411 F: 55 11 285-1426

Segamarchi, Ana Lucia, Ms.
GOVERNO DE SAO PAULO
Sao Paulo - SP, Brazil

Seger, Celso D., Sr.
Biologo
SOC. DE PESQ. EM VIDA SELVAGEM E EDUCACAO AMBIENTAL
Rua Guttemberg 345
Curitiba - PR 80420-030, Brazil
T: 55 41 222-3425 F: 55 41 222-3425

Segovia, James, Mr.
PROUTIST UNIVERSAL
Caixa Postal 12117
Sao Paulo - SP 12117, Brazil
T: 55 11 950-9086 F: 55 11 298-6571

Seligmann, Peter, Mr.
Chairman Of The Board
CONSERVATION INTERNATIONAL
Rua Bueno Brandao 393
Belo Horizonte - MG 31010, Brazil
T: 55 31 226-5145 F: 55 31 222-8429

Sena, Alberto
Editor
ESTADO DE MINAS
R Goias 36
Belo Horizonte - MG, Brazil
T: 55 31 273 2322 F: 55 31 2268070

Sena, Paulo Sergio De, Sr.
UNIVERSIDADE DE TAUBATE
Praca Marcelino Monteiro 63
Taubate - SP 12100, Brazil
T: 11 122327555

Sendra, Claudia
CHICO JR. & ASSOCIADOS
Rio de Janeiro, Brazil

Senna, Euclides Goncalves, Mr.
Presidente
MOV. DEFESA DE PORTO SEGURO
Av. Portugal 368
Porto Seguro - BA 45820/000, Brazil
T: 073288 2488

Sennahill, Telenia Maria De
OSHO SPIRITUAL HEALTH ORGANIZATION
Rua dos Araujos,11-4 Bl. 2/301
Rio de Janeiro - RJ 20.521, Brazil
T: 55 21 264-5790

Sennewald, Stefan, Dr.
Perito
SUREHMA
Rua Eng. Reboucas 1206
Curitiba - PR 80215 100, Brazil
T: 55 41 225 6283

Senra, Ricardo
Editor
REDE GLOBO
R. Lopes Quintas 303
Rio de Janeiro -RJ 2, Brazil
T: 55 21 2947732

Seovino, Ronald Jose Terra, Sr.
Rua Domingues de Sa, 350/902 - Icarai
Niteroi - RJ 24200, Brazil
T: 55 21 710-3455

Serpa, Ricardo
SISTEMA JORNAL DO BRASIL
Av. Brasil 500
Rio de Janeiro -RJ, Brazil
T: 55 21 5854422 F: 55 21 5854428

Serquiz, Daniel, Jr., Mr.
GUARANA TIBIRICA
RUA BENTO LISBOA # 54- 1 Andar
Rio de Janeiro - RJ 22221, Brazil
T: 55 21 205-8963 F: 55 21 556-3195

Serra, Claudia De Pavia
Reporter
RADIOBRAS
Brasilia, Brazil

Serra, Jose Roberto
SISTEMA JORNAL DO BRASIL
Av. Brasil 500
Rio de Janeiro -RJ, Brazil
T: 55 21 5854422 F: 55 21 5854428

Serret, Pedro Ernesto Roldan-Toledo, Sr.
Rua Voluntarios Da Patria 114/607
Rio de Janeiro - RJ, Brazil
T: 55 21 266 6870

Sertie, Jayme Aboim, Sr.
IECAM
Dr. Moreira 250
Manaus - AM 69000, Brazil

Servadei, Massimo, Sr.
Rua Decio Vilares
Rio de Janeiro - RJ, Brazil
T: 55 21 255 8254

Servulo, Milton
REDE RECORD
Av. Miruna 713
Sao Paulo _SP 04084, Brazil
T: 55 11 5429000 F: 55 11 612228

Sette, Barbara Dauster, Ms.
IBASE
Rio de Janeiro - RJ, Brazil

Severo, Maria Isabel Guimaraes, Ms.
GOVERNO DO ESTADO DA BAHIA - SEC. DE EDUCACAO
Ilheus - BA, Brazil
F: 55 73 212-2195

Sevje, Svein, Mr.
Embassy Counsellor
NORWEGIAN DELEGATION TO UNCED
Embaixada Real de Noruega, Caixa Postal 07 0670
Brasilia - DF 70351, Brazil
T: 55 61 243-8720 F: 55 61 242-7989

Sexauer, Michael, Sr.
Representative
HONST H. BAUMANN COMMUNICATION + ART
Rua Pereira da Silva 251/Bl.1/302
Rio de Janeiro - RJ, Brazil
T: 55 21 205-2326 F: 55 21 245-7304

Sforza, Roberto, Mr.
CENTRO DE PESQUISAS E MANEJO DE ABROLHOS
Rua Almirante Pereira Guimares 79/701 Leblon-Rio
Rio de Janeiro - RJ, Brazil
T: 55 21 239-8181

Shadt, Christian, Mr.
Student
R. General Dionisio 63, Botafogo
Rio de Janeiro - RJ, Brazil
T: 55 21 286-0303

Sharp, Sonia Maria G., Sra.
COMUNIDADE BAHA'I
Rua do Cedro 158 - Vila B
Foz Iguacu - PR 85890, Brazil
T: 55 455 735347

Shayani, Farhad, Sr.
Presidente
SOCIEDADE BRASILEIRA DE MEDICOS PELA PAZ
Rua Prof. Clovis Jaguaribe, 195 - Bairro Bom Pastor
Juiz de Fora - MG 36100, Brazil
T: 55 32 211 1698

Shellenberger, Michael, Mr.
Pesquisador
SOCIEDADE DE DEFESA - DIREITOS HUMANOS
Rua Antonio Rayol 190 - Centro
Sao Luis - MA 65000, Brazil
T: 55 85 222-8546

Shiba, Hiroshi, Sr.
Gerente
Rua Paraiba 36
Rio de Janeiro - RJ, Brazil
T: 55 21 273-8346 F: 55 21 293-3894

Shimizu, Kazuo, Mr.
Guia De Turismo
TURISMO
Rua Nascimento Silva, 76/104 - Ipanema
Rio de Janeiro - RJ, Brazil
T: 55 21 227 6649

Shishao, Bao
Redator Internacional
DIARIO DO POVO - CHINA
SHIS QL8 CJ.8 Casa 10
Brasilia - DF, Brazil
T: 55 61 2481855

Sholentzov, Boris, Jr., Sr.
Fazenda Pindorama Cx. Postal 03
Sao Simao 75890, Brazil

Shull, Roger
INTERNATIONAL TELEVISION NEWS
Av. Rio Branco, 25/9 andar
Rio de Janeiro RJ, Brazil
T: 55 21 2335480

Shun, George Hawryly
BRASIL ENERGIA
Rua do Riachuelo, 109/13 andar
Rio de Janeiro - RJ, Brazil
T: 55 21 2241830 F: 55 21 2241731

Shutze, Maria Luize, Sra.
Prof. De Ciencis Biologicas
UNIVERSIDADE SANTA URSULA
Rua Fernando Ferrari, 75
Rio de Janeiro - RJ 22231, Brazil
T: 55 21 551-5542 F: 55 21 551-6446

Siag, Isac Paschoal
PALMARES PRODUCOES E JORNALISMO LTDA
Rio de Janeiro, Brazil

Siano, Luiz Roberto Leven
ASSOCIACAO BRASILEIRA DE DIREITO MARITIMO
Rio de Janeiro, Brazil

Signor, Luiz Augusto, Mr.
ASSOCIACAO BENTO GONCALVENSE DE PROTECAO AO AMBIENTE
Rua Candido Costa 194 Edifio Kemet Sala 301
Bento Goncalves - RS, Brazil
T: 55 54 252-4131 F: 55 54 252-1427

Sil, Ana
Photographer
FREELANCER
R Professor Manoel Ferreira 89/203 Gavea -
Rio de Janeiro - RJ, Brazil
T: 55 21 259 4353

Silva, Ademir Milo Motta Da, Sr.
Environment Dept. Director
FUNDACAO AGUA VIVA - FAVI
Caixa Postal 62
Sideropolis-SC, Brazil
T: 55 484 35 30 81

Silva, Adriana Goncalves Moreira E, Sra.
Reporter
VISAO MUNDIAL BRASIL
Rua Antonio de Albuquerque 788 - Funcionarios
Belo Horizonte - MG 30110, Brazil
T: 55 31 221 8944 F: 55 31 227 2995
TL: 316052

Silva, Adriana Nunes De
Journalist
RADIO GLOBO
Rua do Russel, 434 Gloria
Rio de Janeiro RJ, Brazil
T: 55 21 2058282

Silva, Aida Maria Farias Da, Ms.
Deputada Estadual
ASSEMBLEIA LEGISLATIVA
Rua Apinages 690, Altos
Belem - PA 66000, Brazil
T: 55 91 223-7188

Silva, Airson Medeiros, Sr.
Tesoureiro
ASSOC.INTERAMERICANA. ENG. SANITARIO E AMBIENTAL
Av. Beira Mar, 216 - 11 Andar
Rio de Janeiro - RJ 20021, Brazil
T: 55 21 210-3221 F: 55 21 262-6838

Silva, Alberto De Lemos M. Da
Diretor Responsavel
MOVIMENTO ECOLOGICO
Av.Nilo Pecanha, 12 Gr.801
Rio de Janeiro, RJ, Brazil
T: 021 2424374

Silva, Albimar Alves
REDE GLOBO
R. Lopes Quintas 303 J. Botanico
Rio de Janeiro - RJ, Brazil
T: 55 21 5292000 F: 55 21 2942042

Silva, Alexandre Ferreira Da
Technician
REDE GLOBO
R. Lopes Quintas 303
Rio de Janeiro -RJ 2, Brazil
T: 55 21 2947732

Silva, Alexandre Giovanni P. Lima
JORNAL TRIBUNA DO SERVIDOR PUBLICO
Rio de Janeiro, Brazil

Silva, Alexandre Jose Rebello Da
SISTEMA GLOBO DE RADIO
R. de Russel 434
Rio de Janeiro RJ, Brazil
T: 55 21 2058282 F: 55 21 2054096

Silva, Altima Alves Da, Sra.
Nal. Coordinator
EXECUTIVA NACIONAL DE ASSISTENCIA JURIDICA
Av. Generalissimo Deodoro 235 - Umarizol
Belem - PA 66030, Brazil
T: 55 91 241-5551

Silva, Americo Araujo
Tecnico De Audio
SISTEMA GLOBO DE RADIO
Rua do Russel, 434
Rio de Janeiro - RJ, Brazil
T: 55 21 2058282 F: 55 21 2054096

Silva, Ana Carolina Martins Da, Sra.
2a. Secretaria
GRUPO ECOLOGICO SENTINELA DOS PAMPAS
Rua Morom, 2032 - Centro
Passo Fundo - RS 99050, Brazil

Silva, Ana Lucas Da, Sra.
Acupunturista
INSTITUTO DE CULTURA CHINESA
Estrada do Rio Grande 4885 - Jacarepagua
Rio de Janeiro - RJ, Brazil
T: 55 21 446-7262

Silva, Ana Lucia, Sra.
C E T E S B
Rua Tabuapa 81
Sao Paulo - SP, Brazil

Silva, Ana Maria Rodrigues, Sra.
Presidente
CONFEDERACAO DAS MULHERES DO BRASIL
Rua Avanhadava, 756
Sao Paulo - SP 01306, Brazil
T: 55 11 256-4354 F: 55 11 257-7648

Silva, Ana Virginia, Sra.
Av. Aguia De Haia 2100 Bl.3 Apt.54
Sao Paulo - SP 03694, Brazil

Silva, Antonio Carlos Milla
TVN PRODUCOES ARTISTICAS
Sao Paulo, Brazil

Silva, Antonio Da, Mr.
Cientista
INSTITUTO FLORESTAL DE SAO PAULO
Caixa Postal 1322
Sao Paulo - SP 01059-970, Brazil
T: 55 11 952-8555 F: 55 11 204-8067
TL: 228.77-BR

Silva, Antonio Orlando Alves Da, Sr.
PREFEITURA MUNICIPAL
Praca Maua s/n 1 andar
Santos - SP 11010, Brazil
T: 55 132 342-2093 F: 55 132 345-809
TL: 131048

Silva, Ardemirio, Mr.
ADUNICAMP
Cidade Universitaria Unicamp Caixa Postal 6158
Campinas - SP, Brazil
T: 55 192 39-1148 F: 55 192 39-5229

Silva, Arlete
JORNAL MAIORIA FALANTE
Rio de Janeiro, Brazil

Silva, Benedita Da, Ms.
Fed. Deputy
STATE OF RIO DE JANEIRO
Rio de Janeiro - RJ, Brazil

Silva, Braz Gomes Da
Editor
REDE GLOBO
R. Lopes Quintas 303
Rio de Janeiro -RJ 2, Brazil
T: 55 21 2947732

Silva, Carlos Alberto, Sra.
Diretor De Comunicacao
IGREJA ADVENTISTA DO SETIMO DIA/ASSOCIACAO RIO DE JANEIRO
Rua do Matoso, 97 - Praca da Bandeira
Rio de Janeiro - RJ 20270, Brazil
T: 55 21 266-2841

Silva, Carlos Alberto Da
Reporter
REDE GLOBO
R. Lopes Quintas 303
Rio de Janeiro -RJ 2, Brazil
T: 55 21 2947732

Silva, Carlos Augusto Da
MANCHETE
R. do Russel 804
Rio de Janeiro -RJ 2, Brazil
T: 55 21 2652012 F: 55 21 2855148

Silva, Carlos Delfino Da
REDE RECORD
Av. Miruna 713
Sao Paulo _SP 04084, Brazil
T: 55 11 5429000 F: 55 11 612228

Silva, Carlos Gilberto Pessoa Da
Photographer
JORNAL A SEMENTE
R.Candido Neves 5 Vila Kosmos
Rio de Janeiro -RJ, Brazil
T: 55 21 3521673

Silva, Carlos Nicodemos Oliveira, Sr.
Advogado
ASSOC.BENEFICENTE SAO MARTINHO
Rua Riachuelo,7
Rio de Janeiro - RJ 20231, Brazil
T: 55 21 221 1453 F: 55 21 242 7629

Silva, Carlos Raffo De A. E
Technician
REDE GLOBO
R. Lopes Quintas 303
Rio de Janeiro -RJ 2, Brazil
T: 55 21 2947732

Silva, Carlos Roberto Amorim Da
Producer
REDE GLOBO
R. Lopes Quintas 303
Rio de Janeiro -RJ 2, Brazil
T: 55 21 2947732

Silva, Carlos Roberto Barbosa
SISTEMA GLOBO DE RADIO
R. de Russel 434
Rio de Janeiro RJ, Brazil
T: 55 21 2058282 F: 55 21 2054096

Silva, Celia Christina Correa Da
Secretaria
BCO. INTERAMERICANO DE DESENVOLVIMENTO
Caixa Postal 08738
Brasilia - DF 70312, Brazil
T: 55 61 321-7848 F: 55 61 321-3112
TL: 61-1204

Silva, Celio Teodoro Da
FUND. CULTURAL DE RADIODIFUSAO EDUCATIVA DA REGIAO DOS LAGOS
Av. Jose Bento Ribeiro Dantas 265 Manguinhos 28905
Armacao dos Buzios -, Brazil
T: 55 246 2753848 F: 55 21 2956494

Silva, Cenira De F. E
Jornalista
JORNAL VEGAN
Rua do Richuelo, 220/406
Rio de Janeiro, RJ, Brazil
T: 021 2220496 F: 021 7762108

Silva, Cesar Faria E
Presidente
VEGAN - A VOZ DO VEGETARIANO
Rua do Riachuelo, 220/406
Rio de Janeiro, RJ, Brazil
T: 021 2220496

Silva, Cezar De Faria E
Editor-Diretor
JORNAL VEGAN
Rua do Richuelo, 220/406
Rio de Janeiro, RJ, Brazil
T: 021 2220496 F: 021 7762108

Silva, Clarita Pacheco E, Sra.
Relacoes Puplicas
AMNP-SP
Al. Gabriel Monteiro Da Silva 1465 - Apt. 11
Sao Paulo - SP 01441, Brazil
T: 55 11 852-6012

Silva, Claudia Fernanda Da
UNIVERSIDADE GAMA FILHO
Rua Manuel Vitorino, 625
Rio de Janeiro, Brazil
T: 55 21 2697272

Silva, Claudio Gustavo Da
MADE FOR TV
Rio de Janeiro, Brazil

Silva, Claudio Ornellas Da
ETHOS COMUNICACAO
Rio de Janeiro, Brazil

Silva, Cleber Palma, Sr.
Professor
UNIVERSIDADE DO RIO GRANDE
Rua Eurico Blanchini 655
Rio Grande 96200, Brazil
T: 55 323200

Silva, Clodomir Monteiro Da, Sr.
IECAM
Dr. Moreira 250
Manaus - AM 69000, Brazil

Silva, Clodomir Padilha Alves, Sr.
Representante
CONFEA
SHIS 2I 15-10-16
Brasilia - DF 71635-300, Brazil
T: 55 61 248-0279

Silva, Creso Pessurno Nunes Da
SECRETARIA DE ESTADO DE CULTURA DO RIO DE JANEIRO
Rua Ramalho Ortigao, 9/4 andar, Centro
Rio de Janeiro RJ 2, Brazil
T: 55 21 2427012

Silva, Daphne Adriane, Sra.
Biologa
FEMA- FUNDACAO ESTADUAL MEIO AMBIENTE
Cpa Centro Politico Administrativo
Cuiaba - MT 78070, Brazil
T: 55 125 361-2095 F: 55 125 627-1527

Silva, Delso Correa Da, Sr.
Coordinator
FUNDACAO BIODIVERSITAS
Rua Maria Vaz de Melo, 71
Belo Horizonte - MG 31250, Brazil
T: 55 31 443-2119 F: 55 31 441-7037

Silva, Durvalino
CORREIO FLUMINENSE
Rio de Janeiro, Brazil

Silva, Edmilson Joao
Tecnico De Audio
SISTEMA GLOBO DE RADIO
Rua do Russel, 434
Rio de Janeiro - RJ, Brazil
T: 55 21 2058282 F: 55 21 2054096

Silva, Edson Ap., Sr.
Diretor
SINDICATO DOS TRABALHADORES EM AGUA/ESGOTO E MEIO AMBIENTE DO ESTADO DE SAO PAULO
Rua dom Rodo 34 - Ponte Pequena
Sao Paulo - SP 01109, Brazil
T: 55 11 229-5033 F: 55 11 229-5033.220

Silva, Eduardo Sadicoff
Interprete
UNIVERSIDADE ASPERJ
Av.N.S.Copacabana, 787/402
Rio de Janeiro - RJ, Brazil
T: 55 21 2572726 F: 55 21 2572726

Silva, Elenice Santos
HOSPITAL ADVENTISTA SILVESTRE
Ladeira dos Guararapes, 263 - Santa Tereza
Rio de Janeiro - RJ, Brazil
T: 55 21 265-0212

Silva, Eliezer Alonso Da
TV RIO
Rio de Janeiro, Brazil

Silva, Enivaldo Alves, Mr.
Diretor Do Depto. Est. E Infom
SEC. DO MEIO AMBIENTE CIENCIA E TECNOLOGIA DO DISTRITO FEDERAL
Srtvs. Q. 701,Lote 01,6 Andar C.Empresarial Assis Chateaubriand
Brasilia - DF 70340, Brazil
T: 55 61 321 6842 F: 55 61 321 6791

Silva, Eumar Luiz Da
Technician
REDE GLOBO
R. Lopes Quintas 303
Rio de Janeiro -RJ 2, Brazil
T: 55 21 2947732

Silva, Flavia Martins Lins E
Reporter
JORNAL DA PUC
R Marques de Sao Vicente 351/202 Gavea
Rio de Janeiro - RJ, Brazil
T: 55 21 239 1451 F: 55 21 3224215

Silva, Francisco De Assis Da
Technician
REDE GLOBO
R. Lopes Quintas 303
Rio de Janeiro -RJ 2, Brazil
T: 55 21 2947732

Silva, Frei Jose Alamiro Andrade
FRANCISCANS AT THE UNITED NATIONS
SINFRAJUPE do Brasil, CP 90174
Petropolis - RJ 25655, Brazil
T: 55 242 42 52 47

Silva, Haroldo Gomes Da, Mr.
Coordenador Nacional
PASTORAL OPERARIA - BRASIL
Av. Pres. Kennedy, 1861 - Sala 08 Duque de Caxias
Rio de Janeiro -RJ 25020 - 000, Brazil
T: 021 771 3459 F: 021 772 5465

Silva, Helena Bertho Da, Ms.
Consultora
AUTONOMA
Rua Sao Clemente 45/605 Bloco B
Rio de Janeiro - RJ 22260, Brazil
T: 55 21 226-7017

Silva, Hilda Da
Photographer
REVISTA FLUMINENSE
Rua Republica do Libano, 61 Sala 1120, Centro
Rio de Janeiro RJ, Brazil
T: 55 21 2429809

Silva, Hopheni Da, Sra.
Procurador
ASSEMBLEIA LEGISLATIVA
Praca Joao Clemaco S/N Cidade Alta
Vitoria - ES 29.000, Brazil
T: 55 27 322-0055

Silva, Iara Da Cruz Froes
Editor
REDE GLOBO
R. Lopes Quintas 303
Rio de Janeiro -RJ 2, Brazil
T: 55 21 2947732

Silva, Ilda Lopes Rodrigues Da, Ms.
Coordenadora
REDE PARA INFANCIA E FAMILIA DA AMERICA LATINA E CARIBE - NUCLEO BRASIL
Rua Enes De Souza 100, Casa 3 - Tijuca
Rio de Janeiro - RJ 20521, Brazil
T: 55 21 254-7544 F: 55 21 246-7629

Silva, Indaya Silva, Ms.
GERMEN / PLANURB
R. Codor Jose Alves Ferreira,129/102 - Garcia
Salvador - BA 40145, Brazil
T: 55 71 235 2927 F: 55 71 241 7034
TL: 71-3265

Silva, Ivani Pires Monteiro Da, Sra.
Director
ASSOCIACAO BRASILEIRA DE ECOLOGIA
Av. Nilo Pecanha 12 Gr 801/803 - 8o. andar
Rio de Janeiro - RJ 20020, Brazil
T: 55 21 242-4374 F: 55 21 252-6203

Silva, Jacyra, Ms.
WOMEN'S CLUB DE PETROPOLIS
Rua Locarno 275 - Morim
Petropolis - RJ, Brazil
T: 55 242 24-9483

Silva, Jaime Alves Da
REDE RECORD
Av. Miruna 713
Sao Paulo _SP 04084, Brazil
T: 55 11 5429000 F: 55 11 612228

Silva, Jair Sarmento, Sr.
Presidente
FUNDACAO BRASIL-FRANCA
Sqn-304 Bloco D-Apt. 412
Brasilia - DF 70736, Brazil
T: 55 61 2257438 F: 55 61 2627782

Silva, Janiliete C., Sra.
Instr. Ens. 2-Grau
SENAI - PR
Rua Chile 1678
Curitiba - PR 80220, Brazil
T: 55 41 233-0322 F: 55 41 233-0322
TL: 4135141

Silva, Jaqueline, Ms.
Jornalista
RUA timbiras 1633
Belo Horizonte - MG 30140, Brazil
T: 55 31 224-0782

Silva, Jeannine Leal Fernandes Da
Gerente De Comunicacao
JOAO FORTES ENGENHARIA S/A
Praia de Botafogo 228/18 andar Botafogo
Rio de Janeiro -RJ, Brazil
T: 55 21 5520022/R:211 F: 55 21 5521247

Silva, Joao Custodio Da
MANCHETE
R. do Russel 804
Rio de Janeiro -RJ 2, Brazil
T: 55 21 2652012 F: 55 21 2855148

Silva, Jonny Carlos Da, Sr.
Environment Desk
COMUNIDADE BAHA'I
Caixa Postal 1081
Florianopolis - SC 88000, Brazil
T: 55 482 341123

Silva, Jorge A L Da
MANCHETE
Rio de Janeiro, Brazil

Silva, Jorge Alexandre
Cameraman
MTV - RIO
Praca Nossa Senhora da Gloria 46
Rio de Janeiro -RJ, Brazil
T: 55 21 2855946 F: 55 21 2055394

Silva, Jorge Da
REDE GLOBO
R. Lopes Quintas 303 J.Botanico
Rio de Janeiro RJ 22, Brazil
T: 55 21 2947732

Silva, Jorge Luis Vieira
TV RIO
Rio de Janeiro, Brazil

Silva, Jorge Mauricio M., Mr.
President
IEDA
R. Mamedio Amaral da Silva 258, J. Equatorial
Macapa - AP 68900, Brazil
T: 55 96 223-5720

Silva, Jose Carlos Miguel Da, Mr.
IBAM
Rio de Janeiro - RJ, Brazil

Silva, Jose Claudio Pinheiro Da
Director
JORNAL BRAZIL EXPORT
Av. Franklin Roosevelt 194 Sobre Loja Conj.202
Rio de Janeiro -RJ 2, Brazil
T: 55 21 2629920 F: 55 21 2404705

Silva, Jose De Arymathea L.
JORNAL TRIBUNA DO SERVIDOR PUBLICO
Rio de Janeiro, Brazil

Silva, Jose Domingos Cantanhede De, Mr.
Assessor
SOCIEDADE MARANHAENSE DE DIREITOS HUMANOS
Rua Antonio Rayol 290 - Centro
Sao Luiz - MA 65015-040, Brazil
T: 55 98 221-3838

Silva, Jose Elias Da
Sao Paulo, Brazil

Silva, Jose Francisco Guerra Da, Mr.
CESP
Rua da Consolacao, 1875, 9 andar, Cerq. Cesar
Sao Paulo - SP, Brazil
T: 55 11 258-2165 F: 55 11 256-2101

Silva, Jose Leonardo Rogaciane Da, Mr.
Rua Candido Mendes, 253-307
Rio de Janeiro - RJ 20241, Brazil
T: 55 21 232-2789

Silva, Jose Lourenco Da
Technician
REDE GLOBO
R. Lopes Quintas 303
Rio de Janeiro -RJ 2, Brazil
T: 55 21 2947732

Silva, Jose Marco Do Nascimento
MANCHETE
R. do Russel 804
Rio de Janeiro -RJ 2, Brazil
T: 55 21 2652012 F: 55 21 2855148

Silva, Jose Maria Da
REDE RECORD
Av. Miruna 713
Sao Paulo _SP 04084, Brazil
T: 55 11 5429000 F: 55 11 612228

Silva, Jose Martins, Sr.
ARQUIPELAGO DE FERNANDO DE NORONHA
Rua Barao de Sao Borja, 243 - Boa Vista
Recife - PE 50070, Brazil
T: 55 81 221-3217 F: 55 81 231-1594
TL: 81.2533

Silva, Jose Pereira Da
JORNAL DO COMMERCIO
Rua de Livramento, 189 Centro
Rio de Janeiro, Brazil
T: 55 21 2337690 F: 55 21 5161180

Silva, Joselito Manoel Da
MANCHETE
R. do Russel 804
Rio de Janeiro -RJ 2, Brazil
T: 55 21 2652012 F: 55 21 2855148

Silva, Josinaldo Soares, Sr.
ARQUIPELAGO DE FERNANDO DE NORONHA
Rua Barao de Sao Borja, 243 - Boa Vista
Recife - PE 50070, Brazil
T: 55 81 221-3217 F: 55 81 231-1594
TL: 81.2533

Silva, Julio Cesar Lisboa Da, Sr.
SATELLITE SYSTEMS DO BRASIL
Av. Rio Branco 12/ 4 Andar
Rio de Janeiro - RJ 20090, Brazil
T: 55 21 233-3788 F: 55 21 263-7047

Silva, Julio Cesar M. Fernandes
Reporter
REDE GLOBO
R. Lopes Quintas 303
Rio de Janeiro -RJ 2, Brazil
T: 55 21 2947732

Silva, Lessandra Da
ABVP
Rua Treze de Maio 489, Bela Vista
Sao Paulo SP 01327, Brazil
T: 55 11 2847862

Silva, Leury Da
MANCHETE
R. do Russel 804
Rio de Janeiro -RJ 2, Brazil
T: 55 21 2652012 F: 55 21 2855148

Silva, Lilian Claudia M.Da
Reporter
REVISTA FERROVIARIA
Rua Mexico, 41 Sl.904
Rio de Janeiro - RJ, Brazil
T: 55 21 2401747/1248 e 5320260 F: 55 21 2400139

Silva, Lorena Souza Lima Da, Sra.
Chefe Editoracao
IDESP
Av. Nazare 871
Belem - PA 06035, Brazil
T: 55 91 224-4411 F: 55 91 225-3414
TL: 091 6905

Silva, Lourdes Pereira, Sra.
FLAMA
Rua Marques de Abrantes, 55 - Flamengo
Rio de Janeiro - RJ 22230, Brazil
T: 55 21 265-5690

Silva, Lucas Dias Da
REDE GLOBO
R. Lopes Quintas 303 J.Botanico
Rio de Janeiro RJ 22, Brazil
T: 55 21 2947732

Silva, Luis Antonio Da
Technician
REDE GLOBO
R. Lopes Quintas 303 Jardim Botanico
Rio de Janeiro -RJ 2, Brazil
T: 55 21 2947732

Silva, Luis Fernando Da
REDE GLOBO
R. Lopes Quintas 303 J. Botanico
Rio de Janeiro - RJ, Brazil
T: 55 21 5292000 F: 55 21 2942042

Silva, Luis Lucien Rosa E, Mr.
Engenheiro Florestal
UFMT
Av. Fernando Correa da Costa S/No.
Cuiaba - MT 78060-900, Brazil
T: 55 065 315-8630 F: 55 065 361-7119

Silva, Luis Miguel Justo, Mr.
Coordenador
SUREHMA/SEMA-PR
Rua: Engenheiro Reboucas, 1206
Curitiba - PR 80215 - 100, Brazil
T: 55 41 200 1919 F: 55 41 225 3411

Silva, Luis Virgilio Afonso Da, Sr.
Coordenador De Imprensa
CENTRO ACADEMICO 'XI DE AGOSTO'
Rua Riachuelo 194
Sao Paulo - SP 01007-000, Brazil
T: 55 11 239-3077 F: 55 11 239-3070

Silva, Luiz, Mr.
Administrador
SERVICO SOCIAL DO COMERCIO
Rua Voluntarios Da Patria, 169
Rio de Janeiro - RJ 22270, Brazil
T: 55 21 286-5152 F: 55 21 286-2638

Silva, Luiz Antonio Rodrigues Da
REDE GLOBO
R. Lopes Quintas 303 J. Botanico
Rio de Janeiro - RJ, Brazil
T: 55 21 5292000 F: 55 21 2942042

Silva, Luiz Carlos
Technician
REDE GLOBO
R. Lopes Quintas 303
Rio de Janeiro -RJ 2, Brazil
T: 55 21 2947732

Silva, Luiz Carlos Cafe Da, Mr.
Advodago
PREFEITURA MUNICIPAL DE CAMACARI
Rua Ezequiel Ponde 53/4o andar
Salvador - BA 40210, Brazil
T: 55 71 237-4747 F: 55 71 821-1580

Silva, Luiz Claudio Rodrigues E
HELIO ALONSO
Rua Muniz Barreto, 51
Rio de Janeiro RJ, Brazil
T: 55 21 5515448

Silva, Luiz Otavio Constantino
Technician
REDE GLOBO
R. Lopes Quintas 303
Rio de Janeiro -RJ 2, Brazil
T: 55 21 2947732

Silva, Luiz Romeiro, Sr.
Cultural Director
FEDERACAO BRASILEIRA DE ASSOCIACOES DE ENGENHEIROS - FEBRAE
Av. Rio Branco 124/20o. andar
Rio de Janeiro - RJ 20042, Brazil
T: 55 21 242-2532 F: 55 21 507-1334

Silva, Luiza Monica Assis Da
Journalist
FREELANCER
QNA casa 22 Jagutinga Norte
Brasilia, Brazil
T: 55 61 5623592

Silva, Lurdinalva, Ms.
Produtora De Cacau
SINDICATO TRABALHADORES RURAIS - BA
Rua Duque de Caxias, 345 - Centro
Itabuna - BA 45600, Brazil

Silva, Manuel Antonio Da
AGENCIA GILLVAN FOTO E VIDEO
Av. Treze de Maio 47/1110
Rio de Janeiro - RJ, Brazil
T: 55 21 220-5400

Silva, Manuela Dias Da
FORUM DO FUTURO
Rio de Jaeiro, Brazil

Silva, Marcia Lopes Da
MANCHETE
Rio de Janeiro, Brazil

Silva, Marcos Antonio Lopes Da, Sr.
FUNDACAO MOKITI-OKADA - IGREJA MESSIANICA MUNDIAL
Rio de Janeiro - RJ, Brazil

Silva, Marcos Henrique B. Da
Technician
REDE GLOBO
R. Lopes Quintas 303
Rio de Janeiro -RJ 2, Brazil
T: 55 21 2947732

Silva, Marcus Vinicius Da
92 GLOBAL FORUM
Rio de Janeiro - RJ, Brazil

Silva, Maria Alaecia Morais, Sra.
Tec. Comunicacao
ASS. FUNCIONARIOS - INCRA
Rua Do Futuro 418 Apto 03
Recife - PE 52050, Brazil
T: 55 81 2210149 F: 55 81 2053099

Silva, Maria Claudia Ferreira Da
Reporter Fotografica
FREELANCER
Av Gilberto Amado 1102 Barra da Tijuca
Rio de Janeiro - RJ, Brazil
T: 55 21 399 8258

Silva, Maria De Lourdes Rodrigues Da
Reporter
UNIVERSDADE FEDERAL DO PARA
Rua Augusto Correa S/N
Belem, Brazil
T: 55 21 2299666

Silva, Maria Do Carmo Custodio Da
Secretary
TELENEWS SERVICE LTDA
Av Princesa Isabel 500/1613
Rio de Janeiro RJ 22, Brazil
T: 55 21 546 6153 F: 55 21 542 4223

Silva, Maria Dos Gracal
Editor
REVISTA TOURIM RIO
Rua Evaisto da Veigo, 41 Sala 707
Rio de Janeiro RJ, Brazil
T: 55 21 5331618 F: 55 21 2401112

Silva, Marita B. A. Da
Reporter
AGENCIA ESTADO
Av. Caetano Alvares, 55
Sao Paulo -SP 02598, Brazil
T: 55 11 8562210 F: 55 11 2652297

Silva, Mariza C. Da
INFORMATIVO 'BRASIL-MEIO AMBIENTE E DESENVOLVIMENTO'
Rio de Janeiro, Brazil

Silva, Milton Ferreira Da, Mr.
GOVERNO DO ESTADO DA BAHIA - SEC. DE EDUCACAO
Ilheus - BA, Brazil
F: 55 73 212-2195

Silva, Nadja Regina Peixoto, Sra.
Tecnica De Educacao Ambiental
FUNDACAO SOS MATA ATLANTICA
Rua Manoel da Nobrega, 456
Sao Paulo - SP 04001, Brazil
T: 55 11 887-1195 F: 55 11 885-1680

Silva, Nalu Faria, Ms.
Psicologa
SOF
Rua Eng. Tomas Watety, 204
Sao Paulo - SP 04742, Brazil
T: 55 11 521 9822

Silva, Natalia Da Pena E
JOAO FORTES ENGENHARIA S/A
Praia de Botafogo 228/18 andar Botafogo
Rio de Janeiro -RJ, Brazil
T: 55 21 5520022/R:211 F: 55 21 5521247

Silva, Olivia Soares Da, Sra.
FUNDACAO ESCOLA DE SERVICO PUBLICO RJ - FESP
Av. Carlos Peixoto, 64
Rio de Janeiro - RJ 22290, Brazil

Silva, Orlando Da
Reporter
JORNAL GAZETA DE NOTICIAS
R. Leandro Martins 72 Centro
Rio de Janeiro -RJ, Brazil
T: 55 21 2630435 F: 55 21 2630584

Silva, Osmar G. Da, Sr.
Vice-Presidente Executivo
SHEET METAL AND AIR CONDITIONING CONTRACTOR'S NATIONAL ASSOCIATION
Al. Barao de Piracicaba 799 - 2 Andar
Sao Paulo - SP 01216, Brazil
T: 55 11 221-5366 F: 55 11 220-5481

Silva, Patricia Henriques Da
Recepcionista
BCO. INTERAMERICANO DE DESENVOLVIMENTO
Caixa Postal 08738
Brasilia - DF 70312, Brazil
T: 55 61 321-7848 F: 55 61 321-3112
TL: 61-1204

Silva, Paulo Cesar Da
Correspondent
JORNAL 'O PROGRESSO'
Rua Godofredo Viana, 853, Centro
Imperatriz, Maranhao, Brazil
T: 55 98 7212908

Silva, Paulo Lima
JORNAL TRIBUNA DO SERVIDOR PUBLICO
Rio de Janeiro, Brazil

Silva, Paulo Mauricio Osorio Da
Editor
REDE GLOBO
R. Lopes Quintas 303
Rio de Janeiro -RJ 2, Brazil
T: 55 21 2947732

Silva, Paulo Oliveira Da
Reporter
O DIA
R. Riachuela 359
Rio De Janeiro, RJ 2, Brazil
T: 55 21 272-8000 F: 55 21 507-1228

Silva, Paulo Roberto
Editor
EDITORA ABRIL
R.da Passagem, 123/8 andar
Rio de Janeiro, RJ, Brazil
T: 021 5468282 F: 021 2759347

Silva, Raimunda Gomes Da, Ms.
Diretora
FEDERACAO DOS TRABALHADORES RURAIS
Rua 25 de Agosto S #3
Miracema, Brazil
T: 55 098 721-5551 F: 55 098 721-5551

Silva, Raimunda Mata, Ms.
CECTR
D. Pedro Ii, 616
Imperatriz-MA 65900, Brazil
T: 091 721-6136

Silva, Renato A. Da
MANCHETE
Rio de Janeiro, Brazil

Silva, Renilde, Ms.
Socia
SINDICATO DOS TRABALHADORES RURAIS DE SAO JOAO DO ARAGUAIA
Rua D. Pedro II, 112
S. Joao Araguaia,PA, Brazil

Silva, Ricardo Ferreira Da
Reporter
SISTEMA GLOBO DE RADIO
Rua do Russel, 434
Rio de Janeiro - RJ, Brazil
T: 55 21 2058282 F: 55 21 2054096

Silva, Ricardo Oliveira Da
Audio Operator
RADIO MANCHETE LTDA
Rua do Russel 766 Gloria
Rio de Janeiro, Brazil
T: 55 21 2059895 F: 55 21 2057078

Silva, Ricardo Sardo
JORNAL TRIBUNA DO SERVIDOR PUBLICO
Rio de Janeiro, Brazil

Silva, Rogerio Alves Da
Editor
RADIOBRAS
Brasilia, Brazil

Silva, Rogerio Miguel Santos Da, Mr.
IBAM
Rio de Janeiro - RJ, Brazil

Silva, Ronaldo Gomes Da
JOVEM PAN TV
Sao Paulo, Brazil

Silva, Rosani Alves Da
EDITORA AZUL
Rio de Janeiro, Brazil

Silva, Rui Da
Technician
REDE GLOBO
R. Lopes Quintas 303
Rio de Janeiro -RJ 2, Brazil
T: 55 21 2947732

Silva, Sandra Ariflor Da
Reporter
FUNTEVE - RADIO MEC
Praca da Republica, 141
Rio de Janeiro RJ, Brazil
T: 55 21 2204466

Silva, Sandro Pereira Da
MANCHETE
R. do Russel 804
Rio de Janeiro -RJ 2, Brazil
T: 55 21 2652012 F: 55 21 2855148

Silva, Sebastiao Vieira Da, Mr.
IGREJA
Rua Ladeira Da Capela 9, Xerem, Duque De Caxias
Rio de Janeiro - RJ 25250, Brazil
T: 55 21 779-1503

Silva, Sergio De Azeredo
Chanceler Fundador
UNIVERSIDADE ASPERJ
Av.N.S.Copacabana, 787/402
Rio de Janeiro - RJ, Brazil
T: 55 21 2572726 F: 55 21 2572726

Silva, Sergio Ferreira Da
Audio Operator
RADIO MANCHETE
Rua do Russel 766 Gloria
Rio de Janeiro RJ, Brazil
T: 55 21 4501484 F: 55 21 2057078

Silva, Sergio Ricardo Da
HELIO ALONSO
Rua Muniz Barreto, 51
Rio de Janeiro RJ, Brazil
T: 55 21 5515448

Silva, Sergio Vaz Da
Editor
AGENCIA ESTADO
Av. Caetano Alvares, 55
Sao Paulo -SP 02598, Brazil
T: 55 11 8562210 F: 55 11 2652297

Silva, Sidney Torquato
Technician
REDE GLOBO
R. Lopes Quintas 303
Rio de Janeiro -RJ 2, Brazil
T: 55 21 2947732

Silva, Silvana Bomfim
ICAE
Sao Paulo, Brazil

Silva, Tania Jandira R. Da, Sra.
Psychologist
CERIS
Rua Dr. Julio Ottoni, 571/35
Rio de Janeiro - RJ 20241-400, Brazil
T: 55 21 265-5177 F: 55 21 285-7276

Silva, Telma Salesa Santana Da, Sra.
CREA-RJ
AV. RIO BRANCO 133 - 21o. ANDAR
Rio De Janeiro - RJ, Brazil

Silva, Terezinha Pereira Da, Ms.
Professora Tribo Surui
CIMI
Aldeia Surui - Caixa Postal 01
S.Dom.do Araguaia 69530, Brazil

Silva, Ubira Benedito Da
UNICEF
Rio de Janeiro, Brazil

Silva, Valeria Cristina Barros Da, Sra.
Tec. Ed. Amb.
SOCIEDADE POTIGUAR DE EDUCACAO - SPEL
Rua Floriano Peixoto, 295, Petropolis
Natal - RN 59020, Brazil
T: 55 84 222-2314 F: 55 84 221-6255

Silva, Vencesil Firmino, Sr.
Diretor
CLUBE DE OBSERV. DE AVES - NUCLEO DE MG - COA-MG
R. Aimores, 656
Belo Horizonte - MG 30140, Brazil
T: 55 31222-6367 R

Silva, Vera Genuina P. Da, Ms.
UNICEF
SBS Edificio Seguradoras 13o Andar
Brasilia - DF, Brazil
T: 55 61 224-7145 F: 55 61 224-5035

Silva, Vera Lucia Neri
JORNAL MAIORIA FALANTE
Rio de Janeiro, Brazil

Silva, Vilson Pereira Da
MANCHETE
R. do Russel 804
Rio de Janeiro -RJ 2, Brazil
T: 55 21 2652012 F: 55 21 2855148

Silva, Viviane C. Ferreira Da, Sra.
Professor
UNIVERSIDADE CATOLICA
Colmeia Favo 6 - Rua do Anfiteatro 181
Sao Paulo - SP 05508, Brazil
T: 55 11 887-5204

Silva, Wagner Cerolim
Reporter
REDE GLOBO
R. Lopes Quintas 303
Rio de Janeiro -RJ 2, Brazil
T: 55 21 2947732

Silva, Wagner Rodrigues Da
MANCHETE
R. do Russel 804
Rio de Janeiro -RJ 2, Brazil
T: 55 21 2652012 F: 55 21 2855148

Silva, Waldise Rossy Clea Lima Da, Ms.
CIA VALE DO RIO DOCE
Av, Graca Aranha, 26
Rio de Janeiro - RJ 20.030, Brazil
T: 55 21 272 4401 F: 55 21 220 0593

Silva, Walid Maia Pinto
FLAMA
Praia do Flamengo, 212 - Flamengo
Rio de Janeiro - RJ, Brazil
T: 55 21 236-4111

Silva, Walter Sabino
Tecnico De Audio
SISTEMA GLOBO DE RADIO
Rua do Russel, 434
Rio de Janeiro - RJ, Brazil
T: 55 21 2058282 F: 55 21 2054096

Silva, Wanderlei Novaes Da
Reporter
RADIO ROQUETE PINTO
Av. Erasmo Braga 118/ 11andar Castelo
Rio de Janeiro -RJ 2, Brazil
T: 55 21 2247727 F: 55 21 2524584

Silva, Wellington G.
JORNAL MAIORIA FALANTE
Rio de Janeiro, Brazil

Silva, Yone Da, Ms.
UNITED METHODIST CHURCH OF BRAZIL
Rua Floriano Peixoto, 195
Santo Antonio - PR 86.430, Brazil

Silva Filho, Nelson Marques, Sr.
PREFEITURA MUNICIPAL
Praca Maua s/n 1 andar
Santos - SP 11010, Brazil
T: 55 132 342-2093 F: 55 132 345-809
TL: 131048

Silvas, Luiz Fernando Neves Da
Photographer
IMAGENS DA TERRA
R. das Laranjeiras 203/202
Rio de Janeiro RJ 2, Brazil
T: 55 21 5562502 F: 55 21 2204653

Silveira, Adriana Oliveira De
Reporter
ESTADO DE MINAS
R. Goias 36 Centro
Belo Horizonte -MG 3, Brazil
T: 55 31 2732322 F: 55 31 2268070

Silveira, Claudio Gaiger
ESCOLA DE COMUNICACOES E ARTES U.S.P.
Caixa Postal 64522
Sao Paulo -SP 05497, Brazil
T: 55 11 8339511 F: 55 11 2583575

Silveira, Cristiana, Sra.
Journalist
UNIVERSIDADE DO RIO DE JANEIRO
Rua Raul Pompeia, 154/804
Rio de Janeiro - RJ 22080, Brazil
T: 55 21 247-4217

Silveira, Iracema Miranda Da, Sra.
Tec. Ed. Amb.
SOCIEDADE POTIGUAR DE EDUCACAO - SPEL
Rua Floriano Peixoto, 295, Petropolis
Natal - RN 59020, Brazil
T: 55 84 222-2314 F: 55 84 221-6255

Silveira, Luciano Marcal, Mr.
Eng. Agronomo
COUAMCA
Rua Candido Graffree No.4, Apt. 201
Rio de Janeiro - RJ 22291, Brazil
T: 55 21 295-8577 F: 55 21 767-4183

Silveira, Luis Antonio
MADE FOR TV
Rio de Janeiro, Brazil

Silveira, Luiz Fernando Nunes
Technician
REDE GLOBO
R. Lopes Quintas 303
Rio de Janeiro -RJ 2, Brazil
T: 55 21 2947732

Silveira, Monica Maria F. N. A.
CONFEDERACAO NACIONAL DA INDUSTRIA
Rio de Janeiro, Brazil

Silveira, Vassia Vanessa Da
FORUM DO FUTURO
Rio de Jaeiro, Brazil

Silveira Lobo, Roberto De Alburquerque, Mr.
Engenheiro Naval
ULTRATEC ENGENHARIA S.A.
Rua Sete de Setembro 48/5 Andar
Rio de Janeiro - RJ, Brazil
T: 55 21 224-7432

Silvestre, Jolsiney Caulia
Technician
REDE GLOBO
R. Lopes Quintas 303
Rio de Janeiro -RJ 2, Brazil
T: 55 21 2947732

Silvia, Carneiro, Ms.
Diretora
ESCOLA AMIGOS DO VERDE
Av. Cristovao Colombo, 3437
Porto Alegre - RS 90440, Brazil
T: 55 512 342-0589

Silvio, Bertolami
Editor
DIE WELTWOCHE
Rua Alm. Alexandrino 3179/201
Rio de Janeiro RJ, Brazil
T: 55 21 2853224 F: 55 21 2853224

Simmonds, Marie-Christine, Sra.
rua montecaseros 288
Petropolis - RJ 25600, Brazil
T: 55 242 431606

Simoes, Eduardo Pires, Sr.
Rua Da Assembleia 45/401
Rio de Janeiro - RJ 20011, Brazil
T: 55 21 507 1628

Simoes, Jose Leonardo Da Silva, Mr.
Voluntario
CRUZ VERMELHA BRASILEIRA
Praca Cruz Vermelha 10/12 - 1o. andar
Rio de Janeiro - RJ, Brazil
T: 55 21 232-6325 F: 55 21 242-6760

Simoes, Marcelo
Redator/Editor
TV Rio
R. Miguel de Frias, 57 Estacio
Rio de Janeiro - RJ, Brazil
T: 55 21 2939888 F: 55 21 2938626

Simoes, Maria Tereza Monteiro, Sra.
Diretora Tesoureira
SINDICATO DOS PROFESSORES DE VOLTA REDONDA
Av. Amaral Peixoto n 199, salas 207-210
Volta Redonda - RJ 27280, Brazil
T: 55 243 422593 F: 55 243 422593

Simoes, Oldenize
RADIO NACIONAL - PROGRAMA GINCANA NACIONAL
Rio de Janeiro, Brazil

Simon, Euripia, Sra.
Estudante
U.F.G.
Rua T-33 #188-Apto.904B -S.Bueno
Goiania - GO 74000, Brazil
T: 55 62 251-5264

Simon, Guilherme, Mr.
UERJ
Rua Sao Francisco Xavier 524
Rio de Janeiro - RJ 20550, Brazil
T: 55 21 284-5794 F: 55 21 284-5794

Simon, Zeno, Mr.
AGAPAN
Caixa Postal 1996
Porto Alegre - RS 90001, Brazil
T: 55 512 228-7352

Simons, Monica Osorio, Dr.
PREFEITURA MUNICIPAL DE GUARULHOS
Rua Joao Marcelo Santoni, 295 - Parque Renato Maia
Guarulhos - SP 07110, Brazil
T: 55 11 208-0104 F: 55 11 940-5166

Simplicio, Jose
JOVEM PAN TV
Sao Paulo, Brazil

Singulane, Leila Marta De C., Ms.
Geologa
FEMA - FUNDACAO ESTADUAL MEIO AMBIENTE
CPA Centro Politico Administrativo
Cuiaba - MT 78080, Brazil
T: 55 065 313 2901

Siqueira, Ana Maria, Ms.
WOMEN'S CLUB
Rua Real Grandeza # 99
Rio de Janeiro - RJ, Brazil

Siqueira, Carla Vieira De
Reporter
FREELANCER
R Ipanema 197/1504 Nova Ipanema/Barra da Tijuca
Rio de Janeiro - RJ, Brazil
T: 55 21 325 9996

Siqueira, Claudia Araujo
Editor
BRASIL ENERGIA
Rua Riachueld 909/3 andar
Rio de Janeiro, Brazil
T: 55 21 2241830 F: 55 21 2241730

Siqueira, Elizete Sherring, Sra.
Tecnica De Nivel Superior
BANDES
Av. Princesa Isabel, 14/13 Andar
Vitoria - ES 29010, Brazil
T: 55 27 223-8339 F: 55 27 223-6307
TL: 27-2131

Siqueira, Luis Augusto
Reporter
GLOBO ECOLOGIA
Rua Santa Alexandrina 343/c3 Rio Comprido
Rio de Janeiro - RJ, Brazil
T: 55 21 502-5451 F: 55 21 273-9087

Siqueira, Ricardo Cavalcanti
EDITORA GLOBO S/A
Rio de Janeiro, Brazil

Siqueira, Roberto Sampaio
Technician
REDE GLOBO
R. Lopes Quintas 303
Rio de Janeiro -RJ 2, Brazil
T: 55 21 2947732

Siviero, Simone De Oliveira, Ms.
Assist. Pesquisa
UNIVERSIDADE ESTADUAL DE CAMPINAS
NEPAM-UNICAMP - Caixa Postal 1170 - CEP 13.081
Campinas - SP, Brazil
T: 55 192 398151 F: 55 192 394717
TL: (019)1150

Sixel, Waldir Barcellos
Technician
REDE GLOBO
R. Lopes Quintas 303
Rio de Janeiro -RJ 2, Brazil
T: 55 21 2947732

Skhwob, Liliane
Reporter
JORNAL O GLOBO
Rua Irineu Marinho, 35
Rio de Janeiro, RJ 2, Brazil
T: 021 2722000

Sleman, Souheil
SISTEMA GLOBO DE RADIO
R. de Russel 434
Rio de Janeiro RJ, Brazil
T: 55 21 2058282 F: 55 21 2054096

Slhessarenk, Serys Marly, Sra.
ASSEMBLEIA LEGISLATIVA
Jardim California Q.H. casa 10
Cuiaba - MT 78000, Brazil

Smith, Ruy Torreao, Jr.
IBAMA
Rio de Janeiro, Brazil

Soares, Airton
FORUM DO FUTURO
Rio de Janeiro, Brazil

Soares, Bittencourt
Editor
BRASIL ENERGIA
Rua do Riachuelo, 109/3 andar, Lapa
Rio de Janeiro, Brazil
T: 55 21 2241830 F: 55 21 2241731

Soares, Carlos Alberto Silveira Netto, Sr.
SINDICATO DOS EMPREGADOS DOS ESTABELECIMENTOS BANCARIOS DE PORTO ALEGRE
Rua Gal. Camara 424
Porto Alegre - RS 90000, Brazil
T: 55 512 227-2199 F: 55 512 227-2478

Soares, Carlos Correa
JOAO FORTES ENGENHARIA S/A
Praia de Botafogo 228/18 andar Botafogo
Rio de Janeiro -RJ, Brazil
T: 55 21 5520022/R:211 F: 55 21 5521247

Soares, Celio, Sr.
Coordenador
FUNDACAO ESCOLA DE SERVICO PUBLICO RJ - FESP
Av. Carlos Peixoto, 64
Rio de Janeiro - RJ 22290, Brazil

Soares, Claudio Motta
GLOBO
Rio de Janeiro, Brazil

Soares, Fernando Arnaldo Teixeira, Sr.
Professor
FACULTADES INTEGRADAS BENNETT
Praia de Botafogo 520 apto.1202
Rio de Janeiro - RJ 22250, Brazil
T: 55 21 295-3371 F: 55 21 541-2237

Soares, Jose Paulo Monteiro, Mr.
Director Executive
EDITORA LIBRIS
Av. Rio Branzo 45/1707
Rio de Janeiro - RJ 200 PO, Brazil

Soares, Katia Geo Leite, Sra.
Diretora De Promocoes
GROWING EARTH ASSOCIATION - GEA
Avenida 1, n 78 - Jardim Vitoria
Belo Horizonte - MG 31950, Brazil
T: 55 31 493-2077 F: 55 31 493-3269

Soares, Luis Bezerra
ROLAND HENZE
Rio de Janeiro, Brazil

Soares, Luis Sergio Mamari De, Sr.
Tecnico
CONFEDERACAO NACIONAL DA INDUSTRIA
Av. Nilo Pecanha, 50 sala 3018
Rio de Janeiro - RJ 20044, Brazil
T: 55 21 262 0181 F: 55 21 262 1465
TL: 22634

Soares, Luiz Carlos De Aguir
O DIA
Rua Riachuelo 359, Centro
Rio de Janeiro RJ 2, Brazil
T: 55 21 2728000 F: 55 21 5071228

Soares, M. Beatriz Da Cunha Santos, Ms.
Coordenadora Iafam
INSTTITUTO DE ASSUNTOS DA FAMILIA - SECRETARIA DE SP
Rua Major Natamall 105
Sao Paulo - SP, Brazil
T: 55 11 258-2047

Soares, Osiris De Paula
Jornalista
JORNAL SOL VIAGEM TURISMO
Caixa Postal 172
Maua - SP 09301, Brazil
T: 55 011 450 4854 F: 55 011 477 4881 (Su

Soares, Pedro, Sr.
Articulista
ECOFORCA
Av. Caetano Alvares, 55
Sao Paulo -SP 02598, Brazil
T: 55 11 8562210 F: 55 11 2652297

Soares, Sergio De Araujo
Technician
REDE GLOBO
R. Lopes Quintas 303
Rio de Janeiro -RJ 2, Brazil
T: 55 21 2947732

Soares Filho, Avaldo, Sr.
Professor
UNIVERSIDADE ESTADUAL DO SUDOESTE DA BAHIA
Estrada do Bem Querer, Km 4, s/n - Candeias
Vit. da Conquista-Ba, Brazil
T: 4241143 F: 4222352

Soares Filho, Jose Maria, Sr.
Eng. Florestal
UNIVERSIDADE FEDERAL RURAL DO RIO DE JANEIRO
Rua Oliveira Fausto ,04- Botafogo
Rio de Janeiro 22280, Brazil
T: 5411391

Soares Santos, Terezinha, Ms.
Vice President
A.M.N.P. LONDRINA
Rua Alagoas, 1439
Londrina PR 86020, Brazil
T: 55 222754 F: 55 241677

Soaza, Moses Katias Ferreira De
Journalist
JORNAL FOLHA DE GAIA
RuaArimateia Cisne, 234 Apeadoupo
S. Luis MA, Brazil
T: 55 48 2212008

Sobeh, Ahmad, Sr.
Embaixador
ORGANIZACAO PARA A LIBERTACAO DA PALESTINA - OLP NO BRASIL
SHIS QI 9 Conjunto 6 Casa 2
Brasilia - DF 70625, Brazil
T: 55 61 248-4760 F: 55 61 248-5879

Sobral, Maria Do Carmo Martins, Ms.
Diretora Meio Ambiente
SEPLAN
Rua Setubal, 1600/201
Recife - PE, Brazil
T: 55 81 3411563 F: 55 81 224 5263

Sobral, Maria Helena Leite
Producer
REDE GLOBO
R. Lopes Quintas 303
Rio de Janeiro -RJ 2, Brazil
T: 55 21 2947732

Sobrinho, Denise Cunha Da Silva
Editor
REDE GLOBO
R. Lopes Quintas 303
Rio de Janeiro -RJ 2, Brazil
T: 55 21 2947732

Soca, Ricardo
Correspondent
EL PAIS
Praia do Flamengo 12/413
Rio de Janeiro, Brazil
T: 55 21 2258812 F: 55 21 2258812

Sochaczewski, Jacques
ICAE / UNIRHMA
12 Rajaji Marg
New Delhi 110011, Brazil
T: 91 11 301-4309

Sodre, Luiz Alberto De Souza
Technician
REDE GLOBO
R. Lopes Quintas 303
Rio de Janeiro -RJ 2, Brazil
T: 55 21 2947732

Sohn, Lidio, Mr.
Diretor De Cultura
PREFEITURA MUNICIPAL DE ARIQUEMES
Al. das Orquideas, 2392 - S.4 - Caixa Postal 222
Ariquemes - RO 78932 000, Brazil
T: 55 69 535 3553

Sol, Luis Carlos Barbosa
Editor
REDE GLOBO
R. Lopes Quintas 303
Rio de Janeiro -RJ 2, Brazil
T: 55 21 2947732

Solem, Knut, Mr.
Consul General, Rio
NORWEGIAN DELEGATION TO UNCED
Real Consulado Geral da Noruega, Caixa Postal 436
Rio de Janeiro - RJ 20000, Brazil
T: 55 21 552-1540 F: 55 21 551-9091

Soler, Antonio Carlos Porciuncula, Mr.
Advogado
CENTRO DE ESTUDOS AMBIENTAIS (CEA)
Rua Henrique Bulle 494- Cassino
Rio Grande-RS 96200, Brazil
T: 55 053232-8821

Soler De Leoni, Margaret, Ms.
Manager
SUSILA- HOLISTIC THERAPY CENTRE
Praca Cruz Vermelha, 9/ XX
Rio de Janeiro - RJ 20230, Brazil
T: 55 21 226-6667 F: 55 21 5333643

Som, Luiz
Diretor
FUNDACAO SOS MATA ATLANTICA
R. Manuel da Nobrega 456
Sao Paulo -SP 04001, Brazil
T: 55 11 8871195 F: 55 11 8851680

Sommer, Miriam, Ms.
Medico
ASSOCIACAO MEDICA HOMEOPATICA BRASILEIRA
Rua Tobias da Silva 267 Sala 403
Porto Alegre - RS 90570, Brazil
T: 55 512 331-5138 F: 55 512 342-9788

Son, Luiz, Sr.
Gerente Geral
FUNDACAO SOS MATA ATLANTICA
Rua Manoel da Nobrega, 456
Sao Paulo - SP 04001, Brazil
T: 55 11 887-1195 F: 55 11 885-1680

Sonoda, Shiyuti, Mr.
Membro
KISSO KOSEI-KAI
Av.Aclimacao 502
Sao Paulo - SP, Brazil
T: 55 11 278-9058

Sook, Silvia
GLOBO
Veneral Garzon 28 Apt 802, Jardin Botanico
Rio de Janeiro, Brazil
T: 55 21 2850448

Sorman, Guy-Bernard
Reporter
AGENCIA ESTADO
Av. Caetano Alvares, 55
Sao Paulo -SP 02598, Brazil
T: 55 11 8562210 F: 55 11 2652297

Sorotiuk, Vitorio, Mr.
Diretor Presidente
ITCF
Desembargador Motta 3384
Curitiba - PR 80430, Brazil
T: 55 41 234-1611 F: 55 41 222-1850

Sorrentino, Marcos, Sr.
Diretor
SODEMAP
Rua do Porto 1793
Piracicaba - SP 13400, Brazil
T: 55 194 33 1845

Sousa, Kuillder Gualberto De, Mr.
Estudante
UFG
Av. Laudelino Gomes De Almeida
Goiania - GO CEP: 74.000, Brazil
T: 55 62 225-9534

Sousa, Mara Juliene De
RADIO UNIVERSITARIA
Universidade Federal de Goias - Setor Oeste Av Anhanguera com Alameda das Rosas
Goiania - GO, Brazil
T: 55 62 233-4473/6896

Sousa, Maria De Fatima Pereira, Ms.
Diretora
AMTR
Rua Das Laranjeiras, 1436
Pedreiras-MA 65725000, Brazil
T: 098 642-2061 F: 098 642-2061

Soutinho, Sergio Tinoco Arpon
Radialista
RADIO TUPI
Rua do Livramento, 189/8 andar
Rio de Janeiro, Brazil
T: 55 21 2032421 F: 55 21 2637323

Souto, Elba De Albuquerque, Sra.
Pesquisadora
INSTITUTO DE ECOLOGIA HUMANA
Av. Dom Manuel de Medeiros s/n - Ed. Newton Bancks - 3o. andar
Recife - PE 52071, Brazil
T: 55 81 441-3931 F: 55 81 441-4697

Souto, Joao Alberto Nobre
Camera
TVSBT
Campo de S.Cristovao, 105
Rio de Janeiro - RJ, Brazil
T: 55 21 5803688 / 5800678 F: 55 21 5809685

Souto, Maria Angelica, Sra.
Plan.Ambiental
CENTRO DE RECURSOS AMBIENTAIS
Rua Rio Sao Francisco, 1 - Monte Serrat
Salvador - BA CEP: 40.425, Brazil
T: 55 71 312-7192 F: 55 71 312-7198
TL: 7148 RAMB

Souza, Adelair Vieira De
Motorista/Tecnico
WDR GERMAN TV CHANNEL
Rua Eca de Queiroz, 479 casa 12
Sao Paulo - SP, Brazil
T: 55 11 5721262 F: 55 11 5728959

Souza, Ademilson Satana De
Technician
REDE GLOBO
R. Lopes Quintas 303
Rio de Janeiro -RJ 2, Brazil
T: 55 21 2947732

Souza, Alciene Freitas Bertholdo De, Ms.
Eng. Sanitarista
FUNDACAO ESTADUAL DO MEIO AMBIENTE - MT
Rua das Dalias, 084 Jardim Cuiaba
Cuiaba - MT 78030, Brazil
T: 55 065 321-1618 F: 55 065 321-2171

Souza, Alexandre Oliveira De
UNICEF
Rio de Janeiro, Brazil

Souza, Alfredo Jose D. Castro
Photographer
ESTADO DE MINAS
R. Goias 36 Centro
Belo Horizonte -MG 3, Brazil
T: 55 31 2732322 F: 55 31 2268070

Souza, Angela Mesquita De, Sra.
Member
GRUPO NATIVA - GOIAS/BR
AOS 06-Bloco A- apt.611
Brasilia - DF 70651, Brazil
T: 55 61 234-1410 F: 55 61 233-8763

Souza, Antonio Altayd De
VIDEOARTE RIO
Rua Senradom 45
Rio de Janeiro, Brazil

Souza, Antonio Augusto De
SISTEMA GLOBO DE RADIO
R. de Russel 434
Rio de Janeiro RJ, Brazil
T: 55 21 2058282 F: 55 21 2054096

Souza, Antonio Carlos De
REVISTA ENFERMAGEM CIENTIFICA
Rio de Janeiro, Brazil

Souza, Antonio Francisco, Mr.
Dirigente Sindical
SINDICATO TRAB. AGUA, ESGOTO E MEIO AMBIENTE - SINTAEMA
Rua Arealva, 158
sao jose dos campos 12220, Brazil
T: 0123 23-4472 F: 011 229-5033

Souza, Antonio Rodrigues De, Sr.
INTERNATIONAL YOUNG CHRISTIAN WORKERS
Rua Condessa de Joaquim, 215, Bairro Bela Vista
Sao Paulo - SP 01320, Brazil
T: 55 11 278-0099

Souza, Barbara M., Sra.
SERVICO MULHER MARGINALIZADA
R. Guilherme Maw, 64, casa 20
Sao Paulo - SP 01105, Brazil
T: 55 11 2284955

Souza, Candida-Luz De
TVN PRODUCOES ARTISTICAS
Sao Paulo, Brazil

Souza, Carlos Alberto Machado De
Reporter
FOLHA DE SAO PAULO
Alameda Barao de Limeira 425
Sao Paulo SP, Brazil
T: 55 11 874 2608/874 2602 F: 55 11 2231644

Souza, Claudia Mendonca De, Sra.
Assessora De Com. Social
CENTRO DE INTEGRACAO EMPRESA-ESCOLA - CIEE
Av Maracana, 1524/1528 - Tijuca
Rio de Janeiro - RJ 20511, Brazil
T: 55 21 268-6447
TL: 2134763

Souza, Claudio Heitor
ROLAND HENZE
Rio de Janeiro, Brazil

Souza, Debora, Sra.
Sociologa
CENTRO DE RECURSOS AMBIENTAIS
Rua Rio Sao Francisco, 01 - Monte Serrat
Salvador - BA CEP: 40.425, Brazil
T: 55 71 312-7192 F: 55 71 312-7198
TL: 7148 RAMB

Souza, Edson Belo C., Mr.
Professor
FUNIOESTE/ FACISA
R. 19, Quadra 9, Casa 8, Vila A
Foz Iguacu - PR 85890, Brazil
T: 55 455 22 2211

Souza, Edvaldo, Mr.
Desenhista
SERVICO SOCIAL DO COMERCIO -DEPARTAMENTO NAC
Rua Voluntarios Da Patria 169
Rio de Janeiro - RJ 22270, Brazil
T: 55 21 286-5152 F: 55 21 286-2638

Souza, Efrem Ribeiro De
Reporter
FOLHA DE SAO PAULO
Alameda Barao de Limeira 425
Sao Paulo SP, Brazil
T: 55 11 874 2608/874 2602 F: 55 11 2231644

Souza, Elaine
JORNAL MAIORIA FALANTE
Rio de Janeiro, Brazil

Souza, Euclisia
Ofic. De Proj.
SAVE THE CHILDREN
Rua Cardeal Arcoverde, 142 - Gracas
Recife - PE 52022, Brazil
T: 55 81 231 1263 F: 55 81 231 1263

Souza, Evandro Luiz Pinheiro, Sr.
Av. Mae Luzia, 558 - Laguinho
Macapa - AP 68900, Brazil
T: 55 96 222 2501

Souza, Georgio Antonio De
Tecnico De Audio
SISTEMA GLOBO DE RADIO
Rua do Russel, 434
Rio de Janeiro - RJ, Brazil
T: 55 21 2058282 F: 55 21 2054096

Souza, Gerson Vieira De
EDITORA AZUL
Rio de Janeiro, Brazil

Souza, Gilmar Dias De
RADIO UNIVERSITARIA FMPB
Paraiba, Brazil

Souza, Gisele Soares De, Ms.
Organizadora
PREFEITURA DE NITEROI
Est. Leopoldo Froes 773
Niteroi - RJ, Brazil
T: 55 21 710-2727 F: 55 21 710-8489

Souza, Helena De Miranda Rosa, Sra.
Praia do Flamengo 60/805
Rio de Janeiro - RJ, Brazil
T: 55 21 245-2977

Souza, Helio Macol
CHICO JR. & ASSOCIADOS
Rio de Janeiro, Brazil

Souza, Heloisa Gonzaga De
PREFEITURA MUNICIPAL
Praca Maua s/n 1 andar
Santos - SP 11010, Brazil
T: 55 132 342-2093 F: 55 132 345-809
TL: 131048

Souza, Jose Carlos De Oliveira
Technician
REDE GLOBO
R. Lopes Quintas 303
Rio de Janeiro -RJ 2, Brazil
T: 55 21 2947732

Souza, Jose Gilberto De, Mr.
Director
YMCA OF MINAS GERAIS
Rua Matias Cardoso, 165 Santo Agostinho
Belo Horizonte - MG, Brazil
T: 55 31 337-3032 F: 55 31 337-1153

Souza, Jose Mauro De
Camera
TVSBT
Campo de S.Cristovao, 105
Rio de Janeiro - RJ, Brazil
T: 55 21 5803688 / 5800678 F: 55 21 5809685

Souza, Josefa Magna A. De, Sra.
IECAM
Dr. Moreira 250
Manaus - AM 69000, Brazil

Souza, Juracy Rabelo De
Operator Of Vt
FUNTELDA
Av. Almirante 735/4 andar
Belem PA, Brazil
T: 55 91 2281000

Souza, Jurandy Dos Santos, Sr.
Economista
CENTRO DE RECURSOS AMBIENTAIS
Rua Rio Sao Francisco, 1 - Monte Serrat
Salvador - BA CEP: 40.425, Brazil
T: 55 71 312-7192 F: 55 71 312-7198
TL: 7148 RAMB

Souza, Katia Luane Gomes De
Special Reporter
JORNAL DO COMMERCIO
Rio de Janeiro, Brazil
T: 55 21 2336562 F: 55 21 2538370

Souza, Luciana Maria B. Da Matta, Sra.
Co-Organizadora
IBFAN - INTERNATIONAL BABY FOOD ACTION NETWORK
Rua Sa Ferreira, 152 apto 501, Copacabana
Rio de Janeiro - RJ 22461, Brazil
T: 55 21 227-1282 F: 55 21 270-2193

Souza, Luiz Carlos De
Reporter
DCI - DIARIO COMERCIO e INDUSTRIA
Av.Rio Branco, 177/6 andar
Rio de Janeiro - RJ, Brazil
T: 55 21 2621016 F: 55 21 2200805

Souza, Luiz Fernando Gil De
Reporter
RADIO MANCHETE LTDA
Rua do Russel 766, Gloria
Rio de Janeiro, Brazil
T: 55 21 2855997 F: 55 21 2057078

Souza, Marco Antonio Ramos
Technician
REDE GLOBO
R. Lopes Quintas 303 Jardim Botanico
Rio de Janeiro -RJ 2, Brazil
T: 021 2947732

Souza, Marco Aurelio R. De
Reporter
REDE GLOBO
R. Lopes Quintas 303
Rio de Janeiro -RJ 2, Brazil
T: 55 21 2947732

Souza, Marcos Antonio Parada De
MANCHETE
R. do Russel 804
Rio de Janeiro -RJ 2, Brazil
T: 55 21 2652012 F: 55 21 2855148

Souza, Marcos Aparecido Felix De
Engineer
AMERICA DO SUL PRODUCOES DE DOCUMENTARIOS LTDA (NAMBEI TSUSHIN)
Av. Paulista 509 /11 andar conj.1104 Cerqueira Cesar
Sao Paulo -SP 01311, Brazil
T: 55 11 2886141 F: 55 11 2512065

Souza, Maria A. Borges De
Editor
RADIOBRAS
Brasilia, Brazil

Souza, Maria Alice Mariano
Reporter
CONTEXTO
Av Calogeras 12/63 Centro
Rio de Janeiro - RJ, Brazil
T: 55 21 2204653 F: 55 21 2204653

Souza, Maria Cibele, Sra.
Assessora
PREFEITURA MUNICIPAL DE BELO HORIZONTE
Av. Afonso Pena 1212/ 5 Andar
Belo Horizonte - MG 30130, Brazil
T: 55 31 223 6837

Souza, Maria Das Gracas, Ms.
OSHO SPIRITUAL HEALTH ORGAN. - OSHO
R. Celestino Basilio 26
Rio de Janeiro - RJ, Brazil
T: 55 21 438-7488 F: 55 21 322-0327

Souza, Maria Jose Cadeco, Ms.
R. Senador Vergueiro 103/401 Flamengo
Rio de Janeiro - RJ, Brazil
T: 55 21 556 2649

Souza, Maria Tereza M. De
Editor
RADIOBRAS
Brasilia, Brazil

Souza, Marise Rodrigues De
RADIO CLUBE DE PERNAMBUCO
Rio de Janeiro, Brazil

Souza, Monica Cadeco, Ms.
R. Sen. Vergueiro 103/401
Rio do Sul - SC 22230, Brazil
T: 55 478 5562649

Souza, Noracy Ruiz De, Ms.
Executive Director
INSTITUTO NASCEE DE ECOLOGIAS E TECNOLOGIAS APROPRIADAS
Rua Timoteo da Costa 297/203
Rio de Janeiro - RJ 22450, Brazil
T: 55 21 2941164

Souza, Patricia Costa Coelho De
EDITORA HORA DO POVO S.A.
Rio de Janeiro, Brazil

Souza, Paulo Renato De, Mr.
Dir. De Operacoes
BCO. INTERAMERICANO DE DESENVOLVIMENTO
Caixa Postal 08738
Brasilia - DF 70312, Brazil
T: 55 61 321-7848 F: 55 61 321-3112
TL: 61-1204

Souza, Paulo Roberto De
SISTEMA GLOBO DE RADIO
R. de Russel 434
Rio de Janeiro RJ, Brazil
T: 55 21 2058282 F: 55 21 2054096

Souza, Paulo Roberto Melo
Editor
JORNAL FOLHA DE GAIA
RuaArimateia Cisne, 234 Apeadoupo
S. Luis MA, Brazil
T: 55 48 2212008

Souza, Paulo Sergio Marques De, Mr.
Director
PROJETO RIO PIAVU
Rua Rui Barbosa no.460 ap. 302 Flamengo
Rio de Janeiro - RJ, Brazil
T: 55 21 5517214 F: 55 21 8642860

Souza, Pedro Ramos De, Sr.
FORUM DAS ONG'S BRASILEIRAS
Palais de Nations
Geneve CH 1211, Brazil
F: 41 22 733-6542

Souza, Roberto Ciancio De
Technician
REDE GLOBO
R. Lopes Quintas 303
Rio de Janeiro -RJ 2, Brazil
T: 55 21 2947732

Souza, Rozemburg Rubeiro Da
Reporter
GRUPO ANONIMO DE CULTURA
Rua Paquequer, 270 Casa 15
Rio de Janeiro, Brazil
T: 55 21 2490753

Souza, Sergio A. De Melo E, Mr.
Gerente De Comunicacao
INSTITUTO BRASILEIRO DE PESQUISAS E ESTUDOS AMBIENTAIS - PRONATURA
Rua Monte Alegre 374 - Santa Teresa
Rio de Janeiro - RJ 20240, Brazil
T: 55 21 224-3994 F: 55 21 222-6742

Souza, Sergio Da Rocha
Rua Dias da Cruz, 303/305
Rio de Janeiro - RJ 20720, Brazil
T: 55 21 249-8247

Souza, Severiano Pereira De
MANCHETE
R. do Russel 804
Rio de Janeiro -RJ 2, Brazil
T: 55 21 2652012 F: 55 21 2855148

Souza, Sheila Maria De Iveira De
EDITORA HORA DO POVO S.A.
Rio de Janeiro, Brazil

Souza, Silvana De Paula, Sra.
EXECUTIVA NAC. DOS ESTUDANTES DE BIOLOGIA-ENEB/RJ
R. J.J. Seabra, 15/202 - Jd. Botanico
Rio de Janeiro - RJ, Brazil
T: 55 21259-0139 R

Souza, Silvino Jose Rodrigues, Sr.
Member
FLAMA
Rua Senador Vergueiro, 184/301
Rio de Janeiro - RJ, Brazil
T: 55 21 551-1725

Souza, Telma De, Sra.
PREFEITURA MUNICIPAL
Praca Maua s/n 1 andar
Santos - SP 11010, Brazil
T: 55 132 342-2093 F: 55 132 345-809
TL: 131048

Souza, Tiorbe
Technician
REDE GLOBO
R. Lopes Quintas 303
Rio de Janeiro -RJ 2, Brazil
T: 55 21 2947732

Souza, Wellington Fernandes De
Technician
REDE GLOBO
R. Lopes Quintas 303
Rio de Janeiro -RJ 2, Brazil
T: 55 21 2947732

Souza, Wilton Fernandes De
Technician
REDE GLOBO
R. Lopes Quintas 303
Rio de Janeiro -RJ 2, Brazil
T: 55 21 2947732

Souza Filho, Aldo De Andrade E, Sr.
Coordenador
I.I.U.LEONAR ORR
R Raimundo Correa 43/302 Copacabana
Rio de Janeiro - RJ 22040, Brazil
T: 55 21 256-8901

Souza Neto, Paulo, Sr.
Vereador
CAMARA MUNICIPAL
Rua 4, Ed. Parthenon Center, 9o. Andar, Centro
Goiania - GO 74000, Brazil
T: 55 62 229-0297 F: 55 62 225-4773

Souzarez, Eliana De
Fotografa
EDITORA MULTIMIDIA
Rua Riachuelo, 221 Cob.03
Rio de Janeiro - RJ, Brazil
T: 55 21 2223192 / 2215894

Souzqa, Luiz Augusto Diogo De
Diretor Executivo
EMCENA COMUNICACOES
Rua da Conceicao, 105/601
Rio de Janeiro - RJ, Brazil
T: 55 21 2330862 F: 55 21 2337953

Spencer, David, Mr.
Public Relations
SUMMER INSTITUTE OF LINGUISTICS
SAI/NO, Lote 'D', Bloco 03
Brasilia - DF 70770, Brazil
T: 55 61 272-1200

Spindel, Cheywa, Ms.
Program Coor.
MAC ARTHUR FOUNDATION
Av. Dr. Arnaldo 1976
Sao Paulo - SP 01255, Brazil
T: 55 11 8647500

Staib, Barbara, Ms.
Journalista
Av.Boa Viagem, 2314
Recife - PE 50020, Brazil
T: 55 81 325-4742 F: 55 81 261-0100

Stalin, Jose, Sr.
Coordenador
ASSOCIACAO BRASILEIRA INTERDISCIPLINAR DE AIDS
Rua Lopes Quintas 576 - Jardim Botanico
Rio de Janeiro - RJ 22460, Brazil
T: 55 21 239-5171 F: 55 21 294-5602
TL: 36466 Base Br

Starec, Ana Marlene, Ms.
Presidente
WOMEN'S INTERNATIONAL ZIONIST ORGANIZATION
Rua Cruz Lima 20 C-01
Rio de Janeiro - RJ 22230, Brazil
T: 55 21 265-5390

Stecher, Karl, Mr.
Economista
S A S O P
Travessa Guia Lopes, 50
Salvador - BA 40120, Brazil
T: 55 71 2428520

Stein, Richard F., Mr.
Camera
2DF-TV ALEMA
Av.Delfim Moreira, 662/704
Rio de Janeiro - RJ, Brazil
T: 55 21 2947434 F: 55 21 5122120

Steinbaun, Volff
GOVERNO DO ESTADO DE SAO PAULO
Sao Paulo, Brazil

Steinberg, Gustavo Soares, Sr.
PAZ UNIVERSAL
Av. Bras Leme 1353
Sao Paulo - SP 02511, Brazil
T: 55 11 290-3002 F: 55 11 050-2726

Stepansky, Daisy, Sra.
Tecnico
SERVICO SOCIAL DO COMERCIO
Rua Voluntarios Da Patria 169
Rio de Janeiro - RJ 20720, Brazil
T: 55 21 286 5152 F: 55 21 286 2638

Sterenberg, Leila
Reporter
REVISTA VIA URBANA
Rua Sao Jose, 90/2005
Rio de Janeiro - RJ, Brazil
T: 55 21 2320157 / 2523276 F: 55 21 2400139

Stern, Elizabeth, Ms.
Rua Dr. Tulio Otoni, 367/101 Santa Tereza
Rio de Janeiro - RJ 20241, Brazil
T: 55 21 245-7390

Stern, Marcos, Sr.
IECAM
Dr. Moreira 250
Manaus - AM 69000, Brazil

Steuer, Ana, Ms.
Engenheira
ROY F. WESTON
Av. Sto Amaro 6238/143
Sao Paulo - SP, Brazil
T: 55 11 522-5783 F: 55 11 522-5783

Stop, Alexandre Soares
Fotografo
A STOP FOTOS & VIDEO
Rua Alvaro Alvim 37/1422-1 Cinelandia
Rio de Janeiro - RJ, Brazil
T: 55 21 240 1192 F: 55 21 240 4469

Storti, Raffaella, Ms.
Cooperante
PROGETTO SVILUPPO-CGIL
Rua Dona Brigida, 299 - Vila Mariana
Sao Paulo - SP 04111, Brazil
T: 55 11 575 5155

Straten, Hessel Van, Mr.
Rua Sambaiba 596 #104 Leblon
Rio de Janeiro - RJ 22450, Brazil
T: 55 21 259-8248

Strong, Ivan Quelhas
REDE GLOBO
R. Lopes Quintas 303 J. Botanico
Rio de Janeiro - RJ, Brazil
T: 55 21 5292000 F: 55 21 2942042

Struck, Gabriele, Sra.
Cooperante
CENTRO DAS MULHERES DO CABO
Rua do Amparo, 135/137
Olinda - PE 53110, Brazil
T: 55 81 439-17996

Sturn, Lucio, Sr.
CIOSL - ORIT
Rua 24 de Maio, no 77 cj 902
Sao Paulo - SP 01041, Brazil
T: 55 11 223-2174 F: 55 11 223-2153

Su, Sheila Cherng Fen
Rua Marques de Abrantes 56/1002
Rio de Janeiro - RJ, Brazil
T: 55 21 245-8587

Suarez, Jose Carlos Dos Santos
Technician
REDE GLOBO
R. Lopes Quintas 303
Rio de Janeiro -RJ 2, Brazil
T: 55 21 2947732

Suck, Chae Keon, Sr.
Manager
SUNKYONG
Av. Paulista, 2073 Horsa II - Cj. 44
Sao Paulo - SP 01311, Brazil
T: 55 11 283-34308 F: 55 11 284-6050
TL: 38095

Sugahara, Mari, Ms.
Assistant
KEIDANREN
R. Barros, 467/1504 B Icarai
Niteroi - RJ, Brazil
T: 55 21 710 7824

Suhett, Gilberto, Mr.
Engenheiro
SEC. DE MINAS E ENERGIA
R. Luiz Leopoldo Fernandes Pinheiro 517, sala 1403
Niteroi - RJ 24030, Brazil
T: 55 21 719-7171

Sundaram, Dharani, Mr.
Membro
GERNE
Caixa Postal 998
Cuiaba - MT 78000, Brazil
T: 55 125 315-8751 F: 55 125 361-1119
TL: 65-1317

Susano, Ciro Vergara
Engenheiro
Rua Barata Ribeiro 496/702
Rio de Janeiro - RJ, Brazil

Sutherland, Louise, Ms.
ELETROBRAS
Av. Presidente Vargas, 642/10
Rio de Janeiro - RJ 20079, Brazil
T: 55 21 263-3137 F: 55 21 233-3248

Sutter, Pedro
CHICO JR. & ASSOCIADOS
Rio de Janeiro, Brazil

Suwa, Toshiaki, Sr.
Vice President
CIA DE SEGUROS AMERICA DO SUL YASUDA
Av. Brigadeiro Luiz Antonio, 2020 - 5o andar
Sao Paulo - SP 01318-911, Brazil
T: 55 11 285-1411 F: 55 11 285-1426

Szpilman, Marcelo, Sr.
AQUALUNG CONFECCAO LTDA.
Av. Londres, 174 - Bonsucesso
Rio de Janeiro - RJ 21041, Brazil
T: 55 21 270-1955

Szpiz, Moyses, Mr.
Arquiteto
ARQUITETO LIBERAL
Rua Assis Brasil # 2/304
Rio de Janeiro - RJ, Brazil
T: 55 21 542-1314

Sztyglic, Samuel, Sr.
Diretor Vice-Presidente
SEAERJ
Rua do Russel no.01 - Gloria
Rio de Janeiro - RJ, Brazil
T: 55 21 205-6797 F: 55 21 205-2795

T.Silva, Gustavo
Oficial Comunicacao Assistente
UNICEF
SBS Ed.Seguradoras, 13 andar
Brasilia - DF, Brazil
T: 55 61 2247145 F: 55 61 2245035

Tabach, Marcelo
SISTEMA JORNAL DO BRASIL
Av. Brasil 500
Rio de Janeiro -RJ, Brazil
T: 55 21 5854422 F: 55 21 5854428

Tabajara, Luiz Liberato C.A., Mr.
Pesquisador
P.M.TDAI - ALAMA
Av. Da Igreja 346
Tramandai - RS 95590, Brazil
T: 55 51 661-1322 F: 55 51 661-1762
TL: 515450

Tabak, Fanny, Ms.
Professora
PUC/RIO - NEM
Rua Toneleros 261 Ap 301
Rio de Janeiro - RJ 22030, Brazil
T: 55 21 237 0261 F: 55 21 529 4546

Tabanez, Marlene, Sra.
Pesquisadora
INSTITUTO FLORESTAL S.M.A.
Estacao Experimental de Assis CxPostal 104
Assis - SP 19800, Brazil
T: 0183221066 F: 011204 8067

Tabel, Ruth Melof., Ms.
Lucis Trust
FUNDACAO CULTURAL AVATAR
Rua Ministro Octavio Kelly, 352, c-6
Niteroi - RJ 24220, Brazil
T: 55 21 611-2489

Tabet, Luiz Carlos Gomes, Sr.
Secretaria Da Industria
PREFEITURA MUNICIPAL
Praca Maua s/n 1 andar
Santos - SP 11010, Brazil
T: 55 132 342-2093 F: 55 132 345-809
TL: 131048

Taboada, Adelino Guillen, Sr.
Senior Engineer-Environment
RTZ MINERACAO LTDA.
SCS Quadra 1-Bl. HY Edificio Morro Vermelho, 10o. Andar
Brasilia - DF, Brazil
T: 55 61 321-5545 F: 55 61 321-4534

Tachibana, Koki
JIJI PRESS NEWS AGENCY
Av Paulista 854 - 13o. andar cj 133 Bela Vista
Sao Paulo - SP, Brazil
T: 55 11 285 0025/287 9526 F: 55 11 285 3816

Taddei, Jayme
ASSOCIACAO COMERCIAL DO RIO DE JANEIRO
Rio de Janeiro, Brazil

Tajima, Maria Luz
ICAE
12 Rajaji Marg
New Delhi 110011, Brazil
T: 91 11 301-4309

Tak, Mun Su, Mr.
Manager
POHANG IRON & STEEL COMPANY LTD
R. Lauro Muller 116/3303, Botafogo
Rio de Janeiro - RJ, Brazil
T: 55 21 541-3945 F: 55 21 542-2692

Takahashi, Munetaka, Sr.
Pecuaria
BRAJUSCO AGRO PASTORIL S/A
Rua Marechal Deodoro, 1528
Pereira Barreto - SP 15370, Brazil
T: 55 187 611236 F: 55 187 611236

Takatsu, Iasufumi
Journalist
YOMIURI SHIMBUN
Av Oswaldo Cruz 149/1201
Rio de Janeiro - RJ, Brazil
T: 55 21 551 6187 F: 55 21 552 2279

Takeda, Hiroyuki, Mr.
Funcionario Adm.
EMBAIXADA DO JAPAO
Sqs 313 'G' Apto 605
Brasilia - DF, Brazil
T: 55 61 2452991

Takemoto, Walter Kiyochi, Mr.
CIOSL - ORIT
Rua 24 de Maio, no 77 cj 902
Sao Paulo - SP 01041, Brazil
T: 55 11 223-2174 F: 55 11 223-2153

Tambasco, Francisco Penedo
Editor
REDE GLOBO
R. Lopes Quintas 303
Rio de Janeiro -RJ 2, Brazil
T: 55 21 2947732

Tanaka, Joamello Massaki, Sr.
Tec. Ed. Am.
SOCIEDADE POTIGUAR DE EDUCACAO - SPEL
Rua Floriano Peixoto, 295, Petropolis
Natal - RN 59020, Brazil
T: 55 84 222-2314 F: 55 84 221-6255

Tancredo, Luciana
CHICO JR. & ASSOCIADOS
Rio de Janeiro, Brazil

Tanda, Etsuo, Sr.
BRAJUSCO AGRO PASTORIL S/A
Rua Marechal Deodoro, 1528
Pereira Barreto - SP 15370, Brazil
T: 55 187 61-1236 F: 55 187 61-1236

Tapajos, Ana Maria, Ms.
Coordenador De Estudos E Des.
AGENCIA BRASIL DE COOP.
Palacio do Itamaraty - Anexo 1 - andar 8
Brasilia - DF, Brazil
T: 55 61 2116811

Taranto, William Pereira, Sr.
COMISSAO ESPECIAL LEI 1849/92
Rio de Janeiro - RJ, Brazil
T: 55 21 273 9442

Tardeli, Jose, Sr.
SIEMESP
Av. Eng. Francisco Pita Brito 488
Sao Paulo - SP, Brazil
T: 55 11 542-4222

Tardim, Marcos
Reporter
JORNAL O GLOBO
Rua Irineu Marinho, 35
Rio de Janeiro - RJ, Brazil
T: 55 21 2722000

Tardim, Nilo Jose R.
Diretor
JORNAL CANAL DIRETO
R.Santa Maria, 21
Colatina, Espirito S, Brazil
T: 027 7222203 F: 027 7222203

Targa, Fernando, Mr.
Diretor
GEMPI
Rua MourTO cOELHO, 208
Sao Paulo - SP Cep.: 05417, Brazil
T: 55 11 280-3382 F: 55 11 852-5585

Tarpani, Suely Torres De Andrade, Sr.
PREFEITURA MUNICIPAL
Praca Maua s/n 1 andar
Santos - SP 11010, Brazil
T: 55 132 342-2093 F: 55 132 345-809
TL: 131048

Tate, Patricia, Sra.
KEY MARKETING BRASIL
Av. Ataulfo de Paiva 135 gr. 710
Rio de Janeiro - RJ 22440, Brazil

Tauile, Rauf
Fotografo
REVISTA MANCHETE
Rua do Russel, 804
Rio de Janeiro - RJ, Brazil
T: 55 21 2652012

Tautz, Carlos F
Redator
ECOLOGIA E DESENVOLVIMENTO
Rua da Gloria 122 gr 105/106
Rio de Janeiro - RJ, Brazil
T: 55 21 242 1957/222 1370

Tavares, Ana Maria F.
Reporter
ESTADO DE SAO PAULO
Av. Engenheiro Caetano Alvares 55
Sao Paulo -SP 02598, Brazil

Tavares, Ana Paula
HELIO ALONSO
Rua Muniz Barreto, 51
Rio de Janeiro RJ, Brazil
T: 55 21 5515448

Tavares, Carla Nunes Vieira
HOSPITAL ADVENTISTA SILVESTRE
Ladeira dos Guararapes, 263 - Santa Tereza
Rio de Janeiro - RJ, Brazil
T: 55 21 265-0212

Tavares, Keylah Da Silva, Sra.
Assistente
CAMARA MUNICIPAL
Palacio Pedro Emesto
Rio de Janeiro - RJ, Brazil
T: 55 21 262-6184

Tavares, Lucia Saraiva
HOSPITAL ADVENTISTA SILVESTRE
Ladeira dos Guararapes, 263 - Santa Tereza
Rio de Janeiro - RJ, Brazil
T: 55 21 265-0212

Tavares, Luciana Maria De Azambuja, Ms.
Filer
JOVEM TURISMO
SCLN 202 Bloco B Lojas 7/11
Brasilia - DF, Brazil
T: 55 61 224-5526 F: 55 61 321-0345
TL: 61-4705

Tavares, Mariza
Reporter
JORNAL O GLOBO
Rua Irineu Marinho, 35
Rio de Janeiro - RJ, Brazil
T: 55 21 2722000

Tavares, Paulo S. Do Espirito Santo
Technician
REDE GLOBO
R. Lopes Quintas 303
Rio de Janeiro -RJ 2, Brazil
T: 55 21 2947732

Tavares, Vera Lucia Marques, Sra.
Advogada
SOCIEDADE DE DEFESA DOS DIREITOS HUMANOS
SPDDH- Trav. Barao do Triunfo, 2129 Pedreira
Belem - PA 66093, Brazil
T: 55 91 225-4677 F: 55 91 225-4677

Tavora, Marina De, Sra.
Av. Rui Barbosa 636 - Apt. 401
Rio de Janeiro - RJ 22250, Brazil
T: 55 21 551-9716 F: 55 21 232-3411
TL: 212-2047

Taylor, Alisa, Ms.
Rua Marques de Sao Vicente 30/602
Rio de Janeiro - RJ 22451, Brazil

Taylor, Dyanna
TWENTY FIRST CENTURY INC.
Rio de Janeiro, Brazil

Tedeschi, Alfredo
INTERNATIONAL TELEVISION NEWS
Av. Rio Branco, 25/9 andar
Rio de Janeiro RJ, Brazil
T: 55 21 2335480

Tegagni, Adriana, Ms.
Rua Almte Guilhem 231/301
Rio de Janeiro - RJ, Brazil

Teixeira, Adriana Da Silva
Assistente Promocoes
JORNAL VEGAN
Rua do Richuelo, 220/406
Rio de Janeiro - RJ, Brazil
T: 55 21 2220496 F: 55 21 7762108

Teixeira, Altemir, Sr.
Marketing
VISAO MUNDIAL
Rua Antonio de Albuquerque, 788 - Savassi
Belo Horizonte - MG 30110, Brazil
T: 55 31 221-8944 F: 55 31 227-2995

Teixeira, Daniela, Ms.
UNESCO
SCN -Quadra 2 Lote B
Brasilia - DF 70710, Brazil
T: 55 61 223-8684 F: 55 61 322-4261

Teixeira, Evandro
SISTEMA JORNAL DO BRASIL
Av. Brasil 500
Rio de Janeiro -RJ, Brazil
T: 55 21 5854422 F: 55 21 5854428

Teixeira, Gervasio
Jornalista
REVISTA ECORIO
Av.Pres.Antonio Carlos, 54 Gr.802
Rio de Janeiro - RJ, Brazil
T: 55 21 5330069 F: 55 21 2402658

Teixeira, Gilberto
Director
SOS TERRA (FREELANCER)
R.Monte Libano, 67 apt. 206 Bairro Centro
Teresopolis -RJ 259, Brazil
T: 55 21 2426237 F: 55 21 2426237

Teixeira, Inacio
Fotografo
ASSOCIATED PRESS
Avenida Brasil 500 sala 847
Rio de Janeiro - RJ, Brazil
T: 55 21 580 4442 F: 55 21 580 2309

Teixeira, Jackson
WTN
Rio de Janeiro, Brazil

Teixeira, Jose Luiz Sampaio, Sr.
Projects Supervisor
VISAO MUNDIAL
Rua General Osorio, 1597 - Centro - Caixa Postal 626
Passo Fundo - RS 99100, Brazil

Teixeira, Leliane Neubarth
Reporter
REDE GLOBO
R. Lopes Quintas 303
Rio de Janeiro -RJ 2, Brazil
T: 55 21 2947732

Teixeira, Luiz, Mr.
Diretor
GRUPO QUATRO S.C.
Rua 124, No. 66
Goiania - GO 74000, Brazil
T: 55 62 224-2555 F: 55 62 229-0110

Teixeira, M. L., Ms.
Professora E Diretora
INSTITUTO AURORA
Praia do Flamengo, 66 Salao 914/916 B
Rio de Janeiro - RJ 22228, Brazil
T: 55 21 205 1570

Teixeira, Maria Do Carmo Martins, Sra.
CIOSL - ORIT
Rua 24 de Maio, no 77 cj 902
Sao Paulo - SP 01041, Brazil
T: 55 11 223-2174 F: 55 11 223-2153

Teixeira, Nelio Jorge
Producer
REDE GLOBO
R. Lopes Quintas 303
Rio de Janeiro -RJ 2, Brazil
T: 55 21 2947732

Teixeira, Patricia Castro
ESCOLA DE COMUNICACOES E ARTES U.S.P.
Caixa Postal 64522
Sao Paulo -SP 05497, Brazil
T: 55 11 8339511 F: 55 11 2583575

Teixeira, Paulo Roberto Marques
Reporter Fotografico
FREELANCER
Rua Pernambuco, 874 c/3
Rio de Janeiro - RJ, Brazil
T: 55 21 5946451

Teixeira, Teresa Patricia
ICAE
Rede Mulher C.P.4651
Sao Paulo SP 01061, Brazil
T: 55 11 627050 F: 55 11 8713457

Teixera, Fernando
GRUPO NOVO DE CINEMA E TV
Rua Morechel Niemeyer 24
Rio de Janeiro - RJ, Brazil
T: 55 21 2663637 F: 55 21 2250435

Telles, Maria Amelia De Almeida, Sra.
Tesoureira
UNIAO DE MULHERES DE SAO PAULO
Rua Coracao da Europa 1395
Sao Paulo - SP 01314, Brazil
T: 55 11 362 367

Telles, Sergio Ricardo B.
MANCHETE
R. do Russel 804
Rio de Janeiro -RJ 2, Brazil
T: 55 21 2652012 F: 55 21 2855148

Tenorio, Maria Alice R. De O., Sr.
IECAM
Dr. Moreira 250
Manaus - AM 69000, Brazil

Teotonio, Dorivalda Perreira
Tec Div Eco Humana E Ambiental
MINISTERIO DA SAUDE
QSC 11 casa 12
Brasilia - DF 72015, Brazil
T: 55 61 226-1369

Testa, Regina
Assessora De Imprensa
FORUM BAHIA DE ONG'S
R.Borges dos Reis, 46 Sala J Boulevard Rio Vermelho
Salvador - BA, Brazil
T: 55 71 2377357 F: 55 71 2355836

Thamm, Carola, Sra.
Biologa
SOC. DE PESQ. EM VIDA SELVAGEM E EDUCACAO AMBIENTAL
Rua Gutenberg 345
Curitiba - PR 80420-030, Brazil
T: 55 41 222-3425 F: 55 41 222-3425

Thaumaturgo, Gracas Assis, Sra.
Documentacao
INSTITUTO DE MEIO AMBIENTE DO ACRE
Rua.: Rui Barbosa, 450 - Centro
Rio Branco - AC 69.900, Brazil
T: 55 68 224 5497 F: 55 68 224 5694
TL: 282692

Theobald, Marcelo
SISTEMA JORNAL DO BRASIL
Av. Brasil 500
Rio de Janeiro -RJ, Brazil
T: 55 21 5854422 F: 55 21 5854428

Theodoro, Julia
JORNAL MAIORIA FALANTE
Rio de Janeiro, Brazil

Thibau, Carlos Eugenio, Mr.
Presidente
FAEAB
Rua Pio Porto Menezes 145 Apto 101
Belo Horizonte - MG 30380, Brazil
T: 55 31 344-3623

Thielmann, Beatriz Helena M. Da S.
Reporter
REDE GLOBO
R. Lopes Quintas 303
Rio de Janeiro -RJ 2, Brazil
T: 55 21 2947732

Thiesen, Wilson, Sr.
Presidente
ORG. COOP. BRASILEIRAS
SCS - Ed. Baracat, 4. andar
Brasilia - DF 70309-900, Brazil
T: 55 61 225 0275 F: 55 61 226 8766
TL: 1879

Thimel, Hartmut F. R., Mr.
Av. Nossa Senhora de Copacabana, 7, apto. 901
Rio de Janeiro - RJ, Brazil
T: 55 21 275-0853 F: 55 21 275-8196

Thivierge, Robert, Mr.
Primeiro Secretario
CANADIAN EMBASSY
SES Av. das Nacoes 803 - Lote 16
Brasilia - DF, Brazil
T: 55 61 321-2171 F: 55 61 321-4529
TL: 061-1296 ECANBR

Thomaz, Marcos Dos Santos
Engineer
AMERICA DO SUL PRODUCOES DE DOCUMENTARIOS LTDA (NAMBEI TSUSHIN)
Av. Paulista 509 /11 andar conj.1104 Cerqueira Cesar
Sao Paulo -SP 01311, Brazil
T: 55 11 2886141 F: 55 11 2512065

Thompson, Jason, Mr.
OSHO SPIRITUAL HEALTH ORGAN. - OSHO
R. Celestino Basilio 26
Rio de Janeiro - RJ, Brazil

Thurler, Ana Liesi, Ms.
NUSS
SQN 116 - Bl. 6, apto. 604
Brasilia - DF 70773, Brazil
T: 55 61 272-4557

Tillander, Staffan, Sr.
Embaixada Da Suecia
Ave. Das Nacoes Lote 29
Brasilia - Df, Brazil
T: 55 61 243-1444 F: 55 61 243-1187

Tims, James Edwin, Mr.
Director
CENTRO ECOLOGICO ANA GONZAGA
Rua 56 no. 171 apt. 302 - Nova Guaratiba
Campo Grande - RJ 23045, Brazil
T: 55 21 394-0334

Tinoco, Pedro
EDITORA ABRIL - REVISTA VEJA RIO
Rio de Janeiro, Brazil

Tobinaga, Rosa, Sra.
Director
CMB
Rua Avanhadava no. 756
Sao Paulo - SP 01306, Brazil
T: 55 11 2564354 F: 55 11 2577648

Tokitaka, Alexandre
Fotojornalista
FREELANCER
R Caetes 737/504 Perdizes
Sao Paulo - SP, Brazil
T: 55 11 626 560

Toledo, Artur Ferreira De, Sr.
Coord. Seg.
PIRELLI ADM E SERVICOS LTDA
Alameda Barao De Piracicaba 77
Sao Paulo - SP 01000, Brazil
T: 55 11 823-3946

Tomoya, Inyaku, Sr.
PACIFIC-ASIA RESOURCE CENTER/IBASE
IBASE - R. Vicente de Souza, 29 - Botafogo
Rio de Janeiro - RJ, Brazil
T: 55 21 286-6161 F: 55 21 286-0541

Torloni, Carlos Eduardo Cappelini, Mr.
CESP
Rua da Consolacao, 1875, 9 andar, Cerq. Cesar
Sao Paulo - SP, Brazil
T: 55 11 258-2165 F: 55 11 256-2101

Tornaghi, Anna Maria Margarida, Sra.
Promoter
ANNA MARIA TORNAGHI PROMOCOES
Rua Camon Franco, 46
Rio de Janeiro - RJ, Brazil
T: 55 21 295-4242 F: 55 21 295-3388

Tornaghi, Pedro, Mr.
OSHO SPIRITUAL HEALTH ORGAN. - OSHO
R. Celestino Basilio 26
Rio de Janeiro - RJ, Brazil
T: 55 21 4377488

Torneros, Aparecida
JORNAL DE ESPANA
Rio de Janeiro, Brazil

Tornquist, Carlos Gustavo, Mr.
Agronomist
FUNDACAO GAIA
Rua Jacinto Gomes, 39
Porto Alegre - RS 90040-270, Brazil
T: 55 512 330-5789

Torres, Erico
Assistente
JORNAL BRAZIL EXPORT
Av. Franklin Roosevelt 194 Sobre Loja Conj.202
Rio de Janeiro -RJ 2, Brazil
T: 55 21 2629920 F: 55 21 2404705

Torres, Fernando S., Sr.
CACERES FLORESTAL S/A - CACERES PLANTATION TEAK
R. Guilherme Dumont Vilares 282
Sao Paulo - SP 05640, Brazil
T: 55 11 843-6244 F: 55 11 843-8680

Torres, Graca Maria Rocha, Ms.
Sociologa
PREFEITURA MUNICIPAL DE CAMACARI
Rua Ezequiel Ponde 53/4o andar
Salvador - BA 40210, Brazil
T: 55 71 237-4747 F: 55 71 821-1580

Torres, Joao Batista
Diretor De Readiojornalismo
SISTEMA JORNAL DO BRASIL
Av.Brasil, 500
Rio de Janeiro - RJ, Brazil
T: 55 21 5854422

Tosta, Dazva Soares
Director
BLOCH EDITORES
Rua do Russel 766- 8 andar
Rio de Janeiro, Brazil
T: 55 21 5554524

Tostes, Antonio Oliveira, Sr.
Aux. Admin
ASSOCIACAO DA UNIAO ESTE BRAS. DOS ASD
Rua do Matoso, 97 - Praca da Bandeira
Rio de Janeiro - RJ 20270, Brazil
T: 55 21 273-5422 F: 55 21 273-5422

Tostes, Edinilson Vieira
Assistente
WDR GERMAN TV CHANNEL
Rua Eca de Queiroz, 479 casa 12
Sao Paulo - SP, Brazil
T: 55 11 5721262 F: 55 11 5728959

Tostes, Otavio
Editor
REDE GLOBO
R. Lopes Quintas 303
Rio de Janeiro -RJ 2, Brazil
T: 55 21 2947732

Toucedo, Carmem, Ms.
Psicologa
SEC. DE MINAS E ENERGIA
R. Luiz Leopoldo Fernandes Pinheiro 517, sala 1403
Niteroi - RJ 24030, Brazil
T: 55 21 719-7171

Tour, Simone De La, Sra.
Departamento De Comunicacao
INSTITUTO INTERNACIONAL DE PROJECIOLOGIA
Rua Santo Amaro 4/3 andar - Gloria
Rio de Janeiro - RJ 22211, Brazil
T: 55 21 221-8954 F: 55 21 2218954

Trajber, Rachel, Ms.
Educadora
ICAE/FDE
Rua Pelotas 184 - Apto. 122
Sao Paulo - SP 04012, Brazil
T: 55 11 549-2373

Travares, Luis Fernando
Vt Operator
MANCHETE
Rua do Russel 766/804 4 andar
Rio de Janeiro, Brazil
T: 55 21 2850033

Treiger, Jayme, Sr.
Presidente
LUCIS TRUST ASSOCIATION / FUNDACAO CULTURAL AVATAR
Rua Ministro Otavio Kelly 352, Casa 6
Rio de Janeiro - RJ, Brazil
T: 55 21 611-2489

Treiger, Marilda De Mattos, Sra.
LUCIS TRUST - FUND. CULT. AVATAR
Min. Otavio Kelly 352 c/6
Niteroi - RJ 24230, Brazil
T: 55 21 611-2489

Treijer, Maria Elisa De Mattos, Ms.
FUNDACAO CULTURAL AVATAR
Rua Min. Octavio Kelly 352 c.6
Niteroi - RJ 24220, Brazil
T: 55 21 611-2489

Treuersch, Marlies, Sra.
Biologa
SUREHMA
Rua Engenheiro Reboucas 1206
Curitiba - PR 80210, Brazil
T: 55 41 225 6283

Trevisan, Mauri, Sr.
Diretor
FUNDACAO EDUCACIONAL DE BARRETOS
Av Prof. Roberto Frade Monte, 389
Barretos - SP 14780, Brazil
T: 22 6411 F: 22 6205

Trigueiro, Marta
LEGIAO DA BOA VONTADE
Av. Rudge 700 Bom Retiro
Sao Paulo -SP 01134, Brazil
T: 55 21 2338381/R:223 F: 55 21 2634025

Trindade, Azoilda
JORNAL MAIORIA FALANTE
Rio de Janeiro, Brazil

Trindade, Cairo Assis
Comentarista
JORNAL CONTEMPORANEO
Ladeira dos Tabajaras 126/302 Copacabana
Rio de Janeiro -RJ 2, Brazil
T: 55 21 2565121

Trindade, Denise
Comentarista
JORNAL CONTEMPORANEO
Ladeira dos Tabajaras 126/302 Copacabana
Rio de Janeiro -RJ 2, Brazil
T: 55 21 2565121

Trindade, Luis Gonzaga
Reporter
REDE GLOBO
R. Lopes Quintas 303
Rio de Janeiro -RJ 2, Brazil
T: 55 21 2947732

Trindade, Maria Helena
Photographer
FREELANCER
R.Lopes Quintas 340 casa 114
Rio de Janeiro -RJ, Brazil
T: 55 21 2594839 F: 55 21 2391478

Trindade, Marines, Sra.
Director
CMB
Rua Avanhadava no. 756
Sao Paulo - SP 01306, Brazil
T: 55 11 2564354 F: 55 11 2577648

Trindade, Maristela
Assistente De Fotagrafia
FREELANCER
R.Sao Clemente, 114/705 Bl.II
Rio de Janeiro - RJ, Brazil
T: 55 21 2464871

Trinta, Acely Borges Dos Santos
TVE - FRP
Avienda Gomes Freire 474 2 andar
Rio de Janeiro, Brazil
T: 55 21 2920012 x169

Trinta, Carlos Alberto Da Silva
Reporter
REDE GLOBO
R. Lopes Quintas 303
Rio de Janeiro -RJ 2, Brazil
T: 55 21 2947732

Tripoli, Ricardo, Mr.
Deputado Estadual
ASSEMBLEIA LEGISLATIVA DO ESTADO DE SAO PAULO
Rua Jeronimo da Ugiga, 164 - 2o. Andar
Sao Paulo - SP, Brazil
T: 55 11 884-2050

Tristan, Marilene De
THE MIAMI HERALD
Rio de Janeiro, Brazil

Troian, Fabio, Sr.
ARQUIPELAGO DE FERNANDO DE NORONHA
Rua Barao de Sao Borja, 243 - Boa Vista
Recife - PE 50070, Brazil
T: 55 81 221-3217 F: 55 81 231-1594
TL: 81.2533

Tropardi, Lilian Rose Dos Santos, Sra.
Ass. Proj. Sociologas
FUNDACAO SAMUEL
Rua Americo Brasiliense, 1062
Sao Paulo - SP 04715, Brazil
T: 55 11 521-5577

Tsiipre, Luiz Filipe Figueiredo, Sr.
Coordenador
CADIRJ
R. Real Grandeza 110/402
Rio de Janeiro - RJ 22281, Brazil
T: 55 21 246-6737

Tureck, Claudio Rodolfo, Sr.
Biologist
PREFEITURA M. S. E. DO SUL
Praca Getulio Vargas No. 01
S. Fco. do Sul - SC 89230, Brazil
T: 55 474 440777 F: 55 474 440130

Tuxen, Maria Da Gloria, Ms.
UFRJ
Rua General Critovao Barcelos, 121 Apto 101 Laranjeiras
Rio de Janeiro - RJ 22251, Brazil
T: 55 21 225 2424

Tyschler, Bruno
Sub-Editor
EDITORA ABRIL
R.da Passagem, 123/8 andar
Rio de Janeiro - RJ, Brazil
F: 55 21 2759347

Ubruatau, Pedro
ECOPRESS
Rua Itapolis 1893
Sao Paulo, Brazil
T: 55 11 2586150 F: 55 11 2576793

Uchoa, Cleofas Ismael
FUND. CULTURAL DE RADIODIFUSAO EDUCATIVA DA REGIAO DOS LAGOS
Av. Jose Bento Ribeiro Dantas 265 Manguinhos 28905
Armacao dos Buzios -, Brazil
T: 55 246 2753848 F: 55 21 2956494

Uchoa, Terezinha M. De Menezes, Sra.
Socia
SOCIEDADE NORDESTINA DE ECOLOGIA
Rua Pessoa de Melo, 355 - Madalena
Recife - PE 50710, Brazil
T: 55 81 227-2487 F: 55 81 222-3361

Ugo, Giuseppe, Sr.
Industriale
MOVIMENTO CITADINI E POPOLI DEL MONDO
Av. Alvaro Ramos, 954
Sao Paulo - SP 03330, Brazil
T: 55 11 292-5262 F: 55 11 938736

Ulhoa, Luiz Gonzaga, Sr.
Diretor Executivo
FEDERACAO NACIONAL DOS TRABALHADORES NAS INDUSTRIAS URBANAS - FNTIU
Rua Visconde de Inhauna, 134, 7$ andar - Centro
Rio de Janeiro - RJ 20019, Brazil
T: 55 21 233-1430 F: 55 21 233-3516
TL: 552134090

Unes, Joao Carlos Arruda
RADIO UNIVERSITARIA
Universidade Federal de Goias - Setor Oeste Av Anhanguera com Alameda das Rosas
Goiania - GO, Brazil
T: 55 62 233-4473/6896

Unger, Nancy Mangabeira, Sra.
Membro
CENTRO JOAO XXIII - IBRADES
Rua Bambina, 115
Rio de Janeiro - RJ 22251, Brazil
T: 55 21 286-8522 F: 55 21 266-6157

Uribbe, Washington
Technician
REDE GLOBO
R. Lopes Quintas 303
Rio de Janeiro -RJ 2, Brazil
T: 55 21 2947732

Uteira, Radames De Oliveira
Editor
RC2 MARKETING
Av. Beira, 406, No 1608
Rio de Janeiro, Brazil
T: 55 21 2243043 F: 55 21 1205075

Uzeda, Conrado L., Mr.
FUNDACAO CULTURAL AVATAR
Rua Min. Octavio Kelly 352 c.6
Niteroi - RJ 24220, Brazil
T: 55 21 611-2489

Vaine, Jorge Luiz, Sr.
Engenheiro
SUREHMA
Rua Santo Antonio 239
Curitiba - PR, Brazil
T: 55 41 225 3411

Vainer, Lydia, Sra.
Al. Casa Branca, 1025
Sao Paulo - SP CEP: 01408, Brazil

Val, Marcos
HELIO ALONSO
Rua Muniz Barreto, 51
Rio de Janeiro RJ, Brazil
T: 55 21 5515448

Valadao, Ines
Reporter
RADIO ROQUETE PINTO
Av. Erasmo Braga 118/ 11andar Castelo
Rio de Janeiro -RJ 2, Brazil
T: 55 21 2247727 F: 55 21 2524584

Valadares, Sergio
Representante
OSHO SPIRITUAL HEALTH ORGANIZATION
Rua Celestino Basilio, 26
Rio de Janeiro - RJ, Brazil
T: 55 21 437-7488 F: 55 21 437-7488

Valcarcel, Ricardo
ECOTROPIC
1880 East River Road
Rio de Janeiro - RJ, Brazil
T: 55 21 577-6476 F: 55 21 577-1195

Valdes, Carlos
INTERNATIONAL TELEVISION NEWS
Av. Rio Branco, 25/9 andar
Rio de Janeiro RJ, Brazil
T: 55 21 2335480

Vale, Andrea Nobre Peixoto Do
UNICEF
Rio de Janeiro, Brazil

Vale, Ramildo Soares
MANCHETE
Rua do Russel, 804
Rio de Janeiro, Brazil
T: 55 21 2650033

Vale, William G., Sr.
U F P A
Campus Universitario
Belem - PA 66050, Brazil
T: 55 91 229-0744 F: 55 91 229-1984
TL: 0911013

Vale Dos Santos, Ana Gabriela, Sra.
ABEMA
Av. Jeronimo Monteiro, 240/S. 811/812
Vitoria - ES, Brazil
T: 55 27 322-4193 F: 55 27 222-3725

Valente, Ivan, Mr.
Deputado
ASSEMBLEIA LEGISLATIVA DE SAO PAULO
Av. Pedro Alvares Cabral, s/n - s/5003 -Ibirapuera
Sao Paulo - SP 04094-050, Brazil
T: 55 11 884 4519 F: 55 11 884 3483

Vales, Leise Thereza Teixeira, Sra.
Expositora
IBASE
Rua Vicente de Souza 12- Botafogo
Rio de Janeiro - RJ 22251, Brazil
T: 55 21 286-6161 F: 55 21 286-0541
TL: 21-36466

Valle, Fernanda De Oliveira
Assessora De Imprensa
FUNDACAO BIODIVERSITAS
Brazil
T: 31 4432119 F: 031 4417037

Valle, Martha Denis
Jornalista
PRENSA LATINA
Rua Marechal Mascarenhas de Morais 121/203 Copacabana
Rio de Janeiro - RJ, Brazil
T: 55 21 256-7259

Valle, Nathanael Maranhao, Mr.
CENTRO JOSUE DE CASTRO
Dom Bosco 779
Recife - PE 53000, Brazil
T: 55 81 222-1906

Valva, Andreina, Sra.
SODERNA / UFG
Cx. Postal 12010
Goiania - GO 74.000, Brazil
T: 55 62 205-1160

Van Schaik, Lot
92 GLOBAL FORUM
Rio de Janeiro - RJ, Brazil

Vanoverbeke, Henri
O N U - UNDCP
40 Ondcp
Brasilia - DF, Brazil
T: 55 61 224 1423

Vanucci, Maria De Paula Freitas, Sra.
NUCLEO DE ESTUDOS E ORGANIZACAO DAS MULHERES
Rua 1 Quadra 1 Casa 6
Cuiaba - MT 78000, Brazil
T: 55 125 361-2884

Varella, Jose Augusto
Photographer
AGENCIA ESTADO
Av. Caetano Alvares, 55
Sao Paulo -SP 02598, Brazil
T: 55 11 8562210 F: 55 11 2652297

Varella, Maria Da Graca, Sra.
Advogada
Rua Sao Bento n 8, 21 Andar,
Rio de Janeiro - RJ 20090, Brazil
T: 55 21 253-1269 F: 55 21 233-6352
TL: 22281

Vargas, Edson Dutra, Sr.
CIA. DE CIGARROS SOUZA CRUZ
Rua da Candelaria 66, 10o. Andar
Rio de Janeiro - RJ 20092-900, Brazil
T: 55 21 276-9091

Vargas, Joao Luiz Dos Santos, Sr.
Deputado
ASSEMBLEIA LEGISTATIVA
Pca Marechal Deodoro s/no.
Porto Alegre - RS 90010, Brazil
T: 55 512 221-4260 F: 55 512 228-6740

Vargas, Maria Celia De
ASSOCIACAO COMERCIAL DO RIO DE JANEIRO
Rua da Candelaria 9/12 andar
Rio de Janeiro RJ, Brazil
T: 55 21 2031229 F: 55 21 2637613

Vasconcelles, Luiz Claudio
Executive Producer
RADIO MANCHETE LTDA
Rua do Russel 766, Gloria
Rio de Janeiro, Brazil
T: 55 21 2855997 F: 55 21 2057078

Vasconcellos, Angela, Ms.
Rio de Janeiro - RJ, Brazil
T: 55 21 285-3592

Vasconcellos, Bianca
TVN PRODUCOES ARTISTICAS
Sao Paulo, Brazil

Vasconcellos, Bruno Cattoni
Editor
REDE GLOBO
R. Lopes Quintas 303
Rio de Janeiro -RJ 2, Brazil
T: 55 21 2947732

Vasconcellos, Maria Dulce Queiros, Ms.
FUNDACAO XAPURI
Rua Da Assembleia, # 10 Sala 1621- Centro
Rio de Janeiro - RJ 20119, Brazil
T: 55 21 242-1462 F: 55 21 224-7213

Vasconcellos, Paulo, Mr.
Divulgacao
VERDE PAZ
Rua Barao Da Torre, 213/405
Rio de Janeiro - RJ 22.411, Brazil
T: 55 21 267-1790

Vasconcelos, Aldo Cesar Frota, Sr.
Cultural Director
AMDA
Rua 'E', no. 80, Loteamento Bela Vista
Macae - RJ 28700, Brazil
T: 55 247 62 5110

Vasconcelos, Alexandre Guimaraes, Sr.
Estudante
Rua Dona Amelia, 30/202 Andarai
Rio de Janeiro - RJ 20541, Brazil
T: 55 21 571-7189

Vasconcelos, Consuelo, Sra.
ARQUIPELAGO DE FERNANDO DE NORONHA
Rua Barao de Sao Borja, 243 - Boa Vista
Recife - PE 50070, Brazil
T: 55 81 221-3217 F: 55 81 231-1594
TL: 81.2533

Vasconcelos, Jose T., Sr.
CEPEMAR - CENTRO DE PESQUISA DO MAR
Rua Mexico 31 - Sala 501
Rio de Janeiro - RJ 20031, Brazil
T: 55 21 240-3822 F: 55 21 240-9516

Vasconcelos, Jose Vac. Hone De, Mr.
Medico
S O H E R J - SOCIEDADE DE HOMEOPATIA DO RIO DE JANEIRO
Rua Lauro Muller 128/1402
Rio de Janeiro - RJ, Brazil
T: 55 21 542-6022

Vasconcelos, Maria Tereza Marques
R. Paissandu
Rio de Janeiro - RJ 20000, Brazil
T: 55 21 2257742

Vasconcelos, Marilda Helena Breglio, Sra.
SHIS. Ql. 05 Conjunto 17 Casa 03
Brasilia - DF 71600, Brazil
T: 55 61 248 1117

Vasconellos, Dilu Marie Assis, Ms.
Biologa
Av. Marechal Rondon 300, Bloco 3, Apt. 303
Rio de Janeiro - RJ, Brazil
T: 55 21 574-9173

Vasques, Paulo Roberto De Freitas, Mr.
Secretario
ASSOC. COM. UNIDOS PRO TERRA
Rua Jurubatuba 1019
S. Bernardo Campo-SP, Brazil
T: 55 11 443-3512

Vassilieff, Paulo, Mr.
Instrutor
INSTITUTO INTERNACIONAL DE MOTIVACAO
R. Sete de Setembro, 155/2oAndar
Rio de Janeiro - RJ 20050, Brazil
T: 55 21 242-5990

Vaughan, Scott, Mr.
Coordinator
UNITED NATIONS ENVIRONMENT PROGRAMME
c/o Av Nilo Pecanha, 50 sala 1313 - Instituto Brasil - PNUMA
Rio de Janeiro - RJ, Brazil
T: 55 21 220-1820 F: 55 21 262-4233

Vaz, Ana Luiza De Lira, Sra.
REDE MULHER
Rua Jose Ramalho, 991 - Perdizes
Sao Paulo - SP 05008, Brazil
T: 55 11 262-9407 F: 55 11 871-3457

Vaz, Antonio Carlos Martins
Editor
REDE GLOBO
R. Lopes Quintas 303
Rio de Janeiro -RJ 2, Brazil
T: 55 21 2947732

Vedova, Jose Carlos Della, Sr.
Administrador
CENTRO DE ESTUDOS AVANCADOS DE PROMOCAO SOCIAL
Rua Paulo Barreto 23 - Botafogo
Rio de Janeiro - RJ 22280, Brazil
T: 55 21 226-1519 F: 55 21 266-7897

Vega, Zoilo Gutierrez M. De La
Editor
AGENCIA EFE S/A
Avenida Rio Branco 25/13 andar
Rio de Janeiro -RJ, Brazil
T: 55 21 2534465 F: 55 21 2534861

Veiga, Ana Luiza
BLOCH EDITORES
Rua do Russel, 766/804
Rio de Janeiro, Brazil
T: 55 21 2652012

Veiga, Bruno
Fotografo
TYBA AGENCIA FOTOGRAFICA
Rua Barao do Flamengo 22/503 Flamengo
Rio de Janeiro - RJ, Brazil
T: 55 21 285 4640/265 4839 F: 55 21 222 6742

Veiga, Cassandra Lopes Da, Sra.
Pesquisadora
INSTITUTO DE ECOLOGIA HUMANA
Av. Dom Manuel de Medeiros, s/n - Ed. Newton Bancks 3o. andar
Recife - PE 52071, Brazil
T: 55 81 441-3931 F: 55 81 441-4697

Veiga, Marcos Figueira Da
MADE FOR TV
Rio de Janeiro, Brazil

Veiga, Regina
Editor
FREELANCER
R. Ataulo de Paiva 50 Bl/C2 apt.801
Rio de Janeiro -RJ, Brazil
T: 55 21 2591114

Veira, Marceu
Reporter
O DIA
R. Riachuela 359
Rio De Janeiro, RJ 2, Brazil
T: 55 21 272-8000 F: 55 21 507-1228

Velasco, Renato Menezes
Reporter
REVISTA VISAO
Av.Rio Branco, 177/6 andar
Rio de Janeiro - RJ, Brazil
T: 55 21 2621085 F: 55 21 2200805

Velasco, Valeria De
Diretora
LETTER
Ed Venancio 2000 bloco B50 sala 443 Setor Comercial Sul
Brasilia - DF, Brazil
T: 55 61 226 4695/226 4865/226 F: 55 61 225 6381

Velasques, Loide Barbosa, Ms.
UNITED METHODIST CHURCH OF BRAZIL
Rua Londrina 410-apt.113,Rudge ramos
Sao Bernard.Campo-SP 09735, Brazil

Velho, Luis
fernando mendes 19/71
Rio de Janeiro - RJ 22021, Brazil
T: 55 21 255-1550 F: 55 21 260-2048

Veliz, Grimaldo, Mr.
HERMANN
Rua Visconde De Piraja 608/ 911
Rio de Janeiro - RJ 22410, Brazil

Vellardi, Ana Cristina, Sra.
PREFEITURA MUNICIPAL
Praca Maua s/n 1 andar
Santos - SP 11010, Brazil
T: 55 132 342-2093 F: 55 132 345-809
TL: 131048

Velloso, Agnes, Ms.
Biologa
Rua Raimundo Magalhaes 86
Rio de Janeiro - RJ 22451, Brazil
T: 55 21 274 4559

Velloso, Andre Tomas
Techinician
AGENCIA ESTADO
Av. Caetano Alvares, 55
Sao Paulo -SP 02598, Brazil
T: 55 11 8562210 F: 55 11 2652297

Veloso, Igor
UNIVERSIDADE GAMA FILHO
Rua Manuel Vitorino, 625
Rio de Janeiro, Brazil
T: 55 21 2697272

Veneri, Joao Antonio, Sr.
Assist. Tec. Ii
SENAI - PR
Rua Chile 1678
Curitiba - PR 80220, Brazil
T: 55 41 233-0322 F: 55 41 233-0322
TL: 4135141

Ventura, Irene, Sra.
Secretaria Da Industria
PREFEITURA MUNICIPAL
Praca Maua s/n 1 andar
Santos - SP 11010, Brazil
T: 55 132 342-2093 F: 55 132 345-809
TL: 131048

Veras, Ana Maraia, Sra.
FLAMA
Rua Marques de Abrantes, 55 - Flamengo
Rio de Janeiro - RJ 22230, Brazil
T: 55 21 265-5690

Veras, Concita, Sra.
ARQUIPELAGO DE FERNANDO DE NORONHA
Rua Barao de Sao Borja, 243 - Boa Vista
Recife - PE 50070, Brazil
T: 55 81 221-3217 F: 55 81 231-1594
TL: 81.2533

Veras, Leonardo, Sr.
ARQUIPELAGO DE FERNANDO DE NORONHA
Rua Barao de Sao Borja, 243 - Boa Vista
Recife - PE 50070, Brazil
T: 55 81 221-3217 F: 55 81 231-1594
TL: 81.2533

Veras, Lucia Maria De Siqueira Caval, Sra.
Pesquisadora
INSTITUTO DE ECOLOGIA HUMANA
Av. Dom Manuel de Medeiros, s/n - Ed. Newton Bancks 3o. andar
Recife - PE 52071, Brazil
T: 55 81 441-3931 F: 55 81 441-4697

Veras, Sabina Travassos
FORUM DO FUTURO
Rio de Jaeiro, Brazil

Vercauteren, Agnes, Ms.
Membro
CEQMTR - PR
Rua Almirante Tamandare, 434
Curitiba - PR 80050, Brazil
T: 55 41 264 4455

Vergani, Marcelo
REVISTA FLUMINENSE
Rua Republica do Libano, 61 Sala 1120, Centro
Rio de Janeiro RJ, Brazil
T: 55 21 2429809

Vergara, Maria Do Rosario Botelho, Ms.
Medica
Rua Barata Ribeiro 496 - 1702
Rio de Janeiro - RJ, Brazil

Verginelli, Jose Geraldo, Sr.
Member
FLAMA
Praia do Flamengo, 22
Rio de Janeiro - RJ, Brazil
T: 55 21 273-6458

Vergueiro, Jose Carlos De Aguiar
Motorista
BCO. INTERAMERICANO DE DESENVOLVIMENTO
Caixa Postal 08738
Brasilia - DF 70312, Brazil
T: 55 61 321-7848 F: 55 61 321-3112
TL: 61-1204

Verissimo, Divino Roberto, Sr.
CEC
SHIS Ql 05 conj.17 casa 03
Brasilia - DF 71600, Brazil
T: 55 61 248 1117

Vermelho, Americo
Fotografo
REVISTA VIA URBANA
Rua Sao Jose, 90/2005
Rio de Janeiro - RJ, Brazil
T: 55 21 2320157 / 2523276 F: 55 21 2400139

Vernes, Adao, Sr.
Rua Toledo Barbosa, 474
Sao Paulo - SP 03061, Brazil
T: 55 11 291-2677

Verocal, Iara
GOVERNO DO ESTADO DE SAO PAULO
Sao Paulo, Brazil

Verri, Solange
Camera Assistant
ABIA -ASSOCIACAO BRASILEIRA INTERDISCIPLINAR DE AIDS
R. Lopes Quintas 576 J. Botanico
Rio de Janeiro -RJ 2, Brazil
T: 55 21 2315171 F: 55 21 2945602

Versiani, Claudio
Photographer
EDITORA ABRIL
R.da Passagem, 123/8 andar
Rio de Janeiro, RJ, Brazil
T: 021 5468282 F: 021 2759347

Vesterlond, Valdecir Jose, Sr.
MIDADE - MOVIMENTO INTERNACIONAL DO APOSTOLADO DAS CRIANCAS
Av. Beira Mar no. 48
Barra Velha - SC 88390, Brazil

Vetromille, Vicente De Paula
Reporter
TV EDUCATIVA - RJ
Av. Gomes Freire 474 Centro
Rio de Janeiro -RJ, Brazil
T: 55 21 2426736 F: 55 21 2426740

Veyl, Andreas, Mr.
RIO DE JANEIRO UNIVERSITY
Rua Caio de Melo Franco 45
Rio de Janeiro - RJ 22461, Brazil
T: 55 21 246-9747

Viana, Elisabete Gomes
Reporter
PONTO 3 EUROPEAN JORNALISTA
Av. Serambetilea 1100/apto 203
Rio de Janeiro RJ, Brazil
T: 55 21 4930728

Viana, Jose Carlos
Technician
REDE GLOBO
R. Lopes Quintas 303
Rio de Janeiro -RJ 2, Brazil
T: 55 21 2947732

Viana, Luis Antonio Vieira
MANCHETE
R. do Russel 804
Rio de Janeiro -RJ 2, Brazil
T: 55 21 2652012 F: 55 21 2855148

Viana, Marcelo Bueno
Technician
REDE GLOBO
R. Lopes Quintas 303
Rio de Janeiro -RJ 2, Brazil
T: 55 21 2947732

Viana, Rosemar Das Nupcias
Technician
REDE GLOBO
R. Lopes Quintas 303
Rio de Janeiro -RJ 2, Brazil
T: 55 21 2947732

Viana, Sandro Jose Carneiro
MANCHETE
R. do Russel 804
Rio de Janeiro -RJ 2, Brazil
T: 55 21 2652012 F: 55 21 2855148

Viana, Virigilio, Sr.
Professor
E S A L Q - UNIVERSIDADE DE SAO PAULO
Depto. Ciencias Florestais - ESALQ
Piracicaba, Brazil
T: 0194 33-4124 F: 0194 33-6081

Vianna, Christiane Fernando
Apresentadora
TV RIO
R. Miguel de Frias, 57 Estacio
Rio de Janeiro-RJ 20, Brazil
T: 55 21 2939888 F: 55 21 2938626

Vianna, Lucila Pinsard, Ms.
Supervisora
I.F. - SMA/SP
Rua Do Horto 931
Sao Paulo - SP 01059, Brazil
T: 55 11 952-8995 F: 55 11 204-8067
TL: 22877 SAGEBR

Vianna, Marco
SISTEMA JORNAL DO BRASIL
Av. Brasil 500
Rio de Janeiro -RJ, Brazil
T: 55 21 5854422 F: 55 21 5854428

Viard, Carlos Henrique
Fotografo
REVISTA MANCHETE
Rua do Russel, 804
Rio de Janeiro - RJ, Brazil
T: 55 21 2652012

Vicente, Jose Alberto T.
Reporter
SISTEMA GLOBO DE RADIO
Rua do Russel, 434
Rio de Janeiro - RJ, Brazil
T: 55 21 2058282 F: 55 21 2054096

Vicente, Ma. De Lourdes, Sra.
FUNDACAO ESCOLA DE SERVICO PUBLICO RJ - FESP
Av. Carlos Peixoto, 64
Rio de Janeiro - RJ 22290, Brazil

Vicentini, Ariovaldo
Photographer
AGENCIA ESTADO
Av. Caetano Alvares, 55
Sao Paulo -SP 02598, Brazil
T: 55 11 8562210 F: 55 11 2652297

Vidal, Ula De Andrade, Ms.
Project Coordinator
GLOBAL HARMONY FOUNDATION
Rua General Polidoro 304, casa XI Botafogo
Rio de Janeiro - RJ, Brazil
T: 55 21 2660100 F: 55 21 5372856

Videla, Daniel
Technician
FUND. CULTURAL DE RADIODIFUSAO EDUCATIVA DA REGIAO DOS LAGOS
Av. Jose Bento Ribeiro Dantas 265 Manguinhos 28905
Armacao dos Buzios -, Brazil
T: 55 246 2753848 F: 55 21 2956494

Viegas, Joao Alfredo Noronha
Manager Environment Matters
CONCREMAT ENGENHARIA & TECNOLOGIA
R Euclides da Cunha 106 Sao Cristovao
Rio de Janeiro - RJ, Brazil
T: 55 21 284 4022 F: 55 21 284 2702

Viegas, Silvio Mothe
Reporter Fotografico
REVISTA ISTO E/SENHOR
Av.Almte.Barroso,63/1512
Rio de Janeiro - RJ, Brazil
T: 55 21 2405224 F: 55 21 2402925

Vieira, Alberto Leite, Sr.
PREFEITURA MUNICIPAL
Praca Maua s/n 1 andar
Santos - SP 11010, Brazil
T: 55 132 342-2093 F: 55 132 345-809
TL: 131048

Vieira, Antonio
SOS TERRA (FREELANCER)
R.Monte Libano, 67 apt. 206 Bairro Centro
Teresopolis -RJ 259, Brazil
T: 55 21 2426237 F: 55 21 2426237

Vieira, Beatriz
ESCOLA DE COMUNICACOES E ARTES U.S.P.
Caixa Postal 64522
Sao Paulo -SP 05497, Brazil
T: 55 11 8339511 F: 55 11 2583575

Vieira, Carla, Ms.
CONFEDERACAO NACIONAL IND.
Nilo Pecanha,50/andar 38
Rio de Janeiro - RJ 22250, Brazil
T: 55 21 552-0099 F: 55 21 552-9392

Vieira, Carlos Adauto Virmond, Mr.
Cons. Juridico
APREMA-SC
Rua Marinho Lobo 80, sala 605
Joinville-SC, Brazil
T: 55 474 224874 F: 55 474 224874

Vieira, Cleidionice, Sra.
HOSPITAL PIO X
Av. Goias 360
Ceres - Goias 76300, Brazil
T: 55 62 721-1261

Vieira, Cleonir De Queiros, Ms.
Advogada
Rua Mexico, 168 Sl 601
Rio de Janeiro - RJ, Brazil
T: 55 21 5330500

Vieira, Delfin
Fotografo
JORNAL O GLOBO
Rua Irineu Marinho, 35
Rio de Janeiro - RJ, Brazil
T: 55 21 2722000

Vieira, Douglas
Radialista/Jornalista
IBASE
Rua Vicente de Souza 29 Botafogo
Rio de Janeiro - RJ, Brazil
T: 55 21 286 6161/8642 F: 55 21 286 0541

Vieira, Fernando Antonio
Tecnico
TVSBT
Campo de S.Cristovao, 105
Rio de Janeiro - RJ, Brazil
T: 55 21 5803688 / 5800678 F: 55 21 5809685

Vieira, Humberto Leal, Sr.
Assessor Legislativo
SENADO FEDERAL
SQS 203, Bloco 'C' apt. 204
Brasilia - DF 70233, Brazil
T: 55 61 311-3311

Vieira, Ilza Pereira, Sra.
INTERNACIONAL (AMD)
Av. Mar. Henrique Lott, 31/1302
Rio de Janeiro - RJ, Brazil
T: 55 21 431-1330

Vieira, Ima Celia Vieira, Sra.
MUSEU GOELDI
Museu Goeldi
Belem - PA, Brazil

Vieira, Jessie Jane S
Redator
ECOLOGIA E DESENVOLVIMENTO
Rua da Gloria 122 gr 105/106
Rio de Janeiro - RJ, Brazil
T: 55 21 242 1957/222 1370

Vieira, Joao Pedro Dias
ASSOCIACAO COMERCIAL DO RIO DE JANEIRO
Rio de Janeiro, Brazil

Vieira, Jorge Cysne
EDITORA AZUL
Rio de Janeiro, Brazil

Vieira, Jose Claudio Alfenas Mitraud, Sr.
AMIGOS DO MEIO AMBIENTE DE OURO PRETO - AMA
Rua Dom Silverio, 258
Ouro Preto MG 35400, Brazil
T: 55 31551-3049 R
TL: 55 31551-128

Vieira, Lia
Journalist
REVISTA FLUMINENSE
Rua Republica do Libano, 61 Sala 1120, Centro
Rio de Janeiro RJ, Brazil
T: 55 21 2429809

Vieira, Maercelo Moraes
FUNDACAO ROBERTO MARINHO
Rio de Janeiro RJ, Brazil

Vieira, Maria Candida
Reporter
DCI - DIARIO COMERCIO e INDUSTRIA
Av.Rio Branco, 177/6 andar
Rio de Janeiro - RJ, Brazil
T: 55 21 2621016 F: 55 21 2200805

Vieira, Newmar Pereira
Reporter
REVISTA PORTOS E NAVIOS
Av Rio Branco, 185/sala 1914 Centro
Rio de Janeiro, Brazil
T: 55 21 2203578

Vieira, Quintiliano Machado, Sr.
Deputado Estadual
ASSEMBLEIA LEGISTATIVA
Praca Matriz
Porto Alegre - RS 90010, Brazil
T: 55 512 221-4343

Vieira, Reginaldo Silveira
Technician
REDE GLOBO
R. Lopes Quintas 303
Rio de Janeiro -RJ 2, Brazil
T: 55 21 2947732

Vieira, Rodinei Pintor, Sr.
Balconista
ASSOCIACAO DA UNIAO ESTE BRAS. DOS ASD
Rua do Matoso, 97 - Praca da Bandeira
Rio de Janeiro - RJ 20270, Brazil
T: 55 21 273-5422 F: 55 21 273-5422

Vieira, Rosangela Cardoso
Editora
REVISTA PORTOS E NAVIOS
Av.Rio Branco, 185/1914
Rio de Janeiro - RJ, Brazil
T: 55 21 2203578

Vieira, Susana Camargo, Ms.
Socia-Advogada
PELLICCIARI CAMARGO VIEIRA ADVOGADOS
Prudddente Correa 409
Sao Paulo - SP 01450-030, Brazil
T: 55 11 2106024

Vigliano, Ricardo
CONFEDERACAO NACIONAL DA INDUSTRIA
Rio de Janeiro, Brazil

Vilagrand, Maria Da Graca, Ms.
UNITED METHODIST CHURCH OF BRAZIL
Rua 20 de Setembro, 799
Alegrete - RS 97540, Brazil

Vilar, Ana Guedes, Sra.
Pesquisadora
INSTITUTO DE ECOLOGIA HUMANA
Av. Dom Manuel de Medeiros, s/n - Ed. Newton Bancks 3o. andar
Recife - PE 52071, Brazil
T: 55 81 441-3931 F: 55 81 441-4697

Vilar, Lucrecia Guedes, Sra.
Pesquisadora
INSTITUTO DE ECOLOGIA HUMANA
Av. Dom Manuel de Medeiros, s/n - Ed. Newton Bancks 3o. andar
Recife - PE 52071, Brazil
T: 55 81 441-3931 F: 55 81 441-4697

Vilela, Francisco Eduardo, Sr.
Membro
APOENA
Rua Dr.Gurgel, 860
Presid. Prudente-SP 19015-140, Brazil
T: 55 182 21-4222 F: 55 182 22-6177

Vilhena, Jose Luiz
Reporter
JORNAL O GLOBO
Rua Irineu Marinho, 35
Rio de Janeiro - RJ, Brazil
T: 55 21 2722000

Villago, Carlos Bravo
Operador De Vt
TV Rio
R. Miguel de Frias, 57 Estacio
Rio de Janeiro - RJ, Brazil
T: 55 21 2939888 F: 55 21 2938626

Villani, Rogerio Tunes
MANCHETE
Rio de Janeiro, Brazil

Villar, Roberto
Reporter
RADIO GAUCHA
Av Ipiranga 1075
Porto Alegre - RS 90060, Brazil
T: 55 512 223 6600 F: 55 512 2336600 r 57

Viola, Eduardo J., Sr.
Professor
UNIVERSIDADE FEDERAL DE SANTA CATARINA
Rua Frederico Veras, Trav. 4, no.05
Florianopolis - SC 88040, Brazil
T: 55 482 334543 F: 55 482 344069

Viola, Regina, Sra.
Director
CMB
Rua Avanhadava no. 756
Sao Paulo - SP 01306, Brazil
T: 55 11 2564354 F: 55 11 2577648

Virginio Filho, Elias De Melo, Sr.
EUPASA/UFPB
Patos - PB 58700, Brazil
T: 083 421-3397 F: 083 421-3696

Visconti, Arides
Journalist
HILL AND KNOWLTON
Av. Afranio de Melo Franco 135 /5 andar Leblon
Rio de Janeiro -RJ 2, Brazil
F: 55 21 2394697

Vitale, Paulo Fernando De S. F.
Photographer
AGENCIA ESTADO
Av. Caetano Alvares, 55
Sao Paulo -SP 02598, Brazil
T: 55 11 8562210 F: 55 11 2652297

Vitorio, Cleverson Andreoli, Mr.
Technician
ABEMA
Av. Jeronimo Monteiro, 240 - salas 811/812 - Centro
Vitoria - ES 29045, Brazil
T: 55 27 322-4193 F: 55 27 222-3725

Vizane, Guilherme Marcati
Reporter
REDE GLOBO
R. Lopes Quintas 303
Rio de Janeiro -RJ 2, Brazil
T: 55 21 2947732

Voloch, Bruno
MANCHETE
Rio de Janeiro, Brazil

Volpi, Mario, Sr.
Coordenador Nacional
MOVIMENTO NACIONAL DE MENINOS E MENINAS DE RUA
SDN - CNB 3o. andar, sala L 403
Brasilia - DF 70070, Brazil
T: 55 61 226-9631 F: 55 61 225-1577

Von Frasunkiewicz, Suzanne, Sra.
Teacher
ESCOLA DAS NACOES
SHIS QI 21 Area Especial Lote C-1 - Caixa Postal 7093
Brasilia - DF 71600, Brazil
T: 55 61 366-1800 F: 55 61 366-3377

Von Mann, Isabel Bernadete, Ms.
Diretora
CINEF-EXTERPRESS
Rua Professor Galotti Jr. 299
Sao Paulo - SP, Brazil
F: 55 11 240-1863
TL: 55 11 240-1863

Von Sperling, Eugenia, Sra.
GOVERNO DO ESTADO DO RIO DE JANEIRO
Rio de Janeiro - RJ, Brazil

Vos, Leo De
Programme Officer
UNICEF
SBS Ed.Seguradoras, 13 andar
Brasilia, DF 70072, Brazil
T: 061 2247145 F: 061 2245035

Vozza, Giuseppe Maria, Sr.
Coordinator In Brazil
TERRA NUOVA
Lgo. Sto. Antonio da Barra s/n - Ladeira da Barra
Salvador - BA 40130, Brazil
T: 55 71 247.9707 F: 55 71 2479707

Vregani, Magda
Assistant Director
REVISTA FLUMINENSE
Rua Republica do Libano, 61 Sala 1120, Centro
Rio de Janeiro RJ, Brazil
T: 55 21 2429809

Wachendorfer, Achim Friedrich, Sr.
ILDES/FES
Rua. Lisboa 191
Sao Paulo - SP 05413, Brazil
T: 55 11 282-3900 F: 55 11 280-1916
TL: 1180616

Wagner, Ralf, Mr.
Agronomist
FUNDACAO GAIA
Rua Jacinto Gomes, 39
Porto Alegre - RS 90040-270, Brazil
T: 55 512 330-5789

Wahren, Axel, Mr.
Rua Sen. Correa 66/801
Rio de Janeiro - RJ 22231, Brazil
T: 55 21 245-9225 F: 55 21 245-5537

Wahren, Joana Monat, Ms.
Social Science
Rua Senador Correa, 66/801
Rio de Janeiro - RJ, Brazil
T: 55 21 245-9225 F: 55 21 245-5537

Wajnberg, Salomao, Sr.
Diretor
TELEBRASIL
Av. Pasteur 383 Urca
Rio de Janeiro - RJ 22290, Brazil
T: 55 21 295-4432 F: 55 21 542-4092
TL: 2123057

Walawisky, Diva, Ms.
CONSELHO ESTADUAL DA MULHER - MINAS GERAIS
Rua Comendador Viana 300
Belo Horizonte - MG, Brazil
T: 55 31 223-7750

Walcacer, Theresa
Chefe De Redacao
GLOBO ECOLOGIA
Rua Santa Alexandrina 343/c3 Rio Comprido
Rio de Janeiro - RJ, Brazil
T: 55 21 502-5451 F: 55 21 273-9087

Wanderley, Anibal, Mr.
I E I / U F R J
Av. Pasteur 250
Rio de Janeiro - RJ 22290, Brazil
T: 55 21 295-8891 F: 55 21 541-8148

Wanderley, Laura L. Lavenere, Sra.
Professor
CENTRO PEDAGOGICO - VFMG
Rua Carangola 32, apt. 21
Belo Horizonte - MG 30350, Brazil
T: 55 31 223-4503

Warren, Jonathan Winddance, Sr.
Pesquisa
SOCIAL SCIENCE RESEARCH COUNCIL
Rua Bernardino de Oliveira no. 22
Varre-Sai - RJ 28386, Brazil
T: 55 249 433228

Watabe, Kazumi
Journalist
YOMIURI SHIMBUN
Av Oswaldo Cruz 149/1201
Rio de Janeiro - RJ, Brazil
T: 55 21 551 6187 F: 55 21 552 2279

Watanabe, Celia Misai, Ms.
Assessora Projeto
SINDICATO TRABALHADORES RURAIS - BA
Rua Duque de Caxias, 345 - Centro
Itabuna - BA 45600, Brazil

Watanabe, Kazumi, Mr.
Diretor De Divulgacao
CEFLAN - CENTRO DE ESTUDOS FLORESTAIS DA AMAZONIA
a/c ACTA Av. Dionisio Bentes 210
Tome-Acu - PA 68682, Brazil
T: 55 91 734-1316

Way, Marion W., Mr.
Social Member
IGREJA METODISTA
Rua Afonso Pena, 61 apt.302
Rio de Janeiro - RJ 20270, Brazil
T: 55 21 254-1933

Waymberg, Francis, Sr.
IBFAN - INTERNATIONAL BABY FOOD ACTION NETWORK
Rua Cupertino Durao, 117/301
Rio de Janeiro - RJ 22441, Brazil
T: 55 21 259-5459

Weckbach, Stefan, Mr.
Consul Geral Adjunto
CONSULADO GERAL DA ALEMANHA
Rua: Presidente C. de Campos, 417 - Laranjeiras
Rio de Janeiro - RJ, Brazil
T: 55 21 553 6777

Wee, Choong Hwan, Mr.
General Manager
POHANG IRON & STEEL COMPANY LTD
R. Lauro Muller 116/3303, Botafogo
Rio de Janeiro - RJ, Brazil
T: 55 21 541-3945 F: 55 21 542-2692

Weffort, Djalma, Sr.
Presidente
Associacao Em Defesa Do Rio Parana - Apoena
Rua Dr. Gurgel 860
Presidente Epitacio 19010, Brazil
T: 55 182 33-4222

Weid, Nahyda Von Der
Techinician
ASSOC. PROJETO RODA VIVA
R.Sao Francisco Xavier, 524/12037-F
Rio de Janeiro - RJ, Brazil
T: 55 21 2645196 F: 55 21 2645196

Weinstein, Mauricio, Mr.
Representative Agent
NEWSWEEK INTERNATIONAL
Rua do Russel, 450 sala 801 - Gloria
Rio de Janeiro - RJ, Brazil
T: 55 21 265-4392 F: 55 21 205-6887

Weissenberg, Maria Cecilia Brito
MANCHETE
R. do Russel 804
Rio de Janeiro -RJ 2, Brazil
T: 55 21 2652012 F: 55 21 2855148

Weil, Pierre, Sr.
Presidente
FUNDACAO CIDADE DA PAZ
Granja do Ipe Caixa Postal 09521
Brasilia - DF 70001, Brazil
T: 55 61 553-1885 F: 55 61 553-1202

Wenerte, Claudio Jaime
Cheif Reporter
BLOCH EDITORES
Rua do Russel 766, 8 andar
Rio de Janeiro, Brazil
T: 55 21 5554277

Wenz, Dietmar, Mr.
FUNDACAO KONRAD ADENALIER
Av. Fernando Simoes Barbosa 22/ 703
Recife - PE, Brazil
T: 55 81 326-4545 F: 55 81 326-3288

Werblowsky, Edgard, Sr.
Membro
APOENA
Rua Cuiaba, 1-19
Presid.Epitacio-SP 19470-000, Brazil
T: 55 182 21-4222 F: 55 182 22-6177

Wertheim, Dayse Abrantes
Journalist
FREELANCER
R.das Camelias H/5 Cond.Bleu et Vert Bairro Braunes
Nova Friburgo - RJ, Brazil
T: 55 245 220192 F: 55 245 220192

Wertheim, Peter Howard
Correspondente Do Ewla-Envir.
FREELANCER
R.das Camelias H/5 Cond.Bleu et Vert Bairro Braunes
Nova Friburgo - RJ, Brazil
T: 55 245 220192 F: 55 245 220192

Wery, Peter Zohan
Photographer
WERY PRODUCOES
Rua Diogenes Sampaio 92/301 Humaita
Rio de Janeiro - RJ, Brazil
T: 55 21 246 3272

West, George, Sr.
Amb.Technolog.
WESTCO INTERNACIONAL LTDA.
Caixa Postal 1063
Goiania - GO 74000, Brazil
T: 55 62 225-7559 F: 55 62 223-4700

Wetterling, Fabio Henrique, Sr.
Vendedor
NEWSWEEK INTERNATIONAL
Rua Do Russel, 450 Sala 801
Rio de Janeiro - RJ 22210, Brazil
T: 55 21 265-4392 F: 55 21 205-6887

Wheatley, Jim, Mr.
Director
SUMMER INSTITUTE OF LINGUISTICS
SAI/NO, Lote 'D', Bloco 03
Brasilia - DF 70770, Brazil
T: 55 61 272-1200

Wheatley, Jonathan
CANADIAN BROADCASTING CORPORATION
Rio de Janeiro, Brazil

Whitaker, Luciana
Photographer
FOLHA DE SAO PAULO
Alameda Barao de Limeira 425
Sao Paulo SP, Brazil
T: 55 11 874 2608/874 2602 F: 55 11 2231644

White, Joan, Ms.
Teacher
ESCOLA AMERICANA DE BRASILIA
SGAS 605 Bloco E
Brasilia - DF, Brazil
T: 55 61 2433237

Whyte, Andrew, Sr.
R.Monte Libano 42
Niteroi - RJ 24020, Brazil
T: 55 21 717 6554

Williams, Charlotte, Ms.
FUNSEC
Av. Eng. Roberto Lassance 1375, Vila Isa
Governador Valadares 35040, Brazil
T: 55 332 78-1929 F: 55 332 21-1686

Williams, Lora, Ms.
Student
PURDUE UNIVERSITY
Vila Gianetti 37
Vicosa, MG. 36570, Brazil
T: 031 899-2799

Willian, Jorge
Fotografo
JORNAL O GLOBO
Rua Irineu Marinho, 35
Rio de Janeiro - RJ, Brazil
T: 55 21 2722000

Wilson, James, Mr.
Public Affairs
SUMMER INSTITUTE OF LINGUISTICS
SAI/NO, Lote 'D', Bloco 03
Brasilia - DF 70770, Brazil
T: 55 61 272-1200

Wilson, Thimoteo, Jr.
CIA DE TURISMO DO ESTADO DO RIO - TURISRIO
Rio de janeiro, Brazil

Wirsig, Waldemar
Repr. Do Bid
BCO. INTERAMERICANO DE DESENVOLVIMENTO
Caixa Postal 08738
Brasilia - DF 70312, Brazil
T: 55 61 321-7848 F: 55 61 321-3112
TL: 61-1204

Wohlfender-Gobet, Anne, Ms.
Praia De Botafogo 528, Bl.C - Cobertura
Rio de Janeiro - RJ, Brazil
T: 55 21 295-6810

Wrzosekosemberg, Bernhard
Cinegrafista
WDR GERMAN TV CHANNEL
Rua Eca de Queiroz, 479 casa 12
Sao Paulo - SP, Brazil
T: 55 11 5721262 F: 55 11 5728959

Wynn-Jones, Nigel
RIO LIFE
Est da Gavea 611/Bl 2/304 Sao Conrado
Rio de Janeiro - RJ, Brazil
T: 55 21 322 2290 F: 55 21 322 2990

Xavier, Alex, Mr.
REDES AT
R. Timoteo da Costa 197/401
Rio de Janeiro - RJ, Brazil
T: 55 21 274-7850 F: 55 21 511-2724

Xavier, Helia N., Dr.
INSTITUTO BRASILEIRO DE ADMINISTRACAO MUNICIPAL - IBAM
Edificio Diogo Lordello de Mello. Largo IBAM no.1. Humaita
Rio de Janeiro - RJ 22282, Brazil
T: 55 21 266-6622 F: 55 21 537-1262
TL: 22638 INBM BR

Xavier, Joao
Assistant Producer
ICS INSTITUTO DE COMUNICACAO SINDICAL
Rua Visconde de Inhauma, 134-Sala 715, Centro
Rio de Janeiro RJ 2, Brazil
T: 55 21 2331430 F: 55 21 2333516

Xavier, Joao Carlos Trindade
Technician
REDE GLOBO
R. Lopes Quintas 303
Rio de Janeiro -RJ 2, Brazil
T: 55 21 2947732

Xavier, Neuza Luiza Da Silva
Reporter
FREELANCER
Rua Pedro de Carvalhes, 580/204 Meizi
Rio de Janeiro RJ, Brazil
T: 55 21 5910118

Xavier, Ruy
Reporter
O DIA
R. Riachuela 359
Rio De Janeiro, RJ 2, Brazil
T: 55 21 272-8000 F: 55 21 507-1228

Ximenes, Joao Batista
Reporter
EDITORA ABRIL
R.da Passagem, 123/8 andar
Rio de Janeiro, RJ, Brazil
T: 021 5468282 F: 021 2759347

Yaamaoka, Kunihiro
Journalist
YOMIURI SHIMBUN
Av Oswaldo Cruz 149/1201
Rio de Janeiro - RJ, Brazil
T: 55 21 551 6187 F: 55 21 552 2279

Yabe, Masayuki
Cameraman
AMERICA DO SUL PRODUCOES DE DOCUMENTARIOS LTDA (NAMBEI TSUSHIN)
Av. Paulista 509 /11 andar conj.1104 Cerqueira Cesar
Sao Paulo -SP 01311, Brazil
T: 55 11 2886141 F: 55 11 2512065

Yacoub, Rosie, Ms.
Ed. Bruno de Menezes / 131 TV Apinages / Apt. 1502
Belem - PA 66000, Brazil
T: 55 91 241-7244

Yahiro, Iris, Ms.
ASSOCIACAO POTIGUAR DE EDUCACAO E CULTURA
Rua Floriano Peixoto, 295 Petropolis
Natal - RN 59020, Brazil
T: 55 84 221-6255 F: 55 84 221-6255

Yaish, Patricia
Correspondent
UPI
R. Uruguaiana 54/18 andar
Rio de Janeiro -RJ 2, Brazil
T: 55 21 2244194 F: 55 21 2328295

Yamasaki, Sergio Takashi
ESCOLA DE COMUNICACOES E ARTES U.S.P.
Caixa Postal 64522
Sao Paulo -SP 05497, Brazil
T: 55 11 8339511 F: 55 11 2583575

Yari, Zargham, Sr.
Member
COMUNIDADE BAHA'I BRASIL
Rua Nelson de Figueiredo Jr., 215 - Bairro Antonio Vendas
Campo Grande - MS 79050, Brazil
T: 55 67 725-8269

Yazaki, Kichi
Photographer
FREELANCER
Largo da palvorz 120, Liberdade
Sao Paulo, Brazil
T: 55 11 2787466 F: 55 11 2709041

Yglesias, Edyala
FORUM DO FUTURO
Rio de Janeiro, Brazil

Yokomizo, Nilse Kasue Shimura, Sr.
Eng. Agro.
INSTITUTO FLORESTAL DE SAO PAULO
Rua do Horto 931- CAIXA POSTAL 1322
Sao Paulo - SP 01059-970, Brazil
T: 55 11 952-8555 F: 55 11 204-8067
TL: 22877 SAGR BR

Yokota, Yukihiro
YOMIURI SHIMBUN
Av Oswaldo Cruz 149/1201
Rio de Janeiro - RJ, Brazil
T: 55 21 551 6187 F: 55 21 552 2279

Yokoyama, Satoru, Mr.
Gerente
NKK
Rua da Assembleia, 10 / s.3501 Centro
Rio de Janeiro - RJ, Brazil
T: 55 21 2246472 F: 55 21 2423928

Yolen, Greg
92 GLOBAL FORUM
Rio de Janeiro - RJ, Brazil

Yone Stroh, Paula, Ms.
Diretora
ECOPOLIS
SCLN 305 Bloco 'B' Sala 212
Brasilia - DF 70136, Brazil
T: 55 61 273 9540 F: 55 61 273 9540

Yong, Xu
Reporter
AGENCIA NOTICIOSA XINHUA
Rua Paissandu,7 apto.401
Rio de Janeiro - RJ, Brazil
T: 55 21 285 0590 / 285 06 03

Yoruba, Togo
SERVICO MAIORIA FALANTE - COMUNICACAO RACIAL E POPULAR
Rua da Lapa 200/808
Rio de Janeiro - RJ 20021, Brazil
T: 55 21 507-1462

Yoshida, Takeshi, Sr.
Guia
Avinida Rio Branco, 15615,.1115
Rio de Janeiro - RJ 20040, Brazil
T: 55 21 210-32634 F: 55 21 220-9854
TL: 30605

Yuba, Silvia Sasaoka
JAPAN YOUTH FEDERATION
1. Alianca Caixa Postal 31
Mirandopolis 16830-000, Brazil
T: 0187 78-1247

Yusim, Liliane Sales
Editor
REDE GLOBO
R. Lopes Quintas 303
Rio de Janeiro -RJ 2, Brazil
T: 55 21 2947732

Zalis, Sergio
Reporter Fotografico
ZNZ AGENCIA FOTOGRAFICA
Rua Cosme Velho 355 Cosme Velho
Rio de Janeiro - RJ, Brazil
T: 55 21 225 5805/0066 F: 55 21 265 6567

Zamgrandy, Reynaldo
Assistente De Camera
2DF-TV ALEMA
Av.Delfim Moreira, 662/704
Rio de Janeiro - RJ, Brazil
T: 55 21 2947434 F: 55 21 5122120

Zamora, Maria Del Carmen
MANCHETE
Rua do Russel 804 Flamengo
Rio de Janeiro, Brazil
T: 55 21 2850033

Zanardi Jr., Volney, Sr.
SECRETARIA MUNICIPAL MEIO AMBIENTE PORTO ALEGRE
Av Carlos Gomes 2120
Porto Alegre - RS 90480, Brazil
T: 55 512 334 8489

Zancheti, Silvio Mendes, Sr.
Director
URBANA: PLANEJAMENTOI E PROJETOS
R. Leao Coroado, 45
Recife - PE 50060, Brazil
T: 55 81 222-0707 F: 55 81 231-7221

Zandona Filho, Arion, Sr.
Instr. Ens. 2-Grau
SENAI - PR
Rua Chile 1678
Curitiba - PR 80220, Brazil
T: 55 41 233-0322 F: 55 41 233-0322

Zanetti, Jose Carlos, Sr.
Socio-Gerente
LOCUS-PLKANEJAMENTO E ASSESSORIA EM DESENVOLVIMENTO SOCIO-AMBIENTAL
Trav. Antonio Seabra-Vilamatos/Rio Vermelho
Salvador - BA 40210, Brazil
T: 55 71 247-5936

Zangari, Milton, Sr.
Chefe Setor De Eng. Civil
NESTLE INDL. COML. LTDA
Rua Da Consolacao 896
Sao Paulo - SP 01302, Brazil
T: 55 11 231-8418 F: 55 11 257-3101
TL: 32580

Zanier, Sandra Ely Guasti De Almeida, Sra.
Tecnician
BANDES
Av. Princesa Isabel, 54 - 12o. andar
Vitoria - ES 29010, Brazil
T: 55 27 223-8333 F: 55 27 223-6307
TL: 2131

Zapa, Regina
Editor
SISTEMA JORNAL DO BRASIL
Av. Brasil 500
Rio de Janeiro -RJ, Brazil
T: 55 21 5854422 F: 55 21 5854428

Zappala, Mario, Mr.
Assessor
UNIV. SANTA URSULA
Rua Jornalista Orlando Dantas 59 - Botafogo
Rio de Janeiro - RJ 21231, Brazil
T: 55 21 552-5422 F: 55 21 551-6446

Zarartini, Monica Rolim
Photographer
AGENCIA ESTADO
Av. Caetano Alvares, 55
Sao Paulo -SP 02598, Brazil
T: 55 11 8562210 F: 55 11 2652297

Zatz, Ines Gonzaga, Sra.
Coordenadora
CPCD
SHIS QI 15 - Ch.18
Brasilia - DF 71600, Brazil
T: 55 61 248-0431

Zavareze, Marcelo
Camera Assistant
MTV - RIO
Praca Nossa Senhora da Gloria 46
Rio de Janeiro -RJ, Brazil
T: 55 21 2855946 F: 55 21 2055394

Zeidler, Herbert, Sr.
Engenheiro
SACTES / SEBRAE
Av. Boa Viagem # 2128 Apto. 601 - Reaje
Petrolina - PE 51011, Brazil
T: 55 81 325-1358

Zein, Rabih O., Mr.
ROYAL COMMISSION FOR JUBAIL & YANBU - SAUDI ARABIA
Rua Sambaiba, 190/601
Rio de Janeiro - RJ 22450, Brazil
T: 55 21 239-3329

Zeitone, Rafael Jackes
Jornalista
JORNAL VEGAN
Rua do Richuelo, 220/406
Rio de Janeiro - RJ, Brazil
T: 55 21 2220496 F: 55 21 7762108

Zenga, Luis Alberto
Correspondent
DEUTSCHE PRESSE - AGENTUR
R. Abade Ramos 65 Jardim Botanico
Rio de Janeiro -RJ, Brazil
T: 55 21 2665937 F: 55 21 2860349

Zerbine, Terezinha, Sra.
Director
CMB
Rua Avanhadava no. 756
Sao Paulo - SP 01306, Brazil
T: 55 11 2564354 F: 55 11 2577648

Zhen, Geng Qiu
Reporter
AGENCIA NOTICIOSA XINHUA
Rua Paissandu,7 apto.401
Rio de Janeiro - RJ, Brazil
T: 55 21 285 0590 / 285 06 03

Zhi, Yu Chzng
Reporter
AGENCIA NOTICIOSA XINHUA
Rua Paissandu,7 apto.401
Rio de Janeiro - RJ, Brazil
T: 55 21 285 0590 / 285 06 03

Zhiming, Li
DIARIO DO POVO - CHINA
SHIS QL8 CJ.8 Casa 10
Brasilia - DF, Brazil
T: 55 61 2481855

Zhouri, Andrea Luisa, Sra.
Researcher
UNICAMP
Rua Luzitana, 1770/ap. 25
Campinas - SP 13015, Brazil
T: 55 192 89512

Zoet, Marcia Xavier
Photographer
AGENCIA ESTADO
Av. Caetano Alvares, 55
Sao Paulo -SP 02598, Brazil
T: 55 11 8562210 F: 55 11 2652297

Zoghaib, Neide, Ms.
WOMEN'S CLUB
Rua Real Grandesa # 99
Rio de Janeiro - RJ, Brazil

Zoller, Niki, Sr.
S A C T E S
Rua Vicente Ribeiro 166
Berilo-MG 39640-000, Brazil
T: 2265139

Zorilla, A., Mr.
UNESCO
SCN -Quadra 2 Lote B
Brasilia - DF 70710, Brazil
T: 55 61 223-8684 F: 55 61 322-4261

Zoucas, Robert, Mr.
Planejamento
VERDE PAZ
Av.Nossa Senhora Copacabana,420/906
Rio de Janeiro - RJ 22.020, Brazil
T: 55 21 236-5884

Zucko, Jussara Elizabeth, Sra.
Research Professor
INSTITUTO MOURA LACERDA
R. Padre Euclides 995
Ribeirao Preto - SP, Brazil
T: 55 16 625-7557 F: 55 16 635-7515

Zuntini, Daniela, Sra.
Membro
APOENA
Rua Antonio Lopes de Azevedo, 410
Presid. Prudente-SP 19030, Brazil
T: 55 182 22-4334

Zuquim, Israel, Mr.
AMBIENTAL VIAGENS E EXPEDICOES
Av. Brigadeiro Faria Lima 1684 S/Cj.40
Sao Paulo - SP 01452, Brazil
T: 55 11 814-6191 F: 55 11 211-2239

Zur, Claudia
Photographer
SOS TERRA (FREELANCER)
R.Monte Libano, 67 apt. 206 Bairro Centro
Teresopolis -RJ 259, Brazil
T: 55 21 2426237 F: 55 21 2426237

Zurcher, Piter Rolan
Cameraman
FUNTELDA
Av. Almirante 735/4 andar
Belem PA, Brazil
T: 55 91 2281000

UN AND UNCED

Diallo, Djibril, Mr.
Chief-Special Advisor
UNICEF
SBS Ed. Seguradoras 13o Andar
Brasilia - DF 70072, Brazil
T: 55 61 224-7145 F: 55 61 224-5035
TL: 1181

Figueiredo, Licinio Coelho De O.
UNICEF
Rio de Janeiro, Brazil

Matheus, Angela, Sra.
Assistente De Informacao
UNICEF
SBS Ed.Seguradoras, 13 andar
Brasilia - DF, Brazil
T: 55 61 2247145 F: 55 61 2245035

Padmini, R., Ms.
Chief-Urban Section
UNICEF
SBS Ed. Seguradoras 13o Andar
Brasilia - DF 70072, Brazil
T: 55 61 224-7145 F: 55 61 224-5035
TL: 1181

Pereira, Egberto Kirt
UNICEF
Rio de Janeiro, Brazil

Rios-Kohn, Rebecca, Ms.
Public Affairs Officer, Ny
UNICEF
SBS Edificio Seguradoras 13o Andar
Brasilia - DF, Brazil
T: 55 61 224-7145 F: 55 61 224-5035

Umemoto, Stephen, Mr.
Deputy Regional Dir
UNICEF
SBS Edificio Seguradoras 13o Andar
Brasilia - DF, Brazil
T: 55 61 224-7145 F: 55 61 224-5035

Usher, John, Mr.
Chief-Media Relations
UNICEF
SBS Ed. Seguradoras 13o Andar
Brasilia - DF 70072, Brazil
T: 55 61 224-7145 F: 55 61 224-5035
TL: 1181

ADVISORS

Davies De Freitas, Maria De Lourdes, Ms.
Coordenadora, Programa Nacional Do Meio Ambiente, Instituto Brasileiro...
Brasilia, DF, Brazil
T: 55 61 223 7957 F: 55 61 226 4345

Dias De Moura, Laercio, Fr.
Rector, Pontifica Universidade Catolica : Member, UNCED Brazil Advisory Group
Rio de Janeiro, RJ 22453, Brazil
T: 55 21 276 4507 F: 55 21 274 4197

Do Nascimento, Pele, Edson Arantes, Mr.
UNCED Goodwill Ambassador
Rio de Janeiro, RJ CEP 22290, Brazil
T: 55 21 541 4447 F: 55 21 542 6991

Lorentzen, Erling, Mr.
Chairman, Board Of Directors, Aracruz Celulose S.A.
Rio de Janeiro, RJ 22290, Brazil
T: 55 21 541 6637 F: 55 21 295 7943

PARLIAMENTARIANS / SPIRITUAL LEADERS

Abi-Ramia, Junot, Deputy
Camara dos Deputados, Palacio do Congresso Nacional
70 000 Brasilia, D.F., Brazil
T: 213 51 51
TL: 1163

Angeranel, Tuga, Deputy
Camara dos Deputados, Palacio do Congresso Nacional
70 000 Brasilia, D.F., Brazil
T: 213 51 51
TL: 1163

Arruda, Inacio, Deputy
Camara dos Deputados, Palacio do Congresso Nacional
70 000 Brasilia, D.F., Brazil
T: 213 51 51
TL: 1163

Bezerra, Jose Mendonca, Deputy
Camara dos Deputados, Palacio do Congresso Nacional
70 000 Brasilia, D.F., Brazil
T: 213 51 51
TL: 1163

Boaventura, Maria Aparecida, Deputy
Camara dos Deputados, Palacio do Congresso Nacional
70 000 Brasilia, D.F., Brazil
T: 213 51 51
TL: 1163

Campos, Roberto, Deputy
Camara dos Deputados, Palacio do Congresso Nacional
70 000 Brasilia, D.F., Brazil
T: 213 51 51
TL: 1163

Carvalho, Wanubia, Deputy
Camara dos Deputados, Palacio do Congresso Nacional
70 000 Brasilia, D.F., Brazil
T: 213 51 51
TL: 1163

Correa, Pedro, Deputy
Camara dos Deputados, Palacio do Congresso Nacional
70 000 Brasilia, D.F., Brazil
T: 213 51 51
TL: 1163

Coutinho Jorge, Fernando, Senator
Senado Federal, Palacio do Congresso Nacional
70 000 Brasilia, D. F., Brazil
T: 211 41 41
TL: 1156

De Castro, Aluisio, Deputy
Camara dos Deputados, Palacio do Congresso Nacional
70 000 Brasilia, D.F., Brazil
T: 213 51 51
TL: 1163

De Castro, Manoel, Deputy
Camara dos Deputados, Palacio do Congresso Nacional
70 000 Brasilia, D.F., Brazil
T: 213 51 51
TL: 1163

De Melo, Oswaldo, Councilman
Camara dos Deputados, Palacio do Congresso Nacional
70 000 Brasilia, D.F., Brazil
T: 213 51 51
TL: 1163

De Miguel, Sidney, Deputy
Camara dos Deputados, Palacio do Congresso Nacional
70 000 Brasilia, D.F., Brazil
T: 213 51 51
TL: 1163

Delgado, Paulo, Deputy
Camara dos Deputados, Palacio do Congresso Nacional
70 000 Brasilia, D.F., Brazil
T: 213 51 51
TL: 1163

Duarte, Tuninho, Deputy
Camara dos Deputados, Palacio do Congresso Nacional
70 000 Brasilia, D.F., Brazil
T: 213 51 51
TL: 1163

Duque, Paulo, Deputy
Camara dos Deputados, Palacio do Congresso Nacional
70 000 Brasilia, D.F., Brazil
T: 213 51 51
TL: 1163

Feldmann, Fabio, Deputy
Camara dos Deputados, Palacio do Congresso Nacional
70 000 Brasilia, D.F., Brazil
T: 213 51 51
TL: 1163

Ferreira, Jose Sivuca, Deputy
Camara dos Deputados, Palacio do Congresso Nacional
70 000 Brasilia, D.F., Brazil
T: 213 51 51
TL: 1163

Filho, Sergio Cabral, Deputy
Camara dos Deputados, Palacio do Congresso Nacional
70 000 Brasilia, D.F., Brazil
T: 213 51 51
TL: 1163

Gamma, Aparecida, Deputy
Camara dos Deputados, Palacio do Congresso Nacional
70 000 Brasilia, D.F., Brazil
T: 213 51 51
TL: 1163

Gushiken, Luiz, Deputy
Camara dos Deputados, Palacio do Congresso Nacional
70 000 Brasilia, D.F., Brazil
T: 213 51 51
TL: 1163

Juca, Teresa, Deputy
Camara dos Deputados, Palacio do Congresso Nacional
70 000 Brasilia, D.F., Brazil
T: 213 51 51
TL: 1163

Kamaiura, Sapaim
Xingu
Mato Grosso State, Brazil

Lima, Luiz Henrique, Deputy
Camara dos Deputados, Palacio do Congresso Nacional
70 000 Brasilia, D.F., Brazil
T: 213 51 51
TL: 1163

Lopes, Ney, Deputy
Camara dos Deputados, Palacio do Congresso Nacional
70 000 Brasilia, D.F., Brazil
T: 213 51 51
TL: 1163

Maia, Cesar, Deputy
Camara dos Deputados, Palacio do Congresso Nacional
70 000 Brasilia, D.F., Brazil
T: 213 51 51
TL: 1163

Mendes, Joao, Deputy
Camara dos Deputados, Palacio do Congresso Nacional
70 000 Brasilia, D.F., Brazil
T: 213 51 51
TL: 1163

Mercadaute, Aloisio, Deputy
Camara dos Deputados, Palacio do Congresso Nacional
70 000 Brasilia, D.F., Brazil
T: 213 51 51
TL: 1163

Moraes, Ricardo, Deputy
Camara dos Deputados, Palacio do Congresso Nacional
70 000 Brasilia, D.F., Brazil
T: 213 51 51
TL: 1163

Murao, Jamil, Deputy
Camara dos Deputados, Palacio do Congresso Nacional
70 000 Brasilia, D.F., Brazil
T: 213 51 51
TL: 1163

Nader, Jose, Deputy
Pres.
State Assembly
Camara dos Deputados, Palacio do Congresso Nacional
70 000 Brasilia, D.F., Brazil
T: 213 51 51
TL: 1163

Oliveira, Aloisio, Deputy
Camara dos Deputados, Palacio do Congresso Nacional
70 000 Brasilia, D.F., Brazil
T: 213 51 51
TL: 1163

Pelaes, Fatima, Deputy
Camara dos Deputados, Palacio do Congresso Nacional
70 000 Brasilia, D.F., Brazil
T: 213 51 51
TL: 1163

Penaforte, Mareo, Deputy
Camara dos Deputados, Palacio do Congresso Nacional
70 000 Brasilia, D.F., Brazil
T: 213 51 51
TL: 1163

Portugal, Paulo, Deputy
Camara dos Deputados, Palacio do Congresso Nacional
70 000 Brasilia, D.F., Brazil
T: 213 51 51
TL: 1163

Quieroz, Giovanni, Deputy
Camara dos Deputados, Palacio do Congresso Nacional
70 000 Brasilia, D.F., Brazil
T: 213 51 51
TL: 1163

Salles, H.E. Dom Eugenio Cardinal
Palacio Sao Joaquim
Rua Benjamin Constant 23, Gloria, CEP 20241
Rio de Janeiro, Brazil

Soares, Odacir, Senator
Senado Federal, Palacio do Congresso Nacional
70 000 Brasilia, D. F., Brazil
T: 211 41 41
TL: 1156

Sobel, Henry, Rabbi
Congregacao Israelita Paulista, Rua Antonio Carlos, 653
Sao Paolo 01309, Brazil
T: 256 7811

Starling, Sandra, Deputy
Camara dos Deputados, Palacio do Congresso Nacional
70 000 Brasilia, D.F., Brazil
T: 213 51 51
TL: 1163

Studart, Heineida, Deputy
Camara dos Deputados, Palacio do Congresso Nacional
70 000 Brasilia, D.F., Brazil
T: 213 51 51
TL: 1163

Tavares, Joaquin, Deputy
Camara dos Deputados, Palacio do Congresso Nacional
70 000 Brasilia, D.F., Brazil
T: 213 51 51
TL: 1163

Waichel, Jose Richard, Deputy
Camara dos Deputados, Palacio do Congresso Nacional
70 000 Brasilia, D.F., Brazil
T: 213 51 51
TL: 1163

WISDOMKEEPERS

Alves Filho, Luiz Gonzaga
Spiritual Master
Uniao do Veget.
Av Joaquim Tavora
Cruzeiro do Sul Acre 69980, Brazil
T: 068-322-2121 or 322-2369 F: 068-322-2369

Best, B.
Consultant
Cruzado Dos Menores
Rua Ahelio dos Santos/37
Bento Ribeiro, Rio 21331, Brazil
T: 011-512-252-7292

Cherman, Sheiva K., Sra.
Av. Afranio de Melo Franco 15/4
Rio de Janeiro, Brazil
T: 021-294-2528

BRITISH VIRGIN ISLANDS - GOVERNMENT

Cambers, Gillian, Mr
Head
CONSERVATION & FISHERIES DEPT
PO Box 860 Road Town
Tortola, British Virgin Islands
T: 809 494 5681 2 F: 809 494 4435

BRUNEI DARUSSALAM - GOVERNMENT

Leong, Ivan O.K.
Assistant Solicitor General
BRUNEI DARUSSALAM
Ministry of Law Attorney General'S Chambers
Bandar Seri Begawan 2016, Brunei Darussalam
T: 6732 244872 F: 6732 222300

BULGARIA - GOVERNMENT

Alexeev, Alexey, Mr.
Chef de Cabinet of the President
Permanent Mission of Bulgaria
11 E. 84th Str.
New York, NY 10028, United States of America
T: 212 737 4790 F: 212 472 9865

Bozhanov, Simeon, Mr.
Adviser on Environmental Matters to the President
Permanent Mission of Bulgaria
11 E. 84th Str.
New York, NY 10028, United States of America
T: 212 737 4790 F: 212 472 9865

Ganev, Stoyan, H.E. Mr.
Minister
Foreign Affairs
Permanent Mission to the United Nations, 11 E. 84th Str.
New York, NY 10028, United States of America
T: 212 737 4790 F: 212 472 9865

Ivanov, Lyubomir, Mr.
Minister Plenipotentiary
Ministry of Foreign Affairs
Permanent Mission to the United Nations, 11 E. 84th Str.
New York, NY 10028, United States of America
T: 212 737 4790 F: 212 472 9865

Lazarov, Dimitar, Mr.
Head of Department
Ministry of Foreign Affairs
Permanent Mission to the United Nations, 11 E. 84th Str.
New York, NY 10028, United States of America
T: 212 737 4790 F: 212 472 9865

Nedelchev, Mikhail, Mr.
Member of Parliament
Permanent Mission of Bulgaria
11 E. 84th Str.
New York, NY 10028, United States of America
T: 212 737 4790 F: 212 472 9865

Popov, Zdravko, Mr.
Foreign Political Adviser to the President
Permanent Mission of Bulgaria
11 E. 84th Str.
New York, NY 10028, United States of America
T: 212 737 4790 F: 212 472 9865

Slavenkov, Boyan, Mr.
Economic Adviser to the President
Permanent Mission of Bulgaria
11 E. 84th Str.
New York, NY 10028, United States of America
T: 212 737 4790 F: 212 472 9865

Stoyanov, Valentin, Mr.
Spokesman of the President
Permanent Mission of Bulgaria
11 E. 84th Str.
New York, NY 10028, United States of America
T: 212 737 4790 F: 212 472 9865

Vassilev, Valentin, H.E. Mr.
Minister of the Environment
Permanent Mission of Bulgaria
11 E. 84th Str.
New York, NY 10028, United States of America
T: 212 737 4790 F: 212 472 9865

Zhelev, Zhelyu, H.E. Dr.
President
Permanent Mission of the Republic of Bulgaria
11 E. 84th Str.
New York NY 10028, United States of America
T: 212 737 4790 F: 212 472 9865

NGO

Gerasimov, Svetoslav
UNP
18, Vitosha Ave.
Sofia 1040, Bulgaria

Golemansky, Vassil
BSNR
Post Box 1136, Sofia-1000
Bulgaria

Stepanov, Sergei
General Director
INTERNATIONAL ACADEMY OF ARCHITECTURE - IAA
2 Rouski Bld
1000 Sofia, Bulgaria

PRESS

Apostolov, Andrei
Editor
OFFICIAL PRESS BULGARIA
Sofia, Bulgaria, Bulgaria

Bojilova, Iordanka
Editor
BULGARIA PRESS
Sofia, Bulgaria, Bulgaria

Garelov, Ivan
Editor
OFFICIAL PRESS BULGARIA
Bulgary, Bulgaria

Gueorguiev, Petko
Editor
OFFICIAL PRESS BULGARIA
Sofia, Bulgaria, Bulgaria

Gueorguiev, Sevastiyan
Editor
OFFICIAL PRESS BULGARIA
Bulgary, Bulgaria

Mikhailov, Krassimir
Editor
OFFICIAL PRESS BULGARIA
Bulgary, Bulgaria

Milanov, Petre
Correspondent
DEMOCRACY
134 Rakovsca St.
Sofia, Bulgaria, Bulgaria

Mutafov, Entcho
Editor
OFFICIAL PRESS BULGARIA
Sofia - Bulgaria, Bulgaria

Petrov, Rumian
Correspondent
OFFICIAL PRESS BULGARIA
Sofia, Bulgaria, Bulgaria

Sargavakian, Ani
Correspondent
BULGARIA PRESS
Benkovskistr. 21
Sofia, Bulgaria, Bulgaria

ORGANIZATIONS

Stoilov, Georgi, Prof.
International Academy Of Architecture
2 Rouski Bld
1000 Sofia, Bulgaria
T: (359 2) 87 13 13 F: (359 2) 87 71 65
TL: 23886 iaa ip bg

GLOBAL FORUM

Angnelov, Zlatko
Editor
STANDARD WEEKEND / EQUIPE 10
71 Dimo Hadjidimov St
Sofia 16016, Bulgaria
T: 359 2 545888 F: 359 2 599126

Dimitrov, Ivan
Prof
THEOLOGICAL FACULTY
New Testament Desk
Sofia 1000, Bulgaria

Kurpanov, Yordan, Mr.
F A O / R O M E
26, Rue Murphy
Sofia 1505, Bulgaria
T: 359 2 447412 F: 359 2 885349

Nedialkov, Simeon, Dr.
President
ASSOCIATION OF BULGARIAN ECOLOGISTS
PL Slavekov 7
Sofia 1000, Bulgaria
T: 359 2 88-5349 F: 359 2 88-5340

Stanishev, Georgi, Mr.
INTERNATIONAL ACADEMY OF ARCHITECTURE
Palais de Nations
Geneve CH 1211, Bulgaria
F: 41 22 733-6542

BURKINA FASO - GOVERNMENT

Bazie, Jeremy, M.
Directeur de la pollution et des nuisances
Permanent Mission of Burkina Faso
115 E. 73rd Str.
New York, NY 10021, United States of America
T: 212 288 7515

Brande, Pierre, M.
Conseiller Technique du Ministre de l'Environnement et du Tourisme
Permanent Mission of Burkina Faso
115 E. 73rd Str.
New York, NY 10021, United States of America
T: 212 288 7515

Compaore, Blaise, Mr.
President
Permanent Mission of Burkina Faso
115 E. 73rd. Str.
New York NY 10021, United States of America
T: 212 288 7515

Honadja, Mamadou, M.
Inspecteur des Eaux et Forets
Permanent Mission of Burkina Faso
115 E. 73rd Str.
New York, NY 10021, United States of America
T: 212 288 7515

Kabore, Omer Moussa, M.
Conseiller du President
Charge du monde rural
Permanent Mission to the United Nations, 115 E. 73rd Str.
New York, NY 10021, United States of America
T: 212 288 7515

Konate, Issa Dominique, S.E.M.
Ministre des Relations Exterieures
Permanent Mission of Burkina Faso
115 E. 73rd Str.
New York, NY 10021United States of America, 212 288 7515

Nebie, Moussa B., M.
Deuxieme Conseiller
Mission Permanente aupres des Nations Unies, New York
Permanent Mission to the United Nations, 115 E. 73rd Str.
New York, NY 10021, United States of America
T: 212 288 7515

Ouali, Louis Armand Mihyemba, S.E.M.
Ministre de l'Environnement et du Tourisme
Permanent Mission of Burkina Faso
115 E. 73rd Str.
New York, NY 10021United States of America, 212 288 7515

Ouedrago, Lambert Alexandre
Conseiller des Affaires Etrangeres
Permanent Mission of Burkina Faso
115 E. 73rd Str.
New York, NY 10021, United States of America
T: 212 288 7515

Ouedraogo, Gaetan R., S.E.M.
Ambassadeur Extraordinaire et Plenipotentiaire
Represent Permanent aupres des Nations Unies
Permanent Mission to the United Nations, 115 E. 73rd Str.
New York, NY 10021United States of America, 212 288 7515

Sanago, Adama, M.
Secretaire Permanente du Comite National de lutte contre la deserification
Permanent Mission of Burkina Faso
115 E. 73rd Str.
New York, NY 10021, United States of America
T: 212 288 7515

Some, Laurent Magloire, M.
Directeur du Centre National de semences forestieres
Permanent Mission of Burkina Faso
115 E. 73rd Str.
New York, NY 10021, United States of America
T: 212 288 7515

Sow, M., Conseiller du President en connunication
Permanent Mission of Burkina Faso
115 E. 73rd Str.
New York, NY 10021, United States of America
T: 212 288 7515

Yougbare, Leon, M.
Directeur du Protocole d'Etat
Permanent Mission of Burkina Faso
115 E. 73rd Str.
New York, NY 10021, United States of America
T: 212 288 7515

NGO

Horie, Noboru
AVD
01 BP 4683
Ouagadougou 01, Burkina Faso

Koala, Georgette
ANAR
01 BP 2314
Ouagadougou 01, Burkina Faso

Ouedraogo, Florent
ASSOCIATION VOLUNTAIRE D.
Palais de Nations
Geneve CH 1211, Burkina Faso
F: 41 22 733-6542

Sompougdou, Jean-Mari
ANAR
01 BP 2314
Ouagadougou 01, Burkina Faso

PRESS

Dabire, James
Editor
AIB/CJE
Ouagadougou, Burkina Faso
Burkina Faso

Ouangraolia, Hamado
Editor
SIDWAYA
Provice du Kadiogo Secteur 20
Burkina Faso, Burkina Faso

Ouedraogo, Rene
Reporter Radio/Television
TV NAT. DU BURKINA
B.P. 2530
Ouagadougou, Burkina Faso, Burkina Faso

Ouedraogo, Roch Parfait
Reporter Radio/Television
RADIODIFFUSION NATIONALE
Ouagadougou
Burkina Faso ?, Burkina Faso

Some, Alexis
Editor
SIDWAYA
01 BP 507
Ouagadougou, Burkina Faso, Burkina Faso

Traore, Allassane
Correspondent
LA PRESSE PRES. DU BURKINA
9, Avenue Houari Boumedienne
Secteur 1, Ouagadougou, Burkina Faso, Burkina Faso

Zongo, Norbert
Correspondent
SIDWAYA
01 BP 4809
Ouagadougou, Burkina Faso, Burkina Faso

ORGANIZATIONS

Ouedraogo, Abdou, Mr.
As. Nationale D'action Rurale (ANAR)
01 BP 2314
Ouagadougou 01, Burkina Faso
T: (226) 30 32 53 F: (226) 30 86 17

Ouedraogo, Clementine, Mme
As. for the Promotion of Women in development and Sport
01 BP 2532
Ouagadougou 01, Burkina Faso
T: (226) 313 052

Ouedraogo, S. Florent, Pr.
As. Des Volontaires Pour Le Developpement
01 BP 4683
Ouagadougou 01, Burkina Faso
T: (226) 30 86 16 F: (226) 30 86 17

Zeba, Souleymane, Mr.
Fondation des Amies de la Nature (NATURAMA)
01 BP 6133
Ouagadougou 01, Burkina Faso
T: (226) 303 119 F: (226) 301 351

GLOBAL FORUM

Bado, Hortense, Ms.
ASSISTANCE ECOLOGIQUE
P. 305, BoBo-Dioulasso 01
Burkina Faso
T: 98-2094 F: 97-0802

Hien, Dieudonne, Mr.
Directeur
BUREAU D'ETUDE ET DE LIAISONS
01 BP 1195
Ouagadougou, Burkina Faso
T: 22 6 306025 F: 22 6 306025

Josephine, Ouedraogo, Mr.
AVD
Orbp 4683 Ouagadougou
Ovsgadougou, Burkina Faso
T: 226 30-8616 F: 226 30-8617

Kayende, M. Marcelle Prisque, Ms.
Translater
A.V.D.
01BP 6683 Guagadougou 01
Guagadougou, Burkina Faso
T: 30 4618

Madjri, Jhon, Mr.
CESAO
BP 305
Bobo - Dioulassou, Burkina Faso
T: 22 6 99 04 01 F: 22 6 98 20 42

Ouedraogo, Bernadette, Ms.
GRADE - FRB
PO Box 3656
Ouagadougou, Burkina Faso
TL: 0978-1111

Ouedraogo, Bernard Ledea, Mr.
GROUPMENTS NAAM SIX S
BP 100
Ouahigouya, Burkina Faso
T: 22 6 55 04 11

Samba, Ibrahim M., Mr.
Directeur
OMS/OCP
Ouagadougou, Burkina Faso

Sompougdov, O. Jean Marie, Mr.
Secretaire
ANAR
01 bp 2314 onagadongon
kadiogo, Burkina Faso
T: 303253

BURUNDI - GOVERNMENT

Banyanankiye, Prosper, M.
Directeur General du Plan
Permanent Mission of Burundi
United Nations Plaza
New York, NY 10017, United States of America
T: 212 963 7171 F: 212 963 1921

Bukuru, Melchiade, M.
Deuxieme Conseiller
Mission Permanente aupres des Nations Unies, New York
Permanent Mission to the United Nations, United Nations Plaza
New York, NY 10017, United States of America
T: 212 963 7171 F: 212 963 1921

Nduwimana, Louis, S.E.M.
Ministre de l'Amenagement, du Tourisme et de l'Environnement
Permanent Mission of Burundi
United Nations Plaza
New York, NY 10017, United States of America
T: 212 963 7171 F: 212 963 1921

Ntahuga, Laurent, M.
Directeur General l'Institut National pour l'Environnement et la Conservation de la Nature
Permanent Mission of Burundi
United Nations Plaza
New York, NY 10017, United States of America
T: 212 963 7171 F: 212 963 1921

NGO

Mayshara, Rurenza
ADESKI
D.S. No. 4 Bújumbura
Burundi

Rwumbuguta, Ruhimbika
ADESKI
D.S. No. 4 Bujumbura
Burundi

PRESS

Baragasika, Ildephonse
Editor
RENOUVEAU
Ngagara Q6, No. 367
Bujumbura, Burundi, Burundi

Mbonerane, Albert
Editor
NDONGOZI Y BURUNDI
Rujumbura, Burundi
Afrique Centrale, Burundi

Ntiruhangura, Athanase
Reporter Radio/Television
RADIO-TV NAT.DU BURUNDI
Ngagara Q6, B16, No. 367
B.P. 1300, Bujunbura, Burundi, Burundi

CAMEROON - GOVERNMENT

Balinga, Victor Sunday, Mr
Research Officer/Study Officer
DEPARTMENT OF SCIENTIFIC AND TECHNI
Ministry of Scientific And Technica Research, B. P. 1457
Yaounde, Cameroon
T: 23 77 20 F: 22 26 76

Betah, Samuel Sona
Director Of Geology
MINISTRY OF MINES,WATER AND ENERGY
Ministry of Mines,Water And Power
Yaounde, Cameroon
T: 237 23 39 70 23 3404

Kwuimo, Jacques, Mr
Attache
PREMIER MINISTERE
B.P. 8275
Yaounde, Cameroon
T: 237 23 72 65

Missi, Jean-Paul, Mr
Charge d'Etudes
MINISTERE DE L'URBANISMEX HABITAT
Yaounde-Cameroon-Minuh
Yaounde, Cameroon
T: 237 23 28 47

Nkoulou Ndanga, Appollnaire
Directeur Des Forets
MINISTERE DE L'ENVIRONNEMENT ET DES
Yaounde
Yaounde, Cameroon
T: 237-22-11-06;23-55-4

Ntoupendi, Henri, Mr
Chef De Service Des Institutions Interafricaines
MINISTERE DES RELATIONS
Exterieures
Yaounde, Cameroon

Samba, Dieudonne
Ingenieur Des Mines, Sous-Directeur
MINISTERE DES MINES, DE L'EAU ET EN
Ministere des Mines, De L'Eau Et De L'Energie
Yaounde, Cameroon
T: 23 39 70 OR 23 34 04

Tchana, Mesack
Directeur Adjoint De l'Environnemen
MINISTERE ENVIRONNEMENT ET DES FORE
Br 1877 Yadunde
Yaounde, Cameroon
T: (237) 233581/204893

NGO

Mengolo, Justin
CEDER
B.P. 59 Nanga-Eboko
Cameroon

Momendeng, Symphorien Kevin
CEDER
B.P. 59 Nanga-Eboko
Cameroon

Nguiffo Tene, Samuel Alain, Mr.
Redator
CERCLE INTERNATIONAL POUR LA PROMOTION DE LA CREACION
BP 1256
Bafoussam, Cameroon
T: 237 44 30 55 F: 237 44 51 91

Nkoyon, Jacqueline
IAEPA
B.P. 6935
Yaounde, Cameroon

Younbi, Augustin
IAEPA
B.P. 6935
Yaounde, Cameroon

PRESS

Ngum Tegha Mua, Boniface
Reporter Radio/Television
CRTV
CRTV Yaounde
Cameroon, Cameroon

ORGANIZATIONS

Kenmagne, Jean Blaise, Mr.
Cercle International pour la Promotion de la Creation (CIPCRE)
BP 1256
Bafoussam, Cameroon

Keou, Francois Tiani, Mr.
Enviro-protect - International Association For Environmental Protection In Afri
B.P. 6935
Yaounde, Cameroon
T: (237) 23 54 35 F: (237) 22 06 56

GLOBAL FORUM

Abassi, Elong, Mr.
MAETUR-ARAN
BP.3428
Douala, Cameroon
T: 237 42 61 00
TL: 5725 KN

Ada, Ndeso-Atanga, Ms.
Country Coordinator
PVO-NGO/NRMS PROJECT
BP 422
Yaoude, Cameroon
T: 23 7 20-65-37 F: 23 7 22-18-73

Alpizar, Lydia, Ms.
Delegate
YOUTH 92 WORLD YOUTH PREPARATORY FORUM FOR UNCED
Can B.P. 337 (Dou/Dasa)
Yaounde, Cameroon
T: 237 231-260 F: 237 221-320

Egbe, Besong Esther Batey Nee, Ms.
MOKOK AGHEM
Small Mamfe, Manyu Div.S.W.P. - PO BOX 129
Manfe 34-1272, Cameroon
T: 237 34-1272

Ndhine, Thomas, Mr.
Directeur De La Recherche
MESERES - ORSTOM
BP.18557
Yaounde, Cameroon

Ngangeum, Nicolas Deuton, Mr.
President
INDEPENDANT ET DU RAYSAN (M.P.P.P.)
B. P. 11912,Yaounde
Yaoude, Cameroon
T: 23 7 432140 F: 23 7 235923
TL: (970)8342

Njonga, Bernard, Mr.
SAILD
BP 11855
Yaounde, Cameroon
T: 237 22 46 82 F: 237 22 51 62

Tappa, Louise, Ms.
WORLD COUNCIL OF CHURCHES
BP 6007, New Ball
Douala, Cameroon

Ticha, Samuel Teneng, Mr.
Exec. Secret.
MOUVEMENT POUR LE PROGRES DE L'OUVRIER INDEPANDANT ET DU PAYSAN (M.P.O.P)
B. P. 11912,Yaounde
Yaoude, Cameroon
T: 23 7 432140 F: 23 7 235923
TL: (970)8342

Youmbi, Augustine, Ms.
ENVIRO-PROTECT
Palais de Nations
Geneve CH 1211, Cameroon
F: 41 22 733-6542

CANADA - GOVERNMENT

Bell, John, M.
Counsellor to the Secretary of State for external Affairs on the Environment
Permanent Mission of Canada
866 United Nations Plaza, Suite 250
New York, NY 10017, United States of America
T: 212 751 5600 F: 212 486 1295

Bloomfield, Michael, Mr.
Heritage Foundation
Permanent Mission to the United Nations, 866 United Nations Plaza, Suite 250 New York, NY, 10017
F: 212 751 5600
TL: 212 486 1295

Browes, Pauline, The Hon.
Minister of State
Permanent Mission of Canada
866 United Nations Plaza, Suite 250
New York, NY 10017, United States of America
T: 212 751 5600 F: 212 486 1295

Campeau, Arthur, H.E. Mr.
Ambassador
Prime Minister's Representative on the Environment
Permanent Mission to the United Nations, 866 United Nations Plaza, Suite 250
New York, NY 10017, United States of America
T: 212 751 5600 F: 212 486 1295

Champagne, Michel, M.
Chairperson of the House of Commons
Standing Committee on Forestry and Fisheries
Permanent Mission to the United Nations, 866 United Nations Plaza, Suite 250 New York, NY, 10017
F: 212 751 5600
TL: 212 486 1295

Charest, Jean, The Hon.
Minister of the Environment
Permanent Mission of Canada
866 United Nations Plaza, Suite 250
New York, NY 10017, United States of America
T: 212 751 5600 F: 212 486 1295

Ciaccia, John, The Hon.
Minister of the Environment
Province of Quebec
Permanent Mission to the United Nations, 866 United Nations Plaza, Suite 250
New York, NY 10017, United States of America
T: 212 751 5600 F: 212 486 1295

Clarke, William, H.E. Mr.
Ambassador Extraordinary and Plenipotentiary to Brazil
Permanent Mission of Canada
866 United Nations Plaza, Suite 250
New York, NY 10017, United States of America
T: 212 751 5600 F: 212 486 1295

Clements, Gilbert, The Hon.
Minister of Environment, Province of Prince Edward Island
Representative of the Canadian Council of Ministers of Environment
Permanent Mission to the United Nations, 866 United Nations Plaza, Suite 250
New York, NY 10017, United States of America
T: 212 751 5600 F: 212 486 1295

Cote, Louise, Mr.
Environment Canada
Permanent Mission to the United Nations, 866 United Nations Plaza, Suite 250 New York, NY, 10017
F: 212 751 5600
TL: 212 486 1295

Crayston, John, Mr
Co Ordinator Envir Programmes
INTERNATIONAL CIVIL AVIATION ORGANI
1000 Sherbrooke St West
Montreal Quebec H3A ZRZ, Canada
T: 514 286 6371 F: 514 288 4772

Crosbie, John, The Hon.
Minister of fisheries and Oceans
Permanent Mission of Canada
866 United Nations Plaza, Suite 250
New York, NY 10017, United States of America
T: 212 751 5600 F: 212 486 1295

Cuthbert, John, Mr.
Environment Canada
Permanent Mission to the United Nations, 866 United Nations Plaza, Suite 250 New York, NY, 10017
F: 212 751 5600
TL: 212 486 1295

Dillon, John, Mr.
Business Representative
Permanent Mission of Canada
866 United Nations Plaza, Suite 250
New York, NY 10017, United States of America
T: 212 751 5600 F: 212 486 1295

Doray, Luc, Mr.
Charge De Mission-Affairs Int'l
VILLE DE MONTREAL
333 Rue Saint-Antoine Est Bureau 228 HZX IR9
Montreal, Canada
T: 514 872 0001 F: 514 872 6067

Draimin, Tim, Mr.
Development NGO
Permanent Mission to the United Nations, 866 United Nations Plaza, Suite 250 New York, NY, 10017
F: 212 751 5600
TL: 212 486 1295

Drake, David, Mr.
Ministry of Forestry
Permanent Mission to the United Nations, 866 United Nations Plaza, Suite 250 New York, NY, 10017
F: 212 751 5600
TL: 212 486 1295

Dube', Robert, Mr
Conseiller Principal
ENVIRONNEMENT CANADA
44 la Casse
Aylmer-Quebec, Canada
T: 819 953 6266 F: 819 953 6897

Dufour, Jacques, Ms
Conseilor
INTERGOVERNMENTAL RELATIONS BRANCH
Quebec Ministry of Environment 3900,Marly Street Floor 5TH
Sainte-Foy (Quebec), Canada
T: 418 643 7360 F: 418 644 4598

Filmon, Gary, The Hon.
Premier
Province of Manitoba
Permanent Mission to the United Nations, 866 United Nations Plaza, Suite 250
New York, NY 10017, United States of America
T: 212 751 5600 F: 212 486 1295

Georges,
Boulet
DIRECTEUR DES RELATIONS INTERGOUVER
Nementales Ministere De L'Environne 3900 Rue De Marly 5E Etage
Sainte-Foy (Quebec), Canada
T: 418 643 8209 F: 418 644 4598

Hannah, Erin, Ms.
Youth Representative
Permanent Mission of Canada
866 United Nations Plaza, Suite 250
New York, NY 10017, United States of America
T: 212 751 5600 F: 212 486 1295

Hanson, Arthur J., Mr.
President & Ceo
INTERNATIONAL INSTITUTE FOR SUSTAINABLE DEVELOPMENT
161 Portage Ave. East , 6th floor
Winnipeg - Alberta, Canada
T: 1 204 958-7700 F: 1 204 958-7710

Johnson, Pierre-Marc
Advisor to Maurice Strong
Permanent Mission of Canada
866 United Nations Plaza, Suite 250
New York, NY 10017, United States of America
T: 212 751 5600 F: 212 486 1295

Jolette, Denis, Mr.
Chief of Staff of Ambassador Campeau
Permanent Mission of Canada
866 United Nations Plaza, Suite 250
New York, NY 10017, United States of America
T: 212 751 5600 F: 212 486 1295

Jolin, Pierre, Mr
Dir. Gen
DEPARTMENT OF INTERNATIONAL AFFAIRS
1225 Place Georges V
Quebec, Canada
T: 418 649 2341

Labaj, Annie, Ms.
Nat'L Rep.
CANADIAN AUTOMOBILE WORKERS
205 Placer Court
Willowdale-Ontario, Canada
T: 1 416 497-4110 F: 1 416 495-6552

Landry, Monique, The Hon.
Minister of External Relations and International Development
Permanent Mission of Canada
866 United Nations Plaza, Suite 250
New York, NY 10017, United States of America
T: 212 751 5600 F: 212 486 1295

Langlois, Charles, Mr.
MP
Permanent Mission of Canada
866 United Nations Plaza, Suite 250
New York, NY 10017, United States of America
T: 212 751 5600 F: 212 486 1295

Lautenback, William, Mr.
Urban Advisor
City of Sudbury
Permanent Mission to the United Nations, 866 United Nations Plaza, Suite 250 New York, NY, 10017
F: 212 751 5600
TL: 212 486 1295

Lavoie-Roux, Therese, Senator
Memeber of Senate
Permanent Mission of Canada
866 United Nations Plaza, Suite 250
New York, NY 10017, United States of America
T: 212 751 5600 F: 212 486 1295

Maini, Jagmohan
Assist Deputy Min, Forest Envir
FORESTRY CANADA
351 St. Joseph Blvd. Place Vincent Massey Bldg., 21ST Fl
Ottawa, Ontario K1A 1G5, Canada
T: 819 997 1107 F: 819 953 3642

McDonald, David, The Hon.
MP, Chairperson of the House of Commons Standing Committee on Environment
Permanent Mission to the United Nations, 866 United Nations Plaza, Suite 250
New York, NY 10017, United States of America
T: 212 751 5600 F: 212 486 1295

McDougall, Barbara, The Hon.
Secretary of State for External Affairs
Permanent Mission of Canada
866 United Nations Plaza, Suite 250
New York, NY 10017, United States of America
T: 212 751 5600 F: 212 486 1295

McGrath, Kathleen, Mr.
Special Assistant to Minister Charest
Permanent Mission to the United Nations, 866 United Nations Plaza, Suite 250 New York, NY, 10017
F: 212 751 5600
TL: 212 486 1295

McGraw, Desiree, Mr.
Environment Canada
Permanent Mission to the United Nations, 866 United Nations Plaza, Suite 250 New York, NY, 10017
F: 212 751 5600
TL: 212 486 1295

Mulroney, Brian, The Rt. Hon.
Prime Minister
Canada
866 United Nations Plaza, Suite 250
New York, NY 10017, United States of America
T: 212 751 5600 F: 212 486 1295

Ningark, John, The Hon.
Minister of Renewable Resources
Government of the Northwest Territories
Permanent Mission to the United Nations, 866 United Nations Plaza, Suite 250
New York, NY 10017, United States of America
T: 212 751 5600 F: 212 486 1295

Oberle, Frank, The Hon.
Minister of Forestry
Permanent Mission of Canada
866 United Nations Plaza, Suite 250
New York, NY 10017, United States of America
T: 212 751 5600 F: 212 486 1295

Robertson, Michael Ross, Mr
Senior Director , Environment
PETRO-CANADA
P O Box 2844
Calgary Alberta, Canada
T: 403 296 8824 F: 403 296 3030

Royer, Lucien, Mr.
Labour Representative
Permanent Mission of Canada
866 United Nations Plaza, Suite 250
New York, NY 10017, United States of America
T: 212 751 5600 F: 212 486 1295

Salter, Robert, Dr.
Deputy Minister
Environment Canada
Permanent Mission to the United Nations, 866 United Nations Plaza, Suite 250 New York, NY, 10017
F: 212 751 5600
TL: 212 486 1295

Senecal, Nicole, Ms.
Deputy Minister
Canadian International Development Agency
Permanent Mission to the United Nations, 866 United Nations Plaza, Suite 250
New York, NY 10017, United States of America
T: 212 751 5600 F: 212 486 1295

Simon, Mary, Ms.
Representative
Indigenous People
Permanent Mission to the United Nations, 866 United Nations Plaza, Suite 250 New York, NY, 10017
F: 212 751 5600
TL: 212 486 1295

Simon, Mary, Ms.
Representative
Indigenous People
Permanent Mission to the United Nations, 866 United Nations Plaza, Suite 250
New York, NY 10017, United States of America
T: 212 751 5600 F: 212 486 1295

Slater, Robert, Dr.
Deputy Minister
Environment Canada
Permanent Mission to the United Nations, 866 United Nations Plaza, Suite 250
New York, NY 10017, United States of America
T: 212 751 5600 F: 212 486 1295

U Phae Than Oo,
Charge d' Affaires
EMBASSY OF THE UNION OF MYANMAR
85, Range Road, Apt.No. 902-903, the Sandringham
Ottawa, Ontario, K1N HJC, Canada
T: 2326434 F: 613 2325435

Wells, Clyde, The Hon.
Premier
Province of New Founland
Permanent Mission to the United Nations, 866 United Nations Plaza, Suite 250
New York, NY 10017, United States of America
T: 212 751 5600 F: 212 486 1295

Williams, Claude
Direitevr-Relations Internationals
CANADA MORTGAGE & HOUSING CORPORATI
1560 Sunview
Orleans Ontario KIC5A3, Canada
T: 613 748 2817 F: 613 748 2302

NGO

Adams, Patricia, Ms.
Exec Dir
PROBE INTERNATIONAL
225 Brunswick Avenue
Toronto - Ontario M5S 2M6, Canada
T: 1 416 978-4842 F: 1 416 978-3824

Amaratunga, Carol
INSTITUT CANADIEN DES OCEANS
1236, Henry Street, 5th Floor
Halifax - NS, Canada
T: 1 902 494-3879 F: 1 902 494-1334
TL: UKB 7491

Armstrong, Brenda Russell, Ms.
V. Chair
WESTERN ENVIRONMENT AND DEVELOPMENT TASK FORCE
14th FL- Commerce Place, 400 Burrard St
Vancouver - BC V6C3G2, Canada
T: 1 604 643 1720 F: 1 604 643 1760

Armstrong, Patrick, Mr.
Director
WESTERN ENVIRONMENT AND DEVELOPMENT TASK FORCE
2800 Park Place, 666 Burrard Street
Vancouver, British Columbia V6C 2Z7, Canada

Aslak Nils, Sara, Ms.
Director
WCIP
555 King Edward Ave
Ottawa, Ontario K1N 6N5, Canada

Atherton, Jacqueline
TIES
39077 Point Grey RPO
Vancouver, British Columbia V6R 4P1, Canada

Bertell, Helen Rosalie
IICPH
830 Bathurst Street
Toronto, Ontario M5R 3G1, Canada

Blouin, Glen
CANADIAN FORESTRY ASSOCIATION
203-185 Somerset St. West
Ottawa - Ontario K2P 0J2, Canada
T: 1 613 232-1815 F: 1 613 2324210

Bouige, Daniel
AIC
1130 Sherbrooke Street West, Suite 410, Montreal, Canada

Braulio, Morales
WCIP
555 King Edward Ave
Ottawa, Ontario K1N 6N5, Canada

Broughton, Lynn
CPCU
63 Sparks Street, Ste. 603
Ottawa, Ontario K1P 5A6, Canada

Brugmann, Jeb
ICLEI
City Hall, East Tower, 8th Floor
Toronto, Ontario M5H 2N2, Canada

Carpenter, Les, Mr.
Vice-President
INUIT CIRCUMPOLAR CONFERENCE
650, 32nd Avenue, Suite 404, Lachine
Lachine - P.Q. H8T3k4, Canada
T: 1 514 6373771 F: 1 514 6373146

Carr, Adriane, Ms.
Executive Director
WESTERN CANADA WILDERNESS COMMITTEE
20 Water Street
Vancouver - BC, Canada
T: 1 604 683-8220 F: 1 604 683-8229
E: wc2wild

Coombs, Gerald
IDERA
2524 Cypress Street
Vancouver, BC V6J 3N2, Canada

Davis, Thomas, Mr.
SCIENCE FOR PEACE (CANADA)
University of Toronto
Toronto, Ontario M5S 1A1, Canada

Davison, Joan Mary
POWER
P.O. Box 274
Acton, Ontario L7S 2M4, Canada

Dewachie, Evelyn
ISIN
298 Elgin Street, Suite 105, Ottawa, Ontario K2P 1 M3, Canada

Exmundo, Nilda
Ms.
ICLEI
City Hall, East Tower, 8th Floor
Toronto, Ontario M5H 2N2, Canada
T: 1 416 392 1462 F: 1 416 392 1478

Ferstman, Miriam, Ms.
Chairman
INTERNATIONAL COUNCIL OF JEWISH WOMEN
1110 Finch Ave. West, Suite 518
Downsview, Ontario M3J 2T2, Canada

Hammarskjold, Knut O.H.A.
IIPT
3680 rue de la Montagne
Montreal, Quebec H3G 2A8, Canada

Hernandez, Monica
CYF
55 Parkdale Ave., 3rd Floor
Ottawa, Ontario K1Y 1E5, Canada

Hopkins, Charles
EE
885 Dundas Street East
Toronto, M4M 1R4, Canada

Hoshak, Abdulkader
ICLEI
City Hall, East Tower, 8th Floor
Toronto, Ontario M5H 2N2, Canada

Johnson, Charlie H.
ICPC
Suite 404, 650 32nd Avenue
Lachine, Quebec H8T 3K4, Canada

Jost, Raymond
SIE
Le Cours Le Royer, 48, rue Le Royer Ouest
Montreal, Quebec H2Y 1W7, Canada

Kewell, Caryn, Ms.
Resource Coordinator
SOCIAL JUSTICE COMMITTEE
1857 de Maisonneuve West
Montreal - Quebec, Canada
T: 1 514 933-6797 F: 1 514 933-9517

Kumar, Ranjit, Mr.
Executive Director
FOUNDATION FOR INTERNATIONAL TRAINING - FIT
1262 Don Mills Road, Suite 200
Don Mills-ON M5R 3B7, Canada
T: 1 416 449-8838 F: 1 416 449-8547
TL: 06-986715

Labrie, Louise, Ms.
Researcher
ASSEMBLY OF FIRST NATIONS
47 Clarence Street, Suite 300
Ottawa, Ontario K1N 9KI, Canada

Laplante, Gary, Mr.
Communication Coordinator
INDIGENOUS SURVIVAL INTERNATIONAL
Suite 105 - 298 Elgin St.
Ottawa - Ontario, Canada
T: 1 613 230 3616 F: 1 613 230 3595

Lemonde, Pierre
IELS
445, St. Laurent Blvd, 5th Floor
Montreal, Quebec H2Y 3T8, Canada

Lesnick, Michael
RAFI
750-130 Slater Street
Ottawa, Ontario K1P 6E2, Canada

McKean, Bruce
ICME
360 Albert Street, Suite 1550
Ottawa, Ontario K1R 7X7, Canada

Mercredi, Ovide
AFN
47 Clarence Street, Suite 300
Ottawa, Ontario K1N 9KI, Canada

Murphy, Sophia, Ms.
CANADIAN COUNCIL INTERNATIONAL COOPERATION
1 Nicholas St. # 300
Ottawa - Ontario K1N 7B7, Canada
T: 1 613 236-4547 F: 1 613 236-2188
E: ccic 1992 (oveb)

Oben, Theodore Epey Ayuk, Mr.
YOUTH 92 / CLUB DES AMIS DE LA NATURE
55 Parkdale Ave, 3rd floor
Ottawa - Ontario, Canada
T: 1 613 761 9206 F: 1 613 761 1441

Padbury, Peter
Canadian Council For International Co-operation
1 rue Nicholas Street, Suite 300
Ottawa, Ontario K1N 7B7, Canada

Parker, David
ICME
360 Albert Street, Suite 1550
Ottawa, Ontario K1R 7X7, Canada

Perron, Michel
RLD
425 Sacre-Couer Ouest, Suite 14
Alma, Quebec G8B IM4, Canada

Prescott, Jacques, Mr.
Delegate
FOSEM
8191, Avenue Du Zoo
Charlesbourg- Quebec GIG 4G4, Canada
T: 1 418 622-0313 F: 1 418 646-9239

Quiroz, Ana Maria
ICAE
720 Bathurts Street, Suite 500
Toronto, Ontario M5S 2R4, Canada

Rosales, Rosa
WCIP
555 King Edward Ave
Ottawa, Ontario K1N 6N5, Canada

Rosenberg, Dorothy G.
IICPH
830 Bathurst Street
Toronto, Ontario M5R 3G1, Canada

Rozan, Stella, Ms.
Immediate Past President
INTERNATIONAL COUNCIL OF JEWISH WOMEN
1110 Finch Ave. West, Suite 518
Downsview, Ontario M3J 2T2, Canada

Santos, Ignacio, Sr.
YOUTH 92 / COUNCIL OF EUROPEAN NATIONAL YOUTH COMMITTEES
55 Parkdale Ave, 3rd floor
Ottawa - Ontario, Canada
T: 1 613 761 9206 F: 1 613 761 1441

Seigel, Leila
INTERNATIONAL COUNCIL OF JEWISH WOMEN
1110 Finch Ave. West, Suite 518
Downsview, Ontario M3J 2T2, Canada

Seymoor, Nola Kate
ECO-ED
885 Dundas Street East
Toronto, M4M 1R4, Canada

Sioui, Konrad
AFN
47 Clarence Street, Suite 300
Ottawa, Ontario K1N 9KI, Canada

Snowshoe, Karen
Researcher
INDIGENOUS SURVIVAL INTERNATIONAL
298 Elgin Street, Suite 105, Ottawa, Ontario K2P 1 M3, Canada

Swan, Judith
INSTITUT CANADIEN DES OCEANS
5670 Spring Garden Road, 9th Floor
Halifax, Nova Scotia B3J 1H6, Canada

Thornburn, Geoffrey, Mr.
Economics Adviser
INTERNATIONAL JOINT COMMISSION
100 Metcalpe St., 18th Floor
Ottawa - Ontario K1S 3R1, Canada
T: 1 613 995-0433 F: 1 613 993-5583

Tibando, Ramona, Ms.
Coordinator
WESTERN CANADA WILDERNESS COMMITTEE
20 Water Street
Vancouver - BC, Canada
T: 1 604 683-8220 F: 1 604 683-8229
E: wc2wild

Toole, Brian Edward
ELC
201, 10350-124 Street
Edmonton, Alberta T5N 3V9, Canada

Tulio Mariano, Gonzalez
WCIP
555 King Edward Ave
Ottawa, Ontario K1N 6N5, Canada

Vaillancourt, Jean-Guy Vailian, Mr.
Delegate
FOSEM
8191, Avenue Du Zoo
Charlesbourg- Quebec GIG 464, Canada
T: 1 418 622-0313 F: 1 418 646-9239

Vasconez, Mario
SIE
Le Cours Le Royer, 48, rue Le Royer Ouest
Montreal, Quebec H2Y 1W7, Canada

Verbisky, Richard
Environment Officer
CAMROSE INTERNATIONAL INSTITUTE
5061-50 Street
Camrose, Alberta T4V 1R3, Canada

Viezzer, Moema
ICAE
720 Bathurts Street, Suite 500
Toronto, Ontario M5S 2R4, Canada

Villeneuve, Claude
President
REGION LABORATIORE DU DEVELOPPEMENT DURABLE
425 Sacre-Couer Ouest, Suite 14
Alma, Quebec G8B IM4, Canada

Vio-Grossi, Francisco
ICAE
720 Bathurts Street, Suite 500
Toronto, Ontario M5S 2R4, Canada

Winkler, Earl
IDERA
2524 Cypress Street
Vancouver, BC V6J 3N2, Canada

Winkowski, Sara
Chairperson For Envmmt'L Comm
INTERNATIONAL COUNCIL OF JEWISH WOMEN
1110 Finch Ave. West, Suite 518
Downsview, Ontario M3J 2T2, Canada

Wood, Peregrine, Ms.
Coordinator, Project Learning
FOUNDATION FOR INTERNATIONAL TRAINING - FIT
1262 Don Mills Road, Suite 200
Don Mills-ON M5R 3B7, Canada
T: 1 416 449-8838 F: 1 416 449-8547
TL: 06-986715

Wylynko, Bradley
Consultant
ENVIRONMENTAL LAW CENTRE OF ALBERTA
201, 10350-124 Street
Edmonton, Alberta T5N 3V9, Canada

PRESS

Adams, Leslie
Senior Producer
VISION TV
315 Queen Street East
Toronto M5A 157, Canada
T: 1 416 366-9221 F: 1 416 368-9774

Altieiri, Adhemar
CBC
13 Wilce Dr.
Ajax, ONT, Canada L1T 3J6, Canada

Archambault, Raymond
Reporter Radio/Television
SOCIETE RADIO-CANADA
320 Armstrong
Rosemere, Qc, Canada J7A 1V3, Canada

Bailey, Phyllis E.
Correspondent
CKKCU FM
4072 Northcliffe Ave.
Montreal, Canada, Canada

Barker, Bill
Correspondent
GLOBAL TELEVISION NETWORK
to be completed upon arrival, Canada

Barr, Shane
Correspondent
CHARLATON
1005 River Rd., Kanata
ON R2R 1X7, Canada, Canada

Basile, Sandra
Reporter Radio/Television
SOCIETE RADIO-CANADA
847 Place Simon, St. Laurent
St. Laurent, Qc, Canada H4M 2W3, Canada

Basu, Robindra Nath
Correspondent
CALGARY CABLE TV/FM
78 Edgedale Wat N.W.
Calgary, Alta, Canada T3A 2X9, Canada

Bennett, Julia
Editor
CBC TV - THE JOURNAL
100 Carlton Street, 3rd Floor
Toronto, Canada M5B 1M3, Canada

Berlyn, Judith, Ms.
Correspondent
WESTMOUNT EXAMINER
381 Elm Ave.
Westmount,Qc, Canada H32 1Z4, Canada

Black, Daniel
Reporter Radio/Television
SOCIETE RADIO-CANADA
Outremont, Quebec
H2V 3H7 Canada, Canada

Blount, John Edmund
Correspondent
IDLER
R.Almte.Gomes Pereira 12/201
22291 Rio de Janeiro-RJ, Brazil, Canada

Borgognon, Alain
Reporter Radio/Television
SOCIETE RADIO-CANADA
16 av de la Brunante
Outremont, Qc, Canada H3T 1R4, Canada

Bouchard, Daniel
Reporter Radio/Television
SOCIETE RADIO-CANADA
130 Baker St.
Sudbury, ONT, CAnada P3C2E9, Canada

Bourgault, Marc
Reporter Radio/Television
SOCIETE RADIO-CANADA
4320 de Lorimier
Montreal, Qc, Canada H2H 2B1, Canada

Boyle, Linda
Correspondent
GLOBAL TELEVISION NETWORK
To be completed upon arrival, Canada

Bringue, Gilbert
Reporter Radio/Television
SOCIETE RADIO-CANADA
41 La Casse
Aylmer, Qc, Canada J9H 6H7, Canada

Brosnahan, Maureen P.
Reporter Radio/Television
CBC
2838 Angus Street
Regina, Saskatchewan, Canada 34S IN8, Canada

Bueckert, Dennis
Correspondent
CANADIAN PRESS
2230 McQuaig, Ottawa
ONT K1H 7N9, Canada, Canada

Cameron, William
Reporter Radio/Television
CBC TV - THE JOURNAL
100 Carlton Street, 3rd Floor
Toronto, Canada M5B 1M3, Canada

Cardinal, Mario
Editor
SOCIETE RADIO-CANADA
380 Wiseman
Outremont, Qc, Canada H2V 3J6, Canada

Colton, Michael
Reporter Radio/Television
CBC-NATIONAL RADIO NEWS
#201-484 Church St., Toronto
Toronto, ONT, Canada M4Y 2C7, Canada

Corbeil, Michel
Correspondent
LE SOLEIL
1451 rue Diderot
Sainte-Foy P.Qc, Canada G2E 4S7, Canada

Daponte, Joe
Correspondent
GLOBAL TELEVISION NETWORK
to be completed upon arrival, Canada

De Freitas, Jones
Correspondent
CBC TV
R.Goncalves Fontes 32/102
Rio de Janeiro, Brazil, Canada

De Marco, Jerry
Correspondent
OBITER DICTA-YORK UNIVERSITY
774 Spadina Rd
Toronto, ONT M5P 2X6 Canada, Canada

Desjardins, Jean
Reporter Radio/Television
SOCIETE RADIO-CANADA
10750 Verville, Montreal
Montreal, Qc, Canada H3L 3E9, Canada

Dessureault, Jean Marc
MAGAZINE QUEBEC INTERNATIONAL;
4822, rue Saint Andre
Montreal, Quebec, Canada H2J 3A3, Canada

Devroede, Pierre
Reporter Radio/Television
SOCIETE RADIO-CANADA
6194 rue Terrebonne
Montreal, PQ, Canada H4B 1A3, Canada

Dixon, Donald John
Editor
TELEVISION INTERNATIONAL CONSULTANTS
Ill Frank St.,
Ottawa, Ontario, Canada, Canada

Drache, Natalie
Reporter Radio/Television
GLOBAL TELEVISION NETWORK
857 Beatty SL s/502
Vancouver, Canada, Canada

Eaton, Heather
Correspondent
CATHOLIC NEW TIMES
1778 Bloor St. West, Apt. 18
Toronto - Ontario M6P3K4, Canada
F: 33 1 40 56 07 81

Elston, Suzanne
Correspondent
THOMSON NEWS
1604 Bloor St. Courtice
Ontario, CANADA LIE 2S3, Canada

Foster, Peter
Editor
CANADIAN BUSINESS MAGAZINE
31 Maple Avenue
Toronto, Ontario, Canada, Canada

Francis, Diane
Editor
FINANCIAL POST
2494 Sharon Crescent
Mississauga, ONT, Canada L5C 1Y8, Canada

Francoeur, Louis-Gilles
Correspondent
LE DEVOIR
3836 Girouard (N.D.G.)
Montreal, Qc, Canada HHA 3C7, Canada

Frymire, Mary M.
Editor
CANADIAN BROADCASTING CORPORATION
253 Augusta Ave., Suite 3
Toronto, ONT, Canada M5T 2L8, Canada

Garneau, Lise
Reporter Radio/Television
SOCIETE RADIO-CANADA
560 Ave. Claremount, Westmount
Westmount, Qc, Canada H3Y 2P1, Canada

Germain, Lyne-Marie
Reporter Radio/Television
RADIO-CANADA CHICOUTIMI
258 Du Gran
Chicoutimi G7J 2K2, Canada, Canada

Girard, Martin
Correspondent
VOIR
48 Ave. Joyce Apt 1
Montreal - Quebec H2V 156, Canada
T: 1 514 272.9705 F: 1 514 273.7147

Goldhawk, Dale
Correspondent
CBC-CROSS COUNTRY CHECK UP
1049 Dalewood Drive
Pickering, ONT, Canada L1V 4P6, Canada

Goldrick, Christopher
Reporter Radio/Television
TELEVISION INTERNATIONAL CONSULTANTS
17 Impasse de la Roserate
Quebec, Canada, Canada

Gorrie, Peter
Correspondent
TORONTO STAR
59 Beatrice St.
Toronto, ONT, Canada M6J 2T2, Canada

Guay Smith, Louise
SOCIETE RADIO-CANADA
47 Kirkland Blv
Kirkland, Qc, Canada H9j1N5, Canada

Haberstroh, Thomas
CTV
4 Ava Road, Ottawa
Ontario, Canada K1K 3E4, Canada

Halton, David C.
Correspondent
CBC
5070 Macomb St. N.W.
Washington, DC 20016, USA, Canada

Hippolyte, Martine
Editor
SOCIETE RADIO-CANADA
215 St. Laurent
Montreal, Qc, Canada H2Y 3T9, Canada

Hoff, George
CBC
4604 Albemarle St. N.W.
Washington, DC 20016, USA, Canada

Hughes, Lesley
Correspondent
RADIO CANADA (ENGLISH SERVICE)
47 Morningside Drive
Winnipeg, Manitoba, R3I 4A2, Canada, Canada

Hurko, Marijka
Reporter Radio/Television
CBC
260 Scarlett Road (#313)
Toronto, ONT, Canada M6N 4X6, Canada

Isacsson, Magnus
Reporter Radio/Television
CINEFLICS
5827 Jeanne-Nance 4
Montreal, Canada, Canada

Jean, Michaelle
Reporter Radio/Television
SOCIETE RADIO-CANADA
2247 Quesnel
Montreal, P.Q., Canada H3J 1G3, Canada

Jodoin, Marie-Aimee
Reporter Radio/Television
SOCIETE RADIO-CANADA
585 Lemoyne O, Apt. 1
Longueuil, P.C. Canada J4H 1X5, Canada

Jones, Leslie
Correspondent
CTV
162 Aylmer Ave., Ottawa
Ontario, Canada K15 2Y4, Canada

Julien, Patrice
Reporter Radio/Television
TVED-INFO MONDE
1879 Champlain
Montreal, Canada, Canada

Kines, Wayne
Correspondent
G.T.N.
to be completed upon
arrival, Canada

Lafrance, Pierre-Leon
Reporter Radio/Television
SOCIETE RADIO-CANADA
2750 Bellevue
Brossard, Qc, Canada J4Z 2S3, Canada

Lamb, Marjorie
Reporter Radio/Television
CBC-NEWSWORLD
96 Highland Avenue
M4W 2A6 Toronto, Ontario, Canada, Canada

Laporte, Paul Joseph Auguste
CINEFLICS
R.Benjamin Constante 144 c/1
Rio de Janeiro, Brazil, Canada

Larson, Tasha
Reporter Radio/Television
ACCESS NETWORK/CKUA RADIO
4014 Hopesmore Dr.
Victoria B.C., Canada, Canada

Latouche, Helene
Correspondent
MAGAZINE QUEBEC INTERNATIONAL
1235 Albert Hozeall
Quebec, Canada, Canada

Leblanc, Yvon
Reporter Radio/Television
SOCIETE RADIO-CANADA
224 D'Orleans
St.Lambert, Qc, Canada, Canada

Lemieux, Raymond
Correspondent
LE DEVOIR&QUEBEC SCIENCE
135, rue Grande-Allee Ouest
Quebec G1R 2H2 Canada, Canada

Leprince, Jean-Michel
Reporter Radio/Television
CBC
6510 Pyle Road
Bethesda, MD 20817, USA, Canada

Lewinsky, De Salcedo Antonia E.
Editor
CBC
R. Padre Carapuceiro 478
Recife, Brazil, Canada

Loucks, William
Editor
CBC
1342 Stokley Way
Vienna, Virginia, 22187, USA, Canada

MacDonald, Dorothy A., Ms.
Journalist
VANCOUVER SUN
3461 Point Gray Road
Vancouver - BC, Canada
T: 1 604 733 3509 F: 1 604 731 3710

Macrae, Lillian
Editor
PLANET TODAY
R.R. No. 3, King City
Ontario, Canada L0G 1K0, Canada

Mahoney, Susan
Editor
CBC-CROSS COUNTRY CHECK UP
917 Cherrier
Montreal, Qc, Canada H2L 1J1, Canada

Martin, Chantal
Production Assistant
SOCIETE RADIO CANADA
1400 Est Blvd. Rene Levesque
Montreal -Quebec H2, Canada
T: 1 514 5975618 F: 1 514 5975498

McElhone, Jane
Editor
CBC-CROSS COUNTRY CHECK UP
5238, Rue Chabot
Montreal, Qc, Canada, Canada

McIlroy, Anne
Correspondent
SOUTHAM NEWS
4A-157 James St.
Ottawa, Canada, Canada

McLauchlin, David
Reporter Radio/Television
CBC-NATIONAL RADIO NEWS
114 Cavell Drive, Winnipeg
Manitoba, Canada R3J 1P1, Canada

Meagher, Michael
Editor
SOUTHAM BUSINESS COMMUNICATIONS, INC.
170 Morningside Avenue
Toronto, Ontario, Canada M6S 1E3, Canada

Michaud, Achille
Journalist
SOCIETE RADIO CANADA
1400 Est Blvd Rene Levesque
Montreal - Quebec, Canada
T: 1 514 5974003 F: 1 514 5974095

Moreno, Hector
Journalist
SOCIETE RADIO CANADA
1400 Est Blvd Rene Levesque
Montreal, Canada
T: 1 514 5974003 F: 1 514 5974095

Mowbray, David
Reporter Radio/Television
CBC-QUIRKS AND QUARKS
350 Athlone Ave.
Ottawa, ONT, Canada K1Z 5M4, Canada

Naumetz, Timothy
Correspondent
TORONTO SUN
10 Windsor Avenue
Ottawa, Ontario Canada, Canada

Nelles, Wayne
Correspondent
CANADIAN ECONOMICAL NEWS
2624 W. 8th Avenue
Vancouover, Canada, Canada

Nichols, Mark
Editor
MACLEAN'S MAGAZINE
62 Mitchell Ave.
Toronto, M6J 1B9, Canada, Canada

Noel, Andre
Reporter
LA PRESSE
7 Saint Jacques
Montreal - Quebec, Canada
T: 1 514 7070 F: 1 514 285 6808

O'Brien, Catherine
Correspondent
CATHOLIC TIMES
2471 rue Ryde
Montreal, Qc, H3K 1R5, Canada

O'Connell, James C.
Reporter Radio/Television
CTV
1705 de Sales St., NW, 306
Washington, DC 20036, USA, Canada

Omelus, Michael Barry
Reporter Radio/Television
BROADCAST NEWS LTD.
1001 North Vermont St., #414
Arlington, VA 22201, USA, Canada

Page, Line
Reporter Radio/Television
SOCIETE RADIO CANADA
4965 du Tremblay, Laval
Laval, Qc, Canada H7W 5H8, Canada

Park, Penny
Correspondent
CBC-QUIRKS AND QUARKS
27 Muriel Ave.
Toronto, ONT, Canada M4J 2X8, Canada

Payne, Stephen
Correspondent
TORONTO SUN
8 Maple Crescent
Orangeville, ONT, Canada L9W 1X7, Canada

Peduzzi, Jeanne
Correspondent
RUDY INC.
R. Severiamo de Rezende, 201
Santos Dumont, Minas Gerais 36240,Brazil, Canada

Peeters, Christian
Executive Producer
WEATHER NETWORK/METEOMEDIA
1755 Rene Levesque Blvd East
Montreal - Quebec H2K 4P6, Canada
T: 1 514 597 1700 F: 1 514 597 1591

Plante, Bernard
Reporter Radio/Television
CKAC
38, rue Franche
St. Basile-le-Grand, Canada J3N 1K2, Canada

Pleszczynska, Patricia
Editor
CBC-CROSS COUNTRY CHECK UP
4655 Patricia
Montreal, Qc, Canada H4B 1Z2, Canada

Poulin, Hugues
Journalist
SOCIETE RADIO CANADA
1400 Est Blvd Rene Levesque
Montreal - Quebec, Canada
T: 1 514 5974003 F: 1 514 5974095

Power, Pamela
Reporter Radio/Television
STANDARD BROADCAST NEWS
130 MacLaren St., Apt. 34
Ottawa, ONT K1V 1Z2, Canada, Canada

Proulx, Mario
Journalist
SOCIETE RADIO CANADA
1400 Est Blvd Rene Levesque
Montreal - Quebec, Canada
T: 1 514 5974003 F: 1 514 5974095

Reardon, Sarah
Editor
CBC TV - THE JOURNAL
100 Carlton Street, 3rd Floor
Toronto, Canada M5B 1M3, Canada

Rivard, Jacques A.
Correspondent
SOCIETE RADIO-CANADA
3600 Ave. Du Park (#1608)
Montreal H2X 3R2, Canada, Canada

Rodriguez Anido, Julio
Correspondent
LA TRIBUNE
3649 rue Gervais
Rock Forest, Quebec, Canada J1N 2V3, Canada

Roelofsen, Theresa E.
Correspondent
TORONTO SUN
15 Marilyn Avenue
Ontario, Canada, Canada

Ross, Nicola
Editor
ENVIROLINE
1411-6A St. N.W., Calgary
Alberta, Canada T2M 3G8, Canada

Runnalls, David, Mr.
Editor
EARTH SUMMIT TIMES
275 Slater St. #500
Ottawa - Ontario K1P 5H9, Canada
T: 1 613 234-2296 F: 1 613 235-8237

Rusk, James
Correspondent
GLOBE & MAIL
240 Shaughnessy Blvd
Willowdale, ONT, Canada M2J 1K6, Canada

Rydell, Mary Kate
Editor
CBC
4307 Alton Pl. N.W.
Washington, DC 20016, USA, Canada

Saint-Cyr, Louis
Editor
CBC
5199 Dungannon Road
FairFax, VA 22030, USA, Canada

Sanche, Richard
Editor
SOCIETE RADIO-CANADA
1897 Gauthier
Montreal, Qc, Canada H2K 1A4, Canada

Savory, Judith Eve T.
Correspondent
CBC
3455 West 18th Ave.,Vancouver
British Columbia,Canada, Canada

Saxe, Dianne
Editor
HAZARDOUS MATERIALS MANAGEMENT MAG.
248 Russel Hill Rd.
Toronto, ON MAV 2T2 Canada, Canada

Senecal, Michel
Reporter Radio/Television
SOCIETE RADIO-CANADA
18 de Gaspe
Dollard-des-Ormeaux, Qc. H9B 3G8,Canada, Canada

Serpell, Frances
Correspondent
SUDBURY STAR
130 Baker Street, Sudbury
ONT P3C 2E9, Canada, Canada

Shonfield, Tamara, J.
Reporter Radio/Television
TELEVISION INTERNATIONAL CONSULTANT
Box 41, Alcove
Quebec, Canada, Canada

Skinner, Patricia L.
Editor
CTV
1705 de Sales St., NW, 306
Washington, DC 20036, USA, Canada

Stefason, Mark, Mr.
Correspondent
CBC RADIO CANADA
1194 Sloan Terrace
Victoria, B.C. Canada, Canada

Stevens, Sinclair
Editor
PLANET TODAY
R.R. No. 3, King City
Ontario, Canada L0G 1K0, Canada

Strong, Frederick Maurice
Editor
EARTH SUMMIT TIMES
778 Eastbourne Ave.
Ottawa, Ontario, Canada, Canada

Thivierge, Jean
Reporter Radio/Television
SOCIETE RADIO-CANADA
1252 Lavigerie, Ste-Foy
Qc, Canada G1W 3W7, Canada

Varin, Claire
Correspondent
NORD-SUD, RADIO-QUEBEC
4243 rue Fabre
Montreal, QC H2J 3T5, Canada, Canada

Villedieu, Yanick
Reporter Radio/Television
SOCIETE RADIO-CANADA
4944 Ch. des Plaines
St.Antoine de Tilly, Qc, Canada GOS 2CO, Canada

Vincent, Isabel
Correspondent
GLOBE & MAIL
R.Almte Gomes Pereira, 12/201
22.291 Urca, Rio de Janeiro-RJ, Brazil, Canada

Walker, William
Correspondent
TORONTO STAR
104 Shehyn Lane
Nepean, Ontario K2G 4Y3, Canada, Canada

ORGANIZATIONS

Bartram, Trevor, Mr.
Canadian Institute of International Affairs (CIIA)
15 King's College Circle
Toronto, Ontario M5S 2V9, Canada
T: (1 416) 979 1851 F: (1 416) 979 8575

Bassinghwaite, John, Mr.
Share B.C.
Box 1074
Ucluelet, British Columbia VOR 3AO, Canada
T: (1 604) 726 2002 F: (1 604) 726 1254

Bennet, David, Mr.
Canadian Labour Congress
2841 Riverside Drive
Ottawa, Ontario K1V 8X7, Canada
T: (1 613) 521 3400 F: (1 613) 521 4655

Bernstein, Johannah, Ms.
Canadian Participatory Committee For UNCED
63 Sparks Street, Ste. 603
Ottawa, Ontario K1P 5A6, Canada
T: (1 613) 238 3811 F: (1 613) 594 2948

Blanchard, Lyse, Ms.
Canadian University Service Overseas (CUSO)
135 Rideau Street
Ottawa, Ontario K1N 9K7, Cañada
T: (1 613) 563 1264

Bretteville, J., Mr.
Aupelf - Uref
CP. 6128, Succ. A
Montreal, Quebec H3C 3J7, Canada
T: (1 514) 343 5720 F: (1 514) 343 5783
TL: 055-60955

Brien, Pierre, Mr.
As. Quebecoise Des Organismes De Co-operation Internationale (AQOCI)
801 Sherbrooke East, 400
Montreal, Quebec H2L 1K7, Canada
T: (1 514) 597 2288 F: (1 514) 597 2334

Bull, Harrison, Mr.
International Organisation Of Indigenous Resource Development
c/o Robert Epstein & Ass., 24 Bayswater Avenue
Ottawa, Ontario K1Y 2E4, Canada
T: (1 613) 761 1765 F: (1 613) 761 1388

Clover, D.E., Mr.
International Council for Adult Education
720 Bathurts Street, Suite 500
Toronto, Ontario M5S 2R4, Canada
T: (1 416) 588 1211 F: (1 416) 588 5725

Coutu, Francois, Mr.
United Nations Association In Canada
808-63 Sparks Street
Ottawa, Ontario K1P 5A6, Canada
T: (1 613) 232 5751

Crabbe, Philippe, Dr.
Institute for Research on Environment and Economy
University of Ottawa, 5 Calixa Lavallee
Ottawa, Ontario K1N 6N5, Canada
T: (1 613) 564 7644 F: (1 613) 564 6716

D'Amore, Louis, Mr.
International Institute for Peace through Tourism
3680 rue de la Montagne
Montreal, Quebec H3G 2A8, Canada
T: (1 514) 281 9956 F: (1 514) 848 1099

Farhi, Andree, Ms.
International Council Of Jewish Women
1110 Finch Ave. West, Suite 518
Downsview, Ontario M3J 2T2, Canada
T: (1 116) 665 8251 F: (1 116) 665 8702

Ferretti, Janine, Ms.
Pollution Probe
12 Madison Avenue
Toronto M5R 2S1, Canada
T: (1 416) 926 1907 F: (1 416) 926 1601/926 8174

Gilfillan, Brian, Mr.
Western Environment and Development
2800 Park Place, 666 Burrard Street
Vancouver, British Columbia V6C 2Z7, Canada
T: (1 604) 643 2953 F: (1 604) 687 1612

Janowitz, Martin W., Mr.
Clean Nova Scotia Foundation
PO Box 2528
Station M, Halifax NS B3J 3N5, Canada
T: (1 902) 424 5245 F: (1 902) 424 5334

Lavoie, Mario, Mr.
Canadian Youth Foundation/Youth92
55 Parkdale Ave., 3rd Floor
Ottawa, Ontario K1Y 1E5, Canada
T: (1 613) 761 9206 F: (1 613) 761 1441/722 4829

Lee, C.A., Mr.
Canadian Institute of Forestry
Suite 1005, 151 Slater Street
Ottawa K1P 5H3, Canada
T: (1 613) 234 2242 F: (1 613) 234 6181

Ley, Gary, Mr.
Forest Alliance of British Columbia, The
Suite 210, 1100 Melville Street
Vancouver ,B.C., Canada
T: (1 604) 685 7507 F: (1 604) 685 5373

Little, Janice, Ms.
International Institute For Sustainable Development
212 McDermot Avenue
Winnipeg, Manitoba R3B OS3, Canada
T: (1 204) 958 7700 F: (1 204) 958 7710

MacLachlan, Janet, Ms.
Canadian Public Health Association
1565 Carling, Suite 400
Ottawa, Ontario K1Z 8R1, Canada
T: (1 613) 725 3769 F: (1 613) 725 9826

Monto, Laurie, Ms.
Assembly Of First Nations
47 Clarence Street, Suite 300
Ottawa, Ontario K1N 9KI, Canada
T: (1 613) 236 0673 F: (1 613) 238 5780

Nash, Gary, Mr.
International Council on Metals and the Environment
360 Albert Street, Suite 1550
Ottawa, Ontario K1R 7X7, Canada
T: (1 613) 235 4263 F: (1 613) 235 2865

Paterson, Wayne, Mr.
Whistler Foundation For A Sustainable Environment, The
c/o The University of Victoria Environmental Studies, PO Box 3045
Victoria, British Colombia V8W 3P4, Canada
T: (1 604) 656 9785 F: (1 604) 656 9785

Shebbeare, R.A., Mr.
Council of Forest Industries of British Columbia
1200-555 Burrard Street
Vancouver, British Columbia V7X 157, Canada
T: (1 604) 684 0211 F: 91 604) 687 4930

Simon, Mary, Ms.
Inuit Circumpolar Conference
Suite 404, 650 32nd Avenue
Lachine, Quebec H8T 3K4, Canada
T: (1 514) 637 3771 F: (1 514) 637 3146

Smith, Dan, Mr.
Native Council Of Canada
384 Bank St., 2nd Floor
Ottawa, Ontario K2P 1Y4, Canada
T: (1 613) 238 3511 F: (1 613) 230 6273

Sylvester, Shauna, Ms.
International Development Education Resources Association
2524 Cypress Street
Vancouver, BC V6J 3N2, Canada
T: (1 604) 732 1496

Wareing, Mark, Mr.
Western Canadian Wilderness Committee
20 Water Street
Vancouver BC V6B 1A4, Canada
T: (1 604) 683 8229 F: (1 604) 683 8220

GLOBAL FORUM

Aharatunga, Carol
Director
INTERNATIONAL CENTRE FOR OCEANS DEVELOPMENT
5676 Spring Garden Road, 9th Floor
Halifax - NS, Canada
T: 1 902 4261512 F: 1 902 4268932

Allard, Daniel, Mr.
Program Officer
SECRETARIAT INTERNATIONAL DE L'EAU
48 Rue Le Royer Ovest
Montreal - Quebec H2Y 1W7, Canada
T: 1 514 849 4262 F: 1 514 849 2822

Allard, Don, Mr.
Advisor
INUIT CIRCUMPOLAR CONFERENCE
650, 32nd Avenue, Suite 404, Lachine
Lachine - P.Q. H8T3k4, Canada
T: 1 514 6373771 F: 1 514 6373146

Andrade, Raymundo De, Mr.
Professor
INSTITUT DE RECHERCHE SUR L'ENVIRONNEMENT ET L'ECONOMIE
Department of Sociology, 550 Cumberland, University of Ottawa
Ottawa - Ontario, Canada
T: 1 613 564-2348 F: 1 613 564-9976

Archer, Lynette, Ms.
Sust. Develop. Coordinator
DISABLED PEOPLES' INTERNATIONAL / CENTRO DE VIDA INDEPENDENTE
101-7 Evergreen Place
Winnipeg - Alberta R3L2T3, Canada
T: 1 204 287-8010 F: 1 204 287-8175

Arnold Allaw, Jean, Ms.
Director
FALLS BROOK CENTRE
RR2 Glassville
N.B. E0J 1L0, Canada
T: 506 379 8143

Auguste, Paul Joseph
Technician
CINEFLICS INC
Suite 190, 238 Davenport Road
Toronto, Ontario M5R 1J6, Canada

Avellar, Claudio, Mr.
GLOBE92
601-535 Thurlow Street
Vancouver - BC, Canada
T: 1 604 666-8020 F: 1 604 666-8123

Avirgan, Shanti
YOUTH 92 / EARTH STEWARDS NETWORK
55 Parkdale Ave, 3rd floor
Ottawa - Ontario, Canada
T: 1 613 761 9206 F: 1 613 761 1441

Azak, Stephen, Mr.
Trustee
NISGA'A TRIBAL COUNCIL
Box 231
New Aiyansh-Bc, Canada
T: 604 6332234 F: 63323467

Baily, Phyllis E.
Editor
OMBUDSPRESS OF NORTH AMERICA
4072 Northcliffe Ave
Montreal, Canada
T: 1 514 4861668 F: 1 514 3692233

Balson, David, Mr.
Senior Program Officer
INTERNATIONAL DEVELOPMENT RESEARCH CENTER
P. O. Box 8500
Ottawa - Ontario, Canada
T: 1 613 236-6163 F: 1 613 238-7230
TL: 533753

Barg, Stephan R., Mr.
Treasurer
INTERNATIONAL INSTITUTE FOR SUSTAINABLE DEVELOPMENT
161 Portage Ave. East , 6th floor
Winnipeg - Alberta, Canada
T: 1 204 958-7700 F: 1 204 958-7710

Barron, Barbara, Ms.
Conselor
167 Fairhaven
Montreal H9S 4A6, Canada
T: 1 514 426-0929

Barron, David, Mr.
Vice-President
CANADIAN PULP & PAPER ASSN
1155 Metcalfe Str.
Pointe Claire,Quebec H3B 4T6, Canada
T: 1 418 866-6621 F: 1 418 866-3035

Baumgartner, Sandy, Ms.
Manager
CANADIAN WILDLIFE FEDERATION
2740 Queensview Dr.
Ottawa - Ontario, Canada
T: 1 613 721-2286 F: 1 613 721-0321

Beaulice, Anore, Mr.
Resarcher
MCGILL UNIVERITY / UNCED SECRETARIAT
2020 University, 24th Floor
Montreal H3A.2A5, Canada
T: 1 514 398-7400 F: 1 514 398-4668

Beaulieu, Martin Prieto, Sr.
YOUTH 92 / RED DE LA ECOLOGIA
55 Parkdale Ave, 3rd floor
Ottawa - Ontario, Canada
T: 1 613 761 9206 F: 1 613 761 1441

Beaumier, Josee, Mr.
WORLD CONGRESS FOR EDUCATION & COMMUNICATION ON ENVIRONMENT & DEVELOPMENT
110 Eglinton Ave. West
Toronto - Ontario M4R 1A3, Canada
T: 1 416 482-9212 F: 1 416 482-9601

Belanger, Chris
385 River Ave.
Winnipeg - Alberta, Canada
T: 1 204 7863421 2370600 F: 1 204 4775730

Belanger, Dunat
Audioman
SOCIETE RADIO CANADA
1400 Est Blvd. Rene Levesque
Montreal -Quebec H2, Canada
T: 1 514 5975618 F: 1 514 5975498

Benjamin-Robinson, Mia, Ms.
Executive
PROJECT PLOUGHSHARES
10642 - 84 Avenue
Edmonton - Alberta, Canada
T: 1 403 433-4064 F: 1 403 439-5081

Bischoff, Angela, Ms.
ECOCITY SOCIETY
9523 Jasper Ave.
Edmonton - Alberta T5H 3V2, Canada
T: 1 403 429-3659 F: 1 403 425-5767

Blanchette, Jean-Marie
Cameraman
SOCIETE RADIO CANADA
1400 Est Blvd. Rene Levesque
Montreal -Quebec H2, Canada
T: 1 514 5975618 F: 1 514 5975498

Bohlen, Jim, Mr.
Director
GREENGRASS INSTITUTE
PO Box 39
Denmand Island - BC, Canada
T: 1 604 335 0322 F: 1 604 335 0322

Boivin, Francois, Mr.
SOCIETE D'EDUCATION DU DEVELOPPEMENT INTERNATIONAL DE GROULX
09 Forget
S. Therese J7EIK7, Canada
T: 1 514 435-9208 F: 1 514 430-6024

Bondar, Roberta, Ms.
INTERNATIONAL CENTRE FOR OCEAN DEVELOPMENT
5670 Spring Garden Road
Halifax - NS, Canada
T: 1 902 426-1512 F: 1 902 426-1599
TL: 019-21670

Boulet, Georges, Mr.
GOUVERNEMENT DU QUEBEC
3900 Marly, Sainte-Foy
Quebec G1X 4E4, Canada
T: 1 418 643-8209 F: 1 418 644-4598

Bourque, Jean-Louis, Mr.
UNION QUEBECOISE DE CONSERVATION DE LA NATURE (UQCN)
3770 rue Lacoste
Quebec, Canada
T: 1 418 871-8550
E: G2E 4P8

Boutin, Chantale, Ms.
Coordennatrice
CENTRE DE FORMATION EN ENVIRONNEMENT
455 Fournier street
St-Jerome (Quebec), Canada
T: 1 514 4367402 F: 1 514 4365212

Brascoupe, Simon, Mr.
Coordinator
INDIGENOUS NETWORK OF INDIGENOUS PEOPLES NON-GOVRNMT'L ORGS AND PRACTIONERS
154 Aylmer Avenue
Ottawa - Ontario, Canada
T: 1 613 2317059 F: 1 613 231-5621

Bressette, Beverly, Ms.
ASSEMBLY OF FIRST NATIONS
55 Murray St., 5th Floor
Ottawa - Ontario, Canada
T: 1 613 2360673 F: 1 613 2385780

Brito, Elisabeth, Ms.
INSTITUT CANADIEN DES OCEANS
5670 Spring Garden Road
Halifax - NS, Canada
T: 1 902 426-1512 F: 1 902 426-1599
TL: 019-21670

Brodeur, Marc, Mr.
SOCIETE D'EDUCATION DU DEVELOPPEMENT INTERNATIONAL DE GROULX
09 Forget
S. Therese J7EIK7, Canada
T: 1 514 435-9208 F: 1 514 430-6024

Brooke, Lorraine, Ms.
Advisor
INUIT CIRCUMPOLAR CONFERENCE
650, 32nd Avenue, Suite 404, Lachine
Lachine - P.Q. H8T3k4, Canada
T: 1 514 6373771 F: 1 514 6373146

Brooks, Michael, Mr.
Policy Analist
ONTARIO GOVERNMENT
RM 440 Queen'S Park
Toronto - Ontario M7A1A7, Canada
T: 1 416 325-7175 F: 1 416 325

Brown, Barbara, Ms.
Policy Adviser
GOV. OF CANADA
47 Fulton
Ottawa - Ontario K15 4Y5, Canada
T: 1 819 994-4168 F: 1 819 953-6356

Brown, Christopher, Mr.
Member
SOCIETY FOR INTERNATIONAL DEVELOPMENT
33 Main St.
Aylmer J9H3L2, Canada
T: 1 613 9430899 F: 1 613 9964309

Brown, David, Mr.
Professor
MONTREAL INTERUNIVERSITY GROUP
1826 Parkdale Verdun
Quebec H3A 1G7, Canada
T: 1 418 766-8048

Brunel, Gilles, Mr.
Representative
RESEAU BRESIL-QUEBEC
15, de Castelneau Quest
Montreal, Canada
T: 1 514 270-6089
E: H2R 2W3

Buchan, Robert, Mr.
Member
WESTERN ENVIRONMENT AND DEVELOPMENT TASK FORCE
Commerce Pl., 14th floor, 400 Burrard St.
Vancouver - BC, Canada
T: 1 604 643-1720 F: 1 604 643-1760

Buck, Wilfred
INTERNATIONAL INSTITUTE FOR SUSTAINABLE DEVELOPMENT
161 Portage Ave. E, 6th floor
Winnipeg - Manitoba R38 OY4, Canada
T: 1 204 958-7700 F: 1 204 958-7710

Burton, Thomas, Mr.
President
GREEN SOLUTIONS LTD.
16-3318 Oak St.
Vancouver - BC, Canada
T: 1 416 360-0101 F: 1 416 360-0981

Butterfield, , Mr.& Mrs.
President
SOUTH ISLAND CORP.
Victoria - BC, Canada

Calder, Frank
NISGA'A TRIBAL COUNCIL
Box 231
New Aiyansh V0J1A0, Canada
T: 604 633-2601 F: 604 633-2361

Caliz, Carmen, Ms.
WOMEN'S STUDIES - UNIVERSITY TORONTO
503 Manning Rd.
Toronto - Ontario, Canada
T: 1 416 533-3944

Cameron, Allison, Ms.
MUSKOKA ENVIRONMENTAL EDUCATORS ASSOCIATION
Box 1690, 28 McMurray St.
Bracebridge - ON P1L LS4, Canada
T: 705 645 4496 F: 705 645 7487

Cameron, Bill
CANADIAN BROADCASTING CORPORATION
CBC TV - The Journal, Box 14000 Station A
Toronto - Ontario, Canada
T: 1 416 9757249 F: 1 416 9639036

Campana, Marco, Mr.
INTERNATIONAL INSTITUTE FOR SUSTAINABLE DEVELOPMENT
161 Portage Ave.East, Sixth Floor
Winnipeg - Alberta R3B OY4, Canada
T: 1 204 958-7700 F: 1 204 958-7710

Casey, L., Mr.
Councillor
CITY OF NEPEAN
10 Wrenwood Cr.
Nepean - ON K2G5U3, Canada
T: 1 613 225-7140

Cassara, Ernest
ICAE
12 Rajaji Marg
New Delhi 110011, Canada
T: 91 11 301-4309

Cavoukian, Raffi, Mr.
ENVIRONMENTAL CHILDREN'S ORGANIZATION
12-1604 Balsam St.
Vancouver - BC V6K 3L9, Canada
T: 1 604 682-8698 F: 1 604 682-4291

Chandra, Premesh
YOUTH 92 / NETWORK OF OVERSEAS COLLECTIVES IN AUSTRALIA
55 Parkdale Ave, 3rd floor
Ottawa - Ontario, Canada
T: 1 613 761 9206 F: 1 613 761 1441

Chaudhari, Ashok
INTERNATIONAL INSTITUTE FOR SUSTAINABLE DEVELOPMENT
161 Portage Ave., 6th Floor
Winnipeg - Manitoba R38 OY4, Canada
T: 1 204 958-7700 F: 1 204 958-7710

Chen, Sherman, Mr.
GIFFORD ENGLE ASSOCIATES
26 Victoria Avenue, Canterbury
Victoria - BC 3126, Canada
T: 61 3 830-4726 F: 61 3 836-7134

Chircop, Aldo, Mr.
INSTITUT CANADIEN DES OCEANS
5670 Spring Garden Road
Halifax - NS, Canada
T: 1 902 426-1512 F: 1 902 426-1599
TL: 019-21670

Chitrenky, Gregory J., Mr.
District Representative
ALBERTA TEACHER'S ASSOCIATION
11010 142 street
Edmonton - Alberta, Canada
T: 1 403 4532411 F: 1 403 455-6481

Christiansen-Ruffman, Linda, Ms.
INTERNATIONAL SOCIOLOGICAL ASSOCIATION RC 32
Dept. of Socialogy Saint Mary'S University
Halifax - NS B3H 3C3, Canada
T: 1 902 420-5886 F: 1 902 420-5561

Clarke, Arlene, Ms.
YOUTH 92/ CAN YOUTH WORKING GROUP ON ENV & DEV
55 Parkdale Ave, 3rd floor
Ottawa - Ontario, Canada
T: 1 613 761 9206 F: 1 613 761 1441

Claveau, Guy, Mr.
Professeur
INST. FOR INTER.DEVEL. CO-OPERATION
25, University -
Ontario, Canada
T: 613 564-5930

Clements, Adrienne, Ms.
WORLD YOUNG WOMEN'S CHRISTIAN ASSOCIATION
20 Howthorn Av.
Toronto - Ontario MYW232, Canada
T: 1 416 922-9177 F: 1 416 731 79 38
TL: WYMCA 41 23 32

Clover, Darlene
ICAE
12 Rajaji Marg
New Delhi 110011, Canada
T: 91 11 301-4309

Cohen, Dian, Ms.
INTERNATIONAL INSTITUTE FOR SUSTAINABLE DEVELOPMENT
161 Portage Ave.East, Sixth Floor
Winnipeg - Alberta R3B OY4, Canada
T: 1 204 958-7700 F: 1 204 958-7710

Cohen, Jesse, Mr.
EARTH TRAIN/WORLD LINK GLOBAL YOUTH SUMMIT
Box 299
Jamestown, Newfoundland, Canada
T: 1 709 473-3131 F: 473-5376

Come, Matthew Coon, Mr.
Grand Chief
GRAND COUNCIL OF THE CREES
2 Lakeshore Road
Nemeska - Quebec J0Y 3B0, Canada
T: 819 673 2600

Comeau, Louise A., Ms.
FRIENDS OF THE EARTH - CANADA
251 Laurier Ave. W #701
Ottawa - Ontario KIP556, Canada
T: 1 613 230-3352 F: 1 613 232-4354

Cosbey, Aaron, Mr.
INTERNATIONAL INSTITUTE FOR SUSTAINABLE DEVELOPMENT
161 Portage Ave.East, Sixth Floor
Winnipeg - Alberta R3B OY4, Canada
T: 1 204 958-7700 F: 1 204 958-7710

Cosway, Frank, Mr.
Program Officer
INTERNATIONAL INSTITUTE FOR SUSTAINABLE DEVELOPMENT
161 Portage Ave. East , 6th floor
Winnipeg - Alberta, Canada
T: 1 204 958-7700 F: 1 204 958-7710

Courcelles, Real, Mr.
HYDRO QUEBEC
75 Boulevard Rene Levesque 9
Montreal - Quebec, Canada
T: 1 514 289-3956

Croteau, Sylvain, Mr.
CENTRE DE SANTE FUIVENCE - ENERGIE
2372 Chehin des Patriotes JOG-IPO
St. Ours, Canada
T: 514 1 785369

Cullis, Tara, Ms.
Executive Director
SUZUKI FOUNDATION
2075 W12th Ave.
British Columbia, Canada
T: 1 604 732-4228 F: 1 604 732-0752

Davidson, Joan, Ms.
Representative
P.O.W.E.R.
Rri
Georgetown - ON L7G 454, Canada
T: 1 416 811-6358

Davis, Athena, Ms.
INTERNATIONAL NETWORK FOR ENVIRONMENTAL MANAGEMENT
4639 Hutchinson Ave.
Montreal - Quebec H2V 4D2, Canada
T: 1 514 948-3494

Davis, Kim, Mr.
INTERNATIONAL NETWORK FOR ENVIRONMENTAL MANAGEMENT
3220 Edward Montpetit #1
Montreal - Quebec H3T 1J9, Canada
T: 1 514 343-9369

Deffrasnes, Robert, Mr.
GOVERNMENT DU QUEBEC
5700 4o. AVENUE OUEST
Charlesbourg- Quebec G1H 6R1, Canada
T: 1 418 643-4676

Delabarre, Ken, Mr.
Advisor
INUIT CIRCUMPOLAR CONFERENCE
650, 32nd Avenue, Suite 404, Lachine
Lachine - P.Q. H8T3k4, Canada
T: 1 514 6373771 F: 1 514 6373146

Delanoe, Maryvon
ICAE
12 Rajaji Marg
New Delhi 110011, Canada
T: 91 11 301-4309

Demarco, Jerry, Mr.
ESSA, YORK UNIVERSITY
3rd Floor, Lumbers Building, 2700 Keele Street, York University
North York - ON, Canada
T: 1 416 736-5252 F: 1 416 736-5679

Demers, Claude, Mr.
Research Scientist
HYDRO - QUEBEC
1010 East STE Catherine, 5th Floor
Montreal - Quebec HAL263, Canada
T: 1 514 985-7272 F: 1 514 985-7203

Domagala, John
CANADIAN BROADCASTING CORPORATION
CBC TV - The Journal, Box 14000 Station A
Toronto - Ontario, Canada
T: 1 416 9757249 F: 1 416 9639036

Droucker, Bonnie, Ms.
Representative
SELKIRK COLLEGE
P.O.Box 1200
Castlegar, Canada

Dufour, Jules, Mr.
Delegate
FOSEM
8191, Avenue Du Zoo
Charlesbourg- Quebec GIG 464, Canada
T: 1 418 622-0313 F: 1 418 646-9239

Dumont, Bill, Mr.
Director
WESTERN ENVIRONMENT AND DEVELOPMENT TASK FORCE
14th FL- Commerce Place, 400 Burrard St
Vancouver - BC V6C3G2, Canada
T: 1 604 643 1720 F: 1 604 643 1760

Dunn, Stephen, Mr.
INT'L COORDINATING COMMITTEE FOR RELIGION AND THE EARTH
67 Lewis St
Toronto - Ontario M4M2H2, Canada
T: 1 416 461-5146

Duwache, Evelyn, Ms.
Advisor
INDIGENOUS SURVIVAL INTERNATIONAL
Suite 105 - 298 Elain St.
Ottawa - Ontario, Canada
T: 1 613 230 3616 F: 1 613 230 3595

Dwyer, Bernadette, Ms.
INSTITUT CANADIEN DES OCEANS
5670 Spring Garden Road
Halifax - NS, Canada
T: 1 902 426-1512 F: 1 902 426-1599
TL: 019-21670

Engle, Adam, Mr.
Director
GIFFORD ENGLE ASSOCIATES
26 Victoria Avenue, Canterbury
Victoria - BC 3126, Canada
T: 61 3 830-4726 F: 61 3 836-7134

Enns, Henry, Mr.
Executive Director
DISABLED PEOPLES'S INTERNATIONAL
101-7 Evergreen Place
Winnipeg - Alberta R3POE8, Canada
T: 1 204 2878010 F: 1 204 2878175

Estudillo, Agustin Olivera, Mr.
INTERNATIONAL INSTITUTE FOR SUSTAINABLE DEVELOPMENT
161 Portage Ave.East, Sixth Floor
Winnipeg - Manitoba R3B OY4, Canada
T: 1 204 958-7700 F: 1 204 958-7710

Fauteux, Andree, Ms.
President
ASSOCIATION QUEBECOISE DE GERENTOLOGIE
840 Cite des Jeunes #5
Quebec HWLL, Canada
T: 1 418 7713704

Feldon, Eide, Ms.
Chef
PRIMAVERA ECO 92
14 Perkins St.
Ottawa - Ontario K1R 7G5, Canada
T: 1 613 234 3622 F: 1 613 238-1677

Fenech, Adam, Mr.
ENVIRONMENTAL STUDIES STUDENT ASSOCIATION / YORK UNIVERSITY
3rd Floor, Lumbers Building, 2700 Keele Street, York University
North York - ON, Canada
T: 1 416 736-5252 F: 1 416 736-5679

Fisk, Ronald B., Mr.
Member
INUIT CIRCUMPOLAR CONFERENCE
650, 32nd Avenue, Suite 404, Lachine
Lafayette, BA H8T3k4, Canada
T: 55 73 6373771 F: 55 73 6373146

Forsythe, Kathleen, Ms.
Chairperson
GREEN SOLUTIONS LTD.
703 Rockleights Ave.
Victoria - BC V9A 6H9, Canada
T: 1 604 361-3613 F: 1 604 360-

Francis, Sara K., Ms.
Delegate
GREEN PARTY
82 Broadway Ave.
Ottawa - Ontario K15 2V6, Canada
T: 1 613 238-7730

Fraser, Whit, Mr.
Chairman
CANADIA POLAR COMMISSION
Suite 1710, constitution Square, 360 Albert Street
Ottawa - Ontario K1R 7X7, Canada
T: 1 613 943-8605 F: 1 613 943-8607

Frey, Sigrun, Ms.
LAURENTIAN UNIVERSITY
MSR # 609, Laurentian University
Sudbury - Ontario P36 2C6, Canada
T: 1 705 688-9192

Gallopin, Gilberto, Dr.
INTERNATIONAL INSTITUTE FOR SUSTAINABLE DEVELOPMENT
161 Portage Ave.East, Sixth Floor
Winnipeg - Alberta R3B OY4, Canada
T: 1 204 958-7700 F: 1 204 958-7710

Gardner, Michael, Mr.
UNITED NATIONS ASSOCIATION IN CANADA
6147 Linden St.
Halifax-Nova Scotia B3H 2K7, Canada
T: 902 429-3314 F: 422-5343

Garneau, Roch
Technician
SOCIETE RADIO CANADA
1400 Est Blvd Rene Levesque
Montreal - Quebec, Canada
T: 1 514 5974003 F: 1 514 5974095

Garon, Emmanuel Y., Mr.
Prof: Matematica, Economia
UNIVERSITE DU QUEBEC
300, Allee Des Ursulines
Rimouski - Quebec G5L 3A1, Canada
T: 1 418 724-1583 F: 1 418 724-1525

Geisler, Morgan, Ms.
ENVIRONMENTAL CHILDREN'S ORGANIZATON
202.1592 Southwest Marine Dr.
British Columbia, Canada
T: 1 604 263-8110

Girard, Remy, Mr.
GOVERNMENT DU QUEBEC
5700 4o. AVENUE OUEST
Charlesbourg- Quebec G1H 6F1, Canada
T: 1 418 643-3987

Glentzakos, Demetrios, Mr.
#1510, 8515 - 112 Street
Edmonton - Alberta, Canada
T: 1 403 433-5317 F: 1 403 492-2589

Goldinrosenberg, Dorothy, Ms.
ESSA, YORK UNIVERSITY
3rd Floor, Lumbers Building, 2700 Keele Street, York University
North York - ON, Canada
T: 1 416 736-5252 F: 1 416 736-5679

Gonzalez, Carolina Garcia Ravesi, Ms.
President
GRUPO ECOLOGISTA INFANTIL
15 Allende #301 - Zona Cenaro
Victoria - BC, Canada
T: 1 604 20110 F: 1 604 29792
TL: 31262

Good, Len, Dr.
INTERNATIONAL INSTITUTE FOR SUSTAINABLE DEVELOPMENT
161 Portage Ave.East, Sixth Floor
Winnipeg - Alberta R3B OY4, Canada
T: 1 204 958-7700 F: 1 204 958-7710

Goodstriker, Jason, Mr.
4 WORLDS DEVELOPMENT PROJECT
Box 356
Cardston, Alberta TOK OKO, Canada
T: 1 403 737-2524

Gorley, Alan, R.P.F., Mr.
President
ASSOCIATION OF BC PROFESSIONAL FORESTERS
440-789 West Pender Street
Vancouver - BC, Canada
T: 1 604 687-8027 F: 1 604 678-3264

Gosnell, Joseph, Mr.
Executive Chairman
NISGA'A TRIBAL COUNCIL
Box 231
New Aiyansh V3J 1A0, Canada
T: 1 603 633-2234 F: 1 603 633-2367

Grandbois, Maryse, Dr.
Professeur
UNIVERSITE DU QUEBEC MONTREAL
CP 8888 Succ A
Montreal - Quebec H3C 3P8, Canada

Gray, Corine, Ms.
Executive Director
INUIT CIRCUMPOLAR CONFERENCE
170 Laurier, Suite 504
Ottawa - Ontario, Canada
T: 1 613 563-2642 F: 1 613 565-3089

Gray, Nathaniel James
Cameraman
CINEFLICS INC
Suite 190, 238 Davenport Road
Toronto, Ontario M5R 1J6, Canada

Green, Daniel, Mr.
Co-President
SOCIETE POUR VAINCRE LA POLLUTION
P.O.Box 65, Place D'Armes
Montreal - Quebec H2Y 3E9, Canada
T: 1 514 844-5477 F: 1 514 844-1446

Grondini, Marcelo
ICAE
12 Rajaji Marg
New Delhi 110011, Canada
T: 91 11 301-4309

Gruben, Roger, Mr.
President
INUVIALUIT REGIONAL CORPORATION
PO Box 2120
Inuvik - NWT X0E 0T0, Canada
T: 1 403 979-2737 F: 1 403 979-2135

Gull, Sam W., Mr.
CREE NATION YOUTH COUNCIL
2 Lake Shore Road
Nemaska - Quebec, Canada
T: 1 819 673-2600 F: 1 819 673-2606

Haidazz, Senator Stanley, Mr.
SENATE OF CANADA
215 East Block, Parliament Hill
Ottawa - Ontario KLN OA4, Canada
T: 1 613 992-2713 F: 1 613 995-6789

Halas, Susan, Ms.
23 Riverbend Gate SE. - Calgary
Alberta T2C 3J5, Canada
T: 001 403 236 0247 F: 001 403 290 6200

Hall, Budd
ICAE
12 Rajaji Marg
New Delhi 110011, Canada
T: 91 11 301-4309

Hallman, David, Mr.
SOCIAL ISSUES & JUSTICE/UCC
85 St. Clair Avenue East
Toronto - Ontario M4T 1MB, Canada

Hamilton, Jane, Mr.
ESSA, YORK UNIVERSITY
3rd Floor, Lumbers Building, 2700 Keele Street, York University
North York - ON, Canada
T: 1 416 736-5252 F: 1 416 736-5679

Handley, Joe, Mr.
GOVERMENT OF THE NORTHWEST TERRITORIES
P.O.Box 1320
Yellowknife - NWT XIAZL9, Canada
T: 1 403 873 7420

Hannak, Erin
CANADIAN YOUTH WORKING GROUP ON ENVIRONMENT
Canada
Ottawa - Ontario, Canada
T: 1 613 761-9206 F: 1 613 761-1441

Hart, Paul, Mr.
Professor
UNIVERSITY OF REGINA
Faculty of Education
Regina-Saskatchewan, Canada
T: 1 306 585-4309 F: 1 306 585-4880

Hernandez, Patricia, Ms.
ENVIRONMENTAL CHILDREN'S ORGANIZATON
2305 W3rd Ave.
Vancouver - BC, Canada
T: 1 604 734-0146

Hoffmann, Helga, Ms.
Member
GREEN WEB
R.R.3
Saltsprings-N.S. BOK IPO, Canada
T: 1 9029252514 F: 1 9029252514

Hoyle, Brian, Mr.
Counsel
GIFFORD ENGLE ASSOCIATES
26 Victoria Avenue, Canterbury
Victoria - BC 3126, Canada
T: 61 3 830-4726 F: 61 3 836-7134

Huot, Gilles, Mr.
Tresorier
SECRETARIAT INTERNATIONAL DE L'EAU
48 Le Royer
Montreal - Quebec HZY 1W7, Canada
T: 1 514 849 4262

Husband, Vicky, Ms.
SIERRA CLUB OF WESTERN CANADA
314,620 View St.
Victoria V8WIJ6, Canada
T: 61 3 386 5255 F: 61 3 386 4453

Hutchinson, Roger, Mr.
EMMANUEL COLLEGE
78 Queen's Park Crescent East
Toronto - Ontario M3S 1K7, Canada

Inglis, Julian, Mr.
Director
ECO-ED
135 Hawthorne Avenue
Ottowa - Ontario K150B2, Canada
T: 1 613 232-1249 F: 1 613 990-8818

Isla, Ana
CIVT / FREELANCER
St. George St
Toronto, Canada
T: 1 416 5950909 F: 1 416 5955604

Jacob, Henri, Mr.
President
LE R.E.V.E. INC.
1035, ave. St-Philippe
Dubuisson-Que J9P 4N7, Canada
T: 1 819 738-5261 F: 1 819 825-4215

John, Joshua, Mr.
INSTITUT CANADIEN DES OCEANS
5670 Spring Garden Road
Halifax - NS, Canada
T: 1 902 426-1512 F: 1 902 426-1599
TL: 019-21670

Jolicoeur, Guadalupe, Ms.
Latina America Coordinator
WESTERN CANADA WILDERNESS COMMITTEE
20 Water Street
Vancouver - BC, Canada
T: 1 604 683-8220 F: 1 604 683-8229
E: wc2wild

Kabatoff, Kathleen, Ms.
Representative
SELKIRK COLLEGE
BR#2 Site 19 C-6
Castlegar - BC V1N 3L4, Canada
T: 1 604 4325

Kalevar, Chaitanya K., Mr.
CANADIAN PEACE ALLIANCE
159 Glenholme Avenue
Toronto - Ontario M6E3C5, Canada
T: 1 416 657-8421 F: 1 416 588-5556

Kennedy, Derek
CANADIAN BROADCASTING CORPORATION
CBC TV - The Journal, Box 14000 Station A
Toronto - Ontario, Canada
T: 1 416 9757249 F: 1 416 9639036

Knight, Gloria, Ms.
Hon.
INTERNATIONAL INSTITUTE FOR SUSTAINABLE DEVELOPMENT
161 Portage Ave.East, Sixth Floor
Winnipeg - Alberta R3B OY4, Canada
T: 1 204 958-7700 F: 1 204 958-7710

Koroitamana, Foasa
YOUTH 92/ UNIV. OF SO. PACIFIC STUDENTS ASSOCIATION
55 Parkdale Ave, 3rd floor
Ottawa - Ontario, Canada
T: 1 613 761 9206 F: 1 613 761 1441

Kuptana, Rosemarie, Ms.
Member
INUIT CIRCUMPOLAR CONFERENCE
650, 32nd Avenue, Suite 404, Lachine
Lachine - P.Q. H8T3k4, Canada
T: 1 514 6373771 F: 1 514 6373146

Labelle, Richard, Mr.
INTERNATIONAL INSTITUTE FOR SUSTAINABLE DEVELOPMENT
161 Portage Ave.East, Sixth Floor
Winnipeg - Alberta R3B OY4, Canada
T: 1 204 958-7700 F: 1 204 958-7710

Labrentz, Arnold, Mr.
Sr. Instructor
ARTIC COLLEGE
Box 1132
Ft. Smith - NWT X0E 0P0, Canada
T: 001 403 872-3541 F: 001 403 872-4511

Lachance, Michel, Mr.
Directeur
UNIVERSITE DU QUEBEC
300 Ursulines
Rimouski,Quebec GIL 3A1, Canada
T: 1 418 724-1660 F: 1 418 724-1525

Laperriere, Pierre, Mr.
Conseiller Pedagogique
CENTRE DE FORMATION EN ENVIRONNEMENT
455 Fournier street
St-Jerome (Quebec), Canada
T: 1 514 4367402 F: 1 514 4365212

Lark, John, Mr.
Director
FISHERIES & OCEANS
200 Kent Street
Ottawa - Ontario KIAUE6, Canada
T: 1 613 990-0011

Lazure, Denis, Mr.
Deputado
GOUVERNEMENT DU QUEBEC
2068 Belburne
Quebec J3U 3X4, Canada
T: 1 418 653-6841

Leblanc, Suzie, Ms.
Program Officer
INSTITUT CANADIEN DES OCEANS
5670 Spring Garden Road
Halifax - NS, Canada
T: 1 902 426-1512 F: 1 902 426-1599
TL: 019-21670

Leclerc, Denis
Technician
SOCIETE RADIO CANADA
1400 Est Blvd Rene Levesque
Montreal - Quebec, Canada
T: 1 514 5974003 F: 1 514 5974095

Leitch, Donald, Mr.
INTERNATIONAL INSTITUTE FOR SUSTAINABLE DEVELOPMENT
161 Portage Ave.East, Sixth Floor
Winnipeg - Alberta R3B OY4, Canada
T: 1 204 958-7700 F: 1 204 958-7710

Lennox, Douglas J., Mr.
244 Riverview Blvd - St. Catharines
Ontario L2T 3M8, Canada
T: 416 682 1921

Lethbridge, Kirby, Mr.
Director
NATIVE COUNCIL OF CANADA
BOX491 Stnb
Happy Vl.- Labrador A0P1E0, Canada
T: 1 709 4425

Lingl, Paul, Mr.
Box 291 101-1184 Derman St.
Vancouver - BC V6G2M9, Canada

Littrell, Michael, Mr.
EARTH DAY INTERNATIONAL
794 Fort St.
Victoria - BC V8W 3W2, Canada
T: 1 604 743-7121 F: 1 604 743-3353

Lopes, Lucia, Ms.
Director
ECOLOGICAL RIGHTS ASSOCIATION
203-2160 Haultain Street
Victoria - BC V8R 2L9, Canada
T: 1 604 370-7087 F: 1 604 721-7767

Lorriman, Donna, Ms.
BALLINAFAD RESEARCH
Box 97
Ballinafad - Ontario, Canada
T: 1 416 873-3149 F: 1 416 873-2735

Lumsden, Sheila, Ms.
Youth Coordinator
INUIT CIRCUMPOLAR CONFERENCE
650, 32nd Avenue, Suite 404, Lachine
Lachine - P.Q. H8T3k4, Canada
T: 1 514 6373771 F: 1 514 6373146

Lutes, Mark, Mr.
ESSA, YORK UNIVERSITY
3rd Floor, Lumbers Building, 2700 Keele Street, York University
North York - ON, Canada
T: 1 416 736-5252 F: 1 416 736-5679

Ly, Aoua, Ms.
AQOCI
801 Rue Sherbrooke Est. Suite 400
Montreal - Quebec H2L 1K7, Canada

MacLellan, Peter, Mr.
INSTITUT CANADIEN DES OCEANS
5670 Spring Garden Road
Halifax - NS, Canada
T: 1 902 426-1512 F: 1 902 426-1599
TL: 019-21670

MacLennan, Gail, Ms.
ECO-ED
110 Eglinton Ave West
Toronto - Ontario M4R 1A3, Canada
T: 1 416 482-9212 F: 1 416 482-9601

MacNeill, Jim, Mr.& Mrs.
INTERNATIONAL INSTITUTE FOR SUSTAINABLE DEVELOPMENT
161 Portage Ave.East, Sixth Floor
Winnipeg - Alberta R3B OY4, Canada
T: 1 204 958-7700 F: 1 204 958-7710

Madsen, Ruth, Ms.
Director
THOMPSON INSTITUTE
170 Nicola St.
Kamloops - BC V2C 2PI, Canada
T: 1 828 828-1984 F: 1 828 828 1215

Maele, Dominique Van De, Ms.
CANADIAN COUNCIL INTERNATIONAL COOPERATION
1 Nicholas St. # 300
Ottawa - Ontario K1N 7B7, Canada
T: 1 613 236-4547 F: 1 613 236-2188
E: ccic 1992 (oveb)

Maisonneuve, Jean-Pierre, Mr.
ECO - ED
22 Kew Beach
Toronto - Ontario M4l 1B7, Canada
T: 1 416 628-1090

Marchand, Claude
Technician
SOCIETE RADIO CANADA
1400 Est Blvd Rene Levesque
Montreal - Quebec, Canada
T: 1 514 5974003 F: 1 514 5974095

Marino, Dian
FACULTY OF ENVRIONMENTAL STUDIES - YORK UNIVERSITY
12 Rajaji Marg
New Delhi 110011, Canada
T: 91 11 301-4309

Marr, Helen, Ms.
President
INTERNATIONAL COUNCIL OF JEWISH WOMEN
1110 Finch Ave. W. # 518
Downsview, Canada
T: 1 416 665-8251 F: 1 416 665-8702

Marr, Pamela, Ms.
President'S Assistant
INTERNATIONAL COUNCIL OF JEWISH WOMEN
2 Penwood Crescent
Don Mills-ON, Canada
T: 1 416 4459100 F: 1 416 4459197

Marshall, David, Mr.
Vice-Chairman
GLOBE
Suite 601 - 535 Thurlow Street
Vancouver - BC, Canada
T: 1 604 666-8020 F: 1 604 666-8123

Martinez, Jaime Luna, Mr.
INTERNATIONAL INSTITUTE FOR SUSTAINABLE DEVELOPMENT
161 Portage Ave. E, 6th floor
Winnipeg - Manitoba R38 OY4, Canada
T: 1 204 958-7700 F: 1 204 958-7710

Martins, Carlos C., Mr.
Forest Technologist
CANADIAN EMBASSY
424 Vancouver St.
Victoria - BC, Canada
T: 1 604 382-2047 F: 1 604 387-5999

Mathieu, Bernard, Mr.
Secretaire Associe
MINISTERE DES AFFAIRES INTERNATIONALES QUEBEC
885 Grande Allee
Quebec, Canada

McCrory, Colleen, Ms.
VALHALLA WILDERNESS SOCIETY
Box 224
New Denver V0G 1S0, Canada
T: 358-2449 F: 358-7950

McCurdy, Rev. Archibald
Member
CHRISTIAN PEACE CONFERENCE , NORTH AMERICA
25 Buchan Court
North York - ON M2J 4S9, Canada
T: 1 416 494-2222 F: 1 416 494-9985

McDermott, Lawrence, Mr.
Executive Director
PLENTY CANADA
Rr #3
Lanark - Ontario KOG1KO, Canada
T: 1 613 278-2416 F: 1 613 278-2416

McDowell, Andrew, Mr.
PEACE AND CONFLICT STUDIES SOCIETY UNIVERSITY OF TORONTO
556 Spadina Crescent #3
Toronto - Ontario M5S 2J9, Canada
T: 1 416 924-1890 F: 1 416 971-2027

McGinnis, Mr.Lloyd, Mr.
INTERNATIONAL INSTITUTE FOR SUSTAINABLE DEVELOPMENT
161 Portage Ave.East, Sixth Floor
Winnipeg - Alberta R3B OY4, Canada
T: 1 204 958-7700 F: 1 204 958-7710

McGinnis, Mrs.Lloyd, Ms.
INTERNATIONAL INSTITUTE FOR SUSTAINABLE DEVELOPMENT
161 Portage Ave.East, Sixth Floor
Winnipeg - Alberta R3B OY4, Canada
T: 1 204 958-7700 F: 1 204 958-7710

McLean, Jennifer, Ms.
Int'L. Department
WORLD COUNCIL OF CHURCHES
15 Breadalbane Street
Toronto - Ontario M4Y 2V5, Canada

Mctlron, Anne
Reporter
SOUTHAM NEWS
151 Sparks St. Suite 512
Ottawa - Ontario, Canada
T: 1 613 2360491 F: 1 613 2361788

Menasse, Gewurz Reich, Mr.
211 Wurtemburg S 312
Ottawa - Ontario K1N 8R4, Canada
T: 1 613 563 0211

Mercer, Karen, Ms.
Executive Assistant
NISGA'A TRIBAL COUNCIL
Box 231
New Aiyansh V0J 1A0, Canada
T: 1 603 633-2234 F: 1 603 633-2367

Mercredi, Tony, Mr.
Chief
ASSEMBLY OF FIRST NATIONS
55 Murray Street, 5TH Floor
Ottowa - Ontario KIN5M3, Canada
T: 613 236-0673 F: 613 238-5780

Miller Retzer, Connie, Ms.
Biologist
UNIVERSITY OF VICTORIA
326-1990 West Sixt Ave
Vancouver - BC V6A 4V4, Canada
T: 1 604 73152 81

Mitchell, Donald G., Mr.
961 Aylmer Road
Aylmer - Quebec J9H 5T7, Canada
T: 1 819 778-1081 F: 1 613 592-8352

Moreira, Iara Verocai Dias, Ms.
INSTITUT CANADIEN DES OCEANS
5670 Spring Garden Road
Halifax - NS, Canada
T: 1 902 426-1512 F: 1 902 426-1599
TL: 019-21670

Moses, Ted, Mr.
Ambassador
GRAND COUNCIL OF CREES
24 Bayswatan St.
Ottawa - Ontario, Canada
T: 1 613 761 1655 F: 1 613 327 5464

Muise, Betty, Ms.
IS FIVE FOUNDATION
400 Mount Pleasant Road
Toronto - Ontario, Canada
T: 1 416 4802408 F: 1 416 4802546

Munro, Ingrid, Ms.
INTERNATIONAL INSTITUTE FOR SUSTAINABLE DEVELOPMENT
161 Portage Ave.East, Sixth Floor
Winnipeg - Alberta R3B OY4, Canada
T: 1 204 958-7700 F: 1 204 958-7710

Murray, John, Mr.
Delegate
CANADIAN PEACE ALLIANCE
1001 Williams
Yellowknife - NWT XIA3PI, Canada
T: 1 403 873-3429

Muskego, Rosie, Ms.
Consultant
WINDIGO TRIBAL COUNCIL
PO Box 299 Sioux Lodkout POV 250
Bearskin Lake -Ont. POV IEO, Canada
T: 801 363 2514 F: 801 363 1066

Myers, Heather, Ms.
Manager
GOVERNMENT OF THE NORTHWEST TERRITORIES
PO Box 1320
Yellowknife - NWT, Canada
T: 1 403 873-6693 F: 1 403 873-0114

Myles, Ian, Ms.
YOUTH 92 / WORLD YOUTH PREPARATORY FORUM FOR UNCED
55 Parkdale Ave, 3rd floor
Ottawa - Ontario K1Y 1ES, Canada
T: 1 613 761 9206 F: 1 613 761 1441

Nana, Olivier, Mr.
Member
AQOSI
590 Place de la Fontaine #A Nun's Island
Montreal H3E 167, Canada
T: 1 514 362-1025 F: 1 514 343-2298

Nelson, Brian, Mr.
IS FIVE FOUNDATION
400 Mount Pleasant Road
Toronto - Ontario, Canada
T: 1 416 4802408 F: 1 416 4802546

N'golle, Magnus, Mr.
INSTITUT CANADIEN DES OCEANS
5670 Spring Garden Road
Halifax - NS, Canada
T: 1 902 426-1512 F: 1 902 426-1599
TL: 019-21670

Nirgark, John, Mr.
Min. Of Renewable Resources
GOVERNMENT OF THE NORTHWEST TERRITORIES
PO Box 1320
Yellowknife - NWT, Canada
T: 1 403 873-6693 F: 1 403 873-0114

Noblega, Ludia De
Reporter
LA PRESSE
7, Rua Saint Jacques
Montreal Quebec, Canada

Norrisette, Larry, Mr.
INTERNATIONAL INSTITUTE FOR SUSTAINABLE DEVELOPMENT
161 Portage Ave. E. 6th floor
Winnipeg - Manitoba R38 OY4, Canada
T: 1 204 958-7700 F: 1 204 958-7710

Ochman, Stefan, Mr.
Coordinator
CANADIAN ENVIRONMENTAL NETWORK
PO Box 1289, Stn.B
Ottawa - Ontario, Canada
T: 1 613 563-2078 F: 1 613 563-7236

Orton, David
GREEN WEB
R.R.3
Saltsprings-N.S. BOK IPO, Canada
T: 1 9029252514 F: 1 9029252514

Painchaud, Paul, Mr.
Professor
UNIVERSITE LAVAL - INTERNATIONAL CONSORTIUM FOR THE STUDY OF ENVIRONMENTAL SECURITY
Universite Laval
Quebec GTK 7P4, Canada
T: 1 418 656-2316 F: 1 418 656-7908

Pakozdy, Steve, Mr.
Ceo
INTERNATIONAL CHLORINE INC.
#220 - 1304 - 15 Avenue SW
Calgary - AL T3C OX7, Canada
T: 1 403 228-3686 F: 1 403 541-9157

Palmer, Geoffrey, Mr.
INSTITUT CANADIEN DES OCEANS
5670 Spring Garden Road
Halifax - NS, Canada
T: 1 902 426-1512 F: 1 902 426-1599
TL: 019-21670

Parfitt, Ian, Mr.
Director Of Mapping
WESTERN CANADA WILDERNESS COMMITTEE
20 Water Street
Vancouver - BC, Canada
T: 1 604 683-8220 F: 1 604 683-8229
E: wc2wild

Parisien, Normand, Mr.
Exec.Dir.
TRANSPORT2000
2520 Lionel-Groulx
Montreal - Quebec H3J 1J8, Canada
T: 1 514 932-8008 F: 1 514 932-2024

Parizeau, Marc
CANADIAN BROADCASTING CORPORATION
Montreal, Canada

Parrot, Franck, Mr.
Env Mgr
ICI EXPLOSIVES CANADA
801 Boul. Richelieu
McMasterville - PQ, Canada

Paschen, Syd, Mr.
CANADIANS FOR RESPONSIBLE NORTHERN DEVELOPMENT
11911 Univ. Avennue
Edmonton - Alberta, Canada
T: 1 403 436-4913

Patterson, Cliff, Mr.
Treasurer
ADVENTIST DEVELOPMENT & RELIEF AGENCY
95 Clarence Biesenthal Dr
Oshawa - ON, Canada
T: 1 416 433-8004 F: 1 416 723-1903
TL: 21-0636700466

Paupanekis, A., Mr.
Asst. Exec. Director
CROSS LAKE FIRST NATION
Cross Lake, Manitoba, Canada
T: 1 204 676-2218 F: 1 204 676-2117

Penchaszadeh, Analia, Ms.
YOUTH 92 / PEACE CHILD INTERNATIONAL
55 Parkdale Ave, 3rd floor
Ottawa - Ontario, Canada
T: 1 613 761 9206 F: 1 613 761 1441

Percival, Valerie, Ms.
Coordinator
PEACE AND CONFLICT STUDIES SOCIETY UNIVERSITY OF TORONTO
556 Spadina Crescent #3
Toronto - Ontario M5S 2J9, Canada
T: 1 416 924-1890 F: 1 416 971-2027

Petrovic, Dr., Mr.
INTERNATIONAL CHLORINE INC.
#220 - 1304 - 15 Avenue SW
Calgary T3C OX7, Canada
T: 1 403 228-3686 F: 1 403 541-9157

Pfeifer, Pitse, Ms.
BAFFIN YOUTH COUNCIL
55 Parkdale Ave, 3rd floor
Ottawa - Ontario, Canada
T: 1 613 761 9206 F: 1 613 761 1441

Phillips, Gilliau, Ms.
Board Member
WEED FOUNDATION
736 Bathurst St.
Toronto - Ontario, Canada
T: 1 416 516-2379 F: 1 516 531-6214

Porter, Marilyn, Ms.
Dr.
WOMEN IN SOCIETY
Dept, Sociology Memorial Univ.
St. Johns HIC55, Canada
T: 709 737 7455 F: 709 739 7982

Prazmowski, Andrea
Coordinator For Int'L Educ.
YMCA OF CANADA/OTTAWA OFFICE
180 Argyle Avenue, Suite 309
Ottowa - Ontario K2P 1B7, Canada

Proulx, Denise, Ms.
CANADIAN ENVIRONMENTAL NETWORK
2217 Chemin Principal
St.Joseph du Lac, QB, Canada
T: 1 514 623-0565 F: 1 514 623-4468

Queiroz, Ana Maria
ICAE
12 Rajaji Marg
New Delhi 110011, Canada
T: 91 11 301-4309

Quigg, Michelle, Ms.
ENVIRONMENTAL CHILDREN'S ORGANIZATON
2305 W3rd Av.
Vancouver - BC, Canada
T: 1 604 734-0146

Rahman, Alam C., Mr.
INTERNATIONAL MOVEMENT TO SAVE BANGLADESH
100 Rexdale Blvd, Suite #302 Rexdale
Toronto - Ontario, Canada
T: 1 416 742-9818 F: 1 416 742-9818

Rahman, Habibur M.M., Mr.
President
INTERNATIONAL MOVEMENT TO SAVE BANGLADESH
100 Rexdale Blvd, Suite #302 Rexdale
Toronto - Ontario, Canada
T: 1 416 742-9818 F: 1 416 742-9818

Rahman, Rani L.J.S.
INTERNATIONAL MOVEMENT TO SAVE BANGLADESH
100 Rexdale Blvd, Suite #302 Rexdale
Toronto - Ontario, Canada
T: 1 416 742-9818 F: 1 416 742-9818

Regallet, Gabriel, Mr.
Program Officer
INTERNATIONAL INSTITUTE FOR SUSTAINABLE DEVELOPMENT
161 Portage Ave. East , 6th floor
Winnipeg - Alberta, Canada
T: 1 204 958-7700 F: 1 204 958-7710

Reisler, Sheldon
Cameraman
VISION TV
315 Queen Street East
Toronto - Ontario, Canada
T: 1 416 366-9221 F: 1 416 368-9774

Retzer, Brian, Mr.
Msc Student
UNIVERSITY OF BRITISH COLUMBIA
326 - 1990 West 6th Ave.
Vancouver - BC V6J 4V4, Canada
T: 1 604 731-5281 F: 1 604 822-8645

Richards, Cathy, Ms.
PEACE AND CONFLICT STUDIES SOCIETY UNIVERSITY OF TORONTO
556 Spadina Crescent #3
Toronto - Ontario M5S 2J9, Canada
T: 1 416 924-1890 F: 1 416 971-2027

Richmond, Claude, Mr.
Chairman
WESTERN ENVIRONMENT AND DEVELOPMENT TASK FORCE
14th FL- Commerce Place, 400 Burrard St
Vancouver - BC V6C3G2, Canada
T: 1 604 643 1720 F: 1 604 643 1760

Riquelme, Monica Hernandez, Ms.
YOUTH 92 / BRIGADA UNIVERSITARIA DE SERVICES COMUNITARIOS PARA LA AUTOGESTION
55 Parkdale Ave, 3rd floor
Ottawa - Ontario, Canada
T: 1 613 761 9206 F: 1 613 761 1441

Riverin, Suzanne, Ms.
MUSKOKA ENVIRONMENTAL EDUCATORS ASSOCIATION
58 Brunel Rd.
Huntsville - ON P0A 1K0, Canada
T: 705 789 5594 F: 705 789 2440

Robinson, John, Mr.
INTERNATIONAL INSTITUTE FOR SUSTAINABLE DEVELOPMENT
161 Portage Ave.East, Sixth Floor
Winnipeg - Alberta R3B OY4, Canada
T: 1 204 958-7700 F: 1 204 958-7710

Rose, Elizabeth, Ms.
NORTHERN TELECOM
3 Robert Speck Pkway.
Mississauga - Ont., Canada
T: 1 416 566-3411 F: 1 416 566-3348

Rowan, Marielle, Ms.
56 Monterey Road
Kitchener - Ontario N2B143, Canada
T: 519 744-1374

Roy, Jane, Ms.
Executive Director
FULL - CYCLE LONDON
269, Colborne Street
London - Ontario N6B 2S7, Canada
T: 519 4348225 F: 519 4346201

Russow, Joan, Ms.
Director
ECOLOGICAL RIGHTS ASSOCIATION
203-2160 Haultain Street
Victoria - BC V8R 2L9, Canada
T: 1 604 370-7087 F: 1 604 721-7767

Sandstra, Ted, Mr.
Photographer
CHRISTIAN ENVIRONMENT ALLIANCE
RURAL ROUTE 7
Deer - Alberta T4N 5E1, Canada
T: 1 612 537 3865 F: 1 612 729 2828

Savoie, Donat, Mr.
Advisor
INUIT CIRCUMPOLAR CONFERENCE
650, 32nd Avenue, Suite 404, Lachine
Montreal H8T3k4, Canada
T: 1 514 6373771 F: 1 514 6373146

Sayne, Pamela, Ms.
N.America Reference Centre
WOMEN & SHELTER NETWORK
2163 Queen St. E C-7
Toronto - Ontario M4L 1J1, Canada
F: 1 416 694-5003

Schoon, Jack
CANADIAN BROADCASTING CORPORATION
CBC TV - The Journal, Box 14000 Station A
Toronto - Ontario, Canada
T: 1 416 9757249 F: 1 416 9639036

Senecal, Yves
Technician
SOCIETE RADIO CANADA
1400 Est Blvd Rene Levesque
Montreal - Quebec, Canada
T: 1 514 5974003 F: 1 514 5974095

Shecapio, James A., Mr.
Member
CREE NATION YOUTH COUNCIL
2 Lakeshore Road
Nemaska, Canada
T: 819 673-2600 F: 819 673-2606

Simard, Christian, Mr.
Director
UQCN (UNION QUEBECOISE POUR LA CONS. DE LA NATURE)
160, 76 E Rue Est. Charlesbourg Qc
Quebec G1H 7H6, Canada
T: 1 418 628-9600 F: 1 418 628-9600

Sirros, Christos, Mr.
Ministre
MINISTRE DELEGUE AUX AFF. AUTUCHTONES
875, Grande Allee # 2400
Quebec, Canada

Sitthiraksa, Sinith, Ms.
Student
FACULTY OF ENVIRONMENTAL STUDIES
York University - 4700 Keele ST
North York - ON, Canada
T: 1 416 736-5252 F: 1 416 736-5679

Smith, Louise Guay
Tv Director
SOCIETE RADIO CANADA
1400 est Rene Ceseaque
Montreal, Quebec H2L, Canada
T: 1 514 597-5618 F: 1 514 597-5498

Sonneborn, Carrie Louise, Ms.
GREENHOUSE ACTION AUSTRALIA
145A PigDon St., North Carlton
Victoria - BC, Canada
T: 61 3 380-8491 F: 61 3 481-8850

Sopuck, Robert, Mr.
INTERNATIONAL INSTITUTE FOR SUSTAINABLE DEVELOPMENT
161 Portage Ave.East, Sixth Floor
Winnipeg - Alberta R3B OY4, Canada
T: 1 204 958-7700 F: 1 204 958-7710

Spier, Robbie, Mr.
PEACE AND CONFLICT STUDIES SOCIETY UNIVERSITY OF TORONTO
556 Spadina Crescent #3
Toronto - Ontario M5S 2J9, Canada
T: 1 416 924-1890 F: 1 416 971-2027

Stewart, John, Mr.
INTERNATIONAL CHLORINE INC.
4309 32nd Road South
Arlington - VA 222069, Canada
T: 1 703 379 4280 F: 1 703 845 1660

Still, Gerry, Mr.
Research Manager
B.C. MINISTRY OF FORESTS
31 Bastion Square
Victoria - BC, Canada
T: 1 604 387-6579 F: 1 604 387-0046

Strangway, David, Dr.
INTERNATIONAL INSTITUTE FOR SUSTAINABLE DEVELOPMENT
161 Portage Ave.East, Sixth Floor
Winnipeg - Alberta R3B OY4, Canada
T: 1 204 958-7700 F: 1 204 958-7710

Sunoto, , Mr.
M.E.S.
ENVIRONMENTAL STUDIES STUDENT ASSOCIATION, YORK UNIVERSITY
3rd Floor, Lambert Building, York University, 4700 Keele St.
North York - ON M3J1P3, Canada
T: 1 416 739-1365 F: 1 416 736-5679

Suttie, Vanessa, Ms.
ENVIRONMENTAL CHILDREN'S ORGANIZATON
1158 Forge Walk
British Columbia, Canada
T: 1 604 733-9083

Suzuki, Sarika, Ms.
ENVIRONMENTAL CHILDREN'S ORGANIZATON
2477 Point Grey Road
British Columbia, Canada
T: 1 604 738-6956

Suzuki, Severn, Ms.
ENVIRONMENTAL CHILDREN'S ORGANIZATION
2477 Point Grey Road
Vancouver - BC, Canada
T: 1 604 738-6956

Tanner, Susan, Ms.
FOE
#701 251 Laurier Avenue West
Ottawa - Ontario K1P 5J6, Canada
T: 1 613 230-3352 F: 1 613 232-4354

Tardif, France, Ms.
AQOCI
801 Sherbrooke Est- Suite 400
Montreal - Quebec H2L 1K7, Canada
T: 1 514 597-2334 F: 1 514 597-2288

Taylor, Tamara, Ms.
UNCED
312 Adelaide St. N. 611
Toronto - Ontario, Canada
T: 1 416 408-3258 F: 1 416 408-3255

Thibeault, Eric Normand, Mr.
Directeur-General
ORG CANADIENNE EN GESTION DE PROJECTS POUR LE DEVELOPPEMENT INTERNATIONAL
Room 251, 100 rue Edmonton
Hull J8X 3L7, Canada
T: 1 819 7703921

Thomas, John A., Mr.
Asst. Exec. Director
CROSS LAKE FIRST NATION
Cross Lake, Manitoba, Canada
T: 1 204 676-2218 F: 1 204 676-2117

Titi, Vangile, Ms.
INTERNATIONAL INSTITUTE FOR SUSTAINABLE DEVELOPMENT
161 Portage Ave.East, Sixth Floor
Winnipeg - Alberta R3B OY4, Canada
T: 1 204 958-7700 F: 1 204 958-7710

Tostevin, Elaine, Ms.
Member
UNITED NATIONS ASSOCIATION - CANADA
508 St George St.
London - Ontario N6A 3B8, Canada
T: 519 433 6487

Urs, Thomas P., Mr.
Environmental Studies Member
INTERNATIONAL STUDIES ASSOCIATION
6307 Beaulieu
Montreal - Quebec, Canada
T: 1 514 767-6220 F: 1 514 987-4749

Urud, Anderson Mutang, Mr.
Director
WESTERN CANADA WILDERNESS COMMITTEE
20 Water Street
Vancouver - BC, Canada
T: 1 604 683-8220 F: 1 604 683-8229
E: wc2wild

Valantin, Robert, Mr.
Director
INTERNATIONAL DEVELOPMENT RESEARCH CENTER
P. O. Box 8500
Ottawa - Ontario, Canada
T: 1 613 236-6163 F: 1 613 238-7230
TL: 533753

Vanderzwaag, David
INSTITUT CANADIEN DES OCEANS
5676 Spring Garden Road, 9th Floor
Halifax - NS, Canada
T: 1 902 4261512 F: 1 902 4268932

Vicuna, Francisco Orrego, Mr.
INSTITUT CANADIEN DES OCEANS
5670 Spring Garden Road
Halifax - NS, Canada
T: 1 902 426-1512 F: 1 902 426-1599
TL: 019-21670

Villegas, Milder, Mr.
Gerant
ASS. COMMUNAUTAIRE D'EMPRUNT DE MONTREAL
914 Rachel St.
Montreal - Quebec H2J 2J1, Canada
T: 1 514 525-6628

Wehner, Sharon Elisa, Ms.
Volunteer
TURTLE ISLAND EARTH STEWARDS
#2-1967 W. 1st Ave.
Vancouver - BC, Canada
T: 1 604 736-7078 F: 1 604 737-1636

Whittington, Les
Reporter
SOUTHAM NEWS
151 Sparks St. Suite 512
Ottawa - Ontario, Canada
T: 1 613 2360491 F: 1 613 2361788

Williamson, Judith, Ms.
INTERNATIONAL INSTITUTE FOR SUSTAINABLE DEVELOPMENT
161 Portage Ave. E, 6th floor
Winnipeg - Manitoba R38 OY4, Canada
T: 1 204 958-7700 F: 1 204 958-7710

Zwicky, Lynn, Mr.
Board Member
INTERNATIONAL INSTITUTE FOR SUSTAINABLE DEVELOPMENT
161 Portage Ave. East , 6th floor
Winnipeg - Alberta, Canada
T: 1 204 958-7700 F: 1 204 958-7710

ADVISORS

Cashore, John, The Hon.
Minister of Environment
Province of British Columbia
Permanent Mission to the United Nations, 866 United Nations Plaza, Suite 250
New York, NY 10017, United States of America
T: 212 751 5600 F: 212 486 1295

Hopper, Wilbert H., Mr.
Chairman And CEO
Petro-Canada
Calgary, Alberta, Canada
T: 1 403 296 8000 F: 1 403 296 8385

Strong-Duffin, Kathy, Mrs.
Calgary, Alberta T2P 2MS, Canada
F: 403 268 8291

Suzuki, David, Mr.
Head, The David Suzuki Foundation
Vancouver, BC, Canada
T: 1 604 822 3381 F: 1 604 732 0752/822 6521

PARLIAMENTARIANS / SPIRITUAL LEADERS

Caccia, Charles
MP
The Senate
Ottawa Ontario K1A 0A4, Canada
T: 1 613 996 6111
TL: 053 3139

Haidasz, Stanley, Senator
The Senate Ottawa Ontario K1A 0A4
Canada, 1 613 996 6111
F: 053 3139

Lavois Roux, Threse, Senator
The Senate
Ottawa Ontario K1A 0A4, Canada
T: 1 613 996 6111
TL: 053 3139

MacDonald, David
MP Chariman Of Environment Committee
The Senate
Ottawa Ontario K1A 0A4, Canada
T: 1 613 996 6111
TL: 053 3139

Soranzo, Luis G., Senator
The Senate
Ottawa Ontario K1A 0A4, Canada
T: 1 613 996 6111
TL: 053 3139

Stewart, Christine
MP
The Senate
Ottawa Ontario K1A 0A4, Canada
T: 1 613 996 6111
TL: 053 3139

Wyman, Miriam, Ms.
President
YOUTH 92
11 Model Avenue
North York - ON, Canada
T: 1 416 633-6837 F: 1 416 672-2099

WISDOMKEEPERS

Armstrong, Jeanette
Indigenous/Elder
Eniowkin Center
257 Brunswick St.
Penticton B.C., Canada
T: 604-492-2359 F: 604-493-5302

Bignell, Quentin
Mother Earth Spiritual Camp
General Delivery
Seven Sisters Manitoba, Canada
T: 204-348-7356

Campiou, Fred
Indigenous/Shaman
Friendship center
General delivery
Driftpile Alberta TOG OVO, Canada
T: 403-355-2142 or 523-4511 F: 403-523-3055

Courchene, David
Indigenous/Shaman
Mother Earth Spiritual Camp
Box 121
Pine Falls Manitoba ROE 1MO, Canada
T: 204-348-7356

De La Barre, Kenneth
Youth Advisor
Inuit Circumpolar Conference
177 St. Paul St.
Montreal Quebec, Canada
T: 514-845-7539 F: 514-982-9786

De Laet, Christian
Manitou Development Alternatives
PO Box 40, Victoria Stn
Montreal Quebec H3Z 2V4, Canada
T: 514-487-9154 F: 514-481-0180

CAPE VERDE - GOVERNMENT

Barreto, Edna, Ms.
Director General
State Protocol
Permanent Mission to the United Nations, 27 E. 69th Str.
New York, NY, 10021
F: 212 472 0333
TL: 212 794 1398

Fonseca, Jorge Carlos Almeida, H.E. Dr.
Minister
Foreign Affairs
Permanent Mission to the United Nations, 27 E. 69th Str.
New York, NY, 10021
F: 212 472 0333
TL: 212 794 1398

Jesus, Jose Luis, H.E. Mr.
Ambassador Extraordinary and Plenipotentiary
Permanent Representative of Cape Verde to the United Nations, New York
Permanent Mission to the United Nations, 27 E. 69th Str.
New York, NY 10021, United States of America
T: 212 472 0333 F: 212 794 1398

Mascarenhas, Domingos, Mr.
Head
International Organisations Division,of Ministry of Foreign Affairs
Permanent Mission to the United Nations, 27 E. 69th Str.
New York, NY, 10021
F: 212 472 0333
TL: 212 794 1398

Monteiro, Antonio Mascarenhas, Mr.
President
Permanent Mission of the Republic of Cape Verde
27 E. 69th. Str
New York NY 10021, United States of America
T: 212 472 0333 F: 212 794 1398

Monteiro, Jose Antonio Pinto, H.E. Mr.
Secretary of State
Agriculture
Permanent Mission to the United Nations, 27 E. 69th Str.
New York, NY, 10021
F: 212 472 0333
TL: 212 794 1398

Oliveira, Anibal Chantre, Mr.
Honorary Consul
Rio de Janeiro
Permanent Mission to the United Nations, 27 E. 69th Str.
New York, NY, 10021
F: 212 472 0333
TL: 212 794 1398

Pereira, Carlos Alberto Monteiro, Mr.
Second Secretary
Permanent Mission of Cape Verde
27 E. 69th Str.
New York, NY 10021, United States of America
T: 212 472 0333 F: 212 794 1398

Ribeiro, Maria Teresa Lopes, Mrs.
Diplomatic Counsellor
President
Permanent Mission to the United Nations, 27 E. 69th Str.
New York, NY, 10021
F: 212 472 0333
TL: 212 794 1398

Rocha, Alberto, Mr.
Aide de Camp
President
Permanent Mission to the United Nations, 27 E. 69th Str.
New York, NY, 10021
F: 212 472 0333
TL: 212 794 1398

Santos, Napoleao Bonaparte, Officer
Protocol of the President
Permanent Mission of Cape Verde
27 E. 69th Str.
New York, NY 10021, United States of America
T: 212 472 0333 F: 212 794 1398

Silva, Jose Maria Tavares, Mr.
Executive Secretary
National Committee for UNCED
Permanent Mission to the United Nations, 27 E. 69th Str.
New York, NY, 10021
F: 212 472 0333
TL: 212 794 1398

Tolentino, Jorge, Mr.
Legal Counsellor
President
Permanent Mission to the United Nations, 27 E. 69th Str.
New York, NY, 10021
F: 212 472 0333
TL: 212 794 1398

Vasconcelos, Luis Carlos, Mr.
Press Adviser
Minister of Foreign Affairs
Permanent Mission to the United Nations, 27 E. 69th Str. New York, NY, 10021
F: 212 472 0333
TL: 212 794 1398

CENTRAL AFRICAN REPUBLIC - GOVERNMENT

Bedaya-Nagaro, Cimon, Prof.
Directeur General de l'organisation Inter Africaine pour la Qualite de la Vie
Permanent Mission of Central African Republic
Permanent Mission to the United Nations,386 Park Ave. South, Room 1614
New York, NY 10016, United States of America
T: 212 689 6195

Bingaba, Thierry, S.E.M.
Ministre de leconomie, du Plan, des Statistiques et de la Cooperation Internationale
Permanent Mission of Central African Republic
Permanent Mission to the United Nations,386 Park Ave. South, Room 1614
New York, NY 10016, United States of America
T: 212 689 6195

Docko, Joachim-Galean, S.E.M.
Ambassadeur
Charge du Protocole Presidentiel
Permanent Mission to the United Nations,386 Park Ave. South, Room 1614 New York, NY, 10016
F: 212 689 6195

Geleng, Manuela, Ms.
Programme Officer
UNITED NATIONS DEVELOPMENT PROGRAM
B.P. 872
Bangai, Central African Rep.
T: 236 611977 F: 236 611732
TL: 5268RC

Joachim, Lakoue-Tene, M.
Depute
Permanent Mission of Central African Republic
Permanent Mission to the United Nations,386 Park Ave. South, Room 1614
New York, NY 10016, United States of America
T: 212 689 6195

Kolingba, Andre, Mr.
President
Permanent Mission of the Central African Republic
386 Park Ave. South, Room 1614
New York NY 10016, United States of America
T: 212 689 6195

Kolingba, Mireille, Mme.
Epouse du Chef de letat
Presidente de la Fondation Vie et Espoir
Permanent Mission to the United Nations,386 Park Ave. South, Room 1614 New York, NY, 10016
F: 212 689 6195

Limbassa, Deniele, S.E.M.
Ambassadeur Charge du Protocole Presidentiel
Permanent Mission of Central African Republic
Permanent Mission to the United Nations,386 Park Ave. South, Room 1614
New York, NY 10016, United States of America
T: 212 689 6195

Magba-Totama, Cherubin-Raphael, S.E.M.
Ambassadeur Charge de Mission
Charge de la Presse Presidentielle
Permanent Mission to the United Nations,386 Park Ave. South, Room 1614 New York, NY, 10016
F: 212 689 6195

Mandata-Nguerekata, Gaston, S.E.M.
Secretaire detat a la Recherche Scientifique et Technique
Permanent Mission of Central African Republic
Permanent Mission to the United Nations,386 Park Ave. South, Room 1614
New York, NY 10016, United States of America
T: 212 689 6195

Mbitikon, Raymond, Mr
Ministre
Eaux, Forets, Chasses
Peches Et Du Tourisore Bp 830
Banaui, Central African Republic
T: 61 07 16 61 51 68

Nzessique, Honore, S.E.M.
Secretaire detat
Charte de Mission Particulier du Chief de letat
Permanent Mission to the United Nations,386 Park Ave. South, Room 1614 New York, NY, 10016
F: 212 689 6195

Pagonendji, Marie Solange, Mme.
Representante de Mouvement de la Jeunesse Centraficaine pour l'Environment
Permanent Mission of Central African Republic
Permanent Mission to the United Nations,386 Park Ave. South, Room 1614
New York, NY 10016, United States of America
T: 212 689 6195

Selebangue, Philemon
Directeur General
Ministere Des Eaux
Forets Chasses Peches Et Du Tourisore Bp 830
Banaui, Central African Republic
T: 236 61 2431 F: 236 61 5455

Sohohang-Kombet, Jean-Pierre, S.E.M.
Ambassadeur Extraordinaire et Plenipotentiare aux Etat-Unis
Permanent Mission of Central African Republic
Permanent Mission to the United Nations,386 Park Ave. South, Room 1614
New York, NY 10016, United States of America
T: 212 689 6195

Sopio, Romain, M.
Charge de Mission Coordinateur, Charge du suivi du Comite Preparatoir
Permanent Mission to the United Nations,386 Park Ave. South, Room 1614 New York, NY, 10016
F: 212 689 6195

Tandeau De Marsae, Gabriel Pierre
Conseillor Technique Forestier
MINISTERE DES EAUR ET
FORETS B.P. 830
BANGUI, Central African Republic
T: 61 02 15

Weinstabel, Peter Eduard, Mr
Head/German Forestry Mission
TO CENTRAL AFRICA/GTZ
B.P. 1303 Mission Forestiere Allemande
Bangui, Central African Republic
T: 236 61 48 39 61 5553 F: 236 61 48 39

Yando, Martin, S.E.M.
Ministre
Chef de Cabinet Particupier du Chief de letat
Permanent Mission to the United Nations,386 Park Ave. South, Room 1614
New York, NY 10016, United States of America
T: 212 689 6195

GLOBAL FORUM

Gerard, Michelle, Ms.
FOCSARENA
Bangui, Central African Rep.
T: 61-2924
TL: BP 934 BANGUI-

M'bitikon, Raymond, Mr.
Minister Of Forestry
MINISTRE DES EAUX, FORETS ET TOURISME
Mefcp BP 830
Bangai, Central African Rep.
T: 236 615553 F: 610163
TL: 5305RC

CHAD - GOVERNMENT

Ali, Mahamat, M.
Expert
Permanent Mission of Chad
211 East 43rd Street, Suite 1703
New York, NY 10017, United States of America
T: 986 0980, 490 2072

Badir, Moumi, M.
Aide de Camp
Permanent Mission of Chad
211 East 43rd Street, Suite 1703
New York, NY 10017, United States of America
T: 986 0980, 490 2072

Blausso, Moulna, M.
Protocole
Permanent Mission of Chad
211 East 43rd Street, Suite 1703
New York, NY 10017, United States of America
T: 986 0980, 490 2072

Dansala, Bambe Damsala
Ministre du Developpement Rural
MINISTERE AGRICULTURE
BP 44
Ndtamena, Chad
T: (235) 513470 F: (235) 515119

Denis, Moussa, M.
Directeur de Cabinet du Ministre du Tourisme et de l'Environnement
Permanent Mission of Chad
211 East 43rd Street, Suite 1703
New York, NY 10017, United States of America
T: 986 0980, 490 2072

Laoumai, Madjioudou Laoundam, M.
Conseiller du Premier Ministre
Permanent Mission of Chad
211 East 43rd Street, Suite 1703
New York, NY 10017, United States of America
T: 986 0980, 490 2072

Mugabe, Rombon, M.
Expert
Permanent Mission of Chad
211 East 43rd Street, Suite 1703
New York, NY 10017, United States of America
T: 986 0980, 490 2072

Nayo, Toumar, M.
Directeur de Cabinet du Premier Ministre
Permanent Mission of Chad
211 East 43rd Street, Suite 1703
New York, NY 10017, United States of America
T: 986 0980, 490 2072

Ngaraud, Baulmbate
Conseiller En Environment
MINISTERE DE LA PLANIFICATION ET DU
Ndjamena, Chad
T: 235 51 41 03

Ngaro, Baulbaye, M.
Expert
Permanent Mission of Chad
211 East 43rd Street, Suite 1703
New York, NY 10017, United States of America
T: 986 0980, 490 2072

Oumar, Acheick Ibn, S.E.M.
Ambassadeur Extraordinaire et Plenipotentiaire
Representent Permenant aupres des Nations Unies, New York
Permanent Mission to the United Nations, 211 East 43rd Street, Suite 1703
New York, NY 10017, United States of America
T: 986 0980, 490 2072

Rombon., Ougabet.
Ecoldgiste/Ingenieur Des Eaux-Foret
MINISTERE DUTOURISME ET DE L'ENVIRO
BP 447 Ndjamena/Chad.
Chad
T: 51 50 26 51 50 32

Togoi, Adoum, S.E.M.
Ministre du Tourisme et de l'Environnement
Permanent Mission of Chad
211 East 43rd Street, Suite 1703
New York, NY 10017, United States of America
T: 986 0980, 490 2072

NGO

Dzeringa, Ladukoura
CLUPOV
BP No. 25
Moundou (LOC), Chad

Mbaiherem, Ladukein
CLUPOV
BP No. 25
Moundou (LOC), Chad

Toningar, Jean
ASSADEC
B.P. 01
Sarh (Moyen-Char), Chad

PRESS

Djariri, Maumouna M.
Reporter Radio/Television
TELE TCHAD
Quartier Faxcha, Carre 3
Concession 89, N'Djamena, Chad, Chad

Neldibaye, Moalbaye
Editor
AGENCE TCHADIENNE DE PRESSE
Ndiamena
Chad

Oulatar, Begoto
Editor
NDJAMENA-HEBDO
B.P. 2207
N'Djamena, Chad, Chad

ORGANIZATIONS

Kourayo, Laoukein Mbaiherem, Mr.
Club des Prospecteurs de l'Or Vert (CLUPOV)
BP No. 25
Moundou (LOC), Chad

CHILE - GOVERNMENT

Acuna, Jaime, Sr.
Consejero
Delegacion de Chile ante los Organismos Internationales con sede en Ginebra
Permanent Mission to the United Nations, 809 United Nations Plaza, 4th Floor
New York, NY 10017, United States of America
T: 212 687 7547 F: 212 972 9875

Alvarado, Luis, Sr.
Ministro de la Bienes Nacionales
Permanent Mission of Chile
809 United Nations Plaza, 4th Floor
New York, NY 10017, United States of America
T: 212 687 7547 F: 212 972 9875

Arriagada, Eduardo, Mr.
COMISION ESPECIAL DE DESCONTAMINACION
Teatinos 120 -07812
Santiago, Chile
T: 56 2 6993435 F: 56 2 699 3660

Asenjo, Rafael, Sr.
Secretario Ejecutivo
Comision nacional del Medio Ambiente
Permanent Mission to the United Nations, 809 United Nations Plaza, 4th Floor
New York, NY 10017, United States of America
T: 212 687 7547 F: 212 972 9875

Aylwin Azocar, Don Patricio, S.E.
Presidente
Republica de Chile
809 United Nations Plaza, 4th Floor
New York, NY 10017, United States of America
T: 212 687 7547 F: 212 972 9875

Azocar, Patricio Aylwin, Mr
President
Permanent Mission of Chile
809 United Nations Plaza, 4th floor
New York NY 10017, United States of America
T: 212 687 7547 F: 212 972 9875

Berguno, Jorge, S.E. Sr.
Embajaor, Director de Politica Especial
Ministerio de Relaciones Exteriores
Permanent Mission to the United Nations, 809 United Nations Plaza, 4th Floor
New York, NY 10017, United States of America
T: 212 687 7547 F: 212 972 9875

Bidart, Carlos, Sr.
Director del Servicio Hidrografico y Oceanografico de la Armada
Permanent Mission of Chile
809 United Nations Plaza, 4th Floor
New York, NY 10017, United States of America
T: 212 687 7547 F: 212 972 9875

Briones, Hernan, Sr.
Presidente
Sociedad de Fomento Fabril
Permanent Mission to the United Nations, 809 United Nations Plaza, 4th Floor
New York, NY 10017, United States of America
T: 212 687 7547 F: 212 972 9875

Couve, Subsecretario Andres, Sr.
Servicio National de Pesca
Corporacion de Fomento de la Producion
Permanent Mission to the United Nations, 809 United Nations Plaza, 4th Floor
New York, NY 10017, United States of America
T: 212 687 7547 F: 212 972 9875

Etchegarray, Alberto, Sr.
Ministro de Vivienda y Uranismo
Permanent Mission of Chile
809 United Nations Plaza, 4th Floor
New York, NY 10017, United States of America
T: 212 687 7547 F: 212 972 9875

Fernandez, Pedro
Presidente
Comite Pro Defensa de la Fauna y la Flora
Casilla #14621 Santiago #21
Santiago, Chile
T: (562) 6715681 F: (562) 6715687

Figuerroa, Juan Augustin, S.E.
Ministro de Agricultura
Permanent Mission of Chile
809 United Nations Plaza, 4th Floor
New York, NY 10017, United States of America
T: 212 687 7547 F: 212 972 9875

Garcia, Subsecretaria Alvaro, Sr.
Ministerio de Planificacion y Coordinacion
Permanent Mission of Chile
809 United Nations Plaza, 4th Floor
New York, NY 10017, United States of America
T: 212 687 7547 F: 212 972 9875

Geisse, Guillermo
Secretario Ejecutivo
Centro de Investigatcion y Planificacion del Medio Ambiente
Permanent Mission to the United Nations, 809 United Nations Plaza, 4th Floor
New York, NY 10017, United States of America
T: 212 687 7547 F: 212 972 9875

Gonzalez, Miguel Angel, Sr.
Delegacion Permanente de Chile an la O.N.U.
Permanent Mission of Chile
809 United Nations Plaza, 4th Floor
New York, NY 10017, United States of America
T: 212 687 7547 F: 212 972 9875

Guzman, Antonio
Presidente de la Confederacion de la Produccion y del Comercio
Permanent Mission of Chile
809 United Nations Plaza, 4th Floor
New York, NY 10017, United States of America
T: 212 687 7547 F: 212 972 9875

Jadresic, Alejandro, Sr.
Secretario Ejecutivo
Comite Interministerial de Desarrollo Productivo
Permanent Mission to the United Nations, 809 United Nations Plaza, 4th Floor
New York, NY 10017, United States of America
T: 212 687 7547 F: 212 972 9875

Jimenez, Jorge, S.E.
Ministro de Salud Publica
Permanent Mission of Chile
809 United Nations Plaza, 4th Floor
New York, NY 10017, United States of America
T: 212 687 7547 F: 212 972 9875

Lavados, Jaime
Rector de la Universidad de Chile
Permanent Mission of Chile
809 United Nations Plaza, 4th Floor
New York, NY 10017, United States of America
T: 212 687 7547 F: 212 972 9875

Leon, Emilio, Sr.
Capitan de Fragata, Direccion General Teritorio Maritimo y marina Mercante
Ministerio de Defensa Nacional
Permanent Mission to the United Nations, 809 United Nations Plaza, 4th Floor
New York, NY 10017, United States of America
T: 212 687 7547 F: 212 972 9875

Mackenna, Cecilia, Srta.
Primer Secretario
Ministerio de Relaciones Exteriores
Permanent Mission to the United Nations, 809 United Nations Plaza, 4th Floor
New York, NY 10017, United States of America
T: 212 687 7547 F: 212 972 9875

Manriquez, Gustavo, Sr.
Director General, Direccion General de Aguas
Ministerio de Obras Publicas
Permanent Mission to the United Nations, 809 United Nations Plaza, 4th Floor
New York, NY 10017, United States of America
T: 212 687 7547 F: 212 972 9875

Martinex Sotomayor, Carols, S.E. Sr.
Embajor de Chile en Brasil
Permanent Mission of Chile
809 United Nations Plaza, 4th Floor
New York, NY 10017, United States of America
T: 212 687 7547 F: 212 972 9875

Moya, Juan, Sr.
Director Ejecutivo
Corporacion nacional Forestal, Ministerio de Agricultura
Permanent Mission to the United Nations, 809 United Nations Plaza, 4th Floor
New York, NY 10017, United States of America
T: 212 687 7547 F: 212 972 9875

Ovalle, Jose Manuel, Mr
Ministro Conse Jero
MINISTERIO-RELACIONES EXTERIORES
Morande 471
Santiago, Chile
T: 6982501

Oyarce, Pedro, Sr.
Consejero
Delegacion de Chile ante los Organismos Internacionales con sede en Ginebra
Permanent Mission to the United Nations, 809 United Nations Plaza, 4th Floor
New York, NY 10017, United States of America
T: 212 687 7547 F: 212 972 9875

Sanchez, Vincente, S.E. Sr.
Embajador de Chile en Kenya
Permanent Mission of Chile
809 United Nations Plaza, 4th Floor
New York, NY 10017, United States of America
T: 212 687 7547 F: 212 972 9875

Silva Cimma, Enrique, S.E.
Ministro de Ralaciones Exteriores
Permanent Mission of Chile
809 United Nations Plaza, 4th Floor
New York, NY 10017, United States of America
T: 212 687 7547 F: 212 972 9875

Skoknic, Milenko, Sr.
Primer Secretario consulado General de Chile en Rio de Janeiro
Permanent Mission of Chile
809 United Nations Plaza, 4th Floor
New York, NY 10017, United States of America
T: 212 687 7547 F: 212 972 9875

Solari, Subsecretaria Ricardo, Sr.
Ministerio Secretaria General de la Presidencia
Permanent Mission of Chile
809 United Nations Plaza, 4th Floor
New York, NY 10017, United States of America
T: 212 687 7547 F: 212 972 9875

Somavia, Juan, S.E. Sr.
Enbajador
Representante Permanente de Chile ante las Naciones Unidas
Permanent Mission to the United Nations, 809 United Nations Plaza, 4th Floor
New York, NY 10017, United States of America
T: 212 687 7547 F: 212 972 9875

Tapia, Alfredo, Sr.
Ministro Consejero
Consul General de Chile en Rio de Janeiro
Permanent Mission to the United Nations, 809 United Nations Plaza, 4th Floor
New York, NY 10017, United States of America
T: 212 687 7547 F: 212 972 9875

Toledo De La Maza, Juan C., Sr.
Vicealmirante, Director General del Territorio Maritimo y Marina Mercante
Ministerio de Defensa Nacional
Permanent Mission to the United Nations, 809 United Nations Plaza, 4th Floor
New York, NY 10017, United States of America
T: 212 687 7547 F: 212 972 9875

Tordesilla, Maria Eliana, Srta
Secretaria
Administratriva
Permanent Mission to the United Nations, 809 United Nations Plaza, 4th Floor
New York, NY 10017, United States of America
T: 212 687 7547 F: 212 972 9875

Vial, Juan de Dios, Sr.
Rector
Pontificia Universidad Catolica de Chile
Permanent Mission to the United Nations, 809 United Nations Plaza, 4th Floor
New York, NY 10017, United States of America
T: 212 687 7547 F: 212 972 9875

Videla, Ricardo, Sr.
Departamento Medio Ambiente
Ministerio de Recaciones Exteriores
Permanent Mission to the United Nations, 809 United Nations Plaza, 4th Floor
New York, NY 10017, United States of America
T: 212 687 7547 F: 212 972 9875

NGO

Abogabir Scott, Ximena
ICRA
Casa de la Paz
Antonia Lope de Bello 024, Santiago, Chile

Caquilpan Lincuante, Francisco
OMM
Calle Cautin No 1635 Temuco, Provincia de Cautin I
Casilla 518, Temuco, Chile

Challapa Mamani, Walter Raymundo
FAM
San Martin No. 372, Casilla Postal No. 635
Iquique, Chile

Cuevas Cueto, C.E.
CFV
General Jofre 50, oficina B
Santiago, Chile

Fernandez Bitterlich, Pedro
CODEFF
Santa Filomena 185, Casilla 3675
Santiago, Chile

Izquierdo Huneeus, Antonia
CCN
Av. Portales 3020, Paradero 6 NOS-Casilla 2-D
San Bernardo, Chile

Mamani Mamani, Antonio Hernan
FAM
San Martin No. 372, Casilla Postal No. 635
Iquique, Chile

Morales, Esther E.
IEC
Agustinas 641 Oficina 11
Santiago, Chile

Nahuelcheo Saldana, Yolanda Alicia
OMM
Calle Cautin No 1635 Temuco, Provincia de Cautin I
Casilla 518, Temuco, Chile

Tovarias Ferrada, Marisol
CODEFF
Santa Filomena 185, Casilla 3675
Santiago, Chile

Vergara Larrain, Sergio
ICRA
Casa de la Paz
Antonia Lope de Bello 024, Santiago, Chile

PRESS

Aguirre, Jose Ignacio O.
Reporter Radio/Television
CHILE TV
Los Quelterues 201B
Penalolen, Santiago, Chile, Chile

Araneda, Claudia Andrea N.
Reporter Radio/Television
RTU - CANAL 11
Chile Espana 405, Dpto. 42
Nunoa, Santiago, Chile, Chile

Avetikian, Tamara
Editor
EL MERCURIO
Patagonia, 249/103
Santiago, Chile, Chile

Briceno, Carmen Luz Gutierrez, Ms.
Editor
MINISTERIO DE OBRAS PUBLICAS
Alvarez 660 - B.Depto. 22
Vina del Mar, Chile, Chile

Brodsky, Roberto
Editor
HOY
Alberto Blest Gana 5903 Casa J
La Reina, Santiago, Chile, Chile

Brodzinsky, Sibylla
Editor
AMERICA ECONOMIA
Merced 186 (#11)
Santiago, Chile, Chile

Castillo, Arturo Enrique V.
Reporter Radio/Television
MEGAVISION TV - CANAL 9
Chopin 3182
San Joaquin, Santiago, Chile, Chile

Castillo Pizarro, Jorge Ivan
Editor
LA NACION
Pedro Rico 5944
Penalolen, Santiago, Chile

Castro, Margarita Andrea D.
Reporter Radio/Television
CANAL 13
Buenaventura 1855, Vitacura
Santiago, Chile, Chile

Cavieres, Eliana Gabriela E.
Reporter Radio/Television
TV CHILE
Las Urbinas 145
Depto 15, Santiago, Chile, Chile

Cerda, Rodrigo I.
Editor
RADIO NUEVO MUNDO
Buenos Aires 654
San Bernardo, Chile, Chile

Contreras G., Graciela Del Carmen
Editor
ORBE
Ismael Valdes Vergara, 452/505
Chile

Correa, Ricardo
Reporter Radio/Television
MASS MEDIA TV
V.Castillo 985
Santiago, Chile, Chile

D'Albuquerque Donoso, Lucia Ines
CANAL 13
Av. Damasceno Viera 44, #133-G
Sao Paulo, Brazil, Chile

Ebensperger, Karin Ahrens
Reporter Radio/Television
CANAL 13
Tomas de Figueroa 2306
Santiago, Chile, Chile

Escobar, Paula
Editor
CARAS
Vecinal 181 depto.31
Las Condes, Santiago, Chile, Chile

Gallegulllos, Ximena Eliana D.
CHILE PRESS
Real Audiencia 1320
Santiago, Chile, Chile

Godoy, Hugo
Editor
PAGINA ABIERTA
General Salvo 88
Providencia, Santiago, Chile, Chile

Gomez Duran, Guillermo
Correspondent
DIARIO ULTIMAS NOTICIAS
Secretaria de Prensa
Santiago, Chile, Chile

Guerra, Maria Teresa X. P.
Reporter Radio/Television
TV CHILE
Bulnes 549 S
Santiago, Chile, Chile

Guerra Yercovic, Ana Maria
Correspondent
DIAIRIO LA SEGUNDA
Secretaria de Prensa
Santiago, Chile, Chile

Hayes Frabasile, Barbara
Editor
DIARIO LA NACION
Los Leones 2042
Providencia, Santiago Chile, Chile

Hernandez, Patricio Eduardo P.
Reporter Radio/Television
CANAL 13
Pasaje Los Planetas 3018
Macul., Santiago, Chile, Chile

Hoffmann, Adriana
Correspondent
EL MERCURIO
Alvaro Casanova 613 - Penalolen
Santiago, Chile
T: 56 2 277-2899

Koppenberger Steppen, Lilly
Editor
TELEVISION NACIONAL DE CHILE
Pje Rio Becker 4798
Av. La Florida, Conjunto El Alba,Chile, Chile

Leon, Ricardo Patricio C.
Editor
TV CHILE
Av. Liberador 1712, El Bosque
Santiago, Chile, Chile

Lopez, Jorge Federico S.
Reporter Radio/Television
CANAL 13
El Matico 3786, Vitacura
Santiago, Chile, Chile

Louis Juffe, Andre
Editor
COSAS DE CHILE Y PERU
Alvaro Casanova 291 C-1
La Reina, Santiago, Chile, Chile

Machuca Carvajal, Arturo
Reporter Radio/Television
CHILE PRESS
Los Pinos
Santiago - Chile, Chile

Matus Villalobos,, Julio
Editor
LAS ULTIMAS NOTICIAS
Av. Santa Maria,5542
Santiago, Chile, Chile

Mendoza Prado, Marcelo
Editor
EL CANELO
Antupiren 7388,
Santiago, Chile, Chile

Meza Antognoni, Roberto
EL DIARIO
Compania 1048, Of. 42
Casilla 9893, Santiago, Chile, Chile

Miranda Osorio, Amelia Ester
Editor
DIARIO LA NACION
Agustinas 1269
Santiago, Chile, Chile

Mizala Salces, Gonzalo Javier
Editor
EL DIARIO FINANCIERO
Pasaje Los Cerezos, 851
Santiago, Chile, Chile

Orozco Bass, Pablo
Editor
DIARIO EL MERCURIO
Camino el Alba 10745
Las Condes, Santiago, Chile, Chile

Otarola, Samuel Eduardo V.
Correspondent
TV CANAL 13
Manuel Montt 2629/107
Santiago, Chile, Chile

Pineda Soto, Marcia Margarita
Reporter Radio/Television
CHILE PRESS
Las Carabelas 384, #6
Santiago, Chile, Chile

Rabelo Nunez, Jose
Correspondent
SESOCU
Ministerio Relaciones Exter.
Santiago, Chile, Chile

Reyes Allendes, Fernando
Correspondent
REVISTA HOY
Secretaria de Prensa
Santiago, Chile, Chile

Riera, Maria Eugenia
CHILE PRESS
Merchant Pereira 11-5
Depto. 32,,Santiago, Chile, Chile

Robles, Manola
Reporter Radio/Television
RADIO COOPERATIVA
Gausetla Sur 1071
Santiago, Chile, Chile

Rodriguez, Maria Angelica E.
Reporter Radio/Television
RADIO CHILENA
Castillo Urizar 2890-A
Dpto.21, Macul, Santiago, Chile, Chile

Rojas Gonzalez, Raul
Editor
LA TERCERA DE LA HORA
Vicuna Mackenna 1870
Santiago, Chile, Chile

Rojo, Emilio
Editor
DIARIO LA EPOCA
Villa Los Presidentes,B.100 D
Dpto. 202, Nunos, Chile

Ruz, Veronica Virginia O.
Reporter Radio/Television
RADIO NACIONAL
Jose Manso de Velosco 6973
Santiago, Chile, Chile

Saavedra, Maria Eugenia N.
Reporter Radio/Television
RADIO MINERIA
Poeta Angen Cruchaga 1741A
Dpt. A, La Reina, Santiago, Chile, Chile

Salinas Matalon, Luis Claudio
Editor
EL MERCURIO
Av. Santa Maria, 5542
Las Condes, Santiago, Chile, Chile

Sanchez, Claudio Mariano V.
Reporter Radio/Television
CANAL 13
Napoleon 3038
Depto. 601, Santiago, Chile, Chile

Schmied, Julie
Correspondent
EL DIARIO
Compania 1048 esq.42
Santiago, Chile, Chile

Silva, Samuel
Editor
AMERICA ECONOMIA
Walter Scott 1381
Vitacura, Santiago, Chile, Chile

Torres Zagal, Ivan
Editor
MASS MEDIA TV
Jose Infante 3295
Santiago, Chile, Chile

Ugarte Atala, Marco
Correspondent
DIARIO LA NACION
Secretaria de Prensa
Santiago, Chile, Chile

Vergara, Sergio
Correspondent
EL DIARIO
Mar del Sur, 1052
Santiago, Chile, Chile

Yerkovic, Ana Maria G.
Editor
DIARIO LA SEGUNDA
Jose Joaquin Vallejos 1572
Dpto. 55, Santiago, Chile, Chile

Zarhi, Andres
Reporter Radio/Television
CANAL 13
Los Diamelos 3061 Providencia
Santiago, Chile, Chile

ORGANIZATIONS

Avendano, Elisa, Ms.
Organizacion Mapuche Ad - Mapu
Calle Cautin No 1635 Temuco, Provincia de Cautin IX Region
Casilla 518, Temuco, Chile

Ergas, Rosita, Ms.
Corporacion Futuro Humano
General Jofre No. 50
Santiago, Chile
T: (56 2) 635 4101

Grau, Juan, Sr.
Instituto de Ecologia de Chile
Agustinas 641 Oficina 11
Santiago, Chile
F: (562) 222 2886

Saldivia, Beatriz, Sra
Centro De Investigacion Y Accion En Salud Poblacional (CIASPO)
Casilla 197-13
Santiago, Chile
T: 522 5235/522 2607/522 5196 F: (56 2) 551 9558

Sandoval, Mario, Mr.
Centro De Formacion Y Desarrollo Juvenil (FOLICO)
Fernandez Albano 8117, San Ramon
Casilla 39, Santiago 15, Chile
T: (56 2) 558 9696 F: (56 2) 551 5331

Tovarias, Marisol, Ms.
National Committee For The Defense Of Fauna, Flora And The Environment Chile
Santa Filomena 185, Casilla 3675
Santiago, Chile
T: (56 2) 377 290 F: (56 2) 377 290

GLOBAL FORUM

Aguayo Chamorro, Ivan, Mr.
Adecoop
CENTRO PARA LA AUTOGESTION Y DESARROLLO COOP.
Freire 1062 (Interior)
Concepcion 3364, Chile
T: 041 238707 F: 041 2270006

Alegria, Paula
Periodista
CANAL 4 LA RED
La Capitunia 1200 Las Condes
Santiago, Chile
T: 1 212 8762 F: 2110492

Alvarez, Carlos Canihuante, Mr.
El Sendro 148
Santiago, Chile
T: 56 2 812-1343

Amigo L., Bernardo, Sr.
CODEFF
Santa Filomena 185
Santiago, Chile
T: 56 2 7771607 F: 56 2 377290

Aravena, Jose C., Mr.
President
URRACAS EMAUS
Jose Besa 1481, Quinta Normal
Santiago 5, Chile
T: 34 81 773-5621 F: 34 81 773-8476
E: Casilla 98/Stgo.5

Arenas, Mario Fernando Ahumada, Sr.
CENTRO HUMANISTA PRA EL DESARROLLO CAMMPESINO - CATEV LTDA
Casilla 429, 1 Oriente no. 978
Talca, Chile

Arevalo, Maria, Ms.
FEDERACION NACIONAL
Mujeu Negocios y Proficionales
Santiago, Chile
T: 56 2 671-1228 F: 56 2 671-1228

Astorga Moreno, Sara, Ms.
GRUPO DE ACCION POR EL BIO BIO
R Antonia Lopez De Belklo 075
Santiago, Chile
T: 56 2 375 251 F: 56 2 375 251

Aylwin, Jose, Mr.
Abogado
COMISION CHILENA DERECHOS HUMANOS
Huerfanos 1805
Santiago, Chile
T: 56 2 698-7287 F: 56 2 699-0841

Baquedano, Manuel, Sr.
INSTITUTO DE ECOLOGIA POLITICA
Palais de Nations
Geneve CH 1211, Chile
F: 41 22 733-6542

Boye, Maria Teresa Portilla, Ms.
Medio Ambiente
CAPE HORNN METHANOL
Ricardo Lyon 755 Depto. 62
Santiago, Chile
T: 56 2 251-3437

Brain, Margarita, Ms.
Professora
CODELCO CHUQUICAMATA
Departamento Legal
Chuquicamata, Chile
T: 56 41 324455

Bravo Gallegos, Saul, Sr.
Secretario Geral
COORDINADORA NACIONAL DE ONG'S-CHILE
Calle Cubuas # 776-E
Paneagua-VI Region, Chile
T: 072 23-0980 F: 072 24-1358

Bruce, Alan, Mr.
Estudante
UNIVERSIDAD DE CONCEPCION
Rio Malleco, 2755
Concepcion, Chile
T: 595 31 37-1895

Candia, Rodrigo Cerda, Mr.
Secretario Executivo
INSTITUTO DE ECOLOGIA POLITICA
Londres, 88 Dpto. 20
Santiago, Chile
T: 56 2 632 6780 F: 56 2 391 784

Cardenas Nunes, Juan Carlos, Sr.
Ocean Ecology Cordinator
GREENPEACE AMERICA LATINA
Vicuna Mackena 3 Apto. 2102
Santiago, Chile
T: 56 2 391366 F: 56 2 391268

Castillo, Raul
CANAL 13 TV - CHILE
Ines Matte Urrejola 0848 Providencia
Santiago, Chile
T: 00562 6323214 F: 00562 302516

Castro, M. Andrea
CANAL 13 TV - CHILE
Ines Matte Urrejola 0848 Providencia
Santiago, Chile
T: 00562 6323214 F: 00562 302516

Claro, Amparo
ISIS INTERNATIONAL
Casilla, 2067
Santiago, Chile
T: 56 2 334-582

Co, Christina, Sra.
Coordenadora De Programa
CEAAL
Rafael Canas 218 - Providencia
Santiago, Chile
T: 56 2 235 6710 F: 56 2 235 6256

Cortez, Pedro, Sr.
CORPORACION NACIONAL DEL COBRE DE CHILE, CODELCO-CHILE
Departmento Legal - Div Chuquicamata, Oficinas Generales - Chuquicamata
Chuquicamata, Chile
T: 56 41 322-332 F: 56 41 322-192

Curriao, Jose Antolin, Mr.
Cacique Del.
GRUPO DE ACCION POR EL BIO BIO
Antoniz Lopez De Bello 075
Santiago, Chile
T: 56 2 375 251 F: 56 2 375 251

Diaz, Margarita Andrea Castro
Producer
CANAL 13 TV - CHILE
Ines matte Urrejola 0848 Providencia
Santiago, Chile
T: 00562 6323214 F: 00562 302516

Escobar, Raul
CANAL 13 TV - CHILE
Ines Matte Urrejola 0848 Providencia
Santiago, Chile
T: 00562 6323214 F: 00562 302516

Fernandez, Carlos P. Fuenzayda, Mr.
Engineer
F.M.N.
Santa Beatriz 168- Providencia
Santiago, Chile
T: 34 81 235-2122 F: 34 81 274-0352

Flores, Rene, Mr.
Const. Civil
CETAL
Abtao, 576 - Cerro Concepcion
Valparaiso, Chile
T: 56 32 213360 F: 56 32 214851
TL: 330506 APB VTR-

Fortez, Cecilia, Mr.
INST. ECOLOGIA POLITICA - IEP
Londres 88 Depto.21
Santiago, Chile
T: 56 2 335531

Fuenzalada, Rafael Valenzuela, Sr.
Professor
UNIVERSIDAD CATOLICA VALPARAISO
Eleuterio Ramires #353
Valparaiso, Chile
T: 56 32 230-030 F: 56 32 218-488

Gaete Nunez, Bernardo Casimiro, Sr.
Coord. Gen.
CETAL
Abtao, 576 - Concepcion
Valparaiso, Chile
T: 56 32 213360 F: 56 32 214851
TL: 330506

Gajardo, Jorge, Mr.
Economista
CEDAL
Roberto Prezot, 26
Santiago, Chile
T: 56 2 671-4124

Garreton K., Rodrigo, Mr.
Coord.Audiovisual
GRUPO DE ACCION POR EL BIO BIO
R Antonia Lopez De Bello 075
Santiago, Chile
T: 56 2 375 251 F: 56 2 375 251

Gavilan, Manuel F., Mr.
Rep. Regional
EMAUS SUDAMERICANO
Jose Besa, 1481 - Quinta Normal
Santiago 5, Chile
T: 34 81 773.5621 F: 34 81 773.8476

Gonzalez, Laura Soledad Suarez, Sra.
Coordinator
CEANIM
Coronel Santiago Bueras 182, Region Metropolitana
Santiago, Chile

Gorban, Carolina
Produtora
REDE CANAL 4
Santiago, Chile
F: 56 2 (185) 211049

Grau, Sergio, Mr.
FUNDACION LAHUEN
Orrego Luco 054
Santiago, Chile
T: 56 2 251 7506

Herrera, Gladys Chaves, Sra.
Coordenadora
MUNICIPAL
Catirai 186
Los Angeles VIII Region, Chile
T: 56 43 31-1484

Huenupi, Jose Bernardino H., Mr.
Cacique Major
GRUPO DE ACCION POR EL BIO BIO
R Antonio Lopez De Bello 075
Santiago, Chile
T: 56 2 375 251 F: 56 2 375 251

Izquierdo, Antonia, Sra.
Coord. Del Prog. Ecol.
CANELO DE NOS
Pastor Fernandez Parcela 24- El Arrayan
Santiago, Chile
T: 56 2 859-3262 F: 56 2 859-1405

Jorquera, Eduardo Astorga, Mr.
Director Abogado
CODEFF CHILE (FOE)
Sta Filomena 185
Santiago, Chile
T: 56 2 777 1607 F: 56 2 37 7290

Kanisawa, Michiko, Ms.
Secretario General
NETWORK YOKOHAMA
Rafael Canas 218 Providencia
Santiago, Chile
T: 56 2 235-6710 F: 56 2 235-6250

La Maza Nielandt, Maria Dolores, Ms.
GRUPO DE ACCION POR EL BIO BIO
Antonia Lopez De Bello 075
Santiago, Chile
T: 56 2 37-5251 F: 56 2 37-5251

Lara Quintino, Fernando, Mr.
Coodinador General
CAMBIO EDUCATIVO
Silvina Hurtado 1782 Providencia
Santiago - NO, Chile
T: 56 2 2748765

Larrain, Sergio Vergara, Mr.
Vice-Chairman
INSTITUTO CHILENO DE RECONOCIMIENTO AMBIENTAL
Antonia Lope de Bello 024
Santiago, Chile
T: 56 2 374280 F: 56 2 777 5065

Latorre, Alvaro Erazo, Mr.
Medico
TALLERES PARA LA EMERGENCIA/ARCA. MEDIO AMBIENTE
Sucre 1248 Nunoia
Santiago, Chile
T: 56 2 220-2106

Lebens, John, Mr.
Director
INTERNATIONAL INSTITUTE FOR ENERGY CONSERVATION/LATIN AMERICA OFFICE
Carmencita 110, No. 102, Las Condes
Santiago, Chile
T: 56 2 2332870 F: 56 2 2332870

Leon, Esther Morales, Ms.
INSTITUTO ECOLOGIA DE CHILE
Lira 325 Apto.
Santiago, Chile
T: 56 2 222 3259

Leppe Cartes, Marcelo, Mr.
Estudante
UNIVERSIDAD DE CONCEPCION
Rio Nuble, 5032 - Denavi-Sur
Talcahuano, Chile
T: 41 58-8171

Letelier, Alicia Stuven, Sra.
President
ECOLOGIA OMKINKAI
Avda Soy Vicenta 1870
Los Angeles VIII Region, Chile
T: 56 43 31-1540

Letelier, Sergio, Mr.
Diretor Adjunto
TECA
Seminario 774, Nunoa
Santiago, Chile
T: 56 2 274 5713 F: 56 2 274 5713
E: casilla181-11

Lopez, Maria Teresa, Sra.
Professor
UNIVERSIDAD DE CONCEPCION
El Golf 2681, depto.21
Concepcion VIII Region, Chile
T: 595 31 23-4985 F: 595 31 24-0280

Maldonado, Sergio, Mr.
CIREN-CORFO
Manuel Montt, 1164
Santiago, Chile
T: 56 2 274 9669 F: 56 2 49 6407
TL: 242017

Manriquez, Samuel Palma, Mr.
Executive Director
SERVICO EVANGELICO PARA EL DESARROLLO - SEPADE
Passy 032, Providencia
Santiago, Chile
T: 34 81 346390 F: 34 81 346786

Mansilla, Ricardo
Camarografo
CANAL 4 LA RED
La Capitunia 1200 Las Condes
Santiago, Chile
T: 1 212 8762 F: 2110492

Marcelo, Wilfredo, Mr.
Encarregado Internacional
INSTITUTO DE ECOLOGIA POLITICA
Londres, 88 Apto. 20
Santiago, Chile
T: 56 2 335 531 F: 56 2 391 784

Maturana, Gonzalo, Mr.
CAMARA CHILENA DE LA CONSTRUCCION
Marchant Pereira 10, 3er Piso
Santiago, Chile
T: 56 2 233-1131 F: 56 2 232-7600

Medina, Rodrigo, Mr.
Estudante
UNIVERSIDAD DE CONCEPCION
Lincoyan, 1041
Concepcion, Chile
T: 56 41 24-0468

Meza, Lucy Verdugo, Sra.
Chefe Tecnico
LICEO TECNICO
Marconi 607 depto. 315
Los Angeles VIII Region, Chile
T: 56 43 31-1049

Mintz, Nicole, Ms.
FUNDACION LAHUEN
Orrego Luco 054
Santiago, Chile
T: 56 2 251-7506

Miranda, Renato, Mr.
Pres. Comision De Protec Ma
CAMARA CHILENA DE LA CONSTRUCCION
Marchant Pereira 010 - 3er Piso Providencia
Santiago, Chile
T: 56 2 223-7326 F: 56 2 223-7970

Mora, Alejandra, Ms.
Coordinadora Greenephone
INSTITUTO DE ECOLOGIA POLITICA
Londren 88 Dpto.20
Santiago, Chile
T: 56 2 335-531 F: 56 2 391-784

Mora Mora, Rodrigo, Mr.
Estudante
UNIVERSIDAD DE CONCEPCION
Los 3 Antonios, 1129 - Nunoa
Santiago, Chile
T: 56 2 238-1264

Moriarty, John, Mr.
C O D E F F
Casilla, 3675
Santiago, Chile
T: 56 2 777-1607 F: 56 2 737-7290

Moya, Javier Munoz
Camera Assistant
CANAL 13 TVUC
Ines Matte Urejola 0848 -
Santiago, Chile
T: 56 2 371134 F: 56 2 302516

Munida, Felipe Banados, Mr.
Lawyer
F.M.N.
Agustinas 972
Santiago 407, Chile
T: 56 2 235-2122 F: 56 2 274-0352

Munoz, Javier
CANAL 13 TV - CHILE
Ines Matte Urrejola 0848 Providencia
Santiago, Chile
T: 00562 6323214 F: 00562 302516

Navia, Pedro Cortez, Sr.
CODELCO
Chuquicamata, Depto Legal Chile,
Chuquicamata, Chile
T: 56 41 322-150 F: 56 41 322-332

O'Ryan, Joyce, Ms.
Disenadora
CETAL
Abtao # 576 - Concepcion
Valparaiso, Chile
T: 56 32 251 313 F: 56 32 214 851
TL: 330506 APB VTR

Oliva M., Myriam, Ms.
Vice-Pres.
CONSEJO ECOLOGICO COMUNAL
Mujica,258
Rancagua, Chile
T: 72 241-621

Olmedo, Jose Ignacio Aguvre
TELEVISION NACIONAL DE CHILE
Bellavista 0990
Santiago, Chile
T: 7777454

Orrego, Juan Pablo, Mr.
Coord. General
GRUPO DE ACCION POR EL BIO BIO
Antonio Lopez De Bello 075
Santiago, Chile
T: 56 2 375 251 F: 56 2 375 251

Osorio, Jorge
CEAAL
Rafael Canas 218 - Providencia
Santiago, Chile
T: 56 2 235-6710 F: 56 2 235-6256

Otarola, Gonzalo, Mr.
Coordinador Regional
COMISION CHILENA DERECHOS HUMANOS
Calle Gamero 761
Rancagua, Chile
T: 72 23-9828

Parra Munoz, Luis Alberto, Mr.
Resp. Ar. En.
CETAL
Abtao, 576 - Concepcion - Casilla 197 - V
Valparaiso, Chile
T: 56 32 213360 F: 56 32 214851
TL: 330506 APB VTR

Patrick, Geraldine, Ms.
INSTITUTO DE ECOLOGIA DE CHILE
Bilbao 3775 Depto,1405
Santiago, Chile
T: 56 2 256-6391

Perez, Patricio Eduardo Hernandez
Reporter
CANAL 13 - CHILE
Ines Matte Urrejola 0848
Santiago, Chile
T: 55 21 9803433

Pimo, Vilma Oriana Perez, Sra.
Project Coordinator
TALLER DE VIVENDA SOCIAL
Punisima 370
Santiago, Chile
T: 56 2 374959 F: 56 2 2359091

Prado, Marcelo Mendoza, Mr.
APMA - REVISTA EL CANELO
Monsenor Muller 54, dpto.4 / Providencia
Santiago, Chile
T: 56 2 251-0148 F: 56 2 251-0148
TL: 29480 canel cl

Quinchanao, Eulogia Lorena Huenchunir, Sra.
SODECAM
Daglfinger Strasse 67
Munchen D 8000 Munchen 81, Chile
T: 49 89 936987 F: 49 89 74 88 07

Quintanilla, Victor, Dr.
DEPTO. DE INGENIERIA CIVIL GEOGRAFICA - UNIV. DE CHILE
Universidad de Chile
Santiago, Chile

Robleto, Maria Luisa
INST. ECOLOGIA POLITICA - IEP
Calle Londres 88 Depto.21
Santiago, Chile
T: 56 2 335531

Rodriguez, Patricio Antonio Yanez, Sr.
CENTRO EL CANELO DE NOS
Dagifinger Strasse 67
Munchen D 8000 Munchen 81, Chile
T: 49 89 936987 F: 49 89 74 88 07

Rodriguez, Regina, Sra.
Encargada Publicaciones
ISIS INTERNATIONAL
Casila Postal 2067, Correo Central
Santiago, Chile
T: 56 2 33 4582

Rojas, Nancy, Sra.
DEPTO. DE INGENIERIA CIVIL GEOGRAFICA - UNIV. DE CHILE
Universidad de Chile
Santiago, Chile

Romero, Hugo, Mr.
Professor
UNIV. DE CHILE
Marcoceta 250
Santiago, Chile
T: 56 2 222-6501 F: 56 2 222-2578

Romirez, Oda De Las Mercedes Jilberto, Ms.
Supervisora Tecnico-Pedagoga
MINISTERIO DE EDUCACION
Luis Cruz 1478
Rencagua, Chile
T: 56 72 211222

Salas, Claudio Enrique Fiabane, Mr.
CENTRO EL CANELO DE NOS LTDA
Avenido Portales
San Bernardo 3020, Chile

Schmidt, Jorge Federico Lopez
CANAL 13 TVUC
Jnes Matte Urrejula 0848
Santiago, Chile
T: 56 2 6323214 F: 56 2 302516

Scott, Maria Ximena Abogabir, Ms.
Board Assistant
INSTITUTO CHILENO DE RECONOCIMIENTO AMBIENTAL
Antonia Lope de Bello 024
Santiago, Chile
T: 56 2 374280 F: 56 2 777 5065

Seguel, Elena, Ms.
FEDERACION NACIONAL
Mujeu Negocios y Profissionals
Santiago, Chile
T: 56 2 6711228 F: 56 2 6711228

Sepulveda, Pastor Juan
SERVICIO EVANGELICO PARA EL DESARROLLO
Casilla 238, Correo III
Santiago, Chile

Shallat, Lezak, Ms.
Editor
ISIS INTERNATIONAL
Casila Postal 2067, Correo Central
Santiago, Chile
T: 56 2 33 4582

Simmonas, Luis, Mr.
CAPE HORN METHANOL
Apoquindo 3200 Piso 5
Santiago, Chile
T: 56 2 2316636 F: 56 2 2316638

Suarez, Laura Soledad, Ms.
Social Worker
CEANIM
Santiago, Chile
T: 56 2 330 514

Tapia, Ansellio Angueline Toro
Cameraman
TELEVISION NATIONAL DE CHILE
Beauvista 0990
Santiago, Chile
T: 7770900

Terrazas, Maria Angelica Rovira, Ms.
INSTITUTO DE ECOLOGIA POLITICA
Alberto Baines 1172 Nunoa
Santiago, Chile
T: 56 2 460-0474 F: 56 2 632-5738
TL: 240717

Tolorza, Erika Fuenzalida, Sra.
CORDEM
Erasmo Escala 23821
Santiago, Chile

Urrutia, Jose Antonio, Mr.
Lawyer
URRUTIA & CIA
6653 Hillandale Rd. Chevy Chase Md.
Washington - DC 20815, Chile
T: 1 301 656-9170 F: 1 202 656-9180

Valdebenito, Raul Venegas, Mr.
CENTRO DE EDUCACION Y TECNOLOGIA CET
Santiago, Chile

Vandorsee, Jaime, Mr.
CETAL
Abtao 567 Valparaiso
Valparaiso, Chile
T: 56 32 211500 F: 56 32 214851

Vargas, Patricia, Ms.
Vice-Presidente
ASS.POR LA CALIDAD DE VIDA
Concordia, 2237 - D 43 - Providencia
Santiago, Chile
T: 56 2 233 3706

Vasquez, Domitila Cuadra, Ms.
Biologa
CETAL
Abtao 576
Valparaiso, Chile
T: 56 32 21 3360

Veloso Aguilera, Eleodoro, Sr.
INSTITUTO DE CAPACITACION Y FORMACION JUVENIL
Palais de Nations
Geneve CH 1211, Chile
F: 41 22 733-6542

Verscheure, Hernan, Sr.
Director
CODEFF
Santa Filomena 185
Santiago, Chile
T: 56 2 377290 F: 56 2 377290

Weinstein, Soledad, Sra.
Coordinadora, Centro De Docume
ISIS INTERNATIONAL
Casila Postal 2067, Correo Central
Santiago, Chile
T: 56 2 33 4582

Yanez, Patricio Antonio, Sr.
CENTRO EL CANELO DE NOS
Casilla 2-D
San Bernardino, Chile

Zambrano, Mario Pena, Mr.
C O D E F F
Santa Filomena 185
Santiago, Chile
T: 56 2 377290 F: 56 2 377290

CHINA - GOVERNMENT

Bai, Xianhong
Deputy Director General
DEPARTMENT OF S & T FOR SOCIAL DEVE
54 Sanlihe Road Beijing,100862 SSTC
Beijing, China
T: 80 1847 F: 80 2594

Chang, Dongzhen, Mrs.
Second Secretary
Latin American Department, Ministry of Foreign Affairs
Permanent Mission to the United Nations, 155 W. 66th Str
New York, NY 10023, United States of America
T: 212 870 0300 F: 212 870 0333

Chen, Dazhen, Mr.
Deputy Director
Protocol Department, Ministry of Foreign Affairs
Permanent Mission to the United Nations, 155 W. 66th Str
New York, NY 10023, United States of America
T: 212 870 0300 F: 212 870 0333

Chen, Jian, Mr.
Deputy Director
International Organizations and Conferences Department, Ministry of Foreign Affiars
Permanent Mission to the United Nations, 155 W. 66th Str
New York, NY 10023, United States of America
T: 212 870 0300 F: 212 870 0333

Chen, Junshuo, Mr.
Deputy Director, World Bank Department
Ministry of Finance
Permanent Mission to the United Nations, 155 W. 66th Str
New York, NY 10023, United States of America
T: 212 870 0300 F: 212 870 0333

Chen, Qinchong, Mr.
Attache, Translation and Interpretation Department
Ministry of Foreign Affairs
Permanent Mission to the United Nations, 155 W. 66th Str
New York, NY 10023, United States of America
T: 212 870 0300 F: 212 870 0333

Chen, Zhenjie, Mr.
Deputy Director, Natural Resources Department
Ministry of Forestry
Permanent Mission to the United Nations, 155 W. 66th Str
New York, NY 10023, United States of America
T: 212 870 0300 F: 212 870 0333

Cui, Tiankai, Mr.
Division Chief International Organizations Department
Ministry of Foreign Affairs
Beijing 100701, China
T: 513 0368 F: 513 4521

Deng, Nan, H.E. Mrs.
Vice Minister in Charge of the State Science and Technology Commission
Permanent Mission of China
155 W. 66th Str
New York, NY 10023, United States of America
T: 212 870 0300 F: 212 870 0333

Dong, Furen, Mr.
Vice Chairman
Financial and Economic Committee of the National People's Congress
Permanent Mission to the United Nations, 155 W. 66th Str
New York, NY 10023, United States of America
T: 212 870 0300 F: 212 870 0333

Du, Bilan, Ms.
Deputy Director
Marine Strategy Institute of Bureau of Oceanography
Permanent Mission to the United Nations, 155 W. 66th Str
New York, NY 10023, United States of America
T: 212 870 0300 F: 212 870 0333

Gao, Guangsheng, Mr.
Deputy Division Chief
Land Department, State Planning Commission
Permanent Mission to the United Nations, 155 W. 66th Str
New York, NY 10023, United States of America
T: 212 870 0300 F: 212 870 0333

Gao, Hong, Mrs.
Third Secretary
Translation and Interpretation Department, Ministry of Foreign Affairs
Permanent Mission to the United Nations, 155 W. 66th Str
New York, NY 10023, United States of America
T: 212 870 0300 F: 212 870 0333

Gao, Kexiang, Mr.
Second Secretary
Latin American Department, Ministry of Foreign Affairs
Permanent Mission to the United Nations, 155 W. 66th Str
New York, NY 10023, United States of America
T: 212 870 0300 F: 212 870 0333

Gu, Keping, Mr.
Division Chief, Protocol Department
Ministry of Foreign Affairs
Permanent Mission to the United Nations, 155 W. 66th Str
New York, NY 10023, United States of America
T: 212 870 0300 F: 212 870 0333

Gu, Ming, Mr.
Vice Chairman
Law Committee of the National People's Congress
Permanent Mission to the United Nations, 155 W. 66th Str
New York, NY 10023, United States of America
T: 212 870 0300 F: 212 870 0333

Li, Jianying, Mr.
Deputy Director
Informational Department, Ministry of Foreign Affairs
Permanent Mission to the United Nations, 155 W. 66th Str
New York, NY 10023, United States of America
T: 212 870 0300 F: 212 870 0333

Li, Peng, H.E. Mr.
Premier of the State Council
The People's Republic of China

Liu, Huaqiu, H.E. Mr.
Vice Minister for Foreign Affairs
Permanent Mission of China
155 W. 66th Str
New York, NY 10023, United States of America
T: 212 870 0300 F: 212 870 0333

Liu, Jiang, H.E. Mr.
Vice Minister in Charge of the State Planning Commission
Permanent Mission of China
155 W. 66th Str
New York, NY 10023, United States of America
T: 212 870 0300 F: 212 870 0333

Liu, Shan, Mr.
Former Ambassador to European Community and Belgium
Permanent Mission of China
155 W. 66th Str
New York, NY 10023, United States of America
T: 212 870 0300 F: 212 870 0333

Lu, Congmin, Mr.
Secretary to the Premier
Permanent Mission of China
155 W. 66th Str
New York, NY 10023, United States of America
T: 212 870 0300 F: 212 870 0333

Luo, Jibin, Mr.
Deputy Director
State Meteorological Administration
Permanent Mission to the United Nations, 155 W. 66th Str
New York, NY 10023, United States of America
T: 212 870 0300 F: 212 870 0333

Mai, Guoyan, Mr.
Division Chief, Latin American Department
Ministry of Foreign Affairs
Permanent Mission to the United Nations, 155 W. 66th Str
New York, NY 10023, United States of America
T: 212 870 0300 F: 212 870 0333

Peng, Li, Mr.
Prime Minister
Permanent Mission of the Peoples Republic of China
155 W. 66th. Str.
New York NY 10023, United States of America
T: 212 870 0300 F: 212 870 0333

Qi, Hualyun, Mr.
Director General
Foreign Affairs Office of the State Council
Permanent Mission to the United Nations, 155 W. 66th Str
New York, NY 10023, United States of America
T: 212 870 0300 F: 212 870 0333

Qiu, Xiaoqi, Mr.
Deputy Division Chief
Latin American Department, Ministry of Foreign Affairs
Permanent Mission to the United Nations, 155 W. 66th Str
New York, NY 10023, United States of America
T: 212 870 0300 F: 212 870 0333

Qu, Geping
Administrator
NATIONAL ENVIRONMENTAL PROTECTION O
China No 115 Nanxiaojie,Xizhimennei
Beijing 10035, China
T: 6015641 6015642 F: 6015641 6011194

Shen, Yun'ao, H.E. Mr.
Ambassador Extraordinary and Plenipotentiary
Federative Republic of Brazil
Permanent Mission to the United Nations, 155 W. 66th Str
New York, NY 10023, United States of America
T: 212 870 0300 F: 212 870 0333

Shi, Mingde, Mr.
Deputy Division Chief
West European Department, Ministry of Foreign Affairs
Permanent Mission to the United Nations, 155 W. 66th Str
New York, NY 10023, United States of America
T: 212 870 0300 F: 212 870 0333

Song, Jian, H.E. Mr.
State Councillor and Minister in Charge of the State Science and Technology Commission
Permanent Mission of China
155 W. 66th Str
New York, NY 10023, United States of America
T: 212 870 0300 F: 212 870 0333

Song, Li, Mr.
Division Chief, Law and Treaty Department
Ministry of Foreign Affairs
Permanent Mission to the United Nations, 155 W. 66th Str
New York, NY 10023, United States of America
T: 212 870 0300 F: 212 870 0333

Su, Wei, Mr.
Deputy Division Chief
Law and Treaty Department, Ministry of Foreign Affairs
Permanent Mission to the United Nations, 155 W. 66th Str
New York, NY 10023, United States of America
T: 212 870 0300 F: 212 870 0333

Sun, Lin, Mr.
Director, Treaty and Law Department
Ministry of Foreign Affairs
Permanent Mission to the United Nations, 155 W. 66th Str
New York, NY 10023, United States of America
T: 212 870 0300 F: 212 870 0333

Tang, Yonggui, Mr.
Deputy Director
Latin American Department, Ministry of Foreign Affairs
Permanent Mission to the United Nations, 155 W. 66th Str
New York, NY 10023, United States of America
T: 212 870 0300 F: 212 870 0333

Wan, Bentai, Mr.
Secretary to State Councillor
Permanent Mission of China
155 W. 66th Str
New York, NY 10023, United States of America
T: 212 870 0300 F: 212 870 0333

Wan, Jingchang, Mr.
Deputy Director
Security Guards Bureau, Ministry of Public Security
Permanent Mission to the United Nations, 155 W. 66th Str
New York, NY 10023, United States of America
T: 212 870 0300 F: 212 870 0333

Wang, Gang, Mr.
Third Secretary
Latin American Department, Ministry of Foreign Affairs
Permanent Mission to the United Nations, 155 W. 66th Str
New York, NY 10023, United States of America
T: 212 870 0300 F: 212 870 0333

Wang, Huijun, Mrs.
Staff Member, Protocal Department
Ministry of Foreign Affairs
Permanent Mission to the United Nations, 155 W. 66th Str
New York, NY 10023, United States of America
T: 212 870 0300 F: 212 870 0333

Wang, Song, Mr.
Professor
Chinese Academy of Sciences
Permanent Mission to the United Nations, 155 W. 66th Str
New York, NY 10023, United States of America
T: 212 870 0300 F: 212 870 0333

Wang, Xinxia, Mr.
Second Secretary
Law and Treaty Department, Ministry of Foreign Affairs
Permanent Mission to the United Nations, 155 W. 66th Str
New York, NY 10023, United States of America
T: 212 870 0300 F: 212 870 0333

Wu, Hailong, Mr.
Second Secretary
Permanent Mission of China
155 W. 66th Str
New York, NY 10023, United States of America
T: 212 870 0300 F: 212 870 0333

Wu, Wenchang, Mr.
Secretary to the Premier
Permanent Mission of China
155 W. 66th Str
New York, NY 10023, United States of America
T: 212 870 0300 F: 212 870 0333

Wu, Xinjian, Mr.
Third Secretary
Protocol Department, Ministry of Foreign Affairs
Permanent Mission to the United Nations, 155 W. 66th Str
New York, NY 10023, United States of America
T: 212 870 0300 F: 212 870 0333

Wu, Zhenxiang, Mr.
First Secretary
Permanent Mission of China
Geneva
155 W. 66th Str New York, NY, 10023
F: 212 870 0300
TL: 212 870 0333

Xia, Guang, Mr.
Deputy Division Chief, Political Law Department
National Environmental Protection Agency
Permanent Mission to the United Nations, 155 W. 66th Str
New York, NY 10023, United States of America
T: 212 870 0300 F: 212 870 0333

Xie, Shaoxiong, Mr.
Director Foreign Affairs Department
Ministry of Engergy
Permanent Mission to the United Nations, 155 W. 66th Str
New York, NY 10023, United States of America
T: 212 870 0300 F: 212 870 0333

Xuan, Zengpei, Mr.
Deputy Division Chief
International Cooperation Department, State Science and Technology Commission
Permanent Mission to the United Nations, 155 W. 66th Str
New York, NY 10023, United States of America
T: 212 870 0300 F: 212 870 0333

Yang, Houdi, Mr.
Director, Political Law Department
State Nationalities Affairs Commission
Permanent Mission to the United Nations, 155 W. 66th Str
New York, NY 10023, United States of America
T: 212 870 0300 F: 212 870 0333

Yang, Jiechi, Mr.
Deputy Director
North American and Oceanian Department, Ministry of Foreign Affairs
Permanent Mission to the United Nations, 155 W. 66th Str
New York, NY 10023, United States of America
T: 212 870 0300 F: 212 870 0333

Yang, Jike, Mr.
Member of the Education, Culture and Public Health Committee of the National People's Congress
Permanent Mission of China
155 W. 66th Str
New York, NY 10023, United States of America
T: 212 870 0300 F: 212 870 0333

Yang, Xiuping, Ms.
Deputy Division Chief
Information Department, Ministry of Foreign Affairs
Permanent Mission to the United Nations, 155 W. 66th Str
New York, NY 10023, United States of America
T: 212 870 0300 F: 212 870 0333

Yao, Fei, Mr.
Division Chief
Information Office of State Council
Permanent Mission to the United Nations, 155 W. 66th Str
New York, NY 10023, United States of America
T: 212 870 0300 F: 212 870 0333

Ye, Ruqiu
Deputy Administrator
National Environmental Protection Agency
115 Xizhimennel Nanxiaojie
Beijing 100035, China
T: 86 1 6011180 F: 86 1 6011194

Yi, Chengde, Mr.
Deputy Division Chief
Policy Research Office, Ministry of Foreign Affairs
Permanent Mission to the United Nations, 155 W. 66th Str
New York, NY 10023, United States of America
T: 212 870 0300 F: 212 870 0333

Yu, Qingtai, Mr.
Deputy Division Chief
International Organization Department, Ministry of Foreign Affairs
Permanent Mission to the United Nations, 155 W. 66th Str
New York, NY 10023, United States of America
T: 212 870 0300 F: 212 870 0333

Zeng, Qingcun, Mr.
Professor
Chinese Academy of Sciences
Permanent Mission to the United Nations, 155 W. 66th Str
New York, NY 10023, United States of America
T: 212 870 0300 F: 212 870 0333

Zhang, Hanwu, Mr.
Second Secretary
Latin American Department, Ministry of Foreign Affairs
Permanent Mission to the United Nations, 155 W. 66th Str
New York, NY 10023, United States of America
T: 212 870 0300 F: 212 870 0333

Zhang, Yesui, Counselor
Permanent Mission of China
155 W. 66th Str
New York, NY 10023, United States of America
T: 212 870 0300 F: 212 870 0333

Zhong, Shukong, Mr
Special Adviser On E&D Issues
C/O INTERNATIONAL DEPARTMENT
Foreign Affairs Ministry
Beijing, China
T: 513 4521 BEIJING

Zou, Jingmeng, Mr.
Director
State Meteorological Administration
Permanent Mission to the United Nations, 155 W. 66th Str
New York, NY 10023, United States of America
T: 212 870 0300 F: 212 870 0333

Zou, Yijun, Mr.
Attache, Translation and Interpretation Department
Ministry of Foreign Affairs
Permanent Mission to the United Nations, 155 W. 66th Str
New York, NY 10023, United States of America
T: 212 870 0300 F: 212 870 0333

NGO

Chan, George, Mr.
ANTENA TECHONOLOGY
25-1 Chogro 5 Ga. Chongro-ku 110-125
Guangzhou, China
T: 82 2 741-7961 F: 82 2 7458006

Mao, Yang, Mr.
CFEJ
15A Xiaoxinglongjie Chongwen District
Beijing 100062, China
T: 86 1 701-3772 F: 86 1 701-3772

Qiuxin, Chen, Mr.
UNCED
15A Xiaoxinglongjie Chongwen District
Beijing 100062, China
T: 86 1 701-3772 F: 86 1 701-3772

Shuqin, Liu
CSES
No. 115 Xizhimennei Nanxiaojie
Beijing, China

Yu Cun, Lu
CSES
No. 115 Xizhimennei Nanxiaojie
Beijing, China

PRESS

Bao, Shishao
Correspondent
PEOPLE'S DAILY
SHIS QL 8 CJ8-Casa 10
CEP 71600 Brasilia-DF Brazil, China

Chen, Qiuxin
Correspondent
CHINA FORUM OF ENV. JOURNALISTS
14A Changjiadao Dongzhimennei
Beijing, China, China

Fang, Changping
Correspondent
XINHUA NEWS AGENCY
Tucuman 540, Piso 14 Dept D.
1049 Buenos Aires, Argentina, China

Geng, Qiu Zhan
Correspondent
XINHUA NEWS AGENCY
R. Paissandu 7 Apt.401
22210 Flamengo, Rio de Janeiro-RJ Brazil, China

Gu, Fan, Mr.
Tv Reporter
CHINA CENTRAL TELEVISION
11 Fuxing Road
Beijing 100859, China
T: 86 1 801-1144 F: 86 1 801-2010
TL: 20062

Guo, Linfa
Correspondent
XNA
Huay Ting 21, Beijing
China P.R. China, China

Hu, Jikang
Correspondent
XINHUA NEWS AGENCY
Qta Sinjua, Final Calle
Maracaibo, Caracas, Veenezuela, China

Li, Dong Sheng
Correspondent
CCTV
11 Fuking Rd.
Beijing, P.R. China, China

Li, Zhiming
Correspondent
PEOPLE'S DAILY
SHIS QL 8 - CJ8 Casa 10
71600 Brasilia-DF, Brazil, China

Liu, Dizhong
Editor
CHINA DAILY
Chaoyang District
Beijing, China, China

Liu, Zhenwu, Mr.
Reporter
SCIENCE AND TECHNOLOGY DAILY
No. 15 Funing Road
Beijing, China
T: 86 1 801-5544 F: 86 1 801-5026

Luo, Guanxing
Correspondent
CPBS
2 Fuxingmenwai St.
Beijing, China, China

Peng, Weixiang
Correspondent
CNS
Bai Wan Zhuang S. St.
Beijing, China, China

Wang, Chunrong
Correspondent
XINHUA NEWS AGENCY
Calle Francisco I. Madero, 17
Mexico 20 DF, Mexico, China

Wang, Jing
Correspondent
CCTV
11 Fuxing Rd.
Beijing, China, China

Wang, Junxian
Editor
CHINA CENTRAL TV
11, Fuxing St.
Beijing, P.R.China, China

Wang, Ning
Reporter Radio/Television
CCTV
Bldg. 46-7-20 Nan Li Shi Rd.
Beijing, P.R. China, China

Wang, Yifei
Correspondent
CCTV
3 Yang Fang Dian Rd.
Beijing, P.R. China, China

Wu, Yongheng
Correspondent
XINHUA NEWS AGENCY
R. Paissandu 7 Apt.401
22210 Flamengo, Rio de Janeiro-RJ Brazil, China

Xu, Yong
Correspondent
XINHUA NEWS AGENCY
1740 N 14th St.
Arlington, VA 22209, USA, China

Xu, Zhenglong
Correspondent
CHINA ENVIRONMENT NEWS
15 A Xinxinlongjie Chorgwen D.
Beijing 1000 P.R. China, China

Yang, Mao
Editor
CHINA FORUM OF ENV. JOURNALISTS
B102 Building 5 Fangqunyuan
Fangzhang, Beijing 100075 China, China

Yang, Shiguany
Reporter Radio/Television
CENTRAL PEOPLES' B'CASTING STATION
R 301 Building 9
Fuxingmennandajie, Beijing, China, China

Yu, Chengzhi
Correspondent
XINHUA NEWS AGENCY
40-35 72nd St.
New York, NY 11377, USA, China

Yu-Hui, Juan
Correspondent
RADIO BEIJING
Beijing RPCH
CEP 100866 China, China

Zeng, Hu
Correspondent
XINHUA NEWS AGENCY
25 Xisonhuan Beilu
Beijing, P.R. China, China

Zhang, Ning
Correspondent
XINHUA NEWS AGENCY
Calle Francisco I. Madero, 17
Mexico 20-DF, Mexico, China

GLOBAL FORUM

Chuangwen, Tong, Mr.
GAM GLOBAL 500 AWARD
Yingshang County, Xiaozhangzhuang Village
Yingshang 236001, China
T: 05687 2487

Dedi, Chen, Ms.
Director
ENVIRONMENTAL SCIENCE SOCIETY OF YONG AN CITY
Yong An Environmental Science Society
Yong An - Fujian 366000, China
T: 86 84 33730

Ding, Yihui, Mr.
Researcher
CHINESE METROLOGICAL SOCIETY
No. 46, Brishiqiao Road
Beijing, China
T: 86 1 831-2277 F: 86 1 331-1191

Dinis, Jose Rocha
Director
TRIBUNA DE MACAU
R. da Praia Grande 33 R/C
Macau, China
T: 8 53 378057 F: 8 53 337305

Fang, Chen, Mr.
Director
ENVIRONMENT PROTECTION AGENCY
Maanshan City, Anhui Province
Maanshan, Anhui, China

Feng, Xiuwen, Mr.
Deputy Director
PLANNING AND ECONOMIC COMMITTEE OF JILIN PROVINCE
11 Xin Fa Road
Chang Chun 130054, China
T: 824503 F: 820051

Guangde, Gao, Mr.
CHINA ENVIRONMENT NEWS
15 A Xiaoxinglongjie, Chongwen District
Beijing 100062, China
T: 86 1 701.2033 F: 86 1 701.3772
TL: 222359 Nepa CN

Han, Wenzao, Dr.
CHINA CHRISTIAN COUNCIL
17 Da Jian Yin Yiang
Nanjing, China

Jia-Shun, Chang
GAM GLOBAL 500 AWARD
Xiao Zhangzhuang Village
Yingshang, China
T: 05687 2487

Junxian, Wang, Mr.
Tv Editor
CHINA CENTRAL TELEVISION
11 Fuxing Road
Beijing 100859, China
T: 86 1 801-1144 F: 86 1 801-2010
TL: 20062

Lei, Xia
Editor
CHINA ENVIRONMENTAL NEWS
15 (A) XiaoXinlongjie Chongwen District
Beijing, China
T: 86 1 752478 F: 86 1 7013772

Liu, Shuqin, Mr.
Depart. Director
CHINESE SOCIETY FOR ENVIRONMENTAL SCIENCES
No. 115, Nanxianjie, Xizhimennei
Beijing 100035, China
T: 86 1 602-0069 F: 86 1 601-5641
TL: 222359

Lu, Yucun, Mr.
Secretary General
CHINESE SOCIETY FOR ENVIRONMENTAL SCIENCES
No. 115, Nanzinojie, Xizhimennei
Beijing 100035, China
T: 86 1 601-1184 F: 86 1 601-5641
TL: 222359

Oliveira, Carlos Jose Barbosa De
Redactor
TRIBUNA DE MACAU
R. da Praia Grande 33 R/C
Macau, China
T: 8 53 378057 F: 8 53 337305

Qiang, Suo, Mr.
Env. Manager
ENVIRONMENTAL PROTECTION SOCIETY OF TIBET
No. 61 Jurzhy Zhonglu Road
Lasa, China
T: 891 3362 F: 891 23418

Quixin, Chen, Mr.
CHINESE FORUM OF ENVIRONMENTAL JOURNALISTS
Palais de Nations
Geneve CH 1211, China
F: 41 22 733-6542

Qun, Chen Yu, Dr.
Professor
SHANGHAI ACADEMY OF SOCIAL SCIENCES
7/622 Huauhai Zhong Rd.
Shanghai, China

Shida, Wong, Mr.
Professor
CHINESE ASSOCIATION FOR SCIENCE AND TECHNOLOGY
Xueynan Road
Beijing 100863, China
F: 86 1 832-1914
TL: 20035

Shiguang, Yang
CENTRAL PEOPLE'S BROADCASTING STATION
Beijing, China

Shih-Mao, Li, Mr.
GAM GLOBAL 500 AWARD
xiao zhang zhuang village
china, China
T: 056 872487

Song, Hui-Gang, Mr.
Program Manager
CHINA WILKDLIFE CONSERVATION ASSOCIATION
Ministry of Forestry, Hepingli
Beijing, China
T: 86 1 421-6343 F: 86 1 421-4717

Sun, Qiulai, Mr.
Tv Lighting Engineer
CHINA CENTRAL TELEVISION
11 Fuxing Road
Beijing 100859, China
T: 86 1 801-1144 F: 86 1 801-2010
TL: 20062

Wang, Kaiyue, Mr.
Director
CHINESE SOCIETY OF SCIENCE AND TECHNOLOGY FOR SOCIAL DEVELOPMENT
No. 55 Bei Caigongzhuang
Beijing, China
T: 86 1 841-5521 F: 86 1 801-4854

Wenzao, Han, Dr.
CHINA CHRISTIAN COUNCIL
17 Da Jian Yin Yiang
Nanjing 1211, China

Wil, Liu Zhen
SCIENCE AND TECHNOLOGY DAILY
Beijing, China

Wu, Zong Xin, Mr.
Prof.
INET TSINGHUA UNIVERSITY
P.O.Box 1021
Beijing, China
T: 86 1 259-5230 F: 86 1 256-4177
TL: 22617 QHTSC-CN

Xi, Zha, Mr.
ENVIRONMENTAL PROTECTION SOCIETY OF TIBET
No. 61 Jurzhy Zhonglu Road
Lasa, China
T: 891 3362 F: 891 23418

Xie, Qimei, Mr.
President
UNITED NATIONS ASSOCIATION OF CHINA
71 Nanchizi StreetNo. 61 Jurzhy Zhonglu Road
Beijing, China
T: 86 1 512-0585 F: 86 1 513-1831
TL: 210435

Zhao, Jingzhu, Mr.
Research Professor
RESEARCH CENTER FOR ECO-ENIRONMENTAL SCIENCES - THE CHINESE ACADEMY OF SCIENCES
19 Zhongguancun Road
Beijing 100080, China
T: 86 1 2561872 F: 86 1 2562775

ADVISORS

Huang, Hua, Mr.
Former Foreign Minister Of China
Beijing 100035, China
T: 86 1 602 4553 F: 86 1 601 494

COLOMBIA - GOVERNMENT

Alzate, Fernando, Sr.
Sub-Director
Asuntos Multilaterales, Ministerio de Relaciones Exteriores
Permanent Mission to the United Nations, 140 E. 57th Str., 5th Floor New York, NY, 10022
F: 212 355 7776
TL: 212 371 2813

Arjona, Fabio, Sr.
Corporacion Rios San Juan y San Jorge
Permanent Mission to the United Nations, 140 E. 57th Str., 5th Floor New York, NY, 10022
F: 212 355 7776
TL: 212 371 2813

Casas, Fernado, Sr.
Asesor de la Cancilleria
Permanent Mission of Colombia
140 E. 57th Str., 5th Floor
New York, NY 10022, United States of America
T: 212 355 7776 F: 212 371 2813

Castano, Juanita, Sra.
Ministro Plenipotenciario de la Mision Permanente ante las Naciones Unidas, Nueva York
Permanent Mission of Colombia
140 E. 57th Str., 5th Floor
New York, NY 10022, United States of America
T: 212 355 7776 F: 212 371 2813

De Gaviria, Ana Milena Monoz, Sra.
Presidente de la Junta
Directiva del Instituto de Beinestar Familiar
Permanent Mission to the United Nations, 140 E. 57th Str., 5th Floor New York, NY, 10022
F: 212 355 7776
TL: 212 371 2813

Garces, Fernando, Sr.
Gobernacion del Valle
Permanent Mission of Colombia
140 E. 57th Str., 5th Floor
New York, NY 10022, United States of America
T: 212 355 7776 F: 212 371 2813

Gaviria, Cesar, E.S. Dr.
Presidente
Republica de Colombia
140 E. 57th Str., 5th Floor
New York, NY 10022, United States of America
T: 212 355 7776 F: 212 371 2813

Gaviria, Diana, Srta.
Departamento Nacional de Planeacion
Permanent Mission to the United Nations, 140 E. 57th Str., 5th Floor New York, NY, 10022
F: 212 355 7776
TL: 212 371 2813

Gomez, Ivan Dario, Sr.
Corporacion Rios Negro y Nare
Permanent Mission to the United Nations, 140 E. 57th Str., 5th Floor New York, NY, 10022
F: 212 355 7776
TL: 212 371 2813

Gonzalez, Andres, E.S.
Vice-Ministro
Relaciones Exteriores
Permanent Mission to the United Nations, 140 E. 57th Str., 5th Floor New York, NY, 10022
F: 212 355 7776
TL: 212 371 2813

Gonzalez, Guillermo Alberto, E.S.
Embajador de Colombia ante el gobierno de Brasil
Permanent Mission of Colombia
140 E. 57th Str., 5th Floor
New York, NY 10022, United States of America
T: 212 355 7776 F: 212 371 2813

Hurtado, Juan, Sr.
Representante a la Camara
Permanent Mission of Colombia
140 E. 57th Str., 5th Floor
New York, NY 10022, United States of America
T: 212 355 7776 F: 212 371 2813

Iriarte Diaz Granados, Patricia Lucia, Ms
Press Accredited
Q.A.P. TV NEWS
A.A. 40047
Bogota', Colombia
T: 433 0424 4137700 F: 4136076

Jaramillo, Hector, Sr.
Gerente de Acuavalle
Permanent Mission of Colombia
140 E. 57th Str., 5th Floor
New York, NY 10022, United States of America
T: 212 355 7776 F: 212 371 2813

Lopez, Alfonso, E.S.
Caballero
Ministro de Agricultura
Permanent Mission to the United Nations, 140 E. 57th Str., 5th Floor New York, NY, 10022
F: 212 355 7776
TL: 212 371 2813

Lopez, Jose Horacio, Sr.
Corporacion de Risaralda
Permanent Mission of Colombia
140 E. 57th Str., 5th Floor
New York, NY 10022, United States of America
T: 212 355 7776 F: 212 371 2813

Lopez, Oscar de Jesus, Sr.
Representante a la Camara
Permanent Mission of Colombia
140 E. 57th Str., 5th Floor
New York, NY 10022, United States of America
T: 212 355 7776 F: 212 371 2813

Lozano, Juan, Sr.
Consul General
Rio de Janeiro
Permanent Mission to the United Nations, 140 E. 57th Str., 5th Floor New York, NY, 10022
F: 212 355 7776
TL: 212 371 2813

Melendez, Ricardo, Sr.
Consejero
Mision Permanente ante las Naciones Unidas y los organismos internacionales, Ginebra
Permanent Mission to the United Nations, 140 E. 57th Str., 5th Floor New York, NY, 10022
F: 212 355 7776
TL: 212 371 2813

Montenegro, Armando, E.S.
Jefe
Departamento Nacional de Planificacion
Permanent Mission to the United Nations, 140 E. 57th Str., 5th Floor New York, NY, 10022
F: 212 355 7776
TL: 212 371 2813

Munera, Alfredo, Sr.
Departamento
Nacional de Planeacion
Permanent Mission to the United Nations, 140 E. 57th Str., 5th Floor New York, NY, 10022
F: 212 355 7776
TL: 212 371 2813

Penalosa, Enrique, E.S.
Embajador
Presidente de la Comision
Permanent Mission to the United Nations, 140 E. 57th Str., 5th Floor New York, NY, 10022
F: 212 355 7776
TL: 212 371 2813

Perea, Ezequiel, Sr.
Director
Corpes de la Amazonia
Permanent Mission to the United Nations, 140 E. 57th Str., 5th Floor New York, NY, 10022
F: 212 355 7776
TL: 212 371 2813

Quiceno, Gloria, Sra.
Representante a la Camara
Permanent Mission of Colombia
140 E. 57th Str., 5th Floor
New York, NY 10022, United States of America
T: 212 355 7776 F: 212 371 2813

Rodriguez, Manuel, Sr.
Gerente del Instituto Nacional de los Recursos Naturales y del Medio Ambiente (INDERENA)
Permanent Mission of Colombia
140 E. 57th Str., 5th Floor
New York, NY 10022, United States of America
T: 212 355 7776 F: 212 371 2813

Saenz, Orlando, Sr.
Experto
Colciencias
Permanent Mission to the United Nations, 140 E. 57th Str., 5th Floor New York, NY, 10022
F: 212 355 7776
TL: 212 371 2813

Sanin De Rubio, Noemi, E.S. Dra.
Ministra de Relaciones Exteriores
Permanent Mission of Colombia
140 E. 57th Str., 5th Floor
New York, NY 10022, United States of America
T: 212 355 7776 F: 212 371 2813

Santos, Manuel, E.S.
Ministro de Comercio Exterior
Permanent Mission of Colombia
140 E. 57th Str., 5th Floor
New York, NY 10022, United States of America
T: 212 355 7776 F: 212 371 2813

Sorzano, Luis Guillermo, Sr.
Senador
Colombian Senate
Congreso De Colombia of. 507 B
Bogota, Colombia
T: 57 1 284-4814 F: 57 1 284-8729

Trujillo, Cesar Gaviria, Mr.
President
Permanent Mission of Colombia
140 E. 57th Str., 5th floor
New York NY 10022, United States of America
T: 212 355 7776 F: 212 371 2813

Uribe, Eduardo, Sr.
Jefe
Division Especial de Politica Ambiental y Corporaciones Autonomas Regionales, Departamento Nacional de Planeacion
Permanent Mission to the United Nations, 140 E. 57th Str., 5th Floor New York, NY, 10022
F: 212 355 7776
TL: 212 371 2813

Vallejo, Nancy, Sra.
Secretaria General del INDERENA
Permanent Mission of Colombia
140 E. 57th Str., 5th Floor
New York, NY 10022, United States of America
T: 212 355 7776 F: 212 371 2813

Velsquez, Ramiro
Editor
EL COLOMBIANO-PERIODICO
Carrera 48 #30 Sur-119
Medellin, Colombia
T: 331 5252 F: 331 4858

NGO

Andrade, German Ignacio
FNC
Carrera 12 No. 70-96 Piso 3
Bogota, Colombia

Assis Aymone, Carlos
ALM
Apartado Aereo 39297
Bogota, Colombia

Borrero Navia, Jose, Mr.
Director
F I P M A
Apdo. Aereo 2741/ Calle 2a Oeste # 2-46
Cali, Colombia
T: 57 23 821723 F: 57 23 821723
E: ge02:fipma

Cabrera Rodriguez, Isidoro
UV
Apartado Aereo 25360
Cali, Colombia

Duque, Gabriel
ALM
Apartado Aereo 39297
Bogota, Colombia

Fonseca Barrera, Alegria
FA
Calle 100 - No. 11B-95 - OF 201
Bogota D.E., Colombia

Gomez Arias, Nelson
UV
Apartado Aereo 25360
Cali, Colombia

Gonzalez, Beatriz, Sra.
FUNDACION PARA LA DEFENSA DEL INTERES PUBLICO
Calle 71+5-83
Bogota, Colombia

Grijalba De Rodado, Elizabeth
RS
Calle 37 No. 22-31, Apartado Aereo 89252
Bogota, Colombia

Melandez, Gerardo
UJ
Unidad de Desarrollo Ambiental, Carrera 7 No. 40-6 2, Edificio 9, Santafe de Bogota, Colombia

Michaels Urbina, Maria Luisa, Sra.
Subdirectora
FUNDACION ALMA
Calle 100 no. 11B-95 Of 201
Santa Fe de Bogota, Colombia
T: 57 1 2180156 F: 57 1 2567339

Noriega, Consuelo
UJ
Unidad de Desarrollo Ambiental, Carrera 7 No. 40-6 2, Edificio 9, Santafe de Bogota, Colombia

Ruiz, Juan Pablo
FNC
Carrera 12 No. 70-96 Piso 3
Bogota, Colombia

Szauer, Maria Tereza
FPR
Carrera 10 No. 24-76 Oficina 305, Apartado Aereo 241 438, Santa Fe de Bogota, Colombia

Usme, Sara
ALM
Apartado Aereo 39297
Bogota, Colombia

Valinotti, Elisa
ALM
Apartado Aereo 39297
Bogota, Colombia

PRESS

Aguilar, Jose Alejandro
Director
RADIO VATICANA
A. A. 116
Guadalatara de Bula, Colombia
T: 57 222 71732 F: 57 222 72818

Caballero Velasco, Antonio Jose
Reporter Radio/Television
RADIO CADENA NACIONAL S.A.
Calle 142, No. 18-20
Colombia, Colombia

Cardenas De Hernandez, Martha
Editor
REVISTA ECOLOGICA
Calle 83, 19-10/301
Bogota, Colombia, Colombia

Child Velez, Jorge
Editor
EL ESPECTADOR
Ce 18 A-1-11
Bogota, Colombia, Colombia

Florez, Luis Bernardo
ESTRATEGIA
Carrera 3, No. 22-47, Apt.1202
Santa Fe de Bogota, Colombia, S.A., Colombia

Forero Monsalve, Luz Marina
Reporter Radio/Television
CADENA DE RADIO COLMUNDO
Carrera 32, No. 72-69
Colombia, Colombia

Galindo, Lina Patricia Lozano
Editor
AGENCIA DE NOTICIAS PRENSA VERDE
Banderas, Bloque D-4
Bogota, Colombia, Colombia

Gallo Rojas, Catalina
Reporter Radio/Television
PROGRAMAR TV
Carrera 14, #76-39, Apt. 501
Bogota, Colombia, Colombia

Garcia M., Rafael
Editor
AMAZONIA PUBLICACIONES
Fedecamaras P. Asis
Putumayo, Calle 12N S.Fe Bogota,Colombia, Colombia

Gomez, Ramiro V.
Editor
EL COLOMBIANO
Calle 41, No. 77-35
Medellin, Colombia, Colombia

Hernandez, Omar Ricardo Giraldo
Reporter Radio/Television
PROYECTAMOS TELEVISION
Calle 59, No. 8
Bogota, Colombia, Colombia

Iriarte, Patricia
Reporter Radio/Television
TV CARACOL
Carrera 12 (83-96)
Bogota, Colombia, Colombia

Melendez, Hernan Javier
Reporter Radio/Television
CV-TV INTERNACIONAL
Cra. 4, 87-93
Bogota, Colombia, Colombia

Melendez, Pedro
Editor
CV-TV INTERNACIONAL
Cra. 4, 87-93
Bogota, Colombia, Colombia

Parra Jaimes, William Eduardo
Reporter Radio/Television
NOTICIERO CMI
Diag. 22A 43-65
Bogota, Colombia, Colombia

Perez, Maytee
Editor
CAMARA TV PROD.
El Limon, Edif Castellania
El Cafetal, Venezuela, Colombia

Restrepo C., William
Reporter Radio/Television
CARACOL RADIO & TV
Trans. 33C 119-30 Bogota, Colombia, Colombia

Reyes, Constanza Cubillos
Editor
EL TIEMPO
Calle 39 Bis B No.28-07
Bogota, Colombia, Colombia

Schiavo, Patricia
Editor
CAMARA TV PROD.
Rodrigo Otavio, 171/301
Rio de Janiero, Brazil, Colombia

Vargas, Martha
Editor
CAMARA TV PROD.
SS East Nsth Street
N.Y. NY 10021 Usa, Colombia

Villa, Maria Eugenia
Reporter Radio/Television
PROGRAMAR TV
Calle 139 # 17-90 Int.7
Bogota, Colombia, Colombia

Young, Susanne
Correspondent
CAMARA TV PROD.
P.O. Box, 3351
Jackson, WY USA, Colombia

ORGANIZATIONS

Campo, Sara, Sra
Instituto de Investigaciones Ecologicas y Energeticas (ICECOL)
Apartado Aereo 050326
Barranquilla, Colombia
F: (57 58) 580 083

De Rodado, Elizabeth, Sra
RenaSer
Calle 37 No. 22-31, Apartado Aereo 89252
Bogota, Colombia
T: (57 1) 269 9737 F: (57 1) 268 2454

Escobar, Elsa Matilde, Sra
Fundacion Natura Colombia
Carrera 12 No. 70-96 Piso 3
Bogota, Colombia
T: (57 1) 310 0026 F: (57 1) 210 4515

Fonseca, Alegria, Sra.
Directora
FUNDACION ALMA
Calle 100 no. 11B-95 Of 201
Santa Fe de Bogota, Colombia
T: 57 1 2180156 F: 57 1 2567339

Maldonado, Juan Mayr, Sr.
Fundacion Pro-sierra Nevada De Santa Marta
Edificio Los Bancos of.602
Santa Marta, Colombia
T: (956) 34697 F: (956) 218 3256

Sarmiento, German, Mr.
Fundacion Para La Defensa Del Interes Publico (fundepublico)
Calle 71+5-83
Bogota, Colombia
T: (57 1) 210 4737/249 1891 F: (57 1) 210 4685

Walschburger, Thomas, Mr.
Fundacion Puerto Rastrojo
Carrera 10 No. 24-76 Oficina 305, Apartado Aereo 241 438, Santa Fe de Bogota, Colombia
T: (57 1) 284 9010/284 1951 F: (57 1) 284 3028

GLOBAL FORUM

Aguilar, Pe. Jose Alejandro
Director
INSTITUTO MAYOR CAMPESINO
Apartado 116
Buga (Valle) 1040, Colombia
T: 57 931 71-732 F: 57 1 268-7230

Alviar, Juan Adolfo Bermudez, Sr.
CODACOP
Daglfinger Strasse 67
Munchen D 8000 Munchen 81, Colombia
T: 49 89 936987 F: 49 89 74 88 07

Axdobidora, Roberto Dacsara, Mr.
GLOBAL ENVIRONMENT TEAM
Apartado Aereo 2158
Bucaramanga, Colombia
T: 57 76 430232

Beltran, Haydee, Mr.
Representative
UNIVERSIDAD DE ANTIOQUIA - INER
Apartado Aereo 11780
Medellin 11780, Colombia
F: 57 4 251-1641

Calderon, Carlos E., Mr.
MINISTERIO DE SALUD
Calle 55 ,#10-32
Bogota Cund., Colombia
T: 57 1 255 1204 F: 57 1 235 8577

Camacho, Hugo Lora, Mr.
FUNDACION FES
Calle 5 No. 6-05
Cali, Colombia
T: 57 27 845910 F: 57 27 834706

Caraball, Adelmo, Mr.
GRUPO DE INTEGRACION RURAL
Calle 14 No. 48 A-31
Cali, Colombia
T: 57 27 923-350270

Cardona, Gloria, Ms.
UNIV. DE ANTIOQUIA
Depto Quimica
Medellin, Colombia

Castro, Jose Horacio Lopez, Mr.
CARDER - CORPORACION AUTONOMA REGIONAL DE RISARALDA
Pereira, Colombia
T: 963 354153 F: 963 355501

Dias Ruiz, David, Mr.
Agronom0
CORPORACION SUNA-HISCA
Carrera 27 #73-28
Bogota, Colombia
T: 57 1 310934

Estrada, Julio Cesar, Sr.
ONIC
Apartado Aereo 32395
Santa Fe de Bogota, Colombia
TL: 411101 MLOS VT

Fonseca, Carlos, Mr.
UNIVERSIDAD DE LA SALLE - INGENIERIA AMBIENTAL
Carrera 2 # 10-70
Bogota, Colombia
T: 57 1 3414169

Gaviria, Luz Beatriz, Sra.
Coordenadora
CINEP
K5 No. 33A-08
Bogota, Colombia
T: 57 1 285-8977 F: 57 1 287-9089

Gil, Elsa Leonor Nivia, Sra.
Executive Director
RAPALMIRA
Calle 37 # 29-35
Palmira, Colombia
T: 57 31 52259 F: 57 31 35252
TL: 05761, E: Apartado 1440

Giron, Harold Walther Amezquita, Sr.
HOLON ASESORIA
Carrera 90 #68 g 11
Medellin, Colombia
T: 57 4 264-8263 F: 57 4 234-1703
E: A.A. 12630

Gutierrez, Lester
C V C
Carrera 56 #11-36
Cali 2366, Colombia
T: 57 27 392174

Guzman, Ivan, Mr.
CORNARE
Rio Negro
Antioquia, Colombia
T: 94 271 1859 F: 94 271 1879
TL: 7336 medelli

Hargrove, Thomas, Dr.
CIAT
Apartado Aereo 6743
Cali, Colombia
T: 57 23 675050 F: 57 23 647243

Hincapple, Fabio Arjona, Mr.
Director General
CORPORACION AUTONOMA REGIONAL DE LOS VALES DE SIMUM DE SAN JORGE
A. A. 355
Montenis, Colombia
T: 947 824501

Leon, Juanita, Ms.
Asesora
FUNDEPUBLICO
Cra 5 ap. 61-11
Bogota, Colombia
T: 57 1 681-3576

Lima, Alejandro, Mr.
FUNDACION FES
Calle 5 # 6-05
Cali 5744, Colombia
T: 57 27 822-524 F: 57 27 834-706

Lopes Mejia, Ana Isabel, Sra.
CORNARE
A.A. 185 Rionegro
Rionegro-Antioquia, Colombia
T: 942 71-3874

Mejia, Ignacio
Photographer
EL LATINAMERICANO
Ap. Aero 16364 Bogota
Bogota, Colombia
T: 3415404

Mejia, Mario, Mr.
CEPROID
Apartado Aereo 1188 - Palmira Valle
Colombia, Colombia

Palacio, German Sarmiento, Mr.
FUNDEPUBLICO
Calle 71 - No. 5 - 83
Bogota, Colombia
T: 57 1 210-4647 F: 57 1 210-4685

Pelaez, Mariluz Pelaez, Ms.
CORNARE
Apartado Aereo 185
Rio Negro, Antioquia, Colombia
T: 57 942 715902

Perez, Gabriel Roldan, Mr.
AMBIECOL LTDA
CRA 43 B 16-80, Oficina 602
Medellin, Colombia
T: 57 4 2688010 F: 57 4 2687473

Pineros, Patricia Suarez, Sra.
Co-Director
FOUNDATION 'AMIGOS DE VERDAD'
Tran 16 No. 7049 A.A. 50844
Bogota, Colombia
T: 57 1 217-6550 F: 57 1 255-8636

Prieto Castillo, Patricia, Sra.
ORIVA
Calle 10 #24-155 Colseguros
Cali, Colombia
T: 57 27 589071

Quijano, Rosario Ortiz-, Sra.
FUNDACION PRO-SIERRA NEVADA DE SANTA MARTA
Palais de Nations
Geneve CH 1211, Colombia
F: 41 22 733-6542

Reichel, Elizabeth D., Ms.
Conferencista
UNIVERSIDAD DE LOS ANDES - COLOMBIA
A.A.4976 - Dpto. Antropologia
Bogota AA4976, Colombia
F: 57 1 2841890
TL: 42343 UNAND C0

Rodriguez, Alvaro, Mr.
FUNDEPUBLICO
Cra 5 #61-11 Apto 401
Bogota, Colombia
T: 57 1 255-6092

Rudas, Olga Piedad, Ms.
Coordinator
EOSAT - PROSIS
carrera # 7 # 7469
Colombia

Suremain, Marie Dominique De
ENDA-THIRD WORLD
PO Box 3370
Dakar, Colombia
T: 221 222125 F: 221 222695
TL: 51456ENDATM, E: gn:endadakar/geo2:enda

Tejada, Gisela Delgado, Ms.
FUNDACION GRUPO ECOLOGICO
Carrera 8 # 4-31
Popayan, Colombia
T: 57 28 240712

Triana, Maria Claudia Pablom, Sra.
GIAC
Calle 4 # 1-14
Popayan, Colombia

Usher, Cristina Echavarria, Sra.
Co-Director
ETNOS T.V.
Rra. 50 #60-50
Medellin, Colombia
T: 57 4 245-2277 F: 57 4 311-0888
E: A.A.612

Valencia, Simon, Sr.
ONIC
Apartado Aereo 32395
Santa Fe de Bogota, Colombia
TL: 411101 MLOS VT

Velasco, Luz Marina Cortes, Ms.
FUNDACION GRUPO ECOLOGICO
Carrea 8 #4-31
Popayan, Colombia
T: 57 28 240712

Velasquez, Marta Cecilia, Ms.
Manager
ECO SOCIAL
Carrera 47 no. S7A47
Medellin, Colombia
T: 57 4 242-6889

Waysirsira, Fidel, Mr.
GLOBAL ENVIRONMENT TEAM
Apartado Aereo 2158
Bucaramanga, Colombia
T: 57 76 430232

Wiederkehr, Hans Peter, Mr.
SWISSAID COLOMBIA
Apartado Aereo No. 18456
Bogota, Colombia
T: 57 1 912432764 F: 57 1 912818525
E: GE02:SA-COL

ADVISORS

Marino De Botero, Margarita, Mrs.
Executive Director, El Collegio Verde : Former Member Of Wced
Bogota, Colombia
T: 57 1 243 8545/7512 F: 57 1 342 3237

Pastrana Borrero, Misael, Dr.
Former President Of The Republic Of Columbia
Santa Fe de Bogota, D.C., Colombia
T: 57 1 2 18 83 54/86 76 63 F: 57 1 2 57 11 44

PARLIAMENTARIANS / SPIRITUAL LEADERS

Galvis, Gustavo, Mr.
Senador
Colombian Senate
K 47 # 65-27
Bucaramanga, Colombia
T: 57 7 643-0037

Laserna, Mario, Senator
Senado de la Republica Capitolio Nacional
Bogota, Colombia
T: 2429833
TL: Senin Co 45 582

COMOROS - GOVERNMENT

Chouzour, Sultan, S.E. M.
Ambassadeur des Comores a Paris
Permanent Mission of Comoros
336 East 45th Street, 2nd Floor
New York, NY 10017, United States of America
T: 972 8010, 972 8042 F: 983 4712

Hachim, Said Hassane Said, S.E. M.
Ministre d' Etat, Ministre des Affaires Etrangeres et de la Cooperation (Chef de la Delegation)
Permanent Mission of Comoros
336 East 45th Street, 2nd Floor
New York, NY 10017, United States of America
T: 972 8010, 972 8042 F: 983 4712

Said Abdfifatait, Rahimat
Directica De La Coojeihou
MIUICTERE DES AFFAIRES
B P 428
Comoros
T: 19 269 73 21 13 F: 19 269 73 22 25

Salali, Jean Pierre, Mr
Delegue/Directern General De
B P GY6 Moroni Rep Fed Isl des Comores
Comoros
T: 73 06 08

CONGO - GOVERNMENT

Adouki, Martin, S.E. M.
Ambassadeur Extraordinaire et Plenipotentiaire
Representant Permanent du Congo aupres des Nations Unies, New York
Permanent Mission to the United Nations, 14 East 65th Street
New York, NY 10021, 744 7840, 7841, 7842
T: 744 7975

Bakdua, Batangouna, M.
Chef de Service des Traites et Conventions Internationaux au Ministere charge de la Justice
Permanent Mission of Congo
14 East 65th Street
New York, NY 10021, 744 7840, 7841, 7842
T: 744 7975

Ebondzo, Rigobert, M.
Directeur General des Eaux et Forets
Permanent Mission of Congo
14 East 65th Street
New York, NY 10021, 744 7840, 7841, 7842
T: 744 7975

Issanga-Ngamissimi, Marius, M.
Conseiller a la Cooperation au Ministere charge de la Science et de la Technologie
Permanent Mission of Congo
14 East 65th Street
New York, NY 10021, 744 7840, 7841, 7842
T: 744 7975

Kama, Joseph, M.
Chef de Service de L'Amenagement du Territoire a la Direction Generale du Plan
Permanent Mission of Congo
14 East 65th Street
New York, NY 10021, 744 7840, 7841, 7842
T: 744 7975

Moumba, Pascal, M.
Directeur des Etudes et de la Planification au Ministere charge de l'Agriculture
Permanent Mission of Congo
14 East 65th Street
New York, NY 10021, 744 7840, 7841, 7842
T: 744 7975

Nanga-Maniane, Jean
Directeur General Environnement
MINISTERE ENVIRONNEMENT
BP 958
Brazzaville, Congo
T: 242-833046 F: 242-837150

Ndouane, Dambert Rene
Ministre Charge Environnement
GOUVERNEMENT CONGO
BP 958 Brazzaville
Brazzaville, Congo
T: 242 836772 F: 242-832908

Nkeoua, Gregoire, M.
Conseiller a l'Environnement au Ministere charge de l'Environnement
Permanent Mission of Congo
14 East 65th Street
New York, NY 10021, 744 7840, 7841, 7842
T: 744 7975

Oyo, Pierre, Mr
Directour Du Droit Et De l'Education a l'Environnement
Directeur Generale de l'Environnement
B P 958
Brazzaville, Congo
T: 242 83 30 46 F: 242 83 71 50

Silou, Thomas, M.
Conseiller a l'Industrie
Commerce, Petites et Moyennes Entreprises, Science et Technologie a la Primature
Permanent Mission to the United Nations, 14 East 65th Street New York, NY, 10021
T: 744 7840, 7841, 7842 F: 744 7975

GLOBAL FORUM

Bankounda, Felix, Dr.
Brazzaville, Congo

Mafouari, P.
DIRECTION DES ETUDES ET DE LA PLANIFICATION
Brazzaville, Congo

Makelola, Marie-Agathe, Ms.
DIRECTION GENERAL DE LA RECHERCHE SCIENTIFIQUE
BP 2499te de Ferney
Brazzaville, Congo

Mankou Goma, Jean Claude, Mr.
PROGRAMME DES VOLONTAIRES DE NATION UNIES
9 Rue Ndouna, Qtier Kisito - Route du djoue
Brazzaville, Congo
TL: 5399 Kg

Onanga, Maurice, Mr.
Directeur General
DGRST
Brazzaville, Congo

COOK ISLANDS - GOVERNMENT

Gosselin, J., Mr.
Adviser to the Prime Minister
Permanent Mission of Cook Islands
United Nations Plaza
New York, NY 10017, United States of America
T: 212 963 7171 F: 212 963 1921

Henry, G.A., Hon.
Prime Minister
Cook Islands
United Nations Plaza
New York, NY 10017, United States of America
T: 212 963 7171 F: 212 963 1921

Ingram, P., Ms.
Permanent Mission of Cook Islands
United Nations Plaza
New York, NY 10017, United States of America
T: 212 963 7171 F: 212 963 1921

King, W., Mr.
Permanent Mission of Cook Islands
United Nations Plaza
New York, NY 10017, United States of America
T: 212 963 7171 F: 212 963 1921

Tangiiti, T., Mr.
Bishop
Permanent Mission of Cook Islands
United Nations Plaza
New York, NY 10017, United States of America
T: 212 963 7171 F: 212 963 1921

Tuara, Patricia Ngamata, Ms
Director Of Fisheries Research
MIN. OF MARINE RESOURCES
P.O. Box 85 Rarotonga
Cook Islands
T: 682 28 723 F: 682 29 721

GLOBAL FORUM

Ingram, Pamela, Dr.
President
COOK ISLANDS BUSINESS & PROFESSIONAL WOMENS ASSOC.
P.O. Box 933
Rarotonga Cibpwa, Cook Islands
T: 682 22608

COSTA RICA - GOVERNMENT

Araya, Carlos, E.S. Lic.
Embajador de Costa Rica en Brasil
Permanent Mission of Costa Rica
211 E. 43th Str., Room 903
New York, NY 10017, United States of America
T: 212 986 6373 F: 212 986 6842

Araya, Johnny, Lic.
Ejecutivo Municipal de San Jose
Permanent Mission of Costa Rica
211 E. 43th Str., Room 903
New York, NY 10017, United States of America
T: 212 986 6373 F: 212 986 6842

Bolanos, Manuel Antonio, Sr.
Diputado Asamblea Legislativa
Partido Unidad Social Cristiana
Permanent Mission to the United Nations, 211 E. 43th Str., Room 903
New York, NY 10017, United States of America
T: 212 986 6373 F: 212 986 6842

Bonilla, Alexander, Sr.
Asesor en Asuntos Ambientales
Permanent Mission of Costa Rica
211 E. 43th Str., Room 903
New York, NY 10017, United States of America
T: 212 986 6373 F: 212 986 6842

Borbon, Jose Maria, E.S.
Embajador
Alterno ante las Naciones Unidas, New York
Permanent Mission to the United Nations, 211 E. 43th Str., Room 903
New York, NY 10017, United States of America
T: 212 986 6373 F: 212 986 6842

Boza, Mario, S.E. Sr.
Viceministro de Recursos Naturales
Permanent Mission of Costa Rica
211 E. 43th Str., Room 903
New York, NY 10017, United States of America
T: 212 986 6373 F: 212 986 6842

Bravo, Hernan, E.S. Ing.
Ministro de Recursos Naturales
Energia y Minas
Permanent Mission to the United Nations, 211 E. 43th Str., Room 903
New York, NY 10017, United States of America
T: 212 986 6373 F: 212 986 6842

Budowski, Gerardo, Sr.
Tecnico en Bosques
Permanent Mission of Costa Rica
211 E. 43th Str., Room 903
New York, NY 10017, United States of America
T: 212 986 6373 F: 212 986 6842

Calderon, Rafael Anel, E.S. Lic.
Presidente
Republica de Costa Rica
211 E. 43th Str., Room 903
New York, NY 10017, United States of America
T: 212 986 6373 F: 212 986 6842

Campos, Max, Sr.
Representante del Instituto Meteorologico
Permanent Mission of Costa Rica
211 E. 43th Str., Room 903
New York, NY 10017, United States of America
T: 212 986 6373 F: 212 986 6842

Carvajal, Claudio Vinicio, Dr.
Diputado Asamblea Legislativa
Partido Unidad Social Cristiana
Permanent Mission to the United Nations, 211 E. 43th Str., Room 903
New York, NY 10017, United States of America
T: 212 986 6373 F: 212 986 6842

Casto, Fernando, Sr.
Secretario
Particular del Sr. Presidente
Permanent Mission to the United Nations, 211 E. 43th Str., Room 903
New York, NY 10017, United States of America
T: 212 986 6373 F: 212 986 6842

Chaves, Esau, Sr.
Representative de SIAPAZ
Permanent Mission of Costa Rica
211 E. 43th Str., Room 903
New York, NY 10017, United States of America
T: 212 986 6373 F: 212 986 6842

Conzalez, Marcos, Lic.
Diputado Asamblea Legislativa
Partido Liberacion Nacional
Permanent Mission to the United Nations, 211 E. 43th Str., Room 903
New York, NY 10017, United States of America
T: 212 986 6373 F: 212 986 6842

Cruz Alvarez, Alfredo, Dr.
Diputado Asamblea Legislativa
Partido Unidad Social Cristiana
Permanent Mission to the United Nations, 211 E. 43th Str., Room 903
New York, NY 10017, United States of America
T: 212 986 6373 F: 212 986 6842

De Niehaus, Gabriela, Sra.
Presidente Comision Ecologica deMinisterio de Relaciones Exteriores y Culto
Embajador en Mision Especial
Permanent Mission to the United Nations, 211 E. 43th Str., Room 903
New York, NY 10017, United States of America
T: 212 986 6373 F: 212 986 6842

Dulock, Jose, Sr.
Representante Indigenas
Permanent Mission of Costa Rica
211 E. 43th Str., Room 903
New York, NY 10017, United States of America
T: 212 986 6373 F: 212 986 6842

Eduardo, Peralta Ballester
Presidente
ASOCIACION COSTARRICENSE DE GEOGRAF
Apolo 3616/1000
San Jose, Costa Rica
T: 36 10 82 506 F: 506 31 47 37

Esau, Chaves Aguilar
Director Del Proyecto Siapaz
MINISTERIO DE RECURSOS NATURALES
Apartado 10104-1000
San Jose, Costa Rica
T: 506 57 09 22 F: 506 23 69 63

Fernandez, Mario, Sr.
Asesor del Ministerio de Ralaciones Exteriores
Permanent Mission of Costa Rica
211 E. 43th Str., Room 903
New York, NY 10017, United States of America
T: 212 986 6373 F: 212 986 6842

Hasmus, Erika, Srita.
Representante Juventud 92
Permanent Mission of Costa Rica
211 E. 43th Str., Room 903
New York, NY 10017, United States of America
T: 212 986 6373 F: 212 986 6842

Hernandez, Rita, Sra.
Subdirectora de Tratados
Permanent Mission of Costa Rica
211 E. 43th Str., Room 903
New York, NY 10017, United States of America
T: 212 986 6373 F: 212 986 6842

Jimenez, Norma, Sra.
Consul General en Rio de Janeiro
Permanent Mission of Costa Rica
211 E. 43th Str., Room 903
New York, NY 10017, United States of America
T: 212 986 6373 F: 212 986 6842

Kernick, Ana Maria, Dra.
Consejera
Permanent Mission of Costa Rica
211 E. 43th Str., Room 903
New York, NY 10017, United States of America
T: 212 986 6373 F: 212 986 6842

Levi, Ing. Samuel, Sr.
Consul de Costa Rica en Niteroi
Permanent Mission of Costa Rica
211 E. 43th Str., Room 903
New York, NY 10017, United States of America
T: 212 986 6373 F: 212 986 6842

Luthmer, Rodolfo, Sr.
Assessor del Presidente de la Republica
Permanent Mission of Costa Rica
211 E. 43th Str., Room 903
New York, NY 10017, United States of America
T: 212 986 6373 F: 212 986 6842

Molina, Enrique, Sr.
Embajador en Mision Especial
Permanent Mission of Costa Rica
211 E. 43th Str., Room 903
New York, NY 10017, United States of America
T: 212 986 6373 F: 212 986 6842

Monge, Victor Manuel, Sr.
Subdirector General de Politica Multilateral
Permanent Mission of Costa Rica
211 E. 43th Str., Room 903
New York, NY 10017, United States of America
T: 212 986 6373 F: 212 986 6842

Morgan, Victor H., Sr.
Asesor en Asuntos de Poblacion
Permanent Mission of Costa Rica
211 E. 43th Str., Room 903
New York, NY 10017, United States of America
T: 212 986 6373 F: 212 986 6842

Murillo, Manuel Maria, Sr.
Tecnico en Oceanografia
Permanent Mission of Costa Rica
211 E. 43th Str., Room 903
New York, NY 10017, United States of America
T: 212 986 6373 F: 212 986 6842

Niehaus-Quesada, Bernd, E.S. Lic.
Ministro de Relaciones Exteriores y Culto
Permanent Mission to the United Nations, 211 E. 43th Str., Room 903
New York, NY 10017, United States of America
T: 212 986 6373 F: 212 986 6842

Olsen de Figueres, Karen, Sra.
Diputado Asamblea Legislativa
Partido Luberacion Nacional
Permanent Mission to the United Nations, 211 E. 43th Str., Room 903
New York, NY 10017, United States of America
T: 212 986 6373 F: 212 986 6842

Peralta, Eduardo, Sr.
Colegio de Geografos
Permanent Mission of Costa Rica
211 E. 43th Str., Room 903
New York, NY 10017, United States of America
T: 212 986 6373 F: 212 986 6842

Porras, Ileana, Srita.
Assessora Juridica Ambiental
Permanent Mission of Costa Rica
211 E. 43th Str., Room 903
New York, NY 10017, United States of America
T: 212 986 6373 F: 212 986 6842

Rojas, Emil, Sr.
Representante Juventud 92
Permanent Mission of Costa Rica
211 E. 43th Str., Room 903
New York, NY 10017, United States of America
T: 212 986 6373 F: 212 986 6842

Rojas Lopez, Roberto, E.S. Ing.
Ministro de Comerico Exterior
Permanent Mission of Costa Rica
211 E. 43th Str., Room 903
New York, NY 10017, United States of America
T: 212 986 6373 F: 212 986 6842

Ruiz, Victor M., Dr.
Director Comision Nacional de Emergencias
Permanent Mission of Costa Rica
211 E. 43th Str., Room 903
New York, NY 10017, United States of America
T: 212 986 6373 F: 212 986 6842

Saenz, Melvin, Sr.
Director General Politica Exterior
Embajador en Mision Especial
Permanent Mission to the United Nations, 211 E. 43th Str., Room 903
New York, NY 10017, United States of America
T: 212 986 6373 F: 212 986 6842

Salas, Jose Luis, Mr
Asesor Ministro Recursos Naturales
SECRETARIA DE PLANIFICACION DE RECU
San Jose, Costa Rica
T: 506 25 05 09 34 65 0 F: 506 34 06 51

Sanabria, Rafael, Sr.
Doitadp Asamblea Legislativa
Partido Unidad Social Cristiana
Permanent Mission to the United Nations, 211 E. 43th Str., Room 903
New York, NY 10017, United States of America
T: 212 986 6373 F: 212 986 6842

Tattenbach, Cristian, E.S.
Embajador
Representante Permanente ante las Naciones Unidas, New York
Permanent Mission to the United Nations, 211 E. 43th Str., Room 903
New York, NY 10017, United States of America
T: 212 986 6373 F: 212 986 6842

Trejos, Raul, Sr.
Consejero
Permanent Mission of Costa Rica
211 E. 43th Str., Room 903
New York, NY 10017, United States of America
T: 212 986 6373 F: 212 986 6842

Ulloa, Erick, Sr.
Assessor del Ministerio de Recursos Naturales
Permanent Mission of Costa Rica
211 E. 43th Str., Room 903
New York, NY 10017, United States of America
T: 212 986 6373 F: 212 986 6842

Umana, Alvaro, Dr.
Miembro del Consejo Executivo de UNESCO
Permanent Mission of Costa Rica
211 E. 43th Str., Room 903
New York, NY 10017, United States of America
T: 212 986 6373 F: 212 986 6842

Valerio, Carlos, Sr.
Tecnico en Biodiversidad
Permanent Mission of Costa Rica
211 E. 43th Str., Room 903
New York, NY 10017, United States of America
T: 212 986 6373 F: 212 986 6842

Vargas, Marco Vinicio, Sr.
Director General de Protocolo
Embajador en Mision Especial
Permanent Mission to the United Nations, 211 E. 43th Str., Room 903
New York, NY 10017, United States of America
T: 212 986 6373 F: 212 986 6842

Wohlstein, Harry, Sr.
Director de Recoup
Permanent Mission of Costa Rica
211 E. 43th Str., Room 903
New York, NY 10017, United States of America
T: 212 986 6373 F: 212 986 6842

Yankelewitz, Samuel, Ing.
Director
Camara de Industrias
Permanent Mission to the United Nations, 211 E. 43th Str., Room 903
New York, NY 10017, United States of America
T: 212 986 6373 F: 212 986 6842

NGO

Arce Arias, Esmeralda
CEPESA
Apartado 1257-1002
San Jose, Costa Rica

Arias Rojas, Ronald
CEPESA
Apartado 1257-1002
San Jose, Costa Rica

Brenes, Lilly, Ms.
WORLD ORGANIZATION OF THE SCOUT MOVEMENT
55 Parkdale Ave, 3rd floor
Ottawa - Ontario, Costa Rica
T: 1 613 761 9206 F: 1 613 761 1441

Gutierrez Monzonis, Blanca
AAPA
Apartado 512 Tibas
1100 Costa Rica, Costa Rica

Penon De Arias, Margarita
FAPPPH
Apartado 8-6410
1000 San Jose, Costa Rica

Penon De Cotter, Maria Eugenia
FAPPPH
Apartado 8-6410
1000 San Jose, Costa Rica

Rojas Ortiz, Jose Dualok
ACSCR
Apartado 906-2150
Moravia, San Jose, Costa Rica

Sibar E Sibar, Pablo
ACSCR
Apartado 906-2150
Moravia, San Jose, Costa Rica

Sote Cruz, Miguel
AAPA
Apartado 512 Tibas
1100 Costa Rica, Costa Rica

PRESS

Cespedes, Chavarria William
Reporter Radio/Television
COSTA RICA-PRESS
Apdo. 4334-1000
San Jose, Costa Rica, Costa Rica

Das Neves, Joao Luiz Pereira
Correspondent
UNIVERSIDAD PARA LA PAZ
Apartado 138
Ciudad Colon, Costa Rica, Costa Rica

Gomez, Luis Eduardo Gomez
Correspondent
COSTA RICA-PRESS
C4 Av. 9 # 913, Apto. 4
San Jose, Costa Rica, Costa Rica

Latham, Debra
Reporter Radio/Television
RADIO FOR PEACE INTERNATIONAL
Ciudad Colon
Costa Rica, Costa Rica

Meloche, Francois
Correspondent
MESO AMERICA
140 Chester VMP
H3RIW2 Quebec, Canada, Costa Rica

Moreno, Johnny Fernandez
CAMARA NACIONAL DE RADIO
Apt. 1583-1002, Estudiantes
San Jose, Costa Rica, Costa Rica

Ramirez, Silvia F.
Reporter Radio/Television
CAMARA NACIONAL DE RADIO
Apartado 2582
1000 San Jose, Costa Rica, Costa Rica

Rojas, Olga Quesada
Reporter Radio/Television
TV-COSTA RICA
Lourdes, Montes de Oca
San Jose, Costa Rica, Costa Rica

Tinoco, Grettel
Reporter Radio/Television
MONUMENTAL
San Jose, Costa Rica, Costa Rica

Valverde, Bosco
Editor
LA NACION
Av.1a.calles 3.5
San Jose - Costa Rica, Costa Rica

Zuniga-Vega, Alejandra
Editor
LA NACION
Del Itan, 200 sur y 50 aeste
Montealegre, Zapote, Costa Rica, Costa Rica

ORGANIZATIONS

Gallegos, Ines, Sra
Fundacion de Parques Nacionales
Apartado 1108-1002
San Jose, Costa Rica
T: (506) 201 744 F: (506) 200 939

Harms, Erika, Sra
As. Pro-Conservation Acuatica de Costa Rica (APROCA)
Apartado 1863 (1002), Paseo los Estudiantes
San Jose, Costa Rica
T: (506) 553 365 F: (506) 553 365

Leon, Mario, Sr.
As. De Voluntarios En Investigacion Y Desarrollo Ambientalista (VIDA)
Apartado 7-0350-1000
San Jose, Costa Rica
T: (506) 211 411/364 274 F: (506) 572 273

Rios, Francisco Rivas, Sr.
Consorcio Ejecutor De Proyectos Economicos Y Sociales (CEPESA)
Apartado 1257-1002
San Jose, Costa Rica
T: (506) 34 9905 F: (506) 25 1239

Rojas, Jose Dualok, Sr.
As. Cultural Sejekto de Costa Rica
Apartado 906-2150
Moravia, San Jose, Costa Rica
T: (506) 361 427 F: (506) 402 275

Varela, Vera, Sra
Fundacion Neotropica
Apartado 236-1002
San Jose, Costa Rica
T: (506) 330 003 F: (506) 330 617

GLOBAL FORUM

Aguilar, Esau Chaves, Sr.
Director
MIRENEM
Apartado Postal 10124-1000
San Jose 1000, Costa Rica
T: 50 6 229260 F: 50 6 236963

Barahona, Francisco
UNIVERSIDADE PARA A PAZ DAS NACOES UNIDAS
Escazu, Costa Rica

Blanch, Jose, Mr.
WORLD VISION
PO BOX 133 - 2300 Curridabat
San Jose, Costa Rica
T: 50 6 53-2926 F: 50 6 24-7335

Castro, Rogelio A. Vazquez, Mr.
A M A
San Jose, Costa Rica

Comino, Silvana, Sra.
Advisor
C.A.T.I.E.
Apd. 126
Turrialba, Costa Rica
T: 506 56 64 38 F: 506 56 64 38

Cotter, Maria Eugenia Penon De, Ms.
Executive Director
FUNDACION ARIAS - IPAC
Apdo 8-6410-1000
San Jose, Costa Rica
T: 50 6 55-2885 F: 50 6 55-2244

Curling, Lena Patricia White, Sra.
FUNDACION ARIAS
P.O.Box 276-1001
San Jose, Costa Rica
T: 50 6 55 2955 F: 50 6 55 22 44

Ellis, Frank, Mr.
P.O. Box 3702-1000
San Jose, Costa Rica
T: 50 6 28.1787 F: 50 6 21.9127

Fallas, Douglas B. Campos, Mr.
UNITED NATIONS OF YOUTH
Watersnip 62, 1452 VE, Ilpendam
Alajuela, Costa Rica
T: 506 410698 F: 506 491929

Fernandez, Dora Romanach, Ms.
Assistant
UNIVERSIDAD NACIONAL
Apdp 1350-3000
Heredia, Costa Rica
T: 50 6 377039 F: 50 6 377036

Hall, Julie, Ms.
P.O. Box 3702-1000
San Jose, Costa Rica
T: 50 6 28.1787 F: 50 6 21.9127

Israel, Rodrigo Barahona, Mr.
Attorney
CEDARENA
Apartado Posta 1754 - 1000
San Jose, Costa Rica
T: 50 6 238463 F: 50 6 534750

Jimehez, Enrique, Mr.
Presidente
BOSQUE PUERTO CARILLO
C-25, Avs-6-8 #673
San Jose 131-1000, Costa Rica
T: 50 6 222818 F: 50 6 226613

Labouchelx, Jean, Mr.
Representative
IRAD / FRANCE
san jose
T: 506 255972 F: 506 250940
E: 11

Leon, Lic. Mario
VIDA
Palais de Nations
Geneve CH 1211, Costa Rica
F: 41 22 733-6542

Lopez, Olhedyis Zamora, Ms.
ESC. CIENCIAS AMBIENTALES UNIVERSIDAD NACIONAL SAN JOSE - COSTA RICA
San Juan de Tibas, apartado 8123
San Jose 1000, Costa Rica
T: 506 35-2270 F: 506 37-6465

Mendoza, Rolando, Dr.
WORLD COUNCIL OF CHURCHES
Apartado 1017
Montes de Oca 2050, Costa Rica

Mo, Claudette Lee, Ms.
Professor
REGIONAL WILDLIFE MANAGEMENT PROGRAM FOR MESOAMERICA AND CARIBBEAN
Universidad Nacional, Apartado 1350-3000
Heredia, Costa Rica
T: 50 6 377039 F: 50 6 377036

Montalvo, Jaime
UNIVERSIDADE PARA A PAZ DAS NACOES UNIDAS
Escazu, Costa Rica

Neves, Joao Luiz Pereira Da
UNIVERSIDADE PARA A PAZ DAS NACOES UNIDAS
Escazu, Costa Rica

Solano Sanchez, Jose Alejandro, Mr.
Licenciado
GOBIERNO COSTA RICA
275 Mts Al Sur Caia Costarricense del Seguro Social
San Jose, Costa Rica
T: 50 6 571 309 F: 50 6 267 301
TL: 1287-1002

Soto Cruz, Miguel, Mr.
Director
ARBOFILIA
Tibai - La Florida - Acera 7 Casa - Costa Rica
San Jose, Costa Rica
T: 50 6 355470 F: 50 6 408832

Trejos, Marta, Ms.
Executive Director
RED DE MUJERES Y MEDIO AMBIENTE - CENTRO AMERICA Y CARIBE
P. O. Box 68 C.A. 3355
San Jose, Costa Rica
T: 50 6 244620 F: 50 6 346875

Vargas, Auxiliadora, Ms.
RED DE MUJERES Y MEDIO AMBIENTE - CENTRO AMERICA Y CARIBE
El Caribe y Mexico/CEFEMINA
San Jose, Costa Rica
T: 50 6 24420 F: 50 6 346875

Vaughan, Christopher, Mr.
Director
REGIONAL WILDLIFE MANAGEMENT PROGRAM FOR MESOAMERICA AND CARIBBEAN
Universidad Nacional - Apartado 1350
Heredia, Costa Rica
T: 506 377039 F: 506 377036

Villalobos, Lilly Caridad Brenes, Ms.
WORLD ORGANIZATION OF THE SCOUT MOVEMENT
Calle 5, Avenida 26-28, Casa no.2675, Barrio San Cayatano
San Jose, Costa Rica
T: 50 6 332-6666 F: 50 6 26-1205

ADVISORS

Arias Sanchez, Oscar, H.E. Dr.
S-G, Centro Para La Paz Y La Reconciliacion : Former President Of Costa Rica
San Jose, Costa Rica
T: 506 55 2955 F: 506 55 2244

COTE D'IVOIRE - GOVERNMENT

Akele, Ezan, S.E. M.
Ministre de l'Environment de la Construction et de l'Urbanisme
Permanent Mission of Cote d'Ivoire
866 United Nations Plaza, Room 566
New York, NY 10017, United States of America
T: 212 371 7036 F: 212 935 5347

Amagou, Victor, M.
UVICOCI
Permanent Mission of Cote d'Ivoire
866 United Nations Plaza, Room 566
New York, NY 10017, United States of America
T: 212 371 7036 F: 212 935 5347

Anikpo N'Tame, Emilienne, Mme.
Coordinateur du Programme National d'Action Environnementale
Permanent Mission of Cote d'Ivoire
866 United Nations Plaza, Room 566
New York, NY 10017, United States of America
T: 212 371 7036 F: 212 935 5347

Barouan, Zeze, S.E. M.
Ambassadeur de Cote d'Ivoire au Bresil
Permanent Mission of Cote d'Ivoire
866 United Nations Plaza, Room 566
New York, NY 10017, United States of America
T: 212 371 7036 F: 212 935 5347

Beh, Belo, M.
Ministere
Communication
Permanent Mission to the United Nations, 866 United Nations Plaza, Room 566 New York, NY, 10017
F: 212 371 7036
TL: 212 935 5347

Bile, Boa, M.
Afrique Environnement
Permanent Mission to the United Nations, 866 United Nations Plaza, Room 566 New York, NY, 10017
F: 212 371 7036
TL: 212 935 5347

Brau, , Mme
CI ECOLOGIE
Permanent Mission to the United Nations, 866 United Nations Plaza, Room 566 New York, NY, 10017
F: 212 371 7036
TL: 212 935 5347

Brou, Kouakou N'Guehou Nathurin
Directeur De La Legislation Et De La Documentation
MINISTERE DE LA JUSTICE
B P V 107 Abidjan Cote
Abidjan, Cote d'Ivoire
T: 22 78 26

Brou, Oura, M.
Ministere de l'Agriculture et des Ressources Animales
Permanent Mission of Cote d'Ivoire
866 United Nations Plaza, Room 566
New York, NY 10017, United States of America
T: 212 371 7036 F: 212 935 5347

Cisse, Habout Dramane, M.
Directeur de la Protection de la Nature
Ministere de l'Agriculture et des Ressouces Animales
01 BP 4126
Abidjan 01, Cote d'Ivoire
T: 225 21 91 41 F: 225 21 46 18

Cissoko, Ahmed Souleymane, M.
Sous-Directeur de la Meteorologie, Representant Permanent de la Cote d'Ivoire a l'Organisation Meteorologie Mondiale
Permanent Mission of Cote d'Ivoire
866 United Nations Plaza, Room 566
New York, NY 10017, United States of America
T: 212 371 7036 F: 212 935 5347

Conde, Abdoulaye, M.
RECI
Permanent Mission to the United Nations, 866 United Nations Plaza, Room 566 New York, NY, 10017
F: 212 371 7036
TL: 212 935 5347

Cowpply, Philippe, Pr.
UVICOCI
Permanent Mission of Cote d'Ivoire
866 United Nations Plaza, Room 566
New York, NY 10017, United States of America
T: 212 371 7036 F: 212 935 5347

Deng, Lual Acuek
Division Chief, Environment & Socia
AFRICAN DEVELOPMENT BANK (DBA)
B.P. Box 1387
Abidjan 01, Cote d'Ivoire
T: 225 20 41 99 F: 225 20 49 07

Diarra, Boubakar, M.
Conseiller Technique du Ministre de l'Environnement
Permanent Mission of Cote d'Ivoire
866 United Nations Plaza, Room 566
New York, NY 10017, United States of America
T: 212 371 7036 F: 212 935 5347

Djedje, Bagno Magloire, Colonel
Directeur General Des Eaux Et Foret
BP V 94 Abidjan
Abidjan, Cote d'Ivoire
T: 22 16 29 43 12 87

Djouka, Anzeni, M.
Directeur de l'Eau, Ministere de l'Equipement des Transports et du Tourisme
Permanent Mission of Cote d'Ivoire
866 United Nations Plaza, Room 566
New York, NY 10017, United States of America
T: 212 371 7036 F: 212 935 5347

Ehivet, Kamon Germain, M.
JCE
Permanent Mission to the United Nations, 866 United Nations Plaza, Room 566 New York, NY, 10017
F: 212 371 7036
TL: 212 935 5347

Ekra, Alain, S.E. M.
Ministre de la Sante et de la Protection Sociale
Permanent Mission of Cote d'Ivoire
866 United Nations Plaza, Room 566
New York, NY 10017, United States of America
T: 212 371 7036 F: 212 935 5347

Essis, Kovame
Directeur Du Developpement Minier
Ministere des Mines et de l'Energie
01 BP 2816 Abidjan
Abidjan, Cote d'Ivoire
T: 225 44 29 94 F: 225 44 08 27

Essy, Amara
S.E. M.
Ministre des Affaires Etrangeres
Permanent Mission to the United Nations, 866 United Nations Plaza, Room 566 New York, NY, 10017
F: 212 371 7036
TL: 212 935 5347

Fanny, Inza, M.
Afrique Environnement
Permanent Mission to the United Nations, 866 United Nations Plaza, Room 566 New York, NY, 10017
F: 212 371 7036
TL: 212 935 5347

Gome, Hilaire, M.
Croix Verte
Permanent Mission of Cote d'Ivoire
866 United Nations Plaza, Room 566
New York, NY 10017, United States of America
T: 212 371 7036 F: 212 935 5347

Gomis, Charles Providence, S.E. M.
Ambassadeur de Cote d'Ivoire aux Etats-Unis d'Amerique
Permanent Mission of Cote d'Ivoire
866 United Nations Plaza, Room 566
New York, NY 10017, United States of America
T: 212 371 7036 F: 212 935 5347

Guede, Max, M.
Ministere
Communication
Permanent Mission to the United Nations, 866 United Nations Plaza, Room 566 New York, NY, 10017
F: 212 371 7036
TL: 212 935 5347

Guerrier, Klaus, M.
RECI
Permanent Mission of Cote d'Ivoire
866 United Nations Plaza, Room 566
New York, NY 10017, United States of America
T: 212 371 7036 F: 212 935 5347

Ibrama, Coulibaly, M.
Secretaire General Adjoint de la Caisse Autonome d'Amortissement
Permanent Mission of Cote d'Ivoire
866 United Nations Plaza, Room 566
New York, NY 10017, United States of America
T: 212 371 7036 F: 212 935 5347

Jankovic, Rok Sama
Mme
JCE
Permanent Mission to the United Nations, 866 United Nations Plaza, Room 566 New York, NY, 10017
F: 212 371 7036
TL: 212 935 5347

Kirine, Antoine Kouadio, S.E. M.
Ambassadeur de Cote d'Ivoire a Addis-Adeba
Representant Permanent aupres du PNUE
Permanent Mission to the United Nations, 866 United Nations Plaza, Room 566 New York, NY, 10017
F: 212 371 7036
TL: 212 935 5347

Konan, Lambert Kouassi, S.E. M.
Ministre de l'Agriculture et des Ressouces Animales
Permanent Mission of Cote d'Ivoire
866 United Nations Plaza, Room 566
New York, NY 10017, United States of America
T: 212 371 7036 F: 212 935 5347

Kouadio, Eloi, M.
Chef du Service des Etudes Environnementales et de la Programmation
Permanent Mission of Cote d'Ivoire
866 United Nations Plaza, Room 566
New York, NY 10017, United States of America
T: 212 371 7036 F: 212 935 5347

Kpan, Victor, M.
Ministere
Communication
Permanent Mission to the United Nations, 866 United Nations Plaza, Room 566 New York, NY, 10017
F: 212 371 7036
TL: 212 935 5347

Latto, Daniel, M.
Delegue de la Federation Ivoirienne de Scoutisme
Ministere de la Jeunesse et des Sports
Permanent Mission to the United Nations, 866 United Nations Plaza, Room 566 New York, NY, 10017
F: 212 371 7036
TL: 212 935 5347

Lounes, Ferhat
Vice-President Central Operations
BANQUEAST AFRICAINE DE DEVELOPEMENT
Afdev 01.BP. 1387
Abidjan, Cote d'Ivoire

Mansour, Franck, Dr.
RECI
Permanent Mission to the United Nations, 866 United Nations Plaza, Room 566 New York, NY, 10017
F: 212 371 7036
TL: 212 935 5347

N'Diaye, Zoubakyiron, M.
JCE
Permanent Mission to the United Nations, 866 United Nations Plaza, Room 566 New York, NY, 10017
F: 212 371 7036
TL: 212 935 5347

N'Takpe, N'Cho, M.
Conseiller
Representation Permanente aupres des Nations Unies, Geneve
Permanent Mission to the United Nations, 866 United Nations Plaza, Room 566 New York, NY, 10017
F: 212 371 7036
TL: 212 935 5347

Narcisse, Kouadio, M.
Premier Conseiller, Ambassade de Cote d'Ivoire au Bresil, Charge l'interpretation
Permanent Mission of Cote d'Ivoire
866 United Nations Plaza, Room 566
New York, NY 10017, United States of America
T: 212 371 7036 F: 212 935 5347

Ndiaye, Babacar
President
BANQUEAST AFRICAINE DE DEVELOPPEMENT
Afdev 01. BP. 1387
Abidjan, Cote d'Iyoire

Okossi, Akoa Benoit
Chef Dept Meteorologie Synoptique
ANAM
5 BP 990 Abidjan 5
Abidjan, Cote d'Ivoire
T: 225 27 90 04 F: 225 27 87 71

Ouattara, Largaton, M.
Sous-Directeur de la Cooperation Multilaterale
Ministere des Affaires Estrangeres
Permanent Mission to the United Nations, 866 United Nations Plaza, Room 566 New York, NY, 10017
F: 212 371 7036
TL: 212 935 5347

Oura, Brou
Conseiller Technique
CABINET DE MINISTRE AGRICULTURE ET
BP V 82 ABIDJAN
ABIDJAN, Cote d'Ivoire
T: 22 08 33 POSTE 569

Ouraga, Kako, M.
Directeur General du Plan
Permanent Mission of Cote d'Ivoire
866 United Nations Plaza, Room 566
New York, NY 10017, United States of America
T: 212 371 7036 F: 212 935 5347

Quattara, Alassane D., S.E. M.
Premier Ministre
Cote d'Ivoire
866 United Nations Plaza, Room 566
New York, NY 10017, United States of America
T: 212 371 7036 F: 212 935 5347

Sagou, Nicolas, M.
Ministere
Communication
Permanent Mission to the United Nations, 866 United Nations Plaza, Room 566 New York, NY, 10017
F: 212 371 7036
TL: 212 935 5347

Sangare, Yaya
Charge De Recherches
INSTITUT D'ECOLOGIE TROPICALE
06BP 299 CIDEX01
06 Aridjan, Cote d'Ivoire
T: 21 40 38

Saracino Tagliante, Jeanine, Mme
Directeur de la Sante Publique
Permanent Mission of Cote d'Ivoire
866 United Nations Plaza, Room 566
New York, NY 10017, United States of America
T: 212 371 7036 F: 212 935 5347

Soko, Zabi, M.
Directeur de l'Environnement
Permanent Mission of Cote d'Ivoire
866 United Nations Plaza, Room 566
New York, NY 10017, United States of America
T: 212 371 7036 F: 212 935 5347

Toure, Abdovlaye, M.
Directeur de la Technologie Industrielle
Ministere de l'Industrie et du Commerce
BP V65 Abidjan
Abidjan, Cote d'Ivoire
T: 2 15 59 F: 21 18 7

Toure, Sidya, M.
Directeur de Cabinet du Premier Ministre
Permanent Mission of Cote d'Ivoire
866 United Nations Plaza, Room 566
New York, NY 10017, United States of America
T: 212 371 7036 F: 212 935 5347

Vila, Jose
RECI
Permanent Mission of Cote d'Ivoire
866 United Nations Plaza, Room 566
New York, NY 10017, United States of America
T: 212 371 7036 F: 212 935 5347

NGO

Franck, Mansour Adeoti
RECI
01 BP 1843
Abidjan 01, Cote d'Ivoire

Koffi, Nari
RECI
01 BP 1843
Abidjan 01, Cote d'Ivoire

PRESS

Bailly, Diegou Jerome
Editor
NOTRE TEMPS
BP 276 Cidex III
Abidjan - Riviera, Ivory Coast, Cote d'Ivoire

Ballo, Beh
Editor
AGENCE IVOIRIENNE DE PRESSE (AIP)
04 BP 312
Abidjan, Ivory Coast, Cote d'Ivoire

Kakore, Kouassi
Editor
IVORY COAST PRESS
Abidjan BP v 138
Abidjan, Cote d'Ivoire

N'Guessan, Komenan Raphael
Editor
FRATERNITE MATIN
17 BP 226
Abidjan, Ivory Coast, Cote d'Ivoire

Nouama, Guy Pierre
Editor
LE DEMOCRATE
Cocody Les Deux Plateaux
Abidjan, Ivory Coast, Cote d'Ivoire

Sagou, Ntobou Nicolas
Reporter Radio/Television
RADIODIFFUSION IVOIRIENNE
B.P. V191
Abidjan 01, Cote d'Ivoire

Zeze Guezi, Norberte
Editor
LE VOIE
17 BP 656
Abidjan 17, Ivory Coast, Cote d'Ivoire

ORGANIZATIONS

Ladji, Meite, Mr.
Ligue International Pour La Promotion De La Sante Communautaire (L.I.E.P.S.C.)
01 Boite Postale 1843
Abidjan 01, Cote d'Ivoire
T: (225) 24 0045 F: (225) 32 6242

GLOBAL FORUM

Adeoti, Mansour, Mr.
LIGUE INTERNATIONALE D'ETUDE E DE PROMTION DE LA SANTE COMMUNAUTAIRE
BP 184
Abidjan 01, Cote d'Ivoire
T: 22 5 24 00 45 F: 22 5 24 20 33

Aminata, Traore, Ms.
Regional Coordinator
PROWWESS/AFRICA
06 bp 941
Abidjan, Cote d'Ivoire
T: 22 5 41-5895 F: 22 5 21-1367

Awa, Sylla, Ms.
Consultant
PROWWESS/AFRIQUE
04 bp 498
Abidjan, Cote d'Ivoire
T: 22 5 41-4264

Bouali, Mathieu Bile, Dr.
FACULTE DE DROIT - ABIDJAN
Abidjan, Cote d'Ivoire

Coulibaly, Djeneba
Assistente Social
PROWWESS/ AFRIQUE
06 bp 941 abidjan 06
Abidjan, Cote d'Ivoire
T: 22 5 41-5895

Hilaire, Gome, Mr.
CROIX VERTE
02 BP699
Abidjan, Cote d'Ivoire
T: 22 5 410450

Koffi, N'dri, Mr.
U A D E
01 BP 1843
Abidjan, Cote d'Ivoire
T: 22 5 241443 F: 22 5 241033
TL: 23395

Komenan, Guessan Franck, Mr.
UNION INTERNATIONAL DE LA JEUNESSE SOCIALITE
22 BP. 302 Abitian
Abitian, Cote d'Ivoire
T: 353550 F: 0225 221918

Musa, El Sadig, Mr.
Division Chief
AFRICAN DEVELOPMENT BANK
PO Box V316
Abidjan, Cote d'Ivoire
T: 22 5 20-41-50 F: 22 5 20-49-07

Savadogo, , Ms.
COTE D'IVOIRE ECOLOGIE
15 BP 950
Abidjan, Cote d'Ivoire
T: 22 5 26 83 25 F: 22 5 24 86 44

Traore, Madeleine, Ms.
Resp.Adm
SADEM
Dibo 1843 Abj 01
Abidjan, Cote d'Ivoire
T: 22 5 95-3780 F: 22 5 45-0916
TL: 23395

CROATIA - GOVERNMENT

Greguric, Franjo, H.E. Dr
Prime Minister
Republic of Croatia
551 5th Ave., 35th Floor
New York, NY 10176, United States of America
T: 212 986 9696

Grubisa, Danir, Mr.
Advisor to the Minister for Foreifn Affairs
Permanent Mission of Croatia
551 5th Ave., 35th Floor
New York, NY 10176, United States of America
T: 212 986 9696

Kosty, Tito, Mr.
Advisor
Ministry of Environment
Permanent Mission to the United Nations, 551 5th Ave., 35th Floor New York, NY, 10176
F: 212 986 9696

Movcan, Josip, Mr.
Expert
Ministry of Environment
Permanent Mission to the United Nations, 551 5th Ave., 35th Floor New York, NY, 10176
F: 212 986 9696

Muck, Visnja Jelic, Mrs.
Advisor
Ministry of Environment
Permanent Mission to the United Nations, 551 5th Ave., 35th Floor New York, NY, 10176
F: 212 986 9696

Noblio, Viktor, Dr.
Assistant Minister
Ministry of Environment
Permanent Mission to the United Nations, 551 5th Ave., 35th Floor New York, NY, 10176
F: 212 986 9696

Sasparovic, Franjo, Dr.
Expert
Ministry of Environment
Permanent Mission to the United Nations, 551 5th Ave., 35th Floor New York, NY, 10176
F: 212 986 9696

Separovic, Zvonimir, H.E. Dr.
Minister of Foreign Affairs
Permanent Mission of Croatia
551 5th Ave., 35th Floor
New York, NY 10176, United States of America
T: 212 986 9696

Simoncic, Viktor, Dr.
Assistant Minister
Ministry of Environment
Permanent Mission to the United Nations, 551 5th Ave., 35th Floor New York, NY, 10176
F: 212 986 9696

CUBA - GOVERNMENT

Castro Ruz, Fidel, Mr.
President
Permanent Mission of Cuba
315 Lexington Ave.
New York NY 10016, United States of America
T: 212 689 7215 F: 212 779 1697

PRESS

Barrera Lopez, Nelson
Reporter Radio/Television
EMISORA RADIO RELOJ NACIONAL
Calle 15, 251
Vedado, Cuba, Cuba

Casals Martinez, Melchor
Editor
INSTITUTO CUBANO DEL ARTE E IND.CIN.
Avenida 33 no. 2616
La Habana, Cuba, Cuba

Chao Mesa, Mario Raul
Editor
INSTITUTO CUBANO RADIO/TV
Avenida 31/6608
La Habana, Cuba, Cuba

Contreras Castro, Orlando
Reporter Radio/Television
RADIO REBELDE
Calle 4 no. 214
La Habana, Cuba, Cuba

Fellu Miranda, Santiago Ronald
Editor
SEMANARIO JUVENTUD REBELDE
Trinidad, 23 Vibora Park
Ciudad Habana, Cuba, Cuba

Forteza Martin, Francisco
Editor
PRENSA LATINA
23 No. 201
Vedado, La Habana, Cuba, Cuba

Gonzalez Albrar, Victor M.
Editor
PERIODICO GRANMA
26, 897 Plaza,
Habana, Cuba, Cuba

Gonzalez Guethon, Alberto
Reporter Radio/Television
INSTITUTO CUBANO RADIO/TV
La Habana, Cuba

Iglesias Celestrin, Carlos
Reporter Radio/Television
TELEVISION CUBANA
Avenida Poruenir 624, Lawton
La Habana, Cuba, Cuba

Nunez Dorta, Nestor Pedro
Editor
REVISTA BOHEMIA
Edificio E no. 25
Alamar, La Habana, Cuba

Otero Cabanas, Pedro Manuel
Reporter Radio/Television
RADIO HABANA CUBA
Calle 72, Playa
La Habana, Cuba, Cuba

Robreno Dolz, Gustavo Felix
Editor
PERIODICO GRANMA
Calle H no. 311
La Habana, Cuba, Cuba

Rubio Perez, Gladys Hortensia
Reporter Radio/Television
TELEVISION CUBANA
Calle 22 no. 257, Vedado
La Habana, Cuba, Cuba

GLOBAL FORUM

Ayerra, Carolina Aguilar, Sra.
Member
FMC
Paseo 26 esq. 13 - Vedado
La Habana, Cuba
T: 53 7 37927 F: 53 7 333019

Carcano, Dora, Sra.
FDIM
Calle 20 no 117
La Habana, Cuba
T: 53 7 23-6742

Castro, Sara Maria Smith, Ms.
Directora
INST. CUBANO DE AMISTAD CON LOS PUEBLOS
17 no.301 esq. I Vedado
Havana 6431, Cuba
T: 51 44 327601 F: 51 44 333041
TL: 511234

Fernandes, Regla Dolores, Sra.
Jefa Dept. Politico
OSPARAL
Calle 29 Y C Vedado
Habana, Cuba
T: 305510
TL: 512259

Gomez, Yolanda Ferrer, Sra.
General Secretary
FMC
Paseo 26 esq. 13 - Vedado
La Habana, Cuba
T: 53 7 39931 F: 53 7 333019

Lezcano, Leonor Rodriguez, Sra.
Member
FMC
Paseo 26 esq. 13 - Vedado
La Habana, Cuba
T: 53 7 37927 F: 53 7 333019

Valladares, Luis Miguel Morejon, Dr.
Vice-President
INST. CUBANO DE AMISTAD CON LOS PUEBLOS
17#301 Vedado Habana
Habana, Cuba
T: 32 7601 F: 33 3041
TL: 511-234

PARLIAMENTARIANS / SPIRITUAL LEADERS

Gomes, Yolanda Ferrer, Dep.
Asamblea Nacional del Poder Popular 42 No 2308 Playa
Ciudad Habana, Cuba
T: 22 2535
TL: 051 1172

CURACAO - GOVERNMENT

De Jong-Mercelina, Joelle
Head International Organizations
DIVISION OF THE FOREIGN AFFAIRS OFF
Fort Amsterdam Y
Curacao
T: 00 5999 613933 F: 00 5999 617123

Richardson, Quirino Beltran
Head Of The Environmental Service
Environmental Service
Derouvilleweg
Curacao
T: 5999 626525 F: 5999 626654

CYPRUS - GOVERNMENT

Cleopas, Myrna, Mrs.
Counsellor
Ministry of Foreign Affairs
Permanent Mission to the United Nations, 13 E. 40th Str.
New York, NY, 10016
F: 212 481 6023
TL: 212 685 7316

Dionyssiou, Christakis, Mr.
Aide de Camp
Permanent Mission of Cyprus
13 E. 40th Str.
New York, NY 10016, United States of America
T: 212 481 6023 F: 212 685 7316

Gavrielides, Andreas, H.E. Mr.
Minister
Agriculture and Natural Resources (Deputy Head of Delegation)
Permanent Mission to the United Nations, 13 E. 40th Str.
New York, NY, 10016
F: 212 481 6023
TL: 212 685 7316

Georgiades, Nicos, Mr.
Senior Land Consolidation Officer
Ministry of Agriculture and Natural Resources
Permanent Mission to the United Nations, 13 E. 40th Str.
New York, NY, 10016
F: 212 481 6023
TL: 212 685 7316

Iacovou, George, H.E. Mr.
Minister for Foreign Affairs (Deputy Head of Delegation)
Permanent Mission of Cyprus
13 E. 40th Str.
New York, NY 10016, United States of America
T: 212 481 6023 F: 212 685 7316

Stavrinos, Michalis, Mr.
Second Secretary
Permanent Mission of Cyprus
13 E. 40th Str.
New York, NY 10016, United States of America
T: 212 481 6023 F: 212 685 7316

Stavrou, Patroclos, H.E. Mr.
Under-Secretary
President
Permanent Mission to the United Nations, 13 E. 40th Str.
New York, NY, 10016
F: 212 481 6023
TL: 212 685 7316

Vassiliou, George, H.E. Mr.
President
Permanent Mission of the Republic of Cyprus
13 E. 40th Str.
New York, NY 10016, United States of America
T: 212 481 6023 F: 212 685 7316

PRESS

Jammal, Laila
Correspondent
FALASTINE AL-THAWRA
7104 - 14th Ave.
Brooklyn N.Y. 11228, Cyprus

GLOBAL FORUM

El Guyind, Fatma, Ms.
ENDA
Alard Publishers Ltd.
Limassol 3074, Cyprus
T: 357 5 366494 F: 357 5 369928

Habib, Vera, Ms.
ENDA
Alard Publishers Ltd P.O.Box 3074
Limassol 3074, Cyprus
T: 353 61 366494 F: 353 61 369928
TL: 6251AZ

CZECHOSLOVAKIA - GOVERNMENT

Adamec, Josef, Mr.
Charge d' affaires a.i.
Federative Republic of Brazil and the Republic of Suriname
Permanent Mission to the United Nations, 1109-1111
Madison Avenue New York, NY, 10021
F: 535 8814, 8815, 8818
TL: 772 0586

Bucek, Milan, Mr.
Deputy Chairman
Commission for Economic Strategy of Slovak Republic
Permanent Mission to the United Nations, 1109-1111
Madison Avenue New York, NY, 10021
F: 535 8814, 8815, 8818
TL: 772 0586

Huba, Mikolas, Mr.
Deputy
Slovak National Council
Permanent Mission to the United Nations, 1109-1111
Madison Avenue New York, NY, 10021
F: 535 8814, 8815, 8818
TL: 772 0586

Kara, Jan, Ms
Advisor
FEDERAL COMMITTEE FOR THE ENVIRONME
Slezska 9 120 29 PRAHA2
Praha, Czechoslovakia
T: 42 2 255226 F: 42 2 259042

Miklos, Ladislav
Vice-Minister Of Environment
Slovak Commission for Environment
Hibokd 2
81235 Bratislava, Czechoslovakia
T: 42 7 492553 F: 42 7 311384

Micoch, Slavomir, Mr.
Deputy Minister
Environment of the Czech Republic
Permanent Mission to the United Nations, 1109-1111
Madison Avenue New York, NY, 10021
F: 535 8814, 8815, 8818
TL: 772 0586

Moldan, Bedrich, Mr.
Deputy
Czech National Council
Permanent Mission to the United Nations, 1109-1111
Madison Avenue New York, NY, 10021
F: 535 8814, 8815, 8818
TL: 772 0586

Rynda, Ivan, Mr.
Deputy
Federal Assembly
Permanent Mission to the United Nations, 1109-1111
Madison Avenue New York, NY, 10021
F: 535 8814, 8815, 8818
TL: 772 0586

Sepelak, Pavol, Mr.
Second Secretary
Permanent Mission of Czechoslovakia
1109-1111 Madison Avenue
New York, NY 10021, United States of America
T: 535 8814, 8815, 8818 F: 772 0586

Tirpak, Ivan, Mr.
Minister-Chairman
Slovak Commission for Environment
Permanent Mission to the United Nations, 1109-1111
Madison Avenue New York, NY, 10021
F: 535 8814, 8815, 8818
TL: 772 0586

Vavrousek, Josef, Mr.
Minister-Chairman
Federal Committee for Environment of the Czech and Slovak Federal Republic (Head of Delegation)
Permanent Mission to the United Nations, 1109-1111
Madison Avenue New York, NY, 10021
F: 535 8814, 8815, 8818
TL: 772 0586

Vurm, Karel, Mr.
Czech and Slovak Confederation of Trade Unions
Permanent Mission to the United Nations, 1109-1111
Madison Avenue New York, NY, 10021
F: 535 8814, 8815, 8818
TL: 772 0586

Zlocha, Jan, Mr.
Deputy
Federal Assembly
Permanent Mission to the United Nations, 1109-1111
Madison Avenue New York, NY, 10021
F: 535 8814, 8815, 8818
TL: 772 0586

NGO

Arias Diaz, Jorge
IUS
17th November Street, PO Box 58
110 01 Prague 01, Czechoslovakia

Bernardes, Luis Carlos
IOJ
Celetna 2
Prague 11001, Czechoslovakia

Hollan, Jan
VEC
PO Box 91
601 91 Brno 1, Czechoslovakia

Kundrata, Miroslav
VEC
PO Box 91
601 91 Brno 1, Czechoslovakia

Mpuru, Moeti, Ms.
Secretary General
INTERNATIONAL UNION OF STUDENTS
58, 17th November Street
Prague, Czechoslovakia
T: 42 2 231-2812 F: 42 2 231-6100
TL: 122858

Rollemberg, Armando Sobral
IOJ
Celetna 2
Prague 11001, Czechoslovakia

Salles, Paulo Sergio
IOJ
Celetna 2
Prague 11001, Czechoslovakia

Tewari, Manish
IUS
17th November Street, PO Box 58
110 01 Prague 01, Czechoslovakia

Woods, Zonny
IUS
17th November Street, PO Box 58
110 01 Prague 01, Czechoslovakia

PRESS

Bouda, Pavel
Correspondent
CZECHOSLOVAK TV
Prague 6
Na dlouhem lenu 16, Czechoslovakia, Czechoslovakia

Karel, Wichs
Correspondent
RADIO ESTATAL CHECOSLOVACA
Praga 5 Banacicova, 19
Checoslovaquia, Czechoslovakia

Majstr, Jiri
Correspondent
CZECHOSLOVAK NEWS AGENCY
40 River Road (#19C)
New York, NY 10044, USA, Czechoslovakia

ORGANIZATIONS

Gatinot, Gerard, Mr.
International Organization Of Journalists
Celetna 2
Prague 11001, Czechoslovakia
T: 236 3932/236 4496/236 3480 F: (42 2) 232 0426

Hamernik, Emil, Mr.
World Federation Of Trade Unions
10 Vinohradska
121 47 Prague 2, Czechoslovakia
T: 235 35 65 F: 235 43 93
TL: 121525 WFTU C

Heyrovska, Neela, Ms.
Brontosaurus
Bubenska 6
170 00 Praha 7, Czechoslovakia
T: (42 2) 80 29 10 F: (42 2) 80 29 06

Hrabalova, Nadia, Ms.
Veronica - The Ecological Centre
PO Box 91
601 91 Brno 1, Czechoslovakia
T: (42 5) 227 04

Mekdad, Fayssal, Mr.
International Union Of Students*
17th November Street, PO Box 58
110 01 Prague 01, Czechoslovakia

Urban, Frantisek, Ing.
Czech Union for Nature Conservation
U Michelskeho Lesa 366
14000 Prague 4-KRC, Czechoslovakia
T: (42 2) 2195 4501 F: (42 2) 734 621

GLOBAL FORUM

Binar, Vaclav, Mr.
EARTH TRAIN/ WORLD LINK GLOBAL YOUTH SUMMIT
Wanklova, 6
Olomouc, Czechoslovakia

Ebringer, Libor, Mr.
Professor
COMENIUS UNIVERSITY
Institute of Molecular & Subcellular Biology, Odborarske, nam. 5
Bratislava 811-07, Czechoslovakia
T: 42 7 63454 F: 42 7 729064

Kwiatkowski, Robert, Mr.
INTERNATIONAL UNION OF STUDENTS
58, 17th November Street
Prague, Czechoslovakia
T: 42 2 231-2812 F: 42 2 231-6100
TL: 122858

Mikulas, Huba, Mr.
Dr.
SLOVAK UNION OF NATURE AND LANDSCAPE PROTECTORS
Gorkeho 6, 811 01
Bratislava, Czechoslovakia
T: 42 7 50665 F: 42 7 50665

Oldrich, Vlk, Mr.
Mayor
MUNICIPALITY OF THE TOWN
T.G. Masyryk Square 29
Moravska Trebova, Czechoslovakia
T: 42 462 6573 F: 42 462 6572
E: 57101

Silina, Mara, Ms.
Coordinator East European Prog
FRIENDS OF THE EARTH INTERNATIONAL
PO Box 163
Bratislava 81499, Czechoslovakia
T: 42 7 495264 F: 42 7 495264

Soberova, Yvona
UNITED NATIONS NON-GOV. LIAISON SERVICE - NGOLS
Palais de Nations
Geneve CH 1211, Czechoslovakia
F: 41 22 733-6542

DEMOCRATIC PEOPLE'S REPUBLIC OF KOREA - GOVERNMENT

Kang, Hui Won, H.E. Mr.
Vice-Premier of the Administration Council
Democratic People's Republic of Korea
Permanent Mission to the United Nations, 866 United Nations Plaza, Suite 300
New York, NY 10017, United States of America
T: 212 371 1280 F: 212 371 8873

Kim, Gyong Ho, H.E. Mr.
Ambassador to the Republic of Peru
Permanent Mission of Democratic People's Republic of Korea
866 United Nations Plaza, Suite 300
New York, NY 10017, United States of America
T: 212 371 1280 F: 212 371 8873

Kim, Hyong U, Mr.
Deputy to the Supreme People's Assembly
Adviser to the Institute of International Affairs
Permanent Mission to the United Nations, 866 United Nations Plaza, Suite 300
New York, NY 10017, United States of America
T: 212 371 1280 F: 212 371 8873

Kim, In Guk, Mr.
Director
Bureau for Environment Protection
Permanent Mission to the United Nations, 866 United Nations Plaza, Suite 300
New York, NY 10017, United States of America
T: 212 371 1280 F: 212 371 8873

Li, Gyun Tae, Mr.
Director Department of Internal Organizations
Ministry of Foreign Affairs
Permanent Mission to the United Nations, 866 United Nations Plaza, Suite 300
New York, NY 10017, United States of America
T: 212 371 1280 F: 212 371 8873

Li, Hyong Chol, Mr.
Chief of Section
Institute of Disarmament and Peace
Permanent Mission to the United Nations, 866 United Nations Plaza, Suite 300
New York, NY 10017, United States of America
T: 212 371 1280 F: 212 371 8873

Li, Tae Gun, Mr.
Expert
Ministry of Foreign Affairs
Permanent Mission to the United Nations, 866 United Nations Plaza, Suite 300
New York, NY 10017, United States of America
T: 212 371 1280 F: 212 371 8873

DENMARK - GOVERNMENT

Anderson, Ellen Ostergaard, Ms
Environment Correspondent
POLITIKEN (NATIONAL DAILY)
Raadhus Pladsen 37
Copenhagen DE 1785, Denmark
T: 445 33118511 EXT 326 F: 445 33154117

Backer, Lise, Ms.
Head of Section
Ministry of Energy
Permanent Mission to the United Nations, United Nations Plaza
New York, NY 10017, United States of America
T: 212 963 7174 F: 212 963 1921

Brusgaard, Aage, Mr.
MP
FOLKETINGET
Christiansborg
Copenhagen K1240, Denmark
T: 33 37 55 00 F: 33 32 85 36

Carstensen, John, Mr.
Head of Section
Ministry of the Environment
Permanent Mission to the United Nations, United Nations Plaza
New York, NY 10017, United States of America
T: 212 963 7174 F: 212 963 1921

Carstensen, Kim, Mr.
Head of Conservation Department
World Wide Fund for Nature, Denmark
Permanent Mission to the United Nations, United Nations Plaza
New York, NY 10017, United States of America
T: 212 963 7174 F: 212 963 1921

Christensen, Jens Bjerregaard, Mr
Deputy Director General
The National Forest and Nature Agency
Slotsmarken 13
DK-29770 Horsholm, Denmark
T: 45 45 76 53 76 F: 45 45 76 54 77

Christensen, Poul, Mr.
County Mayor
Association of County Councils in Denmark
Permanent Mission to the United Nations, United Nations Plaza
New York, NY 10017, United States of America
T: 212 963 7174 F: 212 963 1921

Christensen, Torben Malland, H.E. Mr.
Ambassador Extraordinary and Plenipotentiary
Ministry of Foreign Affairs
Permanent Mission to the United Nations, United Nations Plaza
New York, NY 10017, United States of America
T: 212 963 7174 F: 212 963 1921

Dithmer, Torben, H.E. Mr.
Ambassador
Federative Republic of Brazil
Permanent Mission to the United Nations, United Nations Plaza
New York, NY 10017, United States of America
T: 212 963 7174 F: 212 963 1921

Dybkjaer, Lone, Ms.
MP
FOLKETINGET
Christiansborg
Copenhagen K1240, Denmark
T: 33 37 55 00 F: 33 32 85 36

Ellefsen, Svend Aage, Mr.
Minister
The Faroese Home Government
Permanent Mission to the United Nations, United Nations Plaza
New York, NY 10017, United States of America
T: 212 963 7174 F: 212 963 1921

Fischer, Per, Mr.
Councellor
PRIME MINISTERS OFFICE
Prins Jorgens Gaard Ii
DK-1218 Copenhagen, Denmark
T: 3392 3300 F: 3311 1665

Gade, Steen, Mr.
MP
FOLKETINGET
Christiansborg
Copenhagen K1240, Denmark
T: 33 37 55 00 F: 33 32 85 36

Garval, Lis, Ms.
Director of Information
Ministry of Foreign Affairs
Permanent Mission to the United Nations, United Nations Plaza
New York, NY 10017, United States of America
T: 212 963 7174 F: 212 963 1921

Glonborg, Knud, Mr.
Member
Parliament
Permanent Mission to the United Nations, United Nations Plaza
New York, NY 10017, United States of America
T: 212 963 7174 F: 212 963 1921

Helms, Hans Jakob, Mr.
Director
Greenland Home Rule
Permanent Mission to the United Nations, United Nations Plaza
New York, NY 10017, United States of America
T: 212 963 7174 F: 212 963 1921

Hilstrom, Niels-Jorgen, Mr.
Secretary
Danish Confederation of Trade Unions
Permanent Mission to the United Nations, United Nations Plaza
New York, NY 10017, United States of America
T: 212 963 7174 F: 212 963 1921

Hougs, Henrik, Mr.
Deputy Director
Confederation of Danish Industries
Permanent Mission to the United Nations, United Nations Plaza
New York, NY 10017, United States of America
T: 212 963 7174 F: 212 963 1921

Jensen, Lars Georg, Mr.
Officer
International Relations, 92-Group
Permanent Mission to the United Nations, United Nations Plaza
New York, NY 10017, United States of America
T: 212 963 7174 F: 212 963 1921

Jensen, Svend Aage
MP
FOLKETINGET
Christiansborg
Copenhagen K1240, Denmark
T: 33 37 55 00 F: 33 32 85 36

Jonsson, Borge, Mr.
Detective Superintendent
PRIME MINISTER'S OFFICE
Borups Alle 266
Copenhagen DK-2200, Denmark
T: 45 33 923300 F: 45 33 111665

Kjaer, Aggi, Ms.
Doctor of Medicine
The National Council of Women in Denmark
Permanent Mission to the United Nations, United Nations Plaza
New York, NY 10017, United States of America
T: 212 963 7174 F: 212 963 1921

Kjeldsen, H.O.A., Mr.
President
Agricultural Council of Denmark
Permanent Mission to the United Nations, United Nations Plaza
New York, NY 10017, United States of America
T: 212 963 7174 F: 212 963 1921

Kleist, Kuupik, Mr.
Minister
Public Works and Transportation, The Greeland Home Rule
Permanent Mission to the United Nations, United Nations Plaza
New York, NY 10017, United States of America
T: 212 963 7174 F: 212 963 1921

Koester, Veit
Head Of Division
NATIONAL FOREST AND NATURE
Agency Ministry of the Env
Hoersholm, Denmark
T: +4545765376 F: +4545765477

Lidegaard, Bo, Mr.
Head of Section
Ministry of Foreign Affairs
Permanent Mission to the United Nations, United Nations Plaza
New York, NY 10017, United States of America
T: 212 963 7174 F: 212 963 1921

Madsen, Jorgen Lund, Mr.
Head of Division
Agency of Environmental Inspection, City of Copenhagen
Permanent Mission to the United Nations, United Nations Plaza
New York, NY 10017, United States of America
T: 212 963 7174 F: 212 963 1921

Magnussen, Thomas, Mr.
Head of Section
The Faroese Home Government
Permanent Mission to the United Nations, United Nations Plaza
New York, NY 10017, United States of America
T: 212 963 7174 F: 212 963 1921

Maltesen, Ib, Mr.
Economic Adviser
Danish Confederation of Trade Unions
Permanent Mission to the United Nations, United Nations Plaza
New York, NY 10017, United States of America
T: 212 963 7174 F: 212 963 1921

Meilbin, Franz-Michael, Mr.
First Secretary
Royal Danish Embassy Brasilia
Permanent Mission to the United Nations, United Nations Plaza
New York, NY 10017, United States of America
T: 212 963 7174 F: 212 963 1921

Moller, Lone
MP
FOLKETINGET
Christiansborg
Copenhagen K1240, Denmark
T: 33 37 55 00 F: 33 32 85 36

Moller, Per Stig, H.E. Mr.
Minister
Environment (Head of Delegation)
Permanent Mission to the United Nations, United Nations Plaza
New York, NY 10017, United States of America
T: 212 963 7174 F: 212 963 1921

Moltke-Leth, Aase, Ms.
Head of Section
Ministry of Foreign Affairs
Permanent Mission to the United Nations, United Nations Plaza
New York, NY 10017, United States of America
T: 212 963 7174 F: 212 963 1921

Mortensen, Pia, Ms.
Secretary
Members of Parliament
Permanent Mission to the United Nations, United Nations Plaza
New York, NY 10017, United States of America
T: 212 963 7174 F: 212 963 1921

Nehring, Niels-Jorgen, Mr.
Deputy Undersecretary
PRIME MINISTER'S OFFICE
Christiansborg
Copenhagen DK-1218 K, Denmark
T: 45 33 923300 F: 45 33 111665

Olsen, Bent Normann, Mr.
County Counsellor
Board of the Association of County Councils in Denmark
Permanent Mission to the United Nations, United Nations Plaza
New York, NY 10017, United States of America
T: 212 963 7174 F: 212 963 1921

Olsen, Ove Rosing, Mr.
Minister
Health, Environment, Industry and Trade
The Greeland Home Rule
Permanent Mission to the United Nations, United Nations Plaza New York, NY, 10017
F: 212 963 7174
TL: 212 963 1921

Pedersen, John, Mr.
Information Officer
Ministry of the Environment
Permanent Mission to the United Nations, United Nations Plaza
New York, NY 10017, United States of America
T: 212 963 7174 F: 212 963 1921

Plougmann, Ole, Mr.
Deputy Permanent Under-Secretary of State
Ministry of the Environment
Permanent Mission to the United Nations, United Nations Plaza
New York, NY 10017, United States of America
T: 212 963 7174 F: 212 963 1921

Poulsen, Johannes, Mr.
Member
Board of the National Association of Local Authorities
Permanent Mission to the United Nations, United Nations Plaza
New York, NY 10017, United States of America
T: 212 963 7174 F: 212 963 1921

Ronholt, Jens
MP
FOLKETINGET
Christiansborg
Copenhagen K1240, Denmark
T: 33 37 55 00 F: 33 32 85 36

Schou, Hans, Mr.
Head of Section
Ministry of Foreign Affairs
Permanent Mission to the United Nations, United Nations Plaza
New York, NY 10017, United States of America
T: 212 963 7174 F: 212 963 1921

Severinsen, Hanne, Ms.
Member
Parliament
Christiansborg
Copenhagen K1240, Denmark
T: 33 37 55 00 F: 33 32 85 36

Sorensen, Niels Bo, Mr.
Head of Section
Ministry of the Environment
Permanent Mission to the United Nations, United Nations Plaza
New York, NY 10017, United States of America
T: 212 963 7174 F: 212 963 1921

Torp, Ulrik, Mr.
Head of Section
Ministry of the Environment
Permanent Mission to the United Nations, United Nations Plaza
New York, NY 10017, United States of America
T: 212 963 7174 F: 212 963 1921

Wolf, Adam, Mr.
Private Secretary to the Prime Minister
Prime Minister's Office
Christiansborg
Copenhagen DK-1218, Denmark
T: 45 33 923300 F: 45 33 111665

NGO

Ajithakumar, C.T.
IYF
Astridsvej 7-5-1
8220 Braband, Denmark

Aparicio, Teresa
IWGIA
Fiolstraede 10
DK 1171 Copenhagen K, Denmark

Bjerregard, Hans, Mr.
Chairman
DANISH ORGANISATION FOR RENEWABLE ENERGY
Willemoesgade 14,
Copenhagen 2100, Denmark
T: 45 31 4-29091 F: 45 31 4-29095

Brodersen, Hans Jorgen, Mr.
Ass.Prof.
INT.TECH.PLANING, AALBORG UNIVERSITY
R.Gr Cleaner Tech., Dept. of Develop. and Plan, Aalborg University; Fib 2
Aalborg, Denmark
T: 45 98158522 F: 45 98153298

Christensen, Per, Mr.
UA
Fibigerstraede 11
9220 Aalborg, Denmark

Geertsen, Uffe, Mr.
DANISH ORGANIZATION FOR RENEWABLE ENERGY
Borgergade 10-14
1300 Kobenhavn K, Denmark

Jernelov, Arne
WAITRO
c/o Danish Technological Institute, Technology Park, DK-8000 Aarhus, Denmark

Lund, Henrik, Mr.
Researcher
DANISH ORGANISATION FOR RENEWABLE ENERGY
Willemoesgade 14,
Copenhagen 2100, Denmark
T: 45 31 4-29091 F: 45 31 4-29095

Lundberg, Bjorn
WAITRO
c/o Danish Technological Institute, Technology Park, DK-8000 Aarhus, Denmark

Madathiparampil Mathews, Mohan
IYF
Astridsvej 7-5-1
8220 Braband, Denmark

Maegaard, Preben
FRE
Kammersgaardsvej 16, Sdr. Ydby
7760 Hurup Thy, Denmark

Magnusson, Sindre
IYF
Astridsvej 7-5-1
8220 Braband, Denmark

Nigam, Jai Krishna
WAITRO
c/o Danish Technological Institute, Technology Park, DK-8000 Aarhus, Denmark

Ocampo, Daniel
IYF
Astridsvej 7-5-1
8220 Braband, Denmark

Oscar, Alarik Boorn
IYF
Astridsvej 7-5-1
8220 Braband, Denmark

Santa, David
IYF
Astridsvej 7-5-1
8220 Braband, Denmark

PRESS

Andersen, Ellen O.
Journalist
POLITIKEN
Politiken Raadhuspladsen, 1785
Copenhagen, Denmark
T: 45 33 118511 F: 45 33 154117

Andersen, Jan
Correspondent
RITZAUS
Haraldsgade 106.2th
2100 Copenhagen, Denmark, Denmark

Andersen, Marianne K.
Correspondent
JYLLANDS-POSTEN
Molleaparken 28
2800 Lyngby, Denmark, Denmark

Baadsgaard, Trine
Correspondent
NOW
Vejlbygade 26
8240 Risskov, Denmark, Denmark

Bech-Danielsen,
Correspondent
POLITIKEN
Frederikkevej 11.1
DK-2900 Hellerup, Denmark, Denmark

Borch Hansen, Birgitte
Correspondent
NOW
Solystgade 55
8000 Arhus C, Denmark, Denmark

Braager, Ulla
Correspondent
UDVIKLING
Helmer Sogaards Alle 28
DK-2770 Kastrup, Denmark, Denmark

Damkjaer-Nielsen, Ole
Correspondent
NOW
Ingerslev Plads 5 4.tv.
8000 Arhus C, Denmark, Denmark

Dithmer, Claus
Correspondent
NOW
Jaegersborggade 18, 2TV
2200 Kobenhavn N, Denmark, Denmark

Enevoldsen, Layla
Editor
LANDBRUGSRAADET
Aarhus Plads 6
2100 Copenhagen, Denmark, Denmark

Erhardtsen, Birgitte
Journalist
DAGLDADEF BORSEN
Moentengade 19
1114 Copenhagen 4, Denmark
T: 45 33 320102 F: 45 33 122445

Fisker, Henrik
Correspondent
KRISTELIGT DAGBLAD
Hejrebanken 26
DK-4681 Herfolge, Denmark, Denmark

Gadegaard, Anne B.
Correspondent
DANSKE KOMMUNER
Gronnemose Alle 17
2400 Copenhagen NV, Denmark, Denmark

Gudiksen, Kirsten
Correspondent
NOW
Falstersgade 29B, 3TV
8000 Arhus C, Denmark, Denmark

Harboe, Jurgen
Correspondent
KONTAKT
Rungstedhave 21 C
DK 2960 Rungsted, Denmark, Denmark

Hjulmand, Karen
Reporter Radio/Television
DANISH B'CASTING CORP.
Thorvaldsensvej 7 st. tv.
1871 FRB.C, Denmark, Denmark

Horsager, Jan
Correspondent
NOW
Valgaardsvej 9, 2.th.
DK-2500 Valby, Denmark, Denmark

Hougaard, Hanne Birgitte
Correspondent
NOW
Falstersgade 29b, 3TV
8000 Arhus C, Denmark, Denmark

Hoxbroe, Allan
Reporter Radio/Television
DANMARKS RADIO-RADIOAVISEN
Horsebakken 55
2400 Kobenhavn NV, Denmark, Denmark

Huus, Jeppe
Editor
SCANWASTE MAGAZINE
Carstensgade 16
1770 Denmark, Denmark

Jacobsen, Kirsten
Correspondent
NOW
Viborgvej 100
8210 Aarhus N, Denmark, Denmark

Jensen, Charlotte Flyvbjerg
Correspondent
NOW
Nordborggade 21, 5.th
8000 Arhus C, Denmark, Denmark

Johannesen, Justinus
Correspondent
NORDISK FAGPRESSE
Godthabsvaenget 21
3th 2000 Frederiksberg, Denmark, Denmark

Jorgensen, Lotte Ann
Correspondent
NOW
Jerichausgade 5
8000 Arhus C, Denmark, Denmark

Kabel, Lars
Editor
NOW
Birkebakken 22
8520 Lystrup, Denmark, Denmark

Kristensen, Jens Klit Work
Correspondent
NOW
Chr. Waerumsgade 16
st.th., 8000 Aarhus C., Denmark, Denmark

Krohn, Ole
Correspondent
SCAN WASTE
Dalgas Have 40
2000 Frederiksberg, Denmark, Denmark

Kruse, Jane
Editor
ENERGY REVUE
Kallempuej 8
Thisted, Denmark, Denmark

Lassen, Regitre, Ms.
Correspondent
FOLKETIDENDE
Gaaardraekkeves 10
Dk 4840 NR Alsley, Denmark

Leick, Claus
Correspondent
MANUS PRESSEBUREAU
Christiansgade 4
2th 5000 Odense C, Denmark, Denmark

Lidegaard, Martin, Mr.
Editor
INTERNATIONAL HORISANT
Edlevei3
2900 Hellerup, Denmark, Denmark

Lundby, Claus Merrild
Correspondent
NOW
Borglumkillegiet, Room 256
Borglumvej 2, 8240 Risskov, Denmark, Denmark

Lyngso-Petersen, Erik
Editor
ENGINEERING WEEKLY
Folkevej 4
DK 2820 Gentofte; Denmark, Denmark

Maybrom, Nis Brummer
Editor
INTERNATIONAL HORIZON
Drageves 16
4040 Jyllinge, Denmark, Denmark

Mikkelsen, Flemming J.
Correspondent
ERHVERS-BLADET
Menelaos Boulevard 52
DK 2650 Hvidovre, Denmark, Denmark

Nielsen, Elsebeth, Sra.
OUR COMMON FUTURE
Ved Strandeu 18
Copenhagen, Denmark
T: 45 33 5-387-7675

Nielsen, Joergen S.
Editor
DANISH DAILY INFORMATION
Overbys Alle 44
DK-2500 Valby, Denmark, Denmark

Nielsen, Niels Christian
Correspondent
MORGENAVISEN JYLLANDS POSTEN
Raadmandsgade 40 D 5.MF
2200 Copenhagen, Denmark, Denmark

Nielsen, Thorkil G.
Correspondent
ARHUS STIFTSTIDENDE
Samsogade 80
DK-8000 Aarhus C, Denmark, Denmark

Pedersen, Kim Vejrup
Correspondent
NOW
Gudrunsvej 76, 7.th.
8220 Brabrand, Denmark, Denmark

Pedersen, Thomas B.
Correspondent
DET FRI AKTUELT
Kongshaven 63 B
2500 Valby, Denmark, Denmark

Roikjaer, Karin
Correspondent
ATUAGAGDWIUTIT-GREENLANDPOST
Nuniaffik 6 101
3905 Nuussuaq, Greenland, Denmark

Rossen, Lars Henrik
Correspondent
NOW
Samsogade 52 1.sal
8000 Arhus C, Denmark, Denmark

Schoen, Anne G.
Reporter Radio/Television
DANISH B'CASTING CORP.
Gersonsvej 26 st.
2900 Hellerup, Denmark, Denmark

Simonsen, Bo
Correspondent
MIDTJYLLANDS AVIS
Moellevancs Alle 175
8200 Aarhus N., Denmark, Denmark

Skovsbo, Mikkel
Correspondent
HORSENS FOLKEBLAD
Vejlby Vaenge 195
DK-8240 Risskov, Denmark, Denmark

Sonderborg, Birgitte R.
Correspondent
NAESTVED TIDENDE
Oster Farinagsgade 4
2100 Q, Denmark, Denmark

Sorensen, Hanne S.
Correspondent
RITZAUS
Vesterbrogade 135
2nd Floor, 1620 Copenhagen V, Denmark, Denmark

Thomsen, Pia Elers
Correspondent
NOW
Strandvej 12
5700 Svendborg, Denmark, Denmark

Vestergaard, Frede
Editor
WEENDAVISEN BERLINGSKE
Borgager 54
2620 Albertslund, Denmark, Denmark

Vestergaard, Ole
Reporter Radio/Television
TV2-DENMARK
Frederikkevej 14
2900 Hellerup, Denmark, Denmark

Vilhelmsen, Claus
Correspondent
SUN MEDIA
Ringertoften 20 St. TV
2400 Copenhagen, Denmark, Denmark

With, Thomas Klenow
Correspondent
MORGENAVISEN JYLLANDS-POSTEN
R. Eliseu Visconti 403/301
20251-Rio de Janeiro-RJ, Brazil, Denmark

ORGANIZATIONS

Dahl, Jens, Mr.
International Work Group For Indigenous Affairs
Fiolstraede 10
DK 1171 Copenhagen K, Denmark
T: (45) 33 12 47 24 F: (45) 33 14 77 49

Forde, Bjorn, Mr.
Danish Association For International Co-operation
Borgergade 10-14
1300 Kobenhavn K, Denmark
T: (45) 33 32 62 44 F: (45) 33 15 62 43

Karottki, Rene Ben, Mr.
Danish Organisation For Renewable Energy, The OVE
Folkecenter for Renewable Energy, PO Box 208
7760 Hurup, Denmark
T: (45) 97 95 65 55 F: (45) 97 95 65 65

Khare, Shiv, Mr.
World Assembly Of Youth*
Ved Bellahoj 4
2700 Bronshoj Copenhagen, Denmark
T: (45) 316 07770 F: (45) 316 05797

Mathews, Mohan, Mr.
Int'l Youth Federation for Env. Studies and Conservation
Astridsvej 7-5-1
8220 Braband, Denmark
T: 45 38 250 190 F: 45 38 861 941

Mengu, Moses D., Mr.
World Ass. of Industrial and Technological Research Organizations (WAITRO)
c/o Danish Technological Institute, Technology Park, DK-8000 Aarhus, Denmark
T: (45 86) 142 400 F: (45 86) 147 722

GLOBAL FORUM

Andersen, Svend, Mr.
Professor Dr.
CHURCH OF DENMARK
Bystaevnet 16
Lystrup 8520, Denmark
T: 45 86 136-711 F: 45 86 130-490

Baek, John, Mr.
Musician
ALL RIGHTS RESERVED
Falstersgade 7
Arhus 8000 C, Denmark
T: 45 86 191065 F: 45 86 320010

Barrett, Soren Sigurd, Mr.
Actor
ALL RIGHTS RESERVED
Herredsvej 44A
Arhus 8210 V, Denmark
T: 45 86 150708 F: 45 86 320010

Baruah, Monica, Ms.
Personal Sec To Prime Min.
PRIME MINISTER'S OFFICE
Christiansborg
Copenhagen DK-1218, Denmark
T: 45 33 923300 F: 45 33 111665

Birkmose, Dorthe, Ms.
Administrator
ENVIRONMENTAL COURSE, AARHUS UNIVERSITY
Aarhus University, Building 135
Aarhus DK 8000, Denmark
T: 45 612-5177 F: 45 612-5175

Brekke, William Ole, Mr.
Director
COMMEDIA SCHOOL
Snerlevej 5
Gentofte 2820, Denmark
T: 45 31 683389 F: 45 31 31683389

Brikmose, Dorthe, Ms.
Administrator
ENVIRONMENTAL COURSE, AARHUS UNIVERSITY
Building 135
Aarhus DK 8000, Denmark
T: 45 8612-5177 F: 45 8612-5175

Bundesen, Lise Lotte, Ms.
Manager
NOVO NORDISK A/S
Novo Alle
Bagsvaerd 2880, Denmark
T: 45 44448888 F: 45 44445918
TL: 37173

Carlsen, Hanne Norup, Ms.
WOMEN'S INTERNATIONAL LEAGUE FOR PEACE & FREEDOM
Jellerod Have 66
Kokkedal 2980, Denmark
T: 454545 766641 F: 454545

Ch-Danielsen, A.Bec
Journalist
POLITIKEN
Politiken Raadhuspladsen, 1785
Copenhagen, Denmark
T: 45 33 118511 F: 45 33 154117

Christensen, Edith, Ms.
NOVO NORDISK
Mosevangen 12
Kalundborg, Denmark

Christensen, Janni K., Ms.
ALL RIGHTS RESERVED
Norreport 39
Arhus 8000, Denmark
T: 45 86 188488 F: 45 86 320010

Christensen, Jorgen, Mr.
NOVO NORDISK
Mosevangen 12
Kalundborg, Denmark

Christensen, Mailand, Mr.
Ambassador
PRIME MINISTER'S OFFICE
Asiatisuc Plads 2
Copenhagen, Denmark

Duholm, Jytte, Ms.
Consultant
KESSEL KONSULTING
International
Waltham - MA 02154, Denmark
T: 1 617 891-6507 F: 1 617 899-5156

Eppler, Soren, Mr.
ALL RIGHTS RESERVED
Bissensgade 17
Bissensgade DK-8000, Denmark
T: 45 86191809 F: 45 89320010

Esbensen, Torben, Mr.
M Sc
CONSULTING ENGINEER
Mollegade 54-56
Sonderborg DK 6400, Denmark
T: 45 744 22250 F: 45 744 22674

Fisher, Henrik
Journalist
KRISTELIGT DAGBLAD
Fanogade 15 DK - 2100
Coppenhagen, Denmark
T: 45 39 271235 F: 45 39 270800

Fox, Noel
Editor
CADOGAN COMMUNICATIONS LTD
Islands Baygge 3
Copenhagen, Denmark
T: 45 33 571309 F: 45 33 959582

Hansen, Eva Egesborg, Ms.
Executive Administrator
INTERPLAN
Strandgade 27 B(Gammel Dok)
Copenhagen 1401, Denmark
T: 45 959360 F: 45 545010

Hansen, Nikolaj, Mr.
Actor
ALL RIGHTS RESERVED
Fenrisvej 19
Abyhoj 8230, Denmark
T: 45 86 86191809 F: 45 86 89320010

Helsing, Eva Schertenleib, Ms.
SIMU - DANISH COALITION FOR INTERNATIONAL ENVIRONMENT & DEVELOPMENT
Veksovej 4B st
Copenhagen 2700 Brh, Denmark
T: 45 288040

Hyland, Larry, Mr.
ECO ERA
Aalborg, Denmark

Jacobsen, Ulrik, Mr.
IBIS
Store Molle Vej 5, 4th.
Copenhagen 5, Denmark
T: 45 31546963 F: 45 31350696

Jensen, Inger Nyegaard, Ms.
WOMENS INTERNATIONAL LEAGUE FOR PEACE AND FREEDOM - WILPPF
Dronninggaards Alle 10 A, II
Holte 2840, Denmark
T: 45 42420874

Jensen, Mette, Ms.
Head Of Administration
DANISH CENTER FOR HUMAN RIGHTS
38 Studiestradede
Copenhagen DK-1455, Denmark
T: 45 33 911299 F: 45 33 910299

Jensen, Svend W., Mr.
Tech.Director
STORSTROM COUNTY
Parkvej 37
Nykobing F. 4800, Denmark
T: 45 548 23232 F: 45 548 22370

Johansen, Bjarke Steen, Mr.
Organisation-Officer
OUR COMMON FUTURE & DANISH YOUTH COUNCIL
Scherfigsvej 5
Copenhagen 2100 O, Denmark
T: 45 31 298888 F: 45 31 298382

Johansen, Kirsten, Ms.
DANISH YOUTH COUNCIL
Solanger 2
Niva, Denmark
T: 42 24-9959

Kingo, Lise, Ms.
Manager
NOVO NORDISK A/S
Novo Alle
Bagsvaerd 2880, Denmark
T: 45 44448888 F: 45 44445918
TL: 37173

Magnussen, Lisbeth, Ms.
DANISH ORGANIZATION FOR RENEWABLE ENERGY
Willemoesgade 14, Kobenhavn 0
Copenhagen DK 2100, Denmark
T: 45 31 429091 F: 45 31 429005

Mayoh, Anne Kathrine, Ms.
Dancer
ALL RIGHTS RESERVED
Mejlbyvej 270
Hjortshoj 8350, Denmark
T: 45 86-221716 F: 45 86-320010

Meyer, Niels I., Mr.
Professor
TECH. UNIVERSITY OF DENMARK
Physics Laboratory 3, Techn. University of Denmark
Lyngby 2800, Denmark
T: 45 42881611 F: 45 45931669
TL: 37529dthdiadk

Nielsen, Hanne, Ms.
IBIS
Store Molle Vej 5, 4th.
Copenhagen 5, Denmark
T: 45 31546963 F: 45 31350696

Nielsen, Jens C. Kwella, Mr.
Musician
ALL RIGHTS RESERVED
Skandergorvez 6
Arhus 8000 C, Denmark
T: 45 86 149726 F: 45 86 320010

Nyholm, Kristian, Mr.
SIMU - DANISH COALITION FOR INTERNATIONAL ENVIRONMENT & DEVELOPMENT
c/o Ms Borgergade 14, 1300 Copenhagen K
Copenhagen 1300, Denmark
T: 45 33 326244 F: 45 33 156243
TL: 15928MSDK

Oleson, Gunnar Boye, Mr.
International Secretary
DANISH ORGANISATION FOR RENEWABLE ENERGY
Willemoesgade 14,
Copenhagen, Denmark
T: 45 33 42-90-91 F: 45 33 42-90-95

Olsen, Ruth E., Ms.
Coordinator
KULU - WOMEN AND DEVELOPMENT
Fr. Nansens Pl. 6
Oslo 0160, Denmark
T: 47 2 426245 F: 47 2 423205

Oreskov, Claus, Mr.
INTERNATIONAL WORK GROUP FOR INDIGENOUS AFFAIRS (IWGIA)
Fiolstraede 10
Copenhagen K 1171, Denmark
T: 45 33124724 F: 45 33147749

Palshoej, Peter, Mr.
Secretary General
AIF
Teglvaericsgade 27
Copenhagen, Denmark
T: 55 31 29 6066 F: 55 31 20 1019

Petersen, Erik Lyngro
Editor
ENGINEERING WEEKLY OF DENMARK
Stelbetegade 4
DK 1503 Kobonhav, Denmark
T: 45 31 216810 F: 45 31 216701

Rasmussen, Lis, Ms.
Coordinator
KULU - WOMEN IN DEVELOPMENT
Fr. Nansen Pl. 6
Oslo 0160, Denmark
T: 47 2 426245 F: 47 2 423205

Schlueter, Poul Holmskov, Mr.
Prime Minister Of Denmark
PRIME MINISTER'S OFFICE
Christiansborg
Copenhagen DK-1218 K, Denmark
T: 45 33 923300 F: 45 33 111665

Sejersen, Frank, Mr.
INTERNATIONAL WORK GROUP FOR INDIGENOUS AFFAIRS (IWGIA)
Fiolstraede 10
Copenhagen K 1171, Denmark
T: 45 33124724 F: 45 33147749

PARLIAMENTARIANS / SPIRITUAL LEADERS

Glonberg, Knud, Mr
MP
Folketinget Christiansborg Palace DK 1218
Copenhagen K, Denmark
T: 01 11 66 00

Simonsen, Ole, Mr.
MP
Folketinget Christiansborg Palace DK 1218
Copenhagen K, Denmark
T: 01 11 66 00

DJIBOUTI - GOVERNMENT

Awaleh, Mohamed, Mr.
Head of the Agriculture Department
Ministry of Agriculture
Permanent Mission to the United Nations, 866 United Nations Plaza, Suite 4011
New York, NY 10017, United States of America
T: 212 753 3163 F: 212 223 1276

Chehem, Mohamed Moussa, H.E. Mr.
Minister of Agriculture
Permanent Mission of Djibouti
866 United Nations Plaza, Suite 4011
New York, NY 10017, United States of America
T: 212 753 3163 F: 212 223 1276

Hamadou, Barkat Gourad, Mr.
Prime Minister
Permanent Mission of the Republic of Djibouti
866 United Nations Plaza, Suite 4011
New York NY 10017, United States of America
T: 212 753 3163 F: 212 223 1276

Hassan, Ali, Commandant
Aide de Camp of the Prime Minister
Permanent Mission of Djibouti
866 United Nations Plaza, Suite 4011
New York, NY 10017, United States of America
T: 212 753 3163 F: 212 223 1276

Ibrahim, Abdourahman Ahmed, Mr.
Counsellor
Permanent Mission of Djibouti
Permanent Mission to the United States, New York
866 United Nations Plaza, Suite 4011 New York, NY, 10017
F: 212 753 3163
TL: 212 223 1276

Mohamed, Ali, Mr.
Head of the Environmental Department
Ministry of Transport and Tourism
Permanent Mission to the United Nations, 866 United Nations Plaza, Suite 4011
New York, NY 10017, United States of America
T: 212 753 3163 F: 212 223 1276

Mohamed, Dileita, Mr.
Chief of Protocol
Permanent Mission of Djibouti
866 United Nations Plaza, Suite 4011
New York, NY 10017, United States of America
T: 212 753 3163 F: 212 223 1276

Olhaye, Roble, H.E. Mr.
Ambassador Extraordinary and Plenipotentiary
Permanent Representative to the United Nations, New York
Permanent Mission to the United Nations, 866 United Nations Plaza, Suite 4011
New York, NY 10017, United States of America
T: 212 753 3163 F: 212 223 1276

Tani, Islail Houssein, Mr.
Permanent Secretary of Information
Permanent Mission of Djibouti
866 United Nations Plaza, Suite 4011
New York, NY 10017, United States of America
T: 212 753 3163 F: 212 223 1276

Tani, Ismael H, Mr
Secretaire General Information
Boite Postole 032
Djibouti, Djibouti
F: 253 3549 50

DOMINICAN REPUBLIC - GOVERNMENT

Charles, Mary Eugenia, The Hon.
B.A. Prime Minister
Permanent Mission of the Commonwealth of Dominica, 820 2nd Ave.
New York, NY 10017, United States of America
T: 212 949 0853 F: 212 808 4975

Gregoire, Felix, Mr
Director Of Forestry & Wildlife
FORESTRY & WILDLIFE DIVISION
Botauic Garden
Commwealth of Dominica, Dominican Republic
T: 809 448 2401 EXT 282 F: 809 448 7999

Troncoso, Carlos Morales
Vice President
Permanent Mission of the Dominican Republic
144 E. 44th Str., 4th floor
New York NY 10017, United States of America
T: 212 867 0833 F: 212 986 4694

White, Russel, Mr.
Plans Analyst
Metro Toronto and Region Conservation Authority
Permanent Mission to the United Nations, 820 2nd Ave. New York, NY, 10017
F: 212 949 0853
TL: 212 808 4975

NGO

Franco, Pedro
CEP
Av. Mexico No. 143, Apartado 450-2
Santo Domingo, Dominican Republic

Selig, Jose
CENTRO DE EDUCACION POPULAR
Av. Mexico No. 143, Apartado 450-2
Santo Domingo, Dominican Republic

GLOBAL FORUM

Baez, Etzel, Mr.
Vice President
FUNDACION CULTURAL DE CINE Y EDUCACION EDUCATIVOS, INC.
Luis F. Thomen 358
Santo Domingo, Dominican Republic
T: 1 809 563-5263

Bautista, Zoila, Ms.
CEDDESEC
Calle Sanchez W125
Sto. Dgo, Dominican Republic
T: 685 0195

Bujosa, Sagrada, Sra.
Information Dept.
CENT. INV.AC.FEMENINA
Luis F. Thomen #358, Ensanche Quisqueya
Santo Domingo, Dominican Republic
T: 1 809 563-5263 F: 1 809 563-1159

Cabrera, Cesar, Mr.
CREAR
Rio Limpio
Loma de Cabrera, Dominican Republic

Campusano, Mercedes, Ms.
RED DE MUJERES Y MEDIO AMBIENTE - CENTRO AMERICA Y CARIBE
Casimiro Maza H104
Santo Domingo, Dominican Republic
F: 1 809 474

Castellanos, Jose R. Reis, Mr.
SECRETARIA DEL ESTADO DE LA AGRICULTURA
Boca de Mao - Mao-Valverde
Mao-Valverde, Dominican Republic
T: 1809 F: 1809

Cruz, Eridania Reyes, Ms.
C E D D E S E C
Calle Dr. Delgado 255
Santo Domingo, Dominican Republic
T: 1 809 6866186

Espinal, Jose Luis Gigni, Mr.
Program Coordinator
CEDECO
Juan Sanchez Ramirez #60 (altos) Zona Universitaria
Santo Domingo, Dominican Republic
T: 1 809 688-2633 F: 1 809 562-5879

Kees, Rosa Julia, Sra.
FUNCACION LOMA QUITA ESPUELA INC.
Apto. Postal 236, Av. Libertad 44
S. Fran. de Macoris, Dominican Republic

Luciano, Olga, Mr.
Diretoria De Projetos
C O D O P R O C
Mercedes 156 (Altos)
Santo Domingo, Dominican Republic
T: 1 809 689-0427 F: 1 809 687-0304
TL: 2770

Martinez, Radames
CEDECO
Apartado 21880
Santo Domingo, Dominican Republic
T: 1 809 688-2633 F: 1 809 5625879

Mosquea, Rosa Gonzalez, Sra.
FUNDACION LOMA QUITA ESCUELA
Daglfinger Strasse 67
Munchen D 8000 Munchen 81, Dominican Republic
T: 49 89 936987 F: 49 89 74 88 07

Paulino, Catalino Antonio, Sr.
SOCIEDAD PARA EL DESARROLLO INTEGRAL DEL NORESTE - SODIN
Calle Emilio Conde No. 52, Prov. Maria Trinidad Sanchez
Nagua, Dominican Republic

Peralta, Teresa, Sra.
Docmentation Center
CENT. INV.AC.FEMENINA
Luis F. Thomen #358, Ensanche Quisqueya
Santo Domingo, Dominican Republic
T: 1 809 563-5263 F: 1 809 563-1159

Santana, Pedro Franco, Sr.
CENTRO DE EDUCACION POPULAR
Palais de Nations
Geneve CH 1211, Dominican Republic
F: 41 22 733-6542

Shields, Margaret, Ms.
Director
UNITED NATIONS INTERNATIONAL RESEARCH & TRAINING INSTITUTE FOR THE ADVANCEMENT OF WO
Calle Cesar Nicolas Penson 102-A
Santo Domingo, Dominican Republic
T: 18 09 685-2111 F: 18 09 685-2117

Stamiris, Eleni, Ms.
Deputy Director
INTERNATIONAL AND TRAINNING INSTITUTE FOR THE ADVANCEMENT OF WOMEN - INSTRAW
Cesar Nicolas Penson no. 102-A
Santo Domingo 21747, Dominican Republic
T: 1 809 685-2111 F: 1 809 685-2117

Thomen, Antonio, Sr.
Presidente
INSTITUTO DOMINICANO DE BIOCONSERVAICION
Ap. Postal 1149
Santo Domingo, Dominican Republic
T: 1 809 6823-3770 F: 1 809 530 6553

Torres, Robert Adames, Mr.
Program Coordinator
CEDECO
Juan Sanchez Ramirez #60 (altos) Zona Universitaria
Santo Domingo, Dominican Republic
T: 1 809 688-2633 F: 1 809 562-5879

Zevallos, Daniel
ENDA-THIRD WORLD
PO Box 3370
Dakar, Dominican Republic
T: 221 222125 F: 221 222695
TL: 51456ENDATM, E: gn:endadakar/geo2:enda

ECUADOR - GOVERNMENT

Arcos Garcia, Fausto, Sr.
Funcionario del Ecuador en Rio De Janeiro
Permanent Mission of Ecuador
United Nations Plaza
New York, NY 10017, United States of America
T: 212 963 7174 F: 212 963 1921

Balda, Galo, Ing.
Funcionario de PETROPRODUCCION
Permanent Mission of Ecuador
United Nations Plaza
New York, NY 10017, United States of America
T: 212 963 7174 F: 212 963 1921

Bonifaz, Neftali, Ing.
Director Ejecutivo de la Fundacion IDEA
Permanent Mission of Ecuador
United Nations Plaza
New York, NY 10017, United States of America
T: 212 963 7174 F: 212 963 1921

Carrera, Raul, Professor
Area de Ciencias Politicas de la FLASCO
Permanent Mission of Ecuador
United Nations Plaza
New York, NY 10017, United States of America
T: 212 963 7174 F: 212 963 1921

Carrera de la Torre, Luis, Sr.
Secretario Pro Tempore del Tratado de Cooperacion Amazonica
Permanent Mission of Ecuador
United Nations Plaza
New York, NY 10017, United States of America
T: 212 963 7174 F: 212 963 1921

Cordovez, Diego, H.E. Dr.
Ministro de Relaciones Exteriores
Representate Personal del Senor Presidente de la Republica (Jefe de la Delegation)
Permanent Mission to the United Nations, United Nations Plaza New York, NY, 10017
F: 212 963 7174
TL: 212 963 1921

Falconi-Puig, Ricardo, Dr.
Ministre-Conseiller Commercial
MINISTERIO DE INDUSTRIAS COMMERCIO
Integracion Y Pesca
Quito, Ecuador

Herrera, Washington, H.E. Lic.
Embajador
Ecuador en Chile (Jefe de la Delegation)
Permanent Mission to the United Nations, United Nations Plaza New York, NY, 10017
F: 212 963 7174
TL: 212 963 1921

Jara, Fausto, Ing.
Funcionario de PETROAMAZONAS
Permanent Mission of Ecuador
United Nations Plaza
New York, NY 10017, United States of America
T: 212 963 7174 F: 212 963 1921

Lucero, Fernando, Sr.
Funcionario de la Unidad de Proteccion de PETROECUADOR
Permanent Mission of Ecuador
United Nations Plaza
New York, NY 10017, United States of America
T: 212 963 7174 F: 212 963 1921

Mantilla, Maria Eugenia, Srta.
Asistente del Ministro de Relaciones Exteriores
Permanent Mission of Ecuador
United Nations Plaza
New York, NY 10017, United States of America
T: 212 963 7174 F: 212 963 1921

Morillo, Marco, Ing.
Subsecretario de Saneamento Ambiental del IEOS
Permanent Mission of Ecuador
United Nations Plaza
New York, NY 10017, United States of America
T: 212 963 7174 F: 212 963 1921

Navarro, Manuel, Lic.
Jefe de la Unidad de Proteccion Ambiental de PETROECUADOR
Permanent Mission of Ecuador
United Nations Plaza
New York, NY 10017, United States of America
T: 212 963 7174 F: 212 963 1921

Paredes, Rafael, Dr.
Consejero de la Embajada del Ecuador en Brasilia
Permanent Mission of Ecuador
United Nations Plaza
New York, NY 10017, United States of America
T: 212 963 7174 F: 212 963 1921

Paz Y Mino, Guilhermo, Lic.
Subsecretario de Medio Ambiente del Ministerio de Energia y Minas
Permanent Mission of Ecuador
United Nations Plaza
New York, NY 10017, United States of America
T: 212 963 7174 F: 212 963 1921

Perez Arteta, Luis, Ing.
Asesor del Ministerio de Relaciones Exteriores
Permanent Mission of Ecuador
United Nations Plaza
New York, NY 10017, United States of America
T: 212 963 7174 F: 212 963 1921

Racuy, Wilson, Dr.
Presidente de la Fundacion Islas Galapagos
Permanent Mission of Ecuador
United Nations Plaza
New York, NY 10017, United States of America
T: 212 963 7174 F: 212 963 1921

Ramon, Francisco, Dr.
Miembro del Consejo de Administracion de PETROECUADOR
Permanent Mission of Ecuador
United Nations Plaza
New York, NY 10017, United States of America
T: 212 963 7174 F: 212 963 1921

Salazar Sancisi, Juan, H.E. Sr.
Subsecretario de Asuntos Politicos Multilaterales del Ministerio de Relaciones Exteriores
Permanent Mission of Ecuador
United Nations Plaza
New York, NY 10017, United States of America
T: 212 963 7174 F: 212 963 1921

Samanes, Roberto, Dr.
Funcionario del Tratado de Cooperacio Amazonica
Permanent Mission of Ecuador
United Nations Plaza
New York, NY 10017, United States of America
T: 212 963 7174 F: 212 963 1921

Tacle, Moises, Dr.
Secretario General
Planificacion del CONADE
Permanent Mission to the United Nations, United Nations Plaza New York, NY, 10017
F: 212 963 7174
TL: 212 963 1921

Troya, Roberto, Dr.
Director Ejecutivo
Fundacion NATURA
Permanent Mission to the United Nations, United Nations Plaza New York, NY, 10017
F: 212 963 7174
TL: 212 963 1921

Ulloa, Marjorie, Lic.
Tercer Secretario
Mision Permanente ante las Naciones Unidas, Nueva York
Permanent Mission to the United Nations, United Nations Plaza New York, NY, 10017
F: 212 963 7174
TL: 212 963 1921

Vargas, Medardo, Ing.
Funcionario de PETROPRODUCCION
Permanent Mission of Ecuador
United Nations Plaza
New York, NY 10017, United States of America
T: 212 963 7174 F: 212 963 1921

Wachsner, Silvia, Sra.
Consul General del Ecuador en Rio de Janeiro
Permanent Mission of Ecuador
United Nations Plaza
New York, NY 10017, United States of America
T: 212 963 7174 F: 212 963 1921

Zarzar Casis, Alonzo, Professor
Coordinator Of The Masters
FLACSO-SEDE ECUADOR
A.V. America 4000 Casilla 17-11-06362
Quito, Ecuador
T: 452 509 452 603 F: 593 2 459589

NGO

Chavez Vallejo, Gina
AE
Casilla 17-15-246C
Quito, Ecuador

Maciel, Creuza Rosa
SJPLA
Casilla Postal 8667
Guayaquil, Ecuador

Villalonga, Juan Carlos
SJPLA
Casilla Postal 8667
Guayaquil, Ecuador

Yanez, Ivonne, Ms.
Coordinator
ACCION ECOLOGICA
Madrid 340 y Coruna
Quito, Ecuador
T: 59 32 547516 F: 59 32 547516

PRESS

Aguiler Negrete, Francisco
Correspondent
JORNAL EXPRESO
R.Bulhoes Carvalho 596/701
Rio de Janeiro, Brazil, Ecuador

Leon Trujillo, Osvaldo
Editor
ALAI
Calle Andalucia 217
La Floresta, Quito, Ecuador, Ecuador

ORGANIZATIONS

Bonofaz, Neptali, Mr.
Instituto de Estrategia Agropecuarias
Casilla 853 Suc. 12 de Octubre
Quito, Ecuador
T: (593 2) 245 344

Bravo, Elizabeth, Ms.
Accion Ecologica
Casilla 17-15-246C
Quito, Ecuador

Carrasco, Alfredo, Sr.
Fundacion Charles Darwin para las Islas Galapagos (FCD)
Apartado Postal 17-01-3891
Quito, Ecuador
T: (59 32) 244 803/241 573 F: (59 32) 443 935

Cruz, Marcelo E., Dr.
Centro de Invest. y Entrenamiento de la Academia Ecuatoriana de Neurociencias
Ramirez Davalos 136
Quito, Ecuador
T: (593 2) 232 057/527 934 F: (593 2) 250 3004

Curbelo, Neisa, Ms.
General Cordinator
Service Justice And Peace In Latin America
Casilla Postal 8667
Guayaquil, Ecuador
T: (593 4) 201 451 F: (593 4) 203 600

De Benavides, Nancy Hilgert, Sra
Corporacion Ornitologica del Ecuador (CECIA)
Av. de los Shyris 2030 y la Tierra,
Casilla 17-07-9068, Quito, Ecuador
T: (59 32) 244 734 F: (59 32) 244 734

Garzozi, Joseph, Mr.
Fundacion Civica Francisco de Orellana
Albocentro IV, Local 2, PO Box 09-01-5945
Guayaquil, Ecuador
T: (593 4) 241 260 F: (593 4) 235 766

Obando, Gorki, Sr.
Federacion Ecuatoriana de Clubes UNESCO
Juan Murillo y San Gregorio (esquina),
Edificio Dinamed, Piso 6, Ofic. 605, Quito, Ecuador
T: (593 2) 522 382

Rendon, Rodolfo, Mr.
Fundacion Natura
PO Box 253
Quito, Ecuador
T: (593 2) 44 7341 F: (593 2) 43 4449

Silva, Danilo, Mr.
Ecocienca - The Ecuadorian Foundation for Ecologic Studies
Av. 12 de Octobre 959 y Roca, Edificio Mariana de Jesus - Piso 7, Quito, Ecuador
T: (593 2) 502 409/548 752 F: (593 2) 502 409

Vargas, Antonio, Sr.
Organizacion de Pueblos Indigenas de Pastaza
Tnte. Ortiz y Gral. Villamil, Apartado 790
Puyo Pastaza, Ecuador

GLOBAL FORUM

Aroca, Raul Patino, Mr.
Abogado
DIPUTADO ECUADOR
Junin H21 y Cordova, 2o. Piso, Oficina 7
Guayaquil 7544, Ecuador
T: 52 5934445588 F: 52 5934445313

Cano, Andrea, Ms.
CLAI
Casilla 17-08-8522
Quito, Ecuador

Castello, Luis, Mr.
U I C N
Av. Los Shyris 1240-of 601
Quito, Ecuador
T: 59 32 430974 F: 59 32 430974

Gonzalez, Augusto, Sr.
Engineer
CLIRSEN
Of. IGM 42
Cotopaxi, Ecuador

Gonzalez, Oswaldo, Mr.
TIERRA VIVA
Luis Cordero 5-78 Y Hondrato Vazquez
Cuenca, Ecuador
T: 59 37 8211 F: 59 37 833642

Jimenez, Jorge A., Mr.
Manager Of Env. And Publ. Aff.
MAXUS ECUADOR INC.
Edificio Antisana Piso 10, Av. Amazonas 3655 PO Box1717105
Quito, Ecuador
T: 59 32 430720 F: 59 32 430738
TL: 21141

Martinez, Esperanza, Ms.
Coordinadora
OBSERVATORIO SOCIO-AMBIENTAL
Luis Felipe Borja 621
Quito 0544, Ecuador
T: 59 32 550 658 F: 59 32 506 617

Medina, Nelson
Sup.Cuencas
INECEL
P.O.Box 6603 Cci
Quito, Ecuador
T: 59 32 243939 F: 59 32 500824

Porate, Carmelina, Ms.
CONALE
6 de Diciembre y Granados 2553
Quito, Ecuador
T: 59 32 248-930 F: 59 32 442-271

Quirola, Fernando, Mr.
Arquiteto
CENTRO ESTUDOS POBLACIONALES
Rocafuerte,754 Y Paez
Machala-Oro, Ecuador
T: 921200

Rodriguez, Carlos Gustavo, Sr.
Coordinator
INECEL
Apartado, 565 - A
Quito, Ecuador
T: 59 32 543850

Starkoff, Silvia, Ms.
P U C E
Hidalgo De Pinto 612
Quito, Ecuador
T: 59 32 454522

Stewart, William, Mr.
Coordinator Safety And Env.
MAXUS ECUADOR INC.
Edificio Antisana Piso 10, Av. Amazonas 3655 PO Box1717105
Quito, Ecuador
T: 59 32 430720 F: 59 32 430738
TL: 21141

Suarez, Mario Vasconez, Mr.
Vice-Presidente
SIA/REDES/CIUDAD
La Gasca 326, Casilla Postal 1708
Quito 8311, Ecuador
T: 59 32 230192 F: 59 32 500322

T., Osvaldo Leon
Director
AGENCIA ALAI
12 de Octubre 622 y Patria Of. 503 C.P. 1712877
Quito, Ecuador
T: 593 2 505074 F: 593 2 505073

Valenzuela, Jaime, Mr.
IULA
J.Pablo Sanz/ 4612 10 de Agosto
Quito, Ecuador
T: 59 32 435205 F: 29 32 435205

PARLIAMENTARIANS / SPIRITUAL LEADERS

Patino, Raul
Camara Nacional de Representates Palacio Legislativo
Quito, Ecuador
T: 552 769
TL: 2788 ED

EGYPT - GOVERNMENT

Aliababidi, M. Mukhtar
Petroleum Expert Dr
LEAGUE OF ARAB STATES (OAPEC)
PO Box 108 Majlis Al-Shaab
Egypt 11516, Egypt
T: 00202 3542660 F: 00202 3542601

Atef Ebeid, , H.E. Dr.
Minister of Cabinet Affairs
Minister of State for Administrative Development (Head of Delegation)
Permanent Mission to the United Nations, United Nations Plaza New York, NY, 10017
F: 212 963 7174
TL: 212 963 1921

Badr, Marwan, H.E. Dr.
Ambassador
Kenya
Permanent Mission to the United Nations, United Nations Plaza New York, NY, 10017
F: 212 963 7174
TL: 212 963 1921

Diab, Salah, Mr.
Private Sector
Permanent Mission of Egypt
United Nations Plaza
New York, NY 10017, United States of America
T: 212 963 7174 F: 212 963 1921

El Arabi, Nabil A., H.E. Dr.
Ambassador Extraordinary and Plenipotentiary
Permanent Representative to the United States, New York (Head of Delegation)
Permanent Mission to the United Nations, United Nations Plaza New York, NY, 10017
F: 212 963 7174
TL: 212 963 1921

El Eslamboli, Hager, Mr.
Councellor
International Organizations Department, Ministry of Foreign Affairs
Permanent Mission to the United Nations, United Nations Plaza New York, NY, 10017
F: 212 963 7174
TL: 212 963 1921

El Maraghi, Sherif, H.E. Mr.
Ambassador Consul General
Rio de Janeiro
Permanent Mission to the United Nations, United Nations Plaza New York, NY, 10017
F: 212 963 7174
TL: 212 963 1921

El-Tellawy, Mervat, H.E. Mr.
Ambassador, Economic Department
Ministry of Foreign Affairs
Permanent Mission to the United Nations, United Nations Plaza New York, NY, 10017
F: 212 963 7174
TL: 212 963 1921

Hafez, Salah, Mr.
Chairman-in-Charge
Egyptian Environmental Affairs Agency
Permanent Mission to the United Nations, United Nations Plaza New York, NY, 10017
F: 212 963 7174
TL: 212 963 1921

Hasan, Ferkhanda, Dra.
Professor of Geology
American University in Cairo
Permanent Mission to the United Nations, United Nations Plaza New York, NY, 10017
F: 212 963 7174
TL: 212 963 1921

Hassan, Dawlat, Mr.
Minister Plenipotentiary
Permanent Delegation to the United Nations, New York
Permanent Mission to the United Nations, United Nations Plaza
New York, NY 10017, United States of America
T: 212 963 7174 F: 212 963 1921

Latif, Aboul Fetouh Abdel, Dr.
Counsultant
Minister of Cabinet Affairs
Permanent Mission to the United Nations, United Nations Plaza New York, NY, 10017
F: 212 963 7174
TL: 212 963 1921

Mounir, Amr Ahmed, Mr.
Attache, Economic Department
Ministry of Foreign Affairs
Permanent Mission to the United Nations, United Nations Plaza New York, NY, 10017
F: 212 963 7174
TL: 212 963 1921

Mubarak, Housein, Mr.
Second Secretary
Ministry of Foreign Affairs
Permanent Mission to the United Nations, United Nations Plaza New York, NY, 10017
F: 212 963 7174
TL: 212 963 1921

Nour, Mohamed Abdalla, Mr
Executive Director
CENTRE FOR ENV & DEVELOPMENT
Arab Region & Europe Nile Tower Building 21 Giza St 13FL
Giza, Egypt
T: 2025703531 202725785 F: 202 728099

Omar, Sherif, Mr
Member Of The Egyptian
PARLIAMENT
11, Boulos Hanna Street
Kokki, Egypt
T: 202 700721 202349370

Saad, Somaia, Ms.
First Secretary
Ministry of Foreign Affairs
Permanent Mission to the United Nations, United Nations Plaza New York, NY, 10017
F: 212 963 7174
TL: 212 963 1921

NGO

Fetnat, Tolba
ASOSEM
41 Soliman Pasha Street
Cairo, Egypt

Mohamed Fari, Emara
ASOSEM
41 Soliman Pasha Street
Cairo, Egypt

PRESS

Sheidid, Ahmed
Correspondent
AL-AHRAM PRESS
Al-Galaa Street
Cairo, Egypt
T: 5747086 F: 5747126

Wagdi, Riad S.
Correspondent
AL-AHRAM
17 Abo. Al-Mahassen St.
Roxy Heliopolic, Cairo, Egypt, Egypt

ORGANIZATIONS

Emara, Mohamed Farid, Dr.
Arab Society of Occupational Safety and Environmental Health (ASOSEM)
41 Soliman Pasha Street
Cairo, Egypt
T: (20 2) 392 7743

Razzak, Nouri Abdul, Mr.
Afro-Asian People's Solidarity Organisation (AAPSO)*
89 Abdel Aziz Saoud St.
AL Manial Cairo, Egypt
T: (20 2) 362 0462 F: (20 2) 363 7361

GLOBAL FORUM

Adly, Emad, Mr.
Coordinator
ARAB NETWORK FOR ENVIRONMENT & DEVELOPMENT
P. O. Box 2, Magles Elshaab
Cairo, Egypt
T: 20 2 363-8534 F: 20 2 363-8534
TL: 23550 YTS

Allam, Magdy, Mr.
President
ARAB OFFICE FOR YOUTH & ENVIRONMENT
P. O. Box 2, Magles Elshaab
Cairo, Egypt
T: 20 2 363-8534 F: 20 2 363-8534
TL: 23550 YTS

Aziz, Abdel Hady Abdel, Dr.
Cairo, Egypt

Aziza, Youssef A., Dr.
Professor
SCIENTIFIC ASSOCIATION FOR ARAB WOMEN
Central Metallurgical Research and Development Institute, P.O.Box, 87 Helwan
Cairo 790775, Egypt
T: 20 2 790775 F: 20 2 790898
TL: 23116CMRDI

Bishay, Adli M., Dr.
Director General And UNDP Adv.
DESERT DEVELOPMENT CENTER
American Univ. in Cairo 113 Kasr El Aini Str.
Cairo, Egypt
T: 20 2 357-6330 F: 20 2 355-7565
TL: 92224 AUC UN

El Malek, Tharwer Ishak Abd, Mr.
WORLD COUNCIL OF CHURCHES
33, Abdel Halim Ali Street, Imbaba
Giza, Egypt

El-Ebrashi, Nabila, Ms.
SCIENTIFIC ASSOCIATION OF ARAB WOMEN (IPAC)
Nrc-Dokki -Cairo
Cairo, Egypt
T: 20 2 701010

Fawzy, Hossam, Mr.
ENVIRONMENTAL AND REMOTE SENSING CENTER
Cairo, Egypt

Hammouda, Faiza, Ms.
SCIENTIFIC ASSOCIATION OF ARAB WOMEN-IPAC
Nrc - Dokki
Cairo, Egypt
T: 20 2 701010

Hassan, Farkhonda, Ms.
Co-Chair
INTERNATIONAL POLICY ACTION COM.
American University Cairo
Cairo, Egypt
T: 20 2 3542966 F: 20 2 3557565

Kader, Ali Ahmed Abdul, Dr.
Professor
EGYPTIAN CENTRE FOR COMMUNITY & ENVIRONMENT
Cairo University
Giza, Egypt
T: 20 2 728237 F: 20 2 727676

Korashy, Mona, Ms.
Executive Director
EGYPTIAN ASS.FOR PRESERVATION OF ENVIRONMENT
3 Ahmed Pasha St.
Garden City - Cairo, Egypt
T: 20 2 701-952 F: 20 2 3550616
TL: 704385

Mahmoud, Mohammed, Mr.
Board Member
ARAB OFFICE FOR YOUTH & ENVIRONMENT
P. O. Box 2, Magles Elshaab
Cairo, Egypt
T: 20 2 363-8534 F: 20 2 363-8534
TL: 23550 YTS

Nawal, Hassan, Ms.
CENTER FOR EGYPTIAN CIVILIZATION STUDIES
18 Saray El Guezira Street2221 Monica Place - Zamalek
Cairo, Egypt

Neanatawa, Mowir S., Mr.
President
ENVIRONMENTAL QUALITY INTER
3B Bahgat Ali St. Zamalek
Cairo, Egypt
T: 20 2 3417879 F: 20 2 3413331
TL: 23172

Rabie, Mahmoud, Mr.
President
ARAB OFFICE FOR YOUTH & ENVIRONMENT
P. O. Box 2, Magles Elshaab
Cairo, Egypt
T: 20 2 363-8534 F: 20 2 363-8534
TL: 23550 YTS

Reveret, Jean-Pierre, Mr.
Director Department
UNIVERSITE SENGHOR/UREF
1 Midam Arabi, El Mancheya
Alexandria, Egypt
T: 20 3 587-0164 F: 20 3 819-975

Sharawi, Ali Mohammed, Dr.
Vice President
EGYPTIAN CENTRE FOR COMMUNITY & ENVIRONMENT
Cairo University
Giza, Egypt
T: 20 2 728237 F: 20 2 727676

Takla, Leila, Ms.
President
NATIONAL ASSOCIATION FOR PROTECTION OF THE ENVIRONMENT
8 Sad El Aly Dokki
Cairo, Egypt
T: 20 2 705 500 F: 20 2 704 385

Yousef, Aziza, Ms.
SCIENTIFIC ASSOCIATION OF ARAB WOMEN-IPAC
NRC - Dokki
Cairo, Egypt
T: 20 2 790775 F: 20 2 790890

Zaalouk, Malak, Ms.
CENTRE NATIONAL DE RECHERCHES SOCIOLOGIQUES
Palais de Nations
Geneve CH 1211, Egypt
F: 41 22 733-6542

PARLIAMENTARIANS / SPIRITUAL LEADERS

Mahfouz, Mahmoud, Senator
Majlis al Chaab
Cairo, Egypt
T: 23130

Sharif, Omar Mohamed, Dr.
MP
Majlis al Chaab
Cairo, Egypt
T: 23130

EL SALVADOR - GOVERNMENT

Araujo Padilla, Miguel Eduardo, Lic.
Director Ejecutivo de la Secretaria del Medio Ambiente (SEMA)
Permanent Mission of El Salvador
46 Park Avenue
New York, NY 10016, United States of America
T: 212 679 1616

Cabrales, Antonio, E. Ing.
Ministro de Agricultura y Ganaderia
Permanent Mission of El Salvador
46 Park Avenue
New York, NY 10016, United States of America
T: 212 679 1616

Cristiani-Burkard, Alfredo, E.S. Lic.
Presidente
Republica de El Salvador
46 Park Avenue
New York, NY 10016, United States of America
T: 212 679 1616

De Escobar, Rita Cartagena, Deputada
Parlamento
Centro Americano
Permanent Mission to the United Nations, 46 Park Avenue New York, NY, 10016
F: 212 679 1616

Gamero, Julio, Dr.
Presidente de la Comision de Proteccion del Medio Ambiente y Salud Publica de la Asamblea Legislativa
Permanent Mission of El Salvador
46 Park Avenue
New York, NY 10016, United States of America
T: 212 679 1616

Henriquez, Dolores, Deputada
Parlamento
Centro Americano
Permanent Mission to the United Nations, 46 Park Avenue New York, NY, 10016
F: 212 679 1616

Llevano de Marques, Mirna, E. Lic.
Ministro de Planificacion
Permanent Mission of El Salvador
46 Park Avenue
New York, NY 10016, United States of America
T: 212 679 1616

Marques, Jose, Sr.
Asesor del la Secretaria del Medio Ambiente
46 Park Avenue New York, NY, 10016
F: 212 679 1616

Paredes Martell, Roberto, E. Arq.
Vice-Ministro de Vivienda y Desarrollo
Permanent Mission of El Salvador
46 Park Avenue
New York, NY 10016, United States of America
T: 212 679 1616

Prieto, Enrique Garcia, E. Dr.
Secretario Geral do CONAMA
Permanent Mission of El Salvador
46 Park Avenue
New York, NY 10016, United States of America
T: 212 679 1616

Ramos, Fidel, Ing.
de la Direccion de Planificacion Ambiental del SEMA
Permanent Mission of El Salvador
46 Park Avenue
New York, NY 10016, United States of America
T: 212 679 1616

Vasquez Sosa, Lisandro, E. Cnel. y Dr.
Ministro de Salud Publica y Asistencia Social
Permanent Mission of El Salvador
46 Park Avenue
New York, NY 10016, United States of America
T: 212 679 1616

NGO

Rodriguez, Wilfredo
ASCMA
Urbanizacion Buenos Aires 3,
Calle Maquilishuat No. 208, San Salvador, El Salvador

PRESS

Martinez, Nestor
Editor
DIARIO LATINO
Col.Sta.Marta 2,Bl.C, Pasaje 3
Casa 69, San Salvador, El Salvador, El Salvador

ORGANIZATIONS

Ibarra, Ricardo Antonio, Lic.
Fundacion Salvadorena de Recursos Marinos y Limnologicos (FUREMAR)
Apartado Postal No. 2827
San Salvador, El Salvador
T: (503) 741 579

Navarro, Ricardo, Mr.
Salvadorean Center for Appropriate Technology (CESTA)
Apartado Postal 3065-33, Calle Poniente 316
San Salvador, El Salvador
T: (503) 256 746

Rivas, Francisco, Sr.
As. Amigos Del Arbol
Colonia Las Mercedes, Calle los Granados No.9
San Salvador, El Salvador
T: (503) 231 841

Vasquez, Alexander, Mr.
As. Salvadorena de Conservacion del Medio Ambiente
Urbanizacion Buenos Aires 3,
Calle Maquilishuat No. 208, San Salvador, El Salvador
T: (503) 265 514

GLOBAL FORUM

Hozman, Vicky, Ms.
RED DE MUJERES Y MEDIO AMBIENTE - CENTRO AMERICA Y CARIBE
Kmt. 62-1/2 Col. El Mora, Apartado Postal 52
Santa Ana, El Salvador
T: 50 3 410646 F: 50 3 407216

Machado, Melany, Ms.
Calle Nueva #2 Casa #9 Colonia Castelon
San Salvador, El Salvador
T: 235385 F: 242478

Valle, Victor M., Mr.
ASOCIACION 'CAMINEMOS JUNTOS'
Boulevard Maria Cristina 128
San Salvador, El Salvador
T: 503 26-5292 F: 503 25-3166

ESTONIA - GOVERNMENT

Kaasik, Tonis, H.E. Mr.
Minister of the Environment
Permanent Mission of Estonia
9 Rockefeller Plaza, Suite 1421
New York, NY 10020, United States of America
T: 247 0499 F: 262 0893

Luedig, Paul, Mr.
Sec. Secretary
Permanent to the United Nations, New York
Permanent Mission to the United Nations, 9 Rockefeller Plaza, Suite 1421
New York, NY 10020, United States of America
T: 247 0499 F: 262 0893

Luik, Heino, Mr.
Environmental Adviser to the Chairman of the Supreme Council
Permanent Mission of Estonia
9 Rockefeller Plaza, Suite 1421
New York, NY 10020, United States of America
T: 247 0499 F: 262 0893

Miller, Tonu, Mr.
Head
International Relations Department, Ministry of the Environment
Permanent Mission to the United Nations, 9 Rockefeller Plaza, Suite 1421 New York, NY, 10020
F: 247 0499
TL: 262 0893

Neljas, Aap, Mr.
Ministry of Foreign Affairs
Permanent Mission of Estonia
9 Rockefeller Plaza, Suite 1421
New York, NY 10020, United States of America
T: 247 0499 F: 262 0893

Raukas, Anto
Professor, Adviser To The Prime Min
ACADEMY OF SCIENCES, INST. OF GEO
Ee 0105 Entonian Avenue 7
Tallinn, Estonia
T: 7 0142 454182 OFF F: 7 0142 444189

Ruutel, Arnold, H.E. Mr.
Chairman of the Supreme Council of the Republic of Estonia
9 Rockefeller Plaza, Suite 1421
New York, NY 10020, United States of America
T: 247 0499 F: 262 0893

Tarand, Andres, Mr.
Chairman
Environmental Commission of the Supreme Council
Lossi Pl 1A Estonian Parliament
Tallinn Estonia Ee 0001, Estonia
T: 0142 606779 F: 0142 449865

ETHIOPIA - GOVERNMENT

Abdulahi Yusuf, Eskinder, Mr.
Assistant
Permanent Mission of Ethiopia
866 United Nations Plaza, Room 560
New York, NY 10017, United States of America
T: 212 421 1830 F: 212 754 0360

Abebe, Mesfin, H.E. Dr.
Deputy Minister of Agriculture
Environmental Protection and Development
Permanent Mission to the United Nations, 866 United Nations Plaza, Room 560
New York, NY 10017, United States of America
T: 212 421 1830 F: 212 754 0360

Abraham, Mekonnen, Mr.
Department Head
Ministry of Planning
Permanent Mission to the United Nations, 866 United Nations Plaza, Room 560
New York, NY 10017, United States of America
T: 212 421 1830 F: 212 754 0360

Asfaw, Gedion, Mr.
Department Head
Ministry of Planning
Permanent Mission to the United Nations, 866 United Nations Plaza, Room 560
New York, NY 10017, United States of America
T: 212 421 1830 F: 212 754 0360

Asfaw, Zegeye, H.E. Mr.
Minister of Agriculture
Environmental Protection and Development
Permanent Mission to the United Nations, 866 United Nations Plaza, Room 560
New York, NY 10017, United States of America
T: 212 421 1830 F: 212 754 0360

Belete, Yewondwosen, Mr.
Assistant
Permanent Mission of Ethiopia
866 United Nations Plaza, Room 560
New York, NY 10017, United States of America
T: 212 421 1830 F: 212 754 0360

Diallo, Issa B.Y.
Acting Executive Secretary
ECONOMIC COMMISSION FOR AFRICA
P O Box 3001
Addis Ababa, Ethiopia
T: 51 72 00 X266 F: 251 1 51 4416

G/Eziabher, Tewolde Berhan, Dr.
Director
Ministry of Planning
Permanent Mission to the United Nations, 866 United Nations Plaza, Room 560
New York, NY 10017, United States of America
T: 212 421 1830 F: 212 754 0360

Goshu, Haile Mariam, Mr.
Counsellor deputy Permanent Representative to the United Nations, New York
Permanent Mission of Ethiopia
866 United Nations Plaza, Room 560
New York, NY 10017, United States of America
T: 212 421 1830 F: 212 754 0360

Haile Mariam, Eshetu, Mr.
General Manager
National Meteorogolical Institute
Permanent Mission to the United Nations, 866 United Nations Plaza, Room 560
New York, NY 10017, United States of America
T: 212 421 1830 F: 212 754 0360

Imbamba, Simeon Kedogo
Regional Advisor
ECA-MULTIDISCIPLINARY REGIONAL ADVI
United Nations Economic Commission PO Box 3001
Addis Ababa, Ethiopia
T: 517200 F: 251 1 514416

Kol, Sarim
Environmental Affairs Officer
NRD/ UN-ECA
P.O. Box 3004
Addis Ababa, Ethiopia
T: 251 1 517200 F: 251 1 514416

Kouadio-Kirine, Antoine
Ambassadeur, Representant Permanent
DE COTE D'IVOIRE AUPRES DU PNUE
P.O. Box 3558
Addis Abeba, Ethiopia
T: 00 251 1 71 12 13 71 F: 00 251 1 71 21 78

Layene, Tamirat, H.E. Mr.
Prime Minister
Transitional Government of Ethiopia
866 United Nations Plaza, Room 560
New York, NY 10017, United States of America
T: 212 421 1830 F: 212 754 0360

Lemma, Kifle
Head,Legal Services
WATER RESOURCES COMMISSION
P.O. Box 1045
Addis Ababa, Ethiopia
T: 185276

Meloye, Mitiku, Mr.
Assistant
Permanent Mission of Ethiopia
866 United Nations Plaza, Room 560
New York, NY 10017, United States of America
T: 212 421 1830 F: 212 754 0360

Mesfin, Seyoum, H.E. Mr.
Minister of Foreign Affairs
Permanent Mission of Ethiopia
866 United Nations Plaza, Room 560
New York, NY 10017, United States of America
T: 212 421 1830 F: 212 754 0360

Mohammed, Duri, H.E. Dr.
Minister of Planing and Economic Development
Permanent Mission of Ethiopia
866 United Nations Plaza, Room 560
New York, NY 10017, United States of America
T: 212 421 1830 F: 212 754 0360

Mwanza, Peter N.
Director
ECONOMIC COMMISSION FOR
Africa (ECA) PO Box 3001
Addis Ababa, Ethiopia
T: 251 1 517200 F: 251 1 514416

Tafesse, Mekuria, Mr.
Commissioner
Water Resources and Development Commission
Permanent Mission to the United Nations, 866 United Nations Plaza, Room 560
New York, NY 10017, United States of America
T: 212 421 1830 F: 212 754 0360

Teferra, Almaz
Administrative Officer
ECA
Box 3005
Addis Ababa, Ethiopia
T: 51 72 00 EXT. 223 F: 251 1 514416

Teklu, Atakilt, Mr.
Assistant
Permanent Mission of Ethiopia
866 United Nations Plaza, Room 560
New York, NY 10017, United States of America
T: 212 421 1830 F: 212 754 0360

Worede, Melaku, Dr.
General Manager
Plant Genetic Resources Institute
Permanent Mission to the United Nations, 866 United Nations Plaza, Room 560
New York, NY 10017, United States of America
T: 212 421 1830 F: 212 754 0360

NGO

Belehu, Shoatsehaye
IHA
P O Box 6889
Addis Ababa, Ethiopia

Elion-Mboussa, Albert
AFME
Addis Ababa University, Dept. of Biology
PO Box 32896, Addis Ababa, Ethiopia

ORGANIZATIONS

Melake, Kiflemariam, Ms.
African Forum for Mathematical Ecology (AFME)
Addis Ababa University, Dept. of Biology
PO Box 32896, Addis Ababa, Ethiopia

Traore, Gaoussou, Mr.
Reseau Africain Pour Le Developpement (RADEV)
PO Box 60166
Addis Ababa, Ethiopia
T: (251 1) 510 175 F: (251 1) 553 549

GLOBAL FORUM

Adhanom, Teferi Micael, Mr.
CATHOLIC SECRETARIATE
P. O. Box 16277, Asmara/Eritrea/Ethiopia
Eritrea - Ethiopia, Ethiopia

Belahu, Shoatsehaye, Ms.
INTEGRATED HOLISTIC APPROACH URBAN DEVELOPMENT PROJECT
Palais de Nations
Geneve CH 1211, Ethiopia
F: 41 22 733-6542

Berhe, Constantinos, Ms.
LEMLEM ENVIRONMENT & DEVELOPMENT MOVEMENT
Palais de Nations
Geneve CH 1211, Ethiopia
F: 41 22 733-6542

Haile, Fekerte, Ms.
I L O (CIDA)
ILO Office Addis Ababa
Addis Ababa P.O.BOX 3745, Ethiopia
T: 25 11 517320 F: 25 11 513633

Ndjiodi, Prosper, Dr.
Director General
RESEAU AFRICAIN POUR LE DEVELOPEMENT - RADEV
P.O. Box 60233
Addis Ababa, Ethiopia
T: 25 11 51 58 33 F: 25 11 55 35 49
TL: 21951 RADEV

Selassie, Tsehai Berhane, Dr.
ETHIOPIAN ORTHODOX CHURCH
PO Box 7978
Addis Ababa, Ethiopia

Tsehai, Berhane S. Lassi, Dr.
WORLD COUNCIL OF CHURCHES
150 route de Ferney
Geneve 1211, Ethiopia
F: 41 22 791 0361

Wedajo, Berhanu
ETHIOPIAN EVANGELICAL CHURCH MEKANE YESUS
PO Box 2087
Addis Ababa, Ethiopia

EUROPEAN ECONOMIC COMMUNITY - GOVERNMENT

Ambrosio, , Mr.
Permanent Mission of European Economic Community
3 Dag Hammarskjold Plaza, 12 Floor
New York, NY 10017, United States of America
T: 212 371 3804 F: 212 758 2718

Antonine-Gregoire, Jean-Louise, Mr.
Permanent Mission of European Economic Community
3 Dag Hammarskjold Plaza, 12 Floor
New York, NY 10017, United States of America
T: 212 371 3804 F: 212 758 2718

Barbaso, Fabrizio, Mr.
Deputy Head of Cabinet of Commissioner Ripa Di Meana
Permanent Mission of European Economic Community
3 Dag Hammarskjold Plaza, 12 Floor
New York, NY 10017, United States of America
T: 212 371 3804 F: 212 758 2718

Baupin, Denis, Mr.
Assistant to Mr. Onesta
Permanent Mission to the United Nations, 3 Dag Hammarskjold Plaza, 12 Floor New York, NY, 10017
F: 212 371 3804
TL: 212 758 2718

Berends, Helena, Mrs.
Administrator
Permanent Mission of European Economic Community
3 Dag Hammarskjold Plaza, 12 Floor
New York, NY 10017, United States of America
T: 212 371 3804 F: 212 758 2718

Boag, Ian, Mr.
Permanent Mission of European Economic Community
3 Dag Hammarskjold Plaza, 12 Floor
New York, NY 10017, United States of America
T: 212 371 3804 F: 212 758 2718

Bourdeau, Philippe, Dr.
Head of Unit
Permanent Mission of European Economic Community
3 Dag Hammarskjold Plaza, 12 Floor
New York, NY 10017, United States of America
T: 212 371 3804 F: 212 758 2718

Brikhorst, Laurens, Mr.
Director General
Permanent Mission of European Economic Community
3 Dag Hammarskjold Plaza, 12 Floor
New York, NY 10017, United States of America
T: 212 371 3804 F: 212 758 2718

Brusasco-Mackenzie, Margaret, Mrs.
Head of Unit
Permanent Mission of European Economic Community
3 Dag Hammarskjold Plaza, 12 Floor
New York, NY 10017, United States of America
T: 212 371 3804 F: 212 758 2718

Burgues, Julio Garcia, Mr.
Administrator
Permanent Mission of European Economic Community
3 Dag Hammarskjold Plaza, 12 Floor
New York, NY 10017, United States of America
T: 212 371 3804 F: 212 758 2718

Charles, Michel, Mr.
Cabinet of the President
Permanent Mission of European Economic Community
3 Dag Hammarskjold Plaza, 12 Floor
New York, NY 10017, United States of America
T: 212 371 3804 F: 212 758 2718

Christoforides, Nicholas, Mr.
Administrator
Permanent Mission of European Economic Community
3 Dag Hammarskjold Plaza, 12 Floor
New York, NY 10017, United States of America
T: 212 371 3804 F: 212 758 2718

Clarke, Timothy, Mr.
Principal Administrator
Permanent Mission of European Economic Community
3 Dag Hammarskjold Plaza, 12 Floor
New York, NY 10017, United States of America
T: 212 371 3804 F: 212 758 2718

Corcelle, Guy, Mr.
Head of Unit
Permanent Mission of European Economic Community
3 Dag Hammarskjold Plaza, 12 Floor
New York, NY 10017, United States of America
T: 212 371 3804 F: 212 758 2718

Delor, Jacques
President
European Economic Community,
3 Dag Hammarskjold Plaza, 12 Floor
New York, NY 10017, United States of America
T: 212 371 3804 F: 212 758 2718

Delors, Jacques, Mr.
President
Delegation of the Commission of the Europaen Communities
3 Dag Hammarskjold Plaza, 12th floor
New York NY 10017, United States of America
T: 212 371 3804 F: 212 758 2718

Depoortere, Charles, Mr.
Administrator
Permanent Mission of European Economic Community
3 Dag Hammarskjold Plaza, 12 Floor
New York, NY 10017, United States of America
T: 212 371 3804 F: 212 758 2718

Dethomas, Bruno, Mr.
Spokesman of the President
Permanent Mission of European Economic Community
3 Dag Hammarskjold Plaza, 12 Floor
New York, NY 10017, United States of America
T: 212 371 3804 F: 212 758 2718

Dewost, Jean-Louis, Mr.
Director-General
Permanent Mission of European Economic Community
3 Dag Hammarskjold Plaza, 12 Floor
New York, NY 10017, United States of America
T: 212 371 3804 F: 212 758 2718

Dissing, Lisbeth, Mrs.
Administrator
Permanent Mission of European Economic Community
3 Dag Hammarskjold Plaza, 12 Floor
New York, NY 10017, United States of America
T: 212 371 3804 F: 212 758 2718

Faross, Peter, Mr.
Head of Section
Permanent Mission of European Economic Community
3 Dag Hammarskjold Plaza, 12 Floor
New York, NY 10017, United States of America
T: 212 371 3804 F: 212 758 2718

Gabrici, Leonello, Mr.
Permanent Mission of European Economic Community
3 Dag Hammarskjold Plaza, 12 Floor
New York, NY 10017, United States of America
T: 212 371 3804 F: 212 758 2718

Galazans, Jacques Santana, Mr.
Head of Unit
Permanent Mission of European Economic Community
3 Dag Hammarskjold Plaza, 12 Floor
New York, NY 10017, United States of America
T: 212 371 3804 F: 212 758 2718

Gomes, Alexandre Borges, Mr.
Permanent Mission of European Economic Community
3 Dag Hammarskjold Plaza, 12 Floor
New York, NY 10017, United States of America
T: 212 371 3804 F: 212 758 2718

Henningsen, Jorgen, Mr.
Director
Permanent Mission of European Economic Community
3 Dag Hammarskjold Plaza, 12 Floor
New York, NY 10017, United States of America
T: 212 371 3804 F: 212 758 2718

Homan, Maria Helena, Mrs.
Permanent Mission of European Economic Community
3 Dag Hammarskjold Plaza, 12 Floor
New York, NY 10017, United States of America
T: 212 371 3804 F: 212 758 2718

Kent, Martini Delogne, Mrs
Permanent Mission of European Economic Community
3 Dag Hammarskjold Plaza, 12 Floor
New York, NY 10017, United States of America
T: 212 371 3804 F: 212 758 2718

Lamy, Pascal, Mr.
Head of Cabinet of President Delors
Permanent Mission of European Economic Community
3 Dag Hammarskjold Plaza, 12 Floor
New York, NY 10017, United States of America
T: 212 371 3804 F: 212 758 2718

Matthews, Christopher, Mr.
Press Officer
Permanent Mission of European Economic Community
3 Dag Hammarskjold Plaza, 12 Floor
New York, NY 10017, United States of America
T: 212 371 3804 F: 212 758 2718

Matutes, Abel
Commissioner(Deputy Head of Delegation)
Permanent Mission of European Economic Community
3 Dag Hammarskjold Plaza, 12 Floor
New York, NY 10017, United States of America
T: 212 371 3804 F: 212 758 2718

McDonald, Brian, Mr.
Counsellor Economic Affairs
Permanent Mission of European Economic Community
3 Dag Hammarskjold Plaza, 12 Floor
New York, NY 10017, United States of America
T: 212 371 3804 F: 212 758 2718

Meana, Carlo Ripa di
Commissioner(Deputy Head of Delegation)
Permanent Mission of European Economic Community
3 Dag Hammarskjold Plaza, 12 Floor
New York, NY 10017, United States of America
T: 212 371 3804 F: 212 758 2718

Medeiros-Ferandes, Robert, Mr.
Head of Division
Permanent Mission of European Economic Community
3 Dag Hammarskjold Plaza, 12 Floor
New York, NY 10017, United States of America
T: 212 371 3804 F: 212 758 2718

Miguel, Ramon de, Mr.
Head of Cabinet of Commissioner Ripa di Meana
Permanent Mission of European Economic Community
3 Dag Hammarskjold Plaza, 12 Floor
New York, NY 10017, United States of America
T: 212 371 3804 F: 212 758 2718

Morongiu, Antonio, Mr.
Counsellor
Permanent Mission of European Economic Community
3 Dag Hammarskjold Plaza, 12 Floor
New York, NY 10017, United States of America
T: 212 371 3804 F: 212 758 2718

Onesta, Gerard, Mr.
Groupe des Verts-France
Permanent Mission to the United Nations, 3 Dag Hammarskjold Plaza, 12 Floor New York, NY, 10017
F: 212 371 3804
TL: 212 758 2718

Pery, Nicole, Mrs.
First Vice President
Socialist Group, France (Head of European Parliament Delegation)
Permanent Mission to the United Nations, 3 Dag Hammarskjold Plaza, 12 Floor
New York, NY 10017, United States of America
T: 212 371 3804 F: 212 758 2718

Pons, Genevieve, Mrs.
Cabinet of the President
Permanent Mission to the United Nations, 3 Dag Hammarskjold Plaza, 12 Floor New York, NY, 10017
F: 212 371 3804
TL: 212 758 2718

Pooley, Peter, Mr.
Deputy Director-General
Permanent Mission of European Economic Community
3 Dag Hammarskjold Plaza, 12 Floor
New York, NY 10017, United States of America
T: 212 371 3804 F: 212 758 2718

Prat, Juan, Mr.
Director-General
Permanent Mission of European Economic Community
3 Dag Hammarskjold Plaza, 12 Floor
New York, NY 10017, United States of America
T: 212 371 3804 F: 212 758 2718

Robins, Nick, Mr.
Administrator
Permanent Mission of European Economic Community
3 Dag Hammarskjold Plaza, 12 Floor
New York, NY 10017, United States of America
T: 212 371 3804 F: 212 758 2718

Rommel, Thierry, Mr.
Head of Section
Permanent Mission of European Economic Community
3 Dag Hammarskjold Plaza, 12 Floor
New York, NY 10017, United States of America
T: 212 371 3804 F: 212 758 2718

Schmiegelow, Henrik, Mr.
Head of Unit
Permanent Mission of European Economic Community
3 Dag Hammarskjold Plaza, 12 Floor
New York, NY 10017, United States of America
T: 212 371 3804 F: 212 758 2718

Strathopulos, Constantin, Mr.
Administrator
Permanent Mission of European Economic Community
3 Dag Hammarskjold Plaza, 12 Floor
New York, NY 10017, United States of America
T: 212 371 3804 F: 212 758 2718

Tent, Hendrik, Mr.
Deputy Director-General
Permanent Mission of European Economic Community
3 Dag Hammarskjold Plaza, 12 Floor
New York, NY 10017, United States of America
T: 212 371 3804 F: 212 758 2718

Valverde-Lopez, M.
Member of European Parliament Delegation
Permanent Mission of European Economic Community
3 Dag Hammarskjold Plaza, 12 Floor
New York, NY 10017, United States of America
T: 212 371 3804 F: 212 758 2718

Van Depoele, Laurent, Mr.
Head of Unit
Permanent Mission of European Economic Community
3 Dag Hammarskjold Plaza, 12 Floor
New York, NY 10017, United States of America
T: 212 371 3804 F: 212 758 2718

Verhagen, Maxine, Mr.
PPE Group
The Netherlands
Permanent Mission to the United Nations, 3 Dag Hammarskjold Plaza, 12 Floor
New York, NY 10017, United States of America
T: 212 371 3804 F: 212 758 2718

Viegas, Francisco Jaime, Mr.
Administrator
Permanent Mission of European Economic Community
3 Dag Hammarskjold Plaza, 12 Floor
New York, NY 10017, United States of America
T: 212 371 3804 F: 212 758 2718

Villepin, Alban de, Mr.
Head of Section
Permanent Mission of European Economic Community
3 Dag Hammarskjold Plaza, 12 Floor
New York, NY 10017, United States of America
T: 212 371 3804 F: 212 758 2718

Waelput, Alex, Mr.
Cabinet of the President
Permanent Mission of European Economic Community
3 Dag Hammarskjold Plaza, 12 Floor
New York, NY 10017, United States of America
T: 212 371 3804 F: 212 758 2718

Wagenbaur, Rolf, Mr.
Counsellor
Permanent Mission of European Economic Community
3 Dag Hammarskjold Plaza, 12 Floor
New York, NY 10017, United States of America
T: 212 371 3804 F: 212 758 2718

Zepter, Bernhard, Mr.
Counsellor Cabinet of the President
Permanent Mission of European Economic Community
3 Dag Hammarskjold Plaza, 12 Floor
New York, NY 10017, United States of America
T: 212 371 3804 F: 212 758 2718

FIJI - NGO

Lechte, Ruth, Ms.
WORLD YOUNG WOMEN'S CHRISTIAN ASSOCIATION
PO Box 9874
Nadir, Fiji
TL: WYMCA 41 23 32

PRESS

Reddan, Erina
Correspondent
ISLAND BUSINESS PACIFIC MAG.
Rua Buru, 315, Ap. 11
18025 Sao Paulo-SP, Brazil, Fiji

ORGANIZATIONS

Salamonsen, Peter J., Mr.
Pacific Conference of Churches (PCC)
PO Box 208
Suva, Fiji
F: (679) 303 205

GLOBAL FORUM

Fong, Gracie, Ms.
DEVELOPMENT ALTERNATIVES WITH WOMEN FOR A NEW ERA (DAWN)
c/o YMCA Sukuna Park
Suva, Fiji

Vunidilo, Katarina, Ms.
W E D O
24 Tubou Str. , Samabula
Suva, Fiji
T: 679 313900

FINLAND - GOVERNMENT

Ahteela, Reijo, Mr.
Special Adviser
Ministry for Foreign Affairs
Permanent Mission to the United Nations, 866 United Nations Plaza, Second Floor New York, NY, 10017
F: 212 355 2100
TL: 212 759 6156

Donner, Jorn, Mr.
Member of Parliament
Permanent Mission of Finland
866 United Nations Plaza, Second Floor
New York, NY 10017, United States of America
T: 212 355 2100 F: 212 759 6156

Ehrnrooth, Henrik, Mr.
Chief Executive Officer
Permanent Mission of Finland
866 United Nations Plaza, Second Floor
New York, NY 10017, United States of America
T: 212 355 2100 F: 212 759 6156

Harmala, Jukka, Mr.
President and CEO
Enso-Gutzeit Co., Confederation of Finnish Industries
Permanent Mission to the United Nations, 866 United Nations Plaza, Second Floor New York, NY, 10017
F: 212 355 2100
TL: 212 759 6156

Huttunen, Satu, Ms.
Chairperson
Finnish Association for Nature Conservation
Permanent Mission to the United Nations, 866 United Nations Plaza, Second Floor New York, NY, 10017
F: 212 355 2100
TL: 212 759 6156

Kaaria, Tauno, H.E. Mr.
Ambassador
Deputy Permanent Representative to the United Nations, New York
Permanent Mission to the United Nations, 866 United Nations Plaza, Second Floor
New York, NY 10017, United States of America
T: 212 355 2100 F: 212 759 6156

Kaisa Aula, Maria, Ms.
Member of Parliament
Permanent Mission of Finland
866 United Nations Plaza, Second Floor
New York, NY 10017, United States of America
T: 212 355 2100 F: 212 759 6156

Kalela, Aira, Ms.
Special Adviser
Ministry of the Environment
Permanent Mission to the United Nations, 866 United Nations Plaza, Second Floor New York, NY, 10017
F: 212 355 2100
TL: 212 759 6156

Kandolin, Salme, Ms.
Secretary
Environment Committee of Parliament
Permanent Mission to the United Nations, 866 United Nations Plaza, Second Floor New York, NY, 10017
F: 212 355 2100
TL: 212 759 6156

Kankaanniemi, Toimi, H.E. Mr.
Minister for Development Cooperation
Permanent Mission of Finland
866 United Nations Plaza, Second Floor
New York, NY 10017, United States of America
T: 212 355 2100 F: 212 759 6156

Karjalainen-Balk, Leena, Ms.
Adviser
Ministry of the Environment
Permanent Mission to the United Nations, 866 United Nations Plaza, Second Floor New York, NY, 10017
F: 212 355 2100
TL: 212 759 6156

Kilela, Kirsti, Ms.
First Secretary
Ministry of Foreign Affairs
Permanent Mission to the United Nations, 866 United Nations Plaza, Second Floor New York, NY, 10017
F: 212 355 2100
TL: 212 759 6156

Kiuru, Esko, Mr.
Ambassador Ministry of Foreign Affairs
Permanent Mission of Finland
866 United Nations Plaza, Second Floor
New York, NY 10017, United States of America
T: 212 355 2100 F: 212 759 6156

Koivisto, Mauno, Mr.
President
Permanent Mission of Finland
866 United Nations Plaza, Second floor
New York NY 10017, United States of America
T: 212 355 2100 F: 212 759 6156

Korkeaoja, Juha, Mr.
Member of Parliament
Permanent Mission of Finland
866 United Nations Plaza, Second Floor
New York, NY 10017, United States of America
T: 212 355 2100 F: 212 759 6156

Kuosmanen, Sirpa, Ms.
Coordinator
SERVICE CENTRE FOR DEVELOPMENT COOPERATION
Fredrikinkatu 63, A.8
Helsinki 00100, Finland
T: 35 80 694-4233 F: 35 80 694-1786
TL: 123998

Kyla-Harakka-Ruonala, Tellervo, Ms.
Specialist Advisor
Environmental Protection, Confederation of Finnish Industries
Permanent Mission to the United Nations, 866 United Nations Plaza, Second Floor New York, NY, 10017
F: 212 355 2100
TL: 212 759 6156

Lindstrom, U.B., Prof.
Permanent Mission of Finland
866 United Nations Plaza, Second Floor
New York, NY 10017, United States of America
T: 212 355 2100 F: 212 759 6156

Morottaja, Matti, Mr.
Planner
Saami Parliament
Permanent Mission to the United Nations, 866 United Nations Plaza, Second Floor New York, NY, 10017
F: 212 355 2100
TL: 212 759 6156

Muhonen, Juhani, Mr.
Ambassador Extraordinary and Plenipetentiary of Finland to Brazil
Permanent Mission of Finland
866 United Nations Plaza, Second Floor
New York, NY 10017, United States of America
T: 212 355 2100 F: 212 759 6156

Numminen, Asko, Mr.
Director
Ministry of Foreign Affairs
Permanent Mission to the United Nations, 866 United Nations Plaza, Second Floor New York, NY, 10017
F: 212 355 2100
TL: 212 759 6156

Nyroos, Hannele, Ms.
Chief Inspector
Ministry of the Environment
Permanent Mission to the United Nations, 866 United Nations Plaza, Second Floor New York, NY, 10017
F: 212 355 2100
TL: 212 759 6156

Paavilainen, Eero, Prof.
Research Director
The Finnish Forest Research Institute
Permanent Mission to the United Nations, 866 United Nations Plaza, Second Floor New York, NY, 10017
F: 212 355 2100
TL: 212 759 6156

Pietikainen, Sirpa, H.E. Ms.
Minister of the Environment
Permanent Mission of Finland
866 United Nations Plaza, Second Floor
New York, NY 10017, United States of America
T: 212 355 2100 F: 212 759 6156

Ranki, Risto, Mr.
Deputy director-General
Ministry of Trade and Industry
Permanent Mission to the United Nations, 866 United Nations Plaza, Second Floor New York, NY, 10017
F: 212 355 2100
TL: 212 759 6156

Ristimaki, Ilkka, Mr.
Ministry of Foreign Affairs
Under-Secretary of state for the Internatiolnal Development Cooperation
Permanent Mission to the United Nations, 866 United Nations Plaza, Second Floor New York, NY, 10017
F: 212 355 2100
TL: 212 759 6156

Seppanen, Esko, Mr.
Member of Parliament
Permanent Mission of Finland
866 United Nations Plaza, Second Floor
New York, NY 10017, United States of America
T: 212 355 2100 F: 212 759 6156

Sourama, Heikki, Mr.
Financial Secretary
Ministry of Finance
Permanent Mission to the United Nations, 866 United Nations Plaza, Second Floor New York, NY, 10017
F: 212 355 2100
TL: 212 759 6156

Stora, Carl, Mr.
Head
Fishery Section, Administration of the Aland Islands
Permanent Mission to the United Nations, 866 United Nations Plaza, Second Floor New York, NY, 10017
F: 212 355 2100
TL: 212 759 6156

Takala, Irmeli, Ms.
Member of Parliament
Permanent Mission of Finland
866 United Nations Plaza, Second Floor
New York, NY 10017, United States of America
T: 212 355 2100 F: 212 759 6156

Toivola, Trio
Chairman Of The Board
VAISALA OY
PL26
Sf-00421 Helsinki, Finland
T: 358 0 89491 F: 358 0 8949206

Toivola, Yrjo, Mr.
Chairman of the Board
Vaisala Co., Confederation of Finnish Industries
Permanent Mission to the United Nations, 866 United Nations Plaza, Second Floor New York, NY, 10017
F: 212 355 2100
TL: 212 759 6156

Tomila, Esa, Mr.
Director
Environmental and Transport Affairs, Confederation of Finnish Industries
Permanent Mission to the United Nations, 866 United Nations Plaza, Second Floor New York, NY, 10017
F: 212 355 2100
TL: 212 759 6156

Tuomio-Nikula, Petri, Mr.
Director
Ministry of Foreign Affairs
Permanent Mission to the United Nations, 866 United Nations Plaza, Second Floor New York, NY, 10017
F: 212 355 2100
TL: 212 759 6156

Uronen, Reino, Mr.
Secretary General Ministry of Agriculture and Forestry
Permanent Mission of Finland
866 United Nations Plaza, Second Floor
New York, NY 10017, United States of America
T: 212 355 2100 F: 212 759 6156

Vainio, Martti, Mr.
Vice-President
Central Association of Finnish Forest Industries
Permanent Mission to the United Nations, 866 United Nations Plaza, Second Floor New York, NY, 10017
F: 212 355 2100
TL: 212 759 6156

Vayrynen, Paavo, H.E. Dr.
Minister for Foreign Affairs (Deputy Head of Delegation)
Permanent Mission of Finland
866 United Nations Plaza, Second Floor
New York, NY 10017, United States of America
T: 212 355 2100 F: 212 759 6156

Vehvilainen, Anu, Ms.
Special Adviser
Ministry for the Foreign Affairs
Permanent Mission to the United Nations, 866 United Nations Plaza, Second Floor New York, NY, 10017
F: 212 355 2100
TL: 212 759 6156

Wallin, Markku, Mr.
Environmental Secretary
Central Organization of Finnish Trade Unions
Permanent Mission to the United Nations, 866 United Nations Plaza, Second Floor New York, NY, 10017
F: 212 355 2100
TL: 212 759 6156

Westphalen, Kirsti, Ms.
Adviser
Ministry of the Environment
Permanent Mission to the United Nations, 866 United Nations Plaza, Second Floor New York, NY, 10017
F: 212 355 2100
TL: 212 759 6156

Yli-Huttula, Tuomo, Mr.
Special Adviser
Ministry of the Environment
Permanent Mission to the United Nations, 866 United Nations Plaza, Second Floor New York, NY, 10017
F: 212 355 2100
TL: 212 759 6156

NGO

Bacellar, Lesith
WPC
Lonnrotink 25 A 6, KRS
00180 Helsinki, Finland

Kiljunen, Marja-Lisa, Ms.
Secretary General
FINNISH UN ASSOCIATION
The Finnish UN Association, Unioninkatu 45 B
Helsinki 00170, Finland
T: 35 80 135-1521 F: 35 80 135-2173

Niia, Lars, Mr.
NORDIC SAAMI COUNCIL
Sf - 99280 Utsjoki
Helsinki, Finland
T: 35 80 71351 F: 35 80 71353

Seurujarvi, Irja, Ms.
President
NORDIC SAAMI COUNCIL
parivaliakonkuja 2m26
00410 helsink, Finland
T: 358 0 566-2393 F: 358 697 71353

PRESS

Ahonen, Ulla, Ms.
Editor
FINNISH ASSOCIATION FOR NATURE CONSERVATION
Petramiehenk 11 A 8
Helsinki 00100, Finland
T: 35 80 642881 F: 35 80 622 1815

Brax, Anne, Ms.
Editor
SUOMEN LUONTO
Sahajarventie 16
02860 ESPOO, Finland, Finland

Eravalo, Esa
Eorrespondent
KRISTITYN VASTUU
Pihlajatie 4.D.7
SF-02270 Espoo, Finland, Finland

Gronquist, Seija
Editor
HEALTH PROMOTION
Punavuorenkatu 16 D 64
00l50 Helsinki 15, Finland, Finland

Heikkila, Markku
Editor
KALEVA
Myllytie 2 C 21
SF-90500 Oulu, Finland, Finland

Hellen, Tomas
Business & Finance Reporter
HUFVUDSTADSBLADET
Jagaregatan 8 AZ
SF-00150 Helsinki, Finland, Finland

Kanninen, Jaana, Ms.
Reporter Radio/Television
FINNISH B'CASTING-RADIO KATU 5
Radio News, P.O.Box 128
SF-00241 Helsinki, Finland, Finland

Karhu, Matti
Reporter Radio/Television
SUOMEN RADIOTIETO
Ryytimaantie 7 C33
00320 - Helsinki, Finland, Finland

Kinnunen, Helena
Editor
HELSINGIN SANOMAT
Rahapajankatu 3 E 34
SF-00160 Helsinki, Finland, Finland

Korpikallio, Asta
Journalist
SEURA
Maistraatinportti 1 00015 Yhtyneet Kuvalehdet
Finland
T: 358 90 15661 F: 358 90 1496472

Kupari, Ritva, Ms.
Editor
SUOMEN LUONTO
Vesakkotie 7 B 42
00630 Helsinki, Finland

Lavonen, Mika
Reporter Radio/Television
FINNISH B'CASTING CO.
Vainamoisenkatu 17 A 15
SF-00100 Helsinki, Finland, Finland

Oster, Harriet
Managing Editor
TEKNIIKKA & TALOUS
P.O.Box 920, SF-00101
Helsinki, Finland
T: 358 14880350 F: 358 141382

Osterlund, Markus
Reporter Radio/Television
FINNISH B'CASTING CO.
Ulfsbyvagen 7 C 27
00350 Helsinki, Finland, Finland

Rajamaki, Paivi
Correspondent
AAMULEHTI NEWSPAPER
Pohjolankatu 2 A 8
33500 Tampere, Finland, Finland

Rannikko, Tuija - Liisa
Correspondent
HAMEEN SANOMAT
Vanavantie 7, PL 530
13111 Hameenlinna, Finland
T: 358 9171511 F: 358 917151492

Rantala, Juho-Pekka
Correspondent
FINNISH B'CASTING CO. (YLE)
Turkhaudantie 1 C59
00700 Helsinki, Finland, Finland

Rasanen, Eeva-Liisa
Editor
FINNISH NEWS AGENCY
Sammatintie 6-8 L 130
00550 Helsinki, Finland, Finland

Rissa, Kari
Reporter Radio/Television
FINNISH B'CASTING CO.
Nauriskashi 11 I 18
02340 Espoo, Finland, Finland

Rutanen, Reijo
Editor
SUOMEN KUVALEHTI
P1 519 C
07900 Loviisa, Finland, Finland

Saario, Teijo Petter
Reporter Radio/Television
FINNISH B'CASTING CO.
Strandkulla 116 C
06100 Porvoo, Finland, Finland

Tastula, Maarit
Reporter Radio/Television
FINNISH B'CASTING CO.
Helletie 2
33180 Tampere, Finland, Finland

Tommola, Mika
Reporter Radio/Television
MTV NEWS
Telakkakatu 1 D 24
00150 Helsinki, Finland, Finland

Torikka, Jalmari
Editor
ILTALEHTI
Ruusankatu 3A
00250 Helsinki, Finland, Finland

Tuorma, Ismo
Editor
DEMARI-LEHTI
Heinileholeya 1B
01350 Vantaa, Finland, Finland

ORGANIZATIONS

Gaup, Ole, Mr.
Nordic Saami Council
SF-99980 Utsjoki, Finland
T: 9697 71 351 F: 9697 71 353

Stewart, Ray, Mr.
World Peace Council
Lonnrotink 25 A 6, KRS
00180 Helsinki, Finland
T: (358 0) 693 1044 F: (358 0) 693 3703

GLOBAL FORUM

Antero, Rutanen Reijo
Journalist
YHTYNEET KUVALEHDET
Mainstraatinportti 1 00015 YHTYNEETKUVALEHDET Finland
T: 358 15661 F: 358 144076

Bento-Bacellar, Lesith, Ms.
Interpreter
WORLD PEACE COUNCIL
Lonnrotinkatu 25A, 6KRS
Helsinki 00180, Finland
T: 35 80 693-1044 F: 35 80 693-3703
TL: 121680

Gronqvist, Greger
Director
FINNISH BROADCASTING COMPANY
00247 Helsinki, Finland
T: 358 074802313

Haapalinna, Marjut, Ms.
FINNISH SOCIETY OF ENVIRONMENTAL LAW
C/O Helsinki University of Technology
Helsinki 02150 ESPOO, Finland
T: 35 80 4511 F: 35 80 477035

Hamialaina, Petri
Editor
FINNISH BROADCASTING COMPANY
Laivarnhatu
00150 Helsinki 15, Finland
T: 358 0175944

Heikenkeimo, Terrtu, Ms.
WORLD YOUNG WOMEN'S CHRISTIAN ASSOCIATION
Takahuhdinkatu 100 C 17
Tampere 33560, Finland
T: 358 31 561-762
TL: WYMCA 41 23 32

Hollo, Erkki J., Mr.
Professor
FINISH SOCIETY OF ENVIRONMENTAL LAW
c/o Helsinki University of Technology
Espoo SF-02150, Finland
T: 358 4 51-3861 F: 358 4 65-077
TL: 125161 htkksf

Kari, Ahola
Cameraman
FINNISH BROADCASTING COMPANY
Box 196
Tampere 33101, Finland
T: 358 31 640204 F: 358 31 442280

Kiljunen, Kimmo, Mr.
Director
INSTITUTE OF DEVELOPMENT STUDIES/HELSINKI UNIVERSITY
Hameecte 157 00560
Helsinki, Finland
T: 35 80 208 4750 F: 35 80 708 4778

Niemela, Teuvo, Mr.
M.D.
FINNISH ASSOCIATION FOR NATURE CONSERVATION
Ppa 2
Ivalo 99800, Finland
T: 358 697 42626 F: 358 697

O-Pekka, Juho
Journalist
FINNISH BROADCASTING CO
Radio News PO Box 128
Helsinki, Finland
T: 35 80 14804458 F: 35 80 14802465

Ojala, Arja, Ms.
COALITION OF WOMEN ASSOCIATIONS FOR JOINT ACTION NITKIS
Paloniityntie 43
Tampere 33610, Finland
T: 358 31 620436

Palo, Matti, Mr.
Professor
FOREST RESEARCH INSTITUTE
P.O.B. 37
Helsinki 00381, Finland
T: 35 80 556276 F: 35 80 5061484

Palosaari, Meeri, Ms.
Councellor For Environment
COALITION OF WOMEN ASSOCIATIONS FOR JOINT ACTION NITKIS
Confederation of Finish Industries, etelarante 10
Helsinki 00130, Finland
T: 35 80 1809251 F: 35 80 01809345

Pekka, Harvia, Mr.
Forestry Advisor
SUOMEN 4H-LIITTO
Bulevardi 28
Helsinki 00120, Finland
T: 35 80 645133 F: 35 80 604612

Peltonen, Elina, Ms.
Teacher, Master Of Science
COALITION OF WOMENS ASSOCIATION FOR JOINT ACTION NYTKIS
Moykky 64850, Finland
T: 358 62 34108

Saorio, Petteri
Editor
FINNISH BROADCASTING COMPANY
Laivarnhatu
00150 Helsinki 15, Finland
T: 358 0175944

Satu, Huttunen, Ms.
Professor
FINNISH ASSOCIATION FOR NATURE CONSERVATION
Department of Bothanic - University of Oulu
Oulu SF-90570, Finland
T: 35 881 35-3216 F: 35 881 35-3266

Uosukainen, Harry, Mr.
Research Manager
TURVERUUKKI OY
Teknologiantic, 12
Oulu 90570, Finland
F: 35 881 5515520

Vera, Adolfo
Photographer
SEURA
Maistraatinportti 1 00015 Yhtyneet Kuvalehdet
Finland
T: 358 90 15661 F: 358 90 1496472

Vilander, Matti, Mr.
Research Assistant
FINISH SOCIETY OF ENVIRONMENTAL LAW
c/o Helsinki University of Technology
Espoo SF-02150, Finland
T: 358 4 51-3861 F: 358 4 65-077
TL: 125161 htkksf

ADVISORS

Jayawardena, Lal, Dr.
Director, World Institute For Development Economics Research (wider)
00100 Helsinki 10, Finland
T: 35 80 693 841 F: 35 80 693 8548

PARLIAMENTARIANS / SPIRITUAL LEADERS

Aula, Maria Kaisa
MP
Eduskuntatalo 00102 Eduskunta
Helsinki, Finland
T: 4321
TL: 1214 64 ekirj sF

Karhunen, Minna
MP
Eduskuntatalo 00102 Eduskunta
Helsinki, Finland
T: 4321
TL: 1214 64 ekirj sF

Kautto, Tarja
MP
Eduskuntatalo 00102 Eduskunta
Helsinki, Finland
T: 4321
TL: 1214 64 ekirj sF

Korkeajo, Juha
MP
Eduskuntatalo 00102 Eduskunta
Helsinki, Finland
T: 4321
TL: 1214 64 ekirj sF

Rask, Maija
Member of Parliament
Eduskuntatalo 00102 Eduskunta
Helsinki, Finland
T: 4321
TL: 1214 64 ekirj sF

Takai, Irineli
MP
Eduskuntatalo 00102 Eduskunta
Helsinki, Finland
T: 4321
TL: 1214 64 ekirj sF

FRANCE - GOVERNMENT

Arnoult, Erik, M.
Conseiller technique au Cabinet du Ministre des Affaires Estrangeres
Permanent Mission of France
1 Dag Hammarskjold Plaza
New York, NY 10017, United States of America
T: 212 308 5700 F: 212 421 6889

Audouze, Jean, M.
Conseiller technique a la Presidence de la Repubique
Permanent Mission of France
1 Dag Hammarskjold Plaza
New York, NY 10017, United States of America
T: 212 308 5700 F: 212 421 6889

Binde, Jerome, Mr.
Senior Programme Planning Officer
UNESCO (BUREAU OF STUDIES, PLANNING
7 Place De Fontenoy 75700
Paris, France
T: 33 1 45 68 12 08 F: 33 1 43 06 01 08

Bornemann-Blanc, Brigitte, Mme
Cabinet du Ministre de la recherche
Permanent Mission of France
1 Dag Hammarskjold Plaza
New York, NY 10017, United States of America
T: 212 308 5700 F: 212 421 6889

Bouillon, Gabriel
A.E.D. (agence Environnement Develo
34000 Montpellier
Grue Draparnavd, France

Castella, Pierre, M.
conseiller technique au Cabinet du Ministre delegue a la Cooperation et au Developpement
Permanent Mission of France
1 Dag Hammarskjold Plaza
New York, NY 10017, United States of America
T: 212 308 5700 F: 212 421 6889

Chabbal, Robert, M.
Conseiller special aupres du Ministre de la Rechereche et de l'Espace
Permanent Mission of France
1 Dag Hammarskjold Plaza
New York, NY 10017, United States of America
T: 212 308 5700 F: 212 421 6889

Chaubon, Pierre, M.
Conseiller technique au Cabinet du Ministre des Affairs Etrangeres
Permanent Mission of France
1 Dag Hammarskjold Plaza
New York, NY 10017, United States of America
T: 212 308 5700 F: 212 421 6889

Collin, Jean-Francois, M.
Cabinet du Ministre de l'Agriculture et de la Foret
Permanent Mission of France
1 Dag Hammarskjold Plaza
New York, NY 10017, United States of America
T: 212 308 5700 F: 212 421 6889

Cousseran, Jean-Claude, M.
Conseiller diplomatique aupres du Premier Ministre
Permanent Mission of France
1 Dag Hammarskjold Plaza
New York, NY 10017, United States of America
T: 212 308 5700 F: 212 421 6889

Curien, Hubert, S.E. M.
Ministre de la Recherche et de l'Espace
Permanent Mission of France
1 Dag Hammarskjold Plaza
New York, NY 10017, United States of America
T: 212 308 5700 F: 212 421 6889

Debarge, Marcel, S.E. M.
Ministre deleque a la Cooperation et au Developpement
Permanent Mission of France
1 Dag Hammarskjold Plaza
New York, NY 10017, United States of America
T: 212 308 5700 F: 212 421 6889

Dejeant-Pons, Maguelonne
Administrateur-Juriste
CARSEIL DE L'EUROPE-DIVISIAN DEL EU
BP 43 R6 67006
Strasbourg, France
T: (33) 884 2398 F: (33) 884 2780

Dumas, Roland, S.E. M.
Ministre d'Etat
Ministre des Affaires Etrangeres
Permanent Mission to the United Nations, 1 Dag Hammarskjold Plaza
New York, NY 10017, United States of America
T: 212 308 5700 F: 212 421 6889

Faure, Jean-Claude, M.
Directeur du Developpment
Ministere de la Cooperation
Permanent Mission to the United Nations, 1 Dag Hammarskjold Plaza
New York, NY 10017, United States of America
T: 212 308 5700 F: 212 421 6889

Fels, Xavier, M.
Conseiller diplomatique du Ministre de la Recherche et de l'Espace
Permanent Mission of France
1 Dag Hammarskjold Plaza
New York, NY 10017, United States of America
T: 212 308 5700 F: 212 421 6889

Grammont, Andre, M.
Directeur de l'Espace rural et de la foret au Ministere de l'Agriculture et de la Foret
Permanent Mission of France
1 Dag Hammarskjold Plaza
New York, NY 10017, United States of America
T: 212 308 5700 F: 212 421 6889

Hassan, Musa Bin Jaffar, H.E.Dr.
Ambassador Permanent
REPRESENTATIVE TO U.N. UNESCO
1 Rue Miollis
Paris 75015, France
T: 45 6830 52 F: 45 67 57 42

Hermelin, Paul, M.
Directeur de Cabinet du Ministre de l'Industrie et du Commerce Exterieur
Permanent Mission of France
1 Dag Hammarskjold Plaza
New York, NY 10017, United States of America
T: 212 308 5700 F: 212 421 6889

Josselin, Charles, S.E. M.
Secretaire d'Etat a la Mer
Permanent Mission of France
1 Dag Hammarskjold Plaza
New York, NY 10017, United States of America
T: 212 308 5700 F: 212 421 6889

Jouve, Marcel, M.
Conseiller technique au Cabinet du Ministre de l'Environnement
Permanent Mission of France
1 Dag Hammarskjold Plaza
New York, NY 10017, United States of America
T: 212 308 5700 F: 212 421 6889

Kouchner, Bernard, S.E. M.
Ministre de la Sante et de l'Action Humanitaire
Permanent Mission of France
1 Dag Hammarskjold Plaza
New York, NY 10017, United States of America
T: 212 308 5700 F: 212 421 6889

Lafon, Jean-Pierre, M.
Directeur des Nations Unies et des Organisations Internationales au Ministere des Affairs Estrangeres
Permanent Mission of France
1 Dag Hammarskjold Plaza
New York, NY 10017, United States of America
T: 212 308 5700 F: 212 421 6889

Lauvergeon, Anne, Mme
Secretaire General adjoint de la Presidence de la Republique
Permanent Mission of France
1 Dag Hammarskjold Plaza
New York, NY 10017, United States of America
T: 212 308 5700 F: 212 421 6889

Laville, Bettina, Mme
Conseiller pour les affaires d'environnement aupres du Premier Ministre
Permanent Mission of France
1 Dag Hammarskjold Plaza
New York, NY 10017, United States of America
T: 212 308 5700 F: 212 421 6889

Le Danff, Jean-Pierre, M.
Ministere de l'Environnement
Permanent Mission of France
1 Dag Hammarskjold Plaza
New York, NY 10017, United States of America
T: 212 308 5700 F: 212 421 6889

Le Marois, Olivier, M.
Conseiller technique au cabinet du Ministre de l'Environnment
Permanent Mission of France
1 Dag Hammarskjold Plaza
New York, NY 10017, United States of America
T: 212 308 5700 F: 212 421 6889

Le Petit, Annick, Mme
Ministere de l'Environnement
Permanent Mission of France
1 Dag Hammarskjold Plaza
New York, NY 10017, United States of America
T: 212 308 5700 F: 212 421 6889

Levy, Jean, M.
Conseiller technique a la Presidence de la Republique
Permanent Mission of France
1 Dag Hammarskjold Plaza
New York, NY 10017, United States of America
T: 212 308 5700 F: 212 421 6889

Luiz Carlos, De Almeida
World President Elect
INTERNATIONAL REAL ESTATE FEDERATIO
23 Av Bosquet
Paris 75007, France
T: 45 50 45 49 F: 45 50 42 00

Mauroy, Pierre, M.
acien Premier Ministre
Permanent Mission of France
1 Dag Hammarskjold Plaza
New York, NY 10017, United States of America
T: 212 308 5700 F: 212 421 6889

Mergret, Alain, M.
Ministere de l'Environnement
Permanent Mission of France
1 Dag Hammarskjold Plaza
New York, NY 10017, United States of America
T: 212 308 5700 F: 212 421 6889

Mermaz, Louis, S.E. M.
Ministre de l'Agriculture et de la Foret
Permanent Mission of France
1 Dag Hammarskjold Plaza
New York, NY 10017, United States of America
T: 212 308 5700 F: 212 421 6889

Mitterand, Francois, S.E. M.
President de la Republique Francaise
Permanent Mission of France
1 Dag Hammarskjold Plaza
New York, NY 10017, United States of America
T: 212 308 5700 F: 212 421 6889

Mitterrand, Francois, Mr.
President
Permanent Mission of France
1 Dag Hammarskjold Plaza
New York NY 10017, United States of America
T: 212 308 5700 F: 212 421 6889

Pascal, Michel, Mr
Conseiller
DELEGATION FRANGAISE-MINISTER DE L'
84 Rue De Greneppe
75353 Paris CFDEX, France
T: 0033 1 45563297 F: 0033 1 45563360

Philippe, Jean-Pierre, M.
Cabinet du Monistre de l'Agriculture et de la Foret
Permanent Mission of France
1 Dag Hammarskjold Plaza
New York, NY 10017, United States of America
T: 212 308 5700 F: 212 421 6889

Pons, Magali
Consultant International-Geographe
8 We Jeanne Jugan 34000
Montpellier, France
T: (33) 67 72 02 67

Ribaut, Jean Pierre
Head Of The Division Of The Environ
COUNCIL OF EUROPE
BP431 R6
Strasbourg Adex 67006, France
T: 33-88-41-22-56 F: 88-41-27-84

Ripert, Jean, M.
President du Comite intergouvernemental de negociation de la Convention cadre sur les changements climatiques
Permanent Mission of France
1 Dag Hammarskjold Plaza
New York, NY 10017, United States of America
T: 212 308 5700 F: 212 421 6889

Royal, Segolene, S.E. Mme
Ministre de l'Environnement
Permanent Mission of France
1 Dag Hammarskjold Plaza
New York, NY 10017, United States of America
T: 212 308 5700 F: 212 421 6889

Salmon, Jean, M.
Conseiller technique au Cabinet du Ministrede l'Environnement
Permanent Mission of France
1 Dag Hammarskjold Plaza
New York, NY 10017, United States of America
T: 212 308 5700 F: 212 421 6889

Smyser, Connie, Ms
Head Of Energy And Environment Div
INTERNATIONAL ENERGY AGENCY
2 Rue Andre Pascal
Paris 75775, France
T: 331 45 24 99 66 F: 331 45 24 99 88

Stoll, Jean Francois, M.
Conseiller technique au Cabinet du Premier Ministre
Permanent Mission of France
1 Dag Hammarskjold Plaza
New York, NY 10017, United States of America
T: 212 308 5700 F: 212 421 6889

Strauss Kahn, Dominique, S.E. M.
Ministre de l'Industrie et du Commerce Exterieur
Permanent Mission of France
1 Dag Hammarskjold Plaza
New York, NY 10017, United States of America
T: 212 308 5700 F: 212 421 6889

Verbrugge, Geneviwe
Charge'De Mission
MINISTERE DE L'ENVIRONNEMENT
14 Bld on General Lederc
92229 Neuilly Sur Seine Centex, France
T: 33 1 40813494 F: 33 1 40819954

Vidal, Jean, M.
Conseiller diplomatique a la Presidence de la Republique
Permanent Mission of France
1 Dag Hammarskjold Plaza
New York, NY 10017, United States of America
T: 212 308 5700 F: 212 421 6889

Vignal, Henri, M.
Chef du Service des Relations Industrielles et Financieres au Ministere des Affaires Etrangeres
Permanent Mission of France
1 Dag Hammarskjold Plaza
New York, NY 10017, United States of America
T: 212 308 5700 F: 212 421 6889

Wolff, Jacques Herve'
Conseiller Special Communcation
DELEGATION DU SENEGAL
30 Rue Duprintemps
75017 Paris, France
T: 1 42 573175 F: 1 45 26 38 25

NGO

Alberto P. Silva, Carlos
IFIA
28 rue Marbeuf
75008 Paris, France

Almeida, Luiz Carlos De
IREF
23 Avenue Bosquet
75007 Paris, France

Almeida-Moro, Fernanda
ICM
Maison de l'Unesco, 1 rue Miollis
75015 Paris, France

Alquerez, Jose Luis
IUTAO
c/o UNESCO, 1 rue de Miollis
75732 Paris Cedex 15, France

Anache, Mariane
WAMM
251 rue de Vaugirard
75740 Paris Cedex 15, France

Arnaud, Andre
FIA
8 Place de la Concorde
75008 Paris, France

Avolio, Giuseppe
IFAP
21 rue Chaplal
75009 Paris, France

Beck, Marie-Laure
WMM
56 rue de Passy
75016 Paris, France

Berman, William R.
FIA
8 Place de la Concorde
75008 Paris, France

Bideau, Alain
CJC
86 rue Pasteur
69365 Lyon Cedex 07, France

Blackburn, Nigel, Mr.
Director
INTERNTIONAL CHAMBER OF COMMERCE
38 Cours Albert 1er
Paris 75008, France
T: 33 1 49532822 F: 33 1 49532924

Borja, Jordi
UTO
22 rue d'Alsace
92300 Levallois Perret, France

Brault, Maria, Ms.
MOUVEMENT NAT. DE LUTTE POUR L' ENVIRONNEMENT
B. P. 79 - 106 Ave. Jean Lolive
Pantin 93500, France
T: 33 1 484-60414 F: 33 1 484-6665

Carneiro, Fernando L.
IUTAO
c/o UNESCO, 1 rue de Miollis
75732 Paris Cedex 15, France

Ceccon, Claudius
CCFD
4 rue Jean Lantier
75001 Paris, France

Covarrubias, Francisco
WAMM
251 rue de Vaugirard
75740 Paris Cedex 15, France

De Freitas Countinho, Marco Antonio
CIRCS
22 rue Beaunier
75014 Paris, France

Delehedde, Andre, Mr.
Depute
ASSEMBLEE INTERNATIONALE DES PARLEMENTAIRES DE LANGUE FRABCAISE/ AIPLF
235, Bd Saint-Germain - 75007
Paris, France
T: 33 1 47-052687 F: 33 1 45-511147
TL: 202 562f

Deschamps, Lucien, Mr.
Seretary General
SEE (SOCIETE DES ELECTRICIENS ET DES ELECTRONICIENS)
48 rue de la Procession
Paris 75724, France
T: 33 1 44496001 F: 33 1 44496049

Drouin, Richard
UNIPEDE
28 rue Jacques Ibert
75858 Paris Cedex 17, France

Dupuy, Pierre
CIDCE
4 Place du Presidial
87031 Limoges, France

Duquesne, Jean
FIA
8 Place de la Concorde
75008 Paris, France

Gabirault, Olivier
AIPLF
235 Boulevard Saint Germain
75007 Paris, France

Gardner, Emily
ET
Villa Hurle Vent, Villard
01220 Divonne les Bains, France

Giovanni, J. Pierre Dominique
AI
1 rue Joux-Aigues
31000 Toulouse, France

Gordon, Judith Bograd, Dr
ISSC
c/o UNESCO, 1 rue Miollis
75015 Paris, France

Gropp, Beatrice
WMM
56 rue de Passy
75016 Paris, France

Hakim, Rose-Helene
AI
1 rue Joux-Aigues
31000 Toulouse, France

Hamelin, Raymond
IEI
2 Boulevard de la Roche-du-Roi, B.P. 128
73101 Aix-Les-Bains Cedex, France

Hobbs, Jonathan C. A.
ICC
38 Cours Albert 1er
75008 Paris, France

Humberset, Suzanne
CRID
49 rue de la Glaciere
75013 Paris, France

Jasqui, Jacobo
IREF
23 Avenue Bosquet
75007 Paris, France

Julia, Didier
WAMM
251 rue de Vaugirard
75740 Paris Cedex 15, France

Kamto, Maurice
CIDCE
4 Place du Presidial
87031 Limoges, France

Kennedy, Norine
ICC
38 Cours Albert 1er
75008 Paris, France

King, David, Mr.
IFAP
21, Rue Chaptal
Paris 75009, France
T: 33 1 452-60553 F: 33 1 48747212
TL: 281210

Kjeldsen, H.O.A.
IFAP
21 rue Chaplal
75009 Paris, France

La Riviere, J.W.M.
ICSU
51 Boulevard de Montmorency
75016 Paris, France

Lakouetene, Pierri, Mr.
Depute
ASSEMBLEE INTERNATIONALE DES PARLEMENTAIRES DE LANGUE FRABCAISE/ AIPLF
235 Bd Saint-Germain - 75007
Paris, France
T: 33 1 47-052687 F: 33 1 45-511147
TL: 202 562f

Le Saux, Alain
WAMM
251 Rue de Vaugirard
75740 Paris Cedex 15, France

Leger, Guy, Mr.
MOUVEMENT NAT. DE LUTTE POUR L' ENVIRONNEMENT
B. P. 79 - 106 Ave. Jean Lolive
Pantin 93500, France
T: 33 1 484-60414 F: 33 1 484-6665

Lesire-Ogrel, Hubert
UTO
22 rue d'Alsace
92300 Levallois Perret, France

Levy Ep. Seligmann, Francoise
MP
139 Boulevard Victor Hugo
F-93400 Saint-Ouen, France

Mendes De Almeida, Candido
ISSC
c/o UNESCO, 1 rue Miollis
75015 Paris, France

Menon, M. G. K., Mr.
President
ICSU
51 Bd de Montmorency
Paris 75016, France
T: 33 1 4525-0329 F: 33 1 4288-9431
TL: ICSU 630553F

Mentre, Paul
E-7
30 rue de Martignac, 75007 Paris
France

O'Brien, Denise, Ms.
Attache
INTERNTIONAL CHAMBER OF COMMERCE
38 Cours Albert 1er
Paris 75008, France
T: 33 1 49532822 F: 33 1 49532924

Parienti, Raoul
CIRCS
22 rue Beaunier
75014 Paris, France

Paris, Antonio
UTO
22 rue d'Alsace
92300 Levallois Perret, France

Pitaluga, Plinio
WVF
16 rue Hamelin
75116 Paris, France

Poole, Cristina
INTERNATIONAL SOCIAL SCIENCE COUNCIL
c/o UNESCO, 1 rue Miollis
75015 Paris, France

Portella, Joaquim
WVF
16 rue Hamelin
75116 Paris, France

Raul Da Silva, Ady
IUTAO
c/o UNESCO, 1 rue de Miollis
75732 Paris Cedex 15, France

Remond-Gouilloud, Martine
CIDCE
4 Place du Presidial
87031 Limoges, France

Rigg, Tony
IUA
51 rue Raynouard
75015 Paris, France

Rivero Torre, Pedro
UNIPEDE
28 rue Jacques Ibert
75858 Paris Cedex 17, France

Rolin, Andre
UNIPEDE
28 rue Jacques Ibert
75858 Paris Cedex 17, France

Rosswall, Thomas
ICSU
51 Boulevard de Montmorency
75016 Paris, France

Rouher, Jean-Charles, Mr.
ICC
38 COURS ALBERT 1er
Paris 75008, France
T: 33 1 49532817 F: 33 1 49532835

Sall, Khalifa Babacar
Depute
ASSEMBLEE INTERNATIONALE DES PARLEMENTAIRES DE LANGUE FRABCAISE/ AIPLF
235 Boulevard Saint Germain
75007 Paris, France

Scheid Lopes, Alfredo
IFIA
28 rue Marbeuf
75008 Paris, France

Seligmann, Pierre
MOUVEMENT DE LA PAIX
139 Boulevard Victor Hugo
F-93400 Saint-Ouen, France

Stevens, Willian Ross, III
ICC
38 Cours Albert 1er
75008 Paris, France

Suzanne, Hamelin
IEI
2 Boulevard de la Roche-du-Roi, B.P. 128
73101 Aix-Les-Bains Cedex, France

Tetart, Jean-Marie
UTO
22 rue d'Alsace
92300 Levallois Perret, France

Tubiana, Laurence
CRID
49 rue de la Glaciere
75013 Paris, France

Wahl, S.M.
IFAP
21 rue Chaplal
75009 Paris, France

PRESS

Abella, Vilma
Reporter Radio/Television
RFI-LA RADIO MONDIALE
12 rue Boutarel
75006 Paris, France, France

Aldecoa, Rafael Noboa
Correspondent
AGENCE FRANCE PRESSE
Rua Maria Quiteria, No. 60
Rio de Janeiro-RJ, Brazil, France

Alyanak, Leyla
Editor
NEW ROAD
Chemin de Gota d'Or, 1095
Lutry, France

Arroyo, Bruno
Correspondent
ESPACES LATINO AMERICAINS
35, rue Paul Verlaine
69700 Villeurbanne, France, France

Atlan, Jean-Pierre
Reporter Radio/Television
TFI - TV
Paris, France

Barneaud, Dominique M.
Reporter Radio/Television
METAPHORE PRODUCTION
27, Rue Lepic
75018 Paris, France, France

Barrere, Martine
Correspondent
LIBERATION
23 Rue Jean Brunet
9219 Meudar, France, France

Benard, Sylvie
Correspondent
CHAR A VOILE MAGAZINE
179 Rue Henri Fichon 16100
Cognac, France, France

Berna, Marie Jeanne
Editor
VIVA
7, rue du Docteur Dumee
77140 Nemours s/s, France, France

Bessis, Sophie
Editor
MEDIAS FRANCE INTERCONTINENTS/RFI
45 Boulevard Beaumarchais
75003 Paris, France, France

Bittner, Katia
Correspondent
LETTRE DE L'ENVIRONNEMENT
S.Ville des Seblons
Paris, France, France

Bizot, Judithe
Correspondent
LE COURRIER DE L'UNESCO
217 Bd Raspail, 7051
Paris, France, France

Blanchet, Anne-Marie
Reporter Radio/Television
TF-1
4, ar. Jean Monnet, 92130
Issy le Roulineaux, France, France

Blouet, Antoine
Editor
DECISION ENVIRONNEMENT
15, rue Boulay
75017 Paris, France, France

Boisgallais, Anne Sophie
Editor
AGENCE ENVIRONNEMENT DEVELOPPEMENT
68, Bld de Picpus 75012
Paris, France, France

Bon, Michel
Editor
AGENCE REUTER
28 Rue Leopold Bellan
Paris, France, France

Bosc, Eric
Editor
ELYSEE
55 rue de Taubourg
Paris, France, France

Boukhari, Sophie
Editor
MEDIAS FRANCE INTERCONTINENTAUX
25, Av. de Laumiere
Paris, France, France

Bourone, Rene
Editor
AIR ESPATIAL
12 Genewe Koenig
15017 Paris, France, France

Bovillon, Gabriel
Editor
AED
9 Rue Draparnaud 34000
Montpellier, France, France

Breed, Henry, Mr.
Correspondent
THE UNESCO COURRIER
Box 245-B
Troy, N.Y., U.S.A., France

Brigouleix, Bernard
Reporter Radio/Television
EUROPE 1
6, Rue Schoelcher
Paris 24 eme, France, France

Bromberger, Dominique
Correspondent
TF-1
12 rue du Bouguet de Pongehaup
75016-Paris, France, France

Brown, Daniel
Reporter Radio/Television
RFI-LA RADIO MONDIALE
36 bix, rue Ernest Renan
92130 Issy, France, France

Burban, Pierre
LOGISTIQUE PRESS
Mauroul, Olargues, France

Calas, Olivier
Editor
AFP
4844 Western Ave.
Bethesda, MD 20816, USA, France

Calderon-Merchan, Juan
Editor
COURRIER INTERNATIONAL
67 Rue Barpault
75013 Paris, France, France

Camuyrano, Claude
Correspondent
RFI
R. Gilberto Cardoso, 270/1302
Leblon Rio de Janeiro, Brazil, France

Cans, Roger
Editor
LE MONDE
7 rue de l'Odeon
75006 Paris, France, France

Chesneaux, Jean
Correspondent
MONDE DIPLOMATIQUE
8 B. Hopital
75005 Paris, France, France

Chevalet, Michel
Reporter Radio/Television
TF-1
7 rue du Plessis
96620-Paris, France, France

Clauss, Francois
Reporter Radio/Television
UROPE 1
12 bis, rue du Val de Grace
75000 Paris, France, France

Coll, Philippe
Editor
DECISION ENVIRONNEMENT
6, Bd. do Stzasborg
75010 Paris, France, France

Colomer, Jean-Francois
Editor
LA FRANCE AGRICOLE
6 Square Theophile Gautier
75016 Paris, France, France

Contini, Alain
Correspondent
COTE DESFOSSES
18, Rue Mabillon
75000 G. Paris, France, France

Cozzolino, Sylvie
Reporter Radio/Television
RADIO NOSTALGIE
166 Bd. du Montparnasse
Paris, France, France

Crie, Helene
Editor
LIBERATION
185 rue Saint Maur
75010 Paris, France, France

De Chazournes, Noir
Reporter Radio/Television
RADIO MONTE CARLO
3 Bld Bourdon
Paris, France, France

De Miscault, Jean
Editor
LE DAUPHINE LIBERE
5 Cours Jean Javres
Grenoble, France, France

De Noray, Marie Laure
Editor
INTERDEPENDANCES
Le Impasse, Trillot
75011 Paris, France, France

De Oliveira, Claire
Correspondent
AFP
Rua Senador Euzebio, No. 14
Flamengo, Rio de Janeiro-RJ, Brazil, France

De Passoz, Anne
Editor
ADEME
6 Passage Montgallet
75012 Paris, France, France

Demigneux, Jean-Loup
Correspondent
TF-1
3116 Volta Place, NW
Washington, DC 20007, USA, France

Desjardins, Charles
Editor
FRANCE SOIR
41 rue Mazarine
75006 Paris, France, France

Dral, Simon
Reporter Radio/Television
ANTENNE 2
39 Bd. Magenta
Paris, France, France

Draine, Lisa
TF-1
2127 California St., NW
Washington, DC 20008, USA, France

Dutilleux, M. Christian
Correspondent
L'AGEFI
Rua Conde de Lages 54/507
Gloria, Rio de Janeiro CEP 20241 Brazil, France

Escande, Philippe
Editor
LES ECHOS
20 av. d'Ivry
75013-Paris, France, France

Espinosa, Januario
ESPACES LATINO AMERICAINS
15 Av. de l'Europe
69140 Ruilliax, France, France

Faucon, Regis
Correspondent
TF-1
3 rue Traktir
75116-Paris, France, France

Favier, Pierre
Editor
FRANCE PRESS
117 Bvd. Jourdan
Paris, France, France

Ferone De La Selva, Gabriel
Editor
COMBAT NATURE
BP 17
Bourre 41400, France, France

Fontrel, Nathalie
Reporter Radio/Television
RADIO FRANCE
24, Rue de L Abbe Groult
75015 Paris, France

Fouin, Julien
Editor
MARCHES TROPICAUX
1 Rue de la Pelleterie
Paris, France, France

Gaurel, Veronique
Editor
FR 3
10 Residence de Belleville
75019 Paris, France, France

Gignoux, Sabine
Editor
LA CROIX
26 Rue d'Alesia
75014 Paris, France, France

Gigon, Laurence
Editor
VIE PUBLIQUE
97 Avenue de Villiers
75017-Paris, France, France

Giloux, Laurent-Yves
Reporter Radio/Television
EUROPE 2
16, rue du Vienx Colombier
75006 Paris, France, France

Giraud, Marc
Editor
TECHNIQUES, ENVIRONNEMENT, INFORMATIQ
4 rue des Taillandiers
75011, Paris, France, France

Gouverne, Louisette
Editor
AGENCE CENTRAL DE PRESSE
147 Rue du Chemin Vert
75020 Paris, France, France

Goyomard, Yann
Editor
LES REALITES DE L'ECOLOGIE
Croix de la Femme,82 2000
Moissac, France, France

Grenier, Emmanuel
Editor
FUSION
15 rue D'Alsace
92110 Clichy, France, France

Gueneau, Stephane
Reporter Radio/Television
COURRIER DE LA PLANETE
134/11/66, Rue de Curat
3400 Montpellier, France

Guentch, Bruno
Editor
LA GAZETTE
183 Rue de la Paissonniere
75009 Paris, France, France

Guihot, Dominique
Reporter Radio/Television
OFREDIA
36 Ave. de la Republic
Gretail, France, France

Guiraut, Denis H.
Editor
LE MONDE
R.Jose Linhares, 44-501 Leblon
22430 Rio de Janeiro, Brazil, France

Harribey, Jean Marie
Correspondent
ALTERNATIVES ECONOMIQUES
33560 Carbon Blanc
France, France

Ivassenko, Nicole
Editor
AGENCE D'INFORMATION ECOLOGIQUE
7, rue du Docteur Dumel
77140 Nemous s/s, France, France

Joannidis, Marie
Editor
AFP
Mijargou Courgeou Morteguc
au Parclu 61400, France, France

Joly, Paul
Reporter Radio/Television
RTL
56 rue Gay Lussac - 75005
Paris, France

Jouve, Sophie
Reporter Radio/Television
ANTENNE 2
29 rue Clauzel
75009 Paris, France, France

Kohler, Pierre
Reporter Radio/Television
RTL
3 Chemin de la Grenouillere
Paris, France, France

Labi, Philippe
Editor
VSD MAGAZINE
3 Rue Boulanger, 75010
France, France

Laventure, Luc
Editor
RFO
5 Avenue du Redam Poicone
Paris, France, France

Le Genissel, Francois
Reporter Radio/Television
OFREDIA
22 Rue du Petit Musc
75004 Paris, France, France

Le Mounier, Jean F.
Correspondent
AFP
R.Paulo Cesar de Andrade, 232
Laranjeiras, Rio de Janeiro-RJ, Brazil, France

Le Pauder, Danielle
Editor
CENDOTEC
Av.Paulista 1195/251
Sao Paulo, S.P., Brazil, France

Lebrun, Martine
Correspondent
PARIS MATCH
8 R.des Heros Nojeunties 94/30
Nojeunt, France, France

Legris, Pierre
Editor
AGENCE PRESSE ENVIRONNEMENT
30 Rue de Tolbiac
75013 Paris, France, France

Lima, Carlos
Correspondent
AFP
Rua Sao Clemente, No. 114
Botafogo, Rio de Janeiro-RJ, Brazil, France

Lindquist, Jan
Editor
ECOCON
2B Rue de Beau Site
Plerin, France, France

Lopez, Jean
Editor
SCIENCE ET VIE JUNIOR
133 Bd. Pereire
75017 Paris, France

Louarn, Jean J.
Reporter Radio/Television
RFI-LA RADIO MONDIALE
62 rue Damremont
75018 Paris, France, France

Macedo, Sandra
Editor
JOURNAL DU DIMANCHE
34 Rue des Vignoles
75020 Paris, France, France

Mamou-Mani, Alain
Editor
DECISION ENVIRONNEMENT
11 rue Theodore de Bauville
75017 Paris, France, France

Martin-Ferrari, Dominique
Reporter Radio/Television
RADIO FRANCE
74 ure Rambutean
75001 Paris, France, France

Mayer, Sylvie, Ms.
Correspondent
L'HUMANITE
2 Place fu Colonel Fabien
75019, Paris, France, France

Meilion, Jacqueline
Editor
PARISIEN LIBERE
46 Rue Saint-Martin
95300 Pontoise, France, France

Meimon, Marc
Editor
FR 3
112 Avenue de la Republique
75011 Paris, France, France

Mennessier, Marc
Editor
LA CROIX
55 rue Jean Jaures
92170 Vanves, France, France

Miramont, Yves
Editor
LES REALITES DE L'ECOLOGIE
71 Panbomg Lacapelle
82000 Montauban, France, France

Morel, Francoise-Marie
Correspondent
TF-1
17 ave. d'Itali
75013-Paris, France, France

Muracciole, Florence
Editor
LE JOURNAL DU DIMANCHE
22 Rue Etex
Paris, France, France

Nevill, Hugh
Correspondent
AFP
4, rue Nobel
75018, Paris, France, France

Normand, Elisabeth
Editor
SERVICE DE PRESSE
64 Rue de Charonne
Paris, France, France

Olivier, Bergeron
Correspondent
LE NOUVEAU POLITIS
29 bad Odgar Quinet
75014 Paris, France, France

Parizot, Fernand
Editor
AFP
36 rue de la tour d'anverque
75009, Paris, France, France

Peltier, Martin
Editor
LE QUOTIDIEN DE PARIS
7, villa Cronstadt
75019 Paris, France, France

Pennequin, Dominique
Reporter Radio/Television
RTL RADIO
154 rue de Saussure
Paris, France, France

Perrot-Lanaud, Monique
Editor
LE MONDE DE L EDUCATION
47 Rue de Belleville
75019 Paris, France, France

Pinto, Ralph
Reporter Radio/Television
FRANCE INTER
2 Rue Beethoven
Paris, France, France

Piolatto, Laurent
Editor
LE PROGRES
01120 Thil, France

Piro, Patrick
Editor
INTERDEPENDANCES
38 Rue du 11 Novembre
94240 L'Hay-Les-Roses, France, France

Poncet, Dominique
Reporter Radio/Television
FR 3
4 Rue Furtado Hetne
75014 Paris, France, France

Raillard, Henri Georges
Reporter Radio/Television
L'EXPRESS TELECABLE
12 bis Rue du Val de Grace
Paris, France, France

Raymond, Alain
Editor
AFP
24 rue des Patriarches
75005 Paris, France, France

Razaf-Masse, Jacqueline
Editor
MEGASCOPE INTERNATIONAL
14, Bd. Jean Mermoz
9220 Neuilly s/s, France, France

Rebeyrotte, Veronique
Reporter Radio/Television
RADIO FRANCE BESANCON
5, rue du Chateau Rose
25000 Besancon, France, France

Recamier, Francoise
Editor
IMAGES DOC
1 Ruye du Regard
75006 Paris, France, France

Rey, Gloria Helena
Correspondent
AFP
SQS. 108, Bloco 'I', Apt. 403
Brasilia-DF, Brazil, France

Reynard, Delphine
Editor
VSD MAGAZINE
1 rue des Trois Bornes
75011 Paris, France, France

Richard, Evelyne
Editor
PRESS ATTACHE
6 Ave. Del Casse
Paris, France, France

Rivoire, Emmanuella
Editor
DECISION ENVIRONNEMENT
11 rue Charles Telliar
75015 Paris, France, France

Robert, Valerie
Editor
PRESS ELYSEE
55 rue du Faubourg
Paris, France, France

Robin Simon, Annie
Editor
VOLONTAIRES/DEFI
21 Affee des Pressoirs
77400 Pomponne, France

Rollat, Alain
Editor
LE MONDE
3 Rue des Aulnes
Paris, France, France

Roquelle, Sophie
Editor
LE FIGARO
39 rue Davy
75017 Paris, France, France

Royant, Olivier
Editor
PARIS MATCH
16 East 77th St. (#5-B)
New York, NY 10021, USA, France

Ruel, Patrick
Editor
LIBERATION
15 du Temple
Paris, France, France

Schaal, Florence
Correspondent
TF-1
14 bis au des Vignes
92210-St. Cloud, France, France

Schnabacher, Richard
Correspondent
COUSTEAU SOCIETY
2104 Pickwick Lane
Alexandria, VA 22307, USA, France

Schneider, Robert
Editor
OFFICIAL PRESS FRANCE
6 rue Goethe
Paris, France, France

Sermet, Georges
Editor
LYON-MATIN/LE PROGRES
30 Rue Claude Joseph Bonnet
69004 Lyon, France, France

Seroussi, Sophie
Editor
LA TRIBUNE DE L'EXPANSION
11 rue des Cloys
75018-Paris, France, France

Sevilla, Jean-Jacques
Correspondent
LIBERATION
Rua Ferreira Viana 44/701
22210 Rio de Janeiro-RJ, Brazil, France

Soerinho, Nacif
Correspondent
AFP
Rua Haddock Lopo, No. 303
Tijuca, Rio de Janeiro-RJ, Brazil, France

Strazzulla, Jerome
Editor
FIGARO
121 Rue de la Renvion
75020 Paris, France, France

Taillandier, Pascal
Editor
AFP
4606 Cooper Lane
Bethesda, Maryland, U.S.A., France

Terrasson, Francois
Editor
COMBAT NATURE
6 Rue Scipion
75005 Paris, France, France

Thewalt, Beate
Correspondent
AFP
Goebenstr.4
5300 Bonn I, Germany, France

Thomas, Collette
Editor
RFI-LA RADIO MONDIALE
21, rue de le Gource
75016-Paris, France, France

Torre, Isabelle
Reporter Radio/Television
RTL
93 Rue D'Agvege au 92100
Boulogre, Paris, France, France

Trefeu, Roger
Editor
TEMOIGNAGE CHRETIEN
44, rue de Montmorency
75003-Paris, France, France

Treuil, Jean Paul
Reporter Radio/Television
RFO MARTINIQUE
Villa Colebu no. 62, Tivoli
Martinique, France, France

Vadrot, Calude-Marie
Reporter Radio/Television
JOURNAL DU DIMANCHE
57 Rue Saint-Blaise
75020 Paris, France, France

Verdier, Marie
Editor
LA CROIX
13 rue Andre Antoine
75018 Paris, France, France

Verilhac, Yves
Editor
RHORE-ALPES NATURE
2 quai Clermcean 69300
Caluire, France, France

Vilanova, Santiago
Editor
ECRIVAINS POUR LA NATURE
Calle Doctor Duv. 10
Entlo, 1re, 08001, Barcelona, Spain, France

Zillo, Maria Da Graca
Correspondent
SAVOIR C EST POUVOIR
R. Manoel C. Godoi 592
18683-540 Lencois - SP, Brazil, France

ORGANIZATIONS

Blamont, Evelyne, Ms.
International Social Science Council
c/o UNESCO, 1 rue Miollis
75015 Paris, France
T: (33 1) 45 68 25 58 F: (33 1) 43 06 87 98

Carlson, Nils, Mr.
International Union of Architects
51 rue Raynouard
75015 Paris, France
T: (33 1) 45 24 02 78

De Leiris, Denise, Ms.
Centre De Recherche Et D'information Pour Le Developpement
49 rue de la Glaciere
75013 Paris, France
T: (33 1) 43 37 71 81 F: (33 1) 43 37 13 39

De Romain, Eric, Mr.
Environnement Sans Frontiere
1 rue du Faubourg Saint Honore
75008 Paris, France
T: (33 1) 40 76 04 04 F: (33 1) 40 76 04 05

De Vaublanc, Monique, Mme
World Movement Of Mothers*
56 rue de Passy
75016 Paris, France
T: (33 1) 45 20 55 80

Dudu, J., Mrs
World Medical Association
Avenue des Alpes
01210 Ferney-Voltaire, France
T: (33) 50 40 75 75 F: (33) 50 40 59 37
TL: 50 40 59 37

Engleberts, Hans, Mr.
Public Services International
45 Avenue Voltaire
01210 Ferney-Voltaire, France
T: (33) 50 54 64 64 F: (33) 50 40 73 20

Goncet, Odette, Ms.
Secretary General
International Federation for Home Economics (IFHE)
5 Avenue de la Porte Brancion
75015 Paris, France
T: (33 1) 48 42 34 74 F: (33 1) 42 50 09 89

Guerraz, Christian, Mr.
Institut De Recherche Pour L'energie, L'environment Et La Construction
383 Avenue de la Gare
74800 La Roche Sur Foron, France
T: (23) 50 25 96 41

Irvine, Andrew, Mr.
International Real Estate Federation
23 Avenue Bosquet
75007 Paris, France
T: (33 1) 45 50 45 49 F: (33 1) 45 50 42 00
TL: 201 339 f

Kinnon, Frances, Ms.
International Federation Of Agricultural Producers
21 rue Chaplal
75009 Paris, France
T: (33 1) 45 26 05 53 F: (33 1) 48 74 72 12
TL: 281210 ifapf

Laplante, Celine, Mme
As. Internationale Des Parlementaires De Langue Francaise
235 Boulevard Saint Germain
75007 Paris, France
T: (33 1) 47 05 26 87 F: (33 1) 45 51 11 47
TL: 202 562 F

Legay, J.M., Prof.
World Federation Of Scientific Workers
31 rue Barrier
69006 Lyon, France
T: (33) 78 52 75 58

Lucenet, Georges, Mr.
International Union of Producers and Distributors of Electrical Energy (UNIPEDE)
28 rue Jacques Ibert
75858 Paris Cedex 17, France
T: (33 1) 40 42 37 08 F: (33 1) 40 42 60 52

MacCarthy, Geraldine, Ms.
World Union Of Catholic Women's Organisations
20 rue Notre-Dame des Champs
75006 Paris, France
T: (33 1) 45 44 27 65 F: (33 1) 42 84 04 80

Maene, Luc, Mr.
International Fertilizer Industry Association
28 rue Marbeuf
75008 Paris, France
T: (33 1) 42 25 27 07 F: (33 1) 42 25 24 08
TL: 640 481 f

Makins, Christopher, Mr.
Amazon India
1 rue Joux-Aigues
31000 Toulouse, France
T: (33) 61 22 46 70 F: (33) 61 42 54 18

Marton-Lefevre, Julia, Mme
International Council Of Scientific Unions
51 Boulevard de Montmorency
75016 Paris, France
T: (33 1) 45 25 03 29 F: (33 1) 42 88 94 31
TL: ICSU PARIS 016

Pecoux, Pierre, Mr.
International Union Of Technical Associations And Organisations*
c/o UNESCO, 1 rue de Miollis
75732 Paris Cedex 15, France
T: (33 1) 45 66 94 10 F: (33 1) 43 06 29 27

Portes, Elisabeth Des, Ms.
International Council Of Museums*
Maison de l'Unesco, 1 rue Miollis
75015 Paris, France
T: (33 1) 47 34 05 00 F: (33 1) 43 06 78 62

Prieur, Michel, Mr.
Centre International De Droit Compare De L'environnement
4 Place du Presidial
87031 Limoges, France
T: (33) 55 79 19 81 F: (33) 55 33 67 18

Remundt, Art Van, Mr.
Earth Trust
Villa Hurle Vent, Villard
01220 Divonne les Bains, France
T: (33) 50 20 75 33 F: (33) 50 20 27 54

Roche, L., Prof.
Centres*
c/o Centre International de Recherche sur le Cancer
150 Cours Albert Thomas, 69372 Lyon, France
T: (33) 78 74 16 74

Sarrut, Jacques, Mr.
Federation Internationale De L'automobile
8 Place de la Concorde
75008 Paris, France
T: (33 1) 42 65 99 51 F: (33 1) 49 24 98 00

Strent, Kari Berg, Mr.
International Chamber Of Commerce
38 Cours Albert 1er
75008 Paris, France
T: (33 1) 4953 2828 F: (33 1) 4953 2859

Testore, Gianfranca Gabbai, Ms.
United Towns Organisation*
22 rue d'Alsace
92300 Levallois Perret, France
T: (33 1) 47 39 36 86 F: (33 1) 47 39 36 85

Veil, Simone, Ms.
Comite Francais pour l'Environnement
1 rue Bixio
75007 Paris, France
T: (33 1) 40 81 21 22 F: (33 1) 47 38 20 18

Vergerio, Jean-Marie, Mr.
Conseil International De Recherches Culturelles Et Spirituelles
22 rue Beaunier
75014 Paris, France
T: (33 1) 45 39 01 19 F: (33 1) 45 40 49 02

Vernet, Etienne, Mr.
Collectif Environment A Dimension Internationale
32 Avenue Pierre 1er de Serbie
75008 Paris, France
T: (33 1) 46 36 45 25 F: (33 1) 43 49 69 70

Vernet, Etienne, Mr.
ECOROPA
24 rue de l'Hermitage
75020 Paris, France
T: (33 1) 46 36 45 25 F: (33 1) 43 49 69 70

Wourgaft, Serge, Mr.
World Veterans Federation
16 rue Hamelin
75116 Paris, France
T: (33 1) 47 04 33 00 F: (33 1) 47 04 20 84
TL: FMACWVF 643253 F

GLOBAL FORUM

Abbe, Pierre
31 Rue Bergere
Paris 75009, France
T: 33 1 48 01 86 00 F: 33 1 40 22 94 12
TL: 280 859

Alliot, M.
31 Rue Bergere
Paris 75009, France
T: 33 1 48 01 86 00 F: 33 1 40 22 94 12
TL: 280 859

Antonini, Jean-Claude, Mr.
President
CONSEIL NATIONAL DU BRUIT
4 Rue Beffroy
Neuilly f 92200, France
T: 33 1 46400262 F: 33 1 47454843

Armengaud, J.
A.F.E.I.D
31 Rue Bergere
Paris 75009, France
T: 33 1 48 01 86 00 F: 33 1 40 22 94 12

Arnold, Ph.
DEPARTMENT DU HAUT-RHIN
31 Rue Bergere
Paris 75009, France
T: 33 1 48 01 86 00 F: 33 1 40 22 94 12
TL: 280 859

Arroto, Bruno
Photographer
ESPACES LATINO AMERICAIN
7 Rue Louis Adam 69100
Villeurbanne, France
T: 78 689377 F: 78 683408

Audubert, A.
MAIRE DE LA CHAPELLE
31 Rue Bergere
Paris 75009, France
T: 33 1 48 01 86 00 F: 33 1 40 22 94 12
TL: 280 859

Aussaresses, Mme, Ms.
31 Rue Bergere
Paris 75009, France
T: 33 1 48 01 86 00 F: 33 1 40 22 94 12
TL: 280 859

Aviam, Alexandre
31 Rue Bergere
Paris 75009, France
T: 33 1 48 01 86 00 F: 33 1 40 22 94 12
TL: 280 859

Barand, G.
NATURE 19
31 Rue Bergere
Paris 75009, France
T: 33 1 48 01 86 00 F: 33 1 40 22 94 12
TL: 280 859

Barbault, Robert, Mr.
Directeur
LABORATOIRE DE ZOOLOGIE
ENS 46, rue d'Ulm
Paris 75005, France
T: 33 1 4753 1163 F: 33 1 47531212
TL: CNRS 205821

Barbut, Monique
31 Rue Bergere
Paris 75009, France
T: 33 1 48 01 86 00 F: 33 1 40 22 94 12
TL: 280 859

Bardou, Gilles
31 Rue Bergere
Paris 75009, France
T: 33 1 48 01 86 00 F: 33 1 40 22 94 12
TL: 280 859

Barnier, Michel
31 Rue Bergere
Paris 75009, France
T: 33 1 48 01 86 00 F: 33 1 40 22 94 12
TL: 280 859

Barthod, Christian
31 Rue Bergere
Paris 75009, France
T: 33 1 48 01 86 00 F: 33 1 40 22 94 12
TL: 280 859

Beau, Jean Philippe, Mr.
Diretor
CEDRE
BP44
La Coucourde 41260, France
T: 33 54742280 F: 33 54740924

Benoit, Theau, Mr.
O.R.C.A.D.E.S.
12, rue des Carmelites
Poitiers 86000, France
T: 33 49 41 49 11 F: 33 49 52 22 6

Besset, Jean-Paul
AGENCE D'INFORMATION ECOLOGIQUE
Paris, France

Bessieres, Annie
Journalist
ENDA TIERS MONDE
5 Rue des Immeubles Industriels
Paris 75011, France
T: 33 1 43 720909 F: 33 1 43 721681

Biache, Roland, Mr.
LFEEP
3 rue Reeamier
Paris 75351, France
T: 33 1 43589733 F: 33 1 42841334

Bienvault, Pierre
Journalist
ENDA TIERS MONDE
5Rue des Immeubles Industriels
Paris, France
T: 33 1 43 720909 F: 33 1 43 721681

Bismuth, Roger, Mr.
MOUVEMENT NATIONAL DE LUTTE POUR L'ENVIRONMENT
Roumaillac
St. Augustine, France
T: 55 213909

Blasco, Francois, Mr.
Directeur
INSTITUTE VEGETAL DE LA CARTE INTERNATIONALE DU TAPIS
Universite Toulouse 3-39 Allee J Guesde
Toulouse 31062, France
T: 33 61 530235 F: 33 61 556470
TL: CNRS 205821

Bobe, P.
C.F.D.T.
31 Rue Bergere
Paris 75009, France
T: 33 1 48 01 86 00 F: 33 1 40 22 94 12
TL: 280 859

Bockel, Jean-Maria
31 Rue Bergere
Paris 75009, France
T: 33 1 48 01 86 00 F: 33 1 40 22 94 12
TL: 280 859

Bonnet, Isabelle, Ms.
ENDA TIERS MONDE
5 rue des Immeubles Industriels
Paris 75011, France
T: 33 1 43 72 09 09 F: 33 1 43 72 16 81

Borione, Delphine
31 Rue Bergere
Paris 75009, France
T: 33 1 48 01 86 00 F: 33 1 40 22 94 12
TL: 280 859

Bouchardeau, Huguette
31 Rue Bergere
Paris 75009, France
T: 33 1 48 01 86 00 F: 33 1 40 22 94 12
TL: 280 859

Boullon, Gabriel
Editor
AED
30 Rue Rambuteau
75012 Paris, France

Bourgoin, J.F.
VALORGA PROCESS
31 Rue Bergere
Paris 75009, France
T: 33 1 48 01 86 00 F: 33 1 40 22 94 12
TL: 280 859

Bourton, Nicolas, Sr.
BETURE SETAME
Paris, France

Bousseau, Brigitte
31 Rue Bergere
Paris 75009, France
T: 33 1 48 01 86 00 F: 33 1 40 22 94 12
TL: 280 859

Brangier, A.
ASSOCIATION RURAVIVE
31 Rue Bergere
Paris 75009, France
T: 33 1 48 01 86 00 F: 33 1 40 22 94 12
TL: 280 859

Brault, Benoit, Mr.
MOUVEMENT NAT. DE LUTTE POUR L' ENVIRONNEMENT
B. P. 79 - 106 Ave. Jean Lolive
Pantin 93500, France
T: 33 1 484-60414 F: 33 1 484-6665

Brunel,
Mme.
31 Rue Bergere
Paris 75009, France
T: 33 1 48 01 86 00 F: 33 1 40 22 94 12
TL: 280 859

Burchill, Shirley Ann, Ms.
INTERNATIONAL COUNCIL OF SCIENTIFIC UNIONS
53 Rue Du Faubourg Du Temple
Paris 75010, France
T: 33 1 42389817

Burger, P.
C.I.E.P.A.D.
31 Rue Bergere
Paris 75009, France
T: 33 1 48 01 86 00 F: 33 1 40 22 94 12
TL: 280 859

Calas, Francois, Mr.
MEDICINS SANS FRONTIERS
8 Rua St Sabin 75011
Paris, France
T: 33 1 40212929 F: 33 1 480- 6868
TL: (042)214360

Calfas, Peirre
31 Rue Bergere
Paris 75009, France
T: 33 1 48 01 86 00 F: 33 1 40 22 94 12
TL: 280 859

Calliope, Beaud, Ms.
Coodinatrice
GROUPE DE VELELAY
Rue Bonnette
Vezelay, France
T: 33 86 33 2635 F: 33 86 33 2463

Calmat, Alain
31 Rue Bergere
Paris 75009, France
T: 33 1 48 01 86 00 F: 33 1 40 22 94 12
TL: 280 859

Cans, Chantal, Mr.
MINISTERE DE L'ENVIRONNEMENT
Neuilly/Seine, France

Carignon, Alain
31 Rue Bergere
Paris 75009, France
T: 33 1 48 01 86 00 F: 33 1 40 22 94 12
TL: 280 859

Carlier, Jean, Mr.
Vice-President
CEDI -(COLLECTIF ENVIRONNMENT DE DIMENSION INTERNATIONAL)
29, Rue Brancas
Sevres 92310, France
T: 33 46 46262716

Caruette, J.P.
TOTAL
31 Rue Bergere
Paris 75009, France
T: 33 1 48 01 86 00 F: 33 1 40 22 94 12
TL: 280 859

Castano, Marie-Therese, Ms.
HORUS - AL TAISIS
Quartier Boussier
La Coucourde 26740, France
T: 33 7590 01 -01

Catalan, Jean-Charles
31 Rue Bergere
Paris 75009, France
T: 33 1 48 01 86 00 F: 33 1 40 22 94 12
TL: 280 859

Cazala, Francois-Roger
31 Rue Bergere
Paris 75009, France
T: 33 1 48 01 86 00 F: 33 1 40 22 94 12
TL: 280 859

Chabason, Lucien
31 Rue Bergere
Paris 75009, France
T: 33 1 48 01 86 00 F: 33 1 40 22 94 12
TL: 280 859

Chauvin, J.
ECOLE ET NATURE
31 Rue Bergere
Paris 75009, France
T: 33 1 48 01 86 00 F: 33 1 40 22 94 12
TL: 280 859

Chesseriaux, Jean-Michel
31 Rue Bergere
Paris 75009, France
T: 33 1 48 01 86 00 F: 33 1 40 22 94 12
TL: 280 859

Chevallier, Daniel
31 Rue Bergere
Paris 75009, France
T: 33 1 48 01 86 00 F: 33 1 40 22 94 12
TL: 280 859

Ciattoni, Jean Pascal, Mr.
Secretary General
CONSEIL NATIONAL DU BRUIT
4 Rue Beffroy
Neuilly F 92200, France
T: 33 1 464002 F: 33 1 474548

Cingal, Georges, Mr.
FRANCE NATURE ENVIRONMENT
57 Rue Cuvier
Paris, France
T: 33 1 4336 7995 F: 33 1 4336 8467

Clamens, Christine
Journalist
ENDA TIERS MONDE
5Rue des Immeubles Industriels
Paris 75011, France
T: 33 1 43 720909 F: 33 1 43 721681

Clerc, Olivier
Chief Editor
NOUVELLES PRESSES INTERNATIONALES
Vallesvilles 31570, France
T: 33 61 83 8251 F: 33 61 830779

Combrouze, Francis
31 Rue Bergere
Paris 75009, France
T: 33 1 48 01 86 00 F: 33 1 40 22 94 12
TL: 280 859

Constantin, Bruno, Mr.
O.R.C.A.D.E.S.
12, rue des Carmelites
Poitiers 86000, France
T: 33 49 41 49 11 F: 33 49 52 22 6

Costa, Jose Da
ENDA TIERS MONDE
5Rue des Immeubles Industriels
Paris, France
T: 33 1 43 720909 F: 33 1 43 721681

Costa, Ph. Da
SCOUTS DE FRANCE
31 Rue Bergere
Paris 75009, France
T: 33 1 48 01 86 00 F: 33 1 40 22 94 12
TL: 280 859

Couge, Gilles, Mr.
Mission Chief
LIGUE FRANCAISE DE L'ENSEIGNEMENT ET DE L'EDUCATION PERMANENTE - LEEEP
3, rue Recamier - Cedex 07
Paris 75341, France
T: 33 1 4358 9739 F: 33 1 4284 1834

Coumba, Ba, Ms.
Depute
ASSEMBLEE INTERNATIONALE DES PARLEMENTAIRES DE LANGUE FRABCAISE/ AIPLF
235, Bd Saint-Germain - 75007
Paris, France
T: 33 1 47-052687 F: 33 1 45-511147
TL: 202 562f

Courtet, Catherine
MINISTRE DE LA RECHERCHE
31 Rue Bergere
Paris 75009, France
T: 33 1 48 01 86 00 F: 33 1 40 22 94 12
TL: 280 859

Crepeau, Michel
31 Rue Bergere
Paris 75009, France
T: 33 1 48 01 86 00 F: 33 1 40 22 94 12
TL: 280 859

Crettiez, Pierre
31 Rue Bergere
Paris 75009, France
T: 33 1 48 01 86 00 F: 33 1 40 22 94 12
TL: 280 859

Cure, Christian
31 Rue Bergere
Paris 75009, France
T: 33 1 48 01 86 00 F: 33 1 40 22 94 12
TL: 280 859

D'Hieres, Laurent, Mr.
S I E
67RUE St Blaise 75020 Paris
Paris, France
T: 33 1 487-83033 F: 33 1 487-41410

D'Orfeuil, Rouille
31 Rue Bergere
Paris 75009, France
T: 33 1 48 01 86 00 F: 33 1 40 22 94 12
TL: 280 859

Da Costa, Jose, Mr.
ENDA TIERS MONDE
5 rue des Immeubles Industriels
Paris 75011, France
T: 33 1 43 72 09 09 F: 33 1 43 72 16 81

Danff, Jean-Pierre
31 Rue Bergere
Paris 75009, France
T: 33 1 48 01 86 00 F: 33 1 40 22 94 12
TL: 280 859

Dangaix, Denis, Mr.
Mission Chief
LIGUE FRANCAISE DE L'ENSEIGNEMENT ET DE L'EDUCATION PERMANENTE - LEEEP
3, rue Recamier - Cedex 07
Paris 75341, France
T: 33 1 4358 9739 F: 33 1 4284 1834

Darche, Michel, Mr.
CENTRE-SCIENCES
CCSTI de la Region Centre. 72, faubourg Bourgogne
Orleans, France
T: 33 38 77-11-06 F: 33 38 77-11-07

Darndenne, Marie-Louise De Noray, Ms.
Public Relations
INTERDEPENDANCES/CIMADE
176 rue de Grenelle
Paris 75007, France
T: 33 1 6556-9290 F: 33 1 6556-9290

Darnel, Annette
31 Rue Bergere
Paris 75009, France
T: 33 1 48 01 86 00 F: 33 1 40 22 94 12
TL: 280 859

Dartout, R.
CONSEIL REGIONAL ILE DE FRANCE
31 Rue Bergere
Paris 75009, France
T: 33 1 48 01 86 00 F: 33 1 40 22 94 12
TL: 280 859

Daubard, J. P., Mr.
AFVP
le Bois Dufaye
Montlherry 9134W, France
T: 33 69011095 F: 33 6907534

Daufresne, Xavier
31 Rue Bergere
Paris 75009, France
T: 33 1 48 01 86 00 F: 33 1 40 22 94 12
TL: 280 859

Dauzie, Brigitte
31 Rue Bergere
Paris 75009, France
T: 33 1 48 01 86 00 F: 33 1 40 22 94 12
TL: 280 859

De Bure, Patrick, Mr.
Commercial Manager
VOYAGES HAMELIN
31 Rue Bergere
Paris 75009, France
T: 33 1 48 01 86 00 F: 33 1 40 22 94 12
TL: 280 859

De Romain, , Mr.
E-S-F
1 rue du Faugbourg Saint-Honore
Paris 75008, France
T: 33 1 40079090 F: 33 1 40070472

Defontaine, Paul
31 Rue Bergere
Paris 75009, France
T: 33 1 48 01 86 00 F: 33 1 40 22 94 12
TL: 280 859

Dejardin, R.
C.G.T
31 Rue Bergere
Paris 75009, France
T: 33 1 48 01 86 00 F: 33 1 40 22 94 12
TL: 280 859

Delacroix, Philippe
31 Rue Bergere
Paris 75009, France
T: 33 1 48 01 86 00 F: 33 1 40 22 94 12
TL: 280 859

Delefosse, Christian
31 Rue Bergere
Paris 75009, France
T: 33 1 48 01 86 00 F: 33 1 40 22 94 12
TL: 280 859

Delhaye, Annick
Vereador
LES VERTS
30 Rue Edoward Hepriot
Aix-en Provence 13090, France
T: 42 64 2754

Delpy, J.
GENERATION ECOLOGIE CORREZE
31 Rue Bergere
Paris 75009, France
T: 33 1 48 01 86 00 F: 33 1 40 22 94 12
TL: 280 859

Depassoz, Anne
31 Rue Bergere
Paris 75009, France
T: 33 1 48 01 86 00 F: 33 1 40 22 94 12
TL: 280 859

Desert, Bernard
31 Rue Bergere
Paris 75009, France
T: 33 1 48 01 86 00 F: 33 1 40 22 94 12
TL: 280 859

Diallo, Marie-Josephine, Ms.
Functionaire
ASSEMBLEE INTERNATIONALE DES PARLEMENTAIRES DE LANGUE FRABCAISE/ AIPLF
235, Bd Saint-Germain - 75007
Paris, France
T: 33 1 47-052687 F: 33 1 45-511147
TL: 202 562f

Donzier, Ch.
FEDERATION JEUNES POUR LA NATURE
31 Rue Bergere
Paris 75009, France
T: 33 1 48 01 86 00 F: 33 1 40 22 94 12
TL: 280 859

Doumbe-Bille, Spephane, Ms.
Maitre De Conference
FACULTE DE DROIT ET DES SCIENCES ECONOMIQUES DE LIMOGES
Place du Presidial
Limoges 87031, France

Druesne, Daniel
31 Rue Bergere
Paris 75009, France
T: 33 1 48 01 86 00 F: 33 1 40 22 94 12
TL: 280 859

Ducret, Jean-Pierre
31 Rue Bergere
Paris 75009, France
T: 33 1 48 01 86 00 F: 33 1 40 22 94 12
TL: 280 859

Durin, Jean
31 Rue Bergere
Paris 75009, France
T: 33 1 48 01 86 00 F: 33 1 40 22 94 12
TL: 280 859

Dutilleux, Jean-Pierre, Mr.
ECO-FOREST
B.P. 109
Blois 41004, France
T: 33 1 48048386 F: 33 1 48 04 99 53

Ebale, Tsoubassaka, Mr.
Secret. Gen.
ORGANISATION INTERNATIONALE POUR LE DEVELOPPEMENT DE L'AFRIQUE
11, Boulevard Rochechouart
Paris 75009, France
T: 33 1 212131 F: 33 1 4 995 9889
TL: 212 131 F

Edouard, Bertrand, Mr.
Vice-President
CONSEIL REGIONAL
31 Rue Bergere
Paris 75009, France
T: 33 1 48 01 86 00 F: 33 1 40 22 94 12
TL: 280 859

Enaldi, Bahgat
Director
COURRIER DE L'UNESCO
31 Rue Franois Bonvin
Paris 75015, France
T: 45684715 F: 33 1 45669270

Esther, Peter-Davis, Ms.
ECO-CONSEIL
6 Rue d'Arvas
Strasburg 67000, France
T: 33 88 60-6397

Fardeau, Jean-Marie, Mr.
Secretario Nacional
AGIR ICI
17 Place de l'Argonne
Paris 75019, France
T: 33 1 403-50700 F: 33 1 403-50620

Faucon, Michel, Mr.
President
CRID
59 Rue de la Glaciere
Paris 75013, France
T: 33 1 53-37-71-81 F: 33 1 33-37-13-39

Ferrari, Martin, Ms.
FRIENDS OF THE EARTH
30 rue Rambuteau
Paris 75003, France
T: 33 1 48049901 F: 33 1 48049902

Ferray, Claude, Mr.
Mission Chief
LIGUE FRANCAISE DE L'ENSEIGNEMENT ET DE L'EDUCATION PERMANENTE - LEEEP
3, rue Recamier - Cedex 07
Paris 75341, France
T: 33 1 4358 9739 F: 33 1 4284 1834

Fillo, Maria De Janea
Correspondent
SAVOIR C'EST POUVOIR
13 rue Borchardom
75010 Paris, France
T: 33 1 43950992 F: 33 1 42450690

Florian, Dominique Van Egmond, Ms.
President
INSTITUT DE RECHERCHE ET APPLICATION EN AGRICULTURE BIOLOGIQUE
Ferme experimentale - 84870 - Loriol-du-Comtar Siege Social: 20, rue Jacob
Paris 75000, France
T: 33 1 603322

Fontan, Jacques, Mr.
Scientific Advisor
PROGRAMME ENVINNOMENT
CNRS 15 quai Anatole-France
Paris 75700, France
T: 33 1 4753 1163 F: 33 1 47531212
TL: CNRS 205821

Fonterel, Nathalie
FRANCE INTER
31 Rue Bergere
Paris 75009, France
T: 33 1 48 01 86 00 F: 33 1 40 22 94 12
TL: 280 859

Forestier, Hubert Le
31 Rue Bergere
Paris 75009, France
T: 33 1 48 01 86 00 F: 33 1 40 22 94 12
TL: 280 859

Fouin, Joulien
Journalist
TROPICANA ET MEDITERANEANN
75008 Paris, France
T: 43 381447 F: 64 954022

Fouque, Antoinette, Ms.
President
ALLIANCE DES FEMMES POUR LA DEMOCRATISATION
5 Rue de Lille
Paris 75007, France
T: 33 1 45 48 83 80 F: 33 1 42 86 03 08

Fur, G. Le
CONFEDERATION PAYSANNE
31 Rue Bergere
Paris 75009, France
T: 33 1 48 01 86 00 F: 33 1 40 22 94 12
TL: 280 859

Gallet, Bertrand
31 Rue Bergere
Paris 75009, France
T: 33 1 48 01 86 00 F: 33 1 40 22 94 12
TL: 280 859

Gardiol, Freddy, Mr.
ASSEMBLEE INTERNATIONALE DES PARLEMENTAIRES DE LANGUE FRABCAISE/ AIPLF
235, Bd Saint-Germain - 75007
Paris, France
T: 33 1 47-052687 F: 33 1 45-511147
TL: 202 562f

Gardiol, Irene, Ms.
Depute
ASSEMBLEE INTERNATIONALE DES PARLEMENTAIRES DE LANGUE FRABCAISE/ AIPLF
235, Bd Saint-Germain - 75007
Paris, France
T: 33 1 47-052687 F: 33 1 45-511147
TL: 202 562f

Garribba, Sergio, Mr.
Director Of Technology
IEA/OECD
2 Rue A. Pascal
Paris 15775, France
T: 33 1 4524-9960

Garric, Claire
31 Rue Bergere
Paris 75009, France
T: 33 1 48 01 86 00 F: 33 1 40 22 94 12
TL: 280 859

Gautier, Luc
31 Rue Bergere
Paris 75009, France
T: 33 1 48 01 86 00 F: 33 1 40 22 94 12
TL: 280 859

Gerente, Marcel
31 Rue Bergere
Paris 75009, France
T: 33 1 48 01 86 00 F: 33 1 40 22 94 12
TL: 280 859

Gignoux, Celine
JOURNAL LA CROIX
31 Rue Bergere
Paris 75009, France
T: 33 1 48 01 86 00 F: 33 1 40 22 94 12
TL: 280 859

Gilbert, Marc
31 Rue Bergere
Paris 75009, France
T: 33 1 48 01 86 00 F: 33 1 40 22 94 12
TL: 280 859

Gilly, Pierre, Mr.
Unmfr
UNION NATIONALE DES MAISONS FAMILIALES RURALES - UNMER
36 Allee Vivaldi
Paris 75012, France
T: 33 1 4001-1212 F: 33 1 4004-9356

Giraud, Marie, Ms.
MDRI
CNRS 15 quai Anatole-France
Paris 75700, France
T: 33 1 4753 1163 F: 33 1 47531212
TL: CNRS 205821

Giraud, Michel
31 Rue Bergere
Paris 75009, France
T: 33 1 48 01 86 00 F: 33 1 40 22 94 12
TL: 280 859

Glass, Michel, Mr.
IGBP Policy Director
IFREMER
155, Rue J.J. Rousseau
T: 0 0 46482222 F: 0 0 46482224

Godard, Olivier, Mr.
PROGRAMME ENVIRONMENT
CNRS 15 quai Anatole-France
Paris 75700, France
T: 33 1 4753 1163 F: 33 1 47531212
TL: CNRS 205821

Goetschel, P.
ORCHIDEES
31 Rue Bergere
Paris 75009, France
T: 33 1 48 01 86 00 F: 33 1 40 22 94 12
TL: 280 859

Goett, Edel, Ms.
President
PAX INTERNATIONAL
43 Rue De Villez
Bevnecourt 78270, France
T: 93-0286 F: 93-0019

Goff, Y. Le
TOTAL
31 Rue Bergere
Paris 75009, France
T: 33 1 48 01 86 00 F: 33 1 40 22 94 12
TL: 280 859

Goldstein, Michele
31 Rue Bergere
Paris 75009, France
T: 33 1 48 01 86 00 F: 33 1 40 22 94 12
TL: 280 859

Gomez, N.
DECISION ENVIRONMENT
31 Rue Bergere
Paris 75009, France
T: 33 1 48 01 86 00 F: 33 1 40 22 94 12
TL: 280 859

Gorel, Mme
M. ROY
31 Rue Bergere
Paris 75009, France
T: 33 1 48 01 86 00 F: 33 1 40 22 94 12
TL: 280 859

Grammont, M.
31 Rue Bergere
Paris 75009, France
T: 33 1 48 01 86 00 F: 33 1 40 22 94 12
TL: 280 859

Gros, Nicolas
Editor
IPM
SG Bd. Foch
93800 Epinay, France
T: 33 1 42 352325 F: 33 1 42 350426

Grumbach, Marie Claude, Ms.
ALLIANCE DES FEMMES
6 Rue De Vernenil 7500
Paris, France

Gryzick, Antoni
FR3
28 Cours Albert 1er
Paris 75008, France
T: 33 1 40744670 F: 33 1 40744700

Guerineau, D.
EPOUESOU
31 Rue Bergere
Paris 75009, France
T: 33 1 48 01 86 00 F: 33 1 40 22 94 12
TL: 280 859

Guibert, Jean-Jaques
Photographer
ESPACES LATINO AMERICAIN
7 Rue Louis Adam 69100
Villeurbanne, France
T: 78 689377 F: 78 683408

Guileminot, Bernard
31 Rue Bergere
Paris 75009, France
T: 33 1 48 01 86 00 F: 33 1 40 22 94 12
TL: 280 859

Guillon, Edwige, Ms.
GLOBAL 500
LE BOULAY Saint-Calais
Paris 72120, France
T: 33 1 43350741

Guinet, Philippe
FR3
28 Cours Albert 1er
Paris 75008, France
T: 33 1 40744670 F: 33 1 40744700

Guy, P.
CHARENTE NATURE
31 Rue Bergere
Paris 75009, France
T: 33 1 48 01 86 00 F: 33 1 40 22 94 12
TL: 280 859

Gveneau, Stephane, Mr.
Encharged Of The Enviroment Pr
SOLAGRAL
c/o IAM - BP 5056
Montpellier 34033, France
T: 33 67 54-6227

Haeringer, Philippe, Mr.
ORSTOM
Saint- Roman 26410
Chatillon-En-Diois, France
T: 75218066

Hamelin, PH
O.R.S.T.O.M.
31 Rue Bergere
Paris 75009, France
T: 33 1 48 01 86 00 F: 33 1 40 22 94 12
TL: 280 859

Han, Phawon
Journalist
ENDA TIERS MONDE
5Rue des Immeubles Industriels
Paris, France
T: 33 1 43 720909 F: 33 1 43 721681

Helmich, Henny, Mr.
External Coop Program
OECD DEVELOPMENT CENTRE
94, rue Chardon-Lagache
Paris 75016, France
T: 33 1 45248285 F: 33 1 45247943

Herisson, P.
OFFICE INTERNATIONAL DE L'EAU
31 Rue Bergere
Paris 75009, France
T: 33 1 48 01 86 00 F: 33 1 40 22 94 12
TL: 280 859

Herrmann, Ralf, Mr.
ASSOCIATION MAIS INTERNATIONAL
Route des Mauvares 13840 Rognes
Aix-En-Provence, France
T: 33 42 50-1691 F: 33 42 50-1742

Hers, Olivier
31 Rue Bergere
Paris 75009, France
T: 33 1 48 01 86 00 F: 33 1 40 22 94 12
TL: 280 859

Heuraux, Christiane
31 Rue Bergere
Paris 75009, France
T: 33 1 48 01 86 00 F: 33 1 40 22 94 12
TL: 280 859

Heyberger, Agnes
31 Rue Bergere
Paris 75009, France
T: 33 1 48 01 86 00 F: 33 1 40 22 94 12
TL: 280 859

Hidalgo, Victor
Journalist
ENDA TIERS MONDE
5Rue des Immeubles Industriels
Paris, France
T: 33 1 43 720909 F: 33 1 43 721681

Hirtz, J.M.
CONSEIL GENERAL DU GREF
31 Rue Bergere
Paris 75009, France
T: 33 1 48 01 86 00 F: 33 1 40 22 94 12
TL: 280 859

Inra,
LABORATOIRE ECONOMIE ET SOCIOLOGIE
31 Rue Bergere
Paris 75009, France
T: 33 1 48 01 86 00 F: 33 1 40 22 94 12
TL: 280 859

Jean, Ch.
LOIRE VIVANTE
31 Rue Bergere
Paris 75009, France
T: 33 1 48 01 86 00 F: 33 1 40 22 94 12
TL: 280 859

Joly, Pierre-Benoit, Mr.
Charge Du Recherche
INRA
94 Ronte De Bordeaux
Lons 64140, France
T: 33 59 326260 F: 621691

Joly, Th.
F.E.E.E.
31 Rue Bergere
Paris 75009, France
T: 33 1 48 01 86 00 F: 33 1 40 22 94 12
TL: 280 859

Jouvray, D.
ECOLE ET NATURE 01
31 Rue Bergere
Paris 75009, France
T: 33 1 48 01 86 00 F: 33 1 40 22 94 12
TL: 280 859

Kabala, Matuka, Mr.
UNESCO
7 Place Fontenoy
Paris 75700, France

Karam, Roger
31 Rue Bergere
Paris 75009, France
T: 33 1 48 01 86 00 F: 33 1 40 22 94 12
TL: 280 859

Kermabou, Jaques
Journalist
ENDA TIERS MONDE
5Rue des Immeubles Industriels
Paris 75011, France
T: 33 1 43 720909 F: 33 1 43 721681

Kremer, Pascale
Journalist
ENDA TIERS MONDE
5Rue des Immeubles Industriels
Paris, France
T: 33 1 43 720909 F: 33 1 43 721681

Kullenberg, Gunnar, Mr.
IOC OF UNESCO
7 Place Fon Tinoy
Paris 75700, France
T: 33 1 45683983

La Prairie, Herve, Mr.
Director
IFOAM
Menehy Bihan 296406
Daoulas 29460, France
T: 3398258604 F: 3398258976

Labrousse, Jean
31 Rue Bergere
Paris 75009, France
T: 33 1 48 01 86 00 F: 33 1 40 22 94 12
TL: 280 859

Lafonta, Jean-Bernard
31 Rue Bergere
Paris 75009, France
T: 33 1 48 01 86 00 F: 33 1 40 22 94 12
TL: 280 859

Lagalle, C.
R.M.C.
31 Rue Bergere
Paris 75009, France
T: 33 1 48 01 86 00 F: 33 1 40 22 94 12
TL: 280 859

Lagalle, M.
RADIO MONTE CARLO
31 Rue Bergere
Paris 75009, France
T: 33 1 48 01 86 00 F: 33 1 40 22 94 12
TL: 280 859

Lahitte, Bernard, Mr.
LES VERTS
Rue De la Sabliere
Paris 60430, France
T: 33 4481-2538

Lahitte, Nicole, Ms.
LES VERTS
5, Rue De la Sabliere
Paris 60430, France
T: 33 1 44812538

Lalonde, Brice
31 Rue Bergere
Paris 75009, France
T: 33 1 48 01 86 00 F: 33 1 40 22 94 12
TL: 280 859

Lamba, Ndikpo Mata, Mr.
Counseler
ASSEMBLEE INTERNATIONALE DES PARLEMENTAIRES DE LANGUE FRANCAISE - AIPLF
235, BD Saint-Germain
Paris 75007, France
T: 33 1 47 05 26 87 F: 33 1 45 51 11 47
TL: 202 562 f

Lasserre, Pierre, Mr.
Professor
STATION BIOLOGIQUE-ROSCOFF
Plac G. Teissier
Roscoff 29680, France
T: 33 98 292323 F: 33 98 292324

Laupetre, Julien
31 Rue Bergere
Paris 75009, France
T: 33 1 48 01 86 00 F: 33 1 40 22 94 12
TL: 280 859

Laurent, Chamussy
Photographer
SIPA - PRESS
707 Murat 75-76
Paris, France
T: 47 434743

Lauzea, Eric
CONCEIL REGIONAL
31 Rue Bergere
Paris 75009, France
T: 33 1 48 01 86 00 F: 33 1 40 22 94 12
TL: 280 859

Le Deaut, Jean-Yves
31 Rue Bergere
Paris 75009, France
T: 33 1 48 01 86 00 F: 33 1 40 22 94 12
TL: 280 859

Le Soileu, Bernard
OUEST-FRANCE
31 Rue Bergere
Paris 75009, France
T: 33 1 48 01 86 00 F: 33 1 40 22 94 12
TL: 280 859

Lebioda, D.
C.P.I.E. CHARENTES MARITIMES
31 Rue Bergere
Paris 75009, France
T: 33 1 48 01 86 00 F: 33 1 40 22 94 12
TL: 280 859

Lebioda, Denis, Mr.
Mission Chief
LIGUE FRANCAISE DE L'ENSEIGNEMENT ET DE L'EDUCATION PERMANENTE - LEEEP
3, rue Recamier - Cedex 07
Paris 75341, France
T: 33 1 4358 9739 F: 33 1 4284 1834

Lecourtier, Philippe
31 Rue Bergere
Paris 75009, France
T: 33 1 48 01 86 00 F: 33 1 40 22 94 12
TL: 280 859

Legrand, P., Mr.
FNE
57 rue Cervier
Paris 75005, France
T: 33 1 43367995 F: 33 1 43368467

Leila, , Ms.
IFAP
21, Rue Chaptal
Paris 75009, France
T: 33 1 452-60553 F: 33 1 48747212
TL: 281210

Lengagne, Guy
31 Rue Bergere
Paris 75009, France
T: 33 1 48 01 86 00 F: 33 1 40 22 94 12
TL: 280 859

Lepetit, Annick
31 Rue Bergere
Paris 75009, France
T: 33 1 48 01 86 00 F: 33 1 40 22 94 12
TL: 280 859

Leveque, Christian, Mr.
Head Of Mission
PROGRAMME ENVINNOMENT
CNRS 15 quai Anatole-France
Paris 75700, France
T: 33 1 4753 1163 F: 33 1 47531212
TL: CNRS 205821

Levy, P.
UNIVERSITE DE FRANCE
31 Rue Bergere
Paris 75009, France
T: 33 1 48 01 86 00 F: 33 1 40 22 94 12
TL: 280 859

Lewino, F.
LE POINT
31 Rue Bergere
Paris 75009, France
T: 33 1 48 01 86 00 F: 33 1 40 22 94 12
TL: 280 859

Lipietz, Alain, Mr.
AITEC - LES VERTS
22 Rue du 11 Novembre
Villejuif 94800, France
T: 33 1 40778464 F: 33 1 44243857

Longuepee, F.
CONSEIL REGIONAL ILE DE FRANCE
31 Rue Bergere
Paris 75009, France
T: 33 1 48 01 86 00 F: 33 1 40 22 94 12
TL: 280 859

Lopinot, Didier
31 Rue Bergere
Paris 75009, France
T: 33 1 48 01 86 00 F: 33 1 40 22 94 12
TL: 280 859

Luciolli, Esmeralde, Ms.
Head Of Progr.
MEDECINS SANS FRONTIERES
8, Rue Saint Sabin
Paris 75011, France
T: 33 1 402.12929 F: 33 1 480.66868
TL: (042)214360

Lucis, Frank De, Mr.
HORUS - AL TAISIS
Quartier Boussier
La Coucourde 26740, France
T: 33 7590 01 -01

Madelaint, Francois, Mr.
IFREMER
155 Rue J.J. Rousseau
Issy, France
T: 464 82100

Malpel, M.
F.N.S.E.A.
31 Rue Bergere
Paris 75009, France
T: 33 1 48 01 86 00 F: 33 1 40 22 94 12
TL: 280 859

Mamere, Noel
31 Rue Bergere
Paris 75009, France
T: 33 1 48 01 86 00 F: 33 1 40 22 94 12
TL: 280 859

Manceron, Gilles, Mr.
Mission Chief
LIGUE FRANCAISE DE L'ENSEIGNEMENT ET DE L'EDUCATION PERMANENTE - LEEEP
3, rue Recamier - Cedex 07
Paris 75341, France
T: 33 1 4358 9739 F: 33 1 4284 1834

Mani, A. Manou
DECISION ENVIRONMENT
31 Rue Bergere
Paris 75009, France
T: 33 1 48 01 86 00 F: 33 1 40 22 94 12
TL: 280 859

Mapel, M.
31 Rue Bergere
Paris 75009, France
T: 33 1 48 01 86 00 F: 33 1 40 22 94 12
TL: 280 859

Marcaillou, Laurent
Journalist
ENDA TIERS MONDE
5Rue des Immeubles Industriels
Paris 75011, France
T: 33 1 43 720909 F: 33 1 43 721681

Marois, Olivier Le
31 Rue Bergere
Paris 75009, France
T: 33 1 48 01 86 00 F: 33 1 40 22 94 12
TL: 280 859

Massias, J.P.
MAIRE DE VIGNOLS
31 Rue Bergere
Paris 75009, France
T: 33 1 48 01 86 00 F: 33 1 40 22 94 12
TL: 280 859

Masure, Philippe, Mr.
Director
BRGM/48 - ENVIRONMENT DEPARTMENT
BP 6009 - 45060
Orleans 2, France
T: 33 38 643803 F: 33 38 643013
TL: 780258F

Matringe, L.
ASSOCIATION PROT. MARAIS POITEVIN
31 Rue Bergere
Paris 75009, France
T: 33 1 48 01 86 00 F: 33 1 40 22 94 12
TL: 280 859

Maugain, Patricia
31 Rue Bergere
Paris 75009, France
T: 33 1 48 01 86 00 F: 33 1 40 22 94 12
TL: 280 859

Megret, Alain
31 Rue Bergere
Paris 75009, France
T: 33 1 48 01 86 00 F: 33 1 40 22 94 12
TL: 280 859

Mengozzi, R.
MENGOZZI SRL
31 Rue Bergere
Paris 75009, France
T: 33 1 48 01 86 00 F: 33 1 40 22 94 12
TL: 280 859

Merchan, Juan Calderon -
Editor
COURRIER INTERNATIONAL
4 Rue Raoul - Dufy
Paris 70020, France
T: 1 43584949 F: 1 43584900

Metayer, Francoise
31 Rue Bergere
Paris 75009, France
T: 33 1 48 01 86 00 F: 33 1 40 22 94 12
TL: 280 859

Meublat, Guy, Mr.
Head Of Mission
PROGRAMME ENVINNOMENT
CNRS 15 quai Anatole-France
Paris 75700, France
T: 33 1 4753 1163 F: 33 1 47531212
TL: CNRS 205821

Meublat, Mrs
PROGRAMME ENVINNOMENT
CNRS 15 quai Anatole-France
Paris 75700, France
T: 33 1 4753 1163 F: 33 1 47531212
TL: CNRS 205821

Migniac, A.
PEULPLE ET CULTURE
31 Rue Bergere
Paris 75009, France
T: 33 1 48 01 86 00 F: 33 1 40 22 94 12
TL: 280 859

Miran, Patrice, Mr.
BULLE BLEUE
12 Rue Frances de Prebeusse
Paris, France
T: 33 1 45.45.48.76 F: 33 1 45.45.77.31

Moisan, Francois
31 Rue Bergere
Paris 75009, France
T: 33 1 48 01 86 00 F: 33 1 40 22 94 12
TL: 280 859

Monediaire, Gerard, Mr.
Maitre De Conferences
FACULTE DE DROIT DE LIMOGES
Place du Presidial
Limoges, France

Montera, Hyacinthe De
31 Rue Bergere
Paris 75009, France
T: 33 1 48 01 86 00 F: 33 1 40 22 94 12
TL: 280 859

Moran, Stephane
Journalist
ENDA TIERS MONDE
5Rue des Immeubles Industriels
Paris, France
T: 33 1 43 720909 F: 33 1 43 721681

Morselli, L.
MENGOZZI SRL
31 Rue Bergere
Paris 75009, France
T: 33 1 48 01 86 00 F: 33 1 40 22 94 12
TL: 280 859

Mouliade, Pierre
31 Rue Bergere
Paris 75009, France
T: 33 1 48 01 86 00 F: 33 1 40 22 94 12
TL: 280 859

Moyence, F.
A.D.E.J
31 Rue Bergere
Paris 75009, France
T: 33 1 48 01 86 00 F: 33 1 40 22 94 12
TL: 280 859

Muller Orengo, Michelle, Ms.
ALLIANCE DES FEMMES
8 Tenasse du Parc 75019
Paris, France

Naigeon, Christophe
Directeur
AGENCE PERISCOPE MULTIMEDIA
34 bis rue de Dunkerque
Paris, France
T: 33 1 42 85 76 50 F: 33 1 40230220

Nedellec, Vincent, Mr.
MEDECINS SANS FRONTIERES
8, Rue Saint Sabin
Paris 75011, France
T: 33 1 402.12929 F: 33 1 480.66868
TL: (042)214360

N'gabie, Felix, Mr.
Counselor
ORGANISATION INTERNATIONALE POUR LE DEVELOPPEMENT DE L'AFRIQUE
11, Boulevard Rochechouart
Paris 75009, France
T: 33 1 212131 F: 33 1 4 995 9889
TL: 212 131 F

Njine, Thomas, Mr.
ORSTOM
Paris, France

Node-Langlois, Fabrice
Journalist
ENDA TIERS MONDE
5Rue des Immeubles Industriels
Paris 75011, France
T: 33 1 43 720909 F: 33 1 43 7216

Noir, Michel
31 Rue Bergere
Paris 75009, France
T: 33 1 48 01 86 00 F: 33 1 40 22 94 12
TL: 280 859

Nungesser, Roland
DEPUTE VAL DE MARNE
31 Rue Bergere
Paris 75009, France
T: 33 1 48 01 86 00 F: 33 1 40 22 94 12
TL: 280 859

Ogrel, Hubert Lesire
31 Rue Bergere
Paris 75009, France
T: 33 1 48 01 86 00 F: 33 1 40 22 94 12
TL: 280 859

Oliver, Gabirault, Mr.
Depute
ASSEMBLEE INTERNATIONALE DES PARLEMENTAIRES DE LANGUE FRABCAISE/ AIPLF
235, Bd Saint-Germain - 75007
Paris, France
T: 33 1 47-052687 F: 33 1 45-511147
TL: 202 562f

Olivier, D.
CAMPAGNIE GENERALE DES EAUX
31 Rue Bergere
Paris 75009, France
T: 33 1 48 01 86 00 F: 33 1 40 22 94 12
TL: 280 859

Oppeneau, Jean-Claude
31 Rue Bergere
Paris 75009, France
T: 33 1 48 01 86 00 F: 33 1 40 22 94 12
TL: 280 859

Orlianges, Philippe
31 Rue Bergere
Paris 75009, France
T: 33 1 48 01 86 00 F: 33 1 40 22 94 12
TL: 280 859

Ouvrieu, Jean-Bernard
31 Rue Bergere
Paris 75009, France
T: 33 1 48 01 86 00 F: 33 1 40 22 94 12
TL: 280 859

Parfait, Gisele
CEMAGREF
31 Rue Bergere
Paris 75009, France
T: 33 1 48 01 86 00 F: 33 1 40 22 94 12
TL: 280 859

Parizot, M.
A.F.P.
31 Rue Bergere
Paris 75009, France
T: 33 1 48 01 86 00 F: 33 1 40 22 94 12
TL: 280 859

Patat, Daniel
31 Rue Bergere
Paris 75009, France
T: 33 1 48 01 86 00 F: 33 1 40 22 94 12
TL: 280 859

Patrick, Pird
Chief Editor
INTERDEPENDANCES
176 Rue de Grenelle
Paris 75007, France
T: 33 1 45569290 F: 33 1 45569290

Pave, Alain, Mr.
Assistant Director
PROGRAMME ENVIRONMENT
CNRS 15 quai Anatole-France
Paris 75700, France
T: 33 1 4753 1163 F: 33 1 47531212
TL: CNRS 205821

Pave, Mrs. Alain
PROGRAMME ENVINNOMENT
CNRS 15 quai Anatole-France
Paris 75700, France
T: 33 1 4753 1163 F: 33 1 47531212
TL: CNRS 205821

Peitransanti, Yves
31 Rue Bergere
Paris 75009, France
T: 33 1 48 01 86 00 F: 33 1 40 22 94 12
TL: 280 859

Peixoto, O.
B.V.A.
31 Rue Bergere
Paris 75009, France
T: 33 1 48 01 86 00 F: 33 1 40 22 94 12
TL: 280 859

Pelletier, Benjamin, Mr.
FONDATION EUROPEENNE POUR L'ENVIRONNEMENT
80 rue des Gavilliers,
Paris 75003, France
T: 33 1 40-29-06-13 F: 33 1 40-29-05-80

Perrin, M.
S.H.F.
31 Rue Bergere
Paris 75009, France
T: 33 1 48 01 86 00 F: 33 1 40 22 94 12
TL: 280 859

Petit, Claude, Mr.
Vice-President
CONSEIL REGIONAL
31 Rue Bergere
Paris 75009, France
T: 33 1 48 01 86 00 F: 33 1 40 22 94 12
TL: 280 859

Pierre, Mariotte, Mr.
Pdg
3E PETROLEVERT
28 Rue Me Lisbonne
Paris 75008, France
T: 33 1 42894638

Pierret, Christian
31 Rue Bergere
Paris 75009, France
T: 33 1 48 01 86 00 F: 33 1 40 22 94 12
TL: 280 859

Pivot, Agnes, Ms.
Assistant
GROUPE DE RECHERCHES SOCIOLOGIQUES
Universite de Paris X 2, rue de Rouen - Bat G
Nantes 92001, France
T: 33 40 977116 F: 33 40 977155
TL: CNRS 205821

Plane, Patrick, Mr.
Charge De M
OBSERVATOIRE DE L'ENVIRONMENT
B.P. 15
Wervicq-sud, France
T: 00 33 20630610 F: 00 33 20391322

Poirot, Mr.
31 Rue Bergere
Paris 75009, France
T: 33 1 48 01 86 00 F: 33 1 40 22 94 12
TL: 280 859

Pollautre, Danielle
31 Rue Bergere
Paris 75009, France
T: 33 1 48 01 86 00 F: 33 1 40 22 94 12
TL: 280 859

Potie, Louis
31 Rue Bergere
Paris 75009, France
T: 33 1 48 01 86 00 F: 33 1 40 22 94 12
TL: 280 859

Poujade, Robert
31 Rue Bergere
Paris 75009, France
T: 33 1 48 01 86 00 F: 33 1 40 22 94 12
TL: 280 859

Poupon, Christine
COURRIER DE LA PLANETE
Passage Penel
75018 Paris, France
T: 1 42510600 F: 1 42511829

Prats, Alain
31 Rue Bergere
Paris 75009, France
T: 33 1 48 01 86 00 F: 33 1 40 22 94 12
TL: 280 859

Priceu, , Mr.
CIDCE
32 rue Turgot
Limoges 87000, France
T: 33 55 323760 F: 33 55 336718

Prudhomme, Florence, Ms.
ALLIANCE DES FEMMES
47 rue n. d. des champs paris
Paris, France

Quincy, Berengere
31 Rue Bergere
Paris 75009, France
T: 33 1 48 01 86 00 F: 33 1 40 22 94 12
TL: 280 859

Rani, Alain Ramov
Publisher
DECISION ENVIRONMENTAL
122 rue de Provence
Paris, France
T: 33 1 42 65 02 02 F: 33 1 42650064

Rappart, Isabelle
31 Rue Bergere
Paris 75009, France
T: 33 1 48 01 86 00 F: 33 1 40 22 94 12
TL: 280 859

Raunet, Michel, Mr.
C I R A D
B. P. 34032
Montpellier, France

Raynaud, Isabelle
31 Rue Bergere
Paris 75009, France
T: 33 1 48 01 86 00 F: 33 1 40 22 94 12
TL: 280 859

Raynaut, , Dr.
GRID
Universite Bordeaux 2 146 Rue Loe Saignat
Bordeaux 33076, France
T: 33 56 51 85 62 F: 33 56 51 85 66

Renaudienau, Mr., Mr.
31 Rue Bergere
Paris 75009, France
T: 33 1 48 01 86 00 F: 33 1 40 22 94 12
TL: 280 859

Reveille, Therese, Ms.
ALLIANCE DES FEMMES
13 Rue A. Vollar 75011 Paris
Paris, France

Ribeiro, R.
INSTITUTION FRERES DES HOMMES
31 Rue Bergere
Paris 75009, France
T: 33 1 48 01 86 00 F: 33 1 40 22 94 12
TL: 280 859

Rist, Laurence
31 Rue Bergere
Paris 75009, France
T: 33 1 48 01 86 00 F: 33 1 40 22 94 12
TL: 280 859

Riviere, J. W. M. La, Mr.
Secretary General
ICSU
51 Bd de Montmorency
Paris 75016, France
T: 33 1 4525-0329 F: 33 1 4288-9431
TL: 645554F, E: OMN2T:ICSU.PARIS

Rizet, Laurence
Journalist
ENDA TIERS MONDE
5Rue des Immeubles Industriels
Paris 7501, France
T: 33 1 43 720909 F: 33 1 43 721681

Roccasera, Jacques
31 Rue Bergere
Paris 75009, France
T: 33 1 48 01 86 00 F: 33 1 40 22 94 12
TL: 280 859

Rocher, J.
LABORATOIRES YVES ROCHER
31 Rue Bergere
Paris 75009, France
T: 33 1 48 01 86 00 F: 33 1 40 22 94 12
TL: 280 859

Romi, Raphael, Mr.
Professeur
UNIVERSITE DE NANTES
Chimin de la Sensive du Tertre
Nantes, France

Rossi, Eleonora
Producer
L'ARGO
11 Avenue de la Republique
Paris, France
T: 33 1 47 00 15 87

Roussel, Odile
31 Rue Bergere
Paris 75009, France
T: 33 1 48 01 86 00 F: 33 1 40 22 94 12
TL: 280 859

Roux, A.L.
UNIVERSITE C. BERNARD LYON 1
31 Rue Bergere
Paris 75009, France
T: 33 1 48 01 86 00 F: 33 1 40 22 94 12
TL: 280 859

Roy, Gerard
FR3
28 Cours Albert 1er
Paris 75008, France
T: 33 1 40744670 F: 33 1 40744700

Ruellan, Alain, Mr.
Directeur
PROGRAMME ENVINNOMENT
CNRS 15 quai Anatole-France
Paris 75700, France
T: 33 1 4753 1163 F: 33 1 47531212
TL: CNRS 205821

Saffache, Raymond, Mr.
CONSEIL REGIONAL
Boulevard du Verdun Martinique
Fort. de France, France
T: 59 63 00 F: 72 68 10

Samba, Ebrahim, Mr.
ORSTOM
Paris, France

Saugier, Bernard, Mr.
PROGRAMME ENVINNOMENT
CNRS 15 quai Anatole-France
Paris 75700, France
T: 33 1 4753 1163 F: 33 1 47531212
TL: CNRS 205821

Sautier, Denis, Mr.
Member
SOLAGRAL - CIMADE
Avenue du Val de Montferrand - BP 5035
Montpellier 34032, France
T: 33 67 61-58 00 F: 33 67 41-40 15

Schapira, Catherine
Journalist
ENDA TIERS MONDE
5Rue des Immeubles Industriels
Paris 75011, France
T: 33 1 43 720909 F: 33 1 43 721681

Sena, L. De
I.R.F.E.D.
31 Rue Bergere
Paris 75009, France
T: 33 1 48 01 86 00 F: 33 1 40 22 94 12
TL: 280 859

Serre, Georges
31 Rue Bergere
Paris 75009, France
T: 33 1 48 01 86 00 F: 33 1 40 22 94 12
TL: 280 859

Serve, Michel De La
MINISTERE DE LA RECHERCHE ET DE L'ESPACE
31 Rue Bergere
Paris 75009, France
T: 33 1 48 01 86 00 F: 33 1 40 22 94 12
TL: 280 859

Simon, Annie, Ms.
TERRE DES HOMMES
4 rue Franklin
Saint Denis 93200, France
T: 33 5 48090976 F: 33 5 4809 1575

Simon, Daniel
AGENCE DE PHOTOS GAMMA
31 Rue Bergere
Paris 75009, France
T: 33 1 48 01 86 00 F: 33 1 40 22 94 12
TL: 280 859

Simon-Collas, Mireille, Ms.
Assistante
3E PETROLEVERT
28 Rue Me Lisbonne
Paris 75008, France
T: 33 1 42894638

Siopathis, Raymond Max, Mr.
Depute
ASSEMBLEE INTERNATIONALE DES PARLEMENTAIRES DE LANGUE FRABCAISE/ AIPLF
235, Bd Saint-Germain - 75007
Paris, France
T: 33 1 47-052687 F: 33 1 45-511147
TL: 202 562f

Smagghe, Bernard, Mr.
ICAE
12 Rajaji Marg
New Delhi 110011, France
T: 91 11 301-4309

Soares, Adelia, Ms.
EUROPE CONSERVATION
BP 44
La Coucourde 41260, France
T: 33 54 742280 F: 33 54 740924

Socials, C.
FORCE OUVRIERE
31 Rue Bergere
Paris 75009, France
T: 33 1 48 01 86 00 F: 33 1 40 22 94 12
TL: 280 859

Souet, Patrick
31 Rue Bergere
Paris 75009, France
T: 33 1 48 01 86 00 F: 33 1 40 22 94 12
TL: 280 859

Soussana, Jean-Francois, Mr.
SOLAGRAL
7, rue des Chaussetiers
Clermond Ferrand 63000, France

Souza, Antonio Ribeiro De
Photojournalist
GAMMA PRESS IMAGES
104 Blv. Arago
Paris 75014, France
T: 33 144 321212 F: 33 144 321200

Stefani, Arlindo, Mr.
Anthropolo Urbanismo
AUAI-SPRL
22 Rue Hoche
Issy Les Mly 92130, France
T: 00 331 46427889 F: 32 2 6738927

Susana, Delbo
31 Rue Bergere
Paris 75009, France
T: 33 1 48 01 86 00 F: 33 1 40 22 94 12
TL: 280 859

Suzanon, Claude, Mr.
COMITE CULTURE EDUCATION ENVIRONNEMENT - GUYANE
31 Rue Bergere
Paris 75009, France
T: 33 1 48 01 86 00 F: 33 1 40 22 94 12
TL: 280 859

Tanon, Marie-Laure
31 Rue Bergere
Paris 75009, France
T: 33 1 48 01 86 00 F: 33 1 40 22 94 12
TL: 280 859

Tazieff, Haroun
31 Rue Bergere
Paris 75009, France
T: 33 1 48 01 86 00 F: 33 1 40 22 94 12
TL: 280 859

Temple, A.
SOLAGRAL
31 Rue Bergere
Paris 75009, France
T: 33 1 48 01 86 00 F: 33 1 40 22 94 12
TL: 280 859

Terra, M.
WIDE/CIDES
31 Rue Bergere
Paris 75009, France
T: 33 1 48 01 86 00 F: 33 1 40 22 94 12
TL: 280 859

Thibault, Olivier
Journalist
ENDA TIERS MONDE
5Rue des Immeubles Industriels
Paris, France
T: 33 1 43 720909 F: 33 1 43 721681

Thomas, Jean-Jacques
31 Rue Bergere
Paris 75009, France
T: 33 1 48 01 86 00 F: 33 1 40 22 94 12
TL: 280 859

Tillon, Cecile, Ms.
Chairwoman
CITY K/C.
2 Rue De Terrenne
Brest, France
T: 33 98 465733

Tiouka, Felix
ASSOCIATION DES AMERINDENS DE GUYANE
31 Rue Bergere
Paris 75009, France
T: 33 1 48 01 86 00 F: 33 1 40 22 94 12
TL: 280 859

Toutain, Francoise
ENDA
31 Rue Bergere
Paris 75009, France
T: 33 1 48 01 86 00 F: 33 1 40 22 94 12
TL: 280 859

Traulmann, Catherine
31 Rue Bergere
Paris 75009, France
T: 33 1 48 01 86 00 F: 33 1 40 22 94 12
TL: 280 859

Trogrlic, J.F.
C.F.D.T.
31 Rue Bergere
Paris 75009, France
T: 33 1 48 01 86 00 F: 33 1 40 22 94 12
TL: 280 859

Trompette, Charles, Mr.
President
CICDA
3 Chemin des Ecoliers
Cutting 57260, France
T: 33 87865475 F: 33 48515273

Truchot, Claude
31 Rue Bergere
Paris 75009, France
T: 33 1 48 01 86 00 F: 33 1 40 22 94 12
TL: 280 859

Turpin, A.
INERIS
31 Rue Bergere
Paris 75009, France
T: 33 1 48 01 86 00 F: 33 1 40 22 94 12
TL: 280 859

Turpin, Michel
31 Rue Bergere
Paris 75009, France
T: 33 1 48 01 86 00 F: 33 1 40 22 94 12
TL: 280 859

Untermaier, Jean, Mr.
Pjrofesseur
FACULTE DE LYON
15, quai Claude Bernard
Lyon 69000, France

Veret, Patrick, Dr.
COURBE
Siege Social 64, rue Michel Ange
Paris 75116, France
T: 33 1 93019292 F: 33 1 93766971

Veron, J.
C.E.P.E.D.
31 Rue Bergere
Paris 75009, France
T: 33 1 48 01 86 00 F: 33 1 40 22 94 12
TL: 280 859

Verrel, J.L.
CEMAGREF
31 Rue Bergere
Paris 75009, France
T: 33 1 48 01 86 00 F: 33 1 40 22 94 12
TL: 280 859

Viaud, Michele, Ms.
81 Rue Saint-Navr
Paris 75011, France

Victoire, Louis Marie, Ms.
Sociologa
71 rue St Jacques
Paris 75005, France
T: 33 1 432-98652

Vidal, Janine, Ms.
ASSOCIACAO FRANCA/BRASIL
2 Rue Patel Bert
Boulogne 92100, France
T: 1 48 258 404

Vigouroux, Robert
31 Rue Bergere
Paris 75009, France
T: 33 1 48 01 86 00 F: 33 1 40 22 94 12
TL: 280 859

Villaudiere, S.
EUROSPANORAMA
31 Rue Bergere
Paris 75009, France
T: 33 1 48 01 86 00 F: 33 1 40 22 94 12
TL: 280 859

Vrignaud, Rene
31 Rue Bergere
Paris 75009, France
T: 33 1 48 01 86 00 F: 33 1 40 22 94 12
TL: 280 859

Waechter, Antoine
LES VERTS
31 Rue Bergere
Paris 75009, France
T: 33 1 48 01 86 00 F: 33 1 40 22 94 12
TL: 280 859

Wandscheer, Joao C., Mr.
Coordenador
MEDECINS DU MONDE
67 Ave. De la Republique
Paris 75001, France
T: 33 1 4929-1515

Weber, Jacques, Mr.
Recherche Scientifique
IFREMER
155, Rue J.J. Rousseau
Issy 92138, France
T: 90 1 46482100 F: 90 1 46482276

Weber, Jean-Louis
31 Rue Bergere
Paris 75009, France
T: 33 1 48 01 86 00 F: 33 1 40 22 94 12
TL: 280 859

Williams, Sarah, Ms.
Communications Assistant
INTERNTIONAL CHAMBER OF COMMERCE
38 Cours Albert 1er
Paris 75008, France
T: 33 1 49532822 F: 33 1 49532924

Yaker, Farid, Mr.
INTENATIONAL OCEAN INSTITUTE
5 Rue Lucien Et Sacha Guitny
Paris 75020, France
T: 33 1 437-1374 F: 33 1 457-7749

Younes, Talal, Mr.
Executive Director
ICSU/IUBS
51 Bd de Montmorency
Paris 75016, France
T: 33 1 4525-0329 F: 33 1 4288-9431
TL: 645554F, E: OMN2T:ICSU.PARIS

UN AND UNCED

Badran, Adnan
United Nations Educational, Scientific, and Cultural Organization
9, place de Fontenoy
75007 Paris, France
T: 331 45681000 F: 331 45671690

Barbiere, Jean
United Nations Educational, Scientific, and Cultural Organization
9, place de Fontenoy
75007 Paris, France
T: 331 45681000 F: 331 45671690

Barbin, Christina
United Nations Educational, Scientific, and Cultural Organization
9, place de Fontenoy
75007 Paris, France
T: 331 45681000 F: 331 45671690

Bernard-Meunier, Marie
United Nations Educational, Scientific, and Cultural Organization
9, place de Fontenoy
75007 Paris, France
T: 331 45681000 F: 331 45671690

Cauwet, Laurence
United Nations Educational, Scientific, and Cultural Organization
9, place de Fontenoy
75007 Paris, France
T: 331 45681000 F: 331 45671690

Da, Alcion
United Nations Educational, Scientific, and Cultural Organization
9, place de Fontenoy
75007 Paris, France
T: 331 45681000 F: 331 45671690

Damiamian, Jeanne
United Nations Educational, Scientific, and Cultural Organization
9, place de Fontenoy
75007 Paris, France
T: 331 45681000 F: 331 45671690

Di, Francesco
United Nations Educational, Scientific, and Cultural Organization
9, place de Fontenoy
75007 Paris, France
T: 331 45681000 F: 331 45671690

Edwards, Judith
United Nations Educational, Scientific, and Cultural Organization
9, place de Fontenoy
75007 Paris, France
T: 331 45681000 F: 331 45671690

Enriquez, Miguel
United Nations Educational, Scientific, and Cultural Organization
9, place de Fontenoy
75007 Paris, France
T: 331 45681000 F: 331 45671690

Fournier, Francine
United Nations Educational, Scientific, and Cultural Organization
9, place de Fontenoy
75007 Paris, France
T: 331 45681000 F: 331 45671690

Glaser, Gisbert
United Nations Educational, Scientific, and Cultural Crganization
9, place de Fontenoy
75007 Paris, France
T: 331 45681000 F: 331 45671690

Hadley, Malcolm
United Nations Educational, Scientific, and Cultural Organization
9, place de Fontenoy
75007 Paris, France
T: 331 45681000 F: 331 45671690

Hall, Orlando
United Nations Educational, Scientific, and Cultural Organization
9, place de Fontenoy
75007 Paris, France
T: 331 45681000 F: 331 45671690

Kabala, David
United Nations Educational, Scientific, and Cultural Organization
9, place de Fontenoy
75007 Paris, France
T: 331 45681000 F: 331 45671690

Karim, R.
United Nations Educational, Scientific, and Cultural Organization
9, place de Fontenoy
75007 Paris, France
T: 331 45681000 F: 331 45671690

Lyard, Chantal
United Nations Educational, Scientific, and Cultural Organization
9, place de Fontenoy
75007 Paris, France
T: 331 45681000 F: 331 45671690

Marques, Carlos
United Nations Educational, Scientific, and Cultural Organization
9, place de Fontenoy
75007 Paris, France
T: 331 45681000 F: 331 45671690

Missotten, Robert
United Nations Educational, Scientific, and Cultural Organization
9, place de Fontenoy
75007 Paris, France
T: 331 45681000 F: 331 45671690

Modoux, Alain
United Nations Educational, Scientific, and Cultural Organization
9, place de Fontenoy
75007 Paris, France
T: 331 45681000 F: 331 45671690

Moore, Howard, Mr.
United Nations Educational, Scientific, and Cultural Organization
9, place de Fontenoy
75007 Paris, France
T: 331 45681000 F: 331 45671690

Portella, Eduardo
United Nations Educational, Scientific, and Cultural Organization
9, place de Fontenoy
75007 Paris, France
T: 331 45681000 F: 331 45671690

Robertson, Jane
United Nations Educational, Scientific, and Cultural Organization
9, place de Fontenoy
75007 Paris, France
T: 331 45681000 F: 331 45671690

Rodriguez, Pio
United Nations Educational, Scientific, and Cultural Organization
9, place de Fontenoy
75007 Paris, France
T: 331 45681000 F: 331 45671690

Roumain, Evelyne
United Nations Educational, Scientific, and Cultural Organization
9, place de Fontenoy
75007 Paris, France
T: 331 45681000 F: 331 45671690

Saleh, Ahmed
United Nations Educational, Scientific, and Cultural Organization
9, place de Fontenoy
75007 Paris, France
T: 331 45681000 F: 331 45671690

Sasson, Albert
United Nations Educational, Scientific, and Cultural Organization
9, place de Fontenoy
75007 Paris, France
T: 331 45681000 F: 331 45671690

Sawaya, J.
United Nations Educational, Scientific, and Cultural Organization
9, place de Fontenoy
75007 Paris, France
T: 331 45681000 F: 331 45671690

Soneghetti, Ricardo
United Nations Educational, Scientific, and Cultural Organization
9, place de Fontenoy
75007 Paris, France
T: 331 45681000 F: 331 45671690

Szollosi-Nagy, Andras
United Nations Educational, Scientific, and Cultural Organization
9, place de Fontenoy
75007 Paris, France
T: 331 45681000 F: 331 45671690

Tochtermann, Wolfgang
United Nations Educational, Scientific, and Cultural Organization
9, place de Fontenoy
75007 Paris, France
T: 331 45681000 F: 331 45671690

Tornudd, Nina
United Nations Educational, Scientific, and Cultural Organization
9, place de Fontenoy
75007 Paris, France
T: 331 45681000 F: 331 45671690

Von, Bernd
United Nations Educational, Scientific, and Cultural Organization
9, place de Fontenoy
75007 Paris, France
T: 331 45681000 F: 331 45671690

Von, Christina
United Nations Educational, Scientific, and Cultural Organization
9, place de Fontenoy
75007 Paris, France
T: 331 45681000 F: 331 45671690

Wright, Gary
United Nations Educational, Scientific, and Cultural Organization
9, place de Fontenoy
75007 Paris, France
T: 331 45681000 F: 331 45671690

ADVISORS

Kamba, W.J., Prof.
President, International Association Of Universities
75732 Paris, Cedex 15, France
T: 33 1 45 68 25 45 F: 33 1 47 34 76 05

Sachs, Ignacy, Prof.
75270 Paris, France
T: 33 1 49 54 20 85 F: 33 1 45 48 83 53/45 44 93 11

FRENCH GUYANA - GLOBAL FORUM

Aboucher, Emmanuelle, Sra.
ASSOCIATION GUYANE HORIZON VERT
8 Rue Dr. Barrat
Cayenne 97300, French Guyana
T: 59 4 30 90 27

Delannon, Roland, Mr.
Chef Du Service Environment
CONSEIL GENERAL (GUYANE)
4 place Leopold Heder
Cayenne 97300, French Guyana
T: 59 4 315708 F: 59 4 305183

Fencois, Marie Carmelie, Ms.
Librarian
BAHA'I
Route de Desgrad des Cannes
Remire/Montjoly 97300, French Guyana
T: 59 4 35 4018

Fevre, Marcel, Mr.
CFPPA
2 Chemin Gibelin
Matoury 97351, French Guyana

Gutierrez, Luis, Sr.
ASSOCIATION GUYANE HORIZON VERT
9 Rue Dr. Barrat
Cayenne 97300, French Guyana
T: 59 4 30 90 27

Hoen, Maria Das Gracas, Ms.
Vice-President
BAHA'I
43, Rue Felix Ebone
Cayenne, French Guyana
T: 59 4 304300

Jacquemot, Herve, Mr.
ASSOCIATION GUYANE HORIZON VERT
64 Rue Mme Payet
Cayenne 97300, French Guyana
T: 59 4 31 02 96

Lancri, Frederic, Mr.
K.K.C.
31 av Pasteur
Cayenne 97300, French Guyana

Martial, Achille, Mr.
Service Environment
CONSEIL GENERAL - GYUANE
4 Place Leopold Heder
Cayenne 97300, French Guyana
T: 59 4 315708 F: 59 4 305183

Maud, Mirval, Ms.
Adjoint Du Service Env.
CONSEIL GENERAL - GYUANE
4 Place Leopold Heder
Cayenne 97300, French Guyana
T: 59 4 315708 F: 59 4 305183

Mirval, Maud
CONSEIL GENERAL (GUYANE)
4 place Leopold Heder
Cayenne 97300, French Guyana
T: 59 4 315708 F: 59 4 305183

Mongenie, Jean Claude, Mr.
ASSOCIATION GUYANE HORIZON VERT
68 Avenue de la Liberte
Cayenne 97300, French Guyana
T: 49 4 30 78 82

Phinera, Stephan, Mr.
Chef Du Service Environment
CONSEIL GENERAL (GUYANE)
4 place Leopold Heder
Cayenne 97300, French Guyana
T: 59 4 315708 F: 59 4 305183

Roland, Delannon, Mr.
Chef Du Service Environnement
CONSEIL GENERAL - GYUANE
4 Place Leopold Heder
Cayenne 97300, French Guyana
T: 59 4 315708 F: 59 4 305183

Thuriaf, Christian, Mr.
ASSOCIATION GUYANE HORIZON VERT
3 Cite Castor
Cayenne, French Guyana
T: 59 4 31 0392

Wood, Kris, Mr.
ASSOCIATION GUAYANE HORIZON VERT
B.P. 119
St. Laurent du Maron 97320, French Guyana
T: 59 49 34 20 97

GABON - GOVERNMENT

Apangha-Okegue, Herve Theophile, M.
Premier Conseiller l' Ambassade au Bresil
Permanent Mission of Gabon
18 E. 41st Str. 6th Floor
New York, NY 10017, United States of America
T: 212 686 9720 F: 212 689 5769

Bie Eyene, , M.
Directeur des Organisations Internationales et de la Cooperation Multilaterale, Ministere des Affaires Etrangeres de la Cooperation et de la Francophonie
Permanent Mission of Gabon
18 E. 41st Str. 6th Floor
New York, NY 10017, United States of America
T: 212 686 9720 F: 212 689 5769

Biffot, Alba, Mlle
Deuxieme Conseiller
Mission Permanente aupres des Nations Unies, New York
Permanent Mission to the United Nations, 18 E. 41st Str. 6th Floor
New York, NY 10017, United States of America
T: 212 686 9720 F: 212 689 5769

Boguikouma, Bruno, Prof.
Medecin personnel du President de la Republique
Permanent Mission of Gabon
18 E. 41st Str. 6th Floor
New York, NY 10017, United States of America
T: 212 686 9720 F: 212 689 5769

Bokoko, Monique, Mme
Expert
Permanent Mission of Gabon
18 E. 41st Str. 6th Floor
New York, NY 10017, United States of America
T: 212 686 9720 F: 212 689 5769

Bongo, Albertine, Mlle
Expert
Ministere des Affaires Etrangeres de la Cooperation et de la Francophonie
Permanent Mission to the United Nations, 18 E. 41st Str. 6th Floor New York, NY, 10017
F: 212 686 9720
TL: 212 689 5769

Bongo, Ali, M.
Depute
Permanent Mission of Gabon
18 E. 41st Str. 6th Floor
New York, NY 10017, United States of America
T: 212 686 9720 F: 212 689 5769

Bongo, El Hadj Omar, Mr.
President
Permanent Mission of the Gabonese Republic
18 E. 41st. Str. 6th floor
New York NY 10017, United States of America
T: 212 686 9720 F: 212 689 5769

Bongo, Nadine, Mlle
Expert
Ministere des Affaires Extrangeres, de la Cooperation et de la Francophonie
Permanent Mission to the United Nations, 18 E. 41st Str. 6th Floor New York, NY, 10017
F: 212 686 9720
TL: 212 689 5769

Bongo, Pascaline, S.E. Mme
Ministre des Affaires Etrangeres de la Cooperation et de la Francophonie
Permanent Mission of Gabon
18 E. 41st Str. 6th Floor
New York, NY 10017, United States of America
T: 212 686 9720 F: 212 689 5769

Bruneau, Roger, General
Directeur general du Protoocole d Etat
Permanent Mission of Gabon
18 E. 41st Str. 6th Floor
New York, NY 10017, United States of America
T: 212 686 9720 F: 212 689 5769

Capito, Eugene, S.E. M.
Ministre des Eaux et Forets
Permanent Mission of Gabon
18 E. 41st Str. 6th Floor
New York, NY 10017, United States of America
T: 212 686 9720 F: 212 689 5769

Dangue Rewaka, Denis, S.E. M.
Ambassadeur Extraordinaire et Plenipotentiaire
Representant Permanent aupres des Nations Unies, New York
Permanent Mission to the United Nations, 18 E. 41st Str. 6th Floor
New York, NY 10017, United States of America
T: 212 686 9720 F: 212 689 5769

Dossou Naki, , Mme.
Conseiller du President de la Republique aux Affaires africaines et internationales
Permanent Mission of Gabon
18 E. 41st Str. 6th Floor
New York, NY 10017, United States of America
T: 212 686 9720 F: 212 689 5769

Dukuly, Julienne, Mme
Expert
Permanent Mission of Gabon
18 E. 41st Str. 6th Floor
New York, NY 10017, United States of America
T: 212 686 9720 F: 212 689 5769

Etchenda, Fidele, M.
Conseiller a l' Information
Permanent Mission of Gabon
18 E. 41st Str. 6th Floor
New York, NY 10017, United States of America
T: 212 686 9720 F: 212 689 5769

Garba, Mohammed Lawal
Secretary General
AFRICAN TIMBER ORGANISATION
B.P. 1077
Libreville, Gabon
T: 241 73 29 28 F: 241 73 40 30

Gondjout, Laure, Mme
Expert
Permanent Mission of Gabon
18 E. 41st Str. 6th Floor
New York, NY 10017, United States of America
T: 212 686 9720 F: 212 689 5769

Houangni Ambourouet, Marie-Augustine, Hon.
Vice-Presidente de l' Assemblee Nationale
Permanent Mission of Gabon
18 E. 41st Str. 6th Floor
New York, NY 10017, United States of America
T: 212 686 9720 F: 212 689 5769

Lemboumba, Jean-Pierre, M.
Directeur de Cabinet du President de la Republique
Permanent Mission of Gabon
18 E. 41st Str. 6th Floor
New York, NY 10017, United States of America
T: 212 686 9720 F: 212 689 5769

Mabicka, , M.
Secretaire General, du Ministere des Finances du Budget et des Participations
Permanent Mission of Gabon
18 E. 41st Str. 6th Floor
New York, NY 10017, United States of America
T: 212 686 9720 F: 212 689 5769

Mamadou, Ntsoumou
Conseiller Du President De La Repub
CABINET DU PRESIDENT DE LA RE'PUBLI
B.P. 546 Libreville
Libreville, Gabon
T: 241 76 33 85 OU 39 F: 241 77 33 46

Mamboungou, Joseph, S.E. M.
Ambassadeur du Gabon au Bresil
Permanent Mission of Gabon
18 E. 41st Str. 6th Floor
New York, NY 10017, United States of America
T: 212 686 9720 F: 212 689 5769

Mavoungou, Vincent, M.
Directeur adjoint du Cabinet du President de la Republique
Permanent Mission of Gabon
18 E. 41st Str. 6th Floor
New York, NY 10017, United States of America
T: 212 686 9720 F: 212 689 5769

Mavoungou Bongo, Donatien, M.
Conseiller Special du President de la Republique pour Affaires Scientifiques
Permanent Mission of Gabon
18 E. 41st Str. 6th Floor
New York, NY 10017, United States of America
T: 212 686 9720 F: 212 689 5769

Mba, Adrien Nguema, Hon.
Parlementaire
Permanent Mission of Gabon
18 E. 41st Str. 6th Floor
New York, NY 10017, United States of America
T: 212 686 9720 F: 212 689 5769

Mba Bekale, Serge, S.E. M.
Ministre de la Justice
Garde des Sceaux
Permanent Mission to the United Nations, 18 E. 41st Str. 6th Floor New York, NY, 10017
F: 212 686 9720
TL: 212 689 5769

Mba Obame, Andre, M.
Secretaire general adjoint a la Presidence de la Republique
Permanent Mission of Gabon
18 E. 41st Str. 6th Floor
New York, NY 10017, United States of America
T: 212 686 9720 F: 212 689 5769

Medeng, , M.
Expert
Ministere des Eaux et Forets
Permanent Mission to the United Nations, 18 E. 41st Str. 6th Floor New York, NY, 10017
F: 212 686 9720
TL: 212 689 5769

Memiaghe, Gisele, Mme
Secretaire general adjoint Ministere des Affaires estrangeres de la Cooperation et de la Francophonie
Permanent Mission of Gabon
18 E. 41st Str. 6th Floor
New York, NY 10017, United States of America
T: 212 686 9720 F: 212 689 5769

Mololas, , M.
Expert, Ministere du Tourisme de l' Environnment et des Parcs nationaux
Permanent Mission of Gabon
18 E. 41st Str. 6th Floor
New York, NY 10017, United States of America
T: 212 686 9720 F: 212 689 5769

Mongockodji, Pepin, S.E. M.
Ministre du Tourisme de lenvironnement et des Parcs nationaux
Permanent Mission of Gabon
18 E. 41st Str. 6th Floor
New York, NY 10017, United States of America
T: 212 686 9720 F: 212 689 5769

Moucani Muesta, Victor, S.E. M.
Ministre des Petites et Moyennes Entreprises et de l Artisanat
Permanent Mission of Gabon
18 E. 41st Str. 6th Floor
New York, NY 10017, United States of America
T: 212 686 9720 F: 212 689 5769

Ndao Rilogue, , M.
Expert
Ministere des Mines et de l energie
Permanent Mission to the United Nations, 18 E. 41st Str. 6th Floor New York, NY, 10017
F: 212 686 9720
TL: 212 689 5769

Ndong, Joseph, M.
Conseiller juridique du President de la Republique
Permanent Mission of Gabon
18 E. 41st Str. 6th Floor
New York, NY 10017, United States of America
T: 212 686 9720 F: 212 689 5769

Ndong Sima, , M.
Expert
Ministere de la Planification
Permanent Mission to the United Nations, 18 E. 41st Str. 6th Floor New York, NY, 10017
F: 212 686 9720
TL: 212 689 5769

Ndong-Obiang, Joseph, Mr
Conseiller Special Du President De
REPUBLIQUE GABONAISE
BP 3503-Libreville (Gabon)
Libreville, Gabon
T: 241 76.40.21

Nze Obame, , M.
Directeur Amerique au ministere des affaires etrangeres de la cooperation et de la francophonie
Permanent Mission of Gabon
18 E. 41st Str. 6th Floor
New York, NY 10017, United States of America
T: 212 686 9720 F: 212 689 5769

Obeyi, Richard, M.
Expert
Permanent Mission of Gabon
18 E. 41st Str. 6th Floor
New York, NY 10017, United States of America
T: 212 686 9720 F: 212 689 5769

Ona-Ondo, Daniel, M.
Conseiller du President de la Republique
BP 4443 Librevlue
Libreville, Gabon
T: 763379 730417 F: 241 763317

Ondo Methogo, Emmanuel, S.E. M.
Ministre de la Planification de leconomie et de l' Amenagement du Territoire
Permanent Mission of Gabon
18 E. 41st Str. 6th Floor
New York, NY 10017, United States of America
T: 212 686 9720 F: 212 689 5769

Oyono, Simon, S.E. M.
Ministre du Travail des Ressources Humaines et de la Formation Professionnelle
Permanent Mission of Gabon
18 E. 41st Str. 6th Floor
New York, NY 10017, United States of America
T: 212 686 9720 F: 212 689 5769

Ping, Jean, M.
Conseiller personnel du President de la Republique
Permanent Mission of Gabon
18 E. 41st Str. 6th Floor
New York, NY 10017, United States of America
T: 212 686 9720 F: 212 689 5769

Rombonot, Albert, M.
Directeur General, Ministere du Tourisme de l' Environnement et des Parcs Nationaus
Permanent Mission of Gabon
18 E. 41st Str. 6th Floor
New York, NY 10017, United States of America
T: 212 686 9720 F: 212 689 5769

Siawaya, Daniel, M.
Directeur general adjoint du Protocole d' Etat
Permanent Mission of Gabon
18 E. 41st Str. 6th Floor
New York, NY 10017, United States of America
T: 212 686 9720 F: 212 689 5769

Toungui, Paul, S.E. M.
Ministre des Finances du Budget et des Participations
Permanent Mission of Gabon
18 E. 41st Str. 6th Floor
New York, NY 10017, United States of America
T: 212 686 9720 F: 212 689 5769

PRESS

Elle, Mekome Gaston
Reporter Radio/Television
TELEVISION GABONAISE
Libreville, Gabon
Gabon

Mboudy, Serge
Editor
TELEVISION GABONAISE
Libreville, Gabon
Gabon

GLOBAL FORUM

Boussiengui, Raymond Mondjo, Mr.
Directeur
EGLISE ALLIANCE CHRETIENNE GABON
BP 13484 Gros Bouquet
Libreville, Gabon

GAMBIA - GOVERNMENT

Jawara, Dawada K., Mr.
President
Permanent Mission of Gambia
820 2nd. Ave. 9th floor
New York NY 10017, United States of America
T: 212 949 6640 F: 212 808 4975

Touray, Hon. Sajo K.
Minister Of Natural Resources And E
GAMBIA
5 Marina Road
Banjul, Gambia
T: 28545/92658

PARLIAMENTARIANS / SPIRITUAL LEADERS

Touray, Sajo, Minister
House of Representatives Legislature Department Parliament Building Independence Drive
Banjul, Gambia
TL: 2204 PRESOF GV

GAZA STRIP - GLOBAL FORUM

Aby Safeyya, Yousef, Mr.
Chairman
AL AZHAR UNIVERSITY OF GAZA
PO Box 345
Gaza, Gaza Strip

GERMANY - GOVERNMENT

Adam, Birgit, Mrs.
Attachee
Embassy to Brazil
Permanent Mission to the United Nations, 600 Third Ave., 41st Floor New York, NY, 10016
F: 212 856 6200
TL: 212 856 6280

Bauer, Prof. Julian, Mr
President
ECO 2 TERRA
P.O. Box 100
0-3414 Hardegsen, Germany
T: 49 711 447878 F: 49 711 445500

Behlau, Arno, Mr.
Intesu. Coordinator
BOND-FEF GERMANY
Im Rueingarten 7 5300 Bonn
Germany
T: 49 228 40097 0

Bergen, Monika, Mrs.
Official
Federal Ministry for Research and Technology
5300 Bonn, Germany
T: 0049 228 59 3834 F: 0049 228 59 3603

Bohnet, Michael, Dr.
Director
Federal Ministry for Economic Cooperation
Permanent Mission to the United Nations, 600 Third Ave., 41st Floor New York, NY, 10016
F: 212 856 6200
TL: 212 856 6280

Bothmer, H. D. von, Dr.
Official
Federal Ministry for Agriculture
Permanent Mission to the United Nations, 600 Third Ave., 41st Floor New York, NY, 10016
F: 212 856 6200
TL: 212 856 6280

Clouth, Elmar, Mr.
Official
Federal Ministry for Economy
Permanent Mission to the United Nations, 600 Third Ave., 41st Floor New York, NY, 10016
F: 212 856 6200
TL: 212 856 6280

Contius, Stephan, Mr.
Official
Federal Ministry for the Environment
Permanent Mission to the United Nations, 600 Third Ave., 41st Floor New York, NY, 10016
F: 212 856 6200
TL: 212 856 6280

De Almeida Rosa, Francisca Martins, Mrs.
Consulate General
Rio de Janeiro
Permanent Mission to the United Nations, 600 Third Ave., 41st Floor New York, NY, 10016
F: 212 856 6200
TL: 212 856 6280

Dik, Wolfgang, Mr.
First Secretary
Embassy to Brazil
Permanent Mission to the United Nations, 600 Third Ave., 41st Floor New York, NY, 10016
F: 212 856 6200
TL: 212 856 6280

Durkop, Klaus, Mr.
National UNCED-Committee
Permanent Mission to the United Nations, 600 Third Ave., 41st Floor New York, NY, 10016
F: 212 856 6200
TL: 212 856 6280

Engelhardt, Wolfgang, Prof. Dr.
National UNCED-Committee
Permanent Mission to the United Nations, 600 Third Ave., 41st Floor New York, NY, 10016
F: 212 856 6200
TL: 212 856 6280

Enkelmann, Dagmar, Dr.
Member
Federal Parliament
Permanent Mission to the United Nations, 600 Third Ave., 41st Floor New York, NY, 10016
F: 212 856 6200
TL: 212 856 6280

Erl, Willi, Mr.
Director
DED
Uhlandstrasse 3
Berlin, Germany
T: 37 30 365 09102 F: 37 2 365 09211
TL: 308628DED

Erlenbach, Doris, Mrs.
Federal Ministry for the Environment
Permanent Mission to the United Nations, 600 Third Ave., 41st Floor New York, NY, 10016
F: 212 856 6200
TL: 212 856 6280

Fabricius, Brunhilde, Mrs.
National UNCED-Committee
Permanent Mission to the United Nations, 600 Third Ave., 41st Floor New York, NY, 10016
F: 212 856 6200
TL: 212 856 6280

Fassbender, Julia, Mrs.
Federal Press and Information Office
Permanent Mission to the United Nations, 600 Third Ave., 41st Floor New York, NY, 10016
F: 212 856 6200
TL: 212 856 6280

Felter, Franz-Josef, Dr.
Director General
Chancellery
Permanent Mission to the United Nations, 600 Third Ave., 41st Floor New York, NY, 10016
F: 212 856 6200
TL: 212 856 6280

Fiege, Klaus Dieter, Dr.
Member
Federal Parliament
Permanent Mission to the United Nations, 600 Third Ave., 41st Floor New York, NY, 10016
F: 212 856 6200
TL: 212 856 6280

Fischer, Alfred-Herwig, Dr. Ing.
Expert
Permanent Mission of Germany
600 Third Ave., 41st Floor
New York, NY 10016, United States of America
T: 212 856 6200 F: 212 856 6280

Fleckenstein, Kurt, Dr.
National UNCED-Committee
Permanent Mission to the United Nations, 600 Third Ave., 41st Floor New York, NY, 10016
F: 212 856 6200
TL: 212 856 6280

Floter, Udo, Mr.
Consulate General
Rio de Janeiro
Permanent Mission to the United Nations, 600 Third Ave., 41st Floor New York, NY, 10016
F: 212 856 6200
TL: 212 856 6280

Fritzenkotter, Andreas, Mr.
Director
Chancellery
Permanent Mission to the United Nations, 600 Third Ave., 41st Floor New York, NY, 10016
F: 212 856 6200
TL: 212 856 6280

Funke-Schmitt-Rink, Margret, Dr.
Member
Federal Parliament
Permanent Mission to the United Nations, 600 Third Ave., 41st Floor New York, NY, 10016
F: 212 856 6200
TL: 212 856 6280

Gallmeier-Melo, Claudia, Mrs.
Consulate General
Rio de Janeiro
Permanent Mission to the United Nations, 600 Third Ave., 41st Floor New York, NY, 10016
F: 212 856 6200
TL: 212 856 6280

Gallus, Georg, Mr.
Parliamentary State Secretary
Federal Ministry for Agriculture
Permanent Mission to the United Nations, 600 Third Ave., 41st Floor New York, NY, 10016
F: 212 856 6200
TL: 212 856 6280

Ganseforth, Monika, Dr.
Prof.
BUNDESTAG
Bundeshaus
Bonn 5300, Germany
T: 49 228 16-1

Gillmann, Helmut, Prof. Dr.
Personal Office
Chancellor
Permanent Mission to the United Nations, 600 Third Ave., 41st Floor New York, NY, 10016
F: 212 856 6200
TL: 212 856 6280

Gnodtke, Hans-Gunther, Mr.
Counsellor
Permanent Mission of Germany
600 Third Ave., 41st Floor
New York, NY 10016, United States of America
T: 212 856 6200 F: 212 856 6280

Gobel, Lindolfo, Mr.
Consulate General
Rio de Janeiro
Permanent Mission to the United Nations, 600 Third Ave., 41st Floor New York, NY, 10016
F: 212 856 6200
TL: 212 856 6280

Granow, Dietrich, Dr.
Head of Division
Federal Ministry for Foreign Affairs
Permanent Mission to the United Nations, 600 Third Ave., 41st Floor New York, NY, 10016
F: 212 856 6200
TL: 212 856 6280

Grassl, Hartmut, Prof. Dr.
Expert
Permanent Mission of Germany
600 Third Ave., 41st Floor
New York, NY 10016, United States of America
T: 212 856 6200 F: 212 856 6280

Grunbeck, Josef, Mr.
Member
Federal Parliament
Permanent Mission to the United Nations, 600 Third Ave., 41st Floor New York, NY, 10016
F: 212 856 6200
TL: 212 856 6280

Hagedorn, Horst, Prof. Dr.
National UNCED-Committee
Permanent Mission to the United Nations, 600 Third Ave., 41st Floor New York, NY, 10016
F: 212 856 6200
TL: 212 856 6280

Hallensleben, Ekkehard, Dr.
Minister
Embassy to Brazil
Permanent Mission to the United Nations, 600 Third Ave., 41st Floor New York, NY, 10016
F: 212 856 6200
TL: 212 856 6280

Haller, Dieter, Mr.
Counsellor
Embassy to Kenya
Permanent Mission to the United Nations, 600 Third Ave., 41st Floor New York, NY, 10016
F: 212 856 6200
TL: 212 856 6280

Hecker, Martin, Dr.
Head of Division
Federal Ministry for Foreign Affairs
Permanent Mission to the United Nations, 600 Third Ave., 41st Floor New York, NY, 10016
F: 212 856 6200
TL: 212 856 6280

Hemmer, Hans-Rimbert, Prof. Dr.
National UNCED-Committee
Permanent Mission to the United Nations, 600 Third Ave., 41st Floor New York, NY, 10016
F: 212 856 6200
TL: 212 856 6280

Herfeldt, Egon, Mr.
Director General
Federal Ministry for the Environment
Permanent Mission to the United Nations, 600 Third Ave., 41st Floor New York, NY, 10016
F: 212 856 6200
TL: 212 856 6280

Herkenrath, Pralat Norbert, Mr.
National UNCED-Committee
Permanent Mission to the United Nations, 600 Third Ave., 41st Floor New York, NY, 10016
F: 212 856 6200
TL: 212 856 6280

Hipler, Heinz, Mr.
Official
Federal Ministry for Construction
Permanent Mission to the United Nations, 600 Third Ave., 41st Floor New York, NY, 10016
F: 212 856 6200
TL: 212 856 6280

Hoenisch, U., Dr.
Head of Division
Federal Ministry for Agriculture
Permanent Mission to the United Nations, 600 Third Ave., 41st Floor New York, NY, 10016
F: 212 856 6200
TL: 212 856 6280

Hoffmann, Wolfgang, Dr.
Head of Division
Federal Ministry for Foreign Affairs
Permanent Mission to the United Nations, 600 Third Ave., 41st Floor New York, NY, 10016
F: 212 856 6200
TL: 212 856 6280

Holzwarth, Andrea, Mrs.
Consulate General
Rio de Janeiro
Permanent Mission to the United Nations, 600 Third Ave., 41st Floor New York, NY, 10016
F: 212 856 6200
TL: 212 856 6280

Jacob, Roland, Mr.
Head of Division
Enquete-Commission
Permanent Mission to the United Nations, 600 Third Ave., 41st Floor New York, NY, 10016
F: 212 856 6200
TL: 212 856 6280

Joeckle, Kirsten, Mrs.
Consulate General
Rio de Janeiro
Permanent Mission to the United Nations, 600 Third Ave., 41st Floor New York, NY, 10016
F: 212 856 6200
TL: 212 856 6280

Jordans, Hans-Carl von, Mr.
Head of Division
Federal Press and Information Office
Permanent Mission to the United Nations, 600 Third Ave., 41st Floor New York, NY, 10016
F: 212 856 6200
TL: 212 856 6280

Kahl, Harald, Dr.
Member
Federal Parliament
Permanent Mission to the United Nations, 600 Third Ave., 41st Floor New York, NY, 10016
F: 212 856 6200
TL: 212 856 6280

Kausch, Hans Gerd, Dr.
Official
Permanent Mission of Germany
600 Third Ave., 41st Floor
New York, NY 10016, United States of America
T: 212 856 6200 F: 212 856 6280

Kindervater, Christine, Mrs.
Head of Protocol
Federal Ministry for the Environment
Permanent Mission to the United Nations, 600 Third Ave., 41st Floor New York, NY, 10016
F: 212 856 6200
TL: 212 856 6280

Kleiber, Wolfgang, Mr.
Head of Division
Federal Ministry for Construction
Permanent Mission to the United Nations, 600 Third Ave., 41st Floor New York, NY, 10016
F: 212 856 6200
TL: 212 856 6280

Klinger, Verena, Mrs.
Official
Federal Ministry for the Environment
Permanent Mission to the United Nations, 600 Third Ave., 41st Floor New York, NY, 10016
F: 212 856 6200
TL: 212 856 6280

Klopfer, Manfred, Mr.
Consul, Consulate General
Rio de Janeiro
Permanent Mission to the United Nations, 600 Third Ave., 41st Floor New York, NY, 10016
F: 212 856 6200
TL: 212 856 6280

Klotz, Peter, Mr.
Consul
Consulate General
Rio de Janeiro
T: 10016
TL: 212 856 6200, E: 212 856 6280

Knolle, Evelyn, Mrs.
Consulate General
Rio de Janeiro
Permanent Mission to the United Nations, 600 Third Ave., 41st Floor New York, NY, 10016
F: 212 856 6200
TL: 212 856 6280

Kobsa, Stefan, Mr.
Vice Consul Consulate General
Rio de Janeiro
Permanent Mission to the United Nations, 600 Third Ave., 41st Floor New York, NY, 10016
F: 212 856 6200
TL: 212 856 6280

Kohl, Helmut, Mr.
Chancellor
Permanent Mission of Germany
600 Third Ave., 41st. floor
New York NY 10016, United States of America
T: 212 856 6200 F: 212 856 6280

Kohlhaw, Monika, Mrs.
Official
Federal Ministry for Economy
Permanent Mission to the United Nations, 600 Third Ave., 41st Floor New York, NY, 10016
F: 212 856 6200
TL: 212 856 6280

Kundruhn, Gunter, Mr.
Consulate General
Rio de Janeiro
Permanent Mission to the United Nations, 600 Third Ave., 41st Floor New York, NY, 10016
F: 212 856 6200
TL: 212 856 6280

Kurstgens, Norbert, Mr.
Consul
Consulate General in Rio de Janeiro
Permanent Mission to the United Nations, 600 Third Ave., 41st Floor New York, NY, 10016
F: 212 856 6200
TL: 212 856 6280

Kurzinger-Wiemann, Edith, Dr.
Die
Permanent Mission of Germany
600 Third Ave., 41st Floor
New York, NY 10016, United States of America
T: 212 856 6200 F: 212 856 6280

Lamberty, Willi, Mr.
Consulate General
Rio de Janeiro
Permanent Mission to the United Nations, 600 Third Ave., 41st Floor New York, NY, 10016
F: 212 856 6200
TL: 212 856 6280

Lange, Hans-Georg, Mr.
National UNCED-Committee
Permanent Mission to the United Nations, 600 Third Ave., 41st Floor New York, NY, 10016
F: 212 856 6200
TL: 212 856 6280

Lehmann-Zwiener, Simone, Mrs.
Chancellery
Permanent Mission to the United Nations, 600 Third Ave., 41st Floor New York, NY, 10016
F: 212 856 6200
TL: 212 856 6280

Lersner, Heinrich von, Dr.
National UNCED-Committee
Permanent Mission to the United Nations, 600 Third Ave., 41st Floor New York, NY, 10016
F: 212 856 6200
TL: 212 856 6280

Liers, Walter, Mr.
Federal Press and Information Office
Permanent Mission to the United Nations, 600 Third Ave., 41st Floor New York, NY, 10016
F: 212 856 6200
TL: 212 856 6280

Lippold, Klaus W., Dr.
Member
Federal Parliament
Permanent Mission to the United Nations, 600 Third Ave., 41st Floor New York, NY, 10016
F: 212 856 6200
TL: 212 856 6280

Loebsack, Willi, Mr.
GTZ
Permanent Mission of Germany
600 Third Ave., 41st Floor
New York, NY 10016, United States of America
T: 212 856 6200 F: 212 856 6280

Loecker, Petra, Mrs.
Official
Federal Ministry for the Environment
Permanent Mission to the United Nations, 600 Third Ave., 41st Floor New York, NY, 10016
F: 212 856 6200
TL: 212 856 6280

Lottmann, Jurgen, Mr.
KfW
Permanent Mission of Germany
600 Third Ave., 41st Floor
New York, NY 10016, United States of America
T: 212 856 6200 F: 212 856 6280

Martini, Klaudia, H.E. Mrs.
Minister
Environment of the Land of Rheinland-Pfalz
Permanent Mission to the United Nations, 600 Third Ave., 41st Floor
New York, NY 10016, United States of America
T: 212 856 6200 F: 212 856 6280

Matthes, Klaus, Dr.
Counsellor
Embassy to Brazil
Permanent Mission to the United Nations, 600 Third Ave., 41st Floor New York, NY, 10016
F: 212 856 6200
TL: 212 856 6280

Matuschka, Mario Graf von, Dr.
Director
Federal Ministry for Foreign Affairs
Permanent Mission to the United Nations, 600 Third Ave., 41st Floor New York, NY, 10016
F: 212 856 6200
TL: 212 856 6280

Mayr, Heidemarie, Mrs.
Consulate General
Rio de Janeiro
Permanent Mission to the United Nations, 600 Third Ave., 41st Floor New York, NY, 10016
F: 212 856 6200
TL: 212 856 6280

Menden, Werner, Dr. Ing.
Director
Federal Ministry for Research and Technology
Permanent Mission to the United Nations, 600 Third Ave., 41st Floor New York, NY, 10016
F: 212 856 6200
TL: 212 856 6280

Michaelis, Hans, Prof. Dr.
Expert
Permanent Mission of Germany
600 Third Ave., 41st Floor
New York, NY 10016, United States of America
T: 212 856 6200 F: 212 856 6280

Mombaur, Peter Michael, Dr.
Expert
Permanent Mission of Germany
600 Third Ave., 41st Floor
New York, NY 10016, United States of America
T: 212 856 6200 F: 212 856 6280

Muhe, Marlene, Mrs.
Head of the Press Division
Federal Ministry for the Environment
Permanent Mission to the United Nations, 600 Third Ave., 41st Floor New York, NY, 10016
F: 212 856 6200
TL: 212 856 6280

Muller, Edda, Dr.
Director
Federal Ministry for the Environment
Permanent Mission to the United Nations, 600 Third Ave., 41st Floor New York, NY, 10016
F: 212 856 6200
TL: 212 856 6280

Muller, Wolfgang, Dr.
First Secretary
Embassy to Brazil
Permanent Mission to the United Nations, 600 Third Ave., 41st Floor New York, NY, 10016
F: 212 856 6200
TL: 212 856 6280

Mullers, Klaus, Mr.
Consulate General
Rio de Janeiro
Permanent Mission to the United Nations, 600 Third Ave., 41st Floor New York, NY, 10016
F: 212 856 6200
TL: 212 856 6280

Muttelsee, Winfried, Dr.
Head of Division
Federal Ministry for Economy
Permanent Mission to the United Nations, 600 Third Ave., 41st Floor New York, NY, 10016
F: 212 856 6200
TL: 212 856 6280

Mutzelburg, Bernd, H.E. Mr.
Ambassador Extraordinary and Plenipotentiary to the Republic of Kenya
Permanent Mission of Germany
600 Third Ave., 41st Floor
New York, NY 10016, United States of America
T: 212 856 6200 F: 212 856 6280

Neuer, Walter, Dr.
Director, Head
Office of the Federal Chancellor
Permanent Mission to the United Nations, 600 Third Ave., 41st Floor New York, NY, 10016
F: 212 856 6200
TL: 212 856 6280

Ollenschlager, Bernd, Dr.
First Secretary
Embassy to Brazil
Permanent Mission to the United Nations, 600 Third Ave., 41st Floor New York, NY, 10016
F: 212 856 6200
TL: 212 856 6280

Pazlorek, Peter, Dr.
Member
Federal Parliament
Permanent Mission to the United Nations, 600 Third Ave., 41st Floor New York, NY, 10016
F: 212 856 6200
TL: 212 856 6280

Pellet, Volker, Mr.
Consulate General
Rio de Janeiro
Permanent Mission to the United Nations, 600 Third Ave., 41st Floor New York, NY, 10016
F: 212 856 6200
TL: 212 856 6280

Platte, Peter, Mr.
Counsellor
Embassy to Brazil
Permanent Mission to the United Nations, 600 Third Ave., 41st Floor New York, NY, 10016
F: 212 856 6200
TL: 212 856 6280

Power, John, Mr.
Official
Federal Ministry for the Environment
Permanent Mission to the United Nations, 600 Third Ave., 41st Floor New York, NY, 10016
F: 212 856 6200
TL: 212 856 6280

Quennet, Cornelia, Mrs.
Official
Federal Ministry for the Environment
Permanent Mission to the United Nations, 600 Third Ave., 41st Floor New York, NY, 10016
F: 212 856 6200
TL: 212 856 6280

Rami, Bernhard, Dr.
Official
Federal Ministry for Research and Technology
Permanent Mission to the United Nations, 600 Third Ave., 41st Floor New York, NY, 10016
F: 212 856 6200
TL: 212 856 6280

Rauw, M., Mr.
Head of Division
Federal Ministry for Transport
Permanent Mission to the United Nations, 600 Third Ave., 41st Floor New York, NY, 10016
F: 212 856 6200
TL: 212 856 6280

Reinhardt, Erika, Mrs.
Member
Federal Parliament
Permanent Mission to the United Nations, 600 Third Ave., 41st Floor New York, NY, 10016
F: 212 856 6200
TL: 212 856 6280

Repnik, Hans Peter, H.E. Mr
Parliamentary State Secretary
Federal Ministry for Economic Cooperation (Deputy Head of Delegation)
Permanent Mission to the United Nations, 600 Third Ave., 41st Floor New York, NY, 10016
F: 212 856 6200
TL: 212 856 6280

Resinek, Helmut, Mr.
Official
Federal Ministry for Foreign Affairs
Permanent Mission to the United Nations, 600 Third Ave., 41st Floor New York, NY, 10016
F: 212 856 6200
TL: 212 856 6280

Reuter, Edzard, Mr.
President
Permanent Mission of Germany
600 Third Ave., 41st Floor
New York, NY 10016, United States of America
T: 212 856 6200 F: 212 856 6280

Sand, Clemens van de, Dr.
Head of Division
Federal Ministry for Economic Cooperation
Permanent Mission to the United Nations, 600 Third Ave., 41st Floor New York, NY, 10016
F: 212 856 6200
TL: 212 856 6280

Schaback, Susanna, Mrs.
Vice Consul
Consulate General in Rio de Janeiro
Permanent Mission to the United Nations, 600 Third Ave., 41st Floor New York, NY, 10016
F: 212 856 6200
TL: 212 856 6280

Schafhausen, Franz Josef, Mr.
Official
Federal Ministry for the Environment
Permanent Mission to the United Nations, 600 Third Ave., 41st Floor New York, NY, 10016
F: 212 856 6200
TL: 212 856 6280

Scheuermann, Horst, Mr.
Consulate General
Rio de Janeiro
Permanent Mission to the United Nations, 600 Third Ave., 41st Floor New York, NY, 10016
F: 212 856 6200
TL: 212 856 6280

Schipulle, Hans-Peter, Dr.
Head of Division
Federal Ministry for Economic Cooperation
Permanent Mission to the United Nations, 600 Third Ave., 41st Floor New York, NY, 10016
F: 212 856 6200
TL: 212 856 6280

Schmickler, Anja, Mrs.
Federal Press and Information Office
Permanent Mission to the United Nations, 600 Third Ave., 41st Floor New York, NY, 10016
F: 212 856 6200
TL: 212 856 6280

Schmidbauer, Bernd, H.E. Mr.
Minister of State
Chancellery
Permanent Mission to the United Nations, 600 Third Ave., 41st Floor New York, NY, 10016
F: 212 856 6200
TL: 212 856 6280

Schmidt, Konrad, Mr.
Official
Committee for the Environment
Permanent Mission to the United Nations, 600 Third Ave., 41st Floor New York, NY, 10016
F: 212 856 6200
TL: 212 856 6280

Schneider, Reinhard, Mr.
Federal Ministry for the Environment
Permanent Mission to the United Nations, 600 Third Ave., 41st Floor New York, NY, 10016
F: 212 856 6200
TL: 212 856 6280

Schroder, Ralf, Mr.
Official
Federal Ministry for Economic Cooperation
Permanent Mission to the United Nations, 600 Third Ave., 41st Floor New York, NY, 10016
F: 212 856 6200
TL: 212 856 6280

Schuldenzucker, Andreas, Mr.
Official
Federal Press and Information Office
Permanent Mission to the United Nations, 600 Third Ave., 41st Floor New York, NY, 10016
F: 212 856 6200
TL: 212 856 6280

Seeber, Eckhard, Mr.
Personal Staff
Federal Chancellor
Permanent Mission to the United Nations, 600 Third Ave., 41st Floor New York, NY, 10016
F: 212 856 6200
TL: 212 856 6280

Sehn, Marita, Mrs.
Member
Federal Parliament
Permanent Mission to the United Nations, 600 Third Ave., 41st Floor New York, NY, 10016
F: 212 856 6200
TL: 212 856 6280

Stabe, Carola, Mrs.
National UNCED-Committee
Permanent Mission to the United Nations, 600 Third Ave., 41st Floor New York, NY, 10016
F: 212 856 6200
TL: 212 856 6280

Steinberg, Karl-Hermann, Prof. Dr.
National UNCED-Committee
Permanent Mission to the United Nations, 600 Third Ave., 41st Floor New York, NY, 10016
F: 212 856 6200
TL: 212 856 6280

Succow, Michael, Prof. Dr.
Expert
Permanent Mission of Germany
600 Third Ave., 41st Floor
New York, NY 10016, United States of America
T: 212 856 6200 F: 212 856 6280

Tiedeken, Hans, Dr.
National UNCED-Committee
Permanent Mission to the United Nations, 600 Third Ave., 41st Floor New York, NY, 10016
F: 212 856 6200
TL: 212 856 6280

Tittel-Gronefeld, Reinhard, Dr.
First Secretary
Embassy to Brazil
Permanent Mission to the United Nations, 600 Third Ave., 41st Floor New York, NY, 10016
F: 212 856 6200
TL: 212 856 6280

Topfer, Klaus, H.E. Prof. Dr.
Federal Minister
Environment, Nature Conservation and Nuclear Safety (Deputy Head of Delegation)
Permanent Mission to the United Nations, 600 Third Ave., 41st Floor
New York, NY 10016, United States of America
T: 212 856 6200 F: 212 856 6280

U Aung Htoo,
Third Secretary
EMBASSY OF THE UNION OF MYANMAR
112, Schumann Str.
5300 Bonn 1, Germany
T: 0228 210091 F: 0228 219316

Ueberschaer, Christian, Dr.
Head of Division
Chancellery
Permanent Mission to the United Nations, 600 Third Ave., 41st Floor New York, NY, 10016
F: 212 856 6200
TL: 212 856 6280

Uhlmann, Petra, H.E. Mr.
Minister
Environment of the Land of Mecklenburg-Vorpommern
Permanent Mission to the United Nations, 600 Third Ave., 41st Floor
New York, NY 10016, United States of America
T: 212 856 6200 F: 212 856 6280

Vogel, Ansgar Otto, Prof. Dr.
Director General
Federal Ministry for the Environment
Permanent Mission to the United Nations, 600 Third Ave., 41st Floor New York, NY, 10016
F: 212 856 6200
TL: 212 856 6280

Vogel, Dieter, Mr.
State Secretary, Head of the Federal Press and Information Office
Permanent Mission of Germany
600 Third Ave., 41st Floor
New York, NY 10016, United States of America
T: 212 856 6200 F: 212 856 6280

Vollmer, Kirsten, Mrs.
Official
Federal Press and Information Office
Permanent Mission to the United Nations, 600 Third Ave., 41st Floor New York, NY, 10016
F: 212 856 6200
TL: 212 856 6280

Voss, Friedrich, Dr.
KfW
Permanent Mission of Germany
600 Third Ave., 41st Floor
New York, NY 10016, United States of America
T: 212 856 6200 F: 212 856 6280

Wallau, Theodor, H.E. Mr.
Ambassador Extraordinary and Plenipotentiary to the Federative Republic of Brazil
Permanent Mission of Germany
600 Third Ave., 41st Floor
New York, NY 10016, United States of America
T: 212 856 6200 F: 212 856 6280

Wallow, Hans, Mr.
Member
Federal Parliament
Permanent Mission to the United Nations, 600 Third Ave., 41st Floor New York, NY, 10016
F: 212 856 6200
TL: 212 856 6280

Weber, Juliane, Mrs.
Head of the Personal Office
Federal Chancellor
Permanent Mission to the United Nations, 600 Third Ave., 41st Floor New York, NY, 10016
F: 212 856 6200
TL: 212 856 6280

Websky, Michael von, Dr.
Director
Federal Ministry for the Environment
Permanent Mission to the United Nations, 600 Third Ave., 41st Floor New York, NY, 10016
F: 212 856 6200
TL: 212 856 6280

Weckbach, Stefan, Dr.
Deputy Consul General
~~Rio de Janeiro~~
Permanent Mission to the United Nations, 600 Third Ave., 41st Floor New York, NY, 10016
F: 212 856 6200
TL: 212 856 6280

Weinzierl, Hubert, Mr.
National UNCED-Committee
Permanent Mission to the United Nations, 600 Third Ave., 41st Floor New York, NY, 10016
F: 212 856 6200
TL: 212 856 6280

Weiszacker, Ernst Ulrich von, Prof. Dr.
National-Committee
Permanent Mission to the United Nations, 600 Third Ave., 41st Floor New York, NY, 10016
F: 212 856 6200
TL: 212 856 6280

Wenderoth, Jurgen, Mr.
Official
Federal Ministry of Finance
Permanent Mission to the United Nations, 600 Third Ave., 41st Floor New York, NY, 10016
F: 212 856 6200
TL: 212 856 6280

Wermann, Ernst, Mr.
Director
Federal Ministry for Agriculture
Permanent Mission to the United Nations, 600 Third Ave., 41st Floor New York, NY, 10016
F: 212 856 6200
TL: 212 856 6280

Wilhelm, Wendelin, Dr.
Head of Division
Federal Ministry for Economy
Permanent Mission to the United Nations, 600 Third Ave., 41st Floor New York, NY, 10016
F: 212 856 6200
TL: 212 856 6280

Yenioglu, Barbel, Mrs.
Federal Ministry for the Environment
Permanent Mission to the United Nations, 600 Third Ave., 41st Floor New York, NY, 10016
F: 212 856 6200
TL: 212 856 6280

Zimprich, Werner, Dr.
Head of Division
Federal Ministry for Foreign Affairs
Permanent Mission to the United Nations, 600 Third Ave., 41st Floor New York, NY, 10016
F: 212 856 6200
TL: 212 856 6280

NGO

Bremen, Volker Von
GFID
Wielinger Strasse 52
D-8133 Feldafing, Germany

Burhenne, Wolfgang E.
ICEL
Adenauerallee 214
D-5300 Bonn, Germany

Carcano, Dora
WIDF
Dresdener Strasse 43, PO Box 940
O-1020 Berlin, Germany

Fienley, Clare
ANPED
c/o World Ecology, Economy and Development
Siefried-Leopold Str. 53, D-5300 Bonn 3, Germany

Ford, Mary
IARF
IARF, Dreieichstrasse 59
D-6000 Frankfurt 70, Germany

Gast, Hartmut
GFID
Wielinger Strasse 52
D-8133 Feldafing, Germany

Hobbs, Jon
INEM
Sillemstrasse 36
D-2000 Hamburg 20, Germany

Jankowian Felix, Fritz
ICEL
Adenauerallee 214
D-5300 Bonn, Germany

Kreibich, Rolf
SFS
Leithestrasse 37-39
4650 Gelsenkirchen, Germany

Kulessa, Margareta
WORLD ECONOMY ENVIRONMENT AND DEVELOPMENT
Siegfred Leopold Str. 53, D-5300 Bonn 3
Germany

Leal Filho, Walter, Dr.
UH
Universitat Hamburg, Institut fur I.V.E.
Sedanstrasse 19 D-2000, Hamburg 13, Germany

Lingscheid, Raines, Mr.
Rev.
ASSOCIATION OF THE CHURCHES DEVELOPMENT SERVICES - AGKED
Kniebisstr. 29
Stuttgart D-7000 1, Germany
T: 49 711 28-2021 F: 49 711 26-7416

Luz, Bertha
MWIA
5 Herbert-Lewin Strasse
D-5000 Koln 41, Germany

Moll, Peter
SFS
Leithestrasse 37-39
4650 Gelsenkirchen, Germany

Newell, Robin Frank
FGI
Gustav Heinemann Ufer 84-88
D-5000 Cologne 51, Germany

Nieklsch, Manfred, Mr.
DEUTSCHER NATURSCHUTERING
Kalkulbtv.24
Bonn 5300-3, Germany
T: 49 228 44-1505 F: 49 228 44-3643

Ojeda-Castaneda, Jorge
OWLS
Lab. of Biophy.,Instit. of Exp. Audiology,Univ. of Muenster,Kardinal-von-galen-ring10,D-4400 Muenster, Germany

Peszko, Gregorz
ANPED
c/o World Ecology, Economy and Development
Siefried-Leopold Str. 53, D-5300 Bonn 3, Germany

Robinson, Nicholas A.
ICEL
Adenauerallee 214
D-5300 Bonn, Germany

Ruppert, Alfred Josef
MISEREOR
Mozarstrasse 9
D-5100 Aachen, Germany

Sato, Hiroyuki
INEM
Sillemstrasse 36
D-2000 Hamburg 20, Germany

Schepelmann, Philip
BUND
Bundesgeschaftsstelle Im Rheingarten 7
5300 Bonn 3, Germany

Schmitthenner, Ulrich
Pfr
ECUMENICAL SOCIETY FOR JPIC
Am. Schonblick 7
D-7070 Schwabisch Grund, Germany

Schulz, Gunther
Presidente
BRASILIEN INITIATIVE FREIBURG E.V.
In den Weihermatten 27
7800 Freibourg, Germany

Seligmann, Francette
WIDF
Dresdener Strasse 43, PO Box 940
O-1020 Berlin, Germany

Spangenberg, Joachim
BUND FUR UMWELT UND NATURSCHUTZ DEUTSCHLAND
Bundesgeschaftsstelle Im Rheingarten 7
5300 Bonn 3, Germany

Stoltz, Hildegard
MWIA
5 Herbert-Lewin Strasse
D-5000 Koln 41, Germany

Terselic, Vesna, Ms.
ANPED
c/o World Ecology, Economy and Development
Siefried-Leopold Str. 53, D-5300 Bonn 3, Germany

Unmussig, Barbara
DEUTSCHER NATURSCHUTERING
Kalkuhlstr. 24
5300 Bonn 3, Germany

Van Howe, Annette
IARF
IARF, Dreieichstrasse 59
D-6000 Frankfurt 70, Germany

Van Mansvelt, Jan Diek
IFOAM
c/o General Secretariat, Hofgut Imsbach
D-6695 Tholey-Theley, Germany

Wilmsen, Christian, Mr.
KINDERNOTHILFE E.V.
Dusseldorfer Landstrasse 180
Duisburg 4100 28, Germany
F: 49 203 7789118

Winter, Georg
INEM
Sillemstrasse 36
D-2000 Hamburg 20, Germany

Wirtz, Karl Josef
MISEREOR
Mozarstrasse 9
D-5100 Aachen, Germany

Zgaga, Christel, Ms.
Diplom Soziolgin
BUENDNIS 90 DIE GRUENEN
Bundeshaus Hi 201
Bonn 5300, Germany
T: 49 228 16 7622 F: 49 228 16 86662
TL: 885 598

PRESS

Adam-Thalec, Maritta, Ms.
Jornalista
BERLINER ZEITUNG
Karl-Liebkuedit-Stv. 29
Berlin 0-1029, Germany
T: 37 2 327-6605 F: 37 2 327-5255

Angres, Volker
Editor
ZDF GERMAN TV
Sertoriusring 1
6500 Mainz 21, Germany, Germany

Arfs, Jorn
Editor
PREMIERE-TELEVISION
Hermann-Loens-Weg, 58
2000 Hamburg 63, Germany, Germany

Bauer, Julian
Correspondent
MEDIA POINT
P.O. Box 100
Hardegsen, Germany, Germany

Beinhauer, Hagen
Reporter Radio/Television
WESTDEUTSCHER RUNDFUNK KOLN WDR
Jogerstrasse 5c
5064 Rosrath 3, Germany, Germany

Bernd, Hilder
Correspondent
ARD-GERMAN RADIO
Volcom 58
Mexico, DF 01900 Mexico, Germany

Beutler, Martin
Editor
DEUTSCHE WELLE
Reuterstr. 8
B-5300 Bonn 1, Germany, Germany

Bielicki, Jan
Editor
DEUTSCHES ALLGEMEINES SONNTAGSBLATT
Luebecker Str. 82
2000 Hamburg 76, Germany

Boldt, Klaus A.
Editor
EPD NEWS AGENCY
Jahnstrasse 23
D-6000 Frankfurt/Main, Germany, Germany

Borchert, Thomas
Science Editor
HAMBURGER ABENDBLATT
Alsterdorfer Strasse 340
2000 Hamburg 60 Germany, Germany

Brajterman, Diana R.
Correspondent
DEUTSCHE PRESSE AGENTUR (DPA)
Ave. Monsenhor Ascaneo 280/102
22621 Rio de Janeiro-RJ, Brazil, Germany

Brandt, Michael
Correspondent
KOELNER STADT-ANZEIGER
Eifelstrasse 19
5 Koeln I, Germany

Brodehl, Axel
Correspondent
ZEITUNG FUR DARMSTADT
Claudiusweg 16 a 6100 Darmstad
Germany, Germany

Burhenne, Falk
Correspondent
RADIO DEMST BONN/EPL
Santa Clara 340/701
Rio de Janeiro, Brazil, Germany

Cabral, Simons
Reporter Radio/Television
RIAS TV
Waitzetr. 8
1000 Berlin 12, Germany, Germany

Carius, Norbert
Editor
SAARLANDISCHER RUNDFUNK
Im Flurchen 69
D-6601 Scheidt, Germany, Germany

Christiansen, Sabine
Editor
ARD-GERMAN TV
2000 Hamburg 52
Elbchaussee 151, Germany, Germany

Coelho, Maria M.
Correspondent
ARD-FIRST CHANNEL
R.Eca de Queiroz 479 c/1
BR-04011 Sao Paulo-SP, Brazil, Germany

Conze, Gabriele, Ms.
GERMAN TELEVISION WDR
Appellhofplatz 1
Cologne 5000, Germany
T: 49 221 220398 F: 49 221 2203162

Czakainski, Martin
Editor
ENERGIEWIRTSCHAFTLICHE TAGESFRAGEN
Winzermarkstrasse 68
D-4320 Hattingen 17, Germany, Germany

De Graeve, Jurgen
Correspondent
BAYERISCHER RUNDFUNK
Endelhauser Str.32
8000 Munchen 21, Germany, Germany

De Medeiros Valle, Eneas
Correspondent
UVP-REPORT
R. Candido Mendes 335/503
20241 Rio de Janeiro-RJ, Brazil, Germany

De Oliveira, Astrid
Correspondent
DIE TAGESZEITUNG
Av. das Americas 411/111 Bl.2
22631 Rio de Janeiro, RJ Brazil, Germany

De Soliz, Neusa
Editor
DEUTSCHE WELLE
Luxemburger Str.76
5000 Koln 51, Germany, Germany

Dregger, Christa, Ms.
Editor
POTSDAMER
Rosa-Luxemburg Str.39
Belzig 0-1820, Germany
T: 49 161 1732460

Fabian, Wolfgang
Correspondent
STERN
R. Dr.Julio Otoni 188
20241 St.Tereza, Rio de Janeiro, Brazil, Germany

Foste, Wolfgang
Editor
UVP-REPORT
Mollerhof 8
D-4800 Bielefeld 18, FRG, Germany

Fox, Stefan Karl
Reporter Radio/Television
ZDF
Shin QI 04, Conjunto 11
Casa 01, Lago Norte, 71500, Brasilia, DF, Germany

Fox, Veronica
Reporter Radio/Television
ZDF
SHIN QI04 Conj.11 C/01
Lago Norte, 715000 Brasilia-DF, Brazil, Germany

Friedmann, Ronald
Correspondent
NEUES DEUTSCHLAND
0-1195 Berlin
Ludwig-Ulapp Str, 15, Germany

Fruhlich, Thomas
Correspondent
SUDDENTSCHE ZEITUNG
Auf dem Acher 24
D-5305 Alfter, Germany

Fuchs, F.J. Robert
Correspondent
FUCHS-AGENTUR
41, Zeppelinstr.
D-7570 Baden-Baden, Germany, Germany

Gahbauer, Wolfgang J.
Correspondent
SUDWESTFUNK
Est. do Soberbo 911
20531 Rio de Janeiro, Brazil, Germany

Geier, Bernward
Editor
ECOLOGY & FARMING
Hofgut Imsbach
D-6695 Tholey-Theley, Germany, Germany

Gester, Martin
Correspondent
FRANKFURTER ALLGEMEINE ZEITUNG
R.Almte. Guillobel 26
Rio de Janeiro, Brazil, Germany

Glusing, Jens
Correspondent
DER SPIEGEL
Avenida Sao Sebastiao, 157
22291-Rio de Janeiro-RJ, Brazil, Germany

Goerdeler, Carl
Correspondent
DIE ZEIT
Av. Portugal, 986 Apt.20
Rio de Janeiro, Brazil, Germany

Graf, Hohenthal Carl
Correspondent
FRANKFURTER ALLGEMEINE ZEITUNG
Reisenweg 8
5207 Wachtsarg/Villipret, Germany, Germany

Gries, Marie-Luise
Reporter Radio/Television
SAARLANDISCHER RUNDFUNK
Dellengartenstr 3a
6600 Saarbruecken, Germany, Germany

Grosse-Leege, Dirk, Mr.
Chairman
Wirtjchaft + Umwelt
Krumme Stasse 2a
Munster 4400, Germany
T: 49 251 56634 F: 49 251 56850

Gruber, Monika
Editor
RIAS-TV
Konigstr. 95
5300 Bonn 1, Germany, Germany

Hallmann, Stephan
Correspondent
ZDF-GERMAN TV
Apto.Postal 75.882
Caracas 1070 A, Venezuela, Germany

Hatz, Oskar
Editor
PASSAUER NEUE PRESSE
Bruggspergerstr. 36
D-8000 Munchen 90, Germany, Germany

Hellenkemper, Annette
Reporter Radio/Television
WDR GERMAN TV
5 Koln 1
Temtoburgerst. 9, Germany, Germany

Heuer, Christine
Editor
DEUTSCHLANDFUNK
Karthauserstrasse 13
5300 Bonn 1, Germany, Germany

Hochstein, Beatriz
Editor
ZDF-GERMAN TV
Apo.Postal 7.882
Caracas, Venezuela, Germany

Hoering, Uwe
Correspondent
W & E
Euskirchener Str.125
5040 Bruhl FR Germany, Germany

Hofmann, Gerhard L.
Correspondent
ARD-FIRST CHANNEL
Talcahuano 981-Piso 20
J-1013 Buenos Aires, Argentina, Germany

Holtz, Joachim
Correspondent
ZDF-GERMAN TV
2000 Broadway
New York, NY 10023, USA, Germany

Huber, Klaus
Reporter Radio/Television
RIAS
Av. Copacabana, 13691/906
22070 Rio de Janeiro, RJ, Brazil, Germany

Ilse, Christian
Reporter Radio/Television
SAT.1 TELEVISION
Bruderstrasse 19
2000 Hamburg 36, Germany, Germany

Jung, Thomas
Reporter Radio/Television
SUDWESTFUNK
Rheinstr. 41
D-5300 Bonn 2, Germany, Germany

Kaffsack, Hanns-Joehen
Correspondent
DEUTSCHE PRESSE AGENTUR (DPA)
5, chemin le Grenier
1291 Commugny/VD, Germany, Germany

Kaltenbrunner, Thomas
Reporter Radio/Television
ARD-GERMAN TV
7573 Sivzheim
Matthiers-Gruenewald SW 24, Germany, Germany

Kleinmann, Reinhard
Editor
SUDWESTFUNK
Zum Keltenring 1
7570 Baden-Baden, Germany, Germany

Klose, Klose
Editor
RTL PLUS GERMAN TELEVISION
Statthaltehofweg 11
5000 Kohl 40, Germany

Klose, Ulrich
Editor
RTL PLUS TV
Statthalterhofweg 11
5000 Koln 40, Germany, Germany

Knebel, Bernd
Editor
HANNOVERSCHE ALLGEMEINE ZEITUNG
Tegeler Weg 29 D 5205
St. Augustin - Menden, Germany

Kneitz, Gerhard, Dr.
Professor
BUND FUR UMWELT UND NATURSCHUTZ
DEUTSCHLAND
Audev Janneciburg 1, Jast. Fur Zoologie
Bonn 5300, Germany
T: 49 228 735113 F: 49 228 4009740

Koch, Einar
Editor
BUNTE MAGAZIN, SUPER
Im Saud 88
5342 Rheinbreitbad, Germany, Germany

Kruchen, Christine
Editor
ARD-FIRST CHANNEL
Talcahuano 981-20
J-1013 Buenos Aires, Argentina, Germany

Krumrey, Henning
Correspondent
WIRTSCHAFTSWOCHE
Gleiwitzer Str. 12a
5047 Wesseling, Germany, Germany

Krzysztof, Rogala
Correspondent
RADIO FREE EUROPE
8155 Niederhasli
Munich, West Germany, Germany

Kulke, Ulrich
Editor
NATUR
Baderstr. 50
W-8000 Munchen 5, Germany, Germany

Kunath, Wolfgang
Editor
STUTTGARTER ZEITUNG
Reuchlinstr. 15
D-7000 Stuttgart 1, Germany, Germany

Kuntze, Sven Ingolf
Correspondent
ARD-GERMAN TV
107 W 86th St.
New York, NY 10024, USA, Germany

Lang, Andrew Marshall
Reporter Radio/Television
SAT.1 TELEVISION
Harkortstrasse 46
2000 Hamburg 60, Germany, Germany

Langer, Thomas
Editor
FALKTUM
Wolbecker Strasse 9
4400 Minster 1, Germany, Germany

Leinemann, Juergen
Correspondent
SPIEGEL
Bleibtreustrasse 17
W-1000 Berlin 12, Germany, Germany

Machado Dos Santos, Sandra
Reporter Radio/Television
ARD-GERMAN TV
Rua Jose Higino 136/602
20520 Rio de Janeiro, Brazil, Germany

Martin, Dr. Hans-Peter
Editor
DER SPIEGEL
Schnudts Treppe 1
W-2000 Hamburg 55, Germany, Germany

Martin, Sabine
Editor
SUDWESTFUNK
Liebfrauenstr. 13
6800 Mannheim 51, Germany, Germany

Mayer, Stephan Andreas
Correspondent
BAVARIAN TV
Erich-Boger-Str. 15
5300 Bonn 1, Germany, Germany

Mohring, Caroline
Science Editor
FRANKFURTER ALLGEMEINE ZEITUNG
Hellerhofskabe 2-4
Frankfurt, Germany
T: 49 69 75911816 F: 49 69 75911743

Muehe, Maria Magdalena
Correspondent
FEDERAL MINISTRY FOR THE ENVIRONMENT
Koenigsstrasse 62
5300 Bonn 1, Germany, Germany

Muehlenberg, Heidi
Reporter Radio/Television
MITTELDEUTSCHER FUNDFUNK
Plaussiger Strasse 12
Leipzig, Germany, Germany

Muhleisen, Stefan
Correspondent
BILD DER WISSENSCHAFT
c/o H.Saemann, Marquardtstr 25
7000 Stuttgart 1, Germany, Germany

Muller, Ingrid
Editor
DER TAGESSPIEGEL
Dortmunder Str. 13,
W-1000 Berlin 21, Germany, Germany

Muller, Johannes G.
Editor
SUDWESTFUNK
Friedrichbring 55/T
7550 Rastatt, Germany, Germany

Nachtigaeller, Jutta
Correspondent
LATEIN-AMERIKA,ANALYSEN,DATEN
Krohaskamp 39
Hamburg 60, Germany, Germany

Niebuhr, Siegfried
Correspondent
DEUTSCHE PRESSE AGENTUR (DPA)
R. Prof. M. Ferreira 127/802
22451 Rio de Janeiro, Brazil, Germany

Norbisrath, Gudrun
Editor
WESTDEUTSCHE ALLGEMEINE ZEITUNG
Schillerstrasse 28
W-5810 Witten 1, Germany, Germany

Nuhn, Jochen
Correspondent
ARD-GERMAN TV
F. Villapando 110-1303
01020-Mexico-DF, Mexico, Germany

Oetling, Beatriz
Reporter Radio/Television
ARD-GERMAN TV
Av. San Francisco, 311-C2
10810-Mexico-DF, Mexico, Germany

Poenitz, Grace
Correspondent
ALLGEMEINER DEUTSCHER NACHRICHTEN-ADN
Oberweissbacher Strasse 4/1808
O-1142 Berlin, Germany, Germany

Posche-Godt, Ulrike
Editor
STERN
Postfach 11 00 11
Hamburg, Germany, Germany

Puhlmann, Peter
Editor
SUDWESTFUNK
Tiergartenwe 5
7570 Baden-Baden, Germany, Germany

Reiche, Matthias
Reporter Radio/Television
MITTELDEUTSCHER RUNDFUNK
Hendrichplatz 33
Berlin 1156, Germany, Germany

Reiermann, Christian
Correspondent
HANDELSBLATT
Elsa-Brandstromstr, 25
W-5300 Bonn 3, Germany, Germany

Reuter, Gerd
Editor
DEUTSCHE PRESSE AGENTUR (DPA)
5300 Bonn 3, Wedner-Koervers
Str. 18, Germany, Germany

Richter, Andrea
Correspondent
MARKISCHE ALLGEMEINE
Oberaustrasse 107
5300 Bonn 2 Germany, Germany

Riedel, Donata
Editor
TAGESZEITUNG
Lepsius Str. 49, 1000
Berlin 41, Germany, Germany

Rohde, Manfred
Correspondent
ZDF-GERMAN TV
Av. Delfin Moreira 662/704
22441 Rio de Janeiro-C, Brazil, Germany

Rosenberg, Diana C.
Correspondent
ARD-FIRST CHANNEL
Talcahuano 843-5
1013 Buenos Aires, Argentina, Germany

Sadrozinski, Joerg
Editor
ARD-GERMAN TV
Waitz Str.49
D-2000 Hamburg, 52 Germany, Germany

Scarvalone, Maria
ARD-GERMAN TV
18 Fuller Place
Brooklyn, NY 11215, USA, Germany

Schmidt, Gerhardt
Editor
RIAS-TV
Wiesengrund 7
5307 Wachtberg-Villip, Germany, Germany

Schollmeyer, Gerhard
Editor
ARD-FIRST CHANNEL
Kellinghusenstrasse 20
D-2000 Hamburg 20, Germany, Germany

Seltzer, Dagmar
Reporter
SUDDEUTSCHER RUNDFUNK
Daklmannstr. 11
5300 Bonn 1, Germany
T: 004922 8213015

Spindler, Edmund A., Mr.
UUP - FOERDERVEREIN IN HAMM
Oesting Str. 13
Hamburg 4700, Germany
T: 49 2381 52129 F: 49 2381 52195

Staedele, Karl
Correspondent
DEUTSCHE PRESSE AGENTUR (DPA)
Weinzierlstr. 15
8400 Regensburg, Germany, Germany

Stefan, Spruck
Editor
NEWS VISION INTERNATIONAL
Radhamsstrasse, 1
6478 Nioda 1, Germany

Stehkamper, Bettina
Reporter Radio/Television
RIAS TV
1000 Berlin 47,
Arn Hanffgraben 7, Germany, Germany

Storz, Wolfgang Georg
Correspondent
BADISCHE ZEITUNG
Richard Wagner-Strasse 9
5300 Bonn 1, Germany, Germany

Sturm, Josef Christian
Correspondent
BILD ZEITUNG
Rabenkopfstr. 11
D 800 Munchen 90, Germany, Germany

Szymanski, Eva
Correspondent
DEUTSCHE WELLE FERNSEHEN
Im Fluerchen 69
0-6601 Scheide, Germany, Germany

Tenhagen, Herman-Josef
Editor
TAGESZEITUNG
Gertendorfes Weg 8
4236 Hamminkeln 3, Germany, Germany

Thomas, Werner
Correspondent
DIE WELT
434 Castania Av.
Cora C. Fables, FL 33146 USA, Germany

Thuronyi, Ma. Louise
Correspondent
DEUTSCHE ZEITUNG
R.Tonelero, 180 Apt.1003
BR-22030 Rio de Janeiro-RJ, Brazil, Germany

Tuck, Jay
Editor
ARD-GERMAN TV
Fahrenkamp 47
D-2000 Wedel, Germany, Germany

Undheim, Aslaug
NORWEGIAN CAMPAIGN FOR ENVIRONMENT AND DEVELOPMENT
Langesgt 4
Osnabruck, Germany
T: 49 541 362035 F: 49 541 361454

Van Hooven, Jorg
Editor
PRO 7 TELEVISION GMBH
Ainmillerstr. 13
8000 Munchen 40, Germany, Germany

Vogel, Immo
Editor
ARD-SWF
D-7562 Gernsback
Sabacher Weg 53, Germany, Germany

Von Hollander, Nicola
Reporter Radio/Television
NDR-TV/RADIO
Ehrenschildtstr. 19
2000 Hamburg 54, Germany, Germany

Von Moltke, Konrad
Editor
INTERNATIONAL ENVIRONMETAL AFFAIRS
R.R. 2 Box 37A
Norwilche, VT 05055 Germany, Germany

Vorholz, Friedrich
Editor
DIE ZEIT
Sierichstrasse 99,
2000 Hamburg 1, Germany, Germany

Vosskuhl, Thomas
Reporter Radio/Television
PRO 7 TV
Niessengasser
D-5216 Niederkassel I, Germany, Germany

Vulin, Antonio J.
Editor
ARD-FIRST CHANNEL
Taicahuano 981-Piso 20
J-1013 Buenos Aires, Argentina, Germany

Wagner, Eva Maria
Reporter Radio/Television
NATIONAL RADIO: DS-KULTUR/ZDF
Manteuffelstr. 19
1000 Berlin 36, Germany, Germany

Weib, Frieder
Reporter Radio/Television
SAT.1 TELEVISION
Schanzenstrasse 28
2000 Hamburg 36, Germany, Germany

Weidenbach, Thomas
Reporter Radio/Television
WDR GERMAN TV
Wormserstr. 45
5000 Koln 1, Germany, Germany

Weiss, Andreas
RIAS TV
Bismarckstrasse 102
D-1000 Berlin 12, Germany, Germany

Welzel, Klaus
Editor
RHEIN-NECKAR-ZEITUNG
6940 Weinheim, Multring 7
Germany, Germany

Werner, Klaus
Correspondent
WDR GERMAN TV
Annastr. 19
500 Koln 51, Germany, Germany

Wernicke, Christian
Editor
POLITICAL RED., DIE FEIT
Postfach 106820
Speersort 1 D-2000 Hamburg 70, Germany
T: 49 40 32 80214 F: 49 40 32 80505

Wichterich, Christa
Editor
BEITRAGE ZUR FEM. THEORIE UND PRAXIS
D 5040 Bruhl
Euskirchenerstr, 125, Germany, Germany

Wille, Joachim Manfred
Editor
FRANKFURTER RUNDSCHAU
Am Klingenborn 4,
D-6238 Hofheim Taunus, Germany, Germany

Wolf, Rodiger Gauer
Reporter Radio/Television
ARD-GERMAN TV
R. Constantino de Souza, 2097
04605- Sao Paulo-SP, Brazil, Germany

Zagatta, Martin
Correspondent
BAYERISCHER RUNDFUNK
Charcas 3497, Piso 11
1425 Buenos Aires, Argentina, Germany

Zenga Domestico, Luis Alberto
Correspondent
AGENCIA ALEMANA DE PRENSA (DPA)
Joaquin Requena 1258
Montevideo, Uruguay, Germany

ORGANIZATIONS

Bally, G. Vor., Mr.
International Society On Optics Within Life Sciences (owls)
Lab. of Biophy.,Instit. of Exp. Audiology,Univ. of Muenster,Kardinal-von-galen-ring10,D-4400 Muenster, Germany
T: (49 251) 836 861 F: (49 251) 836 812

Bandler, Henry, Mr.
International Council Of Environmental Law
Adenauerallee 214
D-5300 Bonn, Germany
T: (49 228) 2692 231 F: (49 228) 2692 250

Biegert, Claus, Mr.
World Uranium Hearing, The
Praterinsel 4
D-400 Munchen 22, Germany
T: (49 89) 228 5924 F: (49 89) 228 5340

Billen, Gerd, Mr.
Verbraucher Initiative
Breite Str. 51
5300 Bonn, Germany
T: (49 228) 726 3393 F: (49 228) 726 3399

Davis, Troy, Mr.
International Network For Environmental Management
Sillemstrasse 36
D-2000 Hamburg 20, Germany
T: (49 40) 40 8182 F: (49 40) 49 2138

Filho, W.D.S. Leal, Dr.
University of Hamburg/Research Unit on Media and Environment
Universitat Hamburg, Institut fur I.V.E.
Sedanstrasse 19 D-2000, Hamburg 13, Germany
T: (49 40) 4123 3732 F: (49 40) 4123 6247

Haase, G., Dr.
World Association For Element-building And Prefabrication*
Schlueterstrasse 6
D-2000 Hamburg 13, Germany
T: (49 40) 45 1725 F: (49 40) 45 1896

Inden-Heinrich, Helga, Ms.
Deutsche Naturschutzring (DNR)
Kalkuhlstr. 24
5300 Bonn 3, Germany
T: (49 228) 44 1505 F: (49 228) 44 4290

Juras, Alexander, Mr.
Institute For European Environmental Policy
Aloys-Schulte-Strasse 6
D-5300 Bonn 1, Germany
T: (49 228) 21 3810 F: (49 228) 22 1982

Kay, Andrew, Mr.
Pro Regenwald e.V.
Frohschammerstr. 14
D-8000 Munich 40, Germany
T: (49 89) 359 8650 F: (49 89) 359 6622

Mittag, Hans-Christian, Dr.
Naturschutzbund Deutschland
Richard-Straube-Strasse 35
4150 Krefeld, Germany
T: (49 2841) 22960 F: (48 2151) 595 211

Racke, Marcus, Dr.
Federation of German Industries
Gustav Heinemann Ufer 84-88
D-5000 Cologne 51, Germany
T: (49 221) 370 8481 F: (49 221) 370 8730

Stahl, Karin, Ms.
As. Of The Churches' Development Services
Kniebisstrasse 29
7000 Stuttgart 1, Germany
T: (49 711) 282 021 F: (49 711) 267 416

Triems, Brigitte, Ms.
Women's International Democratic Federation
Dresdener Strasse 43, PO Box 940
O-1020 Berlin, Germany
T: (37 2) 275 5028 F: (37 2) 279 3633

Waberzeck, Dagmar, Ms.
Medical Women's International Association
5 Herbert-Lewin Strasse
D-5000 Koln 41, Germany
T: (49 221) 400 4558 F: (49 221) 400 4388

Wahl, Peter, Mr.
Alliance Of Northern People For Environment And Development
c/o World Ecology, Economy and Development
Siefried-Leopold Str. 53, D-5300 Bonn 3, Germany
T: (49 228) 470 806 F: (49 228) 473 682

GLOBAL FORUM

Alfes, Berthold, Mr.
Pre Manager
Wirtjchaft + Umwelt
Krumme Str. 2a
Munster 4400, Germany
T: 49 251 56634 F: 49 251 56850

Avadhuta, A V
Editor
NEW RENAISSANCE MAGAZINE
Weisenauer Weg 4
Mainz, Germany
T: 49 6131131832995 F: 49 6131131 834628

Baerbel, Hoehn, Ms.
Member Parlament
THE GREEN PARTY
Giesbertstr 15
Oberhausen, Germany
T: 49 208 892006 F: 49 208 892006

Bartlau, Cornelia, Ms.
Teacher
ENVIRONMENTAL GROUP
AM Planetarium, 49 0-6900
Jena, Germany
T: 37 78 7825122

Barz, Jana-Antonia, Mr.
EARTH TRAIN/ WORLD LINK GLOBAL YOUTH SUMMIT
Breitscheidstr. 16, 0
Erkner, Germany
T: 4033

Baukloh, Anja, Sra.
Student
CARL DUISBERG GESELLSCHAFT E.V.
Lutzowufer 6-9 D-1000
Berlin 30, Germany
T: 49 30 25482-0 F: 49 30 25482-105
TL: 184628 cdgbl d

Beleites, Johannes, Mr.
EVANGELISCHES STUDIENWERK E.V.
Keplerstrasse 6
Gottingen W-3400, Germany
T: 49 551 42160

Berr, Isabella, Ms.
Photograhper
Zieblandstrasse 10
Munchen, Germany
T: 49 89 272-2930 F: 49 89 272 3930

Binder, Jurgen, Mr.
President
VERKEHRSCLUB DER BUNDERSREPUBLIK DEUTSCHLAND E.V.
Rotebuhlstrasse 84/1
Stuttgart 70000 1, Germany
T: 49 711 62 88 33 F: 49 711 61 89 31

Boelts, Hartmut, Dr.
SPARKASSE MARBURG-BIEDENKOPF
Postfach 2109
Marburg 3550, Germany

Bohlen, Kicki Von
ARTISTS UNITED FOR NATURE E.V.
Frohschammerstrasse 14
Munchen 8000 40, Germany
T: 49 89 359-4874 F: 49 89 359-6556

Bolschowa, Ludimila Iwanowa, Dr.
DNR - BUNDESVERBAND FUR UMWELTSCHUTZ
Kalkuhlstr. 24
Bonn 5300 - 3, Germany
T: 49 228 44 1505 F: 49 228 44 2277
TL: 8861170 dnr d

Brede, Dunja, Ms.
STUDIENWERK VIVIGST
Bleicherstrasse 26
Bremen 2800, Germany
T: 49 421 32 7721

Briggs, Curtis, Mr.
President
ARTISTS UNITED FOR NATURE
Frohschammerstr. 14
Munchen, Germany
T: 49 89 359-4874 F: 49 89 359-6556

Buck, Juergen, Mr.
Org. In Music
INTERNATIONAL FORUM OF CARDIOLOGY
Allemanha 18110 Mumian Bayerladeweg 10
Mumian - Bavaria 8110, Germany
T: 49 99554 F: 49 11808

Cardona, Sylvia Irina, Ms.
TERRE DES HOMMES
PO Box 4126
Osnabruck 4500, Germany
T: 49 541 7-10-10 F: 49 541 70-72-33

Cassens, Hans Jurgen, Mr.
Dipl. Ing.
TECHNISCHE UNIIVERSITAT BERLIN
TU Berlin Sekr.FR2-7 Landschaftsokonomie Franklinstrasse 28/29
Berlin 10 D-1000, Germany
F: 37 2 31421123

Chiparukhin, Alexander, Mr.
Reinhards Str. 47
Berlin 1040, Germany

Clar, Gunter, Dr.
G T Z
Postfach 5180
Eschborn W-6236, Germany

Couze, Gabriele
Editor
WDR INTERNATIONAL
Wormserstrasse 45
Cologne, Germany
T: 49 221 380 826 F: 49 221 376 1567

Dickhaut, Joachim, Mr.
Dipl. Eng.
UVP-AG-ENTWICKLUNGSLAENDER UVP-FOERDERVEREIN
Horner Weg 72
Hamburg 2000, Germany
T: 49 40 21-2683

Dickhaut, Wolfgang, Mr.
Dipl. Ing.
UVP-AG-ENTWICKLUNGSLANDER / UVP - FORDERVEREIN
Mathildenstr. 50
Darmstadt 6100, Germany
T: 00496151163448 F: 00496151163758

Dieke, Gerold, Mr.
Diretor Geral
GTZ
Dag Hammarskjold - weg 1-2 Eschborn
Frankfurt D-6236, Germany
T: 49 69 619-6790 F: 49 69 619-6794
TL: 407 501 - 0 9+2

Dimola, Victor
NORWEGIAN CAMPAIGN FOR ENVIRONMENT AND DEVELOPMENT
Langesgt 4
Osnabruck, Germany
T: 49 541 362035 F: 49 541 361454

Druke, Johannes
Cameraman
Rtl Tv
Aachenerstr 1036
5000 Cologne 40, Germany
T: 49 22 14895580

Durbeck, Mr.
GTE - PROTRADE
W/I/M / Management centre Elbroich AM Falder 4
Dusseldorf D 4000, Germany
T: 49 211 62-7819 F: 49 211 61-4879

Eber, Gerhard, Mr.
YMCA
Lilien Str. 4
Maxhuette 2414, Germany
T: 45 471 20400

Ecker, Joachim, Mr.
Director
YOUTH DEVELOPS FUTURE
Rosa-Luxemburgstr. 39
Belzig 0-1820, Germany
T: 49 161 1732460

Eidam, Gerd, Dr.
Board Member
HRV
Bodekerstr. 80
Hannover 3000, Germany
T: 49 511 390-2200 F: 49 511 390-2222

Fatheuer, Thomas
Director
LATEINAMERIKA - NACHRICHTEN
Gneisen Austr. 2
1000 Berlin 62, Germany
T: 004930 6934029 F: 004930 6926

Feige, Klaus-Dieter, Mr.
FEDERAL PARLIAMENT (BUNDESTAG)
Bundeshaus
Bonn 5300, Germany
T: 49 228 16-1

Fischer, Eric, Mr.
MOVIMENTO CIDADANIA ECOLOGICA
Sprengelstr. 22
Berlin 1000 65, Germany
T: 37 2 453-5205

Fischer, Gudrun, Ms.
Reporter
MULHERES E CIENCIAS
Schleswigerstr. 1
Bremen 2800, Germany
T: 49 421 396-4077

Franeck, Maik, Mr.
EARTH TRAIN/ WORLD LINK GLOBAL YOUTH SUMMIT
Spindelstr. 2
Potsdam, Germany
T: 49 331 77-054

Frohlids, Thomas
Correspondent
SUDDEUTSCHE ZEITUNG
Dahlmannstrasse 11,
Bonn D-5300, Germany
T: 49 228 214016 F: 49 228 219406

Furst, Birgit, Ms.
TULPENWEG 7 7910 NEU-ULM 7
Tulpenweg 7, 7910 Neu-Ulm 7
Germany
T: 7 31 77454

Geissler, Wenzel, Mr.
Evangelisches Studienwerk
Haus Villigst
Schwerte 5840, Germany
T: 49 02304 755 0

Giessner, Klaus, Dr.
Prof.Dr.
KATHOLISCHE UNIVERSITAT EICHSTATT
Lehrstuhl Physiche Geographie Ostenstrasse 26-28
Eichstatt D-8078, Germany
T: 49 84 212-0303 F: 49 84 212-0474

Graeve, Jurgen De
Editor
ARD - RADIO ALEMA
Postfach 20 05 08
D-8000 Munchenz Al, Germany
T: 49 89 590001 F: 49 89 59002639

Guminska, Maria, Dr.
DNR - BUNDESVERBAND FUR UMWELTSCHUTZ
Kalkuhlstr. 24
Bonn 5300 - 3, Germany
T: 49 228 44 1505 F: 49 228 44 2277
TL: 8861170 dnr d

Hamme, Iris, Ms.
VILLIGST
Max-Joseph-Str. 51
Heidelberg, Germany
T: 49 622136 509 F: 49 6221

Hartnagel, Berend, Mr.
EXPO 2000 PLANNING GROUP
Planckstr. 2
Hannover 3000, Germany
T: 49 511 1206991 F: 49 511 1202670

Henn, Klaus Peter, Mr.
Dr.
UVP - ZENTRUM
Bahnhefstr. 85
Bonn 5300, Germany

Hermanns, Ute
Vorbergstr.3 1000 Berlin 62
Berlin 62, Germany

Hokektkal, Carl Graf
Correspondent
FRANKFURTER ALLGEMEINE ZEITUNG
PedallionBouk, Fritz-Schaffer 13
Bonn, Germany
T: 49 228 12289122 F: 49 228 220382

Hollander, Nicola V.
Journalist
NDR- TV ALEMA
Gazellenkamp 57, 2 Hamburg 54
Hamburg, Germany
T: 49 4 4156-0 F: 49 4 4156-5410

Horn, Jakob
GERMAN ADULT EDUCATION ASSOCIATION
12 Rajaji Marg
New Delhi 110011, Germany
T: 91 11 301-4309

Hoss, Willi, Mr.
U F P A
Stuttgart 70 Karl Pfaff St. 22
Stuttgart 7000, Germany

Jaeger, Johann George, Mr.
EVANGELISCHES STUDIENWERK E.V.
Gartnerstrasse 6
Rostock Q-2500, Germany
T: 31 81 453481 F: 31 81 23916

Janecki, Janusz, Dr.
DNR - BUNDESVERBAND FUR UMWELTSCHUTZ
Kalkuhlstr. 24
Bonn 5300 - 3, Germany
T: 49 228 44 1505 F: 49 228 44 2277
TL: 8861170 dnr d

Jourdan, Karl-Heinz, Mr.
7080 Aalen, Rembrandtstr.7
Allen 7080 I, Germany

Jurina, Natalie Petrowna, Dr.
DNR - BUNDESVERBAND FUR UMWELTSCHUTZ
Kalkuhlstr. 24
Bonn 5300 - 3, Germany
T: 49 228 44 1505 F: 49 228 44 2277
TL: 8861170 dnr d

Jutz, Christoph, Mr.
Eisenbahnstr. 34
Ravensburg 7980, Germany
T: 0049751 22853

Kemper, Nina Helen
Researcher
WDR INTERNATIONAL
Wormserstrasse 45
Cologne, Germany
T: 49 221 380 826 F: 49 221 376 1567

Kersting, Felix, Mr.
Technician
GERMAN TELEVISION WDR
Surther Haupt 85
Cologne 5000, Germany
T: 49 221 679606 F: 49 221 68477

Kindernothilfe, E. V., Mr.
Dusseldorfer Landstrasse 180
Duisburg 4100, Germany

Kippenberg, Juliane, Ms.
EVANGELISCHES STUDIENWERK
Haus Villigst
Schwerte 5840, Germany
T: 49 02304 755 0

Klumb, Astrid, Ms.
Student
EV. STUDIENWERU VILLIGST
Grahenstuuiken 5A
Hamburg 2000, Germany
T: 49 40 895128

Kluver, Reymer
Editor
SUDDEUTSCHE ZEITUNG
Redaktion Seite 3 Sendtinger Strasse 80
Munchen, Germany
T: 49 89 709 F: 49 89 2183-787

Kroger, Andreas, Mr.
Student
EV. STUDIENWERK, VILLIGST
Yorckstr. 66 / 1000
Berlin 61, Germany
T: 37 2 786-3942

Lange, Andrea, Ms.
TERRE DES HOMMES / EV. STUDIENWERK E V.
Postfach 4126, Ruppenkampstr 11 a
Osnabruck W4500, Germany
T: 49 541 710-1166

Lange, Anne, Ms.
EVANGELISCHES STUDIENWERK E.V.
Ulmenmarkt 1
Rostock Q-2500, Germany
T: 31 81 453481 F: 31 81 23916

Lange, Eckard, Mr.
UVP-ZENTRUM
Friedrichstr, 92
Berlin, Germany
T: 37 2 2512701 F: 37 2 2516324

Langenscheidt, Florian, Ms.
Member Of The Board
ARTISTS UNITED FOR NATURE E.V.
FrohschammerstraBe 14, 8000
Munchen 40, Germany
T: 49 89 3594874 F: 49 89 3596622

Laubach, Birgit, Ms.
BUENDNIS 90 DIE GRUENEN
Bundeshaus Hi 201
Bonn 5300, Germany
T: 49 228 16 7622 F: 49 228 16 86662
TL: 885 598

Lavalette, Annette, Ms.
Senior Consultant
ARTHUR D. LITTLE INTERNATIONAL
Gustav-Stresemann-Ring 1
Wiesbaden 62000, Germany
T: 49 611 7148-190 297

Leutner, Hans, Dr.
DEUTSCHE FORSCHUNGS GEMEINSCHAFT- DFG
Kennedyallee 40
Bonn 5300, Germany
T: 49 228 885-2346 F: 49 228 885-2221

Lobsack, Lili, Ms.
Director
GTZ
Dag Hammarskjold-weg 1-2 Eschborn 6
Frankfurt D-6236, Germany
T: 49 69 244-1118 F: 49 69 243-4991
TL: 407 501 0 9+2-d

Lopes, Sandra Maria, Ms.
FRAUEN ANSTIFRUNG
Carl-Severing, 24
Bremen 2800 41, Germany
T: 49 421 466879

Loy, Detlef, Mr.
Dipl.Ing.
SENATE OF BERLIN
Georg Wilhelm Str. 18
Berlin 1000, Germany
T: 37 2 891-9458

Lutz, Rudiger, Mr.
Prof.
FUTURE LAB
Begstr. 19 5257
Spriye, Germany
T: 004 50414745

Magalhaes, Antonio Carlos De Melo, Dr.
Doctor Of Theology
MISSIONSAKADEMIE AN DER UNIVERSITAT HAMBURG
Rupertistrasse 67
Hamburg 2000, Germany
T: 49 40 827468 F: 49 40 8226361

Marler, Fred
FREELANCER
6 Hauptmann Str 14
Gelsenkirchen 0-4650, Germany
T: 49 209 397956 F: 49 209 377405

Mastaller, Michael, Mr.
Ph.D.
GESELLSCH F. ORGANISATION
Tournelystr. 16
Stadtbergen D-8901 2, Germany
T: 49 821 43 48 01

Merlitz, Reinhard, Mr.
UVP FORDERVEREIN
Viktoriastrasse 94- Darmstadt
Hessen -RFA 6100, Germany
T: 049 615 23435

Meyer, Roswitha, Ms.
Dr.
UNIVERSITY TUBINGEN
Univ. Tubingen, Chem. Abs. Morgenstelle 22
Tubingen 7400, Germany
T: 49 7071292033 F: 49 7071292033

Monninghoff, Hans, Mr.
Commissioner
CITY OF HANNOVER
P.O. Box 125
Hannover 3000, Germany

Mueller, Anneliese, Ms.
Member Of Board
GERMAN WOMEN'S COUNCIL
Simrockstrasse 5
Bonn 53 Bonn 1, Germany
T: 49 228 223008 F: 49 228 218819

Muhlnickel, Rainer, Mr.
Dipl. Betriebsw
TECHNISCHE UNIIVERSITAT BERLIN
TU Berlin. Sekr.FR2-7 Landschaftsokonomie Franklinstrasse 28/29,
Berlin 10 D-1000, Germany
F: 37 2 31421123

Muller-Rockstroh, Peter, Dr.
Director
CARL DUISBERG GESELLSCHAFT E.V.
Lutzowufer 6-9 D-1000
Berlin 30, Germany
T: 49 30 25482-0 F: 49 30 25482-105
TL: 184628 cdgbl d

Nessler, Nikolaus, Mr.
INSTITUTE GOETHE
Bergerstr. 39
Frankfurt, Germany
T: 49 69 442350 F: 49 69 442350

Neugebauer, Bernd, Dr.
TREES FOR PEOPLE
Graf-Durckheim-Weg 7
Todtmoos D 7865, Germany
T: 49 674 748806 F: 49 674 748807

Neveling, Stefanie, Sra.
Student
CARL DUISBERG GESELLSCHAFT E.V.
Lutzowufer 6-9 D-1000
Berlin 30, Germany
T: 49 30 25482-0 F: 49 30 25482-105
TL: 184628 cdgbl d

Nicolaysen, Anders
NORWEGIAN CAMPAIGN FOR ENVIRONMENT AND DEVELOPMENT
Langesgt 4
Osnabruck, Germany
T: 49 541 362035 F: 49 541 361454

Pawelke, Dr. Rainer, Dr.
Doctor
INTERNATIONAL FORUM OF CARDIOLOGY AND RHAUMATOLOGY
8201 Bad Feilnbach
Germany
T: 0 80668200 F: 0 80661329

Pollmann, Hanne E., Ms.
Member Of Board
GERMAN WOMEN'S COUNCIL
Simrockstrasse 5
Bonn 53 Bonn 1, Germany
T: 49 228 223008 F: 49 228 218819

Porsch, Siglinde, Ms.
Member Of Board
GERMAN WOMEN'S COUNCIL
Simrockstrasse 5
Bonn 53 Bonn 1, Germany
T: 49 228 223008 F: 49 228 218819

Quistorp, Eva M., Ms.
Co-Speaker
WOMEN FOR PEACE AND ECOLOGY
Bundeshaus, HT 108
Bonn 5300 1, Germany
T: 49 228 16 7311 F: 49 228 16 9090

Rautter, Marietta Von
ARTISTS UNITED FOR NATURE E.V.
Frohschammerstrasse 14
Munchen 8000 40, Germany
T: 49 89 359-4874 F: 49 89 359-6556

Riorden, Ellen, Ms.
Schliemann Str. 46
Berlin 1058, Germany

Ruiter, Wim De, Mr.
Coordinator
SMILE EUROPEAN INCENTIVE EXECUTIVES
WIM Management center Elbroich AM Falder 4
Dusseldorf D 4000, Germany
T: 49 211 627-819 F: 49 211 614-879

Sander, Reinhard, Mr.
DEUTSCHER NATURSCHUTERING
Kalkulibtv. 24
Bonn 5300-3, Germany
T: 49 228 44-1505 F: 49 228 44-3643

Sandul, Wlazheslaw A.
DNR - BUNDESVERBAND FUR UMWELTSCHUTZ
Kalkuhlstr. 24
Bonn 5300 - 3, Germany
T: 49 228 44 1505 F: 49 228 44 2277
TL: 8861170 dnr d

Schaller, Andrea
General Manager
MU-LI VERWALTUNGS- UND HANDELSGES. AG
Gnesener Str. 4-6
Munchen D-8000, Germany
T: 49 89 93 00 080 F: 49 89 93 00 859
TL: 522372

Scheffler, Marion, Mr.
Consultant
EXPO 2000 PLANNING
Planckstrasse 2 - 3000
Hannover 3000, Germany
T: 49 511 120-2186 F: 49 511 120-2670

Scheumann, Waltina, Ms.
Dipl. Pol.
TECHNISCHE UNIIVERSITAT BERLIN
Sekr.FR2-7 Landschaftsokonomie Franklinstrasse 28/29
Berlin 10 D-1000, Germany
F: 37 2 31421123

Schielmann, Horst
FREELANCER
6 Hauptmann Str 14
Gelsenkirchen 0-4650, Germany
T: 49 209 397956 F: 49 209 377405

Schmidt, Silke Eva, Ms.
M.A.
EVANGELISCHES STUDIENWERK VILLIGST
Lisztst. 2A D-2000
Hamburg 50, Germany
T: 49 40 880-0837 F: 49 40 822-6361

Scholles, Frank, Mr.
Dipl. Ing.
UVP - FOERDERVEREIN
Destingstr 13
Hamm 4700, Germany
T: 49 238152129 F: 49 238152195

Schulze, Martina, Ms.
U F P A
Ernst August Str. 9
Hamburg 2000, Germany

Silva, Luis, Mr.
INTERNATIONAL WORKERS AID
Marie-Juchacz-Haus, Oppelner Strabe 130
Bonn, Germany
T: 49 228 668-5211 F: 49 228 668-3254
TL: 886954

Souza, Michelle Cristine F., Ms.
WORLD VISION INTERNATIONAL
Adenauer Alle 32
Oberursel, Germany
T: 49 617156074 F: 49 6171 56078

Spruck, Stefan
Communications Director
WORLD VISION INTERNATIONAL
Adenauer Allee 32
Oberursel, Germany
T: 49 6171156074 F: 49 6171156078

Starck, Dorothee, Ms.
VILLIGST
Fleischstr. 77-78
Trier 5500, Germany
T: 49 651 43184

Stawitzki, Frank, Mr.
UVP - FORDEREVBIN HAMM
Oesting Str. 13
Hamburg 4700, Germany
T: 49 2381 52129 F: 49 2381 52195

Steger, Prof Dr. Ulrich, Mr.
Member Of The Managment Board
VOLKSWAGEN
Wolfsburg, Germany

Stibbe, Harald, Ms.
Manager
VOLKSWAGEN
Wolfsburg, Germany

Strunder, Christian, Mr.
GTE - PROTRADE
W/I/M / Management centre Elbroich AM Falder 4
Dusseldorf D 4000, Germany
T: 49 211 62-7819 F: 49 211 61-4879

Szillinsky, Albrecht, Dr.
DEUTSCHE FORSCHUNGSGEMEINSCHAFT
Kennedyeliee 40
Bonn -NRW 5300, Germany
T: 49 228 885247 F: 49 222 82221

Theiemann, Doerte, Ms.
Dipl-Kaufm.
UVP - FOERDERVEREIN
Oesting Str. 13
Hamm, Germany
T: 49 2381 52129 F: 49 2381 52195

Turley, Frank Otto, Mr.
Ecologist
SACTES
Kurfurstenstr, 23
Frankfurt D-6000, Germany
T: 49 69 777-631

Uskow, Arkadij G.
DNR - BUNDESVERBAND FUR UMWELTSCHUTZ
Kalkuhlstr. 24
Bonn 5300 - 3, Germany
T: 49 228 44 1505 F: 49 228 44 2277
TL: 8861170 dnr d

Von Ledebur, Jan, Mr.
IFOAM
6100 Darmstadt
Germany

Von Weizsacker, Ernst U., Dr.
President
WUPPERTAL INSTITUTE
Doppersberg 19 D-5600
Wuppertal 1, Germany
T: 49 202 2492100 F: 49 202 2492208

Weder, Dietrich Jorn
Editor
HESSISCHER RUNDFUNK
Bertramstrasse 8
Frankfurt 6000, Germany
T: 69 1553105 F: 69 1552900

Weindenbach, Thomas
Reporter
WDR INTERNATIONAL
Wormserstrasse 45
Cologne, Germany
T: 49 221 380 826 F: 49 221 376 1567

Weiss, Gerd, Mr.
GERMAN TELEVISION WDR
Surther Hauptstr. 85
Cologne 5000, Germany
T: 49 221 679606 F: 49 221 68477

Wietersheim, Viggo Von
ARTISTS UNITED FOR NATURE E.V.
Frohschammerstrasse 14
Munchen 8000 40, Germany
T: 49 89 359-4874 F: 49 89 359-6556

Wildner, Axel, Mr.
GTE - PROTRADE
W/I/M / Management centre Elbroich AM Falder 4
Dusseldorf D 4000, Germany
T: 49 211 62-7819 F: 49 211 61-4879

Willie, Joachim
Editor
FRANKFURTER RUNDSCHAU
Grosse Eschenheimer Strasse, 18-18,
Frankfurt, Germany
T: 49 69 2199583 F: 49 69 2199521

Winter, Gerd, Mr.
Professeur
UNIVERSITATSALLE GW1
Zentrum fur Europaiche Rechtspolitik
Bremen D 2800 33, Germany

Wittekindt, Elisabeth, Dr.
Member Of Board
GERMAN WOMEN'S COUNCIL
Simrockstrasse 5
Bonn 53 Bonn 1, Germany
T: 49 228 223008

Zillmann, Kerstin, Ms.
TECHNICAL UNIVERSITY OF HAMBURG / FEMINIST ORGANIZATION OF ARCHITECTS
Schulterblatt 84
Hamburg, Germany
T: 49 40 4395568

Zimmerle, Birgit
FAKT
Stephan - Bla Amann Str., 11
Furtwangen, Germany

ADVISORS

Hahn, Carl, Dr.
Chairman, Board Of Management
Volkswagen
3180 Wolfsburg 1, Germany
F: 49 5361 921020

Hauff, Volker, Dr.
Bundesminister A.D., Former Wced Member
5000 Koln, Germany
T: 49 221 382 111 F: 49 221 375 546

PARLIAMENTARIANS / SPIRITUAL LEADERS

Caspers-Merk, Marion, Mrs.
Member
Federal Parliament
Volkskammer der DDR Marx Engels Platz 1020
Berlin, Germany
T: 238 3246
TL: Voka 1152406

GHANA - GOVERNMENT

Abaka, Charlotte, Dr.
Chairperson
National Council on Women and Development
19 E. 47th Str.
New York, NY 10017, United States of America
T: 212 832 1300 F: 212 751 6734

Adzobu, C. Dorm, Dr.
Director of Programmes
Environmental Protection Council
19 E. 47th Str.
New York, NY 10017, United States of America
T: 212 832 1300 F: 212 751 6734

Ahwoi, Kwamena, Hon. Mr.
P.N.D.C. Secretary
Local Government and Environmental Affairs
19 E. 47th Str.
New York, NY 10017, United States of America
T: 212 832 1300 F: 212 751 6734

Annan, Justice D.F., Mr.
Vice President
Permanent Mission of Ghana
19 E. 47th. Str.
New York NY 10017, United States of America
T: 212 832 1300 F: 212 751 6734

Awoonor, Kofi Nyidevu, H.E.
Ambassador Extraordinary and Plenipotentiary
Permanent Representative to the United Nations, New York
19 E. 47th Str.
New York, NY 10017, United States of America
T: 212 832 1300 F: 212 751 6734

Benneh, George, Prof.
vice Chancellor
University of Ghana
19 E. 47th Str.
New York, NY 10017, United States of America
T: 212 832 1300 F: 212 751 6734

Kwami Hamenoo, Michael Charles, H.E. Mr.
Ambassador of Ghana to Brazil
19 E. 47th Str.
New York, NY 10017, United States of America
T: 212 832 1300 F: 212 751 6734

Naadjie, James, Mr.
Counsellor
19 E. 47th Str.
New York, NY 10017, United States of America
T: 212 832 1300 F: 212 751 6734

Osei-Bonsu, , Dr.
National Development Planning Commission
19 E. 47th Str.
New York, NY 10017, United States of America
T: 212 832 1300 F: 212 751 6734

Quaigde, Sam, Mr.
Chief News Editor
Ghana News Agency
19 E. 47th Str.
New York, NY 10017, United States of America
T: 212 832 1300 F: 212 751 6734

Yeboah-Asuamah, Kojo, Mr.
Counsellor
Embassy of Ghana, Brasilia
19 E. 47th Str.
New York, NY 10017, United States of America
T: 212 832 1300 F: 212 751 6734

NGO

Ahadzie, George
GREEN EARTH ORGANIZATION - GEO
PO Box 16641
Accra North, Ghana

Ali, Turabo, Mr.
ALL AFRICA STUDENTS UNION
P.O. Box M274
Accra - North, Ghana
T: 233 21 663 450 F: 233 21 664 293

Drame, Sountoucoung
AASU
P.O. Box M. 274
Accra, Ghana

Konadu Bonsu, Sakeena
ECOG
P.O. Box 168
Trade Fair Site, Accra, Ghana

Opoku Kwasi, William
GEO
PO Box 16641
Accra North, Ghana

Owusu-Mpiani, Kwaku
ECOG
P.O. Box 168
Trade Fair Site, Accra, Ghana

PRESS

Adadeyoh, Sika, Ms.
Reporter
WEEKLY SPECTATOR
P.O. BOX 2638, Accra-North
Accra - North, Ghana
T: 233 21 228282

Quaicoe, Samuel Boniface
Editor
Ghana News Agency
P.O. Box 2118
Accra, Ghana, Ghana

ORGANIZATIONS

Quachey, Lucia, Ms.
As. Of Women In Development (awide)
P.O. Box 7600
Accra, North, Ghana
T: (23321) 222 459
TL: 2340 or 2394 MNJ GH

GLOBAL FORUM

Aboah, Samuel, Mr.
SAMBUS CO. LTD.
Accra, Ghana

Adjei, Kwabena, Mr.
Deputy Secretaty
MINISTRY OF INDUSTRIES SCIENCE AND TECHNOLOGY
Post Office Box M39
Accra - North, Ghana
T: 233 21 66-2533

Amenghe, Clothilda, Ms.
PO Box 7600
Accra - North, Ghana
T: 233 21 222 459 F: 233 21 228 398

Avoka, Chris
Producer
GHANA BROADCASTING CORPORATION (GBC)
PO Box 43
Bolgatanga U-E-R, Ghana
T: 72 30657

Ayiku, Moses, Dr.
Director
TECHNOLOGY TRANSFER CENTRE
c/o P.O. Box M.39
Accra - North, Ghana
T: 233 21 77-3315

Brenu, Kweku Osae, Mr.
President
PLANNED PARENTHOOD ASSOCIATION OF GHANA
P.O.Box 5756
Aura, Ghana
T: 233 21 774 338

Guri, Bernard Y., Mr.
NATIONAL CATHOLIC SECRETARIATE
P.O.Box 9712 - Accra Airport -Ghana - Africa
Ghana - Africa, Ghana
TL: 2471 ncs gh

Parry, Sam Amobil, Mr.
SOCIETY FOR INTER. DEV. COUNCIL FOR INTERNATIONAL ACTION FOR DEVELOPMENT IN AFRICA
P.O.Box 14337
Accra - North, Ghana
F: 233 21 665960
TL: 3033GH

Quarcoo, A.K., Dr.
Rev.
METHODIST CHURCH - GHANA / BOARD OF CHURCH AND SOCIETY
PO Box 403de Ferney
Accra, Ghana

ADVISORS

Ayensu, Edward S., Prof.
President, Panafrican Union For Science And Technology
Accra, Ghana
T: 233 21 775 554 F: 233 21 773 116

Obeng, Letitia E., Dr.
Director, Environment Management Services
Accra, Ghana
T: 233 21 776 248

GREECE - GOVERNMENT

Avramopoulos, Dimitrios, Mr.
First Secretary of Embassy, Spokesman
Ministry of Foreign Affairs
Permanent Mission to the United Nations, 733 3rd Ave., 23rd Floor
New York, NY 10017, United States of America
T: 212 490 6060 F: 212 490 5894

Bakoyannis, Dora, Mrs.
Member of Parliament
Permanent Mission of Greece
733 3rd Ave., 23rd Floor
New York, NY 10017, United States of America
T: 212 490 6060 F: 212 490 5894

Decleris,
Vice President Of The Council Of
COUNCIL OF STATE
82 Fokionos Negri, Athens 113 61 Greece
Athens, Greece
T: 1 82 15 946 F: 1 82 15 946

Denaxas, Evangelos, Mr.
Counsellor of Embassy
Ministry of Foreign Affairs
11 Miaouli Str 15232 Khalandri Athens
Athens, Greece
T: 00301 3619225 F: 00301 36241953609716

Dimitropoulos, Christos, Mr.
Expert-Counsellor
Ministry of Foreign Affairs
Permanent Mission to the United Nations, 733 3rd Ave., 23rd Floor
New York, NY 10017, United States of America
T: 212 490 6060 F: 212 490 5894

Dokianos, Spyridon, H.E. Mr.
Ambassador of Greece to Brasilia
Permanent Mission of Greece
733 3rd Ave., 23rd Floor
New York, NY 10017, United States of America
T: 212 490 6060 F: 212 490 5894

Frangos, Vassilios, Mr.
Deputy Director, General Secretariat for Forests and Natural Environment
Ministry of Agriculture as Representative of Greece to UNCED
Permanent Mission to the United Nations, 733 3rd Ave., 23rd Floor
New York, NY 10017, United States of America
T: 212 490 6060 F: 212 490 5894

Georgiou, George, H.E. Mr.
Ambassador, Head of Protocol
Ministry of Foreign Affairs
Permanent Mission to the United Nations, 733 3rd Ave., 23rd Floor
New York, NY 10017, United States of America
T: 212 490 6060 F: 212 490 5894

Goulandris, Niki, Mrs.
Special Collaborator of the Minister of the Environment
Town Planning and Public Works
Permanent Mission to the United Nations, 733 3rd Ave., 23rd Floor
New York, NY 10017, United States of America
T: 212 490 6060 F: 212 490 5894

Halastanis, Constantin, Mr.
First Secretary of Embassy
Prime Minister's Diplomatic Office
Permanent Mission to the United Nations, 733 3rd Ave., 23rd Floor
New York, NY 10017, United States of America
T: 212 490 6060 F: 212 490 5894

Karamanlis, Achilles, H.E. Mr.
Minister of the Environment
Town Planning and Public Works
Permanent Mission to the United Nations, 733 3rd Ave., 23rd Floor
New York, NY 10017, United States of America
T: 212 490 6060 F: 212 490 5894

Kardassis, George, Mr.
Director-General for the Environment, Ministry of the Environment
Town Planning and Public Works
Permanent Mission to the United Nations, 733 3rd Ave., 23rd Floor
New York, NY 10017, United States of America
T: 212 490 6060 F: 212 490 5894

Kartalis, Konstantinos, Mr.
Special Adviser to the Mayor of Athens
Permanent Mission of Greece
733 3rd Ave., 23rd Floor
New York, NY 10017, United States of America
T: 212 490 6060 F: 212 490 5894

Katramadou, Kouzinia, Mrs.
Ministry of Foreign Affairs
Permanent Mission of Greece
733 3rd Ave., 23rd Floor
New York, NY 10017, United States of America
T: 212 490 6060 F: 212 490 5894

Katsoudas, Dimitrios, Mr.
Secretary-General for Forests and Natural Environment
Ministry of Agriculture
Permanent Mission to the United Nations, 733 3rd Ave., 23rd Floor
New York, NY 10017, United States of America
T: 212 490 6060 F: 212 490 5894

Kypraiou, Artemissia, Mrs.
Director
Prime Minister's Private Office
Permanent Mission to the United Nations, 733 3rd Ave., 23rd Floor
New York, NY 10017, United States of America
T: 212 490 6060 F: 212 490 5894

Liaskas, Konstantinos, Mr.
President of the Technical Chamber of Greece
Permanent Mission of Greece
733 3rd Ave., 23rd Floor
New York, NY 10017, United States of America
T: 212 490 6060 F: 212 490 5894

Mitsotakis, Konstantinos, Mr.
Prime Minister
Permanent Mission of Greece
733 3rd. Ave. 23rd floor
New York NY 10017, United States of America
T: 212 490 6060 F: 212 490 5894

Mourmouri, Athina, Dr.
Ministry of the Environment
Town Planning and Public Works
Permanent Mission to the United Nations, 733 3rd Ave., 23rd Floor
New York, NY 10017, United States of America
T: 212 490 6060 F: 212 490 5894

Nakos, George, Mr
Lab. Director
National Institute of Forest Research
Terma Alkmanos GR-115 28
Athens, Greece
T: 7784240

Neonakis, Ioannis, Mr.
Consul-General of Greece to Rio de Janeiro
Permanent Mission of Greece
733 3rd Ave., 23rd Floor
New York, NY 10017, United States of America
T: 212 490 6060 F: 212 490 5894

Papagrigoriou, Spyridon, Mr.
Adviser
Technical Chamber of Greece
Permanent Mission to the United Nations, 733 3rd Ave., 23rd Floor
New York, NY 10017, United States of America
T: 212 490 6060 F: 212 490 5894

Plakoudi, Athina, Mrs.
Embassy of Greece in Brasilia
Permanent Mission of Greece
733 3rd Ave., 23rd Floor
New York, NY 10017, United States of America
T: 212 490 6060 F: 212 490 5894

Polydoras, Vyron, H.E. Mr.
Deputy Minister of the Presidency.
Government Spokesman
Permanent Mission to the United Nations, 733 3rd Ave., 23rd Floor
New York, NY 10017, United States of America
T: 212 490 6060 F: 212 490 5894

Simantonis, Ioannis, Mr.
Ministry of the Environment
Town Planning and Public Works
17 Amaliados Str
Athens 11523, Greece
T: 6442397 F: 6425300

Skalaios, Helias, Mr.
Vice Mayor of Athens
Permanent Mission of Greece
733 3rd Ave., 23rd Floor
New York, NY 10017, United States of America
T: 212 490 6060 F: 212 490 5894

Staikou, Sophia, Mrs.
Adviser to the Minister of the Environment
Town Planning and Public Works
Harilaou Triuoupi 182
Athens 10178, Greece
T: 01 6446659 F: 01 6427520

Tassioulas, Ionannis, Mr.
Embassy of Greece in Brasilia
Permanent Mission of Greece
733 3rd Ave., 23rd Floor
New York, NY 10017, United States of America
T: 212 490 6060 F: 212 490 5894

Theophanous, George, Mr.
Secretary-General for International Economic Relations
Ministry of National Economy
Permanent Mission to the United Nations, 733 3rd Ave., 23rd Floor
New York, NY 10017, United States of America
T: 212 490 6060 F: 212 490 5894

Tsilas, Loukas, H.E. Mr.
Ambassador, Director
Prime Minister's Diplomatic Office
Permanent Mission to the United Nations, 733 3rd Ave., 23rd Floor
New York, NY 10017, United States of America
T: 212 490 6060 F: 212 490 5894

Zerephos, Christos, Prof.
Special Collaborator of the Minister of the Environment
Town Planning and Public Works
Permanent Mission to the United Nations, 733 3rd Ave., 23rd Floor
New York, NY 10017, United States of America
T: 212 490 6060 F: 212 490 5894

Zikos, Paris, Mr.
Ministry of the Environment
Town Planning and Public Works
147 Patission Str. GR11251
Athens, Greece
T: 8650053 8650334 F: 8647420

PRESS

Karabelias, Georgios
Correspondent
SKY 100.4 FM
106-108 Mavronichali
Athens, Greece, Greece

Panagiotopoulos, Panagiotis
Editor
ANTENNA TV
28 Lycevittey Dtc.
Athenas, Greece, Greece

Serezis, Constantinos
Reporter Radio/Television
E.R. TV
37 Sarantapihou St.
Athenas, Greece, Greece

Tsirigotakis, Charalambolaris
Correspondent
HELLENIC BROADCASTING CORP
39/40 Upper Grosvgnor Street
London W1, UK, Greece

Vassilios, Drimaropoulos
Editor
E.R. TV
49 Lazaraki
Athenas, Greece, Greece

Vembos, Athanassios
Editor
INDUSTRIAL REVIEW
4 Tenedou St.
Athens 112-57, Greece, Greece

Vergis, Dimosthenis
Editor
NATURE NEWS PAPER
Santaroza 1A
Athenas, Greece, Greece

Vretos, Spiros
Editor
ECONOMICOS TACHYDROMOS MAGAZINE
16, Rue Guyton de Morveau
75103 Paris, France, Greece

Xenakis, Sotirios
Reporter Radio/Television
E.R. TV
Thisoo 4 Alimos
Athenas, Greece, Greece

ORGANIZATIONS

Mitsatsos, Dimitris, Mr.
Hellenic Marine Environment Protection Association (HELMEPA)
5 Pergamou Street, N. Smyrni
17121 Athens, Greece
T: (30 1) 934 3088/932 6277 F: (30 1) 935 3845

GLOBAL FORUM

Koutsaktikis, Konstantinos
Photojournalist
ECOLOGICAL NEWSPAPER 'NATURE'
Santaroza 1D Apanedistimiou 6
Athens, Greece
T: 3621140

Persellis, Emmanuel P., Dr.
GREEK ORTHODOX PATRIARCHATE OF ALEXANDRIA
62, C. Varnali St. GR - 144 52 Metamorphosis
Athens, Greece

Venizelos, Lily, Ms.
President
MEDITERRANEAN ASSOCIATION TO SAVE THE SEA TURTLES - MEDASSET
1C Licavitou St.
Athens 10672, Greece
T: 30 1 3624971 F: 30 1 3636415

Vlavianos-Arvanitis, Agni, Dr.
President
BIOPOLITICS INTERNATIONAL ORGANIZATION
10 Tim Vassou
Athens 11521, Greece
T: 30 1 6432 419 F: 30 1 6434 093

GREENLAND - NGO

Egede, Ingmar, Mr.
Council Member
INUIT CIRCUMPOLAR CONFERENCE
PO Box 901, NUUK
Nuuk DK 3900, Greenland
T: 299-27290 F: 299-23001

GRENADA - GLOBAL FORUM

Joseph, Matthias, Mr.
ST. ANDREW PEOPLES / MAROON ASSOC. (SAPMA)
Knox House
St. Georges, Grenada

GUAM - GOVERNMENT

Ada, Joseph F, Mr
Governor Of Guam
Governor'S Office of Guam
Agana 96910, Guam
T: 671 477 9858 8931 F: 671 477 GUAM

Ada, Tricia R.S.
Representative
GOVERNOR'S OFFICE OF GUAM
P.O. Box 6750 Tamuning
Agana 96911, Guam
T: 671 477 9858 F: 671 477 GUAM

Castro, Fred, Mr
Director
GUAM ENVIRONMENTAL PROTECTION AGENC
C/O Governor'S Office
Agona 96910, Guam
T: 671 477 8931 F: 671 477 GUAM

Manibuoan, Marilyn, Ms
Senator
THE GUAM LEGISLATURE
155 Hester St
Agana, Guam
T: (671) 472-3450/3451 F: (671) 472-4090

Ulloa, Anna, Ms
Communications
SEN. MARILYN MANIBUSAN'S OFFICE
the Guam Legislature 155 Hesler St Agana,Guam 96910
Agana, Guam
T: (671) 472-3450/3451 F: (671) 472-4090

GUATEMALA - GOVERNMENT

Aguilar Hecht, Francis Eric, Sr.
Ministro Consejero
Mision Permanente ante las Naciones Unidas, New York
Permanent Mission to the United Nations, 57 Park Avenue
New York, NY 10016, United States of America
T: 212 679 4760 F: 212 685 8741

Archila Serrano, Hilda Mireya, Sra.
Sub-Coordinadora de la Comision de Medio Ambiente
Permanent Mission of Guatemala
57 Park Avenue
New York, NY 10016, United States of America
T: 212 679 4760 F: 212 685 8741

Berger, Francois, Sr.
Assessor de la Presidencia
Permanent Mission of Guatemala
57 Park Avenue
New York, NY 10016, United States of America
T: 212 679 4760 F: 212 685 8741

Berger, Maria Eugenia de, Sra.
Personera del Presidente en cuestiones de Medio Ambiente
Permanent Mission of Guatemala
57 Park Avenue
New York, NY 10016, United States of America
T: 212 679 4760 F: 212 685 8741

Burgos, Amilcar, Sr.
Asesor de la Presidencia de la Republica
Permanent Mission of Guatemala
57 Park Avenue
New York, NY 10016, United States of America
T: 212 679 4760 F: 212 685 8741

Calle S., Juan de Dios, Sr.
Asesor de Contaminacion Ambiental
Permanent Mission of Guatemala
57 Park Avenue
New York, NY 10016, United States of America
T: 212 679 4760 F: 212 685 8741

Castellanos Carillo, Rafael, E.S.
Embajador Extraordinario y Plenipotenciario
Encargado de Negocios ante las Naciones Unidas, New York
Permanent Mission to the United Nations, 57 Park Avenue
New York, NY 10016, United States of America
T: 212 679 4760 F: 212 685 8741

Estuardo Roldan B., Manuel, Sr.
Primero Secretario y Consul
Rio de Janeiro
Permanent Mission to the United Nations, 57 Park Avenue
New York, NY 10016, United States of America
T: 212 679 4760 F: 212 685 8741

Ferrate, Luis, Dr.
Experto en Recursos Naturales Renovables
Permanent Mission of Guatemala
57 Park Avenue
New York, NY 10016, United States of America
T: 212 679 4760 F: 212 685 8741

Ferrate Felice, Antonio, Ing.
Coordinador de la Comision Nacional de Medio Ambiente
Permanent Mission of Guatemala
57 Park Avenue
New York, NY 10016, United States of America
T: 212 679 4760 F: 212 685 8741

Font, Eduardo Hugo, Sr.
Sub-directorde la Direccion
General de Bosques y Vida Silvestre
Permanent Mission to the United Nations, 57 Park Avenue
New York, NY 10016, United States of America
T: 212 679 4760 F: 212 685 8741

Godoy, Juan Carlos, Sr.
Experto de biodiversidad
Permanent Mission to the United Nations, 57 Park Avenue
New York, NY, 10016
F: 212 679 4760
TL: 212 685 8741

Guzman Ovalle, Edwin Fernando, Sr.
Representante del Ministerio de la Defensa Nacional
Permanent Mission of Guatemala
57 Park Avenue
New York, NY 10016, United States of America
T: 212 679 4760 F: 212 685 8741

Lopez, Leonel, Sr.
Representante del Ministro de Energia y Minas
Permanent Mission of Guatemala
57 Park Avenue
New York, NY 10016, United States of America
T: 212 679 4760 F: 212 685 8741

Menedez Park, Gonzalo, E.S. Lic.
Ministro de Relaciones Exteriores
Permanent Mission of Guatemala
57 Park Avenue
New York, NY 10016, United States of America
T: 212 679 4760 F: 212 685 8741

Milian, Francisco, Lic.
Asesor del Medio Amiente
Presidencia de la Republica
Permanent Mission to the United Nations, 57 Park Avenue
New York, NY 10016, United States of America
T: 212 679 4760 F: 212 685 8741

Oribe, Jose, Lic.
Agregado Comercial
Embajada de Guatemala en Washington
Permanent Mission to the United Nations, 57 Park Avenue
New York, NY 10016, United States of America
T: 212 679 4760 F: 212 685 8741

Palaez, Arelis, Sr.
Representante del Ministro de Desarrollo Urbano y Rural
Permanent Mission of Guatemala
57 Park Avenue
New York, NY 10016, United States of America
T: 212 679 4760 F: 212 685 8741

Serrano, Elias Jorge Antonio, Mr.
President
Permanent Mission of Guatemala
57 Park Avenue
New York NY 10016, United States of America
T: 212 679 4760 F: 212 685 8741

Serrano Elias, Jorge, E.S. Ing.
Presidente Constitucional
Republica de Guatemala
57 Park Avenue
New York, NY 10016, United States of America
T: 212 679 4760 F: 212 685 8741

Valdez, Rolando, Sr.
Representante del Ministro de Agricultura
Ganaderia y Alimentacion
Permanent Mission to the United Nations, 57 Park Avenue
New York, NY 10016, United States of America
T: 212 679 4760 F: 212 685 8741

Velasquez Vasquez, Estuardo, Sr.
Director del Instituto Nacional de Sismologia
Metereologica e Hidrologia
Permanent Mission to the United Nations, 57 Park Avenue
New York, NY 10016, United States of America
T: 212 679 4760 F: 212 685 8741

NGO

Pilon De Pacheco, Maria
AGPMA
20 Calle 19-44, zona 10
Guatemala 01010, Guatemala

Vasquez, Arnulfo Zacarias, Sr.
YOUTH 92 / COOPERACION PARA EL DESARROLLO RURAL DE OCCIDENTE
Canton Chuanoj
Totonicapan, Guatemala
T: 32 66 1115 F: 32 66 1115

PRESS

Asturias, Carlos
Reporter Radio/Television
GUATEMALA PRESS
18 Av. A D-32 Z.15 V.H. II
Guatemala, Guatemala

Asturias, Gonzalo
Reporter Radio/Television
GUATEMALA PRESS
14 Av. 2-56 Z.1
Guatemala, Guatemala

GLOBAL FORUM

Centeno, Hector
DEFENSORES DE LA NATURALEZA
Palais de Nations
Geneve CH 1211, Guatemala
F: 41 22 733-6542

Cerezo, Marco, Mr.
Director-General
FUNDACION PARA EL ECODESARROLLO Y LA CONSERVACION
17 Calle B 14-24 Zona 10 Colonia Oakland
Guatemala (City) 01010, Guatemala
T: 502 2 682754 F: 502 2 682754

Hammond, David, Mr.
Director
ALTERTEC
2a. Calle 10-14, Zona 1
Guatemala (City), Guatemala
T: 502 2 082-1343 F: 502 2 24165

Henirson, Lic. Nacida
REPAC/REDES
Palais de Nations
Geneve CH 1211, Guatemala
F: 41 22 733-6542

Monterroso, Tulio, Mr.
Executive Director
ARCAS - WILD LIFE RESCUE AND CONSERVATION ASSOC.
5a. CALLE 9-55 ZONA 1
Guatemala (City), Guatemala
T: 502 2 81647 F: 502 2 538507

Morrison, Sabrina, Ms.
FUNDACION PARA EL ECODESARROLLO Y LA CONSERVACION
14 Calle B Zona 10 Oakland
Guatemala (City) 01010, Guatemala
T: 502 2 682454 F: 502 2 682454

Pacheco, Marta Pilon, Ms.
ASOCIACION GUATEMALTECA PRO-DEFENSA DEL MEDIO AMBIENTE
20 Calle 19-52 Zona 10
Guatemalla C.A., Guatemala
T: 44 681 327 F: 44 372084

Recinos Cueto, Marco Augusto, Mr.
Coord.
ECOTEC
7A,Av. 8-90 Zona 2
Guatemala (City), Guatemala
T: 502 2 22471 F: 502 2 24871

Rivera, Hilda, Ms.
RED DE MUJERES Y MEDIO AMBIENTE - CENTRO AMERICA Y CARIBE
14 Calle 5-57, casa 11
Guatemala (City), Guatemala
T: 502 2 730657 F: 502 2 318769

PARLIAMENTARIANS / SPIRITUAL LEADERS

De Escobar, Rita, Dep.
Congreso de la Republica 9a Av 9 44 Zona 1
Guatemala City, Guatemala
T: 21260 21263

Enriquez, Dolores, Dep.
Congreso de la Republica 9a Av 9 44 Zona 1
Guatemala City, Guatemala
T: 21260 21263

Fernandez, Maria Eugenia Castillo, Dep.
Congreso de la Republica 9a Av 9 44 Zona 1
Guatemala City, Guatemala
T: 21260 21263

Klee, Hector, Dep.
Congreso de la Republica 9a Av 9 44 Zona 1
Guatemala City, Guatemala
T: 21260 21263

Murillo, Pedro, Dep.
Congreso de la Republica 9a Av 9 44 Zona 1
Guatemala City, Guatemala
T: 21260 21263

Pinelo, Jose Eduardo Mendes, Dep.
Congreso de la Republica 9a Av 9 44 Zona 1
Guatemala City, Guatemala
T: 21260 21263

GUINEA - GOVERNMENT

Bah, Maadjou, Mr.
Chef
Division pour Protection de la Nature et de Ressources
Permanent Mission to the United Nations, 140 E. 39th Str.
New York, NY 10016, United States of America
T: 212 687 8115 F: 212 687 8248

Conte, Lanasna, S.E. Le General
President de la Republique
Republique de Guinee
140 E. 39th Str.
New York, NY 10016, United States of America
T: 212 687 8115 F: 212 687 8248

Diallo, Mamadou Saliou, Mr.
Juriste
Administrateur
Permanent Mission to the United Nations, 140 E. 39th Str.
New York, NY 10016, United States of America
T: 212 687 8115 F: 212 687 8248

Goubmassa, Ardoulaye Cherif, Mr.
Director National de l'Environnement
Chef de Mission
Permanent Mission to the United Nations, 140 E. 39th Str.
New York, NY 10016, United States of America
T: 212 687 8115 F: 212 687 8248

Kane, Djiba, Mr.
Coordonnateur du Comite National Preparatoire
Permanent Mission of Guinea
140 E. 39th Str.
New York, NY 10016, United States of America
T: 212 687 8115 F: 212 687 8248

Sylla, Ibrahima, S.E. M.
Ministre des Affaires
Estrangeres et de la Cooperation
Permanent Mission to the United Nations, 140 E. 39th Str.
New York, NY 10016, United States of America
T: 212 687 8115 F: 212 687 8248

Toumany Sako, Dakoum, S.E. M.
Ministre des Ressources Naturelles
Energies et Environnement
Permanent Mission to the United Nations, 140 E. 39th Str.
New York, NY 10016, United States of America
T: 212 687 8115 F: 212 687 8248

Traore, Ahmed Faya, Mr.
Chef de Division
Direction Nationale de la Meteorologie
Permanent Mission to the United Nations, 140 E. 39th Str.
New York, NY 10016, United States of America
T: 212 687 8115 F: 212 687 8248

NGO

Amadon Oury, Barry
UGVD
Siege Labe
BP 67, Guinea

Bah, Ibrahima
VGEMV
Bureaux Institut Recherche Agronomique de Guinee
BP 1861, Conakry, Guinea

Bah, Ousmane
AGUIPER
BP 1217
Conakry, Guinea

Balde, Youssouf
VGEMV
Bureaux Institut Recherche Agronomique de Guinee
BP 1861, Conakry, Guinea

Diallo, Mamadou Bella
ASSOANE
BP 206
Bis Conakry, Guinea

Fadiga, Gaoussou
CPATA
Boite Postale 530
Conakry, Guinea

Henderson, Ima
CPATA
Boite Postale 530
Conakry, Guinea

Mamadon Dulin, Diallo
UGVD
Siege Labe
BP 67, Guinea

Sow, Diamilatou
REDC
B.P. 1382
Conakry, Guinea

Sow, Marcel
REDC
B.P. 1382
Conakry, Guinea

Sylla, O.
GE
BP 4665
Conakry, Guinea

PRESS

Camara, Naby
Editor
RADIO DIFFUSION TV-GUINEENNE
Conakry, Guinea

Dieng, Ibrahim Sory
Editor
QUOTIDIEN NATIONAL GUINEA PRESS
B.P. 341 Horoya
Conakry, Guinea, Guinea

ORGANIZATIONS

Bangoura, Abdel K., Mr.
Guinee-Ecologie
BP 4665
Conakry, Guinea
TL: 22350 Mine Geo GE

Cesse, Alpha Fode, Mr.
As. Guineenne pour le Developpement de l'Artisanat
BP 12237
Conakry, Guinea

Diallo, Abdoulaye Sadio, Mr.
Volontaires Guineens pour l'Environnement 'Mission Verte'
Bureaux Institut Recherche Agronomique de Guinee
BP 1861, Conakry, Guinea
F: (224) 444 456
TL: 22188

Diallo, Mamadou Dulin, Mr.
Union Guineenne des Volontaires du Developpement (UGVD)
Siege Labe
BP 67, Guinea

Diallo, Mamadou Lamine, Mr.
As. des Amies de la Nature et de l'Environnement (ASSOANE)
BP 206
Bis Conakry, Guinea
T: 441 278

GLOBAL FORUM

Claude, Veremen, Mr.
Coordinator Financement
GUINEE ECOLOGIE
BP 3266
Conakry, Guinea
T: 224 4 44-5033 F: 224 4 44-2874
TL: 0995 22280

Sow, Djiamilatou, Mr.
RESEAU D'EDUCATION POUR LE DEVELOPPEMENT EN COOPERATION REDEC
Palais de Nations
Geneve CH 1211, Guinea
F: 41 22 733-6542

Sow, Nene Ousmane, Ms.
W E D O
Mme BP 295
Conakry, Guinea
T: 224 4 44-5001
TL: 22350

GUINEA-BISSAU - GOVERNMENT

Barber, Hilia, Mme
Chef
Departement Europe/Amerique Ministere des Affaires Etrangeres
Permanent Mission to the United Nations, 211 E. 43rd Str. Room 604 New York, NY, 10017
F: 212 661 3977
TL: 212 983 2794

Cabral, Mario, S.E. M.
Ministre du Developpement Rural et de l' Agriculture
Permanent Mission of Guinea-Bissau
211 E. 43rd Str. Room 604
New York, NY 10017, United States of America
T: 212 661 3977 F: 212 983 2794

Cabral, Raul de M., M.
Conseiller
Mission Permanente aupres des Nations Unies, New York
Permanent Mission to the United Nations, 211 E. 43rd Str. Room 604
New York, NY 10017, United States of America
T: 212 661 3977 F: 212 983 2794

Cardoso, Joao, S.E. M.
Secretaire d Etat aux Resources Naturelles
Permanent Mission of Guinea-Bissau
211 E. 43rd Str. Room 604
New York, NY 10017, United States of America
T: 212 661 3977 F: 212 983 2794

Cassama, Cipriano, M.
Directeur General des Forets et de la Chasse
Coordinateur de la Commission Technique du CNE
Permanent Mission to the United Nations, 211 E. 43rd Str. Room 604 New York, NY, 10017
F: 212 661 3977
TL: 212 983 2794

Cassimo, Jose Manuel, M.
Membre de la Comission Technique de la Commission Nationale de l Environment
Permanent Mission of Guinea-Bissau
211 E. 43rd Str. Room 604
New York, NY 10017, United States of America
T: 212 661 3977 F: 212 983 2794

D'Alva, Henrique M., M.
Directeur General du CITA
Permanent Mission of Guinea-Bissau
211 E. 43rd Str. Room 604
New York, NY 10017, United States of America
T: 212 661 3977 F: 212 983 2794

Lopes, Alberto, M.
conseiller Juridique du CNE
Permanent Mission of Guinea-Bissau
211 E. 43rd Str. Room 604
New York, NY 10017, United States of America
T: 212 661 3977 F: 212 983 2794

Mane, Ansumane, S.E. M.
Secretaire d Etat du Plan
Permanent Mission of Guinea-Bissau
211 E. 43rd Str. Room 604
New York, NY 10017, United States of America
T: 212 661 3977 F: 212 983 2794

Proenca, Flavio, S.E. M.
Chef de Cabinet de la Presidence
Permanent Mission of Guinea-Bissau
211 E. 43rd Str. Room 604
New York, NY 10017, United States of America
T: 212 661 3977 F: 212 983 2794

Ribeiro, Rui, M.
Rapporteur National
Permanent Mission of Guinea-Bissau
211 E. 43rd Str. Room 604
New York, NY 10017, United States of America
T: 212 661 3977 F: 212 983 2794

Semedo, Julio, S.E. M.
Ministre des Affaires Etrangeres
Permanent Mission of Guinea-Bissau
211 E. 43rd Str. Room 604
New York, NY 10017, United States of America
T: 212 661 3977 F: 212 983 2794

Spain, Eugenio, M.
Directeur General du Protocole
Permanent Mission of Guinea-Bissau
211 E. 43rd Str. Room 604
New York, NY 10017, United States of America
T: 212 661 3977 F: 212 983 2794

Toure, Boubacar, S.E. M.
Ambassadeur Extraordinaire et Plenipotentiaire
Representant Permanent aupres des Nation Unies, New York
Permanent Mission to the United Nations, 211 E. 43rd Str. Room 604
New York, NY 10017, United States of America
T: 212 661 3977 F: 212 983 2794

Vieira, Joao Bernardo, Mr.
President
Permanent Mission of the Republic of Guinea-Bissau
211 E. 43rd Str. Room 604
New York NY 10017, United States of America
T: 212 661 3977 F: 212 983 2794

PRESS

Abima, Simao Domingos
Editor
NO PINTCHA
Bissau, Guine-Bissau, Guinea-Bissau

Sa, Victorino
Reporter Radio/Television
TV GUINE-BISSAU
Bairro Penha, Apart. 3, bl. 4
Bissau, Guine Bissau, Guinea-Bissau

Soares, Ricardo Semedo
Reporter Radio/Television
RADIO NACIONAL GUINE-BISSAU
Bissau, Av. 14 Novembro
Bissau, Guine-Bissau, Guinea-Bissau

GLOBAL FORUM

Intchasso, Nharebat N., Ms.
WEDO
Mdra, CP 71
Bissau, Guine-Bissau
T: 21 11 34 25 1129

Miranda, Rui, Mr.
Membro Do Secr. Executivo
ALTERNAG
Bairro Belem/Guine-Bissau
Bissau, Guine-Bissau
T: 21-1314 F: 21-1168

Nosolini, Miranda Isabel, Ms.
Presidente
AD (ACCAO P/ DESENVOLVIMENTO)
Bissau, Guine-Bissau
TL: CP606

GUYANA - GOVERNMENT

Chin, Dr. Walter, Mr
Executive Chairman
G.A.H.E.F.
Liliendaal
Greater Georgetown, Guyana
T: 57523 F: 57523

Hoyt, Hugh Desmond, Mr.
President
Permanent Mission of the Republic of Guyana
866 United Nations Plaza, Suite 555
New York NY 10017, United States of America
T: 212 527 3232 F: 212 935 7548

King, Winston Morris, Mr
Executive Chairman
GUYANA NATURAL RESOURCES AGENCY
41 Brickdam And Boyle Place Stabroek
Georgetown, Guyana
T: 592 2 66549 F: 592 2 71211

Wiltshire, Faith, Ms
Director
CARICOM SECRETARIAT
Bank of Guyana Building P O Box 10827
Georgetown, Guyana
T: 592 2 51960 54493 F: 592 2 66091 51964

PRESS

David, Clement
Editor
GUYANA BROADCASTING CORP.
123 Regent Rd. Bourda
Georgetown, Guyana, Guyana

Deolall, Indranie
Correspondent
GUYANA NATIONAL NEWSPAPERS LTD
45 Durban/Palm Sts
Werk-En-Rust, Georgetown, Guyana, Guyana

Detain, Dominique
Reporter Radio/Television
RFO GUYANE
4, Res. Dufournier
97300 Cayenne, Guyana, Guyana

Seeraj, Gladys
Correspondent
GUYANA PUB.COMM AGENCY
236 D'Andrade St.
Georgetown, Guyana, Guyana

Woolford, Enrico
Editor
GUYANA PUBLIC COMMUNICATIONS
252 Stone Ave.
Georgetown, Guyana, Guyana

GLOBAL FORUM

Carberry, Lance, Mr.
Deputy Head
GUYANA NATURAL RESOURCES AGENCY
Brick Dam and Boyle Place
Georgetown, Guyana
T: 592 2 60435 F: 592 2 71211

Clementine Fox, Desrey, Ms.
Researcher
AMERIDIAN RESEARCH UNIT UNIVERSITY OF GUYANA
29 West Road, Sparendaam East Coast Demerara
Georgetown, Guyana
T: 592 2 54841

Dilling, John, Mr.
Representative
NATIONAL SPIRITUAL ASSEMBLY OF THE BAHA'IS OF GUYANA
National Baha'i Centre, 220 Charlotte Street, Bourda
Georgetown, Guyana
T: 592 2 65952 F: 592 2 65952

Sukhraj, Jacinta, Ms.
Ex. Dir. St. Francis Human Dev
CORPUS CHRISTI CARMELITE SISTERS
St. Bernadette's Hostel - 105 Lamaha Street
Georgetown, Guyana
T: 592 2 64851

Walcott, George, Mr.
Consultant
GUYANA DELEGATION TO UNCED
4 Lamaka Gardens
Georgetown 66282, Guyana

Willems, John, Mr.
Chairman
FOREST PRODUCTS ASSOCIATION
Water Street
Georgetown, Guyana

HAITI - GOVERNMENT

Aristide, Jean-Bertrand Pere
President
Permanent Mission of Haiti
801 2nd Ave, Room 300
New York NY 10017, United States of America
T: 212 370 4840 F: 212 661 8698

Aristide, Pere Jean-Bertrand, S.E. le Reverend
President
Republique d Haiti,
801 2nd Ave, Room 300
New York, NY 10017, United States of America
T: 212 370 4840 F: 212 661 8698

Augustin, Marie Roger, Mme
Secretariat du President
Permanent Mission of Haiti
801 2nd Ave, Room 300
New York, NY 10017, United States of America
T: 212 370 4840 F: 212 661 8698

Jean-Baptiste, Gerard E., M.
Consultant/Expert
Permanent Mission of Haiti
801 2nd Ave, Room 300
New York, NY 10017, United States of America
T: 212 370 4840 F: 212 661 8698

Jean-Enard, Turenne, Mme
Charge d Affaires a l Ambassade d Haiti au Bresil
Permanent Mission of Haiti
801 2nd Ave, Room 300
New York, NY 10017, United States of America
T: 212 370 4840 F: 212 661 8698

Jerome, Judith, Mme
Secretariat du Presedent
Permanent Mission of Haiti
801 2nd Ave, Room 300
New York, NY 10017, United States of America
T: 212 370 4840 F: 212 661 8698

Longchamp, Fritz, S.E. M.
Ambassadeur Extraordinaire et Plenipotentiaire
Representant Permanent aupres des Nations Unies (Chef Adjoint de la Delegation)
Permanent Mission to the United Nations, 801 2nd Ave, Room 300 New York, NY, 10017
F: 212 370 4840
TL: 212 661 8698

Monplaisir, Roberte, Mme
Consultant/Expert
Permanent Mission of Haiti
801 2nd Ave, Room 300
New York, NY 10017, United States of America
T: 212 370 4840 F: 212 661 8698

Murat, Jean-Maxime, M.
Ministre Conseiller
Mission Permanente aupres des Nations Unies, New York
Permanent Mission to the United Nations, 801 2nd Ave, Room 300
New York, NY 10017, United States of America
T: 212 370 4840 F: 212 661 8698

Sabalat, Jean-Robert, S.E. M.
Ministre des Affaires Etrangeres et des Cultes
Permanent Mission of Haiti
801 2nd Ave, Room 300
New York, NY 10017, United States of America
T: 212 370 4840 F: 212 661 8698

Targ, Nicholas, M.
Consultant/Expert
Permanent Mission of Haiti
801 2nd Ave, Room 300
New York, NY 10017, United States of America
T: 212 370 4840 F: 212 661 8698

Valcin, Raymond, M.
Ministre Conseiller au Ministere des Affaires Etrangeres
Permanent Mission of Haiti
801 2nd Ave, Room 300
New York, NY 10017, United States of America
T: 212 370 4840 F: 212 661 8698

PRESS

Chauvet, Pierre
Editor
LE NOUVELLISTE
17 Rue Moise
Petion Ville, Haiti, Haiti

Lindor, Deweare Ivann
Editor
HAITI TRIBUNE MAGAZINE
Kenscoff, Rte Fort Jacques 220
Rue de Casernes, 46, Haiti

Lorine, Frantz
Editor
JOOURNAL A.J.H.
Rua Cafois #9 Le Cados
Port-au-Prince, Haiti, Haiti

Victor, Jean Andre
Editor
JORNAL LE NOUVELLISTE
Defuras 31 R.Dubois 3
Port-au-Prince,Haiti, Haiti

GLOBAL FORUM

Abraham, Ernst, Mr.
SERVICE CRETLEN D'HAITI
BP 285
Port-au-Price, Haiti

HOLY SEE - GOVERNMENT

Canuto, Vittorio, Prof.
Permanent Mission of Holy See
20 E. 72nd Str.
New York, NY 10021, United States of America
T: 212 734 2900 F: 212 988 3633

Chagas, Carlos, Mgr. Prof.
Permanent Mission of Holy See
20 E. 72nd Str.
New York, NY 10021, United States of America
T: 212 734 2900 F: 212 988 3633

De Marcellus De Vollmer, Cristina, Mme
Permanent Mission of Holy See
20 E. 72nd Str.
New York, NY 10021, United States of America
T: 212 734 2900 F: 212 988 3633

Dewane, l' abbe Frank, M.
Permanent Mission of Holy See
20 E. 72nd Str.
New York, NY 10021, United States of America
T: 212 734 2900 F: 212 988 3633

Ferreira Da Nova, Abilio, M.
Permanent Mission of Holy See
20 E. 72nd Str.
New York, NY 10021, United States of America
T: 212 734 2900 F: 212 988 3633

Filoni, Fernando, Mgr.
Permanent Mission of Holy See
20 E. 72nd Str.
New York, NY 10021, United States of America
T: 212 734 2900 F: 212 988 3633

Klink, John M., M.
Permanent Mission of Holy See
20 E. 72nd Str.
New York, NY 10021, United States of America
T: 212 734 2900 F: 212 988 3633

Martino, L archeveque Renato R., S.E.
Nonce apostolique, Observateur permanant aupre des Nations Unies (chef Adjoint de la Delegation)
Permanent Mission of Holy See
20 E. 72nd Str.
New York, NY 10021, United States of America
T: 212 734 2900 F: 212 988 3633

McHuch, James T., S.E. Mgr.
Permanent Mission of Holy See
20 E. 72nd Str.
New York, NY 10021, United States of America
T: 212 734 2900 F: 212 988 3633

Mendes de Almeida, Candido Antonio, Dr.
Permanent Mission of Holy See
20 E. 72nd Str.
New York, NY 10021, United States of America
T: 212 734 2900 F: 212 988 3633

Salvador, Roberta, Soeur
Permanent Mission of Holy See
20 E. 72nd Str.
New York, NY 10021, United States of America
T: 212 734 2900 F: 212 988 3633

Santoro, l' abbe Filippo, M.
Permanent Mission of Holy See
20 E. 72nd Str.
New York, NY 10021, United States of America
T: 212 734 2900 F: 212 988 3633

Sodano, Angelo Cardinal, S.E.
Secretary of State
Permanent Observer Mission of the Holy See
20 E. 72nd Str.
New York NY 10021, United States of America
T: 212 734 2900 F: 212 988 3633

HONDURAS - GOVERNMENT

Arriaga Chinchilla, Jose Ricoberto, S.E. Dr.
Embajador de Honduras en Brasil
Permanent Mission of Honduras
866 United Nations Plaza, Suite 417
New York, NY 10017, United States of America
T: 212 752 3370 F: 212 223 0498

Banegas, Esmeralda, Sra.
Encargada de Asuntos Administrativos
Consulado de Rio de Janeiro
Permanent Mission to the United Nations, 866 United Nations Plaza, Suite 417 New York, NY, 10017
F: 212 752 3370
TL: 212 223 0498

Bertrand Anduray, Franklin, S.E. Ing.
Ministro Secretario Ejecutivo de la Comision para Medio Ambiente (Jefe Adjunto de la Delegacion)
Permanent Mission of Honduras
866 United Nations Plaza, Suite 417
New York, NY 10017, United States of America
T: 212 752 3370 F: 212 223 0498

Bogran, Samuel, Ing.
Presidente
Comision Medio Ambiente
del Congreso Nacional
T: 10017
TL: 212 752 3370, E: 212 223 0498

Bruchmann, Cuadalupe, Sra.
Asesora
Casa Presidencial
Permanent Mission to the United Nations, 866 United Nations Plaza, Suite 417 New York, NY, 10017
F: 212 752 3370
TL: 212 223 0498

Budde, Federico, Sr.
Encargado de Asuntos Administrativos
Embajada de Honduras en Brasil
Permanent Mission to the United Nations, 866 United Nations Plaza, Suite 417 New York, NY, 10017
F: 212 752 3370
TL: 212 223 0498

Carias Zapata, Mario, S.E. Dr.
Ministro de Relaciones Exteriores
Permanent Mission of Honduras
866 United Nations Plaza, Suite 417
New York, NY 10017, United States of America
T: 212 752 3370 F: 212 223 0498

De Carias Zapata, Florencia, Sra.
Asesora del Ministerio de Relaciones Exteriores
Permanent Mission of Honduras
866 United Nations Plaza, Suite 417
New York, NY 10017, United States of America
T: 212 752 3370 F: 212 223 0498

Fajardo, Modesto, Sr.
Vice Consul
Consulado de Rio de Janeiro
Permanent Mission to the United Nations, 866 United Nations Plaza, Suite 417 New York, NY, 10017
F: 212 752 3370
TL: 212 223 0498

Ferrera, Juan, Sr.
Presidente del Consejo hondureno de la Empresa Privada
Permanent Mission of Honduras
866 United Nations Plaza, Suite 417
New York, NY 10017, United States of America
T: 212 752 3370 F: 212 223 0498

Funez Padilla, Magaly, Sra.
Consul General
Rio de Janeiro
Permanent Mission to the United Nations, 866 United Nations Plaza, Suite 417 New York, NY, 10017
F: 212 752 3370
TL: 212 223 0498

Kamper, Marta, Sra.
Consejera
Embajada en Argentina
Permanent Mission to the United Nations, 866 United Nations Plaza, Suite 417 New York, NY, 10017
F: 212 752 3370
TL: 212 223 0498

Kattan Arita, Roberto, Lic.
Consejero Comercial
Consulado de Rio de Janeiro
Permanent Mission to the United Nations, 866 United Nations Plaza, Suite 417 New York, NY, 10017
F: 212 752 3370
TL: 212 223 0498

Kawas, Marie J., Sra.
Asociacion de Periodistas Ambientalistas y Agroforestales de Honduras
Permanent Mission to the United Nations, 866 United Nations Plaza, Suite 417 New York, NY, 10017
F: 212 752 3370
TL: 212 223 0498

Martinez, Hector, Lic.
Edecan del Vice-Presidente
Permanent Mission of Honduras
866 United Nations Plaza, Suite 417
New York, NY 10017, United States of America
T: 212 752 3370 F: 212 223 0498

Martinez, Lozano Roberto, Mr.
Vice President
Permanent Mission of Honduras
866 United Nations Plaza, Suite 417
New York NY 10017, United States of America
T: 212 752 3370 F: 212 223 0498

Martinez Lozano, Roberto, S.E. Lic.
Vice Presidente
Republica de Honduras
866 United Nations Plaza, Suite 417
New York, NY 10017, United States of America
T: 212 752 3370 F: 212 223 0498

Nufio de Figueroa, Cristiana, Dra.
Comision Nacional de Medio Ambiente
Permanent Mission of Honduras
866 United Nations Plaza, Suite 417
New York, NY 10017, United States of America
T: 212 752 3370 F: 212 223 0498

Parodi, Newton, Dr.
Consul Honorario de Honduras en Curitiba
Permanent Mission of Honduras
866 United Nations Plaza, Suite 417
New York, NY 10017, United States of America
T: 212 752 3370 F: 212 223 0498

Paz Sierke, Edgaro, Sr.
Consejero de la Embajada en Brasil
Permanent Mission of Honduras
866 United Nations Plaza, Suite 417
New York, NY 10017, United States of America
T: 212 752 3370 F: 212 223 0498

Ponce de Bueso, Alicia Maria, Sra.
Encargada de Asuntos Economicos en el Consulado de Rio de Janeiro
Permanent Mission to the United Nations, 866 United Nations Plaza, Suite 417 New York, NY, 10017
F: 212 752 3370
TL: 212 223 0498

Romero, Elias, Sr.
Asociacion de Periodistas Ambientalistas y Agroforestales de Honduras
Permanent Mission to the United Nations, 866 United Nations Plaza, Suite 417 New York, NY, 10017
F: 212 752 3370
TL: 212 223 0498

Valenzuela, Dolores, Sra.
Asociacion de Periodistas Ambientalistas y Agroforestales de Honduras
Permanent Mission to the United Nations, 866 United Nations Plaza, Suite 417 New York, NY, 10017
F: 212 752 3370
TL: 212 223 0498

Varela, Jorge, Sr.
Fedambiente
Permanent Mission to the United Nations, 866 United Nations Plaza, Suite 417 New York, NY, 10017
F: 212 752 3370
TL: 212 223 0498

Villatoro, Edna H., Sra.
Encargada de Asuntos Culturales
Consulado de Rio de Janeiro
Permanent Mission to the United Nations, 866 United Nations Plaza, Suite 417 New York, NY, 10017
F: 212 752 3370
TL: 212 223 0498

GLOBAL FORUM

Bunch, Roland, Mr.
TREES FOR PEOPLE
Apartado 3385
Tegucigalpa, Honduras

Fullerton, Helena, Sra.
ASOCIACION HONDURENA DE ECOLOGICA REDES
Palais de Nations
Geneve CH 1211, Honduras
F: 41 22 733-6542

Green, Nancy, Ms.
CUPRODEIM
Subida A la Leora
Tegucigalpa, Honduras
T: 50 4 226714

Guevara, Roy, Sr.
OFRANEH
Palais de Nations
Geneve CH 1211, Honduras
F: 41 22 733-6542

Pineda, Jose Leonel Duarte, Sr.
COOPERACION HONDURENO - ALEMANA DE SEGURIDAD ALIMENTARIA
Apto. Postal no. 86
Choluteca, Honduras

HONG KONG - NGO

Bhattarai, Rajan
ASA
511 Nathan Road 1/F, Kowloon,
Hong Kong

Chandran, Premesh
ASA
511 Nathan Road 1/F, Kowloon,
Hong Kong

Fung, Shiu-Wing
CA
Shop 301, Lai Mei House, Lai Kok Estate
Cheung Sha Wan, Kowloon, Hong Kong

Gan, Diong Keng
ASA
511 Nathan Road 1/F, Kowloon,
Hong Kong

Ng, Cho-Nam
CA
Shop 301, Lai Mei House, Lai Kok Estate
Cheung Sha Wan, Kowloon, Hong Kong

Villamor, Veronica
ASA
511 Nathan Road 1/F, Kowloon,
Hong Kong

PRESS

Schwarz, Adam
Correspondent
FAR EASTERN ECONOMIC REVIEW
Wisma Antara 141013
Iakarta Indonesia, Hong Kong

ORGANIZATIONS

Ng, Gordon, Mr.
Conservancy Association, The
Shop 301, Lai Mei House, Lai Kok Estate
Cheung Sha Wan, Kowloon, Hong Kong
T: (852) 815 5805 F: (852) 545 7439

GLOBAL FORUM

Buchanan, James, Dr.
Professor
HONG KONG UNIVERSITY OF SCIENCE AND TECHNOLOGY / CONSERVANCY ASSOCIATION OF HONG KONG
Clear Water Bay, Kowloon
Hong Kong (City), Hong Kong
T: 852 5 7195354 F: 852 5 7195342

Chan, Stephan, Mr.
CONSERVANCY ASSOCIATION
S301, Lai Mei House, Lai Kok Estate, cheung Sha Wan
Hong Kong (City), Hong Kong
T: 852 726781 F: 852 7285538

Cheung, Lister, Ms.
CONSERVANCY ASSOCIATION
S 301, Lai Mei House, Lai Kok Estate, Cheung Sha Wan
Cheung Sha Wan, Hong Kong
T: 652 726781 F: 652 7285538

Cho-Nam, Ng, Mr.
CONSERVANCY ASSOCIATION
S 301, Lai Mei House, Lai Kok Estate, Cheung Sha Wan
Cheung Sha Wan, Hong Kong
T: 652 726781 F: 652 7285538

Hung, Wing-Tat, Mr.
Secretary
CONSERVANCY ASSOCIATION
S301, Lai Mei House, Lai Kok Estate, Cheung Sha Wan
Hong Kong (City), Hong Kong
T: 852 728-6781 F: 852 728-5538

Poon, Peggy, Ms.
Executive Comm. Member
CONSERVANCY ASSOCIATION
S 301, Lai Mei House, Lai Kok Estate, Cheung Sha Wan
Cheung Sha Wan, Hong Kong
T: 652 726781 F: 652 7285538

Samuel, John Victor
CHRISTIAN CONFERENCE OF ASIA
2 Jordan Road
Kowloon, Hong Kong

Shiu-Wing, Fung, Mr.
Chaiman
CONSERVANCY ASSOCIATION
S 301, Lai Mei House, Lai Kok Estate, Cheung Sha Wan
Cheung Sha Wan, Hong Kong
T: 652 726781 F: 652 7285538

Ting-Leung, Ng, Mr.
CONSERVANCY ASSOCIATION
S 301, Lai Mei House, Lai Kok Estate, Cheung Sha Wan
Cheung Sha Wan, Hong Kong
T: 652 726781 F: 652 7285538

HUNGARY - GOVERNMENT

Bago, Sandor, Dr.
Counsellor
Embassy of the Republic of Hungary to Brazil
Permanent Mission to the United Nations, 227 E. 52nd Street New York, NY, 10022
F: 212 752 0209
TL: 212 755 5395

Benyhe, Janos, H.E. Mr.
Ambassador
Federative Republic of Brasil
Permanent Mission to the United Nations, 227 E. 52nd Street New York, NY, 10022
F: 212 752 0209
TL: 212 755 5395

Botos, Balazs, Dr.
Under-General Secretary
Ministre of Industry and Trade
Permanent Mission to the United Nations, 227 E. 52nd Street New York, NY, 10022
F: 212 752 0209
TL: 212 755 5395

Bulla, Miklos, Dr.
Director-General
Ministry of Environment and Regional Policy
Permanent Mission to the United Nations, 227 E. 52nd Street New York, NY, 10022
F: 212 752 0209
TL: 212 755 5395

Csorba, Andras, Dr
Senior Adviser
Ministry for International Economic Relations
Honved U. 13-15
Budapest, Hungary
T: 1116 800 F: 1532 794

Dudas, Mihaly, Mr.
Third Secretary
General Consulate, Sao Paulo
Permanent Mission to the United Nations, 227 E. 52nd Street New York, NY, 10022
F: 212 752 0209
TL: 212 755 5395

Fagaro, Tibor
Director
HUNGARIAN METEOROLOGICAL SERVICE
Kitaibel 1 H-1525 POB 38
Budapest/Hungary, Hungary
T: 361 135 3194 F: 361 135 3194 361 135

Fekete, Gyogy, Mr.
Commercial Secretary
Rio de Janeiro
Permanent Mission to the United Nations, 227 E. 52nd Street New York, NY, 10022
F: 212 752 0209
TL: 212 755 5395

Goncz, Arpud, Mr.
President
Permanent Mission of the Republic of Hungary
227 E. 52nd Str
New York NY 10022, United States of America
T: 212 752 0209 F: 212 755 5395

Gyebnar, Istva'n, Mr.
Senior Adviser
Ministry for Foreign Affairs
BEM RKR 47.
Budapest Ii, Hungary
T: 00361/1568 000 F: 00361/201 93 70

Katona, Tamas, H.E. Dr.
State Secretary
Foreign Affairs (Deputy Head of Delegation)
Permanent Mission to the United Nations, 227 E. 52nd Street New York, NY, 10022
F: 212 752 0209
TL: 212 755 5395

Keresztes K., Sandor, H.E. Mr.
Minister
Environment and Regional Policy (Deputy Head of Delegation)
Permanent Mission to the United Nations, 227 E. 52nd Street New York, NY, 10022
F: 212 752 0209
TL: 212 755 5395

Kiss, Robert, Mr.
Vice-Chairman
Committee for Environment, Hungarian Parliament
Permanent Mission to the United Nations, 227 E. 52nd Street New York, NY, 10022
F: 212 752 0209
TL: 212 755 5395

Lang, Istvan, Dr.
Secretary General
Hungarian Academy of Sciences
Roosevelt Ter
Budapest 1051, Hungary
T: 361 111 9812 F: 361 112 8483

Meszaros, Attila, Mr.
High School Student
Permanent Mission of Hungary
227 E. 52nd Street
New York, NY 10022, United States of America
T: 212 752 0209 F: 212 755 5395

Nagy, Balint, Mr.
First Secretary
Embassy of the Republic of Hungary to Brazil
Permanent Mission to the United Nations, 227 E. 52nd Street New York, NY, 10022
F: 212 752 0209
TL: 212 755 5395

Nathon, Istva'n, Dr.
Director-General for International Economic Relations
Ministry for Foreign Affairs
BEM RKP. 47
Budapest Ii, Hungary
T: 00361/1568 000 F: 00361/201 9370

Nechay, Gabor
Senior Adviser
MINISTRY FOR ENVIRONMENT
Ktm,Otvh Budapest Kolto U 21 H 1121
Budapest, Hungary
T: 36 1 1562133 1756458 F: 36 1 17 57 457

Poos, Miklos, Dr.
Director-General
Ministry of Industry and Trade
Permanent Mission to the United Nations, 227 E. 52nd Street New York, NY, 10022
F: 212 752 0209
TL: 212 755 5395

Rott, Nandor, Dr.
Chairman
Committee for Environment, Hungarian Parliament
Permanent Mission to the United Nations, 227 E. 52nd Street New York, NY, 10022
F: 212 752 0209
TL: 212 755 5395

Szombati, Bela, Mr.
Foreign Policy Adviser to the President
Permanent Mission of Hungary
227 E. 52nd Street
New York, NY 10022, United States of America
T: 212 752 0209 F: 212 755 5395

Tokes, Estvan, Mr.
Director General
Ministry of Environment and Regional Policy
Permanent Mission to the United Nations, 227 E. 52nd Street New York, NY, 10022
F: 212 752 0209
TL: 212 755 5395

Vasarhelyi, Judit, Dr
Vice President
Independent Ecological Center
Mikl'os T'ER1. 'Selyemgombolyito'
Budapest 1035, Hungary
T: 36 1 16 86 229 F: 36 1 180 45 46

NGO

Schmuck, Erzebet, Ms.
NATIONAL SOCIETY OF CONSERVATIONISTS
Palais de Nations
Geneve CH 1211, Hungary
F: 41 22 733-6542

PRESS

Arpad, Juhasz
Reporter Radio/Television
HUNGARIAN TELEVISION
Tarogato Str UJ
H-1026, Hungary, Hungary

Csobor d'Molnarne, Eva
Editor
HUNGARIAN TV
4-1112 Budapest
Budapest, Hungria, Hungary

D'Molnarne, Eua Csobor
Editor
HUNGARIAN TELEVISION
Menyecske V.9
H-112 Budapest, Hungary, Hungary

Feuer, Andras
Correspondent
HUNGARIAN TV
1063 Budapest
Kmetty U.15, Hungary, Hungary

Gabor, Szove
Correspondent
HUNGARIAN TELEVISION
Bimbo Str Jh
H-1022 Budapest, Hungary, Hungary

Harcsa, Bela
Editor
AGRO TV LTD.
Budapest XVII
Baross U. 36, Hungary, Hungary

Horvath, Gabor
Correspondent
NEPSZABADSAG
Bajnou U. 19
1063 Budapest Hungary, Hungary

Kakuk, Gyorgy
Reporter Radio/Television
HUNGARIAN TV
1026 Budapest
Eslter U-2, Hungary, Hungary

Kiss, Peter Dezso
Editor
VASARNAPI HIREK
1138 Budapest
Nepfurdo u.21/F, Hungary, Hungary

Krasznai, Zoltan
Editor
MAGYAR NEMLET/DAILY
Rakocz sit. 411
Budapest VII 1072, Hungary, Hungary

Kurucz, Alexander
Editor
AGRO TV LTD.
1013 Budapest
Attila U. 33, Hungary, Hungary

La Szic, Szabo
Editor
HUNGARIAN TV
Racz Acadar Btr 38
Budapest, Hungary, Hungary

Laszlo, Palok
Reporter Radio/Television
HUNGARINA TELEVISION
Koroypai V 21-21 A
H-1124 Budapest, Hungary, Hungary

Laszlo, Peterfalvi
Editor
HUNGARIAN TV
1134 BP
Devai-u 20/B, Huagary, Hungary

Laszlo, Szabo
Editor
HUNGARINA TELEVISION
Dacz Aladar Bt 38
H-1121 Budapest, Hungary, Hungary

Lokos, Zoltan
Editor
VASARNAPI HIREK
1051 Budapest, Alpari
Gyula u.21, Hungary, Hungary

Molnarne, Eva Csobod
Editor
HUNGARIAN TELEVISION
Menyecske V. 9
H-1112 Budapest, Hungar, Hungary

Palok, Laszlo
Reporter Radio/Television
HUNGARIAN TELEVISION
Korompai v. 21-23/A
H-1124 Budapest, Hungar, Hungary

Rudi, Zoltan
Editor
HUNGARIAN TV
Budapest 1097 Dregeley
Utca 13 V./16, Hungary, Hungary

Sipos, Zoltan
Editor
HUNGARIAN TV BUDAPEST
1019 Logodi 54
Hungary, Hungary

Sotonyi, Jozsef
Editor
AGRO TV LTD.
1162 Budapest Fahej U. 32
Hungary, Hungary

Szabo, Laszlo
Editor
HUNGARIAN TELEVISION
Racz Aladar Str. 38
H-1121 Budapest, Hungar, Hungary

Szoke, Gabor
Correspondent
HUNGARIAN TV
Bimbo Str
Budapest, Hungary, Hungary

Ternovszky, Nora
Reporter Radio/Television
AGRO TV
Postacim 1097
Budapest, Hungary, Hungary

Zilahy, Zsuzsana De
Correspondent
TRANS GLOBAL NETWORK
Budapest 1011, Hungary

GLOBAL FORUM

Basel, Seter
TUDIMANYOS ISMERETTERJESZTO TARSULAT
12 Rajaji Marg
New Delhi 110011, Hungary
T: 91 11 301-4309

Ferenc, Nemeth
Senior Editor
REFORM NEWS MAGAZINE
1027 Budapest , Kiraly u-41
Budapest, Hungary
T: 36 1 1424350 F: 36 1 1224240

Gyorot, Kakuk
Reporter
HUNGARIAN TELEVISION
Szabadsagten 17
Budapest, Hungary
T: 1321589

Imre, Laszlo, Mr.
Chairman
H-ISES
Thechnical University Budapest 1111 Muegyetem RKP.3
Budapest, Hungary
T: 36 1 1812-192 F: 36 1 1666-808
TL: 225931 MUEGY-H

Palugyai, Istvan,
Science Editor
NEPSZABADSAG
Becsi Ut 122-124
Budapest 1034, Hungary
T: 36 1 1686880/303 F: 31 1 1689098

Reti, Ervin
Foreign Editor
ESTI HIRLAP
Blaha Luiza TER 1-3 yentro
Budapest, Hungary
T: 36 1 1185785/1384062 F: 36 1 1384550

Zoltau, Rudi
Editor
HUNGARIAN TELEVISION
Szabadsagten 17
Budapest, Hungary
T: 1321589

ICELAND - GOVERNMENT

Bjornsson, Sveinn, H.E. Mr.
Ambassador of the Federative Republic of Iceland, Secretary-General
Office of the President of Iceland
Permanent Mission to the United Nations, 370 Lexington Ave., 5th Floor New York, NY, 10017
F: 212 686 4100
TL: 212 532 4138

Egudnason, , H.E. Mr.
Minister for the Environment (Deputy Head of Delegation)
Permanent Mission of Iceland
370 Lexington Ave., 5th Floor
New York, NY 10017, United States of America
T: 212 686 4100 F: 212 532 4138

Einarsdottir, Kristin, Mrs.
Member of Parliament
Permanent Mission of Iceland
370 Lexington Ave., 5th Floor
New York, NY 10017, United States of America
T: 212 686 4100 F: 212 532 4138

Eiriksson, Gudmundur, H.E. Mr.
Ambassador, Legal Adviser
Ministry of Foreign Affairs (Deputy Head of Delegation)
Permanent Mission to the United Nations, 370 Lexington Ave., 5th Floor New York, NY, 10017
F: 212 686 4100
TL: 212 532 4138

Finnbogadottir, Vigdis, H.E. Mrs.
President
Republic of Iceland
370 Lexington Ave., 5th Floor
New York, NY 10017, United States of America
T: 212 686 4100 F: 212 532 4138

Gudnason, Eidur, H.E. Mr.
Minister
Environment (Head of Delegation)
Permanent Mission to the United Nations, 370 Lexington Ave., 5th Floor New York, NY, 10017
F: 212 686 4100
TL: 212 532 4138

Guttormsson, Hjorleifur, Mr.
Member
Parliament
Permanent Mission to the United Nations, 370 Lexington Ave., 5th Floor New York, NY, 10017
F: 212 686 4100
TL: 212 532 4138

Johannesson, Magnus, Mr.
Special Assistant
Minister for the Environment
Permanent Mission to the United Nations, 370 Lexington Ave., 5th Floor New York, NY, 10017
F: 212 686 4100
TL: 212 532 4138

Mathiesen, Arni M., Mr.
Member of Parliament
Permanent Mission of Iceland
370 Lexington Ave., 5th Floor
New York, NY 10017, United States of America
T: 212 686 4100 F: 212 532 4138

Ottosson, Jon Gunnar, Dr.
Head of Division
Ministry for the Environment
Permanent Mission to the United Nations, 370 Lexington Ave., 5th Floor New York, NY, 10017
F: 212 686 4100
TL: 212 532 4138

Ringseth, Tom Mario, Mr.
Vice consul, Consultant General
Rio de Janeiro
Permanent Mission to the United Nations, 370 Lexington Ave., 5th Floor New York, NY, 10017
F: 212 686 4100
TL: 212 532 4138

Schram, Gunnar G., Dr.
Professor
University of Iceland
Permanent Mission to the United Nations, 370 Lexington Ave., 5th Floor New York, NY, 10017
F: 212 686 4100
TL: 212 532 4138

Tomasson, Tomas A., H.E. Mr.
Ambassador
Republic of Iceland to Brazil
Permanent Mission to the United Nations, 370 Lexington Ave., 5th Floor New York, NY, 10017
F: 212 686 4100
TL: 212 532 4138

GLOBAL FORUM

Thorgeirsson, Halldor, Mr.
Secretary
INTERNATIONAL BAHA'i COMMUNITY
Hjallabrekka 27
Kopavoghr IS 200, Iceland
T: 354 1 812230 F: 354 1 671101

INDIA - GOVERNMENT

Malhotra, Ajai, Mr
Director (un)
MINISTRY OF EXTERNAL AFFAIRS, NEW
C/O Gen O.P. Malhotra, C-I/12 Lodhi Garden
New Delhi 1100, India
T: 6876153 (OFF) 674779 F: 3010700 OR 301 0680

Rajamani, Raghunathan
Secretary,Government Of India
MINISTRY OF ENVIRONMENT FORESTS
Paryavaran Bhavan, Cgo Complex, Lodi Road, New Delhi-3
New Delhi 1100, India
T: 4360721 F: 4360678

Ramisetti, Murali
Social Worker
MODERN ARCHITECTS FOR RURAL INDIA
16-7-219 Dayanand Colony
Lakshmipuram, Warangal 506013, India
T: 91 8712 23706 F: 91 842 869243

Rao, P.V. Narasimha, Mr.
Prime Minister
Permanent Mission of India
866 United Nations Plaza, Suite 505
New York NY 10017, United States of America
T: 212 751 0900 F: 212 751 1393

Sarabhai, Kartikeya Vikram
Director
CENTRE FOR ENVIRONMENT EDUCATION
Thaltej Tekra
Ahmedabad-380054, India
T: 272 442642 F: 91 272 467342

Singh, Har Swarup, Dr
Governor
UNION TERRITORY GOVERNMENT
Raj Nivas
Pondicherry, India
T: 24051 24050 CODE0413 F: 0413 27575

Sinh, Digvijay, Dr
First Minister Of Environment
IN THE GOVERNMENT OF INDIA
the Palace,
Wankaner 363621, India
T: 02828 324

NGO

Anand, S.
KKWS
PO Box 28
Chiballapur 562 101, India

Arunachalam, Jaya
WORKING WOMEN'S FORUM
55 Shimasena Garden Road, Mylapore
Madras 600 004, India
T: 91 44 74553 F: 91 44 94 4444

Balijepalli V., Subba Rao
AGS
2-2-1133/5/5 New Nallakunta
Hyderabad 500 044, Andra Pradesh, India

Bandhu, Desh, Dr.
Secretaty - General
ASIAN ENVIRONMENTAL SOCIETY
U-112 (3rd floor), Vidhata House, Vikas Marg. Shakarpur
New Delhi 710092, India
T: 91 11 222-3311 F: 91 11 331-9435

Baskar, Joseph Paul, Mr.
PEACE TRUST
Peace Nagar, Near Police Housing Colony, Trichy Ro
Dindigul 624 009, Quaid-e-Milleth District, India

Basu, Aparna, Dr.
Secretary General
ALL INDIA WOMEN'S CONFERENCE
6 Bhagwan Dass Road
New Delhi 110 001, India
T: 91 11 389680

Basu, Biplab B., Mr.
SCHOOL OF FUNDAMENTAL RESEARCH
49/B, Tolly Gunge Road
Calcutta 700026, India
T: 91 33 423348 F: 91 33 286604
TL: 021-7614 TRNSIN

Brajabidhu Singh, Wangkhe Imayum
RDO
Lamsang Bazar
Manipur - 795 146, India

Chainani, Shyam
BOMBAY ENVIRONMENTAL ACTION GROUP
9 St. James Court
Marine Drive, Bombay 400 020, India
T: 91 22 287 3500 F: 91 22 220 187

Chandulal, Parmar
USED
945 Boottee Street
Pune 411 001, India

Chaturvedi, Pradeep
IAAS
55 Kaka Nagar
New Delhi - 110 003, India

Chinnathambi,
TTAP
No. 102A, Railway Station Road, Tirupattur, N.A.
Tamil Nadu-Pin 635 601, India

Daniel, Alexander Vethackan
IIRD
Kanchan Nagar, Nakshatrawadi, PO Box 12
Aurangabad 431 002, Maharashtra, India

Daniel, Evelyn Vasantha
IIRD
Kanchan Nagar, Nakshatrawadi, PO Box 12
Aurangabad 431 002, Maharashtra, India

Dave, Satish
B
15, Sethi Nagar, Ujjain (Madhya Pradesh) - 456 001
India

Doshi Suresh Chandra,
SRD
S-3, Govind Nagar (West)
Jaipur 302 002, India

Gujja, B.
DDS
A6,Meera Apts., Skyline Theatre Lane
Basheerbagh, Hyderabad, A.P., India

Jain, Jambu Kumar
RDYTI
1-K-22 Dabadari
Kota 324 009 (Rajasthan), India

John, Mathew
ICDSS
No. 1, 3rd Cross, Vivekananda Nagar
Bangalore - 33, India

Kannabiran, Vasantha
DDS
A6,Meera Apts., Skyline Theatre Lane
Basheerbagh, Hyderabad, A.P., India

Krishna, Nanditha
CPREEC
1a Eldams Road
Madras 600 018, India

Krishna, Shyamala M.
CEE
Nehru Foundation for Development, Thaltej Tekra
Ahmedabad 380 054, India

Kumar, Prem
ESC
Karuna Sadan, Sector 11-B
Chandigarth 160 011, India

Kumbham, P. Reddy
AGS
2-2-1133/5/5 New Nallakunta
Hyderabad 500 044, Andra Pradesh, India

Latika, Shiva
SSS
Shanti Kunj, 31 Mahipal Nagar
Rae Bareli-229 001 UP, India

Leelavathy,
WWF
55 Bhimasena Garden Road
Mylapore, Madras 600 004, India

Mathur, Ajay, Dr.
Dean, Energy, Eng & Tech
TATA ENERGY RESEARCH INSTITUTE
9 Jorbagh
New Delhi 110003, India
T: 91 11 618-516 F: 91 11 462-1770
TL: 31-61593 Teri I

Mayur, Rashmi
GLOBAL FUTURES NETWORK
73A Mittal Tower, Nariman Point
Bombay 400021, India

Meda, Gurudutt P.
AART
23-1, Meda Nivas
Jangareddigudem-534 447, W.G. Dt. A.P., India

Mehta, Sanat
SAP
Mangalmurti, Ashram Road
Ahmedabad 380 009, Gujarat, India

Mohapatro, Pitabas
OES
N-6/119, Nayapalli, Bhubaneswar
Orissa 751015, India

Muthu, Peraman, Mr.
Ashoka Fellow
TRUST - HELP
L1/984, 43rd Cross St., Thiruvalluvar Nagar, Thiruvanmiyur
Madras 600 041, India
T: 91 44 413948 F: 91 44 944444

Narayan, Ravi
ICYO
F-13, South Extension Part One
New Delhi - 110 049, India

Nayar, Soman
TBGRI
Pacha-Palode P.O.
Thiruvananthapuram - 695 562, Kerala, India

Oza, Gunavantral
ISN
Oza Building, Salatwada
Baroda 390 001, India

Oza, Premlata
ISN
Oza Building, Salatwada
Baroda 390 001, India

Pandey, Rajesh
SOCIETY FOR PARTICIPATORY RESEARCH IN ASIA
42 Tughlakabad Intitutional Area
New Delhi 110 062, India

Parvathi, A. Shiva
JR
2-76 Green House, Suraram Colony Road, I.D.A.
Jeedimetla 500 855, R.R. Dist. (A.P.), India

Patel, Anil
SAP
Mangalmurti, Ashram Road
Ahmedabad 380 009, Gujarat, India

Philip, Mathews
ICDSS
No. 1, 3rd Cross, Vivekananda Nagar
Bangalore - 33, India

Prakash Sj, Cedric
SXSSS
Post Box 4088, Navrangpura
Ahmedabad 380 009 Gujarat, India

Prasad, Manaparambil Koru
KERALA SASTRA SAHITYA PARISHAD
Parishad Bhavan, Guruvayur Road, Trissur, Kerala -
India

Pushpangadan, Palpil
Tropical Botanic Garden & Research Institute
Pacha-Palode P.O.
Thiruvananthapuram - 695 562, Kerala, India

Rai, Ashok K.
CRPA
16 Dakshneshwar, 10- Hailey Road
New Delhi 110 001, India

Rajendra Kumar Nayak,
ILI
Bhagwandas Road
New Delhi 110 001, India

Rama Subramaniam, Attili
IAS
Andhra University
Visakhapatnam - 530 003 (AP), India

Ranade, Smt. Shobhana
ALL INDIA WOMEN'S CONFERENCE
6 Bhagwan Dass Road
New Delhi 110 001, India
T: 91 11 389680

Ribeiro, E.F.N.
AES
U-112 A (3rd Floor), Vidhata House
Vikas Marg, Delhi 110 092, India

Samant, Jay, Mr.
Director
BOMBAY NATURAL HISTORY SOCIETY
Hornbill House, Shaheed - Bhagat Singh Road
Bombay, India
T: 91 22 24-3869

Santhakumar, C. Job
CMAC
5 Damodar Street
Gopalapuram, Madras 600 086, India

Shankar Pandey, Yamuna
SBLKS
Village and Post Gouramafi, Amargarh,
Dist. Pratap Garh, UP, India

Shiv Varan, Shukla
SSS
Shanti Kunj, 31 Mahipal Nagar
Rae Bareli-229 001 UP, India

Soyajl, Banoo
USED
945 Boottee Street
Pune 411 001, India

Sudhirendar, Sharma
EEG
B-415 Ansal Chambers-1
3, Bhikaiji Cama Place, New Dehli-110066, India

Sundaram, Rajlakshmi
CEMS
20F Basant Lok, Vasant Vihar
New Delhi 110 057, India

Sundaram, Sudha
CEMS
20F Basant Lok, Vasant Vihar
New Delhi 110 057, India

Tadepally, Haranath, Mr.
SHIP
SOCIETY FOR HUMAN INTEGRITY AND PROSPERITY
9 - 81 Gandhinagar, Nuzvid 521 201
Andhra Pradesh, India
T: 91 842 8656406

Tandon, Rajesh
SOCIETY FOR PARTICIPATORY RESEARCH IN ASIA
42 Tughlakabad Intitutional Area
New Delhi 110 062, India

Upadhyaya, Ashutosh
EPS
'Vidhya Niketan', 4, Station Plot
Gondal (PIN 360 311), India

Vaish, Ashok
SS
12 Rajendra Nagar, Gariawas Road, Tekri
Udaipur 313 001, India

Vali, S. Mahaboob
JR
2-76 Green House, Suraram Colony Road, I.D.A.
Jeedimetla 500 855, R.R. Dist. (A.P.), India

Verma, Vijendra
B
15, Sethi Nagar, Ujjain (Madhya Pradesh) - 456 001
India

Yelamanchili V., N.B.
SHIP
Door No. 9-81, Gandhi Nagar, Nuzuid 521 201
Krishna District, Andhra Pradesh, India

PRESS

Agarwal, Ajay
Editor
AMAR UJALA
65, New Raja Mandi
Raja Mandi AGRA 282002 India, India

Aguilar, Zuniga Thais
Correspondent
WOMEN'S FEATURE SERVICE
Ave. 7 Calles 13 y 15
San Jose, Costa Rica, India

Anand, Anita
Editor
WOMEN'S FEATURE SERVICE
E-181 Greater Kailasm II
New Delhi 110048, India, India

Ansari, Khalid
Editor
EARTH SÙMMIT TIMES
P.O. Box 51820, Desai
Bombay, India, India

Ansari, Tehzeeb
Correspondent
EARTH SUMMIT TIMES
12, Summerville, B. Desai Rd.
Bombay 400026 India, India

Ashok, Tharuvai Nilakantan
Correspondent
PRESS TRUST OF INDIA
A-9 Samachar Apts.,Mayur Vihar
Phase I Ex1, Delhi 110091, India

Avinash, Acharya
Correspondent
DOORDARSHAN INDIA
Acharya Aginash, Nagar
New Delhi, India, India

Baeza Lopez, Magadalena
WOMEN'S FEATURE SERVICE
Via Fomogosta 45
00192 Rome, Italy, India

Balakrishnan, Lalita
Editor
TIMES OF INDIA
1193 Sect. B/I Vasant Kanj
New Delhi 110030, India, India

Bandyopadhyay, Soumya
Editor
ANANDA BAZAR PATRIKA
B-31 Gulhomar Park
New Delhi 110049, India, India

Bhat, Ishwara P.S.
Correspondent
LIDAYAVAND PUBLICATIONS
Dastata Panaje Puttur
Karnataka India 574259, India

Chandra, Vijay B.
Editor
GOVERNMENT OF INDIA
C-37 Hyderabad Estate
Bombay, India, India

Chaturvedi, L.C.
Editor
TV INDIA
1/7663 Garakh Park
Delhi, India, India

Chaves, Solano Magaly
Correspondent
WOMEN'S FEATURE SERVICE
100 Norte 75 Oest Apt.Cafe#3
San Jose, Costa Rica, India

Chengappa, Raj
Correspondent
INDIA TODAY
M22, Sakt
New Delhi 110 017, India, India

D'Monte, Darryl
THE TIMES OF INDIA
29-B Carter Road
Bandra (West) Bombay 400 050 India, India

Datta, Ella
Correspondent
BUSINESS STANDARD
114C, Selimpore Road
Calcutta, 700031 India, India

Dhar, Mohan Krishen
Correspondent
THE HINDUSTAN TIMES
K.G.Marg
New Delhi, India, India

Foster, Rebecca
Editor
WOMEN'S FEATURE SERVICE
925 Union St. (#5E)
Brooklyn, NY 11215 USA, India

Gupta, Joydeep
Correspondent
SUNDAY MAIL
234 Aravalli Apartments
Alaknanda, New Delhi 110019, India, India

Gupta, Madan Mohan
Editor
DAINIK JAGRAN
B-22 #4
Bungalows, Bhopal, India, India

Gupta, Virendra Kumar
Correspondent
DOOR DARSHAN INDIA
166 Surya Nagar
Gaziabad, India, India

Hedge, Nagesh
Correspondent
PRAJAVANI
MF, 7/12 BDA Flats
Bangalore 560078 India, India

Inderjit, Kaushlsh Poonam
Correspondent
INDIA NEWS AND FEATURE ALLIANCE
115 Golf Links
New Delhi, India, India

Jain, Minu
Correspondent
SUNDAY
I-34 Dangpura Extension
New Delhi 110014, India, India

Kacker, Ranjana
Correspondent
MAYA
E 202 Greater Kailash II
New Delhi 110048, India, India

Kamini, Sawhney
Reporter Radio/Television
INDIA TV
Doordarshan Kendia,
Parlament Street, New Delhi, India

Kapoor, Coomi
Correspondent
DEV FEATURES
C-12, East Nizamuddin
New Delhi, India 110013, India

Kaur, Harminder
Correspondent
NEWSTIME
India

Lakshman, Nirmala
Correspondent
HINDU
c/o The Hindu, Mount Road
Madras 600 002, India, India
T: 91 44 835325

Laul, Mahinder Kumar
Correspondent
UNI
GG 1/64-B, Vikaspuri
New Delhi, India, India

Made, Patricia Ann
Writer
WOMEN'S FEATURE SERVICE
49 Golf Links
New Dehli 40003, India
T: 91114629886 F: 91114626699

Madhok, Sujata
Editor
WOMEN'S FEATURE SERVICE
4/3 Shanti Niketan
New Delhi 110021, India, India

Miglani, Sanjeev
Correspondent
THE STATESMAN
P.25 South Extension
Part-2, New Delhi, India, India

Mirza, Bamboodali Baig
Editor
SIASI MAHAZ DAILY
18-8-192/9 A bess Santoshnagar
Hydrabad-500659, India, India

Mohammed, Ahmedullah
Correspondent
SUNDAY MAIL
92 Hutchins Main Road.
St. Thomas Town, Bangal one, India, India

Morna, Colleen
Correspondent
WOMENS FEATURE SERVICE
Flat One 17 York Terrace East
London, United Kingdom, India

Parasuram, T.V.
Correspondent
PRESS TRUST OF INDIA
4450 S. Park Ave. (#1719)
Chevy Chase, MD 20815, USA, India

Patra, Kalyan
Editor
SRISHTI
94 Neeti Bagh
New Delhi, India, India

Radhakrishnan, Prasannan
Correspondent
MALAYALA MANORAMA
199 Samachar Apts. Mayur Vihar
Phase I Extn., New Delhi 110 092, India, India

Raghunath, Sankaran
Correspondent
VIDESH SANCHAR NIGAM
Sahar Road, Andheri
Bombay 400064, India, India

Rai, Usha
Correspondent
INDIA EXPRESS
161 Kaka Nagar
New Delhi 110003, India, India

Ramamohana Rao, M.
Editor
JANATA
Durga Agraharam
Vijavawada, India, India

Rammanohar, Reddy C.
Correspondent
DECCAN HERALD
2, Cornwell Rd.
Langford Gardens, Bangalore 560025,India, India

Ranu, A.
Editor
BOMBAY U.N.I.
44-Saikripa, Sector 16A,
Vashinagar, New Bombay 400705, India, India

Reddy, Sheela
Editor
WOMEN'S FEATURE SERVICE
E-4 Press Apartments
Patparganj, Delhi, India, India

Sabherwal, O.P.
Editor
INDIA PRESS AGENCY
2800 Netaji Nagar
New Delhi, 110023 , India, India

Sahay, Anand Kishore
Correspondent
THE HINDU
A-30 Friends Golony
New Delhi, India, India

Saini, Bhupal Indar
Correspondent
INDIA RADIO
304 Model Gram
Ludhiana, India, India

Salvi, Gouri
Editor
WOMEN'S FEATURE SERVICE
G-36, Masjid Moth
New Delhi, India, India

Sanatani, Saurabh
Correspondent
INDIAN SCIENCE CRUISER
Geweygasse 1/4/4, A-1190
Vienna, Austria, India

Sawhney, Kamini
Reporter Radio/Television
DOORDARSHAN INDIA
C-124 B Greater Kalish-I
New Delhi, India, India

Seema, Paul
Correspondent
THE TELEGRAPH
G-4 Chander Nagar
Delhi 11051 India, India

Shah, Jasu
Editor
CHEMICAL ENGINEERING WORLD
26 Maker Chambers VI
Bombay, India, India

Sharma, Dinesh Chandra
Correspondent
OBSERVER OF BUSINESS & POLITICS
86, Vasant Apts., Mayur Vimar
Phase I Ext., New Delhi 110 091, Inda, India

Sharma, Mukul
Correspondent
NAVBHARAT TIMES DELHI
222-C, Pocket-F, MIG, GTB
110093 Delhi, India, India

Shepherd, Anne
Correspondent
WOMEN'S FEATURE SERVICE
Flat 1 17 York Terrace East
London NW1 4PT, United Kingom, India

Shukla, Rajiv
Editor
SUNDAY OBSERVER
100 Aeiad Village
New Delhi, India, India

Shyam, Gouind Lal
Correspondent
INDIAN PRESS
B-14 Sakef DDA Fld.
New Delhi, India, India

Singh, Rahul
Correspondent
MIDDAY PUBLICATIONS
2/6Sorab Bharucha Road
Bombay 4-00005, India, India

Sinha, Mahesh Pratap
Correspondent
OFFICIAL PRESS INDIA
4-A Shriji Niketan 5ui-D
Dr. B.A. Road Matunga, Bombay, India, India

Sri, Prema Dhar Sharma
Correspondent
DAINIK ASAM
Sukreswar, Guwahati 781001
Assam, India, India

Sri Renganathan, Thygarajan
Editor
INDIA RADIO
18/2D sector 2
New Delhi, India, India

Sunil, Sethi
Correspondent
THE ECONOMIC TIMES
G-9 South Extension Part II
New Delhi, India, India

Swarup, Hari Har
Correspondent
PRESS TRUST OF INDIA
D-II/310 Vinaymare
New Delhi, India, India

Thanvi, Omprakash
Correspondent
JANSATTA DAILY
2539 Sec 35-C
Chandigarh, India, India

Tharuvai, Nilakantan Ashok
Correspondent
PRESS TRVST
A-9 Samachar Apto.
Dechi-91, India, India

Trehan, P.I.S.
Reporter Radio/Television
INDIA RADIO
B-1/99, Passhim Vihar
Delhi, India, India

ORGANIZATIONS

Agarwal, Anil Kumar, Mr.
Director
Centre For Science And Environment
F-6 Kailash Colony
New Delhi-110048, India
T: (91 11) 643 3394/634 8109 F: (91 11) 644 1711

Ahuja, S.P., Mr.
Centre for Research, Planning and Action
16 Dakshneshwar, 10- Hailey Road
New Delhi 110 001, India
T: (91 11) 332 6955/332 9216

Babu, Harinath, Prof.
Indian Institute of Forest Management
Nehru Nagar
Bhopal 462 003, MP, India

Bagchi, Ajoy, Mr.
People's Commission on Environment and Development India, The
3 Nyaya Marg, Chanakyapuri
New Delhi 110 021, India
T: (91 11) 462 7102 F: (91 11) 462 0102

Chhokar, Kiran B., Dr.
Centre For Environment Education
Nehru Foundation for Development, Thaltej Tekra
Ahmedabad 380 054, India
T: (91 272) 44 2642
TL: 121-6779 CEE IN

Dash, Sriram, Mr.
Gen.Secretary
Jana Seva Parisad
Abhaya Bhawan, Kendrapara
Cuttack 754211, Orissa, India

Dastagir, C.K.M., Mr.
Kwality- Karnataka Welfare Society
PO Box 28
Chiballapur 562 101, India
T: 08156 486

Dhandhania, Kishorilal, Mr.
Sri Aurobindo Society
Pondicherry 605002, India
T: 24382
TL: 0469-221 SAS IN

Haranath, T., Mr.
Academy of Gandhian Studies
2-2-1133/5/5 New Nallakunta
Hyderabad 500 044, Andra Pradesh, India

Haranath, Tadepally, Mr.
Society for Human Integrity and Prosperity
Door No. 9-81, Gandhi Nagar, Nuzuid 521 201
Krishna District, Andhra Pradesh, India

Hegde, Narayan G., Mr.
Baif Development Research Foundation
'Kamdhenu', Senapati Bapat Road
Pune 411 016, India
T: (91 212) 342 621/342 466 F: (91 212) 0145 7283

Jordan, Richard, Mr.
Global Futures Network
73A Mittal Tower, Nariman Point
Bombay 400021, India
T: (91 22) 204 57 58

Kulkarni, Manu N., Mr.
Taralabalu Rural Development Foundation
PO Sirigere-577 541
Dist. Chitradurga, Karnataka, India

Kumar, M.S.R. Prem, Mr.
Youth Charitable Organization
20/14 Urban Bank Street
Yellamanchili 531 055, Vizag Dt. (A.P.), India

Kusum, Bhawani Shanker, Mr.
Society for Rural Development
S-3, Govind Nagar (West)
Jaipur 302 002, India

Manoharan, M.L.E., Mr.
Gramodaya
Post Box 6, Dist: Trichy
Manapparai 621 306, Tamil Nadu, India

Mathew, K.C., Mr.
International Centre For Study And Development
Shangrila Hills, Valakom P.O. 691532
Kerala State, India
T: VALACOM 76

Mehta, Pradeep S., Mr.
General Secretary
Consumer Unity And Trust Society
3-B, Camac Street
Calcutta-700 016, India
T: (91 33) 29 7391 F: (91 33) 29 2815

Misra, Purna Chandra., Mr.
Indian Institute of Youth and Development
Kalinga, Phulbani
Orissa 762 022, India

Moulik, T.K., Mr.
International Institute of Sustainable Development and Management
GHB - Subhashnagar, Chandkheda
Vastrapur, Ahmedabad 382 424, India

Murali, B., Dr.
Development Alternatives
B-32 Institutional Area, Tara Crescent,
New Mehrauli Road, New Delhi-110 016, India
T: (91 11) 66 5370 F: (91 11) 686 6031

Murty, K.S., Prof.
Gondwana Geological Society
c/o University Dept. of Geology,
Law College Compound, Nagpur 440 001, India
T: (91 712) 532 411

Nayak, D.K., Mr.
Village Institution for Social Action
House No. 111-B, Saktinagar M.E. School Lane
Rourkela 769 014, India

Pachauri, R.K., Dr.
Tata Energy Research Institute
7 Jorbagh
New Delhi - 110 003, India
T: (91 11) 462 7651 F: (91 11) 462 1770

Parikh, Jyoti, Dr.
Indira Gandhi Institute for Development Research
Gen. Vaidya Marg, Goregaon (E)
Bombay 400 065, India
T: (91 22) 680 0919 F: (91 22) 680 2752

Patro, S.N., Dr.
Orissa Environmental Society
N-6/119, Nayapalli, Bhubaneswar
Orissa 751015, India

Pattanaik, P.K., Mr.
Orissa State Volunteers And Social Workers Association
Udayapur
Nuagaon - 752083, Puri, Orissa, India
T: 24 NUGN

Pradhan, Manoj K., Mr.
Council of Professional Social Workers
2130/4158 Vivekananda Marg
Bhubaneswar 751002, India

Prakash, Cedric, S.J. Fr.
St. Xavier'S Social Service Society
Post Box 4088, Navrangpura
Ahmedabad 380 009 Gujarat, India
T: (91 272) 495 238

Qureshy, M.N., Mr.
Centre For Science And Technology Of The Non-alligned And Other Developing Coun
C-4/16, Safdarjung Development Area
New Delhi-110016, India
T: (91 11) 686 5623

Reddy, R. Thimma, Mr.
Bangalore Integrated Rural Development Society (BIRDS)
Chandapura Post
Bangalore 562 145, India
T: (91 812) 422 359

Satheesh, P.V., Mr.
Deccan Development Society
A6,Meera Apts., Skyline Theatre Lane
Basheerbagh, Hyderabad, A.P., India
T: (91 842) 231 60

Sharma, Ram Naresh, Mr.
Gram Swarajya bakhtiyarpur
At Post- Salimpur, Via Khuruspur
Dist- Patna (Bihar), India

Sharma, S.K., Mr.
Environment Society of Chandigarth
Karuna Sadan, Sector 11-B
Chandigarth 160 011, India

Singh, Gurmit, Mr.
Environment Preservation Society
'Vidhya Niketan', 4, Station Plot
Gondal (PIN 360 311), India
T: #357 182

Subramaninam, A.R., Prof.
Indian Agrometeorological Society
Andhra University
Visakhapatnam - 530 003 (AP), India
T: (91 891) 64 871 F: (91 891) 50 644

Swarup, Hem Lata, Prof.
All India Women's Studies And Development Organisation
111/98A, Ashok Nagar
Kanpur 208012, India
T: (91 512) 294 130

Vaish, Prerana, Dr.
Sarita Society
12 Rajendra Nagar, Gariawas Road, Tekri
Udaipur 313 001, India

Venkatachalam, R., Mr.
Community Services Guild, The
20, C-Block, 9th Street, Ana Nagar East
Madras 600 102, India
T: (91 44) 616 359

Venugopal, D., Mr.
Save Nilgiris Campaign
22-C, Tirupathy Apartments, Venus Colony
Alwarpet, Madras 600 018, India
T: (91 44) 864 141 EXT. 8 F: (91 44) 825 3395

GLOBAL FORUM

Alagh, Y.K., Dr.
SADVICHAR PARIVAR
Mangaimurti, Ashram Road, Navarangpura
Ahmedabad 380009, India
T: 91 272 405260-90 F: 91 272 44052 54

Alsunbul, Abdulaziz
ALESCO
12 Rajaji Marg
New Delhi 110011, India
T: 91 11 301-4309

Atisha, Tenzin Phuntsok, Mr.
Reasercher
INFORMATION OFFICE OF DALAI LAMA
Gangchen Kyisang -HP
Dharamsala, India
T: 91 2457 F: 91 2357

Babu, Yelananchili V. Nagendra, Mr.
Ship
SOCIETY FOR HUMAN INTEGRITY AND PROSPERITY
9 - 81 Gandhinagar, Nuzvid 521 201
Andhra Pradesh, India
T: 91 842 3215015 F: 91 842 2269021
TL: 614688

Bapna, Ashok, Mr.
Council Member
SOCIETY FOR INTERNATIONAL DEVELOPMENT - SID
334, SSO Building, Secretariat
Jaipur 302005, India
T: 91 141 521109
TL: 0365-2444 SECT

Barka, Lalla Ben
ACORD
12 Rajaji Marg
New Delhi 110011, India
T: 91 11 301-4309

Benjamin, Bennet, Mr.
NCC INDIA - CHRISTIAN COUNCIL LODGE
Post Bag 105, rney
Nagpur 440 001, India

Bhagat, Man Mohan, Mr.
Chairman
EARTH FOUNDATION
253 Bhagat Wing, Claridges, 12 Aurangzeb Road
New Delhi 110011, India
T: 91 11 301-5524 F: 91 11 301-0546
TL: (81)(31)66604

Bhatnagar, P.D.B, Mr.
Secretary
CONSORTIUM OF INDIAN SCIENTISTS FOR SUSTAINABLE DEVELOPMENT
9/1 Institutional Area, Seet Singh Marg, opp INU, East Gate
New Delhi, India
T: 91 11 66-1242

Biksham, Gujja, Mr.
Director
DECCAN DEVELOPMENT SOCIETY
A-6 Meera Apartments, Basheerbagh
Hyderabad 500029, India
T: 91 842 231260 F: 91 842 869243

Bird, Edris
CARCAE
12 Rajaji Marg
New Delhi 110011, India
T: 91 11 301-4309

Borthakur, Shanti Sharma
KRISHNANAGAR SOCIAL WELFARE
Krishnanagar Biswantketan Constructive Board, Krishnanagar Road,
Jorhat, Assam, India

Cadena, Felix
CLASEP
12 Rajaji Marg
New Delhi 110011, India
T: 91 11 301-4309

Coyali, Banoo Jehangir, Ms.
Director
UNITED SOCIO ECONOMIC DEVELOPMENT AND RESEARCH PROGRAMME - UNDARP
945, Boottee St.
Pune 411001, India
T: 91 212 290 30 F: 91 212 444 253

Dava, Sonam
LADAKH ECOLOGICAL DEVELOPMENT GROUP
Palais de Nations
Geneve CH 1211, India
F: 41 22 733-6542

Desai, Ranjan
SELF EMPLOYED WOMEN'S ASSOCIATION - SEWA
SEWA Reception Centre, Bhadra, Ahmedabad
Ahmedabad 380001, India
T: 91 272 39-0577

Dudani, A. T., Mr.
Senior Fellow
SOUTH - SOUTH SOLIDARITY
P-4 Green Park Extension - First floor
New Delhi Pin 110017, India
T: 91 11 66-9370 F: 91 11 656-6720

Geeta, Menon, Ms.
INDIAN SOCIAL INSTITUTE
28 Lodi Road - Institutional Aerea
New Delhi, India
TL: 31-74005

Ghosh, Prodipto, Dr.
TATA ENERGY RESEARCH INSTITUTE
9 Jorbagh
New Delhi 110003, India
T: 91 11 618-516 F: 91 11 462-1770
TL: 31-61593 Teri I

Godbole, Archana Jagdish, Ms.
JEEWAN SANSTHA
B-7, Seema, Shirole Road
Pune 411004, India
T: 91 212 56531

Godbole, Jagdish Rangnath, Mr.
JEEWAN SANSTHA
B-7, Seema, Shirole Road
Pune 411004, India
T: 91 212 56531

Grossi, Francisco Vio
CENTRO EL CANELO DE NOS
12 Rajaji Marg
New Delhi 110011, India
T: 91 11 301-4309

Gunnarsson, Steingrimur, Mr.
CENTRE FOR ENVIRONMENTAL AND MANAGEMENT STUDIES
New Friends Colony
New Delhi, India
T: 91 11 688-4332

Gupta, P.K., Mr.
Professor
NATIONAL FOUNDATION OF INDIAN ENGINEERS
11/6B, Shanti Chambers, Pusa Road
New Delhi 110005, India
T: 91 11 573 1446 F: 91 11 575 5000
TL: 031-77375WCST

Gupta, Priya, Ms.
Environment Reseearcher
CONSUMER EDUCATION AND RESEARCH CENTRE
Thakoreshai Desai Smarak Bhavan Ellisbridge
Ahmedabad, India
T: 91 272 448833 F: 91 272 468661
TL: 0121-6109

Gupta Mayor, Dauji, Dr.
President
SOCIETY FOR RESEARCH AND ENVIRONMENT
Post. Box. 288
Lucknow, India
T: 247 491 F: 91 522 242 061

Harijanhunayan, Akhil Bharat, Mr.
KRISHNANAGAR SOCIAL WELFARE
Krishnanagar Biswantketan Constructive Board, Krishnanagar Road,
Jorhat, Assam, India

Holt, Lilian
ABORIGINAL COMMUNITY COLLEGE
12 Rajaji Marg
New Delhi 110011, India
T: 91 11 301-4309

Inbaraj, Prianayagam Alexander, Mr.
Consultant
COOPERATION AND HUMAN DEVELOPMENT - INDIA
40 Sarvadhana, 413 Annasalai, Tiruvammamach, T.S Dt Tamunadu, India
Tamilnadu, India
T: 91 04175 2123 F: 91 791041252123

Islam, Nazrul, Mr.
President
MALER HAT JUBO SHANGHA
Haimchar
Chandpur, India

Joshi, Sharad, Mr.
Secretary
CECOEDECON
Development Centre, Shilki Dungari Chaksu 303901
Jaipur, India
T: 91 141 51-4950 F: 91 141 51-4950

Joshi, Vasant, Dr.
Director
OSHO SPIRITUAL HEALTH ORGANIZATION
17 Koregaon Park
Poona 411001, India
T: 91 212 66-0963 F: 91 212 66-4181

Joshi, Vidyut, Dr.
SADVICHAR PARIVAR
Mangaimurti, Ashram Road, Navarangpura
Ahmedabad 380009, India
T: 91 272 405260-90 F: 91 272 44052 54

Kananda, John
AALAE
12 Rajaji Marg
New Delhi 110011, India
T: 91 11 301-4309

Khanna, Anjani, Ms.
Research Assoc
CENTRE FOR SCIENCE AND ENVIRONMENT
F-6 Kailash Lolony
New Delhi 110048, India
T: 91 11 643-3394 F: 91 11 643-8109

Khosla, Ashok, Mr.
DEVELOPMENT ALTERNATIVES AND ICSU
B 32 Institutional Area TARA Crescent, New Mehrauli Rd.
New Delhi 110016, India
T: 91 11 66-5370 F: 91 11 686-6031
TL: 31-73216, E: GEO2:TARA

Kumar-d'Souza, Corinne, Ms.
CENTRE FOR DEVELOPMENT STUDIES - INDIA/WEDO
No. 40, Meg Colony
Bagalore, India
T: 812 572568 F: 812 585964

Ladeira, Mariangela
ICAE
12 Rajaji Marg
New Delhi 110011, India
T: 91 11 301-4309

Levine, Lani
Ambassador
OSHO SPIRITUAL HEALTH ORGANIZATION
17 Koregaon Park
Poona 411001, India
T: 91 212 66-0963 F: 91 212 66-4181

Luithui, Luingam
NAGA PEOPLE'S MOVEMENT FOR HUMAN RIGHTS
Palais de Nations
Geneve CH 1211, India
F: 41 22 733-6542

Malhotra, K.C., Dr.
INDIAN STATISTICAL INSTITUTE
203 B.T. Road, Calcutta - 700035
India

Mathew, Ajit Ninan
Cartoonist
INTER PRESS SERVICE
India Today Connaught Place
New Dehli, India
T: 3315801 F: 3315801

Mikkelsen, Kent
DANISH COUNCIL FOR ADULT EDUCATION
12 Rajaji Marg
New Delhi 110011, India
T: 91 11 301-4309

Mishra, Veer Bhadra, Ms.
President
SANKAT MOCHAN FOUNDATION
Tulsi Mandir Tullsi Ghat
Varanasi 221001, India

Mohanan, N., Dr.
ASSOCIATION OF NON-GOVENMENTAL ORGANISATIONS
185 North Avenue
New Delhi 100011, India
T: 91 11 301-0077 F: 91 11 379-2267
TL: 0425-6443 DEVI

Moreira, Tereza
ICAE
12 Rajaji Marg
New Delhi 110011, India
T: 91 11 301-4309

Mueller, Marti, Ms.
Co-Director
CHILDREN AND TREES AUROVILLE
Shanta (via ABRI), Success Rd, Auroville, Tamil Nadu
Auroville 605101, India

Mukhopadhyay, Jill Carr Harris, Ms.
Executive Director
SOUTH - SOUTH SOLIDARITY
P - 4 Green Park Extension - First floor
New Delhi, India
T: 91 11 66-9370 F: 91 11 656-6720

Murali, Ramisetti, Mr.
ACADEMY OF GANDHIAN STUDIES
Action for World Solidarity , 12-13-304, Street No.9-Tarnaka
Secunderabad (A.P.) 0842, India
T: 91 869243

Myles, Raymond M., Mr.
Executive Director
ACTION FOR FOOD PRODUCTION (AFPRO)
25/1A Institutional Area, B - Bloc, Janakpuri
New Delhi 110058, India
T: 91 11 555 5412 F: 91 11 550 0343
TL: 3165899 AFPROIN

Narain, Sunita, Ms.
CENTRE OF SCIENCE AND ENVIRONMENT
F-6 Kailash Colony New Delhi
New Delhi 110048, India
T: 91 11 6433394 F: 91 11 6438109

Niroop, Patiolla, Mr.
DECCAN DEVELOPMENT SOCIETY
A/C Meera Apartments - Huderabad
Andhra Pradesh 500029, India
T: 91 842 212234 F: 91 842 231260

Ogelsby, Leni
UNIVERSITY OF SHEFFIELD
12 Rajaji Marg
New Delhi 110011, India
T: 91 11 301-4309

Oswal, Gautam, Mr.
INTERNATIONAL MAHAVIR JAIN MISSION
Acharya Sushil Ashram C-599, Chetna Marg Defence Colony
New Delhi 110024, India
T: 91 11 462-2729 F: 91 11 463-3622

Pal, Manindra, Mr.
SRI AUROBINDO ASHRAM
Gloria Land, Sri Aurobindo Ashram
Pondicherry, India
T: 91 29017

Panwalker, Pratima, Ms.
TATA INSTITUTE OF SOCIAL SCIENCES
PO Box 8313, Sion Trombay Rd., Deonar
Bombay 40088, India
T: 91 22 523 701 F: 91 22 615 25 27

Panwalker, Virup, Mr.
TATA INSTITUTE OF SOCIAL SCIENCES
PO Box 8313, Sion Trombay Rd., Deonar
Bombay 40088, India
T: 91 22 523 701 F: 91 22 615 25 27

Paranjpye, Vijay, Mr.
Professor
'DURGA' 92/2, ERANDAWE, OPP.
kamla Nehru Park
Pune, India
T: 91 332448 F: 91 212331250

Parikh, Kirit, Mr.
Director
INDIRA GHANDI INSTITUTE OF DEVELOPMENT
Palais de Nations
Geneve CH 1211, India
F: 41 22 733-6542

Parmar, Punamchand Chandulal, Mr.
President
UNITED SOCIO ECONOMIC DEVELOPMENT AND RESEARCH PROGRAMME - UNDARP
945, Boottee St.
Pune 411001, India
T: 91 212 290 30 F: 91 212 444 253

Patel, Adi, Mr.
Professor
I G S S S
28mla Nehru Park
Pune, India
T: 6921923
TL: 31-74005

Patel, C.C., Dr.
SADVICHAR PARIWAR
Ashram Road, Navrangpura
Ahmedabad 380009, India
T: 91 272 405 260 F: 91 272 23056

Patel, Krishnaprasad, Mr.
Convenor
SADVICHAR PARIVAR
Mangaimurti, Ashram Road, Navarangpura
Ahmedabad 380009, India
T: 91 272 405260-90 F: 91 272 44052 54

Patel, Urmilaben, Dr.
SADVICHAR PARIVAR
Mangaimurti, Ashram Road, Navarangpura
Ahmedabad 380009, India
T: 91 272 405260-90 F: 91 272 44052 54

Patil, Major Suresh, Mr.
CHATRAPATI SHIVAJI TREKKING & ENVIRONMENTAL INSTITUTE
14/105 - 106 Agarkar Nagar
Pune 411011, India
T: 91 212 667076 F: 91 212 662059

Patil, Shri Shivajirao G., Mr.
President
NATIONAL FEDERATION OF COOP. SUGAR FACTORIES
83 83 Nehru Place
New Delhi 110 058, India
T: 91 11 641-7481
TL: 031-71090

Paul, Smt. Kunti
ALL INDIA WOMEN'S CONFERENCE
6 Bhagwan Dass Road
New Delhi 110 001, India
T: 91 11 389680

Pimple, Minar, Mr.
HABITAT INTERNATIONAL COALITION
Cordobanes 24, Col. San Jose Insurgentes
Mexico City 03900, India
T: 52 5 6512371 F: 52 5 5453263

Prakash, Sanjeev, Mr.
Coordinator
ECO TIBET - INDIA
G-11 Saket
New Delhi, India
T: 91 11 651038 F: 91 11 646-1914

Proulx, Jacques
DEPARTMENT DE PSYCHOLOGIE FLSH
12 Rajaji Marg
New Delhi 110011, India
T: 91 11 301-4309

Rajput, Pam
Chairperson
WOMEN, POLITICS AND DEVELOPING NATIONS STUDY GROUP 30 OF POLITICAL SCIENCE ASSOCIATIO
Centre for Women's Studies, Panjab Univ. Chandigharh
Chandigarh 160014, India
T: 91 172 314370
TL: 395464 RSIC IN

Ramdas, Lalita
ITFL - NAVY HOUSE
12 Rajaji Marg
New Delhi 110011, India
T: 91 11 301-4309

Ramprasad, Vanaja, Ms.
ASIAN HEALTH NETWORK
839.23 Main, J.P. Nagar, 2nd Phase
Bangalore 0160, India
T: 91 812 644963 F: 91 812 641585

Ranga, Sudhir, Mr.
Chairman
MINORITIES FRONT
Kilacolony, Jhajjar
Haryana 12485, India

Reddy, C. Rammanohar
Assistant Editor
DECCAN HERALD
G6, M.G. Road,
Bangalore 560 025, India
T: 588999 F: 586643

Reddy, Kullisham Purilshatham, Ms.
ACADEMY OF GANDHIAN STUDIES
2-2-1133/5/4 New Nallakunta
Hyderabad 500044, India

Roy, Sanjit (Bunker), Mr.
SOCIAL WORK RESEARCH CENTER
Tilonia 305816 Ajmer Dt. Rajasthan
Tilonia Dt. Rajastha 305816, India

Roychowdhury, Anumita, Ms.
CENTRE FOR SCIENCE AND ENVIRONMENT
Palais de Nations
Geneve CH 1211, India
F: 41 22 733-6542.

Sabnis, Sharad, Dr.
SADVICHAR PARIVAR
Mangaimurti, Ashram Road, Navarangpura
Ahmedabad 380009, India
T: 91 272 405260-90 F: 91 272 44052 54

Satapathy, Annapurna, Ms.
COMMUNITY DEVELOPMENT CENTRE
Ramabhadrapuram Vizianagaram
Andhra Pradesh 532 579, India

Saxena, Mahesh Prasad, Dr.
MADHYA PRADESH STATE COOPERATIVE BANK LTD
Multi Storeyed Building, T.T. Nagar
Bhopal, India
T: 91 755 550221
TL: 0705-293 COOP-I

Sen, Gita, Ms.
DEVELOPMENT ALTERNATIVES WITH WOMEN FOR A NEW ERA (DAWN)
c/o Center for Population & Dev. Stud. Harvard Un. 9 Bow St.
Cambridge - MA 02138, India

Shah, Manubhai, Prof.
Managing Trustee
CONSUMER EDUCATION AND RESEARCH CENTRE
Thakorebhai Desai Smarak Bhavan - near Law College-Ellisbridge
Ahmedabad, India
T: 91 272 448833 F: 91 272 468661
TL: 0121-6109

Shah, Priyavadan, Mr.
ENVIRONMENTAL ACTION GROUP
Opposite Bhagwati Chamber, Gugaret, Vidyapith Ashram Road,Ahmedabad
Gujarat, India

Shah, Sharad, Dr.
GLOBAL FUTURES NETWORK
73 A Mittal Tower, Nariman Point
Bombay 400021, India

Sharma, Kali Das Deva
KRISHNANAGAR SOCIAL WELFARE
Krishnanagar Biswantketan Constructive Board, Krishnanagar Road,
Jorhat, Assam, India

Sharma, Kimti Lal, Mr.
Partner
DKS
B-26, Nizamuddin East
New Delhi, India
T: 91 11 694511 F: 91 11 4621147

Sharma, Sashi Prava
KRISHNANAGAR SOCIAL WELFARE
Krishnanagar Biswantketan Constructive Board, Krishnanagar Road,
Jorhat, Assam, India

Sharma, Vijay Laxmi
KRISHNANAGAR SOCIAL WELFARE
Krishnanagar Biswantketan Constructive Board, Krishnanagar Road,
Jorhat, Assam, India

Shenoy, K.G., Mr.
Exec. Director
GLOBAL FORUM NETWORK
Maker Chamber IV Narimar Point
Bombay 400021, India

Sheth, Pravin, Mr.
SADVICHAR PARIVAR
Mangaimurti, Ashram Road, Navarangpura
Ahmedabad 380009, India
T: 91 272 405260-90 F: 91 272 44052 54

Silva, Denise Ferreira Da, Ms.
OSHO SPIRITUAL HEALTH ORGANIZATION - OSHO
17 Koregaon Park
Pune 411001, India
T: 91 212 4377488 F: 91 212 4377488

Somanath, E
Sub Editor
MALAYALA MANIRAMA
Malayala Manirama Kottayam
Kerala, India
T: 91 565346 F: 91 562479

Srivastava, Anand K., Mr.
Delegate
GLOBAL FUTURES NETWORK
20 b Lands End, Dongarsi Road, Malabar Hill
Bombay 400006, India

Subba Rao, B.V., Mr.
Ags
ACADEMY OF GANDHIAN STUDIES
2-2-1133/5/4, New Nallakunta
Hyderabad 500044, India

Suryanarayana, Veeravilli, Mr.
c/o Prof. S.M. Agrawal, P-3 Gvt. Engg. College Colony
Rewa M.P. 486 002, India
T: 91 76622804 F: 91 76623487

Taware, Gurunath Janardhan, Mr.
Chairman
NATIONAL COOP. AGRICULTURE AND RURAL DEV. BANK'S FEDERATION LTD.
shivshakti, 2nd floor, B.G. Kher Road - Worli
Bombay 400018, India
T: 91 22 493-4349

Thomas, Mary
Professor
EVERGREENS
Parottukonam, Nalanchira PO
Trivundrum 695 015, India

Toth, Mindy, Ms.
Director Of Education
AUROVILLE HOLISTIC HEALING CENTER
Aspiration
Auroville 605 101, India
T: 91 862 123 F: 91 28 132

Trivedi, R.C., Mr.
Manager Of Environment Protect
CONSUMER EDUCATION AND RESEARCH CENTRE
Thakorebhai Desai Smarak Bhavan, near Law College-Ellisbridge
Ahmedabad, India
T: 91 272 448833 F: 91 272 468661
TL: 0121-6109

Tsering, Tempa, Mr.
Secretary
INFORMATION OFFICE OF DALAI LAMA
Gangchen Kyisanng - HP
Dharamsala, India
T: 91 2457 F: 91 2357

Vinayak, Arun
VIKAS ADHYAYAN KENDRA
D-1 Shivdham, 62 Link Road Malad (West)
Bombay 400 064, India
T: 22 6822850 F: 22 2042329/6422

Wangoola, Paul
AALAE
12 Rajaji Marg
New Delhi 110011, India
T: 91 11 301-4309

ADVISORS

Singh, Karan, Dr.
Chairman, The People'S Commission On Environment And Development India
New Delhi 110003, India
T: 91 11 462 7102 F: 91 11 462 0102

Swaminathan, M.S., Dr.
Chairman, M.S. Swaminathan Research Foundation
Madras 600 085, India
T: 91 44 416923/410031 F: 91 44 478148

PARLIAMENTARIANS / SPIRITUAL LEADERS

Gregorios, Paulos Mar, Metropolitan
President
World Council of Churches
Delhi Orthodox Centre 2 Tughlakabad Institution Area
New Delhi 110062, India
T: 6436417 F: 91 11 647 4975

Tulku, Doboom, The Ven. Dr.
Director
Tibet House
1 Institutional Area, Lodi Road
New Delhi 3, India
T: 61 15 15

WISDOMKEEPERS

Corkery, Forrest Philip, Mr.
Co-Director
CHILDREN AND TREES AROVILLE
Shanta (via ABRI), Success Rd, Auroville, Tamil Nadu
Auroville 605101, India

Giordano, Edward, Mr.
Coordinator
Auroville Greenwork Resource Ctr
Isaiambauam
Auroville Tamil Nadu 605101, India
T: 91-413-862290 F: 91-413-28132

Jain, Narenda P.
Former Ambassador Of India To U.N.
International Mahavir Jain Mission
45 South Rajmohalla
Indore M.P., India
T: 011-91-731-491273

INDONESIA - GOVERNMENT

Aman, Dalindra, H.E.
Ambassador Extraordinary and Plenipotentiary
Permanent Representative to the UNEP, Nairobi
Permanent Mission to the United Nations, 325 E. 38th Street
New York, NY 10016, United States of America
T: 212 972 8333 F: 212 972 9780

Brotodiningrat, Soemadi D. M., H.E. Mr.
Ambassador Extraordinary and Plenipotentiary
Permanent Representative to the United Nations, Geneva
Permanent Mission to the United Nations, 325 E. 38th Street
New York, NY 10016, United States of America
T: 212 972 8333 F: 212 972 9780

Daryadi, Lukito, Mr.
Senior Advisor to the Minister of Forestry
Department of Forestry
Manggola Wanabakt Bldg JL Gatot Subroto
Jakarta, Indonesia
T: 62 21 5701118 F: 62 21 5700226

Efransjah, Efransjah, Dr
Official
Ministry of Forestry
P.O.Box 6561 Jkwb
Jakarta, Indonesia
T: 62 21 5701114 F: 62 21 5700226

Jayasusila, I Wayan, Mr.
Second Secretary
Embassy of Indonesia, Brasilia
Permanent Mission to the United Nations, 325 E. 38th Street
New York, NY 10016, United States of America
T: 212 972 8333 F: 212 972 9780

Kismadi, M. S., Mr.
Senior Adviser to the Minister of State for Population and Environment
Permanent Mission of Indonesia
325 E. 38th Street
New York, NY 10016, United States of America
T: 212 972 8333 F: 212 972 9780

Luhulima, Achie Sudiarti
Assistant Minister Iv
STATE MIN FOR ROLE OF WOMEN
Merdeka Barat 15
Jakarta 10110, Indonesia
T: 62 21 3805558 F: 62 21 3805562

Moerdiono, , H.E. Mr.
Minister of State Secretariat
Permanent Mission of Indonesia
325 E. 38th Street
New York, NY 10016, United States of America
T: 212 972 8333 F: 212 972 9780

Naro, H. J., Mr.
Vice Chairman
House of Representatives
Permanent Mission to the United Nations, 325 E. 38th Street
New York, NY 10016, United States of America
T: 212 972 8333 F: 212 972 9780

Nazif, Amru Hydari
Assistant Director
ASEAN SECRETARIAT
Jalan Sisingamangaraja FOA
Kakarta 12110, Indonesia
T: 021 716451 F: 021 7398234

Nitisastro, Widjojo, Mr.
Senior Adviser
Government of the Republic of Indonesia
Permanent Mission to the United Nations, 325 E. 38th Street
New York, NY 10016, United States of America
T: 212 972 8333 F: 212 972 9780

Saleh, Johan S., Mr.
Minister Councellor
Permanent Mission of Indonesia
Geneva
325 E. 38th Street New York, NY, 10016
F: 212 972 8333
TL: 212 972 9780

Salim, Emil, H.E. Mr.
Minister of State
Population and Environment
Permanent Mission to the United Nations, 325 E. 38th Street
New York, NY 10016, United States of America
T: 212 972 8333 F: 212 972 9780

Silalahi, Adian, Mr.
Director for Multilateral Economic Cooperation
Department of Foreign Affairs
JL Taman Pejamibon 6
Jakarta, Indonesia
T: 3848626

Soegandhy, Aca, Mr.
Assistant to the Minister of State for Population and Environment
Permanent Mission of Indonesia
325 E. 38th Street
New York, NY 10016, United States of America
T: 212 972 8333 F: 212 972 9780

Soeharto, , H.E. Mr.
President
Permanent Mission of the Republic of Indonesia
325 E. 38th Street
New York, NY 10016, United States of America
T: 212 972 8333 F: 212 972 9780

Soeriaatmadja, R. E., Mr.
Assistant to the Minister of State for Population and Environment
Permanent Mission of Indonesia
325 E. 38th Street
New York, NY 10016, United States of America
T: 212 972 8333 F: 212 972 9780

Soeriaatmadja, R. I. Rhousdy, H.E. Mr.
Ambassador Extraordinary and Plenipotentiary to the Federative Republic to Brasilia
Permanent Mission of Indonesia
325 E. 38th Street
New York, NY 10016, United States of America
T: 212 972 8333 F: 212 972 9780

Soerlaatmadja, Roehajat Emon
Assistant Minister
MINISTRY OF STATE FOR POPULATION &
Environment, Arthaloka Bldg, J.L. Jendral Sudirman 2
Jakarta, Indonesia
T: 570 3419 583792 F: 62 22 583793

Suganday, Aca
Member Of Delegate
MINISTRY OF POPULATION
And Environment Jalan Merdeka Barat 15B
Jakarta, Indonesia
T: 374 371 F: 62 21 374 307

Sumarwoto, Otto, Mr.
Expert on Environment
Department of Forestry
Permanent Mission to the United Nations, 325 E. 38th Street
New York, NY 10016, United States of America
T: 212 972 8333 F: 212 972 9780

Suryawinata, Benny P., Mr.
Official
Department of Foreign Affairs
Permanent Mission to the United Nations, 325 E. 38th Street
New York, NY 10016, United States of America
T: 212 972 8333 F: 212 972 9780

Sutresna, Nana, H.E. Mr.
Ambassador Extraordinary and Plenipotentiary
Permanent Representative to the United Nations, New York
Permanent Mission to the United Nations, 325 E. 38th Street
New York, NY 10016, United States of America
T: 212 972 8333 F: 212 972 9780

Wibowo, Triyono, Mr.
Second Secretary
Permanent Mission of Indonesia
325 E. 38th Street
New York, NY 10016, United States of America
T: 212 972 8333 F: 212 972 9780

NGO

Moniaga, Sandra, Ms.
FE
Jl. Penjernihan I.. Komp. Keuangan No. 15
Pejompongan, Jakarta 10210, Indonesia

Pramono, Albertus, Mr.
Co2 Campaigner
WALHI(INDONESIAN FORUM FOR ENVIROMENT
J. PENJERNIHAN I KOMP KEVANGAN No15
Jakarta 10210, Indonesia
T: 62 21 583975 F: 62 21 588416
TL: 44672 SWIDJOIA

Witoelar, Erna, Ms.
INTERNATIONAL ORGANIZATION OF CONSUMERS UNION
PO Box 1045
Penang 10830, Indonesia
T: 60 4 3711396 F: 60 4 366506
TL: Ma40164 APIOCU

PRESS

Achmadi, Heri
Correspondent
JANA POS (INDONESIA DAILY)
312 N. George Mason Rd.
Arlington, VA 22203 USA, Indonesia

Bakran, Asmawi
Correspondent
ANTARA NEWS AGENCY
39-18 50th Avenue
Sunnyside, NY 11104, USA, Indonesia

Hartiningsih, Maria
Editor
KOMPAS DAILY
Kompleks BN1 Blok D-S, SL1P1
Jakarta, Barat, Indonesia, Indonesia

Johannes, Wilhelm
Reporter Radio/Television
RCTI
Pandega Duta III/17
Yogyakarta 55281, Indonesia, Indonesia

Pudjomartono, Susanto
Editor
TEMPO MAGAZINE
Batuampar IV/32, Kramatjati
Jakarta Timur, Indonesia, Indonesia

Saksono, Djoko
Reporter Radio/Television
INDONESIA RADIO
Tylyan Indah 4-3/9
Bekasi, Indonesia, Indonesia

Sawitri, Isma
Editor
TEMPO WEEKLY NEWS MAGAZINE
71 Kemanggisan X/3
Jakarta, Indonesia, Indonesia

Sutoyo, Denny
Correspondent
KOMPAS DAILY
Joseph Plaut Weg,3
4930 Detmold, F.R.Germany, Indonesia

Yatim, Debra
Correspondent
JAKARTA POST
Jalah Wijaya Timur 78
Jakarta, Indonesia, Indonesia

ORGANIZATIONS

Zulkarnaren, M., Mr.
Wahana Lingkungan Hidup Indonesia (friends Of The Earth, Indonesia)
Jl. Penjernihan I.. Komp. Keuangan No. 15
Pejompongan, Jakarta 10210, Indonesia
T: (62 21) 586 820/583 975 F: (62 21) 588 416

GLOBAL FORUM

Afiff, Suraya, Ms.
WALHI/FOE INDONESIA
JL Penjernihan I/15 Komp. Kenangen
Jakarta 10240, Indonesia
T: 62 21 586820 F: 62 21 588416

Erari, K.Ph.
Rev.
EVANGELICAL CHRISTIAN CHURCH IN IRIAN JAYA
Jin, Argapura No. 21 Kotak Pos 14 Jayapura
Irian Jaza, Indonesia

Indrayana, Eddy, Mr.
Government-Off
MUNICIPALITY OF SURABAYA
Jalan Kutisari VII 19
Surabaya-East Java, Indonesia
T: 62 31 815255 F: 62 31 46303

Isomo, Sutopo, Mr.
INDONESIAN'S VETERANS ORGANIZATION
Kompe Polri C-1 16 Gatot Moroso
Jakarta 12930, Indonesia
T: 62 21 516431 F: 62 21 584894

Jhamtani, Hira, Ms.
KOMPALINDO
Palais de Nations
Geneve CH 1211, Indonesia
F: 41 22 733-6542

Mokoginta, Dv. H.Ijan, Mr.
MEMBER OF PARLIAMENT R.I.
Wisma DPR F,440 Kalibata
Jakarta, Indonesia
T: 62 21 799-8764

Prawiroatmodjo, Suryo W., Mr.
Leader
ENVIRONMENTAL EDUCATION CENTER
P.O. Box 03 Trawa
Mojorkerto-East Java 61375, Indonesia
T: 62 31 44806 F: 62 31 332537

Silas, Johan, Mr.
Lecturer
INSTITUT TEKNOLOGI 10 NOP.
Jalan Doho, 47
Surabaya-East Jaua 60265, Indonesia
T: 62 31 578157 F: 62 31 578157

Soerjani, Mohamad, Mr.
Prof
UNIVERSITY OF INDONESIA
Jalan Salemba 4
Jakarta 10430, Indonesia
T: 62 21 335929 F: 62 21 330266

Suharta, Herliyani, Ms.
Engineer/Master
INDONESIA ENERGY TECHNOLOGY LABORATORY
Perum Puspiptek Blok 4 H3, Serpong
Tangerang 15310, Indonesia
T: 62 21 756-0904 F: 62 21 756-0914
TL: 61331 BPPT IA

Sutopo, Susilastuti, Ms.
INDONESIAN WOMEN CONGRESS
Kompe Polri C-1 16 Gatot Moroso
Jakarta 12930, Indonesia
T: 62 21 516431 F: 62 21 584894

Wijaya, Hesti, Ms.
Ph.D
RURAL DEVELOPMENT FOUND./WOMEN STUDIES GROUP, BRAWIJAYA UNIVERSITY
Jl Wisma Nenanggal 11/49
Surabaya 60234, Indonesia
T: 62 31 832323 F: 62 31 573537

Zaidi, Zaim, Mr.
SKEPI
Palais de Nations
Geneve CH 1211, Indonesia
F: 41 22 733-6542

Zulkarnaen, Abdul Garuda Nusantara, Mr.
WALHI
Palais de Nations
Geneve CH 1211, Indonesia
F: 41 22 733-6542

PARLIAMENTARIANS / SPIRITUAL LEADERS

Isnomo, Susilastuti Sutopo, Mrs
Dewan Perwakilan Rakyat Republik Indonesia Jalan Janderal Gatot Subroto
Jakata, Indonesia
T: 5801324

Mokoginta, Jan, Mr
MP
Dewan Perwakilan Rakyat Republik Indonesia Jalan Janderal Gatot Subroto
Jakata, Indonesia
T: 5801324

Naro, Jailani, Dr
Vice Speaker
House of Representatives
Dewan Perwakilan Rakyat Republik Indonesia Jalan Janderal Gatot Subroto
Jakata, Indonesia
T: 5801324

IRAQ - GOVERNMENT

Abd, Salah Abdul-Rahman, Mr.
Third Secretary
Ministry of Foreign Affairs
Permanent Mission to the United Nations, 14 E. 79th St. New York, NY, 10021
F: 212 737 4433

Al-Anbari, Abd al-Amir, H.E. Dr.
Ambassador Extraordinary and Plenipotentiary
Permanent Representative to the United Nations, New York (Deputy Head of Delegation)
Permanent Mission to the United Nations, 14 E. 79th St. New York, NY, 10021
F: 212 737 4433

Al-Mukhtar, Qais Tawfiq, H.E. Mr.
Ambassador
Federative Republic to Brazil
Permanent Mission to the United Nations, 14 E. 79th St. New York, NY, 10021
F: 212 737 4433

Alanbari, Abidil-Amir, Dr.
Ambassador
Permanent Representative of Iraq to the United Nations
14 E 79TH Street
New York, NY 10021, United States of America
T: 212 7374433

Aziz, H.E. Taniq, Mr
Deputy Prime Minister Of Iraq (head
Baghdad
Iraq

Aziz, Tariq, H.E. Mr.
Deputy Prime Minister
Republic of Iraq
14 E. 79th St.
New York, NY 10021, United States of America
T: 212 737 4433

Ghadhban, Thamir Abas, Mr.
General Director of Studies
Ministry of Oil
Permanent Mission to the United Nations, 14 E. 79th St. New York, NY, 10021
F: 212 737 4433

Mohammad, Fares Ali, Mr.
Third secretary
Embassy to the Federative Republic of Brazil
Permanent Mission to the United Nations, 14 E. 79th St. New York, NY, 10021
F: 212 737 4433

Polus, Samir A.M., Mr.
Secretary to the Deputy Prime Minister
Permanent Mission of Iraq
14 E. 79th St.
New York, NY 10021, United States of America
T: 212 737 4433

Shakir, Qasim Abdulbaqi, Mr.
Counsellor
Permanent Mission of Iraq
14 E. 79th St.
New York, NY 10021, United States of America
T: 212 737 4433

GLOBAL FORUM

Al-Khateeb, Muhyi, Mr.
Ambassador
INTERNATIONAL UNION OF YOUNG CHRISTIAN DEMOCRATS - UIJDC
Caracas, Iraq
F: 58 2 2610547

IRELAND - GOVERNMENT

Burke, John O., Mr.
Counsellor
Deputy Permanent Representative to the United Nations, New York
Permanent Mission to the United Nations, One Dag Hammarskjold Plaza, 19th Floor
New York, NY 10017, United States of America
T: 212 421 6934 F: 212 223 0926

Clifford, Mary, Ms.
Assistant Principal
Department of Energy
Permanent Mission to the United Nations, One Dag Hammarskjold Plaza, 19th Floor New York, NY, 10017
F: 212 421 6934
TL: 212 223 0926

Connolly, Una, Mr.
Secretary to the Women's Committee
Irish Parliament
Permanent Mission to the United Nations, One Dag Hammarskjold Plaza, 19th Floor New York, NY, 10017
F: 212 421 6934
TL: 212 223 0926

Crean, Paul, Mr.
Honorary Consul-General of Ireland
Rio de Janeiro
Permanent Mission to the United Nations, One Dag Hammarskjold Plaza, 19th Floor New York, NY, 10017
F: 212 421 6934
TL: 212 223 0926

Daly, T.D., Brendan, H.E. Mr.
Minister of State
Department of Foreign Affairs
Permanent Mission to the United Nations, One Dag Hammarskjold Plaza, 19th Floor New York, NY, 10017
F: 212 421 6934
TL: 212 223 0926

De Valera, Sile, Ms., T.D.
Member of the Irish Parliament
Permanent Mission of Ireland
One Dag Hammarskjold Plaza, 19th Floor
New York, NY 10017, United States of America
T: 212 421 6934 F: 212 223 0926

Dirr, Noel, Mr.
Secretary-General
Department of Foreign Affairs
Permanent Mission to the United Nations, One Dag Hammarskjold Plaza, 19th Floor New York, NY, 10017
F: 212 421 6934
TL: 212 223 0926

Duignan, Sean, Mr.
Government Press Secretary
Department of the Taoiseach (Prime Minister)
Permanent Mission to the United Nations, One Dag Hammarskjold Plaza, 19th Floor New York, NY, 10017
F: 212 421 6934
TL: 212 223 0926

Gallan, Niall, Mr.
Assistant Secretary-General
Department of the Environment
Permanent Mission to the United Nations, One Dag Hammarskjold Plaza, 19th Floor New York, NY, 10017
F: 212 421 6934
TL: 212 223 0926

Hanberry, Bertie, Mr.
Private Secretary to the Minister of State
Department of Foreign Affairs
Permanent Mission to the United Nations, One Dag Hammarskjold Plaza, 19th Floor New York, NY, 10017
F: 212 421 6934
TL: 212 223 0926

Harney, T.D., Mary, H.E. Ms.
Minister of State
Department of Environment, Deputy Head of Delegation
Permanent Mission to the United Nations, One Dag Hammarskjold Plaza, 19th Floor New York, NY, 10017
F: 212 421 6934
TL: 212 223 0926

Hayes, Francis Mahon, H.E. Mr.
Ambassador Extraordinary and Plenipotentiary
Permanente Representative to the United Nation, New York
Permanent Mission to the United Nations, One Dag Hammarskjold Plaza, 19th Floor
New York, NY 10017, United States of America
T: 212 421 6934 F: 212 223 0926

Hoey, Michael, Mr.
Counsellor
Department of Foreign Affairs
Permanent Mission to the United Nations, One Dag Hammarskjold Plaza, 19th Floor New York, NY, 10017
F: 212 421 6934
TL: 212 223 0926

Howlin, Brendan, Mr., T.D.
Member of the Irish Parliament
Permanent Mission of Ireland
One Dag Hammarskjold Plaza, 19th Floor
New York, NY 10017, United States of America
T: 212 421 6934 F: 212 223 0926

Kinch, Aidan, Mr.
Private Secretary to the Minister of State
Department of the Environment
Permanent Mission to the United Nations, One Dag Hammarskjold Plaza, 19th Floor New York, NY, 10017
F: 212 421 6934
TL: 212 223 0926

Kirwan, Betty, Ms.
Clerical Assistant
Department of Foreign Affairs
Permanent Mission to the United Nations, One Dag Hammarskjold Plaza, 19th Floor New York, NY, 10017
F: 212 421 6934
TL: 212 223 0926

McCarthy, Brian, Mr.
Principal Officer
Department of the Taoiseach (Prime Minister)
Permanent Mission to the United Nations, One Dag Hammarskjold Plaza, 19th Floor New York, NY, 10017
F: 212 421 6934
TL: 212 223 0926

McGuinness, Ann, Ms.
Assistant Principal
Department of the Environment
Permanent Mission to the United Nations, One Dag Hammarskjold Plaza, 19th Floor New York, NY, 10017
F: 212 421 6934
TL: 212 223 0926

McNamara, Brendan, Mr.
Third Secretary
Department of Foreign Affairs
Permanent Mission to the United Nations, One Dag Hammarskjold Plaza, 19th Floor New York, NY, 10017
F: 212 421 6934
TL: 212 223 0926

Morgan, Donagh, Mr.
Private Secretary to the Taoiseach (Prime Minister)
Permanent Mission of Ireland
One Dag Hammarskjold Plaza, 19th Floor
New York, NY 10017, United States of America
T: 212 421 6934 F: 212 223 0926

Moylan, Mary, Ms.
Principal Officer
Department of the Environment
Permanent Mission to the United Nations, One Dag Hammarskjold Plaza, 19th Floor New York, NY, 10017
F: 212 421 6934
TL: 212 223 0926

Murphy, Ronan, Mr.
Assistant Secretary-General
Department of Foreign Affairs
Permanent Mission to the United Nations, One Dag Hammarskjold Plaza, 19th Floor New York, NY, 10017
F: 212 421 6934
TL: 212 223 0926

Nally, Dermot, Mr.
Secretary to the Government
Permanent Mission of Ireland
One Dag Hammarskjold Plaza, 19th Floor
New York, NY 10017, United States of America
T: 212 421 6934 F: 212 223 0926

O' Donoghue, Brendan, Mr.
Secretary-General
Department of the Environment
Permanent Mission to the United Nations, One Dag Hammarskjold Plaza, 19th Floor New York, NY, 10017
F: 212 421 6934
TL: 212 223 0926

Pinto Vilella, Neyda, Mrs.
Assistant to the Honorary Consul-General
Permanent Mission of Ireland
One Dag Hammarskjold Plaza, 19th Floor
New York, NY 10017, United States of America
T: 212 421 6934 F: 212 223 0926

Reynolds, Albert, Mr.
Prime Minister
Permanent Mission of Ireland
One Dag Hammarskjold Plaza, 19th floor
New York NY 10017, United States of America
T: 212 421 6934 F: 212 223 0926

Reynolds, T.D. Taoiseach, Albert, H.E. Mr.
Prime Minister
Ireland
One Dag Hammarskjold Plaza, 19th Floor
New York, NY 10017, United States of America
T: 212 421 6934 F: 212 223 0926

Rice, Gerry, Mr.
Private Secretary to the Minister for the Environment
Permanent Mission of Ireland
One Dag Hammarskjold Plaza, 19th Floor
New York, NY 10017, United States of America
T: 212 421 6934 F: 212 223 0926

Sartori, Marcia, Ms.
Assistant to the Honorary Consul-General
Permanent Mission of Ireland
One Dag Hammarskjold Plaza, 19th Floor
New York, NY 10017, United States of America
T: 212 421 6934 F: 212 223 0926

Smith, T.D., Michael, H.E. Mr.
Minister for the Environment, Deputy Head of Delegation
Permanent Mission of Ireland
One Dag Hammarskjold Plaza, 19th Floor
New York, NY 10017, United States of America
T: 212 421 6934 F: 212 223 0926

Thompson, June, Ms.
Assistant Principal
Wildlife Service, Office of Public Works
Permanent Mission to the United Nations, One Dag Hammarskjold Plaza, 19th Floor New York, NY, 10017
F: 212 421 6934
TL: 212 223 0926

Valentim, Christina, Mrs.
Assistant to the Honorary Consul-General
Permanent Mission of Ireland
One Dag Hammarskjold Plaza, 19th Floor
New York, NY 10017, United States of America
T: 212 421 6934 F: 212 223 0926

Wallace, Mary, Ms., T.D.
Member of the Irish Parliament
Permanent Mission of Ireland
One Dag Hammarskjold Plaza, 19th Floor
New York, NY 10017, United States of America
T: 212 421 6934 F: 212 223 0926

NGO

Brady, Martin, Mr.
IRELAND'S NATIONAL YOUTH ENVIRONMENTAL ORGANIZATION
11 Cope Street
Dublin 2, Ireland
T: 353 1 679 9673 F: 353 1 711 002

PRESS

Higgins, Michael D.
Reporter Radio/Television
OCEAN FILM PRODUCTIONS
Circular Road, Rahoon
Galway, Ireland, Ireland

McDonald, Frank
Correspondent
IRISH TIMES
6 Casimir Road, Harold's Cross
Dublin 6W, Ireland, Ireland

Ryan, Michael
Correspondent
IRISH NATIONAL RADIO & TV
Killowen, New Ross
Co. Wexford, Ireland, Ireland

Stapleton, Francis
Editor
OCEAN FILM PRODUCTIONS
45 Belgrave Square, Monkstown
Co. Dublin, Ireland, Ireland

Tiernan, Catherine
Editor
OCEAN FILM PRODUCTIONS
49 Larchfield Road, Goatstown
Dublin 14, Ireland, Ireland

Wegimont, Liam, Mr.
Reporter Radio/Television
RTE
31 Turnberry Baldoyle
Dublin 13 Ireland, Ireland

GLOBAL FORUM

Cabot, Penny, Ms.
BANULACHT - WOMEN FOR DEVELOPMENT
I Upper Camden St.
Dublin 2, Ireland

Cahill, Enda
OCEAN FILM PRODUCTIONS
Dublin, Ireland

Cantwell, Aengus, Mr.
Director
WORLD EDUCATION PROJECT
St. Angela's College of Education , Nat'l Univ of Ireland Lough Gill
Sligo, Ireland
T: 35 71 45485 F: 35 71 5

Dickson, Audrey, Ms.
IRISH WOMENS ENVIRONMENTAL NETWORK
3 Asgard Rd., Houth
Dublin, Ireland
T: 353 1 754162

Dixon, Audrey, Ms.
IRISH WOMENS ENVIRONMENTAL NETWORK
3 Asgard Rd., Howth, Co.
Dublin, Ireland
T: 353 1 322923

Dolan, Tony, Mr.
COMHLAMH / TULLAMORE ONE WORLD GROUP
One World Off., Bridge St.
Tullamore-Co.Offaly, Ireland
T: 353 506 5236173 F: 353 506

Eccles, Pauline, Ms.
IRISH COMMISSION FOR JUSTICE & PEACE
109 Booterstown Avenue, Co. Dublin
Dublin, Ireland
T: 353 1 2884 853

Humphries, Mary, Ms.
Head Of Education
CONCERN
I Upper Camden St.
Dublin 2, Ireland
T: 353 1 754162

Keady, William
OCEAN FILM PRODUCTIONS
Dublin, Ireland

McGonigle, Bridget, Ms.
DONEGAL ONE WORLD GROUP
Youth Off., Market Place
Letterkenny-Co.Done., Ireland
T: 353 74 22133

O Gorman, Fergus, Mr.
President
IRISH WILDLIFE FEDERATION
132A East Wall Road
Dublin 3, Ireland

O'Donoghue, John, Mr.
38 College Ave.
Clonnel - Tipperary, Ireland

O'Sullivan, Maeve, Ms.
Spokesperson
EARTHWATCH
Harbour View
Bantry-County Cork, Ireland
T: 353 27 50968 F: 353 27 50545

Riada, Liadh Ni
OCEAN FILM PRODUCTIONS
Dublin, Ireland

Thoele, Cecilia, Ms.
SLIGO WOMEN'S GROUP NGO
Hazelwood
Sligo, Ireland

Walker, Leona, Ms.
EARTHWATCH
10 hillbrook dive drogueda
Louth, Ireland
T: 04138197 F: 04138580

Walsh, Geraldine, Ms.
AN TAISCE - NATIONAL TRUST FOR IRELAND
17 Duke St.,
Dublin 2, Ireland
T: 353 1 6797 558

PARLIAMENTARIANS / SPIRITUAL LEADERS

Barnes, Monica, Ms., T.D.
Member of the Irish Parliament
Permanent Mission of Ireland
One Dag Hammarskjold Plaza, 19th Floor
New York, NY 10017, United States of America
T: 212 421 6934 F: 212 223 0926

Carey, Donal
MP
Mildere Street
Dublin 2, Ireland
T: 789911
TL: 24377 Da11 Ei

Devalera, Cile
MP
Mildere Street
Dublin 2, Ireland
T: 789911
TL: 24377 Da11 Ei

Houlin, Brendan
MP
Mildere Street
Dublin 2, Ireland
T: 789911
TL: 24377 Da11 Ei

ISLAMIC REPUBLIC OF IRAN - GOVERNMENT

Ghasemzadeh, Davood, Dr
Head, Energy Section, Opec
NATIONAL IRANIAN OIL CO
P.O. Box 1863 Taleghani Ave
Tehran, Islamic Republic of Iran
T: 666961 6152599 61528 F: 666961

Habibi, Hassan, Mr.
First Vice President
Permanent Mission of the Islamic Republic of Iran
622 3rd Ave, 34th floor
New York NY 10017, United States of America
T: 212 687 2020 F: 212 867 7086

Rasale, Hassan
NIOC
P.O. Box 1863 Tehran-Iran
Tehran, Islamic Republic of Iran
T: 6152965 F: 2121771212248N10C1R

PRESS

Hossein, Hajipour
Reporter Radio/Television
IRIB
33 Ofogh St.
Tebran, Islamic Republic of Iran

Yahya, Ashjazadeh
Reporter Radio/Television
IRIB
No,114 Saman St.
Tehran, Islamic Republic of Iran

GLOBAL FORUM

Taravati, Hamid, Mr.
M.D.
WORLDWATCH
Noia, Pasteur 161, Mashad, Iran
khorasan, Islamic Republic of Iran
T: 5 71-1942 F: 5 71-1896
TL: 51-2479

ISRAEL - GOVERNMENT

Eirati, David, H.E. Mr.
Ambassador to the Federative Republic of Brazil
Permanent Mission of Israel
800 Second Avenue
New York, NY 10017, United States of America
T: 351 5200 F: 697 6272

Gal-on, Azriel, Mr.
Consul-General
Consulat General, Rio de Janeiro
Permanent Mission to the United Nations, 800 Second Avenue
New York, NY 10017, United States of America
T: 351 5200 F: 697 6272

Loya, Yosef, Prof.
Dean
Faculty of Life Sciences, Tel-Aviv University
Permanent Mission to the United Nations, 800 Second Avenue
New York, NY 10017, United States of America
T: 351 5200 F: 697 6272

Marinov, Uri, H.E. Dr.
Director-General
Ministry of Environment (Head of Delegation)
Permanent Mission to the United Nations, 800 Second Avenue
New York, NY 10017, United States of America
T: 351 5200 F: 697 6272

Oron, Benjamin, Mr.
Director
Division for Political Affairs at the United Nations, Ministry of Foreign Affairs
Permanent Mission to the United Nations, 800 Second Avenue
New York, NY 10017, United States of America
T: 351 5200 F: 697 6272

Pruginin, Amram, Dr.
Deputy Director-General
Ministry of Environment
Permanent Mission to the United Nations, 800 Second Avenue
New York, NY 10017, United States of America
T: 351 5200 F: 697 6272

Schmorak, Dov, Dr.
Deputy Director-General
Ministry of Foreign Affairs
Permanent Mission to the United Nations, 800 Second Avenue
New York, NY 10017, United States of America
T: 351 5200 F: 697 6272

Tenne, Arie, Mr.
Minister-Plenipotentiary
Permanent Mission of Israel
800 Second Avenue
New York, NY 10017, United States of America
T: 351 5200 F: 697 6272

NGO

Katzab, Nuzhat
HECPA
93 Arlozorov St.
Tel Aviv 62098, Israel

Paz, Yehudah
IILD
7 Nehardea Street, P.O. Box 16201
Tel Aviv 64235, Israel

Szir, Igal, Mr.
Representative
HISTADRUTH
Arlozorov 93
Tel-Aviv, Israel
T: 54 1 353550 F: 54 1 3343358

GLOBAL FORUM

Ambra, Abdel-Rahman, Mr.
Engineer
LAND AND WATER ESTABLISHMENT
PO Box 20873
Jerusalem, Israel
T: 972 2 812-364 F: 972 2 812-364

Brill, Peggy, Ms.
ISRAEL WOMEN'S NETWORK / SOCIETY FOR THE PROTECTION OF NATURE IN ISRAEL
1 Caspi Street
Jerusalem 93554, Israel
T: 972 2 735-286

Horowitz, Nitzan
Foreign Affairs Editor
HA'ARETZ /IDF RADIO
23 Shoken St
Tel-Aviv, Israel
T: 972 3 5121212 F: 972 3 810012

Manrike, Jose, Sr.
Fundador E Diretor Presidente
RENASKIGI
Caixa Postal 106.051
Niteroi - RJ 24232, Israel
T: 55 21 711-5251 F: 55 21 616-1355

Shkirat, Khader, Mr.
Advocoy Organization
LAND AND WATER ESTABLISMENT
PO. Box 20.873
Jerusalem, Israel
T: 972 2 81-2364 F: 972 2 81-2364

Tabor, Harry, Mr.
Sci. Director
SCIENTIFIC RESEARCHER FOUNDATION
Box 3745
Jerusalem 91036, Israel
T: 972 2 437470 F: 972 2 513984

Tabor, Vivienne, Ms.
SCIENTIFIC RESEARCHER FOUNDATION
Box 3745
Jerusalem 91036, Israel
T: 972 2 437470 F: 972 2 513984

Zvirin, Ilana, Ms.
ISES - ISRAEL
Mechanical Eng. Technion, I.I.I.
Haifa, Israel
T: 972 4 4-292070 F: 972 4 4-324533

Zvirin, Yoram, Mr.
Professor
ISES - ISRAEL
Mechanical Eng. Technion, I.I.I.
Haifa, Israel
T: 972 4 4-292070 F: 972 4 4-324533

PARLIAMENTARIANS / SPIRITUAL LEADERS

Kelman, Levi, Rabbi
Congregation Kol Haneshama, 57 Harakavet
Jerusalem, Israel

ITALY - GOVERNMENT

Ahmed, Salahuddin, Mr.
Deputy Executive Director
OED/WORLD FOOD PROGRAMME-UN
Via Cristoforo Colombo 426 00145
Rome, Italy
T: 5797 3651/5797 6334 F: 5797 5652

Alessandrini, Alessandro, Mr.
Ministry of Forestry and Agriculture
Permanent Mission to the United Nations, United Nations Plaza New York, NY, 10017
F: 212 963 7174
TL: 212 963 1921

Ambassadeur, Julio Cesar, Lupinacci
Institut Italo-Latino Americain
PIAZZA GUGLIELMO MARCONI,26
Rome, Italy
T: 59097 F: 5974923

Benassai, Sergio
Committee Of Experts On The Transpo
UN-ECON COMM FOR EUROPE
Enea/Disp V. Vitaliano Brancati 48
Rome, Italy
T: 06 5007 2065 F: 06 5007 2807

Bologna, Gianfranco, Mr.
World Wildlife Fund (Italy)
Permanent Mission to the United Nations, United Nations Plaza New York, NY, 10017
F: 212 963 7174
TL: 212 963 1921

Bramati, Luciano
Deputy Manager Environmental Group
ENTE NAZIONALE POR L'ENERGIA ELETRICA
L. Bramati - Enel Groppo Amriente Via Panama 12
Roma CAP001981, Italy
T: 06 85032120 F: 06 85035860

Bresso, Mercedes, Prof.
President
Environment and Work Association
Corso Marelli 497
Milano 20099, Italy
T: 39 2 26223120 F: 39 2 26223130

Cagliari, Gabriele, Mr.
President
National Hydrocarbon Agency
Permanent Mission to the United Nations, United Nations Plaza New York, NY, 10017
F: 212 963 7174
TL: 212 963 1921

Caglioti, Luciano, Professor
Adviser
PROFESSOR ORGANIC CHEMISTRY UNIV RO
Via 9.9. Porro 22 00197
Roma, Italy
T: 39 6 4462729 39 2 62 F: 39 2 62705055

Cassini, Giuseppe, Mr.
Minister
Diplomatic Adviser
Ministry of Environment
T: 10017
TL: 212 963 7174, E: 212 963 1921

Catalano, Antonio, Mr.
Minister
Ministry of Foreign Affairs
Permanent Mission to the United Nations, United Nations Plaza New York, NY, 10017
F: 212 963 7174
TL: 212 963 1921

Cecchetti, Gaetano, Mr.
National Hydrocarbon Agency
Permanent Mission to the United Nations, United Nations Plaza New York, NY, 10017
F: 212 963 7174
TL: 212 963 1921

Clini, Corrado, Mr.
General Director
Ministry of Environment
P.Venezia
Rome 00184, Italy
T: 39 6 7027349 F: 39 6 7027195

Cogliandro, Giuseppe, Mr.
Counsellor, Chief of Cabinet
Ministry of Environment
Permanent Mission to the United Nations, United Nations Plaza New York, NY, 10017
F: 212 963 7174
TL: 212 963 1921

Colella, Alberto, Dr.
Permanent Mission of Italy
United Nations Plaza
New York, NY 10017, United States of America
T: 212 963 7174 F: 212 963 1921

Corrias, Alberto, Mr.
Institute for Industrial Reconstruction
Permanent Mission to the United Nations, United Nations Plaza New York, NY, 10017
F: 212 963 7174
TL: 212 963 1921

Costa, Fabrizio, Mr.
Ministry of Treasury
Permanent Mission to the United Nations, United Nations Plaza New York, NY, 10017
F: 212 963 7174
TL: 212 963 1921

Crawshaw, Bruce, Mr
Policy Analist
OED/WORLD FOOD PROGRAMME-UN
Via Cristoforo Colombo 426 00145
Rome, Italy
T: 5797 3651/5797 6334 F: 5797 5652

De Michelis, Gianni, H.E.
Minister for Foreign Affairs (Head of Delegation)
Permanent Mission of Italy
United Nations Plaza
New York, NY 10017, United States of America
T: 212 963 7174 F: 212 963 1921

Di Palma, Giovannino, Mr.
Ministry of Industry and Commerce
Permanent Mission to the United Nations, United Nations Plaza New York, NY, 10017
F: 212 963 7174
TL: 212 963 1921

Diouf, Bouna Semou
Director External Relations Divisio
IFAD / UNITED NATIONS
Via Del Serafico 107
Rome 00142, Italy
T: 39 6 54592252 F: 39 6 5043463

Fondi, Patrizio, Mr.
First Secretary
Permanent Mission of Italy
United Nations Plaza
New York, NY 10017, United States of America
T: 212 963 7174 F: 212 963 1921

Franceschi, Roberto
Ambassadeur
INSTITUT ITALO-LATINO AMERICAIN
Piazza Guglielmo Marconi,26
Rome, Italy
T: 59091 F: 5974923

Garaguso, Giulio Cesare, H.E. Mr.
Ambassador, Special Adviser
Ministry of Environment
Permanent Mission to the United Nations, United Nations Plaza New York, NY, 10017
F: 212 963 7174
TL: 212 963 1921

Giacomelli, Michele, Mr.
First Secretary
Embassy of Italy to Brazil
Permanent Mission to the United Nations, United Nations Plaza New York, NY, 10017
F: 212 963 7174
TL: 212 963 1921

Gioia, Giuseppe, Mr.
Italian Agricultural Association
Permanent Mission to the United Nations, United Nations Plaza New York, NY, 10017
F: 212 963 7174
TL: 212 963 1921

Golfari, Cesare, Sen.
Member of Parliament
Permanent Mission to the United Nations, United Nations Plaza New York, NY, 10017
F: 212 963 7174
TL: 212 963 1921

Illomei, Saturno, Mr.
Head of Press Office
Ministry of Environment
Permanent Mission to the United Nations, United Nations Plaza New York, NY, 10017
F: 212 963 7174
TL: 212 963 1921

Jacoangeli, Giuseppe, H.E. Mr.
Ambassador
Ministry of Foreign Affairs
Permanent Mission to the United Nations, United Nations Plaza New York, NY, 10017
F: 212 963 7174
TL: 212 963 1921

Le Boffe, Maurilio, Mr.
Italian Industrial Association
Permanent Mission to the United Nations, United Nations Plaza New York, NY, 10017
F: 212 963 7174
TL: 212 963 1921

Malerba, Angelo, Mr.
Ministry of Budget and Financial Planning
Permanent Mission to the United Nations, United Nations Plaza New York, NY, 10017
F: 212 963 7174
TL: 212 963 1921

Manos, Aldo, Mr.
Permanent Mission of Italy
United Nations Plaza
New York, NY 10017, United States of America
T: 212 963 7174 F: 212 963 1921

Marchitelli, Patrizia, Ms.
Permanent Mission of Italy
United Nations Plaza
New York, NY 10017, United States of America
T: 212 963 7174 F: 212 963 1921

Maroni, Roberto, Hon.
Member of Parliament
Permanent Mission to the United Nations, United Nations Plaza New York, NY, 10017
F: 212 963 7174
TL: 212 963 1921

Masci, Pietro, Mr.
Director General
Ministry of Treasury
Permanent Mission to the United Nations, United Nations Plaza New York, NY, 10017
F: 212 963 7174
TL: 212 963 1921

Mattioli, Gianni, Hon.
Member of Parliament
Permanent Mission to the United Nations, United Nations Plaza New York, NY, 10017
F: 212 963 7174
TL: 212 963 1921

Maung, Marlar Sein, Miss
Second Secretary
EMBASSY OF THE UNION OF MYANMAR
Via Vincenzo Bellini, 20, Interno 1
Rome, 00198, Italy
T: 8549374 F: 8413167

Mauro, Francesco, Mr.
National Agency for the Alternative Energies
Permanent Mission to the United Nations, United Nations Plaza New York, NY, 10017
F: 212 963 7174
TL: 212 963 1921

McCusker, Alison, Ms
Deputy Director, Research
CGIAR/INTERNATIONAL BOARD FOR PLANT GENETIC RESOURCES
Via Delle Sette Chiese 142 Box 793 UIDDEW 2606
Rome,Italy 00145, Italy
T: 39 6 574 4719; 61-6 F: 39 6 575 0309

Mongelli, Nicola, Mr.
National Hydrocarbon Agency
Permanent Mission to the United Nations, United Nations Plaza New York, NY, 10017
F: 212 963 7174
TL: 212 963 1921

Ondarsa Linares, Franz
Ambassadeur
INSTITUT IIALO-LATINO AMERICAIN
Piazza Guglielmo Marconi,26
Rome, Italy
T: 59091 F: 5914923

Ortiz, Fernando
Secretaire Tecnico-Scentifihue
INSTITUT ITALO-LATINO AMERICAIN
Piazza Guglielmo Marconi,26
Rome, Italy
T: 59091 F: 5914923

Passanti, Paolo, Mr.
Vice-President
Confindustria
Permanent Mission to the United Nations, United Nations Plaza New York, NY, 10017
F: 212 963 7174
TL: 212 963 1921

Pera, Costanza, Ms.
Director General
Ministry of Environment
Permanent Mission to the United Nations, United Nations Plaza New York, NY, 10017
F: 212 963 7174
TL: 212 963 1921

Pettinari, Paola, Ms.
Ministry of Treasury
Permanent Mission to the United Nations, United Nations Plaza New York, NY, 10017
F: 212 963 7174
TL: 212 963 1921

Pinchera, Giancarlo, Mr.
National Agency for the Alternative Energies
Permanent Mission to the United Nations, United Nations Plaza New York, NY, 10017
F: 212 963 7174
TL: 212 963 1921

Pinchera, Giovanni Carlo, Mr
Senior Adviser For Energy And Envir
ENEA
Viale Regina Margherita 125
Roma 00198, Italy
T: 39 6 8528 2426 F: 39 6 8528 2302

Pratesi, Fulco, Hon.
Permanent Mission of Italy
United Nations Plaza
New York, NY 10017, United States of America
T: 212 963 7174 F: 212 963 1921

Ragno, Aldea, Ms.
Permanent Mission of Italy
United Nations Plaza
New York, NY 10017, United States of America
T: 212 963 7174 F: 212 963 1921

Reitano, Giorgio, H.E. Mr.
Ambassador
Ministry of Foreign Affairs
Permanent Mission to the United Nations, United Nations Plaza New York, NY, 10017
F: 212 963 7174
TL: 212 963 1921

Rondo-Filho, Milton
Secretaire Socio Economirue
INSTITUT ITALO-LATINO AMERICAIN
Piazza Guglielmo Marconi 26
Rome, Italy
T: 5909462 F: 5914923

Ruffolo, Giorgio, H.E.
Minister for Environment
Permanent Mission of Italy
United Nations Plaza
New York, NY 10017, United States of America
T: 212 963 7174 F: 212 963 1921

Salvatici, Silvano, Mr.
Ministry of Forestry and Agriculture
Permanent Mission to the United Nations, United Nations Plaza New York, NY, 10017
F: 212 963 7174
TL: 212 963 1921

Serpi, Paolo, Mr.
First Secretary
Permanent Mission of Italy
United Nations Plaza
New York, NY 10017, United States of America
T: 212 963 7174 F: 212 963 1921

Sesenna, Fernando
Ambassadeur
INSTITUT ITALO-LATINO AMERICAIN
Piazza Guglielmo Marconi,26
Rome, Italy
T: 59091 F: 5914923

Smeraldi, Roberto, Mr.
International NGO Association
Permanent Mission to the United Nations, United Nations Plaza New York, NY, 10017
F: 212 963 7174
TL: 212 963 1921

Tarelli, Antonio, Mr.
Minister
Ministry of Foreign Affairs
Permanent Mission to the United Nations, United Nations Plaza New York, NY, 10017
F: 212 963 7174
TL: 212 963 1921

Tarony, Paolo, H.E. Mr.
Ambassador
Federative Republic of Brazil
Permanent Mission to the United Nations, United Nations Plaza New York, NY, 10017
F: 212 963 7174
TL: 212 963 1921

Traxler, Vieri, H.E. Mr.
Ambassador Extraordinary and Plenipotenciary
Permanent Representative to the United Nations, New York
Permanent Mission to the United Nations, United Nations Plaza
New York, NY 10017, United States of America
T: 212 963 7174 F: 212 963 1921

Villa, Lorenzo, Prof.
Institute of Health
Permanent Mission to the United Nations, United Nations Plaza New York, NY, 10017
F: 212 963 7174
TL: 212 963 1921

Yanga, Thomas, Mr
Alternate Permanent Representative
CAMEROON EMBASSY
Via Siracusa, 6
00161 Rome, Italy
T: 00 396 8554173 85534 F: 00 396 8553060

NGO

Adalberto, Vallega
ENTE
Via Sottoripa 5, Palazzo Serra Gerace
16123 Genova, Italy

Agbedjro, Pierre, Mr.
Assistant
THIRD WORLD ACADEMY OF SCIENCES (TWAS)
c/o Intl Centre for Theoretical Physics - Strada Costiera 11 - P,O,BOX 586
Trieste 34100, Italy
T: 39 40 2240-251 F: 39 40 224-559
TL: 460392 ICTP

Baker, Christoph, Mr.
CAMPAGNA NORD-SUD
V.S. Maria Dell Anima, 30
Rome 00186, Italy
T: 39 6 6832726 F: 39 6 686584

Beaudouezy, Jean Philippe
EC
Via Fusetti 14
20143 Milano, Italy

Bernadi, Lelio
ICE
Corte Suprema di Cassazione, Palazzo di Giustizia
Palazzo Cavour, 1. 00193 Roma, Italy

Bertarelli, Adriana, Ms.
Marketing & Promotions
SOCIETY FOR INTERNATIONAL DEVELOPMENT
Palazzo Civita' del Lavoro - 00144 EUR
Rome, Italy
T: 396 591-7897 F: 396 591-9836
TL: 616484 SID I

Biagini, Bonizella, Ms.
LEGA PER L'AMBIENTE
Via Salaria 280
Rome, Italy
T: 39 6 8841552 F: 39 6 8443505
E: 65:Eco004

Bohn, Marino
CAI
Palazzo San Calisto
V-00120 Vatican City, Italy

Cassani, Robert, Mr.
Executive Director
SOCIETY FOR INTERNATIONAL DEVELOPMENT
Palazzo Civilta del Lavoro
Rome 00144, Italy
T: 39 6 5917897 F: 39 6 5919836
TL: 616484 SID I

Corrado Maria, Dacion
ECES
Via Po 14
10123 Torino, Italy

Cremer, Georg
CAI
Palazzo San Calisto
V-00120 Vatican City, Italy

De C. Godinho, Rui M.
CEMR
Palazzo Cavalli 4090, San Marco
Venezia, Italy

Eccles, Pauline
WIDE
c/o SID, Palazzo Civilta del Lavoro
00144 EUR, Roma, Italy

El-Baz, Farouk
THIRD WORLD ACADEMY OF SCIENCES (TWAS)
c/o Intl Centre for Theoretical Physics - Strada Costiera 11 - P,O,BOX 586
Triestre 34100, Italy
T: 39 40 2240-251 F: 39 40 224-559
TL: 460392 ICTP

Foca, Anna
WIDE
c/o SID, Palazzo Civilta del Lavoro
00144 EUR, Roma, Italy

Gregory, Affonso Felippe
CAI
Palazzo San Calisto
V-00120 Vatican City, Italy

Guttieres, Mario
President
INT'L JURIDICAL ORGANIZATION FOR ENVIRONMENT & DEVELOPMENT
Via Barberni 3
Rome 00187, Italy
T: 39 6 742117 F: 39 6 4745779

Helandri, Giovanna
LA
Via Salaria, 280
00199 Roma, Italy

Langer, Alexander, Mr.
CNS
Via S. Maria dell'Anima
30-00186 Rome, Italy

Nico, Vassalo
ECES
Via Po 14
10123 Torino, Italy

Ostyn, Celine
WIDE - WALING GROUP ON ENVIRONMENT AND DEVELOPMENT
c/o SID, Palazzo Civilta del Lavoro
00144 EUR, Roma, Italy

Pansa, Angelo
Pesquisador
LELIO BASSO INTERNATIONAL FOUNDTION
Via della Dogana Vecchia, 5
00186 Roma, Italy

Pineschi, Laura
CNPDS
Palazzo Comunale delle Scienze Sociali
Piazza Castello, 3; 20121 Milano, Italy

Poelmann, Boudewijn
The Netherlands
IPS
Via Panisperna 207
Rome, Italy
T: 39 6 485 692 F: 39 6 81-7877

Postiglione, Amadeo, Mr.
President
INTERNATIONAL COURT OF THE ENVIRONEMNT
Court of Cassation - Piazza Cavour
Rome 00193, Italy
T: 39 6 6868597 F: 39 6 6874170

Salam, Abdus, Mr.
President & Nobel Laureate
THIRD WORLD ACADEMY OF SCIENCES (TWAS)
c/o Intl Centre for Theoretical Physics - Strada Costiera 11 - P,O,BOX 586
Triestre 34100, Italy
T: 39 40 2240-251 F: 39 40 224-559
TL: 460392 ICTP

Sebek, Victor
CEMR
Palazzo Cavalli 4090, San Marco
Venezia, Italy

Techera, Jorge
CAI
Palazzo San Calisto
V-00120 Vatican City, Italy

Wacker, Corinne
WIDE
c/o SID, Palazzo Civilta del Lavoro
00144 EUR, Roma, Italy

PRESS

Acacia, Dourado
Editor
GBR TV
Via del Sudario 35
00186 Roma, Italy, Italy

Aleotti, Paolo
Reporter Radio/Television
RAI
1 Irving Place, 24D, Village,
New York, N.Y. 10003, U.S.A., Italy

Alonso, Fernandez Enrique
Editor
IPS
Fernandez Crespo 1988 apt.10
Montevideo, Uruguay, Italy

Anyadike, Obinna
Correspondent
IPS
4 Argyle Court, Lubu Rd.
Lusaka, Zambia, Italy

Avila, Diana
Correspondent
IPS
Liedenhof 211
1108 Hm Amsterda, Z.O. Holland, Italy

Awori, Horace
Kenya
IPS
Via Panisperna 207
Rome, Italy
T: 39 6 485 692 F: 39 6 81-7877

Barbi-Cinti, Giovanni
Journalist
RAI - ITALIAN TELEVISION TG1
Via Teulada, 66
Rome, Italy
T: 39 6 372030 F: 39 6 331919

Barendson, Guido
Editor
TG5 CANALE 5
Via delle Madrelari,28
Roma, Italy, Italy

Bassetti, Annamaria
Editor
CBR TELEVISIONE
Via Panama II
Roma, Italy, Italy

Benozzo, Ricardo
ANSA
Av.Eduardo Madero 9840/942
Buenos Aires, Argentina, Italy

Bernardi, Lelio
Editor
AMBIENTE NEWS
Via Tiburtina, 602
00159, Roma, Italy, Italy

Biancacci, Franco
Correspondent
RAI
9C, Plateau de Frontenex
CH-1205 Geneva, Switzerland, Italy

Blackman, Kenneth
Correspondent
IPS
72 Coolshade Drive, Havendale
Kingston 19, Jamaica, W.I., Italy

Blasi, Marina
Editor
CANALE 5 RTI
Via del Caravaggio
12 Milan, Italy, Italy

Bonasso, Miguel Luis
Editor
IPS
Millet 44/101
Mexico - DF, Mexico

Borioni, Anna
Editor
AS e T RISERCHE e TECNOLOGIA AVANZATE
Via Giovani Lanza, 152
Rome, Italy
T: 39 6 4825319 F: 39 6 4820278

Borri, Maria
Editor
CANALE 5 RTI
Via Sanremo
20 Milano, Italy, Italy

Bortolan, Gerardo
Reporter Radio/Television
RAI-TV TG 3
P. Accademia Antiquaria, 29
Roma, Italy, Italy

Bovina, Livia
Editor
CBR TELEVISIONE
Via Bosco degli Arvali 169
Roma, Italy, Italy

Brener, Jaime
Editor
ANSA
R.Nicolau de Souza Querios811/
131 CEP 04105, Sao Paulo, Brazil, Italy

Brighi, Cecilia, Ms.
Responsabile Ambinte
CONFEDERACIONE ITALIANA SINDACATI LAVORATORI
Via Po 21
Rome 00198, Italy
T: 39 6 8473377 F: 39 6 8546076

Broglia, Andrea
Editor
CANALE 5 RTI
Via Donatello
27 Milano, Italy, Italy

Brum, Antonio
Reporter Radio/Television
RAI
Via Andrea Doria, 48
Roma, Italia, Italy

Cadringher, Manuela
Editor
RAI-TG2
Via Giuseppe Palumbo, 26
00195 Rome, Italy, Italy

Cairone, Andrea
Editor
RAI-TG2
100 W. 57th Street
New York, NY 10019, USA, Italy

Candito, Domenico Mimmo
Correspondent
LA STAMPA
Corso Dante 119
Torino, Italy, Italy

Canihuante, Gabriel
Foreign Correspondent
IPS
R. P. Miguelinho 88, Casa 201
20240-Rio de Janeiro-RJ, Brazil, Italy

Caputi, Claudia, Ms.
Journalist
RAI ITALIAN TV TG1
Via Tasso 428/A
Napoli, Italy, Italy

Carabini, Orazio
Editor
IL MONDO MAGAZINE
Via di Villa Torlomia 6
Rome, Italy, Italy

Carazzolo, Barbara
Editor
FAMIGLIA CRISTIANA
Via Giulio Curioni 99
Roma, Italy, Italy

Casanova, Carlamaria
Correspondent
IL RESTO DEL CARLINO
Via Mario Pagano 69
20145 Milano, Italy, Italy

Castilho, Carlos Albano
Editor
IPS
Calle 11y Avda 1y3, Ed.152
1002 San Jose, Costa Rica, Italy

Cattani, Roberto
Editor
ANSA
Av. Sao Luiz 258
Sao Paulo, Brazil, Italy

Ceccanti, Alessio
Correspondent
ECONOMIA E AMBIENTE
Via Vespucci 86
Rome, Italy, Italy

Cerice, Alberto
Correspondent
OASIS
Neyran-Brissogne
Valle D'Aosta,Italy, Italy

Cerquetelli, Moreno
Reporter Radio/Television
RAI TV TG 3
Via della Resistenza, 7
S. Gemini (TR), Italy, Italy

Chiaberge, Riccardo
Editor
CORRIERE DELLA SERA
via Gherardini 3
Milan, Italy, Italy

Cianciullo, Antonio
Editor
LA REPUBBLICA
Via Annia 62
Roma 00184, Italy, Italy

Ciani, Rossana
Editor
RAI-TG2
Via F.S. Benucci, 55
Rome, Italy, Italy

Ciavoni, Carlos
Editor
LA REPUBBLICA
Via Flaminia Nuova 290
00191 roma, Italy, Italy

Cingolani, Federica
Editor
ANSA
Via T.Salvia 55
Rome, Italy, Italy

Clark, Chris
Editor
IPS
144 Purley Oaks Road
Sanderstead,Surrey,CR2 ONS, UK, Italy

Collymore, Yvette
St Vincent
IPS
Via Panisperna 207
Rome, Italy
T: 39 6 485 692 F: 39 6 81-7877

Colomba, Rocco
Editor
AMBIENTE NEWS
Via C. Pirzio Biroli, 48
00189 Roma, Italy

Confalonieri, Luisa
Editor
CANALE 5 RTI
Via Stoppani
75 Milan, Italy, Italy

Cordova, Luis
Correspondent
IPS
3RA Transv. delos dos Caminos
Caracas, Venezuela, Italy

Cortese, Fabio
Reporter Radio/Television
RAI-TV TG3
Via Nemorense, 83
Roma, Italy, Italy

Cossu, Simonetta
Reporter Radio/Television
RAI-TG3
2 Charlton St.
New York, NY 10014, USA, Italy

Coutu, Francois
Correspondent
IPS
8 Joseph-Baker, Hull, Qc
Canada J9A 2M1, United Kingdom, Italy

Da Mata-Machado, Glaucia
Correspondent
TELEMONTECARLO
V. di S. Giovanni, 15
00179-Rome, Italy, Italy

Dantas, Marcos
Correspondent
IPS
R. Viuva Lacerda 128, Apt. 702
Rio de Janeiro-RJ, Brazil, Italy

De Araujo, Luciana
Reporter Radio/Television
TELEMONTECARLO
Via Sestriere, 5
00135 Rome, Italy, Italy

De Lorenzo, Aristide
Editor
IL MATTINO
Via Pagliano 7
Napoli, Milan, Italy

De Marchis, Sergio
Reporter Radio/Television
RAI
Via Sefro 12
Roma 00010 Italy, Italy

De Paiva Pereira, Ricardo Leitao
Editor
TELEMONTECARLO
Via di S. Giovanni
A Porta Latina, 15-Roma, Italy, Italy

De Poli, Marco
Editor
POPOLI
Via S. Marco 26
20121 Milan, Italy, Italy

De Tomassi, Massimiliano
Correspondent
CBR TELEVISIONE
Viale di Villa Pamphili 65
Roma, Italy, Italy

Deleidi, Maria Paola Casasco
Editor
TV SORRISI E CANZONI
Vivai 20-27100
Paia, Italy, Italy

Di Bella, Antonio
Reporter Radio/Television
RAI-TG3
108 Wooster St.
New York, NY USA, Italy

Dourado, Acacia
Editor
CBR TELEVISIONE
Via del Sudario 35
Roma, Italy, Italy

Drago, Tito
Correspondent
IPS
c/ Comercio 4
28007-Madrid, Spain, Italy

Dujisin, Mario
Correspondent
IPS
Rua da Vigia 20, A. 2775
Parede, Portugal, Italy

Exell Pirro, Deirdre
Editor
AMBIENTE NEWS
Via dei Macci, 19
50122 Firenze, Italy, Italy

Fasano, Federico
Editor
IPS
Joaquin Nunez 3067
Montevideo, Uruguay, Italy

Favarato, Gianni
Correspondent
MATTINO DI PADOVA/NUOVA VENEZIA
Via Temanza 2
Padova, Italy, Italy

Federovisky, Sergio
Argentina
IPS
Via Panisperna 207
Rome, Italy
T: 39 6 485 692 F: 39 6 81-7877

Fernandez, Carmen Alicia
Correspondent
IPS
Calle Ayacucho, Res.Say Park
II, Apt.43 - Caracas Venezuela, Italy

Ferraro, Cristina
RAI-TG2
40 E. 94th Street
New York, NY 10019, USA, Italy

Foresta-Martin, Francesco
Editor
CORRIERE DELLA SERA
Via Raffaele Balestra 2/A
00152 Roma, Italy, Italy

Foti, Paola Filomena
Editor
RAI
Via Tarraglia 21
00197 Roma, Italy, Italy

Francescato, Grazia
Editor
AIROME
Via Mattonato 4
00153 ROME, Italy, Italy

Galli, Mario
Editor
MOZART
Conca del Naviglio, 37
Milano, Italy, Italy

Gallucci, Carlo
Science Editor
L'ESPRESSO MAGAZINE
Via PO 12 00198
Rome, Italy
T: 39 6 84781 F: 39 6 8845167

Gardner, Danielle
Correspondent
GIOIA
1150 Fifth Avenue
New York, USA, Italy

Giachetti, Roberto
Reporter Radio/Television
RADIO RADICALE
Via delle Mura Gianicolensi 12
Rome, Italy, Italy

Giancarlo, Summa
Correspondent
LA STAMPA
R. Augusta 434, #22
01304 Sao Paulo, SP, Brazil, Italy

Giannini, Humberto A.
Correspondent
ANSA
Av.Pres.Antonio Carlos 40
(Cobertura)Rio de Janeiro-RJ, Brazil, Italy

Giannotti, Tullio
Editor
ANSA
Via della Dataria 94
Rome, Italy, Italy

Gilioli, Alessandro Maurice
Editor
EUROPEO MAGAZINE
Via A.Rizzoli 4
Milano, Italy, Italy

Giuseppina, Cataldo
Reporter Radio/Television
RAI
Via Brogano 14
Rome, Italy, Italy

Glattbach, John
Editor
IPS
La Pierre, Route de Meribel
01170 Gex, France, Italy

Goncalvez De Araujo, Luciana
Reporter Radio/Television
TELEMONTECARLO
Via Sestriere, 5
00135-Rome, Italy, Italy

Grassi, Riccardo Alberto
Correspondent
IPS
Via del Boschetto 68
00184 Rome, Italy, Italy

Greco, Pietro
Editor
L'UNITA
Via C.Buono 4
Barano D'Ischia, Italy, Italy

Green, Christina
Correspondent
IPS
Apdo. 3834
1000 San Jose, Costa Rica, Italy

Humeres Galvez, Paulina
Correspondent
IPS
Iva del Boschetto 68
00184 Rome, Italy, Italy

Isberto, Ramon R.
Correspondent
IPS
44 Valentine St.,Merville Subd
Paranaque, Metro Manila, Philippines, Italy

Ismartono, Yuli
Correspondent
IPS
22/4 Sikhumvit Soi 18
Bangkok, Thailand, Italy

Jayasekera, Rohan A.
Correspondent
IPS
Kanaalstraat 11
1071 WX Amsterdam, Netherlands, Italy

Kamal, Baher
Correspondent
IPS
V. Constantino 143
00100 Rome, Italy, Italy

Kassam, Amin
Editor
IPS
Burg. Teerstraat 31
1511 BP Oostzaan, Netherlands, Italy

Kennedy, Amber
Correspondent - Terra Viva
IPS (INTERPRESS)
Via Panisperna
Rome, Italy
T: 68 533038 F: 68 533038

Kirk, Alejandro
Correspondent
IPS
Via Bolognesi 25
00152-Rome, Italy, Italy

Koindrekaj, Gjon
RAI
Via Pompei 3
Rome, Italy, Italy

Laruffa, Dario
Reporter Radio/Television
RAI-GR1
Via di Villa Zingone 26
00151-Rome, Italy, Italy

Lasserre, Colette
Correspondent
INTERPRESS SEREVICE
Via Monte Savello 30
Roma, Italy, Italy

Leon, Patricia
Correspondent
IPS
P.O.Box 49-3000
Costa Rica, Italy

Lippolis, Ignazio
Editor
LA GAZZETTA DEL MEZZOGIORNO
Via Olbia, 18
70026 Modugno, Bari, Italy, Italy

Lombardo, Alina
Correspondent
ALIN PRESS AGENCY
Via Pelafio Palafi, 9
20129, Milano, Italy, Italy

Longo, Paolo
Correspondent
RAI
36 Gramercy Park East, 6N
New York, NY USA, Italy

Lubin, Flaminia
RAI-TG2
440 E. 56th Street
New York, NY 10022, USA, Italy

Maccari, Alberto
Reporter Radio/Television
RAI
Via Giannetto Valli, 95
00100 Rome, Italy, Italy

Machado, Glaucia
Editor
TELEMONTECARLO
Via de Giovanni A
Porta Latina, 15 Rome, Italy, Italy

Maglie, Maria Giovanna
Correspondent
RAI-TG2
923 Fifth Avenue
New York, NY 10021, USA, Italy

Maia, Viviane
TELEMONTECARLO
Piazza della Balduina 49
Rome 00136, Italy
T: 396 3304299/265 F: 396 3304257

Malinverno, Giuliano
Correspondent
SIAL
Via Sant Elembardo 1
Milano, Italy, Italy

Mannucci, Anna
Reporter Radio/Television
RADIO POPOLARE
Via Valery 3
Milano 20163, Italy, Italy

Manrique, Salvarrey Gaudin
Editor
IPS
Almte. Brown, 706 piso 10-A
Buenos Aires, Argentina, Italy

Marino, Fiorella
Correspondent
GIORNALE IMPRESA AGRICOLA
V.C. Maiocchi 19 Campitello
Mantova 46010 Italy, Italy

Massidda, Enrico, Ms.
RAI-ITALIAN TELEVISION TG1
Via Teulada 66
Rome, Italy
T: 39 6 3420241

Mastrantonio, Cecilia, Ms.
Editor
LA NUOVA ECOLOGIA
Via Ugo Inchiostri, 95
00143 Roma, Italy, Italy

Masullo, Fernando
Reporter Radio/Television
RAI-GR2(GIORNALE RADIO)
1 Irving Place, Park 23 D
New York, NY 10003, Italy

Matteuzzi, Mauricio
Editor
IL MANIFESTO
Via del Boschetto, 3
00184-Rome, Italy, Italy

Matthew, Mathai Koyapillil S.J.
RADIO VATICANA
St. Joseph's College
Box 315, Tiruchirapalli 620002, India, Italy

McCarthy, John
Correspondent
RADIO VATICANA
24 Bvd. Street, St. Michel
1040 Bruxelles, Belgium, Italy

Melega, Marco
Reporter Radio/Television
RAI
Via dei Colli 8 Fontanelice
Bologna, Italia, Italy

Melnick, Ana
Correspondent
IPS
Rodrigo de Triana 4344
Santiago, Chile, Italy

Mertens, Frederico F.
Editor
IPS
Garibaldi 2579
Montevideo, Uruguay, Italy

Molesti, Romano
Editor
ECONOMIA E AMBIENTE
Via Pratale 64
Roma, Italy, Italy

Monsieur, Christiane
Correspondent
IPS
25 Via F. Bolognesi
00152 Rome Italy, Italy

Montautti Ibarra, Americo
Editor
IPS
Joaquin de Salterain, 1557
Montevideo, Uruguay, Italy

Navarrete, Ma. Isabel
Correspondent
IPS
Calel 12, No. 17
Santafe de Bogota, DC, Colombia, Italy

Niehaus, Ingo
Correspondent
IPS
Apt. 4374
1000 San Jose, Costa Rica, Italy

Ninan, Ajit
Correspondent
IPS
A2/79 Janakpuri
New Delhi, 110058, India, Italy

Orsello, Gian Piero
Editor
L'ITALIA E L'EUROPA
Corso Francia 231
00191 Roma, Italy, Italy

Ortiz, Ana Ma.
Editor
IPS
San Isidro de Coronado
100 Este, 100 Sur, 75 Oeste, Costa Rica, Italy

Osava, Chizuo
Correspondent
IPS
R. Pereira da Silva, 492/104
22221-Rio de Janeiro-RJ, Brazil, Italy

Paladino, Anna
Editor
RAI
Via del Cappellari 35
Rome, Italy, Italy

Pallavidini, Manuel H.
Correspondent
ANSA
Av.Pres.Antonio Carlos 40
(Cobertura)Rio de Janeiro-RJ, Brazil, Italy

Paravagna, Roberto
Correspondent
IL GIORNALE
Via Gorani 5
20123 Milan, Italy, Italy

Pereira, Ricardo L.
Editor
TELEMONTECARLO
Via di s. Giovanni A
Porta Latina, 15, Rome, Italy, Italy

Perlotto, Francesco A.
Correspondent
AVVENIRE
Via Della Stampa 8
36070 Trissino, Italy, Italy

Pieri, Massimo
Correspondent
AS & T
Via Vitorchiano, 23
00189 Rome, Italy, Italy

Pioli, Giampalo
Correspondent
LA NAZIONE-POLIPRESS
300 E 40th St. (#11R)
New York, NY 10016, USA, Italy

Pluviano, Oliviero
Correspondent
ANSA
Av. Sao Luiz, 258 Apt.2009
01046 Sao Paolo-SP, Brazil, Italy

Po, Ruggero
Reporter Radio/Television
RAI
Via Torriglia 18
Roma, Italy, Italy

Prattico, Francesco
Editor
LA REPUBBLICA
via Molveno 105
00135 Rome, Italy, Italy

Quaranta, Paolo
Editor
RAI-TG3
63 East 9th Street, No. 14Y
New York, NY 10003, USA, Italy

Rainieri, Renata
Editor
RAI
Viale Mazzini,33
Roma, Italia, Italy

Rami, Paolo
Correspondent
AISE/L'UMANITA EUROPA
Rua Capitao Salomao, 14/904
Rio de Janeiro-RJ, Brazil, Italy

Ramirez, Lyng-Hou
Correspondent
IPS
Av.Fuerzas Armadas, No.174-B
Caracas, Venezuela, Italy

Rara, Ma. Cristina I.
Correspondent
IPS
67 Gen. Avenue, GSIS Village
Project 8, Quezon City, Philippines, Italy

Robinson, Claude
Correspondent
IPS
26 Whittingham Close
Kingston 10, Jamaica, Italy

Ruggero, Leonardi
Correspondent
DOVE
Via Giovanni, 23
20093 Milano - Italy, Italy

Ruggero, Po
Reporter Radio/Television
RAI-GR2
Via Sisto IV, 5
00167-Rome, Italy, Italy

Sandrow, Valeria
Correspondent
ECONOMIA E AMBIENTE
Via Amerigo Vespucci
Rome, Italy, Italy

Santori, Isa In Fenocchio
Correspondent
RAI
Via Italo Piccagli, 9
00189 Rom, Italy, Italy

Santovincenzo, Ivano
Reporter Radio/Television
TELEMONTECARLO
Via Ugo de Carolis, 90
Rome, Italy, Italy

Sanvoisin, Marco
Editor
AGENZIA GIORNALISTICA ITALIA
Av.Rui Barbosa 280/1402
Rio de Janeiro, Brazil, Italy

Sassano, Marco
Editor
IL GIORNO
Via Aurelia 137
Rome, 00165 Italy, Italy

Savio, Roberto
Editor
IPS
Via Montesavello 30
00184 Rome, Italy, Italy

Semplici, Andrea
Correspondent
ALIN PRESS AGENCY
Via di Mucciana, 6
50026, Firenze, Italy, Italy

Sharma, Yojana
United Kingdom
IPS
Via Panisperna 207
Rome, Italy
T: 39 6 485 692 F: 39 6 81-7877

Sinclair, David
Editor
IPS
Passage d'en Cipria
66300 Thuir, France, Italy

Solana, Luis J.
Editor
IPS
Gob. Francisco Garcia Conde,5
11850 Mexico-DF, Mexico, Italy

Spiezie, Annalisa
TELEMONTECARLO
Piazza della Balduina 49
Rome 00136, Italy
T: 396 3304299/265 F: 396 3304257

Squadroni, Giulia
Correspondent
LA STAMPA
Via Torricelli 30
Milano, Italy, Italy

Stefanat, Bruno
Reporter Radio/Television
RAI
Via Poliziano
Roma, Italy, Italy

Steffen, Hermann
Costa Rica
IPS
Via Panisperna 207
Rome, Italy
T: 39 6 85 692 F: 39 6 81-7877

Stella, Bruno
Editor
ORIZZONTI ALENIA
Fregoli 8
Rome, Italy, Italy

Tallachini, Mariachiara
Correspondent
TEMPO MEDICO
V. Veneto 9/9 Peschiera
Milano, Italy, Italy

Taverniti, Fausto
RAI - RADIO TELEVISIONE ITALIANA/STUDIO IMMAGINE S.R.L.
Rome, Italy

Teodori, Laura Maria
Reporter Radio/Television
RAI
Via A.Doria, 48
Roma, Italia, Italy

Thalif, Deen
Correspondent
IPS
34 Burgher Avenue
Staten Island, New York 10304, USA, Italy

Thompson, Eulalee
Jamaica
IPS
Via Panisperna 207
Rome, Italy
T: 39 6 485 692 F: 39 6 81-7877

Torricelli, Maddalena
Correspondent
AMBIENTE NEWS
Via B. Varchi 39
50132 Florence, Italy, Italy

Treglia, Angela
RAI-TG2
2313 86th Street
Brooklyn, NY 11214, USA, Italy

Trincia, Stefano
Correspondent
IL MESSAGGERO
235 E 18th (#2)
New York, NY 10003, USA, Italy

Urioste, Ricardo
Editor
IPS
2720 Apt.201
Montevideo, Uruguay, Italy

Urruzola, Maria
Editor
IPS
Maldonado 1939 Apt.302
Montevideo, Uruguay, Italy

V. Beek, Jan
Editor
IPS
Oostduinlaan 191
2596 J.L. The Hague, Netherlands, Italy

Valentini, Daniele
Journalist
RAI - ITALIAN TELEVISION TG1
Via Teulada, 66
Rome, Italy
T: 39 6 372030 F: 39 6 331919

Valsania, Marco
Correspondent
IL SOLE 24 ORE
40 McDougal (#11)
New York, NY 10012, USA, Italy

Vega, Alejandro
Editor
ANSA
Av.Pres.Antonio Carlos 40
(Cobertura)Rio de Janeiro-RJ, Brazil, Italy

Vierci, Pablo
Editor
IPS
Lorenzo Perez no.3118
Montevideo, Uruguay, Italy

Villar, Isabel
Editor
IPS
Joaquin Nunez 3067
Montevideo, Uruguay, Italy

Vitello, Carlo
Correspondent
AGENZIA GIORNALISTICA ITALIA
Via Mincio 52
20089 Rozzano, Milano, Italy, Italy

Vittorio, Benedetto
Correspondent
NEW ITALPRESS
Via del Biancospini 13
Milano, Italy, Italy

Vivarelli, Alessandra
Editor
PANDA MAGAZINE
Via Marco Aurelio 5
Italy, Italy

Vizioli, Giorgio
IL SOLE-24 ORE
Via Prina 1
Milan, Italy, Italy

Vlahou, Assimina
Correspondent
TELEMONTECARLO
Clivo di Cinna 205
Rome, Italy, Italy

Wilkinson, Bert
Correspondent
IPS
112 B Laluni St.,Queenstown
Georgetown, Guyana, S.A., Italy

Willson, Richard
Correspondent
IPS
Shepherds Hill
Peaslake, Surrey, GUR 9TB U.K., Italy

Zeledon, Sylvia F.
Correspondent
IPS
Apt.300 Escazu
San Jose, Costa Rica, Italy

Zucconi, Vittorio
Correspondent
LA REPUBBLICA
Rome, Italy
Italy

ORGANIZATIONS

Ausenda, Fabio, Mr.
Europe Conservation
Via Fusetti 14
20143 Milano, Italy
T: (39 2) 5810 3135 F: (39 2) 8940 0649

Farina, Almo, Mr.
International Association for Landscape Ecology
c/o Lunigiana Museum of Natural History,
Fortezza della Brunella, 54011 Aulla, Italy
T: (39 187) 420 374 F: (39 187) 420 727

Harcourt, Wendy, Ms.
WIDE - Women in Development Europe
c/o SID, Palazzo Civilta del Lavoro
00144 EUR, Roma, Italy
T: (39 6) 591 7897 F: (39 6) 591 9836

Hassan, Mohammed, Dr.
Third World Academy Of Sciences
PO Box 486, Miramare, Strada Costura 11
34100 Trieste, Italy
T: (39 40) 224 559

Longo, Justice Giovanni, Mr.
International Association Of Judges
Palazzo di Giustizia, Piazza Cavour
00193 Rome, Italy

Meier, Gerhard, Mr.
Caritas Internationalis
Palazzo San Calisto
V-00120 Vatican City, Italy
T: (39 6) 698 7197 F: (39 6) 698 7237

Melandri, Elisabetta, Ms.
Centro Informazione e Educazione allo Sviluppo
Via Palermo, 36
00184 Roma, Italy
T: (39 6) 474 6246/474 7696 F: (39 6) 486 419

Melandri, Giovanna, Ms.
Lega per l'Ambiente
Via Salaria, 280
00199 Roma, Italy
T: (39 6) 844 2277/ 844 3582 F: (39 6) 844 3504

Sikabonyi, Mary Ellen, Ms.
International Juridical Organization for Environment and Development
Via Barberini 3
00187 Roma, Italy
F: (39 6) 474 5779

Steigerwald, Jutta, Ms.
Campagna Nord-Sud: Biosphere-Human Survival-Debt
Via S. Maria dell'Anima
30-00186 Rome, Italy
T: (39 6) 686 5842 F: (39 6) 686 5842
TL: E geo2:campagna

Williams, Maurice, Mr.
Society For International Development (SID)
Palazzo Civilta del Lavoro
00144 Rome, Italy
T: (39 6) 591 7897 F: (39 6) 591 9836

Zorzetto, Gaetano, Mr.
Council Of European Municipalities And Regions, Working Group On Marine Polluti
Palazzo Cavalli 4090, San Marco
Venezia, Italy
T: (39 41) 270 82 83 F: (39 41) 270 85 95

GLOBAL FORUM

Abram, Mathias, Mr.
Consultant
TERRA NUOVA
Via Urbana 156
Rome 00184, Italy
T: 39 6 4747859 F: 39 6 4747599

Acevedo, Lucio Cabrera De, Mr.
CEISAL - GRUPO DE TRABAJO DE JURISPRUDENCIA
P.zza Castello 11 - Casella postale 81
Sassari 07100, Italy
T: 39 79 237364 F: 39 79 200083

Agnesi, Valerio, Mr.
Professor
UNIVERSITA DEGLI STUDI DI PALERMO
Departamento De Gelogia Corso Tukory 131
Palermo 90.134, Italy
T: 39 91 651-2019 F: 39 91 651-6070

Agostinelli, Agostino, Mr.
Consigliere
REGIONE LOMBARDIA
Via Ugo Bassi 2
Milan 20159, Italy
T: 39 2 6975197 F: 39 2 67655948

Agostino, Agostinelli, Mr.
Con. Reg. Le.
CONCIGLIO REGIONALE LOMBARDIA
Via Ugo Bassi, 2
Milano, Italy
T: 39 2 69751 F: 39 2 6975497

Albertini, Piergiorcio
RAI - RADIO TELEVISIONE ITALIANA/STUDIO IMMAGINE S.R.L.
Rome, Italy

Alcazar, Jose T. Esquinas, Mr.
F A O
Via Delle Terme Di Caracalla
Rome 00100, Italy
T: 39 6 57974986 F: 39 6 57973152

Alessandro, Maurice Gilioli
EUROPEU MAGAZINE
Court of Cassation - Piazza Cavour
Rome 00193, Italy
T: 39 6 6868597 F: 39 6 6874170

Alfieri, Mario, Dr.
LEGA PER L'AMBIENTE
Via Casoretto 8
Milan 20131, Italy
T: 39 2 2857754 F: 39 2 62706355

Allegree, Francesso
Journalist
TERRAMATER
Gubbio, Italy
T: 25 9274606 F: 25 9274211

Allegrucci, Francesco, Mr.
TERRA MATER
Officio Beni Attivito Culturali, Via Matteotti 17
Gubbio 06024, Italy
T: 39 75 927-4604 F: 39 75 927-4720
E: 06024

Amadio, Vittorio, Ms.
LEGA PER L'AMBIENTE
Via Salaria 280
Rome, Italy
T: 39 6 8841552 F: 39 6 8443505
E: 65:Eco004

Ambrogio, Brenna, Mr.
Secretario Nazionale
FIM - CISL
Corso Trieste 36
Rome 00198, Italy
T: 39 6 847-1208 F: 39 6 847-1262

Amendola, Giorgio
RAI - ITALIAN TELEVISION
Rome, Italy

Angelucci, Carlo, Mr.
ENEL
Via Dei Crociferi
Rome 00187, Italy
T: 39 6 678 1280 F: 39 6 678 1348

Antonelli, Gianpiero, Mr.
Dirigente
ASSESSORATO AMBIENTE REGIONE UMBRIA
P.za Partigiani 1
Perugia 06100, Italy
T: 39 75 61167 F: 39 75 5042732

Arevalo, Marco Vinicio Cerezo, Mr.
CEISAL - GRUPO DE TRABAJO DE JURISPRUDENCIA
Piazza Castello 11 - Casella postale 81
Sassari, Italy
T: 39 79 237364 F: 39 79 200083

Arrigoni, Paolo, Mr.
Consigliere
REGIONE LOMBARDIA
Via Ugo Bassi 2
Milan 20159, Italy
T: 39 2 6975197 F: 39 2 67655948

Bacci, Pietro, Mr.
Executive
ENTE NAZIONALE POR L'ENERGIA ELETRICA
Via G.N. Martini 3
Rome 00198, Italy
T: 39 6 85091
TL: 610518

Bachelet, Michel, Mr.
CEISAL - GRUPO DE TRABAJO DE JURISPRUDENCIA
P.zza Castellato 11 - Casella postale 81
Sassari 07100, Italy
T: 39 79 237364 F: 39 79 200083

Balzamonti, Claudio
EMIGRAZIONE
F.I.L.E.F. Via IV Novembre 125/3 piano
Rome 00160, Italy

Baraldi, Gildo, Mr.
President
COCIS
Lungotevere Dei Mellini 39
Rome 00100, Italy
T: 39 6 323-3163 F: 39 6 324-4163

Barbarella, Carla, Ms.
President
CIDIS
Via Della Viola 1
Perugia, Italy
T: 39 75 20895

Barbieri, Giulia, Ms.
CLUB ALPINO ITALIANO
Via Andrea Costa, 7
Milano 20131, Italy
T: 39 2 268-23874

Barboni, Paolo, Mr.
Prefetto
MAYOR OF GUBBIO
Via Matteotti, 17
Gubbio, Italy
T: 39 75 927-4604 F: 39 75 926-4720

Bardanzellu, Francesco, Sr.
Representante
PROVINCIA DI SASSARI
Piazza D'Italia
Sassari, Italy
T: 079 123 2369

Barrera, Pietro, Mr.
LEGA PER L'AMBIENTE
Via Salaria 280
Rome, Italy
T: 39 6 8841552 F: 39 6 8443505
E: 65:Eco004

Battaglia, Luisella, Ms.
Professor
LEGA ITALIANA DIRITI DELL' ANIMALE
Via Richetti 4-12
Genova 16146, Italy
T: 307868

Baumann, W., Mr.
CEISAL - GRUPO DE TRABAJO DE JURISPRUDENCIA
P.zza Castello 11 - Casella postale 81
Sassari, Italy
T: 39 79 237364 F: 39 79 200083

Beek, Jan Van
The Netherlands
IPS
Via Panisperna 207
Rome, Italy
T: 39 6 485 692 F: 39 6 81-7877

Benedetto, Tiziana De, Ms.
Secretaria
REGIONE LOMBARDIA
Via Ugo Bassi 2
Milan 20159, Italy
T: 39 2 6975197 F: 39 2 67655948

Benelli, Carla, Sra.
CENTRO INTERNAZIONALE CROCEVIA
Via Ferraironi 88-G
Rome 00172, Italy
T: 39 6 241-3976 F: 39 6 242-4177

Benigno, Elisa
Via Ludovisi 26
Rome 00187, Italy
T: 39 6 487 0122 F: 39 6 487 0066

Berbenni, Paolo
Scientific Director
ETAS SRL - RIVISTA INQUINAMENTO
4ia Mecenate 89
Milano, Italy
T: 39 02 58084227 F: 39 02 55400388

Bernabucci, Claudio, Mr.
Presidente
MO LI SV
Pza Albania 10
Rome 00153, Italy
T: 39 6 5750941

Bernardini, Alfredo
Via Ludovisi 26
Rome 00187, Italy
T: 39 6 487 0122 F: 39 6 487 0066

Bianchi, Giuseppe, Mr.
Administration Counseler
ENTE NAZIONALE POR L'ENERGIA ELETRICA
Via G.N. Martini 3
Rome 00198, Italy
T: 39 6 85091
TL: 610518

Biondi, Vitaliano
Architetto
Court of Cassation - Piazza Cavour
Rome 00193, Italy
T: 39 6 6868597 F: 39 6 6874170

Boccone, Patrizio, Sr.
Expositor
FLACAM
Via San Giovanni 398
Cagliari - Cagliari 09127, Italy
T: 0039070 498723

Bonacini, Moris, Sr.
Assessore
AMBIENTE REGIONE EMILIA ROMAGNA
P.za Partigiani 1
Perugia 06100, Italy
T: 39 75 61167 F: 39 75 5042732

Bossi, Angela, Ms.
Cons. Reg.
CONCIGLIO REGIONALE LOMBARDIA
Via Ugo Bassi, 2
Milano, Italy
T: 39 2 69751 F: 39 2 6975497

Botero, Luis Santiago, Sr.
F.A.O
Via Della Terme De Caracalla,
Rome, Italy
T: 39 6 579 75088

Bruni, Antonio
RAI - RADIO TELEVISIONE ITALIA
Rome, Italy

Buffaria, Bruno
Via Ludovisi 26
Rome 00187, Italy
T: 39 6 487 0122 F: 39 6 487 0066

Cadringher, Manisela
Editor
RAI TV TG2
Via Tulada, 66
Rome 00195, Italy
T: 319663 F: 386608

Caizzone, Guiseppe, Mr.
Giudice
CORTE SUPREMA DI CASSAZIONE
Court of Cassation - Piazza Cavour
Rome 00193, Italy
T: 39 6 6868597 F: 39 6 6874170

Calsamiglia, Franco, Mr.
Dir. Cord.
CONCIGLIO REGIONALE LOMBARDIA
Via Ugo Bassi, 2
Milano, Italy
T: 39 2 69751 F: 39 2 6975497

Caltabiano, Eleonora, Ms.
CO.P.E.
Via Gabrriello Carnazza 43
Catania 95129, Italy
T: 095 53-6831

Camillo, Piazza, Mr.
CONCIGLIO REGIONALE LOMBARDIA
Via Ugo Bassi, 2
Milano, Italy
T: 39 2 69751 F: 39 2 6975497

Camin, Silvia, Ms.
LEGA PER L'AMBIENTE
Via Ortles 21/2
Bolzano, Italy
T: 471 20-2295

Cammarano, Bruna
RAI - RADIO TELEVISIONE ITALIANA/STUDIO IMMAGINE S.R.L.
Rome, Italy

Cannata, Giuliano, Mr.
Socio
MO LI SV
Pza Albania 10
Rome, Italy
T: 39 6 5750941

Caporizzi, Lucio, Mr.
Dirigente
ASSESSORATO AMBIENTE REGIONE UMBRIA
P.za Partigiani 1
Perugia 06100, Italy
T: 39 75 61167 F: 39 75 5042732

Caracciolo, Francesco
Via Ludovisi 26
Rome 00187, Italy
T: 39 6 487 0122 F: 39 6 487 0066

Carli, Anna, Ms.
CGIL - SINDICATO
Corso d'Italia nr. 25
Rome 00198, Italy
T: 39 6 84761 F: 39 6 884-5683

Carlo, Carmine, Mr.
FILCAMS - CGIL
Via Marconi, 67-2
Bolonha 40122, Italy
T: 39 51 25-3230 F: 39 51 25-1062

Carlo, Monguzzi, Mr.
Con. R. Le.
CONCIGLIO REGIONALE LOMBARDIA
Via Ugo Bassi, 2
Milano, Italy
T: 39 2 69751 F: 39 2 6975497

Carrano, Maurizio
RAI - RADIO TELEVISIONE ITALIANA/STUDIO IMMAGINE S.R.L.
Rome, Italy

Carrano, Paola
RAI - RADIO TELEVISIONE ITALIANA/STUDIO IMMAGINE S.R.L.
Rome, Italy

Carta, Giuseppe, Mr.
Manager
ENTE NAZIONALE POR L'ENERGIA ELETRICA
Via G.N. Martini 3
Rome 00198, Italy
T: 39 6 85091
TL: 610518

Casagrande, Franco Raffi, Mr.
President
TERRA MATER
Piazza Oderist 1
Gubbio, Italy
T: 39 75 922-0693

Castelletti, Lelio, Mr.
TERRA NUOVA
Via Urbana 155
Rome, Italy
T: 39 6 485534 F: 39 6 474594

Castelletti, Osvaldo, Mr.
ACRA
via Breda 54
Milan 20126, Italy
T: 39 2 255-2324 F: 39 2 255-2270
TL: 333187acrami I

Castelli, Gianfranco, Mr.
Deputy Manager
ENTE NAZIONALE POR L'ENERGIA ELETRICA
Via G.N. Martini 3
Rome 00198, Italy
T: 39 6 85091
TL: 610518

Castro, Vania Dias De
RAI - RADIO TELEVISIONE ITALIANA/STUDIO IMMAGINE S.R.L.
Rome, Italy

Catalano, Pierangelo, Mr.
CEISAL - GRUPO DE TRABAJO DE JURISPRUDENCIA
Piazza Castello 11 - Casella Postale 81
Sassari 07100, Italy
T: 39 79 237364 F: 39 79 200083

Cataldo, Pina
RAI - RADIO TELEVISIONE ITALIANA/STUDIO IMMAGINE S.R.L.
Rome, Italy

Cavallo, Anna Rosa, Ms.
Consigliere Regionale
CONSIGLIO REGIONALE DEL LAZIO
Via Della Pisana, 1301
Rome, Italy
T: 39 6 500 0672 F: 39 6 500 0956

Cay, Mario, Mr.
President
TERRA NUOVA
Via Urbana 156
Rome 00184, Italy
T: 39 6 4747859 F: 39 6 4747599

Cevallos, Diego
Ecuador
IPS
Via Panisperna 207
Rome, Italy
T: 39 6 485 692 F: 39 6 81-7877

Chicchella, Francesco, Mr.
Assessore
ASSESSORATO AMBIENTE REGIONE UMBRIA
P.za Partigiani 1
Perugia 06100, Italy
T: 39 75 61167 F: 39 75 5042732

Chiummariello, Miriam, Ms.
CONSIGLIO NAZIONALE DELLE RICERCHE
Viale Cramsei #5
Napoles, Italy
T: 39 81 761 1575 F: 39 81 761 2509
E: 80122

Cignitti, Benedetto, Mr.
Funzionario
CONSIGLIO REGIONALE DEL LAZIO
Via Della Pisana, 1301
Rome, Italy
T: 39 6 500 0672 F: 39 6 500 0956

Clasamiglia, Franco, Mr.
Dirigente
REGIONE LOMBARDIA
Via Ugo Bassi 2
Milan 20159, Italy
T: 39 2 6975197 F: 39 2 67655948

Comi, Nadia, Ms.
ISCOS - CISL
Via Boncompagni 19
Rome, Italy
T: 39 6 481 7100

Corazzi, Ubaldo, Mr.
President Of Comunity
COM. MONTANA ALTO CHIASCO GUBBIO
Via Matteotti, 17
Gubbio, Italy
T: 39 75 927-4604 F: 39 75 927-4720

Corradi, Mario, Mr.
MUNICIPALITY OF MILAN (ITALY)
Piazza Duomo 21
Milano 20100, Italy
T: 39 2 8690017 F: 39 2 8690973

Cortiana, Fiorello, Mr.
Consigliere
REGIONE LOMBARDIA
Via Ugo Bassi 2
Milan 20159, Italy
T: 39 2 6975197 F: 39 2 67655948

Corvi, Corrado, Mr.
Deputy Manager
ENTE NAZIONALE POR L'ENERGIA ELETRICA
Via G.N. Martini 3
Rome 00198, Italy
T: 39 6 85091
TL: 610518

Coulibaly, Aly
Stringer
IPS - INTER PRESS SERVICE
Via Panispena 207
Rome, Italy
T: 39 6 4742497/485692 F: 39 6 740776

Crippa, Federico, Mr.
Parlamentare
PARLAMENTO ITALIANO
Via Gerola 17
Treviglio 24047, Italy
T: 0363302010

Cuminatto, Giorgio, Dr.
ASSOCIAZIONE INTERNAZIONALE MAGNA GRECIA
Via Tibullo 10
Rome 00193, Italy
T: 39 6 686-9758 F: 39 6 654-7791

D'Alfonso, Giovani, Mr.
Consigliere
REGIONE LOMBARDIA
Via Ugo Bassi 2
Milan 20159, Italy
T: 39 2 6975197 F: 39 2 67655948

D'Asaro, Anita
RAI - ITALIAN TELEVISION
Rome, Italy

Dacion, Corrado Maria, Sr.
EUROPEAN CENTER FOR ENVIRONMENTAL STUDIES
ECES - Via Po 14
Turin 10123, Italy

Daghino, Carlo, Mr.
Direttore
ISCOS-CISL PIEMONTE
V. Volturno 11
Torino 10133, Italy
T: 003911 660-4606 F: 003911 660-3575

Dalmasso, Marco, Mr.
Reponsavel Torino
LEGA AMBIENTE
Murazzi Di PO, 52
Torino 10100, Italy
T: 512-5386

Debouck, Daniel, Mr.
INTERNATIONAL BOARD FOR PLANT GENETIC RESOURCES
Via delle Sette Chiese 142
Rome 00145, Italy
T: 39 6 5744719 F: 39 6 5750309
TL: 4900005332IBRUI, E: 157CGI101

Declich, Francesca, Ms.
Social Research
CISP - COMITATO INTERNAZIONALE PER LO SUILUPPO DEI POPOLI
Via Marziale 47
Rome, Italy
T: 39 6 345-4063 F: 39 6 344461

Deen, Thalif
Sri Lanka
IPS
Via Panisperna 207
Rome, Italy
T: 39 6 485 692 F: 39 6 81-7877

Del Monte, Gabriele-Maria, Mr.
CENTRO INTERNAZIONALE DI COOPERAZIONE ALLO SVILUPPO
Viaz Cresangio 82
Rome 00193, Italy
T: 39 6 689 2355 F: 39 6 68307508
TL: 6211100CICS

Diament, Daniel
GIORNALE ARCI NOTIZIE
Via Francesco Carrara 24
Roma, Italy
T: 6 3227791 F: 6 3610858

Dragone, Umberto, Mr.
Director
ENTE NAZIONALE POR L'ENERGIA ELETRICA
Via G.N. Martini 3
Rome 00198, Italy
T: 39 6 85091
TL: 610518

Fabretti, Giuseppe, Mr.
Vice President
NATIONAL CONSUMERS COOPERATIVES ASSOCIATION
Via Panaro 14,
Rome 00199, Italy
T: 39 6 83-09-71 F: 39 6 832-0033

Faldutti, Roberto, Mr.
CIDIS
Via Della Viola 1
Perugia, Italy
T: 39 75 20895

Fatale, Gianpaolo, Mr.
Assessore
ASSESSORATO AMBIENTE REGIONE UMBRIA
P.za Partigiani 1
Perugia 06100, Italy
T: 39 75 61167 F: 39 75 5042732

Fazio, Lorenzo, Mr.
ASSESSORE AMBIENTE PROVINCIA DI BARI
Court of Cassation - Piazza Cavour
Rome 00193, Italy
T: 39 6 6868597 F: 39 6 6874170

Fazzuoli, Anita
RAI - RADIO TELEVISIONE ITALIANA/STUDIO IMMAGINE S.R.L.
Rome, Italy

Fazzuoli, Federico
Anchorman
RAI RADIOTELEVISIONE ITALIANA
Via Pasubio, 2
Rome, Italy
T: 39 6 37264962 F: 39 6 382674

Fazzuoli, Luca
RAI - RADIO TELEVISIONE ITALIA
Rome, Italy

Ferraresi, Giorgio, Mr.
UNIVERSIDADE POLITECNICO MILANO/ ECOPOLIS
Galleria Vittorio Emanuele 11/12
Milan 20133, Italy
T: 39 2 890-0247 F: 39 2 890-0213

Ferraris, Sergio
Photographer
LA NUOVA ECOLOGIA
Via Savoia 37
Rome, Italy
T: 39 6 553777 F: 39 6 416256

Filho, Paulo Santos
RAI - RADIO TELEVISIONE ITALIANA/STUDIO IMMAGINE S.R.L.
Rome, Italy

Fiume, Giorgio Dal, Mr.
Funzionario
CONFEDERAZIONE GENERALE ITALIANA DEL LAVORO
Via Marconi 67/2
Bolonha 40122, Italy
T: 39 51 24 90 51 F: 39 51 25 10 62

Fontanella, Matia, Mr.
FILCAMS - CGIL
Via Marconi, 67-2
Bolonha 40122, Italy
T: 39 51 25-3230 F: 39 51 25-1062

Fraschini, Angela, Ms.
UNIVERSITY OF PAVIA
Strada Nuova 65
Pavia, Italy
T: 0382 38-7407

Galati, Giuseppe, Sr.
IRITECNA
Via Sistina 123
Rome, Italy
T: 39 6 4747633 F: 39 6 48 17 263

Gandus, Nicoletta, Ms.
PRETURA DE MILANO
Via Morosini 31
Milan 20135, Italy
T: 39 2 599-00262

Garcia, Maria Isabel
Colombia
IPS
Via Panisperna 207
Rome, Italy
T: 39 6 485 692 F: 39 6 81-7877

Garcia, Serge Michel, Mr.
Director
FAO FISHERIES DEPARTMENT
Fishery Resources and Environment Division
Rome, Italy
T: 39 6 579-7467 F: 39 6 512-0330

Gasparri, Giancarlo
RAI - RADIO TELEVISIONE ITALIANA/STUDIO IMMAGINE S.R.L.
Rome, Italy

Giallonso, Venera, Ms.
Dirigente
ASSESSORATO AMBIENTE REGIONE UMBRIA
P.za Partigiani 1
Perugia 06100, Italy
T: 39 75 61167 F: 39 75 5042732

Giangirolamo, Franco Di, Mr.
Dirigente
CONFEDERAZIONE GENERALE ITALIANA DEL LAVORO
Via Marconi 67/2
Bolonha 40122, Italy
T: 39 51 24 90 51 F: 39 51 25 10 62

Gioucfiero, Tellarini
Art Critic
FREELANCER
Via Tinati 11
Italy
T: 543 744839

Giovagnoli, Segio
GIORNALE ARCI NOTIZIE
Via Francesco Carrara 24
Roma, Italy
T: 6 3227791 F: 6 3610858

Giovannetti, Alcide, Mr.
Deputy Manager
ENTE NAZIONALE POR L'ENERGIA ELETRICA
Via G.N. Martini 3
Rome 00198, Italy
T: 39 6 85091
TL: 610518

Giovanni, Alemanno, Mr.
Consigliere
CONSIGLIO REGIONALE LAZIO
Via Della Pisana 1301
Rome, Italy
T: 39 6 65000672

Giuliana, Gurisatti, Ms.
Responsavel No Brasil
O.N.G. C.E.V.I.
Via Prampero,43
Gemona - UD 33013, Italy

Giuseppe, Ruggiero
Editor
RAI TV TG2
Via Tulada, 66
Rome 00195, Italy
T: 319663 F: 386608

Giviliano, Malinverno
Correspondent
SERVIZIO INFORMASZIONA AMERICA LATINA
Via Bacilieri, 1/A
27139 Verona, Italy
T: 45 8900329 F: 45 8903199

Gnecchi, Luisa, Sra.
LEGA PER L'AMBIENTE
Via Ortles 21/2
Bolzano, Italy
T: 471 20-2295 F: 471 93-5403

Gomes, Magno Mauricio
TELEMONTECARLO
Piazza della Balduina 49
Rome 00136, Italy
T: 396 3304299/265 F: 396 3304257

Gonzales, Gustavo
Chile
IPS
Via Panisperna 207
Rome, Italy
T: 39 6 485 692 F: 39 6 81-7877

Greco, Roberto
RASSEGNA SINDICALE
Via Dei Frentani 5
Rome 00185, Italy

Guanero, Renato, Mr.
Presidente
COMMISSIONE AMBIENTALE DI BARI
Court of Cassation - Piazza Cavour
Rome 00193, Italy
T: 39 6 6868597 F: 39 6 6874170

Guglielmo, Zanetta
MENABO
FILT-CGIL Via G.B.Morgagni 27A R.Francisco Ottaviano 23/701 RJ-RJ
Rome 0061, Italy
T: 39 6 8496504 tel.Rio 021 52 F: 39 6 8496435

Guidotti, Susanna
Correspondent
SINISTRA ARCOBALENO NEWS
Via Cavour 2
Florence, Italy
T: 55 2387477 F: 55 2381197

Handl, Maria Matilda, Sra.
Osb
MISSIONARY BENEDICTINE SISTERS
Via dei Bevilacqua, 60
Rome 00163, Italy
T: 39 6 6641-9002

Harris, Philip
Editor
SOCIETY FOR INTERNATIONAL DEVELOPMENT
Palazzo Civilta del Lavero
00144 Rome, Italy
T: 5917897 F: 5919836

Hawtin,
INTERNATIONAL BOARD FOR PLANT GENETIC RESOURCES
Via delle Sette Chiese 142
Rome 00145, Italy
T: 39 6 5744719 F: 39 6 5750309
TL: 4900005332IBRUI, E: 157CGI101

Iannelli, Mario, Mr.
CORTE SUPREMA DI CASSAZIONE
Court of Cassation - Piazza Cavour
Rome 00193, Italy
T: 39 6 6868597 F; 39 6 6874170

Ingrao, Renata, Ms.
LEGA PER L'AMBIENTE
Via Salaria 280
Rome, Italy
T: 39 6 8841552 F: 39 6 8443505
E: 65:Eco004

Iuliano, Giuseppe, Mr.
Dipartimento Internazionale
CONF. ITALIANA SINDACATI LAV. - CISL
Via Po,21
Rome, Italy
T: 39 6 8473357 F: 39 6 8413782

Joannilli, Edmondo, Mr.
Executive
ENTE NAZIONALE POR L'ENERGIA ELETRICA
Via G.N. Martini 3
Rome 00198, Italy
T: 39 6 85091
TL: 610518

Kleinmoedig, Mario
The Netherlands
IPS
Via Panisperna 207
Rome, Italy
T: 39 6 485 692 F: 39 6 81-7877

Krombacher, Margareta, Sra.
Rel Sister Of Tutzing
MISSIONARY BENEDICTINE SISTERS
Via dei Bevilacqua, 60
Rome 00163, Italy
T: 39 6 664-19002

Labini, Giovanni Sylos
INTERNATIONAL COURT OF THE ENVIRONEMNT
Court of Cassation - Piazza Cavour
Rome 00193, Italy
T: 39 6 6868597 F: 39 6 6874170

Landi, Giuseppe L., Mr.
Secretario Geral
POLITECNICO INTERNAZIONALE PER LO SVILUPPO INDUSTRIALE ED ECONOMICO
Via S. Marco,13
Iesi (Ancona) 60035, Italy
T: 0039731 64253 F: 0039731 64384
TL: 561007, E: 364

Landini, Paolo, Mr.
Deputy Manager
ENTE NAZIONALE POR L'ENERGIA ELETRICA
Via G.N. Martini 3
Rome 00198, Italy
T: 39 6 85091
TL: 610518

Lanly, Jean-Paul, Mr.
U.N. Officer
F.A.O. (U.N.)
13 via odoardo beccari
Rome 00154, Italy
T: 39 6 57973574 F: 39 6 57975137

Lanzanova, Lucia, Ms.
ICEL
Via Salvini 3
Milan 20122, Italy
T: 39 2 799144 F: 39 2 76000101

Lavitola, Valter, Mr.
CACCIATORI DELL'AMBIENTE
44 Plinio Street
Rome, Italy
T: 39 6 67781 F: 39 6 873927

Leomardi, Ruggro
Coordinator
DRAGOSTIMI-RIZZOLI
Via Gozzi 1A,
Milan, Italy
T: 39 2 70023259

Lilli, Paolo, Mr.
Vice-Prefetto
COM. MONTANA ALTO CHIASCO GUBBIO
Via Matteotti, 17
Gubbio, Italy
T: 39 75 927-4604 F: 39 75 927-4720

Lima, Sirlandio De
TELEMONTECARLO
Piazza della Balduina 49
Rome 00136, Italy
T: 396 3304299/265 F: 396 3304257

Limonet, Renato, Mr.
Assessore
ASSESSORATO AMBIENTE REGIONE VALLE D'AOSTA
Via Cerise 1
Aosta 11100, Italy
T: 39 165 238 097 F: 39 165 238 089

Lingsma, Tjitske
The Netherlands
IPS
Via Panisperna 207
Rome, Italy
T: 39 6 485 692 F: 39 6 81-7877

Lion, Marco, Sr.
Coordinator
FED. VERDI MARCHE
Via Circonvallazione 44
Senigallia - Ancona, Italy
T: 071 598381 F: 071 598458

Lispi, Ugo, Mr.
COMITATO PROMOTORE
Court of Cassation - Piazza Cavour
Rome 00193, Italy
T: 39 6 6868597 F: 39 6 6874170

Lo Curto, Aldo, Mr.
Medico Pesquisador
Via Paradiso 7
Canzo - Como 22035, Italy
T: 39 31 682512

Lodi, Leopoldo
A.M.Director
IDROTECNA
V. Pilsudsky, 124
Rome 00187, Italy
T: 39 6 847-9323 F: 39 6 807-7497

Lojacomo, Giuseppe, Mr.
Editor
EPIDEMIOLOGIA E PREVENZIONE - SCIENTIFIC JOURNAL
p. Trinita dei Pellegrini 91
Rome, Italy
T: 39 6 684-0637 F: 39 6 679-7655

Lomardi, Gabriel, Mr.
MEPES BRASIL
Via Monastero
Parona Verona 37025, Italy
T: 0039 45 941755

Lorenzo, Aristide De
IL MATTINO
Court of Cassation - Piazza Cavour
Rome 00193, Italy
T: 39 6 6868597 F: 39 6 6874170

Lucchetta, Maria Elisa, Ms.
TRIBUNALE NAZIONALE AMBIENTE
Via Elea 4
Rome 00183, Italy
T: 39 6 7009345 F: 39 6 9417041

Luigi, Martinelli, Mr.
Capo Del.
CONCIGLIO REGIONALE LOMBARDIA
Via Ugo Bassi, 2
Milano, Italy
T: 39 2 69751 F: 39 2 6975497

Maggioni, Maurizio, Mr.
Con. Reg. Le.
CONCIGLIO REGIONALE LOMBARDIA
Via Ugo Bassi, 2
Milano, Italy
T: 39 2 69751 F: 39 2 6975497

Magni, Roberto, Sr.
Agronomo-Voluntario
FONDAZIONE GIUSEPPE TOVINI
Av. Copacabana 945/1001
Rio de Janeiro - RJ 22060, Italy
T: 55 21 255-3352 F: 55 21 242-9871
TL: 32628

Magri, Gisberto, Mr.
Consigliere
REGIONE LOMBARDIA
Via Ugo Bassi 2
Milan 20159, Italy
T: 39 2 6975197 F: 39 2 67655948

Mahler, Marie Antoinette, Ms.
FAO OF UNITED NATIONS
Via Dei Pirenei
Rome, Italy
T: 39 6 5917051

Marchis, Sergio De, Mr.
President Of Tv
COM. MONTANA ALTO CHIASCO GUBBIO
Via Matteotti, 17
Gubbio, Italy
T: 39 75 927-4604 F: 39 75 927-4720

Marcialis, Sergio, Mr.
Expositore
FLACAM
Via Peretti II Trav. S/#
Cagliari 09134, Italy
T: 0039070 53-2040 F: 0039070 53-2040

Marco, Iob, Mr.
Responsavel No Brasil
O.N.G. C.E.V.I.
Via Prampero,43
Gemona - UD 33013, Italy

Marconi, Pia, Ms.
MONTEDISON SPA
Foro Buonaparte 31
Milano 20121, Italy
T: 39 2 627-05808 F: 39 2 627-05952

Mario, Ugo Di
RAI - RADIO TELEVISIONE ITALIANA/STUDIO IMMAGINE S.R.L.
Rome, Italy

Maron, Jose Pelaez, Mr.
CEISAL - GRUPO DE TRABAJO DE JURISPRUDENCIA
P.zza Castello 11 - Casella postale 81
Sassari 07100, Italy
T: 39 79 237364 F: 39 79 300083

Marossi, Walter, Mr.
M. Director
CONCIGLIO REGIONALE LOMBARDIA
Via Dante, 12
Milano, Italy
T: 39 2 801071 F: 39 2 340245
E: 20121

Martinelli, Luigi, Mr.
Presidente Consiglio
REGIONE LOMBARDIA
Via Ugo Bassi 2
Milan 20159, Italy
T: 39 2 6975197 F: 39 2 67655948

Marzotto, Lidia, Ms.
LELIO BASSO INTERNATIONAL FOUNDATION FOR THE RIGHTS AND LIBERATION OF PEOPLES
Via Della Dogana Vecchia 5
Rome, Italy
T: 39 6 654-1463 F: 39 6 687-7774

Masella, Maria Grazia, Ms.
TERRA NUEVA
Vio Urbana 156
Rome, Italy
T: 39 6 485534 F: 39 6 4747599

Mastrantoni, Primo, Mr.
Consigliere Regionale
CONSIGLIO REGIONALE DEL LAZIO
Via Della Pisana, 1301
Rome, Italy
T: 39 6 500 0672 F: 39 6 500 0956

Mateu, Monica, Sra.
MBA BOCCONI
Via Broggi 8, apt. 7
Milan 20129, Italy
T: 39 2 295-28274 F: 39 2 583-66623

Maurizi, Mauro, Mr.
RAI TV TGI NEWS
Via Teu lade, 66
Rome 00100, Italy
T: 39 6 314 427

Mazzonna, Claudio
RAI - RADIO TELEVISIONE ITALIANA/STUDIO IMMAGINE S.R.L.
Rome, Italy

Mazzonna, Marco
RAI - ITALIAN TELEVISION
Rome, Italy

Menchini, Piergiorgio, Mr.
Project Director
TERRA NUOVA
Via Urbana 156
Rome 00184, Italy
T: 39 6 4747859 F: 39 6 4747599

Milesi, Gianangelo, Mr.
CESVI
Via Pignolo 50
Bergamo, Italy
T: 39 35 243990 F: 39 35 229096

Minore, Vito
Tv Director
RAI RADIOTELEVISIONE ITALIANA
Via Pasubio, 2
Rome, Italy
T: 39 6 37264962 F: 39 6 382674

Monguzzi, Carlo, Mr.
Consigliere
REGIONE LOMBARDIA
Via Ugo Bassi 2
Milan 20159, Italy
T: 39 2 6975197 F: 39 2 67655948

Montanari, Pinuccia
COMITATO PROMOTORE
Court of Cassation - Piazza Cavour
Rome 00193, Italy
T: 39 6 6868597 F: 39 6 6874170

Montis, Antonella De, Mr.
Court of Cassation - Piazza Cavour
Rome 00193, Italy
T: 39 6 6868597 F: 39 6 6874170

Morezzi, Ettore, Mr.
Director
ENTE NAZIONALE POR L'ENERGIA ELETRICA
Via G.N. Martini 3
Rome 00198, Italy
T: 39 6 85091
TL: 610518

Moriconi, Maurizio
TRIBUNALIE INTERNAZIONALE DELL'AMBIENTE
Court of Cassation - Piazza Cavour
Rome 00193, Italy
T: 39 6 6868597 F: 39 6 6874170

Mosca, Cataldo
CONSIGLIO PROVINCIALE DI BARI
Court of Cassation - Piazza Cavour
Rome 00193, Italy
T: 39 6 6868597 F: 39 6 6874170

Munari, Teresa, Sra.
PRESS
Viale M. Pilsovski
Rome, Italy
T: 39 6 84791

Musto, Paola, Ms.
TRIBUNALE INTERNAZIONALE DEL AMBIENTE
Court of Cassation - Piazza Cavour
Rome 00193, Italy
T: 39 6 6868597 F: 39 6 6874170

Mutschlechner, Elisabeth, Ms.
MONTEDISON SPA
Viale Castrense
Rome 00184, Italy
T: 39 6 70301862

Neri, Luciano, Mr.
Consigliere
ASSESSORATO AMBIENTE REGIONE UMBRIA
P.za Partigiani 1
Perugia 06100, Italy
T: 39 75 61167 F: 39 75 5042732

Nigri, Antonio, Mr.
Dirigente
AMMINISTRAZIONE PROVINCIALE FOGGIA
Pza XX Settembre
Foggia 71100, Italy
T: 39 861 675112 F: 39 861 20007

Notargiovanni, Alessandro, Sr.
CGIL - SINDICATO
Corso d'Italia nr. 25
Rome 00198, Italy
T: 39 6 84761 F: 39 6 8845683

Notargiovanni, Sandro, Mr.
CONF. ITALIANA SINDICATI LAV.
C.So d'Italia 25
Rome 00198, Italy
T: 39 6 84761

O'Mahony, Donal, Mr.
COM. MONTANA ALTO CHIASCO GUBBIO
Via Matteotti, 17
Gubbio, Italy
T: 39 75 927-4604 F: 39 75 927-4720

Okada,
INTERNATIONAL BOARD FOR PLANT GENETIC RESOURCES
Via delle Sette Chiese 142
Rome 00145, Italy
T: 39 6 5744719 F: 39 6 5750309
TL: 4900005332IBRUI, E: 157CGI101

Olori, Toye
Nigeria
IPS
Via Panisperna 207
Rome, Italy
T: 39 6 485 692 F: 39 6 81-7877

Onorati, Antonio, Mr.
President
CENTRO INTERNAZIONALE CROCEVIA
Via Ferraironi 88/6
Rome 00172, Italy
T: 39 6 2413976 F: 39 6 2424177

Orlando, Al, Mr.
Presidente
CATANIAIMMAGINA
V. Le A. Doria, 69
Catania 95123, Italy
T: 095 50-5620 F: 095 50-5620

Ortu, Maurizio, Sr.
MAIS
Ave.Baltimore 298 Esplanada
Nova Iguacu - RJ 26015, Italy
T: 55 21 767-4183 F: 55 21 767-4183

Osava, Mario
Brazil
IPS
Via Panisperna 207
Rome, Italy
T: 39 6 485 692 F: 39 6 81-7877

Paganini, Valeria, Sra.
Director
ISTITUTO NAZIONALE PER IL COMMERCIO ESTERO
Via Liszt 21
Rome, Italy
T: 39 6 59921 F: 39 6 591-0508
TL: 610160 ICERM, E: 00144

Pagliarulo, Giovanni
CORTE SUPREMA
Court of Cassation - Piazza Cavour
Rome 00193, Italy
T: 39 6 6868597 F: 39 6 6874170

Palladino, Anna
RAI - ITALIAN TELEVISION
Rome, Italy

Panaccia, Viviana
INTERNATIONAL COURT OF THE ENVIRONEMNT
Court of Cassation - Piazza Cavour
Rome 00193, Italy
T: 39 6 6868597 F: 39 6 6874170

Paolo, Arrigoni, Mr.
Cons. Re. Le.
CONCIGLIO REGIONALE LOMBARDIA
Via Ugo Bassi, 2
Milano, Italy
T: 39 2 69751 F: 39 2 6975497

Pasolini, Desideria, Ms.
ITALIA NOSTRA
Via N. Porpora 22
Rome, Italy
T: 39 6 8416765 F: 39 6 8844634

Pedemonte, Enrico
Science Editor
L'ESPRESSO MAGAZINE
Via PO 12 001
Roma, Italy
T: 39 6 84781 F: 39 6 8845167

Pelanda, Carlo
TRIBUNALE INTERNAZIONALE DELL'AMBIENTE
Court of Cassation - Piazza Cavour
Rome 00193, Italy
T: 39 6 6868597 F: 39 6 6874170

Pennacchini, Paolo
Court of Cassation - Piazza Cavour
Rome 00193, Italy
T: 39 6 6868597 F: 39 6 6874170

Pezzola, Maria Guiditta
RAI - ITALIAN TELEVISION
Rome, Italy

Piazza, Camillo, Mr.
Consigliere
REGIONE LOMBARDIA
Via Ugo Bassi 2
Milan 20159, Italy
T: 39 2 6975197 F: 39 2 67655948

Picciurro, Guiseppe
RAI - RADIO TELEVISIONE ITALIANA/STUDIO IMMAGINE S.R.L.
Rome, Italy

Piermario, Manca, Sr.
Representante
PROVINCIA DI SASSARI
Piazza D'Italia
Sassari, Italy
T: 079 1232369

Piermatti, Roberto, Mr.
Assessore
ASSESSORATO AMBIENTE REGIONE UMBRIA
P.za Partigiani 1
Perugia 06100, Italy
T: 39 75 61167 F: 39 75 5042732

Pietro, Sanga Mario
Editor
RAI TV TG2
Via Tulada, 66
Rome 00195, Italy
T: 319663 F: 386608

Pinzuti, Alberto
Tv Programer
RAI RADIOTELEVISIONE ITALIANA
Via Pasubio, 2
Rome, Italy
T: 39 6 37264962 F: 39 6 382674

Pirro, Deirdre Exell
CENTRO NAZIONALE RICHERCHE
Court of Cassation - Piazza Cavour
Rome 00193, Italy
T: 39 6 6868597 F: 39 6 6874170

Pnnamperuma, Cyril, Mr.
Fellow
THIRD WORLD ACADEMY OF SCIENCES (TWAS)
c/o Intl Centre for Theoretical Physics - Strada Costiera 11 - P,O,BOX 586
Triestre 34100, Italy
T: 39 40 2240-251 F: 39 40 224-559
TL: 460392 ICTP

Polimeni, Piero, Mr.
Director
CENTRO INTERVENTO REGIONALE COOPEREZ
Via Sanna 43/A
Reggio Calabria 89100, Italy
T: 39 965 812345 F: 39 965 812560

Pota, Eugenia, Ms.
Dirigente
AMMINISTRAZIONE PROVINCIALE FOGGIA
Pza XX Settembre
Foggia 71100, Italy
T: 39 861 675112 F: 39 861 20007

Przewozny, Padre Bernardo, Mr.
CENTRO FRANSCESCANO STUDI AMBIENTALI
Court of Cassation - Piazza Cavour
Rome 00193, Italy
T: 39 6 6868597 F: 39 6 6874170

Quercia, Piero, Mr.
TRIBUNALE NAZIONALE AMBIENTE
Via Elea 4
Rome 00183, Italy
T: 39 6 7009345 F: 39 6 9417041

Raffi, Franco, Mr.
President
COM. MONTANA ALTO CHIASCO GUBBIO
Via Matteotti, 17
Gubbio, Italy
T: 39 75 926-4604 F: 39 75 926-4720

Ranieri, Renata
RAI - ITALIAN TELEVISION
Rome, Italy

Raymond, Ruth D., Ms.
Public Affairs Officer
INTERNATIONAL BOARD FOR PLANT GENETIC RESOURCES - IBPGR
Via Delle Sette Chiese 142
Rome 00145, Italy
T: 39 6 574-4719 F: 39 6 575-0309
E: CGI101 / IBPGR

Ricci, Arnaldo Cesar
Brazil
IPS
Via Panisperna 207
Rome, Italy
T: 39 6 485 692 F: 39 6 81-7877

Riccioni, Roberto, Mr.
Engineer
ENEL
Viale Giulio Cesare 29
Bergamo, Italy
T: 39 35 224 850

Ricco, Lubrano Giovanni Di, Mr.
Giudice
CORTE SUPREMA DI CASSAZIONE
Court of Cassation - Piazza Cavour
Rome 00193, Italy
T: 39 6 6868597 F: 39 6 6874170

Rivolta, Francesco, Mr.
Con. Reg. Le
CONCIGLIO REGIONALE LOMBARDIA
Via Ugo Bassi, 2
Milano, Italy
T: 39 2 69751 F: 39 2 6975497

Rizzi, Giacomo, Mr.
Ingegnere
PRESIDENTE ORDINE INGEGNERE
Palazzo Della Concordia -Roma
Rome, Italy
T: 39 6 591-1790 F: 39 6 591-1795

Rizzi, Paola Tamburrini, Ms.
Assistente
ASSISTENTE UNIVERSITA INGEGN. FISICA TECNICA
Via Eudossiana 18- Dip. Fisica Tecnica
Rome, Italy
T: 39 6 481-9585 F: 39 6 481-2120

Rizzo, Valeria, Ms.
MINISTRY OF ENVIRONMENT-ITALY
P.Venezia
Rome 00184, Italy
T: 39 6 7029210 F: 39 6 7027195

Rocha, Valeria, Ms.
Resp.Depto.
CISS
Via Benedetto D,Acquisto 30
Palermo, Italy
T: 91 39 611-1062 F: 91 39 33-2739

Rolla, Antonio, Mr.
Dirigente
AMMINISTRAZIONE PROVINCIALE FOGGIA
Pza XX Settembre
Foggia 71100, Italy
T: 39 861 675112 F: 39 861 20007

Rosaria, Onida M, Ms.
CESVI
Via Pignolo 50
Bergamo, Italy
T: 39 35 243990 F: 39 35 229096

Rossana, Ciani
Journalist
RAI TV TG2
Via Tulada, 66
Rome 00195, Italy
T: 319663 F: 386608

Ruggero, Michele
UNITA
Via Dei Taurini 19
Rome 00185, Italy
T: 39 6 44490258 F: 39 6 4455305

Ruiz-Gimenez, Guadalupe, Ms.
CEISAL - GRUPO DE TRABAJO DE JURISPRUDENCIA
P.zza Castello 11 - Casella postale 81
Sassari 07100, Italy
T: 39 79 237364 F: 39 79 200083

Rutelli, Francesco, Mr.
Presidente
CAMERA DEI DEPUTATI / GRUPPO VERDI
Piette Montecitorio
Rome 00186, Italy
T: 39 6 6760-2191 F: 39 6 3215491

Santori, Isabella
RAI - RADIO TELEVISIONE ITALIANA/STUDIO IMMAGINE S.R.L.
Rome, Italy

Sarli, Luigi, Mr.
IRITECNA
Yiale Liegi 33
Rome 00137, Italy
T: 39 6 8301848 F: 39 6 8555744

Savlia, Paolo
INTERNATIONAL JURIDICAL ORGANIZATION FOR ENVIRONMENT & DEVELOPMENT
Via Barberni 3
Rome 00187, Italy
T: 39 6 4742117 F: 39 6 4745779

Sbrenna, Giuseppe, Mr.
Regional Counselor
REGIONE UMBRIA/ITALY
Palazzo Cesaroni, Piazza Italia, 2
Perugia, Italy
T: 39 75 696-3236 F: 39 75 696-3573

Scalabrini, Laura, Ms.
Consigliere Regionale
CONSIGLIO REGIONALE DEL LAZIO
Via Della Pisana, 1301
Rome, Italy
T: 39 6 500 0672 F: 39 6 500 0956

Schaluck, Herman, Rev.
Reverendo
FRANCISCANS AT THE UNITED NATIONS
Curia Generalizia, Via Santa Maria Mediatrici, 25
Rome 00165, Italy

Scheele, Reinier J., Mr.
CEISAL - GRUPO DE TRABAJO DE JURISPRUDENCIA
P.zza Castello 11 - Casella postale 81
Sassari 07100, Italy
T: 39 79 237364 F: 39 79 200083

Schiavone, Giuseppe
Vice Presidente
INSTITUTO 'DE GASPERI'
Court of Cassation - Piazza Cavour
Rome 00193, Italy
T: 39 6 6868597 F: 39 6 6874170

Schiavoni, Anna, Ms.
Vice President
AIDOS
Via Del Giubbonari 30
Rome 00186, Italy
T: 39 6 6873214 F: 39 6 6872549

Sequeira, Maricel
Costa Rica
IPS
Via Panisperna 207
Rome, Italy
T: 39 6 485 692 F: 39 6 81-7877

Serena, Palieri Maria, Ms.
LEGA AMBIENTE
Via Degli Scipioni 92/ 00192
Rome, Italy
T: 39 6 3709993 F: 39 6 444-9026

Siniscalco, Domenico, Mr.
Director
ENI ENRICO MATTEI FOUNDATION
Via Sofia 27
Milan 20122, Italy
T: 39 2 52030436 F: 39 2 52036940

Sironi, Virgilio, Mr.
Consigliere
REGIONE LOMBARDIA
Via Ugo Bassi 2
Milan 20159, Italy
T: 39 2 6975197 F: 39 2 67655948

Soares, Walter Luis Cirne
RAI - RADIO TELEVISIONE ITALIANA
Rome, Italy

Somma, Roberto, Mr.
INTERNATIONAL COURT OF THE ENVIRONMENT
Court of Cassation - Piazza Cavour
Rome 00193, Italy
T: 39 6 6868597 F: 39 6 6874170

Stefano, Vincenze De, Mr.
Dirigente
AMMINISTRAZIONE PROVINCIALE FOGGIA
Pza XX Settembre
Foggia 71100, Italy
T: 39 861 675112 F: 39 861 20007

Stepanat, Bruno
RAI - ITALIAN TELEVISION
Rome, Italy

Storelli, Vittorio, Dr.
Via Crispi, 72
Napoles, Italy
T: 39 81 669325 F: 39 81 407850

Stozelli, Vittorio, Mr.
INTERNATIONAL SOLAR ENERGY SOCIETY - ISES
Via Bozmida, 2
Rome, Italy

Summa, Giancarlo
Stringer Para America Latina
L'UNITA
Via dei Taurini 19
Rome, Italy
T: 39 6 44901 F: 39 6 4490328

Swantz, Marja-Liisa, Ms.
SOCIETY FOR INTERNATIONAL DEVELOPMENT
Palazzo Civilta del Lavoro
Rome 00144, Italy
T: 39 6 591-7897 F: 39 6 591-9836

Tartaglione, Nicola, Mr.
Assessore
ASSESSORATO AMBIENTE REGIONE VALLE D'AOSTA
Via Cerise 1
Aosta 11100, Italy
T: 39 165 238 097 F: 39 165 238 089

Tassara, Carlo, Mr.
Deputy Director
CISP - COMITATO INTERNAZIONALE SVILUPPO POPOLI
Viale Medaglie D'Oro, 201
Rome 00136, Italy
T: 39 6 347431 F: 39 6 344461

Tola, Vittoria, Ms.
Consigliere Regionale
CONSIGLIO REGIONALE DEL LAZIO
Via Della Pisańa, 1301
Rome, Italy
T: 39 6 500 0672 F: 39 6 500 0956

Tomasin, Paolo, Mr.
MLAL
Via Palestro 56
Rome, Italy
T: 39 6 446-9658 F: 39 6 494-0635

Torres, Marcio
TELEMONTECARLO
Piazza della Balduina 49
Rome 00136, Italy
T: 396 3304299/265 F: 396 3304257

Torsani, Romano
RAI - RADIO TELEVISIONE ITALIANA/STUDIO IMMAGINE S.R.L.
Rome, Italy

Trevisiol, Roberto, Mr.
I U A V
CA Tron S. Cruce 1957
Venice 30135, Italy
T: 39 41 796-223 F: 39 41 524-0403

Trojani, Carlo Antonio, Mr.
COMITATO PROMOTORE
Court of Cassation - Piazza Cavour
Rome 00193, Italy
T: 39 6 6868597 F: 39 6 6874170

Tudini, Angelica, Ms.
Research Assistant
ENI ENRICO MATTEI FOUNDATION
Via Sofia 27
Milan 20122, Italy
T: 39 2 52030436 F: 39 2 52036940

Tudini, Siniscalco, Sr.
FONDAZIONE ENI ENRICO MATTEI
Via Santa Sofia 27
Milan 20122, Italy

Turroni, Sauro
DEPUTADO PARLAMENTO ITALIANO
Camara de Depotati Monteytoro
Rome, Italy

Vairano, Roberto
RAI - ITALIAN TELEVISION
Rome, Italy

Valente, Marcela
Argentina
IPS
Via Panisperna 207
Rome, Italy
T: 39 6 85 692 F: 39 6 81-7877

Vassallo, Nico, Sr.
EUROPEAN CENTER FOR ENVIRONMENTAL STUDIES
ECES - Via Po 14
Turin 10123, Italy

Vetrali, T., Mr.
TERRA MATER
Piazza Oderist 1
Gubbio, Italy
T: 39 75 922-0693

Vigliar, Paola, Mr.
IRITECNA
Rome 00198, Italy
T: 39 6 84461 F: 39 6 85-7575

Wennik, Karen Nelcimar Oliveira, Sra.
Via Vicenzo Qasini,6 Piazza Delle Bandenere
Milano, Italy
T: 39 2 4044936

White, Jr., T. Kelley, Mr.
Div. Director
FOOD & AGRICULTURAL ORG. OF THE UN (FAO)
via Terme di Caracalla
Rome, Italy
T: 39 6 5797-3462 F: 39 6 5797-3152

Williams, Morris, Mr.
President
SOCIETY FOR INTERNATIONAL DEVELOPMENT
Palazzo Civilta del Lavoro
Rome 00144, Italy
T: 39 6 5917897 F: 39 6 5919836
TL: 616484 SID I

Wolpin, Milton, Mr.
EARTH SAVE
1321 Palms Bl.
Venice 90291, Italy
T: 39 41 392-1685

Zanzi, Andrea, Mr.
FILCAMS - CGIL
Via Marconi, 67-2
Bolonha 40122, Italy
T: 39 51 25-3230 F: 39 51 25-1062

Zapalla, Francesco
TRIBUNALE INTERNAZIONALE DELL'AMBIENTE
Court of Cassation - Piazza Cavour
Rome 00193, Italy
T: 39 6 6868597 F: 39 6 6874170

Zorzolli, Battista G., Mr.
Director
ENTE NAZIONALE POR L'ENERGIA ELETRICA
Via G.N. Martini 3
Rome 00198, Italy
T: 39 6 85091
TL: 610518

UN AND UNCED

Botero, L.S.
Food and Agriculture Organization of the United Nations
Via delle Terme di Caracalla
00100 Rome, Italy
T: 396 57971

Camara, J.
Food and Agriculture Organization of the United Nations
Via delle Terme di Caracalla
00100 Rome, Italy
T: 396 57971

De, H.
Food and Agriculture Organization of the United Nations
Via delle Terme di Caracalla
00100 Rome, Italy
T: 396 57971

Esquinas, J.T.
Food and Agriculture Organization of the United Nations
Via delle Terme di Caracalla
00100 Rome, Italy
T: 396 57971

Garcia, S.M.
Food and Agriculture Organization of the United Nations
Via delle Terme di Caracalla
00100 Rome, Italy
T: 396 57971

Holt, C.M.
Food and Agriculture Organization of the United Nations
Via delle Terme di Caracalla
00100 Rome, Italy
T: 396 57971

Killingsworth, Kay
Food and Agriculture Organization of the United Nations
Via delle Terme di Caracalla
00100 Rome, Italy
T: 396 57971

Mahler, J.P., Mr.
Food and Agriculture Organization of the United Nations
Via delle Terme di Caracalla
00100 Rome, Italy
T: 396 57971

Mann, Wendy, Ms.
Senior Liaison Officer
FOOD AND AGRICULTURE ORGANIZATION OF THE UNITED NATIONS - FAO
Via delle Terme de Caracalla
Rome 00100, Italy
T: 39 6 579 73842 F: 39 6 579 75784
TL: 610181 FAO I

Moreno, R.
Food and Agriculture Organization of the United Nations
Via delle Terme di Caracalla
00100 Rome, Italy
T: 396 57971

Murray, C.H.
Food and Agriculture Organization of the United Nations
Via delle Terme di Caracalla
00100 Rome, Italy
T: 396 57971

Rojas, G.
Food and Agriculture Organization of the United Nations
Via delle Terme di Caracalla
00100 Rome, Italy
T: 396 57971

Rosenegger, P.
Food and Agriculture Organization of the United Nations
Via delle Terme di Caracalla
00100 Rome, Italy
T: 396 57971

Saouma, Eduardo
Food and Agriculture Organization of the United Nations
Via delle Terme di Caracalla
00100 Rome, Italy
T: 396 57971

Sombroek, W.G., Mr.
Food and Agriculture Organization of the United Nations
Via delle Terme di Caracalla
00100 Rome, Italy
T: 396 57971

Syrogianis-Camara, J.
Food and Agriculture Organization of the United Nations
Via delle Terme di Caracalla
00100 Rome, Italy
T: 396 57971

Toros, H.
Food and Agriculture Organization of the United Nations
Via delle Terme di Caracalla
00100 Rome, Italy
T: 396 57971

White, K.
Food and Agriculture Organization of the United Nations
Via delle Terme di Caracalla
00100 Rome, Italy
T: 396 57971

Zehni, M.S.
Food and Agriculture Organization of the United Nations
Via delle Terme di Caracalla
00100 Rome, Italy
T: 396 57971

ADVISORS

Colombo, Umberto, Prof.
President
National Agency for the Alternative Energies
00198 Rome, Italy
T: 39 6 85 281 F: 39 6 85 282/591

Ullsten, Ola, H.E. Mr.
Ambassador Of Sweden To Italy : Former Foreign Minister Of Sweden
I-00100 Roma, Italy
T: 39 6 884 04 41 F: 39 6 440 27 71

PARLIAMENTARIANS / SPIRITUAL LEADERS

Andreini, Elios, Senator
Senato della Republica Palazzo Madama 00100
Roma, Italy
T: 67061

Balocchi, , Dep.
Senato della Republica Palazzo Madama 00100
Roma, Italy
T: 67061

Bonino, Emma, Dep.
Senato della Republica Palazzo Madama 00100
Roma, Italy
T: 67061

Cutrera, Achille, Senator
MP
Senato della Republica Palazzo Madama 00100
Roma, Italy
T: 67061

Golfari, Cesare, Senator
Senato della Republica Palazzo Madama 00100
Roma, Italy
T: 67061

Martin, Diarmud, Fr.
Under Secretary
Pontificia Commissione
Justitia et Pax Piazza S. Calisto, 16
Vatican City, Italy
T: 39 6 698 7191 F: 39 6 698 7205

Napoli, Vito, Dep.
Senato della Republica Palazzo Madama 00100
Roma, Italy
T: 67061

Rastrelli, Antonio, Senator
Senato della Republica Palazzo Madama 00100
Roma, Italy
T: 67061

Rossi, Oreste, Dep.
Senato della Republica Palazzo Madama 00100
Roma, Italy
T: 67061

Talevi, Paulo, Interpreter
Senato della Republica Palazzo Madama 00100
Roma, Italy
T: 67061

JAMAICA - GOVERNMENT

Aldridge, T.E., Dr.
Adviser
Ministry of Tourism and the Environment
Permanent Mission to the United Nations, 2 Dag Hammerskjold Plaza, 15th Floor New York, NY, 10017
F: 212 688 7040
TL: 212 308 3730

Junor, John, Hon.
Minister of Tourism and the Environment
Permanent Mission of Jamaica
2 Dag Hammerskjold Plaza, 15th Floor
New York, NY 10017, United States of America
T: 212 688 7040 F: 212 308 3730

Lewis, R.
Inspector
Permanent Mission of Jamaica
2 Dag Hammerskjold Plaza, 15th Floor
New York, NY 10017, United States of America
T: 212 688 7040 F: 212 308 3730

Lynch, Jacqueline, Miss
Special Assistant to the Prime Minister
Permanent Mission of Jamaica
2 Dag Hammerskjold Plaza, 15th Floor
New York, NY 10017, United States of America
T: 212 688 7040 F: 212 308 3730

McDonald, Franklin, Mr.
Executive Director
Naturial Resources Conservation Authority
Permanent Mission to the United Nations, 2 Dag Hammerskjold Plaza, 15th Floor New York, NY, 10017
F: 212 688 7040
TL: 212 308 3730

Millis, OJ, Don, Hon.
Chairman
Natural Resources Conservation Authority
Permanent Mission to the United Nations, 2 Dag Hammerskjold Plaza, 15th Floor New York, NY, 10017
F: 212 688 7040
TL: 212 308 3730

O'Meally-Nelson, Blossom, Dr.
Permanent Mission of Jamaica
2 Dag Hammerskjold Plaza, 15th Floor
New York, NY 10017, United States of America
T: 212 688 7040 F: 212 308 3730

Patterson, Percival James, Hon.
Q.C., M.P., Prime Minister
Jamaica,
2 Dag Hammerskjold Plaza, 15th Floor
New York, NY 10017, United States of America
T: 212 688 7040 F: 212 308 3730

Royes, Veronica Irene Joy
NGO Delegate For Jamaica
NGO
4 Charlemont Ave
Kingston 6, Jamaica
T: 92 75949

Stone, Alison, Miss
Foreign Service Officer
Permanent Mission of Jamaica
2 Dag Hammerskjold Plaza, 15th Floor
New York, NY 10017, United States of America
T: 212 688 7040 F: 212 308 3730

Thorpe, Patrecia Federicka, Ms.
Graduate Student
UNIVERSITY OF THE WEST INDIES
Department of Government
Mona Kingston 7, Jamaica
T: 1 809 927-2592

NGO

Levy, Horace
CVSS
32 1/2 Duke Street (7th Floor)
Kingston, Jamaica

Smith, David
JCDT
46 Duke Street
Kingston, Jamaica

PRESS

Chaplin, Kenneth
Editor
JAMPRESS
48 A Stilwell Rd.
Kingston, 8, Jamaica, Jamaica

GLOBAL FORUM

Bernard, Marion, Ms.
SISTREN/WEDO
20 Kensington Cres. Kgn 5
Kingston, Jamaica
T: 1 809 908-0895 F: 1 809 926-6990

Cupidon, Janet A., Ms.
Chief Executive Officer
COUNCIL OF VOLUNTARY SOCIAL SERVICES
320 Duke Street, Jamaica National Centre Building, 7th floor
Kingston, Jamaica
T: 1 809 92-29365 F: 1 809 92-25433

Gordon-Gofton, Lorna
ICAE
12 Rajaji Marg
New Delhi 110011, Jamaica
T: 91 11 301-4309

Henry, J. Macdonald, Ms.
Special Representative
FARGUHARSON INSTITUTE OF PUBLIC AFFAIRS
2 Lord Nelson Way
Kingston 10, Jamaica
T: 1 809 926-2321 F: 1 809 927-0795

Ross-Frankson, Joan, Ms.
Communications Consultant
DEVELOPMENT ALTERNATIVES WITH WOMEN FOR A NEW ERA (DAWN)
21 Windsor Avenue
Kingston, Jamaica
T: 1 809 9278701 F: 1 809 9266990

ADVISORS

Seaga, Edward, The Rt. Hon.
PC, MP, Leader Of The Opposition
Kingston 5, Jamaica
T: 1 809 92 99600 F: 1 809 92 98039

JAPAN - GOVERNMENT

Akao, Nobutoshi, H.E. Mr.
Ambassador
Global Environment Affairs
Permanent Mission to the United Nations, 866 United Nations Plaza, 2nd Floor New York, NY, 10017
F: 223 4300
TL: 751 1966

Aoki, Satoshi, Dr.
Office of Environmental Research Agriculture, Forestry and Fisheries Research Council, Ministry of Agriculture, Forestry and Fisheries
Permanent Mission to the United Nations, 866 United Nations Plaza, 2nd Floor
New York, NY 10017, United States of America
T: 223 4300 F: 751 1966

Aoyama, Shin, Mr.
Space Activities Planning Division
Research and Development Bureau, Science and Technology Agency
Permanent Mission to the United Nations, 866 United Nations Plaza, 2nd Floor
New York, NY 10017, United States of America
T: 223 4300 F: 751 1966

Asami, Yoshinori, Mr.
Deputy Director-General
Transport Policy Bureau, Ministry of Transport
Permanent Mission to the United Nations, 866 United Nations Plaza, 2nd Floor
New York, NY 10017, United States of America
T: 223 4300 F: 751 1966

Fujihara, Masahiko, Mr.
Global Environmental Affairs Office, Industrial Location and Environmental Protection Bureau
Permanent Mission to the United Nations, 866 United Nations Plaza, 2nd Floor
New York, NY 10017, United States of America
T: 223 4300 F: 751 1966

Fujiwara, Masahiro, Dr.
Director
Water Supply Division, Water Supply and Environmental Sanitation Department, Environmental Health Bureau, Ministry of Health and Welfare
Permanent Mission to the United Nations, 866 United Nations Plaza, 2nd Floor
New York, NY 10017, United States of America
T: 223 4300 F: 751 1966

Fukushima, Masato, Mr.
Office of Environment Promotion, Minister's Secretariat, Ministry of Agriculture, Forestry and Fisheries
Permanent Mission to the United Nations, 866 United Nations Plaza, 2nd Floor
New York, NY 10017, United States of America
T: 223 4300 F: 751 1966

Fukushima, Tadahiko, Mr.
Director
Lower Secondary School Division, Elementary and Secondary Educational Bureau, Ministry of Education
Permanent Mission to the United Nations, 866 United Nations Plaza, 2nd Floor
New York, NY 10017, United States of America
T: 223 4300 F: 751 1966

Funano, Ryuhei, Mr.
Chief
Overseas Land Improvement Cooperation Office, Design Division, Agricultural Structure Improvement Bureau, Ministry of Agriculture, Forestry and Fisheries
Permanent Mission to the United Nations, 866 United Nations Plaza, 2nd Floor
New York, NY 10017, United States of America
T: 223 4300 F: 751 1966

Goto, Satoru, Mr.
Counsellor in Charge of International Fisheries
Oceanic Fisheries Department, Fisheries Agency
Permanent Mission to the United Nations, 866 United Nations Plaza, 2nd Floor
New York, NY 10017, United States of America
T: 223 4300 F: 751 1966

Gurgulino De Souza, Heitor
Rector
UNITED NATIONS UNIVERSITY
53-70 Jingumae 5-Chome Shibuya-KU
Tokyo 150, Japan
T: 813 3499 2811 F: 813 5467 1201

Hamanaka, Hironori, Mr.
Director
Planning Division, Global Environment Department, Environment Agency
Permanent Mission to the United Nations, 866 United Nations Plaza, 2nd Floor
New York, NY 10017, United States of America
T: 223 4300 F: 751 1966

Hanai, Masamitsu, Dr.
Monuments and Sites Division, Cultural Properties Protection Department, Agency for Cultural Affairs
Permanent Mission to the United Nations, 866 United Nations Plaza, 2nd Floor
New York, NY 10017, United States of America
T: 223 4300 F: 751 1966

Hanazumi, Kazuo, Mr.
Director
Economic Affairs Division, United Nations Bureau, Ministry of Foreign Affairs
Permanent Mission to the United Nations, 866 United Nations Plaza, 2nd Floor New York, NY, 10017
F: 223 4300
TL: 751 1966

Hara, Eiji, Mr.
First Economic Cooperation Division
Coordination Bureau, Economic Planning Agency
Permanent Mission to the United Nations, 866 United Nations Plaza, 2nd Floor
New York, NY 10017, United States of America
T: 223 4300 F: 751 1966

Hashimoto, Masahiro, Mr.
Biochemical Industry Division, Basic Industries Bureau
Permanent Mission to the United Nations, 866 United Nations Plaza, 2nd Floor
New York, NY 10017, United States of America
T: 223 4300 F: 751 1966

Honda, Hiroshi
Senior Economist
THE INSTITUTE OF ENERGY ECONOMICS,
No.10 Mori Building, 1-18-1 Toranomon, Minato-KU
Tokyo 105, Japan
T: 03 (3501) 6625 F: 03 (3501) 7860

Horie, Takeshi
Director
Marine Hydrodynamics Divsion
Port & Harbour Research Institute Ministry Of Transport
1-1 3-Chome, Nagase, Yokosuka 239,, Japan
T: 0468 44 5015 F: 0468 42 9265

Horiuchi, Tetsuo, Mr.
Director
Environment and Ocean Development Division, Transport Policy Bureau
Permanent Mission to the United Nations, 866 United Nations Plaza, 2nd Floor
New York, NY 10017, United States of America
T: 223 4300 F: 751 1966

Hoshino, Kazuaki, Mr.
First Secretary
Embassy of Japan in Kenya
Permanent Mission to the United Nations, 866 United Nations Plaza, 2nd Floor
New York, NY 10017, United States of America
T: 223 4300 F: 751 1966

Hugono, Nobuyoshi, Dr.
Director-General
Communications Research Laboratory, Ministry of Posts and Telecommunications
Permanent Mission to the United Nations, 866 United Nations Plaza, 2nd Floor
New York, NY 10017, United States of America
T: 223 4300 F: 751 1966

Iijima, Takashi, Mr.
Director
Research and Information Office, Global Environment Department, Environment Agency
Permanent Mission to the United Nations, 866 United Nations Plaza, 2nd Floor
New York, NY 10017, United States of America
T: 223 4300 F: 751 1966

Ikkatai, Seiji, Mr.
Coordinator
Health and Welfare Division, Environment Agency
Permanent Mission to the United Nations, 866 United Nations Plaza, 2nd Floor
New York, NY 10017, United States of America
T: 223 4300 F: 751 1966

Imamura, Koji, Mr.
Counsellor
Oceanic Fisheries Department, Fisheries Agency
Permanent Mission to the United Nations, 866 United Nations Plaza, 2nd Floor
New York, NY 10017, United States of America
T: 223 4300 F: 751 1966

Isashiki, Shinichi, Mr.
Director
Global Environmental Affairs Division, United Nations Bureau, Ministry of Foreign Affairs
Permanent Mission to the United Nations, 866 United Nations Plaza, 2nd Floor
New York, NY 10017, United States of America
T: 223 4300 F: 751 1966

Ito, Makoto, Mr.
Cousellor
Embassy of Japan in Brazil, Ministry of Foreign Affairs
Permanent Mission to the United Nations, 866 United Nations Plaza, 2nd Floor New York, NY, 10017
F: 223 4300
TL: 751 1966

Iwamoto, Satoshi
Representative Director
CITIZENS ALLIANCE FOR SAVING THE ATMOSPHERE AND THE EARTH-CASA
1-3-17-813, Tanimachi, Chuo-KU
Osaka 540, Japan
T: 81 6 941 3745 F: 81 6 941 5699

Kadowaki, Kuniyasu, Mr.
Seeds and Seedlings Division, Agricultural Production Bureau, Ministry of Agriculture, Forestry and Fisheries
Permanent Mission to the United Nations, 866 United Nations Plaza, 2nd Floor
New York, NY 10017, United States of America
T: 223 4300 F: 751 1966

Kamigawara, Kenji, Mr.
Global Environmental Affairs Division, United Nations Bureau, Ministry of Foreign Affairs
Permanent Mission to the United Nations, 866 United Nations Plaza, 2nd Floor
New York, NY 10017, United States of America
T: 223 4300 F: 751 1966

Katayama, Shoichiro
Director
OFFICE OF EARTH S & T, R & D BUREAU
S & T Agency 2-1-1 Kasunigasti Chi Yoda-KU
Tokyo 100, Japan
T: 81 3 3581 5271 EX433 F: 81 3 3595 0980

Kato, Hisakazu, Mr.
Director
Office of Oversea's Environmental Cooperation, Global Environment Department, Environment Agency
Permanent Mission to the United Nations, 866 United Nations Plaza, 2nd Floor
New York, NY 10017, United States of America
T: 223 4300 F: 751 1966

Kato, Kazu
Director, Office Of Overseas
ENVIRONMENTAL COOPERATION
1-2-2 Kasumigaseki, Chiyoda-KU, Tokyo 100
Tokyo, Japan
T: 03 3580 4982 F: 03 3504 1634

Kato, Saburo, Mr.
Director
Global Environmental Department, Environment Agency
Permanent Mission to the United Nations, 866 United Nations Plaza, 2nd Floor
New York, NY 10017, United States of America
T: 223 4300 F: 751 1966

Kidokoro, Motoyuki, Mr.
Director
Road Environment Research Section, Planning Division, Road Bureau, Ministry of Construction
Permanent Mission to the United Nations, 866 United Nations Plaza, 2nd Floor
New York, NY 10017, United States of America
T: 223 4300 F: 751 1966

Kikuchi, Kunio, Mr.
Director
Wildlife Division, Nature Conservation Bureau, Environment Agency
Permanent Mission to the United Nations, 866 United Nations Plaza, 2nd Floor
New York, NY 10017, United States of America
T: 223 4300 F: 751 1966

Kikuoka, Eiji, Mr.
General Affairs Division, Industrial Location and Environmental Protection Bureau
Permanent Mission to the United Nations, 866 United Nations Plaza, 2nd Floor
New York, NY 10017, United States of America
T: 223 4300 F: 751 1966

Kobayashi, Hikaru, Mr.
Director
Office of Policy Planning and Research, Planning and Coordination Bureau, Environment Agency
Permanent Mission to the United Nations, 866 United Nations Plaza, 2nd Floor
New York, NY 10017, United States of America
T: 223 4300 F: 751 1966

Kojima, Seiji, Mr.
Director
Research and Programming Division, Economic Cooperation Bureau
Permanent Mission to the United Nations, 866 United Nations Plaza, 2nd Floor New York, NY, 10017
F: 223 4300
TL: 751 1966

Kojima, Toshiro, Mr.
Director
Environmental Impact Assessment Division, Planning and Coordination Bureau, Environment Agency
Permanent Mission to the United Nations, 866 United Nations Plaza, 2nd Floor
New York, NY 10017, United States of America
T: 223 4300 F: 751 1966

Kondo, Kenji, Mr.
Director
Energy and Environmental Affairs Office, Agency of Natural Resources and Energy
Permanent Mission to the United Nations, 866 United Nations Plaza, 2nd Floor
New York, NY 10017, United States of America
T: 223 4300 F: 751 1966

Koyanagi, Yoshihiro, Mr.
Wood Distribution Division, Forest Policy Planning Department, Forestry Agency
Permanent Mission to the United Nations, 866 United Nations Plaza, 2nd Floor
New York, NY 10017, United States of America
T: 223 4300 F: 751 1966

Kuno, Masae, Ms.
Fishery Division Economic Affairs Bureau, Ministry of Foreign Affairs
Permanent Mission to the United Nations, 866 United Nations Plaza, 2nd Floor
New York, NY 10017, United States of America
T: 223 4300 F: 751 1966

Masuda, Hiroya, Mr.
City Planning Division
City Bureau, Ministry of Construction
Permanent Mission to the United Nations, 866 United Nations Plaza, 2nd Floor
New York, NY 10017, United States of America
T: 223 4300 F: 751 1966

Matsubara, Koji, Mr.
Conservation Environment and Ocean Development Division, Transport Policy Bureau
Permanent Mission to the United Nations, 866 United Nations Plaza, 2nd Floor
New York, NY 10017, United States of America
T: 223 4300 F: 751 1966

Matsumura, Takashi, Mr.
Planning Division
Global Environment Department
Permanent Mission to the United Nations, 866 United Nations Plaza, 2nd Floor
New York, NY 10017, United States of America
T: 223 4300 F: 751 1966

Matsumura, Tetsuro
Head, The Office Of Climate Change
PLANNING DIVISION, JAPAN
Meteorological Agency 1-3-4 Otemachi Chiyodaku
Tokyo, 100, Japan
T: 03 3212 8341EXT 2261 F: 8341

Matsuno, Taro, Dr.
Professor
Center for Climate System Research, The University of Tokyo, Ministry of Education
Permanent Mission to the United Nations, 866 United Nations Plaza, 2nd Floor
New York, NY 10017, United States of America
T: 223 4300 F: 751 1966

Mikami, Masahiro, Mr.
Developing Economies Division, Economic Affairs Bureau, Ministry of Foreign Affairs
Permanent Mission to the United Nations, 866 United Nations Plaza, 2nd Floor
New York, NY 10017, United States of America
T: 223 4300 F: 751 1966

Minamikawa, Hideki, Mr.
Director
Office of Public Relations, Minister's Secretariat, Environment Agency
Permanent Mission to the United Nations, 866 United Nations Plaza, 2nd Floor
New York, NY 10017, United States of America
T: 223 4300 F: 751 1966

Miyake, Shotaro, Mr.
Counsellor
Minister's Secretariat, Ministry of Finance
Permanent Mission to the United Nations, 866 United Nations Plaza, 2nd Floor
New York, NY 10017, United States of America
T: 223 4300 F: 751 1966

Miyazawa, Kiichi, H.E. Mr.
Prime Minister
Japan
866 United Nations Plaza, 2nd Floor
New York, NY 10017, United States of America
T: 223 4300 F: 751 1966

Mochizuki, Katsuya, Mr.
Economic Cooperation Division, International Economic Affairs Department, International Trade Bureau
Permanent Mission to the United Nations, 866 United Nations Plaza, 2nd Floor
New York, NY 10017, United States of America
T: 223 4300 F: 751 1966

Mori, Kazuaki, Mr.
Planning Division, Water Supply and Environmental Sanitation Department, Environmental Health Bureau
Permanent Mission to the United Nations, 866 United Nations Plaza, 2nd Floor
New York, NY 10017, United States of America
T: 223 4300 F: 751 1966

Motoki, Akiyoshi, Mr.
Science Division, Science and International Affairs Bureau, Ministry of Education
Permanent Mission to the United Nations, 866 United Nations Plaza, 2nd Floor
New York, NY 10017, United States of America
T: 223 4300 F: 751 1966

Murazumi, Yasushi, H.E. Mr.
Ambassador Extraordinary and Plenipotentiary
Japan to Brazil
Permanent Mission to the United Nations, 866 United Nations Plaza, 2nd Floor New York, NY, 10017
F: 223 4300
TL: 751 1966

Nagame, Ichiro, Mr.
Office of International Forestry Cooperation, Planning Division, Private Forest Department
Permanent Mission to the United Nations, 866 United Nations Plaza, 2nd Floor
New York, NY 10017, United States of America
T: 223 4300 F: 751 1966

Nagatomo, Kenji, Mr.
Fishing Ground Environment Conservation Division Research Department, Fisheries Agency
Permanent Mission to the United Nations, 866 United Nations Plaza, 2nd Floor
New York, NY 10017, United States of America
T: 223 4300 F: 751 1966

Nakajima, Mizue, Ms.
Global Environmental Affairs Division, United Nations Bureau, Ministry of Foreign Affairs
Permanent Mission to the United Nations, 866 United Nations Plaza, 2nd Floor
New York, NY 10017, United States of America
T: 223 4300 F: 751 1966

Nemoto, Norio, Mr.
Director, General Planning and Research Office
General Planning and Policy Division, Minister's Secretariat, Ministry of Posts and Telecommunications
Permanent Mission to the United Nations, 866 United Nations Plaza, 2nd Floor
New York, NY 10017, United States of America
T: 223 4300 F: 751 1966

Nishi, Tatsuo, Mr.
Director
Economic Structural Adjustment Office
Planning Bureau, Economic Planning Agency
Permanent Mission to the United Nations, 866 United Nations Plaza, 2nd Floor New York, NY, 10017
F: 223 4300
TL: 751 1966

Nitta, Takashi
Director-General
JAPAN METEOROLOGICAL AGENCY
1-3-4 Otemachi,
Chiyoda-KE Tokyo 100, Japan
T: +813 3212 8341 EXT 2 F: +813 3212 2453

Nomura, Ichiro, Mr.
International Affairs Division, Oceanic Fisheries Department, Fisheries Agency
Permanent Mission to the United Nations, 866 United Nations Plaza, 2nd Floor
New York, NY 10017, United States of America
T: 223 4300 F: 751 1966

Nomura, Ryuji, Mr.
Economic Structrual Adjustment Office
Planning Bureau, Economic Planning Agency
Permanent Mission to the United Nations, 866 United Nations Plaza, 2nd Floor
New York, NY 10017, United States of America
T: 223 4300 F: 751 1966

Numata, Sadaaki, Mr.
Deputy, Press Secretary, Deputy Director General
Public Information
Permanent Mission to the United Nations, 866 United Nations Plaza, 2nd Floor New York, NY, 10017
F: 223 4300
TL: 751 1966

Oga, Toshiko, Mr.
General Affairs Division
Minister's Secretariat
Permanent Mission to the United Nations, 866 United Nations Plaza, 2nd Floor
New York, NY 10017, United States of America
T: 223 4300 F: 751 1966

Omori, Shigeru, Mr.
Planning Division
Global Environment Department
Permanent Mission to the United Nations, 866 United Nations Plaza, 2nd Floor
New York, NY 10017, United States of America
T: 223 4300 F: 751 1966

Onogawa, Kazunobu, Mr.
Special Advisor to the Director General
Global Environment Department, Environment Agency
Permanent Mission to the United Nations, 866 United Nations Plaza, 2nd Floor
New York, NY 10017, United States of America
T: 223 4300 F: 751 1966

Osawa, Tsutomu, Mr.
First Secretary
Permanent Mission of Japan
Permanent Mission to the United Nations and other International Organizations, Geneva
866 United Nations Plaza, 2nd Floor New York, NY, 10017
F: 223 4300
TL: 751 1966

Ozawa, Fusho, Mr.
Director-General
Forestry Agency, Ministry of Agriculture, Forestry and Fisheries
Permanent Mission to the United Nations, 866 United Nations Plaza, 2nd Floor New York, NY, 10017
F: 223 4300
TL: 751 1966

Saiga, Fumiko, Ms.
Director in Charge of the Preparation for the United Nations Conference on Environment and Development
Ministry of Foreign Affairs
Permanent Mission to the United Nations, 866 United Nations Plaza, 2nd Floor New York, NY, 10017
F: 223 4300
TL: 751 1966

Saito, Mamoru, Mr.
Deputy Vice-Minister for General Policy Coordination, Minister's Secretariat
Ministry of Construction
Permanent Mission to the United Nations, 866 United Nations Plaza, 2nd Floor New York, NY, 10017
F: 223 4300
TL: 751 1966

Saito, Takashi, Mr.
Director
Developing Economies Division, Economic Affairs Bureau, Ministry of Foreign Affairs
Permanent Mission to the United Nations, 866 United Nations Plaza, 2nd Floor New York, NY, 10017
F: 223 4300
TL: 751 1966

Sakano, Masatoshi, Mr.
Office of Environment Promotion, Minister's Secretariat, Ministry of Agriculture, Forestry and Fisheries
Permanent Mission to the United Nations, 866 United Nations Plaza, 2nd Floor
New York, NY 10017, United States of America
T: 223 4300 F: 751 1966

Sasaki, Shuichi, Mr.
Director
Chemical Products Safety Division, Basic Industries Bureau, Ministry of International Trade and Industry
Permanent Mission to the United Nations, 866 United Nations Plaza, 2nd Floor
New York, NY 10017, United States of America
T: 223 4300 F: 751 1966

Seiki, Katsuo, Mr.
Deputy Director-General
Industrial Location and Environment Protection Bureau, Ministry of International Trade and Industry
Permanent Mission to the United Nations, 866 United Nations Plaza, 2nd Floor
New York, NY 10017, United States of America
T: 223 4300 F: 751 1966

Seki, Hikariko, Ms.
Economic Affairs Division, United Nations Bureau, Ministry of Foreign Affairs
Permanent Mission to the United Nations, 866 United Nations Plaza, 2nd Floor
New York, NY 10017, United States of America
T: 223 4300 F: 751 1966

Seki, Soichiro, Mr.
Planning Division
Air Quality Bureau
Permanent Mission to the United Nations, 866 United Nations Plaza, 2nd Floor
New York, NY 10017, United States of America
T: 223 4300 F: 751 1966

Seo, Masaki, H.E. Mr.
Ambassador Extraordinary and Plenipotentiary in Charge of Preparation
United Nations Conference on Environment and Development
Permanent Mission to the United Nations, 866 United Nations Plaza, 2nd Floor New York, NY, 10017
F: 223 4300
TL: 751 1966

Sesaki, Katsumi, H.E. Mr.
Ambassador Extraordinary and Plenipotentiary, Deputy Permanent Representative
United Nations, New York
Permanent Mission to the United Nations, 866 United Nations Plaza, 2nd Floor
New York, NY 10017, United States of America
T: 223 4300 F: 751 1966

Shimada, Michio, Mr.
Director
Oceanic Fisheries Department, Fisheries Agency
Permanent Mission to the United Nations, 866 United Nations Plaza, 2nd Floor
New York, NY 10017, United States of America
T: 223 4300 F: 751 1966

Shimizu, Takahiro, Mr.
Construction and Development Coordination Division Economic Affairs Bureau
Ministry of Construction
Permanent Mission to the United Nations, 866 United Nations Plaza, 2nd Floor New York, NY, 10017
F: 223 4300
TL: 751 1966

Shimizu, Takao, Mr.
Energy Policy Planning Division, Agency of Natural Resources and Energy
Permanent Mission to the United Nations, 866 United Nations Plaza, 2nd Floor
New York, NY 10017, United States of America
T: 223 4300 F: 751 1966

Shimizu, Yasuhiro, Mr.
Central and Cooperation Division
Global Environmental Department
Permanent Mission to the United Nations, 866 United Nations Plaza, 2nd Floor
New York, NY 10017, United States of America
T: 223 4300 F: 751 1966

Shinohara, Naoyuki, Mr.
Director in Charge of Development Issues
International Finance Bureau, Ministry of Finance
Permanent Mission to the United Nations, 866 United Nations Plaza, 2nd Floor
New York, NY 10017, United States of America
T: 223 4300 F: 751 1966

Shoji, Ryuichi, Mr.
Counsellor
Permanent Mission of Japan to the United Nations, New York
Permanent Mission to the United Nations, 866 United Nations Plaza, 2nd Floor
New York, NY 10017, United States of America
T: 223 4300 F: 751 1966

Sugano, Yukio, H.E. Mr.
Minister
Permanent Mission of Japan
866 United Nations Plaza, 2nd Floor
New York, NY 10017, United States of America
T: 223 4300 F: 751 1966

Sugihara, Atsushi, Mr.
Evaluation Division, Economic Cooperation Bureau, Ministry of Foreign Affairs
Permanent Mission to the United Nations, 866 United Nations Plaza, 2nd Floor
New York, NY 10017, United States of America
T: 223 4300 F: 751 1966

Susuki, Hideo, Mr.
Director-General
Industrial Location and Environmental Protection Bureau, Ministy of International Trade and Industry
Permanent Mission to the United Nations, 866 United Nations Plaza, 2nd Floor New York, NY, 10017
F: 223 4300
TL: 751 1966

Susuki, Yasuhiro, Mr.
Office for Special Desease Certification
Environmental Health Department
Permanent Mission to the United Nations, 866 United Nations Plaza, 2nd Floor
New York, NY 10017, United States of America
T: 223 4300 F: 751 1966

Suyama, Akira, Mr.
Deputy Consul-General
Consulate General, Rio de Janeiro
Permanent Mission to the United Nations, 866 United Nations Plaza, 2nd Floor
New York, NY 10017, United States of America
T: 223 4300 F: 751 1966

Suzuki, Yasuji, Mr.
Private Secretary to the Minister
Environmental Health Department
Permanent Mission to the United Nations, 866 United Nations Plaza, 2nd Floor
New York, NY 10017, United States of America
T: 223 4300 F: 751 1966

Takase, Yasuo, Mr.
Deputy Director in Charge of the Preparation
United Nations Conference on Environmental and Development
Permanent Mission to the United Nations, 866 United Nations Plaza, 2nd Floor
New York, NY 10017, United States of America
T: 223 4300 F: 751 1966

Takemura, Kotaro, Mr.
Development Division
River Bureau, Ministry of Construction
Permanent Mission to the United Nations, 866 United Nations Plaza, 2nd Floor
New York, NY 10017, United States of America
T: 223 4300 F: 751 1966

Tamura, Shuji, Mr.
Counsellor
Minister's Secretariat, Environment Agency
Permanent Mission to the United Nations, 866 United Nations Plaza, 2nd Floor
New York, NY 10017, United States of America
T: 223 4300 F: 751 1966

Tanabe, Toshiaki, Mr.
Consul-General
Rio de Janeiro, Brazil
Permanent Mission to the United Nations, 866 United Nations Plaza, 2nd Floor New York, NY, 10017
F: 223 4300
TL: 751 1966

Tanaka, Katsutoshi, Mr.
Office of Environmental Chemicals Safety, Planning Division, Environmental Health Bureau
Permanent Mission to the United Nations, 866 United Nations Plaza, 2nd Floor
New York, NY 10017, United States of America
T: 223 4300 F: 751 1966

Tanaka, Masanori, Mr.
Director
Planning Division, Private Department, Forestry Agency
Permanent Mission to the United Nations, 866 United Nations Plaza, 2nd Floor
New York, NY 10017, United States of America
T: 223 4300 F: 751 1966

Tani, Shuichi, Dr.
Counsellor for Science and Technology
Minister's Secretariat, Ministry of Health and Welfare
Permanent Mission to the United Nations, 866 United Nations Plaza, 2nd Floor
New York, NY 10017, United States of America
T: 223 4300 F: 751 1966

Tsukioka, Hideto, Mr.
Monuments and Sites Division, Cultural Properties Protection Department, Agency for Cultural Affairs
Permanent Mission to the United Nations, 866 United Nations Plaza, 2nd Floor
New York, NY 10017, United States of America
T: 223 4300 F: 751 1966

Tsuruta, Yasunori, Mr.
Director
Office for Environmental Chemicals Safety, Environmental Health Bureau
Permanent Mission to the United Nations, 866 United Nations Plaza, 2nd Floor
New York, NY 10017, United States of America
T: 223 4300 F: 751 1966

Uemura, Nobuo, Mr.
Office of International Cooperation, International Affairs Division, Ministry of Health and Welfare
Permanent Mission to the United Nations, 866 United Nations Plaza, 2nd Floor
New York, NY 10017, United States of America
T: 223 4300 F: 751 1966

Uesaka, Tsuneo, Mr.
Director
Water Resources Investigation Office, National Land Agency
Permanent Mission to the United Nations, 866 United Nations Plaza, 2nd Floor
New York, NY 10017, United States of America
T: 223 4300 F: 751 1966

Usuki, Mitsuo, Mr.
Planning and Coordination Division
Nature Conservation Bureau
Permanent Mission to the United Nations, 866 United Nations Plaza, 2nd Floor
New York, NY 10017, United States of America
T: 223 4300 F: 751 1966

Wakabayashi, Shigeyoshi, Mr.
International Policy Coordination Division
Minister's Secretariat, Ministry of Posts and Telecommunications
Permanent Mission to the United Nations, 866 United Nations Plaza, 2nd Floor
New York, NY 10017, United States of America
T: 223 4300 F: 751 1966

Wakasugi, Ryuhei, Dr.
Director
Environmental Policy Division, Industrial Location and Environmental Protection Bureau
Permanent Mission to the United Nations, 866 United Nations Plaza, 2nd Floor
New York, NY 10017, United States of America
T: 223 4300 F: 751 1966

Watanabe, Hideki, Mr.
International Affairs Division
Economic Affairs Bureau, Ministry of Construction
Permanent Mission to the United Nations, 866 United Nations Plaza, 2nd Floor
New York, NY 10017, United States of America
T: 223 4300 F: 751 1966

Yabunaka, Mitoji, Mr.
Counsellor
Permanent Mission United Nations and other International Organizations, Geneva
Permanent Mission to the United Nations, 866 United Nations Plaza, 2nd Floor New York, NY, 10017
F: 223 4300
TL: 751 1966

Yagihashi, Atsuo, Mr.
Director-General
Planning and Coordination Bureau, Environment Agency
Permanent Mission to the United Nations, 866 United Nations Plaza, 2nd Floor New York, NY, 10017
F: 223 4300
TL: 751 1966

Yamada, Kazuhiko, Mr.
Counsellor
Minister's Secretariat, National Land Agency
Permanent Mission to the United Nations, 866 United Nations Plaza, 2nd Floor
New York, NY 10017, United States of America
T: 223 4300 F: 751 1966

Yamaji, Junichi, Mr.
Counsellor
Minister's Secretariat, Science and Technology Agency
Permanent Mission to the United Nations, 866 United Nations Plaza, 2nd Floor
New York, NY 10017, United States of America
T: 223 4300 F: 751 1966

Yasuda, Takehiko, Mr.
Policy Division
Atomic Energy Bureau
Permanent Mission to the United Nations, 866 United Nations Plaza, 2nd Floor
New York, NY 10017, United States of America
T: 223 4300 F: 751 1966

Yatsu, Ryutaro, Mr.
Planning Division
Global Environment Department
Permanent Mission to the United Nations, 866 United Nations Plaza, 2nd Floor
New York, NY 10017, United States of America
T: 223 4300 F: 751 1966

Yokomaku, Toshiki, Mr.
Deputy Director
Global Environmental Police Bureau, Ministry of Posts and Telecommunications
Permanent Mission to the United Nations, 866 United Nations Plaza, 2nd Floor
New York, NY 10017, United States of America
T: 223 4300 F: 751 1966

NGO

Abiru, Takeshi, Mr.
KEIDANREN
1-9-4 Ohte Machi Chiyoda-KU
Tokyo, Japan
T: 81 3214 1411 F: 81 35255 6233

Aleman Velasco, Miguel
GPH
610 Sangun Kaikan, 2-1-1 Nagato-Cho, Chiyoda-Ku
Tokyo 100 Japan, Japan

Aoki, Yoichi
JAIRI
c/o Mitsubishi Research Institute, Inc., Time & Li
2-3-6 Otemachi, Chiyoda-Ku, Tokyo 100, Japan

Bauer, Conrado E., Mr. & Mrs.
INTERNATIONAL LAKE ENVIRONMENT COMMITTEE FOUNDATION
3-4-22 Kyomachi, Otsu
Shiga 520, Japan
T: 81 775 25-1076 F: 81 775 23-1581
TL: 5464850 ILEC J

Chiba, Yoshiaki, Mr.
JGA
1-15-12 Toranomon, Minato-ku
Tokyo 105, Japan

Goto, Yasuo
Ms.
JFEO
9-4 Otemachi 1-Chome
Chiyoda-Ku, Tokyo 100, Japan

Gotoh, Yasuko
PAPE
Nichiei-Sangyo Bldg. 4f, 2-8-2 Aioi Cho
Akashi City, Hyogo 673, Japan

Hashizume, Shigeyuki
JEA
Toranomon 1 Bldg., 5-8, Toranomon 1-Chome
Minato-Ku, Tokyo 106, Japan

Hattori, Noriyoshi
ICR
Tokyo Suisan Bldg. 4-18, Toyomi-Cho, Chuo-Ku
Tokyo 104, Japan

Hayashi, Katsuhiko
KF
c/o Felissimo Corporation, 1-1-3-2900 Umeda,
Kita-ku, Osaka, Japan

Hirakata, Tatsuo
GIFRFJ
Sengokuyama Annex 107, 5-3-20 Toranomon
Minato-Ku, Tokyo 105, Japan

Iba, Mikako
SCEF
1-2-28, Itsutsubashi, Aobaku, Sendaishi, Miyagiken
Japan

Ida, Keiko, Ms.
Secretary General
JAPAN FEDERATION OF BAR ASSOCIATIONS
1-1-1 Kasumigaseki, Chiyoda-Ku
Tokyo 100, Japan

Imai, Michiko, Ms.
M.D.
JAPAN WOMEN'S GLOBAL ENVIRONMENT NETWORK INTERNATIONAL
3-19-4 Miyasaka, Setagaya-ku
Tokyo 156, Japan
T: 81 3 3439-3599 F: 81 3 3506-8085

Inoue, Yoshio, Mr.
JRCC
1-3-17-813 Tanimachi, Chuo-Ku
Osaka 540, Japan

Iwasaki, Shunsuke, Mr.
92 NGO FORUM JAPAN
4F Kanda Kyoudo Bldg., 2-13-4 Iwamoto-Cho
Tokyo, Japan
T: 81 3 3861-0540 F: 81 3 3861-5048

Kada, Ryohei, Mr.
JFE
1-3-17 Elf Otemae 809
Tanimachi, Chuo-Ku, Osaka 540, Japan

Kaji, Ryuichi, Mr.
ERC
Shin-Maru Bldg., 8th Floor, 5-1,
5-1, Marunouchi 1-chome, Chiyoda-ku, Tokyo 100, Japan

Kaji, Takashi
JEA
Toranomon 1 Bldg., 5-8, Toranomon 1-Chome
Minato-Ku, Tokyo 106, Japan

Kamiya, Masamichi
RISSHO KOSEI-KAI FUND FOR PEACE
11-1 Wada 2-Chome, Suginami-Ku
Tokyo 166, Japan

Kato, Ichiro
EIC
Office Toranomon First Building
1-5-8 Toranomon, Minato-ku, Tokyo 105, Japan

Kawashima, Toshio, Mr.
JRCC
1-3-17-813 Tanimachi, Chuo-Ku
Osaka 540, Japan

Kidahashi, Tsutomu
NEDO
Susshine 60 29 F, 1-1, 3-Chome, Higashi-Ikebukuro
Toshima-Ku, Tokyo 170, Japan

Kinkyo, Hoichi
JAIRI
c/o Mitsubishi Research Institute, Inc., Time & Li
2-3-6 Otemachi, Chiyoda-Ku, Tokyo 100, Japan

Kitamoto, Hiroki, Sr.
Director
CHUGOKU ELECTRIC POWER CO., INC.
4-33 Komachi, Naka-ku
Hiroshima, Japan
T: 81 82 241-0211 F: 81 82 240-7244

Kitsuki, Fumio, Mr.
Director
OISCA - INTERNATIONAL
6-12, Izumi 3-Chome, Suginami-ku
Tokyo, Japan
T: 81 3 3322-5161 F: 81 3 3324-7111
TL: J29397 OISCA

Koganezawa, Takaaki, Mr.
SCEF
1-2-28, Itsutsubashi, Aobaku, Sendaishi, Miyagiken
Japan

Kuno, Katsuji, Mr.
Professor
JAPAN ASSOCIATION ON THE ENVIRONMENTAL STUDIES
3-5-8 Saiwaicho Fuchu-city
Tokyo, Japan
T: 81 423 34 6884 F: 81 423 34 6883

Kuriyama, Koichi
JPC
3-1-1 Shibuya, Shibuya-Ku
Tokyo 150, Japan

Kurosaka, Miwaiko, Ms.
SPF
Sasakawa Hall, 3-12-12 Mita Minato-Ku
Tokyo 108, Japan

Macnow, Alan
ICR
Tokyo Suisan Bldg. 4-18, Toyomi-Cho, Chuo-Ku
Tokyo 104, Japan

Masaki, Shinichi
JGA
1-15-12 Toranomon, Minato-ku
Tokyo 105, Japan

Matsui, Mitsuma
JFTA
No. 7 Rokuban-Cho
Chiyoda-Ku, Tokyo 102, Japan

Matsui, Saburo, Dr.
Professor
INTERNATIONAL LAKE ENVIRONMENT COMMITTEE FOUNDATION
3-4-22 Kyomachi, Otsu
Shiga 520, Japan
T: 81 775 25-1076 F: 81 775 23-1581
TL: 5464850 ILEC J

Miyadera, Hiroshi, Mr.
ERC
Shin-Maru Bldg., 8th Floor, 5-1,
5-1, Marunouchi 1-chome, Chiyoda-ku, Tokyo 100, Japan

Miyamoto, Kenichi
JEC
Dept. of Law, Rikkyo University,
Nishiikebukuro 3-34-1, Toshima-Ku, Tokyo 171, Japan

Miyamoto, Takao
JPC
3-1-1 Shibuya, Shibuya-Ku
Tokyo 150, Japan

Mizutani, Tsutomu, Mr.
Chairman
OSAKA INDUSTRIAL ASSOCIATION
2-8 Honmachi-Bashi Cyuoku
Osaka, Japan
T: 81 6 944-6112 F: 81 6 944-6115

Mizutani, Yoichi, Mr.
JAPAN SCIENTISTS ASSOCIATION
9-16 Yushima 1-Chome, Bunkyo-Ku
Tokyo 113, Japan

Morikawa, Tohru
JFA
Sankaido Bldg., 1-9-13 Akasaka
Minato-Ku, Tokyo 107, Japan

Morishima, Akio
JCGE
Office Toranomon First Bldg., 1-5-8 Toranomon
Minato-Ku, Tokyo 105, Japan

Moriwaki, Kimio, Mr.
JAPAN AIR POLLUTION VICTIMS ASSOCIATION
Sunny City, Shinjuku Gyoen, 10th Floor
Shinjuku 2-1-3, Tokyo-to, 160, Japan

Nakagami, Kenichi
OIA
2-8 Honmachi-Bashi
Chuoku, Osaka 540, Japan

Nakanishi, Junko
JWGEN
3-19-4 Miyasaka, Setagaya-ku
Tokyo, Japan

Namba, Yoshiyuki, Mr.
Managing Director
RESEARCH INSTITUTE OF INNOVATIVE TECHNOLOGY FOR THE EARTH
Shinkyoto Center Blvd. 4F, Shiokoji-dori, Karasumanishi-iru Shimogyo-ku
Kyoto, Japan
T: 81 75 361-3611 F: 81 75 361-3607

Negami, Takuya
JFEO
9-4 Otemachi 1-Chome
Chiyoda-Ku, Tokyo 100, Japan

Nishikawa, Koichi
JCIA
Tokyo Club Building, 2-6 Kasumigaseki 3-Chome
Chiyoda-Ku, Tokyo 100, Japan

Nishimura, Tadayuki, Mr.
Attorney-At-Law
JAPAN FEDERATION OF BAR ASSOCIATIONS
1-1-1 Kasumigaseki, Chiyoda-Ku
Tokyo 100, Japan

Nishimura, Yoshio
JJC
2-14-3 Hirakawacho, Chiyoda-Ku
Tokyo 102, Japan

Noboru, Horie
NEDO
Susshine 60 29 F, 1-1, 3-Chome, Higashi-Ikebukuro
Toshima-Ku, Tokyo 170, Japan

Ogushi, Michimasa
OFCF
Akasaka Twin Tower, 18F
17-22 Akasaka 2, Minato-ku Tokyo 107, Japan

Okada, Nobuhiro
Executive Vice President
JAPAN JUNIOR CHAMBER INC.
2-14-3 Hirakawacho, Chiyoda-Ku
Tokyo 102, Japan

Okinaga, Yasuko, Ms.
Secretary
PEOPLE'S ALLIANCE TO PROTECT THE ENVIRONMENT OF THE JAPANESE INLAND SEA
6-1-34-58 Manabigaoka, Tarumi-ku, Kobe-city
Hyogo-ken, Japan
T: 81 78 784-9271

Okubo, Seishi
ALLBE
2140-581 Oaza Nagakura, Karuizawa-cho,
Nagano 389-01, Japan

Oya, Masao
92NF
Kanda Kyoudo Building 2-13-4 Iwamoto-cho
Tokyo IOI, Japan

Ozaki, Reiko
92NF
Kanda Kyoudo Building 2-13-4 Iwamoto-cho
Tokyo IOI, Japan

Sadamichi, Shigemi, Mr.
Trustee
WATER RESOURCES ENVIRONMENT TECHNOLOGY CENTER
NK. Bldg., 2-14-2, Kojimachi, Chiyoda-ku
Tokyo, Japan
T: 81 3 3263-9921 F: 81 3 3263-9922

Saito, Kenji
FNROJ
2-7 Motoyoyogi-Cho
Shibuya-Ku, Tokyo 151, Japan

Sakanoi, Kazuyuki, Mr.
Guest Researcher
WATER RESOURCES ENVIRONMENT TECHNOLOGY CENTER
NK. Bldg., 2-14-2, Kojimachi, Chiyoda-ku
Tokyo, Japan
T: 81 3 3263-9921 F: 81 3 3263-9922

Sawada, Akiyoshi, Mr.
RISSHO KOSEI-KAI FUND FOR PEACE
2-20-12, Wada, Suginami-ku
Tokyo 166, Japan
F: 81 3 3381-1138

Severo, Larissa
ICETT
1-5 Suwa-Cho, Yokkaichi
Mie 510, Japan

Shibata, Tokue
JEC
Dept. of Law, Rikkyo University,
Nishiikebukuro 3-34-1, Toshima-Ku, Tokyo 171, Japan

Shiragami, Osamu
Sen. Exec. Director
ASAHI GLASS FOUNDATION
Bank of Tokyo Building 12F, Marunouchi 1-4-2, Chiy
Tokyo 100, Japan

Sugai, Hisao
JCIA
Tokyo Club Building, 2-6 Kasumigaseki 3-Chome
Chiyoda-Ku, Tokyo 100, Japan

Suzuki, Ikuo
JFTA
No. 7 Rokuban-Cho
Chiyoda-Ku, Tokyo 102, Japan

Tachibana, Hiroshi
JFEO
9-4 Otemachi 1-Chome
Chiyoda-Ku, Tokyo 100, Japan

Takagi, Isamu
Director
FEDERATION OF ELECTRIC POWER COMPANIES
Keidanren Kaikan, 1-9-4, Ote-Machi, Chiyoda-Ku
Tokyo, Japan

Takagi, Yoshihiro
Director
OVERSEAS FISHERY COOPERATION FOUNDATION
Akasaka Twin Tower, 18F
17-22 Akasaka 2, Minato-ku Tokyo 107, Japan

Takeuchi, Takashi
ALLBE
2140-581 Oaza Nagakura, Karuizawa-cho,
Nagano 389-01, Japan

Tanaka, Norio
GISPRI
7th Floor, Mori Building No. 33, 3-8-21 Toranomon
Minato-Ku, Tokyo 105 Japan, Japan

Taniyama, Tetsuro
JFE
1-3-17 Elf Otemae 809
Tanimachi, Chuo-Ku, Osaka 540, Japan

Vannucci, Marta
ISME
c/o College of Agriculture, University of the Ryukyus, 1 Senbaru Nishihara, Okinawa 903-01, Japan

Velasquez, Mercedes
GPH
610 Sangun Kaikan, 2-1-1 Nagato-Cho, Chiyoda-Ku
Tokyo 100 Japan, Japan

Yamamoto, Norio
GIFRFJ
Sengokuyama Annex 107, 5-3-20 Toranomon
Minato-Ku, Tokyo 105, Japan

Yamamura, Tsunetoshi, Mr.
Representative Director
CITIZENS ALLIANCE FOR SAVING THE ATMOSPHERE AND THE EARTH (CASA)
1-3-17-813 Tanimachi, Chuo-ku, Osaka-city
Osaka, Japan
T: 81 6 941 3745 F: 81 6 941 5699

Yamaoka, Kikuo
FNROJ
2-7 Motoyoyogi-Cho
Shibuya-Ku, Tokyo 151, Japan

Yamazaki, Keiichi
JAPAN AIR POLLUTION VICTIMS ASSOCIATION
Sunny City, Shinjuku Gyoen, 10th Floor
Shinjuku 2-1-3, Tokyo-to, 160, Japan

Yashuhara, Tadashi
EIC
Office Toranomon First Building
1-5-8 Toranomon, Minato-ku, Tokyo 105, Japan

Yasumoto, Nagasaki
JJC
2-14-3 Hirakawacho, Chiyoda-Ku
Tokyo 102, Japan

Yonemoto, Shohei
MACHIDA RESIDENT CONCERN THE ENVIRONMENT
1-6-12 Naruse, Machida-Shi
Tokyo 194, Japan
T: 81 3 6280 F: 81 3 6301

Yonezawa, Kunio
JFA
Sankaido Bldg., 1-9-13 Akasaka
Minato-Ku, Tokyo 107, Japan

Yoshida, Masahito
NATURE CONSERVATION SOCIETY OF JAPAN
Toranomon-Denki Bldg. 4 F, 2-8-1 Toranomon
Minato-Ku, Tokyo 105, Japan

Yoshii, Sumiyuki, Mr.
ICETT
1-5 Suwa-Cho, Yokkaichi
Mie 510, Japan

PRESS

Akasaka, Emely Eiko T.
Correspondent
YOMIURI SHIMBUN
Est. da Gavea 827-1802
Sao Conrado, Brazil, Japan

Akiba, Shun
Correspondent
TV ASAHI
Alejandro Dumas 158
Col. Polanco, Mexico - DF, Mexico

Akio, Sano
Editor
NHK
3-5-25-708 Bessho
Minami-ku, Yokohama-city, Japan, Japan

Akitaka, Murata
Correspondent
KYODO NEWS SERVICE
408 4-5-2 Hakusan
Kawasaki-City Kanagawa Pref. 215 Japan, Japan

Akiyama, Teruto
Correspondent
NIHON KEIZAI SHIMBUN, INC.
18 Ave. Mansions, Finchley Rd.
London, NW3, UK, Japan

Akiyama, Toyohiro
Reporter Radio/Television
TOKYO B'CASTING SYSTEM, INC.
3-4-18-909 Nozawa
Setagaya-Ku, Tokyo, Japan, Japan

Anzai, Naoto
Editor
CHUNICHI SHINBUN
1-6-2 Honkomagome, Bunkyo-Ku
Tokyo, Japan, Japan

Arai, Marie
Translator
ASAHI SHIMBUN
5-3-2 Tsukiji, Chuo-ku R.Sampaio Viana 203 c.13 Paraiso
Sao Paulo -SP 04004, Japan

Araki, Hiroshi
Correspondent
NHK
5-19-1-2-905 Koremasa,
Fuchu-shi, Tokyo, Japan, Japan

Asaba, Masaharu
Editor
YOMIURI SHIMBUN
B-101, 1006-3 Ohfusa
Kashigaya-Ku, Saitama 343, Japan, Japan

Asai, Kenji
Correspondent
TV ASAHI
6600 Millwood Rd.,
Bethesda, MD 20817, USA, Japan

Asai, Takao
Editor
ASAHI SHIMBUN
4-28-1-502 Atago
Tama-shi, Tokyo Japan, Japan

Asato, Vani
Correspondent
TOKYO B'CASTING SYSTEM, INC.
1712 17th Street, NW #4F
Washington, D.C. 20009, USA, Japan

Asaumi, Nobuo
Correspondent
YOMIURI SHIMBUN
2-25-2 Inogata, Komae-Ku
Tokyo 201, Japan, Japan

Athon, Demetra
Correspondent
NIPPON TELEVISION NETWORK
16 W 82nd St. (#4B)
New York, NY 10024, USA, Japan

Atsushi, Honda
Editor
CHUNICHI SHINBUN
1730 Maxwell Ct.
McLean, VA 22101, USA, Japan

Atsuta, Mitsuyoshi
Reporter Radio/Television
FUJI TELEVISION NETWORK, INC.
1525 Shimoyama Hayama
Kanagawa-ken, Japan, Japan

Azumi, Atsushi
Correspondent
NHK JAPAN BROADCASTING CORP.
2-24-18-305 Hyakunin-cho
Shinjuku-ku, Tokyo, Japan, Japan

Batista, Mariko Arai
Correspondent
HOKKAIDO SHIMBUN
Rua Fernado de Noronha, 90
Rio de Janeiro-RJ, Brazil, Japan

Brock, Bruce E.
Correspondent
YOMIURI SHIMBUN
Chatelet Zoshigaya 102,
1-10-2 Zoshigaya, Toshima-Ku,Tokyo,Japan, Japan

Busby, Rose Sakae
Reporter Radio/Television
NHK
R. Apiacas, 886/14 Sumare
Sao Paulo, Brazil, Japan

Chiba, Satoshi
Correspondent
NHK JAPAN BROADCASTING CORP.
2-14-12, B-102 Inogata
Komae-Shi, Tokyo, Japan, Japan

Chiba, Shinya
Correspondent
FUJI TELEVISION NETWORK, INC.
1-4-10-204 Kamijyujyo
Kita-Ku, Tokyo, Japan, Japan

Chien, Yu Ming
Correspondent
NHK JAPAN BROADCASTING CORP.
R.das Laranjeiras 478, Apt.307
22240 Rio de Janeiro, RJ, Brazil, Japan

Chino, Keiko
Correspondent
SANKEI SHIMBUN
220 East, 54th Street, #8-E
New York, NY 10022, Japan

Chitose, Inaba
Editor
CHUNICHI SHINBUN
2-11-5 Tamagawa, Ohtaku
Tokyo 100, Japan, Japan

Clark, Patrick
Reporter Radio/Television
NIPPON TELEVISION NETWORK
142 E 39th St. (#3B)
New York, NY 10016, USA, Japan

Da Silva, Celinda Carvalho
JIJI PRESS
R. Ibira, 27 Riachuelo
RJ, RJ, Brazil, Japan

Ebara, Miki
Correspondent
NHK JAPAN BROADCASTING CORP.
5 Templewood Ave., Hampstead
London NW3 7UY, United Kingdom, Japan

Ebisawa, Ken
Reporter
KYODO NEWS SERVICE
2 2 5 Toranomon Minato-Ku
Tokyo, Japan
T: 81 33 5844111 F: 3-35893683

Egawa, Yasuhiko
Correspondent
NHK JAPAN BROADCASTING CORP.
1-31-18-201 Hanegi
Setagaya-Ku, Tokyo, Japan, Japan

Eguchi, Motonori
Reporter Radio/Television
TV ASAHI
6-24-12-501 Mure Mitaka
Tokyo, Japan, Japan

Fuchigami, Eiji
Correspondent
NHK JAPAN BROADCASTING CORP.
R.Marques de Parana 62,Apt.305
22230, Rio de Janeior, RJ, Brazil, Japan

Fujiki, Tatsuhiro
Correspondent
NHK JAPAN BROADCASTING CORP.
2-16-7-2-106 Azamino
Midori-Ku, Yokohama-Shi, Kanagawa, Japan, Japan

Fujimaki, Hideki
Correspondent
NIHON KEIZAI SHIMBUN, INC.
11 Rue de Javel 75015
Paris, France, Japan

Fujimoto, Hiroshi
Correspondent
FUJI TELEVISION NETWORK, INC.
Rua Jose Getulio, 546
Sao Paulo-SP, Brazil, Japan

Fujimune, Hiroshi
Reporter Radio/Television
FUJI TELEVISION NETWORK, INC.
3-9-27 Minamiazabu
Minato-Ku, Tokyo, Japan, Japan

Fujimura, Katia R.
Correspondent
TV ASAHI
Rua: Mendes Caldeira
285 apto:03, Bras, Sao Paulo, Brazil, Japan

Fujioka, Nobuo
Correspondent
TV ASAHI
236 East 47th Street, #30-B
New York, NY 10017, USA, Japan

Fujioka, Yukio
Correspondent
TV ASAHI
52 Avenue des Champs-Elysees
75008 Paris, France, Japan

Fujita, Yasuko
Editor
NHK JAPAN BROADCASTING CORP.
3-49-15-201 Honmachi
Shibuya-Ku, Tokyo 151, Japan, Japan

Fujiwara, Junji
Editor
CHEMICAL DAILY
102-11-6-2 Takara, Ichikawa-Ku
Chiba Pref. 272-01, Japan, Japan

Fukada, Tsuneo
Editor
NHK
10-35-15-6-402 Ooaza Bushi
Iruma City, Saitama, Japan, Japan

Fukasawa, Paulo
Reporter Radio/Television
NIPPON TELEVISION NETWORK
Av. Kuwahara, 65
Jardim Quarto Centenario, SP, Brazil, Japan

Fukase, Hiromu
Editor
NHK JAPAN BROADCASTING CORP.
340 East 51st Street, #11-4
New York, NY 10022, USA, Japan

Fukuda, Jun
Correspondent
NHK JAPAN BROADCASTING CORP.
3-3-2-206 Sakae-Machi
Higashimurayama-Shi, Tokyo, Japan, Japan

Fukuda, Toshio
Reporter Radio/Television
TV ASAHI
166 High St.
Hastings-on Hudson, New York 10706, Japan

Fukuya, Sei
Editor
NIHON KEIZAI SHIMBUN, INC.
1-33-1, Nakakasai
Edogawa-ku, Tokyo 100, Japan, Japan

Fushimi, Sayuru
Editor
NIHON KEIZAI SHIMBUN, INC.
86-3-2-702, Mizonokuchi
Takatsu-ku, Kawasabi-shi 213, Japan, Japan

Goro, Koide
NHK JAPAN BROADCASTING CORP.
1-34-1-607 Kaitori, Tama-Shi
Tokyo 206, Japan, Japan

Hacidume, Etsuko
Correspondent
NHK
Rua Senador Vergueiro,55/1002
22230 Rio de Janeiro-RJ, Brazil, Japan

Hagihara, Yutaka
Reporter Radio/Television
TOKYO B'CASTING SYSTEM, INC.
1-13-4-210 Tamagawadai
Setagaya-Ku, Tokyo, Japan, Japan

Hagiwara, Kotaro
Reporter Radio/Television
TV ASAHI
Thukimi-dai 130, Hodogaya-Ku
Yokohama-Shi, Kanagawa-Ken, Japan, Japan

Hakamada, Naoki
Correspondent
NIPPON TELEVISION NETWORK
31-2-516 Hon-cho, Wakoh-city
Saitama-Pref., Japan, Japan

Hamada, Masaaki
Correspondent
YOMIURI SHINBUN
2-15-1-515, Masago, Mihama-ku,
Chiba-shi, Chiba, Japan, Japan

Hamatani, Koji
Correspondent
AKAHATA
Nebraska 43-301, Col.Napoles
Mexico - DF, Mexico

Hanamura, Keiko
Reporter Radio/Television
TV ASAHI
Shinjuku-Ku
Kitashinju-Ku 4-7-4-205,Tokyo,Japan, Japan

Handa, Seizo
Editor
WWF-JAPAN
3-10-14-301 Ebisu-minami
Shibuya-Ku, Tokyo, Japan, Japan

Hanyuda, Kenzo
Correspondent
TOKYO B'CASTING SYSTEM, INC.
Al.Joaquim Eugenio de Lima 838
Sao Paulo, Brazil, Japan

Hara, Takeshi
Chief Editor Of Science Dept
MAINICHI SHIMBUN
2-8-1 Wakaba-Cho
Tachikawa-Ku, Tokyo, Japan, Japan

Harada, Katsuhiro
Correspondent
NIHON KEIZAI SHIMBUN
44 Bote Road
Greenwich, CT 06830, USA, Japan

Harano, Joji
Editor
JIJI PRESS
1-13-1-409 Arima Miyamae-ku
Kawasaky-city Kanagawa, Pref, Japan, Japan

Harvey, Mark
NHK JAPAN BROADCASTING CORP.
21A Foyle Road
London SE3 7RQ, United Kingdom, Japan

Hasegawa, Takashi
Correspondent
NHK
4-2-10-332 Kamiaoki,
Kawaguchi-shi, Saitama-ken 333 Japan, Japan

Hashimoto, Goro
Correspondent
YOMIURI SHIMBUN
16-4 Namiki-Cho, Hachioji-Ku
Tokyo 193, Japan, Japan

Hayashi, Jusaburo
Reporter Radio/Television
NIPPON TELEVISION NETWORK
195, Ryder Road
Manhasset, NY 11030, USA, Japan

Hayashi, Rie
Correspondent
NHK JAPAN BROADCASTING CORP.
1-529-505 Kosugimachi
Nakahara-ku, Kawasaki-shi, Japan, Japan

Hide, Sandra
Editor
YOMIURI SHIMBUN
Andre Cavalcante 95/405
Brazil, Japan

Hirai, Fumio
Correspondent
FUJI TV
2311 M Street, N.W.
Washington, D.C., U.S.A., Japan

Hiraka, Kazuyuka
Correspondent
NHK JAPAN BROADCASTING CORP.
3-1-5 Wada, Suginami-Ku
Tokyo 166, Japan, Japan

Hirata, Yoshihiro
Correspondent
NIHON KEIZAI SHIMBUN, INC.
Av. Paulista 1274/36
01310 Sao Paulo, SP, Brazil, Japan

Hiroi, Terumasa
Correspondent
SEIKYO PRESS
Nicolau de S. Queiros 267/33
Sao Paulo, Brazil, Japan

Hironaka, Yoshimichi
Correspondent
YOMIURI SHIMBUN
1371-1 Komaoka, Apt. A401
Yokohama-Shi, Kanagawa 230, Japan, Japan

Hirose, Sadako
Editor
TBS
R.Nigeria, 27
Sao Paulo, Brazil, Japan

Hirotaka, Fukumoto
Correspondent
KYODO NEWS SERVICE
4-6-2 Izumi, Suginami-Ku
Tokyo, Japan, Japan

Hiroyuki, Minoura
Editor
CHUNICHI SHINBUN
44 Flursby Avenue
London N3 1PL, United Kingdom, Japan

Hiroyuki, Moriyama
Correspondent
SHINANO-MAINICHI SHIMBUN
3-4-11 Miwa
Nagano City, Nagano 380 Japan, Japan

Hiroyuki, Otsuka
Correspondent
YOMIURI SHIMBUN
8-8-304 Kashima, Hachioji-Shi
Tokyo, Japan, Japan

Hizawa, Keiichi
Correspondent
KYODO NEWS SERVICE
155 West 70th Street, Apt.12C
New York, NY 10023, USA, Japan

Honda, Masakazu
Correspondent
ASAHI SHIMBUN
1-30-1-203 Kaminomiya
Tsurumi-Ku, Yokohama, Japan, Japan

Honda, Raisuke
Editor
YOMIURI SHIMBUN
5-23-6 Aarima, Kawasaki City
Kanagawa 261, Japan, Japan

Hori, Yasuhiko
Correspondent
TELEVISION TOKYO CHANNEL 12
25 Rockledge Ave.
1014 White Plains, NY 10601, USA, Japan

Horioka, Jun
Editor
NHK JAPAN BROADCASTING CORP.
2-4-1 Kamisaginomiya
Nakano-Ku, Tokyo 165, Japan, Japan

Ichinose, Hideki
Correspondent
KYODO NEWS SERVICE
6-11-33-106, Funabori
Edogawa-ku, Tokyo, Japan, Japan

Idaka, Hiroaki
Correspondent
KYODO NEWS SERVICE
2-2-13 Nakajyujyo, Kita-Ku
Tokyo, Japan, Japan

Ideishi, Tadashi
Correspondent
NHK JAPAN BROADCASTING CORP.
304, 3-36-7 Sakuradai
Nerima-ku, Tokyo 176, Japan, Japan

Iida, Masayuki
Editor
THE YOMIURI SHINBUN
6-18-11 Minami-oi
Shinagawa-ku, Tokyo, Japan, Japan

Iiyama, Masashi
Correspondent
YOMIURI SHIMBUN
Av. Oswaldo Cruz 149/1106
Rio de Janeiro, Brazil, Japan

Ikeda, Motohiro
Correspondent
NIHON KEIZAI SHIMBUN, INC.
Olympic Penta Hotel, Rm. 5041
Olympiskij Prospect 18/1, Moscow, R.Fed., Japan

Ikeda, Teijiro
Editor
NHK JAPAN BROADCASTING CORP.
1-2-24-506 Osaki, Shinagawa-Ku
Tokyo, Japan, Japan

Ikuma, Mikio
Correspondent
YOMIURI SHIMBUN
2141 Winsconsin Ave., Apt.304
Washington, D.C. 20007, USA, Japan

Imagawa, Gen
Editor
NIHON KEIZAI SHIMBUN, INC.
3-1-5-407 Takiyama
Migashi kurume-shi, Tokyo 203, Japan, Japan

Imaguma, Shoji
Correspondent
SEKAI NIPPO
Felicio dos Santos, 230/52
Sao Paulo, Brazil, Japan

Imai, Tamaki
Correspondent
JAPAN BROADCASTING CORPORATION
2-6-5-212, Daizawa
Setagaya-ku, Tokyo, Japan, Japan

Imaizumi, Hiromi
Reporter Radio/Television
NIPPON TELEVISION NETWORK
6-8-9 Takanedai
Funabashi-city, Chiba-Pref., Japan, Japan

Imamura, Shosuke
Correspondent
TV ASAHI
1-12-24 Shinohara Higashi
Kohoku-Ku, Yokohama, Kanagawa,Japan, Japan

Inagaki, Shinji
Editor
SHOGAKUKAN INC.
3-5-7-402 Izumi-cho
Kokubunji-shi, Tokyo, Japan, Japan

Inaki, Setsuko
Correspondent
NIPPON TELEVISION NETWORK
301 W 57th St. (#12D)
New York, NY 10020, USA, Japan

Inoue, Kido
Correspondent
EARTH SUMMIT TIMES
1669 Tadanoumi Hiroshima
Japan, Japan

Inoue, Kyousuke
Correspondent
NHK JAPAN BROADCASTING CORP.
5-30-2-201 Kamisaginomiya
Nakano-Ku, Tokyo, Japan, Japan

Inoue, Masamoto
Editor
JAPAN AGRICULTURAL PRESS
1-15-11 Senegendai-Nishi
Koshigaya-Shi, Saitama-Ken, Japan, Japan

Inoue, Yutaka
Correspondent
NIHON KEIZAI SHIMBUN
Flat 7, 59-61 Cadgan Gardens
London SW3, UK, Japan

Isamu, Hosoda
Editor
PUMPKIN-USHIO SHUPPANSHA
823-7-20-5, Kitasuna
Koutou-Ku, Tokyo, Japan, Japan

Ishi, Hiroyuki
Editor
ASAHI SHIMBUN
2-20-9 Nozawa, Setagaya-Ku
Tokyo, Japan, Japan

Ishiai, Tsutomu
Correspondent
ASAHI SHIMBUN
1-3-5-203 Enoki Motomachi
Sakai-City, Osaka 590, Japan, Japan

Ishibashi, Osamu
Correspondent
HOKKAIDO SHIMBUN
1-17-6-605 Sakai
Musashino, Tokyo, Japan, Japan

Ishii, Takehito
Reporter Radio/Television
FUJI TELEVISION NETWORK, INC.
4-21-13-202 Higashinogawa
Komae-shi, Tokyo 201, Japan, Japan

Ishikawa, Makiko
Correspondent
NHK
1-4-5 Asashigaoka
Yostsukaido-ku, Chiba-Pref. 284, Japan, Japan

Ishikawa, Masahiko
Editor
ASAHI SHIMBUN
#573 5-31 Okaminami-Cho
Hirakata-City, Japan, Japan

Ishikawa, Takaaki
Staff Writer
MAINICHI-SHIMBUN
1-1-1 Chiocla-Ku
Tokyo, Japan
T: 81 3 32120321 F: 81 3 32120635

Ishimura, Eijiro
Correspondent
NHK JAPAN BROADCASTING CORP.
404 Rune, 1-35-47 Kaminagaya
Konan-ku, Yokohama-shi, Japan, Japan

Isomura, Kentaro
Editor
ASAHI SHIMBUN
Susukino 3-2-11-410, Midori-Ku
Yokohama-Shi, Kanagawa, Japan, Japan

Isono, Akihiko
Correspondent
THE MMAINICHI SHINBUN
1-22-8 Shimorenjaku
Mitaka-shi, Tokyo, Japan, Japan

Itano, Yuji
Correspondent
NHK JAPAN BROADCASTING CORP.
5-2-3-904 Hikarigaoka
Nerima-ku, Tokyo, Japan, Japan

Ito, Minoru
Editor
NHK JAPAN BROADCASTING CORP.
2-2-25-203 Odawara
Miyagino-ku, Miyagi-Pref. 983, Japan, Japan

Ito, Yuji
Editor
TOKYO B'CASTING SYSTEM, INC.
4996 Shobu-Machi
Saitama-Pref., Japan, Japan

Iwakiri, Kiyoshi
Editor
TOKYO B'CASTING SYSTEM, INC.
1555-4 Mizuko, Tujimi-Shi
Saitama, Japan, Japan

Iwasaki, Sadaaki
Correspondent
TV ASAHI
12-6, 3 Chome, Oi
Shinagawa-Ku, Tokyo, Japan, Japan

Iwata, Itsuki
Correspondent
YOMIURI SHIMBUN
7-3-14 Honfujisawa
Fujisawa-Ku, Kanagawa 251, Japan, Japan

Jin, Masao
Correspondent
SANKEI SHIMBUN
22-15 Yokododai
Chiba-Shi, Japan, Japan

Jotaro, Nonaka
Reporter Radio/Television
FM SOUND CHIBA CO. LTD.
202.6-32-16 Todoroki
Setagaya-ku, Tokyo, Japan, Japan

Junko, Brasilina Abuno
Reporter Radio/Television
NIPPON TELEVISION NETWORK
R. da Gloria, 654 Apt.36
Rio de Janeiro-RJ, Brazil, Japan

Kaetsu, Hiroshi
Editor
NISHINIPPON SHIMBUN
3-38-9-404 Minami-Hanashima
Matsudo-Ku, Chiba-Pref. 271, Japan, Japan

Kaifu, Kazuo
NHK JAPAN BROADCASTING CORP.
101, 1-5-4 Saginuma
Miyamae-Ku, Kawasaki City, Japan, Japan

Kajiwara, Hideyuki
Correspondent
MAINICHI SHIMBUN
2-28-1-1306 Takashimadaira
Itabashi-Ku, Tokyo 175, Japan, Japan

Kakuma, Takashi
Correspondent
SHOGAKUKAN INC.
5-4-2-211 Minamiaoyama
Minato-ku, Tokyo, Japan, Japan

Kamachi, Akihiro
Correspondent
YOMIURI SHIMBUN
3-11-13 Tokiwa, Urawa-Ku
Saitama 336, Japan, Japan

Kameyama, Michiko Hirose
Correspondent
TBS
Av.Sernambetiba, 3600/bl6/402
Rio de Janeiro, Brazil, Japan

Kamiya, Hiroshi
Correspondent
ASAHI SHIMBUN
3-1-17-1009 Nagahama, Chuo-Ku
Fukuoka-Shi, Japan, Japan

Kamizaki, Masanori
Correspondent
JIJI PRESS LTD.
277-305 Yanokuchi, Inagi-Shi
Tokyo 206, Japan, Japan

Kanazawa, Mitsuru
Editor
MAINICHI FILM PRODUCTIONS
5-7-15 Higashifushimi,Hoya-shi
Tokyo, Japan, Japan

Kanda, Shunsuke
Correspondent
YOMIURI SHIMBUN
5-6-23 Shintakane,Funabashi-Ku
Chiba 251, Japan, Japan

Kano, Tadashi
Reporter Radio/Television
FUJI TELEVISION NETWORK, INC.
1-31-5-301 Kugahara, Ota-ku
Tokyo, Japan, Japan

Kasahara, Toshihiko
Correspondent
MAINICHI SHIMBUN
85 Aoshirocho Ichijyoji
Sakyo-Ku, Kyoto, Japan, Japan

Kasai, Masumi
Editor
TV TOKYO CH 12
1-14-3 Senkawa, Toshima-ku
Tokyo, Japan, Japan

Kashiwa, Yuko
Reporter Radio/Television
FUJI TELEVISION NETWORK, INC.
No.2B, 4-23-16 Daita
Setagaya-ku, Tokyo, Japan, Japan

Kashiwaya, Toshio
Editor
NHK JAPAN BROADCASTING CORP.
428-2 Kozukue-Cho
Kouhoku-Ku, Yokohama 222, Japan, Japan

Kashiyama, Yukio
Correspondent
THE SANKEI SHINBUN
45-11-202 Oyamakanai-cho
Itabashi-ku Tokyo, 173 Japan, Japan

Kasuga, Takayuki
Correspondent
MAINICHI SHIMBUN
4-10-10-105 Atago, Nishi-Ku
Fukuoka-Shi 809, Japan, Japan

Kasuo, Uchiyama
Editor
CHUNICHI SHINBUN
Av. Brigadeiro Luis Antonio
Sao Paulo-SP, Brazil, Japan

Katada, Tomoyuki
Editor
CHUNICHI SHINBUN
12-150 Yahata, Titashi
Aichi, Japan, Japan

Kato, Isao
Correspondent
KAHOKU SIMPO PRESS
14-5 Shirahagi-Chou
Wakabayashi-Ku, Sendai 983, Japan, Japan

Kato, Kazuhiro
Correspondent
TV ASAHI
8-2-5 Kitakarasuyama
Setagaya-Ku, Tokyo 157, Japan, Japan

Kato, Kiyotaka
JIJI PRESS LTD.
2-1-53 Bingohigashi
Kasukabe, Saitama-Ken, Japan, Japan

Kawaguchi, Jun
Correspondent
KYODO NEWS SERVICE, INC.
R. Ministro A. Ribeiro 64/103
Rio de Janeiro, RJ, Brazil, Japan

Kawai, Makoto
Reporter Radio/Television
FUJI TELEVISION NETWORK, INC.
1-22-20 Higashitamagawa
Setagaya-ku Tokyo, Japan, Japan

Kawana, Hideyuki
Correspondent
WEEKLY ENERGY & ENVIRONMENT
1016-33 Sonno-Cho
Chiba-Shi 281, Japan, Japan

Kawanishi, Akira
Correspondent
TV ASAHI
10100 St.Monica Blvd., #2062
Century City, Los Angeles,CA 90067, USA, Japan

Kawauchi, Shiro
Correspondent
TOYAMA
Alameda Jau, #205, Apt. 71
Sao Paulo, SP, Brazil, Japan

Keiji, Ishizaka
Editor
TOKYO B'CASTING SYSTEM, INC.
640 Yanaguchi Inagi-shi
Tokyo, Japan, Japan

Keizaburo, Kubo
Reporter Radio/Television
NAMBEI TSUSHIN
1302 Fukushima 3 Chome 1-49
Fukushima -Ku Osaka, Japan

Kido, Mutsuo
Correspondent
NIHON KEIZAI SHIMBUN, INC
1-3-1, 1-509 Baraki
Ichikawa-Shi, Chiba-Ken, Japan, Japan

Kikuchi, Keiichi
Editor
FUJI TELEVISION NETWORK, INC.
2-5-40-204 Izumi Suginami-Ku
Tokyo, Japan, Japan

Kimoto, Clarice Mariko
Correspondent
TV ASAHI
Rua: Vergueiro, 439 apto.101
Bloco.A, Vergueiro, SP, Brazil, Japan

Kimura, Kyoko
Correspondent
YOMIURI SHIMBUN
2-2-44 Midori-Cho
Koganei-Ku, Tokyo 184, Japan, Japan

Kimura, Mitsuo
Correspondent
KYODO NEWS SERVICE
1800 Old Meadow Rd., Apt. 1102
McLean, VA 22102, USA, Japan

Kishimoto, Ichiro
Editor
FUJI TELEVISION NETWORK, INC.
1-13-4 Kitashinjuku
Shinjuku-ku, Tokyo 169, Japan, Japan

Kiso, Kazuhiko
Editor
TOKYO B'CASTING SYSTEM, INC.
2-27-1 Kamean, Katsushika-ku
Tokyo, Japan, Japan

Kitamori, Eri
Correspondent
NHK JAPAN BROADCASTING CORP.
R.Paissandu 328, Apt. 402
22210 Rio de Janeior, RJ, Brazil, Japan

Kitamura, Kazuya
Correspondent
NIPPON TELEVISION NETWORK
3-18-6 Sakuradai Nerima-Ku
Tokyo, Japan, Japan

Kitamura, Shunzo
Editor
NIPPON TELEVISION NETWORK CORP.
4-5-8 Muraokahigasi
Fujisawa-shi Kanagawa, Japan, Japan

Kitasei, Junichi
Correspondent
TV ASAHI
160 Mt. Airy Rd.
Croton-on-Hudson, NY 10520, USA, Japan

Kiuchi, Yasuhiro
Correspondent
NIHON KEIZAI SHIMBUN, INC.
4-6-4 Toyotamakita, Nerima-Ku
Tokyo 176, Japan, Japan

Kiyokazu, Matsuzawa
Correspondent
TV ASAHI
Rm.A-103, 1769 Kizukisumiyoshi
Nakahara-Ku, Kawasaki-City, Japan, Japan

Kobayashi, Hiroko
Correspondent
SHINANO MAINICHI SHIMBUN
603 Craige Hall, UNC
Chapel Hill, NC 27514, USA, Japan

Kobayashi, Hirotaka
Correspondent
NIPPON TELEVISION NETWORK
5-20-10 Ohsawa, Mitaka-City
Tokyo, Japan, Japan

Kobayashi, Takashi
Correspondent
NIHON KEIZAI SHIMBUN, INC.
124-1-20F Nobacho, Konan-Ku
Yokohama-City, Kanagawa-Pref.223, Japan, Japan

Kobayashi, Toshinori
Reporter Radio/Television
FUJI TELEVISION NETWORK, INC.
2-5-10-401 Shimotakaido
Suginami-Ku, Tokyo, Japan, Japan

Kobayashi, Yasuhiro
Correspondent
ASAHI SHIMBUN
4-6-10 Nagasawa, Tama-Ku
Kawasaki City, Japan, Japan

Kodama, Masashi
Correspondent
HOKKAIDO SHIMBUN
2-5-1 Utsukushigaoka
Midori-ku, Yokohama 225, Japan, Japan

Kodama, Yoko
Editor
JAPAN AGRICULTURAL PRESS
1-25-24 Naka-Cho, Meguro-Ku
Tokyo, Japan, Japan

Kohno, Kenji
Correspondent
NHK
1034 Papermmill Ct. N.W.
Washington, D.C., USA, Japan

Koich, Obata
Editor
CHUNICHI SHINBUN
6503 Machodoc Ct.
Falls Church, VA 22043, USA, Japan

Koiwai, Tadamichi
Correspondent
KYODO NEWS SERVICE
5-17-6-103 Nagayama, Tama-Shi
Tokyo, Japan, Japan

Kojima, Kazuhiko
Editor
CHUNICHI SHINBUN
4-9-6 Kahata, Moriyama-Ku
Nagoya, Japan, Japan

Komaki, Yasuhiro
Correspondent
YOMIURI SHIMBUN
103 Aikopo, 1-14-17 Kyoudou
Setagaya-Ku, Tokyo 156, Japan, Japan

Konishi, Miwako
Reporter Radio/Television
TV ASAHI
2-12-5-201 Tairamachi
Meguro-Ku, Tokyo, Japan, Japan

Kono, Hiroko
Correspondent
YOMIURI SHIMBUN
5-1-11-408 Nishikasai
Edogawa-Ku, Tokyo, Japan, Japan

Kono, Yorihisa
Reporter Radio/Television
NIPPON TELEVISION NETWORK
3-5-18 Sakuradai, Nerima-Ku
Tokyo 176, Japan, Japan

Konya, Masaari
Correspondent
MAINICHI SHIMBUN
Itabashi-Ku
Minami-Cho 37-1-204, Tokyo, Japan, Japan

Kubo, Tetsuo
Editor
FOURIN INC.
202 Kopo-Jyoshin 3-7-15
Kaminagoya, Nishi-ku Nagoya C. Japan, Japan

Kumamura, Takeyuki
Correspondent
NIHON KEIZAI SHIMBUN
8015 Whittier Blvd.
Bethesda, Maryland, U.S.A., Japan

Kumiko, Suzuki
Editor
CHUNICHI SHINBUN
2-11-5 Jamagawa, Ohtaku
Tokyo 100, Japan, Japan

Kuramata, Toshio
Correspondent
NHK
2-52-2 B105 Komazawa,Setagaya
Tokyo , Japan, Japan

Kurasawa, Haruo
Correspondent
NIPPON TELEVISION NETWORK
449-32 Sakusabe-cho
Chiba-city, Chiba-Pref., Japan, Japan

Kure, Fumihiko
Reporter Radio/Television
NIPPON TELEVISION NETWORK
1-13-18-105 Umesato
Suginami-Ku, Tokyo, Japan, Japan

Kuroiwa, Yuji
Reporter Radio/Television
FUJI TV
3-9-2-107 Sakashita
Itabashi-ku, Tokya 174, Japan, Japan

Kurokouchi, Yutaka
Correspondent
YOMIURI SHIMBUN
1-2-8-202 Kami-Osaki
Shinagawa-Ku, Tokyo 141, Japan, Japan

Maeda, Ichiro
Reporter Radio/Television
TOYAMA TV B'CASTING CO. LT
4-5 Shinjoji-Machi
Uoze-Shi, Toyama 939, Japan, Japan

Maeda, Junji
Editor
FUJI TELEVISION NETWORK, INC.
1-302 3-27 Nakadai Itabashi-Ku
Tokyo, Japan, Japan

Maki, Akira
Editor
TOKYO B'CASTING SYSTEM, INC.
4-21-10-602 Kosuge
Katsushika-Ku, Tokyo 124, Japan, Japan

Makino, Kazuhiro
Correspondent
KYODO NEWS SERVICE
3-16 Togoshi 5-Chome
Shinagawa-Ku, Tokyo, Japan, Japan

Maruyama, Taku
Reporter Radio/Television
TOKYO B'CASTING SYSTEM, INC.
1-13-11 Goutakuji, Setagaya-Ku
Tokyo 154, Japan, Japan

Maruyama, Yasuyuki
Correspondent
YOMIURI SHIMBUN
1-12-17 Fujimi, Chiyoda-Ku
Tokyo 102, Japan, Japan

Masanobu, Ohashi
Editor
CHUNICHI SHINBUN
Hinter Hoben 169
5300 Bonn 1, Germany, Japan

Masashi, Hirano
Correspondent
KYODO NEWS SERVICE
2-13-5 Veno Sakuragi
Taito-Ku, Tokyo, Japan, Japan

Mascarenas, Dolly
Correspondent
TV ASAHI
Fresa 151-102, Del Valle
D.F, Mexico, Japan

Masui, Shigeo
Correspondent
YOMIURI SHIMBUN
250 East First Street
Los Angeles, CA 90012, USA, Japan

Matsuda, Keiko
Editor
FUJI TELEVISION NETWORK, INC.
1-34-8-202 Todoroki
Setagaya-ku, Tokyo 158, Japan, Japan

Matsui, Ryuich
Correspondent
KYODO NEWS SERVICE
Izumi 1-1-10-801, Kamamotosi
Kumamotoken 862, Japan, Japan

Matsumoto, Masaya
Editor
FUJI TELEVISION NETWORK, INC.
5-3-8-201 Nishioi Shinagawa-Ku
Tokyo, Japan, Japan

Matsumoto, Tsunao
Correspondent
TOYAMA TV
R. Guatapara 149, #84
U. Guarani, 04310 Sao Paulo, Brazil, Japan

Matsuo, Mitsuhiro
Editor
NHK
436-1,Ichigao,Misori-ku,
Yokohama-shi, Kanagawa, Japan, Japan

Matsuura, Shinichi
Correspondent
HOKKAIDO SHIMBUN
9-4-15 Sumikawa 4-Jou
Minami-Ku, Sapporo City, Hokkaido, Japan, Japan

Matsuyama, Kaori
Reporter Radio/Television
FUJI TV
17, Wakakusa-cho, Minami-ku
Nagoya, Aichi, 457, Japan, Japan

Maurer, Kim
Editor
TOKYO B'CASTING SYSTEM, INC.
200 Rector Place, Apt. #14J
New York, NY 10280, USA, Japan

McArthur, Ian
Correspondent
KYODO NEWS SERVICE
Maruei Haimu Tomikoa 202
Tomioka 2-8-5, Koto-Ku, Tokyo, Japan, Japan

Meshino, Katsuhiko
Correspondent
NIHON KEIZAI SHIMBUN, INC.
3-92 Jianguomen-Wai, Gonyu
Chaoyang, Beijing, China, Japan

Mikitoshi, Kato
Editor
CHUNICHI SHINBUN
23 rue Las Casas
75007 Paris, France, Japan

Minami, Naoki
Correspondent
NHK JAPAN BROADCASTING CORP
8-13-16 Tamagawa-gakuen
Machida City, Tokyo 194, Japan, Japan

Mitsuyama, Toshiyuki
Correspondent
YOMIURI SHIMBUN
1-475-1-303 Maruko-Dori
Kawasaki-Shi, Kanagawa-Ken, Japan, Japan

Miura, Junji
Correspondent
KYODO NEWS SERVICE
6217 Loch Raven Drive
McLean, VA 22101, USA, Japan

Miyakawa, Masaaki
Correspondent
ASAHI SHIMBUN
R. Sampaio Viana 203, Apt.131
CEP. 04004, Paraiso, SP, SP., Brazil, Japan

Miyata, Francisca
Correspondent
YOMIURI SHIMBUN
R. Brasilio de Brito 28, #501
Rio de Janeiro, RJ, Brazil, Japan

Mizuguchi, Masato
Editor
TOKYO B'CASTING SYSTEM, INC.
Seta 5-9-30 Setagya-Ku
Tokyo-To, Japan, Japan

Mizuno, Masayuki
Correspondent
YOMIURI SHIMBUN
88-77 Yaho, Kunitati-Ku
Tokyo 186, Japan, Japan

Mori, Nobuhito
Correspondent
KYODO NEWS SERVICE
1-6-21 Shoan, Suginami-Ku
Tokyo, Japan, Japan

Motoo, Asakuno
Editor
OITA BROADCASTING SYSTEM
11-4 Tsubakigaoka Oita
Japan, Japan

Murao, Jun
Correspondent
YOMIURI SHIMBUN
3-4-18-804 Kitasuna
Kohtoh-Ku, Tokyo 136, Japan, Japan

Muronoi, Takahiro
Correspondent
YOMIURI SHIMBUN
3-17-3-304 Shioyaki
Ichikawa-Ku, Chiba 272-01, Japan, Japan

Nagahama, Takayuki
Correspondent
JIJI PRESS LTD.
235 Bellemeade Ave.
Fort Lee, NJ 07024, USA, Japan

Nagao, Akira
Editor
TOKYO B'CASTING SYSTEM, INC.
1-16-9-101 Nishihara
Shibuya-Ku, Tokyo, Japan, Japan

Nakagawa, Hiroyuki
Correspondent
JIJI PRESS LTD.
265 Van Buren Ave.
Teaneck, NJ 07666, USA, Japan

Nakagawa, Thoshiya
Correspondent
YOMIURI SHIMBUN
3-19-4 Jyosuishinmachi
Kodaira-Ku, Tokyo 187, Japan, Japan

Nakajima, Kenichi
Correspondent
KYODO NEWS SERVICE
137-1 Tamadaira 2-9, Hino-City
Tokyo, Japan, Japan

Nakajima, Mitsunobu
Correspondent
HOKKAIDO SHIMBUN
15-E-106 Kitaterao 3-Chome
Tsurumi-Ku, Kanagawa 230, Japan, Japan

Nakamichi, Masaki
Correspondent
YOMIURI SHIMBUN
1-35-13-402 Statsukigaoka
Chiba, Japan, Japan

Nakamura, Genki
Reporter Radio/Television
NIPPON TELEVISION NETWORK
3-7-14-407 Nakamagome
Ohta-Ku, Tokyo, Japan, Japan

Nakamura, Haruo
Correspondent
TV ASAHI
35-1212 River Drive South
Jersey City, NJ 07310, USA, Japan

Nakamura, Shinichi
Correspondent
KYODO NEWS SERVICE
52-5-306 Matsugaya, Hachioji
Tokyo, Japan, Japan

Nakamura, Yoshitaka
Correspondent
SHOGAKUKAN INC.
1-33-21 Higashigaoka
Meguro-ku, Tokyo, Japan, Japan

Nakano, Hiroshi
Correspondent
FOURIN INC.
3-101-7 Ishibashi-cho
Hekinan city Aichi, Japan, Japan

Nakayama, Shigeo
Editor
FUJI TELEVISION NETWORK, INC.
1-50-17, Gotokuji
Setagaya-ku, Tokyo 154, Japan, Japan

Nakayama, Yoshiaki
Correspondent
MAINICHI SHIMBUN
Sunkijhto 1-202, 3-13-31
Nishiahigumi, Nerima-Ku, Tokyo, Japan, Japan

Nameshida, Takashi
Correspondent
MAINICHI SHIMBUN
1-7-36-307 Fuji-Cho, Hoya-Shi
Tokyo 202, Japan, Japan

Narusawa, Naoki
Correspondent
NIHON KEIZAI SHIMBUN, INC.
H304 Bundeskanzlerplatz 2-10
5-300 Bon 1, Germany, Japan

Neided, Toshimitsu
Correspondent
KYODO NEWS SERVICE
R. das Laranjeiras 466/505
22240 Rio de Janeiro, RJ Brazil, Japan

Nemoto, Seiki
Correspondent
ASAHI SHIMBUN
272-11-104 Yatsuka-cho
Soka-shi, Saitama, Japan, Japan

Neshibuchi, Kenji
Editor
FUJI TELEVISION NETWORK, INC.
3-25-11 Asahi-cho
Kawagoe-shi, Saitama 350, Japan, Japan

Nishihara, Joichi
Correspondent
NHK JAPAN BROADCASTING CORP.
1-7-3-204 Kitamachi
Hoya-shi, Tokyo, Japan, Japan

Nishikawa, Yoshio
Correspondent
NHK JAPAN BROADCASTING CORP.
4-8-23-205 Narimasu
Itabashi-ku, Tokyo, Japan, Japan

Nishimura, Shuichi
Correspondent
MAINICHI SHIMBUN
1-21-3-203 Nezu, Bunkyo-Ku
Tokyo 113, Japan, Japan

Nishimura, Toshinori
Correspondent
JIJI PRESS LTD.
Av. Paulista 854-13
Sao Paulo 01310, Brazil, Japan

Nishimura, Yoko
Reporter Radio/Television
FUJI TELEVISION NETWORK, INC.
1-20-15-602 Otowa
Bunkyo-Ku, Tokyo, Japan, Japan

Nishiyama, Sachiyuki
Photographer
ASAHI SHIMBUN
5-3-2 Tsukiji, Chuo-ku R.Sampaio Viana 203 c.13 Paraiso
Sao Paulo -SP 04004, Japan
T: 55 11 8841263 F: 55 11 8879990

Nitta, Shinzo
Editor
NHK JAPAN BROADCASTING CORP.
1822 Noborito, Tama-Ku
Kawasaki-Shi, Kanagawa 214, Japan, Japan

Nobutoshi, Kamiya
Reporter Radio/Television
OITA BROADCASTING SYSTEM
Iwatamachi Oita City
Japan, Japan

Nomura, Akio
Correspondent
ASAHI SHIMBUN
700 New Hampshire Ave., NW
Washington, DC 20037, USA, Japan

Nonaka, Naoko
Reporter Radio/Television
NIPPON CULTURAL B'CASTING, INC.
1-6-10 Izumihon-cho, Komah-shi
Tokyo 201, Japan, Japan

Norio, Noro
Editor
CHUNICHI SHINBUN
3-10-3, Nishidai, Itabashi-ku
Tokyo 100, Japan, Japan

Noro, Masayuki
Correspondent
ASAHI SHIMBUN
1-13-611 Imazumasago-Cho
Nishinomiya, Japan, Japan

Obara, Mario
Correspondent
ASAHI SHIMBUN
Rua Fernando de Noronha, 296
Sao Paulo, SP., Brazil, Japan

Oda, Takashi
Correspondent
YOMIURI SHIMBUN
1-9-10-302 Takaban
Meguro-Ku, Tokyo 152, japan, Japan

Ogata, Akinobu
Reporter Radio/Television
NHK
3-30-15 Kishiya Tsurumi-ku
Yokohama City, Kanagawa Pre., Japan, Japan

Ogawa, Jun
Correspondent
TOKYO B'CASTING SYSTEM, INC.
G-204, 1-1 Ogura
Kawasaki, Japan, Japan

Ogi, Noriyuki
Correspondent
NHK JAPAN BROADCASTING CORP.
Daizawa 2-6-5-132, Setagata-Ku
Tokyo 155, Japan, Japan

Ohashi, Akiko
Correspondent
MAMORU UTSUMI
41 Shagich-Kamitakano
Sakyo-ku, Kyoto, Japan, Japan

Oiibana, Makoto
Correspondent
JIJI PRESS LTD.
A302, 146-1 Suenaga Takatu-Ku
Kawasaki-Shi, Kanagawa-Ken 213, Japan, Japan

Oishi, Itaru
Correspondent
NIHON KEIZAI SHIMBUN, INC.
3-11-1-504 Botan, Koto-Ku
Tokyo, Japan, Japan

Oishi, Masayasu
Correspondent
MAINICHI SHIMBUN
4-11-2-301 Shimizu, Kita-Ku
Nagoya-City, Aichi-Pref. 462, Japan, Japan

Oiwa, Tomoko
Correspondent
NHK JAPAN BROADCASTING CORP.
H. Tsustujigaoka 2-4-3-403
Chofu, Tokyo 182, Japan, Japan

Oka, Hideo
Correspondent
MAINICHI SHIMBUN
1-6-4-502 Marishoji, Asahi-Ku
Osaka 535, Japan, Japan

Okabe, Toru
Correspondent
NHK
15 Sycamore Lane
Irvington, NY 10533, Japan

Okajima, Shigeyuki
Correspondent
YOMIURI SHIMBUN
1-5-5-204 Komatsugawa
Edogawa-Ku, Tokyo 132, Japan, Japan

Okamoto, Noboru
Correspondent
JIJI PRESS LTD.
Av.Paulista 854-13
Sao Paulo 01310, Brazil, Japan

Okuda, Kayo
Correspondent
FUJI TELEVISION NETWORK, INC.
Estrada Municipal do Perobal
Sao Paulo-SP, Brazil, Japan

Omuta, Toru
Correspondent
ASAHI SHIMBUN
3-16-20 Nishi-Kasai
Edogawa-Ku, Tokyo 134, Japan, Japan

Onishi, Takao
Correspondent
HOKKAIDO SHIMBUN
3-10-24-205 Takashima-Daira
Itabashi-Ku, Tokyo, Japan, Japan

Ono, Masami
Correspondent
ASAHI SHIMBUN
22, Dom 87, Ulitsa Vavilova
Moscow, Russia, Japan

Onose, Masahisa
Reporter Radio/Television
TV ASAHI
2-35-4-102 Shimo-Ishihara
Chofu-Shi, Tokyo, Japan, Japan

Onuma, Chieko
Correspondent
TV ASAHI
1-33-9 Yotsugi Katsushika-Ku
Tokyo, Japan, Japan

Onuma, Erika
Correspondent
ASAHI SHIMBUN
R. Pamplona 163-AP 41
CEP 01405 Sao Paulo, SP, Brazil, Japan

Oshima, Hiroshi
Correspondent
KYODO NEWS SERVICE
20 Edgewood Avenue
Larchmont, NY 10538, USA, Japan

Ota, Hiromichi
Editor
NHK JAPAN BROADCASTING CORP.
3-10-13-423 Miyazaki
Miyamae-ku, Kawasaki-shi, Japan, Japan

Ota, Shin
Reporter Radio/Television
FUJI TELEVISION NETWORK, INC.
1-17-7-203, Kamiuna
Setagaya-ku, Tokyo 154, Japan, Japan

Ota, Tomoko
Correspondent
NHK JAPAN BROADCASTING CORP.
R.Paissandu 328, Apt. 402
22210 Rio de Janeior, RJ, Brazil, Japan

Otsuka, Ryuichi
Correspondent
YOMIURI SHIMBUN
6-26-1-701 Hongo, Bunkyo-Ku
Tokyo 131, Japan, Japan

Otsuka, Takashi
Correspondent
ASAHI SHIMBUN
5-15-6 Matsubara, Setagaya
Tokyo, Japan, Japan

Otsuka, Tomohiko
Correspondent
MAINICHI SHIMBUN
18-1-404 Shimoshinozaki-Cho
Edogawa-Ku, Tokyo 133, Japan, Japan

Oyama, Yasushi
Editor
FUJI TV
1-26-12-A-102 Nukuikitamachi
Koganeishi, Tokyo 184, Japan, Japan

Ozaki, Michio
Correspondent
MAINICHI SHIMBUN
6-6-32 Nakakasai
Edogawa-Ku, Tokyo, Japan, Japan

Ozaki, Newton Yukio
Correspondent
NHK JAPAN BROADCASTING CORP.
R.Andrade Neves 241, Apt. 307
20510 Rio de Janeior, RJ, Brazil, Japan

Ozawa, Kazuo
Correspondent
KYODO NEWS SERVICE
7-31 Misumi-Cho, Chigasaki-Shi
Kanagawa, Japan, Japan

Pimentel, Paulo Roberto Morgira
Editor
FUJI TV
28 Glenmore Road,
NW 3 London, England, Japan

Quinocita, Nair Yoko
Correspondent
TV ASAHI
Rua: Martins, 148 Butanta
Sao Paulo, Brazil, Japan

Ratliff, Timothy Mottat
Correspondent
TV ASAHI
517 Spring Street
SE, Vienna, VA 22180, USA, Japan

Rippei, Tomono
Reporter Radio/Television
TOKYO B'CASTING SYSTEM
1-44-2 Yoyogi
Shibuya-ku, Tokyo, Japan, Japan

Ryuzaki, Takashi
Correspondent
MAINICHI SHIMBUN
1-21-3 Nezu, Bunkyo-ku
Tokyo, Japan, Japan

Sadakazu, Otsuka
Reporter Radio/Television
OITA BROADCASTING SYSTEM
Kanoo Oita City
Japan, Japan

Sadamori, Daiji
Correspondent
ASAHI SHIMBUN
5600 Wisconsin Ave.
Chevy Chase, MD 20815, USA, Japan

Saito, Hirohito
Editor
CHUNICHI SHINBUN
4-16-208 Yamato-Cho
Itabashi-Ku, Tokyo, Japan, Japan

Saito, Lirica
Correspondent
YOMIURI SHIMUBN
R. Raja Gabaglia, 118/301
Grajau, RJ, Brazil, Japan

Saito, Michio
Correspondent
TOKYO B'CASTING SYSTEM, INC.
7316 Churchill Rd.
McLean, Virginia 22101, USA, Japan

Saito, Takahiro
Editor
TOKYO B'CASTING SYSTEM, INC.
C-1314, 30-1 Hisamoto
Takatu-Ku, Kawasaki-City, Japan, Japan

Saito, Yuji
Correspondent
CHUNICHI SHINBUN
4-16-208 Yamato-cho
Itabashi-ku, Tokyo, Japan, Japan

Sakurada, Junko
Reporter Radio/Television
NIPPON TELEVISION NETWORK
5-16-5 Apt. 804, Roppongi
Minato-Ku, Tokyo, Japan, Japan

Sakurai, Mitsuo
Correspondent
KYODO NEWS SERVICE
2-3-7-1011 Minowa-Cho
Kohoku-Ku, Yokohama, Japan, Japan

Sakurai, Yoshiko
Reporter Radio/Television
NIPPON TELEVISION NETWORK
705-2-27-5 Hatsudai
Shibuya-Ku, Tokyo 151, Japan, Japan

Sakurai, Yuki
Editor
TERRA
3-101 Tokyu-Dwell Ka-883
Chiba-pref, Japan 270-01, Japan

Sasada, Naomi
Correspondent
JIJI PRESS LTD.
Av. Paulista 854-13
Sao Paulo 01310, Brazil, Japan

Sasaki, Hiroshi
Correspondent
ASAHI SHIMBUN
405 Yscore, 2-30-4 Higasi
Tutujigaoka, Chouhu-Shi, Tokyo, Japan, Japan

Sato, Hidenori
Correspondent
YOMIURI SHIMBUN
4-2-9-303 Shiohama, Ichikawa
Chiba 272-01, Japan, Japan

Sato, Kenji
Editor
NHK
1-13-15 Goutokuji, Setagaya-ku
Tokyo, Japan, Japan

Sato, Mikio
Correspondent
NHK
2-40-12 Fujigaoka, Midoriku
Yokohama-ku, Tokyo, Japan, Japan

Sato, Noriaki
Correspondent
YOMIURI SHIMBUN
307-4-9-4 Motobuto, Urawa-Ku
Saitama-Pref. 336, Japan, Japan

Sato, Takayuki
Journalist
ASAHI SHIMBUN
5-3-2 Tsukiji, Chuo-ku R.Sampaio Viana 203 c.13 Paraiso
Sao Paulo -SP 04004, Japan
T: 55 11 8841263 F: 55 11 8879990

Sato, Toshio
Editor
JIJI PRESS LTD.
3016-1-1316 Nagatuta-Machi
Midori-Ku, Kanagawa-Ken 227, Japan, Japan

Sato, Yoshimi
Correspondent
SANKEI SHIMBUN
6-20-12-303 Arakawa
Arakawa-Ku, Tokyo 116, Japan, Japan

Satoh, Keiichi
Reporter Radio/Television
NIPPON TELEVISION
4-17-20 Kamikitazawa
Setagaya-ku, Tokyo, 156 Japan, Japan

Sawa, Yuji
Editor
FUJI TELEVISION NETWORK, INC.
4-25-3, Hijirigaoka
Setagaya-ku, Tokyo 154, Japan, Japan

Sawada, Isao
Correspondent
NIPPON TELEVISION NETWORK
Rua Muniz de Souza 1170
01534, Sao Paulo-SP, Brazil, Japan

Segawa, Shiro
Correspondent
MAINICHI SHIMBUN
2-23-53 Oizumigakuen-Cho
Nerima-Ku, Tokyo, Japan, Japan

Seisuke, So
Reporter Radio/Television
OITA BROADCASTING SYSTEM
997 Kumi Oita
Japan, Japan

Sekiguchi, Natsuko
Correspondent
YOMIURI SHIMBUN
R. Garvao Bueno 376, #102
Rio de Janeiro, RJ, Brazil, Japan

Sekikawa, Shuichi
Reporter Radio/Television
TV ASAHI
1-16-13 Takaido Nishi
Saginamitku, Tokyo, Japan, Japan

Serikawa, Yoichi
Correspondent
NIHON KEIZAI SHIMBUN, INC.
3-17-11 Mukaihara, Asaoku
Kawasaki-Ku, Kanagawa-pref., Japan, Japan

Seto, Jyunichi
Correspondent
MAINICHI SHIMBUN
5-1-16-403 Matuba-Cho
Kashiwa-Shi, Chiba-Pre., Japan, Japan

Setsu, Mikiko
Correspondent
ASAHI SHIMBUN
4-13-14 Shirogane, Minato-Ku
Tokyo, Japan, Japan

Shibo, Mitsukazu,
Correspondent
WEEKLY ASAHI
1-44-11 Hatsudai Shibuya-ku
Tokyo, Japan
T: 81 3 33752506 F: 81 3 33752506

Shimada, Toshio
Correspondent
NHK JAPAN BROADCASTING CORP.
4-8-22-322, Shimoochiai
Shinjuku-ku, Tokyo, Japan, Japan

Shimizu, Fumio
Editor
WEEKLY ENERGY & ENVIRONMENT
4-11-28 Midori-Cho
Akishima-Shi, Tokyo 196, Japan, Japan

Shimura, Tomoko
Editor
NATURE CONSERVATION JAPAN
3-22-14-502 Minami
Tokyo, Japan, Japan

Shin-ichi, Torikai
Editor
PUMPKIN-USHIO SHUPPANSHA
3-25-21 Maehara
Koganei-Shi, Tokyo, Japan, Japan

Shinkura, Hiromasa
Correspondent
HOKKAIDO SHIMBUN
10202 Flemming Avenue
Bethesda, MA 20014, USA, Japan

Song, Cecilia
Correspondent
WWF-JAPAN
1-51-18 Wada Suginami-ku
Tokyo, Japan, Japan

Steffen, Alexis
Correspondent
JAPAN TIMES
403 D. Takara Haitsu, 3-544
Ohmiya-Cho, Nara-Shi, Nara-ken, Japan, Japan

Suehiro, Jun
Editor
KUMAMOTO-NICHINICHI SHIMBUN
5-44-2 Chuo, Nakano-Ku
Tokyo 164, Japan, Japan

Suenobu, Yoshimasa
Correspondent
TV ASAHI
257 Westview Ave.
Fort Lee, NJ 07024, USA, Japan

Sueoka, Mariko
Correspondent
TV ASAHI
829-203 Shimosakunobe
Takatsu-Ku, Kawasaki, Japan, Japan

Sugioka, Jun
Editor
NHK JAPAN BROADCASTING CORP.
3-13-8 Sakuradai, Nerima-Ku
Tokyo 176, Japan, Japan

Sugishita, Tsuneo
Correspondent
YOMIURI SHIMBUN
946-7 Junisho, Kamakura-Ku
Kanagawa, Japan, Japan

Suigo, Naoya
Editor
JAPAN TIMES
Oshima 4-6-6, Kawasaki-ku
Kanagawa-ken, 210 Japan, Japan

Susuki, Nobuo
Correspondent
ASAHI SHIMBUN
4-13-3 Setagaya-Ku
Tokyo, Japan, Japan

Suzuki, Masanori
Correspondent
SHOGAKUKAN INC.
2-24-21 Nishiikuta, Tama-ku
Kawasaki-Shi, Kanagawa Ken, Japan, Japan

Suzuki, Norio
Editor
ASAHI SHIMBUN
4-1-12-205, Kirigasaka
Midori-Ku, Yokohama, Japan, Japan

Suzuki, Seiji
Editor
TOKYO B'CASTING SYSTEM, INC.
1-4-5 Midonigaoka, Meguro-Ku
Tokyo 100, Japan, Japan

Suzuki, Suzana Eriko
Translator
ASAHI SHIMBUN
5-3-2 Tsukiji, Chuo-ku R.Sampaio Viana 203 c.13 Paraiso
Sao Paulo -SP 04004, Japan
T: 55 11 8841263 F: 55 11 8879990

Syoji, Ito
Editor
CHUNICHI SHINBUN
3-34-2 Sasage, Kounan-Ku
Yokohama, Kanagawa, Japan, Japan

Tabata, Mami
Correspondent
SHOGAKUKAN INC.
5-4-6-202 Minamiaoyama
Minato-ku, Tokyo, Japan, Japan

Tabata, Tadashi
Reporter Radio/Television
TV ASAHI
Komazawa 4-20-7-201 Setagaya
Tokyo, Japan, Japan

Tachibana, Kouki
Correspondent
JIJI PRESS LTD.
1-17-15 Kinuta, Setagaya-Ku
Tokyo, Japan, Japan

Tadaii, Hugo Yoshii
Correspondent
TOKYO B'CASTING SYSTEM, INC.
Av. Paulista 1471, 5 AND
CJ 503, Sao Paulo, CEP 01311, Brazil, Japan

Tadashi, Miyazaki
Editor
KYODO NEWS SERVICE
2-4-6-301 Susukino, Midori-Ku
Yokohama City, Japan, Japan

Taguchi, Masayuki
Editor
JAPAN AGRICULTURAL PRESS
3-29-8 Minamitanaka, Nerima-ku
Tokyo 177, Japan, Japan

Taichi, Kiko
Correspondent
NHK
5-17-9 Daita Setagaya-ku
Tokyo, Japan, Japan

Tajima, Tutomu
Editor
CHUNICHI SHINBUN
69-11 Manse Street
Forest Hills, NY 11375, USA, Japan

Takahara, Kanji
Editor
TELEVISION TOKYO CHANNEL 12
D-834, 1-1-5 Sekido, Tama City
Tokyo, Japan, Japan

Takahashi, Kumiko
Editor
TOKYO B'CASTING SYSTEM, INC.
2-7-10-205 Sakura Shinmathi
Setagaya-Ku, Tokyo 154, Japan, Japan

Takahashi, Yuichi
Correspondent
NIHON KEIZAI SHIMBUN, INC.
20A Chemin de Velours Conches
Geneva, Switzerland, Japan

Takahashi, Yuichiro
Correspondent
FUJI TELEVISION NETWORK, INC.
Av. Bartolomeu Mitre, 330
Rio de Janeiro-RJ, Brazil, Japan

Takahata, Minako
Correspondent
TOKYO B'CASTING SYSTEM, INC.
333 E. 30th Street
New York, NY 10016, Japan

Takanarita, Toru
Editor
ASAHI SHIMBUN
4-10-14 Sakuragaoka
Setagaya-Ku, Tokyo 156, Japan, Japan

Takao, Nagatake
Editor
CHUNICHI SHINBUN
3018-9 Daitakubo, Urawa-Shi
Saitama, Japan, Japan

Takasaki, Kunio
Correspondent
FUJI TELEVISION NETWORK, INC.
2-42-21-103 Tagara
Nerima-Ku, Tokyo, Japan, Japan

Takashi, Fukushima
Reporter Radio/Television
TOKYO B'CASTING SYSTEM, INC.
1-11-4 Aobadai, Midori-Ku
Yokohama 227, Japan, Japan

Takatsu, Yasufumi
Correspondent
YOMIURI SHIMBUN
101-7-10-3 Tokiwa-cho
Urawa City, Saitama, Japan, Japan

Takeda, Norio
Correspondent
NIHON KEIZAI SHIMBUN, INC.
1-3-21 Sakae-Cho, Asakashi
Saitama 351, Japan, Japan

Takeda, Yorimasa
Correspondent
NHK JAPAN BROADCASTING CORP.
5-1-1 Minami-Odori
Kushiro City, Hokkaido, Japan, Japan

Takehara, Akiko
Editor
TERRA
3-7-16 210 Shibuya-Ku
Sendagaya, Tokyo, Japan, Japan

Takei, Nobusuke
Reporter Radio/Television
NIPPON TELEVISION NETWORK
4-16-5-308 Ken-gun
Kumamoto-city, Kumamoto-pref., Japan, Japan

Takemura, Masahiro
Correspondent
YOMIURI SHIMBUN
3623 38th St. N.W., Apt.101
Washington, D.C. 20016, USA, Japan

Takeuchi, Akihiko
Correspondent
SHOGAKUKAN INC.
30-1 C-611 Hisamoto, Takatu-ku
Kawasaki-shi, Kanagawaken, Japan, Japan

Takeuchi, Keiji
Journalist
ASAHI SHIMBUN
5-3-2 Tsukiji, Chuo-ku R.Sampaio Viana 203 c.13 Paraiso
Sao Paulo -SP 04004, Japan
T: 55 11 8841263 F: 55 11 8879990

Takeuchi, Ken
Correspondent
ASAHI SHIMBUN
2-10-26-110 Kohinata
Bunkyo-Ku, Tokyo, Japan, Japan

Takeuchi, Masatu
Correspondent
NIHON KEIZAI SHIMBUN, INC.
3-28-6-101 Ogikubo
Suginami-Ku, Tokyo 167, Japan, Japan

Takeuchi, Shumei
Correspondent
ASAHI SHIMBUN
2-3-313 Tsutsujigaoka
Midori-Ku, Yokohama, Japan, Japan

Takikawa, Toru
Editor
MAINICHI SHIMBUN
121-1-104 Matudoshinden
Matudo-Ku, Chiba-Pre., Japan, Japan

Takiyama, Susumu
Correspondent
NIHON KEIZAI SHIMBUN, INC.
1727-3 Kanamori, Machida-shi
Tokyo 194, Japan, Japan

Tamura, Hiroko
Reporter Radio/Television
FUJI TELEVISION NETWORK, INC.
2-2-I Koenji-Kita, Suginami-KU
Tokyo 166, Japan, Japan

Tamura, Walter Yukio
Correspondent
TOKYO SHIMBUN
R. Monte Aprazivel 423
Sao Paulo, Brazil, Japan

Tamura, Yuji
Correspondent
HOKKAIDO SHIMBUN
3-4-21-3 Eda-Minami, Midori-Ku
Yokohama 225, Japan, Japan

Tanaka, Hiroaki
Correspondent
NHK JAPAN BROADCASTING CORP.
223, 2-27-8 Shimokodanaka
Nakahara-Ku, Kawasaki-Shi,Kanagawa,Japan, Japan

Tanaka, Narunobu
Editor
NIHON KEIZAI SHIMBUN, INC.
188 East 64th Street
1706, New York, NY 10021, USA, Japan

Taniguchi, Taizo
Editor
MAINICHI SHIMBUN
2-22-10-702 Akabane-Kita
Kita-Ku, Tokyo, Japan, Japan

Taniguchi, Tetsuo
Correspondent
ASAHI SHIMBUN
6-19-8-507 Higashi-Mikuni
Yodogawa-Ku, Osaka City, Japan, Japan

Tanihara, Kazunori
Reporter Radio/Television
NIPPON TELEVISION NETWORK
4-1-10-1103 Kamisaginomiya
Nakano-Ku, Tokyo, Japan, Japan

Tanogashira, Yoichi
Editor
FUJI TELEVISION NETWORK, INC.
84-402, 2-i Sakurazutsumi-cho
Musashino-shi, Tokyo 180, Japan, Japan

Terada, Masaomi
Correspondent
YOMIURI SHIMBUN
2025 Broadway, #27A
New York, NY 10023, USA, Japan

Terasaki, Hisanori
Correspondent
NHK JAPAN BROADCASTING CORP.
1-12-14-111, Saginuma
Miyamae-ku, Kawasaki City, Japan, Japan

Terashima, Mitsuo
Editor
FUJI TELEVISION NETWORK, INC.
1-11-17-602 Hachimancho
Fuchu-shi, Tokyo 183, Japan, Japan

Tokunaga, Hiromitsu
Correspondent
NIPPON TELEVISION NETWORK
16 Dobbs Terrace, Scarsdale
New York, NY 10583, USA, Japan

Torii, Hiroyuki
Editor
NIHON KEIZAI SHIMBUN, INC.
Hhizumigakuenmachi 7-25-6
Nerima-Ku, Tokyo, Japan, Japan

Toshiyuki, Yamagishi
Editor
CHUNICHI SHINBUN
2-30-2 Kashigene, Itabashi-ku
Tokyo 100, Japan, Japan

Totani, Mitsuteru
Correspondent
TV ASAHI
4-27-4 Kitakarasuyama
Setagaya, Tokyo, Japan, Japan

Totsune, Tadashi
Correspondent
NIPPON TELEVISION NETWORK
5-5-44-609 Higashiterao
Tasurumi-Ku, Yokohama-city, K.P, Japan, Japan

Tsuji, Yomei
Editor
ASAHI SHIMBUN
503 1-13-7 Miyazaki
Miyamae-Ku, Kawasak-City 216, Japan, Japan

Tsukui, Naoya
Reporter Radio/Television
TOKYO BROADCASTING SYSTEM
2-33-9-203 Uehara Shibuya-ku
Tokyo, Japan, Japan

Tsutomu, Aoki
NHK
520-20 Sanyu-Machi
Hacioji, Tokyo, Japan, Japan

Tuyuki, Shigeru
Reporter Radio/Television
FUJI TELEVISION NETWORK, INC.
4-7-15 Soshigaya, Setagaya-ku
Tokyo, Japan, Japan

Uchida, Akinori
Correspondent
YOMIURI SHIMBUN
3-13-19-302 Minami-Gyotoku
Ichikawa-Ku, Chiba 272-01, Japan, Japan

Uchida, Masaru
Editor
TOKYO B'CASTING SYSTEM, INC.
3-7-13 Nisi, Kunitachi-Si
Tokyo 186, Japan, Japan

Uchihata, Tsugumasa
Correspondent
SENKEI SHIMBUN
C, 1-2-3 Kichijojikitamachi
Musashino-Ku, Tokyo 180, Japan, Japan

Uchiyama, Mituhiro
Reporter Radio/Television
TV ASAHI
Azamino 3-2-8-202 Midori-Ku
Yokohama-Shi, Kanagawa-Ken, Japan, Japan

Ueda, Tomoki
Correspondent
KYODO NEWS SERVICE
Jardin 13, Col.Tlacopac
San Angel CP 01040, DF, Mexico, Japan

Uemura, Noriko
Correspondent
ASAHI SHIMBUN
R. Fradique Coutinho 1419
Pinheiros, CEP 05416, SP.,SP., Brazil, Japan

Uemura, Takeshi
Correspondent
YOMIURI SHIMBUN
3-239-9 Akayamacho
Koshigaya-Ku, Saitama 343, Japan, Japan

Umemura, Itsuo
Correspondent
NHK JAPAN BROADCASTING CORP.
2-18-12-101 Naname, Meguro-ku
Tokyo 152, Japan, Japan

Wada, Masami
Editor
NIHON KEIZAI SHIMBUN, INC.
42 Breegemont Avenue
Riverside, CT 06878, USA, Japan

Waguri, Yoshikuni
Editor
JAPAN AGRICULTURAL PRESS
7-22 Aoyagi, Soka City
Saitama 340, Japan, Japan

Wakabayashi, Muneo
Editor
TELEVISION TOKYO CHANNEL 12
Room 211, Tsurumaki 3-22-25
Setagaya-ku, Tokyo 154, Japan, Japan

Wakamiya, Toshihiko
Editor
NHK JAPAN BROADCASTING CORP.
9-5-10 Shimorenzyaku
Mitaka-Shi, Tokyo 181, Japan, Japan

Wakayama, Shigeki
Correspondent
ASAHI SHIMBUN
37-4 Mibushyzaku, Nakagyo-Ku
Kyoto City 604, Japan, Japan

Wakita, Tetsushi
Correspondent
NHK JAPAN BROADCASTING CORP.
1421 Greenfield Ave., #5
Los Angeles, CA 90025, USA, Japan

Watanabe, Kojiro
Correspondent
TV ASAHI
6603 Rannoch Road
Md 20817, Usa, Japan

Watanabe, Takashi
Correspondent
JIJI PRESS LTD.
6633 Melrose Drive
McLean, VA 22101, USA, Japan

Watanabe, Terutaka
Reporter Radio/Television
TV ASAHI
4-31-21 Shinishikawa,Midori-Ku
Yokohama, Japan, Japan

Watanabe, Toru
Editor
FUJI TV
510 E. 85th 5c
New York, NY 10028, Japan

Watanabe, Tsutomu
Correspondent
ASAHI SHIMBUN
1-26-5-C208 Takaidonishi
Suginami-ku, Tokyo, Japan, Japan

Watanabe, Yoshiyuki
Correspondent
MAINICHI SHIMBUN
1-1-1-201 Tate, Shiki-Shi
Saitama 353, Japan, Japan

Yabunami, Tadashi
Editor
NHK JAPAN BROADCASTING CORP.
3-1-15 Wada, Suginami-Ku
Tokyo 166, Japan, Japan

Yabutani, Kiyoshi
Editor
TOKYO B'CASTING SYSTEM, INC.
3-24-10 Kitazawa, Setagaya-Ku
Tokyo 156, Japan, Japan

Yajima, Seiji
Correspondent
SANKEI SHIMBUN
R. Valduc., 158
1160 Bruxelles, Belgium, Japan

Yajima, Yasuhiro
Editor
NHK JAPAN BROADCASTING CORP.
1-18-5 Hatogaoka
Hatoyama-machi, Saitama-Ken, Japan, Japan

Yamada, Fumio
Correspondent
KYODO NEWS SERVICE
33 Knox St., Waverley
Johannesburg, South Africa, Japan

Yamada, Michiko
Correspondent
MAINICHI SHIMBUN
3-4-41 Shimorenjya-Ku
Mitaka-Shi, Tokyo, Japan, Japan

Yamada, Tetsuhiro
Correspondent
TV ASAHI
100 UN Plaza (#8C)
New York, NY 10017 USA, Japan

Yamagami, Tsuyoshi
Correspondent
TV ASAHI
30 Blvd., d'Argenson Neuilly
Sur Seime 92200, France, Japan

Yamaguchi, Akirako
Correspondent
YOMIURI SHIMBUN
1-34-10-303 Kichijoji-Honcho
Musashino-Ku, Tokyo 180, Japan, Japan

Yamaguchi, Hiroaki
Editor
CHUNICHI SHINBUN
3-37-36 Sasage, Kounan-Ku
Yokohama, Kanagawa, Japan, Japan

Yamaguchi, Hiroya
Correspondent
YOMIURI SHIMBUN
1-5-205 Makinoharadanchi
435-1 Makinohara, Chiba 270, Japan, Japan

Yamahata, Marcia
Correspondent
YOMIURI SHIMBUN
R. do Bispo 119/Apt. 201
Rio Comprido, RJ, RJ, Brazil, Japan

Yamamoto, Daijiro
Correspondent
YOMIURI SHIMBUN
943-3-113 Kamiodanaka
Kawasaki-Ku, Kanagawa 211, Japan, Japan

Yamamoto, Kenichi
Correspondent
NHK JAPAN BROADCASTING CORP.
4-2-10-233 Kamiaoki
Kawaguchi City, Saitama Pref., Japan, Japan

Yamaoka, Kunihiko
Correspondent
YOMIURI SHIMBUN
7-1-4-106, Fujiwara
Funabashi-shi, Chiba 273, Japan, Japan

Yamashita, Hiroshi
Correspondent
TV ASAHI
Flat 4, 29 Rendington Road
London NW3 7QY, UK, Japan

Yamazaki, Kazuo
Editor
NIKKAN KOGYO SHIMBUN, LTD.
1-8-10 Kudan-Kita, Chiyoda-Ku
Tokyo 102, Japan, Japan

Yano, Shuji
Editor
FUJI TELEVISION NETWORK, INC.
2-1-39-305 Higashioizuni
Nerina-ku, Tokyo 178, Japan, Japan

Yasuhiro, Tanaka
Correspondent
AKAHATA
4615 North Park Ave. (#502)
Chevy Chase, MD 20815 USA, Japan

Yasunori, Komori
Correspondent
MAINICHI BROADCASTING
1-4-28 Manabiguoka
Kobe, Japan, Japan

Yokota, Hiroyuki
Correspondent
YOMIURI SHIMBUN
1-16-1 Suwada, Ichikawa-Ku
Chiba 272, Japan, Japan

Yokota, Mitsutoshi
Correspondent
HOKKAIDO SHIMBUN
3-27 Nakadai
Itabashi-ku, Tokyo, Japan, Japan

Yokoyama, Isao
Editor
NIKKEI BUSINESS PUBL., INC.
1117-8 Matsudo, Matsudo-shi
Chiba, Japan, Japan

Yokoyama, Yuji
Correspondent
SEKAI NIPPO
2529 Fling Hill Road
Vienna, Virginia 22181, USA, Japan

Yoneshima, Shinichi
Correspondent
NHK JAPAN BROADCASTING CORP.
1-29-203 Keyakidaira
Miyamae-ku, Kawasaki City, Pref., Japan, Japan

Yoneyama, Masahiro
Editor
ASAHI SHIMBUN
1-5-33-501 Torikai, Chuo-Ku
Fukuoka-Si 810, Japan, Japan

Yoshida, Fumihiko
Correspondent
ASAHI SHIMBUN
8516 Victory Lant
Potomac, M.D. 20854, USA, Japan

Yoshida, Takafumi
Journalist
ASAHI SHIMBUN
5-3-2 Tsukiji, Chuo-ku R.Sampaio Viana 203 c.13 Paraiso
Sao Paulo -SP 04004, Japan
T: 55 11 8841263 F: 55 11 8879990

Yoshida, Toru
Correspondent
NIHON KEIZAI SHIMBUN, INC.
Ogikubo 5-16-7-310
Suginami-Ku, Tokyo, Japan, Japan

Yoshizaki, Takashi
Editor
TOKYO B'CASTING SYSTEM, INC.
3-24-1 Minami Ikuta, Tama-Ku
Kawasaki City, Japan, Japan

Yukawa, Te Tsuo
Correspondent
TOKYO B'CASTING SYSTEM, INC.
30-1 Hisamoto Takatsu-ku
Kawasaki, Japan, Japan

Yuki, Takahiko
Correspondent
YOMIURI SHIMBUN
1-2-21 Namiki, Kawaguchi-Ku
Saitama, 332, Japan, Japan

Yutaka, Shimizu
Correspondent
FUJI TV
200E. 72nd St. 25
New York, N.Y. 10021, Japan

ORGANIZATIONS

Awaji, Takahisa, Mr.
Japan Environment Council
Dept. of Law, Rikkyo University,
Nishiikebukuro 3-34-1, Toshima-Ku, Tokyo 171, Japan
T: (81 3) 3985 2569 F: (81 3) 3983 0174

Fujimuri, Akiharu, Mr.
Global Industrial And Social Progress Research Institute
7th Floor, Mori Building No. 33, 3-8-21 Toranomon
Minato-Ku, Tokyo 105 Japan, Japan
T: (81 3) 3435 8800 F: (81 3) 3435 8810

Fukasawa, Kazunori, Mr.
New Energy and Industrial Technology Developement Organization (NEDO)
Susshine 60 29 F, 1-1, 3-Chome, Higashi-Ikebukuro
Toshima-Ku, Tokyo 170, Japan
T: (81 3) 3987 9363 F: (81 3) 3981 1744

Hasegawa, Masayo, Ms.
Sasakawa Peace Foundation
Sasakawa Hall, 3-12-12 Mita Minato-Ku
Tokyo 108, Japan
T: (81 3) 3769 2081 F: (81 3) 3769 2090

Hirano, Hiroya, Mr.
Asahi Glass Foundation
Bank of Tokyo Building 12F, Marunouchi 1-4-2, Chiyoda-Ku
Tokyo 100, Japan
T: (81 3) 3285 0591 F: (81 3) 3285 0592

Kaibe, Kohji, Mr.
Federation of Electric Power Companies
Keidanren Kaikan, 1-9-4, Ote-Machi, Chiyoda-Ku
Tokyo, Japan
F: (81 3) 3264 5672

Kawashima, Iyoshi Ike, Mr.
Japan Junior Chamber, Inc.
2-14-3 Hirakawacho, Chiyoda-Ku
Tokyo 102, Japan
F: (81 3) 3265 2409/3234 7183

Kira, Tatuo, Mr.
International Lake Environment Committee Foundation
3-4-22 Kyomachi, Otsu
Shiga 520, Japan
T: (81 775) 251 076 F: (81 775) 231 581/253 278

Kohda, Yoshihiro, Mr.
International Society for Mangrove Ecosystems
c/o College of Agriculture, University of the
Ryukyus, 1 Senbaru Nishihara, Okinawa 903-01, Japan
T: 895 2523/895 2221 EXT. 2876 F: (81 98) 727 9805

Kohno, Masatoshi, Mr.
Rissho Kosei-kai
11-1 Wada 2-Chome, Suginami-Ku
Tokyo 166, Japan
T: (81 3) 3381 1138 F: (81 3) 3382 8455

Nakanishi, Tamako, Mr.
Global Parliamentarians On Habitat
610 Sangun Kaikan, 2-1-1 Nagato-Cho, Chiyoda-Ku
Tokyo 100 Japan, Japan
T: (81 3) 3508 8610 F: (81 3) 3508 8610 E

Nakano, Toshihiro, Mr.
Organisation For Industrial, Spiritual And Cultural Advancement (OISCA)
6-12 Izumi 3-Chome, Suginami-Ku
Tokyo 168, Japan
T: (81 3) 322 5161/4 F: (81 3) 324 7111

Osawa, Shuzo, Mr.
People's Alliance to Protect the Environment of the Japan Inland Sea Area
Nichiei-Sangyo Bldg. 4f, 2-8-2 Aioi Cho
Akashi City, Hyogo 673, Japan
T: (81 78) 912 1193 F: (81 78) 912 0700

Sawama, Yasuo, Mr.
Japan Productivity Center
3-1-1 Shibuya, Shibuya-Ku
Tokyo 150, Japan
T: (81 3) 3409 1111 F: (81 3) 3409 5880

Shimogaito, Hiroshi, Mr.
Citizen's Alliance For Saving The Atmosphere And The Earth (CASA)
1-3-17-813 Tanimachi, Chuo-Ku
Osaka 540, Japan
T: (81 6) 941 3745 F: (81 6) 941 5699

Tanaka, Hideki, Mr.
Petroleum Association of Japan
9-4, 1-Chome, Ohtemachi, Chiyoda-Ku
Tokyo 100, Japan
T: (81 3) 3279 3811 F: (81 3) 3242 5688

Toyoda, Makoto, Mr.
National Alliance of Victims of Minamata Disease & Lawyers
Katsuhata Building 4F, 2-5 Kanda Tausaka-Cho,
Chiyoda-Ku, Tokyo 101, Japan
T: (81 3) 3293 3621 F: (81 3) 3293 3627

Yamaguchi, Tsutomu, Mr.
Research Institute of Innovative Technology for the Earth (RITE)
4th Floor, Shin-Kyoto Center Bldg., Karasumanishi-iru, Shiokouji-dori, Shimogyo-Ku, Kyoto 600, Japan
T: (81 3) 361 3611 F: (81 3) 361 5607

Yamanaka, Masami, Mr.
Japan Chemical Industry Association
Tokyo Club Building, 2-6 Kasumigaseki 3-Chome
Chiyoda-Ku, Tokyo 100, Japan
T: (81 3) 580 0751 F: (81 3) 580 0764

Yuasa, S., Mr.
Japan Scientists Association, The
9-16 Yushima 1-Chome, Bunkyo-Ku
Tokyo 113, Japan
T: (81 3) 3813 2363 F: (81 3) 3813 2363

GLOBAL FORUM

Abe, Osamu, Mr.
JAPANESE SOCIETY OF THE ENVIRONMENTAL EDUCATION
4-406-204, Namiki, Tsukuba-shi, Ibaraki 305
Tsukuba, Japan
T: 81 0298512488 F: 81 0298512988

Abe, Yozo, Mr.
RENGO
3-10-3 Mita, Minato-ku
Tokyo, Japan
T: 81 3 3769-7545 F: 81 3 3454-6206

Abuno, Brasilina Junko
Interpreter
NIPPON TELEVISION NET. CO.
14 Niban-Cho, Chiyoda-Ku
Tokyo, Japan
T: 03 32301960 F: 03 31344044

Agerstrand, Christer
Reporter
NYA NORRLAND
Box 55, 89101 Te Uusborgsv.33 12632 Hagerten
Osaka, Japan
T: 81 6 82380 F: 81 6 83033

Aichi, Ayako, Ms.
Nagata-Cho 2-1, Chiyoda-KU Tokyo
Tokyo, Japan

Akiko, Ohashi
Reporter
Kyoto Shimbun
Karasuma St. Ebisugaira, Nakagyota
Kyote, Japan
T: 81 75 2415930 F: 81 75 2410048

Akira, Yoshino, Mr.
GREEN HEART FOUNDATION
1280 Cigalki
Hondo-Shi - Kumamoto, Japan
T: 81 969 924 2282 F: 81 969 924 2276
E: 863

Akita, Seiko, Ms.
UNITED CHURCH OF CHRIST
4-6-12 Dalzawa, Setagaya-ku
Tokyo 155, Japan

Akiyama, Kazutake, Mr.
GISPRI
3-8-21 Toranomon Minato-Ku
Tokyo, Japan
T: 81 3 3435 8800 F: 81 3 3435 8810

Amano, Reiko, Ms.
92 NGO FORUM JAPAN
4F Kanda Kyyodo Bldg., 2-13-4 Iwamoto-Cho
Tokyo, Japan
T: 81 3 3861-0540 F: 81 3 3861-5048

Amaya, Kazuo, Mr.
Professor
JAPAN ASSOCIATION ON THE ENVIRONMENTAL STUDIES
3-5-8 Saiwaicho Fuchu-city
Tokyo, Japan
T: 81 423 34 6884 F: 81 423 34 6883

Amenomori, Machiko, Mr.
YAMAGISHI ASSOTIATION
2000-10-6-201, Kosugaya-cho, Jakaeku
Yokohama, Japan
T: 045 892 4832

Andoh, Kyoji, Mr.
JAPANESE CONSUMERS' CO-OPERATIVE UNION
Seikyo-kaikan 1-1-13, Sendagaya, Shibuya-ku
Tokyo, Japan
T: 81 3 34979103 F: 81 3 34970722
TL: 2423380 NCOP J

Aokage, Takako, Ms.
NETWORK YOKOHAMA
Yokohama City Office 1-1 Minato-cho
Kanagawa, Japan
T: 81 45 671 3038 F: 81 45 681 7388

Aoki, Terumitsu, Mr.
JAPAN NETWORK FOR ENVIRONMENT & POLLUTION
2-15-22 Shinmachi Akatsuka, Itabashi-ku
Tokyo, Japan
T: 2676

Aoyagi, Tadashi, Sr.
Director Of Ecology Dept.
MITSUBISHI RESEARCH INSTITUTE
2-3-6 Otemach, Chiyoda-ku
Tokyo, Japan
T: 81 3 3277-0546 F: 81 3 3277-0512

Aoyama, Tasuku, Mr.
Student
WASEDA UNIVERSITY
8 Yonbacho Chiyoda-ku, 102
Tokyo, Japan
T: 81 3 3263-9656

Arai, Naoko
Staff Writer
92 NGO FORUM JAPAN NEWS
4 F Kanda Kyodo Bldg, 2-13-4 Iwamotocho, Chiyoda-Ku
Tokushimo-ken, Japan
T: 81 886 3861-0540 F: 81 886 3861-5048

Araki, Teizo, Mr.
Member
OISCA - INTERNATIONAL
6-12, Izumi 3-Chome, Suginami-ku
Tokyo, Japan
T: 81 3 3322-5161 F: 81 3 3324-7111
TL: J29397 OISCA

Arayashiki, Hitoshi, Mr.
Director
OISCA - INTERNATIONAL
6-12, Izumi 3-Chome, Suginami-ku
Tokyo, Japan
T: 81 3 3322-5161 F: 81 3 3324-7111
TL: J29397 OISCA

Arboleda, Amadio A., Mr.
Director
UNITED NATIONS UNIVERSITY PRESS
Toho Seimei Bldg. 15-1 Shibuya 2-chome, Shibuya-ku
Tokyo, Japan
T: 81 3 3499-2811 F: 81 3 3499-2828
TL: J25442

Asai, Kazuko, Mr.
Advisor
JAPAN FOREST TECHNICAL ASSOCIATION
No. 7, Rokubancho chiyoda-ku
Tokyo, Japan
T: 81 3 261-5281 F: 81 3 261-5393

Asai, Masato, Mr.
THE SANITEI SHINBUN
6-34-6 Jingou-Mae Shibuya
Tokyo, Japan
F: 81 3 3243-1800

Asakura, Masayuki, Mr.
NATIONAL ALLIANCE OF VICTIMS OF MINAMATA DISEASE & LAWYERS
6-15-7, Fukazawa, Setagaya-ku
Tokyo, Japan
T: 81 3 3704-6871

Asano, T., Mr.
DELEGATES OF CASA-MEMBER CO-OPS
2-7-22 Tenma, Kita-ku, Osaka-city, PAL co-op
Osaka-tu, Japan
T: 81 6 354 8711 F: 81 6 354 8715

Asaoka, Mie, Ms.
NATIONAL ALLIANCE OF VICTIMS OF MINAMATA DISEASE & LAWYERS
6-15-7, Fukazawa, Setagaya-ku
Tokyo, Japan
T: 81 3 3704-6871

Azuma, Syozo, Mr.
Head
KOMEITO
2-1-2 Syu-giin dai-2 giinkaikan 733 Nagata-che, Tiyoda-ku
Tokyo, Japan
T: 81 3 3508-7320 F: 81 3 3593-7320

Barreto, Ari
Motorista
ASAHI SHIMBUN
5-3-2 Tsukiji, Chuo-ku R.Sampaio Viana 203 c.13 Paraiso
Sao Paulo -SP 04004, Japan
T: 55 11 8841263 F: 55 11 8879990

Bloch, Jayne, Ms.
Vice President
ASAHI GLASS FOUNDATION
Bank of Tokyo Building, 12F Marunouchi 1-4-2, Chiyoda-ku
Tokyo 100, Japan
T: 81 3 285-0591 F: 81 3 285-0592

Chay Lau, Chu, Ms.
Member
OISCA - INTERNATIONAL
6-12, Izumi 3-Chome, Suginami-ku
Tokyo, Japan
T: 81 3 3322-5161 F: 81 3 3324-7111
TL: J29397 OISCA

Chen, Steven K. J., Mr.
CHUNG HUAN FOUNDATION
2-1-4 Sadoharacho, Ichigaya, Shinjuku-ku
Tokyo, Japan
T: 81 3 3266-1598 F: 81 3 3266-9188

Chi-Chuan, Yang, Mr.
Chairman, Oisca Roc Taiwan
OISCA - INTERNATIONAL
6-12, Izumi 3-Chome, Suginami-ku
Tokyo, Japan
T: 81 3 3322-5161 F: 81 3 3324-7111
TL: J29397 OISCA

Chigami, Kunio, Mr.
DELEGATES OF CASA-MEMBER CO-OPS
7-4-15 Tanimachi, Choo-ku, Osaka-city
Osaka-tu, Japan
T: 81 6 762-7220 F: 81 6 762-7296

Cortez, Madgaleno, Mr.
President, Oisca Philippines
OISCA - INTERNATIONAL
6-12, Izumi 3-Chome, Suginami-ku
Tokyo, Japan
T: 81 3 3322-5161 F: 81 3 3324-7111
TL: J29397 OISCA

Doi, Hidetaka, Mr.
DENRYOKU SOREN
2-7-15 Mita, Minato-ku
Tokyo, Japan
T: 81 3 3454-0231

Domoto, Akiko, Ms.
Biodiversity Chairperson
GLOBE INTERNATIONAL
422 Sangiin Kaikan, 2-1-1 Nagatacho Chiyoda-ku
Tokyo, Japan
T: 81 3 3508-8422 F: 81 3 3506-8085

Dya, Kenji, Mr.
Program Specialist
UNITED NATIONS CENTRE FOR REGIONAL DEVELOPMENT
Nagono 1-417-1 Nakamura-ku
Nagoya 450, Japan
T: 81 52 561 9377 F: 81 52 561 9374

Egoshi, Kazunobu, Mr.
NATIONAL ALLIANCE OF VICTIMS OF MINAMATA DISEASE & LAWYERS
6-15-7, Fukazawa, Setagaya-ku
Tokyo, Japan
T: 81 3 3704-6871

Emoto, Akiko, Ms.
DELEGATES OF CASA-MEMBER CO-OPS
Seiwa-bild. Nijo-aaru, Karasuma-beri, Nakagyo-ku, Kyoto-cityshima Tokushima-coop
Kyoto-tu, Japan
T: 81 75 465-8280 F: 81 75 463 1393

Enoki, Akinori, Mr.
DELEGATES OF CASA-MEMBER CO-OPS
1-6-31 Teraikedai, Tondabayashi-cityamura, Kitajima-cho Tokushima Tokushima-coop
Osaka-tu, Japan
T: 81 721 29 3684

Eriguchi, Michiko, Ms.
Member
OISCA - INTERNATIONAL
6-12, Izumi 3-Chome, Suginami-ku
Tokyo, Japan
T: 81 3 3322-5161 F: 81 3 3324-7111
TL: J29397 OISCA

Eto, Saburo, Mr.
Senior Research Fellow
KEIZAI DOYUKAI
4-6, Marunouchi I-Chome, Chiyoda-ku
Tokyo, Japan
T: 81 3 3211-1271 F: 81 3 3213-2946
TL: J32531Keidoyu

Fujimo, Tadashi, Mr.
NATIONAL ALLIANCE OF VICTIMS OF MINAMATA DISEASE & LAWYERS
6-15-7, Fukazawa, Setagaya-ku
Tokyo, Japan
T: 81 3 3704-6871

Fujimoto, Yasuko, Ms.
DELEGATES OF CASA-MEMBER CO-OPS
2-7-22 Tenma, Kita-ku, Osaka-city, PAL Co-op
Osaka-tu, Japan
T: 81 6 354-8711 F: 81 6 354 8715

Fujinaga, Nobuyo, Ms.
Executive Director
JAPAN FOUNDATION FOR THE EARTH
4-10 Miyakojima-Kitadoori, Miyakojima-ku
Osaka, Japan
T: 81 6 921-8752 F: 81 6 921-8752

Fujinara, Yoshinari, Mr.
4-7-57 Asachigaoka Hino-shi
Tokyo, Japan
T: 81 3 425-84 681

Fujino, Yoko, Ms.
NATIONAL ALLIANCE OF VICTIMS OF MINAMATA DISEASE & LAWYERS
6-15-7, Fukazawa, Setagaya-ku
Tokyo, Japan
T: 81 3 3704-6871

Fujita, Machiko, Ms.
JAPAN YOUTH FEDERATION
Nihon Seinen Kan 2F, 15 Kasumigaoka-Cho, Shinjuku
Tokyo, Japan
T: 81 3 3475-2490 F: 81 3 3475-0668

Fujiwara, Motoji, Mr.
KOTSU ROREN
2-20-12 Shiba , Minato-KU
Tokyo, Japan
T: 81 3 3451-7243

Fujiwara, Takeji, Mr.
Attorney-At-Law
JAPAN FEDERATION OF BAR ASSOCIATIONS
5-8-8 Hyakurakuen Nara-city
Nara-ken, Japan
T: 81 742 491893 F: 81 742 365-0458

Fukuda, Masae, Ms.
NATIONAL ALLIANCE OF VICTIMS OF MINAMATA DISEASE & LAWYERS
6-15-7, Fukazawa, Setagaya-ku
Tokyo, Japan
T: 81 3 3704-6871

Fukuda, Yuji, Mr.
Denryoku Soren
DENRYOKU SOREN
2-7-15 Mita. Minato-ku
Tokyo, Japan
T: 81 3 3454-0231 F: 81 3 3798-1470

Fukui, Jhoji, Mr.
NATIONAL ALLIANCE OF VICTIMS OF MINAMATA DISEASE & LAWYERS
6-15-7, Fukazawa, Setagaya-ku
Tokyo, Japan
T: 81 3 3704-6871

Fukui, Shinji
Assistent
SEIKATSU CLUB CONSUMER'S CO-OPERATIVE UNION
Miyasaka 3-2-28 Setagaya-ku
Tokyo 156, Japan
T: 3 3706-0039 F: 3 34279041

Fukushima, Tetsuro, Mr.
THE JAPAN ELECTRICAL MANUFACTURERS ASSOCIATION - JEMA
4-15 Nagata-Cho 2-Chome, Chiyoda-KU
Tokyo 100, Japan
T: 81 3 3581-4845 F: 81 3 3506-0475
TL: 2222619 JEMA J

Furuhashi, Hiroko, Ms.
JAPANESE CONSUMERS' CO-OPERATIVE UNION
Seikyo-kaikan 1-1-13, Sendagaya, Shibuya-ku
Tokyo, Japan
T: 81 3 34979103 F: 81 3 34970722
TL: 2423380 NCOP J

Furukawa, Ayako, Ms.
JAPAN YOUTH FEDERATION
Nihon Seinen Kan 2F, 15 Kasumigaoka-Cho, Shinjuku
Tokyo, Japan
T: 81 3 3475-2490 F: 81 3 3475-0668

Furukawa, Takumi, Ms.
JAPAN YOUTH FEDERATION
Nihon Seinen Kan 2F, 15 Kasumigaoka-Cho, Shinjuku
Tokyo, Japan
T: 81 3 3475-2490 F: 81 3 3475-0668

Furuyama, Yoko, Ms.
JAPANESE MUSICIANS FOR THE GLOBAL ENVIRONMENT
1-48-4-103 Nishiahara, Kita-ku
Tokyo-to, Japan
T: 81 3 5394-2097

Galera, Ricky W., Mr.
General Manager
BRAZIL PROMOTIONS LTD.
3-7-7, Kubota Bld. 3 Fl. Shibuya, Shibuya-ku
Tokyo, Japan
T: 81 3 3499-3737 F: 81 3 499-3736

Garcia, Erusal
Reporter
NHK
2-2-1 Jin-Nan Shibuya-ku
Tokyo, Japan
T: 81 3 54783523 F: 81 3 54783541

Garcia-Franco, Mauricio, Mr.
JAPAN TUNA
201 Upstairs 5-9-11 Shimomeguro, Meguro-Ko
Tokyo 153, Japan
T: 81 3 3710-0261 F: 81 3 3710-0261

Goto, Kazue, Ms.
JAPAN YOUTH FEDERATION
Nihon Seinen Kan 2F, 15 Kasumigaoka-Cho, Shinjuku
Tokyo, Japan
T: 81 3 3475-2490 F: 81 3 3475-0668

Haguchi, Yukio, Mr.
Secretary
PEOPLE'S ALLIANCE TO PROTECT THE ENVIRONMENT OF THE JAPANESE INLAND SEA
58-5 Higashi-Futami, Futami-cho, Akashi-city
Hyogo-ken, Japan
T: 81 78 442-2705

Hamada, Hiroshi, Mr.
Manager
LIAISON ON WATER INFORMATION, A
Tokiwadaira 5-28-5-710 - Matsudo
Chiba 270, Japan
T: 81 473 84-5466 F: 81 473 86-5324

Hamanaka, Nobuo, Mr.
STAFF MEMBER OF CHIBA PREF
Chiba Noba-Shi
Chiba, Japan

Hamazima, Yasuhiro, Mr.
Musician
JAPANESE MUSICIANS FOR THE GLOBAL ENVIRONMENT
2-16-36 Okubo, Shinjuku-ku, Onaku-Center
Tokyo-to, Japan
T: 81 3 3200-0106 F: 81 3 3200-0105

Hani, Kanta
Editor Reporter
A SEED JAPAN
4-3-302, Sakuraoka-Chyo, Shibuyaku
Tokio, Japan
T: 81 3 3476 3252 F: 81 3 3476 5046

Harago, Yuta, Mr.
92 NGO FORUM JAPAN
4F Kanda Kyyodo Bldg., 2-13-4 Iwamoto-Cho
Tokyo, Japan
T: 81 3 3861-0540 F: 81 3 3861-5048

Harako, Eiichiro, Mr.
YMCA OF JAPAN
2-3-18 Nishi-Waseda, Shinjuku-ku
Tokyo 169, Japan

Hasegawa, Tatsuko, Ms.
Secretary
JAPAN AIR POLLUTION VICTIMS ASSOCIATION
1-10-7-304 Nakasato, Kita-ku, Tokyo
Tokyo-to, Japan
T: 81 3 3823 5968

Hashiguchi, Saburo, Mr.
NATIONAL ALLIANCE OF VICTIMS OF MINAMATA DISEASE & LAWYERS
6-15-7, Fukazawa, Setagaya-ku
Tokyo, Japan
T: 81 3 3704-6871

Hashimoto, Yasuo, Mr.
THE JAPAN ELECTRICAL MANUFACTURERS ASSOCIATION - JEMA
4-15 Nagata-Cho 2-Chome, Chiyoda-KU
Tokyo 100, Japan
T: 81 3 3581-4845 F: 81 3 3506-0475
TL: 2222619 JEMA J

Hatakeyama, Minoru, Mr.
NATIONAL ALLIANCE OF VICTIMS OF MINAMATA DISEASE & LAWYERS
6-15-7, Fukazawa, Setagaya-ku
Tokyo, Japan
T: 81 3 3704-6871

Hatano, Hiromi, Mr.
Guide
KINKI NIPPON TOURIST
Daishin Bldg. 6F, 3-8-6 Shimbashi, Minato-KU
Tokyo, Japan
T: 81 3 5473-0540

Hayakawa, Mitsutoshi, Mr.
Attorney-At-Law
LIAISON COUNCIL OF PLAINTIFS AND ATTORNEYS GROUPS IN AIR POLLUTION CASES
2-7-2-201 Nishi tunabashi, Hirakata-city Osaka
Osaka, Japan
T: 81 720 52 3363 F: 81 720 67 1104

Hayama, Keiko, Ms.
JAPANESE CONSUMERS' CO-OPERATIVE UNION
Seikyo-kaikan 1-1-13, Sendagaya, Shibuya-ku
Tokyo, Japan
T: 81 3 34979103 F: 81 3 34970722
TL: 2423380 NCOP J

Hayami, Yu, Ms.
JAPANESE MUSICIANS FOR THE GLOBAL ENVIRONMENT
Tokiwaparesu 1012, 1-1-1 Shimoochiai, Shinjuku-ku
Tokyo, Japan
T: 33603978

Hayashi, Misuko, Ms.
DELEGATES OF CASA-MEMBER CO-OPS
50-3 Hiashi-tsutsuminouchi, Aza, Nakamura, Kitajima-cho Tokushima Tokushima-coop
Tokushimo-ken, Japan
T: 81 886 98-5100

Hayashi, Tatsuo, Mr.
92 NGO FORUM JAPAN
4F Kanda Kyyodo Bldg., 2-13-4 Iwamoto-Cho
Tokyo, Japan
T: 81 3 3861-0540 F: 81 3 3861-5048

Haziame, Adachi, Mr.
DENTSU INC.
6-25-4 Honmachi sibuya.ku
Tokyo, Japan
T: 81 3 337-76939

Hidefira, Neusa Atsuko, Ms.
OVERSEAS FISHERY COOPERATION FOUNDATION
Akasaka Z-117-22
Tokyo, Japan
T: 81 35 85-5382

Hidetoshi, Hayashida, Mr.
TAKENAKA CO. LTD
2-5-14, Minamisuna-ku
Tokyo, Japan
T: 81 3 3640-3161

Higashi, Tatsuo, Mr.
92 JAPAN NGO FORUM
1-1-2-2 Honcho Sapporo Hokkaido
Sapporo, Japan
T: 011 786 0149 F: 011 785 5044

Higashiyama, Masamori, Mr.
NATIONAL ALLIANCE OF VICTIMS OF MINAMATA DISEASE & LAWYERS
6-15-7, Fukazawa, Setagaya-ku
Tokyo, Japan
T: 81 3 3704-6871

Higuchi, Kazuhiro, Mr.
92 NGO FORUM JAPAN
Nakamachi 3-2-14 Musashino City
Tokyo 180, Japan
T: 81 3 371 6325

Higuchi, Shunji, Mr.
92 NGO FORUM JAPAN
c/o Kokuro Headquarter. 1-11-4 Marunouchi - Chiyoda-ku
Tokyo, Japan
T: 81 3 31 4046 F: 81 3 3215 1154

Hillis, Janette, Ms.
Member
GLOBE INTERNATIONAL
422 Sangiin Kaikan, 2-1-1 Nagatacho Chiyoda-ku
Tokyo, Japan
T: 81 3 3508-8422 F: 81 3 3506-8085

Hirabayashi, Tomoki
92 NGO FORUM JAPAN
4F Kanda Kyyodo Bldg., 2-13-4 Iwamoto-Cho
Tokyo, Japan
T: 81 3 3861-0540 F: 81 3 3861-5048

Hirakuri, Sofia Rumiko
Researcher
JAPAN TROPICAL FOREST ACTION NETWORK
801 Shibuya Mansion, Uguisudani-cho7-1 Shibuya-Ku 150
Tokyo, Japan
T: 81 3 5489-4547 F: 81 3 3770-0727

Hiroko, Onishi, Mr.
Committe Member
JAPAN FEDERATION OF BAR ASSOCIATIONS
Jitsugyo-bldg. SF, 5-15-18 Nishi Temma , Kita-ku
Osaka, Japan
T: 81 6 365-7325 F: 81 6 365-7326

Hirose, Michio, Mr.
Director
OISCA - INTERNATIONAL
6-12, Izumi 3-Chome, Suginami-ku
Tokyo, Japan
T: 81 3 3322-5161 F: 81 3 3324-7111
TL: J29397 OISCA

Hirose, Takeo, Mr.
GISPRI
c/o Nomura Tourist Bureau 1-5-3 Nihombashi, Chuo-Ku
Tokyo, Japan
T: 81 3 3281-4561 F: 81 3 3274-2670
TL: J29269

Hishino, Yushimi
A SEED JAPAN
4-3-302, Sakuraoka-Chyo, Shibuyaku
Tokyo, Japan
T: 81 3 3476 3252 F: 81 3 3476 5046

Holliman, Jonathan, Mr.
92 NGO FORUM JAPAN
4F Kanda Kyyodo Bldg., 2-13-4 Iwamoto-Cho
Tokyo, Japan
T: 81 3 3861-0540 F: 81 3 3861-5048

Honami, Sanac, Ms.
JAPANESE CONSUMERS' CO-OPERATIVE UNION
Seikyo-kaikan 1-1-13, Sendagaya, Shibuya-ku
Tokyo, Japan
T: 81 3 34979103 F: 81 3 34970722
TL: 2423380 NCOP J

Hondoh, Takashigo, Mr.
JAPANESE CONSUMERS' CO-OPERATIVE UNION
Seikyo-kaikan 1-1-13, Sendagaya, Shibuya-ku
Tokyo, Japan
T: 81 3 34979103 F: 81 3 34970722
TL: 2423380 NCOP J

Horikawa, Megumi, Ms.
YWCA / JAPAN
2-16-20-301, Sakaeche, Higashimurayama
Tokyo 189, Japan
T: 81 423 97-3499 F: 81 423 97-3499

Hosoda, Isamu, Mr.
Member
KOMEITO
5-20-7-823 Kita-suna Kato-ku
Tokyo, Japan
T: 81 3 3648-9018

Hoyano, Hatsuko
Reporter
SEIKATSU CLUB CONSUMER'S CO-OPERATIVE UNION
Miyasaka 3-2-28 Setagaya-ku
Tokyo 156, Japan
T: 3 3706-0039 F: 3 34279041

Hurusawa, Koyu, Mr.
92 NGO FORUM JAPAN
4F Kanda Kyyodo Bldg., 2-13-4 Iwamoto-Cho
Tokyo, Japan
T: 81 3 3861-0540 F: 81 3 3861-5048

Ichihara, Kyoko, Ms.
NATIONAL ALLIANCE OF VICTIMS OF MINAMATA DISEASE & LAWYERS
6-15-7, Fukazawa, Setagaya-ku
Tokyo, Japan
T: 81 3 3704-6871

Ichikizaki, Yoshiko, Ms.
Musician
JAPANESE MUSICIANS FOR THE GLOBAL ENVIRONMENT
2-16-36 Okubo, Shinjuku-ku
Tokyo-to, Japan
T: 81 3 3200-4497

Idiraw, Jatsuro, Mr.
GISPRI
3-8-21 Toranomon, Minato-Ku
Tokyo, Japan
T: 81 3 3435-8800 F: 81 3 3435-8810

Iizaka, Yoshiaki, Mr.
WORLD CONFERENCE ON RELIGION AND PEACE
c/o Fumon Hall 2-6-1 Wada Suginamii-ka
Tokyo, Japan
T: 81 3 3384-2337 F: 81 3 3383-7993

Ikari, Yosuke, Mr.
Attorney-At-Law
JAPAN FEDERATION OF BAR ASSOCIATIONS
Udagawa-Bldg. 2F, 2-27 Aiai-chu, Naka-ku, Yokohama-city
Kanagawa-ken, Japan
T: 81 45 662-5045

Ikeda, Susumu, Mr.
92 NGO FORUM JAPAN
4F Kanda Kyyodo Bldg., 2-13-4 Iwamoto-Cho
Tokyo, Japan
T: 81 3 3861-0540 F: 81 3 3861-5048

Ikehata, Kinuyo, Ms.
JAPANESE CONSUMERS' CO-OPERATIVE UNION
Seikyo-kaikan 1-1-13, Sendagaya, Shibuya-ku
Tokyo, Japan
T: 81 3 34979103 F: 81 3 34970722
TL: 2423380 NCOP J

Inakatsu, Hideo, Mr.
JAPANESE CONSUMERS' CO-OPERATIVE UNION
Seikyo-kaikan 1-1-13, Sendagaya, Shibuya-ku
Tokyo, Japan
T: 81 3 34979103 F: 81 3 34970722
TL: 2423380 NCOP J

Inoue, Junichi, Mr.
Secretary
PEOPLE'S ALLIANCE TO PROTECT THE ENVIRONMENT OF THE JAPANESE INLAND SEA
6-5-1-920 Nagatenji-cho, Nagata-ku, Kobe-city
Hyogo-ken, Japan
T: 81 78 641-5722 F: 81 78 641-5922

Inoue, Yuichi, Mr.
Director
INSTITUTE OF GLOBAL ENVIRONMENTAL CULTURE
1170-74 Akishinodai - Nara-city
Nara-ken, Japan
T: 81 7442468255 F: 81 7442

Ise, Hiroaki, Mr.
JAPAN SCIENTISTS ASSOCIATION
4-7-15-905 Sendagi, Bunkyo-ku
Tokyo-to, Japan
T: 81 3 3827 8486 F: 81 3 5685 3210

Ishida, Hiroshi, Mr.
95-2, Nagaoka Okayama-shi, Okayama-ken
Okayama 703, Japan
T: 81 862 79-0055

Ishigami, Toshifumi, Mr.
NATIONAL ALLIANCE OF VICTIMS OF MINAMATA DISEASE & LAWYERS
6-15-7, Fukazawa, Setagaya-ku
Tokyo, Japan
T: 81 3 3704-6871

Ishii, Misako, Ms.
92 NGO FORUM JAPAN
4F Kanda Kyyodo Bldg., 2-13-4 Iwamoto-Cho
Tokyo, Japan
T: 81 3 3861-0540 F: 81 3 3861-5048

Ishikawa, Kenji, Mr.
Environmental Engineering Pro.
TAKENAKA COORPORATION
1-13-1, Kachidoki, Chu-o-ku
Tokyo, Japan
T: 81 3 3542 7100 F: 81 3 3536 6309

Ishikawa, Noboru, Mr.
SHINRIN REREN
3-28-7 Otuska, Bunkyo-ku
Tokyo, Japan
T: 81 3 3945-6385 F: 81 3 3945-6496

Ishikawa, Sachihiko, Mr.
Tour Scort
KINKI NIHON, TOURTIST CO. LTD.
1-12-16, Jinnan, Shibuya-ku
Tokyo, Japan
T: 81 3 3464 2851 F: 81 3 5489 2406

Ishizawa, Kiyoshi, Mr.
Executive Producer
GISPRI
c/o Nomura Tourist Bureau 1-5-3 Nihombashi, Chuo-Ku
Tokyo, Japan
T: 81 3 3281-4561 F: 81 3 3274-2670
TL: J29269

Ital, Masaru, Mr.
NATIONAL ALLIANCE OF VICTIMS OF MINAMATA DISEASE & LAWYERS
6-15-7, Fukazawa, Setagaya-ku
Tokyo, Japan
T: 81 3 3704-6871

Itai, Yaeko, Ms.
NATIONAL ALLIANCE OF VICTIMS OF MINAMATA DISEASE & LAWYERS
6-15-7, Fukazawa, Setagaya-ku
Tokyo, Japan
T: 81 3 3704-6871

Ito, Haruo, Mr.
Member
OISCA - INTERNATIONAL
6-12, Izumi 3-Chome, Suginami-ku
Tokyo, Japan
T: 81 3 3322-5161 F: 81 3 3324-7111
TL: J29397 OISCA

Ito, Hidetugu, Mr.
Member
KOMEITO
5-8-3 Asahi-cha Atsugi-shi
Kanagawa, Japan
T: 81 45 28-3193 F: 81 45 28-3193

Ito, Naoaki, Mr.
Professor
TOKYO METROPOLITAN UNIVERSITY
Minami-Osawa, Hachioji
Tokyo 192-033, Japan
T: 81 3 426-77-2807 F: 81 3 426-77-2793

Ito, Sanji, Sr.
Manager Of Ecology
CHUBU ELECTRIC POWER
1-Banchi, Higashi-Shinmachi, Higashi-ku
Nagoya, Japan
T: 38 52 951-8211 F: 38 52 971-1837

Ito, Takaichi, Mr.
OISCA - INTERNATIONAL
6-12, Izumi 3-Chome, Suginami-ku
Tokyo, Japan
T: 81 3 3322-5161 F: 81 3 3324-7111
TL: J29397 OISCA

Ito, Yaeko, Ms.
JAPAN NETWORK FOR ENVIRONMENT & POLLUTION
1-36 Jinpo-cho, Kanda, Chiyoda-ku
Tokyo-to, Japan
T: 81 3 7243-0581 F: 81 3 7243-0574

Itoh, Angela Mayumi
ASAHI SHIMBUN
Tokyo, Japan

Itou, Sakae, Mr.
JAPAN AIR POLLUTION VICTIMS ASSOCIATION
2-6-51 Tenpaku-cho, Minami-ku, Naoya-city
Aichi-ken, Japan
T: 81 52 612 3243

Itoya, Wataru, Mr.
92 NGO FORUM JAPAN
4F Kanda Kyyodo Bldg., 2-13-4 Iwamoto-Cho
Tokyo, Japan
T: 81 3 3861-0540 F: 81 3 3861-5048

Iwamoto, Kazunori, Mr.
Staff
ESCO CO. LTD.
Akjaishi Bldg. 2F, 1-9-3 HIgashi-Azabu Minato-ku
Tokyo, Japan
T: 81 3 3582-7760 F: 81 3 3586-4303

Iwata, Yukio, Mr.
OSAKA CONFEDERATION OF TRADE UNIONS
2-4-38 Ankoki-cho, Takatsuki-city
Osaka-tu, Japan
T: 81 6 890773

Izawa, Masaaki, Mr.
Advisor
JAPAN FOREST TECHNICAL ASSOCIATION
No. 7, Rokubancho chiyoda-ku
Tokyo, Japan
T: 81 3 261-5281 F: 81 3 261-5393

Izumi, Kunihiko, Mr.
Representative Director
CITIZENS ALLIANCE FOR SAVING THE ATMOSPHERE AND THE EARTH (CASA)
1-3-17-813 Tanimachi, Chuo-ku, Osaka-city
Osaka-tu, Japan
T: 81 6 941 3745 F: 81 6 941 5699

Jintoku, Kazuo, Mr.
92 NGO FORUM JAPAN
4F Kanda Kyyodo Bldg., 2-13-4 Iwamoto-Cho
Tokyo, Japan
T: 81 3 3861-0540 F: 81 3 3861-5048

Kajiwara, Toshiyuki, Mr.
Attorney-At-Law
LIAISON COUNCIL OF PLAINTIFS AND ATTORNEYS GROUPS IN AIR POLLUTION CASES
1-10-7 Chuoi, Chiba-city, Chiba Daiichi Law Office
Chiba-ken, Japan
T: 81 472 24 7366 F: 81 472 27 7083

Kakehi, Munenori, Mr.
Attorney-At-Law
JAPAN FEDERATION OF BAR ASSOCIATIONS
4-3-3 Midori-cho, Ashiya-city
Hyogo-ken, Japan
T: 81 7977318952

Kakuta, Yoichi, Mr.
Member
CITIZENS ALLIANCE FOR SAVING THE ATMOSPHERE AND THE EARTH (CASA)
1-3-17-813 Tanimachi, Chuo-ku, Osaka-city
Osaka-tu, Japan
T: 81 6 941 3745 F: 81 6 941 5699

Kamei, Naomi, Ms.
92 NGO FORUM JAPAN
4F Kanda Kyyodo Bldg., 2-13-4 Iwamoto-Cho
Tokyo, Japan
T: 81 3 3861-0540 F: 81 3 3861-5048

Kami, Osamu, Mr.
YAMAGISHI ASSOCIATION
555 Kawahigashi, Iga-cho, Ayaama-gun
Mie-Prof, Japan
T: 81 594 54594 F: 81 594 53329

Kamiya, Yuji
Journalist
ASAHI SHIMBUN
5-3-2 Tsukiji, Chuo-ku R.Sampaio Viana 203 c.13 Paraiso
Sao Paulo -SP 04004, Japan
T: 55 11 8841263 F: 55 11 8879990

Kan, Shingo, Mr.
Director
GISPRI
c/o Nomura Tourist Bureau 1-5-3 Nihombashi, Chuo-Ku
Tokyo, Japan
T: 81 3 3281-4561 F: 81 3 3274-2670
TL: J29269

Kan, Yoshito, Mr.
OSAKA CONFEDERATION OF TRADE UNIONS
2-13-17 Kiyomidai, Kawachinagano-city
Osaka-tu, Japan
T: 81 6 631419

Kanaguchi, Jun
Reporter
KYODO NEWS SERVICE
2 2 5 Toranomon Minato-Ku
Tokyo, Japan
T: 81 33 5844111 F: 3-35893683

Kanasaki, Tsukoko, Ms.
NATIONAL ALLIANCE OF VICTIMS OF MINAMATA DISEASE & LAWYERS
6-15-7, Fukazawa, Setagaya-ku
Tokyo, Japan
T: 81 3 3704-6871

Kanasaki, Yukie, Ms.
DELEGATES OF CASA-MEMBER CO-OPS, PAL CO-OP
2-7-22 Tenma, Kita-ku, Osaka-city
Osaka-tu, Japan
T: 81 6 354-8711 F: 81 6 354 8715

Kanda, Toshiko, Ms.
JAPANESE CONSUMERS' CO-OPERATIVE UNION
Seikyo-kaikan 1-1-13, Sendagaya, Shibuya-ku
Tokyo, Japan
T: 81 3 34979103 F: 81 3 34970722
TL: 2423380 NCOP J

Kaneko, Osamu, Mr.
NATIONAL ALLIANCE OF VICTIMS OF MINAMATA DISEASE & LAWYERS
6-15-7, Fukazawa, Setagaya-ku
Tokyo, Japan
T: 81 3 3704-6871

Karasawa, Kei
Professor
RISTUMEIKAN UNIVERSITY
14-1 Kinugasa Baba-cho Kitaku
Kyoto, Japan
T: 81 75 464-6267

Kariyawasan, Andrew, Mr.& Mrs.
President, Oisca Sri Lanka
OISCA - INTERNATIONAL
6-12, Izumi 3-Chome, Suginami-ku
Tokyo, Japan
T: 81 3 3322-5161 F: 81 3 3324-7111
TL: J29397 OISCA

Kashihata, Naohisa, Mr.
Vice-President
JAPAN JUNIOR CHAMBER, INC.
2-14-3 Hirakawa-Cho Chiyoda-Warp
Tokyo, Japan
T: 81 3 32425601

Kashimura, Chushi
92 NGO FORUM JAPAN
4F Kanda Kyyodo Bldg., 2-13-4 Iwamoto-Cho
Tokyo, Japan
T: 81 3 3861-0540 F: 81 3 3861-5048

Kato, Nishida, Mr.
President
JAPAN CENTER FOR HUMAN ENVIRONMENT
241 Marunouchi Chiyodaku
Tokyo 100, Japan
T: 81 3 3201-1301 F: 81 3 3201-1303

Kato, Tetsuo, Mr.
92 NGO FORUM JAPAN
c/o Kokuro Headquarter, 1 -11-4 Marunouchi - Chiyoda-ku
Tokyo, Japan
T: 81 3 31 4046 F: 81 3 3215 1154

Katsube, Kinichi, Mr.
JAPANESE CONSUMERS' CO-OPERATIVE UNION
Seikyo-kaikan 1-1-13, Sendagaya, Shibuya-ku
Tokyo, Japan
T: 81 3 34979103 F: 81 3 34970722
TL: 2423380 NCOP J

Katsumata, Makoto, Mr.
92 NGO FORUM JAPAN
4F Kanda Kyyodo Bldg., 2-13-4 Iwamoto-Cho
Tokyo, Japan
T: 81 3 3861-0540 F: 81 3 3861-5048

Kawada, Koichiro, Mr.
JAPANESE CONSUMERS' CO-OPERATIVE UNION
Seikyo-kaikan 1-1-13, Sendagaya, Shibuya-ku
Tokyo, Japan
T: 81 3 34979103 F: 81 3 34970722
TL: 2423380 NCOP J

Kawaguchi, Hisataka, Mr.
Programme Coordinator
INTERNATIONAL LAKE ENVIRONMENT COMMITTEE FOUNDATION
3-4-22 Kyomachi, Otsu
Shiga 520, Japan
T: 81 775 25-1076 F: 81 775 23-1581
TL: 5464850 ILEC J

Kawai, Daisuke, Mr.
Nshi-KU Hassamu 1-3-14-7
Sappord, Japan
T: 011 6664520

Kawai, Midorj, Ms.
JAPANESE CONSUMERS' CO-OPERATIVE UNION
Seikyo-kaikan 1-1-13, Sendagaya, Shibuya-ku
Tokyo, Japan
T: 81 3 34979103 F: 81 3 34970722
TL: 2423380 NCOP J

Kawai, Naoki, Mr.
JAPAN NETWORK FOR ENVIRONMENT & POLLUTION
6-14-305 Tsurukawa, Machida-city
Tokyo-to, Japan
T: 81 35 0427-9598

Kawamoto, Kazuko, Ms.
Attorney-At-Law
LIAISON COUNCIL OF PLAINTIFS AND ATTORNEYS GROUPS IN AIR POLLUTION CASES
1-10-7 Chuoi, Chiba-city, Chiba Daiichi Law Office
Chiba-ken, Japan
T: 81 472 24 7366 F: 81 472 27 7083

Kawamura, Hideki, Mr.
WORLD CONFERENCE ON RELIGION AND PEACE
c/o Fumon Hall 2-6-1 Wada Suginamii-ka
Tokyo, Japan
T: 81 3 3384-2337 F: 81 3 3383-7993

Kawamura, Sigeyuki, Mr.
Former Teacher
JAPAN SCIENTISTS ASSOCIATION
5-1-4N404 Asahigooka Kiyose, Tokyo
Tokyo-to, Japan
T: 81 424 91 9963

Kawanabe, Hiroya, Mr.
Professor
DEPT. OF ZOOLOGY, KYOTO UNIVERSITY
Dept. of Zoology, Faculty of Science, Kyoto University
Kyoto 606-01, Japan
T: 81 75 753-4092 F: 81 75 753-4101
TL: J-54-22302

Kawanishi, Nobuko, Ms.
DELEGATES OF CASA-MEMBER CO-OPS
2-7-22 Tenma, Kita-ku, Osaka-city, PAL Co-op
Osaka-tu, Japan
T: 81 6 243 8711 F: 81 6 354 8715

Kawasaki, Takashi, Mr.
MITSUBISHI HEADY INDUSTRIES
7-2- Akasaka Minatoku
Tokyo, Japan
T: 81 3 3796-1445

Kawasaki, Yutaka, Mr.
Folk Singer
JAPAN AIR POLLUTION VICTIMS ASSOCIATION
1560-8-2-2 Makino, Shindachi Sennan-city
Sennan-city, Japan
T: 81 724 83 8736 F: 81 724 52 6486

Kawashima, Jin, Mr.
Director
KEIDANREN-KEIZAI KOHO CENTER
Otemachi Building - Chome 6-1 - Otemachi - Chiyoda-Ku
Tokyo, Japan
T: 81 3 32-011412 F: 81 3 32-011418

Kazuya, Kitamura
Correspondent
NIPPON TELEVISION NET. CO.
14 Niban-Cho, Chiyoda-Ku
Tokyo, Japan
T: 03 32301960 F: 03 31344044

Kenji, Yoshida, Mr.
SENKEI NEWS PRESS
Japan Kanagwaken Kwsaki Ooshima 1-1-5
Kawasaki, Japan
T: 81 44 222-0920

Kenko, Minami, Ms.
Representative
RAINFORST FOUNDATION - JAPAN
#402, 2-7, Sarugaku-cho Sibuya-ku
Tokyo, Japan
T: 81 3 3477-2589 F: 81 3 3477-0977

Kibara, Yusuke Saka
Coordinator
LOVES END HUNGER
Iwasaki 133 Hodogaya Yokohana
Kanagawa, Japan
T: 81 45 331-7868 F: 81 33 769-2090

Kihara, Keikichi
92 NGO FORUM JAPAN
4F Kanda Kyyodo Bldg., 2-13-4 Iwamoto-Cho
Tokyo, Japan
T: 81 3 3861-0540 F: 81 3 3861-5048

Kikushi, Yumi, Ms.
92 NGO FORUM JAPAN
4F Kanda Kyyodo Bldg., 2-13-4 Iwamoto-Cho
Tokyo, Japan
T: 81 3 3861-0540 F: 81 3 3861-5048

Kikutake, Kiyonori, Mr.
President
KIKUTAKE ARCHITECT & ASSOCIATES
3-1-1 Sunshine 60, Higashi-Ikebukuro, Toshima-ku 170
Tokyo, Japan
T: 81 3 3986-6121 F: 81 3 3986-6375

Kikutake, Mitsunori, Mr.
Student
RHODE ISLAND UNIVERSITY
1-11-15 Ohtsuka Bunkyo-ku, 112
Tokyo, Japan
T: 81 3 3947-0830

Kikutake, Yuki, Ms.
1-11-15 Ohtsuka Bunkyo-ku, 112
Tokyo, Japan
T: 81 3 3947-0830

Kimura, Kumiko, Ms.
JAPANESE CONSUMERS' CO-OPERATIVE UNION
Seikyo-kaikan 1-1-13, Sendagaya, Shibuya-ku
Tokyo, Japan
T: 81 3 34979103 F: 81 3 34970722
TL: 2423380 NCOP J

Kimura, Yoko, Ms.
JAPANESE CONSUMERS' CO-OPERATIVE UNION
Seikyo-kaikan 1-1-13, Sendagaya, Shibuya-ku
Tokyo, Japan
T: 81 3 34979103 F: 81 3 34970722
TL: 2423380 NCOP J

Kinoshita, Katsuhira, Mr.
SHINRIN REN
3-2807 Otauka, Bunkyo-ku
Tokyo, Japan
T: 81 3 3945-6385 F: 81 3 3945-6477

Kiriyama, Shoko
JAPANESE CONSUMERS' CO-OPERATIVE UNION
Seikyo-kaikan 1-1-13, Sendagaya, Shibuya-ku
Tokyo, Japan
T: 81 3 34979103 F: 81 3 34970722
TL: 2423380 NCOP J

Kishi, Nobutaka, Mr.
92 NGO FORUM JAPAN
c/oKokuro Headquarter, 1-11-4 Marunouchi - Chiyoda-ku
Tokyo, Japan
T: 81 3 31 4046 F: 81 3 3215 1154

Kishi, Tatsuya, Mr.
JAPAN YOUTH FEDERATION
Nihon Seinen Kan 2F, 15 Kasumigaoka-Cho
Tokyo, Japan
T: 81 3 3475-2490 F: 81 3 3475-0668

Kishimoto, Naoto, Mr.
JAPANESE CONSUMERS' CO-OPERATIVE UNION
Seikyo-kaikan 1-1-13, Sendagaya, Shibuya-ku
Tokyo, Japan
T: 81 3 34979103 F: 81 3 34970722
TL: 2423380 NCOP J

Kishimoto, Tatsuo, Mr.
Bank Officer
THE LONG-TERM CREDIT BANK OF JAPAN
1-2-4 Otemachi Chiyoda
Tokyo 100, Japan
T: 81 3 3211 5111 F: 81 3 3211 6828

Kita, Masaji, Mr.
JAPANESE CONSUMERS' CO-OPERATIVE UNION
Seikyo-kaikan 1-1-13, Sendagaya, Shibuya-ku
Tokyo, Japan
T: 81 3 34979103 F: 81 3 34970722
TL: 2423380 NCOP J

Kitagawa, Haruo
Technician
NIPPON TELEVISION NET. CO.
14 Niban-Cho, Chiyoda-Ku
Tokyo, Japan
T: 03 32301960 F: 03 31344044

Kitamura, Hissho, Sr.
Gerente Geral De Presidencia
YASUDA FIRE & MARINE INS. CO., LTD
1-26-1 Nishishinjuku Shinjukuku
Tokyo 01318-911, Japan
T: 81 3 3349-3044 F: 81 3 3348-6229

Kitaoka, Hideo, Mr.
NATIONAL ALLIANCE OF VICTIMS OF MINAMATA DISEASE & LAWYERS
6-15-7, Fukazawa, Setagaya-ku
Tokyo, Japan
T: 81 3 3704-6871

Kitaouji, Yorimobu, Sr.
Director
YAMAGISHI ASSOCIATION
555 Kawahigashi 519-14
Iga-cho, Japan
T: 059 545 5550 F: 059 545 4870

Kitazawa, Yoshimi, Ms.
JAPANESE CONSUMERS' CO-OPERATIVE UNION
Seikyo-kaikan 1-1-13, Sendagaya, Shibuya-ku
Tokyo, Japan
T: 81 3 34979103 F: 81 3 34970722
TL: 2423380 NCOP J

Koaina, Masashi
Correspondent
THE HOKKAIDO SHIMBUN PRESS
6-1, Ginza, Chuo-ku
Tokyo, Japan
T: 03 35713131 F: 03 35710943

Kobayashi, Isau, Mr.
Manager Of Env.Dev.Dept.
KAJIMA COORPORATION
1-2-7, Motoakasaka, Minato-ku
Tokyo, Japan
T: 81 3 5561 2081 F: 81 3 5561 2089

Kobayashi, Masaoki, Mr.
OSAKA CONFEDERATION OF TRADE UNIONS
Sanhaitsu-Tamiya, C-3, 446-3 Kubo, Kumatori-cho, Sennan-gun
Osaka-tu, Japan
T: 81 6 528487

Kobayashi, Shigeru, Mr.
Manager
GISPRI
c/o Nomura Tourist Bureau 5-3 Nihonbashi 1-Chome Chuo-Ku
Tokyo, Japan
T: 81 3 3281-4561 F: 81 3 3274-2670
TL: J29269

Koido, Iwao, Mr.
JAPANESE CONSUMERS' CO-OPERATIVE UNION
Seikyo-kaikan 1-1-13, Sendagaya, Shibuya-ku
Tokyo, Japan
T: 81 3 34979103 F: 81 3 34970722
TL: 2423380 NCOP J

Kojima, Michkazu, Mr.
INSTITUTE OF DEVELOPING ECONOMIES
6-24-14 Seijo Setagaya-k
Tokyo, Japan
T: 81 3 3482-7803

Kojima, Nobuo, Mr.
Committe Member
JAPAN FEDERATION OF BAR ASSOCIATIONS
14 bldg 3-18-11 Hongo Bunkyo-ku
Tokyo, Japan
T: 81 3 3813-6544 F: 81 3 3814-0252

Kojima, Ryoichiro, Mr.
RENGO
1-10-3 Mita, Minato-ku
Tokyo, Japan
T: 81 3 3769-6546 F: 81 3 3798-4227

Kometani, Daisuke, Mr.
3-57-1, Nagao-higashi-machi
Hirakata, Japan
T: 38 720 58-0008

Komoda, Nobuoo, Mr.
Attorney-At-Law
JAPAN FEDERATION OF BAR ASSOCIATIONS
1-14-10 Ichiban-cho Matsuyama-city
Ehime-ken, Japan
T: 81 899 32-1666 F: 81 899 72-1748

Komoda, Tetsu, Mr.
Committe Member
JAPAN FEDERATION OF BAR ASSOCIATIONS
Daiichi-Hoshi-bldg 805, 3-3-2 Ayase Adachi-ku
Tokyo, Japan
T: 81 3 3620-8994

Komori, Chosei, Mr.
Former Lecturer
JAPAN SCIENTISTS ASSOCIATION
Takao Park Heights B-410, Hatsuzawamachi 1231-19, Hachioji
Tokyo, Japan
T: 81 426 65 7128

Konno, Takehiro, Mr.
JAPAN YOUTH FEDERATION
Nihon Seinen Kan 2F, 15 Kasumigaoka-Cho
Tokyo, Japan
T: 81 3 3475-2490 F: 81 3 3475-0668

Konno, William Hiroki
ASAHI SHIMBUN
Tokyo, Japan

Kosaka, Tomonori, Mr.
Jga
JAPAN FISHERIES ASSOCIATION
9-13 Akaasaka I Minatokh
Toquio, Japan
T: 03 3585 6683 F: 03 3585 2337

Koshiishi, Nabuo, Mr.
Project Manager
JOHSON ENTERPRISE
3-7-7, Kubota Bld. 3 Fl. Shibuya, Shibuya-ku
Tokyo, Japan
T: 81 3 3499-3737 F: 81 3 499-3736

Kubo, Iku, Ms.
Parliamentaire
JAPAN WOMEN'S GLOBAL ENVIRONMENT NETWORK INTERNATIONAL
3-19-4 Miyasaka, Setagaya-ku
Tokyo 156, Japan
T: 81 3 3439-3599 F: 81 3 3506-8085

Kubo, Keizaburo, Mr.
Membership
EARTH DAY NETWORK JAPAN
3-1-49-1302, Fukushima, Fukushima-ku
Osaka, Japan
T: 81 6 451 4435

Kubota, Masahiko, Mr.
JAPAN YOUTH FEDERATION
Nihon Seinen Kan 2F, 15 Kasumigaoka-Cho, Shinjuku
Tokyo, Japan
T: 81 3 3475-2490 F: 81 3 3475-0668

Kumada, Nobuyuki, Mr.
Staff
ESCO CO. LTD.
Akaishi Bldg. 2F 1-9-3 Higashi-Azabu Minato-ku
Tokyo, Japan
T: 81 3 3582-7760 F: 81 3 3586-4303

Kunimoto, Mikio, Mr.
JAPANESE CONSUMERS' CO-OPERATIVE UNION
Seikyo-kaikan 1-1-13, Sendagaya, Shibuya-ku
Tokyo, Japan
T: 81 3 34979103 F: 81 3 34970722
TL: 2423380 NCOP J

Kunimune, Naoko, Ms.
NATIONAL ALLIANCE OF VICTIMS OF MINAMATA DISEASE & LAWYERS
6-15-7, Fukazawa, Setagaya-ku
Tokyo, Japan
T: 81 3 3704-6871

Kunito, Shunji, Sr.
Director Of Ecology Dep.
FEDERATION OF ELECTRIC POWER COMPANIES
1-9-4, Otemachi, Keidaren Kaikan, Chiyoda-ku
Tokyo, Japan
T: 81 3 3279-2188 F: 81 3 3241-1826

Kurimolo, Akira, Mr.
JAPANESE CONSUMERS' CO-OPERATIVE UNION
Seikyo-kaikan 1-1-13, Sendagaya, Shibuya-ku
Tokyo, Japan
T: 81 3 34979103 F: 81 3 34970722
TL: 2423380 NCOP J

Kuriyama, Kohichi, Mr.
Corporate Planning Dept.
IDEMITSU KOSAN CO., LTD
3-1-1, Marunouchi, Chiyooda-ku
Tokyo, Japan
T: 81 3 3213 9342 F: 81 3 3211-3521

Kuroda, Fumio, Mr.
JAPAN TSACHERS' UNION
Kyoiku-kaikan, 2-6-2 Hitotsubashi, Chiyoda-ku
Tokyo, Japan
T: 81 3 3265-2171 F: 81 3 3230-0172

Kuroda, Yoichi, Mr.
Coordinator
JAPAN TROPICAL FOREST ACTION NETWORK
801 Shibuya Mansion, Uguisudani-cho7-1 Shibuya-Ku 150
Tokyo, Japan
T: 81 3 3770-6308 F: 81 3 3770-0727

Kurosawa, Yasuju, Mr.
JAPANESE CONSUMERS' CO-OPERATIVE UNION
Seikyo-kaikan 1-1-13, Sendagaya, Shibuya-ku
Tokyo, Japan
T: 81 3 34979103 F: 81 3 34970722
TL: 2423380 NCOP J

Kurumagi, Yoko, Ms.
Membership
PEOPLE'S ALLIANCE TO PROTECT THE ENVIRONMENT OF THE JAPANESE INLAND SEA
5-3-812 Tanaka-cho, Higashinada-ko, Kobe-city
Hyogo-ken, Japan
T: 81 78 451-5550

Kusano, Chiyomi, Ms.
JAPAN YOUTH FEDERATION
Nihon Seinen Kan 2F, 15 Kasumigaoka-Cho, Shinjuku
Tokyo, Japan
T: 81 3 3475-2490 F: 81 3 3475-0668

Kyumon, Tarobe, Mr.
Director
JAPAN RACHEL CARSON COUNSIL
988 Mochio Kanan-cho Minami-Kawachi-gun
Osaka, Japan
T: 81 721 93-2845

Lechifata, Kugumara
Staff Writer
THE SANKEI SHIMBUN
7-2 Otemachi Chiyoda-ku
Tokyo, Japan
T: 81 3 32758786 F: 81 3 32758765

Loures, Maria Aparecida Limongi, Sra.
92 NGO FORUM JAPAN
4F Kanda Kyyodo Bldg., 2-13-4 Iwamoto-Cho
Tokyo, Japan
T: 81 3 3861-0540 F: 81 3 3861-5048

Machimura, Sueyoshi, Mr.
Member Of Central Committee
OISCA - INTERNATIONAL
6-12, Izumi 3-Chome, Suginami-ku
Tokyo, Japan
T: 81 3 3322-5161 F: 81 3 3324-7111
TL: J29397 OISCA

Maeda, Ayumi
92 NGO FORUM JAPAN
4F Kanda Kyyodo Bldg., 2-13-4 Iwamoto-Cho
Tokyo, Japan
T: 81 3 3861-0540 F: 81 3 3861-5048

Maeda, Keiko, Ms.
WORLD YOUNG WOMEN'S CHRISTIAN ASSOCIATION
1-1-20 Kamitsutsuidari Chuo-ku
Kobe 651, Japan
T: 81 78 231-6201 F: 81 78 231-6692
TL: WYMCA 41 23 32

Maeda, Masahisa, Mr.
Director
INTER GROUP
6-1-119,Satsukkigaoka-Minami
Suita, Japan
T: 3728048 F: 3726127

Majima, Tokujiro, Mr.
Chairman
SRID OF JAPAN
Srid-C/O Idc. Kyofuku.Bldg. Tamioka. 2-9-11. Koutou-KU.
Tokyo 135, Japan
T: 81 3 3630-7820 F: 81 3 3630-7880

Makimura, Hisako, Mr.
Director
INSTITUTE OF GLOBAL ENVIRONMENTAL CULTURE
4-1-6 Hiramatsu, Nara-city
Nara-ken, Japan
T: 81 7 44-3604

Makoto, Numata, Mr.
President
NATURE CONSERVATION SOCIETY OF JAPAN
Toranomon Denki-bldg, 2-8-1, Toranomon, Minatoku
Tokyo, Japan
T: 81 3 35034896 F: 81 3 35920496

Manaki, Akio, Mr.
NATIONAL ALLIANCE OF VICTIMS OF MINAMATA DISEASE & LAWYERS
6-15-7, Fukazawa, Setagaya-ku
Tokyo, Japan
T: 81 3 3704-6871

Masaaki, Nakamura, Mr.
JAPAN TUNA
2-3-22 Kudankita Chiyodaku
Tokyo 102, Japan
T: 81 3 3264-6167 F: 81 3 3234-7455

Masahiko, Konno, Mr.
2-87-C5 Toyo Ohmiya
Ohmiya, Japan
T: 048 6685648

Masuya, Tadashi, Sr.
Manager
KANSAI ELECTRIC POWER CO.
3-3-22 Nakanoshima, Kita-ku
Osaka, Japan
T: 81 6 441-8821 F: 81 6 441-8821

Masuyama, Akemi, Ms.
JAPAN YOUTH FEDERATION
Nihon Seinen Kan 2F, 15 Kasumigaoka-Cho, Shinjuku
Tokyo, Japan
T: 81 3 3475-2490 F: 81 3 3475-0668

Matsuara, Shinici
Correspondent
THE HOKKAIDO SHIMBUN PRESS
6-1, Ginza, Chuo-ku
Tokyo, Japan
T: 03 35713131 F: 03 35710943

Matsuba, Kenzo, Mr.
Attorney-At-Law
JAPAN FEDERATION OF BAR ASSOCIATIONS
326-9 Chishaku-cho, Yokkaichi-city
Mie-ken, Japan
T: 81 593 262476

Matsuda, Kimiko, Ms.
THE HUMAN ECOLOGY INSTITUTE
3-13-3 Sakuragaoka Tama City Tokyo
Tokyo, Japan

Matsumoto, Hideki, Mr.
ARTIST
Kubota Blvd. 3 Fl. Shibuya, Shibuya-Ku 3 - 7 - 7 150
Tokyo, Japan
T: 81 3 499 3737 F: 81 3 3499 3736

Matsumoto, Tsukio, Mr.
NATIONAL ALLIANCE OF VICTIMS OF MINAMATA DISEASE & LAWYERS
6-15-7, Fukazawa, Setagaya-ku
Tokyo, Japan
T: 81 3 3704-6871

Matsumura, Kunimori, Mr.
92 NGO FORUM JAPAN
4F Kanda Kyyodo Bldg., 2-13-4 Iwamoto-Cho
Tokyo, Japan
T: 81 3 3861-0540 F: 81 3 3861-5048

Matsumura, Makoto, Mr.
Manager
GISPRI
c/o Nomura Tourist Bureau 1-5-3 Nihombashi, Chuo-Ku
Tokyo, Japan
T: 81 3 3281-4561 F: 81 3 3274-2670
TL: J29269

Matsuno, Nobuo, Mr.
NATIONAL ALLIANCE OF VICTIMS OF MINAMATA DISEASE & LAWYERS
6-15-7, Fukazawa, Setagaya-ku
Tokyo, Japan
T: 81 3 3704-6871

Matsushima, Tsuneharu, Mr.
NATIONAL ALLIANCE OF VICTIMS OF MINAMATA DISEASE & LAWYERS
6-15-7, Fukazawa, Setagaya-ku
Tokyo, Japan
T: 81 3 3704-6871

Miho, Kazumi, Ms.
Secretary
EARTH DAY NETWORK JAPAN
1-3-17-813 Tanimachi, Chuo9-ku
Osaka, Japan
T: 81 6 941 3745 F: 81 6 941 5699

Mimura, Satoshi
A SEED JAPAN
4-3-302, Sakuraoka-Chyo, Shibuyaku
Tokyo, Japan
T: 81 3 3476 3252 F: 81 3 3476 5046

Minami, Hirotaka, Sr.
Director Of Planning Dept.
ELECTRIC POWER DEVELOPMENT CO.
6-15-1, ginza, Chuo-ku
Tokyo, Japan
T: 81 3 3546-2211 F: 81 3 3546-0531

Minamigata, Atsushi, Mr.
5-9-4 Soyama-cho Kita-ku
Kobe, Japan
T: 81 78 594-2011 F: 81 78 594-8909

Ming, Chu, Mr.
Dir Of The Board, Roc Taiwan
OISCA - INTERNATIONAL
6-12, Izumi 3-Chome, Suginami-ku
Tokyo, Japan
T: 81 3 3322-5161 F: 81 3 3324-7111
TL: J29397 OISCA

Minobe, Yuki, Ms.
52--26 Matsumidai
Nara-ken, Japan
T: 81 74424-662555 F: 81 7442

Mitani, Takeya, Mr.
OLYMPUS OPTICAL CO., LTD.
E104 2-11 Nishiturrugaok Ohimachi Irumagun Saitamaken
Tokyo, Japan

Mitsuyoshi, Hitoshi, Mr.
JAPAN YOUTH FEDERATION
Nihon Seinen Kan 2F, 15 Kasumigaoka-Cho, Shinjuku
Tokyo, Japan
T: 81 3 3475-2490 F: 81 3 3475-0668

Mittwer, Henry, Mr.
Monk
TENRYO-JI ZEN MONASTERY
Ukyo-ku Saga Tenryu-ji
Kyoto 616, Japan
T: 81 75 881-5236 F: 81 75 881-5236

Miyama, Chikahiro, Mr.
Kokuiryou 3-2-33 Takaokouri Nada-ku
Kobe, Japan
T: 81 78 861-4333

Miyamoto, Hiroshi, Mr.
JAPANESE CONSUMERS' CO-OPERATIVE UNION
Seikyo-kaikan 1-1-13, Sendagaya, Shibuya-ku
Tokyo, Japan
T: 81 3 34979103 F: 81 3 34970722
TL: 2423380 NCOP J

Miyata, Ikuro, Mr.
JAPANESE CONSUMERS' CO-OPERATIVE UNION
Seikyo-kaikan 1-1-13, Sendagaya, Shibuya-ku
Tokyo, Japan
T: 81 3 34979103 F: 81 3 34970722
TL: 2423380 NCOP J

Miyata, Kayoko, Ms.
JAPANESE CONSUMERS' CO-OPERATIVE UNION
Seikyo-kaikan 1-1-13, Sendagaya, Shibuya-ku
Tokyo, Japan
T: 81 3 34979103 F: 81 3 34970722
TL: 2423380 NCOP J

Miyata, Manabu, Mr.
NATIONAL ALLIANCE OF VICTIMS OF MINAMATA DISEASE & LAWYERS
6-15-7, Fukazawa, Setagaya-ku
Tokyo, Japan
T: 81 3 3704-6871

Miyawaki, Shigeki, Mr.
Manager
GISPRI
c/o Nomura Tourist Bureau 1-5-3 Nihombashi, Chuo-Ku
Tokyo, Japan
T: 81 3 3281-4561 F: 81 3 3274-2670
TL: J29269

Miyazaki, Norio, Mr.
NATIONAL ALLIANCE OF VICTIMS OF MINAMATA DISEASE & LAWYERS
6-15-7, Fukazawa, Setagaya-ku
Tokyo, Japan
T: 81 3 3704-6871

Miyazawa, Kanako, Ms.
JAPANESE CONSUMERS' CO-OPERATIVE UNION
Seikyo-kaikan 1-1-13, Sendagaya, Shibuya-ku
Tokyo, Japan
T: 81 3 34979103 F: 81 3 34970722
TL: 2423380 NCOP J

Mizuhiko, Kiyohara, Mr.
Professor
UNIVERSITY TOKAI
Minamisawa 2-2-11-27 , Minamiku
Sapporo - Hokkaido, Japan
T: 0115713204 F: 0115713204

Mizuno, Takeo, Mr.
Attorney-At-Law
JAPAN FEDERATION OF BAR ASSOCIATIONS
14-5 Nishikakiuchi, Terado-cho, Muko-city
Kyoto-tu, Japan
T: 81 675 364-3224 F: 81 75 364-3226

Mizutani, Shoko, Mr.
RISSHO KOSEI-KAI FUND FOR PEACE
2-20-12, Wada, Suginami-ku
Tokyo 166, Japan
F: 81 3 3381-1138

Mochizuki, Yuko, Ms.
JAPAN YOUTH FEDERATION
Nihon Seinen Kan 2F, 15 Kasumigaoka-Cho
Tokyo, Japan
T: 81 3 3475-2490 F: 81 3 3475-0668

Mori, Kenji
Photographer
MAINICHI SHINBUM
1-1-1 Hito Stubashi , Chiyodaku
Tokyo, Japan
T: 03 3212 0640 F: 03 3287 1865

Mori, Norikazu, Mr.
NATIONAL ALLIANCE OF VICTIMS OF MINAMATA DISEASE & LAWYERS
6-15-7, Fukazawa, Setagaya-ku
Tokyo, Japan
T: 81 3 3704-6871

Morimoto, Machiko
JAPANESE CONSUMERS' CO-OPERATIVE UNION
Seikyo-kaikan 1-1-13, Sendagaya, Shibuya-ku
Tokyo, Japan
T: 81 3 34979103 F: 81 3 34970722
TL: 2423380 NCOP J

Morita, , Mr.
NATIONAL ALLIANCE OF VICTIMS OF MINAMATA DISEASE & LAWYERS
6-15-7, Fukazawa, Setagaya-ku
Tokyo, Japan
T: 81 3 3704-6871

Morita, Akira, Mr.
Attorney-At-Law
JAPAN FEDERATION OF BAR ASSOCIATIONS
3-5-24 Hiyoshi, Minato-kita, Yokohama-city
Kanagawa-ken, Japan
T: 81 45 561-4980

Morita, Chiharu, Ms.
1000-258 Kuriyama Yotsukaido-shi Chyiba-ken
Yotsukaido 7284, Japan
T: 81 43 421-0435

Moriya, Yoshihiko, Sr.
Manager Of Planning Dept.
FEDERATION OF ELECTRIC POWER COMPANIES
1-9-4 Otemachi, Chiyoda-ku
Tokyo, Japan
T: 81 3 3279-2340

Moriyama, Hiroyuki, Mr.
Advisor
JAPAN FOREST TECHNICAL ASSOCIATION
No. 7, Rokubancho chiyoda-ku
Tokyo, Japan
T: 81 3 261-5281 F: 81 3 261-5393

Moriyasu, Tsuneo, Mr.
Chief, Creative Division
MOVIE 'THE EARTH' SCREENING COMMITTEE
C/O Sanpord CO., Ltd., 1-7-27, Juso Honmachi, Yodogawa-KU
Osaka, Japan
T: 81 6 302-8826 F: 81 6 300-0170

Motonami, Yoshinori, Mr.
CITIZENS GROUP FOR THE ENVIRONMENT
2-1-37-101 Imafukuhigashi, Joto-ku
Osaka, Japan
T: 81 6 9330565

Mukaida, Eiko, Ms.
NETWORK YOKOHAMA
Yokohama City Office 1-1 Minato-Cho
Kanagawa, Japan
T: 81 45 671 3038 F: 81 45 681 7388

Murakami, Shoji, Mr.
NATIONAL ALLIANCE OF VICTIMS OF MINAMATA DISEASE & LAWYERS
6-15-7, Fukazawa, Setagaya-ku
Tokyo, Japan
T: 81 3 3704-6871

Murata, Satoshi, Mr.
OVERSEAS FISHERY COOPERATION FOUNDATION
Tokyo
Tokyo, Japan

Murayama, Mitsonobu, Mr.
NATIONAL ALLIANCE OF VICTIMS OF MINAMATA DISEASE & LAWYERS
6-15-7, Fukazawa, Setagaya-ku
Tokyo, Japan
T: 81 3 3704-6871

Murayama, Takehiko, Mr.
92 NGO FORUM JAPAN
4F Kanda Kyyodo Bldg., 2-13-4 Iwamoto-Cho
Tokyo, Japan
T: 81 3 3861-0540 F: 81 3 3861-5048

Nagai, Iku, Ms.
3-20-8 onodai Osakasayamari Nada-ku
Osaka, Japan

Nagamoto, Koichi, Mr.
RISSHO KOSEI-KAI FUND FOR PEACE
2-20-12, Wada, Suginami-ku
Tokyo 166, Japan
F: 81 3 3381-1138

Nagatake, Hiroko, Ms.
JAPAN YOUTH FEDERATION
Nihon Seinen 2F, 15 Kasumigaoka-Cho, Shinjuku
Tokyo, Japan
T: 81 3 3475-2490 F: 81 3 3475-0668

Nagatake, Takao -
Editor
THE CHUNICH SHIMBUN
2-3-13 Konan - Minate-Ka
Tokyo, Japan
T: 03 34912211

Nakabayashi, Hiroharu, Mr.
TOSHIBA
3-1-18 Fufimidai Nerimaku
Tokyo, Japan
T: 81 3 5241-4972

Nakagawa, Shuji, Mr.
THE JAPAN ELECTRICAL MANUFACTURERS ASSOCIATION - JEMA
4-15 Nagata-Cho 2-Chome, Chiyoda-KU
Tokyo 100, Japan
T: 81 3 3581-4845 F: 81 3 3506-0475
TL: 2222619 JEMA J

Nakai, Yasuyuki, Mr.
Attorney-At-Law
JAPAN FEDERATION OF BAR ASSOCIATIONS
2-19-2 Kita-Yamato, Ikoma-city
Hara-ken, Japan
T: 81 9 7437-3465

Nakamura, Kazuo, Mr.
NATIONAL ALLIANCE OF VICTIMS OF MINAMATA DISEASE & LAWYERS
6-15-7, Fukazawa, Setagaya-ku
Tokyo, Japan
T: 81 3 3704-6871

Nakamura, Masaaki, Mr.
3-25-4 Kamiikbukuro Toshima
Tokyo, Japan
T: 81 3 3916-3209

Nakamura, Masako, Ms.
JAPANESE CONSUMERS' CO-OPERATIVE UNION
Seikyo-kaikan 1-1-13, Sendagaya, Shibuya-ku
Tokyo, Japan
T: 81 3 34979103 F: 81 3 34970722
TL: 2423380 NCOP J

Nakamura, Munekazu, Mr.
Assistant Manager
GISPRI
c/o Nomura Tourist Bureau 1-5-3 Nihombashi, Chuo-Ku
Tokyo, Japan
T: 81 3 3281-4561 F: 81 3 3274-2670
TL: J29269

Nakamura, Tetsuya, Mr.
JAPANESE CONSUMERS' CO-OPERATIVE UNION
Seikyo-kaikan 1-1-13, Sendagaya, Shibuya-ku
Tokyo, Japan
T: 81 3 34979103 F: 81 3 34970722
TL: 2423380 NCOP J

Nakamura, Yoko, Ms.
JAPANESE CONSUMERS' CO-OPERATIVE UNION
Seikyo-kaikan 1-1-13, Sendagaya, Shibuya-ku
Tokyo, Japan
T: 81 3 34979103 F: 81 3 34970722
TL: 2423380 NCOP J

Nakamura, Yoshihiko, Mr.
Attorney-At-Law
EARTH DAY JAPAN
5F Uni-yodoyabashi Bild, 3-1-14 Kitahama, Chuo-ku, Osaka
Osaka, Japan
T: 81 6 231 7110 F: 81 6 231 3114

Nakanishi, Yasuko, Ms.
Executive Director
JAPAN FOUNDATION FOR THE EARTH
6-9-45 Mikageyamate, Higashinada-ku, Kobe-city
Hyogo, Japan
T: 81 78 841-8720 F: 81 78 822-3422

Nakano, Etsuko, Ms.
Vice Chairman
OISCA - INTERNATIONAL
6-12, Izumi 3-Chome, Suginami-ku
Tokyo, Japan
T: 81 3 3322-5161 F: 81 3 3324-7111
TL: J29397 OISCA

Nakano, Ricko, Ms.
JAPANESE CONSUMERS' CO-OPERATIVE UNION
Seikyo-kaikan 1-1-13, Sendagaya, Shibuya-ku
Tokyo, Japan
T: 81 3 34979103 F: 81 3 34970722
TL: 2423380 NCOP J

Nakatani, Takeshi, Mr.
Advisor
JAPAN FOREST TECHNICAL ASSOCIATION
No. 7, Rokubancho chiyoda-ku
Tokyo, Japan
T: 81 3 261-5281 F: 81 3 261-5393

Nakatomi, Masayoshi, Mr.
Chairman
OISCA - INTERNATIONAL
6-12, Izumi 3-Chome, Suginami-ku
Tokyo, Japan
T: 81 3 3322-5161 F: 81 3 3324-7111
TL: J29397 OISCA

Nakayama, Toshiaki, Mr.
LIAISON COUNCIL OF PLAINTIFS AND ATTORNEYS GROUPS IN AIR POLLUTION CASES
4-10-12 Chuo, Chiba-city
Chiba-ken, Japan
T: 81 472 25 4567 F: 81 472 25 1507

Nakayama, Yuji, Mr.
NATIONAL ALLIANCE OF VICTIMS OF MINAMATA DISEASE & LAWYERS
6-15-7, Fukazawa, Setagaya-ku
Tokyo, Japan
T: 81 3 3704-6871

Nakazawa, Yukio, Mr.
NATIONAL ALLIANCE OF VICTIMS OF MINAMATA DISEASE & LAWYERS
6-15-7, Fukazawa, Setagaya-ku
Tokyo, Japan
T: 81 3 3704-6871

Naor, Gideon, Mr.
Sec-Gen Oisca Israel
OISCA - INTERNATIONAL
6-12, Izumi 3-Chome, Suginami-ku
Tokyo, Japan
T: 81 3 3322-5161 F: 81 3 3324-7111
TL: J29397 OISCA

Negoro, Akio, Mr.
THE JAPAN ELECTRICAL MANUFACTURERS ASSOCIATION - JEMA
4-15 Nagata-Cho 2-Chome, Chiyoda-KU
Tokyo 100, Japan
T: 81 3 3581-4845 F: 81 3 3506-0475
TL: 2222619 JEMA J

Nezu, Mitsuo, Mr.
GISPRI
c/o Nomura Tourist Bureau 1-5-3 Nihombashi, Chuo-Ku
Tokyo, Japan
T: 81 3 3281-4561 F: 81 3 3274-2670
TL: J29269

Nhigula, George M., Mr.
Chairman, Oisaca Tanzania
OISCA - INTERNATIONAL
6-12, Izumi 3-Chome, Suginami-ku
Tokyo, Japan
T: 81 3 3322-5161 F: 81 3 3324-7111
TL: J29397 OISCA

Niikura, Toshiko
92 NGO FORUM JAPAN
4F Kanda Kyyodo Bldg., 2-13-4 Iwamoto-Cho
Tokyo, Japan
T: 81 3 3861-0540 F: 81 3 3861-5048

Nishi, Yuji, Mr.
DELEGATES OF CASA-MEMBER CO-OPS
2-2-15 Minamihana daguchi-cho, Sakai-city, Izumi Co-op
Osaka-tu, Japan
T: 81 32 722-3111 F: 81 32 722 5074

Nishikawa, Eiichi, Mr.
JAPAN SCIENTISTS ASSOCIATION
2-1173-7 Mukogaoka-cha, Ueneshiba, Sakai-city
Osaka, Japan
T: 81 722 77 7130

Nishimura, Fuji, Ms.
CITIZENS GROUP FOR THE ENVIRONMENT
2-3-30 Uenomaru, Akashi-city
Hyogo-ken, Japan
T: 81 78 9182333

Nishioka, Tatsu, Mr.
92 NGO FORUM JAPAN
4F Kanda Kyyodo Bldg., 2-13-4 Iwamoto-Cho
Tokyo, Japan
T: 81 3 3861-0540 F: 81 3 3861-5048

Nishizawa, Jun, Sr.
Senior Manager
MITSUBISHI COORPORATION
2-6-3 Marunouchi, Chiyoda-ku Higashi-ku
Tokyo, Japan
T: 81 3 3210-6095 F: 81 3 3210-8159

Niwano, Kichiya, Mr.
JAPANESE CONSUMERS' CO-OPERATIVE UNION
Seikyo-kaikan 1-1-13, Sendagaya, Shibuya-ku
Tokyo, Japan
T: 81 3 34979103 F: 81 3 34970722
TL: 2423380 NCOP J

No, Hiran
KYODO NEWS SERVICE
2-2-5 Torakomon, Minato-Ku
Tokyo, Japan
T: 81 3 4111 F: 81 3 3683

Noda, Tomosuke, Mr.
92 NGO FORUM JAPAN
1-38-3 Shimoarata
Kagoshima, Japan
T: 992 57 5255

Nonaka, Shigeo, Mr.
NATIONAL ALLIANCE OF VICTIMS OF MINAMATA DISEASE & LAWYERS
6-15-7, Fukazawa, Setagaya-ku
Tokyo, Japan
T: 81 3 3704-6871

Nonaka, Toru, Mr.
JAPAN NETWORK FOR ENVIRONMENT & POLLUTION
Tokyo-kokko, Tsusahshe-nai, 1-3-1 Kasumigaseki, Chiyoda-ku
Tokyo-to, Japan
T: 81 3 3501-1511 F: 81 3 3580-4847

Noro, Hiroshi, Mr.
Attorney-At-Law
JAPAN FEDERATION OF BAR ASSOCIATIONS
2nd-Kinenbashi Bldg. 4-16-16 Osu, Naka-ku, Naoya-city
Aichi-ken, Japan
T: 81 52 241-7613 F: 81 52 241-1055

Nouda, Tetsuya, Mr.
JAPANESE CONSUMERS' CO-OPERATIVE UNION
Seikyo-kaikan 1-1-13, Sendagaya, Shibuya-ku
Tokyo, Japan
T: 81 3 34979103 F: 81 3 34970722
TL: 2423380 NCOP J

O, Hirano, Mr.
KYODO NEWS SERVICE
2-2-5 Torakomon, Minato-Ku
Tokio, Japan
T: 03 3584 4111 F: 03 3589 3683

Obara, Hideo, Mr.
Director General
NATURE CONSERVATION SOCIETY OF JAPAN
Toranomon Denki-bldg, 2-8-1, Toranomon, Minatoku
Tokyo, Japan
T: 81 3 35034896 F: 81 3 35920496

Ochi, Hiroshi, Sr.
General Manager
CHUBU ELECTRIC POWER
1-Banchi, Higashi-Shinmachi, Higashi-ku
Nagoya, Japan
T: 38 52 951-8211 F: 38 52 971-1837

Ochiai, Ichiro, Mr.
NATIONAL ALLIANCE OF VICTIMS OF MINAMATA DISEASE & LAWYERS
6-15-7, Fukazawa, Setagaya-ku
Tokyo, Japan
T: 81 3 3704-6871

Odaka, Ushimatsu, Mr.
Attorney-At-Law
LIAISON COUNCIL OF PLAINTIFS AND ATTORNEYS GROUPS IN AIR POLLUTION CASES
4-3-13 Tendai, Chiba-city
Chiba-ken, Japan
T: 81 472 52 1875 F: 81 472 25 9313

Ogawa, Fumiko, Mr.
NATIONAL ALLIANCE OF VICTIMS OF MINAMATA DISEASE & LAWYERS
6-15-7, Fukazawa, Setagaya-ku
Tokyo, Japan
T: 81 3 3704-6871

Ogawa, Kazuo, Mr.
JAPANESE CONSUMERS' CO-OPERATIVE UNION
Seikyo-kaikan 1-1-13, Sendagaya, Shibuya-ku
Tokyo, Japan
T: 81 3 34979103 F: 81 3 34970722
TL: 2423380 NCOP J

Ogawa, Yozou, Mr.
JAPANESE CONSUMERS' CO-OPERATIVE UNION
Seikyo-kaikan 1-1-13, Sendagaya, Shibuya-ku
Tokyo, Japan
T: 81 3 34979103 F: 81 3 34970722
TL: 2423380 NCOP J

Ohizumi, Ai, Ms.
JAPANESE CONSUMERS' CO-OPERATIVE UNION
Seikyo-kaikan 1-1-13, Sendagaya, Shibuya-ku
Tokyo, Japan
T: 81 3 34979103 F: 81 3 34970722
TL: 2423380 NCOP J

Ohno, Hiroshi, Mr.
JIDOSHA SOREN
Kokuryu-shibakoen Bldg., 2-6-15 Shibakoen, Minato-ku
Tokyo, Japan
T: 81 3 3434-7641 F: 81 3 3454-7428

Ohno, Koichi, Mr.
92 NGO FORUM JAPAN
4F Kanda Kyyodo Bldg., 2-13-4 Iwamoto-Cho
Tokyo, Japan
T: 81 3 3861-0540 F: 81 3 3861-5048

Ohsawa, Masahiko, Mr.
Doctor
ECOLOGICAL SOCIETY OF JAPAN
Laboratory of Ecology,
Chiba 261, Japan
T: 81 473 251 1111 F: 81 473 256 5793

Ohsawa, Toshikazu, Mr.
JAPANESE CONSUMERS' CO-OPERATIVE UNION
Seikyo-kaikan 1-1-13, Sendagaya, Shibuya-ku
Tokyo, Japan
T: 81 3 34979103 F: 81 3 34970722
TL: 2423380 NCOP J

Ohya, Masao, Mr.
JAPANESE CONSUMERS' CO-OPERATIVE UNION
Seikyo-kaikan 1-1-13, Sendagaya, Shibuya-ku
Tokyo, Japan
T: 81 3 34979103 F: 81 3 34970722
TL: 2423380 NCOP J

Oka, Yasutoshi, Mr.
Manain Director
INSTITUTE OF GLOBAL ENVIRONMENTAL CULTURE
1-1-2-920 Takeshirodai, Sakai-city
Osaka, Japan
T: 81 6 912977

Okamoto, Takayuki, Mr.
Secretary General
ORGANIZATION FOR INDUSTRIAL SPIRITUAL AND CULTURAL ADVANCEMENT-INTERNATIONAL
6-12 - Izumi 3-Chome Saginami-ku
Tokyo 168, Japan
T: 81 33 3322-5161 F: 81 33 3324-7111
TL: J29397 OISCA

Okamura, Etsuji, Mr.
Membership
EARTH DAY NETWORK JAPAN
4-7-22-619 Nanko-naka, Suminee-ku
Osaka, Japan
T: 81 6 613 0941

Okamura, Jun, Mr.
NATIONAL ALLIANCE OF VICTIMS OF MINAMATA DISEASE & LAWYERS
6-15-7, Fukazawa, Setagaya-ku
Tokyo, Japan
T: 81 3 3704-6871

Okano, Mitsuki, Mr.
Assistant Manager
GISPRI
c/o Nomura Tourist Bureau 1-5-3 Nihombashi, Chuo-Ku
Tokyo, Japan
T: 81 3 3281-4561 F: 81 3 3274-2670
TL: J29269

Okawa, Akira, Mr.
STAFF OF OKAYAMA PREF.
Okayama, Japan

Okubo, Seiji, Mr.
92 NGO FORUM JAPAN
4F Kanda Kyyodo Bldg., 2-13-4 Iwamoto-Cho
Tokyo, Japan
T: 81 3 3861-0540 F: 81 3 3861-5048

Okuda, Sachiko, Ms.
F C C RESEARCH INSTITUTE INC.
2-7-1 Nishishinjuku, Shinjuku
Tokyo, Japan
T: 81 3 3344-43980

Okui, Akiko, Ms.
1-4-3-501 Tamagawa Den-enchofu Setagaya-ku
Tokyo, Japan
T: 81 3 3721-6402

Okumura, Keiichiro, Mr.
NATIONAL ALLIANCE OF VICTIMS OF MINAMATA DISEASE & LAWYERS
6-15-7, Fukazawa, Setagaya-ku
Tokyo, Japan
T: 81 3 3704-6871

Okuno, Saburo
Photographer
ASAHI SHIMBUN
5-3-2 Tsukiji, Chuo-ku R.Sampaio Viana 203 c.13 Paraiso
Sao Paulo -SP 04004, Japan
T: 55 11 8841263 F: 55 11 8879990

Okutomi, Kiyoshi, Mr.
Director
NATURE CONSERVATION SOCIETY OF JAPAN
Toranomon Denki-bldg, 2-8-1, Toranomon, Minatoku
Tokyo, Japan
T: 81 3 35034896 F: 81 3 35920496

Onishi, Susumu, Mr.
Composer
JAPANESE MUSICIANS FOR THE GLOBAL ENVIRONMENT
2-16-36 Okubo, Shinjuku-ku, Onaku-Center
Tokyo-to, Japan
T: 81 3 3200-0106 F: 81 3 3200-0105

Onozuka, Harukichi, Mr.
JAPAN SCIENTISTS ASSOCIATION
3-20-5 Adachi, Adachi-ku
Tokyo, Japan
T: 81 3 3886 7597

Onuki, Daisuke
92 NGO FORUM JAPAN
4F Kanda Kyyodo Bldg., 2-13-4 Iwamoto-Cho
Tokyo, Japan
T: 81 3 3861-0540 F: 81 3 3861-5048

Orita, Naoaki, Mr.
Director
92 NGO FORUM JAPAN
C/O Kokuro Headquarter, 1-11-4 Marunouchi
Chiyoda-ku - Tokyo, Japan
T: 03 3 31 4046 F: 03 32151154

Osakabe, Takeshi, Mr.
DELEGATES OF CASA-MEMBER CO-OPS
2-7-22 Tenma, Kita-ku, Osaka-city, PAL Co-op
Osaka-tu, Japan
T: 81 6 354 8711 F: 81 6 354 8715

Osaki, Kozo, Mr.
Manager
TOYO ENGINEERING CORP.
3-2-5, Kasumigaseki, Chiyoda-kku
Tokyo, Japan
T: 81 3 3592 7711 F: 81 3 3593 0749

Ota, Yukari, Mr.
DELEGATES OF CASA-MEMBER CO-OPS
2-2-15 Minamihana daguchi-cho, Sakai-city, Izumi Co-op
Osaka-tu, Japan
T: 81 3 722-3111 F: 81 32 722-5074

Otsuka, Eliza
Translator
ASAHI SHIMBUN
5-3-2 Tsukiji, Chuo-ku R.Sampaio Viana 203 c.13 Paraiso
Sao Paulo -SP 04004, Japan
T: 55 11 8841263 F: 55 11 8879990

Otsuka, Hidetoshi, Sr.
Sub-Manager Of Dev. Dept.
MITSUBISHI CORPORATION
2-6-3 Marunouchi, Chiyoda-ku
Tokyo, Japan
T: 81 3 3210-6095 F: 81 3 3210-8159

Otsuka, Kochi, Mr.
39 Futaba-cho Himeji-shi
Hyogo, Japan
T: 81 792 84-04133

Ozaki, Toshiyuki, Mr.
NATIONAL ALLIANCE OF VICTIMS OF MINAMATA DISEASE & LAWYERS
6-15-7, Fukazawa, Setagaya-ku
Tokyo, Japan
T: 81 3 3704-6871

Pramanik, M.A.H., Dr.
Professor
UNIVERSITY OF TOKYO / INSTITUTE OF INDUSTRIAL SCIENCE
7-22-1 Roppongi, Minato-Ku
Tokyo 106, Japan
T: 81 3 34026231 F: 81 3 34024165

Ryoei, Nakadera, Mr.
GLOBAL ENVIRONMENTAL FORUM
1-9-7 Azabudai Minato- KU
Tokyo, Japan
T: 81 3 55619735

Saijo, Nobuyo, Ms.
Musician
JAPANESE MUSICIANS FOR THE GLOBAL ENVIRONMENT
4-9-4-303 Ayase, Adachi-ku
Tokyo-to, Japan
T: 81 3 3628-0710

Saito, Koichi, Mr.
RISSHO KOSEI-KAI FUND FOR PEACE
2-20-12, Wada, Suginami-ku
Tokyo 166, Japan
F: 81 3 3381-1138

Sakai, Kanji, Mr.
General Manager
OBAYASHI CORPORATION
2-3, Kanda Tsukasa-cho, Chiyoda-ku
Tokyo, Japan
T: 81 3 5256 4844 F: 81 3 5256 4498

Sakai, Kenichi, Mr.
Secretary
JAPAN AIR POLLUTION VICTIMS ASSOCIATION
2-3-14 Matsunami, Chiba-city
Chiba-ken, Japan
T: 81 472 52 0377 F: 81 472 52 6486

Sakai, Tatsuo, Mr.
4-5 Chuocho 2 Chome Meguro-KU
Tokyo 152, Japan
T: 81 3 3715-1111 F: 81 3 3715-8085

Sakakibara, Yusuke, Mr.
Coordinator
AIUEO (LOVES END HUNGER)
Iwasaki 133 Hodogaya
Yokohama Kanagawa, Japan
T: 81 45 331-7868 F: 81 45 3769-2090

Sakamaki, Nobuhiro, Mr.
Chiyoda Corporation
ECO-LIFE CENTER INEM
12-1 Tsurumichuo 2- Chome - Tsurumi-KU
Yokohama 230, Japan
T: 045 521 1231 F: 045 506 4204

Sakami, Sho, Mr.
THE JAPAN ELECTRICAL MANUFACTURERS ASSOCIATION - JEMA
4-15 Nagata-Cho 2-Chome, Chiyoda-KU
Tokyo 100, Japan
T: 81 3 3581-4845 F: 81 3 3506-0475
TL: 2222619 JEMA J

Sakamoto, Yoshiko, Ms.
Vice-Secretary General
OSAKA LIAISON COMMITTEE OF CONSUMERS ORGANIZATIONS
1-3-17-813 Tanimachi, Chuo-ku
Osaka, Japan
T: 81 6 941-3745 F: 81 6 941-5699

Saruva, Takaaki, Mr.
JAPANESE CONSUMERS' CO-OPERATIVE UNION
Seikyo-kaikan 1-1-13, Sendagaya, Shibuya-ku
Tokyo, Japan
T: 81 3 34979103 F: 81 3 34970722
TL: 2423380 NCOP J

Sasai, Kojiro
Cameraman
NIPPON TELEVISION NET. CO.
14 Niban-Cho, Chiyoda-Ku
Tokyo, Japan
T: 03 32301960 F: 03 31344044

Sasaki, Kiyoshi, Mr.
Staff
OITA PREFECTURAL GOVERNMENT
Mitsuyoshidai
Oita, Japan
T: 683494

Sasaki, Rica, Ms.
J T B SUPPORT INTERNATIONAL
1-25-22 Nakamachi Megure-KM
Tokyo 153, Japan
T: 81 3 3793-1547

Sato, Kazuo
92 NGO FORUM JAPAN
4F Kanda Kyyodo Bldg., 2-13-4 Iwamoto-Cho
Tokyo, Japan
T: 81 3 3861-0540 F: 81 3 3861-5048

Sato, Mitsuo, Mr.
JAPAN NETWORK FOR ENVIRONMENT & POLLUTION
Giyugaoka-m, 103, 7-4-13 Okusawa, Setagaya-ku
Tokyo-to, Japan
T: 81 3 3701-6770

Sato, Shuji, Mr.
JAPAN PRODUCTIVITY CENTER
3-1-1 Shibuya, Shibuya-ku
Tokyo 150, Japan
T: 81 3 3409-1111 F: 81 3 3409-5880
TL: J23296(JPCTOKYO

Sato, Tsutomu, Mr.
Member
OISCA - INTERNATIONAL
6-12, Izumi 3-Chome, Suginami-ku
Tokyo, Japan
T: 81 3 3322-5161 F: 81 3 3324-7111
TL: J29397 OISCA

Sato, Yosuke, Mr.
JAPAN JUNIOR CHAMBER
Yokohama Kanagawa, Japan
T: 81 45 243-6171 F: 81 45 243-6333

Sato, Yousuke, Mr.
JAPAN NETWORK FOR ENVIRONMENT & POLLUTION
218-4 Yoyogi, Shibuya-ku
Tokyo-to, Japan
T: 81 3 3379 1421 F: 81 3 3379 1427

Sekiguchi, Katsutoshi, Mr.
NATIONAL ALLIANCE OF VICTIMS OF MINAMATA DISEASE & LAWYERS
6-15-7, Fukazawa, Setagaya-ku
Tokyo, Japan
T: 81 3 3704-6871

Sekiguchi, Takashi, Mr.
YAMAGISHI ASSOCIATION
555 Kawahigashi, Iga-cho, Ayaama-gun
Mie-Prof, Japan
T: 81 594 54594 F: 81 594 53329

Sekimoto, Kiyoyasu, Mr.
DELEGATES OF CASA-MEMBER CO-OPS
83-2 Yoshibara, Higashi-Osaka-city, Osaka-block
Osaka-tu, Japan
T: 81 729 61 7282 F: 81 729 61 7283

Sema, Toru, Sr.
Director Of Ecology Dept.
CENTRAL RESEARCH INSTITUTE OF ELCTIRC POWER INDUSTRY
1-6-1 Otemach, chiyoda-ku
Tokyo, Japan
T: 81 3 3205-6601 F: 81 3 3287-2880

Senba, Shigekatsu, Mr.
NATIONAL ALLIANCE OF VICTIMS OF MINAMATA DISEASE & LAWYERS
6-15-7, Fukazawa, Setagaya-ku
Tokyo, Japan
T: 81 3 3704-6871

Seo, Jiro, Mr.
92 NGO FORUM JAPAN
4F Kanda Kyyodo Bldg., 2-13-4 Iwamoto-Cho
Tokyo, Japan
T: 81 3 3861-0540 F: 81 3 3861-5048

Shibata, Akane, Ms.
NATIONAL ALLIANCE OF VICTIMS OF MINAMATA DISEASE & LAWYERS
6-15-7, Fukazawa, Setagaya-ku
Tokyo, Japan
T: 81 3 3704-6871

Shibata, Jyoji, Mr.
NATIONAL ALLIANCE OF VICTIMS OF MINAMATA DISEASE & LAWYERS
6-15-7, Fukazawa, Setagaya-ku
Tokyo, Japan
T: 81 3 3704-6871

Shibata, Naoyo, Ms.
Secretary Of Overseas Relation
SAINFOREST FOUNDATION - JAPAN
#402, 2-7, Saruqaku-cho, Shibuya
Tokyo 150, Japan
T: 81 3 3477-2589 F: 81 3 3477-0966

Shibatsuji, Masaki, Mr.
Senior Assistant
SHOWA SHELL. SEKIYO K.K.
3-2-5, Kasumigaseki, Chiyoda-ku
Tokyo, Japan
T: 81 3 3503 4372 F: 81 3 3581 5701

Shijmogatto, Hiroshi, Mr.
Representative Director
CITIZENS ALLIANCE FOR SAVING THE ATMOSPHERE AND THE EARTH (CASA)
1-3-17-813 Tanimachi, Chuo-ku, Osaka-city
Osaka, Japan
T: 81 6 941 3745 F: 81 6 941 5699

Shimada, Motomasa, Mr.
Advisor
JAPAN FOREST TECHNICAL ASSOCIATION
No. 7, Rokubancho chiyoda-ku
Tokyo, Japan
T: 81 3 261-5281 F: 81 3 261-5393

Shimagami, Motoko, Ms.
92 NGO FORUM JAPAN
4F Kanda Kyyodo Bldg., 2-13-4 Iwamoto-Cho
Tokyo, Japan
T: 81 3 3861-0540 F: 81 3 3861-5048

Shimamura, Masatochi, Mr.
RISSHO KOSEI-KAI FUND FOR PEACE
2-20-12, Wada, Suginami-ku
Tokyo 166, Japan
F: 81 3 3381-1138

Shimasaki, Chitose, Mr.
Advisor
JAPAN FOREST TECHNICAL ASSOCIATION
No. 7, Rokubancho Chiyoda-ku
Tokyo, Japan
T: 81 3 261-5281 F: 81 3 261-5393

Shimasaki, Katsuaki, Mr.
Advisor
JAPAN FOREST TECHNICAL ASSOCIATION
No. 7, Rokubancho Chiyoda-ku
Tokyo, Japan
T: 81 3 261-5281 F: 81 3 261-5393

Shimizu, Kumiko, Ms.
JAPANESE CONSUMERS' CO-OPERATIVE UNION
Seikyo-kaikan 1-1-13, Sendagaya, Shibuya-ku
Tokyo, Japan
T: 81 3 34979103 F: 81 3 34970722
TL: 2423380 NCOP J

Shimizu, Tetsunosuke, Mr.
Ghuide
KINKI NIPPON TOURIST
Daishin Bldg, 6F, 3-8-6 Shimbashi, Minato-KU
Tokyo, Japan
T: 81 3 5473-0540 F: 81 3 5473-0541

Shimokawa, Nobuaki, Mr.
NATIONAL ALLIANCE OF VICTIMS OF MINAMATA DISEASE & LAWYERS
6-15-7, Fukazawa, Setagaya-ku
Tokyo, Japan
T: 81 3 3704-6871

Shinba, Michiko, Ms.
JAPANESE CONSUMERS' CO-OPERATIVE UNION
Seikyo-kaikan 1-1-13, Sendagaya, Shibuya-ku
Tokyo, Japan
T: 81 3 34979103 F: 81 3 34970722
TL: 2423380 NCOP J

Shinbo, Dahihachi, Mr.
Student
WASEDA UNIVERSITY
2-2-7 Higashi-Horikiri, Katsushika-ku
Tokyo, Japan
T: 81 3 3601-4820

Shinkai, Tadataka, Mr.
JICHIRO
6-1 Chiyoda-ku
Tokyo, Japan
T: 81 3 3263-0272 F: 81 3 5210-7422

Shinohara, Fumio, Mr.
Director
KINKI NIPPON TOURIST
2-3-6 Otemachi, Chiyoda-KU
Tokyo, Japan
T: 81 3 3277-0546 F: 81 3 3277-0512

Shiota, Naoshi, Mr.
NATIONAL ALLIANCE OF VICTIMS OF MINAMATA DISEASE & LAWYERS
6-15-7, Fukazawa, Setagaya-ku
Tokyo, Japan
T: 81 3 3704-6871

Shirai, Masaaki, Mr.
Attorney-At-Law
JAPAN FEDERATION OF BAR ASSOCIATIONS
1-33-4 Hakasato, Kita-ku, Tokyo
Tokyo-to, Japan
T: 81 3 3822-1098

Shirota, Naohiko, Mr.
JAPAN NETWORK FOR ENVIRONMENT & POLLUTION
3-77-5 Nakasato, Kiyose-city
Tokyo-to, Japan
T: 81 424 931-7566

Shoji, Ito
Editor
THE CHUNICH SHIMBUN
2-3-13 Konan - Minate-Ka
Tokyo, Japan
T: 03 34912211

Sin, Christopher, Mr.
Chairman, Oisca Hong Kong
OISCA - INTERNATIONAL
6-12, Izumi 3-Chome, Suginami-ku
Tokyo, Japan
T: 81 3 3322-5161 F: 81 3 3324-7111
TL: J29397 OISCA

Sinha, D. N., Mr.
President, Oisca India
OISCA - INDIA
6-12, Izumi 3-Chome, Suginami-ku
Tokyo, Japan
T: 81 3 3322-5161 F: 81 3 3324-7111
TL: J29397 OISCA

Sonoda, Shoto, Mr.
NATIONAL ALLIANCE OF VICTIMS OF MINAMATA DISEASE & LAWYERS
6-15-7, Fukazawa, Setagaya-ku
Tokyo, Japan
T: 81 3 3704-6871

Sonoyama, Yutaka, Mr.
THE JAPAN ELECTRICAL MANUFACTURERS ASSOCIATION - JEMA
4-15 Nagata-Cho 2-Chome, Chiyoda-KU
Tokyo 100, Japan
T: 81 3 3581-4845 F: 81 3 3506-0475
TL: 2222619

Souza, Heitor Gurgulino De, Mr.
Rector
UNITED NATIONS UNIVERSITY
2871 Madison Av. Suite 908
New York - NY, Japan
T: 1 212 340-4150 F: 1 212 297-0967

Suekawa, Chihoko, Ms.
JAPANESE CONSUMERS' CO-OPERATIVE UNION
Seikyo-kaikan 1-1-13, Sendagaya, Shibuya-ku
Tokyo, Japan
T: 81 3 34979103 F: 81 3 34970722
TL: 2423380 NCOP J

Suenobu, Hidekatsu, Mr.
Member
OISCA - INTERNATIONAL
6-12, Izumi 3-Chome, Suginami-ku
Tokyo, Japan
T: 81 3 3322-5161 F: 81 3 3324-7111
TL: J29397 OISCA

Sueyoshi, Setsuji, Mr.
General Secretary
DAI-NIPPON INK & CHEMICALS WORKERS UNION
3-4-8, Hacchobori, Chu-o-ku
Tokyo, Japan
T: 81 3 3553 4300 F: 81 3 3555 2657

Sugawara, Momoko, Ms.
Musician
JAPANESE MUSICIANS FOR THE GLOBAL ENVIRONMENT
1-21 Yakaba Shinjuku-ku
Tokyo-to, Japan
T: 81 3 3355-5438 F: 81 3 3355-5400

Sugawara, Yasushiko, Mr.
KOTSUKO SOREN
1-2-1 Kasumigasaki, Chiyoda-ku
Tokyo, Japan
T: 81 3 3506-1395 F: 81 3 3580-8038

Sugaya, Mitsuhiro, Mr.
JAPAN YOUTH FEDERATION
Nihon Seinen 2F, 15 Kasumigaoka-Cho, Shinjuku
Tokyo, Japan
T: 81 3 3475-2490 F: 81 3 3475-0668

Sugihara, Michio, Mr.
DELEGATES OF CASA-MEMBER CO-OPS
430-7 Ota, Wakayama-city, Wakayama Co-op
Wakayama-ken, Japan
T: 81 734 73 8653 F: 81 734 74 8649

Sugihara, Takuji, Mr.
Chief Of Kansai Office
UNITED LEAGUE FOR DEMOCRACY / NEWSPAPER 'MINSHUTOITSU'
5-2-1-201 Shibahara-chyo
Toyonaka 560, Japan
T: 81 6 855-7681 F: 81 6 855-2148

Sugii, Kyoko, Ms.
Secretary
OISCA - INTERNATIONAL
6-12, Izumi 3-Chome, Suginami-ku
Tokyo, Japan
T: 81 3 3322-5161 F: 81 3 3324-7111
TL: J29397 OISCA

Sugio, Naoya
Staff Writer
THE JAPAN TIMES
4-5-4, Shibaura, Minato - ku,
Tokyo 180, Japan
T: 81 3 34522099 F: 81 3 34535456

Sugiyama, Hiroshi, Mr.
TETSUKO ROREN
4-7 Daido-cho, Minami-ku, Nagoya-shi
Aichi, Japan
T: 81 52 611-9445 F: 81 52 612-6772

Sumi, Kazuo, Mr.
Chairman
JAPAN ODA STUDY GROUP
713-1-636 Marakodori Nakahara-Ku Kawasaki-Shi
Kawasaki 211, Japan
T: 81 44 434-2650

Sumi, Taeko, Ms.
Staff
STAFF MEMBER OF KANAGAWA
Yokohama
Kanagawa, Japan
T: 81 45 783-2731

Sumino, Setsuko, Ms.
Membership
EARTH DAY NETWORK JAPAN
201-3-13-8 Kitano Mitaka-city, Tokyo
Tokyo, Japan
T: 81 3 41 6157 F: 81 3 41 6157

Suzuki, Kazuo, Mr.
THE JAPAN ELECTRICAL MANUFACTURERS ASSOCIATION - JEMA
4-15 Nagata-Cho 2-Chome, Chiyoda-KU
Tokyo 100, Japan
T: 81 3 3581-4845 F: 81 3 3506-0475
TL: 2222619 JEMA J

Suzuki, Keiko, Ms.
NATIONAL ALLIANCE OF VICTIMS OF MINAMATA DISEASE & LAWYERS
6-15-7, Fukazawa, Setagaya-ku
Tokyo, Japan
T: 81 3 3704-6871

Suzuki, Makoto, Mr.
Lecturer
JAPAN SCIENTISTS ASSOCIATION
Daito-Apartments Kitayasiki 201, Kitayasiki 63,. Asiya, Kota-cho
Nakata-gun, Japan
T: 81 564 63 0494

Suzuki, Mamoru, Mr.
Attorney-At-Law
LIAISON COUNCIL OF PLAINTIFS AND ATTORNEYS GROUPS IN AIR POLLUTION CASES
1-10-7 Chuo, Chiba-city, Chiba Daiichi Law Office
Chiba-ken, Japan
T: 81 472 24 7366 F: 81 472 27 7083

Suzuki, Takahiro, Mr.
NATIONAL ALLIANCE OF VICTIMS OF MINAMATA DISEASE & LAWYERS
6-15-7, Fukazawa, Setagaya-ku
Tokyo, Japan
T: 81 3 3704-6871

Suzuki, Takashi, Mr.
92 NGO FORUM JAPAN
c/o Kokuro Headquarter, 1-11-4 Marunouchi - Chiyoda-ku
Tokyo, Japan
T: 81 3 31 4046 F: 81 3 3215 1154

Suzuki, Toshisaburo, Mr.
3-5-1 Shibuya Shibuyaku
Tokyo, Japan
T: 81 3 34065436 F: 81 3 34065867

Suzuki, Tsuneya, Mr.
NATIONAL ALLIANCE OF VICTIMS OF MINAMATA DISEASE & LAWYERS
6-15-7, Fukazawa, Setagaya-ku
Tokyo, Japan
T: 81 3 3704-6871

Syoji, Shigenori, Mr.
JAPANESE CONSUMERS' CO-OPERATIVE UNION
Seikyo-kaikan 1-1-13, Sendagaya, Shibuya-ku
Tokyo, Japan
T: 81 3 34979103 F: 81 3 34970722
TL: 2423380 NCOP J

Tabata, Katsumine, Mr.
JAPAN YOUTH FEDERATION
Nihon Seinen Kan 2F, 15 Kasumigaoka-Cho, Shinjuku
Tokyo, Japan
T: 81 3 3475-2490 F: 81 3 3475-0668

Tachi, Hiromichi, Mr.
JAPAN NETWORK FOR ENVIRONMENT & POLLUTION
4-18-33 Naka-cho, Koanei-city
Tokyo, Japan
T: 5472 5841 F: 5472 5845

Takagaki, Haruo, Mr.
JAPAN YOUTH FEDERATION
Nihon Seinen Kan 2F, 15 Kasumigaoka-Cho, Shinjuku
Tokyo, Japan
T: 81 3 3475-2490 F: 81 3 3475-0668

Takagi, Yoshiyuki, Mr.
Assistant Manager
GISPRI
c/o Nomura Tourist Bureau 1-5-3 Nihombashi, Chuo-Ku
Tokyo, Japan
T: 81 3 3281-4561 F: 81 3 3274-2670
TL: J29269

Takahashi, Isao, Mr.
Attorney-At-Law
LIAISON COUNCIL OF PLAINTIFS AND ATTORNEYS GROUPS IN AIR POLLUTION CASES
4-10-12 Chuo Chiba-city, Chiba Chuo Law Office
Chiba-ken, Japan
T: 81 472 25 4567 F: 81 472 25 1507

Takahashi, Kaori, Mr.
Youth Representative
A SEED JAPAN
4-3-302, Sakuraoku-Chyo, Shibuyaky
Tokyo, Japan
T: 81 3 3476-3252 F: 81 3 3476-5040

Takahashi, Keiichi, Mr.
THE JAPAN ELECTRICAL MANUFACTURERS ASSOCIATION - JEMA
4-15 Nagata-Cho 2-Chome, Chiyoda-KU
Tokyo 100, Japan
T: 81 3 3581-4845 F: 81 3 3506-0475
TL: 2222619 JEMA J

Takahashi, Masaichi, Mr.
JAPANESE CONSUMERS' CO-OPERATIVE UNION
Seikyo-kaikan 1-1-13, Sendagaya, Shibuya-ku
Tokyo, Japan
T: 81 3 34979103 F: 81 3 34970722
TL: 2423380 NCOP J

Takahashi, T., Mr.
Attorney-At-Law
JAPAN SCIENTISTS ASSOCIATION
4-4 Naritanishimachi Neyaawashi
Osaka, Japan
T: 81 720 34 0799 F: 81 720 34 0799

Takahashi, Tadashi, Mr.
CONSUMER ACTION NETWORK
29-2 Ueta, Shimizuyama, Toyohashi-city
Aichi-ken, Japan
T: 81 52 25-5467 F: 81 52 25-5467

Takahashi, Yoshiyuki
92 NGO FORUM JAPAN
4F Kanda Kyyodo Bldg., 2-13-4 Iwamoto-Cho
Tokyo, Japan
T: 81 3 3861-0540 F: 81 3 3861-5048

Takaki, Kayoko, Ms.
Trustee
JAPAN FOUNDATION FOR THE EARTH
2-1-4 Daiwa-higashi Kawanishi-city
Hyogo, Japan
T: 81 78 94-1005

Takakuwa, Eimatsu, Mr.
Advisor
KOMEITO
Room 416, Sangiin-giin-kaikan, Nagata0cho, Chiyoda-ku
Tokyo, Japan
T: 81 3 2508-8416 F: 81 3 3508-8416

Takano, Hideo, Mr.
NATIONAL ALLIANCE OF VICTIMS OF MINAMATA DISEASE & LAWYERS
6-15-7, Fukazawa, Setagaya-ku
Tokyo, Japan
T: 81 3 3704-6871

Takemoto, Etsuko, Ms.
Head Sales Manager
NIPPON LIFE
Koenji City House #302 3-21-9 Minami Koenji
Tokyo, Japan
T: 81 3 3288-2091 F: 81 3 3288-2065

Takenaka, Toshihiko, Mr.
NATIONAL ALLIANCE OF VICTIMS OF MINAMATA DISEASE & LAWYERS
6-15-7, Fukazawa, Setagaya-ku
Tokyo, Japan
T: 81 3 3704-6871

Takeuchi, Akiko, Ms.
JAPANESE CONSUMERS' CO-OPERATIVE UNION
Seikyo-kaikan 1-1-13, Sendagaya, Shibuya-ku
Tokyo, Japan
T: 81 3 34979103 F: 81 3 34970722
TL: 2423380 NCOP J

Takeuchi, Masanori, Mr.
BEYOND THE EARTH SUMMIT
702, 35-9, I-Chome, Den-en-chofu, Ohta-ku
Tokyo 145, Japan
T: 81 3 3722-1716

Takiguchi, Rie, Mr.
Attorney-At-Law
CO-OPS
1-10-7 Chuoi, Chiba-city, Chiba Daiichi Law Office
Chiba-ken, Japan
T: 81 472 24 7366 F: 81 472 27 7083

Takimoto, Maki, Ms.
JAPAN YOUTH FEDERATION
Nihon Seinen Kan 2F, 15 Kasumigaoka-Cho, Shinjuku
Tokyo, Japan
T: 81 3 3475-2490 F: 81 3 3475-0668

Takuya, Negami, Mr.
KEIDANREN
1-9-4 Ohtemachi, Chiyodaku
Tokyo, Japan
T: 81 3 279-1411

Tamai, Yoku, Mr.
FUJI TELEVISION NETWORK
20 Higashitakasago-cho Urawa-Shi Saidama-Prof
Saidama, Japan

Tamura, Toro, Mr.
THE JAPAN ELECTRICAL MANUFACTURERS ASSOCIATION - JEMA
4-15 Nagata-Cho 2-Chome, Chiyoda-KU
Tokyo 100, Japan
T: 81 3 3581-4845 F: 81 3 3506-0475
TL: 2222619 JEMA J

Tanabe, Akira, Sr.
Gerente
TOKYO ELECTRIC POWER CO., INC.
1-1-3 Uchisaiwaichou Chiyodaku
Tokyo, Japan
T: 81 3 3501-8111

Tanaka, Fumiko, Ms.
NATIONAL ALLIANCE OF VICTIMS OF MINAMATA DISEASE & LAWYERS
6-15-7, Fukazawa, Setagaya-ku
Tokyo, Japan
T: 81 3 3704-6871

Tanaka, Kanji, Mr.
JAPAN NETWORK FOR ENVIRONMENT & POLLUTION
1599-2 Kamishizu, Sakura-city
Chiba-ken, Japan

Tanaka, Miyoko, Ms.
DELEGATES OF CASA-MEMBER CO-OPS
430-7 Ota, Wakayama-city, Wakayama Co-op
Wakayama-ken, Japan
T: 81 734 74 8653 F: 81 734 74 8649

Tanaka, Noriko, Ms.
CONSUMER ACTION NETWORK
1-107-609, Chiyogaoka, Chikusa-ku, Nagoya-city
Aichi-ken, Japan
T: 81 52 775-7176

Taniai, Teruyuki, Mr.
3-19-3, Kisabe-Nishi
Katano, Japan
T: 81 720 92-8482

Tanida, Ichiro, Mr.
JAPAN YOUTH FEDERATION
Nihon Seinen Kan 2F, 15 Kasumigaoka-Cho, Shinjuku
Tokyo, Japan
T: 81 3 3475-2490 F: 81 3 3475-0668

Tanigawa, Akio
Photographer
ASAHI SHIMBUN
5-3-2 Tsukiji, Chuo-ku R.Sampaio Viana 203 c.13 Paraiso
Sao Paulo -SP 04004, Japan
T: 55 11 8841263 F: 55 11 8879990

Taniguchi, Minoru, Mr.
GISPRI
c/o Nomura Tourist Bureau 1-5-3 Nihombashi, Chuo-Ku
Tokyo, Japan
T: 81 3 3281-4561 F: 81 3 3274-2670
TL: J29269

Tatsukawa, Momoc, Ms.
JAPANESE CONSUMERS' CO-OPERATIVE UNION
Seikyo-kaikan 1-1-13, Sendagaya, Shibuya-ku
Tokyo, Japan
T: 81 3 34979103 F: 81 3 34970722
TL: 2423380 NCOP J

Tatsumi, Furuse, Mr.
JAPAN TUNA
2-3-22 Kudankita Chiyodaku
Tokyo 102, Japan
T: 81 3 3264-6167 F: 81 3 3234-7455

Teixeira, Antonio Coutinho
Motorista
ASAHI SHIMBUN
5-3-2 Tsukiji, Chuo-ku R.Sampaio Viana 203 c.13 Paraiso
Sao Paulo -SP 04004, Japan
T: 55 11 8841263 F: 55 11 8879990

Terada, Takehiko, Mr.
Attorney-At-Law
JAPAN FEDERATION OF BAR ASSOCIATIONS
Asahi-Build. 65 Yanaginobanba-kado Oike-deri Nakagyo-ku, Kyoto-city
Kyoto-tu, Japan
T: 81 75 211-6361 F: 81 75 231-1189

Tomoki, Hirabayashi, Mr.
Member
KOMEITO
2-42-3-602 Minamioutuka - Toshima-ku
Tokyo, Japan
T: 81 3 3946-6325 F: 81 3 3355--453

Tonegawa, Haruo, Mr.
Professor
JAPAN ASSOCIATION ON THE ENVIRONMENTAL STUDIES
3-5-8 Saiwaicho Fuchu-city
Tokyo, Japan
T: 81 423 34 6884 F: 81 423 34 6883

Torikai, Shin-Ichi, Mr.
Head
KOMEITO
3-25-21 Maehara Koganei-shi
Tokyo, Japan
T: 81 3 83-9834 F: 81 3 28-3888

Toshimitsu, Neide
KYODO NEWS SERVICE
2 2 5 Toranomon Minato-Ku
Tokyo, Japan
T: 81 33 5844111 F: 3-35893683

Toyoda, Fumiko, Ms.
NATIONAL ALLIANCE OF VICTIMS OF MINAMATA DISEASE & LAWYERS
6-15-7, Fukazawa, Setagaya-ku
Tokyo, Japan
T: 81 3 3704-6871

Toyota, Yoshiko
92 NGO FORUM JAPAN
4F Kanda Kyyodo Bldg., 2-13-4 Iwamoto-Cho
Tokyo, Japan
T: 81 3 3861-0540 F: 81 3 3861-5048

Tsuchiya, Takeo, Mr.
Staff
STAFF OF KANAGAWA-PREF.
Yokohama
Yokohama Kanagawa, Japan
T: 81 45 966-2503

Tsuda, Atsuko, Ms.
CITIZENS GROUP FOR THE ENVIRONMENT
1-12-55-606 Esaka-cho, Suita-city
Osaka, Japan
T: 81 6 3858049

Tsugumi, Kenichi, Mr.
Representative
JAPAN TROPICAL FOREST ACTION NETWORK (JATAN)
Sanjui kari machi tei 2-3 Kanazawa shi Ishikawaken
Kanazawa, Japan
T: 81 762 98.6858

Tsujimoto, Kiyomi, Mr.
92 NGO FORUM JAPAN
4F Kanda Kyyodo Bldg., 2-13-4 Iwamoto-Cho
Tokyo, Japan
T: 81 3 3861-0540 F: 81 3 3861-5048

Tsuzuki, Eiza, Ms.
Interpreter
JAPAN CONSUMERS' COOPERATIVE UNION
4-1-13 Sendagaya, Shibuya
Tokyo 151, Japan
T: 81 3 349-9124

Tundisi, J., Dr.
Professor
INTERNATIONAL LAKE ENVIRONMENT COMMITTEE FOUNDATION
3-4-22 Kyomachi, Otsu
Shiga 520, Japan
T: 81 775 25-1076 F: 81 775 23-1581
TL: 5464850 ILEC J

Uchida, Toshio, Mr.
STAFF OF SAITAMA PREF.
Fukaya - Shi, Saitama, Japan

Uchikawa, Hiroshi, Mr.
NATIONAL ALLIANCE OF VICTIMS OF MINAMATA DISEASE & LAWYERS
6-15-7, Fukazawa, Setagaya-ku
Tokyo, Japan
T: 81 3 3704-6871

Uchino, Tetsuya, Mr.
Managing Director
ASAHI GLASS FOUNDATION
Bank of Tokyo Bldg. 12F, Marunouchi 1-4-2, Chiyoda-Ku
Tokyo, Japan
T: 81 3 3285-0591 F: 81 3 3285-0592

Uchiyama, Hideo, Mr.
NATIONAL ALLIANCE OF VICTIMS OF MINAMATA DISEASE & LAWYERS
6-15-7, Fukazawa, Setagaya-ku
Tokyo, Japan
T: 81 3 3704-6871

Udo, Masao, Mr.
NATIONAL ALLIANCE OF VICTIMS OF MINAMATA DISEASE & LAWYERS
6-15-7, Fukazawa, Setagaya-ku
Tokyo, Japan
T: 81 3 3704-6871

Ueda, Masuo, Mr.
92 NGO FORUM JAPAN
4F Kanda Kyyodo Bldg., 2-13-4 Iwamoto-Cho
Tokyo, Japan
T: 81 3 3861-0540 F: 81 3 3861-5048

Ueda, Shiro, Mr.
JAPANESE CONSUMERS' CO-OPERATIVE UNION
Seikyo-kaikan 1-1-13, Sendagaya, Shibuya-ku
Tokyo, Japan
T: 81 3 34979103 F: 81 3 34970722
TL: 2423380 NCOP J

Ueki, Hisako, Ms.
Attorney-At-Law
JAPAN FEDERATION OF BAR ASSOCIATIONS
5-8-15 Nishitenma, Kita-ku, Osaka-city
Osaka, Japan
T: 81 6 363-0146 F: 81 6 363-4788

Ueki, Minoru F., Dr.
President, Oisca Palau
OISCA - INTERNATIONAL
6-12, Izumi 3-Chome, Suginami-ku
Tokyo, Japan
T: 81 3 3322-5161 F: 81 3 3324-7111
TL: J29397 OISCA

Ui, Jun, Mr.
President
ASIAN ENVIRONMENTAL SOCIETY
981 Kokuba
Naha, Japan
T: 81 98 8531194

Ui, Noriko, Ms.
ASIAN ENVIRONMENTAL SOCIETY
981 Kokuba
Naha, Japan
T: 81 98 8531194

Umeshima, Aiko, Ms.
NATIONAL ALLIANCE OF VICTIMS OF MINAMATA DISEASE & LAWYERS
6-15-7, Fukazawa, Setagaya-ku
Tokyo, Japan
T: 81 3 3704-6871

Umeshima, Kiichi, Mr.
NATIONAL ALLIANCE OF VICTIMS OF MINAMATA DISEASE & LAWYERS
6-15-7, Fukazawa, Setagaya-ku
Tokyo, Japan
T: 81 3 3704-6871

Ura, Nakamu, Mr.
KYODO NEWS SERVICE
2-2-5 Torakomon, Minato-Ku
Tokio, Japan
T: 03 3584 4111 F: 03 3589 3683

Utagawa, Manabu, Mr.
JAPAN NETWORK FOR ENVIRONMENT & POLLUTION
Zentsusan Tsusansho-nai, 1-3-1 Kasumigaseki, Chiyeda-ku
Tokyo-to, Japan
T: 81 3 3580-1707 F: 81 3 3580-4847

Wada, Akira, Mr.
KOMEITO
17-1, Minamimotomachi Sinjuku-ku
Tokyo, Japan
T: 81 3 335-3011

Wada, Osamu, Mr.
Director Int'L. Div.
ASAHI GLASS FOUNDATION
Bank of Tokyo Bldg. 12F, Marunouchi 1-4-2, Chiyoda-Ku
Tokyo, Japan
T: 81 3 3285-0591 F: 81 3 3285-0592

Wada, Takeshi, Mr.
Professor
RESEARCH COUNCIL ON GLOBAL ENVIRONMENT
6-15-13 Hiyeshidai, Takatsuki-city
Osaka-tu, Japan
T: 81 6 87 8243

Wakabayashi, Saeko, Ms.
JAPANESE CONSUMERS' CO-OPERATIVE UNION
Seikyo-kaikan 1-1-13, Sendagaya, Shibuya-ku
Tokyo, Japan
T: 81 3 34979103 F: 81 3 34970722
TL: 2423380 NCOP J

Wang, Bobo, Mr.
CHUNG HUAN FOUNDATION
2-1-4 Sadoharacho, Ichigaya, Shinjuku-ku
Tokyo, Japan
T: 81 3 3266-1598 F: 81 3 3266-9188

Wara, Ikuko, Ms.
THE JAPAN ELECTRICAL MANUFACTURERS ASSOCIATION - JEMA
4-15 Nagata-Cho 2-Chome, Chiyoda-KU
Tokyo 100, Japan
T: 81 3 3581-4845 F: 81 3 3506-0475
TL: 2222619 JEMA J

Watanabe, Akira, Mr.
Associate Professor
JAPAN SCIENTISTS ASSOCIATION
9-2 Aza-Dotenmae Hiyano Iizaka-machi
Fiukushima, Japan
T: 81 245 42 8992

Watanabe, Junji, Mr.
JAPANESE CONSUMERS' CO-OPERATIVE UNION
Seikyo-kaikan 1-1-13, Sendagaya, Shibuya-ku
Tokyo, Japan
T: 81 3 34979103 F: 81 3 34970722
TL: 2423380 NCOP J

Watanabe, Kimio, Mr.
NATIONAL ALLIANCE OF VICTIMS OF MINAMATA DISEASE & LAWYERS
6-15-7, Fukazawa, Setagaya-ku
Tokyo, Japan
T: 81 3 3704-6871

Watanabe, Mistuko, Ms.
NETWORK YOKOHAMA
Yokohama City Office 1-1 Minato-cho
Yokohama Kanagawa, Japan
T: 81 45 671-3038 F: 81 45 681-7388

Watanabe, Tadashi, Mr.
Director
OISCA - INTERNATIONAL
6-12, Izumi 3-Chome, Suginami-ku
Tokyo, Japan
T: 81 3 3322-5161 F: 81 3 3324-7111
TL: J29397 OISCA

Watanabe, Yuko, Ms.
KOMEITO
17-1, Minamimotomachi Sinjuku-KU
Tokyo, Japan
T: 81 3 335-3011

Watanuki, Reiko, Mr.
92 NGO FORUM JAPAN
4F Kanda Kyyodo Bldg., 2-13-4 Iwamoto-Cho
Tokyo, Japan
T: 81 3 3861-0540 F: 81 3 3861-5048

Winkle, Dwight Vamn, Mr.
RESEARCH COUNCIL ON GLOBAL ENVIRONMENT
1-3-17-813 Tanimachi, Chuo-ku
Osaka, Japan
T: 81 6 941 3745 F: 81 6 941 3745

Wsiu-Hwn, Yang Lie, Ms.
Alternate Director
OISCA - INTERNATIONAL
6-12, Izumi 3-Chome, Suginami-ku
Tokyo, Japan
T: 81 3 3322-5161 F: 81 3 3324-7111
TL: J29397 OISCA

Yabe, Yoshitada, Mr.
Director
OKAYAMA-PREF
5-18-11 Saidaji
Okaya-City, Japan
T: 086253 9128

Yabuuchi, Takao, Mr.
TETSUKO ROREN
Hibiya-Kokusai Bldg., 2-2-3 Uchisaiwai-cho, Chiyoda-ku
Tokyo, Japan
T: 81 3 3597-4772 F: 81 3 3597-4775

Yagi, Kenzo, Mr.
92 NGO FORUM JAPAN
4F Kanda Kyyodo Bldg., 2-13-4 Iwamoto-Cho
Tokyo, Japan
T: 81 3 3861-0540 F: 81 3 3861-5048

Yagi, Mariyo, Ms.
NAWALOGY - OMAME
1-23-21 Hilidaira
Ohtsu - Shisa 520, Japan
T: 81 775 290158 F: 81 775 29 0418

Yagi, Toshihiko, Mr.
JAPAN SCIENTISTS ASSOCIATION
3-436 Keyamacho-kita
Tottori, Japan
T: 81 857 28 2350

Yagi, Toshiyuki, Mr.
92 NGO FORUM JAPAN
4F Kanda Kyyodo Bldg., 2-13-4 Iwamoto-Cho
Tokyo, Japan
T: 81 3 3861-0540 F: 81 3 3861-5048

Yamachika, Shigeru, Mr.
NATIONAL ALLIANCE OF VICTIMS OF MINAMATA DISEASE & LAWYERS
6-15-7, Fukazawa, Setagaya-ku
Tokyo, Japan
T: 81 3 3704-6871

Yamada, Kazuya, Mr.
JAPAN NETWORK FOR ENVIRONMENT & POLLUTION
781 Ura-takao-che, Hachioji-city
Tokyo-to, Japan
T: 81 3 63-7955 F: 81 3 62-8115

Yamada, Naohiko, Mr.
GISPRI
c/o Nomura Tourist Bureau 1-5-3 Nihombashi, Chuo-Ku
Tokyo, Japan
T: 81 3 3281-4561 F: 81 3 3274-2670
TL: J29269

Yamamoto, Katsuhiko, Mr.
OSAKA CONFEDERATION OF TRADE UNIONS
27-1-208 Fujiaoka-che, Suita-city
Osaka-tu, Japan
T: 81 6 789-6552

Yamamoto, Kotaro, Mr.
AVANCE CO., LTD.
5-39-7 Jingumae, Shibuya-tu
Tokyo, Japan
T: 81 3 5466-6772

Yamamoto, Sunao, Mr.
Sgi
SOKA GAKKAI INTERNACIONAL
32 Shimanomachi Shinjuku - Ku
Tokyo, Japan
T: 81 3 353-7248 F: 81 3 353-5145

Yamamoto, Wakaba, Ms.
92 NGO FORUM JAPAN
1-25-25 Nakamachi, Meguro-Ku
Tokyo, Japan
T: 81 3 3714 7241

Yamane, Ikuo, Mr.
DELEGATES OF CASA-MEMBER CO-OPS
2-2-15 Minamihana daguchi-cho, Sakai-city, Izumi Co-op
Osaka-tu, Japan
T: 81 32 722-3111 F: 81 32 722-5074

Yamashiro, Hideki, Mr.
JAPAN JUNIOR CHAMBER, INC.
2-8-25 Hiranuma. Nishiku
Yokohama Kanagawa, Japan
T: 81 45 3216090 F: 81 45 3216091

Yamazaki, Hiroyuki, Mr.
Director
JAPAN YOUTH FEDERATION
Nihon Seinen Kan 2F, 15 Kasumigaoka-Cho, Shinjuku
Tokyo, Japan
T: 81 3 3475-2490 F: 81 3 3475-0668

Yamazaki, Kazukiko, Mr.
Vice President
KIKUTAKE ARCHITECT & ASSOCIATES
3-1-1 Sunshine 60, Higashi-Ikebukuro, Toshima-ku 170
Tokyo, Japan
T: 81 3 3986-6121 F: 81 3 3986-6375

Yamazaki, Kei, Mr.
Director General
INTERNATIONAL LAKE ENVIRONMENT COMMITTEE FOUNDATION
3-4-22 Kyomachi, Otsu
Shiga 520, Japan
T: 81 775 25-1076 F: 81 775 23-1581
TL: 5464850 ILEC J

Yasuda, Walter, Mr.
Enviroment
HIDEKOKUSAKUCHO
Osaka-Shi Minato-KU Benten 3-20-11
Osaka, Japan
T: 81 6 576 4839

Yasui, Narumi, Ms.
JAPAN YOUTH FEDERATION
Nihon Seinen Kam 2F, 15 Kasumigaoka-Cho, Shinjuku
Tokyo, Japan
T: 81 3 3475-2490 F: 81 3 3475-0668

Yasuyuki, Tada, Mr.
Musician
GREEN HEART FOUNDATION
1280 Cigalki
Hondo-Shi - Kumamoto, Japan
T: 81 969 24 2282 F: 81 969 24 2276

Yokota, Katsumi, Mr.
JAPANESE CONSUMERS' CO-OPERATIVE UNION
Seikyo-kaikan 1-1-13, Sendagaya, Shibuya-ku
Tokyo, Japan
T: 81 3 34979103 F: 81 3 34970722
TL: 2423380 NCOP J

Yokoyama, Manaby, Mr.
JAPANESE CONSUMERS' CO-OPERATIVE UNION
Seikyo-kaikan 1-1-13, Sendagaya, Shibuya-ku
Tokyo, Japan
T: 81 3 34979103 F: 81 3 34970722
TL: 2423380 NCOP J

Yonetani, Tetsuji, Mr.
NATIONAL ALLIANCE OF VICTIMS OF MINAMATA DISEASE & LAWYERS
6-15-7, Fukazawa, Setagaya-ku
Tokyo, Japan
T: 81 3 3704-6871

Yoneyama, Tetsuo, Mr.
Member
OISCA - INTERNATIONAL
6-12, Izumi 3-Chome, Suginami-ku
Tokyo, Japan
T: 81 3 3322-5161 F: 81 3 3324-7111
TL: J29397 OISCA

Yoshihiro, Otaki, Mr.
DENTSU INC.
Chuoku Tsakuda 2-2-8-1203
Tokyo 104, Japan
T: 81 3 556-00756 F: 81 3 556-00756

Yoshikawa, Koji, Mr.
JAPANESE CONSUMERS' CO-OPERATIVE UNION
Seikyo-kaikan 1-1-13, Sendagaya, Shibuya-ku
Tokyo, Japan
T: 81 3 34979103 F: 81 3 34970722
TL: 2423380 NCOP J

Yoshiki, Higashide, Mr.
1-13-8 Kiboubbaoka Chikkusa-ku
Nagoya, Japan
T: 81 52 762-3989

Yu Huang, Ching, Mr.
Member
OISCA - INTERNATIONAL
6-12, Izumi 3-Chome, Suginami-ku
Tokyo, Japan
T: 81 3 3322-5161 F: 81 3 3324-7111
TL: J29397 OISCA

Yuichi, Fujimoto, Mr.
Director
THE JAPAN ELECTRICAL MANUFACTURERS ASSOCIATION - JEMA
4-15 Nagata-Cho 2-Chome, Chiyoda-KU
Tokyo 100, Japan
T: 81 3 3581-4845 F: 81 3 3506-0475
TL: 2222619 JEMA J

Yuri, Kayano, Ms.
RAINFOREST FOUNDATION - JAPAN
#402, 2-7, Sarugakucho Shibuyaku
Tokyo 221, Japan
T: 81 3 3477-2589 F: 81 3 3477-0966

Yuriko, Sakai
SOCIETY AGAINST THE NAGARA RIVER ESTUARY
241, 3-308, 117 Zenbo Chu Asahiku Yokahana Chi Kanagaw
Yokohama Kanagawa, Japan
T: 81 45 361-7917

ADVISORS

Aichi, Kazuo, Mr.
Member Of Parliment
Tokyo 100, Japan
F: 81 3 3504 1634

Fujiwara, Nobuo, Dr.
Managing Director, Fujiwara Associates Inc.
Tokyo, Japan
T: 03 3239 5617 F: 03 3239 5618

Hashimoto, Ryutaro, H.E. Mr.
Member Of The House Of Representatives
Tokyo 100, Japan
F: 81 3 3504 1634

Kaifu, Toshiki, The Hon.
Former Prime Minister : Member Of Parliament, Japan
Tokyo 100, Japan
F: 81 3 3504 1634

Kawai, Saburo, Mr.
President, Sasakawa Peace Foundation
Tokyo 108, Japan
T: 81 3 3769 2081 F: 81 3 3709 2090

Nakano, Yoshiko, Ms.
President, Int'L Org. For Industrial, Spiritual And Cultural Advancement (OISCA)
Tokyo 168, Japan
T: 81 3 322 5161/5164 F: 81 3 3324 7111

Okita, Saburo, H.E. Dr.
Chairman, Institute For Domestic And International Policy Studies
Tokyo 100, Japan
T: 81 3 3508 2251 F: 81 3 3508 2100

Takahashi, Kazuo, Mr.
Program Director, The Sasakawa Peace Foundation
Tokyo 108, Japan
T: 81 3 3760 2084 F: 81 3 3769 2090

Takeshita, Noboru, The Hon.
Former Prime Minister Of Japan
Tokyo 100, Japan
T: 81 3 3581 3351 F: 81 3 3504 1634

Yamashita, Hideaki, Mr.
Chairman Of The Board, The Asahi Glass Foundation
Tokyo 100, Japan
T: 81 3 3285 0591 F: 81 3 3285 0592

Yazaki, Katsuhito, Mr.
S-G, Kyoto Forum
Osaka 530, Japan
T: 81 6 344 6121 F: 81 6 345 7132

PARLIAMENTARIANS / SPIRITUAL LEADERS

Deguchi, Kyotaro, The Rev.
The Oomoto Foundation
Kameoka, Kyoto, 621, Japan
T: 817712 2 0410 F: 81 7712 2 6725

Nakashini, Tamako
MP
Shugi in 7 1 Nagatacho 1 Chiyodaku 100
Tokyo, Japan
T: 03 581 5111

Sonoda, Minoru, Rev.
Chief Priest Of Chichibu Jinja, Professor Of Kyoto University
Kyoto, Japan
T: 81 494 22 0262 F: 81 494 24 5596

JORDAN - GOVERNMENT

Abandah, Ali Ibrahim, Dr.
Director-General
Department of Metereology
P.O. Box 341011
Amman, Jordan
T: 6 894409 F: 6 894409

Abdel Jaber, Tayseer
Usg Executive Secretary
UN ECON. & SOCIAL COMMISSION FOR WE
P O Box 927115
Amman, Jordan
T: 009626 694351 F: 009626 694980

Akasheh, Talal, Dr
Director Of Environment
HIGHER COUNCIL FOR SCIENCE & TECHNO
Box 36 A1-Jubeiha
Amman, Jordan
T: 009626 835 280 F: 840589

Al Akashah, Talal, Dr.
Higher Council for Science and Technology
Permanent Mission to the United Nations, 866 United Nations Plaza, Room 550-552 New York, NY, 10017
T: 752 0135, 0136 F: 826 0830

Al-Share, Dr. Saleh
Director-Environment Department
MINISTRY OF MUNICIPALS AND ENVIRONM
P.O. Box 1799
Amman, Jordan

Al-Shunaq, Issam, Dr.
Royal Society for the Protection of Nature
Permanent Mission to the United Nations, 866 United Nations Plaza, Room 550-552 New York, NY, 10017
T: 752 0135, 0136 F: 826 0830

Ash-Shari, Saleh, Dr.
Director of Department of Environment
Ministry of Municipal Rural Affairs and Environment
Permanent Mission to the United Nations, 866 United Nations Plaza, Room 550-552 New York, NY, 10017
T: 752 0135, 0136 F: 826 0830

Bino, Murad, Dr.
Royal Scientific Society
Permanent Mission to the United Nations, 866 United Nations Plaza, Room 550-552 New York, NY, 10017
T: 752 0135, 0136 F: 826 0830

Gedeon, Dr Raja Hanna
Director
DEPT. OF LABS & ENVIRONMENT
P.O. Box 2412 Water Authority of Jordan
Amman, Jordan
T: 711361 711362 F: 712275

Inahab, Mohammad, Mr
Oic, Environment & Human Settlement
UNITED NATIONS-ESCWA DIVISION
P O Box 927115
Amman, Jordan
T: 9626 694 3511 EXT601 F: 9626 694981 2

Jadoun, Raja, Dr.
Ministry of Water and Irrigation
Permanent Mission to the United Nations, 866 United Nations Plaza, Room 550-552 New York, NY, 10017
T: 752 0135, 0136 F: 826 0830

Mawla, Naif, H.E. Mr.
Ambassador
Federative Republic of Brazil
Permanent Mission to the United Nations, 866 United Nations Plaza, Room 550-552 New York, NY, 10017
T: 752 0135, 0136 F: 826 0830

Othman, Khalil L., H.E. Mr.
Ambassador, Director of International Organizations and Conference Department
Ministry of Foreign Affairs
Permanent Mission to the United Nations, 866 United Nations Plaza, Room 550-552 New York, NY, 10017
T: 752 0135, 0136 F: 826 0830

Shafa-Amri, Hussein, Eng.
Ministry of Planning
Permanent Mission to the United Nations, 866 United Nations Plaza, Room 550-552 New York, NY, 10017
T: 752 0135, 0136 F: 826 0830

Shamaylah, Jamal, Mr.
Second Secretary
Embassy, Brasilia
Permanent Mission to the United Nations, 866 United Nations Plaza, Room 550-552 New York, NY, 10017
T: 752 0135, 0136 F: 826 0830

Shannag, Esam
Professor
JORDAN DELEGATION
Biology Dept /Yarmouk University
Irbid, Jordan
T: 962 2 271100 EXT2494 F: 962 2 271100 EXT3181

Shatanawi, Muhammad Rashid, Dr.
Director; Professor
University of Jordan
Water & Environment Research Center University of Jordan
Amman, Jordan
T: 962 6 846940 F: 962 6 832318

Tubaishat, Abdel-Razzaq, H.E. Dr.
Minister
Municipal Rural Affairs and Environment
P.O. Box 1799
Amman, Jordan
T: 641 393 F: 672 135

Zein, Ahmad Mukhtar, Mr.
Honorary Consul
Rio de Janeiro
Permanent Mission to the United Nations, 866 United Nations Plaza, Room 550-552 New York, NY, 10017
T: 752 0135, 0136 F: 826 0830

PRESS

Abdel-Hadi, Umayce
Correspondent
SAWT AL KUWAIT NEWSPAPER
3701 Georgem. Dr. 813
Fallschurch, Va. USA, Jordan

Dawodieh, Mohammad
Editor
AL-HASAD
P.O. Box 591
Amman, Jordan, Jordan

Tell, Sufyan
Editor
JSCEP
P.O. Box 644, Jubeiha
Amman, Jordan, Jordan

ORGANIZATIONS

Yassin, El Sayed, Mr.
Arab Thought Forum
PO Box 925 418
Amman, Jordan
T: (962 6) 678 707/678 708 F: (962 6) 675 325

GLOBAL FORUM

Abu Rmeileh, Nawal B., Mr.
President
AL-NASSER LADIES CENTRE
Aman, Jordan
T: 96 26 644-505 F: 96 26 618-505
E: P.O. Box 4600

Badran, Laila, Ms.
Consult.Min.
JORDANIAN SOCIETY FOR ENVIRONMENT CONTROL & MINISTRY OF HEALTH
P.O.Box 921941
Amman, Jordan
F: 96 26 606-606

Mouasher, Anis, Mr.
WORLD COUNCIL OF CHURCHES
P. O. Box 1387
Amman, Jordan

KAZAKHSTAN - GOVERNMENT

Bras Varquez Pareira, Silvio, Mr.
Adviser Minister of Ecology
Permanent Mission of Kazakhstan
136 E. 67th Str.
New York, NY 10021, United States of America
T: 212 861 4900 F: 212 628 0252

Medvedev, Svyatoslav, H.E. Mr.
Minister Ecology (Head of Delegation)
Permanent Mission of Kazakhstan
136 E. 67th Str.
New York, NY 10021, United States of America
T: 212 861 4900 F: 212 628 0252

PRESS

Maksimovna, Marina Petrovna, Ms.
Correspondent
SOVETY KAZAKHSTAN (NEWSPAPER)
Ul. Aiteke B. 67
Alma-Ata, Kazakhstan
T: 7 2 63-7731 F: 7 2 63-1207
TL: 251232PBT SU

GLOBAL FORUM

Azimov, Sergei, Mr.
FILMMAKERS' UNION OF KAZAKHSTAN
20 ul Sovetskaia
Alma-Ata 480100, Kazakhstan

Nezhina, Tamara, Ms.
Tour Agent Adv.
ALTYN ALMA TOURS
200 Tchaikovsky Street, Suite 14
Alma-Ata, Kazakhstan
T: 7 2 63-7731 F: 7 2 63-1207
TL: 251232PBT SU

Ramos, Gilbert M., Dr.
South American Advisor
FILMMAKERS' UNION OF KAZAKHSTAN
ul. Sovetskaia, 20
Alma-Ata 480100, Kazakhstan
T: 47 2 470788 F: 47 2 470507

Uzakov, Akhmet
FILMMAKER'S UNION OF KAZAKHSTAN
1880 East River Road
Tucson - AZ, Kazakhstan
T: 1 602 577-6476 F: 1 602 577-1195

Zullo, Lesley-Anne, Ms.
Us Rep/Int'L Advisor
FILMMAKERS' UNION OF KAZAKHSTAN
ul Sovetskaia, 20
Alma-Ata 480100, Kazakhstan
T: 7 2 611851 F: 7 2 613062
TL: 251244 ORLANSU

KENYA - GOVERNMENT

Adala, Ochieng, H.E. Mr.
Ambassador Extraordinary and Plenipotentiary
Permanent Representative to United Nations, New York
Permanent Mission to the United Nations, 866 UN Plaza, Room 486
New York, NY 10017, United States of America
T: 212 421 4740 F: 212 486 1985

Alopaeus-Stahl, Dorrit M
Counsellor,Deputy Permanent De
EMBASSY OF SWEDEN NAIROBI
PO Box 30600
Nairobi, Kenya
T: 254 2 229042 F: 254 2 218908

Amadi, Roselyn V, Ms
Senior State Counsel
ATTORNEY GENERALS CHAMBERS
PO Box 40112
Nairobi, Kenya
T: 330167

Atandi, S., Ms.
Protocol Officer
Ministry of Foreign Affairs and International Cooperation
Permanent Mission to the United Nations, 866 UN Plaza, Room 486 New York, NY, 10017
F: 212 421 4740
TL: 212 486 1985

Bett, F.K., Mr.
Deputy Comptroller
State House
Permanent Mission to the United Nations, 866 UN Plaza, Room 486 New York, NY, 10017
F: 212 421 4740
TL: 212 486 1985

Chesereck, Enock, Mr.
Office of the President
Permanent Mission to the United Nations, 866 UN Plaza, Room 486 New York, NY, 10017
F: 212 421 4740
TL: 212 486 1985

Habimana, Cyprien
Ambassadeur
AMBASSA DE DU RWANDA
International House P.O. Box 48579
Nairori, Kenya
T: 00 254 2 33 43 41 F: 00 254 2 33 63 65

K'Omudho, B.O., Mr.
Permanent Mission of Kenya
866 UN Plaza, Room 486
New York, NY 10017, United States of America
T: 212 421 4740 F: 212 486 1985

Kahindi, Z., Mr.
State House
Permanent Mission to the United Nations, 866 UN Plaza, Room 486 New York, NY, 10017
F: 212 421 4740
TL: 212 486 1985

Kathurima, H.M., Mr.
Chief of Protocol
Ministry of Foreign Affairs and International Cooperation
Permanent Mission to the United Nations, 866 UN Plaza, Room 486 New York, NY, 10017
F: 212 421 4740
TL: 212 486 1985

Khimulu, Paschal, Dr.
Consultant
UNEP
P.O. Box 73751
Nairobi, Kenya
T: 254 2 504455 F: 254 2 521933

Kiai, Sylvester, Mr
Senior Chemist
PO Box 62678
Nairobi, Kenya
T: 608252

Kiilu, R.M., H.E. Mr.
Ambassador Extraordinary and Plenipotentiary
Permanent Representative to UNEP
Permanent Mission to the United Nations, 866 UN Plaza, Room 486 New York, NY, 10017
F: 212 421 4740
TL: 212 486 1985

Kiptanui, A.K., Mr.
Private Secretary
Comptroller State House
Permanent Mission to the United Nations, 866 UN Plaza, Room 486 New York, NY, 10017
F: 212 421 4740
TL: 212 486 1985

Koece, Korir, Dr.
Permanent Mission of Kenya
866 UN Plaza, Room 486
New York, NY 10017, United States of America
T: 212 421 4740 F: 212 486 1985

Koech Dr., Michael Kipkorir
Senior Lecturer And Dean
FACULTY OF ENVIRONMENTAL STUDIES
Kenyatta University Box 30943
Nairobi, Kenya
T: 254 2 81901 12 EXT 3

Koinange, W.K., Dr.
Permanent Secretary
Office of the Vice-President and Ministry of Finance
Permanent Mission to the United Nations, 866 UN Plaza, Room 486 New York, NY, 10017
F: 212 421 4740
TL: 212 486 1985

Komen, Joel, Mr.
Office of the President
Permanent Mission to the United Nations, 866 UN Plaza, Room 486 New York, NY, 10017
F: 212 421 4740
TL: 212 486 1985

Korir, Tirop Yegor, Mr.
State House
Permanent Mission to the United Nations, 866 UN Plaza, Room 486 New York, NY, 10017
F: 212 421 4740
TL: 212 486 1985

Kosgei, S. J., Dr.
Permanent Secretary
Ministry of Foreign Affairs and International Cooperation
Permanent Mission to the United Nations, 866 UN Plaza, Room 486 New York, NY, 10017
F: 212 421 4740
TL: 212 486 1985

Kulei, J.C., Mr.
State House
Permanent Mission to the United Nations, 866 UN Plaza, Room 486 New York, NY, 10017
F: 212 421 4740
TL: 212 486 1985

Lagat, Geoffrey, Mr.
Office of the President
Permanent Mission to the United Nations, 866 UN Plaza, Room 486 New York, NY, 10017
F: 212 421 4740
TL: 212 486 1985

Leakey, P., The Hon.
M.P.
Minister Environment and Natural Resources
Permanent Mission to the United Nations, 866 UN Plaza, Room 486 New York, NY, 10017
F: 212 421 4740
TL: 212 486 1985

Looremeta, S., Mr.
Office of the President
Permanent Mission to the United Nations, 866 UN Plaza, Room 486 New York, NY, 10017
F: 212 421 4740
TL: 212 486 1985

Mboya, P., H.E. Mrs.
Ambassador Extraordinary and Plenipotentiary
Permanent Representative to HABITAT
Permanent Mission to the United Nations, 866 UN Plaza, Room 486 New York, NY, 10017
F: 212 421 4740
TL: 212 486 1985

Moi, Daniel Toroitich arap, The Hon.
C.G.H., M.P., President
Republic of Kenya
866 UN Plaza, Room 486
New York, NY 10017, United States of America
T: 212 421 4740 F: 212 486 1985

Msosa, Aggrey Joe
Diplomat (adviser)
MALAWI MISSION TO UNEP
P O Box 30453
Nairobi, Kenya
T: 221174 F: 331297

Mugo, Vincent R., Mr.
Permanent Mission of Kenya
866 UN Plaza, Room 486
New York, NY 10017, United States of America
T: 212 421 4740 F: 212 486 1985

Muliro, J., The Hon.
M.P.
Assistant Minister
Foreign Affairs and International Cooperation
T: 10017
TL: 212 421 4740, E: 212 486 1985

Murunga, Bernard, Mr.
Office of the President
Permanent Mission to the United Nations, 866 UN Plaza, Room 486 New York, NY, 10017
F: 212 421 4740
TL: 212 486 1985

Mutisya, D.R., Mr.
Chief Accountant
Ministry of Foreign Affairs and International Cooperation
Permanent Mission to the United Nations, 866 UN Plaza, Room 486 New York, NY, 10017
F: 212 421 4740
TL: 212 486 1985

Mwanzia, P.M., Dr.
Head, Americas Division
Ministry of Foreign Affairs and International Cooperation
Permanent Mission to the United Nations, 866 UN Plaza, Room 486 New York, NY, 10017
F: 212 421 4740
TL: 212 486 1985

Ndiangui, Ndegwa, Mr.
Permanent Mission of Kenya
866 UN Plaza, Room 486
New York, NY 10017, United States of America
T: 212 421 4740 F: 212 486 1985

Ndolo Ayah, Wilson, The Hon.
M.P.
Minister Foreign Affairs and International Cooperation (Deputy Head of Delegation)
Permanent Mission to the United Nations, 866 UN Plaza, Room 486 New York, NY, 10017
F: 212 421 4740
TL: 212 486 1985

Njihia, F.K., Mr.
Protocol Officer
Ministry of Foreign Affairs and International Cooperation
Permanent Mission to the United Nations, 866 UN Plaza, Room 486 New York, NY, 10017
F: 212 421 4740
TL: 212 486 1985

Njiru, Lee, Mr.
Head
Presidential Press Unit
Permanent Mission to the United Nations, 866 UN Plaza, Room 486 New York, NY, 10017
F: 212 421 4740
TL: 212 486 1985

Nyagah, Verity, Mrs.
Permanent Mission of Kenya
866 UN Plaza, Room 486
New York, NY 10017, United States of America
T: 212 421 4740 F: 212 486 1985

Oduor-Noah, Elizabeth Atieno, Ms
Ecologist/Urban Planner
NATIONAL ENVIRONMENT SECRETARIAT
P.O. Box 67839 Nairobi
Nairobi, Kenya
T: 229261 X 141 F: 338272

Okeyo, M.G., Mr.
Permanent Secretary
Ministry of Environment and Natural Resources
Permanent Mission to the United Nations, 866 UN Plaza, Room 486 New York, NY, 10017
F: 212 421 4740
TL: 212 486 1985

Okinda, A., Mr.
Office of the President
Permanent Mission to the United Nations, 866 UN Plaza, Room 486 New York, NY, 10017
F: 212 421 4740
TL: 212 486 1985

Ontita, Charles, Mr.
Office of the President
Permanent Mission to the United Nations, 866 UN Plaza, Room 486 New York, NY, 10017
F: 212 421 4740
TL: 212 486 1985

Onyonka, Z.T., The Hon. Dr.
M.P.
Minister Planning and National Development
Permanent Mission to the United Nations, 866 UN Plaza, Room 486 New York, NY, 10017
F: 212 421 4740
TL: 212 486 1985

Philip Richard Okanda, Owade
Senior Legal Officer
MINISTRY OF FOREIGN AFFAIRS
P O Box 30551
Nairobi, Kenya
T: 334433

Raichenah, Gerald Kiambi, Mr.
Under Secretary
Ministry of Foreign Affairs and International Cooperation
P.O.Box 30551
Nairobi, Kenya
T: 334433

Ramadhan, Hassan, Mr.
State House
Permanent Mission to the United Nations, 866 UN Plaza, Room 486 New York, NY, 10017
F: 212 421 4740
TL: 212 486 1985

Reuben Vincent, Mucto
Ag. Director
NATIONAL ENVIRONMENT
Secretariat-PO Box 47746
Nairobi, Kenya
T: 213774

Sang, Stephen, Mr.
Office of the President
Permanent Mission to the United Nations, 866 UN Plaza, Room 486 New York, NY, 10017
F: 212 421 4740
TL: 212 486 1985

Silverstein, D.M., Dr.
Office of the President
Permanent Mission to the United Nations, 866 UN Plaza, Room 486 New York, NY, 10017
F: 212 421 4740
TL: 212 486 1985

Sitieney, A.K., Lt. Col.
Aide de camp to His Excellency the President
Permanent Mission of Kenya
866 UN Plaza, Room 486
New York, NY 10017, United States of America
T: 212 421 4740 F: 212 486 1985

Tolle, E.M., Ms.
Head, International Organizations and Conference Divison
Ministry of Foreign Affairs and International Cooperation
Permanent Mission to the United Nations, 866 UN Plaza, Room 486 New York, NY, 10017
F: 212 421 4740
TL: 212 486 1985

Toroitich Arap Moi, Daniel, Mr.
President
Permanent Mission of the Republic of Kenya
866 UN Plaza, Room 486
New York NY 10017, United States of America
T: 212 421 4740 F: 212 486 1985

Toure, Fatoumaata, Ms.
Official
Ministry of Foreign Affairs and International Cooperation
Permanent Mission to the United Nations, 866 UN Plaza, Room 486 New York, NY, 10017
F: 212 421 4740
TL: 212 486 1985

Towett, David, Mr.
State House
Permanent Mission to the United Nations, 866 UN Plaza, Room 486 New York, NY, 10017
F: 212 421 4740
TL: 212 486 1985

Wanguru, S., Mr.
Permanent Mission of Kenya
866 UN Plaza, Room 486
New York, NY 10017, United States of America
T: 212 421 4740 F: 212 486 1985

NGO

Awori, Achoka
KEEO
PO Box 48197
Nairobi, Kenya

Chege, Miriam W., Ms.
GREEN BELT MOVEMENT OF KENYA
P.O. Box 67545
Nairobi, Kenya

Kiwanuka, Ruth, Ms.
GREEN BELT MOVEMENT OF KENYA
P.O. Box 67545
Nairobi, Kenya

Muchura, Justina Elizabeth, Ms.
Editor
EAST AFRICAN WILD LIFE SOCIETY
Museum Hill Centre, PO Box 20110
Nairobi, Kenya
T: 254 2 748170 F: 254 2 746868

Muntemba, Shimwaayi
Executive Director
ENVIRONMENT LIAISON CENTRE INTERNATIONAL - KENYA
P.O.Box 72461
Nairobi, Kenya
T: 254 2 56-2015 F: 254 2 56-2172
TL: 23240 ELC KE

Rotich, Nehemiah K. Arap, Mr.
Executive Director
EAST AFRICAN WILD LIFE SOCIETY
PO Box 20110
Nairobi, Kenya
T: 254 2 748170 F: 254 2 746868

Sharma, Ravi
FELCI
PO Box 72461
Nairobi, Kenya

Shete, Mwamba H.A., Mr.
Assistant Director
EAST AFRICAN WILD LIFE SOCIETY
P.O.Box 20110
Nairobi, Kenya
T: 254 2 748-170 F: 254 2 746-868

Wilfred, Nyangena
ACTS
Saint George's House, Parliament Road
P.O.Box 45917 Nairobi, Kenya

PRESS

Bhardwaj, Prabha, Ms.
Yoard Member
WORLDWIDE NETWORK
1331 H Street, Suite 903
Washington - DC 20005, Kenya
T: 1 202 347-1514 F: 1 202 347-1524

Igambi, Herman
Editor
KENYA PRESS
P.O. Box 30958
Nairobi, Kenya, Kenya

Lokorio, John
Editor
KENYA BROADCASTING CORPORATION
P.O. Box 40530
Nairobi, Kenya, Kenya

Munro, Robert
Correspondent
NATION NEWSPAPER
Box 41479
Nairobi, Kenya, Kenya

Mwangi, John
Correspondent
KENYA BROADCASTING CORPORATION
P.O. Box 40530
Nairobi, Kenya, Kenya

Ndumbu, Abel
Editor
DEVELOPMENT HORIZONS
P.O. Box 58933
Nairobi, Kenya, Kenya

Ngunia, Judith
Editor
KENYA BROADCASTING CO.
P. O. Box 40530
Nairobi, Kenya, Kenya

Rotich, Joseph
Editor
KENYA BROADCASTING CORPORATION
P.O. Box 40530
Nairobi, Kenya, Kenya

Salim, Mohamed
Reporter Radio/Television
KENYA BROADCASTING CO.
P.O. Box 40530
Nairobi, Kenya, Kenya

ORGANIZATIONS

Arum, Gilbert, Mr.
Kenya Energy & Environment Organizations
PO Box 48197
Nairobi, Kenya
T: (254 2) 749 747 F: (254 2) 749 382

Juma, Calestous, Dr.
African Centre For Technology Studies
Saint George's House, Parliament Road
P.O.Box 45917 Nairobi, Kenya
T: (254 2) 74 40 47/74 40 95 F: (254 2) 74 39 95

Kahihu, Ndungu, Mr.
Kenya Scouts Association
PO Box 41422
Nairobi, Kenya
T: (25 42) 568 111 F: (25 42) 230 876/569 795

Maathai, Wangari, Prof.
Green Belt Movement, The
P.O. Box 67545
Nairobi, Kenya
T: (254 2) 504 264

Malaho, Silvanus A.B., Mr.
Kenya Volunteer Development Services
PO Box 310
Bungoma, Kenya
F: (254) 337 20235

Mann, Erica, Ms.
Council For Human Ecology - Kenya
PO Box 20360
Nairobi, Kenya
T: (254 2) 720 399

Muchiru, Simon, Mr.
African NGOs Environment Network (ANEN)
PO Box 53844
Nairobi, Kenya
T: (25 42) 28 138 F: (25 42) 335 108

Phiri, Seko H., Ms.
African Association For Literacy And Adult Education
Finance House, 6th Floor Loita Street
331512 Nairobi, Kenya
T: (254 2) 222 391 F: (254 2) 340 849

Tindimubona, Alex R., Dr.
African Academy of Sciences, The
PO Box 14798
Nairobi, Kenya
T: (25 42) 802 182/802 176 F: (25 42) 802 185

GLOBAL FORUM

Akinyi, Elizabeth, Ms.
AAWORD - KENYA
P.O.Box 51195
Nairobi, Kenya
T: 254 2 277111

Aremo, James, Mr.
KENYA FREEDOM FROM HUNGER COUNCIL
Palais de Nations
Geneve CH 1211, Kenya
F: 41 22 733-6542

Aweri, Achoka, Mr.
KENGO
Palais de Nations
Geneve CH 1211, Kenya
F: 41 22 733-6542

Brevik, Tore J., Sr.
Chief Information
UNEP
Box 30552
Nairobi, Kenya
T: 254 2 23 0084 F: 254 2 22 6831
TL: 22068 UNEP KE

Brown, Joan Martin, Ms.
Chief Washi. Off.
UNEP
Box 30552
Nairobi, Kenya
T: 254 2 23 0084 F: 254 2 22 6831
TL: 22068 UNEP KE

Cege, Lydia W., Ms.
KENYA INSTITUTE OF ORGANIC FARMING
P.O.Box 34972
Nairobi, Kenya
T: 254 2 732487

Fayemi, Jean-Marie, Mr.
ENVIRONMENT LIAISON CENTRE INTERNATIONAL
P. O. Box 72461
Nairobi, Kenya
T: 254 2 562015

Fitter, Perin, Ms.
National Coordinator
KENYA YOUTH ASSOCIATION COUNCIL
P.O.Box 1690
Kisumu, Kenya
T: 254 35 44240

Garau, Maia, Ms.
P.O. Box 24350
Nairobi, Kenya
T: 254 2 88-2235

Karekezi, Stephen, Mr.
Executive Secretary
FIND & AFREPREN
PO Box 30979, Ole Odume Road - Kiumani
Nairobi, Kenya
T: 254 2 566032 F: 254 2 561464
TL: 25516 FWD NET

Keya, Willian, Mr.
DIOCESE OF NAKURU-AGRICULTURAL TEAM
P.O.Box 938 -
Nakuru, Kenya

Lamba, Davinder, Mr.
Executive Director
MAZINGIRA INSTITUTE
PO Box 14564
Nairobi, Kenya
T: 254 2 442577 F: 254 2 442592

Lee-Smith, Diana, Ms.
Secretary
WOMEN AND SHELTER NETWORKERS
PO Box 14564
Nairobi, Kenya
T: 254 2 740284 442577 F: 254 2 442592

Liamine, Nathalie, Ms.
Liaison Officer
ELCI/CLEI
PO Box 72461
Nairobi, Kenya
T: 254 2 562015 F: 254 2 562175

Lumbama, Noel, Mr.
Information Manager
ENVIRONMENT LIAISON CENTRE INTERNATIONAL
PO Box 72461
Nairobi, Kenya
T: 254 2 562015 F: 254 2 562175
TL: 23240 ELCKE

M©Wereria, Godfrey, Mr.
SOUTHERN NETWORK FOR DEVELOPMENT SONED
Palais de Nations
Geneve CH 1211, Kenya
F: 41 22 733-6542

Macharia, Dianah, Ms.
Information Manager
ENVIRONMENT LIAISON CENTRE INTERNATIONAL
PO Box 72461
Nairobi, Kenya
T: 254 2 562015 F: 254 2 562175
TL: 23240 ELCKE

Mathai, Wangari, Dr.
Coordinator
GREEN BELT MOVEMENT OF KENYA
P. O. Box 67545
Nairobi, Kenya
T: 254 2 504264

Mbaya, V. Beaman, Dr.
GREENBELT MOVEMENT
Palais de Nations
Geneve CH 1211, Kenya
F: 41 22 733-6542

M'mwereria, Godfrey K., Mr.
Co-Ordinator- Africa Regions
SOUTHERN NETWORKS FOR DEVELOPMENT
P.O.Box. 14205
Nairobi, Kenya
T: 254 2 445893 F: 254 2 445894

Munro, Bob, Mr.
Chairman
MATHARE YOUTH SPORTS ASSOCIATION (MYSA)
Box 41479
Nairobi, Kenya
T: 254 2 741642 F: 254 2 762178

Mwangi, Wagaki, Ms.
Editor - Econews Africa
KENYA CONSUMER ORGANIZATION
PO Box 21136
Nairobi, Kenya
T: 254 2 226-028 F: 254 2 729-447

Mwereria, Godfrey, Dr.
Corrdinator
SONED- AFRICAN REGION
P.O.Box 14205
Nairobi, Kenya

Ng'Weno, Hilary B
Executive Director
AFRICAN CENTRE FOR COMMUNICATIONS & DEVELOPMENT
5th Floor,Prudential Assurance Building, Wabera St PO Box 20190
Nairobi, Kenya
T: 254 2 218043 F: 254 2 222555

Njoroge, John P. W., Mr.
Director
KENYA INSTITUTE OF ORGANIC FARMING
P.O.Box 34972
Nairobi, Kenya
T: 254 2 732487

Obonyo, Stephen O
Editor
SAUTI YA GUSII
PO Box 763
Kisii, Kenya
T: 254 0381/21038

Okech-Owiti, Maurice
UNIVERSITY OF NAIROBI
12 Rajaji Marg
New Delhi 110011, Kenya
T: 91 11 301-4309

Olindo, Perez, Mr.
AFRICAN WILDLIFE FOUNDATION
PO9 Box 48177
Nairobi, Kenya
T: 254 2 710-367

Sylvain, Marie Ange, Sra.
Coordinator
UNITED NATIONS ENVIRONMENT PROGRAMME
PO Box 30552
Nairobi, Kenya
T: 254 2 23 0084 F: 254 2 22 6831
TL: 22068 UNEP KE

Wachiira, Kamoji, Mr.
GREEN BELT MOVEMENT OF KENYA
c/o Resource Renewal Institute, FT. Mason Center Bldg A
San Francisco - CA, Kenya
T: 1 415 928-32774 F: 1 415 928-6529

Wafula, A. Okhoya, Mr.
Programme Officer
AFRICAN ASSOCIATION OF PUBLIC ADMINISTRATION AND MANAGEMENT
P.O. Box 48677 Nairobi
Nairobi, Kenya

Yimbo, Silas, Mr.
Director
AEROSPACE SURVEY
Nairobi, Kenya

UN AND UNCED

Aga, Gugu
United Nations Enviroment Programme
P.O. Box 30552
Nariobi, Kenya
T: 254-2 226886 F: 254-2 228890

Belmont, Francoise
United Nations Enviroment Programme
P.O. Box 30552
Nariobi, Kenya
T: 254-2 226886 F: 254-2 228890

Boyd, Dale
United Nations Enviroment Programme
P.O. Box 30552
Nariobi, Kenya
T: 254-2 226886 F: 254-2 228890

Carvalho, Elizabeth
United Nations Enviroment Programme
P.O. Box 30552
Nariobi, Kenya
T: 254-2 226886 F: 254-2 228890

Lizarraga-Reyes, Jose
United Nations Enviroment Programme
P.O. Box 30552
Nariobi, Kenya
T: 254-2 226886 F: 254-2 228890

Mattos, Haroldo
United Nations Enviroment Programme
P.O. Box 30552
Nariobi, Kenya
T: 254-2 226886 F: 254-2 228890

Olembo, Reuben
United Nations Enviroment Programme
P.O. Box 30552
Nariobi, Kenya
T: 254-2 226886 F: 254-2 228890

Osman, Salih
United Nations Enviroment Programme
P.O. Box 30552
Nariobi, Kenya
T: 254-2 226886 F: 254-2 228890

Woollaston, Philip
United Nations Enviroment Programme
P.O. Box 30552
Nariobi, Kenya
T: 254-2 226886 F: 254-2 228890

KIRIBATI - GOVERNMENT

Teannaki, Teatao, Mr.
President
Kiribati

KUWAIT - GOVERNMENT

Al-Abdali, Fatima Khudayer, Dr.
Associate Research Scientist
Kuwait Institute for Scientific Research
Permanent Mission to the United Nations, 321 E. 44th Str.
New York, NY 10017, United States of America
T: 212 973 4300 F: 212 370 1733

Al-Ahmad Al-Jaber Al-Sabah, Jaber, Amir
Permanent Mission of the State of Kuwait
321 E. 44th Str.
New York NY 10017, United States of America
T: 212 973 4300 F: 212 370 1733

Al-Awadhi, Maryam Abdul Karim, Mrs.
Assistant Under Secretary for Technical Co-operation Affairs
Ministry of Planning
Permanent Mission to the United Nations, 321 E. 44th Str.
New York, NY 10017, United States of America
T: 212 973 4300 F: 212 370 1733

Al-Faily, Rida Yousef, Mr.
Assistant Under Secretary for Television Affairs
Ministry of Information
Permanent Mission to the United Nations, 321 E. 44th Str.
New York, NY 10017, United States of America
T: 212 973 4300 F: 212 370 1733

Al-Fetouh, Mohammad Yehia A. Abou, Mr.
Legal Adviser
Ministry of Public Health
Permanent Mission to the United Nations, 321 E. 44th Str.
New York, NY 10017, United States of America
T: 212 973 4300 F: 212 370 1733

Al-Fouzan, Abdulwahab S., H.E. Dr.
Minister of Public Health
Chairman of the Environmental Protection Council
Permanent Mission to the United Nations, 321 E. 44th Str.
New York, NY 10017, United States of America
T: 212 973 4300 F: 212 370 1733

Al-Ghais, Faisal Rashed, H.E. Mr.
Ambassador of the State of Kuwait to Brazil
Permanent Mission of Kuwait
321 E. 44th Str.
New York, NY 10017, United States of America
T: 212 973 4300 F: 212 370 1733

Al-Hassam, Mohammad Abou, H.E. Mr.
Ambassador Extraordinary and Plenipotentiary
Permanent Representative to the United Nations, New York
Permanent Mission to the United Nations, 321 E. 44th Str.
New York, NY 10017, United States of America
T: 212 973 4300 F: 212 370 1733

Al-Hassan, Jassim Mohammad, Prof.
Professor of Biochemistry, Head of Biochemistry, Faculty of Science
Kuwait University
Permanent Mission to the United Nations, 321 E. 44th Str.
New York, NY 10017, United States of America
T: 212 973 4300 F: 212 370 1733

Al-Meshan, Meshal Abdulla, Dr.
Head
Environment Health Department College of Health Science
Permanent Mission to the United Nations, 321 E. 44th Str.
New York, NY 10017, United States of America
T: 212 973 4300 F: 212 370 1733

Al-Nakib, Ahmad Mohammad S., Mr.
Deputy Director General
Authority for Agricultural Department
Permanent Mission to the United Nations, 321 E. 44th Str.
New York, NY 10017, United States of America
T: 212 973 4300 F: 212 370 1733

Al-Omani, Sulaiman Nisf Jassim, Mr.
Under Secretary
Ministry of Oil
Permanent Mission to the United Nations, 321 E. 44th Str.
New York, NY 10017, United States of America
T: 212 973 4300 F: 212 370 1733

Al-Razoki, Seham, Ms.
Assistant Under Secretary, Economic Affairs
Ministry of Petroleum
Permanent Mission to the United Nations, 321 E. 44th Str.
New York, NY 10017, United States of America
T: 212 973 4300 F: 212 370 1733

Al-Sabah, Sabah Abdallah, Mr.
Department of International Organizations
Ministry of Foreign Affairs
Permanent Mission to the United Nations, 321 E. 44th Str.
New York, NY 10017, United States of America
T: 212 973 4300 F: 212 370 1733

Al-Sabah, Sheikh Jaber Al-Ahmad Al-Jaber, His Highness
Amir
Kuwait
321 E. 44th Str.
New York, NY 10017, United States of America
T: 212 973 4300 F: 212 370 1733

Al-Saif, Ali Yousef, Dr.
Assistant Under Secretary for Public Health Affairs
Ministry of Public Health
Permanent Mission to the United Nations, 321 E. 44th Str.
New York, NY 10017, United States of America
T: 212 973 4300 F: 212 370 1733

Al-Sarawi, Mohammad, Dr.
Professor, Faculty of Science
Kuwait University
Permanent Mission to the United Nations, 321 E. 44th Str.
New York, NY 10017, United States of America
T: 212 973 4300 F: 212 370 1733

Al-Shuaiby, Khaled Mubarak, Mr.
Civil Aviation Authority
Permanent Mission to the United Nations, 321 E. 44th Str.
New York, NY 10017, United States of America
T: 212 973 4300 F: 212 370 1733

Al-Wagayyan, Wagayyan Y.F., Mr.
Director
Minister's Office Ministry of Public Health
Permanent Mission to the United Nations, 321 E. 44th Str.
New York, NY 10017, United States of America
T: 212 973 4300 F: 212 370 1733

Ali, Mahmoud Adulrahim, Dr.
Deputy Director, Environment Protection Department
Ministry of Public Health
Permanent Mission to the United Nations, 321 E. 44th Str.
New York, NY 10017, United States of America
T: 212 973 4300 F: 212 370 1733

Hussein, Nezar M., Mr.
Deputy Director General, Research/Life and Environmental Sciences
Kuwait Institute for Scientific Research
Permanent Mission to the United Nations, 321 E. 44th Str.
New York, NY 10017, United States of America
T: 212 973 4300 F: 212 370 1733

Mourady, Abdallah, Mr.
Director, International Organizations Department
Ministry of Foreign Affairs
Permanent Mission to the United Nations, 321 E. 44th Str.
New York, NY 10017, United States of America
T: 212 973 4300 F: 212 370 1733

Ramadan, Mohammad, Dr.
Director
Kuwait Institute for Energy Research
Permanent Mission to the United Nations, 321 E. 44th Str.
New York, NY 10017, United States of America
T: 212 973 4300 F: 212 370 1733

PRESS

Al-Jaser, Adnan
Correspondent
KUWAIT NEWS AGENCY
Street 16 # 51
Al-Safat, Kuwait, Kuwait

Al-Rewaished, Rashed
Editor
KUWAIT NEWS AGENCY
Al-Jabreya St. #7
Kuwait, Kuwait

Jasem, Arti
Correspondent
KUWAIT NEWS AGENCY
P.O.Box 2602
Al-Safat, Kuwait, Kuwait

GLOBAL FORUM

Al-Meer, Adnan Y., Mr.
Chairman
SCIENCE CLUB
P.O.Box 23259
Kuwait 13093, Kuwait
T: 965 75 476-2562 F: 965 75 476-2567

Almeer, Adnan, Mr.
Chairman
KUWAIT SCIENCE CLUB
PO Box 23259
Kuwait, Kuwait
T: 965 75 476-2562 F: 965 75 476-2567

Mohammad, Dawood S., Mr.
General Secretary
SCIENCE CLUB
P.O.Box 23259 -Safat
Kuwait 13093, Kuwait
T: 61 75 4762562 F: 61 75 4762567
TL: 31949Nadalm

Sulaiman-Ahmad, Dawood, Mr.
General-Secretary
KUWAIT SCIENCE CLUB
PO Box 23259
Safat, Kuwait
T: 965 476-2561 F: 965 476-2567

PARLIAMENTARIANS / SPIRITUAL LEADERS

Al Bishir, Badar Nasir, Mr
MP
National Assembly PO Box 716 Safat
Kuwait, Kuwait
T: 439031
TL: MAJLIS 46000 KT

Al Enizi, Khalaf Dmeitheer, Mr
MP
National Assembly PO Box 716 Safat
Kuwait, Kuwait
T: 439031
TL: MAJLIS 46000 KT

Al Omar, Abdullah Omar, Dr
MP
National Assembly PO Box 716 Safat
Kuwait, Kuwait
T: 439031
TL: MAJLIS 46000 KT

Alweheib, Tawfiq Saud, Mr
MP
National Assembly PO Box 716 Safat
Kuwait, Kuwait
T: 439031
TL: MAJLIS 46000 KT

KYRGYSTAN - GLOBAL FORUM

Suleimenov, Olzas Omarovich, Mr.
President
Nevada-Semipalatinskk Anti-Nuclear
Prospekt Abia, 1 - Movement
Alma-Ata Republic 48100, Kyrgystan
T: 7 327 63-3648 F: 7 327 63-6634
TL: 251232PBT SU

LAOS PEOPLE'S DEMOCRATIC REPUBLIC - GOVERNMENT

Kittikhoun, Alounkeo, M.
Directeur
Departement des Organisations Internationales, Ministere des Affaires Etrangeres, (Chef-Adjoint de la Delegation)
Permanent Mission to the United Nations, 820 Second Avenue, Suite 400
New York, NY 10017, United States of America
T: 986 0227 F: 750 0039

Meuangnalat, Pho, M.
Chef de Cabinet
Ministere de la Science et de la Technologie
Permanent Mission to the United Nations, 820 Second Avenue, Suite 400
New York, NY 10017, United States of America
T: 986 0227 F: 750 0039

Nanthavong, Soull, S.E. M.
Ministre de la Science la Technologie (Chef de la Delegation)
Permanent Mission of Laos People's Democratic Republic
820 Second Avenue, Suite 400
New York, NY 10017, United States of America
T: 986 0227 F: 750 0039

Phommachanh, Kanika, Mme
Ministre Conseillere
Permanent Mission of Laos People's Democratic Republic
820 Second Avenue, Suite 400
New York, NY 10017, United States of America
T: 986 0227 F: 750 0039

Savathvong, Silavanh, M.
Directeur Adjoint
Departement des Forets et de l' Environnement, Ministere de l' Agriculture et des Forets
Permanent Mission to the United Nations, 820 Second Avenue, Suite 400
New York, NY 10017, United States of America
T: 986 0227 F: 750 0039

GLOBAL FORUM

Sumbath, , Dr.
RICE INTEGRATED FARMING SYSTEM
Palais de Nations
Geneve CH 1211, Laos People's Democratic Republic
F: 41 22 733-6542

LATVIA - GOVERNMENT

Baumanis, Aivars, H.E. Mr.
Ambassador Extraordinary and Plenipotentiary
Permanent Representative to the United Nations, New York
Permanent Mission to the United Nations, 115 W. 183rd Str.
New York, NY 10453, United States of America
T: 212 933 4500 F: 212 562 7434

Dinduns, Maris, Mr.
Head
Foreign Relations Department, Environmental Protection Committee
Permanent Mission to the United Nations, 115 W. 183rd Str.
New York, NY 10453, United States of America
T: 212 933 4500 F: 212 562 7434

Emsis, Indulis, H.E. Mr.
Chairman
Environmental Protection Committee
Permanent Mission to the United Nations, 115 W. 183rd Str.
New York, NY 10453, United States of America
T: 212 933 4500 F: 212 562 7434

Gorbunovs, Anatolijs, H.E. Mr.
Chairman of the Supreme Council
Republic of Latvia
Permanent Mission to the United Nations, 115 W. 183rd Str.
New York, NY 10453, United States of America
T: 212 933 4500 F: 212 562 7434

Krikis, Andris, Dr.
Head of Coordination Department
Environmental Protection Committee
Permanent Mission to the United Nations, 115 W. 183rd Str.
New York, NY 10453, United States of America
T: 212 933 4500 F: 212 562 7434

Petersone, Karina, Mr.
Assistant to the Chairman
Supreme Council of the Republic of Latvia
Permanent Mission to the United Nations, 115 W. 183rd Str.
New York, NY 10453, United States of America
T: 212 933 4500 F: 212 562 7434

Treimanis, Mara, Ms.
Minister Counsellor
Permanent Mission of Latvia
115 W. 183rd Str.
New York, NY 10453, United States of America
T: 212 933 4500 F: 212 562 7434

Ulme, Arvids, Mr.
President
Environmental Protection Club
Permanent Mission to the United Nations, 115 W. 183rd Str.
New York, NY 10453, United States of America
T: 212 933 4500 F: 212 562 7434

Ziedins, Aivars, Mr.
Counsellor
Permanent Mission of Latvia
115 W. 183rd Str.
New York, NY 10453, United States of America
T: 212 933 4500 F: 212 562 7434

LEBANON - GOVERNMENT

Al-Daher, Rachid
Ex-Ambassadeur Aupres Des Nations
AVOCAT INTERNATIONALISTE
President De L'Universite Pour la
Beyrouth, Lebanon
T: B/ 966 1 200012 13 F: 200014

Arsanios, Youssef, H.E. Mr.
Ambassador
Ministry of Foreign Affairs
Permanent Mission to the United Nations, 866 United Nations Plaza, Room 531-533 New York, NY, 10017
F: 355 5460, 5461
TL: 838 2819

Boueiz, Farez, H.E. Mr.
Minister for Foreign Affairs (Head of Delegation)
Permanent Mission of Lebanon
866 United Nations Plaza, Room 531-533
New York, NY 10017, United States of America
T: 355 5460, 5461 F: 838 2819

Chidiac, Ghazi, H.E. Mr.
Ambassador Federative Republic of Brazil
Permanent Mission of Lebanon
866 United Nations Plaza, Room 531-533
New York, NY 10017, United States of America
T: 355 5460, 5461 F: 838 2819

Daher, Rachid Daher
Lawyer-Former Ambassador
RUE ECOLE EBRINE
Lm Samaha 5' Et
Beyrouth, Lebanon
T: 961 1 200012 13

El-Khouli, Mohamed, Dr.
Expert on Environment
Permanent Mission of Lebanon
866 United Nations Plaza, Room 531-533
New York, NY 10017, United States of America
T: 355 5460, 5461 F: 838 2819

Fawaz, Mohamad
Directeur General De L Urbanisme
MINISTERE DES TRAVAUX PUBLICS ET DE
Corniche Mazraa-In Saab
Beirut, Lebanon
T: 01 865713 01 301919

Ghanem, Fouad El-Khoury, Mr.
Consul General
Rio de Janeiro
Permanent Mission to the United Nations, 866 United Nations Plaza, Room 531-533 New York, NY, 10017
F: 355 5460, 5461
TL: 838 2819

Hobeika, Samir, H.E. Mr.
Ambassador, Director of Immigration
Ministry of Foreign Affairs
Permanent Mission to the United Nations, 866 United Nations Plaza, Room 531-533
New York, NY 10017, United States of America
T: 355 5460, 5461 F: 838 2819

Jarjoui, Milad, Prof.
Expert on Environment
Permanent Mission of Lebanon
866 United Nations Plaza, Room 531-533
New York, NY 10017, United States of America
T: 355 5460, 5461 F: 838 2819

Mallat, Hiam
Professor Of Environmental Law
UNDA BEIRUT
POB 1352 Limassol - Cyprus
Beyrouth, Lebanon
T: 00 357 5 539146 F: 00 357 5 339084

Saad, Fouad, Dr.
Ministry of Agriculture
Permanent Mission to the United Nations, 866 United Nations Plaza, Room 531-533
New York, NY 10017, United States of America
T: 355 5460, 5461 F: 838 2819

GLOBAL FORUM

Ghougassian, Boghos, Mr.
Coordinator
MECTAT
P.O. Box 113-5474
Beirut, Lebanon
T: 34-1332
TL: 41224-LE

Khawlie, Mohamad, Mr.
Chairman
AMERICAN UNIVERSITY OF BEIRUT
Geology Department - A.U.B.
Beirut, Lebanon
T: 350000
TL: 2080/LE

Mikirdltsian, Mikirdits, Dr.
ARMENIAN PRELACY OF LEBANON
P. O. Box 409
Beirut, Lebanon

Nasr, Julinda Abu, Dr.
Director, Women'S Studies
BEIRUT UNIVERSITY COLLEGE
Beirut, Lebanon

PARLIAMENTARIANS / SPIRITUAL LEADERS

Maouad, Nayla, Dep.
Assemblee Nationale
Beyrouth, Lebanon
T: 388175
TL: 23596 ASS NAT LE

LESOTHO - GOVERNMENT

Ramaema, Elias, Mr.
Prime Minister
Permanent Mission of the Kingdom of Lesotho
204 E. 39th Str.
New York NY 10016, United States of America
T: 212 661 1690 F: 212 682 4388

PRESS

Let'Soara, Rankolopane
Editor
CHRISTIAN COUNCIL OF LESOTHO
P.O. Box 2055
Maseru 102 - Lesotho, Lesotho

GLOBAL FORUM

Matela, Lipalesa, Ms.
PLENTY LESOTHO
Palais de Nations
Geneve CH 1211, Lesotho
F: 41 22 733-6542

Sello, Caleb, Mr.
LESOTHO COUNCIL OF NGO'S
Palais de Nations
Geneve CH 1211, Lesotho
F: 41 22 733-6542

LIBERIA - GOVERNMENT

Reeves, Z. Moulai B.
Senior Economist(Deputy Ministe)
MINISTRY OF PLANNING & ECONOMIC AFF
P.O. Box 9016
Monrovia, Liberia
T: 221 532 222 086

PARLIAMENTARIANS / SPIRITUAL LEADERS

Kuyon, Bismarck N., Mr
MP
The Senate Capitol
Monrovia, Liberia
T: 225210 OR 225006

LIBYA ARAB JAMAHIRIYA - GOVERNMENT

Abdulsami Sherif, Abdul Gader
Professor Of Toxicology
AND CHAIRMAN OF LIBYAN
Environment Society P.O. Box 14147
Tripoli, Libya Arab Jamahiriya
T: FAX 39307 F: 39307 905989

Al-Aujali, Ali Suleiman, H.E. Mr.
Ambassador
Secretary of the Libyan People's Bureau in Brazil
Permanent Mission to the United Nations, 309-315 East 48th Street New York, NY, 10017
F: 752 5772
TL: 593 4784

Al-Mohrak, Yusif Ibrahim, Dr.
Secretary
Technological Centre for Environment Preservation (Deputy Head of Delegation)
Permanent Mission to the United Nations, 309-315 East 48th Street New York, NY, 10017
F: 752 5772
TL: 593 4784

Al-Sebai, Mustafa, Mr.
Strategic Industries Secretariat
Permanent Mission to the United Nations, 309-315 East 48th Street New York, NY, 10017
F: 752 5772
TL: 593 4784

Al-Shridy, Mahamad Moamar, Mr.
Treaty and Legal Affairs Administration
Permanent Mission to the United Nations, 309-315 East 48th Street New York, NY, 10017
F: 752 5772
TL: 593 4784

Attaihi, Jadallah Azuz, H.E. Mr.
Secretary
Strategic Industries (Head of Delegation)
Permanent Mission to the United Nations, 309-315 East 48th Street New York, NY, 10017
F: 752 5772
TL: 593 4784

Attiari, Mahamada, Eng.
Petrol Research Centre
Permanent Mission to the United Nations, 309-315 East 48th Street New York, NY, 10017
F: 752 5772
TL: 593 4784

El-Swei, Mohamed Abdalla, Mr.
Libyan People's Bureau in Brazil
Permanent Mission to the United Nations, 309-315 East 48th Street New York, NY, 10017
F: 752 5772
TL: 593 4784

Elhouderi, Ali Ahmed, H.E. Dr.
Ambassador Extraordinary and Plenipotentiary
Permanent Representative of the United Nations, New York
Permanent Mission to the United Nations, 309-315 East 48th Street
New York, NY 10017, United States of America
T: 752 5772 F: 593 4784

Elmehrik, Yusef
Director General
TECHNICAL CENTRE FOR ENVIRONMENT PR
P.O. Box 3545,Tripoli,Libya
Tripoli, Libya Arab Jamahiriya
T: 48452 45795 48668

Shaban, Attahir Umran Bin, Mr.
International Organizations Administration
Permanent Mission to the United Nations, 309-315 East 48th Street New York, NY, 10017
F: 752 5772
TL: 593 4784

PRESS

Abi Habib, Vera
Correspondent
Libyan Arab Jamahiriya
Cyprus, Libya Arab Jamahiriya

Al-Hariri, Rania
Correspondent
LIBYAN ARAB JAMAHIRIYA
Cyprus
Libya Arab Jamahiriya

Hamed, Farag Saad
Reporter Radio/Television
LIBYAN ARAB JAMAHIRIYA
Tripoly, Libya Arab Jamahiriya

GLOBAL FORUM

Alharill, Nidal Ahmed, Mr.
ENDA
Al Aide Publisher
Tipoly, Libya Arab Jamahiriya
F: 218 5 366494
TL: 218 5 369928, E: 6251 AZ

Kdai, Omar, Mr.
P.O.Box 4814
Tripoli 4814, Libya Arab Jamahiriya
T: 218 21 49294

LIECHTENSTEIN - GOVERNMENT

Nascher, Felix, Mr.
Head
National Office for Forestry
Permanent Mission to the United Nations, 405 Lexington Avenue, 43rd Floor
New York, NY 10174, United States of America
T: 599 0220 F: 599 0064

LITHUANIA - GOVERNMENT

Juceryte, Dalia, Ms.
Representative
Lithuanian Greens Movement
Permanent Mission to the United Nations, 41 W. 82nd Str.
New York, NY 10024, United States of America
T: 212 877 4552 F: 212 595 8301

Kiskis, Inesis, Mr.
Foreign Relations Division
Environmental Protection Department
Permanent Mission to the United Nations, 41 W. 82nd Str.
New York, NY 10024, United States of America
T: 212 877 4552 F: 212 595 8301

Landsbergis, Vytautas, H.E. Mr.
President of the Supreme Council
Republic of Lithuania
41 W. 82nd Str.
New York, NY 10024, United States of America
T: 212 877 4552 F: 212 595 8301

Simenas, Jonas, Mr.
Member of Parliament, Chairman
Environmental Protection Committee
Permanent Mission to the United Nations, 41 W. 82nd Str.
New York, NY 10024, United States of America
T: 212 877 4552 F: 212 595 8301

Suziedelis, Darius, Mr.
Counsellor
Permanent Mission of Lithuania
41 W. 82nd Str.
New York, NY 10024, United States of America
T: 212 877 4552 F: 212 595 8301

Vaisvilla, Zigmas, H.E. Mr.
Deputy Prime Minister
Permanent Mission of Lithuania
41 W. 82nd Str.
New York, NY 10024, United States of America
T: 212 877 4552 F: 212 595 8301

Vebra, Evaldas, Mr.
Director
Lithuanian Environmental Protection Department
Permanent Mission to the United Nations, 41 W. 82nd Str.
New York, NY 10024, United States of America
T: 212 877 4552 F: 212 595 8301

LUXEMBOURG - GOVERNMENT

Bodry, Alex, S.E. M.
Ministre de l' Amenagement du Territoire et de l' Environnement
Permanent Mission of Luxembourg
801 2nd Ave.
New York, NY 10017, United States of America
T: 212 370 9850 F: 212 922 1685

Feyder, Jean, S.E. M.
Ambassadeur Extraordinaire et Plenipotentiaire
Representant Permanent aupres des Nations Unies, New York
Permanent Mission to the United Nations, 801 2nd Ave.
New York, NY 10017, United States of America
T: 212 370 9850 F: 212 922 1685

Franck, Claude, M.
Conseiller de Direction
Ministere de l' Environnement
Permanent Mission to the United Nations, 801 2nd Ave. New York, NY, 10017
F: 212 370 9850
TL: 212 922 1685

Haine, Henri, M.
Attache d' administration
Ministere de l' Environnement
Permanent Mission to the United Nations, 801 2nd Ave. New York, NY, 10017
F: 212 370 9850
TL: 212 922 1685

Heinen, Georges, M.
Ministere du Tresor
Permanent Mission of Luxembourg
801 2nd Ave.
New York, NY 10017, United States of America
T: 212 370 9850 F: 212 922 1685

Kollwelter, Rene, M.
President
Commission de l' Environnement, Chambre des Deputes
Permanent Mission to the United Nations, 801 2nd Ave. New York, NY, 10017
F: 212 370 9850
TL: 212 922 1685

Santer, Jacques, S.E. M.
Premier Ministre
Permanent Mission of Luxembourg
801 2nd Ave.
New York NY 10017, United States of America
T: 212 370 9850 F: 212 922 1685

PRESS

Garcia, Robert, Mr.
Editor
GRENGESPOUN
4, rue Hubert Clement
L-3444 Dudelange, Luxembourg, Luxembourg

GLOBAL FORUM

Ballabio, Rosella, Ms.
Interprete
EUROPEAN PARLIAMENT
9, Rue de Kroentgeshof
Contern L-5312, Luxembourg
T: 00 352 359630

Degreve, Rita Raum, Ms.
EURECO
Luxembourg, Luxembourg

Earley, Elizabeth, Ms.
Member
SAVE THE RAIN FOREST FOUNDATION
BP 1147
Luxembourg L-1011, Luxembourg
T: 35 2 48-5050 F: 35 2 48-1385

Gerardin, Bernard, Mr.
Representative
EUROPEAN INVESTMENT BANK
100 Boulevard Konrad Adenaver
Luxembourg 2950, Luxembourg
T: 35 2 437-97439 F: 35 2 437-704
TL: 3530 BNKE

Ryckevorsel, Maurits Van, Mr.
Member
SAVE THE RAIN FOREST FOUNDATION
6, Rue Heine
Luxembourg L-1011, Luxembourg
T: 35 2 48-5050 F: 35 2 48-1385

Turmes, Claude, Mr.
Campaigner
FOE LUXEMBOURG
39, r. Fort Neigpperj L-2230
Luxembourg, Luxembourg
T: 35 2 400427 F: 35 2 405849

Zamaron, Alix, Ms.
Delegue General
ECOROPA / INSTITUTE SCHUMAN POUR L'EUROPE
Esvscherange
Luxembourg L9747, Luxembourg
T: 35 2 91407 F: 35 2 91405

Zamaron, Bernard, Mr.
Deleque General
ECOROPA / INSTITUTE SCHUMAN POUR L'EUROPE
Esvscherange
Luxembourg L9747, Luxembourg
T: 35 2 91407 F: 35 2 91405

MACAU - GLOBAL FORUM

Tsui, Lawrence
ICAE
12 Rajaji Marg
New Delhi 110011, Macau
T: 91 11 301-4309

MADAGASCAR - GOVERNMENT

Rabary, Victor, M.
Directeur General de l'Office National de l'Environnement
Permanent Mission of Madagascar
801 2nd Ave, Suite 404
New York, NY 10017, United States of America
T: 212 986 9491 F: 212 986 6271

Rabevazaha-Rabekoriana, Celine
Economiste, Secretaire General
A.P.E.M.
Rue Rajakoba Augustin
Anhananarivo 101, Madagascar
T: 217 6 3

Rakotonarivo, Georges Henri, Mr.
Chef Du Service Des Ecosystemes Forestiers
DIRECTION DES EAUX ET FORET
B.P. 243
Antananarivd, Madagascar

Rakotovahiny, Emmanuel, S.E. M.
Ministre d'Etat a l'Agriculture et au developpement rural (Chef de Delegation Adjoint)
Permanent Mission of Madagascar
801 2nd Ave, Suite 404
New York, NY 10017, United States of America
T: 212 986 9491 F: 212 986 6271

Ralimanga, Viviane Sosoa, Ms
Coordinatrice Des Aides
OFFICE NATIONAL DE L'ENVIRONNEMENT
Lot IG161A-Ambalavaoisotry Anta
Antananarivo 101, Madagascar
T: 25999 F: 30693

Ranaivoharisoa, Lala, Mme.
Conseiller du Premier Ministre en matiere d'Environnement
Permanent Mission of Madagascar
801 2nd Ave, Suite 404
New York, NY 10017, United States of America
T: 212 986 9491 F: 212 986 6271

Razanamasy, Guy Willy, S.E. M.
Premier Ministre et Chef du Gouvernement
Republique democratique de Madagascar
801 2nd Ave, Suite 404
New York, NY 10017, United States of America
T: 212 986 9491 F: 212 986 6271

GLOBAL FORUM

Albignac, Roland, Mr.
UNESCO/PNUD
Pnnd BP 1348
Tananarive, Madagascar
T: 35-359 F: 35-370

Rabenirina, Remi Joseph, Mr.
EKLESIA EPISKOPALY MALAGSY
Eveche Anglican, Ambohimanoro
Antananarivo 101, Madagascar

Rabevazaha, Celine, Ms.
ENDA TIERS MONDE
VU 142 Ambohipotsy
Antananarivo, Madagascar

Rajeriarison, , Ms.
Professor University
GLOBAL 500
32 Cite des Professeurs
Tananarine, Madagascar

Rakotozafy, Jean Martin, Mr.
C A P R
P.P. 40,301 - Fianarantsoa
Madagascar, Madagascar

Ramaroharinosy, William, Mr.
COMODE
BP 3875
Antananarivo, Madagascar
T: 26 12 40525

Ranaivoson, Charles, Mr.
President
PVO-NGO/NRMS PROJECT
BP 8357
Antananarivo, Madagascar
T: 26 12 34 1 93 F: 26 12 40 5 61

MALAWI - GOVERNMENT

Chilingulo, Richard, Mr
Chief Industrial Dev. Officer
MINISTRY OF TRADE AND INDUSTRY
P.O. Box 30366
Lilongwe 3, Malawi
T: 732711

Juma, Arnold Mwanapwa
Senior Economist (energy)
DEPT OF ECONOMIC PLANNING
And Development P.O. Box 30136
Lilongwe 3, Malawi
T: 265 731 777 F: 265 731 669

Kawonga, Harned Penjan
Deputy Secetary
MINISTRY OF FINANCE MARAW GOVERNMEN
P O Box 30049
Lilongwe 3, Malawi
T: 731311 734 175

Matenje, Stephen Dick Tennyson, Mr
Parliamentary Draftsman
MINISTRY OF JUSTICE
Private BAG 333
Lilongwe 3, Malawi
T: (265) 731 533OFFICE F: (265) 731 776

Mtonakutha, Sadwick Limbikani
Economist
GOVERNMENT
Economic Planning And Development P.O. Box 30136, Lildngwe 3, Malawi
Lildngwe, Malawi
T: 731777 F: 265 731669

Sefu, Leonard Donald, Mr
Wildlife Officer
DEPT. OF NATIONAL PARKS & WILDLIFE
P.O. Box 30131
Lilongwe 3, Malawi
T: 265 723 566 676 F: 265 723 089

GLOBAL FORUM

Mkolesia, P., Mr.
MALAWI COUNCIL FOR SOCIAL WELFARE SERVICES
Palais de Nations
Geneve CH 1211, Malawi
F: 41 22 733-6542

Nandolo, E., Mr.
CHRISTIAN SERVICE COMMITTEE
Palais de Nations
Geneve CH 1211, Malawi
F: 41 22 733-6542

Ndisale, Jessie, Ms.
PAN AFRICAN GREEN BELT MOV.
Muluwuuzi Sec. School, PO Box 138
Zomba, Malawi
T: 523053

MALAYSIA - GOVERNMENT

Abdul, Jabar Mohamad, Director
External Information Division
Ministry of Foreign Affairs
Permanent Mission to the United Nations, 140 E. 45th Str., 43rd Floor New York, NY, 10017
F: 212 986 6310
TL: 212 490 8576

Abdul, Malek Hanafiah, H.E. Dato Dr.
Political Secretary
Permanent Mission of Malaysia
140 E. 45th Str., 43rd Floor
New York, NY 10017, United States of America
T: 212 986 6310 F: 212 490 8576

Abdul, Ranman Jamal, Director
Regional Development Economic Planning Unit
Prime Minister's Department
Permanent Mission to the United Nations, 140 E. 45th Str., 43rd Floor New York, NY, 10017
F: 212 986 6310
TL: 212 490 8576

Abdul Aziz bin Ismail, Dato, H.E.
Principal Private Secretary to the Prime Minister
Permanent Mission of Malaysia
140 E. 45th Str., 43rd Floor
New York, NY 10017, United States of America
T: 212 986 6310 F: 212 490 8576

Abdul Taib Mahmud, Tan Sri Datuk Patinggi, The Hon.
Chief Minister of Sarawak
Permanent Mission of Malaysia
140 E. 45th Str., 43rd Floor
New York, NY 10017, United States of America
T: 212 986 6310 F: 212 490 8576

Abdullah Gin Ahmad Badawi, Datuk, The Hon.
Minister for Foreign Affairs (Deputy Head of Delegation)
Permanent Mission of Malaysia
140 E. 45th Str., 43rd Floor
New York, NY 10017, United States of America
T: 212 986 6310 F: 212 490 8576

Affifuddin bin Haji Omar, Dato, The Hon.
Member of Parliament
Permanent Mission to the United Nations, 140 E. 45th Str., 43rd Floor New York, NY, 10017
F: 212 986 6310
TL: 212 490 8576

Ahmad, Mokhtar Selat, Mr.
Charge d'Affaires, Embassy of Malaysia in Venezuela
Permanent Mission of Malaysia
140 E. 45th Str., 43rd Floor
New York, NY 10017, United States of America
T: 212 986 6310 F: 212 490 8576

Ali Abul Hassan Sulaiman, Datuk, H.E.
Director General
Economic Planning Unit
Prime Minister's Department
T: 10017
TL: 212 986 6310, E: 212 490 8576

Amdan, Mat Din, Mr.
Senior Federal Counsel
Attorney General's Chamber's
Permanent Mission to the United Nations, 140 E. 45th Str., 43rd Floor New York, NY, 10017
F: 212 986 6310
TL: 212 490 8576

Amha, Buang, Principal Assistant Secretary
Ministry of Primary Industries
Permanent Mission of Malaysia
140 E. 45th Str., 43rd Floor
New York, NY 10017, United States of America
T: 212 986 6310 F: 212 490 8576

Ariffin bin Raja Sulaiman, Dato Raja, The Hon.
Deputy Minister
Prime Minister's Department
Permanent Mission to the United Nations, 140 E. 45th Str., 43rd Floor New York, NY, 10017
F: 212 986 6310
TL: 212 490 8576

Bin Mohamed, Mahathir, Mr.
Prime Minister
Permanent Mission of Malaysia
140 E. 45th Str., 43rd floor
New York NY 10017, United States of America
T: 212 986 6310 F: 212 490 8576

Bofy, Park Seng, Mr.
Third Secretary
Embassy to Brazil
Permanent Mission to the United Nations, 140 E. 45th Str., 43rd Floor New York, NY, 10017
F: 212 986 6310
TL: 212 490 8576

Clifford, Herbert, Mr.
Deputy Secretary General, Ministry of Finance
Permanent Mission of Malaysia
140 E. 45th Str., 43rd Floor
New York, NY 10017, United States of America
T: 212 986 6310 F: 212 490 8576

Danabalan, V., Mr.
Secretary General
Ministry of Science, Technology and Environment
Permanent Mission to the United Nations, 140 E. 45th Str., 43rd Floor New York, NY, 10017
F: 212 986 6310
TL: 212 490 8576

David, Kala, Mr.
Representative of Penan Community
Permanent Mission of Malaysia
140 E. 45th Str., 43rd Floor
New York, NY 10017, United States of America
T: 212 986 6310 F: 212 490 8576

Fauziah, Mohd. Taib, Ms.
Principal Assistant Secretary, Ministry of Foreign Affairs
Permanent Mission of Malaysia
140 E. 45th Str., 43rd Floor
New York, NY 10017, United States of America
T: 212 986 6310 F: 212 490 8576

Gooi, Hoe Hin, Mr.
Private Secretary to the Minister of Primary Industries
Permanent Mission of Malaysia
140 E. 45th Str., 43rd Floor
New York, NY 10017, United States of America
T: 212 986 6310 F: 212 490 8576

Hamidah, Mohd. Yusof, Ms.
Special Officer to the Prime Minister
Permanent Mission of Malaysia
140 E. 45th Str., 43rd Floor
New York, NY 10017, United States of America
T: 212 986 6310 F: 212 490 8576

Hashim, Makaruddin, Mr.
Press Secretary to the Prime Minister
Permanent Mission of Malaysia
140 E. 45th Str., 43rd Floor
New York, NY 10017, United States of America
T: 212 986 6310 F: 212 490 8576

Hassan, Sui, Mr.
Representative of Penan Community
Permanent Mission of Malaysia
140 E. 45th Str., 43rd Floor
New York, NY 10017, United States of America
T: 212 986 6310 F: 212 490 8576

Hayati, Ismail, Ms.
Assistant Secretary
Ministry of Foreign Affairs
Permanent Mission to the United Nations, 140 E. 45th Str., 43rd Floor New York, NY, 10017
F: 212 986 6310
TL: 212 490 8576

Hussein, Haniff
Counsellor
Permanent Mission of Malaysia
140 E. 45th Str., 43rd Floor
New York, NY 10017, United States of America
T: 212 986 6310 F: 212 490 8576

Ibrahim Ali, , The Hon.
Member of Parliament
Permanent Mission to the United Nations, 140 E. 45th Str., 43rd Floor New York, NY, 10017
F: 212 986 6310
TL: 212 490 8576

Jalaludin, Bahaudin, Mr.
Special Assistant to the Prime Minister
Permanent Mission of Malaysia
140 E. 45th Str., 43rd Floor
New York, NY 10017, United States of America
T: 212 986 6310 F: 212 490 8576

Jamal, Abdul Rahman, Mr
Director
ECONOMIC PLANNING UNIT PMS DEPT
Jalan Dato Onn 50502 Kilhmphr
Malaysia
T: 603 293 3414 F: 603 291 4268

James, Dawos Mamit, Senior Forestry Officer
Forestry Department, Sarawak
Permanent Mission of Malaysia
140 E. 45th Str., 43rd Floor
New York, NY 10017, United States of America
T: 212 986 6310 F: 212 490 8576

Jee, Guan Keen, Mr.
Communication Officer
Sarawak
Permanent Mission to the United Nations, 140 E. 45th Str., 43rd Floor New York, NY, 10017
F: 212 986 6310
TL: 212 490 8576

Jini, Wat, Mr.
Private Secretary to the Minister of Science, Technology and Environment
Permanent Mission of Malaysia
140 E. 45th Str., 43rd Floor
New York, NY 10017, United States of America
T: 212 986 6310 F: 212 490 8576

Kamarulzaman, Zainal, Mr.
Secretary to the Minister for Foreign Affairs
Permanent Mission of Malaysia
140 E. 45th Str., 43rd Floor
New York, NY 10017, United States of America
T: 212 986 6310 F: 212 490 8576

Law Hieng Ding, , The Hon.
Minister of Science, Technology and Environment
Permanent Mission of Malaysia
140 E. 45th Str., 43rd Floor
New York, NY 10017, United States of America
T: 212 986 6310 F: 212 490 8576

Leo, Chai, H.E. Datuk
Director
Sarawak Forestry Department
Permanent Mission to the United Nations, 140 E. 45th Str., 43rd Floor New York, NY, 10017
F: 212 986 6310
TL: 212 490 8576

Lim Keng Yaik, Dato Seri, The Hon.
Minister of Primary Industries
Permanent Mission of Malaysia
140 E. 45th Str., 43rd Floor
New York, NY 10017, United States of America
T: 212 986 6310 F: 212 490 8576

Looi Cheok Hun, , Mr. H.E.
Ambassador Extraordinary and Plenipotentiary to Brasil
Permanent Mission of Malaysia
140 E. 45th Str., 43rd Floor
New York, NY 10017, United States of America
T: 212 986 6310 F: 212 490 8576

M.M. Sathiah, Dato, H.E.
Deputy Secretary General II
Ministry of Foreign Affairs
Permanent Mission to the United Nations, 140 E. 45th Str., 43rd Floor New York, NY, 10017
F: 212 986 6310
TL: 212 490 8576

Mahathir bin Mohamad, Dato's Seri, The Hon.
Prime Minister
Malaysia
140 E. 45th Str., 43rd Floor
New York, NY 10017, United States of America
T: 212 986 6310 F: 212 490 8576

Michael Lisa Kaya, , The Hon.
Member of Parliament
Permanent Mission to the United Nations, 140 E. 45th Str., 43rd Floor New York, NY, 10017
F: 212 986 6310
TL: 212 490 8576

Mohamad Ali, Mr.
ADC the Prime Minister
Permanent Mission of Malaysia
140 E. 45th Str., 43rd Floor
New York, NY 10017, United States of America
T: 212 986 6310 F: 212 490 8576

Mohamed Abdul Kadir, Syed Syed, Dr.
Deputy Under Secretary
Ministry of Finance
Permanent Mission to the United Nations, 140 E. 45th Str., 43rd Floor New York, NY, 10017
F: 212 986 6310
TL: 212 490 8576

Mustafa, Ong, Mr.
Second Secretary
Embassy to Brazil
Permanent Mission to the United Nations, 140 E. 45th Str., 43rd Floor New York, NY, 10017
F: 212 986 6310
TL: 212 490 8576

Noordin, Mohd. Bachik, Mr.
Personal Secretary to the Prime Minister
Prime Minister's Department
Permanent Mission to the United Nations, 140 E. 45th Str., 43rd Floor New York, NY, 10017
F: 212 986 6310
TL: 212 490 8576

Ong Tee Kiat, , The Hon.
Deputy Speaker
House of Representatives
Permanent Mission to the United Nations, 140 E. 45th Str., 43rd Floor New York, NY, 10017
F: 212 986 6310
TL: 212 490 8576

Rahim, Dahlan, Mr.
Director
Prime Minister's Department
Permanent Mission to the United Nations, 140 E. 45th Str., 43rd Floor New York, NY, 10017
F: 212 986 6310
TL: 212 490 8576

Razali Esmail, Tan Sri, H.E.
Ambassador Extraordinary and Plenipotentiary
Permanent Mission of Malaysia
140 E. 45th Str., 43rd Floor
New York, NY 10017, United States of America
T: 212 986 6310 F: 212 490 8576

Rozali, Mohamed Ali, Dr.
Assistant Director General, Institute of Strategic and International Studies
Permanent Mission of Malaysia
140 E. 45th Str., 43rd Floor
New York, NY 10017, United States of America
T: 212 986 6310 F: 212 490 8576

Samsudin, Marsop, Mr.
Director
Industry Division, Ministry of International Trade and Industry
Permanent Mission to the United Nations, 140 E. 45th Str., 43rd Floor New York, NY, 10017
F: 212 986 6310
TL: 212 490 8576

Sudirman Ahmad, Haji, Mr.
Director
Prime Minister's Department
Permanent Mission to the United Nations, 140 E. 45th Str., 43rd Floor New York, NY, 10017
F: 212 986 6310
TL: 212 490 8576

Tan, K.K.
C.E.O
MTIDC
9B 9TH Floor Bangnnal Arab Malaysia Taleq Raja Chulan 50200 Knala Lumpa
Kuala Lumpar, Malaysia
T: 603 2323999 F: 603 2386376

Tan, Meng Leng, Mr.
Deputy Director General
Ministry of Science, Technology and Environment
Permanent Mission to the United Nations, 140 E. 45th Str., 43rd Floor New York, NY, 10017
F: 212 986 6310
TL: 212 490 8576

Thanarajasingam, S., Mr.
Under Secretary (Economic), Ministry of Foreign Affairs
Permanent Mission to the United Nations, 140 E. 45th Str., 43rd Floor New York, NY, 10017
F: 212 986 6310
TL: 212 490 8576

Thang, Hooi Chiew, Mr.
Director
Forestry Department, Ministry of Primary Industries
Permanent Mission to the United Nations, 140 E. 45th Str., 43rd Floor
New York, NY 10017, United States of America
T: 212 986 6310 F: 212 490 8576

Ting, Wen Lian, Ms.
Ambassador Extraordinary and Plenipotentiary to Rome and FAO
Permanent Mission of Malaysia
140 E. 45th Str., 43rd Floor
New York, NY 10017, United States of America
T: 212 986 6310 F: 212 490 8576

Wong Kum Choon, Tan Sri Dato, H.E.
Secretary General
Ministry of Primary Industries
Permanent Mission to the United Nations, 140 E. 45th Str., 43rd Floor New York, NY, 10017
F: 212 986 6310
TL: 212 490 8576

Zainal, Abidin Abdul Hamid, Mr.
Prime Minister's Department
Permanent Mission to the United Nations, 140 E. 45th Str., 43rd Floor New York, NY, 10017
F: 212 986 6310
TL: 212 490 8576

Zulkifli Abdul Rahman, Dato, Director
Ministry of Home Affairs
Permanent Mission of Malaysia
140 E. 45th Str., 43rd Floor
New York, NY 10017, United States of America
T: 212 986 6310 F: 212 490 8576

NGO

Chee, Yoke Ling
SAM
43 Salween Road
10050 Penang, Malaysia

Shiva, Vandana
TWN
87 Cantonmont Road
10250 Penang, Malaysia

PRESS

Guan Oo, Kuah
Editor
BERNAMA
120 Jalan Desa Mesra, T.D.
Kuala Lumpur 58100, Malaysia, Malaysia

Jabar, Abdul Mohamad
GOVERNMENT PRESS
Ministry Foreign Affairs
Kuala Lumpur, Malaysia

Kamarulshaid, Mior
Editor
BERITA HARIAN
31 Lorong Kanan Satu, Kampun P
Kuala Lumpur, Malaysia, Malaysia

Khoo, Swee See
Correspondent
THIRD WORLD NETWORK
39 Sdok, Tembaga Satu,
Island Park 11600, Penang, Malaysia, Malaysia

Mathews, Philip
Editor
ASIA-PACIFIC
18 Lengkok Zaaba, Taman Tun
60000 Kuala Lumpur, Malaysia, Malaysia

Mustapa, Ismail Bin
Editor
RADIO TV MALAYSIA
77 Medan Athinahapan Dua
60000 Kuala Lumpur, Malaysia, Malaysia

Poole, Ah Lek
Correspondent
NEWSPAPER SIN CHEW JIT POH
3694 Jalan 18/57
Selangor, Malaysia, Malaysia

Raghavan, Chakravarthi
Editor
THIRD WORLD NETWORK
2 Rue des Barbues
CH-1207 Geneva, Switzerland, Malaysia

Salleh, Nordin Salleh
Editor
RADIO TV MALAYSIA
1G, Jalan Pandan Indah 1/16
55100 Kuala Lumpur, Malaysia, Malaysia

ORGANIZATIONS

How, Agnes, Ms.
As. of Development Research and Training Institutes of Asia and the Pacific
c/o APDC, Pesiaran Duta, PO Box 12224
50770 Kuala Lumpur, Malaysia
T: (60 3) 254 8088 F: (60 3) 255 0316

Khor, Martin, Mr.
Third World Network
87 Cantonmont Road
10250 Penang, Malaysia
T: (60 4) 37 3511 F: (60 4) 36 8106

Ling, Chee Yoke, Ms.
Sahabat Alam Malaysia
43 Salween Road
10050 Penang, Malaysia
T: (60 4) 376930 F: (60 4) 368106

Malek, Hasan Bin, Mr.
Asian Youth Council
52 International Youth Centre, Jalan Tenteram
Bandar Tun Raza, Cheras 56000 Kuala Lumpur, Malaysia
T: (60 3) 971 7373 F: (60 3) 971 2296

GLOBAL FORUM

Badariah, Arshad, Ms.
ENVIRONMENTAL ANALYSIS, MALAYSIA
Kuala Lumpur, Malaysia
T: 60 3 689-3444

Barrera, Cristina, Ms.
INTERNATIONAL ORGANIZATION OF CONSUMERS UNIONS - IOCU
P.O.Box 1045
Penang 10830, Malaysia
T: 60 4 371-396/371-318
TL: MA 40164 APIOCU, E: GE02IOCU (GEONET)

Chin, Saik Yoon, Mr.
THIRD WORLD NETWORK
87 Cantonment Road
Penang 10250, Malaysia
T: 60 4 373511 F: 60 4 368106
TL: MA40989 CAPPG

Dall, Rinta Ak, Ms.
INDIGENOUS PROGRAM
c/o Rainforst Information Centre, P. O. Box 368
Lismore - N.S.W. 2480, Malaysia
T: 61 66 218505 F: 61 66 222339

Davison, Geoffrey, Mr.
President
W W F MALAYSIA
Locked Bag No. 911, Jalan Sultan PO
Petaling Jaya 46990, Malaysia
F: 60 9 7565594

George, Susanna, Ms.
DEVELOPMENT ALTERNATIVES WITH WOMEN FOR A NEW ERA (DAWN)
c/o Asia & Pacific Dev. Centre 50770
Kuala Lumpur, Malaysia

Hiratsuka, Shuei, Mr.
INTERNATIONAL ORGANIZATION OF CONSUMERS UNION
PO Box 1045
Penang 10830, Malaysia
T: 60 4 3711396 F: 60 4 366506
TL: Ma40164 APIOCU

Jalong, Thomas, Mr.
SAHABAT ALAM MALAYSIA
43 Salween Road
Penang 10050, Malaysia
T: 60 4 376930 F: 60 4 375705
TL: MA40989 CAPPG

Jantan, Wahab, Mr.
MALAYSIA ENVIRONMENTAL GROUP
168 Petaling Jaya
Selangor, Malaysia

Lajo, Paul Joseph, Mr.
INDIGENOUS PROGRAM
c/o Rainforst Information Centre, P. O. Box 368
Lismore - N.S.W. 2480, Malaysia
T: 61 66 218505 F: 61 66 222339

Lee, Wee You
Journalist
TV3 MALAYSIA
18, Jacan Liku
Bangsar, Malaysia
T: 2826066

Limat, Yon
Journalist
TV3 MALAYSIA
18, Jacan Liku
Bangsar, Malaysia
T: 2826066

Mudir, Abdul Rahim
Cameraman
TV3 MALAYSIA
18, Jacan Liku
Bangsar, Malaysia
T: 2826066

Nang, Wong Sing, Mr.
Lawyer
INSTITUTE PENGAJARAN KOMUNITI
P.O.Box 8
Sibu, Sarawak 96007, Malaysia
T: 60 84 333367 F: 60 84 333368

Nasir, Gamal
Journalist
UTUSAN MALAYSIA
Jalan Chan Sow Lin,
55200 Kuala Lumpur, Malaysia
T: 603 2217055 F: 603 2227876

Omar, Adhari
Cameraman
TV3 MALAYSIA
18, Jacan Liku
Bangsar, Malaysia
T: 2826066

Ong, Mustapha, Mr.
Observer
ENVIROMENTAL GROUP
45 JLN. SS 21/21, Damansara Letama, Petaling Jaya
Selangor, Malaysia

Sindu, Patrick, Mr.
President
CONSUMERS' ASSOCIATION OF SABAH (CASH)
PO Box 14859, 88855 Kota Kinabalu, Sabah, East Malaysia
Kota Kinabalu, Malaysia
T: 60 88 234616 F: 60 88 214144

Souza, Denise De, Ms.
Observer
ENVIROMENTAL GROUP
45 JLN. SS 21/21, Damansara Letama, Petaling Jaya
Selangor, Malaysia

Teck Ghee, Lim, Mr.
Professor
UNIVERSITY OF MALAYA
Kuala Lumpur, Malaysia

Teoh, Cheng Hai, Mr.
GOLDEN HOPE
Prang Besar Research Station
Kajang - Selangor 43009, Malaysia
T: 3 825-7597 F: 825-7598

Tungang, Yahuda, Mr.
INDIGENOUS PROGRAM
c/o Rainforst Information Centre, P. O. Box 368
Lismore - N.S.W. 2480, Malaysia
T: 61 66 218505 F: 61 66 222339

Wan, Mujan, Ms.
INDIGENOUS PROGRAM
c/o Rainforst Information Centre, P. O. Box 368
Lismore - N.S.W. 2480, Malaysia
T: 61 66 218505 F: 61 66 222339

Yusoff, Hamidah, Ms.
ENVIRONMENTAL ANALYSIS MALAYSIA
Kuala Lumpur, Malaysia
T: 60 3 689-3444

PARLIAMENTARIANS / SPIRITUAL LEADERS

Ali, Ibrahim, Mr
MP
Dewan Negara Bangunan Parlimen Parliament House
Kuala Lumpur 10 02, Malaysia
T: 2321955

Kaya, Michael Lisa
MP
Dewan Negara Bangunan Parlimen Parliament House
Kuala Lumpur 10 02, Malaysia
T: 2321955

Kiat, Ong Tea
MP
Dewan Negara Bangunan Parlimen Parliament House
Kuala Lumpur 10 02, Malaysia
T: 2321955

Omar, Afifuddin Haji, Dr
MP
Dewan Negara Bangunan Parlimen Parliament House
Kuala Lumpur 10 02, Malaysia
T: 2321955

MALDIVES - GOVERNMENT

Gayoom, Maumoon Abdul, Mr.
President
Permanent Mission of the Republic of the Maldives
820 2nd Ave., Suite 800c
New York NY 10017, United States of America
T: 212 599 6195 F: 212 972 3970

PRESS

Rafeeq, Ali
Editor
HAVEERY
H. Neel Villa, Male 20-06
Rep. of Maldives, Maldives

MALI - GOVERNMENT

Berthe, Abou Lamine
Directeur Adjoint
DIRECTION NATIONALE EAUX ET FORETS
BP 275 B
Bamako, Mali
T: 2258 50 225973

Dembele, Aly, M.
Directeur
National de l'Hydraulique et de l'Energie
BP 66
Bamako, Mali
T: 223 222588 222589

Diallo, Moulaye, M.
Conseiller Technique
Ministre de l'Agricultrue, de l'Elevage et de l'Environnement
Permanent Mission to the United Nations, 111 East 69th Street New York, NY, 10021
F: 737 4150, 794 1311
TL: 472 3778

Haidara, Lansiri Nana, Mme
Secretariat d'Etat a l'action sociale et a la Promotion Feminine
Permanent Mission of Mali
111 East 69th Street
New York, NY 10021, United States of America
T: 737 4150, 794 1311 F: 472 3778

Kaba, Sory Ibrahima, M.
Directeur
National de la Sante
Permanent Mission to the United Nations, 111 East 69th Street New York, NY, 10021
F: 737 4150, 794 1311
TL: 472 3778

Konare, Kaliba
Director
METEOROLOGICAL DEPARTMENT
BE 237 Bamako
Bamako, Mali
T: 223 222101 F: 223 222101/224151

Kone, Yassoungo, M.
Premier Conseiller
Ambassade du Mali aux Etats-Unis d'Amerique
Permanent Mission to the United Nations, 111 East 69th Street New York, NY, 10021
F: 737 4150, 794 1311
TL: 472 3778

Sissoko, Sountoucoumba, M.
Counseiller
Technique au Ministere du Plan et de la Cooperation Internationale
Permanent Mission to the United Nations, 111 East 69th Street New York, NY, 10021
F: 737 4150, 794 1311
TL: 472 3778

Toure, Mohamed Alhousseyni, S.E. M.
Ambassadeur du Mali au Bresil avec residence a Washington D.C. (Chef de la Delegation)
Permanent Mission of Mali
111 East 69th Street
New York, NY 10021, United States of America
T: 737 4150, 794 1311 F: 472 3778

NGO

Abdel Kader Ladji, Thera
CS
B.P. 3225,
Bamako, Mali

Aboubacar, Fane
CS
B.P. 3225,
Bamako, Mali

PRESS

Dansoko, Abdel Kader
Editor
L'ESSOR
Rue 578 Lafiabougon
Mali, Mali

Dansoko, Cheik
Editor
L'ESSOR
Rue 578 49
Lafiabouguli Bomaks, Mali, Mali

Maiga, Ousmane M.
Correspondent
L'AMAP
Bamalho, Rep. of Mali
Mali

Sy, A. Saloum
Reporter Radio/Television
RADIO & TV MALI
Karofina Nord
Bamako Mali, Mali

Traore, Modibo B.
Reporter Radio/Television
RTM
BP 171 Bamako
Mali, Mali

ORGANIZATIONS

MacDonald, Ted, Mr.
Carrefour Sahel
B.P. 3225,
Bamako, Mali

GLOBAL FORUM

Abd-El Kader Ladji, Thera, Mr.
President
CARREFOUR SAHEL
BP 3225
Bamako, Mali

Ba, Mariam, Ms.
Presidente
GUAMINA
BP 2744 Bamako
Bamako, Mali
T: 2 23 22-8083 F: 2 23 22-8083

Coulibaly, Daba, Mr.
STOP SAHEL
Bamako, Mali
F: 2 23 22 23 59

Dakouo, Jean, Mr.
Country Coordinator
PVO-NGO/NRMS PROJECT
BP 1721
Bamako, Mali
T: 2 23 22-80-83 F: 2 23 22-23-59

Dicko, Modibo, Mr.
CRAT
Bamako, Mali
F: 2 23 22 02 07

Keita, Maria, Ms.
GROUPMENTS PAYSANS CCA
s/c AFVP BP1721
Bamako, Mali
T: 2 23 22 33 28

Maimouna, Traore, Ms.
Aurore
JOURNALISTE
bp 3150
Bamako, Mali
T: 2 23 22-6922 F: 2 23 22-6922

Thera, Abd-El Kader, Mr.
CARREFOUR SAHEL
B. P. 3225
Bamako, Mali
F: 2 23 22 23 59

Traore, Ibrahima, Mr.
Premier Secretaire
PARTI ECOLOGISTE POUR L'INTEGRATION
Ecole-Normale-Supemieure Bp 241
Bamako, Mali
T: 2 23 222189

Traore, Maimouna, Ms.
WORLDWIDE NETWORK
1331 H Street, Suite 903
Washington - DC 20005, Mali
T: 1 202 347-1514 F: 1 202 347-1524

Yattara, Inamoud, Mr.
Teacher
A M C F E - ASSOC. MALIENNE POUR LA CONSERVATION DE LA FAUNE ET DE SON ENV.
B.P.2921
Bamako, Mali
T: 2 23 22-2189

MALTA - GOVERNMENT

Fenech-Adami, Edward, Mr.
Prime Minister
Permanent Mission of Malta
249 E. 35th. Str.
New York NY 10016, United States of America
T: 212 725 2345 F: 212 779 7097

NGO

Mann Borgese, Elisabeth
IOI
PO Box 524
Valetta, Malta

Mifsud, Joe, Mr.
President
INTERNATIONAL YOUNG NATURE FRIENDS
138 Triq. L-Antinni
Mosta, Malta
T: 356 247855 F: 356 240717

Singh, Bhagwat
IOI
PO Box 524
Valetta, Malta

Yaker, Layachi
IOI
PO Box 524
Valetta, Malta

PRESS

Bondi, Lou
Reporter Radio/Television
MALTA NATIONALIST PRESS
Qasam Street, Swieqi,
Malta, Malta

Bugeja, Nazzareno
Reporter Radio/Television
XANDIR MALTA
Triq. Il-Hobbejza
Rabat, Malta, Malta

Vella, Charles
Editor
MALTA LABOUR PRESS
Block H, Flat 8
Hamrun, Malta, Malta

GLOBAL FORUM

Micallef, Robert, Mr.
PEACE ORGAN. FOR PEOPLE OF EUROPE AND MED.
10 Valletta Buildings
Valletta, Malta
T: 238434 F: 238788

Ragonesi, Rudolf, Sr.
Dr
MOVIMENT GHALL-AMBJENT
P.O.Box 13
Valletta, Malta
T: 356 23-5306 F: 356 22-4379

PARLIAMENTARIANS / SPIRITUAL LEADERS

Warter, Carlos, Sir
MP
House of Representatives
Valletta, Malta
T: 22294
TL: TX MANUAL 403 1100 MODMLT MT

MARSHALL ISLANDS - GOVERNMENT

Aisea, Tony, Mr.
Administrative Officer
Kawjalein Atoll Development Authority
Permanent Mission to the United Nations, Dag Hammerskjold Plaza, 7th Floor
New York, NY 10017, United States of America
T: 212 702 4850 F: 212 207 9888

Balos, Henchi, The Hon.
Minister of Health and Environment
Permanent Mission of Marshall Islands
Dag Hammerskjold Plaza, 7th Floor
New York, NY 10017, United States of America
T: 212 702 4850 F: 212 207 9888

Crawford, Martha, Ms.
Special Consultant
Environmental Protection Authority
Permanent Mission to the United Nations, Dag Hammerskjold Plaza, 7th Floor
New York, NY 10017, United States of America
T: 212 702 4850 F: 212 207 9888

Danz, Gregory, Mr.
Legal Counsel to the President and Cabinet
Permanent Mission of Marshall Islands
Dag Hammerskjold Plaza, 7th Floor
New York, NY 10017, United States of America
T: 212 702 4850 F: 212 207 9888

Heine, Carl L., The Hon.
Ambassador Extraordinary and Plenipotentiary
Permanent Representative to the United Nations, New York
Permanent Mission to the United Nations, Dag Hammerskjold Plaza, 7th Floor
New York, NY 10017, United States of America
T: 212 702 4850 F: 212 207 9888

Helgenberger, Kazuo, Mr.
General Manager
Environmental Protection Authority
Permanent Mission to the United Nations, Dag Hammerskjold Plaza, 7th Floor
New York, NY 10017, United States of America
T: 212 702 4850 F: 212 207 9888

Inok, Fountain, Mr.
Treasurer
Kwajalein Atoll Development Authority
Permanent Mission to the United Nations, Dag Hammerskjold Plaza, 7th Floor
New York, NY 10017, United States of America
T: 212 702 4850 F: 212 207 9888

Kabua, Amata, Mr.
President
Permanent Mission of the Republic of the Marshall Islands
1 Dag Hammerskjold Plaza, 7th floor
New York NY 10017, United States of America
T: 212 702 4850 F: 212 207 9888

Kabua, Imata, The Hon.
Senator
Representative Kwajalein Atoll
Permanent Mission to the United Nations, Dag Hammerskjold Plaza, 7th Floor
New York, NY 10017, United States of America
T: 212 702 4850 F: 212 207 9888

Kijiner, Tom D., The Hon.
Minister for Foreign Affairs
Permanent Mission of Marshall Islands
Dag Hammerskjold Plaza, 7th Floor
New York, NY 10017, United States of America
T: 212 702 4850 F: 212 207 9888

Lowe, David, Mr.
Legal Counsel
Kwajalein Atoll Development Authority
Permanent Mission to the United Nations, Dag Hammerskjold Plaza, 7th Floor
New York, NY 10017, United States of America
T: 212 702 4850 F: 212 207 9888

Mathew, Amenta, Ms.
Clerk of the Cabinet
Permanent Mission of Marshall Islands
Dag Hammerskjold Plaza, 7th Floor
New York, NY 10017, United States of America
T: 212 702 4850 F: 212 207 9888

Predo, Fred J., Mr.
Special Assistant to the President
Permanent Mission of Marshall Islands
Dag Hammerskjold Plaza, 7th Floor
New York, NY 10017, United States of America
T: 212 702 4850 F: 212 207 9888

Riklon, Johnsay, Mr.
Executive Director
Kwajalein Atoll Development Authority
Permanent Mission to the United Nations, Dag Hammerskjold Plaza, 7th Floor
New York, NY 10017, United States of America
T: 212 702 4850 F: 212 207 9888

MARTINIQUE - NGO

Chartol, Paul-Henri
ASSOCIATION DE SAUVEGARDE DU PATRIMOINE MARTINIQUAIS
Immeuble Canavalia-Residence du Square, Place d'Armes
97232 Lamentin., Martinique

Ega, Jean-Luc, Mr.
ASPM
Immeuble Canavalia-Residence du Square, Place d'Armes
Lamentin 97232, Martinique
T: 596 515884

GLOBAL FORUM

Alcinador, Paul
Vice President
CONSEIL GENERAL
Godissard - Lot. Azalees - Apt. 2-B1
Fort de France 972, Martinique
T: 596 552600 F: 596 701920

Canavalia, Ivana, Ms.
RHCHARTOL
Place D'Armes 97232
Lamentin, Martinique

Catorc, Rodrigues, Mr.
MAIRIE DE F. DE F.
Fort - de - France
Fort de France, Martinique

Glondu, Patrice, Mr.
Tourism - Environnement
CONSEIL REGIONAL
Boulevard de Verdun Martinique
Fort de France, Martinique
T: 59 6 616177 F: 59 62 612272

Montoussamy, Carmelien, Mr.
REGION MARTINIQUE (FRANCE)
Rue de Verdun
Fort de France, Martinique
T: 59 6300 F: 72 6810

Nosel, Jose, Mr.
Directeur
PARC NATUREL REGIONAL MARTINIQUE
Domaine De Tivoli BP 437
Fort De France, Martinique
T: 19 596 644259 F: 19 596 647227

Pamar, Assita
RCHARTOL
Place D'Armes 97232
Lamentin, Martinique

MAURITANIA - GOVERNMENT

Ainina, Mohamed Vall Ould, S.E. M.
Ambassadeur
Aupres des Etats-Unis d' Amerique
Permanent Mission to the United Nations, 211 E. 43rd. Str., Suite 2000 New York, NY, 10017
F: 212 986 7963
TL: 212 986 8419

Boubacar, Sidi Mohamed Ould, Mr.
Prime Minister
Permanent Mission of the Islamic Republic of Mauritania
211 E. 43rd. Str., Suite 2000
New York NY 10017, United States of America
T: 212 986 7963 F: 212 986 8419

Boulebatt, Mohamed Aby Ould, M.
Coordinateur de la Desertification (MDRE)
Permanent Mission of Mauritania
211 E. 43rd. Str., Suite 2000
New York, NY 10017, United States of America
T: 212 986 7963 F: 212 986 8419

Chah, Mohamed Lemine Ould, M.
Attache de Presse
Cabinet du Premier Ministre
Permanent Mission to the United Nations, 211 E. 43rd. Str., Suite 2000 New York, NY, 10017
F: 212 986 7963
TL: 212 986 8419

Dahi, Mohamed Mahmould, M.
Directeur
Genie Rural
Permanent Mission to the United Nations, 211 E. 43rd. Str., Suite 2000 New York, NY, 10017
F: 212 986 7963
TL: 212 986 8419

Deddach, Mahfoudh Ould, S.E. M.
Ministre
Developpement Rural et de l'Environnement (Chef Adjoint de la Delegation)
Permanent Mission to the United Nations, 211 E. 43rd. Str., Suite 2000 New York, NY, 10017
F: 212 986 7963
TL: 212 986 8419

El Ghaoth, Mohamed Mahmoud Ould, M.
Conseiller
Mission Permanente aupres des Nations Unies, New York
Permanent Mission to the United Nations, 211 E. 43rd. Str., Suite 2000
New York, NY 10017, United States of America
T: 212 986 7963 F: 212 986 8419

Elbarra, M'Barka Mint, Mme
Chargee de Mission
Ministere du Developpement Rural et de l'Environnement (MDRE)
Permanent Mission to the United Nations, 211 E. 43rd. Str., Suite 2000 New York, NY, 10017
F: 212 986 7963
TL: 212 986 8419

Ghadi, Ahmed Ould, M.
Directeur
Adjoint du Protocole
Permanent Mission to the United Nations, 211 E. 43rd. Str., Suite 2000 New York, NY, 10017
F: 212 986 7963
TL: 212 986 8419

Kane, Aissata, Mme
Representante
ONG Nationales
Permanent Mission to the United Nations, 211 E. 43rd. Str., Suite 2000 New York, NY, 10017
F: 212 986 7963
TL: 212 986 8419

Maaouya, Mohamed Ould, S.E. M.
Directerur
Cabinet du Premier Ministre
Permanent Mission to the United Nations, 211 E. 43rd. Str., Suite 2000 New York, NY, 10017
F: 212 986 7963
TL: 212 986 8419

Merzoug, Dahmoud Ould, M.
Directeur
Protection de la Nature
Permanent Mission to the United Nations, 211 E. 43rd. Str., Suite 2000 New York, NY, 10017
F: 212 986 7963
TL: 212 986 8419

Moustapha, Abderrahmane Ould, M.
Conseiller
Cabinet du Premier Ministre
Permanent Mission to the United Nations, 211 E. 43rd. Str., Suite 2000 New York, NY, 10017
F: 212 986 7963
TL: 212 986 8419

Ould Boubacar, Sidi Mohamed, S.E. M.
Premier Ministre
Republique Islamique de Mauritanie
211 E. 43rd. Str., Suite 2000
New York, NY 10017, United States of America
T: 212 986 7963 F: 212 986 8419

Salem, Cheikhna Ould Mohamed, M.
Fonctionnaire
MDRE
Permanent Mission to the United Nations, 211 E. 43rd. Str., Suite 2000 New York, NY, 10017
F: 212 986 7963
TL: 212 986 8419

Soueidatt, Zeine Ould, Lieutenant
Aide de Camp
Premier Ministre
Permanent Mission to the United Nations, 211 E. 43rd. Str., Suite 2000 New York, NY, 10017
F: 212 986 7963
TL: 212 986 8419

NGO

Ahmed, A.
CANPE
Boite Postale 620
Novakchott, Mauritania

Ahmed Salem, O.B.
CANPE
Boite Postale 620
Novakchott, Mauritania

Aidara, Cheikhna
ASNPE
BP 5275
Nouakchott, Mauritania

Saw, Aliou
ASNPE
BP 5275
Nouakchott, Mauritania

MAURITIUS - GOVERNMENT

Bheenick, R., Mr.
Director
Ministry of Economic Planning and Development
Permanent Mission to the United Nations, 211 East 43rd St., 15th Floor New York, NY, 10017
F: 949 0190, 0191
TL: 697 3829

Kasenally, A.S., The Hon.
Minister of the Environment (Head of Delegation)
Permanent Mission of Mauritius
211 East 43rd St., 15th Floor
New York, NY 10017, United States of America
T: 949 0190, 0191 F: 697 3829

Peerthum, Satteeanund, H.E. Dr.
Ambassador Extraordinary and Plenipotentiary
Permanent Representative to the United Nations, New York
Permanent Mission to the United Nations, 211 East 43rd St., 15th Floor
New York, NY 10017, United States of America
T: 949 0190, 0191 F: 697 3829

Prayag, R., Mr.
Director
Department of the Environment
Permanent Mission to the United Nations, 211 East 43rd St., 15th Floor New York, NY, 10017
F: 949 0190, 0191
TL: 697 3829

Venkatasamy, V., Mr.
Adviser to the Minister
Ministry of Environment and Quality of Life
Permanent Mission to the United Nations, 211 East 43rd St., 15th Floor New York, NY, 10017
F: 949 0190, 0191
TL: 697 3829

NGO

Beenah Devi, Seebaluck
MCD
c/o PO Box 1124
Port Louis, Mauritius

Naik, Rajkarrun Thumba
MFPA
30 Sir Seewoosagur Ramgoolam Street
Port Louis, Mauritius

Rajendranath, Awotar
MCD
c/o PO Box 1124
Port Louis, Mauritius

Ramgoolam, Kishore C.
MAUCESCO
Isidore-Rose
Quatre-Coco, Mauritius

ORGANIZATIONS

Awotar, R., Mr.
Maudesco-council For Development, Environmental Studies And Conservation
c/o PO Box 1124
Port Louis, Mauritius
F: (230) 208 5607

Gunput, Chooramun, Mr.
Action For Development
P.O Box 1042
Port Louis, Mauritius
T: (230) 208 4822 F: (230) 208 2634

Hurrynag, Hemsing, Mr.
Mauritius Council Of Social Service
Astor Court
Port Louis, Mauritius
T: (230) 212 0242 F: (230) 208 6370

Koenig, Monique, Dr.
Friends of the Environment
25 St. Georges Street
Port Louis, Mauritius
T: (230) 212 6783/212 2524 F: (230) 208 8414

Oodit, Geeta, Ms.
Mauritious Family Planning Association
30 Sir Seewoosagur Ramgoolam Street
Port Louis, Mauritius

GLOBAL FORUM

Acuna, Maria Antonietta, Ms.
FLACAM
Via Dewe Serra 18
Quartucciu 09044, Mauritius
T: 230 884813

Awatar, Rajen, Mr.
COUNCIL FOR DEVELOPMENT, ENVIRONMENTAL STUDIES & CONSERVATION
Palais de Nations
Geneve CH 1211, Mauritius
F: 41 22 733-6542

Deepchand, L., Mr.
MAURITIUS COUNCIL OF SOCIAL SERVICE MACOSS
Palais de Nations
Geneve CH 1211, Mauritius
F: 41 22 733-6542

Ramlallah, Beekrumsing
President (and Editor In Chief
MAURITIUS UNION OF JOURNALISTS
23 Bourbon St
Port Louis, Mauritius
T: 230 2121313/ 2121743 F: 230 4409 Beech

MEXICO - GOVERNMENT

Acevedo Danache, Gerardo, Lic.
Consulado General de Mexico en Rio
SRE
Permanent Mission to the United Nations, 2 United Nations Plaza, 28th Floor New York, NY, 10017
F: 212 752 0220
TL: 212 688 8862

Alcalde, Celia, Lic.
Secretaria
SRE
Permanent Mission to the United Nations, 2 United Nations Plaza, 28th Floor New York, NY, 10017
F: 212 752 0220
TL: 212 688 8862

Alvarado Gonzales, Jose Luis, Lic.
Consulado General en Rio De Janeiro
SRE
Permanent Mission to the United Nations, 2 United Nations Plaza, 28th Floor
New York, NY 10017, United States of America
T: 212 752 0220 F: 212 688 8862

Arizpe, Lourdes, Ms.
Executive Committee
INTERNATIONAL SOCIAL SCIENCE COUNCI
Campestre 54
Mexico D.F. 01060, Mexico
T: 52 5 5 48 34 19 F: 52 5 5 48 34 19

Arriola Aguirre, Ma. de los Angeles, Lic.
Consulado General en Rio de Janerior
SRE
Permanent Mission to the United Nations, 2 United Nations Plaza, 28th Floor New York, NY, 10017
F: 212 752 0220
TL: 212 688 8862

Bugeda, Beatriz, Lic
Asesora de la Coordinacion de Recursos Naturales y Medio Ambiente
SRE
Permanent Mission to the United Nations, 2 United Nations Plaza, 28th Floor New York, NY, 10017
F: 212 752 0220
TL: 212 688 8862

Calderon Bartheneuf, Jose Luis, Ing.
Subdirector General Administracion del Agua
CONAGUA
Permanent Mission to the United Nations, 2 United Nations Plaza, 28th Floor New York, NY, 10017
F: 212 752 0220
TL: 212 688 8862

Camacho Solis, Manuel, Lic.
Jefe
Departmento del Distrito Federal
Permanent Mission to the United Nations, 2 United Nations Plaza, 28th Floor New York, NY, 10017
F: 212 752 0220
TL: 212 688 8862

Cervera, Antonio, Lic.
Coordinador de Asesores del Subsecretario de Asuntos Financieros Internacionales
SHCP
Permanent Mission to the United Nations, 2 United Nations Plaza, 28th Floor New York, NY, 10017
F: 212 752 0220
TL: 212 688 8862

Colosio, Luis Donaldo, Lic.
Secretario de Desarrollo Social
Permanent Mission of Mexico
2 United Nations Plaza, 28th Floor
New York, NY 10017, United States of America
T: 212 752 0220 F: 212 688 8862

De Alba, Edmundo, Dr.
Coordinator Ejecutivo
Comision Nacional de Ecologia, SEDESOL
Permanent Mission to the United Nations, 2 United Nations Plaza, 28th Floor New York, NY, 10017
F: 212 752 0220
TL: 212 688 8862

Del Amo, Silvia, Dra.
Vocal Ejecutive del Programa de Accion Forestal Tropical
SARH
Permanent Mission to the United Nations, 2 United Nations Plaza, 28th Floor New York, NY, 10017
F: 212 752 0220
TL: 212 688 8862

Farres, Oliver, Lic.
Consul General En Rio de Janeiro
SRE
Permanent Mission to the United Nations, 2 United Nations Plaza, 28th Floor New York, NY, 10017
F: 212 752 0220
TL: 212 688 8862

Glender, Roberto, Lic.
Asesor de la Coordinacion de Recursos Naturales Y Medio Ambiente
SRE
Permanent Mission to the United Nations, 2 United Nations Plaza, 28th Floor New York, NY, 10017
F: 212 752 0220
TL: 212 688 8862

Gilly Hernandez, Emilio Jose, Lic.
Consolado General en Rio de Janeiro
SRE
Permanent Mission to the United Nations, 2 United Nations Plaza, 28th Floor New York, NY, 10017
F: 212 752 0220
TL: 212 688 8862

Gomez Leon, Alejandra G., Lic.
Asesora
SRE
Permanent Mission to the United Nations, 2 United Nations Plaza, 28th Floor New York, NY, 10017
F: 212 752 0220
TL: 212 688 8862

Gonzales Castillo, German
Mayor
SEDESOL
Permanent Mission to the United Nations, 2 United Nations Plaza, 28th Floor New York, NY, 10017
F: 212 752 0220
TL: 212 688 8862

Kawage Vera, Camilo, Lic.
Director de Ceremonial
Direccion General del Protocolo, SRE
Permanent Mission to the United Nations, 2 United Nations Plaza, 28th Floor New York, NY, 10017
F: 212 752 0220
TL: 212 688 8862

Luna Parra, Angelica, Lic. Ma.
Directora General de Concertacion Social, SEDESOL
Permanent Mission of Mexico
2 United Nations Plaza, 28th Floor
New York, NY 10017, United States of America
T: 212 752 0220 F: 212 688 8862

Mateos, Juan Antonio, S.E. Sr.
Embajador en Kenya y Representante ante el Pnuma, SRE
Permanent Mission of Mexico
2 United Nations Plaza, 28th Floor
New York, NY 10017, United States of America
T: 212 752 0220 F: 212 688 8862

Mendez Corona, C. Antonio, Consulado General en Rio de Janeiro
SRE
Permanent Mission of Mexico
2 United Nations Plaza, 28th Floor
New York, NY 10017, United States of America
T: 212 752 0220 F: 212 688 8862

Montano, Jorge, S.E. Dr.
Embajador Extraprdinario Y Plenipotenciario
Represenante de Mexico ante las Naciones Unidas, SRE, New York
Permanent Mission to the United Nations, 2 United Nations Plaza, 28th Floor
New York, NY 10017, United States of America
T: 212 752 0220 F: 212 688 8862

Munguia, Norma, Lic.
Asesora
SRE
Permanent Mission to the United Nations, 2 United Nations Plaza, 28th Floor New York, NY, 10017
F: 212 752 0220
TL: 212 688 8862

Onate, Ma. de la Luz, Lic.
Director General del Protocolo
SRE
Permanent Mission to the United Nations, 2 United Nations Plaza, 28th Floor New York, NY, 10017
F: 212 752 0220
TL: 212 688 8862

Onate, Santiago, Lic.
Procurado Federal de Protection del Albiente, SEDESOL
Permanent Mission to the United Nations, 2 United Nations Plaza, 28th Floor New York, NY, 10017
F: 212 752 0220
TL: 212 688 8862

Pacheco, Manuel, C.
SEDESOL
Permanent Mission to the United Nations, 2 United Nations Plaza, 28th Floor New York, NY, 10017
F: 212 752 0220
TL: 212 688 8862

Ponce Nava, Diana, Lic.
Coordinadora Adjunta del Programa de Cooperacion Internacional para la Ecologia el Medio Ambiente de Mexico
SEDESOL
Permanent Mission to the United Nations, 2 United Nations Plaza, 28th Floor New York, NY, 10017
F: 212 752 0220
TL: 212 688 8862

Quadri, Gabriel, Ing.
Director de Planeacion Ecologica
DDF
Permanent Mission to the United Nations, 2 United Nations Plaza, 28th Floor
New York, NY 10017, United States of America
T: 212 752 0220 F: 212 688 8862

Ramos, Jeronimo, Lic.
Director General de Asuntos Pesqueros Internacionales
SEPESCA
Permanent Mission to the United Nations, 2 United Nations Plaza, 28th Floor New York, NY, 10017
F: 212 752 0220
TL: 212 688 8862

Renorio Maranon, Jose Carlos, Ing.
Subdirector Programa de Cooperacion Internacional para la Ecologia y el Ambiente de Mexico
SEDESOL
Permanent Mission to the United Nations, 2 United Nations Plaza, 28th Floor New York, NY, 10017
F: 212 752 0220
TL: 212 688 8862

Resendiz, Rafael, Lic.
Oficial Mayor
SEDESOL
Permanent Mission to the United Nations, 2 United Nations Plaza, 28th Floor New York, NY, 10017
F: 212 752 0220
TL: 212 688 8862

Reyes Lujan, Sergio, Fis.
Instituto Nacional de Ecologia, SEDESOL
Permanent Mission to the United Nations, 2 United Nations Plaza, 28th Floor New York, NY, 10017
F: 212 752 0220
TL: 212 688 8862

Rozental, Andres, S.E. Sr.
Embajador
Subsecretario de Relationes Exteriores, SRE
Permanent Mission to the United Nations, 2 United Nations Plaza, 28th Floor New York, NY, 10017
F: 212 752 0220
TL: 212 688 8862

Ruiz-Cabanas Izquierdo, Miguel
Ministro del Servicio Exterior Mexicano
SER
Permanent Mission to the United Nations, 2 United Nations Plaza, 28th Floor New York, NY, 10017
F: 212 752 0220
TL: 212 688 8862

Salinas De Gortari, Carlos, S.E. Sr.
Presidente Constitucional
Estados Unidos Mexicanos
2 United Nations Plaza, 28th Floor
New York, NY 10017, United States of America
T: 212 752 0220 F: 212 688 8862

Solana, Fernando, Sr.
Sr.
Secretario de Relaciones Exteriores
Permanent Mission to the United Nations, 2 United Nations Plaza, 28th Floor New York, NY, 10017
F: 212 752 0220
TL: 212 688 8862

Torres Mendoza, Jorge, Lic.
Consulado Genera en Rio de Janeiro
SRE
Permanent Mission to the United Nations, 2 United Nations Plaza, 28th Floor New York, NY, 10017
F: 212 752 0220
TL: 212 688 8862

Yazquez Arroyo, Carmen Yolando, C.
Consulado General en Rio de Janeiro
SRE
Permanent Mission to the United Nations, 2 United Nations Plaza, 28th Floor New York, NY, 10017
F: 212 752 0220
TL: 212 688 8862

NGO

Alatorre, Gerardo
ERA
Calle Escuela Naval Militar No. 420
Colonia Reforma, Oaxaca, Oax., Mexico

Arias Chavez, Jose
AEC
Apartado Postal 21-898
04000 Mexico, Mexico

Corte Gonzales, Bertha C.
AEC
Apartado Postal 21-898
04000 Mexico, Mexico

Espriu, Martha Patricia
CE
Magdalena 22-404
Mexico D.F. 03100, Mexico

Ferber-Aridijis, B.
GC
Sierra Jiutepec No. 155-B
Lomas Barrilaco 11010, Mexico DF, Mexico

Guzman Ortega, Susana
INAINE
Coladiolas 56, Col. Ciudad Jardin, Delegacion
Coyoacan, Mexico D.F. c.p. 04370, Mexico

Hermann, Hans
PPY
Calle 38 No. 21 (x17) Dept.E-1, Royal Campestre V,
Col. Campestre, 97120 Merida, Yucatan, Mexico

Lamba, Davinder
HIC
Cordobanes N. 24, Col. San Jose Insurgentes
Mexico 03900 D.F., Mexico

Lee Smith, Diana
HIC
Cordobanes N. 24, Col. San Jose Insurgentes
Mexico 03900 D.F., Mexico

Madrid-Zubiran, Sergio
ESTUDIOS RURALES Y ASESORIA
Calle Escuela Naval Militar No. 420
Colonia Reforma, Oaxaca, Oax., Mexico

Reyes, Francisco
ERA
Calle Escuela Naval Militar No. 420
Colonia Reforma, Oaxaca, Oax., Mexico

Rodriguez, Anibal Severino
CE
Magdalena 22-404
Mexico D.F. 03100, Mexico

Romani, Julio
ERA
Calle Escuela Naval Militar No. 420
Colonia Reforma, Oaxaca, Oax., Mexico

Rubello, Leticia, Ms.
Secretary
GRUPO ECOLOGISTA DEL MAYAB, A. C.
PO Box 479
Cancun - QR 77500, Mexico
T: 52 988 46944 F: 52 988 46944

Ruiz, Fernando
ERA
Calle Escuela Naval Militar No. 420
Colonia Reforma, Oaxaca, Oax., Mexico

Sosa Ferreira, Ana Pricila
GEM
Apartado Postal 479
Cancun, Quintana Roo 77500, Mexico

Van Putten, Han
HIC
Cordobanes N. 24, Col. San Jose Insurgentes
Mexico 03900 D.F., Mexico

Yanez-Arancibia, Alejandro, Mr.
Director Scientifico
EPOMEX PROGRAM / ECOMAR A.C.
Program of Ecology, Fisheries and Oceanonagraphy-Gulf of Mexico
Campeche 24030, Mexico
T: 52 981 11600 F: 52 981 65954

PRESS

Acosta, Carlos Gerardo
Editor
PROCESO
Axolahua No. 94
Mexico - DF, Mexico

Aguilar Flores, Jorge Arturo
CEPROPIE/TV
Cantaro 33-C-305
Col. Villacoapa, Mexico - DF, Mexico

Alday, Ricardo
Correspondent
NOTIMEX
155 E 31st Apt. 19D
New York, NY, USA, Mexico

Alvarez Diaz De, Domingo Javier
Reporter Radio/Television
TELEVISA
Eucken 6, #1 Colonia Anzures
Mexico, Mexico

Aponte, Guadalupe David
Editor
EL UNIVERSAL
I. 1124-3, Col. Reforme Abril
Mexico - DF, Mexico

Bardacke, Theodore
Correspondent
EL FINANCIERO
Apartado Postal 22-146
Mexico _ DF 14000, Mexico

Barragan Lannoy, Roberto
MEXICAN PRESS
Plaza Palomas No. 15
Mexico - DF, Mexico

Becerra, Bertha
Editor
EL SOL DE MEXICO
Rio Hudson No. 27-202
Mexico - DF, Mexico

Breton, Elizabeth
RADIO MIL
Los Pinos,
Mexico - DF, Mexico

Calixto Hernandez, Pablo
MEXICAN PRESS
Distrito Federal No. 12
Col. A. Obregon, Mexico - DF, Mexico

Calleja Bello, Maria Eugenia
Editor
EPOCA
Carracci No. 29, Col. Mixcoac
Mexico - DF, Mexico

Camacho, Jose
Reporter Radio/Television
CANAL 6 GUADALAJARA
Av. Del Sur 2003
Guadalajara, Jalisco, Mexico, Mexico

Carmona, Teresita
EL FINANCIERO
Avenida 565, 79 Unidad Aragon
Secc. 2, Mexico-DF, Mexico, Mexico

Carreon, Victor
Reporter Radio/Television
RADIO PROGRAMAS DE MEXICO
Victoria,#7,Col.S.Lucas Patoni
54100 Mexico, DF, Mexico

Castaneda Millan, Juan Francisco
Reporter Radio/Television
TELEVISA
Lux Savinon 926
Col.Del. Valle, Mexico, DF, Mexico

Castellanos, Rocio
MEXICAN PRESS NOVEDADES
Los Pinos
Mexico - DF, Mexico

Cervera Gomez, Hector
CEPROPIE/TV
Desierto de los Leones, 4036
Mexico - DF, Mexico

Chavela Ruiz, Onofre
MEXICAN PRESS
Av. del Rosal No. 290
Col. M. de Rosas, Mexico - DF, Mexico

Chavez Lopez, Alejandro
MEXICAN PRESS
Dr. Garcia Diego No. 139
Mexico - DF, Mexico

Chihuaque Gerardo, Jacobo
MEXICAN PRESS
Belisario Dominguez No. 375
Mexico - DF, Mexico

Conde, Estefano
Correspondent
CENTRO DE INVESTIGACIONES SOBRE LIBRE
Retorno Julieta 50
Mexico,D.F. 11000, Mexico

Contreras Torres, Alvaro Octavio
MEXICAN PRESS
Ave. Tepetlapa and 12 Edif.
Mexico - DF, Mexico

Cordero Guerrero, Pedro Manuel
MEXICAN PRESS
Dinares No. 118
Mexico - DF, Mexico

Cortes, Lidia C. Barron
Reporter Radio/Television
CANAL 11
Privada 5 de Mayo, #9
SN. Cristobal Ecatepec, Mexico, Mexico

Cruz, Rosaura
IMER
Los Pinos
Mexico, DF, Mexico

Cruz De Gante, Maria Rosaura
Reporter Radio/Television
INSTITUTO MEXICANA DE LA RADIO
Odesa No. 715 Dept. C
Col. Portales, Mexico DF, Mexico

Cruz Sanchez, Armando
Editor
SUMMA
Calz. San Juan de Aragon
Mexico - DF, Mexico

Del Pozo, Carlos
MEXICAN PRESS
Los Pinos
Mexico - DF, Mexico

Denti, Jorge Luis
Editor
NOTICERO ALA
Jaina, 29
Mexico -DF, Mexico

Diaz Otero, Elizabeth
Reporter Radio/Television
GRUPO ACIR
Praga 27
Tlalnepantla, Mexico - DF, Mexico

Dominguez Olmos, Benjamin
PUNTO Y APARTE
Los Pinos
Mexico - DF, Mexico

Echeverria Cota, Monica
MEXICAN PRESS
Cerro 2, Conejos 146
Mexico - DF, Mexico

Fernandez, Maria Angeles
MEXICAN PRESS EL HERALDO
Los Pinos
Mexico - DF, Mexico

Flores, Daniel
Reporter Radio/Television
TELEVISA
Av. Chapultepec 35-101
Mexico - DF, Mexico

Flores, Gerardo
Editor
EL ECONOMISTA
F. de Diana 88
Mexico - DF, Mexico

Franco, Eleazar
UNIVERSAL
Los Pinos
Mexico - DF, Mexico

Franco, Fernando
MEXICAN PRESS
Los Pinos
Mexico - DF, Mexico

Franco Sanchez, Jose Luciano
Editor
NOVEDADES
Cuauhtemoc 130
07000 Mexico - DF, Mexico

Freitas Tavares, Flavio Aristides
Correspondent
EXCELSIOR
Av. del Libertador 4780/9B
Buenos Aires, Cap. Federal Argentina, Mexico

French Iduarte, Leonardo
MEXICAN PRESS
Soledad No. 12, Contreras
Mexico - DF, Mexico

Gallardo, Cristina Ma.
MEXICAN PRESS
Minerva 202, Col. Florida
Mexico - DF, Mexico

Gallegos Ramirez, Ma. Elena
Editor
LA JORNADA
Av. Mexico 11-201
Col. Condesa, Mexico - DF, Mexico

Ganem Rodriguez, Rita
Reporter Radio/Television
TELEVISA
Mun. Libre 185, Col. Portales
Mexico - DF, Mexico

Garcia, Luis Alberto
Editor
REVISTA VIVA
Viaducto M. Aleman 91-2
Mexico, 7 D.F., Mexico, Mexico

Garcia, Veronica Cruz
Reporter Radio/Television
IMER-MEXICAN RADIO
Vallarta 178, 4 Seccion
Jardines de Casanueva, Ecatepec,Mexico, Mexico

Garcia Miranda, Gaston
Editor
CADENA RASA
A. Molina Enriquez 4233
Col. Viaducto, Mexico - DF, Mexico

Garduno, Roberto
CANAL 13
Los Pinos
Mexico - DF, Mexico

Garduno Espinoza, Roberto Alejandro
Editor
IMEVISION
Unidad Xochimilco Nativitas
Xochimilco, Mexico - DF, Mexico

Gomez Granados, Angel
MEXICAN PRESS
Bosques de Chiapas No. 69
Tlalnepantla, Mexico - DF, Mexico

Gomez Serrano, Maricela
Editor
MEXICAN PRESS
Callejon San Fernando No. 18
Mexico - DF, Mexico

Gonzalez, Hector
MEXICAN PRESS
Los Pinos
Mexico - DF, Mexico

Gonzalez Martinez, Isidro
MEXICAN PRESS
Portal No. 46
Mexico - DF, Mexico

Granados Granados, Agustin
Reporter Radio/Television
TELEVISA
Galicia 256, Col Alamos
Delegacion Juarez, Mexico, DF, Mexico

Guevara, Patricia
Editor
PERIODICO EXCELSIOR
Hilario Perez de Leon, 37
Mexico - DF, Mexico

Guzman, Susana
Reporter Radio/Television
RADIO RED
Edif. F-2, Entr.1, Depto. 43
Unidad Plateros, Mexico-DF, Mexico 01480, Mexico

Hernandez, Ada
Reporter Radio/Television
EL HERALDO DE MEXICO
Zempoala Norte 208
03020 Mexico - DF, Mexico

Hernandez, Evangelina
Correspondent
LA JORNADA
Vito Alessio 34 Col. Jacaranda
Mexico, Mexico

Hernandez, Francisco
MEXICAN PRESS
Los Pinos
Mexico - DF, Mexico

Hernandez Bazaldua, Tomas
MEXICAN PRESS
Coahuila No. 122-124-H-201
Col. Roma, Mexico - DF, Mexico

Hernandez Gomez, Arturo
Editor
MEXICAN PRESS
America No. 1206
Col. P. Nacional, Mexico - DF, Mexico

Hernandez Tellez, Evangelina
Correspondent
LA JORNADA NEWSPAPER
Vito Alessio R # 34
Mexico - DF, Mexico

Herrera, Araceli
FOTOGRAFA SIEMPRE
Los Pinos
Mexico - DF, Mexico

Herrera Arguelles, Hector
Reporter Radio/Television
RADIO FORMULA
Cerro de la Luna 219
Mexico - DF, Mexico

Hiriat, Pablo
MEXICAN PRESS
Los Pinos
Mexico - DF, Mexico

Jerezano, Rafael
RADIO RED
Los Pinos
Mexico - DF, Mexico

Juarez Montes De Oca, Elizabeth
Reporter Radio/Television
STEREO 100
Fuente de Trevi No. 29
Col.L. de Tecamachalo, Mexico - DF, Mexico

Kudimov, Yuri
Correspondent
NEW TIMES
Sierra Nevada, #725
Col.Lomas de Chapultepec, 11000, Mexico, Mexico

Lang, Patricia C.
Editor
PERIODICO UNOMASUNO
R. Tigris 57-2, Col.Cuauhtemoc
06500-Mexico - DF, Mexico

Lara Guadalupe, Celio
MEXICAN PRESS
Calle Homero M-1, L-14
Mexico - DF, Mexico

Leon Santiago, Hector
MEXICAN PRESS
Gen. Francisco Ramirez 28-11
Mexico - DF, Mexico

Limon, Dante
EPOCA
Los Pinos
Mexico - DF, Mexico

Lomas Maldonado, Emilio E.
Editor
LA JORNADA
Periferico Sur 7654
Tlalpan, Mexico - DF, Mexico

Lopez, Domingo
MEXICAN PRESS
Los Pinos
Mexico - DF, Mexico

Lopez, Laura
Correspondent
TIME MAGAZINE
Fle. de Pescador 147, Tecamach
Tecamachalco Naucalpan 53950, Mexico

Lopez Doriga Velandia, Joaquin
Editor
EL HERALDO DE MEXICO
Crater 460, Col. Pedregal
Mexico - DF, Mexico

Lopez Lorenzo, Lozada
MEXICAN PRES
Edif. 76-502
Mexico - DF, Mexico

Lopez Pastran, Domingo
MEXICAN PRESS
Av. Los Alfanjes 67
Mexico - DF, Mexico

Lopez Portillo Vargas, Guillermo
Reporter Radio/Television
NUCLEO RADIO MIL
Nino Jesus 89-5
14090 Mexico, Mexico

Lopez-Doriga, Joaquin
EL HERALDO DE MEXICO
Los Pinos
Mexico - DF, Mexico

Luhnow, David
Editor
NEWS
Taine #227, Piso 2, Colonia
Polanco, DF Mexico, 11560, Mexico

Luna, Tonatiuh
MEXICAN PRESS
Los Pinos
Mexico - DF, Mexico

Luna D., Jose M.
MEXICAN PRESS
Los Pinos
Mexico - DF, Mexico

Macias Garcia, Javier
Editor
EL NACIONAL
Av. Morelos 595, Edif. 'W'
Mexico - DF, Mexico

Margolis, Therese Ann
Editor
THE NEWS
Tenancingo 38-302
Col. Condesa, Mexico - DF, Mexico

Marquez, Raul
MEXICAN PRESS
Los Pinos
Mexico - DF, Mexico

Martinez, Javier
MEXICAN PRESS
Los Pinos
Mexico - DF, Mexico

Martinez, Salvador
Editor
EL EXCELSIOR
Presa 206, Casa 4
10200 Mexico - DF, Mexico

Maussan Flota, Jose Jaime
Reporter Radio/Television
TELEVISA
Zamora 169-1 Col
Condesa, Mexico, Mexico

Medellin, Jorge Alejandro
Editor
IMPACTO
Insurgentes Sur 441
Col. Tlalcoligia, Mexico - DF, Mexico

Medina Acosta, Rolando
Reporter Radio/Television
TELEVISA
Calle 5, No. 92
Col. I. Zaragoza, Mexico - DF, Mexico

Medina Barranco, Patricia Veronica
Reporter Radio/Television
TV-UNAM
Melchor Ocampo 346-102
06500-Cuahtemoc, Mexico, Mexico

Montes De Oca D., Fernando Jose
Reporter Radio/Television
CANAL 6 GUADALAJARA
Garibaldi 1810
L. de Guevara, Mexico

Mora Tavares, Carlos Eduardo
Editor
EPOCA
Ave. Mexico 147-10
06110-Mexico-DF, Mexico

Moreno, Julio
MEXICAN TV
Los Pinos
Mexico - DF, Mexico

Muniz Vazquez, Constantino
MEXICAN PRESS
Estrella No. 168-E-303
Mexico - DF, Mexico

Navarro, Fernando Jose Becker
Reporter Radio/Television
TV-UNAM
Puerto Mexico 55-2
06700-Mexico-DF, Mexico, Mexico

Navarro Patino, Raul
MEXICAN PRESS
Calle 1, Eje Satelite No. 15
Tlalnepantla, Mexico - DF, Mexico

Nieto Martinez, Juan
Editor
LA PRENSA
Retama 77, Col. La Joya
Mexico - DF, Mexico

Noguez Arana, Victor Manuel
MEXICAN PRESS
Porfirio Diaz No. 19
Mexico - DF, Mexico

Oliva Cuevas, Marco Antonio
Editor
MEXICAN PRESS
Ignacio Mariscal No. 29
Mexico - DF, Mexico

Ortega, Gabriela R.
Editor
PERIODICO EL NACIONAL
Tuy No. 15-5 Col. Postal
CP 03440, Mexico - DF, Mexico

Ortiz De Zarate, Rafael Jerezano
Reporter Radio/Television
RADIO RED
Benjamin Franklin 30-13
Col. Escandon, Mexico - DF, Mexico

Paez Montoya, Victor Manuel
CEPROPIE/TV
Calle Rif. 904
Col. Portales, Mexico - DF, Mexico

Pardinas, Juan Ernesto, Mr.
Primera Cerrada de Minerva, 64-1B Co. Florida
Mexico City C.P. 01030, Mexico
T: 52 5 524.2148 F: 52 5 534.6847

Patino, Angelica
CANAL 11
Los Pinos,
Mexico - DF, Mexico

Paulino, Diego
MEXICAN PRESS
Los Pinos
Mexico - DF, Mexico

Pelaez, Ana Cristina
Reporter Radio/Television
TELEVISA
Rio Tiber No. 48-504
Col. Cuauhtemoc, Mexico - DF, Mexico

Perez, Leticia
MEXICAN PRESS
Los Pinos
Mexico - DF, Mexico

Perez Martinez, Alfredo
MEXICAN PRESS
Calle San Ignacio de Loyola 70
Mexico - DF, Mexico

Perez Orozco, Leticia
CEPROPIE/TV
Calle las Cuestas No. 10
07279 Mexico - DF, Mexico

Perez Silva, Ciro
Reporter Radio/Television
STEREO REY
Las Flores 56
Coyoacan, Mexico - DF, Mexico

Quadra, Gabriel
Correspondent
EL NACIONAL
Beta 85, Robero de Terreros
Mexico - DF, Mexico

Ramirez, Ricardo
MEXICAN PRESS
Los Pinos
Mexico - DF, Mexico

Ramirez C., Enrique
MEXICAN PRESS EL DIA
Los Pinos
Mexico - DF, Mexico

Ramirez Mendoza, Diego Raul
Reporter Radio/Television
TELEVISA
Av. Ejercito Nacional, #340
Col. Chapultepec Morales, Mexico,DF, Mexico

Ramos Aguilar, Fernando
Reporter Radio/Television
CANAL 6 GUADALAJARA
Nacar 2456 Res. Vic
Guadalajara, Jal., Mexico

Rangel, Valente Felipe
Editor
TELEVISA/ECO
Av.Chaultepec 28
Mexico - DF, Mexico

Repper Jaramillo, Luis
Reporter Radio/Television
RADIO CENTRO
F. de Verona No. 5
Tultitlan, Mexico - DF, Mexico

Resendez, Gabriela
Reporter Radio/Television
TELEVISA/ECO
Av.Chapultepec 28
Mexico - DF, Mexico

Resendez, Gerardo
PRESIDENCIA DE MEXICO
Residencial Oficial Los Pinos
Mexico - DF, Mexico

Resendez Garza, Gerardo Rafael
MEXICAN PRESS
Circuito Fuentes del Pedregal
Mexico - DF, Mexico

Reyes Razo, Miguel
Editor
EXCELSIOR
Ricalde 742-30
Mexico - DF, Mexico

Reyna Munoz, Roman
MEXICAN PRESS
Batalla de Santa Maria 5340-A
Mexico - DF, Mexico

Rivera Pina, Ruben
MEXICAN PRESS
Edificio 63-501
Tlalnepantla, Mexico - DF, Mexico

Rivera Tripp, Valentin
Editor
MEXICAN PRESS
Div. del Norte, 2981
Coyoacan, Mexico - DF, Mexico

Robles Vazquez, Daniel
Reporter Radio/Television
RAIDO EDUCACION
Calle Caballero M-189
San Nicolas, Mexico - DF, Mexico

Rodas Dominguez, Enrique
MEXICAN PRESS
Calle Palomas No. 720
Mexico - DF, Mexico

Rodriguez, Francisco
EL SOL DE MEXICO
Los Pinos
Mexico - DF, Mexico

Rodriguez, Luiz Alberto
Editor
LA JORNADA
Atenas 113
Mexico - DF, Mexico

Rodriguez, Marcos
MEXICAN PRESS
Los Pinos
Mexico - DF, Mexico

Rodriguez, Rita Ganem
Reporter Radio/Television
TELEVISA
Municipio Libre 185
Col. Portales, Mexico, DF, Mexico

Rodriguez Hernandez, Marcos
MEXICAN PRESS
Ave. 625 No. 84
Mexico - DF, Mexico

Romero, David
Editor
OVACIONES
Feria 834-45, Col. Jardines
Mexico - DF, Mexico

Ruge, Tlahoga, Ms.
Reporter Radio/Television
TELEVISA-CAD.DE LAS AMERICAS
Jaime Nuno #77
Mexico 01020 DF, Mexico, Mexico

Ruiz Posada, Gonzalo
Correspondent
EL NUEVO AMANECER DE QUERETANO
Juarez 48 Sur
Queretano QRO. Mexico, Mexico

Saad Sotomayor, Patricia
Editor
EXCELSIOR
Privada Florida # 16
Coyocan, Mexico, Mexico

Salazar Gonzalez, Eduardo
Correspondent
TELEVISA
Sierra Candela 39
Lomas de Chapultepec, Mexico, DF, Mexico

Samaniego Reyes, Fidel
Editor
EL UNIVERSAL
Guadalupe I, Ramirez 1008
Mexico - DF, Mexico

Sanchez Espino, Antonio
Editor
MEXICAN PRESS
Ramales No. 33, El Tejocote
Mexico - DF, Mexico

Santillan Velazquez, Enrique
MEXICAN PRESS
Dr. Atl. No. 30
Col. B. Dominguez, Mexico - DF, Mexico

Sixto, Gerardo
MEXICAN PRESS
Los Pinos
Mexico - DF, Mexico

Solorzano Zinser, Javier
Reporter Radio/Television
IMEVISION
Periferico Sur 4121
CP 14141 Mexico, Mexico

Sosa Calderon, Alberto
Reporter Radio/Television
TELEVISA
Joaquin Mucel, #19-2
Col. Huizachal, Mexico, DF, Mexico

Soto Ortiz, Luis
Editor
EL FINANCIERO
Norte 70, 8015 Col. S. Miron
Mexico-DF, Mexico, Mexico

Sotomayor, Walter Carlos
Correspondent
NOTIMEX
SHIN QI7 Conj.1 Casa 3
70515 Brasilia-DF, Brazil, Mexico

Sotomayor Orellana, Enrique
MEXICAN PRESS
Paseo de la Reforma No.27-13
Mexico - DF, Mexico

Tavares, Flavio A. Freitas
Correspondent
EXCELSIOR
Av. Libertador 4780, #9B
Buenos Aires 1426, Argentina, Mexico

Teheran, Jorge
MEXICAN PRESS EL ECONOMISTA
Los Pinos
Mexico - DF, Mexico

Tellez Flores, Armando
Editor
EL DIARIO DE MEXICO
Av. Diagonal San Antonio
Mexico - DF, Mexico

Thompson, Charlotte R.
Editor
MEXICO DESCONOCIDO MAGAZINE
Liverpool No. 164, Zona Rosa
Mexico City, Mexico, Mexico

Tirado Aguilar, Carlos
MEXICAN PRESS
Alberto Gozales 118
Mexico - DF, Mexico

Torio Ramos, Carlos
Editor
EL NORTE DE MONTERREY
Sheskespeare 183
Mexico - DF, Mexico

Torres, Victor Manuel
Editor
UNO MAS UNO
Redencion 150, Edif. 'D'
Xochimilco, Mexico - DF, Mexico

Trad, Elias
MEXICAN PRESS
Montes Urales 832
Mexico - DF, Mexico

Trejo, Martha
NOTIMEX-
Los Pinos
Mexico - DF, Mexico

Trejo Hernandez, Jose Jesus
MEXICAN PRESS
Carrizo No. 11
Mexico - DF, Mexico

Trejo Sanchez, Ricardo
Reporter Radio/Television
NUCLEO RADIO MIL
Rio Tuxpan 3
Ecatepec, Mexico - DF, Mexico

Trevedan, Jesus Mario
MEXICAN PRESS
Los Pinos,
Mexico - DF, Mexico

Trevino, Javier
MEXICAN PRESS
Retorno 24 de Camen Aviles
Mexico - DF, Mexico

Urban Pastran, Alberto
MEXICAN PRESS
M. de los Reyes M. 17, Lote 21
Zaragoza, Mexico - DF, Mexico

Valadez, Rene
MEXICAN PRESS
Los Pinos
Mexico - DF, Mexico

Varela, Jose Pedro
Correspondent
TELEVISA
R.Paulo Cesar Andrade,106/303
Rio de Janeiro, Brazil, Mexico

Vega Saldana, Jorge Carlos
Editor
OVACIONES
Lerdo 17, Depto. 402-A
06300 Mexico - DF, Mexico

Vega Sotelo, Daniel
MEXICAN PRESS
Av. Marina Nacional 200
Mexico - DF, Mexico

Velazquez, Cesar
PRESIDENCIA DE MEXICO
Residencia Oficial Los Pinos
Mexico - DF, Mexico

Viola, Giuseppe
Editor
ECONOMIA NACIONAL
Maria Del Rey 1517
Puerto Vallarca - Mèxico, Mexico

ORGANIZATIONS

Barba, Regina, Sra
As. Ecologica Coyoacan, a.c.
Apartado Postal 21-898
04000 Mexico, Mexico
T: (52 5) 544 7172 F: (52 5) 532 5639

Cipres, Alfonso, Mr.
Movimiento Ecologista Mexicano
Calzada de Guadalupe 120, Edificio 5, 004 Colonia
Ex. Hacienda Coapa, 14300 Mexico D.F., Mexico
T: (52 5) 678 0487 F: (52 5) 673 6635

Gonzalez, Luis, Sr.
Pronatura Peninsula de Yucatan
Calle 38 No. 21 (x17) Dept.E-1, Royal Campestre V,
Col. Campestre, 97120 Merida, Yucatan, Mexico
F: (52 99) 268 423

Guerra, Luis Manuel, Sr.
Instituto Autonomo De Investigaciones Ecologicas (INAINE A.C.)
Coladiolas 56, Col. Ciudad Jardin, Delegacion
Coyoacan, Mexico D.F. c.p. 04370, Mexico
T: (52 5) 689 0473 F: (52 5) 689 5972

Guerrero, Fernando, Mr.
Grupo Ecologista del Mayab a.c.
Apartado Postal 479
Cancun, Quintana Roo 77500, Mexico
T: (52 988) 45 365 F: (52 988) 49 857

Ortiz, Enrique, Mr.
Habitat International Coalition
Cordobanes N. 24, Col. San Jose Insurgentes
Mexico 03900 D.F., Mexico
T: (52 5) 651 6807 F: (52 5) 593 5194

Severino, Anibal, Mr.
President
CULTURA ECOLOGICA
Magdalena 22-404
Mexico City 03100, Mexico
T: 52 5 2775864 F: 52 5 6873085

GLOBAL FORUM

Aguirre, Beatriz, Ms.
N O V A
Galindo Y Villa #1 CTO,Geografos
Naucalpan 53100, Mexico
T: 55626654 F: 53265740

Ahumada, Aurelio, Mr.
Coordinator
E C O S O L A R
Eugenia 1510 - C. Narvarte
Mexico City D.F. 03600, Mexico
T: 52 5 687 5474 F: 52 5 682 0507

Alvarez, Marcella, Ms.
P.D.
GESTION DE ECOSISTEMAS A.C.
Crater 21, Jardines del Pedregal
Mexico City 01900, Mexico
T: 52 5 568-2905 F: 52 5 568-5933

Ambriz, Raul A., Mr.
Operations Control Manager
CEMENTOS DE CHIHUAHUA
Apdo Postal 241 CP 31110
Chihuahua, Mexico
T: 52 14 17-5100 F: 52 14 17-0104

Amiga, Jacobo Jasqui, Mr.
Un Representative
FEDERACION INTERNACIONAL DE PROFESIONES INMOBILIARIAS - FIABCI
Gaeuss 12 - 502 Col. Anzures
Mexico City 11590, Mexico
T: 52 5 250-2699 F: 52 5 254-0871

Amor, Antonieta Tejeda De, Sra.
Secretaria
PRIVADA ESCUELA DE GUERRA
no. 10 San Jeronimo Lidice
Mexico City 10200, Mexico
T: 52 5 681-1432 F: 52 5 575-5891

Aridjis, Betty Ferber De, Ms.
International Coordinator
GRUPO DE LOS CIEN
Sierra Jiutepec 155-B Lomas Barrilaco
Mexico City 11010, Mexico
T: 52 5 540-7379 F: 52 5 520-3577

Armenta, Maria De Jesus, Ms.
TELEVICION DE MEXICO
Cda Acamapixtel M.11.L 10, Col.Paraiso C.P.09230
Mexico City, Mexico
T: 52 5 604-2306 F: 52 5 605-5566

Arzate, Enrique Rlico, Sra.
DIPUTADO FEDERAL MOVIMIENTO MUNICIPALISTA
Juarez 309
Paseo el Alto - GFO, Mexico

Azispe, Lourdes, Dr.
INSTITUTO DE INVESTIGACIONES ANTROPOLOGICAS - UNAM
Ciudad Universitaria, Coyoacan
Mexico City 04510, Mexico
T: 52 5 548-3667 F: 52 5 548-3667

Barocio, Octavio, Mr.
Director
MANANTIAL DE ALTERNATIVAS
Nubes 211
Mexico City D.F. 01900, Mexico
T: 52 5 568 7191 F: 52 5 568 5070

Bedia Estrada, Josue, Mr.
MOVIMIENTO ECOLOGISTA MEXICANO DE LEON A.C.
Av. Pradera 204
Leon, Mexico
T: 52 471 14229

Behn, Mercedes, Ms.
Ecologist
PRESCOTT COLLEGE FOR ENV STUDIES
CPA de Presa 150-6, San Jeronimo Lidice
Monte Aprazivel, SP 10200, Mexico
T: 55 172 6531

Behn, Veronica, Ms.
Ecologist
PRESCOTT COLLEGE FOR ENV STUDIES
CPA de Presa 150-6, San Jeronimo Lidice
Mexico City 10200, Mexico
T: 52 5 95-65-31

Benavides, Marta
Diretora
MEDEPAZ
Apdo 57-317 / Cap. 06501
Mexico City D.F. 06501, Mexico
T: 52 5 651 4915

Briggs, Christopher F., Mr.
Director
BRITISH COUNCIL
Cromwell House
East Haddam - CT RH19 3DE, Mexico
T: 1 203 322 400

Bueno, Agustina, Sra.
SOCIEDAD DE ECOLOGISTAS
10. de Maio #2606
Chihuahua, Mexico
T: 52 14 152814

Bueno, Juan, Sr.
Director De Produccion
SOCIEDAD DE ECOLOGISTAS
Cortez de Monroy, 3102 - Parques San Felipe
Chihuahua 31240, Mexico
T: 52 14 44552

Bustamante, Homero Blas, Mr.
Representante
CONFEDERACION CARIBENA LATINOAMERICANA DE ESTUDIANTES AGRONOMIA
Km 38.5 Carr. Mex-Tex, UACh.
Chapingo, Texcoco 56230, Mexico
T: 91 595 693-5019 F: 91 595 42200

Cachoua, Adriana O. De
Officer
INAINE
Gladiolas 56
Mexico City D.F. 04370, Mexico
T: 52 5 689-6885

Cachoua, Victor, Mr.
Officer
INAINE
Gladiora 54
Mexico City D.F. 04370, Mexico
T: 52 5 689-6885 F: 52 5 689-5972

Campana, Luis Miguel Flores, Sr.
UNIVERSIDAD AUTONOMA DE SINALOA
Claussen s/n, A.P. 610
Mazatlan - Sinaloa, Mexico
T: 52 69 82-8656 F: 52 69 81-0156

Carabias, Julia, Ms.
U N A M
Faculdad De Ciencias
Mexico City D.F. 04510, Mexico
T: 52 5 568-6949 F: 52 5 548-8186

Carmona, Maricarmen, Ms.
ENLACE AMBIENTAL ACADEMIA MEX DE DERECHO ECOLOGICO
Irapuato 28-A
Mexico City, Mexico

Carrillo, Alberto, Mr.
CONGRESISTA MEXICANO CONGRESSO DE LA UNION
Mexico City, Mexico

Carrion, Diego, Mr.
HABITAT INTERNATIONAL COALITION
Cordobanes 24, Col. San Jose Insurgentes
Mexico City 03900, Mexico
T: 52 5 6512371 F: 52 5 5453263

Caso, Affonso, Sr.
Director
NORTH AMERICA ENVIROMENTAL FUND
Insonge Fessun 197 3 Piso 13
Mexico City, Mexico
T: 52 5 3332566 F: 52 5 325 6009

Castillo, German R Mendez Del, Mr.
Veterinarian/Ecologist
PO Box 396, Ave 35 Sur # 110
Cozumel 77600, Mexico
T: 52 987 21232 F: 52 987 21545

Castillo, Pedro Reyes, Mr.
Director General
CENTRO DE ECO DESARROLLO
Altadena # 8 Col. Napoles
Mexico City D.F., Mexico
T: 52 5 523-1802 F: 52 5 669-0481

Ceballos, Jose, Mr.
Board Member
FCOC
Cordobanes 24, Col. San Jose Insurgentes
Mexico City 03900, Mexico
T: 52 5 6512371 F: 52 5 5453263

Cespedes, Mario, Mr.
Board Member
FCOC
Cordobanes 24, Col. San Jose Insurgentes
Mexico City 03900, Mexico
T: 52 5 6512371 F: 52 5 5453263
TL: 1776454 MAFOME

Chapa, Omar, Mr.
FRENTE EMILIANO ZAPATA EN DEFESA DEL EJIDO
Cordobanes 24, Col. San Jose Insurgentes
Mexico City 03900, Mexico
T: 52 5 6512371 F: 52 5 5453263

Chavez, Jose Arias, Sra.
PACTO LATINO AMERICANO DE ACCION ECOLOGICA
Palais de Nations
Geneve CH 1211, Mexico
F: 41 22 733-6542

Claveran, Jorge Gonzalez, Mr.
Director For America
INTERNATIONAL ACADEMY OF ARCHITECTURE
Georgia 163-13
Mexico City 03810, Mexico
T: 52 5 687-3347 F: 52 5 669-3931

Collado, Pablo, Sr.
GBM
Reforma 382-2
Mexico City 0066, Mexico
T: 52 5 207-0202

Corrales, Pedro Leon, Mr.
Asesor
UNION DE PUEBLOS DE MANANTLAN
Espanola, 2864, Jardines de la Cruz
Guadalara -Jalisco, Mexico
T: 36 6115222

Cortes, Julio Romani, Sr.
ESTUDIOS RURALES Y ASESORIA
Palais de Nations
Geneve CH 1211, Mexico
F: 41 22 733-6542

Covarrubias, Francisco, Mr.
Vice-President
METROPOLIS
B. De Acacias 65
Mexico City D.F. 11.700, Mexico
T: 52 5 596-3920 F: 52 5 596-8694

Cruz, Varonica Garcia
Reporter
INSTITUTO MEXICANODE
Margaritas #18 Colonia Florida
Mexico
T: 5246151 F: 5342491

De La Vega, Guadalupe
Pres. Comunitario
FED.MEXICANA DE ASS. DESALUD Y DESARROLLO
Plutarco E Calles 744 Norte
Chihuahua, Mexico
T: 52 14 166161 F: 52 14 566535

Delamo, Silvia, Ms.
Vocal Consejera
S A R H
Mexico City, Mexico

Diaz, Marcelino, Mr.
Comisionario
HABITAT INTERNATIONAL COALITION
Cordobanes 24, Col. San Jose Insurgentes
Mexico City 03900, Mexico
T: 52 5 6512371 F: 52 5 5453263

Domingo, Rosalinda Jinich, Sra.
Diretora
MUSEO DE LA ISLA DE COZUMEL
Av. Rafael E. Melgar Esq. Con 6 Norte Apdo Postal 35
Cozumel 77600, Mexico
T: 52 987 215-45 F: 52 987 215-45

Duran, Martha Magdalena Aguiar, Ms.
PATRONATO CERRO DE LA SILLA
Ortega Y Gasset 2810, Col. Contry la Silla
Monterrey - NL 67170, Mexico
T: 52 83 571328

Duran, Reynalda, Ms.
PATRONATO CERRO DE LA SILLA
Ortega Y Gassey 2810, Silla, Col. Contry la Silla
Monterrey - NL 67170, Mexico
T: 52 83 571328

Encinas, Alejandro, Mr.
Dep.Federal
CONGRESSISTA MEXICANO CONGRESSO DE LA UNION
Mexico City, Mexico

Ensastiga, Leopoldo, Mr.
Secretario
UCP - UNION DE COLONIAS POPULARES
Copevi. A.C. 1o de Mayo 151, Col. San Pedro de los Pinos, C.P. 03800
Mexico City, Mexico
T: 52 5 515-9627 F: 52 5 271-4119

Escobar, Saul, Mr.
Secretario Tecnico
MOVIMIENTO NACIONAL MUNICIPALISTA
Copevi. A.C. 1o de Mayo 151, Col. San Pedro de los Pinos, C.P. 03800
Mexico City, Mexico
T: 52 5 515-9627 F: 52 5 271-4119

Escudero, Natalia, Ms.
Secretary General
PARTIDO ECOLOGISTA DE MEXICO
Fte. de Medusa 9, Tecamachalco
Mexico City 53950, Mexico
T: 52 5 251-3662 F: 52 5 531-1866

Espindola, Hermilo Salas, Mr.
LA MADALENA CONTRERAS - SUBDELEGACION DE ECOLOGIA, DDF
A. Alvaro Obreo No 20, Col Barranca Seca C.P. 10580
Mexico City, Mexico
T: 52 5 645-4963 F: 52 5 645-4966

Fernandez, Adolfo, Mr.
GOBIERNO DEL ESTADO DE MORELOS
Palacio De Gobierno
Cuernavaca, Mexico
T: 91 73 123348

Flores, Luis, Sr.
ESC. CIENCIAS DEL MAR
A.P. 610 Mazatlan
Sinaloa, Mexico
T: 52 69 82-86-56 F: 52 69 81-01-56

Frenk, Gerardo Alatorre, Mr.
Investigador
GRUPO DE ESTUDIOS AMBIENTALES
Allende # 7 Col. Sta Ursula Coapa
Mexico City D.F. 04650, Mexico
T: 52 5 684 0253 F: 52 5 684 0377

Friedland, Peter, Mr.
HABITAT INTERNATIONAL COALITION
Cordobanes 24, Col. San Jose Insurgentes
Mexico City 03900, Mexico
T: 52 5 6512371 F: 52 5 5453263

Garcia, Maria Del Rayo, Ms.
COLECTIVO ECOLOGISTA JALISCO
Alpes # 1276A Col. Independencia S.L.
Guadalajara, Mexico
T: 52 36 516442 F: 52 36 516442

Garcia Campos, Helio M., Sr.
Miembro/Tesorero
FOROMEX/RIAEV
Alfaro #55 Centro
Xalapa - Ver 91000, Mexico
T: 52 281 792 57 F: 52 281 638 48

Garcia Rizzo, Elida, Sra.
GOBIERNO MEXICO
Monterrey - NL, Mexico
T: 52 83 470270 F: 52 83 441179

Garduno, Daniel, Mr.
Mexico City, Mexico

Garduno, Hector, Mr.
COMISSION NACIONAL DE AGUA MEXICO
Mexico City, Mexico

Gaye, Malick, Mr.
Board Member
HABITAT INTERNATIONAL COALITION
Cordobanes 24, Col. San Jose Insurgentes
Mexico City 03900, Mexico
T: 52 5 6512371 F: 52 5 5453263

Godas, Magali Daltabuit, Ms.
CENTRO REGIONAL DE INVESTIGACIONES MULTIDISCIPLINARIAS
Av. Universidad s/n Circuito II, Col. Chamilpa
Cuernavaca, Mexico
T: 52 731 13-0555 F: 52 731 17-5981

Godoy, Lucas, Mr.
Dirigente
XACHIMILCO
Cordobanes 24, Col. San Jose Insurgentes
Mexico City 03900, Mexico
T: 52 5 6512371 F: 52 5 5453263

Gonzales, Pedro Vazques, Sr.
ASSEMBLEA NACIONAL URBANO POPULAR
Bernardo Reyes 6350 nte. Popular de Mexico (ANAMUP)
Monterrey - NL, Mexico
T: 52 83 264-4359 F: 52 83 264-4376

Gonzalez, Bertha Corte, Sra.
FORD MEXICANO
Alcazar Nom. 6-1, Jardines del Sur, Xochimilco
Mexico City D.F. 16050, Mexico
T: 52 5 653-0580 F: 52 5 534-2030

Gonzalez, Emilio Becerra, Mr.
Presidente
UNION CIVICA SALMANTINA
Andres Delgado #200
Salamanca-Guanajuato, Mexico
T: 91 464 82030

Gonzalez, Vincente Andrade
Cameraman
IMEVISION
Periferico Sur 4121
Mexico
T: 6455705

Gonzalez Torres, Jorge, Mr.
PARTIDO ECOLOGISTA DE MEXICO
Volcan 10 esquina con caseada - Pedregal
Pedregal (Mexico), Mexico
T: 52 595.8090 F: 52 594.4522

Gottfried, Carlos, Mr.
President
MONARCA AC
Av. Constituyentes 345
Mexico City D.F. 11830, Mexico
T: 52 5 515-9910 F: 52 5 604-8525
TL: 1774313 PCECHE

Gritti, Margarita Soria, Sra.
President
A.S.P.A., AC.
Almada # 35 Col. Centro
Navolato - SI, Mexico
T: 52 672 70355 F: 52 672 70401

Guerrero, Gabriela, Ms.
Dep. Federal
CONGRESISTA MEXICANA CONGRESSO DE LA UNION
Congresso De la Union
Mexico City, Mexico

Gutierrez, Alberto Anaya, Mr.
ANAMUP - FCOC
Bernado Reyes 6350
Monterrey, NL, Mexico
T: 731056 F: 736730

Gutierrez, Blanca, Dr.
Deputy
CLUB AMISTUR MEXICO A.C.
P.O. Box 61094 C.P.
Mexico City 06600, Mexico

Gutierrez, Jorge Gaston, Sr.
COLECTIVO ECOLOGISTA JALISCO
Palais de Nations
Geneve CH 1211, Mexico
F: 41 22 733-6542

Guzman, Lucero Jimenez, Ms.
Sid
Rio Elba 32-104 Cuannnhtemoc
Mexico City D.F. 06500, Mexico
T: 52 5 286-9111

Hernandez, Edgar, Mr.
TECNOLOGICO DE MONTERREY
Eucaliptos # 55 Izcalli Del Bosque
Naucalpan 53270, Mexico
T: 53732827 F: 53265740

Hernandez, Edmundo, Mr.
MOV. ECOLOGISTA BAJA CALIFORNIA
Fuentas Brotantes I-6, 602 Col. Hidalgo, Del. Tlalpan
Mexico City D.F., Mexico
T: 52 5 553-9702 F: 52 5 553-0808

Herrmann, Hans, Mr.
Director General
PRONATURA
Av. Nuevo Lem # 144, Col. Hipodromo Condasa
Mexico City D.F. 06100, Mexico
T: 52 5 286-9642 F: 52 5 286-9480

Hoyos Cerna, Emilio De, Dr.
MOVIMENTO ECOLOGISTA MEXICANO
5 Mayo 450 SUR - Acono
Coahulla, Mexico
T: 8772-11-03 F: 8772-11-03

Huerta, Manuel, Mr.
Diputado Federal
CAMERA DE DIPUTADOS
Copevi. A.C. 1o de Mayo 151, Col. San Pedro de los Pinos, C.P. 03800
Mexico City, Mexico
T: 52 5 515-9627 F: 52 5 271-4119

Huerta, Rosalinda, Ms.
Regidor
DIPUTADA LOCAL VERACRUZ
Copevi. A.C. 1o de Mayo 151, Col. San Pedro de los Pinos, C.P. 03800
Mexico City, Mexico
T: 52 5 515-9627 F: 52 5 271-4119

Hume, Patricia, Sra.
CULTURA ECOLOGICA
Palais de Nations
Geneve CH 1211, Mexico
F: 41 22 733-6542

Jaime, Jose Luis, Mr.
Coordinador Maestria
AYUNTAMIENTO DE TOLUCA, ESTADO DE MEXICO
Copevi. A.C. 1o de Mayo 151, Col. San Pedro de los Pinos, C.P. 03800
Mexico City, Mexico
T: 52 5 515-9627 F: 52 5 271-4119

Jasoui, Jacobo, Sr.
Representative To Un
FIABCI-FEDERACION INTERNACIONAL DE PROFESIONES IMOBILIARIAS
Gauss 12 - 502 Col. Anzures
Mexico City 11590, Mexico
T: 52 5 250-3699 F: 52 5 254-0871

Jimenez, Ernesto, Mr.
Assistant
ASAMBLEA DE REPRESENTANTES DEL DISTRITO FEDERAL
Copevi. A.C. 1o de Mayo 151, Col. San Pedro de los Pinos, C.P. 03800
Mexico City, Mexico
T: 52 5 515-9627 F: 52 5 271-4119

La Cueva, Carlos Meade De, Mr.
Coordenador De Comunicacao
ORGANIZACAO DE EJIDOS PRODUTORES FORESTALES DE LA ZONA MAYA, S.C.
Av. Lazaro Cardenas s/n - Edif. Prodemay, A.P. #8
F.Carrillo,Pto.Quint 77200, Mexico
T: 52 983 4-0307 F: 52 983 4-0307

Lasse, Rolando De, Dr.
Asesor
SECRETARIA DE AGRICULTURA Y RECURSOS HIDRAULICOS
Providencia no. 706 Col. del Valle CP 03100
Mexico City, Mexico
T: 52 5 543-6055 F: 52 5 280-3196

Legorreta, Jorge, Sr.
FRENTE EMILIANO ZAPATA DEFENSA EJIDAL
Sr. Gregorio, Xochimilco
Xochimilco, Mexico
T: 5744117

Leon, Miguel A., Mr.
Dep.Federal
CONGRESSISTA DE UNION
Congresso De la Union
Mexico City, Mexico

Licea, Carlos, Mr.
COLEGIO DE INGIENIEROS ARQUITECTOS
Pacifico # 350 Edif E - 304 Coxoacau Mexico
Mexico City, Mexico
T: 52 5 689 2830 F: 52 5 549 4946

Limon, Armida, Sra.
SOCIEDAD DE ECOLOGISTAS
Cortez de Monroy, 3102 - Parques San Felipe
Chihuahua 31240, Mexico
T: 52 14 44552

Limon Dominguez, Geraldo Arturo, Sr.
Presidente
SOCIEDAD DE ECOLOGISTAS
Cortez de Monroy 3102 Parques San Felipe
Chihuahua 31240, Mexico
T: 52 14 144552

Lombera, Rocio, Ms.
Codirectora
CODEVI-CENTRO OPERACIONAL DE VIVIENDA Y POBLAMIENTO
Copevi. A.C. 1o de Mayo 151, Col. San Pedro de los Pinos, C.P. 03800
Mexico City, Mexico
T: 52 5 515-9627 F: 52 5 271-4119

Longoria, Martin, Mr.
Board Member
FCOC
Cordobanes 24, Col. San Jose Insurgentes
Mexico City 03900, Mexico
T: 52 5 6512371 F: 52 5 5453263

Lopez, Francisco Sanchez, Ms.
FORO REGIONAL ECOLOGISTA VALLE DE MEXICO
Georgia, 30 - Esq. Nebraska (bajos) - Col. Napoles
Mexico City 03810, Mexico
T: 52 5 687-0889 F: 52 5 543-1103

Lopez-Portillo, Guillermo
Reporter
NUCLEO RADIO MIL
Iusangentes Sur 1870 Col Florida
01030 Mexico, Mexico
T: 915 6612032 F: 915 6612032

Lopezllera, Luis, Sr.
Presidente
TLALOC 40-3
Tlaloc 40-3
Mexico City D.F. 11370, Mexico
T: 52 5 5664265 F: 52 5 5921989

Lucero Marquez, Alvaro F., Sr.
ECOLINE
Federacao Mexicana de Futbol # 14-301
Mexico City 14370, Mexico
T: 671-2664

Luevano, Alejandro, Mr.
Diputado Federal
CAMARA DE DIPUTADOS
Copevi. A.C. 1o de Mayo 151, Col. San Pedro de los Pinos, C.P. 03800
Mexico City, Mexico
T: 52 5 515-9627 F: 52 5 271-4119

Lujan, Berta, Sra.
ASOCIACION ECOLOGICA DE COYOACAN
Palais de Nations
Geneve CH 1211, Mexico
F: 41 22 733-6542

Lungo, Mario, Mr.
Member
HABITAT INTERNATIONAL COALITION
Cordobanes 24, Col. San Jose Insurgentes
Mexico City 03900, Mexico
T: 52 5 6512371 F: 52 5 5453263

Macias, Rolando Delasse, Mr.
Mexico City, Mexico

Madrigal M., Juan Manuel, Mr.
FORO MEXICANO
Americas #62 - A
Uruapan - Michoacan, Mexico
T: 91 452 41077 F: 91 452 33374

Mandrogin, Angeles
Reporter
EL HERALDO DE MEXICO
Mexico
T: 5787022 F: 5789824

Marina, Luz Maria Delgado De, Ms.
Commissioner
FORO REGIONAL ECOLOGISTA VALLE DE MEXICO
Georgia, 30 - Esq. Nebrasca (bajos) - Col. Napoles
Mexico City 03810, Mexico
T: 52 5 687-0889 F: 52 5 543-1103

Martin Arrieta, Cristina, Ms.
Assistant To Gen Coordinator
DEVELOPMENT ALTERNATIVES WITH WOMEN FOR A NEW ERA (DAWN)
c/o WAND, UWI School of Continuing Studies, Pinelands
St Michael, Mexico
T: 1 809 4269288 F: 1 809 4294854

Martinez, Marcela Rodriguez, Sra.
FORO REGIONAL ECOLOGISTA VALLE DE MEXICO
Georgia, 30 - Esq. Nebraska (bajos) - Col. Napoles
Mexico City 03810, Mexico
T: 52 5 687-0889 F: 52 5 543-1103

Mastretta, Veronica, Ms.
Pueblaverdeac
PUEBLA VERDE AC.
Av. Bugambilias 5316
Puebla, Mexico
T: 52 22 44 5234 F: 52 22 44 5234

Medina, Ofelia, Ms.
Fundadora
SALUD NINOS INDIOS
Afonso Pruneda, 12
Mexico City D.F. 04360, Mexico
T: 52 5 659 4742 F: 52 5 659 5113

Menchu, Rigoberta, Sra.
REPRESENTACION INTERNACIONAL DEL COMITE DE UNIDAD CAMPESINA
Apartado Postal 20475, Colonia San Angel
Mexico City D.F. 01000, Mexico

Mendoza, Jose Roberto Fong, Sr.
Director
MAESTRO
Colon 739 otc. CP.80000
Culiancan, Mexico
T: 52 671 127216 F: 52 671 172400

Merlos, Rocio, Mr.
Coodinador De Programa
AYUNTAMIENTO DE TOLUCA, ESTADO DE MEXICO
Copevi. A.C. 1o de Mayo 151, Col. San Pedro de los Pinos, C.P. 03800
Mexico City, Mexico
T: 52 5 515-9627 F: 52 5 271-4119

Moguel, Julio, Mr.
MOVIMIENTO NACIONAL MUNICIPALISTA - MNM
Av. Torres 58 Ed.A - 404, Keryes - Coyoacan
Mexico City, Mexico

Mojica, Armando, Mr.
Member
TALLER 'ESPACIO VERDE' A.C.
Arista, 22 C.P. 62000 Centro
Cuernavaca - Morelos 62000, Mexico
T: 91 73 12 15 60 F: 91 73 12 69 36
TL: 886

Mondragon, Luis Enrique Sanchez
Consuejo Colegiado
GRUPO ECOLOGISTAS TLALPAN
Juarez Oriente No. 15 - San Andres Tlalpan
Mexico City, Mexico
T: 52 5 849-0073 F: 52 5 849-0091

Morgon Colon, Humberto, Sr.
CONSEJO POPULAR JUVENIL
Paso Chacabuco #40-B, Col. la Mexicana, Sta. Fe
Mexico City D.F. 01260, Mexico
T: 52 5 259-7895

Munoz, Mindahi C. Bastida
YOUTH 92
55 Parkdale Ave, 3rd floor
Ottawa - Ontario, Mexico
T: 1 613 761 9206 F: 1 613 761 1441

Najera, Raquel Gutierrez, Ms.
Asesor
UNION DE PUEBLOS DE MANANTLAN
Espanola no.2864 - Jardines de la Cruz
Guadalara- Jalisco, Mexico
T: 36 6115222

Nemi, Juan Antonio, Mr.
Dep.Federal
CONGRESISTA MEXICANO CONGRESSO DE LA UNION
Mexico City, Mexico

Nunes, Carlos, Sr.
Diretor
IMDEC
Pino 2237 - Colonia del Fresmo
Guadalajara, Mexico

Olguin, Eugenia J., Ms.
Dr.
INSTITUTE OF ECOLOGY
KM 2.5 Antigua Carretera A Coatepec. Apdo. Postal 63
Xalapa - Ver, Mexico
T: 52 281 8-60-00 F: 52 281 8-68-09

Olguin, Mari, Ms.
WEDO - CASA DE LA MUJER ROSARIOS CASTELLANOS
Constitucion # 30
Oaxaca 68000, Mexico
T: 52 954-6427 F: 52 954-6427

Ondarza, Raul, Mr.
UNIVERSIDAD NACIONAL DE MEXICO
Olivar De Los Padres 941
Mexico City D.F. 01780, Mexico
T: 52 5 595-4195

Ortega, Jose Ignacio Felix-Diaz, Sr.
Jefe Del Dept.De Medio Ambien.
UNIVERSIDAD AUTONOMA METROPOLITANA AZCAPOTZALCO
Nebraska 209 Depto.3 Col. Napoles CP03810
Mexico City, Mexico
T: 52 5 687-0727 F: 52 5 394-8396

Ortega, Susana Guzman, Ms.
Media
INAINE
Gladiolas # 56 - Colonia Ciudad Jardin
Mexico City D.F., Mexico
T: 52 5 549-2642 F: 52 5 689-5972

Ortiz-Monasterio, Martha, Ms.
Apdo 28
Tepoztlan 62520, Mexico

Palazuelos, Eduardo, Mr.
Pedriata
HOSPITAL ABC
Sur 136 Exq. Observatorio 107
Mexico City D.F. 01120, Mexico
T: 52 5 2723585 F: 52 5 2722155

Parra, Maria Angelica Luna, Ms.
Membro
MOVIMIENTO CIUDADANO DE UNE-PRI
Pareo De la Reforma 2410-6
Mexico City 06030, Mexico
T: 52 5 596-6591

Pena, Ofelia Perez, Ms.
Env. Educ. Coord.
IMDEC
Pino 2237 Col del Fresno
Guadalajara, Mexico
T: 52 36 104536 F: 52 36 110714

Peralta, Joaquin Esteva, Sr.
CENTRO DE ESTUDIOS SOCIALES Y ECOLOGICOS - CESE
Navarrete 50
Patzcuaro, Michoacan 61600, Mexico
T: 52 562 20852 F: 52 562 20852

Peralta, Rosa Martha, Ms.
Cordinator
PROEZA, A.C.
Torrente 4-801 Col-Las Aguilas
Mexico City 01710, Mexico
T: 52 5 61-51-97-92

Perez, Adrian Javier Lopez, Mr.
Delegado
ASOCIACION NACIONAL DE ESTUDIANTES DE LAS CIENCIAS AGROPECUARIAS (ANECA)
Km 38.5 Carr.
Chapingo, Texcoco 56230, Mexico
T: 91 595 4-22-00 F: 91 595 4-22-00

Perez, Maria Elena, Ms.
CONSEJO MEXICANO DE LOS 500 DE RESISTENCIA
Fco Field y Jurado 51
Mexico City, Mexico
T: 52 5 539-0015 F: 52 5 539-0015

Picado, Enrique, Mr.
Board Member
FCOC
Cordobanes 24, Col. San Jose Insurgentes
Mexico City 03900, Mexico
T: 52 5 6512371 F: 52 5 5453263

Ponge-Nava, Diana, Ms.
Dir Of International Coop
SECRETARIA DE DESARROLHO SOCIAL
P Hacienda 85 503
Mexico City 04250, Mexico

Preciado Cueva, Antonio, Mr.
ASOCIACION ECOLOGICA COYOACAN, A.C.
San Lucas Coyoacan, 18
Mexico City, Mexico
T: 52 5 549-05-74 F: 52 5 532-56-39

Provencio, Enrique, Mr.
PROGRAMA DE APROVECHAMIENTO DE RECURSOS
Argos 101 Predegal 2
Mexico City D.F. 1072O, Mexico
T: 52 5 5689021 F: 52 5 5488186

Putten, Han Van, Mr.
President
HABITAT INTERNATIONAL COALITION
Cordobanes 24, Col. San Jose Insurgentes
Mexico City 03900, Mexico
T: 52 5 6512371 F: 52 5 5453263

Reyes, Javier, Mr.
Coordinator
CESE-CENTRO DE ESTUDOS SOCIALES E ECOLOGICOS
Copevi. A.C. 1o de Mayo 151, Col. San Pedro de los Pinos, C.P. 03800
Mexico City, Mexico
T: 52 5 515-9627 F: 52 5 271-4119

Rincon, Froylan, Mr.
CIAT
Apartado Aereo 6743
Cali, Mexico
T: 57 23 675050 F: 57 23 647243

Rio Uscanga, Felipe, Mr.
UP WITH PEOPLE
Bernal Diaz del Castillo 141 esq. Colon
Veracruz, Mexico

Riojas, Guadalupe, Ms.
Org. Eventos Especiater
INAINE
Gladielas 56
Mexico City D.F. 03020, Mexico
T: 52 5 689-0473 F: 52 5 689-5972

Robles, Fidel, Mr.
Board Member
FCOC
Cordobanes 24, Col. San Jose Insurgentes
Mexico City 03900, Mexico
T: 52 5 6512371 F: 52 5 5453263

Rojo, Jorge, Sr.
ROTARY INTERNATIONAL
Calle 13 G4 Club De Golf Mexico
Mexico City 14620, Mexico
T: 52 5 6891405 F: 52 5 5443803

Romero, Gustavo, Mr.
Director
FOSOVI-FONDO SOLIDARIO PARA LA VIVIENDA
Copevi. A.C. 1o de Mayo 151, Col. San Pedro de los Pinos, C.P. 03800
Mexico City, Mexico
T: 52 5 515-9627 F: 52 5 271-4119

Rooy, Cornelius Kees Van, Sr.
Professor/Investigador
VAM
Av. San Pablo no. 180 Co. Reynosa Tamps
Mexico City 02200, Mexico
T: 52 5 382-5000 F: 52 5 394-8396

Royes, Javier, Sr.
Coordinador
CENTRO DE ESTUDIOS SODIALES Y ECOLOGICOS
Av. 1o. de Mayo 151, Col San Pedro de los Pinos
Mexico City 03800, Mexico
T: 52 5 515-9627 F: 52 5 271-4119

Ruge, Reinhart, Mr.
President
WORLD CONSTITUTION AND PARLIAMENT ASC.
Apartado 607
Cuernavaca, Mexico
T: 52 739 50003 F: 52 739 51264

Salazar, Hilda, Ms.
FORO MEXICANO
Vesubio 57 Col. Alpes
Mexico City D.F., Mexico
T: 52 5 658 9690 F: 52 5 660 8994

Salinas, Fernanda Paz, Ms.
CENTRO REGIONAL DE INVESTIGACIONES MULTIDISCIPLINARIAS
Av. Universidad s/n Circuito II, Col. Chamilpa
Cuernavaca, Mexico
T: 52 731 13-0555 F: 52 731 17-5981

Sanches, Daniel, Mr.
Gegrafo
UNE
Edif. I-6 Depto. 602, Fuentas Brotantes, Tlalpan
Mexico City D.F. 14410, Mexico
T: 52 5 705-1303

Sanchez, Ernesto Aloma, Mr.
Secretary General
WORLD CONSTITUTION AND PARLIAMENT ASSOCIATION
Ave. Revolucion #25
Tepotztlan - MO 62520, Mexico
T: 52 739 50003 F: 52 739 51264

Sanguinetti, Ana, Sra.
I F B P W
Presidente Mazarik 29
Mexico City D.F., Mexico
T: 52 5 250-7883

Schuchmann, Helena, Ms.
I N A I N E
Gladiolla 56
Cidade Do Mexico, Mexico

Senaual, Corta
Reporter
SERVICIOS INFOMATIVOS Y DE ORIENTACION
Enrique Rebsamen, 545
Narvarte, Mexico
T: 5436184 F: 6065883

Sosa, Pricila, Sra.
GEMA
Plaza Galerias, Local 13. Aptado Postal 479
Cancun - QR, Mexico
T: 52 988 469 44 F: 52 988 469 44

Soto, Cecilia, Ms.
Dep.Federal
CONGRESSISTA MEXICANA
Congresso De la Union
Mexico City, Mexico

Tellez, Evagelina Hernandez
Correspondent
LA JORNADA
Balderas #68, Centro
Mexico
T: 5 5181764 F: 5 5212763

Travesi, Carolina Garcia, Sra.
President
CLUB ECOLOGISTA
15 Allende 301, Centro
Victoria - Tam, Mexico

Travesi, Laura Gonzalez, Sra.
Child Ecologist
15 Allende 301, Centro
Victoria - Tam, Mexico

Tum, Rigoberta Menchu
Miembro Directiva
COMITE DE UNIDAD CAMPESINA
Calle Patricio Sanz 449 - Col Del Valle
Mexico City, Mexico
T: 52 5 523-0492 F: 52 5 523-0492

Uriarte, Jose Luis Soria, Sr.
Active Member
A.S.P.A., AC.
Almada # 35 Col. Centro
Navolato - SI, Mexico
T: 52 672 70355 F: 52 672 70401

Valencia, Miguel, Mr.& Mrs.
Commissioner
FORO REGIONAL ECOLOGISTA VALLE DE MEXICO
Georgia, 30 - Esq. Nebraska (bajos) - Col. Napoles
Mexico City 03810, Mexico
T: 52 5 687-0889 F: 52 5 543-1103

Valle, Lenica, Ms.
Asesor
SECRETARIA DE AGRICULTURA Y RECURSOS HIDRAULICOS
Providencia no. 706 Col. del Valle CP 03100
Mexico City, Mexico
T: 52 5 543-6055 F: 52 5 280-3196

Velazques, Enrique, Sr.
Delegado
FOROMEX
Copevi. A.C. 1o de Mayo 151, Col. San Pedro de los Pinos, C.P. 03800
Mexico City, Mexico
T: 52 5 515-9627 F: 52 5 271-4119

Velazques, Socorro, Ms.
Coordinadora Area Urbana
HABITAT INTERNATIONAL COALITION
Cordobanes 24, Col. San Jose Insurgentes
Mexico City 03900, Mexico
T: 52 5 6512371 F: 52 5 5453263

Ylatalo, Riitta, Ms.
HABITAT INTERNATIONAL COALITION
Cordobanes 24, Col. San Jose Insurgentes
Mexico City 03900, Mexico
T: 52 5 6512371 F: 52 5 5453263

ADVISORS

De La Madrid Hurtado, Miguel, H.E. Mr.
Former President Of Mexico
04010 Mexico, D.F., Mexico
T: 525 658 4459/554 7784 F: 525 658 7979

PARLIAMENTARIANS / SPIRITUAL LEADERS

Armenta, Alberto Carillo
Senado Xicotencatl 9
Mexico 1, DF, Mexico
T: 521 00 20
TL: 017 60172 CSNME

Corrales, Miguel Angel Leon
Senado Xicotencatl 9
Mexico 1, DF, Mexico
T: 521 00 20
TL: 017 60172 CSNME

Dib, Juan Antonbio Nemi
Senado Xicotencatl 9
Mexico 1, DF, Mexico
T: 521 00 20
TL: 017 60172 CSNME

Esponda, Blanca, Dep.
Senado Xicotencatl 9
Mexico 1, DF, Mexico
T: 521 00 20
TL: 017 60172 CSNME

Gonzales, Cecilia Soto, Dep.
Senado Xicotencatl 9
Mexico 1, DF, Mexico
T: 521 00 20
TL: 017 60172 CSNME

Oliveros, Gabriela Guerrero
Senado Xicotencatl 9
Mexico 1, DF, Mexico
T: 521 00 20
TL: 017 60172 CSNME

Rodriguez, Alejandro Encinas
Senado Xicotencatl 9
Mexico 1, DF, Mexico
T: 521 00 20
TL: 017 60172 CSNME

Velasco, Miguel Aleman, Senator
Senado Xicotencatl 9
Mexico 1, DF, Mexico
T: 521 00 20
TL: 017 60172 CSNME

WISDOMKEEPERS

Aridjis, Homero & Betty
Grupo de los Cien
Sierra Jiutepec 155-B
Lomas Barrilaco, Mexico City 11010, Mexico
T: 52-5-540-7379 F: 52-5-520-3577

MICRONESIA - GOVERNMENT

Fritz, Jack, The Hon.
Speaker
Federated States of Micronesia Congress
Permanent Mission to the United Nations, 820 2nd Ave., Suite 204
New York, NY 10017, United States of America
T: 212 697 8370 F: 212 697 8295

Iehsi, Ieske, Mr.
Chief of Staff, Aide de camp
Office of the President
Permanent Mission to the United Nations, 820 2nd Ave., Suite 204
New York, NY 10017, United States of America
T: 212 697 8370 F: 212 697 8295

Marehalau, Jessie B., H.E. Mr.
Ambassador Extraordinary and Plenipotentiary
Permanent Representative to the United Nations, New York
Permanent Mission to the United Nations, 820 2nd Ave., Suite 204
New York, NY 10017, United States of America
T: 212 697 8370 F: 212 697 8295

Moses, Resio S., The Hon.
Secretary of External Affairs
Permanent Mission of Micronesia
820 2nd Ave., Suite 204
New York, NY 10017, United States of America
T: 212 697 8370 F: 212 697 8295

Oltar, Bailey, H.E. Mr.
President
Federated States of Micronesia
820 2nd Ave., Suite 204
New York, NY 10017, United States of America
T: 212 697 8370 F: 212 697 8295

Phelan, Maureen, Ms
Assistant Attorney General
OFFICE OF ATTORNEY GENERAL
Box PS-105, Palikir
Pohnpei FM 96941, Micronesia
T: 691 320 2644 F: 691 320 2234

Ragimar-Subolmar, J., Mr.
Director
Department of External Affairs
Permanent Mission to the United Nations, 820 2nd Ave., Suite 204
New York, NY 10017, United States of America
T: 212 697 8370 F: 212 697 8295

Rospel, Elieser, Mr.
Office of Public Information
Office of the President
Permanent Mission to the United Nations, 820 2nd Ave., Suite 204
New York, NY 10017, United States of America
T: 212 697 8370 F: 212 697 8295

Stovall III, James T., Mr.
Envoy Extraordinary and Minister Plenipotentiary
Deputy Permanent Representative to the United Nations, New York
Permanent Mission to the United Nations, 820 2nd Ave., Suite 204
New York, NY 10017, United States of America
T: 212 697 8370 F: 212 697 8295

MONACO - GOVERNMENT

Briand, Frederic, Mr
Director General
C.I.E.S.M.
16 Bvd De Suisse
MI 98000 Monag, Monaco
T: (33) 93 30 38 79 F: (33) 92 16 11 95

Fautrier, Bernard, S.E. M.
Conseiller de Gouvernement pour les Travaux Publics et les Affaires Sociales (Chef de la Delegation)
Permanent Mission of Monaco
845 3rd Ave., 19th Floor
New York, NY 10022, United States of America
T: 212 759 5227 F: 212 754 9320

Franzi, Etienne, M.
Delegue Permanent Adjoint aupres des Organismes Internationaux
Vice President du Bureau des Parties Contractantes a la Convention de Barcelone pour la protection de la Mer Mediterranee contre la pollution
Permanent Mission to the United Nations, 845 3rd Ave., 19th Floor
New York, NY 10022, United States of America
T: 212 759 5227 F: 212 754 9320

Rainier III, Son Altesse Serenissime, Prince
Head of State
Permanent Observer Mission of Monaco
845 3rd Ave., 19th Floor
New York NY 10022, United States of America
T: 212 759 5227 F: 212 754 9320

Van Klaveren, Patrick, M.
Chef du Service de l' Environnement
Permanent Mission of Monaco
845 3rd Ave., 19th Floor
New York, NY 10022, United States of America
T: 212 759 5227 F: 212 754 9320

MONGOLIA - GOVERNMENT

Adyasuren, Tsohiogyn, Mr.
Chief of Department
State Committe for Environmental Protection
Permanent Mission to the United Nations, 6 E. 77th Str. New York, NY, 10021
F: 212 861 9460
TL: 212 861 9464

Batjargal, Zambyn, H.E. Mr.
Minister of State
Committee for Environmental Protection
(Deputy Head for Delegation)
T: 10021
TL: 212 861 9460, E: 212 861 9464

Buuvei, Balhaajavn, Mr.
Personal assistant to the Prime Minister
Permanent Mission of Mongolia
6 E. 77th Str.
New York, NY 10021, United States of America
T: 212 861 9460 F: 212 861 9464

Byambasuren, Dashiin, H.E. Mr.
Prime Minister
Permanen Mission of Mongolia
6 E. 77th Str.
New York, NY 10021, United States of America
T: 212 861 9460 F: 212 861 9464

Erdenechuluun, Luvsangiin, H.E. Mr.
Ambassador Extraordinary and Plenipotentiary
Permanent Mission of Mongolia
6 E. 77th Str.
New York, NY 10021, United States of America
T: 212 861 9460 F: 212 861 9464

Munhtuvshingyn, Ganbat, Mr.
Chairman
Mongolian Green Party
Ulaanbaata C.P.O.B-1089
Ulaanbaatar, Mongolia
T: 23871

Zambyn, Batjargal
Minister
MINISTRY FOR NATURE AND ENVIRONMENT
Khudaldaany Gudamj 5 Batgal Yam
Ulaanbaatar, Mongolia
F: TEX 79343 MACNE MH

MONROVIA - PARLIAMENTARIANS / SPIRITUAL LEADERS

Campbell, Ishmael P.
MP
The Senate Capitol, Monrovia
F: 221840

MOROCCO - GOVERNMENT

Abouzaid, Houssain, Dr
Directeur De La Qualite De l'Eau
ONEP
BP Rabat - Chellah
Rabat, Morocco
T: 2127 758892 F: 2127 752377

Abouzaid, Lahcen, M.
Directeur du Laboratoire de Controle de la Qualite des Eaux
Permanent Mission of Morocco
767 3rd Ave, 30th Floor
New York, NY 10017, United States of America
T: 212 421 1580 F: 212 980 1512

Achergui, Mohamed, M.
President
Conseil Municipal de Mohammedia
Permanent Mission to the United Nations, 767 3rd Ave, 30th Floor New York, NY, 10017
F: 212 421 1580
TL: 212 980 1512

Agoumi, Ali, M.
Directeur de la Recherche et de la Formation continue de l' Ecole Hassania des Travaux Publics
767 3rd Ave, 30th Floor New York, NY, 10017
F: 212 421 1580
TL: 212 980 1512

Akhmiss, Mustapha, M.
Secretaire General du Ministere de la Sante Publique
Permanent Mission of Morocco
767 3rd Ave, 30th Floor
New York, NY 10017, United States of America
T: 212 421 1580 F: 212 980 1512

Alaoui El Mdarhri, El Mostafa
Charge d'Etude
MINISTERE DE L'ENERGIE
Et des Mines
Morocco
T: 212 7 774715 F: 212 7 772115

Alaoui Mdaghri, Driss, S.E.M.
Ministre
Ministere de l'Energie et des Mines
Morocco
T: 212 7 773675 F: 212 7 772115

Benmoussa, Abderrahim, M.
Directeur de la Cooperation Economique Multilaterale au Ministere d' Etat Charge des Affaires Etrangeres et de la Cooperation
Permanent Mission of Morocco
767 3rd Ave, 30th Floor
New York, NY 10017, United States of America
T: 212 421 1580 F: 212 980 1512

Bensari, Ahmed, Mr.
Directeur
Meteorologie Nationale
BP 8106 Casablanca Oasis Morocco Casablanca Oasis,
F: 903330 902178
TL: 900742

Cheikh, Biadi Lah Mohamed, M.
Depute
President de la Commission Infrastructure, Poste et Telecommunications
Permanent Mission to the United Nations, 767 3rd Ave, 30th Floor New York, NY, 10017
F: 212 421 1580
TL: 212 980 1512

Dadi Essakalli, Zahia, Mme
Inspecteur General
Ministere du Tourisme
Permanent Mission to the United Nations, 767 3rd Ave, 30th Floor New York, NY, 10017
F: 212 421 1580
TL: 212 980 1512

Elkhaier, Abdelhak, M.
Directeur de Developpement
Office Cherifien des Phosphates
Permanent Mission to the United Nations, 767 3rd Ave, 30th Floor New York, NY, 10017
F: 212 421 1580
TL: 212 980 1512

Fassi Fihri, Lahcen, M.
Chef de la Division des Amenagements et des Equipements
Ministere du Tourisme
Permanent Mission to the United Nations, 767 3rd Ave, 30th Floor New York, NY, 10017
F: 212 421 1580
TL: 212 980 1512

Fassi-Fihri, El Houcine
Repris Eut Out
ADMINISTRATION DU TOUISME
42 Rue Oulad Fyane
Aviation Rarat, Morocco
T: 212 7 70 12 80

Fassi-Fihri, Zineb
Chargee d'Etudes
MINISTERE DE L'ENERGIE ET des Hines - Ag Dal Rabat
Rabat, Morocco
T: 77 47 15 F: 77 47 21

Fikri, Mohamed, M.
President
Conseil Municipal de Mechraa Belksiri
Permanent Mission to the United Nations, 767 3rd Ave, 30th Floor New York, NY, 10017
F: 212 421 1580
TL: 212 980 1512

Filali, AbDellatif, S.E. M.
Ministre d'Etat Charge
Affaires Etrangeres et de la Cooperation (Chef Adjoint de la Delegation)
Permanent Mission to the United Nations, 767 3rd Ave, 30th Floor New York, NY, 10017
F: 212 421 1580
TL: 212 980 1512

Frouji, Abdellah, M.
Ministere l'Education Nationale
Permanent Mission of Morocco
767 3rd Ave, 30th Floor
New York, NY 10017, United States of America
T: 212 421 1580 F: 212 980 1512

Guessab, Driss, M.
Directeur des Affaires Rurales, Ministere de l' Interieur et de l' Information
Permanent Mission of Morocco
767 3rd Ave, 30th Floor
New York, NY 10017, United States of America
T: 212 421 1580 F: 212 980 1512

Hilali, M'hamed, M.
Charge d' Etudes au Ministere
Charge des Affaires Economiques et de la Privatisation
Permanent Mission to the United Nations, 767 3rd Ave, 30th Floor New York, NY, 10017
F: 212 421 1580
TL: 212 980 1512

Jellali, Mohammed
Director Of Water Research & Planni
ADMINISTRATION DE L'HYDRAULIQUE
Rue Hassan Benchekroun Agdal
Rabat, Morocco
T: 212 7 77 86 90 F: 212 7 77 60 81

Kemmou, Mohamed, M.
President
Conseil Municipal Hay El Hassani, Vice-President de la Communaute Urbaine de Casablanca
Permanent Mission to the United Nations, 767 3rd Ave, 30th Floor New York, NY, 10017
F: 212 421 1580
TL: 212 980 1512

Kerbid, Mohamed, M.
Secretaire General
Ministere de l'Energie et des Mines
Permanent Mission to the United Nations, 767 3rd Ave, 30th Floor New York, NY, 10017
F: 212 421 1580
TL: 212 980 1512

Khalili, Abdelhamid, M.
Rapporteur-Adjoint
Commission Parlementaire de l'Habitat, de l'Amenagement du Territoire et de l'Environnement
Permanent Mission to the United Nations, 767 3rd Ave, 30th Floor New York, NY, 10017
F: 212 421 1580
TL: 212 980 1512

Lakhimiri, Abdeljabbar, M.
Chef de Service a la Direaction Generale de l'Urbanisme et de l'Amenagement du Teritoire et de l'Environnement
Ministere de l'Interieur et de l'Information
Permanent Mission to the United Nations, 767 3rd Ave, 30th Floor New York, NY, 10017
F: 212 421 1580
TL: 212 980 1512

Layachi, Bani, Mme
Directeur de l' Amenagement du Territoire et de l' Environnement, Ministere de l' Interieur et de l' Information
Permanent Mission of Morocco
767 3rd Ave, 30th Floor
New York, NY 10017, United States of America
T: 212 421 1580 F: 212 980 1512

M'Daghri Alaoui, Mustapha, M.
Directeur de Cabinet du Ministre de l' Energie et des Mines
Permanent Mission of Morocco
767 3rd Ave, 30th Floor
New York, NY 10017, United States of America
T: 212 421 1580 F: 212 980 1512

Maliti, M'hamed
Chef Division Environnement
Ministere de l' Interieur
Villa 218 El Harboura Plaze
Temara, Morocco
T: 212 7 7476 10

Mohamed, Son Altesse Royale le Prince Heritier Sidi
Morocco
767 3rd Ave, 30th Floor
New York, NY 10017, United States of America
T: 212 421 1580 F: 212 980 1512

Mohamed-Cheikh, Biadillah
Professeur De Medecine
DEPUTE (IPU)
Lot Almanar-WG-Villa
Maryam Casablanca, Morocco
T: 365046 22

Nouiji, Ahmed, M.
Directeur du Centre des Etudes Recherches Demographiques
Ministere du Plan
Permanent Mission to the United Nations, 767 3rd Ave, 30th Floor New York, NY, 10017
F: 212 421 1580
TL: 212 980 1512

Nouri, Chakib, M.
Chef du Service de l'Information et des Relations publiques a la Direction Generale de l'Urbanisme et de l'Amenagement du Territoire National et de l'Environnement
Ministere de l'Interieur et de l'Information
Permanent Mission to the United Nations, 767 3rd Ave, 30th Floor New York, NY, 10017
F: 212 421 1580
TL: 212 980 1512

Reffouh, Larbi, S.E. M.
Ambassadeur
Sa Majeste le Roi au Bresil
Permanent Mission to the United Nations, 767 3rd Ave, 30th Floor New York, NY, 10017
F: 212 421 1580
TL: 212 980 1512

Senoussi, Badredine, M.
President
Conseil Municipal de Rabat-Youssoufla
Permanent Mission to the United Nations, 767 3rd Ave, 30th Floor New York, NY, 10017
F: 212 421 1580
TL: 212 980 1512

Snoussi, Maana, M.
President de la Commission Parlementaire de l'Habitat de l'Amenagement du Territoire et de l'Environnement
Permanent Mission of Morocco
767 3rd Ave, 30th Floor
New York, NY 10017, United States of America
T: 212 421 1580 F: 212 980 1512

Tagrite, Lahcen, M.
Directeur General
Ministere de l' Interieur
Rabat, Morocco
T: 766658 59

Tamouh, Mohamed, M.
Chef de Service de la Protection de l'Environnement
Ministere du Commerce et de l'Industrie
Permanent Mission to the United Nations, 767 3rd Ave, 30th Floor New York, NY, 10017
F: 212 421 1580
TL: 212 980 1512

Tijani, Rhamni, M.
Directeur de la Marine Marchande, Ministere des Peches Maritimes et de la Marine Marchande
Permanent Mission of Morocco
767 3rd Ave, 30th Floor
New York, NY 10017, United States of America
T: 212 421 1580 F: 212 980 1512

Zerouali, Abdelhaj, M.
Chef De Laboratore Pollution
DIR AMENAGEMENT TERRITOIRE ET ENVIR
M. Interieur
Rabat, Morocco
T: 7666 59 768423

Zouagui, Mohamed, M.
de la Direction Generale des Collectivites Locales
Ministere de l'Interieur et de l'Information
Permanent Mission to the United Nations, 767 3rd Ave, 30th Floor New York, NY, 10017
F: 212 421 1580
TL: 212 980 1512

NGO

Alaoui, Fatima, Ms.
FORUM MAGHREBIN POUR L'ENVIRONNEMENT ET DE DEVELOPPEMENT - FMED
2 rue Zahla - B.P. 403
Rabat, Morocco
T: 72 746 2127 F: 72 746 2127

Benjelloun, Ahmed
FMPED
2 rue Zahla, BP 403
Rabat, Morocco

GLOBAL FORUM

Abdellah, Baudahrain, Mr.
Professor
UNIVERSITE MOHAMMDE V
36 Rue du Point du Jour Bourgogne
CasaBlanca 01, Morocco
T: 212 2 27 96 77

Dinia, Hayat
RIBAT AL FATH ASSOCIATION
Rue Madani Bel Housni Route de Zaer
Rabat, Morocco
T: 212 7 75610608

Diouri, Azzedine, Mr.
Chief Engineer
LABORATOIRE PUBLIC D'ESSAIS ET D'ETUDES
25 Rue D'Azilal - BP 13389
CasaBlanca, Morocco
T: 212 2 307510 F: 212 2 301550
TL: 45046

Fatima, Alaoui
FORUM MAGHREBIN POUR L'ENVIRONNEMENT ET LE DEVELOPEMENT
BP 403
Rabat, Morocco
T: 212 7 727 406 F: 212 7 727 406

Hakimi, Ahmed, Mr.
General Director
LABORATOIRE PUBLIC D'ESSAIS ET D'ETUDES
25 Rue D'Azilal - BP 13389
CasaBlanca, Morocco
T: 212 2 307510 F: 212 2 301550
TL: 45046

Hassar, Fatima, Sra.
Leader
LIGUE MAROCAINE PROTECTION DE L'ENFANCE
3, RueBhotess al av Tzi
Rabat, Morocco
T: 212 7 759675 F: 212 7 759680

Hayat, Dinia, Ms.
Doctor
RIBAT ELFATH ASSOCIATION
Rue Madami Ben El Housinia
Rabat, Morocco
T: 212 7 75-6406 F: 212 7 75-6409
TL: 32993M

Jabry, El Hadj, Mr.
Chief Engineer
LABORATOIRE PUBLIC D'ESSAIS ET D'ETUDES
25 Rue D'Azilal - BP 13389
CasaBlanca, Morocco
T: 212 2 307510 F: 212 2 301550
TL: 45046

Kaouakbi, Khedija, Ms.
ASSOCIATION POUR LE DROIT DE L'ENVIRONNEMENT
BP 5137
Rabat, Morocco
T: 212 7 75 80 23

Laalou, Habiba Amal, Ms.
BOUREGREG ASSOCIATION
Rue Gued Fddahar Bettana
Sale, Morocco
T: 212 7 780787 F: 212 7 781104

Laovina, Abdellah, Mr.
Professor
COMMITTEE INTER. ASSOCIATIONS
Rue Madami Ben El Housinia
Rabat, Morocco
T: 212 7 75-6106 F: 212 7 75-6409
TL: 32993M

Mostapha, El Haiba, Mr.
LABORATOIRE PUBLIC D'ESSAIS ET D'ETUDES
25 Rue D'Azilal - BP 13389
CasaBlanca, Morocco
T: 212 2 307510 F: 212 2 301550
TL: 45046

Naciri, Mohamed, Mr.
INSTITUT AGRONOMIQUE ET VETERINAIRE HASSAN II
5, Zenket Jagir Ibn Hayane
Rabat, Morocco
T: 212 7 77 17 59 F: 212 7 77 98 85
TL: 5725 KN

Sbai, Nafissa, Ms.
FONDS D'APPUI ONG
Rabat, Morocco

Skalli, Zahia, Ms.
President
MOVEMENT NATIONAL DES ECOLOGISTS
7 Rue Abdelonahed Monrrakouchi-Orangers
Rabat, Morocco
T: 212 7 72-7552

Zahia, Essakalli Dadi, Ms.
Presidente
MOUVEMENT NATIONAL DES ECOLOGISTES
7 Rue Abdelouahed El Marrakchi Les Orangers
Rabat, Morocco
T: 212 7 76 4150 F: 212 7 76 0574

Zemmana, Mohamed, Mr.
FES SAIS
BP 529
Rabat, Morocco
T: 212 7 76-2540

Zniber, Fouzi, Mr.
ASSOCIATION BOUREGREG
Rabat, Morocco

MOZAMBIQUE - GOVERNMENT

Afonso, Pedro Comissario, H.E. Dr.
Ambassador Extraordinary and Plenipotentiary
Permanent Representative to the United Nations, New York
Permanent Mission to the United Nations, 70 E. 79th Street
New York, NY 10021, United States of America
T: 212 517 4550 F: 212 734 3083

Baquete, Dulcineia
Technician
National Comission Of Environment
Aj. Acordes De Lusaka,No2115 C.P.1310-Maputo
Maputo, Mozambique
T: 460503

Baquete, Evaristo, Mr.
Director
Food Hygiene and Water Laboratory, Ministry of Health
Permanent Mission to the United Nations, 70 E. 79th Street New York, NY, 10021
F: 212 517 4550
TL: 212 734 3083

Bata, Roque, Mr.
Director for Administration and Finance President's Office
Permanent Mission of Mozambique
70 E. 79th Street
New York, NY 10021, United States of America
T: 212 517 4550 F: 212 734 3083

Candeeiro, Leonardo, Mr.
Secretary General
Mozambican Youth
Permanent Mission to the United Nations, 70 E. 79th Street New York, NY, 10021
F: 212 517 4550
TL: 212 734 3083

Chissano, Joaquim Alberto, H.E. Mr.
President
Permanent Mission of Mozambique
70 E. 79th Str.
New York NY 10021, United States of America
F: 212 517 4550
TL: 212 734 3083

Couto, Mia, Mr.
Environmental Expert of the Division of Environment
Permanent Mission of Mozambique
70 E. 79th Street
New York, NY 10021, United States of America
T: 212 517 4550 F: 212 734 3083

Ferraz, Bernardo, Mr. Dr.
General Director
Division of Environment
Permanent Mission to the United Nations, 70 E. 79th Street New York, NY, 10021
F: 212 517 4550
TL: 212 734 3083

Ferreira, Sergio, Mr.
Director
National Institute of Meteorology, Ministry of Transports and Communications
Permanent Mission to the United Nations, 70 E. 79th Street New York, NY, 10021
F: 212 517 4550
TL: 212 734 3083

Infante, Florencio, Mr.
Economic Adviser
President's Office
Permanent Mission to the United Nations, 70 E. 79th Street New York, NY, 10021
F: 212 517 4550
TL: 212 734 3083

Kachamila, John William, H.E. Mr.
Minister
Mineral Resources
Permanent Mission to the United Nations, 70 E. 79th Street New York, NY, 10021
F: 212 517 4550
TL: 212 734 3083

Mafulana, Arminda, Mrs.
Secretary of the President
Permanent Mission of Mozambique
70 E. 79th Street
New York, NY 10021, United States of America
T: 212 517 4550 F: 212 734 3083

Mandlate, Felix, Mr.
President
Institute for Rural Development
Permanent Mission to the United Nations, 70 E. 79th Street New York, NY, 10021
F: 212 517 4550
TL: 212 734 3083

Manhica, Virgilio, Mr.
Officer of Protocol
Permanent Mission of Mozambique
70 E. 79th Street
New York, NY 10021, United States of America
T: 212 517 4550 F: 212 734 3083

Mathusse, Hilario, Mr.
Secretary General
Mozambican Journalists
Permanent Mission to the United Nations, 70 E. 79th Street New York, NY, 10021
F: 212 517 4550
TL: 212 734 3083

Mocumbi, Pascoal Manuel, H.E. Dr.
Minister
Foreign Affairs
Permanent Mission to the United Nations, 70 E. 79th Street New York, NY, 10021
F: 212 517 4550
TL: 212 734 3083

Molane, Arlindo, Mr.
National Director of Industry
Ministry of Industry and Energy
Permanent Mission to the United Nations, 70 E. 79th Street New York, NY, 10021
F: 212 517 4550
TL: 212 734 3083

Mota, Helena, Mrs.
National Director of Fishing, Secretary of State for Fishing
Permanent Mission of Mozambique
70 E. 79th Street
New York, NY 10021, United States of America
T: 212 517 4550 F: 212 734 3083

Muchangos, Aniceto dos, H.E. Mr.
Minister
Education
Permanent Mission to the United Nations, 70 E. 79th Street New York, NY, 10021
F: 212 517 4550
TL: 212 734 3083

Nunes Junior, Jose Miguel, Mr.
National Director of Protocol
Permanent Mission of Mozambique
70 E. 79th Street
New York, NY 10021, United States of America
T: 212 517 4550 F: 212 734 3083

Nuvunga, Milagre, Mr.
National Deputy Director of Forestry and Wild Fauna
Ministry of Agriculture
Permanent Mission to the United Nations, 70 E. 79th Street New York, NY, 10021
F: 212 517 4550
TL: 212 734 3083

Pelembe, Abrao, Mr.
Officer of Protocol
Permanent Mission of Mozambique
70 E. 79th Street
New York, NY 10021, United States of America
T: 212 517 4550 F: 212 734 3083

Sales, Tomas, Mr.
Second Secretary
Permanent Mission of Mozambique
70 E. 79th Street
New York, NY 10021, United States of America
T: 212 517 4550 F: 212 734 3083

Ventura, Flora, Ms.
Member of 'Group de Trabalho Ambiental'
Permanent Mission of Mozambique
70 E. 79th Street
New York, NY 10021, United States of America
T: 212 517 4550 F: 212 734 3083

Zambeze, Amelia, Mrs.
Associacao Mocombicana da Mulher Rural
P. O. Box 1892
Maputo, Mozambique
F: 25 81 491628

Zandamela, Alexandre da Conceicao, Mr.
Director
Division of International Organization and Conferences, Ministry of Foreign Affairs
Permanent Mission to the United Nations, 70 E. 79th Street New York, NY, 10021
F: 212 517 4550
TL: 212 734 3083

PRESS

Cossa, Hilario Cesar Carlos
Editor
JORNAL NOTICIAS
Rua de Hulene
Maputo, Mozambique, Mozambique

Lauriciano, Gil Fabiao
Correspondent
MEDIA FAX MOZAMBIQUE
Av. Salvador Allende, 93
Maputo, Mozambique, Mozambique

Mausse, Jaime Zaquere
Reporter Radio/Television
RADIO MOZAMBIQUE
Maputo, Mozambique

Torohato, Artur
PRESIDENCIA DA REPUBLICA
Maputo, Mozambique

GLOBAL FORUM

Andrade, Ximena, Ms.
Geographer
WOMEN AND ENVIRONMENT GROUP - MOZAMBIQUE
Universidde Eduardo Mondlana - Centro de Estudos Africanos
Maputo, Mozambique
T: 25 81 491896 F: 25 81 491896
TL: P.O. Box 1635

Candieiro, Leonardo, Mr.
ORGANIZATION OF MOZAMBICAN YOUTH
P. O. Box 2998
Maputo, Mozambique
T: 25 81 492541 F: 25 81 490161

Caulo, Carlos Alberto, Mr.
Federacao Internacional M.C. Juridica
Av. 24 De Julho, 458
Maputo, Mozambique
T: 25 81 491121 F: 25 81 492068
TL: 243

Lopes, Henrique, Mr.
Director
AFRICA INTERNATIONAL PROGRAM FOR EXCHANGE OF TECHNICAL INFORMATION
Av. Vladimir Lenine 570, 2o. PO Box 1574
Maputo, Mozambique
T: 25 81 431029
TL: 6-555 UEM CB MO

Mata, Eusebia, Ms.
Geographer
WOMEN AND ENVIRONMENT GROUP - MOZAMBIQUE
NORAD - Maputo - Rua de Kassuende, 118
Maputo, Mozambique
T: 25 81 491585 F: 25 81 491591
TL: P.O. Box 828

Matusse, Hilario Manuel Eugenio, Sr.
Secretario Geral
SINDICATO DE JORNALISTAS MOCAMBIQUE
Av 24 De Julho, 231
Maputo, Mozambique
T: 25 81 2581492031

Mejia, Margarita, Ms.
WOMEN AND ENVIRONMENT GROUP - MOZAMBIQUE
NORAD - Maputo - Rua de Kassuende, 118
Maputo, Mozambique
T: 25 81 491585 F: 25 81 491591
TL: P.O. Box 828

Mussagy, Aidate, Ms.
Biologist
WOMEN AND ENVIRONMENT GROUP - MOZAMBIQUE
Universidade Eduardo Mondlane - Faculdade de Biologia
Maputo, Mozambique
T: 25 81 492142

MYANMAR - GOVERNMENT

U Khin Maung Win,
Director
PUBLIC RELATIONS AND INFORMATION
Ministry of Foreign Affairs, Pyay Road
Yangon, Myanmar
T: 22844 F: 22940

U Kyaw Tint Swe,
Director/Secretary Of The National
ECONOMIC DIVISION/OFFICE OF THE NAT
Ministry of Foreign Affairs, Pyay Road
Yangon, Myanmar
T: 22844 F: 22950

U Ohn Gyaw, , H.E.
Minister For Foreign Affairs/Chairm
MINISTRY OF FOREIGN AFFAIRS/OFFICE
Pyay Court, Pyay Road
Yangon, Myanmar
T: 22844 F: 22950

NAMIBIA - GOVERNMENT

Nujoma, Sam, Mr.
President
Permanent Mission of Namibia
135 E. 36th Str.
New York NY 10016, United States of America
T: 212 685 2003 F: 212 685 1561

PRESS

Stoldt, Gitta
Reporter Radio/Television
NAMIBIAN B'CASTING CORP.
P.O.Box 1593
Windhoek 900, Namibia, Namibia

Swartz, Jackson
Editor
NEW ERA
CL-1- A. Mashego Str.
Katutura, Namibia, Namibia

GLOBAL FORUM

Gitta, Stoudt
Chief Reporter
NAMIBIAN BROADCASTING CORP.
NBC Box 321 Windhoek
Namibia
T: 61 215811 F: 61 216579

Koenraad, Ligthelm Willem
Cameraman
NAMIBIAN BROADCASTING CORP.
NBC Box 321 Windhoek
Namibia
T: 61 215811 F: 61 216579

Munjanu, Oloff, Mr.
NAMIBIA DEVELOPMENT TRUST
Palais de Nations
Geneve CH 1211, Namibia
F: 41 22 733-6542

NAURU - GOVERNMENT

Dowiyogo, Bernard, Mr.
President
Nauru
Nauru

Keke, Leo D, Mr
Presidential Counsel
EXECUTIVE OFFICE OF THE PRESIDENT
P O Box 284
Republic of Nauru, Nauru
T: 674 3360 3300 F: 674 3361 3422

NEPAL - GOVERNMENT

Bhattarai, Tara Nath, Ms
Chief Environment Division
MINISTRY OF FORESTS AND ENVIRONMENT
Babar Mahal
Kathmandu, Nepal
T: 977 1 224891

Dhakal, Hon Ble Birmani
State Minister Of Forests Environmt
MINISTRY OF FORESTS AND ENVIRONMENT
Babar Mahal
Kathmandu, Nepal
T: 2 20160

Koirala, Girija Prasad
Prime Minister
Permanent Mission of Nepal
820 2nd. Ave, Suite 202
New York NY 10017, United States of America
T: 212 370 4188 F: 212 953 2038

Wagley, Mohan
Under Secretary
MINISTRY OF FOREST AND ENVIORNMENT
Babar Mahal
Kathmandu, Nepal

NGO

Bhattarai, Binod
NFEJ
PO Box 5143
Kathmandu, Nepal

Gautam, Achul
NICU
PO Box 5841
Pulchowk, Lalitpur, Kathmandu, Nepal

Ghimire, Ganesh, Mr.
Gen. Secretary
COALITION NEPAL
PO Box 4067
Kathmandu, Nepal
T: 977 1 527781 F: 977 1 521506

Sharma, Bharat, Mr.
Coordinator
NEPAL ENVORONMENTAL CONSERVATION GROUP
Palais de Nations
Geneve CH 1211, Nepal
F: 41 22 733-6542

Shrestha, Prachet
ECCA
PO Box 3923
Kathmandu, Nepal

Tuladhar, Jyoti
CWD
P.O. Box 3637
Kamaldi, Kathmandu, Nepal

PRESS

Adhikary, Hari
Editor
DESHANTER WEEKLY
Shanti Goreto, Chabahil 7
Kathmandu, Nepal, Nepal

Beaudry, Colette
Correspondent
INDEPENDENT WEEKLY
31 Farm Lane
Levittown, NY 11756, USA, Nepal

Pokharel, Krishna
Correspondent
DESHANTER WEEKLY
P.O Box 3228
Kathmandu, Nepal, Nepal

ORGANIZATIONS

Chitrakar, Anil, Mr.
Environmental Camps for Conservation Awareness
PO Box 3923
Kathmandu, Nepal

GLOBAL FORUM

Adhikari, Rabindra, Mr.
NEPAL NATIONAL FEDERATION OF STUDENTS
Post Box 2737
Kathmandu, Nepal
T: 977 1 410126 F: 977 1 227431
TL: 2314Raitra NP

Shrestha, Ramesh Babu, Mr.
General Secretary
NEPAL NATIONAL FEDERATION OF STUDENTS
PO Box 379 Kamaladj.
Kathmandu, Nepal
T: 977 1 5-24630 F: 977 1 5-22633

NETHERLANDS - GOVERNMENT

Alders, J.G.M., H.E. Mr.
Minister of Housing
Physical Planning and Environment
Permanent Mission to the United Nations, 711 3rd. Ave ., 9th Floor
New York, NY 10017, United States of America
T: 212 697 5547 F: 212 370 1954

Alvarez, M.M., Mr.
Administrative Officer
Ministry of Foreign Affairs
Permanent Mission to the United Nations, 711 3rd. Ave ., 9th Floor
New York, NY 10017, United States of America
T: 212 697 5547 F: 212 370 1954

Boer-Wijnberg, M.H.A., Ms.
Member
Parliament
Permanent Mission to the United Nations, 711 3rd. Ave ., 9th Floor New York, NY, 10017
F: 212 697 5547
TL: 212 370 1954

Curet, E.A., Mr.
Department of Energy Affairs
Ministry of Economic Affairs, Aruba
Permanent Mission to the United Nations, 711 3rd. Ave ., 9th Floor
New York, NY 10017, United States of America
T: 212 697 5547 F: 212 370 1954

De Jong, A.M., Mr.
Telecommunications Facilities Service
Ministry of Foreign Affairs
Permanent Mission to the United Nations, 711 3rd. Ave ., 9th Floor
New York, NY 10017, United States of America
T: 212 697 5547 F: 212 370 1954

De Jongh, P.E., Mr.
Deputy Director-General for Environment Protection
Ministry of Housing, Physical Planning and Environment
Permanent Mission to the United Nations, 711 3rd. Ave ., 9th Floor
New York, NY 10017, United States of America
T: 212 697 5547 F: 212 370 1954

De Loenen, C., Mr.
Information Officer
Ministry of Foreign Affairs
Permanent Mission to the United Nations, 711 3rd. Ave ., 9th Floor
New York, NY 10017, United States of America
T: 212 697 5547 F: 212 370 1954

De Merez Oyens, W., Mr.
Consul-General
Kingdom of the Netherlands, Rio de Janeiro
Permanent Mission to the United Nations, 711 3rd. Ave ., 9th Floor
New York, NY 10017, United States of America
T: 212 697 5547 F: 212 370 1954

Deeker, R.H., Mr.
Head International Water Policy Division
Ministry of Transport, Public Works and Water Management
Permanent Mission to the United Nations, 711 3rd. Ave ., 9th Floor
New York, NY 10017, United States of America
T: 212 697 5547 F: 212 370 1954

Dekker, Robert Heuk, Mr
Director International Water Policy
MINISTRY OF TRANSPORT, PUBLIC WORKS
Management P.O. Box 20906
NL-2500 Ex the Hague, Netherlands
T: 31 70 3744141 F: 31 70 3744335

Douma, W., Ms.
Directorate for International Environmental Affairs
Ministry of Housing, Physical Planning and Environment
Permanent Mission to the United Nations, 711 3rd. Ave ., 9th Floor
New York, NY 10017, United States of America
T: 212 697 5547 F: 212 370 1954

Enthoven, M.E.E., Mr.
Director-General for Environmental Protection
Ministry of Housing, Physical Planning and Environment
Permanent Mission to the United Nations, 711 3rd. Ave ., 9th Floor
New York, NY 10017, United States of America
T: 212 697 5547 F: 212 370 1954

Falad, A.A., Mr.
Major-Domo
Netherlands Embassy, Brasilia
Permanent Mission to the United Nations, 711 3rd. Ave ., 9th Floor
New York, NY 10017, United States of America
T: 212 697 5547 F: 212 370 1954

Flaes, R., H.E. Mr.
Ambassador Extraordinary and Plenipotentiary
Federative Republic of Brazil
Permanent Mission to the United Nations, 711 3rd. Ave ., 9th Floor
New York, NY 10017, United States of America
T: 212 697 5547 F: 212 370 1954

Gabor, J., H.E. Mr.
Deputy Minister of Agriculture
Nature Management and Fisheries
Permanent Mission to the United Nations, 711 3rd. Ave ., 9th Floor
New York, NY 10017, United States of America
T: 212 697 5547 F: 212 370 1954

Hamburger, A., Mr.
Minister Plenipotentiary
Permanent Mission, New York
Permanent Mission to the United Nations, 711 3rd. Ave ., 9th Floor
New York, NY 10017, United States of America
T: 212 697 5547 F: 212 370 1954

Hernandez, Raymond, Mr.
Member
Parliament
Permanent Mission to the United Nations, 711 3rd. Ave ., 9th Floor New York, NY, 10017
F: 212 697 5547
TL: 212 370 1954

Ijserinkhuijsen, I.S., Ms.
Administrative Officer
Netherlands Embassy, Brasilia
Permanent Mission to the United Nations, 711 3rd. Ave ., 9th Floor
New York, NY 10017, United States of America
T: 212 697 5547 F: 212 370 1954

Jonkheer, , H.E. Mr.
Minister of State and Advisor to the Secretary-General
Conference on Environment and Development
Permanent Mission to the United Nations, 711 3rd. Ave ., 9th Floor
New York, NY 10017, United States of America
T: 212 697 5547 F: 212 370 1954

Kakebeeke, W.J., Prof.
Director for International Environmental Affairs
Ministry of Housing, Physical Planning and Environment
Permanent Mission to the United Nations, 711 3rd. Ave ., 9th Floor
New York, NY 10017, United States of America
T: 212 697 5547 F: 212 370 1954

Lamers, J.G., Prof.
Deputy Legal Adviser
Ministry of Foreign Affairs
Permanent Mission to the United Nations, 711 3rd. Ave ., 9th Floor
New York, NY 10017, United States of America
T: 212 697 5547 F: 212 370 1954

Lubbers, R.F.M., H.E. Mr.
Prime Minister and Minister
General Affairs of the Kingdom of the Netherlands
711 3rd. Ave., 9th Floor
New York, NY 10017, United States of America
T: 212 697 5547 F: 212 370 1954

Mazairac, L.P.J., H.E. Mr.
Ambassador Extraordinary and Plenipotentiary
Permanent Representative to the United Nations Environmental Programme, Nairobi
Permanent Mission to the United Nations, 711 3rd. Ave ., 9th Floor
New York, NY 10017, United States of America
T: 212 697 5547 F: 212 370 1954

Metz, B., Mr.
Deputy Director Air Quality
Ministry of Housing, Physical Planning and Environment
Permanent Mission to the United Nations, 711 3rd. Ave ., 9th Floor
New York, NY 10017, United States of America
T: 212 697 5547 F: 212 370 1954

Nolet, G.P., Mr.
Directorate of Foreign Financial Relations
Ministry of Finance
Permanent Mission to the United Nations, 711 3rd. Ave ., 9th Floor
New York, NY 10017, United States of America
T: 212 697 5547 F: 212 370 1954

Pietermaat, D.F.W.T., Mr.
Directorate-General of Energy
Ministry of Economic Affairs
Permanent Mission to the United Nations, 711 3rd. Ave ., 9th Floor
New York, NY 10017, United States of America
T: 212 697 5547 F: 212 370 1954

Pronk, J.P., H.E. Mr.
Minister
Development Cooperation
Permanent Mission to the United Nations, 711 3rd. Ave ., 9th Floor
New York, NY 10017, United States of America
T: 212 697 5547 F: 212 370 1954

Reijnders, Y.S.H., Ms.
Assistant-Chancellor
Netherlands Consulate-General, Rio de Janeiro
Permanent Mission to the United Nations, 711 3rd. Ave ., 9th Floor
New York, NY 10017, United States of America
T: 212 697 5547 F: 212 370 1954

Roos, I.G., Mr.
Economic Cooperation Department
Ministry of Foreign Affairs
Permanent Mission to the United Nations, 711 3rd. Ave ., 9th Floor
New York, NY 10017, United States of America
T: 212 697 5547 F: 212 370 1954

Schlingemann, F., Mr.
Head of the Environment Programme
Directorate-General International Cooperation, Ministry of Foreign Affairs
Permanent Mission to the United Nations, 711 3rd. Ave ., 9th Floor
New York, NY 10017, United States of America
T: 212 697 5547 F: 212 370 1954

Schmitz Filho, R., Mr.
Major-Domo
Netherlands Consulate General, Rio de Janerio
Permanent Mission to the United Nations, 711 3rd. Ave ., 9th Floor
New York, NY 10017, United States of America
T: 212 697 5547 F: 212 370 1954

Schroder, C., Mr.
Head of the Staff Office Environment Affairs
Ministry of Economic Affairs
Permanent Mission to the United Nations, 711 3rd. Ave ., 9th Floor
New York, NY 10017, United States of America
T: 212 697 5547 F: 212 370 1954

Slingenberg, M.H., Mr
Department of Nature, Forests, Landscape and Wildlife
Ministry of Agriculture, Nature Management and Fisheries
Permanent Mission to the United Nations, 711 3rd. Ave ., 9th Floor
New York, NY 10017, United States of America
T: 212 697 5547 F: 212 370 1954

Slingenberg, Max
Senior Executive Officer
MINISTRY OF AGRICULTURE NATURE MANA
PO Bo 20401
the Hague 2500EK, Netherlands
T: 0 70 3792934 F: 0 70 3478228

Smeeman, J.H.D., Mr.
Consul
Kindgom of the Netherlands, Rio de Janeiro
Permanent Mission to the United Nations, 711 3rd. Ave ., 9th Floor
New York, NY 10017, United States of America
T: 212 697 5547 F: 212 370 1954

Smits, S.J.H., Mr.
First Secretary
Permanent Mission, Geneva
Permanent Mission to the United Nations, 711 3rd. Ave ., 9th Floor
New York, NY 10017, United States of America
T: 212 697 5547 F: 212 370 1954

Te Veldhuis, A.J., Mr.
Member
Parliament
Permanent Mission to the United Nations, 711 3rd. Ave ., 9th Floor New York, NY, 10017
F: 212 697 5547
TL: 212 370 1954

Tommel, D.K.J., Mr.
Member
Parliament
Permanent Mission to the United Nations, 711 3rd. Ave ., 9th Floor New York, NY, 10017
F: 212 697 5547
TL: 212 370 1954

Van Dijck, L., Ms.
Child Representative
Permanent Mission of Netherlands
711 3rd. Ave ., 9th Floor
New York, NY 10017, United States of America
T: 212 697 5547 F: 212 370 1954

Van Donkersgoed, P.M., Mr.
Second Secretary
Netherlands Embassy, Brasilia
Permanent Mission to the United Nations, 711 3rd. Ave ., 9th Floor
New York, NY 10017, United States of America
T: 212 697 5547 F: 212 370 1954

Van Gennip, J.J.A.M., Mr.
Member
Parliament
Permanent Mission to the United Nations, 711 3rd. Ave ., 9th Floor New York, NY, 10017
F: 212 697 5547
TL: 212 370 1954

Van Giezen, M.J.H.A., Mr.
Spokesperson of the Minister of Housing
Physical Planning and Environment
Permanent Mission to the United Nations, 711 3rd. Ave ., 9th Floor
New York, NY 10017, United States of America
T: 212 697 5547 F: 212 370 1954

Van Ginkel, C.J., Mr.
Head of External Relations Bureau
Ministry of Housing, Physical Planning and Environment
Permanent Mission to the United Nations, 711 3rd. Ave ., 9th Floor
New York, NY 10017, United States of America
T: 212 697 5547 F: 212 370 1954

Van Lennep, E., Mr.
Permanent Mission of Netherlands
711 3rd. Ave ., 9th Floor
New York, NY 10017, United States of America
T: 212 697 5547 F: 212 370 1954

Van Schaik, R.J., H.E. Mr.
Ambassador Extraordinary and Plenipotentiary
Permanent Representative to the United Nations, New York
Permanent Mission to the United Nations, 711 3rd. Ave ., 9th Floor
New York, NY 10017, United States of America
T: 212 697 5547 F: 212 370 1954

Van Spaendonck, E., Mr.
Assistant to H.E. Jonkheer
Permanent Mission of Netherlands
711 3rd. Ave ., 9th Floor
New York, NY 10017, United States of America
T: 212 697 5547 F: 212 370 1954

Van der Kooi, H., Mr.
Deputy Head Environment Coordination, Department for the Environment
Quality and Nutrition, Ministry of Agriculture, Nature Management and Fisheries
Permanent Mission to the United Nations, 711 3rd. Ave ., 9th Floor
New York, NY 10017, United States of America
T: 212 697 5547 F: 212 370 1954

Van der Ploeg, J.H., Mr.
Deputy General-Director
Netherlands Information Service, Spokesperson of the Prime Minister
Permanent Mission to the United Nations, 711 3rd. Ave ., 9th Floor
New York, NY 10017, United States of America
T: 212 697 5547 F: 212 370 1954

Verbeek, B., Ms
Administrative Officer, Directorate-General for International Environmental Affairs
Ministry of Housing, Physical Planning and Environment
Permanent Mission to the United Nations, 711 3rd. Ave ., 9th Floor
New York, NY 10017, United States of America
T: 212 697 5547 F: 212 370 1954

Verberne, M.M., Ms.
International Organizations Department
Ministry of Foreign Affairs
Permanent Mission to the United Nations, 711 3rd. Ave ., 9th Floor
New York, NY 10017, United States of America
T: 212 697 5547 F: 212 370 1954

Vereecken, S.M., Mr.
Information Officer
Ministry of Foreign Affairs
Permanent Mission to the United Nations, 711 3rd. Ave ., 9th Floor
New York, NY 10017, United States of America
T: 212 697 5547 F: 212 370 1954

Verheij, H.Th.H, Mr.
Directorate for International Environmental Affairs
Ministry of Housing, Physical Planning and Environment
Permanent Mission to the United Nations, 711 3rd. Ave ., 9th Floor
New York, NY 10017, United States of America
T: 212 697 5547 F: 212 370 1954

Verspaget, J.M., Mr.
Member
Parliament
Permanent Mission to the United Nations, 711 3rd. Ave ., 9th Floor New York, NY, 10017
F: 212 697 5547
TL: 212 370 1954

Waller-Hunter, J.H., Ms.
Director Strategic Planning
Ministry of Housing, Physical Planning and Environment
Permanent Mission to the United Nations, 711 3rd. Ave ., 9th Floor
New York, NY 10017, United States of America
T: 212 697 5547 F: 212 370 1954

Weitering, D., Mr.
Child Representative
Permanent Mission of Netherlands
711 3rd. Ave ., 9th Floor
New York, NY 10017, United States of America
T: 212 697 5547 F: 212 370 1954

Willems, Wilbert J, Mr
Member Of Parliament
Twerde Kamer 20018
2500 Ea Den Haag, Netherlands
T: 070 3183030 F: 070 3182685

Willemsen, F.H., Mr.
Directorate of Legislation
Ministry of Justice, Aruba
Permanent Mission to the United Nations, 711 3rd. Ave ., 9th Floor
New York, NY 10017, United States of America
T: 212 697 5547 F: 212 370 1954

Wulsman, G.P.I.M., Mr.
Advisor
Prime Minister's Office
Permanent Mission to the United Nations, 711 3rd. Ave ., 9th Floor
New York, NY 10017, United States of America
T: 212 697 5547 F: 212 370 1954

Zijlstra, K., Mr.
Member
Parliament
Permanent Mission to the United Nations, 711 3rd. Ave ., 9th Floor New York, NY, 10017
F: 212 697 5547
TL: 212 370 1954

NGO

Abraham, Martin, Mr.
INTERNATIONAL ORGANIZATION OF CONSUMERS UNION
9 Emmastraat
2595 EG The Hague, Netherlands

Bas, Arts
ASD
P.O. Box 19245
3501 De Utrecht, Netherlands

Berghuizen, Andrea, Ms.
ALLIANCE FOR SUTAINABLE DEVELOPMENT
PO Box 19245
Utrecht 3501 DE, Netherlands
T: 31 30 36 79 95 F: 31 30 36 79 98

Blydorp, Janneke
PB
PO Box 19245
3501 De Utrecht, Netherlands

Booth, Andrew
GI
Keizersgracht 176
1016 DW Amsterdam, Netherlands

Breivik, Roy, Mr.
WFM
Leliegracht 21
1016 GR Amsterdam, Netherlands

Corbett, Helen
CPHRPN
Post Box 85875
2508 CN The Hague, Netherlands

Firebrale, James
IOCU
9 Emmastraat
2595 EG The Hague, Netherlands

Grattapaglia, Ursula, Sra.
Director
UNIVERSALA ESPERANTO-ASOCIO
Nieuwe Binnenweg 176
3015 BJ Rotterdam, Netherlands

Gyari, Kasur Lodi
CPHRPN
Post Box 85875
2508 CN The Hague, Netherlands

Hallo, Ralph, Mr.
STICHTING NATUUR EN MILIEU
Donkerstraat 17
Utrecht 3511 KB, Netherlands
T: 31 30 331328 F: 31 30 331311

Harding, Thomas
EYFA
PO Box 566
Sitard, Netherlands

Hesselink, F. K., Mr.
Managing Director
SME, INSTITUTE FOR ENVIRONMENTAL COMMUNICATION
P. O. Box 13030
Utrecht 3507 LA, Netherlands
T: 31 30 713734 F: 31 30 714369

Holliman, Jonathan
FEI
PO Box 19199
1000 GD Amsterdam, Netherlands

Kappers, Vera
ASD
P.O. Box 19245
3501 De Utrecht, Netherlands

Kennedy, Danny
EYFA
PO Box 566
Sitard, Netherlands

Lees, Andrew
FEI
PO Box 19199
1000 GD Amsterdam, Netherlands

Mohr, Anne Tamara, Ms.
BOTH ENDS
Damrak 28-30
Amsterdam 1012 Lj, Netherlands
T: 31 20 62 30 826 F: 31 20 62 08 049
E: GEO2:BOTHENDS

Morgan, Eka, Ms.
Info. Officer
FRIENDS OF THE EARTH INTERNATIONAL
P.O.Box 19199
Amsterdam 1000, Netherlands
T: 31 20 20-6221369 F: 31 20 20-6275287
TL: 202 562f

Mueller, Nicole
EUROPEAN YOUTH FOREST ACTION
PO Box 566
Sitard, Netherlands

Orlebeke, Suzan
ASD
P.O. Box 19245
3501 De Utrecht, Netherlands

Pace, William
WFM
Leliegracht 21
1016 GR Amsterdam, Netherlands

Pereira De Souza, Delio
UEA
Nieuwe Binnenweg 176
3015 BJ Rotterdam, Netherlands

Roberts, John
WFM
Leliegracht 21
1016 GR Amsterdam, Netherlands

Romme, Margarita, Ms.
Representative
IFPE
Javastraat, 40 A
The Hague 2585 AT, Netherlands
T: 31 70 360-3318 F: 31 70 360-3346

Roy, Chandra K., Ms.
Director
CENTRE FOR PROMOTION OF HUMAN RIGHTS AND PROTECTION OF NATURE
PO Box 85878
The Hague 2508 CN, Netherlands
T: 31 70 360-3318 F: 31 70 360-3346

Sanchez, Vicente
IWT
Damrak 83
1012 LN Amsterdam, Netherlands

Schravesande, Hans
IODC
P O Box 151
3700 AD Zeist, Netherlands

Schwartzberg, Joseph
WORLD FEDERALIST ASSOCIATION
Leliegracht 21
1016 GR Amsterdam, Netherlands

Van Damme, Johannes Hans
IWSC
P O Box 93190, 2509 AD The Hague
Netherlands

Van Koolwyk, Theo
BE
Damrak 28-30
1012 L Amsterdam, Netherlands

Van Norden, Arthur
IWT
Damrak 83
1012 LN Amsterdam, Netherlands

Veening, Wouter
Coordinator
NETHERLANDS COMMITTE FOR IUCN
Plantage Middenlaan 2B
1018 DD Amsterdam, Netherlands

Wilson, Roger
GREENPEACE INTERNATIONAL
Keizersgracht 176
1016 DW Amsterdam, Netherlands

Zagema, Bertram
EYFA
PO Box 566
Sitard, Netherlands

Zwetsloot, Gerard
IODC
P O Box 151
3700 AD Zeist, Netherlands

PRESS

Cavalcanti Araujo, Alvaro
Reporter Radio/Television
RADIO NEDERLAND
Tobias Ajjerlaan 229
1111JT-Diemen, Netherlands, Netherlands

De Miguel Garcia, Constantino
Editor
RADIO NEDERLAND WERELDOMROEP
Karel Doormanlaan 70
1403 TP, Bussum, Netherlands, Netherlands

Harkema, Harm
Editor
GEMEENSCHAPPELIJKE PERS DIENST
Zorgvlietstr 201
2513 HA Denhaag, Netherlands, Netherlands

Holtwijk, Ineke
Correspondent
DE VOLKSKRANT
Rua do Russel, 300 Apt.701
22210 Rio de Janeiro-RJ, Brazil, Netherlands

Hoogenboom, Annemieli
IPS
Van Eeghenstraat 77
1071 Ex Amsterdam, Netherlands
T: 31 20 6640817 F: 31 20 6626416

Knoppers, Rijkert
Editor
NRC HANDELSBLAD
Redemptoristenpad 42, 5211 XR
's Hertogenbosch, The Netherlands, Netherlands

Lindenkamp, Frans
Correspondent
ZUID OOST PERS
R.Barata Ribeiro 260 Apt.65
01308 Sao Paulo-SP, Brazil, Netherlands

Maharaj, Niala
Reporter Radio/Television
RADIO NETHERLANDS
Zocherstraat 27/3
1054 LK Amsterdam, Netherlands

McKay, David
Editor
LANDSCAPE AND URBAN PLANNING
5259 Cole Lane
College Station, TX 77845, USA, Netherlands

Nijdam, Pieter
Editor
DE TELEGRAAF
Lauriergracht 5
Amsterdam, Netherlands, Netherlands

Peters, Patrick F.H.J., Mr.
Editor
LIBERAL FORUM MAGAZINE
Postbus 2730
Utrecht, Netherlands

Pilgram, A.
Reporter Radio/Television
DUTCH KRO-RADIO
Kolfmakersteeg 17,
2311 VE Leidem, Netherlands, Netherlands

Romeijn, Willem
Correspondent
DE TELEGRAAF
P.O.Box 277
18130 Sao Roque, SP, Brazil, Netherlands

Scheele, Huub
Correspondent
GREEN TREE NEWS
Ouertoom447-B
1054 KG Amsterdam, The Netherlands, Netherlands

Schoenmakers, Jan
Editor
FINANCIAL ECONOMIC MAGAZINE
Lomanstraat 44
1075 RC Amsterdam, Netherlands, Netherlands

Schurink, Johannes
Correspondent
ALGEMEEN DAGBLAD
CP 137
Sao Roque, S.P., Brazil, Netherlands

Schwartz, Jacobus W.B.
Editor
TROUW
Abbenesstraat 16 II
1059 TD Amsterdam, Netherlands, Netherlands

Stapel, Allard, Mr.
Press Officer
WWF - NETHERLANDS
Postbus 7
Zeist 3700 AA, Netherlands
T: 31 34 404-22164 F: 31 34 404-12064

Steketee, Hans
Editor Foreign Desk
NRC HANDELSBLAD
Postbus 824
Rotterdam, Netherlands
T: 31 10 4067319 F: 31 10 4066967

Thielen, Jan
Correspondent
NOS RADIO
Estados Unidos 780
1005 Buenos Aires, Argentina, Netherlands

Van Der Boon, Ferenc
Correspondent
HET FINANCIEELE DAGBLAD
1e Spechtstraat 15
Utrecht, Netherlands, Netherlands

Van Der Colk, Rudolf Johannes
Editor
AURO TV
Bezuidenhoutseweg 67
2500 EB The Hague, Netherlands, Netherlands

Van Der Willigen, Marc Anthony
Correspondent
NETHERLANDS BROADCASTING CO.
Rua Goitacazes 104/201
Rio de Janeiro, Brazil, Netherlands

Van Kleef, Michiel A.
Editor
MILIEU MAGAZINE
Botenmakersstraat 135
1506 TD Zaandam, Netherlands, Netherlands

Van Seeters, Piet
Editor
DE VOLKSKRANT
Nyenheim 11-16
3704 Ag Zeist, Netherlands, Netherlands

Weckseler, Henk
Correspondent
VEREINIGING '31'
Dorpsstraat 12-A
6721 NJ Beunelan, Netherlands, Netherlands

Wijninga, Derk
Editor
ALGEMEEN NEDERLANDS PERSBUREA (ANP)
Paulus van Asperenpad 15
3151 Ja Hook of Holland,The Netherlands, Netherlands

ORGANIZATIONS

Abrams, Isabel, Ms.
Caretakers Of The Environment International
Secretariat - Nassauplein 8
1815 GM Alkmaar, Netherlands
T: (1 708) 251 8935 F: (312) 440 9426

Atwood, David, Mr.
International Fellowship Of Reconciliation*
Spoorstraat 38
1815 BK Alkmaar, Netherlands
T: (31 72) 12 30 14 F: (31 72) 15 11 02

Bilderbeek, Simone, Ms.
Netherlands National Committee For Iucn
Plantage Middenlaan 2B
1018 DD Amsterdam, Netherlands
T: (31 20) 626 1732 F: (31 20) 627 9349

Dirx, Jan, Mr.
Alliance For Sustainable Development, Netherlands
P.O. Box 19245
3501 De Utrecht, Netherlands
T: (31 30) 367995 F: (31 30) 367998

Henselmans, Jan, Mr.
Stichting Natuur En Milieu
Donkerstraat 17
3511 KB Utrecht, Netherlands
T: (31 30) 331 328 F: (31 30) 331 311

Hyvarinen, Yoy, Ms.
Greenpeace International*
Keizersgracht 176
1016 DW Amsterdam, Netherlands
T: (31 20) 5236555 F: (31 20) 5236500
TL: 18775 GPINT NL

Koolwyck, Theo Van, Mr.
Both Ends
Damrak 28-30
1012 L Amsterdam, Netherlands
T: (31 20) 623 0823 F: (31 20) 620 8049
TL: E GEO2:BOTHENDS

Kranendonk, Marie, Ms.
Landelijk Milieu Overleg (LMO - Dutch Environmental Forum)
Donkerstraat 17
3511 KB Utrecht, Netherlands
T: (31 30) 331 328 F: (31 30) 331 311

Nic, Eileen, Ms.
International Organization Of Consumer's Unions*
9 Emmastraat
2595 EG The Hague, Netherlands
T: (212) 663 6378 F: (212) 749 3829

Oudejans, Judi, Ms.
World Federalist Movement
Leliegracht 21
1016 GR Amsterdam, Netherlands
T: (31 20) 22 7502 F: (31 20) 20 8727

Pakulski, Jan, Mr.
Youth For Development And Cooperation
Overschiestraat 9
1062 HN Amsterdam, Netherlands
T: (31 20) 14 25 10 F: (31 20) 17 55 45

Parfait, Francois, Mr.
International Federation for Housing and Planning
43 Wassenaarseweg
2596 CG, The Hague, Netherlands
T: (31 70) 324 4557 F: (31 70) 328 2085

Schlanger, John, Mr.
Inter Press Service
Van Eeghensor 77
1071 Ex-Amsterdam, Netherlands
T: (31 20) 66 40 817 F: (31 20) 67 19 701

Van Den Schoor, Finche, Mr.
Platform Brazilia 92
PO Box 19245
3501 De Utrecht, Netherlands
T: (31 30) 367997 F: (31 30) 367998

GLOBAL FORUM

Altchouler, Igor, Mr.
ASEI / WISE
Geldershoofd 928
Amsterdam 1103 BG, Netherlands
T: 31 20 699-3377 F: 31 20 6262051

Altchouler, Raissa Molastova, Ms.
Board Member
ASEI / WISE
Geldershoofd 928
Amsterdam 1103 BG, Netherlands
T: 31 20 699-3377 F: 31 20 6262051

Anderson, Patrick, Mr.
GREENPEACE INTERNATIONAL
176 Keizergracht
Amsterdam 1016 DW, Netherlands
T: 31 20 523-6555 F: 31 20 523-6500

Andriessen, Gijs Jurriaan
Video-Director - Journalist
GIJS ANDRIESSEN VIDEO PRODUCTIONS
Badhuisweg 179
Denhaag 2597JP, Netherlands
T: 70 3559789 F: 55 21 2311521 (Rio

Arts, B.J.M., Mr.
PEACE RESEARCH CENTRE
Muntweg 109
Nijmegen 6532 TH, Netherlands
T: 31 80 56 72 32

Bakker, Nicolaas, Mr.
ADO
Buitenom 216, 2512 Xd
The Hague, Netherlands

Bijfleveld, May, Ms.
UNIVERSALA ESPERANTO-ASOCIO
R. Santa Alexandrina 481/606
Rio de Janeiro - RJ, Netherlands
T: 55 21 273-8108

Boelsma, E., Dr.
ZONTA INTERNATIONAL
59 Lokhorst
Leiderdorp 2352 KE, Netherlands
T: 31 71 290 419

Boschman, H., Ms.
Drs.Coordinator
WEMOS
Minahassastraat 1 Postbus 40066
Amsterdam, Netherlands
T: 31 20 66 53 115 F: 31 20 66 84 085

Brakel, Manus Van, Mr.
FRIENDS OF THE EARTH INTERNATIONAL
P.O.Box 19199
Amsterdam 1000, Netherlands
T: 31 20 622-1366 F: 31 20 627-5287

Bruno, Kenny, Mr.
GREENPEACE INTERNATIONAL
176 Keizergracht
Amsterdam 1016 DW, Netherlands
T: 31 20 523-6555 F: 31 20 523-6500

Buitenkamp, Maria, Ms.
Climate Changer
FRIENDS OF THE EARTH INTERNATIONAL
P.O.Box 19199
Amsterdam 1000, Netherlands
T: 31 20 6221366 F: 31 20 6275287
TL: 202 562f

Burger, Willem, Mr.
INSTITUTE OF SOCIAL STUDIES
PO Box 90733
The Hague 2509 LS, Netherlands
T: 31 70 3510100 F: 31 70 3549851

Cavalcanti, Alvaro
Reporter
RADIO NEDERLAND
Witte Kruislaan 55
Hilversum 1200 JG, Netherlands
T: 31 35 724211 F: 31 35 724352

Colk, Ruud Van Der
Director
HAN PEEKEL PRODUKTIES B.V.
Insulindellan 28
1217 HL Hilversum, Netherlands
T: 31 35 218241 F: 31 35 218984

Damen, Pieter, Sr.
CEBEMO
Rhijngeestersstraatweg, 40
Oegstgeest 2340 AB, Netherlands
T: 46 171 159159 F: 46 171 175391

Dankelman, Irena, Ms.
NOVIB
Amaliastraat 7
Haya 2514 JC, Netherlands
T: 31 70 342.1621 F: 31 70 361.4461

Detiger-Huber, Elinore, Ms.
Founder Director
DETIGER FOUNDACION / FUNDADION VICTORIA / INT'L COUNCIL OF WISE WOMEN
P.O.1023 Wassenaar, the Netherlands 2240AB
Wassenaar 2240AB, Netherlands
T: 31 17 5177271 F: 31 17 5177518
E: 4406817588

Dijksta, Maria Elizabeth, Ms.
Student
Markt 9a
Wageningen 6701cx, Netherlands
T: 0837 14198

Dillen, R. B. Van, Ms.
GRONTMIJ
De Holle Bilt 1, P. O. Box 203
De Bilt 3730 AE, Netherlands
T: 31 30 207430 F: 31 30 210730
TL: 70279

Dinteren, Lambert Van, Mr.
ORTHDOX PEACE FELLOWSHIP
Kolpingstraat 63
Nijmegen NL-6532 WJ, Netherlands

Figari, Andrea, Ms.
GREENPEACE INTERNATIONAL
176 Keizergracht
Amsterdam 1016 DW, Netherlands
T: 31 20 523-6555 F: 31 20 523-6500

Garcia, Constantino Miguel
Reporter
RADIO NEDERLAND
Witte Kruislaan 55
Hilversum 1200 JG, Netherlands
T: 31 35 724211 F: 31 35 724352

Gendt, Rien Van, Mr.
Executive Director
BERNARD VAN LEER FOUNDATION
Eisenhowerlaan 156
Haya, Netherlands
T: 31 70 351-2040 F: 31 70 350-2373
E: 33678BVLFH

Gianni, Matthew, Mr.
GREENPEACE INTERNATIONAL
176 Keizergracht
Amsterdam 1016 DW, Netherlands
T: 31 20 523-6555 F: 31 20 523-6500

Harold, Mike, Mr.
GREENPEACE INTERNATIONAL
176 Keizergracht
Amsterdam 1016 DW, Netherlands
T: 31 20 523-6555 F: 31 20 523-6500

Hazeu, Marcel, Mr.
Student
Tarthorst 855
Wageningen, Netherlands
T: 31 8370-12098

Hermsen, Wil
Cameraman
HAN PEEKEL PRODUKTIES B.V.
Insulindellan 28
1217 HL Hilversum, Netherlands
T: 31 35 218241 F: 31 35 218984

Heybrock, Ferry
Press Officer
WWF - NETHERLANDS
Postbus 7
Zeist 3700 Aa, Netherlands
T: 31 34 404-22164 F: 31 34 404-12064

Hoedt, Selma, Ms.
Acquisition Editor
MARTINUS NIJHOFF PUBLISHERS
Spuiboulevard 50 P.O. Box 163
Dordrecht 3300, Netherlands
T: 31 78 334-204 F: 31 78 334-254
TL: 29245

Hoogmoed, Marinus S., Dr.
NATIONAL MUSEUM OF NATURAL HISTORY
Postbus 9517
Lieden 2300RA, Netherlands
T: 31 71 143844 F: 31 71 133344

Huizenga, Cornie, Mr.
IDEM CONSUL
Rynlann 66
Utrecht, Netherlands
T: 31 30 800217

Hummel, John, Mr.
INTERNATIONAL FORESTRY STUDENTS SYMPOSIUM FOUNDATION 91
Market Q A
Wageningen 6707 CX, Netherlands
T: 31 0837014198

Johan, , Mr.
GREENPEACE INTERNATIONAL
176 Keizergracht
Amsterdam 1016 DW, Netherlands
T: 31 20 523-6555 F: 31 20 523-6500

Kant, Mrs, Ms.
Drs
WEMOS
Kijkduinstraat 43II
Amsterdam 1055 XR, Netherlands
T: 31 20 682-8330

Karliner, Josh, Mr.
GREENPEACE INTERNATIONAL
176 Keizergracht
Amsterdam 1016 DW, Netherlands
T: 31 20 523-6555 F: 31 20 523-6500

Keijzer, Marc, Mr.
Student
UNIVERSITY OF WAGENINGEN
Troelstraweg 173
Wageningen 6702 AK, Netherlands
T: 31 8370 22680

Kester, Boris, Mr.
UNIVERSITY OF AMSTERDAM
V. Tuyll V. Serooskerkenweg, 52 li
Amsterdam, Netherlands
T: 31 20 673-2642

Kleef, Michiel A Van
Editor In Chief
MILIEU MAGAZINE
PO Box 532
Amsterdam, Netherlands
T: 31 20 6383011 F: 31 20 6382871

Koesman, Wouter
Sound
HAN PEEKEL PRODUKTIES B.V.
Insulindellan 28
1217 HL Hilversum, Netherlands
T: 31 35 218241 F: 31 35 218984

Koning, Jan-Kees, Mr.
E.C. Member
WORLD FEDERALIST THE NETHERLANDS
Weerstraat 28-B 4001 LD
Tiel, Netherlands
T: 31 344023305 F: 31 3440206208727

Krommendijk, Bernardo, Sr.
CEBEMO
Rhijngeestersstraatweg, 40
Oegstgeest 2340 AB, Netherlands
T: 46 171 159159 F: 46 171 175391

Larrain, Sara, Ms.
GREENPEACE INTERNATIONAL
176 Keizergracht
Amsterdam 1016 DW, Netherlands
T: 31 20 523-6555 F: 31 20 523-6500

Leggett, Jeremy, Mr.
GREENPEACE INTERNATIONAL
176 Keizergracht
Amsterdam 1016 DW, Netherlands
T: 31 20 523-6555 F: 31 20 523-6500

Maccow, Charles F., Mr.
ENVIROMENTAL HEALTH
Philipsburg Box 924 Stmaarten
Netherlands

Mahdi, Raya Issara Nunez, Ms.
ROYAL TROPICAL INSTITUTE
Badhuisweg 185
Den Haag, Netherlands
T: 31 70 350-2128

Mansvelt, J.D. Van, Dr.
IFOAM
Sterkenburgerlaan 10
Doorn 3q4iBD, Netherlands
T: 31 343020269 F: 31 837084995

Marel, Ljudmilla Van Der, Ms.
President
CLEAN AIR FOUNDATION
Ketensaan 15
Oosterhout 4903RW, Netherlands
T: 31 881816112252 F: 31 881816111335
TL: 54949pent

Martone, Francesco, Mr.
GREENPEACE INTERNATIONAL
176 Keizergracht
Amsterdam 1016 DW, Netherlands
T: 31 20 523-6555 F: 31 20 523-6500

Meijer, C. M., Ms.
NOUVELLES FRONTIERES HOLLAND B.V.
Van Baerlestraat 3
Amsterdam 1071 AL, Netherlands

Mersky, Marcie, Ms.
GREENPEACE INTERNATIONAL
176 Keizergracht
Amsterdam 1016 DW, Netherlands
T: 31 20 523-6555 F: 31 20 523-6500

Meyer, Maria B.L., Ms.
OBSERVER RELATIV IUCN
Weteringschans 112 II
Amsterdam 1017 MT., Netherlands
T: 31 20 623-0894

Nijpels, Ed., Dr.
Chairman
WWF - NETHERLANDS
Postbus 7
Zeist 3700 AA, Netherlands
T: 31 34 422 1649 F: 31 34 412 0643

Nypels, Maarten
WORLD INFORMATION ON ENERGY - AMSTERDAM
PO Box 18185
Amsterdam 1001 ZB, Netherlands
T: 31 20 639-2681 F: 31 20 639-1379

Osborn, Paul., Mr.
THE ONE FOUNDATION
H. Gorterhof 102
1422JS Uithoorn, Netherlands
TL: GE02:ONE

Pannocchia, Cristina, Ms.
WISE- AMSTERDAM
PO Box 18185
Amsterdam 1001 ZB, Netherlands
T: 31 20 639-2681 F: 31 20 639-1379

Parren, Marc, Mr.
TROPENBOS FOUNDATION
Sumatralstraat 18
Wageningen 6707 EG, Netherlands
T: 31 0837616854

Paulis, Paulis Olf, Mr.
GROEN AMSTERDAM/DE GROENEN (GREEN PARTY HOLLAND)
c/o Stadhuis (City Hall), Amstel 1
Amsterdam 1011PN, Netherlands
F: 31 20 552-3426

Pinxteren, Bert Van, Mr.
Coordinator
FRIENDS OF THE EARTH INTERNATIONAL
PO Box 19199
Amsterdam 1000 GD, Netherlands
T: 31 20 6221369 F: 31 20 6275287

Pluta-Charkiewicz, Eua, Ms.
I PREFER TO BE (POLAND)
C/O Iss, Postbus 90733
the Magne NL-2597LS, Netherlands
T: 31 70 351-0346 F: 31 70 354-9851

Podea, Titus, Dr.
INTERNATIONAL FELLOWSHIP OF RECONCILIATION
Speerstraat 38
Alkmaar NL-181510016, Netherlands
T: 31 72 123014 F: 31 72 3792004

Praag, Michael Van Walt Van, Dr.
Director
CENTRE FOR PROMOTION OF HUMAN RIGHTS AND PROTECTION OF NATURE
PO Box 85878
The Hague 2508 CN, Netherlands
T: 31 70 360-3318 F: 31 70 360-3346

Rogers, Maria, Ms.
GREENPEACE INTERNATIONAL
176 Keizergracht
Amsterdam 1016 DW, Netherlands
T: 31 20 523-6555 F: 31 20 523-6500

Runhaar, Job, Mr.
De Wittenkade, 3#
Amsterdam, Netherlands

Schoor, T. Van Der, Ms.
Secretary
PLATFORM BRASILIE92
Postbus 19245
Utrecht 3501 DE, Netherlands
T: 31 30 36 79 97 F: 31 30 36 79 98

Schravesande, J.H., Mr.
WORKING-GROUP CHURCH AND ENVIRONMENT OF THE COUNCIL OF CHURCHES IN THE NETHERLANDS
CZWD, PO Box 200
Leusden 3830 AE, Netherlands
T: 31 33 96 05 37 F: 31 33 94 87 07
TL: 79494 GKN NL

Schuurman, Johanna, Ms.
CRITICAL AGRICULTURAL COUNCIL
Nassaulaan 19
Allmaar, Netherlands
T: 072 118543

Schwartz, J.
Editor
TROUW
Parkstraat 22
Den Haag, Netherlands
T: 09-317-03466300 F: 09-317-03656911

Smits, Paul, Mr.
RAWOO
PO Box 90734
The Hague 2509 LS, Netherlands
T: 31 70 35 10 581 F: 31 70 35 10 513
TL: 33565 nufic

Soetendorp, Sira, Ms.
Rabbi
GLOBAL FORUM OF SPIRITUAL AND PARLIAMENTARY LEADERS
856 L. V. Meerdervoolt
Den Haag 2564AS, Netherlands
T: 31 70 325-3441

Spinelli, Claudio, Mr.
Delegate
UNIVERSALA ESPERANTO-ASOCIO
R. 8, casa 172
Ilha- Rio 21931, Netherlands
T: 55 21393-1539

Stevenson, Peg, Ms.
GREENPEACE INTERNATIONAL
176 Keizergracht
Amsterdam 1016 DW, Netherlands
T: 31 20 523-6555 F: 31 20 523-6500

Thielen, Johannes E. M.
Correspondent
NOS - RADIO
Hiluersum, Netherlands
T: 35 7998822

Tongeren, P. Van
Amsterdam, Netherlands

Van De Veen, H.
Editor
ONZE WERELD
PO Box 1050
Amsterdam 1000 BB, Netherlands
T: 31 20 5518560 F: 31 20 6251288

Van Der Beek, Paul Jcm, Mr.
CMDC-WORLD ENERGY COALITION-DUTCHT CHAPTER
Rustenburger Str 124-I
Amsterdam 1073 GH, Netherlands
T: 31 20 675 7423 F: 31 20 675 8881

Van Geloven, Yonke, Ms.
Participant
WAGENINGEN UNIVERSITRY
Herenstraat 29
Wageningen 6701 PH, Netherlands
T: 31 087370-18656

PARLIAMENTARIANS / SPIRITUAL LEADERS

Boer-Wynberg, Mrs.
Eerste Kamer der Staten General Binnenhof 1 a
2500EA s'Gravenhage
Netherlands, 64.48.00
F: 32653 PARL NL

De Veldhuis, A.J.
Eerste Kamer der Staten General Binnenhof 1 a
2500EA s'Gravenhage, Netherlands
T: 64.48.00
TL: 32653 PARL NL

Soetendorp, Awraham, Rabbi
Albert Einstein Institute, Princessegracht 26
The Hague 2564AS, Netherlands
T: 070 656 893 F: 31/10 4047 4639

Terpstra, Erica, Mrs.
MP
Eerste Kamer der Staten General Binnenhof 1 a
2500EA s'Gravenhage, Netherlands
T: 64.48.00
TL: 32653 PARL NL

Tommel, D.
Eerste Kamer der Staten General Binnenhof 1 a
2500EA s'Gravenhage, Netherlands
T: 64.48.00
TL: 32653 PARL NL

Van Gennip, J.A.M
MP
Eerste Kamer der Staten General Binnenhof 1 a
2500EA s'Gravenhage, Netherlands
T: 64.48.00
TL: 32653 PARL NL

Verspaget, J, Mrs.
Eerste Kamer der Staten General Binnenhof 1 a
2500EA s'Gravenhage, Netherlands
T: 64.48.00
TL: 32653 PARL NL

Zylstra, R.
MP
Eerste Kamer der Staten General Binnenhof 1 a
2500EA s'Gravenhage, Netherlands
T: 64.48.00
TL: 32653 PARL NL

NEW CALEDONIA - NGO

D'Auzon, Jean-Louis, Mr.
President
ASSOCIATION POUR LA SAUVAGARDE DE LA NATURE NEO-CALEDONIENNE
37 Rue Geroges Clemenceau / BP 1772
Noumea, New Caledonia
T: 6 87 283275 F: 6 87 283275

GLOBAL FORUM

Bone, Charles, Mr.
P.C.R.C.
Palais de Nations
Geneve CH 1211, New Caledonia
F: 41 22 733-6542

NEW GUINEA - GOVERNMENT

Konabe, Chawi
Deputy Director
PNG FOREST RESEARCH INSTITUTE
PO Box 314,Lae
Papua, New Guinea
T: 675 42 4197 F: 675 42 4357

Nadarajah, Tahereh, Ms
Research Associate
NATIONAL RESEARCH INSTITUTE
P.O. Box 5854 Boroko
Papua, New Guinea
T: 675 260 300 F: 675 260 213

GLOBAL FORUM

Sconadrawu, Mary, Ms.
EAST SEEPIK COUNCIL OF WOMEN
Palais de Nations
Geneve CH 1211, New Guinea
F: 41 22 733-6542

NEW ZEALAND - GOVERNMENT

Adank, J., Mr.
Second Secretary
Permanent Mission of New Zealand
United Nations Plaza
New York, NY 10017, United States of America
T: 212 963 7174 F: 212 963 1921

Atairangikaahu, Te, Dame
Patron
National Maori Congress, Maori Queen
Permanent Mission to the United Nations, United Nations Plaza New York, NY, 10017
F: 212 963 7174
TL: 212 963 1921

Barrett, T., Mr.
Ministry of External Relations and Trade
Permanent Mission to the United Nations, United Nations Plaza New York, NY, 10017
F: 212 963 7174
TL: 212 963 1921

Barton, E., Mrs.
Permanent Mission of New Zealand
United Nations Plaza
New York, NY 10017, United States of America
T: 212 963 7174 F: 212 963 1921

Blakeley, R., Dr.
Secretary
Environment
Permanent Mission to the United Nations, United Nations Plaza New York, NY, 10017
F: 212 963 7174
TL: 212 963 1921

Caygill, D., Hon.
Member
Parliament
Permanent Mission to the United Nations, United Nations Plaza New York, NY, 10017
F: 212 963 7174
TL: 212 963 1921

Collier, T., Mr.
Ministry of Commerce
Permanent Mission to the United Nations, United Nations Plaza New York, NY, 10017
F: 212 963 7174
TL: 212 963 1921

Conroy, M., Mr.
Ministry of External Relations and Trade
Permanent Mission to the United Nations, United Nations Plaza New York, NY, 10017
F: 212 963 7174
TL: 212 963 1921

Kettle, P., Mr.
Ministry of Agriculture and Fisheries
Permanent Mission to the United Nations, United Nations Plaza New York, NY, 10017
F: 212 963 7174
TL: 212 963 1921

Mead, A., Ms.
National Maori Congress
Permanent Mission to the United Nations, United Nations Plaza New York, NY, 10017
F: 212 963 7174
TL: 212 963 1921

Mohamed, N., Mr.
Ministry for the Environment
Permanent Mission to the United Nations, United Nations Plaza New York, NY, 10017
F: 212 963 7174
TL: 212 963 1921

O'Brien, T.C., H.E. Mr.
Ambassador Extraordinary and Plenipotentiary
Permanent Representative to the United Nations, New York (Deputy Head of Delegation)
Permanent Mission to the United Nations, United Nations Plaza New York, NY, 10017
F: 212 963 7174
TL: 212 963 1921

Palmer, Sir Geoffrey, Rt. Hon.
Permanent Mission of New Zealand
United Nations Plaza
New York, NY 10017, United States of America
T: 212 963 7174 F: 212 963 1921

Paterson, R., Ms.
Ministry of External Relations and Trade
Permanent Mission to the United Nations, United Nations Plaza New York, NY, 10017
F: 212 963 7174
TL: 212 963 1921

Payton, D., Mr.
Ministry of External Relations and Trade
Permanent Mission to the United Nations, United Nations Plaza New York, NY, 10017
F: 212 963 7174
TL: 212 963 1921

Reedy, T., Dr.
National Maori Congress
Permanent Mission to the United Nations, United Nations Plaza New York, NY, 10017
F: 212 963 7174
TL: 212 963 1921

Reeves, S., Mr.
New Zealand NGO/UNCED Committee
Permanent Mission to the United Nations, United Nations Plaza New York, NY, 10017
F: 212 963 7174
TL: 212 963 1921

Rider, P., Mr.
Permanent Mission to the Office of the United Nations, Geneva
Permanent Mission to the United Nations, United Nations Plaza New York, NY, 10017
F: 212 963 7174
TL: 212 963 1921

Shand, D., Ms.
New Zealand NGO/UNCED Committee
Permanent Mission to the United Nations, United Nations Plaza New York, NY, 10017
F: 212 963 7174
TL: 212 963 1921

Shirley, K., Mr.
New Zealand Forest Owner's Association
Permanent Mission to the United Nations, United Nations Plaza New York, NY, 10017
F: 212 963 7174
TL: 212 963 1921

Smith, N., Mr.
Member
Parliament
Permanent Mission to the United Nations, United Nations Plaza New York, NY, 10017
F: 212 963 7174
TL: 212 963 1921

Storey, W.R., H.E. Hon.
Minister
Environment (Head of Delegation)
Permanent Mission to the United Nations, United Nations Plaza New York, NY, 10017
F: 212 963 7174
TL: 212 963 1921

Vaughan, B., Ms.
Executive Assistant
Minister for the Environment
Permanent Mission to the United Nations, United Nations Plaza New York, NY, 10017
F: 212 963 7174
TL: 212 963 1921

Wije-Wardana, Don, Mr
Chief Economist
Ministry of Forestry
P O Box 1610
Wellington, New Zealand
T: 64 4 4721 569 F: 64 4 4722 314

NGO

Holdaway, Jim, Mr.
NZN/WORLD WIDE FUND FOR NATURE
35 Ocean View Road
Auckland, New Zealand
T: 9 4183242 F: 9 4802808

Hope, Gillian
NZN
PO Box 10 123
Wellington, New Zealand

Minhinnick, Nganeko
AKM
Pouaka 250
Waiuku, South Auckland, Aotearoa, New Zealand

Senituli, Lopeti
PCR
C.P.O. Box 3148
Auckland, Aotearoa, New Zealand

Small, Dolores Loretta
NZNHF
PO Box 10 123
Wellington, New Zealand

Springett, Delyse, Ms.
Executive Director
NEW ZEALAND NATURAL HERITAGE FOUNDATION
c/o Massey University
Palmerston North, New Zealand
T: 64 6 356-9099 F: 64 6 356-7286

Tangiora, Pauline, Ms.
WOMENS INTERNATIONAL LEAGUE PEACE FREEDOM / M.W.WL NZ MAORI COUNCIL
PO Box 33 Mahia Hawkobay
Aotearoa, New Zealand
T: 64 6 837-5816

Temaru, Oscar
PCR
C.P.O. Box 3148
Auckland, Aotearoa, New Zealand

Uregei, Louis Kotra
PCR
C.P.O. Box 3148
Auckland, Aotearoa, New Zealand

PRESS

Lessing, Judy, Ms.
Correpondent
RADIO NEW ZEALAND NEWS
c/o 1422 Beverly Rd Brooklyn
New York - NY, USA 11226, New Zealand
T: 1 212 4692767 F: 1 212 4692808

ORGANIZATIONS

Dell, Miriam, Dame
New Zealand Ngo/unced Liaison Committee
PO Box 10 123
Wellington, New Zealand
T: (64 4) 828 125 F: (64 4) 828 129

Mahuika, Apirana, Mr.
National Maaori Congress
c/o 26 Brandon Street, PO Box 5236
Wellington, New Zealand

Thompson, , Ms.
Awora Ki Manuka
Pouaka 250
Waiuku, South Auckland, Aotearoa, New Zealand
T: (6485) 59120

Wehipeihana, Maria, Ms.
South Pacific Regional Indigenous Science Network, The
48 Winara Avenue
Waikanae, New Zealand
T: (64 58) 32 932 F: (64 06) 368 0440

GLOBAL FORUM

Bosselmann, Klaus, Mr.
Prof. Dir.
UNIVERSITY OF AUCKLAND
9 Eden Crescent, Private Bay
Auckland 92019, New Zealand
T: 64 9 3737999 F: 64 9 3737440

Dolan, Josie, Mr.
NETWORK OF SISTERS BROTHERS & PRIESTS FOR JUSTICE
36 Stenhope Cres, Corstorphine Otepoti
Dunedin, New Zealand
T: 64 3 487-9748

Healy, Peter, Mr.
NETWORK FOR JUSTICE
305 Te Atatu Road, Te Atatu South, Tomaki Makauran
Auckland, New Zealand
T: 64 9 834-3366

Hennessy, Wayne, Mr.
COAL RESEARCH
Box 31-244
Wellington, New Zealand
T: 64 4 566-2289 F: 64 4 566-7737

Holdaway, Ann, Ms.
Wwf
WORLD WIDE FUND FOR NATURE
35 Ocean View Rd.
Auckland, New Zealand
T: 64 9 4183242 F: 64 9 4802808

Kettle, Peter, Mr.
Dr
MAF
Pobox 2526
Wellington, New Zealand
T: 64 4 472-0367 F: 64 4 474-4163
TL: NZ MAF 31532

Mead, Aroha, Ms.
Convenor, Int'L Commitee
NATIONAL MAORI CONGRESS
9 Boubon Tce- Karori
Wellington, New Zealand
T: 64 4 476-2205 F: 64 4 499-2388

Reedy, Tamati, Mr.
Convenor, Int'L Committee
NATIONAL MAORI CONGRESS
26 Brandon ST. Central House
Wellington, New Zealand
T: 64 4 499-1417 F: 64 4 499-1418

Robertson, Graeme Leonard, Mr.
NEW ZEALAND INSTITUTE OF ARCHITECTS
Uni.of Auckland, Private Bag 92019
Auckland, New Zealand
T: 64 9 373-7599 F: 64 9 373-7410

Shand, Diana, Ms.
NATIONAL NGO UNCED COMMITTEE
311 Montreal St.
Christchurch, New Zealand
T: 64 3 650546 F: 64 3 772387

Small, Delle, Ms.
NATIONAL NGO UNCED COMMITTEE
175 Fitzgerald Ave.
Christchurch, New Zealand
T: 64 3 792-185 F: 64 3 792-185

Smith, Nick, Mr.
Doctor
NZ GOVERMENT
43 Ellis St.,Brightwater
Nelson-South Island, New Zealand
T: 64 3 5423341 F: 64 3 5423351

Storey, Lorraine, Ms.
GOVERNMENT
Parliament Buildings
Wellington, New Zealand
T: 64 4 471-9157 F: 64 4 471-2954

Storey, Rob, Mr.
Minister
GOVERNMENT
Parliament Buildings
Wellington, New Zealand
T: 64 4 471-9157 F: 64 4 471-2954

Taylor, Prue
Prof Dir
UNIVERSITY OF AUCKLAND
9 Eden Crescent, Private Bay
Auckland 92019, New Zealand
T: 64 9 3737999 F: 64 9 3737440

Vaughan, Beth, Ms.
GOVERNMENT
Paliament Building
Wellington, New Zealand
T: 64 4 471-9157 F: 64 4 471-2954

Weiss, Diana
OVERSEAS DEVELOPMENT NETWORK
32 Douglas St.
Gisborne, New Zealand
T: 64 6 867-2790

PARLIAMENTARIANS / SPIRITUAL LEADERS

Caygill, David
MP
House of Representatives Parliament Buildings
Wellington, New Zealand
T: 749 199
TL: 31042

Smith, Rick
MP
House of Representatives Parliament Buildings
Wellington, New Zealand
T: 749 199
TL: 31042

Sotrey, Rob
MP
House of Representatives Parliament Buildings
Wellington, New Zealand
T: 749 199
TL: 31042

NICARAGUA - GOVERNMENT

Arguello-Guardian, Gilda, Ms
Presidenta
FUNDACION MUJERES POIRA EL MEDIOAMN
Villa Fontana Norte #72
Managua, Nicaragua
'T: 505 2 679763 F: 505 2 679763

Barrios de Chamorro, Violeta, MS
President
Permanent Mission of Nicaragua
820 2nd Ave., 8th floor
New York NY 10017, United States of America
T: 212 490 7997 F: 212 286 0815

Gutierrez-Huete, Claudio
Director General
INSTITUTONICARAGUENSE DE ESTUDIOS
Apartado 2110
Managua, Nicaragua
T: 505 2 41890 505 2 44 F: 505 2 41890

Incer-Barquero, Jaime
Ministro
MINISTERIO DEL AMBIENTE Y REC NATUR
KM 12.5 Corretera Norte
Managua, Nicaragua
T: 505 2 631271 631273 F: 505 2 631274

Picado Trana, Freddy J., Mr
Director Tecnico Meteorologia
INETER
Managua-Nicaragua, Apartado 2110 Telex-Fax 41890
Managua, Nicaragua
T: 44739 666093 666058 F: 41890 31610

Tercero Talayera, Victor, Mr
Director Gral Planificauon
MINISTERIO DE REWRSOS NATURALES
Apdo 3263 Managua/Nic
Managua, Nicaragua
T: 664848 613110 15 F: 503 2 627134

Terez-Paguaga, Patricio
Vice Ministro
MINISTERIO DEL AMBIENTE Y REC NAT.
KM 12.5 Carretera Norte
Mawagua, Nicaragua
T: 505 2 631343 F: 505 2 631274

NGO

Espinosa Rojas, Eudoro
MAN
Apartado A-99
Managua, Nicaragua

Guillen, Jaime Rafael
REDESCA
Altamira d'Este, Casa No. 774
Managua, Nicaragua

Ruiz Melendez, Antonio
MAN
Apartado A-99
Managua, Nicaragua

ORGANIZATIONS

Espinosa, Eudoro, Ing.
Movimiento Ambientalista Nicaraguense
Apartado A-99
Managua, Nicaragua
T: (50 52) 74835 F: (50 52) 674 021

Montiel, Juan Jose, Lic.
REDES Centro America
Altamira d'Este, Casa No. 774
Managua, Nicaragua

GLOBAL FORUM

Amador, Carlos
General Secretary
ASOCIACION CRISTIANA DE JOVENES DE NICARAGUA
0-IV-4, de la Iglesia Pio X, 400 metros al sur y 50 arriba
Managua, Nicaragua

Castrillo, Sandra, Ms.
THIRD WORLD NETWORK
Casilla 2715
Managua, Nicaragua

Castro U., Maria Del Carmen, Ms.
C E S A D E
Altamora C. Los Madronos No. 65
Managua, Nicaragua
T: 505 2 780847 F: 505 2 70950

Garcia E., Ma. Lourdes, Sra.
REDES
Palais de Nations
Geneve CH 1211, Nicaragua
F: 41 22 733-6542

Hafner, Winfried, Mr.
Ingeniero
Granada, Nicaragua
T: 505 55 221 172

Lorio, Margarita, Ms.
RED DE MUJERES Y MEDIO AMBIENTE - CENTRO AMERICA Y CARIBE
Apartado A-99
Managua, Nicaragua
T: 505 2 74836

Membreno, Lic. Conny
GUNATT-LEON
Palais de Nations
Geneve CH 1211, Nicaragua
F: 41 22 733-6542

Mercado, Pedro, Mr.
Licenciado
MOVIMIENTO AMBIENTALISTA NICARAGUENSE
Man Apartado A-99, Bello Horizonte, L--1--39
Managua, Nicaragua
T: 505 2 74835 F: 505 2 74836
TL: A-99

Moreira, Roberto Munoz, Sr.
Manager-Projects Formulation
FACTS
Semaforos Telcor Villa Fontana 300 mts. norte 20 mts. este.
Managua, Nicaragua
T: 505 2 74773 F: 505 2 675670
TL: 2361 FACTS NK, E: 2458, Zona Post. 5

Morgan, Gloria Calero, Ms.
Cihuatlampa
ONG NICARAGUA
Apdo EC-22
Managua, Nicaragua
T: 505 2 43189

Munoz, Roberto, Sr.
FUNDACION AUGUSTO C. SANDINO
Apartado 2458, Zona Postal 5
Managua, Nicaragua
F: 505 2 675670
TL: 2361 FACS NK

Narvaez, Juan Ruiz, Mr.
Directivo Regional
ASOCODE
Man Apartado A-99, Bello Horizonte, L--1--39
Managua, Nicaragua
T: 505 2 74835 F: 505 2 74836

Picado, Jaime, Mr.
Ing. Agronomo
MOVIMIENTO AMBIENTALISTA NICARAGUENSE
Man Apartado A-99, Bello Horizonte, L--1--39
Managua, Nicaragua
T: 505 2 74835 F: 505 2 74836

Querol, Daniel, Sr.
THIRD WORLD NETWORK
Casilla 2715
Managua, Nicaragua

Rojas, Eudoro Espinosa, Mr.
Ing. Agronomo
MOVIMIENTO AMBIENTALISTA NICARAGUENSE
Man Apartado A-99, Bello Horizonte, L--1--39
Managua, Nicaragua
T: 505 2 74835 F: 505 2 74836

Romero, Rayneiro, Mr.
MOVIMIENTO AMBIENTALISTA NICARAGUENSE
Man Apartado A-99, Bello Horizonte, L--1--39
Managua, Nicaragua
T: 505 2 74835 F: 505 2 74836
TL: A-99

Ruiz, Antonio, Mr.
Sociologo
MOVIMIENTO AMBIENTALISTA NICARAGUENSE
Parque Memorial Sandino Managua
Managua, Nicaragua
T: 505 2 74835 F: 505 2 74836
TL: A-99

Ruiz, Juan, Mr.
ASSOCODE
Parque las Palmas 3045 ,Suv
Managua, Nicaragua
T: 505 2 660632 F: 505 2 662318

Sanders, Pedro Mercado, Mr.
CONSEJO REGIONAL AUTONOMO
Puerto Cabezas
Puerto Cabezas, Nicaragua

Shogreen, Rev. Andy
MORAVIAN CHURCH IN NICARAGUA
Puerto Cabezas
Zelaya, Nicaragua

Toruno, Rainiero Jose Romero, Mr.
CENTRO DE INVESTIGACIONES Y DOCUMENTACION DE LA COSTA ATLANTICA
Reparto Pancasan
Managua, Nicaragua
T: 505 2 70983 F: 505 2 673006
E: P-183

NIGER - GOVERNMENT

Boukar, Abba Malam, S.E. M.
Ministre l'Agriculture et de l'Elevage
Permanent Mission of Niger
417 East 50th Street
New York, NY 10022, United States of America
T: 421 3260 F: 753 6931

Dankolodo, Dan Dicko, M.
Physicien
Centre Regional de l'Energie Solaire
Permanent Mission to the United Nations, 417 East 50th Street New York, NY, 10022
F: 421 3260
TL: 753 6931

Elhadji Mahamane, Mahamane Lawali
Directeur Projet Energie Ii-Volet O
DIRECTION-DE L'ENVIRONNEMENT
BP 578 Niamey 1
Niamey, Niger
T: 227 733393/227 72219 F: 227 7347 09

Garba, Adamou, Mr.
RASSEMBLENT POUR UN SAHEL VERT
BP 12 515 Niamey (Niger)
Niamey, Niger
T: 22 7 74 11 25 F: 22 7 72 23 36
TL: 5358 NI

Hadjo, Abdoulmoumine, Mme
Directrice des Organisations et Conferences Internationales
Ministere des Affaires Etrangeres et de la Cooperation
Permanent Mission to the United Nations, 417 East 50th Street New York, NY, 10022
F: 421 3260
TL: 753 6931

Hassane, Abdou, S.E. M.
Ministre l'Hydraulique et de l'Environnement (Chef de la Delegation)
Permanent Mission of Niger
417 East 50th Street
New York, NY 10022, United States of America
T: 421 3260 F: 753 6931

Hassane, Mamoudou, M.
Juriste
Ministere de l'Agriculture et de l'Elevage
Permanent Mission to the United Nations, 417 East 50th Street New York, NY, 10022
F: 421 3260
TL: 753 6931

Kombo, Adamou, M.
Deuxieme Vice-President
Haut Conseil de la Republique
Permanent Mission to the United Nations, 417 East 50th Street New York, NY, 10022
F: 421 3260
TL: 753 6931

Mahamane, Elhadj Lawali, M.
Directeur du Projet Energie II
Ministere de l'Hydraulique et de l'Environnement
Permanent Mission to the United Nations, 417 East 50th Street New York, NY, 10022
F: 421 3260
TL: 753 6931

GLOBAL FORUM

Keita, Mariama, Ms.
ASAFED
P.O.Box 12010
Niamey, Niger
T: 22 7 735820 F: 22 7 732855
TL: 5269NI

Younoussi, Boukary, Mr.
SEAG
BP 12675
Niamey, Niger
T: 22 7 73 35 27

NIGERIA - GOVERNMENT

Adejuwon, Adeniran, Mr
Assistant Director
NATIONAL PLANNING COMMISSION
the Presidency, Federal Secretariat Ikoyi
Logos, Nigeria
T: 01 687614

Adesiyan, Olufemi Zaccheus
Deputy Director Of Forestry
FEDERAL DEPT OF FORESTRY
PMB 135, Wuse Zone 1
Abuja, Nigeria
T: 09 5232659

Agah, Yonov Frederick
Asst. Director (external Trade)
MULTILATERAL TRADE DIVISION
Federal Ministry of Commerce And Tourism
Garki-Abnja, Nigeria
T: 09 2341792

Aina, Evans, Olufunso
Director - General
ENVIRONMENTAL PROTECTION AGENCY
P.O. Box 2831
Lagos, Nigeria
F: 680071

Babangida, Ibrahim B., Mr.
President
Permanent Mission of Nigeria
733 3rd Ave., 15th Floor
New York NY 10017, United States of America
T: 212 953 9130 F: 212 697 1970

Bassey, James Obong, Mr
Director Water Supply & Quality Con
FED. MIN. OF AGRIC, WATER RESOURCES
Area 1, Block A, PMB 159, Garki,
Abuja, Nigeria
T: 234 09 2342734 F: 234 09 2342509

Bawuro Bapetel, Yahaya
Executive Director
ADAMAWA STATE ENVIRONMENTAL AGENCY
PMB 2170 Yola; Adamawa State
Adamawa State, Nigeria

Belo, Ganiyu Akanbi
Executive Secretary
NIGERIAN EDUCATIONAL RESEARCH&DEV.
P.M.B. 91 Fed Capital Territory
Abuja, Nigeria
T: 09 8821614

Eluwa, Nnenna Achama Eluwa
Assistant Director Women Employment
NATIONAL DIRECTORATE OF EMPLOYMENT
Plot 1623 Saka Jojo Street P.O. Box 71649 Victoria Island
Lagos, Nigeria

Mahmood, Sadiq Muhammad
Director Of Dams & Reservoirs
FEDERAL MIN OF AGRIC & WATER RESOUR
P.M.B. 159 Garki, Abuja
Abuja, Nigeria
T: 234 09 2342324 F: 09 2342509

Odu, Godfrey Odu
Correspondent
NIGERIAN TELEVISION
Authority-News Directorate
Lagos, Nigeria
T: 610057

Omotosho, David Bola, Dr
Agm
FEDERAL ENVIRONMENTAL PROTECTION AG
P.M.B. 12620
Lagos, Nigeria
T: (234-1) 680308 F: (234-1) 687600

Osibanjo; Prof., Oladele
Adviser
FEDERAL ENVIRONMENTAL PROTECTION AG
Department of Chemistry University of Ibadan
Ibadan, Nigeria
T: 022 400550 EXT 2555: F: 01 611531

Yaye, Abdoulmoumine Hadjio
Directrice Organisations L.
CONFERENCES INTERNATIONALES
Maec BP 396
Niamey, Nigeria
T: 73 42 18 - 72 3031

NGO

Adeniran, Adedoja
FADU
Brown House (2nd Floor) Monatan Bus-Stop
P.M.B. 56 Agodi, Ibadan, Nigeria

Anadu, P. A.
NCF
PO Box 74638
Victoria Island, Lagos, Nigeria

Chigbolu, E.C.
CWAN
No.2, Afunbiowo Street, P.M.B. 809
Akure, Ondo State, Nigeria

Hassan, Bukar
CAZS
University of Maiduguri, P.M.B. 1069
Maiduguri, Borno State, Nigeria

Isaiah, Ogundele
FADU
Brown House (2nd Floor) Monatan Bus-Stop
P.M.B. 56 Agodi, Ibadan, Nigeria

Japhet Ukpong, Samuel
SIRF
58 Ekpo Abasi St., PO Box 1130
Calabar, Nigeria

Kamarakafego, Pauulu, Mr.
INT'L. PAN AFRICA MOVEMENT
2-8 Calcutta Crescent, P O Box 610, APAPA
Lagos, Nigeria

Nascimento, Edialeda Salgado
PAM
2-8 Calcutta Crescent, P O Box 610, APAPA
Lagos, Nigeria

Obot, Sunday Udo, Mr.
National President
NATIONAL CONSUMERS NIGERIA
28 E Yellow Duke St., P.O. Box 851
Calabar -CRS, Nigeria
T: 234 87 222252

Oseji, Humphrey
SIRF
58 Ekpo Abasi St., PO Box 1130
Calabar, Nigeria

Ser, Julie Shown, Ms
President
COUNTRY WOMEN ASSOCIATION OF NIGERIA (COWAN) - WEDO
No.2, Afunbiowo Street, P.M.B. 809
Akure, Ondo State, Nigeria

PRESS

Abdulmalik, Ibrahim
Editor
NEWSLINE
Paiko Road, P.M.B. 10
Minna, Nigeria, Nigeria

Akingbade, Tunde
Correspondent
AFRICAN SCIENCE MONITOR
63 DuroJane St.
Lagos, Nigeria, Nigeria

Akinremi, Folarin Olufemi
Editor
PEOPLE'S EPITATH
P.O. Box 5571, Totoro P.O.
Abeokuta, Ogun State, Nigeria, Nigeria

Amupitan, Ola
Editor
REPUBLIC NEWSPAPERS LTD.
13 Adepegba St, Ilupeju Estate
Lagos State, Nigeria, Nigeria

Awojuola, T. Yemi
Reporter Radio/Television
FEDERAL RADIO CO. OF NIGERIA
P. O Box 3702
Lagos, Nigeria, Nigeria

Ekpu, Ray
Editor
NEWSWATCH
49 Isaac John Street
GRA, Ikeja, Lagos, Nigeria, Nigeria

Kingsley, Agha
Editor
NIGERIAN TV AUTHORITY NEWS
Blk 4, No. 1, Bar Beach Towers
Victoria Island, Lagos, Nigeria, Nigeria

Olukoshi, Jacob
Editor
NAN/ NEWS AGENCY OF NIGERIA
Third Ave., X Close
Block 1, Flat 11, Festac Town, Nigeria, Nigeria

Tijani, Adebisi
Editor
DAILY TIMES OF NIGERIA
New Isher Road, PMB 21340
Agidongbi, Ikeja, Lagos, Nigeria, Nigeria

ORGANIZATIONS

Adedoja, Ade, Mr.
Farmers Development Union (FADU)
Brown House (2nd Floor) Monatan Bus-Stop
P.M.B. 56 Agodi, Ibadan, Nigeria

Ajayi, Akinola, Mr.
Youth Alliance of Nigeria
7, Dotun Isijola Street, Aka-Okokomaiko
PO Box 3171 Surulere, Lagos, Nigeria

Amokeodo, Ore, Mr.
Nigerian Conservation Foundation
PO Box 74638
Victoria Island, Lagos, Nigeria
T: (2341) 683 563/686 163/687 385

Gadzama, N.M., Prof.
Centre for Arid Zone Studies
University of Maiduguri, P.M.B. 1069
Maiduguri, Borno State, Nigeria

Moma, Fatma E., Ms.
UNCED92 African NGOs Co-ordinating Committee
PO Box 3063
Lagos, Nigeria
T: (23 41) 612 091

Nweke, Chike, Mr.
Green Environment Movement
18 Montgomery Street, Sabo Yaba-Lagos
GPO Box 5867, Lagos, Nigeria
T: (234 1) 865 914 F: (234 1) 860 161

Ogunleye, Adeleke, Mr.
Country Women Association of Nigeria
No.2, Afunbiowo Street, P.M.B. 809
Akure, Ondo State, Nigeria
F: (234) 231 633

Salau, Ademola T., Sr.
Chairman
NEST (THE NIGERIAN ENVIRONMENTAL STUDY/ACTION TEAM)
P.M.B. 5297 -18B Aare Avenue -New Bodija
Ibadan - Oyo, Nigeria
TL: 31123cusong

Samuel, Eke, Mr.
Africa Harvest Mission
P.O. Box 514, Nkwogwu
Mbaise, Imo State, Nigeria

Ukpong, Samuel Japhet, Dr.
Sajju Institute and Research Foundation
58 Ekpo Abasi St., PO Box 1130
Calabar, Nigeria

GLOBAL FORUM

Abubakar, Atiku
NIGERIAN BRANCH OF THE WORLD CONSTITUTION AND PARLIAMENT ASSOCIATION
5, Godwin Okigbo St.
Lagos, Nigeria
T: 234 41 835722

Ademidun, Fao Le Grace, Ms.
Project Coordinator
UNFPA/IPAC NIGERIAN CHAPTER
PO Box 531
Akure, Nigeria
T: 234 34 230486 F: 234 34 231633

Aina, Tade
NEST (THE NIGERIAN ENVIRONMENTAL STUDY/ACTION TEAM)
P.M.B.5297 -18B Aare Avenue -New Bodija
Ibadan, Nigeria
TL: 31123cusong

Akandu, Lewis Ngozi, Mr.
BUDGET OFFICE
Room 746, PH II, Fed Secretariat
Lagos Ikoyi, Nigeria
T: 234 1 682682

Amakiri, Mildred A., Dr.
Professor
UNIVERSITY
Rivers State University of Science and Technology
Port-Harcourt, Nigeria
T: 234 84 334838

Amujogu, Anthony, Mr.
NIGERIAN BRANCH OF THE WORLD CONSTITUTION AND PARLIAMENT ASSOCIATION
5, Godwin Okigbo St.
Lagos, Nigeria
T: 234 41 835722

Anekwe, Ezekiel, Mr.
NIGERIAN BRANCH OF THE WORLD CONSTITUTION AND PARLIAMENT ASSOCIATION
5, Godwin Okigbo St.
Lagos, Nigeria
T: 234 41 835722

Awosika, Chief Winifred A., Ms.
28 Opebi Road, P.H.B. 21614 - Ikeja
Lagos, Nigeria
T: 234 1 964524

Awosika, Winifred, Ms.
GLOBAL FORUM OF SPIRUTUAL AND PARLIAMENTARY LEADER ON HUMAN SURVIVAL
28 Opebi Road P.M.B. 21614
Lagos, Nigeria
T: 234 1 964-529

Azu, K, Mr.
President
EARTHSEARCH
11 Ochi Street, PO Box 1230
Enugu, Nigeria
T: 234 4201330481 F: 234 4201332059
TL: 51445/51300

Bell-Gam, Winston, Mr.
Publisher/Editor
INTERNATIONAL JOURNAL OF ENVIRONMENTAL MANAGEMENT AND DEVELOPMENT
PMB 6171
Portharcourt, Nigeria
T: 234 84 334940
TL: 611205NG, E: BOX 267

Chigbolu, Elizabeth Ngozi, Ms.
President
COUNTRY WOMEN ASSOCIATION OF NIGERIA (COWAN)/WEDO
1 Umutei Quarters, P.O. Box 93
Illah, Nigeria

Chilagorom, Eric, Mr.
President
EARTHSEARCH
11 Ochi Street, PO Box 1230
Enugu, Nigeria
T: 234 4201330481 F: 234 4201332059
TL: 51445/51300

Dappa-Biriye, Atama, Ms.
RIVERS CHIEFS AND PEOPLES CONFERENCE
12 Rickett Street
London SW6 1RU, Nigeria
T: 44 71 386-0266 F: 44 71 386-0275

Dappa-Biriye, H.R.H. Harold, Dr.
President
RIVERS CHIEFS AND PEOPLES CONFERENCE
12 Rickett Street
London SW6 1RU, Nigeria
T: 44 71 386-0266 F: 44 71 386-0275

Edoga, Elizabeth
BETTER LIFE PROGRAMME
Government House
Enugu, Nigeria
T: 234 4201 F: 234 4201

Eguriase, Samuel, Mr.
NIGERIAN ENVIRONMENTAL SOCIETY
P.O.Box 9187, Marina,Lagos 13 D'Alberto Road, Palm Grove Estate
Lagos, Nigeria
T: 234 1 880241

Falomo, Adejoju, Ms.
FRIENDS OF THE ENVIRONMENT - NIGERIA
Lagos, Nigeria

Faoye, Grace Ademidun
COUNTRY WOMEN ASSOCIATION OF NIGERIA-UNFPA
PO Box 2430
Akure, Nigeria
T: 234 34 231857

Fubara, Dagogo M., Mr.
Professor
RIVERS CHIEFS AND PEOPLES CONFERENCE
12 Rickett Street
London SW6 1RU, Nigeria
T: 44 71 386-0266 F: 44 71 386-0275

Hanson, Elkanah, Mr.
Reverend
RIVERS STATE CHIEFS AND PEOPLE'S CONFERENCE
2 Anokuru St. Diobu, Mile 1
Port-Harcourt, Nigeria
T: 234 84 330388

Idoniboye-Obu, B., Dr.
Chief
RIVERS CHIEFS AND PEOPLES CONFERENCE
12 Rickett Street
London SW6 1RU, Nigeria
T: 44 71 386-0266 F: 44 71 386-0275

Ilegbune, Theresa O., Ms.
Dep. of Public & Private Law, Faculty of Law, University of NigeriaEnugu Campus
Enugu, Nigeria
T: 234 4201337278 F: 234 4201

Ilo, Gogo Rose, Ms.
Co-Chairwoman
IPAC / WEDO (NIGERIA CHAPTER)
10, Lafiya Road
Nassarawa, Kano, Nigeria
T: 234 64 6171

Itakura, Sumiko, Ms.
Nest
NIGERIAN ENVIRONMENTAL STUDY / ACTION TEAM
18B Aave Avenue , New Bodija
Ibadan 5297, Nigeria

Itakura, Yutaka, Mr.
NIGERIAN ENVIRONMENTAL SOCIETY
13,D'Alberto Road Palmg Rove Estate P.O.BOX9187 Marina
Lagus, Nigeria
T: 01 88024

Job, Margaret Yob, Ms.
State Secretary
COWAN - COUNTRY WOMEN ASSOCIATION OF NIGERIA
C/O Box 6504 - Anglo - JOS - Plateau State
Jos - Plateau State, Nigeria

Kilanko, Glory Afi, Ms.
Chariperson
NIGERIA LABOUR CONGRESS
29, Olajuwon St., Off. Ojuelegba Rd. PO Box 620 Yaba
Lagos, Nigeria
T: 234 1 835571 F: 234 1 835578

Lahadi, Tseayo, Ms.
NIGERIAN ENVIRONMENTAL STUDY/ACTION TEAM (NEST)
186 AARE Avenue, IBADAN
Ibadan, Nigeria
T: 234 22 41-6244

Madu, T I, Mr.
President
EARTHSEARCH
11 Ochi Street, PO Box 1230
Enugu, Nigeria
T: 234 4201330481 F: 234 4201332059
TL: 51445/51300

Madunagu, Bene, Ms.
WOMEN IN NIGERIA
c/o WAND, UWI School of Continuing Studies, Pinelands
St Michael, Nigeria
T: 1 809 4269288 F: 1 809 4294854

Mazi, Bernardshaw, Mr.
NIGERIAN BRANCH OF THE WORLD CONSTITUTION AND PARLIAMENT ASSOCIATION
5, Godwin Okigbo St.
Lagos, Nigeria
T: 234 41 835722

Moka, G.U.
Prince
NIGERIAN BRANCH OF THE WORLD CONSTITUTION AND PARLIAMENT ASSOCIATION
5, Godwin Okigbo St.
Lagos, Nigeria
T: 234 41 835722

Muhammed, Ajoke, Ms.
MURTALA MUHAMMED MEMORIAL BOTANICAL GARDENS
4 Tafawa Ralena Square, Race Course
Lagos, Nigeria
T: 632-413

Ngene, Amaka, Ms.
NIGERIAN BRANCH OF THE WORLD CONSTITUTION AND PARLIAMENT ASSOCIATION
5, Godwin Okigbo St.
Lagos, Nigeria
T: 234 41 835722

Nwodo, Doroth, Ms.
BETTER LIFE PROGRAMME
Government House Enugu
Enugu, Nigeria
T: 234 4201 F: 234 4201

Obadiah, Jesseca Di, Ms.
Treasurer
COWAN - COUNTRY WOMEN ASSOCIATION OF NIGERIA
PO Box 6504, Anglo-Jos
Jos - Plateau State, Nigeria

Obidike, Bob J.N., Mr.
NIGERIAN BRANCH OF THE WORLD CONSTITUTION AND PARLIAMENT ASSOCIATION
5, Godwin Okigbo St.
Lagos, Nigeria
T: 234 41 835722

Odukoya, Segun A., Dr.
Deputy Councilor
SOCIETY OF NIGERIAN DOCTORS FOR THE WELFARE OF MANKIND
9 Sam Shonibare St., s/ Lere
Lagos, Nigeria
T: 234 41 870265 F: 234 41 667884

Ogunleye, Chief Bisi, Ms.
National Coordinator
COUNTRY WOMEN ASSOCIATION OF NIGERIA (COWAN)/WEDO
2 Afunbiowo Street P.M.B. 809
Akure, Nigeria
T: 234 34 231-945 F: 234 34 231-633

Ogunsanya, Adenrele, Ms.
NIGERIAN BRANCH OF THE WORLD CONSTITUTION AND PARLIAMENT ASSOCIATION
5, Godwin Okigbo St.
Lagos, Nigeria
T: 234 41 835722

Okafor, Josephine, Ms.
NIGERIAN BRANCH OF THE WORLD CONSTITUTION AND PARLIAMENT ASSOCIATION
5, Godwin Okigbo St.
Lagos, Nigeria
T: 234 41 835722

Okali, David
NEST (THE NIGERIAN ENVIRONMENTAL STUDY/ACTION TEAM)
P.M.B.5297 -18B Aare Avenue -New Bodija
Ibadan, Nigeria
TL: 31123cusong

Okolo-Ayeni, Victoria, Ms.
govt house, ENUGU STATE
Enugu, Nigeria
T: 234 4201338570 F: 234 4201

Oladele, Olori Folasayo Olamike, Ms.
State Coordinator
COUNTRY WOMEN ASSOCIATION OF NIGERIA (COWAN)/WEDO
c/o St. John's Anglican Church, P.O. BOX 32, Osum State
Ipetumodu, Nigeria

Olowokere, B.A., Ms.
COUNTRY WOMEN ASSOCIATION OF NIGERIA (COWAN)/WEDO
PO Box 2430
Akure, Nigeria
T: 234 34 231857

Onyekakeyah, Luke O., Mr.
President
EARTHSEARCH
11 Ochi Street GPO Box 1230
Enugu, Nigeria
T: 234 4201330481 F: 234 4201332059
TL: 51445/51300

Onyirimba, E, Mr.
President
EARTHSEARCH
11 Ochi Street, PO Box 1230
Enugu, Nigeria
T: 234 4201330481 F: 234 4201332059
TL: 51445/51300

Ozakunle, Onayemi Ozarotimi, Mr.
President
NIGERIAN ENVIRONMENT NETWORK
27a Ibadan Road, Ijebu-ODE
Ogun State, Nigeria
T: 234 37 430329

Paduka, Joanna, Ms.
FRIENDS OF THE ENVIRONMENT - NIGERIA
22 Simpson St. Box 10627
Lagos 20117, Nigeria
T: 234 1 633-988 F: 234 1 611-424

Salau, Oreoluwa, Ms.
NIGERIAN ENVIRONMENTAL STUDY/ACTION TEAM (NEST)
186 Aare Avenue, New Bodija
Ibadan, Nigeria
T: 234 22 416244
TL: CUSO NG 31123

Tunji-Olagunju, Omobola, Ms.
FRIENDS OF THE ENVIRONMENT - NIGERIA
22 Simpson St. Box 10627
Lagos 20117, Nigeria
T: 234 1 633-988 F: 234 1 611-424

Utulu, Evelyn, Ms.
NIGERIAN BRANCH OF THE WORLD CONSTITUTION AND PARLIAMENT ASSOCIATION
5, Godwin Okigbo St.
Lagos, Nigeria
T: 234 41 835722

Uwechue, Justina, Ms.
Vice President
UNITED NATIONS ASSOCIATION OF NIGERIA
P.O. Box 662
Lagos, Nigeria
T: 234 1 61-8465

Yonkat, Jessica, Ms.
State Coordinator
COWAN (COUNTRY WOMEN ASSOCIATION OF NIGERIA)
c/o Box 6504 - Anglo
Jos - Plateau State, Nigeria

NORWAY - GOVERNMENT

Brundtland, Gro Harlem, MS
Prime Minister
Permanent Mission of Norway
825 3rd. Ave, 39th floor
New York NY 10022, United States of America
T: 212 421 0280 F: 212688 0554

NGO

Chatfield, John
IULA
PO Box 1378
Vika, N-0114, Oslo 1, Norway

Dale, Caroline, Ms.
Project Mgr.
IEB/ICC
P.O. Box 301
Lysaker 1324, Norway
T: 47 2 581800 F: 47 2 581875

Droyer, Annelise, Ms.
Freelancer
EDUCATORS FOR PEACE
Haunabakken 29
Oslo 0874, Norway
T: 47 2 236630

Eskeland, Kristin, Ms.
Coordinator Voice Of Children
NORWEGIAN CAMPAIGN FOR ENVIRONMENT & DEVELOPMENT
Langes Gt.4
Oslo 0165, Norway
T: 47 2 362035 F: 47 2 361454

Gaenger, Hasse, Mr.
Adviser
WORLDVIEW INTERNATIONAL FOUNDATION
Welhavensgt 1
Oslo 0166, Norway
T: 47 2 114980 F: 47 2 114988

Gulichsen, Karin
CIWI
Fr. Nansens Plass 6
0160 Oslo 1, Norway

Halonon, Leif, Mr.
NORDIC SAAMI COUNCIL
N-9520 Vautokeino
Finermanda, Norway
T: 47 84 56686

Hauge, Wenche, Ms.
Chef Of Information
NORTH/SOUTH COALITION
Boks 23, Vindern
Oslo 0319, Norway
T: 47 2 451818 F: 47 2 451810

Haugestad, Anne Kristine, Ms.
President
FUTURE IN OUR HANDS
Skoleveien 37
Fjellstrand 1458, Norway
T: 47 9 918538 F: 47 9 918538

Hersvik, Rune, Mr.
Dir External Rel
WORLDVIEW INTERNATIONAL FOUNDATION
Strandkaien 2
Stavanger 4005, Norway
T: 47 45 525767 (2)114980 F: 47 45 526480

Instefjord, Gunnstein, Mr.
Project Secretary
CHURCH OF NORWAY MAGAZINE - COUNCIL ON FOREIGN RELATIONS
P.B. 5816 Hegdehaugen
Oslo 0308, Norway
T: 47 2 694044 F: 47 2 565864

Korsrud, Mariette
NCED
Langesgt.
0165 Oslo 1, Norway

Lem, Steinar, Mr.
Head Info
FUTURE IN OUR HANDS
PO Box 4743 Sofienberg
0506 Oslo, Norway

Lynggard, Trine
CIWI
Fr. Nansens Plass 6
0160 Oslo 1, Norway

Malkenes, Runar, Mr.
Inform.Officer
NORWEGIAN CAMPAIGN FOR ENVIRONMENT & DEVELOPMENT
Langesgt 4
Oslo 0165, Norway
T: 47 2 362035 F: 47 2 361454

Nicolaisen, Betty
WPN
Box 389
7001 Trondheim, Norway

Nordland, Eva
NORWEGIAN PEACE SOCIETY
Eikstubben 6 b
1343 Eiksmarka, Norway

Nordland, Odd, Ms.
Professor
NORWEGIAN PEACE SOCIETY
Bjerkasen 19
Blommenholm 1310, Norway
T: 47 2 548482 F: 47 2 24032940329

Siljeholm, Jorn
Spec Adv
NORWEGIAN CAMPAIGN FOR ENVIRON & DEV
c/o IN-sambandet, Langes gl.4
Oslo 0165, Norway
T: 47 2 362035 F: 47 2 361454

Willums, Jan-Olaf, Mr.
Executive Assistant
IEB/ICC
P.O. Box 301
Lysaker 1324, Norway
T: 47 2 581800 F: 47 2 581875

PRESS

Agnets, Haaland
Reporter Radio/Television
NORWEGIAN TV
Skarspnogt 2
0271 Oslo, Norway, Norway

Bentzrod, Sveinung Berg
Journalist
AFTENPOSTEN
Postboks 1178 Sentrum
Oslo 0107, Norway
T: 47 2 863000 F: 47 2 863423

Bolstad, Gunnar, Mr.
Editor
FOE-NORWAY
P.O. Box 2113 , Grunerlokka
Oslo 0505, Norway
T: 47 2 715520 F: 47 2 715640

Borsum, Morten
Correspondent
TONSBERGS BLAD
Hertug Guttormsgate 25
N-3100 Tonsberg, Norway, Norway

Dahle, Kjell
Editor
SENTERPRESSENS OSLOREDAKSSON
Akersbakken 19
0172 Oslo, Norway

Edvin, Borgen E.
Correspondent
NORWEGIAN TV
Las Hurtewsias 2822
Dept. 1002, Santiago, Chile, Norway

Eldem, Liv
Correspondent
OSTLANDETS BLAD
Holteveine 25
1410 Kolboto, Norway, Norway

Erling, Borgen
Correspondent
NORWEGIAN TV
Las Hortensias, 2822
Santiago, Chile, Norway

Finnseth, Asle
Correspondent
VART LAND
Herslebs Gate 37
0578 Oslo, Norway, Norway

Haaland, Agnete
Reporter Radio/Television
NRK TELEVISION NORWAY
Skarpsnost 2,
0271 Oslo, Norway, Norway

Haaland, Leif
Editor
TEKNISK UBEBLAD
Skaugveien 42
1911 Flateby, Norway, Norway

Haugsgjerd, Hilde
Correspondent
DAGBLADET
Grimstadgt 15, 0566
Oslo, Norway, Norway

Heiberg, Turid
Editor
DIN HORSEL
Myklegardsgt. 1
0656 Oslo, Norway, Norway

Heiding, Lita
VIOQ VART-HEALTH ENVIRONMENT MAG.
Solveien 131
1169 Oslo 11, Norway, Norway

Hess, Anne M.
Journalist
HESS PRESS
Ostadalsv.79
0753 Oslo 7, Norway
T: 47 23 45055 F: 47 23 49520

Hjorthol, Lars Martin
Correspondent
NORWEGIAN NEWS AGENCY (NTB)
Tiriltunga 244
1259 Oslo, Norway, Norway

Jortveit, Bjorne
NORWEGIAN CAMPAIGN FOR ENVIRONMENT AND DEVELOPMENT
Langesgt 4
Oslo 0165, Norway
T: 47 2 362035 F: 47 2 361454

Kalvemo, Johs
Reporter Radio/Television
NORWEGIAN B'CASTING CORP.
Gimlevegen 56 D
9000 Tromso, Norway, Norway

Lyche, Anders, Mr.
Correspondent
BELLONA MAGAZINE
Sognsueien 6B
0451-Oslo, Norway, Norway

Magnum, Anders
Correspondent
BERGENS TIDENDE
Bispengsgaten 17
N-5018 Bergen, Norway, Norway

Marthinsen, Tom
Correspondent
DAGENS NAERINGSLIV
Flaktveitv.518
5095 Flaktvett, Norway, Norway

Mathismoen, Ole
Correspondent
AFTENPOSTEN
GML Drammensvei 187C
1310 Blommenholm, Norway, Norway

Mellqvist, Lars Erik
Correspondent
FOLKEVETT
Ostliveien 6
N-1415 Oppegard, Norway, Norway

Nass Avo, Lilly
Correspondent
NYTID
Sondregate 12
0550 Oslo, Norway, Norway

Nicholaysen, Anders
Reporter Radio/Television
NORWEGIAN B'CASTING CORP.
Rolf Wickstromsv. 9, N-0873
Oslo, Norway, Norway

Oddekalv, Kurt
Correspondent
BERGENSAVISEN
Helcubin 5210
Kalandseid, Norway, Norway

Omdahl, Jan
Editor
BELLONA MAGAZINE
Vibesgate 16-3
0356-Oslo, Norway, Norway

Parmann, Georg
NORWEGIAN CAMPAIGN FOR ENVIRONMENT AND DEVELOPMENT
Langesgt 4
Oslo 0165, Norway
T: 47 2 362035 F: 47 2 361454

Rokseth, Olav, Mr.
Editor
MAGASINET FOLKEVETT
Akersbakken 13
0172 Oslo, Norway, Norway

Rygh, Bjorn
Correspondent
CHR. MICHELSEN INSTITUTE
Coahuila 156-26, Col. Roma
06700 Mexico - DF, Norway

Sagflaat, Erik
NORWEGIAN CAMPAIGN FOR ENVIRONMENT AND DEVELOPMENT
Langesgt 4
Oslo 0165, Norway
T: 47 2 362035 F: 47 2 361454

Silva, Carlos Alfredo M.
Correspondent
VART LAND
R.Visc.Piraja 207/805
Rio de Janeiro, Brazil, Norway

Strandhagen, Sverre
Correspondent
DAGENS NAERINGSLIV
Niels Juelsgt 47
0259 Oslo, Norway, Norway

Talsnes, Stal
Correspondent
VERDENS GANG
Drammensveien 45
0271 Oslo, Norway, Norway

Thomassen, Carsten
Correspondent
DAGENS NAESTINGSLIV
Marselisgate, 29
055 Oslo, Norway

Toenset, Snorre
Reporter Radio/Television
NORWEGIAN B'CASTING CORP.
Helgesens Gate 79C
0563 Oslo 5, Norway, Norway

Vatne, Karl Johan
Correspondent
SUNNMORSPOSTEN
Myrabakken
6010 Aalesund, Norway, Norway

Vermes, Thomas
Correspondent
KLASSEKAMPEN
Normannsgate 61,
N-0655 Oslo, Norway, Norway

Willich, Wenche
Correspondent
NYTID
Sognsveien 50
N-0851 Oslo, Norway, Norway

ORGANIZATIONS

Bakkevig, Trond, Mr.
Church of Norway - Council on Foreign Relations
PO Box 5816 Hegdehaugen
N-0308 Oslo 3, Norway
T: (47 2) 694 044 F: (47 2) 565 864

Borgen, Jan, Mr.
International Union Of Local Authorities
PO Box 1378
Vika, N-0114, Oslo 1, Norway
T: 31 70 32 44 032 F: 31 70 32 46 916

Enge, Elin, Ms.
Director
NORWEGIAN CAMPAIGN FOR ENVIRONMENT & DEVELOPMENT
Langes Gt. 4
Oslo 0165, Norway
T: 47 2 362035 F: 47 2 361454

Larsen, Terje, Mr.
Project Alternative Future
Sognsveien 70
0855 Oslo 3, Norway
T: (47 2) 18 1170 F: (47 2) 18 2077

Schmuck, Elizabeth, Ms.
North-South Coalition
Diakonveien 14, PO Box 23
Vinderen, N-0319 Oslo, Norway
T: (47 2) 451 811 F: (47 2) 541 810

GLOBAL FORUM

Akervik, Anne, Ms.
Member Of Parliament, A
NORWEGIAN DELEGATION TO UNCED
Karl Johans gt. 22
Oslo N-0026, Norway
T: 47 2 313183 F: 47 2 313861

Ananthakrishnan, S., Mr.
ANTI-RACIST CENTRE
Hjorteveien 4
Oslo 0852, Norway
T: 47 2 468 123 F: 47 2 170 561

Baadsvik, Karl, Mr.
Director General
NINA-NORWEGIAN INSTITUTE FOR NATURE RESEARCH
Tungalsletta 2
Trondheim N-7005, Norway
T: 47 7 580677 F: 47 7 915433

Bendiksen, Randi, Ms.
Dir. Gen. Min. Foreign Affairs
NORWEGIAN DELEGATION TO UNCED
Postboks 8013 Dep.
Oslo N-0030, Norway
T: 47 2 343912 F: 47 2 349560

Bergensen, Helge Ole, Mr.
Research Fellow
FRIDTJOF NANSEN INSTITUTE/GREEN GLOBE YEARBOOD
Fridtjof Nansen Institute, PO Box 326
Lysaker N-1324, Norway
T: 47 4 562924 F: 47 2 562924

Berner, Yrsa, Ms.
ASSOCIATED COUNTRY WOMEN OF THE WORLD
Kajaveien 33
As N1430, Norway
T: 47 10 994-1249 F: 47 10 994-8502

Berntsen, H.E. Thorbjorn
Minister Of The Environment
NORWEGIAN DELEGATION TO UNCED
Postboks 8013 Dep.
Oslo N-0030, Norway
T: 47 2 345700 F: 47 2 349560

Bjorheim, Petter, Mr.
Member Of Parliament, Frp
NORWEGIAN DELEGATION TO UNCED
Postboks 8114 Dep.
Oslo N-0032, Norway
T: 47 2 313203 F: 47 2 313856

Borgen, Erlince
Reporter
NRK - TV
Nrk
Oslo 3, Norway
T: 02 459050 F: 02 457178

Brother, Mona, Ms.
Head Of Division
NORWEGIAN DELEGATION TO UNCED
Postboks 8114 Dep.
Oslo N-0030, Norway
T: 47 2 343534 F: 47 2 349580

Bryn, Kare, Mr.
Dir. Gen. Min. For. Affairs
NORWEGIAN DELEGATION TO UNCED
Postboks 8114 Dep.
Oslo N-0032, Norway
T: 47 2 343602 F: 47 2 349580

Bye, Minati, Ms.
GREENS NORWAY
Grensen 8
Oslo 0159, Norway
T: 47 2 242-9758

Caviedes, Sergi Alcaino
Cameraman
TV NORWAY
Oslo 3, Norway

Christiansen, Arne, Mr.
Leader
LAND CONSOLIDATING OFFICE
Tretjonnvn. 16
Kristiansand 4633, Norway
T: 47 42 90-953

Christiansen, Roar
NORWEGIAN CAMPAIGN FOR ENVIRONMENT AND DEVELOPMENT
Langesgt 4
Oslo 0165, Norway
T: 47 2 362035 F: 47 2 361454

Dybdal, Birgit, Ms.
GRANDMOTHERS AGAINST NUCLEAR WEAPONS
Svaleveien 15
Oslo 1182, Norway
T: 47 2 678514

Eidheim, Idunn, Ms.
Asst. Dir. Gen.
NORWEGIAN DELEGATION TO UNCED
Postboks 8013 Dep.
Oslo N-0030, Norway
T: 47 2 345992 F: 47 2 349580

Erik, Furuluno Svein, Mr.
Photographer
WORLD CITY DIRCOVERY AS
Uilevn Alle 41-0311
Oslo, Norway
T: 47 2 505100 F: 47 2 524778

Faremo, H.E. Grete
Minister For Dev. Cooperation
NORWEGIAN DELEGATION TO UNCED
Postboks 8114 Dep.
Oslo N-0030, Norway
T: 47 2 343900 F: 47 2 349580

Fjortoft, Ragnhild S., Ms.
Public Relation
UNIFEM-NORGE
Welhavensgt 1
Oslo, Norway
T: 47 2 114980 F: 47 2 114988
TL: 74023, E: WIF OSLO 115058

Fjukstad, Sissel Hartmark, Ms.
FUTURE IN OUR HANDS
Arenfeldtsvei 39
Kristiansund, Norway
T: 47 42 91390

Forseth, Elisabeth, Ms.
Project Coordinator
FAFO- INTERNATIONAL
Helgesensgt. 12, 0553
Oslo, Norway
T: 47 2 37-9980 F: 47 2 37-5956

Furulund, Svein Erik
Photographer
WORLD CITY DISCOVERY AS
Ullern Alle 41
Oslo, Norway
T: 47 2 505100 F: 47 2 524778

Gaski, Harald, Mr.
Professor
UNIVERSITY OF TROMSO
Tromso N-9037
Tromso, Norway
T: 47 83 44000 F: 47 83 44239

Graham, Elizabeth, Ms.
Senior Clerk
NORWEGIAN DELEGATION TO UNCED
Postboks 8114 Dep.
Oslo N-0030, Norway
T: 47 2 343621 F: 47 2 349580

Grann, Odd
WORLD CITY DISCOVERY AS
Ullern Alle 41 PO Box 50 Ovre Ullen
Oslo 0311, Norway
T: 47 2 505-100 F: 47 2 524-778

Gravir, Wetil
Editor
NORWEGIAN BROADCASTING
NRK
Oslo 3, Norway
T: 47 2 489050 F: 47 2 458471

Gronn, Erik, Mr.
Asc. Professor
NORWEGIAN SCHOOL OF MANAGEMENT
Postboks 580, Elias Smithsvei 15
Sandvika N-1301, Norway
T: 47 2 470788 F: 47 2 470570

Hansen, Ragnar, Mr.
ASSOCIATION OF NORWEGIAN VISUAL ART
Formannsvei 38A
Bergen, Norway
T: 47 5 327454

Hansen, Stein, Mr.
Adviser
NORWEGIAN DELEGATION TO UNCED
Ornevelen 46 a.
Bekkestua N-1346, Norway
T: 47 2 247856 F: 47 2 247856

Hareide, Dag, Mr.
General Secretary
FRIENDS OF THE EARTH, NORWAY
PO Box 2113 - Grunerlokka
Oslo 0505, Norway
T: 47 2 71.55.20 F: 47 2 71.56.40

Hauge, Frederic, Mr.
Leader
ENVIRONMENTAL FOUNDATION BELLONA
Box 8874 Youngs Torget
Oslo 1, Norway
T: 47 2 38241 F: 47 2 383862

Henriksen, Arent M., Mr.
Sec.Gen.Norw.Fishermen'S Asso.
NORWEGIAN DELEGATION TO UNCED
Pir-Senteret
Trondheim N-7005, Norway
T: 47 7 529680 F: 47 7 534742

Hofseth, Paul, Mr.
Special Adviser
NORWEGIAN DELEGATION TO UNCED
Postboks 8013 Dep.
Oslo N-0030, Norway
T: 47 2 345690 F: 47 2 349564

Holbye, Geir, Mr.
Asst. Dir. Conf. Norw. Bus.
NORWEGIAN DELEGATION TO UNCED
Middelthuns gt. 27
Oslo N-0368, Norway
T: 47 2 965000 F: 47 2 965185

Holthe, Ole Kristian, Mr.
Dep. Dir., Gen.
NORWEGIAN DELEGATION TO UNCED
Postboks 8114 Dep.
Oslo N-0030, Norway
T: 47 2 343600 F: 47 2 349580

Holvik, Oystein, Mr.
COMMUNITY (KRISTIANSAND)
Kristiansand Kommune
Kristiansund, Norway
T: 47 42 75000

Hrabar, Dean, Mr.
Project Manager
IEB/ICC
P.O. Box 301
Lysaker 1324, Norway
T: 47 2 581800 F: 47 2 581875

Jordre, Ivar, Mr.
Student
LATIN AMERICA GROUP
Kronstadveien 21B
Solheimsvik 5037, Norway

Kapoor, Nita, Ms.
MIRA CENTRE
P. O. Box 1749 - Vika
Oslo 0121, Norway
T: 47 2 116920

Kloster, Knut V., Mr.
Chairman
GAIASHIP FOUNDATION
Ullern Alle 41
Oslo, Norway
T: 47 2 525 100 F: 47 2 524 778

Knudsen, Henrik, Mr.
Sleipwervn 46
Larvik 3250, Norway

Korsrud, Marie Henriette, Ms.
FISHERMEN'S WIVES ASSOCIATION
Toppoy
Reine 8390, Norway
T: 47 88 92345 F: 47 88 92432

Kristoffersen, Bjorn, Mr.
STATOIL
PG 300 Statoil
Stavanger 400, Norway
T: 47 45 808080

Kullervd, Lars, Mr.
Geologist
GRID ARENDAL
PO Box 1602
Arendal 4801, Norway
T: 47 41 35650 F: 47 41 35050

Lager, Ragnhild, Ms.
Info Officer
UN - ASSOCIATION OF NORWAY
Ostre Strandgt 5
Kristiansund 4610, Norway
T: 47 42 24900 F: 47 42 24824

Larsen, Mona Eline, Ms.
Exec. Officer
NORWEGIAN DELEGATION TO UNCED
Postboks 8013 Dep.
Oslo N-0030, Norway
T: 47 2 345962 F: 47 2 349564

Leiro, Jostein, Mr.
Sr. Exec. Officer
NORWEGIAN DELEGATION TO UNCED
Postboks 8114 Dep.
Oslo N-0032, Norway
T: 47 2 343617 F: 47 2 349580

Lena, Steinar
Head Of Information
FUTUREICH AUR VARDS
Tozgt 35
0133 Oslo, Norway
T: 2201045 F: 427204788

Lindebotten, John
NORWEGIAN CAMPAIGN FOR ENVIRONMENT AND DEVELOPMENT
Langesgt 4
Oslo 0165, Norway
T: 47 2 362035 F: 47 2 361454

Lindsoe, Anne Brit, Ms.
Sr. Exec. Officer, Prime Min.
NORWEGIAN DELEGATION TO UNCED
Postboks 8001 Dep.
Oslo N-0030, Norway
T: 47 2 344033 F: 47 2 349501

Linggard, Trine
Journalist
CENTER FOR INTERNATIONAL WOMEN'S ORGANIZATION
0160 Oslo, Norway
T: 47 2426205 F: 47 2423205

Lopez, Noemi Meyer, Ms.
MIRA CENTRE
P. O. Box 1749 - Vika
Oslo 0121, Norway
T: 47 2 116920

Lovold, Lars, Mr.
Executive Director
NOWEGIAN RAINFOREST FOUNDATION
PO Box 2113 - Grunerlokka
Oslo 0505, Norway
T: 47 2 71.55.20 F: 47 2 71.56.40

Lund, Hauk, Mr.
Press Officer
IEB/ICC
P.O. Box 301
Lysaker 1324, Norway
T: 47 2 581800 F: 47 2 581875

Lunde, Leiv, Mr.
Project Secretary
CHURCH OF NORWAY MAGAZINE - COUNCIL ON FOREIGN RELATIONS
P.B. 5816 Hegdehaugen
Oslo 0308, Norway
T: 47 2 694044 F: 47 2 565864

Lundstol, Ottar, Mr.
Manager
GAIASHIP FOUNDATION
Ullern Alle 41
Oslo, Norway
T: 47 2 525 100 F: 47 2 524 778

Magga, Ole Henrik, Mr.
President, Sami Parliament
NORWEGIAN DELEGATION TO UNCED
Postboks 254
Kautokeino M-9520, Norway
T: 47 84 56253 F: 47 84 66949

Meisfjord, Arna
INTERNATIONAL KUINNELIGA FOR FRED OG FRIHET - WILPF NORWAY
Lovenskioldsvei 12, 1342 Jar
Oslo, Norway
T: 47 2 531300 F: 47 2 531300

Molvaer, Reidulf, Mr.
Program Director
INTERNATIONAL PEACE RESEARCH INSTITUTE
Prio, Fuglehauggata 11
Oslo N-0260, Norway
T: 47 2 55-7150 F: 47 2 55-8422
TL: 72400 Fotex N

Museus, Lena, Ms.
Project Man
GAIASHIP FOUNDATION
Ullern Alle 41
Oslo, Norway
T: 47 2 525 100 F: 47 2 524 778

Naerstad, Aksel, Mr.
Leader
RED ELECTION ALLIANCE
Hvalstadvn. 36B
Hvalstad 1364, Norway
T: 47 2 384250 F: 47 2 384602

Naess, Lilly
NORWEGIAN CAMPAIGN FOR ENVIRONMENT AND DEVELOPMENT
Langesgt 4
Oslo 0165, Norway
T: 47 2 362035 F: 47 2 361454

Nilsen, Knut Erik, Mr.
Manager
ENVIRONMENTAL FOUNDATION BELLONA
Box 8874 Youngs Torget
Oslo 1, Norway
T: 47 2 38241 F: 47 2 383862

Nordbo, Eldrid, Mr.
Special Adviser
NORWEGIAN DELEGATION TO UNCED
Postboks 8013 Dep.
Oslo N-0030, Norway
T: 47 2 345991 F: 47 2 349564

O'Sullivan, Brenda May, Ms.
Konsulent
NORWEGIAN SOCIETY FOR CONSERVSATION OF NATURE
Slattum Tevr 331
Nittedal 1480, Norway
T: 47 952 77-7048 F: 47 952 77-2270

O'Sullivan, Karin Jayne, Ms.
NORWEGIAN SOCIETY FOR CONSERVATION OF NATURE (NNV)
Rodstua
Toyenhaugen 1487, Norway
T: 47 952 77-6066 F: 47 952 77-2270

Ole, Mathismoen, Mr.
Journalist
AFTENPOSTEN NEWSPAPER
Postboks 1178, Sentrum
Oslo 0107, Norway
T: 47 2 863000 F: 47 2 864035

Olsen, Terje, Mr.
Lecturer
BODO LIERERHOGSKOLE
Amtm Worsoesgt 28
Bodo N-8000, Norway
T: 47 81 81560 F: 47 81 83560

Olsson, Nils, Mr.
ISFIT
Herman Krags Veg 25-14
Trondheim 7035, Norway
T: 47 7 588147 F: 47 7 591045

Overaa, Severin J., Mr.
Regional Manager
UN ASSOCIATION OF NORWAY
Olav Kyrresgt 31
Bergen 5014, Norway
T: 47 5 316789 F: 47 5 321689

Overli, Jan M, Mr.
Dir Environment Prof Env/Energ
STATOIL
PG 300 Statoil
Stavanger 400, Norway
T: 47 45 808080

Paludo, Suetelane, Ms.
ISFIT
Oslov 18-B
Trondheim 7018, Norway
T: 47 7 523781 F: 47 7 591045

Paulsen, Arve, Mr.
INTERNATIONAL PEACE RESEARCH
Fuglemauggt 11, 0260
Oslo, Norway
T: 47 2 557150 F: 47 2 558422

Quinteros, William, Mr.
FUTURE IN OUR HANDS
Stobakken 18
Sogne, Norway
T: 47 42 97990

Rahiff, Elsebet, Sra.
ASSOCIATION OF NORWEGIAN VISUAL ART
Fredlundbakken 3, 5032 Minde
Bergen, Norway
T: 47 5 287 412

Rahiff, Sixten, Sr.
ASSOCIATION OF NORWEGIAN VISUAL ART
Fredlundbakken 3, 5032 Minde
Bergen, Norway
T: 47 5 287412

Refsnes, Laila, Ms.
Exec. Officer
NORWEGIAN DELEGATION TO UNCED
Postboks 8001 Dep.
Oslo N-0030, Norway
T: 47 2 345991 F: 47 2 349564

Rostvik, Harald N., Mr.
President
SUN LAB ORGANIZATON (SOLAR ENERGY)
Steingaten 87
Stavanger 4024, Norway
T: 47 45 33442 F: 47 45 52402

Rune, Haaland, Mr.
Research Fellow
ENVIRONMENTAL FOUNDATION BELLONA
Box 8874 Youngs Torget
Oslo 1, Norway
T: 47 2 38241 F: 47 2 383862

Saddler, Heather, Ms.
VOICE OF THE CHILDREN
c/o Norwegian Campaign fo Env and Dev, Langes gt 4
Oslo 0165, Norway
T: 47 2 362035 F: 47 2 361454

Salimi, Fakhra, Ms.
MIRA CENTRE
P. O. Box 1749 - Vika
Oslo 0121, Norway
T: 47 2 116920

Schei, Peter Johan, Mr.
Director General
NORWEGIAN DELEGATION TO UNCED
Postboks 8013 Dep.
Oslo N-0030, Norway
T: 47 2 345966 F: 47 2 349564

Skagestad, Odd Gunnar, Mr.
Asst. Director General
NORWEGIAN DELEGATION TO UNCED
Postboks 8118 Dep.
Oslo N-0032, Norway
T: 47 2 346450 F: 47 2 349585

Skjevik, Erlend, Mr.
NORWEGIAN CAMPAIGN FOR ENVIROMENT AND DEVELOPMENT
Langes Gate 4
Oslo 0165, Norway
T: 47 2 36-2035 F: 47 2 36-1454

Skrent, Kari Berg, Ms.
Executive Assistant
IEB/ICC
P.O. Box 301
Lysaker 1324, Norway
T: 47 2 581800 F: 47 2 581875

Slatto, Torhild, Ms.
Head Of Department
ROYAL NORWEGIAN SOCIETY FOR RURAL DEVELOPMENT
PO Box 115, 2013 Skjetten
Skedsmo, Norway
T: 47 6 84-5000 F: 47 6 84-5130

Snorre, Toenset
NORWEGIAN BROADCASTING CORPORATION
Oslo, Norway

Sohr, Mauricio Gorget
Audio
TV NORWAY
Oslo 3, Norway

Solheim, Erik, Mr.
Member Of Parliament, Sv
NORWEGIAN DELEGATION TO UNCED
Karl Johans gt. 22
Oslo N-0026, Norway
T: 47 2 313228 F: 47 2 313857

Sorbotten Tveide, Anne, Ms.
Teacher
NORGES BONDEKVINNELAG
4760 Birkeland
Arendal, Norway
T: 47 41 76163

Spinnangr, Kjell
KRISTIANSAND RIOS MILJOGRUPPE
v/ Sissel Hartmark Fjukstad / Arenfeldtsvei 39
Kristiansund 4630, Norway

Stephens, Sharon, Dr.
Fulbright Professor
NORWEGIAN CENTRE FOR CHILD
N-7055 Dragvoll
Trondheim, Norway
T: 47 7 596243 F: 47 7 596240

Strandli, Bjorn, Mr.
Prize Winner
NORWEGIAN DELEGATION TO UNCED
Stotvigveien 40
Larkollen N-1560, Norway
T: 47 9 263244

Stubsjoen, Magne, Mr.
Dir. Gen. Min. Agriculture
NORWEGIAN DELEGATION TO UNCED
Postboks 8007 Dep.
Oslo N-0030, Norway
T: 47 2 349300 F: 47 2 349557

Tandberg, Jon Eivind, Mr.
Asst. Dir. Gen. Min. Finance
NORWEGIAN DELEGATION TO UNCED
Postboks 8008 Dep.
Oslo N-0030, Norway
T: 47 2 344322 F: 47 2 349505

Titlestad, Jan, Mr.
Board Member
NORWEGIAN SOCIETY FOR CONSERVATION OF NATURE
Postboks 1201
Bergen N-5001, Norway
T: 47 5 313-851 F: 47 5 31-5392

Tito, Mr.
NORWEGIAN BROADCASTING CORPORATION
Oslo, Norway

Tveitdal, Svein, Mr.
Director
GRID ARENDAL
PO Box 1602
Arendal 4801, Norway
T: 47 41 35650 F: 47 41 35050

Tveter, Oystein, Mr.
Coordinator
NORTH/SOUTH COALITION
Boks 23, Vindern
Oslo 0319, Norway
T: 47 2 451818 F: 47 2 451810

Undheim, Liv, Ms.
Sec. To Workers' Union
NORWEGIAN DELEGATION TO UNCED
Youngsgt. 11
Oslo N-0181, Norway
T: 47 2 401050 F: 47 2 401743

Utheim, Bjornar, H.E.
Ambassador, Min. For. Affairs
NORWEGIAN DELEGATION TO UNCED
Postboks 8114 Dep.
Oslo N-0032, Norway
T: 47 2 343619 F: 47 2 349580

Venger, Amund, Mr.
Sec. Gen. Norw. Farmers' Union
NORWEGIAN DELEGATION TO UNCED
Postboks 9354 Dep.
Oslo N-0032, Norway
T: 47 2 173400 F: 47 2 173668

Wiik, Flavio Braune, Mr.
NORWEGIAN CAMPAIGN FOR ENVIROMENT AND DEVELOPMENT
Langes Gate 4
Oslo 0165, Norway
T: 47 2 36-2035 F: 47 2 36-1454

Wilhjelm, Hanne, Ms.
Editor
WOMEN'S FRONT
Nadderudlia 15A, 1340 Bekkestua
Oslo 0160, Norway
T: 47 2 345932 F: 47 2 423205

Wisloff, Gunn-Bjorg
Board Member
CENTER FOR INTERNATIONAL WOMEN'S ISSUES
Oslo, Norway

PARLIAMENTARIANS / SPIRITUAL LEADERS

Aakervik, Anne, Ms.
MP
Stortinget Karl Johansgate 22
Oslo 1, Norway
T: (02) 31.30.50
TL: 72732 stort n

Bernander, John, Mr.
MP
Stortinget Karl Johansgate 22
Oslo 1, Norway
T: (02) 31.30.50
TL: 72732 stort n

OMAN - GOVERNMENT

Al Hosny, He Shaikh Amer Bin Shuwain
Minister For Environment And Regional Municipalities
P.O. BOX 323
Muscat.
Sultanate of Oman, Oman
T: 696464 - 696469

Al Moosawi, Moosa Bin Jaffar Bin Ali
Acting Director Agcc And Inter Orgn
MIN OF REGIONAL MUNIC AND ENVIR
PO Box 323
Muscat, Oman
T: 604981 696444 EXT311 F: 602320

Al-Abri, Said Bin Salman, Mr.
Legal Adviser
Minister's Office, Ministry of Regional Municipalities and Environment
Permanent Mission to the United Nations, 866 United Nations Plaza, Suite 450 New York, NY, 10017
F: 355 3505, 3506, 3507
TL: 644 0070

Al-Battashi, Ali Bin Thabit, Mr.
Director-General
Council of Corporations, Ministry of Petroleum and Minerals
Permanent Mission to the United Nations, 866 United Nations Plaza, Suite 450 New York, NY, 10017
F: 355 3505, 3506, 3507
TL: 644 0070

Al-Farsi, Ibrahim Bin Mohammed, Mr.
First Secretary
Permanent Mission of Oman
Geneva
T: 10017
TL: 355 3505, 3506, 3507, E: 644 0070

Al-Harmi, Abdul Rahim Bin Salem, Mr.
Assistant Director-Gen for Meteorological Affairs
Ministry of Communication
Seeb Airport P.O. Box 1021
Sultanate of Oman, Oman
T: 519361 F: 510122

Al-Hosni, Sheikh Amer Bin Shwain, H.E. Mr.
Minister for Regional Municipalities and Environment (Head of Delegation)
Permanent Mission of Oman
866 United Nations Plaza, Suite 450
New York, NY 10017, United States of America
T: 355 3505, 3506, 3507 F: 644 0070

Al-Kisouqi, Lutfi Abdul Kader, Mr.
Expert on Natural Conservation of the Environment
Ministry of Regional Municipalities and Environment
Permanent Mission to the United Nations, 866 United Nations Plaza, Suite 450 New York, NY, 10017
F: 355 3505, 3506, 3507
TL: 644 0070

Al-Maamary, Ahmed Bin Rashid, Mr.
Director-General
Coordination Department, Development Council
Permanent Mission to the United Nations, 866 United Nations Plaza, Suite 450 New York, NY, 10017
F: 355 3505, 3506, 3507
TL: 644 0070

Al-Musawi, Moosa Bin Jafar Bin Ali, Dr.
Director
Department of the Gulf Council Affairs and International Organizations, Ministry of Regional Municipalities and Environment
Permanent Mission to the United Nations, 866 United Nations Plaza, Suite 450 New York, NY, 10017
F: 355 3505, 3506, 3507
TL: 644 0070

Al-Riyami, Said Bin Juma, Mr.
Director of the Minister's Office
Ministry of Regional Municipalities and Environment
Permanent Mission to the United Nations, 866 United Nations Plaza, Suite 450 New York, NY, 10017
F: 355 3505, 3506, 3507
TL: 644 0070

Al-Said, Barghash Bin Ghaleb, Mr.
Director-General
Water Resources Assessment, Ministry of Water Resources
Permanent Mission to the United Nations, 866 United Nations Plaza, Suite 450 New York, NY, 10017
F: 355 3505, 3506, 3507
TL: 644 0070

Al-Shanfari, Saleh Bin Mohammed Bin Amer, Mr.
Deputy Director-General of Agriculture, Livestock and Fisheries for Dhofar Province
Ministry of Agriculture and Fisheries
Permanent Mission to the United Nations, 866 United Nations Plaza, Suite 450 New York, NY, 10017
F: 355 3505, 3506, 3507
TL: 644 0070

Al-Sheryani, Ahmed Bin Ali
Director General For Environmental Ministry Of Regional Municipalities
P.O. BOX 323
Muscat
Sultanate of Oman, Oman
T: 00968 696441 F: 00968 602320

Al-Zadgaly, Suleiman Bin Lashkaran, Mr.
Second Secretary
Department of Conferences and International Organizations, Ministry of Foreign Affairs
P O. Box 252
Muscat- Sultanate of Oman, Oman
T: 968 699 500 F: 968 698 321

Alshanfari, Saleh
Director General At Ministry Of Agr
DIRECTORATE GEN. OF AG. LIVESTOCK A
P.O. Box 18475
Salalah, Oman
T: 294421 F: 293919

Ba-Omar, Taher Bin Abdulrahman, Dr.
Lecturer
Faculty of Science, Sultan Qaboos University
P.O. Box 32486
Muscat, Oman
T: 515441

Desougi, Lutfi A., Dr
Expert Nature Reserves
MINISTRY OF REGIONAL MUNCIPALITIES AND ENVIRONMENT
P O Box 323
Muscat SULTANATE OF OMAN, Oman
T: 696444 312

Hassan, Moosa Bin Jafar Bin, H.E. Mr.
Ambassador Extraordinary and Plenipotentiary
Permanent Representative to UNESCO, Paris
Permanent Mission to the United Nations, 866 United Nations Plaza, Suite 450 New York, NY, 10017
F: 355 3505, 3506, 3507
TL: 644 0070

PRESS

Al Naby, Salim
Editor
OMAN NEWS PAPER
P.O. Box 6002
Muscat, Oman, Oman

Al-Mantheri, Faisal
Correspondent
OMAN NEWS AGENCY
P.O. Box 9406
Sultanase of Oman, Oman

PAKISTAN - GOVERNMENT

Banuri, Tariq J.
SUSTAINABLE DEVELOPMENT POLICY INST
Box 2342
Islamabad, Pakistan
T: 92 51 211097 F: 92 51 822313

Hassan, Parvez
Chairman, Commission On Environment
IUCN-THE WORLD CONSERVATION UNION
Paaf Building 7D Kashmir Egerton Ro
Lahore, Pakistan
T: 92 42 360800 TO 3608 F: 92 42 303825+303826

Nawaz, Sharif Mohamed, Mr.
Prime Minister
Permanent Mission of Pakistan
8 E. 65th Str.
New York NY 10021, United States of America
T: 201 879 8600 F: 212 744 7348

Qureshi, Zulfiqar Ali
Alternative Representative
ENVIRONMENT MINISTRY
Additional Secretary Incharge Envir Onment And Urban Affairs Division
Islamabad, Pakistan
T: 813373 851105 F: 9251 824122

U Pe Thein Tin, , H.E.
Ambassador
EMBASSY OF THE UNION OF MYANMAR
12/1, Street No. 13, Sector F-7/2
Islamabad, Pakistan
T: 822460 F: 92 51 820123

NGO

Ahmad, Syed
CCMA
515, Street 18, G-8/1
Islambad, Pakistan

Hamdani, Vigar A.
WMC
Site 9/A, B1. 7
Gulshan Iqbal, Karachi-75300, Pakistan

Husain, Syed
CCMA
515, Street 18, G-8/1
Islambad, Pakistan

Ibtehajuddin Khuram, Shah
FPAP
3-A Temple Road
Lahore 54000, Pakistan

Kango, Allahdino
RDF
RDF Centre (I&T), Mauve Area, G9/1, PO Box 1170
Islamabad, Pakistan

Khan, Kausar S., Ms.
SHIRKAT GAH WOMEN'S RESOURCE CENTRE
Palais de Nations
Geneve CH 1211, Pakistan
F: 41 22 733-6542

Khan, Muhammad
SCSP
B-7, Sheet No. 25 Model Colony
Karachi 75100, Pakistan

Malik, Sadiq
RDF
RDF Centre (I&T), Mauve Area, G9/1, PO Box 1170
Islamabad, Pakistan

Mushtaq Ghani,
APYF
Haleem Building, 4-Mozang Road
Lahore, Pakistan

Nazim, Mohammad
ENERGY AND ENVIRONMENT SOCIETY OF PAKISTAN
123-J, Model Toen
Lahore 54700, Pakistan

Sardar Khan, Malik
WMC
Site 9/A, B1. 7
Gulshan Iqbal, Karachi-75300, Pakistan

Shahid, Yasmin
Senior Director
FAMILY PLANNING ASSOCIATION OF PAKISTAN
3-A Temple Road
Lahore 54000, Pakistan

Tanveer, Arif
SCOPE
511, Al-Amin Tower (NIPA Chowrangi), University Road, Gulshan-e-Iqbal, Karachi 75300, Pakistan

Tariq Haleem,
APYF
Haleem Building, 4-Mozang Road
Lahore, Pakistan

PRESS

Abbasi, Niaz
Correspondent
DAILY NAWA-I-WAQT
1601 W 5th St., Apt. B2
Brooklyn, NY 11223, USA, Pakistan

Abdus Sayeed, Khan Q.
Editor
PPI
70 F. Mazhar Qayyum Roid
Lalazar, Rawalpindi, Pakistan

Ahmed, Jalil
Correspondent
ASSOCIATED PRESS OF PAKISTAN
A-26 Bl.4-A
Gulshan-E-IQBAL, Karachi-47, Pakistan, Pakistan

Bhatti, Razia
Editor
NEWSLINE MAGAZINE
A 101, Clifton Gardens Fl.1
Kehkashan, Clifton, Karachi, Pakistan, Pakistan

Bukhari, Syed Nazir
Correspondent
PAKISTAN BROADCASTING CORPORATION
Syed Nazir Bukhari, H. No.603
G-10/4, Islamabad, Pakistan, Pakistan

Hanif, Khalid
Correspondent
DAILY JANG
496-C, Sector G-9/3
Islamabad, Pakistan

Hashimi, Ehsan Qadir
Correspondent
PAKISTAN TELEVISION CORPORATION
House No. 392, St. No. 31
G-8/2, Islamabad, Pakistan, Pakistan

Khalid, Hafiz Abdul
Correspondent
MARKIZ
Khyber Plaza, Blue Area
Islamabad, Pakistan, Pakistan

Malik, Tahir
Correspondent
DAILY NAWA-I-WAQT
H No.L315 - Model Town Ext
Lahore, Pakistan, Pakistan

Mehmood, Rao Khalid
Correspondent
PAKISTAN LAHORE
P-1044 B Said Pur Road
Asghar Mall Chowk Rawalpindi, Pakistan, Pakistan

Niaz, Anjum
Editor
DAWN PAKISTAN
Raleigh House #3
Karachi, Pakistan, Pakistan

Qureshi, Abdul Wadood
Correspondent
MARKIZ
Street No. 7 Qasimabar
Rawalpindi, Pakistan, Pakistan

Sohail, M.Z.
Correspondent
PAKISTAN TELEVISION CORPORATION
House No. 303, St. No. 16
F-10/2, Islamabad, Pakistan, Pakistan

Zaidi, Meesaq Hussain
Correspondent
PAKISTAN BROADCASTING CORPORATION
Q/290, Saidpuri Gate
Tehshil, Rawalpindi, Pakistan, Pakistan

ORGANIZATIONS

Arif, Tanveer, Mr.
Society for Conservation and Protection of Environment (SCOPE)
511, Al-Amin Tower (NIPA Chowrangi), University Road, Gulshan-e-Iqbal, Karachi 75300, Pakistan

Ciszak, E., Mr.
World Muslim Congress
Site 9/A, B1. 7
Gulshan Iqbal, Karachi-75300, Pakistan

Malik, M. Sadik, Dr.
Rural Development Foundation
RDF Centre (I&T), Mauve Area, G9/1, PO Box 1170
Islamabad, Pakistan
T: (92 51) 858 972 F: (92 51) 851 078

Mumtaz, Khawar, Mr.
Shirkat Gah
18-A, Mian Mir Road, PO Mughalpura
Lahore 15, Pakistan
T: (92 42) 874 951

GLOBAL FORUM

Ahmad, S. Atiq, Mr.
President
COMMUNITY CAPITAL MANAGEMENT ASSOCIATION
House 185 - Street 74 - G9/3
Islamabad, Pakistan
T: 92 51 858-629 F: 92 51 256-061

Choudhry, Tariq Haleem, Mr.
President
ALL PAKISTAN YOUTH FEDERATION
Haleem Bldg., 4 Mozang Road
Lahore, Pakistan
T: 92 42 66231 211153 F: 92 42 223974
TL: CTOLHPK-4494,

Dogar, Nasir
Administrator
JOURNALISTS' RESOURCE CENTRE FOR THE ENVIRONMENT
1-Bath Island Road
Karditas, Pakistan
T: 30 441 573079 F: 30 441 530976

Ghani, Mushtaq Ahmed, Mr.
Secretary General
ALL PAKISTAN YOUTH FEDERATION
Haleem Bldg., 4 Mozang Road
Lahore, Pakistan
T: 92 42 66231 211153 F: 92 42 223974
TL: CTOLHPK-4494,

Habib, Begum Meherafroze, Ms.
Vice President
ALL PAKISTAN WOMENS ASSOCIATION
67 - B, Garden Road
Karachi, Pakistan
T: 92 2 712991

Hai, Begum Jahan Ara, Ms.
Chairperson
ALL PAKISTAN WOMENS ASSOCIATION
67 - B, Garden Road
Karachi, Pakistan
T: 92 712991

Hidayatullah, Charmaine, Ms.
Chairperson
ALL PAKISTAN WOMENS ASSOCIATION
67 - B, Garden Road
Karachi, Pakistan
T: 92 2 712991

Hussain, Saneeya
Director
JOURNALISTS' RESOURCE CENTRE FOR THE ENVIRONMET
1-Bath Island Road
Karditas, Pakistan
T: 30 441 573079 F: 30 441 530976

Jagerhorn, Martina, Ms.
House 23, Street 1, Sector G-6/3
Islamabad, Pakistan
T: 92 5 815845

Jawad, Begum Masudah
Executive Vice President
ALL PAKISTAN WOMENS ASSOCIATION
67 - B, Garden Road
Karachi, Pakistan
T: 92 21 712991

Kermani, Begum Meher, Ms.
Chairperson
ALL PAKISTAN WOMENS ASSOCIATION
67 - B, Garden Road
Karachi, Pakistan
T: 92 2 712991

Naveed, Ahmed, Mr.
Gen. Sec. Crescent Youth Org.
ALL PAKISTAN YOUTH FEDERATION
Haleem Bldg., 4 Mozang Road
Lahore, Pakistan
T: 92 42 66231 211153 F: 92 42 223974
TL: CTOLHPK-4494,

Pasbaan, Yasmin Sharid, Ms.
FAMILY PLANNING ASSOCIATION OF PAKISTAN
Palais de Nations
Geneve CH 1211, Pakistan
F: 41 22 733-6542

Rashid, Ishrat, Ms.
Member
SIND GOVT EMPLOYEES COOPERATIVE HOUSING SOCIETY LTD
558-A Gulshan-e-Iqbal
Karachi, Pakistan
T: 92 21 460543 F: 92 21 469092

Rashid, Nargis, Ms.
Member Executive Committee
SIND GOVT EMPLOYEES COOPERATIVE HOUSING SOCIETY LTD
558-A Gulshan-e-Iqbal
Karachi, Pakistan
T: 92 21 460543 F: 92 21 469092

Rizwanullan, Colonel Muhammad, Mr.
Projects Director
SIND GOVT EMPLOYEES COOPERATIVE HOUSING SOCIETY LTD
558-A Gulshan-e-Iqbal
Karachi, Pakistan
T: 92 21 460543 F: 92 21 469092

Shah, Hadi H., Mr.
SIND GOVT EMPLOYEES COOPERATIVE HOUSING SOCIETY LTD
558-A Gulsman-e-Iqbal
Karachi, Pakistan
T: 92 2 460543 F: 92 2 469092

Shaukat, Tajwar, Ms.
Vice President
ALL PAKISTAN WOMENS ASSOCIATION
67 - B, Garden Road
Karachi, Pakistan
T: 92 712991

Siddique, Bashir Ahmed Khan, Mr.
Chairman
SIND GOVT EMPLOYEES COOPERATIVE HOUSING SOCIETY LTD
558-A Gulsman-e-Iqbal
Karachi, Pakistan
T: 92 2 460543 F: 92 2 469092

Taseer, Usaf Zai, Mr.
Secretary
SIND GOVT EMPLOYEES COOPERATIVE HOUSING SOCIETY LTD
558-A Gulshan-e-Iqbal
Karachi, Pakistan
T: 92 21 460543 F: 92 21 469092

Warriach, Shahid, Mr.
Vice Chairman
ALL PAKISTAN YOUTH FEDERATION
Haleem Bldg., 4 Mozang Road
Lahore, Pakistan
T: 92 42 66231 211153 F: 92 42 223974
TL: CTOLHPK-4494,

PARLIAMENTARIANS / SPIRITUAL LEADERS

Khan, Anwar Saifullah, Min.
Senate
Islamabad, Pakistan
T: 823734

Panezai, Noor Johan, Dr.1
Vice Pres.
Senate
Islamabad, Pakistan
T: 823734

Zaki, Akram, Senator
Senate
Islamabad, Pakistan
T: 823734

PALESTINE - GLOBAL FORUM

Abu-Kishk, Rula, Ms.
Engineer
PALESTINE HYDROLOGY GROUP
PO Box 20873
Jerusalem, Palestine
T: 972 2 812-364 F: 972 2 812-364

Issa, Shawqi
Lawyer
LAND AND WATER ESTABLISHMENT
PO Box 20873
Jerusalem, Palestine
T: 972 2 812-364 F: 972 2 812-364

Koresh, Ibrahim, Mr.
G.U.P.S.
R Palestine
Tulkarn, Palestine
T: 237 702 F: 766 344

PANAMA - NGO

Alfaro, Teofrida, Ms.
ANCON
Apartado 1387
Panama 1, Panama

Archibold, Guilhermo
AEKP
Apartado Postal 2012
Paraiso, Ancon Panama, Panama

Chang, Rene
CECA
Playa El Agallito, Distrito de Chitre
Provencia de Herrera, Panama

Eligio, Alvarado
FDY
Apartado 2072, Balboa
Republica de Panama, Panama

Ferrer, Maximiliano
KUNA
Via Espana, Edif. Brasilia, Piso 1, Ofic. 9-A
P O Box 536, Panama 1, Panama

Geodisio, Castillo
FDY
Apartado 2072, Balboa
Republica de Panama, Panama

Martinez, Atencio Lopez
KUNA
Via Espana, Edif. Brasilia, Piso 1, Ofic. 9-A
P O Box 536, Panama 1, Panama

Mendez De Sierra, Ana
CECA
Playa El Agallito, Distrito de Chitre
Provencia de Herrera, Panama

Smith, Teofrida Alfaro
AEKP
Apartado Postal 2012
Paraiso, Ancon Panama, Panama

ORGANIZATIONS

Bonamico, Glenda, Ms.
Grupo de Tecnologia Apropriada
Apartado 8046, Zona 7
Panama, Panama
T: (507) 617 353 F: (507) 618 072

Castillo, Geodisio, Sr.
Fundacion Dobbo-Yala
Apartado 2072, Balboa
Republica de Panama, Panama
T: (507) 276 022 EXT. 330 F: (507) 625 942

Rovi, Julio Ivan, Sr.
Fundacion Para El Desarrollo De Los Jovenes (FUNDEJOVEN)
Apartado 87-1740, Zona 7
Panama, Panama
T: (507) 611 520 F: (507) 291 051

Telesca, Antonio, Lic.
As. Nacional Para La Conservacion De La Naturaleza - Panama (ancon)
Apartado 1387
Panama 1, Panama
T: (507) 648 100 F: (507) 641 836

GLOBAL FORUM

Eleta, Avila Fany, Ms.
KUNA
Veracruz Baniada Kashkuna 70
Panama, Panama
T: 507 63-40-27 F: 507 69-35-14

Rojas, Itzel-Damaris, Ms.
RED DE MUJERES Y MEDIO AMBIENTE - CENTRO AMERICA Y CARIBE
Fundacion de Parques Nacionales y Medio Ambiente (Panama)
Panama, Panama
T: 507 626234 F: 507 640371
E: 6-6623 EL DORADO

Soto, Teofilo Ku, Sr.
Representative
FUNDACAO MELO
Casilla Postal 218
Panama 9, Panama
T: 507 238316
TL: 2278

PARLIAMENTARIANS / SPIRITUAL LEADERS

Graegi, Hermann, Deputy
Consejo Nacional de Legislacion
Panama City, Panama
T: 62 6447

PAPUA NEW GUINEA - GOVERNMENT

Kula, Gaikovina
First Assistant Secretary Nature Co
DEPARTMENT OF ENVIRONMENT & CONSERV
P O Box 6601
Boroko, Papua New Guinea
T: 675 271791 F: 675 271044

Taylor, Meg, Ms
Ambassador
EMBASSY OF PAPUA NEW GUINEA
1615 New Hampshire Ave
Washington DC, United States of America
T: 202 745 3680 F: 202 745 3679

Wilmot, John Mark
Scientific Advisor
DEPARTMENT OF ENVIRNOMENT & CONSERV
P O Box 6601
Boroko, Papua New Guinea
T: 675 271763 F: 675 271900

NGO

Solon, Mark
MIN
P.O. Box 571
Goroka, Eastern Highlands Province, Papua New Guinea

GLOBAL FORUM

Atmeyor, Daniel, Mr.
OK TEDI/ FLY RIVER ENECO ASSOCIATION
P.O.Box 527 Tabubil
Western Province, Papua New Guinea

Dagi, Rex, Mr.
Chairman
ENECO ASSOCIATION
Cl - P.O.Box 527 Tabubil
Kiunga- Western, Papua New Guinea

Gabara, Dair, Mr.
Lawyer
OK TEDI/ FLY RIVER ENECO ASSOCIATION
P.O.Box 527 - Tabubil
Kiunla - Western, Papua New Guinea
T: 254422 F: 259682

Kanupa, Nalaba, Mr.
Commitee Member
OK TEDI/ FLY RIVER ENECO ASSOCIATION
Eneco P.O.Box 521
Port Moresby, Papua New Guinea

Kirokim, Blui, Mr.
OK TEDI/ FLY RIVER ENECO ASSOCIATION
P.O.Box 527 Tabubil
Kiunga - Western, Papua New Guinea

Martin, Semnebe, Mr.
OK TEDI/ FLY RIVER ENECO ASSOCIATION
P.O.Box 527 Tabubin
Western PNG, Papua New Guinea
T: 659 077 F: 250 682

Soondrawu, Mary, Ms.
EAST SEPIK COUNCIL OF WOMEN
P.O.Box 75 Wewak East Sepik Province Papua
Port Moresby, Papua New Guinea
T: 6 75 862827 F: 6 75 862131

Uako, Barnabas, Mr.
Commitee
OK TEDI/ FLY RIVER ENECO ASSOCIATION
Fly River Provincial Government P.O.Box 521
Daru - Western, Papua New Guinea

PARAGUAY - GOVERNMENT

Rodriquez, Andres, Mr.
President
Permanent Mission of Paraguay
211 E. 43rd Str., Room 1202
New York NY 10017, United States of America
T: 212 687 3490 F: 212 818 1282

NGO

Baldwin, Angelica, Sra.
ROAP
25 de Mayo 1618
Asuncion, Paraguay

Diaz Pena, Elias
ROAP
25 de Mayo 1618
Asuncion, Paraguay

Farina Insfran, Gladys Maria
CECTEC
Casilla Correo 1730
Asuncion, Paraguay

Garcia Gomez, Mirtha
SOB
25 de Mayo 1618, Casilla de Correos 1380
Asuncion, Paraguay

Rivas, Oscar
SOB
25 de Mayo 1618, Casilla de Correos 1380
Asuncion, Paraguay

Wehrle Rivarola, Andres Teodoro
CECTEC
Casilla Correo 1730
Asuncion, Paraguay

PRESS

Arguello Duo, Luis Manuel
Reporter Radio/Television
RADIO CARITAS
Sto. Domingo y 4ta. (Capiata)
Asuncion, Paraguay, Paraguay

Cespedes, Mauro Ramon
Correspondent
ZP 16 RADIO PARQUE
Edificio APE Department
Ciudad del Este, Paraguay, Paraguay

Chavez, Celso
Editor
NOTICIAS EL DIARIO
Artigas y Dr. Prieto, No 145
Asuncion, Paraguay, Paraguay

Escobar, Franco Desiderio
Editor
DIARIO ABC COLOR
Cnel. Rogelio R. Benitez 1148
Fdo. de la Mora, Paraguay, Paraguay

Ferreira Fleitas, Felix Vidal
Reporter Radio/Television
CANAL 13
Cnel. Lirio, 1956
entre 24 y 25 (sajonia)Asuncion, Paraguay

Goossen Martens, Teresa Renate
Reporter Radio/Television
RADIO NANDUTI
Bruselas 1696
Asuncion, Paraguay, Paraguay

Gregor Hellmann, Jose Alfredo
Editor
NOTICIAS EL DIARIO
Teniente Farina 3.919
Radio Operadores del Chaeo, Paraguay, Paraguay

Insaurralde, Nestor
Editor
DIARIO HOY
Palma Loma 272
Luque, Asuncion, Paraguay, Paraguay

Leon Da Costa, M. Antonio
Reporter Radio/Television
RADIO PRIMERO DE MARZO
Cap. Lezcano c/ G. Centurion
Asuncion, Paraguay, Paraguay

Lopez, Rogelio
Editor
DIARIO ULTIMA HORA
Ponce de Leon 126
Fdo. de la Mora, Asuncion, Paraguay, Paraguay

Mendez Grimaldi, Idilio
Editor
REVISTA CAMPO ABIERTO
Tte. Soliz 1056
Asuncion, Paraguay, Paraguay

Miranda, Risoleta Cardoso
Reporter Radio/Television
TV CULTURA DO PARA
Tr. Mauriti, 1006/504
Belem, Para, Paraguay

Ortiz, Bareiro Valois G.
Reporter Radio/Television
RADIO NACIONAL DEL PARAGUAY
Av. 1 da Marzo V. Porvener
Lambare, Paraguay, Paraguay

Ruiz Diaz, Juan Carlos
Editor
REVISTA CAMPO ABIERTO
Manuel Dominguez 597
Asuncion, Paraguay, Paraguay

Sanchez, Isidoro Gonzalez
Reporter Radio/Television
ZP 16 RADIO PARQUE
Bario Juan Pablo Segundo
Ciudad del Este, Paraguay, Paraguay

Yaluff, Guillermo
Reporter Radio/Television
SNT - CANAL 9
14 de Mayo, 1720
Asuncion, Paraguay, Paraguay

GLOBAL FORUM

Acevedo, Julio Cesar Cabanas, Mr.
President
ASOCIACION MEDIO AMBIENTE
Lazaro De Rivera, 1407
Asuncion, Paraguay

Beaufort, Humberto Ruben, Sr.
Student
UNA
Kennedy 506 Esq. cerro Cora
Assuncion, Paraguay
T: 595 21 27-646

Chaves, Sara Rachel, Ms.
Director
ONE MILLION TREES
Benitez Vera, 791 Y Sillo
Asuncion, Paraguay
T: 59 21-203897 F: 221 447 806

Escobar, Manuelita, Sra.
Coordenadora
ALTER VIDA
Artigas 960 c/Cusmanich - Casilla Correo 2334
Asuncion, Paraguay
T: 207246 F: 444572

Espinola, Miryan Teresa Caballero De, Sra.
Alter Vida
CENTRO DE ESTUDIOS Y FORMACION PARA EL ECODESARROLLO
Artigas 960
Assuncion 2334, Paraguay

Garcia, Mirtha, Sra.
SOBREVIVENCIA
Palais de Nations
Geneve CH 1211, Paraguay
F: 41 22 733-6542

Ihle, Armin, Mr.
Director
COMITE DE IGLESIAS
Gral. Diaz 429
Assuncion, Paraguay
T: 595 21 495 161 F: 595 21 443 932
TL: 2085

Ihle, Sonia, Ms.
Miembro
COMITE DE IGLESIAS
Gral. Diaz 429
Asuncion, Paraguay
T: 595 21 495 161 F: 595 21 443 932

Lloret, Maria Teresa, Ms.
FORO SIGLO XXI
Boqueron No.559
Asuncion, Paraguay
T: 595 21 24088 F: 595 21 24722

Mabel, Ovando
Alas Paraguayas 810
Assuncion, Paraguay
T: 595 21 605056

Marini, Tereza Gonzalez
ASOCIACION CRISTIANA DE JOVENES (YMCA) DE PARAGUAY
Calle Luis A. de Herera 962, Casilla 672
Asuncion, Paraguay

Mena, Ramona Concepcion Alcaraz, Sra.
FACULTAD DE INGENIERIA AGRONOMICA
Ruta Mcal. Estigarribia, Km.11.5
San Lorenzo, Paraguay

Mendoza, Rocio, Sra.
JORNADAS POPULARES DE DISCUSION ECOLOGICA
Palais de Nations
Geneve CH 1211, Paraguay
F: 41 22 733-6542

Mongelos, Ricardo Fabian Yorg, Sr.
Estudante
FACULDADE DE ECONOMIA
Ruta Mariscal Estiparibia km 11 - Campus Universitario
San Lorenzo, Paraguay
F: 595 21 50-8475

Pena, Elias Diaz, Sr.
REDES, C.C. 1380
Palais de Nations
Geneve CH 1211, Paraguay
F: 41 22 733-6542

Peterkin, John David, Mr.
Coordinator
ONE MILLION TREES
Benitez Vera 791 Esp. Lillo
Asuncion, Paraguay
T: 59 521-23427 F: 214-47800

Roche, Rich Ann, Ms.
Investigator
CERI
Casilla de Correo 1986
Asuncion, Paraguay
T: 595 21 406-002

Ruiz, Luis Alberto Costa, Mr.
AUDIO IMAGEN Y MEDIO AMBIENTE
Independencia Nacional 948
ITA - Central, Paraguay

Schmeling, Irene, Ms.
Coordinator
UNITED NATIONS - PARAGUAY
Pouhon 25, B-4920
Aywaille, Paraguay
T: 32 86 434116 F: 32 86 823108

Valinotti Bareiro, Elisa, Ms.
Sec Ejecutiva De Paraguay
A.LA.MA.
Dr. M. Dominguez 587
Asuncion, Paraguay
T: 449364

Yubi, Carlas Zenaida Bocado, Sra.
Student
UNA
Kennedy 506 Esq. cerro Cora
Assuncion, Paraguay
T: 595 21 27-646

Yubi, Elizabeth Bogado, Sra.
Student
UNA
Kennedy 506 Esq. cerro Cora
Assuncion, Paraguay
T: 595 21 27-646

PERU - GOVERNMENT

Barrantes Takata, Carmen Luisa, Ms.
INSTITUTO SABER/ DIARIO LA REPUBLICA
Los Alamos 204, Residencial San Felipe, Jesus Maria
Lima, Peru
F: 51 14 335678

Barton Eckfeldt, Michelle, Ms
Head Environmental Education Chapte
SIRENA COMUNICACION PARA LA PAZ
General Borgono 449 'A' Miraflores
Lima 18, Peru
T: 466887

Fujimori, Alberto, Mr.
President
Permanent Mission of Peru
820 2nd Ave., Suite 1600
New York NY 10017, United States of America
T: 212 687 3336 F: 212 972 6975

Moromisato Miasato, Doris, Ms
Directora
COMYC (COMUNICACION) CULTURA PARA
General Borgono 449'A' Miraflores
Lima 18, Peru
T: 00 51 14 466887 F: 00 51 14 676661

Mutal, Sylvio S
Chief Technical Advisor And Regiona
UNDP/UNESCO
CASILLA 4480 (4480)
Lima, Peru
T: 51 14 402401 F: 51 14 402484

NGO

Bottger, Miriam
PT
Av. Esteban Campodonico No. 208, Urb. Santa Catalina, La Victoria, Lima 13, Peru

Castano, Maite
ICSA
Apartado Postal 2054
Correo Central, Lima 100, Peru

Dulois, Maylis
ICSA
Apartado Postal 2054
Correo Central, Lima 100, Peru

Espinoza Llanos, Roberto
COICA
Jiron Almagro 614
Lima 11, Peru

Fernandez Velasquez, Bernardo
ULI
Av. Javier Prado Este S/N
Monterrico, Lima 33, Peru

Grefa, Valerio Almoya
COICA
Jiron Almagro 614
Lima 11, Peru

Guerrero, Raul
DESCO
Leon de la Fuente No. 110
Lima 17, Peru

Guevarra Linares, Nestor Abelardo
ICMA
Manco Capac 473, Oficina 2
Miraflores, Lima, Peru

Koechlin Von Stein, Pedro
I_PS
478 Camino Real, 8o Piso
San Isidro, Lima 27, Peru

Kun Bodanszky, Pedro Pablo
HPS
Luis N. Saenz 449, Jesus Maria
Lima 11, Peru

Landeo Del Pino, Liliam Astrid
CAAAP
Apartado 14-0166
Lima 14, Peru

Leo, Diana Mariella
APCN
Parque Jose de Acosta 187
Magdalena, Lima 17, Peru

Linares Meneses, Felicitas Estelas
ICMA
Manco Capac 473, Oficina 2
Miraflores, Lima, Peru

Luna Espinoza, Ali Iram
HPS
Luis N. Saenz 449, Jesus Maria
Lima 11, Peru

Manrique, Jorge
CAME
Jiron Arequipa No. 120, 3er piso B
Puno, Peru

Obregon, Luis A., Mr.
President
ANDESTUDIO / ESTUDIO Y PROMOCION DE PROYECTOR MULTIDICIPLINAREIOS DE DESAROLLO
Casilla 270027
Lima 27, Peru
F: 51 14 41 53 257350

Orozco Huaytalla, Adrian
ICSA
Apartado Postal 2054
Correo Central, Lima 100, Peru

Palao, Juan, Sr.
CONSEJO ANDINO DE MANEJO ECOLOGICO
Palais de Nations
Geneve CH 1211, Peru
F: 41 22 733-6542

Pessoa Mendoza, Vicente
COICA
Jiron Almagro 614
Lima 11, Peru

Puertas Pulgar, Marita
CAAAP
Apartado 14-0166
Lima 14, Peru

Pulgar-Vidal, Manuel
SPDA
Plaza Arrospide No. 9
San Isidro, Lima 27, Peru

Ramirez Jara, Eleuterio
ICSA
Apartado Postal 2054
Correo Central, Lima 100, Peru

Rosas Moscoso, Fernando
ULI
Av. Javier Prado Este S/N
Monterrico, Lima 33, Peru

Ruiz Pereyra, Gustavo
I_PS
478 Camino Real, 8o Piso
San Isidro, Lima 27, Peru

Sanchez, Silvia
APCN
Parque Jose de Acosta 187
Magdalena, Lima 17, Peru

Surber Devoto, Conrado, Mr.
ANDESTUDIO
Casilla 270 027, Lima 27 (Correo San Isidro)
Peru

Turpo, Fortunato
CJPI
Urbanizacion Dolores 6-20
Paucarpata, Apartado No 230, Arequipa, Peru

PRESS

Aliaga Paricahua, Pedro Pablo
Correspondent
PERUVIAN PRESS
R.Vitor Meirelles 505
Maracana,Rio de Janeiro, Brazil, Peru

Angeles, Maria Luisa
Correspondent
RADIORED
Av. San Borja Norte 883
Lima 41, Peru, Peru

Baella Herrera, Alfonso
Editor
OFFICIAL PRESS PERU
Lima, Peru

Bittencourt, Silvia Helena Guimaraes
Correspondent
RIDAT
Av. RAinha Elizabeth, 653
Rio de Janeiro, Brazil, Peru

Cannor, Patrick
Correspondent
EL COMERCIO
Jiron Miro Quesada, No.300
Lima, Peru, Peru

Castro Rivera, Eva
Correspondent
OFFICIAL PRESS PERU
Lima, Peru

Delta Parodi, Monica Cecilia
Reporter Radio/Television
PANAMERICANA TV
Baca Flor 168 San Isidro
Lima, Peru, Peru

Goycochea Ramirez, Julia
Reporter Radio/Television
ASOCIACION VIDEO Y CULTURA
Las Orquideas E-20 P. Libre
Lima, Peru, Peru

Jara, Humberto
Editor
EXPRESO
Lima, Peru, Peru

Lazarte, Armando
Editor
EXPRESO
Garcia Bedoya, 185
Lima, Peru

Madalengoitia, Laura
Editor
SUPLEMENTO ECOLOGIA Y DESARROLLO
Las Begonias 1503 R. S. Felipe
LIma II, Peru, Peru

Meier Miro-Quesada, Martha E., Ms
Head Ecology Section
'EL COMERCIO' NEWSPAPER
Jiron A. Miro-Quesada #300
Lima 1, Peru
T: 00 51 14 287660 EXT F: 00 51 14 310810

Quesada, Maqui Miro
Correspondent
EXPRESO
Jiron Ica, No. 646
Lima, Peru, Peru

Rojas Samanez, Alvaro
Editor
DIARIO EXPRESSO
Melchormalo 183
Surco, Peru, Peru

Saenz A., Luz Aurea
Reporter Radio/Television
LA REPUBLICA
Monte de las Caobas 795
Monterrico Sur. Lima, Peru, Peru

Trejo, Sonia
Correspondent
EXPRESSO
21 Stuyvesant Oval Apt. M-F
New York, NY 10009, USA, Peru

ORGANIZATIONS

Andaluz, Antonio, Sr.
Proterra
Av. Esteban Campodonico No. 208, Urb. Santa Catalina, La Victoria, Lima 13, Peru
T: (51 14) 723 800/703 371 F: (51 14) 417 001

Arrieta, Gerardo Gonzales, Mr.
Inter-American Housing Union
Santa Teresita No. 253, San Isidro, PO Box 18-1366
Lima 27, Peru
T: (51 14) 418 384/418 380 F: (51 14) 415 659

Caillaux, Jorge, Sr.
Sociedad Peruana de Derecho Ambiental
Plaza Arrospide No. 9
San Isidro, Lima 27, Peru
T: (51 14) 400 549/224 033 F: (51 14) 703 111

Correa, Marcel Rubio, Sr.
DESCO - Centro De Estudios Y Promocion Del Desarrollo
Leon de la Fuente No. 110
Lima 17, Peru
T: (51 14) 62 71 93 F: (51 14) 61 73 09

Davies, Maria De Lourdes, Ms.
Instituto de Desarrollo y Medio Ambiente (IDMA)
Apartado Postal 110 384
Lima 11, Peru

Ikanan, Evaristo Nugkuag, Mr.
Coordinadora De Las Organizaciones Indigenas De La Cuenca Amazonica
Jiron Almagro 614
Lima 11, Peru
T: (51 14) 631 983 F: (51 14) 631 983

Iturregui, Juan Augusto Palao, Sr.
Consejo Andino de Manejo Ecologico
Jiron Arequipa No. 120, 3er piso B
Puno, Peru
T: (51 54) 351 619 F: (51 54) 351 616

Koechlin, Peter, Mr.
Sociedad Pachamama
478 Camino Real, 8o Piso
San Isidro, Lima 27, Peru
T: (51 14) 428 946/428 951 F: (51 14) 411 990

Leo, Mariella L., Ms.
As. Peruana para la Conservacion de la Naturaleza
Parque Jose de Acosta 187
Magdalena, Lima 17, Peru
T: (51 14) 616 316 F: (51 14) 424 585

Meneses, Estela Linares, Sra
Instituto de Comunicacion y Medio Ambiente (ICMA)
Manco Capac 473, Oficina 2
Miraflores, Lima, Peru
T: (51 14) 478 024 F: (51 14) 478 024

Ramirez, Eleuterio, Sr.
Indian Council of South America
Apartado Postal 2054
Correo Central, Lima 100, Peru

Sarasana, Cesar, Sr.
Confederacion de Nacionalidades Amazonicas del Peru (CONAP)
Brigadier Pumacahua No. 974 Jesus Maria
Lima, Peru
T: (51 14) 238 391

Turpo, Aurellano, Sr.
Comision Juridica de los Pueblos de Integracion Tawantinsuyana
Urbanizacion Dolores 6-20
Paucarpata, Apartado No 230, Arequipa, Peru
T: (51 54) 238 383/228 231

GLOBAL FORUM

Abraham, Rosario, Ms.
Jr. Ica 646
Lima, Peru
T: 51 14 313985 F: 51 14 318314
TL: 25307

Acosta Vargas, Gladys, Ms.
CENTRO DE LA MUJER PERUANA FLORA TRISTAN
Parque Hernan Velarde # 42
Lima, Peru
T: 51 14 330694 F: 51 14 339060
TL: GZ02: FLORA

Agurto Saenz, Tomas, Mr.
UNIVERSIDAD RICARDO PALMA
Av. Benavides, 5400 - Surco
Lima, Peru
T: 51 14 236535

Alaba, Marco
CIED
las Mongolias 2741 Lince
Lima 110104, Peru
T: 51 14 421766

Arce, Carlos F. Aranda, Mr.
Director
SOUTHERN PERU COPPER CORPERATION
ave. caminos dez inca 171, chacarilla del estanque surquillo
Lima, Peru
T: 51 14 368813 F: 51 14 365599

Arias, Rina Flores, Ms.
Una-Peru
ONG ADELPA-LIVOS-AREAS VERDES, ETC.
URB San Jose B-5 - Umacollo
Arequipa, Peru
T: 51 54 212031 F: 51 54 234476

Barrantes, Carmen, Ms.
INSTITUTO DE COMUNICATION Y MEDIO AMBIENTE
Rijeca, Peru

Beat, Weiser, Mr.
SWISS TELEVISION
San Jorge 111 Monterrico
Lima 33, Peru
T: 51 14 369179

Benavides, Cesar, Mr.
Ong Secretario
ASOCIACION PERUANO SUECA
Aporimac 224/505- Lima 1 Peru
Lima, Peru
T: 51 14 49-2157 F: 51 14 41-2794
TL: 14-0140-LIMA

Bernardo, Fernandez Valasquez
UNIVERSIDAD DE LIMA
Av. Javier Prado Este s/n, Monterrico
Lima, Peru
T: 51 14 356552 F: 51 14 356552

Calero Merino, Mariano
UNIVERSIDADE NACIONAL DE PIURA
Rijeca, Peru

Carcausto, Jesus Antonio Subia, Sr.
Consultant
ASSOC. CIVIL LABOR
Jr. Abtao, no.607
Ilo, Peru
T: 51 781816 F: 51 781208

Celeste, Cambria, Ms.
CENTRO FLORA TRISTAN
Parque Mernan Velarde 42
Lima, Peru
T: 51 14 331457 F: 51 14 339060

Chipoco Malmborg, Teresa, Sra.
Presidenta
C.E.R.
Parque Antequera 815-3
Lima 27, Peru
T: 51 14 423 096

Correa, Alcides Pelayo Chavarry, Mr.
Member
EQUIPO DERECHO AMBIENTAL - TALLER DE DERECHO
Pontificia Universidad Catolica del Peru - Av. Universitaria s/n - San Miguel
Lima, Peru
T: 51 14 611785

Cugler, Alexandra, Ms.
President
IDEFE - INSTITUTO DE ESTUDIOS DE FACTIBILIDAD ECOLOGICA
Bartolome Trujillo 656, San Antonio, Miraflores
Lima 18, Peru
T: 54 14 469339 F: 353 61 486667

Davey, Carlos Melo, Sr.
Vice President
OCCIDENTAL PETROLEUM CORPORATION OF PERU
Casilla 6226
Lima 100, Peru
T: 51 14 42-8110 F: 51 14 42-8140 42-
TL: 0077, E: 01:YNS020

Diaz Palacios, Julio E., Mr.
O N G - LABOR
Jr. Abtao 607 - 2doPisu
Ilo, Peru
T: 51 54 781816 F: 51 54 781208

Drabbe, Eddy, Ms.
Una-Peru
ONG ADELPA-LIVOS-AREAS VERDES, ETC.
URB San Jose B-5 - Umacollo
Arequipa, Peru
T: 51 54 212031 F: 51 54 234476

Estela Castro, Julio, Mr.
EDAC/CIEA
Jiron Amalia Puga 465
Casamarca, Peru
T: 51 44 923125 F: 923429

Farro, Maria Helena Foronda, Ms.
INSTITUTO NATURA
Chimbote, Peru

Fernandez, Hilda Marlene Valdivia, Sra.
ASSOC. CIVIL LABOR
Jr. Abtao, no.607
Ilo, Peru
T: 51 781816 F: 51 781208

Fernando, Rosas Moscoso
UNIVERSIDAD DE LIMA
Av. Javier Prado Este s/n, Monterrico
Lima, Peru
T: 51 14 378491 F: 51 14 356552

Gomez, Boris, Sr.
Gerente
MANU NATURE TOURS
Ave. Sol, 582
Cuzco, Peru
T: 51 84 224-384 F: 51 84 234-793

Gonzales, Cesar Juan Sotomayor, Sr.
CONVENIO PERU-ALEMANA PARA CULTIVOS ANDINOS - COPACA
Casilla Postal 807, Avenida el Sol 817
Cuzco, Peru

Guerrero, Luis, Sr.
Director
CIED
Las Magnolias 2741
Lince - Lima 110104 (110485), Peru
T: 428747 F: 51-14-421766

Hernandez, Gaby
ECO INTEGRAL
An+onio Portugal 757, Urb. Los Ficos
Santa Anita, Peru

Hernandez, Zoila
INSTITUTO DE COMUNICATION Y MEDIO AMBIENTE
Rijeca, Peru

Hoz, Enrique La, Sr.
Secretario Executivo
RINAP
Parque 27 de Noviembre 430 Oficina 2A San Isidoro
Lima, Peru
T: 51 14 419942 F: 51 14 419942

Huaman, Josefina, Ms.
WOMEN AND SHELTER NETWORK
Cordobanes 24, Col. San Jose Insurgentes
Mexico City 03900, Peru
T: 52 5 6512371 F: 52 5 5453263

Koechlin, Pedro G., Mr.
Presidente
SOCIEDAD PACHAMAMA
Av Camino Real 479
Lima 27, Peru
T: 51 14 41-6031 F: 51 14 41-1990

Lavado, Demetrio Carranza, Mr.
Engineer
INSTITUTO PERUANO DE PATERNIDAD RESPONSABLE
Gregorio Escobedo 115 Jesus Maria
Lima, Peru
T: 51 14 63-9823 F: 51 14 63-5965

Legonia, Carmen
INSTITUTO DE COMUNICATION Y MEDIO AMBIENTE
Rijeca, Peru

Linares, Felicitas, Ms.
INSTITUTO DE COMUNICATION Y MEDIO AMBIENTE
Lima, Peru

Llanos, Gabriela, Ms.
President
AFAC
Av. General Pezet 961-San Isindro
Lima 27, Peru
T: 51 14 40-5233 F: 51 14 40-5233

Llanos, Roberto Espinoza, Sr.
COICA
Larco Herrera, 1057
Lima, Peru
T: 051 14 619228 F: 051 14 619228

Llosa, Gonzalo, Sr.
EKODES CONSULTORES
Avda. Nicolas de Rivera 544
Lima 27, Peru
T: 51 14 408256 F: 51 14 426407

Loyaga De Ludena, Martha Elena, Ms.
UNIVERSIDAD RICARDO PALMA
Av. Los Tallanes,1408 - Wrd. Ing. La Castellana - Surco
Lima, Peru

Ludena Luque, Guillermo, Mr.
UNIVERSIDAD RICARDO PALMA
Av. Benavides 5400
Lima, Peru

Luna, Boris Gomez, Sr.
Gerente
MANU NATURE TOURS
Av. Sol, 582
Cuzco, Peru
T: 51 84 224 384 F: 51 84 234 793

Luna, Henry Gomez, Sr.
Sales Representative
MANU NATURE TOURS
Av. Sol, 582
Cuzco, Peru
T: 51 84 224 384 F: 51 84 234 793

Mauleon, Cecelia Oleo, Ms.
DEVELOPMENT ALTERNATIVES WITH WOMEN FOR A NEW ERA (DAWN)
c/o Flora Tristan Plaza Hernan Velarde 42
Lima, Peru

Miro-Quesada, Martha Meier
Head Ecology Section
EL COMERCIO
Jiron Antonio Miro-Quesada 300
Lima, Peru
T: 51 14 287 660 F: 51 14 310810

Morales, Gunter Morales Arnao, Mr.
Member
SOCIEDAD PERUANA DE DERECHO AMBIENTAL
Hernando de Soto no.250 Salamanca
Lima 03, Peru
T: 51 14 351 746

Mutal, Nina, Ms.
UNESCO
Av. Jucen Perez
Lima, Peru
T: 51 14 223-3846 F: 51 14 402484

Peralta, Elias Carreno, Sr.
Presidente
ADARI
Avenida Sol 346 Oficina 209
Cuzco, Peru
T: 51 84 228 825 F: 51 84 221 1408

Pino, Liliam A. Landeo Del Pino, Sra.
CAAAP
Av. Gonzales Irada 626
Lima 17-1, Peru
T: 51 14 615223 F: 51 14 638846

Plenge, Inigo Aspillaga, Mr.
Vice President
OCCIDENTAL PETROLEUM CORPORATION OF PERU
Casilla 6226
Lima 100, Peru
T: 51 14 428110 F: 51 14 8140
E: 01:YNS020

Pulzar, Marita Angelica Puertas, Ms.
Coordinator
EQUIPO DERECHO AMBIENTAL - TALLER DE DERECHO
Pontificia Universidad Catolica del Peru - Av. Universitaria s/n - San Miguel
Lima, Peru
T: 51 14 611785

Ribeyro, Piedad
INSTITUTO DE COMUNICATION Y MEDIO AMBIENTE
Rijeca, Peru

Rodriguez, Hugo Rolando Quito, Mr.
INSTITUTO NATURA
Chimbote, Peru

Rodriquez, Maria Elvira Luna De, Ms.
INDEPENDIENTE
Mza. A Lote 3 Urb. Santa Laura Chorrillos
Lima, Peru
T: 51 14 677406

Rojas, Danilia, Ms.
INSTITUTO NACIONAL DE DESARROLLO
Lima, Peru

Ruiz, Gustavo, Mr.
TRI/YALE
C. de Alcantara 117
Lima, Peru
T: 51 14 904-730 F: 51 14 425-855

Sallat, Jacques, Sr.
CENTRO DE ESTUDIOS 'BARTOLOME DE LAS CASAS'
Carlos Alaysa y Roel 2626
Lima 14, Peru
T: 51 14 429992 F: 51 14 427894

Sarmiento, Juan Roberto Ramirez-Otarola, Mr.
INDUSTRIA ALIMENTARIA S.A.
Fundo La Tablada-Huacho Av. Peralvillo 2710
Huacho - Dep. Lima, Peru

Suarez, Manuel d'Ornellas, Dr.
Director
DIARIO EXPRESSO
Jr. Ica 646
Lima, Peru
T: 51 14 313985 F: 51 14 318314
TL: 25307

Torres, Amelia
FUNDACION PERUANA PARA LA CONSERVACION DE LA NATURALEZA
Lima, Peru

Vadillo, Jesus Percy Alvarado, Sr.
PRO-PAIS
Av. Aviacion #2552 - Distrito de Santa Borja
Lima, Peru
T: 51 14 751213

Vargas, Felix, Mr.
EMPRESA DE SERVICIOS MUNICIPALES
Rijeca, Peru

Watson, Deborah, Ms.
SISTERS OF MERCY
Apartado 295
Puno, Peru
T: 51 54 351184 F: 51 54 351574

Yangua, Jose Lirio, Sr.
CONFEDERACION DE NACIONALIDADES AMAZONICAS DEL PERU-CONAP
Palais de Nations
Geneve CH 1211, Peru
F: 41 22 733-6542

Zevallos, Dioni Alberto Guevara, Mr.
Promotor Desarrollo Ambiental
ASSOC. CIVIL LABOR
Jr. Abtao, no.607
Ilo, Peru
T: 51 781816 F: 51 781208

PARLIAMENTARIANS / SPIRITUAL LEADERS

Ulloa, Manuel, Senator
Senado Congreso Nacional
Lima, Peru

PHILIPPINES - GOVERNMENT

Basu, Chiranjib Kumar
Head/Environment
COLOMBO PLAN STAFF COLLEGE CPSC
PO Box 7500, Airmail Distribution 1300 Pasay City Metromanila
Manila, Philippines
T: 632 631 0992 5 F: 632 631 0996

Birosel, Roger Cuaresma, Mr
Science & Technology Adviser
PHILIPPINE SENATE
28 Yulo Plaza Don Bosco Cor - Pasona Tamo
Makati, Metromanica, Philippines
T: 8193494 8152689 F: 407637

De Jesus, Agnes, Ms
Manager
ENVIRONMENTAL MANAGEMENT DIVISION E
Merritt Road, Fort Bonifacio, Metro
Manila, Philippines
T: 87 87 62 OR 815 89 6 F: 632 815 2747

Ordonez, Gina
Correspondent
NGO PROJECT FINDER
AA-202 Galeria Magallanes Magallanes Village
Makati,M. Manila, Philippines
T: 833 6741 F: 833 6741

NGO

Anda, Grizelda
CFI
Rm. 406 Victoria Condominium, 41 Annapolis Street, Grenhills, San Juan, Metro Manila 1502, Philippines

Araneta, Patricia
Ms.
GRP
3rd Floor, Liberty Buildings, 835 Pasay Road
Makati, Metro Manila, Philippines

Caringas, Yul
NFIPP
Rm. 701, WEB-JET Building, 64 Quezon Ave. Cor. BMA
St. Quezon City,, Philippines

Chebat, Maria Asuncion, Ms.
TWMA
41 Rajah Matanda, Project 4
Quezon City 1109, Philippines

Corpuz, Victoria T.
CWERC
P.O. Box 7691, Garcom-Baguio (752)
Dapo 1300, Domestic Road, Pasay City, Philippines

Danant, Joelle
PCD
MCC PO Box 740
Makati, Metro Manila 1299, Philippines

De Dios, Aurora
CFI
Rm. 406 Victoria Condominium, 41 Annapolis Street, Grenhills, San Juan, Metro Manila 1502, Philippines

De Guzman-Castillo, Raquel
IBON
PO Box SM-447
Manila 1008, Philippines

Gamo, Eileen
GCI
Suite 421 Equitable Bank II, Ortigas Avenue
Greenhills, San Juan, Metro Manila, Philippines

Goerin, Carmelita
NFIPP
Rm. 701, WEB-JET Building, 64 Quezon Ave. Cor. BMA
St. Quezon City,, Philippines

Korten, David C., Dr.
Fellow
PEOPLE CENTERED DEVELOPMENT FORUM
MCC P.O.Box 740, Makati
Metro Manila, Philippines
T: 63 832 866-413 F: 63 832 818-6488
TL: 53283 FFMAIL PN

Lubi, Elisa Tita, Ms.
Int. Relations Officer
GABRIELA
35 Scout Delgado St
Quezun City, Philippines
T: 2 63 99-8034 F: 2 63 922-9567

Manubay, Mary Ann
IBON
PO Box SM-447
Manila 1008, Philippines

Morales, Horacio, Mr.
RURAL RECONSTRUCTION MOVEMENT PRRM
Kayumanggi Press Building
940 Quezon City, Philippines

Olalia, Diwata, Ms.
PHILIPPINE WOMEN'S UNIVERSITY
Taft Avenue
Manila, Philippines

Ouano, Ely Anthony
ENCI
9th Floor Mondragon Building, 324 Sen. Gil Puyat Ave., Makati, Metro Manila 1200, Philippines

Perpinan, Norma (Mary Soledad)
TWMA
41 Rajah Matanda, Project 4
Quezon City 1109, Philippines

Regpala, Maria Elena
CWERC
P.O. Box 7691, Garcom-Baguio (752)
Dapo 1300, Domestic Road, Pasay City, Philippines

Serrano, Isagani R., Mr.
Vice President
PHILIPPINE RURAL RECONSTRUCTION MOVEMENT(PRRM)
Kayumanggi Press Bldg. , 940 Quezon Ave.
Quezon City, Philippines
T: 63 2 985563/991866 F: 63 2 997919

Tiagan, Luz
G
No. 35, Scout Delgado Street
Roces District, Quezon City, Philippines

PRESS

Amor, Adlai
Correspondent
DEPTHNEWS ASIA
Ch. de la Chavanne, 9
CH-1196 Gland, Switzerland, Philippines

Arquiza, Yasmin
EARTH STATION
24 Maalalahanin St.
Quezon City, Philippines, Philippines

Baylon, Gloria Jane
Correspondent
CEBU NEWSTIME DAILY
Kitasuna 6-19-11-102, Koto-ku
Tokyo 136, Japan, Philippines

Bulong, Mario
Correspondent
GUMIL
Suite 203, Erlene Bldg,22
Sen.Gil Puyat Ave., Makati,Manila, Phils, Philippines

Calso, Alfredo
Correspondent
GUMIL
Suite 203, Erlene Bldg, No.22
Sen.Gil Puyat Ave. Makati, Manila, Phils, Philippines

Claudio, Corazon
Editor
OFFICIAL PRESS PHILLIPINES
30 Macapuno, 87
Valle Verde I, Pasig, Manila, Philippines

Fawthrop, Tom
Reporter Radio/Television
EASTERN HORIZONS TV
2/20 Russell Avenue
Pasay, Manila 1300, Philippines, Philippines

Icamina, Paul
Editor
DEPTHNEWS
3rd Fl.S&L Bldg.1500 Roxas Blv
Manila, Philippes, Philippines

Libanio, Jose Carlos De A.
Reporter Radio/Television
DEPTHNEWS RADIO
SQS 107 Bl.B/303
Brasilia, D.F., Brazil, Philippines

Maceda, Sonia
Correspondent
ELCOREPORTS NEWS & PHOTO AGENCY
8 Free Press Street
West Triangle, Quezon City, Philippines, Philippines

Makabenta, Leah P.
Correspondent
IPS
217 Cityland 7, Alfaro St.
Salcedo Village, Makati, Philippines

Manabat, Flordeliza
Correspondent
GUMIL
Suite 203, Erlene Bldgs, No.22
Sen.Gil Puyat Ave., Makati, Manila,Phils, Philippines

Navalta, Frankie
Correspondent
DZCV
8 Free Press St.
West Triangle, Quezon City, Philippines, Philippines

Ordonez, Regina
Correspondent
CORDILLERA NEWS AGENCY
AA-202 Galeria Magallanes Vil
M. MNL, Philippines, Philippines

Paat, Anastacio
Correspondent
GUMIL
Suite 203, Erlene Bldg,22
Sen.Gil Puyat Ave.,Makati,Manila, Phils, Philippines

Taguar, Danilo
Correspondent
GUMIL
Suite 203, Erlene Bldg, No.22
Sen.Gil Puyat Ave.,Makati, Manila, Phils, Philippines

Villadiego, Rita
PHILIPPINE DAILY INQUIRER
870-B Jhocson St.
Sampaloc, Manila, Philippines, Philippines

ORGANIZATIONS

Camacho, Leonarda N., Ms.
Metro Manila Coucil of Women Balikatan Movement Inc.
215 Regent Park, 207 Santolan Road, Greenhills
Quezon City, Philippines
T: (63 2) 721 0767/693 0494

Gonsalves, Julian F., Mr.
International Institute of Rural Reconstruction
Rm. 38, Elena Apartments, 512R, Salas St.
Ermita, Manila, Philippines
T: (63 2) 582 659 F: (63 2) 522 2494

Hilario, Ernesto, Mr.
Council For People's Development
Databanking & Publications Dept., PO Box SM 359
Santa Mesa, Manila, Philippines
T: (63 2) 921 1531 F: (63 2) 921 1531

Kalaw, Maximo, Mr.
Green Forum Philippines
3rd Floor, Liberty Buildings, 835 Pasay Road
Makati, Metro Manila, Philippines
T: (63 2) 85 1110 F: (63 2) 818 3207

Lopez, Victoria M., Ms.
SIBAT - Wellspring of Science and Technology
PO Box 375
CPO Manila, Philippines
T: (63 2) 994 781 F: (63 2) 994 781

Perpinan, Maria Soledad, Sister
Third World Movement Against the Exploitation of Women
41 Rajah Matanda, Project 4
Quezon City 1109, Philippines
T: (63 2) 786 469 F: (63 2) 921 5662

Pineda, Susan, Ms.
Green Coalition Inc., The
Suite 421 Equitable Bank II, Ortigas Avenue
Greenhills, San Juan, Metro Manila, Philippines
T: (63 2) 773 419/721 3147 F: (63 2) 493 029

Quizon, Antonio, Mr.
Asian Ngo Coalition (ANGOC)
47 Matrinco Building, 2178 Pasong Tamo Street
Makati, Metro Manila 1200 Philippines, Philippines
T: (63 2) 816 3033 F: (63 2) 815 1198

Ramos, Ricardo B., Mr.
Environmental Network Center Inc. (ENCI)
9th Floor Mondragon Building, 324 Sen. Gil
Puyat Ave., Makati, Metro Manila 1200, Philippines
T: (63 2) 818 1506/815 3202

Rarallo, Encarnacion, Ms.
Philippine Women's University, The
Taft Avenue
Manila, Philippines
T: (63 2) 522 3383/502 733

Santos, Filipina Del Rosario, Ms.
Appropriate Technology Center for Rural Development
63 Magnolia Street
Roxas District, Quezon City, Philippines
T: (63 2) 989 728 F: (63 2) 521 7225

Tinker, Jon, Mr.
Process Inc.
54 Estrella St.
Makati, Metro Manila, Philippines
T: 817 5825

Velazquez, Maria Aida, Sr.
Lingkod Tao-Kalikasan Foundation Inc. (In the Service of The Human Earth Commun.)
PO Box 2734
Manila 1099, Philippines
T: (63 2) 591 185 F: (63 2) 8180107

GLOBAL FORUM

Alvarado, Wilhelmino R., Mr.
PHILIPPINE GREEN CAMPAIGN
Box 8721 Adc-Mia
Manila, Philippines
F: 63 2 817-9742

Alvarez, Cecile, Ms.
EARTH SAVERS/POPULATION COMMUNICATIONS INTERNATIONAL/ITI
ninoy aquiro park,quezon blvd
Manila, Philippines
T: 63 2 494911-3 F: 63 2 407637

Anasco, Mari Elinor G., Ms.
Asst Coord
GREEN FORUM PHILIPPINES
3/F, Liberty Building, 835 Pasay Road, Makati
Metro Manila, Philippines
T: 63 832 85-11-10 F: 63 832 818-3207

Ancheta, Armando A., Mr.
PHILIPPINE GREEN CAMPAIGN
Box 8721 Adc-Mia
Manila, Philippines
F: 63 2 817-9742

Anita, Celdran, Mr.
GREEN FORUM PHILIPPINES
3/F, Liberty Building, 835 Pasay Road, Makati
Metro Manila, Philippines
T: 63 832 85-11-10 F: 63 832 818-3207

Antonio, Benito S., Sr.
INTERNATIONAL CRUSADE FOR FILIPINO GREATNESS & SOCIAL PROGRESS, INC.
8 Free Press St. West Triangle
Quezon City, Philippines
T: 63 2 972341 F: 63 2 8179742
TL: 63199 ETPIMO PN

Antonio, Quizon, Mr.
Exec Dir
ASIAN NGO COALITION
4/F, Matrinco Bldg, Pasong Tamo, Makati
Metro Manila, Philippines
T: 63 832 816-3033 F: 63 832 815-1198

Bagadion Jr., Benjamin C., Mr.
Professor
ASIAN INSTITUTE OF MANAGEMENT
123 Paseo de Roxas, Makati (MCC P.O. Box 898, Makati)
Metro Manilla, Philippines
T: 63 2 874011-19 F: 63 2 817-9240
TL: 63778 AIMPN

Balazo, Carmen, Ms.
Community Worker
LAKAS - AYTA PEOPLES ALLIANCE OF ZAMBALES
c/o Rainforst Information Centre, P. O. Box 368
Lismore - N.S.W. 2480, Philippines
T: 61 66 218505 F: 61 66 222339

Bascug, Laurentino, Mr.
Chairman
CONGRESS FOR A PEOPLE'S AGRARIAN REFORM (CPAR)
71 Kalayaan St. Diliman
Quezon City, Philippines

Benitez, Giovanna, Ms.
PHILIPPINE WOMEN'S UNIVERSITY
Taft Avenue
Manila, Philippines

Benitez, Jose, Mr.
PHILIPPINE WOMEN'S UNIVERSITY
Taft Avenue
Manila, Philippines

Briones, Leonor, Dr.
FREEDOM FROM DEBT COALITION
College of Business Administration, Univ of the Phillipines, Diliman
Quezon City, Philippines

Cabalic, Palawig, Mr.
Chairperson
LAKAS - AYTA PEOPLES ALLIANCE OF ZAMBALES
c/o Rainforst Information Centre, P. O. Box 368
Lismore - N.S.W., Philippines
T: 61 66 218505 F: 61 66 222339

Camara, Philip, Mr.
FOUNDATION COMMUNITY ORGANIZATION AND MANAGEMENT TECHNOLOGY
Rm 204, Raman Condominium, Pasang Tamo, Makati
Metro Manila, Philippines
T: 63 832 818-1933 F: 63 832 816-4423

Caringas, Julieto, Mr.
Secretary General
KAMP
Kamp KM 701 Wes-Jet Bldg # 04 Quezon Av Cor Bma 57, Quezn City Phil
Quezun City, Philippines
T: 2 63 7120451

Carnate, Jesus C., Mr.
Executive Director
ASIA-PACIFIC CULTURAL & ECONOMIC DEVELOPMENT FOUNDATION LTD
Rm. 203, Erlene Bldg., No. 22 Scn. Cil J. Puyat Avenue
Makati, Philippines
T: 63 2 862-193 F: 63 2 815-0756

Catalan, Alberto M., Sr.
ASIA-PACIFIC CULTURAL & ECONOMIC DEVELOPMENT FOUNDATION, INC.
Room 203, Erlene Building NO. 22 Sen, Gil J. Puyat Avenue
Makati, Philippines
T: 632 972341 F: 632 8179724

Celdran, Anita, Ms.
Program Officer
GREEN FORUM - PHILIPPINES
3rd Floor Liberty Bldg. 835 Pasay Rd. Makati
Metro Manila, Philippines
T: 63 632 851110 F: 63 632 8183207

Claparols, Antonio, Mr.
ECOLOGICAL SOCIETY OF THE PHILIPPINES
Philippines (Esp)???
Manila, Philippines
T: 63 2 631-7351 F: 63 2 631-7357

Claudio, Isaac T., Sr.
ASIA-PACIFIC CULTURAL & ECONOMIC DEVELOPMENT FOUNDATION, INC.
Room 203, Erlene Building NO. 22 Sen, Gil J. Puyat Avenue
Makati, Philippines
T: 632 972341 F: 632 8179724

Collado Jr., Gaudencio A., Mr.
PHILIPPINE GREEN CAMPAIGN
Box 8721 Adc-Mia
Manila, Philippines
F: 63 2 817-9742

Concepcion, Wilberto P., Sr.
ASIA-PACIFIC CULTURAL & ECONOMIC DEVELOPMENT FOUNDATION, INC.
Room 203, Erlene Building NO. 22 Sen, Gil J. Puyat Avenue
Makati, Philippines
T: 632 972341 F: 632 8179724

Cunanan, Jose Pepito, Mr.
NATIONAL COUNCIL OF CHURCHES IN THE PHILIPPINES
879 Edsa
Quezon City, Philippines
T: 63 2 924-0224 F: 63 2 967076

Daoanay, Liza, Ms.
PHILIPPINE RURAL RECONSTRUCTION MOVEMENT
940 Kayumanggi Press Bldg, Quezon Ave
Quezon City, Philippines
T: 63 2 991866 F: 63 2 99-79-19

Del Rosario, Claudine, Ms.
Intern/Volunteer
ASSOCIATION OF FOUNDATIONS
3/F, Liberty Building, 835 Pasay Road, Makati Tel. 854757
Metro Manila, Philippines
T: 63 832 85-11-10 F: 63 832 818-3207

Delgado, Nelson M., Sr.
INTERNATIONAL CRUSADE FOR FILIPINO GREATNESS & SOCIAL PROGRESS, INC.
8 Free Press St. West Triangle
Quezon City, Philippines
T: 63 2 972341 F: 63 2 8179742
TL: 63199 ETPIMO PN

Duhaylungsod, Noel, Mr.
PHILIPPINE ENVIRONMENTAL ACTION NETWORK
109 Scout Fuentebella St
Quezon City, Philippines
T: 63 2 922-8515

Escudero, Danilo S., Sr.
ASIA-PACIFIC CULTURAL & ECONOMIC DEVELOPMENT FOUNDATION, INC.
Room 203, Erlene Building NO. 22 Sen, Gil J. Puyat Avenue
Makati, Philippines
T: 632 972341 F: 632 8179724

Faustino, Benjamin R., Sr.
ASIA-PACIFIC CULTURAL & ECONOMIC DEVELOPMENT FOUNDATION, INC.
Room 203, Erlene Building NO. 22 Sen, Gil J. Puyat Avenue
Makati, Philippines
T: 632 972341 F: 632 8179724

Fuentes, Orlando D., Sr.
INTERNATIONAL CRUSADE FOR FILIPINO GREATNESS & SOCIAL PROGRESS, INC.
8 Free Press St. West Triangle
Quezon City, Philippines
T: 63 2 972341 F: 63 2 8179742
TL: 63199 ETPIMO PN

Gasgonia, Donna, Ms.
Trustee
PANLIPI
TCG Bldg., Unit I, E. Gaches cor. Palanca Sts.
BF Homes Paranagua, Philippines
T: 63 2 802-3541 F: 842-5511

Gatmaitan, Rodolfo M., Mr.
Executive Director
PHILIPPINE GREEN CAMPAIGN
Free Press/Liwayway Streets, West Triangle
Quezon City, Philippines
T: 63 2 972341 F: 63 2 8179742
TL: 63199ETPIMO

Gatmaytan, Augustus, Mr.
LEGAL RIGHTS CENTER
Palais de Nations
Geneve CH 1211, Philippines
F: 41 22 733-6542

Gertes, Benjamin, Mr.
Manager Director
PHILIPPINE CENTER FOR APP. TECHNOLOGY AND TRAINING
224 Diego Silana Str.
Batangas City 4200, Philippines
T: 724-1609

Gonzales, Rosanna O., Ms.
PHILIPPINE GREEN CAMPAIGN
Box 8721 Adc-Mia
Manila, Philippines
F: 63 2 817-9742

Guardino, Cresencia, Ms.
POOK KALINISAN PHILIPPINES/ WEDO
4517 Surigao St. BO. Pitogo
Makati, Philippines
T: 63 2 815-0285

Honculada, Jurgette, Ms.
PILIPINA
28 Marunong Central District
Quezon City, Philippines
T: 63 2 922-7746 F: 63 2 521-7225

Hunziker, Olalia Diwata, Ms.
PHILIPPINE RURAL RECONSTRUCTION MOVEMENT
Kayumanggi Press Bldg. 940 Quezon Avenue
Quezon City, Philippines
T: 63 2 722-7093 F: 63 2 522-4002

Ignacio, Me-Ann, Ms.
PHILSSAORUM PHILIPPINES
c/o HASIK Office, 2-A Malusog cor Matiwasay Sts
Quezon City, Philippines
T: 63 2 921-8207

Ito, Tomas, Mr.
Council Of Elders
SBM
Kidapawan, Cotatato
Cotatato, Philippines

Jesus, Edilberto C. De, Mr.
Associate Dean Of Research
ASIAN INSTITUTE OF MANAGEMENT
123 Paseo de Roxas, Makati, (MCC PO Box 898)
Metro Manila, Philippines
T: 63 832 87 40 11 F: 63 832 817924
TL: 63778aimpn

Jugatan, Ben, Mr.
LAKAS - AYTA PEOPLES ALLIANCE OF ZAMBALES
c/o Rainforst Information Centre, P. O. Box 368
Lismore - N.S.W. 2480, Philippines
T: 61 66 218505 F: 61 66 222339

King, Belen, Ms.
HARIBON FOUNDATION
Suite 901, Richbelt Towers, 17 Anapolis St, Greenhills, San Juan
Metro Manila, Philippines
T: 63 832 706611 F: 63 832 7226357

Laguerder, Jean C., Ms.
Asst. Vice President
PHILIPPINE RURAL RECONSTRUCTION MOVEMENT(PRRM)
Kayumanggi Press Bldg. , 940 Quezon Ave.
Quezon City, Philippines
T: 63 2 985563/995576/980079 F: 63 2 997919

Lesaco, Alfonso L., Sr.
ASIA-PACIFIC CULTURAL & ECONOMIC DEVELOPMENT FOUNDATION, INC.
Room 203, Erlene Building NO. 22 Sen, Gil J. Puyat Avenue
Makati, Philippines
T: 632 972341 F: 632 8179724

Lim-Perez, Adelaida, Ms.
CORDILLERA NEWS AGENCY
27 Yangco Road
Baguio, Philippines
T: 63 74 4427314

Lorenzo, Andres B., Sr.
ASIA-PACIFIC CULTURAL & ECONOMIC DEVELOPMENT FOUNDATION, INC.
Room 203, Erlene Building NO. 22 Sen, Gil J. Puyat Avenue
Makati, Philippines
T: 632 972341 F: 632 8179724

Lozano, Rommel V., Sr.
INTERNATIONAL CRUSADE FOR FILIPINO GREATNESS & SOCIAL PROGRESS, INC.
8 Free Press St. West Triangle
Quezon City, Philippines
T: 63 2 972341 F: 63 2 8179742
TL: 63199 ETPIMO PN

Lucas, Francis
PHILDHRRA
59-C Salvador St, Loyola Heights
Quezon City, Philippines
T: 63 2 887538 F: 63 2 961335

Mallari, Ricardo O., Sr.
INTERNATIONAL CRUSADE FOR FILIPINO GREATNESS & SOCIAL PROGRESS, INC.
8 Free Press St. West Triangle
Quezon City, Philippines
T: 63 2 972341 F: 63 2 8179742
TL: 63199 ETPIMO PN

Manalo, Orlando B., Sr.
INTERNATIONAL CRUSADE FOR FILIPINO GREATNESS & SCOCIAL PROGRESS
8 Free Press St. West Triangle
Quezon City, Philippines
T: 63 2 97-23-41 F: 63 2 81-79-42
TL: 63199 ETPIMO PN

Maza, Liza, Ms.
Secretary General
GABRIELA
35 Scont Delgado
Quezon City, Philippines
T: 63 2 99-8034 F: 63 2 922-9567

Melzatamuk, Sinabadam Ka Baqpbo, Ms.
General Secretary
ERA ESPANA
006 Plaridd St. Lud
Cotatato, Philippines

Mondonedo, Nora Linda, Ms.
CORDILLERA NEWS AGENCY
27 Yangco Road
Baguio, Philippines
T: 63 74 442-7314

Morales Jr., Horacio R., Mr.
President
PHILIPPINE RURAL RECONSTRUCTION MOVEMENT(PRRM)
Kayumanggi Press Bldg. , 940 Quezon Ave.
Quezon City, Philippines
T: 63 2 980079 F: 63 2 997919

Olegario, Arturo Jr, Mr.
CONGRESS FOR A PEOPLE'S AGRARIAN REFORM (CPAR)
71 Kalayaan Ave., Diliman
Quezon City, Philippines

Oliveros-Vistro, Teresita, Ms.
Deputy Sec.Gen.
AMIHAN-GABRIELA
113-E 13th Ave. Cubao
Quezon City, Philippines
T: 63 2 721-3941 F: 63 2 721-3941

Orpilla, Amado C., Mr.
Executive Director
INTERNATIONAL CRUSADE FOR FILIPINO GREATNESS & SOCIAL PROGRESS
Rm. 203 Erlene Bldg., No. 22 Sen. Gil J. Puyat Ave.
Makati, Philippines
T: 632 86-2193 F: 632 815-0756

Orpilla, Edwina R., Ms.
PHILIPPINE GREEN CAMPAIGN
Box 8721 Adc-Mia
Manila, Philippines
F: 63 2 817-9742

Orpilla, Gregorio C., Sr.
INTERNATIONAL CRUSADE FOR FILIPINO GREATNESS & SOCIAL PROGRESS, INC.
8 Free Press St. West Triangle
Quezon City, Philippines
T: 63 2 972341 F: 63 2 8179742
TL: 63199 ETPIMO PN

Orpilla, Marissa Zabala
Vice Pres Environ Studies
THIRD WORLD DEVELOPMENT STUDIES FOR BETTER ASIAN LIFE
Suite 403, PM Building, No 26 N. Domingo Street
Quezon City, Philippines
T: 63 2 707-561 F: 63 2 796-228

Orpilla, Virgilio R., Mr.
PHILIPPINE GREEN CAMPAIGN
Box 8721 Adc-Mia
Manila, Philippines
F: 63 2 817-9742

Paredes, Jim, Mr.
GREEN EARTH MOVEMENT
113 B. Gonzales St, Loyola Heights
Quezon City, Philippines
T: 63 2 997404 F: 63 2 997404

Perlas, Nicanor, Mr.
CENTER FOR ALTERNATIVE DEVELOPMENT INITIATIVE
110 Scout Rallos St.
Quezon City, Philippines
T: 63 2 993986 F: 63 2 997608

Punzealan, Marvin E., Sr.
INTERNATIONAL CRUSADE FOR FILIPINO GREATNESS & SOCIAL PROGRESS, INC.
8 Free Press St. West Triangle
Quezon City, Philippines
T: 63 2 972341 F: 63 2 8179742
TL: 63199 ETPIMO PN

Quiambao, Conrado C., Sr.
INTERNATIONAL CRUSADE FOR FILIPINO GREATNESS & SOCIAL PROGRESS, INC.
8 Free Press St. West Triangle
Quezon City, Philippines
T: 63 2 972341 F: 63 2 8179742
TL: 63199 ETPIMO PN

Ragragio, Concheta, Ms.
GREEN FORUM PHILIPPINES
3/F, Liberty Building, 835 Pasay Road, Makati
Metro Manila, Philippines
T: 63 832 85-11-10 F: 63 832 818-3207

Ramos, Lilia, Ms.
I S W / APPROTECH ASIA
Psdcenter,Cor.Real Magallanes Sts,Intramurus
Manila, Philippines
T: 63 2 47-99-18 F: 63 2 47-99-18
TL: 23313 NPCA PH

Ravanera, Roel, Mr.
ASIAN NGO COALITION
4/F, Matrinco Bldg, Pasong Tamo, Makati
Metro Manila, Philippines
T: 63 832 816-3033 F: 63 832 815-1198

Ribo, F. Roy, Mr.
PEASANT MOVEMENT OF THE PHILIPPINES- CPAR
Rm. 210 Kaimo Bldg.,101 Quezon Avenue Corner Sto. Domingo St.
Quezon City, Philippines
T: 63 2 7410032 F: 63 2 7118238

Riego, Gerardo G., Sr.
ASIA-PACIFIC CULTURAL & ECONOMIC DEVELOPMENT FOUNDATION, INC.
Room 203, Erlene Building NO. 22 Sen, Gil J. Puyat Avenue
Makati, Philippines
T: 632 972341 F: 632 8179724

Rosales, Gaudencio B., Mr.
Bishop
DIOCESE DE MALAYBALAY
Box 102 Bukidnon, Malaybay
Philippines, Philippines
F: 774212

Roxas, Sixto K., Mr.
FOUNDATION COMMUNITY ORGANIZATION AND MANAGEMENT TECHNOLOGY
Rm 204, Raman Condominium, Pasong Tamo, Makati
Metro Manila, Philippines
T: 63 832 818-1933 F: 63 832 816-4423

Sambajon, Rodolfo, Mr.
NACFAL NATIONWIDE COALITION OF FISHERFOLK
89-I west avenue
Quezon City, Philippines
T: 63 2 965198
TL: P.O.BOX 10223

Sanchez, Carlito A., Sr.
INTERNATIONAL CRUSADE FOR FILIPINO GREATNESS & SOCIAL PROGRESS, INC.
8 Free Press St. West Triangle
Quezon City, Philippines
T: 63 2 972341 F: 63 2 8179742
TL: 63199 ETPIMO PN

Santiago, Ramil B., Sr.
INTERNATIONAL CRUSADE FOR FILIPINO GREATNESS & SOCIAL PROGRESS, INC.
8 Free Press St. West Triangle
Quezon City, Philippines
T: 63 2 972341 F: 63 2 8179742
TL: 63199 ETPIMO PN

Santos, Roman G., Sr.
INTERNATIONAL CRUSADE FOR FILIPINO GREATNESS & SOCIAL PROGRESS, INC.
8 Free Press St. West Triangle
Quezon City, Philippines
T: 63 2 972341 F: 63 2 8179742
TL: 63199 ETPIMO PN

Serrano, Gilbert B., Sr.
INTERNATIONAL CRUSADE FOR FILIPINO GREATNESS & SOCIAL PROGRESS, INC.
8 Free Press St. West Triangle
Quezon City, Philippines
T: 63 2 972341 F: 63 2 8179742
TL: 63199 ETPIMO PN

Soliman, Corazon J., Mr.
CONGRESS FOR PEOPLES AGRARIAN REFORM
71 Kalayaan Ave.
Quezon City, Philippines
T: 63 2 922-2668 F: 63 2 922-2668

Soliman, Hector
HARIBON FOUNDATION
Suite 901, Richbelt Towers, 17 Annapolis St, Greenhills, San Juan
Metro Manila, Philippines
T: 63 832 706611 F: 63 832 722-3657

Sta Ana, Filomeno III, Mr.
Secretary-General
FREEDOM FROM DEBT COALITION
FDC, UP Solair Bldg., UP Campus Diliman
Quezon City 1101, Philippines
T: 63 2 976061-69 4808

Tanada, Rebecca Karen N., Ms.
Coordinator
WOMEN'S ACTION NETWORK FOR DEVELOPMENT
111-B Scout Lozano Street
Quezon City, Philippines
T: 63 2 984803 F: 63 2 984803

Tauli-Corpuz, Victoria, Ms.
CORDILLERA WOMEN'S EDUCATION AND RESOURCE CENTER
Palais de Nations
Geneve CH 1211, Philippines
F: 41 22 733-6542

Ticzon, Lucia Pavia, Ms.
Coordenation
ISIS INTERNATIONAL MANILA
85 A East Maya Phil Am Life Homs
Quezon City, Philippines
T: 63 2 993292 F: 63 2 997512

Valencia, Ariel S., Sr.
INTERNATIONAL CRUSADE FOR FILIPINO GREATNESS & SOCIAL PROGRESS, INC.
8 Free Press St. West Triangle
Quezon City, Philippines
T: 63 2 972341 F: 63 2 8179742
TL: 63199 ETPIMO PN

Velasquez, Aida
Sister
LINGKOD TAO KALIKASAN
2470 del Carmen St, Malate
Manila, Philippines
T: 63 2 59-11-85

Villanueva, Lourdes, Ms.
Treasurer/1st V.P.
NCSD/S.T.O.P.
2002 Jesus Pandacan
Manila, Philippines
T: 63 2 572830 F: 63 2 922-2860

Villaruel, Adrian L., Sr.
INTERNATIONAL CRUSADE FOR FILIPINO GREATNESS & SOCIAL PROGRESS, INC.
8 Free Press St. West Triangle
Quezon City, Philippines
T: 63 2 972341 F: 63 2 8179742
TL: 63199 ETPIMO PN

Villoria, Alice, Ms.
KALIKASAN PUBLICATIONS
LG-1 Cityland Bldg, Herrera cor Esteban Sts, Legaspi Village, Makati
Metro Manila, Philippines
T: 63 832 85-94-46 F: 63 832 817-7574

Walpole, Peter, Mr.
INSTITUTE ON CHURCH AND SOCIAL ISSUES
Box 102 Bukidnon
Malaybay, Philippines
F: 774212

ADVISORS

Benitez, Helena, Dr.
President, The Philippine Women'S University
Manila, Philippines
F: 632 522 4002/0143

WISDOMKEEPERS

Cortez, Magdaleno B.
OISCA International
23 Pukalaw St. Tierra
Dura Homes, Diliman Quezon City, Philippines
T: 8122520

Geenan, Jim
Plan Baguio
140 Navy Base Road
Baguio 5423, Philippines
T: 63-74-442-5423 F: 63-74-442-2091

POLAND - NGO

Van Der Marel, Ludmila
CAF
Ul. Jakuba Kubickiego 17/23
02-954 Warsaw, Poland

PRESS

Ejerman, Agnieszka
Reporter Radio/Television
POLISH RADIO
Bonifacego 83
02-945 Warsaw, Poland, Poland

Tryc-Ostrowska, Malgorzata
Correspondent
JOURNAL RZECZPOSPOLITA
Synow Putku 9M15
01-354 Varsovic, Poland, Poland

Walczak, Krzysztof
Correspondent
ZYCIE WARSZAWY
02-647 Warszawa
UL.Tagiqre'A 4/26 Warsaw, Poland, Poland

ORGANIZATIONS

Gerhardt, Andrzei, Mr.
Centre For Environment And Development
c/o Centre for Environmental Protection
5/11 Krucza Str., 00548 Warszava, Poland
T: (48 22) 213 670 F: (48 22) 295 263

Manczarski, Piotr, Mr.
International Environmental Service Centre
Ul. Czackiego 3/5
00 - 043 Warsaw, Poland
T: (48 22) 267 461 EXT.317 F: (48 22) 273 843

GLOBAL FORUM

Adam, Gula, Dr.
POLISH ECOLOGICAL CLUB
ul. jPoselska 20/15
Krakow, Poland
T: 48 22 273271 F: 48 22 273271

Czajkowski, Przemyslaw, Mr.
SOEM / THE CENTRE FOR ENVIRONMENT AND DEVELOPMENT / ANPED
Szara Str. K/34
Warsaw 00420, Poland
T: 48 22 296433 F: 48 22 296433

Gula, Adam, Dr.
POLISH ECOLOGICAL CLLUB
ul. Poselska 20/15
Krakow, Poland
T: 48 22 273271 F: 48 22 273271

Ostrarska, Antonia, Ms.
Professor
INSTITUTE OF PHILOSOPHY & SOCIOLOGY
Polish Academy of Science, Nawy Swiat 72
Warsaw P 00330, Poland

Piotrowska, Krystyna, Ms.
Professor
CENTRO DE ESTUDIOS LATINO AMERICANOS - CESLA , UNIVERSIDAD DE VARSOVIA
Dobra 56
Warsaw 00312, Poland
T: 48 22 266081 F: 48 22 267520
TL: 825439vwpl

Szymanczak, Michal, Mr.
National Program Director
POLSKA YMCA
Konopnickiej 6
Warsaw 00-590, Poland

PARLIAMENTARIANS / SPIRITUAL LEADERS

Rzymelha, Jan
MP
Sejm Wiejska 6/8 00 489
Warsaw, Poland
T: 287.001
TL: 812544 Sejm pl

PORTUGAL - GOVERNMENT

Cavaco, Silva Anibal, Mr.
Prime Minister
Permanent Mission of Portugal
I777 3rd. Ave., 27 Floor
New York NY 10017, United States of America
T: 212 759 9444 F: 212 355 1124

Mendes, Jose Casimiro
Head Of Division
INST. NAC. METEOROLOGIA
Rua C Do Aeroporto
Lisbon, Portugal
T: 357 1 8483961 F: 351 1 802370

Soares, Mario Nobre, Mr.
President
Permanent Mission of Portugal
777 3rd. Ave., 27 Floor
New York NY 10017, United States of America
T: 212 759 9444 F: 212 355 1124

NGO

Domingos, Jose
IDEA
Apartado 21.529
1134 Lisbon Codex, Portugal

Jardim, Agostinho
OIKOS
Ave. Visconde Valmor 35, 3.DT
1000 Lisboa, Portugal

Silva, Maria
QANCN
R. Trinidade Coelho, 1 C-4
P-4000 Porto, Portugal

Tavares, Luis
IDEA
Apartado 21.529
1134 Lisbon Codex, Portugal

PRESS

Aguiar, Aldina
Reporter Radio/Television
RTP
R. Prof.Eduardo Araujo
1600 Lisboa, Portugal, Portugal

Belisario De Souza, Joao Cesar
Correspondent
JORNAL DE NOTICIAS
Estrada Dona Castorina, 105
22460, Rio de Janeiro-RJ, Brazil, Portugal

Bosco Jardim, Joao
Correspondent
PUBLICO
Rua Eduardo Guinle 59, Apto706
22260 Botafogo, Rio de Janeiro Brazil, Portugal

Capinha, Antonio Julio Almeida
Reporter Radio/Television
RADIO COMERCIAL
R.Viveiro Lte.2
Estoril, Portugal, Portugal

Cardoso, Rui
Reporter Radio/Television
EXPRESSO DE PORTUGAL
R.Escolas Gerais 96
Lisboa, Portugal, Portugal

Castanheira, Henrique
Correspondent
CEREGRA EDITORA, LDA.
Rua Professor Sousa da Camara
Lisboa, Portugal, Portugal

Da Silva, Luiz Sergio F.
Correspondent
LUSA-AGENCIA DE INFORMATION
R. Timoteo da Costa, 445/603
22450-Rio de Janeiro-RJ, Brazil, Portugal

De Matos, Helena Ma. F.
Reporter Radio/Television
CORREIO DA MANHA RADIO
Pct. Bernardo Santareno, No.5
Drt. 2700, Amadora, Portugal, Portugal

De Salles Freaza, Iza
Correspondent
EXPRESSO
R. Jose Linhares 14/109
22430 Lesblon, Rio de Janeiro, Brazil, Portugal

Delgado, Jose
Editor
REVISTA SABADO
R.Americo Durao 2 Dpt.1900
Lisboa, Portugal, Portugal

Figueiredo, Mariza
Correspondent
JORNAL CORREIO DA MANHA
Br.Assuncao, Zona C Lt.7-ID
Cascais, Portugal, Portugal

Figueiredo, Sergio
Editor
EXPRESSO
Rua Antonio Galvao, No. 9
2780 Ogiras, Portugal, Portugal

Fragoso Pereira, Julio Cesar
Reporter Radio/Television
RDP
R. Dr. Manuel de Arriaga, 35
1 DTO, Queluz, Portugal, Portugal

Gasperini, Valeria
Correspondent
RADIO NOVA
Rua Correia de Faria 258
03461 Sao Paulo-SP, Brazil, Portugal

Granado, Antonio
Reporter Radio/Television
PUBLICO
Av.Colombano B.Pinheiro 70/ 3
Lisboa, Portugal, Portugal

Guerra, Paulo
Reporter Radio/Television
TSF RADIO JORNAL
R. Gones da Silva, 1 R/C DTO
100 Lisboa, Portugal, Portugal

Marinho, Antonio Luis
Editor
LUSA
R. da Junqueira, 390 1 esq.
1300 Lisboa, Portugal, Portugal

Medeiros, Lourenco
Reporter Radio/Television
RTP
R.D.Filipa de Vilalua 6
Lisboa, Portugal, Portugal

Mendes, Carlos Jose
Reporter Radio/Television
RADIODIFUSAO PORTUGUESA-E.P.
R. Crisostomo Figueiredo 10
Dto 1700, Lisboa, Portugal, Portugal

Moita, Carla
Reporter Radio/Television
RADIO RENASCENCA
Av.Seculo 56/I Dpt.
Samora Correia, Portugal, Portugal

Narciso, Carlos
Reporter Radio/Television
RADIOTELEVISAO PORTUGUESA
R. da Graca 60 - 1100
Lisboa, Portugal, Portugal

Neto, Mario Fernando
Reporter Radio/Television
CORREIO DA MANHA RADIO
R.Educardo Macedo 5/I Dept.
Antonio dos Cavaleiros, Portugal, Portugal

Panciarelli, Antonio
Editor
CEREGRA EDITORA, LDA.
Av.Columbano Bordado Pinheiro
59-6 D, 1000 Lisboa, Portugal, Portugal

Pimentel, Marina
Editor
RADIO RENASLENCA
Av. C. Barjona de Freitas, 20
1500 Lisbon, Portugal, Portugal

Pinheiro, Zelia
Editor
O INDEPENDIENTE
rua Almada Negreiros, Lote 460
Lisboa, Portugal, Portugal

Pinto, Vitor
Reporter Radio/Television
RADIO NOVA
Praceta Diogo Macedo 65
Vila Nova de Gaia, Portugal

Ramos Pitarma, Marina
Reporter Radio/Television
RTP
R.Ruy Telo 20
Almada, Portugal, Portugal

Redmont, M. Manuela P.
Correspondent
DIARIO DE NOTICIAS
Via Alessandro Poerio
59 Rome 00152, Italy, Portugal

Santos Moita, Carla Isabel
Editor
RADIO RENASCENCA
Rua Ivens, 14
1200 Lisboa, Portugal, Portugal

Serrinha, Domingos
Correspondent
ANTENA 1
Av.Amaralina 994/304
Salvador BASEP, Portugal, Portugal

Silva, Maria De Fatima M.
Editor
AGENCIA LUSA
R.Teixeira Aguillar L-19
Parede, Portugal, Portugal

Soares, Dina
Editor
JORNAL SEMANARIO
R. Dr. Silva Teles, 7
1000 Lisboa, Portugal, Portugal

Varela, Reinaldo Da Silva
Correspondent
RTP
R. Anita Garibaldi 29, #602
Copacabana, 22041 Rio de Janeiro, Brazil, Portugal

ORGANIZATIONS

Goncalves, A. Jardim, Mr.
Oikos - Cooperacao E Desenvolvimento
Ave. Visconde Valmor 35, 3.DT
1000 Lisboa, Portugal
T: (35 11) 76 4719 F: (35 11) 793 9791

GLOBAL FORUM

Afonso, Ana Alexandra Rosa Escaleira, Sra.
Socia
ASSOCIACAO PORTUGUESA DE ENGENHEIROS DO AMBIENTE
R. Cidade da Horta, 16-B, sala 22
Lisboa 1000, Portugal
T: 351 1 352-0305

Afonso, Paula Cristina Rodrigues, Sra.
Socia
ASSOCIACAO PORTUGUESA DE ENGENHEIROS DO AMBIENTE
R. Cidade da Horta, 16-B, sala 22
Lisboa 1000, Portugal
T: 351 1 352-0305

Agostinho, Jorge Manuel, Mr.
Vereador
ASSOC. MUN. OESTE
Av. Independencia Nacional, 8-1o.
Caldas da Rainha, Portugal
T: 844213 F: 831561

Andrade Silva, Maria Ines, Ms.
Guia Turismo
FUNDACAO DOM MANUEL II
Rua dos Duques de Braganca,10
Lisboa 1200, Portugal
T: 351 1 3423705 F: 351 1 3477537
TL: 13-655Ruy-vaz-p

Antunes, Manuel De Azevedo, Mr.
Representative
CONFEDERACAO PORTUGUESA
Olho de Boi
Almada 2800, Portugal
T: 351 1 274-5244

Avelino, Alberto, Mr.
Presidente
ASSOC. MUN. OESTE
Av. Independencia Nacional, 8-1o.
Caldas da Rainha, Portugal
T: 844213 F: 831561

Baptista, Viriato Augusto, Sr.
National Secretary
U.G.T. - PORTUGAL
Rua de Buenos Aires, 11
Lisboa 1200, Portugal

Barata, Jose Filipe Leal Dias Fialho, Sr.
Socio
ASSOCIACAO PORTUGUESA DE ENGENHEIROS DO AMBIENTE
R. Cidade da Horta, 16-B, sala 22
Lisboa 1000, Portugal
T: 351 1 352-0305

Barros, Cristina Maria Handem De, Ms.
Environment Sector Worker
SUL-ASSOCIACAO DE COOPERACAO PARA O DESENVOLVIMENTO
Rua Carlos Mardel 139 - 1o. esq
Lisboa 1900, Portugal
T: 351 1 8487482

Bastos, Miguel Dias De Souza Pestana, Sr.
QUERCUS
Rua Goncalves Zarco, 6f 6o.d
Lisboa 1400, Portugal
T: 351 1 301-3760

Benavente, Joao Francisco Gomes, Mr.
Presidente
ASSOC. MUN. OESTE
Av. Independencia Nacional, 8-1o.
Caldas da Rainha, Portugal
T: 844213 F: 831561

Bernardo, Paula Cristina Correia, Sra.
Socio
ASSOCIACAO PORTUGUESA DE ENGENHEIROS DO AMBIENTE
R. Cidade da Horta, 16-B, sala 22
Lisboa 1000, Portugal
T: 351 1 352-0305

Bogalho, Antonio Lopes, Mr.
Presidente
ASSOC. MUN. OESTE
Av. Independencia Nacional, 8-1o.
Caldas da Rainha, Portugal
T: 844213 F: 831561

Cachado, Jose Julio Teixeira, Mr.
Vereador
ASSOC. MUN. OESTE
Av. Independencia Nacional, 8-1o.
Caldas da Rainha, Portugal
T: 844213 F: 831561

Caeiro, Sandra Sofia F. Da Silva, Sra.
ASSOCIACAO PORTUGUESA DE ENGENHEIROS DO AMBIENTE
R. Cidade da Horta, 16-B, sala 22
Lisboa 1000, Portugal
T: 351 1 352-0305

Caetano, Jose Manuel, Sr.
President
C.P.C. / F.P.C.T
Av. Migyel Bombarda - 147 - 2a dt.
Lisboa 1000 LX, Portugal
T: 351 1 3156086 F: 351 1 3561253

Caixinhas, Raul Domingos, Sr.
Dr
DIRECAO GERAL DOS RECURSOS NATURAIS
Av. Alm. Gago Coutinho, 30
Lisboa 1000, Portugal
T: 351 1 8470080 F: 351 1 807726

Canoa, Antonio, Sr.
ASSOC. MUN. OESTE
Av. Independencia Nacional, 8-1o.
Caldas da Rainha, Portugal
T: 844213 F: 831561

Carraca, Emilia Maria Rocha Duque, Sra.
Socio
ASSOCIACAO PORTUGUESA DE ENGENHEIROS DO AMBIENTE
R. Cidade da Horta, 16-B, sala 22
Lisboa 1000, Portugal
T: 351 1 352-0305

Carvalho, Mario, Mr.
Presidente
ASSOC. MUN. OESTE
Av. Independencia Nacional, 8-1o.
Caldas da Rainha, Portugal
T: 844213

Castanha, Cristina, Ms.
GREEN BELT MOVEMENT OF KENYA
c/o Resource Renewal Institute, FT. Mason Center Bldg A
San Francisco - CA, Portugal
T: 1 415 928-32774 F: 1 415 928-6529

Castilho, Carlos
Correspondent
JORNAL PUBLICO
R Almicar Cabral LT 1
Lisboa, Portugal
T: 351 1 7599271 F: 351 1 7587373

Charneco, Isabel, Ms.
Lawyer
FEDERATION INTERNATIONALE DES FEMMES DES CARRIERES JURIDIQUE
R. Passos Manuel, 4 - 2 Dto.
Lisboa 1100, Portugal
T: 351 1 3524593 F: 351 1 3153750

Coimbra, Claudio Telmo Do Carmo Marque, Sr.
ASSOCIACAO PORTUGUESA DE ENGENHEIROS DO AMBIENTE
R. Cidade da Horta, 16-B, sala 22
Lisboa 1000, Portugal
T: 351 1 352-0305

Correia, Rui, Sr.
Presidente Comite Executivo
SUL - ASSOCIACAO DE COOPERACAO PARA O DESENVOLVIMENTO
Travessa Maria da Fonte, 5 - Esgueira
Aveiro 3800, Portugal
T: 351 34 315-292

Crisostomo, Nuno M.C., Sr.
Member
AGIR-AQUI/ A CHAMA
Rua Vila Catio 399 R/C Esq.
Lisboa 1800, Portugal
T: 351 1 851-6396 F: 351 1 848-0488

Custodio, Paula Cristina Alves, Sra.
ASSOCIACAO PORTUGUESA DE ENGENHEIROS DO AMBIENTE
R. Cidade da Horta, 16-B, sala 22
Lisboa 1000, Portugal
T: 351 1 352-0305

Dias, Ana Cristina Vasconcelos, Sra.
ASSOCIACAO PORTUGUESA DE ENGENHEIROS DO AMBIENTE
R. Cidade da Horta, 16-B, sala 22
Lisboa 1000, Portugal
T: 351 1 352-0305

Dias, Sofia Alexandra Sousa, Sra.
ASSOCIACAO PORTUGUESA DE ENGENHEIROS DO AMBIENTE
R. Cidade da Horta, 16-B, sala 22
Lisboa 1000, Portugal
T: 351 1 352-0305

Eloy, Antonio
Coordenador Suplemento Ambient
GAZETA DAS CALDAS
2500 Caldas da Rainha
Portugal
T: 351 1 3479599 F: 351 1 3479635

Fava, Sofia Mesquita, Sra.
ASSOCIACAO PORTUGUESA DE ENGENHEIROS DO AMBIENTE
R. Cidade da Horta, 16-B, sala 22
Lisboa 1000, Portugal
T: 351 1 352-0305

Fernandes, Eduardo De Oliveira, Sr.
DEPARTOMENTA ENG. MECANICA - UNIVERSIDADE DO PORTO
Rua dos Bragas 4099
Porto Codex, Portugal
T: 351 2 312-476 F: 351 2 319-280
TL: 27323 FEUP P

Ferreira, Eduardo Jose Rebelo, Mr.
Vereador
ASSOC. MUN. OESTE
Av. Independencia Nacional, 8-1o.
Caldas da Rainha, Portugal
T: 1 403 844213 F: 1 403 831561

Ferreira, Maria Teresa Lourenco, Ms.
Engineer
DIRECCAO GERAL DOS RECURSOS NATURAIS
Almirante Gago Coutinho, 30/16o. piso
Lisboa 1000, Portugal
T: 351 1 847-0080 F: 351 1 848-0933

Fibueiredo, Mariza
Journalist
JORNAL CORREIO DA MANHA
Rua Mouzinho da Silveira, 27
Lisbon, Portugal
T: 35 11 526553 F: 36 11 533726

Figueiredo, Armenio De, Mr.
Coordenador - Ambiente
CGTP - IN
Rua Victor Cordon, 1 - 2o.
Lisboa, Portugal
T: 351 1 3472181 F: 351 1 347 2189
TL: 16795

Figueiredo, Maria Manuel Reis De, Ms.
Engineer
COMISSAO DE COORDENACAO REGIONAL
R. Rainha D. Estefania, 251
Porto 4100, Portugal
F: 351 29 600.5401

Freitas, Fernando, Sr.
Vereador
ASSOC. MUN. OESTE
Av. Independencia Nacional, 8-1o.
Caldas da Rainha, Portugal
T: 844213 F: 831561

Geraldes, Joao Eduardo, Mr.
Assessor
CAMARA MUNICIPAL DE LISBOA
Praca do Municipio
Lisboa 1194 Lisboa Codex, Portugal
T: 351 1 347 6879 F: 351 1 346 7548

Godinho, Rui Manuel, Mr.
Vice Presidente
CAMARA MUNICIPAL DE LISBOA
Praca do Municipio
Lisboa 1194 Lisboa Codex, Portugal
T: 351 1 3479677 F: 351 1 346 7548

Goncalves, Agostinho Jardim, Sr.
Secret. Geral
OIKOS
AV. VISCONDE VALMOR 35 - 3o. DTo.
Lisboa 1500, Portugal
T: 351 1 793-9790 F: 351 1 793-9791

Goncalves, Maria Graca Pereira, Sra.
ASSOCIACAO PORTUGUESA DE ENGENHEIROS DO AMBIENTE
R. Cidade da Horta, 16-B, sala 22
Lisboa 1000, Portugal
T: 351 1 352-0305

Heine, Rita Luiza Da Silva Niza, Sra.
ASSOCIACAO PORTUGUESA DE ENGENHEIROS DO AMBIENTE
R. Cidade da Horta, 16-B, sala 22
Lisboa 1000, Portugal
T: 351 1 352-0305

Jardim, Joao Bosco
Correspondent
PUBLICO
Rua Amilcar Cabral, Lote 1
Lisbon, Portugal
T: 351 17599135 F: 351 17587373

Lopes, Ana Maria, Sra.
Socia
ASSOCIACAO PORTUGUESA DE ENGENHEIROS DO AMBIENTE
R. Cidade da Horta, 16-B, sala 22
Lisboa 1000, Portugal
T: 351 1 352-0305

Lopes, Rita Silva, Ms.
Engineer
DIRECCAO GERAL DOS RECURSOS NATURAIS
Almirante Gago Coutinho, 30/16o. piso
Lisboa 1000, Portugal
T: 351 1 847-0080 F: 351 1 848-0933

Martinho, Ana Paula, Sra.
ASSOCIACAO PORTUGUESA DE ENGENHEIROS DO AMBIENTE
R. Cidade da Horta, 16-B, sala 22
Lisboa 1000, Portugal
T: 351 1 352-0305

Martinho, Isabel, Sra.
NORTH-SOUTH CENTRE
Avenida da Liberdade, 299 - 4
Lisboa P-1200, Portugal
T: 351 1 52-2903 F: 351 1 53-1329

Martinho, Maria Da Graca Madeira, Sra.
President
ASSOCIACAO PORTUGUESA DE ENGENHEIROS DO AMBIENTE
R. Cidade da Horta, 16-B, sala 22
Lisboa 1000, Portugal
T: 351 1 352-0305

Martins, Zelia Cristina Vitorino, Sra.
Socia
ASSOCIACAO PORTUGUESA DE ENGENHEIROS DO AMBIENTE
R. Cidade da Horta, 16-B, sala 22
Lisboa 1000, Portugal
T: 351 1 352-0305

Marujo, Miguel., Mr.
General Secretary
ALL AFRICA CONFERENCE OF CHURCHES
PO Box 14205
Portugal

Meira, Rita, Sr.
Socio
ASSOCIACAO PORTUGUESA DE ENGENHEIROS DO AMBIENTE
R. Cidade da Horta, 16-B, sala 22
Lisboa 1000, Portugal
T: 351 1 352-0305

Monterroso, Luis, Mr.
ASSOC. MUN. OESTE
Av. Independencia Nacional, 8-1o.
Caldas da Rainha, Portugal
T: 844213 F: 831561

Moreira, Antonio, Sr.
ASSOC. MUN. OESTE
Av.. Independencia Nacional, 8-1o.
Caldas da Rainha, Portugal
T: 844213 F: 831561

Moura, Carlos, Sr.
Socio
ASSOCIACAO PORTUGUESA DE ENGENHEIROS DO AMBIENTE
R. Cidade da Horta, 16-B, sala 22
Lisboa 1000, Portugal
T: 351 1 352-0305

Moura, Evelina Brigite Pires De, Sra.
Socia
ASSOCIACAO PORTUGUESA DE ENGENHEIROS DO AMBIENTE
R. Cidade da Horta, 16-B, sala 22
Lisboa 1000, Portugal
T: 351 1 352-0305

Nunes, Luis Miguel A. E. Fernandes, Sr.
Socio
ASSOCIACAO PORTUGUESA DE ENGENHEIROS DO AMBIENTE
R. Cidade da Horta, 16-B, sala 22
Lisboa 1000, Portugal
T: 351 1 352-0305

Nunes, Pedro Da Silva Vicente, Sr.
Socio
ASSOCIACAO PORTUGUESA DE ENGENHEIROS DO AMBIENTE
R. Cidade da Horta, 16-B, sala 22
Lisboa 1000, Portugal
T: 351 1 352-0305

Oliveira, Jorge Manuelvassalo, Mr.
Presidente
ASSOC. MUN. OESTE
Av. Independencia Nacional, 8-1o.
Caldas da Rainha, Portugal
T: 844213 F: 831561

Oliveira, Luisa Maria Horta Fernanades, Sra.
Socio
ASSOCIACAO PORTUGUESA DE ENGENHEIROS DO AMBIENTE
R. Cidade da Horta, 16-B, sala 22
Lisboa 1000, Portugal
T: 351 1 352-0305

Oliveira, Manuel Alberto Barbosa De, Sr.
Secretario Nacional
U.G.T. - PORTUGAL
Rua de Buenos Aires, 11
Lisboa 1200, Portugal
T: 351 1 600247 F: 351 1 397-4612
TL: 15581

Partidario, Maria Rosario, Ms.
Engenheira
CEPGA
Rua dos Contrabandistas no. 19
Lisboa 1300, Portugal
T: 351 1 3953852 F: 351 1 3953852

Pedro, Alvaro, Sr.
Pres. Camara
ASSOC. MUM. OESTE
Av. Independencia Nacional, 8-1o.
Caldas da Rainha, Portugal
T: 844213

Peneda, Maria Constanca, Dr.
LNETI-DEII - LABORATORIO NAL. DE ENGENHARIA E TECNOLOGIA INDUSTRIAL
Azinhaga dos Lameiros a Estrada do Paco do Lumiar
Lisboa 1699, Portugal
T: 351 1 716-6141 F: 351 1 716-0901
TL: 424 86 LNETI P

Pereira, , Sr.
Presidente
ASSOC. MUN. OESTE
Av. Independencia Nacional, 8-1o.
Caldas da Rainha, Portugal
T: 844213 F: 831561

Pereira, Maria Miguel Henrique, Sra.
Socia
ASSOCIACAO PORTUGUESA DE ENGENHEIROS DO AMBIENTE
R. Cidade da Horta, 16-B, sala 22
Lisboa 1000, Portugal
T: 351 1 352-0305

Pinho, Joaquim Neves Correia De, Sr.
Medico Do Trabalho
Rua Oliveira Junior 248 3o. Dir
Sao Joao da Madeira 3700, Portugal
T: 351 56-26090

Pinto, Carla Isabel De Sousa, Sra.
Socia
ASSOCIACAO PORTUGUESA DE ENGENHEIROS DO AMBIENTE
R. Cidade da Horta, 16-B, sala 22
Lisboa 1000, Portugal
T: 351 1 352-0305

Pinto, Jose Geraldes, Sr.
Administrador
ASSOC. MUM. OESTE
Av. Independencia Nacional, 8-1o.
Caldas da Rainha, Portugal

Quadros, Silvia Alexandra Bettencourt, Sra.
Socia
ASSOCIACAO PORTUGUESA DE ENGENHEIROS DO AMBIENTE
R. Cidade da Horta, 16-B, sala 22
Lisboa 1000, Portugal
T: 351 1 352-0305

Ramos, Tomas Augusto Barros, Sr.
Socio
ASSOCIACAO PORTUGUESA DE ENGENHEIROS DO AMBIENTE
R. Cidade da Horta, 16-B, sala 22
Lisboa 1000, Portugal
T: 351 1 352-0305

Rodrigues, Valdemar Jose C. B., Sr.
Socio
ASSOCIACAO PORTUGUESA DE ENGENHEIROS DO AMBIENTE
R. Cidade da Horta, 16-B, sala 22
Lisboa 1000, Portugal
T: 351 1 352-0305

Rosmaninho, Luis, Mr.
COMITE COORDENADOR 'BRASIL 92'
Rua de S. Domingos A. Lapa 26
Lisboa 1200, Portugal
T: 351 1 346-4447 F: 351 1 396-4629

Salgueiro, Mario Andre E. De Oliveira, Sr.
ASSOCIACAO PORTUGUESA DE ENGENHEIROS DO AMBIENTE
R. Cidade da Horta, 16-B, sala 22
Lisboa 1000, Portugal
T: 351 1 352-0305

Santos, Antonio Manoel Pinheiro, Mr.
Vereador
ASSOC. MUN. OESTE
Av. Independencia Nacional, 8-1o.
Caldas da Rainha, Portugal
T: 1 403 844213 F: 1 403 831561

Santos, Manuel Antonio Araujo, Sr.
National Secretary
U.G.T. - PORTUGAL
Rua de Buenos Aires, 11
Lisboa 1200, Portugal
T: 351 1 600247 F: 351 1 3974612
TL: 15581

Santos, Maria
Deputada
GLOBE - GLOBAL LEGISLATION ORGANISATION FOR A BALANCED ENVIRONMENT
Largo Leam Maemet, No. 1 - 6
Lisboa, Portugal
T: 351 1 578031 F: 351 1 540004

Santos, Paula Cristina Ventura Dos Sa, Sra.
Socia
ASSOCIACAO PORTUGUESA DE ENGENHEIROS DO AMBIENTE
R. Cidade da Horta, 16-B, sala 22
Lisboa 1000, Portugal
T: 351 1 352-0305

Santos, Pedro Manuel Ribeiro Simoes D, Mr.
Executive Committee
SUL-ASSOCIACAO DE COOPERACAO PARA O DESENVOLVIMENTO
Av Defensores de Chaves 67 - 6o. esq
Lisboa 1000, Portugal
T: 351 1 763181

Santos, Susana Cristina Nunes C., Sra.
Socia
ASSOCIACAO PORTUGUESA DE ENGENHEIROS DO AMBIENTE
R. Cidade da Horta, 16-B, sala 22
Lisboa 1000, Portugal
T: 351 1 352-0305

Simas, Maria Joao Bittencourt, Sra.
Socia
ASSOCIACAO PORTUGUESA DE ENGENHEIROS DO AMBIENTE
R. Cidade da Horta, 16-B, sala 22
Lisboa 1000, Portugal
T: 351 1 352-0305

Sousa, Adilson De Oliveira
Assistant
RADIOTELEVISAO PORTUGUESA
Av. Soutvero 797
7000 Lisbon, Portugal
T: 55 21 7931774

Souza, Joao Cesar Bellario De
Correspondent
JORNAL DE NOTICIAS
Rua Goncalo Cristovao, 195 - 4052
Porto, Portugal
T: 2081331 F: 2006330

Tamagnini, Carla Isabel Marcelo Matos, Sra.
Socia
ASSOCIACAO PORTUGUESA DE ENGENHEIROS DO AMBIENTE
R. Cidade da Horta, 16-B, sala 22
Lisboa 1000, Portugal
T: 351 1 352-0305

Teixeira, Denise Maria Dos Santos, Mr.
UNIVERSIDADE DE AVEIRO
Aveiro, Portugal

Teixeira, Maria Margarida C. G. Ribau, Sra.
Socia
ASSOCIACAO PORTUGUESA DE ENGENHEIROS DO AMBIENTE
R. Cidade da Horta, 16-B, sala 22
Lisboa 1000, Portugal
T: 351 1 352-0305

Telhado, Ana Maria Martins, Ms.
Engenheira
DIRECAO GERAL DOS RECURSOS NATURAIS
Av. Alm. Gago Coutinho, 30
Lisboa 1000, Portugal
T: 351 1 8470080 F: 351 1 807726

Tuta, Carlos Alberto Santos, Mr.
Presidente
ASSOC. MUN. OESTE
Av. Independencia Nacional, 8-1o.
Caldas da Rainha, Portugal
T: 844213 F: 831561

Urbano, Ana Paula Da Silva, Sr.
Socia
ASSOCIACAO PORTUGUESA DE ENGENHEIROS DO AMBIENTE
R. Cidade da Horta, 16-B, sala 22
Lisboa 1000, Portugal
T: 351 1 352-0305

PARLIAMENTARIANS / SPIRITUAL LEADERS

Castro, Isabel, Sra.
MP
Assembly of the Republic Fargo das Cortes 1200
Lisbon, Portugal
T: 66 01 41
TL: 1304LARP

QATAR - GOVERNMENT

Abdul Mohsin Al-Mahmoud, Abdul Rahman Mahmoud
Head Of Maritime Adm.
DEPT. OF PORTS, MARITIME AFF.
PO Box 1645
Doha, Qatar
T: 974 833805 F: 974 871577

Al-Midfa, Abdul Aziz
General Secretary
ENVIRONMENT PROTECTION COMMITTEE
PO Box #3799
Doha, Qatar
T: 415246 F: 415246

REPUBLIC OF KOREA - GOVERNMENT

Chang, Byong Wan, Mr.
Director
Science Development Planning Division, Economic Planning Board
Permanent Mission to the United Nations, 866 United Nations Plaza, Suite 300 New York, NY, 10017
F: 212 371 1280
TL: 212 371 8873

Chang, Seung-Woo, Mr.
Director General
International Policy Coordination Office, Economic Planning Board
L Chungangdong, Kwachonsiz
Kyunggido, Republic of Korea
T: 82 2 503 3134

Choi, Min Hyu
Director/General,Forest Management
FORESTRY RESEARCH INSTITUTE
#207 Chungyangri,Seoul,Korea(South)
Seoul, Republic of Korea
T: 966 8968 F: 967 5101

Choi, Seok Young, Mr.
Deputy Director
Science and Environment Division, Ministry of Foreign Affairs
Permanent Mission to the United Nations, 866 United Nations Plaza, Suite 300 New York, NY, 10017
F: 212 371 1280
TL: 212 371 8873

Chung, Rae Kwon, Mr.
Director
Science and Environment Division, Ministry of Foreign Affairs
Permanent Mission to the United Nations, 866 United Nations Plaza, Suite 300 New York, NY, 10017
F: 212 371 1280
TL: 212 371 8873

Chung, Won Shik, H.E. Mr.
Prime Minister
Permanent Mission of the Republic of Korea
866 United Nations Plaza, Suite 300
New York NY 10017, United States of America
T: 212 371 1280 F: 212 371 8873

Cmol, Young Myong, Mr.
Senior Member
Presidential Commission on the 21-st Century
Permanent Mission to the United Nations, 866 United Nations Plaza, Suite 300 New York, NY, 10017
F: 212 371 1280
TL: 212 371 8873

Han, Chul Soo, H.E. Mr.
Ambassador Extraordinary and Plenipotentiary to Brazil
Permanent Mission of Republic of Korea
866 United Nations Plaza, Suite 300
New York, NY 10017, United States of America
T: 212 371 1280 F: 212 371 8873

Han, Moon Hi
Senior Research Fellow
GENETIC ENGINEERING RESEARCH INSTIT
P.O. Box 17, Daeduk Science Town Daejeon, Korea (Republic of)
Daejeon, Republic of Korea
T: 042 860 4120 F: 042 861 1759

Kim, Hyung Cheull, Mr.
Assistant Minister
Planning and Management Office, Ministry of Environment
Permanent Mission to the United Nations, 866 United Nations Plaza, Suite 300 New York, NY, 10017
F: 212 371 1280
TL: 212 371 8873

Kim, Ok Jo, Mr.
Secretary General to the Prime Minister
Permanent Mission of Republic of Korea
866 United Nations Plaza, Suite 300
New York, NY 10017, United States of America
T: 212 371 1280 F: 212 371 8873

Kim, Wan Joong, Mr.
Assistant Director
South America Division, Ministry of Foreign Affairs
Permanent Mission to the United Nations, 866 United Nations Plaza, Suite 300 New York, NY, 10017
F: 212 371 1280
TL: 212 371 8873

Kwon, E Hyock, H.E. Mr.
Minister for Environment
Permanent Mission of Republic of Korea
866 United Nations Plaza, Suite 300
New York, NY 10017, United States of America
T: 212 371 1280 F: 212 371 8873

Kwon, In Hyuck, H.E. Mr.
Ambassador for Environmental Affairs
Ministry of Foreign Affairs
Permanent Mission to the United Nations, 866 United Nations Plaza, Suite 300 New York, NY, 10017
F: 212 371 1280
TL: 212 371 8873

Lee, Jae Hoon, Mr.
Director
Multilateral Cooperation Division, Ministry of Commerce and Industry
Permanent Mission to the United Nations, 866 United Nations Plaza, Suite 300 New York, NY, 10017
F: 212 371 1280
TL: 212 371 8873

Lee, Keun Sik, Mr.
Senior Protocol Secretary to the Prime Minister
Permanent Mission of Republic of Korea
866 United Nations Plaza, Suite 300
New York, NY 10017, United States of America
T: 212 371 1280 F: 212 371 8873

Lee, Suk Jo, Mr.
Counsellor
Permanent Mission of Republic of Korea
Geneva
T: 10017
TL: 212 371 1280, E: 212 371 8873

Li, Tae Gun
Expert
P.O. Box 44
Pyongyang, Republic of Korea

Lim, Dai Yong, Mr.
Director
South America Division, Ministry of Foreign Affairs
Permanent Mission to the United Nations, 866 United Nations Plaza, Suite 300 New York, NY, 10017
F: 212 371 1280
TL: 212 371 8873

Lim, Soo Young, Mr.
Counsellor
Embassy to Brazil
Permanent Mission to the United Nations, 866 United Nations Plaza, Suite 300 New York, NY, 10017
F: 212 371 1280
TL: 212 371 8873

Min, Byung Hwan, Mr.
Assistant Protocol Secretary
Office of Prime Minister
Permanent Mission to the United Nations, 866 United Nations Plaza, Suite 300 New York, NY, 10017
F: 212 371 1280
TL: 212 371 8873

Nam, Jeong Pan, Mr.
Senior Press Secretary
Office of Prime Minister
Permanent Mission to the United Nations, 866 United Nations Plaza, Suite 300 New York, NY, 10017
F: 212 371 1280
TL: 212 371 8873

Noh, Boo Ho, Mr.
Director
Global Environment Division, Ministry of Environment
Permanent Mission to the United Nations, 866 United Nations Plaza, Suite 300 New York, NY, 10017
F: 212 371 1280
TL: 212 371 8873

Roe, Chang Hee, H.E. Mr.
Vice-Minister for Foreign Affairs
Permanent Mission of Republic of Korea
866 United Nations Plaza, Suite 300
New York, NY 10017, United States of America
T: 212 371 1280 F: 212 371 8873

Shin, Kook Ho, Mr.
Protocol Secretary
Office of Prime Minister
Permanent Mission to the United Nations, 866 United Nations Plaza, Suite 300 New York, NY, 10017
F: 212 371 1280
TL: 212 371 8873

Son, Jeong Soo, Mr.
Director
International Cooperation Division, Ministry of Agriculture, Forestry and Fisheries
#2 Gov'T Complex
Kwocheon, Republic of Korea
T: 82 2 503 7227 F: 82 2 503 7249

Yoon, Dae Soo, Mr.
Director
Technology Cooperation Division II, Ministry of Science and Technology
Permanent Mission to the United Nations, 866 United Nations Plaza, Suite 300 New York, NY, 10017
F: 212 371 1280
TL: 212 371 8873

Yu, Myung Hwan, Mr.
Deputy Director-General for American Affairs
Ministry of Foreign Affairs
Permanent Mission to the United Nations, 866 United Nations Plaza, Suite 300 New York, NY, 10017
F: 212 371 1280
TL: 212 371 8873

NGO

Chung, Sung Chul
CCEJ
25-1 Chongro S-GA
Chongro-Gu, Seoul, Korea 110-125, Republic of Korea

Lee, Kyong Jae
Director
KOREA ANTI-POLLUTION MOVEMENT ASSOCIATION
100 Chungsin-Kong Boduk B/D 4 Jongro-Ku
Seoul, Republic of Korea
T: 82 2 7444549 F: 82 2 7411240

Lee, Sijae
KAPMA
Choong Shan-Dong Boduk B/D 4
Jongro-ku, 110 490 Seoul, Republic of Korea

Yoo, Jae Hyun, Ms.
CITIZEN'S COALITION FOR ECONOMIC JUSTICE
25-1 Chogro 5 Ga. Chongro-ku 110-125
Seoul, Republic of Korea
T: 82 2 741-7961 F: 82 2 7458006

PRESS

Byun, Suk-Chan
Reporter Radio/Television
KBS
101, 58 Chulsan-dong, Woo Sung
Gwangmyung City, Gyungi-do, Korea, Republic of Korea

Byung-Chul, Min
Reporter Radio/Television
KBS
Sindonga Apt. 2-1502
Bangbae-Dong, Seocho-ku, Seoul, Korea, Republic of Korea

Chi-Jeong, Park
Correspondent
THE KWANGJU DAILY NEWSPAPER
194 Kakhwa-Dong
Kwangju, Korea, Republic of Korea

Cho, Sung Boo
Correspondent
YON HAP NEWS AGENCY
Maure 1560 2R
Argentina, Republic of Korea

Choi, Bo Gik
Correspondent
CHOSHUN ILBO
Bun-Dong Apt.112-1104
1-120 Bong Gu, Seoul Korea, Republic of Korea

Choi, Ham Il
Reporter Radio/Television
SEOUL B'CASTING SYSTEM
24P-3 Youn Ham Dong Mapo Gu
Seoul, Korea, Republic of Korea

Chun, Sang-Hoon
Managing Editor
THE ENVIRONMENT & POLLUTION DAILY
Seoul Building 4Fl #105-96 Gong 1, Mapo-Ku
Seul, Republic of Korea
T: 82 2 7064956 F: 82 2 7192068

Dong-Hun, Lee
Correspondent
NAEWAY ECONOMIC DAILY
Jookong Apt. 1902-1104
Chang-dong, Dobong-ku, Seoul, Korea, Republic of Korea

Hwang, In Sun
Correspondent
SEOUL JEONG-JE SHIU MOON
Bun Dong Apt. 112-1310
Bun Dong Du Bong Ku, Seoul, Republic of Korea

Jeon, Jae Chol
Correspondent
MBC TV
1463 John Street
Fort Lee, NJ 07024, Republic of Korea

Kim, Cha-Ung
Correspondent
DONG-A ILBO
13 Arthur Ct.
Closter, NJ 07624, USA, Republic of Korea

Kim, In-Ki
Reporter Radio/Television
SEOUL B'CASTING SYSTEM
755 Yoksam-Dong Kangnam Gu
Seoul, Korea, Republic of Korea

Kim, Seung Young
Correspondent
CHOSHUN ILBO
23 Princeton Street
Closter, NJ 07624, USA, Republic of Korea

Kim, Soo-Jong
Correspondent
HANKOOK ILBO & THE KOREA TIMES
863 Harristown Rd.
Glen Rock, NJ 07452, USA, Republic of Korea

Koo, Heui-Cheon
Correspondent
KOOKMIN ILBO
Sinahn Apt. 1-306,Bangwha-dong
Kangsu-ku, Seoul, Korea, Republic of Korea

Lee, Daewoo
Correspondent
MBC TV
213 N. Pleasant Avenue
Ridgewood, NJ 07450, Republic of Korea

Lee, Woo Ho
Reporter
MBC - TV
31 Yoido dong
Seoul, Republic of Korea
T: 82 2 7812571 F: 82 2 7824144

Lim, Choon Woong
Correspondent
SEOUL SHINMUN DAILY NEWS
465 5th Street, #2
Palisade Park, NJ 07650, USA, Republic of Korea

Oh, Myung Ock C.
Correspondent
ENVIRONMENT&POLLUTION DAILY
#206-1401 Hyundai Apt.
Apkujung-dong, Kongnan-Ku, Seoul, Korea, Republic of Korea

Park, Joon
Correspondent
JOONG-ANG DAILY NEWS
713 Hillcrest Road
Washington Twp., NJ 07675, USA, Republic of Korea

Sum-Ho, Jeoung
Reporter Radio/Television
SEOUL B'CASTING SYSTEM
Yoongdungpo-Gu 82 H2
Seoul, Korea, Republic of Korea

Yi, John U.
Correspondent
MBC TV
92 Waterside Drive
Little Ferry, NJ 07643, Republic of Korea

ORGANIZATIONS

Kyung-Suk-So, , Rev.
Citizen's Coalition For Economic Justice
25-1 Chongro S-GA
Chongro-Gu, Seoul, Korea 110-125, Republic of Korea
T: (82 2) 741 7961 F: (82 2) 745 8006

GLOBAL FORUM

Ahn, Jung Son
Deputy Of Women'S Committee
KOREA ANTI-POLLUTION MOVEMENT ASSOCIATION
100 Chungsin-Kong Boduk B/D 4 Jongro-Ku
Seoul, Republic of Korea
T: 001 82 2 7444549 F: 001 82 2 7411240

Ahn, Key Hee, Mr.
THE DEMOCRATIC LIBERAL
#1202-1403, Kaepotong, Kang Nangu
Seoul, Republic of Korea
T: 82 2 574 4522

Bai, Soon Ja, Ms.
Secty. Board Of Trustees
NATIONAL YWCA OF KOREA
1-3 First Street Myung Dong
Seoul, Republic of Korea
T: 82 2 774-97027 F: 82 2 774-9724

Cha, Chul-Whan, Mr.
Dean Of Public Housing
COLLEGE OF MEDICINE KOREA UNIVERSITY
496-5 Pyungchang-dong, Chongro-gu Seoul Medical Technology
Seoul, Republic of Korea
T: 82 2 927-7224

Chae-Shik, Rho, Mr.
Vice-Chaiman
KOREAN ENVIRONMENTAL PRESERVATION ASSOCIATION
#45, 4-ka Namdaemun-ro, Chung-ku
Seoul, Republic of Korea
T: 82 2 042-868-2604 F: 82 2 042-861-1700

Chang, Won Suk, Mr.
CITIZEN'S COALITION FOR ECONOMIC JUSTICE
25-1 Chongro 5 Ga. Chongro-ku 110-125
Seoul, Republic of Korea
T: 82 2 741-7961 F: 82 2 745-8006

Chang, Woo-Zoo
KOREA ANTI-POLLUTION MOVEMENT ASSOCIATION
100 Chungsin-Dong Boduk B/D 4
Seoul Jongro Ku, Republic of Korea
T: 82 2 744 4549 F: 82 2 741 1240

Chasang, Ku, Mr.
Regional Director
NGO'S COMMITTEE FOR UNCED KOREA
Chungsin 100 Bobug Bd, 4th Jongro
Seoul 110-490, Republic of Korea
T: 82 2 741 2118 F: 82 2 741 1240

Cho, Yun-Syng, Mr.
President
THE KOREAN SOCIETY FOR SANITATION
280 - 17 Pulkwang-dong, bunpyung-Gu
Seoul, Republic of Korea

Choe, Buong Mo
Law Consultant
KOREA ANTI-POLLUTION MOVEMENT ASSOCIATION
100 Chungsin-Kong Boduk B/D 4 Jongro-Ku
Seoul, Republic of Korea
T: 001 82 2 7444549 F: 001 82 2 7411240

Choi, Byung Soo, Mr.
Artist
KOREA ANTI-POLLUTION MOVEMENT ASSOCIATION
Cung Sin Dong 100 Budug Bd 4th Pl, Jongro Ku
Seoul 110400, Republic of Korea
T: 82 2 744-4549 F: 82 2 741-1240

Choi, Byung-Mo, Mr.
Law Consultant
NGO'S COMMITTEE FOR UNCED KOREA
Chungsin 100 Bobug Bd, 4th Jongro
Seoul 110-490, Republic of Korea
T: 82 2 741 2118 F: 82 2 741 1240

Choi, Do Young
Advisor
KOREA ANTI-POLLUTION MOVEMENT ASSOCIATION
100 Chungsin-Kong Boduk B/D 4 Jongro-Ku
Seoul, Republic of Korea
T: 001 82 2 7444549 F: 001 82 2 7411240

Choi, Ye Yong
Finance Committee
KOREA ANTI-POLLUTION MOVEMENT ASSOCIATION
100 Chungsin-Kong Boduk B/D 4 Jongro-Ku
Seoul, Republic of Korea
T: 001 82 2 7444549 F: 001 82 2 7411240

Choi, Yul
Chairman
KOREA ANTI-POLLUTION MOVEMENT ASSOCIATION
100 Chungsin-Kong Boduk B/D 4 Jongro-Ku
Seoul, Republic of Korea
T: 001 82 2 7444549 F: 001 82 2 7411240

Chongho, Ha, Mr.
Regional Director
NGO'S COMMITTEE FOR UNCED KOREA
Chungsin 100 Bobug Bd, 4th Jongro
Seoul 110-490, Republic of Korea
T: 82 2 741 2118 F: 82 2 741 1240

Chul-Kyu, Kang, Mr.
CITIZEN'S COALITION FOR ECONOMIC JUSTICE
25-1 Chongro 5 Ga, Chongro-ku 110-125
Seoul, Republic of Korea
T: 82 2 714-7961 F: 82 2 745-8006

Chun, Oh Chung-Ock
Journalist
ENVIRONMENT & POLLUTION DAILY
Seoul Building 4Fl #105-96 Gong 1, Mapo-Ku
Seoul, Republic of Korea
T: 82 2 7064956 F: 82 2 7192068

Chun,Oh, Myung-Ock
Journalist
THE ENVIRONMENT & POLLUTION DAILY
Seoul Building 4Fl #105-96 Gong 1, Mapo-Ku
Seul, Republic of Korea
T: 82 2 7064956 F: 82 2 7192068

Chungryang, Kang, Mr.
Regional Director
NGO'S COMMITTEE FOR UNCED KOREA
Chungsin 100 Bobug Bd, 4th Jongro
Seoul 110-490, Republic of Korea
T: 82 2 741 2118 F: 82 2 741 1240

Gilseung, Yang, Mr.
Personal Adviser
NGO'S COMMITTEE FOR UNCED KOREA
Chungsin 100 Bobug Bd, 4th Jongro
Seoul 110-490, Republic of Korea
T: 82 2 741 2118 F: 82 2 741 1240

Ha, Jong Ho
Regional Director
KOREA ANTI-POLLUTION MOVEMENT ASSOCIATION
100 Chungsin-Kong Boduk B/D 4 Jongro-Ku
Seoul, Republic of Korea
T: 001 82 2 7444549 F: 001 82 2 7411240

Hacksun, Lee, Sr.
Professional Adisor
NGO'S COMMITTEE FOR UNCED KOREA
Chungsin 100 Bobug Bd, 4th Jongro
Seoul 110-490, Republic of Korea
T: 82 2 741 2118 F: 82 2 741 1240

Ham, Saivet Ho
Cameraman
MBC - TV
31 Yoido dong
Seoul, Republic of Korea
T: 82 2 7812571 F: 82 2 7824144

Han, Gleyang, Mr.
KOREA ANTI-POLLUTION MOVEMENT ASSOCIATION
100 Chungsin-Dong Boduk B/D 4
Seoul Jongro Ku, Republic of Korea
T: 82 2 744 4549 F: 82 2 741 1240

Han, Myung Sook
Director
KOREA ANTI-POLLUTION MOVEMENT ASSOCIATION
100 Chungsin-Kong Boduk B/D 4 Jongro-Ku
Seoul, Republic of Korea
T: 001 82 2 7444549 F: 001 82 2 7411240

Heo, Hoon Soon
Agricultural Coordinator
KOREA ANTI-POLLUTION MOVEMENT ASSOCIATION
100 Chungsin-Kong Boduk B/D 4 Jongro-Ku
Seoul, Republic of Korea
T: 001 82 2 7444549 F: 001 82 2 7411240

Heulgyun, Choi, Mr.
Professional Advisor
NGO'S COMMITTEE FOR UNCED KOREA
Chungsin 100 Bobug Bd, 4th Jongro
Seoul 110-490, Republic of Korea
T: 82 2 741 2118 F: 82 2 741 1240

Hoon, Oh, Mr.
Professional Advisor
NGO'S COMMITTEE FOR UNCED KOREA
Chungsin 100 Bobug Bd, 4th Jongro
Seoul 110-490, Republic of Korea
T: 82 2 741 2118 F: 82 2 741 1240

Im, Nack Pyeong
Regional Director
KOREA ANTI-POLLUTION MOVEMENT ASSOCIATION
100 Chungsin-Kong Boduk B/D 4 Jongro-Ku
Seoul, Republic of Korea
T: 001 82 2 7444549 F: 001 82 2 7411240

Janet, Won, Mr.
Director
BAEDAL ENVIRONMENTAL INSTITUTE
102-2 Jungchon-Dong, Jung-Gy
Taejon - South Korea, Republic of Korea
T: 042 257 6853 F: 042 257 6854

Jin, Hwang, Mr.
Regional Director
NGO'S COMMITTEE FOR UNCED KOREA
Chungsin 100 Bobug Bd, 4th Jongro
Seoul 110-490, Republic of Korea
T: 82 2 741 2118 F: 82 2 741 1240

Jong-Sung, Yoo, Mr.
CITIZEN'S COALITION FOR ECONOMIC JUSTICE
25-1 Chogro 5 Ga. Chongro-ku 110-125
Seoul, Republic of Korea
T: 82 2 741-7961 F: 82 2 7458006

Joo, Kwang Jin
Director
KOREA ANTI-POLLUTION MOVEMENT ASSOCIATION
100 Chungsin-Kong Boduk B/D 4 Jongro-Ku
Seoul, Republic of Korea
T: 001 82 2 7444549 F: 001 82 2 7411240

Jung-Sik, Kim, Mr.
Manager
TECHNOLOGY DEVELOPMENT DEPARTMENT/KOREA GENERAL CHEMICAL CORPORATION
80 Susong-Dong. Chongro-Gu
Seoul, Republic of Korea
T: 82 2 723-0524 F: 82 2 735-5605

Kang, Moon Kyu
Director Of YWCA
KOREA ANTI-POLLUTION MOVEMENT ASSOCIATION
100 Chungsin-Kong Boduk B/D 4 Jongro-Ku
Seoul, Republic of Korea
T: 001 82 2 7444549 F: 001 82 2 7411240

Kim, Dong Young
Researcher
KOREA ANTI-POLLUTION MOVEMENT ASSOCIATION
100 Chungsin-Kong Boduk B/D 4 Jongro-Ku
Seoul, Republic of Korea
T: 001 82 2 7444549 F: 001 82 2 7411240

Kim, Heung-Sik, Mr.
ASSOCIATION OF CHUNG-NAM ANTI-POLLUTION MOVEMENT
102-3 Jeong-RI Jo-Chi-Won-UP
Yeon-Ki-Kun, Republic of Korea
T: 415 865-0416 F: 865-0416

Kim, Hyun Ja, Ms.
Board Member
NATIONAL YWCA OF KOREA
Yuwen Apt. 202-1109, tangsan Dong 5Ka, Yungdung Po Ku
Seoul, Republic of Korea
T: 82 2 333-2771 F: 82 2 338-1983

Kim, Jung Wk
Advisor
KOREA GENERAL CHEMICAL CORPORATION
100 Chungsin-Kong Boduk B/D 4 Jongro-Ku
Seoul, Republic of Korea
T: 001 82 2 7444549 F: 001 82 2 7411240

Kim, Kyeong Ah
Researcher
KOREA ANTI-POLLUTION MOVEMENT ASSOCIATION
100 Chungsin-Kong Boduk B/D 4 Jongro-Ku
Seoul, Republic of Korea
T: 001 82 2 7444549 F: 001 82 2 7411240

Kim, Mikyong
Cultural Dept.
KOREA ANTI-POLLUTION MOVEMENT ASSOCIATION
100 Chungsin-Kong Boduk B/D 4 Jongro-Ku
Seoul, Republic of Korea
T: 001 82 2 7444549 F: 001 82 2 7411240

Kim, Misook
KOREA ANTI-POLLUTION MOVEMENT ASSOCIATION
100 Chungsin-Dong Boduk B/D 4
Seoul Jongro Ku, Republic of Korea
T: 82 2 744 4549 F: 82 2 741 1240

Kim, Myung Suk
Finance
KOREA ANTI-POLLUTION MOVEMENT ASSOCIATION
100 Chungsin-Kong Boduk B/D 4 Jongro-Ku
Seoul, Republic of Korea
T: 001 82 2 7444549 F: 001 82 2 7411240

Kim, Sang Jong, Mr.
Researcher
KOREA ANTI-POLLUTION MOVEMENT ASSOCIATION
Cung Sin Dong 100 Budug Bd 4th Pl, Jongro Ku
Seoul 110400, Republic of Korea
T: 82 2 744-4549 F: 82 2 741-1240

Kim, Soeg Chang
Regional Director
KOREA ANTI-POLLUTION MOVEMENT ASSOCIATION
100 Chungsin-Kong Boduk B/D 4 Jongro-Ku
Seoul, Republic of Korea
T: 001 82 2 7444549 F: 001 82 2 7411240

Kim, Sook He, Ms.
President
NATIONAL YWCA OF KOREA
1-3 First Street Myung Dong, Choong Ku
Seoul, Republic of Korea
T: 82 2 774-9702 F: 82 2 774-9724

Kim, Sung-Soo, Mr.
NATIONAL COUNCIL OF YMCA'S OF KOREA
C.P.O. Box 1056
Seoul, Republic of Korea

Kim, Young Man
Cultural Dept.
KOREA ANTI-POLLUTION MOVEMENT ASSOCIATION
100 Chungsin-Kong Boduk B/D 4 Jongro-Ku
Seoul, Republic of Korea
T: 001 82 2 7444549 F: 001 82 2 7411240

Koh, Chul Hwan, Mr.
Leader Of Research Commt.
KOREA ANTI-POLLUTION MOVEMENT ASSOCIATION
Cung Sin Dong 100 Budug Bd 4th Pl, Jongro Ku
Seoul 110400, Republic of Korea
T: 82 2 744-4549 F: 82 2 741-1240

Koo, Ja Song
Professional Advisor
KOREA ANTI-POLLUTION MOVEMENT ASSOCIATION
100 Chungsin-Kong Boduk B/D 4 Jongro-Ku
Seoul, Republic of Korea
T: 001 82 2 7444549 F: 001 82 2 7411240

Kwon, Heonyl
Activist
KOREA ANTI-POLLUTION MOVEMENT ASSOCIATION
Cung Sin Dong 100 Budug Bd 4th Pl, Jongro Ku
Seoul 110400, Republic of Korea
T: 82 2 744-4549 F: 82 2 741-1240

Kwon, Sook-Pyo, Mr.
Attending Professor
COLLEGE OF MEDICINE YONSEI UNIVERSITY
300-77 Echon-dong Yongsan-gu
Seoul, Republic of Korea
T: 82 2 495-5578

Lee, Chul
Professional Advisor
KOREA ANTI-POLLUTION MOVEMENT ASSOCIATION
100 Chungsin-Kong Boduk B/D 4 Jongro-Ku
Seoul, Republic of Korea
T: 001 82 2 7444549 F: 001 82 2 7411240

Lee, Duck Hee
Deputy Of Research Commt.
KOREA ANTI-POLLUTION MOVEMENT ASSOCIATION
100 Chungsin-Kong Boduk B/D 4 Jongro-Ku
Seoul, Republic of Korea
T: 82 2 7444549 F: 82 2 7411240

Lee, Duk Seung, Mr.
Secretary
SEOUL YMCA IN LOREA
9TH 2ND St Jong-Ro
Seoul, Republic of Korea
T: 82 2 732-2941 F: 82 2 734-8003

Lee, Hak Sun
Professional Advisor
KOREA ANTI-POLLUTION MOVEMENT ASSOCIATION
100 Chungsin-Kong Boduk B/D 4 Jongro-Ku
Seoul, Republic of Korea
T: 001 82 2 7444549 F: 001 82 2 7411240

Lee, Hye Sook, Ms.
International Coordinator
CITIZEN'S COALITION FOR ECONOMIC JUSTICE
25-1 Chongro 5 Ga, Chong ro-ku 110-125
Seoul, Republic of Korea
T: 82 2 741-7961 F: 82 2 745-8006

Lee, In Sig
Kan Regional Director
KOREA ANTI-POLLUTION MOVEMENT ASSOCIATION
100 Chungsin-Kong Boduk B/D 4 Jongro-Ku
Seoul, Republic of Korea
T: 001 82 2 7444549 F: 001 82 2 7411240

Lee, Jin Ah, Ms.
CITIZEN'S COALITION FOR ECONOMIC JUSTICE
25-1 Chogro 5 Ga. Chongro-ku 110-125
Seoul, Republic of Korea
T: 82 2 741-7961 F: 82 2 7458006

Lee, Min
Cameraman
MBC - TV
31 Yoido dong
Seoul, Republic of Korea
T: 82 2 7812571 F: 82 2 7824144

Lee, See Jae
Director Of Vocational Commt.
KOREA ANTI-POLLUTION MOVEMENT ASSOCIATION
100 Chungsin-Kong Boduk B/D 4 Jongro-Ku
Seoul, Republic of Korea
T: 001 82 2 7444549 F: 001 82 2 7411240

Lee, Seog Tae
Law Consultant
KOREA ANTI-POLLUTION MOVEMENT ASSOCIATION
100 Chungsin-Kong Boduk B/D 4 Jongro-Ku
Seoul, Republic of Korea
T: 001 82 2 7444549 F: 001 82 2 7411240

Lee, Seon, Mr.
Professor
KYUNG-HEE UNIVERSITY
51, Heoki-Dong, Dongdaemoon-ku
Seoul, Republic of Korea
T: 82 2 9610482 F: 82 2 9665925

Lee, Seung Hoon, Mr.
CITIZEN'S COALITION FOR ECONOMIC JUSTICE
25-1 Chogro 5 Ga. Chongro-ku 110-125
Seoul, Republic of Korea
T: 82 2 741-7961 F: 82 2 7458006

Lee, Seung Sil
Editorial Dept.
KOREA ANTI-POLLUTION MOVEMENT ASSOCIATION
100 Chungsin-Kong Boduk B/D 4 Jongro-Ku
Seoul, Republic of Korea
T: 001 82 2 7444549 F: 001 82 2 7411240

Lee, Sug-Tae, Mr.
Law Consultant
NGO'S COMMITTEE FOR UNCED KOREA
Chungsin 100 Bobug Bd, 4th Jongro
Seoul 110-490, Republic of Korea
T: 82 2 741 2118 F: 82 2 741 1240

Lee, Sung-Moo, Mr.
Professor
YONSEI UNIVERSITY
#3-401, Shindong A Apto Yong-san-ku,
Seoul, Republic of Korea
T: 82 2 798 3915

Lee, Yong Bae
Cultural Dept.
KOREA ANTI-POLLUTION MOVEMENT ASSOCIATION
100 Chungsin-Kong Boduk B/D 4 Jongro-Ku
Seoul, Republic of Korea
T: 001 82 2 7444549 F: 001 82 2 7411240

Lim, Jin Teak
Director
KOREA ANTI-POLLUTION MOVEMENT ASSOCIATION
100 Chungsin-Kong Boduk B/D 4 Jongro-Ku
Seoul, Republic of Korea
T: 001 82 2 7444549 F: 001 82 2 7411240

Min, Byung-Chul, Mr.
Reporter
KRB (KOREAN BROADCASTING SYSTEM)
Science Nrews Div., News Department, KRB, #18 Yoido-dong
Seoul, Republic of Korea
T: 82 2 781-4183 F: 82 2 781-4183

Min, Tae Ik
Professional Advisor
KOREA ANTI-POLLUTION MOVEMENT ASSOCIATION
100 Chungsin-Kong Boduk B/D 4 Jongro-Ku
Seoul, Republic of Korea
T: 001 82 2 7444549 F: 001 82 2 7411240

Moon, Young Mee, Ms.
International Coordinator
CITIZEN'S COALITION FOR ECONOMIC JUSTICE
25-1 Chongro 5 Ga, Chong ro-ku 110-125
Seoul, Republic of Korea
T: 82 2 741-7961 F: 82 2 745-8006

Myung, Ock Chun Oh, Ms.
Journalist
ENVIRONMENT AND POLLUTION DAILY
Seoul Building 4 Fl, # 105-96, Gongdok1 -Dong, Mapo-Ku
Seoul, Republic of Korea
T: 82 2 706-4954-7 F: 82 2 719-2068
TL: kitsel k29271-2

Nackpyung, Im, Mr.
Regional Director
NGO'S COMMITTEE FOR UNCED KOREA
Chungsin 100 Bobug Bd, 4th Jongro
Seoul 110-490, Republic of Korea
T: 82 2 741 2118 F: 82 2 741 1240

Oh, Heung San
Director
KOREA ANTI-POLLUTION MOVEMENT ASSOCIATION
100 Chungsin-Kong Boduk B/D 4 Jongro-Ku
Seoul, Republic of Korea
T: 001 82 2 7444549 F: 001 82 2 7411240

Ohm, Dae Woo
Regional Director
KOREA ANTI-POLLUTION MOVEMENT ASSOCIATION
100 Chungsin-Kong Boduk B/D 4 Jongro-Ku
Seoul, Republic of Korea
T: 001 82 2 7444549 F: 001 82 2 7411240

Park, Chang Bin
Guidance Committee
KOREA ANTI-POLLUTION MOVEMENT ASSOCIATION
100 Chungsin-Kong Boduk B/D 4 Jongro-Ku
Seoul, Republic of Korea
T: 001 82 2 7444549 F: 001 82 2 7411240

Park, Chung Hee, Ms.
President
NATIONAL YWCA OF KOREA
1-1 First St. Myung-dong, Chung-ku, Korea
Seoul, Republic of Korea
T: 82 2 779-7567 F: 82 2 774-8984

Park, Hee Jung
Cultural Dept.
KOREA ANTI-POLLUTION MOVEMENT ASSOCIATION
100 Chungsin-Kong Boduk B/D 4 Jongro-Ku
Seoul, Republic of Korea
T: 001 82 2 7444549 F: 001 82 2 7411240

Park, Hye Sook
Cultural Dept.
KOREA ANTI-POLLUTION MOVEMENT ASSOCIATION
100 Chungsin-Kong Boduk B/D 4 Jongro-Ku
Seoul, Republic of Korea
T: 001 82 2 7444549 F: 001 82 2 7411240

Park, Sang Chul
Regional Staff
KOREA ANTI-POLLUTION MOVEMENT ASSOCIATION
100 Chungsin-Kong Boduk B/D 4 Jongro-Ku
Seoul, Republic of Korea
T: 001 82 2 7444549 F: 001 82 2 7411240

Park, Yong Il
Law Consultant
KOREA ANTI-POLLUTION MOVEMENT ASSOCIATION
100 Chungsin-Kong Boduk B/D 4 Jongro-Ku
Seoul, Republic of Korea
T: 001 82 2 7444549 F: 001 82 2 7411240

Park, Young Sook
Representative
KOREA ANTI-POLLUTION MOVEMENT ASSOCIATION
100 Chungsin-Kong Boduk B/D 4 Jongro-Ku
Seoul, Republic of Korea
T: 001 82 2 7444549 F: 001 82 2 7411240

Rhee, Jin Yong, Ms.
Director
NATIONAL YWCA OF KOREA
1-3 First St. Myung Dong
Seoul, Republic of Korea
T: 82 2 774-9702 F: 82 2 774-9724

Rho, Young-Sook, Ms.
San 56-1 Sinlim-Dong Kwanag-KU
Seoul, Republic of Korea
T: 82 2 880-5650 F: 82 2 773-7092

Rho, Yung-Hee, Mr.
Professor
GRADUATE SCHOOL OF ENVIRONMENT - SEOUL NATIONAL UNIVERSITY
San 56-1 Sinlim-dong Kwanag-ku
Seoul, Republic of Korea
T: 82 2 880-5650 F: 82 2 773-7092

Ryu, Jae-Sung, Mr.
Environment Manager
DOOSAN GROUP OF COMPANIES
Chung-gu Apt. 7-206 Hagae-dong, 133BI Nowon-gu
Seoul, Republic of Korea
T: 82 2 759-6950 F: 82 2 753-0563

Sang, Hoon Chun, Mr.
Managing Editor
ENVIRONMENT AND POLLUTION DAILY
Seoul Building 4 Fl, # 105-96, Gongdok1 -Dong, Mapo-Ku
oul, Republic of Korea
2 706-4954-7 F: 82 2 719-2068
el k29271-2

, Kim, Mr.
ALITION FOR ECONOMIC JUSTICE
Ga, Chongro-ku 110-125
Korea
82 2 745-8006

Sangchul, Park, Mr.
Regional Staff
NGO'S COMMITTEE FOR UNCED KOREA
Chungsin 100 Bobug Bd, 4th Jongro
Seoul 110-490, Republic of Korea
T: 82 2 741 2118 F: 82 2 741 1240

Sangoon, Lee, Mr.
Professional Adisor
NGO'S COMMITTEE FOR UNCED KOREA
Chungsin 100 Bobug Bd, 4th Jongro
Seoul 110-490, Republic of Korea
T: 82 2 741 2118 F: 82 2 741 1240

Seog, Dong Il
Advisor
KOREA ANTI-POLLUTION MOVEMENT ASSOCIATION
Cung Sin Dong 100 Budug Bd 4th Pl, Jongro Ku
Seoul 110400, Republic of Korea
T: 82 2 744-4549 F: 82 2 741-1240

Seung, Lee Duk, Mr.
YMCA OF KOREA
15 Breadalbane Street
Toronto - Ontario M4Y 2V5, Republic of Korea

Seung-Woo, Nam, Mr.
CITIZEN'S COALITION FOR ECONOMIC JUSTICE
25-1 Chogro 5 Ga. Chongro-ku 110-125
Seoul, Republic of Korea
T: 82 2 741-7961 F: 82 2 7458006

Seungho, Han, Mr.
Professional Adisor
NGO'S COMMITTEE FOR UNCED KOREA
Chungsin 100 Bobug Bd, 4th Jongro
Seoul 110-490, Republic of Korea
T: 82 2 741 2118 F: 82 2 741 1240

Shin, Chung-Sup, Mr.
Chairman
KOREA GENERAL CHEMICAL CORPORATION
80 Susong-dong, Chogro-ku
Seoul 110140, Republic of Korea
T: 82 2 7355602 F: 82 2 7355605

Shin, Eung-Bai, Mr.
Professor
HANYANG UNIVERSITY
Dept of Civil Eng. Hanyang University
Seoul 133-791, Republic of Korea
T: 82 2 2918089

Soon-Young, Kim, Ms.
CITIZEN'S COALITION FOR ECONOMIC JUSTICE
25-1 Chogro 5 Ga. Chongro-ku 110-125
Seoul, Republic of Korea
T: 82 2 741-7961 F: 82 2 7458006

Suckchang, Kim, Mr.
Regional Director
NGO'S COMMITTEE FOR UNCED KOREA
Chungsin 100 Bobug Bd, 4th Jongro
Seoul 110-490, Republic of Korea
T: 82 2 741 2118 F: 82 2 741 1240

Suk, Dong Il
Advisor
KOREA ANTI-POLLUTION MOVEMENT ASSOCIATION
100 Chungsin-Kong Boduk B/D 4 Jongro-Ku
Seoul, Republic of Korea
T: 001 82 2 7444549 F: 001 82 2 7411240

Sung-Chul, Chung, Mr.
Advisor
CITIZEN'S COALITION FOR ECONOMIC JUSTICE
25-1 Chongro 5 Ga, Chongro-ku 110-125
Seoul, Republic of Korea
T: 82 2 714-7961 F: 82 2 745-8006

Sungo Moo, Lee, Mr.
Professor
YONSEI UNIVERSITY
shindonga, Apt 3-4401 Seobinggo-dong Yongaan-gu
Seoul, Republic of Korea
T: 82 2 361-2751

Sungtea, Lee, Mr.
Professional Advisor
NGO'S COMMITTEE FOR UNCED KOREA
Chungsin 100 Bobug Bd, 4th Jongro
Seoul 110-490, Republic of Korea
T: 82 2 741 2118 F: 82 2 741 1240

Taeick, Min, Mr.
Professional Advisor
NGO'S COMMITTEE FOR UNCED KOREA
Chungsin 100 Bobug Bd, 4th Jongro
Seoul 110-490, Republic of Korea
T: 82 2 741 2118 F: 82 2 741 1240

Wa, Miung
Reporter Producer
MBC - TV
31 Yoido dong
Seoul, Republic of Korea
T: 82 2 7812571 F: 82 2 7824144

Wee, Eui Hwan
Coordinator
KOREA ANTI-POLLUTION MOVEMENT ASSOCIATION
100 Chungsin-Kong Boduk B/D 4 Jongro-Ku
Seoul, Republic of Korea
T: 001 82 2 7444549 F: 001 82 2 7411240

Wi, Eui Hoan
Coordinator
KOREA ANTI-POLLUTION MOVEMENT ASSOCIATION
100 Chungsin-Kong Boduk B/D 4 Jongro-Ku
Seoul, Republic of Korea
T: 001 82 2 7444549 F: 001 82 2 7411240

Won, Kyung Sun, Mr.
CITIZEN'S COALITION FOR ECONOMIC JUSTICE
25-1 Chogro 5 Ga. Chongro-ku 110-125
Seoul, Republic of Korea
T: 82 2 741-7961 F: 82 2 7458006

Woonjin, Yang, Mr.
Reional Director
NGO'S COMMITTEE FOR UNCED KOREA
Chungsin 100 Bobug Bd, 4th Jongro
Seoul 110-490, Republic of Korea
T: 82 2 741 2118 F: 82 2 741 1240

Yeo-Chang, Youn, Mr.
CITIZEN'S COALITION FOR ECONOMIC JUSTICE
25-1 Chogro 5 Ga. Chongro-ku 110-125
Seoul, Republic of Korea
T: 82 2 741-7961 F: 82 2 7458006

Yoo-Hwan, Lee, Mr.
International Coordinator
CITIZEN'S COALITION FOR ECONOMIC JUSTICE
25-1 Chongro 5 Ga, Chong ro-ku 110-125
Seoul, Republic of Korea
T: 82 2 741-7961 F: 82 2 745-8006

Yoon, Je Yong
Researcher
KOREA ANTI-POLLUTION MOVEMENT ASSOCIATION
Cung Sin Dong 100 Budug Bd 4th Pl, Jongro Ku
Seoul 110400, Republic of Korea
T: 82 2 744-4549 F: 82 2 741-1240

Yoon, Jong Hyun
Law Consultant
KOREA ANTI-POLLUTION MOVEMENT ASSOCIATION
100 Chungsin-Kong Boduk B/D 4 Jongro-Ku
Seoul, Republic of Korea
T: 001 82 2 7444549 F: 001 82 2 7411240

Yoon, Yeo Kyeong
Professional Advisor
KOREA ANTI-POLLUTION MOVEMENT ASSOCIATION
100 Chungsin-Kong Boduk B/D 4 Jongro-Ku
Seoul, Republic of Korea
T: 001 82 2 7444549 F: 001 82 2 7411240

Youngsook, Park, Ms.
Representative
NGO'S COMMITTEE FOR UNCED KOREA
Chungsin 100 Bobug Bd, 4th Jongro
Seoul 110-490, Republic of Korea
T: 82 2 741 2118 F: 82 2 741 1240

Yu, Keun Bae
Director
KOREA ANTI-POLLUTION MOVEMENT ASSOCIATION
100 Chungsin-Kong Boduk B/D 4 Jongro-Ku
Seoul, Republic of Korea
T: 001 82 2 7444549 F: 001 82 2 7411240

Yul, Choi, Mr.
Chief Executive
NGO'S COMMITTEE FOR UNCED KOREA
Chungsin 100 Bobug Bd, 4th Jongro
Seoul 110-490, Republic of Korea
T: 82 2 741 2118 F: 82 2 741 1240

PARLIAMENTARIANS / SPIRITUAL LEADERS

Choi, Joo-Sub
MP
National Assembly 1 1 Yeoidodong Yeongdungpo ku
Seoul, Republic of Korea
T: 782 3351
TL: NAL ROK K25849

Kim, Dong Gill
MP
National Assembly 1 1 Yeoidodong Yeongdungpo ku
Seoul, Republic of Korea
TL: NAL ROK K25849

Park, Jun Byung
MP
National Assembly 1 1 Yeoidodong Yeongdungpo ku
Seoul, Republic of Korea
T: 782 3351
TL: NAL ROK K25849

Sonn, Se-Il
MP
National Assembly 1 1 Yeoidodong Yeongdungpo ku
Seoul, Republic of Korea
T: 782 3351
TL: NAL ROK K25849

Suh, Sang Mok
MP
National Assembly 1 1 Yeoidodong Yeongdungpo ku
Seoul, Republic of Korea
T: 782 3351
TL: NAL ROK K25849

REPUBLIC OF MOLDOVA - GOVERNMENT

Dediu, Ion, Mr.
Head of Environment Protection
MINISTRY OF ENVIRONMENT
Stefan Cel Mare Ave
73 277001 Kishinau, Republic of Moldova
T: 22 61 61 F: 23 38 06

Scobioala, Veaceslav, Mr.
Expert
Permanent Mission of Republic of Moldova
573-577 3rd Ave.
New York, NY 10017, United States of America
T: 212 682 3274 F: 212 682 9746

Snegur, Mircea, H.E. Mr.
President
Republic of Moldova
573-577 3rd Ave.
New York, NY 10017, United States of America
T: 212 682 3274 F: 212 682 9746

Tiu, Nicolae, H.E. Mr.
Minister for Foreign Affairs
Permanent Mission of Republic of Moldova
573-577 3rd Ave.
New York, NY 10017, United States of America
T: 212 682 3274 F: 212 682 9746

REUNION - NGO

Paternoster, Monique
CJBM
Domaine des Colimacons,
F-97436 Saint Leu, Reunion

Valck, Dominique
CJBM
Domaine des Colimacons,
F-97436 Saint Leu, Reunion

ROMANIA - GOVERNMENT

Iliescu, Ion, Mr.
President
Permanent Mission of Romania
573 3rd. Ave.
New York NY 10016, United States of America
T: 212 682 3273 F: 212 682 9746

NGO

Bucuroiu, Vasile
FYR
4-6 Dem I Dobrescu Street
Bucharest 70119, Romania

Georgescu, Tudor
FYR
4-6 Dem I Dobrescu Street
Bucharest 70119, Romania

PRESS

Bunescu, Aristotel
Correspondent
TINETETUL LIBER.
Str. Snagdu 52
Bucharest, Romania, Romania

Jurca, Emil
Editor
LIBERTATEA
Rue Bibescu Voda Str. 19
Bucharest, Romania, Romania

Stoica, Dan
Editor
RADIO-ROMANIA
Str. Veteranilo 23
Bucharest 6, Romania, Romania

Torcatoriu, Bogdan Stefan
Editor
LUMEA
Str. Cap. Bulugea Ch. Nr. 14
Bucharest, Romania, Romania

GLOBAL FORUM

Iancu, Gheorghe, Mr.
Parlement de la Roumanie - Chambre des Deputes, 13 Septiembre, Secteur 5
Bucharest, Romania
T: 40 148091 F: 40 148091

Maiorescu, Toma George, Mr.
M.E.R. ECOLOGIST MOVEMENT OF ROMANIA
Bucurest, 2, 11 Al. Philippide Str.
Bucharest, Romania
T: 40 11293 F: 40 104858

Uliescu, Marilena, Ms.
Directeur
INSTITUT DES RECHERCHES JURIDIQUES
Bucharest, Romania

PARLIAMENTARIANS / SPIRITUAL LEADERS

Bienhu, Matcian, Senator
Marea Adunare Nationala Aleea Marii Adunari Nationale, No. 5
Bucarest
Romania, 16 21 50

Verestoy, Attila
Parloument of Roumaric Bd.Mihai Brauv N2196 Bl.200/28
Bucuresh, Romania
T: 00400 222927 F: 00400 158056

Weber, Peter
MP President
Environmental Committee
Marea Adunare Nationala Aleea Marii Adunari Nationale, No. 5
Bucarest, Romania
T: 16 21 50

RUSSIAN FEDERATION - GOVERNMENT

Alekseyev, S. A., Mr.
Director,Kronots Biosphere Park
Co-Chairman of the Regional Society of Ecological Protection and Greenpeace, Kamchatka Branch
Permanent Mission to the United Nations, 136 E. 67th Str.
New York, NY 10021, United States of America
T: 212 861 4900 F: 212 628 0252

Bolshova, L. I., Ms.
Chief Expert, Main Department for International Cooperation
Ministry for Ecology and Natural Resources, Member of the Central Council of Russian Society for Conservation of Nature
Permanent Mission to the United Nations, 136 E. 67th Str.
New York, NY 10021, United States of America
T: 212 861 4900 F: 212 628 0252

Burliay, Ya. A., Mr.
Director, Department of Central and Latin American
Ministry of Foreign Affairs
Permanent Mission to the United Nations, 136 E. 67th Str.
New York, NY 10021, United States of America
T: 212 861 4900 F: 212 628 0252

Danilov-Danilian, V.I., H.E. Mr.
Minister for Ecology and Natural Resources
Permanent Mission of Russian Federation
136 E. 67th Str.
New York, NY 10021, United States of America
T: 212 861 4900 F: 212 628 0252

Fedorov, A. V., Mr.
Adviser to the Vice President
Permanent Mission of Russian Federation
136 E. 67th Str.
New York, NY 10021, United States of America
T: 212 861 4900 F: 212 628 0252

Gaponenko, Oksana, Ms.
Project Manager
RUSSIAN ASSOCIATION FOR UN
Russia 119899 Moscow Lenin Hills, MSU Economics Departme
Moscow, Russian Federation
F: 007 095 9390877

Gavrilov, I. T., Mr.
Adviser to the Vice President
Permanent Mission of Russian Federation
136 E. 67th Str.
New York, NY 10021, United States of America
T: 212 861 4900 F: 212 628 0252

Glaziev, S. Yu., H.E. Mr.
First Deputy Minister
External Economic Relations
Permanent Mission to the United Nations, 136 E. 67th Str.
New York, NY 10021, United States of America
T: 212 861 4900 F: 212 628 0252

Golovatiy, A.I., H.E. Mr.
Deputy Minister of Finance
Permanent Mission of Russian Federation
136 E. 67th Str.
New York, NY 10021, United States of America
T: 212 861 4900 F: 212 628 0252

Gudyma, A. M., Mr.
Chief of Section, Main Department for International Cooperation
Ministry for Ecology and Natural Resources
Permanent Mission to the United Nations, 136 E. 67th Str.
New York, NY 10021, United States of America
T: 212 861 4900 F: 212 628 0252

Isaev, A.S., Mr.
Director
International Institute of Forestry
Permanent Mission to the United Nations, 136 E. 67th Str.
New York, NY 10021, United States of America
T: 212 861 4900 F: 212 628 0252

Kasumian, E.V., Mr.
Head of Directorate
Ministry for Ecology and Natural Resources
Permanent Mission to the United Nations, 136 E. 67th Str.
New York, NY 10021, United States of America
T: 212 861 4900 F: 212 628 0252

Kazakov, Yu. E., Mr.
Head, Main Directorate for International Cooperation
Ministry for Ecology and Natural Resources
Permanent Mission to the United Nations, 136 E. 67th Str.
New York, NY 10021, United States of America
T: 212 861 4900 F: 212 628 0252

Kimstach, V.P., Mr.
Deputy Chairman
Committee for Hydrometeorology and Environment Monitoring
Permanent Mission to the United Nations, 136 E. 67th Str.
New York, NY 10021, United States of America
T: 212 861 4900 F: 212 628 0252

Kokeyev, M.E., Mr.
Deputy Head, Directorate for International Science and Technology Cooperation
Ministry of Foreign Affairs
Permanent Mission to the United Nations, 136 E. 67th Str.
New York, NY 10021, United States of America
T: 212 861 4900 F: 212 628 0252

Konygin, E. A., Mr.
Deputy Director, Main Department for International Cooperation
Ministry for Ecology and Natural Resources
Permanent Mission to the United Nations, 136 E. 67th Str.
New York, NY 10021, United States of America
T: 212 861 4900 F: 212 628 0252

Koptiug, V.A., Mr.
Vice President
Academy of Sciences
Permanent Mission to the United Nations, 136 E. 67th Str.
New York, NY 10021, United States of America
T: 212 861 4900 F: 212 628 0252

Kosov, N. N., Mr.
Assistant to the Vice President
Permanent Mission of Russian Federation
136 E. 67th Str.
New York, NY 10021, United States of America
T: 212 861 4900 F: 212 628 0252

Kransov, V.N., Mr.
Chief of the Secretariat of the Vice President
Permanent Mission of Russian Federation
136 E. 67th Str.
New York, NY 10021, United States of America
T: 212 861 4900 F: 212 628 0252

Makarov, N. V., Mr.
Academician-Secretary
Bioenergy and Ecology Section, Academy of Sciences
Permanent Mission to the United Nations, 136 E. 67th Str.
New York, NY 10021, United States of America
T: 212 861 4900 F: 212 628 0252

Parfenov, B.F., Mr.
Head
Department of Ecology and Use of Natural Resources of the Government Administration
Permanent Mission to the United Nations, 136 E. 67th Str.
New York, NY 10021, United States of America
T: 212 861 4900 F: 212 628 0252

Ponizova, Oega
Researcher
MOSCOW STATE UNIVERSITY
Russia, Moscow, 127521 Theremetievskaya St 7 37-61
Moscow, Russian Federation
T: 7 095 219 60 15 F: 007 095 9390877

Routskoy, Aleksandr, H.E. Mr.
Vice President
Russian Federation
136 E. 67th Str.
New York, NY 10021, United States of America
T: 212 861 4900 F: 212 628 0252

Rudnev, V. A., Mr.
Director-General
Soyuzkhimexport
Permanent Mission to the United Nations, 136 E. 67th Str.
New York, NY 10021, United States of America
T: 212 861 4900 F: 212 628 0252

Taranenko, V. N., Mr.
Chief
Security Service of the Vice President
Permanent Mission to the United Nations, 136 E. 67th Str.
New York, NY 10021, United States of America
T: 212 861 4900 F: 212 628 0252

Tashchev, D. E., Mr.
Deputy Chief
Service of Protocol of the Administration of the Vice President
Permanent Mission to the United Nations, 136 E. 67th Str.
New York, NY 10021, United States of America
T: 212 861 4900 F: 212 628 0252

Tpaev, Alexandr
Director International Forest Insti
Novocheriomushkinska St
69 Moscow 117418, Russian Federation
T: 332 86 52

Vorfolomeev, V.P., Mr.
Chairman
Committee for Ecology and Use of Natural Resources of the Supreme Council of the Russian Federation
Permanent Mission to the United Nations, 136 E. 67th Str.
New York, NY 10021, United States of America
T: 212 861 4900 F: 212 628 0252

Yablokov, A.V., Mr.
State Counsellor
Russian Federation on the Policy in the Field of Environment and Health Care
Permanent Mission to the United Nations, 136 E. 67th Str.
New York, NY 10021, United States of America
T: 212 861 4900 F: 212 628 0252

Yakovlev, V. G., Mr.
Member of the Supreme Council
Committee on International Affairs
Permanent Mission to the United Nations, 136 E. 67th Str.
New York, NY 10021, United States of America
T: 212 861 4900 F: 212 628 0252

Yurina, H. P., Ms.
Deputy Chairperson
International Association Ecology and Peace
Permanent Mission to the United Nations, 136 E. 67th Str.
New York, NY 10021, United States of America
T: 212 861 4900 F: 212 628 0252

NGO

Ernsteins, Raimonds
ULEC
19 Rainis Blv.
Riga 226 098, Latvia, Russian Federation

Evgenjavitch, R.V.
IB
Akademgorodok
660 036 Krasnoyarsk, Russian Federation

Georgievitch, D.A.
IB
Akademgorodok
660 036 Krasnoyarsk, Russian Federation

Kokashvili, Vladimir, Mr.
INTERNATIONAL PEACE FOUNDATIONS ASSOCIATION
10, Prechistenka str.
Moscow 119889, Russian Federation
T: 7 95 201-74-77 F: 7 95 202-37-32
TL: 411489

Leonid, Pavlov
IAMEFE
30-2-338 Kronchtadsky Boulevard
125499 Moscow, Russian Federation

Luzhbin, Alexey
CEB
116, Bolshevistskaya Street
Perm, Russian Federation

Lyubimov, Alexander
CEB
116, Bolshevistskaya Street
Perm, Russian Federation

Nikolai, Mikhailov
IAMEFE
30-2-338 Kronchtadsky Boulevard
125499 Moscow, Russian Federation

PRESS

Babinsky, Alexander
Correspondent
RUSSIAN INFORMATION AGENCY
Molodeznaja 5
Moscow, Russia, Russian Federation

Beldinsky, Sergey Igorevich
Correspondent
RUSSIAN NATIONAL TV
Dom 3, Korpus 2, Kvartira 319
Litovsky Bulvar, Moscow, Russia, Russian Federation

Beloshapko, Viktor Grigoryevich
Correspondent
NOVOSTI
R. Almirante Guilhem 141/201
Leblon, 22440, Rio de Janeiro, Brazil, Russian Federation

Cheporov, Edgar
Correspondent
LITERARY GAZETTE
30 W 63rd St. (#11U)
New York, NY 10023, USA, Russian Federation

Chistiacov, Vladimir
Correspondent
NOVOSTI
Moscou, Russia, Russian Federation

Egorov, Viatcheslav
Editor
MOSCOW NEWS
Komintern St. 46-20
Moscow, Russia, Russian Federation

Kourgouzov, Andrei
Correspondent
TASS NEWS AGENCY
SQS-114, RL.b, #408
70257 Brazilia, DF, Brazil, Russian Federation

Mineev, Alexander
Correspondent
TASS
Borisovski 22
Moscow 115563, Russia, Russian Federation

Nikishin, Leonard
Editor
MOSCOW NEWS WEEKLY
4/65 Pavlova St., Kaliningzad
141070-Moscow Region, Russia Federation, Russian Federation

Pesliak, Alexandre M.
Correspondent
TV Russia
Krupskaya
Moscow, Russia, Russian Federation

Rudenko, Victor
Editor
ECOS MAGAZINE
Koroleva St. 8-1-232
Moscow, Russia, Russian Federation

Tcheporov, Edgard
Correspondent
ZITERARY GAZETTE
30 W. 63 St., II U
New York, N.Y., U.S.A., Russian Federation

Tchistov, Vadim
Correspondent
TASS NEWS AGENCY
SQS 114, BL.b, #605
70377 Brasilia, DF, Brazil, Russian Federation

ORGANIZATIONS

Aksenov, Vladimir, Mr.
International Peace Foundations Associations (previously Soviet Peace Fund)
10 Kropotkinskaya Str.
Moscow 119889, Russian Federation
T: (7 095) 201 7477 F: (7 095) 200 1208

Koptyug, V.A., Prof.
International Union Of Pure And Applied Chemistry
Leninskii Prospekt 14
SU-117901 Moscow, Russian Federation
T: (7 095) 237 7004
TL: 411964 ANS SU

Kouznetsov, V., Mr.
International Consortium 'ecological Reconstruction'
8B Sadovaja Str.
Moscow 103379, Russian Federation
T: (7 095) 209 3271 F: (7 095) 200 9271

Lascorin, Boris, Mr.
Ecoforum For Peace - Ussr
Prospect Marksa, 11
Moscow, Russian Federation

Marchuk, G.I., Dr.
Academy of Sciences of the USSR
Leninskii Prospekt 14
Moscow V-71, Russian Federation
T: (7 095) 237 37 42 F: (7 095) 230 27 41
TL: 411 964 ANS SU/Moscow 23 05 91

GLOBAL FORUM

Afanassiev, Vladimir, Mr.
KOLA SAAMI ASSOCIATION
183012 Murmansk, St. Lenina R.95-1
Murmansk, Russian Federation
T: 5-14-78

Cheparukhin, Alexander, Mr.
President
GREENWAVE INTERNATIONAL
115612 Kashirskoe Shosse 53-5-527
Moscow, Russian Federation
T: 7 95 344-8240 F: 7 95 344-8240

Deptosova, Ludmila, Ms.
KAMENSK - UZALSKY ECOLOGICAL COMMITTEE
623408 Sverdlovsk region, 26 Dzerzhinsky st, 12 flat
Kamensk-Uzalsky, Russian Federation
T: 7 2-5745

Enikolopov, Nickolai, Mr.
RUSSIAN ACADEMY OF SCIENCES
10, Prechistenka str.
Moscow 119889, Russian Federation
T: 7 95 201-74-77 F: 7 95 202-37-32
TL: 411489

Gouchtcho, Iouri, Mr.
INTERNATIONAL ASS. 'PERSONALITY ECOLOGY PEACE'
Novojasenevsky PR. 14-2-492
Moscow, Russian Federation
T: 7 95 421-7836 F: 7 95 310-7061

Goudyma, Alexander
MINISTRY OF ECOLOGY INTERNATIONAL COOPERATION DEPARTMENT
Palais de Nations
Geneve CH 1211, Russian Federation
F: 41 22 733-6542

Gwmerov, Airat, Mr.
SCIENTIFIC ASSOC.
St. Gotvald,11
Moscow, Russian Federation
T: 7 95 251-4692 F: 7 95 263-0400

Kharitonenko, Nick, Mr.
Editor-In-Chief
MAGAZINE 'GREEN CROSS'
1st Sovietski per. 5-F3, Mitichi
Moscow 141000, Russian Federation
T: 7 95 481-0966

Kovrizhenko, Gregory, Mr.
UNITED NATIONS ASSOCIATION
Prospekt Mira 36
Moscow 129010, Russian Federation
T: 7 95 280-3358 F: 7 95 200-4250

Kozeltsev, Michael, Mr.
MOSCOW STATE UNIVERSITY
Moscow, Russian Federation

Kozhernikov, Piotr, Mr.
ECOLOGICAL ASSOCIATION 'DELTA'
194295, Khudozhnikov pr., 9-2-285
St. Petersburg, Russian Federation
T: 7 511-3896 F: 7 552-6952

Kozlov, Yuri, Mr.
Professor
UNIVERSITY FOR FRIENDSHIP AMONG THE PEOPLES
Miklujo - Maklaia 6
Moscow 117198, Russian Federation
T: 7 95 434-6641 F: 7 95 433-7385

Loukina, Natalya Vladimirovna, Ms.
Engineer
NEW SYSTEMS
22/11 Kasetvilla Tower, Soi Tanpuying Pahon, Ngarmvongvarn Road
Bangkok 10900, Russian Federation
T: 66 2 579-7608 F: 66 2 579-8944
TL: 87373

Manushin, Edvard
ICAE
12 Rajaji Marg
New Delhi 110011, Russian Federation
T: 91 11 301-4309

Mikhailov, Sergei, Mr.
UNITED NATIONS ASSOCIATION
Prospekt Mira 36
Moscow 129010, Russian Federation
T: 7 95 280-3358 F: 7 95 200-4250

Mironova, Natalia I., Ms.
THE COMMITTE ON NUCLEAR SAFETY
c/p Bellona, PB 8874, Youngtorget
Oslo, Russian Federation
T: 47 2 382-410 F: 47 2 383-862

Petrov, Valery Anatoljevich
RUSSIAN ECOLOGICAL CENTER
Taganrogskaja St. 10/21-25 Moscow 109386
Moscow, Russian Federation
T: 7 95 2038234 F: 7 95 2005263
E: 121069, a/ja 175

Petrova, Lyudmila Nikolaevna, Ms.
RUSSIAN ECOLOGICAL CENTER
Taganrogskaja St. 10/21-25 Moscow 109386
Moscow, Russian Federation
T: 7 95 2038234 F: 7 95 2005263
E: 121069, a/ja 175

Riazantsev, Yan Vladimirovich, Mr.
Engineer
NEW SYSTEMS
22/11 Kasetvilla Tower, Soi Tanpuying Pahon, Ngarmvongvarn Road
Bangkok 10900, Russian Federation
T: 66 2 579-7608 F: 66 2 579-8944
TL: 87373

Sangui, Vladimir, Mr.
President
ASSOCIATION PEOPLE OF THE NORTH OF RUSSIA
121248 Moscow Nac.T. Shevtschenko, 3-3-119
Moscow, Russian Federation
T: 7 95 243 4159 F: 7 95 243 4159

Semenko, Marina, Ms.
Chairman Gv Ikar
TADJIK DEPATMENT OF SOCIO-ECOLOGICAL UNION
Kasimduskaya st 1/2-81
Moscow, Russian Federation
T: 7 95 3263509
E: 734058

Skuratovskaja, Larissa, Dr.
RUSSIAN WOMEN UNION
26 Boku Commissary Street, 3-3-340
Moscow 117571, Russian Federation
T: 7 95 433-5557 F: 7 95 433-5557

Sosnovsky, Anatoly, Mr.
INTERNATIONAL PEACE FOUNDATIONS ASSOCIATION
10, Prechistenka str.
Moscow 119889, Russian Federation
T: 7 95 201-74-77 F: 7 95 202-37-32
TL: 411489

Srapenyants, Rigo, Mr.
General Director
PUBLIC FUND REVIVAL OF AMENTA
Krasikova Str.32, CEMI
Moscow 117418, Russian Federation
T: 70 95 4293900 F: 70 95 3107015
TL: 411076 TSEMT SU

Stanis, Elena V., Ms.
Professor
UNIVERSITY FOR FRIENDSHIP AMONG THE PEOPLES
Miklujo - Maklaia 6
Moscow 117198, Russian Federation
T: 7 95 434 66 41 F: 7 95 433 73 85

Timochek, Vladimir, Mr.
Int'L. Secretary
UNIVERSITY FOR FRIENDSHIP AMONG THE PEOPLES
Miklujjo - Maklaia 6
Moscow 117198, Russian Federation
T: 7 95 434 66 41 F: 7 95 433 73 85
TL: 411768RTCNT SU

Veriguine, Vitali Yeugenyevich, Mr.
NEW SYSTEMS
22/11 Kasetvilla Tower, Soi Tanpuying Pahon, Ngarmvongvarn Road
Bangkok 10900, Russian Federation
T: 66 2 579-7608 F: 66 2 579-8944
TL: 87373

Zarezov, Kirill, Mr.
RAINBOW (NGO)
lenin hills economic dept. apt. 484
Moscow 119899, Russian Federation
T: 7 95 939-3555 F: 7 95 939-0877

PARLIAMENTARIANS / SPIRITUAL LEADERS

Jeaev, Alxandr
MP
Sovet Soiuza Soviet of the Union USSR Supreme Soviet Kremlin
Moscow, Russian Federation
T: 224 05 10

Vladimir, Vorfolomeev
MP
Sovet Soiuza Soviet of the Union USSR Supreme Soviet Kremlin
Moscow, Russian Federation
T: 224 05 10

RWANDA - GOVERNMENT

Bahigiki, Emmanuel, Mr.
Consultant
Duhamic-Adri
B.P 1458
Kigali, Rwanda
T: 250 76776 F: 250 72217

Habiyambeke, Thaddee
Directeur General Des Forets
MINISTERE DE L'AGRICULTURE ET
De L'Elevage B P 621
Kigali, Rwanda
T: 250 82756 F: 250 82103

Hakizimana, Jean De Diev, Ms
Chef Decabinet Relation Publigue
MINISTERE DEL ENVIRONNEMENT ET DU T
BP 2378
Kigali, Rwanda
T: 250 774 5

Kabanza, Fraterne
Journaliste Secretaire General De L
PRESSE RWANDAISE/ASSOCIATION RWANDA
Le Development Local (ARDEL) B.P.31
Kibuye, Rwanda
T: 250 68186

Mugorewera, Droeela
Charge Du Guier Et Evaluation
MINISTERE DE L'ENVIRONMENT ET DU TO
B.P 2378
Kigali, Rwanda
T: 250 77283

Ruhumuliza, Gaspard
Ministre
MINISTERE DEL'ENVIRONNEMENT
BP. 2378
Kigali, Rwanda
T: 250 CHAD F: 77415
TL: 250 74834

Rukangira, Ernest, Mr
Directeur
MINISTERE DEL ENVIRONNEMENT ET OLU
Tourisme BP 2378
Kigali, Rwanda
T: 250 72237 774 5

Sisi, Jean Damascene
Chef De Section Pub And Propa Gande
OFFICE RWANSAIS DU TOURISME ET
des Parcs Nationaux BP 905
Kigali, Rwanda
T: 250 765 4 250 84666 F: 250 765 2

Uwilingiyimana, Juvenal
Directeur
OFFICE DU TOURISME ET DES PARES NAT
B.P. 905 Kigali
Kigali, Rwanda
T: 250 76514 76514 F: 250 76512 250 76514

Winterbottom, Robert
Environmental Advisor
MINISTRY OF ENVIRONMENT AND TOURISM
P.O. Box 28 C/O Usaid Natural Resources Managem
Kigali, Rwanda
T: 250 772 83

GLOBAL FORUM

Drocella, Mugorewera, Ms.
ASSOCIATION RWANDAISE DES FEMMES POUR L'INVIRONNEMENT
P.O.Box 2378
Kigali, Rwanda
T: 250 77-283

Oreste, Habyarimana Mupanda, Mr.
Executive Secretary
A.R.D.I.
B.P. 1295
Kigali, Rwanda
T: 250 73961 F: 250 73961

SAINT KITTS AND NEVIS - GOVERNMENT

Edwards, Erstein, Mr
Permanent Secetary
Ministry of Foreign Affairs
Government Headquarters P.O.Box 186
Basseterre, Saint Kitts
T: 809 465 2521 F: 809 465 5202

Herbert, William Valentine, Dr
Ambassador
UNITED NATIONS
Frigate Bay Road
Basseterre, Saint Kitts
T: 809 465 2031 F: 809 465 5202

Petty, Eugene, Mr
Permanent Seretary
MINISTRY OF DEVELOPMENT
Government Headquarters P.O. Box 186
Basseterre, Saint Kitts
T: 809 465 2521 F: 809 465 5202

Simmonds, Kennedy, Dr. The Rt. Hon.
Prime Minister and Minister, Foreign Affairs of Saint Kitts and Nevis
Government
Government Headquarters P.O.Box 186
Basseterre, Saint Kitts
T: 809 465 2521 F: 809 465 5202

SAINT LUCIA - GOVERNMENT

Flemming, Charles Stephen, H.E. Dr.
Ambassador Extraordinary and Plenipotentiary
GOVERNMENT OF SAINT LUCIA
P.O. Box 1412 Castries
Saint Lucia, Saint Lucia
T: 809 452 8490 F: 809 452 8490

Lansiquot, Romanus, The Hon.
Minister of Tourism
Public Utilities and National Mobilisation
Permanent Mission to the United Nations, 820 Second Avenue,Deputy Permanent Representative to the United Nations, New York Suite 900
New York, NY 10017, 697 9360, 9361
T: 370 7867

Sands, Philippe, Mr.
Adviser
Permanent Mission of Saint Lucia
820 Second Avenue, Suite 900
New York, NY 10017, 697 9360, 9361
T: 370 7867

Singh, Naresh, Dr.
Executive Director
Caribbean Environmental Health Institute
Box 1111
Castries, Saint Lucia
T: 809 452 2501 F: 809 453 2721

Springer, Cletus, H.E. Mr.
Permanent Secretary, Ministry of Tourism
Public Utilities and National Mobilisation
Permanent Mission to the United Nations, 820 Second Avenue, Suite 900
New York, NY 10017, 697 9360, 9361
T: 370 7867

SAINT VINCENT - GOVERNMENT

Allen, Hon Alpian R.O.
Parliamentary Secretary
MINISTRY OF HEALTH & THE ENVIRONMEN
Kingstown St. Vincent & the
Grenadines, Saint Vincent
T: 71898

Browne, Bentley Ashford, Mr
Town-Planner
CENTRAL PLANNING DIVISION
P.O. Box 608 Kingstown
Saint Vincent
T: 809 45 71746 F: 809 45 62430

GLOBAL FORUM

Peacocke, Nan, Ms.
Newsletter Editor
WOMEN & DEVELOPMENT UNIT (WAND)
UWI School of Continuing Studies, Pinelands
St Michael, Saint Vincent
T: 1 809 4366312 F: 1 809 4274397

SAMOA - GOVERNMENT

Alesana, Tofilau Eti, The Hon.
Prime Minister
Permanent Mission of Western Samoa
820 2nd Ave., Suite 800D
New York NY 10017, United States of America
T: 212 599 6196 F: 212 972 3970

Laulu, Fiu Mataese Elisala, Mr.
Director
Lands and Environment
Permanent Mission to the United Nations, United Nations Plaza New York, NY, 10017
F: 212 963 7171
TL: 212 963 1921

Pati, Faasootauloa, Hon.
Minister
Lands and Environment
Permanent Mission to the United Nations, United Nations Plaza New York, NY, 10017
F: 212 963 7171
TL: 212 963 1921

Sanerivi, Leaupepe Muliaumaseali, Mr.
Attorney General
Permanent Mission of Samoa
United Nations Plaza
New York, NY 10017, United States of America
T: 212 963 7171 F: 212 963 1921

Sesega, Samuelu, Mr.
Principal Conservation Officer
Lands and Environment
Permanent Mission to the United Nations, United Nations Plaza New York, NY, 10017
F: 212 963 7171
TL: 212 963 1921

Sua, Mose Pouvi, Mr.
Secretary for Foreign Affairs
Permanent Mission of Samoa
United Nations Plaza
New York, NY 10017, United States of America
T: 212 963 7171 F: 212 963 1921

GLOBAL FORUM

Matatumua, Lemamea 'ata, Mr.
President
OLE SIOSIOMAGA SOCIETY
PO Box 5774
Apia, Samoa
T: 685 21993 F: 685 21993

Westhead, Lisa, Ms.
SAMOA NGO ASSOCIATION SANGO
Palais de Nations
Geneve CH 1211, Samoa
F: 41 22 733-6542

SANTA LUCIA - GLOBAL FORUM

Jules, Didacus
ICAE
12 Rajaji Marg
New Delhi 110011, Santa Lucia
T: 91 11 301-4309

SAO TOME AND PRINCIPE - GOVERNMENT

Da Cunha, Lisboa Trovoada, M.A.
President
Permanent Mission of Sao Tome and Principe
801 2nd Ave., Suite 603
New York NY 10017, United States of America
T: 212 697 4211212 687 8389

NGO

Pinto Da Costa, Henrique
ECOSTP
P O Box 484
Sao Tome, Sao Tome and Principe

SAUDI ARABIA - GOVERNMENT

Akhdar, Fawaz, Mr
Director Of General Resources Dept
MINISTRY OF PLANNING
P.O. Box (358)
Riyadh 11138, Saudi Arabia
T: 9661 4049278 F: 9661 4010385

Al-Athel, Saleh
President
KACST
P.O. Box 6086
Riyadh 11442, Saudi Arabia
T: 966 1 4883467 F: 966 1 4883756

Al-Haizan, Mohammed, Mr
International Economic
RELATIONS MINISTRY OF FINANCE AND NATIONAL ECONOMY
Rigadn 11177, Saudi Arabia
T: 4050000 / 1156 F: 425 8788

Al-Jubeir, Maher
Advisor,Office Of The Minister
MINISTRY OF PETROLEUM & MINERAL RES
Kingdom of Saudi Arabia 4660 Kenmore Dr.NW
Washington DC 20007, United States of America
T: 202 337 2833 F: 202 337 3282

Arab, Asem, Mr
Assistant Deputy Minister
MINISTRY OF PLANNING
P.O. Box (358)
Riyadh 11182, Saudi Arabia
T: 9661 4022692 F: 9661 4010385

Mahamed, Zeid
Prof Asstc Riyadh
ARAB SECURITY STUDIES TRAINING
P O Box 6830 Riyadh
Riyadh, Saudi Arabia
T: 009661 246 3444 F: 009661 246 4713

NGO

Al Hammad, Mohamed
AUDI
PO Box 6892
Riyadh, Saudi Arabia

PRESS

Abdulwahab, Well
Correspondent
ASHARQ AL AWSAT
108 Huddleston Road
London N7 OEG, UK, Saudi Arabia

Al Kahtani, Mejarri M.M.
Editor
SAUDI TV
P.O.Box 57351
Riyadh 11573, Saudi Arabia, Saudi Arabia

Al Samman, Jamil M.A.B.
Correspondent
SAUDI TV
P.O.Box 18143
Riyadh 11415, Saudi Arabia, Saudi Arabia

Al-Mosibeeh, Saud Salih
Editor
OKAZ NEWSPAPER
P.O. Box 60629
Riyad 11555, Saudi Arabia

Algain, Abdulmomen, Mr.
Correspondent
SAUDI PRESS AGENCY
P.O.Box 58702
Riyadh 11515, Saudi Arabia, Saudi Arabia

ORGANIZATIONS

Hamamd, M.A. Al, Dr.
Arab Urban Development Institute (AUDI)
PO Box 6892
Riyadh, Saudi Arabia
T: (96 61) 441 8100/441 8235
TL: 403566 AUDI S.J.

GLOBAL FORUM

Abotteen, Saleh, Mr.
Program Manager
ROYAL COMMISSION FOR JUBAIL & YANBU
Jubail Ind. City, Saudi Arabia
T: 966 3 341-1603 F: 966 3 341-1238

Al-Gain, Abdulrazaq, Mr.
ROYAL COMMISSION FOR JUBAIL & YANBU
PO Box 5964
Riyadh 11432, Saudi Arabia
T: 966 1 479-4445
TL: 401386

Al-Mazini, Khalid, Mr.
ROYAL COMMISSION FOR JUBAIL & YANBU
P.O. Box 5964
Riyadh 11432, Saudi Arabia
T: 966 1 477-4445 F: 966 1 477-5404

Al-Muhanna, Mohammed, Mr.
Deputy Director General
ROYAL COMMISSION FOR JUBAIL & YANBU
PO Box 5964
Riyadh 11432, Saudi Arabia
T: 966 1 4794445 F: 966 1 341-9891
TL: 401386

Al-Oyoni, Yahiya, Mr.
Department Director
ROYAL COMMISSION FOR JUBAIL & YANBU
PO Box 5964
Riyadh 11432, Saudi Arabia
T: 966 1 479-4445 F: 966 1 341--9891
TL: 401386

Al-Rasheed, Faisal M., Mr.
Director, Communities Affairs
ROYAL COMMISSION FOR JUBAIL & YANBU
PO Box 5964
Riyadh 11432, Saudi Arabia
T: 966 1 479-4445 F: 966 1 477-5404
TL: 401386

Dahlawi, Muneer, Mr.
Environment
ROYAL COMMISSION FOR JUBAIL & YANBU
PO Box 5964
Riyadh 11432, Saudi Arabia
T: 966 1 479-4445 F: 966 1 477-5404
TL: 401386

Dakkak, Abdulbadea M.K., Mr.
President
DAKKAK TRADING GROUP
PO Box 5221
Jeddah, Saudi Arabia
T: 96 62 669-2326

Dakkak, Ali Mahd Jamil, Mr.
Gen Manager
DAKKAK TRADING GROUP
PO Box 5221
Jeddah, Saudi Arabia
T: 96 62 6601142

Kahjani, Mejarri M Al
News Editor
SAUDI YV
P.O.Box 57137
Riyadh 11574, Saudi Arabia
T: 014014440 F: 014036085

Khoja, Abdoulrhman
NATIONAL WILDLIFE RESEARCH CENTER- NWRC
P.O. Box 1086
Taif, Saudi Arabia
T: 02 745-5188 F: 02 745-5176

Koshak, Abdulkader, Mr.
Gen. Secretary
ORGANIZATION OF ISLAMIC CAPITALS & CITIES (OICC)
P. O. Box 9009
Makkari, Saudi Arabia
T: 966 2 665-5896 F: 966 2 665-7516

Mousli, Adel, Mr.
Diretor
ROYAL COMMISSION FOR JUBAIL & YANBU
P.O.B. 5964
Riyadh 11432, Saudi Arabia
T: 966 1 479 4445 F: 966 1 477 5404

Obaid, Saleh
SAUDI YV
P.O.Box 57137
Riyadh 11574, Saudi Arabia
T: 014014440 F: 014036085

Samman, Jamil Al
Correspondent
SAUDI YV
P.O.Box 57137
Riyadh 11574, Saudi Arabia
T: 014014440 F: 014036085

Serhan, Khaled Al
Cameraman
SAUDI YV
P.O.Box 57137
Riyadh 11574, Saudi Arabia
T: 014014440 F: 014036085

Zahrani, Abdullah, Mr.
Okaz
N.W.L.C.-OKAZ
Taif. Saudi Arabia
Saudi Arabia
T: 0 2 745 5184 F: 0 2 745 5184

SENEGAL - GOVERNMENT

Ba, Seydou
Conseiller Juridique Du Premier
MINITRE-PRIMATURE
Building Administratif
Dakar, Senegal
T: 22 48 70 F: 221 229413

Cisse, Keba Birane
Ambassador-Permanent Rep To The U N
PERMANENT MISSION OF SENEGAL
238 East 68TH Street
New York 10021 NY, United States of America
T: 1 212 517 9030

Diene, Mame Ndiobo
Inginieur Agronome
MINISTERE DU DEVELOPEMENT RURAL ET
B P L005 Dakar
Dakar, Senegal
T: 234216 233578 237785 F: 237596

Diouf, Abdou, Mr.
President
Permanent Mission of Senegal
238 E. 68th Str.
New York NY 10021, United States of America
T: 212 517 9030 F: 212 737 7461

Jiaw, Dieynaba
Ingenneus Chemist
MINITERE DE L'INJUSTRIE
BP 1238
Dakar 320730, Senegal
F: 00221 225594

Mbaye, Babacar Carlos, Mr
Ambass Conseiller Diplom Du Pres
PRESIDENCE DE LA REPUBLIQUE
BP 4023
Dakar, Senegal
T: 237984 F: 232840

Seck, Mansour
Delegue.
MINISTERE DE L'EQUIPEMENT DE TRANSP
B.P. 4014
Dakar/Senegac, Senegal
T: 231 2385 60

Toure, Fatimata Dia
Directrice Adjointe Del Enviromente
MINISTERE DU TOURISME ET DE P'ENVIR
23 Rue Calmette BP
Dakar, Senegal
T: BUR 8.1 0725 DOM 345 F: 221 22 9413

NGO

Ben Abdallah, Taoufik
ENDA
B.P. 3370
Dakar, Senegal

Daff, Amadou
CONGAD
Km. 1, Avenue Cheikh Anta Diop
BP 4109, Dakar, Senegal

Damoda, Diop
ABACED
Villa No. 2170 Dieuppeul 1, BP 12135
Colobane, Dakar, Senegal

Diong, Ibrahima Cheikh, Mr.
RADI
B.P. 12085
Dakar, Senegal

Diop, Makhtar
CONGAD
Km. 1, Avenue Cheikh Anta Diop
BP 4109, Dakar, Senegal

Diouf, Aminata
ABACED
Villa No. 2170 Dieuppeul 1, BP 12135
Colobane, Dakar, Senegal

Ka, Abasse
AJED
Parcelles Assaines Unite 10 No. 143
BP 12 035, Dakar, Senegal

Ly, Bineta Sarr
FAVDO
B.P. 12085
Dakar, Senegal

Mazambani, David, Mr.
ENDA
B.P. 3370
Dakar, Senegal

Sall, Ihrahima
FAVDO
B.P. 12085
Dakar, Senegal

Seck, Abdoul
CONGAD
Km. 1, Avenue Cheikh Anta Diop
BP 4109, Dakar, Senegal

Sokona, Youba
ENDA-THIRD WORLD
PO Box 3370
Dakar, Senegal
T: 221 222125 F: 221 222695
TL: 51456ENDATM, E: gn:endadakar/geo2:enda

PRESS

Ben Arrous, Michel
Editor
SUD-HEBDO
BP 21141 Dakar Ponty
Dakar, Senegal, Senegal

Diakhate, Cheukh
Reporter Radio/Television
LE SOLEI
Louga, r.de Coki r.de Millan
Dakar, Senegal

Drame, Alioune
Correspondent
LE SOLEIL
Le Soleil
Dakar, Senegal, Senegal

Gueye, Babacar
Editor
OFFICIAL PRESS SENEGAL
Dakar, Senegal, Senegal

Ndieguene,, Mamadou Lo
Reporter Radio/Television
OFFICIAL PRESS SENEGAL
Ouagon Niayes, 2429
Dakar, Senegal

ORGANIZATIONS

Bugnicourt, Jacques, Mr.
Environnement Et Developpement Du Tiers-monde (ENDA)
B.P. 3370
Dakar, Senegal
T: (221) 22 42 29 F: (221) 22 26 95

Diop, Daoudo, Mr.
As. Des Bacheliers Pour L'emploi Et Le Developpement (abaced)
Villa No. 2170 Dieuppeul 1, BP 12135
Colobane, Dakar, Senegal
T: (221) 257 056 F: (221) 257 132

Kane, Thierno Hameth, Mr.
Federation des Associations du Fouta pour le Developpement
Sicap Liberte II Villa 1382
BP 3865, Dakar, Senegal
T: (221) 257 233 F: (221) 242 431

Ndiaye, Abdou El Mazide, Mr.
Forum Of African Voluntary Development Organizations
B.P. 12085
Dakar, Senegal
T: (221) 22 44 95 F: (221) 22 15 44
TL: 1304 SG

Ndiaye, Fatime Christiane, Ms.
Conseil des Organisations Non-Gouvernementales d'Appui au Developpement (CONGAD)
Km. 1, Avenue Cheikh Anta Diop
BP 4109, Dakar, Senegal
T: (221) 214 720

Ndiaye, Mazide, Mr.
Reseau Africain Pour Le Developpement Integre
B.P. 12085
Dakar, Senegal
T: (221) 255562/63 F: (221) 255564

Ndiaye, Yaya, Mr.
As. des Jeunes pour l'Education et le Developpement
Parcelles Assaines Unite 10 No. 143
BP 12 035, Dakar, Senegal
T: (221) 350 320

Sow, Fatoumata, Ms.
As. des Professionelles Africaines de la Communication (APAC)
BP 4234
Dakar, Senegal
T: (221) 210 815 F: (221) 220 0442

GLOBAL FORUM

Abasse, Ka, Mr.
AJED
B P 12035
Dakar, Senegal
T: 221 1 350320

Abdallah, Taoufik Ben
ENDA-THIRD WORLD
PO Box 3370
Dakar, Senegal
T: 221 222125 F: 221 222695
TL: 51456ENDATM, E: gn:endadakar/geo2:enda

Arrous, Ben
Journalist
SUD-HERDO
B.P.4130
Dakar, Senegal
T: 1 221 225393 F: 1 221 225290

Ba, Soukenya
FDEA
12 Rajaji Marg
New Delhi 110011, Senegal
T: 91 11 301-4309

Bassirou, Kane Amadou, Mr.
Member
CONGAD
Dakar, Senegal
T: 221 1 214720 F: 221 1 257132

Correa, Marie Louise, Ms.
CONFEDERATION SENEGALAISE DE SCOUTISME
95 Rue Mousse Diop
Dakar, Senegal
T: 221 1 22 00 59

Diallo, Amadou, Mr.
ENDA TIERS MONDE / ATELIER AUDIO-VISUEL
BP 3370
Dakar, Senegal
T: 221 1 22 26 95

Enda, T.M.
Gen. Secretary
BUGNICOURT
B. P. 3370
Dakar, Senegal
T: 221 222125 F: 221 222695
TL: 51456 ENDA TMSG

Engelhard, Philippe
ENDA-THIRD WORLD
PO Box 3370
Dakar, Senegal
T: 221 222125 F: 221 222695
TL: 51456ENDATM, E: gn:endadakar/geo2:enda

Guibbert, Jean Jacques
ENDA-THIRD WORLD
PO Box 3370
Dakar, Senegal
T: 221 222125 F: 221 222695
TL: 51456ENDATM, E: gn:endadakar/geo2:enda

Hamida, Essma Ben, Mr.
ENDA-THIRD WORLD
PO Box 3370
Dakar, Senegal
T: 221 222125 F: 221 222695
TL: 51456ENDATM, E: gn:endadakar/geo2:enda

Ki-Zerbo, Jacqueline, Ms.
ENDA TIERS MONDE
BP.3370
Dakar, Senegal
T: 221 1 22 42 29

Ly, Habib, Mr.
Directeur General De La Reche.
ISRA
BP.2241
Dakar, Senegal

Ly, Nohamadou, Mr.
ISRA
BP 3120
Dakar, Senegal
T: 221 1 220375 F: 221 1 223413
TL: GU1786

Moctar, Fall, Mr.
Rerpes.A Dakar
A.S.E.S.C.A.W
Ronkh
Dakar, Senegal
T: 00221 321264 F: 00221 321407
TL: BP 3588

Ndeye Gnima, Diop, Ms.
U J T / P D S
H L M GUEDIALOAYA VILLA No. 13
Stockholm 11383, Senegal
T: 46 8 150600 F: 46 8 153226

Ndiaye, Raphael
ENDA-THIRD WORLD
PO Box 3370
Dakar, Senegal
T: 221 222125 F: 221 222695
TL: 51456ENDATM, E: gn:endadakar/geo2:enda

Ndione, Emmanuel
ENDA-THIRD WORLD
PO Box 3370
Dakar, Senegal
T: 221 222125 F: 221 222695
TL: 51456ENDATM, E: gn:endadakar/geo2:enda

Niang, Babacar, Mr.
PDG-CAFAL
Dakar
Dakar, Senegal
T: 221 1 342979 F: 221 1 344177
TL: 314180, E: 2044

Niang, Mame Thiam, Ms.
Secretary
PDG-CAFAL
Dakar, Senegal
T: 221 1 340117 F: 221 1 344177
TL: 314180

Remal, Mahmoud
ENDA-THIRD WORLD
PO Box 3370
Dakar, Senegal
T: 221 222125 F: 221 222695
TL: 51456ENDATM, E: gn:endadakar/geo2:enda

Sambe Gorgui, Papa, Mr.
E.L.C. / A.R.E.P.S.
7 Rue Mouhamet V Dakar
Dakar, Senegal
T: 221 1 221798

Seck, Moussa, Mr.
Agro-Biologistic
ENDA TIERS MONDE
BP.3370
Dakar, Senegal
T: 221 1 22 42 29

Sissoko, Amadou, Mr.
FONGS
BP 260
Thies, Senegal
T: 22 1 51 12 37

Soumare, Mohamed
ENDA-THIRD WORLD
PO Box 3370
Dakar, Senegal
T: 221 222125 F: 221 222695
TL: 51456ENDATM, E: gn:endadakar/geo2:enda

Sy, El-Hadj, Mr.
ICASO - AFRIQUE
Dakar, Senegal
T: 221 1 22 42 29 F: 221 1 22 26 95

Tadesse, Zenebeworke, Ms.
DEVELOPMENT ALTERNATIVES WITH WOMEN FOR A NEW ERA (DAWN)
Dawn c/o Codesria BP 3304
Dakar, Senegal

Tamba, Yancouba, Mr.
General Secretary
ASSOCIATION FEDERATIVE DES AGRICULTEURS DU DEPT DE SEDHIOU
Rue 11, Plle n$68/A - Grand-Dakar
Dakar, Senegal

Tanor Fall, Thiendella, Mr.
General Manager
A S A C E
28, Rue des Essarts X Place de l'Independance - B.P. 3939
Dakar, Senegal
T: 221 1 224234 F: 221 1 213611
TL: 51509 SG

Thiam, Abou, Mr.
PRONAT - ENDA TIERS MONDE
BP 3370
Dakar, Senegal
T: 221 1 22 42 29 F: 221 1 22 26 95

Vincke, Pierre, Dr.
Head Of Project
SENEGAL APRES-BARRAGES
C/O Ambassade de BELGIRUE P.O. BOX 524
Dakar, Senegal
T: 221 1 22-1688 F: 221 1 21-6345

ADVISORS

Diop, Mamadou, Mr.
Mayor Of Dakar : Pres. Du Comite De La Communaute Urbaine De Dakar
Dakar, Senegal
T: 221 23 89 08 F: 221 21 37 35

SEYCHELLES - GOVERNMENT

Marengo, Marc, Mr.
First Counsellor
Permanent Mission of Seychelles
820 Second Avenue, Room 900F
New York, NY 10017, United States of America
T: 687 9766, 9767 F: 922 9177

Mascarenhas, J., Mr.
Principal Secretary
Environment Department
Permanent Mission to the United Nations, 820 Second Avenue, Room 900F
New York, NY 10017, United States of America
T: 687 9766, 9767 F: 922 9177

St. Jorre, Danielle de, H.E. Mrs.
Minister for Planning and External Relations
Permanent Mission of Seychelles
820 Second Avenue, Room 900F
New York, NY 10017, United States of America
T: 687 9766, 9767 F: 922 9177

SIERRA LEONE - GOVERNMENT

Bangali, Andrew Cibebay, Mr
Acting Permanent Representative
PERMANENT MISSION OF SIERRA LEONE
245 East 49TH Street
New York, New York 10017, United States of America
T: 212 688 1656 6884985 F: 212 688 4924

NGO

Kamara, Alieu Braimah, Mr.
Organizer/Coordinator
PORT LOKO UNITED NATIONS STUDENTS ASSOCIAION
P.O. Box 788
Freetown, Sierra Leone

SINGAPORE - GOVERNMENT

Foo, Kim Boon
State Counsel
Attorney General's Chambers
1 Coleman Street # 10 -00 Th Adelphi
Singapore 0617, Singapore
T: 3308960 F: 3308984

NGO

Check Woo, Foo
SABS
68 Lorong 16 Geylang, No. 04-03 Association Buildi
Singapore 1439, Singapore

PRESS

Ghosh, Nirmal
Correspondent
BUSINESS TIMES
5000 E Marine Parade Road
13-20 Singapore 1544, Singapore

ORGANIZATIONS

Menon, Vijay, Mr.
Asian Mass Communication Research And Information Centre (AMIC)
39 Newton Road
Singapore 1130, Singapore
T: (65) 251 5106 F: (65) 253 4535

Wee, Y.C., Prof.
Nature Society, The, Singapore
Dept. of Botany, National University of Singapore,
Lower Kent Ridge Road, Singapore 0511, Singapore
T: (65) 772 2858 F: (65) 779 5671

SLOVENIA - GOVERNMENT

Bricelj, Mitja, H.E. Mr.
Deputy Minister for Environment and Regional Planning
Permanent Mission of Slovenia
United Nations Plaza
New York, NY 10017, United States of America
T: 212 963 7171 F: 212 963 1921

Drnovsek, Janez, Mr.
President
Permanent Mission of Slovenia
1 Dag Hammerskjold Plaza, 7th floor
New York NY 10017, United States of America
T: 212 702 4884 F: 212 207 9888

Golob, Ignac, H.E. Mr.
Ambassador Extraordinary and Plenipotentiary
Permanent Representative to the United Nations, New York
Permanent Mission to the United Nations, United Nations Plaza
New York, NY 10017, United States of America
T: 212 963 7171 F: 212 963 1921

Jazbinsek, Miha, H.E. Mr.
Minister for Environment and Regional Planning
Permanent Mission of Slovenia
United Nations Plaza
New York, NY 10017, United States of America
T: 212 963 7171 F: 212 963 1921

Kramar, Marjan, Mr.
Head
Cabinet of the President of the Government
Permanent Mission to the United Nations, United Nations Plaza
New York, NY 10017, United States of America
T: 212 963 7171 F: 212 963 1921

Rupel, Dimitrij, H.E. Dr.
Minister for Foreign Affairs
Permanent Mission of Slovenia
United Nations Plaza
New York, NY 10017, United States of America
T: 212 963 7171 F: 212 963 1921

PRESS

Dobnikar, Romana Seruga
Correspondent
DELO
Marinkov Trg. 2,
Ljubljana, Slovenia, Slovenia

Sotensek, Darko
Reporter Radio/Television
RTV SLOVENIJA
Av.Sernambetiba 3300 Bl.7/1101
Rio de Janeiro, Brazil, Slovenia

SOLOMON ISLANDS - GOVERNMENT

Agorau, Transform, Mr.
Legal Counsel
Ministry of Foreign Affairs and Trade Relations
Permanent Mission to the United Nations, 820 2nd Ave., Suite 800A
New York, NY 10017, United States of America
T: 212 599 6193 F: 212 972 3970

Ariki, Mike, Mr.
Director of Meteorology
Minister of Tourism and Aviation
Permanent Mission to the United Nations, 820 2nd Ave., Suite 800A
New York, NY 10017, United States of America
T: 212 599 6193 F: 212 972 3970

Bugotu, Francis, H.E. Mr.
Amabssador Extraordinary and Plenipotentiary
Permanent Representative to the United Nations
Permanent Mission of the Solomon Islands, 820 2nd Ave., Suite 800A
New York, NY 10017, United States of America
T: 212 599 6193 F: 212 972 3970

Daokalia, Paul, Mr.
Special Secretary to Prime Minister
Office ofthe Prime Minister
Permanent Mission to the United Nations, 820 2nd Ave., Suite 800A
New York, NY 10017, United States of America
T: 212 599 6193 F: 212 972 3970

Isa, Henry, Mr.
Environmental Officer
Ministry of Natural Resources
Permanent Mission to the United Nations, 820 2nd Ave., Suite 800A
New York, NY 10017, United States of America
T: 212 599 6193 F: 212 972 3970

Mamaloni, Solomon S., Mr.
Prime Minister
Permanent Mission of the Solomon Islands
820 2nd. Ave, Suite 800A
New York NY 10017, United States of America
T: 212 599 6193 F: 212 972 3970

Misros, Lensay, Mr.
Head
Policy Evaluation Unit
Permanent Mission to the United Nations, 820 2nd Ave., Suite 800A
New York, NY 10017, United States of America
T: 212 599 6193 F: 212 972 3970

Ngele, Victor, The Hon.
Minister for Natural Resources
Permanent Mission of Solomon Islands
820 2nd Ave., Suite 800A
New York, NY 10017, United States of America
T: 212 599 6193 F: 212 972 3970

GLOBAL FORUM

Daga, Mali, Ms.
WORLD YOUNG WOMEN'S CHRISTIAN ASSOCIATION
PO Box 52
Munda, Solomon Islands
T: 67 7 61089 F: 67 7 61150
TL: WYMCA 41 23 32

SOUTH AFRICA - GOVERNMENT

Sangweni, Stanislaus S.
Professor / Director
SCHOOL OF RURAL COMMUNITY DEVELOPME
University of Natal Box 375 Pietermaritzburg 3200
Pietermaritzburg, South Africa
F: 27 331 955495

NGO

Cooper, David Harris, Mr.
EMG-ENVIRONMENT MONITORING GROUP
2156 Upper Buitenkant St.
Vredehoek, 8001, Cape Town, South Africa

Sekobe, Gopolang
NACTU
P.O. Box 10928
Johannesburg, South Africa

Smith, Brian
Minister
METHODIST CHURCH OF SOUTHERN AFRICA
114 Rissik Street
Braamfontein 2017, Johannesburg, South Africa

PRESS

Arnaud, Jean-Michel
Reporter Radio/Television
RADIO RSA
56, Cross Street, Pres. Ridge
Johannesburg, 2194 South Africa, South Africa

Barrett, Karen L.
Correspondent
BUSHCALL
108 Aberdeen St.
Melville 2092, RSA, South Africa

Botha, Johann
Correspondent
DIE BURGER
9 Rue Le Havre, Montclair
Somerset West, South Africa, South Africa

Holt-Biddle, David
Reporter Radio/Television
RADIO RSA
5 Chichester Street
Westdene, Johannesburg,South Africa 2092, South Africa

L'Ange, Jacqueline
Correspondent
FEMINA MAGAZINE
16 Ayres Rod, Rosebank
Cape Town 7700, South Africa, South Africa

Van Der Merwe, Christian
Correspondent
VRYE WEEKBLAD
P.O. Box 91872
Auckland Park 2092, South Africa, South Africa

Yeld, John
Correspondent
ARGUS NEWSPAPER
11 Mons Avenue, Newlands
7700 South Africa, South Africa

GLOBAL FORUM

Burns, Derrick, Mr.
INDUSTRIAL ENVIRONMENTAL FORUM
P.O.Box 35
Cape Town, South Africa
T: 27 21 64-1647

Burns, Diana, Ms.
ENGEN
P.O.Box 35
Cape Town, South Africa
T: 27 21 64-1647

Cook, Cecil, Mr.
Managing Director
TRANSKEI APPROPRIATE TECHNOLOGY UNIT
P.O. Box 225
Umtata - Transkei, South Africa
T: 27 471 24383.4.5 F: 27 471 24385

Dembele, Alaine, Mr.
DIOCESE DE SAN - ACTION SOCIALE
BP 48-San
Mali, South Africa
F: 226306025

Diallo, Hamid, Mr.
INADES-FORMATION, ANTENNE DE MAROUA
B.P. 167, Maroua/cameron/Afrique
Camero, South Africa
T: 28-2930 F: 44-0641

Fewell, Dorothy, Ms.
GENMIN
Box 61820 Marshalltown
Johannesburg, South Africa
T: 27 021 3763218

Fewell, P.T., Mr.
GENMIN
PO Box 61820, Marshlltown
Johannesburg, South Africa
T: 27 11 3763218 F: 27 11 8337132
TL: 4-85839 SA

Golding, Dave, Mr.
SOUTHERN AFRICAN COUNCIL FOR CATHOLIC SOCIAL SERVICE-SACCSS
P.O.Box 16277
Doornfontein, South Africa
T: 0116143429 F: 0116143429

Greeff, Johan, Mr.
General Manager
CHAMBER OF MINES
Box 809
Johannesburg, South Africa
T: 27 11 838 8211 F: 27 11 8341884

Holim, D., Mr.
Professor
UNIVERSITY PRETORIA
Dept. Architecture
Pretoria - Transvaal, South Africa
T: 27 12 420-2550 F: 27 12 432-185

Holm, Dieter, Dr.
Professor
UNIVERSITY OF PRETORIA
Dept. Architecture, University of Pretoria
Pretoria, Transvaal, South Africa

Kachiliko, Joseph, Mr.
MISSSIONARIES OF AFRICA
P.O. Box 410153
Kasama, South Africa
T: 221569 F: 221410

Lingen, Carl David Van Der, Mr.
Ashley Cottage, 7 Quarterdeck Rd. Kalk Bay
Cape Town 7975, South Africa
T: 27 21 788-7677 F: 27 21 439-9345

Makhayd, Muzi
FREELANCER
Johanesburg, South Africa

Mashinini, Gladys Khangwayini, Ms.
ECO-LINK
P. O. Box 727
Whiteriver, South Africa
T: 27 131132120 F: 27 131133287

Ngcukana, Cunningham
General Secretary
NATIONAL COUNCIL OF TRADE UNIONS
Po Box 10928
Johannesburg 2000, South Africa
T: 27 11 29 80 31 F: 27 11 333 7625

Rensburg, Bridget Van, Ms.
YOUTH 92 / WE CREATE OUR FUTURE
55 Parkdale Ave, 3rd floor
Ottawa - Ontario, South Africa
T: 1 613 761 9206 F: 1 613 761 1441

Usdin, Martine, Ms.
PROJECT EARTH
25 Riverside 18 Lane Rd Lyndhurst 2192
Johannesburg 2192, South Africa
T: 27 11 8826566

Wynberg, Rachel, Ms.
Dept of Env'l. & Geog. Science - Univ. of Cape Town - Rondebosch
Cape Town 7700, South Africa
T: 27 21 788-7677 F: 27 21 650-2784

SPAIN - GOVERNMENT

Gonzalez, Don Felipe, Mr.
President
Permanent Mission of Spain
809 United Nations Plaza, 6th floor
New York NY 10017, United States of America
T: 212 661 1050

NGO

Alarcon Y Puerto, Antoni
COBS
c/ Bailen 20 40 20
08010 Barcelona, Spain

Anzo, Javier
ADT
c/o Juan Pradillo, 26 lo
28039 Madrid, Spain

Arias Diaz, Montserrat
AVS
Clot, 39, 2o
08018 Barcelona, Spain

Castroviejo Bolibar, Javier
AAD
Panama 6, Apartado Correos 2182
41080, Sevilla, Spain

Cendrero, Antonio
UIP
Plaza Mayor 10
37002 Salamanca, Spain

Da Cruz, Humberto
ADT
c/o Juan Pradillo, 26 lo
28039 Madrid, Spain

Garcia Gallo, Antonio
COBS
c/ Bailen 20 40 20
08010 Barcelona, Spain

Lopez Vazquez, Luis
FM
Crta. Pozuelo a Majadahonda, km. 3,500
28220 Majadahonda, Madrid, Spain

Marti, Felix, Mr.
Director
CENTRE UNESCO DE CATALUNYA
Mallorca 285
Barcelona 08037, Spain
T: 34 3 4071716 F: 34 3 4575851
TL: 98314

Martin Mateo, Ramon
ADAE
Universidad de Alicante, Depto. de Estudio Juridicos del Estado y de la Empresa, Alicante, Spain

Martinez Camarero, Carlos
AEDENAT
c/ Campomanes 13, 20 Izquierda, 28013 Madrid
Spain

McPherson, Janice
UAM
Ciudad Universitaria de Cantoblanco
28049 Madrid, Spain

Perez Royo, Javier
AAD
Panama 6, Apartado Correos 2182
41080, Sevilla, Spain

Portabella, Jordi, Mr.
CENTRE UNESCO DE CATALUNYA
Mallorca 285
Barcelona 08037, Spain
T: 34 3 4071716 F: 34 3 4575851
TL: 98314

Real Ferrer, Gabriel
ADAE
Universidad de Alicante, Depto. de Estudio Juridicos del Estado y de la Empresa, Alicante, Spain

Ruiz, Juan P.
UAM
Ciudad Universitaria de Cantoblanco
28049 Madrid, Spain

Ruiz Sanchez, Jose
AVS
Clot, 39, 2o
08018 Barcelona, Spain

Santamarta Florez, Jose
AEDENAT
c/ Campomanes 13, 20 Izquierda, 28013 Madrid
Spain

Sola, Maria De Los Angeles
FSM
Doctor Esquerdo, 125-3
28007 Madrid, Spain

Vitoria Ruiz, Victor
FSM
Doctor Esquerdo, 125-3
28007 Madrid, Spain

Vozmediano Y Gomez-Feu, Jesus
SB
Apartado Correos 10.043, 41080 Sevilla
Spain

PRESS

Acosta, Rafael C.
Editor
EFE-NEWS AGENCY
Super Quadra Norte, 211
405 Brasilia-DF, Brazil, Spain

Alvarez Rodriguez, Faustino
ANTENA 3 TV
C/Cartagena, 11-3
Madrid, Spain, Spain

Andreu, Papiol Joan
Reporter Radio/Television
CATALUNYA/TV-3
Gran de Gracia 53, 1-2
08012 Barcelona, Spain, Spain

Anon, Maria G.
Correspondent
LA VOZ DE GALICIA
Apartado 500
1508, La Coruna, Spain, Spain

Aroca, Santiago
Editor
EL MUNDO
C. Sanchez Pacheco, 61
Madrid, Espana, Spain

Arroba, Jose Ma. P.
Reporter Radio/Television
CADENA SER
Lucienete No.7 DCHA.
28005 Madrid, Spain, Spain

Astiz, Carlos F.
Reporter Radio/Television
CANAL +
Sta. Genoveva, 52
Madrid, Spain, Spain

Ayuso, Javier
Editor
EL PAIS
c/Quintana 24
28008 Madrid, Spain, Spain

Balgorria, Juana E.
Correspondent
TELEVISION ESPANHOLA
J.Ingenieros 2805
Prov., Buenos Aires, Argentina, Spain

Balcells, Jord Bigas I
Editor
INTEGRAL
Mendel 6 08034
Barcelona, Spain, Spain

Baltanas, Reyes
Reporter Radio/Television
ANTENA 3 TV
C/Torpedero Tucuman 17
Madrid, Spain, Spain

Barrero, Maria Jose H.
Editor
DIARIO YA
Avda de Betanzos, 68
Madrid, Spain, Spain

Becerril, Jesus G.
Editor
EFE-NEWS AGENCY
C/ Cava Baja, 8
28005-Madrid, Spain, Spain

Besson, Blaya J.
Editor
CATALUNYA/TV-3
Joan Prom 266
C. Panollers, Barcelona, Spain, Spain

Blanco, Carlos
CATALUNYA/TV-3
C/Marques de Sentmenat, #20
Ato. 2, Barcelona, Spain, Spain

Blanes, Jose Manuel G.
TV-E
C/O'Donnell 77
Madrid, Spain, Spain

Borrero, Miguel A.F.
Reporter Radio/Television
CANAL SUR TELEVISION
c/El Mirador, 17-2 C
Tomares, Sevilla, Spain, Spain

Cadavid, Carlos M.
Editor
EFE-NEWS AGENCY
R. J. de Castilhos, 40
Copacabana, Rio de Janeiro-RJ, Brazil, Spain

Capella, Josep
Editor
AVUI
Pilar Cavero 3 2oC
28027 Madrid, Spain, Spain

Castellot, Almudena G.
Reporter Radio/Television
CANAL +
c/ E. Calderon, 7 1C
28020-Madrid, Spain, Spain

Castillo Giron, Jose M.
Reporter Radio/Television
ANTENA 3 TV
C/Julio Palacios, 4-12B
Madrid, Spain, Spain

Castrinon, Elena H.
Reporter Radio/Television
TELEMADRID
C/Maqneda, #120
Spain

Cembrero, Ignacio
Correspondent
EL PAIS
C. Don Ramon de la Cruz, 4
28001 Madrid, Espana, Spain

Cerrillo, Antonio
Editor
LA VANGUARDIA
Aragon 158, 5 2
Barcelona, Spain, Spain

Comas, Jose
Correspondent
EL PAIS
Florida 1065 Tercero A
1005 Buenos Aires, Argentina, Spain

Contreras, Edgar H.
EFE-NEWS AGENCY
Avenida Rio Branco, 25
Rio de Janeiro-RJ, Brazil, Spain

De Elvira, Magdalena R.
Editor
EL PAIS
Fermin Caballero 58
28034, Madrid, Spain, Spain

De La Fuente, Ana P.
Reporter Radio/Television
TELEMADRID
C/Lopez de Maro 17
Spain

De La Fuente, Teresa P.
Reporter Radio/Television
TV-E
C/Jeronima Llorente 4, dup.20B
28039 Madrid, Spain, Spain

De La Vega, Zoilo G.
Correspondent
EFE-NEWS AGENCY
R. Rino Levi, 375, Apto. 720
B. da Tijuca, Rio de Janeiro-RJ, Brazil, Spain

Del Cerro, Carmen
Correspondent
RADIO NACIONAL DE ESPANA
Ocana, Madrid
Spain

Dominguez, Pilar
Correspondent
EFE-NEWS AGENCY
50 Lexington Ave., Apt. 25A
New York, NY 10010, USA, Spain

Elzabuirre Ciriza, Josean
ANTENA 3 TV
C/Jaboneria, 6-1G
Madrid, Spain, Spain

Ercasi, Claudia G.
EFE-NEWS AGENCY
Av. Rio Branco, 25/13 andar
Rio de Janeiro-RJ, Brazil, Spain

Espinosa, Ana Paula
Correspondent
La Gaceta De Los Negocios
Rua Felinto de Almeida,86
Cosme Velho Rio de Janeiro, Brazil, Spain

Fernandez Bayo, Ignacio
Editor
DIARIO 16
c/ Santa Tecla, 34
Madrid, Spain, Spain

Figueroa, Francisco R.
Correspondent
EFE-NEWS AGENCY
C/Manuel Gonzalez Olachea 201
San Isidro, Lima-27, Peru, Spain

Foraste, Xavier G.
Reporter Radio/Television
CATALUNYA RADIO
Zurbano 26, #6-D
28010 Madrid, Spain, Spain

Gallego-Diaz, Soledad
Editor
EL PAIS
Gen. Cabrera 27
Madrid, Spain, Spain

Galvan, Ricardo V.
Correspondent
EL PAIS
Praia do Flamengo 12 (#413)
22210, Rio de Janeiro, RJ, Brazil, Spain

Gancedo, Pilar
Reporter Radio/Television
ANTENA 3 TV
Jorge Juan 15
28001 Madrid, Spain, Spain

Garcia, Alberto G.
Editor
DIARIO EL SOL
Garellano 21. 2A
28039 Madrid, Spain, Spain

Garcia, Javier T.
Editor
EFE-NEWS AGENCY
C/Esperanza 1, 2oB
(LAS ROZAS) Madrid, Spain, Spain

Garcia Becerril, Jesus
Editor
AGENCIA EFE
C Cava Baja,8
Madrid,Spain 28005, Spain

Garcia Miranda, Amanda
Correspondent
EFE-NEWS AGENCY
Ercilla, 48 2E
28005 Madrid, Spain, Spain

Giordano, Gerardo M.
Editor
EFE-NEWS AGENCY
R. Bulhoes de Carvalho, 489
Copacabana, Rio de Janeiro-RJ, Brazil, Spain

Gonzalez, Jose Maria
Reporter Radio/Television
ANTENA 3 TV
C/Maria Zayas 14
San Sebastian de Los Reyes, Madrid,Spain, Spain

Gonzalez Vega, Geraldo L.
Correspondent
EFE-NEWS AGENCY
Calle 32, No.112
Miramar, La Habana, Cuba, Spain

Gonzalez Yuste, Juan
Correspondent
EL PERIODICO DE CATALUNYA
Don Ramon de la Cruz 68
28001 Madrid, Spain, Spain

Gregory Roig, Javier
Reporter Radio/Television
CADENA SER
Calle Uruguay No. 39- 2o. D
28820 Coslada, Madrid, Spain, Spain

Gutierrez, Roldan J.
Editor
CATALUNYA/TV-3
Princep de Vergara, 205 2, 2A
28002 Madrid, Sapin, Spain

Gutierrez Roldan, Jordi
Editor
TV CATLUNYA
Princep de Vergara, 205
Madrid, 28002, Espana, Spain

Hermana, Luis Angel
Editor
EL PERIODICO DE CATALUNYA
Duran Y Borrell 9,AT
08023 Barcelona, Spain, Spain

Hernandez, Ma.Esperanza
Reporter Radio/Television
ANTENA 3 DE RADIO SA
c/Luciente, 7
Madrid, Spain, Spain

Iborra Gonzalez, Olga
Reporter Radio/Television
ANTENA 3 TV
C/Dracena, 42-1 izda
Madrid, Spain, Spain

Juri, Daniel
Reporter Radio/Television
TV-E
Lafuente 890
Buenos Aires, Argentina, Spain

Leroux, Anne
Correspondent
EFE-NEWS AGENCY
311 E. 38th St., Apt. 8F
New York, NY 10016, USA, Spain

Lizarraga, Ricardo J.
Reporter Radio/Television
ONDA CERO RADIO
C/S Olano 31-2-B
Prado de Somosaguas, Madrid, Spain, Spain

Llaurado, B. Esther
Reporter Radio/Television
CATALUNYA/TV-3
Frederic Rahola, 64, 2, 2
08032 Barcelona, Spain, Spain

Lopez Jordan, Jesus
Reporter Radio/Television
TELEMADRID
Sector Eswijores, #29 3-C
Tres Canios, Madrid, Spain, Spain

Lopez Nicolas, Luis
Editor
ABC-MADRID
Avda C. Herrera Oria 171-14D
Madrid 28034, Spain, Spain

Malpica Chica, Isabel
TV-E
C/O'Donnell 77
Madrid, Spain, Spain

Martinez, Antonio
Editor
EFE-NEWS AGENCY
Via Argentina, 60
Panama, Rep. de Panama, Spain

Menendez-Moran, Sofia R.
Editor
ECOLOGIA Y SOCIEDAD
Lilas, 11
28036 Madrid, Spain

Merino, Enrique
Correspondent
EFE-NEWS AGENCY
3003 Van Nese Street, NW
Washington, D. C. 20008 USA, Spain

Montanes, Cid Jose Francisco
Reporter Radio/Television
EBRE INFORMES MIGJORN
C/Doctor Vila, 7
43500 Tortosa, Spain, Spain

Moral, Javier G.
Reporter Radio/Television
RADIO NACIONAL DE ESPANA
Huergo 210, 17 B
Buenos Aires, Argentina, Spain

Morandini, Norma
Correspondent
CAMBIO 16
Avda del Libertador no.4780
1429 Buenos Aires, Argentina, Spain

Moreno Lacave, Emilio
Reporter Radio/Television
ANTENA 3 DE RADIO SA
R.de Santana, 150 Grs.102/103
20.230 Rio de Janeiro-RJ, Brazil, Spain

Nicolas, Fernando O.
Reporter Radio/Television
TELEMADRID
C/Villaamil 21, 1-B
28039 Madrid, Spain, Spain

Odoriz, Silvia
Correspondent
EFE-NEWS AGENCY
Sadovo-Samotiochnaya
UL, 12-24, Moscow, Russia, Spain

Pallole, Manuel A.
Reporter Radio/Television
RADIO POPULAR-CADENA COPE
c/Playa de Samil, 3
Collado-Villalba, 28460, Madrid, Spain, Spain

Paz, Rosa
Editor
EL SOL
Orellana 9
Madrid, Spain, Spain

Pena Marino, Ana
Reporter Radio/Television
CADENA SER
Brescia, S. 9D.
Madrid, Espana, Spain

Pepe Leira, Alberto
Editor
EFE-NEWS AGENCY
C/Maldonado 1395
Montevideo, Uruguay, Spain

Ramirez, Javier A.
Editor
RADIO NACIONAL DE ESPANA
Avda. Italia 26 - Pozuelo
Madrid, Spain, Spain

Reales, Luis
Editor
LA VANGUARDIA
Maestro Nrolan, 9-11
08021 Barcelona, Spain, Spain

Rios, Jose Luis
TV-E
C/O'Donnell 77
Madrid, Spain, Spain

Rius, Davi Eugenio
Reporter Radio/Television
CATALUNYA/TV-3
C/Girona #21, 9 Piso
Granollons, Barcelona, Spain, Spain

Rivera, Alicia
Editor
EL PAIS
Camino Nuevo, 109
La Moraleja 28109 Madrid, Spain, Spain

Rodriguez, Anna C.
Reporter Radio/Television
CATALUNYA RADIO
Spain

Roque, Francisco
Correspondent
EFE-NEWS AGENCY
4600 S. Four Mile Ryan Dr.#115
Arlington, VA 22204, USA, Spain

Ros, Montserrat E.
Reporter Radio/Television
CATALUNYA RADIO
Abat Escarre 3
08100 Mollet del Valles, Barcelona,Spain, Spain

Rubio Paillole, Manuel Alberto
Reporter Radio/Television
RADIO POPULAR CADENA
C/Playa de Samil, 3. 2B
Collado, Villalba, 28400 Madrid, Spain, Spain

Salvador, Ma.Delmar C.
TV-E
C/O'Donnell 77
Madrid, Spain, Spain

Sampedro, Amalia S.
Reporter Radio/Television
TV-E
C/O'Donnell 77
Madrid, Spain, Spain

Sanchez, Joaquin
Editor
RADIO NACIONAL DE ESPANA
Galileo, 71
Madrid, Spain, Spain

Sanchez Sampedro, Amalia
Reporter Radio/Television
TVE
C. O'Donnell, 77
Madrid, Espana, Spain

Sandoval, Jose Maria Montero
Reporter Radio/Television
CANAL SUR TELEVISION
c/Diego de Merlo 8, 1 I2P
41003 Sevilla, Spain, Spain

Sole, Laura E.
Reporter Radio/Television
CATALUNYA/TV-3
C/Fructuos Gelabert 5, 3rd 4th
St. Joan Despi, Barcelona, Spain, Spain

Soler, Granal A.
Editor
CATALUNYA/TV-3
Zurband, 7 4th P14
28010 Maurio, Spain, Spain

Suarez Hernandez, Maria Esperanza
Reporter Radio/Television
ANTENA 3 DE RADIO, S.A.
C/ Luciente,7
Madrid,Spain, Spain

Toharia, Manuel
Reporter Radio/Television
CADENA SER
Ginzo de Limia 55
28034 Madrid, Spain, Spain

Torres, Patricio
TV-E
C/O'Donnell 77
Madrid, Spain, Spain

Tovar Garcia, Javier
Editor
EFE-NEWS AGENCY
c/ Esperanza 1
Madrid, Spain, Spain

Ubeda, Carulla A.
Reporter Radio/Television
CATALUNYA/TV-3
Crtra. Kabassada, 44B
Barcelona 08023, Spain, Spain

Vazquez, Ignacio C.
Correspondent
EL PAIS
C.Don Ramon de la cruz, 40
28001 Madrid, Spain, Spain

Vidal Coy, Jose L.
Correspondent
DIARIO 16
Albasan 14
28037 Madrid, Spain, Spain

Zamora, Juan C.
Editor
EFE-NEWS AGENCY
General Yague 30
28020 Madrid, Spain, Spain

ORGANIZATIONS

Alcantara, Andres, Mr.
Silvema
C. Mariblanca No. 21 bajo
Malaca 29012, Spain
F: (34 952) 229 595

Anso, Javier, Mr.
Comision Justicia y Paz de Espana
Francisco Silvela 77bis, la dcha.
28028 Madrid, Spain
T: (34 1) 261 1214

Cooper, David, Mr.
Genetic Resources Action International (GRAIN)
Jonqueres 16, 6e D
E-08003 Barcelona, Spain
T: (34 3) 310 59 09 F: (34 3) 310 59 52
TL: 99767 CIDOB E

De Senillosa, Ignacio, Mr.
Fundacion Intermon
c/o Roger de Lluria, 15
08010 Barcelona, Spain
T: (93) 301 2936 F: (34 3) 301 2221

De Vinuesa, Maya Garcia, Ms.
Fundacion Santa Maria
Doctor Esquerdo, 125-3
28007 Madrid, Spain
T: (34 1) 573 0299/409 2872 F: (34 1) 504 0135

Mateo, Ramon Martin, Mr.
As. de Derecho Ambiental Espanol
Universidad de Alicante, Depto. de Estudio Juridi-dicos del Estado y de la Empresa, Alicante, Spain
T: (34 96) 590 3400 EXT. 3342 F: (34 96) 590 3464/590 3571

Munn, Wanda, Ms.
Ass. de Naturalistas del Sureste
Ricardo Gil, 25-3
30002-Murcia, Spain
T: 219 862

Pablo, Fernando, Mr.
Fundacion Mapre
Crta. Pozuelo a Majadahonda, km. 3,500
28220 Majadahonda, Madrid, Spain
T: (34 1) 626 31 39

Ruiz, Jose, Mr.
As. Vida Sana Para El Fomento De La Cultura Y El Desarrollo Biologicos
Clot, 39, 2o
08018 Barcelona, Spain
T: ((34 3) 245 0661

GLOBAL FORUM

Alarcon, A., Sr.
Coordinador De Medio Ambiente
COLEGIO OFICIAL DE BIOLOGOS
C. Bailen 20 4o. 2a.
Barcelona 08010, Spain
T: 34 3 265-2393 F: 34 3 265-2691

Artes Perez, Juan, Mr.
DIAGONAL 372 - 2o.
Barcelona 08037, Spain
T: 34 3 207-6633 F: 34 3 459-0122

Assuncion, Maria Del Mar, Ms.
ALTERNATIVA SOLIDARIA - PLENTY
Aurora 31
Madrid 28035, Spain

Ayala, Ines., Ms.
Ejecutiva Confederal
UGT UNION GENERAL DE TRABAJADORES
C/ Hortaleza 88
Madrid - ID 28004, Spain
T: 34 1 589 7600 F: 34 1 589 7603

Barbuda, Cassio Everaldo Meyer, Mr.
Coordenador Geral
ASSOC. IBEROAMERICANA DE DIREITO AMBIENTAL
Ferraz, 3 - 3o. Derecha
Madrid 28008, Spain
T: 34 1 559-5134 F: 34 1 541-7573

Boada, Lluis, Mr.
PREFEITURA DE BARCELONA
Pasaje Foraste 15
Barcelona 08022, Spain
T: 34 3 417-0797

Borja, Jordi, Mr.
Delegado Cooperacion
AYUNTAMIENTO BARCELONA
Pza Sant Jaume
Barcelona 08002, Spain
T: 34 3 4027443 F: 34 3 4027453

Canals, Puri, Ms.
DEPANA
C/ Arago 281, 2, 08009
Barcelona 08009, Spain
T: 34 3 215-1484 F: 34 3 487-1560

Carbonell, Lydia, Ms.
Professor
CONFERENCE DES EGLISES PROTESTANTES DES PAYS LATINS D'EUROPE
Calle Supervia 52
Zaragosa E-50006, Spain

Catala, Josep, Mr.
Delegate
AJUNTAMENT DE CABRILS (CATALONIA)
Emilia Carles
Cabrils - Barcelona, Spain
T: 34 3 753-2410 F: 34 3 753-2410

Cendon, Rosa Gomez, Ms.
Mapere
Echegaray 25 - 2nd Planta
Madrid, Spain
T: 34 91 581 5340

Coma, Antoni Garcia, Mr.
Director Of Water Treatment Ag
MINISTRY OF ENVIRONMENT
Department de Medi Ambient. Passeig de Grrrrracia, 94
Barcelona 08008, Spain
T: 34 3 487-2234 F: 34 3 487-3049

Costa, Miquel-Didac Pinero
ALTERNATIVA VERDA
Passeio del Mar 2
Li Escala 17130, Spain
T: 34 72 972 772011

Cotta, Rose, Sra.
Global 500
SOROPTIMIST
dr. carulla 31-33
Barcelona, Spain
T: 34 3 211 1532 F: 34 3 212 8576

Cruz, Humberto Da, Sr.
AMIGOS DE LA TIERRA
Juan Pradillo 26
Madrid, Spain
T: 34 1 311 2186 F: 34 1 311 4874

Cunha, Ester Piro Da
LA REGION INTERNACIONAL
Madrid, Spain

Farran Pinto, Margarita, Sra.
Presidenta
AMAZONAS
Calle Gerona no. 35 3o.1o.
Barcelona, Spain
T: 34 3 412-1134

Freijo Lizan, Jose Lluis, Sr.
Portavoz Relaciones Interna
LOS VERDES - UNION VERDE
Puigi Cadagalch 246 - Bajos
Barcelona ZIP: 08303, Spain
T: 34 3 798-7717 F: 34 3 798-7717

Gallo, Antonio Garcia, Mr.
Delegado
COLEGIO DE BIOLOGIA
Bailen, 20 4o 2o
Barcelona 08010, Spain
T: 34 3 265-2398 F: 34 3 265-2691

Gonzalez, Albert Vilalta I., Mr.
Minister
MINISTRY OF ENVIRONMENT
Department de Medi Ambient. Passeig de Gracia, 94
Barcelona 08008, Spain
T: 34 3 487-2234 F: 34 3 487-3049

Gortari Solbes, Javier, Mr.
AV. BAJA NAVARRA, 16 - 4o.
Pamplona 31004, Spain
T: 34 48 247-996

Iriarte, Jose M. Ordonez, Mr.
Coordinator
SCHOOL OF PUBLIC HEALTH
c/ General Oraa 39
Madrid 28006, Spain
T: 34 1 5642499 F: 34 1 4116696

Isern Aranda, Joan Josep, Mr.
GENERALITAT CAT.
Via Laietana, 69
Barcelona, Spain
T: 34 3 318-7204

Javier, Anso, Mr.
Secretario General
JUSTICIA Y PAZ-ESPANA
Francisco Silvela 77 Bis-1
Madrid 28028, Spain
T: 34 1 561 1214 F: 34 1 576 3890

Landa, Jose Allende, Mr.
Professor
UNIVERSIDAD DEL PAIS VASCO
Fac Ciencias Economicas, Auda Lebiendakaru Agirre, 83
Bilbao 48015, Spain
T: 34 4 4696390 F: 34 4 4475154

Loperena, Demetrio, Mr.
Professor
UNIVERSIDAD DEL PAIS VASCO
Fac Derecho P. M. de Lardizabal 2
S Sebastian 20009, Spain
T: 34 43 21 03 00 F: 34 43 21 94 04

Marino, Ana M. Paula, Sra.
Periodista
CADENA SER
Gran Via 32
Madrid 28010, Spain
T: 34 1 347-0700 F: 34 1 347-0779

Martin, Katia Cocera, Sra.
Ecologa
IZQUIERDA UNIDA
General Rodrigo 6
Madrid 28003, Spain
T: 34 1 91 5534909 F: 34 1 91 5349747

Martinez-Ruiz, Carlos, Mr.
AMAZONIA ASSEMBLEA SOLIDARITAT
C/St Vicens, 3 , 2
Barcelona 08001, Spain
T: 34 3 441-5887

Massachs, Gloria Compte, Ms.
XILONE - ONG ESPANA
Felipe Gil no. 05
Barcelona 08023, Spain
T: 34 3 212-0390 F: 34 3 265-7175

Miguelez, Aquilino, Mr.
ALTERNATIVA SOLIDARIA - PLENTY
Aurora 31
Madrid 28035, Spain

Miranda, Isabel
FEDERACION ESPANOLA DE UNIVERSIDADES POPULARES
12 Rajaji Marg
New Delhi 110011, Spain
T: 91 11 301-4309

Nadal Reimat, Eugenio, Mr.
Ingeniero
FUNDACION JOAQUIN COSTA
Manuel Bescos no.1, 4o.C
Huesca, Aragon, Spain
T: 34 74 222521

Novo, Maria
ICAE
12 Rajaji Marg
New Delhi 110011, Spain
T: 91 11 301-4309

Orcoyen, Cristina Garcia, Ms.
Secretary General
WWF - SPAIN
Adena, Santa Engracia 6
Madrid 28010, Spain
T: 34 1 308-2309 F: 34 1 308-3293

Pajares-Miravalles, Jorge, Mr.
Research Journalist
CENTRAL AMERICAN EDUCATIONAL FUND
carrer bonaire 8
mataro, Spain
T: 93 798 9720

Panmocchia, M. C.
INTEGRAL
Passeiz Margall 371
Barcelona, Spain
T: 33 581810 F: 33 583053

Picazo, Herminio, Sr.
COLEGIO DE BIOLOGIA
C/Bailen 20 4th 2nd
Barcelona, Spain
T: 34 3 265 2393 F: 34 3 265 2691

Plaza, Guillermo Garlick
Cameraman
CANAL SUR TELEVISION
Carretera San Juan de Aznalfarache/Tomares Km 1.3 San Juan de Aznalfarache
Sevilla, Spain
T: 34 54 4550759 F: 34 54 4550711

Pomar Sasot, J. Manuel, Mr.
Presidente
FUNDACION M-II
FERNANDO EL CATOLICO 65 - 4o. ESC DOHA
Zaragosa 50006, Spain
T: 34 76 356663

Provansal, Kenty Richardson I., Mr.
International Affairs
MINISTRY OF ENVIRONMENT
Department de Medi Ambient. Passeig de Grrrrracia, 94
Barcelona 08008, Spain
T: 34 3 487-2234 F: 34 3 487-3049

Puerto, Antoni Alarconi, Mr.
Delegado
COLEGIO DE BIOLOGIA
Bailen, 20 4o 2o
Barcelona 08010, Spain
T: 34 3 265-2398 F: 34 3 265-2691

Puig, Josep, Mr.
ALTERNATIVA VERDA
Eduard Toda 98 1 2
Barcelona 08031, Spain
T: 34 3 4272449

Quadras Sans, Miguel De, Sr.
RED INTERCULTURAL DE INNOVACION SOCIAL
Ronda General Mitre 165.C
Barcelona 08022, Spain
T: 34 3 417 4160

Rafa, Miguel, Mr.
Projects Director
DEPANA
c/Arago 281, 2
Barcelona 08009, Spain
T: 34 3 215-1484 F: 34 3 487-1560

Ramos, Herminia Merino
Productora De Informativos
CANAL SUR TELEVISION
Carretera San Juan de Aznalfarache/Tomares Km 1.3 San Juan de Aznalfarache
Sevilla, Spain
T: 34 54 4550759 F: 34 54 4550711

Ruiz, Victor Vitoria, Mr.
Biologo
FUNDACION SANTA MARIA
DR. ESQUERDO 125 - 3o.
Madrid 28007, Spain
T: 34 1 91 5730299 F: 34 1 1 5040135

Sala, Lloret, Sr.
CAMARA OFICIAL DE COMERCIO,INDUSTRIA Y NAVEGACION DE ALICANTE
Pascual Perez 36
Alicante, Spain

Sanchez, Hector Roque
Editor/Director
BRECHA- INDEPENDIENTE
25 de Mayo 73 -P.A. Oficinas 5/6/7 Cordoba
Cordoba Capital 500, Spain
T: 51 37924 F: 51 241405

Sanchez, Jose Ruiz, Mr.
ASOCIACION VIDA SANA
Clot 39
Barcelona, Spain
T: 34 3 245-0661 F: 34 3 265-2445

Santos, Marian Sola, Ms.
Biologa
FUNDACION SANTA MARIA
DR. ESQUERDO 125 - 3o.
Madrid 28007, Spain
T: 34 1 91 5730299 F: 34 1 5040135

Segura, Alfons, Mr.
Asesor Cooperacion
AYUNTAMIENTO BARCELONA
Pza Sant Jaume
Barcelona 08002, Spain
T: 34 3 402 7443 F: 34 3 402 7453

Steiner, Mary, Ms.
Counsellor
COUNSEL FOR BETTER ENERGY TRANSFER
Cost Azul, Paseo Maritimo 7 07014
Palma de Mallorca 07014, Spain
T: 34 71 971-231940

Torra, Monserrat De Vehi I, Dr.
General Secretary
MINISTRY OF ENVIRONMENT
Department de Medi Ambient. Passeig de Gracia, 94
Barcelona 08008, Spain
T: 34 3 487-2234 F: 34 3 487-3049

Trias, Pilar Heras, Ms.
Profes
DEPARTAMENTO DE TEORIA E HISTORIA DE LA EDUCACION
Universidad Barcelona, Facultat de Pedagogia
Barcelona 08028, Spain
T: 34 3 317-1047 F: 34 3 334-9193

Varillas, Benigno
Editor
QUERCUS (MAGATIRE)
La Pedriza, 1 E-28002
Madrid, Spain
T: 1 341 3262202 F: 1 341 5192194

SRI LANKA - GOVERNMENT

Amaratunga, Gunasiri
Chairman
CENTRAL ENVIRONMENTAL AUTHORITY
Parisara Mavata Maligawatta Colombo lo
Sri Lanka
T: 446750 F: 446749

Wijetunga, D.B., Mr.
Prime Minister
Permanent Mission of Sri Lanka
630 3rd Ave., 20th floor
New York NY 10017, United States of America
T: 212 986 7040 F: 212 986 1838

NGO

Algama, Ravi
EF
29 Siripa Road
Colombo 5, Sri Lanka

Dharman, Wickremaratne
SLEJF
No. 5, 4th Lane
Nawala, Rajagiriya, Sri Lanka

Fernando, Ranjen
WNPSS
Chaitiya Road
Marine Drive, Fort, Sri Lanka

Fjortoft, Arne, Mr.
Secr General
WORLDVIEW INTERNATIONAL FOUNDATION
8 Kinross Avenue
Colombo 4, Sri Lanka

Gunaratne, K.D.A.
SLEJF
No. 5, 4th Lane
Nawala, Rajagiriya, Sri Lanka

Indasena, Jayasingha
SSE
National Secretariat, Anuradhapura Road
Tawala Farm, Tawala, Sri Lanka

Jayasena, Hewage, Mr.
Hony General Secretary
WILDLIFE & NATURE PROTECTION
Chaitiya Rd.
Colombo 1, Sri Lanka
T: 94 1 25248 F: 94 1 580721
TL: 21537 Metaliy C

Serasinghe, Iranganie
TSSL
50/7C Siripa Road
Colombo 5, Sri Lanka

Vitarana, Kalmini
EF
29 Siripa Road
Colombo 5, Sri Lanka

PRESS

Wanigasundara, Mallika
Correspondent
DAILY NEWS-SRI LANKA
A3/1, Government Flats
Hospital Road, Dehiwala, Sri Lanka, Sri Lanka

ORGANIZATIONS

De Silva, Lalanath, Mr.
Environmental Foundation Ltd.
29 Siripa Road
Colombo 5, Sri Lanka

De Zoysa, Uchita, Mr.
Public Campaign On Environment And Development
50/7C, Siripa Road
Colombo 5, Sri Lanka
T: (94 1) 582 439 F: (94 1) 546 518

Gunawardene, Nalaka, Ms.
Sri Lanka Environmental Journalists Forum
No. 5, 4th Lane
Nawala, Rajagiriya, Sri Lanka
T: (94 1) 562 215/563 068 F: (94 1) 698 315

Guwardene, Nalaka, Mr.
Worldview International Foundation
8 Kinross Avenue
Colombo 4, Sri Lanka
T: (94 1) 58 9225/58 3109 F: (94 1) 58 9225

Jayasingha, Indasena, Mr.
Sama Sevaya
National Secretariat, Anuradhapura Road
Tawala Farm, Tawala, Sri Lanka
T: (94 25) 6266 F: (94 1) 440 950

Kathirgamanathan, Nagules, Dr.
South-East Asia Centre for Environment and Development (SEACED)
UNAJU Secretariat
Citadal Velanai NP, Sri Lanka

Vitarana, Kamini M., Mr.
Ruk Rakaganno (The Tree Society of Sri Lanka)
50/7C Siripa Road
Colombo 5, Sri Lanka
F: (941) 588 804

GLOBAL FORUM

Amarasingha, Ranjan
Vice President
NATIONAL JOURNALISTS GUILD OF SRI LANKA
15/1 First Lane Gothami Road
Colombo 08, Sri Lanka
T: 94 1 564545 F: 94 1 500544

Barroso, Rafael S., Sr.
Director
PROUT RESEARCH INSTITUTE
15 Lewis Place
Negombo, Sri Lanka

Hemantha, Withanage Don, Mr.
ENVIRONMENTAL FOUNDATION LTD
Palais de Nations
Geneve CH 1211, Sri Lanka
F: 41 22 733-6542

Parnajith, Sudeepa
Treasurer
NATIONAL JOURNALISTS GUILD OF SRI LANKA
15/1 First Lane Gothami Road
Colombo 08, Sri Lanka
T: 94 1 564545 F: 94 1 500544

Perera, B. C.
Vice President
NATIONAL JOURNALISTS GUILD OF SRI LANKA
15/1 First Lane Gothami Road
Colombo 08, Sri Lanka
T: 94 1 564545 F: 94 1 500544

Sirimanna, Bandula
President
NATIONAL JOURNALISTS GUILD OF SRI LANKA
15/1 First Lane Gothami Road
Colombo 08, Sri Lanka
T: 94 1 564545 F: 94 1 500544

Thera, Keeranthidiya Pragnasekera, Mr.
JATHIKA SAMPATH SUREKEEME BHIKKU SANGVIHANAYA
Palais de Nations
Geneve CH 1211, Sri Lanka
F: 41 22 733-6542

Varnakulasuriya, Keerthi
Secretary
NATIONAL JOURNALISTS GUILD OF SRI LANKA
15/1 First Lane Gothami Road
Colombo 08, Sri Lanka
T: 94 1 564545 F: 94 1 500544

Weerasekara, Wimal
Assistant Secretary
NATIONAL JOURNALISTS GUILD OF SRI LANKA
15/1 First Lane Gothami Road
Colombo 08, Sri Lanka
T: 94 1 564545 F: 94 1 500544

Wickremarathe, Dharman
General Secretary
SRI LANKA ENVIRONMENTAL JOURNALIST FORUM
P O Box 26,434/4
Sri Jayawardenapur, Sri Lanka
T: 94 1 562215 F: 94 1 698315

Zoysa, Uchita De, Mr.
PUBLIC CAMPAIGN ON ENVIRONMENT AND DEVELOPMENT
Palais de Nations
Geneve CH 1211, Sri Lanka
F: 41 22 733-6542

SUDAN - GOVERNMENT

Abdel Nour, Hassan Osman
General Manager
FORESTS NATIONAL CORPORATION
P.O. Box 658
Khartoum, Sudan
T: 447664

Ali, Mohamed Abdulla, Mr
Secretary General
HIGHER COUNCIL FOR ENVIR. & NAT. RE
Ministry of Agriculture
Khartoum, Sudan

Elkhalifa, Mamoun Daovd, Prof
Director
INSTITUTE OF ENVIRONMENTAL
Studies University of Khartoum
Khartoum, Sudan
T: 80993

NGO

Badri, Balghis
SCIENTIFIC ASSOCIATION FOR WOMEN'S STUDIES
PO Box 167
Omdurman, Sudan
T: 249 11 54163

El Tayeb, Somaya
SCIENTIFIC ASSOCIATION FOR WOMEN'S STUDIES
PO Box 167
Omdurman, Sudan
T: 249 11 54163

Elsirag Ahmed, Samia
ES
PO Box 321
Khartoum, Sudan

Jamal, Alawiyya Mohamed, Mr.
ENVIRONMENTALISTS SOCIETY
Institute of Environmental Studies, University of Khartoum, P.O. Box 321
Khartoum 80993, Sudan

Mohamed Babiker, Elham Elgaily
NRPG
PO Box 822
Khartoum, Sudan

PRESS

Gelon, Michael
Editor
NEW HORIZON NEWSPAPER
P.O.Box 2651
Khartoum, Sudan, Sudan

ORGANIZATIONS

Elkhalig, Ragaa Abd, Ms.
Babiker Badri Scientific Association For Women Studies
PO Box 167
Omdurman, Sudan
T: 53363-54409

Gamal, Arif, Mr.
Natural Resources Protection Group
PO Box 822
Khartoum, Sudan

Taha, Izat M., Mr.
Environmentalists Society
PO Box 321
Khartoum, Sudan
T: (249 11) 80 993

GLOBAL FORUM

Ahmed, Samia Elsirac, Ms.
ENVIRONMENTALISTS SOCIETY
Institute of Environmental Studies, University of Khartoum, P.O. Box 321
Khartoum, Sudan

Ali, Osman Mirghani, Ms.
Finance Secretary
SUDANESE ENVIRONMENTAL CONSERVATION SOCIETY
Institute of Environmental Studies, University of Khartoum
Khartoum, Sudan
T: 249 11 8-0963
TL: 22738

Badri, Babiker, Ms.
SCIENTIFIC ASSOCIATION FOR WOMEN STUDIES
Palais de Nations
Geneve CH 1211, Sudan
F: 41 22 733-6542

El-Mograby, Asim Ibrahim, Mr.
Professor
SUDANESE ENVIRONMENTAL CONSERVATION SOCIETY
PO Box 44266
Khartoum, Sudan
T: 249 11 80963
TL: 22738

Elgaily, Elham, Ms.
NATURAL RESOURCES PROTECTION
PO Box 822
Khartoum, Sudan
T: 249 11 445978

Hammour, Fatima Abd Elaal, Ms.
NATIONAL POPULATION COMMITTEE - NPC
St.47. P.O. Box 3995
Khartoum 2, Sudan
T: 249 11 44-3641
E: 3995

Hijazi, Naila, Ms.
W E D O
University of Khartown
Khartown, Sudan

Mostafa, Aisha, Ms.
Coordinator/Consultant
INSTITUTE ENVIRONMENTAL STUDIES
Hotel Paysandu
Khartoum BOX 321, Sudan
T: 249 11 80 933 F: 249 11 355 7565
TL: 92224 AVCUN

Sanousi, Magda M. El, Ms.
Executive Director
BABIKER BEDRI SCIENTIFIC ASSOCIATION FOR WOMEN'S STUDIES
P.O.Box 167
Omdurman, Sudan
T: 249 11 53-363
TL: 24259 MAZEN SD

Seed, Mirghani Tag El, Mr.
Associate Professor
SUDANESE ENVIRONMENTAL CONSERVATION SOCIETY
Institute of Environmental Studies, University of Khartoum
Khartoum, Sudan
T: 249 11 80963
TL: 22738

Zeid, Hashim Abu
MINISTRY OF EDUCATION
12 Rajaji Marg
New Delhi 110011, Sudan
T: 91 11 301-4309

SURINAME - GOVERNMENT

Venetiaan, Runaldo Ronald, Mr.
President
Permanent Mission of Surinam
866 United Nations Plaza, Suite 320
New York NY 10017, United States of America
T: 212 826 0660 F: 212 980 7029

PRESS

Imambaks, Saira-Banu
Reporter Radio/Television
OFFICIAL PRESS SURINAME
Weg Raar Peperpot Bl.201
Suriname, Suriname

GLOBAL FORUM

Cederbourn-Ritfeld, Elfriede, Ms.
NATIONAL COUNCIL OF WOMEN
Van Aalsstraat 13
Paramaribo, Suriname
T: 5 97 497714

Werners-Samsom, Christine, Ms.
Nickeirestraat 4
Paramaribo, Suriname
T: 5 97 499-560

SWAZILAND - GOVERNMENT

Allen, Irma, Ms
Director In-Service Education & Tra
MINISTRY OF EDUCATION SWAZILAND
P.O. Box 135
Mbabane, Swaziland
T: 09268 42376 F: 09268 54328

Chiya, Enock, Mr.
Swaziland Council of Churches
Permanent Mission to the United Nations, 866 United Nations Plaza, Suite 420
New York, NY 10017, United States of America
T: 212 371 8910

Dlamini, C., H.E. Mr.
High Commissioner to Nairobi
Permanent Mission of Swaziland
866 United Nations Plaza, Suite 420
New York, NY 10017, United States of America
T: 212 371 8910

Dlamini, Fikile, Ms.
Attorney Generals Office
Permanent Mission to the United Nations, 866 United Nations Plaza, Suite 420
New York, NY 10017, United States of America
T: 212 371 8910

Dlamini, Mbuso, Mr.
Ministry of Housing and Township Development
Permanent Mission to the United Nations, 866 United Nations Plaza, Suite 420
New York, NY 10017, United States of America
T: 212 371 8910

Dlamini, Nomathemba, Ms.
Ministry of Economic Planning
Permanent Mission to the United Nations, 866 United Nations Plaza, Suite 420
New York, NY 10017, United States of America
T: 212 371 8910

Dlamini, O. Z., Mr.
Member of Parliament
Permanent Mission of Swaziland
866 United Nations Plaza, Suite 420
New York, NY 10017, United States of America
T: 212 371 8910

Dlamini, Obed, Mr.
Prime Minister
Permanent Mission of Swaziland
866 United Nations Plaza, Suite 420
New York NY 10017, United States of America
T: 212 371 8910

Dlamini, Richard M., Mr.
Aide de Camp
Permanent Mission of Swaziland
866 United Nations Plaza, Suite 420
New York, NY 10017, United States of America
T: 212 371 8910

Dlamini, Sicelo, Mr.
Private Secretary to the Prime Minister
Permanent Mission of Swaziland
866 United Nations Plaza, Suite 420
New York, NY 10017, United States of America
T: 212 371 8910

Dlamini, Solomon, Hon.
Minister of Economic Planning and Development
Permanent Mission of Swaziland
866 United Nations Plaza, Suite 420
New York, NY 10017, United States of America
T: 212 371 8910

Dlamini, Stephen, Mr.
Ministry of Natural Resources Energy
Permanent Mission to the United Nations, 866 United Nations Plaza, Suite 420
New York, NY 10017, United States of America
T: 212 371 8910

Dlamini, Timothy L. L., H.E. Dr.
Ambassador Extraordinary and Plenipotentiary
Permanent Representative to the United Nations, New York
Permanent Mission to the United Nations, 866 United Nations Plaza, Suite 420
New York, NY 10017, United States of America
T: 212 371 8910

Dlamini, Wilson, Mr.
Ministry of Agriculture
Permanent Mission to the United Nations, 866 United Nations Plaza, Suite 420
New York, NY 10017, United States of America
T: 212 371 8910

Mabuzu, Harry, Mr.
Chief Protocol Officer
Ministry of Foreign Affairs
Permanent Mission to the United Nations, 866 United Nations Plaza, Suite 420
New York, NY 10017, United States of America
T: 212 371 8910

Magagula, Dumisile, Ms.
Ministry of Economic Planning
Permanent Mission to the United Nations, 866 United Nations Plaza, Suite 420
New York, NY 10017, United States of America
T: 212 371 8910

Mamba, A. V., H.E. Mr.
Ambassador Extraordinary and Plenipotentiary to the United States of America
Permanent Mission of Swaziland
866 United Nations Plaza, Suite 420
New York, NY 10017, United States of America
T: 212 371 8910

Masina, Greenman, Dr.
Family Life Association
Permanent Mission to the United Nations, 866 United Nations Plaza, Suite 420
New York, NY 10017, United States of America
T: 212 371 8910

Mdluli, Sethabile E., Ms.
First Secretary
Permanent Mission of Swaziland
866 United Nations Plaza, Suite 420
New York, NY 10017, United States of America
T: 212 371 8910

Mkhabela, Lolo, Ms.
Ministry of Housing and Township Development
Permanent Mission to the United Nations, 866 United Nations Plaza, Suite 420
New York, NY 10017, United States of America
T: 212 371 8910

Nkambule, Lindi, H.E. Ms.
Personal Secretary to Ambassador
Permanent Mission of Swaziland
866 United Nations Plaza, Suite 420
New York, NY 10017, United States of America
T: 212 371 8910

Nkambule, M. N., Mr.
Principal Secretary
Ministry of Natural Resources and Energy
Permanent Mission to the United Nations, 866 United Nations Plaza, Suite 420
New York, NY 10017, United States of America
T: 212 371 8910

Nkambule, Noah Mtsenjwa, Mr
Principal Secretary
MINISTRY OF NATURAL RESOURCES & ENE
P.O. Box 57
Mbabane, Swaziland
T: 44707

Nkambule, Solomon, Mr.
Ministry of Transport and Communications
Permanent Mission to the United Nations, 866 United Nations Plaza, Suite 420
New York, NY 10017, United States of America
T: 212 371 8910

Shabangu, Isaac, Mr.
Principal Secretary
Ministry of Finance
Permanent Mission to the United Nations, 866 United Nations Plaza, Suite 420
New York, NY 10017, United States of America
T: 212 371 8910

Shabangu, Paul, Mr.
Principal Secretary
Prime Minister's Office
Permanent Mission to the United Nations, 866 United Nations Plaza, Suite 420
New York, NY 10017, United States of America
T: 212 371 8910

Tshabalala, Senzenjani, Hon.
Minister of Natural Resources and Energy
Permanent Mission of Swaziland
866 United Nations Plaza, Suite 420
New York, NY 10017, United States of America
T: 212 371 8910

NGO

Malles, Teresa
YONGE NAWE CONSERVATION CLUBS OF SWAZILAND
Box 2061
Mbabane, Swaziland

PRESS

Dlamini, Lavumisa
SWAZI TV
PO Box 2032
Mbabane, Swaziland, Swaziland

Simelane, Percy N.
Reporter Radio/Television
RADIO SWAZILAND
101 Parkview Heights, Allisier
Mbabane, Swaziland, Swaziland

SWEDEN - GOVERNMENT

Bildt, Carl, Mr.
Prime Minister
Permanent Mission of Sweden
I Dag Hammerskjold Plaza, 46th Floor
New York NY 10017, United States of America
T: 212 751 5900 F: 212 832 0389

Gustaf, Carl XVI
King
Pernament Mission of Sweden
1 Dag Hammerskjold Plaza, 46th Floor
New York NY 10017, United States of America
T: 212 751 5900 F: 212 832 0389

NGO

Bandh, Cecilia
YMCAS
Box 2054
S- 103 12 Stockholm, Sweden

Bertilsson, Cecilia, Ms.
International Secretary
SWEDISH NATIONAL YOUTH COUNCIL
Box 49005
Stockholm, Sweden
T: 46 8 7850690 F: 46 8 6537497

Eder, Josef Boorn Henrik
CP
Fjallgatan 23a
S-116 28 Stockholm, Sweden

Ferrez-Weinberg, Marc
JV
Fjallgaten 23A
S-11628 Stockholm, Sweden

Gabelic, Aleksander, Mr.
International Secretary
SWEDISH NATIONAL YOUTH COUNCIL
Box 49005
Stockholm, Sweden
T: 46 8 7850690 F: 46 8 6537497

Gan, Lin
UL
Box 2017
S-220 02 Lund, Sweden

Hellstrom, Annette, Ms.
Head Of Env. Dept.
FEDERATION OF SWEDISH FARMERS
Lrf
Stockholm 10533, Sweden
T: 46 8 787-5370 F: 46 8 248119

Kristoferson, Lars, Mr.
Dr.
STOCKHOLM ENVIRONMENT INSTITUTE
Box 2142
Stockholm 10314, Sweden
T: 46 8 723-0260 F: 46 8 723-0348
TL: 19580 seis

Pirak, Katarina, Ms.
International Secretary
SWEDISH NATIONAL YOUTH COUNCIL
Box 49005
Stockholm, Sweden
T: 46 8 7850690 F: 46 8 6537497

Strom, Jakob
CP
Fjallgatan 23a
S-116 28 Stockholm, Sweden

Van Pinxteren, Bert
JV
Fjallgaten 23A
S-11628 Stockholm, Sweden

PRESS

Agren, Helen
Correspondent
MINISTRY OF THE ENVIRONMENT
Arrendevagen 5
17546 Jarfalla, Sweden, Sweden

Akerman, Lars
Reporter Radio/Television
SWEDISH TV2
Lovasvagen 22
Bromma, Sweden, Sweden

Anders, Thoren
Editor
SKOGSINDUSTRIERNA
Vapnovagen 47
S-302 45 Halmstad, Sweden, Sweden

Andersson, Thomas B.
Correspondent
LAND MAGAZINE
Oslogatan 33
S-752 64 Uppsala, Sweden, Sweden

Baltscheffsky, Susanna
Reporter
VETENSKAPS JOURNALISTERNA
Soldatvagen 40
Sollentuna 191 73, Sweden
T: 46 8 6436423 F: 46 8 6440306

Beckman, Mats
Reporter Radio/Television
SWEDISH TV4 NEWS
Valbergavogan 3
S-17563 Jorfolla, Sweden, Sweden

Billinghurst, Kevin
Editor
TOMORROW - THE GLOBAL ENVIRONMENT MAGAZINE
Kungsgaten 27
Stockholm, Sweden
T: 46 8 24 34 80 F: 46 8 24 08 09

Bjorkluud, Leif
Reporter Radio/Television
SWEDISH NATIONAL RADIO
Grytagard PL 317
74082 Orsuudsbro, Sweden, Sweden

Bondeson, Karl J.
Correspondent
GOTEBORGS-POSTEN
Korngangen 7
S-442, 39 Kungalv, Sweden, Sweden

Borjesson, Sven A.
Reporter Radio/Television
SWEDISH NATIONAL RADIO
Seminariev 25
352 38 Voxjo, Sweden, Sweden

Dahmen, David, Mr.
Editor
WORLD HORIZONS MAGAZINE
Zetterlunds Vag 138
S-186 51 Vallentuna, Sweden, Sweden

Decker-Molina, Carlos
Editor
SWEDISH NATIONAL RADIO
Surbrunnsgatan 8 nb
11421 Stockholm, Sweden, Sweden

Ekdal, Niklas
Editor
EXPRESSEN
Lundagatan 36:8,
117 27 Stockholm, Sweden, Sweden

Ekeblad, Urban
Correspondent
ALTERNATIVET
Gotgatan 118
118 62 Stockholm, Sweden, Sweden

Englund, Else
Reporter Radio/Television
RADIO SWEDEN
Soldatgatan 6
212 33 Malmoe, Sweden, Sweden

Eriksson, Magnus
Correspondent
GAUDEAMUS
Myggdalswagen 55, 135
Tyreso, Sweden, Sweden

Falck, Karin
Reporter Radio/Television
SWEDISH TELEVISION
Enebygatan 29A
601 18 Norrkoping, Sweden, Sweden

Frolich, Maud
Correspondent
ENVIRONMENT IN SWEDEN
Krikonstigen 38 S-19632
Kungsagen, Sweden, Sweden

Gahrton, Per
Correspondent
MILJOMAGASINET ALTERNATIVET
Henriksdalsringen 35
S-13132 Nacqa, Sweden, Sweden

Ganning, Bjorn
Editor
AMBIO-JOURNAL OF HUMAN ENV.
S-16354 Spanga
Sweden, Sweden

Gunnar, Vagerstam
Editor
SWEDISH TELEVISION
Muraregatan 14A
652 28 Karlstad, Sweden, Sweden

Gustavsson, Ingrid
Reporter Radio/Television
SWEDISH NATIONAL RADIO
Hesselmans vag 3, 752
Uppsala, Sweden, Sweden

Haglind, Carl Jurgen
Correspondent
SAF
Professorsgatan 5A, 223
63 Lund, Sweden, Sweden

Haldin, Jan
Correspondent
GOTEBORGS-POSTEN
Kabelgatan 15B
S-414, 57 Goteborg, Sweden, Sweden

Hammarstrom, Tommy
Editor
EXPRESSEN
Knutsbog Dunker
64032 Malmkoping, Sweden, Sweden

Hanneberg, Peter
Editor
ENVIRO MAGAZINE-SNV
Beckasinvagen 45 B
S-131 44 Nacka, Sweden, Sweden

Hard, Berit
Editor
TILL LANS
Luthagsesplanader 11
75225 Uppsala, Sweden, Sweden

Hasselgren, Maria K.
Correspondent
BRODERSKAP
Kantorsgatan 46
75424 Uppsala Sweden, Sweden

Hoglin, Renee
Correspondent
KOMMUNALARBETAREN
Gotgatan 118
118 62 Stockholm, Sweden, Sweden

Holmgren, Lars, Mr.
Correspondent
INTERNATIONALEN
Karraviken Medevi
S-59199 Motala, Swedem, Sweden

Jansen, Per
Reporter Radio/Television
SWEDISH EDUCATIONAL B'CASTING CORP.
Kavesta 69072
Skollersta, Sweden, Sweden

Johansson, Bo Bertil
Correspondent
NFLA-NOTICIAS LATINO-AMERICANAS
Rua Paissandu 156/707
Rio de Janeiro, Brazil, Sweden

Kamp, Kristina
Correspondent
AFTONBLADET
Oruaroddsvag 6
S-11254 Stockholm, Sweden, Sweden

Karlsson, Kalle
Editor
MEDIA EXPRESS
5-12646 Hagerssen
Sweden, Sweden

Karlsson, Lars-Ingmar
Editor
DAGENS NYHETER
Bergengatang, 6tr
S-16435 Kista, Sweden, Sweden

Karlsson, Srig
Reporter Radio/Television
RADIO NORRBOTTEN
Ostra Vilaugen 30A
95161 Luiea, Sweden, Sweden

Karvonen, Anita
Reporter Radio/Television
SWEDISH NATIONAL RADIO
Anders Reimers Vas 4, III.TR
11750 Stockholm, Sweden, Sweden

Kessler, Elisabeth, Ms.
Editor
AMBIO-JOURNAL OF HUMAN ENV.
Gubbkarrsbacken 29
Bromma S-161 51 Sweden, Sweden

Kind, Berit, Ms.
Publ Editor
ROYAL SWEDISH ACADEMY OF SCIENCES
Ambio Box 50005
Stockholm, Sweden
T: 46 8 150430 F: 46 8 166251

Kirwin, James
Correspondent
TOMORROW MAGAZINE
Kvarnosvagen 13
18451 Osterskor, Sweden, Sweden

Kjeang, Are, Mr.
Energy Adviser
SOCIETY OF ENERGY COUNSELLING IN SWEDEN
c/o Are Kjean , Falkenberg Community
Falkenberg S-31180, Sweden
T: 46 3 8-6040 F: 46 3 1-3340

Laquist, Anne, Ms.
Editor
SWEDISH ASSOCIATION FOR NATURE CONSERVATION
Box 4625
11691 Stockholm, Sweden
T: 46 8 7026500 F: 46 8 7022702

Laurell, My
Correspondent
UPSALA NYA TIDNING
Slagugglan 22
S-742 36 Osthammar, Sweden, Sweden

Liljedahl, Ola
Correspondent
EXPRESSEN
155 E 29th St. ap. 330
New York, NY 10016, USA, Sweden

Lindevall, Bo, Mr.
Editor
SKOGEN (THE FOREST)
Haggkullev.81
184 37 Akersberga, Sweden, Sweden

Lindstrom, Per
Correspondent
CENTRAL SWEDISH NEWS AGENCY
Tappgrand 109, III
S-122 50 Enskede, Sweden, Sweden

Lockne, Gunilla
Editor
SUNDSVALLS TIDNING
Larsbodavagen 38
12341 Farsta, Sweden, Sweden

Midman, Tadia Ednond
Editor
TIDNINNGEN NU
Urbergsvagen 4
Stockholm, Sweden, Sweden

Nilsson, Annika
Reporter Radio/Television
SWEDISH NATIONAL RADIO
Atlestigen 7
S-141 41 Huddinge, Sweden, Sweden

Nyberg, Svante, Mr.
Lic Ph
NATURAL RESOURCES MANAGEMENT INSTITUTE
Stockholm University S-106 91
Stockholm, Sweden
T: 46 8 164215 F: 46 8 155987

Palmeus, Lennart
Correspondent
DAGENS INDUSTRI
R. Ubaldino Amaral, 40, C-01
20231-Rio de Janeiro-RJ, Brazil, Sweden

Sahlim, Mats
Correspondent
LIBERAL YOUTH OF SWEDEN
G Maynig 15
113 83 Stulen, Sweden, Sweden

Samuelsson, Marianne
Correspondent
MILJOMAGASINET
Pl. 5505 S-44194 Alingsas
Sweden, Sweden

Sjoberg, Claes
Editor In Chief
TOMORROW - THE GLOBAL ENVIRONMENT MAGAZINE
Kungsgaten 27
Stockholm, Sweden
T: 46 8 24 34 80 F: 46 8 24 08 09

Spens, Elizabeth
Correspondent
HALSA
Berguddsvagen 5
S-182 33 Danderyd, Sweden, Sweden

Strandberg, Hans
Correspondent
SVENSKA DAGBLADET
Vallingbyuagen 87
16221 Vallingby, Sweden, Sweden

Svensson, Peter
Correspondent
EXPRESSEN
Upplandsgatan 33
S-11328 Stockholm, Sweden, Sweden

Thibaud, Ewa
Correspondent
DAGENS NYHETER
300 E 56th St. (#11G)
New York, NY 10022, Sweden

Thoren, Anders
Editor
SKOGSINDUSTRIERNA
Vapnovagen 47
S-302 45 Halmstad, Sweden, Sweden

Thulin, Asa
Correspondent
MINISTRY OF THE ENVIRONMENT
Bordsvagen 52
12246 Enskede, Sweden, Sweden

Vagerstam, Gunnar
Editor
SWEDISH TELEVISION
Muraregatan 14A
852 28 Karlstad, Sweden, Sweden

Von Schmeling, Astrid
Correspondent
TOMORROW MAGAZINE
Boukholmeg 37
116 63 Stockholm, Sweden, Sweden

ORGANIZATIONS

Daleus, Lennart, Mr.
Swedish Campaign For Environment And Development 92, The
Box 15050
S-104 65 Stockholm, Sweden
T: (46 8) 644 6671 F: (46 8) 641 8876

Hammar, Marcus, Mr.
Swedish United Nations Association
Svenska FN-Forbundet, Box 15115
S-10465 Stockholm, Sweden
T: (46 8) 449 835 F: (46 8) 418 876

Landholm, Ulla, Ms.
YWCA - YMCA of Sweden
Box 2054
S- 103 12 Stockholm, Sweden

Rosemarin, Arno, Dr.
Stockholm Environment Institute
Box 2142
S-104 13 Stockholm, Sweden
T: (46 8) 723 0260 F: (46 8) 723 0348

Tandberg, Olof G., Dr.
Royal Swedish Academy of Sciences
Box 50005
104 05 Stockholm, Sweden
F: (46 8) 150 430

GLOBAL FORUM

Agerstrom, Magnus, Mr.
NATLIKAN AB
Markvardsq.5
Stockholm, Sweden

Ahlgren, Sven-Olov, Mr.
Photographer
Box 39
Knutby S-740 12, Sweden
T: 46 174 40085

Ahnberg, Annika, Ms.
SWEDISH LEFT PARTY
Nydalav 129
Sodertaye, Sweden
T: 46 8 786-4000

Alarik, Oscar, Mr.
FALTBIOLOGERNA/IYF
Lugnesberg, S-755 91
Uppsala, Sweden
T: 46 18 754-4640 F: 46 18 754-5365

Algerman, Lars, Mr.
Science Editor
SWEDISH TELEVISION, CHANNEL 2 NEWS
Lovasvogen 22
Brommo S-161 35, Sweden
T: 46 8 261686 F: 46 8 261686

Allard, Goran, Mr.
STIFTETS SKOGAR
Box 1334 - 62124 Visby
Visby, Sweden

Anderson, Thomas, Mr.
Journalist
LAND MAGAZINE
Stockholm 10533, Sweden
T: 46 8 787-5100 F: 46 8 211421

Andersoon, Anette
Photographer
VETENSKAPS JOURNALISTERNA
Soldatvagen 40
Sollentuna 191 73, Sweden
T: 46 8 6436423 F: 46 8 6440306

Andersson, Anders-P, Mr.
CREATURES OF THE SWEDISH FOREST
St. Larsv. 18
Lund 22270, Sweden
T: 46 46 13 4222 F: 46 46 18 9726

Angren, Sven Olaf
Photographer
Freelancer
Box 39 74012
Knutby, Sweden
T: 46 17440085

Anner, Carina, Ms.
Periodista
T U F F
Lovstigenio 13550 Tyneso
Stockholm, Sweden
T: 46 8 712-4232

Arrhenius, Erik, Mr.
Prof.
NATURAL RESOURCES MANAGEMENT INSTITUTE
Stockholm University S-106 91
Stockholm, Sweden
T: 46 8 164215 F: 46 8 155987

Asander, Maj-Britt
UN ASSOCIATION OF SWEDEN
Larmgatan 6A 90322
Umea, Sweden
T: 46 90 117605

Astrand, Kerstin, Ms.
FONO/ESAM AB
Mariehemvagen 6, BOX 6554
Umea, Sweden
T: 46 90 122598 F: 46 90 163719

Astrom, Anita, Ms.
Psychologist
UN ASSOCIATION OF SWEDEN
Svidjev. 21, 904 40 Roback
Umea, Sweden
T: 46 90 22259

Astrom, Lage, Mr.
Ing.
UN ASSOCIATION
Svidjerigen 21 90440 Roback
Umea 90440, Sweden
T: 46 90 22259

Backman, Ida-Lota, Ms.
Director
HIWAK PRODUCTIONS THEATRE
Box 73
Karesuando, Sweden
T: 46 9 812 0242

Bamdh, Cecilia, Ms.
Y M C A - SWEDEN
Box 2054
Stockholm S-10312, Sweden
T: 46 8 145330 F: 46 8 217522

Bengtsson, Tommy, Mr.
Dr.
ROYAL SWEDISH ACADEMY OF SCIENCES
Box 50005, S-104 05
Stockholm, Sweden
T: 46 8 8-15 04 30 F: 46 8 8-15 56 70

Bentsson, Ingemund, Mr.
SWEDISH DELEGATION TO UNCED
Varberg
Varberg, Sweden

Berglof, Lina, Ms.
BSC
NATURAL RESOURCES MANAGEMENT INSTITUTE
Stockholm University S-106 91
Stockholm, Sweden
T: 46 8 164215 F: 46 8 155987

Bjork, Tord, Mr.
Ba
FONO/ESAM AB
Mariehemvagen 6, BOX 6554
Umea, Sweden
T: 46 90 122598 F: 46 90 122170

Bjorklund, Leif, Ms.
Producer
SWEDISH NATIONAL RADIO COMPANY
Science Department - Swedish National Radio Company
Stockholm S-105 10, Sweden
T: 46 8 784-1727 F: 46 8 784-1500

Bjorn, Gronholm, Mr.
Student
NORDIC CENTER YOUTH
Bergsgatan 7B S/10422
Stockholm 10422, Sweden
T: 46 8 652-9951 F: 46 8 652-6440

Bokalders, Varis, Mr.
Architect
STOCKHOLM ENVIRONMENT INSTITUTE
Box 2142
Stockholm 10314, Sweden
T: 46 8 723-0260 F: 46 8 723-0348
TL: 19580 sei

Boschi, Ana Maria
UNIVERSIDAD NACIONAL DEL COMAHUE
Box 6508
Stockholm 11383, Sweden
T: 46 8 150600 F: 46 8 153226

Brown, Amy
Associate Editor
TOMORROW - THE GLOBAL ENVIRONMENT MAGAZINE
Kungsgaten 27
Stockholm, Sweden
T: 46 8 24 34 80 F: 46 8 24 08 09

Brundin, Gunnar, Mr.
FONO/ESAM AB
Mariehemvagen 6, BOX 6554
Umea, Sweden
T: 46 90 122598 F: 46 90 122170

Bruun, Christina, Ms.
SWEDISH UNITED NATIONS ORGANIZATION
Stureplan 6, 11435
Stockholm 11435, Sweden
T: 46 8 679-9500 F: 46 8 611-0160

Carlsson, Reidar, Mr.
MINISTRY OF THE ENVIRONMENT AND NATURAL RESOURCES
Fredsgatan 8
Stockholm S-103 33, Sweden
T: 46 8 763-1000 F: 46 8 219628

Carlsson, Sylvia, Ms.
Grasloksgatan 18 754 46
Uppsala, Sweden

Charles, Norman, Mr.
Actor
HIWAK
Box 73
Karesuando, Sweden

Dahl, Anna, Ms.
WORLD YOUNG WOMEN'S CHRISTIAN ASSOCIATION
Badstrandsvagen 26
Stockholm 11265, Sweden
T: 46 8 753 89 27 F: 46 8 753 60 48
TL: WYMCA 41 23 32

Edman, Stefan, Mr.
Vice President
SWEDISH SOCIETY FOR NATURE CONSERVATION
Box 4625
Stockholm 5-11691, Sweden
T: 46 8 7026500 F: 46 8 7020855

Eeg-Olofsson, Mia Karolina, Ms.
THE INTERNATIONAL YOUTH RED CROSS/SWEDISH YOUTH FOR UNCED
Skogsgatan 75
Linkoping 58257, Sweden
T: 46 13 154159

Ek, Aina, Ms.
SWEDISH FORESTRY ASSOCIATION
Stockholm, Sweden
F: 46 8 755-8602

Ek, Ingvar, Mr.
SWEDISH FORESTRY ASSOCIATION
Stockholm, Sweden
F: 46 8 755-8602

Ekblad, Permilla, Ms.
CREATURES OF THE SWEDISH FOREST
St. Larsv. 18
Lund 22270, Sweden
T: 46 46 13 4222 F: 46 46 18 9726

Engberg, Peter
FOLKBILDNINGSFORBUNDET
12 Rajaji Marg
New Delhi 110011, Sweden
T: 91 11 301-4309

Erickson, Carina, Ms.
Administrator
SWEDISH CAMPAIGN FOR ENVIRONMENT & DEVELOPMENT 92
P.O.Box 15050
Stockholm 10465, Sweden
T: 46 8 644-6671 F: 46 8 641-8876

Ericsson, Sven, Mr.
NATLIKAN
Markvardsgatan 5
Stockholm 11353, Sweden

Eriksson, Bernt, Mr.
CREATURES OF THE SWEDISH FOREST
St. Larsv. 18
Lund 22270, Sweden
T: 46 46 13 4222 F: 46 46 18 9726

Eriksson, Ingela, Ms.
SWEDISH COMPANY FOR ENVIRONMENT AND DEVELOPMENT F-92
Box 15050
Stockholm 10465, Sweden
T: 46 8 644-6671 F: 46 8 641-8876

Essen, Charlotte Von, Ms.
FONO/ESAM AB
Mariehemvagen 6, BOX 6554
Umea, Sweden
T: 46 90 122893 F: 46 90 163719

Fedrizzi, Beatriz, Ms.
FCSB
Osbygatan 19
Malmo 21443, Sweden
T: 46 40 233900

Folkesson, Helena, Mr.
LIBERAL YOUTH OF SWEDEN
Box 6508
Stockholm 11383, Sweden
T: 46 8 150600 F: 46 8 153226

Forsberg, Meri Helena, Ms.
SWEDISH UNITED NATIONS ORGANIZATION
Stureplan 6, 11435
Stockholm 11435, Sweden
T: 46 8 679-9500 F: 46 8 611-0160

Francke, Ulrika, Ms.
President
CITY OF STOCKHOLM
Karrgrand 53, S-162 46 Vallingby
Stockholm, Sweden

Fries, Ingrid, Ms.
Teacher Ba
FORESTERS ASSOCIATION
Backvagen 7
Upplands Vasby 19440, Sweden
T: 46 18 671983

Fries, Joran, Mr.
Professor
FORESTERS ASSOCIATION
Backvagen 7
Upplands Vasby 19440, Sweden
T: 46 18 671983

Froberg, Odd Jonas Axel, Mr.
Student
CAMPAIGN FOR THE EARTH
Leljevagen 3
Kungsbacka, Sweden
T: 46 30 201113 F: 46 30 141410

Ganning, Biurn
Science Editor
ROYAL SWEDISH ACADEMY OF SCIENCES
Ambio Box 50005
Stockholm S-104 05, Sweden
T: 46 8 150430 F: 46 8 166251

Ganrton, Per, Mr.
MILSE MAGASINET ALTERNATIVET
Henriusdalsringea 35
Stockholm S-10172, Sweden
T: 46 8 615-2408 F: 46 8 615-2363

Granstrom, Jens
Representative
LINGILIN
Polhemsgatan 21
11230 Stockholm, Sweden
T: 46 8 6509887 F: 46 8 6442590

Gustafsson, Ingrid
SWEDISH NATIONAL RADIO COMPANY
Science Department S-105 10
Stockholm, Sweden
T: 46 8 7841727 F: 46 8 7841500

Gustavsson, Tomas, Mr.
SWEDISH UNITED NATIONS ORGANIZATION
Stureplan 6, 11435
Stockholm 11435, Sweden
T: 46 8 679-9500 F: 46 8 611-0160

Haglind, Jorgen
Correspondent
SAF- TIDNINGEN
Professoragatan 5A
Lund 22363, Sweden
T: 46 46 146750 F: 46 46 114709

Hakanson, Hakan, Mr.
Director
SKOGSVARDSSTYRELSEN
Box 50123
Lund, Sweden
T: 46 46 920-10090 F: 46 46 920-12588

Hallman, Katrin, Ms.
MINISTRY OF THE ENVIRONMENT AND NATURAL RESOURCES
Fredsgatan 8
Stockholm S-103 33, Sweden
T: 46 8 763 1000 F: 46 8 21 9628

Hamilton, Raoul, Mr.
Count
OVERSHOLMS GODS
Ovesholms Castle
Kristianstad 29192, Sweden
T: 46 44 239096 F: 46 44 239086

Hedman, Gunnel, Mr.
MINISTRY OF THE ENVIRONMENT AND NATURAL RESOURCES
S-103 33 Prodsgatan 8
Stockholm, Sweden
T: 46 8 763-1000 F: 46 8 219628

Hellstrom, Erik, Mr.
Landscape Architect
ORG. FOR LANDSCAPE ARKITEKTS
EDS Gatanv. 3
Vanersborg 46234, Sweden
T: 46 521 60540 F: 46 521 17985

Hjelte, Hans, Mr.
Forest Biologist
DOMAN SKOG
Box 810
Lund, Sweden

Holnigsen, Lars
Correspondent
ENVIRONMENTAL FEDERATION - ENVIRONMENTAL MAGAZINE
Bona FHS
Motala, Sweden
T: 46 141 233923 F: 46 141 976008

Hoog, Leif, Mr.
SWEDISH UNITED NATIONS ORGANIZATION
Stureplan 6, 11435
Stockholm 11435, Sweden
T: 46 8 679-9500 F: 46 8 611-0160

Hulth, Mats, Mr.
Vice Mayor
CITY OF STOCKHOLM
Stockolm City Hall, S-105 35
Stockholm, Sweden
T: 46 8 785-9230 F: 46 8 785-9876

Hultin, Torsten, Mr.
HUMANISTIC FORUM
Syster Emmas Gata II
Gothenburg 43024, Sweden
T: 46 31 821231 F: 46 31 336618

Hunhammar, Sven, Mr.
STOCKHOLM ENVIRONMENT INSTITUTE
Box 2142
Stockholm 10314, Sweden
T: 46 8 723-0260 F: 46 8 723-0348
TL: 19580 seis

Ihse, Margareta, Ms.
Vice President
SWEDISH SOCIETY FOR NATURE CONSERVATION
Box 4625
Stockholm 5-11691, Sweden
T: 46 8 7026500 F: 46 8 7020855

Janse, Per
SWEDISH NATIONAL RADIO COMPANY
Science Department S-105 10
Stockholm, Sweden
T: 46 8 7841727 F: 46 8 7841500

Jensen, Tove, Ms.
Instructor
HUNGER PROJECT
Love Almquistsvag 42
Sollentuna 19144, Sweden
T: 46 8 964834 F: 46 8 336618

Jertfelt, Pia, Ms.
FUTURE EARTH
Fjallg. 28A
Stockholm, Sweden
T: 46 8 642-6225

Johannisson, Eva, Ms.
NATIONAL DEFENSE RESEARCH ESTABLISHMENT/FOA
17290 Sundryberg
Stockholm, Sweden
T: 46 8 663.1500 F: 46 8 733.5240

Johansson, Mats, Mr.
Student
BONA FOLLEHIGHSCHOOL/ENVIRONMENTAL FEDERATION
Bona FHS
Motala S-591 97, Sweden
T: 46 141 21100 F: 46 141 21051

Johansson, Olof, Mr.
DOMAN SKOG
Stockholm, Sweden
F: 46 8 755-8602

Jonsson, Mrs Sibylla
UN ASSOCIATION OF SWEDEN
Storgatan, 23
Vilhelmina 91233, Sweden
T: 10-727

Jonsson, Roger, Mr.
ENVIRONMENTAL FEDERATION/BONA FHSK
Bona FHSK
Motala S-59197, Sweden
T: 46 141 21100 F: 46 141 21051

Junker, Kerstin, Ms.
Director Of Education
ZONTA INTERNATIONAL
Munkvagen 26
Smedjebalken 77751, Sweden
T: 46 240 39622 F: 46 240 39622

Junker, Norbert, Mr.
ZONTA INTERNATIONAL
Munkvagen 26
Smedjebalken 77751, Sweden
T: 46 240 39622 F: 46 240 39622

Kampfie, Bengt, Mr.
STIFTETS FOREST
Box 1314
Uppsala 75143, Sweden
T: 46 18 753-0390 F: 46 18 755-8602

Karlsson, Sylvia, Ms.
Student
SWEDISH BAHA'I COMMUNITY
Grasloksgatan 18
Uppsala, Sweden
T: 46 18 250506 F: 46 760 73885

Kauppi, Monica, Ms.
SWEDISH DENTAL PATIENT ORGANIZATION
Tegelbruksv. 2, 12632 Hagersten
Stockholm, Sweden
T: 46 8 184086 F: 46 8 6450086

Kemper, Karin, Ms.
International Secretary
SWEDISH NATIONAL YOUTH COUNCIL
Box 49005
Stockholm, Sweden
T: 46 8 7850690 F: 46 8 6537497

Kessler, Ann
AMBIO-ROYAL SWEDISH ACADEMY OF SCIENCES
Box 50005
Stockholm S - 104 05, Sweden
T: 46 8 15-0430 F: 46 8 16-6251

Kristensen, Torben, Mr.
General Manager
SCANWATER AB
Box 34
Varberg S-43221, Sweden
T: 46 340 78040 F: 46 340 83714

Kristoffersson, Nils Olof, Mr.
DEVELOPMENT FORUM OF THE SWEDISH CHURCHES
Gotgatan 3 nb
Uppsala S-73 15, Sweden
T: 46 18 54-9770 F: 46 18 50-5107

Kronvall, Stellan, Mr.
MINISTRY OF THE ENVIRONMENT AND NATURAL RESOURCES
Fredsgatan 8
Stockholm S-103 33, Sweden
T: 46 8 763-1000 F: 46 8 21 96 18

Kvick, Nils, Mr.
SWEDISH DENTAL PATIENT ORGANIZATION
Vallavagen 5
Stockholm, Sweden

Lackenbauer, Helene, Ms.
Theologian
DEVELOPMENT FORUM OF THE SWEDISH CHURCHES
Gotgatan 3 nb
Uppsala S - 753 15, Sweden
T: 46 18 549770 F: 46 18 505107

Landberg, Hans, Mr.
Professor
ROYAL SWEDISH ACADEMY OF SCIENCES
Box 50005, S-104 05
Stockholm, Sweden
T: 46 8 8-15 04 30 F: 46 8 8-15 56 70

Landborn, Gunnar, Mr.
Secretary General
SWEDISH SOCIETY FOR NATURE CONSERVATION
Box 4625
Stockholm 5-11691, Sweden
T: 46 8 7026500 F: 46 8 7020855

Larson, Hildegun, Ms.
UN ASSOCIATION OF SWEDEN
Korpralso 50A
Umea 90432, Sweden

Larsson, Kjell-Arne, Mr.
SWEDISH SOCIETY FOR NATURE CONSERVATION
P.O. Box 1807
Uppsala S-75148, Sweden
T: 46 18 10-6152

Larsson, Margareta, Ms.
CREATURES OF THE SWEDISH FOREST
St. Larsv. 18
Lund 22270, Sweden
T: 46 46 13 4222 F: 46 46 18 9726

Larsson, Peter, Mr.
Expert
SOC. DEM. PARTY
Stockholm, Sweden
T: 46 8 786-4347 F: 46 8 211524

Larsson Holmberg, Birgitta, Ms.
Teacher
UN ASSOCIATION OF SWEDEN
9050 S- 68500 Torsby
Torsby, Sweden
T: 560 31238

Lembeig, Ulla, Ms.
VORRUNAL ARBEFLAREW
Henriusdalsringen 35
Stockholm S-10172, Sweden
T: 46 8 6152408 F: 46 8 6152363

Levin, Titti, Mr.
WORLDWIDE NETWORKING FOR INNER PEACE
Gibraltargatan 20
Gothenburg 41132, Sweden
T: 46 31 168759

Lilja, Gunnel, Ms.
Professor
CITY OF STOCKHOLM
S-105 35 City Hall
Stockholm, Sweden
T: 46 8 785-9120 F: 46 8 650-8668

Lilja, Torsten, Mr.
Town Clerk
CITY OF STOCKHOLM
S-105 35 City Hall
Stockholm, Sweden
T: 46 8 785-9120 F: 46 8 650-8668

Lindahl-Kiessling, Kerstin, Ms.
Professor
ROYAL SWEDISH ACADEMY OF SCIENCES
Box 50005, S-104 05
Stockholm, Sweden
T: 46 8 8-15 04 30 F: 46 8 8-15 56 70

Lindgren, Kerstin, Ms.
CREATURES OF THE SWEDISH FOREST
St. Larsv. 18
Lund 22270, Sweden
T: 46 46 13 4222 F: 46 46 18 9726

Lindhe, Jenny, Ms.
BONA FHSK
Bona FHSK
Motala S-59197, Sweden
T: 46 141 21100 F: 46 141 21051

Lindne, Jenny
Editor
MOTALA NEWSPAPER
Industrigatan 9 Box 945
Motala 59760, Sweden
T: 0141 16390

Lindzvist, Hans, Mr.
Potitcien
COUNTY COUNSIL STOCKHOLM
Backregen 8
Stockholm 13900, Sweden
T: 46 8 737-2500

Ljungberg, Lars Ove, Mr.
Secretary Gen.
DEVELOPMENT FORUM OF THE SWEDISH CHURCHES
Gotgatan 3 nb
Uppsala S - 753 15, Sweden
T: 46 18 549770 F: 46 18 505107

Lofgren, Lars, Mr.
Member
GREEN PARTY SWEDEN
Ennvvveommen
Stockholm, Sweden
T: 46 8 650-5590

Lundberg, Hans, Mr.
Environmental Secretary
ROYAL SWEDISH ACADEMY OF SCIENCES
Box 50005, S-104 05
Stockholm, Sweden
T: 46 8 8-15 04 30 F: 46 8 8-15 56 70

Lunden, Jan-Ake, Mr.
SODRA SKOGSAGARNA
Stockholm, Sweden

Lunden, Sven, Mr.
Chairman
HUMANISTISKT FORUM
Mjorn 1306
Grabo 44006, Sweden
T: 46 30242281 F: 46 30242560

Malmberg, Ole, Mr.
Chief District Medical Officer
SWEDISH DOCTORS FOR ENVIRONMENT/NORDIC SOCIETY FOR HUMAN ECOLOGY
Handskmakaregatan 2C
Kristianstad S-291 54, Sweden
T: 46 44 310537 F: 46 44 311945

Mansson, Tommy, Mr.
Dep. Director
ENVIRONMENTAL ADVISORY COUNCIL
Ministry of Environment of Sweden
Stockholm S-103 33, Sweden
T: 46 8 763-2059 F: 46 8 204331

Martins, Anna, Ms.
BONA FHSK
Bona FHSK
Motala S-59197, Sweden
T: 46 141 21100 F: 46 141 21051

Mauritzon, Dagmar, Ms.
SWEDISH UNITED NATIONS ORGANIZATION
Stureplan 6, 11435
Stockholm 11435, Sweden
T: 46 8 679-9500 F: 46 8 611-0160

Maxe, Gote, Mr.
BONA FELKHOGSKALA
Bona FHSK
Motala S-59197, Sweden
T: 46 141 21100 F: 46 141 21051

Molander, Karin, Ms.
DEVELOPMENT FORUM OF THE SWEDISH CHURCHES
Enerhogen 37
Hjalteby 44074, Sweden
T: 46 31 66 78 90 F: 46 31 82 07 13

Molander, Sverker, Dr.
DEVELOPMENT FORUM OF THE SWEDISH CHURCHES
Gotgatan 3 nb
Uppsala S-73 15, Sweden
T: 46 18 54-9770 F: 46 18 50-5107

Niklasson, Laila, Ms.
SWEDISH UNITED NATIONS ORGANIZATION
Stureplan 6, 11435
Stockholm 11435, Sweden
T: 46 8 679-9500 F: 46 8 611-0160

Nilsson, Lars J., Mr.
M.Sc.
UNIVERSITY OF LUND - DEESS
Gerdagatan 13, S-22362
Lund, Sweden
T: 46 46 104833 F: 46 46 108644

Nord, Par
Editor
RIKSRADION
S-851 80
Sundsvall, Sweden
T: 46 60 15 44 40 F: 46 60 11-09 27

Nordanstig, Gunner, Mr.
Project Leader'
NATIONAL BOARD OF FORESTRY
Stogsvardsstyreisen - Box 1213
Visby 62123, Sweden
T: 46 36 49818000 F: 46 36 498-8595

Nordmalm, Pelle, Mr.
Information Officer
SWEDISH SOCIETY FOR NATURE CONSERVATION
Box 4625
Stockholm 5-11691, Sweden
T: 46 8 7026500 F: 46 8 7020855

Nystrom, Bruno, Mr.
SWEDISH UNITED NATIONS ORGANIZATION
Stureplan 6, 11435
Stockholm 11435, Sweden
T: 46 8 679-9500 F: 46 8 611-0160

Ohman, May-Britt
LIBERAL YOUTH OF SWEDEN
Box 6508
Stockholm 11383, Sweden
T: 46 8 150600 F: 46 8 153226

Olausson, Birgitta, Ms.
SWEDISH UNITED NATIONS ORGANIZATION
Stureplan 6, 11435
Stockholm 11435, Sweden
T: 46 8 679-9500 F: 46 8 611-0160

Ornstedt, Svea
SWEDISH UNITED NATIONS ORGANIZATION
Stureplan 6, 11435
Stockholm 11435, Sweden
T: 46 8 679-9500 F: 46 8 611-0160

Osterberg, Arne, Mr.
City Commissioner
CITY OF GOTHENBURG
S-40482
Gothenburg, Sweden
T: 46 31 611000 F: 46 31 138886

Owerholm, Cecelia
TOMORROW - THE GLOBAL ENVIRONMENT MAGAZINE
Kungsgaten 27
Stockholm, Sweden
T: 46 8 24 34 80 F: 46 8 24 08 09

Paulsson, Valfrid, Mr.
SWEDISH DELEGATION TO UNCED
Tornbyvagen 1, S-19441
Upplands Vasby, Sweden

Pereira, Denilza De Souza, Sra.
Guide
CREATURES OF THE SWEDISH FOREST
St. Larsv. 18
Lund 22270, Sweden
T: 46 46 13 4222 F: 46 46 18 9726

Persson, Annette, Ms.
CREATURES OF THE SWEDISH FOREST
St. Larsv. 18
Lund 22270, Sweden
T: 46 46 13 4222 F: 46 46 18 9726

Persson, Sture, Mr.
MINISTRY OF THE ENVIRONMENT AND NATURAL RESOURCES
Fredsgatan 8
Stockholm S-103 33, Sweden
T: 46 8 763-1000 F: 46 8 219628

Petre, Hans, Mr.
Chief Forester
COUNTRY FORESTRY BOARD
Falhemssvagen 52
Falun, Sweden
T: 46 23 11416 F: 46 23 19297

Petre, Ingrid, Ms.
Chief Forester
COUNTRY FORESTRY BOARD
Falhemssvagen 52
Falun, Sweden
T: 46 23 11416 F: 46 23 19297

Pieterek, Ulla, Ms.
CREATURES OF THE SWEDISH FOREST
St. Larsv. 18
Lund 22270, Sweden
T: 46 46 13 4222 F: 46 46 18 9726

Planhem, Marie, Ms.
SWEDISH UNITED NATIONS ORGANIZATION
Stureplan 6, 11435
Stockholm 11435, Sweden
T: 46 8 679-9500 F: 46 8 611-0160

Ramel, Claes, Mr.
ROYAL SWEDISH ACADEMY OF SCIENCES
Box 50005, S-104 05
Stockholm, Sweden
T: 46 8 8-15 04 30 F: 46 8 8-15 56 70

Ratti-Carbonari, Graciela Hilda, Ms.
Lawyer
UNIVERSITY OF LUND
PO Box 207
Lund S-221 00, Sweden
T: 46 46 108088 F: 46 46 104720
TL: 33533 LINIVER S

Rosenmarin, Arno
Information Manager
STOCKHOLM ENVIRONMENT INSTITUTE
Box 2142
Stockholm 103 14, Sweden
T: 46 8 7230260 F: 46 8 7230348

Rosslin, Linda, Ms.
CREATURES OF THE SWEDISH FOREST
St. Larsv. 18
Lund 22270, Sweden
T: 46 46 13 4222 F: 46 46 18 9726

Rudback, Goran T., Mr.
Bsc
NATURAL RESOURCES MANAGEMENT INSTITUTE
Stockholm University S-106 91
Stockholm, Sweden
T: 46 8 164215 F: 46 8 155987

Rudolfsson, Martin, Mr.
BONA FELKHOGSKALA
Bona FHSK
Motala S-59197, Sweden
T: 46 141 21100 F: 46 141 21051

Sandbu, Magnus D., Mr.
Llm
RAOUL WALLENBERG INSTITUTE
S:t Annegatan 4
Lund 223-50, Sweden
T: 46 46 104311 F: 46 46 104445

Schedin, Sven O., Mr.
Consultant
Fyrmastaregrngen 16
Gothenburg 5-41318, Sweden
T: 46 31 112404 F: 46 31 114946

Schmeling, Astrid Von
Editorial Assistant
TOMORROW - THE GLOBAL ENVIRONMENT MAGAZINE
Kungsgaten 27
Stockholm, Sweden
T: 46 8 24 34 80 F: 46 8 24 08 09

Segelstroem, Inger, Ms.
SWEDISH UNITED NATIONS ORGANIZATION
Stureplan 6, 11435
Stockholm 11435, Sweden
T: 46 8 679-9500 F: 46 8 611-0160

Silveira, Serge, Mr.
Manager
OVERSHOLMS GODS
Ovesholms Castle
Kristianstad 29192, Sweden
T: 46 44 239096 F: 46 44 239086

Simma-Charles, Asa, Ms.
Actress
HIWAK
Box 73
Karesuando, Sweden

Sindre, Magnusson, Mr.
FALTBIOLOGERNA/IYF
Lungnesberg
Uppsala 75591, Sweden
T: 46 18 360446

Skoglund, Gunnar, Mr.
F.P. Landstdi Rail
COUNTY COUNSIL STOCKHOLM
Livholnsci 21
Stockholm, Sweden
T: 46 8 189057 F: 46 8 189057

Soderling, Louise, Ms.
HUNGERPROJECT
Love Almquistsvag 42
Sollentuna, Sweden
T: 46 8 964834 F: 46 8 336618

Steen, Peter, Mr.
NATIONAL DEFENSE RESEARCH ESTABLISHMENT/FOA
17290 Sundryberg
Stockholm, Sweden
T: 46 8 663.1500 F: 46 8 733.5240

Stenberg, Lilian, Ms.
CREATURES OF THE SWEDISH FOREST
St. Larsv. 18
Lund 22270, Sweden
T: 46 46 13 4222 F: 46 46 18 9726

Stenberg, Torbjorn, Mr.
CREATURES OF THE SWEDISH FOREST
St. Larsv. 18
Lund 22270, Sweden
T: 46 46 13 4222 F: 46 46 18 9726

Strandberg, Margit
SWEDISH UNITED NATIONS ORGANIZATION
Stureplan 6, 11435
Stockholm 11435, Sweden
T: 46 8 679-9500 F: 46 8 611-0160

Strid, Johan, Mr.
International Secretary
SWEDISH NATIONAL YOUTH COUNCIL
Box 49005
Stockholm, Sweden
T: 46 8 7850690 F: 46 8 6537497

Sunden, Christina, Ms.
CREATURES OF THE SWEDISH FOREST
St. Larsv. 18
Lund 22270, Sweden
T: 46 46 13 4222 F: 46 46 18 9726

Svard, Krister, Mr.
STOCKHOLM ENVIRONMENT INSTITUTE
Box 2142
Stockholm 10314, Sweden
T: 46 8 723-0260 F: 46 8 723-0348
TL: 19580 seis

Svedin, Uno, Mr.
Dr.
ROYAL SWEDISH ACADEMY OF SCIENCES
Box 50005, S-104 05
Stockholm, Sweden
T: 46 8 8-15 04 30 F: 46 8 8-15 56 70

Svenson, Kerstin, Ms.
City Commisioner
CITY OF GOTHENBURG
S-40482
Gothenburg 40482, Sweden
T: 46 31 611048 F: 46 31 138886

Swisher, Joel, Mr.
Engineer
ENVIRONMENT AND ENERGY SYSTEMS
Lund University / Gerdagatan 13
Lund 22362, Sweden
T: 46 46 108645 F: 46 46 108644

Syvajarvi, Aulis, Mr.
SWEDISH UNITED NATIONS ORGANIZATION
Stureplan 6, 11435
Stockholm 11435, Sweden
T: 46 8 679-9500 F: 46 8 611-0160

Tenne, Susanne, Ms.
CREATURES OF THE SWEDISH FOREST
St. Larsv. 18
Lund 22270, Sweden
T: 46 46 13 4222 F: 46 46 18 9726

Tihbaud, Eva
DAGENS NYHETER
Stockholm, Sweden
T: 46 8 7381059 F: 46 8 6564722

Vikberg, Margareta, Ms.
Teacher
UN ASSOCIATION OF SWEDEN
Foreningsgatan 41
Vilhelmina, Sweden
T: 094012301

Warholm, Evy, Ms.
UNIFEM SWEDEN/SWEDISH UNITED NATIONS ORGANIZATION
Stureplan 6, 11435
Stockholm 11435, Sweden
T: 46 8 679-9500 F: 46 8 611-0160

Warlsson, Lars Ingmar
Science Editor
DAGENS NYHETER
Stockholm S-105 15, Sweden
T: 46 8 7381059 F: 46 8 6564722

Watte, Laura, Ms.
CREATURES OF THE SWEDISH FOREST
St. Larsv. 18
Lund 22270, Sweden
T: 46 46 13 4222 F: 46 46 18 9726

Watte, Renate, Mr.
CREATURES OF THE SWEDISH FOREST
St. Larsv. 18
Lund 22270, Sweden
T: 46 46 13 4222 F: 46 46 18 9726

Weinberg, Marc, Mr.
Project Coordenator
FUTURE EARTH
Fjallgatan 23A
Stockholm S-11628, Sweden
T: 46 8 6426225 F: 46 8 6426261

Wennerberg, Pia, Ms.
BONA FELKHOGSKALA
Bona FHSK
Motala S-59197, Sweden
T: 46 141 21100 F: 46 141 21051

Westman, Ulf, Mr.
FONO/ESAM AB
Mariehemvagen 6, BOX 6554
Umea, Sweden
T: 46 90 122893 F: 46 90 163719

Wijkman, Anders, Mr.
Dir. Gen
S A R E C
Stockholm, Sweden
T: 46 8 791 2120

Williamson, Philip, Mr.
Director
IGBP
The Royal Swedish Academy of Sciences PO Box 50005, S-10405
Stockholm, Sweden
T: 46 8 166448 F: 46 8 166405
TL: 17509

Wilson, Deborah, Ms.
Researcher
ENVIRONMENTAL AND ENERGY SYSTEMS STUDIES
Lund University, Gerdagatan 13
Lund S-233 62, Sweden
T: 46 46 104684 F: 46 46 108644
E: debw@seldc52.bitnet

UN AND UNCED

Hammar, Magnus, Mr.
UNITED NATIONS ASSOCIATION IN STOCKHOLM
PO Box 15115
S-104 65 Stockholm, Sweden

Lundgren, Marten, Mr.
UN ASSOCIATION OF SWEDEN
Strandvagen 8c, 91234
Vilhelmina, Sweden
T: 094010755

ADVISORS

Chadwick, Michael J., Prof.
Director, Stockholm Environmental Institute
S-103 14 Stockholm, Sweden
T: 46 8 723 0260 F: 46 8 723 0348

PARLIAMENTARIANS / SPIRITUAL LEADERS

Bjorling, Fiona, Ms.
Riksdagen S-100 12
Stockholm, Sweden
T: 08/786 40 00
TL: 13126 Parliment S

Deleus, Lennart
Riksdagen S-100 12
Stockholm, Sweden
T: 08/786 40 00
TL: 13126 Parliment S

Ericsson, Dan
Riksdagen S-100 12
Stockholm, Sweden
T: 08/786 40 00
TL: 13126 Parliment S

Fremling, Lennart
Riksdagen S-100 12
Stockholm, Sweden
T: 08/786 40 00
TL: 13126 Parliment S

Persson, Goran
Riksdagen S-100 12
Stockholm, Sweden
T: 08/786 40 00
TL: 13126 Parliment S

Smith, Karin Svensson
Riksdagen S-100 12
Stockholm, Sweden
T: 08/786 40 00
TL: 13126 Parliment S

Theorin, Naj Veirr
Riksdagen S-100 12
Stockholm, Sweden
T: 08/786 40 00
TL: 13126 Parliment S

Winden, Krister
Riksdagen S-100 12
Stockholm, Sweden
T: 08/786 40 00
TL: 13126 Parliment S

SWITZERLAND - GOVERNMENT

Al-Eshaiwi, Mussaid M.S.
Saudi Arabia Attache For Gatt
K.SAUDIARABIA MISSION GENEVE
46 Ave Blanc
Geneve 1202 CH, Switzerland
T: 738 3030 F: 738 3134

Al-Farsy, Ibrahim Mohammed Sharif
First Secretary
PERMANENT MISSION OF THE SULTANATE of Oman Chemin du Petit-Saconnex 28 B.
Geneva CH1209, Switzerland
T: 022 733 73 20 F: 022 740 10 75

Arda, Mehmet
Officer-In-Charge, General Studies
SECTION COMMODITIES DIVISION.UNCTAD
UNCTAD, Palais des Nations
Geneva 10 CH1211, Switzerland
T: 4122 734 6011 5790 F: 4122 734 2754

Ballaman, Jean-Pierre, M.
Consul
Consulat general de Suisse, Rio de Janeiro
Permanent Mission to the United Nations, 757 3rd. Ave., 21st Floor New York, NY, 10017
F: 212 421 1480
TL: 212 751 2104

Blankart, Franz, S.E. M.
Secretaire d'Etat
Directeur de l'Office federal des affairs economiques exterieures, Department federal de l'economie publique, (Chef Adjoint de la Delegation)
Permanent Mission to the United Nations, 757 3rd. Ave., 21st Floor New York, NY, 10017
F: 212 421 1480
TL: 212 751 2104

Boehlen, Bruno, S.E. M.
Conseiller scientifique
Permanent Mission of Switzerland
757 3rd. Ave., 21st Floor
New York, NY 10017, United States of America
T: 212 421 1480 F: 212 751 2104

Cameroni, Marco, M.
Departement federal des affaires etrangeres
Permanent Mission to the United Nations, 757 3rd. Ave., 21st Floor New York, NY, 10017
F: 212 421 1480
TL: 212 751 2104

Chave, Olivier, M.
Departement federal des affaires etrangeres
Permanent Mission to the United Nations, 757 3rd. Ave., 21st Floor New York, NY, 10017
F: 212 421 1480
TL: 212 751 2104

Clemencon, Raymond, M.
Departement federal de l'interieur
Permanent Mission to the United Nations, 757 3rd. Ave., 21st Floor New York, NY, 10017
F: 212 421 1480
TL: 212 751 2104

Cotti, Flavio, Mr.
Federal Councillor
Permanent Observer Mission of Switzerland
757 3rd. Ave., 21st. Floor
New York NY 10017, United States of America
T: 212 421 1480 F: 212 751 2104

Ducret, Claude G., M.
Departement federal des affaires etrangeres
Permanent Mission to the United Nations, 757 3rd. Ave., 21st Floor New York, NY, 10017
F: 212 421 1480
TL: 212 751 2104

Eigensatz, Gabriela, Mme
Departement federal des affaires etrangeres
Permanent Mission to the United Nations, 757 3rd. Ave., 21st Floor New York, NY, 10017
F: 212 421 1480
TL: 212 751 2104

Erkman, Suren
Science & Technology Editor
JOURNAL DE GENEVE
P. O. Box
Geneva 11 1211, Switzerland
T: 41 22 3115511 F: 41 22 3115520

Francis, Pascale Morand, Mme
Permanent Mission of Switzerland
757 3rd. Ave., 21st Floor
New York, NY 10017, United States of America
T: 212 421 1480 F: 212 751 2104

Gatan, Leslie
First Secretary
PHILIPPINE MISSION
47 Avenue Blanc
1202 Geneva, Switzerland
T: 022 731 8320 F: 022 731 6888

Giovannini, Jean-Francois, M.
Directeur suppleant de la Direction de la cooperation au developpement et de l'aide humanitaire
Departement federal des affaires etrangeres
Permanent Mission to the United Nations, 757 3rd. Ave., 21st Floor New York, NY, 10017
F: 212 421 1480
TL: 212 751 2104

Haeberli, Christian, M.
Departement federal de l'economie publique
Permanent Mission to the United Nations, 757 3rd. Ave., 21st Floor New York, NY, 10017
F: 212 421 1480
TL: 212 751 2104

Haegi, Claude, M.
Conseiller d'Etat de la Republique et Canton de Geneve
Permanent Mission of Switzerland
757 3rd. Ave., 21st Floor
New York, NY 10017, United States of America
T: 212 421 1480 F: 212 751 2104

Hilber, Rudolf, M.
Consul general
Consulat general de Suisse, Rio de Janeiro
Permanent Mission to the United Nations, 757 3rd. Ave., 21st Floor New York, NY, 10017
F: 212 421 1480
TL: 212 751 2104

Husy, Stephan, M.
Departement federal des affaires etrangeres
Permanent Mission to the United Nations, 757 3rd. Ave., 21st Floor New York, NY, 10017
F: 212 421 1480
TL: 212 751 2104

Imboden, Nicolas, S.E. M.
Ambassadeur
Delegue aux accords commerciaux de l'Office federal des affaires economiques exterieures, Departement federal de l'economie publique
Permanent Mission to the United Nations, 757 3rd. Ave., 21st Floor New York, NY, 10017
F: 212 421 1480
TL: 212 751 2104

Joray, Jacqueline, Mme
Ambassade de Suisse au Bresil
Permanent Mission to the United Nations, 757 3rd. Ave., 21st Floor New York, NY, 10017
F: 212 421 1480
TL: 212 751 2104

Juen, Christoph, M.
Conseiller
Permanent Mission of Switzerland
757 3rd. Ave., 21st Floor
New York, NY 10017, United States of America
T: 212 421 1480 F: 212 751 2104

Kohn, Michael, Mr
Chairman
ENERGY COMMISSION ICC
P.O.B 4813
Zurich 8022, Switzerland
T: 01 281 1616 F: 01 202 1583

Kreisel, Wilfried, Mr.
Director
WORLD HEALTH ORGANIZATION
Geneva 1211 27, Switzerland
T: 0041-22-7913582/3/4 F: 0041-22-788 4218

Krieg, Catherine, S.E. Mme
Ambassadeur
Ambassade de Suisse au Bresil
Permanent Mission to the United Nations, 757 3rd. Ave., 21st Floor New York, NY, 10017
F: 212 421 1480
TL: 212 751 2104

Ksentini, Fatna. Zohra
Expert. Member Sub.Commission
CENTER FOR HUMAN RIGHTS
United Nations Geneva
Geneva, Switzerland
T: 41 22 7346011 F: 7339879

Kunz, Kurt, M.
Ambassade de Suisse au Bresil
Permanent Mission to the United Nations, 757 3rd. Ave., 21st Floor New York, NY, 10017
F: 212 421 1480
TL: 212 751 2104

Linn Locher, Monika, Mme
Departement federal de l'interieur
Permanent Mission to the United Nations, 757 3rd. Ave., 21st Floor New York, NY, 10017
F: 212 421 1480
TL: 212 751 2104

Longet, Rene, M.
Conseiller
Permanent Mission of Switzerland
757 3rd. Ave., 21st Floor
New York, NY 10017, United States of America
T: 212 421 1480 F: 212 751 2104

Lopez Ehavaee, Mario
Second Secretary
MISSION DU PERU AUPREJ N J
63 Rue De Lawanne
Geneva 1202, Switzerland
T: 41 22 731 1130 F: 41 22 731 1168

Mercier, Michel Jean
Director
IPCS/WHO
Av. Appia 1211 Geneva 27
Geneva, Switzerland
T: (22) 797-35-88 F: (22) 788-13-49

Messerli, Bruno, M.
Conseiller
Permanent Mission of Switzerland
757 3rd. Ave., 21st Floor
New York, NY 10017, United States of America
T: 212 421 1480 F: 212 751 2104

Moeri, Jacqueline, Mme
Departement federal de l'interieur
Permanent Mission to the United Nations, 757 3rd. Ave., 21st Floor New York, NY, 10017
F: 212 421 1480
TL: 212 751 2104

Moglia, Ana M., Ms
First Secretary
PERMANENT MISSION OF ARGENTINA
Ao, Route De C'Aeroport 1215 Geneva As
Geneva, Switzerland
T: 798 19 52

Morand Francis, Pascale D
Senior Advisor
INTERNATIONAL AFFAIRS/FEDERAL OFFIC E HALLWY1STROSSE 4
Berna 3003, Switzerland
T: 41 31 61 68 62 F: 41 31 43 81 37

Pauletto, Christian, M.
Departement federal de l'economie publique
Permanent Mission to the United Nations, 757 3rd. Ave., 21st Floor New York, NY, 10017
F: 212 421 1480
TL: 212 751 2104

Pestana-Macedo, Carlos R., Mr
Second Secretary
PERMANENT MISSION OF VENEZUELA
Mission du Venezuela.18-A.CH.Franqo Lehmann.1218.Grand-Saconnex.Geneve.
Geneve, Switzerland
T: 7982621 F: 7985877

Peytrignet, Gerard, Mr
Head Of Icrc Mission In Brazil
INTERNATIONAL COMMITTEE OF THE RED
ICRC, 19 Av De la Paix 'Merci Faire Suirre'
1202 Geneva, Switzerland
T: 55 61 2480250 248384 F: 55 61 2485173

Risch, Christian, M.
Departement federal de l'interieur
Permanent Mission to the United Nations, 757 3rd. Ave., 21st Floor New York, NY, 10017
F: 212 421 1480
TL: 212 751 2104

Schmid, Wilhelm, M.
Vice-directeur de l'Office federal de l'environnement, des forets et du paysage
Departement federal de l'interieur
Permanent Mission to the United Nations, 757 3rd. Ave., 21st Floor New York, NY, 10017
F: 212 421 1480
TL: 212 751 2104

Staehelin, Fritz, Mr.
Ambassadeur
Directeur de la Direction de la cooperation au developpement et de l'aide humanitaire, Department federal des affaires etrangeres
Permanent Mission to the United Nations, 757 3rd. Ave., 21st Floor New York, NY, 10017
F: 212 421 1480
TL: 212 751 2104

Suedi, Khamis Juma
Director
WORLD INTECLECTUAL PROPERTY ORGANIN
34 Clr. des Colombehes. 1211 Geneve, Suisse 20
Geneva, Switzerland
T: (022) 730 9535 F: 733 5428

Takala, Jukka
Head Of CIS
CIS/INTERNAT. LABOUR ORGANISATION
ILO/CIS
Geneva 22 CH-1211, Switzerland
T: 41 22 799 6740 F: 41 22 798 6253

Vchida, Kosaku
Scientist
DIV. OF ENVIRONMENTAL HEALTH, WHO
20 Ave. Appia, CH-1211
Geneva 27, Switzerland
T: 022 791 3601 F: 022 791 0746

Vossenaar, Rene
Senior Economic Affairs Officer
UNCTAD
Palais des Nations 1211 CH
Geneva 1211 CH, Switzerland
T: 41 22 7346011 EX5679 F: 41 22 8342953

NGO

Aram, Muthukumaraswamy, Dr.
WCRP INTERNATIONAL
14, Chemin Auguste-Vibert, 1218 Grand Saconnex
Geneve, Switzerland
F: 41 22 731-0034

Arosemena, Pablo
IFBW
20 route de Pre-Bois
1215 Geneva 15 Airport, Switzerland

Backstrand, Goran
LRCRCS
PO Box 372
1211 Geneva 19, Switzerland

Banda, Moses, Mr.
WORLD UNIV. SERVICE
5, Chemin des Iris
Geneva 1216, Switzerland
T: 41 22 798-8711 12 F: 41 22 798-0829

Barrat, Olga
IFUW
37 Quai Wilson
CH-1202 Geneva, Switzerland

Baskar, Paul
IFTH
22 rue Michel Chauvet
1208 Geneva, Switzerland

Bayon, Ricardo
COCF
Palais Wilson, 52 rue des Paquis
CH-1201 Geneve, Switzerland

Beguin, Renata
HP
1 rue de Varembe, Centre International
1211 Geneva, Switzerland

Berner, Y.
ACWW
149 D, route de Ferney
1218 Grand-Saconnex GE, Switzerland

Berry, Victori, Ms.
WORLD COUNCIL OF CHURCHES
150 route de Ferney
Geneve 1211, Switzerland
F: 41 22 791 0361

Beukes, Edel Havin
WILPF
1 rue de Varembe, CP 28
1211 Geneva, Switzerland

Binns, Beth, Ms.
Vice Director
INTERNATIONAL ACADEMY OF THE ENVIRONMENT
4, Ch. de Conches
Conches - Geneva 1231, Switzerland
T: 41 22 789-1311 F: 41 22 7892538

Bitar, Fouad
UCIP
37-39 Rue de Vermont, C.P, 197
CH-1211 Geneva, Switzerland

Boggian, Alessandro
ISDE
Vicolo Gesora 7
6929 Gravesano, Switzerland

Boquet, Genevieve
UTANSC
26 rue l'Athenee, Case Postale 370
CH-1211 Geneva 12, Switzerland

Boquet, Pierre
UTANSC
26 rue l'Athenee, Case Postale 370
CH-1211 Geneva 12, Switzerland

Bosshardt, Frank W., Mr.
Executive Assistant To The Cha
BUSINESS COUNCIL FOR SUSTAINABLE DEVELOPMENT
Hurdnerstrasse 10, PO Box 1474
Hurden 8640, Switzerland
T: 41 55 484824

Brakemeier, Gottfried
LWF
150 route de Ferney, Case Postale 2100
Geneva, Switzerland

Brandao, Angela
LLLI
Alte Landstrasse 89
8802 Kilchberg, Switzerland

Brugger, Ernst A., Mr.
Director For Regional Affairs
BUSINESS COUNCIL FOR SUSTAINABLE DEVELOPMENT
Haus Inseli
Niederurnen 8867, Switzerland
T: 41 58 211750 F: 41 58 211709

Burhenne, Guilmin
IUCN
Avenue du Mont-Blanc
CH-1196 Gland, Switzerland

Carbonell, Lydia
CEC
PO Box 2100, 150 route de Ferney
CH-1211 Geneva 2, Switzerland

Carpentier De Gourdon, Come
SAES
Seefeldstrasse 301, Postfach
CH-8034 Zurich, Switzerland

Castro, Emilio
WCC
150 route de Ferney
1211 Geneva 20, Switzerland

Chazerand, Odile, Ms.
Program Officer/Environment
INTER-PARLIAMENTARY UNION
Place du Petit-Saconnex - Bp 438
1211 Geneve 19 Switzerland,
T: 022 7344150 F: 022 7333141
TL: 414 217 IPU CH

Condell, Yvonne C., Ms.
INTERNATIONAL FEDERATION OF UNIV WOMEN
37 Quai Wilson
CH-1202 Geneva, Switzerland

Costa, Pedro
IUCN
Avenue du Mont-Blanc
CH-1196 Gland, Switzerland

Costa Ribeiro, Carlos
IFCE
PO Box 86
1000 Lausanne 12-Chailly, Switzerland

Cropper, Angela, Ms.
Head-Governance
IUCN-THE WORLD CONSERVATION UNION
Av Mont Blanc, CH - 1196
Gland, Switzerland
T: 41 22 649 232 F: 41 22 642 926

Dallari, Dalmo
ICJ
109 route de Chene / PO Box 120
CH-1224 Chene-Bougeries GE, Switzerland

Dallari, Sueli
ICJ
109 route de Chene / PO Box 120
CH-1224 Chene-Bougeries GE, Switzerland

De Boulloche, Carola
ICW
12 Chemin Vert-Pre
1231 Conches GE, Switzerland

De Haes, Charles
WWFN
Avenue du Mont-Blanc
CH-1196 Gland, Switzerland

Dobbie, Leoni
IATA
IATA Centre, 33 route de l'aeroport
CH-1215 Geneva 15 Airport, Switzerland

Dyer, Simon, Mr.
ITA
2 Quai Gustave Ador
1207 Geneva, Switzerland

Ettlin, Jost
HP
1 rue de Varembe, Centre International
1211 Geneva, Switzerland

Faulkner, Hon J. Hugh, Mr.
Executive Director
BUSINESS COUNCIL FOR SUSTAINABLE DEVELOPMENT
World Trade Center Building, Route de l'Aeroport 10, PO Box 365
Geneve 1215, Switzerland
T: 41 22 7883202 F: 41 22 7883211

Frazer, Garth
IFC
Palais Wilson, 52 rue des Paquis
1201 Geneva, Switzerland

Fundanga, Caleb, Mr.
President
WORLD UNIVERSITY SERVICE
5 Chemin Des Iris
Geneve 1216, Switzerland
T: 41 22 798-8711 F: 41 22 798-0829
TL: 415537WUSH

Giovannini, Bernard, Mr.
Director
INTERNATIONAL ACADEMY OF THE ENVIROMENT
4 CH. De Conches
1231 Conches, Switzerland
T: 4122 789-1311 F: 4122 789-2538

Gravina, Braz Alberto
IFCE
PO Box 86
1000 Lausanne 12-Chailly, Switzerland

Gyger, Pia, Ms.
Director
ST. KATHARINA - WERK
Postfaon Holeestrasse 123
Basel CH-4015, Switzerland
T: 41 61 301-2323 F: 41 61 301-1714

Halldor, Thorgeirsson
BIC
15 route des Morillons
1218 Grand-Saconnex-GE, Switzerland

Halle, Mark, Mr.
Director-Development
IUCN - THE WORLD CONSERVATION UNION
Av Mont Blanc, CH - 1196
Gland, Switzerland
T: 41 22 649 232 F: 41 22 642 926

Han Ju, Kim Farley
BIC
15 route des Morillons
1218 Grand-Saconnex-GE, Switzerland

Hartley, Nigel
WUS
5 Chemin des Iris
1216 Geneva, Switzerland

Heini, Claude
HP
1 rue de Varembe, Centre International
1211 Geneva, Switzerland

Heini, Erich, Mr.
Director For Public Affairs
BUSINESS COUNCIL FOR SUSTAINABLE DEVELOPMENT
Hurdnerstrasse 10, PO Box 1474
Hurden 8640, Switzerland
T: 41 55 484824

Hellman, Marion Felix
IFBW
20 route de Pre-Bois
1215 Geneva 15 Airport, Switzerland

Hesse Steel, Elaine
WYWCA
37 Quai Wilson
1201 Geneva, Switzerland

Horst, Morich
IFBW
20 route de Pre-Bois
1215 Geneva 15 Airport, Switzerland

Howard, Phyllis C.
President
ASSOCIATED COUNTRY WOMEN OF THE WORLD
149 D, route de Ferney
1218 Grand-Saconnex GE, Switzerland

Hurtado, Maria Elena, Ms.
NCU
15 Chemin Rojoux
1231 Conches GE, Switzerland

Iisaka, Yoshiaki, Mr.
Professor
WCRP INTERNATIONAL
14 Chemin Auguste Vilbert
1218 Grand-Saconnex GE, Switzerland

Kang, Moon Kyu
WAYMCA
37 Quai Wilson
1201 Geneva, Switzerland

Kapfer, Paul, Mr.
WORLD ORGANIZATION OF THE SCOUT MOVEMENT
PO Box 241
1211 Geneva 4, Switzerland

Karlsson, Gunnar A., Mr.
IFBWW
INTERNATIONAL FEDERATION OF BUILDING AND WOOD WORKERS
P.O. Box 733
Geneva, Switzerland
T: 41 22 788-0888 F: 41 22 788-0716
TL: 415327flitb

Kloetzli, Stefan, Mr.
WPF
Engehaldenstrasse 4 P, Postfach
CH-3001 Berne, Switzerland

Kuypers, Ursula, Ms.
SKWB
Holeestrasse 123
CH-4015 Basel, Switzerland

Laing, Sara, Ms.
WORLD CITIZENS ASSEMBLY
PO Box 128
1211 Geneva 20, Switzerland

Larson, Rebecca
LWF
150 route de Ferney, Case Postale 2100
Geneva, Switzerland

Lawrence, Arturo
BIC
15 route des Morillons
1218 Grand-Saconnex-GE, Switzerland

Lovell, Jean, Ms.
WORLD YOUNG WOMEN'S CHRISTIAN ASSOCIATION/BARBADOS
37 Quai Wilson
1201 Geneva, Switzerland

Marcus, Lars
ICA
15 route des Morillons
1218 Grand-Saconnex GE, Switzerland

Markham, Adam
WORLD WILDLIFE FUND - WWF
Rue du Mont Blanc
Gland, Switzerland
T: 41 22 64 95 25 F: 41 22 64 58 79

Marroquim, Pajucara
LLLI
Alte Landstrasse 89
8802 Kilchberg, Switzerland

Mauser, Bruno
STP
Bruggerstrasse 78
CH-5400 Baden, Switzerland

McAllister, Ian
LRCRCS
PO Box 372
1211 Geneva 19, Switzerland

McCammon, Antony, Mr.
FOUNDATION FOR ENVIRONMENTAL CONSERVATION
7 Chemin Tavernay
1218 Grand-Saconnex-GE, Switzerland

Montant, Francoise, Ms.
Programme Manager
CENTRE FOR APPLIED STUDIES IN INTERNATIONAL NEGOTIATIONS
11a, Av. de la Paix
Geneva 1202, Switzerland
T: 41 22 734-8950 F: 41 22 733-6444

Mucke, Peter
IFTH
22 rue Michel Chauvet
1208 Geneva, Switzerland

Munro, David, Mr.
Consultant
IUCN - THE WORLD CONSERVATION UNION
Av Mont Blanc, CH - 1196
Gland, Switzerland
T: 41 22 649 232 F: 41 22 642 926

Muntasser, Mohamed
CMDC
Kellerweg 38
CH-8055 Zurich, Switzerland

Murthy, Rama
CMDC
Kellerweg 38
CH-8055 Zurich, Switzerland

Norup Carlsen, Hanne
WILPF
1 rue de Varembe, CP 28
1211 Geneva, Switzerland

Nussbaumer, Werner
ISDE
Vicolo Gesora 7
6929 Gravesano, Switzerland

Ouedraogo, Bernadette
IFTH
22 rue Michel Chauvet
1208 Geneva, Switzerland

Pancini, Paola
ISDE
Vicolo Gesora 7
6929 Gravesano, Switzerland

Paull-Borja, Laura
NCU
15 Chemin Rojoux
1231 Conches GE, Switzerland

Pernato, Ellen
COCF
Palais Wilson, 52 rue des Paquis
CH-1201 Geneve, Switzerland

Pregon, Noel
IFTH
22 rue Michel Chauvet
1208 Geneva, Switzerland

Rasool, Ebrahim, Mr.
WORLD CONFERENCE ON RELIGION PEACE
14 Chemin Auguste Vilbert
1218 Grand-Saconnex GE, Switzerland

Renfer, Sonja M.
ZONTA
22 Avenue du Chateau
CH-1008 Prilly, Switzerland

Rexsodihardjo, W. Soengeng
ISTF
Ch. du Reposoir 5-7
1260 Nyon, Switzerland

Rodrigues, Roberto
ICA
15 route des Morillons
1218 Grand-Saconnex GE, Switzerland

Romizi, Roberto
ISDE
Vicolo Gesora 7
6929 Gravesano, Switzerland

Saint Laurent, Carole
WWFN
Avenue du Mont-Blanc
CH-1196 Gland, Switzerland

Sarathy, Partha, Mr.
Chair IUCN Commission
WORLD CONSERVATION UNION ON EDUCATON AND COMMUNICATION- IUCN
Av.du Mont-Blanc
Gland CH-1196, Switzerland
T: 41 22 364-9232 F: 41 22 364-2926

Schneider, Rudolf
INSTITUTE FOR PLANETARY SYNTHESIS
15 rue de Varembe
CH-1211 Geneva 20, Switzerland

Seitz, Jurgen, Mr.
Systems Manager
IUCN - THE WORLD CONSERVATION UNION
Av Mont Blanc, CH - 1196
Gland, Switzerland
T: 41 22 649 232 F: 41 22 642 926

Sotto, Jose Andres, Mr.
Director
IFC FOR INDEPENDENT SECTORS FOR UNCED
52, Rue de Paquis
Geneve 1201, Switzerland
T: 41 22 732-7117 F: 41 22 738-5046
TL: 412436-csen ch

Sowunmi, Adebisi
WORLD COUNCIL OF CHURCHES
150 route de Ferney
1211 Geneva 20, Switzerland

Strom, Turid, Ms.
Chairwoman
INTERNATIONAL COOPERATIVE ALLIANCE (ICA)
15, Route des Morillons - Geneva
Grand-Saconnex 1218, Switzerland
T: 41 22 798.4121 F: 41 22 798.4122

Stromberg, Kristen
LWF
150 route de Ferney, Case Postale 2100
Geneva, Switzerland

Szemere, Juan Ivan
IPSY
15 rue de Varembe
CH-1211 Geneva 20, Switzerland

Taylor, John B., Mr.
Secretary-General
WORLD CONFERENCE ON RELIGION AND PEACE/INTERNATIONAL
14 ch. Auguste-vilbert 1218 Grand-Saconnex
Geneve, Switzerland
T: 41 22 798-5162 F: 41 22 791-0034

Techeste, Ahderom
BIC
15 route des Morillons
1218 Grand-Saconnex-GE, Switzerland

Tissot, Luc
SAES
Seefeldstrasse 301, Postfach
CH-8034 Zurich, Switzerland

Toivola, Anja
LRCRCS
PO Box 372
1211 Geneva 19, Switzerland

Tomaz, Wonghon
WCOTP
5 Avenue du Moulin
1110 Morges, Switzerland

Ulatowska, Lisinka
WCA
PO Box 128
1211 Geneva 20, Switzerland

Van Der Weid, Denis
AT
10 rue grenus
CH-1201 Geneve, Switzerland

Van Griethuysen, Pascal
CMDC
Kellerweg 38
CH-8055 Zurich, Switzerland

Vischer, Lukas
CEC
PO Box 2100, 150 route de Ferney
CH-1211 Geneva 2, Switzerland

Wee, Paul
LUTHERAN WORLD FEDERATION
150 route de Ferney, Case Postale 2100
Geneva, Switzerland

PRESS

Abranoff, Simone
EBU
R.Clodomiro Amazonas 960/171
Sao Paulo, S.P., Brazil, Switzerland

Achermann, Urlich
Correspondent
RADIO SUIGA INTERNATIONAL
Gicicometticbraste
3000 Bern, Switzerland
T: 31 439222

Banderet, Luc
Correspondent
DRS-SWISS B'CASTING CORP.
R. Sabara 318, ap. 143
Higienopolis 01239, SP, Brazil, Switzerland

Bertolami, Silvio
Editor
DIE WELTWOCHE
R.Almte.Alexandrino 3179/201
Rio de Janeiro, Brazil, Switzerland

Bischofberger, Tonia
Editor
SWISS B'CASTING CORP.
Am Bollwerk 1
CH-4102 Binningen, Switzerland, Switzerland

Born, Martin
Editor
TELEMUNDI
Ch. Planta 3
1223 Cologny, Geneva Switzerland, Switzerland

Buchi, Andres Bruno
Reporter Radio/Television
SONNTAGS ZEITUNG
Englischviertelstr.50
8032 Zurich CH, Switzerland, Switzerland

De Azevedo Irving, Marta
Correspondent
NEUE ZURCHER ZEITUNG
Rua Marques de Olinda, 90
Botafogo, Rio de Janeiro-RJ, Brazil, Switzerland

Dettling, Erwin
Correspondent
DER BULID
Berlia, Switzerland
T: 41 31 311211

Dompka, Victoria
Correspondent
NEW SCIENTIST
Rue la Potern 2
1260 Nyon, Switzerland, Switzerland

Everaert, Catharina
EBU
3120 R Street NW (#208)
Washington, DC 20007, USA, Switzerland

Fluhler, Johannes
Correspondent
ECOMED
CH-8044
Zurich, Swiss, Switzerland

Fritsch, Bruno
Editor
ECOMED
Aussichtsstrasse 13,
CH-8704 Herrliberg, Swiss, Switzerland

Galland, Bertil
Editor
LE NOUVEUAU QUTIDIEN
3 Chermin des Fleurettes
Ch 1006 Lausanne, Suiza, Switzerland

Gallavresi, Lucilla
Correspondent
CORRIERE DEL TICINO
Via Disciplini 18
20123 Milano, Italy, Switzerland

Glogger, Beat
Editor
SWISS TV
Hohestr. 71
8702 Zolgikon, Switzerland, Switzerland

Gygi, Beat
Editor
NEUE ZURCHER ZEITUNG
Kalcheggweg 21
CH-3006 Bern, Switzerland, Switzerland

Hammerli, Fredy
Editor
CASH MAGAZINE
Krauchthal Strasse 60
CH-3065 Bolligen,Switzerland, Switzerland

Hart, Klaus
Correspondent
NEUE ZURCHER ZEITUNG
Rua Taylor 39, Apt. 905
20421, Rio de Janeiro-RJ, Brazil, Switzerland

Henderson, Susan
EBU
4240 Alton Place, NW
Washington, DC 20016, USA, Switzerland

Hottinger, Jessica
Reporter Radio/Television
RADIO SWISS INTERNATIONAL
Giacomettistr. 6
3006 Bern, Switzerland, Switzerland

Hufschmid, Peter
Editor
TAGES-ANZEIGER ZURICH
25 Tavelweg
3006 Bern, Switzerland, Switzerland

Joutet, Francois
Correspondent
INFOSUD/LAUSANNE
36 Route de Coppet
CH-1291 Commugny, Switzerland, Switzerland

Labeviere, Richard Etienne
Correspondent
SWISS TV
83, Boulevard de la Corniche
Thonon Savoie, Switzerland, Switzerland

Lietti, Anna
Editor
LE NOUVEAU QUOTIDIEN
Terreaux 9
1003 Lausanne, Switzerland, Switzerland

Ludin, Walter
Correspondent
ITE
Wesemlin stralle 42
Ch 6000 Luzem 10, Swiss, Switzerland

Magistra, Michel
Correspondent
SWISS TV
Helide 6815
Switzerland, Switzerland

Maurice, Antoine
Editor
JOURNAL DE GENEVE
5, rue des Granges
1204 Geneve, Switzerland, Switzerland

Meyer, Pierre
Editor
LA TRIBUNE DE GENEVE
53, Route de Chene
1208 Geneve, Switzerland, Switzerland

Mueller, Thomas
Editor
BASLER ZEITUNG
Vogesenstr. 100
4056 Basel,Switzerland, Switzerland

Pfluger, August
Correspondent
DER ZEIT PUNKT
P.O. Box 3207
2500 Biel 3 Switzerland, Switzerland

Razali, Anne
Editor
INGOSUD
1523 GRanges Rarnaud
Geneve, Switzerland, Switzerland

Rime, Michel
Editor
24 HEURES
39 rue des Bains
1205 Geneve, Switzerland, Switzerland

Schiesser, Walter
Editor
NEUE ZURCHER ZEITUNG
Zurichstrasse 35
CH-8700 Kusnacht, Switzerland, Switzerland

Schwarzstein, Jacques
Correspondent
SCHWEIZERISCHE DEPESCHENAGENTUR AG
R.Alda Garrido 583/102
Rio de Janeiro, Brazil, Switzerland

Stoecklin, Stefan
Correspondent
BASLER ZEITUNG
Clarahofweg 16
4058 Basel, Switzerland, Switzerland

Tavares, Armando
EBU
CH. De Vleurgat 282 A
9050 Bruxelles, Belgium, Switzerland

Van Hooland, Jacques
EBU
Rue Veydt 27
1050 Bruxelles, Belgium, Switzerland

Vellasco, Helena A.R.
EBU
Av.Henrique Dodsworth 64/305-B
Rio de Janeiro, Brazil, Switzerland

Warne, Peter
Correspondent
SWISS BUSINESS
Florastr. 40
4057 Basel, Switzerland, Switzerland

Zbinden, Jurg
Editor
SWISS ILLUSTRATED
Bern, Brunngasse, Switzerland, Switzerland

ORGANIZATIONS

Aecherli, Willy, Mr.
International Association Against Noise
Hirschenplatz 7
CH-6004 Luzern, Switzerland
T: 041 22 00 15

Ahiaba, Beatriz, Ms.
INTERNATIONAL INDIGENOUS COMMISSION
Centre for Our Common Future Palais Wilson, 52 rue des Paquis
Geneve 1201, Switzerland
T: 41 22 732-7750 F: 41 22 738-5046

Arturo, Lawrence, Mr.
Bahai International Community
15 route des Morillons
1218 Grand-Saconnex-GE, Switzerland
T: (41 22) 798 5400 F: (41 22) 798 65 77

Asp, Ulf, Mr.
International Federation Of Building And Woodworkers
20 route de Pre-Bois
1215 Geneva 15 Airport, Switzerland
T: (41 22) 788 0888 F: (41 22) 788 0716

Ballantyne, Edith, Ms.
Women's International League For Peace And Freedom*
1 rue de Varembe, CP 28
1211 Geneva, Switzerland
T: (41 22) 733 6175 F: (41 22) 740 1063

Beguin, Fred, Mr.
Hunger Project, The
1 rue de Varembe, Centre International
1211 Geneva, Switzerland
T: (41 22) 734 6519 F: (41 22) 733 9203

Bell, Kirk, Mr.
International Organisation For Migration
17 route des Morillons
CH-1211 Geneva 19, Switzerland
T: (41 22) 717 911 F: (41 22) 798 6150

Bichsel, Anne, Ms.
Swiss Aid Agencies Coalition
Giesereiweg 9
3007 Bern, Switzerland
T: (41 31) 455 412 F: (41 31) 451 411

Blasco, Delmar, Mr.
International Council Of Voluntary Agencies
13 rue Gautier
1201 Geneva, Switzerland
T: (41 22) 732 66 00 F: (41 22) 738 99 04
TL: 412586 ICVA CH

Bloom, Renate, Ms.
World Federation Of Methodist Women
103 bis, route de Thonon
CH-1222 Vesenaz GE, Switzerland
T: (41 22) 52 15 36 F: (41 22) 752 4104

Boyd, Doreen, Ms.
World Young Women's Christian Association (World YWCA)
37 Quai Wilson
1201 Geneva, Switzerland
T: (41 22) 732 3100 F: (41 22) 731 7938

Bridel, Danielle, Ms.
Zonta International
22 Avenue du Chateau
CH-1008 Prilly, Switzerland
T: (41 21) 635 41 47

Chau, Andre Van, Dr.
International Catholic Migration Commission
37-39 rue de Vermont
1211 Geneva, Switzerland
T: (41 22) 733 41 50 F: (41 22) 734 79 29
TL: 28 100 ICMC CH

Chessex, Cyril, Mr.
Rotary International
La Maison du Bailli
1816 Chailly-Montreux, Switzerland
T: (41 21) 964 65 18

Choa, Diana, Ms.
Food for the Hungry International
108 Route de Suisse
1290 Versoix GE, Switzerland
T: (41 22) 755 1444 F: (41 22) 755 1686

Cohen, Maria Luisa, Ms.
Assisi Nature Council
3 Avenue de Jaman
1005 Lausanne, Switzerland
T: (41 21) 207 043 F: (41 21) 230 736

Dalton, Sue, Ms.
International Union For The Conservation Of Nature
Avenue du Mont-Blanc
CH-1196 Gland, Switzerland
T: (41 22) 64 91 14 F: (41 22) 64 29 26
TL: 419 605 IUCN CH

Davies, Dorothy, Ms.
International Federation Of University Women
37 Quai Wilson
CH-1202 Geneva, Switzerland
T: (41 22) 731 2380 F: 941 22) 738 0440

De Boccard, Jeannine, Mme
International Council Of Women
12 Chemin Vert-Pre
1231 Conches GE, Switzerland
T: (41 22) 47 02 28 F: (41 22) 46 79 21

De Puymege, Alison, Mme
Universities
10 Conseil-General
CH-1211 Geneva 4, Switzerland
T: (41 22) 29 26 44 F: (41 22) 29 28 21

Dieng, Adama, Mr.
International Commission Of Jurists
109 route de Chene / PO Box 120
CH-1224 Chene-Bougeries GE, Switzerland
T: (41 22) 49 35 45 F: (41 22) 49 31 45
TL: 418 531 ICJ CH

Erazo, Zimena, Mr.
World University Service*
5 Chemin des Iris
1216 Geneva, Switzerland
T: (41 22) 798 8711 F: (41 22) 798 0829

Flache, Stanislas, Dr.
World Federation For Mental Health
9C Plateau de Frontenex
1208 Geneva, Switzerland
T: (41 22) 736 8435 F: (41 22) 736 3058

Frado, Dennis, Mr.
Lutheran World Federation
150 route de Ferney, Case Postale 2100
Geneva, Switzerland
T: (41 22) 791 61 11 F: (41 22) 791 03 61

Gamma, Anna, Dr.
St. Katharina Werk Basel
Holeestrasse 123
CH-4015 Basel, Switzerland
T: (41 61) 38 2323/302 4903 F: (41 61) 38 1714

Gebhardt, Gunther, Mr.
World Conference On Religion And Peace
14 Chemin Auguste Vilbert
1218 Grand-Saconnex GE, Switzerland
T: (41 22) 798 51 62 F: (41 22) 791 00 34

Gilbert-Miguet, Barry, Mr.
International Public Relations Association*
Case Postale 2100
CH-1211 Geneva 2, Switzerland
T: (41 22) 791 0550 F: (41 22) 788 0336

Granberg-Michaelson, Wesley, Mr.
World Council Of Churches
150 route de Ferney
1211 Geneva 20, Switzerland
T: (41 22) 791 61 11 F: (41 22) 791 03 61
TL: 23423 OIK CH

Grob, Gustav, Mr.
World Circle Of The Consensus (CMDC)
Kellerweg 38
CH-8055 Zurich, Switzerland
T: (01) 463 9252/463 0226 F: (01) 463 0252

Guy, Andre, Mr.
Vague, La
20 rue Henri Mussard
1208 Geneve, Switzerland
T: (41 22) 736 34 17 F: (41 22) 66 24 75

Gysi, Marshall, Mr.
International Federation Of Consulting Engineers
PO Box 86
1000 Lausanne 12-Chailly, Switzerland
T: (41 21) 653 5003 F: (41 21) 653 5432

Hagmajer, Marek, Mr.
World Federation Of United Nations Associations
Palais des Nations
1211 Geneva 10, Switzerland
T: (41 22) 733 07 30 F: (41 22) 733 48 38
TL: 289696 UNO CH

Harris, David, Mr.
Business Council For Sustainable Development
10 route de l'Aeroport
1215 Geneva 15, Switzerland
T: (41 22) 788 3202 F: (41 22) 788 3211

Heiner, Howard H., Mr.
International Seed Trade Federation
Ch. du Reposoir 5-7
1260 Nyon, Switzerland
T: (41 22) 619 977/619 914 F: (41 22) 619 219

Holzer, Marcus, Mr.
European Confederation Of Agriculture
Postfach 87
5200 Brugg, Switzerland
T: (41 56) 41 31 77 F: (41 56) 41 44 77

Kadjar-Hamouda, Eylah, Mr.
International Federation Terre Des Hommes
22 rue Michel Chauvet
1208 Geneva, Switzerland
T: (41 22) 736 33 72

Koch, Joanna, Mrs
La Leche League International
Alte Landstrasse 89
8802 Kilchberg, Switzerland
T: (41 1) 715 19 46 F: (41 1) 202 30 93

Kyte, Rachel, Ms.
Centre For Our Common Future
Palais Wilson, 52 rue des Paquis
CH-1201 Geneve, Switzerland
T: (41 22) 732 7117 F: (41 22) 738 5046
TL: 27910 CH

Laughton, John, Mr.
Foundation For Environmental Conservation
7 Chemin Tavernay
1218 Grand-Saconnex-GE, Switzerland
T: (41 22) 798 2383 F: 941 22) 798 2384

Lingner, Klaus, Mr.
International Organization For Standardization
Case Postale 56
CH-1211 Geneva 20, Switzerland
T: (41 22) 749 01 11 F: (41 22) 733 34 30
TL: 41 22 05 iso ch

Lonn, Jan, Mr.
International Youth And Student Movement For The United Nations*
16 Avenue Jean Trembley
Geneva, Switzerland
T: (41 22) 798 5850 F: (41 22) 733 4838

Moreillon, Jacques, Mr.
World Organization Of The Scout Movement
PO Box 241
1211 Geneva 4, Switzerland
T: (41 22) 204 233 F: (41 22) 781 2053

Noll, Rudiger, Mr.
Conference Of European Churches
PO Box 2100, 150 route de Ferney
CH-1211 Geneva 2, Switzerland
T: (41 22) 791 6111 F: (41 22) 791 0361

Nyaga, Richard, Mr.
International Air Transport Association
IATA Centre, 33 route de l'aeroport
CH-1215 Geneva 15 Airport, Switzerland
T: (41 22) 799 25 25
TL: 415586

Olivastri, Beatrice, Ms.
International Facilitating Committee, The
Palais Wilson, 52 rue des Paquis
1201 Geneva, Switzerland
T: (41 22) 732 7117 F: (41 22) 737 5046

Pitt, David, Mr.
International Peace Research Institute
34 Boulevard du Pont d'Arve
CH-1205 Geneva, Switzerland
T: (41 22) 601 452

Poletti, Luigi B., Mr.
United Towns Agency For North-south Cooperation
26 rue l'Athenee, Case Postale 370
CH-1211 Geneva 12, Switzerland
T: (41 22) 46 71 71 F: (41 22) 46 78 81

Ritchie, Cyril, Mr.
Ngo Committee On Unicef
15 Chemin Rojoux
1231 Conches GE, Switzerland
T: (41 22) 733 6717 F: (41 22) 784 3533

Robel, Blaise, Mr.
World Confederation Of Labour*
1 rue de Varembe, Case Postale 122
1211 Geneva 20, Switzerland
T: (41 22) 733 66 88 F: (41 22) 733 47 85

Robert, Christian Nils, Prof.
World Lawyers Association For The Protection Of Environment
Universite de Geneve, Faculte de Droit, 5 rue St. Ours
1211 Geneve 4, Switzerland
T: (1) 45 00 85 55 F: (1) 45 01 27 54

Schneider, , Ms.
World Citizen's Assembly
PO Box 128
1211 Geneva 20, Switzerland
T: (41 22) 733 8876 F: (41 22) 733 6649

Schumperli, Catherine, Ms.
Declaration De Berne, La
Chemin de Boston 11
Lausanna CCP 10-10813, Switzerland
T: (41 21) 24 54 17

Shepherd, Gordon, Mr.
World Wide Fund For Nature
Avenue du Mont-Blanc
CH-1196 Gland, Switzerland
T: (41 22) 64 9111 F: (41 22) 64 4238

Strahl, Alexander, Mr.
Airport Association Council International
P.O. Box 125
1215 Geneva 15 Airport, Switzerland
T: (41 22) 798 41 41 F: (41 22) 788 09 09
TL: 427 130 THX CH

Taylor, Thomas, Mr.
International Peace Bureau*
41 route de Zurich
CH-1201 Geneva, Switzerland
T: (41 22) 731 6429 F: (41 22) 731 6429

Thordarson, Bruce, Mr.
International Co-operative Alliance
15 route des Morillons
1218 Grand-Saconnex GE, Switzerland
T: (41 22) 798 41 21 F: (41 22) 798 41 22
TL: 41 56 20 ICA CH

Towsley, Lona, Ms.
World Confederation Of Organisations Of The Teaching Profession
5 Avenue du Moulin
1110 Morges, Switzerland
T: (41 21) 801 7467 F: (41 21) 801 7469

Vaerno, Grethe, Dr.
Associated Country Women Of The World
149 D, route de Ferney
1218 Grand-Saconnex GE, Switzerland
T: (41 22) 798 90 19

Veith, Werner K., Dr.
International Federation of Industrial Energy Consumers
7 Chemin des Tattes
CH-1222 Vesenaz GE, Switzerland
F: (41 22) 752 5624

Walker, Peter, Mr.
League Of Red Cross & Red Crescent Societies
PO Box 372
1211 Geneva 19, Switzerland
T: (41 22) 734 55 80 F: (41 22) 733 03 95
TL: 412 133 LRC CH

Ward, John, Mr.
International Touring Alliance
2 Quai Gustave Ador
1207 Geneva, Switzerland
T: (41 22) 735 2727 F: (41 22) 735 2326

Weid, Denis Von Den, Mr.
Antenna Technologie
10 rue grenus
CH-1201 Geneve, Switzerland
T: (41 22) 731 8036 F: (41 22) 731 9786

Ziegler, Aila, Ms.
Society for Threatened Peoples
Bruggerstrasse 78
CH-5400 Baden, Switzerland
T: (41 56) 221 787 F: (41 56) 221 809

GLOBAL FORUM

Abdel Malek, Ishak, Mr.
WORLD COUNCIL OF CHURCHES
150 Rte de Ferney
Geneva, Switzerland

Abu Nasr, Julenda, Mr.
WORLD COUNCIL OF CHURCHES
150 route de Ferney
Geneve 1211, Switzerland
F: 41 22 791 0361

Adams, Sally
WORLD WILDLIFE FUND - WWF
Rue du Mont Blanc
Gland, Switzerland
T: 41 22 64 95 25 F: 41 22 64 58 79

Ahumada, Patricia V., Ms.
WORLD YOUNG WOMEN'S CHRISTIAN ASSOCIATION / CHILE
37, Quai Wilson
Geneva 1201, Switzerland
T: 41 22 732 31 00 F: 41 22 731 79 38
TL: WYMCA 41 23 32

Alho, Celina, Ms.
WORLD WILDLIFE FUND - WWF
Rue du Mont Blanc
Gland, Switzerland
T: 41 22 64 95 25 F: 41 22 64 58 79

Alho, Cleber J. R., Mr.
WORLD WILDLIFE FUND - WWF
Rue du Mont Blanc
Gland, Switzerland
T: 41 22 64 95 25 F: 41 22 64 58 79

Almeida, Enea De Stuz, Ms.
WORLD COUNCIL OF CHURCHES
150 Rte De Ferney
Geneva, Switzerland

Altman, Walter, Mr.
Rev.
WORLD COUNCIL OF CHURCHES
150 Rte de Ferney
Geneva, Switzerland

Antoine, Maurice, Mr.
Redator
JOURNAL DE GENEVE ET GAZETTE DE LAUSANNE
5-7 General Dufour Boite Postale 439 1211 Geneve 11
Geneve 1211, Switzerland
T: 41 22 280350 F: 41 22 283450
TL: 422-214

Barsoum, Nadine, Ms.
WORLD YOUNG WOMEN'S CHRISTIAN ASSOCIATION / EGYPT
37, Quai Wilson
Geneva 1201, Switzerland
T: 41 22 732 31 00 F: 41 22 731 79 38
TL: WYMCA 41 23 32

Beat, Glogger, Mr.
SWISS TELEVISION
Peruselistr. 1-4 Redaktion MTV
Zurich 8052, Switzerland
T: 41 1 2055939 F: 41 1 2055810

Berham, Wedajo, Mr.
WORLD COUNCIL OF CHURCHES
150 route de Ferney
Geneve 1211, Switzerland
F: 41 22 791 0361

Bernard, Giovannini, Mr.
INTERNATIONAL ACADEMY OF THE ENVIRONMENT
4, Ch. de Conches
Conches - Geneva 1231, Switzerland
T: 41 22 789-1311 F: 41 22 7892538

Bizuru, Joyce, Ms.
WORLD YOUNG WOMEN'S CHRISTIAN ASSOCIATION / TANZANIA
37, Quai Wilson
Geneva 1201, Switzerland
T: 41 22 732 31 00 F: 41 22 731 79 38
TL: WYMCA 41 23 32

Boccard, Jeanne-Marie De, Ms.
Vice-President
INTERNATIONAL COUNCIL OF WOMEN
12 Chemin Vert-Pre - CH 1231 Conches
Geneve, Switzerland
T: 41 22 47-0228 F: 41 22 46-7921

Boulloche, Carola De, Ms.
INTERNATIONAL COUNCIL OF WOMEN
07 Route de Pressy 1253 Vandoeuvres
Geneve, Switzerland
T: 41 22 750-1438

Bretel, Valery Aida Georgette Aliaga, Sr.
WORLD YOUNG WOMEN'S CHRISTIAN ASSOCIATION / BOLIVIA
37, Quai Wilson
Geneva 1201, Switzerland
T: 41 22 732 31 00 F: 41 22 731 79 38
TL: WYMCA 41 23 32

Buechi, Andres, Mr.
Journalist
SONNTAGSZEITUNG
c/o Mondorama Zurich AG - Central 2
Zurich, Switzerland
T: 41 1 262-3320 F: 41 1 262-2306
TL: 817801

Burhenne, Francoise, Ms.
Director - Env. Law Centre
IUCN - THE WORLD CONSERVATION UNION
Av Mont Blanc, CH - 1196
Gland, Switzerland
T: 41 22 649 232 F: 41 22 642 926

Burhenne, Wolfgang, Mr.
Legal Adviser
IUCN - THE WORLD CONSERVATION UNION
Av Mont Blanc, CH - 1196
Gland, Switzerland
T: 41 22 649 232 F: 41 22 642 926

Cesar, Ely Eser, Ms.
WORLD COUNCIL OF CHURCHES
150 Rte de Ferney
Geneva, Switzerland

Cha, Kyung-Hwan, Mr.
WORLD COUNCIL OF CHURCHES
150 Rte de Ferney
Geneva, Switzerland

Chakarian, Janet, Ms.
GATT
154 VE De Lausanne
Geneva 1211, Switzerland
T: 41 22 739-5565

Charity, Sandra
WORLD WILDLIFE FUND - WWF
Rue du Mont Blanc
Gland, Switzerland
T: 41 22 64 95 25 F: 41 22 64 58 79

Cheda, Cornelio
Casa Girasole - 6673 Maggia
Ticino 6673, Switzerland
T: 093 87-1318

Correia, Evanilza De Barros, Ms.
Rev.
WORLD COUNCIL OF CHURCHES
150 Rte de Ferney
Geneva, Switzerland

Cotrim, Maria Leticia, Ms.
IRED
Rua Varembe No.3
Geneva, Switzerland
T: 41 22 737-1716

Cundiff, Ron, Mr.
Singer
SINGER-COMPOSER ENVIRONMENTAL SONGS
C.P. 478
Geneva 1214, Switzerland
T: 41 22 340-2425

Dahinden, Justus, Mr.
Professor
IAA INTERNATIONAL ACADEMY OF ARCHITECTURE
Hotel Rio Atlantico
Zurich, Switzerland
T: 244-5056

Dawson, Anna, Ms.
Programme Manager
INTERNATIONAL ACADEMY OF THE ENVIRONMENT
4, Ch. de Conches
Conches - Geneva 1231, Switzerland
T: 41 22 789-1311 F: 41 22 7892538

Devoto, Maria, Ms.
WORLD YOUNG WOMEN'S CHRISTIAN ASSOCIATION
37, Quai Wilson
Geneva 1201, Switzerland
T: 41 22 732 31 00 F: 41 22 731 79 38
TL: WYMCA 41 23 32

Dlamini, Sarah, Ms.
COORDINATING ASSEMBLY OF NGO'S
Palais de Nations
Geneve CH 1211, Switzerland
F: 41 22 733-6542

Dos Santos, Claudio, Mr.
WORLD COUNCIL OF CHURCHES
150 Rte de Ferney
Geneva, Switzerland

Efthimiou, Miltiades, Mr.
WORLD COUNCIL OF CHURCHES
150 Rte de Ferney
Geneva, Switzerland

Egli, Ruedi
WORLD WILDLIFE FUND - WWF
Rue du Mont Blanc
Gland, Switzerland
T: 41 22 64 95 25 F: 41 22 64 58 79

Eglin, Richard, Mr.
Director
GATT
Rue de Lausanne 154
Geneva, Switzerland
T: 41 22 739-5148

Ekerman, Suren
Editor
JOUNAL DE GENEVE et GAZETTE DE LOUSANE
5-7 Genaral Defour P.O.Box 439
Geneve, Switzerland
T: 41 22 280350 F: 41 22 283450

Embren, Kaj, Mr.
INTERNATIONAL COOPERATIVE ALLIANCE (ICA)
15, Route des Morillons - Geneva
Grand-Saconnex 1218, Switzerland
T: 41 22 798.4121 F: 41 22 798.4122

Esparza, Maria G.G., Ms.
WORLD YOUNG WOMEN'S CHRISTIAN ASSOCIATION / MEXICO
37, Quai Wilson
Geneva 1201, Switzerland
T: 41 22 732 31 00 F: 41 22 731 79 38
TL: WYMCA 41 23 32

Fenger, Pernille, Ms.
INTERNATIONAL LABOUR OFFICE
4, Route des Morillons
Geneva 1211, Switzerland
T: 41 22 799-7177 F: 41 22 798-8685

Ferreira, Jorge Luis Dominguez, Mr.
WORLD COUNCIL OF CHURCHES
150 Rte De Ferney
Geneva, Switzerland

Fortes, Marcio, Mr.
BUSINESS COUNCIL FOR SUSTAINABLE DEVELOPMENT
Hurdnerstrasse 10, PO Box 1474
Hurden 8640, Switzerland
T: 41 55 473377 F: 41 55 484824

Fouad, Bitar, Mr.
Expert
UNION CATHOLIQUE INTERNATIONAL DE LA PRESSE (UCIP)
10,rue des Lilas
Geneve 1202, Switzerland
T: 41 22 7336219 F: 41 22 7331051
TL: 412946 ucip ch

Gavrila, Lucian, Mr.
Prof.
WORLD COUNCIL OF CHURCHES
150 Rte de Ferney
Geneva, Switzerland

Ghai, Dharam, Mr.
UNRISD
Palais des Nations,1211 Geneva
Geneva, Switzerland
T: 41 22 798 8400 F: 41 22 7400791

Giovannini, Maria-Rosa, Ms.
WORLD COUNCIL OF CHURCHES
150 route de Ferney
Geneve 1211, Switzerland
F: 41 22 791 0361

Goldstein, Wendy, Ms.
Coordinator
I.U.C.N.
Av. du Mont Blanc
Geneva, Switzerland
T: 41 22 364-8360 F: 41 22 364-8325

Goukwayo, Helen, Ms.
WORLD YOUNG WOMEN'S CHRISTIAN ASSOCIATION - UGANDA
37, Quai Wilson
Geneva 1201, Switzerland
T: 41 22 732 31 00 F: 41 22 731 79 38
TL: WYMCA 41 23 32

Grandmaison, Roberto, Mr.
Padre
WORLD COUNCIL OF CHURCHES
150 Rte de Ferney
Geneva, Switzerland

Granfar, Ramin, Mr.
Interv.
BAHA'I INTERNATIONAL COMMUNITY
9A Plateau de Frintenex
Geneve 1208, Switzerland
T: 41 22 7986577

Griethuysen, Pascal Van, Mr.
WORLD CIRCLE OF THE CONSENSUS
Chemin de Champ-Rond 28
Lausanne CH-1010, Switzerland
T: 41 21 653-0084 F: 41 21 237615

Grob, Monika, Ms.
WORLD COUNCIL OF CHURCHES
150 route de Ferney
Geneve 1211, Switzerland
F: 41 22 791 0361

Guzman, R.A. De, Dr.
Special Assistant
WORLD METEOROLOGICAL ORGANIZATION
Case Postale NO.2300
Geneva CH 1211, Switzerland
T: 41 22 730-8200 F: 41 22 734-2326

Haemmerli, Fredy, Mr.
Editor
CASH MAGAZINE
P. O. Box 3000
Berna 3000 7, Switzerland
T: 41 31 213833 F: 41 31 213840

Haes, Charles De, Mr.
WORLD WILDLIFE FUND - WWF
Rue du Mont Blanc
Gland, Switzerland
T: 41 22 64 95 25 F: 41 22 64 58 79

Hajjar, Raja, Mr.
Dr.
WORLD COUNCIL OF CHURCHES
150 Rte de Ferney
Geneva, Switzerland

Helfer, Marisa, Ms.
Economista
Schlettstadterstr, 50
Basileia, Switzerland

Hellmann, Marion F., Mr.
Ifbww
INTERNATIONAL FEDERATION OF BUILDING AND WOOD WORKERS
P.O. Box 733
Geneva, Switzerland
T: 41 22 788-0888 F: 41 22 788-0716
TL: 415327fitb

Herrera, Tina, Ms.
WORLD YOUNG WOMEN'S CHRISTIAN ASSOCIATION
37, Quai Wilson
Geneva 1201, Switzerland
T: 41 22 732 31 00 F: 41 22 731 79 38
TL: WYMCA 41 23 32

Hill, Tony, Mr.
Coordinator
UNITED NATIONS NGO LIAISON SERVICE
Palais de Nations
Geneve CH 1211, Switzerland
F: 41 22 733-6542

Holguin, Adolfo, Sr.
FUNDACION NATURA
Palais de Nations
Geneve CH 1211, Switzerland
F: 41 22 733-6542

Humberg, Mario Ernesto, Mr.
WORLD WILDLIFE FUND - WWF
Rue du Mont Blanc
Gland, Switzerland
T: 41 22 64 95 25 F: 41 22 64 58 79

Inder, Brigid, Ms.
WORLD YOUNG WOMEN'S CHRISTIAN ASSOCIATION
37, Quai Wilson
Geneva 1201, Switzerland
T: 41 22 732 31 00 F: 41 22 731 79 38
TL: WYMCA 41 23 32

Jager, Doris, Ms.
WORLD COUNCIL OF CHURCHES
150 Rte de Ferney
Geneva, Switzerland

James, Patricia, Ms.
WORLD YOUNG WOMEN'S CHRISTIAN ASSOCIATION / GRENADA
37, Quai Wilson
Geneva 1201, Switzerland
T: 41 22 732 31 00 F: 41 22 731 79 38
TL: WYMCA 41 23 32

Jefferies, Anna, Ms.
Executive Secretary
BUSINESS COUNCIL FOR SUSTAINABLE DEVELOPMENT
World Trade Center, 10 route de l'Aeroport, Case Postale 365
Geneva 1215, Switzerland
T: 41 22 788-3202 F: 41 22 788-3211

Jesus, A. Nascimento, Ms.
WORLD COUNCIL OF CHURCHES
150 Rte de Ferney
Geneva, Switzerland

John, Erica, Ms.
WORLD YOUNG WOMEN'S CHRISTIAN ASSOCIATION / ST. VINCENT
37, Quai Wilson
Geneva 1201, Switzerland
T: 41 22 732 31 00 F: 41 22 731 79 38
TL: WYMCA 41 23 32

Joshua, Frank T., Mr.
Economist
UNCTAD
Palais de Nations, 1211 Geneva 10
Geneva 1211, Switzerland
T: 41 22 734 60 11 F: 41 22 733 65 42
TL: 41 29 62

Karsay, Eszter, Ms.
WORLD COUNCIL OF CHURCHES
150 Rte de Ferney
Geneva, Switzerland

Khalid, Mansour, Dr.
Chairman
CENTRE FOR OUR COMMON FUTURE
c/o Palais Wilson 52, Rue des Paquis
Geneve 1201, Switzerland
T: 41 22 732-7117

Kidane, Fekrou, Mr.
Advisor
INTERNATIONAL OLYMPIC COMMITEE
CH-1007 Lausanne
Lausanne, Switzerland
T: 41 21 6216111 F: 41 21 253818

Klebnikoff, Marie C. De, Ms.
BUSINESS COUNCIL FOR SUSTAINABLE DEVELOPMENT
World Trade Center Building, Route de l'Aeroport 10, PO Box 365
Geneve 1215, Switzerland
T: 41 22 7883202 F: 41 22 7883211

Klebnikoff, Serge De, Mr.
BUSINESS COUNCIL FOR SUSTAINABLE DEVELOPMENT
World Trade Center Building, Route de l'Aeroport 10, PO Box 365
Geneve 1215, Switzerland
T: 41 22 7883202 F: 41 22 7883211

Kohler, Larry, Mr.
INTERNATIONAL LABOUR OFFICE
4, Route des Morillons
Geneva 1211, Switzerland
T: 41 22 799-7456 F: 41 22 798-8685

Koukoura, Dimitra, Dr.
WORLD COUNCIL OF CHURCHES
150 Rte de Ferney
Geneva, Switzerland

Kraav, Anne, Ms.
WORLD YOUNG WOMEN'S CHRISTIAN ASSOCIATION / FINLAND
37, Quai Wilson
Geneva 1201, Switzerland
T: 41 22 732 31 00 F: 41 22 731 79 38
TL: WYMCA 41 23 32

Kwok, Pui Lan
Prof.
WORLD COUNCIL OF CHURCHES
150 Rte de Ferney
Geneva, Switzerland

Lauhonen, Harri
WORLD WILDLIFE FUND - WWF
Rue du Mont Blanc
Gland, Switzerland
T: 41 22 64 95 25 F: 41 22 64 58 79

Lawrence, Richard, Mr.
Rev.
WORLD COUNCIL OF CHURCHES
150 Rte de Ferney
Geneva, Switzerland

Lenares, Denise, Ms.
WORLD YOUNG WOMEN'S CHRISTIAN ASSOCIATION / BELIZE
37, Quai Wilson
Geneva 1201, Switzerland
T: 41 22 732 31 00 F: 41 22 731 79 38
TL: WYMCA 41 23 32

Lesaffre, Dominique, Mr.
RAFAD
Case Postale 117
Geneve 1211, Switzerland
T: 41 22 733 50 73 F: 41 22 740 00 11

Leuenberger, Nicolas, Mr.
PARTI ECOLOGISTE GENEVE
110, Route De Vandoeuvres 1253, Vandoeuvres
Geveva, Switzerland
T: 19 41 22 750-2335

Limstrand, Siv, Mr.
WORLD COUNCIL OF CHURCHES
150 Rte de Ferney
Geneva, Switzerland

Lopez, Fermina, Ms.
WORLD COUNCIL OF CHURCHES
150 Rte de Ferney
Geneva, Switzerland

Lyonette, Kevin, Mr.
WORLD WILDLIFE FUND - WWF
Rue du Mont Blanc
Gland, Switzerland
T: 41 22 64 95 25 F: 41 22 64 58 79

Mailliet, Michele, Ms.
Audio Visual
UNEP-IUCC
Palais des Nations
Geneva CM-1211, Switzerland
T: 41 22 789-4062 F: 41 22 789-4073

Maneno, Caroline, Ms.
WORLD YOUNG WOMEN'S CHRISTIAN ASSOCIATION
37, Quai Wilson
Geneva 1201, Switzerland
T: 41 22 732 31 00 F: 41 22 731 79 38
TL: WYMCA 41 23 32

Marras, Finola, Ms.
Secretary
INTERNATIONAL CO-OPERATIVE ALLIANCE
15 Route Des Morillons 1218 G-D Saconnex
Geneve, Switzerland
T: 44 22 798-4121 F: 44 22 798-4122

McSweeney, Brenda, Ms.
Executive Coodinator
U.N. VOLUNTEERS
86, Ave. L. Casai
Geneva, Switzerland
T: 41 22 788-2455

Menegueti, Mercio, Mr.
Rev.
WORLD COUNCIL OF CHURCHES
150 Rte de Ferney
Geneva, Switzerland

Meta, Brigitta, Ms.
Observer, Networker, Consultan
M.E.C.C.A.; E.R.F.; FOE; A.C.F.; WINDSTAR
c/o Fam. Glauser, Rutistr. 14
Eschenbach 8733, Switzerland
T: 41 11 55 86 17 27

Missbach, Andreas, Mr.
Neugasse 60
Zurich 8005, Switzerland

Montanez, Wilfredo, Mr.
WORLD COUNCIL OF CHURCHES
150 Rte de Ferney
Geneva, Switzerland

Morelli, Dom Mauro, Mr.
Bispo
WORLD COUNCIL OF CHURCHES
150 Rte de Ferney
Geneva, Switzerland

Morgan, Richie, Mr.
WORLD COUNCIL OF CHURCHES
150 Rte de Ferney
Geneva, Switzerland

Morich, Horst, Mr.
Ifbww
INTERNATIONAL FEDERATION OF BUILDING AND WOOD WORKERS
P.O. Box 733
Geneva, Switzerland
T: 41 22 788-0888 F: 41 22 788-0716
TL: 415327fitb

Mullings, Pauline, Ms.
WORLD YOUNG WOMEN'S CHRISTIAN ASSOCIATION / JAMAICA
37, Quai Wilson
Geneva 1201, Switzerland
T: 41 22 732 31 00 F: 41 22 731 79 38
TL: WYMCA 41 23 32

Mutambirwa, James, Dr.
WORLD COUNCIL OF CHURCHES
150 route de Ferney
Geneve 1211, Switzerland
F: 41 22 791 0361

Neto, Reynaldo, Mr.
WORLD COUNCIL OF CHURCHES
150 Rte de Ferney
Geneva, Switzerland

Ocola, Jessica M. De, Ms.
WORLD YOUNG WOMEN'S CHRISTIAN ASSOCIATION / PERU
37, Quai Wilson
Geneva 1201, Switzerland
T: 41 22 732 31 00 F: 41 22 731 79 38
TL: WYMCA 41 23 32

Oever, Pietronella Van Den, Ms.
Head, Social Sciences Div.
IUCN
Ave. du Mont Blanc
Gland CH 1196, Switzerland
T: 41 22 364-8360 F: 41 22 364-8325

Oit, Betty
Uniced Communications Coordina
WWF - WORLDWIDE FUND FOR NATURE
Avenue de Mont-Blanc, CH-1196 Gland
Geneva, Switzerland
T: 41 22 56441563 F: 41 22 36165358

Oliveira, Rafael Soares De, Mr.
WORLD COUNCIL OF CHURCHES
150 Rte De Ferney
Geneva, Switzerland

Oshigue, Fernando
WORLD COUNCIL OF CHURCHES
150 Rte de Ferney
Geneva, Switzerland

Paarlberg, John, Mr.
WORLD COUNCIL OF CHURCHES
150 Rte deFerney
Geneva, Switzerland

Palma, Marta, Ms.
WORLD COUNCIL OF CHURCHES
150 route de Ferney
Geneve 1211, Switzerland
F: 41 22 791 0361

Palma, Samuel, Mr.
WORLD COUNCIL OF CHURCHES
150 Rte de Ferney
Geneva, Switzerland

Paschoaletto, Lourdes, Ms.
WORLD COUNCIL OF CHURCHES
150 Rte de Ferney
Geneva, Switzerland

Passah, Leila, Ms.
WORLD YOUNG WOMEN'S CHRISTIAN ASSOCIATION
37, Quai Wilson
Geneva 1201, Switzerland
T: 41 22 732 31 00 F: 41 22 731 79 38
TL: WYMCA 41 23 32

Pellerano, Miguel
WORLD WILDLIFE FUND - WWF
Rue du Mont Blanc
Gland, Switzerland
T: 41 22 64 95 25 F: 41 22 64 58 79

Peteru, Clark, Mr.
Director
O'LE SIOSIOMAGA SOCIETY
Palais de Nations / box 5774 Apia
Geneve CH 1211, Switzerland
T: 685 21993 F: 41 22 733-6542

Pinder, Cherrylee, Ms.
WORLD YOUNG WOMEN'S CHRISTIAN ASSOCIATION / BAHAMAS
37, Quai Wilson
Geneva 1201, Switzerland
T: 41 22 732 31 00 F: 41 22 731 79 38
TL: WYMCA 41 23 32

Pret, Marie-Claire, Ms.
Independent Consultant
WORLD COUNCIL OF CHURCHES
P. O. Box 2100 - 150 Route de Ferney
Geneve 1211, Switzerland
T: 41 22 791-6323 F: 41 22 791-0361
TL: 415730

Primavesi, Mark, Mr.
WORLD COUNCIL OF CHURCHES
150 Rte De Ferney
Geneve, Switzerland

Reed, David
WORLD WILDLIFE FUND - WWF
Rue du Mont Blanc
Gland, Switzerland
T: 41 22 64 95 25 F: 41 22 64 58 79

Rey, Maria C.D., Ms.
WORLD YOUNG WOMEN'S CHRISTIAN ASSOCIATION / URUGUAY
37, Quai Wilson
Geneva 1201, Switzerland
T: 41 22 732 31 00 F: 41 22 731 79 38
TL: WYMCA 41 23 32

Ribeiro, Claudio, Mr.
Rev.
WORLD COUNCIL OF CHURCHES
150 Rte de Ferney
Geneva, Switzerland

Robinson, Nick, Mr.
Deputy Chair - Com. On E. Law
IUCN - THE WORLD CONSERVATION UNION
Av Mont Blanc, CH - 1196
Gland, Switzerland
T: 41 22 649 232 F: 41 22 642 926

Saint-Laurent, Carole
WORLD WILDLIFE FUND - WWF
Rue du Mont Blanc
Gland, Switzerland
T: 41 22 64 95 25 F: 41 22 64 58 79

Sami Manorama, Ruth, Ms.
WORLD COUNCIL OF CHURCHES
150 Rte de Ferney
Geneva, Switzerland

Sanchez, Marina, Sra.
WORLD YOUNG WOMEN'S CHRISTIAN ASSOCIATION-COLOMBIA
37, Quai Wilson
Geneva 1201, Switzerland
T: 41 22 732 31 00 F: 41 22 731 79 38
TL: WYMCA 41 23 32

Santos, Aureo, Mr.
WORLD COUNCIL OF CHURCHES
150 Rte de Ferney
Geneva, Switzerland

Schatt, Rahera, Ms.
BUSINESS COUNCIL FOR SUSTAINABLE DEVELOPMENT
Hurdnerstrasse 10, PO Box 1474
Hurden 8640, Switzerland
T: 41 55 484824

Schmid, Walter, Mr.
Eibenstr. 6
Zurich 8045, Switzerland

Schmidth, Ervino, Mr.
WORLD COUNCIL OF CHURCHES
150 Rte de Ferney
Geneva, Switzerland

Schmittfull, Hildegard, Ms.
ST. KATHARINAWERK BASEL
Holee Str. 123
Basel 4015, Switzerland
T: 41 61 301-2323

Schneider, Alice Boainain, Ms.
INSTITUTE FOR PLANETARY SYNTHESIS
P. O. Box 128
Geneve 1211, Switzerland
T: 41 22 733 88 76 F: 41 22 733 66 49
TL: 427993

Schuller, Marilia, Ms.
WORLD COUNCIL OF CHURCHES
150 route de Ferney
Geneve 1211, Switzerland
F: 41 22 791 0361

Silva, Jaider Batista, Mr.
WORLD COUNCIL OF CHURCHES
150 Rte de Ferney
Geneva, Switzerland

Soares, Rafael Oliveira, Mr.
WORLD COUNCIL OF CHURCHES
150 Rte de Ferney
Geneva, Switzerland

Sof, Marita, Ms.
WORLD YOUNG WOMEN'S CHRISTIAN ASSOCIATION / SURINAME
37, Quai Wilson
Geneva 1201, Switzerland
T: 41 22 732 31 00 F: 41 22 731 79 38
TL: WYMCA 41 23 32

Steel, Elaine
Staff
WORLD YOUNG WOMEN'S CHRISTIAN ASSOCIATION
37, Quai Wilson
Geneva 1201, Switzerland
T: 41 22 732 31 00 F: 41 22 731 79 38
TL: WYMCA 41 23 32

Suren, Erkman, Mr.
Redator
JOURNAL DE GENEVE ET GAZETTE DE LAUSANNE
5-7 General Dufour Boite Postale 439 1211 Geneve 11
Geneve 1211, Switzerland
T: 41 22 280350 F: 41 22 283450
TL: 422-214

Szekely, Francisco, Mr.
Professor Envir Management
INTERNATIONAL INSTITUTE FOR MANAGEMENT
23 Chemin de Bellerive, PO Box 915
Lausanne 1001, Switzerland
T: 41 21 6180111 F: 41 21 266725

Taran, Patrick, Mr.
WORLD COUNCIL OF CHURCHES
P. O. Box 2100 - 150 Route de Ferney
Geneve 1211, Switzerland
T: 41 22 791-6323 F: 41 22 791-0361
TL: 415730

Thiollay, Jean Marc
WORLD WILDLIFE FUND - WWF
Rue du Mont Blanc
Gland, Switzerland
T: 41 22 64 95 25 F: 41 22 64 58 79

Trisoglio, Alex, Mr.
Associate
INTERNATIONAL ACADEMY OF THE ENVIRONMENT
4, Ch. de Conches
Conches - Geneva 1231, Switzerland
T: 41 22 789-1311 F: 41 22 7892538

Troya, Roberto
WORLD WILDLIFE FUND - WWF
Rue du Mont Blanc
Gland, Switzerland
T: 41 22 64 95 25 F: 41 22 64 58 79

Tucker, Lynda, Ms.
CAMPAIGN AT THE EARTH
Hearts Bad,Vt
Zurich, Switzerland

Varela, Lic. Vera
NEOTROPICA
Palais de Nations
Geneve CH 1211, Switzerland
F: 41 22 733-6542

Vianna, Anna Lucia, Ms.
WORLD WILDLIFE FUND - WWF
Rue du Mont Blanc
Gland, Switzerland
T: 41 22 64 95 25 F: 41 22 64 58 79

Wahi, Shri Mohan, Mr.
Trustee
WORLD FARMERS TIMES FOUNDATION
Talstrasse 39
Zurich, Switzerland
T: 41 1 211-8812 F: 41 1 211-8812

Waldelich, Otto, Mr.
WORLD COUNCIL OF CHURCHES
150 Rte de Ferney
Geneva, Switzerland

Walter, Ludin
Editor
ITE REVUE
Wesemlinstrasse, 42
CH 6000 Luzern 10, Switzerland
T: 41 41 362155 F: 41 41 362041

Weber, Thomas, Mr.
ST. KATHARINA - WERK
Postfach
Basel CH-4015, Switzerland
T: 41 61 301-2323 F: 41 61 301-1714

Weid, Bernardette Von Der, Ms.
Representative
WORLD ASSOCIATION FOR GIRL GUIDES AND GIRLS SCOUTS
8 Ch. E. Cuchet Albaret, Vandoeuvre
Geneva 1235, Switzerland
T: 41 22 750-1926

Weid, Von Der, Mr.
President
ANTENNA TECHNOLOGY
10 Rue de Grenus
Geneve, Switzerland
T: 41 22 731-8036 F: 41 22 731-9786

Willey, , Mr.
UNCTAD
Palais de Nations
Geneve 1211, Switzerland
T: 41 22 734 60 11 F: 41 22 733 65 42
TL: 41 29 62

Wohler, Jens
President
BLUE CLUE
Nevengasse 39, Postfach
3009 Bern, Switzerland
T: 41 37 219209 F: 41 37 219703

Wong, Peony, Ms.
WORLD COUNCIL OF CHURCHES
150 Rte de Ferney
Geneva, Switzerland

Z'graggen, Andreas, Mr.
BELLERIVE FOUNDATION
4 Rue Munier Romilly, 1206
Geneve, Switzerland
T: 41 22 468866 F: 41 22 479159
TL: 429835muro

UN AND UNCED

Acolant, Florence
United Nations Conference on Environment and Development
160 Route de Florissant, P.O. Box 80
CH-1231 Conches, Switzerland
T: 41-22 7891676 F: 41-22 789-3536

Afonso, Carlos
United Nations Conference on Environment and Development
160 Route de Florissant, P.O. Box 80
CH-1231 Conches, Switzerland
T: 41-22 7891676 F: 41-22 789-3536

Alikhan, Sajid
United Nations Conference on Environment and Development
160 Route de Florissant, P.O. Box 80
CH-1231 Conches, Switzerland
T: 41-22 7891676 F: 41-22 789-3536

Alleman, Angela
United Nations Conference on Environment and Development
160 Route de Florissant, P.O. Box 80
CH-1231 Conches, Switzerland
T: 41-22 7891676 F: 41-22 789-3536

Alvarez, Ms.
United Nations Conference on Environment and Development
160 Route de Florissant, P.O. Box 80
CH-1231 Conches, Switzerland
T: 41-22 7891676 F: 41-22 789-3536

Andrade, Helena
United Nations Conference on Environment and Development
160 Route de Florissant, P.O. Box 80
CH-1231 Conches, Switzerland
T: 41-22 7891676 F: 41-22 789-3536

Antoine, Serge
United Nations Conference on Environment and Development
160 Route de Florissant, P.O. Box 80
CH-1231 Conches, Switzerland
T: 41-22 7891676 F: 41-22 789-3536

Anukam, Lawrence
United Nations Conference on Environment and Development
160 Route de Florissant, P.O. Box 80
CH-1231 Conches, Switzerland
T: 41-22 7891676 F: 41-22 789-3536

Arias, Margarita
United Nations Conference on Environment and Development
160 Route de Florissant, P.O. Box 80
CH-1231 Conches, Switzerland
T: 41-22 7891676 F: 41-22 789-3536

Assuncao, Lucas
United Nations Conference on Environment and Development
160 Route de Florissant, P.O. Box 80
CH-1231 Conches, Switzerland
T: 41-22 7891676 F: 41-22 789-3536

Ayad, M.N.
United Nations Conference on Environment and Development
160 Route de Florissant, P.O. Box 80
CH-1231 Conches, Switzerland
T: 41-22 7891676 F: 41-22 789-3536

Azevedo, Valeria
United Nations Conference on Environment and Development
160 Route de Florissant, P.O. Box 80
CH-1231 Conches, Switzerland
T: 41-22 7891676 F: 41-22 789-3536

Baptista, Marcelo
United Nations Conference on Environment and Development
160 Route de Florissant, P.O. Box 80
CH-1231 Conches, Switzerland
T: 41-22 7891676 F: 41-22 789-3536

Barcena, Alicia
United Nations Conference on Environment and Development
160 Route de Florissant, P.O. Box 80
CH-1231 Conches, Switzerland
T: 41-22 7891676 F: 41-22 789-3536

Beaulieu, Andre
United Nations Conference on Environment and Development
160 Route de Florissant, P.O. Box 80
CH-1231 Conches, Switzerland
T: 41-22 7891676 F: 41-22 789-3536

Beekman, G.
World Meteorological Organization
41, avenue G. Motta
1211 Geneva 20, Switzerland
T: 41-22 730811 F: 41-22 7242326

Benedick, Richard
United Nations Conference on Environment and Development
160 Route de Florissant, P.O. Box 80
CH-1231 Conches, Switzerland
T: 41-22 7891676 F: 41-22 789-3536

Berlinck, D.
United Nations Conference on Environment and Development
160 Route de Florissant, P.O. Box 80
CH-1231 Conches, Switzerland
T: 41-22 7891676 F: 41-22 789-3536

Berthelot, Yves
United Nations Conference on Trade and Development
8-14, avenue de la Paix, Palais des Nations
1211 Geneva 10, Switzerland
T: 41-22 7346011 F: 41-22 7336542

Blanco, Lucas
United Nations Conference on Environment and Development
160 Route de Florissant, P.O. Box 80
CH-1231 Conches, Switzerland
T: 41-22 7891676 F: 41-22 789-3536

Borghi, Christina
United Nations Conference on Environment and Development
160 Route de Florissant, P.O. Box 80
CH-1231 Conches, Switzerland
T: 41-22 7891676 F: 41-22 789-3536

Boulharouf, Rajeb
United Nations Conference on Environment and Development
160 Route de Florissant, P.O. Box 80
CH-1231 Conches, Switzerland
T: 41-22 7891676 F: 41-22 789-3536

Braganca, P.
World Meteorological Organization
41, avenue G. Motta
1211 Geneva 20, Switzerland
T: 41-22 730811 F: 41-22 7242326

Britto, Flavio
United Nations Conference on Environment and Development
160 Route de Florissant, P.O. Box 80
CH-1231 Conches, Switzerland
T: 41-22 7891676 F: 41-22 789-3536

Brose, David
United Nations Conference on Environment and Development
160 Route de Florissant, P.O. Box 80
CH-1231 Conches, Switzerland
T: 41-22 7891676 F: 41-22 789-3536

Bulwa, Beatrice
United Nations Conference on Environment and Development
160 Route de Florissant, P.O. Box 80
CH-1231 Conches, Switzerland
T: 41-22 7891676 F: 41-22 789-3536

Campos, Ricardo
United Nations Conference on Environment and Development
160 Route de Florissant, P.O. Box 80
CH-1231 Conches, Switzerland
T: 41-22 7891676 F: 41-22 789-3536

Capanema, Andrea
United Nations Conference on Environment and Development
160 Route de Florissant, P.O. Box 80
CH-1231 Conches, Switzerland
T: 41-22 7891676 F: 41-22 789-3536

Castello, Mauricio
United Nations Conference on Environment and Development
160 Route de Florissant, P.O. Box 80
CH-1231 Conches, Switzerland
T: 41-22 7891676 F: 41-22 789-3536

Chadwick, Josephine
United Nations Conference on Environment and Development
160 Route de Florissant, P.O. Box 80
CH-1231 Conches, Switzerland
T: 41-22 7891676 F: 41-22 789-3536

Chandiramani, Lata
United Nations Conference on Environment and Development
160 Route de Florissant, P.O. Box 80
CH-1231 Conches, Switzerland
T: 41-22 7891676 F: 41-22 789-3536

Clements, Millard
United Nations Conference on Environment and Development
160 Route de Florissant, P.O. Box 80
CH-1231 Conches, Switzerland
T: 41-22 7891676 F: 41-22 789-3536

Coan, Brian
United Nations Conference on Environment and Development
160 Route de Florissant, P.O. Box 80
CH-1231 Conches, Switzerland
T: 41-22 7891676 F: 41-22 789-3536

Cole, Seraju
United Nations Conference on Environment and Development
160 Route de Florissant, P.O. Box 80
CH-1231 Conches, Switzerland
T: 41-22 7891676 F: 41-22 789-3536

Cordeiro, Cesar
United Nations Conference on Environment and Development
160 Route de Florissant, P.O. Box 80
CH-1231 Conches, Switzerland
T: 41-22 7891676 F: 41-22 789-3536

Couchene, David
United Nations Conference on Environment and Development
160 Route de Florissant, P.O. Box 80
CH-1231 Conches, Switzerland
T: 41-22 7891676 F: 41-22 789-3536

Cousteau, Mrs.
United Nations Conference on Environment and Development
160 Route de Florissant, P.O. Box 80
CH-1231 Conches, Switzerland
T: 41-22 7891676 F: 41-22 789-3536

Crichton, Jacinth
World Association of Former United Nations Internes and Fellows
16, avenue Jean Trembley
1209 Geneva, Switzerland
T: 41-22 7985850 F: 41-22 7910827

Cunha, Bruno
United Nations Conference on Environment and Development
160 Route de Florissant, P.O. Box 80
CH-1231 Conches, Switzerland
T: 41-22 7891676 F: 41-22 789-3536

D'Amico, Pierina
World Meteorological Organization
41, avenue G. Motta
1211 Geneva 20, Switzerland
T: 41-22 730811 F: 41-22 7242326

Da, Gerson
United Nations Conference on Environment and Development
160 Route de Florissant, P.O. Box 80
CH-1231 Conches, Switzerland
T: 41-22 7891676 F: 41-22 789-3536

Dadzie, Kenneth
United Nations Conference on Trade and Development
8-14, avenue de la Paix, Palais des Nations
1211 Geneva 10, Switzerland
T: 41-22 7346011 F: 41-22 7336542

Davidson, Marion
United Nations Conference on Environment and Development
160 Route de Florissant, P.O. Box 80
CH-1231 Conches, Switzerland
T: 41-22 7891676 F: 41-22 789-3536

De, Ana
United Nations Conference on Environment and Development
160 Route de Florissant, P.O. Box 80
CH-1231 Conches, Switzerland
T: 41-22 7891676 F: 41-22 789-3536

De, Anne
United Nations Conference on Environment and Development
160 Route de Florissant, P.O. Box 80
CH-1231 Conches, Switzerland
T: 41-22 7891676 F: 41-22 789-3536

De, Edmund
United Nations Conference on Environment and Development
160 Route de Florissant, P.O. Box 80
CH-1231 Conches, Switzerland
T: 41-22 7891676 F: 41-22 789-3536

De, J.C.
World Meteorological Organization
41, avenue G. Motta
1211 Geneva 20, Switzerland
T: 41-22 730811 F: 41-22 7242326

De, P.C.
World Meteorological Organization
41, avenue G. Motta
1211 Geneva 20, Switzerland
T: 41-22 730811 F: 41-22 7242326

De, R.
World Meteorological Organization
41, avenue G. Motta
1211 Geneva 20, Switzerland
T: 41-22 730811 F: 41-22 7242326

De, R.A.
World Meteorological Organization
41, avenue G. Motta
1211 Geneva 20, Switzerland
T: 41-22 730811 F: 41-22 7242326

Debray, Martine
United Nations Conference on Environment and Development
160 Route de Florissant, P.O. Box 80
CH-1231 Conches, Switzerland
T: 41-22 7891676 F: 41-22 789-3536

Dejene, Alemneh
United Nations Conference on Environment and Development
160 Route de Florissant, P.O. Box 80
CH-1231 Conches, Switzerland
T: 41-22 7891676 F: 41-22 789-3536

Dengo, Manuel B., Mr.
World Meteorological Organization
41, avenue G. Motta
1211 Geneva 20, Switzerland
T: 41-22 730811 F: 41-22 7242326

Dennis, Ameena
United Nations Conference on Environment and Development
160 Route de Florissant, P.O. Box 80
CH-1231 Conches, Switzerland
T: 41-22 7891676 F: 41-22 789-3536

Desai, A
United Nations Conference on Environment and Development
160 Route de Florissant, P.O. Box 80
CH-1231 Conches, Switzerland
T: 41-22 7891676 F: 41-22 789-3536

Desai, Nitin
United Nations Conference on Environment and Development
160 Route de Florissant, P.O. Box 80
CH-1231 Conches, Switzerland
T: 41-22 7891676 F: 41-22 789-3536

Di, Tiziana
United Nations Conference on Environment and Development
160 Route de Florissant, P.O. Box 80
CH-1231 Conches, Switzerland
T: 41-22 7891676 F: 41-22 789-3536

Diallo, Arbo
United Nations Conference on Environment and Development
160 Route de Florissant, P.O. Box 80
CH-1231 Conches, Switzerland
T: 41-22 7891676 F: 41-22 789-3536

Dias, Laercio
United Nations Conference on Environment and Development
160 Route de Florissant, P.O. Box 80
CH-1231 Conches, Switzerland
T: 41-22 7891676 F: 41-22 789-3536

Diperna, Paula
United Nations Conference on Environment and Development
160 Route de Florissant, P.O. Box 80
CH-1231 Conches, Switzerland
T: 41-22 7891676 F: 41-22 789-3536

Do, Pele
United Nations Conference on Environment and Development
160 Route de Florissant, P.O. Box 80 CH-1231 Conches,
F: 41-22 7891676
TL: 41-22 789-3536

Dodds, Michael
World Federation of United Nations Associations
16, avenue Jean Trembley
1209 Geneva, Switzerland
T: 41-22 7985850 F: 41-22 7910827

Dos, Ismael
United Nations Conference on Environment and Development
160 Route de Florissant, P.O. Box 80
CH-1231 Conches, Switzerland
T: 41-22 7891676 F: 41-22 789-3536

Duguid, Sharon
United Nations Conference on Environment and Development
160 Route de Florissant, P.O. Box 80
CH-1231 Conches, Switzerland
T: 41-22 7891676 F: 41-22 789-3536

Dunkley, David
United Nations Conference on Environment and Development
160 Route de Florissant, P.O. Box 80
CH-1231 Conches, Switzerland
T: 41-22 7891676 F: 41-22 789-3536

Eguchi-Sirvain, Yoko
United Nations Conference on Environment and Development
160 Route de Florissant, P.O. Box 80
CH-1231 Conches, Switzerland
T: 41-22 7891676 F: 41-22 789-3536

Ellyard, Heather
United Nations Conference on Environment and Development
160 Route de Florissant, P.O. Box 80
CH-1231 Conches, Switzerland
T: 41-22 7891676 F: 41-22 789-3536

Espindola, Roberto
United Nations Conference on Environment and Development
160 Route de Florissant, P.O. Box 80
CH-1231 Conches, Switzerland
T: 41-22 7891676 F: 41-22 789-3536

Evans, Gayle
United Nations Conference on Environment and Development
160 Route de Florissant, P.O. Box 80
CH-1231 Conches, Switzerland
T: 41-22 7891676 F: 41-22 789-3536

Faby, Jean-Claude
United Nations Conference on Environment and Development
160 Route de Florissant, P.O. Box 80
CH-1231 Conches, Switzerland
T: 41-22 7891676 F: 41-22 789-3536

Farwell, Edith
United Nations Conference on Environment and Development
160 Route de Florissant, P.O. Box 80
CH-1231 Conches, Switzerland
T: 41-22 7891676 F: 41-22 789-3536

Faucher, George
United Nations Conference on Environment and Development
160 Route de Florissant, P.O. Box 80
CH-1231 Conches, Switzerland
T: 41-22 7891676 F: 41-22 789-3536

Fauci, George
United Nations Conference on Environment and Development
160 Route de Florissant, P.O. Box 80
CH-1231 Conches, Switzerland
T: 41-22 7891676 F: 41-22 789-3536

Ferrey, Aura
United Nations Conference on Environment and Development
160 Route de Florissant, P.O. Box 80
CH-1231 Conches, Switzerland
T: 41-22 7891676 F: 41-22 789-3536

Figueira, Saliel
United Nations Conference on Environment and Development
160 Route de Florissant, P.O. Box 80
CH-1231 Conches, Switzerland
T: 41-22 7891676 F: 41-22 789-3536

Figueiras, Silvia
United Nations Conference on Environment and Development
160 Route de Florissant, P.O. Box 80
CH-1231 Conches, Switzerland
T: 41-22 7891676 F: 41-22 789-3536

Figueiredo, Astrid
United Nations Conference on Environment and Development
160 Route de Florissant, P.O. Box 80
CH-1231 Conches, Switzerland
T: 41-22 7891676 F: 41-22 789-3536

Figueroa, Godofredo
United Nations Conference on Environment and Development
160 Route de Florissant, P.O. Box 80
CH-1231 Conches, Switzerland
T: 41-22 7891676 F: 41-22 789-3536

Fonda, Jane
United Nations Conference on Environment and Development
160 Route de Florissant, P.O. Box 80
CH-1231 Conches, Switzerland
T: 41-22 7891676 F: 41-22 789-3536

Fordyce, Urma
United Nations Conference on Environment and Development
160 Route de Florissant, P.O. Box 80
CH-1231 Conches, Switzerland
T: 41-22 7891676 F: 41-22 789-3536

Fram, Steven
United Nations Conference on Environment and Development
160 Route de Florissant, P.O. Box 80
CH-1231 Conches, Switzerland
T: 41-22 7891676 F: 41-22 789-3536

Fruttig, Marcelo
United Nations Conference on Environment and Development
160 Route de Florissant, P.O. Box 80
CH-1231 Conches, Switzerland
T: 41-22 7891676 F: 41-22 789-3536

Garcia, Carmen
United Nations Conference on Environment and Development
160 Route de Florissant, P.O. Box 80
CH-1231 Conches, Switzerland
T: 41-22 7891676 F: 41-22 789-3536

Gardner, Richard, Mr.
United Nations Conference on Environment and Development
160 Route de Florissant, P.O. Box 80
CH-1231 Conches, Switzerland
T: 41-22 7891676 F: 41-22 789-3536

Garfunkel, Elizabeth
United Nations Conference on Environment and Development
160 Route de Florissant, P.O. Box 80
CH-1231 Conches, Switzerland
T: 41-22 7891676 F: 41-22 789-3536

Garton, Andrew
United Nations Conference on Environment and Development
160 Route de Florissant, P.O. Box 80
CH-1231 Conches, Switzerland
T: 41-22 7891676 F: 41-22 789-3536

Garza, B
United Nations Conference on Environment and Development
160 Route de Florissant, P.O. Box 80
CH-1231 Conches, Switzerland
T: 41-22 7891676 F: 41-22 789-3536

Gauthier, Yvette, Ms.
World Association of Former United Nations Internes and Fellows
16, avenue Jean Trembley
1209 Geneva, Switzerland
T: 41-22 7985850 F: 41-22 7910827

Gazarian, Jean
United Nations Conference on Environment and Development
160 Route de Florissant, P.O. Box 80
CH-1231 Conches, Switzerland
T: 41-22 7891676 F: 41-22 789-3536

Ghai, Aniket
United Nations Conference on Environment and Development
160 Route de Florissant, P.O. Box 80
CH-1231 Conches, Switzerland
T: 41-22 7891676 F: 41-22 789-3536

Given, Anne
United Nations Conference on Environment and Development
160 Route de Florissant, P.O. Box 80
CH-1231 Conches, Switzerland
T: 41-22 7891676 F: 41-22 789-3536

Goodman, Gordon
United Nations Conference on Environment and Development
160 Route de Florissant, P.O. Box 80
CH-1231 Conches, Switzerland
T: 41-22 7891676 F: 41-22 789-3536

Gordon, Anita
United Nations Conference on Environment and Development
160 Route de Florissant, P.O. Box 80
CH-1231 Conches, Switzerland
T: 41-22 7891676 F: 41-22 789-3536

Gordon, Bruce
United Nations Conference on Environment and Development
160 Route de Florissant, P.O. Box 80
CH-1231 Conches, Switzerland
T: 41-22 7891676 F: 41-22 789-3536

Gorre, Eirah
World Meteorological Organization
41, Avenue G. Motta
1211 Geneva 20, Switzerland
T: 41-22 730811 F: 41-22 7242326

Gosovic, Branislav
United Nations Conference on Environment and Development
160 Route de Florissant, P.O. Box 80
CH-1231 Conches, Switzerland
T: 41-22 7891676 F: 41-22 789-3536

Granados, Asa
United Nations Conference on Environment and Development
160 Route de Florissant, P.O. Box 80
CH-1231 Conches, Switzerland
T: 41-22 7891676 F: 41-22 789-3536

Gray, Fabienne
United Nations Conference on Environment and Development
160 Route de Florissant, P.O. Box 80
CH-1231 Conches, Switzerland
T: 41-22 7891676 F: 41-22 789-3536

Greenberg, Melvyn
United Nations Conference on Environment and Development
160 Route de Florissant, P.O. Box 80
CH-1231 Conches, Switzerland
T: 41-22 7891676 F: 41-22 789-3536

Greenberg, Sandi
United Nations Conference on Environment and Development
160 Route de Florissant, P.O. Box 80
CH-1231 Conches, Switzerland
T: 41-22 7891676 F: 41-22 789-3536

Griffiths, Sally
United Nations Conference on Environment and Development
160 Route de Florissant, P.O. Box 80
CH-1231 Conches, Switzerland
T: 41-22 7891676 F: 41-22 789-3536

Grojsgold, Alexandre
United Nations Conference on Environment and Development
160 Route de Florissant, P.O. Box 80
CH-1231 Conches, Switzerland
T: 41-22 7891676 F: 41-22 789-3536

Gupte, Pranay
United Nations Conference on Environment and Development
160 Route de Florissant, P.O. Box 80
CH-1231 Conches, Switzerland
T: 41-22 7891676 F: 41-22 789-3536

Hadj-Sadok, Tahar
United Nations Conference on Environment and Development
160 Route de Florissant, P.O. Box 80
CH-1231 Conches, Switzerland
T: 41-22 7891676 F: 41-22 789-3536

Helmer, Richard
United Nations Conference on Environment and Development
160 Route de Florissant, P.O. Box 80
CH-1231 Conches, Switzerland
T: 41-22 7891676 F: 41-22 789-3536

Hill, Kevin
United Nations Conference on Environment and Development
160 Route de Florissant, P.O. Box 80
CH-1231 Conches, Switzerland
T: 41-22 7891676 F: 41-22 789-3536

Htun, Nay
United Nations Conference on Environment and Development
160 Route de Florissant, P.O. Box 80
CH-1231 Conches, Switzerland
T: 41-22 7891676 F: 41-22 789-3536

Hyttinen, Lars
United Nations Conference on Environment and Development
160 Route de Florissant, P.O. Box 80
CH-1231 Conches, Switzerland
T: 41-22 7891676 F: 41-22 789-3536

Jampies, Sylvia
United Nations Conference on Environment and Development
160 Route de Florissant, P.O. Box 80
CH-1231 Conches, Switzerland
T: 41-22 7891676 F: 41-22 789-3536

Jawetz, Pincas
World Association of Former United Nations Internes and Fellows
16, avenue Jean Trembley
1209 Geneva, Switzerland
T: 41-22 7985850 F: 41-22 7910827

Johnson, Pierre
United Nations Conference on Environment and Development
160 Route de Florissant, P.O. Box 80
CH-1231 Conches, Switzerland
T: 41-22 7891676 F: 41-22 789-3536

Jung, Eileen
United Nations Conference on Environment and Development
160 Route de Florissant, P.O. Box 80
CH-1231 Conches, Switzerland
T: 41-22 7891676 F: 41-22 789-3536

Kakabadse, Yolanda
United Nations Conference on Environment and Development
160 Route de Florissant, P.O. Box 80
CH-1231 Conches, Switzerland
T: 41-22 7891676 F: 41-22 789-3536

Kalirai, S
United Nations Conference on Environment and Development
160 Route de Florissant, P.O. Box 80
CH-1231 Conches, Switzerland
T: 41-22 7891676 F: 41-22 789-3536

Kaplan, Steve
United Nations Conference on Environment and Development
160 Route de Florissant, P.O. Box 80
CH-1231 Conches, Switzerland
T: 41-22 7891676 F: 41-22 789-3536

Keil, Marilia
United Nations Conference on Environment and Development
160 Route de Florissant, P.O. Box 80
CH-1231 Conches, Switzerland
T: 41-22 7891676 F: 41-22 789-3536

Khosla, Iqbal
United Nations Conference on Environment and Development
160 Route de Florissant, P.O. Box 80
CH-1231 Conches, Switzerland
T: 41-22 7891676 F: 41-22 789-3536

Kittani, Ismat
United Nations Conference on Environment and Development
160 Route de Florissant, P.O. Box 80
CH-1231 Conches, Switzerland
T: 41-22 7891676 F: 41-22 789-3536

Krishnamurthy, P.
United Nations Conference on Environment and Development
160 Route de Florissant, P.O. Box 80
CH-1231 Conches, Switzerland
T: 41-22 7891676 F: 41-22 789-3536

Kushner, Louis
United Nations Conference on Environment and Development
160 Route de Florissant, P.O. Box 80
CH-1231 Conches, Switzerland
T: 41-22 7891676 F: 41-22 789-3536

Langel, Marcia
United Nations Conference on Environment and Development
160 Route de Florissant, P.O. Box 80
CH-1231 Conches, Switzerland
T: 41-22 7891676 F: 41-22 789-3536

Langer, Berthold
United Nations Conference on Environment and Development
160 Route de Florissant, P.O. Box 80
CH-1231 Conches, Switzerland
T: 41-22 7891676 F: 41-22 789-3536

Larangeira, Manuel
United Nations Conference on Environment and Development
160 Route de Florissant, P.O. Box 80
CH-1231 Conches, Switzerland
T: 41-22 7891676 F: 41-22 789-3536

Lassonde, Louise
United Nations Conference on Environment and Development
160 Route de Florissant, P.O. Box 80
CH-1231 Conches, Switzerland
T: 41-22 7891676 F: 41-22 789-3536

Lawson, B
United Nations Conference on Environment and Development
160 Route de Florissant, P.O. Box 80
CH-1231 Conches, Switzerland
T: 41-22 7891676 F: 41-22 789-3536

Lawson, Jean-Louis
United Nations Conference on Environment and Development
160 Route de Florissant, P.O. Box 80
CH-1231 Conches, Switzerland
T: 41-22 7891676 F: 41-22 789-3536

Leger, Ron
United Nations Conference on Environment and Development
160 Route de Florissant, P.O. Box 80
CH-1231 Conches, Switzerland
T: 41-22 7891676 F: 41-22 789-3536

Leland, William
United Nations Conference on Environment and Development
160 Route de Florissant, P.O. Box 80
CH-1231 Conches, Switzerland
T: 41-22 7891676 F: 41-22 789-3536

Lemaresquier, Thierry, Mr.
Manager
UNDP-United Nations Development Programme
16, avenue Jean Trembley
1209 Geneva, Switzerland
T: 41-22 7985850 F: 41-22 7987524

Lescano, Hector
United Nations Development Programme
16, avenue Jean Trembley
1209 Geneva, Switzerland
T: 41-22 7985850 F: 41-22 7987524

Liburd, Charles
United Nations Conference on Environment and Development
160 Route de Florissant, P.O. Box 80
CH-1231 Conches, Switzerland
T: 41-22 7891676 F: 41-22 789-3536

Lim, Frances
United Nations Conference on Environment and Development
160 Route de Florissant, P.O. Box 80
CH-1231 Conches, Switzerland
T: 41-22 7891676 F: 41-22 789-3536

Lindner, Warren, Mr.
United Nations Conference on Environment and Development
160 Route de Florissant, P.O. Box 80
CH-1231 Conches, Switzerland
T: 41-22 7891676 F: 41-22 789-3536

Lockhart, C
United Nations Conference on Environment and Development
160 Route de Florissant, P.O. Box 80
CH-1231 Conches, Switzerland
T: 41-22 7891676 F: 41-22 789-3536

Lund, Alan
United Nations Conference on Environment and Development
160 Route de Florissant, P.O. Box 80
CH-1231 Conches, Switzerland
T: 41-22 7891676 F: 41-22 789-3536

Lyons, Tara
United Nations Conference on Environment and Development
160 Route de Florissant, P.O. Box 80
CH-1231 Conches, Switzerland
T: 41-22 7891676 F: 41-22 789-3536

MacDonald, Malcolm
United Nations Conference on Environment and Development
160 Route de Florissant, P.O. Box 80
CH-1231 Conches, Switzerland
T: 41-22 7891676 F: 41-22 789-3536

Magalhaes, Janair
United Nations Conference on Environment and Development
160 Route de Florissant, P.O. Box 80
CH-1231 Conches, Switzerland
T: 41-22 7891676 F: 41-22 789-3536

Marois, Maurice
United Nations Conference on Environment and Development
160 Route de Florissant, P.O. Box 80
CH-1231 Conches, Switzerland
T: 41-22 7891676 F: 41-22 789-3536

Martin, Stanley
United Nations Conference on Environment and Development
160 Route de Florissant, P.O. Box 80
CH-1231 Conches, Switzerland
T: 41-22 7891676 F: 41-22 789-3536

Matsushita, Kazuo
United Nations Conference on Environment and Development
160 Route de Florissant, P.O. Box 80
CH-1231 Conches, Switzerland
T: 41-22 7891676 F: 41-22 789-3536

Matsushita, T
United Nations Conference on Environment and Development
160 Route de Florissant, P.O. Box 80
CH-1231 Conches, Switzerland
T: 41-22 7891676 F: 41-22 789-3536

Matte-Baker, Andrea
United Nations Conference on Environment and Development
160 Route de Florissant, P.O. Box 80
CH-1231 Conches, Switzerland
T: 41-22 7891676 F: 41-22 789-3536

McClellan, Monique
United Nations Conference on Environment and Development
160 Route de Florissant, P.O. Box 80
CH-1231 Conches, Switzerland
T: 41-22 7891676 F: 41-22 789-3536

McDaniel, B
United Nations Conference on Environment and Development
160 Route de Florissant, P.O. Box 80
CH-1231 Conches, Switzerland
T: 41-22 7891676 F: 41-22 789-3536

Melendez, Giselle
United Nations Conference on Environment and Development
160 Route de Florissant, P.O. Box 80
CH-1231 Conches, Switzerland
T: 41-22 7891676 F: 41-22 789-3536

Meynet, Virginie
United Nations Conference on Environment and Development
160 Route de Florissant, P.O. Box 80
CH-1231 Conches, Switzerland
T: 41-22 7891676 F: 41-22 789-3536

Miller, Marcy
United Nations Conference on Environment and Development
160 Route de Florissant, P.O. Box 80
CH-1231 Conches, Switzerland
T: 41-22 7891676 F: 41-22 789-3536

Miranda, Evaristo
United Nations Conference on Environment and Development
160 Route de Florissant, P.O. Box 80
CH-1231 Conches, Switzerland
T: 41-22 7891676 F: 41-22 789-3536

Monaghan, Michael
United Nations Conference on Environment and Development
160 Route de Florissant, P.O. Box 80
CH-1231 Conches, Switzerland
T: 41-22 7891676 F: 41-22 789-3536

Mooney, Pauline
United Nations Conference on Environment and Development
160 Route de Florissant, P.O. Box 80
CH-1231 Conches, Switzerland
T: 41-22 7891676 F: 41-22 789-3536

Moraes, Luis
United Nations Conference on Environment and Development
160 Route de Florissant, P.O. Box 80
CH-1231 Conches, Switzerland
T: 41-22 7891676 F: 41-22 789-3536

Mostafa, Intisar
United Nations Conference on Environment and Development
160 Route de Florissant, P.O. Box 80
CH-1231 Conches, Switzerland
T: 41-22 7891676 F: 41-22 789-3536

Mwagiru, Wanjiku
United Nations Conference on Environment and Development
160 Route de Florissant, P.O. Box 80
CH-1231 Conches, Switzerland
T: 41-22 7891676 F: 41-22 789-3536

Nakamuru, Tomoko
World Association of Former United Nations Internes and Fellows
16, avenue Jean Trembley
1209 Geneva, Switzerland
T: 41-22 7985850 F: 41-22 7910827

Nattrass, Brian
United Nations Conference on Environment and Development
160 Route de Florissant, P.O. Box 80
CH-1231 Conches, Switzerland
T: 41-22 7891676 F: 41-22 789-3536

Nenu, Luciana
United Nations Conference on Environment and Development
160 Route de Florissant, P.O. Box 80
CH-1231 Conches, Switzerland
T: 41-22 7891676 F: 41-22 789-3536

Neves, Marco
United Nations Conference on Environment and Development
160 Route de Florissant, P.O. Box 80
CH-1231 Conches, Switzerland
T: 41-22 7891676 F: 41-22 789-3536

O'Connor, Kris
United Nations Conference on Environment and Development
160 Route de Florissant, P.O. Box 80
CH-1231 Conches, Switzerland
T: 41-22 7891676 F: 41-22 789-3536

Obasi, G.O. P., Mr.
Secretary-General
WORLD METEOROLOGICAL ORGANIZATION
Case Postale NO.2300
Geneva CH 1211, Switzerland
T: 41 22 730-8200 F: 41 22 734-2326

Odabashian, Elisa
United Nations Conference on Environment and Development
160 Route de Florissant, P.O. Box 80
CH-1231 Conches, Switzerland
T: 41-22 7891676 F: 41-22 789-3536

Oliveira, Alessandra
United Nations Conference on Environment and Development
160 Route de Florissant, P.O. Box 80
CH-1231 Conches, Switzerland
T: 41-22 7891676 F: 41-22 789-3536

Ortiz-Gandarillas, Erwin
United Nations Conference on Environment and Development
160 Route de Florissant, P.O. Box 80
CH-1231 Conches, Switzerland
T: 41-22 7891676 F: 41-22 789-3536

Osseiran, Nada
United Nations Conference on Environment and Development
160 Route de Florissant, P.O. Box 80
CH-1231 Conches, Switzerland
T: 41-22 7891676 F: 41-22 789-3536

Paganelli, Antonio
United Nations Conference on Environment and Development
160 Route de Florissant, P.O. Box 80
CH-1231 Conches, Switzerland
T: 41-22 7891676 F: 41-22 789-3536

Park, Anne
United Nations Conference on Environment and Development
160 Route de Florissant, P.O. Box 80
CH-1231 Conches, Switzerland
T: 41-22 7891676 F: 41-22 789-3536

Pasztor, Janos
United Nations Conference on Environment and Development
160 Route de Florissant, P.O. Box 80
CH-1231 Conches, Switzerland
T: 41-22 7891676 F: 41-22 789-3536

Peart,
United Nations Conference on Environment and Development
160 Route de Florissant, P.O. Box 80
CH-1231 Conches, Switzerland
T: 41-22 7891676 F: 41-22 789-3536

Peregrino, Mr.
United Nations Conference on Environment and Development
160 Route de Florissant, P.O. Box 80
CH-1231 Conches, Switzerland
T: 41-22 7891676 F: 41-22 789-3536

Pernetta, John
United Nations Conference on Environment and Development
160 Route de Florissant, P.O. Box 80
CH-1231 Conches, Switzerland
T: 41-22 7891676 F: 41-22 789-3536

Peter, Ian
United Nations Conference on Environment and Development
160 Route de Florissant, P.O. Box 80
CH-1231 Conches, Switzerland
T: 41-22 7891676 F: 41-22 789-3536

Plant, Glen
United Nations Conference on Environment and Development
160 Route de Florissant, P.O. Box 80
CH-1231 Conches, Switzerland
T: 41-22 7891676 F: 41-22 789-3536

Poitier-Mutal, Aziyade
United Nations Conference on Environment and Development
160 Route de Florissant, P.O. Box 80
CH-1231 Conches, Switzerland
T: 41-22 7891676 F: 41-22 789-3536

Pottorf, Darryl
United Nations Conference on Environment and Development
160 Route de Florissant, P.O. Box 80
CH-1231 Conches, Switzerland
T: 41-22 7891676 F: 41-22 789-3536

Prasek, Dariusz
United Nations Conference on Environment and Development
160 Route de Florissant, P.O. Box 80
CH-1231 Conches, Switzerland
T: 41-22 7891676 F: 41-22 789-3536

Qian, Hao
United Nations Conference on Environment and Development
160 Route de Florissant, P.O. Box 80
CH-1231 Conches, Switzerland
T: 41-22 7891676 F: 41-22 789-3536

Ramos, Michael
United Nations Conference on Environment and Development
160 Route de Florissant, P.O. Box 80
CH-1231 Conches, Switzerland
T: 41-22 7891676 F: 41-22 789-3536

Ray, Carole
United Nations Conference on Environment and Development
160 Route de Florissant, P.O. Box 80
CH-1231 Conches, Switzerland
T: 41-22 7891676 F: 41-22 789-3536

Read, B
United Nations Conference on Environment and Development
160 Route de Florissant, P.O. Box 80
CH-1231 Conches, Switzerland
T: 41-22 7891676 F: 41-22 789-3536

Read, John
United Nations Conference on Environment and Development
160 Route de Florissant, P.O. Box 80
CH-1231 Conches, Switzerland
T: 41-22 7891676 F: 41-22 789-3536

Ribeiro, Flavio
United Nations Conference on Environment and Development
160 Route de Florissant, P.O. Box 80
CH-1231 Conches, Switzerland
T: 41-22 7891676 F: 41-22 789-3536

Roberts, Kirk
United Nations Conference on Environment and Development
160 Route de Florissant, P.O. Box 80
CH-1231 Conches, Switzerland
T: 41-22 7891676 F: 41-22 789-3536

Rocha, Antonio
United Nations Conference on Environment and Development
160 Route de Florissant, P.O. Box 80
CH-1231 Conches, Switzerland
T: 41-22 7891676 F: 41-22 789-3536

Rodriguez, Mercedes
United Nations Conference on Environment and Development
160 Route de Florissant, P.O. Box 80
CH-1231 Conches, Switzerland
T: 41-22 7891676 F: 41-22 789-3536

Rodriguez, Rafael
United Nations Conference on Environment and Development
160 Route de Florissant, P.O. Box 80
CH-1231 Conches, Switzerland
T: 41-22 7891676 F: 41-22 789-3536

Rogers, Murray
United Nations Conference on Environment and Development
160 Route de Florissant, P.O. Box 80
CH-1231 Conches, Switzerland
T: 41-22 7891676 F: 41-22 789-3536

Rosen, Marni
United Nations Conference on Environment and Development
160 Route de Florissant, P.O. Box 80
CH-1231 Conches, Switzerland
T: 41-22 7891676 F: 41-22 789-3536

Ryan, Janette
United Nations Conference on Environment and Development
160 Route de Florissant, P.O. Box 80
CH-1231 Conches, Switzerland
T: 41-22 7891676 F: 41-22 789-3536

Sampaio, Marcos
United Nations Conference on Environment and Development
160 Route de Florissant, P.O. Box 80
CH-1231 Conches, Switzerland
T: 41-22 7891676 F: 41-22 789-3536

Sand, Peter
United Nations Conference on Environment and Development
160 Route de Florissant, P.O. Box 80
CH-1231 Conches, Switzerland
T: 41-22 7891676 F: 41-22 789-3536

Sandoval, Lila
United Nations Conference on Environment and Development
160 Route de Florissant, P.O. Box 80
CH-1231 Conches, Switzerland
T: 41-22 7891676 F: 41-22 789-3536

Sarto, Vilson
United Nations Conference on Environment and Development
160 Route de Florissant, P.O. Box 80
CH-1231 Conches, Switzerland
T: 41-22 7891676 F: 41-22 789-3536

Scaramuzza, Carlos
United Nations Conference on Environment and Development
160 Route de Florissant, P.O. Box 80
CH-1231 Conches, Switzerland
T: 41-22 7891676 F: 41-22 789-3536

Schak, Sabine
United Nations Conference on Environment and Development
160 Route de Florissant, P.O. Box 80
CH-1231 Conches, Switzerland
T: 41-22 7891676 F: 41-22 789-3536

Schmidt, Erick
United Nations Conference on Environment and Development
160 Route de Florissant, P.O. Box 80
CH-1231 Conches, Switzerland
T: 41-22 7891676 F: 41-22 789-3536

Serwadda, Ronah
United Nations Conference on Environment and Development
160 Route de Florissant, P.O. Box 80
CH-1231 Conches, Switzerland
T: 41-22 7891676 F: 41-22 789-3536

Severino, A
United Nations Conference on Environment and Development
160 Route de Florissant, P.O. Box 80
CH-1231 Conches, Switzerland
T: 41-22 7891676 F: 41-22 789-3536

Shah, Mahendrah
United Nations Conference on Environment and Development
160 Route de Florissant, P.O. Box 80
CH-1231 Conches, Switzerland
T: 41-22 7891676 F: 41-22 789-3536

Shibuya, H
United Nations Conference on Environment and Development
160 Route de Florissant, P.O. Box 80
CH-1231 Conches, Switzerland
T: 41-22 7891676 F: 41-22 789-3536

Sitek, Dorothy
United Nations Conference on Environment and Development
160 Route de Florissant, P.O. Box 80
CH-1231 Conches, Switzerland
T: 41-22 7891676 F: 41-22 789-3536

Sonntag, Nicholas
United Nations Conference on Environment and Development
160 Route de Florissant, P.O. Box 80
CH-1231 Conches, Switzerland
T: 41-22 7891676 F: 41-22 789-3536

Spencer, Claire
United Nations Conference on Environment and Development
160 Route de Florissant, P.O. Box 80
CH-1231 Conches, Switzerland
T: 41-22 7891676 F: 41-22 789-3536

Stanton, Michael
United Nations Conference on Environment and Development
160 Route de Florissant, P.O. Box 80
CH-1231 Conches, Switzerland
T: 41-22 7891676 F: 41-22 789-3536

Steady, Filomina
United Nations Conference on Environment and Development
160 Route de Florissant, P.O. Box 80
CH-1231 Conches, Switzerland
T: 41-22 7891676 F: 41-22 789-3536

Steiger, Alessandra
United Nations Conference on Environment and Development
160 Route de Florissant, P.O. Box 80
CH-1231 Conches, Switzerland
T: 41-22 7891676 F: 41-22 789-3536

Steiger, Diana
United Nations Conference on Environment and Development
160 Route de Florissant, P.O. Box 80
CH-1231 Conches, Switzerland
T: 41-22 7891676 F: 41-22 789-3536

Strong, H
United Nations Conference on Environment and Development
160 Route de Florissant, P.O. Box 80
CH-1231 Conches, Switzerland
T: 41-22 7891676 F: 41-22 789-3536

Strong, K
United Nations Conference on Environment and Development
160 Route de Florissant, P.O. Box 80
CH-1231 Conches, Switzerland
T: 41-22 7891676 F: 41-22 789-3536

Strong, Maurice
United Nations Conference on Environment and Development
160 Route de Florissant, P.O. Box 80
CH-1231 Conches, Switzerland
T: 41-22 7891676 F: 41-22 789-3536

Takahashi, Tadao
United Nations Conference on Environment and Development
160 Route de Florissant, P.O. Box 80
CH-1231 Conches, Switzerland
T: 41-22 7891676 F: 41-22 789-3536

Teakle, Christopher
United Nations Conference on Environment and Development
160 Route de Florissant, P.O. Box 80
CH-1231 Conches, Switzerland
T: 41-22 7891676 F: 41-22 789-3536

Ten, Kerry
United Nations Conference on Environment and Development
160 Route de Florissant, P.O. Box 80
CH-1231 Conches, Switzerland
T: 41-22 7891676 F: 41-22 789-3536

Thacher, Peter
United Nations Conference on Environment and Development
160 Route de Florissant, P.O. Box 80
CH-1231 Conches, Switzerland
T: 41-22 7891676 F: 41-22 789-3536

Tiso, Marco
United Nations Conference on Environment and Development
160 Route de Florissant, P.O. Box 80
CH-1231 Conches, Switzerland
T: 41-22 7891676 F: 41-22 789-3536

Tran, My
United Nations Conference on Environment and Development
160 Route de Florissant, P.O. Box 80
CH-1231 Conches, Switzerland
T: 41-22 7891676 F: 41-22 789-3536

Tun, Thandar
United Nations Conference on Environment and Development
160 Route de Florissant, P.O. Box 80
CH-1231 Conches, Switzerland
T: 41-22 7891676 F: 41-22 789-3536

Ul, Mrs.
United Nations Conference on Environment and Development
160 Route de Florissant, P.O. Box 80
CH-1231 Conches, Switzerland
T: 41-22 7891676 F: 41-22 789-3536

Urquhart, Charles
United Nations Conference on Environment and Development
160 Route de Florissant, P.O. Box 80
CH-1231 Conches, Switzerland
T: 41-22 7891676 F: 41-22 789-3536

Valent, Dante
United Nations Conference on Environment and Development
160 Route de Florissant, P.O. Box 80
CH-1231 Conches, Switzerland
T: 41-22 7891676 F: 41-22 789-3536

Van, Andreas
United Nations Conference on Environment and Development
160 Route de Florissant, P.O. Box 80
CH-1231 Conches, Switzerland
T: 41-22 7891676 F: 41-22 789-3536

Van, Lot
United Nations Conference on Environment and Development
160 Route de Florissant, P.O. Box 80
CH-1231 Conches, Switzerland
T: 41-22 7891676 F: 41-22 789-3536

Vasilyev, Andrey
United Nations Conference on Environment and Development
160 Route de Florissant, P.O. Box 80
CH-1231 Conches, Switzerland
T: 41-22 7891676 F: 41-22 789-3536

Vavilov, Andrey
United Nations Conference on Environment and Development
160 Route de Florissant, P.O. Box 80
CH-1231 Conches, Switzerland
T: 41-22 7891676 F: 41-22 789-3536

Veras, Ernesto
United Nations Conference on Environment and Development
160 Route de Florissant, P.O. Box 80
CH-1231 Conches, Switzerland
T: 41-22 7891676 F: 41-22 789-3536

Viana, Helio
United Nations Conference on Environment and Development
160 Route de Florissant, P.O. Box 80
CH-1231 Conches, Switzerland
T: 41-22 7891676 F: 41-22 789-3536

Vickers, Bill
United Nations Conference on Environment and Development
160 Route de Florissant, P.O. Box 80
CH-1231 Conches, Switzerland
T: 41-22 7891676 F: 41-22 789-3536

Vilela, Miriam
United Nations Conference on Environment and Development
160 Route de Florissant, P.O. Box 80
CH-1231 Conches, Switzerland
T: 41-22 7891676 F: 41-22 789-3536

Villas, Marcos
United Nations Conference on Environment and Development
160 Route de Florissant, P.O. Box 80
CH-1231 Conches, Switzerland
T: 41-22 7891676 F: 41-22 789-3536

Visser, Koenraad
United Nations Conference on Environment and Development
160 Route de Florissant, P.O. Box 80
CH-1231 Conches, Switzerland
T: 41-22 7891676 F: 41-22 789-3536

Voronov, Anatoly
United Nations Conference on Environment and Development
160 Route de Florissant, P.O. Box 80
CH-1231 Conches, Switzerland
T: 41-22 7891676 F: 41-22 789-3536

Voss, Annemarie
United Nations Conference on Environment and Development
160 Route de Florissant, P.O. Box 80
CH-1231 Conches, Switzerland
T: 41-22 7891676 F: 41-22 789-3536

Warwar, Alexandre
United Nations Conference on Environment and Development
160 Route de Florissant, P.O. Box 80
CH-1231 Conches, Switzerland
T: 41-22 7891676 F: 41-22 789-3536

Wasambla, Gladys
United Nations Conference on Environment and Development
160 Route de Florissant, P.O. Box 80
CH-1231 Conches, Switzerland
T: 41-22 7891676 F: 41-22 789-3536

Wheeler, Joseph
United Nations Conference on Environment and Development
160 Route de Florissant, P.O. Box 80
CH-1231 Conches, Switzerland
T: 41-22 7891676 F: 41-22 789-3536

Wheeler, V
United Nations Conference on Environment and Development
160 Route de Florissant, P.O. Box 80
CH-1231 Conches, Switzerland
T: 41-22 7891676 F: 41-22 789-3536

Willecke, Valeria
United Nations Conference on Environment and Development
160 Route de Florissant, P.O. Box 80
CH-1231 Conches, Switzerland
T: 41-22 7891676 F: 41-22 789-3536

Williams, Stan
United Nations Conference on Environment and Development
160 Route de Florissant, P.O. Box 80
CH-1231 Conches, Switzerland
T: 41-22 7891676 F: 41-22 789-3536

Wilson, Paul
United Nations Conference on Environment and Development
160 Route de Florissant, P.O. Box 80
CH-1231 Conches, Switzerland
T: 41-22 7891676 F: 41-22 789-3536

Yerg, Martin
World Meteorological Organization
41, avenue G. Motta
1211 Geneva 20, Switzerland
T: 41-22 730811 F: 41-22 7242326

Zentilli, Bernardo
United Nations Conference on Environment and Development
160 Route de Florissant, P.O. Box 80
CH-1231 Conches, Switzerland
T: 41-22 7891676 F: 41-22 789-3536

ADVISORS

Golubev, Genady, Prof.
Asst. D-G, Conservation Programme, Iucn : Faculty Of Geography, Moscow Univ.
CH-1996 Gland, Switzerland

Hentsch, Leonard, Mr.
Board Member, Centre For Our Common Future
1245 Collonge-Bellerive, Switzerland
T: 21 90 11

Holdgate, Martin, Dr.
Director General
IUCN - THE WORLD CONSERVATION UNION
Av Mont Blanc, CH-1196 Gland, Switzerland
T: 41 22 649 232 F: 41 22 642 926

Mathur, M.G., Mr.
Former Deputy D-G Of The Gatt
1261 Chavannes-de-Bogis, Switzerland
T: 022 776 24 60

Nerfin, Marc, Mr.
President, International Foundation For Development Alternatives
CH-1260 Nyon, Switzerland
T: 022 61 82 81 F: 022 61 05 25

Schmidheiny, Stephen, Dr.
Chairman, Business Council For Sust. Dev. : Principle Adviser To UNCED
CH-8640 Hurden, Switzerland
T: 055 47 33 77 F: 055 48 41 17

SYRIAN ARAB REPUBLIC - GOVERNMENT

Abosamra, Fouad
Dr.
H.I.A.S.T.
P.O. Box 7028
Damascus, Syrian Arab Republic
T: 772 603 /U/5

Al-Ahmad, Farouk, Dr.
Director Of Forests
Ministry of Agriculture
Damascus-Syrian Arab Republic,
F: 233868 OR 810221

Al-Mounajed, Abdul Hameed, H.E.
Minister of State for Environmental Affairs
Permanent Mission of Syrian Arab Republic
United Nations Plaza
New York, NY 10017, United States of America
T: 212 963 7171 F: 212 963 1921

Al-Shawaf, Saadallah
Director Of Water Pollution Control
MINISTRY OF IRRIGATION
P O Box 4451
Damascus, Syrian Arab Republic
T: 210175

ISSA, Mouhydeen, H.E. Dr.
Deputy Minister of Higher Education
Permanent Mission of Syrian Arab Republic
United Nations Plaza
New York, NY 10017, United States of America
T: 212 963 7171 F: 212 963 1921

Issa, Mohiddin, Dr
Vice Minister For Higher Education
Ministry for Higher Edu Damascus
Damascus, Syrian Arab Republic

Kaddour, Nasser, H.E. Mr.
Minister of State for Foreign Affairs
Permanent Mission of Syrian Arab Republic
United Nations Plaza
New York, NY 10017, United States of America
T: 212 963 7171 F: 212 963 1921

Khaddam, Abdul Halim, Mr.
Vice President
Permanent Mission of Syria
820 2nd. Ave., 10th Floor
New York NY 10017, United States of America
T: 212 661 1313 F: 212 983 4439

Monajad, Abdulhamad
Minister Of Invironment
MINISTERY OF INVIRONMENT
Minister of Invironment
Damascus, Syrian Arab Republic
T: 335645

Othaman, Ibraheem, Dr.
Chief
Prevention and Safety Department Atomic Energy Commission
Permanent Mission to the United Nations, United Nations Plaza New York, NY, 10017
F: 212 963 7171
TL: 212 963 1921

Othman, Ibrahim
Head Of Department
ATOMIC ENERGY COMMISSION
P.O. Box 6091
Damascus, Syrian Arab Republic
T: 668784/668114 F: 411422

Rifai, Ghassoub, H.E. Dr.
Ambassador Extraordinary and Plenipotentiary to Brazil
Permanent Mission of Syrian Arab Republic
United Nations Plaza
New York, NY 10017, United States of America
T: 212 963 7171 F: 212 963 1921

Samra, Fouad Abou, Dr.
Chief
Labratory of Environmental Studies, Higher Institute for Applied Sciences and Technology
Permanent Mission to the United Nations, United Nations Plaza New York, NY, 10017
F: 212 963 7171
TL: 212 963 1921

TAIWAN - PRESS

Chung, Che-Fang
Reporter Radio/Television
CHINESE TV SYSTEM
1273 National Press Bldg.
Washington, D.C., U.S.A., Taiwan

Huang, Wan-Chu, Dr.
Editor
NEWS REPORTMENT
333, Sec. 2 Tunhwa Srd.
Taipei, Taiwan, Taiwan

Lu, Li-Teh
Editor
CHINA TIMES
8floor, 111, Sec. 2 Cheng-kuo
RD Taipei, Taiwan, Taiwan

Lui, Li-Hui
Reporter Radio/Television
TAIWAN TV ENTERPRISE
3F. 7 Alley 7 Lane 12 Pa Te Rd
Taipei, Taiwan ROC, Taiwan

Mei-Hua, Cheng
Reporter Radio/Television
BROADCASTING CORP. OF CHINA
2F. N.41 Alley 5, Lane 70
Sec.6 Chungsiao, E.R. Taipei, Taiwan, Taiwan

GLOBAL FORUM

Chan, Kow-Lung, Mr.
Professor
NATIONAL TAIWAN UNIVERSITY
3F, No. 60-5, Chang Hsin Street
Taipei, Taiwan
T: 886 2 968-6733 F: 886 2 968-3235

Chen, Kai-Hung, Ms.
Assistant
LEGISLATIVE YUAN
5 FL. No. 54, Section 1, CUNG HSIAO E Road
Taipei, Taiwan
T: 886 2 397-5206 F: 886 2 397-9766

Chen, Larry L.G., Dr.
I.E.P.
Bldg. 64, 195, Chung Hsing Rd. Section 4, Chutung
Hsin-chu, Taiwan
T: 88 6 35-966100 F: 88 6 35-943680

Chen, Ming-Yih, Mr.
Professor
NATIONAL CHUNG-HSING UNIVERSITY
Dept. of Botany
Taichung 402, Taiwan
T: 886 42 285-1333 F: 886 42 287-4740

Chen, Nan Hwa, Dr.
I.E.P.
Bldg. 64, 195, Chung Hsing Rd. Section 4, Chutung
Hsin-chu, Taiwan
T: 88 6 35-966100 F: 88 6 35-943680

Chen, Yieche Grace, Dr.
I.E.P.
Bldg. 64, 195, Chung Hsing Rd. Section 4, Chutung
Hsin-chu, Taiwan
T: 88 6 35-966100 F: 88 6 35-943680

Cheng, Mei-Hua
Reporter
BROADCASTING CORPORATION OF CHINA
Jen-Ai Road, Tapei
Taiwan
T: 886 2 721-6530 F: 886 2 771-3787

Chi, Liang Yu
Reporter
UNITED DAILY NEWS
555 Chung-Hsiao E. Rd. Sec 4
Tapei, Taiwan
T: 886 2 7631482 F: 886 2 7646893

Chiu, Jong-Jen, Dr.
I.E.P.
Bldg. 64, 195, Chung Hsing Rd. Section 4, Chutung
Hsin-chu, Taiwan
T: 88 6 35-966100 F: 88 6 35-943680

Deng, Chen-Chung, Dr.
I.E.P.
Bldg. 64, 195, Chung Hsing Rd. Section 4, Chutung
Hsin-chu, Taiwan
T: 88 6 35-966100 F: 88 6 35-943680

Ding, Tian-Kwei, Mr.
Director
BEAUTIFUL TAIWAN FOUNDATION
16F-1, 93 Hsin-Hai Road
Taipei, Taiwan
T: 886 2 362-1823 F: 886 2 362-1825

Fan, Richard K.S., Dr.
I.E.P.
Bldg. 64, 195, Chung Hsing Rd. Section 4, Chutung
Hsin-chu, Taiwan
T: 88 6 35-966100 F: 88 6 35-943680

Fang, Jay, Mr.
Coordinator
EARTHDAY TAIWAN
2F-1, No. 28, Sec. 1, Shing-Sheng South Road
Taipei, Taiwan
T: 886 2 778-0277 7816878 F: 886 2 781-6873

Hsien, Yein-Rui, Dr.
I.E.P.
Bldg. 64, 195, Chung Hsing Rd. Section 4, Chutung
Hsin-chu, Taiwan
T: 88 6 35-966100 F: 88 6 35-943680

Hsu, I-Shia, Ms.
Assistant
HEALTH, WELFARE AND ENVIRONMENTAL FOUNDATION
5 FL. No. 54, Section 1, CUNG HSIAO E Road
Taipei, Taiwan
T: 886 2 397-5206 F: 886 2 397-9766

Huang, Chen Lai-Hong, Ms.
Student
HOMEMAKERS' UNION & FOUNDATION
4FL, 2, Lane 160, SGC 3, Ting-Chou Rd.
Taipei, Taiwan
T: 886 2 368-6211 F: 886 2 368-6213

Huang, Min-Ho
HEALTH, WELFARE AND ENVIRONMENTAL FOUNDATION
5 FL. No. 54, Section 1, CUNG HSIAO E Road
Taipei, Taiwan
T: 886 2 397-5206 F: 886 2 397-9766

Huang, Ta-Jen, Dr.
I.E.P.
Bldg. 64, 195, Chung Hsing Rd. Section 4, Chutung
Hsin-chu, Taiwan
T: 88 6 35-966100 F: 88 6 35-943680

Huei, Liu Li
Reporter
TAWAIN TV LTD
1 Pa Teh Road
Taipei, Taiwan
T: 7579617 F: 7579618

Kao, Chih-Wen, Dr.
Legislator
LEGISLATIVE YUAN
5 FL. No. 54, Section 1, CUNG HSIAO E Road
Taipei, Taiwan
T: 886 2 397-5206 F: 886 2 397-9766

Kao, Jones, Mr.
Deputy Director
CHINESE PETROLEUM CORP.
83 Ching Hwa Rd. Sec. 1
Taipei, Taiwan
T: 886 2 311-3429 F: 886 2 331-7718

Lee, Li-Hsun, Ms.
Reportertor
CHINA TIME EXPRESS
5-7, Lan 52, Sec. 2, Ho-ping W. Road
Taipei, Taiwan
T: 886 2 305-3013

Lee, Ping, Ms.
WORLD YOUNG WOMEN'S CHRISTIAN ASSOCIATION
24-2 Lane 381 WU
Hsin-chu, Taiwan
F: 88 6 383-1340
TL: WYMCA 41 23 32

Lee, Shiaw-Tzong, Dr.
I.E.P.
Bldg. 64, 195, Chung Hsing Rd. Section 4, Chutung
Hsin-chu, Taiwan
T: 88 6 35-966100 F: 88 6 35-943680

Li, Ching-Chang, Mr.
Director
HEALTH, WELFARE AND ENVIRONMENTAL FOUNDATION
5 FL. No. 54, Section 1, CUNG HSIAO E Road
Taipei, Taiwan
T: 886 2 397-5206 F: 886 2 397-9766

Li-Teh, Lu
Reporter
CHINA TIMES
132 Da-li Street
Taipei, Taiwan
T: 886 2 3081111 F: 886 2 3069050

Lin, Ada Wen-Shung Ma, Ms.
General Director
I.E.P.
Bldg 32. 195 Chung Hsing Rd., sec 4
Hsin-chu ROC, Taiwan
T: 886 35 986173 F: 886 35 957177

Lin, Edgar J. Y., Mr.
Professor Of Biology
TUNG-HAI UNIVERSITY
Tung-Hai Road
Taichung, Taiwan
T: 565 42 3590249 F: 565 42 3595622

Lin, Ho, Mr.
Professor
NATIONAL TAIWAN UNIVERSITY
4F, No.2, Lane 160, Sec. 3, Ting-Chou Rd.
Taipei, Taiwan
T: 886 2 368-6211 F: 886 2 368-6213

Liu, Jimmy, Mr.
2 Alley 31 Lane 421 Sec 2 - Chih-Shan Rd. Shih-Lin Dist.
Taipei, Taiwan
T: 886 2 841 1022

Liu, Wen-Chao, Mr.
Professor
NATIONAL TAIWAN UNIVERSITY
3F, No. 60-5, Chang Hsin Street
Taipei, Taiwan
T: 886 2 968-6733 F: 886 2 968-3235

Lu, Jhy-Ming, Dr.
I.E.P.
Bldg. 64, 195, Chung Hsing Rd. Section 4, Chutung
Hsin-chu, Taiwan
T: 88 6 35-966100 F: 88 6 35-943680

Petrucelli, Linda, Ms.
GLOBAL ENVIRONMENT TEAM
53 Chung Hwa, 3rd Road, Floor 5, Apt. 1
Kaohsiung 800, Taiwan

Tang, Hsiao-Yu, Mr.
Chief
RESOURCES CONSERVATION DIVISION FORESTRY DEPT. COUNCIL OF AGRICULTURE
37 Nanhai Rd.
Taipei, Taiwan
T: 886 2 312-4058 F: 886 2 312-5857

Tsai, Pei-Shun, Mr.
Deputy Director - General
I.E.P.
No. 2, 1st. Section, Hanchow South Road
Taipei, Taiwan
T: 886 2 341-7634 F: 886 2 341-4281

Wang, Juju C. S., Ms.
Assoc. Professor/Director
NATIONAL TSING-HUA UNIERSITY
101, Sec. 2, Kuang-Fu Rd.
Hsin-chu, Taiwan
T: 886 72251833 F: 886 72369135

Wang, Shing, Mr.
I.E.P.
Bldg. 64, 195, Chung Hsing Rd. Section 4, Chutung
Hsin-chu, Taiwan
T: 88 6 35-966100 F: 88 6 35-943680

Wu, Mei-Ling Lee, Ms.
Chariman
HOMEMAKERS' UNION & FOUNDATION
4FL, 2, Lane 160, SGC 3, Ting-Chou Rd.
Taipei, Taiwan
T: 886 2 368-6211 F: 886 2 368-6213

Yang, Jih Chang
Vice President
I.E.P.
Bldg. 64, 195, Chung Hsing Rd. Section 4, Chutung
Hsin-chu, Taiwan
T: 88 6 35-966100 F: 88 6 35-943680

Yang, Sen Hong, Mr.
Director
EARTHDAY TAIWAN
2F No. 28 Sec. 1, Shing-Sheng South Road
Taipei, Taiwan
T: 886 2 778-0277 6716878 F: 886 2 781-6893

Yeh, Jiunn-Rong, Mr.
Assoc. Professor/Director
NATIONAL TSING-HUA UNIVERSITY
21 Hsu-Chow Road
Taipei 2, Taiwan
F: 886 2 365-2891 341-7047
TL: 886 2 341-0495

Yih, Hong-Ting, Dr.
I.E.P.
Bldg. 64, 195, Chung Hsing Rd. Section 4, Chutung
Hsin-chu, Taiwan
T: 88 6 35-966100 F: 88 6 35-943680

Zhen, Chinlong, Mr.
Research Fellow
CHUNG-HUA INSTITUTION FOR ECONOMIC RESEARCH
75 Chang Hsing St.
Taipei, Taiwan
T: 886 2 735-6006 F: 886 2 735-06035

TAJIKISTAN - GLOBAL FORUM

Atakhanov, Khamid, Mr.
Chairman Sg. Ikar
TADJIK DEPATMENT OF SOCIO-ECOLOGICAL UNION
Tadjikistan Dushanbe Gissarskaya Str. 15 fl 27
Dushanbe, Tajikistan
T: 350543 F: 233754
E: 734058

THAILAND - GOVERNMENT

Attavipach, Pricha, Mr.
Director-General
Department of Industrial Works, Ministry of Indusry
Permanent Mission to the United Nations, 351 E. 52nd Str.
New York, NY 10022, United States of America
T: 212 754 2230 F: 212 754 2535

Boonpradit, Wattana, Miss
Attache
Embassy to Brazil
Permanent Mission to the United Nations, 351 E. 52nd Str.
New York, NY 10022, United States of America
T: 212 754 2230 F: 212 754 2535

Buranasin Rtn, Sutee, Captain
Royal Aide-de-Camp
Permanent Mission of Thailand
351 E. 52nd Str.
New York, NY 10022, United States of America
T: 212 754 2230 F: 212 754 2535

Chandraprabha, Krisana, Mr.
Counsellor
Permanent Mission of Thailand
351 E. 52nd Str.
New York, NY 10022, United States of America
T: 212 754 2230 F: 212 754 2535

Charoonluk, Benja, Miss
Third Secretary
Embassy to Brazil
Permanent Mission to the United Nations, 351 E. 52nd Str.
New York, NY 10022, United States of America
T: 212 754 2230 F: 212 754 2535

Chiranond, Karn, Mr
Counsellor
Department of Treaties and Legal Affairs
Ministry of Foreign Affairs
Bangkok 10400, Thailand
T: 245 5669 F: 245 9373 245 9903

Choompoopet, Wilaiwan, Mrs.
First Secretary
Royal Thai Embassy, Brasilia
Permanent Mission to the United Nations, 351 E. 52nd Str.
New York, NY 10022, United States of America
T: 212 754 2230 F: 212 754 2535

Chulabhorn Mahidol, Her Royal Highness Princess, Professor Dr.
President, Chulabhorn Research Institute
Special Representative of His Majesty the King of Thailand
351 E. 52nd Str.
New York, NY 10022, United States of America
T: 212 754 2230 F: 212 754 2535

Didyasarin, Veerayuth, Group Captain
Vice President
Chulabhorn Research Institute, Special Adviser
Permanent Mission to the United Nations, 351 E. 52nd Str.
New York, NY 10022, United States of America
T: 212 754 2230 F: 212 754 2535

Israsena, Birath, Mr.
Director-General
Department of International Organizations, Ministry of Foreign Affairs
Permanent Mission to the United Nations, 351 E. 52nd Str.
New York, NY 10022, United States of America
T: 212 754 2230 F: 212 754 2535

Jirapaet, Krirk-Krai, Mr.
Deputy Permanent Secretary
Ministry of Commerce
Permanent Mission to the United Nations, 351 E. 52nd Str.
New York, NY 10022, United States of America
T: 212 754 2230 F: 212 754 2535

Kajavong, Santas, Mr.
Second Secretary
Royal Thai Embasssy, Brasilia
Permanent Mission to the United Nations, 351 E. 52nd Str.
New York, NY 10022, United States of America
T: 212 754 2230 F: 212 754 2535

Karim, Rezaul
Chief, Environment Section
UNITED NATIONS ECONOMIC AND SOCIAL
U.N. Building Rajdamavern Avenue
Bangkok, Thailand
T: 282 9161 EXT 1614 F: 282 9602

Kraiwixien, Tanin, H.E. Mr.
Privy Counsellor
Vice President, Chulabhorn Research Institute
Permanent Mission to the United Nations, 351 E. 52nd Str.
New York, NY 10022, United States of America
T: 212 754 2230 F: 212 754 2535

Kullavanijaya, Pichitr
Dep Permanent Scty Of Def;Ass Dir
CHULABHORN RESEARCH INSTITUTE
Lak Si,
Bangkok, Thailand
T: 510 9022 F: 066 2 519 5805

Kullawanich, Pichitr, General
Associate Vice-President for Special Projects
Chulabhorn Research Institute
Permanent Mission to the United Nations, 351 E. 52nd Str.
New York, NY 10022, United States of America
T: 212 754 2230 F: 212 754 2535

Kulwat, Suchada, Ms.
Second Secretary
Permanent Mission of Thailand
351 E. 52nd Str.
New York, NY 10022, United States of America
T: 212 754 2230 F: 212 754 2535

Mahidol, Chulahorn
Princess
Permanent Mission of Thailand
351 E. 52nd Str.
New York NY 10022, United States of America
T: 212 754 2230 F: 212 754 2535

Na Nagara Rtn, Suriya, Captain
Royal Physician
Permanent Mission of Thailand
351 E. 52nd Str.
New York, NY 10022, United States of America
T: 212 754 2230 F: 212 754 2535

Pakkasem, Phisit, Dr.
Secretary-General
Office of the National Economic and Social Development Board
Permanent Mission to the United Nations, 351 E. 52nd Str.
New York, NY 10022, United States of America
T: 212 754 2230 F: 212 754 2535

Plamphongsant, Suphavit
Senior Environmental Expert
OFFICE OF ENVIRONMENTAL POLICY AND
60/1 501 Phibun Wattana Rama Vi Road
Bangkok, Thailand
T: 66 2 279 4769 F: 66 2 271 3226

Pibulsonggram, Nitya, H.E. Mr.
Ambassador Extraordinary and Plenipotentiary
Permanent Representative to the United Nations, New York
Permanent Mission to the United Nations, 351 E. 52nd Str.
New York, NY 10022, United States of America
T: 212 754 2230 F: 212 754 2535

Pibulsonggram, Patcharin, Mrs.
Executive Secretary
Her Royal Highness Princess Chulabhorn
Permanent Mission to the United Nations, 351 E. 52nd Str.
New York, NY 10022, United States of America
T: 212 754 2230 F: 212 754 2535

Prachimdhit, Suchat, Dr.
Deputy Director-General
Department of International Organizations, Ministry of Foreign Affairs
Permanent Mission to the United Nations, 351 E. 52nd Str.
New York, NY 10022, United States of America
T: 212 754 2230 F: 212 754 2535

Rojanapaibulya, Anek, Mr.
Inspector-General
Ministry of Interior
Permanent Mission to the United Nations, 351 E. 52nd Str.
New York, NY 10022, United States of America
T: 212 754 2230 F: 212 754 2535

Ruchirawat, Mathuros, Associate Professor Dr.
Associate Vice-President for Scientific Affairs
Chulabhorn Research Institute
Permanent Mission to the United Nations, 351 E. 52nd Str.
New York, NY 10022, United States of America
T: 212 754 2230 F: 212 754 2535

Sakornvanasak, Pinyo, Mr.
Second Secretary
Embassy to Brazil
Permanent Mission to the United Nations, 351 E. 52nd Str.
New York, NY 10022, United States of America
T: 212 754 2230 F: 212 754 2535

Sarikahputi, Yookti, Mr.
Permanent Secretary
Ministry of Agriculture and Cooperatives
Permanent Mission to the United Nations, 351 E. 52nd Str.
New York, NY 10022, United States of America
T: 212 754 2230 F: 212 754 2535

Satayavivad, Jutamaad, Associate Professor Dr.
Chief, Office of Scientific Affairs
Chulabhorn Research Institute
Permanent Mission to the United Nations, 351 E. 52nd Str.
New York, NY 10022, United States of America
T: 212 754 2230 F: 212 754 2535

Snidvongs, Kasem, Mr.
Permanent Secretary
Ministry of Science, Technology and Environment
Permanent Mission to the United Nations, 351 E. 52nd Str.
New York, NY 10022, United States of America
T: 212 754 2230 F: 212 754 2535

Sri Sa-Arn, Wichitr, Professor Dr.
Vice President, Chulabhorn Research Institute
Permanent Secretary, Ministry of University Affairs
Permanent Mission to the United Nations, 351 E. 52nd Str.
New York, NY 10022, United States of America
T: 212 754 2230 F: 212 754 2535

Srinilta, M. Sawarng
Deputy Permanent Scretary
Ministry of Transport and Communications
Radcha-Damnoen Nork Ave., Bangkok 10100, Thailand
Bangkok, Thailand
T: 2813220 2829595

Suphamongkhon, Kantathi, Dr.
Counsellor, Department of International Organizations
Ministry of Foreign Affairs, Executive Secretary to Her Royal Highness Princess Chulabhorn
Permanent Mission to the United Nations, 351 E. 52nd Str.
New York, NY 10022, United States of America
T: 212 754 2230 F: 212 754 2535

Suphapodok, Arthorn, Mr.
Secretary-General, Office of the National Environment Board
Ministry of Science, Technology and Environment
Permanent Mission to the United Nations, 351 E. 52nd Str.
New York, NY 10022, United States of America
T: 212 754 2230 F: 212 754 2535

Thananan, Anurak, H.E. Mr.
Ambassador Extraordinary and Plenipotentiary
Royal Thai Embassy, Brasilia
Permanent Mission to the United Nations, 351 E. 52nd Str.
New York, NY 10022, United States of America
T: 212 754 2230 F: 212 754 2535

Weerawedphisai, Poowadol, Mr.
Second Secretary, Department of International Organizations
Ministry of Foreign Affairs
Permanent Mission to the United Nations, 351 E. 52nd Str.
New York, NY 10022, United States of America
T: 212 754 2230 F: 212 754 2535

Yingvanasiri, Thanit, Mr.
Agricultural Counsellor
Permanent Mission of Thailand
351 E. 52nd Str.
New York, NY 10022, United States of America
T: 212 754 2230 F: 212 754 2535

NGO

Munyan, Shirley B.
PAN PACIFIC & SOUTHEAST ASIA WOMEN'S ASSOCIATION
2234 New Petchburi Road
Bangkok, Thailand

Niamskul, Chalinee, Ms.
Director, International Rela.
INTERNATIONAL BOARD FOR SOIL RESARCH AND MANAGEMENT - IBSRAM
PO Box 9-109 Bandkhen
Bangkok 10900, Thailand
T: 66 2 579-7590 F: 66 2 561-1230

Pynappel, Leoni
PSAWAI
2234 New Petchburi Road
Bangkok, Thailand

PRESS

Laird, John Stuart
Correspondent
NATIONAL BUSINESS REVIEW
P.O.Box 16
Hua Hin 77110, Thailand, Thailand

Sukin, Kamol
Correspondent
MANAGER
93/16 Soi Kerdsap Bangkhunnon
Bankoknoi, 10700, Bangkok, Thailand, Thailand

Thaitawat, Nusara
BANGKOK POST
c/o News Dept., Bangkok Post
968 Rama IV Rd.,Bangkok 10500, Thailand, Thailand

Thongtham, Charunee
Correspondent
BANGKOK POST
10/1 SOi Prasartsook
Bangkok 10120 Thailand, Thailand

Usher, Ann Danaiya
Correspondent
NATION
954 West 7th Avenue
Vancouver, BC V5Z 1C3, Canada, Thailand

ORGANIZATIONS

Latham, Marc, Dr.
International Board for Soil Research and Management
PO Box 9-109
Bangkhen, Bnagkok 10900, Thailand
T: (66 2) 579 7590/579 4012 F: (66 2) 561 1230

GLOBAL FORUM

Chutima, Gawin
SVITA FOUNDATION NATIONAL NGO FOCAL POINT
22/11 Kasetvilla Toewr, Soi Tanpuying Pahon, Ngarmvongvarn
Bankok, Thailand
T: 66 2 579-7608 F: 66 2 579-8944
TL: 87373

Deetes, Tuenjai, Ms.
Secretary General
HILL AREA DEVELOPMENT FOUNDATION
PO BOX 11 - Mae Chan
Chiengrai 57110, Thailand
T: 66 53 713988 F: 66 02 513-3031

Permongsacharoen, Witoon, Mr.
PROJECT FOR ECOLOGICAL RECOVERY
Palais de Nations
Geneve CH 1211, Thailand
F: 41 22 733-6542

Sabur, Abdul, Mr.
ASIAN CULTURAL FORUM ON DEVELOPMENT ACFOD
Palais de Nations
Geneve CH 1211, Thailand
F: 41 22 733-6542

Seeharat, Geerasak, Mr.
SVITA FOUNDATION NATIONAL NGO FOCAL POINT
22/11 Kasetvilla Toewr, Soi Tanpuying Pahon, Ngarmvongvarn
Bankok, Thailand
T: 66 2 579-7608 F: 66 2 579-8944
TL: 87373

Sharp, Tim
Editor / Publisher
FREELANCER
PO Box 4 Doi Saket
Chiang Mai, Thailand
T: 66 53 495022 F: 66 53 212219

Silpa-Anan, Surat
MINISTRY OF EDUCATION
12 Rajaji Marg
New Delhi 110011, Thailand
T: 91 11 301-4309

Sudara, Suraphol, Dr.
Vice-President
SVITA FOUNDATION NATIONAL NGO FOCAL POINT
22/11 Kasetvilla Tower, Soi Tanpuying Pahon, Ngarmvongvarn Road
Bangkok 10900, Thailand
T: 66 2 579-7608 F: 66 2 579-8944
TL: 87373

Vidal, Melba J., Ms.
CIDSE VIETNAM
G.P.O. Box 2914, General Post Office
Bangkok 54971, Thailand
T: 411101 MLOS VT

Wattanasiritham, Paiboon
FOUNDATION FOR THAILAND RURAL RECONSTRUCTION
Palais de Nations
Geneve CH 1211, Thailand
F: 41 22 733-6542

Yevgenyevich, Veriguine Vitali, Mr.
Engineer
SVITA FOUNDATION NATIONAL NGO FOCAL POINT
22/11 Kasetvilla Tower, Soi Tanpuying Pahon, Ngarmvongvarn Road
Bangkok 10900, Thailand
T: 66 2 579-7608 F: 66 2 579-8944
TL: 87373

TOGO - GOVERNMENT

Abotchi, N'koley Koffi
Ministre Du Developpement Rural
MINISTERE DU DEV RURAL
BP 385
Lome, Togo
T: 21 04 82 21 75 85 F: 21 87 92

Binguitcha - Fare, Kpandta Ismail
Ministre De l'Environnement
MINISTERE DE L'ENVIRONNEMENT
BP 4825
Lome, Togo
T: 228 21 3078 F: 228 21 0333

Cheaka, Aboudou Toure
Ministre Des Affaires Etrangeres
MINISTERE DES AFFAIRES ETRANGERES
BP. 900
Lome, Togo
T: 228 213601

Degonzague, Adaku Sittu
Representante
HAUT CONSEIL DELA REPUBLIGUE
Boite Postale 327
Lome, Togo
T: 21 83 00

Emoe, Felix Komla
Ingenieur-Conseiller Technique
MINISTERE DE L'ENVIRONNEMENT
BP 4825
LOME, Togo
T: BUREAU 215658 F: 21 03 33 TG

Kambia, Essobeheyi
BP 80162
Lome, Togo
T: 21 20 39 21 04 82 F: 21 87 92

Lawson, Latevi Akpe
Conseiller
MINISTERE DU PLAN
B.P. 1667
Lome, Togo
T: 228 21 5010 F: 228 21 3751

Tengue, Kodjo Mawuenyega
Ingenieur Forestier-Coordonnateur
P.A.F.T. - TOGO/MDR
B.P. 341
Lome, Togo
T: 228 21 30 96 F: 228 21 87 92

Typamm, Ayi Messan
Secretaire Des A.E.
MINISTERE DES A.E.C.
B.P. 900
Lome, Togo
T: 21 29 11 13

NGO

Adika, Samuel Koku
RAAILD
B.P. 458
Kpalime, Togo

Adjalla, Roger
ASAFED
rue des Hydrocarbures BP 3907
Lome, Togo

Ayessaki, Boukari
V
B.P. 47
Dapaong, Togo

Gle, Seth Kossi
RAAILD
B.P. 458
Kpalime, Togo

Niang, Mody
ASAFED
rue des Hydrocarbures BP 3907
Lome, Togo

Razanamparany, Lucie
ASAFED
rue des Hydrocarbures BP 3907
Lome, Togo

PRESS

De Gonzague, Adaku
Editor
AVENIR
14 Avenue des Kondona
lokoin Forever, Lome, Togo, Togo

ORGANIZATIONS

Anika, Anani, Mr.
Barvensco
B.P. 1288
Lome, Togo

Attignou, Koffi, Mr.
As. Africaine D'education Pour Le Developpement*
rue des Hydrocarbures BP 3907
Lome, Togo
T: (228) 21 63 16

Gle, K.A., Mr.
Reseau Africain d'Appui aux Initiatives Locales de Developpement
B.P. 458
Kpalime, Togo
T: 410 180

Okorie, Okorie O., Dr.
International Agency for Rural Industrialization (INARI)
58 rue Bocco Agegee- Lom Nava
BP 2809, Lome, Togo

Todzro, Mensah T., Mr.
Les Amis de la Terre
BP 20190
Lome, Togo
T: (228) 214 965

GLOBAL FORUM

Ajavon, Ayikoe Oswald, Mr.
President
ASSOCIATION POUR L'ENVIRONMENT EN AFRIQUE (AEA)
P.O.Box 3522
Lome, Togo
T: 228 210114 F: 228 210543

Kluvia, Seth Mateli, Mr.
Director
YMCA OF TOGO
UCJG/YMCA, B.P. 02
Lome, Togo
T: 228 214669 F: 228 212967
TL: 5327 CONGATE

TONGA - GLOBAL FORUM

Fifita, Netatua, Ms.
WOMEN'S ORGANIZATION
P.O.Box 5
Nukualofa, Tonga
T: 676 23611 F: 676 23216

Finau, Bishop Patelisio
CATHOLIC DIOCESE OF TONGA
P. O. Box 1
Nuku'alofa, Tonga

TRINIDAD AND TOBAGO - GOVERNMENT

Applewhite, Clyde Carlton, Mr
Director
UNITED NATIONS ECONOMIC COMMISSION
P.O.Box 1113 22-24 St. Vincent St
Port-of-Spain, Trinidad, Trinidad and Tobago
T: 809 623 5595 5428 F: 809 623 8485

Charles, Hollis
Adviser On Environment
CARIBBEAN COMMUNITY SECRETARIAT
PO Box 1113
Port of Spain, Trinidad, Trinidad and Tobago
T: 809 623 2280 F: 809 625 2568

Daniel Paul, Judy
Legal Officer
ENVIRONMENT DIVISION
Ministry of Planning & Development Whitehall St Clair
Trinidad, Trinidad and Tobago
T: 1 809 628 5233 F: 1 809 628 5233

NGO

Dumas, Regina, Sra.
CARIBBEAN NETWORK FOR INTEGRATED RURAL DEVELOPMENT
40 Eastern Main Road
St. Augustine, Trinidad and Tobago

Shand, Eden
CFCA
P O Box 679, 12 Ferndale Terrace
St. Ann's, Port of Spain, Trinidad and Tobago

Shand, Gillian
CFCA
P O Box 679, 12 Ferndale Terrace
St. Ann's, Port of Spain, Trinidad and Tobago

Wiltshire, Rosina, Ms.
Research Coordinator
DEVELOPMENT ALTERNATIVES WITH WOMEN FOR A NEW ERA (DAWN)
c/o WAND, UWI School of Continuing Studies, Pinelands
St Michael, Trinidad and Tobago
T: 1 809 4269288 F: 1 809 4294854

GLOBAL FORUM

Ramesar, Esmond
UNIVERSITY OF THE WEST INDIES
12 Rajaji Marg
New Delhi 110011, Trinidad and Tobago
T: 91 11 301-4309

Ramesar, Mrs, Ms.
UNIVERSITY OF WEST INDIES
12 Rajaji Marg
New Delhi 110011, Trinidad and Tobago
T: 91 11 301-4309

Sitahal, Harold
Rev
THEOL. REFLECTION & CONSCIENTIZATION - CAREBBEAN CONFERENCE OF CHURCHES
PO Box 876
Port-of-Spain, Trinidad and Tobago

Talma, Grace, Ms.
Chairman
FAMILY PLANNING OF TRINIDAD & TOBAGO
79 Western Main Road
Port-of-Spain, Trinidad, Trinidad and Tobago
T: 1 809 622-7452 F: 1 809 622-7452

TUNISIA - GOVERNMENT

Abdellah, Hichem, M.
Conseiller l' Ambassade au Bresil
Permanent Mission of Tunisia
United Nations Plaza
New York, NY 10017, United States of America
T: 212 963 7174 F: 212 963 1921

Bovali, Naceur, M.
Secretaire l' Ambassade au Bresil
Permanent Mission of Tunisia
United Nations Plaza
New York, NY 10017, United States of America
T: 212 963 7174 F: 212 963 1921

Ennabli, Mohamed, M.
President et Directeur General de l' Agence Nationale de Protection de l' Environnement
Permanent Mission of Tunisia
United Nations Plaza
New York, NY 10017, United States of America
T: 212 963 7174 F: 212 963 1921

Hentati, Adel, M.
Expert l' Agence National de Protection de l' Environnement
Permanent Mission of Tunisia
United Nations Plaza
New York, NY 10017, United States of America
T: 212 963 7174 F: 212 963 1921

Jebali, Salah, S.E. M.
Ministre de l' Environnement l' Amenagement du Territoire
Permanent Mission of Tunisia
United Nations Plaza
New York, NY 10017, United States of America
T: 212 963 7174 F: 212 963 1921

Jomaa, Ghazi, M.
Secretaire des Affaires Etrangeres
Permanent Mission of Tunisia
United Nations Plaza
New York, NY 10017, United States of America
T: 212 963 7174 F: 212 963 1921

Kanoun, Houda, Mme
Depute
Permanent Mission of Tunisia
United Nations Plaza
New York, NY 10017, United States of America
T: 212 963 7174 F: 212 963 1921

Mestiri, Zinelabidine, S.E. M.
Ambassadeur Extraordinaire et Plenipotnetiaire au Bresil
Permanent Mission of Tunisia
United Nations Plaza
New York, NY 10017, United States of America
T: 212 963 7174 F: 212 963 1921

Mlika, Mehdi, M.
President et Directeur General de l' Office National de l' Assainissement
Permanent Mission of Tunisia
United Nations Plaza
New York, NY 10017, United States of America
T: 212 963 7174 F: 212 963 1921

Souissi, Ahmed, M.
Directeur General de l' Environnement
Permanent Mission of Tunisia
United Nations Plaza
New York, NY 10017, United States of America
T: 212 963 7174 F: 212 963 1921

Tebourbi, Habib, M.
Secretaire des Affaires Etrangeres
Permanent Mission of Tunisia
United Nations Plaza
New York, NY 10017, United States of America
T: 212 963 7174 F: 212 963 1921

Yahia, Habib Ben, S.E. M.
Ministre Affaires Etrangeres (Chef de la Delegation)
Permanent Mission of Tunisia
United Nations Plaza
New York, NY 10017, United States of America
T: 212 963 7174 F: 212 963 1921

NGO

Abrougui, Mohamed Ali
ATPNE
12 rue Tantaoui El Jawhari
El Omrane, 1005 Tunis, Tunisia

Azouz, Mokhtar
APEN
87 Place du 13 Janvier 52
7000 Bizerte, Tunisia

Azzouz, Ahmed
APEN
87 Place du 13 Janvier 52
7000 Bizerte, Tunisia

Debbiche, Lanouar
FTDC
11 rue d'Irak
1002 Tunis, Tunisia

Ernez, Amor
ATPNE
12 rue Tantaoui El Jawhari
El Omrane, 1005 Tunis, Tunisia

PRESS

Bouzid, Dorra
Editor
LE RENOUVEAU
Apt. C42, Cite Sprie
La Marsa 2070, Tunis, Tunisia, Tunisia

ORGANIZATIONS

Ali, Abrougui Mohamed, Dr.
As. Tunisienne pour la Protection de la Nature et de l'Environnement (ATPNE)
12 rue Tantaoui El Jawhari
El Omrane, 1005 Tunis, Tunisia
T: (216 1) 288 141 F: (216 1) 797 295

Ellouze, Abdelhafidh, Mr.
Fondation Tunisienne de Developpement Communautaire
11 rue d'Irak
1002 Tunis, Tunisia
T: (216 1) 283 116/782 513
TL: 14.689 (SCFDC TN)

Mokhtar, Azouz, Dr.
As. de Protection de l'Environnement et de la Nature du Gouvernorat de Bizerte
87 Place du 13 Janvier 52
7000 Bizerte, Tunisia
T: (216 2) 32 403//31 406 F: (216 2) 37 650

GLOBAL FORUM

Attia, Habib, Mr.
CNRS
Tunis, Tunisia
T: 216 1 27 60 88

Bouraoui, Soukaina, Mr.
Proffesseur
FACULTE DES SCIENCES JURIDIQUES
Tunis 11 2049, Tunisia

Chabbi, Morched, Mr.
Charcheur-Urbaniste
ENDA TIERS MONDE
24, rue Ilb Charaf
Tunis 1002, Tunisia
T: 216 1 792 343

Cracknel, Michael, Mr.
ENDA INTER ARABE
6 Impasse de la mer rouge
Ariana 2080, Tunisia
T: 216 1 718 340 F: 216 1 718 340

Dorra, Bouzid
Reporter
RTT LE RENOUVEU
Aff Cye Cite Spric
La Marsa 2070 Tunis, Tunisia
T: 216 1 746 536/240 585 F: 216 1 746 446/788

Ghaddoum, Mongi, Mr.
FONDATION TUNISIENNE DE DEVELOPPEMENT COMMUNAUTAIRE
Palais de Nations
Geneve CH 1211, Tunisia
F: 41 22 733-6542

Mabrouk, Ridha, Mr.
NADI AL BASSAR - NORTH AFRICAN CENTER FOR SIGHT AND VISUAL SCIENCES
9, Boulevard Bab Menara
Tunis 1088, Tunisia
T: 216 1 561737 F: 216 1 352321
TL: BT140.70 TN

TURKEY - GOVERNMENT

Akad, Ozger, Mr.
Under-Secretary
Ministry of Environment
Permanent Mission to the United Nations, 821 United Nations Plaza, 1st Floor
New York, NY 10017, United States of America
T: 212 949 0150 F: 212 949 0086

Aksoy, Oktay, H.E.
Ambassador, Director General
Department of International Economic Organizations, Ministry of Foreign Affairs
Permanent Mission to the United Nations, 821 United Nations Plaza, 1st Floor
New York, NY 10017, United States of America
T: 212 949 0150 F: 212 949 0086

Akyurek, B. Dogancan, H.E. Mr.
Minister of Environment
Permanent Mission of Turkey
821 United Nations Plaza, 1st Floor
New York, NY 10017, United States of America
T: 212 949 0150 F: 212 949 0086

Alper, Deniz, Mr.
Special Adviser of the Minister
Ministry of Environment
Permanent Mission to the United Nations, 821 United Nations Plaza, 1st Floor
New York, NY 10017, United States of America
T: 212 949 0150 F: 212 949 0086

Alptuna, Akin, Mr.
Minister Plenipotentiary, Deputy Director General
Department of International Economic Organizations
Permanent Mission to the United Nations, 821 United Nations Plaza, 1st Floor
New York, NY 10017, United States of America
T: 212 949 0150 F: 212 949 0086

Baran, Fikret T., Mr.
Deputy Director General for Public Finance
Permanent Mission of Turkey
821 United Nations Plaza, 1st Floor
New York, NY 10017, United States of America
T: 212 949 0150 F: 212 949 0086

Cakmakci, Akin, Mr.
Under-Secretary
Ministry of Industry and Trade
Sanayive Ticaret Bakanligi
Maltepe Ankara, Turkey
T: 90 4 2293811 2291979 F: 2312208 2308705

Cerrahoglu, Saeed Rabb, Mr.
President
Association of the Protection of the Natural Life
Permanent Mission to the United Nations, 821 United Nations Plaza, 1st Floor
New York, NY 10017, United States of America
T: 212 949 0150 F: 212 949 0086

Culha, Durukal, Mr.
Director General of Technical Research and Implementation
Ministry of Public Works and Housing
Permanent Mission to the United Nations, 821 United Nations Plaza, 1st Floor
New York, NY 10017, United States of America
T: 212 949 0150 F: 212 949 0086

Demirel, Suleyman, H.E. Mr.
Prime Minister
Republic of Turkey
821 United Nations Plaza, 1st Floor
New York, NY 10017, United States of America
T: 212 949 0150 F: 212 949 0086

Erdogan, Senol, Dr
Deputy Undersecretary
Ministry of Agriculture and Rural Affairs
Namik Kemal MAH. 3.Cadde No 1/1
Ankara 06650, Turkey
T: 4181300 F: 4170237

Karaca, Hayrettin, Mr.
Recipient of the '1991 Global 500 Award'
Permanent Mission of Turkey
821 United Nations Plaza, 1st Floor
New York, NY 10017, United States of America
T: 212 949 0150 F: 212 949 0086

Karadag, Esra Fatma, Mrs.
Deputy Expert
Ministry of Environment
Permanent Mission to the United Nations, 821 United Nations Plaza, 1st Floor
New York, NY 10017, United States of America
T: 212 949 0150 F: 212 949 0086

Kochas, Suleyman, Mr.
General Coordinator of the Mediterranean Countries Youths for Environment and Services
Permanent Mission of Turkey
821 United Nations Plaza, 1st Floor
New York, NY 10017, United States of America
T: 212 949 0150 F: 212 949 0086

Mater, Baris, Prof. Dr.
President
Organization of the Protection of Environment and Greenification of Turkey
Permanent Mission to the United Nations, 821 United Nations Plaza, 1st Floor
New York, NY 10017, United States of America
T: 212 949 0150 F: 212 949 0086

Pazarci, Huseyin
Chief Legal Adviser
FOREIGN MINISTRY OF TURKEY
T C Disisleri Bakanligi Hukuk Musavirligi
Balgert Anakara, Turkey
T: 4 212 89 46

Polat, Mehmet, Mr.
Head of Department of Sectors, General Directorate of Economic Planning, Under-Secretariat of the State Planning Organization
Prime Minister
Permanent Mission to the United Nations, 821 United Nations Plaza, 1st Floor
New York, NY 10017, United States of America
T: 212 949 0150 F: 212 949 0086

Sabuncu, Hilmi, Mr.
Expert, General Directorate of Social Planning, Under-Secretariat of the State Planning Organization
Prime Ministry
Permanent Mission to the United Nations, 821 United Nations Plaza, 1st Floor
New York, NY 10017, United States of America
T: 212 949 0150 F: 212 949 0086

Saygilioglu, Nevzat, Mr.
Deputy Director General, General Derectorate of Revenues
Ministry of Finance and Customs
Permanent Mission to the United Nations, 821 United Nations Plaza, 1st Floor
New York, NY 10017, United States of America
T: 212 949 0150 F: 212 949 0086

Sen, Mujgan, Mrs.
Head of Department of Energy
Ministry of Energy and Natural Resources
Permanent Mission to the United Nations, 821 United Nations Plaza, 1st Floor
New York, NY 10017, United States of America
T: 212 949 0150 F: 212 949 0086

Sipahioglu, Sengun, Mr.
Head of Department of Weather Forecasting, General Directorate of State Meteorological Affairs
Prime Ministry
Permanent Mission to the United Nations, 821 United Nations Plaza, 1st Floor
New York, NY 10017, United States of America
T: 212 949 0150 F: 212 949 0086

Soyslemez, Turan, Mr.
Permanent Mission of Turkey
821 United Nations Plaza, 1st Floor
New York, NY 10017, United States of America
T: 212 949 0150 F: 212 949 0086

Surucu, Feyza, Mrs.
Head of Department
Ministry of Tourism
Ministry of Tourism Kat 4 Bahcelivelér
Ankara, Turkey
T: 90 4 2128378 F: 90 4 2128397

Talu, Nuran, Mrs.
Head of Department of International Relations
Ministry of Environment
Permanent Mission to the United Nations, 821 United Nations Plaza, 1st Floor
New York, NY 10017, United States of America
T: 212 949 0150 F: 212 949 0086

Tanlay, Fuat, Mr.
Counsellor
Permanent Mission of Turkey
821 United Nations Plaza, 1st Floor
New York, NY 10017, United States of America
T: 212 949 0150 F: 212 949 0086

Tarlan, Tahsin, H.E. Mr.
Ambassador
Permanent Mission of Turkey
821 United Nations Plaza, 1st Floor
New York, NY 10017, United States of America
T: 212 949 0150 F: 212 949 0086

User, Erol, Mr.
Member of the Board of Directors
Turkish National Committee of the International Chamber of Commerce
Permanent Mission to the United Nations, 821 United Nations Plaza, 1st Floor
New York, NY 10017, United States of America
T: 212 949 0150 F: 212 949 0086

Yavuz, Huseyin, Mr.
Deputy Head of Department of Evaluation and Planning, General Directorate of State Water Works
Ministry of Public Works and Housing
DSI Genel Mudurlugu Etud VE Pland D Bsk Yrd
06100 Ankara, Turkey
T: 4 4183420 4 418 3406 F: 4 417 13 78

Yuksel, Sukruye, Mrs.
Head of Department of Environmental Health, General Directorate of Principal Health Service
Ministry of Health
Permanent Mission to the United Nations, 821 United Nations Plaza, 1st Floor
New York, NY 10017, United States of America
T: 212 949 0150 F: 212 949 0086

Yuksel, Yavuz, Mr.
Deputy Under-Secretary
Ministry of Forestry
Permanent Mission to the United Nations, 821 United Nations Plaza, 1st Floor
New York, NY 10017, United States of America
T: 212 949 0150 F: 212 949 0086

PRESS

Alp, Sahap
Editor
TRT
Beykoz Sok 32/20 Kavaklidere
Ankara, Turkey, Turkey

Cekirge, Yusuf Fatih
Editor
TURKEY PRESS
Cinnah Cad. 8 Kavaklidere
Ankara, Turkey, Turkey

Celebioglu, Meral
TV Reporter
TURKISH RADIO AND TELEVISION (TRT)
D-1 Blok D:704 Oran-Ankara Turkey
Ankara, Turkey
T: 90 4 428 1650 F: 468 0440

Cemal, Hazan
Editor
SABAH GAZETESI
Sabath Gazetesi
Stambul, Turkey

Dimli, Serdar
Reporter Radio/Television
TRT
Gokdere sok.11/1 Yukari
Ayranci, Ankara, Turquey, Turkey

Inanc, Gulper
Reporter Radio/Television
TRT
Getim Emec Bulvari Cevre S.
95/6 Balgat, Ankara, Turquey, Turkey

Polat, Yilmar
Correspondent
ANATOLIAN NEWS AGENCY
6810 Supreme Court
Springfield, VA 22150, Turkey

Toprak, Cengiz N.
Correspondent
TURKIYE
821 UN Plaza
New York, NY 10017, USA, Turkey

Yavuz, Turan
Correspondent
MILLIYET NEWSPAPER
NPB suite 898
Washington, D.C., U.S.A., Turkey

Yener, Yucel
Editor
TRT
Yudum sokak 4/3 Kucukesat
Ankara, Turkey, Turkey

GLOBAL FORUM

Behrend, Hartmut, Mr.
G.T.Z.
Turgutlu Sok. 19/17 Ankara,Turkiye
Ankara, Turkey
T: 90 4 436 0726 F: 90 4 434 2945

Erim, Aydan, Ms.
Vice-President
TURKISH CHAMBER OF ARCHITECTS
Konur Socak # 4/2 Kizilay
Ankara 06690, Turkey
T: 90 4 417-3727 F: 90 4 417-0361

Ersoy, Fureya, Ms.
Ali Kaya Suk No. 8, Levent
Istanbul 80720, Turkey
T: 90 1 269-6680 F: 90 1 269-6699

Kocbas, Suleyman, Mr.
Dr.
SASA - TURKEY
380 Ata Turk Cad., Suite #103
Alsancak Izmir 35220, Turkey
T: 90 51 217974 F: 90 51 631552

Ozkan, Noyan
Lawyer/Environmentalist
Sehit Fethibey Caddesi, No.:36
Izmir, Turkey
T: 90 51 130790 F: 90 51 130790
TL: 52584

Sonmez, Cidem, Ms.
Director
ULUSLARARASI MIRAS
1388 STR. No.3, SUITE #9
Alsancak Izmir 35220, Turkey
T: 90 51 225877 F: 90 51 631552

TUVALU - GOVERNMENT

Paeniu, Bikenibeu, Mr
Prime Minister
Tuvalu
Tuvalu

UGANDA - GOVERNMENT

Acere, T., Mr.
Commissioner for Environment
Ministry of Water, Energy, Minerals and Environment Protection
Permanent Mission to the United Nations, 336 E. 45th Str. New York, NY, 10017
F: 212 949 0110
TL: 212 687 4517

Byaruhanga, Emmamuel, Mr.
Commissioner for Housing
Ministry of Lands, Housing and Urban Development
Permanent Mission to the United Nations, 336 E. 45th Str. New York, NY, 10017
F: 212 949 0110
TL: 212 687 4517

Etuket, David, Mr.
First Secretary
Permanent Mission of Uganda
336 E. 45th Str.
New York, NY 10017, United States of America
T: 212 949 0110 F: 212 687 4517

Gwage, Philip, Mr.
Assistant Director
Meteorology Department, Ministry of Water, Energy, Minerals and Environment Protection
Permanent Mission to the United Nations, 336 E. 45th Str. New York, NY, 10017
F: 212 949 0110
TL: 212 687 4517

Kagimba, J., Dr.
Presidential Adviser
Permanent Mission of Uganda
336 E. 45th Str.
New York, NY 10017, United States of America
T: 212 949 0110 F: 212 687 4517

Kagoro, M., Mr.
Deputy Commissioner
Forestry
Permanent Mission to the United Nations, 336 E. 45th Str. New York, NY, 10017
F: 212 949 0110
TL: 212 687 4517

Kajura, Henry M., The Hon.
Minister
Water, Energy, Minerals and Environment Protection
Permanent Mission to the United Nations, 336 E. 45th Str. New York, NY, 10017
F: 212 949 0110
TL: 212 687 4517

Kamugisha, Wycliffe, Mr.
Presidential Aide
Permanent Mission of Uganda
336 E. 45th Str.
New York, NY 10017, United States of America
T: 212 949 0110 F: 212 687 4517

Kamunanwire, Perezi K., H.E. Prof.
Ambassador Extraordinary and Plenipotentiary
Permanent Representative to the United Nations
Permanent Mission to the United Nations, 336 E. 45th Str. New York, NY, 10017
F: 212 949 0110
TL: 212 687 4517

Kavuma, J.N., Mrs.
Chief Environment Officer
Permanent Mission of Uganda
336 E. 45th Str.
New York, NY 10017, United States of America
T: 212 949 0110 F: 212 687 4517

Kavuma, Sam, Mr.
Presidential Adviser
Permanent Mission of Uganda
336 E. 45th Str.
New York, NY 10017, United States of America
T: 212 949 0110 F: 212 687 4517

Kivengere, Hope, Ms.
Presidential Adviser
Permanent Mission of Uganda
336 E. 45th Str.
New York, NY 10017, United States of America
T: 212 949 0110 F: 212 687 4517

Kyambadde, Amelia, Mrs.
Presidential Adviser
Permanent Mission of Uganda
336 E. 45th Str.
New York, NY 10017, United States of America
T: 212 949 0110 F: 212 687 4517

Kyeyune, Alexie, Mrs.
Foreign Service Officer
Ministry of Foreign Affairs
Permanent Mission to the United Nations, 336 E. 45th Str. New York, NY, 10017
F: 212 949 0110
TL: 212 687 4517

Lwanga-Lunyiingo, , Prof.
Presidential Adviser
Permanent Mission of Uganda
336 E. 45th Str.
New York, NY 10017, United States of America
T: 212 949 0110 F: 212 687 4517

Magazi, Joan, Mrs.
Principal Private Secretary
State House
Permanent Mission to the United Nations, 336 E. 45th Str. New York, NY, 10017
F: 212 949 0110
TL: 212 687 4517

Muduuli, M., Mrs.
Principal Economist
Ministry of Finance and Economic Planning
Permanent Mission to the United Nations, 336 E. 45th Str. New York, NY, 10017
F: 212 949 0110
TL: 212 687 4517

Mukassa-Ssali, Paul, Mr.
Presidential Adviser
Permanent Mission of Uganda
336 E. 45th Str.
New York, NY 10017, United States of America
T: 212 949 0110 F: 212 687 4517

Museveni, Yoweri Kaguta, H.E. Mr.
President
Republic of Uganda
336 E. 45th Str.
New York, NY 10017, United States of America
T: 212 949 0110 F: 212 687 4517

Mutebile, Tumusiime, Mr.
Permanent Secretary
Ministry of Finance and Economic Planning
Permanent Mission to the United Nations, 336 E. 45th Str. New York, NY, 10017
F: 212 949 0110
TL: 212 687 4517

Muyingo, Steven, Mr.
Presidential Adviser
Permanent Mission of Uganda
336 E. 45th Str.
New York, NY 10017, United States of America
T: 212 949 0110 F: 212 687 4517

Ocaya-Lakidi, J.A., Mrs.
Secretary
Environment Protection
Permanent Mission to the United Nations, 336 E. 45th Str. New York, NY, 10017
F: 212 949 0110
TL: 212 687 4517

Rutaroh, Joan, Mrs.
Assistant Commissioner
Ministry of Agriculture and Animal Industry
Permanent Mission to the United Nations, 336 E. 45th Str. New York, NY, 10017
F: 212 949 0110
TL: 212 687 4517

Ssemogerere, Paul K., The Hon. Dr.
Second Deputy Prime Minister and Minister for Foreign Affairs
Permanent Mission of Uganda
336 E. 45th Str.
New York, NY 10017, United States of America
T: 212 949 0110 F: 212 687 4517

Tomusange, Joseph, H.E.
High Commissioner
Kenya
Permanent Mission to the United Nations, 336 E. 45th Str. New York, NY, 10017
F: 212 949 0110
TL: 212 687 4517

Wadenya, C., Mr.
Senior Environment Officer
Permanent Mission of Uganda
336 E. 45th Str.
New York, NY 10017, United States of America
T: 212 949 0110 F: 212 687 4517

NGO

Bigirwenkya, Jane
UWTPM
PO Box 10351
Kampala, Uganda

Mubiru, Ruth
UGANDA WOMEN TREE PLANTING MOVEMENT
PO Box 10351
Kampala, Uganda

GLOBAL FORUM

Mubbala, Susan, Ms.
Country Coordinator
PVO-NGO/NRMS PROJECT
PO Box 9007
Kampala, Uganda
T: 256 41 24-24-29 F: 256 41 23-17-43

PARLIAMENTARIANS / SPIRITUAL LEADERS

Magezi, D.S., Dr.
National Assembly Parliament House P.O. Box 7178
Kampala, Uganda
T: 34340
TL: 61093 & 61095

UKRAINE - GOVERNMENT

Humenny, Ihor V., Mr.
First Secretary
Ministry of Foreign Affairs
Permanent Mission to the United Nations, United Nations Plaza
New York, NY 10017, United States of America
T: 212 963 7171 F: 212 963 1921

Kostenko, Yuri, Mr.
Deputy Head
Standing Commission of Verkhovna Rada of Ukraine on Ecology and Natural Resources Management
Permanent Mission to the United Nations, United Nations Plaza
New York, NY 10017, United States of America
T: 212 963 7171 F: 212 963 1921

Maidannyk, Oleksandr I., Mr.
Chief, Intergovernmental Relations Department
Ministry of Environmental Protection
Permanent Mission to the United Nations, United Nations Plaza
New York, NY 10017, United States of America
T: 212 963 7171 F: 212 963 1921

Mysko, Bohdan, Mr.
Adviser to the President of Ukraine
Permanent Mission of Ukraine
United Nations Plaza
New York, NY 10017, United States of America
T: 212 963 7171 F: 212 963 1921

Plioushch, Ivan, Mr.
Vice President
Permanent Mission of the Ukraine
136 E. 67th Str.
New York NY 10021, United States of America
T: 212 535 3418 F: 212 288 5361

Scherbak, Yuri M., H.E. Mr.
Minister of Environmental Protection of Ukraine
Deputy Head of the Delegation
Permanent Mission to the United Nations, United Nations Plaza
New York, NY 10017, United States of America
T: 212 963 7171 F: 212 963 1921

Yavorivskiy, Volodymyr O., Mr.
Head, Standing Commission of Verkhovna Rada of Ukraine on Chornobyl Disaster
Deputy Head of Delegation
Permanent Mission to the United Nations, United Nations Plaza
New York, NY 10017, United States of America
T: 212 963 7171 F: 212 963 1921

NGO

Koseltsev, Mikhail
SPAN
P.B. 54
Moscow 113105, Ukraine

PRESS

Glazovoi, Andrei
Editor
ZELENY SVIT
254032, Tolstoi St. 25 Apt.44
Kiev-32 Ukraine, Ukraine

GLOBAL FORUM

Vadimov, Vadim M., Mr.
Professor
Golovka, 18 - Flat 44
Poltava 314037, Ukraine
T: 1-10-89

UNITED ARAB EMIRATES - GOVERNMENT

Abdulla, Abdul Razak, Mr.
Director
Fisheries Research Center, Ministry of Agriculture and Fisheries
Permanent Mission to the United Nations, 747 Third Avenue, 36th Floor
New York, NY 10017, 371 0480
T: 371 4923

Al Absy, Mohammad, Mr.
First Secretary, International Organization & Conference Department
Ministry of Foreign Affairs
Permanent Mission to the United Nations, 747 Third Avenue, 36th Floor
New York, NY 10017, 371 0480
T: 371 4923

Al Dhahery, Salem, Dr.
Environmental Advisor
Ministry of Health
Permanent Mission to the United Nations, 747 Third Avenue, 36th Floor
New York, NY 10017, 371 0480
T: 371 4923

Al Hossain, Ali, Mr.
Third Secretary
Embassy to Brazil
Permanent Mission to the United Nations, 747 Third Avenue, 36th Floor
New York, NY 10017, 371 0480
T: 371 4923

Al Khameri, Hassan, Mr.
Head of Civil Aviation and Communication Division
Ministry of Communication
Permanent Mission to the United Nations, 747 Third Avenue, 36th Floor
New York, NY 10017, 371 0480
T: 371 4923

Al Muhaaideb, Abdul Wahab, H.E. Dr.
Assistant Under Secretary, Deputy Chairmanof Higher Environmental Committee
Ministry of Health
Permanent Mission to the United Nations, 747 Third Avenue, 36th Floor
New York, NY 10017, 371 0480
T: 371 4923

Al Mutawa, Abdulla, Mr.
Director of Water Production
Ministry of Electricity & Water
Permanent Mission to the United Nations, 747 Third Avenue, 36th Floor
New York, NY 10017, 371 0480
T: 371 4923

Al Noaimi, Rashid Abdullah, H.E. Mr.
Minister for Foreign Affairs
Permanent Mission of United Arab Emirates
747 Third Avenue, 36th Floor
New York, NY 10017, 371 0480
T: 371 4923

Al Numairy, Saad, Dr.
Chief in Chargeof Environmental Polution & General Secretary of the Higher Environmental Committee
Ministry of Health
Permanent Mission to the United Nations, 747 Third Avenue, 36th Floor
New York, NY 10017, 371 0480
T: 371 4923

Al Qasimi, Jassim, Mr.
Third Secretary
Economic Relations Department
Permanent Mission to the United Nations, 747 Third Avenue, 36th Floor
New York, NY 10017, 371 0480
T: 371 4923

Al Shair, Hamdan, Mr.
Assistant president of the Environment Protection
Permanent Mission of United Arab Emirates
747 Third Avenue, 36th Floor
New York, NY 10017, 371 0480
T: 371 4923

Al Shamsi, Abdul Azis Bin Rahama, H.E. Mr.
Ambassador Extraordinary and Plenipotentiary
Federal Republic of Brazil
Permanent Mission to the United Nations, 747 Third Avenue, 36th Floor
New York, NY 10017, 371 0480
T: 371 4923

Al Suwaidi, Ali Thani, Mr.
Minister Plenipotentiary
Permanent Mission of United Arab Emirates
747 Third Avenue, 36th Floor
New York, NY 10017, 371 0480
T: 371 4923

Al Zaabi, Saeed Obid, Mr.
Counsellor, Protocal Department
Ministry of Foreign Affairs
Permanent Mission to the United Nations, 747 Third Avenue, 36th Floor
New York, NY 10017, 371 0480
T: 371 4923

Al-Zaabi, Abdulla Bin Ali, H.E. Mr.
Ambassador
Ministry of Foreign Affairs
Permanent Mission to the United Nations, 747 Third Avenue, 36th Floor
New York, NY 10017, 371 0480
T: 371 4923

Ebrahim, Ahmed Abdulla, H.E. Mr.
Director of H.E.
Minister of Foreign Affairs
Permanent Mission to the United Nations, 747 Third Avenue, 36th Floor
New York, NY 10017, 371 0480
T: 371 4923

Hadi, Abdul Rahman, Mr.
First Secretary
The Legal Affairs & Studies Department, Ministry of Foreign Affairs
Permanent Mission to the United Nations, 747 Third Avenue, 36th Floor
New York, NY 10017, 371 0480
T: 371 4923

Majid, Ahmed, Mr.
Director of Petroleum Department
Ministry of Petroleum & Mineral Resources
Permanent Mission to the United Nations, 747 Third Avenue, 36th Floor
New York, NY 10017, 371 0480
T: 371 4923

Missmar, Salem, Mr.
Assistant Director of the Environment
Permanent Mission of United Arab Emirates
747 Third Avenue, 36th Floor
New York, NY 10017, 371 0480
T: 371 4923

Orman, Mohammad, Mr.
First Secretary
Permanent Mission to United Nations, Geneva
Permanent Mission to the United Nations, 747 Third Avenue, 36th Floor
New York, NY 10017, 371 0480
T: 371 4923

Samhan, Mohammed Jasim, H.E. Mr.
Ambassador Extraordinary and Plenipotentiary
Permanent Representative to the United Nations, New York
Permanent Mission to the United Nations, 747 Third Avenue, 36th Floor
New York, NY 10017, 371 0480
T: 371 4923

Youssef, Amine Mohamed, Dr.
Assistant director of Environment Laboratory
Permanent Mission of United Arab Emirates
747 Third Avenue, 36th Floor
New York, NY 10017, 371 0480
T: 371 4923

Yousuf, Amin Mohamed, Mr
Head Of Environmental Affairs
FOOD AND ENV CONTROL CENTRE
PO Box 3111
Abu Dhabi, United Arab Emirates
T: 725008 F: 785961

PRESS

Karkouti, Mohammed Khaled
Correspondent
AL-IHIHAD NEWSPAPER
165 More Close, St. Paul's Crt
C. Gardens, London W14, United Kingdom, United Arab Emirates

UNITED KINGDOM - GOVERNMENT

Ashworth, Patrick, Mr.
Second Secretary, UNCED Coordinator
British Consulate-General, Rio de Janeiro
Permanent Mission to the United Nations, Permanent Mission to the United Nations, 845 3rd. Ave., 10th Floor
New York, NY 10022, United States of America
T: 212 745 0200 F: 212 745 0316

Bearpark, Andrew, Mr.
Information Officer
Overseas Development Administration
Permanent Mission to the United Nations, 845 3rd. Ave., 10th Floor
New York, NY 10022, United States of America
T: 212 745 0200 F: 212 745 0316

Bennett, Andrew, Mr.
Under-Secretary
Overseas Development Administration
Permanent Mission to the United Nations, Permanent Mission to the United Nations, 845 3rd. Ave., 10th Floor
New York, NY 10022, United States of America
T: 212 745 0200 F: 212 745 0316

Bone, Roger, Mr.
Assistant Under-Secretary
Foreign and Commonwealth Office
Permanent Mission to the United Nations, Permanent Mission to the United Nations, 845 3rd. Ave., 10th Floor
New York, NY 10022, United States of America
T: 212 745 0200 F: 212 745 0316

Brenton, Anthony, Mr.
Counsellor, Head of Environment Science and Energy Department
Foreign and Commonwealth Office
Permanent Mission to the United Nations, Permanent Mission to the United Nations, 845 3rd. Ave., 10th Floor
New York, NY 10022, United States of America
T: 212 745 0200 F: 212 745 0316

Burke, Tom, Mr.
Special Adviser to the Secretary of State
Deparment of Environment
Permanent Mission to the United Nations, Permanent Mission to the United Nations, 845 3rd. Ave., 10th Floor
New York, NY 10022, United States of America
T: 212 745 0200 F: 212 745 0316

Chakrabarti, Suma, Mr.
Private Secretary to the Minister
Overseas Development Administration
Permanent Mission to the United Nations, Permanent Mission to the United Nations, 845 3rd. Ave., 10th Floor
New York, NY 10022, United States of America
T: 212 745 0200 F: 212 745 0316

Chalker, Lynda, Baroness
Minister for Overseas Development
Permanent Mission of United Kingdom
Permanent Mission to the United Nations, 845 3rd. Ave., 10th Floor
New York, NY 10022, United States of America
T: 212 745 0200 F: 212 745 0316

Chatfield, John, Mr.
Local Government Representative
Permanent Mission of United Kingdom
Permanent Mission to the United Nations, 845 3rd. Ave., 10th Floor
New York, NY 10022, United States of America
T: 212 745 0200 F: 212 745 0316

Coates, Barry, Mr
Head Of Development Policy
World Wide Fund For Nature WWF-UK
Weyside Park Catteshall Lane
GODALMING SURREY, England GU7 1XR, United Kingdom
T: 44-483-426 444 F: 44-483-426 409

Cohen, Shimon, Mr.
Consultant To Malaysia
LOWE BELL COMMUNICATIONS
7 Hartford Street Mayfair
London W1T7DY, United Kingdom
T: 071 495 4044 F: 071 629 1277

Deyes, Maya, Mrs.
Assistant Coordinator
Rio de Janerio
Permanent Mission to the United Nations, Permanent Mission to the United Nations, 845 3rd. Ave., 10th Floor
New York, NY 10022, United States of America
T: 212 745 0200 F: 212 745 0316

Dodds, Felix Michael, Mr
Head Of Campaigns
UNITED NATIONS ASSOCIATION
3 Whitehall Court
London, United Kingdom
T: 71 930 2931 F: 71 930 5893

Ennals, The Lord, The Rt. Hon.
UN Association
Permanent Mission of United Kingdom
Permanent Mission to the United Nations, 845 3rd. Ave., 10th Floor
New York, NY 10022, United States of America
T: 212 745 0200 F: 212 745 0316

Feeney, Patricia, Ms.
OXFAM-UK
274 Banbury RD
Oxford, United Kingdom
T: 44 865 311 311 F: 44 865 312 600

Garner-Winship, Stephen, Mr.
HM Counsel
Rio de Janerio
Permanent Mission to the United Nations, Permanent Mission to the United Nations, 845 3rd. Ave., 10th Floor
New York, NY 10022, United States of America
T: 212 745 0200 F: 212 745 0316

Gee, Jack, Mr.
Information Officer
Department of Environment
Permanent Mission to the United Nations, Permanent Mission to the United Nations, 845 3rd. Ave., 10th Floor
New York, NY 10022, United States of America
T: 212 745 0200 F: 212 745 0316

Godfrey, Ian, Mr.
Technical Engineer
Permanent Mission of United Kingdom
Permanent Mission to the United Nations, 845 3rd. Ave., 10th Floor
New York, NY 10022, United States of America
T: 212 745 0200 F: 212 745 0316

Godson, Anthony, Mr.
First Secretary
United Permanent to the United Nations, Geneva
Permanent Mission to the United Nations, Permanent Mission to the United Nations, 845 3rd. Ave., 10th Floor
New York, NY 10022, United States of America
T: 212 745 0200 F: 212 745 0316

Hammond, Mark, Mr.
Principal, Global Atmosphere Division
Department of Environment
Permanent Mission to the United Nations, Permanent Mission to the United Nations, 845 3rd. Ave., 10th Floor
New York, NY 10022, United States of America
T: 212 745 0200 F: 212 745 0316

Handley, Peter, Mr.
Principal, Environment Protection
International Division, Department of Environment
Permanent Mission to the United Nations, Permanent Mission to the United Nations, 845 3rd. Ave., 10th Floor
New York, NY 10022, United States of America
T: 212 745 0200 F: 212 745 0316

Hannay, David, H.E. Sir
Ambassador Extraordinary and Plenipotentiary
Permanent Representative to the United Nations, New York
Permanent Mission to the United Nations, Permanent Mission to the United Nations, 845 3rd. Ave., 10th Floor
New York, NY 10022, United States of America
T: 212 745 0200 F: 212 745 0316

Harman, John, Mr.
Local Government Representative
Permanent Mission of United Kingdom
Permanent Mission to the United Nations, 845 3rd. Ave., 10th Floor
New York, NY 10022, United States of America
T: 212 745 0200 F: 212 745 0316

Hill, Julie, Ms.
Green Alliance
Permanent Mission of United Kingdom
Permanent Mission to the United Nations, 845 3rd. Ave., 10th Floor
New York, NY 10022, United States of America
T: 212 745 0200 F: 212 745 0316

Holt, Denise, Mrs.
First Secretary
British Embassy, Brasilia
Permanent Mission to the United Nations, Permanent Mission to the United Nations, 845 3rd. Ave., 10th Floor
New York, NY 10022, United States of America
T: 212 745 0200 F: 212 745 0316

Horn, Margaret, Ms.
Second Secretary, Environment Science & Energy Department
Assistant Delegation Secretary
Permanent Mission to the United Nations, 845 3rd. Ave., 10th Floor
New York, NY 10022, United States of America
T: 212 745 0200 F: 212 745 0316

Howard, Michael, The Rt. Hon.
Secretary of the State
Permanent Mission of United Kingdom
Permanent Mission to the United Nations, 845 3rd. Ave., 10th Floor
New York, NY 10022, United States of America
T: 212 745 0200 F: 212 745 0316

Ireton, Barry, Mr.
Under-Secretary and Principal Finance Officer
Overseas Development Administration
Permanent Mission to the United Nations, Permanent Mission to the United Nations, 845 3rd. Ave., 10th Floor
New York, NY 10022, United States of America
T: 212 745 0200 F: 212 745 0316

Johnson, Vicky, Ms.
Action Aid
Permanent Mission of United Kingdom
Permanent Mission to the United Nations, 845 3rd. Ave., 10th Floor
New York, NY 10022, United States of America
T: 212 745 0200 F: 212 745 0316

Kelly, Patricia, Ms.
HM Consul-General
Rio De Janeiro
Permanent Mission to the United Nations, Permanent Mission to the United Nations, 845 3rd. Ave., 10th Floor
New York, NY 10022, United States of America
T: 212 745 0200 F: 212 745 0316

Kent, Mark, Mr.
Information Secretary
British Embassay, Brasilia
Permanent Mission to the United Nations, Permanent Mission to the United Nations, 845 3rd. Ave., 10th Floor
New York, NY 10022, United States of America
T: 212 745 0200 F: 212 745 0316

Lea, David, Mr.
Assistant General Secretary
TUC
Permanent Mission to the United Nations, Permanent Mission to the United Nations, 845 3rd. Ave., 10th Floor
New York, NY 10022, United States of America
T: 212 745 0200 F: 212 745 0316

Lowcock, Mark, Mr.
Principal, Natural Resources & Environment Department
Overseas Development Administration
Permanent Mission to the United Nations, Permanent Mission to the United Nations, 845 3rd. Ave., 10th Floor
New York, NY 10022, United States of America
T: 212 745 0200 F: 212 745 0316

Major, John, The Rt. Hon.
Prime Minister and the First Lord of the Treasury
United Kingdom of Great Britain and Northern Ireland
Permanent Mission to the United Nations, Permanent Mission to the United Nations, 845 3rd. Ave., 10th Floor
New York, NY 10022, United States of America
T: 212 745 0200 F: 212 745 0316

Martin, John, Mr.
Counsellor
Permanent Mission of United Kingdom
Permanent Mission to the United Nations, 845 3rd. Ave., 10th Floor
New York, NY 10022, United States of America
T: 212 745 0200 F: 212 745 0316

McConnell, Fiona, Ms.
Assistant Secretary, Environment Protection
International Division, Department of Environment
Permanent Mission to the United Nations, Permanent Mission to the United Nations, 845 3rd. Ave., 10th Floor
New York, NY 10022, United States of America
T: 212 745 0200 F: 212 745 0316

McGlone, Alistair, Mr.
Legal Adviser
Department of Environment
Permanent Mission to the United Nations, Permanent Mission to the United Nations, 845 3rd. Ave., 10th Floor
New York, NY 10022, United States of America
T: 212 745 0200 F: 212 745 0316

Newington, Michael, Mr.
HM Ambassador
Federative Republic of Brazil
Permanent Mission to the United Nations, Permanent Mission to the United Nations, 845 3rd. Ave., 10th Floor
New York, NY 10022, United States of America
T: 212 745 0200 F: 212 745 0316

O'Neill, Stephen, Mr.
Information Officer
Department of Environment
Permanent Mission to the United Nations, Permanent Mission to the United Nations, 845 3rd. Ave., 10th Floor
New York, NY 10022, United States of America
T: 212 745 0200 F: 212 745 0316

Osborn, Derek, Mr.
Deputy Secretary
Department of Environment
Permanent Mission to the United Nations, Permanent Mission to the United Nations, 845 3rd. Ave., 10th Floor
New York, NY 10022, United States of America
T: 212 745 0200 F: 212 745 0316

Ramsaran, Clare, Ms.
UNCED FOR YOUTH
UNA, 3 Whitehall Court
London SW1A 2EL, United Kingdom
T: 44 71 925-2614 F: 44 71 930-5893

Sandbrook, Richard, Mr.
UNEP-UK
Permanent Mission of United Kingdom
Permanent Mission to the United Nations, 845 3rd. Ave., 10th Floor
New York, NY 10022, United States of America
T: 212 745 0200 F: 212 745 0316

Sheridan, William, Mr.
Head of International Relations Division
Forestry Commission
Permanent Mission to the United Nations, Permanent Mission to the United Nations, 845 3rd. Ave., 10th Floor
New York, NY 10022, United States of America
T: 212 745 0200 F: 212 745 0316

Smith, Stephen, Mr.
First Secretary
Environment Science & Energy Department, Delegation Secretary
Permanent Mission to the United Nations, 845 3rd. Ave., 10th Floor
New York, NY 10022, United States of America
T: 212 745 0200 F: 212 745 0316

Storey, Neil, Mr.
Technical Engineer
Permanent Mission of United Kingdom
Permanent Mission to the United Nations, 845 3rd. Ave., 10th Floor
New York, NY 10022, United States of America
T: 212 745 0200 F: 212 745 0316

Thornton, Ian, Professor
The Royal Society
Permanent Mission of United Kingdom
Permanent Mission to the United Nations, 845 3rd. Ave., 10th Floor
New York, NY 10022, United States of America
T: 212 745 0200 F: 212 745 0316

Townend, Anna, Ms
Designer/Environmentalist
PARC - PEOPLE AGAINST THE RIVER CRO
22 Coleraine Rd Blackheath London SE37PQ
London, United Kingdom
T: 081 858 3844 071 435 F: 071 435 9032

Ward, Philip, Mr.
Private Secretary to the Secretary of State
Department of Environment
Permanent Mission to the United Nations, Permanent Mission to the United Nations, 845 3rd. Ave., 10th Floor
New York, NY 10022, United States of America
T: 212 745 0200 F: 212 745 0316

Watson, Gavin, Mr.
Director
Environment Protection Command, Department of Environment
Permanent Mission to the United Nations, Permanent Mission to the United Nations, 845 3rd. Ave., 10th Floor
New York, NY 10022, United States of America
T: 212 745 0200 F: 212 745 0316

Wynn, Grahaam, Mr.
RSPB
Permanent Mission of United Kingdom
Permanent Mission to the United Nations, 845 3rd. Ave., 10th Floor
New York, NY 10022, United States of America
T: 212 745 0200 F: 212 745 0316

Yarnell, Christopher, Mr.
Assistant Director, Environment Unit
Department of Trade and Industry
Permanent Mission to the United Nations, Permanent Mission to the United Nations, 845 3rd. Ave., 10th Floor
New York, NY 10022, United States of America
T: 212 745 0200 F: 212 745 0316

NGO

Black, David, Mr.
PEOPLE AGAINST THE RIVER CROSSING
62 Burnell Av.
Welling DA16 3HW, United Kingdom
T: 81 854-4763 F: 81 278-9460

Blom, Ulla-Liza
IFBPW
Studio 16, Cloisters Business Centre, 8 Battersea Park Road, London SW8 4BG, United Kingdom

Brenu, Kineku Osae
IPPF
Regent's College, Inner Circle, Regent's Park
London NW1 4NS, United Kingdom

Caldas, Tadeu, Mr.
Director
IFOAM - ORGANIC AGRICULTURE
Parracombe, Chapel Lane Forest Row
Sussex RH185OX, United Kingdom
T: 44 243 282-3235 F: 44 243 282-4949

Calvert, Peter
UG
69 Oakfield Avenue
Glasgow G12 8HQ, United Kingdom

Chishti, Saadia
CHEC
57/58 Stanhope Gardens
London SW7 5RF, United Kingdom

Clark, Lorne
WTTC
WTTC 50 Victoria Street, Floor 5
London SW1H ONH, United Kingdom

Cury, Marta Bittar, Ms.
INTERNATIONAL FEDERATION OF BUSINESS AND PROFESSIONAL WOMEN
Studio 16, Cloisters Business Centre, 8 Battersea Park Road, London SW8 4BG, United Kingdom

Da Gama, Louis
BA
1 Queen Anne's Gate
London SW1H 9BT, United Kingdom

Davidson, Joan, Ms.
Representative
OXFAM UK
274 Banbury Road
Oxford OX2 7DZ, United Kingdom

Davies, William
WCIA
Temple of Peace, Cathays Park
Cardiff CF1 3AP, United Kingdom

Faulks, John, Mr.
ACPS
57 Duke Street Grosvenor Square
London W1M 5DH, United Kingdom

Feeney-Gryus, Tricia
OF
274 Banbury Road
Oxford OX2 7DZ, United Kingdom

Forbes, Gerald Innes, Mr.
Doctor
ENVIRONMENTAL HEALTH UNIT
Common Services Agency, Trinity Park House,
South Trinity Road, Edinburgh EH5 3ST, United Kingdom

Gadotti, Moacir
ICEA
Lyng Hall, Blackberry Lane
Coventry CV2 3JS, United Kingdom

Gibson, Craig
FINDHORN FOUNDATION
Cluny Hill College, FORRES IV 36 ORD
Scotland, United Kingdom

Gosling, David, Mr.
COUNCIL OF CHURCHES FOR BRITAIN AND IRELAND
Iner-Church House
35-41 Lower Marsh, London SE1 7RL, United Kingdom

Hale, Monica, Ms.
IEEM
36 Kingfisher Court, Hambridge Road
Newbury, Berkshire RG14 5SJ, United Kingdom

Hale, Robert
IEEM
36 Kingfisher Court, Hambridge Road
Newbury, Berkshire RG14 5SJ, United Kingdom

Hemming, John, Dr.
Director
ROYAL GEOGRAPHICAL SOCIETY
1 Kensington Gore
London SW7 2AR, United Kingdom
T: 44 71 589-0648 F: 44 71 584-4447

Honari, Morteza
CHEC
57/58 Stanhope Gardens
London SW7 5RF, United Kingdom

Howard, Charles
RBG
Richmond, Surrey, TW9 3AB
United Kingdom

Howard, Eric Sevan
WCMC
219 Huntingdon Road
Cambridge CB3 ODL, United Kingdom

Jenkins, Nicola Caroline, Ms.
VTIC
8 John Adam Street
London WC2N 6EZ, United Kingdom

Johnson, Stanley, Mr.
IFAW
Tubwell House, New Road
Crowborough, East Sussex TN6 2HQ, United Kingdom

Juniper, Tony, Mr.
FRIENDS OF THE EARTH-UK
26-28 Underwood St.
London N1 7JQ, United Kingdom
T: 44 71 490-1555 F: 44 71 490-0881

Leach, Gerald, Mr.
STOCKHOLM ENVIRONMENT INSTITUTE
3 Tanza Road
London NW3 2UA, United Kingdom
T: 44 71 794-6228 F: 44 71 431-5480

Leggett, Jeremy, Dr.
GLOBAL SECURITY PROGRAMME
Free School Lane,University of Cambridge
Cambridge CB23RQ, United Kingdom
T: 44 223 334509 F: 44 223 335065

Lima, Marcos De Almeida
SDI
Anugraha, Wick Lane, Engelfield Green
Surrey TW20 OXN, United Kingdom

Locke, Brian
BACR
The Court House, Bisley, Stroud
Gloucestershire GL6 7AA, United Kingdom

Malta, Flavio
CHEC
57/58 Stanhope Gardens
London SW7 5RF, United Kingdom

Mayo, Ed
NEF
Universal House, 2nd Floor, 88-94 Wentworth Street
London E1 7SA, United Kingdom

McCorry, Denise, Ms.
Science Coordinator
CORAL CAY CONSERVATION
Sutton Business Centre, Restmor Way
Wallington-Surrey SM6 7AH, United Kingdom
T: 44 81 669-0011 F: 44 81 773-9656

Mertens, Dirk
IFIAG
22-23 Great Tower Street
London EC3R 5AQ, United Kingdom

North, Sachlan, Ms.
Coordinator
SUSILA DHARMA INTERNATIONAL
Anugraha, Wick Lane, Engelfield Green
Surrey TW20 OXN, United Kingdom

Palmer, Joy, Dr.
COEE
University of Reading, London Road
Reading RG1 5AQ, United Kingdom

Passos, Angela Suzanna A. Dauche De Castro
WAGGGS
Olave Centre, 12c Lyndhurst Road
London NW3 5PQ, United Kingdom

Pearce, Greg
ILA
Charles Clore House, 17 Russel Square
London WC1B 5DR, United Kingdom

Pellew, Robin
WCMC
219 Huntingdon Road
Cambridge CB3 ODL, United Kingdom

Pineiro East, May
F_GF
18 Well Walk
London NW3 ILD, United Kingdom

Posada Salazar, Alvaro
WSPA
Park Place, 10 Lawn Lane
London SW8 1UD, United Kingdom

Posey, Edward
F_GF
18 Well Walk
London NW3 ILD, United Kingdom

Raines, Peter, Mr.
CORAL CAY CONSERVATION
Sutton Business Centre, Restmor Way
Wallington-Surrey, United Kingdom
T: 44 81 669-0011 F: 44 81 773-9656

Read, Anthony
OIIEPF
25-28 Old Burlington St.
London W1X 1L, United Kingdom

Rustad, Alf
IWSA
1 Queen Anne's Gate
London SW1H 9BT, United Kingdom

Sagel, Lois
SOI
87 Gilisson Road
Cambridge CB1 2HG, United Kingdom

Silva, Laura Muri
WAGGGS
Olave Centre, 12c Lyndhurst Road
London NW3 5PQ, United Kingdom

Singhvie, L.M.
CHEC
57/58 Stanhope Gardens
London SW7 5RF, United Kingdom

Stephens, Rachel
COUNCIL OF CHURCHES FOR BRITAIN AND IRELAND
Iner-Church House
35-41 Lower Marsh, London SE1 7RL, United Kingdom

Stirton, Charles Howard
Professor
ROYAL BOTANIC GARDENS, KEW
Richmond, Surrey, TW9 3AB
United Kingdom

Sutherland, Will
NEW ECONOMICS FOUNDATION
Universal House, 2nd Floor, 88-94 Wentworth Street
London E1 7SA, United Kingdom

Sutton, Alison, Researcher
ANTI-SLAVERY INTERNATIONAL
180 Brixton Road
London SW9 6AT, United Kingdom

Tavares Ramos, Maria Lucia
WAGGGS
Olave Centre, 12c Lyndhurst Road
London NW3 5PQ, United Kingdom

Taylor, Clifford James, Mr.
Campaign Officer
PEOPLE AGAINST THE RIVER CROSSING (PARC)
72 Gerdia Road, New Eltham
London SE9335J, United Kingdom
T: 44 81 850-7037 F: 44 71 492-2910

Thomson, Koy Charles
IIED
3 Endsleigh St.
London WC1H 0DD, United Kingdom

Toni, Ana
ACTION AID
Tapstone Road, Chard
Somerset TA20 2AB, United Kingdom

Unnithan, T.K. Narayanan
CHEC
57/58 Stanhope Gardens
London SW7 5RF, United Kingdom

Von Der Weid, Bernadette
WAGGGS
Olave Centre, 12c Lyndhurst Road
London NW3 5PQ, United Kingdom

Watson, Fiona
SURVIVAL INTERNATIONAL
310 Edgware Road
London W2 1DY, United Kingdom

Wiedemann, Rudolf
ICEA
Lyng Hall, Blackberry Lane
Coventry CV2 3JS, United Kingdom

Williams, Nicole
WCI
Vicarage House, 58/60 Kensington Church Street
London W8 4DB, United Kingdom

Winberg, Margarethe
SIW
Maritime House, Old Town, Clapham
London SW4 0JW, United Kingdom

Wood, Howard
FINDHORN FOUNDATION
Cluny Hill College, FORRES IV 36 ORD
Scotland, United Kingdom

Worcester, Tracy
F_GF
18 Well Walk
London NW3 ILD, United Kingdom

PRESS

Adams, Tani
Correspondent
AGENCIA LATINOAMERICANA
59 Marne Ave. SF
CA 94127 Usa, United Kingdom

Akhurst, James
Editor
ITN-INDEPENDENT TELEVISION NEWS
15 Copperfields, Kemsing
Sevenoaks, Kent TN15 6QG, United Kingdom, United Kingdom

Alagiah, George
Correspondent
BBC TELEVISION NEWS
11 Norcott Road
London N16 7BJ, UK, United Kingdom

Allen-Mills, Anthony B.
Correspondent
SUNDAY TIMES
Uhlandstr 162
100 Berlin 15, Germany, United Kingdom

Anstiss, Paul
Correspondent
BRITISH SATELLITE NEWS
Flat 2, 32 Fontenoy Road
London SW12 9LU, United Kingdom, United Kingdom

Arden, John
Editor
BBC TELEVISION NEWS
Via Gallia 64-15
00183 Rome, Italy, United Kingdom

Arden-Clarke, Charles
Correspondent
GUARDIAN
Ch de Fin D'elez 1261
Muids, Switzerland, United Kingdom

Aref, Mohammed
Science And Technology Editor
AL-HAYAT
22 Newborough Green
New Malden, Surrey KT3 5H, UK, United Kingdom

Bartle, Alison
Editor
INTL.WATER POWER&DAM CONSTRUCTION
17, Crystal Palace Rd.
London SE22 9EX, United Kingdom, United Kingdom

Bellinger, Christopher
Editor
BBC TELEVISION NEWS
24 Summerfield Road
Ealing, London W5 1ND, United Kingdom, United Kingdom

Berke, Ronni
Editor
WTN
254-24 Walden Ave.
Great Neck, NY 11020, USA, United Kingdom

Bernstein, Anita
Correspondent
THE BRAZILIAN GAZETTE
Av. Niemeyer 805/704
Rio de Janeiro, Brazil, United Kingdom

Bestia, Richard
Correspondent
SKY NEWS
94 Vanburgh Court, Wincott St.
London SE11 4NR, United Kingdom, United Kingdom

Boulton, Thomas A.B.
Correspondent
SKY NEWS
1 Dunollie Road
London NW5 2XN, United Kingdom, United Kingdom

Breach, Ian
Correspondent
BBC
9 Park Place East
Sunderland, SR2 8EE, UK, United Kingdom

Briston, Anne, Ms.
Reporter Radio/Television
BBC-WORLD SERVICE
69, Elsinore Road
Forest Hill, London SE23 25H, UK, United Kingdom

Brown, Amanda
Correspondent
PRESS ASSOCIATION
177 Crookston Road, Eltham
London SE9 1YG, United Kingdom, United Kingdom

Brown, Julia
Correspondent
GREEN MAGAZINE
16 Ennerdale House
Newnton Close, London N4 2RP, UK, United Kingdom

Brown, Paul
Correspondent
GUARDIAN
25 Rothschild Road
Leighton Buzzard, Beds LU7 7SY, UK, United Kingdom

Buie, Elizabeth A.
News Reporter
THE GLASGOW HERALD
195 Albion Street
Glasgow, United Kingdom
T: 44 41 5526255 F: 44 41 5522288

Byrne, Louise
Correspondent
OBSERVER
Chelsea Bridge House, Queenstown Rd.
London, United Kingdom
T: 44 71 3503544 F: 44 71 6275570

Cabral, Tim
Editor
BBC-WORLD SERVICE
27 Hope Park
Bromley Kent BR13RG, UK, United Kingdom

Cahen, Eva
Correspondent
WTN/TCTV
24 Rue Ledic
75018 Paris, France, United Kingdom

Cairncross, Frances
Correspondent
ECONOMIST
6 Canonbury Lane
London, N1, UK, United Kingdom

Campbell, Alastair
Correspondent
DAILY MIRROR
31 Oak Village
London NW5 4QN, United Kingdom, United Kingdom

Cannon, Howard
Correspondent
SUPER CHANNEL
14 Canark Square, Limeharbor
London, United Kingdom
T: 44 71 499418 F: 44 71 4189419

Carritt, Hugh Anthony B.
Correspondent
REUTERS
Kloosterlaan 17
1970 Wezembeek, United Kingdom, United Kingdom

Caspari, Andrew
Reporter Radio/Television
BBC-RADIO 4 TODAY PROGRAMME
17F John Spencer Square
Islington, London, N1, UK, United Kingdom

Cayon, Margarita Maria
Editor
VISNEWS
Rua Barao de Guaratiba 218/304
22211 Rio de Janeiro-RJ, Brazil, United Kingdom

Cecil, Jesica Mary
Reporter Radio/Television
BBC
10 Dorlcote Road, London
United Kingdom, United Kingdom

Chapoval, Tova
Correspondent
REUTERS
R. Pres. Alfonson Lopes 20/807
22071 Rio de Janeiro-RJ, Brazil, United Kingdom

Clover, Charles R.H.
Correspondent
DAILY TELEGRAPH
15 Wilkes st.
London E1 6QF, UK, United Kingdom

Comenda, Sidnei
BBC-WORLD SERVICE
R.Manoel de Soveral 77 Apt.84B
02040 Sao Paulo-SP, Brazil, United Kingdom

Connelly, Frances Anne, Ms.
Editor
FIELD-FOUNDATION FOR INT. ENV. LAW AND DEVELOPMENT
King's College London, Manresa Road
London SW3 6LX, United Kingdom
T: 44 71 352-8123 F: 44 71 351-6435

Couri, George
Reporter Radio/Television
TELEVISION TRUST FOR THE ENVIRONMENT
18 Eurmalus St. Mosman
2088 Australia, United Kingdom

Cowell, John Adrian
Reporter Radio/Television
I TV
Al Anson Rd.,
London N1 OAB, UK, United Kingdom

Cox, Moira
Correspondent
BBC
77 Islip Rd.
Oxford OX2 75P, England, United Kingdom

Cramb, Auslan, Mr.
Environment Corr.
SCOTSMAN
20 North Bridge
Edimburg EH1 1YT, United Kingdom
T: 44 31 243-3366 F: 44 31 226-7420

Crossland, John William
Editor
WTN/TCTV
38 Princes Ave., Petts Wood
Orsington, Kent BP5 1Q5, United Kingdom, United Kingdom

Crosson, Judith H.
Correspondent
REUTERS
3329 E Bayard Ave. (#1108)
Denver, Colorado 80209, USA, United Kingdom

Dafforn, Katherine
Reporter Radio/Television
BBC
1 Central Avenue, Hanham
Bristol, United Kingdom

Darch, Colin
Correspondent
SOCIALIST
Pompeu Loreiro 114/3
Rio de Janeiro, Brazil, United Kingdom

Dawson, Judith Helen
Correspondent
SKY NEWS
21 Boltons Court, 216 B. Road
London SW5 0BZ, United Kingdom, United Kingdom

Del Carpio, Luis
Correspondent
VISNEWS
4067 Wimbledon Dr.
Cooper City, Florida, USA, United Kingdom

Depledge, Joanna
Reporter Radio/Television
GOING LIVE,BBC
32 Wheelys Rd.Edgbaston
Birmingham,U.K., United Kingdom

Dickson, Adrian David
Correspondent
REUTERS
Haddock Lobo 1435 (#92)
Sao Paulo-SP, Brazil, United Kingdom

Dilks, Robin
Editor
BBC-WORLD SERVICE
17 The Ave. Kew Richmond
Surrey 7W9 2AL, United Kingdom, United Kingdom

Dimbleby, Jonathan
Correspondent
BBC
1 Horbury Crescent
London W1l 3NF, United Kingdom

Dowd, Mark
Correspondent
BBC
51 Cranhurst Rd.
London , U.K., United Kingdom

Dowdney, Mark
Editor
DAILY MIRROR
29 Alder Road, Mortlake
London SW14 8ER, United Kingdom, United Kingdom

Downing, Malcolm
Editor
BBC-RADIO NEWS&CURRENT AFFAIRS
18 Somerfield Road
London N42JJ, UK, United Kingdom

Duggan, Clare
Reporter Radio/Television
ITN
87 St. Agustines Rd.
London NW1, UK, United Kingdom

Durra, Yasar
Editor
WTN/TCTV
3 Holland Park Court
London W 14 8DN, United Kingdom, United Kingdom

Emert, Harold
Correspondent
DAILY EXPRESS
Miguel Lemos 51/604
Copacabana, Rio de Janeiro Brazil, United Kingdom

Ennals, David
Correspondent
NEW STATESMAN
16 Ingnam Road
London N2 9QA United Kingdom, United Kingdom

Fabra, Adriana Aguilar, Ms.
Research Fellow
FIELD-FOUNDATION FOR INT. ENV. LAW AND DEVELOPMENT
King's College London, Manresa Road
London SW3 6LX, United Kingdom
T: 44 71 352-8123 F: 44 71 351-6435

Fidler, Amanda
Correspondent
BBC
161 St.Albans Ave.
London, U.K., United Kingdom

Ford, Simon
Correspondent
BBC
9A Elsynge Rd.
London, U.K., United Kingdom

Frampton, Robin
Editor
BUTTERWORTH-HEINEMANN
40 Rockhampton Road
Kingston, Surrey-KT2 5, UK, United Kingdom

Frier, Angela
Producer
ITN-INDEPENDENT TELEVISION NEWS
200 Grams
INN London WCl, United Kingdom
T: 44 71 8333000 F: 44 71 4304019

Gaillard, Claudia
Editor
VISNEWS
Ave. Eduardo Modero 942
Buenos Aires, Argentina, United Kingdom

Gaillard, Marina
Correspondent
VISNEWS
328 North Miami Ave.
Miami, FL 33128, USA, United Kingdom

Gamini, Gabriella
Correspondent
TIMES
Amenabar 2322, &C
Buenos Aires, Argentina, United Kingdom

Gaunt, Jeremy
Correspondent
REUTERS
662 Grove St.
Upper Montclair, NJ 07043, USA, United Kingdom

Ghazi, Polly
Correspondent
OBSERVER
Flat 3, 55 Highbury Park
London N5 1TJ, United Kingdom, United Kingdom

Ghaznavi, Dawood
Correspondent
THE GUARDIAN
144-D Defense Society
Lahhore Pakistan, United Kingdom

Gilz, Sergio
WTN
4 Sarre Road
London NW2 3SL, UK, United Kingdom

Glover, Margaret
Correspondent
FRIEND
3 Manor Gardens, Larkhall Rise
London SW4 6JZ, UK, United Kingdom

Gluchowski, Krzysztof
Correspondent
POLISH DAILY
R. P. Coutinho Frois 81 C01
Barra de Tijuca, Rio de Janeiro, Brazil, United Kingdom

Godlee, Fiona
Correspondent
BRITISH MEDICAL JOURNAL
35 a Addington Sq.
SES 7LB, London, United Kingdom

Gonzales, Elizabeth
Reporter Radio/Television
VISNEWS
Nunes de Balboa 550
Lima 18, Peru, United Kingdom

Greig, Gordon
Correspondent
DAILY MAIL
Flat 303, Howard House
Dolphin Square, London SW1V 3PF, UK, United Kingdom

Gribben, Claire
Reporter Radio/Television
INDEPENDENT RADIO NEWS
7 Abbey Gardens
London NW 8 9AS, UK, United Kingdom

Grice, Andrew
Correspondent
SUNDAY TIMES
40 Crothall Close
London N13 4BN, United Kingdom, United Kingdom

Grout-Smith, Tim
Reporter Radio/Television
BBC-WORLD SERVICE
Church Path Lane End, High
Wycombe, Bucks, UK HP143HD, United Kingdom

Gunn, Sueli
Reporter Radio/Television
BBC-WORLD SERVICE-BRAZILIAN SECTOR
R. Gen. Barbosa Lima,57 (#303)
22011 Rio de Janeiro, Brazil, United Kingdom

Gunness, Christopher
Reporter Radio/Television
BBC
245 E 44th St.
New York, NY 10017, USA, United Kingdom

Haar, Monica
Correspondent
SUPER CHANNEL
16 Abbey Lodge Park Rd.
Regents Park, London NW8 7RJ, United Kingdom

Hampson, Judith
Correspondent
SOUTH MAGAZINE
77 Parliament Mill
London NW3 2TM, United Kingdom, United Kingdom

Hannah, Karen
Reporter Radio/Television
BBC
10 Ryebank Mews
Manchesteer M21 1LY, England, United Kingdom

Harold, Michael
Correspondent
SUPER CHANNEL
Primrose Cottage
Sweeney, Mnt. United Kingdom, United Kingdom

Harrabin, Roger
Reporter Radio/Television
BBC RADIO NEWS
46 Spencer Rise
London NW5 1AP, UK, United Kingdom

Harris, Amanda
Editor
BBC
5, Nasmyth Street
London W6 OHA, UK, United Kingdom

Hedges, Stephen
Reporter Radio/Television
BBC-WORLD SERVICE&GLOBAL CONCERS
25 Heath Drive
London SW20 9BE, UK, United Kingdom

Hendrick, Richard
Editor
SUPER CHANNEL
Box 105, Orford
New Hampshire, NH 03777, USA, United Kingdom

Hickey, Frederick
ITN
49 Sherwood Way, West Wickham
Kent BR4 9PB, United Kingdom, United Kingdom

Hinds, Anne
Editor
BBC
The Cottage, 13 Girdlers Road
London W14, UK, United Kingdom

Hinnigan, Deborah
Editor
BBC
46 Prince's Court, Shoot Up Hl
London NW2, United Kingdom, United Kingdom

Holland, John
Editor
SUPER CHANNEL
13 Wolfington Rd.
London SE 27 OJF, UK, United Kingdom

Homewood, Brian
Correspondent
NEW SCIENTIST
Av.Copacabana 1250/505
Rio de Janeiro, Brazil, United Kingdom

Hrris, Jeremy
Correspondent
BBC
5412 Surrey St.
Chevy Chase, Maryland, U.S.A., United Kingdom

Huber, Phylis
Editor
BBC
Av. Niemeyer 174/302
Rio de Janeiro, Brazil, United Kingdom

Hughes, David
Correspondent
SUNDAY TIMES
9 Southfield Gardens
Twickenmam, United Kingdom, United Kingdom

Hughes, Mark
Editor
ITN-CHANNEL 4 DAILY
Flat 5, 4A Downshire Hill
London NW3 1NR, United Kingdom, United Kingdom

Hunt, John
Correspondent
FINACIAL TIMES
Chesil, Westwood Ave., Woodham
Weybridge, Surrey, UK, United Kingdom

Hurst, Peter
Correspondent
MORNING STAR
20A Bishopsthorpe Rd.
London, SE26 4NY, UK, United Kingdom

Hurtado, Maria Elena
Correspondent
LATIN AMERICAN NEWSLETTERS
95 Devonshire Road
London, UK, United Kingdom

Ivens, Katherine
Editor
STARBIRD SATELLITE SERVICES
49 Maresfield SDNS
London NW3 5TE, United Kingdom, United Kingdom

Jay, Xanthe
Correspondent
RADIO NORTH SOUND
Braehead House,79,Balgownie Rd
Aberdeen, Scotland, UK, United Kingdom

Jones, David
Editor
TODAY NEWSPAPER
5 Lynwood Drive
Worcester Park, Surrey, United Kingdom, United Kingdom

Jones, George
Correspondent
DAILY TELEGRAPH
57, Gowrie Road
London, SW11 5NN, United Kingdom, United Kingdom

Jones, Hywel
Correspondent
ITN-CHANNEL 4 DAILY
Flat 6, 50 Compayne Gardens
London NW6 3RY, United Kingdom, United Kingdom

Kao, Thomas Kuang Po
Correspondent
WTN/TCTV
731 S. Michillinda Ave.
Pasadena, CA 91107, USA, United Kingdom

Kavanagh, Trevor
Correspondent
SUN
15 Denham Road
Epsom, Surrey, KT13AA United Kingdom, United Kingdom

Keane, Sean
Editor
STARBIRD SATELLITE SERVICES
'Wakanui' Mt. Avenue, Hutton
Essex, CM13 2NZ, United Kingdom, United Kingdom

Kelleher, Kristine
Editor
ITN
2501 Calvert Street, NW #605
Washington, DC 20008, USA, United Kingdom

Kinch, Diana Constance Sylvia
Editor
METAL BULLETIN PLC
R. Maestro Francisco Braga,230
22041, Rio de Janeiro-RJ, Brazil, United Kingdom

King, Katherine
Correspondent
REUTERS
R. Caconde 125 (#91)
01425 Sao Paulo-SP, Brazil, United Kingdom

Kirby, Alex
Reporter Radio/Television
BBC
Ashness Road
London, SW11, UK, United Kingdom

Knights, Melanie
Assistant Producer
BBC TELEVISION
Old Rectory Cottage,
Bletsoe, Bedford MIC, United Kingdom
T: 44 234 782230

Lamb, Christina
Correspondent
FINANCIAL TIMES
R. Sao Sebastiao 165 Apt.201-S
Rio de Janeiro-RJ, Brazil, United Kingdom

Lamb, Robert
Correspondent
WORLD MAGAZINE & SUNDAY TIMES
13 AE Wortfort Koad
Lewes, E Sussex, Bn7 ISP UK, United Kingdom

Lascelles, David
Correspondent
FINANCIAL TIMES
60 Talbot Road
London N6 4RA, UK, United Kingdom

Lean, Geoffrey
Correspondent
OBSERVER
25 Woodhurst Lane
Exted, Surrey RH8 9HN, United Kingdom, United Kingdom

Lewington, Charles
Correspondent
SUNDAY EXPRESS
Basement Flat, 26 R. Crescent
London W11 48N, England, UK, United Kingdom

Leyland, Jean
Assistant Reporter
IPS
91 Embassy Court, Kings Road
Brighton BN1 2PY, United Kingdom
T: 44 273 779391

Lithgow, Adrian
Correspondent
MAIL ON SUNDAY
7, Burnaby Street
London SW10 OPR, United Kingdom, United Kingdom

Lord, Arthur
Producer
ITN-INDEPENDENT TELEVISION NEWS
200 Grams
INN London WCI, United Kingdom
T: 004471 8333000 F: 004471 43040

Lutley, Martin Neil
Editor
SKY NEWS
132 Brixton Hill Crt
Brixton, London SW2 1QZ, United Kingdom, United Kingdom

Lye, Geoff, Mr.
Correspondent
STAR NEWSPAPERS
Dolphin
Wigginton, Oxona, United Kingdom, United Kingdom

Macaskill, Ian
Correspondent
NEWS OF THE WORLD
25 Sunny Bank
Epsom, Surrey United Kingdom KT81 704, United Kingdom

Macintire, Donald
Correspondent
INDEPENDENT ON SUNDAY
Top Flat 33, Leinster Square
London W2, England UK, United Kingdom

Maddox, Bronwen
Correspondent
FINANCIAL TIMES
39 Shepherds Bush Road
London W6 7LU, UK, United Kingdom

Markham, Daniel
WTN
3636 Fieldston Road (#2B)
Riverdale, NY 10463, USA, United Kingdom

Marti, Miguel
Reporter Radio/Television
SUPER CHANNEL
1048 La Buaria
Moravia, Costa Rica, United Kingdom

Martins, Alejandra
Reporter Radio/Television
BBC
20 Wilton Avenue
Chiswick, London W4, UK, United Kingdom

Mathew, Don
Correspondent
SURVEYOR MAGAZINE
38 Reeve Street
Lowestoft, Suffolk, NR32 1UP, UK, United Kingdom

Matthews, Eamonn
Reporter Radio/Television
BBC
26 Norroy Road
London SW15, United Kingdom

McCarthy, Michael
Correspondent
TIMES
26 Wolseley Road, Hornsey
London N8 8RP, United Kingdom, United Kingdom

McDonald, Trevor
Reporter Radio/Television
ITN
18 Sheen Gate Gardens
London SW14, United Kingdom, United Kingdom

McGinty, Lawrence
Correspondent
ITN-INDEPENDENT TELEVISION NEWS
200 Grams
INN London WCI, United Kingdom
T: 44 71 8333000 F: 44 71 4304019

Mesquita, Cristiana
Reporter Radio/Television
WTN
R. Nina Rodrigues 93 Apt.101
22461 Rio de Janeiro-RJ, Brazil, United Kingdom

Miller, Richard
Editor
PROJECT EARTH
32 Andridge Rd. Villas
London, UK, United Kingdom

Milne, Roger Stuart
Correspondent
LBC NEWSTALK RADIO
13 Bunces Close, Eton Wick
Bergshire, United Kingdom, United Kingdom

Mishori, Ranit
Editor
WTN
57 W 10th St.
New York, NY 10011, USA, United Kingdom

Morgan, James
Reporter Radio/Television
BBC-WORLD SERVICE
20 Rectory Road
London, SW13 ODT, UK, United Kingdom

Murphy, Jonathan, Mr.
Reporter Radio/Television
BBC
102 Cator Lane, Chilwell
Notts, NG94BB, United Kingdom, United Kingdom

Mylrea, Paul
Correspondent
REUTERS
Onofre Jarpa 9788, Casa B
La Reina, Santiago, Chile, United Kingdom

Nadkarni, Maurice
Reporter Radio/Television
BBC
R. Tavares Bastos 414/66
Catete, 22221 Rio de Janeiro, Brazil, United Kingdom

Neale, Gregory
Correspondent
SUNDAY TELEGRAPH
204 Westbourne Park Road
London W11 1EP, England, United Kingdom, United Kingdom

Neely, Bill
Correspondent
ITN-INDEPENDENT TELEVISION NEWS
200 Grams
INN London WCI, United Kingdom
T: 44 71 8333000 F: 44 71 4304019

Nelson, Daniel
Editor
PS MAGAZINE/PANOS FEATURES
136 Southgate Road
London N.1, United Kingdom, United Kingdom

Norris, Alexander
Correspondent
GEMINI NEWS SERVICE
9 White Lion St.
London, United Kingdom
T: 44 71 2277584 (Rio) F: 44 71 8375118 Lond

Oakley, Robert
Correspondent
TIMES
Quarry House 14, Lynwood Road
Epson Surrey K717 4LD, United Kingdom, United Kingdom

Ohara, Naoki
SUPER CHANNEL
2-55-18 Toyoda,Hino-City
Tokyo - Japan, United Kingdom

Oishi, Yoko
WTN
2-55-4 Kamitakada, Nakano-ku
Tokyo 164, Japan, United Kingdom

Olatunji, Lardner
Correspondent
WEST AFRICA MAGAZINE
572 Marlborough Road
Brooklyn, NY 11226, USA, United Kingdom

Olt, Betty
Correspondent
WORLD WIDE FUND FOR NATURE
Sous les Vignes A 25
Ch-1196 Gland, Switzerland, United Kingdom

Oox, Moira
Editor
BBC-WORLD SERVICE
77 Islip Road
Oxford OX2 7SP, United Kingdom, United Kingdom

Ortega, Luz
Correspondent
WTN Corp.
2303 Turnbull Ave.
Bronx, New York, U.S.A., United Kingdom

Ouseley, Rachel
Editor
GREEN ENERGY MATTERS
7 Stag Lane, Buckhurst Hill
Essex IG9 5TD, United Kingdom, United Kingdom

Parsons-O'Mahony, Katherine
Journalist/Coordinator
FUJI TELEVISION
1st Floor, 15-16 New Burlington St.
London WIX IFF, United Kingdom
T: 44 71 4942591 F: 44 71 4396206

Patten, Emer
News Producer
SUPER CHANNEL
Melrose House, 14 Lanark Square
London, United Kingdom
T: 44 71 4189418 F: 44 71 4189421

Pearce, Frederick
Editor
NEW SCIENTIST MAGAZINE
42 Bramford Road
London SW18 1AP, UK, United Kingdom

Pedraza, Patricia
Reporter Radio/Television
WTN
56 E 87th St. (#5B)
New York, NY 10128, USA, United Kingdom

Pickard, Christopher
Correspondent
MOVING PICTURES INT'L
Estrada da Gavea 611/Bl.2/304
Sao Conrado, Rio de Janeiro-RJ, Brazil, United Kingdom

Platt, Nicholas Jr.
Correspondent
REUTERS
67 Willow St.
Brooklyn, NY 11201, USA, United Kingdom

Polaine, Guy
Editor
ENVIRONMENT STRATEGY EUROPE
11 Monson Road
London NW10 5UR, United Kingdom, United Kingdom

Porritt, Jonathon
Correspondent
BBC WILDLIFE MAGAZINE
9, Lypiatt Terrace,Lypiatt Rd.
Cheltenham, Glos., GL50 25X, UK, United Kingdom

Powell, George
Reporter Radio/Television
BBC-RADIO NEWS-OUTLOOK
37 Sutherland Square
London SE17 3EE, UK, United Kingdom

Pratt, Kerrie-Anne
Correspondent
SUPER CHANNEL
63 Ridge Rd.
Crouch End, London N8, UK, United Kingdom

Pritchard, Thomas Owen
Reporter Radio/Television
BBC
124 Penrhos Rd.
Bangor, Wales, United Kingdom, United Kingdom

Qassim, Ali
Correspondent
NEWSPAPER PUBLISHING PLC
Las Carabelas 382 - 47
Santiago de Chile, United Kingdom

Ramalho, Gabriel P.
Correspondent
SUPER CHANNEL
68A Caulfield Road
East Ham, London E6 2EN, UK, United Kingdom

Rawet, Ilan Alberto
Reporter Radio/Television
VISNEWS
R.Jose Linhares 47/202
Rio de Janeiro, Brazil, United Kingdom

Rddriguez, Carlos Enrique
Correspondent
VISNEWS
Via 7, no. 82-62 apto. 401
Bogota, Colombia, United Kingdom

Redhead, Brian
Reporter Radio/Television
BBC
71 Thomas More House
Barbican, London EC2Y 8AB, UK, United Kingdom

Reiss, Charles
Correspondent
EVENING STANDARD
Belsize Lane
London NW35AN, England, United Kingdom

Revill, Evejoanne
Correspondent
THE MAIL ON SUNDAY
1, Colville Terrace
London W11 2BE, UK, United Kingdom

Reynolds, Paul
Correspondent
BBC
6 Prince Arthur Road
London NW 3 6AV, UK, United Kingdom

Richards, Jenny
Editor
TV TRUST FOR THE ENVIRONMENT
81 Charlotte St.
London W1P, United Kingdom, United Kingdom

Rin, Jonathan Da
Editor
KAPOK GROUP
20 Redburn St.
London SW 3, United Kingdom, United Kingdom

Robinson, John
Editor
STARBIRD SATELLITE SERVICES
86 Nevill Rd, Stoce Newington
London N16 0SX, United Kingdom, United Kingdom

Robiou, Alan
Editor
BBC
137 Princes Gardens
London W3 0LS, UK, United Kingdom

Rocha, Jan
Reporter Radio/Television
BBC-WORLD SERVICE
R. Prof. Joao Arruda 209
05012 Sao Paulo-SP, Brazil, United Kingdom

Rowlands, Ian, Mr.
Lecturer In Env. And Dev.
LONDON SCHOOL OF ECONOMICS AND POLITICAL SCIENCE
Houghton St.
London Wc2AE, United Kingdom
T: 44 71 405-7686 F: 44 71 242-0392
TL: 24655 LSELEON G

Rowley, John
Editor
EARTHWATCH
60 Twisden Road
London NW5 1DN, England UK, United Kingdom

Rush, Julian
News Producer
BBC TELEVISION NEWS
TV Centre Wood Lane
London WIR 7RJ, United Kingdom
T: 44 81 7438000 F: 44 81 5767957

Ryan, Sean
Correspondent
SUNDAY TIMES
Winhaven, Churchway,Westilglet
Newbury, Dorkshire, RG16 OAH, UK, United Kingdom

Sadler, Richard
Correspondent
YORKSHIRE POST
P.O. Box 168
Willington, Street, Leeds LS1 1RF, UK, United Kingdom

Salama, Fawzia
Editor
AL SHARQ AL AWSAT
59 Ullswater Crescent
London SW15 3RG, England UK, United Kingdom

Salles, Flavio
Correspondent
EUROPEAN CHEMICAL NEWS
Av.Ataufo de Paiva,135 S.808
22440 Rio de Janeiro-RJ, Brazil, United Kingdom

Salt, Julian
Correspondent
ENVIRONMENT NEWS
Flat 8, 2 Cecil Ave.
Bradford, B07 3BP, United Kingdom, United Kingdom

Samphier, Tony
Editor
BBC
7 Baumont Crt. Streatham Place
SW2 4PY, United Kingdom, United Kingdom

Schoon, Nicholas
Correspondent
INDEPENDENT
35 Annandale Road
London SE10 ODD, United Kingdom, United Kingdom

Scott, Jennifer
Editor
BUTTERWORTH-HEINEMANN
25 Hart St
Oxford, OX 6BN, UK, United Kingdom

Sharifi, Tara
Correspondent
ACCOUNTANCY AGE MAGAZINE
9 Lindfiled GDNS
London NW3 6PX, England, United Kingdom

Sharpe-Newton, Geraldine
Correspondent
WORLD WIDE FUND FOR NATURE
29 Albert Mansions
Albert Bridge Road, London SW11, UK, United Kingdom

Sheather, Michelle
Correspondent
SUPER CHANNEL
Tichelsstraat 5
Amsterdam, The Netherlands, United Kingdom

Siddiqi, Jameela Asma
Correspondent
WTN/TCTV
8 South Rise,St.George's Field
London W2 2YD, United Kingdom, United Kingdom

Simms, Andrew
Correspondent
NEW INTERNATIONALIST
126 Magdalen Road
Oxford, United Kingdom, United Kingdom

Simpson, John
Correspondent
BBC
96 Cornwall Gardens
London, SW 7, UK, United Kingdom

Smith, Alistair P.
Editor
FOOD MATTERS WORLDWIDE
38 Exchange St
Norwich, United Kingdom
T: 44 603 761645 F: 44 603 765434

Sorrentino, Bruno Leopoldo
Correspondent
TELEVISION TRUST FOR ENVIRONMENT
96 Brondwsbury RD
London NW6 6RX, United Kingdom, United Kingdom

Southworth, June
Correspondent
DAILY MAIL
33 Epsom Road
Guildford Sussels, GU13LA, UK, United Kingdom

Steen, Nicola
Correspondent
ROYAL INSTITUTE
38 Margate Rd.
London SW2 5DT, United Kingdom, United Kingdom

Stewart, Alastair
Presenter
ITN-INDEPENDENT TELEVISION NEWS
200 Grams
INN London WCI, United Kingdom
T: 44 71 8333000 F: 44 71 4304019

Stewart, Kirsty
Reporter Radio/Television
SKY NEWS
35 Sussex Gardens, Highgate
London N6 4LY, United Kingdom, United Kingdom

Stewart, Sarah
Correspondent
CHRISTIAN AID
Herald House, Lambs passage
Bunhill Row, London EC1Y 8TQ, UK, United Kingdom

Stewart, William
Correspondent
DAILY EXPRESS
3 Harold Road Hornsey
London N8 7DE, United Kingdom, United Kingdom

Storey, David
Correspondent
REUTERS
124 Iffley Road
London W6 OPE, United Kingdom, United Kingdom

Stott, Michael James
Correspondent
REUTERS
Av. Joao Luis Alves 268
Rio de Janeiro-RJ, Brazil, United Kingdom

Stourton, Edward
Reporter Radio/Television
ITN
73 Hendham Road
London SW17 7DH, United Kingdom, United Kingdom

Sullivan, Robert E.
WTN
1 Baptist Church Road
Yorktown Heights, NY 10598, USA, United Kingdom

Tapper, Richard
Correspondent
ECOS MAGAZINE
26 Perryn Road, East Acton
London, UK, United Kingdom

Thomas, Jerry
Editor
SUPER CHANNEL
Melrose House 14 Lamark Syvare
Limeharbour, London E14900, UK, United Kingdom

Throssell, Elizabeth
Correspondent
BBC-WORLD SERVICE
R. Itambe 316, Apto. 101
Sao Paulo-SP, Brazil, United Kingdom

Tidmarsh, John A.
Correspondent
BBC-WORLD SERVICE
23 Gloucester St.
London, SW1V 2DB, UK, United Kingdom

Tinson, Susan
Editor
ITN
56 Portobello Road
London W11 3DL, United Kingdom, United Kingdom

Tisoall, Simon
Correspondent
THE GUARDIAN
4801 Bending Lane
Washington, D.C.,U.S.A., United Kingdom

Titchen, Richard
Editor
SUPER CHANNEL
8 Malden Road
London NW5, UK, United Kingdom

Toksvig, Nicholas
Correspondent
WTN/TCTV
19, The Street, Capel
Surrey, RH5 5LD, United Kingdom, United Kingdom

Townley, Alistair, Mr.
GREEN MAGAZINE
Northern & Shell Building, PO Box 381, Mill Harbour
London G14 9TW, United Kingdom
T: 44 71 9855090 F: 44 71 5383690

Turner, James N.
Editor
WTN
229 1/2 E 81th St.
New York, NY 10028, USA, United Kingdom

Turner, Rikki
Correspondent
INDEPENDENT
Rua Simao Alvares 923
05417-Sao Paulo-SP, Brazil, United Kingdom

Veitch, Andrew
Correspondent
ITN
Old Police Hse., Bull Lane,
Eccles, Aylesford, Kent ME 20 7HT, UK, United Kingdom

Vergara, Marc
Reporter Radio/Television
BBC-WORLD SERVICE-FRENCH SERVICE
15 Consort Road
London SE15 2PH United Kingdom, United Kingdom

Vidal, John
Correspondent
GUARDIAN
40 Redcliffe Rd.
London SW10 9NJ, United Kingdom, United Kingdom

Vieira, Beth
Reporter Radio/Television
BBC-WORLD SERVICE
Alameda Casa Branca, 791
01408-Sao Paulo-SP, Brazil, United Kingdom

Ward, Halina, Ms.
Staff Lawyer
FIELD-FOUNDATION FOR INT. ENV. LAW AND DEVELOPMENT
King's College London, Manresa Road
London SW3 6LX, United Kingdom
T: 44 71 352-8123 F: 44 71 351-6435

Wason, John
Correspondent
SUPER CHANNEL
315 1/2, North Agassiz
Flagstaff, AZ,8600L, USA, United Kingdom

Wastell, David
Correspondent
SUNDAY TELEGRAPH
22a Huddleston Road
London N7 OAC, United Kingdom, United Kingdom

Webb, Caroline
Correspondent
INT'L B'CASTING TRUST
21 Kynaston Road
London N16 0EA, United Kingdom, United Kingdom

Weir, Fiona
Correspondent
EARTH MATTERS
30 B Albany Road
London N4 4RL, England, UK, United Kingdom

Weli, Abdulwahab K.
Journalist
ASHARQ ALAWSAT
Saudi Research & Marketing 184 High Holborn
London WCIV7AP, United Kingdom
T: 071 8318181 F: 071 8312310

Wells, Andrew Keith
Correspondent
LIFETIME MAGAZINE
Yr Ocar Coed-Y-Parc
NR. Bethersda, Guynedd, LL57, United Kingdom

Welsh, Cheryl
Correspondent
CHRISTIAN AID
B'Ment Flat, 170 Richmond Rd.
Mackney E8 3MN London, United Kingdom, United Kingdom

Welsh, Paul
Reporter Radio/Television
BBC-WORLD SERVICE
Flat 4, 31 Eaton Rise
Ealing, London W5, United Kingdom, United Kingdom

West, Ruth, Ms.
Reporter Radio/Television
NEXUS TV
The Chalfont, Blomfield Road
London W9 England, United Kingdom

Wheeler, Paul
Correspondent
ASAHI SHIMBUN
Rowan House, Harmers Hill
East Sussex, United Kingdom, United Kingdom

Whittaker, Mark
Reporter Radio/Television
BBC
1 Western Gardens Ealing
London W5 3RS, UK, United Kingdom

Wilenius, Paul
Correspondent
TODAY NEWSPAPER
37 Wesrelot Road
Swindon, Wiltshire UK, United Kingdom

Wilson, Andrew M.
Reporter Radio/Television
SUPER CHANNEL
4 Ashness Road
London SW11 GRY, UK, United Kingdom

Wilson, Dolores
Reporter Radio/Television
WTN
33 East End Ave. (#1C)
New York, NY 10028, USA, United Kingdom

Wilson, Jennifer
Editor
WTN
39 Hargrave Road
London N.19 5SH, UK, United Kingdom

Wolfe-Mcnaught, Karen
Editor
SKY NEWS
5 Cardigan Road, Richmond Hill
Surrey TW1Q 6BJ, United Kingdom, United Kingdom

Woollcombe, David
Correspondent
RESURGENCE
Much Hadham Herts
United Kingdom, United Kingdom

Wright, Denis
Reporter Radio/Television
BBC
R.Gal. Glicerio 85/801
Rio de Janeiro, Brazil, United Kingdom

ORGANIZATIONS

Bourillon, Christopher, Mr.
World Coal Institute
Vicarage House, 58/60 Kensington Church Street
London W8 4DB, United Kingdom
T: (44 71) 937 4600 F: (44 71) 376 0453

Briggs, Julia, Ms.
ActionAid
Tapstone Road, Chard
Somerset TA20 2AB, United Kingdom
T: (44 460) 62972 F: (44 460) 67191

Caldwell, Neil, Mr.
Campaign for the Protection of Rural Wales
Ty Gwyn 31 High Street, Welshpool
Powys SY21 7JP, United Kingdom
T: (44 938) 552 525/556 212 F: (44 938) 552 741

Cameron, James, Mr.
Centre For International Environmental Law (CIEL)
3 Gray's Inn Place, Gray's Inn
London WC1R 5EA, United Kingdom
T: (44 71) 831 8441 F: (44 71) 831 8479

Catherines, Fiona, Ms.
Royal Geographical Society
Kensington Gore
London SW7 2AR, United Kingdom
T: (44 71) 589 5466 F: (44 71) 584 4447

Clegg, Michael, Mr.
Energy, Technology And The Environment (ETE 21)
17/19 Barter Street
London WC1A 2AQ, United Kingdom
T: (44 71) 404 0221 F: (44 21) 405 7429

Davies, W.R., Mr.
Welsh Centre for International Affairs
Temple of Peace, Cathays Park
Cardiff CF1 3AP, United Kingdom
T: (44 222) 228 549/395 664

Davis, Derek, Mr.
World Energy Council*
Smug Oak Lane, Bracket Wood, St. Albans
Herts. AL2 3UE, United Kingdom
T: (44 923) 893 989 F: (44 923) 894 173

Daysh, Zena, Ms.
Commonwealth Human Ecology Council*
57/58 Stanhope Gardens
London SW7 5RF, United Kingdom
T: (44 71) 373 6761 F: (44 710 244 7470

De Lupis, Ingrid, Ms.
Institute for Studies in Environmental Law
1 Hare Court, Temple
London EC4Y 2BE, United Kingdom
T: (44 71) 353 3171/589 041 F: (44 71) 225 2929/583 9127

Dirickx, J., Mr.
International Water Supply Association
1 Queen Anne's Gate
London SW1H 9BT, United Kingdom
T: (44 71) 222 8111 F: (44 71) 222 7243

Forbes, G.I., Dr.
Environmental Health (Scotland) Unit
Common Services Agency, Trinity Park House,
South Trinity Road, Edinburgh EH5 3ST, United Kingdom
T: (44 41) 946 7120 F: (44 41) 945 3892

Francis, Solveig, Ms.
Housewives in Dialogue
PO Box 287
London NW6 5QU, United Kingdom
T: (44 81) 837 7509 F: (44 81) 531 0571

Gomilny, Jake, Mr.
European Local Environmental Information Clearinghouse (INFOLIFE)
Unit 109, 23 New Mount Street
Manchester M4 4DE, United Kingdom
T: (44 61) 953 4094 F: (41 61) 953 4001

Goriup, Paul, Mr.
Institute of Ecology and Environmental Management
36 Kingfisher Court, Hambridge Road
Newbury, Berkshire RG14 5SJ, United Kingdom
T: (44 635) 37 715 F: (44 635) 550 230

Griver, Celia, Ms.
World Association Of Girl Guides And Girl Scouts
Olave Centre, 12c Lyndhurst Road
London NW3 5PQ, United Kingdom
T: (44 71) 794 1181 F: (44 71) 431 3764

Haslegrave, Marianne, Ms.
International Federation Of Business And Professional Women*
Studio 16, Cloisters Business Centre, 8 Battersea
Park Road, London SW8 4BG, United Kingdom
T: (44 71) 738 8323 F: (44 71) 622 8528

Havard, John, Dr.
Commonwealth Medical Association*
c/o BMA House, Tavistock
London WC1H 9JP, United Kingdom
T: (44 71) 383 6266 F: (44 71) 383 6233

Holloway, Jennie, Ms.
Advisory Committee On Protection Of The Sea
57 Duke Street Grosvenor Square
London W1M 5DH, United Kingdom
T: (44 71) 499 0704 F: (44 71) 493 3092

Holmberg, Johan, Mr.
International Institute For Environment & Development
3 Endsleigh St.
London WC1H 0DD, United Kingdom
T: (44 71) 388 2117 F: (44 71) 388 2826
TL: 261681 EASCAN G

Imboden, Christoph, Mr.
International Council For Bird Preservation
32 Cambridge Road, Girton
Cambridge CB3 OPJ, United Kingdom
T: (44 223) 277 318 F: (44 223) 277 200

Lanchbery, John, Mr.
Verification Technology Information Centre
8 John Adam Street
London WC2N 6EZ, United Kingdom
T: (44 71) 925 0867 F: (44 71) 925 0861

Langer, S.S., Mr.
Royal Society of Chemistry, The
Burlington House, Picadilly
London WIV OBN, United Kingdom
T: (44 71) 437 8656 F: (44 71) 734 1227

Leith, Helen, Ms.
Friends Of The Earth (uk)
26-28 Underwood Street
London N1 7JQ, United Kingdom
T: (44 71) 490 1555 F: (44 71) 490 0881

Lemlin, John, Mr.
International Petroleum Industry Environmental Conservation Association
2nd Floor, Monmouth House, 87-93 Westbourne Grove
London W2 4UL, United Kingdom
T: (44 71) 221 2026 F: (44 71) 229 4948

Lipman, Geoffrey H., Mr.
World Travel and Tourism Council
WTTC 50 Victoria Street, Floor 5
London SW1H ONH, United Kingdom
T: (44 71) 222 1955 F: (44 71) 222 4983

Lovejoy, Derek, Prof.
International Federation Of Landscape Architects
8-11 Denbigh Mews
London SW1V 2HQ, United Kingdom
T: (44 71) 828 6392 F: (44 71) 630 6958

McKenzie, J.C., Mr.
World Federation Of Engineering Organisations
1-7 Great George Street
London SW1P 3AA, United Kingdom
T: (44 71) 222 7722 F: (44 71) 222 0812

Medawan, Jean, Ms.
Margaret Pyke Trust
15 Bateman Buildings, Soho Square
London W1V 5TW, United Kingdom
T: (44 71) 734 9351

Munson, Abby, Ms.
Global Security Programme, The
Faculty of Social & Political Sciences, University of Cambridge, Free School Lane, Cambridge CB2 3RQ, United Kingdom
T: (44 223) 334 509 F: (44 223) 335 065

Myers, Dorothy, Ms.
Oxfam
274 Banbury Road
Oxford OX2 7DZ, United Kingdom
T: (44 865) 31 22 79 F: (44 865) 57 612

O'Brien, Fred, Mr.
International Federation of Environmental Health
Chadwick House, 48 Rushworth Street
London SE1 0QT, United Kingdom
T: (44 71) 928 6006

O'Sullivan, Michael, Mr.
World Society for the Protection of Animals
Park Place, 10 Lawn Lane
London SW8 1UD, United Kingdom
T: (44 71) 793 0540 F: (44 71) 793 0208

Paul, Helena, Ms.
Gaia Foundation, The
18 Well Walk
London NW3 ILD, United Kingdom
T: (44 71) 435 5000 F: (44 71) 431 0551

Puri, Sunetra, Ms.
International Planned Parenthood Federation
Regent's College, Inner Circle, Regent's Park
London NW1 4NS, United Kingdom
T: (44 71) 486 0741 F: (44 71) 487 7950
TL: 919573 IPEPEE C

Roberts, Glyn, Mr.
Tools For Self-reliance
Netley Marsh
Southhampton SO4 2GY, United Kingdom
T: (44 703) 869 697 F: (44 703) 868 544

Roddick, Jacqueline, Dr.
University of Glasgow /Institute of Latin American Studies
69 Oakfield Avenue
Glasgow G12 8HQ, United Kingdom
T: (44 41) 339 8855 EXT. 5025 F: (44 41) 330 4808

Rodriguez-Jonas, Maria, Ms.
Socialist International Women
Maritime House, Old Town, Clapham
London SW4 0JW, United Kingdom
T: (44 71) 627 4449 F: (44 71) 720 4448
TL: 261735 SISEC G

Sattar, Danyal, Mr.
New Economics Foundation
Universal House, 2nd Floor, 88-94 Wentworth Street
London E1 7SA, United Kingdom
T: (44 71) 377 5696 F: (44 71) 247 4725

Senior, Philippa S., Ms.
BioIndustry Association
1 Queen Anne's Gate
London SW1H 9BT, United Kingdom
T: (44 71) 222 2809 F: (44 71) 222 8876

Smith, Alex, Mr.
International Association Of Ports And Harbors
c/o British Ports Federation, Victoria House,
Vernon Place, London WC1B 4LL, United Kingdom
T: (44 71) 242 1200 F: (44 71) 405 1069
TL: 295741 BRITPA G

Thornback, Jane
UK TROPICAL FOREST FORUM
c/o ROYAL BOTANIC GARDEN
Kew, Richmond, Surrey TW9 3AB, United Kingdom
T: (081)332-6299 F: (081)332-6294

Verheyen, Wilbert, Mr.
Susila Dharma International
Anugraha, Wick Lane, Engelfield Green
Surrey TW20 OXN, United Kingdom
T: (44 784) 436 450 F: (44 784) 430 596
TL: 928116 ANUGRA G

Zilka, Michael, Mr.
David Shepherd Conservation Foundation, The
PO Box 123
Godalming, Surrey GU8 4JS, United Kingdom
T: (44 48) 632 576 F: (44 48) 632 253

GLOBAL FORUM

Adamou, Adam
Training Officer
INTERNATIONAL CENTRE FOR CONSERVATION EDUCATION
Greenfield House Guiting Power
Cheltenham, Glos. Gl54 5T2, United Kingdom
T: 44 242 850777 F: 44 242 850705

Agudelo, Patricia Peralta, Sra.
Researcher
UNIVERSIDADE DE NOTTINGHAM
Geography Dept. Uni.Park
Nottingham NG7 2RD, United Kingdom
T: 44 602 484848 F: 44 602 421676

Aldaraca, Norma Munguia, Ms.
Research Fellow
FIELD-FOUNDATION FOR INT. ENV. LAW AND DEVELOPMENT
King's College London, Manresa Road
London SW3 6LX, United Kingdom
T: 44 71 352-8123 F: 44 71 351-6435

Asogwa, Cecile, Ms.
WORLD COUNCIL OF CHURCHES
NO. 4 Montagu Terrace
Edimburgo EH3 5QX, United Kingdom

Aspinwall, Rod, Mr.
Chairman
ASPINWALL & CO.
Walford Manor, Shrewsbury
Shrewbury, United Kingdom
T: 939 261144 F: 939 261146

Bal, Eugeny, Mr.
President
COMMITEE FOR SAVING AZOV SEA
Lenin pz 1 341032
Maziupol, United Kingdom
T: 7 33 70 18
TL: 115 156

Ball, Sarah
Producer
GRANADA TELEVISION
34 The Colonnades Albert Dock
Liverpool, United Kingdom
T: 44 51 55 709 6338 F: 44 51 55 709 5259

Barakat, Sultan Zuhdi, Mr.
POSTWAR RECONSTRUCTION AND DEVELOPMENT
Institute of Advanced Architectural Studies, University of York, King's Manor
York Y012EP, United Kingdom
T: 44 433968 F: 44 433949

Barker, Andrew
Reporter
SUNDERLAND ECHO
Penntwell Ind Est, Penntwell, Sunderland, Tyne and Wear
SR4 9ER United Kingdom, United Kingdom
T: 91 5343011 F: 91 5345975

Barrett, Scott, Mr.
Professor
UNCTAD/LONDON BUSINESS SCHOOL
Sussex Place, Regents Park
London NWI 4SA, United Kingdom
T: 44 71 262-5050 F: 44 71 724-7875

Begum Hussain, Nazmi, Ms.
UNIVERSITY OF LONDON UNION
Malet Street London Lucie 7HY
London, United Kingdom
T: 44 71 5809551 F: 44 71 4364604

Benefield, Jim, Mr.
Manager Audio Visual Services
WASTE MANAGEMENT INTERNATIONAL - WMI
Windsor House, 55-56 St. James's Street
London SW1A 1LA, United Kingdom
T: 44 71 493-2554 F: 44 71 499-8684

Benford, Roy
Sound
BBC TELEVISION NEWS
TV Centre Wood Lane
London WIR 7RJ, United Kingdom
T: 44 81 7438000 F: 44 81 5767957

Bennet, R., Ms.
INFOLIFE
23 New Mount Street
Manchester M4 4DE, United Kingdom
T: 44 61 9534094 F: 44 61 9534001
TL: 6105942ELE, E: GeoNet MCR1:ELEICH

Bennett, Rosey
Group Editor
BUTTERWORTH HEINEMANN
Linacre House Jordan Hill
Oxford, United Kingdom
T: 44 865 310366 F: 44 865 311268

Berridge, Richard
Sound Recordist
ITN-INDEPENDENT TELEVISION NEWS
200 Grams
INN London WCI, United Kingdom
T: 44 71 8333000 F: 44 71 4304019

Berstein, Jodie, Ms.
V.P. Environmental Policy
WASTE MANAGEMENT INTERNATIONAL - WMI
Windsor House, 55-56 St. James's Street
London SW1A 1LA, United Kingdom
T: 44 71 493-2554 F: 44 71 499-8684

Binyon, Felicity, Ms.
Participant
CAMPAIGN FOR THE EARTH
9 Buckinsham St.
Oxford OXI4CH, United Kingdom
T: 44 865 245243

Bird, Ian, Mr.
Vp & Corp. Sec. Gov'T Affairs
WASTE MANAGEMENT INTERNATIONAL - WMI
Windsor House, 55-56 St. James's Street
London SW1A 1LA, United Kingdom
T: 44 71 493-2554 F: 44 71 499-8684

Bird, Marcela, Ms.
Anthropologist
WASTE MANAGEMENT INT.
55/56 St. James St.
London, United Kingdom
T: 44 71 4998684

Blackler, Zoe, Ms.
ECO - ACTION
124 Montaga Manstons
London W1A 1LE, United Kingdom

Blair, Lorna
STRATHCLYDE REGIONAL CONCIL
Strathclyde House, 20 India St., Charing Cross
Glasgow G2 4PF, United Kingdom
T: 44 41 204-2900 F: 44 41 227-2870
TL: 77428

Blowers, Andrew, Mr.
Professor
TOWN AND COUNTRY PLANNING ASSOCIATION
17 Carlton House Terrace
London SW1Y 5AS, United Kingdom
T: 44 71 930 8904 F: 44 71 930 3280

Bourne, Richard, Mr.
Special Advisor
COMMONWEALTH SECRETARIAT
Marlborough Movie, Pall Mall
London SW1Y 5P7, United Kingdom
T: 44 71 839-3411

Bright, Peter E., Mr.
Shell International Petroleum Co. Ltd.
Shell Centre
London SEI 7NA, United Kingdom
T: 44 71 934-5528 F: 44 71 934-5555
TL: 919651 SHELL G

Britto, Alberto Campus
Sound
BBC TELEVISION NEWS
TV Centre Wood Lane
London WIR 7RJ, United Kingdom
T: 44 81 7438000 F: 44 81 5767957

Brown, Bart, Mr.
Associate
FIELD-FOUNDATION FOR INT. ENV. LAW AND DEVELOPMENT
King's College London, Manresa Road
London SW3 6LX, United Kingdom
T: 44 71 352-8123 F: 44 71 351-6435

Brown, Bill, Mr.
Dir. Environmental Affairs
WASTE MANAGEMENT INTERNATIONAL - WMI
Windsor House, 55-56 St. James's Street
London SW1A 1LA, United Kingdom
T: 44 71 493-2554 F: 44 71 499-8684

Bruce, David, Mr.
STRATHCLYDE REGIONAL CONCIL
Strathclyde House, 20 India St., Charing Cross
Glasgow G2 4PF, United Kingdom
T: 44 41 204-2900 F: 44 41 227-2870
TL: 77428

Buntrock, Dean, Mr.
Chairman Of Board And Ceo
WASTE MANAGEMENT INTERNATIONAL - WMI
Windsor House, 55-56 St. James's Street
London SW1A 1LA, United Kingdom
T: 44 71 493-2554 F: 44 71 499-8684

Butler, Michael, Mr.
Board Director
CHURCH OF ENGLAND / SUSSEX
30 Bigwood Av.
Hove, Sussex 3BNIEN, United Kingdom
T: 02 73 29023

Cameron, Sarah, Ms.
INDEPENDENT FILM MAKER
Flat 44 Brandram's Wharf, 127 Rotherhithe St.
London SE16 4NF, United Kingdom
T: 44 71 252-1969

Cando, Philippe, Mr.
Director
FIELD-FOUNDATION FOR INT. ENV. LAW AND DEVELOPMENT
King's College London, Manresa Road
London SW3 6LX, United Kingdom
T: 44 71 352-8123 F: 44 71 351-6435

Castner, Ed
Cameraman
ITN-INDEPENDENT TELEVISION NEWS
200 Grams
INN London WCI, United Kingdom
T: 44 71 8333000 F: 44 71 4304019

Christie, Christina, Ms.
Horticultural.
ASPINWALL & CO.
Walford Manor, Shrewsbury
Shrewsbury, United Kingdom
T: 939 261144 F: 939 261146

Christoforou, Elizabeth, Ms.
FRIENDSHIP WITHOUT FRONTIERS
15 Holyhead Road Bangor Gwynedd
Bangor - Wales LL57 2eg, United Kingdom
T: 44 248 601-095 F: 44 248 601-095

Clark, Micheal, Mr.
Director
ROCKWERK
43 Harmer Green Lane Y- Digswell
Digswell - Welling N AL6 0AP, United Kingdom
T: 44 43 871 4071

Clark, Paul Wilson, Mr.
MANDATE FOR LIFE ON EARTH
211 Piccadilly
London W12 8LH, United Kingdom
T: 44 71 740-1554 F: 44 71 740-6470

Cohen, Jeff
Technician
BBC RADIO NEWS
Broadcasting House Portland Place
London WIA IAA, United Kingdom
T: 44 81 7438000 F: 44 81 5767957

Corbett, Malcolm, Mr.
POPTEL-GEONET COMMUNICATIONS AND INFORMATION NETWORK SPACE
25 Downham Road
London N1 5AA, United Kingdom
T: 44 71 249-2948 F: 44 71 254-1102
TL: 051918023GEONET, E: GEO2:SUPPORT (Geonet

Cowell, Adrian N., Ms.
Producer
CENTRAL TV
46 Anson Rd.
London NZ0 AB, United Kingdom
T: 44 71 6091240

Crombie, Russ
Editor
BBC TELEVISION NEWS
TV Centre Wood Lane
London WIR 7RJ, United Kingdom
T: 44 81 7438000 F: 44 81 5767957

Cunn, Suell
Brazilian Service
BBC WORLD SERVICE
Room 828 CB Bush House, Aldwych
London, United Kingdom
T: 44 71 257 2771 F: 44 71 240 2541

Cunningham, Sue
TIMES - LONDON
London, United Kingdom

Dafforn, Jenny
Producer
BBC TELEVISION NEWS
TV Centre Wood Lane
London WIR 7RJ, United Kingdom
T: 44 81 7438000 F: 44 81 5767957

Davis, Linda, Mr.
Editor
ECOLOGIST MAGAZINE
Agriculture House, Bath Rd - Sturminster Newton
Dorset, United Kingdom
T: 44 258 860-022 F: 44 258 737-48

De Rothschild, Anne, Ms.
GRAESSER - BIOTECHNA LIMITED
81 Wimpole Street
London WIM7DR, United Kingdom
T: 44 71 486-5140 F: 44 71 486-0819

Dillon, Justin S., Mr.
CENTER FOR EDUCATIONAL STUDIES, KING'S COLLEGE LONDON
Cornwall House Annex, Waterloo Road
London SE1 8TX, United Kingdom
T: 44 71 872-3096 F: 44 71 872-3172
E: 01:YNS020

Donnelly, Richard
Video Editor
ITN-INDEPENDENT TELEVISION NEWS
200 Grams
INN London WCI, United Kingdom
T: 44 71 8333000 F: 44 71 4304019

Dorgham, Mohammed A, Dr.
Editor In Chief
INTERNATIONAL JOURNAL OF SUSTAINABLE DEVELOPMENT
17 Beeward Close,The Leys, Wolverton Hill
Milton Keynes, United Kingdom
T: 44 908 314248 F: 44 908 221509

Dyer, Simon, Mr.
Director General President
AUTOMOBILE ASSOCIATION
Fanum House, Basing View
Basingstoke-Hampshire RG21 2EC, United Kingdom
T: 44 256 49-3080 F: 44 256 49-2077

Early, Daphne I. K., Ms.
UNITED NATIONS ASSOCIATION, OXFORD
Yatsen Bears'Hill
Oxford, United Kingdom
T: 44 865 75356

Edwards, Mark
President
STILL PICTURES
199A Shooters Hill Rd
London SE3 8UL, United Kingdom
T: 44 81 8588307 F: 44 81 8582049

Falkman, Edwin G., Mr.
President
WASTE MANAGEMENT INTERNATIONAL - WMI
Windsor House, 55-56 St. James's Street
London SW1A 1LA, United Kingdom
T: 44 71 493-2554 F: 44 71 499-8684

Farookhi, Intiaz, Mr.
Chief Executive
LEICESTER CITY COUNCIL
New Walk Centre
Leicester, United Kingdom

Faulkner, Julian, Mr.
EARTH TRAIN/WORLDLINK/GLOBAL YOUTH SUMMIT
Grange,King's School
Canterbury -Kent GT12E5, United Kingdom

Ferreira, Virginia, Ms.
Research Assistant
FIELD-FOUNDATION FOR INT. ENV. LAW AND DEVELOPMENT
King's College London, Manresa Road
London SW3 6LX, United Kingdom
T: 44 71 352-8123 F: 44 71 351-6435

Ferrey, Jo Espinoza, Mr.
Senior Programme Officer
INTERNATIONAL MARITIME ORGANIZATION
4 Albert Embankment
London SE1 3210, United Kingdom
T: 44 71 7357611 F: 44 71 5873210
TL: 23588

Finney, C.
Researcher
FOOD MATTERS WORLDWIDE
38 Exchange St
Norwich, United Kingdom
T: 44 603 761645 F: 44 603 765434

Fitch, Brian, Mr.
INTERNATIONAL NFZ
165 Waldegrave Road
Brighton BNI 6GJ, United Kingdom
T: 44 273 553 983 F: 44 273202102

Footman, Bill
Producer
ITN-INDEPENDENT TELEVISION NEWS
200 Grams
INN London WCI, United Kingdom
T: 004471 8333000 F: 004471 43040

France, Nicky
Editor
ENERGY POLICY/UTILITIES POLICY/IAEE NEWSLETTER
101 Trafalgar Road Horsham
Sussex, United Kingdom
T: 44 243 211160 F: 44 243 66738

Fraser, Alan
Cameraman
GRANADA TELEVISION
34 The Colonnades Albert Dock
Liverpool, United Kingdom
T: 44 51 55 709 6338 F: 44 51 55 709 5259

Frost, Bill
Video Editor
ITN-INDEPENDENT TELEVISION NEWS
200 Grams
INN London WCI, United Kingdom
T: 44 71 8333000 F: 44 71 4304019

Garcia, Dylan, Mr.
Photographer
PANOS PICTURES/PANOS INSTITUTE
9 White Lion Street
London NI 9PD, United Kingdom
T: 44 71 278-1111 F: 44 71 278-0345
TL: 4419293 PanosG

Gennaro, Pietro, Dr.
Chair - Waste Management
WASTE MANAGEMENT INTERNATIONAL - WMI
Windsor House, 55-56 St. James's Street
London SW1A 1LA, United Kingdom
T: 44 71 493-2554 F: 44 71 499-8684

Ghosh, Pallab
Producer
BBC RADIO NEWS
Broadcasting House Portland Place
London WIA IAA, United Kingdom
T: 44 81 7438000 F: 44 81 5767957

Gibbs, Andrew, Mr.
INTERNATIONAL BUSINESS IN THE COMMUNITY
5 Cleveland Place
London, United Kingdom
T: 44 71 925-2933 F: 44 71 321-6480

Girardet, Herbert, Mr.
Co-Organizer
EARTH PARLIAMENT
93 Cambridge Gardens
London W1067E, United Kingdom
T: 44 71 81969 6375 F: 44 71 81969 6375

Gontry, Bob
Producer
ITN-INDEPENDENT TELEVISION NEWS
200 Grams
INN London WCI, United Kingdom
T: 004471 8333000 F: 004471 43040

Gonzalez, Alex, Mr.
Vp - Latin America
WASTE MANAGEMENT INTERNATIONAL - WMI
Windsor House, 55-56 St. James's Street
London SW1A 1LA, United Kingdom
T: 44 71 493-2554 F: 44 71 499-8684

Gormley, Antony, Mr.
Artist
GOETHE INSTITUTE
13, South Villas
London, United Kingdom
T: 44 71 482-7383 F: 44 71 639-1303

Goswami, Rita, Ms.
FRIENDSHIP WITHOUT FRONTIERS - LIFE FOUNDATION SCHOOL OF THERAPEUTICS
15 Holyhead Road
Bangor - Wales LL57 2ES, United Kingdom
T: 44 248 601095 370076 F: 44 248 601095

Granich, Sarah
Editor
TIEMPO
47 Wellington Rd
Nottingham, United Kingdom
T: 44 602 610854 F: 44 602 610854

Gray, John, Mr.
STRATHCLYDE REGIONAL CONCIL
Strathclyde House, 20 India St., Charing Cross
Glasgow G2 4PF, United Kingdom
T: 44 41 204-2900 F: 44 41 227-2870
TL: 77428

Griffiths, David, Mr.
CORPORATE PROFILES SERVICES LIMITED
12 Rickett St.
London SW6 1RU, United Kingdom
T: 44 71 386-0266 F: 44 71 386-0275

Grubb, Michael, Mr.
Professor
UNCTAD/RIIA
Chatman House, 10 St. James Square
London GW1X 4LE, United Kingdom
T: 44 71 957-5700 F: 44 71 957-5710

Guedes, Olga, Ms.
Professor
LOUGHBOROUGH UNIVERSITY - DEPT. SOCIAL SCIENCES
L'borough - Leics
L'borough-Leics, United Kingdom
T: 44 0509263171 F: 44 0509238277

Guy, David Eric, Mr.
Colonel
SALVATION ARMY
Salvation Army International Headquarters, 101 Queen Victoria Street
London, United Kingdom
T: 44 71 236-5222

Hailes, Julia, Ms.
Director
SUSTAINABILITY
The Director's Mall 91-97 Freston Road
London, United Kingdom
T: 44 71 243-1277 F: 44 71 243-0367

Hall, Anthony, Mr.
Dr.
LONDON SCHOOL OF ECONOMICS
Houghton St.
London WC2 A 2AE, United Kingdom
T: 44 71 955-7354 F: 44 71 955-7415

Hall, D. O., Mr.
Prof.
KING'S COLLEGE LONDON
Campden Hill Road
London W8 7AH, United Kingdom
T: 44 71 333-4371 F: 44 71 937-7783
TL: 8954102 BBSLONG

Hall, David, Mr.
Director
TOWN AND COUNTRY PLANNING ASSOCIATION
17 Carlton House Terrace
London SW1Y 5AS, United Kingdom
T: 44 71 9308904 F: 44 71 9309280

Halmos, Tony, Mr.
Adviser
GOVERNMENT OF BOTSWANA
C/O 5 - 11 Theobalds Road
London, United Kingdom
T: 44 71 413 3043

Hams, Tony, Mr.
Env. Adviser
UK LOCAL GOVERNMENT
Arndale House, Arndale Centre
Luton LU1 2TS, United Kingdom
T: 44 582 451166 F: 44 582 412525

Handl, Gunther, Mr.
Associate
FIELD - FOUNDATION FOR INTERNATIONAL ENVIRONMENTAL LAW AND DEVELOPMENT
King's College London, Manresa Road
London SW3 6LX, United Kingdom
T: 44 71 352-8123 F: 44 71 351-6435

Hardie, Jacqueline, Ms.
LONDON BOROUGH OF ENFIELD
23 Helmsley, Cleveland Rd., South Woodford
London, United Kingdom
T: 44 71 5803709

Hardie, Philip James, Mr.
LONDON BOROUGH OF ENFIELD
23 Helmsley, Cleveland Rd., South Woodford
London, United Kingdom
T: 44 71 5803709

Harris, Errol E., Dr.
ONE WORLD WCPA
High Wray House
Ambleside-Gumbria LA 22 OJQ, United Kingdom
T: 44 5394-32107

Harvey, Gillian, Mr.
Project Coordinator
TREE OF LIFE PROJECT
30 Swinton Street
London WCIX 9NX, United Kingdom
T: 44 71 833-0252 F: 44 71 833-0405

Hay, Annette, Ms.
UNA OF GREAT BRITAIN AND NOTHERN IRELAND
3 Whitehall Court
London SW1A 2EL, United Kingdom
T: 44 71 930-2931 - 2 F: 44 71 930-5893
TL: 837883 SPEND G

Haygood, Leah, Ms.
Manager Environmental Affairs
WASTE MANAGEMENT INTERNATIONAL - WMI
Windsor House, 55-56 St. James's Street
London SW1A 1LA, United Kingdom
T: 44 71 493-2554 F: 44 71 499-8684

Haynes, Charlotte, Ms.
Director
RAINFOREST FOODS
2 Rylett RD
London W12 9SS, United Kingdom
T: 44 71 740 4574 F: 44 71 740 4574

Hensmall, Ian
Editor Outlook Mg. / Co-Editor
PLANET NEWS/OUTLOOK MAGAZINE
87 Kirkstall Rd
London, United Kingdom
T: 44 71 671 7920

Herbert, Adele L., Ms.
Coord. Corp. Presentations
WASTE MANAGEMENT INTERNATIONAL - WMI
Windsor House, 55-56 St. James's Street
London SW1A 1LA, United Kingdom
T: 44 71 493-2554 F: 44 71 499-8684

Herbert, Duncan
Editor
BBC TELEVISION NEWS
TV Centre Wood Lane
London WIR 7RJ, United Kingdom
T: 44 81 7438000 F: 44 81 5767957

Hill, Regina Nogueira, Ms.
Coordinator
DINEFWR GREEN GROUP
Llwyn Onn - Trapp - Llandeilo
Dyfed,Wales SA196TR, United Kingdom
T: 5 58 82-3515

Hockman, Deborah, Ms.
Asst. Dir. Lab Operations
WASTE MANAGEMENT INTERNATIONAL - WMI
Windsor House, 55-56 St. James's Street
London SW1A 1LA, United Kingdom
T: 44 71 493-2554 F: 44 71 499-8684

Holloway, Lester
WORLD DEVELOPMENT MOVVEMENT
25 Beetlive Place
London, United Kingdom
T: 44 71 737-6215

Holtzlaw, Lilian, Ms.
SUN GROUP OF COMPANIES
11 New Burlington Place
London WIXIFA, United Kingdom
T: 44 71 4960408 F: 44 71 4960814
TL: 27189

Horberry, John, Mr.
ENVIRONMENTAL RESOURCES LTD
106 Gloucester Place
London W1H 3DB, United Kingdom

Howard, Michael
SUPER CHANNEL
14 Canark Square, Limeharbor
London, United Kingdom
T: 44 71 499418 F: 44 71 4189419

Hunter, Debra, Ms.
SUN GROUP OF COMPANIES
11 New Burlington Place
London WIXIFA, United Kingdom
T: 44 71 4960408 F: 44 71 4960814
TL: 27189

Hyndman, Martin, Mr.
Publisher
CHAPMAN & HALL
2-6 Boundary Row
London SEI 8HN, United Kingdom
T: 44 71 865-0066 F: 44 71 522-9623

Insley, Jane Elizabeth, Ms.
SOROPTIMIST INTERNATIONAL
Flat 1, 20 Chesterton Road
London W10 5LK, United Kingdom

Jacobie, Jackie, Ms.
Manager Special Projects
WASTE MANAGEMENT INTERNATIONAL - WMI
Windsor House, 55-56 St. James's Street
London SW1A 1LA, United Kingdom
T: 44 71 493-2554 F: 44 71 499-8684

Jennings, Vernon, Dr.
Director Of Consultancy
SUSTAINABILITY
The People's Hall, 91-97 Freston Road
London Wii 4BD, United Kingdom
T: 44 71 2431277 F: 44 71 2430364

Jesh, L., Mr.
Dr.
FRAKLIN COMPANY CONSULTANTS
192 Franklin Road
Birminghan, United Kingdom
F: 44 21 214598206

Jones, Christina, Ms.
Trainee Solicitor
FIELD - FOUNDATION FOR INTERNATIONAL LAW AND DEVELOPMENT
King's College London, Manresa Road
London SW3 6LX, United Kingdom
T: 44 71 352-8123 F: 44 71 351-6435

Jones, John, Mr.
FRIENDSHIP WITHOUT FRONTIERS - LIFE FOUNDATION SCHOOL OF THERAPEUTICS
15 Holyhead Road
Bangor - Wales LL57 2ES, United Kingdom
T: 44 248 601095 370076 F: 44 248 601095

Jopling, John, Mr.
Trustee
ASHTAN TRUST
1 Chamberlain Street
London NW1 8X13, United Kingdom
T: 44 71 722-3710 F: 44 71 722-3959

K.K., Tan, Mr.
LOWE BELL
7 Hartford St.
London, United Kingdom

Karpan, Natalia, Ms.
GREEN HELP FOUNDATION FOR HELP TO VICTIMS OF ECODISASTERS
13 Malozhitomirskaya Str., Fl.16
Kiev 252001, United Kingdom
T: 7 2256473 F: 7 4844580

Kelly, Mick
Editor
TIEMPO
47 Wellington Rd
Nottingham, United Kingdom
T: 44 602 610854 F: 44 602 610854

Kelly, Ross
Reporter
GRANADA TELEVISION
34 The Colonnades Albert Dock
Liverpool, United Kingdom
T: 44 51 55 709 6338 F: 44 51 55 709 5259

Kennedy, James, Mr.
STRATHCLYDE REGIONAL CONCIL
Strathclyde House, 20 India St., Charing Cross
Glasgow G2 4PF, United Kingdom
T: 44 41 204-2900 F: 44 41 227-2870
TL: 77428

Kerslake, Louise
Editor
BBC TELEVISION NEWS
TV Centre Wood Lane
London WIR 7RJ, United Kingdom
T: 44 81 7438000 F: 44 81 5767957

Kerslake, Simon
FUJISANKEI COMMUNICATIONS INTERNATIONAL INC.
London, United Kingdom

Khemka, Shiv, Mr.
Director
SUN GROUP OF COMPANIES
11 New Burlington Place
London WIXIFA, United Kingdom
T: 44 71 4960408 F: 44 71 4960814
TL: 27189

Khemka, Uday, Mr.
Director
SUN GROUP OF COMPANIES
11 New Burlington Place
London WIXIFA, United Kingdom
T: 44 71 4960408 F: 44 71 4960814
TL: 27189

Kippert, Jashana, Ms.
Staff Member
FINDHORN FOUNDATION
The Park, Findhorn
Forres 1V360TZ, United Kingdom
T: 44 309 690751 F: 44 309 691301

Kobiliatskij, Anatolij, Mr.
GREEN HELP
24 Vunacharskogo st., apt 100
Kiev 97 252097, United Kingdom
T: 7 517-7158 F: 7 517-7158

Kohn, Roger, Mr.
Senior Ext Rel Officer
INTERNATIONAL MARITIME ORGANIZATION
4 Albert Embankment
London SE1 3210, United Kingdom
T: 44 71 7357611 F: 44 71 5873210
TL: 23588

Kumar, Satish, Mr.
Director /editor
SCHUMACHER COLLEGE /RESURGENCE MAGAZINE
The Old Postem, Dartington
Totnes - Devon TQ9 6EA, United Kingdom
T: 44 803 865934 F: 44 803 866899

Lahellec, Annie, Ms.
POPTEL - GEONET
134 rainha guilhermina/25 downham kd
London, United Kingdom
T: 44 71 2492948

Lanchester, Barry
Camera
BBC TELEVISION NEWS
TV Centre Wood Lane
London WIR 7RJ, United Kingdom
T: 44 81 7438000 F: 44 81 5767957

Leach, Penelope, Dr.
EPOCH WORLDWIDE
3 Tanza Road
London NW3 2UA, United Kingdom
T: 44 71 435-9025 F: 44 71 431-5480

Lowe, Julie, Ms.
ASHTAN TRUST
7 Chamberlain Street
London NW1 8XB, United Kingdom
T: 44 71 722-3710 F: 44 71 722-3959

Lunden, Jaana, Ms.
Architect
POSTWAR RECONSTRUCTION AND DEVELOPMENT
Institute of Advanced Architectural Studies, University of York, King's Manor
York Y012EP, United Kingdom
T: 44 433968 F: 44 433949

Magaldi, Francisco
Camera
BBC TELEVISION NEWS
TV Centre Wood Lane
London WIR 7RJ, United Kingdom
T: 44 81 7438000 F: 44 81 5767957

Marsh, June, Ms.
ASHTAN TRUST
7 Chamberlain Street
London NW1 8XB, United Kingdom
T: 44 71 722-3710 F: 44 71 722-3959

Martin, John
Cameraman
ITN-INDEPENDENT TELEVISION NEWS
200 Grams
INN London WCI, United Kingdom
T: 44 71 8333000 F: 44 71 4304019

May, Charles A., Mr.
Dr.
THAMES POLYTECHNIC
School of Engineering, Wellington St., Woolwich
London, United Kingdom
T: 44 71 316-8622 F: 44 71 316-8605

McCauley, Mark
Sound
BBC TELEVISION NEWS
TV Centre Wood Lane
London WIR 7RJ, United Kingdom
T: 44 81 7438000 F: 44 81 5767957

McDonagh, Sean, Mr.
COLUMBANS
Dalgan, Navaw - Ireland
United Kingdom
T: 353 46 21325 F: 353 46 22799

McGuinty, David J., Mr.
Solicitor
GOULDENS
22 Tudor Street
London EC4Y0JJ, United Kingdom
T: 44 71 583-7777 F: 44 71 583-3051
TL: 21520

McHarg, Allison
STRATHCLYDE REGIONAL CONCIL
Strathclyde House, 20 India St., Charing Cross
Glasgow G2 4PF, United Kingdom
T: 44 41 204-2900 F: 44 41 227-2870
TL: 77428

McHutchon, Graham
Technician
BBC RADIO NEWS
Broadcasting House Portland Place
London WIA IAA, United Kingdom
T: 44 81 7438000 F: 44 81 5767957

McNally, Tony, Mr.
Assistant
GREEN SHOW
85 Moor Str. CO
Coventry, United Kingdom

McNeill, Sabine, Ms.
Netwroker
GLOBAL HEART
21-A Gold Hurst Terrace
London NW6 3HB, United Kingdom

McPherson, Andrew
STRATHCLYDE REGIONAL CONCIL
Strathclyde House, 20 India St., Charing Cross
Glasgow G2 4PF, United Kingdom
T: 44 41 204-2900 F: 44 41 227-2870
TL: 77428

Miller, David, Mr.
COOPERS & LYBRAND
Plumtree Court
London EL4, United Kingdom
T: 71 212-8217

Monaghan, Aileen
STRATHCLYDE REGIONAL CONCIL
Strathclyde House, 20 India St., Charing Cross
Glasgow G2 4PF, United Kingdom
T: 44 41 204-2900 F: 44 41 227-2870
TL: 77428

Monetti, Myriam, Ms.
Translator
WASTE MANAGEMENT INTERNATIONAL - WMI
Windsor House, 55-56 St. James's Street
London SW1A 1LA, United Kingdom
T: 44 71 493-2554 F: 44 71 499-8684

Motley, Tony, Mr.
Envirnmental Consultant
WASTE MANAGEMENT INTERNATIONAL - WMI
Windsor House, 55-56 St. James's Street
London SW1A 1LA, United Kingdom
T: 44 71 493-2554 F: 44 71 499-8684

Mushet, Julie, Ms.
SCHUMACHER COLLEGE
The Old Postem, Dartington
Totnes - Devon TQ9 6EA, United Kingdom
T: 44 803 865934 F: 44 803 866899

Nadkarni, Bob
News Producer
BBC TELEVISION NEWS
TV Centre Wood Lane
London WIR 7RJ, United Kingdom
T: 44 81 7438000 F: 44 81 5767957

Nagorny, Alexey, Mr.
PUBLIC SCIENTIFIC ORGANIZATION
Nizhnedneprovskija St 4 fl 26
Zaporozhye, United Kingdom

Nath, Krishna Dulal Deb, Mr.
I.E.D.S.-F.O.E.
c/o M.A. Razzali, 32 High Street
Stanford-le-Stope Essex SS17 0HQ, United Kingdom

Nevens, Betty
Producer
ITN-INDEPENDENT TELEVISION NEWS
200 Grams
INN London WCI, United Kingdom
T: 004471 8333000 F: 004471 43040

Newton, Geraldine Sharpe, Ms.
Director
WWF - UK
Panda House, Weyside Park, goldalming
Godalming-Surrey GU7 1XR, United Kingdom
T: 44 44 834-26444 F: 44 44 834-26409

Ni Riain, Noirin, Ms.
ASHTAN TRUST
7 Chamberlain Street
London NW1 8XB, United Kingdom
T: 44 71 722-3710 F: 44 71 722-3959

Ninoseki, Yoshio
Journalist
FUJI TELEVISION
1st Floor, 15-16 New Burlington St.
London WIX IFF, United Kingdom
T: 44 71 4942591 F: 44 71 4396206

O'Donnell, Mary, Ms.
Chairperson
SKIBBEREEN ENVIRONMENTAL ACTION SCHEME
Ballydehob
Co. Cork, Ireland, United Kingdom
T: 44 353 28-31223 F: 44 353 28-22267

O'Neil, William, Mr.
Secretary General
INTERNATIONAL MARITIME ORGANIZATION
4 Albert Embankment
London SE1 3210, United Kingdom
T: 44 71 7357611 F: 44 71 5873210
TL: 23588

O'Reilly, Francis
Conference Editor
SCIENCE TECHNOLOGY DEVELOPMENT FORUM
72 Hainault Court , Forest Rise, Whipps Cross
London, United Kingdom
T: 44 71 5211415 F: 44 71 7646003

Orr, Patricia
Editor
TIMBER GROWER UK
28, Rosary Gdns.
London SW7 4NT, United Kingdom
T: 44 71 3733406

Orr, Patrick
Correspondent
COMMONWEALTH TRUST
18 Northumberland Ave
London W1, United Kingdom
T: 44 71 8283961 F: 44 71 6309750

Orr, Paul
STRATHCLYDE REGIONAL CONCIL
Strathclyde House, 20 India St., Charing Cross
Glasgow G2 4PF, United Kingdom
T: 44 41 204-2900 F: 44 41 227-2870
TL: 77428

Osbourne, Carol, Ms.
PROJECT EARTH
140 Tottenham Cr. Road
London, United Kingdom
T: 44 71 945 6965

Parker, Philip
Senior Editor
GAIA BOOKS
20 High Street, Stroud
Gloucestershire GL5, United Kingdom
T: 44 453 752985 F: 44 453 752987

Patel, Mamta
Enviroment Reporter
EUROPEAN CHEMICAL NEWS
Quadrant House,
Sutton, Surrey SM2 8, United Kingdom
T: 44 81 6523153 F: 44 81 6523375

Poole, George
Camera
BBC TELEVISION NEWS
TV Centre Wood Lane
London WIR 7RJ, United Kingdom
T: 44 81 7438000 F: 44 81 5767957

Powell, Geri, Ms.
Manager- Media Relations
WASTE MANAGEMENT INTERNATIONAL - WMI
Windsor House, 55-56 St. James's Street
London SW1A 1LA, United Kingdom
T: 44 71 493-2554 F: 44 71 499-8684

Prescott, Tony
Director
GRANADA TELEVISION
34 The Colonnades Albert Dock
Liverpool, United Kingdom
T: 44 51 55 709 6338 F: 44 51 55 709 5259

Primavesi, Anne, Ms.
WORLD COUNCIL OF CHURCHES
Manor Lane
Newbury - Berks. RG13 2QX, United Kingdom

Puri, Shamlal
NEWSLINK AFRICA LTD
London, United Kingdom

Quinn, Kerrie-Anne
Producer
SUPER CHANNEL
Melrose House, 14 Lanark Square
Limeharbour, London, United Kingdom
T: 44 71 4189418 F: 44 71 418942

Radford, Andrew, Mr.
Coordinator
BABY MILK ACTION
23 St. Adrew's Street
Cambridge, United Kingdom
T: 44 223 464-420 F: 44 223 464-419

Rajan, Mukund, Mr.
Student
OXFORD UNIVERSITY
Worcester College, Oxford University
Oxford, United Kingdom

Randles, David
Engineer
ITN-INDEPENDENT TELEVISION NEWS
200 Grams
INN London WCI, United Kingdom
T: 44 71 8333000 F: 44 71 4304019

Ribot, Jesse, Mr.
M.I.T.
885 mass. ave 16
Cambridge 02139, United Kingdom
T: 44 223 3543070

Roberts, Ian, Mr.
Director
LEICESTER ENVIRONMENT CITY
Town Hall
Leicester England, United Kingdom

Rodgerson, Gillian
International News Editor
GAY TIMES MAGAZINE
Worldwide House , Ground Floor 116-134 Bayham St
London NW1 OBA, United Kingdom
T: 44 71 4822576 F: 44 71 2840329

Rose, Greg, Mr.
Staff Lawyer
FIELD - FOUNDATION FOR INTERNATIONAL LAW AND DEVELOPMENT
King's College London, Manresa Road
London SW3 6LX, United Kingdom
T: 44 71 352-8123 F: 44 71 351-6435

Rubin, Frances, Ms.
OXFAM
20 Hanfield Rd.
Oxford OX2 6TT, United Kingdom
T: 44 865 510659 F: 44 865 313780

Rughani, Dee, Mr.
LEAP TRAINING
38 South Park Crescent, Ilford, Essex IG1 1XU
London, United Kingdom

Russell, Bob, Mr.
Manager Environmental Affairs
WASTE MANAGEMENT INTERNATIONAL - WMI
Windsor House, 55-56 St. James's Street
London SW1A 1LA, United Kingdom
T: 44 71 493-2554 F: 44 71 499-8684

Seddon-Brown, Bill, Mr.
WASTE MANAGEMENT INTERNATIONAL - WMI
Windsor House, 55-56 James's Street
London SW1A 1LA, United Kingdom
T: 44 71 493-2554 F: 44 71 499-8684

Simmonds, Julian
THE SUNDAY TELEGRAPH
London, United Kingdom

Sjolin, Karin, Ms.
Programme Officer
INTERNATIONAL MARITIME ORGANIZATION
4 Albert Embankment
London SE1 3210, United Kingdom
T: 44 71 7357611 F: 44 71 5873210
TL: 23588

Smyth, John, Mr.
Professor
SEEC
Glenpark
Johnstone, Scotland PA5 051, United Kingdom
T: 44 505 20219 F: 44 505 20219

Soulsby, Peter, Mr.
Leader
LEICESTER CITY COUNCIL
New Walk Centre
Leicester, United Kingdom
T: 0533 526043 F: 0533 555726

Surjadi, Laura, Ms.
Reserve Official Rep. For Sdi
SUSILA DHARMA INTERNATIONAL - SDI
9 Albert St. Slough
Berks SL1 2EH, United Kingdom
T: 44 753 55 22 43 F: 44 753 55 46 10

Swan, Robert, Mr.
Unep Ambassador
PROJECT EARTH
140 Tottenham Cr.Road
London, United Kingdom
T: 44 71 945 6965

Tait, Joyce, Ms.
Professor
STRATHCLYDE GRADUATE BUSINESS SCHOOL
130, Rottenrow
Glasgow 9409E, United Kingdom
T: 44 41 553 6014 F: 44 41 552 2501
TL: 77472UNSWB

Tanzarella, Filomena, Ms.
Dr.
CEMP
24 Spital
Arberdeen, United Kingdom
T: 224 63-3663

Terrins, Paul
Cameraperson
FUJI TELEVISION
1st Floor, 15-16 New Burlington St.
London WIX IFF, United Kingdom
T: 44 71 4942591 F: 44 71 4396206

Thamage, Thobeka Mjolo, Ms.
Teaching Assistant
FIELD - FOUNDATION FOR INTERNATIONAL LAW AND DEVELOPMENT
King's College London, Manresa Road
London SW3 6LX, United Kingdom
T: 44 71 352-8123 F: 44 71 351-6435

Thompson, Alan
Cameraman
ITN-INDEPENDENT TELEVISION NEWS
200 Grams
INN London WCI, United Kingdom
T: 44 71 8333000 F: 44 71 4304019

Thompson, Koy, Mr.
Liaison Officer - Northern Ngo
INTERNATIONAL INSTITUTE FOR ENVIRONMENT AND DEVELOPMENT
3 Endsleigh St.
London WC1HODD, United Kingdom
T: 44 71 388-2117 F: 44 71 388-2826
TL: 261681

Thompson, Nigel
Cameraman
ITN-INDEPENDENT TELEVISION NEWS
200 Grams
INN London WCI, United Kingdom
T: 44 71 8333000 F: 44 71 4304019

Throssell, Liz
Reporter
BBC RADIO NEWS
Broadcasting House Portland Place
London WIA IAA, United Kingdom
T: 44 81 7438000 F: 44 81 5767957

Tilson, Sue
Producer
ITN-INDEPENDENT TELEVISION NEWS
200 Grams
INN London WCI, United Kingdom
T: 44 71 8333000 F: 44 71 4304019

Timberlake, Lloyd, Mr.
Director - External Affairs
INTERNATIONAL INSTITUTE FOR ENVIRONMENT AND DEVELOPMENT
3 Endsleigh St.
London WC1HODD, United Kingdom
T: 44 71 388-2117 F: 44 71 388-2826
TL: 261681

Timms, Alwyn, Mr.
ASHTAN TRUST
7 Chamberlain Street
London NW1 8XB, United Kingdom
T: 44 71 722-3710 F: 44 71 722-3959

Ullstein, Bartholomeu, Mr.
Diretor
BANSON
3 Turville Street
London E27MR, United Kingdom
T: 44 71 729 7315 F: 44 71 729 7870

Unsworth, Tony
Sound
GRANADA TELEVISION
34 The Colonnades Albert Dock
Liverpool, United Kingdom
T: 44 51 55 709 6338 F: 44 51 55 709 5259

Vasquez-Velasquez, Debora, Ms.
Represetative
LEEDS POLYTECHNIC STUDENTS UNION
Welfare Office , City Site , Calverley Street
Leeds - W.Yorkshire LS15 8AU, United Kingdom
T: 44 532 430171 F: 44 532 973

Vergara, Dan
Journalist
BBC WORLD SERVICE
15B Consort Road
London, United Kingdom
T: 44 71 2572549

Vickers, Hilary, Ms.
Editorial Assistant Policy Jou
BUTTERWORTH HEINEMANN
Linacre House Jordan Hill
Oxford, United Kingdom
T: 44 865 310366 F: 44 865 311268

Walker, Keith
STRATHCLYDE REGIONAL CONCIL
Strathclyde House, 20 India St., Charing Cross
Glasgow G2 4PF, United Kingdom
T: 44 41 204-2900 F: 44 41 227-2870
TL: 77428

Walsh-Evans, Louise, Ms.
Fundraising & Public Relations
TREES FOR LIFE/FINDHORN FOUNDATION
The Findhorn Foundation, The Park, Findhorn, Moray IV36 OTZ
Findhorn, United Kingdom
T: 44 309 691292

Wason, John
SUPER CHANNEL
14 Canark Square, Limeharbor
London, United Kingdom
T: 44 71 499418 F: 44 71 4189419

Waterworth-Owen, Gary, Mr.
INEM/EEEMA
Ivy House Farm
Twyford, United Kingdom
T: 44 664 184-0069 F: 44 664 1123711

Watson, Graham M., Mr.
General Manager
MANDATE FOR LIFE ON EARTH
211 Piccadilly
London W12 8LH, United Kingdom
T: 44 71 740-1554 F: 44 71 740-6470

Werksman, Jake, Mr.
Staff Attorney
FIELD-FOUNDATION FOR INT. ENV. LAW & DEVELOPMENT
King's College London, Manresa Road
London SW3 6LX, United Kingdom
T: 44 71 352-8123 F: 44 71 351-6435

Whalley, Lilian, Ms.
FRIENDSHIP WITHOUT FRONTIERS - LIFE FOUNDATION SCHOOL OF THERAPEUTICS
15 Holyhead Road
Bangor - Wales LL57 2ES, United Kingdom
T: 44 248 601095 370076 F: 44 248 601095

Wilkinson, Peter
Cameraman
ITN-INDEPENDENT TELEVISION NEWS
200 Grams
INN London WCI, United Kingdom
T: 44 71 8333000 F: 44 71 4304019

Wilkinson, Wendy
Reporter
ITN-INDEPENDENT TELEVISION NEWS
200 Grams
INN London WCI, United Kingdom
T: 004471 8333000 F: 004471 43040

Woodrow, Bill, Mr.
ARTE AMAZONAS
44 Comberwell Grove
London, United Kingdom
T: 44 71 701 3763

Wregglesworth, Victoria, Ms.
PROJECT EARTH
140 Tottenham Cr. Road
London, United Kingdom

Wynne, Graham, Mr.
Director-Conservation
Royal Society For The Protection Of Birds
Rspb. The Lodge
Sandy Beds, United Kingdom
T: 44 767 680551 F: 44 767 692365

Yamin, Farhana, Ms.
Staff Lawyer
FIELD-FOUNDATION FOR INT. ENV. LAW AND DEVELOPMENT
King's College London, Manresa Road
London SW3 6LX, United Kingdom
T: 44 71 352-8123 F: 44 71 351-6435

Young, Kenny, Mr.
Trustee
EARTH LOVE FUND
Lazy Moon
Upper Brailes - Oxon OX155BA, United Kingdom
T: 44 60885269

UN AND UNCED

Edwards, David Thiele, Mr.
International Maritime Organization
4, Albert-Embankment
London, SE 1, 7 SR, United Kingdom
T: 4471 7357611 F: 4471 58732100

Labastida, F.
International Maritime Organization
4, Albert-Embankment
London, SE 1, 7 SR, United Kingdom
T: 4471 7357611 F: 4471 58732100

Nauke, M.
International Maritime Organization
4, Albert-Embankment
London, SE 1, 7 SR, United Kingdom
T: 4471 7357611 F: 4471 58732100

Voskresensky, K.
International Maritime Organization
4, Albert-Embankment
London, SE 1, 7 SR, United Kingdom
T: 4471 7357611 F: 4471 58732100

Wonham, J.
International Maritime Organization
4, Albert-Embankment
London, SE 1, 7 SR, United Kingdom
T: 4471 7357611 F: 4471 58732100

ADVISORS

De Rothschild, Edmund, Mr.
London EC4P 4DU, United Kingdom
T: 44 71 280 5000 F: 44 71 929 1643

Ramphal, Shridath, The Hon. Sir
Former S-G Of The Commonwealth Secretariat
London W9 1HR, United Kingdom
T: 44 71 266 3409 F: 44 71 286 2302

PARLIAMENTARIANS / SPIRITUAL LEADERS

Ennals, Lord
House of Lords Palace of Westminster
London SWI OPW, United Kingdom
T: 219.3000
TL: 881 3033

UNITED REPUBLIC OF TANZANIA - GOVERNMENT

Chande, A., Mr.
Permanent Mission to the United Nations
205 E. 42nd Str., 13th Floor
New York, NY 10017, United States of America
T: 212 972 9160 F: 212 682 5232

Chokola, P., Mr.
Permanent Mission to the United Nations
205 E. 42nd Str., 13th Floor
New York, NY 10017, United States of America
T: 212 972 9160 F: 212 682 5232

Daraja, , Mr.
Permanent Mission to the United Nations
205 E. 42nd Str., 13th Floor
New York, NY 10017, United States of America
T: 212 972 9160 F: 212 682 5232

Diria, Ahmed Hassan, The Hon.
Minister
Foreign Affairs and International Cooperation
Permanent Mission to the United Nations
205 E. 42nd Str.,13th Floor New York, NY, 10017
F: 212 972 9160
TL: 212 682 5232

Feruzi, J., Mr.
Permanent Mission to the United Nations
205 E. 42nd Str.,13th Floor
New York, NY 10017, United States of America
T: 212 972 9160 F: 212 682 5232

Ihema, S., Mr.
Permanent Mission to the United Nations
205 E. 42nd Str., 13th Floor
New York, NY 10017, United States of America
T: 212 972 9160 F: 212 682 5232

Isike, H., Mr.
Permanent Mission to the United Nations
205 E. 42nd Str., 13th Floor
New York, NY 10017, United States of America
T: 212 972 9160 F: 212 682 5232

Issa, Abdulrahman
Director
DEPARTMENT OF ENVIRONMENT
P.O. Box 811
Zanzibar, United Republic of Tanzania
T: 054 30269 32288 3098 F: 054 30647 31252

Jabir, M.H., Mr.
Permanent Mission to the United Nations
205 E. 42nd Str., 13th Floor
New York, NY 10017, United States of America
T: 212 972 9160 F: 212 682 5232

Jamal, A., Mr.
Permanent Mission to the United Nations
205 E. 42nd Str., 13th Floor
New York, NY 10017, United States of America
T: 212 972 9160 F: 212 682 5232

Johari, T., Col.
Permanent Mission to the United Nations
205 E. 42nd Str., 13th Floor
New York, NY 10017, United States of America
T: 212 972 9160 F: 212 682 5232

Kaboyoka, , Lt. Col.
Permanent Mission to the United Nations
205 E. 42nd Str., 13th Floor
New York, NY 10017, United States of America
T: 212 972 9160 F: 212 682 5232

Kamuhanda, S., Mr.
Permanent Mission to the United Nations
205 E. 42nd Str., 13th Floor
New York, NY 10017, United States of America
T: 212 972 9160 F: 212 682 5232

Kamukala, Godfrey L., Mr
Director-General
NATIONAL ENVIROMENT MANAGEMENT COUN
P.O. Box 63154
Dar-es-Salaam, United Republic of Tanzania
T: 34603

Kombe, I.H., Mr.
Permanent Mission to the United Nations
205 E. 42nd Str., 13th Floor
New York, NY 10017, United States of America
T: 212 972 9160 F: 212 682 5232

Korsheni, , Mr.
Permanent Mission to the United Nations
205 E. 42nd Str., 13th Floor
New York, NY 10017, United States of America
T: 212 972 9160 F: 212 682 5232

Lusinde, J.M., Mr.
Permanent Mission to the United Nations
205 E. 42nd Str., 13th Floor
New York, NY 10017, United States of America
T: 212 972 9160 F: 212 682 5232

Mallya, F., Mr.
Permanent Mission to the United Nations
205 E. 42nd Str., 13th Floor
New York, NY 10017, United States of America
T: 212 972 9160 F: 212 682 5232

Masilingi, W., Mr.
Permanent Mission to the United Nations
205 E. 42nd Str., 13th Floor
New York, NY 10017, United States of America
T: 212 972 9160 F: 212 682 5232

Mbwana, S., Mr.
Permanent Mission to the United Nations
205 E. 42nd Str., 13th Floor
New York, NY 10017, United States of America
T: 212 972 9160 F: 212 682 5232

Meteleka, T., Mr.
Permanent Mission to the United Nations
205 E. 42nd Str., 13th Floor
New York, NY 10017, United States of America
T: 212 972 9160 F: 212 682 5232

Mgumia, A. Y., The Hon.
Minister
Tourism, Natural Resources and Environment
Permanent Mission to the United Nations
205 E. 42nd Str., 13th Floor New York, NY, 10017
F: 212 972 9160
TL: 212 682 5232

Mkanga, P.J., Mr.
Permanent Mission to the United Nations
205 E. 42nd Str., 13th Floor
New York, NY 10017, United States of America
T: 212 972 9160 F: 212 682 5232

Mugurusi, E., Mr.
Permanent Mission to the United Nations
205 E. 42nd Str., 13th Floor
New York, NY 10017, United States of America
T: 212 972 9160 F: 212 682 5232

Mwazyunga, F., Mr.
Permanent Mission to the United Nations
205 E. 42nd Str., 13th Floor
New York, NY 10017, United States of America
T: 212 972 9160 F: 212 682 5232

Mwinyi, Hassan, The Hon.
President
United Republic of Tanzania
205 E. 42nd Str., 13th Floor
New York, NY 10017, United States of America
T: 212 972 9160 F: 212 682 5232

Nasser, Suleiman Mohammed
Terrestrial Ecologist
DEPT OF ENVIRONMENT
P.O. Box 811 Zanzibar
Tanzania,, United Republic of Tanzania
T: 054 30269 F: 054 30647

Ngaiza, P., Mr.
Permanent Mission to the United Nations
205 E. 42nd Str., 13th Floor
New York, NY 10017, United States of America
T: 212 972 9160 F: 212 682 5232

Nijiku, P., Mr.
Permanent Mission to the United Nations
205 E. 42nd Str., 13th Floor
New York, NY 10017, United States of America
T: 212 972 9160 F: 212 682 5232

Nyakyi, Antony, H.E. Mr.
Ambassador Extraordinary and Plenipotentiary
Permanent Mission to the United Nations
205 E. 42nd Str., 13th Floor
New York, NY 10017, United States of America
T: 212 972 9160 F: 212 682 5232

Nyenzi, B., Dr.
Permanent Mission to the United Nations
205 E. 42nd Str., 13th Floor
New York, NY 10017, United States of America
T: 212 972 9160 F: 212 682 5232

Nyenzi, Buruhan Salum, Dr.
Meteorologist
MINISTRY OF TRANSPORT
And Communications P.O. Box 3056
Dar-ES-Salaam, United Republic of Tanzania
T: 255 51 44438 F: 255 51 32591

Rajab, Salum, The Hon.
Minister
Water, Energy and Minerals
Permanent Mission to the United Nations
205 E. 42nd Str., 13th Floor New York, NY, 10017
F: 212 972 9160
TL: 212 682 5232

Rupia, Paul, H.E. Mr.
Ambassador
Permanent Mission to the United Nations
205 E. 42nd Str., 13th Floor
New York, NY 10017, United States of America
T: 212 972 9160 F: 212 682 5232

Rutabanzibwa, Patrick, Mr.
Senior Executive Engineer
MINISTRY OF WATER,ENERGY AND MINERALS
PO Box 2000
Dares Salaam, United Republic of Tanzania
T: 31433 F: 37139

Suleiman, K.M., Mr.
Permanent Mission to the United Nations
205 E. 42nd Str., 13th Floor
New York, NY 10017, United States of America
T: 212 972 9160 F: 212 682 5232

Suleiman, S.A., Mr.
Permanent Mission to the United Nations
205 E. 42nd Str., 13th Floor
New York, NY 10017, United States of America
T: 212 972 9160 F: 212 682 5232

NGO

Juma, S.
TES
PO Box 1309
Dar Es Salaam, United Republic of Tanzania

PRESS

Chokala, P.
Correspondent
UNITED REPUBLIC OF TANZANIA PRESS
205 E.42nd Street
New York, N.Y., U.S.A., United Republic of Tanzania

Kalemera, Rose
Editor
DAILY NEWS PAPERS
P.O.Box 9033 Dar-Es
Salaami, UNITED REPUBLIC OF TANZANIA, United Republic of Tanzania

Kuandika, Givic
Correspondent
BUSINESS TIMES
P.O. Box 71439
Dar-Es-Salaam, UNITED REPUBLIC OF TANZANIA, United Republic of Tanzania

Liana, Jacqueline
Editor
UHURU/MZALENDO NEWSPAPERS
Box 9221
Dar Es Salaam, UNITED REPUBLIC OF TANZANIA, United Republic of Tanzania

Msham, Jibrahim Mohammed
Editor
FAMILY MIRROR NEWSPAPER
P. O. Box 6804
Dares Salaam, United Republic of Tanzania

ORGANIZATIONS

Chomba, Hussien J., Mr.
Tanzania Environmental Society
PO Box 1309
Dar Es Salaam, United Republic of Tanzania

Kisangure, Daudi Chacha, Mr.
Tarime Rural Development Trust Fund
PO Box 140
Tarime, United Republic of Tanzania
T: (255) 137

GLOBAL FORUM

Iguandiga, Givic
News Editor
BUSINESS TIMES
Box 71439 Dar - EG
Salarm, United Republic of Tanzania
T: 46074 F: 21773

Kiwasila, Hilda, Ms.
INSTITUTE OF RESOURCES ASSESSMENT
Palais de Nations
Geneve CH 1211, United Republic of Tanzania
F: 41 22 733-6542

Obilla, Mussa R., Mr.
TANGO
Palais de Nations
Geneve CH 1211, United Republic of Tanzania
F: 41 22 733-6542

Ole-Nangoro, Benedict, Mr.
Develop. Worker
ADDO
P.O.Box. 3044
Arusha, United Republic of Tanzania
T: 255 57 2410 F: 255 57 8263

Rwechungura, Cleophas C
Editor
UKULIMA WA KISASA MAGAZINE
PO Box 2308
Dar Es Salaam, United Republic of Tanzania
T: 51 29047

Rweyemanu, Salvator N B
News Editor
UHURU AND MZALENDO NEWSPAPERS
PO Box 9221
Dar Es Salaam, United Republic of Tanzania
T: 255 51 3564/46860

Shaba, Marie, Ms.
TAMWA/WEDO
P.O.Box 6143
Dar Es Salaam, United Republic of Tanzania
T: 255 51 32181

ADVISORS

Nyerere, Mwalimu Julius K., Mr.
Chairman, The South Center : Former President of Tanzania
Dar-es-Salaam, United Republic of Tanzania
T: 255 51 46924/38525/38526 F: 255 51 44144/38527 : 789 85 31

UNITED STATES OF AMERICA - GOVERNMENT

Aronson, Bernard W, The Hon.
Assistant Secretary for InterAmerican Affairs
Department of State
The Capitol
Washington, D.C. 20510, United States of America
T: 202 224 3121

Bannon, Maureen, Lieut. Com.
USN, Office of the Representative for Oceans Policy
Office of the Secretary, Department of Defense
The Capitol
Washington, D.C. 20510, United States of America
T: 202 224 3121

Biniaz, Susan, Ms.
Assistant Legal Adviser for Oceans
International Environmental and Scientific Affairs, Office of the Legal Adviser, Department of State
The Capitol
Washington, D.C. 20510, United States of America
T: 202 224 3121

Bliss-Guest, Patricia, Ms.
Associate Director for International Law and Policy
Council on Environmental Quality, Executive Office of the President
The Capitol
Washington, D.C. 20510, United States of America
T: 202 224 3121

Bohlen, Curtis, The Hon.
Assistant Secretary for Oceans and International Environmental and Scientific Affairs
Department of State
The Capitol
Washington, D.C. 20510, United States of America
T: 202 224 3121

Bonilla, Luis A, Dr
Chairman Puerto Rico's Advisory
OFFICE OF LEGISLATIVE SERVICES
P O Box 3986
SAN JUAN, Puerto Rico 00904, United States of America
T: 809 723 4112 F: 809 724 2428

Bush, George, Mr.
President
White House Office
1600 Pennsylvania Ave
Washington DC 20500, United States of America
T: 202 456 1414 F: 202 456 2461

Carter, Nancy O'Neal, Ms
Coordinator For Population Affairs
US Dept of State
des/CP RM 7825
Washington 20520, United States of America
T: 202 647 3472 F: 202 647 0774

Chaffee, John H., Senator
U.S. Senate, The Capitol
Washington, D.C. 20510, United States of America
T: 202 224 3121

Connors, Donald L., Mr
Chairman Board Of Directors
ENVIRONMENTAL BUSINESS COUNCIL INC.
53 State St. Exchange Place
Boston Mass 02109, United States of America
T: 617 227 5020 F: 617 227 7566

Davis, J. Michael, The Hon.
Assistant Secretary for Conservation and Renewable Energy
Department of Energy
The Capitol
Washington, D.C. 20510, United States of America
T: 202 224 3121

Davis, Trevor
Alternate NGO Representative
U.C.C.-O.C.I.S.
243 Injun Hollow Rd
Haddam Nec, Conn 06424, United States of America
T: 203 525 9171 F: 249 7916

Deland, Michael R., The Hon.
Chairman, Council on Environmental Quality
Executive Office of the President
The Capitol
Washington, D.C. 20510, United States of America
T: 202 224 3121

Delello, Michael S., Mr.
Special Assistant
Office of the Secretary, Department of Energy
The Capitol
Washington, D.C. 20510, United States of America
T: 202 224 3121

Dernbach, John C., Mr.
Special Assistant
PENNSYLVANIA DEPT OF ENVIRONMENTAL RESOURCES
2410 Green St.
Harrisburg-PA, United States of America
T: 1 717 787-9871 F: 1 717 787-1904

Drake, Susan, Ms
UN Programs Manager
Department of State
2201 C St NW Oeslenv RM 4325
Washington DC 20520, United States of America
T: 202 647 9266 F: 202 647 5947

Drucker, Milton K., Mr.
Office of International Commodities, Energy, Resources and Food Policy
Bureau of Economic and Business Affairs, Department of State
The Capitol
Washington, D.C. 20510, United States of America
T: 202 224 3121

Durnil, Gordon K, D
United States Chairman
INTERNATIONAL JOINT COMMISSION
1250 23RD Street NW
Washington DC, United States of America
T: 202 736 9000 F: 202 736 9015

Giddings, Val L., Mr.
Office of Biotechnology Biologics and Environmental Protection
Animal and Plant Health Inspection Service, Department of Agriculture
The Capitol
Washington, D.C. 20510, United States of America
T: 202 224 3121

Gilman, Benjamin A., Congressman
House of Representatives, The Capitol
Washington, D.C. 20515, United States of America
T: 202 224 3121

Gore, Jr., Albert, Senator
U.S. Senate, The Capitol
Washington, D.C. 20510, United States of America
T: 202 224 3121

Grady, Robert, The Hon.
Associate Director for Natural Resources, Energy and Science, Office of Management and Budget
Executive Office of the President
The Capitol
Washington, D.C. 20510, United States of America
T: 202 224 3121

Graham, Bob, Senator
U.S. Senate, The Capitol
Washington, D.C. 20510, United States of America
T: 202 224 3121

Green, Bill, Congressman
House of Representatives, The Capitol
Washington, D.C.

Hayashi, Moritaka
Principal Officer
DIV FOR OCEAN AFFAIRS AND THE LAW
DC2-432, United Nations
New York. NY, United States of America
T: 212 963 3990 F: 212 963 5847

Hecht, Alan D., Mr.
Deputy Assistant Administrator for International Activities
Environmental Protection Agency
The Capitol
Washington, D.C. 20510, United States of America
T: 202 224 3121

Hertel, Dennis M., Congressman
House of Representatives, The Capitol
Washington, D.C. 20515, United States of America
T: 202 224 3121

Johnson, Twig, Mr.
Ph.D, Director, Office of Environmental and Natural Resources
Bureau of Research and Development, Agency for International Development
The Capitol
Washington, D.C. 20510, United States of America
T: 202 224 3121

Keller, Ed, Mr
Air Quality Engineer
SANTA BARBARA AIR POLLUTION CONTROL
26 Castilian Dr. B-23
Goleta, CA 93117, United States of America
T: 805 961 8885 F: 805 961 8801

Kindall, Mark P., Mr.
Office of the International Activities
Environmental Protection Agency
The Capitol
Washington, D.C. 20510, United States of America
T: 202 224 3121

Klien, Stephen, Mr.
U.S. UNCED Coordination Center
Bureau of Oceans and International Environmental and Scientific Affairs, Department of State
The Capitol
Washington, D.C. 20510, United States of America
T: 202 224 3121

Kovar, Jeffrey D., Mr.
Attorney-Adviser
Office of Legal Adviser, Departmentof State
The Capitol
Washington, D.C. 20510, United States of America
T: 202 224 3121

Lake, William, Mr.
Consultant to the Administrator
Environmental Protection Agency
The Capitol
Washington, D.C. 20510, United States of America
T: 202 224 3121

Larsen, Gary L., Mr.
U.S. UNCED Coordination Center
Bureau of Oceans and International Environmental and Scientific Affairs, Department of State
The Capitol
Washington, D.C. 20510, United States of America
T: 202 224 3121

Lautenberg, Frank R., Senator
U.S. Senate, The Capitol Washington, D.C., 20510
F: 202 224 3121

Lovejoy, Thomas E., The Hon.
Assistant Secretary for External Affairs
Smithsonian Institution, Washington, D.C.
The Capitol
Washington, D.C. 20510, United States of America
T: 202 224 3121

McGuinness, John P., Mr.
Office of Technical Specialized Agencies
Bureau of International Organization Affairs, Department of State
The Capitol
Washington, D.C. 20510, United States of America
T: 202 224 3121

Miller, Dennis F, Mr.
Science Advisor
OFFICE OF TECHNOLOGY DEVELOPMENT
Environmental Restoration And Waste Management,U.S.Department of Energy
Washington D.C. 20585, United States of America
T: 202 586 3022 F: 202 586 8872

Miller, George, Congressman
House of Representatives, The Capitol
Washington, D.C. 20515, United States of America
T: 202 224 3121

Miller, John, Congressman
House of Representatives, The Capitol
Washington, D.C. 20515, United States of America
T: 202 224 3121

Moore, Jonathan, The Hon.
Ambassador
United States Representative to the Economic and Social Council of the United Nations, New York
The Capitol
Washington, D.C. 20510, United States of America
T: 202 224 3121

Perkins, Edward J., The Hon.
Ambassador Extraordinary Plenipotentiary
Permanent Representative United Nations, New York
The Capitol
Washington, D.C. 20510, United States of America
T: 202 224 3121

Reilly, William K., The Hon.
Administrator
Environmental Protection Agency
The Capitol
Washington, D.C. 20510, United States of America
T: 202 224 3121

Reinstein, Robert A., Mr.
Deputy Assistant Secretary for Environment
Bureau of Oceans and International Environmental and Scientific Affairs, Department of State
The Capitol
Washington, D.C. 20510, United States of America
T: 202 224 3121

Rohena, Santos, Sr.
Natural Resources Secretary
DEPARTAMENTO RECURSOS NATURALES
Apdo Correo 5887
San Juan, Puerto Rico, United States of America
T: 1 809 7233090 F: 1 809 7234255

Roskens, Ronald W., The Hon.
Administrator
Agency for International Development
The Capitol
Washington, D.C. 20510, United States of America
T: 202 224 3121

Ryan, Robert J., The Hon.
Coordinator for United Nations Conference on Environment and Development preparations
Bureau of Oceans and International Environmental and Scientific Affairs, Department of State
The Capitol
Washington, D.C. 20510, United States of America
T: 202 224 3121

Savage, Eleanor W., Ms.
Director, Office of Ecology, Health and Conservation
Bureau of Oceans and International Environmental and Scientific Affairs, Department of State
The Capitol
Washington, D.C. 20510, United States of America
T: 202 224 3121

Schmitz, John P., Mr.
Deputy Counsel to the President
Executive Office of the President
The Capitol
Washington, D.C. 20510, United States of America
T: 202 224 3121

Scully, R. Tucker, Mr.
Director, Office of Ocean Affairs
Bureau of Oceans and International Environmental and Scientific Affairs, Department of State
The Capitol
Washington, D.C. 20510, United States of America
T: 202 224 3121

Sirmon, Jeff M., Mr.
Deputy Chief for International Forestry
United States Forest Service, Department of Agriculture
The Capitol
Washington, D.C. 20510, United States of America
T: 202 224 3121

Steever, Zell, Mr.
U.S. UNCED Coordination Center
Bureau of Oceans and International Environmental and Scientific Affairs, Department of State
The Capitol
Washington, D.C. 20510, United States of America
T: 202 224 3121

Strachan, Linda, Ms.
Special Assistant for Congressional Relations
Bureau of Oceans and International Environmental and Scientific Affairs, Department of State
The Capitol
Washington, D.C. 20510, United States of America
T: 202 224 3121

Waldrop III, Neal A., Mr.
Office of Global Change
Bureau of Oceans and International Environmental and Scientific Affairs, Department of State
The Capitol
Washington, D.C. 20510, United States of America
T: 202 224 3121

Young, Frank E., Dr.
Deputy Assistant Secretary for Health
Public Health Service, Department of Health and Human Services
The Capitol
Washington, D.C. 20510, United States of America
T: 202 224 3121

Young, Michael K., The Hon.
Deputy Under Secretary for Economic and Agricultural Affairs
Department of State
The Capitol
Washington, D.C. 20510, United States of America
T: 202 224 3121

Zoellick, Robert, The Hon.
Under Secretary for Economic and Agricultural Affairs
Department of State
The Capitol
Washington, D.C. 20510, United States of America
T: 202 224 3121

NGO

Ackermann, Jean-Philippe, Mr.
JUNIOR CHAMBER INTERNATIONAL
PO Box 140577, Coral Gables
Fl 33114-0577, United States of America

Adams, John, Mr.
Executive Director
NATURAL RESOURCES DEFENSE COUNCIL (NRDC)
40 West 20th Street
New York - NY 10011, United States of America
T: 1 212 727-2700 F: 1 212 727-177
TL: 4900010562

Alcala, Maria Jose
PGA
211 East 43rd Street, Suite 1604
New York, NY 10017, United States of America

Almeida, Waldemar
ILSI
1126 Sixteenth Street NW
Washington DC 20036, United States of America

Anderson, Bruce
EARTH DAY USA
P O Box 470, Peterborough, NH 03458
United States of America

Anderson, Nancy W., Ms.
North American Committee Chair
WORLD CONVERSATION UNION
Lincoln Filene Center, Tufts University
Medford - MA 02155, United States of America
T: 1 617 381-3451 F: 1 617 627-3401

Armstrong, Susan
EDA
P O Box 707
Newtown, PA 18941, United States of America

Balbach, John, Mr.
Fellow
FOUNDATION FOR GLOBAL COMMUNITY
222 High Street
Palo Alto - CA 94301, United States of America
T: 1 415 328-7756 F: 1 415 328-7785

Baldwin, John, Mr.
NAAEE
Suite 400, 1255 23rd Street, NW
Washington DC 20037, United States of America

Barratt-Brown, Elizabeth, Ms.
Attorney
NATURAL RESOURCES DEFENSE COUNCIL
1350 New York Avenue, NW #300
Washington - DC 20005, United States of America
T: 1 202 783-7800 F: 1 202 783-5917
TL: 4900010562

Bartholomew, Joy A.
ISEE
PO Box 1589
Solomons, MD 20688, United States of America

Bauer, Joanne
CCEIA
Merrill House, 170 East 64th Street
New York, NY 10021-7478, United States of America

Beebe, Spencer B., Mr.
President
ECOTRUST
10 SW Ash Street
Portland - OR, United States of America
T: 1 503 2276225 F: 1 503 2221517

Benjamin, Richard
ANSO
555 North Kensington Avenue
La Grange Park, Illinois 60525, United States of America

Bereano, Philip
WBAC
20319-92nd Ave. West
Edmonds, Washington 98020, United States of America

Bergmann, Edda
CBJO
823 United Nations Plaza
New York, NY 10017, United States of America

Berkebile, Robert
AIA
1735 New York Avenue, NW; Washington DC 20006-5295
United States of America

Bhagat, Shantilal
CBGB
1451 Dundee Avenue
Elgin, Illinois 60120-1694, United States of America

Blackwelder, Brent
FEUSA
218 D Street, SE
Washington DC 20003, United States of America

Boff, Leonardo
FC
135 Remsen Street
Brooklyn, New York 11201, United States of America

Boom, Brian
NYBG
Bronx
NY 10458-5126, United States of America

Borges, Jose Roberto, Mr.
Amazon Campaign Coordinator
RAINFOREST ACTION NETWORK
301 Broadway Suite A
San Francisco CA 94133, United States of America

Bortnyk, George P.
EERG
13938-85th Terrace North
Seminole, Florida 34646, United States of America

Bowen, Elizabeth
PSR
1000 Sixteenth Street, NW, Suite 810
Washington DC 20036, United States of America

Bowles, Ian
CONSERVATION INTERNATIONAL
1015 18th Street, NW, Suite 1000
Washington DC 20036, United States of America

Bramble, Barbara J., Ms.
Director Of Int'L. Programs
NATIONAL WILDLIFE FEDERATION
1400 - 16th st. NW
Washington - DC, United States of America
T: 1 202 797-6601 F: 1 202 797-5486

Brennan, Andrew
INSEE
Dept. of Philosophy, Colorado State University
Fort Collins, CO 80523, United States of America

Brenner, Beverley
UDWC
1010 S. Flower St., Suite 500
Los Angeles, CA 90015, United States of America

Broisman, Emma R., Ms.
Associate
CENTER FOR THE STUDY OF WOMEN AND SOCIETY
City University of New York, 33 West 42nd Street
New York - NY, United States of America
T: 1 212 6422954 F: 1 212 6422642

Brown, Donald A., Mr.
Director Board
EARTH ETHICS RESARCH GROUP
301 Chestnut St., 4th floor
Harrisburg-PA 17101, United States of America
T: 1 717 787-9368 F: 1 717 787-9379

Brown, Foster
WHRC
13 Church Street, Woods Hole
Massachusetts 02543, United States of America

Brown, Lester R., Mr.
President
WORLDWATCH INSTITUTE
1776, Massachusetts Ave., N.W.
Washington - DC, United States of America
T: 1 202 452-1999 F: 1 202 296-7365

Buschbacher, Robert, Dr.
WORLD WILDLIFE FUND - USA
1250 Twenty-Fourth Street, NW
Washington DC 20037-1175, United States of America

Cagle, Fred
LCF
2255 avenida de la Playa, Suite 14
La Jolla, CA 92037-3287, United States of America

Callicott, J. Baird
INSEE
Dept. of Philosophy, Colorado State University
Fort Collins, CO 80523, United States of America

Calmell Del Solar, Eduardo
PGA
211 East 43rd Street, Suite 1604
New York, NY 10017, United States of America

Cameron Porter, Kathryn
COI
1015 18th Street, NW, Suite 1000
Washington DC 20036, United States of America

Campbell, Laura
ABCNY
42 West 44th Street
New York, NY 10036-6690, United States of America

Campbell, Martha, Ms.
POPULATION COMMUNICATIONS INTERNATIONAL
6200 E. Charlow Dr. PO 111
Englewood - CO, United States of America
T: 1 303 303-2402202 F: 1 303 303-9628

Cargile, Jo Luck, Ms.
Director, Global Services
HEIFER PROJECT INTERNATIONAL
1015 So. Louisiana St.
Little Rock - AR 72202, United States of America
T: 1 501 376-6836 F: 1 501 376-8906
TL: 4949415HEIFER

Carpenter, George, Mr.
Chairman
GLOBAL ENVIRONMENTAL MANAGEMENT INITIATIVE
6110 Center Hill Rd.
Cincinnati - OH 45224, United States of America
T: 1 513 634-7768 F: 1 513 634-6998

Carroll, William J.
AAES
1111 Nineteenth Street, NW, Suite 608
Washington, D.C. 20036-3690, United States of America

Carter, Curtis, Mr.
EEI
701 Pennsylvania Ave., NW
Washington DC 20004, United States of America

Chappell, Thomas, Mr.
CERES
711 Atlantic Avenue
Boston, MA 02111, United States of America

Chasek, Pamela
COL
1709 New York Avenue NW, Suite700
Washington DC 20006, United States of America

Chivian, Eric, Dr.
Professor
PHYSICIANS FOR SOCIAL RESPONSIBILITY
1000 Sixteenth Street, NW, Suite 810
Washington DC 20036, United States of America

Christie, Patrick, Mr.
Univ.Michigan
School of Natural Resources
Ann Harbor, MI 48109-1115, United States of America

Chu, Chia Yu
ICFA
1525 Wilson Boulevard, Suite 500
Arlington, VA 22209, United States of America

Chu, Chung Lan
ICFA
1525 Wilson Boulevard, Suite 500
Arlington, VA 22209, United States of America

Chu, Ming
ICFA
1525 Wilson Boulevard, Suite 500
Arlington, VA 22209, United States of America

Chuse, Patricia, Ms.
CENTER FOR INTERNATIONAL COOPERATION
Rt, 1, Box 220
Castleton 22716, United States of America
T: 1 703 9375219 F: 1 703 9374037

Clark, John
SAF
4 Librairy Court, SE
Washington DC 20003, United States of America

Clugston, Richard, Mr.
Director
HUMANE SOCIETY OF THE U.S./CENTER FOR RESPECT OF LIFE & ENVIRONMENT
2100 L Street NW
Washington - DC 200037, United States of America
T: 1 202 452-1100 F: 1 202 778-6132

Collins, Susie Skelly
EARTH STATION FOUNDATION
1000 Ponce de Leon Blvd., Suite 203
Coral Gables, Florida, United States of America

Colloredo Mansfield, Susanna
LFC
Tufts University, Medford, MA 02155
United States of America

Comitta, Carolyn, Ms.
WORLD INFORMATION TRANSFER INC
274 Madison Ave, Suite 601
New York - NY 10016, United States of America
T: 1 212 2730253 F: 1 212 3792004

Cook, Kenneth A.
CREIP
1718 Connecticut Avenue N.W. Suite 300
Washington D.C. 20009, United States of America

Costanza, Robert, Mr.
President
INTERNATIONAL SOCIETY FOR ECOLOGICAL ECONOMICS
P.O. Box 1589
Solomons-MD 20688, United States of America
T: 410 326-0794 F: 410 326-6342

Court, Eryl
UUUNO
777 U.N. Plaza, Suite 7D
New York, NY 10017, United States of America

Cowart, Marthena
USCNU
300 Broadway, Suite 39
San Francisco CA 94133, United States of America

Craemer, Kevin John
JCI
PO Box 140577, Coral Gables
Fl 33114-0577, United States of America

Creelman, Lynn Williams, Ms.
Research Ass.
UCLA BIOTECHNOLOGY RESEARCH CENTER
405, Hilgard Ave.
Los Angeles 90024, United States of America
T: 310 206 6818 F: 310 825 8440

Cudlipp, Kate
DW
1244 19th Street NW
Washington DC 20036, United States of America

D'Amato, Anthony
ASIL
2223 Massachusetts Ave., NW
Washington DC 20008-2864, United States of America

Da Silva Amaral, P. P.
SGI
7 East 15th Street
New York, NY 10003, United States of America

Dart Lancaster, Star
ESPI
5 Mead Circle, Lexington
Massachusetts 02173, United States of America

Davies, Rennard G.
CEF
2120 13th Street
Boulder, Colorado 80302-4802, United States of America

Davis, William B., Mr.
President
U.S. ASSOCIATION FOR THE CLUB OF ROME
11800 Ambleside Drive, Potomac
Potomac - MD 20854-2104, United States of America
T: 11 301 4243944 F: 11 301 7389042

Dillon-Ridgley, Dianne, Ms.
YWCA OF THE U.S.A., NATIONAL BOARD
726 Broadway
New York - NY 10003, United States of America
T: 1 212 614-2700 F: 1 212 677-9716

Dobson, Tracy
MICHIGAN STATE UNIVERSITY
207 Center for International Programs, Michigan St
East Lansing, MI 48824-1035, United States of America

Dorsey, Michael, Mr.
STUDENT ENVIRONMENTAL ACTION COALITION
Campus Y Building, University of North Carolina
Chapel Hill, NC 27514, United States of America

Drew, Wallace, Mr.
NUCLEAR AGE PEACE FOUNDATION
1187 Coast Village Road, Suite 123
Santa Barbara, CA 93108, United States of America

Driscoll, John, Mr.
ENVIRONMENTAL BUSINESS COUNCIL
53 State Street, Suite 3400
Boston, Massachusetts 02109, United States of America

Duke, Robin Chandler, Ms.
National Chair
POPULATION CRISIS COMMITTEE
1120 19th Street, NW, Suite 550
Washington - DC 20036, United States of America
T: 1 202 659 1833 F: 1 202 293 1795
TL: 440450

Dunlap, Riley
GALLUP INTERNATIONAL INSTITUTE
47 Hulfish Street, P.O. Box 140
Princeton - NJ 08542, United States of America
T: 1 609 921-6200 F: 1 609 924-0228

Eblen, Ruth A.
RDCHE
100 East 85th Street
New York, NY 10028, United States of America

Eblen, William R.
RDCHE
100 East 85th Street
New York, NY 10028, United States of America

Efthimiou, Milton
GOA
177 East 75th Street
New York, NY 10021, United States of America

Eisenberg, Warren Wolff
CBJO
823 United Nations Plaza
New York, NY 10017, United States of America

Elswick, Linda, Ms.
Director
ORGANIC FARMERS ASSOCIATION COUNCIL
P. O. Box 213
Putney - VT, United States of America
T: 1 802 387-2429 F: 1 802 387-4001

Engle, Dawn G.
CBJO
823 United Nations Plaza
New York, NY 10017, United States of America

Epstein, Paul, Dr.
INTERNATIONAL PHYSICIANS FOR THE PREVENTION OF NUCLEAR WAR
126 Rogers Street
Cambridge - MA, United States of America
T: 1 617 868-5050 F: 1 617 868-2560
TL: 4430017

Featherstone, Roger
SAF
4 Librairy Court, SE
Washington DC 20003, United States of America

Fernandez Porto, Jorge
MIPR
PO Box 363 728
San Juan, Puerto Rico 00936-3728, United States of America

Ferraz, Luciana
BKWSU
c/o Gayatri Naraine, 866 UN Plaza, Room 582
New York, NY 10017, United States of America

Fischer, Dietrich
EAAR
475 Riverside Drive, Room 828
New York, NY 10115, United States of America

Fitzgerald, John, Mr.
Chair International Committee
DEFENDERS OF WILDLIFE
1244 19th Street NW
Washington - DC 20036, United States of America
T: 1 202 2294967 F: 1 202 8333349

Flad, Ethan, Mr.
ANGLICAN CONSULTATIVE COUNCIL
815 Second Avenue
New York - NY, United States of America
T: 1 212 922-5222 F: 1 212 490-6684

Flavin, Christopher, Mr.
Vice-President For Research
WORLDWATCH INSTITUTE
1776, Massachusetts Ave., N.W.
Washington - DC, United States of America
T: 1 202 452-1999 F: 1 202 296-7365

Forero, Enrique
NYBG
Bronx
NY 10458-5126, United States of America

Fornos, Werner, Mr.
President
POPULATION INSTITUTE
110 Maryland Ave. NW
Washington - DC 20002, United States of America
T: 1 202 544-3300 F: 1 202 544-0068

Forster, Thomas, Mr.
Director
ORGANIC FARMERS ASSOCIATION COUNCIL
P. O. Box 213
Putney - VT, United States of America
T: 1 802 387-2429 F: 1 802 387-4001

Fox, Margie
LCF
2255 Avenida de la Playa, Suite 14
La Jolla, CA 92037-3287, United States of America

Francis, Douglas T.
UACL
Speedway and Montain
Tucson, AZ 85721, United States of America

Freeman, Helen, Ms.
Acting Chairperson
WORLDWIDE NETWORK
1331 H Street, Suite 903
Washington - DC 20005, United States of America
T: 1 202 347-1514 F: 1 202 347-1524

Freeman, Jim
CIELUS
1621 Connecticut Ave., NW, Suite 300
Washington DC 20009, United States of America

French, Hilary
WI
1776 Massachusetts Ave., NW
Washington DC 20036, United States of America

Funtomiez, Silvio
ISEE
PO Box 1589
Solomons, MD 20688, United States of America

Gallup, George H. Jr., Mr.
Chairman
GEORGE H. GALLUP INTERNATIONAL INSTITUTE
47 Hulfish Street Princeton
Princeton - NJ 08542, United States of America
T: 1 609 921-6200 F: 1 609 924-0228

Gauger, Eugene, Mr.
UNITARIAN UNIVERSALIST UN OFFICE
777 U.N. Plaza, Suite 7D
New York, NY 10017, United States of America

Genshlea, Catherine
USCNU
300 Broadway, Suite 39
San Francisco CA 94133, United States of America

Getz, Arthur
PE
6911 Richmond Highway, Suite 308
Alexandria, Virginia 22306, United States of America

Girard, Wendy Alicia
ISI
1274 Calle de Comercio
Sante Fe, New Mexico 87501, United States of America

Gittinger, J. Price
IDTI
1747 Pennsylvania Ave. NW, Suite 950
Washington, DC 20006, United States of America

Glauthier, T.J.
WWFUSA
1250 Twenty-Fourth Street, NW
Washington DC 20037-1175, United States of America

Gordon, Michael
NSJ
209 East 10th Street, No. 10
New York, NY 10003, United States of America

Gottschalk, Loretta, Ms.
Professor
CENTER OF CONCERN
3700 13th St. NE
Washington - DC 20017, United States of America
T: 1 202 635-2757 F: 1 202 832-9494

Graumer, Denise
ED
P O Box 470, Peterborough, NH 03458
United States of America

Gray, David, Mr.
EARTH LIGHT NETWORK INC.
8111 Lagos de Campo Blvd.
Ft. Lauderdale - FL 33321, United States of America
T: 1 305 721-8585

Greenberg, Dean, Mr.
PROGRESS AGENCY
One Union Square West 201
New York, NY 10003, United States of America

Grogan, Peter
GTF
PO Box 7490
Boulder, CO 80306-7490, United States of America

Gross, Mayne M.
POI
110 Maryland Avenue NE
Washington DC 20002, United States of America

Gupta, Seshadri, Mr.
Executive Vice President
INTERNATIONAL ORGANIZATION FOR WORLD PEACE, DISARMAMENT, DEVELOPMENT AND HUMAN RIGHTS
314 67th St.
Brooklyn - NY 11220, United States of America
T: 1 718 921-6352 F: 1 212 371-9020
TL: 66560 MCIUN

Haas, Peter*
ISA
216 Herald R. Clark Building, Brigham Young University, Provo, Utah 84602, United States of America

Hair, Jay D., Dr.
President
NATIONAL WILDLIFE FEDERATION
1400 Sixteenth Street, NW
Washington - DC 20036-2266, United States of America
T: 1 202 797-6842 F: 1 202 797-6866

Hakkarinen, Charles, Dr.
EPRI
3412 Hillview Avenue
Palo Alto, CA 94304, United States of America

Halleck, Andrea
ENCOR
1218 Third Avenue, Suite 1515, Seattle, WA
United States of America

Hansen, Richard, Mr.
ENERSOL ASSOCIATES INC.
1 Summer Street
Somerville - MA, United States of America
T: 1 617 628-3550 F: 1 617 617-623-5845

Harding, Thomas B., Mr.
Vice-President
IFOAM - INT'L. FEDERATION OF ORGANIC AGRICULTURE MOVEMENTS
c/o AgriSystems Int'l - 125 W. 7th St.
Wind Gap - PA 18091, United States of America
T: 1 215 863-6700 F: 1 215 863-4622

Heffernan, Elizabeth, Ms.
Coordinator Peace Trees
EARTH STEWARDS NETWORK
PO Box 10697 - Bainbridge Island
Washington - VA 98110, United States of America
T: 1 703 842-7986 F: 1 703 842-8918

Hegstrand, John P.
KSE
620 Mendelssohn, Suite 145
Plymouth, Minnesota 55427, United States of America

Hellinger, Stephen, Mr.
Executive Director
DEVELOPMENT GAP
1400 I Street, NW, Suite 520
Washington - DC 20005, United States of America
T: 1 202 898-1566 F: 1 202 898-1612

Hendryx, Nona
MERI
235 West 76th Street
New York, NY 10023, United States of America

Henry, Elizabeth
HSUS
2100 L Street, NW
Washington DC 20037, United States of America

Henson, Dave, Mr.
HIGHLANDER RESEARCH AND EDUCATION CENTER
1939 Highlander Way
New Market - TN 37820, United States of America
T: 1 615 933 3443 F: 1 615 933 3424

Hernandez, Silvia
PGA
211 East 43rd Street, Suite 1604
New York, NY 10017, United States of America

Heyduk, Daniel, Mr.
Director
INSTITUE OF INTERNATIONAL EDUCATION
809 UN Plaza
New York - NY, United States of America
T: 1 212 984-5332 F: 1 212 984-5393
TL: TRT 175977, E: TCN 1513

Holden, John B., Mr.
WORLD FEDERALIST ASSOCIATION
418 7th Street, SE
Washington DC 20003, United States of America

Hollingshead, Laura, Ms.
NEW YORK ZOOLOGICAL SOCIETY/WILDLIFE CONS. INT'L
185th St. & Southern Blvd., Bronx Zoo, Bronx
New York - NY 10460, United States of America
T: 1 212 220-5090 F: 1 212 364-7963

Hoover, W. Scott
NEW ENGLAND TROPICAL CONSERVATORY
PO Box 93, 905 Hancock Road
Williamstown, MA 01267, United States of America

Hopkins, Graham
PU
1354 Montague St., NW. P O Box 56466
Washington DC 20011 P, United States of America

Howell, Christopher L.
Journalis/Producer
UNITED NATIONS ASSOCIATION
723 E Green St.
Pasadena - CA, United States of America
T: 1 818 4638350 F: 1 818 3032391

Humphrey Iii, Hubert H., Mr.
NATIONAL ASSOCIATION OF ATTORNEYS GENERAL
Hall of the States, 444 North Capitol Street
Washington DC 20001, United States of America

Hung, James C., Mr.
SEAC
Campus Y Building, University of North Carolina
Chapel Hill, NC 27514, United States of America

Irwin, Paul G., Mr.
President
HUMANE SOCIETY OF THE U.S.
2100 Street, N.W.
Washington - DC 20037, United States of America
T: 1 202 4521100 F: 1 202 7786132

Iwu, Maurice M.
RA
270 Lafayette Street, Suite 512
New York, NY 10012, United States of America

Jaeger, Dale, Ms.
AMERICAN PLANNING ASSOCIATION
1776 Massachusetts Ave., NW
Washington DC 20036, United States of America

Jodry, Barbara R., Ms.
SOLAR BOX COOKERS INTERNATIONAL
1724 Eleventh Street
Sacramento - CA 95814, United States of America
T: 1 916 444-6616 F: 1 916 447-8689

Kane, Charles
YMCA
101 North Wacker Drive
Chicago, Illinois 60606-7386, United States of America

Kass, Stephen L., Mr.
ABCNY
42 West 44th Street
New York, NY, 10036-6690, United States of America

Kaufman, Steven, Mr.
ENERSOL ASSOCIATES INC.
1 Summer Street
Sommerville, MA 02143, United States of America

Kelly, Patrick, Mr.
YMCA OF THE USA / INTERNATIONAL DIVISION
101 N. Wacker Drive
Chicago - IL IL60606, United States of America
T: 1 312 9970031 F: 1 312 497-9424
TL: 4979424

Kimerling, Judith
WIPACED
c/o Women's Foreign Policy Council, 845 Third Aven
New York NY 10022, United States of America

Kraft, Pamela
NSJ
209 East 10th Street, No. 10
New York, NY 10003, United States of America

Krizek, Bryan, Mr.
President
PARTNERS IN THE ENVIRONMENT
6911 Richmond Hwy., Ste. 308
Alexandria-VA, United States of America
T: 1 703 765-8463 F: 1 703 765-8426

Krupp, Fred, Mr.
Executive Director
ENVIRONMENTAL DEFENSE FUND
257 Park Avenue South
New York - NY 10010, United States of America
T: 1 212 5052100 F: 1 212 5052375

Kurzman, Nancy White
CCCUN
301 East 45th St. (No. 20B)
New York, NY 10017, United States of America

Kurzman, Peter
CCCUN
301 East 45th St. (No. 20B)
New York, NY 10017, United States of America

La Borde, Roger
MF
Box 250, Crestone
Colorado 81131, United States of America

Laird, Sarah
RA
270 Lafayette Street, Suite 512
New York, NY 10012, United States of America

Lamb, George
MF
Box 250, Crestone
Colorado 81131, United States of America

Lamont, Tara, Ms.
DAMIEN FOUNDATION
Russ Building, Suite 1010, 235 Montgomery Street
San Francisco, CA 94104, United States of America

Lancaster, Justin, Mr.
Executive Director
ENVIRONMENTAL SCIENCE AND POLICY INSTITUTE
5 Mead Circle
Lexington - MA 02173, United States of America
T: 1 617 674-1147 F: 1 617 674-2556

Lange, Wanda
WPU
PO Box 10869
Eugene, OR 97440, United States of America

Langman, James E., Mr.
Assistant To Chair
EARTH ISLAND INSTITUTE
300 Broadway, Suite 28
San Francisco - CA 94133, United States of America
T: 1 415 788-3666 F: 1 415 788-7324
TL: 6502829302

Langston, Amelia Nan, Ms.
Public Education Coodinator
CENTER FOR PSYCHOLOGICAL STUDIES IN THE NUDEAR ASSOCIATION
45 Francis Avenue
Cambridge - MA 02138, United States of America
T: 1 617 493-4735 F: 1 617 497-0122

Langston, Judy Ann C.
CENTER FOR PSYCHOLOGICAL STUDIES IN THE NUDEAR ASSOCIATION
1493 Cambridge Street
Cambridge, MA 02139, United States of America

Lashof, Daniel, Mr.
Scientist
NATURAL RESOURCES DEFENSE COUNCIL (NRDC)
1350 New York Ave., NW Suite 300
Washington - DC, United States of America
T: 1 202 783-7800 F: 1 202 783-5917
TL: 4900010562

Lazaro, Margo
GFM
112 Jordan Avenue, San Anselmo, CA 94960
United States of America

Lee White, Sharon
CEF
2120 13th Street
Boulder, Colorado 80302-4802, United States of America

Leet, Glen, Mr.
President
TRICKLE UP PROGRAM
54 Riverside Dr
New York - NY 10024, United States of America
T: 1 212 362-7958 F: 1 212 877-7464

Lerner, Michael, Mr.
President
COMMONWEAL
1517 Swann Str., NW
Washington DC 20009, United States of America

Lesh, Donald R., Mr.
President
GLOBAL TOMORROW COALITION
1325 G Street N.W. Suite 1010
Washington - DC 20005-3104, United States of America
T: 1 202 6284016 F: 1 202 6284018

Lilley, Floy
ANSO
555 North Kensington Avenue
La Grange Park, Illinois 60525, United States of America

Lim, Gill-Chin, Mr.
Dean
MICHIGAN STATE UNIVERSITY
207 International Centre, MSU
E Lansing - Mi 48824, United States of America
T: 517 359-2350 F: 353-7254

Long, Beverly A.
LWV
1730 M Street, NW
Washington DC 20036, United States of America

Lopez Montanez, Wilfredo
MIPR
PO Box 363 728
San Juan, Puerto Rico 00936-3728, United States of America

Lowenthal, Abrao
CBJO
823 United Nations Plaza
New York, NY 10017, United States of America

Lowrie, Diane G., Ms.
Vice-President
GLOBAL TOMORROW COALITION
1325 G Street N.W. Suite 1010
Washington - DC 20005-3104, United States of America
T: 1 202 6284016 F: 1 202 6284018

Lutz, Robert
NHI
Cronkhite Beach, Building 1055, Golden Gate National Recreation Area, Sausalito, CA 94965-2609, United States of America

Lyons, Chief Oren Joheqisho
CIC
Route 1, Box 220
Castleton, VA 22716, United States of America

Macan, Dominick H.T.
FOIN
91 Gregory Lane, No. 21
Pleasant Hill, CA 94523, United States of America

Malone, James
HF
214 Massachusetts Avenue, NE
Washington DC 20002-4999, United States of America

Mangelsdorf, Fred
GCC
c/o Edison Electric Institute, 701 Pennsylvania Av e., NW Washington DC 20004, United States of America

Mankin, William E., Mr.
International Committee
SEIRRA CLUB
1283 Briacliff Rd., #2
Atlanta - GA 30306, United States of America
T: 1 404 881-1532

Manno, Jack
GLU
State University College at Buffalo, Cassety Hall, 1300 Elmwood Av., Buffalo. NY 14222, United States of America

Marcus, Genevieve
EXCI
9872 Easton Drive
Beverly Hills, CA 90210-1418, United States of America

Marshal, Helga, Iaia
IAIA
P O Box 70
Belhaven, NY 27810, United States of America

Marshall, David W.I.
IAIA
P O Box 70
Belhaven, NY 27810, United States of America

Martinez, Sofia, Ms.
SOUTH WEST ORGANIZING PROJECT
211 10th Street SW
Albuquerque, New Mexico 87102, United States of America

Matheus, Carlos
GGII
47 Hulfish Street
Princeton, New Jersey, United States of America

Mathias, Erich D., Mr.
GLOBAL ENVIRONMENT TEAM
UCBWM-Africa, 475 Riverside Dr.
New York - NY 10115, United States of America
T: 1 212 870-2701

Maxman, Susan, Ms.
President-Elect
AMERICAN INSTITUTE OF ARCHITECTS
1735 New York Ave. N.W.
Washington - DC 20006-5292, United States of America
T: 1 202 626-7319 F: 1 202 626-7426

Mayo, Kristina
MF
Box 250, Crestone
Colorado 81131, United States of America

McCleary, Joel, Mr.
President
INSTITUTE FOR ASIAN DEMOCRACY
P.O. Box 524
The Plains - VA 22171, United States of America
T: 1 703 754-8713 F: 1 703 754-1475

McCollough, Charles
UCC
700 Prospect Avenue
Cleveland OH 44115, United States of America

McComie, Elia Garcia, Ms.
Board Member
WORLDWIDE NETWORK
1331 H Street, Suite 903
Washington - DC 20005, United States of America
T: 1 202 347-1514 F: 1 202 347-1524

McDonough, Lee, Ms.
CENTER FOR INTERNATIONAL ENVIRONMENTAL LAW-US
1621 Connecticut Avenue, Suite 300
Washington - DC 20009-1076, United States of America
T: 1 202 332-4840 F: 1 202 332-4865

McDonough, William
WCAR
188 Wittenburg Road
Bearsville, NY 12409, United States of America

McGuinness, Patrick
ICFA
1525 Wilson Boulevard, Suite 500
Arlington, VA 22209, United States of America

McIntwre,
TFGU
2129 13th Street
Boulder, CO 80302, United States of America

Mejia, Louis
President
EARTH STATION FOUNDATION
1000 Ponce de Leon Blvd.Suite 203
Coral Gables - FL, United States of America
T: 1 305 4438652 F: 1 305 8885161

Messenger, Michael, Mr.
WORLD VISION INTERNATIONAL
919 W. Huntington Drive
Monrovia, CA 91016, United States of America

Meyerson, Frederick A. B., Sr.
Representative
RENE DUBOS CENTER FOR HUMAN ENVIRONMENTS
142 East 19TH Street
New York - NY 10003, United States of America
T: 1 212 254-7578 F: 1 212 254-7578

Mikhailkov, Victor
NF
3601 C Street, Suite 798, anchorage, AK 99503
United States of America

Minger, Terry, Mr.
CENTER FOR RESOURCE MANAGEMENT
262 South 200 West
Salt Lake City - UT 84101, United States of America
T: 1 801 3220530 F: 1 801 3283457

Mintzer, Irving M, Mr.
Project Director
STOCKHOLM ENVIRONMENT INSTITUTE
9514 Garwood St
Silver Spring - MD 20901, United States of America
T: 1 301 587-8714 F: 1 301 587-8716

Mittermeier, Russell, Dr.
CONSERVATION INTERNATIONAL
1015 18th Street, NW, Suite 1000
Washington DC 20036, United States of America

Moraes, Jose Mauro
ILSI
1126 Sixteenth Street NW
Washington DC 20036, United States of America

Moshe, Nancy
ICCRE-INTERNATIONAL COORDINATING COMMITTEE ON RELIGION AND THE EARTH
55 Marcourt Drive
Chappaqua, NY 10514, United States of America

Mueller, Peter
EPRI
3412 Hillview Avenue
Palo Alto, CA 94304, United States of America

Munoz, Marya
SCT
Condominium El Centro I, Ofic. 607
Hato Rey, Puerto Rico 00918, United States of America

Mutiso, Reuben
ARC-PEACE
c/o ADPSR, 225 Lafayette Street
New York NY 10012, United States of America

Myers, Colleen
CERES
711 Atlantic Avenue
Boston, MA 02111, United States of America

Myers, P.
FF
320 East 43rd Street
New York, NY 10023, United States of America

Newman, Arnold, Mr.
Director
INTERNATIONAL SOCIETY FOR THE PRESERVATION OF THE TROPICAL RAINFOREST
3931 Camino de la Cumbra
Sherman Oaks - CA 91423, United States of America
T: 1 818 7882002 F: 1 818 9903333

Ninio, Alberto
Staff Attorney
ENVIRONMENTAL LAW INSTITUTE
1616 P Street, NW
Washington DC 20036, United States of America

Nolan, Walker, Mr.
Exec. Vice President
EDISON ELETRIC INSTITUTE
701 Pennsylvania Avenue, N.W.
Washington - DC 20004, United States of America
T: 1 202 508-5460 F: 1 202 508-5403

Nolte, Thomas, Mr.
Techniocal Manager
GALLUP INTERNATIONAL INSTITUTE
47 Hulfish Street, P.O. Box 140
Princeton - NJ 08542, United States of America
T: 1 609 921-6200 F: 1 609 924-0228

Norse, Elliott A., Mr.
Chief Scientist
CENTER FOR MARINE CONSERVATION
15806 NE 47th Court
Redmond - WA 98052-5208, United States of America
T: 1 206 883-8914 F: 1 206 883-3017

Norton, Gale
NAAG
Hall of the States, 444 North Capitol Street
Washington DC 20001, United States of America

Nottinghan, Gregory Philip, Mr.
CHURCH WORLD SERVICE OF WITNESS
2838 A El Lago North Drive
Indianapolis - IND, United States of America
T: 1 317 243-5173 F: 1 317 243-5092

Ofosu-Amaah, Waafas, Ms.
Managing Director
WORLDWIDE NETWORK
1331 H Street, Suite 903
Washington - DC 20005, United States of America
T: 1 202 347-1514 F: 1 202 347-1524

Olds, Glenn A.
NF
3601 C Street, Suite 798, Anchorage, AK 99503
United States of America

Oliveira, Celso Luiz Claro De
COS
870 Greenbrier Circle, Suite 402
Chesapeake, Virginia 23320, United States of America

Omland, Mari
FRIENDS WORLD COMMITTEE FOR CONSULTATION
247E 48th Street
New York, 10017, United States of America

Oppenheimer, Michael
CAPE
801 Pennsylvannia Avenue,SE
Washington DC 20003, United States of America

Ortiz, Alma D. Torres, Ms.
Environmental Coordinator
UNIVERSITY OF PUERTO RICO
PO Box 22785 U.P.R. Station
Rio Piedras, Puerto Rico, United States of America
T: 1 809 7650000 F: 1 809 7538995

P. Mafra, J. Humberto
DF
Russ Building, Suite 1010, 235 Montgomery Street
San Francisco, CA 94104, United States of America

Pannuti, Luiz H.
CENTER OF CONCERN
3700 13th Street,NE
Washington DC 20017, United States of America

Paolozzi, Alicia Spaulding, Ms.
Vice President
INTERNATIONAL COUNCIL OF WOMEN
30 West 61
New York - NY 10017, United States of America
T: 1 212 757-4840

Parker, Vanter, Mr.
Vice President
SIERRA CLUB LEGAL DEFENSE FUND
180 Montgomery St., suite 1400
San Francisco - CA 94104, United States of America
T: 1 415 627-6700 F: 1 415 627-6740

Pat Kenoyer, S. L.
LC
635 E. 12th, 6K
New York, NY 10009, United States of America

Paul, Shaun
UM
School of Natural Resources
Ann Harbor, MI 48109-1115, United States of America

Peacock, Susan
UCC
700 Prospect Avenue
Cleveland OH 44115, United States of America

Pearlman, Donald H., Mr.
CLIMATE COUNCIL
Suite 700, 2550 Main Street NW
Washington - DC 20037, United States of America
T: 1 202 457-6601 F: 1 202 457-6315

Pearlman, Shirley A.
CLIMATE COUNCIL
Suite 700, 2550 Main Street NW
Washington - DC 20037, United States of America
T: 1 202 457-6601 F: 1 202 457-6315

Pesinova, Veronica, Ms.
UCLA BIOTECHNOLOGY RESEARCHE CENTER
405 Hilgard Ave
Los Angeles - CA 90024, United States of America
T: 1 213 2066818 F: 1 213 8258440

Peters, James R., Jr., Mr.
Trainning Coordinator
OREGON STATE UNIVERSITY
400 Snell Hall, Office of Int'l Research and Development
Corvallis - OR, United States of America
T: 1 503 737-6408 F: 1 503 737-3447

Peterson, Glenda Marion
KSE
620 Mendelssohn, Suite 145
Plymouth, Minnesota 55427, United States of America

Petiot, Patrick
SEIN
c/o King, 7C, 321 W. 24th St.
New York, NY 10011, United States of America

Pezzi, Terry Ann, Ms.
President
TERRA CHRISTA COMMUNICATIONS
1880 East River Road
Tucson - AZ, United States of America
T: 1 602 577-6476 F: 1 602 577-1195

Philpott, Julia Ann, Ms.
Policy Analyst
INSTITUTE FOR TRANSPORTATION & DEVELOPMENT POLICY
1787 Columbia Road, N.W.
Washington - DC 20009, United States of America
T: 1 202 387-1434 F: 1 202 387-1450

Phoenix, Arlyn
SEN
426 6th Avenue
Brooklyn, NY 11215, United States of America

Plant, Donathan
GEMI
1828 L Street, NW, Suite 711
Washington DC 20036, United States of America

Pokorny, Bradley, Mr.
Editor 'one Country'
BAHA'I INTERNATIONAL COMMUNITY
866 UN Plaza, Suite 120
New York - NY 10017, United States of America
T: 1 212 7563500 F: 1 212 7563573
E: 152:CNM057

Pollard, Robert, Mr.
Project Director
INTERNATIONAL SYNERGY INSTITUTE
801 Homestead St.
Baltimore, Md., United States of America
T: 1 301 243-2131 F: 1 301 243-7357

Porter, Catherine, Ms.
US CITIZENS NETWORK ON UNCED
300 Broadway-suite 39
San Francisco - CA, United States of America
T: 1 415 956-6162 F: 1 415 956-0241

Postel, Sandra, Ms.
Vice-President For Research
WORLDWATCH INSTITUTE
1776, Massachusetts Ave., N.W.
Washington - DC, United States of America
T: 1 202 452-1999 F: 1 202 296-7365

Prestley, Peter
AMERICAN BAR ASSOCIATION
750 North Lake Shore Drive
Chicago, IL 60611, United States of America

Price, Charlotte
WIPACED
c/o Women's Foreign Policy Council, 845 Third Aven
New York NY 10022, United States of America

Princen, Thomas
GREAT LAKES UNITED
State University College at Buffalo, Cassety Hall,
1300 Elmwood Av., Buffalo. NY 14222, United States of America

Quigley, John
FC
135 Remsen Street
Brooklyn, New York 11201, United States of America

Rabben, Linda, Dr.
RAINFOREST FOUNDATION
270 Lafayette St.
New York - NY, United States of America
T: 1 212 431-9098 F: 1 212 431-9197

Ramage, Patrick R., Mr.
NACRE
5 Thomas Circle, NW
Washington DC 20005, United States of America

Ramp, Floyd
WFA
418 7th Street, SE
Washington DC 20003, United States of America

Randall, Robert
RRI
1727 Massachusetts Avenue, NW
Washington DC 20036, United States of America

Register, Richard
FFAI
Post Box 4347
Arcata, CA 95521, United States of America

Reid, Walter V., Mr.
Vice-President
WORLD RESOURCES INSTITUTE
1709 New York Avenue, N.W
Washington - DC 20006, United States of America
T: 1 202 638-6300 F: 1 202 638-0036
TL: 64414 WRI WASH

Renner, Magda, Ms.
FRIENDS OF THE EARTH - ADFG
1880 East River Road
Tucson - AZ, United States of America
T: 1 602 577-6476 F: 1 602 577-1195

Reyes, Cecile Joaquim, Ms.
Joint Secretary
GLOBAL FORUM OF SPIRITUAL AND PARLIAMENTARY LEADERS ON HUMAN SURVIVAL
304 East 45th St. Room 476
New York - NY, United States of America
T: 1 212 953-7947 F: 1 212 557-20961

Ris, Howard, Mr.
Executive Director
UNION OF CONCERNED SCIENTISTS
26 Church St.
Cambridge - MA 02238, United States of America
T: 1 617 547-5552 F: 1 617 864-9405

Roberts, Malcolm
Northern Forum
3601 C Street, Suite 798, Anchorage, AK 99503
United States of America
T: 1 907 465-3500 F: 1 907 463-3454

Rodrigues, Luiz, Sr
Env. Educator
REEF RELIEF
PO Box 430
Key West, Florida 33041, United States of America

Rogers, Frank
MI
P.O. Box 566
Cambridge, MA 02238, United States of America

Rosenberg Zalk, Sue
CUNY
Health Center
535 East 80th St. New York, NY 10021, United States of America

Rowland, Sandra, Ms
APPROPRIATE TECHNOLOGY INTERNATIONAL
1331 H Street, NW
Washington DC 20005, United States of America
T: 202 8792900 F: 202 6284622
TL: 64661 ATI

Roxanne, Kremer
ISPTR
3302 North Burton Avenue
Rosemead, CA 91770, United States of America

Ruddy, Timothy
ISA
216 Herald R. Clark Building, Brigham Young University, Provo, Utah 84602, United States of America

Saito, Roberto Y.
SGI
7 East 15th Street
New York, NY 10003, United States of America

Sandford, Betty
IAVE
PO Box 27095
Los Angeles, CA 90027-0095, United States of America

Schearer, S. Bruce, Mr.
Executive Director
SYNERGOS INSTITUTE
100 East 85th St.
New York - NY 10028, United States of America
T: 1 212 517-4900 F: 1 212 517-4815

Schlitt, Carol
CDIL
PO Box 53149
Washington DC 20009-3149, United States of America

Schmidt, Martha
TIILO
25 West Ninth Street, Suite 3B, New York, NY 10011
United States of America

Schneider, Franklin
WPU
PO Box 10869
Eugene, OR 97440, United States of America

Schroyer, Trent
TOES
PO Box 12003
Austin, TX 78711, United States of America

Sharma, Shyam
PU
1354 Montague St., NW. P O Box 56466
Washington DC 20011 P, United States of America

Shaw, James
CEA
PO Box 47504
Plymouth, MN 55447-0504, United States of America

Shlaes, John, Mr.
Executive Director
GLOBAL CLIMATE COALITION
1331 Pennsylvania Avenue, NW, 15th floor
Washington - DC 20004, United States of America
T: 1 202 637-3162 F: 1 202 638-1043

Shoen, Sophia
Executive Director
GLOBAL OBJECTIVES, INC.
3653 N 6th Avenue No. C-10
Phoenix, Arizona 85013, United States of America

Shropshire, Steve, Mr.
Special Assistant To G.Hickel
STATE OF ALASKA, NORTHERN FORUM
3601 'C' St., Suite 724
Anchorage - Alaska 99503, United States of America
T: 1 907 465-2500 F: 1 907 463-3841

Siglin, Douglas
EESI
122c Street NW, Suite 700
Washington DC 20001, United States of America

Sims, William
TFGU
2129 13th Street
Boulder, CO 80302, United States of America

Sive, David
HC
P.O. Box 299 148 Cream Hill Road
West Cornwall, CT 06796, United States of America

Smith, Elliot
CHF
PO Box 91280
Washington DC 20090-1280, United States of America

Smith, Robert L.
EXCI
9872 Easton Drive
Beverly Hills, CA 90210-1418, United States of America

Smith, Turner, Mr.
Attorney
ENVIRONMENTAL LAW INSTITUTE
1616 P. Street, N.W., Suite 200
Washington - DC 20036, United States of America
T: 1 202 939-3853 F: 1 202 328-5002

Snyder, Mary-Jane, Ms.
Board
POPULATION INSTITUTE
107 2nd Ave, N. E.
Washington - DC 20002, United States of America
T: 1 202 544-3300

Soleri, Paolo
ARCP
c/o ADPSR, 225 lafayette Street
New York NY 10012, United States of America

Springuel, Natalie
SAF
4 Libraiy Court, SE
Washington DC 20003, United States of America

Springuel, Rene H.
IDTI
1747 Pennsylvania Ave. NW, Suite 950
Washington, DC 20006, United States of America

Stabinsky, Doreen
WBAC
20319-92nd Ave. West
Edmonds, Washington 98020, United States of America

Stang, Betsy
WCAR
188 Wittenburg Road
Bearsville, NY 12409, United States of America

Steger, Will
AAEF
Gilbert Building No. 200, 413 Wacouta Street
St. Paul, Minnesota 55101, United States of America

Stephenson, Carolyn, Ms.
INTERNATIONAL STUDIES ASSOCIATION
Political Science Depto. - University of Hawaii of Manoa
Honolulu - HI, United States of America
T: 1 808 956-8195 F: 1 808 956-6877

Stovall, Robert Kirk
Director Of Programs
JUNIOR CHAMBER INTERNATIONAL
PO Box 140577, Coral Gables
Fl 33114-0577, United States of America
T: 1 305 446-7608 F: 1 305 442-0041

Strock, James
CMC
1725 De Sales Street, NW
Washington D.C. 20036, United States of America

Swart, Lydia
CDIL
PO Box 53149
Washington DC 20009-3149, United States of America

Tenney, Anne
CENTER FOR ENVIRONMENTAL INFORMATION
46 Prince Street
Rochester NY 14607-1016, United States of America

Terry, Charles
ICIS
121 Avenue of The Americas
New York, NY 10013, United States of America

Thomas, Urs
ISA
216 Herald R. Clark Building, Brigham Young
University, Provo, Utah 84602, United States of America

Thompson, Gregory
WVI
919 W. Huntington Drive
Monrovia, CA 91016, United States of America

Thorhaug, Anitra
USACR
11800 Ambleside Drive
Potomac, MD 20854-2104, United States of America

Tian-Kwei, Ding
CCEIA
Merrill House, 170 East 64th Street
New York, NY 10021-7478, United States of America

Tinker, Catherine
TIILO
25 West Ninth Street, Suite 3B, New York, NY 10011
United States of America

Treadwell, Mead
NF
3601 C Street, Suite 798, Anchorage, AK 99503
United States of America

Tsalis, Tula, Ms.
TOES
PO Box 12003
Austin, TX 78711, United States of America

Tzelios, Menelaos
IFPE
11-25 30th Avenue
Long Island City, NY 11102, United States of America

Uribe, Gonzalo
IALC
300 22nd Street, Oak Brook
Illinois 60521-8842, United States of America

Vachta, Kerry
SEAC
Campus Y Building, University of North Carolina
Chapel Hill, NC 27514, United States of America

Vallet, Ronald E., Mr.
ECSS
1100 West 42nd Street, Suite 335, Indianapolis, In
diana 46208, United States of America

Van De Woestyne, Geert
SEN
426 6th Avenue
Brooklyn, NY 11215, United States of America

Van Nort, Marjorie
GO
3653 N 6th Avenue No. C-10
Phoenix, Arizona 85013, United States of America

Vannatta, Paula
SCEC
PO Box 32305
Los Angeles, CA 90032, United States of America

Veith, Cheryl
EDA
P O Box 707
Newtown, PA 18941, United States of America

Veloso, Mario
Executive Secretary
GENERAL CONFERENCE OF S.D.A.
12501 Old Columbia Pike
Silver Spring, Maryland 20904-1608, United States of America
F: 1 301 680 6090

Viereck, John Alexis
NPF
1835 K Street, NW, Suite 610
Washington DC 20006, United States of America

Vineyard, Phyllis
Director
POPULATION INSTITUTE
110 Maryland Avenue NE
Washington DC 20002, United States of America

Vogelsberg, Tony
AFEAS
The West Tower - Suite 400
1333 H Street NW, Washington DC 20005, United States of America

Wallace, Richard, Mr.
Research Assistant
OVERSEAS DEVELOPMENT COUNCIL
1717 Massachusetts Ave. NW
Washington - DC 20036, United States of America
T: 1 202 234-0871 F: 1 202 745-0067

Warner, Virginia
SHARARUNA FOUNDATION
PO Box 130
The Plains, Virginia 22171, United States of America

Watkins, Anthony
ARCP
c/o ADPSR, 225 lafayette Street
New York NY 10012, United States of America

Weiss, Edith Brown
ASIL
2223 Massachusetts Ave., NW
Washington DC 20008-2864, United States of America

Westra, Laura
INSEE
Dept. of Philosophy, Colorado State University
Fort Collins, CO 80523, United States of America

Westwood, David
WORLD VISION INTERNATIONAL
919 West Huntington Drive
Monrovia -CA 91016, United States of America
T: 1 818 3017756 F: 1 818 3582548

Wilcox, Bruce
ISD
3000 Sand Hill Road, Building 1,Suite 102
Menlo Park, CA 94025, United States of America

Wilke, Richard
NORTH AMERICAN ASSOCIATON FOR ENVIRONMENT EDUCATION
Suite 400, 1255 23rd Street, NW
Washington DC 20037, United States of America

Williams, Larry
SC
730 Polk Street
San Francisco, CA 94109, United States of America

Wilmot, Christopher, Mr.
ZERO POPULATION GROWTH
1400 16th street NW suite320
Washington - DC 20036, United States of America
T: 1 202 3322200 F: 1 202 3322302

Wilson, Dwight, Mr.
Director, Global Service Prog.
EARTH STEWARDS NETWORK
PO Box 10697 - Bainbridge Island
Washington - VA 98110, United States of America
T: 1 206 842-7986 F: 1 206 842-8918

Wise, Donna, Ms.
V.P. Policy Affa
WORLD RESOURCES INSTITUTE
1709 New York Ave., NW, Suite 700
Washington - DC, United States of America
T: 1 202 638-6300 F: 1 202 638-0036
TL: 64414WRIWASH

Wittbecker, Alan
MI
P.O. Box 566
Cambridge, MA 02238, United States of America

Wittwer, Nancy
LC
635 E. 12th, 6K
New York, NY 10009, United States of America

Young, Roy
GTF
PO Box 7490
Boulder, CO 80306-7490, United States of America

Zanes, Anne, Ms.
Member Of Board Of Directors
COMMUNICATIONS COORDINATIVE COMM. FOR UN
290 Riverside Dr. - Apt.15A
New York - NY, United States of America
T: 1 212 749-6297

Zwerver, John Harry
GLOBAL FAMILY
112 Jordan Ave
San Anselmo,CA 94960, United States of America

PRESS

Abbasi, Daniel
Correspondent
EARTH SUMMIT TIMES
30 Cat Rock Road
Cos Cob, CT 06342, USA, United States of America

Abrams, Isabel
Correspondent
GENERAL LEARNING CORP.
2216 Schiller Avenue
Wilmette, IL 60091, USA, United States of America

Abramson, Rudy
Correspondent
LOS ANGELES TIMES
11806 Tree Fern Court
Reston, VA 22091, USA, United States of America

Abs, Susan
Reporter Radio/Television
MONITOR RADIO
2-2254 Cornwall Avenue
Vancouver, BC V6K 155, Canada, United States of America

Achenbach, Joel
Correspondent
WASHINGTON POST
5422 Cathedral Avenue, NW
Washington, D.C. 20016, USA, United States of America

Adam, Glenn
Editor
GREENWIRE
2702 Wisconsin Avenue, NW
Washington, DC 20007 USA, United States of America

Adams, Sarah Lynn
Correspondent
EAGLE TIMES
1514 17th Street, N.W.
Washington, D.C. U.S.A., United States of America

Aglow, Robert
ABC NEWS
30 Prescott Ave.
Bronxville, NY 10708 USA, United States of America

Ahmad, Nermin
Correspondent
INTERNATIONAL REPORT
3020 Dent Pl. NW (#14W)
Washington, DC 20007, USA, United States of America

Akers, Dee, Mr.
Correspondent
INSIDE ENVIRONMENT
132 Wildwood Lane
Kenbrucky, USA 40223, United States of America

Akers, Lilialyce
Correspondent
COURIER JOURNAL
132 Wildwood Lane
Louisville, KY 40223, USA, United States of America

Alboum, Kenneth
Correspondent
DOLLARS AND SENSE
705 Centre St. Boston,
Ma. 02130, USA, United States of America

Allen, Frank Edward
Editor
WALL STREET JOURNAL
185 Ladderback Lane
Devon, Pennsylvania 19333 USA, United States of America

Allen, John
Correspondent
LAHAINA NEWS
Sunspace Ranch
Oralle, Arizona, USA, United States of America

Allen, Marlon
Editor
INT'L ENVIRONMENT REPORTER
1924 15th St., N.W.
Washington, DC 20009, USA, United States of America

Alves Pinto, Daniela K.
Correspondent
META
Ataulfo de Paiva 983/703
Rio de Janeiro, Brazil, United States of America

Alvin, Jones
Reporter Radio/Television
ICTV
22610 Francais Irwis Blvd.
Cambrim, USA, United States of America

Amanpour, Christiane
Reporter
CNN BROADCASTING SYSTEM
One CNN Center, Box 105366
Atlanta - GA, United States of America
T: 1 404 8371918 F: 1 404 8274292

Ambrose, Larry
Correspondent
INTERMOUNTAIN JEWISH NEWS
3970 Zenobia
Denver, CO 80212 USA, United States of America

Anderson, Judith
Editor
21ST CENTURY NEWS
6655 North Canyon Crest Drive
Tucson, AZ 85718, USA, United States of America

Andrade, Walter
Reporter Radio/Television
USIA/TV MARTI
9431 Lee Highway
Fairfax, VA 22031, USA, United States of America

Arango, Victor M.
Correspondent
INTERNATIONAL COMMUNITY TV
1726 12th Ave.
Coralville, IA, U.S.A., United States of America

Araujo, Alan
Reporter Radio/Television
21ST CENTURY NEWS
Av. Atlantica 586/402
Rio de Janeiro, Brazil, United States of America

Aresti, Lore
Correspondent
FINAL CALL NEWSPAPER
Av.Pipioltepec 51
Mexico - DF, United States of America

Askari, Emilia
Correspondent
DETROIT FREE PRESS
321 W. Lafayette
Detroit, Michigan 48226, United States of America

Assis Rodrigues, Carlos Otavio
EARTH SUMMIT TIMES/JB
Trav. Julio Froes, 141
Engh, Niterio, Rio de Janeiro,Brazil, United States of America

Atkinson, Darren
Reporter Radio/Television
MOONDRAGON PICTURES, INC
11 Altamont Road
Willowdale, ONT, Canada M2M 1S5, United States of America

Atlas, Terry
Correspondent
CHICAGO TRIBUNE
3123 N.17th St.
Arlington, VA 22201 USA, United States of America

Avirgan, Shanti
Correspondent
MESOAMERICA
Apdo 518-1007 San Jose
Costa Rica, United States of America

Azcuy, Ana M.
Reporter Radio/Television
UNIVISION
6401 Riviera Drive
Coral Gables, FL 33146, USA, United States of America

Baca, James
Reporter Radio/Television
HISPANIC RADIO NETWORK, INC.
3128 La Mancha Place, NW
Albuquerque, NM 87104, USA, United States of America

Bach, Rudi
Reporter Radio/Television
OREGON PUBLIC BROADCAST
2701 2nd Ave N. Seattle
WA 98109, Usa, United States of America

Bailey, Pamela Michelle
Correspondent
PACIFIC NEWS SERVICE
56 Marin Av.
Sausalito, CA 94765, United States of America

Bailey, Ronald, Mr.
Correspondent
REASON MAGAZINE
2350 Highland Ave.
Falls Church, VA 22046, USA, United States of America

Balbin, Maribel
Reporter Radio/Television
RADIO MIAMI INTERNATIONAL/AHORA
14200 Alamanda Ave.
Miami, FL, USA, United States of America

Ballard, Melissa
Producer
CNN BROADCASTING SYSTEM
One CNN Center, Box 105366
Atlanta - GA, United States of America
T: 1 404 8371918 F: 1 404 8274292

Banks, David O.
Correspondent
COAL VOICE
P.O. Box 74
Dunn Loring, VA 22027, USA, United States of America

Barber, Carolyn
Reporter Radio/Television
GLOBALVISION
33 Gold Street, #124
New York, N.Y. 10038, U.S.A., United States of America

Barnd, William Jeffrey
Reporter Radio/Television
WBFF-TV FOX 45
8465-4D Oakton Lane
Ellicoot City, MD 21043, United States of America

Barnes, Cristine
Correspondent
THE WORLD PAPER
18 Samoset Ave.
Barrington RI 02806, United States of America

Barrett, Minna
Reporter Radio/Television
INTERNATIONAL COMMUNITY TV
35 Knights Bridge Rd. I-7
Great Necxk, New York, U.S.A., United States of America

Basaria, Abdul Malik
Correspondent
WORLD POPULATION NEWS SERVICE
6902 Hutchison St.
Falls Church VA 22222222043, United States of America

Bascom, Jon
Correspondent
ABC RADIO
1748 S. Street, N.W.
Washington, D.C. 20009, USA, United States of America

Bassett, Carol Ann
Reporter Radio/Television
WISCONSIN PUBLIC TELEVISION
P.O.Box 1509
#ucson, Arizona, 85702 USA, United States of America

Becker, Daniel, Mr.
Editor
SIERRA MAGAZINE
3324 19th St., NW
Washington, DC 20010, USA, United States of America

Beers, Thomas David
CNN/NETWORK EARTH
210 Degress St.
Atlanta, GA 30303, USA, United States of America

Berenyi, John Andrew
Editor
GOVERNMENTAL ADVISORY ASSOCIATES,INC
64 E 86th St.
New York, NY 10028 USA, United States of America

Berg, Jordana
Editor
WTN
R. Almirante Alexandrino, 3780
Rio de Janeiro, Brazil, United States of America

Berle, Peter A.A.
Editor
ADUBON MAGAZINE
530 East 86 Street
New York, N.Y. 10028, United States of America

Bernthal, Ron
Reporter Radio/Television
MIDDLE SUNDAY RECORD/WJFF NPR
P.O.Box D
Hurleyville, NY 12747, USA, United States of America

Bierbauer, Charles
Correspondent
CNN
9408 Duxford Ct.
Potomac, Maryland, U.S.A., United States of America

Bishop, Robert John
Editor
CIESIN
2853 W. Jolly Rd.
Okemos, MI 48864 USA, United States of America

Black, Naomi, Ms.
Senior Editor
COLORS MAGAZINE
50 West 17th St., 12th floor
New York - NY 10011, United States of America
T: 1 212 645-2126 F: 1 212 645-7789

Blair, Russel
Correspondent
EARTH SUMMIT TIMES
502 Park Ave.
New York, N.Y., U.S.A., United States of America

Blanco, Luis
Editor
UNIVISION
10281 N.W. 56 St.
Miami, FL 33178, USA, United States of America

Boddy, Patricia
Correspondent
WEDO
2007 46TH Street
Des Maines, Iowa, U.S.A., United States of America

Bohn, Kevin
CNN
4529 S. 28th St.
Arlington, Virginia, U.S.A., United States of America

Bourin, Len
Reporter Radio/Television
ABC NEWS
555 NE 34th St.
Miami, FL 33137 USA, United States of America

Brady, Monica
Reporter Radio/Television
USIA/VOA
R. Sta. Clara 277/101
Copacabana, Rio de Janeiro-RJ, Brazil, United States of America

Brandao, Gloria
Editor
CNN
Av. Min.Afranio Costa, 21
Rio de Janeiro-RJ, Brazil, United States of America

Bransilver, Maud
USIA/TV MARTI
2250 Clarendon Blvd., 806
Arlington, VA 22201, USA, United States of America

Bronstein, Scott
Correspondent
ATLANTA JOURNAL
124 Glendale Ave.
Decatur, GA 30030, USA, United States of America

Brooke, Elizabeth
Correspondent
NEW YORK TIMES
R. Leoncio Correa 123
Rio de Janeiro, Brazil, United States of America

Brooke, James
Correspondent
NEW YORK TIMES
Ave. Visconde de Piraja
550/2012-CEP 22410 Ipanema, RJ-RJ,Brazil, United States of America

Brooks, Sam W.
Editor
ABC
4501 N. 17th St.
Arlington VA, USA, United States of America

Bruce, James
Correspondent
JOURNAL OF COMMERCE
B. Luiz Antonio 3004/9
Sao Paulo, Brazil, United States of America

Bruner, Karla
Correspondent
TIME MAGAZINE
Rua Prof. Joao Arruda 197
Perdizes, Sao Paulo SP 05012, Brazil, United States of America

Buck, Randall Dean
Correspondent
DAILY BEE
9220 Lower Pack River Road
Sandpoint, Idaho, 83864, USA, United States of America

Burdett, Harld N.
Editor
WORLD POPULATION NEWS SERVICE
107 - Second street NE
Washington DC 20002, United States of America
T: 1 202 544 330

Burgess, Philip
Correspondent
ROCKY MOUNTAIN NEWS
1625 Broadway, Suite 600
Denver, CO. 80202, USA, United States of America

Burkhardt, Bruce
Reporter Radio/Television
CNN/TBS
909 Los Angeles Ave., NE
Atlanta, GA 30306, USA, United States of America

Bustamante, Paola
Reporter
NEWSWEEK
Rua Ramon Frang No. 100
22290-Urca, Brazil, United States of America

Camargo, Zeca
Reporter Radio/Television
MTV
Praca Buritama 48
Apt. 01, Rio de Janeiro, Brazil, United States of America

Cameron, Martha
Editor
EARTH SUMMIT TIMES
696 Pres. St.
Brooklyn, NY 11215, USA, United States of America

Cappo, Joseph C.
Editor
CRAIN COMMUNICATIONS, INC
823 Monticello Place
Evanston, Illinois 60201 USA, United States of America

Carlson, Eric, Mr.
Correspondent
WOMEN'S INTERNATIONAL NETWORK
P.O.Box 3425, 16016 via Dicha
Rancho Santa Fe, CA 92067, USA, United States of America

Carneiro, Daniel
Editor
AIRLINES MAGAZINE
R. Conde de Bonfim 682/201
Rio de Janeiro, Brazil, CEP20520, United States of America

Carrasco, Lorenzo
Correspondent
EXECUTIVE INTELLIGENCE REVIEW
Av. Graca Aranha 327 s.1108
Rio de Janeiro-RJ, Brazil, United States of America

Carter, John M.
Editor
HEARST MAGAZINE
20 Ridge Road
Bronxville, New York,USA, United States of America

Carty, Winthrop P.
Editor
POPULATION TODAY
3900 Watson Place-A, N.W.
Washington, DC 20016 USA, United States of America

Case, Patricia
Editor
DISCIPLE
956 Ellenberger Parkway W. Dr.
Indianapolis, IND 46219, USA, United States of America

Casselman, Marilyn
Correspondent
EARTH SUMMIT TIMES
P.O. Box 343
Taos Nm 87571, United States of America

Castillo, Sharon J.
USIA/TV MARTI
4601 N. Park Ave., No. 617
Chevy Chase, MDA 20815, USA, United States of America

Chang, Albert
Correspondent
KBS
4112 Chess Dr.
Anchorage Ak 99508 USA, United States of America

Chang, Williamson
Correspondent
HONOLULU WEEKLY
969 Ahuwale Place
Honolulu. Hawaii 96821, USA, United States of America

Chappie, Damon
Editor
BNA-DAILY ENVIRONMENT REPORT
5325 Buxton Ct.
Alexandria, VA 22310 USA, United States of America

Chapple, Stephen
Correspondent
LOS ANGELES TIMES MAGAZINE
735 Union St.
San Francisco, CA 94133, USA, United States of America

Christensen, John
Correspondent
PACIFIC NEWS SERVICE
6185 Franktown Road
Carson City, Nevada 89704, United States of America

Christian, Diji
Correspondent
TIMELINE
101 Hillborn Ave 3
Vallejo, CA 94590, USA, United States of America

Cismowski, Dorothy
Editor
CHANNEL ONE
222 Marguerite Avenue
Mill Valley, CA 94941, USA, United States of America

Cochran, John Jerome
Correspondent
NBC
3147 O Street NW
Washington Dc, 20007 USA, United States of America

Colgrove, Nana
Correspondent
SOJOURNER PRODUCTIONS/META
225 W 86th St.
New York, NY 10024, USA, United States of America

Collins, Carole J.L.
Un Correspondent
NATIONAL CATHOLIC REPORTER
115 E Armour Blvd
Kansas City, MO 6411, United States of America
T: 1 816 5310538 F: 1 212 7499335

Collins, Sharon
Correspondent
CNN/NETWORK EARTH
855 Piedmont Ave. (#1D)
Atlanta, GA 30308 USA, United States of America

Compton, Ann
Reporter Radio/Television
ABC
5009 Tilden St. NW
Washington, DC 20016 USA, United States of America

Connell, Joan
Correspondent
NEWHOUSE NEWS SERVICE
8003 Newdale Road
Bethesda, MD 20814, USA, United States of America

Content, Robert
Correspondent
FOUNDATION FOR GLOBAL COMMUNITY
1746 Bayview Dr. Hermosa B.
Ca, USA, United States of America

Cook, Diane
Editor
CAMARA PRODUCTIONS
170 Marion Street
Denver, CO 80218, USA, United States of America

Cookson, Michael
Correspondent
GLOBAL 500 NEWSLETTER(UNEP)
13 Regent St.,
MT Waverly 3149, United States of America

Cooper, Marc
Correspondent
VILLAGE VOICE
22800 Calabash
Woodland Hills, CA 91364, USA, United States of America

Cooper, Richard T.
Editor
LOS ANGELES TIMES
210 South Lee St.
Alexandria, VA 22314, USA, United States of America

Covarrubias, Alvaro E.
Correspondent
GLOBAL FAMILY NEWS NETWORK
Diagonal Paraguay 360 Dept.521
Santiago, Chile, United States of America

Coxton, Angela
Correspondent
WORLDLINK
407 Beirut Ave.
Pacific Palisades, CA 90272, USA, United States of America

Craatty, Carol
CNN
10201 B. Willow Mist Ct.
Oakton, Virginia, U.S.A., United States of America

Craig, James Dexter
Reporter Radio/Television
21ST CENTURY NEWS
1810 East Blackldge, No. 104
Tucson, AZ 85719, USA, United States of America

Crosby, William
Reporter Radio/Television
WISCONSIN PUBLIC TELEVISION
1700 N. Wentworth Rd.
Tucson, Arizona, 85749 USA, United States of America

Crutcher, Anh
Correspondent
MEDIA LINK
PO Box 2455
Amagansett New York, United States of America

Cuauhtemoc, Perez
Correspondent
TIMELINE
820 Humboldt Rd
Brisbane, CA 94005, USA, United States of America

Cullison, Paula
Correspondent
TODAY'S ARIZONA WOMAN
13058 N Surrey Circle
Phoenix, AZ 85029, United States of America

Cuomo, Kerry Kennedy
Reporter Radio/Television
CNN/TBS
122 Grosvenor St.
Douglaston, NY 11363 USA, United States of America

Curwood, Stephen T.
Editor
LIVING ON EARTH/NAT'L PUB.RADIO
26 Chillon St.
Cambridge, MA 02138, USA, United States of America

Daines, William Kay
Editor
INTERNATIONAL TRADE CALENDAR
1226 Ranleigh Road
McLean, Virginia 22101, United States of America

Danger, Lisa
Correspondent
SIERRA
730 Polk Street
San Francisco -CA 9, United States of America
T: 415-9235608 F: 415-7764868

David S., Helvarg
Correspondent
KRON TV
140 Buchanan
Sausalito, California 94965 USA, United States of America

Day, William Charles
Correspondent
ENVIRONMENT NEWS REPORT
11 Birch LANE
O6830 Greenwich,CT, USA, United States of America

De Carrasco, Silvia
Correspondent
EXECUTIVE INTELLIGENCE REVIEW
Av. Graca Aranha 327 S.1108
Roi de Janeiro-RJ, Brazil, United States of America

De La Lande, Bob
Correspondent
SEVEN NETWORK AUSTRALIA
10445 Wilshire Blvd 604
Los Angeles 90045, USA, United States of America

De Melo, Ayla Tiago
Reporter Radio/Television
WNET-13
54 Barrow St.
New York, NY 10014, USA, United States of America

De Molina, Raul Gomez
Correspondent
RADIO MIAMI INTERNATIONAL/AHORA
8788 SW 12St.#106
Miami, FL, USA, United States of America

De Saram, Asokamala
Correspondent
WORLD POPULATION NEWS
26/2 Lauries Rd.
Sri Lanka, United States of America

De Saram, John H.
Correspondent
WORLD POPULATION NEWS
46 Jl. Setiabakti 9
Kuala Lumpur, Malaysia, United States of America

De Tristan, Marilene
Correspondent
MIAMI HERALD
Ro.Raul Pompeia 131/1011
Copacabana, Rio de Janeiro-RJ, Brazil, United States of America

Del Mastro, Salvatore
ABC NEWS
112 Lighthouse Drive
Freehold, NJ 07728, USA, United States of America

Delaski, Kathleen
Reporter Radio/Television
ABC
2110 Patty LA
Vienna, VA 22182, United States of America

Delgado, Oscar
Correspondent
NBC NEWS
Roma 13, #202
Mexico-City, Mix., United States of America

Demaio, Don
Editor
EARTH SUMMIT TIMES
147 Farragut Pkwy
Hastings-on-Hudson, NY 10706, USA, United States of America

Devroy, Ann
Correspondent
WASHINGTON POST
3215 MacKinley N.W.
Washington, USA, United States of America

Dillon, Arlene M.
Reporter Radio/Television
CBS
2534 E. So. Walter Reed Dr.
Arlington, VA 22206 USA, United States of America

Dillon, Samuel
Correspondent
MIAMI HERALD
Av. Portugal 248, Urca
22291 Rio de Janeiro-RJ, Brazil, United States of America

Diringer, Elliot
Staff Writer
SAN FRANCISCO CHRONICLE
827 Broadway, Suite 340
Oakland, California, United States of America
T: 1 510 465-0514 F: 510 834-5556

Do Nascimento, Olimpio Jose
Correspondent
KSQQ
Rua Monsenhor Jeronimo 400
Rio de Janeiro, Brazil, United States of America

Dobbs, Francis
Reporter Radio/Television
WORLD BANK FILM & VIDEO UNIT
2401 S. Dinwiddie St.
Arlington, VA 22206, USA, United States of America

Dolan, Maura E.
Correspondent
LOS ANGELES TIMES
1150 N Mar Vista Ave.
Pasadena, CA 91104, USA, United States of America

Donald, B. Marti
Correspondent
EARTH SUMMIT TIMES
163 Prospeg
Brooklyn no. 11215, United States of America

Dougherty, Jill
Correspondent
CNN
2475 Virginia Ave., N.W.
Washington, D.C., U.S.A., United States of America

Doyle, Lawrence
Correspondent
CBS
4770 Biscayne Blvd., No. 1170
Miami, FL 33137, USA, United States of America

Doyle, Michael William
Correspondent
GOLDEN DOME PRODUCTIONS
50857 Safari Drive
Granger, IN 46530, USA, United States of America

Drowne, Roger H.
Correspondent
STOWE REPORTER
3164 Nebrask Valley Road
Stowe Vermont, USA 05672, United States of America

Duffy, Michael
Correspondent
TIME MAGAZINE
315 5th Street, N.E.
Washington, D.C., U.S.A., United States of America

Dugdale, Joel
Reporter Radio/Television
21ST CENTURY NEWS
2756 Arcadia, Lake
Hauasu City, AZ 86403, United States of America

Dumanoski, Dianne
Correspondent
BOSTON GLOBE
107 Park St.
Newton, MA 02158, USA, United States of America

Dunleavey, Mary Patricia
Correspondent
THE NEW YORK TIMES
327 E. 12 St., 16
NY, 10003, Usa, United States of America

Dunne, Cory, Ms.
Correspondent
ASPEN TIMES
1220,Mtn. View Drive
Aspen, CO, USA 81611, United States of America

Edgerton, Patricia Jones, Ms.
Correspondent
GREENWIRE
Timber Creek Farm - Garth Road
Charlottsville - VA 22901, United States of America
T: 1 804 977-3870

Ehlers, Jake
Reporter Radio/Television
VH-1
Harrison NY 10524
New York, USA, United States of America

Einreinhofer, William
Editor
WNET-13
904 Hudson St.
Hoboken, NJ 07030, USA, United States of America

Elsner, Peter
Correspondent
NEWSDAY
2801 Ponce Deleon Blvd.
Suite 531, Coral Gables, FL.33134, USA, United States of America

Engelhart, Gita
Correspondent
Hollywood Reporter
R. Cupertino Durao 45/305
2241 Leblon, Rio de Janeiro, Brazil, United States of America

Ensign, Thomas D.
Correspondent
GUARDIAN
182 E. 7th St., 5N
New York, NY 10009, USA, United States of America

Esserman, Lauren
Editor
EARTH SUMMIT TIMES
398 Pacific St.
New York, USA, United States of America

Evenson, Dean
Reporter Radio/Television
KVOS
2501 College St.
Bellingham, WA 98225, United States of America

Fabricio, Roberto
Editor
SUN-SENTINEL
6835 Queen Palm Terrace
Miami Lakes, Florida, USA, United States of America

Fainaru, Steven
Correspondent
BOSTON GLOBE
535 W 111th St. (#1)
New York, NY 10025, USA, United States of America

Fay, Haussman
INDUSTRIAL WORLD
114 E 84th St.
New York, NY 10028 USA, United States of America

Finley, Bruce
Correspondent
DENVER POST
1507 Culebra
Colorado Spring, USA 80907, United States of America

Fisher, Lawrence
Correspondent
EARTH SUMMIT TIMES
120 E 36th St. (#5C)
New York, NY 10016, USA, United States of America

Fisher, Lianne
Correspondent
EARTH SUMMIT TIMES
120 E 36th St. (#5C)
New York, NY 10016, USA, United States of America

Flatow, Ira
Reporter Radio/Television
NATIONAL PUBLIC RADIO
99 Club Rd.
Stamford, CT 06905, USA, United States of America

Foley, Kevin
Reporter Radio/Television
RADIO FREE EUROPE
7701 Lafayette Forest Dr., #2
Annandale, VA 22033, USA, United States of America

Foster, Richard
Editor
BRAZIL WATCH
SQN 316 Bloco E Apto 119
70000 Brasilia, DF Brazil, United States of America

Franco, Lucrecia
Correspondent
UNIVISION NEWS
Marquis de Sao Vicente 67/402
22451 Gavea, Rio de Janeiro-RJ, Brazil, United States of America

Franco, Maria Luiza Leite
RAI - TV CHANNEL 3
New York, United States of America

Frank, Phillip J.
Editor
CNN/TBS
3479 Rockhaven CIR.NE
Atlanta, GA 30324 USA, United States of America

Fraser, C. Gerald
Correspondent
EARTH SUMMIT TIMES
788 Columbus Ave. (#17E)
New York, NY 10025, USA, United States of America

Freeman, Alan
Editor
ENS WORLD NEWS
5200 Boaronacic
Ventnor NJ 08406, USA, United States of America

Freeman, Jack
Correspondent
EARTH SUMMIT TIMES
287 Woodside Ave.
Ridgewood, NJ 07450, United States of America

Froehlich, Vera
Correspondent
AP
Frankfurter Str. 40
6368 Bad Vilbel, Germany, United States of America

Fry, Gordon
Correspondent
AMC JOURNAL
113 4th St. NE
Washington, D.C., 20002, United States of America

Fuss, Robert
Correspondent
NBC
3009 Cedar Hill Rd.
Falls Church, Virginia, U.S.A., United States of America

Galanternick, Mery
Correspondent
NEW YORK TIMES
Ave. Visconde de Piraja
550/2012-CEP22410, Ipanema, RJ-RJ,Brazil, United States of America

Gamble, Christopher, Mr.
UPI/GLOBAL ENVIRONMENT TEAM
1346 S. Roxbury Rd.,
Salt Lake City - UT 84108, United States of America
T: 1 801 583-6446

Garcia, Dorothy Linda
Correspondent
AUDUBON
5449 23rd St. NW
Washington D.C., USA, United States of America

Garcia, Sidiny
Reporter Radio/Television
USIA/VOA
R. Dr. Olinto Magalhaes, 20
Vidigal, Rio de Janeiro-RJ, Brazil, United States of America

Gelbspan, Ross
Correspondent
BOSTON GLOBE
247 Kent St.
Brookline, MA 02146, USA, United States of America

Gerson, Noel Lillian
Correspondent
EARTH SUMMIT TIMES
5114 Fulton Street, NW
Washington DC 20016, United States of America
T: 1 202 6860697

Getches, David H.
Editor
COLORADO JOURNAL OF INT. LAW + POLICY
627 Pine SY
Boulden Colorado 80302, USA, United States of America

Gil, Enrique
Editor
HISPANIC RADIO NETWORK, INC.
25724 Stafford CNY Road
Stevenson, Ranch, CA 91381, United States of America

Gilbert, Anne
Correspondent
IMAGINE
60 Gramercy Park
New York NY 10010, USA, United States of America

Gildred, Kathleen
Correspondent
KPFK CENTURY CABLE
2716 Marquette Dr.
Topanga, CA 90290 Los Angeles, USA, United States of America

Gillette, Christopher
Reporter Radio/Television
NIGHTLY BUSINESS REPORT/WPBT
3575 West Glencoe
Coconut Grove, FL. 33133, USa, United States of America

Girshman, Peggy
Deputy Science Editor
NATIONAL PUBLIC RADIO
2025 M St - NW
Washington - DC, United States of America
T: 1 202 8222784 F: 1 202 8222203

Goldensohn, Marty
Reporter Radio/Television
MARKETPLACE
635 Wyndham Road
Teaneck, NJ 07666, USA, United States of America

Goldin, Gregory
Correspondent
L.A. WEEKLY
816 South Stanley Avenue
Los Angeles, CA 90036, USA, United States of America

Goldman, Benjamin
Correspondent
IN THESE TIMES
530 E. 18th Street, #34
New York, NY 10009, USA, United States of America

Goldman, Joseph
Correspondent
MC GRAW-HILL GROUP
Belgrand 225
8B, United States of America

Goldman, Lois M.
Correspondent
IN THESE TIMES
530 E 13th St. 34
New York, NY 10009, USA, United States of America

Goldman, Thomas
Reporter Radio/Television
NATIONAL PUBLIC RADIO
40004 St. N.W. #201
Washington, DC 20007, United States of America

Goldman, Tom, Mr
Radio Producer / Reporter
NATIONAL PUBLIC RADIO
2025 M. St. N.W.
Washington, DC, United States of America
T: 202 822 2175 F: 202 822 2203

Goler, Wendell
Correspondent
AP RADIO
1503 Woodman Ave.
Silver Spring, Maryland, U.S.A., United States of America

Gonzalez, Ada
Reporter Radio/Television
EARTH SUMMIT TIMES
4415 Palmarito St. C.Gables
Florida, USA, United States of America

Gould, Lance
Correspondent
EARTH SUMMIT TIMES
111 Shearwater Court #53
Jersey City, NJ 07305, USA, United States of America

Graber, Dean
Correspondent
RELIGIOUS NEWS SERVICE
Rua 2 de Dezembro 131/1001
22220 Rio de Janeiro-RJ, Brazil, United States of America

Graca, Marita
Correspondent
AUSTRALIAN B'CASTING CORP.
Rua Santa Glafira 635
Sao Conrado, Rio de Janeiro, Brazil, United States of America

Graham, Sheila
Editor
PLAIN TRUTH MAGAZINE
40633 Maestro Lane
Palmdale, CA 93551, USA, United States of America

Granados, Godofredo
Editor
USIA/RADIO MARTI
10823 Warwick Ave.
Fairfax VA. 22030, USA, United States of America

Grant, Andrea Joan
Correspondent
GLOBAL PRODUCTION
1881 9th. St. Suite200
Boulder,Colorado, United States of America

Grasser, John
Correspondent
COAL VOICE
6701 Melville Place
Chevy Chase, WD 20815 USA, United States of America

Gray, Nathan
Correspondent
IMAGINE
60 Gramercy Park
New York NY 10010, USA, United States of America

Green, Paula
Correspondent
JOURNAL OF COMMERCE
Two Tudor City, Apt. 6PS
New York, NY 10017, USA, United States of America

Greene, Richard
Correspondent
KTWV RADIO
17 Northstar St. 204
Marina del Rey, United States of America

Greenfield, Laura B.
Reporter Radio/Television
CITY TV OF SANTA MONICA
617 Bay St.
Santa Monica, CA 90405, USA, United States of America

Grossfeld, Stan
Editor
THE BOSTON GLOBE
46 Monmouth St.
Boston, Mass., U.S.A., United States of America

Gubar, Thomas R.
Editor
ABC NEWS
235 Prospect Ave. (#11F)
Hackensack, NJ 07061 USA, United States of America

Gudenkauf, Anne
Editor
NATIONAL PUBLIC RADIO
2025 M St. NW
Washington DC 20036, United States of America
T: 001 202 8222784 F: 001 202 8222

Guilfoyle, Jean M.
Correspondent
GLOBAL FAMILY NEWS NETWORK
2840 Duvall Rd.
Woodbine, Md. 21797, USA, United States of America

Guinle, Maria Celina
Reporter Radio/Television
MTV
R. Sao Gualter 1578
Rio de Janeiro, RJ, Brazil, United States of America

Gunst, Kathy
Correspondent
PARENTING MAGAZINE
126 Old Fields Road
South Berwick, MA 03908, USA, United States of America

Guruswamy, Lakshman
Editor
ARIZONA JOURNAL OF INT'L & C. LAW
750 Camino Alberta
Tucson, Arizona, USA, United States of America

Gutfeld, Rose
Correspondent
WALL STREET JOURNAL
1418 Montague St., NW
Washington, D.C. 20011, United States of America

Gutierrez, Alberto
Reporter Radio/Television
RADIO MIAMI INTERNATIONAL
4773SW 154 Court
Miami, FL, USA, United States of America

Guzman, Armando
Correspondent
UNIVISION NEWS
2902 27th St. North
Arlington, VA 22207 USA, United States of America

Hackin, Steven
Correspondent
GLOBE PRESS
827 4. Street
Santa Monica, CA, 90403 USA, United States of America

Hadlock, Charles
Correspondent
MIDDLESEX NEWS
Sandy Pond Road
Lincoln, MA 01779, USA, United States of America

Hager, Mary
Correspondent
NEWSWEEK
1607 8m Pl
McLean, VA 22101 USA, United States of America

Hahnen, Richard
Correspondent
THE NEWS SUN
2417 N. Lewis Avenue
Waukegan, IL 60087-3058 USA, United States of America

Haines-Stiles, Geof
33RD PRODUCTIONS, INC.
41 Rowan Road
Summit, NJ 07901 USA, United States of America

Hall, Janice
Correspondent
EXPANSION MANAGEMENT
1903 Gross St. #1
Boulde, CO, USA, United States of America

Hall, Savario Charles
Correspondent
SECRETARIAT NEWS
325 E 92St. Apt. 3C
N.Y. New York, United States of America

Hamann, Jack
Reporter Radio/Television
CNN/TBS
953 W. Emerson St.
Seattle, WA 98119 USA, United States of America

Hamilton, G. William III
Correspondent
ROTARIAN MAGAZINE
7000 Charles Street
St. Augustine, FL 32084, USA, United States of America

Handler, Jessica A.
CNN/NETWORK EARTH
1140 Woodland Ave. S.E.
Atlanta, GA 30310, USA, United States of America

Harkavy, Angela
Correspondent/Reporter
NAT'L WILDLIFE FEDERATION
1400 16TH St NW, #502
Washington DC 20036, United States of America
T: 202 797 6609 F: 202 797 5486

Harris, Jeremy
Correspondent
BBC
5412 Surrey St. Chevy Chase,
Md 20815, Usa, United States of America

Harris, Richard
Correspondent
NATIONAL PUBLIC RADIO
323 Ethan Allen Ave.
Takoma Park, MD 20912, USA, United States of America

Harrison, Anne Hunter
Correspondent
UPI
Av. Belgrano 271, 1 piso,
Capital Federal, Buenos Aires, Argentina, United States of America

Harrison, Emmett B.
Correspondent
CHEMICAL WEEK MAGAZINE
3201 N.Vermont St.
Arlington, Virginia, U.S.A., United States of America

Hartigan, Neil F.
Editor
TFI
1120 W. Albion Ave.
Chicago, ILL, USA, United States of America

Harvey, Richard
Correspondent
DAILY PROGRESS
P.O.Box 9038-340
Charlottsville, VA, USA, United States of America

Harvey, Thomas
Correspondent
SUN-SENTINEL
301 Hendricks Isle No.6
Ft. Lauderdale, Florida, 33301, USA, United States of America

Hashim, Zuraidah
Correspondent
SOJOURNER PRODUCTIONS/META
2016 N. Adams No, 305
Arlington, VA 22201, USA, United States of America

Hathaway, Bonnie
Editor
TIMELINE
RHI Box 381
New Ipswich, NH 03071-9635, USA, United States of America

Heissenbuttel, John Fredrick
Correspondent
TREE FARMER MAGAZINE
5803 Cannon Lane
Alexandria, VA 20036, USA, United States of America

Hendrick, Thomas Thurston
Correspondent
FOX NEWS
5211 Massachusetts Avenue
Bethesda, Maryland 20810, USA, United States of America

Herbert, Mark
Reporter Radio/Television
KFJC RADIO
1901 Jardin Drive
Mountain View, CA 94040, USA, United States of America

Herbst, Christian
Correspondent
SUPER COMPUTING REVIEW
1400 N. Meade St. (#304)
Arlington, VA 22209, USA, United States of America

Hermanson, Sylvia
Correspondent
THE BRISTOL PRESS
163 Church St.
Bristol,C.T. 06010, United States of America

Hernandorena, Diego
Reporter Radio/Television
21ST CENTURY NEWS
Praia do Flamengo, 274/901
Rio de Janeiro, Brazil, United States of America

Hertsgaard, Mark
Correspondent
NEW YORKER MAGAZINE
2516 Parkview Rd.
Baltimore, MD 21207, USA, United States of America

Hileman, Bette
Senior Editor
CHEMICAL & ENGINEERING NEWS
1155 16TH St. N.W.
Washington, D.C. 20036, United States of America
T: 202 872 4583 F: 202 872 8727

Hilker, Colleen
Research Manager
IMPACT ENVIROMENTAL REPORTS
747 Front St.
San Francisco CA 94, United States of America
T: 1 415 3972876 F: 1 415 4214982

Hill, Jay
Editor
NEWSWEEK
235 W 102nd St. (#7E)
New York, NY 10025, USA, United States of America

Hill, Prudence
Reporter Radio/Television
GLOBALVISION
329 E. 9th Street
New York, NY 10003, USA, United States of America

Hirsh, Stephen
Reporter Radio/Television
ABC
4415 39Th St. NW
Washingto, DC, 20016 USA, United States of America

Hogue, Cheryl
Correspondent
BNA-DAILY ENVIRONMENT REPORT
4128 Wexford Drive,
Kensington, MD 20895, USA, United States of America

Holland, Steve
Correspondent
REUTERS
906 S. Asaph St.
Alexandria, Virginia, U.S.A., United States of America

Holloway, Marguerite
Editor
SCIENTIFIC AMERICAN
358 West End Ave. (#3B)
New York, NY 10024, USA, United States of America

Holman, Richard
Editor
THE WALL STREET TRANSCRIPT
61 East 77th Street
New York, N.Y., U.S.A., United States of America

Homsy, George C.
Reporter Radio/Television
LIVING ON EARTH/NAT'L PUB.RADIO
118 Josephine Ave.
Somerville, MA 02144, USA, United States of America

Honores, Hugo Coya
Correspondent
UPI
R. Emilio Berla, 105, 2 andar
22.061 Rio de Janeior, RJ, Brazil, United States of America

Horner, Lisa Emma
Reporter Radio/Television
21ST CENTURY NEWS
1616 North Treat Avenue
Tucson, AZ 85716, USA, United States of America

Hosken, Fran, Ms.
WOMEN'S INTERNATION NETWORK
187 Grant St
Lexington - MA 02173, United States of America
T: 1 617 862-943

Hostetler, Timothy
Correspondent
SECRETARIAT NEWS
211 N. 34th.St.
Bellington, WA 98225, United States of America

Hubbard, Lela
Correspondent
HONOLULU WEEKLY
99-402 Aina Hts Road
Honolulu, Hawaii 96701, USA, United States of America

Hudson, Adrian, Mr.
Editor
CALIFORNIA NATURAL GAS BULLETIN
480 Summit Spgs. Rd.
Woodside, CA 94062, United States of America
T: 415 851-4363 F: 415 851 5473

Hudson, Richard
Editor
GLOBAL REPORT
218 E 18th St.
New York, NY 10003, USA, United States of America

Huffman, Charles Alan
Correspondent
GOLDEN DOME PRODUCTIONS
1004 Wheatley Ct.
South Bend, IN 46614, USA, United States of America

Hull, Fritz
Correspondent
NEW AGE
195 Second St.
Langley Washington, USA, United States of America

Hull, Vivienne
Correspondent
THE SOUTH WHIDBEY RECORD
6296 South Brightan Beach Road
Washington, 98236 USA, United States of America

Hume, Alexander B.
Reporter Radio/Television
ABC
1000 29th. At. NW 310
Washington Dc, USA, United States of America

Hume, Brit
Correspondent
ABC NEWS
3100 N Street, N.W.
Washington, D.C., U.S.A., United States of America

Huml, Peter
Reporter Radio/Television
CNN
4 Square Perronet
92200 Neuilly Sur Seine, Paris, France, United States of America

Hunt, Susan
Correspondent
DISCOVERY CHANNEL
5217 Sherrier Pl., NW
Washington, DC 20016, USA, United States of America

Ichiyanagi, Kristine
Editor
EARTH SUMMIT TIMES
1866 Fell Street
San Francisco, CA 94117, USA, United States of America

Iker, Samuel R.
Reporter Radio/Television
USIA/VOA
5214 Norway Drive
Chevy Chase, Maryland, USA, United States of America

Ingendaay, Elizabeth H.
Correspondent
KNIGHT-RIDDER FINANCIAL NEWS
R. Prof. Joao Arruda 197
05012 Sao Paulo, SP, Brazil, United States of America

Jackson, Barbara H.
Correspondent
SECRETARIAT NEWS
1903 9St.
Anacortes, WA 98221, United States of America

Jaffe, Mark
Correspondent
PHILADELPHIA INQUIRER
115 Harrogat Rd.
Wynnewodd, PA 19096, USA, United States of America

Jagger, Bianca
Correspondent
HBO/CINE ART
530 Park Ave.
New York, N.Y., U.S.A., United States of America

Janiot, Patricia
Reporter Radio/Television
CNN
2921 Lenox Rd. NE, Apt. 10
Atlanta, GA 30324, USA, United States of America

Jarvis, Michaela
Correspondent
DETROIT NEWS
587 Valle Vista Ave.
Oakland, CA 94610 USA, United States of America

Jasper, William F.
NEW AMERICAN MAGAZINE
6730 Somersworth Dr.
Citrus Heights, CA 95621, USA, United States of America

Jeffrey, A. Kramer
Correspondent
ELECTRIC PERSPECTIVES
4604 Lawn CT
Fairfax VA 22032, United States of America

Johnson, Mark C.
Reporter Radio/Television
MOONDRAGON PICTURES, INC.
1088 Grandview Ave.
Fullerton, CA 92635, USA, United States of America

Johnston, Henry
Correspondent
CBS
Ave.Rainha Elisabete 86/601
Rio de Janeiro, Brazil, United States of America

Jones, Artur Eric
Correspondent
CHANNEL ONE
222 Marguerite Ave.
Mill Valley, CA 94941, USA, United States of America

Kamm, Thomas
Correspondent
WALL STREET JOURNAL
Av Beira Mar 406/1104
20021, Rio de Janeior, RJ, Brazil, United States of America

Kanamine, Linda
Correspondent
USA TODAY
2475 Virginia Ave., NW
Washington, DC 20037, United States of America

Karan, Brian
Reporter Radio/Television
THE WORLD PAPER
71 Driftwood Dr.Port.Washingto
New York 11050, United States of America

Kasindorf, Martin
Correspondent
NEWSDAY
Washington, DC 20008, USA, United States of America

Katz, Ian Mitchell
Reporter Radio/Television
USIA/VOA
Arenales 961, Piso 11, No. 53
Buenos Aires, Argentina, United States of America

Kavits, Philip B., Mr.
Director Radio/TV
NATIONAL WILDLIFE FEDERATION
1400 16thh St. NW
Washington - DC 20036, United States of America
T: 1 202 797-6852 F: 1 202 939-3320

Kay, Jane
Environmental Writer
SAN FRANCISCO EXAMINER
11 5Th Street
San Francisco - CA, United States of America
T: 1 415 7778704 F: 1 415 7772525

Keiko, Matsushita
Reporter Radio/Television
DOLPHIN PRODUCTIONS
140 E 7th St. (#3K)
New York, NY 10009, USA, United States of America

Keller, Hans J., Mr.
Editor
WHO IS WHO
184 Shelton Cove
Waynesville, NC 28786 USA, United States of America

Kelly, David
Reporter Radio/Television
WNHV RADIO
19 School Street
Hanover, NH 03755, USA, United States of America

Kepp, Michael Carl
Correspondent
FAIRCHILD NEWS SERVICE
Rua Alvaro Ramos 235/302
22280, Rio de Janeiro-RJ, Brazil, United States of America

Kerry, Kurt
Correspondent
WCAX TV-3
32 Shore Acres
Col. Chester, Vt. 05446, USA, United States of America

Kheel, Theodore W.
Editor
EARTH SUMMIT TIMES
407 W. 246 New York NY
Usa, United States of America

Khosravi, Parisa
Reporter Radio/Television
CNN
2321 Windmere CT.
Marietta, GA 30062, United States of America

Killebrew, Martha, Ms.
Correspondent
ALAMEDA TIMES STAR
555 Pierce St. B-607
Albany, CA 94706, USA, United States of America

Kim, Hyung Laf
Correspondent
KOREA BROAD. SYST.
44 Woodcrest Dr.
Livingston, NJ 07039, United States of America

Kipnis, Hilery Joy
Correspondent
SOJOURNER PRODUCTIONS/META
511 Warrenton DR.
Silver Spring, MD 20904, USA, United States of America

Kline, Jeff
Editor
HISPANIC RADIO NETWORK, INC.
207 Montoya Street
Santa Fe, NW 87501, USA, United States of America

Kliver, Reymer
Editor
SUEDDEUTSCHE ZEITUNG
Liebigstrasse 19
8000 Muenich 22, Germany, United States of America

Knapp, Peggy
Reporter Radio/Television
CNN/TBS
P.O. Box 1348
Guerneville, CA 95446 USA, United States of America

Knickerbocker, Brad
Correspondent
CHRISTIAN SCIENCE MONITOR
188 Scenic Drive
Ashland, Oregon 97520 USA, United States of America

Knoller, Mark
Correspondent
CBS
4515 Willard Ave.
Chevy Chase, MD 20815 USA, United States of America

Knutsson, Gunilla
Reporter Radio/Television
CNN
19 Rue Brezin 75014
Paris, France, United States of America

Kolt, Robert Thomas
Editor
CIESIN
Jolley Rd.
Okemos, Michigan USA, United States of America

Kong, Michael
Editor
ECO-VISION
2 Mott Street
New York, NY 10013, USA, United States of America

Kopco, Patricia A.
Editor
NBC NEWS
5363 La Gorce Drive
Miami, Florida, USA, United States of America

Kottary, Sailesh
Correspondent
EARTH SUMMIT TIMES
S.Krishna Sadan S.T. Road
Mahim, Bombay-16, India, United States of America

Kramer, Marty
CNN
1706 1/2 Summit Pl., N.W.
Washington, D.C., United States of America

Kroll, Keeva
Reporter Radio/Television
OREGON PUBLIC BROADCASTING
14555 Stone Ave. N.
Seattle, WA 98133, USA, United States of America

Kuhon, Albert
Bureau Chief (U.S.)
SUARA PEMBARUAN DAILY
2950 Van Ness St. NW #612
Washington DC 20008-1116, United States of America
T: (202) 686-4654 F: (202) 686-4655

Kundert, Glen
Correspondent
WASHINGTON LETTER
1129 N. Illinois,St. Arlington
Va. 22205, USA, United States of America

Kuo, Christina
Correspondent
PLAIN TRUTH MAGAZINE
1625 Fremont Ave., No. 7
S. Pasadena, California, USA, United States of America

Kurt, Kerry H.
Reporter Radio/Television
WCAX-TV
32 Shore Acres
Colchester, UT 05446, United States of America

Lahive, James P., Mr.
Editor
ENVIRONMENT WATCH-LATIN AMERICA
2 Pearl St., Apt. 7
Charlestown, MA 02129, USA, United States of America

Lakshman, Buruswamy
Editor
ARIZONA JOURNAL OF INT'L & C. LAW
750 Camino Alberca
Tucson,Arizona 85718, USA, United States of America

Langlois, Dale
Reporter Radio/Television
MOONDRAGON PICTURES, INC.
8660 SW 212th St. (#211)
Miami, FL 33189, USA, United States of America

Lauer, Peggy, Ms.
Correspondent
SAN FRANCISCO EXAMINER
6 Tamal Vista Drive
California, USA, United States of America

Lawler, Andrew
Correspondent
SPACE NEWS
1026 D St., NE
Washington, DC 20002, USA, United States of America

Lawrence, Joseph
Correspondent
VOGUE
278 6th Ave. 2, Brooklyn
Usa 11215, United States of America

Ldewenstein, Louis
Reporter Radio/Television
KALW -FM
3858 Kackson St.
San Francisco, CA. USA, United States of America

Leach, Jim
Correspondent
TIMELINE
3252 Hillview Lane
Lafayette, CA 94549, United States of America

Leal, Kathryn Jason
Correspondent
EDITORA BLOCH
480 Park Avenue
New York 10001, USA, United States of America

Lee, Elizabeth, Ms.
Associate Editor
POPULATION REFERENCE BUREAU
1875, Connecticut Ave, N.W. Suite 520
Washington - DC, United States of America
T: 1 202 483-1100 F: 1 202 328-3937

Lee, Jessica
Correspondent
USA TODAY
581 17th Street, S.E.
Washington, D.C., U.S.A., United States of America

Lee, Kyu-Young
Correspondent
EARTH SUMMIT TIMES
45 Topland Rd.
Harbdale, NY 10530, USA, United States of America

Lefevere, Patricia Ann
Correspondent
RELIGIOUS NEWS SERVICE
24 Mackay Dr.
Tenafly, NJ 07670-2420, United States of America
T: 1 201 568-5821

Lehman, Stanley Howard
Correspondent
ASSOCIATED PRESS
Av. Brasil 500, Sala 847
Sao Gristovao, Rio de Janeiro, RJ,Brazil, United States of America

Leitao, Marcellus Eugenio
Reporter Radio/Television
USIA/VOA
Rua Sta. Clara 431, Apt. 903
22041 Rio de Janeiro-RJ, Brazil, United States of America

Levy, Marc
Correspondent
ENVIRONMENT MAGAZINE
Center for Science And Internationa 79 Kennedy Street
Cambridge, Mass 02138, United States of America
T: 617 495 1416 F: 617 495 8963

Lewan, Theodore J.
Correspondent
AP
Av. Brasil 500, Sala 847
Sao Gristovao, Rio de Janeiro, RJ,Brazil, United States of America

Lewis, Paul
Correspondent
NEW YORK TIMES
35 Sutton Place (#14F)
New York, NY 10022, USA, United States of America

Lewis, Sunny P.
Editor
ENVIRONMENT NEWS SERVICE
2488A Waiomao Rd.
Honolulul, United States of America

Light, Timothy
Correspondent
WORLDPAPER
4078 East Gull Lake Dr.
Hickory Corners, MI 49060, USA, United States of America

Linden, Eugene
Correspondent
TIME MAGAZINE
139 West 82nd Street, #6F
New York, NY 10024, USA, United States of America

Lino, Geraldo Luis S.
Correspondent
21ST CENTURY NEWS
R. Haddock Lobo, 145/805
20260 Rio de Janeiro-RJ, Brazil, United States of America

Lionel, David, Mr.
Correspondent
LIONEL TV PRODUCTIONS
3329 Brookside Dr.
Malibu, CA 90265 USA, United States of America

Little, Paul Elliott
Correspondent
IN THESE TIMES
SCLN 209 Bloco B, sala 210
Brasilia, DF, Brazil, United States of America

Lloyd, Robin
Correspondent
NBC NEWS
6215 Kennedy Dr.
Bethesda, MD. 20815, USA, United States of America

Lobe, James R.
Correspondent
IPS
8404 Garland Ave.
Takoma Park, MD 20912, United States of America

Loftis, Randy Lee
Correspondent
DALLAS MORNING NEWS
634 Sherwood Dr.
Richardson, TX 75080, USA, United States of America

Long, William
Correspondent
LOS ANGELES TIMES
Av.Sao Sebastiao 157, Urca
22291 Rio de Janeiro-RJ, Brazil, United States of America

Lopes, Valeria R.
Correspondent
MIAMI HERALD
R.Julio de Castilhos,55/206
Copacabana, Rio de Janeiro-RJ, Brazil, United States of America

Lorina, Luiz Eduardo
Editor
UNIVISION NEWS
Jardim Botanico 464 Apt.302
222461 Rio de Janeiro-RJ, Brazil, United States of America

Lowe, Justin
Editor
EARTH ISLAND JOURNAL
3638 Anza St.
San Francisco, CA, USA, United States of America

Lustig, Charles
Editor
ABC NEWS
14 Carlyle Court
Robbinsville, NJ 08691, USA, United States of America

Lyons, Jane, Ms.
Project Director
NATIONAL AUDUBON SOCIETY
2525 Wallingwood, Suite 1505
Austin, Texas 78746, United States of America
T: 512 327 3249 F: 512 327 1264

MacLeod, Kirsteen
Correspondent
FINANCIAL POST
3 Baird Avenue
Toronto, ONT M4J 1H1, Canada, United States of America

Macintyre, James
Correspondent
CAMARA PRODUCTIONS
2129 13th St.
Boulder, CO 80302, USA, United States of America

Mackenzie, Robert
Reporter Radio/Television
KTVU (FOX)
1740 Broadway
San Francisco, CA, United States of America

Macklowe, Benjamin
Correspondent
NEW ENGLAND
1133 Fifth Ave.
NY 10128, Usa, United States of America

Maer, Peter
Reporter Radio/Television
WESTWOOD ONE/NBC RADIO
4503 Wakefield Drive
Annandale, VA 22003, USA, United States of America

Maffitt, Peter
Editor
TRIDENT SYNDICATE, INC.
3465 Wickersham
Houston, TX. 77027, USA, United States of America

Majot, Juliette
Correspondent
SAN FRANCISCO BAY GUARDIAN
2442 Prince St.
Berkeley, CA 94705, USA, United States of America

Mano, Toru
Correspondent
IMAGINE
60 Gramercy Park
New York NY 10010, USA, United States of America

Mardon, Mark
Editor
SIERRA MAGAZINE
377 Hermann St.
San Francisco, CA 94117 USA, United States of America

Margolis, George M.
Correspondent
NEWSWEEK
Ave. Sao Sebastiao 157
22291 Rio de Janeiro-RJ, Brazil, United States of America

Mariel, Gustav
Correspondent
UNIVISION NEWS
16400 Henderson Pass # 721
San Antonio, Texas, U.S.A., United States of America

Marshall, Paul
Correspondent
SEVEN NETWORK AUSTRALIA
601 Pennsylvania Ave.
Washington DC 20004, USA, United States of America

Marti, Donald B.
Correspondent
EARTH SUMMIT TIMES
163 Prospect Park W. # 3R
Brooklyn, N.Y. U.S.A., United States of America

Martin, P. Joy
Reporter Radio/Television
KBCO
2720 Broadway
Santa Monica, CA, USA, United States of America

Martins, Joana Andrea
Reporter Radio/Television
EARTH SUMMIT TIMES
Rua Ministro Godoy 836, 4053
Sao Paulo, Brazil, United States of America

Mason, Douglas
Correspondent
BLOOMINGTON VOICE
571 Woodruff Pl. East Dr.
Indianapolis, IN 46201 USA, United States of America

Matlack, Carolyn B., Ms.
Correspondent
THE POPULATION INSTITUTE
Rt.11, Box 576
Mooresville, NC 28115, United States of America
T: 1 704 664-7627

Mattesky, Thomas
Reporter Radio/Television
CBS
800 South Saint ASaph St. 217
Alexandria VA 22314 USA, United States of America

Mayer, Paul
Correspondent
CHRISTIANITY & CRISIS
546 Park Ave.
East Orange, N.J. 07017 USA, United States of America

Maynard, Edward, Mr.
GLOBAL ENVIRONMENT TEAM
696 South Madison Avenue
Pasadena - CA 91106, United States of America
T: 1 818 796-2951 F: 1 818 792-2121

McAlary, David
Science Correspondent
VOICE OF AMERICA NEWS
330 Independence Ave.,S.W.
Washington,D.C.,20547-0001, United States of America
T: 202 619 0713 F: 202 619 2400

McClintock, Sandra
Editor
WHO IS WHO
8271 N. Oracle Road, #244
Tuscon, AZ 85741, USA, United States of America

McCluskey, Ian
Correspondent
TIME MAGAZINE
Rua Prof. Gaotao Bahiana 429
Apto. 402 Copacabana 22071 RJ Brazil, United States of America

McCoy, Charles
Correspondent
WALL STREET JOURNAL
215 Francisco, # 3
San Francisco, CA 94133, United States of America

McCoy, Michael
Assignment Editor
EARTH SUMMIT TIMES
784 Columbus Ave 156
New York NY 10025, United States of America
T: 1 212 6639238

McCullough, Edmund
Editor
AP
Montevideo 1740, Apt. 10C
Buenos Aires, Argentina, United States of America

McDonald, Greg
Correspondent
HOUSTON CHRONICLE
3100 Connecticut Ave.,N.W.#433
Washington, D.C., U.S.A., United States of America

McDonald, Mary S.
Reporter Radio/Television
21ST CENTURY NEWS
R. Lopes Quintas, 120 Bl.2/104
Rio de Janeiro, Brazil, United States of America

McFarlane-Taylor, Joann
Editor
CNN
1080 Carroll Street
Brooklyn, N.Y., USA, United States of America

McGrath, James
Correspondent
MISSOULA INDEPENDENT
1626 Cooper
Missoula, MT 59802, USA, United States of America

McManus, Doyle
Correspondent
LOS ANGELES TIMES
7802 Old Chester Road
Bethesda, MD 20817, USA, United States of America

McPherson, Mary M.
Correspondent
SCOTTSDALE PROGRESS
720 N. 82nd St. (#E-13)
Scottsdale, AZ 85257, USA, United States of America

McQuillan, Laurence
Correspondent
REUTERS
13726 Middlevale Lane
Silver Spring, Maryland, U.S.A., United States of America

Mead, Walter
Fellow
LOS ANGELES TIMES
147, Royal St.
Los Angeles - CA, United States of America
T: 1 213 504-5248648 F: 1 213 504-5291105

Mederos, Jorge Washington, M.
Correspondent
AP
Av. Brasil 500, Sala 847
Sao Gristovao, Rio de Janeiro, RJ,Brazil, United States of America

Meenan, Kyle
Reporter Radio/Television
ESPN
4560 W. Sahara, #2005
Las Vegas, NV 89102, USA, United States of America

Meersmith, Thomas
Reporter
MINNESOTA PUBLIC RADIO
45 east 7th st,
St. Paul, MN, United States of America
T: 1 612 2901474 F: 1 612 2901295

Meidlinger, Leo James
Editor
ABC NEWS
9313 Telfer Court
Vienna, VA 22182, USA, United States of America

Meisler, Stanley
Correspondent
LOS ANGELES TIMES
5403 Cromwell Dr.
Bethesda, MD 20816, USA, United States of America

Melnicove, Margo
Reporter Radio/Television
CHRISTIAN SCIENCE MONITOR
245 Lake Shore Drive
Wayland, MA. 01778, USA, United States of America

Melvarg, David
Correspondent
KRON TV
140 Buchanan Drive
Sausalito, California 94965, USA, United States of America

Mena, Staci Matlock
Editor
SONORAN JOURNAL
2965 N. Bear Canyon Rd.
Tucson, Arizon, USA, United States of America

Menagh, Melanie
Correspondent
OMNI MAGAZINE
476 Ninth Street
Brooklyn, NY 11215, USA, United States of America

Mendonca Lima, Dora
Editor
RAI 3
Avenida Epitacio Pessoa 1704
Rio de Janeiro, Brazil, United States of America

Meng, Hsuan
Correspondent
WORLD JOURNAL
141-07 Zoav Whitestone
New York NY, United States of America
T: 1 718 7468889 F: 1 718 7466509

Michaels, Julia
Correspondent
CHRISTIAN SCIENCE MONITOR
R. das Barcas, 115
04641 Sao Paulo-SP, Brazil, United States of America

Michaud, Rudolph
Editor
ODYSSEY PRODUCTIONS
351 W 4th Street
New York, NY 10014, USA, United States of America

Michaux, Claude M.
Correspondent
JET PROPULSION LABORATORY
34 Rue Scheffer
Paris, France, United States of America

Miklaszewski, James Alan
Correspondent
NBC
19837 Meredith Drive
Derwood, MD, 20855 USA, United States of America

Miller, Richard Bruce
Correspondent
CAPE COD TIMES/OTTOWAY NEWS SERVICE
Rua Sta. Clara 277/101
22041-Rio de Janeiro-RJ, Brazil, United States of America

Mills, Stephen
Correspondent
OMNI MAGAZINE
476 Ninth Street
Brooklyn, NY 11215, USA, United States of America

Mines, Steven, Mr
Journalist
2116 Guadalupe St #125
Austin, Texas 78705, United States of America
T: 512 370 9578

Mirabella, Marina
Reporter
CNN BROADCASTING SYSTEM
One CNN Center, Box 105366
Atlanta - GA, United States of America
T: 1 404 8371918 F: 1 404 8274292

Mische, Ann, Ms.
Correspondent
SOUTH-NORTH NEWS SERVICE
427 E.11th Street # 33
New York, N.Y., United States of America

Mische, Patricia M., Ms.
Press, National Catholic Reporter
C/O GLOBAL EDUCATION ASSOCIATES
500 Riverside Dr, #456
New York, NY, United States of America
T: 212 870 3290 F: 212 870 2055

Molineaux, Diana
Editor
USIA/RADIO MARTI
3001 Veazey Terrace N.W.
Washington, DC 2008-5455, USA, United States of America

Molinski, Michael
Correspondent
TAMPA TRIBUNE
Rua Caconde 125, Apto 91
01425 Sao Paulo, SP Brazil, United States of America

Molloy, Christina Alves
Correspondent
THE GUARDIAN
523 West 112th St., Apt. 54
New York, USA, United States of America

Monsalvo, Bernabe
Correspondent
THE PLAIN TRUTH
RR3 Box 19
Gladewater, Tx. 75647 USA, United States of America

Moss, Douglas E.
Editor
E,THE ENVIRONMENTAL MAGAZINE
23 Grove Point Road
Westport, CT 06880, USA, United States of America

Mott, Richard
Correspondent
JOURNAL OF ENV. LAW & LITIGATION
5719 Thrid St.
Arz, VA, 22204, USA, United States of America

Mueller, Don
Correspondent
TIMELINE
603 21st.
Huntington Beach, CA 92648, United States of America
T: 1 714 9603354

Muhammad, Ishmael Rahman
Correspondent
FINAL CALL NEWSPAPER
818 Maple Road
Flossmoor, Illinois 60422, United States of America

Muhammad, Tynnetta A.
Correspondent
FINAL CALL NEWSPAPER
734 W. 79th Street
Flossmoor, Illinois, U.S.A., United States of America

Muller, Helmut
Editor
EARTH MONITOR
433 N. Park Drive
Arlington, VA, 22203 USA, United States of America

Murdock, Janet Lynn
Correspondent
USIA/RADIO MARTI
6343, Wingate Street,
Alexandria, VA 22312, USA, United States of America

Murray, Frank
Correspondent
THE WASHINGTON TIMES
14308 Delcastle Dr.
Bowie,Maryland, U.S.A., United States of America

Myers, John Peterson
Correspondent
GREENWIRE
Rte 2 Box 360 A
Greenwire, USA, United States of America

Nash, Stephen
Correspondent
ATLANTIC NEWS-FEATURES
9702 Gayton Rd.
Richmond, VA USA, United States of America

Nathan, Donald
Correspondent
MOSCOU MAGAZINE
405 W. 57 St.
New York City, NY USA, United States of America

Nelson, Jack
LOS ANGELES TIMES
4 Wynkoop Court
Bethesda, MD 20817, USA, United States of America

Nevins, Elizabeth
NBC NEWS
2669 S. Walter Reed Drive
Arlington, VA. 22206, USA, United States of America

Newman, Lucia
Reporter Radio/Television
CNN
Los Espanoles 1986
Providence, Santiago, Chile, United States of America

Newport, David B.
Editor
FLORIDA ENVIRONMENTS PUBL.
3416 N.W. 49th Terrace
Gainesville, FL 32606, USA, United States of America

Nichols, Richard
Correspondent
PHILADELPHIA INQUIRER
208 Chestnut Ave.
Narberth, PA 19072, USA, United States of America

Nickel, Janice
Correspondent
WORLD VISION INTERNATIONAL
1106 Okoboji
Arcadia, California, U.S.A., United States of America

Nielsen, John
Reporter Radio/Television
NATIONAL PUBLIC RADIO
2001 Kalovama Rd. #8 NW
Washington, DC 20009, USA, United States of America

Nielson, Daniel L., Mr.
Editor In Chief
JOURNAL OF ENVIRONEMT & DEVELOPMENT
9500 Gilman Drive, Graduate School of Int'l Relations & Pacific Studies
La Jolla - CA, United States of America
T: 1 619 534-7617 F: 1 619 534-3939

Nogueira, Tania
Correspondent
UNIVERSITY OF ILLINOIS PRESS
R. Monte Alegre 1159, Apt.22
Sao Paulo, SP, Brazil, United States of America

Noronha, Vivian
Correspondent
EARTH SUMMIT TIMES
183-11 Hillside Ave.
Jamaica Estates, NY 11432, USA, United States of America

Norrris, Peggy, Ms.
Correspondent
GLOBAL FAMILY NEWS NETWORK
7845 - E Airpark Rd.
Gaithersburg - MD, United States of America
T: 1 301 670 7884 F: 1 301 869 7363

O'Connor, Rory Edward
Reporter Radio/Television
GLOBALVISION
420 Riverside Drive
New York, NY 10025, USA, United States of America

O'Donnell, Mogan
News Correspondent
PHILADELPHIA INQUIRER
400 Blood St
Philedelphia, PA 19, United States of America
T: 1 215 18542400

O'Rourke, Patrick Jake
Correspondent
ROLLING STONE MAGAZINE
3133 Connecticut Ave., NW, 801
Washington, DC 20008, USA, United States of America

Obata, Koichi
Correspondent
THE TOKYO SHIMBUN
6503 Machodoc Ct.
Falls Church, Virginia, U.S.A., United States of America

Obert, Albert D.
Editor
ABC NEWS
5 Foxden Court
Rockville, Maryland USA, United States of America

Oehl, Ronald
Correspondent
EARTH SUMMIT TIMES
791 Park Ave
New York, NY, 10021, USA, United States of America

Oehl, Sarah
Correspondent
EARTH SUMMIT TIMES
791 Park Avenue
New York, USA, United States of America

Okie, Martha E.
Correspondent
EARTH SUMMIT TIMES
333 Old Long Ridge Rd.
Stamford, Connecticut, U.S.A., United States of America

Olactregui, Demerio
Reporter
CNN - SPANISH
One CNN Center
Atlanta, GA, United States of America
T: 1 404 8274034 F: 1 404 8273319

Oliver, Joan Duncan
Correspondent
GOOD HEALTH MAGAZINE/NY TIMES
315 East 68th Street
New York, NY 10021 USA, United States of America

Oliver, Lisa Elaine
Editor
CNN/TBS
645 North Highland Ave. N.E.
Atlanta, GA 30306 USA, United States of America

Orelt, Angel
Correspondent
CITY GUIDE (USA)
322 West 57th St., No. 25T
New York, NY 10019, USA, United States of America

Ortiz, Robert
Reporter Radio/Television
MOONDRAGON PICTURES, INC.
10505 SW 210th Terrace
Miami, FL 33187, USA, United States of America

Osorio, Carlos Roberto
Correspondent
CNN - TURNER BROADCASTING SYSTEM
One CNN Center, Box 105366
Atlanta, GA 30335, United States of America
T: 404 8371918 F: 404 8274292

Paen, Alexander L.
Correspondent
KABC TV
11800 Goshen Avenue
Los Angeles, CA 90049, USA, United States of America

Palma, Sonia Maria
Correspondent
SOJOURNER/PRODUCTIONS/META
Av. A. de Melo Franco,365/1203
Rio de Janeiro, Brazil, United States of America

Parke, Juan William
Correspondent
TIMES-PICAPURRE
1900 Pine St.
San Juan, Puerto Rico, United States of America

Parker-Ambrose, Jane
Correspondent
INTERMOUNTAIN JEWISH NEWS
3970 Zenobia St.
Denver, Colorado, USA, United States of America

Parson, Edward A.
Correspondent
ENVIRONMENT MAGAZINE
50A Roberts Rd.
Cambridge, MA 02138 USA, United States of America

Pasquale Mancini, Dr. Antonio Angelo
Editor
THE OVERSEAS NEWS AGENCY
R. Prof. G. Bahiana, 58/101
Rio de Janeiro, Brazil, United States of America

Passo, Mario
Correspondent
USIA/RADIO MARTI
R. Alberto Victor 15
24020-Niteroi-RJ, Brazil, United States of America

Pearlman, Nancy
Reporter Radio/Television
HOPE PRODUCTIONS
P.O.Box 351419
Los Angeles, USA, United States of America

Penney, Alexandra
Editor
EARTH SUMMIT TIMES
435 East 52nd Street
N.Y.C. N.Y. 10017, Usa, United States of America

Pereira Das Neves, Joao Luiz
Correspondent
UNIVERSITY FOR PEACE
Av. Portugal, 818/502
Urca 22291, Rio de Janeiro, RJ, Brazil, United States of America

Perkins, Paul
Correspondent
IMAGINE
60 Gramercy Park
New York, NY 10010 USA, United States of America

Peters, Jill
Reporter Radio/Television
WNET-13 EARTH TECH 92
238 Walnut Street
Peekskill, NY 10566, USA, United States of America

Petersen, Barbara
Correspondent
CENTER FOR ART & EARTH TIMES
704 Stemboat Road
Greenwich, CT 06830, USA, United States of America

Petit, Charles
Correspondent
SAN FRANCISCO CHRONICLE
1007 Cragmont Ave.
Berkeley, Ca 94708 USA, United States of America

Pettie, Marion D.
Correspondent
GLOBAL PRESS
3918 W St. N.W.
Washington, DC 20007 USA, United States of America

Pidcoke, Heather
Reporter Radio/Television
HOPE PRODUCTIONS
696 S.Madison Ave.
Pasadena, California, U.S.A., United States of America

Piel, Eleanor Jackson
Editor
EARTH SUMMIT TIMES
1115 Fifth Av.
New York, NY 10128, USA, United States of America

Piel, Gerard
Editor
EARTH SUMMIT TIMES
1115 Fifth Ave.
New York, N.Y., U.S.A., United States of America

Pierre, James Granderson
Editor
FINAL CALL NEWSPAPER
8946 S.Blackstone
Chicago, Illinois, U.S.A., United States of America

Pinilla, Carlos A.
Editor
CNN
2683 Camille Dr., N.E.
Atlanta, GA 30319, USA, United States of America

Pinkston, Randall
Correspondent
CBS
8104 Flower Ave.
Takoma Park MD 20912 USA, United States of America

Plesser, Andy
Reporter Radio/Television
EARTH SUMMIT TIMES
14 Niweva Ave.
NY 11023, Usa, United States of America

Plocek, Tracy
Correspondent
NEW TEXAS MAGAZINE
1508 Richcreek Road
Austin, Texas 78757 USA, United States of America

Poe, Melissa
Reporter Radio/Television
21ST CENTURY NEWS
1002 Grassiane Lane
Nashville TN, United States of America

Poe, Patricia
Reporter Radio/Television
21ST CENTURY NEWS
1002 Grasland Lane
Nashuville TN 372200, United States of America

Popovich, Luke
Editor
TREE FARMER
3311 Highland Pl.NW
Washington, DC, USA, United States of America

Potter, Mark Edward
Correspondent
ABC NEWS
8995 Northeast 4th Ave.
El Portal, FL 33138 USA, United States of America

Powers, Bertram A.
Correspondent
EARTH SUMMIT TIMES
10 E 10 St. NY
New York, USA, United States of America

Powers, Martha
INTERNATIONAL REPORT
2805 St. Paul Street
Baltimore, MD 21218, USA, United States of America

Preston, Julia
Correspondent
WASHINGTON POST
Av Portugal 248-501
22291, Rio de Janeiro, RJ Brazil, United States of America

Principe, Leonide
Editor
AMAZON MULTIMEDIA STOCK
R.Benjamin Silva 38
Manaus, Amazonas, United States of America

Prothap, Chatterjee
Reporter Radio/Television
PACIFICA RADIO
1545 Dwight Way
Berkeley, CA 94703, USA, United States of America

Pruna, Andres F.
Reporter Radio/Television
MOONDRAGON PICTURES, INC.
8660 SW 212th Street (#211)
Miami, FL 33189, USA, United States of America

Puyol, Fernando
Correspondent
TIMELINE
28546 Thackeray
Hayward, CA 94544, USA, United States of America

Raabe, Mark, Mr.
CENTER FOR INETERNATIONAL ENVIRONMENTAL LAW - USA
1621 Connecticut Ave., NW, Suite 300
Washington - DC 20009, United States of America
T: 1 202 332-4840 F: 1 202 332-4865

Rabel, Edward
Correspondent
NBC NEWS
507 Scenic Way
Great Falls, VA. 22066, USA, United States of America

Rabine, Mark
Correspondent
SOCIAL JUSTICE
60 Dearborn St, #8
San Francisco, CA 94110, United States of America

Radford, Jeff
Editor
CORRALES COMMENT
164 Calle Conejo
Corrales, NM 87048, USA, United States of America

Raeburn, Paul
Correspondent
AP
15 Randolph Place
Ridgewood, New Jersey, U.S.A., United States of America

Ramos, Jorge Gilberto
Correspondent
UNIVISION
6821 Pallazzo Coral Gables
Florida 33148, USA, United States of America

Randle, David
Correspondent
UNITED PRESS INTERNATIONAL
9844 Glendover Way
Sandy, UT 84092, United States of America

Rathbun, Richard
Correspondent
TIMELINE
575 Kellogg Ave.
Palo Alto, CA 94301, USA, United States of America

Raum, H.Thomas
Correspondent
AP
5657 Rayburn Av.
Alexandria, Virginia, U.S.A., United States of America

Ray, Terrence
Reporter Radio/Television
ABC NEWS
12626 Tinonium Terrace
North Potomac, Maryland, U.S.A., United States of America

Real De Azua, Christine
Correspondent
ADUBON MAGAZINE
4502 Elm St.
Chevy Chase MD 20815, United States of America
T: 1 202 547-9009 F: 1 202 547-9022

Reed, Paul Bradford
Reporter Radio/Television
EARTH NEWS
1200 10th Street
Manhattan Beach, CA 90266 USA, United States of America

Reed, Susan
Correspondent
PEOPLE MAGAZINE
Brooklyn NY 11215
New York, USA, United States of America

Rees, John
Editor
INTERNATIONAL REPORT
2805 St. Paul Street
Baltimore, MD 21218, USA, United States of America

Regalado, Tomas Pedro
Correspondent
VIVA AMERICA
2424 S.W. 20th St.
Miami, Florida, U.S.A., United States of America

Register, Larry
Correspondent
CNN
1089 Club Trace
Atlanta, GA 30319, USA, United States of America

Reisner, Hiram
Editor
BUSINESS PUBLISHERS, INC.
1545 18th Street, N.W., #909
Washington, D.C. 20036, United States of America

Ricciardi, Lorenzo
Editor
GLOBAL COMMUNICATIONS CONSERVATION
105 East 16th Street
New York, NY 10003, United States of America

Richman, Gary Dale
Correspondent
AP
Av. Brasil 500, Sala 847
Sao Cristovao, Rio de Janeiro, RJ,Brazil, United States of America

Richmond, Bret
Correspondent
ENTERPRISE NEWS
660 Alger Dr. Palo Alto
CA, Usa 94306, United States of America

Richmond, William
Correspondent
ENTERPRISE NEWS
660 Alger Dr. Palo Alto
CA, Usa, United States of America

Rick, Sandra
Correspondent
INSIGHT FEATURES NEWS SERVICE
11258 S. Champlain Ave.
Chicago Il, 60628, USA, United States of America

Riley, Ann Ruth
Correspondent
NEW ENGLAND
204 Broderhouse Blvd. 2
Somerville, USA, United States of America

Riotta, Giovanni
Correspondent
CORRIERE DELLA SERA
25 Central Park W.
New York, N.Y., U.S.A., United States of America

Robinson, Eugene
Correspondent
WASHINGTON POST
Corrientes, 456/181
Buenos Aires, Argentina, United States of America

Robinson, Linda
Correspondent
US NEWS & WORLD REPORT
1915 Brickell Ave.
Miami, FL 33129 USA, United States of America

Roche, Robert M.
Reporter Radio/Television
WGBH-TV
560 Bedford St. (#C5)
Abington, MA 02351, United States of America

Rockland, Dr. David B.
Correspondent
TIMES MIRROR MAGAZINE
802 Dale Drive
Silver Spring, MD 20910, USA, United States of America

Rodriguez, Carmen Maria
Reporter Radio/Television
USIA NEWS SERVICE
260 Richmond Terrace
New York 10301, USA, United States of America

Rodriguez, Helen
Correspondent
PACIFIC NEWS SERVICE
508 2ND Av.
San Francisco, CA, USA, United States of America

Rodriguez, Raphael
Editor
CNN
2707 Norseman Drive, Apt. 2
Smyma, GA 30080, USA, United States of America

Rogers, Adam
Editor
EARTH NEWS
5126 Clareton Drive, Suite 200 Agoura Hills
California, United States of America
T: 51 14 5978733 F: 51 14 5971607

Rogers, Norman L.
Reporter Radio/Television
CAL.POLYTECHNIC STATE UNV.
8 Quail Ridge Dr.
Atascadero, CA 93422 USA, United States of America

Rovin, Gary B.
Correspondent
COMMUNITY NEWSPAPER
16017 S.W. 74 Place
Miami, Florida 33157, USA, United States of America

Roza, Nara
Correspondent
TELEMUNDO
2836 Sw 34 Ave.
Miami - Fla. 33133, United States of America

Rubenstein, Theodore
TURNER BROADCASTING SYSTEM / CNN
One CNN Centre
Atlanta - GA, United States of America
T: 1 404 827-3973 F: 1 404 827-1015

Rudolph, John
Reporter Radio/Television
CHRISTIAN SCIENCE MONITOR
126 Old Fields Road
S. Berwick, ME. 03908, USA, United States of America

Ruiz, Orlando
Editor
CNN
710 Winston Rd.
Marietta, GA 30060, USA, United States of America

Russo, Kurt
Correspondent
BELLINGHAM HERALD
2405 W. Shore Drive
Lummi Island, Washington USA, United States of America

Rutherford, John
NBC NEWS
5810 Bent Twig Road, NcLean
VA 22101, Usa, United States of America

Sallesh, Kottary
Correspondent
EARTH SUMMIT TIMES
S. Krishna Sadan, S.T. Road
Bombay, 16, India, United States of America

Sanchez Triana, Angela Ines
Correspondent
ENVIRONMENT WATCH-L.A.
Transversal 119-57, apt. 202
Bogota, Colombia, United States of America

Sand, Faith
Editor
HOPE PRODUCTIONS
696 S.Madison Av.
Pasadena, Calif., U.S.A., United States of America

Savage, J.A.
Correspondent
SAN FRANCISCO BAY GUARDIAN
2117 5th St.
Berkeley, CA 94710, USA, United States of America

Sawa, Bozena
Correspondent
RELAX MAGAZINE
34 Hillside Av.. apt.5T
New York, N.Y., U.S.A., United States of America

Scarborough, Katherine Tracy Alvord
Correspondent
GREENPEACE MAGAZINE
P.O.Box 943
Occidental, CA 95465, USA, United States of America
T: 1 707 874 2052

Schechter, Daniel
Correspondent
GLOBALVISION
248 West 23rd Street
New York, N.Y., U.S.A., United States of America

Scherr, S Jacob, Mr.
Special Correspondent
AMICUS JOURNAL
Natural Resources Defense Council B50 New York, Ave NW
Washington, DC 20005, United States of America
T: 202 783 7800 F: 202 783 5917

Schmulh, Peter, Mr.
Correspondent
LOS ANGELES TIMES
927 Toulouse 57, apt 3
New Orleans - LA, United States of America
T: 1 504 524-8648 F: 1 504 529-1105

Schramm, Dorothy
Correspondent
BURLINGTON HAWK EYE
2700 South Main St.
Burlington, Iowa 52601, United States of America

Schumacher, Edward
Correspondent
IMAGINE
60 Gramercy Park
New York NY 10010, USA, United States of America

Scott, Fraser
Reporter Radio/Television
WGBH-TV
118 Java St.
Brooklyn, NY 11222, USA, United States of America

Seem, Thomas H.
Reporter Radio/Television
CBS
4601 Herend Place
Fairfax VA 22032 USA, United States of America

Selendy, Janine
Correspondent
TRIBUTE
148 C. Hill, West
Cornwell, CT, USA, United States of America

Sender, Stuart
Reporter Radio/Television
GLOBALVISION
305 W 106th St.
New York, NY 10025, USA, United States of America

Serafin, Barry
Reporter
ABC NEWS
47 West 66th street
New York, NY 10023-6, United States of America
T: 001 212 4562800

Serradas, Mary
Correspondent
CREW, INC.
347 West 55th St, 6B
New York, NY 10019, USA, United States of America

Settipani, Frank
Reporter Radio/Television
CBS
21 Fox Run
Murray Hill, NJ 07974, USA, United States of America

Severinghaus, Jeffrey Peck
Correspondent
BENNINGTON BANNER
P.O. Box 718
Palisades,New York 10964 USA, United States of America

Shabecoff, Philip
Correspondent
GREENWIRE
C/O Maurice Strong
United States of America

Shen, Sinyan
Editor
WORLD RESOURCE REVIEW
1 Heritage Plaza
Wrudridge IL, 60517, United States of America

Shepard, Gertrude
Correspondent
EARTH SUMMIT TIMES
67-18 L 195 Lane
Fresh Meadows, N.Y. 11365, USA, United States of America

Shepard, Richard
Correspondent
EARTH SUMMIT TIMES
67-87 L 195 Lane
Fresh Meadows, NY 11365, USA, United States of America

Sherrow, M.T.
Correspondent
LAHAINA NEWS
20 Alaeloa #35
Lahania, Maui, Hawaii 96761, United States of America

Shumaker, Anne Williams, Ms.
Advocate
THE POPULATION INSTITUTE
270 East Main Street
Centre = NY 35960, United States of America
T: 1 205 927-5856 F: 1 2055927-5568

Silva, Eduardo G.
Correspondent
FOLHA DO BRASIL
Av. Monteiro Lobato, 6
Cumbica, Guarulhos,SP 07180, Brazil, United States of America

Silva, Teresa C.A.
Editor
BRAZILIAN TIMES/OCA MAGAZINE
610 W 150th St. (#3F)
New York, NY 10031 USA, United States of America

Silverstein, Kenneth Mark
Correspondent
ASSOCIATED PRESS
Av. Brasil 500, Sala 847
Sao Cristovao, Rio de Janeiro, RJ,Brazil, United States of America

Simon, Scott
Reporter Radio/Television
THIRTEEN/WNET
2450 Virginia Ave. NW
Washington, DC 20037, United States of America

Sinai, Ruth
Correspondent
THE ASSOCIATED PRESS
3315 Wisconsin Ave. # 704
Washington, D.C., U.S.A., United States of America

Slater, Chuck
Editor
EARTH SUMMIT TIMES
28 Dorothy St.
Staten Island, NY 10314, USA, United States of America

Sloan, Lucy
Correspondent
COMMERCIAL FISHERIES NEWS
188 Frankunst.
Cambridge, USA, United States of America

Smiroldo, Diane
Correspondent
WASHINGTON LETTER
7743 Invershan Dr. 195
Falls Church, UA 22042, USA, United States of America

Smith, Fred Lee
Correspondent
NATIONAL REVIEW
2039 New Hampshire Ave.# 206
Washington, D.C., U.S.A., United States of America

Smith, Geri Louise
Correspondent
BUSINESS WEEK MAGAZINE
R. Mundo Novo 701/301
Botafogo, 22251, Rio de Janeiro, Brazil, United States of America

Smith, Haruko
Correspondent
EARTH SUMMIT TIMES
3 Roque Moraes Ct.
Mill Valley CA, 94941 USA, United States of America

Smith, Katie R.
Correspondent
KTAR
7079 E. Sunnyvile Rd.
Pv A2 85253 Usa, United States of America

Smith, Melinda
Editor
UNIVISION NEWS
2252 Cathedral Ave.NW (#1)
Washington, DC 20008 USA, United States of America

Smith, Nancy
Correspondent
EARTH SUMMIT TIMES
501 E 81th
Nyc 10128, USA, United States of America

Smith, Taro
Correspondent
EARTH SUMMIT TIMES
3 Roque Moraes CT.
Mill Valley, USA, United States of America

Smith, Timothy Adam
Correspondent
EARTH SUMMIT TIMES
RD1 Box 262
Nm, Usa, United States of America

Snow, Crocker
Editor
WORLDPAPER
144 Tapsfield Road
Ipswich, MA 01938, United States of America

Sodetani, Wendy Naomi
Correspondent
HONOLULU WEEKLY
3814 Kaimuki Avenue
Honolulu, Hawaii USA, United States of America

Solares, Andre
Correspondent
RADIO MIAMI INTERNATIONAL
1435 SW 82 Place
Miami, FL, USA, United States of America

Solares-Olmo, Pedro L.
Reporter Radio/Television
RADIO MIAMI INTERNATIONAL/AHORA
1210 W. 6 Terrace
Hialeah, FL, USA, United States of America

Solom, Rachel
Reporter Radio/Television
MICHIGAN PUBLIC RADIO
P.O. 8318
Ann Arbor, MI 48103, United States of America

Sorensen, Marjorie M.
Correspondent
OUR TOWN
210 East 72nd Street
New York, NY 10021, USA, United States of America

Soto, Roberto
USIA/TV MARTI
3107 South St., NW
Washington, DC 20007, USA, United States of America

Spencer, Susan R.
Correspondent
CBS
2705 Dumbarton ST. NW
Washington, DC 20007 USA, United States of America

Spreyer, Frederick
Correspondent
THE OIL DAILY
552 La Guardia Place
New York,N.Y. 10012, USA, United States of America

Spring, Abigail, Ms.
Editor
WORLD WILDLIFE FUND NEWS
1868 Columbia Rd., NW, No. 110
Washington, DC 20009, USA, United States of America

Stanley, John R.G.
Correspondent
EARTH SUMMIT TIMES
7407 Denton Road
Bethesda, Maryland, USA 20814, United States of America

Stapleton, Sally
AP
101 W 87th St. (#9H)
New York, NY 10024, USA, United States of America

Staril, Robert
Editor
PACCA
1833 Lamont N.W.
Washington, D.C., U.S.A., United States of America

Steiner, Robert
Correspondent
SCIENCES
250 W 76th (#1B)
New York, NY 10023, USA, United States of America

Stern, Marc, Mr.
Managing Editor
JOURNAL OF ENVIRONMENT & DEVELOPMENT
Mail Code 0519, Univ. of CA,S.Diego 9500 Gilman Drive
La Jolla - CA 0519, United States of America
T: 1 619 534-7617 F: 1 619 534-3939

Steuart, Elizabeth
NBC NEWS
4418 Garrison Street, NW
Washington, DC, USA, United States of America

Stevens, William K.
Correspondent
NEW YORK TIMES
33 Tuers Place
Montclair, NJ 07043, USA, United States of America

Stilkind, Jerry
Correspondent
USIA NEWS SERVICE
2801 Cortland Place, N.W.
Washington, DC 20008, USA, United States of America
T: 1 202 619 4157 F: 1 202 619-6267

Stone, Florence
Consultant
EARTHWATCH
1228 1/2 31st St. N.W. 410
Washington - DC 20007, United States of America
T: 1 202 3422564 F: 1 202 3420751

Stone, Roger David
Correspondent
WILDLIFE CONSERVATION MAG.
3403 'O' St., NW
Washington, DC 20007, USA, United States of America

Strand, Jennifer
Editor
CAMARA PRODUCTIONS
84 Ontario Court
Boulder, CO 80303, USA, United States of America

Straus, Ellen S.
Correspondent
EARTH SUMMIT TIMES
1321 4th St. S.W.
Washington D.C., 20024 USA, United States of America

Straus, Peter
Reporter Radio/Television
EARTH SUMMIT TIMES
1321 4th St. SW
Washington D.C. 20024 USA, United States of America

Strauss, Marilyn
UNIVISION
300 Galen Drive, #301
Key Biscayne, Florida 33140, USA, United States of America

Studebaker, Scott
Correspondent
INT'L ENVIRONMENT REPORTER
Gal. Ribeiro DA Costa 178/1001
22010 Rio de Janeiro, Brazil, United States of America

Swain, Virginia
Correspondent
UN OBSERVER AND INTERNATIONAL REPORT
Box 921
Old Lyme, Ct 06371, United States of America

Swart, Lidwina C.
Reporter Radio/Television
DCTV
2144 California, St.
NW 504 Washington, D.C. 20008, USA, United States of America

Szlemko, Hope
Correspondent
SGI-USA WORLD TRIBUNE
5252 Cahuenga Blvd (#7)
North Hollywood, CA 92601, USA, United States of America

Tajibnapis, Zulkarnain H.
Correspondent
USIA/VOA-EAST ASIA/PACIFIC
6559 Rockland Dr.
Clifton, Va 22024, USA, United States of America

Tajima, Tstomu
TOKYO/CHINICHI SHIMBUN
69-11, Manse St.
Forest Hills NY 11375, United States of America
T: 1 212 9691870 F: 1 212 9691843

Tanzer, Michael David
Correspondent
MONTHLY REVIEW
221 W 82nd St.
New York, NY 10024, USA, United States of America

Taylor, Adrian
Correspondent
CBS
716 Gibbon St. Alexandria
VA, 22314 Usa, United States of America

Terjen,
Reporter Radio/Television
CHRISTIAN BROADCASTING NETWORK
1333 Starling Court
Virginia, 23451, United States of America

Ternes, Alan P.
Editor
NATURAL HISTORY MAGAZINE
333 Central Park West
New York, NY 10024, USA, United States of America

Terrel, Gaynell
Correspondent
HOUSTON POST
10615 Meadowglen (#302)
Houston, TX 77042 USA, United States of America

Terrence, Ray
Reporter Radio/Television
ABC
12626 Timonium Terrace
N. Potomac, Maryland, 20878, USA, United States of America

Thomas, Evelyn
Reporter Radio/Television
CBS
129 U Street NW
Washington, Dc USA, United States of America

Thomas, Kit
Correspondent
WORLDLINK
407 Beirut Avenue
California 90272, USA, United States of America

Thomson, Peter P.
Editor
LIVING ON EARTH/NAT'L PUB.RADIO
26 Sixth St.
Cambridge, MA 02141, USA, United States of America

Tillotson, Mary
Correspondent
CNN
8212 Custer Rd.
Bethesda, Md., U.S.A., United States of America

Toshiyuki, Kora
Editor
DOLPHIN PRODUCTIONS
Takatsuku Sulenaga 909
Kawasakichi, Kanagawaken, Japan, United States of America

Triay, Andres Pedro
Reporter Radio/Television
CBS
4770 Biscayne Blvd., No. 1170
Miami, FL 33137, USA, United States of America

Tunnell, Douglas
Correspondent
CBS
4770 Biscayne Blvd.
Miami, FL 33137, USA, United States of America

Turner, Paul Warren
Correspondent
INDIANA DAILY STUDENT
120 Kingston Place, Apt. E-49
Bloomington, IN 47408, USA, United States of America

Tyson, Rae
Correspondent
USA TODAY
13033 Compton Road
Clifton, VA 22024, USA, United States of America

Ulatowska, Elizabeth A.
Reporter Radio/Television
RADIO PEACE INTERNATIONAL
170-40 Henley Rd. 1B
Jamica States, United States of America

Unger, Bruce
Reporter Radio/Television
WWBT-TV CHANNEL 12
119 Fairway Lane
Ashland, VA 23005, USA, United States of America

Usdin, Steve
Editor
GLOBAL PRESS
1110 N. Arlington Blvd.
Arlington, VA 22209, USA, United States of America

Vaisman, Cecilia
Reporter Radio/Television
NATIONAL PUBLIC RADIO
HC32 Box 557, Groom Creek Rte.
Prescott, Arizona 86303, USA, United States of America

Valdez, William
Reporter Radio/Television
USIA/RADIO MARTI
5601 Seminary Rd. Apt.1110-N
Falls Church VA.22041,USA, United States of America

Van Hof, Josephine
Reporter Radio/Television
ONE WORLD ENVIRONMENTAL TV
Zwanenzude 18, Leiden
Netherlands, United States of America

Varni, Allan, Mr.
UNCED-Team Member
FOUNDATION FOR GLOBAL COMMUNITY
222 High St.
Palo Alto - CA, United States of America
T: 1 415 328-7756 F: 1 415 328-7785
TL: 856588

Vasconcellos, Mary Ann
Correspondent
THE COUNTY TELEGRAM TRIBUNE
2363 Leona Avenue
San Luis Obispo, CA 93401, United States of America

Vasquez, Juan
Correspondent
CBS
5545 S.W. 81 Terr.
Miami, FL 33143, USA, United States of America

Veale, Noelle Alejandra
Reporter Radio/Television
CBS
602 Linda Vista Vve.
Pasadena, CA 91105, USA, United States of America

Verhagen, Frans
Reporter Radio/Television
QUEENS PUBLIC ACCESS TV
97-37 63rd Rd., 15E
Forest Hills North, NY 11374, USA, United States of America

Vermeire, Donna
Correspondent
SOJOURNER PRODUCTIONS/META
152 W 94th St., No. 4
New York, NY 10028, USA, United States of America

Wacker, Ronnie
Correspondent
SEA FRONTIERS
2455 Masseu Point Road
Cutchoque, NY 11935, USA, United States of America

Wagner, Michael
Correspondent
INTERIORS&ARCHITECTURE MAGAZINES
230 W 105th St.
New York, NY 10025, USA, United States of America

Walden, Shelton Harrison
Chief Announcer/Public Affairs
WBAI RADIO/PACIFICA
65 West 96th St. #24C
New York - NY, United States of America
T: 1 212 6623385 F: 1 212 5645359

Walker, Wendy
Editor
CNN
4893 MacArthur Blvd., N.W.
Washington, D.C. 20007, United States of America

Wall, Timothy Joseph
CNN
P.O.Box 105366
Atlanta, GA 30348-5366, USA, United States of America

Wallace, Scott B.
Correspondent
PRODIGY SERVICES CO.
508 W 112th St. (#1A)
New York, NY 10025 USA, United States of America

Walsh, Kenneth
Correspondent
US NEWS & WORLD REPORT
6400 Tulsa Lane
Bethesda, Maryland, U.S.A., United States of America

Walsh, Patricia
Correspondent
UPI
8504 Flower Ave.
Takoma Park, MD 20912, USA, United States of America

Walsh, Simon
Correspondent
BLOOMBERG
1620D Corcoran St. NW
Washington, Dc 20009, USA, United States of America

Wapner, Paul
Correspondent
TIKKUN MAGAZINE
2122 Decatur Place, N.W. 39
Washington, D.C., U.S.A., United States of America

Warner, Craig Marshall
Reporter Radio/Television
WESTWOOD ONE/NBC RADIO
9329 Limestone Place
College Park, Maryland 20740, USA, United States of America

Wayne, Gresham
Correspondent
AMC JOURNAL
105 13th St. N.W.
Washington D.C., USA, United States of America

Weaver, Dana
Correspondent
WORLD POPULATION NEW SERVICES
503 Taylor Columbia
Usa, United States of America

Webb, Geoff
Correspondent
HISPANIC RADIO NETWORK, INC.
Rt. 14, Box 210A
Santa Fe, NM 87505, USA, United States of America

Weisbard, Charles
Correspondent
FOLHA DO BRASIL
31-31 29th St. (#3F)
Astoria, New York, NY 11106, USA, United States of America

Weiss, David
Reporter Radio/Television
WTN
690 Fort Washington Ave.
New York, NY 10040, USA, United States of America

Weisskopf, Michael
Correspondent
WASHINGTON POST
2817 Bellevue Terrace, N.W.
Washington, D.C. 20007, USA, United States of America

Weissman, Robert, Mr.
MULTINATIONALS AND DEVELOPMENT CLEARINGHOUSE
PO Box 19405
Washington - DC 20036, United States of America
T: 1 202 3878030 F: 1 202 2345176
TL: 6502633455

Wells, Janet
Team Member
FOUNDATION FOR GLOBAL COMMUNITY
3641 Tanglewood Dr.
Atlanta - GA 30339, United States of America
T: 1 404 438121

Welna, David
Reporter
NATIONAL PUBLIC RADIO
2025 M St. NW
Washington - DC, United States of America
T: 1 202 8222784 F: 1 202 8222203

Wenden-Verhagen, Anita
Reporter Radio/Television
QUEENS PUBLIC ACCESS TV
97-37 6rd Rd., 15E
Forest Hills North, NY 11374, United States of America

Whitbeck, Harris
Correspondent
CNN
3481 Lakeside Dr., Apt. 2501
Atlanta, GA 30326, USA, United States of America

White, Bruce
Correspondent
HEARST MAGAZINES
141 West Lafayette Avenue
Baltimore, Maryland 21217, USA, United States of America

White, Lisa
Correspondent
HEARST MAGAZINES
141 West Lafayette Avenue
Beltimore, Maryland 21217, USA, United States of America

Wilkinson, Loren Earl, Dr.
Correspondent
CHRISTIANITY TODAY
Rural Route 2, Galiano Island
British Colombia, Canada V0N 1PO, United States of America

Wilkinson, Wendla
NBC NEWS
1627 Roxanna Road, NW
Washington, DC. 20012, USA, United States of America

Wilkison, Roger
Correspondent
USIA/VOA
R. Timoteo da Costa 815
Rio de Janeiro-RJ, Brazil, United States of America

Willis, Nathaniel
Correspondent
CHICAGO CITY COLLEGE/KENNEDY KING
400 E 41th St. #24075
Chicago, Ill., U.S.A., United States of America

Wilson, Howard
Correspondent
ASPEN DAILY NEWS
407 Park Ave 'B'
Aspen, Co 81611, USA, United States of America

Wines, Stephen Michael
Correspondent
THE NEW YORK TIMES
4501 S.Chelsea Lane
Bethesda, Maryland, U.S.A., United States of America

Winn, Randolph
Correspondent
GLOBAL PRESS
3918 W St. NW
Washington, D.C. 20007, USA, United States of America

Winslow, Williams
Correspondent
RELIGION NEWS SERVICE
50 E 89 St.
New York NY USA, United States of America

Winton, Anthony
Reporter Radio/Television
AP RADIO
1579 Meridian Ave.
Miami Beach, FL 33139 USA, United States of America

Wolf, Mary Kathleen
Editor
CIESIN
861 Hintz,
Owosso, MI 48867 USA, United States of America

Wolfson, Paula
Reporter Radio/Television
UDA
1833 Calif. Sd., N.W. #304
Washington, D.C. U.S.A., United States of America

Wood, Philip
TERRA CHRISTA COMMUNICATIONS
1880 East River Road
Tucson - AZ, United States of America
T: 1 602 577-6476 F: 1 602 577-1195

Woodruff, Paul
Correspondent
ENR
744 S. Warren Ave.
Malvern, PA 19355, USA, United States of America

Worcman, Nira Broner
Correspondent
ENVIRONMENT MAGAZINE
116 Pinehurst Ave., Apt. M22
New York, NY 10033, USA, United States of America

Wright, Linda
C-SPAN
4230 Chesapeake St., N.W.
Washington, DC 20016, United States of America

Yamada, Hidzki
Correspondent
NHK-JAPN
2230 California St., N.W.
Washington, D.C. U.S.A., United States of America

Yates, Angus
Executive Producer
DISCOVERY CHANNEL
7700 Wisconsin Ave. 7th Floor
Bthesda M.D. 20814 -, United States of America
T: 1 301 9861999 F: 1 301 9864822

Yazaki, Hitomi
Correspondent
EARTH SUMMIT TIMES
1675 Bergen Blvd.
Fort Lee, NJ 07024 USA, United States of America

Yazigi, Monique
Correspondent
NEW YORK TIMES
PO Box 20222
New York, NY 10017, USA, United States of America

Younger, James P.
Correspondent
SAN FRANCISCO EXAMINER
1110 Hastings Ranch Drive
Pasadena, CA 91107 USA, United States of America

Zimmer, Carole
Reporter Radio/Television
NBC/MUTUAL
39 Plaza St.
Brooklyn, NY 11217, USA, United States of America

Zirato, Nina E.
Reporter Radio/Television
FOX NEWS
5331 Dianco way
Culver City, CA 90230, USA, United States of America

ORGANIZATIONS

Abzug, Bella, Ms.
Women's International Policy Action Committee On Environment And Development
c/o Women's Foreign Policy Council, 845 Third Avenue, 15th Floor
New York NY 10022, United States of America
T: (1 212) 759 7982 F: (1 212) 759 8647

Adams, Lauri J., Ms.
Director
SIERRA CLUB LEGAL DEFENSE FUND
180 Montgomery St., Suite 1400
San Francisco - CA 94104, United States of America
T: 1 415 627-6700 F: 1 415 627-6740

Adriance, Peter, Mr.
National Spiritual Assembly of the Bahais of the United States
1606 New Hampshire Avenue NW
Washington DC 20009, United States of America
T: (1 202) 265 8830 F: (1 202) 265 6012

Archibold, Guillermo, Sr.
Antartica Project
707 D Street, S.E.
Washington, D.C. 20003, United States of America
T: (1 202) 544 0236 F: (1 202) 544 8453

Astley, Doreen, Ms.
Synergos Institute, The
100 East 85th Street
New York, NY 10028, United States of America
T: (1 212) 517 4900 F: (1 212) 517 4815

Barnes, James, Mr.
Friends Of The Earth (usa)
218 D Street, SE
Washington DC 20003, United States of America
T: (1 202) 544 2600 F: (1 202) 543 4710

Barney, Gerald O., Mr.
Institute For 21st Century Studies
1611 North Kent Street, Suite 610
Arlington VA 22209, United States of America
T: (1 703) 841 0048 F: (1 703) 525 1744

Barsh, Russel, Mr.
Four Directions Council
4733 17th Avenue N.E.,No. 37
Seattle, Washington 98105, United States of America
T: (1 206) 527 9527 F: (1 206) 527 9527

Bass, Susan, Ms.
Director, Inter-American Program
Environmental Law Institute, USA
1616 P Street, NW
Washington DC 20036, United States of America
T: (1 202) 328 5150 F: (1 202) 328 5002

Best, Barbara H., Ms.
Tides Foundation/Adopt a Tree Project
1388 Sutter, 10th Floor
San Francisco, CA 94109, United States of America

Blades, Christopher Scott
21st Century News
1880 East River Road, Suite 210
Tuscon, Arizona 85718, United States of America
T: (1 602) 577 1137 F: (1 602) 577 1195

Bragonier, Penelope, Ms.
Center for Psychological Studies in the Nuclear Age
1493 Cambridge Street
Cambridge, MA 02139, United States of America
T: (1 617) 497 1553 F: (1 617) 497 0122

Brown, Jessica, Ms.
Quebec- Labrador Mission Foudation Inc./Atlantic Center for the Environment
39 South Main Street
Ipswich, Massachusetts 01938, United States of America
T: (1 508) 356 0038 F: (1 508) 356 7322

Campello, Silvana, Ms.
Nature Conservancy, The
1815 North Lynn Street
Arlington, Virginia 22209, United States of America
T: (1 703) 841 5300 F: (1 703) 841 1283/841 4880

Christ, Michael, Mr.
International Physicians for the Prevention of Nuclear War
126 Rogers Street
Cambridge, MA 02142, United States of America
T: (1 617) 868 5050 F: (1 617) 868 2560

Cicchelli, , Rev.
Anglican Consultative Council*
815 Second Avenue
New York NY 10017, United States of America
T: (1 212) 922 5163 F: (1 212) 687 1336

Clark, Deborah, Ms.
Institute for Sustainable Development
3000 Sand Hill Road, Building 1,Suite 102
Menlo Park, CA 94025, United States of America
T: (1 415) 854 5510 F: (1 415) 854 2330

Cohen, Albert G., Rev.
Southern California Ecumenical Council/Ecology Task Force
PO Box 32305
Los Angeles, CA 90032, United States of America
T: (1 818) 796 2951 F: (1 818) 792 2121

Cohen, Albert G., Rev.
United Church Of Christ, USA
700 Prospect Avenue
Cleveland OH 44115, United States of America
T: (1 216) 736 2174

Collett, Stephen, Mr.
Friends World Committee For Consultation*
247E 48th Street
New York, 10017, United States of America
T: (1 212) 682 2745 F: (1 212) 983 0034

Commins, Stephen, Mr.
World Vision International
919 W. Huntington Drive
Monrovia, CA 91016, United States of America
T: (1 818) 303 8811 F: (1 818) 301 7786

Coulter, Robert, Mr.
Indian Law Resource Centre*
508 Stuart Street
Helena, Montana 59601, United States of America
T: (1 406) 449 2006 F: (1 406) 449 2031

Cox, Larry, Mr.
Rainforest Foundation, The
270 Lafayette Street, Suite 1205
New York NY 10012, United States of America
T: (1 212) 431 9098 F: (1 212) 431 9197

Craig, Bob, Mr.
Keystone Centre, The
Box 606, Keystone
Colorado 80435, United States of America
T: (1 303) 468 5822

Crosby, Faye, Ms.
Society for the Psychological Study of Social Issues
PO Box 1248
Ann Arbor, MI 48106-1248, United States of America
T: (1 313) 662 9130 F: (1 313) 662 5607

Crumbley, Cathy, Ms.
Coolidge Center for Environmental Leadership
1675 Massachusetts Avenue
Cambridge, Massachusetts 02138-1836, United States of America
T: (1 617) 864 5085 F: (1 617) 864 6503

Cywinski, Robert, Mr.
International Association Of Lions Clubs
300 22nd Street, Oak Brook
Illinois 60521-8842, United States of America
T: (1 708) 571 54 66 F: (1 708) 571 88 90
TL: 297236 lion ur

Dann, Melissa, Ms.
WorldWIDE Network
1331 H. Street, NW, Suite 903
Washington DC 20005, United States of America
T: (1 202) 347 1514 F: (1 202) 347 1524

Davies, Peter, Mr.
American Council For Voluntary International Action
200 Park Avenue South
New York N.Y. 10003, United States of America
T: (1 212) 777 8210 F: (1 212) 995 2942

Davis, Hope White, Ms.
World Association Of Former United Nations Internes And Fellows*
2 United Nations Plaza, 11th Floor
New York NY 10017, United States of America
T: (1 212) 963 3110

Dawkins, Kristin, Ms.
Institute For Agriculture And Trade Policy
1313 Fifth Street SE, Suite 303
Minneapolis MN 55414, United States of America
T: (1 612) 379 5980 F: (1 612) 379 5982

Dawkins, Kristin, Ms.
International Environmental Negotiation Network (IENN)
Harvard Law School, 512 Pound Hall
Cambridge MA 02138, United States of America
T: (1 617) 495 1684 F: (1 617) 495 7818

Durbak, Christine, Dr.
World Information Transfer, Inc.
274 Madison Avenue, Suite 601
New York, NY 10016, United States of America
T: (1 212) 696 2037 F: (1 212) 532 1775

Everts, Ted, Mr.
As. of Space Explorers USA
35 White Street
San Francisco, CA 94109, United States of America
T: (1 415) 931 0585

Fitzgerald, Anne, Ms.
Development Group for Alternative Policies, Inc., The (DGAP)
1400 I Street, NW, Suite 520
Washington DC 20005, United States of America
T: (1 202) 898 1566 F: (1 202) 898 1612

Gabriel, Bruce, Mr.
Centre For Development Of International Law
PO Box 53149
Washington DC 20009-3149, United States of America
T: (1 202) 462 6251 F: (1 202) 462 4941

Gacek, Christopher, Mr.
Heritage Foundation, The
214 Massachusetts Avenue, NE
Washington DC 20002-4999, United States of America
T: (1 202) 546 4400 F: (1 202) 544 2260

Gebert, Andrew, Mr.
Soka Gakkai International
7 East 15th Street
New York, NY 10003, United States of America
T: (1 212) 727 7715 F: (1 212) 727 7712

Goreau, Thomas, Dr.
Global Coral Reef Alliance
324 North Bedford Road, Chappaqua
New York 10514, United States of America
T: (1 914) 238 8788 F: (1 914) 238 8768

Graham, Kennedy, Dr.
Parliamentarians for Global Action
211 East 43rd Street, Suite 1604
New York, NY 10017, United States of America
T: (1 212) 687 7755 F: (1 212) 687 8409

Gregory, Michael, Mr.
Appropriate Technology International (ATI)
1331 H Street, NW
Washington DC 20005, United States of America
T: (1 202) 879 2905 F: (1 202) 628 4622

Gregory, Michael, Mr.
Arizona Toxics Information Inc. (ATI)
PO Box 1896
Bisbee, Arizona 85603, United States of America
T: (1 602) 432 7340 F: (1 602) 432 7340

Griffel, Andrew, Mr.
American Jewish World Service
15 West 26th Street, 9th Floor
New York, NY 10010, United States of America
T: (1 212) 683 1161 F: (1 212) 683 5187

Guertin, Donald L., Mr.
Atlantic Council, The
1616 H Street, NW
Washington DC 20006, United States of America
T: (1 202) 347 9353 F: (1 202) 737 5163

Hagan, Patrick, Mr.
International Society for Ecological Economics (ISEE)
PO Box 1589
Solomons, MD 20688, United States of America
T: (1 301) 326 4281 F: (1 301) 326 6342

Hajost, Scott, Mr.
Consortium For Action To Protect The Earth
801 Pennsylvannia Avenue,SE
Washington DC 20003, United States of America
T: (1 202) 547 9009 F: (1 202) 547 9022

Hajost, Scott, Mr.
Environmental Defense Fund
257 Park Avenue South
New York, NY 10010, United States of America
T: (1 212) 505 2100

Hargrove, John Lawrence, Mr.
American Society of International Law, The
2223 Massachusetts Ave., NW
Washington DC 20008-2864, United States of America
T: (1 202) 797 7133

Harper, Conrad K., Mr.
As. of the Bar of the City of New York, The
42 West 44th Street
New York, NY 10036-6690, United States of America
T: (1 212) 382 6700 F: (1 212) 768 8116

Harris, Ladonna, Ms.
Americans For Indian Opportunity, AIO
3508 Garfield Street, NW
Washington DC 20007, United States of America
T: (1 202) 338 8809 F: (1 202) 965 0898

Hartke, Jan A., Mr.
Earthkind USA
2100 L Street, NW
Washington DC 20037, United States of America
T: (1 202) 452 1100 F: (1 202) 778 6132

Hayes, Randall, Mr.
Rainforest Action Network
301 Broadway Suite A
San Francisco CA 94133, United States of America
T: (1 415) 398 4404 F: (1 415) 398 2732

Hermanson, Judith A., Ms.
Cooperative Housing Foundation
PO Box 91280
Washington DC 20090-1280, United States of America
T: (1 301) 587 4700 F: (1 301) 587 2626

Higby, Annette, Ms.
Centre For Rural Affairs
P.O. Box 405
Walthill, Nebraska 68067, United States of America
T: (1 402) 846 5428

Higginson, Charles, Mr.
Council on Ocean Law
1709 New York Avenue NW, Suite700
Washington DC 20006, United States of America
T: (1 202) 347 3766 F: (1 202) 638 0036

Hirsch, Ralph, Mr.
International Federation Of Pedestrians
3500 Race Street
Philadelphia PA 19104-2440, United States of America
T: (1 215) 386 1270 F: (1 215) 386 1270

Hirshbein, Nancy, Ms.
Zero Population Growth
1400 16th Street, NW, Suite 320
Washington DC 20036, United States of America
T: (1 202) 332 2200 F: (1 202) 332 2302

Hollist, W. Ladd, Mr.
International Studies Association
216 Herald R. Clark Building, Brigham Young University, Provo, Utah 84602, United States of America
T: (1 801) 378 5459 F: (1 801) 378 7075

Horowitz, Michael, Mr.
Institute For Development Anthropology
99 Collier Street, PO Box 2207
Binghamton NY 13902, United States of America
T: (1 607) 772 6244 F: (1 607) 773 8993

Huddle, Norie, Ms.
Center for New National Security
1016 Prince Street, No. 5
Alexandria, VA 22314, United States of America
T: (1 703) 549 2862 F: (1 703) 549 2862

Hug, James E., Mr.
Climate Council
2550 M Street, NW Suite 700, Washington DC, 20037
United States of America
T: (1 202) 457 6501 F: (1 202) 457 6000

Huttner, Susanne, Dr.
University of California/Systemwide Biotechnology Research & Education Program
405 Hilgard Avenue UCLA
Los Angeles, CA 90024-1570, United States of America
T: (1 310) 206 6814 F: (1 310) 206 3747

Jepsen, Patricia, Ms.
Centre for International Cooperation
Route 1, Box 220
Castleton, VA 22716, United States of America
T: (1 703) 937 5219 F: (1 703) 937 4039

Karim, A., Mr.
Natural Resources Defence Council*
40 W 20 St.
New York, NY 10011, United States of America
T: (1 212) 727 2700 F: (1 212) 609 452 0284

Knecht, Robert, Prof.
International Coastal And Ocean Organisation
c/o College of Marine Studies, University of Delaware, Newark DE 19716, United States of America
T: (1 302) 451 8086 F: (1 302) 292 3668

Knecht, Robert W., Prof.
International Coastal & Ocean Organisation
Center for the Study of Marine Policy, University of Delaware
Newark - DE 19711, United States of America
T: 1 302 451 8086 F: 1 302 292 3668
TL: 19716

Kramer, Jeff, Mr.
Manager
Edison Electric Institute
701 Pennsylvania Ave., NW
Washington DC 20004, United States of America
T: (1 202) 508 5460 F: (1 202) 508 5403

Kremer, Roxanne, Ms.
International Society for the Preservation of the Tropical Rainforest
3302 North Burton Avenue
Rosemead, CA 91770, United States of America
T: (1 818) 572 7273 F: (1 818) 572 9521

Kresse, Joseph, Mr.
Foundation for Global Community
222 High Street
Palo Alto, CA 94301-1097, United States of America
T: (1 415) 328 7756 F: (1 415) 328 7785

Krieger, David, Mr.
Nuclear Age Peace Foundation
1187 Coast Village Road, Suite 123
Santa Barbara, CA 93108, United States of America
T: (1 805) 965 3443 F: (1 805) 568 0466

Lamey, W. Daniel, Mr.
Junior Chamber International
PO Box 140577, Coral Gables
Fl 33114-0577, United States of America
T: (1 305) 446 7608 F: (1 305) 442 0041
TL: ITT 441084 JAYCEES

Langer, Molly, Ms.
Unitarian Universalist United Nations Office
777 U.N. Plaza, Suite 7D
New York, NY 10017, United States of America
T: (1 212) 986 5165

Laurenti, Jeffrey, Mr.
United Nations Association Of America
485 Fifth Avenue
New York NY 10017-6104, United States of America
T: (1 212) 697 3232 F: (1 212) 682 9185

Leet, Mildred Robbins, Ms.
President
Trickle Up Program
54 Riverside Drive
New York NY 10024-6509, United States of America
T: (1 212) 362 7958 F: (1 212) 877 7464

Lerner, Stephen, Mr.
Commonweal
1517 Swann Str., NW
Washington DC 20009, United States of America
T: (1 202) 232 5548 F: (1 202) 667 6255

Louis, Andre, Mr.
International Development Conference
Suite 1100, 1401 New York Ave, NW
Washington, DC 20005-2160, United States of America
T: (202) 638 3111 F: (202) 638 1374

Mafra, Humberto, Mr.
Damien Foundation
Russ Building, Suite 1010, 235 Montgomery Street
San Francisco, CA 94104, United States of America
T: (1 415) 421 7555 F: (1 415) 421 0712

Magowan, Runa, Ms.
Gateway Pacific Foundation
c/o Aquatech, Ghirardelli sq. 900 N. Point
San Francisco, CA 94109, United States of America
T: (1 415) 673 2793 F: (1 415) 673 9152

Mai, Lorraine M., Ms.
International Center for Integrative Studies
121 Avenue of The Americas
New York, NY 10013, United States of America
T: (1 212) 941 9090 F: (1 212) 941 0714

Malaspina, Alex, Mr.
International Life Sciences Institute*
1126 Sixteenth Street NW
Washington DC 20036, United States of America
T: (1 202) 659 0074 F: (1 202) 659 3859

Manohar, G.S. Abraham, Mr.
Center for Marine Conservation
1725 De Sales Street, NW
Washington D.C. 20036, United States of America
T: (1 202) 429 5609

March, Ruth, Ms.
International Association for Volunteer Effort
PO Box 27095
Los Angeles, CA 90027-0095, United States of America
T: (1 213) 467 6443 F: (1 310) 459 1022

Martin, Dan, Mr.
International Co-ordinating Committee On Religion And The Earth
55 Marcourt Drive
Chappaqua, NY 10514, United States of America

Matsumura, Akio, Mr.
Executive Director
GLOBAL FORUM OF SPIRITUAL AND PARLIAMENTARY LEADERS ON HUMAN SURVIVAL
304 East 45th St. Room 476
New York - NY, United States of America
T: 1 212 953-7947 F: 1 212 557-20961

Maughan, Janet, Ms.
Ford Foundation
320 East 43rd Street
New York, NY 10023, United States of America
T: (1 212) 573 5138 F: (1 212) 297 0969

May, Richard, Mr.
American Planning Association
1776 Massachusetts Ave., NW
Washington DC 20036, United States of America
T: (1 202) 872 0611

McCloskey, Maxine E., Ms.
Defenders Of Wildlife
1244 19th Street NW
Washington DC 20036, United States of America
T: (1 202) 659 9510 F: (1 202) 833 3349

McCloskey, Michael, Mr.
Sierra Club
730 Polk Street
San Francisco, CA 94109, United States of America
T: (1 415) 776 2211 F: (1 415) 776 0350

Menth, Paul R., Mr.
Chairman
Christian Environmental Alliance
PO Box 47504
Plymouth, MN 55447-0504, United States of America
T: (1 612) 537 3865

Mercieca, Carlos, Dr.
International Association of Educators for World Peace
P.O. Box 3282, Mastin Lake Station
Huntsville, Alabama 35810-0282, United States of America
T: (1 205) 534 5501/851 5341 F: (1 205) 851 9157

Merrick, Thomas W., Mr.
Population Reference Bureau, Inc.
1875 Connecticut Ave., NW, Suite 520
Washington DC 20009-5728, United States of America
T: (1 202) 483 1100 F: (1 202) 328 3937

Metcalf, Robert, Mr.
Solar Box Cookers International
1724 11th Street
Sacramento CA 95819, United States of America
T: (1 916) 444 6616 F: (1 916) 447 8689

Meyer, Alden, Mr.
Union of Concerned Scientists
1616 P Street, NW, Suite 310
Washington DC 20036, United States of America
T: (1 202) 332 0900 F: (1 202) 332 0905

Meyer, Gerard S., Mr.
French-american Environmental Technology Centre
Suite 3140, One Oxford Centre
Pittsburg, PA 15219, United States of America
T: (1 412) 391 2911 F: (1 412) 391 2912

Meyn, Marianne, Ms.
Mision Industrial de Puerto Rico, Inc.
PO Box 363 728
San Juan, Puerto Rico 00936-3728, United States of America
T: (1 809) 765 4303/767 0820 F: (1 809) 754 6462

Miller, Kenton, Mr.
World Resources Institute
1709 New York Avenue, NW
Washington DC 20006, United States of America
T: (1 202) 638 6300 F: (1 202) 638 0036
TL: 64414 WRIWASH

Miller, Kristine, Ms.
Great Lake United
State University College at Buffalo, Cassety Hall,
1300 Elmwood Av., Buffalo. NY 14222, United States of America
T: (1 716) 886 0142 F: (1 716) 886 0303

Milliken, Christine T., Ms.
National Association of Attorneys General
Hall of the States, 444 North Capitol Street
Washington DC 20001, United States of America
T: (1 202) 628 0435 F: (1 202) 347 4882

Moore, Julia A., Ms.
Physicians for Social Responsability
1000 Sixteenth Street, NW, Suite 810
Washington DC 20036, United States of America
T: (1 202) 785 3777 F: (1 202) 785 3942

Morehouse, Ward, Mr.
Council on International and Public Affairs
777 United Nations Plaza
New York, NY 10017, United States of America
T: (1 212) 972 9877

Morgan, Jim, Mr.
ARC Peace
c/o ADPSR, 225 lafayette Street
New York NY 10012, United States of America
F: (1 212) 914 9679

Moses, Ted, Mr.
Global Climate Coalition
c/o Edison Electric Institute, 701 Pennsylvania Av e., NW Washington DC 20004, United States of America
T: (202) 637 3161 F: (202) 508 5403

Mueller, Christine, Ms.
National Wildlife Federation
1400 Sixteenth Street, NW
Washington DC 20036-2266, United States of America
T: (1 202) 797 5444 F: (1 202) 797 5486

Naraine, Gayatri, Ms.
Brahma Kumaris World Spiritual University
c/o Gayatri Naraine, 866 UN Plaza, Room 582
New York, NY 10017, United States of America
T: (1 212) 688 1335

Nitze, William A., Mr.
Alliance to Save the Energy, The
1725 K Street, NW, Suite 914
Washington DC 20006-1401, United States of America
T: (1 202) 857 0666 F: (1 202) 331 9588

Norris, Tyler, Mr.
National Civic League
1445 Market Street, Suite 300
Denver, CO 80202-1728, United States of America
T: (1 303) 571 4343 F: (1 303) 571 4404

Nunez, Antonio, Dr.
Together Foundation for Global Unity
2129 13th Street
Boulder, CO 80302, United States of America
T: (1 303) 444 9567 F: (1 303) 444 7512

O'Grady, Mary, Ms.
Sacharuna Foundation
PO Box 130
The Plains, Virginia 22171, United States of America
T: (1 703) 754 8713 F: (1 703) 754 1475

Olsen, Lance, Mr.
Great Bear Foundation
P.O. Box 2699
Missoula, MT 59806, United States of America
T: (1 406) 721 3009

Paprocki, Charles R., Mr.
Proutist Universal
1354 Montague St., NW. P O Box 56466
Washington DC 20011 P, United States of America
T: (1 202) 829 2278

Paprocki, Chuck, Mr.
Progress Agency Inc., The
One Union Square West 201
New York, NY 10003, United States of America
T: (1 212) 989 6900

Patico, Alexander, Mr.
Institute of International Education
1400 K Street, NW
Washington DC 20005-2403, United States of America
T: (1 202) 898 0600 F: (1 202) 842 1219
TL: E DIALCOM141:TCN1505

Petesch, Patti, Ms.
Overseas Development Council, The
1717 Massachusetts Avenue, NW
Washington DC 20036, United States of America
T: (1 202) 234 8701 F: (1 202) 645 0067

Porter, Gareth, Mr.
Environmental And Energy Study Institute
122c Street NW, Suite 700
Washington DC 20001, United States of America
T: (1 202) 628 1400 F: (1 202) 628 1825

Ramakrishna, Kilaparti, Mr.
Woods Hole Research Centre, The
13 Church Street, Woods Hole
Massachusetts 02543, United States of America
T: (1 508) 540 9900 F: (1 508) 540 9700

Rice, Andrew, Mr.
U.S. Citizens' Network On UNCED
300 Broadway, Suite 39
San Francisco CA 94133, United States of America
T: (1 415) 956 6162 F: (1 415) 956 0241

Robinson, John, Dr.
New York Zoological Society/Wildlife Conservation International
Bronx
New York 10460, United States of America
T: (1 212) 220 6864 F: (1 212) 220 7114
TL: 428279 nyzwci

Rochte, Tim, Mr.
National Association of Service and Conservation Corps International Committee
PO Box 1380
San Luis Obispo, CA 93406, United States of America
T: (1 805) 549 3700 F: (1 805) 549 3711

Rolston, Holmes, Dr.
International Society For Environmental Ethics
Dept. of Philosophy, Colorado State University
Fort Collins, CO 80523, United States of America
T: (1 303) 491 5328/491 6315 F: (1 303) 491 0528

Rosborough, Brian, Mr.
Earthwatch
680 Mt. Auburn Street
Watertown, MA 02172, United States of America
T: (1 617) 926 8200 F: (1 617) 926 8532

Roshan, S., Dr.
International Health Awareness Network
300 E Street, Suite 19D
New York, NY 10022, United States of America
T: (1212)644 9304/(1201)378 2225 F: (1 201) 378 9774

Rossi, Gianfranco, Mr.
General Conference Of Seventh-day Adventists*
12501 Old Columbia Pike, Silver Spring
Maryland 20904-1608, United States of America
T: (1 301) 680 6000 F: (1 301) 680 6090

Russell, Virginia, Ms.
American Bar Association
750 North Lake Shore Drive
Chicago, IL 60611, United States of America
T: (1 312) 988 5000 F: (1 312) 988 6281

Sanna, Lucy, Ms.
Electric Power Research Institute
3412 Hillview Avenue
Palo Alto, CA 94304, United States of America
T: (1 415) 855 2000 F: (1 415) 855 2954

Schneider, Richard, Mr.
World Peace University/Radio for Peace International
PO Box 10869
Eugene, OR 97440, United States of America
T: (1 503) 741 1794

Schwabacher, Richard, Mr.
Cousteau Society Inc.
870 Greenbrier Circle, Suite 402
Chesapeake, Virginia 23320, United States of America
T: (1 804) 523 9335 F: (1 804) 523 2747

Schwartz, Michelle, Ms.
Natural Heritage Institute
Cronkhite Beach, Building 1055, Golden Gate
National Recreation Area, Sausalito, CA 94965-2609, United States of America
T: (1 415) 331 1232 F: (1 415) 331 7761

Scott, Tom B., Mr.
International Union of Housing Finance Institutions
11 East Wacker Drive
Chicago, Illinois 60601-4680, United States of America
T: (1 312) 856 0482 F: (1 312) 856 0435

Shaw, Susan D., Ms.
Marine Environmental Research Institute
235 West 76th Street
New York, NY 10023, United States of America
T: (1 212) 721 6243 F: (1 212) 721 6282

Sherwood, Diane, Dr.
North American Coalition On Religion And Ecology, The
5 Thomas Circle, NW
Washington DC 20005, United States of America
T: (1 202) 462 2591 F: (1 202) 234 0307

Smith, Kevin, Brother
Franciscans
135 Remsen Street
Brooklyn, New York 11201, United States of America
T: (1 718) 858 8217 F: (1 718) 858 8306

Smith, Mary Ann, Sister
Coordination in Development, Inc. (CODEL)
475 Riverside Drive, Room 1842
New York,NY 10115, United States of America
T: (1 212) 870 3000 F: (1 212) 870 3545

Sohn, Louis B., Prof.
Commission to Study the Organization of Peace
George Washington University, 720 20th Street, NW
Washington DC 20052, United States of America

Spaulding, Tom, Mr.
YMCA Of The U.S.A.
101 North Wacker Drive
Chicago, Illinois 60606-7386, United States of America
T: (1 312) 977 0031 F: (1 312) 977 9063

Speidel, Joseph, Mr.
Population Crisis Committee
1120 19th Street, NW, Suite 550
Washington DC 20036, United States of America
T: (1 202) 659 1833 F: (1 202) 293 1795
TL: 440450

Spivy-Weber, Francis, Ms.
Audubon Society*
801 Pennsylvannia Ave, SE
Washington DC 20003, United States of America
T: (1 202) 547 9009 F: (1 202) 547 9022

Stangel, Paul J., Mr.
International Fertilizer Development Center
PO Box 2040
Muscle Shoals, Alabama 35662, United States of America
T: (1 205) 381 6600 F: (1 205) 381 7408

Stetson, Marnie, Ms.
Worldwatch Institute
1776 Massachusetts Ave., NW
Washington DC 20036, United States of America
T: (1 202) 452 1999 F: (1 202) 296 7365

Strickland, Stephen, Mr.
National Peace Foundation
1835 K Street, NW, Suite 610
Washington DC 20006, United States of America
T: (1 202) 223 1770 F: (1 202) 223 1718

Strong, Hanne, Mrs
Manitou Foundation
Box 250, Crestone
Colorado 81131, United States of America
T: (1 719) 256 4266 F: (1 719) 256 4266

Tagge, Anne Catherine, Ms.
Campbell Institute
37 Avon Road
Wellesley, Massachusetts 02181, United States of America
T: (1 617) 237 1597 F: (1 617) 237 2842

Thalman, Dart, Mr.
Experiment in International Living, The
Kipling Road, Brattleboro
Vermont 05301, United States of America
T: (1 802) 257 7751 F: (1 802) 254 6674

Thorndike, Elisabeth, Ms.
Centre For Environmental Information
46 Prince Street
Rochester NY 14607-1016, United States of America
T: (1 716) 271 3550 F: (1 716) 271 0606

Torres, Alma, Ms.
Universidad de Puerto Rico - Recinto de Rio Piedras
Apartado 23300, Estacion U.P.R.
San Juan, Puerto Rico 00931-3300, United States of America
T: (1 809) 754 8532

Valentine, Mark, Mr.
Earth Island Institute
300 Broadway, Suite 28
San Francisco, CA 94133, United States of America
T: (1 415) 788 3666 F: (1 415) 788 7324

Wedigg, L.J., Mr.
International Coalition Of Fisheries Associations
1525 Wilson Boulevard, Suite 500
Arlington, VA 22209, United States of America
T: (1 703) 524 8880 F: (1 703) 524 4619

Yeager, Bruce, Mr.
Foster Parents Plan International
804 Quaker Lane, East Greenwich
Rhode Island 02818, United States of America
T: (1 401) 826 2500 F: (1 401) 826 2680

Zaelke, Durwood, Mr.
Center for International Environmental Law - US, (CIEL-US)
1621 Connecticut Ave., NW, Suite 300
Washington DC 20009, United States of America

Zajickova, Zdena, Dr.
World Safety Organisation
305 East Market Street, PO Box 518
Warrensburg, MO 64093, United States of America
T: (1 816) 747 3132 F: (1 816) 747 2647

GLOBAL FORUM

Abioudon, Abigun Ade, Mr.
Expert
OOSA / ONU
Room 3260 United Nations Plaza
New York - NY, United States of America
T: 1 212 963-7255 F: 1 212 963-7998
TL: 420544

Abrams, Daniel
YMCA EARTH ACTION CORPS
909 4th Av
Seattle - WA, United States of America
T: 1 206 382-5013 F: 1 206 382-7283

Abrams, Marna
YMCA EARTH ACTION CORPS
909 4th Av
Seattle - WA, United States of America
T: 1 206 382-5013 F: 1 206 382-7283

Abranson, Neil
OUT OF EPOCA FILMS
N. Cahuenga Blvd.
Hollywood -CA 90038, United States of America
T: 1 213 4694405 F: 1 213 4695010

Ackerman, Pablo
Visual Designer
DOVE & CO./MMH
4759 15th Ave NE
Seattle - WA 98105, United States of America
T: 1 206 344-6321

Adams, Barbara, Ms.
UN NGLS
Palais de Nations
Geneve CH 1211, United States of America
F: 41 22 733-6542

Admussen, Betty Jane, Ms.
UNITED METHODIST CHURCH
5604 N. Oaktree Lane
Kansas City - MO 64118, United States of America

Affolter, Jim, Mr.
Professor
CORNELL UNIVERSITY
Cornell Plantations
Ithaca-NY 14850, United States of America
T: 1 617 255-2018 F: 1 617 255-2404

Agle, Andrew, Mr.
Director Of Operations
GLOBAL 2000
Carter Presidential Center, One Copenhill
Atlanta - GA 30307, United States of America
T: 1 404 872 3848 F: 1 404 874 5515
TL: 49611071

Agyepong, Irene, Ms.
Icour Project
1717 Massachusetts Ave. NW, 2nd floor
Washington - DC 20036, United States of America
T: 1 202 667-1142 F: 1 202 332-4496
TL: 440384

Ahdrom, Techeste, Mr.
Director Un Office
BAHA'I INTERNATIONAL COMMUNITY
866 UN Plaza, Suite 120
New York - NY 10017, United States of America
T: 1 212 7563500 F: 1 212 7563573
E: 152:CNM057

Aiken, Robbie, Mr.
Manager
PINNACLE WEST CAPITAL CORP.
1735 I St., Nw.#916
Washington - DC 20006, United States of America
T: 1 202 508-5460 F: 1 202 508-5403

Ainbinder, David, Dr.& Mrs.
EARTH EMERGE & SEE
RD 8 Box 311
Middletown-NY, United States of America
T: 1 914 342-5411 F: 1 914 342-3252

Ainbinder, Faith, Ms.
EARTH EMERGE & SEE
277 Hudson St.
Cornwall-on-Hudson-N 12520, United States of America
T: 1 914 342-5411 F: 1 914 342-3252

Akida, Amina, Ms.
GREEN BELT MOVEMENT OF KENYA
c/o Resource Renewal Institute, FT. Mason Center Bldg A
San Francisco - CA, United States of America
T: 1 415 928-32774 F: 1 415 928-6529

Alam, Muhammad Nurul
General Secretary
WORLD SPIRITUAL ASSEMBLY
314-67 St. Apt.2R Brooklyn
New York - NY 11220, United States of America
T: 1 212 921 6352 F: 1 212 645 6866
TL: 6656MIC UN

Alamdari, Soraya, Ms.
Staff
EARTH TRAIN/WORLDLINK/GLOBAL YOUTH SUMMIT
22746 Mulholland Drive
Woodland Hills - CA 91364, United States of America
T: 1 818 591-7010 F: 1 213 225-3250

Alario, Celia, Ms.
SUN MICROSYSTEMS / GEOSPHERE PROJECT
1367 Oakwood
Topanga-CA, United States of America
T: 1 310 455-3448 F: 1 818 905-5841

Alexander, Eric, Mr.
Program Officer
GLOBAL EDUCATION ASSOCIATION
475 Riverside Dr.
New York - NY 10115, United States of America
T: 1 212 870 3290 F: 1 212 870 2055

Alexander, Michelle, Dr.
WORLD INFORMATION TRANSFER INC
274 Madison Ave, Suite 601
New York - NY 10016, United States of America
T: 1 212 2730253 F: 1 212 3792004

Ali, Mohamed
MINISTRY OF PLANNING AND ENVIRONMENT
Ghazee Building Ameer Ahmed Madu
Male Maldives Island, United States of America
T: 1 602 577-6476 F: 1 602 577-1195

Alldredge, Tom
Techniciancer
C - SPAN
400 N.Capitol St.,NW #650
Washington, DC 20001, United States of America
T: 202 6264648 F: 202 7376226

Allen, Anna H, Ms.
President
PAN PACIFIC & SOUTH EAST ASIA WOMEN'S ASSOCIATION OF THE USA INC
PO Box 171
Point Pleasant - PA 18950-0171, United States of America
T: 1 215 297-5118 F: 1 215 348-9751

Allen, Glenn, Mr.
35 Union Pk No. 1
Boston - MA, United States of America
T: 1 617 482-5104 F: 1 617 482-5104

Allen, Janet, Ms.
Public Speaker
COALTION TO STOP FOOD IRRADIATION
26015 Redbluff Drive
Calabasas - CA 91302, United States of America
T: 1 818 878 1906

Allen, Kari, Ms.
GLOBAL ENERGY NETWORK INT'L
PO Box 81565
San Diego - CA 92138, United States of America
T: 1 619 595-0139 F: 1 619 595-0403

Allen, Linda Bawden, Ms.
Executive Producer
HOMETOWN FILMS
8306 Wilshire Blvd Suite 254
Beverly Hills - CA, United States of America
T: 1 213655 9403 F: 1 213655 2434

Allen, Meredith E, Ms.
OBSERVER
PO Box 171
Point Pleasant - PA 18950-0171, United States of America
T: 1 215 297-5118 F: 1 215 348-9751

Almaraz, Lisa, Ms.
INT'L SOCIETY FOR THE PRESERVATION OF THE RAINFOREST
1187 Coast Village Dr. #192
Santa Barbara - CA 93108, United States of America

Almeida, Carla Brito De, Ms.
Editor
UNDP/UNITED NATIONS
1 un Plaza #1936
New York - NY, United States of America
T: 1 212 906-5324 F: 1 212 906-5364

Almeida, Leopoldo Goncalves P.A., Mr.
Assistant
GLOBAL FORUM OF SPIRITUAL AND PARLIAMENTARY LEADERS ON HUMAN SURVIVAL
304 East 45th St. Room 476
New York - NY, United States of America
T: 1 212 953-7947 F: 1 212 557-20961

Almond, Peter O., Mr.
Chairman Board Of Directors
COMMONWEAL
PO Box 316
Bolinas - CA, United States of America
T: 1 415 8680970 F: 1 415 8682230

Alston, Dana, Ms.
PANOS INSTITUTE
1717 Mass. Ave., NW
Washington - DC 20003, United States of America
T: 1 202 483-0044 F: 1 202 483-3095

Alves, Alexandra, Ms.
Ass. To The Program Director
COUNCIL ON FOREIGN RELATIONS
86-19 Britton Ave.
Elmhurst - NY 11373, United States of America
T: 1 718 5659075 F: 1 212 8611916

Alves, Gabriela
TERRA CHRISTA COMMUNICATIONS
1880 East River Road
Tucson - AZ, United States of America
T: 1 602 577-6476 F: 1 602 577-1195

Alves, Tania
TERRA CHRISTA COMMUNICATIONS
1880 East River Road
Tucson - AZ, United States of America
T: 1 602 577-6476 F: 1 602 577-1195

Amberger, Dave, Mr.
GLOBAL YOUTH SUMMIT
10575 Waheo Lake
Waheo Lake - MI, United States of America
T: 1 616 816-7576 F: 1 616 846-2536

Amin, Mahindra, Mr.
TERRA CHRISTA COMMUNICATIONS
1880 East River Road
Tucson - AZ, United States of America
T: 1 602 577-6476 F: 1 602 577-1195

Amon, Amelia, Ms.
INDUSTRIAL DESIGNERS SOCIETY OF AMERICA
116 E 27th Street - 12 H
New York - NY Cep.:20016, United States of America
T: 1 212 447 5515 F: 1 212 779 9589

Amos, Zoe, Ms.
Student
INTERNATIONAL HONORS PROGRAM OF GLOBAL ECOLOGY
11 Carpenter Rd.
Hopewell Jct. - NY, United States of America
T: 1 914 221-7863

Anderson, Carolyn, Ms.
GLOBAL FAMILY
112 Jordan Ave.
San Anselmo CA 94960, United States of America

Anderson, David, Mr.
President
GAIA CORPORATION
PO Box 26138
Fresno - CA CA 93720, United States of America
T: 1 209 433-1516

Anderson, Imarra, Ms.
President
TWENTY-FIRST CENTURY NEWS INC
1880 E River Road, Suite 210
Tucson - AZ 85718, United States of America
T: 1 602 577 1137 F: 1 602 577-1195

Anderson, Jerry, Mr.
Marketing Director
RAINFOREST TV NETWORK
4521 Campus, Suite 275
Irvine - CA 92715, United States of America
T: 1 714 497-9151

Anderson, Lean
YMCA EARTH ACTION CORPS
909 4th Av
Seattle - WA, United States of America
T: 1 206 382-5013 F: 1 206 382-7283

Anderson, Leilani, Ms.
Owner
GLOBAL IDEAS
6846 San Pedro #112
San Antonio - TX 78216, United States of America
T: 1 512 7343652 F: 1 512 7372507

Anderson, Paul S., Mr.
ILLINOIS STATE UNIVERSITY
Dept. of Geography-Geology
Normal-IL 61761, United States of America
T: 1 309 438-7649 F: 1 309 438 5310

Anderson, Sharon, Ms.
TERRA CHRISTA COMMUNICATIONS
1880 East River Road
Tucson - AZ, United States of America
T: 1 602 577-6476 F: 1 602 577-1195

Andrade, Christa, Sra.
WORLD COUNCIL OF CHURCHES
Urb. Riverside L-10
San German, Puerto Rico 00683, United States of America

Andrade, Fernando
Journalist
SABA PRESS PHOTOS
150 East 35th St Suite 940
New York NY 10016, United States of America
T: 1 212 6795454 F: 1 212 6795491

Andres, Catherine Evans, Ms.
President
HARMONIOUS HORIZONS
2737-2 Pacific Heights Rd.
Honolulu - HI 96813, United States of America
T: 1 808 5244686

Anele Helges, O.P., Ms.
Member
DOMINICAL WORLD ORDER TASK FORCE
100 LaSalle St., 3-D
New York - NY 10027, United States of America
T: 1 6630267

Angelis, Fusako De, Mr.
PLUTONIUM FREE FUTURE
P.O. BOX 140 , 2018 Shattuck Avenue
Berkeley - CA 947090, United States of America
T: 1 510 5407645 F: 1 510 5407645

Angelo, Homer, Mr.
FORESTA INSTITUTE FOR OCEAN & MOUNTAIN STUDIES
PO Box 41567
Tucson - AZ 85717, United States of America
T: 1 602 8816174 F: 1 602 3229850

Anglin, Genevieve, Ms.
33 Meadow Lane
Dubuque - IA 50010, United States of America
T: 1 319 292-3275

Anglin, Robert, Mr.
2916 West St., apt.4,
Ames - IA 50010, United States of America
T: 1 515 292-3275

Anglin, Tanya, Ms.
33 Meadow Lane
Dubuque - IA 50010, United States of America
T: 1 319 292-3275

Antrim, Lance, Mr.
CIESIN
1111 Army Novy Drive
Arlington - VA, United States of America
T: 1 703 979 9124

Anwar, Dr. Husayn
Representative
BAHA'I INTERNATIONAL COMMUNITY
866 United Nations Plaza , Suite 120
New York - NY 10017-1811, United States of America
T: 1 212 7563500 F: 1 212 7563573
E: 152:cnm057

Aquino, Mauricio
ECOTROPIC
1880 East River Road
Tucson - AZ, United States of America
T: 1 602 577-6476 F: 1 602 577-1195

Arango, Manuel, Mr.
FUNDACION UNIVERSO VEINTIUNO A.C
1880 East River Road
Tucson - AZ, United States of America
T: 1 602 577-6476 F: 1 602 577-1195

Araujo, Mae, Ms.
Executive Director
AMAZONIA AIDS & HEALTH PROJECT
3180, 18th. Street at Folsom
San Francisco - CA 94110, United States of America
T: 1 415 476-3990 F: 1 415 476-3655

Arbess, Saul
INDIGENOUS PEOPLES MEDIA CENTER
77 Main Street
Newark - DE, United States of America
T: 1 302 3650414 F: 1 302 3650864

Archbold, Anne Elizabeth
Journalist
MARYKNOLL MAGAZINE LATIN AMERICA PRESS
Maryknoll Missiones
Maryknoll NY 10501, United States of America
T: 1 914 9417634

Arnault, Jean
UN DEPARTMENT OF PUBLIC INFORMATION
New York, United States of America

Aronson, Melissa, Ms.
Professor
UNA / USA
21711 So.McBride
Escalan-CA 95320, United States of America
T: 1 209 838-1869 F: 1 209 667-3358

Arrowsmith, John
Director, Foreign News Coverag
ABC NEWS
47 West 66th street
New York - NY, United States of America
T: 1 212 4562800

Arsky, Oleg, Mr.
GEOSPHERE PROJECT, THE
146 Entrada Drive
Santa Monica - CA 90402, United States of America
T: 1 310 459-4342 F: 1 310 459-8299

Askew, Cleo
HIGHLANDER CENTER
1939 Highlander Way
New Market - TN 37820, United States of America
T: 1 615 933 3443 F: 1 615 933 3424

Atkinson, Marjorie
Team Member
FOUNDATION FOR GLOBAL COMMUNITY
24475 Paseo Privado
Salinas - CA 93908, United States of America
T: 1 408 4841349

Atkinson, Sunny, Ms.
EARTH SOCIETY
3600 S. Harbor Blvd. Suite 247
Channel Islands, United States of America
F: 1 805 9840593

Atta, Pabla Alessandra
USA TODAY
1865 Brickell Ave A1102
Miami FL 33129, United States of America
T: 1 305 3666006 F: 1 305 8582594

Atta, Paula Van, Ms.
GLOBAL ENVIRONMENT TEAM
2534 Ganesha Ave.
Altadena - CA 91001, United States of America
T: 1 818 791-1335

Attiga, Ali Ahmed, Mr
Asg/Mndp
130 W 79 St Apt 11C
New York NY, United States of America
T: 0 212 906 5450 H F: 212 906 5487

Audwyn, Ondine Wilhelm, Ms.
Student
INTERNATIONAL HONORS PROGRAM - BARD COLLEGE
1211 Poplar Hill Road
Baltimore, Md. 21210, United States of America
T: 1 301 323-4668

Auer, Brad, Mr.
YALE UNIVERSITY
477 Prospect St.
New Haven - CT 06511, United States of America
T: 1 203 436-3536

Augustine, David, Mr.
EARTH TRAIN/WORLD LINK GLOBAL YOUTH SUMMIT
1884 E. Street
Hayward - CA 94541, United States of America
T: 1 510 538-3530

Avadhuta, Dada Satyeshvarananda
Chief Secretary
PROUTIST UNIVERSAL
1354, Montague St. NW
Washington DC 20011, United States of America

Ayala, Juan O., Mr.
Ayudante Especial
DEPARTAMENTO RECURSOS NATURALES
Pta. de Tierra San Juan
San Juan, Puerto Rico 00956, United States of America
T: 1 809 723-3040

Bachula, Gary, Mr.
Director
CIESIN
2250 Pierce Rd
Saginal, MI 48710, United States of America
T: 1 517 791 7371 F: 1 517 790 0735

Baek, Joanne, Ms.
35 Village Park
Amherst - MA 01002, United States of America
T: 1 413 549-1084

Baez, Shad, Mr.
Student
LAUREL SPRINGS SCHOOL
PO Box 1440
Ojai - CA 93024, United States of America
T: 1 805 646-2473 F: 1 805 646-0186

Bagga, Daljeet
GLOBAL FUTURES NETWORK
1880 East River Road
Tucson - AZ, United States of America
T: 1 602 577-6476 F: 1 602 577-1195

Bailey, Ken
MIRANDA SMITH PRODUCTIONS INC.
251 West 30th Street #16-W
New York - NY, United States of America
T: 1 212 5632370 F: 1 212 5632372

Baker, Ben, Mr.
UP WITH PEOPLE
3103 North Campbell Av.
Tucson - AZ 85719, United States of America

Baker, Lynn, Ms.
TERRA CHRISTA COMMUNICATIONS
1880 East River Road
Tucson - AZ, United States of America
T: 1 602 577-6476 F: 1 602 577-1195

Baker, Madison, Ms.
PEACEABLE REALM FOUNDATION
4329 Brookshire Circle
Santa Rosa - CA 95405, United States of America

Baker, Paul, Mr.
TERRA CHRISTA COMMUNICATIONS
1880 East River Road
Tucson - AZ, United States of America
T: 1 602 577-6476 F: 1 602 577-1195

Ball, Jocelyn, Ms.
Landscape Architect
UNA PASADENA
4798 Sullivan St #203
Ventura, United States of America
T: 1 805 6509516

Ballerstaedt, Maria Esther, Ms.
CASA DE LA MUJER
c. 21 de Mayo 541
Santa Cruz - CA, United States of America
T: 1 408 32-4931

Ballew, Cathy, Ms.
Project Coordinator
LUMMI INDIAN BUSSINESS COUNCIL
2470 Kwina Road
Bellighan 982265, United States of America
T: 206384 5288 F: 206384 5521

Balmaceda, Mary, Mr.
WORLD INFORMATION TRANSFER INC
274 Madison Ave, Suite 601
New York - NY 10016, United States of America
T: 1 212 2730253 F: 1 212 3792004

Banker, Deborah, Ms.
President
FRIENDS OF ONE WORLD
Ste. 1-316 23715 W. Malibu Rd.
Malibu - CA 90265, United States of America
T: 1 310 700-1294 F: 1 310 591-0854

Baptista, Sandra, Ms.
BROWN UMIVERSITY - CENTER FOR ENVIRONMENTAL STUDIES
PO Box 5523
Providence - RI 02912, United States of America
T: 401 521-6202 F: 401 863-3503
TL: st802218@BROWNU

Barat, Stephan Mustapha
RAI - TV CHANNEL 3
New York, United States of America

Barber, Charles, Mr.
Associate Forests & Biodiversi
WORLD RESOURCES INSTITUTE
1709 New York Avenue, N.W
Washington - DC 20006, United States of America
T: 1 202 638-6300 F: 1 202 638-0036
TL: 64414 WRI WASH

Barber, Lois, Ms.
Advisor
EARTHACTION
30 Cottage St.
Amherst - MA 01002, United States of America
T: 1 413 549-4285 F: 1 413 549-0544

Barber, Sarah, Ms.
Staff
EARTH TRAIN/WORLDLINK/GLOBAL YOUTH SUMMIT
12974 Robin Road
Nevada City - CA 95959, United States of America
T: 1 916 272 8141

Barbosa, Alexandre Sodre, Mr.
ECOTROPIC
1880 East River Road
Tucson - AZ, United States of America
T: 1 602 577-6476 F: 1 602 577-1195

Barill, Terryn, Ms.
Scientist
PRACTICAL SOLUTIONS
686 S. Arroyo Parkway # 265
Pasadena-CA 91105, United States of America
T: 818 304-4756 F: 818 247-9558

Barker, David, Mr.
Tech. Assistance Coordinator
WINROCK INTERNATIONAL
1611 N.Kent St., Suite 600
Arlington - VA, United States of America
T: 1 703 525-9430 F: 1 703 649-1106
TL: 6491106 WIDC, E: DIALCOM YI: ICN 408

Barnes, Christine, Ms.
HORIZONS COUNSELING
P.O. Box 35893
Tucson - AZ 85740, United States of America

Barni, Alex, Mr.
GEOSPHERE PROJECT, THE
146 Entrada Drive
Santa Monica - CA 90402, United States of America
T: 1 310 459-4342 F: 1 310 459-8299

Barrera, Ana Maria, Ms.
GLOBAL ENVIRONMENT TEAM
696 South Madison Avenue
Pasadena - CA 91106, United States of America
T: 1 818 796-2951 F: 1 818 792-2121

Barroso, Carmen, Ms.
Diretor
MAC ARTHUR FOUND.
Chicago - IL, United States of America

Bartone, Carl, Mr.
Environment Specialist
URBAN MANAGEMENT PROGRAM / WORLD BANK
1818 H Street NW
Washington - DC 20433, United States of America
T: 1 202 4733861 F: 1 202 4771391

Barzaghi, Jacques, Mr.
WE THE PEOPLE
643 E. Channel Rd Santa Monica
Santa Monica - CA 90402, United States of America
T: 1 310 454 9905 F: 1 310 449 1976

Basi, Raghbir S., Mr.
Advisor
WORLD PEACE UNIVERSITY
c/o Maharajas 328 'G' Street
Anchorage - Alaska, United States of America
T: 1 907 272-2233 F: 1 907 272-7597

Bassett, Elizabeth, Ms.
Press Coordinator
GLOBAL FORUM OF SPIRITUAL AND PARLIAMENTARY LEADERS ON HUMAN SURVIVAL
304 East 45th St. Room 476
New York - NY, United States of America
T: 1 212 953-7947 F: 1 212 557-20961

Baun, Kyle
UP WITH PEOPLE
3103 North Campbell Av.
Tucson - AZ 85719, United States of America

Baure, Joanne, Ms.
CARNEGIE COUNCIL ON ETHICS AND INTERNATIONAL AFFAIRS
170 E. 64th St.
New York - NY 10021, United States of America
T: 1 212 838-4120 F: 1 212 752-2432

Bexter, Tricia Ann
TURNER BROADCASTING SYSTEM / CNN
One CNN Centre
Atlanta - GA, United States of America
T: 1 404 827-3973 F: 1 404 827-1015

Bearcomesout, Charles, Mr.
NATIVE ACTION
PO Box 682
Lame Deer - MT 59043, United States of America
T: 1 406 477-6392

Bearcomesout, Dalberta, Ms.
NATIVE ACTION
PO Box 682
Lame Deer - MT 59043, United States of America
T: 1 406 477-6392

Beaton, Andrew, Mr.
GLOBAL ENVIRONMENT TEAM
81 Waterman Ave.
Rumson - NJ 07760, United States of America
T: 1 908 741-7411

Beck, Robert, Mr.
Director
EDISON ELETRIC INSTITUTE
701 Pennsylvania Avenue, N.W.
Washington - DC 20004, United States of America
T: 1 202 508-5460 F: 1 202 508-5403

Bedoya, Kathleen, Ms.
219 East 25th St, Apt 2F
New York - NY 10010, United States of America
T: 1 212 512-1097 F: 1 212 412-5698

Bedrossian, Vanessa C., Ms.
Head Orgaizer
LION COALITION FOR THE ENVIRONMENT & PEOPLE
710 Leland 1s
St. Louis, MO, United States of America
T: 1 314 8633302

Beesley, Tina, Ms.
UP WITH PEOPLE
3103 North Campbell Av.
Tucson - AZ 85719, United States of America

Belcon, Patricia, Ms.
CENTER FOR THE STUDY OF WOMEN & SOCIETY
City University of New York, 33 West 42nd St.
New York - NY 10036, United States of America

Belding, Laura, Ms.
UP WITH PEOPLE
3103 North Campbell Av.
Tucson - AZ 85719, United States of America

Bell, Perry
GLOBAL ENVIRONMENT TEAM
696 South Madison Avenue
Pasadena - CA 91106, United States of America
T: 1 818 796-2951 F: 1 818 792-2121

Bella, Antonio Di
RAI - TV CHANNEL 3
New York, United States of America

Bello, Waldemar F., Sr.
Executive Director
INSTITUTE FOR FOOD AND DEVELOPMENT POLICY
140 Ninth St.
San Francisco - CA, United States of America
T: 1 415 864-8555 F: 1 415 864-3909

Bellomo, Angelo J., Mr.
ICF KAISER ENGINEERS
10 Universal Plaza # 2400
Universal City, CA, United States of America
T: 818 509-3123 F: 818 509-3137

Bennet, Joan, Ms.
Director Of Public Affairs
HILL & KNOWLTON PACIFIC COMMUNICATIONS, INC.
820 Mililani St., Suite 400
Honolulu - HI, United States of America
T: 1 808 543-3523 F: 1 808 537-1531

Bennett, Sherrin, Ms.
Executive Producer
EARTH TRAIN / GLOBAL YOUTH SUMMIT
1505 Bridgeway
Sausalito - CA 94965, United States of America
T: 1 415 331-4073 F: 1 415 331-0825

Benson, Lesley, Ms.
TEACHER LAUSD
1441 West Olympic Boulevard
Los Angeles - CA, United States of America
T: 1 213 736-1094 F: 1 213 380-3769

Benson, Robert, Mr.
Professor
LOYOLA UNIVERSITY LAW SCHOOL
1441 West Olympic Boulevard
Los Angeles - CA, United States of America
T: 1 213 736-1094 F: 1 213 380-3769

Bentley, Shannon
YMCA EARTH ACTION CORPS
909 4th Av
Seattle - WA, United States of America
T: 1 206 382-5013 F: 1 206 382-7283

Benton, Inka, Sra.
FOUNDATION FOR GLOBAL COMMUNITY
251 Bret Harte Rd.
San Rafael - CA 94901, United States of America
T: 1 415 461-1468

Beredjick, Nick, Mr.
Advisor
ENEL
194-06 87th Road Hollis
New York - NY 11423, United States of America
T: 1 212 479 2153

Berenholtz, Jim, Mr.
Consultant
EARTH DAY INTERNATIONAL
231-A Fairview St.
Laguna Beach - CA 92651, United States of America
T: 1 714 797-9618

Bergstrom, Kirk, Mr.
Executive Director
EARTH TRAIN/WORLDLINK/GLOBAL YOUTH SUMMIT
P.O. Box 480483
Los Angeles - CA 90048, United States of America
T: 1 213 273-2636

Berman, Paul, Mr.
Consultant
MEGA - CITIES
1185 Keeler Ave.
Berkeley - CA, United States of America
T: 1 510 843-8574

Bermingham, John R., Mr.
President
COLORADO POPULATION COALITION
601 Franklin Street
Denver - CO 80218, United States of America
T: 1 303 322-8290

Best, Laura, Ms.
Int'L Coordinator
WORLD ALLIANCE FOR BREASTFOOD ACTION (WABA)
34-11 146th St.
Flushing-NY 11354, United States of America
T: 1 718 886-4423 F: 1 718 886-4423

Beyer, Curtis, Mr.
GLOBAL ENVIRONMENT TEAM
5108 Cimirron Lane
Culver City - CA 90230, United States of America
T: 1 310 839-8131

Beynon, Alaea, Ms.
FRIENDS OF THE AMAZON FOREST
47 Little Nahant Road
Nahant - MA 01908, United States of America
T: 001 617 593-4329 F: 001 617 593-4329

Beynon, Rex, Mr.
FRIENDS OF THE AMAZON FOREST
P.O.Box 625
Cambridge - MA 02140, United States of America

Bhansal, Tarachand
MAHAVIR AGRICULTURE SYSTEMS -INDIA
1880 East River Road
Tucson - AZ, United States of America
T: 1 602 577-6476 F: 1 602 577-1195

Bharat, Parekh, Dr.
President
GANTECH CORPORATION
66 East 9th Street suite 1811
St. Paul - MN 55101, United States of America

Bhattacharya, Bhaswati, Ms.
TERRA CHRISTA COMMUNICATIONS
1880 East River Road
Tucson - AZ, United States of America
T: 1 602 577-6476 F: 1 602 577-1195

Biaggi, Cristina, Ms.
Member
WEDO - WOMEN'S ENVIRONMENT & DEVELOPMENT ORGANIZATION
1 Ludlow Lane
Palisades - NY 10964, United States of America
T: 1 914 359-5898 F: 1 201 767-6471

Biehn, Patricia M., Ms.
NORTH AMERICAN COALITION ON RELIGION AND ECOLOGY
8702 Mossford Dr.
Hunlington Beach-CA, United States of America
T: 1 714 963-7754 F: 1 714 962-1716

Bihun, Ulana
WORLD INFORMATION TRANSFER INC
274 Madison Ave, Suite 601
New York - NY 10016, United States of America
T: 1 212 2730253 F: 1 212 3792004

Bird, Rachel A., Ms.
GEOSPHERE PROJECT, THE
146 Entrada Drive
Santa Monica - CA 90402, United States of America
T: 1 310 459-4342 F: 1 310 459-8299

Birk, Mia, Ms.
Programme Associate
INTERNATIONAL INSTITUTE FOR ENERGY CONSERVATION
420 C Street N.E.
Washington - DC, United States of America
T: 1 202 546-3388 F: 1 202 546-6978
TL: 249114 IIEC UR

Bishop, Judy, Ms.
GLOBAL ENERGY NETWORK INT'L
PO Box 81565
San Diego - CA 92138, United States of America
T: 1 619 595-0139 F: 1 619 595-0403

Bishop, Kristina, Ms.
THE COLLEGE EXPLORATION
1431 Towston Road
Vienna - VA 22182, United States of America
T: 1 703 255-2672 F: 1 703 255-2792

Bishop, Lawrence, Mr.
Hon. Director
INT'L SOC PRESERVATION OF TROPICAL RAIN FOREST
3302 N. Burton Avenue
Rosemead - CA 91770, United States of America
T: 1 818 572-7273 F: 1 818 572-9521

Bjanesoy, Kjartan
UP WITH PEOPLE
3103 North Campbell Av.
Tucson - AZ 85719, United States of America

Blackford, Loren, Ms.
Consult
UN NGLS
Palais de Nations
Geneve CH 1211, United States of America
F: 41 22 733-6542

Blackman, Laura A., Ms.
Attorney
72 Park Terrace West E85
New York - NY, United States of America
T: 1 212 930-0214 F: 1 212 930-0196

Blackstone, Wendy, Ms.
Composer
WORLD CITY - BLACHSTONE GAIA LJ
59 W 10 Street
New York - NY 10011, United States of America
T: 1 212 228-4091

Blades, Scott, Mr.
21ST. CENTURY NEWS
1880 East River Road
Tucson - AZ, United States of America
T: 1 602 577-6476 F: 1 602 577-1195

Blair, James G., Mr.
Special Assistant
UNITED STATES (TEAM) DELEGATION
3134 Carleton St. # C
San Diego - CA, United States of America
T: 1 619 430-2478 F: 1 619 725-2208

Bland, George, Mr.
Research Director
TREE TECHNOLOGY INTERNATIONAL, INC.
440 Park Avenue South 13th floor
New York - NY, United States of America
T: 1 212 685-8400 F: 1 212 689-3701

Blankenship, Thomas, Mr.
Vice-Presidente
PEACE BY PEACE INTERNATIONAL
7904 110TH SE
Renton, WA 98056, United States of America
T: 206 948-5086 F: 916 389-2390

Blobaum, Roger, Mr.
Asst. Director
WORLD SUSTAINABLE AGRICULTURE ASSOCIATION
1331 Pennsylvania Ave., NW, Suite 907, North Tower
Washington - DC 20004, United States of America
T: 1 202 347-0633 F: 1 202 662-7035

Bloom, Gretchen, Ms.
President
SOCIETY FOR INTERNATIONAL DEVELOPMENT
1401 New York Ave 1100
Washington - DC, United States of America
T: 1 202 7850831 F: 1 202 2233865

Bloomer, Julie, Ms.
CENTER FOR INTRNATIONAL COOPERATION
1820 Westridge Rd
Los Angeles - CA 90049, United States of America
T: 1 213 476-6326 F: 1 213 476-7000

Blum, Bev, Mr.
Executive Director
SOLAR BOX COOKERS INTERNATIONAL
1724 11th Street
Sacramento - CA 95814, United States of America
T: 1 916 444-6616 F: 1 916 447-8689

Boatman, Marise, Ms.
Producer Manager
INTERNATIONAL SYNERGY INSTITUTE
508 Price St. #2B
Alexandria-VA 22314, United States of America
T: 1 703 548-6425 F: 1 703 652-5217

Bodansky, Daniel, Mr.
UNIVERSITY OF WASHINGTON
1002 58st. AUC NE
Seattle - WA 98115, United States of America
T: 1 206 543 7491 F: 1 206 5435671

Bodarth, Daniel Michael
Assistant Professor
UNIVERSITY OF WASHINGTON LAW SCHOOL
School of Law JB-20 University of Washington
Seattle WA 98115, United States of America
T: 206 543 7491 F: 206 543 5671

Bodmer, Richard, Dr.
Post-Doctoral Fellow
UNIVERSITY OF FLORIDA
Dept. Zoology, University of Florida
Gainesville - FL 32605, United States of America
T: 1 904 373-3186

Bodzin, Steven
FREELANCE JOURNALIST
3362A 24th St.
San Francisco - CA 94110, United States of America
T: 1 415 285-1929

Boehme, Eric Alexander, Mr.
President
GLOBAL EDUCATION ASSOCIATE
475 Riverside Drive, Suite 456
New York - NY 10115, United States of America
T: 1 212 870-3290 F: 1 212 870-2055

Bogart, Kirstin, Ms.
FOUNDATION FOR GLOBAL COMMUNITY
222 High Street
Palo Alto - CA 94301, United States of America

Boggs, Carl, Mr.
EARTHSAVE
1251 Wellesley
Los Angeles - CA 90025, United States of America
T: 1 213 826-0671

Bograd-Gordon, Judith, Dr
Rep To Nfo Commitee Ch State Of Wom
IN'L SOCIAL SCIENCE COUNCIL & UNIV
Univ. of New Haven Dept of Sociology
West Haven, CT 06316, United States of America
T: 203 288 7815 F: 203 937 0756

Bonds, Nedra, Ms.
HIGHLANDER CENTER
1939 Highlander Way
New Market - TN 37820, United States of America
T: 1 615 933 3443 F: 1 615 933 3424

Bondy, Magda, Ms.
SOCIAL WORKER
12-3 Grandda Crescent
White Plains - NY 100603, United States of America
T: 1 914 946-9383

Bondy, Rob
Tv Cameraman
AUST BROADCASTING COMMISSION
2030 M Street NW 702
Washington - DC, United States of America
T: 1 202 466 6575 F: 1 202 775 8291

Bones, David, Mr.
GREEN (GLOBAL RIVERS ENVIRONMENTAL EDUCATION NETWORK)
University of Michigan School of Natural Resouces -Dana Building 430E
Ann Arbor - MI, United States of America
T: 1 313 7645171 F: 1 313 9362195

Boni, Dennis P.
Cameraman meta inc
Permanent Mission of United States of America
2100 N 14th St
Arlington VA, United States of America
T: 1 703 243360

Bookser, Mary, Ms.
EARTH CONNECTION
5820 Bender Bd. 145233
Cincinnati - OH, United States of America
T: 1 513 922-1468

Booth, Dorraine, Ms.
INDIGENOUS PEOPLES MEDIA CENTER
77 Main Street
New York - NY, United States of America
T: 1 212 3650414 F: 1 212 3650864

Boren, Ann, Ms.
Member
WORLDWIDE NETWORK
2610 31st.St. N.W. - Network
Washington - DC 20008, United States of America
T: 1 202 342-8727

Boren, Sarah, Ms.
Staff
THE POPULATION INSTITUTE
107 Second Street, NE
Washington - DC 20002, United States of America
T: 1 202 544-3300 F: 1 202 544-0068

Boshart, David, Mr.
WILDLIFE ART INTERNATIONAL
11000-33 Metro PNWY
Fort Myers - FL 33912, United States of America
T: 1 813 936-7518 F: 1 813 936-2788

Boule, Nathalie, Ms.
UP WITH PEOPLE
3103 North Campbell Av.
Tucson - AZ 85719, United States of America

Bowles, Eugene, Mr.
TEMPLE OF UNDERSTANDING
St. Emeric Church / 185 Avenue D
New York - NY 10009, United States of America
T: 1 212 353-0001 F: 1 212 353-0170

Bowles, Margaret, Ms.
TEMPLE OF UNDERSTANDING
St. Emeric Church / 185 Avenue D
New York - NY 10009, United States of America
T: 1 212 353-0001 F: 1 212 353-0170

Bowman, Margaret, Ms.
BOILINGSTONE
P. O. Box 12199
San Antonio - TX 78212, United States of America
T: 1 512 236-9617 F: 1 512 735-7033

Brahmananda, Dada, Mr.
Chief Secretary
PROUTIST UNIVERSAL
1354 Montague St., NW, P.O. Box 56466
Washington - DC 20040, United States of America
T: 1 202 829-2278 F: 1 202 829-0462
TL: 150255668

Brandat, Gloria De Pablo Pardo
Editor
CNN
1 CNN Center
Atlanta, GA 30303, United States of America
T: 55 21 2595407

Brannigan, Bill, Mr.
Information Officer
THE WORLD BANK
1818 H St NW
Washington - DC 20433, United States of America
T: 1 202 473-1784 F: 1 202 676-0578

Brauns, Andrei
Cameraman
CNN BROADCASTING SYSTEM
One CNN Center, Box 105366
Atlanta - GA, United States of America
T: 1 404 8371918 F: 1 404 8274292

Breed, Hellen Illick, Dr.
Advisor
WIPAC
R.D.3, Box 245-B
Troy - NY 12180, United States of America
T: 1 518 279-3255 F: 1 518 584-5330

Bresenham, David, Mr.
President
WOFFORD COLLEGE
PO Box 267 Wofford College
Spartanburg 29303-3663, United States of America
T: 303 597-4010 F: 063 803 597-4019

Bresnahan, Chris
Audio
WGBH-TV
125 Western Ave.
Boston -Massachusett, United States of America
T: 1 617 4495830 F: 1 617 4446450

Bretz, Laurie, Ms.
Spanish Editor
INTERNATIONAL SYNERGY INSTITUTE
2740 Eliot Circle
Westminister, United States of America
T: 1 303 426-8963 F: 1 303 292-2645

Bridges, Helice, Ms.
Exec. Director
DIFFERENCE MAKERS, INT'L
P.O.Box 2115
Del Mar - CA 92014, United States of America

Broderson, Peggie, Ms.
UP WITH PEOPLE
3103 North Campbell Av.
Tucson - AZ 85719, United States of America

Brodt, Alexander, Mr.
CAMPAIGN FOR THE EARTH
2120 13th Street
Boulder - CO 80302, United States of America
T: 1 303 444-3800 F: 1 303 444-3811

Brooks, Stephen Raymond
Tape Editor
CNN BROADCASTING SYSTEM
One CNN Center, Box 105366
Atlanta - GA, United States of America
T: 1 404 8371918 F: 1 404 8274292

Brosnan, Peter
Tv Editor
AUST BROADCASTING COMMISSION
2030 M Street NW 702
Washington - DC, United States of America
T: 1 202 466 8575 F: 1 202 775 8291

Brown, Judie, Ms.
Co-Chairman-Board Of Directors
ICRI - INTERNATIONAL CHILD RESOURCE INSTITUTE
1810 Hopkins St.
Berkeley - CA 94707, United States of America
T: 1 510 644-1000 F: 1 510 525-4106

Brown, Michael, Mr.
Project Director
PVO-NGO/NRMS PROJECT
1250, 24th Street, NW- Suite 500
Washington - DC, United States of America
T: 1 202 293-4800 F: 1 202 223-6971

Brown, Noel, Mr.
Director
UNEP
2 Un Plaza
New York - NY, United States of America
T: 1 212 980-3392 F: 1 212 963-7341

Brown, Rachel Noel, Ms.
Student
INTERNATIONAL HONORS PROGRAM - BARD COLLEGE
11 Morrislea Court
Baltimore, Md. 21234, United States of America
T: 1 301 823-8785

Brown, Sarah, Ms.
YMCA EARTH ACTION CORPS
909 4th Av
Seattle - WA, United States of America
T: 1 206 382-5013 F: 1 206 382-7283

Bruce, Carolyn, Ms.
UP WITH PEOPLE
3103 North Campbell Av.
Tucson - AZ 85719, United States of America

Bruhwiler, Erik, Mr.
GEOSPHERE PROJECT, THE
146 Entrada Drive
Santa Monica - CA 90402, United States of America
T: 1 310 459-4342 F: 1 310 459-8299

Bruner, Tina
UNITED NATIONS ASSOCIATION
Abbott Road 4522
Dallas - TX 75205, United States of America

Brunner, Alison, Ms.
Staff
EARTH TRAIN/WORLDLINK GLOBAL SUMMIT
1638 Daphne Dr.
San Jose 95129, United States of America
T: 408 255 6063

Bruss, Jane, Ms.
President
CROSS CULTURAL COMMUNICATINS
6319 Barrie Road
Minneapolis - MN, United States of America
T: 1 612 920-8951 F: 1 612 920-8951

Bryce, Gabriole Van, Ms.
57 W. 58th Street
New York - NY, United States of America
T: 1 212 751-2951 F: 1 800 762-2187

Brym, , Ms.
INSTITUTE FOR ECONOMIC ANALYSIS
New York University, 269 Mercer Street Im 205
New York - NY, United States of America
T: 1 212 988-7480 F: 1 212 995-4465

Brym, Maia, Ms.
Student
171 Blackberry Dr.
Stanford 06903, United States of America
T: 203 329-2533

Buescher, Kay, Ms.
UNITED METHODIST CHURCH
925 S.E. Malden
Portland - OR, United States of America

Bullard, Robert, Mr.
Professor
UNIVERSITY OF CALIFORNIA
Watkins Hall
Riverside-CA 92521, United States of America
T: 1 714 787-5444 F: 1 714 787-3330

Burn, John
Publisher
UNIFO PUBLISHERS
PO Box 3858
Sarasota - FL, United States of America
T: 1 813 953 7900 F: 1 813 953 4999

Burns, Sarah, Ms.
Ngo Liaison
WORLD RESOURCES INSTITUTE
1709 New York Avenue NW
Washington - DC 2006, United States of America
T: 1 202 662-2559 F: 1 202 638-0036

Burris, Juanita Salvador, Ms.
ASIAN-AMERICAN WOMEN'S NETWORK & HYDE PARK ENVIRONMENTAL ACTION COALITION
6107 South Woodlawn Avenue
Chicago - IL 60637, United States of America
T: 1 312 955-0291 F: 1 312 363-4880

Burtt, Michael C., Mr.
Priest
ORDER OF MELCHIZEDEK
PO Box 331791
Miami - Florida 33233, United States of America

Busch, Adam, Mr.
YMCA EARTH ACTION CORPS
909 4th Ave.
Seattle - WA, United States of America
T: 1 206 352-5013 F: 1 206 382-7283

Butler, Geoffrey, Mr.
Member
WORLD THAT WORKS FOR EVERY ONE, THE
PO Box 20506
Juneau - Alaska, United States of America
T: 1 907 465-3879

Butler, John
WORLD WILDLIFE FUND - WWF
Rue du Mont Blanc
Gland, United States of America
T: 41 22 64 95 25 F: 41 22 64 58 79

Butter, James, Mr.
Musician
PARLIAMENTARY GLOBAL FORUM
P.O.Box 68
Litchfield, United States of America
T: 203 567-8796 F: 203 567-4276

Byrne, Warren, Mr.
288 Cozy Beach Avenue
East Haven - CT 06512, United States of America
T: 1 203 468-8609 F: 1 203 432-5942

C.Bedrossian, Vanessa
Head Orgaizer
LION COALITION FOR THE ENVIRONMENT & PEOPLE
710 Leland 1s
St.Louis, MO 63130, United States of America
T: 314 8633302

Caballero, Ma. Cristina C., Ms.
President
DIALOGUE ON DIVERSITY
6041 Forest Hollow Lane
Springfield-VA, United States of America
T: 1 703 644-0169 F: 1 703 644-6709

Cage, Henry
INDIGENOUS PEOPLES MEDIA CENTER
77 Main Street
New York - NY 10083, United States of America
T: 1 212 3650414 F: 1 212 3650864

Caldwell, William, Dr.
V.Presid.
UNA COASTLINE CHAPTER
5553B - Rayo del Sol (Laguna Hills)
Laguna Hills Zip.: 92653, United States of America
T: 714 472-6262

Campbell, Duncan, Mr.
CAMPAIGN FOR THE EARTH
2120 13th Street
Boulder - CO 80302, United States of America
T: 1 303 444-3800 F: 1 303 444-3811

Campbell, Nancy, Ms.
Fund-Administrator
TIDES FOUNDATION
P.O. Box 121
Kentfield, CA 94914, United States of America
T: 415 924-8644 F: 415 924-5304

Canaletich, Tony, Mr.
ECO ERA
213 Richardson Drive
Mill Valley - CA, United States of America
T: 1 415 721-9665 F: 1 415 381-8028

Cancian, Antonio Joao, Mr.
MISSAO EVANGELICA KAIWA
Caixa Postal 04
Downey - CA 79800, United States of America

Candelas, Fernanda
ECOTROPIC
1880 East River Road
Tucson - AZ, United States of America
T: 1 602 577-6476 F: 1 602 577-1195

Canfield, Christopher, Mr.
CERRO GORDO TOWN FORUM
Dorena Lake, Box 569,
Cottage Grove, OR 97424, United States of America
T: 1 503 942 7720

Cannon, Laura-Lea, Ms.
Vice President
ENVIRONMENTAL MARKETING SERVICES, INC.
4295 Eldorado Springs Drive
Boulder - CO 80303, United States of America
T: 1 303 4942210 F: 1 303 4942210

Capellini, Steven, Mr.
ORGANICA PRESS
3859 Kumquat Ave.
Miami - Florida, United States of America
T: 1 305 4419252 F: 1 305 4419252

Caplan, Maia, Ms.
WHARTON SCHOOL - U. OF PENNSYLVANIA
3514 Lancaster Ave, 413
Philadelphia - PA, United States of America
T: 1 215 898-3018 F: 1 215 573-2006

Caprio, Theresa
TEMPLE OF UNDERSTANDING
St. Emeric Church / 185 Avenue D
New York - NY 10009, United States of America
T: 1 212 353-0001 F: 1 212 353-0170

Card, Margaret Ann, Ms.
SO.CALIF. ECOLOGY TASK FORCE
9203 Cord Ave.
Downey - CA 90240, United States of America
T: 1 310 861-3207

Cardamone, Anne, Ms.
CENTER FOR INTERNATIONAL COOPERATION
712 Barnes Ave
Syracuse - NY 13207, United States of America
T: 1 315 492 4704

Cardo, Fabio, Ms.
TERRA PAR LATDA
Rodovia Campinas/Capivari km 23,5 -cp 51
Monterey - CA, United States of America
T: 1 408 79-1311 F: 1 408 79-2161

Cargas, James P., Mr.
CENTER FOR INTERNATIONAL ENVIRONMENTAL LAW-US
1621 Connecticut Avenue, Suite 300
Washington - DC 20009-1076, United States of America
T: 1 202 332-4840 F: 1 202 332-4865

Caris, Chris
Producer
CBS NEWS
524 W 57th St
New York NY 10019, United States of America
T: 1 212 5051911 F: 1 212 5051179

Carlsson, Chris
Editor
PROCESSED WORLD
1095 Market Street suite 209
San Francisco CA 941, United States of America
T: 1 415 6262160 F: 1 415 6262685

Carmichael, Judy, Ms.
Director
WOMEN FOR A MEANINGFUL SUMMIT / GLOBAL NETWORK 2000
Suite 180, 6 University Dr.
Amherst - MA 01002, United States of America
T: 1 413 549-1071 F: 1 413 549-6401

Carney, Patricia, Ms.
Director
EARTH SAVE
706, Frederick St
Santa Cruz, United States of America
T: 408 423 4069 F: 408 458 0255

Carr, Caroline, Ms.
BROWN FOR PRESIDENT
309 Mariemont Dr.
Lexington - KY 40505, United States of America
T: 606 299 2829 F: 606 299 3227

Carroll, Daniel, Mr.
MG TRADING
516 S. 4TH St.
Las Vegas - NV 89101, United States of America
T: 1 702 229 2245

Carroll, Frank, Mr.
ENERGY & RESOURCES GROUP
6618 Tremont Street
Oakland - CA 94609, United States of America
T: 1 510 547-4269

Carter, Nicholas
Journalist
SMITHSONIAN TROPICAL RESEACH INSTITUTE
Unit 0946, APO AA 0948
United States of America
T: 1 507 525539 F: 1 507 625942

Cartwright, Lisa, Ms.
Professor
UNIVERSITY OF ROCHESTER
380 Wellington Ave
Rochester - NY 14619, United States of America
T: 1 212 436-8059 F: 1 212 432-

Caruso, Judy, Ms.
CENTER FOR INTERNATIONAL COOPERATION
PO Box 102
Washington - VA 22747, United States of America
T: 1 703 9375219 F: 1 703 9374039

Carvalho, Nanci Bunte De, Sra.
Project Coordinator
INTERNATIONAL CHILD RESOURCE INSTITUTE - ICRI
1819 Hopkins St.
Berkeley - CA 94707, United States of America
T: 1 510 664-1000 F: 1 510 525-4106

Cassara, Beverly
RFD 2
12 Rajaji Marg
New Delhi 110011, United States of America
T: 91 11 301-4309

Cassinelli, Nicole Albertal, Ms.
Adm. Assistant
CULTURAL SURVIVAL
53 A Church Street
Cambridge - MA 02138, United States of America
T: 1 617 495-2562 F: 1 617 495-1396

Castenada, Theresa, Ms.
Secretary
PERMANENT MISSION OF VANUATU TO THE U.N
416 Convent Avenue
New York - NY 10031, United States of America
T: 1 212 577-6476 F: 1 212 577-1195

Castillo, Carlos, Mr.
WORLD ASSN OF FORMER U.N. INTERNES AND FELLOWS INC (WAFUNIF)
c/o Ms Hope P White-Davis, United Nations, Room DC2-1118, 2 UN Plaza
New York - NY, United States of America
T: 1 212 9633110 F: 1 212 9634116

Castro, Aurelia, Ms.
GRUPO JANSTAY
Ameghino 1341 Marquezado
San Juan, Puerto Rico, United States of America

Castro, Flavia
Television Production Speciali
CONSERVATION INTERNATIONAL TV PRODUCTION
1015 18Th St., NW
Washington - DC 20036, United States of America
T: 1 202 4295660 F: 1 202 8875188

Castro Neves, Oscar, Mr.
PARLIAMENTARY GLOBAL FORUM
Box 68
Litenfield, United States of America

Cavalieri, L.F., Dr.
COUNCIL FOR RESPONSIBLE GENETICS AND BIOTECHNOLOGY WORKING GROUP
735 Anderson Hill Road ns
Purchase - NY 10577-1400, United States of America
T: 1 914 251-6643 F: 1 914 251-6635

Cave, Christine
Administrative Assistant
NATIONAL PUBLIC RADIO
2025 M St NW
Washington DC 20036, United States of America
T: 1 202 8222000 F: 1 202 8222329

Cedarface, Reg, Mr.
Director
PARTNERS IN THE ENVIRONMENT
6911 Richmond Hwy., Ste. 308
Alexandria-VA, United States of America
T: 1 703 765-8463 F: 1 703 765-8426

Cerqueira, Eneida, Ms.
INT'L ASSN. FOR ECOLOGICAL ECONOMICS
Solomons
Solomons- Maryland, United States of America
T: 021 252 3001 F: 021 252 3001

Cervantes, Luz, Mr.
GLOBAL HEART PROJECT
395 Precitn Ave
San Francisco - CA 94110, United States of America
T: 1 415 285-2287

Ceyssens, Guido, Mr.
UP WITH PEOPLE
3103 North Campbell Av.
Tucson - AZ 85719, United States of America

Chadwick, Bruce P., Mr.
Pesquisas De Ciencia Pol.
COLUMBIA UNIVERSITY
400, Central Park West
New York - NY, United States of America
T: 1 212 865-8628 F: 1 212 865-8628
E: BPC1@CUNIXB.CC.

Chaikovsky, Helen, Ms.
Vp For Financial Assistant
ENVIRONMENTAL ENTERPRISES ASSISTANCE FUND
1611 N. Kent Street, Suite 600
Arlington - VA 22209, United States of America
T: 1 703 522-5928 F: 1 703 522-6450

Chaitkio, Stuart
URBAN ECOLOGY
P.O.Box 10144
Berkeley - CA 94709, United States of America
T: 1 510 549-1724 F: 1 415 7031981

Chakhanov, Moukhtar, Mr.
President
ARAL-ASIA-KAZAKHSTAN INT'L. PUB.CMTE.
24 Sawmill Road
Norwalk - CT 06851, United States of America

Chan, Erika, Ms.
Student
CHILDREN'S ALLIANCE FOR THE PROTECTION OF THE ENV.
Austin, United States of America

Chapman, Frederica, Ms.
DIANA OF THE DEER MASKS
325 Town Farm RD
Wind Gap - PA 03086, United States of America

Chappell, Chris, Mr.
CERES / TOM'S OF MAINE
P.O. Box 710
Kennebunk - ME, United States of America
T: 1 207 985-2944 F: 1 207 985-3982

Chappell, Sarah, Ms.
CERES / TOM'S OF MAINE
P.O. Box 710
Kennebunk - ME, United States of America
T: 1 207 985-2944 F: 1 207 985-3982

Charette, Bill
Cameraman
WGBH-TV
125 Western Ave.
Boston -Massachusett, United States of America
T: 1 617 4495830 F: 1 617 4446450

Chasnoff, Debra
Interviewer
NORTHCUTT PRODUCTIONS
2966 Diamond St. #171
San Francisco - CA, United States of America
T: 1 415 3377044 F: 1 415 3370630

Chatterjee, Prothap
Reporter
PACIFICA RADIO
505 Eighth St
New York, NY 10018, United States of America
T: 1 718 7891847

Chawla, Prabhu
Executive Editor
GLOBAL EDITION
1875 Connecticut Ave, NW Suite 520
Washington DC 20009, United States of America
T: 1 202 483 1100 F: 1 202 328 39

Checon, Amarilho, Mr.
Padre
PONTIFICA UNIVERSIDADE CATOLICA DO RIO DE JANEIRO
1880 East River Road
Tucson - AZ, United States of America
T: 1 602 577-6476 F: 1 602 577-1195

Chew, Cory, Ms.
TOGETHER FOUNDATION
2129 13th St.
Boulder - CO 80302, United States of America
T: 1 303 444 9567 F: 1 303 444 7512

Chivello, Roya, Ms.
EARTH PLUS - PARADIGM
3015 Main Street Suite 340
Santa Monica - CA 90405, United States of America
T: 1 310 452-9075 F: 1 310 450-1069

Choy, Glenn, Mr.
Administrator
NATIVE HAWAIIAN ENVIRONMENTAL DEFENSE
2515 Alaula Way
Honolulu - HI 96817, United States of America
T: 1 808 988-4445

Christensen, Jon, Mr.
FORESTA INSTITUTE FOR OCEAN & MOUNTAIN STUDIES
PO Box 41567
Tucson - AZ 85717, United States of America
T: 1 602 8816174 F: 1 602 3229850

Christian, Dorothy I., Ms.
FOUNDATION FOR GLOBAL COMMUNITY AND CITIZENS NETWORK FOR UNCED
101 Hillborn Ave 3
Valleys - CA 94590, United States of America
T: 1 707 6444384

Chung, Tina
CHINESE TV SYSTEM
1273 National Press Building
Washington -DC 2004, United States of America
T: 1 202 6628950 F: 1 202 6628865

Cicin-Sain, Biliana, Dr.
INTERNATIONAL COASTAL & OCEAN ORGANIZATION
Center for the Study of Marine Policy, University of Delaware
Newark, NJ 19716, United States of America
T: 1 302 831-8086 F: 1 302 831-3608

Clark, Ana, Sra.
Executive Director
GLOBAL WOMAN PROJECT
5010 Wiseonsin Ave. N.W.
Washington - DC 20016, United States of America
T: 1 202 966-8207 F: 1 202 363-4600

Clark, Donald B., Mr.
GLOBAL ENVIRONMENT TEAM
12 Terrace Ave.
Nutley - NJ 07110, United States of America
T: 1 201 667-0079

Clark, Francis, Dr.
Secretary General
INTER-RELIGIOUS FEDERATION FOR WORLD PEACE
4 West 43rd Street, 5th Floor
New York - NY 10036, United States of America
T: 1 212 8696023 F: 1 212 8696424

Clark, Susan, Ms.
GLOBAL ENVIRONMENT TEAM
2153 N. 64th St.
Seattle - WA 98103, United States of America
T: 1 206 524-9052

Clarke, Catherine K., Ms.
Director
ONENESS - CENTER FOR PURPOSEFUL LIVING
2444 Carnation Lane
Chesapeake - VA 23325, United States of America
T: 1 804 523-4842

Claxton, Philander P., Mr.
Director
WORLD POPULATION SOCIETY
5837 Osceola Ct.
Bethesda - MD, United States of America
T: 1 202 775-9680 F: 1 202 775-9694

Clements, Phyllis, Ms.
Professor
A.P.C. / NEW YORK UNIVERSITY
635 East Building
New York - NY 10003, United States of America
T: 1 212 9985495

Clover, Charles W., Mr.
Owner
CLOVER CREATIONS UNLIMITED
230 Karls Dell Rd.
Scotts Valley - CA 95066, United States of America
T: 1 408 439-0605

Coan, Donald W., Mr.
SOLAR BOX COOKERS INTERNATIONAL
1724 Eleventh St.
Sacramento - CA 95814, United States of America
T: 1 916 444-6616 F: 1 916 447-8689

Cock-Calvo, Gabriel, Mr.
Consultant
ESRI - EXPOSITOR
380 New York St.
Redlands - CA 92373, United States of America
T: 1 714 793-2853 F: 1 714 793-5953

Cofield, Philip, Mr.
GLOBAL ENVIRONMENT TEAM
California Institute of Technology
Pasadena - CA 91125, United States of America
T: 1 818 796-2951 F: 1 818 792-2121

Cohen, Martin
Camera Asst.
OUT OF EPOCA FILMS
N. Cahuenga Blvd.
Hollywood -CA 90038, United States of America
T: 1 213 4694405 F: 1 213 4695010

Cohen, Michael, Mr.
Division Chief
URBAN MANAGEMENT PROGRAM / WORLD BANK
1818 H Street NW
Washington - DC 20433, United States of America
T: 1 202 4733861 F: 1 202 4771391

Coleman, Paul
CAMPAIGN FOR THE EARTH
1880 East River Road
Tucson - AZ, United States of America
T: 1 602 577-6476 F: 1 602 577-1195

Colon, Alice, Ms.
UNIVERSIDAD DE PUERTO RICO
Concordia Gardens, apt. 2D
Rio Piedras, Puerto Rico 00924, United States of America

Condon, Camy
NEW MEXICO CONF. OF CHURCHES/CENTER FOR WILDLIFE LAW
235 Mezcal NW
New Mexico 87105, United States of America
T: 1 505 2056

Connell, Noreen, Ms.
WOMEN'S ENVIRONMENTAL & DEVELOPMENT ORGANIZATION - WEDO
151 W74 St. Apt. 3C
New York - NY 10023, United States of America
T: 1 212 724-7156

Connor, James V., Mr.
Director
OUTREACH
200 East Building, 239 Greene St.
New York - NY 10003, United States of America
T: 1 212 475-1993 F: 1 212 995-4049

Connor, Joseph E., Mr.
Chairman
PRICE WATERHOUSE WORLD FIRM
1285 Avenue of The Americas
New York - NY 10019, United States of America
T: 1 212 2373923

Connors, Margaret S., Ms.
CONNORS LANDSCAPE DESIGN
P. O. Box G
Duxbury 02331, United States of America
T: 1 617 934-5200 F: 1 617 934-0697

Constantine, Mark, Mr.
WORLD BANK
1818 H St. NW
Washington - DC 20433, United States of America
T: 1 202 458-1943

Contact, Robert, Mr.
FOUNDATION FOR GLOBAL COMMUNITY
222 High Street
Palo Alto - CA 94301, United States of America

Content, Bob, Mr.
UNCED-Team Member
FOUNDATION FOR GLOBAL COMMUNITY
222 High Street
Palo Alto - CA, United States of America
T: 1 415 328-7756 F: 1 415 328-7785
TL: 856588

Cook, Jeffrey, Mr.
President
PLEA INTERNATIONAL
Passive and Low Energy School of Architecture, A.S.U.
Tempe - AZ, United States of America
T: 1 602 965-3536 F: 1 602 965-0968

Cooke, Charlotte, Ms.
Nuclear Educator
CONCERNED CITIZENS FOR NUCLEAR SAFETY
412 W. San Francisco St.
Santa Fe, NM 87501, United States of America

Cooper, Douglas, Mr.
61 Country Ridge Circle Rye Brook
New York - NY 10573, United States of America
T: 1 212 937-3715 F: 1 212 921-0521

Cooper, Stephanie, Ms.
ICF INCORPORATED / BROWN UNIVERSITY
9300 Lec Highway
Fairfax - VA 22031-1207, United States of America
T: 1 703 9343660 F: 1 703 9343590

Cooper-Sheehan, Christine, Ms.
A WORLD THAT WORKS FOR EVERYONE
PO Box 2982
Seward - Alaska 99664, United States of America
T: 1 907 224-8773

Copeland, David, Mr.
Founder
VALUE EARTH
Non White Horse Pike Suite L-2
Abselon, United States of America
T: 609 641 2400 F: 609 272 157

Cordaro, Sharron V., Ms.
Member
INTERNATIONAL HONORS PROGRAM / GLOBAL ECOLOGY
3040 Dumbarton St. NW
Washington - DC 20007, United States of America
T: 1 202 684-7887

Correia, Clemisom, Mr.
Environmental Engineer
CH2M HILL
2020 SW 4th St., 2nd Floor
Portland - OR, United States of America
T: 1 503 224-9190 F: 1 503 295-4446

Corrigan, Kevin, Mr.
INDIGENOUS PEOPLES MEDIA CENTER
77 Main Street
New York - NY, United States of America
T: 1 212 3650414 F: 1 212 3650864

Cortese, Anthony D., Mr.
Dean
TUFTS UNIVERSITY/CEM
Curtis Hall, 474 Boston Avenue
Medford - MA, United States of America
T: 1 617 627.3452 F: 1 617 627.3480

Cota-Robles, Patricia, Ms.
NEW AGE STUDIES FOR HUMANITIES PURPOSE
P.O.Box 41883
Tucson - AZ, United States of America
T: 1 602 577-6476 F: 1 602 577-1195

Cota-Robles, Victoria, Ms.
NEW AGE STUDY OF HUMANITYIES PURPOSE
P.O.Box 41883
Tucson - AZ 85717, United States of America

Cournand, Marie-Claire, Dr.
TEMPLE OF UNDERSTANDING
St. Emeric Church / 185 Avenue D
New York - NY 10009, United States of America
T: 1 212 353-0001 F: 1 212 353-0170

Cowherd, Alexander, Mr.
CENTER FOR INTERNATIONAL ENVIRONMENTAL LAW
411 W. 115th St. #44
New York - NY, United States of America

Cox, Dallas W., Jr., Mr.
111 West Port Plaza Suite 610
St. Louis, MO, United States of America
T: 1 314 434-1111 F: 1 314 434-6778

Cox, Steven, Mr.
Director
E O S A T
4300 Forbes
Lanham - MD 20854, United States of America
T: 1 301 552-0557

Cox Jr., Dallas W., Mr.
111 West Port Plaza Suite 610
St. Louis, MO, United States of America
T: 1 314 4341111 F: 1 314 4346778

Crabtree, Robert, Mr.
Res. Ecologist
YELLOWSTONE NATIONAL PARK
271 Quinn Creek Rd.
Bozeman - Montana 59715, United States of America
T: 1 406 587-4079 F: 1 406 994-3091

Craig, Chip, Mr.
21ST. CENTURY NEWS
1880 East River Road
Tucson - AZ, United States of America
T: 1 602 577-6476 F: 1 602 577-1195

Craighead, Derek, Mr.
President
CRAIGHEAD WILDLIFE WILDLANDS INSTITUTE
5200 Upper Miller Creek Road
Missoula - MT 59803, United States of America
T: 1 406 251-3867 F: 1 406 251-5069

Crane, Jody, Ms.
President
NEW SOLUTIONS
7 Nelson Street
Stamford - CT 06902, United States of America
T: 1 203 353-9660 F: 1 203 975-1717

Cravinho, Cristiane Mattos, Ms.
Assistant
GLOBAL FORUM OF SPIRITUAL AND PARLIAMENTARY LEADERS ON HUMAN SURVIVAL
304 East 45th St. Room 476
New York - NY, United States of America
T: 1 212 953-7947 F: 1 212 557-20961

Crazy Horse, Roy
Chief
POW HATAN RENAPE NATION
Rankokus Indian Reservation
Rancocas - NJ, United States of America
T: 1 609 261-4747 F: 1 609 261-7310

Criswell, David R., Mr.
UNIVERSITY OF HOUSTON
Institute Space Systems
Houston - TX 77024, United States of America
T: 1 713 486-5019 F: 1 713 486-5019

Crockett, David W., Mr.
Councilman
CITY OF CHATTANOOGA, TENNESSEE
Room 111 - City Hall
Chatanooga, TN, United States of America
T: 1 615 757-5196 F: 1 615 757-4857

Cronwell, Mare, Ms.
GREEN (GLOBAL RIVERS ENVIRONMENTAL EDUCATION NETWORK)
University of Michigan School of Natural Resouces -Dana Building 430E
Ann Arbor - MI, United States of America
T: 1 313 7645171 F: 1 313 9362195

Crosby, Sky Williams
E.C.O PRODUCTIONS
1700 N. Wentworth
Tucson - AZ, United States of America
T: 1 602 577-6476 F: 1 602 577-1195

Croxton, Randolph, Mr.
Boardmember
AMERICAN INSTITUTE OF ARCHITECTS
1735 New York Avenue N.W.
Washington - DC, United States of America
T: 1 202 794-2285

Cullen, Michael
NATIONAL PUBLIC RADIO
2025 M St NW
Washington DC 20036, United States of America
T: 1 202 8222000 F: 1 202 8222329

Cullen, Tim, Mr.
WORLD BANK
1818 H. Street N.W.
Washington - DC 20433, United States of America

Cullison, Pamela, Ms.
HOPE PUBLISHING HOUSE
13058 N. Surrey Circle
Phoenix - AZ 85029, United States of America

Culver, Katey, Ms.
Representative
NASHVILLE PEACE & JUSTICE CENTER
716 GEORGETOWN Dr.
Nashville - TN 37205, United States of America
T: 615 352 6037

Cummings, Karen Armstrong-, Ms.
Director Of Administration
KENTUCKY NATURAL RESOURCES CABINET
4th floor, Capital Plaza Tower
Frankfort - KY, United States of America
T: 1 502 564-7320 F: 1 502 564-2043

Curry, Donna, Dr.
EMISSARY FOUNDATION
5669 N. Country RD.29
Loveland, CO 80538, United States of America
T: 577-6476 F: 577-1195

Curtin, Leah, Ms.
Editor
NURSING MANAGEMENT JOURNAL
3734 Glenway
Cincinnati - OH 45205, United States of America

Curtis, Clifton E., Mr.
Lobbyist
GREENPEACE INTERNATIONAL
1436 U St. NW
Washington DC, United States of America
T: 1 202 4621177 F: 1 202 4624507

Curtis, Mike, Mr.
Sr. Engineer
RADIAN CORP
1420 S. Meadows
Austin - TX, United States of America
T: 1 512 8361077 F: 1 512 8364517

Czeplowodzki, Leopoldo F., Mr.
Architect
CARIBBEAN UNIVERSITY
Resid. Torre del Norte 1718,Ave. Universidad # 4
R. Piedras, Puerto Rico 00925, United States of America

D'Amico, Carol, Ms.
Director/Teacher
KIDS AGAINST POLLUTION
PO Box 699 / 3 Kiley Court
Provincetown - MA 02657, United States of America
T: 1 508 487-1782

D'Amico, Simone, Ms.
KIDS AGAINST POLLUTION
PO Box 699 / 3 Kiley Court
Provincetown - MA 02657, United States of America
T: 1 508 487-1782

D.Beers, Thomas
CNN BROADCASTING SYSTEM
One CNN Center, Box 105366
Atlanta - GA, United States of America
T: 1 404 8371918 F: 1 404 8274292

D'Marie, May, Ms.
SOLAR BOX COOKERS INTERNATIONAL
1724 Eleventh St.
Sacramento - CA 95814, United States of America
T: 1 916 444-6616 F: 1 916 447-8689

Da Silva, Vera, Ms.
CAMPAIGN FOR THE EARTH
13th St.
Boulder - CO, United States of America
T: 1 303 444-3800

Da-Rin, Silvio
WORLD BANK
Washington DC, United States of America

Daher, Joseane
Photographer
AMERICAN MUSEUM OF NATURAL HISTORY
79 Central Park West
New York NY, United States of America
T: 1 212 7695886

Daley, Devonne L., Ms.
DE PAUL UNIVERSITY
25 E Jackson
Chicago - IL 60604, United States of America
T: 1 312 362-5258 F: 1 312 362-6908

Dall, Eugenio
TERRA CHRISTA COMMUNICATIONS
1880 East River Road
Tucson - AZ, United States of America
T: 1 602 577-6476 F: 1 602 577-1195

Daly, Herman, Dr.
WORLD BANK
1818 H Street, NW
Washington - DC 20433, United States of America

Danant, Joelle, Ms.
Coordinator
AMERICAN FORUM FOR GLOBAL EDUCATION
45 John Street, Suite 908
New York - NY 10038, United States of America
T: 1 212 7328606 F: 1 212 7914132

Davia, Jerry L., Mr.
Representative
HUNGER PROJECT
One Madison Ave.
New York - NY, United States of America

Davidson, Art, Mr.
Director
ENDANGERED PEOPLES PROJECT
2200 Gambell
Anchorage - Alaska, United States of America
T: 1 907 653-7675

Davidson, Eric A., Mr.
Assistant Scientist
WOODS HOLE RESEARCH CENTER
PO Box 296
Woods Hole - MA 02543, United States of America
T: 1 508 540-9900 F: 1 508 540-9700

Davies, Kathleen, Ms.
Executive Producer
HOMETOWN FILMS
8306 Wilshire Blvd Suite 254
Beverly Hills - CA, United States of America
T: 1 655 9403 F: 1 655 2434

Davies, Phyllis, Ms.
HOPE PUBLISHING HOUSE
5005 Jesperson Rd.
San Luis Obispo - CA 93401, United States of America
T: 1 805 544-2072

Davis, Garry, Mr.
Coordinator
WORLD SERVICE AUTHORITY
Suite 1106 - Continental Bldg. 1012 14th. Street NW
Washington - DC, United States of America
T: 1 202 638-2662 F: 1 202 638-0638

Davis, Gloria, Mr.
Representative
WORLD BANK
1818 H St. NW
Washington - DC 20433, United States of America
T: 1 202 458-1943

Davis, Lee, Mr.
CARE
660 First Avenue
New York - NY 10016, United States of America
T: 1 212 686-3110 F: 1 212 686-3675

Davis, Rennie, Mr.
Coordinator
CAMPAIGN FOR THE EARTH
2120 13th Street
Boulder - CO 80301, United States of America

Davis, Shelton, Mr.
WORLD BANK
1818 H St. NW
Washington - DC 20433, United States of America
T: 1 202 458-1943

Davis, Wendy, Ms.
P.O. Box 1203
Boulder - CO 80306, United States of America
F: 1 303 938 8463

Dawson, Marian Rapp, Ms.
GLOBAL ENVIRONMENT TEAM
696 South Madison Avenue
Pasadena - CA 91106, United States of America
T: 1 818 796-2951 F: 1 818 792-2121

Dawson, Melanie, Ms.
UP WITH PEOPLE
3103 North Campbell Av.
Tucson - AZ 85719, United States of America

De Menar, Bernard, Ms.
WORLD INFORMATION TRANSFER INC
274 Madison Ave, Suite 601
New York - NY 10016, United States of America
T: 1 212 2730253 F: 1 212 3792004

Deiro, Ana Luica
ABC NEWS
New York, United States of America

Delgado, Sharon, Ms.
GLOBAL ENVIRONMENT TEAM
P. O. Box 235
Newcastle - CA 95658, United States of America
T: 1 916 663-2250

Deling, Wang, Ms.
Exec. Director
THE EARTH FOUNDATION
P.O.Box 1198
New York - NY 10027-1198, United States of America
T: 1 212 330 9015 F: 1 212 769 3847

Della Cava, Ralph, Mr.
COLUMBIA UNIVERSITY
Latin American Institute
New York - NY, United States of America

Denes, Agnes, Ms.
Participant
GLOBAL FORUM OF SPIRUTAL AND PARLIMENTARY LEADERS
304 East 45th St., 4th Floor
New York - NY, United States of America
T: 1 212 953-7947 F: 1 212 557-2061

Depuyt, Diane, Ms.
Assoc. Director
SACRED EARTH NETWORK
220 West 21st Street, Suite 7A
New York - NY, United States of America
T: 1 212 9242295 F: 1 212 9242295

Deus, Vincente Manna, Mr.
Political Science/ Video Arts
NEW COLLEGE OF CALIFORNIA
677 - 4th. Avenue
San Francisco - CA 94118-3910, United States of America
T: 1 415 221-8197

Devenport, Deborah S., Ms.
ATLANTA GREENS
2494 Williams Lane, apt.5
Decatur - GA, United States of America
T: 1 404 712-9558
E: POLBT 374 @ EMUVM1

Dewing, Adrian, Mr.
Chairman Of The Board
NATURAL RESOURCES DEFENSE COUNCIL (NRDC)
40 West 20th Street
New York - NY 10011, United States of America
T: 1 212 727-2700 F: 1 212 727-1773
TL: 4900010562

Dharlo, Rinchen
Representative
OFFICE OF TIBET
241 East 32nd St
New York - NY 10016, United States of America
T: 1 212 2135010 F: 1 212 7799245

Diamond, Nancy K., Dr.
Agroforestry
UNITED STATES AGENCY FOR INTERNATIONAL DEVELOPMENT
Office of Economic & Institutional Development SA-18, room 622C
Washington - DC 20523-1814, United States of America
T: 1 703 875-4737 F: 1 703 875-4949

Dias, Nadia De Souza
4235 Crosswood dr.
Buktonsville 20866, United States of America
T: 301 604 1821

Dickhut, Kathleen, Ms.
1769 W Arthur Ave
Chicago - IL, United States of America
T: 1 312 427-4256 F: 1 312 427-4261

Digiacomo, Karin, Ms.
MANITOU FOUNDATION
1880 East River Road
Tucson - AZ, United States of America
T: 1 602 577-6476 F: 1 602 577-1195

Dikgang, Eureka, Ms.
Nurse
CENTRE FOR DEVELOPMENT AND POPULATION ACTIVITIES - CEDPA
1717 Massachusetts Ave. NW, 2nd floor
Washington - DC 20036, United States of America
T: 1 202 667-1142 F: 1 202 332-4496
TL: 440384

Dimartino, Michael, Mr.
TERRA CHRISTA COMMUNICATIONS
1880 East River Road
Tucson - AZ, United States of America
T: 1 602 577-6476 F: 1 602 577-1195

Dingee, Charles, Mr.
Director
WORLD GAME INSTITUTE
3508 Market Street
Philadelphia - PA 19104, United States of America
T: 1 215 387-0227 F: 1 215 387-3009

Dinwiddie, Parker S., Mr.
President
THE TIRELESS EFFORT
433 Town Center Ste. 71 B
California 94925, United States of America
T: 415 924-4700 F: 415 924-4749

Dion, Mark, Mr.
ARTE AMAZONAS
519 W. 143 Street
New York - NY, United States of America
T: 1 212 238 0928

Diss, Sylvia, Ms.
Coord. Spec.
NORTH AMERICAN COALITION ON RELIGION AND ECOLOGY - CORE-INT'L
5 Thomas Circle
Washington - DC, United States of America
T: 1 204 462-2591

Dluhy, Nina
Producer
OUT OF EPOCA FILMS
N. Cahuenga Blvd.
Hollywood -CA 90038, United States of America
T: 1 213 4694405 F: 1 213 4695010

Dmytrow, Marianne, Ms.
President
PHONE KING INT'L LTD
388 Van Duzer St.
Staten Island - NY, United States of America
T: 1 718 720-3300 F: 1 718 720-6832

Do Nascimento, Heidi, Ms.
MCCORMICK SEMINARY
5555 Woodlawn
Chicago - IL, United States of America

Doderer, Minnette, Ms.
State Rep.
UNITED NATIONS ASSOC. SA.
2008 Dunlap Ct
Iowa City, Iowa, United States of America
T: 1 319 337-9282

Dolan, Father Luis M.
TEMPLE OF UNDERSTANDING
St. Emeric Church / 185 Avenue D
New York - NY 10009, United States of America
T: 1 212 353-0001 F: 1 212 353-0170

Domenici, Nicole
Photographer
INTERIORS +ARQUITECTURE MAGAZINES
1515 Broadway 39th Floor
New York -NY 10036, United States of America
T: 1 212 7491735 F: 1 212 3914157

Donnenfeld, Eleanor L., Ms.
Trustee
PAN PACIFIC & SOUTHEAST ASIA WOMEN'S ASSOCIATION
160 East 84th St. # 3D
New York - NY 10028, United States of America
T: 1 212 988-1084

Dower, Roger, Mr.
Director
WORLD RESOURCES INSTITUTE
1709 New York Ave., N.W.
Washington - DC, United States of America
T: 1 202 662-2544 F: 1 202 638-0036

Dowling, Michael, Mr.
Member
INTERNATIONAL LIFE SCIENCE INST.
609 Greystone Park
Atlanta - GA 30324, United States of America
T: 1 404 6077775 F: 1 404 5251964

Downes, David, Mr.
General Counsel
CENTER FOR INTERNATIONAL ENVIRONMENTAL LAW-US
1621 Connecticut Avenue, Suite 300
Washington - DC 20009-1076, United States of America
T: 1 202 332-4840 F: 1 202 332-4865

Doyle, Walter, Mr.
President
TREE TECHNOLOGY INTERNATIONAL, INC.
440 Park Ave. South 13th floor
New York - NY, United States of America
T: 1 212 685-8400 F: 1 212 689-3701

Draines, Holli, Ms.
SO. CALIF. ECOLOGY TASK FORCE
197 Blue Hills Ave.
Hartford - CT 06112, United States of America
T: 1 203 525-6926

Drummond, Jose
INDIGENOUS PEOPLES MEDIA CENTER
77 Main Street
New York - NY, United States of America
T: 1 212 3650414 F: 1 212 3650864

Duarte, Ana Gabriela Lopes
Technician
CNN BROADCASTING SYSTEM
One CNN Center, Box 105366
Atlanta - GA, United States of America
T: 1 404 8371918 F: 1 404 8274292

Dubash, Navroz, Mr.
Policy Analyst
ENVIRONMENTAL DEFENSE FUND
1616 P street, NW, suite 150
Washington - DC 20036, United States of America
T: 1 202 3873500 F: 1 202 2346049

Dubes, Cameron, Dr.
Founder
CHERNOBYL'S CHILDREN
351 Buena Vista Ave. East # 605E
San Francisco - CA 94117, United States of America
T: 1 415 552-1676

Duchin, Faye, Dr.
INSTITUTE FOR ECONOMIC ANALYSIS
New York University, 269 Mercer Street lm 205
New York - NY, United States of America
T: 1 212 988-7480 F: 1 212 995-4465

Ducker, Lakin
UP WITH PEOPLE
3103 North Campbell Av.
Tucson - AZ 85719, United States of America

Dudek, Dan, Mr.
Senior Economist
ENVIRONMENTAL DEFENSE FUND
257 Park Ave., South
New York - NY 10010, United States of America
T: 1 212 505-2100 F: 1 212 505-2375

Duff, Craig
Coordinating Producer
TURNER BROADCASTING SYSTEM / CNN
One CNN Centre
Atlanta - GA, United States of America
T: 1 404 827-3973 F: 1 404 827-1015

Duffy, Beth Anne, Ms.
NACRE
2520 Highway Al. NE
Solon - Iowa 52333-9783, United States of America
T: 1 319 351-8854

Dugundji, Elenna, Ms.
39 Albert Ave.
Belmont - MA 02178, United States of America
T: 1 617 484-9272

Duke, Angier Biddle, Mr.
Participant
GLOBAL FORUM SPIRITUAL AND PARLIMENTARY LEADERS
304 East 45th St., 4th Floor
New York - NY, United States of America
T: 1 212 953-7947 F: 1 212 557-2061

Dunham, Lucile, Ms.
Member
INTERNATIONAL FEDERATION OF BUSINESS AND PROFESSIONAL WOMEN
131 Sunset Circle #58,
Benicia - CA 94510, United States of America
T: 1 707 747-0843

Duning, Xiao, Mr.
Professor
INSTITUTE OF FORESTRY AND SOCIAL SCIENCE
1880 East River Road
Tucson - AZ, United States of America
T: 1 602 577-6476 F: 1 602 577-1195

Dunn, Kim, Ms.
1252 Pender View Tr.
Fairfax - VA 22030, United States of America
T: 703 385-1413

Dunstan, Denise, Ms.
INDIGENOUS PEOPLES MEDIA CENTER
77 Main Street
New York - NY, United States of America
T: 1 212 3650414 F: 1 212 3650864

Dunstan, Ruby
INDIGENOUS PEOPLES MEDIA CENTER
77 Main Street
New York - NY, United States of America
T: 1 212 3650414 F: 1 212 3650864

Durden, Maria, Ms.
WORLD YOUNG WOMEN'S CHRISTIAN ASSOCIATION
816 E. Prospect apt.no.01
Seattle - WA 98102, United States of America
T: 1 206 324-4426
TL: WYMCA 41 23 32

Dusseau, Charles, Mr.
Commissioner
METRO - DADE COUNTY
111 N.W. First St.
Miami - Florida, United States of America
T: 1 305 375-5117

Ebenezer, Job S, Mr.
Director Office Of Environment
EVANGELICAL LUTHERAN CHURCH
8765 West Higgins Road
Chicago - IL, United States of America
T: 1 312 3802708 F: 1 312 3801465

Echaniz, Marcelo
TERRA CHRISTA COMMUNICATIONS
1880 East River Road
Tucson - AZ, United States of America
T: 1 602 577-6476 F: 1 602 577-1195

Eddy, Edith T., Ms.
Senior Research Ass
COMMONWEAL
P.O. Box 316
Bolinas - CA, United States of America
T: 1 415 968-0970 F: 1 415 968-2230

Eddy, Jack, Mr.
Chief Scientist
CIESIN
2250 Pierle Rd.
Saginau - Mi 48710, United States of America
T: 1 517791 7371 F: 1 517790 0735

Edgerton, Lynn, Mr.
ENVIRONMENTAL LAW INSTITUTE
1616 P.Street, N.W., Suite 200
Washington - DC 20036, United States of America
T: 1 202 939-3853 F: 1 202 328-5002

Edwards, Don, Mr.
PANOS INSTITUTE
1717 Massachusetts Ave NW Suite 301
Washington - DC 20036, United States of America
T: 1 202 483-0043 F: 1 202 483-3059

Edwards, Michael J., Mr.
Director
GLOBAL SUSTAINABLE DEVELOPMENT SYSTEMS
1110 Reed St.
Safety Harbor - FL, United States of America
T: 1 813 725-1286 F: 1 813 725-2268

Egawa, Chenoa
INDIGENOUS PEOPLES MEDIA CENTER
77 Main Street
New York - NY, United States of America
T: 1 212 3650414 F: 1 212 3650864

Egger, Manfredo, Mr.
Naturalist
INT'L SOCIETY FOR THE PRESERVATION OF THE RAINFOREST
3302 N. Burton Ave.
Rosemead - CA 91770, United States of America

Ehrhardt, Sylvia, Ms.
PARTNERS IN THE ENVIRONMENT
6911 Richmond Hwy., Ste. 308
Alexandria-VA, United States of America
T: 1 703 765-8463 F: 1 703 765-8426

Ehrhardt, Walter, Mr.
PARTNERS IN THE ENVIRONMENT
6911 Richmond Hwy., Ste. 308
Alexandria-VA, United States of America
T: 1 703 765-8463 F: 1 703 765-8426

Eidlin, Richard, Mr.
TERRA CHRISTA COMMUNICATIONS
1880 East River Road
Tucson - AZ, United States of America
T: 1 602 577-6476 F: 1 602 577-1195

El-Ashry, Mohamed, Mr.
Representative
WORLD BANK
1818 H St. NW
Washington - DC 20433, United States of America
T: 1 202 458-1943

Elgin, Lysa, Ms.
TERRA CHRISTA COMMUNICATIONS
4830 University Blvd. East C4
Tuscaloosa Al 35404, United States of America
T: 1 602 577-6476 F: 1 602 577-1195

Elliott, Cynthia, Ms.
MIAMI-DADE COMUNITY COLLEGE
Miami-dade Community College
Miami - Florida 33176, United States of America
T: 1 305 237-2158 F: 1 305 237-2928

Ellis, William, Mr.
Ceo
NORTHEAST UTILITIES
107 Selden St.
Berlin - CT 06037, United States of America
T: 1 508-5460 F: 1 508-5403

Elmendorf, Mary, Ms.
Antropologist
UNA (USA) & AAUN
Lido key, 535 S. Blvd of Presidents
Sarasota - FL 34236, United States of America
T: 1 813 3884184 F: 1 813 3883534

Emenhiser, Don, Mr.
Consultant
GREENHEART AMERICA
Box 946
Corvallis - OR 97339, United States of America
T: 1 503 754-6435

Engelman, Bob, Mr.
Dir Pop./envrnmtl Prog
POPULATION CRISIS COMMITTEE
1120 19th Street, NW, Suite 550
Washington - DC 20036, United States of America
T: 1 202 659 1833 F: 1 202 293 1795
TL: 440450

Enote, James E., Mr.
Project Leader
PUEBLO OF ZUNI - CONSERVATION PROJECT
PO Box 339
Zuni-NM, United States of America
T: 1 505 782-4481 F: 1 505 782-2700

Ensign, Tod, Mr.
Founder
ONE WORLD NOW
1510 Lyric Center, 440 Louisiana
Houston - TX, United States of America
T: 1 713 222-2288

Enzer, Steve, Mr.
GEOSPHERE PROJECT, THE
146 Entrada Drive
Santa Monica - CA 90402, United States of America
T: 1 310 459-4342 F: 1 310 459-8299

Eppler, Dianne, Ms.
Director Of Int'L Programs
AMERICAN WIND ENERGY ASSN.
777 N. Capitol St., NE, #805
Washington - DC, United States of America
T: 1 202 408-8988 F: 1 202 408-8536

Epstein, Sally, Ms.
Board Member
POPULATION INSTITUTE
5620 Oregon Ave. NW
Washington - DC 20015, United States of America
T: 1 202 363-3664

Ernst, John, Mr.
Ecologist
1253 E. Madison
Colorado Springs,CO, United States of America

Ethelston, Sally, Ms.
Director Of Media Relations
POPULATION CRISIS COMMITTEE
1120 19th Street, NW, Suite 550
Washington - DC 20036, United States of America
T: 1 202 659 1833 F: 1 202 293 1795
TL: 440450

Evans, Daniel, Mr.
GLOBE UNITED STATES
507 Saward Square SE#4
Washington - DC 20003, United States of America
T: 1 202 546-3457 F: 1 202 225-9095

Evans, Jodie, Ms.
WE THE PEOPLE
643 E. Channel Rd Santa Monica
Santa Monica - CA 90402, United States of America
T: 1 310 454 9909 F: 1 310 449 1976

Evans, Kara, Ms.
UNIVERSITY OF CALIFORNIA AT SANTA CRUZ
113 Peyton St.
Santa Cruz - CA, United States of America
T: 1 408 429-6152

Eyster, Dar
111 West Port Plaza Suite 610
St. Louis, MO, United States of America
T: 1 314 434-1111 F: 1 314 434-6778

Fakerang, Stefan, Mr.
Representative
TP AMERICAS
1100 Peachtree Street
Atlanta - GA, United States of America
T: 1 404 815-3400

Falicov, Yael M.
YOUTH ALLIANCE FOR SOCIAL SURVIVAL
2227 Ward St.
Berkley, CA, United States of America
T: 1 415 704-0559 F: 1 415 642-7551

Falloux, Francois, Mr.
WORLD BANK
1818 H St. NW
Washington - DC 20433, United States of America
T: 1 202 458-1943

Fang, William, Mr.
General Counsel
EDISON ELETRIC INSTITUTE
701 Pennsylvania Avenue, N.W.
Washington - DC 20004, United States of America
T: 1 202 508-5460 F: 1 202 508-5403

Farago, Stephanie, Ms.
Jornalist
CAMPAIGN FOR THE EARTH/PRESS
6510 W. 6th St.
Los Angeles - CA 90048, United States of America
T: 1 213 655 8310 F: 1 213 655 8556

Farby, Edward
AFTER IMAGE INTERNATIONAL
New York, United States of America

Farrell, Amy, Ms.
Production Assistan
WORLD PEACE PRAYER SOCIETY
22 Sheridan Drive
Pawling-NY, United States of America
T: 1 914 855-5877

Farrell, Christopher, Mr.
Music Director
WORLD PEACE PRAYER SOCIETY
22 Sheridan Drive
Pawling-NY, United States of America
T: 1 914 855-5877

Fass, Francine Susan, Ms.
Assistant To Us Representative
FILMMAKERS' UNION OF KAZAKHSTAN
2 N Pasture Rd
Westport - CT, United States of America
T: 1 203 4544980 F: 1 203 4544628
TL: 495-0063

Faundeen, John, Mr.
Scientist
UN/GRID
2213 Wellington
Sioux Falls 57106, United States of America
T: 605 361-2283 F: 594-6587

Fazzolo, Jussara
GLOBAL WOMAN PROJECT
Washington DC 20016, United States of America

Featherstone-Witty, Venetia, Ms.
RAINFOREST ACTION NETWORK
4153 Vantage Ave
Studio City 91607, United States of America
T: 818 506 0133

Feleus, Laurens
UP WITH PEOPLE
3103 North Campbell Av.
Tucson - AZ 85719, United States of America

Fenton, Betsy, Ms.
YMCA EARTH ACTION CORPS
909 4th Av
Seattle - WA, United States of America
T: 1 206 382-5013 F: 1 206 382-7283

Fenton, Patrick, Mr.
Assistant
COLORADO INTERNATIONAL TRADE OFFICE
1625 Broadway, Suite 680
Denver, Colorado 80202, United States of America
T: 1 303 892 3850 F: 1 303 892 3820

Ferguson, David, Mr.
GLOBAL ENVIRONMENT TEAM
70A Devon Ct. #3ite 4224
Edwardville-IL 62025, United States of America
T: 1 618 656-0502

Ferguson, Ken, Mr.
Director
EOSAT
1300 Forbes Boulevard
Lanham - MD 20706, United States of America
T: 1 301 552-0571 F: 1 301 552-5476

Ferguson, Laura, Ms.
HIGHLANDER CENTER
1939 Highlander Way
New Market - TN 37820, United States of America
T: 1 615 933 3443 F: 1 615 933 3424

Fernandez, Jorge, Mr.
MISION INDUSTRIAL DE PUERTO RICO - INC
Aptdo 363728
San Juan, Puerto Rico 000936-3728, United States of America

Fernando, Rita, Ms.
Chief
UNITED NATIONS POSTAL ADMINISTRATION - NY
No. 2 UN Plaza
New York - NY 10017, United States of America
T: 1 212 963-1502 F: 1 212 963-2975

Ferre, Angele, Ms.
S.A.F.E. CONSULTING FOR THE EARTH
Signal Mt Lodge, Grandteton National Park, P.O.Box 50
Moran - Wyoming 83013, United States of America
T: 1 307 543-2831

Ferreira, Flavio
WORLD BANK
Washington DC, United States of America

Ferriera, Marcela, Ms.
CITY COLLEGE OF NEW YORK
138th St and Convent Ave, Rm 6/109
New York - NY, United States of America
T: 1 212 650-6909 F: 1 212 650-6970

Fiedler, Eugene G., Mr.
Business Development
RUBY DOME INC ENVIRONMENTAL
3766 Bradview Way
Sacramento - CA, United States of America
T: 1 916 368-4700 F: 1 916 368-4704

Fields, Bonnie, Ms.
Director
WORLD COMMUNITY FOUNDATION
866 United Nations Plaza, Suite 438
New York - NY, United States of America
T: 1 212 752-0672 F: 1 212 752-0685

Filemyr, Ann, Ms.
Professor Of Journalism
ANTIOCH COLLEGE
795 Livermore Street
Yellow Springs,Ohio 45387, United States of America
T: 1 513 767-1086 F: 1 513 767-1891

Findley, Rober W., Dr.
Professor Of Law
LOYOLA LAW SCHOOL
1441 West Olympic Blvd
Los Angeles - CA 90015, United States of America

Fink, Christopher, Mr.
BLUE GREEN PRODUCTIONS
731 North Street
Santa Rosa - CA 95405, United States of America
T: 1 707 579-4297 F: 1 707 344-1515

Fisher, Frances, Ms.
CENTER FOR THE STUDY OF WOMEN AND SOCIETY
26 Higgins Street
Alston - MA 02134, United States of America

Fisher, Mary Pat, Ms.
Representative
GOBIND SADAN
657 Chaffeeville Road
Storrs - CT 06268, United States of America
T: 1 203 4290710

Fisher, Mike, Mr.
GLOBAL ENERGY NETWORK INT'L
PO Box 81565
San Diego - CA 92138, United States of America
T: 1 619 595-0139 F: 1 619 595-0403

Fisher, Padraic, Mr.
Dir. Fund Raising/Develop.
INDONESIAN CULTURAL FOUNDATION
60 E. 42nd Street
New York - NY 10165, United States of America
T: 1 212 989-1709 F: 1 212 944-9050

Fitzgibson, Theresa, Ms.
EARTH DAY ALLIANCE / FRIENDS PEACE COMMITTEE
1515 Cherry St.
Philadelphia - PA 19102, United States of America
T: 1 215 241-7233 F: 1 215 567-2046

Flanders, Stephanie
SCHOOL OF THEOLOGY AT CLAREMONT
1325 N. College Ave. PO Box 107
Claremont-CA 91711-3919, United States of America
T: 1 714 399-5706 F: 1 714 626-7062

Fleming, Tracy Ann
Tape Editor
CNN BROADCASTING SYSTEM
One CNN Center, Box 105366
Atlanta - GA, United States of America
T: 1 404 8371918 F: 1 404 8274292

Fletcher, Susan, Ms.
Board Member
U.S LIBRARY OF CONGRESS
9906 Wildwood Road
Kensington - MD 20895, United States of America
T: 1 202 707-1231 F: 1 202 707-6745

Flores, Clara
Latin America Sales Rep.
UNITED NATIONS PUBLICATIONS
Hotel Barra Bela
New York - NY 1P017, United States of America

Foer, Paul
Writer
FREELANCER
PO Box 5024
Annapolis, United States of America
T: 1 410 269 6289 F: 1 410 269 6289

Fontes, Angela Maria Belmiro, Ms.
CALIFORNIA WORLD FOUNDATION
P.O.Box 587
Half Moon Bay - CA, United States of America
T: 1 726 262216

Ford, Carolyn, Ms.
Director
JOLIET DIOCESE
St. Isidore Church 427 W. Army Trail Road
Bloomingdale -IL 60108, United States of America

Ford, Mary Anne, Ms.
UNITARIAN UNIVERSALIST UN OFFICE
505 N. Lake Shore Dr.
Chicago - IL, United States of America
T: 1 312 527-4964 F: 1 312 527-4964

Forden, Clark, Mr.
Secretary
PROUTIST UNIVERSAL INC, NORTH AMERICA
1354 Montegue Street, N.W.
Washington - DC 20011, United States of America
T: 1 202 829-2278 F: 1 202 829-0462

Forkan, Patricia
Executive Vice Presidente
HUMANE SOCIETY OF THE UNITED STATES
2100 L Street - NW
Washington - DC 20037, United States of America
T: 1 202 452-1100 F: 1 202 778-6132

Fortner, Rosanne, Ms.
Professor
OHIO STATE UNIVERSITY
2021 Coffey Rd.
Columbus-OH 43210, United States of America
T: 1 614 292-9826 F: 1 614 292-7162

Foster, Tonia Shoumatoff, Ms.
Editor / Coordinator
WORLD PEACE PRAYER SOCIETY
800 Third Avenue, 37th flr.
New York - NY 10022, United States of America
T: 1 212 755 4755 F: 1 212 935 1389

Fouracre, Anthony, Mr.
Chief Unpa
UNITED NATIONS
1 UN Plaza
New York - NY 00017, United States of America
T: 1 212 963-0932 F: 1 212 963-4148

Fowlor, Richard
Correspondent
VOICE OF AMERICA
330 Independence Ave SW
Washington DC 20001, United States of America
T: 1 202 6197225 F: 1 202 6197113

Francis, Tyler, Mr.
TERRA CHRISTA COMMUNICATION
1524 E 8TH Street
Tuscon, AZ 85719, United States of America
T: 1 602 628-8477

Franklin, Bernard B., Dr.
PHYSICIANS FOR SOCIAL REPONSIBILITY
6200 Wilshire Boulevard, Suite 800
Los Angeles - CA, United States of America
T: 1 213 930-1200 F: 1 213 930-1205

Fraser, Scott
Producer
WGBH-TV
125 Western Ave.
Boston -Massachusett, United States of America
T: 1 617 4495830 F: 1 617 4446450

Frederick, Pia, Ms.
UP WITH PEOPLE
3103 North Campbell Av.
Tucson - AZ 85719, United States of America

Freeman, James S., Mr.
Lega Interna
CENTER FOR INTERNATIONAL ENVIRONMENTAL LAW - US
1621 Connecticut Av., NW. Suite 300
Washington - DC 20009-1076, United States of America
T: 1 202 332-4840 F: 1 202 332-4865

Freeman, Mike, Mr.
UP WITH PEOPLE
3103 North Campbell Av.
Tucson - AZ 85719, United States of America

Frelinghuysen, Barrett, Ms.
Staff
NATURAL RESOURCES DEFENSE COUNCIL (NRDC)
40 West 20th St.
New York - NY 10011, United States of America
T: 1 212 783-7800 F: 1 212 783-5917
TL: 4900010562

Frempong, Richard A., Dr.
President
ATALNTA MGMT INSTITUTE
2175 Park Lake Dr. Ne, Suite 220
Atlanta - GA, United States of America
T: 1 404 934-3031 F: 1 404 934-3241
TL: 9102406054

Frett, John J., Mr.
Director
TOWER CONFERENCE MANAGEMENT CO.
800 Roosevelt Rd., Bldg. E-408
Glen Ellyn, Il 601367, United States of America
T: 1 708 469-3373 F: 1 708 469-7477

Freymann, John, Mr.
Senior Analyst
THE FUTURE GROUP
One Thomas Circle
Washington - DC 20005, United States of America
T: 1 202 775-9680

Friedlander, Eva, Ms.
UNITED NATIONS DEVELOPMENT FUND FOR WOMEN
304 East 45th St.
New York - NY 10017, United States of America
T: 1 212 9066400 F: 1 212 9066705

Fry, Kirby S., Mr.
Activist
INDEPENDENT CONSERVATION
6431 Belmont St.
Houston - TX, United States of America
T: 1 713 668-5775 F: 1 713 960-8215

Fujimoto, Isao, Mr.
Professor
UNIVERSITY OF CALIFORNIA, DAVIS
University of California - Dept. of Applied Bahavioral Sciences
Davis - CA, United States of America
T: 1 916 752-5660

Fulford, Jeanne E., Ms.
GLOBAL ENVIRONMENT TEAM
226 Seaview Ave
Swansea - MA 02777, United States of America
T: 1 508 678-3881

Fulgencio, Nuria, Ms.
TEMPLE OF UNDERSTANDING
St. Emeric Church / 185 Avenue D
New York - NY 10009, United States of America
T: 1 212 353-0001 F: 1 212 353-0170

Fuller, Carlos, Mr.
METEOROLOGICAL DEPARTMENT
1880 East River Road
Tucson - AZ, United States of America
T: 1 602 577-6476 F: 1 602 577-1195

Fuller, Richard, Mr.
President
GREAT FOREST INC.
Eleven Penn Plaza
New York - NY, United States of America
T: 1 212 967-4757 F: 1 212 279-6839

Funaba, Masatomi, Dr.
Professor/Advisor
HIROSHIMA UNIVERSITY / KUMAMOTO PREFECTURE
1225 Observatory Dr
Madison - WI 53706, United States of America
T: 1 608 2630446 F: 1 608 2332660

Futrell, J. William, Mr.
President
ENVIRONMENTAL LAW INSTITUTE
1616 P Street, N.W., Suite 200
Washington - DC 20036, United States of America
T: 1 202 939-3853 F: 1 202 328-5002

Gabel, Medard, Mr.
Executive Director
WORLD GAME INSTITUTE
3508 Market Street
Philadelphia - PA 19104, United States of America
T: 1 215 387-0227 F: 1 215 387-3009

Gale, Janice M., Ms.
Director
BERKELEY ENVIRONMENTAL EDUCATION RESEARCH DEV. GROUP
1185 Keith Ave
Berkeley - CA 94708, United States of America
T: 1 510 843-9993 F: 1 510 526-7695

Gall, Peter, Mr.
Chief, Public Affairs,
UNITED NATIONS DEVELOPMENT PROGRAMME
1 UN Plaza,
New York - NY 100017, United States of America
T: 1 212 906-5312 F: 1 212 905-5364

Gallardo, Hernan, Mr.
SALVAM - VIDA Y ECOLOGIA
P.O.Box 7463
New York - NY 11374, United States of America
T: 1 718 361-5732

Gallup, Kingsley, Ms.
GALLUP INTERNATIONAL INSTITUTE
47 Hulfish St., Po Box 140
Princeton - NJ 08542, United States of America
T: 1 609 921-6200 F: 1 609 924-0228

Gamble, Kathleen, Ms.
GLOBAL ENVIRONMENT TEAM
1346 S. Roxbury Rd.,
Salt Lake City - UT 84108, United States of America
T: 1 801 583-6446

Gamburg, Angela
Tape Operator
NORTHCUTT PRODUCTIONS
2966 Diamond St. #171
San Francisco - CA, United States of America
T: 1 415 3377044 F: 1 415 3370630

Ganti, Shankar, Dr.
Physician
INTERNATIONAL EYE FOUNDATION
744, Palisade Ave.
Yonilers-NY 10703, United States of America
T: 78 44 1980 F: 78 44 1980

Gaona, Noberto, Ms.
UP WITH PEOPLE
3103 North Campbell Av.
Tucson - AZ 85719, United States of America

Garcia, Martha P., Ms.
GLOBAL ENVIRONMENT TEAM
696 South Madison Avenue
Pasadena - CA 91106, United States of America
T: 1 818 796-2951 F: 1 818 792-2121

Gardner, Richard N., Mr.
Prof. Of Int. Law
COLUMBIA UNIVERSITY
435 West 116th Street
New York - NY 10027, United States of America
T: 1 212 880-4923 F: 1 212 557-8136

Garrett, Michael
UN DEPARTMENT OF PUBLIC INFORMATION
New York, NY, United States of America

Garris, Judy, Ms.
Environmental Director
BROWN FUR PRESIDENT
2121 Coverfield Bldg
Santa Monica - CA, United States of America
T: 1 310 449-1993

Garrison, John, Mr.
Representative
INTER-AMERICAN FOUNDATION
231 Gundev Dr.
Falls Church - VA 22046, United States of America
T: 703 841-3890 F: 703 841-0973

Gaumer, Denise, Ms.
EARTH DAY USA
1880 East River Road
Tucson - AZ, United States of America
T: 1 602 577-6476 F: 1 602 577-1195

Gauntlett, Suzanne, Ms.
President
THE BARBARA GAUNTLET FOUNDATION
International Square 1825 Eye St., N.W. suite 400
Washington - DC, United States of America
T: 1 202 429-2055 F: 1 202 466-3973

Geller, Howard, Mr.
Executive Director
AMERICAN COUNCIL FOR AN ENERGY-EFFICIENT ECONOMY
1001 Connecticut Ave. NW #801
Washington - DC 20036, United States of America
T: 1 202 429-8873 F: 1 202 429-2248

Genshlea, Katie, Ms.
US CITIZENS NETWORK ON UNCED
300 Broadway-suite 39
San Francisco - CA, United States of America
T: 1 415 956-6162 F: 1 415 956-0241

Gerace, Robert C., Mr.
Sales Manager
BOWKER A&I PUBLISHING CO. - BUTTERSWOTH HEINEMANN
515 East 85 St. - Suite 10D
New York - NY 10028, United States of America
T: 1 212 734-3855 F: 1 212 996-9003

Germain, Adrienne, Ms.
INTERNATIONAL WOMEN'S HEALTH COALITION
24 East 21ST St.
New York - NY 10010, United States of America
T: 1 212 979-8500 F: 1 212 979-9009

Gershon, David, Mr.
GLOBAL ACTION PLAN
84 Yerry Hill Road
Woodstock NY, United States of America

Gershowitz, Paul, Mr.
WORLD PEACE PRAYER SOCIETY
518 Beekman Rd.
Hopewell Jct. - NY 12533, United States of America
T: 1 212 227-7898 F: 1 212 935-1389

Gerzon, Daniel, Mr.
Master Of Environmental Manag.
DUKE UNIVERSITY SCHOOL OF THE ENVIRONMENT
2401 West Club Blvd. Apt 4
Durham, N.C. 27705, United States of America
T: 1 919 286 3396 F: 1 919 684 8741

Getter, Howard, Mr.
Executive Director
AMERICAN COUNCIL FOR AN ENERGY-EFFICIENT ECONOMY
1001 Connecticut Ave. NW - # 801
Washington - DC, United States of America
T: 1 202 4298873 F: 1 202 4292248

Gilbert, David, Mr.
Diretor Media
AMAZONIA AIDS AND HEALTH
993 CAROLINA St.
San Francisco - CA 95060, United States of America
T: 1 415 4763990 F: 1 415 4763655

Gillespie, Kate, Ms.
GLOBAL YOUTH SUMMIT
186 Sierra View
Pasadena - CA 91105, United States of America
T: 1 818 449-9505

Gillespie, Robert W., Mr.
POPULATION COMMUNICATION
1489 E. Colorado Bvd., Suite 202
Pasadena - CA 91106, United States of America
T: 1 818 793-4750 F: 1 818 793-4791

Gilley, Cynthia, Ms.
Director Of Public Policy
POPULATION CRISIS COMMITTEE
1120 19th Street, NW, Suite 550
Washington - DC 20036, United States of America
T: 1 202 659 1833 F: 1 202 293 1795
TL: 440450

Giniger, Lauren, Ms.
Env.Monitor-Systm.Anal.
TERRA CHRISTA COMMUNICATIONS
P.O.Box 2630
Santa Cruz - CA 95603, United States of America

Gips, Terry, Mr.
President
INTERNATIONAL ALLIANCE FOR SUSTAINABLE AGRICULTURE
1701 University Avenue SE
Minneapolis - MN, United States of America
T: 1 612 331-1099 F: 1 612 379-1527

Gladstein, Clifford, Mr.
Vp
PS ENTERPRISES
309 Santa Monica Blvd, Suite 322
Santa Monica 90401, United States of America
T: 1 310 393-3703 F: 1 310 393-7012

Glaeser, Carsten, Mr.
Manager
NEW YORK CITY DEPARTMENT OF PARKS
152-18 Union TPKE. #4M
Flushing-NY, United States of America
T: 1 718 9690857 F: 1 718 9658989

Glaeser, Nedenia, Ms.
Manager
NEW YORK CITY DEPARTMENT OF PARKS
152-18 Union TPKE. #4M
Flushing-NY, United States of America
T: 1 718 9690857 F: 1 718 9658989

Glass, Arthur, Mr.
Owner
PHONE KING INT'L LTD
388 Van Duzer St.
Staten Island - NY, United States of America
T: 1 718 720-3300 F: 1 718 720-6832

Glazer, Teri, Ms.
5901 SW 107th St.
Miami - Florida 33156, United States of America
T: 1 305 685-6743 F: 1 305 696-1906

Gleazer, Edmund
ICAE
12 Rajaji Marg
New Delhi 110011, United States of America
T: 91 11 301-4309

Gleckman, Harris, Mr.
Chief, Environment Unit
UN/DESP
Room DC2-1202, United Nations
New York - NY 10017, United States of America
T: 1 212 963-3160 F: 1 212 963-2146

Glenn, A. Adam
Managing Editor
GREENWIRE
282 N Washington St
Falls Church Virgini, United States of America
T: 1 703 237 5130 F: 1 703 237 91

Gobster, Paul, Mr.
1769 W Arthur Ave
Chicago - IL, United States of America
T: 1 312 427-4256 F: 1 312 427-4261

Goderkart, Anne
Editor
NATIONAL PUBLIC RADIO
2025 M St. NW
Washington - DC, United States of America
T: 1 202 8222784 F: 1 202 8222203

Gohr, Tiffany
Video Tape Logger
TURNER BROADCASTING SYSTEM / CNN
One CNN Centre
Atlanta - GA, United States of America
T: 1 404 827-3973 F: 1 404 827-1015

Goldfarb, Brian, Mr.
UNIVERSITY OF ROCHESTER
380 Wellington Ave
Rochester - NY 14619, United States of America
T: 1 212 436-8059 F: 1 212 432-

Goldsmith, Steve, Mr.
Regional Director
ENVIRONMENTAL FEDERATION OF CALIFORNIA
1242 24th
Santa Monica - CA 90404, United States of America
T: 1 310 828-5137 F: 1 310 829-9164
E: 90404

Goldstein, Natalie
Environment Writer
FREELANCER
2911 Route 209
Kingston NY 12401, United States of America
T: 1 914 331 7392

Gomez, Marsha A., Ms.
INDIGENOUS WOMEN'S NETWORK
13621 FM 2769
Austin - TX 78726, United States of America
T: 1 512 258-3880 F: 1 512 472-1043

Gomez-Echeverri, Luis, Mr.
Director Environment Group
NACOES UNIDAS
23 Philips Lane
Darien - CT, United States of America
T: 1 203 961-9592 F: 1 242 906-5057

Gonzales, Liz, Ms.
AMERICAN BAPTIST CHURCH
P. O. Box 851
Valley Forge - PA 19482-0851, United States of America

Gonzalez S.C., Paula
Co-Director
EARTH CONNECTION
5820 Bender Rd. 45233
Cincinnati - OH, United States of America
T: 1 513 922-1468

Gordon, Jesse, Mr.
Consultant
KESSEL KONSULTING
International
Waltham - MA 02154, United States of America
T: 1 617 891-6507 F: 1 617 899-5156

Gourlay, Mary Ann, Ms.
Director Of Communications
ALLIANCE TO SAVE ENERGY
1725 K St. NW Suite 509
Washington - DC 20006, United States of America
T: 1 202 857 -0666 F: 1 202 331 -9588

Gowan, Petra Mac, Ms.
YMCA EARTH ACTION CORPS
909 4th Av
Seattle - WA, United States of America
T: 1 206 382-5013 F: 1 206 382-7283

Grant, Debbie, Ms.
Ass. Conf. Org.
FORMER ASSOCIATION OF FORMER U.N. INTERNES AND FELLOWS
United Nations, Room DC II - 1118
New York - NY 10017, United States of America
T: 1 212 963 3110

Gravir, Ketil, Mr.
Board Member
WORLDWATCH INSTITUTE
1776, Massachusetts Ave., N.W.
Washington - DC, United States of America
T: 1 202 452-1999 F: 1 202 296-7365

Gray, Susan, Ms.
GENERAL BOARD OF CHURCH AND SOCIETY - UNITED METHODIST CHURCH
606 Greenwood SE. Ne
Grand Rapids - MI 49506, United States of America
T: 1 616 458-2750 F: 1 616 459-0191

Greedberg, Dean
President
PROGRESS AGENCY INC
1 Union square West 201
New York -NY 10003, United States of America
T: 1 212 9246550 F: 1 212 6456866

Green, Danielle, Ms.
4106 N Ashland, Rear
Chicago - IL 60613, United States of America
T: 1 312 525-4447

Green, Eileen
INDIGENOUS PEOPLES MEDIA CENTER
77 Main Street
New York - NY, United States of America
T: 1 212 3650414 F: 1 212 3650864

Greenmun, Aaron, Mr.
YMCA EARTH ACTION CORPS
909 4th Av
Seattle - WA, United States of America
T: 1 206 382-5013 F: 1 206 382-7283

Gregorius, Metropolitan Paulos Mar, Mr.
INTER-RELIGIOUS FEDERATION FOR WORLD PEACE
4 West 43rd Street, 5th Floor
New York - NY 10036, United States of America
T: 1 212 8696023 F: 1 212 8696424

Grey, Mara
Musician
DOVE & CO./MMH
4759 15th Ave NE
Seattle - WA 98105, United States of America
T: 1 206 344-6321

Grigg, Alice, Ms.
Trade Specialist
COLORADO INTERNATIONAL TRADE OFFICE
1625 Broadway, Suite 680
Denver, Colorado 80202, United States of America
T: 1 303 892 3850 F: 1 303 892 3820

Grijalva, Christina, Ms.
GROWING CONNECTIONS
2123 E. Grant Rd.
Tucson - AZ 85719, United States of America
T: 1 602 3257909 F: 1 602 3257961

Gron, Christian, Mr.
UP WITH PEOPLE
3103 North Campbell Av.
Tucson - AZ 85719, United States of America

Gross, Vicki, Ms.
Executive Director
RIVERKEEPER
PO Box 130 Castle Rock Field Station
Garrison - NY 10524, United States of America
T: 1 914 4244149 F: 1 914 4244150

Grossman, Art, Mr.
Representative
DADE COUNTY WOMEN'S COALITION FOR A HEALTHY PLANET
20100 West Country Club Drive
Aventura-FL 33180, United States of America
T: 1 305 932-0042 F: 1 305 651-7291

Grossman, Ginger, Ms.
Representative
DADE COUNTY WOMEN'S COALITION FOR A HEALTHY PLANET
20100 West Country Club Drive
Aventura-FL 33180, United States of America
T: 1 305 932-0042 F: 1 305 651-7291

Gubuan, Darlene Felisa, Ms.
Inter
CHURCH WOMEN UNITED - UN OFFICE
New York, NY, United States of America

Gucovsky, Michael, Mr.
UNDP- UNITED NATIONS DEVELOPMENT PROGRAMME
12 SE 72nd St.
New York - NY, United States of America
T: 1 212 906-5049 F: 1 212 906-6998

Gudzij, Michael, Mr.
CENTER FOR INTERNATIONAL ENVIRONMENTAL LAW - US
1621 Connecticut Av., Suite 300
Washington - DC 2009-1076, United States of America
T: 1 202 332-4840 F: 1 202 332-4865

Gulhati, Kaval, Ms.
Co-Founder
CENTRE FOR DEVELOPMENT AND POPULATION ACTIVITIES - CEDPA
1717 Massachusetts Ave. NW, 2nd floor
Washington - DC 20036, United States of America
T: 1 202 667-1142 F: 1 202 332-4496
TL: 440384

Gwin, Todd, Mr.
UP WITH PEOPLE
3103 North Campbell Av.
Tucson - AZ 85719, United States of America

Haas, Peter M., Mr.
Professor
UNIVERSITY OF MASSACHUSSETS & HARVARD UNIVERSITY
203 State Street
Northampton - MA 01060, United States of America
T: 1 413 584-4881 F: 1 413 545-4902

Haden, Steven T., Mr.
Associate
TREE TECHNOLOGY INTERNATIONAL, INC.
440 Park Ave South
New York - NY 10016, United States of America
T: 1 212 685-8400 F: 1 212 689-3701

Hadlock, Joanne, Ms.
Psychologist
BENTLEY COLLEGE
747 Main Street
Concord-Massachusset 01742, United States of America
T: 001 508 371-0109

Hagerty, Tina, Ms.
YMCA EARTH ACTION CORPS
909 4th Av
Seattle - WA, United States of America
T: 1 206 382-5013 F: 1 206 382-7283

Hahm, Michael, Mr.
UNITED METHODIST CHURCH
38 W. 32nd. Street, 1108
New York - NY 1001, United States of America

Haimbaugh, George D. Jr., Mr.
Professor
UNIVERSITY OF SOUTH CAROLINA SCHOOL OF LAW
Main and Greene Street
Columbia - SC, United States of America
T: 1 803 777-7788 F: 1 803 777-9405

Halbert, Herschel
TEMPLE OF UNDERSTANDING
St. Emeric Church / 185 Avenue D
New York - NY 10009, United States of America
T: 1 212 353-0001 F: 1 212 353-0170

Hallunbak, Hans, Mr.
Exec. Dir.
KV MARKETING INC.
12 Huntville Rd.
Katonah - NY 10536, United States of America
T: 1 914 2327566

Hamer, Lois V., Ms.
WOMEN'S INTERNATIONAL LEAGUE FOR PEACE AND FREEDOM
821 S. Walnut St.
Anaheim - CA, United States of America
T: 1 714 774-2389 F: 1 714 239-6276

Hamilton, William
Reporter
ROTARIAN MAGAZINE
201 Owens Ave
St. Augustine FL 32, United States of America
T: 1 904 4710440 F: 1 904 4719897

Hancock, Helene, Ms.
UNA PASADENA CALIFORNIA
677 West Calif. Blvd.
Pasadena - CA 91105, United States of America
T: 1 818 793-3925

Hancock, Lowell, Mr.
UNA PASADENA CALIFORNIA
677 West California Blvd.
Pasadena - CA 91105, United States of America
T: 1 818 793-3925

Hancock, Vicki, Ms.
Assistant Director
ASSOCIATION FOR SUPERVISION AND CURRICULUM DEVELOPMENT
1250 N. Pitt St.
Alexandria-VA 22314, United States of America
T: 1 703 5499110 ext519 F: 1 703 549-3891

Haney, Trey
Cameraman
CNN BROADCASTING SYSTEM
One CNN Center, Box 105366
Atlanta - GA, United States of America
T: 1 404 8371918 F: 1 404 8274292

Hanley, Kathy, Ms.
Leader
GIRLS SCOUTS U.S.A.
34 Legend Lane
Ballston Spa - NY 12020, United States of America
T: 1 518 583-0481

Hansen, Bo, Mr.
UP WITH PEOPLE
3103 North Campbell Av.
Tucson - AZ 85719, United States of America

Hansen, Kathryn W., Ms.
Board Member
NATIONAL COUNCIL OF RETURNED PEACE CORPS VOLUNTEER
Box 246
Orange City - IA 51041, United States of America
T: 1 712 737.2700 F: 1 712 7373696

Hansen, Michael W., Mr.
Associate Expert
UNCTC
2 United Nations Plaza, Rm DC 2 1110
New York - NY 10017, United States of America
T: 1 212 963-3119 F: 1 212 963-2146

Hansen, Thomas P., Mr.
GLOBAL ENVIRONMENT TEAM
696 South Madison Avenue
Pasadena - CA 91106, United States of America
T: 1 818 796-2951 F: 1 818 792-2121

Hanson, Jaydee, Mr.
GENERAL BOARD OF CHURCH AND SOCIETY - UNITED METHODIST CHURCH
100 Maryland Ave Ne
Washington - DC 20002, United States of America
T: 1 202 488-5650 F: 1 202 488-5639

Haq, A Gary A., Sr.
LANCASTER UNIVERSITY
S Ezngsdale Rd, Lancaster

Harrison, Laura, Ms.
GLOBAL HEART
10200 Pine Hurst
Austin - TX 787474, United States of America

Hart, Stanley, Mr.
UNA -USA, SIERRA CLUB
3104 Mt. Curve Ave.
Altadena - CA 91001, United States of America
T: 1 818 791-9348 F: 1 818 791-9348

Hartley, Elda, Ms.
TEMPLE OF UNDERSTANDING
St. Emeric Church / 185 Avenue D
New York - NY 10009, United States of America
T: 1 212 353-0001 F: 1 212 353-0170

Hartman, Mitra
GLOBAL ENVIRONMENT TEAM
Biology Dept. 216-76, Cal Tech University
Pasadena - CA 91125, United States of America
T: 1 818 356-6824

Harvey, Hal, Mr.
Executive Director
THE ENERGY FOUNDATION
75 Federal St.
San Francisco - CA 94107, United States of America
T: 1 415 546-7400 F: 1 415 546-1794

Harvey, Tom
Senior Correspondent
SUN-SENTINEL
200 East Las Olas Blvd
Ft. Lauderdale - FL, United States of America
T: 1 305 356 4632 F: 1 305 356 4559

Harwell, Christine, Ms.
Director
NORTH-SOUTH CENTER UNIVERSITY OF MIAMI
University of Miami, P. O. Box 248123 / 1500 MONZA AVE.
Coral Gables - FL 33146, United States of America
T: 1 305 361-4168 F: 1 305 361-4077

Hasse, John, Mr.
RR1 87c Lincoln Road
Mullica Hill - NJ 08062, United States of America
T: 1 609 478-6527

Hastings, Lisa, Ms.
GLOBAL ENVIRONMENT TEAM
696 South Madison Avenue
Pasadena - CA 91106, United States of America
T: 1 818 796-2951 F: 1 818 792-2121

Havlick, Spenser, Mr.
Professor
UNIVERITY OF COLORADO
C.B. 314
Boulder - CO 80309, United States of America
T: 1 303 492-0664 F: 1 303 492-6163

Hawk, Homer, Mr.
HOPE PUBLISHING HOUSE
5800 Prospect Ave.
Plymouth - OH 44865, United States of America
T: 1 419 687-3951

Hawken, Paul E., Mr.
221 Hillside
Minneapolis - MN 90941, United States of America
T: 1 612 383-9653

Hawkings, Donna, Mr.
AMERICAN COUNCIL FOR AND ENERGY-EFFICIENT ECONOMY
1001 Connecticut Avenue N.W., Suite 801
Washington - DC 20036, United States of America
T: 1 202 429-8873 F: 1 202 429-2248

Hawkins, Ann Patrick, Ms.
Professor
UNIVERSITY OF OREGON
International Studies, Prince Lucien Campbell Hall
Eugene - OR 97403, United States of America
T: 1 503 346-5052 F: 1 503 346-3660

Hawley, Lisa, Ms.
President
STARLIGHT PRODUCTIONS, INC.
496 Kokomo Road
Haiku - IL 06708, United States of America
T: 1 808 572-4188

Hayenga, Mary, Ms.
GENERAL BOARD OF CHURCH AND SOCIETY - UNITED METHODIST CHURCH
Route 1
Andover-SD 57422, United States of America
T: 605 298-5617

Haynie, Alan, Mr.
SEAS - STUDENTS FOR ENVIRONMENTAL ACTION AT STANFORD
P.O.Box 11602
Stamford - CT 94309, United States of America
T: 001 415 723 4444 F: 001 415 725 6566

Hazday, Ronald, Mr.
President
ENVIRONMENTAL PLASTICS OF PUERTO RICO
1407, Ashford Ave, Suite 502
San Juan, Puerto Rico 00907, United States of America
T: 1 809 725 2726 F: 1 809 725 2726

Heart, Carolene, Ms.
Reverend
ART OF LIVING CENTER
10339 Delmar Cr.
Tampa - FL 33624, United States of America
T: 1 813 779-2203

Helges, O.P., Anele, Ms.
Member
DOMINICAN WORLD ORDER TASK FORCE
100 LaSalle St., 3-D
New York - NY 10027, United States of America
T: 1 212 663-0267

Helmuth, Christopher, Mr.
US CITIZENS NETWORK
1841 Santa Rosa Ave
Pasadena - CA, United States of America
T: 1 818 3989750 F: 1 818 4493735

Hempel, Lamont C., Mr.
Professor Of Env. Politics And
CLAREMONT GRADUATE SCHOOL
170 E. 10th Street
Claremont-CA 91711, United States of America
T: 1 714 621-8171 F: 1 714 621-8039

Henderson, Sean
UP WITH PEOPLE
3103 North Campbell Av.
Tucson - AZ 85719, United States of America

Hendrickson, Laurel, Ms.
Director Of Operations
L. A. SYSTEMS, INC.
4521 Campus, Suite 275
Irvine - CA 92715, United States of America
T: 1 714 497-9151 F: 1 714 497-1187

Hennin, Elizabeh, Ms.
Interpreter
URBAN ECOLOGY
1532 Ward St.
Berkeley - CA, United States of America
T: 1 510 486-1477

Henry, Ashley, Ms.
Program Coordinator
HUMANE SOCIETY OF THE U.S./CENTER FOR RESPECT OF LIFE & ENVIRONMENT
2100 L Street NW
Washington - DC 200037, United States of America
T: 1 202 452-1100 F: 1 202 778-6132

Hernandes, Carlos, Dr.
GLOBAL FUTURES NETWORK
52 Mercer Street
Jersey City 07302, United States of America

Hernandez, Ayala Cinthia, Ms.
Tec. Ambiental
DEPARTAMENTO RECURSOS NATURALES
Pta. de Tierra San Juan,
San Juan, Puerto Rico 00956, United States of America
T: 1 809 723-3090 F: 1 809 722-2785

Herrera, Lupita, Ms.
UP WITH PEOPLE
3103 North Campbell Av.
Tucson - AZ 85719, United States of America

Herrera, Sylvia, Ms.
PODER
2514 Mamie
Brownsville-TX 78520, United States of America

Herzog, Linda Ruth, Ms.
GROWING CONNECTIONS
2123 E. Grant Rd.
Tucson - AZ 85719, United States of America
T: 1 602 3257909 F: 1 602 3257961

Hickel, H.E.Walter, Mr.
Governor
ALASKA
PO Box 110001
Juneau - Alaska 99811, United States of America
T: 1 907 465-3500

Hickey, Chimene, Ms.
UNA SANTA BARBARA & PASADENA
6657 El Colegio Rd #48
Goleta - CA, United States of America
T: 1 805 685-5232 F: 1 805 685-7912
E: 4649ch@ucsbxa.ucsb.edux

Hildebrandt, Denise, Ms.
UP WITH PEOPLE
3103 North Campbell Av.
Tucson - AZ 85719, United States of America

Hill, Joyce, Ms.
UNITED METHODIST CHURCH
16 June Avenue
Norwalk - CT 06850, United States of America

Hinrichs, Johnny, Mr.
HIGHLANDER CENTER
1939 Higlander Way
New Market - TN 37820, United States of America
T: 1 615 933 3443 F: 1 615 933 3424

Hinson, Jacquelyn, Ms.
GLOBAL ENVIRONMENT TEAM
241 So. Ave. 57, #110
Los Angeles - CA 90042, United States of America
T: 1 213 254-7813

Hintzke, Teresa A., Ms.
Nat. 1st Vice President
PAN PACIFIC & SOUTHEAST ASIA WOMEN'S ASSOCIATION
416 Maple
Winnetka - IL 60093, United States of America
T: 1 708 446-4998 F: 1 708 446-4765
TL: J-=54-22302

Hiroko, Atkins, Ms.
Guia De Turismo
TURISMO
Av. Nilo Pecanha, 50, s/2101
Robbinsville - NJ, United States of America
T: 1 609 532 2455

Hixon, Kate
UN DEPARTMENT OF PUBLIC INFORMATION
New York, NY, United States of America

Hocevar, Carl, Mr.
Program Manager
USAID/EPIC PROGRAM
1611 N.Kent St., Suite 206
Arlington - VA 22209, United States of America
T: 1 703 524-1171 F: 1 703 841-0689

Hoffman, Sandra, Ms.
Policy Advisor
WELLSTART
3333 K Street, NW - Suite 101
Washington - DC 20007, United States of America
T: 1 202 298-7979 F: 1 202 298-7988

Hogan, Jean, Ms.
TERRA CHRISTA COMMUNICATIONS
1880 East River Road
Tucson - AZ, United States of America
T: 1 602 577-6476 F: 1 602 577-1195

Holden, Mary Catherin, Ms.
WORLD FEDERALIST ASSOCIATION
530 N St.
Washington - DC 20024, United States of America
T: 1 202 554 4984

Holl, Kevin, Mr.
Advisor
WE THE PEOPLE
60 Balceta
San Francisco - CA 94127, United States of America

Holley, Betty, Ms.
SUSTAINABLE ENV'L/EDUCATION AND DEVELOP. FORUM OF NORTH CAROLINA
14-E Hunt Club Road
Greensboro, NC 27410, United States of America
T: 1 919 855.0156

Holley, Jason E., Mr.
GLOBAL ENVIRONMENT TEAM
222 1941 Hall, Princeton U.
Princeton - NJ 08544, United States of America
T: 1 609 258-8143

Holley Jr., Rev. Solomon J., Mr.
Founder/Director
SUSTAINABLE ENV'L/EDUCATION AND DEVELOP. FORUM OF NORTH CAROLINA
14-E Hunt Club Road
Greensboro, NC 27410, United States of America
T: 1 919 855.0156

Holm, Amy Elizabeth, Ms.
696 Stone Canyon Rd.
Los Angeles - CA 90077, United States of America
T: 1 213 440-2655 F: 1 213 440-2654

Hope, William, Mr.
Vice-President
COCA COLA COMPANY - USA
No 1 Coca Cola Plaza
Atlanta - GA 30301, United States of America
T: 1 404 676 2314 F: 1 404 515 2133

Hopkins, Frank, Mr.
Representative
WORLD CONGRESS OF POETS
329 4th St
Carlstadt - NJ 07072, United States of America
T: 1 201 933-1765 F: 1 212 243-3288

Hopper, Donald
TURNER BROADCASTING SYSTEM / CNN
One CNN Centre
Atlanta - GA, United States of America
T: 1 404 827-3973 F: 1 404 827-1015

Horner, Sarah, Ms.
UP WITH PEOPLE
3103 North Campbell Av.
Tucson - AZ 85719, United States of America

Horrell, Dana, Mr.
GLOBAL ENVIRONMENT TEAM
United States of America
T: 86 267-9652

Horsley, Kris, Ms.
UNCED Staff Coordinator
POPULATION INSTITUTE
110 Maryland Ave. NW
Washington - DC 20002, United States of America
T: 1 202 544-3300 F: 1 202 544-0068

Howe, Annete Van, Ms.
Board Of Directors
FLORIDA WOMEN'S CONSORTIUM
2100 S. Ocean Dr., 16K
Ft. Lauderdale - FL, United States of America
T: 1 305 7632581 F: 1 305 5257310

Howe, Charles
Baha'I Newsreel
MEDIA SERVICES - BAHA'I CENTER
1233 Central Street
Evanston, Illinois 6, United States of America
T: 1 708 2561361 F: 1 708 2561372

Howe, Thomas
Baha'I Newsreel
MEDIA SERVICES - BAHA'I CENTER
1233 Central Street
Evanston, Illinois 6, United States of America
T: 1 708 2561361 F: 1 708 2561372

Hudgens, Thomas, Mr.
Vice President
ASSOCIATION TO UNITE THE DEMOCRACIES
1506 Pennsylvania Ave SE
Washington - DC 20003, United States of America
T: 1 202 5445152 F: 1 202 5443742

Hudson, Helen, Ms.
Administrative Assistant
CENTER FOR WAR/PEACE STUDIES
218 East 18th Street
New York - NY 10003, United States of America
T: 1 212 4751077 F: 1 212 2606384

Huffman, Sandra, Dr.
Policy Advisor
WELLSTART
3333 K Street NW Suite 101
Washington - DC 20007, United States of America
T: 1 202 298-7979 F: 1 202 298-7988

Huffsmith, Barbara, Ms.
A WORLD THAT WORKS FOR EVERYONE
P.O. Box 240683
Anchorage - Alaska 99524, United States of America
T: 1 907 2670885 F: 1 907 276-4429

Hunt, Douglas B., Mr.
Convener
CHESAPEAKE CENTER
1820 Sanford Road
Wheaton - MD 20723, United States of America
T: 1 301 890-4172 F: 1 301 755-3234

Hunt, Elizabeth H., Ms.
Secretary Board Of Directors
CENTER FOR INTERNATIONAL COOPERATION
PO Box 444
Washington - DC 22747, United States of America
T: 1 202 675-1071

Hunter, Jennifer, Ms.
GLOBAL ENVIRONMENT TEAM
696 South Madison Avenue
Pasadena - CA 91106, United States of America
T: 1 818 796-2951 F: 1 818 792-2121

Huq, Shireen, Ms.
NARIPOKKHO
1717 Massachusetts Ave. NW, 2nd floor
Washington - DC 20036, United States of America
T: 1 202 667-1142 F: 1 202 332-4496
TL: 440384

Huslid, H.E. Martin
Ambassador, N.Y.
NORWEGIAN DELEGATION TO UNCED
825 Third Ave., 18th floor
New York - NY, United States of America

Hutchinson, Peter, Mr.
TEMPLE OF UNDERSTANDING
St. Emeric Church / 185 Avenue D
New York - NY 10009, United States of America
T: 1 212 353-0001 F: 1 212 353-0170

Huthsteiner, George M.D., Mr.
President
U.N.A., U.S.A. SAN FERNANDO VALLEY, CA
4537 Gable Drive,
Encino-CA, United States of America
T: 1 818 342-6195

Hutton, Bruce, Mr.
Dean
GRADUATE SCHOOL OF BUSINESS/UNIVERSITY OF DENVER
Graduate School of Business, 2020 S. Race St.
Denver - CO, United States of America
T: 1 303 871-3416

Ingham, Fredericka A.
Freelance
COPLEY NEWS SERVICE
3050 Rue D'Orleans
San Diego CA 92110, United States of America
T: 1 609 2236643 F: 1 609 2236643

Irish, Jeffrey S.
SCHOOL OF THEOLOGY AT CLAREMONT
505 W. 9th St.
Claremont-CA 91711-3919, United States of America
T: 1 714 621-5254

Irizarr, Maria Margarita, Ms.
Diretora Executiva
AUTORIDAD ACUODUTOS E ALCANTARILLADOS
Av. Barbosa 604
San Juan, Puerto Rico, United States of America
T: 1 809 756-7422 F: 1 809 754-3092

Isabel, Rosa Maria
COOPERATIVA AGROEXTRATIVISTA DE XAPURI
1880 East River Road
Tucson - AZ, United States of America
T: 1 602 577-6476 F: 1 602 577-1195

Isely, Margaret, Ms.
WORLD CONSTITUTION AND PARLIAMENT ASSOCIATION
1480 Hoyt St., Suite 31
Lakewood - CO 80215, United States of America
T: 1 303 233-3548 526-0463 F: 1 303 526-2185

Isely, Philip, Mr.
Secretary General
WORLD CONSTITUTION AND PARLIAMENT ASSOCIATION
1480 Hoyt St., Suite 31
Lakewood - CO 80215, United States of America
T: 1 303 233-3548 526-0463 F: 1 303 526-2185

Isom, John, Mr.
Research Associate
ECO TIBET - CALIFORNIA
347 Dolores st., Room 206
San Francisco - CA 94110, United States of America
T: 1 415 241-9197 F: 1 415 626-0865

Ives, Diane, Ms.
TIDES FOUNDATION
1388 Sutter Street # 1010
San Francisco - CA 94109, United States of America
T: 1 415 771-4308 F: 1 415 771-0535

Jackson, Courtney, Mr.
Student
LAUREL SPRINGS SCHOOL
PO Box 1440
Ojai - CA 93024, United States of America
T: 1 805 646-2473 F: 1 805 646-0186

Jacobs, Daniel, Mr.
Professor
AMERICAN UNIVERSITY
2450 Virginia Avenue
Washington, Virginia 20037, United States of America

Jacobson, Micah, Mr.
Staff
EARTH TRAIN/WORLDLINK/GLOBAL YOUTH SUMMIT
3220 Sacramento St.
San Francisco - CA 94115, United States of America
T: 1 415 931 1290

Jacqz, Jane, Ms.
Senior Advisor
UNDP- UNITED NATIONS DEVELOPMENT PROGRAMME
One UN Plaza, Room 1958
New York - NY 10017, United States of America
T: 1 212 906-6076 F: 1 212 906-5313

Jaeger, Rob, Mr.
Appraiser
APPRAISAL INSTITUTE
3170 Lake Ranch Circle
Gainesville, United States of America
T: 1 904 534-7024 F: 1 904 534-3836

James, Sarah
INDIGENOUS PEOPLES MEDIA CENTER
77 Main Street
New York - NY, United States of America
T: 1 212 3650414 F: 1 212 3650864

Jarvie, Leslie, Sra.
Director Of Environmental
RAINFOREST ACTION NETWORK
c/o 6420 Fairmount Ave.
El Cerrito-CA 94530, United States of America
T: 1 510 527-8100 F: 1 510 525-7710
TL: 335-429

Jarvis, Tracey, Ms.
UP WITH PEOPLE
3103 North Campbell Av.
Tucson - AZ 85719, United States of America

Jefferson, Richard, Mr.
LUMMI TRIBE
4315 Lummi Share Road
Ferndale 98.248, United States of America
T: 206647-6150 F: 206384-5521

Jenkins, Al, Mr.
Team Member
FOUNDATION FOR GLOBAL COMMUNITY
14284 Eastridge
Whittier - CA, United States of America
T: 1 310 6936748

Jenkins, Nancy
Team Member
FOUNDATION FOR GLOBAL COMMUNITY
14284 Eastridge
Whittier - CA, United States of America
T: 1 310 6936748

Jhirad, David J., Mr.
Director
U.S. AGENCY FOR INTERNATIONAL DEVELOPMENT
Red/Ei 508, Sa-18
Washington - DC 20523-1810, United States of America
T: 1 703 875-4047 F: 1 703 875-4053

Jipp, Peter, Sr.
Pesquisado
WOODS HOLE RESEARCH CENTER
P.O.Box 296
Woods Hole - MA, United States of America
T: 1 508 540-9900

John, Dove
Director
DOVE & CO./MMH
4759 15th Ave NE
Seattle - WA 98105, United States of America
T: 1 206 344-6321

Johnson, Carolyn, Sra.
Bm
UNITED METHODIST CHURCH
2550 Yeager Road 19-2
West Lafayette - IN 47906, United States of America

Johnson, Charles, Mr.
TERRA CHRISTA COMMUNICATIONS
1880 East River Road
Tucson - AZ, United States of America
T: 1 602 577-6476 F: 1 602 577-1195

Johnson, Hazel, Mr.
Founder/Ceo
PEOPLE FOR COMMUNITY RECOVERY
13116 South Ellis Avenue
Chicago - IL, United States of America
T: 1 312 468-1645 F: 1 312 468-8105

Johnson, Iaw, Mr.
Administrator
GLOBAL ENVIRONMENT FACILITY
c/o World Bank
Washington - DC, United States of America
T: 1 202 673 1053 F: 1 202 477 0551

Johnson, Kersten, Ms.
EAST-WEST CENTER (IEDP)
1777 East-West Road
Honolulu - HI, United States of America
T: 1 808 944-7525 F: 1 808 044-7970

Johnson, Norma, Ms.
INTERFAITH CONFERENCE / SENSE
5908 Dale Lane
Greendale-WI, United States of America

Johnson, Randall H, Mr.
Director
INTERNATIONAL SOCIETY FOR THE PRESERVATION OF THE TROPICAL RAINFOREST
3302 North Burton Avenue
Rosemead - CA 91770, United States of America
T: 1 818 5727273 F: 1 818 5729521

Johnston, Suzanna H., Ms.
Chief Sales Section
UNITED NATIONS PUBLICATIONS
U.N. Plaza - Room DC-853
New York - NY 10017, United States of America
T: 1 212 963-8302 F: 1 212 963-3489

Jonas, Susanne, Ms.
Correspondent
PACIFIC NEWS SERVICE/UNIVERSITY OF CALIFORNIA AT SANTA CRUZ
3311 Mission St. # 189
San Francisco - CA 94110, United States of America
T: 1 415 826-8338 F: 1 415 459-3125

Jones, Jolie, Ms.
President
TAKE IT BACK FOUNDATION
111 N. Hollywood Way
Burbank CA 91505, United States of America
T: 001 818 569-0552 F: 001 818 569-0539

Jones, Nell, Ms.
Trustee
GEORGIA WILDLIFE FEDERATION
4211 Conway, Valley Road
Atlanta - GA 30327, United States of America
T: 1 404 233 4343 F: 1 404 416 1521

Jordan, Richard T.
Correspondent
GLOBAL EDUCATION ASSOCIATES
26 Gramercy Park IE
New York -NY 10003, United States of America
T: 1 212 6739398 F: 1 212 5055233

Jorgensen, Erik E., Mr.
Director
CENTRE FOR PEACE AND THE ENVIRONMENT
PO Box 393
Santa Barbara - CA 93102, United States of America
T: 1 805 9657505

Jorgensen, Karen, Ms.
Env. Management Advisor
UNITED NATIONS DEVELOPMENT PROGRAM - UNDP
Room DC1-2088, One UN Plaza, United Nations
New York - NY 10017, United States of America
T: 1 212 906-5008 F: 1 212 906-5365

Jorre, Danielle St., Ms.
PERMANENT MISSION OF SEYCHELLES TO THE U.N.
820, 2nd Avenue Rm. 900f
New York - NY 10017, United States of America
T: 1 212 444-3800 F: 1 212 444-3811

Joseph, Niki, Ms.
Coordinator
HAITI IN DEVELOPMENT
Box 769
New York - NY 10185, United States of America
T: 1 212 465-4683 F: 1 212 249-400
E: TAS 0 994

Joshi, Chandni, Ms.
Regional Officer
UNITED NATIONS DEVELOPMENT FUND FOR WOMEN
304 East 45th Street
New York - NY 10017, United States of America
T: 1 212 906-6440

Judge, Michael, Mr.
WORLD COMMUNITY FOUNDATION
866 United Nations Plaza, Suite 438
New York - NY, United States of America
T: 1 212 752-0672 F: 1 212 752-0685

Jung, Elba, Ms.
GLOBAL ENERGY NETWORK INTERNATIONAL
221 B Solon Dr.
Ballwin 63121, United States of America
T: 314 230-0586

Junior, Reynaldo Zangradi
RAI - TV CHANNEL 3
New York, United States of America

Kackloudis, Nick S., Mr.
President
ECOLOGIA INC.
3690 Runnymede Blvd.
Cleveland Hts. - OH 44121, United States of America
T: 1 216 291-6909

Kahare, Paul Njoru, Mr.
GREEN BELT MOVEMENT OF KENYA
c/o Resource Renewal Institute, FT. Mason Center Bldg A
San Francisco - CA, United States of America
T: 1 415 928-32774 F: 1 415 928-6529

Kahill, Emil, Mr.
Field Rep.
ONE WORLD NOW
1510 Lyric Center, 440 Louisiana
Houston - TX, United States of America
T: 1 713 222-2288

Kahn, Les, Mr.
Musician
PARLIAMENTARY GLOBAL FORUM
374 Whispering Pines
Boulder - CO, United States of America
T: 1 303 567-8796

Kale, P.M., Dr.
ENGINEERING SERVICES PROJECT
1880 East River Road
Tucson - AZ, United States of America
T: 1 602 577-6476 F: 1 602 577-1195

Kaminsky, Leib, Mr.
GREEN (GLOBAL RIVERS ENVIRONMENTAL EDUCATION NETWORK)
University of Michigan School of Natural Resouces -Dana Building 430E
Arcata, United States of America
T: 1 707 7645171 F: 1 707 9362195

Kane, A. Charles, Mr.
YMCA OF THE USA / INTERNATIONAL DIVISION
101 N. Wacker Drive
Chicago - IL IL60606, United States of America
T: 1 312 9970031 F: 1 312 497-9424
TL: 4979424

Kane, Jack L., Mr.
Vice President
INTERNATIONAL MARKETING
27707 Mountain Meadow Road #50
Escondido - CA 92026, United States of America
T: 1 619 7497849

Kane, Richard Scott
Sound Recordist
META INC
2100 N 14th
Arlingon VA, United States of America
T: 1 703 2433608

Kapadia, Farzana, Mr.
WORLD INFORMATION TRANSFER INC
274 Madison Ave, Suite 601
New York - NY 10016, United States of America
T: 1 212 2730253 F: 1 212 3792004

Kaplan, Brian
Reporter
EPIIC
Miner Hall, Tufts University
Medford MA 02155, United States of America
T: 1 617 3813314 F: 1 617 3813940

Karel, Rogers, Ms.
POPULATION INSTITUTE
209 ponderosa dr.
Alamosa 81101, United States of America
T: 719 589-7211 F: 719 589-7522

Karjalainen, Piia, Ms.
UP WITH PEOPLE
3103 North Campbell Av.
Tucson - AZ 85719, United States of America

Karnavy, Faye, Ms.
Staff
EARTH TRAIN/WORLDLINK/GLOBAL YOUTH SUMMIT
3220 Sacramento St.
San Francisco - CA 94115, United States of America
T: 1 415 931 1290

Kashiwaya, Itaru, Mr.
Graduate Student
DEPARTMENT OF SOCIOLOGY, UNIVERSITY OF TSUKUBA, JAPAN
Yaguchi-So, Kami-Hiroaka 382
Tucson - AZ 305, United States of America
T: 1 602 0298 57 2611
E: .336

Kasimoff, Kyril, Mr.
GLOBAL ENVIRONMENT TEAM
337 N. Larchmont Blvd.
Los Angeles - CA 90004, United States of America
T: 1 213 255-6620

Kathan, Joyce C., Ms.
GLOBAL ENVIRONMENT TEAM
229 Cheshire Rd
Prospect - CT 91106, United States of America
T: 1 203 758-4606

Kaufman, Herbert
ECOTROPIC
1880 East River Road
Tucson - AZ, United States of America
T: 1 602 577-6476 F: 1 602 577-1195

Kaufmann, Frank, Dr.
Executive Director
INTER-RELIGIOUS FEDERATION FOR WORLD PEACE
4 West 43rd Street, 5th Floor
New York - NY 10036, United States of America
T: 1 212 8696023 F: 1 212 8696424

Kawano, Masami
UP WITH PEOPLE
3103 North Campbell Av.
Tucson - AZ 85719, United States of America

Kay, Alan F., Mr.
President
AMERICANS TALK SECURITY FOUNDATION
907 6th St. SW, #602C
Washington - DC 20024, United States of America
T: 1 202 488 5889 F: 1 202 488 8024

Keane, Patrice, Ms.
AMERICAN SOCIETY FOR PSYCHICAL RESEARCH
s west 73 rd st. new york
New York - NY, United States of America
T: 1 212 799 5050 F: 1 212 496 2497

Kelley, Florence M., Ms.
President
UNA/USA - HAWAII DIVISION
46-284 Kamehameha Hwy
Kaneohe - HI 96744, United States of America
T: 1 808 247-1550 F: 1 808 595-4468
E: 144:bhn037

Kelley, John H., Mr.
Exec. Comm.
ONE WORLD NOW
1510 Lyric Center, 440 Lousiana
Houston - TX 77002, United States of America
T: 1 713 222-2288 F: 1 713 222-0319

Kellman, Barry, Mr.
Professor
DE PAUL UNIVERSITY
25 E Jackson Blvd
Chicago - IL 60604, United States of America
T: 1 312 362-5258 F: 1 312 362-6908

Kelly, Erle, Mr.
2 Park Ave
Long Beach - CA 90803, United States of America
T: 1 310 434-4645

Kelly, Karen, Ms.
PROJECTS FOR GLOBAL HARMONY
P.O.Box 705
Goffstown NY 03045, United States of America

Kelly, Kristine, Ms.
2 Park Ave
Long Beach - CA 90803, United States of America
T: 1 310 434-4645

Kelly, Patricia
Vice-President For Communications
CONSERVATION INTERNATIONAL TV PRODUCTION
1015 18Th St., NW
Washington - DC, United States of America
T: 1 202 4295660 F: 1 202 8875188

Kelly, Sheila, Ms.
Executive Committee
U.S. CITIZENS NETWORK ON UNCED
1444 N.W. Woodbine Way
Seattle - WA 98177, United States of America
T: 1 206 3657892 F: 1 206 3653212

Kemp, Peggy
TURNER BROADCASTING SYSTEM / CNN
One CNN Centre
Atlanta - GA, United States of America
T: 1 404 827-3973 F: 1 404 827-1015

Kennedy, Joan, Ms.
500 Westcott St.
Syracuse - NY 13210, United States of America
T: 1 315 422-0709

Kenoyer, S.L., Pat, Ms.
Loretto Ngo-Un Rep.
SISTERS OF LORETTO
635 E. 12th St., Apt. 6K
New York - NY 10009 - 3603, United States of America
T: 1 212 674 25 42

Kersting, Fritz, Mr.
RELIGIAO FE BAHA'I
Wm 3 Dalia Los Angeles
CAROLINA, Puerto Rico 00979, United States of America
T: 809 789 3578 F: 9575

Kerttula, Elizabeth J., Ms.
ALASKA DEPARTMENT OF LAW
1105 Mendenhall Penn Road
Juneau - Alaska 99801, United States of America
T: 1 907 790-2023 F: 1 907 465-5274

Keshner, Ethel, Ms.
WORLD CITIZENS ASSEMBLY
301 E 21 St
New York - NY 10010, United States of America
T: 1 212 674-5467

Ketchum, Robert Glenn, Mr.
Curator Of Photography
NATIONAL PARK FOUNDATION
696 Stone Canyon Rd.
Los Angeles - CA 90077, United States of America
T: 1 213 440-2655 F: 1 213 440-2654

Khalid, Kahlid R.O., Mr.
MULTINATIONALS AND DEVELOPMENT CLEARINGHOUSE
PO Box 19405
Washington - DC 20036, United States of America
T: 1 202 553-5777 F: 1 202 234-5176

Khan, Malik, Mr.
Counselor
WORLD MUSLIM CONGRESS
4281 SW 15 St.
Miami - Florida 33134, United States of America
T: 1 305 446-7372 F: 1 305 444-2001

Khan, Salma, Ms.
Member
WOMAN FOR WOMAN OF BANGLADESH
Eisenhower Felloships 256 South 16th st.
Philadelphia - PA 19102, United States of America
T: 1 215 546-1738 F: 1 215 546-4567

Khan, Uzma, Ms.
UNITED NATIONS DEVELOPMENT FUND FOR WOMEN
304 East 45th St.
New York - NY 10017, United States of America
T: 1 212 9066400 F: 1 212 9066705

Kilday, Gregory
TURNER BROADCASTING SYSTEM / CNN
One CNN Centre
Atlanta - GA, United States of America
T: 1 404 827-3973 F: 1 404 827-1015

Kim-Farley, Han Ju, Mr.
Representative
BAHA'I INTERNATIONAL COMMUNITY
866 UN Plaza, Suite 120
New York - NY 10017, United States of America
T: 1 212 7563500 F: 1 212 7563573
E: 152:CNM057

King, Carol
Cameraman
FREELANCER
212 Riverside Drive, 7B
New York, NY 10025, United States of America
T: 1 212 6626858

Kiran, Stordalen, Ms.
Manager
AVEDA CORPORATION
4000 Pheasant Ridge Drive
Minneapolis - MN, United States of America
T: 1 612 783-4000 F: 1 612 783-4110

Kittel, Tim, Mr.
Asst Project Scientist
UNIVERSITY CORP FOR ATMOS RESEARCH
1850 Table Mesa Drive
Boulder - CO, United States of America
T: 1 303 4971611 F: 1 303 4971137
TL: 989764

Kleeschulte, Sonja, Ms.
UP WITH PEOPLE
3103 North Campbell Av.
Tucson - AZ 85719, United States of America

Kmiani, Debora, Ms.
Assoc. Publisher
E - THE ENVIRONMENTAL MAGAZINE
P.O.Box 5098
Westport - CT 06881, United States of America
T: 1 203 854-5559 F: 1 203 866-0602

Knowles, Melanie K., Ms.
Member Senate Rep.
OVERSEES DEVELOPMENT NETWORK
Ocmr Box 2770 Oberlin College
Oberlin - OH 44074, United States of America
T: 1 216 774 5003 F: 1 216 775 8666

Knox, Adelle B., Ms.
UNA PASADENA
1290 Meadowbrook
Altadena - CA 91001, United States of America
T: 1 818 794-5929

Knox, Robert B., Mr.
UNA PASADENA
1290 Meadowbrook
Altadena - CA 91001, United States of America
T: 1 818 794-5929

Knudson, Barbara, Ms.
Professor
UNIVERSITY OF MINNESOTA - INSTITUTE OF INTERNATIONAL STUDIES
1823 SE Franklin Ave.
Minneapolis - MN 55414, United States of America
T: 1 612 378-2634

Kols, Dennis, Mr.
Science & Technology Mgmt.
MODEL EVALUATION CONSORTIUM FOR CLIMATE ASSESSMENT
2511N 124th Street - suite 205
Brookfield - WI 53005, United States of America
T: 1 414 785-5946 F: 1 414 785-5950

Kolmodin, Ada I., Ms.
Observer
PPSEAWA - PAN PACIFIC AND SOUTH EAST ASIAN WOMEN'S ASSOCIATION
6805 W. Highland Ave.
Chicago - IL, United States of America
T: 1 312 774-2262

Komer, Daniel I., Mr.
CENTER FOR HUMAN ECOLOGY
P.O. Box 5210
Austin 78763, United States of America
T: 1 512 443-4042 F: 1 512 443-0973

Kondylis, Katherine, Ms.
UNIVERSITY OF COLORADO
113 Peyton St.
Santa Cruz - CA, United States of America
T: 1 408 429-6152

Kopf, Petra C., Ms.
Consultant
NORWEGIAN TRADE COUNCIL
800 Third Avenue
New York - NY 10022, United States of America
T: 1 212 4219210 F: 1 212 8380374

Kordestani, Kameron, Mr.
Participant
EARTH TRAIN/ WORLD LINK GLOBAL YOUTH SUMMIT
1012 Redmond Court
San Jose - CA 95120, United States of America
T: 4089976408

Koski, Scott, Mr.
Director
MID PENINSULA ENVIRONMENTAL EDUCATION ALLIANCE
2448 Watson Court
Palo Alto - CA, United States of America
T: 1 415 497-4737 F: 1 415 494-1946

Kovich, Matt, Mr.
YMCA EARTH ACTION CORPS
909 4th Av
Seattle - WA, United States of America
T: 1 206 382-5013 F: 1 206 382-7283

Kramer, Mark A. M., Mr.
Participant
BAHA'I INTERNATIONAL COMMUNITY
Yellow Springs -Ohio
Yellow Springs,Ohio 45387, United States of America
T: 1 513 767 7079 F: 1 513 285 5754

Kranish, Arthur
Editor
TRENDS PUBLISHING INC
1079 National Press Bldg
Washington - DC, United States of America
T: 1 202 393 0031 F: 1 202 393 1732

Krasney, Martin N, Mr.
Senior Research Associate
COMMONWEAL
PO Box 316
Bolinas - CA, United States of America
T: 1 415 8680970 F: 1 415 8682230

Krasnow, Sandy, Ms.
The Carlyle Hotel 35 E. 76th Street
New York - NY 10003, United States of America
T: 1 212 744-1600 F: 1 212 439-5746

Krech, M. Chapin, Mr.
HOPE PUBLISHING HOUSE
1537 Saragosa Ave.
Coral Gables - FL 33134, United States of America
T: 1 305 661-4343

Kreh, Katherine A., Ms.
PLAN IT EARTH PRODUCTIONS
P.O.Box 15858,
West Palm Beach - FL 33416, United States of America
T: 1 407 966-7890 F: 1 407 641-5757

Kreitler, Peter Gwillim, Mr.
Exec Director
EARTH SERVICE INCORPORATED
2450 Broadway Avenue, Suite 560
Santa Monica - CA, United States of America
T: 1 310 829-9190

Krishnan, Krish R., Mr.
DECCAN DEVELOPMENT SOCIETY (DDS)
1617, Sixth Place
McLean - VA 22101, United States of America
T: 1 703 761-1466 F: 1 703 893-2123

Krizek, Rita, Ms.
PARTNERS IN THE ENVIRONMENT
6911 Richmond Hwy., Ste. 308
Alexandria-VA, United States of America
T: 1 703 765-8463 F: 1 703 765-8426

Kroeger, Ken, Mr.
TERRA CHRISTA COMMUNICATIONS
1880 East River Road
Tucson - AZ, United States of America
T: 1 602 577-6476 F: 1 602 577-1195

Kuczynski, Terry, Ms.
GENERAL BOARD OF CHURCH & SOCIETY /UNITED METHODIST CHURCHES
24 Church Street -P.O.Box 1187
Hardwick - VT, United States of America
T: 1 802 888-4902 F: 1 914 684-6874

Kuhn, Thomas, Mr.
President
EDISON ELETRIC INSTITUTE
701 Pennsylvania Avenue, N.W.
Washington - DC 20004, United States of America
T: 1 202 508-5460 F: 1 202 508-5403

Kuhne, K. John, Mr.
Co-Director
UNIVERSITY OF THE NATIONS
75-5851 Kuakini Hwy
Kailua-Kona - HI 96740, United States of America
T: 1 808 326-7228 F: 1 808 329-2387
TL: 510-601-4522

Kumar, Acharya Sushil, Mr.
INTERNATIONAL MAHAVIR JAIN
Siddhachalam 65,Mud Pond Rd
Blairstown, NJ, United States of America
T: 1 908 362- 9793 F: 1 908 362- 9649

Kupec, Anthony, Mr.
Research Assistant
NORTH-SOUTH CENTER UNIVERSITY OF MIAMI
1500 Monza Avenue
Coral Gables - FL 33146, United States of America
T: 1 305 284 6868 F: 1 305 284 6370

Kupec, Ivy, Ms.
NORTH-SOUTH CENTER UNIVERSITY OF MIAMI
1500 Monza AVe.
Miami - Florida 33146-3027, United States of America
T: 1 305 284 6868 F: 1 305 2846370

Kyle, Paula, Ms.
CENTER FOR INTERNATIONAL COOPERATION
5523 Sherrier Plasce NW
Washington - DC 20016, United States of America
T: 1 202 244 0994 F: 1 202 937 4039

La Salle, Amy Elizabeth, Ms.
Coordinator
HEALING ARTS TEAM
2756 Arcadia Drive
Lake Havasu City -AZ 86403, United States of America
T: 1 602 453-2730

Laan, Rosemary C. Van Der, Ms.
Coordinator
A WORLD THAT WORKS FOR EVERYONE
6405 Colgate Drive
Anchorage - Alaska 99504, United States of America
T: 1 907 337-5580 F: 1 907 337-8816

Lagerros, Dorothy, Dr.
GLOBAL ENVIRONMENT TEAM
507 Chapel Av.
Ashland-WI 54806, United States of America
T: 1 715 682-6532

Lakes Jr., George, Mr.
SCHOOL OF THEOLOGY AT CLAREMONT
1325 N. College Ave. PO Box 107
Claremont-CA 91711-3919, United States of America
T: 1 714 399-5706 F: 1 714 626-7062

Lambe, Tony, Mr.
Translation Coordinator
INTERNATIONAL SYNERGY INSTITUTE
80-66 Cornish Av.
Elmhurst - NY, United States of America
T: 1 718 473-7841

Landgraf, John L., Mr.
Anthropologist
UNA (USA) & AAUN
535 South Boulevard of Presidents
Sarasota - FL 20037, United States of America
T: 1 813 388-1184 F: 1 813 388-3534

Landoff, Lawrence, Mr.
Student
PO Box 1440
Ojai - CA 93024, United States of America
T: 1 805 646-2473 F: 1 805 646-0186

Landvatter, Jeffrey Dennis, Mr.
Adviser
STATE OF ALASKA
PO Box 110001
Juneau - Alaska 99811, United States of America
T: 1 907 465 3500

Lane, Donna M., Ms.
Vice-President
PLAN IT EARTH PRODUCTIONS
P.O.Box 15858,
West Palm Beach - FL 33416, United States of America
T: 1 407 966-7890 F: 1 407 641-5757

Laporte, Christine
Director
SACHARUNA FOUNDATION
P.O. Box 130
The Plains - VA, United States of America
T: 1 703 754-8713 F: 1 703 754-1475

Larrance, Deven
UP WITH PEOPLE
3103 North Campbell Av.
Tucson - AZ 85719, United States of America

Larsen, Janet J., Dr.
Professor Emertus
UNIVERSITY OF FLORIDA
805 N.W. 20th Terrace
Gainesville 32603, United States of America
T: 1 904 3966689

Larson, Nancy, Ms.
CAMPAIGN FOR THE EARTH
1441 Mariposa Av.
Boulder - CO 80302, United States of America
T: 1 303 444-8002 F: 1 303 444-3811

Larson, Robert, Mr.
SOLAR BOX COOKERS INTERNATIONAL
1724 Eleventh St.
Sacramento - CA 95814, United States of America
T: 1 916 444-6616 F: 1 916 447-8689

Larsson, Alice, Ms.
Partner, Co-Founder
WORLD INTUACTION
P.O. Box 3241 / 54650 Tahquitz View Drive
Idyllwild-CA, United States of America
T: 1 714 659-9858 F: 1 714 659-3761

Lascot, Zann, Ms.
CAMPAIGN FOR THE EARTH
39 Spring Canyon P.O. Box 711
Bisbee - AZ 85603, United States of America
T: 818 999 0369

Lathan, Anthony Michell, Mr.
Board Member
OREGON UNITED NATIONS ASSOCIATION
Portland Chapter United Nations Association 121 S.W. Salmon St. Room 320
Portland, Oregon 972, United States of America
T: 001 503 2251157

Lattimer, Tim, Mr.
UNCED-Team Member
FOUNDATION FOR GLOBAL COMMUNITY
222 High St.
Palo Alto - CA, United States of America
T: 1 415 328-7756 F: 1 415 328-7785
TL: 856588

Lawrence, Bob
INDIGENOUS PEOPLES MEDIA CENTER
77 Main Street
New York - NY, United States of America
T: 1 212 3650414 F: 1 212 3650864

Lawrence, Kathy L., Ms.
Director
ORGANIC FARMERS ASSOCIATION COUNCIL
P. O. Box 213
Putney - VT, United States of America
T: 1 802 387-2429 F: 1 802 387-4001

Le Compte, Aimee, Ms.
Admin. Assist.
WORLD WISE
1817 T St. #3
Sacramento - CA, United States of America
T: 1 916 446-5944 F: 1 916 739-6951
TL: 6503859849

Le Moigne, Guy, Mr.
WORLD BANK
1818 H St. NW
Washington - DC 20433, United States of America
T: 1 202 458-1943

Leach, Brad, Mr.
Economist
NEW YORK MERCANTILE EXCHANGE
Washington - DC, United States of America
F: 1 202 785-5351

Lear, Jon, Mr.
CENTER FOR RESOURCE MANAGEMENT
262 South 200 West
Salt Lake City - UT, United States of America
T: 1 801 3220530 F: 1 801 3283457

Leavett, Jackie, Ms.
UNIVERSITY OF CALIFORNIA - LOS ANGELES
1317 Perloff, 41 Hillgard Avenue
Los Angeles - CA 90024, United States of America

Leblanc, Alice, Ms.
Staff Economist
ENVIRONMENTAL DEFENSE FUND
257 Park Ave., South
New York - NY 10010, United States of America
T: 1 212 505-2100 F: 1 212 505-2375

Lecompte, Aimee, Ms.
Administrative Director
WORLDWISE
401 Sw Miguel Way
Sacramento - CA 95814, United States of America
T: 1 916 456-9205

Ledbetter, Brownie W., Ms.
WOMEN'S ENVIRONMENT AND DEVELOPMENT ORGANIZATION
103 West Capitol Ave, Suite 1115
Little Rock - AR, United States of America
T: 1 501 376-7913 F: 1 501 374-3935

Lee, Boying
SCHOOL OF THEOLOGY AT CLAREMONT
1325 N. College Ave. 275
Claremont-CA 91711-3919, United States of America
T: 1 714 399-5706 F: 1 714 626-7062

Lee Keller, Edward, Ms.
Director
OIL AND GASTEK INC.
301 Rosewood Street
Ventura,CA 93001, United States of America
T: 805 648 3743 F: 805 961 8801

Legler, Heidi
INTERFAITH CONFERENCE / SENSE
5908 Dale Lane
Greendale-WI, United States of America

Leitmann, Joe, Mr.
Urban Planner
URBAN MANAGEMENT PROGRAM / WORLD BANK
1818 H Street NW
Washington - DC 20433, United States of America
T: 1 202 4733861 F: 1 202 4771391

Lemley, Bruno, Mr.
WORLD INFORMATION TRANSFER INC
274 Madison Ave, Suite 601
New York - NY 10016, United States of America
T: 1 212 2730253 F: 1 212 3792004

Leonard, Amber, Ms.
Project Manager
STOCKHOLM ENVIRONMENT INSTITUTE
9514 Garwood St
Silver Spring - MD 20901, United States of America
T: 1 301 587-8714 F: 1 301 587-8716

Lepisto, Gretchen, Ms.
UP WITH PEOPLE
3103 North Campbell Av.
Tucson - AZ 85719, United States of America

Lerner, Peter Ted
TURNER BROADCASTING SYSTEM / CNN
One CNN Centre
Atlanta - GA, United States of America
T: 1 404 827-3973 F: 1 404 827-1015

Levin, Penny, Ms.
EAST-WEST CENTER (IEDP)
1777 East-West Road
Honolulu - HI, United States of America
T: 1 808 944-7525 F: 1 808 044-7970

Levitt, David M., Mr.
WORLD CONFERENCE ON RELIGION AND PEACE
515 East 72nd Street, apt 30 D
New York - NY 10021, United States of America
T: 212 288-5367 F: 212 988-2046

Levitt, Norma U., Ms.
International President
WORLD CONFERENCE ON RELIGION AND PEACE
515 East 72nd Street, apt 30 D
New York - NY 10021, United States of America
T: 1 212 288-5367 F: 1 212 988-2046

Lewis, Tammy, Ms.
Student
UNIVERSITY OF CALIFORNIA, DAVIS
1828 Laurel Lane
Sacramento - CA 95691, United States of America
T: 1 916 371-0920

Lierop, Robert Van, Dr.
Ambassador
GLOBAL HEART/ VANUATU
1880 East River Road
Tucson - AZ, United States of America
T: 1 602 577-6476 F: 1 602 577-1195

Liley-Winter, Cher, Ms.
Musician
PARLIAMENTARY GLOBAL FORUM
P.O.Box 68
Litchfield, United States of America
T: 203 567-8796 F: 203 567-4272

Lima, Gail, Ms.
EARTH EMERGE & SEE
277 Hudson St.
Cornwall-on-HudsonNY 12520, United States of America

Lima, Lauro, Mr.
Representative Of Unesco
PAN PACIFIC & SOUTHEAST ASIA WOMEN'S ASSOCIATION
1 Lexington Avenue
New York - NY, United States of America
T: 1 212 228-5307 F: 1 212 473-0942

Linda, Andy
Cameraman Operator
NORTHCUTT PRODUCTIONS
2966 Diamond St. #171
San Francisco - CA, United States of America
T: 1 415 3377044 F: 1 415 3370630

Lindahl, Lena, Ms.
Coordinator
GLOBE
409 3rd St. SW #204
Washington - DC 20003, United States of America
T: 1 202 863 0153 F: 1 202 479 9447

Lindquist, Russell, Dr.
UNITED NATIONS OF THE US
126 River Alto Canal
Long Beach - CA 90803, United States of America
T: 1 310 433-7231 F: 1 310 433-6318

Lini, Walter H., Mr.
Father
GLOBAL HEART/ VANUATU
1880 East River Road
Tucson - AZ, United States of America
T: 1 602 577-6476 F: 1 602 577-1195

Linz, Helen, Ms.
TEMPLE OF UNDERSTANDING
St. Emeric Church / 185 Avenue D
New York - NY 10009, United States of America
T: 1 212 353-0001 F: 1 212 353-0170

Linz, Werner, Mr.
TEMPLE OF UNDERSTANDING
St. Emeric Church / 185 Avenue D
New York - NY 10009, United States of America
T: 1 212 353-0001 F: 1 212 353-0170

Lippert, Merrilee, Ms.
Press
CAMPAIGN FOR THE EARTH
50 Graeser Acres
St. Louis, MO 63146, United States of America
T: 1 314 434-6580 F: 1 314 567-3107

Litenski, Chaim
UN DEPARTMENT OF PUBLIC INFORMATION
Room S-1037D
New York _NY 10017, United States of America
T: 1 212 9638842 F: 1 212 9636914

Little, Kathryn, Ms.
YMCA EARTH ACTION CORPS
909 4th Av
Seattle - WA, United States of America
T: 1 206 382-5013 F: 1 206 382-7283

Livengood, David L., Mr.
Environmental Specialis / Res.
UNIVERSITY OF OREGON
1230 Franklin Blvd.
Eugene - OR, United States of America
T: 1 503 346-3197 F: 1 503 726-0816

Lloveras, Ramon, Mr.
Abogado
ESTADO LIBRE ASOCIADO DE PUERTO RICO
Rey Luis 433 la Villa De Torrimar
Guaynabo 00969, United States of America
T: 809 759 3115 F: 809 767 3943

Lloyd, Judy Amber, Ms.
EARTH SAVE- LOS ANGELES
1880 East River Road
Tucson - AZ, United States of America
T: 1 602 577-6476 F: 1 602 577-1195

Loaney, Jenifer, Ms.
UP WITH PEOPLE
3103 North Campbell Av.
Tucson - AZ 85719, United States of America

Locke, Kevin, Mr.
Representative
BAHA'I INTERNATIONAL COMMUNITY
866 UN Plaza, Suite 120
New York - NY 10017, United States of America
T: 1 212 7563500 F: 1 212 7563573
E: 152:CNM057

Locke, Patricia, Ms.
BAHA'I INTERNATIONAL COMMUNITY
866 UN Plaza, Suite 120
New York - NY 10017, United States of America
T: 1 212 7563500 F: 1 212 7563573
E: 152:CNM057

Locke, Waniya, Ms.
BAHA'I INTERNATIONAL COMMUNITY
866 UN Plaza, Suite 120
New York - NY 10017, United States of America
T: 1 212 7563500 F: 1 212 7563573
E: 152:CNM057

Locklin, Ken, Mr.
APPROPRIATE TECHNOLOGY INTERNATIONAL
1331 H St. NW
Washington - DC, United States of America
T: 1 202 8792900 F: 1 202 6284622
TL: 64661 ATI

Lodico, Yvonne, Ms.
INTERNATIONAL INSTITUTE FOR LAW, SCIENCE AND TECHNOLOGY
2871 Madison Av. Suite 908
New York - NY, United States of America
T: 1 212 340-4150 F: 1 212 297-0967

Loewenstein, Louis K., Mr.
Member Board Of Directors
UNITED STATES ASSOCIATION FOR THE CLUB OF ROME
3858 Jackson Street
San Francisco - CA, United States of America
T: 1 415 387-2600 F: 1 1415

Logan, Ronald, Mr.
PROUTIST UNIVERSAL
1354 Montague St., NW, P.O. Box 56466
Washington - DC 20040, United States of America
T: 1 202 829-2278 F: 1 202 829-0462
TL: 150255668

Logreira, Margaret, Ms.
Administrative Assistant
CENTER FOR THE STUDY OF WOMEN AND SOCIETY
City University of New York, 33 West 42nd Street
New York - NY, United States of America
T: 1 212 6422954 F: 1 212 6422642

Lonergan, Dennis
INDIGENOUS PEOPLES MEDIA CENTER
77 Main Street
New York 10083, United States of America
T: 914 3650414 F: 914 3650864

Long, Donna, Ms.
Photographer
JERRY BROWN PRESIDENTIAL
120 Seabreeze Ave.
Palm Beach, FL 33480, United States of America
T: 1 407 659 5670 F: 1 407 659 3191

Longley, Robert, Mr.
Partner, Co-Founder
WORLD INTUACTION
2696 Centennial Court
Alexandria-VA 22311, United States of America
T: 1 703 820-7826 F: 1 703 659-3761

Longstreth, Molly, Ms.
UNITED THEOLOGICAL SEMINARY
1810 Harvard BLVD.
Dayton, OH, United States of America
T: 1 513 2785817 F: 1 513 2781218

Looye, Johanna W., Ms.
CORNELL /UFRJ
School of Planning, University of Cincinnati
Cincinnati - OH, United States of America
T: 1 513 556-0216 F: 1 513 556-3288

Lopez, Guadalupe, Ms.
Prog. Associate
UNITARIAN UNIVERSALIST SERVICE COMMITTEE
130 Prospect Street
Cambridge - MA 02139, United States of America
T: 1 617 868.6600 F: 1 617 868.7102

Lousma, Jack, Mr.
President
CIESIN
1968 Green Road
Ann Arbor - MI, United States of America
T: 1 313 775 6600

Low, Patricia, Ms.
INTERNATIONAL EDUCATION CENTER
309 Charles Street Ave. #E
Baltimore, Md. 21204, United States of America

Lowthian, Douglas, Mr.
119 Brainerd Rd, Apt 15
Allston - MA 02134, United States of America
T: 1 617 232-7424

Lowthian, Elizabeth, Ms.
119 Brainerd Rd, Apt 15
Allston - MA 02134, United States of America
T: 1 617 232-7424

Lucca, Liliana De
UN DEPARTMENT OF PUBLIC INFORMATION
New York, NY, United States of America

Luckett, Dolores Garcia, Ms.
Int'L Sales Coordinator
ERDAS INC.
2801 Buford Highway, Suite 300
Atlanta - GA 30329, United States of America
T: 1 404 248-9000 F: 1 404 248-9400

Lulic, Shirley J., Ms.
Vice-President
PPSEWA / DELEGATE
3440 Golfview Dr., 105
Eagan - MN 55123, United States of America
T: 1 612 454-0128

Lunceford, Miguel, Mr.
Research Associate
DAEDALUS EDUCATION FOUNDATION
12702 Via Cortina
Del Mar - CA, United States of America
T: 1 619 7930411 F: 1 619 7930523

Lund, Heidi, Ms.
UP WITH PEOPLE
3103 North Campbell Av.
Tucson - AZ 85719, United States of America

Lundberg, Jan Christian, Mr.
President
FOSSIL FUELS POLICY ACTION INSTITUTE
761 8th street P.O. Box 4347
Arcata - CA 95521, United States of America
T: 1 707 826-7775 F: 1 707 822-7007

Lundin, Anders, Mr.
SWEDISH CAMPAIGN FOR ENVIRONMENT AND DEVELOPMENT-92
C/O Swedish Embassy, 600 New Hampshire Ave., N.W.
Washington - DC, United States of America

Lundin, Michael, Mr.
SWEDISH CAMPAIGN FOR ENVIRONMENT AND DEVELOPMENT-92
C/O Swedish Embassy, 600 New Hampshire Ave, NW
Washington - DC, United States of America

Lutzenberger, Jose, Mr.
GLOBAL HEART
1880 East River Road
Tucson - AZ, United States of America
T: 1 602 577-6476 F: 1 602 577-1195

Lynn, Nina, Ms.
CAMPAIGN FOR THE EARTH
2120 13th Street
Boulder - CO 80302, United States of America
T: 1 303 444-3800 F: 1 303 444-3811

Lyons, Stephen, Mr.
President
GREEN HEART AMERICA
5715 SE Taylor Street
Portland - OR 97401, United States of America
T: 1 503 239-4344 F: 1 503 239-4344

Macgowan, Paul D.
Videographer
WCAX - TV (CBS)
PO Box 608
Burlington VT 0540, United States of America
T: 1 802 6586300 F: 1 802 6586300

Machado, Beatriz
TERRA CHRISTA COMMUNICATIONS
1880 East River Road
Tucson - AZ, United States of America
T: 1 602 577-6476 F: 1 602 577-1195

Mackenzie, Ross, Mr.
Director Religion Dept
CHAUTAUQUA INSTITUTION
Chautauqua Institution
Chautauqua - NY 14722, United States of America
T: 1 718 3576242 F: 1 718 3579014

Macnow, Alan, Mr.
Exec. Director
FRIENDS OF WHALERS
321 East 53TH Street
New York - NY 10022, United States of America
T: 1 212 688-5580 F: 1 212 688-5857

Macs, Dina, Ms.
Intl Liason
YMCA EARTH ACTION CORPS
909 4th Av
Seattle - WA, United States of America
T: 1 206 382-5013 F: 1 206 382-7283

Madera, Oscar, Mr.
Staff
EARTH TRAIN/WORLDLINK/GLOBAL YOUTH SUMMIT
3220 Sacramento Street
San Francisco - CA 94115, United States of America
T: 1 415 931-1290

Maeda, Hitomi, Ms.
Director
HONNOKI USA
300 Broadway, Suite 39
San Francisco - CA, United States of America
T: 1 415 392-3151 F: 1 415 392-3151

Magalhaes, Enaldo
TERRA CHRISTA COMMUNICATIONS
1880 East River Road
Tucson - AZ, United States of America
T: 1 602 577-6476 F: 1 602 577-1195

Magnuson, Nicole, Ms.
Consultant
4135 Gilbert St.
Oakland - CA 94611, United States of America
T: 1 510 6541930 F: 1 510 643-8683
E: UEXTBONN2GARNET.BERKELEY

Maguire, Andrew
President
APPROPRIATE TECHNOLOGY INTERNATIONAL
1331 H St. NW
Washington - DC, United States of America
T: 1 202 8792900 F: 1 202 6284622
TL: 64661 ATI

Mahar, Dennis, Mr.
WORLD BANK
1818 H Street
New York - NY 20433, United States of America

Mahdi, Sulaiman
HIGHLANDER CENTER
1939 Highlander Way
New Market - TN 37820, United States of America
T: 1 615 933 3443 F: 1 615 933 3424

Mahinpey, Pari
PEACE 12 12
P.O.Box 78813
Tucson - AZ, United States of America
T: 1 602 577-6476 F: 1 602 577-1195

Mainhardt, Heike, Ms.
GREEN (GLOBAL RIVERS ENVIRONMENTAL EDUCATION NETWORK)
University of Michigan School of Natural Resouces -Dana Building 430E
Aracaju - SE, United States of America
T: 55 7645171 F: 55 9362195

Malte, , Ms.
GEOSPHERE PROJECT, THE
146 Entrada Drive
Santa Monica - CA 90402, United States of America
T: 1 310 459-4342 F: 1 310 459-8299

Maldonado Ojeda, Pedro A., Mr.
Presidente
JUNTA DE CALIDAD AMBIETNTAL DE PR
Ave. Ponce De Leon 431,, Hato Rey
San Juan, Puerto Rico, United States of America
T: 1 809 767-8056 F: 1 809 767-8348

Malebanye, Chieftaness Mashoaepane
GREEN BELT MOVEMENT OF KENYA
c/o Resource Renewal Institute, FT. Mason Center Bldg A
San Francisco - CA, United States of America
T: 1 415 928-32774 F: 1 415 928-6529

Malik, Masood K., Mr.
Advisor
SEED
1709 Foxhall Road N.W.
Washington - DC 20007, United States of America
T: 1 202 337-2770 F: 1 202 337-3545

Malone, Marjorie, Ms.
Faculty
UNIVERSITY OF CALIFORNIA AT LOS ANGELES
PO Box 1737
El Prado - NW, United States of America

Malone, Rosalie A., Ms.
SIMA XI, THE SCIENTIFIC RESEARCH SOC
99 Alexander Drive, PO Box 13975, Research Triangle Park
Research Triangle Pk 27709, United States of America
T: 1 5494691 F: 1 5490090

Mandi, Malati, Ms.
UNITED NATIONS DEVELOPMENT FUND FOR WOMEN
304 East 45th St.
New York - NY 10017, United States of America
T: 1 212 9066400 F: 1 212 9066705

Manley, Susane, Ms.
GREEN PARTY OF CALIFORNIA
8745-F Woodman Way
Sacramento - CA 95826, United States of America
T: 1 916 368-0245

Mansfield, Susanna Colloredo, Ms.
Board Member
LINCOLN FILENE CENTER TUFTS UNIVERSITY
Winthrop St. / So
Hamilton - MA 01982, United States of America
T: 1 508 468-3310 F: 1 508 468-3490

Marcello, Richard, Mr.
L.A. Ca. Representative
HELP END MARIJUANA PROHIBITION - HEMP
5632 Vannuys Blvd. 310
Vannuys CA 91401, United States of America
T: 392-1806

Marcus, David Tex
Bureau Chief
DALLAS NEWS
508 Young Street
Dallas TX 75265, United States of America
T: 1 214 9778089 F: 1 214 9778089

Marden, Mark, Mr.
EARTH ISLAND INSTITUTE
300 Broadway, Suite 28
San Francisco - CA 94133, United States of America
T: 1 415 788-3666 F: 1 415 788-7324
TL: 6502829302

Margarida-Lloveras, Teresa, Ms.
Customer Service Manager
PROCTER & GAMBLE
Rey Luis 433, la Villa De Torrimar
GUAYNABO, Puerto Rico 00969, United States of America
T: 1 809 721-7070 F: 1 809 729-7310

Margulis, Sergio, Mr.
Representative
WORLD BANK
1818 H St. NW
Washington - DC 20433, United States of America
T: 1 202 458-1943

Marks, Claudia, Ms.
TEMPLE OF UNDERSTANDING
St. Emeric Church / 185 Avenue D
New York - NY 10009, United States of America
T: 1 212 353-0001 F: 1 212 353-0170

Marks, Leonard, Mr.
TEMPLE OF UNDERSTANDING
St. Emeric Church / 185 Avenue D
New York - NY 10009, United States of America
T: 1 212 353-0001 F: 1 212 353-0170

Marlow, Nancy Harrer, Ms.
Conference Coordinator
INTERNATIONAL SYNERGY INSTITUTE
19 A Randall St.
Annapolis, United States of America
T: 1 410 280-9083

Marques, Victor
TURNER BROADCASTING SYSTEM / CNN
One CNN Centre
Atlanta - GA, United States of America
T: 1 404 827-3973 F: 1 404 827-1015

Marsden, Claire, Ms.
TERRA CHRISTA COMMUNICATIONS
1880 East River Road
Tucson - AZ, United States of America
T: 1 602 577-6476 F: 1 602 577-1195

Marsh, Langdon, Mr.
ENVIRONMENTAL LAW INSTITUTE
1616 P.Street, N.W., Suite 200
Washington - DC 20036, United States of America
T: 1 202 939-3853 F: 1 202 328-5002

Martin-Brown, J., Ms.
Director
U.N.E.P.
5508 39th Street NW
Washington - DC, United States of America
T: 1 202 289-8956 F: 1 202 289-4267

Martinez, Neftali Garcia, Mr.
Director
SERVICIOS CIENTIFICOS Y TECNICOS
Cond El Centro I Of 607
San Juan, Puerto Rico 00918, United States of America
T: 1 809 7598787 F: 1 809 7676757

Martino, Maria De, Ms.
EARTH ANGELS
7240 Briarcliff Rd. - Route 3
Dora - AL 35064, United States of America
T: 1 205 648-3426

Martins, Michael De, Mr.
President
FOUR CORNERS PROJECT
22 & San Francisco St. #438
Flagtaff - AZ, United States of America
T: 1 602 773-9669 F: 1 602 774-9074

Mascarenhas, Andrea Marques, Ms.
Intern
CULTURAL SURVIVAL
53 A Church Street Cambridge, MA
Cambridge - MA 02138, United States of America
T: 1 617 495-2562 F: 1 617 495-1396

Mastro, Sal Del
ABC NEWS
47 West 66th St
New York NY 10023-6, United States of America
T: 1 212 4562800

Matarazzo, Trisha, Ms.
Participant
EARTH TRAIN/WORLD LINK GLOBAL YOUTH SUMMIT
11140 Arco Dr.
Escondido - CA 92026, United States of America
T: 1 619 741-9523 F: 1 619 741-9573

Mathews, Dan, Mr.
P E T A
4928 Wyaconda Rd.
Rockville-MD 20852, United States of America
T: 1 301 770-7444 F: 1 301 770-8969

Mathews, Jessica T., Dr.
Vice President
WORLD RESOURCES INSTITUTE
1709 New York Ave., NW, Suite 700
Washington - DC, United States of America
T: 1 202 638-6300 F: 1 202 638-0036
TL: 64414WRIWASH

Matlock-Merla, Staci
Editor
SONORON JOURNAL
Box 258, 8963 E. Taigue Verde Rd.
Tucson AZ 85749, United States of America
T: 1 602 7493931

Matos, Milly, Sra.
Asesora Del Presidente
CAMARA DE REPRESENTANTES DE PUERTO RICO
Apartado 2228 Capitolio
San Juan, Puerto Rico, United States of America
T: 1 809 724-4464 F: 1 809 723-4711

Matsumura, Maki, Ms.
Assistant
GLOBAL FORUM OF SPIRITUAL AND PARLIAMENTARY LEADERS ON HUMAN SURVIVAL
304 East 45th St. Room 476
New York - NY, United States of America
T: 1 212 953-7947 F: 1 212 557-20961

Mattos, Maril M., Sra.
Pesquisadora
WOODS HOLE RESEARCH CENTER
P.O.Box 296 Woods Hole
Woods Hole - MA, United States of America
T: 1 508 540-9900

Mauldin, Corrie
BOILINGSTONE
P.O.BOX 12199
San Antonio - TX 78212, United States of America
T: 512 7369617 F: 512 7357033

Mavroudi, Maria, Ms.
Professor
NEW YORK UNIVERSITY
General Studies Program ,239 Greeve St. - Room 810
New York - NY 10012, United States of America
T: 1 212 998 5696 F: 1 212 995 4193

Mawlawi, Farouk
UN DEPARTMENT OF PUBLIC INFORMATION
New York, NY, United States of America

Max, Lisa, Ms.
President
BETTERWORLD INC.
2501 E. Commercial Blvd. Suite 201
Ft. Lauderdale - FL 33808, United States of America
T: 1 305 776-3390 F: 1 305 776-9005

Maxx, Suzanne
Producer
BROWN FOR PRESIDENT/WORLD TEAM
3345 Oak Glen Dr.
Los Angeles - CA 90056, United States of America

Mc Daniel, Robert, Mr.
Member
ILSI - INT'L LIFE SCIENCE INSTITUTE
P.O.Drawer 1734
Atlanta - GA 30301, United States of America
T: 1 404 676-1212 F: 1 404 676-6275

McAlevey, Jane, Ms.
HIGHLANDER CENTER
1939 Highlander Way
New Market - TN 37820, United States of America
T: 1 615 933 3443 F: 1 615 933 3424

McAlister, Roy E., Mr.
President
AMERICAN HYDROGEN ASSOCIATION
219 S. sieta Ln, suite 101
Temple - AZ 85281, United States of America
T: 1 602 921-0443 F: 1 602 967-6601

McCabe, Ron, Mr.
President And Project Coordina
RAINFOREST ALLIANCE INFORMATION NETWORK
923 East John St. #302
Seattle - WA 98102, United States of America
T: 1 206 3247163 F: 1 206 5621727

McCandliss, John
INDIGENOUS PEOPLES MEDIA CENTER
77 Main Street
New York 10083, United States of America
T: 914 3650414 F: 914 3650864

McCann, Nancy Wilson, Ms.
GLOBAL ENVIRONMENT TEAM
696 South Madison Avenue
Pasadena - CA 91106, United States of America
T: 1 818 796-2951 F: 1 818 792-2121

McCarthy, Patricia L., Ms.
DE PAUL UNIVERSITY
25 E Jackson
Chicago - IL 60604, United States of America
T: 1 312 362-5258 F: 1 312 362-6908

McCarthy, Thomas, Mr.
LEMNA CORPORATION
St. Paul - MN, United States of America
T: 1 612 688-0876 F: 1 612 688-8813

McClary, Eric, Mr.
FORESTA INSTITUTE FOR OCEAN & MOUNTAIN STUDIES
PO Box 41567
Tucson - AZ 85717, United States of America
T: 1 602 8816174 F: 1 602 3229850

McClean, Oronde, Mr.
GLOBAL ENVIRONMENT TEAM
696 South Madison Avenue
Pasadena - CA 91106, United States of America
T: 1 818 796-2951 F: 1 818 792-2121

McClellan, Jo Eva, Sra.
Bm
UNITED METHODIST CHURCH
807 Main Street, Box 248
Palco - KS 67657, United States of America

McClugston, Richard
Reporter
INDIGEONOUS PEOPLE'S MEDIA CENTER
Crle 2100 L St NW
Washington DC 20037, United States of America
T: 1 202 4521100 F: 1 202 7786132

McCully, Shannon, Ms.
GLOBAL ENVIRONMENT TEAM
P.O. Box 9285
Monterey - CA 93942, United States of America
T: 1 408 646-9768

McDonald, David C., Mr.
SIERRA CLUB - MINNESOTA
2200 First National Bank Bldg.
St. Paul - MN 55101, United States of America
T: 1 612 222-6212 F: 1 612 223-6450

McDonald, Norm, Mr.
Director
NORTHERN TELECOM
Miami Lakes Drive
Miami Lakes - FL, United States of America
T: 1 305 526-8483

McElroy, Susan, Ms.
Coordinator
WORLDWIDE NETWORK
1331 H Street, Suite 903
Washington - DC 20005, United States of America
T: 1 202 347-1514 F: 1 202 347-1524

McGreal, Shirley, Ms.
Chairwoman
INTERNATIONAL PRIMATE PROTECTION LEAGUE
POB 766 Protection League
Summerwell - SC 29484, United States of America
T: 803 871 2280 F: 803 871 7988

McIvor, Daniel
Cameraman
CNN BROADCASTING SYSTEM
One CNN Center, Box 105366
Atlanta - GA, United States of America
T: 1 404 8371918 F: 1 404 8274292

McJunkin, James, Mr.
HOPE PUBLISHING HOUSE
4411 E. 175th St.
Cleveland - OH 44128, United States of America
T: 1 216 283-4365

McKenney, Joan, Ms.
Field Rep
ONE WORLD NOW
1510 Lyric Center, 440 Louisiana
Houston - TX 77002, United States of America
T: 1 713 221-2288

McKeon, Isadora Zoe, Ms.
SO.CALIF. ECOLOGY TASK FORCE
9203 Cord Ave.
Downey - CA 90240, United States of America
T: 1 310 861-3207

McLoed, Teri, Ms.
UP WITH PEOPLE
3103 North Campbell Av.
Tucson - AZ 85719, United States of America

McMahon, James P., Mr.
BOILINGSTONE
P.O.Box 12199
San Antonio - TX 18212, United States of America
T: 1 512 7369617 F: 1 512 7357033

McNamara, Kristin
TEMPLE OF UNDERSTANDING
St. Emeric Church / 185 Avenue D
New York - NY 10009, United States of America
T: 1 212 353-0001 F: 1 212 353-0170

McNamara, Patrick, Mr.
ORGANISATIONAL TRANSFORMATION NETWORK/THE BUSINESS NETWORK
19645 Detroit Road, Suite 155
Rocky River, OH, United States of America
T: 1 216 333 4812

McNealey, Connie, Ms.
Staff
SAVE OUR CUMBERLAND MOUNTAINS
Rt. 1, Box 375
Jellico - TN 37762, United States of America
T: 1 784 9374

McNeil, Mary, Ms.
Public Affairs
URBAN MANAGEMENT PROGRAM / WORLD BANK
1818 H Street NW
Washington - DC 20433, United States of America
T: 1 202 4733861 F: 1 202 4771391

McPike, Barbara, Ms.
PEACEABLE REALM FOUNDATION
4329 Brookshire Circle
Santa Rosa - CA 95405, United States of America

Meegan, Anne, Ms.
ZONTA INTERNATIONAL
557 W.Randolph Street
Chicago - IL 60661-2206, United States of America
T: 1 312 930-5848 F: 1 312 930-0951

Meighan, Matt, Mr.
PROGRESS AGENCY
855 Orman Dr.
Boulder - CO 80303, United States of America
T: 1 303 4994672

Meisen, Peter, Mr.
Director
GLOBAL ENERGY NETWORK INT'L
PO Box 81565
San Diego - CA 92138, United States of America
T: 1 619 595-0139 F: 1 619 595-0403

Melli, Frank
PROGRESS AGENCY INC
1 Union square West 201
New York -NY 10003, United States of America
T: 1 212 9246550 F: 1 212 6456866

Meltzer, Erica
UN DEPARTMENT OF PUBLIC INFORMATION
New York, NY, United States of America

Mendonza, Roberto, Mr.
Coord. Committee
GREENS / GREEN PARTY U.S.A.
2 Florence St.
Boston - MA, United States of America
T: 1 617 327-1638 F: 1 617 357-4410

Menge, Mike, Mr.
Director
ALASKA DEPARTMENT OF ENVIRONMENT CONVERVATION
1105 Mendenhall Penn Road
Juneau - Alaska 99801, United States of America
T: 1 907 790-2023 F: 1 907 465-5274

Menotti, Victor A., Mr.
Info Coordinator
RAINFOREST ACTION NETWORK
450 Sansome St. Suite 700
San Francisco - CA, United States of America
T: 1 415 398-4404 F: 1 415 398-2732

Mensah, Christian
PERMANENT MISSION OF VANUATU TO THE U.N
416 convent Ave.
New York - NY 10031, United States of America
T: 1 212 577-6476 F: 1 212 577-1195

Merlin, Mark, Mr.
UNIVERSITY OF HAWAII
General Science Department, Dean Hall 2
Honolulu - HI 96822, United States of America
T: 1 808 956 6038 F: 1 808 956 4745

Merritt, Justine, Ms.
HOPE PUBLISHING HOUSE
235 White Ave.
Grand Junction - CO 81501, United States of America
T: 1 303 243-1110

Messer, Bonnie, Ms.
7327 S. Newport Way
Englewood - CO 80112, United States of America
T: 1 303 744-1287

Messer, Donald E., Mr.
UNITED METHODIST CHURCH OF BRAZIL
2201 S.University Blvd.
Denver, Colorado 80210, United States of America

Messer, Kent, Ms.
7327 S. Newport Way
Englewood - CO 80112, United States of America
T: 1 303 744-1287

Meyer, Kay
NEW AGE STUDY OF HUMANITYIES PURPOSE
9010 E. Kenyon
Tucson - AZ 85710, United States of America

Meyers, Chuck, Mr.
TERRA CHRISTA COMMUNICATIONS
1880 East River Road
Tucson - AZ, United States of America
T: 1 602 577-6476 F: 1 602 577-1195

Meyers, Erik, Mr.
General Counsel
ENVIRONMENTAL LAW INSTITUTE
1616 P. Street, N.W., Suite 200
Washington - DC 20036, United States of America
T: 1 202 939-3853 F: 1 202 328-5002

Michael, Andy, Mr.
WORLD SPIRITUAL ASSEMBLY
State Capitol, Room 6026
Sacramento - CA 95814, United States of America
T: 1 916 445-4253

Michael, Donald, Mr.
Senior Research Associate
COMMONWEAL
PO Box 316
Bolinas - CA, United States of America
T: 1 415 8680970 F: 1 415 8682230

Michelle, Mark De, Mr.
Ceo
ARIZONA PUBLIC SERVICE COMPANY
400 North 5th St.
Phoenix - AZ, United States of America
T: 1 602 508-5460 F: 1 602 508-5403

Migdalia, Del Valle, Ms.
Specialist
DEPARTMENT OF NATURAL RESOURCES
Miramar
San Juan, Puerto Rico, United States of America
T: 1 809 723-3090 F: 1 809 721-5984

Miller, Anita M, Ms.
Executive Director
CALIFORNIA WOMEN LAWYERS
3028 Huntington Road
Sacramento - CA, United States of America
T: 1 916 4413703 F: 1 916 4415723

Miller, Jacqueline, Ms.
WORLD SPIRITUAL ASSEMBLY
687 Spindrift Way
Half Moon Bay - CA 94019, United States of America
T: 1 415 712-1222

Miller, Richard Gordon, Mr.
Ngo Delegate
FORESTA INSTITUTE FOR OCEAN & MOUNTAIN STUDIES
PO Box 41567
Tucson - AZ 85717, United States of America
T: 1 602 8816174 F: 1 602 3229850

Miller, Vernice, Ms.
U.S CITIZENS NETWORK
1880 East River Road
Tucson - AZ, United States of America
T: 1 602 577-6476 F: 1 602 577-1195

Milley, Jennifer, Ms.
UP WITH PEOPLE
3103 North Campbell Av.
Tucson - AZ 85719, United States of America

Milligan, Nancy, Ms.
GLOBAL ENVIRONMENT TEAM
2266 S. Columbine St.
Denver - CO 80210, United States of America
T: 1 303 733-1454

Milliken, Irene, Ms.
GLOBAL ENVIRONMENT TEAM
2 Northgate Village
Media - PA 91106, United States of America

Mills, Deborah, Ms.
796 A Juniper Walk
Goleta - CA, United States of America
T: 1 805 968-8493

Mills, Julia, Ms.
Correspondent
DELAWARE VALLEY RAINFOREST ACTION GROUP
3707 Elmhurst Avenue
Bristol - PA 19007, United States of America
T: 1 215 788-9422

Minton, Jonas, Mr.
Manager
CALIFORNIA DEPT. OF WATER RESOURCES
1416 - 9TH Str., Room 804
Sacramento - CA 95814, United States of America
T: 1 916 653-9167 F: 1 916 653-4275

Mische, Gerald, Mr.
President
GLOBAL EDUCATION ASSOCIATES
475 Riverside Drive, Suite 456
New York - NY 10115, United States of America
T: 1 212 8703290 F: 1 212 8702055

Mische, Jerry, Mr.
GLOBAL EDUCATION ASSOCIATION
1880 East River Road
Tucson - AZ, United States of America
T: 1 602 577-6476 F: 1 602 577-1195

Mitchell, Konny Light, Ms.
Director
MITCOT- DAWLWE INTERNATIONAL, INC./ SOURCE CONSULTANTS INTERNATIONAL
2285 Peach Tree Rd NE Suite 210
Atlanta - GA, United States of America
T: 1 404 351-6204 F: 1 404 351-6083

Miura, Eutah, Mr.
JAPAN ENVIROMENT EXCHANGE
2425 B. Channing Way
Berkley, CA, United States of America
T: 510 5491206 F: 510 549 1206

Mocuim, Melony
Audio Technician
NORTHCUTT PRODUCTIONS
2966 Diamond St. #171
San Francisco - CA, United States of America
T: 1 415 3377044 F: 1 415 3370630

Moeira, Ana Cristina Leme
Journalist
SMITHSONIAN TROPICAL RESEARCH INSTITUTE
Unit 0946, APO AA 0948
United States of America
T: 1 507 525539 F: 1 507 625942

Moir, Matthew, Mr.
Student
LAUREL SPRINGS SCHOOL
PO Box 1440
Ojai - CA 93024, United States of America
T: 1 805 646-2473 F: 1 805 646-0186

Moitoza, Kaaren
Musician
DOVE & CO./MMH
4759 15th Av. NE
Seattle - WA 98105, United States of America
T: 1 206 344-6321

Monin, Noel, Mr.
YMCA EARTH ACTION CORPS
909 4th Av
Seattle - WA, United States of America
T: 1 206 382-5013 F: 1 206 382-7283

Monte, Gilberto, Mr.
Producer
HOMETOWN FILMS
8306 Wilshire Blvd Suite 254
Beverly Hills - CA, United States of America
T: 1 655 9403 F: 1 655 2434

Monteiro, Fernando
Cameraman
WORLD MEDIA ASSOCIATION
1012 14th Street, NW
Washington DC 63826, United States of America
T: 1 202 6382662 F: 1 202 6380638

Montez, Ceasar, Mr.
Assistant
TRADITIONAL HOPI & DINEH ELDERS
P.O. Box 1178
WINSLOW- Arizona 86047, United States of America
T: 602 289 4968 F: 602289 4968

Montgomery, Katy, Ms.
Reporter
INTERNATIONAL RIVERS NETWORK
1847 Berkeley Way
Berkeley - CA, United States of America
T: 1 510 848-1155 F: 1 510 848-1008

Mooneyhan, Dee, Ms.
A.A
UNEP-GRIP
1104 W.57th St. 302A
Sioux Falls-S.Dakota 57106, United States of America
T: 001 605 594-6107 F: 001 605 594-6589

Mooneyhan, Wayne, Mr.
Manager
UNEP-GRID
1104 W 57th ST 302 A
Sioux Falls-S.Dakota 57106, United States of America
T: 001 605 5946107 F: 001 605 5946589

Moonstar, Joy, Ms.
Global Resource Specialist
AVEDA
4000 Pheasant Ridge Drive
Minneapolis - MN 55434, United States of America
T: 1 612 783-4110 F: 1 800 283-3224

Moore, Calvin, Mr.
716 Georgetown
Nashville - TN, United States of America
T: 615352 6047

Moore, Christopher T., Dr.
WILDERNESS MEDICAL SOCIETY
c/o Aiken Clinic, 4066 Hixson Clinic
Chahanooga - TN 37415, United States of America
T: 1 615 875-0999 F: 1 615 875-0896

Moore, Marjorie, Ms.
WOMEN'S ENVIRONMENT AND ECONOMIC DEU
850 Third Ave.
New York - NY 10010, United States of America
T: 1 212 481-4355

Moore-Kochlacs, Peter, Mr.
SO.CALIF. ECOLOGY TASK FORCE
1725 Sunny Park
Redlands - CA 92374, United States of America
T: 1 714 794-8898

Moorhead, Erica S., Ms.
Associate
ECO - SPHERE - JB PRODUCTIONS
4836 Loughbord Rd
Washington - DC 20016, United States of America
T: 1 202 3627735 F: 1 202 3625625

Moraes, Luzia, Sra.
GLOBAL HEART
1880 East River Road
Tucson - AZ, United States of America
T: 1 602 577-6476 F: 1 602 577-1195

Moreria, Wilma, Ms.
HIGHLANDER CENTER
1939 Highlander Way
New Market - TN 37820, United States of America
T: 1 615 933 3443 F: 1 615 933 3424

Morgan, David
Tv Correspondent
AUST BROADCASTING COMMISSION
2030 M Street NW 702
Washington - DC, United States of America
T: 1 202 466 8575 F: 1 202 775 8291

Morgan, Marlys, Ms.
Council Member
ART OF LIVING CENTER
516 W.Cherry, 9
Flagstaff, United States of America
T: 1 602 779-2203

Morgan, Maurene M., Ms.
IA. DIVISION, U.N. ASSOCIATION- USA
20 E. Market St
Iowa City, Iowa 52245, United States of America
T: 1 319 337-7290

Morgan, Tom, Mr.
Director
PISCATAQUA PLANNING
P.O. Box 4121
Portsmouth - NH 03801, United States of America
T: 1 603 4309283

Moriarty, Burke, Mr.
Vice President
INTERNATIONAL CHLORINE INC.
4309 32nd Road South
Arlington - VA 222069, United States of America
T: 1 703 379-4280 F: 1 703 845-1660

Morimoto, Sara, Ms.
Support
RAINFORST ACTION NETWORK
450 Sausome, Suite 700
San Francisco - CA 94111, United States of America
T: 1 415 398-4404 F: 1 415 398-2732
TL: 151276475

Morris, Langdon, Mr.
GLOBAL YOUTH SUMMIT
209 Lasso Circle
San Ramon - CA 94583, United States of America
T: 1 510 830-1083

Morrison, Dorothy, Ms.
NACRE
11 Collins Av.
Clearfield - PA 16830, United States of America
T: 1 814 765-1491

Morrison, Sherman, Mr.
Dir. Of Administration
NORTH AMERICAN COALITION ON RELIGION AND ECOLOGY - CORE-INT'L
5 Thomas Circle - NW
Washington - DC 20005, United States of America
T: 1 202 462-2591 F: 1 202 234-0307

Moseley, Jennifer, Ms.
UP WITH PEOPLE
3103 North Campbell Av.
Tucson - AZ 85719, United States of America

Mosley, Marilyn, Ms.
Director
LAUREL SPRINGS SCHOOL
PO Box 1440
Ojai - CA 93024, United States of America
T: 1 805 646-2473 F: 1 805 646-0186

Mosley, Ramaa, Ms.
LAUREL SPRINGS SCHOOL
P. O. Box 1440
Ojai - CA 93024, United States of America
T: 1 805 646-2473 F: 1 805 646-0186

Moss, Deborah Kamlani, Ms.
Associate Publisher
E MAGAZINE
P.O.Box 5098
Westport - CT 06881, United States of America
T: 1 203 8545559 F: 1 203 8660602

Mouche, Carol
Editor
ENVIRONMENTAL PROTECTION MAGAZINE
225 North New Rd
Waco Texas 76710, United States of America
T: 1 817 7515163 F: 1 817 7769018

Mpofu, Shahadat
Editor
GLOBAL EDITION
1875 Connecticut Ave, NW Suite 520
Washington - DC, United States of America
T: 1 202 483 1100 F: 1 202 328 3937

Mudd, Alexandra R, Ms.
Senior Research Associate
COMMONWEAL
PO Box 316
Bolinas - CA, United States of America
T: 1 415 8680970 F: 1 415 8682230

Mueller, Mary, Ms.
2221 Monica Place
Sarasota - FL 33580, United States of America

Muhlenberg, Nicholas, Mr.
President
ENVIROMENTAL PRODUCTIONS INC./ PRESIDENT
326 S. Swedley St
Philadelphia - PA 19103, United States of America
T: 1 215 735 7103

Muir, Ruth, Ms.
Representative
HOWEL CITY AAUW
6 Glendale Court
Iowa City, Iowa, United States of America
T: 1 319 337-7387

Muller, Robert, Mr.
Fmr. Asst. Sec. Gnl. To U.N
GLOBAL HEART
1880 East River Road
Tucson - AZ, United States of America
T: 1 602 577-6476 F: 1 602 577-1195

Munasinghe, Mohan, Mr.
Div. Chief
WORLD BANK
1818 H. Street NW
Washington - DC 20433, United States of America
T: 1 202 473-3204 F: 1 202 477-0968

Muniz, Maria Cecilia C. P.
Reporter
INT. INST. OF LAW, SCIENCE AND TECHNOLOGY
271 Madison Ave Suite 908
New York NY, United States of America
T: 1 212 3404150 F: 1 212 2970967

Munroe, Ervine, Ms.
YMCA EARTH ACTION CORPS
909 4th Av
Seattle - WA, United States of America
T: 1 206 382-5013 F: 1 206 382-7283

Munyan, Winthrop R., Mr.
PPSEAWA
1 Lexington Ave.
New York - NY 10010, United States of America
T: 1 212 228-5307 F: 1 212 473-0942

Musabbir, Abdul, Mr.
President
BANGLADESH FRONT INC.
55 First Avenue, Suite # 9
New York - NY 10003, United States of America
T: 1 212 505-5710 F: 1 212 243-8644

Musselwhite, David G., Mr.
President
THE CORPORATE ECOLOGY GROUP
316 Irving Place
Lookout Mt. - TN 37350, United States of America
T: 1 615 828-1328

Mussi, Marco Aurelio Jaikh, Sr.
Assistente
GLOBAL FORUM OF SPIRITUAL AND PARLIAMENTARY LEADERS ON HUMAN SURVIVAL
304 East 45th St. Room 476
New York - NY, United States of America
T: 1 212 953-7947 F: 1 212 557-20961

Musslewhite, Benton, Mr.
President
ONE WORLD NOW
1510 Lyric Center, 440 Lousiana
Houston - TX 77002, United States of America
T: 1 713 222-2288 F: 1 713 222-0319

Myers, Carol
Producer
EARTH STATION FOUNDATION
1000 Ponce de Leon Blvd.Suite 203
Coral Gables, FL 331, United States of America
T: 001 305 4438652 F: 001 305 8885

Myers, Daniel R., Mr.
Representante
SIERRA CLUB
3306, Swann Ave.
Tampa-Florida 33609, United States of America
T: 8767624

Myers, Michael S., Mr.
CITY OF AUSTIN
206 E. 9th Street
Austin 78701, United States of America
T: 1 512 499-3508 F: 1 512 499-2859

Myhre, Liv Irene, Ms.
Editor
WORLD CITY DISCOVERY
261 Newtown Trunpike
Weston - CT, United States of America
T: 1 203 226-1642 F: 1 203 226-1642

Nahulu, Linda B., Dr.
Honolulu, Hawaii, United States of America

Nakamaru, Kaoru, Ms.
Chair Person
INTERNATIONAL AFFAIRS INSTITUTION FOR WORLD PEACE
245 E. 63rd Street # 405
New York - NY, United States of America
T: 1 212 752-6493

Nakamaru, Tomoko, Ms.
WORLD ASSOCIATION OF FORMER UN INTERNS AND FELLOWS, INC.
245 E 63rd St.
New York - NY 10017, United States of America
T: 1 212 752.6493 F: 1 212 308.7852

Narayanan, Jaya, Ms.
YMCA EARTH ACTION CORPS
909 4th Av
Seattle - WA, United States of America
T: 1 206 382-5013 F: 1 206 382-7283

Narby, Jeremy
Correspondent
PROCESSED WORLD
1095 Market Street suite 209
San Francisco CA 941, United States of America
T: 1 415 6262160 F: 1 415 6262685

Narine, Gyatri
Producer
EARTH STATION FOUNDATION
1000 Ponce de Leon Blvd.Suite 203
Coral Gables, FL 331, United States of America
T: 001 305 4438652 F: 001 305 8885

Nascimento, Monica, Sra.
Assistant Marketing
OXFORD UNIVERSITY PRESS
200 Madison Avenue - 9th floor
New York - NY 10016, United States of America
T: 1 212 679-0300..7125 F: 1 212 725-2972

Nashban, Nicole, Ms.
Secretary
WORLD PEACE PRAYER SOCIETY
518 Beekman Rd.
Hopewell Jct. - NY 12533, United States of America
T: 1 212 227-7898 F: 1 212 935-1389

Nazareth, Premila, Ms.
Ass. Trans. Corp. Officer
UNITED NATIONS
2 United Nations Plaza, Rm 1230
New York - NY 10017, United States of America
T: 1 212 963-7631 F: 1 212 963-2146

Ndiritu, Josephine Wangari, Mr.
GREEN BELT MOVEMENT OF KENYA
c/o Resource Renewal Institute, FT. Mason Center Bldg A
San Francisco - CA, United States of America
T: 1 415 928-32774 F: 1 415 928-6529

Ndisale, J.E., Ms.
GREEN BELT MOVEMENT OF KENYA
c/o Resource Renewal Institute, FT. Mason Center Bldg A
San Francisco - CA, United States of America
T: 1 415 928-32774 F: 1 415 928-6529

Neal, Rosanne, Ms.
Administrative Assistant
CHRISTIAN ENVIRONMENT ALLIANCE
3110 RALEIGH AVE So #103
Minneapolis - MN 55416, United States of America
T: 1 612 926 3828 F: 1 612 729 2828

Nelson, Maryann, Ms.
Director
SIERRA CLUB
10 Gore St.
Boston - MA 02120, United States of America
T: 1 617 442 0123

Nelson, Pete
Editor In Chief
GREENWIRE
282 North Wasington St. Falls Church
Virginia 22046, United States of America
T: 1 703 2375130 F: 1 703 2375149

Nepstad, Dan, Dr.
Assistant Scientist
WOODS HOLE RESEARCH CENTER
13 Church Street - P.O. Box 296
Woods Hole - MA 02543, United States of America
T: 1 508 540-9900 F: 1 508 540-9700

Nevarrez, David, Mr.
Photographer
GLOBAL HEART
210 W. 70th St. #1507
New York - NY 10023, United States of America

Nevers, Jim, Mr.
RANCHO BUENA VISTA HIGH SCHOOL
1601 Longhorn Drive
Vista 92083, United States of America

Nevin, John, Mr.
GLOBAL FUTURES NETWORK
339 N. Fullerton
Upper Montclair-NJ 07043, United States of America

Newcombe, Kenneth, Mr.
WORLD BANK
1818 H St. NW
Washington - DC 20433, United States of America
T: 1 202 458-1943

Newell, Leslie Brooke, Ms.
President
UNITED EARTH
1225 Park Avenue
New York - NY 10128, United States of America
T: 1 212 517-7776

Newkirk, Grace
SCHOOL OF THEOLOGY AT CLAREMONT
1325 N. College Ave. PO Box 107
Claremont-CA 91711-3919, United States of America
T: 1 714 399-5706 F: 1 714 626-7062

Newman, Lorrenda, Ms.
UP WITH PEOPLE
3103 North Campbell Av.
Tucson - AZ 85719, United States of America

Nicholls, Joel, Mr.
UP WITH PEOPLE
3103 North Campbell Av.
Tucson - AZ 85719, United States of America

Nichols, Sheila, Ms.
UP WITH PEOPLE
3103 North Campbell Av.
Tucson - AZ 85719, United States of America

Nicholson, Patrick, Mr.
N-VIRO ENERGY SYSTEMS
3450 W. Central Ave.
Toledo - Ohio 43606, United States of America
T: 1 619 535-6374 F: 1 619 535-7008

Nielson, Dennis
Audio Technician
NATIONAL PUBLIC
2025 M St NW
Washington DC 20036, United States of America
T: 1 202 8222480

Nienhueser, Helen, Ms.
A WORLD THAT WORKS
2561 Lovejoy Drive
Anchorage - Alaska 99508, United States of America
T: 1 907 277-9330

Nishimura, Niro, Mr.
UP WITH PEOPLE
3103 North Campbell Av.
Tucson - AZ 85719, United States of America

Njehu, Lilian Wanjiru Njoroge, Ms.
GREEN BELT MOVEMENT OF KENYA
c/o Resource Renewal Institute, FT. Mason Center Bldg A
San Francisco - CA, United States of America
T: 1 415 928-32774 F: 1 415 928-6529

Nobel, Claes, Mr.
Founder-Chairman
UNITED EARTH
700 East Daisy Ln
Milwaukee-WI. 5327, United States of America
T: 414 351-2737 F: 414 351-4493

Nobel, Margaretha, Ms.
Director
UNITED EARTH
700 East Daisy Ln
Milwaukee-WI. 53217, United States of America
T: 414 351-2737 F: 414 351-4493

Nocella, Susan
INTERFAITH CONFERENCE / SENSE
5908 Dale Lane
Greendale-WI, United States of America

Noffs, David, Mr.
President
LIFE EDUCATION CENTER FOUNDATION
P. O. Box 1454
Elmhurst - IL 60126-2127, United States of America
T: 1 708 530-8999 F: 1 708 530-7241

Noonan, Kate
Writer Producer
FREELANCER
2142 N. Cahuenga Blvd.
Hollywood-C.A. 90038, United States of America
T: 1 213 4694405 F: 1 213 4695010

Norderhaug, Anne-Li, Ms.
WORLDWATCH INSTITUTE
1776, Massachusetts Ave., N.W.
Washington - DC, United States of America
T: 1 202 452-1999 F: 1 202 296-7365

Norderhaug, Magnar, Mr.
Executive Director
WORLDWATCH INSTITUTE
1776, Massachusetts Ave., N.W.
Washington - DC, United States of America
T: 1 202 452-1999 F: 1 202 296-7365

Nordland, Jeannie, Mr.
CAMPAIGN FOR THE EARTH
2120 13th Street
Boulder - CO 80302, United States of America
T: 1 303 444-3800 F: 1 303 444-3811

Norelli, Gianfranco, Mr.
Director
EAST-WEST PRODUCTIONS, NEW YORK
71 Vandam Street
New York - NY 10013, United States of America
T: 1 212 463-9908 F: 1 212 675-6344

Nort, Marjorie Van, Ms.
Executive Assistant
GLOBAL OBJECTIVES, INC.
3653 N. Central Ave., Suite C-10
Phoenix - AZ, United States of America
T: 1 602 230-2357 F: 1 602 264-5030

North, Shanti, Ms.
Observer
FRIENDS OF THE AMAZON FOREST
GA, United States of America
T: 1 912 354-6644

Northcutt, Patsy
Owner, Producer
NORTHCUTT PRODUCTIONS
2966 Diamond St. #171
San Francisco - CA, United States of America
T: 1 415 3377044 F: 1 415 3370630

Nottingham, Dawn Marie, Ms.
2838 A El Lago North Drive
Indianapolis - IND, United States of America
T: 1 317 243-5173 F: 1 317 243-5092

Novar, Patricia, Ms.
COMMUNICATIONS COORDINATIVE COMM. FOR UN
290 Riverside Dr. - Apt.15A
New York - NY, United States of America
T: 1 212 749-6297

Nungester, Norma, Ms.
HIGHLANDER CENTER
1939 Highlander Way
New Market - TN 37820, United States of America
T: 1 615 933 3443 F: 1 615 933 3424

Nussbaum, Larry, Mr.
STUDENT ENVIRONMENTAL ACTION COALITION (SEAC)
403 Barnard Ave
Cedarhurst - NY 11516, United States of America
T: 1 516 374-6047

Nyman, Carie, Ms.
YMCA EARTH ACTION CORPS
909 4th Av
Seattle - WA, United States of America
T: 1 206 382-5013 F: 1 206 382-7283

O'Conner, Geoffrey
Producer
CBS NEWS
524 W. 57th Street
New York NY 10025, United States of America
T: 1 212 5051911 F: 1 212 5051179

O'Hanlon, Barbara, Ms.
THE FUTURES GROUP
One Thomas Circle, NW
Washington - DC, United States of America

O'Hara, Sabine U., Dr.
GENERAL BOARD OF CHURCH & SOCIETY /UNITED METHODIST CHURCHES
362 State Street
Albany - NY 122100, United States of America
T: 1 518 4369319 F: 1 518 427-6705

O'Keefe, Hsu, Ms.
Economist
9 Gristmill rd.
Randolf NJ, United States of America
T: 01 201 989-0661

O'Reilly, Nancy, Ms.
Reverend
ART OF LIVING CENTER
15 E. Cherry Ave. Suite #203
Flagtaff - AZ 86004, United States of America
T: 1 602 779-2203

Obert, Douglas
ABC NEWS
47 West 66th street
New York - NY, United States of America
T: 1 212 4562800

Ocloo, Esther, Ms.
Member
WOMEN FOREGN POLICY COUNCIL
845 Third Avenue 15 Floor N 10022
New York - NY, United States of America
T: 1 212 759 7983

Okada, Nancy, Ms.
RALPH BUNCHE INSTITUTE ON THE UN
33 West 42nd St.
New York - NY 10036, United States of America

Okwesa, Geraldine, Ms.
Teacher
DISTRICT OF COLUMBIA SCHOOLS
8th Oglethorpe Street N.W PAUL JUNIOR HIGH SCHOOL
Washington - DC 30022, United States of America
T: 1 202 576-6190 F: 1 202 576-6196

Oliveira, Jose Alberto, Sr.
SERENGUEIRO DO SERINGAL CACHOEIRA
1880 East River Road
Tucson - AZ, United States of America
T: 1 602 577-6476 F: 1 602 577-1195

Oliveira, Letitia, Ms.
WORLD BANK
1818 H St. NW
Washington - DC 20433, United States of America
T: 1 202 458-1943

Oliveira, Osmar, Sr.
SIND.DOS TRABALHADORES RURAIS DE XAPURI
1880 East River Road
Tucson - AZ, United States of America
T: 1 602 577-6476 F: 1 602 577-1195

Oliveira, Siloni, Ms.
DAMIEN FOUNDATION
477 Summit Ave.
Mill Valley - CA 94941, United States of America

Oliver, Patsy
NATIONAL TOXICS CAMPAIGN
1939 Highlander Way
New Market - TN 37820, United States of America
T: 1 615 933 3443 F: 1 615 933 3424

Orts, Eric Wills, Mr.
WHARTON SCHOOL - U. OF PENNSYLVANIA
3514 Lancaster Ave, 413
Philadelphia - PA, United States of America
T: 1 215 898-3018 F: 1 215 573-2006

Osa, Richard, Mr.
SCIENCE & TECHNOLOGY MANAGEMENT, INC.
2511 N. 124th Street, Suite 205
Brookfield - WI, United States of America
T: 1 414 785-5946 F: 1 414 785-5950

Osbourne, Molly L., Dr.
SUSIAL DHARMA INTERNATIONAL
319 SE Gilham Avenue
Portland - OR 97215, United States of America
T: 1 503 253-7601 F: 1 503 494-6670

Otani, Mr., Mr.
INTERNATIONAL AFFAIRS INSTITUTION FOR WORLD PEACE
245 E. 63rd Street # 405
New York - NY, United States of America
T: 1 212 752-6493

Otani, Ms., Ms.
INTERNATIONAL AFFAIRS INSTITUTION FOR WORLD PEACE
245 E. 63rd Street # 405
New York - NY, United States of America
T: 1 212 752-6493

Otee, Kamala, Ms.
PROUT RESEARCH INSTITUTE
4095 Jackdaw
San Diego - CA 92103, United States of America
T: 1 619 260-1989 F: 1 619 287-2508

Overberg, Debra, Ms.
GREENS / WG / USA
308 Wood Rose CT
Windsor 95492, United States of America
T: 1 707 838-2926

Overend, Ralph, Mr.
N R E L
1617,Cole Boulevard
Golden 80401, United States of America
T: 1 303 231 1450 F: 1 303 231 1352

Owens, Owen, Mr.
GLOBAL ENVIRONMENT TEAM
PO Box 851
Valley Forge - PA 19482, United States of America

Packard, Linda, Ms.
RAINFOREST ACTION NETWORK
37 Windjammer
Irvine - CA 92714, United States of America
T: 1 714 786-6554

Paden, Mary, Ms.
Senior Associate
WORLD RESOURCES INSTITUTE
1709 New York Ave.
Washington - DC 20006, United States of America
T: 1 202 662-2573

Page, William R., Mr.
HARVARD MEDICAL SCHOOL
BOX 482, Rd 2
Orleans - VT, United States of America
T: 1 802 525-3655 F: 1 802 525-3988

Paige, Bernice, Ms.
GLOBAL HEART
1880 East River Road
Tucson - AZ, United States of America
T: 1 602 577-6476 F: 1 602 577-1195

Paige, Ralph, Mr.
FED. OF SOUTHERN CO-OPS.
100 Edgewood Ave. Ste. 814
Atlanta - GA 30303, United States of America

Palacio, Maria Teresa
MIAMI ANNIVERSARY COMMITTEE
8730 sw 51 street
Miami - Florida 33165, United States of America
T: 1 305 598 7494 F: 1 305 598 7496

Palma, Renee De
Producer
OUT OF EPOCA FILMS
N. Cahuenga Blvd.
Hollywood -CA 90038, United States of America
T: 1 213 4694405 F: 1 213 4695010

Palmer, Andrew, Mr.
Political Dir
AMERICAN OCEANS CAMPAIGN
235 Pennsylvania Ave, SE, 3rd Flr
Washington - DC, United States of America
T: 1 202 544-3526 F: 1 202 544-5625

Palmer, Patricia, Ms.
GLOBAL ENVIRONMENT TEAM
140 Prospect Av.
Long Beach - CA 90803, United States of America
T: 1 310 439-1654

Papadopoulos, George, Mr.
Representative
WORLD BANK
1818 H St. NW
Washington - DC 20433, United States of America
T: 1 202 458-1943

Paprocki, David, Mr.
PROUT. INTERNATIONAL
663 W. Buckingham
Chicago - IL, United States of America
T: 1 312 929-8491 F: 1 312 929-8491

Paredes, Viviane, Ms.
Development Director
AMAZONIA AIDS & HEALTH PROJECT
3180 18th Street, Suite 102
San Francisco - CA 94110, United States of America
T: 1 415 476-3990 F: 1 415 476-3655

Park, Hee Sook
TEMPLE OF UNDERSTANDING
St. Emeric Church / 185 Avenue D
New York - NY 10009, United States of America
T: 1 212 353-0001 F: 1 212 353-0170

Parke, Beth
Senior Producer And Host
CONSIDER THE ALTERNATIVES RADIO
5808 Green Street
Philadelphia - PA, United States of America
T: 1 215 848 4100 F: 1 215 844-5988

Parker, Carole, Ms.
Congressional Staff
GLOBE
403 Cannon House Office Building
Washington - DC 20515, United States of America
T: 1 202 225-2271

Parker, Elissa, Ms.
Director
ENVIRONMENTAL LAW INSTITUTE
1616 P. Street, N.W., Suite 200
Washington - DC 20036, United States of America
T: 1 202 939-3853 F: 1 202 328-5002

Parker, Paul, Mr.
Vice-President
CENTER FOR RESOURCE MANAGEMENT
262 South 200 West
Salt Lake City - UT, United States of America
T: 1 801 3220530 F: 1 801 3283457

Parris, Tom, Mr.
Senior Analyst
CIESIN
1825 K st. NW
Washington - DC 20006, United States of America
T: 1 202 775 6600 F: 1 202 775 6622

Passos, Veronica T. Da R., Ms.
WOODS HOLE RESEARCH CENTER
P. O. Box 296
Woods Hole - MA 02543, United States of America
T: 1 508 540-9900 F: 1 508 540-9700

Paterson, Janet, Ms.
UNIVERSITY OF IOWA
94 Chatham Street
Chatham - NJ, United States of America
T: 1 201 635-8085

Paterson, Kurtis G., Mr.
Biochemistry Reseracher
UNIVERSITY OF IOWA
Dept. Civil and Environmental Engineering, The University of Iowa
Iowa City - IA, United States of America
T: 1 319 335-5663 F: 1 319 335-5777

Patrick, Charles, Mr.
President
ASSOCIATION TO UNITE THE DEMOCRACIES
1506 Pennsylvania Ave SE
Washington - DC 20003, United States of America
T: 1 202 5445152 F: 1 202 5443742

Patton, Sharyle M, Ms.
Associate Director
COMMONWEAL
PO Box 316
Bolinas - CA, United States of America
T: 1 415 8680970 F: 1 415 8682230

Paulk, Daryl
UP WITH PEOPLE
3103 North Campbell Av.
Tucson - AZ 85719, United States of America

Pautzman, Anne, Ms.
HUMANE SOCIETY
1185 Keeler Ave
Berkeley 94708, United States of America
T: 501 8431246

Paz, Lisia, Sra.
Asesor Auxiliar Asuntos
OFICINA DEL GOBERNADOR
Calle Estado 667 Apt. 4
Miramar, Puerto Rico, United States of America
T: 1 809 725-1967 F: 1 809 724-4916

Peabody III, N. S., Mr.
Chief Of Party
WINROCK INTERNATIONAL
1611 N.Kent St., Suite 600
Arlington - VA 22209, United States of America
T: 1 703 525-9430 F: 1 703 516-0481
TL: 6491106 WIDC, E: DIALCOM YI: ICN 408

Pearce, Vin
GLOBAL ENERGY NETWORK INT'L
PO Box 81565
San Diego - CA 92138, United States of America
T: 1 619 595-0139 F: 1 619 595-0403

Pearlberg, Rev. John D.
REFORMED CHURCH IN AMERICA
475 Riverside Drive
New York - NY 10015, United States of America

Pearlmutter, Howard, Mr.
President, Founder
SOFTWEAVER; EARTHEART
18424 Harbin Springs Road; Box 1600
Middletown - CA, United States of America
T: 1 707 987-9000 F: 1 707 987-0616
TL: D0390 APPLELINK

Pearson, David, Mr.
GLOBAL ENVIRONMENT TEAM
63 S. Main St.
Rochester - NH 03867, United States of America
T: 1 603 332-1121

Peavey, Fran, Ms.
SANKAT MOCHAN FOUNDATION
3181 Mission St. #30
San Francisco - CA 94110, United States of America
T: 1 510 428-0240 F: 1 510 601-5683

Peck, John E., Mr.
STUDENT ENVIRONMENTAL ACTION COALITION
12299 Sauk River Rd.
St. Cloud, United States of America
T: 612 685 3365

Pederson, Kusumita, Ms.
Participant
GLOBAL FORUM OF SPIRUTUAL AND PARLIMENTARY LEADERS
304 East 4th St., 4th Floor
New York - NY, United States of America
T: 1 212 953-7947 F: 1 212 557-2061

Pelter, Ken, Mr.
CAMPAIGN FOR THE EARTH
1880 East River Road
Tucson - AZ, United States of America
T: 1 602 577-6476 F: 1 602 577-1195

Pena, Susana, Ms.
UP WITH PEOPLE
3103 North Campbell Av.
Tucson - AZ 85719, United States of America

Pereira, Alberto Carlos Lourenco
Grad. Student
UNIV. CALIFORNIA, LOS ANGELES
3174 Sawtelle Blvd. # 2
Los Angeles - CA 90066, United States of America
T: 1 213 398-7782 F: 1 213 398-7782

Pereira, Cassio A., Sr.
Pesquisador
WOODS HOLE RESEARCH CENTER
P.O.Box 296
Woods Hole - MA, United States of America
T: 1 508 540-9900

Perez, Clarissa, Ms.
Member
INDIGENOUS WOMEN'S NETWORK
35 Bridge St.
Eureka Springs -AR, United States of America
T: 1 501 253-9545

Perez, Cuauhtemoc
Team Member
FOUNDATION FOR GLOBAL COMMUNITY
222 High St.
Palo Alto - CA 94301, United States of America
T: 1 415 3287756 F: 1 415 3287785

Perez, Reinaldo
Cameraman
CNN BROADCASTING SYSTEM
One CNN Center, Box 105366
Atlanta - GA, United States of America
T: 1 404 8371918 F: 1 404 8274292

Perfetti, Patricia, Dr.
Professor
UNIVERSITY OF TENNESSEE AT CHATTANOOGA
Dept. of Biology & Environmental Science, 615 McCallie
Chattanooga - TN, United States of America
T: 615 755-4794 F: 615 755-4061

Persily, Meredith, Ms.
BROWN UNIVERSITY
P.O.Box 3536
Providence - RI 02912, United States of America
T: 1 401 421-3270

Peters, Aurora, Ms.
YMCA EARTH ACTION CORPS
909 4th Av
Seattle - WA, United States of America
T: 1 206 382-5013 F: 1 206 382-7283

Peters, Josh D., Mr.
STUDENT ENVIRONMENTAL ACTION COALITION - SEAC
VSTEP, UVM, Billings Center
Burlington - VT 05405, United States of America
T: 1 802 656.4484 F: 1 802 656.7719

Petersen, Sam, Ms.
Publicity And Promotion Manag.
SIERRA CLUB BOOKS
100 Bush Street 13th Floor
San Francisco - CA 94104, United States of America
T: 1 415 291-1196 F: 1 415 291-1602

Petit, M. D.
GLOBAL PRESS
Washington DC, United States of America

Petruzziello, Armando, Mr.
Executive Director
EUROPEAN ENVIRONMENT BUSINESS
189 Bedford Street, 02173
Lexington - MA 02173, United States of America
T: 1 617 862-9332 F: 1 617 863-9405

Phares, Robert W., Mr.
Principal Investigator
MONDEX
333 Old Mill Rd., # 318
Santa Barbara - CA 93110, United States of America
T: 1 805 964-6034 F: 1 805 964-3337

Philleo, Wendy, Ms.
Special Ass. To Managing Dir.
WORLDWIDE NETWORK
1331 H. Street, Suite 903
Washington - DC 20005, United States of America
T: 1 202 347-1514 F: 1 202 347-1524

Phillips, Jodi, Ms.
UP WITH PEOPLE
3103 North Campbell Av.
Tucson - AZ 85719, United States of America

Phoenix, Heart, Ms.
SACRED EARTH NETWORK
c/o Phoenix 3232 SW 35 Blvd suite 121
Gainesville - FL, United States of America
T: 1 904 466-0222 F: 1 904 466-0221

Pidcoke, Heidi, Ms.
GLOBAL ENVIRONMENT TEAM
1230 Hayes St. #5
San Francisco - CA 91106, United States of America
T: 1 415 928-1276

Pierce, Catherine, Ms.
Member
POPULATION INSTITUTE
Washington - DC, United States of America

Piere, Lisa, Ms.
Env. Scient.
UNITED NATIONS ASSOCIATION
255 W. State St.
Pasadena - CA 91105, United States of America
T: 1 818 585-6041 F: 1 818 585-6336

Pierro, Christina Del, Ms.
GLOBAL ENVIRONMENT TEAM
PO Box 216
Middlebury-VT 05753, United States of America
T: 1 802 462-2668

Piestrup, Sarah, Ms.
Staff
EARTH TRAIN/WORLDLINK/GLOBAL YOUTH SUMMIT
3220 Sacramento
San Francisco - CA 94115, United States of America
T: 1 415 931 1290

Pimenta-Neves, Antonio, Mr.
Representative
WORLD BANK
1818 H St. NW
Washington - DC 20433, United States of America
T: 1 202 458-1943

Pinheiro, Jose, Mr.
UNIVERSITY OF ARIZONA
1402 E. Manlove St. apt. 26
Tucson - AZ 85719, United States of America
T: 1 602 577-6476 F: 1 602 577-1195

Pinto, Marco Antonio F.
WGBH-TV
125 Western Ave.
Boston -Massachusett, United States of America
T: 1 617 4495830 F: 1 617 4446450

Pitcher, W. Alvin, Mr.
GLOBAL ENVIRONMENT TEAM
6107 S. Woodlawn
Chicago - CA, United States of America
T: 1 312 643-4141

Plucknett, Donald, Mr.
Scientific Advisor
C G I A R
1818 H St. N.W.
Washington - DC 20433, United States of America
T: 1 202 473-8934 F: 1 202 334-8750

Podea, Melody, Ms.
U.N.Rep.
INTERNATIONAL FELLOWSHIP OF RECONCILIATION
301 Richards St.
Iowa City - IA 52246, United States of America
T: 1 319 338-044723014

Poe, Trish, Ms.
KIDS FOR A CLEAN ENVIRONMENT (FACE)
P.O.Box 158254
Nashville TN 37215, United States of America

Polaner, Steven, Mr.
Llewllyn Pk
West Orange -Nj 07052, United States of America
T: 201 7310109

Pollock, Donald, Dr.
Professor
STATE UNIVERSITY OF NEW YORK/BUFFALO
Dept of Anthropology, SUNY/BUFFALO, 380 MFAC, Ellicott Complex
Buffalo - NY 14261, United States of America
T: 1 716 636 3080 F: 1 716 636 3808

Pollock, Ed, Mr.
PERFORMING AND FINE ARTISTS FOR WORLD PEACE
PO Box 312
Hanalei-Kavai, HAWAII 96714, United States of America
T: 1 808 826-9251

Pollock, Marilyn, Ms.
PERFORMING AND FINE ARTISTS FOR WORLD PEACE
PO Box 312
Hanalei-Kavai, HAWAII 96714, United States of America
T: 1 808 826-9251

Polston, Pamela
Director Of Development
TOWARD FREEDOM
209 College St
Burlington VT 05401, United States of America
T: 1 802 658 2523 F: 1 802 658 3738

Porter, Mary Marvin, Ms.
VIDEO JOURNALISM
Malibu - CA, United States of America

Porto, Jorge Fernandez, Sr.
HIPR
Navarro # 53, 1er. piso
Hato Rey, Puerto Rico, United States of America
T: 1 767-0820 F: 1 754-6462

Posner, Joseph, Sr.
6117 Benhurst Rd.
Baltimore, Md. 21209, United States of America
T: 1 301 358-9097

Possiel, William J., Mr.
Brazil Program Director
NATURE CONSERVANCY
1815 N. Lynn St.
Arlington - VA 22209, United States of America
T: 1 703 8414820 F: 1 703 8414880

Potenza, , Ms.
PEACEABLE REALM FOUNDATION
4329 Brookshire Circle
Santa Rosa - CA 95405, United States of America

Powell, Allison, Ms.
PERMANENT MISSION OF VANUATU TO THE U.N
416 Convent Ave.
New York - NY 10031, United States of America
T: 1 212 577-6476 F: 1 212 577-1195

Powell, James E., Mr.
Deputy Director
ALASKA DEPARTMENT OF ENVIRONMENT CONVERVATION
1105 Mendenhall Penn Road
Juneau - Alaska 99801, United States of America
T: 1 907 790-2023 F: 1 907 465-5274

Powers, Kay, Ms.
UNITED NATIONS ASSOCIATION
20 East Market Street
Iowa City, Iowa, United States of America
T: 1 319 337-7290 F: 1 319 335-2951

Pozzi, John, Mr.
PROGRESS AGENCY
126 SE 3 St
Hallardale - Fl, United States of America
T: 1 305 458-7941

Praag, Nick Van
Freelance Journalist
GEF
1523 28th St NW
Washington DC 20007, United States of America
T: 1 202 3382930 F: 1 202 4770551

Prajean, Marion, Mr.
UP WITH PEOPLE
3103 North Campbell Av.
Tucson - AZ 85719, United States of America

Pratt, Jane, Ms.
Representative
WORLD BANK
1818 H St. NW
Washington - DC 20433, United States of America
T: 1 202 473-1865

Price, Helen T., Ms.
IPAC/WEDO
37 Garlen Road
Katonah - NY 10536, United States of America
T: 1 914 232-8380

Price, Philip, Mr.
24373 La Masina Court
Calabasas - CA 91302, United States of America
T: 1 818 886-0942

Price, Victoria, Ms.
24373 La Masina Court
Calabasas - CA 91302, United States of America
T: 1 818 886-0942

Proietto, Javier, Mr.
WORLD ASSN OF FORMER U.N. INTERNES AND FELLOWS INC (WAFUNIF)
c/o Ms Hope P White-Davis, United Nations, Room DC2-1118, 2 UN Plaza
New York - NY, United States of America
T: 1 212 9633110 F: 1 212 9634116

Propper, Diana, Ms.
Business Manager
CULTURAL SURVIVAL
53-A Church Street
Cambridge - MA 02138, United States of America
T: 1 617 495-2562 F: 1 617 495-1396

Prout, Christine Ellen, Ms.
SO.CALIF. ECOLOGY TASK FORCE
9203 Cord Ave.
Downey - CA 90240, United States of America
T: 1 310 861-3207

Prout, Shanen Reid, Mr.
SO.CALIF. ECOLOGY TASK FORCE
9203 Cord Ave.
Downey - CA 90240, United States of America
T: 1 310 861-3207

Prout, Traverson Austin, Mr.
SO.CALIF. ECOLOGY TASK FORCE
9203 Cord Ave.
Downey - CA 90240, United States of America
T: 1 310 861-3207

Qadir, Salma A., Ms.
Board Of Directors
OVERSEAS DEVELOPMENT NETWORK
149 Columbia Heights
Brooklyn - NY 1120, United States of America
T: 1 718 703-2020 F: 1 718 791-4132

Quay, Beth, Ms.
Director Envirn. Tech. Affairs
THE COCA-COLA COMPANY
One Coca-Cola Plaza
Atlanta - GA 30313, United States of America
T: 1 404 676-6629

Queiroz, Maria De Conceicao, Sra.
Chief
UNITED NATIONS DEVELOPMENT PROGRAMME
One UN Plaza - Room DC1-1926
New York - NY, United States of America
T: 1 212 906-5302 F: 1 212 906-5364

Quere-Messing, Marie Louise
UN DEPARTMENT OF PUBLIC INFORMATION
New York, NY, United States of America

Quintero, Maria
UP WITH PEOPLE
3103 North Campbell Av.
Tucson - AZ 85719, United States of America

Qureshi, Ata, Dr
Director
GLOBAL ENVIRONMENT PROGRAMS
P.O. Box 12073
Washington DC 20005, United States of America
T: 202 547 0104 F: 202 547 0111

Quirino, Joseph
YMCA EARTH ACTION CORPS
909 4th Av
Seattle - WA, United States of America
T: 1 206 382-5013 F: 1 206 382-7283

Radwanski, Therezinha, Ms.
ECOTROPIC
1880 East River Road
Tucson - AZ, United States of America
T: 1 602 577-6476 F: 1 602 577-1195

Rainey, Robert
TURNER BROADCASTING SYSTEM / CNN
One CNN Centre
Atlanta - GA, United States of America
T: 1 404 827-3973 F: 1 404 827-1015

Ramadan, Yolanda, Ms.
WHAT'S NEXT SEMINARS
230 East 44TH St., Apt. 102
New York - NY 10017, United States of America
T: 1 212 687-4832 F: 1 609 894-2537

Ramirez, Carlos, Sr.
Ayudante Del Presidente
CAMARA DE DIPUTADOS - PUERTO RICO
Appo. 2228 Capitolio
San Juan, Puerto Rico, United States of America
T: 1 809 723 4711

Ramos, Mario, Mr.
Observer
UNIVERSITY OF RICHMOND
68 Parkshine Ave
Iseekonk, United States of America
T: 508 399 6755

Ramp, Marjorie, Ms.
SO. CALIF. ECOLOGY TASK FORCE
225 Hollywood St.
Oberlin - OH 44074, United States of America
T: 1 216 774-1049

Randel, Sally, Ms.
President
ENVIRONMENTAL MARKETING SERVICES, INC.
4295 Eldorado Springs Drive
Boulder - CO 80303, United States of America
T: 1 303 4942210

Randle, Kenneth, Mr.
HOPE PUBLISHING HOUSE
9844 Glendover Way
Sandy - UT 84092, United States of America
T: 1 801 572-6304

Raposo, Evelyne, Ms.
EARTH TRAIN/WORLDLINK/GLOBAL YOUTH SUMMIT
470 HAO St.
Honolulu - HI 96821, United States of America
T: 1 808 3732667

Rasberry, Salli, Ms.
Writer
GREEN HEART AMERICA
13055 Bodega Hwy
Freestone - CA 95472, United States of America
T: 001 707 823-2207

Raven, Tamra, Ms.
INTERNATIONAL COUNCIL OF WOMEN
New York, NY, United States of America

Raye, Marina, Ms.
HIGH PERFORMANCE TRAINING
P.O.Box 6250
Colorado SPRINGS CO 80934, United States of America

Rebik, Lisa, Ms.
UP WITH PEOPLE
3103 N. Campbell Ave.
Tucson - AZ 85718, United States of America
T: 1 602 577-6476 F: 1 602 577-1195

Rechelbacher, Horst, Mr.
Ceo
AVEDA CORPORATION
400 Pheasant Ridge Drive
Minneapolis - MN 55434, United States of America
T: 1 612 783-4000 F: 1 612 783-4110

Reed, Bradford, Mr.
1200 10th Street
Manhattan Beach - CA 90266, United States of America
T: 1 310 379 6285 F: 1 310 379 0967

Regan, Mary Ann
Journalist
MARYKNOLL MAGAZINE LATIN AMERICA PRESS
Maryknoll Missiones
Maryknoll NY 10501, United States of America
T: 1 914 9417634

Regnier, Andre
TERRA CHRISTA COMMUNICATIONS
1880 East River Road
Tucson - AZ, United States of America
T: 1 602 577-6476 F: 1 602 577-1195

Rehlander, John, Mr.
UP WITH PEOPLE
3103 North Campbell Av.
Tucson - AZ 85719, United States of America

Reid, Judith A., Ms.
Teacher
ANCHORAGE SCHOOL DISTRICT (SELF)
2807 Sunflower
Anchorage - Alaska, United States of America
T: 1 907 337-8052 F: 1 907 276-7177

Reilly, Charles, Mr.
Doctor
FUNDACAO INTERAMERICANA
901NO Stuart Street
Arlington - VA 22203, United States of America
T: 1 703 841-3809

Reimann, Cornelia Camila, Ms.
DAMIEN FOUNDATION
477, Sumit Ave.
San Francisco - CA, United States of America
T: 1 415 381-2778

Reister, Kurt
Sound Tech
CNN BROADCASTING SYSTEM
One CNN Center, Box 105366
Atlanta - GA, United States of America
T: 1 404 8371918 F: 1 404 8274292

Resersen, Kusumita, Ms.
Joint Secretary
GLOBAL FORUM OF SPIRITUAL AND PARLIAMENTARY LEADERS ON HUMAN SURVIVAL
304 East 45th St. Room 476
New York - NY, United States of America
T: 1 212 953-7947 F: 1 212 557-20961

Reveron, Carmen, Ms.
Professor-Advisor-Research
COLEGIO UNIVERSITARIO DEL ESTE
Apartado 2010
Carolina, Puerto Rico, United States of America
T: 758-7171

Rexford, Burton, Mr.
Chairman
ALASKA ESKIMO WHALING COMMISSION
Box 570, Barrow
Barrow-AK 99723, United States of America
T: 1 907 852-2392 F: 1 907 852-2303

Reyes, Carmen, Ms.
Project Director
CHILDREN'S EARTH FUND
40 West 20th Street, 11th Floor
New York - NY 10010, United States of America
T: 1 212 7274488 F: 1 212 7271773

Reyes, Jose M., Mr.
Assoc.
CHILDREN'S EARTH FUND
30 Rockefeler Plaza
New York - NY 10.112, United States of America
T: 1 212 649-5819

Riad, Wagdi
Environmnet Editor
GLOBAL EDITION
1875 Connecticut Ave, NW Suite 520
Washington - DC 20009, United States of America
T: 1 202 483 1100 F: 1 202 328 3937

Ribalow, Meir Z., Mr.
Assistant
GLOBAL FORUM OF SPIRITUAL AND PARLIAMENTARY LEADERS ON HUMAN SURVIVAL
304 East 45th St. Room 476
New York - NY, United States of America
T: 1 212 953-7947 F: 1 212 557-20961

Ricciardi, Mirella, Ms.
Vice-Pres.
RICCIARDI FAMILY FOUNDATION
105 East 16TH Street #9S
New York - NY 10003, United States of America

Rich, Bruce, Mr.
Chairman, International Prog.
ENVIRONMENTAL DEFENSE FUND
1616 P street, NW, Suite 150
Washington - DC 20036, United States of America
T: 1 202 3873500 F: 1 202 2346049

Richards, Jinni Lue, Ms.
Executive Director
INTERNATIONAL SYNERGY INSTITUTE
1274 Calle de Comercio, Suite One
Santa Fe, NM 87501, United States of America
T: 1 505 471-4535 F: 1 505 988-5809

Richie, Melissa, Ms.
UP WITH PEOPLE
3103 North Campbell Av.
Tucson - AZ 85719, United States of America

Richman, Barbara T
Managing Editor
ENVIRONMENT
1319 18th St NW
Washington DC 20036-, United States of America
T: 1 202 296 6267 F: 1 202 2965149

Richman, Stacy
Deputy Bureau Chief
CBS NEWS
4770 Biscayne Blvd. Florida
Miami - Florida, United States of America
T: 1 305 5455000 F: 1 305 5450307

Richmond, Dorothy, Ms.
TERRA CHRISTA COMMUNICATIONS
1195 E. Edison
Tucson - AZ, United States of America
T: 1 602 577-6476 F: 1 602 577-1195

Richmond, Franko, Ms.
EARTH EMERGE & SEE
277 Hudson St.
Cornwall-on-HudsonNY 12520, United States of America

Rickenbrode , Robert K., Mr.
ECOLOGY TASK FORCE
Mailbox #869 Caltech
Pasadena - CA, United States of America
T: 1 818 356-6180

Riddle, Martyn, Mr.
WORLD BANK
1818 H St. NW
Washington - DC 20433, United States of America
T: 1 202 458-1943

Riley, Kate, Ms.
CAMPAIGN FOR THE EARTH
P.O.Box 250
Hurleyville 92747, United States of America

Riley, Matthew, Mr.
President
HORIZONS COUNSELING
P.O. Box 35893
Tucson - AZ 85740, United States of America

Rimar, Brian, Mr.
Enviromental Protection Specia
LIST ENVIRONMENTAL PROTECTION AGENCY
P.O. Box 2022
Denver, Colorado 80201, United States of America
T: 1 303 293-1673 F: 1 303 293-1488

Rindge, Karen, Ms.
Population Program Coordinator
NATIONAL WILDLIFE FEDERATION
1400 16th St., NW
Washington - DC 20036, United States of America
T: 1 202 939-3311 F: 1 202 797-5486

Rios, Astrid E., Ms.
CITY COLLEGE STUDENT N.Y. - INTERNATIONAL STUDIES
1212 8th Ave. apt.03
Brooklyn - NY 11215, United States of America
T: 1 718 832-2467

Rivers, Joyce M., Ms.
WORLD THAT WORKS FOR EVERYONE
2741 W 42nd Place
Anchorage - Alaska, United States of America
T: 1 907 248-2909

Roberts, Wilma, Ms.
UNITED METHODIST CHURCH
3777 Independece Avenue, 16J
Riverdale - NY 11201, United States of America

Robertson, Adinah, Ms.
ROBERT SUN COMPANY
240, Great Circle Road, #344
Tennessee, United States of America
T: 615 251-0680 F: 615 251-0682

Robertson, Wayne, Mr.
ROBERT SUN COMPANY
240, Great Circle Road, #344
Tennessee, United States of America
T: 615 251-0680 F: 615 251-0682

Robinson, Jerry
President & Editorial Director
CARTOONISTS & WRITERS SYNDICATE
67 Riverside Drive
New York NY 10024, United States of America
T: 1 212 3639256 F: 1 212 5954218

Roche, Maria Cristina O.
Translator
WGBH-TV
125 Western Ave.
Boston -Massachusett, United States of America
T: 1 617 4495830 F: 1 617 4446450

Rochedo, Santos, Mr.
Secretario De Rec Naturales
DEPT. RECURSOS NATURALES
Apdo Correo 5887,
San Juan, Puerto Rico, United States of America
T: 1 809 723-3090 F: 1 809 723-4255

Rodgers, R. Jean
HOPE PUBLISHING HOUSE
P.O. Box 291
Gridley - IL 61744, United States of America
T: 1 309 747-2541

Rodrigues, Gomercindo, Sr.
COOPERATIVA AGROEXTRAVISTA DE XAPURI
1880 East River Road
Tucson - AZ, United States of America
T: 1 602 577-6476 F: 1 602 577-1195

Rodriguez, Benjamin
HOPE PRODUCTIONS
Pasadena, CA, United States of America

Rodriguez, Rafael
Cameraman
CNN SPANISH
1 CNN Center
Atlanta - GA, United States of America
T: 1 404 8816727 F: 1 404 8273319

Rodriguez, Robert A., Esq., Mr.
Attorney
27-16 Seymour Ave.
Bronx - NY, United States of America
T: 1 718 381-6247

Roemer, Dana
GLOBAL ENVIRONMENT TEAM
2334 Chalet Gardens Rd. #1
Madison - WI 53711, United States of America
T: 1 608 271-6573

Roen, Nancy, Ms.
Vicepresident
FLORIDA POWER AND LIGHT COMPANY
PO Box 078768
Palm Beach, FL 33407, United States of America
T: 1 508-5460 F: 1 508-5403

Roesner, Lori
INTERNATIONAL UNION OF YOUNG CHRISTIAN DEMOCRATS - UIJDC
Caracas, United States of America
F: 58 2 2610547

Rogers, Cornish
CLARENCE SCHOOL OF THEOLOGY
1717 Sierra Madre Vella Av.
Pasadena - CA 91107, United States of America

Rogers, Sharon, Ms.
ENVIRONMENTAL COMMUNITY CONSULTANTS
17 Trembley Lane
Wright City - MA 63390, United States of America
T: 1 314 745-3980

Rogers Jr., Frank, Mr.
SCHOOL OF THEOLOGY AT CLAREMONT
1325 N. College Ave. PO Box 107
Claremont-CA 91711-3919, United States of America
T: 1 714 399-5706 F: 1 714 626-7062

Romero, Yvonne Mejia, Ms.
HUXLEY COLLEGE OF ENVIRONMENT STUDIES
Western Washington University, 516 High Street
Bellingham - WA 98225, United States of America
T: 206676 3520 F: 206647 6274

Roo, Macon
KIDS FOR A CLEAN ENVIRONMENT (F.A.C.E.)
P.O.Box 158254
Nashville TN 37215, United States of America

Roos, Andreas, Mr.
UP WITH PEOPLE
3103 North Campbell Av.
Tucson - AZ 85719, United States of America

Rose, Adam, Mr.
Professor
UNCTAD/PENNSYLVANIA STATE UNIVERSITY
222 Walker Building, University Park
University Park - PA 16802, United States of America
T: 814 865-2549 F: 814 863-7433

Rosen, Michael
Editor
ABC NEWS
47 West 66th street
New York - NY, United States of America
T: 1 212 4562800 F: 1 212 4562795

Rosenberg, Barbara, Dr.
FEDERATION OF AMERICAN SCIENCTISTS
735 Anderson Hill Road ns
Purchase - NY 10577-1400, United States of America
T: 1 914 251-6643 F: 1 914 251-6635

Rosenthal, Joyce, Ms.
Director
EPEA - HAMBURG ENVIRONMENTAL INSTITUTE
116 E. 27 St.
New York - NY 10016, United States of America
T: 1 212 685 4962

Rosier, Tara, Ms.
International Director
EC 2000
1717 Stewart St.
Santa Monica - CA 74074, United States of America
T: 1 310 453-7007

Rothschild, Brian, Mr.
UP WITH PEOPLE
3103 North Campbell Av.
Tucson - AZ 85719, United States of America

Rubiner, Sinde
GLOBAL FAMILY
P.O.Box 50063
Tucson - AZ 85703, United States of America
T: 1 602 577-6476 F: 1 602 577-1195

Ruiz-Valdepena, Jorge, Mr.
Professor
ESRI - EXIBITOR/EXPOSITOR
380 New York st.
Redlands - CA 92373, United States of America
T: 1 714 793-2853 F: 1 714 793-5953

Rupp, Judith, Ms.
Representative Represent
NATURAL WORLD
2120 Via Mar Valle
Del Mare 92014, United States of America
T: 619 792 0765

Russell, Gray, Mr.
N. J. ENVIRONMENTAL FEDERATION
14 South Park Street
Montclair 07042, United States of America
T: 701 783-5112 F: 201 783-5247

Russell, Robert H., Mr.
President
NEAHTAWANTA RESEARCH AND EDUCATION CENTER
1308 Neahtawnata Rd.
Traverse City-MI 49684, United States of America
T: 1 616 2237315 F: 1 616 2237557

Russell, Walter, Mr.
WORLD POLICY INSTITUTE
927 Toulouse 57, apt 3
New Orleans - LA, United States of America
T: 1 504 524-8648 F: 1 504 529-1105

Russian, Hilary, Ms.
YMCA EARTH ACTION CORPS
909 4th Av
Seattle - WA, United States of America
T: 1 206 382-5013 F: 1 206 382-7283

Ryan, Hope, Ms.
EARTH EMERGE & SEE
277 Hudson St.
Cornwall-on-Hudson-N 12520, United States of America
T: 1 914 342-5411 F: 1 914 342-3252

Rymjanov, Oraz
Board Secretary
FILMMAKERS' UNION OF KAZAKHSTAN
24 Sawmill Road.
Norwalk - CT 06851, United States of America

S.Abrams, Isabel
Science, Health + Environment
FREELANCER
2216 Schiller
Wilmington - Ohio, United States of America
T: 1 919 2518935

Sa, Luiz Carlos
WTN
Broadway 1955
New York NY, United States of America
T: 1 212 3624410

Sacca, Michael F.
Cameraman
WCAX - TV
PO Box 608
Burlington VE 05489, United States of America
T: 1 802 6586300 F: 1 802 6586300

Sadotomo, Tetsu, Mr.
Visiting Scholar
MERSHON CENTER OF THE OHIO STATE UNIVERSITY
199 West 10th Avenue
Columbus-OH 43201, United States of America
T: 1 614 2921681 F: 1 614 2922407

Salas, Gilbert
Camera Asst.
OUT OF EPOCA FILMS
N. Cahuenga Blvd.
Hollywood -CA 90038, United States of America
T: 1 213 4694405 F: 1 213 4695010

Salk, Judy, Ms.
Executive Editor
R.R. BOWKER/REED REFERENCE PUBLISHING
Serials/A&I Departments - 121 Chanlon Rd
New Providence, NY 07974, United States of America
T: 1 908 464-6800 F: 1 908 665-6688

Salleh, Ariel, Ms.
WOMENS' ENVIRONMENT AND DEVELOPMENT ORGANIZATION
737 east building
NW York 10003, United States of America
T: 212 9985635

Saltzman, Bob, Mr.
GLOBAL HEART
1443 Adelsas Ave.
Encinitas CA, United States of America

Saltzman, Catherine, Ms.
GLOBAL HEART
1443 Adelsas Ave.
Encinitas CA 92024, United States of America

Samson, Michael J., Mr.
Economic Consultant
COOPERATIVA CENTROAMERICANA
149 Columbia Heights
Brooklyn - NY 1120, United States of America
T: 1 718 703-2020 F: 1 718 791-4132

Samuels, Helen, Ms.
Director
CAMPAIGN FOR THE EARTH
2265 S Westwood Blvd.
Los Angeles - CA 90064, United States of America
T: 1 213 960-7888

Sander, Mr., Mr.
Ex. Managing Director
UNCTAD/KIDDER, PEABODY & CO.
20 Exchange Place
New York - NY 10005, United States of America
T: 1 212 510-4493 F: 1 212 510-4493

Sanders, Brian, Mr.
INTERNATIONAL STUDIES ASSOCIATION
1246 Walker Ave. #108
Walnut Creek-Ca 94596, United States of America
T: 1 510 933-4256

Sanders, Marcia
Programme Assistant
INTERNATIONAL WOMEN'S TRIBUNE CENTRE
777 United Nations Plaza
New York - NY 1007, United States of America
T: 1 212 6878633 F: 1 212 6612704

Sandor, Richard, Mr.
Ex Managing Director
UNCTAD/KIDDAR, PEABODY & CO.
20 Exchange place
New York 10005, United States of America
T: 212 510 4419 F: 212 510 4493

Sandvick, Carrie, Ms.
YMCA EARTH ACTION CORPS
909 4th Av
Seattle - WA, United States of America
T: 1 206 382-5013 F: 1 206 382-7283

Sanford, Beth
Pres.Pasadena Chapter
UNITED NATIONS ASSOCIATION
723 E Green St.
Pasadena - CA, United States of America
T: 1 818 4638350 F: 1 818 3032391

Sankaran, K., Dr.
THANE MUNICIPAL CORPORATION
1880 East River Road
Tucson - AZ, United States of America
T: 1 602 577-6476 F: 1 602 577-1195

Sansone, Marie, Ms.
Asst Attorney General
ALASKA DEPARTMENT OF LAW
c/o Elizabeth Kerttula, 1105 Mendenhall Penn Road
Juneau - Alaska 99801, United States of America
T: 1 907 790-2023 F: 1 907 465-5274

Santoro, Rosangela
TERRA CHRISTA COMMUNICATIONS
1880 East River Road
Tucson - AZ, United States of America
T: 1 602 577-6476 F: 1 602 577-1195

Santos, Adriana Dos Santos E, Sra.
Recepcionista
GLOBAL FORUM OF SPIRITUAL AND PARLIAMENTARY LEADERS ON HUMAN SURVIVAL
304 East 45th St. Room 476
New York - NY, United States of America
T: 1 212 953-7947 F: 1 212 557-20961

Santos, Angelo
Pesquisador
SMITHSONIAN INSTITUTION
Nhb/Mrc 180
Washington - DC 20560, United States of America
T: 1 202 786-2821 F: 1 202 786-2934

Santos, Gilberto Gomes Dos, Mr.
PONITIFICA UNIVERSIDADE CATOLICA DO RIO DE JANEIRO
1880 East River Road
Tucson - AZ, United States of America
T: 1 602 577-6476 F: 1 602 577-1195

Santos, Lydia, Ms.
Secretary
GLOBAL FORUM OF SPIRITUAL AND PARLIAMENTARY LEADERS ON HUMAN SURVIVAL
304 East 45th St. Room 476
New York - NY, United States of America
T: 1 212 953-7947 F: 1 212 557-20961

Santos, Maria Cecilia MacDowell Dos, Ms.
Pesquisadora
RAINFOREST ACTION NETWORK
San Some St.
San Francisco - CA 94111, United States of America
T: 1 415 3984404

Sarenas, Patricia M., Ms.
CEDPA KAHAYAG FDN/ PILIPINA/WAND
1717 Massachusetts Ave. NW, 2nd floor
Washington - DC 20036, United States of America
T: 1 202 667-1142 F: 1 202 332-4496
TL: 440384

Sarti, Riccardo, Mr.
President
ENVIRONMENTAL DEFENSE TECHNOLOGIES INC.
c/o 1737 H Street N.W.
Washington - DC, United States of America
T: 1 202 3317760 F: 1 202 3310463

Satouris., Elizabeth, Ms.
WORLDWIDE INDIGENOUS SCIENCE NETWORK
1880 East River Road
Tucson - AZ, United States of America
T: 1 602 577-6476 F: 1 602 577-1195

Sauni, Miriama, Mr.
WEDO
PO Box 3230
Pago Pago 96299, United States of America
T: 684 633-7212 F: 684 633-7212

Savitsky, Basil, Mr.
Directorthe Strom Thurmond Ins
TITUTE OF GOVERNMENT AND PUBLIC AFFAIRS
Strom Thurmond Institute, Clemson University
Clemson - SC 29634-5130, United States of America
T: 1 803 656-0214 F: 1 803 656-4780

Savitt, Chuck, Mr.
President
ISLAND PRESS
1718 Connecticut Ave. Suite 300
Washington - DC, United States of America
T: 1 202 232-7033 F: 1 202 234-1328

Sayres, Sandra, Mr.
Professor
THE COOPER UNION
Fac. of Humanities 141, Cooper Square
New York - NY 10003, United States of America
T: 1 212 353-4394 F: 1 212 4398

Scaglione, Cynthia M., Ms.
UNITED NATIONS ASSOCIATION
9201 Fair Oaks Ave, 807
Dallas - TX, United States of America
T: 1 214 6552106 (work) F: 1 214 5288516

Schaffer, Marilyn, Ms.
Director, Int'L Cntr. Ed. &tec
UNIVERSITY OF HARTFORD
200 Bloomfield Ave.
W. Hartford - CT 06117, United States of America
T: 203 243-4277 F: 203 242-7002

Schaffrath, Sara, Ms.
ANANDA MARGA (PROVT)
243 Bedford Road
New Boston - NH 03070, United States of America
T: 1 603 487-2199

Schartz, Michelle, Ms.
NATURAL HERITAGE INSTITUTE
1055 Fort Cronkhite
Sausalito - CA, United States of America
T: 1 415 3311232 F: 1 415 3311232

Schaub, Nancy, Ms.
UNCED-Team Member
FOUNDATION FOR GLOBAL COMMUNITY
222 High Street
Palo Alto - CA, United States of America
T: 1 415 328-7756 F: 1 415 328-7785
TL: 856588

Schiavo, Marcio
ECOTROPIC
1880 East River Road
Tucson - AZ, United States of America
T: 1 602 577-6476 F: 1 602 577-1195

Schipper, Lee, Mr.
Dr.
STOCKHOLM ENVIRONMENT INSTITUTE
International Studies, LBL, Room 3125 Bldg 90
Berkeley - CA 94720, United States of America
T: 1 510 486-5057 F: 1 510 486-6996
TL: 9103662067

Schmaltz, Mary Sue, Ms.
YOKOHAMA WOMEN'S FORUM
820 Greenst., apt.5A
Iselin - NY 08830, United States of America
T: 1 212 636-9143 F: 1 212 313-2654

Schmidt, Deborah Beaumont, Ms.
UNITED METHODIST CHURCH
426 Monroe Avenue
Helena - MT 59601, United States of America

Schmitz, Tosh Wulff, Mr.
FERTILITY OF THE EARTH
Dixon - NW 87527, United States of America
T: 1 505 579 4650

Schneider, Dennis, Mr.
R And D Director
MICRO-BAC INTERNATIONAL
9607 Gray Blvd
Austin - TX, United States of America
T: 1 512 837-1145 F: 1 512 837-5968

Schneider, Martin L., Mr.
1710 South Tyler
Amarillo - TX 79102, United States of America
T: 1 806 354-9331

Schnoor, Jerald L., Mr.
Professor
UNIVERSITY OF IOWA
1136 Engineering Bldg.
Iowa City, Iowa, United States of America
T: 1 319 335-5649 F: 1 319 335-5777

Schub, Nancy, Ms.
Team Member
FOUNDATION FOR GLOBAL COMMUNITY
E.1437 - 26Th
Spokane - WA 99203, United States of America
T: 1 509 5346431

Schwartz, Eric, Mr.
Staff
EARTH TRAIN/WORLDLINK/GLOBAL YOUYTH SUMMIT
3220 Sacramento
San Francisco - CA 94115, United States of America
T: 1 415 931-1290

Schwartz, Jonathan, Mr.
Director
INTERLOCK/LEARNTECH
19 Arrow St.
Cambridge - MA 02138, United States of America
T: 1 617 354-3595 F: 1 617 547-4711

Schwartzman, Stephan, Mr.
Anthropologist
ENVIRONMENTAL DEFENSE FUND
1616 P street, NW, suite 150
Washington 20036, United States of America
T: 202 3873500 F: 202 2346049

Schwarz, Gail, Mr.
Press
CAMPAIGN FOR THE EARTH
50 Graeser Acres
St. Louis, MO 63146, United States of America
T: 1 314 434-6580 F: 1 314 547-3107

Schwind, Ronald E., Mr.
Treasurer
A WORLD THAT WORKS FOR EVERYONE
HC02 Box 7251
Palmer - Alaska 01501, United States of America
T: 1 907 745-6243 F: 1 907 271-3260

Schwung, Stefan, Mr.
UP WITH PEOPLE
3103 North Campbell Av.
Tucson - AZ 85719, United States of America

Scotto, Erin Lynn, Ms.
SO.CALIF. ECOLOGY TASK FORCE
9203 Cord Ave.
Downey - CA 90240, United States of America
T: 1 310 861-3207

Scotto, Hazel Nila, Ms.
SO.CALIF. ECOLOGY TASK FORCE
9203 Cord Ave.
Downey - CA 90240, United States of America
T: 1 310 861-3207

Scroggs, Jenny, Ms.
YMCA EARTH ACTION CORPS
909 4th Av
Seattle - WA, United States of America
T: 1 206 382-5013 F: 1 206 382-7283

Selfa, Theresa, Ms.
Phd Candidate
CORNELL UNIVERSITY / DEPT. OF RURAL SOCIETY
132 Warren Hal
Ithaca-NY 14853, United States of America
T: 1 617 256-0460 F: 1 617 255-9984

Sellman, Buzz, Mr.
CONSORTIUM FOR INTERNATIONAL EARTH SCIENCE
1968 Green Rd., PO Box 134003
Ann Arbor - MI 48113, United States of America
T: 1 313 471-4673 F: 1 313 663-6622

Seniloni, Ratu Manasa, Mr.
Ambassador
PERMANENT MISSION TO THE REPUBLIC OF FIJI TO U.N.
1 United Nations Plaza 26TH Fl.
New York - NY 10017, United States of America
T: 1 212 444-3800 F: 1 212 444-3811

Seth, Anand, Mr.
Representative
WORLD BANK
1818 H St. NW
Washington - DC 20433, United States of America
T: 1 202 458-1943

Shaifer, Norman
Co-Ordinator
INDIGENOUS PEOPLES MEDIA CENTER
77 Main Street
New York - NY 10083, United States of America
T: 1 212 3650414 F: 1 212 3650864

Shakhavov, Mukhtar
FILMMAKER'S UNION OF KAZAKHSTAN
1880 East River Road
Tucson - AZ, United States of America
T: 1 602 577-6476 F: 1 602 577-1195

Shanley, Patricia, Sra.
Pesquisadora
WOODS HOLE RESEARCH CENTER
P.O.Box 296
Woods Hole - MA, United States of America
T: 1 508 540-9900

Sharp, Barbara, Ms.
ERDAS INC.
2801 Buford Highway, Suite 300
Atlanta - GA 30329, United States of America
T: 1 404 248-9000 F: 1 404 248-9400

Sharp, William, Mr.
ERDAS INC.
2801 Buford Highway, Suite 300
Atlanta - GA 30329, United States of America
T: 1 404 248-9000 F: 1 404 248-9400

Shaw, Vera C., Ms.
NACRE
10 Stiles Terrace
Newton Centre - MA 02159, United States of America
T: 1 617 332-2640

Sheehan, Jeffrey, Mr.
A WORLD THAT WORKS FOR EVERYONE
PO Box 2982
Seward - Alaska 99664, United States of America
T: 1 907 224-8773

Sheehey, Christopher B., Mr.
Associate
HUNTON & WILLIAMS
200 Park Avenue, 43rd Floor
New York - NY, United States of America
T: 1 212 3091184 F: 1 212 3091100

Shelkaev, Yurri, Mr.
FILMMAKER'S UNION OF KAZAKHSTAN
1880 East River Road
Tucson - AZ, United States of America
T: 1 602 577-6476 F: 1 602 577-1195

Shemandach, Leon
INDIGENOUS PEOPLES MEDIA CENTER
77 Main Street
New York 10083, United States of America
T: 914 3650414 F: 914 3650864

Shepperd, Lisa, Ms.
Research Assistant
F S E C
300 State Road 401
Cape Canaveral - Fl 32920, United States of America
T: 1 407 783-0300

Sher, Ferida, Ms.
CENTRE FOR DEVELOPMENT AND POPULATION ACTIVITIES - CEDPA
1717 Massachusetts Ave. NW, 2nd floor
Washington - DC 20036, United States of America
T: 1 202 667-1142 F: 1 202 332-4496
TL: 440384

Sherman, Scott, Mr.
Official Representative
ECOLOGY LAW QUARTERLY
Boalt Hall School of Law, UC Berkeley
Berkeley - CA 94720, United States of America
T: 1 510 643 F: 1 510 643 6171

Sherrow, Teri
CAMPAIGN FOR THE EARTH
1880 East River Road
Tucson - AZ, United States of America
T: 1 602 577-6476 F: 1 602 577-1195

Shilling, John, Mr.
Advisor
WORLD BANK
1818 H St. NW
Washington - DC 20433, United States of America
T: 1 202 458-1943

Shilling, Kaitlin, Ms.
Advisor
WORLD BANK
1839 Wyoming Ave. NW
Washington - DC 20009, United States of America
T: 1 202 462-5438

Shin, Hue Jai Chang, Ms.
TEMPLE OF UNDERSTANDING
St. Emeric Church / 185 Avenue D
New York - NY 10009, United States of America
T: 1 212 353-0001 F: 1 212 353-0170

Shinozaki, Akira, Mr.
Director
KEIDANREN
145 W57 St.
New York - NY 10019, United States of America
T: 1 212 489 6206 F: 1 212 489 6211

Shortsleeve, Kevin, Mr.
Volunteer Intern
LEARN TECH/INTERLOCK MEDIA
19 Arrow St.
Cambridge - MA 02238, United States of America
T: 1 617 354 3595 F: 1 617 547 4711

Shotkin, Bruce Mortimer, Mr.
President
PANOPOLIS CORPORATION
101 West Sixth Street, Suite 402
Austin 78701, United States of America
T: 1 512 476-3785 F: 1 512 476-3786

Shoumatoff, Kon, Mr.
Coordinator
WORLD PEACE PRAYER SOCIETY
800 Third Ave., 37th Floor
New York - NY 10017, United States of America
T: 1 212 755.4755 F: 1 212 935.1389

Showen, Mary Ann, Ms.
UNA - USA ORLANDO, FL CHAPTER
44 Riverview Lane
Cocoa Beach - FL, United States of America
T: 1 407 784-0641

Shrestha, Chandrika, Ms.
UNITED NATIONS DEVELOPMENT FUND FOR WOMEN
304 East 45th Street
New York - NY 10017, United States of America
T: 1 212 906-6440

Shull, Thomas, Mr.
HEARTLAND INSTITUTE
645 Griswold Ste. 2525
Detroit - MI 48226-4202, United States of America
T: 1 313 961-1950 F: 1 313 961-6186

Siemers, Paula, Ms.
HIGHLANDER CENTER
1939 Highlander Way
New Market - TN 37820, United States of America
T: 1 615 933 3443 F: 1 615 933 3424

Silberling, Louise, Ms.
Estudante
CORNELL UNIVERSITY
Warren Hall - Rural Sociology
Ithaca-NY ZIP: 14853, United States of America
F: 1 617 255-9984

Silen, Francisco, Mr.
Relacion Publica
AUTORIDAD ACUODUTOS E ALCANTARILLADOS
Av. Barbosa 604
San Juan, Puerto Rico, United States of America
T: 1 809 756-2222 F: 1 809 754-3092

Sills, Charles F., Mr.
Executive Director
THE BARBARA GAUNTLET FOUNDATION
International Square 1825 Eye St., N.W. suite 400
Washington - DC, United States of America
T: 1 202 429-2055 F: 1 202 466-3973

Sills, Joe
UN DEPARTMENT OF PUBLIC INFORMATION
Room S-1037D
New York _NY 10017, United States of America
T: 1 212 9638842 F: 1 212 9636914

Silva, Jose Vital Da
HOPE PRODUCTIONS
Pasadena, CA, United States of America

Silva, Julia, Sra.
CENTRO DOS TRABALHADORES DA AMAZONIA
1880 East River Road
Tucson - AZ, United States of America
T: 1 602 577-6476 F: 1 602 577-1195

Silva, Margarida, Ms.
QUERCUS-PORTUGAL
993 Dryden Road # 8
Ithaca-NY 14850, United States of America
T: 1 607 277-7821
E: WT3@CORNELLC.CIT.CORNELL.

Silva, Vera Lucia Da
HOPE PRODUCTIONS
Pasadena, CA, United States of America

Silver, Diane, Ms.
GREEN (GLOBAL RIVERS ENVIRONMENTAL EDUCATION NETWORK)
University of Michigan School of Natural Resouces -Dana Building 430E
Ann Arbor - MI, United States of America
T: 1 313 7645171 F: 1 313 9362195

Silvers, Barbara, Ms.
GLOBAL HEART
1880 East River Road
Tucson - AZ, United States of America
T: 1 602 577-6476 F: 1 602 577-1195

Silverstein, Duane, Mr.
Executive Director
GOLDMAN ENVIRONMENTAL FOUNDATION
One Lombard St., # 303
San Francisco - CA, United States of America
T: 1 415 788-9090 F: 1 415 788-7890

Simwanza, Elizabeth, Ms.
CENTER FOR THE STUDY OF WOMEN AND SOCIETY
420 West 118th St.
New York - NY 10027, United States of America

Sindab, Jean, Ms.
VNCCCUSA
475 Riverside Drive, Room 572
New York - NY 10015-0050, United States of America

Sindair, Tom, Mr.
Public Info. Manager
INST. FOR ENVIRONMENT STUDIES, UNIV. OF WISCONSIN
550 North Park St., 30A Science Hall
Madison - WI 53706, United States of America
T: 1 608 263 5599 F: 1 608 262 2273

Singh, Kathleen T, Ms.
Alternate Rep Un Ecosoc
PAN PACIFIC & SOUTHEAST ASIA WOMEN'S ASSOCIATION INTERNATIONAL
182 Lincoln Place
Brooklyn - NY 11217, United States of America
T: 1 6220058

Singh, Meera, Ms.
Programme Ass.
INTERNATIONAL WOMEN'S TRIBUNE CENTRE
777 UN Plaza
New York - NY 10017, United States of America
T: 1 212 687-8633 F: 1 212 661-2704

Singh, Rajwant, Dr.
GURU GOBIND SINGH FOUNDATION
Sikh Center, 13814 Travilan Road
N. Potomac - MD 13814, United States of America
T: 1 301 384-3576

Sinsheimer, Peter, Mr.
Student
UNIV. OF CALIFORNIA, LOS ANGELES
2721/2 3rd St #A
Santa Monica - CA 90405, United States of America
T: 1 310 396 6084

Sirri, Betty, Ms.
UNA-USA-PASADENA
757 Galaxy Hts.
La Canada - CA 91011, United States of America
T: 818 790-5904

Sirri, Norri, Mr.
UNA-USA-PASADENA
757, Galaxy Heights Dr.
La Canada 91011, United States of America
T: 818790 5904

Siry, Barbara, Ms.
ROLLINS COLLEGE
271 Sandlewood Trail
Winter Park - FL 32789, United States of America
T: 1 407 6296564

Siry, Joseph, Mr.
Professor
ROLLINS COLLEGE
271 Sandlewood Trail
Winter Park - FL 32789, United States of America
T: 1 407 6296564

Sites, Betsi, Ms.
CAMPAIGN FOR THE EARTH
P. O. Box 447
Santa Cruz - CA 95061, United States of America

Siudy, George S., Mr.
GLOBAL ENVIRONMENT TEAM
4455 Torrance Blvd Ste 379
Torrance - CA 90503, United States of America
T: 1 213 542-0685

Sklar, Scott, Mr.
Executive Director
UNITED STATES EXPORT COUNCIL FOR RENEWABLE ENERGY - US/ECRE
PO Box 10095
Arlington - VA, United States of America
T: 1 202 408-0665 F: 1 202 408-8536

Sloan, Garnsey, Ms.
STUDENT ENVIRONMENT ACTION COALITION (SEAC)
PO BOx 1232, Wheaton College
Norton - MA 02766, United States of America
T: 1 508 285 7720 F: 1 508 285 2908

Sloan, Julia, Ms.
PETA - PEOPLE FOR THE ETHICAL TREATMENT OF ANIMALS
735 Webster St.
San Francisco - CA 94117, United States of America
T: 1 415 292-6975

Smith, Grace, Ms.
Coordinator
TRADITIONAL HOPI & DINEH ELDERS
P.O.Box:1178
Winslow 86047, United States of America
T: 1 602 289 4968 F: 1 602 289 4968

Smith, Jay M., Mr.
THE ACADEMY
2722 Benvenue
Berkeley - CA, United States of America
T: 1 510 4549-0605

Smith, Jennifer, Ms.
GREEN (GLOBAL RIVERS ENVIRONMENTAL EDUCATION NETWORK)
University of Michigan School of Natural Resouces -Dana Building 430E
Ann Arbor - MI, United States of America
T: 1 313 7645171 F: 1 313 9362195

Smith, Kari, Ms.
Specialist
CALIFORNIA ENERGY COMMISSION
2643 17TH St.
Sacramento - CA 95818, United States of America
T: 1 916 654-4568

Smith, Kent, Mr.
Consultant
ESRI - EXPOSITOR
380 New York St.
Redlands - CA 92373, United States of America
T: 1 714 793-2853 F: 1 714 793-5953

Smith, Miranda, Ms.
President
MIRANDA SMITH PRODUCTIONS INC.
251 West 30th Street #16-W
New York - NY, United States of America
T: 1 212 5632370 F: 1 212 5632372

Smith, Skye, Ms.
TERRA CHRISTA COMMUNICATIONS
1880 East River Road
Tucson - AZ, United States of America
T: 1 602 577-6476 F: 1 602 577-1195

Smolla, Rodney, Mr.
Director
BILL OF RIGHTS INSTITUTE
william and mary law school
Willansburg -VA, United States of America
T: 804 221 3808 F: 804 221 3775

Sniderman, Julia, Ms.
1846 W Berkau
Chicago - IL, United States of America
T: 1 312 868-0418

Snyder, Ken, Mr.
Consultants
NATIONAL AUDUBON SOCIETY
c/o Fran Spruy-Weber 666 Pennsylvania Av. SE
Washington - DC 20003, United States of America
T: 1 202 547-9009 F: 1 202 547-9022

Soares, Wanda
TERRA CHRISTA COMMUNICATIONS
1880 East River Road
Tucson - AZ, United States of America
T: 1 602 577-6476 F: 1 602 577-1195

Soderling, Ragnar
CAMPAIGN FOR THE EARTH
2120 13th Street
Boulder - CO 80302, United States of America
T: 1 303 444-3800 F: 1 303 444-3811

Sodre, Maria Tereza
YMCA EARTH ACTION CORPS
909 4th Av
Seattle - WA, United States of America
T: 1 206 382-5013 F: 1 206 382-7283

Soltani, Atossa, Ms.
RAINFOREST ACTION NETWORK
1647 Ocean Front Walk #15
Santa Monica - CA, United States of America
T: 1 310 458-2068 F: 1 310 393-2154

Somplatsky-Jarman, William, Mr.
Associate Environmental Justic
PRESBYTERIAN CHURCH USA
100 Witherspoon St
Louisille - KE, United States of America
T: 1 502 5695809 F: 1 502 5695018

Sonobe, Eiichi
Correspondent
KYODO NEWS SERVICE
4312 Spencer St
Torrance CA 90503, United States of America
T: 1 310 3715534 F: 1 213 6803547

Soto, Thomas, Mr.
Owner/President
PS ENTERPRISES
309 Santa Monica Blvd, Suite 322
Santa Monica - CA 90401, United States of America
T: 1 310 393-3703 F: 1 310 393-7012

Soucie, Gary, Mr.
Senior Editorial Staff
NATIONAL GEOGRAPHIC MAGAZINE
1145 17th St NW
Washington - DC 20036, United States of America
T: 1 202 8577176 F: 1 202 4295767

Souleimenov, Oljas
President
NEVADA-SEMIPALATINSK ANTI-NUCLEAR MOVEMENT
24 Sawmill Road
Norwalk - CT 06851, United States of America

Souza, Maria Amalia, Ms.
UP WITH PEOPLE
3103 North Campbell Av.
Tucson - AZ 85719, United States of America

Souza, Marineusa De
HOPE PRODUCTIONS
Pasadena, CA, United States of America

Souza, Paulo Roberto Silva E, Mr.
COOPEAMA
1880 East River Road
Tucson - AZ, United States of America
T: 1 602 577-6476 F: 1 602 577-1195

Souza, Suelly Machado Lima De
Producer
CBS NEWS
4770 Biscayne Bolevard
Miami FL 37170, United States of America
T: 1 305 5455000 F: 1 305 5450307

Sozen, Joyce, Ms.
Representative
WORLD CONGRESS OF POETS
35 Grove, #GM
New York - NY 10014, United States of America
T: 1 212 255-5677 F: 1 212 243-3288

Sparr, Pamela, Ms.
UNITED METHODIST CHURCH OF BRAZIL
6714 Fairfax Road
Chevy Chase - MD 20815, United States of America

Spencer, Alice, Ms.
CERES / TOM'S OF MAINE
P.O. Box 710
Kennebunk - ME, United States of America
T: 1 207 985-2944 F: 1 207 985-3982

Sperry, Steve, Mr.
ERDAS INC.
2801 Buford Highway, Suite 300
Atlanta - GA 30329, United States of America
T: 1 404 248-9000 F: 1 404 248-9400

Speth, James Gustave, Mr.
President
WORLD RESOURCES INSTITUTE
1709 New York Ave., NW, Suite 700
Washington - DC, United States of America
T: 1 202 638-6300 F: 1 202 638-0036
TL: 64414WRIWASH

Spielman, Linda, Ms.
Member
U.S. CITIZENS NETWORK
1209 Windsor #5
Austin - TX, United States of America
T: 1 512 4787635

Spitz, Michael D., Mr.
President
LIFE FORCE
27232 102nd Dr. PO Box 1223
Stanwood, United States of America
T: 1 206 629-2040 F: 1 206 629-4069

Spring, Fiona
Student
LAUREL SPRINGS SCHOOL
PO Box 1440
Ojai - CA 93024, United States of America
T: 1 805 646-2473 F: 1 805 646-0186

Srilaa, , Ms.
TERRA CHRISTA COMMUNICATIONS
1880 East River Road
Tucson - AZ, United States of America
T: 1 602 577-6476 F: 1 602 577-1195

Stack, Victoria, Ms.
SIERRA CLUB
Harbor Ridge Dr.
Newport Beach - CA 92660, United States of America
T: 1 214 760-1814

Stanistreet, Shaun, Mr.
President
COMMITTEE OF THE WHALE
3600 South Harbor Blvd. #247
Channel Islands - CA, United States of America
T: 1 805 984-2734 F: 1 805 984-2734

Starbird, Ellen, Ms.
Economist
THE FUTURES GROUP
One Thomas Circle, Suite 600
Washington - DC, United States of America

Startari, Wayne, Mr.
8410 S.W. 154th Circle Ct. # 620
Miami - Florida 33193, United States of America
T: 1 305 388-1308

Stefani, Michelle, Ms.
KIDS AGAINST POLLUTION
PO Box 699 / 3 Kiley Court
Provincetown - MA 02657, United States of America
T: 1 508 487-1782

Stefani, Nancy, Ms.
KIDS AGAINST POLLUTION
PO Box 699 / 3 Kiley Court
Provincetown - MA 02657, United States of America
T: 1 508 487-1782

Stegenga, James, Dr.
Prof. Int'L Relations
PARDUE UNIVERSITY
Dept. Political Science
West Lafayette - IN 47907, United States of America
T: 1 317 494-4170

Stein, Edith Carol
President
ENVIRONMENTAL DATA RESEARCH INSTITUTE
797 Elmwood Ave
Rochester - NY, United States of America
T: 1 212 473-3090 F: 1 212 473-0968

Stern, Stephanie
INTERFAITH CONFERENCE / SENSE
5908 Dale Lane
Greendale-WI, United States of America

Stevens, Catherine Suarez, Ms.
Marketing Director
WILDLIFE ART INTERNATIONAL
11000-33 Metro PNWY
Fort Myers - FL 33912, United States of America
T: 1 813 936-7518 F: 1 813 936-2788

Stewart, Jim, Mr.
Communications
CAMPAIGN FOR THE EARTH
2120 13TH St.
Boulder - CO 80302, United States of America
T: 1 303 4943800 F: 1 303 4443811

Stiffarm, John
TERRA CHRISTA COMMUNICATIONS
1880 East River Road
Tucson - AZ, United States of America
T: 1 602 577-6476 F: 1 602 577-1195

Stirling, Catherine, Ms.
UP WITH PEOPLE
3103 North Campbell Av.
Tucson - AZ 85719, United States of America

Stone, Jean
HIGHLANDER CENTER
1939 Highlander Way
New Market - TN 37820, United States of America
T: 1 615 933 3443 F: 1 615 933 3424

Strachan, Deirdre, Mr.
Vice President
PATHFINDER INTERNATIONAL
Nene Galen Street, Suite 217
Watertown - MA 02172, United States of America
T: 1 617 924-7200 F: 1 617 924-3833
TL: 681-7095

Strasburg, Katherine, Ms.
Senior Research Associate
COMMONWEAL
PO Box 316
Bolinas - CA, United States of America
T: 1 415 8680970 F: 1 415 8682230

Strozier, Robert, Mr.
KPMG - PEAT MARWICK
3 Chestnut Ridge Road, Building 2
Montvale-NJ, United States of America
T: 1 201 307-7278 F: 1 201 307-7103

Stuart, Marilyn B., Ms.
President
POPULATION EDUCATION COMMITTEE
2425 Colorado Avenue, Suite 204
Santa Monica - CA, United States of America
T: 1 310 449-7744 F: 1 310 449-7741

Stull, Lyyn
President
PROJECT EARTH
921 SW Davenport
Portland - OR, United States of America
T: 1 503 228 6022 F: 1 503 2281574

Stylzer, Sergio
TERRA CHRISTA COMMUNICATIONS
1880 East River Road
Tucson - AZ, United States of America
T: 1 602 577-6476 F: 1 602 577-1195

Subak, Susan, Ms.
Information Manager
STOCKHOLM ENVIRONMENT INSTITUTE
Box 2142
Stockholm 103 14, United States of America
T: 46 8 7230260 F: 46 8 7230348

Sugarman, Jim, Mr.
Director
MULTINATIONALS AND DEVELOPMENT CLEARINGHOUSE
PO Box 19405
Washington - DC 20036, United States of America
T: 1 202 3878030 F: 1 202 2345176
TL: 6502633455

Sugasti, Alex, Mr.
TERRA CHRISTA COMMUNICATIONS
1880 East River Road
Tucson - AZ, United States of America
T: 1 602 577-6476 F: 1 602 577-1195

Sullivan, Katy Marie, Ms.
Admin. Assist.
SIERRA CLUB LEGAL DEFENSE FUND
180 Montgomery St, Suite 1400
San Francisco - CA, United States of America
T: 1 415 627-6700 F: 1 415 627-6740

Sultana, Bibi, Ms.
UNITED NATIONS DEVELOPMENT FUND FOR WOMEN
304 East 45th St.
New York - NY 10017, United States of America
T: 1 212 9066400 F: 1 212 9066705

Sumbung, Titi, Ms.
CENTRE FOR DEVELOPMENT AND POPULATION ACTIVITIES - CEDPA
1717 Massachusetts Ave. NW, 2nd floor
Washington - DC 20036, United States of America
T: 1 202 667-1142 F: 1 202 332-4496
TL: 440384

Suzuki, Sakae, Mr.
GLOBAL ENVIRONMENT TEAM
California Institure of Technology
Pasadena - CA 91125, United States of America
T: 1 818 356-6163

Swanson, Alfred B., Dr.
Chairman
ALTERNATIVE METHODS FOR INTERNATIONAL STABILITY (AMIS)
1900 Wealthy Street Suite 290
Grand Rapids - MI, United States of America
T: 1 616 774-0191 F: 1 616 774-8280

Swartz, Beth Ann, Ms.
Artist
GLOBAL HEART
1880 East River Road
Tucson - AZ, United States of America
T: 1 602 577-6476 F: 1 602 577-1195

Swift, Susan, Ms.
APPROPRIATE TECHNOLOGY INTERNATIONAL
1331 H St. NW
Washington - DC, United States of America
T: 1 202 8792900 F: 1 202 6284622
TL: 64661 ATI

Swisher, Randall, Mr.
Executive Director
AMERICAN WIND ENERGY ASSOC.
777 N. Capitol Street, N.E. Suite 805
Washington - DC, United States of America
T: 1 202 4088988 F: 1 202 4088536

Sygen, Vic, Mr.
LEMNA CORPORATION
1408 Northland Dr. 310
St. Paul - MN 55120, United States of America
T: 1 612 688-0836 F: 1 612 688-8813

Sylva, Carlos
TERRA CHRISTA COMMUNICATIONS
1880 East River Road
Tucson - AZ, United States of America
T: 1 602 577-6476 F: 1 602 577-1195

Syverson, Michelle, Ms.
Exec. Director
ENVIRONMENTAL NEWS NETWORK
1442-A Walhut Street - Suite 81
Berkeley - CA Zip.: 94709, United States of America
T: 1 510 524 0795

Talbot, John, Mr.
Management Director
GREEN SHOW
A-H House Upper Gong St.
Birmingham, United States of America

Talukder, Abdul Mannan, Mr.
Executive Director
INTERNATIONAL ORGANIZATION FOR WORLD PEACE, DISARMAMENT, DEVELOPMENT AND HUMAN RIGHTS
314-67 St. Apt.2R Brooklyn
New York - NY 11220, United States of America
T: 1 212 921 6352 F: 1 212 645 6866
TL: 6656MIC UN

Tanahashi, Kazuaki, Mr.
Steering Committee
PLUTONIUM FREE FUTURE
P.O. BOX 140 , 2018 Shattuck Avenue
Berkeley - CA 947090, United States of America
T: 1 510 5407645 F: 1 510 5407645

Tanaka, Yasuhiro
Bureau Chief Of Washington
AKAMATA
Suite 918 National Press Building
Washington - DC, United States of America
T: 1 202 3935238 F: 1 202 3935239

Taqueti, Claudemiro Fae, Mr.
President
THE AMAZONIA FOUNDATION
77 Town Hill Ave.
Danbury, 06810, United States of America
T: 203 778-2033 F: 203 778-4916

Tavares, John, Mr.
UP WITH PEOPLE
3103 North Campbell Av.
Tucson - AZ 85719, United States of America

Taylor, Joann Mcfarlane
Editor
CNN
1 CNN Center
Atlanta, GA 30303, United States of America
T: 55 21 2595407

Taylor, Martha, Ms.
QUAKER UNITED NATIONS
1613 Lawrence Road
Lawrenceville - NJ 08648, United States of America
T: 1 609 771-8160 F: 1 609 924-3556

Taylor-Snow, Dianne, Ms.
Secretary
INTERNATIONAL PRIMATE PROTECTION LEAGUE
POB 766
Summerwell - SC 29484, United States of America
T: 209 275 2834 F: 209 275 2834

Tdc, Carlos Humberto
Photographer/ Correspondent
CONTACT PRESS IMAGES
116 East 27th ST 8th floor
New York - NY, United States of America
T: 1 212 481 6910 F: 1 212 4816909

Tendick, Barton, Mr.
President-Harbor Chapter
UNA PASADENA
955 Ashbridge Lane
Harbor City - CA 90710, United States of America
T: 1 310 530-3656

Terjen, Judy Bend, Ms.
President
ECO - SPHERE - JB PRODUCTIONS
1333 Starling Court
Virginia Beach 23451, United States of America
T: 1 804 4220660 F: 1 804 4912090

Thacher, Shaw, Mr.
Programmer
ESRI - EXIBITOR
380 New York St.
Redlands - CA 92373, United States of America
T: 1 714 793-2853 F: 1 714 793-5953

Thalman, C. Dart, Dr.
Associate Professor
SCHOOL FOR INTERNATIONAL TRAINING / EXPERIMENT IN INTERNATIONAL LIVING
Kipling Road
Brattleberg-VT VT 05301, United States of America
T: 1 802 257 7751 F: 1 802 254 6674

Theiler, Julio, Sr.
FACULDADE DE CIENCIAS HIDRICAS
Cc 495 Parje El Doro
Santa Fe, NM, United States of America
T: 1 505 042 21384 F: 1 505 042 52494

Theon, John, Dr.
Chief
NASA - USA
2871 Madison Av. Suite 908
New York - NY, United States of America
T: 1 212 340-4150 F: 1 212 297-0967

Tholkes, Steve, Mr.
UP WITH PEOPLE
3103 North Campbell Av.
Tucson - AZ 85719, United States of America

Thomas, Christopher
Photographer
SIERRA
730 Polk Street
San Francisco -CA 9, United States of America
T: 415-9235608 F: 415-7764868

Thompson, June, Dr.
Staff
EARTH TRAIN/WORLDLINK/GLOBAL YOUTH SUMMIT
40 Deerpath Drive
Portola Valley - CA 94028, United States of America
T: 1 415 851-4293 F: 1 415 834-2275

Thompson, Odell, Mr.
UNITED METHODIST CHURCH OF BRAZIL
1740 Sixth Street
Beloit - WI 53511, United States of America

Thompson, Peter P.
Producer
LIVING ON EARTH/NATIONAL PUBLIC RADIO
55 Wheeler Street
Cambridge, MA 02138, United States of America
T: 617 8688810 F: 617 8688659

Thompson, Willis H., Mr.
Professor
SAN DIEGO STATE UNIVERSITY
Natural Science Dept., S.D.S.U.
San Diego - CA, United States of America
T: 1 619 594-6594 F: 1 619 454-7761

Thorne, Elizabeth, Ms.
UNIVERSITY OF CALIFORNIA AT SANTA CRUZ
2091 Upper Scenic Dr.
Felton - CA, United States of America
T: 1 408 3355234

Tietenberg, Tom, Mr.
Professor
UNCTAD/COLBY COLLEGE
Maine
Maine 04901, United States of America
T: 207 872-3143 F: 207 872-3555

Tilousi, Carletta
SOUTH WEST NETWORK FOR ECONOMIC & ENVIRONMENTAL JUSTICE
1939 Highlander Way
New Market - TN 37820, United States of America
T: 1 615 933 3443 F: 1 615 933 3424

Timpson, Sarah, Ms.
Director
PNUD
One United Nations Plaza
New York - NY 10017, United States of America
T: 1 212 906-6090

Tisbert, Ronald, Mr.
L.A. Ca. Representative
HELP END MARIJUANA PROHIBITION - HEMP
5632 Vannuys Bl. 310 California
T: 392-1806

Tomalsquim, Mauricio
ECOTROPIC
1880 East River Road
Tucson - AZ, United States of America
T: 1 602 577-6476 F: 1 602 577-1195

Topalian, Teny, Dr.
Assistant Professor
CALIFORNIA STATE UNIVERSITY - LONG BEACH
College of Natural Sciences 1250 Bellflower Blvd.
Long Beach - CA 90840, United States of America
T: 1 310 485-5220 F: 1 310 495-0656

Torres, Padre Henry L. Beauchamp, Sr.
Delegado
MAYAGUEZANOS POR LA SALUD Y EL AMBIENTE
Calle Claudio Carrero #293 El Mani
Mayaguez, Puerto Rico, United States of America
T: 1 809 831-6544

Totten, Michael, Mr.
Senior Associate
INTERNATIONAL INSTITUTE FOR ENERGY CONSERVATION
420 C Street NE
Washington - DC, United States of America
T: 1 202 546-3388 F: 1 202 546-6978

Toure, Nkenge, Ms.
GLOBAL ECONOMY PROJECT
1007 Maryland Ave. NE #302
Washington - DC 20002, United States of America
T: 1 202 546-8459

Tourrelles, Francisco A., Mr.
Vice-President
NATURE CONSERVANCY
1815 North Lynn Street
Arlington - VA 22209, United States of America
T: 1 703 841-4861 F: 1 703 841-8796

Towler Iii, William, Mr.
GLOBAL ENVIRONMENT TEAM
160 Westminster Rd.
Rochester - NY 14607, United States of America
T: 1 716 244-9073

Travesi, Carolina, Ms.
CAMPAIGN FOR THE EARTH
1880 East River Road
Tucson - AZ, United States of America
T: 1 602 577-6476 F: 1 602 577-1195

Traylor, Martha, Ms.
Dr.
INTERNATIONAL FACILITATING
1613 Lawrence Rd
Lawrenceville - NJ 08648, United States of America
T: 1 609 924 4536 F: 1 609 924 3556

Trinta, Joaozinho, Sr.
ESPACO FLOR DO AMANHA
1880 East River Road
Tucson - AZ, United States of America
T: 1 602 577-6476 F: 1 602 577-1195

Truebody, Miriam, Ms.
CENTRE FOR DEVELOPMENT AND POPULATION ACTIVITIES - CEDPA
1717 Massachusetts Ave. NW, 2nd floor
Washington - DC 20036, United States of America
T: 1 202 667-1142 F: 1 202 332-4496
TL: 440384

Truppel, Wagner, Dr.
QUERCUS-PORTUGAL
993 Dryden Road # 8
Ithaca-NY 14850, United States of America
T: 1 607 277-7821
E: WT3@CORNELLC.CIT.CORNELL.

Trzil, Jan Lee, Ms.
Member
U.S. CITIZENS NETWORK
8246 SE Rhone St.
Portland - OR, United States of America
T: 1 503 771-4275 F: 1 503 273-4822

Tsomo, Tsering, Ms.
Research Associate
ECO TIBET - CALIFORNIA
347 Dolores st., Room 206
San Francisco - CA 94110, United States of America
T: 1 415 241-9197 F: 1 415 626-0865

Tucker, Cora, Ms.
NATIONAL TOXICS CAMPAIGN
1939 Highlander Way
New Market - TN 37820, United States of America
T: 1 615 933 3443 F: 1 615 933 3424

Tuddenham, Peter, Mr.
Executive Director
THE COLLEGE EXPLORATION
1431, Towlston Road
Vienna 22182, United States of America
T: 703255 2672 F: 703255 2792

Turchyn, Luba
WORLD INFORMATION TRANSFER INC
274 Madison Ave, Suite 601
New York - NY 10016, United States of America
T: 1 212 2730253 F: 1 212 3792004

Turkovic, Robert, Mr.
NORTHERN TELECOM
5505 NW 60 Dr.
Coal Springs - FL 33067, United States of America

Twine, Francine Winddance Twine, Sra.
Pesquisa
UN. BERKELEY - ANTHROPOLOGY
Rua Bernardino de Oliveira no. 22
Varre-Sai - RJ 28386, United States of America
T: 55 249 433228

Tyler, Robert R., Mr.
OMNIGRAPHICS
Penobscot Blvd.
Detroit - MI, United States of America
T: 1 313 961-1340

Ucatowsha, Lisinka, Dr.
Journalist
RADIO FOR PEACE INTERNATIONAL
170 - 40 Henley Road #1B
Jamaica Estates NY, United States of America
T: 1 718 6585872 F: 1 718 6585158

Ulrych, Andre
ENVIRONMENTAL SERVICES GROUP
1880 East River Road
Tucson - AZ, United States of America
T: 1 602 577-6476 F: 1 602 577-1195

Ulrych, Judi, Ms.
ENVIRONMENTAL SERVICES GROUP
1880 East River Road
Tucson - AZ, United States of America
T: 1 602 577-6476 F: 1 602 577-1195

Underwood, Margo L., Ms.
Past President
IOWA AAUW - AMERICAN ASSOCIATION OF UNVERSITY WOMEN
2 Post Road
Mason City-IA 50401, United States of America
T: 1 515 4248600

Underwood, Paula, Ms.
CENTER FOR INTERNATIONAL COOPERATION
P.O Box 216
San Anselmo 944979, United States of America
T: 1 703 9375219 F: 1 703 9374039

Urb, Tarmo, Mr.
GLOBAL HEART
2308 Alma Ave,
Manhattan Beach - CA 90366, United States of America

Urb, Thomas, Mr.
GLOBAL HEART
2308 Alma Ave.
Manhattan Beach - CA 90266, United States of America

Uriate, Maria Isabel Ibarrola, Ms.
CENTRE FOR DEVELOPMENT AND POPULATION ACTIVITIES - CEDPA
1717 Massachusetts Ave. NW, 2nd floor
Washington - DC 20036, United States of America
T: 1 202 667-1142 F: 1 202 332-4496
TL: 440384

Urmantaev, Akim
FILMMAKER'S UNION OF KAZAKHSTAN
1880 East River Road
Tucson - AZ, United States of America
T: 1 602 577-6476 F: 1 602 577-1195

Valenzuela, Itala, Sra.
Sr. Technical Adviser
CENTRE FOR DEVELOPMENT AND POPULATION ACTIVITIES - CEDPA
1717 Massachusetts Ave. NW, 2nd floor
Washington - DC 20036, United States of America
T: 1 202 667-1142 F: 1 202 332-4496
TL: 440384

Valladares, Guilerme, Mr.
UNIVERSITY OF CALIFORNIA
319 Anapamu # 13
Santa Barbara - CA 93101, United States of America
T: 1 805 564 8434

Valle, Doris, Ms.
Directora Oficina Comunicacao
JUNTA DE CALIDAD AMBIENTAL
Calle 2 # 42 Ure Villa Marisla
Verdi Isla, Puerto Rico 00979, United States of America
T: 809 728-2261 F: 809 762-8348

Van Daam, John, Mr.
President
CALIFORNIA VIDEO PRODUCTIONS
10 Liberty Ship Way # 108C
Sausalito - CA 94965, United States of America
T: 1 415 332-5300 F: 1 415 332-1319

Vanderzee, Darrel, Mr.
UNEP-GRID
Ekos Data Center
Stoux Falls - SD 57198, United States of America
T: 605 594 6854 F: 605594 6589

Vargas, German, Sr.
Environemtal Ed. Consultant
INTERLOCK/LEARNTECH
55 Chapel Street
Newton Centre - MA 02160, United States of America
T: 1 617 354-3595 F: 1 617 547-4711

Vasconcelos, Mary Ann
THE COUNTY TELEGRAM-TRIBUNE
San Luis Obispo CA, United States of America

Vazao, Aldina, Ms.
BROWN UNIVERSITY
Box 5729 Brown University
Providence 02912, United States of America
T: 401 521 6438 F: 401 863 1270

Vazquez, Jaime, Sr.
INTERNATONAL SECRETARY OF NUCLEAR FREE ZONES
280 Groul St.
Jersey City ZIP: 07302, United States of America
T: 201 547-5204 F: 201 547-5230

Veach, Kevin C., Mr.
Enviromental Engineer
4725 35 Avenue S.
Minneapolis - MN, United States of America
T: 1 612 721-8903

Veit, John, Mr.
WORLD ASSOCIATION OF FORMER UN INTERNS FELLOW
138th St and Convent Ave, Rm 6/109
New York - NY, United States of America
T: 1 212 650-6909 F: 1 212 650-6970

Velasquez, Linda, Ms.
UNA PASADENA
223 Museum DR
Los Angeles - CA, United States of America
T: 1 213 223.0043

Velez, Mavirdia, Ms.
Dr.
U OF M - PRSO - SOLIDARITY
118 N. Thayer #2
Ann Arbor - MI 48823, United States of America
T: 1 313 995-0355

Venkataraman, M.C., Mr.
MTNL
1880 East River Road
Tucson - AZ, United States of America
T: 1 602 577-6476 F: 1 602 577-1195

Vidal, Nilda
TERRA CHRISTA COMMUNICATIONS
1880 East River Road
Tucson - AZ, United States of America
T: 1 602 577-6476 F: 1 602 577-1195

Vilas, Franklin E., Dr.
Chairman Environmental Comm.
EPISCOPAL CHURCH
94 Chatham Street
Chatham - NJ, United States of America
T: 1 201 635-8085

Villarosa, Lori, Sra.
Writer
MOTT FOUNDATION
Flint,MI, United States of America

Viscardi, Tadeu
TERRA CHRISTA COMMUNICATIONS
1880 East River Road
Tucson - AZ, United States of America
T: 1 602 577-6476 F: 1 602 577-1195

Vittachi, Varindra Tarzie, Ms.
Press
GLOBAL FORUM OF SPIRUTUAL AND PARLIMENTARY LEADERS
304 East 45th St., 4th Floor
New York - NY, United States of America
T: 1 212 953-7947 F: 1 212 557-2061

Vogel, Joseph Henry, Mr.
Director
POLYGLOTT
816 East Laurel St.
Hattiesburg - MS 39401, United States of America
T: 1 601 583-9452 F: 1 601 584-8906
E: vogel@ct6usm.bitnet

Wachholz, Chris, Mr.
Director Development
DIVERS ALERT NETWORK
3101 Petty Rd, Suite 1300
Durham - NC 27707, United States of America
T: 1 919 490.6055 F: 1 919 490 6630

Wade, Isabel, Ms.
URBAN RESOURCE SYSTEMS
1783 Buenavista West
San Francisco - CA 94117, United States of America

Wade, Susan, Ms.
TEMPLE OF UNDERSTANDING
St. Emeric Church / 185 Avenue D
New York - NY 10009, United States of America
T: 1 212 353-0001 F: 1 212 353-0170

Wakeley, Carol, Ms.
Leader
GIRLS SCOUTS U.S.A.
34 Legend Lane
Ballston Spa - NY 12020, United States of America
T: 1 518 583-0481

Wall, Michael, Mr.
Program Director
EARTH TRAIN/WORLDLINK/GLOBAL YOUTH SUMMIT
3220 Sacramento St.
San Francisco - CA 94115, United States of America
T: 1 415 931 1290

Walla, Grace, Ms.
CENTRE FOR DEVELOPMENT AND POPULATION ACTIVITIES - CEDPA
1717 Massachusetts Ave. NW, 2nd floor
Washington - DC 20036, United States of America
T: 1 202 667-1142 F: 1 202 332-4496
TL: 440384

Wallace, Michael
SCHOOL OF THEOLOGY AT CLAREMONT
1325 N. College Ave.#372
Claremont-CA 91711-3919, United States of America
T: 1 714 399-5706 F: 1 714 626-7062

Walsh, Bryan, Mr.
Monsignor
ARCHDIOCESE OF MIAMI
9401 Biscayne Blvd.D
Miami - Florida 33138, United States of America
T: 1 305 577-6476 F: 1 305 577-1195

Ward, Linda A., Ms.
GLOBAL ENVIRONMENT TEAM
6549 Sunnyside Ave.
Seattle - WA, United States of America
T: 1 206 523-4309

Ward, Robert E., Mr.
GLOBAL ENVIRONMENT TEAM
696 South Madison Avenue
Pasadena - CA 91106, United States of America
T: 1 818 796-2951 F: 1 818 792-2121

Ward, Timothy, Mr.
English Outreach
INTERNATIONAL SYNERGY INSTITUTE
6800 Breeze Wood Ter.
Rockville-MD, United States of America
T: 1 301 984-1523

Wason, Bill
ENVIRONMENTAL SOLUTIONS
1949 McAllister St.
San Francisco - CA 94115, United States of America
T: 1 415 474-5423 F: 1 415 474-5423

Waterfall, Julia, Ms.
UP WITH PEOPLE
3103 North Campbell Av.
Tucson - AZ 85719, United States of America

Watson, Beverly, Ms.
Intern
WORLD RESOURCE INSTITUTE
1709 New York Avenue
Washington - DC 20004, United States of America
T: 1 202 638-6300 F: 1 202 638-0034

Watson, Leon, Mr.
Phd
US CITIZENS NETWORK ON UNCED
P.O. Box 4701
East Lansing, United States of America
T: 1 517 355-5964 F: 1 517 336-2296

Watson, Paul, Mr.
Captain
SEA SHEPHERD CONSERVATION SOCIETY
1314 @nd St.
Santa Monica - CA 90401, United States of America
T: 1 310 394-3198 F: 1 310 394-0360

Watts, Cheryl A.
SCHOOL OF THEOLOGY
5601 W. Burnfield St.
Los Angeles - CA 90016, United States of America
T: 1 213 751-8724

Wawro, James, Mr.
Chair - Intl
AMERICAN BAR ASSOICATION
750 North Lake Shore Drive
Chicago, IL 60611, United States of America
T: 1 213 612-2500 F: 1 213 612-2554

Weatherspoon, Jacquelyne K., Ms.
U.N. Observer
CHURCH WOMEN UNITED
53 Court St.
Exeter - NH, United States of America
T: 1 603 778-9493

Weatherspoon, Russell D., Mr.
Instructor
PHILLIPS EXETER ACADEMY
53 Court St. - Exeter
New Hampshire 03833, United States of America

Weaver, Dennis, Mr.
KIDS FOR A CLEAN ENVIRONMENT (F.A.C.E)
13867 County Rd. #1
Ridgeway CO 81432, United States of America

Weaver, Gerry, Ms.
LOVE IS FEEDING EVERYONE (L.I.F.E)
13867 County Road #1
Ridgway CO 81432, United States of America

Weber, Linda R., Dr.
Assistant Professor
STATE UNIVERSITY OF NEW YORK INSTITUTE OF TECHNOLOGY AT UTICA/ROME
Department of Sociology, PO Box 3050
Utica - NY 13504-3050, United States of America
T: 1 315 792-7324 F: 1 315 792-7222

Weichbrodt, Randall, Mr.
10 Boulevard Dr. #101
Danbury - CT 06810, United States of America
T: 1 203 730-8924

Weiss, Arnie, Mr.
GLOBAL ENERGY NETWORK INT'L
PO Box 81565
San Diego - CA 92138, United States of America
T: 1 619 595-0139 F: 1 619 595-0403

Weiss, Mara Noble, Mr.
Intern
YOUTH ALLIANCE FOR SOCIAL SURVIVAL
2701 Benvenue # 5
Berkeley - CA 94705, United States of America
T: 1 510 704-0604 F: 1 510 549-9960

Weld, Jamie
UP WITH PEOPLE
3103 North Campbell Av.
Tucson - AZ 85719, United States of America

Wells, Rebecca
INDIGENOUS PEOPLES MEDIA CENTER
77 Main Street
New York - NY, United States of America
T: 1 212 3650414 F: 1 212 3650864

Wemhoff, Daniel M., Mr.
ADVOGADO INTERNACIONAL
1111 Arlington Blvd. #B03
Arlington - VA, United States of America
T: 1 703 538-9117

Wescott, William F.
Senior Consultant
ARTHUR D.LITTLE, INC.
Acorn Park Cambridge
Massachusetts 02140-, United States of America
T: 617 8645770 F: 617 8648719

Weston, Burns H., Mr.
Professor
UNIVERSITY OF IOWA COLLEGE OF LAW
Iowa City, Iowa, United States of America
T: 1 319 335-9106 F: 1 319 335-9019

Whaley, Kevin, Mr.
POPULATION INSTITUTE
Biopnysies/Jenkins, John Wopkins U., 34th +Charles
Baltimore, Md. 21218, United States of America

Whelan, Carolyn, Ms.
EARTH ISLAND INSTITUTE
300 Broadway, Suite 28
San Francisco - CA 94133, United States of America
T: 1 415 788-3666 F: 1 415 788-7324
TL: 6502829302

Whitaker, Paul, Mr.
UP WITH PEOPLE
3103 North Campbell Av.
Tucson - AZ 85719, United States of America

White, C. Dale, Mr.
Bishop
GENARAL BOARD OF CHURCH & SOCIETY /UNITED METHODIST CHURCHES
252 Bryant Ave.
White Plains - NY, United States of America
T: 1 914 684-6922 F: 1 914 684-6874

White, Jeanne Marie, Ms.
Coordinator
MEDITATION GROUPS INC
PO Box 566
Ojai - CA, United States of America
T: 1 805 6465508 F: 1 805 6465520

White, P.L
TERRA CHRISTA COMMUNICATIONS
1880 East River Road
Tucson - AZ, United States of America
T: 1 602 577-6476 F: 1 602 577-1195

White, Phil, Mr.
Coordinator
COMUNICA
PO Box 245
Ojai - CA, United States of America
T: 1 805 6531722 F: 1 805 6537260

White, Sasha, Ms.
CAMPAIGN FOR THE EARTH
2120 13th Street
Boulder - CO 80302, United States of America
T: 1 303 444-3800 F: 1 303 444-3811

Whitehead, Brad, Mr.
Associate
MCKINSEY AND CO.
130, E. 9th. Street, Suite 2400
Cleveland - OH 44114, United States of America
T: 1 216 861 9300 F: 1 216 861 9315

Wiener, Jonathan, Mr.
Policy Council
US OFFICE OF SCIENCE AND TRCHNOLOGY
Old Executive Office Building 431
Washington - DC 20500, United States of America
T: 1 202 3953638 F: 1 202 3953719

Wiley, Laura K., Ms.
Administrative Assistant
INTERNATIONAL COUNCIL OF WOMEN - ICW
777 U.N. Plaza, 7th Floor
New York - NY 10017, United States of America
T: 1 212 697-1278 F: 1 212 972-0164

Willard, Monica, Ms.
GLOBAL ENVIRONMENT TEAM / THE RIBBON INTERNATIONAL
3 Harbor Ct., Centerport, NY
Centerport 11721, United States of America
T: 1 516 754-1008

Willey, Zach, Mr.
Economist
ENVIRONMENTAL DEFENSE FUND
257 Park Ave. South
New York - NY 10010, United States of America
T: 1 212 505-2100 F: 1 212 505-2375

Williams, Aubrey, Mr.
WORLD BANK
1818 H. St, N.W.
Washington - DC 20433, United States of America

Williams, Benjamin B., Mr.
GLOBAL OBJECTIVES
16 Henchman Street Apt. 3M
Boston - MA 02113, United States of America
T: 1 617 367-6321 F: 1 617 367-3167

Williams, Neville, Mr.
President
SOLAR ELECTRIC LIGHT FUND
1739 Connecticut Ave NW
Washington - DC 20009, United States of America
T: 1 202 2347265 F: 1 202 2347265

Williams, Ray, Mr.
YMCA EARTH ACTION CORPS
909 4th Av
Seattle - WA, United States of America
T: 1 206 382-5013 F: 1 206 382-7283

Williams, Raymond
Chairman
INDIGENOUS PEOPLES MEDIA CENTER
77 Main Street
New York 10083, United States of America
T: 914 3650414 F: 914 3650864

Williamson, Jan, Ms.
GEOSPHERE PROJECT, THE
146 Entrada Drive
Santa Monica - CA 90402, United States of America
T: 1 310 459-4342 F: 1 310 459-8299

Williamson, Theresa D., Ms.
STUDENTS FOR ENVIRONMENTAL ACTION
3919 Oliver St.
Chevy Chase - MD 20315, United States of America
T: 1 654-5312

Wilson, Larry, Mr.
HIGHLANDER RESERCH AND EDUCATION CENTER
1939 Highlander Way
New Market - TN 37820, United States of America
T: 1 615 933 3443 F: 1 615 933 3424

Wilson, Michelle, Ms.
Staff
EARTH TRAIN/WORLDLINK GLOBAL YOUTH SUMMIT
Earth Train, 1212 Preservation Park Way
Oakland - CA 94612, United States of America
T: 1 510 834 2272 F: 1 510 834 2275

Winslow, Bill, Mr.
Production Assistant
VISION TV (CANADA)
475 Riverside, 16th floor
New York - NY, United States of America
T: 1 212 932-1236

Winston, Roland, Mr.
Chairman
UNIVERSITY OF CHICAGO, DEPARTMENT OF PHYSICS
5640 S. Ellis Ave.
Chicago - IL 60637, United States of America
T: 1 312 702-7756 F: 1 312 702-2045

Winter, Francis De, Mr.
Board Member
INTERNATIONAL. SOC. EN. SOC.
1401 Laurent St.
Santa Cruz - CA, United States of America
T: 1 408 425-1211 F: 1 408 425-8533

Winter, Paul, Mr.
Misician
PARLIAMENTARY GLOBAL FORUM
Box 68
Litchfield, United States of America
T: 203 567-8796 F: 203 567-4276

Wise, Steve, Mr.
CIESIN
2250 Pierce Rd.
Saginaw, Michigan 48710, United States of America
T: 1 517 791 7371 F: 1 517 790 0735

Wister, Effie, Ms.
RICCIARDI FAMILY FOUNDATION
105 East 16TH Street #9S
New York - NY 1003, United States of America

Witherite, Lou, Ms.
Director Int'L Programmes
UNITARIAN UNIVERSALIST SERVICE COMMITTEE
130 Prospect Street
Cambridge - MA 02139, United States of America
T: 1 617 868.6600 F: 1 617 868.7102

Woestyne, Geert Van De, Mr.
SAGRED EARTH NETWORK
c/o Phoenix 3232 SW 35 Blvd suite 121
Gainesville, United States of America
T: 1 904 466-0222 F: 1 904 466-0221

Wolcott, Nancy, Ms.
TEMPLE OF UNDERSTANDING
St. Emeric Church / 185 Avenue D
New York - NY 10009, United States of America
T: 1 212 353-0001 F: 1 212 353-0170

Wold, Chris A., Mr.
Atty
CENTER FOR INTENATIONAL ENVIRONMENTAL LAW - US
1621 Connecticut Av., Suite 300
Washington - DC 20009-2076, United States of America
T: 1 202 332-4840 F: 1 202 332-4865

Wood, Susan, Ms.
INTERNATIONAL WOMEN'S HEALTH COALITION
24 East 21ST St.
New York - NY 10010, United States of America
T: 1 212 979-8500 F: 1 212 979-9009

Woodruf, Paul H., Mr.
President
ENVIRONMENTAL RESOUCES MANAGEMENT, INC.
855 Springdale Drive
Exton-PA, United States of America
T: 1 215 524-3501 F: 1 215 524-8980

Woodwell, George M., Dr.
WOODS HOLE RESEARCH CENTER
13 Church Street - P.O. Box 296
Woods Hole - MA 02543, United States of America
T: 1 508 540-9900 F: 1 508 540-9700

Worth, Tim, Mr.
Senator
SENATOR COLORADO
1880 East River Road
Tucson - AZ, United States of America
T: 1 602 577-6476 F: 1 602 577-1195

Wright, Lloyd, Mr.
Environmental Expert
US EPA
United Nations Plaza 2, Room DC21110
New York - NY 10017, United States of America
T: 1 212 963-3119 F: 1 212 963-2146

Wright, Tom
Editor
ADVANCED TELEVISION PUBLISHING
4955 Castana 1#
Lakewood - CO, United States of America
T: 1 303 5311589 F: 1 303 4874187

Wuerth, Soren
Member
WORLD THAT WORKS FOR EVERY ONE, THE
PO Box 20506
Juneau - Alaska, United States of America
T: 1 907 465-3879

Wurster, Charles, Mr.
Trustee
ENVIRONMENTAL DEFENSE FUND
257 Park Av. South
New York - NY 10010, United States of America
T: 1 212 5052100 F: 1 212 5052375

Wyman, Michael, Mr.
President
WORLD SPIRITUAL ASSEMBLY
135 Macdougal St., Apt.2B
New York - NY, United States of America
T: 1 212 995-0532 F: 1 212 645-6866
TL: 66560

Wynn, Vanessa, Ms.
UP WITH PEOPLE
3103 North Campbell Av.
Tucson - AZ 85719, United States of America

Yamada, Jewelle, Ms.
Ass. Supervisor
HILL AND KNOWLTON
420 Lexington Ave.
New York - NY, United States of America
T: 1 212 967-5600 F: 1 212 697-6478

Yang, K. P., Ms.
ZIGEN FUND - FOR GRASSROOTS DEVELOPMENT
United States of America
T: 279 2674 F: 663 7784

Yap, Lorene, Ms.
WORLD BANK
1818 H St. NW
Washington - DC 20433, United States of America
T: 1 202 458-1943

Yardley, Miriam, Ms.
WOMEN'S ENVIRONMENTAL AND DEVELOPMENT ORGANIZATION
2053 Kula St.
Honolulu - HI 96817, United States of America
T: 1 808 595-6201

Yarus, Matthew, Mr.
President
TIMEWEAVE
1427 Pearl, # 2
Boulder - CO 80302, United States of America
T: 1 303 939 8463 F: 1 303 938 8463

Ybarra, Debi, Ms.
TERRA CHRISTA COMMUNICATIONS
1880 East River Road
Tucson - AZ, United States of America
T: 1 602 577-6476 F: 1 602 577-1195

York, Alexandra, Ms.
T.O.E.S.
172 East 61ST St., Apt. 4-B
New York - NY, United States of America
T: 1 212 416-8484 F: 1 212 416-6007

Young, Daniel
TURNER BROADCASTING SYSTEM / CNN
One CNN Centre
Atlanta - GA, United States of America
T: 1 404 827-3973 F: 1 404 827-1015

Young, Gary M, Mr.
Director
INTERNATIONAL RELIEF FRIENDSHIP FOUNDATION
c/o IRFF 4 W 43rd St
New York - NY, United States of America
T: 1 212 869-2614 F: 1 212 869-4268

Young, Katherine, Ms.
24 West Hilton Ave.
Redlands, CALIFORNIA 92373, United States of America
T: 1 714 793-5057

Young, Peter, Mr.
UP WITH PEOPLE
3103 North Campbell Av.
Tucson - AZ 85719, United States of America

Zalk, Sue Rosenberg, Ms.
Director
CENTER FOR THE STUDY OF WOMEN AND SOCIETY
City University of New York, 33 West 42nd Street
New York - NY, United States of America
T: 1 212 6422954 F: 1 212 6422642

Zalph, Ruth, Ms.
Delegate Or Representative
WILPF
525 Sandpiper Way
Boca Raton - FL 33431, United States of America
T: 1 487 3927146 F: 1 487 395-3850850

Zandburgher, Karin, Ms.
Assistant
GLOBE
409 3rd St. SW # 204
Washington - DC 20003, United States of America
T: 1 202 8630153 F: 1 202 4799447

Zarco, Ray, Mr.
160 West 24th St.
New York - NY, United States of America
T: 1 212 809-5099 F: 1 212 809-5497

Zissis, Cecelia, Ms.
Associate Professor Emeritus
PURDUE UNIVERSITY
629 Rose Street
West Lafayette - IN, United States of America
T: 1 317 7431014

Zucker, Julie, Ms.
Professor
UNIVERSITY OF MASSACHUSSETS & HARVARD UNIVERSITY
203 State Street
Northampton - MA 01060, United States of America
T: 1 413 584-4881 F: 1 604 545-4902

Zurbach, Matthew, Mr.
Member
UNA - USA / PASADENA CHAPTER
30102 Hillside Terrace
S.Juan Capistrano-CA 92675, United States of America
T: 1 714 493-5174

Zurbach, Robert L., Mr.
Board Member
UNA/USA, PASADENA (CA) CHAPTER
620 Arbor Street
Pasadena - CA 91105, United States of America
T: 1 818 792-8607

Zuzworsky, Rose, Ms.
NACRE
256 Etna St. - Brooklyn
New York - NY 11208, United States of America
T: 1 212 235-4574 F: 1 212 277-6723

UN AND UNCED

Adams, Thomas
United Nations
2 UN Plaza
New York, NY, United States of America
T: 212 9637174 F: 212 9631921

Adotevi, Stanislav
UNICEF
2 UN Plaza
New York, NY, United States of America
T: 212 9637174 F: 212 9631921

Allen, Dennis
United Nations
2 UN Plaza
New York, NY, United States of America
T: 212 9637174 F: 212 9631921

Anderson, Jim
United Nations
2 UN Plaza
New York, NY, United States of America
T: 212 9637174 F: 212 9631921

Arcos, Manuel
United Nations
2 UN Plaza
New York, NY, United States of America
T: 212 9637174 F: 212 9631921

Astolfi, Beti
UNIFEM
2 UN Plaza
New York, NY, United States of America
T: 212 9637174 F: 212 9631921

Azimi, Nassrine
UNITAR
2 UN Plaza
New York, NY, United States of America
T: 212 9637174 F: 212 9631921

Bajracharva, Deepak
UNICEF
2 UN Plaza
New York, NY, United States of America
T: 212 9637174 F: 212 9631921

Bartelmus, Peter
United Nations
2 UN Plaza
New York, NY, United States of America
T: 212 9637174 F: 212 9631921

Barton, Kenneth
United Nations
2 UN Plaza
New York, NY, United States of America
T: 212 9637174 F: 212 9631921

Beza, Sterie
International Monetary Fund
700 19th Street, N.W.
Washington, D.C. 20431, United States of America
T: 202 6237000 F: 202 6236391

Bezold, Willy, Mr.
UNICEF
2 UN Plaza
New York, NY, United States of America
T: 212 9637174 F: 212 9631921

Bhatia, Doris
International Monetary Fund
700 19th Street, N.W.
Washington, D.C. 20431, United States of America
T: 202 6237000 F: 202 6236391

Bhatia, Rattan
International Monetary Fund
700 19th Street, N.W.
Washington, D.C. 20431, United States of America
T: 202 6237000 F: 202 6236391

Birchem, Regina, Ms.
Credentialed Observor To UNCED
Women's Int'l League For Peace And Freedom
R.D.1, Box 134
Irwin , Pennsylvania 15642, United States of America
T: 412 446 3157 F: 412 838 4203

Bobish, Frank
United Nations
2 UN Plaza
New York, NY, United States of America
T: 212 9637174 F: 212 9631921

Bourne, Lynden
United Nations
2 UN Plaza
New York, NY, United States of America
T: 212 9637174 F: 212 9631921

Boutros-Ghali, Boutros, H.E.
Secretary-General of the United Nations
Office of the Secretary-General
S-3800, United Nations
New York, NY 10017, United States of America
T: 212 9635012 F: 212 9632155

Brauning, Roberto
International Monetary Fund
700 19th Street, N.W.
Washington, D.C. 20431, United States of America
T: 202 6237000 F: 202 6236391

Bruno, Dino
United Nations
2 UN Plaza
New York, NY, United States of America
T: 212 9637174 F: 212 9631921

Burke, Phillip
United Nations
2 UN Plaza
New York, NY, United States of America
T: 212 9637174 F: 212 9631921

Cabal, Elizabeth
United Nations
2 UN Plaza
New York, NY, United States of America
T: 212 9637174 F: 212 9631921

Cabuay, Joseph
United Nations
2 UN Plaza
New York, NY, United States of America
T: 212 9637174 F: 212 9631921

Caldas, Paulo
UNICEF
2 UN Plaza
New York, NY, United States of America
T: 212 9637174 F: 212 9631921

Camdessus,
International Monetary Fund
700 19th Street, N.W.
Washington, D.C. 20431, United States of America
T: 202 6237000 F: 202 6236391

Capeling-Alakija, Sharon
UNIFEM
2 UN Plaza
New York, NY, United States of America
T: 212 9637174 F: 212 9631921

Carvalho, Eduardo
UNICEF
2 UN Plaza
New York, NY, United States of America
T: 212 9637174 F: 212 9631921

Caufield, Walter
United Nations
2 UN Plaza
New York, NY, United States of America
T: 212 9637174 F: 212 9631921

Chaozhu, Ji
United Nations
2 UN Plaza
New York, NY, United States of America
T: 212 9637174 F: 212 9631921

Chowdury, Anwarul
UNICEF
2 UN Plaza
New York, NY, United States of America
T: 212 9637174 F: 212 9631921

Colavito, Francesco
United Nations
2 UN Plaza
New York, NY, United States of America
T: 212 9637174 F: 212 9631921

Colomes, Alejandro
UNICEF
2 UN Plaza
New York, NY, United States of America
T: 212 9637174 F: 212 9631921

Cooper, Tania
UNFPA
2 UN Plaza
New York, NY, United States of America
T: 212 9637174 F: 212 9631921

Crossan, Thomas
United Nations
2 UN Plaza
New York, NY, United States of America
T: 212 9637174 F: 212 9631921

Cunha, Alana
UNICEF
2 UN Plaza
New York, NY, United States of America
T: 212 9637174 F: 212 9631921

Curran, Marybeth
United Nations
2 UN Plaza
New York, NY, United States of America
T: 212 9637174 F: 212 9631921

Da, Marcos
UNICEF
2 UN Plaza
New York, NY, United States of America
T: 212 9637174 F: 212 9631921

Davies, William
UNICEF
2 UN Plaza
New York, NY, United States of America
T: 212 9637174 F: 212 9631921

De, Eliomar
UNICEF
2 UN Plaza
New York, NY, United States of America
T: 212 9637174 F: 212 9631921

De, Jose
UNICEF
2 UN Plaza
New York, NY, United States of America
T: 212 9637174 F: 212 9631921

De, Leo
UNICEF
2 UN Plaza
New York, NY, United States of America
T: 212 9637174 F: 212 9631921

De Soto, Alvaro
Senior Politcal Advisor to the Secretary-General
Office of the Secretary-General
S-3840, United Nations
New York, NY 10017, United States of America
T: 212 9635034 F: 212 9632155

Del Castillo, Graciand
Senior Ecomonic Affairs Officer
Office of the Secretary-General
S-3810A, United Nations
New York, NY 10017, United States of America
T: 212 9635033 F: 212 9632155

Dias, Tania
UNICEF
2 UN Plaza
New York, NY, United States of America
T: 212 9637174 F: 212 9631921

Earley, Steven
United Nations
2 UN Plaza
New York, NY, United States of America
T: 212 9637174 F: 212 9631921

Eastman, Kenneth
United Nations
2 UN Plaza
New York, NY, United States of America
T: 212 9637174 F: 212 9631921

Fajgenbaum, Jose
International Monetary Fund
700 19th Street, N.W.
Washington, D.C. 20431, United States of America
T: 202 6237000 F: 202 6236391

Ferrari, Mario, Mr.
UNICEF
2 UN Plaza
New York, NY, United States of America
T: 212 9637174 F: 212 9631921

Ferreira, Marcos
UNFPA
2 UN Plaza
New York, NY, United States of America
T: 212 9637174 F: 212 9631921

Folkes, Anthony
United Nations
2 UN Plaza
New York, NY, United States of America
T: 212 9637174 F: 212 9631921

Garcia, William
United Nations
2 UN Plaza
New York, NY, United States of America
T: 212 9637174 F: 212 9631921

Garnes, Alfred
United Nations
2 UN Plaza
New York, NY, United States of America
T: 212 9637174 F: 212 9631921

Glattbach, Jack
UNICEF
2 UN Plaza
New York, NY, United States of America
T: 212 9637174 F: 212 9631921

Gomes, Keyla
UNICEF
2 UN Plaza
New York, NY, United States of America
T: 212 9637174 F: 212 9631921

Grant, Ellan
UNICEF
2 UN Plaza
New York, NY, United States of America
T: 212 9637174 F: 212 9631921

Grant, James
UNICEF
2 UN Plaza
New York, NY, United States of America
T: 212 9637174 F: 212 9631921

Grenville-Wood, Geoffrey
UNITED NATIONS ASSOCIATION IN CANADA
485 5th Avenue
New York, NY 10017, United States of America
T: 212 9636870 F: 212 9634556

Grimm, Dennis
United Nations
2 UN Plaza
New York, NY, United States of America
T: 212 9637174 F: 212 9631921

Guimaraes, Antonio
UNICEF
2 UN Plaza
New York, NY, United States of America
T: 212 9637174 F: 212 9631921

Gustavo, Lopez Ospina
Director Unesco Office New York
UNESCO
Two United Nations Plaza DC 2 Room 920
New York NY 10017, United States of America
T: 212 9635978 F: 212 3555627

Guthrie, Victor
United Nations
2 UN Plaza
New York, NY, United States of America
T: 212 9637174 F: 212 9631921

Halpa, Achim
UNITAR
2 UN Plaza
New York, NY, United States of America
T: 212 9637174 F: 212 9631921

Hart, John
United Nations
2 UN Plaza
New York, NY, United States of America
T: 212 9637174 F: 212 9631921

Haxton, David
UNICEF
2 UN Plaza
New York, NY, United States of America
T: 212 9637174 F: 212 9631921

Hedlund, Inga-Lisa, Ms
Senior Environment Adviser
UNDP
C/O Undp One UN Plaza, Room Dci-2092
New York, NY 10017, United States of America
T: 212 906 5059 9804529 F: 212 906 5365

Herencia, Salvador
UNICEF
2 UN Plaza
New York, NY, United States of America
T: 212 9637174 F: 212 9631921

Hinds, Berold
United Nations
2 UN Plaza
New York, NY, United States of America
T: 212 9637174 F: 212 9631921

Ignaciuk, Stanley
United Nations
2 UN Plaza
New York, NY, United States of America
T: 212 9637174 F: 212 9631921

Jackson, Hyclis
United Nations
2 UN Plaza
New York, NY, United States of America
T: 212 9637174 F: 212 9631921

Jasper, Alphonso
United Nations
2 UN Plaza
New York, NY, United States of America
T: 212 9637174 F: 212 9631921

Jesus, Ubira
UNICEF
2 UN Plaza
New York, NY, United States of America
T: 212 9637174 F: 212 9631921

Jinvit, Vajirayu
United Nations
2 UN Plaza
New York, NY, United States of America
T: 212 9637174 F: 212 9631921

Jolly, Alison, Ms.
UNICEF
2 UN Plaza
New York, NY, United States of America
T: 212 9637174 F: 212 9631921

Jolly, Richard
UNICEF
2 UN Plaza
New York, NY, United States of America
T: 212 9637174 F: 212 9631921

Joseph, Kenzie
United Nations
2 UN Plaza
New York, NY, United States of America
T: 212 9637174 F: 212 9631921

Juliao-Pockels, Gustavo
United Nations
2 UN Plaza
New York, NY, United States of America
T: 212 9637174 F: 212 9631921

Kaempfert, Volker
United Nations
2 UN Plaza
New York, NY, United States of America
T: 212 9637174 F: 212 9631921

Kaji, Hideki
United Nations
2 UN Plaza
New York, NY, United States of America
T: 212 9637174 F: 212 9631921

Kayayan, Agop
UNICEF
2 UN Plaza
New York, NY, United States of America
T: 212 9637174 F: 212 9631921

Kearins, Michael
United Nations
2 UN Plaza
New York, NY, United States of America
T: 212 9637174 F: 212 9631921

Khalil, Moataz
United Nations
2 UN Plaza
New York, NY, United States of America
T: 212 9637174 F: 212 9631921

Kinzo, Mary
UNFPA
2 UN Plaza
New York, NY, United States of America
T: 212 9637174 F: 212 9631921

Koo, Samuel
UNICEF
2 UN Plaza
New York, NY, United States of America
T: 212 9637174 F: 212 9631921

Kuhl, Corinna
UNFPA
2 UN Plaza
New York, NY, United States of America
T: 212 9637174 F: 212 9631921

Lambertie, John
United Nations
2 UN Plaza
New York, NY, United States of America
T: 212 9637174 F: 212 9631921

Leal, Joseph
United Nations
2 UN Plaza
New York, NY, United States of America
T: 212 9637174 F: 212 9631921

Levin, Herbert
United Nations
2 UN Plaza
New York, NY, United States of America
T: 212 9637174 F: 212 9631921

Levy, Derrick
United Nations
2 UN Plaza
New York, NY, United States of America
T: 212 9637174 F: 212 9631921

Lima, Neusa
UNICEF
2 UN Plaza
New York, NY, United States of America
T: 212 9637174 F: 212 9631921

Lima, Ricardo
Columbia University Student
U.N. SUPPORT STAFF-GTN
898 Park Avenue
New York, NY 10021, United States of America
T: 1 212 734 8329 F: 1 212 744 1132

Lopes, Gilberto
UNICEF
2 UN Plaza
New York, NY, United States of America
T: 212 9637174 F: 212 9631921

Lopes, Patricia
UNICEF
2 UN Plaza
New York, NY, United States of America
T: 212 9637174 F: 212 9631921

Maass, Robert
UNICEF
2 UN Plaza
New York, NY, United States of America
T: 212 9637174 F: 212 9631921

Maeurer, Anthony
United Nations
2 UN Plaza
New York, NY, United States of America
T: 212 9637174 F: 212 9631921

Magnoni, Graciela
UNICEF
2 UN Plaza
New York, NY, United States of America
T: 212 9637174 F: 212 9631921

Marshall, Alexander
UNFPA/POPULI
2 UN Plaza
New York, NY, United States of America
T: 212 9637174 F: 212 9631921

Matheus, Euler
UNICEF
2 UN Plaza
New York, NY, United States of America
T: 212 9637174 F: 212 9631921

McNeill, Charles, Dr.
Global Environment Facility
BUREAU FOR AFRICA, UNDP
One UN Plaza
New York, NY, United States of America
T: 212-906-5960 F: 212-906-5423

Medina, Ivan
United Nations
2 UN Plaza
New York, NY, United States of America
T: 212 9637174 F: 212 9631921

Mera, Ruby, Ms.
Photographer
UNICEF
445 East 77th Street , apt. 5K
New York - NY 10021, United States of America
T: 1 212 326-7286

Moore, Phyllis
United Nations
2 UN Plaza
New York, NY, United States of America
T: 212 9637174 F: 212 9631921

Moreira, Branca
UNIFEM
2 UN Plaza
New York, NY, United States of America
T: 212 9637174 F: 212 9631921

Moreira, Eduardo
UNICEF
2 UN Plaza
New York, NY, United States of America
T: 212 9637174 F: 212 9631921

Myers, William
UNICEF
2 UN Plaza
New York, NY, United States of America
T: 212 9637174 F: 212 9631921

Nellor, David
International Monetary Fund
700 19th Street, N.W.
Washington, D.C. 20431, United States of America
T: 202 6237000 F: 202 6236391

Ngarambe, Gunvor
UNITED NATIONS ASSOCIATION IN STOCKHOLM
485 5th Avenue
New York, NY 10017, United States of America
T: 212 9636870 F: 212 9634556

Nichols, Kenneth
United Nations
2 UN Plaza
New York, NY, United States of America
T: 212 9637174 F: 212 9631921

Noble, Nileema
Gef Co-Ordinator, Rbap
UNDP, NEW YORK
1 UN Plaza New York
New York NY 10017, United States of America
T: (212) 906-6575 F: (212) 906-5825

Novichenko, Igor
United Nations
2 UN Plaza
New York, NY, United States of America
T: 212 9637174 F: 212 9631921

O'Haire, Hugh
Editor
UNFPA/POPULI
49-01 39th Ave.
Sunnyside, NY 11104, USA, United States of America
T: 212 9637174 F: 212 9631921

Ottoni, Cristiano
UNFPA
2 UN Plaza
New York, NY, United States of America
T: 212 9637174 F: 212 9631921

Pastizzi-Ferencic, Dunja
United Nations
2 UN Plaza
New York, NY, United States of America
T: 212 9637174 F: 212 9631921

Paul, Dorothy
UNITED NATIONS ASSOCIATION OF AMERICA
485 5th Avenue
New York, NY 10017, United States of America
T: 212 9636870 F: 212 9634556

Peterson, Jeannie
UNFPA
2 UN Plaza
New York, NY, United States of America
T: 212 9637174 F: 212 9631921

Piers, Ronald
United Nations
2 UN Plaza
New York, NY, United States of America
T: 212 9637174 F: 212 9631921

Powell, Leo
United Nations
2 UN Plaza
New York, NY, United States of America
T: 212 9637174 F: 212 9631921

Prince, Marcia
United Nations
2 UN Plaza
New York, NY, United States of America
T: 212 9637174 F: 212 9631921

Pronove, Gao
UNITAR
2 UN Plaza
New York, NY, United States of America
T: 212 9637174 F: 212 9631921

Puentes, Hernan
International Monetary Fund
700 19th Street, N.W.
Washington, D.C. 20431, United States of America
T: 202 6237000 F: 202 6236391

Quarterman, Charles
United Nations
2 UN Plaza
New York, NY, United States of America
T: 212 9637174 F: 212 9631921

Rivera, Elvis
United Nations
2 UN Plaza
New York, NY, United States of America
T: 212 9637174 F: 212 9631921

Roach, Stanley
United Nations
2 UN Plaza
New York, NY, United States of America
T: 212 9637174 F: 212 9631921

Rosario, Kenneth
United Nations
2 UN Plaza
New York, NY, United States of America
T: 212 9637174 F: 212 9631921

Sazanami, Hidehiko
United Nations
2 UN Plaza
New York, NY, United States of America
T: 212 9637174 F: 212 9631921

Schiefer, N.
United Nations
2 UN Plaza
New York, NY, United States of America
T: 212 9637174 F: 212 9631921

Schwieger, Norbert
UNICEF
2 UN Plaza
New York, NY, United States of America
T: 212 9637174 F: 212 9631921

Sessions, Kathryn
UNITED NATIONS ASSOCIATION OF AMERICA
485 5th Avenue
New York, NY 10017, United States of America
T: 212 9636870 F: 212 9634556

Sigurdsson, Gudmundur
United Nations
2 UN Plaza
New York, NY, United States of America
T: 212 9637174 F: 212 9631921

Silva, Gustavo
UNICEF
2 UN Plaza
New York, NY, United States of America
T: 212 9637174 F: 212 9631921

Simpson, Jenaro D., Mr.
Deputy Division Chief
INTERNATIONAL MONETARY FUND
700 19th St.
Washington - DC 20431, United States of America
T: 1 202 6236516 F: 1 202 6236220

Smith, Muriel
UNITED NATIONS ASSOCIATION IN CANADA
485 5th Avenue
New York, NY 10017, United States of America
T: 212 9636870 F: 212 9634556

Smith-Barbour, Printhea
United Nations
2 UN Plaza
New York, NY, United States of America
T: 212 9637174 F: 212 9631921

Souza, Alexandre
UNICEF
2 UN Plaza
New York, NY, United States of America
T: 212 9637174 F: 212 9631921

Spencer, Richard
UNICEF
2 UN Plaza
New York, NY, United States of America
T: 212 9637174 F: 212 9631921

Spiers, Ronald
United Nations
2 UN Plaza
New York, NY, United States of America
T: 212 9637174 F: 212 9631921

Steneker, Sietake
UNFPA
2 UN Plaza
New York, NY, United States of America
T: 212 9637174 F: 212 9631921

Stevenson, Maurice
United Nations
2 UN Plaza
New York, NY, United States of America
T: 212 9637174 F: 212 9631921

Sweeney, John
United Nations
2 UN Plaza
New York, NY, United States of America
T: 212 9637174 F: 212 9631921

Taylor, Jayah
United Nations
2 UN Plaza
New York, NY, United States of America
T: 212 9637174 F: 212 9631921

Teymour, Aly
United Nations
2 UN Plaza
New York, NY, United States of America
T: 212 9637174 F: 212 9631921

Thomas, Kuttipurathu
United Nations
2 UN Plaza
New York, NY, United States of America
T: 212 9637174 F: 212 9631921

Thunqvist, Kerstin
UNITED NATIONS ASSOCIATION IN STOCKHOLM
485 5th Avenue
New York, NY 10017, United States of America
T: 212 9636870 F: 212 9634556

Vartholomatos, Dennis
United Nations
2 UN Plaza
New York, NY, United States of America
T: 212 9637174 F: 212 9631921

Vazquez, Gwendolyn
United Nations
2 UN Plaza
New York, NY, United States of America
T: 212 9637174 F: 212 9631921

Vazquez-Aldana, Louis
United Nations
2 UN Plaza
New York, NY, United States of America
T: 212 9637174 F: 212 9631921

Volpe, Vincent
United Nations
2 UN Plaza
New York, NY, United States of America
T: 212 9637174 F: 212 9631921

Walcott, Al
United Nations
2 UN Plaza
New York, NY, United States of America
T: 212 9637174 F: 212 9631921

Wall, Karl
United Nations
2 UN Plaza
New York, NY, United States of America
T: 212 9637174 F: 212 9631921

Ward, Richard
United Nations
2 UN Plaza
New York, NY, United States of America
T: 212 9637174 F: 212 9631921

White-Davis, Hope P., Ms.
President
WORLD ASSN OF FORMER U.N. INTERNES AND FELLOWS INC (WAFUNIF)
United Nations, Room DC2-1118, 2 UN Plaza
New York - NY, United States of America
T: 1 212 9633110 F: 1 212 9634116

Yevoly, Peter
United Nations
2 UN Plaza
New York, NY, United States of America
T: 212 9637174 F: 212 9631921

Yin, Yin
UNICEF
2 UN Plaza
New York, NY, United States of America
T: 212 9637174 F: 212 9631921

ADVISORS

Anderson, Robert O., Mr.
Roswell, NM 88201, United States of America
T: 1 505 625 8700 F: 1 505 625 6825

Cousteau, Jacques-Yves, Capt.
New York, NY 10022, United States of America
T: 212 826 2940 F: 212 751 4349

Coward, E. Walter, Mr.
Director, Rural Poverty And Resources Program, The Ford Foundation
New York, NY 10017, United States of America
T: 1 212 573 4615 F: 1 212 297 0969

Denver, John, Mr.
UNCED Goodwill Adviser
Aspen, CO 81612, United States of America
T: 1 303 925 6352 F: 1 303 920 1919

Earle, Sylvia, Ms.
Advisor To The Administrator, U.S. Department Of Commerce
Washington, DC 20230, United States of America
F: 1 202 408 9674/377 8203

Fuller, Kathryn S., Ms.
President, Worldwide Fund For Nature (WWF)
Washington, DC 20037-1175, United States of America
T: 1 202 293 4800 F: 1 202 293 9211

Hickel, Walter J., The Hon.
Governer, State Of Alaska
Juneau, AK, United States of America

Leeburg, Louis G., Mr.
Vice-President, Finance, The Fetzer Institute
Kalamazoo, MI 49009, United States of America
T: 1 616 375 2000 F: 1 616 372 2163

Malone, Thomas F., Dr.
Director
The Sigma Xi Center
Research Triangle Park, NC 27709, United States of America
T: 1 919 549 4691 F: 1 919 549 0090

Muskie, Edmund S., The Hon.
Chairman, UNA-USA Project On The Global Environment
New York, NY 10017-6104, United States of America
T: 1 212 697 3232 F: 1 212 682 9185

Prince, Gregory, Mr.
President, Hampshire College
Amherst, MA 01002, United States of America
T: 1 413 549 4600 x521 F: 1 413 549 0707

Pyle, Barbara, Ms.
Environmental Policy Department, Turner Broadcasting System, Inc.
Atlanta, GA 30348-5366, United States of America
T: 1 404 827 3183 F: 1 404 827 4292

Rauschenberg, Robert, Mr.
Artist Of The Official Earth Summit Poster
Captive, FL 33924, United States of America
F: 1 813 472 5447

Reid, John, Mr.
Chief Environment Officer, Coca-Cola Company
Atlanta, GA 30301, United States of America
T: 1 404 676 2371

Sawhill, John C., Mr.
President And Ceo, The Nature Conservancy
Arlington, VA 22209, United States of America
T: 1 703 841 5300 F: 1 703 841 1283

Speth, Gus, Mr.
President, World Resources Institute
Washongton, DC 20006, United States of America
T: 202 638 6300 F: 202 638 0036

Topping, Seymour, Mr.
Editor
New York Times
New York, NY 10036, United States of America
T: 1 212 556 1978 F: 1 212 556 3722

Turner, Ted, Mr.
Chairman Of The Board, Turner Broadcasting System, Inc.
Atlanta, GA 30348-530, United States of America
F: 1 404 827 3107/1575

Ul Haq, Mahbub, Mr.
Special Advisor To Administrator, UNDP
New York, NY 10017, United States of America
T: 1 212 906 5764 F: 1 212 906 5778

Van Agt, Andreas, Mr.
Ambassador, Head Of Delegation Of The Comission Of The European Community
Washington, DC 20037, United States of America
T: 1 202 862 9500 F: 1 202 429 1766

PARLIAMENTARIANS / SPIRITUAL LEADERS

Arafat, Iman M. Bashar
6629 Johnny Cake Rd.
Baltimore, MD. 21207, United States of America
T: 301 744 1532 F: 301 747 4869

Baucus, Max, Senator
U.S. Senate, The Capital
Washington, D.C. 20510, United States of America
T: 202 224 3121
TL: 89 554

Beilenson, Anthony C., Congressman
House of Representatives, The Capital
Washington, D.C. 20515, United States of America
T: 202 224 3121
TL: 89 554

Bingham, Sally Grover, Rev.
Chair
Diocesan Commission on the Environment
Episcopal Diocese of California 7 Laurel St.
San Francisco, CA. 94118, United States of America

Blaz, Ben G., Congressman
House of Representatives, The Capital
Washington, D.C. 20515, United States of America
T: 202 224 3121
TL: 89 554

Chaffee, John H., Sen.
U.S. Senate, The Capital
Washington, D.C. 20510, United States of America
T: 224 3121

Collins, Cardiss, Congresswoman
House of Representatives, The Capital
Washington, D.C. 20515, United States of America
T: 202 224 3121

Davis, Robert, Congressman
House of Representatives, The Capital
Washington, D.C. 20515, United States of America
T: 202 224 3121

Fascell, Dante B., Cong.
Chairman
House of Representatives, The Capital
Washington, D.C. 20515, United States of America
T: 224 3121

Getahoun, Rebequa, Ms.
Deputy Director/Environment
BAHA'I INTERNATIONAL COMMUNITY
866 UN Plaza, Suite 120
New York - NY 10017, United States of America
T: 1 212 7563500 F: 1 212 7563573
E: 152:CNM057

Gore, Albert, Jr., Senator
U.S. Senate, The Capital
Washington, D.C. 20510, United States of America
T: 224 3121

Hamdani, Viqar, Dr.
5 Stuyvesant Oval, #4H
New York, N.Y. 10009, United States of America

Helges, Anele E., Dr. O.P.
International Communications Facilitator
Education forAll Global Network
United Nations, DC2 Room 1124
New York, N.Y. 10017, United States of America

Heschel, Susannah, Professor
Department of Religion
Case Western Reserve University, 10900 Euclid Avenue
Cleveland, OH 44106-7112, United States of America
T: 216 368 2741 F: 216 368 4681

James, Jewell Praying Wolf
c/o Lummi Indian Nation, 2616 Kuina Road
Belingham, WA. 98226, United States of America

Kerry, John, Sen.
U.S. Senate, The Capital
Washington, D.C. 20510, United States of America
T: 202 224 3121
TL: 89 554

Lyons, Oren, Chief
Faith Keeper (Onandaga)
Box 200, Onandaga Nation
Nedro, N.Y. 13120, United States of America

Maharaj, H.H. Acharya Sushil Kumar
International Mahavir Jain Mission
Siddhachalam, 65 Mud Pond Road
Blairstown, N.J. 07825, United States of America

Martin, Daniel, The Rev. Dr.
Director
International Coordinating Cttee on Religion and The Earth
P.O. Box 67
Greenwich, Connecticut 06831 0767, United States of America
T: 914 238 5032 F: 914 238 5032

Meyer, Marshall, Rabbi
Temple B'nai Jeshurun, 270 West 89th St.
New York, N.Y. 10024, United States of America

Palaniswami, Sivasiva, The Rev.
Saiva Siddhanta Church
Kadavul Hindu Temple 107 Kaholalole Rd.
Kapaa, Hawaii 95746, United States of America
T: 808 822 0080 F: 808 822 4351

Pell, Claiborne, Senator
U.S. Senate, The Capital
Washington, D.C. 20510, United States of America
T: 202 224 3121
E: 89 554

Pelosi, Nancy, Congresswoman
House of Representatives, The Capital
Washington, D.C. 20515, United States of America
T: 202 224 3121

Pertell, Dennis, Cong.
House of Representatives, The Capital
Washington, D.C. 20515, United States of America
T: 224 3121

Piyatissa, K., The Ven. Dr.
New York Buddhist Vihara
84 32 124th St.
Kew Gardens, N.Y. 11415, United States of America

Porter, John E., Congressman
House of Representatives, The Capital
Washington, D.C. 20515, United States of America
T: 202 224 3121

Pressler, Larry, Senator
U.S. Senate, The Capital
Washington, D.C. 20510, United States of America
T: 202 224 3121

Rasmussen, Larry, Dr.
Reinhold Nieburh Professor Of Social Ethics
Union Theological Seminary
3041 Broadway at Reinhold Niebuhr Place
New York, N.Y. 10027, United States of America
T: 212 662 7100

Reeves, Paul, The Most Rev. Sir
Anglican Observer At The United Nations
815 Second Avenue
New York, N.Y. 10017, United States of America
T: 922 5163 F: 687 1336

Roshi, Maezumi, The Rev.
Zen Center of Los Angeles
923 South Normandi Avenue
Los Angeles, CA. 90006, United States of America
T: 213 387 2351 F: 2130 387 2377

Scheuer, James H., Congressman
House of Representatives, The Capital
Washington, D.C. 20515, United States of America
T: 202 224 3121

Shenandoah, Leon, Chief
Tadodaho (chief Of Chiefs)
Six Nations of the Iroquois Confederacy
c/o Chief Oren Lyons, Box 200, Onandaga Nation
Nedro, New York 13120, United States of America

Sikorski, Gerry, Congressman
House of Representatives, The Capital
Washington, D.C. 20515, United States of America
T: 202 224 3121

Solarz, Stephen J., Cong.
House of Representatives, The Capital
Washington, D.C. 20515, United States of America
T: 224 3121

Subramuniyaswami, H.H. Sivya, Mr.
Saiva Siddhanta Church
Kadavul Hindu Temple 107 Kaholalole Rd.
Kapaa, Hawaii 95746, United States of America
T: 808 822 0080 F: 808 822 4351

Symms, Steve, Senator
U.S. Senate, The Capital
Washington, D.C. 20510, United States of America
T: 202 224 3121

Torricelli, Robert G., Congressman
House of Representatives, The Capital
Washington, D.C. 20515, United States of America
T: 202 224 3121

Unsoeld, Jolene, Congresswoman
House of Representatives, The Capital
Washington, D.C. 20515, United States of America
T: 202 224 3121

Wellstone, Paul, Senator
U.S. Senate, The Capital
Washington, D.C. 20510, United States of America
T: 202 224 3121

Wirth, Timothy E., Senator
U.S. Senate, The Capital
Washington, D.C. 20510, United States of America
T: 202 224 3121

Yrado The First, Nana Boakyewa, Mme.
152-18 Union Turnpike, #80
Flushing, N.Y. 11367, United States of America
T: 718 380 5283

WISDOMKEEPERS

Aiko, Gary Puniwai Keawe
47-001 Okana Rd.
Kaneohe Hawaii 96744, United States of America

Aiko, Laura Kaaiohelo
47-001 Okana Rd.
Kaneohe Hawaii 96744, United States of America

Anderson, Tracy
Campaign for the Earth/Keystone Ctr.
Box 204
Breckenridge CO 80424, United States of America
T: 303-453-0562 F: 303-444-1286

Barnes, John-Rupert
Institute of Cultural Affairs
3416 Newton St.
Mount Ranier MD 20712, United States of America
T: 301-864-3755

Bray, David
Seneca Nation/Haudenosaunee
PO Box 126
via Versailles New York 14168, United States of America
T: 91-434-43383

Bronfman, Jeffrey
Ecologist
Threshold Foundation
176 Valley Drive
Sante Fe New Mexico 87501, United States of America
T: 505-988-5924 F: 505-982-5029

Brower, David
Chairman
Earth Island Institute
300 Broadway, Ste 28
San Francisco CA 94113, United States of America
T: 415-788-3666 F: 415-788-7324

Brown, Suzanne
Director
Daedulus Foundation
12702 Via Cortina #210B
Del Mar CA 92014, United States of America
T: 619-793-0411

Chappell, Joy
Joyful Sound Co.
2010 S. Broadway
Pittsburg Kansas 66720, United States of America
T: 316-231-3188 F: 316-231-3198

Chappell, Walter
INTERNATIONAL TECHNOLOGY INSTITUTE
Box 1171
Pittsburg Kansas 66762, United States of America
T: 316-231-3188 F: 316-231-3198

Conroy, Donald
Consortiumon on Religion and Ecology Int'l/NACRE
5 Thomas Cr. N.W.
Washington, D.C. 20005, United States of America
T: 202-462-2591 F: 202-234-0307

Currier, Lavinia
Sacharuna Foundation
Roland Farm
The Plains Virginia 22171, United States of America
T: 703-754-8713 F: 703-754-1475

Cymerman, Shabari, Ms.
Gaia Institute
5484 Trimble Rd.
Atlanta Georgia 30342, United States of America
T: 404-256-9002 F: 404-252-9160

De La Vista, Rio
Manitou Foundation
Box 1379
Pagosa Springs Colorado 81147, United States of America
T: 303-731-5742

Giglio, Nick, Mr.
President
Emissary Fndn
5569 N. Cr 29
Loveland CO 80538, United States of America
T: 303-679-4222 F: 303-679-4233

Heffernan, Liz
Earth Stewards
PO Box 10697
Bainbridge Island Washington 98110, United States of America
T: 206-842-87986

Henderson, Hazel
author
Permanent Mission of United States of America
200 Zoratoa Ave.
Anastasia Island Florida 32084, United States of America
T: 904-829-3140 F: 904-826-0325

Herbertson, Gary
Vice President
Interfaith Council for the UN
4300 Beuflower
Lakewood CA 90713, United States of America
T: 310-429-5640 F: 310-496-1486

Hetzel, Rattana
Author/Inner And Outer Ecology
Global Vision
1833 Myrtle Ave.
San Diego CA 92103, United States of America
T: 619-296-5107

Huddle, Norie, Ms.
Green Communications/Center for New National Security
1016 Prince Street, #5
Alexandria VA 22314, United States of America
T: 703-549-2862 F: 703-5492862

Impallomeni, Vicki, Ms.
Reef Relief/Earth Week FLA Keys 93
23 Key Haven Terr.
Key West Florida 33040, United States of America
T: 305-294-9731 F: 305-293-9515

Jennings, John
Int'l Sister Forest Assoc.
PO Box 2156
Portland OR 97208, United States of America
T: 503-231-0332 F: 503-235-6000

Johnson, Huey
President
Resource Renewal Institute
San Francisco, United States of America

Johnson, Vincent
Indigenous/Chief
Onondaga Nation
Buffalo Hill Road
Lafayette New York 13084, United States of America
T: 315-677-7786

Kanegis, Arthur
Producer
OOOPS! Movie Productions
105 Kamino Teresa
Sante Fe New Mexico 87501-4703, United States of America
T: 505-982-8882

Kay, Allan
Americans Talk Issues Fndn
907 - 6 St. SW
Washington, D.C. 20024, United States of America
T: 202-488-5889 F: 2-2-488-8024

Keawe Aiko, Genoa L.
2283 Tantalus
United States of America

URUGUAY - GOVERNMENT

Lacalle, Luis Alberto, Mr.
President
Permanent Mission of Uruguay
747 3rd. Ave., 21st. Floor
New York NY 10017, United States of America
T: 212 752 8240 F: 212 593 0935

NGO

Alves Teixeira, Alvaro
ARPEL
Javier de Viana 2345
11200 Montevideo, Uruguay

Bequio, Beatriz
AMULP
Canelones 929
Montevideo, Uruguay

Brussoni, Aldo
ARPEL
Javier de Viana 2345
11200 Montevideo, Uruguay

Filion, J. Paul
ARPEL
Javier de Viana 2345
11200 Montevideo, Uruguay

Fontes, Dora
GB
Loc. Grupo Bosque, Cantegril Country Club
Punta del Este, Uruguay

Israel, Sergio, Mr.
CIPFE
Canelones 1164
11100Montevideo, Uruguay
T: 598 2 922363 F: 598 2 98-5959

McCully, Patrick, Mr.
ITM
Miguel del Corro 1461
Montevideo 11200, Uruguay

Prieto, Ruben
REDES
Milan 4115
Montevideo, Uruguay

Recine, Juan Miguel Moyano
ARPEL
Javier de Viana 2345
11200 Montevideo, Uruguay

Splay Memoryroig, Graziela
GB
Loc. Grupo Bosque, Cantegril Country Club
Punta del Este, Uruguay

Valverde, Marta
AMULP
Canelones 929
Montevideo, Uruguay

Von Sanden, Cecilia
CIPFE
Canelones 1164
11100 Montevideo, Uruguay

PRESS

Aharonian, Aram Ruben
Editor
ASIN
Av. La Salle Colinas de los
Caobos, Caracas, Venezuela, Uruguay

Barboza, Jose Luis
Reporter Radio/Television
URUGUAY PRESS
Remigio Gastellanos 721
Melo, Uruguay, Uruguay

Bocage, Alberto
Correspondent
REPORTE SNS
Vilardego 1307
Montevideo, Uruguay, Uruguay

Bottero, Monica
Editor
BUSQUEDA
Edil. Hugo Prato 2249, Ap. 503
Montevideo, Uruguay, Uruguay

Caimi, Walter
Correspondent
LA JUVENTUD
Nina 1617
CP. 12.500 Montevideo, Uruguay, Uruguay

Gustavo, Guida
Reporter Radio/Television
CANAL 5 SODRE TV
Avda. tomas Giribaldi
Montevideo, Uruguay, Uruguay

Leal Rovira, Griselda
Reporter Radio/Television
RADIO PANAMERICANA
Ciudadela 1166 Apt.401
Montevideo, Uruguay, Uruguay

Lees, Martin
Reporter Radio/Television
SAETZ TV
Concenvion 1162, ap. 002
Montevideo, Uruguay, Uruguay

Magarinos, De Mello Mateo J.
Reporter Radio/Television
RADIO CARVE
Echevarriarza 3396
Montevideo,Uruguay 11300, Uruguay

Olascoaga, Paulo
Editor
SAETA TV
Av. 18 de Julio 2033, 602
Montevideo, Uruguay, Uruguay

Pascale, Graziano
Reporter Radio/Television
MONTE CARLO TV CANAL 4
J.B. Blango 3280/201
Uruguay, Uruguay

Pauletti, Carlos
Correspondent
URUGUAY PRESS
Ramon Fernandez 230
Montevideo, Uruguay, Uruguay

Portella, Lucia Helena De Souza
Correspondent
SISTEMA NACIONAL DE TV
Rua Jose de Brito 47
Rio de Janeiro, Brazil, Uruguay

Sampognaro, Pilar
Editor
BUSQUEDA
Scoseria 2677
Montevideo, Uruguay, Uruguay

Saucero, Jose Pedro
Reporter Radio/Television
CX8 RADIO SARANDI
Dr.Alejandro Gallinal 1553
Montevideo, Uruguay, Uruguay

Sorhuet, Hernan
Editor
TRAFFIC SUDAMERICA
C. Roxlo 1496/301
Montevideo 11200, Uruguay
T: 598 2 49-3384 F: 598 2 49-3384

Vignola, Jorge
Reporter Radio/Television
RADIO CARVE
Luis A. de Herrera 2682, #503
Montevideo, Uruguay, Uruguay

Wolf, Annabell
Reporter Radio/Television
RADIO CARVE
Mercedes 973
Montevideo, Uruguay, Uruguay

Xalambri, Ricardo
Correspondent
CONO SUR PRESS
Solano Lopez 1352/12
Montevideo, Uruguay, Uruguay

ORGANIZATIONS

Balparda, Maria De Los Angeles, Ms.
As. de Mujeres Uruguayas Lourdes Pintos
Canelones 929
Montevideo, Uruguay

Bissio, Roberto, Sr.
Instituto del Tercer Mundo
Miguel del Corro 1461
Montevideo 11200, Uruguay
T: (5982) 496 192 F: (5982) 419 222
TL: E ax:chasque geo2:chasque

Peixoto, Jorge, Fr.
Centro de Investigacion y Promocion Franciscano y Ecologico
Canelones 1164
11100 Montevideo, Uruguay
T: (5982) 907 648/922 362 F: (5982) 985 959

Teixeira, Alvaro Alves, Mr.
Asistencia Reciproca Petrolera Estatal Latino Americana (ARPEL)
Javier de Viana 2345
11200 Montevideo, Uruguay
T: (598 2) 40 6993 F: (598 2) 23 7023

Vescovi, Enrique, Sr.
Grupo Bosque
Loc. Grupo Bosque, Cantegril Country Club
Punta del Este, Uruguay
T: (59 842) 32 440 F: (59 842) 21 349

Villareal, Alberto, Mr.
REDES (Red De Ecologia Social)
Milan 4115
Montevideo, Uruguay
T: (5982) 38 16 40

GLOBAL FORUM

Aguirre, Raul Mesa, Mr.
Montevideo, Uruguay

Aren, Rita Stella, Ms.
Professora De Biologia
ED. SECUNDARIA
Canelones,863
Montevideo, Uruguay
T: 598 2 914538

Avelino, Tulio
ESCUELA NACIONAL DE ECOLOGIA APLICADA - ENEA
18 de Julio 1235
Montevideo, Uruguay
T: 598 2 90 16 16 F: 598 2 90 67 39

Barreiro, Jorge, Mr.
NGONET
Casilla Correo 1539
Montevideo 11000, Uruguay
T: 598 2 49 61 92 F: 598 2 41 92 22
E: org.uy/geo2:ngonet

Behr, Howard, Mr.
Pcv
PEACE CORPS URUGUAY
Plaza Independencia 822, 4o Piso
Montivideo, Uruguay
T: 598 2 904123 F: 598 2 923479

Behrens, Chris, Mr.
Volunteer-Biologist
PEACE CORPS URUGUAY
Plaza Independencia 822, 4o Piso
Montevideo, Uruguay
T: 598 2 904123 F: 598 2 923479

Bernardi, Rafael, Sr.
GRUPO ESTUDIANTIL PRO MEDIO AMBIENTE
S. Vazquez 1140
Montevideo 11300, Uruguay
T: 598 2 783679

Blenker, Roberto, Mr.
Sub-Director
PEACE CORPS
Plaza Independencia 822, 4 piso
Montevideo, Uruguay
T: 598 2 90-4123 F: 598 2 92-3479

Bonno, Sra.
AMULP
Canelones 929
Montevideo, Uruguay
T: 598 2 903422 F: 598 2 904961

Broumand, Mihandokht R., Ms.
BAHA'I COMMUNITY OF URUGUAY
Brenda 5751
Montevideo 11400, Uruguay
T: 598 2 601944 F: 598 2 617282

Burnett, Enrique, Mr.
GRUPO BOSQUE
Cantegril Country Club Local 04
Punta del Leste, Uruguay
T: 596 42 32440 F: 596 42 32440

Cairolo, Carlos, Mr.
CIPFE
Montevideo, Uruguay

Canclini, Jose Pedro Lopez, Mr.
Vereador
JUNTA DEPARTAMENTAL DE MONTEVIDEO
25 de Mayo, 609
Montevideo, Uruguay
T: 598 2 952126 F: 598 2 963614

Carambula, Antonio, Mr.
Coordinator
PAZ Y BIEN
Canelones 1164
Montevideo, Uruguay
T: 598 2 922363 F: 598 2 98-5959

Cardoso, Aura Luz, Ms.
Med. Pediatra
DEFENSA AMBIENTAL DE LEZICA Y COLON
Hector Miranda 87
Sta Lucia, Uruguay
T: 03346402

Cardozo, Ana Maria
Secretaria
P. SOCIALISTA
Soriano 1218
Montivideo, Uruguay

Cardozo, Jorge Orestes De Leon, Mr.
Professor Biologia
DEPTO DE BIOLOGIA - C.E.D.O.
Hugo Prato, 2269 Y Azevedo Dias
Montevideo 12000, Uruguay
T: 598 2 413467

Ceplikas, Lilian, Ms.
COTIDIANO MUJER
Jackson1270 s.s. 101
Montevideo 11100, Uruguay
T: 598 2 40-3709 F: 598 2 49-5651

Cervini, Raul, Sr.
Professor
UNIVERSIDAD CATOLICA URUGUAY
Plaza Independencia 811 - PB
Montevideo, Uruguay
T: 598 2 921515 F: 598 2 925454
TL: 22379 YR

Chew, Tracy, Ms.
Volunteer
PEACE CORPS URUGUAY
Plaza Independencia 822, 4o Piso
Montivideo, Uruguay
T: 598 2 904123 F: 598 2 923479

Clavito, Lilian, Ms.
GARDEN CLUB OF URUGUAY
Maldonado Punta Del Este - Chalet Andor 24 Mansa
Punta del Leste, Uruguay
T: 596 42 25-369 F: 596 42 41-887

Clermont, Ecilda Isabel Lasa, Ms.
Prof. De Biologia
EDUCACION SECUNDARIA
Brenda 5901
Montevideo, Uruguay
T: 598 2 617104

Coirolo, Carlos, Sr.
CIPFE
Canelones 1164
Montevideo 11100, Uruguay
T: 598 2 922362 F: 598 2 985959

Engstrom, Charlene, Ms.
Volunteer
PEACE CORPS URUGUAY
Plaza Independencia 822, 4o Piso
Montevideo, Uruguay
T: 598 2 904123 F: 598 2 923479

Escofet, Adrian, Sr.
GRUPO ESTUDIANTIL PRO MEDIO AMBIENTE
S. Vazquez 1140
Montevideo 11300, Uruguay
T: 598 2 783679

Evia, Graciela, Ms.
C.L.A.E.S.
Casilla Correo 13000
Montevideo 11700, Uruguay
T: 598 2 783132 F: 598 2 921117
TL: 23391 UY

Farriols, Isabel
ESCUELA NACIONAL DE ECOLOGIA APLICADA - ENEA
18 de Julio 1235
Montevideo, Uruguay
T: 598 2 90 16 16 F: 598 2 90 67 39

Furman, Charna, Ms.
UNIDAD DE VIVIENDA
Br. Artigas 1031
Montevideo, Uruguay
T: 598 2 806950 F: 598 2 807433

Galeano, Pablo, Mr.
REDES/ FRIENDS OF THE EARTH - URUGUAY
Millan 4113
Montevideo 12900, Uruguay
T: 598 2 356265 F: 598 2 381640

Galecki, Jennifer A., Ms.
Volunteer
PEACE CORPS URUGUAY
Plaza Independencia 822, 4o Piso
Montivideo, Uruguay
T: 598 2 904123 F: 598 2 923479

Garcia Mattes, Mabel, Ms.
GRUPO BOSQUE
Garaudi, 678- Ap. 805
Punta del Leste, Uruguay
T: 596 42 20-758

Garrido, Lydia Maria Luzardo, Sra.
Directivo
ADIPE
Avda. Gorleto 993
Punta del Leste 20100, Uruguay
T: 596 42 43239-43187 F: 596 42 45679

Garrone, Maria Del Lujan Banchero, Sra.
INSTITUTO DE PROMOCION ECONOMICO SOCIAL DEL URUGUAY
Colonia 2069
Montevideo, Uruguay

Gascue, Susana, Ms.
Educadora
HORNEROS POR LA VIDA 'EMAUS'
Caraguatay 2086
Montevideo 11800, Uruguay
T: 598 2 584222 F: 598 2 565239

Gil, Cecilia, Sra.
AMULP
Canelones 929
Montevideo, Uruguay
T: 598 2 903422 F: 598 2 904961

Goldflus, Mirta, Ms.
Vice President
INTERNATIONAL COUNCIL OF JEWISH WOMEN
Vina del Mar 6824
Montivideo, Uruguay
T: 598 2 60-31-05

Gomara, Susana, Ms.
Vice-Pres.
GARDEN CRUB
Chaletsineu - Parada 11
Punta del Leste, Uruguay
T: 596 42 8-1910

Gonzalez, Ileana, Ms.
CEATECO
18 De Julio 2294- Apto. 801
Montevideo, Uruguay
T: 598 2 49-2837

Grana, Adriana
ESCUELA NACIONAL DE ECOLOGIA APLICADA - ENEA
18 de Julio 1235
Montevideo, Uruguay
T: 598 2 90 16 16 F: 598 2 90 67 39

Gross, Daniel J., Mr.
Volunteer
PEACE CORPS URUGUAY
Plaza Independencia 822, 4o Piso
Montivideo, Uruguay
T: 598 2 904123 F: 598 2 923479

Gudynas, Eduardo, Mr.
Academic Coordinator
M.F.A.L
Canelones 1164
Montevideo, Uruguay
T: 598 2 922363 F: 598 2 98-5959

Johnson, Jennie, Ms.
Volunteer
PEACE CORPS URUGUAY
Plaza Independencia 822, 4o Piso
Montivideo, Uruguay
T: 598 2 904123 F: 598 2 923479

Katz, Dana, Ms.
Volunteer
PEACE CORPS URUGUAY
Plaza Independencia 822, 4o Piso
Montevideo, Uruguay
T: 598 2 904123 F: 598 2 923479

Lara, Michael, Mr.
Natural Resources Volunteer
PEACE CORPS URUGUAY
Plaza Independencia 822, 4o Piso
Montevideo, Uruguay
T: 598 2 904123 F: 598 2 923479

Mahlau, Kristen Verner, Mr.
Volunteer
PEACE CORPS URUGUAY
Plaza Independencia 822, 4o Piso
Montevideo, Uruguay
T: 598 2 904123 F: 598 2 923479

Marquez, Claudia
ESCUELA NACIONAL DE ECOLOGIA APLICADA - ENEA
18 de Julio 1235
Montevideo, Uruguay
T: 598 2 90 16 16 F: 598 2 90 67 39

Martinez, Olga Margot Garcia
CENTRO INVESTIGACIONES
Luis Bature Berros SI 86
Montevideo, Uruguay
T: 598 2 35-0837

Mautone, Marcelo, Sr.
Rep. Gestiones
GRUPO ETO-ECOLOGICO DE DURAZNO
Durazno
Durazno, Uruguay

Mazuncalde, Susana, Ms.
ALTERNATIVAS
Amsterdam 1404 Apt. 802
Montevideo, Uruguay
T: 598 2 692819

Mazzei, Alex, Mr.
Professor
FUNDACION PLEMUU
Uruguay, 1555
Montevideo, Uruguay
T: 598 2 417470 F: 598 2 480556

Mello, Mateo Magarinos De, Mr.
President
COMMISSION INTER - AMERICAINE DU DROIT DE L'ENVIRONNEMENT
Rue Echvarriarze 3396
Montevideo, Uruguay

Michetti, Luis Alberto Santos, Sr.
Chemical Engineer
Blanca del Tabare 2980/102
Montevideo 11300, Uruguay
T: 598 2 704-3168 F: 598 2 98-5282

Mieres, Bruno
ESCUELA NACIONAL DE ECOLOGIA APLICADA - ENEA
18 de Julio 1235
Montevideo, Uruguay
T: 598 2 90 16 16 F: 598 2 90 67 39

Mieres, Homero, Mr.
ESCUELA NACIONAL DE ECOLOGIA APLICADA - ENEA
18 de Julio 1235
Montevideo, Uruguay
T: 598 2 90 16 16 F: 598 2 90 67 39

Miles, Fernando Perez, Mr.
CIPFE
Canelones 1164
Montevideo, Uruguay
T: 598 2 922363 F: 598 2 98-5959

Milicua, Dalila, Ms.
CONSEJO NACIONAL DE MUJERES
Andes 1433/107
Montivideo, Uruguay
T: 598 2 913611

Mortezayani, Oliver, Sr.
Membro
COMUNIDADE BAHA©I
Joaquin Pequena 1090
Montevideo, Uruguay

Nansen, Karin, Ms.
REDES/ FRIENDS OF THE EARTH - URUGUAY
Millan 4113
Montevideo 12900, Uruguay
T: 598 2 356265 F: 598 2 381640

Navarrete, Maria Severina, Ms.
President
CACPLA
Somme 1612
Montevideo, Uruguay
T: 598 2 783447 F: 598 2 792108

Nicholson, Patrick D., Mr.
Volunteer
PEACE CORPS URUGUAY
Plaza Independencia 822, 4o Piso
Montevideo, Uruguay
T: 598 2 904123 F: 598 2 923479

Nocetti Cotelo, Ana Maria, Sra.
Diretora
FUNDACION PLEMUU
Uruguai, 1555
Montevideo, Uruguay
T: 598 2 417470 F: 598 2 480556
E: 480556

Ortiz, Cesar, Mr.
RED URUGUAYA DE ONG'S AMBIENTALISTAS
Millan 4113
Montevideo 12900, Uruguay
T: 598 2 356265 F: 598 2 381640

Paradell, Lidia, Ms.
GRUPO BOSQUE
Cantegril Country Club Local 04
Maldonado 20100, Uruguay
T: 598 423-2440 F: 598 423-2440

Parikh, Jagdish
NGONET
Casilla Correo 1539
Montevideo 11000, Uruguay
T: 598 2 496192 F: 598 2 419222

Perasso, Mario, Mr.
ESCUELA NACIONAL DE ECOLOGIA APLICADA - ENEA
18 de Julio 1235
Montevideo, Uruguay
T: 598 2 90 16 16 F: 598 2 90 67 39

Pereyra, Ricardo, Mr.
ENCUENTRO ECOLOGICO 'CAD'
Lezica 6375
Montevideo, Uruguay
T: 598 2 328077

Perez, Alejandro Gil, Sr.
Directivo
C.E.U.T.A.
Santiago do Chile, 1183
Montivideo, Uruguay
T: 598 2 914004 F: 598 2 914004

Perez, Raul, Mr.
GRUPO MADRE TIERRA
Santanjer 2069
Montevideo 11400, Uruguay
T: 598 2 587 489

Perone, Janet, Ms.
NGONET
Casilla Correo 1539
Montevideo 11000, Uruguay
T: 598 2 496192 F: 598 2 419222

Philips, Gillian, Ms.
NGONET
Casilla Correo 1539
Montevideo 11000, Uruguay
T: 598 2 496192 F: 598 2 419222

Porta, Carlos, Sr.
SOCIEDAD DE CONSERVACION DEL MEDIO AMBIENTE
Palais de Nations
Geneve CH 1211, Uruguay
F: 41 22 733-6542

Portillo, Mabel, Ms.
ESCUELA NACIONAL DE ECOLOGIA APLICADA - ENEA
18 de Julio 1235
Montevideo, Uruguay
T: 598 2 90 16 16 F: 598 2 90 67 39

Reed, Kevin, Mr.
Volunteer Biologist
PEACE CORPS URUGUAY
Plaza Independencia 822, 4o Piso
Montevideo, Uruguay
T: 598 2 904123 F: 598 2 923479

Ribeiro, Silvia, Ms.
FUTURE EARTH NETWORK TIERRA DEL FUTURO
Millan 4113
Montevideo 12900, Uruguay
T: 598 2 356265 F: 598 2 381640

Rice, Peter, Mr.
Volunteer
PEACE CORPS URUGUAY
Plaza Independencia 822, 4o Piso
Montevideo, Uruguay
T: 598 2 904123 F: 598 2 923479

Rivera, Daniel A., Mr.
Volunteer
PEACE CORPS URUGUAY
Plaza Independencia 822, 4o Piso
Montevideo, Uruguay
T: 598 2 904123 F: 598 2 923479

Rodriguez, Laura, Sra.
AMULP
Canelones 929
Montevideo, Uruguay
T: 598 2 903422 F: 598 2 904961

Romero, Hector, Sr.
GRUPO ESTUDIANTIL PRO MEDIO AMBIENTE
S. Vazquez 1140
Montevideo 11300, Uruguay
T: 598 2 783679

Rosier, Jean-Charles, Sr.
Florencio Sanchez, 1512
Rivera 40000, Uruguay

Saldivia, Jorge, Mr.
Educador
HORNEROS POR LA VIDA 'EMAUS'
Caraguatay 2086
Montevideo 11800, Uruguay
T: 598 2 297383 F: 598 2 908097

Sanchez, Elsie F., Ms.
Volunteer
PEACE CORPS URUGUAY
Plaza Independencia 822, 4o Piso
Montevideo, Uruguay
T: 598 2 904123 F: 598 2 923479

Sande, Maria Cristina, Ms.
THIRD WORLD NETWORK
Jackson, 1136
Montevideo, Uruguay
T: 598 2 496192 F: 598 2 404119

Sanden, Cecilia Von, Ms.
Environmental Coordinator
CIPFE
Canelones 1164
Montevideo, Uruguay
T: 598 2 922363 F: 598 2 98-5959

Shields, Melba, Ms.
Zoologist
PEACE CORPS URUGUAY
Plaza Independencia 822, 4o Piso
Montevideo, Uruguay
T: 598 2 904123 F: 598 2 923479

Sigillito, Magela, Ms.
NGONET
Casilla Correo 1539
Montevideo 11000, Uruguay
T: 598 2 496192 F: 598 2 419222

Simpson, Elizabeth Kelly, Ms.
Envmmt Education Specialist
PEACE CORPS URUGUAY
Plaza Independencia 822, 4o Piso
Montevideo, Uruguay
T: 598 2 904123 F: 598 2 923479

Soler, Dany, Ms.
Secretary Geral
INTERNAT. FELINE FUND.
C. Roxlo 1496/301
Montevideo 11200, Uruguay
T: 598 2 493384 F: 598 2 493384

Stephenson, Amy Marie, Ms.
Volunteer
PEACE CORPS URUGUAY
Plaza Independencia 822, 4o Piso
Montevideo, Uruguay
T: 598 2 904123 F: 598 2 923479

Tinger Jr., John, Mr.
Volunteer
PEACE CORPS URUGUAY
Plaza Independencia 822, 4o Piso
Montivideo, Uruguay
T: 598 2 904123 F: 598 2 923479

Tron, Maria Emilia, Ms.
CONSEJO NACIONAL DE MUJERES
Andes 1433/107
Montivideo, Uruguay
T: 598 2 913611

Tron, Silvia, Ms.
Jornalist
CONSEJO NACIONAL DE MUJERES
Andes 1433/107
Montivideo, Uruguay
T: 598 2 913611

Trotter, Nathan, Mr.
Volunteer
PEACE CORPS URUGUAY
Plaza Independencia 822, 4o Piso
Montivideo, Uruguay
T: 598 2 904123 F: 598 2 923479

Vazquez, Nelida
ESCUELA NACIONAL DE ECOLOGIA APLICADA - ENEA
18 de Julio 1235
Montevideo, Uruguay
T: 598 2 90 16 16 F: 598 2 90 67 39

Vila, Gabriel Martinez, Sr.
Directivo
C.E.U.T.A.
Santiago do Chile, 1183
Montivideo, Uruguay
T: 598 2 914004 F: 598 2 914004

Vilarreal, Alberto, Mr.
FRIENDS OF THE EARTH INTERNATIONAL
Millan 4113
Montevideo, Uruguay
T: 598 2 356-265 F: 598 2 381-640

Villalba-Macias, Juan S., Mr.
Director
TRAFFIC SUDAMERICA
C. Roxlo 1496/301
Montevideo 11200, Uruguay
T: 598 2 49-3384 F: 598 2 49-3384

Whipple, Karen, Ms.
Volunteer
PEACE CORPS
Plaza Independencia 822, 4o Piso
Montivideo, Uruguay
T: 598 2 904123 F: 598 2 923479

Wilshusen, Peter, Mr.
PEACE CORPS URUGUAY
Plaza Independencia 822, 4o Piso
Montivideo, Uruguay
T: 598 2 904123 F: 598 2 923479

Zabaleta, Maria Angelica
ESCUELA NACIONAL DE ECOLOGIA APLICADA - ENEA
18 de Julio 1235
Montevideo, Uruguay
T: 598 2 90 16 16 F: 598 2 90 67 39

UN AND UNCED

Peirano, Miguel, Mr.
NGONET
Casilla Correo 1539
Montevideo 11000, Uruguay
T: 598 2 496192 F: 598 2 419222

UZBEKISTAN - GLOBAL FORUM

Chernogaev, Evgeni A., Mr.
ECOLOGIST
Nuratan Reserve, Severo-Vostok-2, 43-59
Tashkent 700124, Uzbekistan
T: 371 2 635-100 F: 371 2 413-904
TL: 116298 MAKET SU

Oralov, Valery V., Mr.
ECOLOGIST
Nuratan Reserve, Severo-Vostok-2, 43-59
Tashkent 700124, Uzbekistan
T: 371 2 635-100 F: 371 2 413-904
TL: 116298 MAKET SU

Tsaruk, Ole E., Mr.
ECOLOGIST
laidar pr. 11-a fl.10
Tashkent 700115, Uzbekistan
T: 371 2 913-925 F: 371 2 413-904
TL: 116298 MAKET SU

Zagribelny, Alexander, Mr.
Chairman, Ed Of 'oasis'
GREEN MOVEMENT CENTRE/EDITOR OF NEWSPAPER 'OASIS'
Lundcharsky str. 42-2
Djambul Kazakhstan, Uzbekistan
T: 70-52483

VANUATU - GOVERNMENT

Carlot, Maxime
Prime Minister
Permanent Mission of Vanuatu
416 Convent Ave.
New York NY 10031, United States of America
T: 212 926 3311 F: 212 926 4131

GLOBAL FORUM

Lini, Fr. Walter, Mr.
Parliamentarian
TERRA CHRISTA
Ohlen Area
Port Vila, Vanuatu
T: 678 22229 F: 678 22910

Lini, Mary, Ms.
House Wife
TERRA CHRISTA
Ohlen Area
Port Vila, Vanuatu
T: 678 22229 F: 678222910

Marae, Clarence, Mr.
Attorney
TERRA CHRISTA
Ohlen Area
Port Vila, Vanuatu
T: 22755 F: 22910

Siwatibau, Suliana, Ms.
WORLD COUNCIL OF CHURCHES
c/o ESCAP - P. O. Box 503
Port Vila, Vanuatu

VENEZUELA - GOVERNMENT

Cisneros, Elia, Ms
Delegation - Uzla
NGO REP IN GOBERMENT
Aptdo 70519 Los Ruices C
Caracas, Venezuela
T: 582 916449/924333 F: 582 926905

Garcia Prince, Evangelina, Ms
Ministra Para La Promocion De La
GOBIERNO DE VENEZUELA
Esq. Jesuitas. Torre Bandagro. Piso
Caracas 1010, Venezuela
T: 02 818214 818211 987 F: 02 818215

Velasquez De Visbal, Carmentuisa
Directora De Relaciones Internatl
MINISTERIO DEL AMBIENTE (NARNR
Torre Sur Piso 18 Oficina 1805 Cen To Simon Bolivar,El Silencio
Caracas, Venezuela
T: 58 2 413298 4834638 F: 58 2 4832445

NGO

Bustamante-Rivas, Eugenia
AMIGRANSA
Apartado Postal 50460
Caracas 1050-A, Venezuela

Diaz, Maria Elena
FEA
Apartado Postal No. 456
Merida 5101, Venezuela

Flores, Celestino
FORJA
Residencias 'Paraguachi', Edificio 17, Apart. 207
Pisa 2. Sector UD3, Parroguia Caricuao, Caracas, Venezuela

Garcia-Scarton, Alicia
AMIGRANSA
Apartado Postal 50460
Caracas 1050-A, Venezuela

Hernandez Arocha, Armando
FP
PO Box 70934
Caracas 1071A, Venezuela

Lascoutx, Alfredo
FEA
Apartado Postal No. 456
Merida 5101, Venezuela

Martinez, Alejandro Ubieda
PACRV
ones Interparlamentarias/Secretaria Ejecutiva
Caracus, Venezuela

Moya, Jose
General Coordinator
FORJA DE VENEZUELA
Residencias 'Paraguachi', Edificio 17, Apart. 207
Pisa 2. Sector UD3, Parroguia Caricuao, Caracas, Venezuela
T: 58 2 431-4437 F: 58 2 431-4437

Palacio, Francisco
PACRV
ones Interparlamentarias/Secretaria Ejecutiva
Caracus, Venezuela

Penaranda, Maria T.
PACRV
ones Interparlamentarias/Secretaria Ejecutiva
Caracus, Venezuela

Rodriguez, Jose Curiel
PACRV
ones Interparlamentarias/Secretaria Ejecutiva
Caracus, Venezuela

PRESS

Barrios, Asdrubal
Correspondent
DIARIO EL NACIONAL
Av. A La Carlota
Caracas, Venezuela, Venezuela

Diaz, Lina
Reporter Tv
SERV. INF. AMERICA LATINA - ECC
Av. Leman 91
Brussels, Belgium, Venezuela

Juhler, Peter
COMUNIDAD EUROPEA
Res. Sierra Nevada, Apto. 42A
Chulavista, Caracas, Venezuela, Venezuela

Las Heras, Txomin
Correspondent
AGENCIA VENEZOLANA DE NOTICIAS
Alameda Sarutaia, 73 Apt. 26
01403-Sao Paulo-SP, Brazil, Venezuela

Martins, Luiz Renato
Correspondent
PRESS COOMMUNAUTE EUROPEENE
R.Monte Alegre 1741
Sao Paulo, S.P., Brazil, Venezuela

Perez, Alvaro
Correspondent
COMUNIDAD EUROPEA
Res. San Fernando, Santa Paula
Caracas, Venezuela, Venezuela

Smith, Alexander
THE DAILY JOURNAL
Caracas, Venezuela

Socorro Morales, Marco Tulio
Correspondent
EL DIARIO DE CARACAS
Av. Principal Urb. San Luis
Apto.6-2, Caracas, Venezuela, Venezuela

ORGANIZATIONS

Berti, Arturo Luis, Sr.
International Rural Housing Association
Apartado Postal 16224
Caracas 1011-A, Venezuela
T: (58 2) 574 2586

Bustamante, Maria Eugenia, Ms.
Sociedad de Amigos en Defensa de la Gran Sabana (AMIGRANSA)
Apartado Postal 50460
Caracas 1050-A, Venezuela
T: (58 2) 662 4487 F: (58 2) 284 9377

De L. Acedo De Sucre, Maria, Sra
Fundacion Para La Defensa De La Naturaleza (FUDENA)
Ave. Principal Los Cortijos de Lourdes c/2a. Transversal, Centro
Empresarial Senderos,Piso 6, Oficina 611-A, Caracas, Venezuela
T: (58 2) 238 29 30 F: (58 2) 239 65 47

Hernandez, Jose I. Moya, Prof.
Federacion De Organizaciones Y Juntas Ambientalistas De Venezuela (FORJA)
Residencias 'Paraguachi', Edificio 17, Apart. 207
Pisa 2. Sector UD3, Parroguia Caricuao, Caracas, Venezuela
T: (58 2) 431 4437 F: (58 2) 408 1464

Lara, Maria Margarita, Ms.
Fundacion Polar
PO Box 70934
Caracas 1071A, Venezuela
T: (58 2) 203 3111 F: (58 2) 203 3722

Martinez, Rafael Elino, Mr.
President
Fundacion Ambiental Francisco Tamayo
Avenida Diaz Moreno, Edif. Oficentro 108, Piso 5
Oficina 5-B, Esquina San Jose, Valencia, Venezuela
T: 22 43 59 F: 57 80 40

Nunez, Miguel Angel, Sr.
Instituto para la Produccion e Investigacion de la Agricultura Tropical (IPIAT)
Apartado 84 Merida
Merida 5101-A, Venezuela
T: (58 74) 523 779 F: (58 74) 527 402

GLOBAL FORUM

Arocha, Armando Hernandez, Sr.
Coordinator Of Environ. Proj.
FUNDACION POLAR
Cal. Hans Neumann con. 2da Av. de los Cortijos de Lourdes, piso 1
Caracas, Venezuela
T: 58 2 203-3762 F: 58 2 203-3722
TL: 21628 FUNPO VC

Astudillo, Neddy A., Sra.
Student
UCV
Apartado 69321 Altamira
Caracas 1062, Venezuela
T: 56 2 32-7139 F: 56 2 284-5475

Bello, Mercedes, Sra.
INTERNATIONAL UNION OF YOUNG CHRISTIAN DEMOCRATS - UIJDC
Caracas, Venezuela
F: 58 2 2610547

Bhat, Keshava, Mr.
MOVIMIENTO DE REEDUCACION NATURISTA
av. TRUCUJILIO, 12-39 MARIPEREZ
Caracas, Venezuela
T: 58 2 781-5962

Bracho, Frank
Adviser
INTERNATIONAL FACILITATING/FUNDAVIME/AMIGRANSA
Caracas, Venezuela

Briceno, Ybelice L., Sra.
Student
UCV
Apartado 69321 Altamira
Caracas 1062, Venezuela
T: 56 2 32-7139 F: 56 2 284-5475

Brino, Kathy, Ms.
TOGETHER FOUNDATION
Cc. Passeo las Mercedes - Nivel Oficinas 2 Piso - Of.1-2
Caracas, Venezuela
T: 58 2 924-333/916-449 F: 58 2 926-905

Bufalo, Enzo Dal, Sr.
Professor
U.C.V.
Calle Geminis Edf. Dom Emillo 41 Santa Paula
Caracas 1067, Venezuela
T: 58 2 987-4449 F: 58 2 285-5129

Bustamante, M. Eugenia, Ms.
AMIGRANSA
Apartado Postal 50460
Caracas 1050 A, Venezuela
T: 58 2 987-6716 F: 58 2 284-9377

Camelo, Nelson, Mr.
Calle Avila # 1
Aragua, Venezuela
T: 04383-3518 F: 04383-0813

Carranza, Zoraya, Sra.
ASO MOSHIMA, ASOCAURIMARE
Calle A-1, Ata Eran Montana, Urb. Caurimare
Caracas, Venezuela
T: 58 2 985--803

Cisneros, Claudia, Ms.
TOGETHER FOUNDATION
Cc. Passeo las Mercedes - Nivel Oficinas 2 Piso - Of.1-2
Caracas, Venezuela
T: 58 2 924-333/916-449 F: 58 2 926-905

Cisneros, Mariela, Ms.
TOGETHER FOUNDATION
Cc. Passeo las Mercedes - Nivel Oficinas 2 Piso - Of.1-2
Caracas, Venezuela
T: 58 2 924-333/916-449 F: 58 2 926-905

Cisneros, Oswaldo, Mr.
TOGETHER FOUNDATION
Cc. Passeo las Mercedes - Nivel Oficinas 2 Piso - Of.1-2
Caracas, Venezuela
T: 58 2 924-333/916-449 F: 58 2 926-905

Daza-Paris, Livia, Sra.
CAONABO NUEVAS FORMAS DE LA DANCA EN LAS AMERICAS
Apdo Correo 4002, Carmelitas
Caracas, Venezuela
T: 58 2 671 5443 F: 58 2 545 2129

Dubois, Mark, Mr.
TOGETHER FOUNDATION
Cc. Passeo las Mercedes - Nivel Oficinas 2 Piso - Of.1-2
Caracas, Venezuela
T: 58 2 924-333/916-449 F: 58 2 926-905

Espinoza, Haydee Josefina Diaz, Ms.
Coordenadora Nacional Legisti
ASSOCIACION DE MUJERES AMBIENTALISTAS DE VENEZUELA (AMAVEN)
Av Paez cm Calle C. - Say-Park IV ,2 piso 23 BW Paraiso
Caracas, Venezuela
T: 58 2 442 9187 F: 58 2 782 2598

Fabres Cordero, Maria, Ms.
ECOVITAL
3 Entrada El Convente, Edf. El Sauce, apto 11-C, El Marques
Caracas, Venezuela
T: 58 2 225889

Fernandez, Armando Chirinos
UIJDC - UNIAO INTERNACIONAL DE JOVENS DEMOCRATAS CRISTAOS
Av. Primero de Altamira Sur Resid. Taurizano 82
Caracas, Venezuela
T: 00582 2630065 F: 00582 2631992

Forwood, William
Director Of Information
EUROPEAN COMMUNITY
Oficiana de Informacao para America Latina, Cee
Las Mercedes, Caraca, Venezuela
T: 58 2 915133 F: 58 2 911114

Garcia, Alicia, Sra.
AMIGRANSA
Palais de Nations
Geneve CH 1211, Venezuela
F: 41 22 733-6542

Garcia, Estevez, Mr.
FUNDACION CIEPE
Caracas, Venezuela
T: 58 2 781-8219 F: 58 2 781-9387

Garcia-Guardilla, Maria Pilar, Ms.
GRIDIA
Aportado Postal 89-000 - Universidad Simon Bolivar
Caracas 1080, Venezuela
T: 58 2 9799085 F: 58 2 9621695

Gomez, Natacha M., Sra.
Student
UCV
Apartado 69321 Altamira
Caracas 1062, Venezuela
T: 58 2 32-7139 F: 58 2 284-5475

Gonzalez, Luis Jose Marcano, Sr.
FUNDACION SERVICIO PARA EL AGRICULTOR
Torre Cavendes, Av Eco de Miranda, Office 806
Caracas, Venezuela

Hernandez, Armando, Mr.
Engineer
FUNDACION POLAR
Edf. Fundacion Polar, Piso 1. Los Cortijos de Lourdes
Caracas, Venezuela
T: 56 2 2395622 F: 56 2 2392704

Hernandez Prada, Coromoto, Ms.
F.L.E.A.
FUNDACION LA ERA AGRICOLA
Av. dos Lora entre 18 y 19, Edif. Cuinas P.B. 1
Merida, Venezuela
T: 58 74 527402 F: 58 74 522795
TL: APDO Postas 456, E: 5101 Venezuela

Izcaray, Flor G., Sra.
Student
UCV
Apartado 69321 Altamira
Caracas 1062, Venezuela
T: 58 2 32-7139 F: 58 2 284-5475

Jones, Kathryn, Ms.
TOGETHER FOUNDATION
Cc. Passeo las Mercedes - Nivel Oficinas 2 Piso - Of.1-2
Caracas, Venezuela
T: 58 2 924-333/916-449 F: 58 2 926-905

Kennedy, Matt, Mr.
TOGETHER FOUNDATION
Cc. Passeo las Mercedes - Nivel Oficinas 2 Piso - Of.1-2
Caracas, Venezuela
T: 58 2 924-333/916-449 F: 58 2 926-905

Lander, Edgardo, Sra.
Director
UCV
Apartado 69321 Altamira
Caracas 1062, Venezuela
T: 58 2 32-7139 F: 58 2 284-5475

Linares, Yamelis, Ms.
FUNEDICUR
AV. BICENTENARIO No. 10-87
Barinas, Venezuela
T: 58 73 24910

Lopez, Argelia Laya, Sra.
Coordinadra General
ASSOCIACION MUJER Y AMBIENTE - AMAVEN
Esquina de Jesuitas-Torre Bandagro - Piso 2
Caracas, Venezuela
T: 58 2 782-2598 F: 58 2 782-2598

Martinez, Renilda, Sra.
ORGANIZACION INDIGENA MOVIMENTO INDIO POR LA IDENTIDAD NACIONAL Y CONIVE
urb-el naranjal avis j-1 n.50 a-46
Maracaibo, Venezuela
T: 58 61 428996

Mata Escobar, Andreina, Sra.
FUNDACION LA ERA AGRICOLA
Av. dos Lora Entre 18 Y 19 EDF. Cuinas P.B.1
Merida, Venezuela
T: 58 74 527402 F: 58 74 522795

McIntyre, Jeff, Mr.
TOGETHER FOUNDATION
Cc. Passeo las Mercedes - Nivel Oficinas 2 Piso - Of.1-2
Caracas, Venezuela
T: 58 2 924-333/916-449 F: 58 2 926-905

McIntyre, Jim, Mr.
TOGETHER FOUNDATION
Cc. Passeo las Mercedes - Nivel Oficinas 2 Piso - Of.1-2
Caracas, Venezuela
T: 58 2 924-333/916-449 F: 58 2 926-905

Meijer-Werner, Eugenia Frota De, Ms.
Delegado
PARTIDO ECOLOGICO VENEZOLANO
Calle la Ceiba - Quinta Y Unam - Prados Del Este
Caracas, Venezuela
T: 58 2 978-1671 F: 58 2 979-2821
TL: APARTADO 61810

Molina, Yolanda, Sra.
Coordinator
FUSAGAI
Fusagri Curimagua
Coro, Venezuela

Montoya, Ruben, Sr.
CESAP
Apartado Postal 4240
Caracas 1010-A, Venezuela

Moreno, Zoraida
AMIGRANSA-SOC. AMIGOS EN DEFESA GRAN SABANA
Centro Comercial Los Chaguaramos, Piso 20, Ofic 20-03, Los Chaguaramos
Caracas, Venezuela
T: 58 2 662-4487 F: 58 2 662-4492

Nava, Alexander Luzardo, Sr.
Presidente
FUNDACION ECODESARROLLO
Torre Phelps piso 17, oficina 'C', Plaza Venezuela
Caracas, Venezuela
T: 58 2 793-0989 F: 58 2 793-0989

Nieves, Anibal Espejo, Sr.
Student
UCV
Apartado 69321 Altamira
Caracas 1062, Venezuela
T: 56 2 32-7139 F: 56 2 284-5475

Nunes, Luis, Sr.
Director
INTERNATIONAL UNION OF YOUNG CHRISTIAN DEMOCRATS - UIJDC
Caracas, Venezuela
F: 58 2 2610547

Nunez, Zaida, Ms.
TOGETHER FOUNDATION
Cc. Passeo las Mercedes - Nivel Oficinas 2 Piso - Of.1-2
Caracas, Venezuela
T: 58 2 924-333/916-449 F: 58 2 926-905

Ortiz, Alexis, Mr.
Ong
ECOVITAL
Sexta Transversal Santa Eduviyis, Ed. Masparro, Apto. 33
Caracas 1010, Venezuela
T: 58 2 283-2552 F: 58 2 414558

Ovalles, Omar
HABITAT
12 Rajaji Marg
New Delhi 110011, Venezuela
T: 91 11 301-4309

Oyarzabal, Nelson, Mr.
CROQUIS
Calle Orodeza Castillo Edif. Ferma apto 36 SAB Grande Caracas
Caracas, Venezuela
T: 58 2 782 7441

Palacios, Joseph, Mr.
TOGETHER FOUNDATION
Cc. Passeo las Mercedes - Nivel Oficinas 2 Piso - Of.1-2
Caracas, Venezuela
T: 58 2 924-333/916-449 F: 58 2 926-905

Paredes, Enrique, Sr.
Institutional Relations Chief
FUNDAMAYO
Av. Diaz Moreno, Edf. Oficentro 108, Piso 5 Of.5-B
Valencia - Caraboro, Venezuela
T: 58 41 577543 F: 58 41 578040
TL: 41414

Petit, Ana Beatriz Perez Osuna De
UIJDC - UNIAO INTERNACIONAL DE JOVENS DEMOCRATAS CRISTAOS
Av. Primero de Altamira Sur Resid. Taurizano 82
Caracas, Venezuela
T: 00582 2630065 F: 00582 2631992

Ramon, Aveledo Hostos, Sr.
SOCIEDAD VENEZUELANA DE CIENCIAS NATUAIS
El Marquez, Calle Arichunz
Caracas, Venezuela
T: 58 2 21-7780 F: 58 2 21-4278

Rino, Fiume, Mr.
GLOBAL EDUCATION
Gram Jabana - S. Margarita
Bolivar, Venezuela

Rodriguez, Jon Paul, Sr.
Director Of Programs
PROVITA
Apdo. 47552
Caracas 1041-A, Venezuela
T: 58 2 576-2828 F: 58 2 576-1579

Romero, Liccia, Ms.
IPIAT
Apartado Postal 17514
Caracas 1015-A, Venezuela
T: 58 2 572-8595 F: 58 2 541-7717

Ruiz, Gustavo
INSTITUTO DE COMUNICACION Y MEDIO AMBIENTE
Caracas, Venezuela

Runfola M., Jose Wilmer, Mr.
FUNDACION LA ERA AGRICOLA
Av. 2 Entre 19 # 18-41 Edificio Cuinas Planta Baja #1
Merida, Venezuela
T: 58 074 527402 F: 58 074 522795

Salvatierra, Marisela
CIRCULO DE PERIODISTAS CIENTIFICOS
Caracas, Venezuela

Sanchez, Antonio
UIJDC - UNIAO INTERNACIONAL DE JOVENS DEMOCRATAS CRISTAOS
Av. Primero de Altamira Sur Resid. Taurizano 82
Caracas, Venezuela
T: 00582 2630065 F: 00582 2631992

Sanchez, Eveling A., Sra.
Student
UCV
Apartado 69321 Altamira
Caracas 1062, Venezuela
T: 56 2 32-7139 F: 56 2 284-5475

Sanchez, Magaly, Ms.
INSTITUTO DE URBANISMO - U.C.V.
Facultad de Arquitectura - U-C-U
Caracas, Venezuela
T: 58 2 662 5717 F: 58 2 662 1316

Sepulveda, Maytee, Ms.
TOGETHER FOUNDATION
Cc. Passeo las Mercedes - Nivel Oficinas 2 Piso - Of.1-2
Caracas, Venezuela
T: 58 2 924-333/916-449 F: 58 2 926-905

Silva, Lisa E., Sra.
Student
UCV
Apartado 69321 Altamira
Caracas 1062, Venezuela
T: 56 2 32-7139 F: 56 2 284-5475

Silveira, Pedro M., Sr.
Student
UCV
Apartado 69321 Altamira
Caracas 1062, Venezuela
T: 58 2 32-7139 F: 58 2 284-5475

Simancas, Francisco, Mr.
Profesor
CEFI - ULA
Av. Universidad - Paramillo
San Cristobal, Venezuela
T: 562184 F: 563411

Simms, Bill, Mr.
TOGETHER FOUNDATION
Cc. Passeo las Mercedes - Nivel Oficinas 2 Piso - Of.1-2
Caracas, Venezuela
T: 58 2 924-333/916-449 F: 58 2 926-905

Soares, Angelina, Ms.
AMAVEN
Lagunelica - Calle Rio Chico - Edo. Miranda
Miranda, Venezuela
T: 321323

Toro S, Ligia, Sra.
Abogado
AMAVEN - VENEZUELA
Ave Caroni 9 Ta
Caracas, Venezuela
T: 58 2 316914 F: 58 2 316910

Trujillo, Rosa, Ms.
Coordinadora
GEMA
Los Caobos - Tomelos Caobos B. Apto 171B
Caracas, Venezuela
T: 58 2 571 7648 F: 58 2 571 7648

Valencia, Judith, Sra.
Teacher
UCV
Apartado 69321 Altamira
Caracas 1062, Venezuela
T: 58 2 32-7139 F: 58 2 284-5475

Vargas, Martha, Ms.
TOGETHER FOUNDATION
Cc. Passeo las Mercedes - Nivel Oficinas 2 Piso - Of.1-2
Caracas, Venezuela
T: 58 2 924-333/916-449 F: 58 2 926-905

Vicentini, Maria Luisa, Sra.
INTERNATIONAL UNION OF YOUNG CHRISTIAN DEMOCRATS - UIJDC
Caracas, Venezuela
F: 58 2 2610547

Villasmil, Marcos, Sr.
General Secretary
INTERNATIONAL UNION OF YOUNG CHRISTIAN DEMOCRATS - UIJDC
Caracas, Venezuela
F: 58 2 2610547

Young, Suzanne, Ms.
TOGETHER FOUNDATION
Cc. Passeo las Mercedes - Nivel Oficinas 2 Piso - Of.1-2
Caracas, Venezuela
T: 58 2 924-333/916-449 F: 58 2 926-905

PARLIAMENTARIANS / SPIRITUAL LEADERS

Gervasi, Oscar Celli
Senado, Congreso de la Republica
Caracas, Venezuela
T: 4836011
TL: 21252 CCASSVC

Gonzales, Eudoro, Sen.
Senado, Congreso de la Republica
Caracas, Venezuela
T: 4836011
TL: 21252 CCASSVC

Guillen, Edgar Gomez
Senado, Congreso de la Republica
Caracas, Venezuela
T: 4836011
TL: 21252 CCASSVC

Laguado, Gabriel Nino
Senado, Congreso de la Republica
Caracas, Venezuela
T: 4836011
TL: 21252 CCASSVC

Martinez, Pedro Escarra
Senado, Congreso de la Republica
Caracas, Venezuela
T: 4836011
TL: 21252 CCASSVC

VIET NAM - GOVERNMENT

Le Quy An, , Mr.
Vice-Chairman
National Commission for Science, (Deputy Head for Delegation)
Permanent Mission to the United Nations, 20 Waterside Plaza New York, NY, 10010
F: 212 679 3779
TL: 212 686 8534

Nguyen Khanh, , H.E. Mr.
Vice Chairman of the Council of Ministers
Socialist Republic of Viet Nam
Permanent Mission to the United Nations, 20 Waterside Plaza New York, NY, 10010
F: 212 679 3779
TL: 212 686 8534

Nguyen Minh Thong, , Mr.
Second Secretary
Permanent Mission of Viet Nam
20 Waterside Plaza
New York, NY 10010, United States of America
T: 212 679 3779 F: 212 686 8534

Nguyen Thi Hoi, Mrs., Deputy-Head
Department for International Organizations Ministry of Foreign Affairs
Permanent Mission of Viet Nam
20 Waterside Plaza
New York, NY 10010, United States of America
T: 212 679 3779 F: 212 686 8534

Pham Sanh Chau, , Mr.
Expert
Ministry of Foreign Affairs
Permanent Mission to the United Nations, 20 Waterside Plaza New York, NY, 10010
F: 212 679 3779
TL: 212 686 8534

Trinh Van Thu, , Mr.
Deputy-Head
Department for Meteorology and Hydrology
Permanent Mission to the United Nations, 20 Waterside Plaza New York, NY, 10010
F: 212 679 3779
TL: 212 686 8534

GLOBAL FORUM

The Tua, Dao, Mr.
INSTITUT NATIONAL DES SCIENCES AGRONOMIQUES
Hanoi, Viet Nam
T: 42 52 650

Tranh, Nguyen Ngoc, Mr.
ENDA TIERS MONDE
Ho Chi Minhville, Viet Nam
T: 84 4 263-127

YEMEN - GOVERNMENT

Abdulghani, Adulaziz, Mr.
Vice President
Permanent Mission of the Republic of Yemen
866 United Nations Plaza, Room 435
New York NY 10017, United States of America
T: 212 355 1730 F: 212 750 9613

Al-Ashtal, Abdalla Saleh, H.E. Mr.
Ambassador Extraordinary and Plenipotentiary
Permanent Representative to the United Nations, New York
Permanent Mission to the United Nations, 866 United Nations Plaza, Room 435
New York, NY 10017, United States of America

Al-Guneid, Hussein Alawi, Mr.
Secretary
Council for the Environment Protection
Permanent Mission to the United Nations, 866 United Nations Plaza, Room 435
New York, NY 10017, United States of America

Al-Hamadani, Mohssen Ali, H.E. Engr.
Minister of State
President of the Council for the Environment Protection
Permanent Mission to the United Nations, 866 United Nations Plaza, Room 435
New York, NY 10017, United States of America

Al-Moqaleh, Abdul Karim Saleh, Mr.
Director
General Office of the Minister of State for Environment Protection
Permanent Mission to the United Nations, 866 United Nations Plaza, Room 435
New York, NY 10017, United States of America

Al-Thaur, Ali, H.E. Mr.
Member of Parliament
Permanent Mission of Yemen
866 United Nations Plaza, Room 435
New York, NY 10017, United States of America

Aqlan, Ahmed Abdul-Raheem, Mr.
Minister Plenipotentiary
Ministry for Foreign Affairs
Permanent Mission to the United Nations, 866 United Nations Plaza, Room 435
New York, NY 10017, United States of America

Baesa, Abdul Aziz, Mr.
Director International Organizations Department
Ministry of Foreign Affairs
Permanent Mission to the United Nations, 866 United Nations Plaza, Room 435
New York, NY 10017, United States of America

Ghanem, Farag Bin, H.E.
Minister of Planning and Development
Permanent Mission to the United Nations, 866 United Nations Plaza, Room 435
New York, NY 10017, United States of America

Ghani, Abdul Aziz Abdul, H.E. Mr.
Head of Delegation
Member of the Presidential Council
Permanent Mission to the United Nations, 866 United Nations Plaza, Room 435
New York, NY 10017, United States of America

Ismail, Zarina Mohammad, Ms.
Director
Technical Secretariat Branch
Environment Council, Aden
Permanent Mission to the United Nations, 866 United Nations Plaza, Room 435 New York, NY, 10017

Jokabul-Khaleq, Jazim, Mr.
Director, Protocol Department
Ministry of Foreign Affairs
Permanent Mission to the United Nations, 866 United Nations Plaza, Room 435
New York, NY 10017, United States of America

Noaman, Izat Amin, Mr.
Director
Office of the President Council
Permanent Mission to the United Nations, 866 United Nations Plaza, Room 435
New York, NY 10017, United States of America

Sadeq, Ahmed, Dr.
Permanent Secretary
Ministry of Planning and Development
Permanent Mission to the United Nations, 866 United Nations Plaza, Room 435
New York, NY 10017, United States of America

PRESS

Saleh, Bashrahil Ahmed
AL-THAWRY
Bab-Assalam House No.56
Sanaa, Yemen Republic, Yemen

YUGOSLAVIA - GOVERNMENT

Blagojevic, Vladislav, Mr.
Counsellor
Embassy to the Federative Republic of Brazil
Permanent Mission to the United Nations, 854 Fifth Avenue New York, NY, 10021
T: 879 8700 F: 879 8705

Bulatovic, Momir, H.E. Mr.
President
Republic of Montenegro
854 Fifth Avenue
New York, NY 10021, United States of America
T: 212 879 8700 F: 212 879 8705

Buric, Mihajlo, H.E. Dr.
Minister
Environment Ptotection of the Republic of Montenegro
Permanent Mission to the United Nations, 854 Fifth Avenue New York, NY, 10021
T: 879 8700 F: 879 8705

Ivanovic, Miodrag, Mr.
Adviser at the Presidency
Republic of Montenegro
Permanent Mission to the United Nations, 854 Fifth Avenue New York, NY, 10021
T: 879 8700 F: 879 8705

Martins Costa, Roberto Guimaraes, Mr.
Honorary Consul
Rio de Janeiro
Permanent Mission to the United Nations, 854 Fifth Avenue New York, NY, 10021
T: 879 8700 F: 879 8705

Mijovic, Dejan, Mr.
Adviser
Agency of the Republic of Montenegro for Economic Restructuring and Foreign Investment
Permanent Mission to the United Nations, 854 Fifth Avenue New York, NY, 10021
T: 879 8700 F: 879 8705

Perovic, Ljubisa, Mr.
Assistant Foreign Minister
Republic of Montenegro
Permanent Mission to the United Nations, 854 Fifth Avenue
New York, NY, 10021
T: 879 8700 F: 879 8705

Rajkovic, Ranko, Mr.
Director General
Hidro-Metorological Institute of Montenegro
Permanent Mission to the United Nations, 854 Fifth Avenue
New York, NY, 10021
T: 879 8700 F: 879 8705

Samardzic, Nikola, H.E. Dr.
Minister for Foreign Affairs
Republic of Montenegro
Permanent Mission to the United Nations, 854 Fifth Avenue
New York, NY, 10021
T: 879 8700 F: 879 8705

Tabas, Ljiljana, Ms.
Minister Plenipotentiry
Federal Secretariat for Foreign Affairs
Permanent Mission to the United Nations, 854 Fifth Avenue
New York, NY, 10021
T: 879 8700 F: 879 8705

Todorovic, Pavle, H.E. Dr.
Minister
Environment of the Republic of Serbia
Permanent Mission to the United Nations, 854 Fifth Avenue
New York, NY, 10021
T: 879 8700 F: 879 8705

Trajkovic, Miodrag, H.E. Dr.
Ambassador to Brazil, (Deputy Head of Delegation)
Permanent Mission of Yugoslavia
854 Fifth Avenue
New York, NY 10021, 879 8700
T: 879 8705

Vujanac-Borovnica, Sofija, Mrs.
Under-Secretary
Federal Secretariat for Development Environmental Division
Mladinskih Bri Gada 1
Belgrade 11000, Yugoslavia
T: 011 199128 F: 011 2222 909

Zupanjevac, Dragan, Mr.
Consellor
Permanent Mission of Yugoslavia
854 Fifth Avenue
New York, NY 10021, 879 8700
T: 879 8705

PRESS

Barjaktarevic, Milica
Reporter Radio/Television
BELGRADE TV
Terazije 38
11000 Belgrade, Yugoslavia, Yugoslavia

Popovic-Milic-Strkalj, Biljana
Editor
TV BELGRADE
B.Yerkovic 151
Belgrade, Yugoslavia, Yugoslavia

Tafra-Vlahovic, Majda
Correspondent
VECERNJI LIST
Trg Hrvatskih velikana 9
Zagreb, Croatia, Yugoslavia, Yugoslavia

GLOBAL FORUM

Tokarev, Mihail, Mr.
Dem - Ecologist Movement Of Macedonia
Partzanski Odredi 24, Box 303
Yugoslavia
T: 91234-974 F: 91 238-010

ZAIRE - GOVERNMENT

Amule, Maddy, Mr.
Director de Gestion des Ressources Naturelles Renovelables
Permanent Mission of Zaire
United Nations Plaza
New York, NY 10017, United States of America
T: 212 963 7174 F: 212 963 1921

Baliene, Banamuhere, Mr.
Conseiller Technique
Permanent Mission of Zaire
United Nations Plaza
New York, NY 10017, United States of America
T: 212 963 7174 F: 212 963 1921

Baliene, Saloman Bonamulere, Mr
Conseiller
MINISTERE DE L ENVIRONMENT
BP 3516
Kinshasa, Zaire
T: 33250

Bihini, Won Wa Musiti
Directeur P.P.M
INSTITUT ZAIROIS POUR LA CONSERVATI
De la Nature B.P. 868
Kinshasa I, Zaire

Bope, Mikobi, Mr.
Chef de Division de Pre-Investissement
Permanent Mission of Zaire
United Nations Plaza
New York, NY 10017, United States of America
T: 212 963 7174 F: 212 963 1921

Chyrorod Kadumba, Isabelle, Ms
Directeue
B.P
Kiushasa, Zaire

Dio, Kudiwu Kengila, Mr.
Ambassadeur Directeur-Chef de la Francophonie
Permanent Mission of Zaire
United Nations Plaza
New York, NY 10017, United States of America
T: 212 963 7174 F: 212 963 1921

Eale, Ekaka Brigitte, Ms
Expert
ETUDES ET PLANIFICATION MINISTERE
BP 2348 Kinshasa I
Kinshasa, Zaire
T: 33251 33252 EXT 242

Eale, Ey'Amba Ea Nsongabyele
Secretaire General
MINISTERE DE L'ENVIRONNEMENT ET CON
12348 Kinshasa I
Kinshasa, Zaire
T: 33251 33252 EXT 237

Ekaka, Eale, Mlle
Expert au Bureau d'Etudes
Permanent Mission of Zaire
United Nations Plaza
New York, NY 10017, United States of America
T: 212 963 7174 F: 212 963 1921

Guma, Mlle
Secretaire
Steno-Dactylo du Ministre
Permanent Mission to the United Nations, United Nations Plaza New York, NY, 10017
F: 212 963 7174
TL: 212 963 1921

Ilanga, Lofonga
Chef De Bureau/Expert
MINISTERE DE L'ENVIRONMENT
12348 Kini.
Kinshasa, Zaire
T: 33250 33251 ET 33252

Kabala, Mwanambuyu, Mr.
Chef de Division au Programme
National d'Assainissement
Permanent Mission to the United Nations, United Nations Plaza
New York, NY 10017, United States of America
T: 212 963 7174 F: 212 963 1921

Kabemba, Mwenbo, Mr.
Directeur
Scientifique a l'Institut Zairois pour la Conservation de la Nature
Permanent Mission to the United Nations, United Nations Plaza
New York, NY 10017, United States of America
T: 212 963 7174 F: 212 963 1921

Kabeya Mukenyi, Roger
Directeur
PROTECTION DE L'ENVIRONNEMENT ETABL
B.P. 790
Kinshasa/Limete, Zaire
T: 243 12 33251 52

Kamakanga, Ngole Kampenga, Mr.
Ambassadeur pres la Republique Federative du Bresil
Permanent Mission of Zaire
United Nations Plaza
New York, NY 10017, United States of America
T: 212 963 7174 F: 212 963 1921

Kamina, Mukenga, Mr.
Conseiller au Cabinet du Ministre
Permanent Mission of Zaire
United Nations Plaza
New York, NY 10017, United States of America
T: 212 963 7174 F: 212 963 1921

Kasulu, Seya Makonga, Ms
Chef De Division
ETUDES ET PLANIFICATION MINISTERE D
BP 12348 Kinshasa I 15 Avenue des Cliniques Kinshasa Go
Kinshasa, Zaire
T: 33251 33252 EXT 242

Khabouji, Lukabu, Mr.
Ministre-Conseiller a la Mission Permanente a New York
Permanent Mission of Zaire
United Nations Plaza
New York, NY 10017, United States of America
T: 212 963 7174 F: 212 963 1921

Kipoy, Kamashi, Mr.
Conseiller a l'Ambassade a Brasilia
Permanent Mission of Zaire
United Nations Plaza
New York, NY 10017, United States of America
T: 212 963 7174 F: 212 963 1921

Lofonga, Ilanga, Mr.
Expert au Bureau d'Etudes
Permanent Mission of Zaire
United Nations Plaza
New York, NY 10017, United States of America
T: 212 963 7174 F: 212 963 1921

Londoale, Louya, Mr.
Ambassadeur Directeur
Chef de Service des Organisations Internationales
Permanent Mission to the United Nations, United Nations Plaza
New York, NY 10017, United States of America
T: 212 963 7174 F: 212 963 1921

Longange, Bomina N'Somi, Mr.
Ambassadeur a Addis-Abebe
Permanent Mission of Zaire
United Nations Plaza
New York, NY 10017, United States of America
T: 212 963 7174 F: 212 963 1921

Lonsa, Kanda, Mr.
Directeur de la Cooperation Multilaterale
Permanent Mission of Zaire
United Nations Plaza
New York, NY 10017, United States of America
T: 212 963 7174 F: 212 963 1921

Luamba, Lua Nsembo
Professeur
GOUVERNEMENT
B.P. 3032
Kinshasa/Gombe, Zaire

Lukabu, Khabouji
Ministre Counsellor
ZAIRE MISSION TO UN
767 Third Ave-25 Fl
New York, NY 10017, United States of America
T: 212 754 1966 68 F: 212 754 1970

Lumuka, Ntonfo, Mr.
Directeur de la Programmation
Formation et Relations Internationales
Permanent Mission to the United Nations, United Nations Plaza
New York, NY 10017, United States of America
T: 212 963 7174 F: 212 963 1921

Lumuka Maniole, Leon Ntondo, Mr
MINISTERE DEL ENVIRONMENT
Et De la Consruation Dela Nature BP 12348
Kinshasa, Zaire
T: 31252 32252

Lutete, Vangu, Mr.
Coordennateur
National du PAFT/Zaire
Permanent Mission to the United Nations, United Nations Plaza
New York, NY 10017, United States of America
T: 212 963 7174 F: 212 963 1921

Mandango, Madragule Angoyo
Prof. Dr
ASSOCIATION DES BOTANISTES DU ZAIRE
B.P. 4125
Kinshasa,li, Zaire

Mankoto, Ma Mbaele Le
President De Le Ave General
INSTITUT ZATROIS TOUR LA CONSERNATI
De la Nature IZCN BP 868 Kinshasa I
Kinshasa, Zaire
T: 33250 33251

Mansoni, Be Dy-Di-, Mr.
President
Delegue General de l'Institur des Jardins Zoologiques et Botaniques du Zaire
Permanent Mission to the United Nations, United Nations Plaza
New York, NY 10017, United States of America
T: 212 963 7174 F: 212 963 1921

Manzjia, Bolndombe, Mr.
Conseiller a l'Ambassade a Brasilia
Permanent Mission of Zaire
United Nations Plaza
New York, NY 10017, United States of America
T: 212 963 7174 F: 212 963 1921

Matomina, Lusanda
Expert
Permanent Mission of Zaire
United Nations Plaza
New York, NY 10017, United States of America
T: 212 963 7174 F: 212 963 1921

Mbaelele, Mankoto-Ma, Mr.
President
Delegue General de l'Institute Zairois pour la Conservation de la Nature
Permanent Mission to the United Nations, United Nations Plaza
New York, NY 10017, United States of America
T: 212 963 7174 F: 212 963 1921

Mbuba, Labu
Professeur
MINISTERE DE L'ENVIRONNEMENT
BP 12348 Kinshasa I
Kinshasa, Zaire
T: 33251 33252 EXT 215

Mbumba, Labu, Mr.
Secretaire Particulier du Ministre
Permanent Mission of Zaire
United Nations Plaza
New York, NY 10017, United States of America
T: 212 963 7174 F: 212 963 1921

Mfulunkatu, Lumba, Dr.
Premier Secretaire a l'Ambassade a Brasilia
Permanent Mission of Zaire
United Nations Plaza
New York, NY 10017, United States of America
T: 212 963 7174 F: 212 963 1921

Mitonga, Zongwe K., Mr.
Conseiller au Cabinet du Ministre
Permanent Mission of Zaire
United Nations Plaza
New York, NY 10017, United States of America
T: 212 963 7174 F: 212 963 1921

Momi, Lingosa, Mr.
Conseiller
Permanent Mission of Zaire
United Nations Plaza
New York, NY 10017, United States of America
T: 212 963 7174 F: 212 963 1921

Mpelenda, Mumbandu, Dr.
Expert a Delegation Speciale a la Francophonie
Permanent Mission of Zaire
United Nations Plaza
New York, NY 10017, United States of America
T: 212 963 7174 F: 212 963 1921

Muembo, Kabemba
Directeur Scientifique
INSTITUT ZAI ROIS POUR
la Conservation De la Nature B.P. 868
Kinshasa, Zaire

Mukenga, Kamina
Conseiller Du Ministre
MINISTERE DES RELATIONS EXTERIEURES
Kinshasa, Zaire
T: 21118

Mulumba, Kabeya Mukendi-Wa-, Directeur
des Etablissements Humains
Permanent Mission of Zaire
United Nations Plaza
New York, NY 10017, United States of America
T: 212 963 7174 F: 212 963 1921

Musitti, Baihini Won-, Mr.
Directeur du Parc President Mobutu
Permanent Mission of Zaire
United Nations Plaza
New York, NY 10017, United States of America
T: 212 963 7174 F: 212 963 1921

Myuavua Makuala Ne Kungu, Maurice, Mr
Directeur Charge d'Etudes
MINISTE'RE DU BUDGET DE LA RE'P DU
Kinshasa, Zaire
T: 33014-33490-33239-33

Nouma Nyamobembo, Annie, Ms
Diplomah
Kinshasa/Gombe
Kinshasa, Zaire

Nsongabyele, Kale Ey'Amba Ea, Mr.
Secretaire General a l'Environnement et Conservation de la Nature
Permanent Mission of Zaire
United Nations Plaza
New York, NY 10017, United States of America
T: 212 963 7174 F: 212 963 1921

Nwanambuyu, Kabala
Chef De Division
PROGRAMME NATIONAL D'ASSAINISSEMENT
Kinshasa, Zaire
T: 77691

Ohoto-Omana Losokola, Emilienne, Ms
Docteur En Biologie (professeur)
FACULTE DES SCIENCES
B. P. 13220 Kinshasa I
Kinshasa, Zaire

Phoba, Massanga, Mr.
Conseiller Coordenateur
Permanent Mission of Zaire
United Nations Plaza
New York, NY 10017, United States of America
T: 212 963 7174 F: 212 963 1921

Pulunyo, Bura, S.E. Prof.
Ministre de l'Environnement et de la Conservation de la Nature
Permanent Mission of Zaire
United Nations Plaza
New York, NY 10017, United States of America
T: 212 963 7174 F: 212 963 1921

Seya, Kasulu, Mr.
Chef de Division et Expert au Bureau d'Etudes
Permanent Mission of Zaire
United Nations Plaza
New York, NY 10017, United States of America
T: 212 963 7174 F: 212 963 1921

Vangu- Lutete, Clenent
Coordonnateur National/Paft
MINISTERE ENVIRONNENENT
B.P. 4951 Kin/Gombe Zaire ZAI/90/007 B.P. 7248
Kini, Zaire

Yansomwe, , Mr.
Directeur au Ministere
United Nations Plaza New York, NY, 10017
F: 212 963 7174
TL: 212 963 1921

Zengeya, Bagbeni Adeito, S.E. Mr.
Ministre des Relations Exterieures
Permanent Mission of Zaire
United Nations Plaza
New York, NY 10017, United States of America
T: 212 963 7174 F: 212 963 1921

ORGANIZATIONS

Bagan, Senkoko Kishar, Mr.
Centre D'appui Au Developpement Global Rural Integre Du Nord-kivu (ADRINOK)
B.P. 170
Goma, Zaire

Lende, Kipupila Nza'Kayasunduako, Sr.
As. D'entr'aide Medico-sociale
B.P. 593
Kinshasha 23, Zaire

GLOBAL FORUM

Is'Etambela, Ishomba, Mr.
Executive Secretary
C.V. AVEQUA MBANDAKA
c/o CNONGD-Zaire PO Box 5744 Kinshasa
Kinshasa -Kinshasa, Zaire
T: 26707

Kabarhuza, Hamuli, Mr.
Executive Secretary
CONSEIL NATIONAL DES ONG DE DEVELOPPEMENT DU ZAIRE
Cnongd PO Box 5744
Kinshasa, Zaire
T: 243 26707

Kasongo, Shimba, Mr.
Director
CRONGD/SHABA
c/o CNONG P.O. Box 5744
Kinshasa - Saba, Zaire
T: 26707

Kitoko, Omekongo, Mr.
Vice-President
VDA
c/o CNONGD/Zaire PO Box 5744 Kinshasa
Kinshasa - Kinshasa, Zaire
T: 26707

Lukagyu, Bitomwa, Mr.
Ingenieur Agronome
GROUPE TECHNIQUE D'ENCADREMENT REGIONAL
B.P. 165 Cyangugu - Rwanda, Boulevard du Lac no. 24 Ibanda
Bukavu, Zaire
T: 2431 F: 2431

Lukasu, Dianda, Mr.
President
CLUB DE LA BONTE
B.P.235, Avenue Mobutu, Quarter 53
Kamina Ville, Zaire

Luonga, Mutchokv, Mr.
President
FAITH IN ACTION TO ZAIRE
Av. Place Kunda , 1 - Bagira-Bukavu PO Box 263
Cyangugu, Zaire
T: 25 3-7173 F: 25 3-7173

Lutonga, Kabeya, Mr.
Administrator
GTER/KASAI-ORIENTAL
c/o CNONGA/Zaire PO Box 5744 Kinshasa
Kinshasa - Kinshasa, Zaire
T: 26707

Mbala, Kasombo, Mr.
CRONGD/KASAI OCCIDENTAL
C/O Cnonga/Zaire PO Box 5744 Kinshasa
Kinshasa, Zaire
T: 24312 26 707 F: 24312

Mboho, Kisusu, Mr.
President
CRONGD/NORD KIVU
PO Box 5744 C/O Cnongo/Zaire
Goma - Nord-Kivu, Zaire

Mumbenga, Mukwa, Mr.
President
GTER/BANDUNDU
PO Box 5744 C/O Cnonga
Kikwit - Bandundu, Zaire

Nenbalemba, Loseke, Mr.
Vice-President
LIZADHO
c/o CNONGA-Zaire PO Box 5744
Kinshasa - Kinshasa, Zaire
T: 26707

Niemba, Mabuelama, Mr.
Executive Secretary
BEDI/BOMA
C/O Cnonga/Zaire
Boma, Zaire
T: 91 22 26 707

Rurenza, Mayshara, Mr.
Social Entertainer
ASSOCIATION ADESKI
Adeski Uvira c/ D.S. no. 4 Bujumbura Via Republique du Burundi
Uvira, Zaire
T: 2431 F: 2431

Semacumu, Gakuru, Mr.
Vide-President
BDD/KISANGANI
C/O Cnonga/Zaire PO Box 5744 Kinshasa
Kinshasa, Zaire
T: 24312 26 707 F: 24312

Wa Numbe, Cissa, Mr.
FERCOOP/SUD KIVU
c/o CNONGD-Zaire
Kinshasa - S/Kivu, Zaire
T: 26707

Zombo, Maurice Musema, Mr.
Director
RESEAU D'APPUI AU DEVELOPPEMENT
d'Activites Rurales (RADAR) BP 10203
Kinshasa, Zaire
T: 24312 F: 24312

PARLIAMENTARIANS / SPIRITUAL LEADERS

Madrabule, Mandango, Prof.
MP
Conseil Legislatif, Palais de la Nation, B. P. 3955
Kinshasa/Gombe, Zaire
T: 31.650
TL: 21319

ZAMBIA - GOVERNMENT

Banda, M., Mr.
Acting Director
Natural Resources Department
Permanent Mission to the United Nations, United Nations Plaza
New York, NY 10017, United States of America
T: 212 963 7174 F: 212 963 1921

Chabala, Isaiah Z., Mr.
Minister Counsellor, Deputy Permanent Representative
Permanent Mission of Zambia
United Nations Plaza
New York, NY 10017, United States of America
T: 212 963 7174 F: 212 963 1921

Chimuka, Augustine N., Mr.
Permanent Secretary
Ministry of Environment and Natural Resources
Permanent Mission to the United Nations, United Nations Plaza
New York, NY 10017, United States of America
T: 212 963 7174 F: 212 963 1921

Chipete, G., Dr.
Director
Meteorologiacal Department
Permanent Mission to the United Nations, United Nations Plaza
New York, NY 10017, United States of America
T: 212 963 7174 F: 212 963 1921

Chipungu, P. N., Dr.
Director
National Environmental Council
Permanent Mission to the United Nations, United Nations Plaza
New York, NY 10017, United States of America
T: 212 963 7174 F: 212 963 1921

Kunda, Humphrey B., Mr.
Deputy Permanent Secretary
Ministry of Foreign Affairs
Permanent Mission to the United Nations, United Nations Plaza
New York, NY 10017, United States of America
T: 212 963 7174 F: 212 963 1921

Mbumwae, Louis, Mr.
Director
Energy and Water Development
Permanent Mission to the United Nations, United Nations Plaza
New York, NY 10017, United States of America
T: 212 963 7174 F: 212 963 1921

Munsanje, J. M., Mr.
Zambia Council for Social Development
Permanent Mission to the United Nations, United Nations Plaza
New York, NY 10017, United States of America
T: 212 963 7174 F: 212 963 1921

Mushota, L., Mrs.
Non-Governmental Organisations Coordinating Committee
Permanent Mission to the United Nations, United Nations Plaza
New York, NY 10017, United States of America
T: 212 963 7174 F: 212 963 1921

Mwilinga, D., Mrs.
Senior Legal Draftsman
National Commission for Development Planning
Permanent Mission to the United Nations, United Nations Plaza
New York, NY 10017, United States of America
T: 212 963 7174 F: 212 963 1921

Nooma, Monica Chundama, Mrs.
Senior Economist
National Commission for Development Planning
Permanent Mission to the United Nations, United Nations Plaza
New York, NY 10017, United States of America
T: 212 963 7174 F: 212 963 1921

Phiri, J., Dr.
World University Service of Zambia
Permanent Mission to the United Nations, United Nations Plaza
New York, NY 10017, United States of America
T: 212 963 7174 F: 212 963 1921

Silangwa, S. L., Dr.
Chairman
National Environmental Council
Permanent Mission to the United Nations, United Nations Plaza
New York, NY 10017, United States of America
T: 212 963 7174 F: 212 963 1921

Walubita, Keli S., H.E. Hon.
MP
Minister for Environment and Natural Resources
Permanent Mission to the United Nations, United Nations Plaza
New York, NY 10017, United States of America
T: 212 963 7174 F: 212 963 1921

Zukas, Simon, H.E. Hon.
MP
Deputy Minister Office of the President
Permanent Mission to the United Nations, United Nations Plaza
New York, NY 10017, United States of America
T: 212 963 7174 F: 212 963 1921

NGO

Jere, Harrington E.
HUZA
5082 Fukwe Close, PO Box 50141, Code 15101
Ridgeway, Lusaka, Zambia

Mushota, Lillian, Ms.
ZAMBIA ALLIANCE OF WOMEN
Palais de Nations
Geneve CH 1211, Zambia
F: 41 22 733-6542

PRESS

Chisupa, Katongo
Correspondent
ZAMBIA B'CASTING CORP.
PO Box R.W. 50620
Lusaka, Zambia, Zambia

Malama, Godfrey
Editor
TIMES NEWSPAPERS LTD.
Plot 11037 Northmead
Lusaka, Zambia, Zambia

Mwanangombe, Lewis
Correspondent
ZANA/ZAMBIA NEWS AGENCY
P.O.Box 30007
Lusaka, Zambia, Zambia

ORGANIZATIONS

Bwamga, Dominic, Mr.
National Centre for Environmental Education (NCEE)
PO Box 36316
ZA- 10101 Lusaka, Zambia
T: (2603) 323 126
TL: ZA 24100

Molotsi, Prisca, Ms.
Zambia Alliance of Women (ZAW)
PO Box 51068
Lusaka, Zambia
F: (260 1) 221 888

Muchelemba, Josephine B., Ms.
Environment, Population and Development Services
PO Box 35239
Lusaka, Zambia

Munsanje, J.N., Mr.
Zambia Council for Social Developpement (ZCSD)
ZCSD Centre
10.1 km Lusaka - Mongu Road, Zambia
T: (260 1) 210 542

GLOBAL FORUM

Jerre, Harrington, Mr.
HUMAN SETTLEMENTS OF ZAMBIA
Palais de Nations
Geneve CH 1211, Zambia
F: 41 22 733-6542

Mumba, Wanga, Ms.
ENVIRONMENT AND POPULATION CENTER
PO Box 35614
Lusaka, Zambia
T: 260 1 251-870

PARLIAMENTARIANS / SPIRITUAL LEADERS

Banda, H.D., Mr.
MP
National Assembly, Parliament Building, P.O. Box 31299
Lusaka, Zambia
T: 218991
TL: Parly ZA 44390

Chiti, D.K., Mr.
(secretary To The Delegation)
National Assembly, Parliament Building, P.O. Box 31299
Lusaka, Zambia
T: 218991
TL: Parly ZA 44390

Tumbila, R.M., Mr.
MP
National Assembly, Parliament Building, P.O. Box 31299
Lusaka, Zambia
T: 218991
TL: Parly ZA 44390

ZIMBABWE - GOVERNMENT

Chitauro, J.J., Mr.
Permanent Secretary
Ministry of Energy and Water Resources and Development
Permanent Mission to the United Nations, 128 E. 56th Str.
New York, NY 10022, United States of America
T: 212 980 9511 F: 212 755 4188

Garwe, E.R.M., Mr.
Chief of Protocol
Permanent Mission of Zimbabwe
128 E. 56th Str.
New York, NY 10022, United States of America
T: 212 980 9511 F: 212 755 4188

Jokonya, T.J.B., Dr.
Senior Secretary
Foreign Affairs
Permanent Mission to the United Nations, 128 E. 56th Str.
New York, NY 10022, United States of America
T: 212 980 9511 F: 212 755 4188

Makado, J., Mr.
Director Department of Agriculture
Technical and Extension Services
Permanent Mission to the United Nations, 128 E. 56th Str.
New York, NY 10022, United States of America
T: 212 980 9511 F: 212 755 4188

Mashingaidze, E.K., Dr.
Permanent Secretary
President's Office
Permanent Mission to the United Nations, 128 E. 56th Str.
New York, NY 10022, United States of America
T: 212 980 9511 F: 212 755 4188

Moyo, M., Mr.
Deputy Secretary
Ministry of Environment and Tourism
Permanent Mission to the United Nations, 128 E. 56th Str.
New York, NY 10022, United States of America
T: 212 980 9511 F: 212 755 4188

Msengezi, Winston T., Mr.
Counsellor
Permanent Mission of Zimbabwe
128 E. 56th Str.
New York, NY 10022, United States of America
T: 212 980 9511 F: 212 755 4188

Muchada, M.M., Mr.
Deputy Secretary
President's Office
Permanent Mission to the United Nations, 128 E. 56th Str.
New York, NY 10022, United States of America
T: 212 980 9511 F: 212 755 4188

Mugabe, Robert Gabriel, Mr.
President
Permanent Mission of the Republic of Zimbabwe
128 E. 56th. Str.
New York NY 10022, United States of America
T: 212 980 9511 F: 212 755 4188

Mugage, Robert Gabriel, H.E. Cde.
President
The Republic of Zimbabwe
128 E. 56th Str.
New York, NY 10022, United States of America
T: 212 980 9511 F: 212 755 4188

Mundangepfupfu, E.T., Mr.
Permanent Secretary
Ministry of Environment and Tourism
Permanent Mission to the United Nations, 128 E. 56th Str.
New York, NY 10022, United States of America
T: 212 980 9511 F: 212 755 4188

Murangari, D.E.H., Mr.
Permanent Secretary
Ministry of Mines
Permanent Mission to the United Nations, 128 E. 56th Str.
New York, NY 10022, United States of America
T: 212 980 9511 F: 212 755 4188

Murerwa, H.M., Hon. Dr.
Minister
Environment and Tourism
Permanent Mission to the United Nations, 128 E. 56th Str.
New York, NY 10022, United States of America
T: 212 980 9511 F: 212 755 4188

Ngara, P., Mr.
Director
Department of Meteorological Services
Permanent Mission to the United Nations, 128 E. 56th Str.
New York, NY 10022, United States of America
T: 212 980 9511 F: 212 755 4188

Sam, , Mr.
Director
Zimbabwe Environmental Research Organization
Permanent Mission to the United Nations, 128 E. 56th Str.
New York, NY 10022, United States of America
T: 212 980 9511 F: 212 755 4188

Shamuyarira, N.M., Hon. Dr.
Minister
Foreign Affairs
Permanent Mission to the United Nations, 128 E. 56th Str.
New York, NY 10022, United States of America
T: 212 980 9511 F: 212 755 4188

NGO

Katerere, Yem
ZERO
PO Box 5338
Harare, Zimbabwe

Mundondo, Iephias
SAHRF
P.O. Box 430
Kwekwe, Zimbabwe

Ngcobo, Archibald
SAHRF
P.O. Box 430
Kwekwe, Zimbabwe

PRESS

Bwititi, Makuwerere
ZIMBABWE INTER-AFRICA NEWS AG.
P.O. Box 8166
Causeway, Harare, Zimbabwe, Zimbabwe

Dakwa, Palnos
Reporter Radio/Television
OFFICIAL PRESS ZIMBABWE
8 Biddulah Road
Harare, Zimbabwe, Zimbabwe

Mpofu, Stephen
Editor
CHRONICLE
103 Beresford Close
Killarney, Bulawayo, Southern Africa, Zimbabwe

ORGANIZATIONS

Moyo, Sam, Mr.
Zero (regional Network Of Energy/environmental Experts)
PO Box 5338
Harare, Zimbabwe
T: (2634) 732 858

GLOBAL FORUM

Antonio, Edward, Dr.
UNIVERSITY OF ZIMBABWE
Dept. of Religious Studies - PO Box 167 - Mt. Pleasant
Harare, Zimbabwe

Chitekuteku, J., Mr.
GLENFOREST TRAINING CENTER
Palais de Nations
Geneve CH 1211, Zimbabwe
F: 41 22 733-6542

Gambanga, John
Features Editor
HERALD
PO Box 4152
Harare, Zimbabwe
T: 263 795771

Gove, Charles, Mr.
Z.E.R.O. (FOCAL POINT-SUTHERN AFRICA)
Box 5338
Harare, Zimbabwe
T: 263 4 791 333 F: 263 4 732 858

Khumalo, Sikhangezile, Ms.
Inyathi Cland PB J5113
Bulamayo, Zimbabwe

Machingauta, Rudo Chitiga, Ms.
Director
IRED DEVELOPMENT INNOVATION
PO Box 8242 Causeway
Harare, Zimbabwe
T: 263 4 796853 F: 263 4 722421
TL: 220552MMT

Made, Ambrose, Mr.
Manager
REALTIME
Harare, Zimbabwe

Mazambani, David, Mr.
ARDA
14 Belvedere Rd. PO Box 3492
Harare, Zimbabwe
T: 263 4 37 773 F: 263 4 72 29 04

Mbanga, Wilf
Chief Executive
COMMUNITY NEWSPAPERS
PO Box 6520
Harare, Zimbabwe
T: 793707 F: 794336

McCracken, Sylvia, Ms.
Youth Environmental Specialist
YOUTH CONSERVATION CLUBS
P.O.Box, 8317 - Causeway
Harare, Zimbabwe
T: 263 4 738 459

Mhlanga, Liberty, Dr.
Exec. Director
ENVIRONMENTAL & DEVELOPMENT AGENCY
P.O.Box 3492, Mount Pleasant
Harare, Zimbabwe

Mooyo-Mhlanga, Khethiwe, Ms.
DEVELOPMENT ALTERNATIVES WITH WOMEN FOR A NEW ERA (DAWN)
c/o Africa 2000 Network 60 Selous Avenue PO Box 4775
Harare, Zimbabwe

Nelson, Jane, Ms.
Consultant
23 Herbert Chitepo Ave.
Harare, Zimbabwe
T: 263 4 724 638

Ngobese, Peter, Mr.
Z.E.R.O. (FOCAL POINT-SUTHERN AFRICA)
Box 5338
Harare, Zimbabwe
T: 263 4 791 333 F: 263 4 732 858

Pilixs, Aeae Esi Amanor
Ips Africa Editor
IPS - INTER PRESS SERVICE
PO Box 6050
Harare, Zimbabwe
T: 263 4 706466/ 706430

Stevens, Jeffrey, Mr.
Hon Secretary
SOLAR ENERGY SOCIETY OF ZIMBABWE
PO Box MP119 Mr Pleasant
Harare, Zimbabwe
T: 263 4 34908

Tabaiba, Clever, Mr.
METHODIST CHURCH IN ZIMBAWE
PO Box 8298 Causeway
Harare, Zimbabwe

PARLIAMENTARIANS / SPIRITUAL LEADERS

Kwidini, D.J., Min.
The Senate, Parliament of Zimbabwe
Harare, Zimbabwe
T: 700181
TL: 4064

NOTES

DIRECTORY

LISTINGS BY ORGANIZATION

2

2 ABC RADIO / 2SER-FM, Hitchman, Kylien, Ms. 90
2 SER FM RADIO, Philpott, Leanne 90
2 SER FM RADIO, Thorman, Pelham 90
21ST CENTURY NEWS, Anderson, Judith 359
21ST CENTURY NEWS, Araujo, Alan 359
21ST CENTURY NEWS, Craig, James Dexter 361
21ST CENTURY NEWS, Dugdale, Joel 361
21ST CENTURY NEWS, Hernandorena, Diego 362
21ST CENTURY NEWS, Horner, Lisa Emma 362
21ST CENTURY NEWS, Lino, Geraldo Luis S. 364
21ST CENTURY NEWS, McDonald, Mary S. 364
21ST CENTURY NEWS, Poe, Melissa 365
21ST CENTURY NEWS, Poe, Patricia 365
21st Century News, Blades, Christopher Scott 368
21ST. CENTURY NEWS, Blades, Scott, Mr. 372
21ST. CENTURY NEWS, Craig, Chip, Mr. 375
24 HEURES, Rime, Michel 333
2DF-TV ALEMA, Bruns, Jam 139
2DF-TV ALEMA, Cavalcante, Guido 142
2DF-TV ALEMA, Depner, Uta 146
2DF-TV ALEMA, F.Stein, Richard 148
2DF-TV ALEMA, Ferreira, Mario 150
2DF-TV ALEMA, Hallmam, Stephan 156
2DF-TV ALEMA, Hockstein, Beatriz 156
2DF-TV ALEMA, Jung, Johaann 158
2DF-TV ALEMA, Keppler, Alexandre 159
2DF-TV ALEMA, Nunes, Paulo Ricardo 172
2DF-TV ALEMA, Stein, Richard F., Mr. 190
2DF-TV ALEMA, Zamgrandy, Reynaldo 195
2SER-FM / GREEN LEFT MAGAZINE, Gilleatt, Brad, Mr. 90

3

33RD PRODUCTIONS, INC., Haines-Stiles, Geof 362
33RD PRODUCTIONS/NEON RIO, Rodrigues, Agnes Consula J. 123
3E PETROLEVERT, Pierre, Mariotte, Mr. 235
3E PETROLEVERT, Simon-Collas, Mireille, Ms. 235

4

4 WORLDS DEVELOPMENT PROJECT, Goodstriker, Jason, Mr. 204

9

92 GLOBAL FORUM, Avelleda, Ana Maria 134
92 GLOBAL FORUM, Belloc, Beth 136
92 GLOBAL FORUM, Cardoso Filho, Gerson 140
92 GLOBAL FORUM, Cesar, Aluisio 142
92 GLOBAL FORUM, Ewart-White, Fraser 148
92 GLOBAL FORUM, Farias, Marcelo Cesar Bastos De 149
92 GLOBAL FORUM, Felix, Teresa Cristina 149
92 GLOBAL FORUM, Hawrylyshyn, Olgha Dierdre Tibytiba 156
92 GLOBAL FORUM, Mamede, Silvia 164
92 GLOBAL FORUM, Mbanefo, Sandra 166
92 GLOBAL FORUM, Medeiros, Alessandra 166
92 GLOBAL FORUM, Medeiros, Maria Angelica F. 166
92 GLOBAL FORUM, Palermo, Renata 174
92 GLOBAL FORUM, Rodriguez, Karla Bruner 180
92 GLOBAL FORUM, Sabra, Georgina 181
92 GLOBAL FORUM, Silva, Marcus Vinicius Da 187
92 GLOBAL FORUM, Van Schaik, Lot 192
92 GLOBAL FORUM, Yolen, Greg 195
92 JAPAN NGO FORUM, Higashi, Tatsuo, Mr. 275
92 NGO FORUM JAPAN, Iwasaki, Shunsuke, Mr. 267
92 NGO FORUM JAPAN, Amano, Reiko, Ms. 274
92 NGO FORUM JAPAN, Harago, Yuta, Mr. 275
92 NGO FORUM JAPAN, Hayashi, Tatsuo, Mr. 275
92 NGO FORUM JAPAN, Higuchi, Kazuhiro, Mr. 275
92 NGO FORUM JAPAN, Higuchi, Shunji, Mr. 275
92 NGO FORUM JAPAN, Hirabayashi, Tomoki 275
92 NGO FORUM JAPAN, Holliman, Jonathan, Mr. 275
92 NGO FORUM JAPAN, Hurusawa, Koyu, Mr. 275
92 NGO FORUM JAPAN, Ikeda, Susumu, Mr. 275
92 NGO FORUM JAPAN, Ishii, Misako, Ms. 275
92 NGO FORUM JAPAN, Itoya, Wataru, Mr. 276
92 NGO FORUM JAPAN, Jintoku, Kazuo, Mr. 276
92 NGO FORUM JAPAN, Kamei, Naomi, Ms. 276
92 NGO FORUM JAPAN, Kashimura, Chushi 276
92 NGO FORUM JAPAN, Kato, Tetsuo, Mr. 276
92 NGO FORUM JAPAN, Katsumata, Makoto, Mr. 276
92 NGO FORUM JAPAN, Kihara, Keikichi 276
92 NGO FORUM JAPAN, Kikushi, Yumi, Ms. 276
92 NGO FORUM JAPAN, Kishi, Nobutaka, Mr. 276
92 NGO FORUM JAPAN, Loures, Maria Aparecida Limongi, Sra. 277
92 NGO FORUM JAPAN, Maeda, Ayumi 277
92 NGO FORUM JAPAN, Matsumura, Kunimori, Mr. 277
92 NGO FORUM JAPAN, Murayama, Takehiko, Mr. 278
92 NGO FORUM JAPAN, Niikura, Toshiko 278
92 NGO FORUM JAPAN, Nishioka, Tatsu, Mr. 278
92 NGO FORUM JAPAN, Noda, Tomosuke, Mr. 278
92 NGO FORUM JAPAN, Ohno, Koichi, Mr. 279
92 NGO FORUM JAPAN, Okubo, Seiji, Mr. 279
92 NGO FORUM JAPAN, Onuki, Daisuke 279
92 NGO FORUM JAPAN, Orita, Naoaki, Mr. 279
92 NGO FORUM JAPAN, Sato, Kazuo 279
92 NGO FORUM JAPAN, Seo, Jiro, Mr. 279
92 NGO FORUM JAPAN, Shimagami, Motoko, Ms. 279
92 NGO FORUM JAPAN, Suzuki, Takashi, Mr. 280
92 NGO FORUM JAPAN, Takahashi, Yoshiyuki 280
92 NGO FORUM JAPAN, Toyota, Yoshiko 281
92 NGO FORUM JAPAN, Tsujimoto, Kiyomi, Mr. 281
92 NGO FORUM JAPAN, Ueda, Masuo, Mr. 281
92 NGO FORUM JAPAN, Watanuki, Reiko, Mr. 281
92 NGO FORUM JAPAN, Yagi, Kenzo, Mr. 281
92 NGO FORUM JAPAN, Yagi, Toshiyuki, Mr. 281
92 NGO FORUM JAPAN, Yamamoto, Wakaba, Ms. 281
92 NGO FORUM JAPAN NEWS, Arai, Naoko 274
92NF, Oya, Masao 267
92NF, Ozaki, Reiko 267

A

A CLASSE OPERARIA, Barros, Jefferson 104
A CRITICA, Bonelli, Arlete 105
A D I C O, Guimaraes, Maria Benita Fernandes, Ms. 156
A D U F U - S. SINDICAL, Dansa, Claudia Valeria De Assis, Sra. 146
A E FORO VERDE, Arias, Elsa Perez, Sra. 80
A E FORO VERDE, Ferrero, Alejandra, Sra. 83
A E FORO VERDE, Hamilton, Joseph Andrew, Mr. 84
A E FORO VERDE, Olivier, Altamirano Diego, Mr. 85
A E FORO VERDE, Olivier, Maria Luisa Scorpati De, Sra. 85
A E FORO VERDE, Olivier, Santiago R., Mr. 85
A E M A B, Estevez, Diana, Ms. 82
A E M A B, Galafassi, Guido, Mr. 83
A GOTA DE ESPERANCA, Descomps, Gregeire 146
A GOTA DE ESPERANCA, Freneaux, Pascal 152
A GOTA DE ESPERANCA, Germano, Ines 153
A GOTA DE ESPERANCA, Lorud, Christophe 162
A L L B E, Furusho, Maria Yoshie, Ms. 152
A M A, Castro, Rogelio A. Vazquez, Mr. 216
A M C F E - ASSOC. MALIENNE POUR LA CONSERVATION DE LA FAUNE ET DE SON ENV., Yattara, Inamoud, Mr. 289
A MAO QUE SEMEA, Fonseca, Ivan Ciaret M., Mr. 151
A NOTICIA, Castro, Daniele De O. 107
A NOTICIA, Dantas, Sergio 109
A NOTICIA, Diniz, Henrique Augusto De O. 110
A NOTICIA, Dos Santos, Sandro Roberto 111
A NOTICIA, Duarte, Jose Luis L. 111
A NOTICIA, Ferreira, Nucia Maira Da S. 112
A NOTICIA, Ferreira, Roberto P. 112
A NOTICIA, Oliveira, Nilandi C. 120
A NOTICIA, Silverio, Marcia Luisa 126
A NOTICIA, Vieira, Wilma Lucia 128
A S A C E, Tanor Fall, Thiendella, Mr. 322
A SEED JAPAN, Hani, Kanta 275
A SEED JAPAN, Hishino, Yushimi 275
A SEED JAPAN, Mimura, Satoshi 277
A SEED JAPAN, Takahashi, Kaori, Mr. 280
A STOP FOTOS & VIDEO, Amorim, Adilson Barbosa De 132
A STOP FOTOS & VIDEO, Stop, Alexandre Soares 190
A UNIAO, Pinto Neto, Francisco 122
A WORLD THAT WORKS, Nienhueser, Helen, Ms. 385
A WORLD THAT WORKS FOR EVERYONE, Cooper-Sheehan, Christine, Ms. 374
A WORLD THAT WORKS FOR EVERYONE, Huffsmith, Barbara, Ms. 379
A WORLD THAT WORKS FOR EVERYONE, Laan, Rosemary C. Van Der, Ms. 381
A WORLD THAT WORKS FOR EVERYONE, Schwind, Ronald E., Mr. 389
A WORLD THAT WORKS FOR EVERYONE, Sheehan, Jeffrey, Mr. 389
AAD, Castroviejo Bolibar, Javier 323
AAD, Perez Royo, Javier 323
AAEE, Ilett, Annie 90
AAEE, Robottom, Ian 90
AAEF, Steger, Will 359
AAES, Carroll, William J. 356
A.A.G.G., Godinho, Marco, Mr. 154
AALAE, Kananda, John 252
AALAE, Wangoola, Paul 253
AAMULEHTI NEWSPAPER, Rajamaki, Paivi 226
AAPA, Gutierrez Monzonis, Blanca 215
AAPA, Sote Cruz, Miguel 215
AAPNE, Brahimi, Mohammed 78
AAPNE, Derbal, Hammouche 78
AAPPP - APANDE, De Sa'Earp Hertz, Alfredo, Mr 100
AART, Meda, Gurudutt P. 250
AASU, Drame, Sountoucoung 244
AAWORD - KENYA, Akinyi, Elizabeth, Ms. 284
ABACED, Damoda, Diop 321
ABACED, Diouf, Aminata 321
ABAPORU EMPREENDIMENTOS, Magalhaes, Sergio Faria 164
ABAPORU EMPREENDIMENTOS, Mello, Regina Lara Silveira 166
ABC, Brooks, Sam W. 360
ABC, Compton, Ann 360
ABC, Delaski, Kathleen 361
ABC, Hirsh, Stephen 362
ABC, Hume, Alexander B. 363
ABC, Terrence, Ray 367
ABC NEWS, Aglow, Robert 359
ABC NEWS, Bourin, Len 360
ABC NEWS, Del Mastro, Salvatore 361
ABC NEWS, Gubar, Thomas R. 362
ABC NEWS, Hume, Brit 363
ABC NEWS, Lustig, Charles 364
ABC NEWS, Meidlinger, Leo James 364
ABC NEWS, Obert, Albert D. 365
ABC NEWS, Potter, Mark Edward 365
ABC NEWS, Ray, Terrence 366
ABC NEWS, Serafin, Barry 366
ABC NEWS, Arrowsmith, John 371
ABC NEWS, Deiro, Ana Luica 375
ABC NEWS, Mastro, Sal Del 383
ABC NEWS, Obert, Douglas 386
ABC NEWS, Rosen, Michael 388
ABC RADIO, Bascom, Jon 360
ABC-MADRID, Lopez Nicolas, Luis 324
ABCNY, Campbell, Laura 356
ABCNY, Kass, Stephen L., Mr. 357
ABEMA, Buege, Tania Regina, Ms. 139
ABEMA, Mendonca Soares, Tania Lucia, Sra. 167
ABEMA, Moreira, Teresa Maria, Sra. 169
ABEMA, Vale Dos Santos, Ana Gabriela, Sra. 192
ABEMA, Vitorio, Cleverson Andreoli, Mr. 194
ABEMA, Bressau, Almir, Jr., Mr. 90
ABEMA, Martins, Nilo, Mr. 91
ABES, Arnt, Clarice, Ms. 133
ABES, Britto, Evandro Rodrigues De, Mr. 139
ABES, Fiori, Heloisa Helena Pinheiro, Sra. 151
ABES, Fonseca, Edmilson, Mr. 151
ABES, Guerrero, Claudio, Sr. 155
ABES, Inacio, Antonio Rogerio Da Silva, Sr. 157
ABES, Machado, Anna Virginia Muniz, Ms. 163
ABES, Oliveira, Marco Antonio Souza De, Sr. 172
ABES, Pereira, Luiz Otavio Mota, Mr. 175
ABES, Pires, Jadson De Araujo, Sr. 176
ABES, Pulcherio, Maria Isabel Capistrano, Sra. 177
ABES, Reis, Aloisio Clovis Dos, Sr. 178
ABES, Ribeiro, Vasti De Moura, Sra. 179
ABES, Santiago, Emiliano, Sr. 182
ABES, Santo, Joper Padrao Do Espirito, Mr. 182
ABES, Santos Neto, Acylino Jose Dos, Sr. 183
ABESA, Mota Pereira, Luiz Otavio 101
ABESA, Padrao Do Espirito Santo, Joper 101
ABIA -ASSOCIACAO BRASILEIRA INTERDISCIPLINAR DE AIDS, H., Margarete 156
ABIA -ASSOCIACAO BRASILEIRA INTERDISCIPLINAR DE AIDS, Leite, Flavio 160
ABIA -ASSOCIACAO BRASILEIRA INTERDISCIPLINAR DE AIDS, Pereira, Miguel 175
ABIA -ASSOCIACAO BRASILEIRA INTERDISCIPLINAR DE AIDS, Verri, Solange 193
ABIM AGENCY, Carden, Elizabeth Flanigen 106
ABIM AGENCY, Ray, Dixy Lee 122
ABIO - ASSOCIACAO DE AGRICULTORES BIOLOGICOS DO ESTADO DO RIO DE JANEIRO, Rousseau, Leon, Sr. 181
ABORIGINAL COMMUNITY COLLEGE, Holt, Lilian 252
ABORIGINAL GOVERNMENT OF AUSTRALIA, Isaacs, Clarence, Mr. 90
ABPOLAR - ASSOC. BRASIL DE PREVENCAS A POLUICAS DO AR, Junior, Renato Guimaraes 158
ABRACAVE, Machado, Marco Aurelio A.C., Sr. 163
ABRAPIA, Filho, Lauro Monteiro, Sr. 151
ABRAPIA, Marques, Telma Ferracioli, Sra. 165
ABRIL TV, Costa, Paula Gasparini A. 108
ABVP, Azevedo, Patricia De 134
ABVP, Byington, Maria 139
ABVP, Cabral, Adilson Vaz 139
ABVP, Camara, Breno Pimentel 139
ABVP, Castro, Marcia Correa E. 141
ABVP, Santos, Fernanda Reznik 182
ABVP, Silva, Lessandra Da 186
ABVP - ASSOCIACAO BRASILEIRA DE VIDEO POPULAR, Mejia, Alberto Lopez 166
ACADEMIA/ORDEM JORNALISTAS BRASIL, Rinski, Wolf Lobo 123
Academy of Gandhian Studies, Haranath, T., Mr. 251
ACADEMY OF GANDHIAN STUDIES, Murali, Ramisetti, Mr. 252
ACADEMY OF GANDHIAN STUDIES, Reddy, Kullisham Purilshatham, Ms. 253
ACADEMY OF GANDHIAN STUDIES, Subba Rao, B.V., Mr. 253
Academy of Sciences, Koptiug, V.A., Mr. 319
ACADEMY OF SCIENCES, INST. OF GEO, Raukas, Anto 224
Academy of Sciences of the USSR, Marchuk, G.I., Dr. 319
ACAE, Valente De B. Barre, Fernando 102
Acao Democratica Feminina Gaucha,/friends Of The Earth, Brazil, Aunvalle, Angela Escostiguy, Ms. 129
ACAO ECOLOGICA - ASSOCIACAO CAMPINEIRA DE ACAO ECOLOGICA, Gordon, Flavio, Sr. 101
ACCESS NETWORK/CKUA RADIO, Larson, Tasha 201
ACCION ECOLOGICA, Yanez, Ivonne, Ms. 222
Accion Ecologica, Bravo, Elizabeth, Ms. 222
ACCION VERDE, Macchiavello, Gabriel, Mr. 85
ACCOUNTANCY AGE MAGAZINE, Sharifi, Tara 349
ACIE, Neto, Isias Nunes Pereira 171
ACM, Paterson, John 90
ACORD, Barka, Lalla Ben 252
ACPG, Araujo, Ariel 79
ACPG, Inet, Inet Jose Maria 79
ACPS, Faulks, John, Mr. 346
ACRA, Castelletti, Osvaldo, Mr. 260
ACRA, Garbellini, Paolo, Mr. 99
ACSCR, Rojas Ortiz, Jose Dualok 215
ACSCR, Sibar E Sibar, Pablo 215
ACTION AID, Toni, Ana 346
Action For Development, Gunput, Chooramun, Mr. 290
ACTION FOR FOOD PRODUCTION (AFPRO), Myles, Raymond M., Mr. 252
ActionAid, Briggs, Julia, Ms. 349
ACTS, Wilfred, Nyangena 283
A.CU.FO.M., Carrasco, Dino David Parra, Mr. 81
ACWW, Berner, Y. 331
AD (ACCAO P/ DESENVOLVIMENTO), Nosolini, Miranda Isabel, Ms. 247
ADAB-ASSN. OF DEVELOPMENT AGENCIES IN BANGLADESH, Abed, Fazle Hasan, Mr. 93
ADAB-ASSN. OF DEVELOPMENT AGENCIES IN BANGLADESH, Ahmed, Qazi Faruque, Mr. 93
ADAB-ASSN. OF DEVELOPMENT AGENCIES IN BANGLADESH, Ahmed, Yasmin, Ms. 93
ADAB-ASSN. OF DEVELOPMENT AGENCIES IN BANGLADESH, Hasan, F.R. Mahmood, Mr. 93
ADAB-ASSN. OF DEVELOPMENT AGENCIES IN BANGLADESH, Kabir, Khushi, Ms. 93
ADAE, Martin Mateo, Ramon 323
ADAE, Real Ferrer, Gabriel 323
ADAMAWA STATE ENVIRONMENTAL AGENCY, Bawuro Bapetel, Yahaya 303
ADARI, Peralta, Elias Carreno, Sr. 311
ADDO, Ole-Nangoro, Benedict, Mr. 354
A.D.E.J, Moyence, F. 234
ADEJA, Andrade, Ernani, Sr. 132
ADEME, De Passoz, Anne 229
ADESKI, Mayshara, Rurenza 198
ADESKI, Rwumbuguta, Ruhimbika 198
ADFG / AMIGOS DA TERRA, Castro, Giselda Escosteguy, Sra. 101
ADFG / AMIGOS DA TERRA, Farias, Ben Hur, Sr. 149
ADFG / AMIGOS DA TERRA, Monticelli, Gislene, Sra. 168
ADIPE, Garrido, Lydia Maria Luzardo, Sra. 397
ADITTEP, Santos, Arlinda Messias Dos, Ms. 182
Adjoint du Protocole, Ghadi, Ahmed Ould, M. 289
ADMINISTRACAO DO DISTITO ESTUDUAL DE FERNANDO DE NARONHA, Andrade, Ana Clara Marinho De 132
Administrateur, Diallo, Mamadou Saliou, Mr. 246
Administration de la Trasorerie, Ministere des Finances, Gots, Franz, M. 95
ADMINISTRATION DE L'HYDRAULIQUE, Jellali, Mohammed 296
ADMINISTRATION DU TOUISME, Fassi-Fihri, El Houcine 296
Administrativa, Tordesilla, Maria Eliana, Srta. 208
ADN - ALUMNOS DEFENSORES DE NATUR, Mabel, Tosi Liliana, Ms. 84
ADNUMA, Valle, Guillermo Del, Mr. 88
ADO, Bakker, Nicolaas, Mr. 300
ADRA, Britton, Walter, Sr. 139
ADRA, Schulz, Luis, Sr. 184
ADRA, Santos, Fernando Pacheco Dos, Mr. 78
ADRA, Viegas, Walter, Sr. 78
ADT, Anzo, Javier 323
ADT, Da Cruz, Humberto 323
ADUBON MAGAZINE, Berle, Peter A.A. 360
ADUBON MAGAZINE, Real De Azua, Christine 366
ADUNICAMP, Silva, Ardemirio, Mr. 185
ADVANCED TELEVISION PUBLISHING, Wright, Tom 393
ADVENTIST DEVELOPMENT & RELIEF AGENCY, Patterson, Cliff, Mr. 205
Adviser to the Institute of International Affairs, Kim, Hyong U, Mr. 219
Advisory Committee On Protection Of The Sea, Holloway, Jennie, Ms. 350
ADVOGADO INTERNACIONAL, Wemhoff, Daniel M., Mr. 392
AE, Chavez Vallejo, Gina 222
AEC, Arias Chavez, Jose 291
AEC, Corte Gonzales, Bertha C. 291
AED, Bovillon, Gabriel 229
AED, Boullon, Gabriel 231
AEDENAT, Martinez Camarero, Carlos 323
AEDENAT, Santamarta Florez, Jose 323
AEKP, Archibold, Guilhermo 308
AEKP, Smith, Teofrida Alfaro 308
AEP-DF ASSEMBLEIA ECOLOGICA PERMANENTE / FENAMAN, Lenox, Roberto, Mr. 161
AEROSPACE SURVEY, Yimbo, Silas, Mr. 284
AEROTERRA S.A., Viola, Carlos, Mr. 88
AES, Ribeiro, E.F.N. 250
AESP - AUSTRALIANS FOR AN ECOLOGICALLY SUSTAINABLE POPULATION, O'Connor, Mark, Mr. 91
AFAC, Llanos, Gabriela, Ms. 311
AFEA, Chacon De Assis, Jose 101
AFEA, Tarcsay, Ronaldo M. 102
AFEAS, Vogelsberg, Tony 359
A.F.E.I.D, Armengaud, J. 231
Affaires Etrangeres et de la Cooperation (Chef Adjoint de la Delegation), Filali, AbDellatif, S.E. M. 296
AFME, Elion-Mboussa, Albert 224
AFN, Mercredi, Ovide 200
AFN, Sioui, Konrad 200
AFP, Calas, Olivier 229

B

C

D

E

F

FUJI TELEVISION NETWORK, INC., Fujimune, Hiroshi 268
FUJI TELEVISION NETWORK, INC., Ishii, Takehito 269
FUJI TELEVISION NETWORK, INC., Kano, Tadashi 269
FUJI TELEVISION NETWORK, INC., Kashiwa, Yuko 269
FUJI TELEVISION NETWORK, INC., Kawai, Makoto 269
FUJI TELEVISION NETWORK, INC., Kikuchi, Keiichi 270
FUJI TELEVISION NETWORK, INC., Kishimoto, Ichiro 270
FUJI TELEVISION NETWORK, INC., Kobayashi, Toshinori 270
FUJI TELEVISION NETWORK, INC., Maeda, Junji 270
FUJI TELEVISION NETWORK, INC., Matsuda, Keiko 270
FUJI TELEVISION NETWORK, INC., Matsumoto, Masaya 270
FUJI TELEVISION NETWORK, INC., Nakayama, Shigeo 271
FUJI TELEVISION NETWORK, INC., Neshibuchi, Kenji 271
FUJI TELEVISION NETWORK, INC., Nishimura, Yoko 271
FUJI TELEVISION NETWORK, INC., Okuda, Kayo 271
FUJI TELEVISION NETWORK, INC., Ota, Shin 271
FUJI TELEVISION NETWORK, INC., Sawa, Yuji 272
FUJI TELEVISION NETWORK, INC., Takahashi, Yuichiro 272
FUJI TELEVISION NETWORK, INC., Takasaki, Kunio 272
FUJI TELEVISION NETWORK, INC., Tamura, Hiroko 272
FUJI TELEVISION NETWORK, INC., Tanogashira, Yoichi 272
FUJI TELEVISION NETWORK, INC., Terashima, Mitsuo 273
FUJI TELEVISION NETWORK, INC., Tuyuki, Shigeru 273
FUJI TELEVISION NETWORK, INC., Yano, Shuji 273
FUJI TV, Hirai, Fumio 268
FUJI TV, Kuroiwa, Yuji 270
FUJI TV, Matsuyama, Kaori 270
FUJI TV, Oyama, Yasushi 271
FUJI TV, Pimentel, Paulo Roberto Morgira 271
FUJI TV, Watanabe, Toru 273
FUJI TV, Yutaka, Shimizu 273
FUJISANKEI COMMUNICATIONS INTERNATIONAL INC., Kerslake, Simon 352
FULBRIGHT FOUNDATION REGIONAL RESEARCH, Alger, Keith, Mr. 130
FULL - CYCLE LONDON, Roy, Jane, Ms. 205
FUMBRASIL, Fontes, Roldao De Siqueira, Sr. 151
FUMBRASIL, Lima, Ana Cristina Siqueira, Sra. 161
FUMDACON NATURA (ARGENTINA), Risiga, Amilcar Hugo 79
FUNAI, Barros Cruz, Luiz Eduardo Monteiro De, Mr. 136
FUNAI, Fernandes, Annemarie 149
FUNAI, Leitao, Wilma Marques, Ms. 160
FUNAM, Gregoret, Maria Silvina, Sra. 83
FUNAM- ENVIROMENT DEFENCE FOUNDANTION, Aun, Elisa 80
Funatura, Padua, Maria Teresa Jorge, Ms. 129
FUNCACION LOMA QUITA ESPUELA INC., Kees, Rosa Julia, Sra. 222
FUNCACION PARA LA DEFENSA DELA VIDA - FEDEVI, Guevara, Irma 84
FUND. BRASIL - FRANCA PARA ECO, Sarmento, Jair Da Silva, Mr. 184
FUND. CULTURAL DE RADIODIFUSAO EDUCATIVA DA REGIAO DOS LAGOS, Araripe, Alex 133
FUND. CULTURAL DE RADIODIFUSAO EDUCATIVA DA REGIAO DOS LAGOS, Cardoso, Fernando Rodrigues 140
FUND. CULTURAL DE RADIODIFUSAO EDUCATIVA DA REGIAO DOS LAGOS, Carvalho, Marcelo Velloso De 141
FUND. CULTURAL DE RADIODIFUSAO EDUCATIVA DA REGIAO DOS LAGOS, Chagas, Marcelo 142
FUND. CULTURAL DE RADIODIFUSAO EDUCATIVA DA REGIAO DOS LAGOS, Chagas, Renata Monteiro Das 142
FUND. CULTURAL DE RADIODIFUSAO EDUCATIVA DA REGIAO DOS LAGOS, Correia, Berenice Maria 143
FUND. CULTURAL DE RADIODIFUSAO EDUCATIVA DA REGIAO DOS LAGOS, Jensen, Elisabeth 158
FUND. CULTURAL DE RADIODIFUSAO EDUCATIVA DA REGIAO DOS LAGOS, Lebre, Fernando Ribeiro 160
FUND. CULTURAL DE RADIODIFUSAO EDUCATIVA DA REGIAO DOS LAGOS, Lima, Rachel Cristina Linhares De 161
FUND. CULTURAL DE RADIODIFUSAO EDUCATIVA DA REGIAO DOS LAGOS, Margulies, Andre Jorge 164
FUND. CULTURAL DE RADIODIFUSAO EDUCATIVA DA REGIAO DOS LAGOS, Medeiros, Gustavo Henrique 166
FUND. CULTURAL DE RADIODIFUSAO EDUCATIVA DA REGIAO DOS LAGOS, Mello, Pedro Paulo Thiago De 166
FUND. CULTURAL DE RADIODIFUSAO EDUCATIVA DA REGIAO DOS LAGOS, Menezes, Sidine Antunes De 167
FUND. CULTURAL DE RADIODIFUSAO EDUCATIVA DA REGIAO DOS LAGOS, Neves, Alice Martins Baeta 171
FUND. CULTURAL DE RADIODIFUSAO EDUCATIVA DA REGIAO DOS LAGOS, Oldemburgo, Cristina 172
FUND. CULTURAL DE RADIODIFUSAO EDUCATIVA DA REGIAO DOS LAGOS, Ornellas, Erica 173
FUND. CULTURAL DE RADIODIFUSAO EDUCATIVA DA REGIAO DOS LAGOS, Quintela, Maria Fernanda Alves P. 177
FUND. CULTURAL DE RADIODIFUSAO EDUCATIVA DA REGIAO DOS LAGOS, Sabino, Bernardo 181
FUND. CULTURAL DE RADIODIFUSAO EDUCATIVA DA REGIAO DOS LAGOS, Santos, Fernando Dos 182
FUND. CULTURAL DE RADIODIFUSAO EDUCATIVA DA REGIAO DOS LAGOS, Santos, Luis Claudio Dos 183
FUND. CULTURAL DE RADIODIFUSAO EDUCATIVA DA REGIAO DOS LAGOS, Santos, Maria Francisca M. Dos 183
FUND. CULTURAL DE RADIODIFUSAO EDUCATIVA DA REGIAO DOS LAGOS, Schawrnopf, Paulo Inacio 184
FUND. CULTURAL DE RADIODIFUSAO EDUCATIVA DA REGIAO DOS LAGOS, Silva, Celio Teodoro Da 186
FUND. CULTURAL DE RADIODIFUSAO EDUCATIVA DA REGIAO DOS LAGOS, Uchoa, Cleofas Ismael 192
FUND. CULTURAL DE RADIODIFUSAO EDUCATIVA DA REGIAO DOS LAGOS, Videla, Daniel 194
FUND. ECOLOGIA UNIVERSAL, Rissi, Luis Antonio, Mr. 86
FUND. ECOLOGICA VALE DO GUAPORE RONDONIA, Oliveira, Erundina Barbosa De, Sra. 172
Fund. Jorge Esteban Roulet/Inst. de Estudios e invest. sobre el Medio Ambiente, Canziani, Osvaldo, Mr. 79
FUND. NEHUEN, Bercovich, Patricia, Sra. 80
FUND. NEHUEN, Martinez, Carlos, Sr. 85
FUNDACAO AGUA VIVA - FAVI, Silva, Ademir Milo Motta Da, Sr. 185
FUNDACAO AMAZONAS FOREVER GREEN, Mantovani, Ivolin, Sr. 101
FUNDACAO AMAZONAS FOREVER GREEN, Gindry, Reny, Mr. 154
FUNDACAO AMAZONAS FOREVER GREEN, Ramos, Adriana Oliveira, Sra. 177
FUNDACAO BIBLIOTECA NACIONAL, Gauz, Valeria, Ms. 153
FUNDACAO BIODIVERISITAS, Paiva, Maria Cristina Ribeiro, Ms. 173
FUNDACAO BIODIVERSITAS, Bernardes, Aline Tristao, Sra. 137
FUNDACAO BIODIVERSITAS, Costa, Leonora Pires, Sra. 144
FUNDACAO BIODIVERSITAS, Fernandes, Glaucia Mendes, Ms. 149
FUNDACAO BIODIVERSITAS, Fonseca, Ana Maria Paiva Da, Sra. 151
FUNDACAO BIODIVERSITAS, Leite, Yuri, Mr. 160
FUNDACAO BIODIVERSITAS, Machado, Ricardo Bonfim, Mr. 163
FUNDACAO BIODIVERSITAS, Santos, Ilmar Bastos, Sr. 183
FUNDACAO BIODIVERSITAS, Silva, Delso Correa Da, Sr. 186
FUNDACAO BIODIVERSITAS, Valle, Fernanda De Oliveira 192
FUNDACAO BRASIL - FRANCA, Cunha, Ricardo Lisboa Da, Sr. 145
FUNDACAO BRASIL 2000, Rodrigues, Roberto Maia E Carmem 123
FUNDACAO BRASIL FRANcA, Madureira, Mario Luiz, Mr. 163
FUNDACAO BRASIL FRANCA PARA O DESENVOLVIMENTO, Borba, Nestor Da Costa, Mr. 137
FUNDACAO BRASILEIRA PARA A CONSERVACAO DA NATUREZA, Amazonas, Patricia O., Sra. 132
FUNDACAO BRASIL-FRANCA, Ferrari, Antenor, Mr. 149
FUNDACAO BRASIL-FRANCA, Franco, Marilia Carmem, Ms. 152
FUNDACAO BRASIL-FRANCA, Malheiros, Tadeu Fabricio, Sr. 164
FUNDACAO BRASIL-FRANCA, Pacios, Valentim Manduca, Sr. 173
FUNDACAO BRASIL-FRANCA, Silva, Jair Sarmento, Sr. 186
Fundacao Brasilliera Para E Conservacao Da Natureza (FBCN), Costa, Jairo, Sr. 129
FUNDACAO CIDADE DA PAZ, Well, Pierre, Sr. 195
FUNDACAO CULTURAL AVATAR, Araujo, Tania Goncalves De, Ms. 133
FUNDACAO CULTURAL AVATAR, Fonseca, Iva Marques Da, Ms. 151
FUNDACAO CULTURAL AVATAR, Oiticica, Vera, Sra. 172
FUNDACAO CULTURAL AVATAR, Tabel, Ruth Melof., Ms. 190
FUNDACAO CULTURAL AVATAR, Treijer, Maria Elisa De Mattos, Ms. 192
FUNDACAO CULTURAL AVATAR, Uzeda, Conrado L., Mr. 192
FUNDACAO CULTURAL RIOGRANDENSE, Gollo, Nestor Jose 154
FUNDACAO DA ARVORE E DA CRIANCA / G.L.M.E.E.S., Fiorot, Antonio Wilson, Sr. 151
FUNDACAO DO MEIO AMBIENTE E TECNOLOGIA DE RORAIMA, Fonseca, Alexandre Alberto Henklain, Sr. 151
FUNDACAO DOM MANUEL II, Andrade Silva, Maria Ines, Ms. 314
FUNDACAO EDUCACIONAL DE BARRETOS, Trevisan, Mauri, Sr. 192
FUNDACAO ESCOLA DE SERVICO PUBLICO RJ - FESP, Barbosa, Rizio, Sr. 135
FUNDACAO ESCOLA DE SERVICO PUBLICO RJ - FESP, Clutilde, Maria, Sra. 143
FUNDACAO ESCOLA DE SERVICO PUBLICO RJ - FESP, Silva, Olivia Soares Da, Sra. 187
FUNDACAO ESCOLA DE SERVICO PUBLICO RJ - FESP, Soares, Celio, Sr. 188
FUNDACAO ESCOLA DE SERVICO PUBLICO RJ - FESP, Vicente, Ma. De Lourdes, Sra. 194
FUNDACAO ESQUEL, Moura, Adriana Maria Magalhaes De 169
FUNDACAO ESTADUAL DO MEIO AMBIENTE, Lopes, Eulinda De Campos, Ms. 162
FUNDACAO ESTADUAL DO MEIO AMBIENTE - MT, Espindola, Albimarcia Das Neves 148
FUNDACAO ESTADUAL DO MEIO AMBIENTE - MT, Souza, Alcilene Freitas Bertholdo De, Ms. 189
FUNDACAO FE E ALEGRIA, Brasileiro, Tula Viera, Ms. 138
FUNDACAO FE E ALEGRIA, Ribeiro, Carlos Dimas Martins, Mr. 179
FUNDACAO FE E ALEGRIA, Rozemblatt, Flavio 181
FUNDACAO FLORESTAL, Costa, Aparecido Antonio Carvalho, Mr. 144
FUNDACAO FORD, Maubury-Lewis, Bioru, Mr. 166
FUNDACAO GAIA, Tornquist, Carlos Gustavo, Mr. 192
FUNDACAO GAIA, Wagner, Ralf, Mr. 194
FUNDACAO GILBERTO FREYRE, Lody, Raul 162
FUNDACAO GRUPO ESQUEL, Glantz, Michael H., Mr. 101
FUNDACAO INTERAMERICANA, Reilly, Charles, Mr. 387
FUNDACAO IPLANCE, Carvalho, Maria Teresa, Ms. 141
FUNDACAO KONRAD ADENALIER, Wenz, Dietmar, Mr. 195
FUNDACAO MELO, Soto, Teofilo Ku, Sr. 308
FUNDACAO MOKITI-OKADA - IGREJA MESSIANICA MUNDIAL, Pereira, Angela Peixoto, Sra. 175
FUNDACAO MOKITI-OKADA - IGREJA MESSIANICA MUNDIAL, Silva, Marcos Antonio Lopes Da, Sr. 187
FUNDACAO MOVIMENTO ONDAZUL, Galvao, Claudia Valeria, Sra. 153
FUNDACAO MUSEU DA IMAGEM E SOM, Gaurio, Jalten De Paula 153
Fundacao Museu Do Homem Americano, Pessis, Anne-Marie, Ms. 129
FUNDACAO NACIONAL DE ACAO ECOLOGICA, Cunha, Icaro, Sr. 145
FUNDACAO NACIONAL DE SAUDE, Marrocos, Ricardo Paysano, Sr. 165
FUNDACAO NACIONAL DO INDIO, Rocha, Maria Da Conceicao Militao, Sra. 179
FUNDACAO NACIONAL SAUDE-FNS, Da Silva, Jose Dibamar, Sr. 145
FUNDACAO OSWALDO CRUZ, Grimaldi Filho, Gabriel, Sr. 155
FUNDACAO OSWALDO CRUZ, Santa Rosa, Paulo, Sr. 182
FUNDACAO PARA A CONSERVACAO E A PRODUCAO FLORESTAL DO ESTADO DE SAO PAULO, Rocha, Ligia Moreira, Ms. 179
FUNDACAO PARA CONSERVACAO DA NATUREZA, Barros, Francisco Anselmo Gomes De, Sr. 135
FUNDACAO PARA CONSERVACAO DA NATUREZA, Brenacher, Maria Helena, Sra. 138
FUNDACAO PARA CONSERVACAO DA NATUREZA, Sampaio, Francisco De Assis, Sr. 182
FUNDACAO PARQUE TECNOLOGICO - PB, Favero, Luiz A., Mr. 149
FUNDACAO PAU BRASIL, Alvim, Paulo T., Mr. 131
FUNDACAO PLANETA OCEANO, Couro, Maria Celia 144
FUNDACAO PRO TAMAR, Marcovaldi, Maria Angela, Sra. 164
FUNDACAO PROJETO SANITARISTA, Alcantara, Sandra Cristina, Sra. 130
FUNDACAO PROJETO SANITARISTA, Maciel, Jacques Antonio Alvim, Sr. 163
FUNDACAO PROJETO SANITARISTA, Maciel, Tereza Maria Cristina Bastos, Sra. 163
FUNDACAO PROJETO SANITARISTA, Ortiz, Eduardo, Sr. 173
FUNDACAO PROJETO SANITARISTA, Ortiz, Francisco, Sr. 173
FUNDACAO PROJETO SANITARISTA, Ortiz, Sonia Regina Carboni, Sra. 173
FUNDACAO PRO-NATUREZA (FUNATURA), Andrade, Fernando Antonio Thome, Sr. 132
FUNDACAO PRO-NATUREZA (FUNATURA), Antas, Paulo De Tarso Zuquim, Sr. 132
FUNDACAO PRO-NATUREZA (FUNATURA), Maury, Cilulia Maria, Sra. 166
FUNDACAO PRO-NATUREZA (FUNTARA), Santo, Cesar Victor Do Espirito, Sr. 182
FUNDACAO ROBERTO MARINHO, Cardim, Roberto 106
FUNDACAO ROBERTO MARINHO, Falcao, Joaquim 111
FUNDACAO ROBERTO MARINHO, Fonseca, Joberto 112
FUNDACAO ROBERTO MARINHO, Ramos, Margarida 122
FUNDACAO ROBERTO MARINHO, Souza, Gilberto De 126
FUNDACAO ROBERTO MARINHO, Britto, Francisco Eugenio Calmon De 139
FUNDACAO ROBERTO MARINHO, Vieira, Maercelo Moraes 194
FUNDACAO ROQUETTE-PINTO, Avelar, Lucila Vasconcelos De 134
FUNDACAO ROQUETTE-PINTO, Guimaraes, Marilia De Lacerde 156
FUNDACAO ROQUETTE-PINTO, Matrianha, Carlos Borges 166
FUNDACAO ROQUETTE-PINTO, Oliveira, Maria Aparecida De 172
FUNDACAO SAMUEL, Tropardi, Lilian Rose Dos Santos, Sra. 192
FUNDACAO SOS MATA ATLANTICA, Alves, Liane C. A. 131
FUNDACAO SOS MATA ATLANTICA, Araujo, Washington Luiz De, Sr. 133
FUNDACAO SOS MATA ATLANTICA, Dias, Ines De Souza 147
FUNDACAO SOS MATA ATLANTICA, Diniz, Iliria Maria, Sra. 147
FUNDACAO SOS MATA ATLANTICA, Hamburger, Diana Sarita, Sra. 156
FUNDACAO SOS MATA ATLANTICA, Hirota, Marcia Makiko, Sra. 156
FUNDACAO SOS MATA ATLANTICA, Klabin, Roberto L., Sr. 159
FUNDACAO SOS MATA ATLANTICA, Lourencao, Otavio Cunha, Sr. 162
FUNDACAO SOS MATA ATLANTICA, Mantovani, Mario Cesar, Sr. 164
FUNDACAO S.O.S. MATA ATLANTICA, Pen, Jeane, Sra. 174
FUNDACAO SOS MATA ATLANTICA, Silva, Nadja Regina Peixoto, Sra. 187
FUNDACAO SOS MATA ATLANTICA, Som, Luiz 189
FUNDACAO SOS MATA ATLANTICA, Son, Luiz, Sr. 189
FUNDACAO S.O.S. PIRACICABA, Amaral, Luiz Fernando, Mr. 132
FUNDACAO S.O.S. PIRACICABA, Goncalves, Clesio Oliveira, Mr. 154
FUNDACAO SOS RIO PIRACICABA, Azevedo, Andre Louis Tenuta, Sr. 134
FUNDACAO SOS RIO PIRACICABA, Azevedo, Maria Do Rosario Rates, Ms. 134
FUNDACAO S.O.S. RIO PIRACICABA, Oliveira, Acir Silva, Sr. 172
FUNDACAO SOS RIO PIRACICABA, Penha, Eri Pimenta Da 174
FUNDACAO UNIVERSIDADE PATAGONICA, Romero, Carlos, Sr. 87
FUNDACAO UNIVERSIDADE PATAGONICA, Woslawski, Mauricio, Mr. 88
FUNDACAO UNIVERSO E VIDA-UNIDA, Ferreira Carneiro, Jaemio, Mr. 150
FUNDACAO VITORIA AMAZONICA, Miller, Carlos, Mr. 101
Fundacao Vitoria Amazonica, Lacerda, Leonard, Mr. 129
FUNDACAO VITORIA AMAZONICA, Fonseca, Claudio Ruy Vasconcelos Da, Sr. 151
FUNDACAO XAPURI, Biar, Paulo, Sr. 137
FUNDACAO XAPURI, Vasconcellos, Maria Dulce Queiros, Ms. 193
FUNDACION ALBERT SCHWEITZER, Recla, Nelida E., Sra. 86
FUNDACION ALMA, Fonseca, Alegria, Sra. 213
FUNDACION ALMA, Michaels Urbina, Maria Luisa, Sra. 213
Fundacion Ambiental Francisco Tamayo, Martinez, Rafael Elino, Mr. 399
FUNDACION AMBIENTE Y RECURSOS NATURALES, Bilbao, Jorge, Mr. 80
FUNDACION AMBIENTE Y RECURSOS NATURALES, Frazier, Jeff L., Mr. 83
FUNDACION AMBIENTE Y RECURSOS NATURALES, Tarat, Pedro, Mr. 88
Fundacion Ambiente y Recursos Naturales (FARN), Cano, Guillermo J., Mr. 79
FUNDACION AMBIENTIS, Antoniow, Marta Victoria, Ms. 80
FUNDACION AMBIENTIS, Domingo, Alberto Virdo, Mr. 82
FUNDACION AMBIENTIS, Virdo, Alberto Nicolas, Mr. 88
FUNDACION ARGENTINA DE ECOLOGIA - FADE, Canduglia Pivorano, Jose Augusto, Sr. 81
FUNDACION ARGENTINA DE ECOLOGIA - FADE, Gitto, Santiago G., Sr. 83
FUNDACION ARGENTINA DE ECOLOGIA - FADE, Pagnotta, Leonardo 86
FUNDACION ARIAS, Curling, Lena Patricia White, Sra. 216
FUNDACION ARIAS - IPAC, Cotter, Maria Eugenia Penon De, Ms. 216
FUNDACION AUGUSTO C. SANDINO, Munoz, Roberto, Sr. 303
FUNDACION BARILOCHE, Abaleron, Carlos Alberto, Mr. 79
FUNDACION BARILOCHE, Mendez, Maria Susana, Ms. 85
FUNDACION BIOMA, Matarasso, Horacio F., Mr. 85
FUNDACION CEPA, Reimondi, Ivan, Sr. 86
Fundacion Charles Darwin para las Islas Galapagos (FCD), Carrasco, Alfredo, Sr. 222
FUNDACION CIEPE, Garcia, Estevez, Mr. 399
Fundacion Civica Francisco de Orellana, Garzozi, Joseph, Mr. 222
FUNDACION CONASED, Luy, Amaila M., Sra. 84
FUNDACION CONASED/FUNDACION DE ESTUDIOS MUNICIPALES Y SOCIALES, Machado, Pablo, Sr. 85
FUNDACION CULTURAL DE CINE Y EDUCACION EDUCATIVOS, INC., Baez, Etzel, Mr. 222
FUNDACION DE ACCION ECOLOGICA, Valle, Genovesi Gabriela Del, Sra. 88
Fundacion de Parques Nacionales, Gallegos, Ines, Sra 216
FUNDACION DESARROLLO ECOLOGICO, Acuna, Ruben Marcelo, Mr. 79

G

J

K

L

M

N

O

P

Q

R

S

T

U

V

W

X

Y

Z

The Companion Book To Who Is Who At The Earth Summit

WHO IS WHO

IN SERVICE TO THE EARTH

People • Projects • Organizations • Key Words

&

41 VISIONS OF A POSITIVE FUTURE BY

Marilyn Ferguson, Lester R. Brown, Hazel Henderson, Joseph A. Califano Jr., Riane Eisler, Edgar D. Mitchell, Jean Houston, and others.

First Edition

Publisher: VisionLink Education Foundation, North Carolina

Editor-In-Chief: Hans J. Keller

Co-editor: Daniel Maziarz

ORDER FORM

WHO IS WHO

AT THE EARTH SUMMIT

Mail To:

Last Name ______________________ First Name ______________________

Organization/Company ______________________

Street or P.O. Box ______________________ City ______________________

State ______________ Zip Code ______________ Country ______________

PLEASE SEND ME _____ COPIES OF WHO IS WHO AT THE EARTH SUMMIT

(The price is $30.00 plus shipping and handling $2.50 per book, $5.00 outside USA.) Total Cost $__________.

My check* for $__________ is enclosed, or bill my credit card ☐ Visa ☐ Master ☐ Amex

In The Name Of ______________________

Card No. ☐☐☐☐ ☐☐☐☐ ☐☐☐☐ ☐☐☐☐ Expiration Date ☐ ☐☐

Make Check Payable To:

VisionLink Education Foundation
130 Biodome Drive
Waynesville, NC 28786 USA
704 926-2200 / FAX: 704 926-9041

* Check must be in US dollars, drawn on a US bank with electronic encoding.

Mail To:

Last Name ______________________ First Name ______________________

Organization/Company ______________________

Street or P.O. Box ______________________ City ______________________

State ______________ Zip Code ______________ Country ______________

PLEASE SEND ME _____ COPIES OF WHO IS WHO AT THE EARTH SUMMIT

(The price is $30.00 plus shipping and handling $2.50 per book, $5.00 outside USA.) Total Cost $__________.

My check* for $__________ is enclosed, or bill my credit card ☐ Visa ☐ Master ☐ Amex

In The Name Of ______________________

Card No. ☐☐☐☐ ☐☐☐☐ ☐☐☐☐ ☐☐☐☐ Expiration Date ☐ ☐☐

Make Check Payable To:

VisionLink Education Foundation
130 Biodome Drive
Waynesville, NC 28786 USA
704 926-2200 / FAX: 704 926-9041

* Check must be in US dollars, drawn on a US bank with electronic encoding.

3 1/2" and 5" disks and labels available. Please inquire.

CD-ROM ORDER FORM
WHO IS WHO AT THE EARTH SUMMIT AND WHO IS WHO IN SERVICE TO THE EARTH

The complete content of both publications on one CD-ROM

Mail To:

Last Name ______________________ First Name ______________________

Organization/Company ______________________

Street or P.O. Box ______________________ City ______________________

State ______________ Zip Code ______________ Country ______________

PLEASE SEND ME _____ CD-ROMS AT $80.00

(Shipping and handling $2.50 per CD, $5.00 outside USA.) Total Cost $__________.

My check* for $__________ is enclosed, or bill my credit card ☐ Visa ☐ Master ☐ Amex

In The Name Of ______________________

Card No. ☐☐☐☐ ☐☐☐☐ ☐☐☐☐ ☐☐☐☐ Expiration Date ☐ ☐☐

Make Check Payable To:

VisionLink Education Foundation
130 Biodome Drive
Waynesville, NC 28786 USA
704 926-2200 / FAX: 704 926-9041

* Check must be in US dollars, drawn on a US bank with electronic encoding.

Mail To:

Last Name ______________________ First Name ______________________

Organization/Company ______________________

Street or P.O. Box ______________________ City ______________________

State ______________ Zip Code ______________ Country ______________

PLEASE SEND ME _____ CD-ROMS AT $80.00

(Shipping and handling $2.50 per CD, $5.00 outside USA.) Total Cost $__________.

My check* for $__________ is enclosed, or bill my credit card ☐ Visa ☐ Master ☐ Amex

In The Name Of ______________________

Card No. ☐☐☐☐ ☐☐☐☐ ☐☐☐☐ ☐☐☐☐ Expiration Date ☐ ☐☐

Make Check Payable To:

VisionLink Education Foundation
130 Biodome Drive
Waynesville, NC 28786 USA
704 926-2200 / FAX: 704 926-9041

* Check must be in US dollars, drawn on a US bank with electronic encoding.

3 1/2" and 5" diskettes available. Please inquire.

NOTES

NOTES

NOTES

NOTES